GLENCOE MATHEMATICS

Teacher Wraparound Edition

Mathematics

Applications and Concepts

Course 2

Bailey
Day
Frey
Howard
Hutchens
McClain
Moore-Harris
Ott
Pelfrey
Price
Vielhaber
Willard

McGraw Hill Glencoe

New York, New York   Columbus, Ohio   Chicago, Illinois   Peoria, Illinois   Woodland Hills, California

Glencoe

The McGraw-Hill Companies

The Standardized Test Practice features in this book were aligned and verified by The Princeton Review, the nation's leader in test preparation. Through its association with McGraw-Hill, The Princeton Review offers the best way to help students excel on standardized assessments.

The Princeton Review is not affiliated with Princeton University or Educational Testing Service.

Send all inquiries to:
Glencoe/McGraw-Hill
8787 Orion Place
Columbus, OH 43240-4027

ISBN: 0-07-829633-1 (Student Edition)
ISBN: 0-07-829634-X (Teacher Wraparound Edition)

2 3 4 5 6 7 8 9 10 043/027 10 09 08 07 06 05 04 03

Contents in Brief

UNIT 1 Decimals, Algebra, and Statistics

1 Decimal Patterns and Algebra

2 Statistics: Analyzing Data

UNIT 2 Integers and Algebra

3 Algebra: Integers

4 Algebra: Linear Equations and Functions

UNIT 3 Fractions

5 Fractions, Decimals, and Percents

6 Applying Fractions

UNIT 4 Proportional Reasoning

7 Ratios and Proportions

8 Applying Percent

9 Probability

UNIT 5 Geometry and Measurement

10 Geometry

11 Geometry: Measuring Two-Dimensional Figures

12 Geometry: Measuring Three-Dimensional Figures

Authors

Rhonda Bailey
Mathematics Consultant
Mathematics by Design
DeSoto, Texas

Roger Day, Ph.D.
Associate Professor
Illinois State University
Normal, Illinois

Patricia Frey
Director of Staffing and Retention
Buffalo City Schools
Buffalo, New York

Arthur C. Howard
Mathematics Teacher
Houston Christian High School
Houston, Texas

Deborah T. Hutchens, Ed.D.
Assistant Principal
Great Bridge Middle School
Chesapeake, Virginia

Kay McClain, Ed.D.
Assistant Professor
Vanderbilt University
Nashville, Tennessee

Beatrice Moore-Harris
Mathematics Consultant
League City, Texas

Jack M. Ott, Ph.D.
Distinguished Professor of Secondary Education Emeritus
University of South Carolina
Columbia, South Carolina

Ronald Pelfrey, Ed.D.
Mathematics Specialist
Appalachian Rural Systemic Initiative
Lexington, Kentucky

Jack Price, Ed.D.
Professor Emeritus
California State Polytechnic University
Pomona, California

Kathleen Vielhaber
Mathematics Specialist
Parkway School District
St. Louis, Missouri

Teri Willard, Ed.D.
Adjunct Instructor and Mathematics Consultant
Montana State University
Bozeman, Montana

Contributing Authors

USA TODAY
The USA TODAY Snapshots®, created by USA TODAY®, help students make the connection between real life and mathematics.

Dinah Zike
Educational Consultant
Dinah-Might Activities, Inc.
San Antonio, Texas

Content Consultants

Each of the Content Consultants reviewed every chapter and gave suggestions for improving the effectiveness of the mathematics instruction.

Mathematics Consultants

L. Harvey Almarode
Curriculum Supervisor, Mathematics K–12
Augusta County Public Schools
Fishersville, VA

Claudia Carter, MA, NBCT
Mathematics Teacher
Mississippi School for Mathematics and Science
Columbus, MS

Carol E. Malloy, Ph.D.
Associate Professor, Curriculum Instruction, Secondary Mathematics
The University of North Carolina at Chapel Hill
Chapel Hill, NC

Melissa McClure, Ph.D.
Mathematics Instructor
University of Phoenix On-Line
Fort Worth, TX

Robyn R. Silbey
School-Based Mathematics Specialist
Montgomery County Public Schools
Rockville, MD

Leon L. "Butch" Sloan, Ed.D.
Secondary Mathematics Coordinator
Garland ISD
Garland, TX

Barbara Smith
Mathematics Instructor
Delaware County Community College
Media, PA

Reading Consultant

Lynn T. Havens
Director
Project CRISS
Kalispell, MT

ELL Consultants

Idania Dorta
Mathematics Educational Specialist
Miami–Dade County Public Schools
Miami, FL

Frank de Varona, Ed.S.
Visiting Associate Professor
Florida International University College of Education
Miami, FL

Teacher Reviewers

Each Teacher Reviewer reviewed at least two chapters of the Student Edition, giving feedback and suggestions for improving the effectiveness of the mathematics instruction.

Royallee Allen
Teacher, Math Department Head
Eisenhower Middle School
San Antonio, TX

Dennis Baker
Mathematics Department Chair
Desert Shadows Middle School
Scottsdale, AZ

Rosie L. Barnes
Teacher
Fairway Middle School–KISD
Killeen, TX

Charlie Bialowas
Math Curriculum Specialist
Anaheim Union High School District
Anaheim, CA

Stephanie R. Boudreaux
Teacher
Fontainebleau Jr. High School
Mandeville, LA

Dianne G. Bounds
Teacher
Nettleton Junior High School
Jonesboro, AR

Susan Peavy Brooks
Math Teacher
Louis Pizitz Middle School
Vestavia Hills, AL

Karen Sykes Brown
Mathematics Educator
Riverview Middle School
Grundy, VA

Kay E. Brown
Teacher, 7th Grade
North Johnston Middle School
Micro, NC

Renee Burgdorf
Middle Grades Math Teacher
Morgan Co. Middle
Madison, GA

Kelley Summers Calloway
Teacher
Baldwin Middle School
Montgomery, AL

Carolyn M. Catto
Teacher
Harney Middle School
Las Vegas, NV

Claudia M. Cazanas
Math Department Chair
Fairmont Junior High
Pasadena, TX

David J. Chamberlain
Secondary Math Resource Teacher
Capistrano Unified School District
San Juan Capistrano, CA

David M. Chioda
Supervisor Math/Science
Marlboro Township Public Schools
Marlboro, NJ

Carrie Coate
7th Grade Math Teacher
Spanish Fort School
Spanish Fort, AL

Toinette Thomas Coleman
Secondary Mathematics Teacher
Caddo Middle Career & Technology School
Shreveport, LA

Linda M. Cordes
Math Department Chairperson
Paul Robeson Middle School
Kansas City, MO

Polly Crabtree
Teacher
Hendersonville Middle School
Hendersonville, NC

Dr. Michael T. Crane
Chairman Mathematics
B.M.C. Durfee High School
Fall River, MA

Tricia Creech, Ph.D.
Curriculum Facilitator
Southeast Guilford Middle School
Greensboro, NC

Lyn Crowell
Math Department Chair
Chisholm Trail Middle School
Round Rock, TX

B. Cummins
Teacher
Crestdale Middle School
Matthews, NC

Debbie Davis
8th Grade Math Teacher
Max Bruner, Jr. Middle School
Ft. Walton Beach, FL

Diane Yendell Day
Math Teacher
Moore Square Museums Magnet Middle School
Raleigh, NC

Wendysue Dodrill
Teacher
Barboursville Middle School
Barboursville, WV

Judith F. Duke
Math Teacher
Cranford Burns Middle School
Mobile, AL

Carol Fatta
Math/Computer Instructor
Chester Jr. Sr. M.S.
Chester, NY

Cynthia Fielder
Mathematics Consultant
Atlanta, GA

Georganne Fitzgerald
Mathematics Chair
Crittenden Middle School
Mt. View, CA

Jason M. Fountain
7th Grade Mathematics Teacher
Bay Minette Middle School
Bay Minette, AL

Sandra Gavin
Teacher
Highland Junior High School
Cowiche, WA

Ronald Gohn
8th Grade Mathematics
Dover Intermediate School
Dover, PA

Larry J. Gonzales
Math Department Chairperson
Desert Ridge Middle School
Albuquerque, NM

Shirley Gonzales
Math Teacher
Desert Ridge Middle School
Albuquerque, NM

Paul N. Hartley, Jr.
Mathematics Instructor
Loudoun County Public Schools
Leesburg, VA

Deborah L. Hewitt
Math Teacher
Chester High School
Chester, NY

Steven J. Huesch
Mathematics Teacher/Department Chair
Cortney Jr. High
Las Vegas, NV

Sherry Jarvis
8th Grade Math/Algebra 1 Teacher
Flat Rock Middle School
East Flat Rock, NC

Teacher Reviewers *continued*

Mary H. Jones
Math Curriculum Coordinator
Grand Rapids Public Schools
Grand Rapids, MI

Vincent D.R. Kole
Math Teacher
Eisenhower Middle School
Albuquerque, NM

Ladine Kunnanz
Middle School Math Teacher
Sequoyah Middle School
Edmond, OK

Barbara B. Larson
Math Teacher/Department Head
Andersen Middle School
Omaha, NE

Judith Lecocq
7th Grade Teacher
Murphysboro Middle School
Murphysboro, IL

Paula C. Lichiello
7th Grade Math and Pre-Algebra Teacher
Forest Middle School
Forest, VA

Michelle Mercier Maher
Teacher
Glasgow Middle School
Baton Rouge, LA

Jeri Manthei
Math Teacher
Millard North Middle School
Omaha, NE

Albert H. Mauthe, Ed.D.
Supervisor of Mathematics (Retired)
Norristown Area School District
Norristown, PA

Karen M. McClellan
Teacher & Math Department Chair
Harper Park Middle
Leesburg, VA

Ken Montgomery
Mathematics Teacher
Tri-Cities High School
East Point, GA

Helen M. O'Connor
Secondary Math Specialist
Harrison School District Two
Colorado Springs, CO

Cindy Ostrander
8th Grade Math Teacher
Edwardsville Middle School
Edwardsville, IL

Michael H. Perlin
8th Grade Mathematics Teacher
John Jay Middle School
Cross River, NY

Denise Pico
Mathematics Teacher
Jack Lund Schofield Middle School
Las Vegas, NV

Ann C. Raymond
Teacher
Oak Ave. Intermediate School
Temple City, CA

M.J. Richards
Middle School Math Teacher
Davis Middle School
Dublin, OH

Linda Lou Rohleder
Math Teacher, Grades 7 & 8
Jasper Middle School
Jasper, IN

Dana Schaefer
Pre-Algebra & Algebra I Teacher
Coachman Fundamental Middle School
Clearwater, FL

Donald W. Scheuer, Jr.
Coordinator of Mathematics
Abington School District
Abington, PA

Angela Hardee Slate
Teacher, 7th Grade Math/Algebra
Martin Middle School
Raleigh, NC

Mary Ferrington Soto
7th Grade Math
Calhoun Middle School-Ouachita Parish Schools
Calhoun, LA

Diane Stilwell
Mathematics Teacher/Technology Coordinator
South Middle School
Morgantown, WV

Pamela Ann Summers
K–12 Mathematics Coordinator
Lubbock ISD–Central Office
Lubbock, TX

Marnita L. Taylor
Mathematics Teacher/Department Chairperson
Tolleston Middle School
Gary, IN

Susan Troutman
Teacher
Dulles Middle School
Sugar Land, TX

Barbara C. VanDenBerg
Math Coordinator, K–8
Clifton Board of Education
Clifton, NJ

Mollie VanVeckhoven-Boeving
7th Grade Math and Algebra Teacher
White Hall Jr. High School
White Hall, AR

Mary A. Voss
7th Grade Math Teacher
Andersen Middle School
Omaha, NE

Christine Waddell
Teacher Specialist
Jordan School District
Sandy, UT

E. Jean Ware
Supervisor
Caddo Parish School Board
Shreveport, LA

Karen Y. Watts
9th Grade Math Teacher
Douglas High School
Douglas, AL

Lu Wiggs
Supervisor
I.S. 195
New York, NY

Teacher Advisory Board and Field Test Schools

Teacher Advisory Board

Glencoe/McGraw-Hill wishes to thank the following teachers for their feedback on *Mathematics: Applications and Concepts*. They were instrumental in providing valuable input toward the development of this program.

Katie Davidson
Legg Middle School
Coldwater, MI

Lynanne Gabriel
Bradley Middle School
Huntersville, NC

Kathleen M. Johnson
New Albany-Plain Local Middle School
New Albany, OH

Ronald C. Myer
Indian Springs Middle School
Columbia City, IN

Mike Perlin
John Jay Middle School
Cross River, NY

Reema Rahaman
Brentwood Middle School
Brentwood, MO

Diane T. Scheuber
Elizabeth Middle School
Elizabeth, CO

Deborah Sykora
Hubert H. Humphrey Middle School
Bolingbrook, IL

DeLynn Woodside
Roosevelt Middle School,
Oklahoma City Public Schools
Oklahoma City, OK

Field Test Schools

Glencoe/McGraw-Hill wishes to thank the following schools that field-tested pre-publication manuscript during the 2002–2003 school year. They were instrumental in providing feedback and verifying the effectiveness of this program.

Knox Community Middle School
Knox, IN

Roosevelt Middle School
Oklahoma City, OK

Brentwood Middle School
Brentwood, MO

Elizabeth Middle School
Elizabeth, CO

Legg Middle School
Coldwater, MI

Great Hollow Middle School
Nesconset, NY

Student Advisory Board

The Student Advisory Board gave the authors, editorial staff, and design team feedback on the lesson design, content, and covers of the Student Editions. We thank these students for their hard work and creative suggestions in making ***Mathematics: Applications and Concepts*** more student friendly.

Front Row: Joey Snyder, Tiffany Pickenpaugh, Craig Hammerstein

Back Row: Brittany Yokum, Alex Johnson, Cimeone Starling, Kristina Smith, Kate Holt, Ben Ball

Teacher Handbook

Table of Contents

Dynamic Instructional Strategies T2

Time-Saving Teacher Resources T4

Reading and Writing T6

Foldables™ T8

Meeting Student Needs T9

Intervention T10

Assessment T12

MathScape T13

USA TODAY® T14

Internet Resources T15

Staff Development T16

Program Validation T18

Pacing T20

NCTM Principles and Standards T21

Teacher Handbook

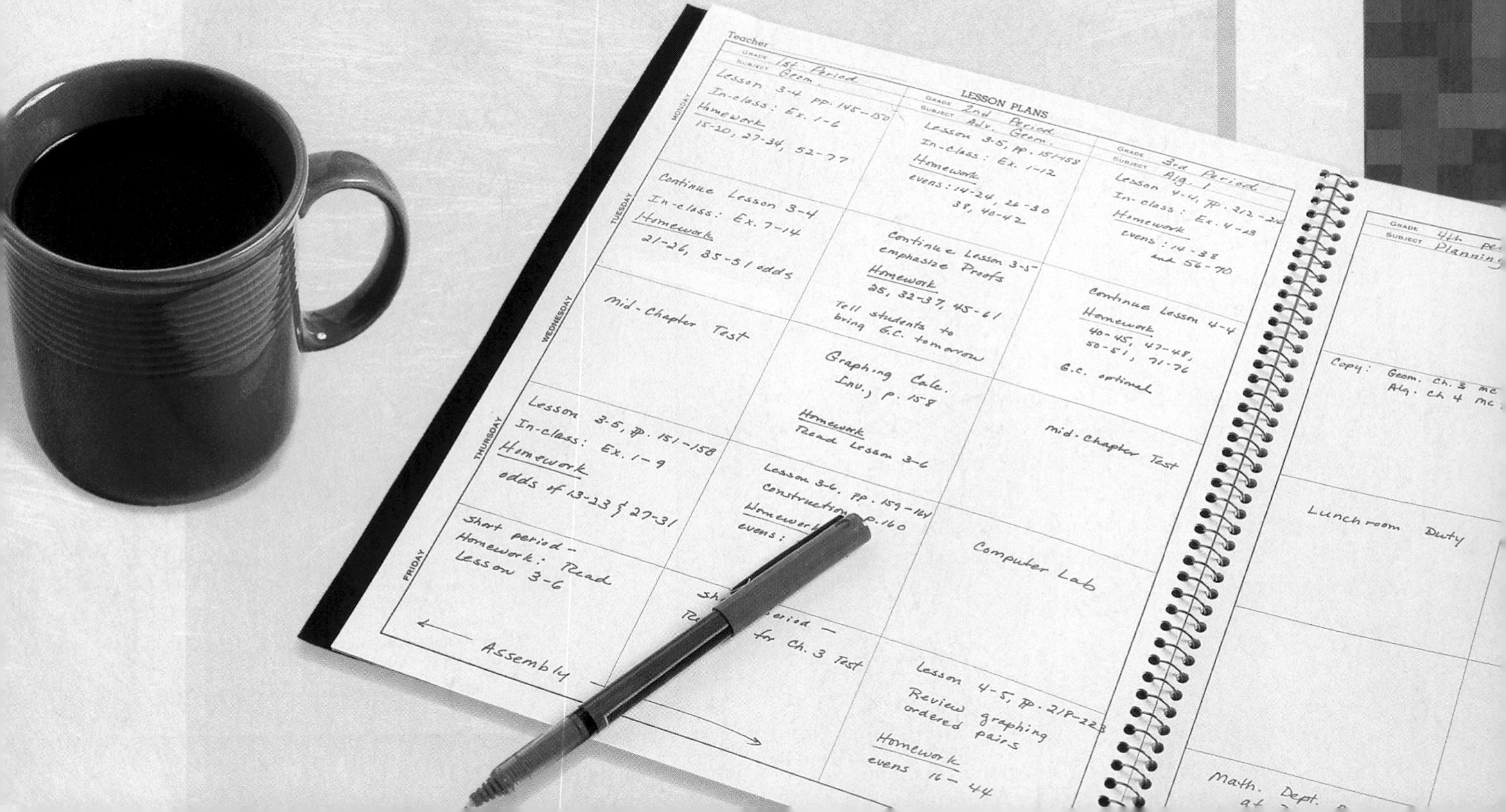

Designed to be in more

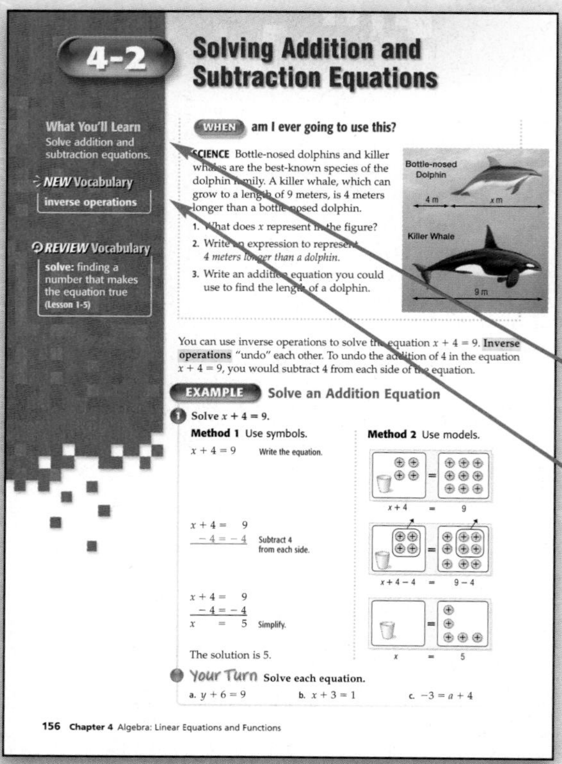

4-2 Solving Addition and Subtraction Equations

What You'll Learn
Solve addition and subtraction equations.

NEW Vocabulary
inverse operations

REVIEW Vocabulary
solve: finding a number that makes the equation true (Lesson 1-5)

WHEN am I ever going to use this?

SCIENCE Bottle-nosed dolphins and killer whales are the best-known species of the dolphin family. A killer whale, which can grow to a length of 9 meters, is 4 meters longer than a bottle-nosed dolphin.

1. What does x represent in the figure?
2. Write an expression to represent *4 meters longer than a dolphin.*
3. Write an addition equation you could use to find the length of a dolphin.

You can use inverse operations to solve the equation $x + 4 = 9$. **Inverse operations** "undo" each other. To undo the addition of 4 in the equation $x + 4 = 9$, you would subtract 4 from each side of the equation.

EXAMPLE Solve an Addition Equation

1 Solve $x + 4 = 9$.

Method 1 Use symbols.

$x + 4 = 9$ Write the equation.

$x + 4 = 9$, $-4 = -4$ Subtract 4 from each side.

$x + 4 = 9$, $-4 = -4$, $x = 5$ Simplify.

The solution is 5.

Method 2 Use models.

$x + 4 = 9$

$x + 4 - 4 = 9 - 4$

$x = 5$

Your Turn Solve each equation.

a. $y + 6 = 9$ b. $x + 3 = 1$ c. $-3 = a + 4$

156 **Chapter 4** Algebra: Linear Equations and Functions

The consistent instructional strategy of the lessons strengthens student learning—from the very beginning of each lesson where students see "What You'll Learn," to the very end of each lesson where they'll have a chance to prepare for the next lesson.

What You'll Learn previews the topics to come and engaging questions incorporate problem-based learning in your classroom.

New and Review Vocabulary terms are listed at the beginning of each lesson for easy reference. New terms are also highlighted in yellow as they occur in the lesson. Link to Reading features relate mathematical terms to everyday meanings. Math Symbols are listed in the lesson in which they are introduced.

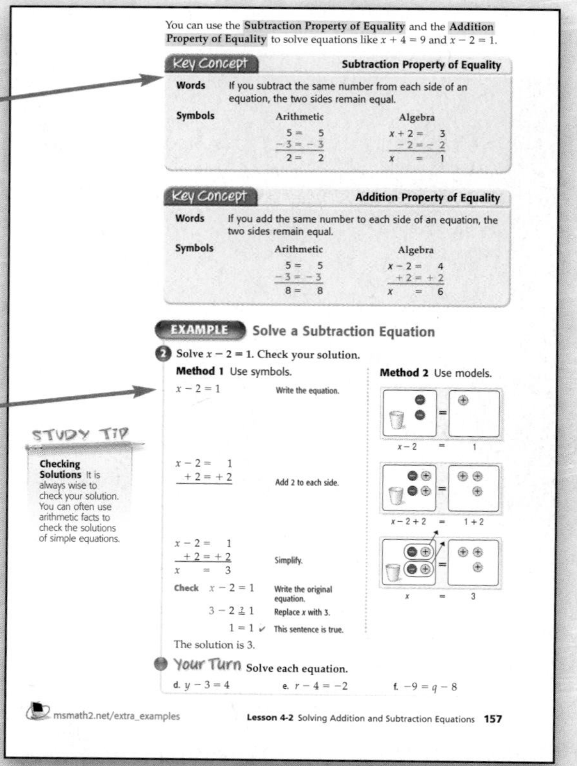

You can use the **Subtraction Property of Equality** and the **Addition Property of Equality** to solve equations like $x + 4 = 9$ and $x - 2 = 1$.

Key Concept — **Subtraction Property of Equality**

Words If you subtract the same number from each side of an equation, the two sides remain equal.

Symbols

Arithmetic	Algebra
$5 = 5$	$x + 2 = 3$
$-3 = -3$	$-2 = -2$
$2 = 2$	$x = 1$

Key Concept — **Addition Property of Equality**

Words If you add the same number to each side of an equation, the two sides remain equal.

Symbols

Arithmetic	Algebra
$5 = 5$	$x - 2 = 4$
$-3 = -3$	$+2 = +2$
$8 = 8$	$x = 6$

EXAMPLE Solve a Subtraction Equation

2 Solve $x - 2 = 1$. Check your solution.

Method 1 Use symbols.

$x - 2 = 1$ Write the equation.

$x - 2 = 1$, $+2 = +2$ Add 2 to each side.

$x - 2 = 1$, $+2 = +2$, $x = 3$ Simplify.

Check $x - 2 = 1$ Write the original equation.

$3 - 2 \stackrel{?}{=} 1$ Replace x with 3.

$1 = 1$ ✓ This sentence is true.

The solution is 3.

Method 2 Use models.

$x - 2 = 1$

$x - 2 + 2 = 1 + 2$

$x = 3$

STUDY TIP

Checking Solutions It is always wise to check your solution. You can often use arithmetic facts to check the solutions of simple equations.

Your Turn Solve each equation.

d. $y - 3 = 4$ e. $r - 4 = -2$ f. $-9 = q - 8$

msmath2.net/extra_examples

Lesson 4-2 Solving Addition and Subtraction Equations 157

Key Concept boxes use words, symbols, and models to illustrate new rules, properties, and definitions, so students can build their reading skills as they build their math skills. **Concept Summary** boxes provide a concise overview of key topics.

Completely worked-out examples with clear explanations are paralleled by the Guided Practice and Practice and Applications exercises that follow. Your Turn exercises are included as appropriate after skill examples. Examples often include strategies for problem-solving and taking standardized tests.

effective,

ways than one.

Skill and Concept Check

You can use this portion of the exercises in class to ensure that all students understand the concepts.

- **Writing Math** exercises give students opportunities to define, describe, and explain the mathematical concepts they've just learned.
- **Guided Practice** presents a representative sample of the exercises in the Practice and Applications section. A key is provided in the Teacher Wraparound Edition that correlates the exercises with appropriate examples.
- **Application** problems give students the opportunity to use the skills they have learned in a real-world setting.

Practice and Applications

- **Skill Exercises** correspond to the Guided Practice exercises and are structured so that students practice the same concepts whether they are assigned odd- or even-numbered problems. *Homework Help* refers students to examples as they complete these exercises.
- **Applications** provide problem-based learning opportunities in both real-life and mathematical situations.
- **Critical Thinking** exercises require students to explain, make conjectures, and prove mathematical relationships.

Standardized Test Practice and Mixed Review

- **Standardized Test Practice** questions provide students with ongoing opportunities to sharpen their test-taking skills.
- **Mixed Review** includes spiraled, cumulative exercises from the two previous lessons as well as earlier lessons.
- **Getting Ready for the Next Lesson** exercises give students the chance to review prerequisite skills for the coming lesson. A reference is provided should students need additional help.

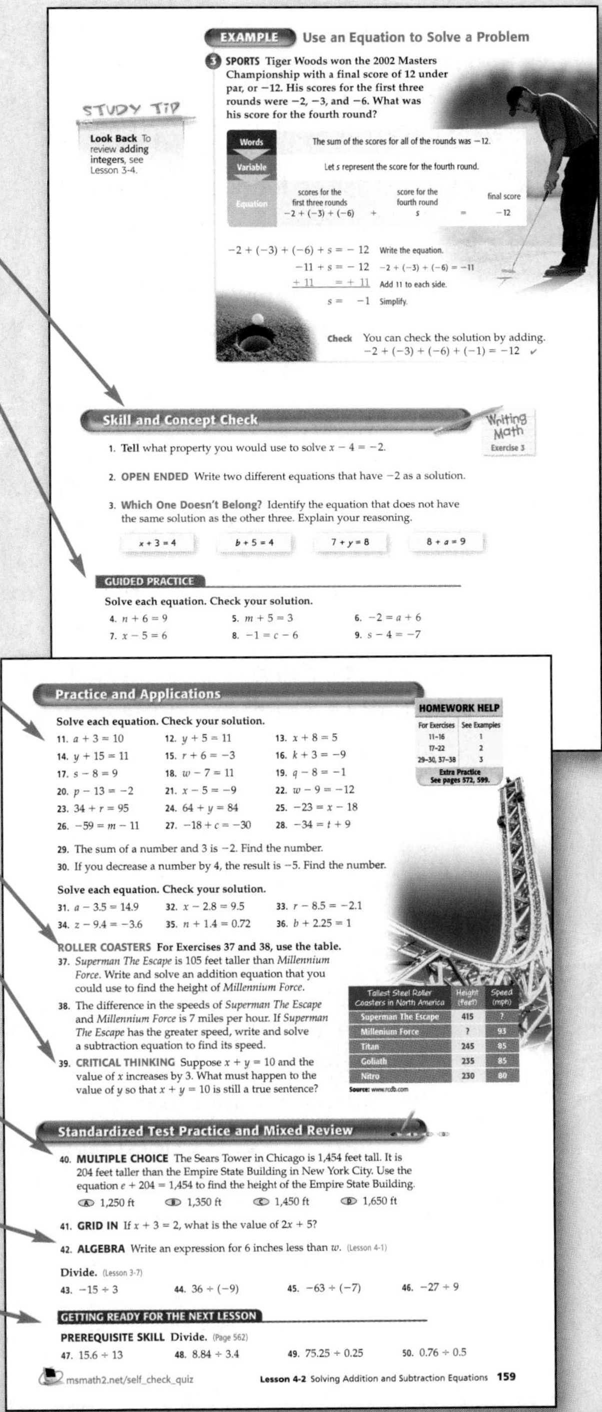

EXAMPLE Use an Equation to Solve a Problem

3 **SPORTS** Tiger Woods won the 2002 Masters Championship with a final score of 12 under par, or −12. His scores for the first three rounds were −2, −3, and −6. What was his score for the fourth round?

STUDY TIP

Look Back To review adding integers, see Lesson 3-4.

Words: The sum of the scores for all of the rounds was −12.

Variable: Let s represent the score for the fourth round.

Equation: scores for the first three rounds $-2 + (-3) + (-6)$ + score for the fourth round s = final score -12

$-2 + (-3) + (-6) + s = -12$ Write the equation.

$-11 + s = -12$ $-2 + (-3) + (-6) = -11$

$+11 \quad = +11$ Add 11 to each side.

$s = -1$ Simplify.

Check You can check the solution by adding. $-2 + (-3) + (-6) + (-1) = -12$ ✓

Skill and Concept Check

Writing Math Exercise 3

1. **Tell** what property you would use to solve $x - 4 = -2$.
2. **OPEN ENDED** Write two different equations that have −2 as a solution.
3. **Which One Doesn't Belong?** Identify the equation that does not have the same solution as the other three. Explain your reasoning.

$x + 3 = 4$ | $b + 5 = 4$ | $7 + y = 8$ | $8 + a = 9$

GUIDED PRACTICE

Solve each equation. Check your solution.

4. $n + 6 = 9$ 5. $m + 5 = 3$ 6. $-2 = a + 6$
7. $x - 5 = 6$ 8. $-1 = c - 6$ 9. $s - 4 = -7$

Practice and Applications

Solve each equation. Check your solution.

11. $a + 3 = 10$ 12. $y + 5 = 11$ 13. $x + 8 = 5$
14. $y + 15 = 11$ 15. $r + 6 = -3$ 16. $k + 3 = -9$
17. $s - 8 = 9$ 18. $w - 7 = 11$ 19. $q - 8 = -1$
20. $p - 13 = -2$ 21. $x - 5 = -9$ 22. $w - 9 = -12$
23. $34 + r = 95$ 24. $64 + y = 84$ 25. $-23 = x - 18$
26. $-59 = m - 11$ 27. $-18 + c = -30$ 28. $-34 = t + 9$

HOMEWORK HELP	
For Exercises	See Examples
11–16	1
17–22	2
29–30, 37–38	3

Extra Practice See pages 572, 599.

29. The sum of a number and 3 is −2. Find the number.
30. If you decrease a number by 4, the result is −5. Find the number.

Solve each equation. Check your solution.

31. $a - 3.5 = 14.9$ 32. $x - 2.8 = 9.5$ 33. $r - 8.5 = -2.1$
34. $z - 9.4 = -3.6$ 35. $n + 1.4 = 0.72$ 36. $b + 2.25 = 1$

ROLLER COASTERS For Exercises 37 and 38, use the table.

37. *Superman The Escape* is 105 feet taller than *Millennium Force*. Write and solve an addition equation that you could use to find the height of *Millennium Force*.
38. The difference in the speeds of *Superman The Escape* and *Millennium Force* is 7 miles per hour. If *Superman The Escape* has the greater speed, write and solve a subtraction equation to find its speed.
39. **CRITICAL THINKING** Suppose $x + y = 10$ and the value of x increases by 3. What must happen to the value of y so that $x + y = 10$ is still a true sentence?

Tallest Steel Roller Coasters in North America	Height (feet)	Speed (mph)
Superman The Escape	415	?
Millenium Force	?	93
Titan	245	85
Goliath	235	85
Nitro	230	80

Source: www.rcdb.com

Standardized Test Practice and Mixed Review

40. **MULTIPLE CHOICE** The Sears Tower in Chicago is 1,454 feet tall. It is 204 feet taller than the Empire State Building in New York City. Use the equation $e + 204 = 1{,}454$ to find the height of the Empire State Building.
Ⓐ 1,250 ft Ⓑ 1,350 ft Ⓒ 1,450 ft Ⓓ 1,650 ft
41. **GRID IN** If $x + 3 = 2$, what is the value of $2x + 5$?
42. **ALGEBRA** Write an expression for 6 inches less than w. (Lesson 4-1)

Divide. (Lesson 3-7)

43. $-15 \div 3$ 44. $36 \div (-9)$ 45. $-63 \div (-7)$ 46. $-27 \div 9$

GETTING READY FOR THE NEXT LESSON

PREREQUISITE SKILL Divide. (Page 562)

47. $15.6 \div 13$ 48. $8.84 \div 3.4$ 49. $75.25 \div 0.25$ 50. $0.76 \div 0.5$

msmath2.net/self_check_quiz **Lesson 4-2** Solving Addition and Subtraction Equations **159**

Accomplish more

Glencoe Mathematics: Applications and Concepts provides so many resources for lesson planning and teaching that you can create a complete, customized course in mathematics quickly...and easily.

This is where you start.

In addition to teaching suggestions, additional examples, and answers, the **Teacher Wraparound Edition** provides a guide for all of the print and software materials available for each lesson.

FAST FILE _Chapter Resource Masters_ contain all of the core supplements you'll need to begin teaching a chapter of *Glencoe Mathematics: Applications and Concepts*. Each chapter booklet features convenient tabs for easy filing.

FAST FILE

- The **Family Letter** and **Family Activity** informs your students' families of the requirements of the chapter and helps them understand how the mathematics students are learning is applicable to real life.
- **Vocabulary Builder** helps students locate and define key vocabulary words from the chapter.
- **Study Guide and Intervention** summarizes key concepts and provides practice.
- **Practice: Skills** provides ample exercises to help students develop computational skills, lesson by lesson.
- **Practice: Word Problems** mimics the verbal problems in each lesson at an average level.
- **Reading to Learn Mathematics** provides students with various reading strategies to master the mathematics presented in each lesson.
- **Enrichment** activities extend students' knowledge and widen their appreciation of how mathematics relates to the world around them.
- **Assessment** options for each chapter include six forms of chapter tests, assessment tasks, quizzes, mid-chapter test, cumulative review, and standardized test practice.

Reading and Writing

WebQuests, Projects, and Interdisciplinary Investigations include teaching notes and answers for the WebQuest Interdisciplinary projects as well as other long-term projects.

Reading and Writing in the Mathematics Classroom features suggestions for including reading as an integral part of the mathematics curriculum as well as differentiated approaches for English learning and inclusion.

Teaching Mathematics with Foldables™ offers guidelines for using Foldables interactive study organizers in your class.

FOLDABLES™ Study Organizer

More information on options for reading and writing is available on pages T6–T7.

Applications

Classroom Math Games include tools for easy use of the Game Zone games in the Student Edition. Also included are additional games designed to help students improve their problem solving, logic, and computational skills. (available online)

Calculator and Spreadsheet Masters include activities to incorporate scientific and graphing calculators and spreadsheets into your mathematics classroom.

Science and Mathematics Lab Manual includes lab activity masters and teaching suggestions for integrating science into your mathematics classroom. (available online)

Real-Life Career Masters features activities that show how mathematics relates to various careers.

What's Math Got to Do with It? videos connect mathematics to real life in an exciting, student-friendly format.

than you'd ever imagine

in less time than you'd ever believe

Assessment and Intervention

5-Minute Check Transparencies with Standardized Test Practice for each lesson evaluates what students learned in the previous lesson. Each transparency also includes a standardized test practice question.

Key Teaching Transparencies provide key concepts and worked-out examples for key lessons.

Closing the Gap for Absent Students provides an easy-to-use summary of all the materials you have covered in the chapter in a format that can be posted or distributed to students who have missed class.

DAILY INTERVENTION

Guide to Daily Intervention offers suggestions for daily assessment and tips on how to help students succeed.

Math Skills Maintenance includes worksheets to review basic mathematics skills.

Staff Development

Answer Key Transparencies provide answers to Student Edition exercises.

Solutions Manual includes completely worked-out solutions for all exercises in the Student Edition.

Using the Internet in the Mathematics Classroom provides guidelines for using the Internet as well as resources available on the Internet.

Teaching Mathematics with Manipulatives features activities and teaching suggestions to help you present mathematics concepts with manipulatives and hands-on materials.

Technology Support for Teachers

TeacherWorks

TeacherWorks All-in-One Lesson Planner and Resource Center CD-ROM includes a lesson planner and interactive Teacher Edition, so you can customize lesson plans and reproduce classroom resources quickly and easily.

Answer Key Maker with Solutions Manual software allows you to customize answer keys for your assignments from the Student Edition exercises.

INTERACTIVE CHALKBOARD
PowerPoint® Presentations

Interactive Chalkboard CD-ROM includes fully worked-out examples, Bellringer Activities, 5-Minute Check Transparencies, and Your Turn problems in a customizable Microsoft PowerPoint® format.

And more...Additional technology products and Internet resources for students, teachers, and parents are discussed on pages T6–T7, T11, and T14–T15.

HELP your students become fluent

Glencoe Mathematics: Applications and Concepts makes it easy for you to incorporate constructive reading and writing strategies into every class you teach.

In the Student Edition

Foldables™ Study Organizers at the beginning of each chapter provide students with tools for organizing what they are reading and studying.

Study Skill activities include strategies that provide students with a variety of methods to read, study, and comprehend the mathematics they are learning.

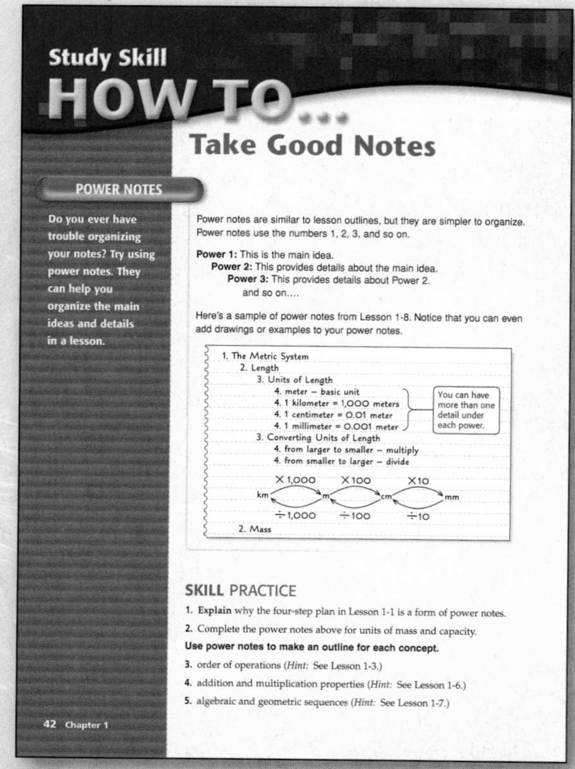

Study Skill

HOW TO...

Take Good Notes

POWER NOTES

Do you ever have trouble organizing your notes? Try using power notes. They can help you organize the main ideas and details in a lesson.

Power notes are similar to lesson outlines, but they are simpler to organize. Power notes use the numbers 1, 2, 3, and so on.

Power 1: This is the main idea.
Power 2: This provides details about the main idea.
Power 3: This provides details about Power 2.
and so on....

Here's a sample of power notes from Lesson 1-8. Notice that you can even add drawings or examples to your power notes.

SKILL PRACTICE

1. **Explain** why the four-step plan in Lesson 1-1 is a form of power notes.
2. Complete the power notes above for units of mass and capacity.

Use power notes to make an outline for each concept.

3. order of operations (*Hint:* See Lesson 1-3.)
4. addition and multiplication properties (*Hint:* See Lesson 1-6.)
5. algebraic and geometric sequences (*Hint:* See Lesson 1-7.)

42 Chapter 1

Writing Math questions in every lesson give students opportunities to define, describe, and explain the mathematical concepts they've just learned.

Reading Math tips appear throughout each chapter to help students learn and use the language of math.

Vocabulary terms are listed at the beginning of each lesson and highlighted when defined.

The **Vocabulary and Concept Check** in each Study Guide and Review summarizes and reviews key terms.

Key Concepts are illustrated using **Words, Symbols,** and **Models** as appropriate. This approach improves reading comprehension by using multiple representations.

WebQuest Interdisciplinary Projects give students the chance to work through long-term projects to develop their research and creative writing skills.

In the Teacher Wraparound Edition

Foldables™ Follow-Up provide suggestions to help students create Foldables that are thorough and effective.

Modeling, Speaking, and **Writing** activities in the Assess step require students to summarize what they have learned by responding to open-ended prompts.

in the Language of MATHEMATICS.

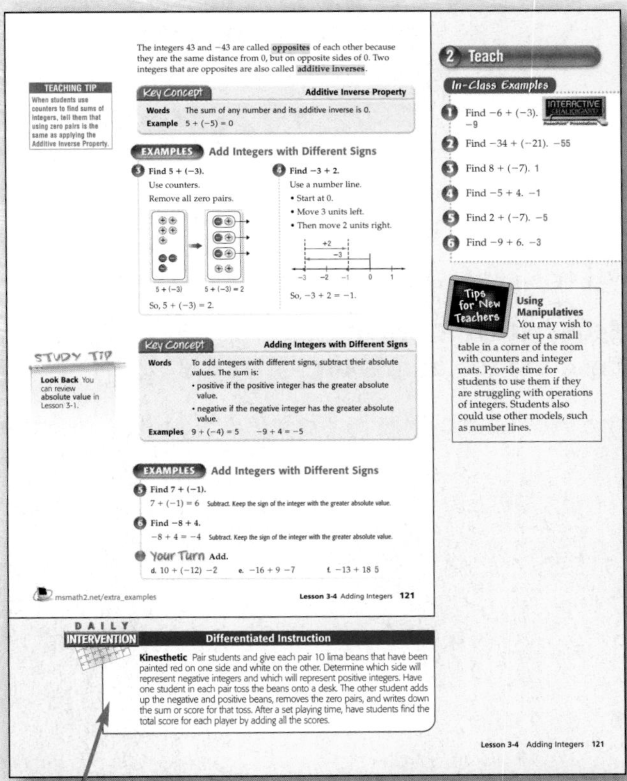
The integers 43 and −43 are called **opposites** of each other because they are the same distance from 0, but on opposite sides of 0. Two integers that are opposites are also called **additive inverses**.

TEACHING TIP
When students use counters to find sums of integers, tell them that using zero pairs is the same as applying the Additive Inverse Property.

Key Concept — Additive Inverse Property
Words: The sum of any number and its additive inverse is 0.
Example: $5 + (-5) = 0$

EXAMPLES Add Integers with Different Signs

3 Find $5 + (-3)$.
Use counters.
Remove all zero pairs.
$5 + (-3)$ — $5 + (-3) = 2$
So, $5 + (-3) = 2$.

4 Find $-3 + 2$.
Use a number line.
• Start at 0.
• Move 3 units left.
• Then move 2 units right.
So, $-3 + 2 = -1$.

STUDY TIP
Look Back You can review absolute value in Lesson 3-1.

Key Concept — Adding Integers with Different Signs
Words: To add integers with different signs, subtract their absolute values. The sum is:
• positive if the positive integer has the greater absolute value.
• negative if the negative integer has the greater absolute value.
Examples: $9 + (-4) = 5$ $-9 + 4 = -5$

EXAMPLES Add Integers with Different Signs

5 Find $7 + (-1)$.
$7 + (-1) = 6$ Subtract. Keep the sign of the integer with the greater absolute value.

6 Find $-8 + 4$.
$-8 + 4 = -4$ Subtract. Keep the sign of the integer with the greater absolute value.

Your Turn Add.
d. $10 + (-12)$ −2 e. $-16 + 9$ −7 f. $-13 + 18$ 5

msmath2.net/extra_examples

Lesson 3-4 Adding Integers 121

2 Teach

In-Class Examples
1 Find $-6 + (-3)$. −9
2 Find $-34 + (-21)$. −55
3 Find $8 + (-7)$. 1
4 Find $-5 + 4$. −1
5 Find $2 + (-7)$. −5
6 Find $-9 + 6$. −3

Tips for New Teachers
Using Manipulatives
You may wish to set up a small table in a corner of the room with counters and integer mats. Provide time for students to use them if they are struggling with operations of integers. Students also could use other models, such as number lines.

DAILY INTERVENTION — Differentiated Instruction
Kinesthetic Pair students and give each pair 10 lima beans that have been painted red on one side and white on the other. Determine which side will represent negative integers and which will represent positive integers. Have one student in each pair toss the beans onto a desk. The other student adds up the negative and positive beans, removes the zero pairs, and writes down the sum or score for that toss. After a set playing time, have students find the total score for each player by adding all the scores.

Lesson 3-4 Adding Integers 121

Differentiated Instruction features help students at all points on the learning spectrum develop their reading, writing, and comprehension skills.

Unlocking Misconceptions features point out preconceived ideas that students may have that are not necessarily true. Often, you can head off a global misconception that students may have.

Flexible Grouping features outline opportunities for using heterogeneous groups for instruction or intervention.

Recommended Outside Reading provides suggestions for both fiction and nonfiction literature that relates to the mathematics in each chapter.

English Language Learners Strategies suggest lesson activities that help English Language Learners grasp content.

For more information on teaching to English Language Learners, see p. T9.

Technology Support

StudentWorks™ includes the entire text, formatted like the hardbound book, so students can study from just about anywhere—no book required. With StudentWorks, students can also access six different worksheets for each lesson as well as online study tools.

msmath2.net/vocabulary_review is a Glencoe site that provides online study tools for reviewing the vocabulary of each chapter.

Vocabulary PuzzleMaker creates crossword, jumble, and word search puzzles online using vocabulary lists that you can customize.

Additional Resources

Chapter Resource Masters

- Vocabulary Builder
- Reading to Learn Mathematics (both available in a workbook)

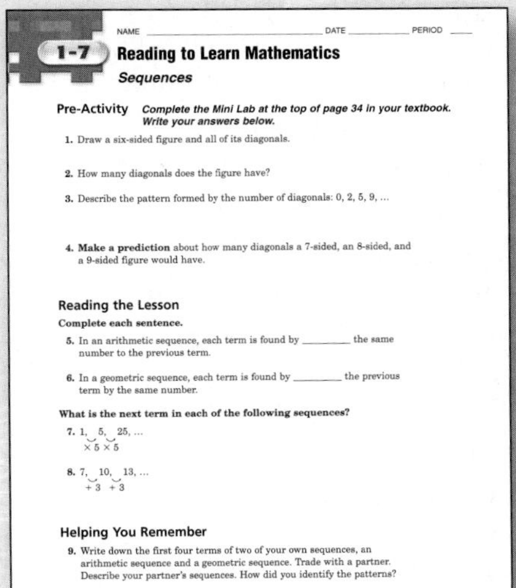
NAME ______ DATE ______ PERIOD ____

1-7 Reading to Learn Mathematics
Sequences

Pre-Activity *Complete the Mini Lab at the top of page 34 in your textbook. Write your answers below.*

1. Draw a six-sided figure and all of its diagonals.
2. How many diagonals does the figure have?
3. Describe the pattern formed by the number of diagonals: 0, 2, 5, 9, …
4. **Make a prediction** about how many diagonals a 7-sided, an 8-sided, and a 9-sided figure would have.

Reading the Lesson
Complete each sentence.
5. In an arithmetic sequence, each term is found by ______ the same number to the previous term.
6. In a geometric sequence, each term is found by ______ the previous term by the same number.

What is the next term in each of the following sequences?
7. 1, 5, 25, … (×5 ×5)
8. 7, 10, 13, … (+3 +3)

Helping You Remember
9. Write down the first four terms of two of your own sequences, an arithmetic sequence and a geometric sequence. Trade with a partner. Describe your partner's sequences. How did you identify the patterns?

© Glencoe/McGraw-Hill 34 *Mathematics: Applications and Concepts, Course 2*

Teaching Mathematics With Foldables™

Reading and Writing in the Mathematics Classroom

WebQuests, Projects, and Interdisciplinary Investigations

For more information on these products, see pp. T4–T5.

DISCOVER how a simple sheet of paper can CHANGE the way your students THINK about math... forever.

Students love Foldables™ because they're fun. Teachers love them because they're effective.

Foldables are easy-to-make, three-dimensional interactive graphic organizers that students create out of simple sheets of paper. These unique hands-on tools for learning and reviewing were created exclusively for *Glencoe Mathematics: Applications and Concepts* by teaching specialist Dinah Zike.

Building Prereading Skills

At the beginning of each chapter, students construct one of a variety of Foldables. Each Foldable helps students create an interactive strategy for organizing what they read and observe. As they work through each chapter, students add more detail to their Foldable until they have created a comprehensive, interactive snapshot of the key concepts and vocabulary of the chapter.

Reading and Writing

Each foldable helps students practice basic reading and writing skills, find and report main ideas, organize information, review key vocabulary terms, and more.

Review and Reinforcement

The completed Foldable is a comprehensive overview of the chapter concepts—perfect for preparing for chapter, unit, and even end-of-course tests.

Assessment

Foldables present an ideal opportunity to probe the depth of your students' understanding of chapter concepts. You'll get detailed feedback on what your students know and any misconceptions they have.

Staff Development

Teaching Mathematics with Foldables™ equips teachers to extend the use of Foldables in their classrooms by exploring the different Foldable formats and providing suggestions for using them throughout the mathematics curriculum.

Give students control, so that they leave your classroom with *knowledge and power* over their own learning.

PROJECT CRISS[SM] Study Skill

Project CRISS℠ [**CR**eating **I**ndependence through **S**tudent-owned **S**trategies] is a research-based staff development program created to help students better organize, understand and retain course information. In short, students receiving the CRISS method of instruction will "LEARN HOW TO LEARN".

CRISS strategies are designed to develop thoughtful and independent readers and learners.

To enhance student learning, CRISS employs several concepts drawn from cognitive psychology.

- Students must be able to integrate new information with prior knowledge.
- Students need to be actively involved in their own learning by discussing, writing, and organizing information.
- Students must self-monitor to identify which strategies are the most effective for their own learning.

These behaviors need to be taught by content teachers to maximize student learning.

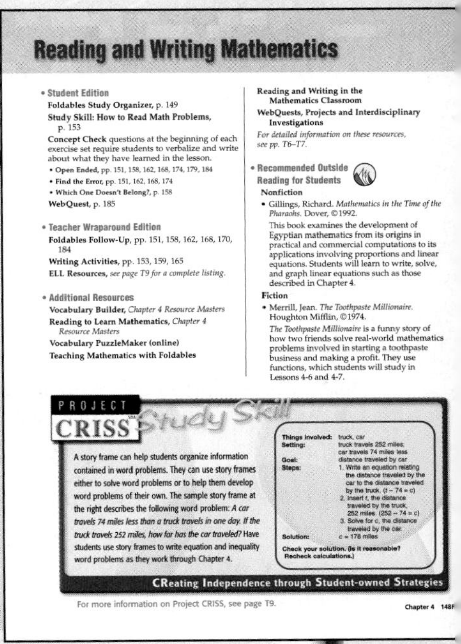

Reading and Writing Mathematics

- **Student Edition**
  - **Foldables Study Organizer,** p. 149
  - **Study Skill: How to Read Math Problems,** p. 153
  - **Concept Check** questions at the beginning of each exercise set require students to verbalize and write about what they have learned in the lesson.
    - **Open Ended,** pp. 151, 158, 162, 168, 174, 179, 184
    - **Find the Error,** pp. 151, 162, 168, 174
    - **Which One Doesn't Belong?,** p. 158
  - **WebQuest,** p. 185
- **Teacher Wraparound Edition**
  - **Foldables Follow-Up,** pp. 151, 158, 162, 168, 170, 184
  - **Writing Activities,** pp. 153, 159, 165
  - **ELL Resources,** *see page T9 for a complete listing.*
- **Additional Resources**
  - **Vocabulary Builder,** *Chapter 4 Resource Masters*
  - **Reading to Learn Mathematics,** *Chapter 4 Resource Masters*
  - **Vocabulary PuzzleMaker (online)**
  - **Teaching Mathematics with Foldables**
  - **Reading and Writing in the Mathematics Classroom**
  - **WebQuests, Projects and Interdisciplinary Investigations**

For detailed information on these resources, see pp. T6–T7.

- **Recommended Outside Reading for Students**
  - **Nonfiction**
    - Gillings, Richard. *Mathematics in the Time of the Pharaohs.* Dover, ©1992.
      This book examines the development of Egyptian mathematics from its origins in practical and commercial computations to its applications involving proportions and linear equations. Students will learn to write, solve, and graph linear equations such as those described in Chapter 4.
  - **Fiction**
    - Merrill, Jean. *The Toothpaste Millionaire.* Houghton Mifflin, ©1974.
      The Toothpaste Millionaire is a funny story of how two friends solve real-world mathematics problems involved in starting a toothpaste business and making a profit. They use functions, which students will study in Lessons 4-6 and 4-7.

PROJECT CRISS[SM] Study Skill

A story frame can help students organize information contained in word problems. They can use story frames either to solve word problems or to help them develop word problems of their own. The sample story frame at the right describes the following word problem: *A car travels 74 miles less than a truck travels in one day. If the truck travels 252 miles, how far has the car traveled?* Have students use story frames to write equation and inequality word problems as they work through Chapter 4.

Things involved: truck, car
Setting: truck travels 252 miles; car travels 74 miles less
Goal: distance traveled by car
Steps:
1. Write an equation relating the distance traveled by the car to the distance traveled by the truck. $(t - 74 = c)$
2. Insert t, the distance traveled by the truck, 252 miles. $(252 - 74 = c)$
3. Solve for c, the distance traveled by the car.

Solution: $c = 178$ miles

Check your solution. (Is it reasonable? Recheck calculations.)

CReating **I**ndependence through **S**tudent-owned **S**trategies

For more information on Project CRISS, see page T9.

Chapter 4 148F

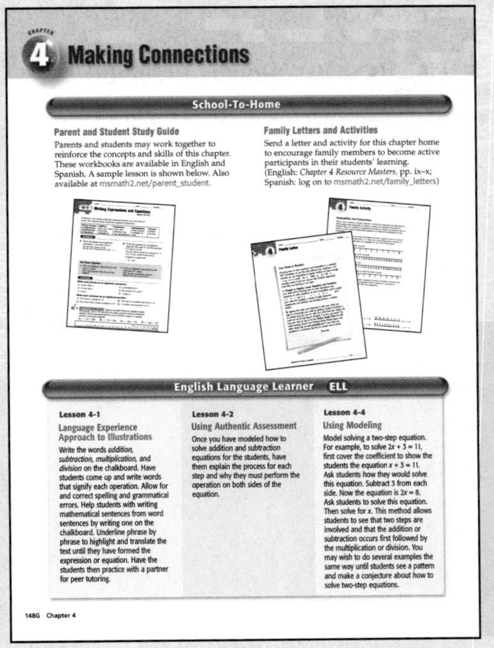

Chapter 4 Making Connections

School-To-Home

Parent and Student Study Guide
Parents and students may work together to reinforce the concepts and skills of this chapter. These workbooks are available in English and Spanish. A sample lesson is shown below. Also available at msmath2.net/parent_student.

Family Letters and Activities
Send a letter and activity for this chapter home to encourage family members to become active participants in their students' learning. (English: *Chapter 4 Resource Masters,* pp. ix–x; Spanish: log on to msmath2.net/family_letters)

English Language Learner ELL

Lesson 4-1
Language Experience Approach to Illustrations
Write the words *addition, subtraction, multiplication,* and *division* on the chalkboard. Have students come up and write words that signify each operation. Allow for and correct spelling and grammatical errors. Help students with writing mathematical sentences from word sentences by writing one on the chalkboard. Underline phrase by phrase to highlight and translate the text until they have formed the expression or equation. Have the students then practice with a partner for peer tutoring.

Lesson 4-2
Using Authentic Assessment
Once you have modeled how to solve addition and subtraction equations for the students, have them explain the process for each step and why they must perform the operation on both sides of the equation.

Lesson 4-4
Using Modeling
Model solving a two-step equation. For example, to solve $2x + 3 = 11$, first cover the coefficient to show the students the equation $x + 3 = 11$. Ask students how they would solve this equation. Subtract 3 from each side. Now the equation is $2x = 8$. Ask students to solve this equation. Then solve for x. This method allows students to see that two steps are involved and that the addition or subtraction occurs first followed by the multiplication or division. You may wish to do several examples the same way until students see a pattern and make a conjecture about how to solve two-step equations.

148G Chapter 4

Implementing CRISS Strategies

Project CRISS Study Skills were developed with leaders from Project CRISS to facilitate the teaching of each chapter of *Glencoe Mathematics: Applications and Concepts*. These strategies appear in the interleaf of the Teacher Wraparound Edition for each chapter.

For more information on Project CRISS℠, visit www.projectcriss.com.

English Language Learners ELL

English language learners may need specialized help in overcoming a language barrier to learn mathematics. Hands-on activities, modeling, working in flexible groups, and vocabulary building activities are particularly helpful to ELL students. Suggested strategies appear in the interleaf of the Teacher Wraparound Edition for each chapter.

With these TOOLS, you'll always know

Whether you need daily intervention resources integrated right into the program, or supplemental materials for after school and summer school programs, *Glencoe Mathematics: Applications and Concepts* puts it all right at your fingertips!

Diagnosis

Glencoe's *Diagnostic and Placement Tests* help you identify the key mathematical objectives that students are struggling with so you can make course placement decisions more effectively. A list of intervention resources is provided for each Glencoe program including *Glencoe Mathematics: Applications and Concepts.*

Prerequisite Skills

Students often struggle in mathematics because they have not mastered the prerequisite skills needed to be successful. *Glencoe Mathematics: Applications and Concepts* provides several opportunities to check student skills and determine if they need additional review and practice.

- The **Prerequisite Skills** at the beginning of every chapter help students identify and practice skills they'll need for each new concept.
- Additional **prerequisite skills** practice is provided at the end of each lesson and includes page references to help students get extra review whenever they need it. More prerequisite skill practice appears in the Student Handbook section at the back of the Student Edition.
- The **Math Skills Maintenance** masters provide extra practice on basic skills needed for success in mathematics.

Daily Intervention Opportunities

Guide to Daily Intervention offers suggestions for using Glencoe materials to intercept students who are having difficulties and prescribe a system of reinforcement to promote student success.

The **Chapter Resource Masters** include several types of worksheets that can be used for daily intervention in each lesson. For a description of each worksheet, see page T4.

- **Study Guide and Intervention***
- **Practice: Skills***
- **Reading to Learn Mathematics**

Each of these types of worksheets is available as a consumable workbook.

**These workbooks are also available in Spanish.*

The **Student Edition** contains additional problems in the Student Handbook to help students master each lesson.

- **Extra Practice** provides additional, immediate practice with the concepts from each lesson.
- **Mixed Problem Solving** includes numerous verbal problems for students to reinforce their problem-solving skills.

who needs EXTRA HELP.

And you'll be able to *DELIVER* it.

Technology Resources for Intervention

In addition to print resources, Glencoe offers a variety of timesaving technology tools to help students build their math skills more effectively.

MathPASS: Tutorial Plus CD-ROM provides an interactive, self-paced tutorial for a complete middle school mathematics curriculum. The lessons are correlated directly to *Glencoe Mathematics: Applications and Concepts.* Each lesson, or concept, includes a pretest, tutorial, guided practice, and posttest. Students' answers to the pretests automatically determine whether they need the tutorial for each concept, so students can take responsibility for their own learning—without taking teacher time for grading.

Additional Teacher Resources

The following materials are available to help you determine which students need intervention and allow you to develop strategies for giving students the help they need. For a description of each feature, see page T5.

- **5-Minute Check Transparencies with Standardized Test Practice**
- **Daily Intervention** features in the Teacher Wraparound Edition
- **Closing the Gap for Absent Students**

Online Study Tools include comprehensive review and intervention tools that are available anytime, anyplace simply by logging on to

msmath2.net

Self-check quizzes are available for every lesson, and immediate feedback helps students check their progress and find specific pages and examples in the Student Edition whenever they need extra review. These Online Study Tools also include extra examples, chapter tests, standardized test practice, and vocabulary review.

Give ASSESSMENT the extra attention it deserves, without the extra prep time.

***Glencoe Mathematics: Applications and Concepts* gives you all the tools you need to prepare students for success—including Standardized Test Practice prepared exclusively for Glencoe by The Princeton Review and the powerful ExamView® Pro CD-ROM.**

Student Edition

Glencoe has partnered with *The Princeton Review* to ensure that all of our materials meet state and national standards. Every lesson contains Standardized Test Practice questions, and every chapter contains a completely worked-out standardized test example as well as two full pages of Standardized Test Practice with Test-Taking Tips.

Chapter Study Guide and Review provides Vocabulary and Concept review—a Glencoe exclusive—and Lesson-by-Lesson Exercises and Examples—all at the point of use for students.

A **Mid-Chapter Practice Test** and a **Practice Test** for each chapter provide the variety of practice questions students need to succeed on tests.

Teacher Wraparound Edition

An **Open-Ended Assessment** activity is provided in each lesson in the margin of the Teacher Wraparound Edition.

Teacher Classroom Resources

5-Minute Check Transparencies with Standardized Test Practice provide full-size transparencies with questions covering the previous lesson or chapter. Standardized Test Practice questions are also included.

Assessment Options in the Chapter Resource Masters These assessment resources are available for each chapter.

- 6 Chapter Tests
- Extended-Response Assessment with Scoring Rubric
- Vocabulary Test and Review—a Glencoe exclusive!
- 4 Quizzes
- Mid-Chapter Test
- Cumulative Review
- 2-page Standardized Test Practice

Unit Tests, Semester Tests, and a **Final Test** are also available at point of use in the Chapter Resource Masters.

Technology Support

Use the networkable **ExamView® Pro Test Bank CD-ROM** to:

- Create **Multiple versions** of tests
- Create **Modified** tests for *Inclusion* students with one mouse click
- **Edit** existing questions and **Add** your own questions.
- Build tests aligned with state standards using built-in **State Curriculum Correlations**
- Change **English** tests to **Spanish** with one mouse click and vice versa

MindJogger Videoquizzes present chapter-by-chapter review sessions in a game show format.

Online Study Tools

- Extra Examples
- Self-Check Quizzes
- Vocabulary Review
- Chapter Test Practice
- Standardized Test Practice
- NAEP 2005 Practice

A Comprehensive, *research-based* curriculum that *engages* students and builds mathematical understanding.

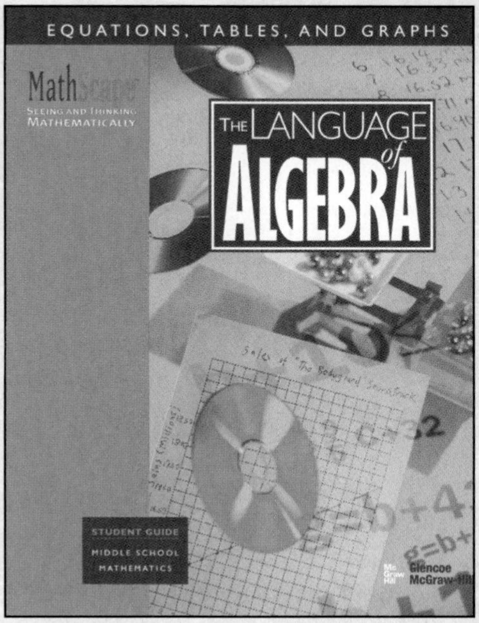

This unique, comprehensive curriculum encourages students to learn mathematics by doing mathematics, by using and connecting mathematical ideas, and by actively increasing their understanding. *MathScape* was developed by the Seeing and Thinking Mathematically Project at the Education Development Center.

- Twenty-one units with rigorous mathematics in engaging projects and investigations that may be used to enhance chapters in *Glencoe Mathematics: Applications and Concepts.*
- The program combines classroom strategies, group activities, cooperative learning situations, and homework.
- MathScape emphasizes mathematical thinking in addition to equations and problems.

The interleaf pages of each chapter in the Teacher Wraparound Edition highlight units from *MathScape* that are appropriate for that chapter. The chart below summarizes the mathematical focus of each unit of MathScape.

	Unit	Title	Mathematical Focus
Course 1	1	What Does the Data Say?	Graphs and Averages
	2	The Language of Numbers	Inventing and Comparing Number Systems
	3	From Zero to One and Beyond	Fractions, Decimals, and Percents
	4	Designing Spaces	Visualizing, Planning, and Building
	5	Number Powerhouse	Operating with Fractions, Decimals, and Percents
	6	Gulliver's Worlds	Measuring and Scaling
	7	Patterns in Numbers and Shapes	Using Algebraic Thinking
Course 2	1	Buyer Beware	Rates, Ratios, Percents, and Proportions
	2	Chance Encounters	Probability in Games and Simulations
	3	Making Mathematical Arguments	Generalizing about Numbers
	4	From the Ground Up	Modeling, Measuring, and Constructing Homes
	5	The Language of Algebra	Equations, Tables, and Graphs
	6	Getting Down to Business	Functions and Spreadsheets
	7	Getting in Shape	Exploring Two-Dimensional Figures
Course 3	1	Looking Behind the Numbers	Correlations, Rankings, and Permutations
	2	Mathematics of Motion	Distance, Speed, and Time
	3	Shapes and Space	Thinking Three-Dimensionally
	4	What Comes Next?	Modeling and Predicting
	5	Exploring the Unknown	Writing and Solving Equations
	6	Roads and Ramps	Slopes, Angles, and Ratios
	7	Family Portraits	Comparing Function Families

Introducing our new partner

USA TODAY® Education

USA TODAY Snapshots®

This is the same up-to-date data you know so well. But now, in an exclusive partnership with Glencoe/McGraw-Hill, USA TODAY® Education has brought its powerful one-of-a-kind perspective and dynamic content to the pages of *Glencoe Mathematics: Applications and Concepts*. USA TODAY Snapshots® explode off the page to make mathematics topics come alive with current, relevant data.

www.education.usatoday.com USA TODAY® K–12 Education's website offers resources and interactive features connected to each day's newspaper. *Experience Today*, USA TODAY®'s daily lesson plan, is available on the site and delivered daily to subscribers. This plan provides instruction for integrating USA TODAY® graphics and key editorial features into your mathematics classroom.

Stay current with additional charts and graphs with USA TODAY®. Log on to www.education.usatoday.com, or call USA TODAY at (800) 757-TEACH.

6. **SCHOOLS** The line graph shows public school teachers' salaries for the past few years. If the trend continues, about what will the average annual salary be in 2005?

7. **RESEARCH** Use the Internet or another source to find a real-life example of a line graph. Write a description of what the graph displays and extend the graph to show where the data will be in the future.

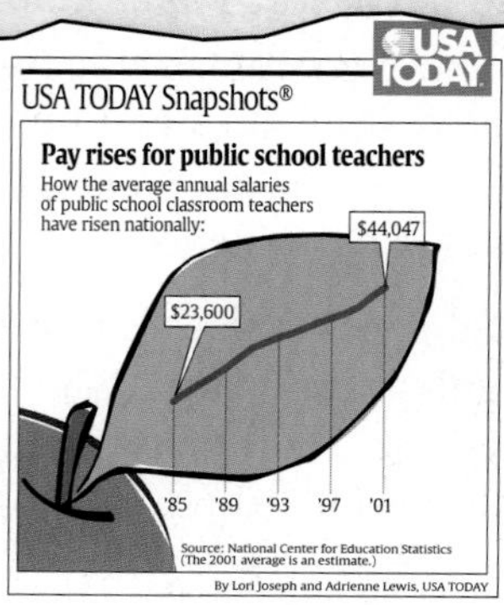

OLYMPICS For Exercises 8 and 9, use the table below. It shows the number of nations that participated in the Summer Olympics from 1948 to 2000.

Year	Number of Nations	Year	Number of Nations
1948	59	1976	92
1952	69	1980	80
1956	72	1984	140
1960	83	1988	159
1964	93	1992	169
1968	112	1996	197
1972	121	2000	199

Source: www.factmonster.com

8. Make a line graph of the data.
9. Predict the number of nations that will participate in the 2008 Summer Olympics.

For E
relati
requi
10. M
h
11. D
12. P

62 Chapt

Practice and Applications

Write each percent as a decimal.

12. 27%
13. 70%
14. 6%
15. 1%
16. 18.5%
17. 2.2%
18. $15\frac{1}{2}$%
19. $30\frac{1}{4}$%

HOMEWORK HELP	
For Exercises	See Examples
12–19, 34–37	1, 2, 3, 4
20–33	5, 6, 7, 8

Extra Practice See pages 576, 600.

Write each decimal as a percent.

20. 0.95
21. 0.08
22. 0.17
23. 0.6
24. 0.675
25. 0.145
26. 0.012
27. 0.7025

TENNIS For Exercises 28–32, use the graphic at the right. Write each decimal as a percent for the following players.

28. Monica Seles
29. Martina Hingis
30. Venus Williams
31. Jennifer Capriati
32. Based on her winning percentage, which player wins 413 out of every 500 games played?

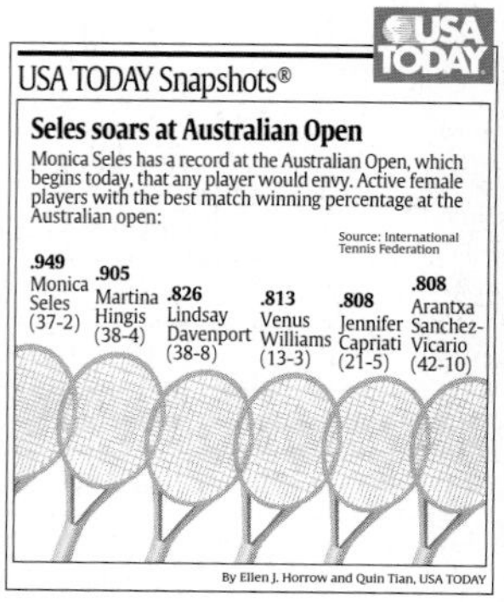

222 **Chapter 5** Fractions, Decimals, and Percents

The INTERNET: One TOOL. Endless possibilities.

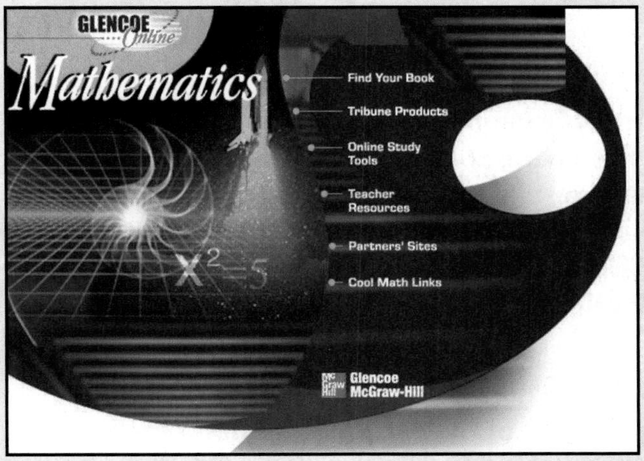

Many of your students may already be familiar with the Internet, but may not have discovered the full potential of this powerful research tool. With *Glencoe Mathematics: Applications and Concepts*, your students can use the Internet to build their math skills. And you can access a wide variety of resources to help you plan classes, extend lessons, even meet professional development requirements.

For Students

Online Study Tools, referenced on the Student Edition pages, are keyed specifically to *Glencoe Mathematics: Applications and Concepts.*

- msmath2.net/extra_examples features additional fully worked-out examples.
- msmath2.net/self_check_quiz allows students to check their progress in each lesson.
- msmath2.net/vocabulary_review lets students check their vocabulary comprehension.
- msmath2.net/chapter_test provides additional practice in test taking.
- msmath2.net/standardized_test simulates questions that appear on standardized and proficiency tests.

Other Online Resources

- msmath2.net/webquest offers an online research project.
- msmath2.net/data_update features links to updated statistical data presented in exercises.
- msmath2.net/careers offers information about career opportunities.
- msmath2.net/other_calculator_keystrokes provides keystroke instructions for various calculators to accompany graphing calculator activities and exercises in the Student Edition.

For teachers

Powerful tools to make your job easier

- Classroom Math Games
- Key Concepts: Mathematical Background and Teaching Notes
- Science and Mathematics Lab Manual
- Real-Life Career Masters
- Problem of the Week Activities
- Family Letters and Family Activities, Spanish versions
- USA TODAY® K–12 Education daily lesson plans
- Bulletin Board
- Activities, Games, Puzzles, and Software Links
- State and National Resources
- The Princeton Review

Staff Development sites

- NCTM links
- Teaching Today link
- McGraw-Hill Learning Network link
- Cooperative learning suggestions
- Using the Internet in the Mathematics Classroom

For parents—

Sites to help parents get involved with their child's learning

- Parent and Student Study Guide Workbook
- Involving Parents and Community in the Mathematics Classroom

It's Staff Development,

As professional development continues to take on greater importance for educators across the country, teachers are constantly looking for easy-to-use tools to help them stay abreast of current trends and issues. At Glencoe, we know how valuable your time is, so we've developed a variety of staff development tools to help you meet your district's requirements.

Teacher Wraparound Edition

Continuity of Instruction and Mathematical Background found at the beginning of each chapter gives you an overview of the mathematics in each lesson. Information about prior knowledge as well as future connections lets you see the how the mathematics your students are learning fits into the whole of their education.

Tips for New Teachers offers helpful suggestions for such things as classroom management, assessment, teaching techniques, and more.

Teaching Tips can be found on the reduced student pages at point of use.

Teacher Classroom Resources

Glencoe Mathematics Staff Development Series is a series of publications that allows you to stay current with issues that affect your teaching effectiveness. The series is intended to help you implement new mathematics strategies and enhance your classroom performance.

Available in print:

- *Using the Internet in the Mathematics Classroom*
- *Reading and Writing in the Mathematics Classroom*
- *Teaching Math with Manipulatives*
- *Teaching Mathematics with Foldables™*

Available online at
math.glencoe.com

- *Key Concepts: Mathematical Background and Teaching Notes*
- *Graphing Calculators in the Mathematics Classroom*
- *Cooperative Learning in the Mathematics Classroom*
- *Alternative Assessment in the Mathematics Classroom*
- *Involving Parents and the Community in the Mathematics Classroom*

made convenient.

Technology Support

At math.glencoe.com, you'll find:

- a Staff Development site that addresses current issues in education.
- a Bulletin Board that allows teachers to discuss issues and ideas with colleagues.
- a State and National Resources site that links to math and math education resources, nationally and by state.

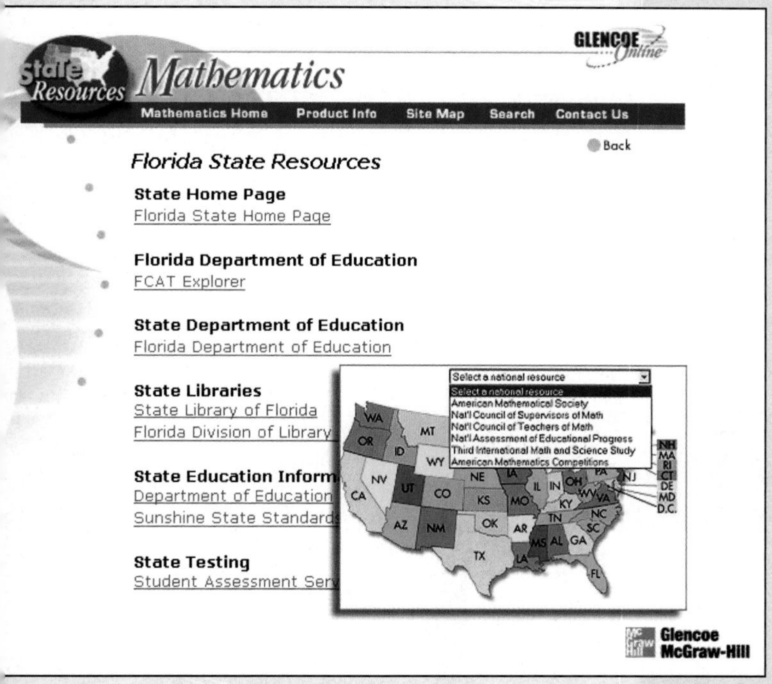

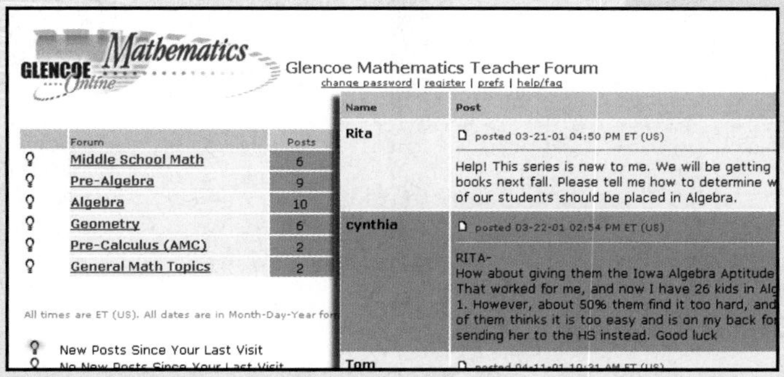

McGraw-Hill Professional Development

McGraw-Hill Professional Development is dedicated to providing outstanding professional development for teachers and school administrators through innovative training, resources, and services.

Mathematics, YES! Online is a web-based professional development program that is self-paced and flexible. The content units are based on major concepts in mathematics curriculum and are standards-based. Actual classroom videos provide the springboard for discussion and activities. University credit for *Mathematics, YES!* can be achieved at school or home at your convenience and pace.

The interleaf pages of each chapter in the Teacher Wraparound Edition highlight units from *Mathematics, YES!* that are appropriate for that chapter.

Mathematics, YES! Table of Contents

Unit 1: Number Sense
Unit 2: Patterns and Functions
Unit 3: Integers and Order of Operations
Unit 4: Fractions, Decimals, and Percentages
Unit 5: Coordinate Pairs
Unit 6: Ratio and Proportional Reasoning
Unit 7: Equations
Unit 8: Geometry and Measurement of Plane Figures
Unit 9: Geometry and Measurement of Solid Figures
Unit 10: Probability, Data Analysis, and Statistics

For more information on *Mathematics, YES! Online*, contact McGraw-Hill Professional Development at (866) 501-6997 or by e-mail at mhpdonline.com.

Glencoe Mathematics Programs

Glencoe's Mathematics programs are the product of ongoing classroom and educational research activities involving students, teachers, curriculum supervisors, administrators, parents, and college-level mathematics educators, mathematicians, and researchers.

Glencoe programs are currently used by millions of students and tens of thousands of teachers. A key reason why Glencoe mathematics programs help to ensure success in the mathematics classroom is because Glencoe author teams are a combination of practicing classroom teachers, curriculum supervisors, and college professors.

SOUND

Prior to the publication of any Glencoe Mathematics program, the following initial research is completed.

- Analyze national and state changes and trends such as graduation requirements, standardized test exams, and the latest NCTM and NAEP reports.
- Incorporate the most current and applicable educational research where reported results show significant improvement on student learning and achievement.
- Evaluate returns from independently contracted mailings and telephone surveys.
- Review all comments and correspondences from teachers, parents, and students pertaining to appropriate prior editions. The teachers' comments help Glencoe to build in staff development support, which makes the programs easy to implement from the first day of use.

Survey Results Responding *Very Satisfied* or *Satisfied* with Glencoe *Middle School Mathematics**

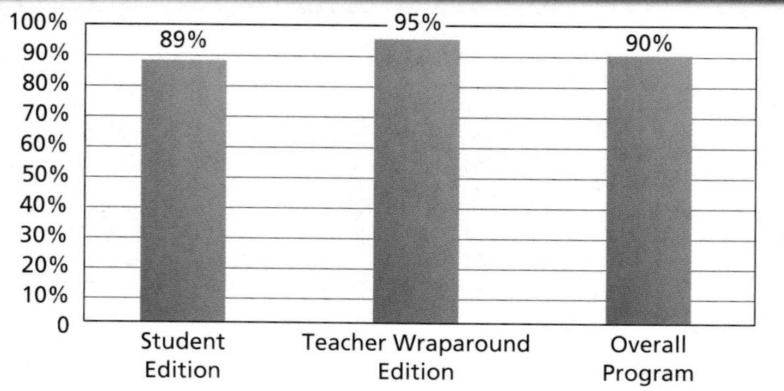

*Ratings for all other middle school math programs were 73%, 84%, and 74%, respectively.

Source: 2001 Middle School Longitudinal Study

Research-based and tested *to ensure success*

PROVEN

Prior to the publication of *Glencoe Mathematics: Applications and Concepts*, the following extensive research was conducted using manuscript and pre-publication versions of the program.

- Glencoe conducts nationwide discussion groups involving mathematics teachers, department chairpersons, supervisors, educational learning specialists, and students.
- Glencoe conducts face-to-face interviews with mathematics teachers.
- Reviewers and consultants reviewed Glencoe manuscripts for accuracy, content development, and thoroughness. The Princeton Review, a leader in test expertise, reviews the Glencoe assessment to ensure test validity.
- Glencoe conducts Learner Verification Research in which students and teachers use pre-publication manuscripts in the classroom. The results, compiled by an independent contractor, for the *Mathematics: Applications and Concepts* are summarized below.

Top-Line Results

- The research indicates that test scores significantly increased among students using *Mathematics: Applications and Concepts*, ©2004.
- Roughly three out of four students earned higher scores after using the Glencoe program.
- The program was equally effective with:
  - boys and girls; and
  - minority and non-minority students.
- Overall, the gap between the average pre-test score and a perfect score closed by 27%. Stated differently, on average, scores increased 29% after students used the Glencoe program.

In addition to pre-publication field testing, Glencoe engages in full year post-publication testing through an independent contractor.

EFFECTIVE

What do teachers say? Here are some testimonials from Learner Verification Research teachers.

- *great emphasis placed on vocabulary, writing, and problem solving*
- *word problems and applications are very up-to-date*
- *user-friendly*
- *takes material at a perfect pace*
- *allows for individualization by teacher or student*

For more details on Glencoe's research, please contact us at math.glencoe.com.

Planning YOUR Mathematics *Course*

Glencoe Mathematics: Applications and Concepts and the accompanying support materials allow you to create a mathematics course that meets the needs of each class of students. The charts shown on these two pages offer general suggestions for pacing your students through the book. Pacing for both standard class periods and block schedule class periods is given.

The total number of days in each level of pacing is less than the typical 180-day school year and 90-day semester to allow for flexibility in planning due to testing, school cancellation, or shortened class periods.

You may want to use one of the Chapter Test 2C provided in the *Chapter Resource Masters* as a pretest to determine how well your students are prepared for each chapter. If you find that they are well prepared, consider using the Study Guide and Review at the end of the chapter as a one-day lesson and proceed to the next chapter.

Year-Long Schedule
45–50-minute periods

Grading Period	Chapter	Days
1	1	13
	2	14
	3	12
2	4	12
	5	12
	6	13
3	7	15
	8	11
	9	12
4	10	18
	11	13
	12	13
	Total	158

Block Schedule
90-minute periods

Chapter	Days
1	7
2	7
3	6
4	6
5	6
6	7
7	8
8	6
9	6
10	9
11	7
12	7
Total	82

Daily Planning

TeacherWorks ***All in One Lesson Planner and Resource Center***

This CD-ROM includes a lesson planner and an Interactive Teacher Edition enabling you to customize an entire course of study to meet your specific needs.

NCTM Standards for School Mathematics

The Standards portion of the *NCTM Principles and Standards for School Mathematics* center upon ten areas of mathematics curriculum development. The number assigned to each standard is for easy reference and is not part of each standard's official title.

Instructional programs from pre-kindergarten through grade 12 should enable all students to—

1 Number and Operations

- Understand numbers, ways of representing numbers, relationships among numbers, and number systems
- Understand the meaning of operations and how they relate to each other
- Compute fluently and make reasonable estimates

Lessons: 1-1–1-9, 2-4, 2-5, 3-1–3-7, 4-1–4-7, 5-1–5-8, 6-1–6-9, 7-1–7-8, 8-1–8-6, 9-1–9-7, 10-1–10-9, 11-1–11-8, 12-1–12-6

2 Algebra

- Understand patterns, relations, and functions
- Represent and analyze mathematical situations and structures using algebraic symbols
- Use mathematical models to represent and understand quantitative relationships
- Analyze change in various contexts

Lessons: 1-3–1-7, 1-9, 3-1–3-7, 4-1–4-4, 4-6, 4-7, 5-1–5-8, 6-5, 6-6, 6-8, 6-9, 7-3–7-5, 7-7, 7-8, 8-2–8-6, 9-3–9-5, 9-7, 10-1–10-9, 11-1–11-8, 12-1–12-6

3 Geometry

- Analyze characteristics and properties of two- and three-dimensional geometric shapes and develop mathematical arguments about geometric relationships
- Specify locations and describe spatial relationships using coordinate geometry and other representational systems
- Apply transformations and use symmetry to analyze mathematical situations
- Use visualization, spatial reasoning, and geometric modeling to solve problems

Lessons: 3-3, 4-6, 4-7, 6-8, 6-9, 11-3–11-8, 12-3

4 Measurement

- Understand measurable attributes of objects and the units, systems, and processes of measurement
- Apply appropriate techniques, tools, and formulas to detemine measurements

Lessons: 1-8, 2-4, 2-6, 6-2–6-4, 6-7–6-9, 7-1–7-4, 9-1–9-4, 11-4–11-8

5 Data Analysis and Probability

- Formulate questions that can be addressed with data and collect, organize, and display relevant data to answer them
- Select and use appropriate statistical methods to analyze data
- Develop and evaluate inferences and predictions that are based on data
- Understand and apply basic concepts of probability

Lessons: 2-1–2-8, 8-3

Instructional programs from pre-kindergarten through grade 12 should enable all students to—

6 Problem Solving

- Build new mathematical knowledge through problem solving
- Solve problems that arise in mathematics and in other contexts
- Apply and adapt a variety of appropriate strategies to solve problems
- Monitor and reflect on the process of mathematical problem solving

Lessons: 1-1–1-9, 2-2, 3-1–3-7, 4-1–4-7, 5-1–5-8, 6-1–6-9, 7-2–7-5, 7-7, 7-8, 8-1–8-6, 9-2–9-5, 9-7, 10-1–10-9, 11-1–11-8, 12-1–12-6

7 Reasoning and Proof

- Recognize reasoning and proof as fundamental aspects of mathematics
- Make and investigate mathematical conjectures
- Develop and evaluate mathematical arguments and proofs
- Select and use various types of reasoning and methods of proof

Lessons: 1-1–1-9, 2-2, 2-7, 2-8, 3-1–3-7, 4-1–4-7, 6-1–6-9, 8-1–8-6, 11-1–11-8, 12-1–12-6

8 Communication

- Organize and consolidate their mathematical thinking through communication
- Communicate their mathematical thinking coherently and clearly to peers, teachers, and others
- Analyze and evaluate the mathematical thinking and strategies of others
- Use the language of mathematics to express mathematical ideas precisely

Lessons: 1-1–1-9, 3-1–3-7, 4-1–4-7, 6-1–6-9, 8-1–8-6, 12-1–12-6

9 Connections

- Recognize and use connections among mathematical ideas
- Understand how mathematical ideas build on one another to produce a coherent whole
- Recognize and apply mathematics in contexts outside of mathematics

Lessons: 1-1–1-9, 2-6, 2-8, 3-1–3-7, 4-1–4-7, 5-1–5-8, 6-1–6-9, 7-1–7-8, 8-1–8-6, 9-1–9-7, 10-1–10-9, 11-1–11-8, 12-1–12-6

10 Representation

- Create and use representations to organize, record, and communicate mathematical ideas
- Select, apply, and translate among mathematical representations to solve problems
- Use representations to model and interpret physical, social, and mathematical phenomena

Lessons: 1-1–1-9, 2-1–2-8, 3-1–3-7, 4-1–4-7, 5-1–5-8, 6-1–6-9, 7-1–7-8, 8-1–8-6, 9-1–9-7, 10-1–10-9, 11-1–11-8, 12-1–12-6

NAEP 2005 Standards

The National Assessment of Educational Progress (NAEP) is the leading national assessment of mathematics and many other academic subjects for American students. For a correlation to the standards to be tested on the NAEP 2005 test, see the interleaf pages in the Teacher Wraparound Edition for each chapter. For a listing of the NAEP 2005 standards, visit **math.glencoe.com**.

UNIT 1

Decimals, Algebra, and Statistics

Decimals, Algebra, and Statistics 2

Chapter 1 **Decimal Patterns and Algebra** 4

WebQuest INTERDISCIPLINARY PROJECT
- Math and Science **3**

1-1 A Plan for Problem Solving **6**
1-2 Powers and Exponents **10**
1-3 Order of Operations **14**
1-4 **Algebra:** Variables and Expressions **18**
1-5a **Problem-Solving Strategy:** Guess and Check **22**
1-5 **Algebra:** Equations **24**
Mid-Chapter Practice Test **28**
The Game Zone: Evaluating Expressions **29**
1-6 **Algebra:** Properties **30**
1-7 Sequences **34**
1-7b **Hands-On Lab:** Exploring Sequences **37**
1-8 **Measurement:** The Metric System **38**
Study Skill: How to Take Good Notes **42**
1-9 Scientific Notation **43**
Study Guide and Review **46**
Practice Test **49**
Standardized Test Practice **50**

Lesson 1-2, p. 13

Prerequisite Skills
- Getting Started 5
- Getting Ready for the Next Lesson 9, 13, 17, 21, 27, 33, 36, 41

FOLDABLES™ Study Organizer 5

Study Skills
- Study Tips 7, 11, 15, 25, 31, 38, 44
- Homework Help 9, 12, 16, 20, 26, 32, 36, 40, 45

Reading and Writing Mathematics
- Link to Reading 30, 34
- Reading Math 10, 18, 19, 39
- Writing Math 8, 11, 16, 20, 26, 32, 35, 37, 40, 44

Standardized Test Practice
- Multiple Choice 9, 13, 17, 21, 23, 27, 28, 33, 41, 45, 49, 50
- Short Response/Grid In 9, 13, 21, 27, 28, 33, 36, 41, 45, 51
- Extended Response 51
- Worked-Out Example 25

USA TODAY Snapshots 6, 17

UNIT 1

Chapter 2 Statistics: Analyzing Data 52

2-1 Frequency Tables 54
2-2a **Problem-Solving Strategy:** Use a Graph 58
2-2 Making Predictions 60
2-3 Line Plots 64
2-4 Mean, Median, and Mode 69
2-4b **Hands-On Lab:** Are You Average? 73
Mid-Chapter Practice Test 74
The Game Zone: Finding the Mean 75
2-5 Stem-and-Leaf Plots 76
2-6 Box-and-Whisker Plots 80
2-6b **Graphing Calculator Investigation:** Box-and-Whisker Plots 84
2-7 Bar Graphs and Histograms 85
2-7b **Spreadsheet Investigation:** Multiple-Line and Multiple-Bar Graphs 90
2-8 Misleading Statistics 92
Study Guide and Review 96
Practice Test 99
Standardized Test Practice 100

Prerequisite Skills
- Getting Started 53
- Getting Ready for the Next Lesson 57, 63, 68, 72, 79, 83, 89

FOLDABLES™ Study Organizer 53

Study Skills
- Study Tips 65, 70, 81, 86
- Homework Help 56, 62, 67, 72, 78, 83, 88, 94

Reading and Writing Mathematics
- Link to Reading 64, 69
- Reading Math 61, 81
- Writing Math 56, 61, 66, 71, 73, 78, 83, 87, 93

Standardized Test Practice
- Multiple Choice 57, 59, 63, 68, 72, 74, 79, 83, 89, 95, 99, 100
- Short Response/Grid In 63, 72, 74, 79, 83, 89, 95, 101
- Extended Response 101
- Worked-Out Example 87

USA TODAY Snapshots 62

WebQuest INTERDISCIPLINARY PROJECT 95

Lesson 2-2a, p. 59

UNIT 2

Integers and Algebra 102

WebQuest INTERDISCIPLINARY PROJECT
- Math and Geography **103**

Chapter 3 Algebra: Integers 104

3-1 Integers and Absolute Value **106**
3-2 Comparing and Ordering Integers **109**
3-3 The Coordinate Plane **112**
Mid-Chapter Practice Test **116**
The Game Zone: Graphing Points on a Coordinate Plane **117**
3-4a Hands-On Lab: Adding Integers **118**
3-4 Adding Integers **120**
Study Skill: How to Take Good Notes **125**
3-5a Hands-On Lab: Subtracting Integers **126**
3-5 Subtracting Integers **128**
3-6a Problem-Solving Strategy: Look for a Pattern **132**
3-6 Multiplying Integers **134**
3-7 Dividing Integers **138**
Study Guide and Review **142**
Practice Test **145**
Standardized Test Practice **146**

Lesson 3-5, p. 129

Prerequisite Skills
- Getting Started 105
- Getting Ready for the Next Lesson 108, 111, 115, 124, 131, 137

FOLDABLES Study Organizer 105

Study Skills
- Study Tips 106, 112, 120, 121, 129, 135
- Homework Help 108, 111, 114, 123, 130, 136, 140

Reading and Writing Mathematics
- Link to Reading 120
- Reading Math 107, 113, 118, 127, 139
- Writing Math 107, 110, 114, 119, 122, 127, 130, 136, 140

Standardized Test Practice
- Multiple Choice 108, 111, 115, 116, 124, 131, 133, 137, 141, 145, 146
- Short Response/Grid In 108, 111, 115, 116, 124, 131, 137, 141, 147
- Extended Response 147
- Worked-Out Example 110

UNIT 2

Chapter 4 Algebra: Linear Equations and Functions 148

4-1 Writing Expressions and Equations 150

Study Skill: How to Read Math Problems 153

4-2a Hands-On Lab: Solving Equations Using Models 154

4-2 Solving Addition and Subtraction Equations 156

4-3 Solving Multiplication Equations 160

4-4a Problem-Solving Strategy: Work Backward 164

4-4 Solving Two-Step Equations 166

Mid-Chapter Practice Test 170

The Game Zone: Solving Equations 171

4-5 Inequalities 172

4-6a Hands-On Lab: Functions and Graphs 176

4-6 Functions and Linear Equations 177

4-7 Lines and Slope 182

Study Guide and Review 186

Practice Test 189

Standardized Test Practice 190

Prerequisite Skills
- Getting Started 149
- Getting Ready for the Next Lesson 152, 159, 163, 169, 175, 181

FOLDABLES Study Organizer 149

Study Skills
- Study Tips 155, 157, 158, 167, 173, 178, 183
- Homework Help 152, 159, 162, 168, 174, 180, 184

Reading and Writing Mathematics
- Link to Reading 182
- Writing Math 151, 155, 158, 162, 168, 174, 176, 179, 184

Standardized Test Practice
- Multiple Choice 152, 159, 163, 165, 169, 170, 175, 181, 185, 189, 190
- Short Response/Grid In 159, 169, 170, 185, 191
- Extended Response 191
- Worked-Out Example 183

USA TODAY Snapshots 175, 180

WebQuest INTERDISCIPLINARY PROJECT 185

Lesson 4-3, p. 163

UNIT 3

Fractions

Fractions 192

Chapter 5 Fractions, Decimals, and Percents 194

5-1a **Hands-On Lab:** Exploring Factors 196
5-1 Prime Factorization 197
5-2a **Problem-Solving Strategy:** Make an Organized List 201
5-2 Greatest Common Factor 203
5-3 Simplifying Fractions 207
5-4 Fractions and Decimals 210
Mid-Chapter Practice Test 214
The Game Zone: Finding Factors 215
5-5 Fractions and Percents 216
5-6 Percents and Decimals 220
5-7 Least Common Multiple 224
5-8 Comparing and Ordering Rational Numbers 227
Study Guide and Review 232
Practice Test 235
Standardized Test Practice 236

WebQuest **INTERDISCIPLINARY PROJECT**
- Math and Health **193**

Lesson 5-4, p. 210

Prerequisite Skills
- Getting Started 195
- Getting Ready for the Next Lesson 200, 206, 209, 213, 219, 223, 226

FOLDABLES Study Organizer 195

Study Skills
- Study Tips 198, 204, 207, 211, 221, 225
- Homework Help 199, 205, 209, 212, 218, 222, 226, 230

USA TODAY Snapshots 209, 222

Reading and Writing Mathematics
- Link to Reading 210
- Reading Math 217
- Writing Math 196, 199, 205, 208, 212, 218, 222, 225, 229

Standardized Test Practice
- Multiple Choice 200, 202, 206, 209, 213, 214, 219, 223, 226, 231, 235, 236
- Short Response/Grid In 200, 206, 209, 213, 214, 219, 223, 226, 231, 237
- Extended Response 237
- Worked-Out Example 229

UNIT 3

Chapter 6 Applying Fractions 238

6-1 Estimating with Fractions 240
6-2 Adding and Subtracting Fractions 244
6-3 Adding and Subtracting Mixed Numbers 248
6-3b **Problem-Solving Strategy:** Eliminate Possibilities 252
6-4 Multiplying Fractions and Mixed Numbers 254
6-5 **Algebra:** Solving Equations 258
Mid-Chapter Practice Test 262
The Game Zone: Multiplying Fractions 263
6-6 Dividing Fractions and Mixed Numbers 264
6-7 **Measurement:** Changing Customary Units 267
6-8 **Geometry:** Perimeter and Area 270
6-9a **Hands-On Lab:** Circumference 274
6-9 **Geometry:** Circles and Circumference 275
Study Guide and Review 278
Practice Test 281
Standardized Test Practice 282

Prerequisite Skills
- Getting Started 239
- Getting Ready for the Next Lesson 243, 247, 251, 257, 261, 266, 269, 273

FOLDABLES Study Organizer 239

Study Skills
- Study Tips 240, 241, 242, 255, 259, 271, 274, 276
- Homework Help 243, 246, 250, 256, 260, 266, 269, 272, 277

Reading and Writing Mathematics
- Link to Reading 258
- Writing Math 242, 246, 250, 256, 260, 266, 268, 272, 274, 276

Standardized Test Practice
- Multiple Choice 243, 247, 251, 253, 257, 261, 262, 269, 273, 277, 281, 282
- Short Response/Grid In 257, 261, 266, 269, 273, 277, 283
- Extended Response 283
- Worked-Out Example 259

USA TODAY Snapshots 261

WebQuest INTERDISCIPLINARY PROJECT 277

Lesson 6-2, p. 245

UNIT 4

Proportional Reasoning

Proportional Reasoning 284

Chapter 7 Ratios and Proportions 286

WebQuest INTERDISCIPLINARY PROJECT
- Math and Recreation 285

7-1 Ratios 288
7-2 Rates 292
7-2b **Hands-On Lab:** Rate of Change 296
7-3 Solving Proportions 297
7-3b **Hands-On Lab:** Wildlife Sampling 301
7-4a **Problem-Solving Strategy:** Draw a Diagram 302
7-4 Scale Drawings 304
7-4b **Spreadsheet Investigation:** Scale Drawings 309
Mid-Chapter Practice Test 310
The Game Zone: Solving Proportions 311
7-5 Fractions, Decimals, and Percents 312
7-6 Percents Greater Than 100% and Percents Less Than 1% 316
7-7 Percent of a Number 319
7-8a **Hands-On Lab:** Using a Percent Model 322
7-8 The Percent Proportion 323
Study Guide and Review 326
Practice Test 329
Standardized Test Practice 330

Lesson 7-8, p. 323

Prerequisite Skills
- Getting Started 287
- Getting Ready for the Next Lesson 291, 295, 300, 308, 315, 318, 321

FOLDABLES Study Organizer 287

Study Skills
- Study Tips 289, 293, 296, 305, 306, 313, 314, 316, 320, 324
- Homework Help 290, 294, 299, 307, 315, 318, 321, 325

Reading and Writing Mathematics
- Reading Math 317
- Writing Math 290, 294, 296, 299, 301, 306, 314, 317, 320, 322, 325

Standardized Test Practice
- Multiple Choice 291, 295, 300, 303, 308, 315, 318, 321, 325, 329, 330
- Short Response/Grid In 300, 308, 310, 315, 321, 325, 331
- Extended Response 331
- Worked-Out Example 293

USA TODAY Snapshots 320

UNIT 4

Chapter 8 Applying Percent

Chapter 8 Applying Percent 332

8-1 Percent and Estimation 334
8-1b **Problem-Solving Strategy:** Reasonable Answers 338
8-2 The Percent Equation 340
8-3a **Hands-On Lab:** Sampling 344
8-3 Using Statistics to Predict 345
Mid-Chapter Practice Test 348
The Game Zone: Percent of a Number 349
8-4 Percent of Change 350
8-5 Sales Tax and Discount 354
8-6 Simple Interest 358
8-6b **Spreadsheet Investigation:** Simple Interest 361
Study Guide and Review 362
Practice Test 365
Standardized Test Practice 366

Prerequisite Skills
- Getting Started 333
- Getting Ready for the Next Lesson 337, 343, 347, 353, 357

FOLDABLES™ Study Organizer 333

Study Skills
- Study Tips 335, 340, 341, 344, 355
- Homework Help 337, 342, 347, 352, 356, 360

Reading and Writing Mathematics
- Link to Reading 340
- Reading Math 359
- Writing Math 336, 342, 346, 352, 356

Standardized Test Practice
- Multiple Choice 337, 339, 343, 347, 348, 353, 357, 360, 365, 366
- Short Response/Grid In 347, 353, 357, 367
- Extended Response 367
- Worked-Out Example 355

USA TODAY Snapshots 343, 345

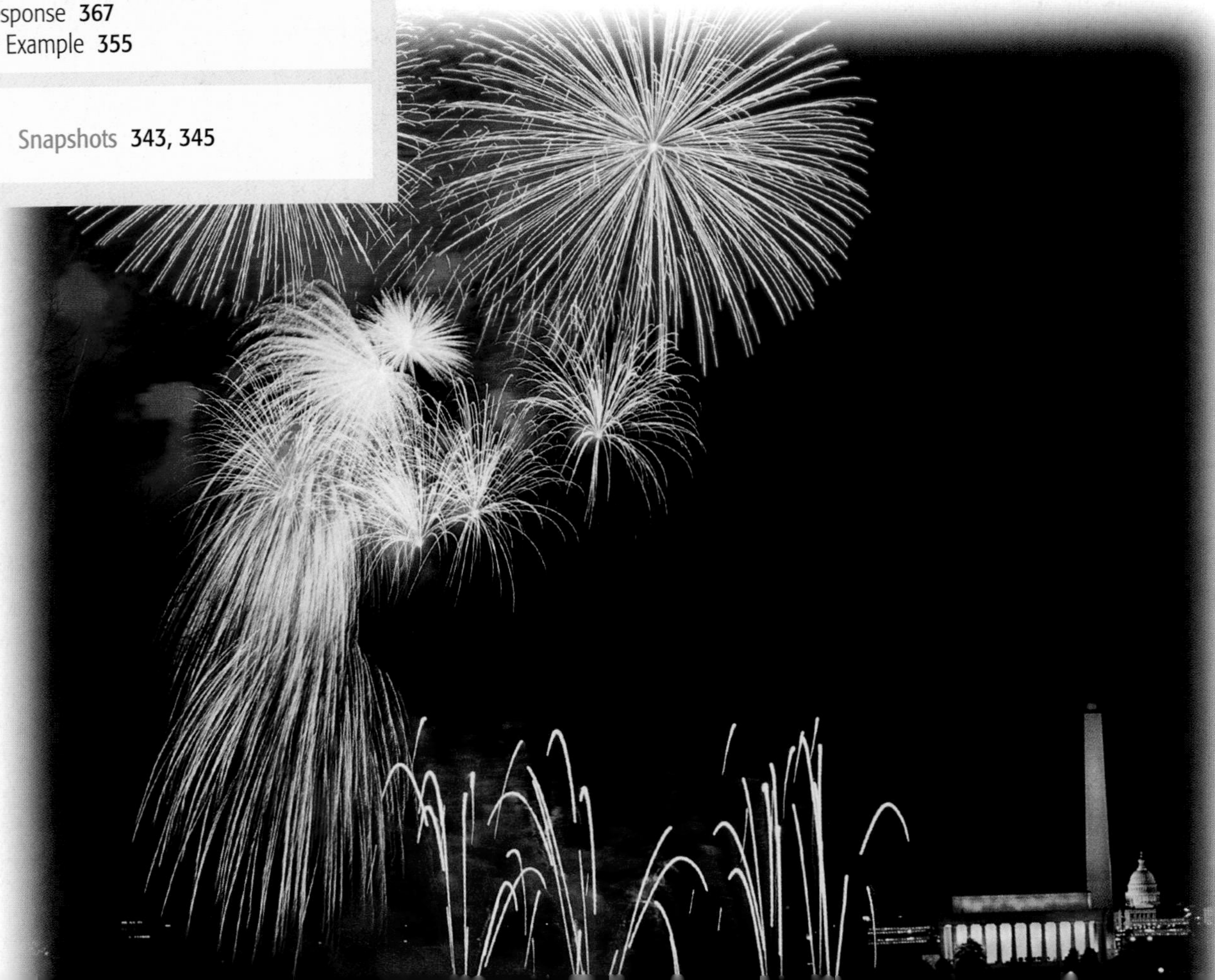

Lesson 8-1, p. 334

UNIT 4

Chapter 9 Probability

Chapter 9 **Probability** **368**

9-1 Simple Events **370**
9-2 Tree Diagrams **374**
9-3 The Fundamental Counting Principle **378**
9-4 Permutations **381**
Mid-Chapter Practice Test **384**
The Game Zone: Probability **385**
9-5a Hands-On Lab: Exploring Combinations **386**
9-5 Combinations **387**
9-6a Problem-Solving Strategy: Act It Out **391**
9-6 Theoretical and Experimental Probability **393**
9-6b Hands-On Lab: Experimental Probability **397**
9-7 Independent and Dependent Events **398**
Study Guide and Review **402**
Practice Test **405**
Standardized Test Practice **406**

Prerequisite Skills
- Getting Started 369
- Getting Ready for the Next Lesson 373, 377, 380, 383, 390, 396

FOLDABLES™ Study Organizer 369

Study Skills
- Study Tips 371, 382, 399
- Homework Help 372, 376, 380, 383, 389, 395, 400

Reading and Writing Mathematics
- Link to Reading 387
- Reading Math 398
- Writing Math 372, 376, 379, 382, 386, 389, 395, 397, 399

Standardized Test Practice
- Multiple Choice 373, 377, 380, 390, 392, 396, 401, 406
- Short Response/Grid In 377, 380, 383, 384, 390, 396, 401, 405, 407
- Extended Response 407
- Worked-Out Example 379

USA TODAY Snapshots 396

WebQuest INTERDISCIPLINARY PROJECT 401

Lesson 9-4, p. 382

UNIT 5

Geometry and Measurement

Geometry and Measurement 408

Chapter 10 Geometry 410

10-1a **Hands-On Lab:** Measuring Angles 412
10-1 Angles 413
10-1b **Hands-On Lab:** Constructing and Bisecting Angles .. 416
10-2 Making Circle Graphs 418
10-3 Angle Relationships 422
10-3b **Hands-On Lab:** Constructing Parallel Lines 426
10-4 Triangles 428
10-4b **Hands-On Lab:** Constructing Triangles 432
10-5 Quadrilaterals 434
Mid-Chapter Practice Test 438
The Game Zone: Finding Squares 439
10-6 Similar Figures 440
10-7a **Problem-Solving Strategy:** Use Logical Reasoning 444
10-7 Polygons and Tessellations 446
10-8 Translations 451
10-8b **Spreadsheet Investigation:** Dilations 455
10-9 Reflections 456
10-9b **Hands-On Lab:** Rotations 460
Study Guide and Review 462
Practice Test 465
Standardized Test Practice 466

WebQuest INTERDISCIPLINARY PROJECT
- Math and History **409**

Lesson 10-9, p. 458

Prerequisite Skills
- Getting Started 411
- Getting Ready for the Next Lesson 415, 421, 425, 431, 437, 443, 450, 454

FOLDABLES Study Organizer 411

Study Skills
- Study Tips 429, 432, 433, 441, 447, 451, 452, 457
- Homework Help 415, 420, 424, 430, 436, 442, 449, 453, 458

USA TODAY Snapshots 421, 425

Reading and Writing Mathematics
- Link to Reading 413, 451
- Reading Math 428, 434, 441, 446, 447, 460
- Writing Math 412, 414, 417, 420, 424, 427, 430, 433, 435, 442, 448, 453, 458, 461

Standardized Test Practice
- Multiple Choice 415, 421, 425, 431, 437, 438, 443, 445, 450, 454, 459, 465, 466
- Short Response/Grid In 415, 421, 425, 431, 438, 443, 450, 454, 459, 467
- Extended Response 467
- Worked-Out Example 441

UNIT 5

Chapter 11 Geometry: Measuring Two-Dimensional Figures

Geometry: Measuring Two-Dimensional Figures 468

11-1 Squares and Square Roots 470
Study Skill: How to Make a Web 474
11-2 Estimating Square Roots 475
11-3a Hands-On Lab: The Pythagorean Theorem 478
11-3 The Pythagorean Theorem 479
11-4 Area of Parallelograms 483
Mid-Chapter Practice Test 486
The Game Zone: Finding Square Roots 487
11-5a Hands-On Lab: Triangles and Trapezoids 488
11-5 Area of Triangles and Trapezoids 489
11-6 Area of Circles 493
11-7a Problem-Solving Strategy: Solve a Simpler Problem 496
11-7 Area of Complex Figures 498
11-8 Area Models and Probability 501
Study Guide and Review 504
Practice Test 507
Standardized Test Practice 508

Prerequisite Skills
- Getting Started 469
- Getting Ready for the Next Lesson 473, 477, 482, 485, 492, 495, 500

FOLDABLES™ Study Organizer 469

Study Skills
- Study Tips 476, 481, 484
- Homework Help 472, 477, 482, 485, 491, 494, 500, 503

Reading and Writing Mathematics
- Link to Reading 498
- Reading Math 470, 471, 480, 490
- Writing Math 472, 476, 478, 481, 484, 488, 491, 494, 499, 502

Standardized Test Practice
- Multiple Choice 473, 477, 482, 485, 486, 492, 495, 497, 500, 503, 507, 508
- Short Response/Grid In 473, 477, 482, 485, 486, 492, 495, 500, 503, 509
- Extended Response 509
- Worked-Out Example 499

USA TODAY Snapshots 492

Lesson 11-5, p. 490

UNIT 5

Chapter 12 Geometry: Measuring Three-Dimensional Figures 510

12-1a **Hands-On Lab:** Building Three-Dimensional Figures 512
12-1 Drawing Three-Dimensional Figures 514
12-1b **Problem-Solving Strategy:** Make a Model 518
12-2 Volume of Rectangular Prisms 520
12-2b **Spreadsheet Investigation:** Similar Solids 523
12-3 Volume of Cylinders 524
Mid-Chapter Practice Test 528
The Game Zone: Three-Dimensional Figures 529
12-4a **Hands-On Lab:** Nets and Surface Area 530
12-4 Surface Area of Rectangular Prisms 532
12-4b **Hands-On Lab:** Changes in Volume and Surface Area 536
12-5 Surface Area of Cylinders 538
12-6 Precision and Measurement 542
Study Guide and Review 546
Practice Test 549
Standardized Test Practice 550

Prerequisite Skills
- Getting Started 511
- Getting Ready for the Next Lesson 517, 522, 527, 535, 541

FOLDABLES™ Study Organizer 511

Study Skills
- Study Tips 515, 521
- Homework Help 516, 522, 526, 534, 540, 544

Reading and Writing Mathematics
- Link to Reading 520
- Writing Math 513, 515, 521, 525, 531, 533, 537, 539, 543

Standardized Test Practice
- Multiple Choice 517, 519, 522, 527, 535, 541, 545, 549, 550
- Short Response/Grid In 522, 528, 535, 551
- Extended Response 551
- Worked-Out Example 525

USA TODAY Snapshots 544

WebQuest INTERDISCIPLINARY PROJECT 545

Lesson 12-2, p. 522

HOW TO...

Use Your Math Book

Why do I need my math book?

Have you ever been in class and not understood all of what was presented? Or, you understood everything in class, but at home, got stuck on how to solve a couple of problems? Maybe you just wondered when you were ever going to use this stuff?

These next few pages are designed to help you understand everything your math book can be used for . . . besides homework problems!

BEFORE YOU READ

Have a Goal

- What information are you trying to find?
- Why is this information important to you?
- How will you use the information?

Have a Plan

- Read *What You'll Learn* at the beginning of the lesson.
- Look over photos, tables, graphs, and opening activities.
- Locate boldfaced words and read their definitions.
- Find Key Concept and Concept Summary boxes for a preview of what's important.
- Skim the example problems.

Have an Opinion

- Is this information what you were looking for?
- Do you understand what you have read?
- How does this information fit with what you already know?

IN CLASS

During class is the opportunity to learn as much as possible about that day's lesson. Take advantage of this time! Ask questions about things that you don't understand, and take notes to help you remember important information.

To help keep your notes in order, try making a Foldables Study Organizer. It's as easy as 1-2-3! Here's a Foldable you can use to keep track of the rules for addition, subtraction, multiplication, and division.

FOLDABLES™ Study Organizer

Operations Make this Foldable to help you organize your notes. Begin with a sheet of 11″ × 17″ paper.

STEP 1

Fold
Fold the short sides toward the middle.

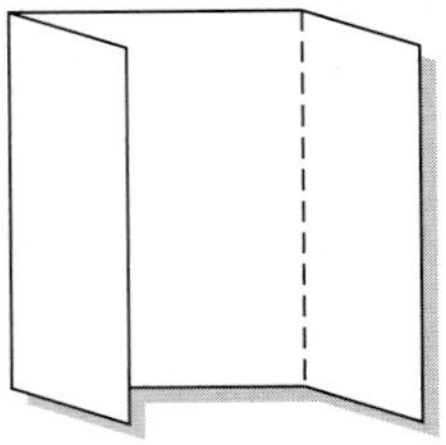

STEP 2

Fold Again
Fold the top to the bottom.

STEP 3

Cut
Open. Cut along the second fold to make four tabs.

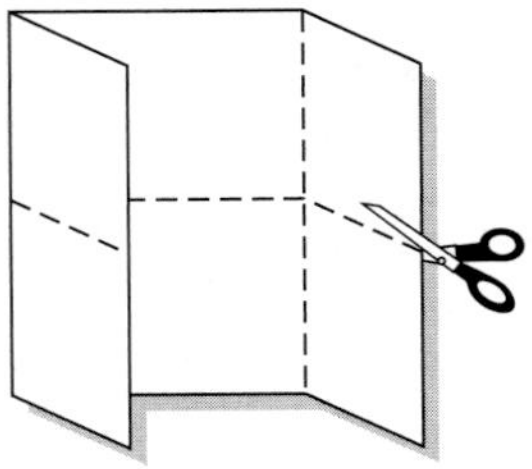

STEP 4

Label
Label each of the tabs as shown.

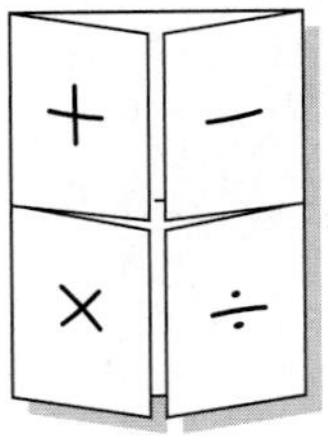

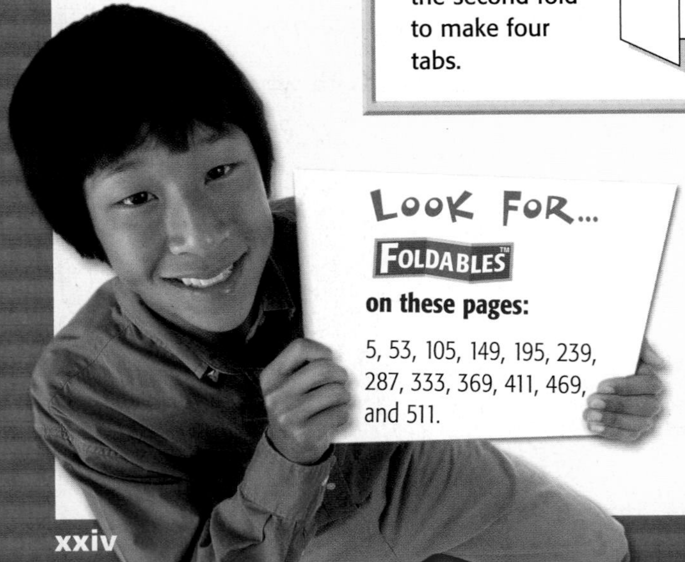

DOING YOUR HOMEWORK

Regardless of how well you paid attention in class, by the time you arrive at home, your notes may no longer make any sense and your homework seems impossible. It's during these times that your book can be most useful.

- Each lesson has example problems, solved step-by-step, so you can review the day's lesson material.
- A Web site has extra examples to coach you through solving those difficult problems.
- Each exercise set has Homework Help boxes that show you which examples may help with your homework problems.
- Answers to the odd-numbered problems are in the back of the book. Use them to see if you are solving the problems correctly. If you have difficulty on an even problem, do the odd problem next to it. That should give you a hint about how to proceed with the even problem.

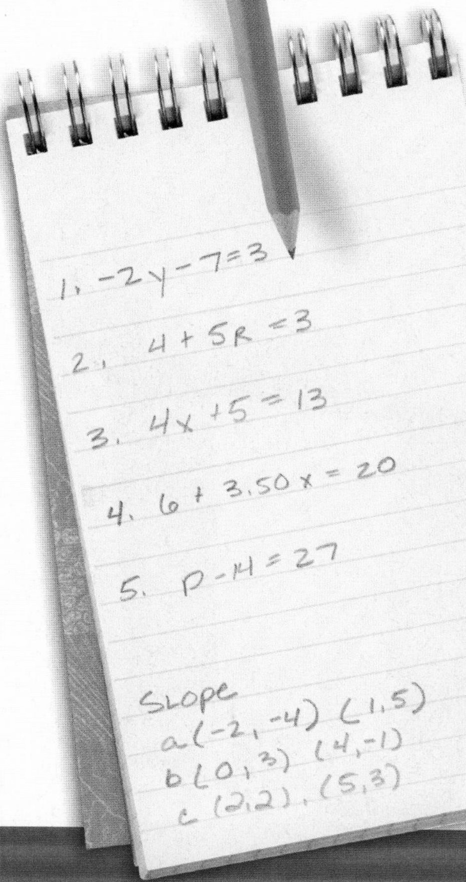

HOW TO...
Use Your Math Book

BEFORE A TEST

Admit it! You think there is no way to study for a math test! However, there *are* ways to review before a test. Your book offers help with this also.

- Review all of the new vocabulary words and be sure you understand their definitions. These can be found on the first page of each lesson or highlighted in yellow in the text.
- Review the notes you've taken on your Foldable and write down any questions that you still need answered.
- Practice all of the concepts presented in the chapter by using the chapter Study Guide and Review. It has additional problems for you to try as well as more examples to help you understand. You can also take the Chapter Practice Test.
- Take the self-check quizzes from the Web site.

LOOK FOR...

The Web site with self-check quizzes on these pages in Chapter 1: 9, 13, 16, 21, 27, 33, 36, 38, and 45.

The Study Guide and Review for Chapter 1 on page 46.

Scavenger Hunt

Let's Get Started

To help you find the information you need quickly, use the Scavenger Hunt below to learn where things are located in each chapter.

1. What is the title of Chapter 1? **Decimal Patterns and Algebra**

2. How can you tell what you'll learn in Lesson 1-1?
   Sample answer: There is a statement in the "What You'll Learn" section at the beginning of Lesson 1-1, in the left-hand column, that will tell me what I will learn in the lesson.

3. Sometimes you may ask, "When am I ever going to use this?" Name a situation that uses the concepts from Lesson 1-2.
   Sample answer: The lesson opener shows an example of how computing power in a computer chip can be modeled by a power of 2.

4. What is the key concept presented in Lesson 1-3?
   The key concept explains how to follow the order of operations.

5. How many Examples are presented in Lesson 1-3? **6**

6. What is the web address where you could find extra examples?
   msmath2.net/extra_examples

7. In Lesson 1-4, there is a paragraph that mentions it is common to use the first letter of a word as a variable. What is the main heading above that paragraph? **Reading Math**

8. There is a Real-Life Career mentioned in Lesson 1-4. What is it? **Athletic Trainer**

9. List the new vocabulary words that are presented in Lesson 1-5.
   equation, solution, solving an equation, defining the variable

10. List the math symbols that are presented in Lesson 1-5.
    $=$ is equal to; $\neq$ is not equal to

11. Suppose you're doing your homework on page 32 and you get stuck on Exercise 12. Where could you find help? **Sample answer: I can look at Examples 1 and 2 on page 30 for help.**

12. What is the web address that would allow you to take a self-check quiz to be sure you understand the lesson? **msmath2.net/self_check_quiz**

13. On what pages will you find the Study Guide and Review? **pages 46–48**

14. Suppose you can't figure out how to do Exercise 25 in the Study Guide on page 47. Where could you find help?
    Sample answer: I can review Examples 1-4 in Lesson 1-4.

15. You complete the Practice Test on page 49 to study for your chapter test. Where could you find another test for more practice? **I can look online at msmath2.net/chapter_test.**

UNIT 1 Notes

Introduction

In this unit, students will explore problem solving using a four-step plan. They will use tools such as decimals, algebra, and statistics for problem solving. They will examine powers, exponents, scientific notation, and the order of operations. In addition, they will use their knowledge of algebraic variables, expressions, and properties to solve equations.

Students will use frequency tables, line plots, and measures of central tendency to analyze data and make predictions. They also will learn to interpret statistical data through the use of stem-and-leaf plots, box-and-whisker plots, bar graphs, and histograms. Finally, students will combine these skills as they analyze how data can be manipulated and displayed in a misleading fashion.

Assessment Options

Unit 1 Test Pages 127–128 of the *Chapter 2 Resource Masters* may be used as a test or a review for Unit 1. This assessment contains both multiple-choice and short-response items.

UNIT 1

Decimals, Algebra and Statistics

Chapter 1

Decimal Patterns and Algebra

Chapter 2

Statistics: Analyzing Data

Your study of math requires analysis of data, much of which will be in decimal form. In this unit, you will learn how to evaluate algebraic expressions involving decimals, as well as analyze and graph data.

What's Math Got to Do With It? Real-Life Videos

What's Math Got to Do With It? Real-Life Videos engage students, showing them how math is used in everyday situations. Use Video 1 to discuss how stem-and-leaf plots, median, and mean are used in animal care and retail sales.

WebQuest INTERDISCIPLINARY PROJECT — MATH and SCIENCE

LIONS, TIGERS, AND BEARS

Are you ready to join a team of animal experts? As part of your application to be a Zoo's new coordinator, you must complete several challenging tasks. You'll make decisions about what animals to purchase for the zoo based on financial information provided to you. You'll gather specific data about the animals you choose, including their weight and expected lifespan. Finally, you'll present your findings to the hiring committee. So pack up your gear and don't forget your algebra tool kit. This adventure is going to be wild!

Log on to msmath2.net/webquest to begin your WebQuest.

WebQuest INTERDISCIPLINARY PROJECT

Lions, Tigers, and Bears

A WebQuest is an online project in which students do research on the Internet, gather data, and make presentations using word processing, graphing, page-making, or presentation software. In each chapter, students advance to the next step in their WebQuest. At the end of Chapter 2, the project culminates with a presentation of their findings.

Teaching suggestions and sample answers are available in the WebQuests, Projects, and Interdisciplinary Investigations.

Math and Science

You can use this WebQuest Interdisciplinary Project with your students' science teacher to make the connection from mathematics to the science your students are studying.

More Interdisciplinary Connections

You may wish to share these suggestions with your students' other teachers.

Math and Art

Choose several animals of interest to you and portray them using painting, drawing, or sculpture. Your depictions should be accurate in terms of habitat and the relative size of the animals to their surroundings.

Math and Social Studies

Research the history of zoos in the United States and around the world and write a report. Include a discussion of how animal parks around the world coordinate with one another to breed animals and aid their conservation.

CHAPTER 1

Chapter Overview

Decimal Patterns and Algebra

	LESSON OBJECTIVES	NCTM Standards	NAEP 2005	State/Local Objectives
1-1	**A Plan for Problem Solving** *(pp. 6–9)* • Solve problems using the four-step plan.	1, 6–10	D1a, N3g, N2c, A1a	
1-2	**Powers and Exponents** *(pp. 10–13)* • Use powers and exponents.	1, 6–10	N1d, N3f, N1j, N5f	
1-3	**Order of Operations** *(pp. 14–17)* • Evaluate expressions using the order of operations.	1, 2, 6–10	N3a, N3f, N5e	
1-4	**Algebra: Variables and Expressions** *(pp. 18–21)* • Evaluate simple algebraic expressions.	1, 2, 6–10	A1a, A1b, A3a, A3b	
1-5	**Algebra: Equations** *(pp. 22–27)* **1-5a Problem-Solving Strategy:** Solve problems using the guess and check strategy. • Solve equations using mental math.	1, 2, 6–10	N3g, N2b, A4a, A4b, A3a, N3d	
1-6	**Algebra: Properties** *(pp. 30–33)* • Use addition and multiplication properties to solve problems.	1, 2, 6–10	N5e, N3a, A2b	
1-7	**Sequences** *(pp. 34–37)* • Recognize and extend patterns for sequences. **1-7b Hands-On Lab:** Explore patterns in sequences using paper folding.	1, 2, 6–10	A1a, A1b, A1c	
1-8	**Measurement: The Metric System** *(pp. 38–41)* • Change metric units of length, capacity, and mass.	1, 4, 6–10	M2b	
1-9	**Scientific Notation** *(pp. 43–45)* • Write numbers greater than 100 in scientific notation and in standard form.	1, 2, 6–10	N1f, N1d, N1j	

Key to NCTM Standards:

1=Number & Operations, 2=Algebra, 3=Geometry, 4=Measurement, 5=Data Analysis & Probability, 6=Problem Solving, 7=Reasoning & Proof, 8=Communication, 9=Connections, 10=Representation

NAEP 2005 Standards: See math.glencoe.com

Pacing suggestions for the entire year can be found on page T20.

Chapter Resource Manager

Time-Saving Tool
TeacherWorks™
All-In-One Planner and Resource Center
See page T5.

FAST FILE

Chapter 1 Resource Masters: Study Guide and Intervention	Practice (Skills and Word Problems)	Reading to Learn Mathematics	Enrichment	Assessment	Applications*	Parent and Student Study Guide Workbook	5-Minute Check Transparencies	Key Teaching Transparencies	MathPASS: Tutorial Plus (lessons)	Teaching Mathematics with Manipulatives	Optional Materials
1	2–3	4	5			1-1	1-1	1	1, 11		
6	7–8	9	10		CM 1–4 CS 13	1-2	1-2			31	
11	12–13	14	15	61	CS 14	1-3	1-3		4		
16	17–18	19	20		CS 15	1-4	1-4	2		32–33	
21	22–23	24	25	61, 63		1-5	1-5	3	38		
26	27–28	29	30			1-6	1-6				
31	32–33	34	35	62		1-7	1-7		25	34	calculator, colored pencils
36	37–38	39	40		SM 25–28	1-8	1-8	4	56		
41	42–43	44	45	47–60, 62, 64–66	RC 15 CS 16	1-9	1-9		26		

*** Key to Abbreviations:** *CS = Calculator and Spreadsheet Masters, RC = Real-Life Career Masters*
CM = Classroom Math Games (online), SM = Science and Math Lab Manual (online)

INTERACTIVE CHALKBOARD

PowerPoint® Presentations

This CD-ROM is a customizable Microsoft® PowerPoint® presentation that includes:

- In-Class Examples
- Links to Online Study Tools
- 5-Minute Check Transparencies
- Bellringers

Math Skills Maintenance

Use the Math Skills Maintenance Masters to keep your students' basic math skills fresh. Students can review skills one at a time or in a combined format. The Masters are also available online at msmath2.net/maintenance.

CHAPTER 1 Mathematical Content and Teaching Strategies

Continuity of Instruction

Building on Prior Knowledge

Prior to this course, students should have learned to add, subtract, multiply, and divide decimals, fractions, and mixed numbers. They have used the order of operations. They have also been introduced to basic algebra problems, such as solving simple equations, and they have experienced working through simple word problems.

Learning in This Chapter

Students improve their problem-solving skills by learning the four-step plan for problem solving. They learn how to use exponents to simplify the multiplication of same terms. Students are reintroduced to variables and use them to evaluate algebraic expressions and then to write algebraic equations. They learn properties of addition and multiplication. Finally, they learn about the metric system and how to write numbers in scientific notation.

Making Future Connections

The four-step plan can be applied to any problem that students may encounter, whether in other classes or in day-to-day life. Exponents and scientific notation are used frequently in the sciences. The tools of algebra are at the foundation of the rest of this course and future math courses. Knowledge of sequences teaches a form of logical thinking that can be applied to computer programming as well as many other areas of life that involve patterns.

Mathematical Background

1-1 A Plan for Problem Solving

The problem-solving plan presented in this lesson became popular after Hungarian mathematician George Polya published his book *How to Solve It* (1957). Polya's book outlined four general steps that can be applied to most problems. Encouraging students to use this plan in math as well as in their other subjects will help them to become problem solvers and critical, independent thinkers.

1-2 Powers and Exponents

Powers and exponents are essential for the expression of both very large and very small numbers. As students progress in math and science, they will need to use exponents to write figures in scientific notation.

1-3 Order of Operations

Point out to students that the order of operations provides structure for mathematical expressions. Without understanding the order of operations, the phrase "$3 + 7 \times 4$" could be interpreted to mean 40 or 31. With practice, the order of operations will become second nature for students. Point out to students that any number raised to the zero power, x^0, equals 1. Demonstrate this by showing them that since any number divided by itself equals 1, and $\frac{a^n}{a^n} = a^{(n-n)} = a^0$, that $a^0 = 1$.

Teacher to Teacher

Renee B. Burgdorf
Morgan County Middle School
Madison, GA

"I introduce the order of operations by writing a mathematical expression on the chalkboard. Then I ask students to evaluate the expression. The variety of answers shows students that an order must be followed. Students *really* enjoy deciding who is correct. Then I proceed to teach the lesson."

Renee B. Burgdorf

1-4 Algebra: Variables and Expressions

While students should be familiar with variables, be sure they understand the vocabulary and the different elements of algebra. The work students do as they evaluate expressions is important background for solving linear equations.

1-5 Algebra: Equations

Solving simple equations is the basis for all advanced algebra. Once students can correctly apply one of the four basic operations to arrive at the solution of an equation, they are prepared for more complex algebraic concepts.

1-6 Algebra: Properties

Applying the Distributive, Commutative, Associative, and Identity properties will help students solve more difficult algebraic equations. Using the properties is often necessary to set up and solve word problems.

1-7 Sequences

Pattern recognition is a skill that applies to a wide variety of disciplines, from the analysis of an artist's style to building neural networks in the field of artificial intelligence. This lesson helps develop students' pattern recognition abilities as they encounter sequences of numbers.

1-8 Measurement: The Metric System

Developed in France during the French Revolution, the metric system is based on units of ten. The system made measurements and calculations much easier than other, older measurement systems. The system of weights and measures used by scientists worldwide, the International System of Units (SI), is based on the metric system.

Team Teaching

Inform the other teachers on your team that your classes are studying the metric system. Suggestions for curriculum integration are:

Science: units of measurement

Social Studies: history of the integration of the metric system into the U.S. economy

Language Arts: use of the metric prefixes in other words

1-9 Scientific Notation

Scientific notation enables students to express and manipulate very large and very small numbers. Scientific notation is based on the fact that multiplying and dividing by powers of ten simply involves moving the decimal place, for example, $6.3 \times 10^4 = 63{,}000$.

McGraw-Hill Professional Development

Mathematics, YES! is a self-paced, web-based professional development program. Units are based on major mathematical concepts and are standards-based. Actual classroom videos provide a springboard for discussion and activities. Units appropriate for this chapter:

- Unit 1, Number Relations
- Unit 2, Patterns and Functions
- Unit 8, Geometry and Measurement of Plane Figures

Key Concepts

Additional mathematical information and teaching notes are available at

msmath2.com/key_concepts.

The section appropriate for this chapter is:

- Scientific Notation

DAILY INTERVENTION and Assessment

	Type	Student Edition	Teacher Resources	Technology/Internet
INTERVENTION	Ongoing	Prerequisite Skills, pp. 5, 13, 21, 36, 41 Mid-Chapter Practice Test, p. 28	5-Minute Check Transparencies Quizzes 1–4, *CRM*, pp. 61–62 Mid-Chapter Test, *CRM*, p. 63 Study Guide and Intervention, *CRM*, pp. 1, 6, 11, 16, 21, 26, 31, 36, 41	MathPASS: Tutorial Plus msmath2.net/self_check_quiz msmath2.net/extra_examples
	Mixed Review	pp. 9, 13, 17, 21, 27, 33, 36, 41, 45	Cumulative Review, *CRM*, p. 64	
	Error Analysis	Find the Error, pp. 16, 26, 40	Find the Error, *TWE*, pp. 16, 26, 40 Unlocking Misconceptions, *TWE*, pp. 19, 43 Tips for New Teachers, *TWE*, p. 7	
ASSESSMENT	Standardized Test Practice	pp. 9, 13 17, 21, 23, 25, 27, 28, 33, 36, 41, 45, 50–51	Standardized Test Practice, *CRM*, pp. 65–66	Standardized Test Practice CD-ROM msmath2.net/standardized_test
	Open-Ended Assessment	Writing Math, pp. 8, 11, 16, 20, 26, 32, 35, 37, 40, 44 Open Ended, pp. 8, 11, 16, 20, 26, 28, 32, 35, 40, 44 Standardized Test, p. 51	Modeling: *TWE*, pp. 21, 33, 42 Speaking: *TWE*, pp. 23, 45 Writing: *TWE*, pp. 9, 17 Extended Response Assessment, *CRM*, p. 59	
	Chapter Assessment	Study Guide, pp. 46–48 Practice Test, p. 49	Multiple-Choice Tests (Forms 1, 2A, 2B), *CRM*, pp. 47–52 Free-Response Tests (Forms 2C, 2D, 3), *CRM*, pp. 53–58 Vocabulary Test/Review, *CRM*, p. 60	ExamView® Pro Test Bank (see below) MindJogger Videoquizzes msmath2.net/vocabulary_review msmath2.net/chapter_test

* **Key to Abbreviations:** *TWE* = Teacher Wraparound Edition, *CRM* = Chapter Resource Masters

Additional Intervention

Skills Intervention for Middle School Mathematics workbook and CD-ROM includes 88 skills with pretests, tutorials, guided practice, and posttests for diagnosis and remediation.

Guide to Daily Intervention booklet details each intervention resource available and guides teachers in using them as intervention tools everyday to assist students who may be having difficulty.

ExamView® Pro

Use ExamView® Pro Test Bank CD-ROM to:

- **Create Multiple versions of tests.**
- **Create Modified tests for Inclusion students with one mouse click.**
- **Edit existing questions and Add your own questions.**
- **Build tests aligned with state standards using built-in State Curriculum Correlations.**
- **Change English tests to Spanish with one mouse click and vice versa.**

Reading and Writing Mathematics

- **Student Edition**

Foldables Study Organizer, p. 5

Study Skill: How to Take Good Notes, p. 42

Concept Check questions at the beginning of each exercise set require students to verbalize and write about what they have learned in the lesson.

- **Open Ended,** pp. 8, 11, 16, 20, 26, 28, 32, 35, 40, 44
- **Find the Error,** pp. 16, 26, 40
- **Which One Doesn't Belong?,** pp. 11, 20

Reading Math Tip, pp. 10, 18, 19, 39

- **Teacher Wraparound Edition**

Foldables Follow-Up, pp. 5, 8, 11, 16, 20, 28, 40, 46

Writing Activities, pp. 9, 17

Differentiated Instruction, (Verbal/Linguistic), p. 15

ELL Resources, *see page T9 for a complete listing.*

- **Additional Resources**

Vocabulary Builder, *Chapter 1 Resource Masters*

Reading to Learn Mathematics, *Chapter 1 Resource Masters*

Vocabulary PuzzleMaker (online)

Teaching Mathematics with Foldables

Reading and Writing in the Mathematics Classroom

WebQuests, Projects, and Interdisciplinary Investigations

For detailed information on these resources, see pp. T6-T7.

- **Recommended Outside Reading for Students**

Nonfiction

- *Powers of Ten,* Encyclopedia Britannica Educational Corporation, ©1998.

  This book is part of the Mathematics in Context series, from the Encyclopedia Britannica Educational Corporation. The student book covers powers of ten as well as exponents and scientific notation. Teacher materials are also available.

Fiction

- Isdell, Wendy. *A Gebra Named Al: A Novel.* Free Spirit Publishing, ©1993.

  The story of a girl who is having problems learning algebra. She travels to the Land of Mathematics and meets a "gebra named Al." She travels through the land encountering many algebraic concepts, finally finding the equation she needs to return home.

PROJECT CRISS[SM] Study Skill

A problem-solution graphic structure can be used by students to develop a step-by-step plan for solving a problem. Students begin by writing details of the problem. Next, they write steps that are necessary to solve the problem. Finally, they solve the problem and write out their result. The graphic structure at the right applies the four-step plan presented in Chapter 1 to solving a problem containing an exponent. Have students develop a similar structure to help them solve sequence problems and other types of problems they encounter in this chapter.

Problem: Evaluate the expression 4×6^3.

Steps to Solution

Step	Result
1. Explore	1. The exponent means the 6 is used as a factor 3 times.
2. Plan	2. Write out all the factors and multiply.
3. Solve	3. $4 \times 6^3 \times 4 \times 6 \times 6 \times 6 = 864$
4. Examine	4. $4 \times 6^3 = 864$ is a reasonable answer.

End Result: $4 \times 6^3 = 864$

CReating Independence through Student-owned Strategies

For more information on Project CRISS, see page T9.

CHAPTER 1 Making Connections

School-To-Home

Parent and Student Study Guide

Parents and students may work together to reinforce the concepts and skills of this chapter. These workbooks are available in English and Spanish. A sample lesson is shown below. Also available at msmath2.net/parent_student.

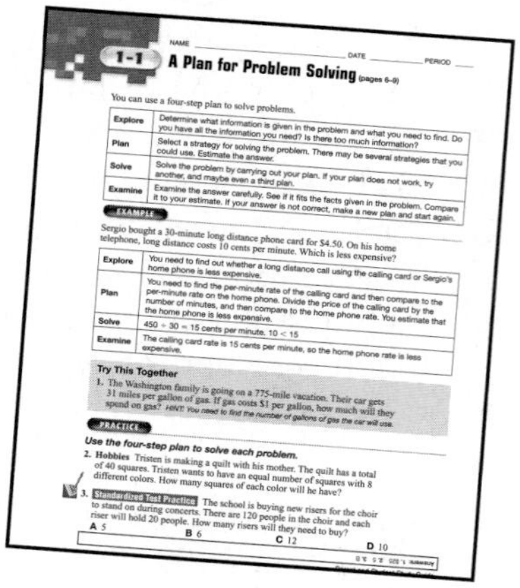

Family Letters and Activities

Send a letter and activity for this chapter home to encourage family members to become active participants in their students' learning. (English: *Chapter 1 Resource Masters*, pp. ix–x; Spanish: log on to msmath2.net/family_letters)

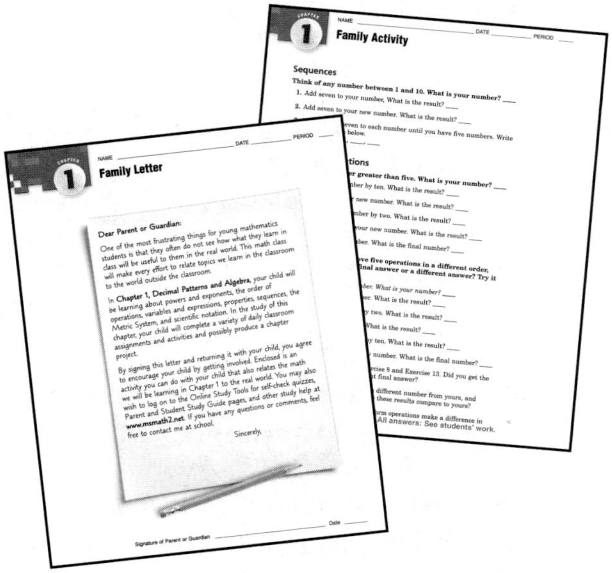

English Language Learner ELL

Lesson 1-4

Language Experience Approach to Illustrations

Write $2x + 3$ on the chalkboard and label each piece. Explain that 2 represents the coefficient. The variable x represents a value that we do not know, and its value varies for different problems. The number 3 is a constant because it does not change. It remains as a value of 3.

Lesson 1-5

Peer Tutoring

Ask English language learners to sit with an English-speaking or bilingual student and read the math problems. When defining a variable, always have students read the problem and then ask what it is that is asked of us.

Lesson 1-6

Higher-Level Thinking

Make a poster to display all of the properties discussed in the lesson. Ask students to study the definitions and examples of each and ask a series of questions: What are the similarities between the Commutative and Associative Properties? What do they have in common? How is the Distributive Property different from the Commutative and Associative Properties?

MathScape™

This project was supported, in part, by the **National Science Foundation**

MathScape is a mathematics curriculum for grades 6–8 developed by the Seeing and Thinking Mathematically Project at the Education Development Center.

Generalizing About Numbers

Making Mathematical Arguments

Summary Students will understand how to add, subtract, multiply, and divide integers. Using the order of operations agreement, students will evaluate expressions with exponents. Students will look for patterns as a problem-solving method.

USA TODAY Education Online Lesson Planner

USA TODAY® Education's Online site offers resources and interactive features connected to each day's newspaper. *Experience Today,* USA TODAY's® daily lesson plan, is available on the site and delivered daily to subscribers. This plan provides instruction for integrating USA TODAY® graphics and key editorial features into your mathematics classroom. Log on to **education.usatoday.com**.

USA TODAY Snapshots® in Chapter 1	
Lesson	**Snapshot**
1-1	N.C., Minn. lead turkey production
1-3	Time spent preparing for testing

Quick Review Math Handbook

is Glencoe's mathematical handbook for students. The Hot Words section includes a glossary of terms while the Hot Topics section consists of explanations of key mathematical concepts. An exercise set is included to check students' understanding of the concepts. This valuable resource can be used as a reference in the classroom or for home study.

Lesson	Hot Topics Section	Lesson	Hot Topics Section
1-1	1-5	**1-5**	6-1
1-2	3-1	**1-6**	1-2
1-3	1-3	**1-8**	8-1
1-4	6-1	**1-9**	2-2

Chapter 1 Notes

Key Chapter Objectives

Lesson 1-1 Solve problems using the four-step plan.

Lesson 1-4 Evaluate simple algebraic expressions.

Lesson 1-6 Use addition and multiplication properties to solve problems.

Lesson 1-7 Recognize and extend patterns for sequences.

Lesson 1-9 Write numbers greater than 100 in scientific notation and in standard form.

"What does exercising have to do with math?"

Point out to students that this is only one of many applications of the math that they are studying. Others are provided in the introduction to each lesson.

Vocabulary Builder, pp. vii–viii ELL

Chapter 1

NAME ________________ DATE ________ PERIOD ____

Reading to Learn Mathematics

Vocabulary Builder

This is an alphabetical list of new vocabulary terms you will learn in Chapter 1. As you study the chapter, complete each term's definition or description. Remember to add the page number where you found the term. Add this page to your math study notebook to review vocabulary at the end of the chapter.

Vocabulary Term	Found on Page	Definition/Description/Example
algebra		
algebraic [al-juh-BRAY-ihk] expression		
arithmetic [air-ith-MEH-tik] sequence		
base		
coefficient		
constant		
cubed		
defining the variable		
equation [ih-KWAY-zhuhn]		
equivalent expressions		
evaluate		
exponent		
exponential form		

Chapter 1 Decimal Patterns and Algebra

"What does exercising have to do with math?"

A 13-year-old has a minimum training heart rate of about 124 beats per minute. You can use the expression **0.6(220 − *a*)**, where ***a*** stands for a person's **age** to find a person's minimum training heart rate. Other real-life situations can also be modeled using expressions and variables.

You will solve problems about your health in Lesson 1-4.

Vocabulary Builder

For a thorough vocabulary list with pronunciations of new words, give students the Vocabulary Builder worksheets from the *Chapter 1 Resource Masters.* Encourage them to complete the definition of each term as they progress through the chapter.

GETTING STARTED

Take this quiz to see whether you are ready to begin Chapter 1. Refer to the lesson or page number in parentheses if you need more review.

Vocabulary Review

State whether each sentence is *true* or *false*. If *false*, replace the underlined word to make a true sentence.

1. A decimal point separates the ones place and the <u>tenths</u> place. (Page 555) **true**
2. When <u>multiplying</u> or subtracting decimals, you must first line up the decimal points. (Page 555) **false; adding**

Prerequisite Skills

Add. (Page 559)

3. 89.3 + 16.5 **105.8**
4. 7.9 + 32.45 **40.35**
5. 54.25 + 6.39 **60.64**
6. 10.8 + 2.6 **13.4**

Subtract. (Page 559)

7. 24.6 − 13.3 **11.3**
8. 9.1 − 6.6 **2.5**
9. 30.55 − 2.86 **27.69**
10. 17.4 − 11.2 **6.2**

Multiply. (Page 560)

11. 4 × 7.7 **30.8**
12. 9.8 × 3 **29.4**
13. 2.7 × 6.3 **17.01**
14. 8.5 × 1.2 **10.2**

Divide. (Page 562)

15. 32.6 ÷ 4 **8.15**
16. 10.6 ÷ 2 **5.3**
17. 5.5 ÷ 5 **1.1**
18. 17.84 ÷ 4 **4.46**

Multiply or divide. (Page 562)

19. 2.45 × 1,000 **2,450**
20. 87.3 ÷ 100 **0.873**
21. 0.61 × 100 **61**
22. 10 ÷ 1,000 **0.01**

FOLDABLES™ Study Organizer

Decimal Patterns and Algebra Make this Foldable to help you organize your notes. Begin with ten sheets of notebook paper.

STEP 1 **Staple**
Staple the ten sheets together to form a booklet.

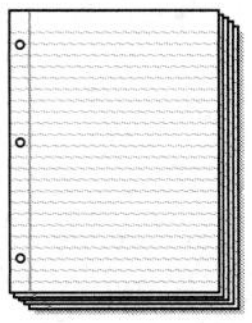

STEP 2 **Cut Tabs**
On the second page, make the top tab the width of the white space. On the third page, make the tab 2 lines longer, and so on.

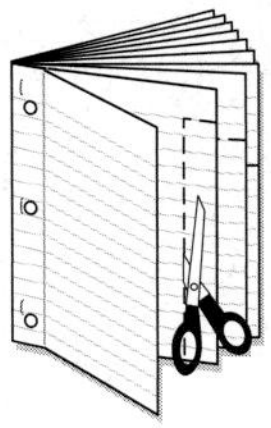

STEP 3 **Label**
Write the chapter title on the cover and label each tab with the lesson number.

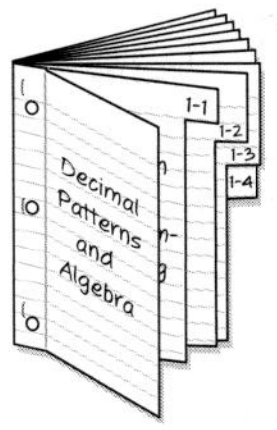

Reading and Writing As you read and study each lesson, write examples on the corresponding page.

GETTING STARTED

This section provides a review of the basic concepts needed before beginning Chapter 1. Page references are included for additional student help.

Prerequisite Skills review skills needed in this chapter.

Exercises	Needed for Lesson	Skill
3–6	1-3	Add Decimals
7–10	1-3	Subtract Decimals
11–14	1-3	Multiply Decimals
15–18	1-3	Divide Decimals
19–22	1-8	Multiply or Divide by a Power of Ten

For more prerequisite skills, see:

- Getting Ready for the Next Lesson, at the end of each lesson
- Prerequisite Skills in the Student Handbook on pp. 554–563
- *Math Skills Maintenance Masters*

Each chapter opens with **Vocabulary Review** and **Prerequisite Skills** practice for lessons in the chapter. See the list above for more Prerequisite Skills practice.

FOLDABLES™ Study Organizer

For more information about Foldables, see *Teaching Mathematics with Foldables.*

This Foldable is a study guide for student writing about decimal patterns and algebra. Students use their Foldables to take notes, define terms, record concepts, and write examples. At the end of each lesson, ask students to write a summary of the lesson, or write in their own words what the lesson was about. Some students might like to paraphrase key points. Summaries are useful for condensing data, and paraphrasing is a good memory aid.

Foldables™ are a unique way to enhance students' study skills.

1-1 Lesson Notes

Lesson Resources

Chapter 1 Resource Masters
- Study Guide and Intervention, p. 1
- Practice: Skills, p. 2
- Practice: Word Problems, p. 3
- Reading to Learn Mathematics, p. 4
- Enrichment, p. 5

Key Teaching Transparencies, 1

MathPASS Tutorial CD-ROM, 1, 11

Parent and Student Study Guide Workbook, 1-1

FAST FILE

1 Focus

5-Minute Check 1-1

INTERACTIVE CHALKBOARD PowerPoint® Presentations

Use as a quiz over prerequisite skills. (Also available as a transparency)

Bellringer

Communication Ask students to determine the two months in which the most members of the class were born. What questions do they need to ask? Have them compile a survey that they might use. Ask students to list the steps necessary to solve the problem and any questions to be answered during the process.

Reading to Learn Mathematics, p. 4 **ELL**

NAME ______________ DATE ________ PERIOD ____

1-1 Reading to Learn Mathematics

A Plan for Problem Solving

Pre-Activity *Read the introduction at the top of page 6 in your textbook. Write your answers below.*

1. Do you have all of the information necessary to solve this problem? See students' work.
2. Explain how you would solve this problem. Then solve it. Sample answer: I would add the number of turkeys produced in the top five states. 46 + 43.5 + 27 + 24 + 24 = 164.5 million
3. Does your answer make sense? Explain. Sample answer: I can compare my answer to the estimate of 50 + 40 + 30 + 25 + 25 or 170 million. So, the total of 164.5 million makes sense.
4. What can you do if your first attempt at solving the problem does not work? Sample answer: I would start over and make a new plan. Then, I would try to solve the problem again. If my answer is not close to the estimate, I would check my

1-1 A Plan for Problem Solving

What You'll Learn
Solve problems using the four-step plan.

The **Lesson Resources** lists all of the resources available for the lesson, including workbooks, blackline masters, transparencies, and technology.

WHEN am I ever going to use this?

ANIMALS The graphic shows the top five turkey-producing states in a recent year. What is the total number of turkeys produced in the top five states? **3–4. See margin.**

1. Do you have all of the information necessary to solve this problem?
2. Explain how you would solve this problem. Then solve it.
3. Does your answer make sense? Explain.
4. What can you do if your first attempt at solving the problem does not work?

1. See students' work.
2. Sample answer: I would add the number of turkeys produced in the top five states. 46 + 43.5 + 27 + 24 + 24 = 164.5 million

In mathematics, there is a *four-step plan* you can use to help you solve math problems.

1. Explore
- Determine what information is given and what you need to find.
- Do you have all the information to solve the problem
- Is there any information given that is not needed?

2. Plan
- Select a strategy for solving the problem. There may be several that you can use.
- It is usually helpful to estimate the answer.

3. Solve
- Carry out your plan.
- If your plan doesn't work, keep trying until you find one that does work.
- Make sure your solution contains appropriate units o labels.

4. Examine
- Does your answer make sense given the facts in the problem?
- Is your answer reasonable compared to your estimate
- If your answer is not correct, make a new plan and start again.

ELL notations indicate items that can assist English-Language Learners. Additional ELL strategies can be found at the beginning of each chapter.

Answers (When am I ever going to use this?)

3. Sample answer: I can compare my answer to the estimate of 50 + 40 + 30 + 25 + 25, or 170 million. So, the total of 164.5 million makes sense.
4. Sample answer: I would start over and make a new plan. Then, I would try to solve the problem again. If my answer is not close to the estimate, I would check my arithmetic.

6 Chapter 1 Decimal Patterns and Algebra

EXAMPLE Use the Four-Step Plan

1 TELEVISION Color TV sets were first mass-produced in 1954. In that year, 5,000 color TVs were produced. How many years passed before at least 100 times as many were produced?

Year	Color TVs Produced	Year	Color TVs Produced
1954	5,000	1960	120,000
1955	20,000	1961	140,000
1956	90,000	1962	450,000
1957	85,000	1963	750,000
1958	80,000	1964	1,500,000
1959	90,000		

Source: *Sams Photofact Book on Color TV*

Explore *What are you trying to find?*
The number of years since 1954 that 100 times as many TVs were produced.

What information do you need to solve the problem?
From the table, you know the number of TVs produced each year from 1954 to 1964.

Plan You can estimate the number of years that passed before at least 100 times 5,000 TVs, or 500,000 TVs were produced. This happened between 1962 and 1963. So, approximately 10 years had passed.

Solve $100 \times 5{,}000$ TVs $= 500{,}000$ TVs

In 1963, 750,000 TVs were produced, which is more than 500,000 TVs. Since $1963 - 1954 = 9$, nine years passed before production was at least 100 times greater than the number of sets produced in 1954.

Examine *Is your answer reasonable?*
Look at the data in the table and count the number of years from 1954 to 1963. Also, compare your answer to the estimate.

Your Turn **Solve using the four-step plan.**

a. How many times more TVs were produced in 1960 than in 1954?

b. Between which two years did TV production increase the most?

a. 24 b. 1963–1964

EAL-LIFE MATH

LEVISION The average
erican watches more than
ours of TV per day.
rce: A.C. Nielsen Company

TEACHING TIP
ell students that they
ill practice using each
rategy throughout the
ook.

Problems can be solved using different operations or strategies.

Key Concept **Problem-Solving Strategies**

guess and check	use a graph
look for a pattern	work backward
make an organized list	eliminate possibilities
draw a diagram	estimate reasonable answers
act it out	use logical reasoning
solve a simpler problem	make a model

STUDY TIP

Four-Step Plan
No matter what strategy you use, you can always use the four-step plan to solve a problem.

msmath2.net/extra_examples

2 Teach

In-Class Example

INTERACTIVE CHALKBOARD
PowerPoint® Presentations

1 SPENDING A can of soda holds 12 fluid ounces. A 2-liter bottle holds about 67 fluid ounces. If a pack of six cans costs the same as a 2-liter bottle, which is the better buy? **pack of six cans**

Tips for New Teachers

Intervention By having students solve problems using different strategies, you are teaching them not to give up on a problem. Have students solve the same problem in two different ways to compare the processes.

New teachers, or teachers new to teaching mathematics, may especially appreciate the **Tips for New Teachers.**

DAILY INTERVENTION **Flexible Grouping**

Problem Solving Use the think-pair-share technique to choose problem-solving strategies. Have students read a word problem. Then pair them up and have them discuss which strategy would work the best for that problem. The pair then presents its results to the class. For each new problem, regroup students into different pairs.

Daily Intervention notes help you help students when they need it most. **Flexible Grouping** suggestions provide a variety of grouping options.

In-Class Example

2 **POPULATION** For every 100,000 people in the United States, there are 5,750 radios. For every 100,000 people in Canada, there are 323 radios. Suppose Sheamus lives in Des Moines, Iowa, and Alex lives in Windsor, Ontario. Both cities have about 200,000 residents. About how many more radios are there in Sheamus's city than in Alex's city? **10,854 radios**

INTERACTIVE CHALKBOARD
PowerPoint® Presentations

3 Practice/Apply

FOLDABLES™ Follow-Up Encourage students to summarize the four-step problem-solving plan on the first section of their Foldables. Students might include an example of how they used this plan to solve a problem.

Foldables™ Follow-Ups remind you to encourage students to add to their Foldables as they work through the chapter and use them to review for the chapter test.

About the Exercises...

Guided Practice Key

Exercises	Examples
4	1
5	2

Odd/Even Assignments
Exercises 6–11 are structured so that students practice the same concepts whether they are assigned odd or even problems.

EXAMPLE Use a Strategy in the Four-Step Plan

2 **SPORTS In a basketball tournament, 128 teams play in the first round. Then there are 64 teams left to play in the second round, 32 teams left to play in the third round, and so on. How many rounds does it take to determine a winner?**

Explore You know the number of teams at the start of the tournament and how many teams play in each of the first three rounds.

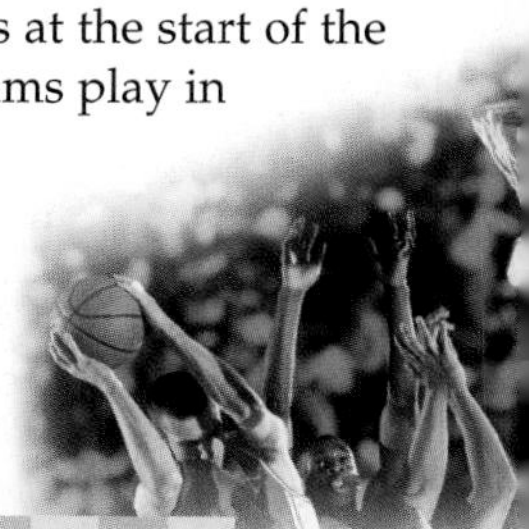

Plan You can look for a pattern by organizing the information in a table. Then continue the pattern until there is one team left.

Solve

Round	1	2	3	4	5	6	7
Number of Teams	128	64	32	16	8	4	2

÷2 ÷2

So, a winner is determined in 7 rounds.

Examine You could complete the diagram below to show that the answer is reasonable.

Round 1 → 128 teams: 64 teams win, 64 teams lose
Round 2 → 64 teams: 32 teams win, 32 teams lose
Round 3 → 32 teams: 16 teams win, 16 teams lose

Your Turn

c. How many rounds would it take to determine a winner if 512 teams participated in the tournament? **9 rounds**

Skill and Concept Check

Writing Math
Exercises 1 & 2

1. **Explain** why it is important to plan before solving a problem.
2. **Describe** what to do if your solution to a problem seems unrealistic.
3. **OPEN ENDED** Write a real-life problem that can be solved by adding 79 and 42. **See students' work.**

1–2. See margin.

Writing Math identifies exercises in which students can practice their communication skills.

GUIDED PRACTICE

Use the four-step plan to solve each problem.

4. **MONEY** To attend the class picnic, each student will have to pay $5.00 for transportation and $6.50 for food. If there are 300 students in the class, how much money will be collected for the picnic? **$3,450**
5. **PATTERNS** If Isaac receives an E-mail every 20 minutes during the workday, how many E-mails would he expect to receive between 8:00 A.M. and noon? **13 E-mails**

Answers

1. Sample answer: It helps to organize your thoughts and focus on how to approach solving the problem.
2. Sample answer: Try solving the problem again using a different strategy.

★indicates increased difficulty

Practice and Applications

Use the four-step plan to solve each problem.

6. **VIDEO RENTALS** A video store took in $5,400 in video rentals during July. January sales are expected to be double that amount. If videos rent for $4, how many video rentals are expected in January? **2,700**

7. **RACING** The Indianapolis Motor Speedway has 3,200,000 bricks that lie beneath the 2.5 miles of track. Find how many bricks lie under one mile of track at the race track. **1,280,000**

8. **COOKING** Mr. Sanchez is serving Cajun fried turkey at 3:00 P.M. The 15-pound turkey has to remain in the fryer 3 minutes for every pound and must cool for at least 45 minutes before it is carved. What is the latest time he can start frying? **1:30 P.M.**

★9. **GEOMETRY** Draw the next two figures in the pattern.

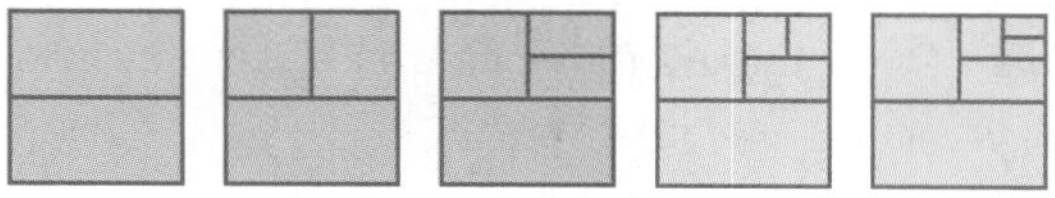

10. **MULTI STEP** Kishi wants to buy a DVD player that costs $250 with tax. So far, she has saved $145. If she saves $5 every week, in how many weeks will she be able to purchase the DVD player? **21 weeks**

11. **COLLECTIONS** In 1999, Dustin started collecting 6 die-cast cars every year. In 2001, his brother Logan started collecting 9 cars per year. In what year did Dustin and Logan have the same number of cars? **2004**

12. **CRITICAL THINKING** Use the digits 1, 2, 3, 4, and 5 to form a two-digit and a three-digit number so that their product is the least product possible. Use each digit only once. **245 × 13**

HOMEWORK HELP

For Exercises	See Examples
6–9	1
10–11	2

Extra Practice
See pages 564, 596.

Students can use **Homework Help** references when they complete the exercises at home. **Extra Practice** for every lesson is provided in the back of the book.

Standardized Test Practice and Mixed Review

13. **SHORT RESPONSE** Jeannie has $86 to spend on birthday gifts for her mother. Which three items from the table could Jeannie purchase for her mother, not including tax? **Sample answer: sweater, gloves, and chocolates**

Item	Cost
sweater	$29.50
gloves	$22.75
purse	$32.00
chocolates	$15.00
movie passes	$27.75

14. **MULTIPLE CHOICE** In a field near Duns, Scotland, a reproduction of the painting *Sunflowers* was created with 250,000 plants and flowers. The "painting" covered a 46,000-square foot area. What is a reasonable number of plants and flowers per square foot? **A**

Ⓐ 5 Ⓑ 50 Ⓒ 100 Ⓓ 500

GETTING READY FOR THE NEXT LESSON

BASIC SKILL Multiply.

15. 10 × 10 **100**
16. 3 × 3 × 3 **27**
17. 5 × 5 × 5 × 5 **625**
18. 2 × 2 × 2 × 2 × 2 **32**

msmath2.net/self_check_quiz

Getting Ready for the Next Lesson exercises target specific skills students will need for the next lesson.

4 Assess

Open-Ended Assessment

Writing Have students write a detailed paragraph explaining how to use the four-step plan to solve a problem. Ask them to include an example.

Study Guide and Intervention, p. 1

NAME ________ DATE ________ PERIOD ____

1-1 Study Guide and Intervention
A Plan for Problem Solving

Four-Step Problem-Solving Plan

When solving problems, it is helpful to have an organized plan to solve the problem. The following four steps can be used to solve any math problem.

1. **Explore** – Get a general understanding of the problem.
2. **Plan** – Make a plan to solve the problem and estimate the answer.
3. **Solve** – Carry out your plan.
4. **Examine** – Check the reasonableness of your answer.

EXAMPLE 1 **Use the four-step plan to solve the problem.**
RECREATION A canoe rental store along the Illinois River in Oklahoma has 30 canoes that it rents on a daily basis during the summer season. If canoes rent for $15 per day, how much money can the store collect for canoe rentals during the month of July?

Explore You know that they rent 30 canoes per day for $15 each. You need to determine the total amount of money that can be collected during the month of July.

Plan First, find the total amount of money that can be collected each day by finding the product of 30 and 15. Next, multiply the previous result by 31, the number of days in July. You can estimate this result by 30. 30 × 15 × 30 = 13,500

Solve Since 30 × $15 = $450, the canoe rental store can collect $450 in rental fees each day. This means the total amount of money that could be collected during the month of July is $450 × 31 or $13,950.

Examine Is your answer reasonable? The answer is close to the estimate of $13,500.

Lesson 1–1

EXERCISES

Use the four-step plan to solve each problem.

1. **MONEY** Colin works for his dad during summer vacation. His dad pays him $5.20 per hour and he works 20 hours per week. How much will Colin earn during his 8-week summer vacation? $832.00
2. **BOOKS** A student assistant in the school library is asked to shelve 33 books. If he puts away 9 books the first hour and then 6 books each hour after that, how long will it take him to shelve all 33 books? 5 hours
3. **SHOPPING** Alicia bought a $48 sweater on sale for $25 and a $36 purse on sale for $22. How much did Alicia save? $37
4. **MAIL** It cost Ramon $3.73 to mail a package to his grandmother. The post office charged $2.38 for the first pound and 45 cents for each additional pound. How much did the package weigh? 4 pounds

Practice: Skills, p. 2

NAME ________ DATE ________ PERIOD ____

1-1 Practice: Skills
A Plan for Problem Solving

Answer these questions about the four-step problem-solving plan.

1. During which step do you ask if your answer makes sense?
Step 4: Examine
2. During which step do you make a new plan if your first plan doesn't work?
Step 4: Examine
3. During which step do you make a strategy for solving the problem?
Step 2: Plan
4. During which step do you ask yourself, "What do I need to find out?"
Step 1: Explore

Choose one of the following to describe how you would plan to solve each problem. Do not solve the problems.

A. Use only one operation, such as addition or multiplication.
B. Use a combination of operations, such as division and addition.
C. Use a different strategy.

5. **MONEY** Julia opened a savings account with a deposit of $36. She then deposited $5 per week for one month. If she then withdrew $9.50, how much is left in her savings account? B
6. In how many different patterns can 3 rose bushes, 2 sunflowers, and 5 tulip plants be planted in a garden? C
7. Use the four-step plan to solve Exercise 5.
   - **A.** Explore
   I need to find how much is left in the savings account. First, I need to find the total deposits.
   - **B.** Plan
   Find the total deposits. Add them to the first deposit, then subtract the withdrawal.
   - **C.** Solve
   Since there are 4 weeks in a month, the total deposits are 4 × 5 = 20. 36 + 20 = 56. Subtract $9.50. 56 − 9.50 = $46.50.
   - **D.** Examine
   The difference between the deposits and withdrawal is about $10. This means the account should be about $10 more than $36, so the answer is reasonable.

Practice: Word Problems, p. 3

NAME ________ DATE ________ PERIOD ____

1-1 Practice: Word Problems
A Plan for Problem Solving

MAGAZINES **For Exercises 1 and 2, use the table that shows the costs of several popular magazines.**

Costs of Popular Magazines		
Magazine	**Cost of Yearly Subscription**	**Cost of a Single Copy**
Teen World	$9.98 (12 issues)	$3.25
Soccer World	$19.97 (6 issues)	$4.99
Book Nation	$19.98 (12 issues)	$2.99
TV Weekly	$46.28 (52 issues)	$1.95

1. How much could you save by buying *Teen World* with a yearly subscription rather than 12 single copies? $29.02
2. Which of the magazines saves you the most money by purchasing a yearly subscription instead of an equivalent number of single copies? How much will you save? TV Weekly; $55.12
3. **BICYCLING** Adriana can ride her bicycle 6 miles in one hour. How long will it take her to ride 15 miles? 2.5 h
4. **BASKETBALL** At Johnson Middle School an average of 500 people attended each of the 15 home basketball games. If admission was $3 per person, about how much money was collected in all? $22,500
5. **THEATER** A local theater has floor seating, balcony seating, and box seating. If the theater contains 2,500 seats with 425 seats in the balcony and 215 box seats, how many seats are on the floor? 1,860 seats
6. **POPCORN** Janelle plans to buy three boxes of popcorn at the movies for herself and two friends. If each box costs $1.95, how much change will she receive when she pays with a ten-dollar bill? $4.15

1-2 Lesson Notes

Lesson Resources

Chapter 1 Resource Masters
- Study Guide and Intervention, p. 6
- Practice: Skills, p. 7
- Practice: Word Problems, p. 8
- Reading to Learn Mathematics, p. 9
- Enrichment, p. 10

Calculator and Spreadsheet Masters, p. 13
Classroom Math Games, pp. 1–4
Parent and Student Study Guide Workbook, 1-2
Teaching Mathematics with Manipulatives, p. 31

FAST FILE

1 Focus

5-Minute Check 1-2
Use as a quiz or review of Lesson 1-1. (Also available as a transparency)

INTERACTIVE CHALKBOARD
PowerPoint® Presentations

Bellringer
Hands-On Activity Give students a pile of 1 × 1 × 1 cubes. Have them stack the unit cubes to make larger cubes. Have students complete the first three rows of the table shown in the bottom margin. Ask them how the numbers relate. The total number of cubes needed to build a cube equals the number of cubes per side times the same number.
(Continued in bottom margin.)

Reading to Learn Mathematics, p. 9 ELL

NAME ______ DATE ______ PERIOD ____
1-2 Reading to Learn Mathematics
Powers and Exponents

Pre-Activity *Read the introduction at the top of page 10 in your textbook. Write your answers below.*
1. How is doubling shown in the table? multiplying by 2
2. If the pattern continued, how much storage space would be available by year 6? 64 units
3. What is the relationship between the number of 2s and the year? The number of 2s is the same as the year.

Reading the Lesson
4. What is the difference between a power and an exponent? An exponent tells how many times the base is used as a factor, and a power is a number expressed using an exponent.

1-2 Powers and Exponents

What You'll Learn
Use powers and exponents.

NEW Vocabulary
factors
exponent
base
powers
squared
cubed
evaluate
standard form
exponential form

New Vocabulary words are listed at the beginning of the lesson and are highlighted in yellow at point of use.

WHEN am I ever going to use this?

TECHNOLOGY Computer engineer Gordon E. Moore made this observation in 1964. Moore's Law says the amount of available storage space on a computer chip doubles every year. 1. multiplying by 2

Year	Amount of Storage Space (u
1	2
2	2 × 2
3	2 × 2 × 2
4	2 × 2 × 2 × 2

1. How is doubling shown in the table?
2. If the pattern continued, how much storage space would be available by year 6? 64 units
3. What is the relationship between the number of 2s and the year? The number of 2s is the same as the year.

Two or more numbers that are multiplied together to form a product are called **factors**. The amount of storage space on a computer chip in year 4 can be written using only factors of 2. When the same factor is used, you may use an exponent to simplify the notation.

$$16 = 2 \cdot 2 \cdot 2 \cdot 2 = 2^4$$

The centered dots indicate multiplication.
common factors
The **exponent** tells how many times the base is used as a factor.
The **base** is the common factor.

Numbers expressed using exponents are called **powers**. You can say that 2^4 is a power of 2. This and the powers 5^2 and 4^3 are read as follow

Symbols	Words
5^2	five to the second power or five **squared**
4^3	four to the third power or four **cubed**
2^4	two to the fourth power

READING Math
First Power When a number does not have an exponent, it is understood to be 1. For example, $8 = 8^1$.

EXAMPLES Write Powers as Products

Write each power as a product of the same factor.

1 7^5
The base is 7. The exponent 5 means that 7 is used as a factor five times.
$7^5 = 7 \cdot 7 \cdot 7 \cdot 7 \cdot 7$

2 3^2
The base is 3. The exponent 2 means the 3 is used as a factor twice.
$3^2 = 3 \cdot 3$

Have students predict the next two rows in the table. Finally, have students build the larger cube to check their answers.

Cubes per Side	Total Cubes Used
1	1
2	4
3	9
4	16
5	25

TEACHING TIP

xplain to students that e expressions 4^3 and · 3 have different alues.
$^3 = 4 \cdot 4 \cdot 4 = 64$
$\cdot 3 = 12$

You can **evaluate**, or find the value of, powers by multiplying the factors. Numbers written without exponents are in **standard form**.

EXAMPLES Write Powers in Standard Form

Evaluate each expression.

3 2^5

$2^5 = 2 \cdot 2 \cdot 2 \cdot 2 \cdot 2$

$= 32$

4 4^3

$4^3 = 4 \cdot 4 \cdot 4$

$= 64$

Your Turn **Evaluate each expression.**

a. 10^2 100 b. 7^3 343 c. 5^4 625

TUDY TIP

Technology You can use a calculator to compute large exponents. To find 4^5, press 4 [^] 5 [ENTER]. The display shows **1024**.

Numbers written with exponents are in **exponential form**.

EXAMPLE Write Numbers in Exponential Form

5 **Write $3 \cdot 3 \cdot 3 \cdot 3$ in exponential form.**

3 is the base. It is used as a factor 4 times. So, the exponent is 4.

$3 \cdot 3 \cdot 3 \cdot 3 = 3^4$

Your Turn **Write each product in exponential form.**

d. $5 \cdot 5 \cdot 5$ 5^3 e. $12 \cdot 12 \cdot 12 \cdot 12 \cdot 12 \cdot 12$ 12^6

Skill and Concept Check

Writing Math
Exercises 1 & 4

1. **Explain** what *five to the fourth power* means. See margin.
2. **Write** 7^5 in words. seven to the fifth power
3. **OPEN ENDED** Write one number in exponential form and another number in standard form. Sample answer: exponential form: 4^2; standard form: 125
4. **Which One Doesn't Belong?** Identify the number that cannot be written as a power with an exponent greater than 1. Explain your reasoning.

4 9 16 50

50; 50 cannot be expressed as a power with an exponent of 2: $4 = 2^2$, $9 = 3^2$, $16 = 4^2$

GUIDED PRACTICE

Write each power as a product of the same factor.

5. 6^2 $6 \cdot 6$
6. 4^4 $4 \cdot 4 \cdot 4 \cdot 4$
7. 8^5 $8 \cdot 8 \cdot 8 \cdot 8 \cdot 8$

Evaluate each expression.

8. 3^4 81
9. 5^5 3,125
10. 10^3 1,000

Write each product in exponential form.

11. $5 \cdot 5 \cdot 5 \cdot 5 \cdot 5 \cdot 5$ 5^6
12. $1 \cdot 1 \cdot 1 \cdot 1$ 1^4
13. Evaluate *eleven to the third power.* 1,331

Students practice comparing and contrasting when they identify **Which One Doesn't Belong?**

msmath2.net/extra_examples

2 Teach

In-Class Examples

INTERACTIVE CHALKBOARD PowerPoint® Presentations

Write each power as a product of the same factor.

1. 8^4 $8 \cdot 8 \cdot 8 \cdot 8$
2. 4^6 $4 \cdot 4 \cdot 4 \cdot 4 \cdot 4 \cdot 4$

Evaluate each expression.

3. 8^3 512
4. 6^4 1,296
5. Write $9 \cdot 9 \cdot 9 \cdot 9 \cdot 9 \cdot 9$ in exponential form. 9^6

In-Class Examples, included for every example in the Student Edition, parallel the examples in the text.

3 Practice/Apply

FOLDABLES™ Follow-Up Remind students to summarize what they learn about powers and exponents on the second section of their Foldables. They might differentiate between the terms *power* and *exponent* and record information on how to read powers as presented on page 10.

Which One Doesn't Belong? In Exercise 4, have students factor each number and compare the factors. Some of the numbers will have the same factors and can be written in exponential form.

Answer

1. Five is used as a factor four times, or $5 \cdot 5 \cdot 5 \cdot 5 = 5^4$.

DAILY INTERVENTION **Differentiated Instruction**

Logical Have students compare multiplying and exponents. To do this, suggest they write out a short series of products and a short series of exponents of the same number. For example, they might compare "$3 \times 1 = 3, 3 \times 2 = 6, \ldots, 3 \times 6 = 18$" with "$3^1 = 3, 3^2 = 9, \ldots, 3^6 = 729$." Ask them to write a sentence that explains how their totals increase in each case.

About the Exercises...

Guided Practice Key

Exercises	Examples
5–7	1, 2
8–10, 13	3, 4
11–12	5

Odd/Even Assignments

Exercises 14–37 and 42–53 are structured so that students practice the same concepts whether they are assigned odd or even problems.

Alert! Exercises 42–47 require a calculator.

Alert! Exercise 60 requires research on the Internet or another source.

About the Exercises helps you plan to get the most from the exercise set.

Answers

41. Sample answer: A number taken to the third power is the same as the volume of a cube, or the amount of space inside a cube.
48. $1^{14}, 17^3, 6^5, 4^{10}$
49. $6^3, 15^2, 3^5, 2^8$
50. $7^2, 5^3, 2^{11}, 4^6$
62. Sample answer: $3^3 = 27, 3^2 = 9, 3^1 = 3, 3^0 = 1$; Each power is three times greater than the next.

★ indicates increased difficulty

Practice and Applications

HOMEWORK HELP

For Exercises	See Examples
14–19	1, 2
20–31, 38, 42–47, 54	3, 4
32–37, 39, 57, 59, 61	5

Extra Practice See pages 564, 596.

Write each power as a product of the same factor.

14. 1^5 $1 \cdot 1 \cdot 1 \cdot 1 \cdot 1$
15. 4^2 $4 \cdot 4$
16. 3^8 $3 \cdot 3 \cdot 3 \cdot 3 \cdot 3 \cdot 3 \cdot 3 \cdot 3$
17. 9^3 $9 \cdot 9 \cdot 9$
18. 10^4 $10 \cdot 10 \cdot 10 \cdot 10$
19. 11^3 $11 \cdot 11 \cdot 11$

Evaluate each expression.

20. 2^3 8
21. 2^6 64
22. 4^3 64
23. 5^4 625
24. 1^{10} 1
25. 10^1 10
26. 3^8 6,561
27. 4^6 4,096
28. 10^5 100,000
29. 12^4 20,736
30. 20^4 160,000
31. 50^3 125,000

Write each product in exponential form.

32. $3 \cdot 3$ 3^2
33. $7 \cdot 7 \cdot 7 \cdot 7$ 7^4
34. $1 \cdot 1 \cdot 1$ 1^3
35. $6 \cdot 6 \cdot 6 \cdot 6 \cdot 6$ 6^5
36. $4 \cdot 4 \cdot 4 \cdot 4 \cdot 4 \cdot 4 \cdot 4$ 4^7
37. $13 \cdot 13 \cdot 13 \cdot 13 \cdot 13$ 13^5
38. Find the value of *6 cubed.* 216

★ 39. Express 256 in exponential form as a power of 2. 2^8

GEOMETRY For Exercises 40 and 41, use the figures below.

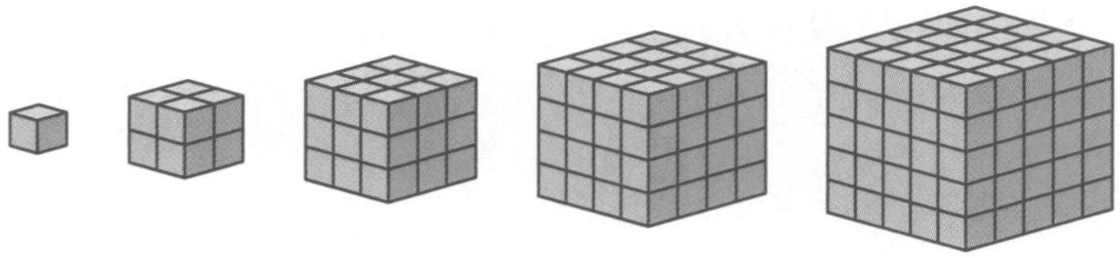

40. Find the number of unit cubes that make up each large cube. Write your answers using exponents. $1^3, 2^3, 3^3, 4^3, 5^3$
41. Why do you think the expression 2^3 is sometimes read as *2 cubed*? See margin.

Use a calculator to determine whether each sentence is *true* or *false*.

42. $3^7 > 7^3$ true
43. $14^4 = 19^2$ false
44. $927 > 9^3$ true
45. $3^2 = 2^3$ false
46. $6^3 < 4^4$ true
47. $3^4 < 9^2$ false

Order the following numbers from least to greatest. 48–50. See margin.

48. $6^5, 1^{14}, 4^{10}, 17^3$
49. $2^8, 15^2, 6^3, 3^5$
50. $5^3, 4^6, 2^{11}, 7^2$
51. $6^4, 3^6, 5^5, 8^3$ — $8^3, 3^6, 6^4, 5^5$
52. $3^7, 10^3, 8^2, 2^{10}$ — $8^2, 10^3, 2^{10}, 3^7$
53. $6^5, 9^4, 11^3, 5^5$ — $11^3, 5^5, 9^4, 6^5$
54. **POPULATION** There are approximately 7^{10} people living in the United States. About how many people is this? Sample answer: 282,000,000 people
55. Explain why a *quadrillion*, 10^{15}, is usually written in exponential form instead of standard form. Sample answer: It is easier and less error-prone to write 10^{15} than writing out 15 zeros.

★ 56. Tell whether the following statement is *true* or *false*. Explain your reasoning. *The number 64 can be written in exponential form in only one way.* False; 64 can be written as 8^2 or 4

★ 57. Use exponents to write $5 \cdot 5 \cdot 5 \cdot 5 \cdot 4 \cdot 4 \cdot 4$ in the shortest form possible. $5^4 \cdot 4^3$

Every effort is made to show the **Answers** to exercises (1) on the Student Edition page, or (2) in the margin of the Teacher Wraparound Edition. Answers that do not fit can be found in the Answer Appendix at the end of each chapter.

12 Chapter 1 Decimal Patterns and Algebra

★ 58. **MULTI STEP** Use a calculator to express 1,679,616 in exponential form as a power of 6. 6^8

★ 59. **POPULATION** The population of Fort Worth, Texas, is approximately 534,000. Find a number less than 534,000 and a number greater than 534,000 that can be expressed in exponential form. Sample answer: $390{,}625 < 534{,}000 < 823{,}543$, or $5^8 < 534{,}000 < 7^7$

60. **RESEARCH** Our numbering system is based on the powers of 10. The ancient Mayan system is based on another power. Use the Internet or another source to write a paragraph about the Mayan numbering system. See students' work.

★ 61. **PHYSICAL SCIENCE** The speed of sound varies as it travels through different substances. At ordinary temperatures, sound travels through water at a rate of 5,000 feet per second. Write 5,000 using exponents and the factors 5 and 10. $5^1 \cdot 10^3$

62. **CRITICAL THINKING** Based on the number pattern shown at the right, write a convincing argument that any number, except 0, raised to the 0 power equals 1. See margin.

$2^4 = 16$
$2^3 = 8$
$2^2 = 4$
$2^1 = 2$
$2^0 = ?$

Standardized Test Practice and Mixed Review

63. **MULTIPLE CHOICE** Which is *two to the sixth power* written in standard form? C

   (A) 26 (B) $6 \cdot 6$
   (C) 64 (D) $2 \cdot 2 \cdot 2 \cdot 2 \cdot 2 \cdot 2$

64. **GRID IN** How many zeros does the value of 10^8 have? 8

65. **RACING** The graph shows which numbered-cars have the most wins at the Indianapolis 500. How many more wins did the number 3 car have than the number 14 car? (Lesson 1-1) 3

Source: *Indianapolis 500 Record Book*

66. **PRODUCTION** A machine on a production line fills 8 soft drink cans per minute. How many cans does it fill in 8 hours? (Lesson 1-1) 3,840 cans

67. *True* or *False*? In the four-step plan, the *Solve* step comes last. (Lesson 1-1) false

GETTING READY FOR THE NEXT LESSON

PREREQUISITE SKILL Add, subtract, multiply, or divide. (Pages 559, 560, 562)

68. $13.4 + 8.7$ 22.1
69. $13.6 - 10.2$ 3.4
70. 1.5×6.8 10.2
71. $8.6 - 5.53$ 3.07
72. $36 \div 3$ 12
73. $1.7 + 32.28$ 33.98
74. 3.2×6 19.2
75. $24.5 \div 7$ 3.5

4 Assess

Higher-Level Thinking

Ask students to determine the steps necessary to evaluate $2^2 \cdot 3^3 \cdot 4^4$. Then evaluate the expression. First they should calculate all of the powers. Then they should multiply the products: $4 \cdot 27 \cdot 256 = 27{,}648$.

There is a Study Guide and Intervention, Practice: Skills, Practice: Word Problems, Reading to Learn Mathematics, and Enrichment Master for every lesson. These masters can be found in the Chapter Resource Masters.

Study Guide and Intervention, p. 6

NAME ____________ DATE ________ PERIOD ____

1-2 Study Guide and Intervention

Powers and Exponents

$3^4 = 3 \cdot 3 \cdot 3 \cdot 3 = 81$ (Exponent: 4; Base: 3; common factors: $3 \cdot 3 \cdot 3 \cdot 3$)

The **exponent** tells you how many times to use the **base** as a factor.

EXAMPLE 1 Write 6^3 as a product of the same factor.

The base is 6. The exponent 3 means that 6 is used as a factor 3 times.
$6^3 = 6 \cdot 6 \cdot 6$

EXAMPLE 2 Evaluate 5^4.

$5^4 = 5 \cdot 5 \cdot 5 \cdot 5$
$= 625$

EXAMPLE 3 Write $4 \cdot 4 \cdot 4 \cdot 4 \cdot 4$ in exponential form.

The base is 4. It is used as a factor 5 times, so the exponent is 5.
$4 \cdot 4 \cdot 4 \cdot 4 \cdot 4 = 4^5$

EXERCISES

Write each power as a product of the same factor.

1. 7^3 $7 \cdot 7 \cdot 7$
2. 2^7 $2 \cdot 2 \cdot 2 \cdot 2 \cdot 2 \cdot 2 \cdot 2$
3. 9^2 $9 \cdot 9$
4. 15^4 $15 \cdot 15 \cdot 15 \cdot 15$

Evaluate each expression.

5. 3^5 243
6. 7^3 343
7. 8^4 4,096
8. 5^3 125

Write each product in exponential form.

9. $2 \cdot 2 \cdot 2 \cdot 2$ 2^4
10. $7 \cdot 7 \cdot 7 \cdot 7 \cdot 7 \cdot 7$ 7^6
11. $10 \cdot 10 \cdot 10$ 10^3
12. $9 \cdot 9 \cdot 9 \cdot 9 \cdot 9$ 9^5
13. $12 \cdot 12 \cdot 12$ 12^3
14. $5 \cdot 5 \cdot 5 \cdot 5$ 5^4
15. $6 \cdot 6 \cdot 6 \cdot 6 \cdot 6$ 6^5
16. $1 \cdot 1 \cdot 1 \cdot 1 \cdot 1 \cdot 1 \cdot 1 \cdot 1$ 1^8

Practice: Skills, p. 7

NAME ____________ DATE ________ PERIOD ____

1-2 Practice: Skills

Powers and Exponents

Write each power as a product of the same factor.

1. 11^2 $11 \cdot 11$
2. 3^4 $3 \cdot 3 \cdot 3 \cdot 3$
3. 2^5 $2 \cdot 2 \cdot 2 \cdot 2 \cdot 2$
4. 9^3 $9 \cdot 9 \cdot 9$
5. 15^3 $15 \cdot 15 \cdot 15$
6. 4^3 $4 \cdot 4 \cdot 4$
7. 1^6 $1 \cdot 1 \cdot 1 \cdot 1 \cdot 1 \cdot 1$
8. 17^4 $17 \cdot 17 \cdot 17 \cdot 17$
9. 3^7 $3 \cdot 3 \cdot 3 \cdot 3 \cdot 3 \cdot 3 \cdot 3$
10. 8^6 $8 \cdot 8 \cdot 8 \cdot 8 \cdot 8 \cdot 8$

Evaluate each expression.

11. 9^2 81
12. 8^2 64
13. 8^3 512
14. 2^4 16
15. 2^5 32
16. 6^3 216
17. 3^4 81
18. 3^5 243
19. 9^3 729
20. 11^2 121
21. 4^7 16,384
22. 12^3 1,728
23. 1^9 1
24. 10^4 10,000
25. 20^4 160,000
26. 2^6 64

Write each product in exponential form.

27. $12 \cdot 12$ 12^2
28. $10 \cdot 10 \cdot 10$ 10^3
29. $4 \cdot 4 \cdot 4 \cdot 4 \cdot 4$ 4^5
30. $9 \cdot 9 \cdot 9 \cdot 9$ 9^4
31. $15 \cdot 15 \cdot 15 \cdot 15 \cdot 15$ 15^5
32. $6 \cdot 6 \cdot 6 \cdot 6 \cdot 6 \cdot 6 \cdot 6 \cdot 6$ 6^8

Practice: Word Problems, p. 8

NAME ____________ DATE ________

1-2 Practice: Word Problems

Powers and Exponents

1. **SPACE SHUTTLE** The cost of each flight of the Space Shuttle is about \$10,000,000. Write this amount in exponential form. 10^7 dollars

2. **ANIMALS** The African bush elephant is the largest land animal and weighs about 8 tons. Write this amount in exponential form. 2^3 tons

3. **VOLUME** To find the volume of a rectangular box you multiply the length times the width times the height. In a cube all sides are the same length. If the cube has length, width, and height of 6 inches, write the volume as a product. Then write it in exponential form. $6 \times 6 \times 6$; 6^3 in^3

4. **SCIENCE** A certain type of cell doubles every hour. If you start with one cell, at the end of one hour you would have 2 cells, at the end of two hours you have 4 cells, and so on. The expression $2 \times 2 \times 2 \times 2 \times 2$ tells you how many cells you would have after five hours. Write this expression in exponential form; then evaluate it. 2^5 cells; 32 cells

5. **MATH** Write 625 using exponents in as many ways as you can. 5^4; 25^2

6. **PREFIXES** Many prefixes are used in mathematics and science. The prefix giga in gigameter represents 1,000,000,000 meters. Write this prefix as a power of ten. 10^9 meters

7. **LIBRARY** The school library contains 9^4 books. How many library books are in the school library? 6,561 books

8. **HOT DOGS** The concession stand at the county fair sold 6^3 hot dogs on the first day. How many hot dogs did they sell? 216 hot dogs

1-3 Lesson Notes

Lesson Resources

Chapter 1 Resource Masters
- Study Guide and Intervention, p. 11
- Practice: Skills, p. 12
- Practice: Word Problems, p. 13
- Reading to Learn Mathematics, p. 14
- Enrichment, p. 15

Calculator and Spreadsheet Masters, p. 14

MathPASS Tutorial CD-ROM, 4

Parent and Student Study Guide Workbook, 1-3

FAST FILE

1 Focus

5-Minute Check 1-3

INTERACTIVE CHALKBOARD PowerPoint® Presentations

Use as a quiz or review of Lesson 1-2. (Also available as a transparency)

Bellringer

Problem Solving Write $5 + 3 \cdot 4$ on the chalkboard and ask students to evaluate it. 17 Ask them which operation they did first and why. multiplication; order of operations If any students mistakenly found a solution of 60, have them enter the entire expression on a scientific calculator and solve. Explain that it matters in what order the operations are solved.

(Continued in bottom margin.)

Reading to Learn Mathematics, p. 14 ELL

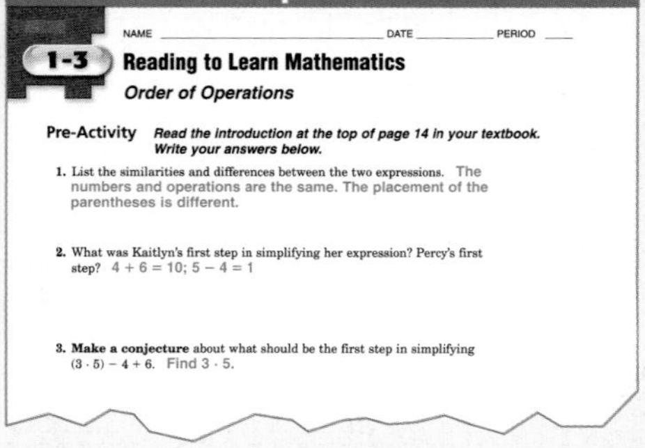
NAME ______ DATE ______ PERIOD ____

1-3 Reading to Learn Mathematics

Order of Operations

Pre-Activity *Read the introduction at the top of page 14 in your textbook. Write your answers below.*

1. List the similarities and differences between the two expressions. The numbers and operations are the same. The placement of the parentheses is different.
2. What was Kaitlyn's first step in simplifying her expression? Percy's first step? $4 + 6 = 10$; $5 - 4 = 1$
3. **Make a conjecture** about what should be the first step in simplifying $(3 \cdot 5) - 4 + 6$. Find $3 \cdot 5$.

1-3 Order of Operations

What You'll Learn
Evaluate expressions using the order of operations.

NEW Vocabulary
- numerical expression
- order of operations

"When am I ever going to use this?" The lesson openers answer that question.

WHEN am I ever going to use this?

GAMES Kaitlyn and Percy are each writing as many different expressions as possible using the numbers rolled on a number cube. In the first round, 3, 5, 4, and 6 were rolled in order.

Kaitlyn	Percy
$3 \cdot 5 - (4 + 6) = 3 \cdot 5 - 10$ $= 15 - 10$ or 5	$3 \cdot (5 - 4) + 6 = 3 \cdot 1 + 6$ $= 3 + 6$ or 9

1. List the similarities and differences between the two expressions. The numbers and operations are the same. The placement of the parentheses is different.
2. What was Kaitlyn's first step in simplifying her expression? Percy's first step? $4 + 6 = 10$; $5 - 4 = 1$
3. **Make a conjecture** about what should be the first step in simplifying $(3 \cdot 5) - 4 + 6$. Find $3 \cdot 5$.

The expressions $3 \cdot 5 - (4 + 6)$ and $3 \cdot (5 - 4) + 6$ are **numerical expressions**. Mathematicians have agreed upon steps called the **order of operations** to find the values of such expressions. These rules ensure that numerical expressions have only one value.

Key Concept — **Order of Operations**

1. Do all operations within grouping symbols first.
2. Evaluate all powers before other operations.
3. Multiply and divide in order from left to right.
4. Add and subtract in order from left to right.

You can use the order of operations to evaluate numerical expressions.

EXAMPLES Evaluate Expressions

Examples illustrate all of the concepts taught in the lesson and closely mirror the exercises in the Guided Practice and Applications sections.

1 Evaluate $5 + (12 - 3)$.

$5 + (12 - 3) = 5 + 9$ Subtract first since $12 - 3$ is in parentheses.
$= 14$ Add 5 and 9.

2 Evaluate $8 - 3 \cdot 2 + 7$.

$8 - 3 \cdot 2 + 7 = 8 - 6 + 7$ Multiply 3 and 2.
$= 2 + 7$ Subtract 6 from 8.
$= 9$ Add 2 and 7.

Your Turn **Evaluate each expression.**

a. $39 \div (9 + 4)$ 3
b. $10 + 8 \div 2 - 6$ 8

Your Turn exercises allow students to practice concepts in a classroom setting.

Then have students solve the problems at the right using a scientific calculator to deduce which operations need to be solved first in each problem.

$3 + 4 \cdot 5$	23
$3 + 4 - 5$	2
$5 - 4 + 3$	4
$9 - 4 \cdot 2$	1
$5 - 9 \div 3$	2
$8 \cdot 3 \div 4$	6
$6 \cdot 7 - 4$	38
$10 \div 2 + 2$	7

EXAMPLES Evaluate Expressions with Powers

3 **Evaluate 15×10^3.**

$15 \times 10^3 = 15 \times 1{,}000$ Find the value of 10^3.

$= 15{,}000$ Multiply.

4 **Evaluate $36 \div (1 + 2)^2$.**

$36 \div (1 + 2)^2 = 36 \div 3^2$ Add 1 and 2 inside the parentheses.

$= 36 \div 9$ Find the value of 3^2.

$= 4$ Divide.

Your Turn Evaluate each expression.

c. 3×10^4 **30,000** d. $(5 - 1)^3 \div 4$ **16**

STUDY TIP

Powers If an exponent lies outside a grouping symbol, complete the operations within the grouping symbol before applying the power.

In addition to using the symbols × and ·, multiplication can be indicated by using parentheses.

$2(3 + 5)$ means $2 \times (3 + 5)$ or $2 \cdot (3 + 5)$.

$(4 - 2)3$ means $(4 - 2) \times 3$ or $(4 - 2) \cdot 3$.

TEACHING TIP

ell students that they an change the order of perations on a alculator by using the arentheses keys.

EXAMPLE Evaluate an Expression

5 **Evaluate $14 + 3(7 - 2) - 2 \cdot 5$.**

$14 + 3(7 - 2) - 2 \cdot 5$

$= 14 + 3(5) - 2 \cdot 5$ Subtract 2 from 7.

$= 14 + 15 - 2 \cdot 5$ Multiply from left to right, $3 \cdot 5 = 15$.

$= 14 + 15 - 10$ Multiply from left to right, $2 \cdot 5 = 10$.

$= 29 - 10$ Add from left to right, $14 + 15 = 29$.

$= 19$ Subtract 10 from 29.

EAL-LIFE MATH

EO GAMES A team of 0 engineers and hnicians built the first eo-game console in 66.

rce: www.geekcomix.com

EXAMPLE Use an Expression to Solve a Problem

6 **VIDEO GAMES Evita is buying a video game station, three extra controllers, and four new video games. What is the total cost?**

Item	Quantity	Unit Cost
game station	1	$180.00
controller	3	$24.95
game	4	$35.99

cost of game station + number of controllers × cost of controller + number of games × cost of game

$180 + 3 × $24.95 + 4 × $35.99

$= 180 + 74.85 + 143.96$ Multiply from left to right.

$= 398.81$ Add.

So, the total cost is $398.81.

Check Check the reasonableness of the answer by estimating. The cost is about $180 + (25 \times 3) + (40 \times 4) = 180 + 75 + 160$, or $415. So, the solution of $398.81 is reasonable.

msmath2.net/extra_examples

2 Teach

In-Class Examples

INTERACTIVE CHALKBOARD PowerPoint® Presentations

1 Evaluate $27 - (18 + 2)$. **7**

2 Evaluate $15 + 5 \cdot 3 - 2$. **28**

3 Evaluate 12×10^4. **120,000**

4 Evaluate $28 \div (3 - 1)^2$. **7**

5 Evaluate $32 - 5(4 + 2) + 8 \cdot 2$. **18**

6 **VIDEO GAMES** Use the table shown in Example 6 in the Student Edition. Taylor is buying two video game stations, five extra controllers, and ten games. What is the total cost? **$844.65**

Tips for New Teachers

Teaching Strategies One common mnemonic for helping students remember the order of operations goes, "*P*lease *e*xcuse *m*y *d*ear *A*unt *S*ally." The first letters of each word stand for *parentheses, exponents, multiply, divide, add,* and *subtract*.

DAILY INTERVENTION **Differentiated Instruction** **ELL**

Verbal/Linguistic Have students work in pairs. Have one student make up a numerical expression and describe it to the other student. The other student has to write it correctly. Then both students evaluate the expression. Make sure students use a variety of operations and include parentheses in their notations.

Daily Intervention notes help you help students when they need it most. **Differentiated Instruction** suggestions are keyed to eight commonly accepted learning styles.

3 Practice/Apply

Skill and Concept Check exercises are intended to be completed in class to ensure that students understand the concepts in the lesson. The **Guided Practice** exercises are representative of the exercises used for homework or seatwork.

FOLDABLES™ Follow-Up Encourage students to summarize the order of operations on the third section of their Foldables, and explain its importance. Students might include an example illustrating how they use the order of operations to evaluate expressions.

DAILY INTERVENTION **FIND THE ERROR** In Exercise 3, encourage students to use parentheses to clarify the order of operations. In Yutaka's evaluation, have students compare the results of evaluating the multiplication before the division and after the division. Remind them that multiplication and division are evaluated in order from left to right.

About the Exercises...

Guided Practice Key

Exercises	Examples
4–9	1, 2
10–11, 13	3, 4
12, 14	5

Odd/Even Assignments
Exercises 15–48 are structured so that students practice the same concepts whether they are assigned odd or even problems.

Skill and Concept Check

Writing Math Exercise 3

1. **Identify** the operation that should be done first in each expression.
   a. $9 \div 3 + (14 - 7)$ $14 - 7$ b. $3 + 24 \div 3 \cdot 4$ $24 \div 3$
2. **OPEN ENDED** Write an expression containing five numbers that is evaluated by first multiplying. Sample answer: $2 + 3(4) - 6 \div 2$
3. **FIND THE ERROR** Yutaka and Cynthia are evaluating $16 - 24 \div 6 \cdot 2$. Who is correct? Explain your reasoning. See margin.

Yutaka
$16 - 24 \div 6 \cdot 2$
$= 16 - 24 \div 12$
$= 16 - 2$ or 14

Cynthia
$16 - 24 \div 6 \cdot 2$
$= 16 - 4 \cdot 2$
$= 16 - 8$ or 8

GUIDED PRACTICE

Evaluate each expression.

4. $11 - (3 \cdot 2)$ 5
5. $25 \div (9 - 4)$ 5
6. $8 - 4 + 3.7$ 7.7
7. $2 + 5 \cdot 5$ 27
8. $14 \div 2 \cdot 6$ 42
9. $8 \cdot 5 - 4 \cdot 3$ 28
10. 4×10^2 400
11. $3.5 \times 5 + 6^2$ 53.5
12. $7 + 4(5.6 - 2) - 9$ 12.4
13. Evaluate $(16 \div 4)^3 - 6$. 58
14. Find the value of $(6 + 8) \div (10 - 8)$. 7

TEACHING TIP Watch for students who incorrectly add first in Exercise 6, or students who incorrectly multiply first in Exercise 8.

★ indicates increased difficulty

Practice and Applications

HOMEWORK HELP

For Exercises	See Examples
15–28	1, 2
29–36, 43–45	3, 4
37–40, 46–48	5
41–42, 50	6

Extra Practice See pages 564, 596.

Evaluate each expression.

15. $(1 + 8) \times 3$ 27
16. $10 - (3 + 4)$ 3
17. $(25 \div 5) + 8$ 13
18. $(11 - 2) \div 9$ 1
19. $3 \cdot 2 + 7$ 13
20. $15 \div 3 + 4$ 9
21. $12 + 6.6 \div 3$ 14.2
22. $18 - 3 \cdot 6$ 0
23. $8 - 7.2 + 5$ 5.8
24. $28 \div 7(5)$ 20
25. $(17 + 3) \div (4 + 1)$ 4
26. $(6 + 5) \cdot (8 - 6)$ 22
27. $21 \div 3 \times 2 - 4$ 10
28. $35 \div 5 + 56 \div 7$ 15
29. $2 \times 9 - 4^2$ 2
30. $24 \div 3 + 5^3$ 133
31. $7 + (8 - 7 + 2)^4$ 88
32. $(2 + 10)^2 \div 4$ 36
33. 6×10^2 600
34. 18×10^3 18,000
35. 1.95×10^2 195
36. 3.7×10^4 37,000
37. $6 + 2(9.4 - 1)$ 22.8
38. $3(4.5 + 7.2) - 5 \cdot 4$ 15.1

★ 39. $72 \div 3 - 5(8.8 - 6) + 9$ 19
★ 40. $9 \div 3 \cdot 14(10 - 8) - 60$ 24

41. **GEOMETRY** The distance around a geometric figure is called its *perimeter.* Write a numerical expression to find the perimeter of the figure at the right. Then evaluate the expression. Sample answer: $2(7) + 2(9.5) + 6$; 39 cm

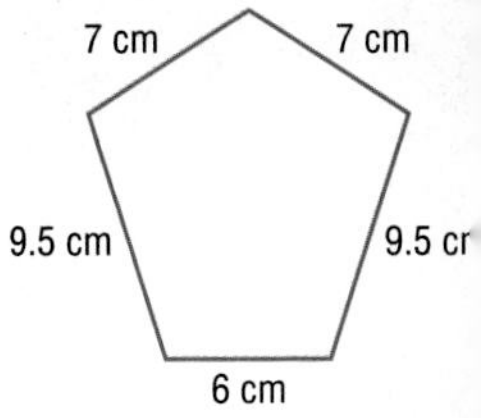

42. **MARATHONS** On Mondays, Wednesdays, and Thursdays, Jacob trains for a marathon for 3.5 hours. On Tuesdays and Fridays, he trains for 2 hours, and on Saturdays, he trains for 4.5 hours. How many hours does Jacob train per week? 19 h/wk

Answers

3. Cynthia; the first step is to do the division, $24 \div 6$. Yutaka incorrectly multiplied 6 and 2 first.

51. 16; The calculator followed the order of operations. Ruby incorrectly added $2 + 2$ first when she should have divided $12 \div 2$ first.

52. Sample answer:
Enter 12 [÷] 2 [+] 2 [×] 5 [ENTER].
If the calculator displays 16, it follows the order of operations. If it displays 40, it evaluated the expression from left to right and does not follow the order of operations.

Find the value of each expression.

43. 4.5×10^2 450 **44.** 3.08×10^4 30,800 **45.** 2.965×10^5 296,500

★**46.** $3 \cdot 4(5 - 3.8) + 2.7$ 17.1 ★**47.** $7.1 \times 9 - (4 + 3) + 1$ 57.9 ★**48.** $(6 + 1)^2 - 2.5(3)$ 41.5

49. Insert parentheses to make $72 \div 9 + 27 - 2 = 0$ true. $72 \div (9 + 27) - 2 = 0$

50. **MONEY MATTERS** Luke is ordering nine reams of paper, four boxes of pens, and two rolls of tape. Use the table to find the cost of his order. $68.09

Item	Cost
ream of paper	$3.95
box of pens	$7.49
roll of tape	$1.29

CRITICAL THINKING Some calculators are programmed to follow the order of operations. 51–52. See margin.

51. Ruby evaluates $12 \div 2 + 2 \cdot 5$ and gets 15. When she checks using her calculator, she gets 16. Which is correct? Explain.

52. Explain how you could tell whether your calculator follows the order of operations.

Standardized Test Practice and Mixed Review

53. **MULTIPLE CHOICE** Which is the first step in evaluating $10 - 3 \cdot 2 + 4(15 - 5)$? D

Ⓐ $10 - 3$ Ⓑ $3 \cdot 2$ Ⓒ $2 + 4$ Ⓓ $15 - 5$

54. **MULTIPLE CHOICE** Anthony has 2 boxes of cookies containing 24 cookies each and 3 packages of brownies containing 15 brownies each. Which expression *cannot* be used to find the total number of items he has? G

Ⓕ $3 \times 15 + 2 \times 24$ Ⓖ $5 \times (24 + 15)$

Ⓗ $2(24) + 3(15)$ Ⓘ $15 + 15 + 15 + 24 + 24$

55. **TESTING** Use the graphic to estimate how much longer educators spend preparing students in grades 6–8 for tests than students in grades 9–12. (Lesson 1-1) Sample answer: 30 − 20, or about 10 days longer

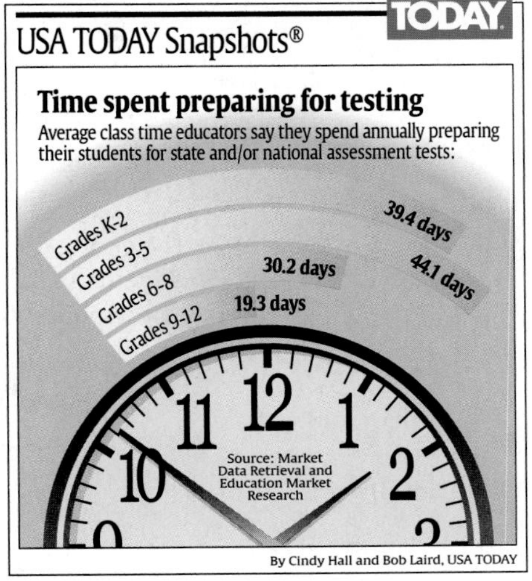

By Cindy Hall and Bob Laird, USA TODAY

Evaluate each expression. (Lesson 1-2)

56. 2^4 16 **57.** 11^2 121

58. 3^6 729 **59.** 4^5 1,024

Write each power as a product of the same factor. (Lesson 1-2)

60. 3^4 $3 \cdot 3 \cdot 3 \cdot 3$ **61.** 8^5 $8 \cdot 8 \cdot 8 \cdot 8 \cdot 8$

62. 6^2 $6 \cdot 6$ **63.** 4^6 $4 \cdot 4 \cdot 4 \cdot 4 \cdot 4 \cdot 4$

64. 1^4 $1 \cdot 1 \cdot 1 \cdot 1$ **65.** 9^1 9

GETTING READY FOR THE NEXT LESSON

BASIC SKILL **Divide.**

66. $45 \div 15$ 3 **67.** $90 \div 5$ 18 **68.** $110 \div 10$ 11 **69.** $52 \div 4$ 13

msmath2.net/self_check_quiz

Glencoe's exclusive partnership with USA TODAY® provides actual USA TODAY Snapshots® that illustrate mathematical concepts.

4 Assess

Open-Ended Assessment

Writing Have students take turns reading complex expressions for the rest of the class to write. Have some students write their interpretations on the chalkboard. Discuss ways of clarifying the order of operations.

Study Guide and Intervention, p. 11

NAME ______ DATE ______ PERIOD ____

1-3 Study Guide and Intervention

Order of Operations

Use the **order of operations** to evaluate numerical expressions.

1. Do all operations within grouping symbols first.
2. Evaluate all powers before other operations.
3. Multiply and divide in order from left to right.
4. Add and subtract in order from left to right.

EXAMPLE 1 **Evaluate $(10 - 2) - 4 \cdot 2$.**

$(10 - 2) - 4 \cdot 2 = 8 - 4 \cdot 2$ Subtract first since $10 - 2$ is in parentheses.
$= 8 - 8$ Multiply 4 and 2.
$= 0$ Subtract 8 from 8.

EXAMPLE 2 **Evaluate $8 + (1 + 5)^2 \div 4$.**

$8 + (1 + 5)^2 \div 4 = 8 + 6^2 \div 4$ First, add 1 and 5 inside the parentheses.
$= 8 + 36 \div 4$ Find the value of 6^2.
$= 8 + 9$ Divide 36 by 4.
$= 17$ Add 8 and 9.

EXERCISES

Evaluate each expression.

1. $(1 + 7) \times 3$ 24 **2.** $28 - 4 \cdot 7$ 0 **3.** $5 + 4 \cdot 3$ 17

4. $(40 \div 5) - 7 + 2$ 3 **5.** $35 \div 7(2)$ 10 **6.** 3×10^3 3,000

7. $45 \div 5 + 36 \div 4$ 18 **8.** $42 \div 6 \times 2 - 9$ 5 **9.** $2 \times 8 - 3^2 + 2$ 9

10. $5 \times 2^2 + 32 \div 8$ 24 **11.** $3 \times 6 - (9 - 8)^3$ 17 **12.** 3.5×10^2 350

Practice: Skills, p. 12

NAME ______ DATE ______ PERIOD ____

1-3 Practice: Skills

Order of Operations

Evaluate each expression.

1. $9 - 3 + 4$ 10 **2.** $8 + 6 - 5$ 9 **3.** $12 \div 4 + 5$ 8

4. $25 \times 2 - 7$ 43 **5.** $36 \div 9(2)$ 8 **6.** $6 + 3(7 - 2)$ 21

7. $3 \times 6.2 + 5^2$ 43.6 **8.** $(1 + 11)^2 \div 3$ 48 **9.** $12 - (2 + 8)$ 2

10. $15 - 24 \div 4 \cdot 2$ 3 **11.** $(4 + 2) \cdot (7 + 4)$ 66 **12.** $(3 \cdot 18) \div (2 \cdot 9)$ 3

13. $24 \div 6 + 4^2$ 20 **14.** $3 \times 8 - (9 - 7)^3$ 16 **15.** $9 + (9 - 8 + 3)^4$ 265

16. $3 \times 2^2 + 24 \div 8$ 15 **17.** $(15 \div 3)^2 + 9 \div 3$ 28 **18.** $(52 \div 4) + 5^3$ 138

19. 26×10^3 26,000 **20.** 7.2×10^2 720 **21.** $5 \times 4^2 - 3 \times 2$ 74

22. $24 \div 6 \div 2$ 2 **23.** $13 - (6 - 5)^5$ 12 **24.** $(8 - 3 \times 2) \times 6$ 12

25. $(11 \cdot 4 - 10) \div 2$ 17 **26.** $10 \div 2 \times (4 - 3)$ 5 **27.** 1.82×10^5 182,000

28. $35 \div 7 \times 2 - 4$ 6 **29.** $2^5 + 7(9 - 1)$ 88 **30.** $12 + 16 \div (3 + 1)$ 16

Practice: Word Problems, p. 13

NAME ______ DATE ______ PERIOD ____

1-3 Practice: Word Problems

Order of Operations

1. **FOOTBALL** The middle school team scored three field goals worth three points each and two touchdowns worth six points each, including extra points. Write a numerical expression to find the team's score. Then evaluate the expression. 3(3) + 2(7); 23 points

2. **BOOKS** Juan goes to the schoolbook fair where paperback books are $1.50 and hardback books are $3.00. Juan buys 5 paperback and 2 hardback books. Write a numerical expression to find how much Juan paid for the books. Then evaluate the expression. 5(1.5) + 2(3); $13.50

3. **GEOMETRY** The perimeter of a hexagon is found by adding the lengths of all six sides of the hexagon. For the hexagon below write a numerical expression to find the perimeter. Then evaluate the expression. 4(5) + 2(8); 36

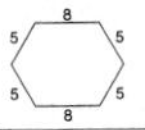

4. **MONEY** Aisha bought school supplies consisting of 6 spiral notebooks costing $0.39 each, 2 packages of pencils at $0.79 each, and a 3-ring binder for $1.99. Write an expression to find the total amount Aisha spent on school supplies. Then evaluate the expression. $6 \times 0.39 + 2 \times 0.79 + 1.99$; $5.91

5. **REASONING** Use the order of operations and the digits 2, 4, 6, and 8 to create an expression with a value of 2. Sample answer: $4^2 - (8 + 6)$

6. **NUMBER SENSE** Without parentheses, the expression $8 + 30 \div 2 + 4$ equals 27. Place parentheses in the expression so that it equals 13; then 23. $8 + 30 \div (2 + 4)$; $(8 + 30) \div 2 + 4$

7. **MONEY** Tyrone bought 5 postcards at $0.55 each and a set of postcards for $1.20. Write an expression to find the total amount Tyrone spent on postcards. Then evaluate the expression. 5(0.55) + 1.2; $3.95

8. **DINING** Mr. Firewalks took his family out to eat. They ordered 3 meals costing $8.99 each, 2 sodas at $1.50 each, and 1 glass of tea for $1.25. Write an expression to find the total amount the Firewalks family spent on dinner before taxes and tip. Then evaluate the expression. 3(8.99) + 2(1.5) + 1.25; $31.22

1-4 Lesson Notes

Lesson Resources

Chapter 1 Resource Masters
- Study Guide and Intervention, p. 16
- Practice: Skills, p. 17
- Practice: Word Problems, p. 18
- Reading to Learn Mathematics, p. 19
- Enrichment, p. 20

Calculator and Spreadsheet Masters, p. 15

Key Teaching Transparencies, 2

Parent and Student Study Guide Workbook, 1-4

Teaching Mathematics with Manipulatives, pp. 32–33

FAST FILE

1 Focus

5-Minute Check 1-4

INTERACTIVE CHALKBOARD PowerPoint® Presentations

Use as a quiz or review of Lesson 1-3. (Also available as a transparency)

Bellringer

Reading Mathematics Read the following sentence. *How alert you are is variable and may depend on how much sleep you get.* Ask students what the word *variable* means in this context. **likely to change** Ask for any other definitions of variable. Explain that in math, *variable* means having no fixed quantitative value.

(Continued in bottom margin.)

Reading to Learn Mathematics, p. 19 **ELL**

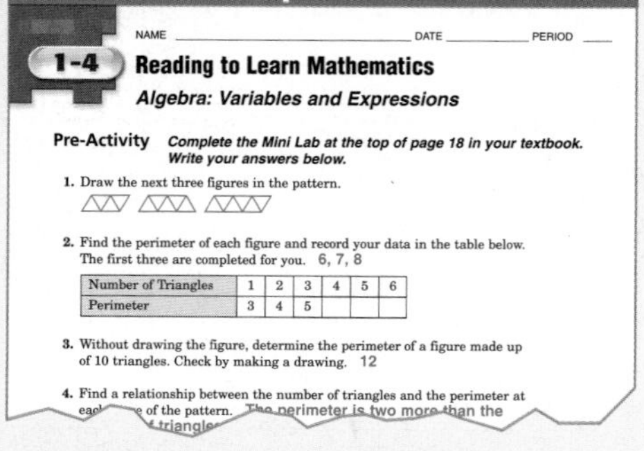
NAME ______ DATE ______ PERIOD ______

1-4 **Reading to Learn Mathematics**

Algebra: Variables and Expressions

Pre-Activity *Complete the Mini Lab at the top of page 18 in your textbook. Write your answers below.*

1. Draw the next three figures in the pattern.

2. Find the perimeter of each figure and record your data in the table below. The first three are completed for you. 6, 7, 8

Number of Triangles	1	2	3	4	5	6
Perimeter	3	4	5			

3. Without drawing the figure, determine the perimeter of a figure made up of 10 triangles. Check by making a drawing. 12

4. Find a relationship between the number of triangles and the perimeter at each of the pattern. The perimeter is two more than the number of triangles.

1-4 Algebra: Variables and Expressions

What You'll Learn
Evaluate simple algebraic expressions.

NEW Vocabulary
- variable
- algebraic expression
- algebra
- term
- coefficient
- constant

REVIEW Vocabulary

evaluate: find the value (Lesson 1-2)

Review Vocabulary lists terms and definitions presented earlier in the course.

HANDS-ON Mini Lab

Materials
- isometric dot paper

Work with a partner.

The pattern below is made up of triangles, each with side lengths of 1.

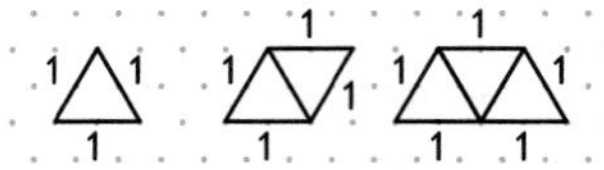

1. Draw the next three figures in the pattern. **See margin.**
2. Find the perimeter of each figure and record your data in a table like the one shown below. The first three are completed for you.

Number of Triangles	1	2	3	4	5	6
Perimeter	3	4	5	?	?	?

6 7 8

3. Without drawing the figure, determine the perimeter of a figure made up of 10 triangles. Check by making a drawing. **12**
4. Find a relationship between the number of triangles and the perimeter at each stage of the pattern.

The perimeter is two more than the number of triangles.

In the Mini Lab, you found that the perimeter of the figure is two more than the number of triangles. You can use a placeholder, or **variable**, to represent the number of triangles.

number of triangles → $n + 2$ → perimeter of figure

The expression $n + 2$ is called an **algebraic expression** because it contains variables, numbers, and at least one operation. The branch of mathematics that involves expressions with variables is called **algebra**.

READING Math

Variables The letter x is often used as a variable. It is also common to use the first letter of a word as the variable. For example, use t for *t*ime or n for *n*umber.

EXAMPLE Evaluate an Expression

1 **Evaluate $n + 3$ if $n = 4$.**

$n + 3 = 4 + 3$ Replace n with 4.

$= 7$

Your Turn **Evaluate each expression if $c = 8$.**

a. $c + 2$ **10** b. $c - 3$ **5** c. $15 - c$ **7**

Ask students why they think variables might be important in math. **They can be used whenever you do not know a number or when the number could change based on the situation.**

Answer (Hands-On Mini Lab)

1. 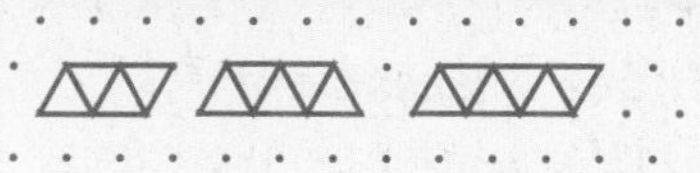

The following symbols are used for multiplication and division with variables. Note that the multiplication sign is usually omitted.

$6d$	means	$6 \times d$ or $6 \cdot d$	$9st$	means	$9 \times s \times t$
ef	means	$e \times f$	$\frac{c}{d}$	means	$c \div d$

When plus or minus signs separate an algebraic expression into parts, each part is a **term**. The numerical factor of a term that contains a variable is called a **coefficient**. A term that does not contain a variable is called a **constant**.

8 is the coefficient of $8x$. → $8x + y + 3$ ← constant

1 is the coefficient of y because $1y = y$.

ading Math
atures help
udents learn and
e the language
mathematics.

EADING Math

Multiplication In Example 2, the $8w$ means 8 times w, but 85 does not mean 8 times 5. So, there are parentheses around the 5 in 8(5).

EAL-IFE CAREERS

w Does an Athletic iner Use Math?

letic trainers use the mula $(220 - a) \times 0.6$, ere a is a person's age, find their minimum ning heart rate.

Online Research
information about a career an athletic trainer, visit: math2.net/careers

EXAMPLES Evaluate Expressions

2 Evaluate $8w - 2v$ if $w = 5$ and $v = 3$.

$8w - 2v = 8(5) - 2(3)$ Replace w with 5 and v with 3.

$= 40 - 6$ Use the order of operations.

$= 34$ Subtract 6 from 40.

3 Evaluate $\frac{mn}{6}$ if $m = 8$ and $n = 12$.

$\frac{mn}{6} = \frac{(8)(12)}{6}$ Replace m with 8 and n with 12.

$= \frac{96}{6}$ The fraction bar is like a grouping symbol.

$= 16$ Divide.

4 Evaluate $y^2 + 2$ if $y = 3$.

$y^2 + 2 = 3^2 + 2$ Replace y with 3.

$= 9 + 2$ or 11 Use the order of operations.

Your Turn **Evaluate each expression if $a = 4$ and $b = 3$.**

d. $9a - 6b$ 18 e. $\frac{ab}{2}$ 6 f. $a^2 + 5$ 21

EXAMPLE Use an Expression to Solve a Problem

5 HEALTH Use the formula at the left to find Kaylee's minimum training heart rate if she is 17 years old.

$(220 - a) \times 0.6 = (220 - 17) \times 0.6$ Replace a with 17.

$= 203 \times 0.6$ or 121.8 Use the order of operations.

Kaylee's minimum training heart rate of 121.8 beats per minute is rounded up to 122 because her heart cannot beat 0.8 times.

msmath2.net/extra_examples

2 Teach

In-Class Examples

INTERACTIVE CHALKBOARD PowerPoint® Presentations

1 Evaluate $t - 4$ if $t = 6$. 2

2 Evaluate $5x + 3y$ if $x = 7$ and $y = 9$. 62

3 Evaluate $\frac{rs}{4}$ if $r = 7$ and $s = 12$. 21

4 Evaluate $5 + a^2$ if $a = 5$. 30

5 **PHYSICS** The final speed of a falling object is found by using the expression $v + 9.8t$, where v is the speed when you begin timing and t is the length of time the object falls. Find the final speed when the object starts falling at 3 meters per second and falls for 2 seconds. 22.6 m/s

DAILY INTERVENTION Unlocking Misconceptions

Variables Make sure students understand there is nothing special about which letters are used for variables. It is common to use letters signifying the object being replaced, such as using a for age. However, this is merely a convenience. Some students may feel that some letters should not be used, such as x and y. Have students practice writing expressions with different variables to become accustomed to using a variety of letters.

Daily Intervention notes help you help students when they need it most. **Unlocking Misconceptions** suggestions help you analyze common errors so you can help students avoid these trouble spots.

3 Practice/Apply

FOLDABLES™ Follow-Up Remind students to use their Foldables to summarize what they learn about variables and expressions. On the fourth section, they might include the definitions of key terms such as *evaluate*, or record and evaluate an example of a simple algebraic expression.

Which One Doesn't Belong? In Exercise 3, make sure students realize that the similarities or differences are not based on the variable being used. You may wish to rewrite all the expressions using the same variables.

About the Exercises...

Guided Practice Key

Exercises	Examples
4, 6	1
5, 7	2
8–9, 11–12	3
10, 13	4

Odd/Even Assignments
Exercises 14–37 and 41–42 are structured so that students practice the same concepts whether they are assigned odd or even problems.

Answer

38.

Number of Games	Amount Earned
1	\$16
2	\$32
3	\$48
4	\$64
5	\$80

Skill and Concept Check

Writing Math Exercise 3

1. **Identify** the coefficient and constant in the expression $2x - 6$.
2. **OPEN ENDED** Write an algebraic expression that involves multiplication and addition. **Sample answer: $3x + 7$** **1. coefficient: 2; constant: -6**
3. **Which One Doesn't Belong?** Identify the expression that does not have the same characteristic as the other three. Explain your reasoning.

$4x$	ab	$r \cdot 8$	$\frac{x}{y}$

$\frac{x}{y}$ is a division expression; the others are multiplication expressions.

GUIDED PRACTICE

Evaluate each expression if $a = 3$, $b = 5$, and $c = 4$.

4. $a + 7$ **10**
5. $4b$ **20**
6. $b + c$ **9**
7. $6c - a$ **21**
8. $\frac{c}{4}$ **1**
9. $\frac{5c + b}{5}$ **5**
10. $4b^2$ **100**
11. $\frac{3c^2}{a}$ **16**
12. What operations are used in the algebraic expression $\frac{xyz}{2}$? **multiplication and division**
13. Evaluate $n^2 + 1$ if $n = 2$. **5**

★ indicates increased difficulty

Practice and Applications

HOMEWORK HELP

For Exercises	See Examples
14–19	1
20–22	2
23–25	3
26–35	4
38–40, 43–45	5

Extra Practice See pages 565, 596.

Evaluate each expression if $d = 8$, $e = 3$, $f = 4$, and $g = 1$.

14. $d + 9$ **17**
15. $10 - e$ **7**
16. $4f$ **16**
17. $8g$ **8**
18. $f - e$ **1**
19. $d + f$ **12**
20. $10g - 6$ **4**
21. $8 + 5d$ **48**
22. $7e + 2d$ **37**
23. $\frac{d}{5}$ **1.6**
24. $\frac{16}{f}$ **4**
25. $\frac{5d - 25}{5}$ **3**
26. e^2 **9**
27. $\frac{d^2}{2}$ **32**
28. $6f^2$ **96**
29. $11.9d - 10e$ **65.2**
30. $\frac{2f}{g^2}$ **8**
31. $\frac{2.5f + 5}{3}$ **5**
32. $7e^2 + 5$ **68**
33. $\frac{(5 + g)^2}{2}$ **18**
34. Evaluate $9h + h^2$ if $h = 3.5$. **43.75**
35. For what value of x is the expression x^2 equal to 9? **3**

Determine whether each statement is *sometimes, always,* or *never* true. Explain.

★ 36. The expressions $x - 3$ and $y - 3$ represent the same value. **Sometimes; $x - 3$ and $y - 3$ represent the same value only when $x = y$.**

★ 37. The expression ab means the same as the expression $a \times b$. **Always; ab means a is multiplied by b.**

BASEBALL For Exercises 38–40, use the following information.
Charlie earns \$16 per game for umpiring Little League baseball games.

38. Make a table that shows how much money Charlie will earn for umpiring 1, 2, 3, 4, and 5 baseball games. **See margin.**
39. Write an expression to find how much money Charlie will earn for any number of games. Let n be the number of games he umpires. **$16n$**
40. Charlie umpires 18 games during the summer. How much will he earn? **\$288**

Complete each table.

★ **41.**

Quarts (q)	Gallons $\left(\frac{q}{4}\right)$
4	1
6	1.5
8	? 2
10	? 2.5
12	? 3
14	? 3.5

★ **42.**

Dollars (d)	Quarters ($4d$)
1	4
2	8
3	? 12
4	? 16
5	? 20
6	? 24

43. HEALTH The expression $110 + \frac{A}{2}$, where A stands for a person's age, is used to estimate a person's normal systolic blood pressure. Estimate the normal systolic blood pressure for a 16-year old. about 118

BOATING For Exercises 44 and 45, use the information below and at the right.
The Lerman family is vacationing at the lake and wants to rent a boat.

44. Write an expression showing the cost of renting a boat for h hours. $45h + 65$

45. How much would it cost if they rented the boat for 2, 3, or 4 hours? $155; $200; $245

★ **46. WRITE A PROBLEM** Write an algebraic expression with the variable x that has a value of 3 when evaluated. Sample answer: $5x - 37$ if $x = 8$

47. CRITICAL THINKING Find values of x and y so that the value of $7x + 2$ is greater than the value of $3y + 23$. Sample answer: $x = 5$, $y = 2$

Standardized Test Practice and Mixed Review

48. MULTIPLE CHOICE Which expression could be used to find the cost of buying b baseball bats at $75 each and g baseball gloves at $98 each? A

Ⓐ $75b + 98g$ Ⓑ $b + g$ Ⓒ $75 + 98$ Ⓓ $75b \div 98g$

49. SHORT RESPONSE Juan opened a bank account one month with $40 and deposited $15 each month after that. He was able to save $100 by December. In which month did he open the bank account? August

Evaluate each expression. (Lesson 1-3)

50. $6(5) - 2$ 28 **51.** $9 + 9 \div 3$ 12 **52.** $4 \cdot 2(8 - 1)$ 56 **53.** $(17 + 3) \div 5$ 4

54. Write 2^5 as a product of the same factor. (Lesson 1-2) $2 \cdot 2 \cdot 2 \cdot 2 \cdot 2$

GETTING READY FOR THE NEXT LESSON

PREREQUISITE SKILL Determine whether each sentence is *true* or *false*. (Lesson 1-3)

55. $15 - 2(3) = 9$ true **56.** $20 \div 5 \times 4 = 1$ false **57.** $4^2 + 6 \cdot 7 = 154$ false **58.** $24 \div (7 - 3) = 6$ true

4 Assess

Open-Ended Assessment

Modeling Have students model simple expressions using counting blocks or tiles. They should choose one of the objects as the variable. For example, if pencils are the variable, 3 blocks + 2 pencils = $3 + 2a$.

Study Guide and Intervention, p. 16

NAME ________ DATE ________ PERIOD ____

1-4 Study Guide and Intervention
Algebra: Variables and Expressions

To evaluate an algebraic expression you replace each variable with its numerical value, then use the order of operations to simplify.

EXAMPLE 1 **Evaluate $6x - 7$ if $x = 8$.**

$6x - 7 = 6(8) - 7$ Replace x with 8.
$= 48 - 7$ Use the order of operations.
$= 41$ Subtract 7 from 48.

EXAMPLE 2 **Evaluate $5m - 3n$ if $m = 6$ and $n = 5$.**

$5m - 3n = 5(6) - 3(5)$ Replace m with 6 and n with 5.
$= 30 - 15$ Use the order of operations.
$= 15$ Subtract 15 from 30.

EXAMPLE 3 **Evaluate $\frac{ab}{3}$ if $a = 7$ and $b = 6$.**

$\frac{ab}{3} = \frac{(7)(6)}{3}$ Replace a with 7 and b with 6.
$= \frac{42}{3}$ The fraction bar is like a grouping symbol.
$= 14$ Divide.

EXAMPLE 4 **Evaluate $x^3 + 4$ if $x = 3$.**

$x^3 + 4 = 3^3 + 4$ Replace x with 3.
$= 27 + 4$ Use the order of operations.
$= 31$ Add 27 and 4.

EXERCISES

Evaluate each expression if $a = 4$, $b = 2$, and $c = 7$.

1. $3ac$ 84 **2.** $5b^3$ 40 **3.** abc 56
4. $5 + 6c$ 47 **5.** $\frac{ab}{8}$ 1 **6.** $2a - 3b$ 2
7. $\frac{b^4}{4}$ 4 **8.** $c - a$ 3 **9.** $20 - bc$ 6
10. $2bc$ 28 **11.** $ac - 3b$ 22 **12.** $6a^2$ 96
13. $7c$ 49 **14.** $6a - b$ 22 **15.** $ab - c$ 1

Practice: Skills, p. 17

NAME ________ DATE ________ PERIOD ____

1-4 Practice: Skills
Algebra: Variables and Expressions

Evaluate each expression if $w = 2$, $x = 3$, $y = 5$, and $z = 6$.

1. $2w$ 4 **2.** $y + 5$ 10 **3.** $9 - z$ 3
4. $x + w$ 5 **5.** $3 + 4z$ 27 **6.** $6y - 5$ 25
7. y^2 25 **8.** $y - x$ 2 **9.** $\frac{z}{2}$ 3

Evaluate each expression if $m = 3$, $n = 7$, and $p = 9$.

10. $m + n$ 10 **11.** $12 - 3m$ 3 **12.** $5p$ 45
13. $3.3p$ 29.7 **14.** $3.3p + 2$ 31.7 **15.** $2p + 3.3$ 21.3
16. $20 + 2n$ 34 **17.** $20 - 2n$ 6 **18.** $\frac{n}{7}$ 1
19. n^2 49 **20.** $6m^2$ 54 **21.** $\frac{p^2}{3}$ 27
22. $1.1 + n$ 8.1 **23.** $p - 8.1$ 0.9 **24.** $3.6m$ 10.8
25. $3n - 2m$ 15 **26.** $3m - n$ 2 **27.** $2.1n + p$ 23.7
28. $\frac{m^2}{p}$ 1 **29.** $\frac{2.5m + 2.5}{5}$ 2 **30.** $\frac{(n + 2)^2}{3}$ 27

Practice: Word Problems, p. 18

NAME ________ DATE ________ F

1-4 Practice: Word Problems
Algebra: Variables and Expressions

1. FIELD TRIP The seventh grade math classes are going on a field trip. The field trip will cost $7 per student. Write an expression to find the cost of the field trip for s students. What is the total cost if 26 students go on the trip? $7s$; $182

2. SOCCER Jason earns $20 per game as a referee in youth soccer games. Write an expression to find how much money Jason will earn for refereeing any number of games. Let n represent the number of games Jason has refereed. How much will he earn for refereeing 6 games? $20n$; $120

3. PROFIT The expressions $c - e$, where c stands for the money collected and e stands for the expenses, is used to find the profit from a basketball concession. If $500 was collected and expenses were $150, find the profit for the concession. $350

4. SAVINGS Kata has a savings account that contains $230. She decides to deposit $5 each month from her monthly earnings for baby-sitting after school. Write an expression to find how much money Kata will have in her savings account after x months. Let x represent the number of months. Then find out how much she will have in her account after 1 year. $5x + 230$; $290

5. MONEY Mr. Wilson has $2,500 in his savings account and m dollars in his checking account. Write an expression that describes the total amount that he has in both accounts. $2,500 + m$

6. ANIMALS Write an expression to represent the total number of legs on h horses and c chickens. How many legs are there in 5 horses and 6 chickens? $4h + 2c$; 32 legs

7. T-SHIRTS The band wants to order T-shirts. The T-shirts cost $15 each plus a shipping fee of $10. Write an expression to find the total cost of c T-shirts. $15c + 10$

8. TEMPERATURE The expression $\frac{9}{5}C + 32$, where C stands for temperature in degrees Celsius, is used to convert Celsius to Fahrenheit. If the temperature is 20 degrees Celsius, find the temperature in degrees Fahrenheit. 68°F

1-5a Lesson Notes

1 Focus

Bellringer

Reading Mathematics As students read the introduction to the lesson, have them assign variables for the values that they are looking for. Students should use their own words to say what information is given and needed. Have them write a single sentence that summarizes what they know and what they need. **$7.50 times the number of adult tickets plus $4 times the number of student tickets = $263.50.**

2 Teach

In-Class Example

INTERACTIVE CHALKBOARD
PowerPoint® Presentations

Solve. Use the guess and check strategy.

CONCESSIONS The concession stand at the school play sold lemonade for $0.50 and cookies for $0.25. They sold 7 more lemonades than cookies, and they made a total of $39.50. How many lemonades and cookies were sold?

55 lemonades, 48 cookies

1-5a Problem-Solving Strategy

A Preview of Lesson 1-5

What You'll Learn
Solve problems using the guess and check strategy.

Guess and Check

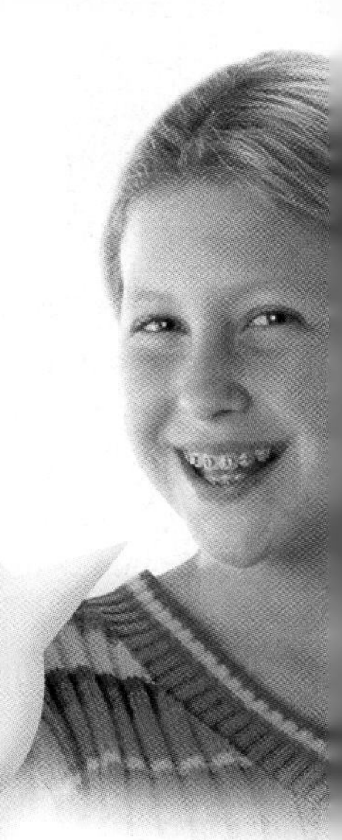

The total ticket sales for the school play were $263.50. Amber, how many adult and student tickets were sold?

Adult tickets are $7.50 and student tickets are $4.00. Also, twice as many students bought tickets as adults. Let's **guess and check** to find out.

Explore	We know adult tickets are $7.50, student tickets are $4.00, and twice as many students bought tickets than did adults.		
Plan	Let's make a guess and check if it is correct. Then adjust the guess until we get the correct answer.		
Solve	Make a guess.		
	10 adults, 20 students	7.5(10) + 4(20) = $155	*too low*
	Adjust the guess upward.		
	20 adults, 40 students	7.5(20) + 4(40) = $310	*too high*
	Adjust the guess downward slightly.		
	18 adults, 36 students	7.5(18) + 4(36) = $279	*still too high*
	Adjust the guess downward again.		
	17 adults, 34 students	7.5(17) + 4(34) = $263.50	*correct*
	There were 17 adult and 34 student tickets sold.		
Examine	Seventeen adult tickets cost $127.50, and 34 student tickets cost $136. Since $127.50 + $136 = $263.50 and 34 tickets are twice as many as 17 tickets, the guess is correct.		

Analyze the Strategy

1. **Explain** why you must keep a careful record of each of your guesses and their results in the *solve* step of the problem-solving plan. **1–2. See margin.**
2. **Write** a problem that could be solved by guess and check. Then write the steps you would take to find the solution to your problem.

The **Problem-Solvi Strategy** lessons provide an opportunity for mixed problem solving.

DAILY INTERVENTION

Differentiated Instruction

Visual/Spatial Have students hang butcher block paper or large packing paper on the wall. Use this paper to make a table to chart the guess and check strategy. Have students work on checking reasonable answers rather than guessing randomly.

pply the Strategy

olve. Use the guess and check strategy.

. **CAR WASH** The band sponsored a car wash to help pay for uniforms. They charged $4 for a car and $6 for an SUV. During the first hour, they washed 16 vehicles and earned $78. How many of each type of vehicle did they wash? **9 cars and 7 SUVs**

4. **SOUVENIRS** The Pike's Peak souvenir shop sells standard-sized postcards in packages of 5 and large-sized postcards in packages of 3. If Juan bought 16 postcards, how many packages of each did he buy? **two 5-card packages and two 3-card packages**

ixed Problem Solving

olve. Use any strategy.

. **MUSIC** Flor is burning a CD for her friend Derek. The CD will hold 30 minutes of music. Which songs should she select from the list at the right to record the maximum time on the CD without going over? **See margin.**

SONG	TIME
A	3 min 30 s
B	6 min 0 s
C	4 min 45 s
D	5 min 30 s
E	5 min 6 s
F	4 min 15 s
G	4 min 30 s
H	3 min 48 s

. **NUMBER THEORY** A number is squared, and the result is 256. Find the number. **16**

. **PATTERNS** Copy the table below. Extend the pattern to complete the table.

Input	1	2	3	4	5	6	7
Output	1	8	27	?	?	?	?

64, 125, 216, 343

. **MULTI STEP** The Wicked Twister roller coaster can accommodate about 1,000 passengers per hour. The coaster has 8 vehicles that each carry 4 passengers. Approximately how many runs are made in 1 hour? **31**

. **PHYSICAL SCIENCE** Telephone calls travel through optical fibers at the speed of light, which is 186,000 miles per second. A millisecond is 0.001 of a second. How far can your voice travel over an optical line in 1 millisecond? **186 mi**

10. **MONEY MATTERS** Syreeta needs to buy four markers to make posters for a science project. She has $4. Does she have enough money if the cost of each marker, including tax, is 89¢? Explain. **See margin.**

SNOWFALL For Exercises 11 and 12, use the graph below.

Source: www.usatoday.com

11. What is the total average snowfall for October through April? **42.4 in.**

12. Between which two consecutive months is the difference in snowfall the least? **February and March**

13. **STANDARDIZED TEST PRACTICE** Kelly has $2.80 in change in her purse. If the change is made up of an equal number of nickels, dimes, and quarters, how many of each does she have? **D**

   Ⓐ 4 Ⓑ 5 Ⓒ 6 Ⓓ 7

You will use the guess and check strategy in the next lesson.

3 Practice/Apply

About the Exercises...

Problem-Solving Strategies
Exercises 5–13 are structured so that students have the opportunity to practice many different problem-solving strategies. You may wish to remind them of some of the strategies they have studied.

- use the four-step plan (p. 6)
- guess and check ✓

Answers

1. Sample answer: You need to keep track of what numbers you have already guessed, so that you do not make the same guess twice. You also need to know what numbers produce answers that are too large or too small, so you can make better guesses.
2. Sample answer: Julia scored 16 points in a basketball game. She missed 18 of her 28 attempts. All of her shots were either one- or two-point shots. How many one- and two-point shots did Julia make?
5. Songs B, C, D, E, G, and H run for a total of 29.65 minutes, or 29 minutes and 39 seconds.
10. Yes, because 4 × $0.89 = $3.56, which is less than $4.

4 Assess

Open-Ended Assessment

Speaking Have students explain how to solve a problem using the guess and check strategy.

1-5 Lesson Notes

Lesson Resources

Chapter 1 Resource Masters
- Study Guide and Intervention, p. 21
- Practice: Skills, p. 22
- Practice: Word Problems, p. 23
- Reading to Learn Mathematics, p. 24
- Enrichment, p. 25

Key Teaching Transparencies, 3
MathPASS Tutorial CD-ROM, 38
Parent and Student Study Guide Workbook, 1-5

1 Focus

5-Minute Check 1-5

INTERACTIVE CHALKBOARD PowerPoint® Presentations

Use as a quiz or review of Lesson 1-4. (Also available as a transparency)

Bellringer

Problem Solving Emily received her $4 allowance and decided to go to the ice cream store for a treat. How much change would she receive from each of the following purchases?

a. banana split $3.25 **$0.75**

b. ice cream cone $1.50 **$2.50**

Write a rule to describe how you found the change.
allowance − cost of treat = change

(Continued in bottom margin.)

Let t represent the cost of the treat and c represent the amount of change. Rewrite the rule using numbers, variables, and an equals sign.
$4 - t = c$

Reading to Learn Mathematics, p. 24 — ELL

NAME ______ DATE ______ PERIOD ____

1-5 Reading to Learn Mathematics
Algebra: Equations

Pre-Activity *Read the introduction at the top of page 24 in your textbook. Write your answers below.*

1. How many losses did each team have? Complete the table.

Women's National Basketball Association, 2001		
Team	Wins	Losses
Cleveland	22	10
New York	21	11
Miami	20	12
Charlotte	18	14
Orlando	13	19
Indiana	10	22

2. Write a rule to describe how you found the number of losses.
32 = wins + losses

3. Let ... the number of wins and ℓ represent the number of losses. ... rule using ...

1-5 Algebra: Equations

What You'll Learn
Solve equations using mental math.

NEW Vocabulary
- equation
- solution
- solving an equation
- defining the variable

MATH Symbols
- = is equal to
- ≠ is not equal to

WHEN am I ever going to use this?

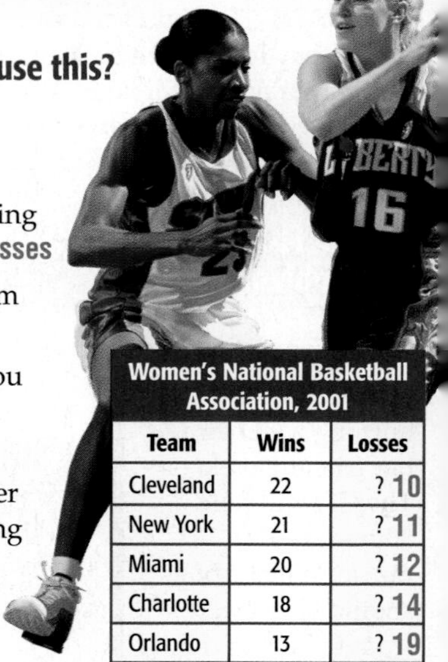

BASKETBALL The table shows the number of wins for six WNBA Eastern Division teams after playing 32 games each. **2. 32 = wins + losses**

1. How many losses did each team have? **See table.**
2. Write a rule to describe how you found the number of losses.
3. Let w represent the number of wins and ℓ represent the number of losses. Rewrite your rule using numbers, variables, and an equals sign. **Sample answer: $32 = w + \ell$**

Women's National Basketball Association, 2001		
Team	Wins	Losses
Cleveland	22	? **10**
New York	21	? **11**
Miami	20	? **12**
Charlotte	18	? **14**
Orlando	13	? **19**
Indiana	10	? **22**

Source: cnn.si.com

An **equation** is a sentence in mathematics that contains an equals sign. The equals sign tells you that the expression on the left is equivalent to the expression on the right.

$4 + 3 = 8 - 1$ $\qquad 3(4) = 24 \div 2$ $\qquad 17 = 13 + 2 +$

An equation that contains a variable is neither true nor false until the variable is replaced with a number. The **solution** of an equation is a number that makes the sentence true.

The process of finding a solution is called **solving an equation**. Some equations are easy to solve using mental mathematics.

TEACHING TIP
Tell students that they will learn how to solve different types of equations in Chapter 4.

EXAMPLE Solve an Equation Mentally

1 Solve $14 + t = 18$ mentally.

$14 + t = 18$ Write the equation.
$14 + 4 = 18$ You know that 14 + 4 is 18.
$18 = 18$ Simplify.

The solution is 4.

Your Turn **Solve each equation mentally.**

a. $p - 5 = 20$ **25** b. $8 = y \div 3$ **24** c. $7h = 56$ **8**

EXAMPLE Graph the Solution of an Equation

2 Graph the solution of the equation in Example 1.

Locate the point named by the solution on a number line. Then draw a dot at the solution, 4.

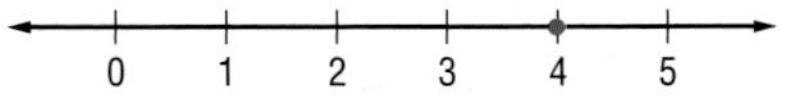

When you write an equation that represents a real-life problem, you are *modeling* the problem. First, choose a variable to represent one of the unknowns. This is called **defining the variable**.

EXAMPLE Write an Equation to Solve a Problem

3 FOOD The total cost of a hamburger, fries, and soft drink is $5.50. If the fries and drink cost $2.50 together, what is the cost of the hamburger?

Words	The cost of a hamburger, fries, and soft drink is $5.50.
Variable	Let h represent the cost of the hamburger.
Equation	$h + 2.50 = 5.50$

$h + 2.50 = 5.50$ Write the equation.

$3 + 2.50 = 5.50$ Replace h with 3 to make the equation true.

$5.50 = 5.50$ Simplify.

The number 3 is the solution of the equation. So, the cost of the hamburger is $3.00.

EXAMPLE Find a Solution of an Equation

4 MULTIPLE-CHOICE TEST ITEM

What value of x is a solution of $x + 14.6 = 30.2$?

(A) 14.9 (B) 15.2 (C) 15.6 (D) 16.1

Read the Test Item

Substitute each value for x to determine which makes the left side of the equation equivalent to the right side.

Solve the Test Item

Replace x with 14.9.	Replace x with 15.2.	Replace x with 15.6.
$x + 14.6 = 30.2$	$x + 14.6 = 30.2$	$x + 14.6 = 30.2$
$14.9 + 14.6 \stackrel{?}{=} 30.2$	$15.2 + 14.6 \stackrel{?}{=} 30.2$	$15.6 + 14.6 \stackrel{?}{=} 30.2$
$29.5 \neq 30.2$ false	$29.8 \neq 30.2$ false	$30.2 = 30.2$ true ✔

The value 15.6 makes the equation true. So, the answer is C.

Words/Variable/Equation entries help students translate between word problems and equations.

STUDY TIP

Defining the Variable Use the first letter of the word you are defining as a variable. For example: h stands for the cost of a hamburger.

A B C D **Standardized Test Practice**

Test-Taking Tip The Princeton Review

Solutions
Since an equation with one variable has only one solution, the value of 16.1, or choice D, does not need to be checked.

TEACHING TIP
Point out to students that they are using the guess-and-check strategy here.

msmath2.net/extra_examples

2 Teach

In-Class Examples

INTERACTIVE CHALKBOARD PowerPoint® Presentations

1 Solve $\frac{x}{2} = 3$ mentally. **6**

2 Graph the solution of the equation in In-Class Example 1.

3 ENTERTAINMENT An adult paid $18.50 for herself and two students to see a movie. If the two student tickets cost $11 together, what is the cost of the adult ticket? **$7.50**

4 MULTIPLE-CHOICE TEST ITEM What value of x is a solution of $12 - x = 5$? **C**

A 5 **B** 6 **C** 7 **D** 8

A Standardized Test Practice example is worked out for students.

Tips for New Teachers

Classroom Management When grouping students, you can separate them by their developmental reading ability, by table groups, by learning style, alphabetically, or randomly.

DAILY INTERVENTION **Differentiated Instruction**

Kinesthetic Have students set up simple equations such as $n + \$0.10 = \$0.25$. Then have them use actual coins to model the equation and solve for n. If students are in groups of threes, they can take turns writing the equation, modeling it, and solving for n. Encourage students to challenge each other with difficult equations.

3 Practice/Apply

DAILY INTERVENTION **FIND THE ERROR**

For Exercise 3, survey the class to see who thinks Joshua is correct and who thinks Ivan is correct. Have students test both values for w to check the answer. Ask students whether they can figure out what Joshua did wrong. **He subtracted 25 from 50 rather than adding.**

About the Exercises...

Guided Practice Key

Exercises	Examples
4–9	1
10	2
13	3
11–12	4

Odd/Even Assignments
Exercises 14–41 are structured so that students practice the same concepts whether they are assigned odd or even problems.

Alert! In Exercise 44, point out to students that calculated answers are not always possible in reality. In this case, 179 episodes in about 10 years means a calculated average of 17.9 episodes a year; however, the true number of episodes in any one year must be a whole number.

Alert! The Data Update after Exercise 45 requires access to the Internet.

Skill and Concept Check

Writing Math
Exercises 1 & 3

1. **Explain** what it means to solve an equation.
Sample answer: Find the value of the variable that makes a true sentence.
2. **OPEN ENDED** Write an example of an equation that is easily solved using mental mathematics and an equation that is not as easily solved using mental mathematics. **Sample answer: $4 + t = 9$ and $4.35 + t = 9.02$**
3. **FIND THE ERROR** Joshua and Ivan had a contest to see who could solve $w - 25 = 50$ the fastest. Their solutions are shown below. Who is correct? Explain. **Ivan; $75 - 25 = 50$ is a true statement.**

Joshua	Ivan
w = 25	w = 75

Find the Error exercises help students identify and address common errors before they occur.

GUIDED PRACTICE

Solve each equation mentally.

4. $y - 18 = 20$ **38**
5. $5 + a = 22$ **17**
6. $w \div 4 = 7$ **28**
7. $\frac{r}{9} = 6$ **54**
8. $75 = w + 72$ **3**
9. $69 = 3f$ **23**

10. Graph the solution of $x + 5 = 11$. **See margin.**

Name the number that is the solution of the given equation.

11. $x + 15 = 19$; 4, 5, 6 **4**
12. $13k = 80.6$; 5.5, 5.8, 6.2 **6.2**

13. **BASKETBALL** Jason scored 14 more points than David in the last basketball game. If David scored 9 points, how many did Jason score? **23 points**

★ **indicates increased difficulty**

Practice and Applications

HOMEWORK HELP

For Exercises	See Examples
14–31	1
32–35	2
40–46	3
36–39	4

Extra Practice
See pages 565, 596.

Solve each equation mentally.

14. $b + 7 = 18$ **11**
15. $8 + x = 15$ **7**
16. $y - 15 = 71$ **86**
17. $a - 18 = 20$ **38**
18. $25 - 19 = n$ **6**
19. $24 + 39 = m$ **63**
20. $12 \cdot 5 = s$ **60**
21. $n = \frac{30}{6}$ **5**
22. $b - 42 = 84$ **126**
23. $7t = 77$ **11**
24. $3d = 99$ **33**
25. $22 + h = 42$ **20**
26. $20 = y \div 5$ **100**
27. $16 = \frac{u}{4}$ **64**
28. $\frac{z}{7} = 12$ **84**
★29. $18j = 360$ **20**
★30. $m - 36 = 123$ **159**
★31. $108 = 9b$ **12**

Graph the solution of each equation on a number line. **32–35.** **See margin.**

32. $w + 3 = 10$
33. $x - 7 = 11$
34. $12y = 48$
35. $11 = 132 \div z$

Name the number that is the solution of the given equation.

36. $13.4 \cdot 9 = h$; 117.8, 118.7, 120.6 **120.6**
37. $n \div 10 = 4$; 20, 30, 40 **40**
38. $9.9 + r = 24.2$; 12.7, 14.3, 16.3 **14.3**
39. $c - 8 = 17$; 15, 25, 35 **25**

40. **ORNAMENTS** Last year, Desiree had 25 glass ornaments. This year, she has 38 ornaments. Solve $25 + n = 38$ to find how many new ornaments Desiree collected. **13**

Answers

10. $x = 6$

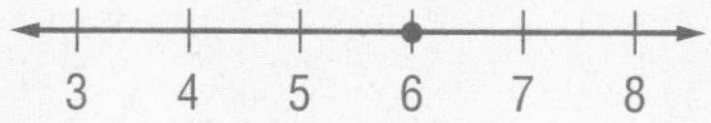

32. $w = 7$

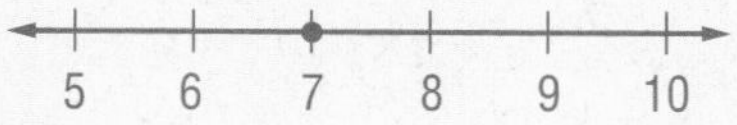

33. $x = 18$

34. $y = 4$

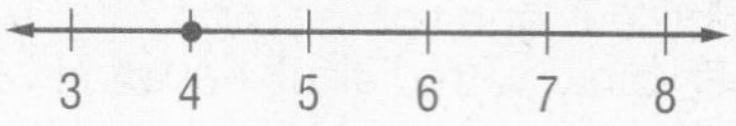

35. $z = 12$

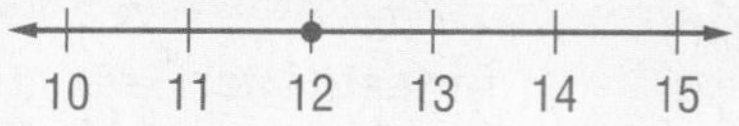

26 Chapter 1 Decimal Patterns and Algebra

★ **41. MONEY** Maria was paid \$9 per hour and earned \$67.50. How many hours did she work? Use the equation $67.50 = 9h$, where h is hours worked. **7.5 h**

WHALES For Exercises 42 and 43, use the information below.
Each winter, Humpback whales migrate about 1,500 miles north to the Indian Ocean. However, scientists tracked one whale that migrated 5,000 miles in one season. **42. $5,000 - 1,500 = n$ or $5,000 = 1,500 + n$**

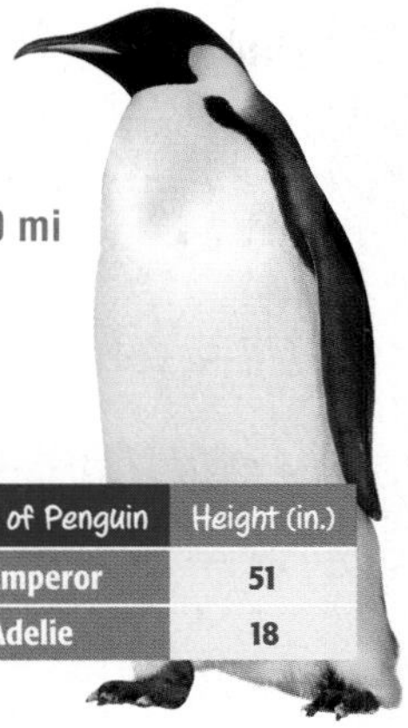

42. Write an equation that can be used to find how many more miles the whale migrated compared to the usual migration.

43. How much farther did the whale travel than the usual migration? **3,500 mi**

44. MULTI STEP A long-running TV show aired 179 episodes from October 1951 to September 1961. About how many episodes were filmed each year? **about 18 episodes per year**

45. PENGUINS On average, how much taller are Emperor penguins than Adelie penguins? Write an equation and solve. **$t = 51 - 18$ or $51 = t + 18$; 33 in.**

Type of Penguin	Height (in.)
Emperor	51
Adelie	18

Data Update How do the heights of other types of penguins compare to Adelie penguins? Visit msmath2.net/data_update to learn more.

46. BASKETBALL Michael Jordan returned to the NBA as a member of the Washington Wizards in 2001. During one game, he scored 51 of the team's 107 points. How many points did the rest of the team score? **$51 + p = 107$; 56 points**

47. CRITICAL THINKING Consider the equation $0 \times a = b$. Find the values of a and b. **Sample answer: $b = 0$; a is any number**

Standardized Test Practice and Mixed Review

48. MULTIPLE CHOICE Choose the solution of the equation $\frac{48}{c} = 4 + 2$. **D**

Ⓐ 2 Ⓑ 4 Ⓒ 6 Ⓓ 8

49. GRID IN Use the table to determine how much farther in air miles Phoenix is from New Orleans than Memphis. **958**

From	To	Distance (air mi)
Memphis, TN	New Orleans, LA	358
Phoenix, AZ	New Orleans, LA	1,316

50. ALGEBRA Evaluate $3a + b^2$ if $a = 2$ and $b = 3$. (Lesson 1-4) **15**

Evaluate each expression. (Lesson 1-3)

51. $11 \cdot 6 \div 3 + 9$ **31** **52.** $5 \cdot 13 - 62$ **3** **53.** $1 + 2(8 - 5)^2$ **19**

54. DINING Four hundred sixty people are scheduled to attend a banquet. If each table seats 8 people, how many tables are needed? (Lesson 1-1) **58 tables**

GETTING READY FOR THE NEXT LESSON

BASIC SKILL Multiply.

55. $2 \cdot (4 + 10)$ **28** **56.** $(9 \cdot 1) \cdot 8$ **72** **57.** $(5 \cdot 3)(5 \cdot 2)$ **150** **58.** $(6 + 8) \cdot 12$ **168**

msmath2.net/self_check_quiz

4 Assess

Higher-Level Thinking

Ask students to outline the steps needed to solve for x if $y + 1 = 2$ and $x + y = 5$. Then have them find the value of x. **First solve for $y = 1$, then substitute the value of y into $x + y = 5$ $(x + 1 = 5)$, and solve for x; 4**

Study Guide and Intervention, p. 21

NAME ______ DATE ______ PERIOD ____

1-5 Study Guide and Intervention
Algebra: Equations

- An **equation** is a sentence in mathematics that contains an equals sign.
- The **solution** of an equation is the value that when substituted for the variable makes the equation true.

EXAMPLE 1 **Solve $23 + y = 29$ mentally.**

$23 + y = 29$ Write the equation.
$23 + 6 = 29$ You know that 23 + 6 is 29.
$29 = 29$ Simplify.
The solution is 6.

EXAMPLE 2 **What value of x is a solution of $x + 8 = 26$?**
A. 16 **B.** 17 **C.** 18 **D.** 19

Substitute each value for x to determine which makes the left side of the equation equivalent to the right side.

Replace x with 16.	Replace x with 17.	Replace x with 18.
$x + 8 = 26$	$x + 8 = 26$	$x + 8 = 26$
$16 + 8 = 26$	$17 + 8 = 26$	$18 + 8 = 26$
$24 \neq 26$ false	$25 \neq 26$ false	$26 = 26$ true

The value of 18 makes the equation true. So, the answer is C.

EXERCISES

Solve each equation mentally.

1. $k + 7 = 15$ 8 **2.** $g - 8 = 20$ 28 **3.** $6y = 24$ 4
4. $\frac{a}{3} = 9$ 27 **5.** $\frac{x}{6} = 9$ 54 **6.** $8 + r = 24$ 16
7. $12 \cdot 8 = h$ 96 **8.** $n \div 11 = 8$ 88 **9.** $48 \div 12 = x$ 4
10. $h - 12 = 24$ 36 **11.** $19 + y = 28$ 9 **12.** $9f = 90$ 10

Name the number that is the solution of the given equation.

13. $27 - h = 10$; 7, 17, 27 17 **14.** $n \div 11 = 4$; 44, 55, 66 44

Practice: Skills, p. 22

NAME ______ DATE ______ PERIOD ____

1-5 Practice: Skills
Algebra: Equations

Solve each equation mentally.

1. $a + 7 = 16$ 9 **2.** $12 + x = 21$ 9 **3.** $4d = 60$ 15
4. $15 = \frac{u}{3}$ 45 **5.** $\frac{b}{7} = 12$ 84 **6.** $13 \cdot 3 = y$ 39
7. $8 + r = 17$ 9 **8.** $27 - 12 = m$ 15 **9.** $h - 22 = 67$ 89
10. $27 + 15 = n$ 42 **11.** $36 + a = 96$ 60 **12.** $99 \div d = 3$ 33
13. $6t = 66$ 11 **14.** $25 = y \div 4$ 100 **15.** $b - 25 = 120$ 145
16. $n \div 5 = 10$ 50 **17.** $4y = 48$ 12 **18.** $5t = 40$ 8
19. $50 \cdot d = 150$ 3 **20.** $w + 61 = 65$ 4 **21.** $88 \div k = 2$ 44

Graph the solution of each equation on a number line.

22. $y + 16 = 21$
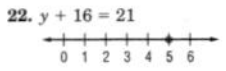
23. $3f = 18$

24. $13 - x = 10$

Name the number that is the solution of the given equation.

25. $4 + a = 21$; 15, 16, 17 17 **26.** $2f = 48$; 20, 22, 24 24
27. $y - 17 = 30$; 37, 47, 57 47 **28.** $48 \div 12 = h$; 3, 4, 5 4

Practice: Word Problems, p. 23

NAME ______ DATE ______ PERIOD ____

1-5 Practice: Word Problems
Algebra: Equations

1. GAS MILEAGE Mr. Moseley's car has a 20-gallon gas tank. It took 14 gallons of gas to fill his tank. Use the equation $14 + g = 20$ to find the number of gallons g that he had before he filled his tank with gas. 6 gal	**2. PAINTING** Latisha earned \$5 an hour painting for her dad. If she made \$40 last week, use $5h = 40$ to find how many hours h she painted. 8 h
3. LUMBER Mrs. Garcia had a piece of board that was 15 feet long. She cut off 6.5 feet. Use the equation $6.5 + \ell = 15$ to determine how much of the board ℓ she has left. 8.5 ft	**4. MAGAZINES** Mahpee was selling magazine subscriptions. He earned \$5 plus \$0.50 for each subscription he sold. If Mahpee earned \$25, use the equation $25 = 5 + 0.50n$ to find the number of subscriptions n he sold. 40 subscriptions
5. TIRES The cost of a car tire is \$45 plus \$10 per order regardless of the number of tires purchased. If Mrs. Sato places an order for \$190, use the equation $45t + 10 = 190$ to find the number of tires t she purchased. 4 tires	**6. AREA** If the area of a rectangle is 30 square centimeters and the length is 6 centimeters, use the equation $30 = 6w$ to find the width w of the rectangle. 5 cm
7. SUPPLIES The Jones Middle School had \$4,000 to spend on office supplies. They had already spent \$1,250. Use the equation $1,250 + d = 4,000$ to find how much money d the school had left for other supplies. \$2,750	**8. PENCILS** Mi-Leng spent 90 cents on 6 pencils. Use the equation $90 = 6c$ to find the cost c of each pencil. 15 cents or \$0.15

Chapter 1 Mid-Chapter Practice Test

Assessment Option

Mid-Chapter Test (Lessons 1-1 through 1-5) is available on p. 63 of the *Chapter 1 Resource Masters*.

ExamView® Pro Test Bank CD-ROM

Use the networkable ExamView® Pro Test Bank CD-ROM to:

- Create **Multiple versions** of tests.
- Create **Modified** tests for **Inclusion** students with one mouse click.
- **Edit** existing questions and **Add** your own questions.
- Build tests aligned with state standards using built-in **State Curriculum Correlations.**
- Change **English** tests to **Spanish** with one mouse click and vice versa.

FOLDABLES™ Follow-Up

Before students complete the Mid-Chapter Practice Test, encourage them to review the notes, definitions, concepts, and main ideas they recorded on the first five tabs of their Foldables.

Chapter 1 Mid-Chapter Practice Test

The Mid-Chapter Practice Test reviews skills and concepts presen in previous lesso

Vocabulary and Concepts

1. **List** the steps in the four-step problem-solving plan. (Lesson 1-1) **explore, plan, solve, and examine**
2. **OPEN ENDED** Write an algebraic expression and identify any coefficients, variables, and constants. (Lesson 1-4) **Sample answer: $7m - 2$; coefficient: 7; variable: m; constant: -2**

Skills and Applications

3. **SPORTS** A baseball stadium holds 20,000 people. If 3,650 people can be seated in the bleachers, how many seats are available in the rest of the stadium? (Lesson 1-1) **16,350 seats**
4. Write $2 \cdot 2 \cdot 2 \cdot 2 \cdot 2 \cdot 2 \cdot 2$ in exponential form. (Lesson 1-2) **2^7**

Evaluate each expression. (Lessons 1-2 and 1-3)

5. 4^5 **1,024**
6. 9^6 **531,441**
7. 3^{10} **59,049**
8. $25 - (3 + 2 \cdot 5)$ **12**
9. $(2^3 + 4) + 3^2$ **21**
10. $\frac{2(7-3)}{2^2}$ **2**

Evaluate each expression if $x = 12$, $y = 4$, and $z = 8$. (Lesson 1-4)

11. $x - 5$ **7**
12. $3y + 10z$ **92**
13. $\frac{yz}{2}$ **16**
14. $\frac{(y+8)^2}{x}$ **12**

Solve each equation mentally. (Lesson 1-5)

15. $x + 16 = 23$ **7**
16. $36 = 3y$ **12**
17. $\frac{65}{z} = 5$ **13**

18. **GEOMETRY** The perimeter of a rectangle is given by the expression $2(\ell + w)$ where ℓ = length and w = width. Find the perimeter of the rectangle. (Lesson 1-4) **15.4 ft**

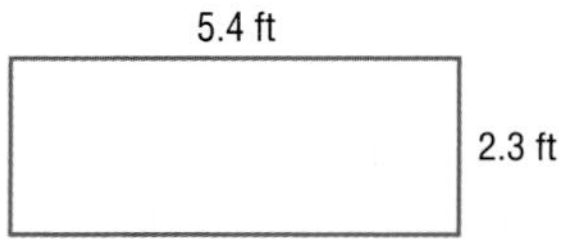

Standardized Test Practice

19. **MULTIPLE CHOICE** A cycling club is planning a 1,800-mile trip. They can average 15 miles per hour for 6 hours each day. How many days will it take them to complete the trip? (Lesson 1-1) **A**

    Ⓐ 20 days Ⓑ 21 days
    Ⓒ 22 days Ⓓ 23 days

20. **GRID IN** The expression $\frac{w}{30}$, where w is a person's weight in pounds, is used to find the approximate number of quarts of blood in the person's body. How much blood does a 120-pound person have? (Lesson 1-4) **4 qt**

The Game Zone

A Place To Practice Your Math Skills

Students will enjoy using **The Game Zone** math games to develop concepts, review, and utilize problem-solving strategies.

Express Yourself!

GET READY!

Players: two, three, or four
Materials: scissors, 18 index cards cut in half, one number cube

GET SET!

- On each card, write a different expression containing only one variable. The coefficients and constants are to be values less than or equal to ten and any exponents should be less than four. Four examples are shown at the right.

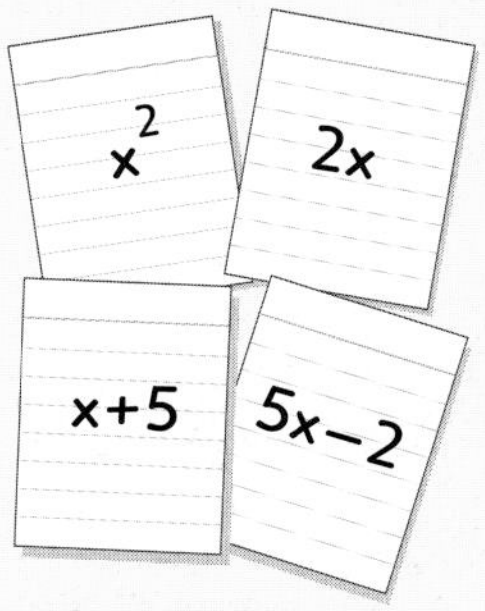

GO!

- Deal all cards to the players. The dealer then rolls the number cube. The number rolled is the value of the variable for the first round.
- The player to the left of the dealer puts a card of his or her choice faceup on the table, evaluates the expression on the card, and announces its value. Play continues until all players have placed one card on the table. This is the end of the first round. The person whose card has the greatest value wins all of the cards for that round.
- The player to the left of the dealer rolls the number cube. This is the value of the variable for the next round. Play continues until all the cards are played.
- **Who wins?** The person who has the most cards at the end of the game is the winner.

Possible Expressions for Cards

x^2	$2x + 4$	$4x$	$9x + 1$
$2x$	$4x - 6$	$3x + 2$	$3x$
$2x + 6$	$9x$	$x + 5$	$3x - 1$
$3x + 1$	$10 + x$	$5x - 2$	$4x - 2$
$2x^2$	$5x$	$x - 1$	$x + 7$
$2x - 2$	$4x + 1$	$7 - x$	$6x$
$7x$	$1 + 3x$	$5x + 5$	$3x^2$

The Game Zone

Glencoe Mathematics Classroom Manipulative Kit
- scissors
- number cubes

GETTING READY

Objective Evaluate expressions.

MANAGEMENT TIPS

Recommended Time
30 minutes

Tips
- Consider having the cards precut and the expressions written in order to save time. A list of possible expressions for the cards is given in the bottom margin.
- Remind students of the meaning of the terms *coefficient* and *constant*.
- Determine whether you are going to allow calculators or rely on mental math. Set that rule at the beginning of the game.
- As a variation for two players, keep the cards face down. Each player selects a card from the top of the pile. Roll a number cube to determine the value of the variable for each round. The player whose card has the greater value wins the cards.
- You may wish to use the game as an assessment tool for the concepts the students have been learning in Lessons 1-1 through 1-5.

For an additional game for this chapter, visit: msmath2.net/gamezone.

1-6 Lesson Notes

Lesson Resources

Chapter 1 Resource Masters
- Study Guide and Intervention, p. 26
- Practice: Skills, p. 27
- Practice: Word Problems, p. 28
- Reading to Learn Mathematics, p. 29
- Enrichment, p. 30

Parent and Student Study Guide Workbook, 1-6

FAST FILE

1 Focus

5-Minute Check 1-6

INTERACTIVE CHALKBOARD
PowerPoint® Presentations

Use as a quiz or review of Lesson 1-5. (Also available as a transparency)

Bellringer

Reading Mathematics Have students copy the table in the bottom margin into their notebooks, leaving plenty of space to write. In the middle column, have them find the dictionary definition for each word. As you go through the lesson, have them add the lesson definition (as a property) to their tables and relate the lesson definition to the dictionary definition.

(Continued in bottom margin.)

Reading to Learn Mathematics, p. 29 **ELL**

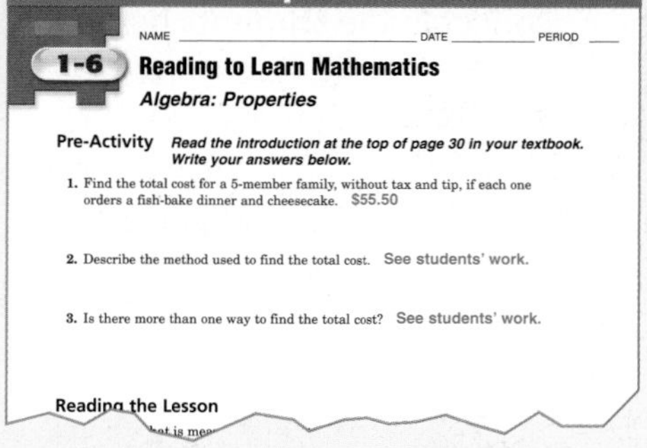
NAME ____________ DATE ________ PERIOD ____

1-6 Reading to Learn Mathematics
Algebra: Properties

Pre-Activity *Read the introduction at the top of page 30 in your textbook. Write your answers below.*

1. Find the total cost for a 5-member family, without tax and tip, if each one orders a fish-bake dinner and cheesecake. $55.50

2. Describe the method used to find the total cost. See students' work.

3. Is there more than one way to find the total cost? See students' work.

Reading the Lesson

1-6 Algebra: Properties

What You'll Learn
Use addition and multiplication properties to solve problems.

NEW Vocabulary
- equivalent expressions
- properties

LINK To Reading
Everyday Meaning of Distribute: to divide among several, as in distribute a deck of cards

Key Concept boxes highlight definitions, formulas, and other important ideas. Multiple representations—words, symbols, examples, models—reach students of all learning styles.

TEACHING TIP
Tell students that the Distributive Property also works when the number is multiplied by the sum. $(b + c)a = b(a) + c(a)$

WHEN am I ever going to use this?

RESTAURANTS Land-Ho! Fish Market is having a Friday night special.

1. Find the total cost for a 5-member family, without tax and tip, if each one orders a fish-bake dinner and cheesecake. **$55.50**
2. Describe the method used to find the total cost.
3. Is there more than one way to find the total cost?

2–3. See students' work.

Here are two ways to find the total cost of the dinner:

$5 \times (\$8.95 + \$2.15)$ — number of people; cost of 1 dinner and 1 cheesecake

$5(\$8.95) + 5(\$2.15)$ — cost of 5 dinners; cost of 5 cheesecakes

The expressions $5(\$8.95 + \$2.15)$ and $5(\$8.95) + 5(\$2.15)$ are **equivalent expressions** because they have the same value, $55.50. This shows how the **Distributive Property** combines addition and multiplication.

Key Concept — Distributive Property

Words To multiply a sum by a number, multiply each addend of the sum by the number outside the parentheses.

Symbols

Arithmetic	Algebra
$3(4 + 6) = 3(4) + 3(6)$	$a(b + c) = a(b) + a(c)$

EXAMPLES Use the Distributive Property

Use the Distributive Property to write each expression as an equivalent expression. Then evaluate the expression.

1 **5(3 + 2)**

$$5(3 + 2) = 5 \cdot 3 + 5 \cdot 2$$
$$= 15 + 10 \quad \text{Multiply.}$$
$$= 25 \quad \text{Add.}$$

2 **(7 + 4)3**

$$(7 + 4)3 = 7 \cdot 3 + 4 \cdot 3$$
$$= 21 + 12 \quad \text{Multiply.}$$
$$= 33 \quad \text{Add.}$$

New Words	Definitions	
	Everyday	Math
distributive		
commutative		
associative		
identity		

EAL-LIFE MATH

STORY Each rider on the
ny Express would travel
0 miles and change
rses every 10–20 miles.
rce: http://members.tripod.com

EXAMPLE Use the Distributive Property

3 **HISTORY The riders for the Pony Express mail service carried mail from St. Joseph, Missouri, to Sacramento, California, in eight days. On average, the riders covered 250 miles each day. About how far did the Pony Express riders travel over the eight-day period?**

You can find how many miles the riders traveled over the eight-day period by finding 8×250. You can use the Distributive Property to multiply mentally.

$8(250) = 8(200 + 50)$	Rewrite 250 as 200 + 50.
$= 8(200) + 8(50)$	Distributive Property
$= 1{,}600 + 400$	Multiply.
$= 2{,}000$	Add.

So, the riders on the Pony Express traveled about 2,000 miles over an eight-day period.

In algebra, **properties** are statements that are true for any number or variable.

STUDY TIP

Properties The *Commutative Property* shows how the terms move and change places.

The *Associative Property* shows how the parentheses move and change places.

Key Concept — **Properties**

Commutative Property	The order in which two numbers are added or multiplied does not change their sum or product.
Associative Property	The way in which three numbers are grouped when they are added or multiplied does not change their sum or product.
Identity Property	The sum of an addend and zero is the addend. The product of a factor and one is the factor.

EXAMPLES Identify Properties

Name the property shown by each statement.

4 $6 + (2 + 7) = (6 + 2) + 7$ Associative Property of Addition

5 $15 \times 10 = 10 \times 15$ Commutative Property of Multiplication

6 $4 \times 1 = 4$ Identity Property of Multiplication

7 $4(6 + 8) = 4(6) + 4(8)$ Distributive Property

Your Turn **Name the property shown by each statement.**

a. $1 \times (3 \times 4) = (1 \times 3) \times 4$ Associative (×)

b. $a + 0 = a$ Identity (+)

msmath2.net/extra_examples

DAILY INTERVENTION **Flexible Grouping**

Properties Separate students into groups of four. Then assign each student one of the properties in the lesson. Have the students meet with the other students assigned the same property to coach each other on that property. If you have a very large class, divide these groups into two. Then the students rejoin their original groups and coach the other students on the property they studied.

2 Teach

In-Class Examples

INTERACTIVE CHALKBOARD PowerPoint® Presentations

Use the Distributive Property to write each expression as an equivalent expression. Then evaluate the expression.

1 $8(5 + 7)$ **96**

2 $(2 + 9)6$ **66**

3 **VACATIONS** Mr. Harmon has budgeted \$150 per day for his hotel and meals during his vacation. If he plans to spend six days on vacation, how much will he spend? **\$900**

Name the property shown by each statement.

4 $7 = 1 \times 7$ **Identity Property of Multiplication**

5 $24 + 5 = 5 + 24$ **Commutative Property of Addition**

6 $7 + 0 = 7$ **Identity Property of Addition**

7 $(11 \times 4) \times 8 = 11 \times (4 \times 8)$ **Associative Property of Multiplication**

Tips for New Teachers

Teaching Strategies When studying mathematical properties, ask students to identify the difference between the left side and the right side of an equation. Once they are certain about the difference, it will be easier for them to name the property.

3 Practice/Apply

FOLDABLES™ Follow-Up Remind students to summarize what they learn about properties. On the sixth section, they might explain how the properties for addition and multiplication of whole numbers are the same for the addition and multiplication of fractions.

NUMBER SENSE
For Exercise 3, You may want to have students write similar, simple problems to help explain their answers.

About the Exercises...

Guided Practice Key

Exercises	Examples
4–6	1, 2
7–10	4–7

Odd/Even Assignments
Exercises 11–16, 19–26, and 31–38 are structured so that students practice the same concepts whether they are assigned odd or even problems.

Skill and Concept Check

Writing Math Exercise 3

1. Rewrite $p(q) + p(r)$ using the Distributive Property. $p(q + r)$
2. **OPEN ENDED** Write an equation that illustrates the Associative Property of Addition. Sample answer: $(5 + z) + 9 = 5 + (z + 9)$
3. **NUMBER SENSE** *True* or *False*? Explain your answer. $(11 + 18) \times 5 = 11 + 18 \times 5$ false; $(11 + 18) \times 5 = 145$; $11 + 18 \times 5 = 101$

GUIDED PRACTICE

Use the Distributive Property to write each expression as an equivalent expression. Then evaluate the expression.

4. $7(4 + 3)$ $7(4) + 7(3)$; 49
5. $(10 + 8)2$ $(10)2 + (8)2$; 36
6. $5(6 - 2)$ $5(6) - 5(2)$; 20

Name the property shown by each statement.

7. $5 + 4 = 4 + 5$ Commutative (+)
8. $(2 \times 3) \times 7 = 2 \times (3 \times 7)$ Associative (×)
9. $6(9 + 3) = 6(9) + 6(3)$ Distributive
10. $5 \times 1 = 5$ Identity (×)

★ indicates increased difficulty

Practice and Applications

HOMEWORK HELP

For Exercises	See Examples
11–16	1, 2
39, 40	3
19–30	4, 5, 6, 7

Extra Practice See pages 565, 596.

Use the Distributive Property to write each expression as an equivalent expression. Then evaluate the expression. 11–16. See margin.

11. $2(6 + 7)$
12. $(3 + 8)4$
13. $(11 + 3)8$
14. $2(5 - 4)$
15. $7(8 - 6)$
16. $6(12 + 5)$
17. Which expression is easier to evaluate mentally: 3(452) or 3(400 + 50 + 2)? Explain. See margin.
18. Is $5 + 15 = 5(1 + 3)$ a true equation? Explain your reasoning. Yes, the equation is true by the Distributive Property.

Name the property shown by each statement. 19–26. See margin.

19. $1 \times (5 \times 9) = (1 \times 5) \times 9$
20. $4 + b = b + 4$
21. $m \times n = n \times m$
22. $8 + 0 = 8$
23. $(1 + r) + s = 1 + (r + s)$
24. $(2.5 + 7)e = 2.5e + 7e$
25. $x(y + z) = xy + xz$
26. $19 \times 1 = 19$

TRAVEL For Exercises 27 and 28, use the table at the right. 27. $239 + 531 = 531 + 239$

27. Write a sentence comparing the mileage from Jacksonville to Charleston to Norfolk and the mileage from Norfolk to Charleston to Jacksonville.
28. Name the property that is shown in the sentence. Commutative (+)

From	To	Driving Distance (mi)
Jacksonville, FL	Charleston, SC	239
Charleston, SC	Norfolk, VA	531

COIN COLLECTING Mrs. Jackson has collected 152 rare coins.

29. If she does not add any coins to her collection, write a sentence that represents this situation. $152 + 0 = 152$
30. Name the property that is illustrated. Identity (+)

Answers

11. $2(6) + 2(7)$; 26
12. $(3)4 + (8)4$; 44
13. $(11)8 + (3)8$; 112
14. $2(5) - 2(4)$; 2
15. $7(8) - 7(6)$; 14
16. $6(12) + 6(5)$; 102
17. Sample answer: The second expression, because it is easy to multiply $3 \cdot 400$, $3 \cdot 50$, and $3 \cdot 6$ mentally, and then add.
19. Associative Property of Multiplication
20. Commutative Property of Addition
21. Commutative Property of Multiplication
22. Identity Property of Addition
23. Associative Property of Addition
24. Distributive Property
25. Distributive Property
26. Identity Property of Multiplication

Use properties to rewrite each expression as an equivalent expression without parentheses.

★ 31. $(y + 1) + 4$ **$y + 5$** ★ 32. $2 + (x + 4)$ **$x + 6$** ★ 33. $4(8b)$ **$32b$** ★ 34. $(3a)2$ **$6a$**

★ 35. $2(x + 3)$ **$2x + 6$** ★ 36. $4(2 + b)$ **$8 + 4b$** ★ 37. $6(c + 1) - c$ **$5c + 6$** ★ 38. $3(f + 4) + 2f$ **$5f + 12$**

39. **MENTAL MATH** Use one of the properties to find 6(48) mentally. Which property did you use? **$6(40 + 8) = 6(40) + 6(8) = 240 + 48 = 288$; Distributive Property**

40. **BASKETBALL** Use mental math to find the total number of NBA draftees that have come from the schools in the graph. **44**

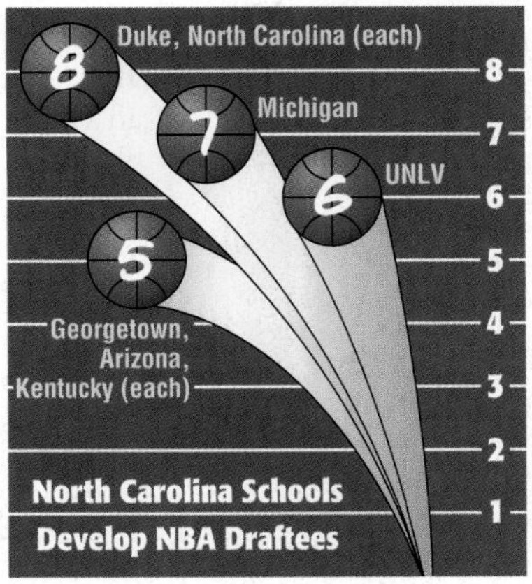

Source: nbcsports.com

41. **WRITE A PROBLEM** Write about a real-life situation that can be solved using the Distributive, Commutative, Associative, or Identity Properties. Then use the property to solve the problem. **Answers will vary.**

42. **CRITICAL THINKING** A *counterexample* is an example showing that a statement is not true. Provide a counterexample to the following statement. *Division of whole numbers is associative.* **See margin.**

Standardized Test Practice and Mixed Review

43. **MULTIPLE CHOICE** Rewrite $p \times (q \times r)$ using the Associative Property. **A**

Ⓐ $(p \times q) \times r$ Ⓑ $p \times q \times r$

Ⓒ $(p + q) + r$ Ⓓ $p \times q \times r \times 1 = p \times q \times r$

44. **SHORT RESPONSE** Rewrite $6(9 + 8)$ using the Distributive Property. **$6(9) + 6(8)$**

Name the number that is the solution of the given equation. (Lesson 1-5)

45. $x + 12 = 20$; 8, 9, 10 **8**

46. $7.3 = t - 4$; 10.3, 11.3, 12.3 **11.3**

47. $35.5 = 5n$; 5.1, 7.1, 9.1 **7.1**

48. $z \div 6 = 18$; 88, 98, 108 **108**

LIFE SCIENCE For Exercises 49 and 50, use the information below.
It is believed that a dog ages 7 human years for every calendar year. (Lesson 1-4)

49. Write an expression for determining a dog's age in human years. Let y represent the number of calendar years the dog has lived. **$7y$**

50. Find the human age of a dog that has lived for 12 calendar years. **84 yr**

51. Evaluate $(14 - 9)^4$. (Lesson 1-3) **625**

GETTING READY FOR THE NEXT LESSON

BASIC SKILL Find the next number in each pattern.

52. 2, 4, 6, 8, ? **10**

53. 10, 21, 32, 43, ? **54**

54. 1.4, 2.2, 3.0, 3.8, ? **4.6**

55. 4, 8, 12, 16, ? **20**

56. 64, 32, 16, 8, ? **4**

57. 0.4, 1.2, 3.6, 10.8, ? **32.4**

4 Assess

Open-Ended Assessment

Modeling Have students use base-ten blocks or counters to illustrate the Commutative, Associative, Distributive, and Identity Properties. For example, $5 \cdot 3 = 3 \cdot 5$; $3 \cdot (4 \cdot 5) = (3 \cdot 4) \cdot 5$; and $3 \cdot (4 + 5) = 3 \cdot 4 + 3 \cdot 5$.

Answer

42. Sample answer: Since $24 \div (12 \div 2) = 4$ and $(24 \div 12) \div 2 = 1$, $24 \div (12 \div 2) \neq (24 \div 12) \div 2$.

Study Guide and Intervention, p. 26

NAME ______ DATE ______ PERIOD ____

1-6 Study Guide and Intervention

Algebra: Properties

Property	Arithmetic	Algebra
Distributive Property	$5(3 + 4) = 5(3) + 5(4)$	$a(b + c) = a(b) + a(c)$
Commutative Property of Addition	$5 + 3 = 3 + 5$	$a + b = b + a$
Commutative Property of Multiplication	$5 \times 3 = 3 \times 5$	$a \times b = b \times a$
Associative Property of Addition	$(2 + 3) + 4 = 2 + (3 + 4)$	$(a + b) + c = a + (b + c)$
Associative Property of Multiplication	$(4 \times 5) \times 6 = 4 \times (5 \times 6)$	$(a \times b) \times c = a \times (b \times c)$
Identity Property of Addition	$5 + 0 = 5$	$a + 0 = a$
Identity Property of Multiplication	$5 \times 1 = 5$	$a \times 1 = a$

EXAMPLE 1 **Use the Distributive Property to write 6(4 + 3) as an equivalent expression. Then evaluate the expression.**

$6(4 + 3) = 6 \cdot 4 + 6 \cdot 3$ Apply the Distributive Property.
$= 24 + 18$ Multiply.
$= 42$ Add.

EXAMPLE 2 **Name the property shown by each statement.**

$5 \times 4 = 4 \times 5$ Commutative Property of Multiplication
$12 + 0 = 12$ Identity Property of Addition
$7 + (6 + 3) = (7 + 6) + 3$ Associative Property of Addition

EXERCISES

Use the Distributive Property to write each expression as an equivalent expression. Then evaluate the expression.

1. $5(7 + 2)$ $5(7) + 5(2)$; 45
2. $4(9 + 1)$ $4(9) + 4(1)$; 40
3. $2(6 + 7)$ $2(6) + 2(7)$; 26

Name the property shown by each statement.

4. $9 \times 1 = 9$ Identity Property (×)
5. $7 \times 3 = 3 \times 7$ Commutative Property (×)
6. $(7 + 8) + 2 = 7 + (8 + 2)$ Associative Property (+)
7. $6(3 + 2) = 6(3) + 6(2)$ Distributive Property
8. $15 + 12 = 12 + 15$ Commutative Property (+)
9. $1 \times 20 = 20$ Identity Property (×)
10. $(9 \times 5) \times 2 = 9 \times (5 \times 2)$ Associative Property (×)
11. $3 = 0 + 3$ Identity Property (+)

Practice: Skills, p. 27

NAME ______ DATE ______ PERIOD ____

1-6 Practice: Skills

Algebra: Properties

Use the Distributive Property to write each expression as an equivalent expression. Then evaluate the expression.

1. $3(5 + 1)$ $3(5) + 3(1)$; 18
2. $(2 + 7)5$ $2(5) + 7(5)$; 45
3. $(10 + 2)7$ $10(7) + 2(7)$; 84
4. $2(9 - 8)$ $2(9) - 2(8)$; 2
5. $4(10 - 2)$ $4(10) - 4(2)$; 32
6. $6(13 + 4)$ $6(13) + 6(4)$; 102

Name the property shown by each statement.

7. $2 \times (3 \times 7) = (2 \times 3) \times 7$ Associative Property (×)
8. $6 + 3 = 3 + 6$ Commutative Property (+)
9. $3(9 - 7) = 3(9) - 3(7)$ Distributive Property
10. $18 \times 1 = 18$ Identity Property (×)
11. $7 \times 2 = 2 \times 7$ Commutative Property (×)
12. $6 + (1 + 4) = (6 + 1) + 4$ Associative Property (+)
13. $7 + 0 = 7$ Identity Property (+)
14. $0 + 12 = 12$ Identity Property (+)
15. $625 + 281 = 281 + 625$ Commutative Property (+)
16. $(12 \times 18) \times 5 = 12 \times (18 \times 5)$ Associative Property (×)
17. $2(8 + 2) = 2(8) + 2(2)$ Distributive Property
18. $(15 + 11) + 9 = 15 + (11 + 9)$ Associative Property (+)
19. $(6 + r) + s = 6 + (r + s)$ Associative Property (+)
20. $(4 \times 8) \times a = 4 \times (8 \times a)$ Associative Property (×)
21. $p \times 1 = p$ Identity Property (×)
22. $a + 5 = 5 + a$ Commutative Property (+)
23. $y \times 3 = 3 \times y$ Commutative Property (×)
24. $b + 0 = b$ Identity Property (+)
25. $(x + y) + z = x + (y + z)$ Associative Property (+)
26. $6(200 + 50) = 6(200) + 6(50)$ Distributive Property

Practice: Word Problems, p. 28

NAME ______ DATE ______ PERIOD ____

1-6 Practice: Word Problems

Algebra: Properties

1. **MUSIC** Mr. Escalante and Mrs. Turner plan to take their music classes to a musical revue. Tickets cost $6 each. Mr. Escalante's class needs 22 tickets, and Mrs. Turner's class needs 26 tickets. Use the Distributive Property to write a sentence to express how to find the total cost of tickets in two ways. $6(22) + 6(26) = 6(22 + 26)$

2. **SAVINGS** Mrs. Perez was looking at her bank account statement. She noticed that her beginning balance was $500, and she had added nothing to her account. What was the ending balance on her statement? What property did you apply? $500; Identity Property of Addition

3. **ADDITION** Mr. Brooks was working on addition using dominos with a group of 1st graders. When picking the domino with 3 dots on one end and 5 dots on the other, some students read. "3 plus 5 equal 8" while others read it as "5 plus 3 equals 8." What property were these children using? Explain. Commutative Property of Addition; the order was different.

4. **AREA** Aleta noticed that for the rectangle below she could either multiply 2 times 3 or 3 times 2 to get its area of 6 square inches. What property allows her to do this? Commutative Property of Multiplication

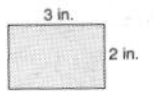

5. **NUMBER CUBES** Students in Mr. Rivas' class were practicing their multiplication skills by rolling three 6-sided number cubes. Wapi rolled a 2, a 3, and a 5 on his roll. He multiplied the three numbers as follows using the order of operations: $(2 \times 3) \times 5 = 30$. Write another way Wapi could have performed the multiplication without changing the order of the numbers. State the property you used. $2 \times (3 \times 5) = 30$; Associative Property of Multiplication

6. **FACTS** Bik was working on memorizing her multiplication facts. She noticed that anytime she multiplied a number by 1, she got the same number she started with. What property allows this to be true? Identity Property of Multiplication

7. **MONEY** Mei was trying to figure out the cost of 4 boxes of cereal for $2.25 each. Write a sentence to show Mei an easy way to do her calculations. What property did you apply to help her? $4(2.25) = 4(2.00 + 0.25)$; Distributive Property

8. **WALKING** Jacob walked 3 blocks to Ping's house, then 5 blocks to Jamal's house. Write a sentence to show that the distance from Ping's to Jamal's is the same as the return walk home. Name the property illustrated in your sentence. $3 + 5 = 5 + 3$; Commutative Property of Addition

1-7 Lesson Notes

Lesson Resources

Chapter 1 Resource Masters
- Study Guide and Intervention, p. 31
- Practice: Skills, p. 32
- Practice: Word Problems, p. 33
- Reading to Learn Mathematics, p. 34
- Enrichment, p. 35

MathPASS Tutorial CD-ROM, 25
Parent and Student Study Guide Workbook, 1-7

FAST FILE

1 Focus

5-Minute Check 1-7

INTERACTIVE CHALKBOARD
PowerPoint® Presentations

Use as a quiz or review of Lesson 1-6. (Also available as a transparency)

Bellringer

Hands-On Activity Have students use counters to represent the sequence 2, 4, 6, 8, ... Have them group together 2 counters, then add 2 more counters to make 4, and so on. Then ask students to predict the next number in the sequence and explain their predictions. **10; Each pile is made by adding 2 counters to the previous pile.**

Reading to Learn Mathematics, p. 34 **ELL**

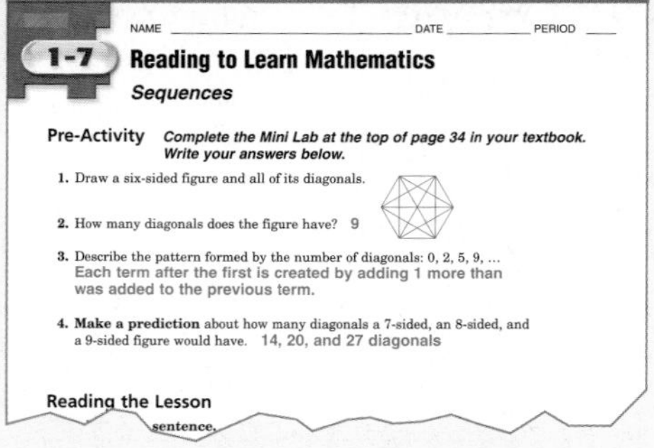
NAME ______ DATE ______ PERIOD ____

1-7 Reading to Learn Mathematics
Sequences

Pre-Activity *Complete the Mini Lab at the top of page 34 in your textbook. Write your answers below.*

1. Draw a six-sided figure and all of its diagonals.
2. How many diagonals does the figure have? 9
3. Describe the pattern formed by the number of diagonals: 0, 2, 5, 9, ... Each term after the first is created by adding 1 more than was added to the previous term.
4. **Make a prediction** about how many diagonals a 7-sided, an 8-sided, and a 9-sided figure would have. 14, 20, and 27 diagonals

Reading the Lesson
sentence.

1-7 Sequences

What You'll Learn
Recognize and extend patterns for sequences.

NEW Vocabulary
- sequence
- term
- arithmetic sequence
- geometric sequence

LINK To Reading
Everyday Meaning of Sequence: a connected series, as in a sequence of events

HANDS-ON Mini Lab

Materials
- straightedge

Work with a partner.

A *diagonal* connects two nonconsecutive vertices in a figure, as shown at the right.

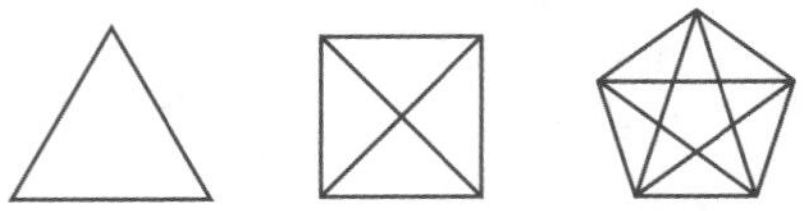

3 sides, 0 diagonals | 4 sides, 2 diagonals | 5 sides, 5 diagonals

1. Draw a six-sided figure and all of its diagonals. **See margin.**
2. How many diagonals does the figure have? **9**
3. Describe the pattern formed by the number of diagonals: 0, 2, 5, 9, … .
4. **Make a prediction** about how many diagonals a 7-sided, an 8-sided, and a 9-sided figure would have. **14, 20, and 27 diagonals**

3. Each term after the first is created by adding 1 more than was added to the previous term.

A **sequence** is an ordered list of numbers. Each number in a sequence is called a **term**.

8, 11, 14, 17, 20, …
+3 +3 +3 +3

In an **arithmetic sequence**, each term is found by *adding* the same number to the previous term.

3, 6, 12, 24, 48, …
×2 ×2 ×2 ×2

In a **geometric sequence**, each term is found by *multiplying* the previous term by the same number.

EXAMPLES Describe Patterns in Sequences

Describe the pattern in each sequence and identify the sequence as *arithmetic*, *geometric*, or *neither*.

1 **8, 13, 18, 23, …**

8, 13, 18, 23, …
+5 +5 +5

Each term is found by adding 5 to the previous term. This sequence is arithmetic.

2 **1, 4, 16, 64, …**

1, 4, 16, 64, …
×4 ×4 ×4

Each term is found by multiplying the previous term by 4. This sequence is geometric.

Your Turn **Describe the pattern in each sequence and identify the sequence as *arithmetic*, *geometric*, or *neither*.**

a. 4, 12, 36, 108, … **× 3; geometric**
b. 1, 2, 4, 7, 11, …

b. Add 1 more than was added to the previous term; neither.

Answer (Hands-On Mini Lab)

1.

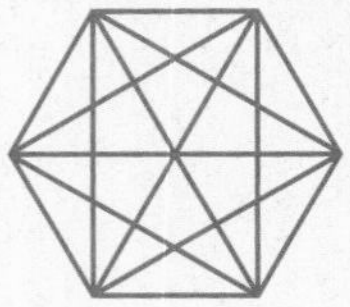

Answer

1. **Sample answer: Each term in an arithmetic sequence is found by adding or subtracting the same number to or from the previous term. Each term in a geometric sequence is found by multiplying or dividing the previous term by the same number.**

If you know the pattern of a sequence, you can use it to determine the terms.

EXAMPLES Determine Terms in Sequences

Write the next three terms of each sequence.

3 **0, 13, 26, 39, ...**

0, 13, 26, 39, ... Each term is 13 greater than the previous term.
+13 +13 +13

Continue the pattern to find the next three terms.

39 + 13 = 52   52 + 13 = 65   65 + 13 = 78

The next three terms are 52, 65, and 78.

4 **0.16, 0.8, 4, 20, 100, ...**

0.16, 0.8, 4, 20, ... Each term is 5 times the previous term.
×5 ×5 ×5

Continue the pattern to find the next three terms.

20 × 5 = 100   100 × 5 = 500   500 × 5 = 2,500

The next three terms are 100, 500, and 2,500.

Your Turn **Write the next three terms of each sequence.**

c. 0.5, 1.5, 2.5, 3.5, ... **4.5, 5.5, 6.5** d. 0.5, 1, 2, 4, ... **8, 16, 32**

Skill and Concept Check

Writing Math
Exercises 1–3

1. **Compare and contrast** arithmetic and geometric sequences.
1–2. See margin.
2. **Explain** how you would find the next term in the sequence 14, 22, 30, 38, 46, ...
3. **OPEN ENDED** Write five terms of an arithmetic sequence and describe the rule for finding the terms. **See students' work.**

GUIDED PRACTICE

Describe the pattern in each sequence and identify the sequence as *arithmetic*, *geometric*, or *neither*.

4. 0, 9, 18, 27, ... **+ 9; arithmetic**
5. 200, 202, 206, 212 **See margin.**
6. 36, 144, 576, 2,304, ... **× 4; geometric**

Write the next three terms of each sequence.

7. 4, 12, 36, 108, ... **324, 972, 2,916**
8. 22, 33, 44, 55, ... **66, 77, 88**
9. 5, 5.4, 5.8, 6.2, ... **6.6, 7, 7.4**
10. Find the missing terms in the sequence 26, 33, 40, __?__, 54, 61, __?__, ... **47; 68**
11. Create a sequence of four terms in which the first term is 70 and the next terms are found by multiplying 0.1. **70, 7, 0.7, 0.07**

DAILY INTERVENTION — Differentiated Instruction

Auditory/Musical Have the class think of ways to demonstrate arithmetic and geometric sequences with sound. **Sample answers: Demonstrate arithmetic sequence by clapping out the number of each term, pausing briefly between terms (clap, clap-clap, clap-clap-clap, . . .). Demonstrate geometric sequence with volume. Start with a very low volume, double it, then double it again. The volume quickly increases.**

2 Teach

In-Class Examples

INTERACTIVE CHALKBOARD PowerPoint® Presentations

Describe the pattern in each sequence and identify the sequence as *arithmetic*, *geometric*, or *neither*.

1. 3, 6, 12, 24, ... **× 2; geometric**
2. 7, 11, 15, 19, ... **+ 4; arithmetic**

Write the next three terms of each sequence.

3. 5, 14, 23, 32, ... **41, 50, 59**
4. 0.2, 1.2, 7.2, 43.2, ... **259.2, 1555.2, 9331.2**

3 Practice/Apply

About the Exercises...

Guided Practice Key

Exercises	Examples
4–6	1, 2
7–11	3, 4

Odd/Even Assignments
Exercises 12–23 are structured so that students practice the same concepts whether they are assigned odd or even problems.

Answers

2. Sample answer: First, find what number is added or multiplied to each term to get the next term.
5. Add 2 more than what was added to the previous term; neither.
17. Multiply by 1 more than was multiplied by the previous term; neither.
19. 20.6, 24.6, 28.6
20. 204.8, 1,638.4, 13,107.2
26. Add 2 to the first term, add 3 to the second term, add 4 to the third term, and so on.

Study Guide and Intervention, p. 31

NAME ______________ DATE ________ PERIOD ____

1-7 Study Guide and Intervention

Sequences

An **arithmetic sequence** is a list in which each term is found by *adding* the same number to the previous term. 2, 5, 8, 11, 14, ...
+3 +3 +3 +3

A **geometric sequence** is a list in which each term is found by *multiplying* the previous term by the same number. 2, 6, 18, 54, ...
+3 +3 +3

EXAMPLE 1 **Describe the pattern in the sequence and identify the sequence as *arithmetic, geometric,* or *neither.* 4, 8, 12, 16, ...**

4, 8, 12, 16, ...
+4 +4 +4

Each term is found by adding 4 to the previous term. This sequence is arithmetic.

EXAMPLE 2 **Describe the pattern in the sequence and identify the sequence as *arithmetic, geometric,* or *neither.* 1, 5, 25, 125, ...**

1, 5, 25, 125, ...
×5 ×5 ×5

Each term is found by multiplying the previous term by 5. This sequence is geometric.

EXAMPLE 3 **Write the next three terms of the sequence 20, 40, 80, 160, ...**

20, 40, 80, 160, ... Each term is two times the previous term.
×2 ×2 ×2

Continue the pattern to find the next three terms.
160 × 2 = 320 320 × 2 = 640 640 × 2 = 1,280
The next three terms are 320, 640, and 1,280.

EXERCISES

Describe the pattern in each sequence and identify the sequence as *arithmetic, geometric,* or *neither.*

1. 2, 4, 6, 8, ... +2; arithmetic
2. 6, 12, 24, 48, ... ×2; geometric
3. 1, 2, 4, 8, ... ×2; geometric
4. 4, 7, 10, 13, ... +3; arithmetic
5. 1, 1, 2, 3, 5, ... neither
6. 4, 4, 4, 4, ... ×1; geometric

Write the next three terms of each sequence.

7. 3, 6, 9, 12, ... 15, 18, 21
8. 2, 6, 18, 54, ... 162; 486; 1,458
9. 0.2, 0.4, 0.8, 1.6, ... 3.2, 6.4, 12.8
10. 10, 20, 30, 40, ... 50, 60, 70
11. 1, 4, 16, 64, ... 256; 1,024; 4,096
12. 4.5, 6.5, 8.5, 10.5, ... 12.5, 14.5, 16.5

Practice: Skills, p. 32

NAME ______________ DATE ________ PERIOD ____

1-7 Practice: Skills

Sequences

Describe the pattern in each sequence and identify the sequence as *arithmetic, geometric,* or *neither.*

1. 3, 6, 12, 24, ... ×2; geometric
2. 1, 3, 5, 7, ... +2; arithmetic
3. 1, 2, 6, 24, ... neither
4. 0, 7, 14, 21, ... +7; arithmetic
5. 2, 5, 8, 11, ... +3; arithmetic
6. 5, 15, 45, 135, ... ×3; geometric
7. 0.3, 1.5, 7.5, 37.5, ... ×5; geometric
8. 1, 10, 100, 1,000, ... ×10; geometric
9. 7, 7, 7, 7, ... ×1; geometric
10. 0.5, 2, 8, 32, ... ×4; geometric
11. 3, 7, 11, 15, ... +4; arithmetic
12. 9, 18, 36, 72, ... ×2; geometric
13. 11, 22, 44, 88, ... ×2; geometric
14. 11, 22, 33, 44, ... +11; arithmetic

Write the next three terms of each sequence.

15. 3, 6, 9, 12, ... 15, 18, 21
16. 2, 4, 8, 16, ... 32, 64, 128
17. 7, 10, 13, 16, ... 19, 22, 25
18. 3, 15, 75, 375, ... 1,875; 9,375; 46,875
19. 1, 3, 9, 27, ... 81, 243, 729
20. 7, 12, 17, 22, ... 27, 32, 37
21. 5, 7, 9, 11, ... 13, 15, 17
22. 5, 15, 25, 35, ... 45, 55, 65
23. 21, 42, 63, 84, ... 105, 126, 147
24. 6, 24, 96, 384, ... 1,536; 6,144; 24,576
25. 0.5, 1.0, 1.5, 2.0, ... 2.5, 3.0, 3.5
26. 1.7, 1.9, 2.1, 2.3, ... 2.5, 2.7, 2.9
27. 0.5, 1.5, 4.5, 13.5, ... 40.5, 121.5, 364.5
28. 2, 8, 32, 128, ... 512; 2,048; 8,192

Practice: Word Problems, p. 33

NAME ______________ DATE ________ PERIOD ____

1-7 Practice: Word Problems

Sequences

1. **NUMBERS** The multiples of two form a sequence as follows: 2, 4, 6, 8, 10, 12, 14, 16, ... What type of sequence do you see? What about the multiples of three? Four? Five? Arithmetic; the multiples of any number would result in an arithmetic sequence.

2. **OLYMPICS** The summer Olympics occur every four years. If the last summer Olympics happened in 2000, when are the next three times that it will occur? What type of sequence do the Olympic years form? 2004, 2008, 2012; arithmetic

3. **BABY-SITTING** Tonya charges $3.50 per hour to baby-sit. The sequence $3.50, $7.00, $10.50, $14.00, ... represents how much she charges for each subsequent hour. For example, $10.50 is the third term that represents how much she charges for 3 hours. What are the next three terms in the sequence? How much does she charge for 7 hours of baby-sitting? $17.50, $21.00, $24.50; $24.50

4. **RECTANGLES** Suppose you start with 1 rectangle and then divide it in half. You now have 2 rectangles. You divide each of these in half, and you have 4 rectangles. The sequence for this division is 1, 2, 4, 8, 16, . . . rectangles after each successive division. What type of sequence results? geometric

5. **BACTERIA** Three bacteria are in a dish. Each hour the number of bacteria multiplies by four. If at the end of the first hour there are 12 bacteria, how many bacteria are there at the end of the next three hours? What type of sequence results? 48, 192, 768; geometric

6. **ENROLLMENT** The enrollment at Grove Middle School is expected to increase by 40 students each year for the next 5 years. If their current enrollment is 600 students, find their enrollment after each of the next 5 years. 640, 680, 720, 760, 800

7. **SALARY** Mrs. Malone's current salary is $1,500. She expects it to increase $100 per year. Write the first 6 terms of a sequence that represents her salary. The first term should be her current salary. What does the sixth term represent? $1,500, $1,600, $1,700, $1,800, $1,900, $2,000; her salary after 5 years or at the beginning of the sixth year

8. **FIBONACCI** The Fibonacci sequence is named after Leonardo Fibonacci who first explored it. Look at the Fibonacci sequence below and describe its pattern. Is it arithmetic, geometric or neither? 1, 1, 2, 3, 5, 8, 13, 21, 34, ... Each term is found by adding the two previous terms; it is neither arithmetic nor geometric.

★ indicates increased difficulty

Practice and Applications

HOMEWORK HELP	
For Exercises	See Examples
12–17, 26	1, 2
18–25	3, 4
Extra Practice See pages 566, 596.	

Describe the pattern in each sequence and identify the sequence as *arithmetic, geometric,* or *neither.*

12. 1, 6, 36, 216, ... × 6; geo.
13. 19, 31, 43, 55, ... + 12; arith.
14. 2, 14, 98, 686, ... × 7; geo.
15. 3.5, 10.5, 31.5, 94.5, ... × 3; geo.
16. 2.0, 3.1, 4.2, 5.3, ... + 1.1; arith.
★17. 1, 2, 6, 24, 120, ... See margin p. 35.

Write the next three terms of each sequence. 19–20. See margin p. 35.

18. 33, 38, 43, 48, ... 53, 58, 63
19. 4.6, 8.6, 12.6, 16.6, ...
20. 0.05, 0.4, 3.2, 25.6, ...
21. 0.1, 0.4, 0.7, 1.0, ... 1.3, 1.6, 1.9
22. 3, 12, 48, 192, ... 768, 3,072, 12,288
★23. 125, 25, 5, 1, ... 0.2, 0.04, 0.008

24. **INSECTS** Every 17 years, billions of insects called *cicada* emerge from hibernation. The last time they emerged was in 1996. Find the next three years in which they will come out of hibernation. 2013, 2030, 2047

GEOMETRY **For Exercises 25 and 26, use the following information.**
Numbers that can be represented by a triangular arrangement of dots are called *triangular numbers*. The first five triangular numbers are shown below.

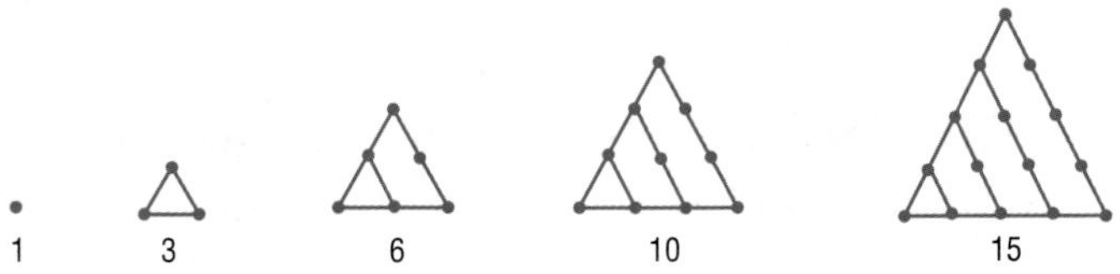

25. Write a sequence formed by the first eight triangular numbers. 1, 3, 6, 10, 15, 21, 28, 36

★26. Write a rule for generating the sequence. See margin p. 35.

27. **CRITICAL THINKING** You can find any term in an arithmetic sequence by using the expression $a + (n - 1)d$, where a is the first term, d is the difference between each pair of consecutive terms, and n is the position in the sequence. Find the eleventh term in the sequence 6, 13, 20, 27, 34, … . 76

Standardized Test Practice and Mixed Review

Time (h)	Height (in.)
0	5
1	27
2	49
3	71

28. **SHORT ANSWER** The table shows the height of a bamboo plant after each number of hours. How tall will the plant be after 6 hours? 137 in.

29. **GRID IN** Find the next term in the sequence 44, 58, 72, 86, … . 100

Name the property shown by each statement. (Lesson 1-6)

30. $8 + 4.1 = 4.1 + 8$ Comm. (+)
31. $3(n + 6) = 3n + 3 \cdot 6$ Distributive

32. Solve $30 + y = 50$ mentally. (Lesson 1-5) 20

GETTING READY FOR THE NEXT LESSON

PREREQUISITE SKILL **Multiply or divide.** (Pages 560, 562)

33. 52 × 10 520
34. 400 ÷ 1,000 0.4
35. 32 ÷ 10 3.2
36. 0.31 × 100 31

msmath2.net/self_check_quiz

4 Assess

Higher-Level Learning

Ask students how they could solve the following problem: If a cow produces its first female calf at age two years and after that produces another single female calf every year, how many female calves are there after 12 years, assuming none die? Every female calf born also reproduces at the same rate. Create tree diagrams for several years, examine the total number of cows for each year to discern a pattern, then continue that pattern to year 12. sequence: 1, 1, 2, 3, 5, 8, 13, 21, 34, 55, 89, 144; pattern: Each term is the sum of the two previous terms.

1-7b HANDS-ON LAB

A Follow-Up of Lesson 1-7

'hat You'll Learn
:plore patterns in
quences using paper
lding.

aterials:
:alculator
piece of paper
ecording table
:olored pencils

nds-On Mini
s and
nds-On Labs
e manipulatives
d models to help
udents learn key
ncepts.

Exploring Sequences

INVESTIGATE *Work with a partner.*

Number of Folds	Layers of Paper	Fraction of Paper *Not* Shaded
1	2	$\frac{1}{2}$
2	4	$\frac{1}{4}$
3	8	
4		
⋮		

STEP 1 Fold a piece of paper in half. Make a table like the one at the right and record the number of layers of paper.

STEP 2 Shade one side of the folded paper.

STEP 3 Open the piece of paper. Record the fractional part of the paper that is *not* shaded. Refold the paper.

STEP 4 Fold your paper in half again so that the unshaded side is on the outside. Record the number of layers of paper.

STEP 5 Shade one side of the folded paper.

STEP 6 Open the piece of paper. Record the fractional part of the paper that is *not* shaded. Completely refold the paper.

STEP 7 Continue this folding, shading, and recording process for five folds.

Writing Math

Work with a partner. **3, 5. See margin.**

1. **Examine** the sequence of numbers in the "Layers" column of your table. Is the sequence arithmetic or geometric? Then write a rule to find the next three terms. **geometric; rule: × 2; 64, 128, 256**
2. **Examine** the sequence of numbers in the "Fraction" column of your table. Is the sequence arithmetic or geometric? Then write a rule to find the next three terms.
3. Assume you could continue the paper-folding process indefinitely. Suppose your unfolded piece of paper is 0.002 inch thick. Add a column to your table and find the thickness of the paper for the first five folds.
4. How many folds would it take until the paper is as tall as you?
5. **Explain** the relationship between the number of layers and the fraction of the shaded region.

geometric; rule: $\frac{1}{2}$; $\frac{1}{64}$, $\frac{1}{128}$, $\frac{1}{256}$

Sample answer: 14 folds is 32.768 in., 15 folds is 65.536 in., 16 folds is 131.072 in.

1-7b HANDS-ON LAB
Lesson Notes

Lesson Resources

Teaching Mathematics with Manipulatives
- student recording sheet 1-7b

GETTING STARTED

Teaching Tip Recommend that students use colored pencils for their shading so that they can keep shaded portions separate. They could also color-code their tables.

TEACH

- Make sure students are careful not to tear the tissue paper when shading.
- Encourage students to be precise in their folding. The more precise they are, the better their results.

ASSESS

Observe as students solve the **Writing Math** exercises. They should:

- correctly determine the rules for the columns of the table, and
- determine the number of folds needed to match their height.

Discuss **Exercise 5**. Students should demonstrate a full understanding of arithmetic sequences.

Answers

3.

Number of Folds	Layers of Paper	Fraction of Paper *Not* Shaded	Thickness of Folded Paper
1	2	1/2	0.004
2	4	1/4	0.008
3	8	1/8	0.016
4	16	1/16	0.032
5	32	1/32	0.064

5. Sample answer: The "Layer" column represents powers of 2, 2^n. The "Fraction" column represents the powers of the reciprocal of 2, $\frac{1}{2^n}$.

1-8 Lesson Notes

Lesson Resources

Chapter 1 Resource Masters
- Study Guide and Intervention, p. 36
- Practice: Skills, p. 37
- Practice: Word Problems, p. 38
- Reading to Learn Mathematics, p. 39
- Enrichment, p. 40

FAST FILE

Key Teaching Transparencies, 4
MathPASS Tutorial CD-ROM, 56
Parent and Student Study Guide Workbook, 1-8
Science and Math Lab Masters, pp. 25–28

1 Focus

5-Minute Check 1-8

INTERACTIVE CHALKBOARD PowerPoint® Presentations

Use as a quiz or review of Lesson 1-7. (Also available as a transparency)

Bellringer

Communication Ask students what they know about the metric system. Then discuss the following history of the metric system.

The International System of Units (SI) is a modernized version of the metric system. The metric system was developed during the French Revolution.

(Continued in bottom margin.)

Reading to Learn Mathematics, p. 39 ELL

NAME ________ DATE ________ PERIOD ____

1-8 Reading to Learn Mathematics
Measurement: The Metric System

Pre-Activity *Complete the Mini Lab at the top of page 38 in your textbook. Write your answers below.*

1. Find three other objects. Use a ruler to find and record the width of all five objects to the nearest millimeter and tenth of a centimeter. Add your measurements to the table. See students' work.
2. Compare the measurements of each object, and write a rule that describes how to convert from millimeters to centimeters. Divide by 10.
3. Use a meterstick to measure the length of your classroom in meters. **Make a conjecture** about which operation you would use to convert this measure to centimeters. Explain. Multiplication; there is a greater number of smaller units than larger units in a measure.

1-8 Measurement: The Metric System

What You'll Learn
Change metric units of length, capacity, and mass.

NEW Vocabulary
- **meter**
- **metric system**
- **gram**
- **kilogram**
- **liter**

Study Tips offer students helpful information about the topics they are studying.

HANDS-ON Mini Lab

Materials
- centimeter ruler
- computer diskette
- CD case
- meterstick

Work with a partner.

The widths of two objects are shown below.

Object	Width (millimeters)	Width (centimeters)
computer diskette	? 89	8.9
CD case	? 144	14.4

1. Find three other objects. Use a ruler to find and record the width of all five objects to the nearest millimeter and tenth of a centimeter. Add your measurements to the table. See students' work
2. Compare the measurements of each object, and write a rule that describes how to convert from millimeters to centimeters.
3. Use a meterstick to measure the length of your classroom in meters. **Make a conjecture** about which operation you would use to convert this measure to centimeters. Explain.

2. Divide by 10.

The **meter** (m) is the base unit of length in the **metric system**. A meter is about the distance from the floor to a doorknob, or a little more than a yard. One kilometer is equivalent to 1,000 meters, and 1 meter is equivalent to 100 centimeters and also to 1,000 millimeters.

Prefix	Meaning in Words	Meaning in Numbers
kilo-	thousands	1,000
centi-	hundredths	0.01
milli-	thousandths	0.001

Metric prefixes indicate the decimal place-value position of the measurement.

MULTIPLY to convert from larger units to smaller units.

kilometer km → ×1,000 → **meter m** → ×100 → centimeter cm → ×10 → millimeter mm

DIVIDE to convert from smaller units to larger units.

millimeter → ÷10 → centimeter → ÷100 → meter → ÷1,000 → kilometer

STUDY TIP

Mental Math To multiply or divide by a power of ten, you can move the decimal point right or left.

3. Multiplication; there is a greater number of smaller units than larger units in a measure.

EXAMPLES Convert Units of Length

Complete.

1 **4.5 m = __?__ cm**

To convert from meters to centimeters, multiply by 100.

$4.5 \times 100 = 450$

4.5 m = 450 cm

2 **2,600 m = __?__ km**

To convert from meters to kilometers, divide by 1,000.

$2,600 \div 1,000 = 2.6$

2,600 m = 2.6 km

Its use was legalized in the United States in 1866. In 1902, proposed congressional legislation requiring the U.S. government to use the metric system exclusively was defeated by a single vote.

EADING Math

Prefixes The kilogram is the only base unit in the metric system that has a prefix as part of its name.

TEACHING TIP

plain to students that ass and weight are fferent measures. An tronaut in space has e same mass as on rth, but has less eight in space.

The **gram** (g) measures *mass*, the amount of matter in an object. A paper clip has a mass of about one gram. The base unit of mass in the metric system is the **kilogram**, which is equivalent to 1,000 grams. One gram is equivalent to 1,000 milligrams.

EXAMPLES Convert Units of Mass

Complete.

3 **8.43 kg = ___?___ g**

To convert from kilograms to grams, multiply by 1,000.

$8.43 \times 1{,}000 = 8{,}430$

8.43 kg = 8,430 g

4 **500 mg = ___?___ g**

To convert from milligrams to grams, divide by 1,000.

$500 \div 1{,}000 = 0.5$

500 mg = 0.5 g

Your Turn Complete.

a. 23.5 g = ___?___ mg **23,500**

b. 9,014 g = ___?___ kg **9.014**

The **liter** (L) is widely used and accepted as the metric measure for capacity. *Capacity* is the amount of dry or liquid material an object can hold. Some soft drinks are sold in one-liter bottles. Kiloliter, liter, and milliliter are related in a manner similar to kilometer, meter, and millimeter.

EXAMPLES Convert Units of Capacity

Complete.

5 **3.2 L = ___?___ mL**

To convert from liters to milliliters, multiply by 1,000.

$3.2 \times 1{,}000 = 3{,}200$

3.2 L = 3,200 mL

6 **1,640 L = ___?___ kL**

To convert from liters to kiloliters, divide by 1,000.

$1{,}640 \div 1{,}000 = 1.64$

1,640 L = 1.64 kL

Your Turn Complete.

c. 0.75 kL = ___?___ L **750**

d. 8,800 mL = ___?___ L **8.8**

EAL-LIFE MATH

TRIC SYSTEM The tric Act of 1866 made use of the metric tem legal in the ted States.

rce: lamar.colostate.edu

The following table is a summary of how to convert measures in the metric system.

Concept Summary — **Converting Units**

	Larger Units → Smaller Units (multiply by a power of 10)	Smaller Units → Larger Units (divide by a power of 10)
Units of Length	1 km = 1,000 m 1 m = 100 cm 1 cm = 10 mm	1 mm = 0.1 cm 1 cm = 0.01 m 1 m = 0.001 km
Units of Mass	1 kg = 1,000 g 1 g = 1,000 mg	1 mg = 0.001 g 1 g = 0.001 kg
Units of Capacity	1 kL = 1,000 L 1 L = 1,000 mL	1 mL = 0.001 L 1 L = 0.001 kL

msmath2.net/extra_examples

2 Teach

In-Class Examples

INTERACTIVE CHALKBOARD PowerPoint® Presentations

Complete.

1. 28 cm = ?̲ mm **280**
2. 438 cm = ?̲ m **4.38**
3. 72 g = ?̲ mg **72,000**
4. 202 g = ?̲ kg **0.202**
5. 2 L = ?̲ mL **2,000**
6. 2.4 kL = ?̲ L **2,400**

Tips for New Teachers

Using Applications Students may be interested to see some recipes from other countries that use metric measurements. You can find these in cookbooks at a library or on the Internet.

DAILY INTERVENTION — **Differentiated Instruction**

Naturalist Have students take a meterstick outside with them. Tell them to record the following information before going outside: 1 kilometer = 1,000 metersticks; 1 decimeter = $\frac{1}{10}$ of a meterstick; 1 centimeter = $\frac{1}{100}$ of a meterstick; and 1 millimeter = $\frac{1}{1{,}000}$ of a meterstick. They should find and record two natural things that approximate each length. The kilometer category will require estimation.

3 Practice/Apply

FOLDABLES™ Follow-Up Remind students to use their Foldables to summarize what they learn about the units of length, capacity, and mass in the metric system. On the eighth section, they might include definitions of key terms such as *meter* or the meanings of metric prefixes.

DAILY INTERVENTION **FIND THE ERROR** In Exercise 3, remind students to check their conversions to see whether they are reasonable. Is it reasonable to have a smaller number for grams than for milligrams? **no**

About the Exercises...

Guided Practice Key

Exercises	Examples
4, 7	1, 2
5, 8	3, 4
6, 9, 10	5, 6

Odd/Even Assignments
Exercises 11–32, 34–37, and 39–40 are structured so that students practice the same concepts whether they are assigned odd or even problems.

Alert! Exercise 42 requires research on the Internet or another source.

Answers

31. 0.02 km, 3,000 cm, 50 m
32. 430,000 mg, 0.05 kg, 32 g
33. 0.06 L, 660 mL, 6.6 kL

Skill and Concept Check

Writing Math Exercises 1 & 3

1. **Explain** why it makes sense to multiply when changing from a larger unit to a smaller unit. **There are a greater number of smaller units.**
2. **OPEN ENDED** List a set of measurement references, made up of everyday objects, that will help you remember the basic units in the metric system. **See students' work.**
3. **FIND THE ERROR** Hunter and Arturo are converting 45.7 milligrams to grams. Who is correct? Explain.

Hunter	Arturo
45.7 mg = 0.0457 g	45.7 mg = 45,700 g

Hunter; Arturo multiplied 45.7 by 1,000; he should have divided by 1,000.

GUIDED PRACTICE

Complete. **5. 45,000**

4. 550 m = ? km **0.55**
5. 45 g = ? mg
6. 1,460 mL = ? L **1.46**
7. 3.7 m = ? cm **370**
8. 3,800 mg = ? g **3.8**
9. 2.34 kL = ? L **2,340**
10. How many milliliters are in 0.04 liter? **40 mL**

★**indicates increased difficulty**

Practice and Applications

HOMEWORK HELP

For Exercises	See Examples
11–41	1–6

Extra Practice See pages 566, 596.

Complete.

11. 720 cm = ? m **7.2**
12. 983 mm = ? m **0.983**
13. 64 kg = ? g **64,000**
14. 7.5 g = ? mg **7,500**
15. 925 mg = ? g **0.925**
16. 345 mL = ? L **0.345**
17. 0.85 cm = ? mm **8.5**
18. 3.2 m = ? cm **320**
19. 9.1 L = ? mL **9,100**
20. 73.2 g = ? mg **73,200**
21. 43 L = ? kL **0.043**
22. 130.5 kL = ? L **130,500**
23. 997 g = ? kg **0.997**
24. 0.046 kL = ? L **46**
25. 82.1 g = ? kg **0.0821**
26. 0.03 m = ? mm **30**
27. 57,000 mL = ? L **57**
28. 0.22 mm = ? cm **0.022**

MONUMENTS For Exercises 29 and 30, use the information about the Washington Monument at the right.

29. What is the height of the monument in kilometers? **0.169 km**
30. What is the height in centimeters? in millimeters? **16,900 cm; 169,000 mm**

★31. Order 0.02 km, 50 m, and 3,000 cm from least to greatest length.
★32. Order 0.05 kg, 32 g, and 430,000 mg from greatest to least mass.
★33. Order 660 mL, 0.06 L, and 6.6 kL from least to greatest capacity.
31–33. See margin.

Complete.

★34. 0.05 km = ? mm **50,000**
★35. 0.93 km = ? cm **93,000**
★36. 23,000 mg = ? kg **0.023**
★37. 40,000 mL = ? kL **0.04**

38. **TRACK AND FIELD** Refer to the table. Convert the distance of each event to kilometers.

Event	Number of Kilometers
10,000-meter run	? 10
110-meter hurdles	? 0.11
1,600-meter relay	? 1.6
3,000-meter steeplechase	? 3

39. **MULTI STEP** Heather needs a 2.5-meter pole for a birdfeeder that she is building. How many centimeters will she need to cut off a 3-meter pole in order to use it for the birdfeeder? **50 cm**

40. **MULTI STEP** Margarita is running the school concession stand. The fruit punch machine holds 15 liters. Each drink cup holds 500 milliliters of punch. How many cups of punch can Margarita sell before the fountain is empty? **30 cups**

41. **BAKING** Evan is making peach cobblers for the fair. The recipe requires 900 grams of peaches per cobbler. If Evan has 4.5 kilograms of peaches, how many cobblers can he make? **5 cobblers**

42. **RESEARCH** Use the Internet or another source to investigate relationships between metric units of mass and capacity. Explain the relationships and give examples of how they are used in real life. **See students' work.**

CRITICAL THINKING The metric prefix *giga* refers to something that is one billion times larger than the base unit. The metric prefix *nano* refers to something that is one billion times smaller than the base unit.

43. How many meters are in 3.9 gigameters? **3,900,000,000 m**

44. How many meters are in 3.9 nanometers? **0.0000000039 m**

Standardized Test Practice and Mixed Review

45. **MULTIPLE CHOICE** Choose the best estimate for the mass of a golf ball. **C**

Ⓐ 4.5 mg Ⓑ 4.5 g Ⓒ 45 g Ⓓ 45 kg

46. **GRID IN** What is the mass in grams of a 200-milligram dosage of medication? **0.2**

Describe the pattern in each sequence and identify the sequence as *arithmetic, geometric,* or *neither*. (Lesson 1-7) **50. square numbers; neither**

47. 27, 36, 45, 54, … **+ 9; arithmetic**

48. 6, 36, 216, 1,296, … **× 6; geometric**

49. 1.1, 2.2, 3.3, … **+ 1.1; arithmetic**

50. 1, 4, 9, 16, …

Use the Distributive Property to write each expression as an equivalent expression. Then evaluate the expression. (Lesson 1-6)

51. 5(9 + 7) **5(9) + 5(7); 80**

52. (12 + 4)4 **(12)4 + (4)4; 64**

53. 8(7 − 2) **8(7) − 8(2); 40**

54. (6 − 5)10 **(6)10 − (5)10; 10**

GETTING READY FOR THE NEXT LESSON

PREREQUISITE SKILL **Multiply.** (Lesson 1-3)

55. $8 \cdot 10^2$ **800**

56. $5 \cdot 10^3$ **5,000**

57. $4 \cdot 10^5$ **400,000**

58. $2.7 \cdot 10^3$ **2,700**

59. $65 \cdot 10^2$ **6,500**

60. $1.35 \cdot 10^4$ **13,500**

msmath2.net/self_check_quiz

4 Assess

Higher-Level Thinking

Discuss with students why they think the United States has not changed over to using the metric system. Whenever there has been a national push, there has been a resistance to the change.

Study Guide and Intervention, p. 36

NAME ____________ DATE ________ PERIOD ____

1-8 Study Guide and Intervention

Measurement: The Metric System

The table below is a summary of how to convert measures in the metric system.

	Larger Units → Smaller Units	Smaller Units → Larger Units
Units of Length (meter)	km to m – multiply by 1,000 m to cm – multiply by 100 m to mm – multiply by 1,000 cm to mm – multiply by 10	mm to cm – divide by 10 mm to m – divide by 1,000 cm to m – divide by 100 m to km – divide by 1,000
Units of Mass (kilogram)	kg to g – multiply by 1,000 g to mg – multiply by 1,000	mg to g – divide by 1,000 g to kg – divide by 1,000
Units of Capacity (liter)	kL to L – multiply by 1,000 L to mL – multiply by 1,000	mL to L – divide by 1,000 L to kL – divide by 1,000

EXAMPLE 1 **Complete. 62 cm = ___ m**

To convert from centimeters to meters, divide by 100.
62 ÷ 100 = 0.62
62 cm = 0.62 m

EXAMPLE 2 **Complete. 2.6 kL = ___ L**

To convert from kiloliters to liters, multiply by 1,000.
2.6 × 1,000 = 2,600
2.6 kL = 2,600 L

EXERCISES

Complete.

1. 650 cm = ___ m 6.5
2. 57 kg = ___ g 57,000
3. 751 mg = ___ g 0.751
4. 8.2 L = ___ mL 8,200
5. 52 L = ___ kL 0.052
6. 892 mm = ___ m 0.892
7. 121.4 kL = ___ L 121,400
8. 0.72 cm = ___ mm 7.2
9. 67.3 g = ___ kg 0.0673
10. 5.2 g = ___ mg 5,200
11. 0.05 m = ___ mm 50
12. 2,500 mg = ___ g 2.5
13. 32 mm = ___ cm 3.2
14. 96 m = ___ cm 9,600

Practice: Skills, p. 37

NAME ____________ DATE ________ PERIOD ____

1-8 Practice: Skills

Measurement: The Metric System

Complete.

1. 660 m = ___ km 0.66
2. 5.7 m = ___ cm 570
3. 543 mL = ___ L 0.543
4. 23.7 g = ___ mg 23,700
5. 529 mg = ___ g 0.529
6. 2,640 mL = ___ L 2.640
7. 4.32 kL = ___ L 4,320
8. 75.4 mg = ___ g 0.0754
9. 8,300 mg = ___ g 8.3
10. 7.3 m = ___ cm 730
11. 250.3 kL = ___ L 250,300
12. 799 g = ___ kg 0.799
13. 8.5 cm = ___ mm 85
14. 450 kg = ___ g 450,000
15. 7.3 L = ___ mL 7,300
16. 6,140 L = ___ kL 6.140
17. 3,500 m = ___ km 3.5
18. 89 km = ___ m 89,000
19. 26.8 mm = ___ cm 2.68
20. 750 m = ___ km 0.75
21. 4.8 m = ___ cm 480
22. 95 g = ___ mg 95,000
23. 389 mm = ___ m 0.389
24. 56 L = ___ kL 0.056
25. 0.32 mm = ___ cm 0.032
26. 39.1 g = ___ kg 0.0391

Practice: Word Problems, p. 38

NAME ____________ DATE ________ PERIOD ____

1-8 Practice: Word Problems

Measurement: The Metric System

1. RUNNING Each morning Carlos runs 1.5 kilometers. How many meters did he run? 1,500 m

2. AVIATION A helicopter was flying 800 meters above the ground. How many kilometers above the ground was it flying? 0.8 km

3. SODA A soda can contains 355 milliliters of liquid. How many liters of liquid does it contain? 0.355 L

4. CONSTRUCTION The ceilings of most classrooms are about 2.5 meters above the floor. How many centimeters high is the ceiling? 250 cm

5. FENCING Gerri's garden is 1,270 centimeters around the edges. How many meters of fencing material does she need to enclose her garden? 12.7 m

6. GARDENING Mr. Chou's lawn sprinker sprays about 150 liters of water each hour. How many kiloliters of water does it spray? 0.15 kL

7. NUTRITION For 11- to 14-year-olds, the Recommended Dietary Allowance (RDA) for protein is about 60 grams daily. How many milligrams do they need daily? 60,000 mg

8. MEASUREMENT A measure of one pound is equivalent to about 454 grams. How many kilograms are in one pound? How many milligrams? 0.454 kg; 454,000 mg

Study Skill

Objective Organize notes for more effective studying.

GETTING STARTED

Taking Notes Before starting this lesson, have students review notes they have taken in the past. Ask them to assess the way their notes are organized. (They will probably indicate their notes reflect the order in which the material was presented.) After teaching power notes, have them review their notes again, indicating how the notes could be reorganized.

TEACH

- Students might be forced to work quickly during a class or lecture if a good deal of new information is being presented. This situation can lead to notes that are messy and difficult to follow. Encourage students to recopy notes immediately (or as soon as they can) after class. The material will still be fresh in their minds, and they can use the power notes format to create well-organized, legible notes.
- After students organize their notes, encourage them to highlight any concepts that are still unclear. Highlighting will make it easy to see areas where they need to ask questions or do extra work.
- Encourage students to use their power notes when studying for the chapter test. They can cover their notes with a piece of paper and uncover one power at a time. Students can quiz themselves on each power and uncover the notes to check.

Study Skill HOW TO... Take Good Notes

POWER NOTES

Do you ever have trouble organizing your notes? Try using power notes. They can help you organize the main ideas and details in a lesson.

Study Skill pages offer practical suggestions for taking notes, learning vocabulary, and reading word problems.

Power notes are similar to lesson outlines, but they are simpler to organiz
Power notes use the numbers 1, 2, 3, and so on.

Power 1: This is the main idea.
  Power 2: This provides details about the main idea.
    Power 3: This provides details about Power 2.
      and so on....

Here's a sample of power notes from Lesson 1-8. Notice that you can eve
add drawings or examples to your power notes.

1. The Metric System
  2. Length
    3. Units of Length
      4. meter – basic unit
      4. 1 kilometer = 1,000 meters
      4. 1 centimeter = 0.01 meter
      4. 1 millimeter = 0.001 meter

You can have more than one detail under each power.

    3. Converting Units of Length
      4. from larger to smaller – multiply
      4. from smaller to larger – divide

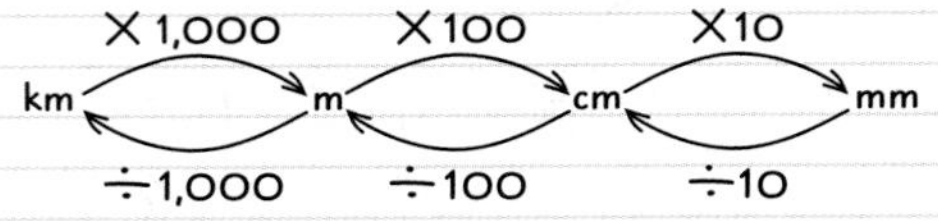

  2. Mass

SKILL PRACTICE

1. **Explain** why the four-step plan in Lesson 1-1 is a form of power notes.
2. Complete the power notes above for units of mass and capacity.

Use power notes to make an outline for each concept.

3. order of operations (*Hint:* See Lesson 1-3.) **2–5. See students' work.**
4. addition and multiplication properties (*Hint:* See Lesson 1-6.)
5. algebraic and geometric sequences (*Hint:* See Lesson 1-7.)

1. Sample answer: The four-step plan is Power 1, the steps are Power and the explanations of the steps are Power

ASSESS

Open-Ended Assessment

Modeling Have students create a template that they can use in the future to remind them how power notes are organized.

1-9 Scientific Notation

What You'll Learn
rite numbers greater
an 100 in scientific
otation and in
andard form.

EW Vocabulary
ientific notation

EVIEW Vocabulary
andard form:
imbers written
ithout exponents
esson 1-2)

ching Tips
the reduced
ident page
er practical
iggestions for
ur lessons.

TEACHING TIP
ll students that
ientific notation is used
express very large or
ry small numbers.

WHEN am I ever going to use this?

ENTERTAINMENT The Mall of America in Bloomington, Minnesota, is the largest shopping and entertainment center in the United States. The table shows two features of the mall. **1. 13,000; 4.2 million**

	Standard Form	Abbreviated Form
Number of Parking Spaces	?	13 thousand
Area of Mall (square feet)	4,200,000	?

1. Complete the table.
2. Write both numbers as the product of a number and a power of ten. **Sample answer: $13 \times 1{,}000$; $4.2 \times 1{,}000{,}000$**
3. Rewrite both powers of ten using exponents.

3. Sample answer: $1{,}000 = 10^3$; $1{,}000{,}000 = 10^6$

When you work with very large numbers, it is difficult to keep track of the place value. You can write numbers like 4.2 million in **scientific notation** by using a power of ten.

$$4.2 \text{ million} = 4.2 \times 1{,}000{,}000$$
$$= 4.2 \times 10^6 \qquad 10^6 = 1{,}000{,}000$$

Key Concept — **Scientific Notation**

Words A number is expressed in scientific notation when it is written as the product of a number and a power of ten. The number must be greater than or equal to 1 and less than 10.

Examples $60{,}000 = 6.0 \times 10^4$ $\qquad$ $3.5 \text{ billion} = 3.5 \times 10^9$

In Lesson 1-3, you used the order of operations to evaluate expressions like 5.2×10^5. You can also use the order of operations to write a number in standard form.

EXAMPLE Write a Number in Standard Form

1 Write 5.2×10^5 in standard form.

$5.2 \times 10^5 = 5.2 \times 100{,}000$ $\qquad 10^5 = 100{,}000$

$= 520000$ $\qquad$ Move the decimal point 5 places to the right.

$= 520{,}000$

Your Turn **Write each number in standard form.**

a. 7.2×10^4 **72,000** $\qquad$ b. 3.8×10^9 **3,800,000,000** $\qquad$ c. 1.0×10^8 **100,000,000**

DAILY INTERVENTION — **Unlocking Misconceptions**

Scientific Notation Some students may mistakenly use a whole number greater than 9 when writing scientific notation. Remind them that there should only be one digit to the left of the decimal point.

1-9 Lesson Notes

Lesson Resources

Chapter 1 Resource Masters
- Study Guide and Intervention, p. 41
- Practice: Skills, p. 42
- Practice: Word Problems, p. 43
- Reading to Learn Mathematics, p. 44
- Enrichment, p. 45

Calculator and Spreadsheet Masters, p. 16
MathPASS Tutorial CD-ROM, 26
Parent and Student Study Guide Workbook, 1-9
Real-Life Career Masters, p. 15

FAST FILE

1 Focus

5-Minute Check 1-9
Use as a quiz or review of Lesson 1-8. (Also available as a transparency)

INTERACTIVE CHALKBOARD
PowerPoint® Presentations

Bellringer
Hands-On Activity Have students measure various items around the room in centimeters. Then have them measure larger items, such as the length of a wall, hallway, or chalkboard. Discuss the fact that writing large lengths in centimeters can be cumbersome. What would be an easier way to write them without converting to meters or kilometers? **Use powers of ten.**

Reading to Learn Mathematics, p. 44 **ELL**

NAME ______ DATE ______ PERIOD ____

1-9 Reading to Learn Mathematics
Scientific Notation

Pre-Activity *Read the introduction at the top of page 43 in your textbook. Write your answers below.*

1. Complete the table.

	Standard Form	Abbreviated Form
Number of Parking Spaces	13,000	13 thousand
Area of Mall (square feet)	4,200,000	4.2 million

2. Write both numbers as the product of a number and a power of ten.
Sample answer: $13 \times 1{,}000$; $4.2 \times 1{,}000{,}000$

3. Rewrite both powers of ten using exponents.
Sample answer: $1{,}000 = 10^3$; $1{,}000{,}000 = 10^6$

Lesson

2 Teach

In-Class Examples

INTERACTIVE CHALKBOARD
PowerPoint® Presentations

1. Write 3.4×10^6 in standard form. **3,400,000**

2. Write 428,000 in scientific notation. **4.28×10^5**

3. **ASTRONOMY** Pluto's maximum distance from Earth is about 4.6 billion miles. One mile is equal to 5,280 feet. Find the approximate maximum distance of Pluto from Earth in feet. Round to the nearest hundredth. **2.43×10^{13} ft**

3 Practice/Apply

NUMBER SENSE
In Exercise 3, have students write both numbers in standard form in order to compare them more easily. Point out that 6 is greater than 3 and that 10^6 has 3 more zeros than 10^3.

About the Exercises...

Guided Practice Key

Exercises	Examples
4–6	1
7–9	2

Odd/Even Assignments
Exercises 10–43 are structured so that students practice the same concepts whether they are assigned odd or even problems.

TEACHING TIP
Tell students that the decimal part of a number written in scientific notation is often rounded to the hundredths place.

To write a number in scientific notation, move the decimal to the right of the first nonzero number. Then find the power of ten.

$13{,}000 = 1.3 \times 10^4$

Move the decimal point to the right of the first nonzero digit.

The decimal point was moved 4 places.

EXAMPLE Write a Number in Scientific Notation

2 Write 23,500,000 in scientific notation.

$23{,}500{,}000 = 2.35 \times 10{,}000{,}000$ Move the decimal point 7 places to find a number between 1 and 10.

$= 2.35 \times 10^7$

Your Turn **Write each number in scientific notation.**

d. 38,000,000 **3.8×10^7**
e. 19,000 **1.9×10^4**
f. 86.1 billion **8.61×10^{10}**

When you use a calculator to compute large numbers, the numbers are often displayed in scientific notation.

STUDY TIP

Technology To enter 3.42×10^{10} on a calculator, press 3.42 [EE] 10 [ENTER]. The screen would look like this: **3.42$_{\times 10^{10}}$**

EXAMPLE Compute with Large Numbers

3 RESERVOIRS Lake Francis Case in South Dakota contains 5.7 billion cubic meters of water. Lake Mead in Nevada has about 6 times as much water. Find the approximate amount of water in Lake Mead.

To find the amount of water in Lake Mead, multiply 5.7 billion by 6.

5.7 billion = 5,700,000,000

Enter 5700000000 [×] 6 [ENTER].

The number **3.42$_{\times 10^{10}}$** on the calculator represents 3.42×10^{10}.

So, Lake Mead has about 3.42×10^{10} cubic meters of water.

Skill and Concept Check

Writing Math
Exercises 1 & 3

1. **Explain** why scientific notation is used with large numbers. **See margin.**
2. **OPEN ENDED** Write a number in standard form and then in scientific notation. **Sample answer: 8,100,000,000; 8.1×10^9**
3. **NUMBER SENSE** Which number describes the population of Chicago, 3.0×10^6 or 3.0×10^3? Explain your reasoning. **See margin.**

Number Sense exercises help students with mental math

GUIDED PRACTICE

Write each number in standard form.

4. 5.1×10^4 **51,000**
5. 1.0×10^5 **100,000**
6. 8.05×10^6 **8,050,000**

Write each number in scientific notation.

7. 70,000 **7.0×10^4**
8. 510,000 **5.1×10^5**
9. 17,500,000 **1.75×10^7**

msmath2.net/extra_examples

Answers

1. Sample answer: It is easier to keep track of the place value.
3. 3.0×10^6; Sample answer: Chicago has a population of about 3 million people: $3.0 \times 10^6 = 3{,}000{,}000$.
10. 4,200
11. 73,000
12. 6,900,000
13. 338,000
14. 110,000,000
15. 8,980,000
16. 51,600
17. 798,000
18. 10,500,000
19. 990,000,000
20. 53,300,000,000
21. 80,600,000
24. 4.6×10^4
25. 1.98×10^4
26. 1.69×10^5
27. 9.97×10^6
28. 3.72×10^6
29. 5.05×10^5

★ indicates increased difficulty

Practice and Applications

HOMEWORK HELP	
For Exercises	See Examples
10–21	1
24–35, 42	2
22, 43	3

Extra Practice
See pages 566, 596.

Write each number in standard form. **10–21. See margin.**

10. 4.2×10^3
11. 7.3×10^4
12. 6.9×10^6
13. 3.38×10^5
14. 1.1×10^8
15. 8.98×10^6
16. 5.16×10^4
17. 7.98×10^5
18. 1.05×10^7
19. 9.90×10^8
20. 5.33×10^{10}
21. 8.06×10^7

22. Triple the number 435,000 and write it in scientific notation. **1.305×10^6**
23. Order the numbers 53,400, 4.98×10^6, 5.03×10^5, and 4,980,100 from least to greatest. **53,400, 5.03×10^5, 4.98×10^6, 4,980,100**

Write each number in scientific notation. **24–35. See margin.**

24. 46,000
25. 19,800
26. 169,000
27. 9,970,000
28. 3,720,000
29. 505,000
30. 607,000
31. 23,600,000
32. 7 million
33. 1.1 million
34. 40 billion
35. 60.8 billion

NUMBER NAMES For Exercises 36 and 37, use the table.

Power of 10	Name
10^6	million
10^9	billion
10^{12}	trillion
10^{15}	quadrillion
10^{18}	quintillion

★36. Write 13 quintillion in scientific notation. **1.3×10^{19}**
★37. Write the name of the number 5.43×10^{16}. **54.3 quadrillion**

Replace each ● with <, >, or = to make a true sentence.

38. 5,000 ● 5.0×10^3 =
39. 600 ● 6.0×10^1 >
40. 81,900 ● 8.19×10^5 <
41. 1 million ● 1.1×10^6 <

42. **GEOGRAPHY** The Mariana Trench is the deepest part of the Pacific Ocean with a depth of about 36,000 feet. Write this depth in scientific notation. **3.6×10^4**
43. **EARTH SCIENCE** A light-year, the distance light travels in one year, is about 5,880,000,000,000 miles. How many miles is 50 light years? **2.94×10^{14} mi**

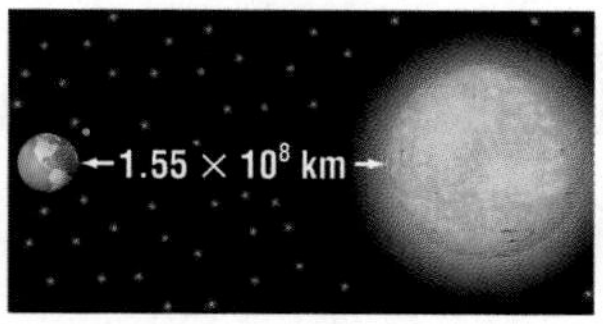

44. **CRITICAL THINKING** The distance from Earth to the Sun is 1.55×10^8 kilometers. If the speed of light is 3×10^5 kilometers per second, about how long does it take for light to travel from the Sun to Earth? (*Hint*: distance = rate × time) **$516.\overline{6}$ s or 517 s or about 9 min**

Standardized Test Practice and Mixed Review

45. **MULTIPLE CHOICE** Asia has an area of 44,579,000 square kilometers. Write 44,579,000 in scientific notation. **B**
    (A) 4.4579×10^6 (B) 4.4579×10^7 (C) 4.4579×10^8 (D) 4.4579×10^9
46. **SHORT RESPONSE** Write the standard form of 5.56×10^{12}. **5,560,000,000,000**

Complete. (Lesson 1-8)

47. 1.01 kg = __?__ g **1,010**
48. 36 mL = __?__ L **0.036**
49. 2.33 km = __?__ m **2,330**

50. **SOCCER** The World Cup is played every four years. If the World Cup was played in 2002, when are the next three years that it will be played? (Lesson 1-7) **2006, 2010, 2014**

msmath2.net/self_check_quiz

Answers

30. 6.07×10^5
31. 2.36×10^7
32. 7.0×10^6
33. 1.1×10^6
34. 4.0×10^{10}
35. 6.08×10^{10}

4 Assess

Open-Ended Assessment

Speaking Have students practice saying numbers written in standard form and in scientific notation.

Study Guide and Intervention, p. 41

NAME ____________ DATE ________ PERIOD ____

1-9 Study Guide and Intervention
Scientific Notation

A number is in scientific notation when it is written as the product of a number and a power of ten. The number must be greater than or equal to 1 and less than 10.
- To write a number in standard form, you apply the order of operations. First evaluate the power of ten and then multiply.
- To write a number in scientific notation, move the decimal point to the right of the first nonzero number. Then, find the power of ten by counting the number of places moved.

EXAMPLE 1 **Write 6.1×10^3 in standard form.**

$6.1 \times 10^3 = 6.1 \times 1{,}000$ $10^3 = 1{,}000$
$= 6.100$ Move the decimal point 3 places to the right.
$= 6{,}100$

EXAMPLE 2 **Write 62,500 in scientific notation.**

$62{,}500 = 6.250 \times 10{,}000$ Move the decimal point 4 places to get a number between 1 and 10.
$= 6.25 \times 10^4$

EXERCISES

Write each number in standard form.

1. 7.25×10^2 725
2. 2.5×10^3 2,500
3. 9.95×10^5 995,000
4. 8.80×10^4 88,000
5. 3.18×10^6 3,180,000
6. 6.12×10^3 6,120

Write each number in scientific notation.

7. 325 3.25×10^2
8. 9,210 9.21×10^3
9. 200 2.0×10^2
10. 5,120 5.12×10^3
11. 561 5.61×10^2
12. 1,230 1.23×10^3
13. 21,300 2.13×10^4
14. 53,000 5.3×10^4
15. 8,930 8.93×10^3

Practice: Skills, p. 42

NAME ____________ DATE ________ PERIOD ____

1-9 Practice: Skills
Scientific Notation

Write each number in standard form.

1. 3.1×10^2 310
2. 2.3×10^3 2,300
3. 9.86×10^2 986
4. 3.25×10^4 32,500
5. 6.10×10^5 610,000
6. 7.87×10^4 78,700
7. 2.2×10^2 220
8. 4.27×10^3 4,270
9. 1.06×10^7 10,600,000
10. 2.11×10^5 211,000
11. 4.82×10^4 48,200
12. 5.55×10^{10} 55,500,000,000

Write each number in scientific notation.

13. 230 2.3×10^2
14. 300 3.0×10^2
15. 720 7.2×10^2
16. 2,790 2.79×10^3
17. 5,000 5.0×10^3
18. 8,800 8.8×10^3
19. 37,000 3.7×10^4
20. 26,300 2.63×10^4
21. 52,100 5.21×10^4
22. 120,000 1.2×10^5
23. 361,000 3.61×10^5
24. 989,000 9.89×10^5
25. 5,000,000 5.0×10^6
26. 82,100,000 8.21×10^7
27. 51,000,000 5.1×10^7

Replace each ● with <, >, or = to make a true sentence.

28. 3,000 ● 3.0×10^3 =
29. 520 ● 5.2×10^1 >
30. 8,800 ● 8.8×10^4 <
31. 659,000 ● 6.59×10^5 =

Practice: Word Problems, p. 43

NAME ____________ DATE ________ PERIOD ____

1-9 Practice: Word Problems
Scientific Notation

1. **ICEBERG** The biggest iceberg ever seen that had split off the Ross Ice Shelf in Antarctica happened during the spring of 2000. It was estimated to weigh about 4,000,000,000,000 tons. Write its weight in scientific notation. 4.0×10^{12} tons
2. **PLANETS** The average distance between Earth and the Sun is 92,960,000 miles. Write this distance in scientific notation. 9.296×10^7 mi
3. **POPULATION** The population of Texas was about 20,900,000 in the 2000 U.S. census. Write this number in scientific notation. 2.09×10^7
4. **BASEBALL** Willie Mays hit 6.6×10^2 career home runs. Write 6.6×10^2 in standard form. 660
5. **EARTH** The mass of Earth is about 6,580,000,000,000,000,000,000 tons. Write this value in scientific notation. 6.58×10^{21} tons
6. **LIGHT** The speed of light is about 2.998×10^8 meters per second or about 1.86×10^5 miles per second. Write these values in standard form. 299,800,000 m/s; 186,000 mi/s
7. **WEALTH** In 1999, the wealthiest man in the world's assets were estimated at about $90,000,000,000. Write this value in scientific notation. $\$9.0 \times 10^{10}$
8. **POPULATION** In 2001, the population of the world was about 6,160,000,000. The projected population of the world for the year 2025 is about 7,840,000,000. Write these values in scientific notation. 6.16×10^9; 7.84×10^9

CHAPTER 1 Study Guide and Review

Vocabulary Assessment

- This alphabetical list of vocabulary terms in Chapter 1 includes a page reference where each term was introduced.
- A vocabulary review/test for Chapter 1 is available on p. 60 of the *Chapter 1 Resource Masters*.

Vocabulary PuzzleMaker

ELL The online Vocabulary PuzzleMaker improves students' mathematics vocabulary using four puzzle formats — crossword, scramble, word search using a word list, and word search using clues. Students can work on a computer screen or from a printed handout.

MindJogger Videoquizzes

ELL MindJogger Videoquizzes provide an alternative review of concepts presented in this chapter. Students work in teams in a game show format to gain points for correct answers. The questions are presented in three rounds.

Round 1 Concepts (5 questions)
Round 2 Skills (4 questions)
Round 3 Problem Solving (4 questions)

ExamView® Pro Test Bank CD-ROM

Use the networkable ExamView® Pro Test Bank CD-ROM to:

- Create **Multiple versions** of tests.
- Create **Modified** tests for **Inclusion** students with one mouse click.
- **Edit** existing questions and **Add** your own questions.
- Build tests aligned with state standards using built-in **State Curriculum Correlations.**
- Change **English** tests to **Spanish** with one mouse click and vice versa.

CHAPTER 1 Study Guide and Review

Vocabulary and Concept Check

algebra (p. 18)	exponent (p. 10)	powers (p. 10)
algebraic expression (p. 18)	exponential form (p. 11)	properties (p. 31)
arithmetic sequence (p. 34)	factors (p. 10)	scientific notation (p. 43)
base (p. 10)	geometric sequence (p. 34)	sequence (p. 34)
coefficient (p. 19)	gram (p. 39)	solution (p. 24)
constant (p. 19)	kilogram (p. 39)	solving an equation (p. 24)
cubed (p. 10)	liter (p. 39)	squared (p. 10)
defining the variable (p. 25)	meter (p. 38)	standard form (p. 11)
equation (p. 24)	metric system (p. 38)	term (pp. 19, 34)
equivalent expressions (p. 30)	numerical expression (p. 14)	variable (p. 18)
evaluate (p. 11)	order of operations (p. 14)	

Choose the letter of the term that best matches each phrase.

1. a number that makes an equation true b
2. numbers expressed using exponents e
3. two or more numbers that are multiplied together to form a product d
4. rules used to ensure that numerical expressions have only one value c
5. tells how many times the base is used as a factor f

a. algebra
b. solution
c. order of operations
d. factors
e. powers
f. exponent

Lesson-by-Lesson Exercises and Examples

1-1 A Plan for Problem Solving (pp. 6–9)

Use the four-step plan to solve each problem.

6. **CELLULAR PHONES** Sherita's service charges a monthly fee of \$20.00 plus \$0.15 per minute. One monthly bill is \$31.25. How many minutes did Sherita use during the month? 75 min
7. **CAR RENTAL** ABC Car Rental charges \$25 per day to rent a mid-sized car plus \$0.20 per mile driven. Mr. Ruiz rents a mid-sized car for 3 days and drives a total of 72 miles. Find the amount of Mr. Ruiz's bill. \$89.40

Example 1 **Use the four-step plan to solve the problem.**

GARDENING A bag of mulch covers 25 square feet of garden space. Taylor uses 7 bags of mulch to cover her garden. How large is Taylor's garden?

Explore Taylor used 7 bags of mulch, each covering 25 square feet.
Plan Multiply 25 by 7.
Solve $25 \cdot 7 = 175$
Taylor's garden is 175 square feet.
Examine Add 25 seven times.

msmath2.net/vocabulary_revie

Key concepts from the lesson, one or two examples, and several practice problems are included in the **Lesson-by-Lesson Exercises and Examples**.

1-2 Powers and Exponents (pp. 10–13)

Evaluate each expression.

8. 3^5 243
9. 7^9 40,353,607
10. 2^8 256
11. 18^2 324
12. 9^4 6,561
13. 10^4 10,000
14. 20^3 8,000
15. 100^1 100
16. Write $15 \cdot 15 \cdot 15$ in exponential form. 15^3

Example 2 **Evaluate 4^5.**

The base is 4. The exponent 5 means that 4 is used as a factor 5 times.

$4^5 = 4 \cdot 4 \cdot 4 \cdot 4 \cdot 4$
$= 1{,}024$

1-3 Order of Operations (pp. 14–17)

Evaluate each expression.

17. $24 - 8 + 3^2$ 25
18. $48 \div 6 + 2 \cdot 5$ 18
19. $5 \cdot 9 - (2 + 3)^2$ 20
20. $(14 - 4) \div 2$ 5
21. $9 + 3(7 - 5)^3$ 33
22. 12×10^4 120,000
23. $42 \div 6 - 63 \div 9$ 0
24. $15 + 9 \div 3 - 7$ 11

Example 3 **Evaluate $24 - (8 \div 4)^4$.**

$24 - (8 \div 4)^4 = 24 - 2^4$ Divide first since $8 \div 4$ is in parentheses.
$= 24 - 16$ Find the value of 2^4.
$= 8$ Subtract.

1-4 Algebra: Variables and Expressions (pp. 18–21)

Evaluate each expression if $a = 10$, $b = 4$, and $c = 8$.

25. $b + 7$ 11
26. $7b$ 28
27. $(a - b)^2$ 36
28. $ab \div c$ 5
29. $3b^2 + c$ 56
30. $\frac{(b + c)^2}{3}$ 48
31. **PRODUCTION** The cost of producing T-shirts is given by the expression $350 + 0.82x$, where x is the number of T-shirts produced. Find the cost of producing 750 T-shirts. $965

Example 4 **Evaluate $2m^2 - 5n$ if $m = 4$ and $n = 3$.**

$2m^2 - 5n = 2(4)^2 - 5(3)$ Replace m with 4 and n with 3.
$= 2(16) - 5(3)$ Find the value of 4^2.
$= 32 - 15$ Multiply.
$= 17$ Subtract.

1-5 Algebra: Equations (pp. 24–27)

Solve each equation mentally.

32. $h + 9 = 17$ 8
33. $5 = c - 3$ 8
34. $32 - 19 = w$ 13
35. $8a = 56$ 7
36. $31 - y = 8$ 23
37. $\frac{t}{9} = 12$ 108
38. $100 = 20g$ 5
39. $p - 49 = 18$ 67

Example 5 **Solve $14 = 5 + x$ mentally.**

$14 = 5 + x$ Write the equation.
$14 = 5 + 9$ You know that $5 + 9 = 14$.
$14 = 14$ Simplify.
The solution is 9.

FOLDABLES™ Study Organizer

For more information about Foldables, see *Teaching Mathematics with Foldables.*

Have students look through the chapter to make sure they have included examples in their Foldables for each lesson or topic in the chapter.

Encourage students to refer to their Foldables while completing the Study Guide and Review and to use them in preparing for the Chapter Test.

Answers

44. geometric; ×3; 405, 1,215, 3,645
45. arithmetic; +1.2; 6.8, 8.0, 9.2
46. neither; add 2 more than was added to the previous term; 21, 31, 43
53. 5.9×10^4
54. 8.46×10^7
55. 3.24×10^5
56. 1.0×10^{10}
57. 1.03×10^6
58. 3.33×10^5
59. 9.1×10^6
60. 6.0×10^9

1-6 Algebra: Properties (pp. 30–33)

Name the property shown by each statement.

40. $3 + 8 = 8 + 3$ Commutative (+)
41. $3 \cdot (2 \cdot 7) = (3 \cdot 2) \cdot 7$ Associative (×)
42. $6(1 + 4) = 6(1) + 6(4)$ Distributive
43. $29 \cdot 1 = 29$ Identity (×)

Example 6 **Name the property shown by the statement $12 + (8 + 2) = (12 + 8) + 2$.**

This statement shows that the way in which three numbers are grouped when they are added does not change their sum. This is an example of the Associative Property of Addition.

1-7 Sequences (pp. 34–36)

Describe the pattern in each sequence and identify the sequence as *arithmetic*, *geometric*, or *neither*. Then find the next three terms. 44–46. See margin.

44. 5, 15, 45, 135, …
45. 2.0, 3.2, 4.4, 5.6, …
46. 1, 3, 7, 13, …

47. **CONTESTS** The prizes for first, second, and third places in a spelling bee are \$1,000, \$500, and \$250, respectively. If this pattern continues, what would the fourth, fifth, and sixth prizes be? \$125, \$62.50, \$31.25

Example 7 **Describe the pattern in the sequence 2, 4, 8, 16, 32, …, and identify the sequence as *arithmetic*, *geometric*, or *neither*. Then find the next three terms.**

This is a geometric sequence created by multiplying the previous term by 2.

2 4 8 16 32 64 128 256

×2 ×2 ×2 ×2 ×2 ×2 ×2

The next three terms are 64, 128, and 256.

1-8 Measurement: The Metric System (pp. 38–41)

Complete.

48. 3.6 km = __?__ m 3,600
49. 29 L = __?__ mL 29,000
50. 237 mg = __?__ g 0.237
51. 7 mL = __?__ L 0.007
52. 3,200 cm = __?__ m 32

Example 8 **Complete.**

6,750 m = __?__ km

Divide 6,750 by 1,000.

$6{,}750 \div 1{,}000 = 6.75$

So, 6,750 meters is equivalent to 6.75 kilometers.

1-9 Scientific Notation (pp. 43–45)

Write each number in scientific notation.

53. 59,000
54. 84,600,000
55. 324,000
56. 10,000,000,000
57. 1,030,000
58. 333,000
59. 9.1 million
60. 6 billion

53–60. See margin.

Example 9 **Write 23,100,000 in scientific notation.**

Move the decimal point 7 places to find a number between 1 and 10.

$23{,}100{,}000 = 2.31 \times 10^7$

48 Chapter 1 Decimal Patterns and Algebra

CHAPTER 1

Practice Test

Vocabulary and Concepts

1. **List** the order of operations used to find the value of an expression.
2. **Explain** the difference between an arithmetic sequence and a geometric sequence. **1–2. See margin.**

Assessment Options lists the tests that are available in the Chapter Resource Masters.

Skills and Applications

Evaluate each expression.

3. 3^5 **243**
4. 15^4 **50,625**
5. $18 - 3 + 5$ **20**
6. $8 + (12 \div 3)^3$ **72**

Evaluate each expression if $x = 12$, $y = 5$, and $z = 3$.

7. $7 - x \div z$ **3**
8. $yz + 23$ **38**
9. $(y - z)^5$ **32**
10. $\frac{xz}{y + 13}$ **2**

Solve each equation mentally.

11. $9 + m = 16$ **7**
12. $d - 14 = 37$ **51**
13. $23 = \frac{92}{t}$ **4**
14. $6x = 126$ **21**

Name the property shown by each statement.

15. $2 + (3 + 7) = (2 + 3) + 7$ **Associative (+)**
16. $5(w + 2) = 5w + 10$ **Distributive**

Complete.

17. 5.45 m = __?__ cm **545**
18. 27 mL = __?__ L **0.027**
19. 8,200 g = __?__ kg **8.2**

20. **CAPACITY** The table lists the capacity of three different containers. Determine which container has the largest capacity. **container #2**

Container	Capacity
#1	5400 mL
#2	0.008 kL
#3	4.78 L

21. **ANIMALS** Irene left both her dog and cat at a kennel for 3 nights. The kennel charges \$8 per night for the dog and \$5 per night for the cat. Find Irene's bill. **\$39**

Write each number in scientific notation.

22. 42,300 4.23×10^4
23. 3,780,000 3.78×10^6
24. 502,000,000 5.02×10^8

Standardized Test Practice

25. Ms. Carter runs a pizza parlor. Her daily costs are \$125 for rent, employee wages, and utilities, plus \$2 for every pizza. If n represents the number of pizzas the pizza parlor makes one day, which expression represents Ms. Carter's total cost for that day? **A**

Ⓐ \$125 + $2n$ Ⓑ $125n$ + \$2 Ⓒ \$$125n + 2n$ Ⓓ \$127

msmath2.net/chapter_test

Portfolio Suggestion

Introduction Various ways of using symbols to express numbers have been discussed in this chapter, including algebraic expressions, sequence notation, and scientific notation.

Ask Students to include in their Portfolios examples and descriptions of each type of mathematical notation they find when reviewing the chapter.

CHAPTER 1 Practice Test

Assessment Options

Vocabulary Test A vocabulary review/test for Chapter 1 can be found on p. 60 of the *Chapter 1 Resource Masters.*

Chapter Tests There are six Chapter 1 Tests and an Extended Response Assessment task available in the *Chapter 1 Resource Masters.*

Chapter 1 Tests			
Form	**Type**	**Level**	**Pages**
1	MC	basic	47–48
2A	MC	average	49–50
2B	MC	average	51–52
2C	FR	average	53–54
2D	FR	average	55–56
3	FR	advanced	57–58

MC = multiple-choice questions
FR = free-response questions

Extended Response Assessment

Extended Response tasks for Chapter 1 can be found on p. 59 of the *Chapter 1 Resource Masters.* A sample scoring rubric for these tasks appears on p. A31.

Answers

1. 1. Do all operations within grouping symbols first. 2. Evaluate all powers before other operations. 3. Multiply and/or divide in order from left to right. 4. Add and/or subtract in order from left to right.
2. Sample answer: Each term in an arithmetic sequence is found by adding the same number to the previous term. Each term in a geometric sequence is found by multiplying the previous term by the same number.

Chapter 1 Standardized Test Practice

These two pages contain practice questions in the various formats that can be found on the most frequently given standardized tests.

A student recording sheet for these two pages can be found on page A1 of the *Chapter 1 Resource Masters.*

Standardized Test Practice
Student Recording Sheet, p. A1

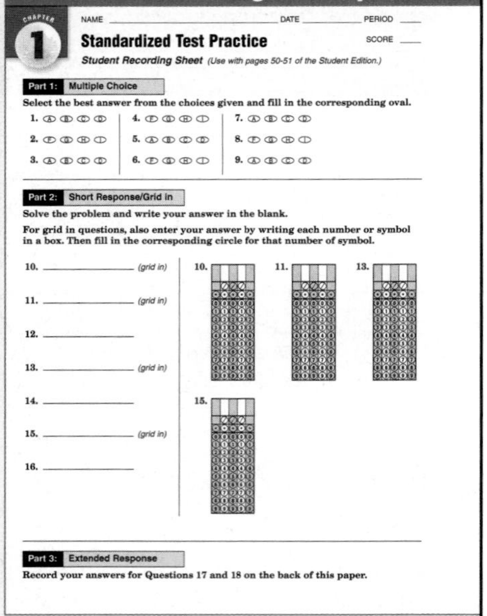
NAME ________ DATE ________ PERIOD ____

Chapter 1 **Standardized Test Practice** SCORE ____

Student Recording Sheet (Use with pages 50-51 of the Student Edition.)

Part 1: Multiple Choice

Select the best answer from the choices given and fill in the corresponding oval.

1. A B C D 4. F G H I 7. A B C D
2. F G H I 5. A B C D 8. F G H I
3. A B C D 6. F G H I 9. A B C D

Part 2: Short Response/Grid In

Solve the problem and write your answer in the blank.

For grid in questions, also enter your answer by writing each number or symbol in a box. Then fill in the corresponding circle for that number of symbol.

10. ______ (grid in)
11. ______ (grid in)
12. ______
13. ______ (grid in)
14. ______
15. ______ (grid in)
16. ______

Part 3: Extended Response

Record your answers for Questions 17 and 18 on the back of this paper.

Additional Practice

See pp. 65–66 of the *Chapter 1 Resource Masters* for additional standardized test practice.

Standardized Test Practice exercises were created to closely parallel those on actual state proficiency tests. Rubrics are provided for assessment.

Chapter 1 Standardized Test Practice

PART 1 Multiple Choice

Record your answers on the answer sheet provided by your teacher or on a sheet of paper.

1. The strategy of guess and check would be *most* useful in solving which problem? (Lesson 1-1) **D**
   - (A) About how long does it take to travel 62 miles at 58 miles per hour?
   - (B) What is the area of a rectangle if it has a length of 4.5 centimeters and a width of 9 centimeters?
   - (C) It costs \$3 to cross the town bridge to enter town, but it is free of charge to cross it to leave town. How much does a person who makes 150 trips across the bridge into town and 50 trips across the bridge to leave town spend in bridge tolls?
   - (D) The product of two consecutive even numbers is 728. Find the two numbers.

2. Ciro earns \$8 per hour plus tips cleaning rooms at Sandman's Inn. Which expression represents Ciro's earnings for a week in which he works 36 hours and collects \$25 in tips? (Lesson 1-3) **F**
   - (F) $(36 \times 8) + 25$
   - (G) $(36 + 25) \times 8$
   - (H) $(8 \times 25) + 36$
   - (I) $(36 \div 8) + 25$

3. To estimate the number of miles a thunderstorm is from you, count the number of seconds between the thunder and the lightning. Then divide the number of seconds by 5. Which expression can be used to estimate the distance? (Lesson 1-4) **D**
   - (A) $5 \div s$
   - (B) $s - 5$
   - (C) $5s$
   - (D) $s \div 5$

4. Which value of p makes $64 \div p = 8$ a true sentence? (Lesson 1-5) **H**
   - (F) 2
   - (G) 2^2
   - (H) 2^3
   - (I) 3^2

5. Which is the graph of the solution of $x - 15 = 33$? (Lesson 1-5) **A**
   - (A) 

   - (B) 

   - (C) 

   - (D) 

6. Identify the property shown by the sentence $16 + (4 + 7) = (16 + 4) + 7$. (Lesson 1-6) **G**
   - (F) Commutative Property of Addition
   - (G) Associative Property of Addition
   - (H) Associative Property of Multiplication
   - (I) Identity Property of Multiplication

7. Which is the next term of the sequence 6, 12, 18, 24, … ? (Lesson 1-7) **B**
   - (A) 28
   - (B) 30
   - (C) 36
   - (D) 48

8. Every morning, Morgan swims 0.5 kilometer. How many meters does she swim? (Lesson 1-8) **I**
   - (F) 5 m
   - (G) 50 m
   - (H) 100 m
   - (I) 500 m

9. Which is 5,300,000 written in scientific notation? (Lesson 1-9) **C**
   - (A) 0.53×10^6
   - (B) 53×10^5
   - (C) 5.3×10^6
   - (D) 5.3×10^7

The Princeton Review **Log On** for Test Practice

The Princeton Review offers additional test-taking tips and practice problems at their web site. Visit princetonreview.com or review.com

ExamView® Pro Test Bank CD-ROM

Special banks of standardized test questions similar to those on the TIMSS and NAEP grade 8 tests can be found on this CD-ROM.

RT 2 Short Response/Grid In

cord your answers on the answer sheet ovided by your teacher or on a sheet of per.

What value is represented by $(1 \times 10^4) + (7 \times 10^3) + (3 \times 10^2) + (0 \times 10^1)$? (Prerequisite Skill, p. 561) **17,300**

The graph shows the number of World Series Championships for selected teams from 1903–2001. How many total championships have been won by these teams? (Lesson 1-1) **50**

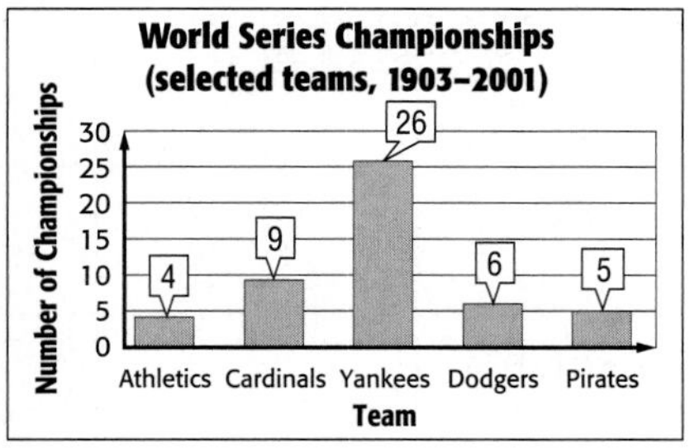

Source: World Almanac, 2001

Write x^6 as a product of the same factor. (Lesson 1-2) **$x \cdot x \cdot x \cdot x \cdot x \cdot x$**

What is the value of $2^2 + 4^3$? (Lesson 1-3) **68**

Tyler works at the library 3 times as many hours as Lugo and Lydia combined. Write an algebraic expression that represents the number of hours Tyler works per week if h represents the number of hours Lugo and Lydia work combined. (Lesson 1-4) **$3h$**

The town of Holden's water tank can hold 10 kiloliters. The town of Jasper's water tank can hold half the amount of Holden's tank. How many liters of water can Jasper's water tank hold? (Lesson 1-8) **5,000**

TEST-TAKING TIP

Question 13 Fill in the answer bubbles below each numeral completely on the answer grid-in sheet. That way your answer will be clear.

16. Four countries and their approximate populations in 2001 are shown in the table. Which country has the least population? (Lesson 1-9) **Croatia**

Country	Population, 2001
Honduras	6.406×10^6
Australia	1.936×10^7
Croatia	4.334×10^6
Denmark	5.353×10^6

Source: www.factmonster.com

PART 3 Extended Response

Record your answers on a sheet of paper. Show your work.

17. Figures 1, 2, and 3 form a pattern made of equal-sized squares. Each side of the square in Figure 1 is 2 units in length. (Lesson 1-7)

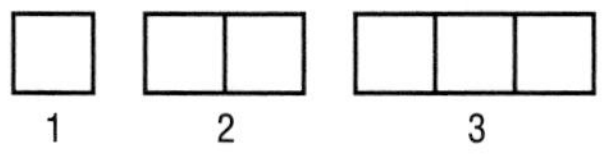

a. Extend the pattern to find the perimeters of Figures 3–8. **See table.**

b. Describe the sequence of perimeters and identify the sequence as *arithmetic*, *geometric*, or *neither*. **+4; arithmetic**

Figure	Perimeter
1	8
2	12
3	? **16**
4	? **20**
5	? **24**
6	? **28**
7	? **32**
8	? **36**

c. The perimeter of a term above is given by the expression $4f + 4$, where f is the figure number in the pattern. What is the perimeter of figure 26? **108**

18. Marlene made her special punch by mixing 4 liters of ginger ale and 2,650 milliliters of cranberry juice. (Lesson 1-8)

a. How many liters are in the punch bowl when the punch is complete? **6.65**

b. About how many 200-mL servings of punch can be served? **33**

Aligned and verified by The Princeton Review

Standardized Test Practice

Rubrics The following rubrics are sample scoring devices for short response and extended response questions.

Short Response

Points	Description
2	The student demonstrates a **thorough understanding** of the mathematics of the task. The response may contain minor flaws that do not detract from the demonstration of a thorough understanding.
1	The student has provided a response that is only **partially correct**.
0	The student has provided a **completely incorrect** solution or no response at all.

Extended Response

Points	Description
4	The student demonstrates a **thorough understanding** of the mathematics of the task. The response may contain minor flaws that do not detract from the demonstration of a thorough understanding.
3	The student demonstrates an **understanding** of the mathematics of the task. The response is essentially correct and demonstrates an essential but less than thorough understanding of the mathematics.
2	The student demonstrates only a **partial understanding** of the mathematics of the task. Although the student may have used the correct approach to a solution or may have provided a correct solution, the work lacks an essential understanding of the underlying mathematical concepts.
1	The student demonstrates a **very limited understanding** of the mathematics of the task. The response is incomplete and exhibits many flaws.
0	The student provides a **completely incorrect** solution or no response at all.

A more detailed 4-point rubric is provided on page A2 of the *Chapter 1 Resource Masters*.

CHAPTER 2

Chapter Overview

Statistics: Analyzing Data

	LESSON OBJECTIVES	NCTM Standards	NAEP 2005	State/Local Objectives
2-1	**Frequency Tables** *(pp. 54–57)* • Organize and interpret data in a frequency table.	5, 10	D1a, D2e	
2-2	**Making Predictions** *(pp. 58–63)* **2-2a Problem-Solving Strategy:** Solve problems by using graphs. • Make predictions from graphs.	5, 6, 7, 10	D1a, D1b, D2e	
2-3	**Line Plots** *(pp. 64–68)* • Construct and interpret line plots.	5, 10	D1b, D1e, D2a, D2b, D2c	
2-4	**Mean, Median, and Mode** *(pp. 69–73)* • Find the mean, median, and mode of a set of data. **2-4b Hands-On Lab:** Use mean, median, mode, and range to describe a set of data.	1, 4, 5, 10	D2a, D2b, D2c, D2d, D3a	
2-5	**Stem-and-Leaf Plots** *(pp. 76–79)* • Construct and interpret stem-and-leaf plots.	1, 5, 10	D1a, D1c, D2a, D1e	
2-6	**Box-and-Whisker Plots** *(pp. 80–84)* • Construct and interpret box-and-whisker plots. **2-6b Graphing Calculator Investigation:** Use a graphing calculator to make box-and-whisker plots.	4, 5, 9, 10	D1a, D2a, D2b, D2c, D1e, D1c	
2-7	**Bar Graphs and Histograms** *(pp. 85–91)* • Construct and interpret bar graphs and histograms. **2-7b Spreadsheet Investigation:** Use a spreadsheet to make a multiple-line graph and a multiple-bar graph.	5, 7, 10	D1a, D1b, D1c, D1e	
2-8	**Misleading Statistics** *(pp. 92–95)* • Recognize when statistics and graphs are misleading.	5, 7, 9, 10	D1a, D1d, D1e, D2a, D2b	

Key to NCTM Standards:

1=Number & Operations, 2=Algebra, 3=Geometry, 4=Measurement, 5=Data Analysis & Probability, 6=Problem Solving, 7=Reasoning & Proof, 8=Communication, 9=Connections, 10=Representation

NAEP 2005 Standards: See math.glencoe.com

Pacing suggestions for the entire year can be found on page T20.

Chapter Resource Manager

Time-Saving Tool
TeacherWorks™
All-In-One Planner and Resource Center
See page T5.

Chapter 2 Resource Masters: Study Guide and Intervention	Practice (Skills and Word Problems)	Reading to Learn Mathematics	Enrichment	Assessment	Applications*	Parent and Student Study Guide Workbook	5-Minute Check Transparencies	Key Teaching Transparencies	MathPASS: Tutorial Plus (lessons)	Teaching Mathematics with Manipulatives	Optional Materials
67	68–69	70	71		RC 16	2-1	2-1	5			
72	73–74	75	76	121	SM 29–32	2-2	2-2	6		37	
77	78–79	80	81		CS 17	2-3	2-3	7			
82	83–84	85	86	121, 123	CM 5–6 CS 18	2-4	2-4			38	markers, ruler
87	88–89	90	91			2-5	2-5	8			
92	93–94	95	96	122	CS 19	2-6	2-6		58	39–40	
97	98–99	100	101		SM 81–84, 85–88	2-7	2-7				
102	103–104	105	106	107–120, 122, 124–128	CS 20	2-8	2-8		57	41	

*** Key to Abbreviations:** *CS = Calculator and Spreadsheet Masters, RC = Real-Life Career Masters, CM = Classroom Math Games (online), SM = Science and Math Lab Manual (online)*

INTERACTIVE CHALKBOARD

PowerPoint® Presentations

This CD-ROM is a customizable Microsoft® PowerPoint® presentation that includes:

- In-Class Examples
- Links to Online Study Tools
- 5-Minute Check Transparencies
- Bellringers

Math Skills Maintenance

Use the Math Skills Maintenance Masters to keep your students' basic math skills fresh. Students can review skills one at a time or in a combined format. The Masters are also available online at msmath2.net/maintenance.

CHAPTER 2 Mathematical Content and Teaching Strategies

Continuity of Instruction

Building on Prior Knowledge

In previous courses, students have collected data and created graphs to display their findings. Among these graphs have been bar graphs and line graphs. Students also have organized information into tables and charts. Students should have basic skills performing mathematical operations with nonnegative fractions and decimals as well as a basic understanding of the order of operations.

Learning in This Chapter

Students acquire numerous skills and tools for collecting, organizing, summarizing, and analyzing data. They use frequency tables, find scales and intervals, interpret line and bar graphs and scatter plots, and analyze graphs to make predictions. Students also make line plots, stem-and-leaf plots, box-and-whisker plots, bar graphs, and histograms. They work with three statistical measures of central tendency–the mean, median, and mode. Students also examine ways in which statistics and graphs can be misleading.

Making Future Connections

Analyzing data from real-life situations shows students how valuable a role mathematics can play in understanding the world around them. Interpreting data and predicting from data give students the power to make informed decisions as citizens and consumers. The study of graphs, including line graphs and scatter plots, will give students a strong foundation for the study of linear equations and functions in future mathematics courses.

Mathematical Background

2-1 Frequency Tables

In this lesson, students learn how to organize data in a frequency table. For numerical data, they must examine the data to choose a data scale that includes the least and greatest data and has equal intervals. For non-numerical data, categories are used without scales and intervals. Students will learn to display frequency data in graphical form in Lesson 2-7.

2-2 Making Predictions

Students begin this lesson by learning how to solve problems using a graph. Different graphs are used to display different kinds of information for different purposes. For example, bar graphs are useful for making comparisons. Circle graphs represent percents of a whole. In this lesson, students will use line graphs and scatter plots to display trends and make predictions. Students will learn about other data displays in the rest of the chapter and use them to solve problems.

Team Teaching

Inform the other teachers on your team that your classes are learning how to make predictions from graphs. Suggestions for curriculum integration are:

Earth Science: fossil fuel depletion, land fill use

Economics: predicting standards of living, economic changes

Geography: population growth, global warming, el Niño patterns

Life Science: predicting the extinction of species

2-3 Line Plots

In previous courses, students gained experience plotting numbers on number lines. In this lesson, they learn to plot entire sets of numerical data on a number line called a line plot. While frequency tables show what intervals have the most data, line plots show where individual data points are clustered and where gaps exist in the data. Students learn visually what the range of a set of data means.

2-4 Mean, Median, and Mode

Students broaden their skills analyzing data sets. They learn to find measures of central tendency—mean, median, and mode. All of these measures find the average of a set of data. To find the median, students must first order the data, either from least to greatest or from greatest to least. Point out that a set of data may have one mode, more than one mode, or no mode.

Teacher to Teacher

Bob Thompson
Thomas Jefferson Middle School
Port Washington, WI

"I discuss how accurately the mean describes a set of data by asking the students to estimate the average age for the people in the classroom. Most students will answer an average student's age. But when my age is included in the data, the mean is higher."

Bob Thompson

2-5 Stem-and-Leaf Plots

In this lesson, students learn yet another way to organize data. A stem-and-leaf plot shows all of the data in a set from least to greatest, and makes it easy to locate the upper and lower extremes and the median of the data. A stem-and-leaf plot is a numerical version of a histogram that is placed on its side. Students will study histograms in Lesson 2-7.

McGraw-Hill Professional Development

Mathematics, YES! is a self-paced, web-based professional development program. Units are based on major mathematical concepts and are standards-based. Actual classroom videos provide a springboard for discussion and activities. Unit appropriate for this chapter:

- **Unit 10, Statistics, Data Analysis, and Probability**

2-6 Box-and-Whisker Plots

Box-and-whisker plots plot upper and lower extremes of a data set, its median value, and the values halfway between the extremes and the median. These values are the lower and upper quartiles, so named because, with the median, they divide the data into four equal parts. Collectively, these values are measures of distribution because they show how the data are spread out. Students learn to analyze the distribution of data by examining the relative lengths of the box and whiskers in the plot.

2-7 Bar Graphs and Histograms

Students review their knowledge of bar graphs from previous courses and learn to make a special type of bar graph called a histogram. A histogram arranges data from a frequency table into evenly sized bars. Because each bar could be said to be a thin bar for each value in the interval, such as 11–20, and since there are no gaps between intervals, bars on a histogram have no space between them.

2-8 Misleading Statistics

This lesson applies the knowledge of data displays and measures of central tendency studied in this chapter to the analysis of graphs and statistics used to persuade consumers and voters. Colorful graphs have become widely used in newspapers, magazines, on television, and the Internet. Thus, students' ability to evaluate graphs has great importance in everyday life as well as the classroom. Students should understand that a measure used to represent average—mean, mode, or median—may not fairly represent a particular set of data.

Key Concepts

Additional mathematical information and teaching notes are available at

msmath2.com/key_concepts.

The sections appropriate for this chapter are:

- **Mean, Median, and Mode**
- **Stem-and-Leaf Plots**
- **Box-and-Whisker Plots**

CHAPTER 2

DAILY INTERVENTION and Assessment

	Type	Student Edition	Teacher Resources	Technology/Internet
INTERVENTION	Ongoing	Prerequisite Skills, pp. 53, 57, 63, 79, 83, 89 Mid-Chapter Practice Test, p. 74	5-Minute Check Transparencies Quizzes 1–4, *CRM*, pp. 121–122 Mid-Chapter Test, *CRM*, p. 123 Study Guide and Intervention, *CRM*, pp. 67, 72, 77, 82, 87, 92, 97, 102	MathPASS: Tutorial Plus msmath2.net/self_check_quiz msmath2.net/extra_examples
INTERVENTION	Mixed Review	pp. 57, 63, 68, 72, 79, 83, 89, 95	Cumulative Review, *CRM*, p. 124	
INTERVENTION	Error Analysis	Find the Error, pp. 66, 71, 78	Find the Error, *TWE*, pp. 66, 71, 78 Unlocking Misconceptions, *TWE*, pp. 59, 80	
ASSESSMENT	Standardized Test Practice	pp. 59, 63, 68, 72, 74, 79, 83, 89, 95, 99, 100–101	Standardized Test Practice, *CRM*, pp. 125–126	Standardized Test Practice CD-ROM msmath2.net/standardized_test
ASSESSMENT	Open-Ended Assessment	Writing Math, pp. 56, 61, 66, 71, 73, 78, 82, 87, 93 Open-Ended, pp. 56, 61, 66, 71, 78, 82, 87, 91, 93 Standardized Test, p. 101	Modeling: *TWE*, pp. 59, 88 Speaking: *TWE*, pp. 68, 95 Writing: *TWE*, pp. 57, 79 Extended Response Assessment, *CRM*, p. 119	
ASSESSMENT	Chapter Assessment	Study Guide, pp. 96–98 Practice Test, p. 99	Multiple-Choice Tests (Forms 1, 2A, 2B), *CRM*, pp. 107–112 Free-Response Tests (Forms 2C, 2D, 3), *CRM*, pp. 113–118 Vocabulary Test/Review, *CRM*, p. 120	ExamView® Pro Test Bank (see below) MindJogger Videoquizzes msmath2.net/vocabulary_review msmath2.net/chapter_test

* **Key to Abbreviations:** *TWE* = Teacher Wraparound Edition, *CRM* = Chapter Resource Masters

Additional Intervention

Skills Intervention for Middle School Mathematics workbook and CD-ROM includes 88 skills with pretests, tutorials, guided practice, and posttests for diagnosis and remediation.

Guide to Daily Intervention booklet details each intervention resource available and guides teachers in using them as intervention tools everyday to assist students who may be having difficulty.

ExamView® Pro

Use ExamView® Pro Test Bank CD-ROM to:

- **Create Multiple versions of tests.**
- **Create Modified tests for Inclusion students with one mouse click.**
- **Edit existing questions and Add your own questions.**
- **Build tests aligned with state standards using built-in State Curriculum Correlations.**
- **Change English tests to Spanish with one mouse click and vice versa.**

Reading and Writing Mathematics

• Student Edition

Foldables Study Organizer, p. 53

Concept Check questions at the beginning of each exercise set require students to verbalize and write about what they have learned in the lesson.

- **Open Ended,** pp. 56, 61, 66, 71, 73, 78, 82, 87, 91, 93
- **Find the Error,** pp. 66, 71, 78
- **Which One Doesn't Belong?,** p. 61

Reading Math Tip, pp. 61, 81

WebQuest, p. 95

• Teacher Wraparound Edition

Foldables Follow-Up, pp. 62, 71, 74, 94

Writing Activities, pp. 57, 79

Differentiated Instruction, (Verbal/Linguistic), p. 55

ELL Resources, *see page T9 for a complete listing.*

• Additional Resources

Vocabulary Builder, *Chapter 2 Resource Masters*

Reading to Learn Mathematics, *Chapter 2 Resource Masters*

Vocabulary PuzzleMaker (online)

Teaching Mathematics with Foldables

Reading and Writing in the Mathematics Classroom

WebQuests, Projects, and Interdisciplinary Investigations

For detailed information on these resources, see pp. T6–T7.

• Recommended Outside Reading for Students

Nonfiction

- Huff, Darrell. *How to Lie with Statistics.* W.W. Norton & Company, ©1954, 1993.

  This classic book, illustrated by Irving Geis, warns readers of methods used to persuade an audience with statistics. Many techniques of manipulating statistics are explained, including the use of scale, presentation, and sample bias.

Fiction

- Scieszka, Jon and Smith, Lane. *Math Curse.* Viking, ©1995.

  This book tells the funny story of a student who begins to see everything she encounters throughout the day as a math problem. Many questions arise which are related to different topics in mathematics, including the topic of *average.*

PROJECT CRISS℠ Study Skill

Vocabulary maps elaborate the meanings of vocabulary words by using multiple details. They are a useful way for students to organize their ideas while taking notes for a chapter. As students work through Chapter 2, have them complete vocabulary maps similar to the one shown at the right for vocabulary words related to statistics topics they encounter. You may wish to have them work in cooperative groups to make large versions of the maps that can be displayed on a bulletin board.

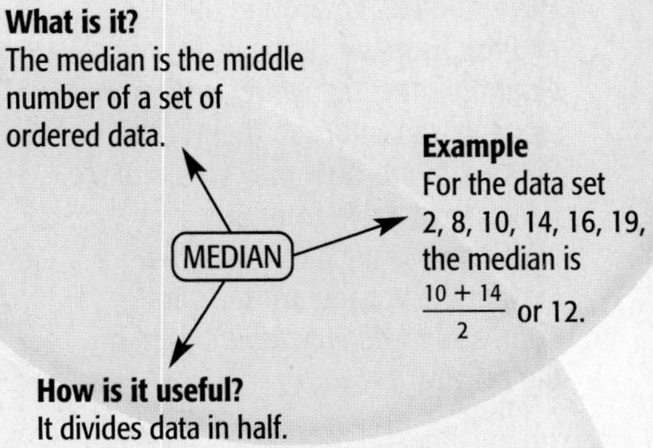

CReating Independence through Student-owned Strategies

For more information on Project CRISS, see page T9.

CHAPTER 2 Making Connections

School-To-Home

Parent and Student Study Guide

Parents and students may work together to reinforce the concepts and skills of this chapter. These workbooks are available in English and Spanish. A sample lesson is shown below. Also available at msmath2.net/parent_student.

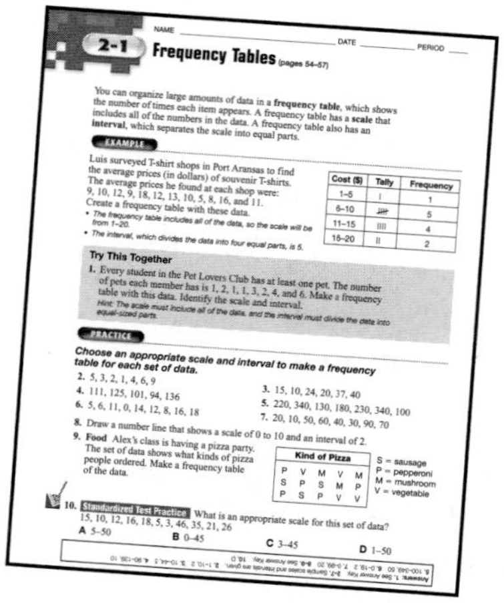

NAME ______ DATE ______ PERIOD ____

2-1 Frequency Tables (pages 54–57)

You can organize large amounts of data in a **frequency table**, which shows the number of times each item appears. A frequency table has a **scale** that includes all of the numbers in the data. A frequency table also has an **interval**, which separates the scale into equal parts.

EXAMPLE

Luis surveyed T-shirt shops in Port Aransas to find the average prices (in dollars) of souvenir T-shirts. The average prices he found at each shop were: 9, 10, 12, 9, 18, 12, 13, 10, 5, 8, 16, and 11. Create a frequency table with these data.

- *The frequency table includes all of the data, so the scale will be from 1–20.*
- *The interval, which divides the data into four equal parts, is 5.*

Cost ($)	Tally	Frequency
1–5	I	1
6–10	~~IIII~~	5
11–15	IIII	4
16–20	II	2

Try This Together

1. Every student in the Pet Lovers Club has at least one pet. The number of pets each member has is 1, 2, 1, 1, 3, 2, 4, and 6. Make a frequency table with this data. Identify the scale and interval.
Hint: The scale must include all of the data, and the interval must divide the data into equal-sized parts.

PRACTICE

Choose an appropriate scale and interval to make a frequency table for each set of data.

2. 5, 3, 2, 1, 4, 6, 9
3. 15, 10, 24, 20, 37, 40
4. 111, 125, 101, 94, 136
5. 220, 340, 130, 180, 230, 340, 100
6. 5, 6, 11, 0, 14, 12, 8, 16, 18
7. 20, 10, 50, 60, 40, 30, 90, 70
8. Draw a number line that shows a scale of 0 to 10 and an interval of 2.
9. **Food** Alex's class is having a pizza party. The set of data shows what kinds of pizza people ordered. Make a frequency table of the data.

Kind of Pizza				
P	V	M	V	M
S	P	S	M	P
P	S	P	V	V

S = sausage
P = pepperoni
M = mushroom
V = vegetable

10. **Standardized Test Practice** What is an appropriate scale for this set of data? 15, 10, 12, 16, 18, 5, 3, 46, 35, 21, 26
A 5–50 **B** 0–45 **C** 3–45 **D** 1–50

Answers: 1. See Answer Key. 2–7. Sample scales and intervals are given. 2. 1–10; 2 3. 10–44; 5 4. 90–139; 10 5. 100–349; 50 6. 0–19; 2 7. 0–99; 20 8–9. See Answer Key. 10. D

Family Letters and Activities

Send a letter and activity for this chapter home to encourage family members to become active participants in their students' learning. (English: *Chapter 2 Resource Masters*, pp. ix–x; Spanish: log on to msmath2.net/family_letters)

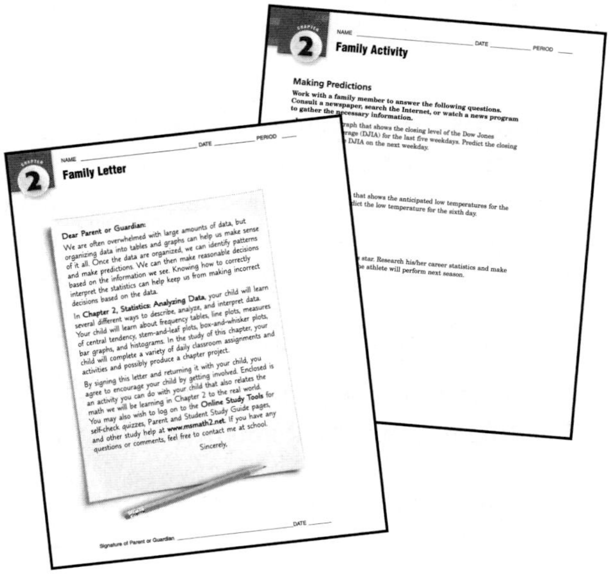

NAME ______ DATE ______ PERIOD ____

2 Family Letter

Dear Parent or Guardian:

We are often overwhelmed with large amounts of data, but organizing data into tables and graphs can help us make sense of it all. Once the data are organized, we can identify patterns and make predictions. We can then make reasonable decisions based on the information we see. Knowing how to correctly interpret the statistics can help keep us from making incorrect decisions based on the data.

In **Chapter 2, Statistics: Analyzing Data**, your child will learn several different ways to describe, analyze, and interpret data. Your child will learn about frequency tables, line plots, measures of central tendency, stem-and-leaf plots, box-and-whisker plots, bar graphs, and histograms. In the study of this chapter, your child will complete a variety of daily classroom assignments and activities and possibly produce a chapter project.

By signing this letter and returning it with your child, you agree to encourage your child by getting involved. Enclosed is an activity you can do with your child that also relates the math we will be learning in Chapter 2 to the real world. You may also wish to log on to the **Online Study Tools** for self-check quizzes, Parent and Student Study Guide pages, and other study help at **www.msmath2.net**. If you have any questions or comments, feel free to contact me at school.

Sincerely,

Signature of Parent or Guardian ______ DATE ______

NAME ______ DATE ______ PERIOD ____

2 Family Activity

Making Predictions

Work with a family member to answer the following questions. Consult a newspaper, search the Internet, or watch a news program to gather the necessary information.

...raph that shows the closing level of the Dow Jones ...rage (DJIA) for the last five weekdays. Predict the closing ... DJIA on the next weekday.

... that shows the anticipated low temperatures for the ...dict the low temperature for the sixth day.

... star. Research his/her career statistics and make ...e athlete will perform next season.

English Language Learner ELL

Lesson 2-1

Using Applications

Have students go around the class and ask as many students as possible what their favorite candy is. Then students can return to their desks. Show students how they would make a frequency table from the information they have collected. Explain the vocabulary *statistics*, *data*, and *frequency table* based on the activity.

Lesson 2-4

Reading and Writing

Have students write an explanation in their own words, and if necessary in their own language, of how to find the mean, median, mode, and range of a set of data. Have a bilingual student discuss the explanations to check for accuracy. Have students color-code the mean, median, mode, and range of a given set of data with the terms defined earlier. For example, "The term mean is highlighted in pink and so is the mean of the set of data."

Lesson 2-5

Using Models

Draw a tree with branches and leaves so that students can make a connection between real life and a stem-and-leaf plot. Next, demonstrate how the plot is drawn. Add to the tree model by drawing a stem with a leaf attached and place the digit in the tens place on the stem and the digit in the ones place on the leaf. Have the students read the numbers from the stem-and-leaf plot. Give students working in groups a stem-and-leaf plot. Ask them to list all the data from the plot and find the measures of central tendency.

MathScape™

This project was supported, in part, by the **National Science Foundation**

MathScape is a mathematics curriculum for grades 6–8 developed by the Seeing and Thinking Mathematically Project at the Education Development Center.

Functions and Spreadsheets

Getting Down to Business

Summary Students will plot data points on a coordinate grid and explore relationships among points on a graph. Tables are used to record, organize, interpret, and analyze information.

USA TODAY Education Online Lesson Planner

USA TODAY® Education's Online site offers resources and interactive features connected to each day's newspaper. *Experience Today,* USA TODAY's® daily lesson plan, is available on the site and delivered daily to subscribers. This plan provides instruction for integrating USA TODAY® graphics and key editorial features into your mathematics classroom. Log on to **education.usatoday.com**.

USA TODAY Snapshots® in Chapter 2	
Lesson	**Snapshot**
2-2	Pay raises for public school teachers

Quick Review Math Handbook

is Glencoe's mathematical handbook for students. The Hot Words section includes a glossary of terms while the Hot Topics section consists of explanations of key mathematical concepts. An exercise set is included to check students' understanding of the concepts. This valuable resource can be used as a reference in the classroom or for home study.

Lesson	Hot Topics Section	Lesson	Hot Topics Section
2-1	4-2	**2-5**	4-2
2-2	4-2, 4-3	**2-6**	4-2, 9-2
2-3	4-2	**2-7**	4-2, 9-4
2-4	4-4	**2-8**	4-1, 4-3

CHAPTER 2

Notes

Key Chapter Objectives

Lessons 2-1, 2-3, 2-5, 2-6, and 2-7 Organize and interpret data in tables and graphs.

Lesson 2-2 Make predictions from graphs.

Lesson 2-4 Find the mean, median, and mode of a set of data.

Lesson 2-8 Recognize when statistics and graphs are misleading.

"What do bridges have to do with math?"

Point out to students that this is only one of many applications of the math that they are studying. Others are provided in the introduction to each lesson.

Vocabulary Builder, pp. vii–viii **ELL**

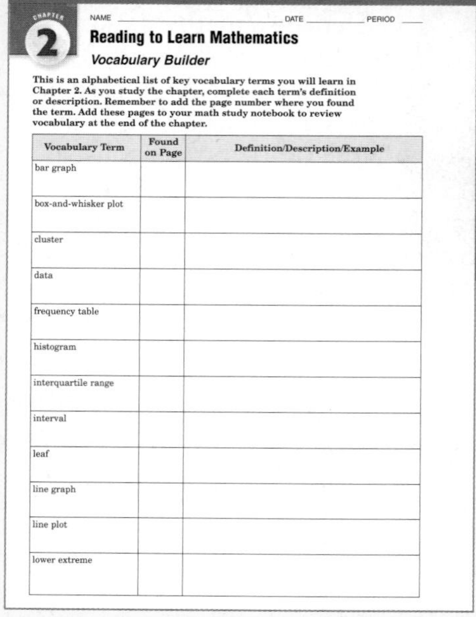

CHAPTER 2 NAME ________ DATE ________ PERIOD ____

Reading to Learn Mathematics

Vocabulary Builder

This is an alphabetical list of key vocabulary terms you will learn in Chapter 2. As you study the chapter, complete each term's definition or description. Remember to add the page number where you found the term. Add these pages to your math study notebook to review vocabulary at the end of the chapter.

Vocabulary Term	Found on Page	Definition/Description/Example
bar graph		
box-and-whisker plot		
cluster		
data		
frequency table		
histogram		
interquartile range		
interval		
leaf		
line graph		
line plot		
lower extreme		

CHAPTER 2

Statistics: Analyzing Data

"What do bridges have to do with math?"

Twenty-two of the longest suspension bridges in North America are 1,500 feet to 4,260 feet long. You can use a box-and-whisker plot to summarize the data about bridges. In mathematics, you will use tables, graphs, and diagrams to organize lists of numbers that describe real-life events.

You will solve problems by organizing data in box-and-whisker plots in Lesson 2-6.

52 **Chapter 2** Statistics: Analyzing Data

Vocabulary Builder

For a thorough vocabulary list with pronunciations of new words, give students the Vocabulary Builder worksheets from the *Chapter 2 Resource Masters*. Encourage them to complete the definition of each term as they progress through the chapter.

GETTING STARTED

Take this quiz to see whether you are ready to begin Chapter 2. Refer to the page number in parentheses if you need more review.

Vocabulary Review

Choose the correct term to complete each sentence.

1. 173.02 and 37.125 have the same digit in the (hundreds, <u>hundredths</u>, tenths) place. (Page 555)
2. The number 0.02 is (less than, equal to, <u>greater than</u>) 0.002. (Page 556)
3. The numbers 10.88, 10.089, and 1.08 are in order from (least to greatest, <u>greatest to least</u>). (Page 556)

Prerequisite Skills

Name the place value of each underlined digit. (Page 555)

4. $7.\underline{9}5$ **tenths**
5. $\underline{2},064$ **thousands**
6. $48.0\underline{5}$ **hundredths**
7. $16\underline{3}.3$ **ones**

Order each set of decimals from least to greatest. (Page 556)

8. 96.2, 96.02, 95.89 **95.89, 96.02, 96.2**
9. 5.61, 5.062, 5.16 **5.062, 5.16, 5.61**
10. 22.02, 22, 22.012 **22, 22.012, 22.02**
11. 3.98, 39.8, 0.398 **0.398, 3.98, 39.8**

Order each set of decimals from greatest to least. (Page 556)

12. 74.65, 74.67, 74.7 **74.7, 74.67, 74.65**
13. 1.26, 1.026, 10.26 **10.26, 1.26, 1.026**
14. 3.304, 3.04, 3.340 **3.340, 3.304, 3.04**
15. 68.9, 69.79, 68.99 **69.79, 68.99, 68.9**

Add or divide. (Pages 559 and 562)

16. 88.21 + 4.75 + 62.4 + 39.60 **194.96**
17. 215 ÷ 2.5 **86**
18. 5.9 + 28.1 + 9.2 **43.2**
19. 333 ÷ 0.6 **555**
20. 51 ÷ 1.7 **30**

FOLDABLES™ Study Organizer

Statistics Make this Foldable to help you organize information about analyzing data. Begin with eight sheets of notebook paper.

STEP 1 Fold
Fold 8 sheets of paper in half along the width.

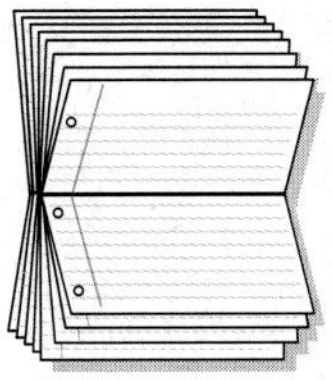

STEP 2 Cut
Cut a 1″ tab along the left edge through one thickness.

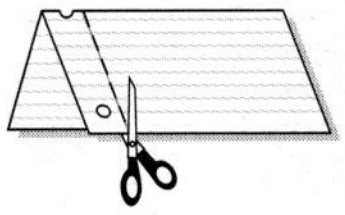

STEP 3 Glue and Label
Glue the 1″ tab down. Write the lesson number and title on the front tab.

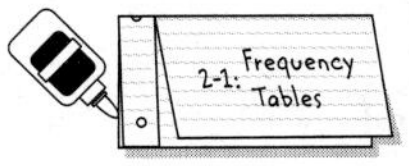

STEP 4 Repeat and Staple
Repeat Steps 2 and 3 for the remaining sheets. Staple them together on the glued tabs to form a booklet.

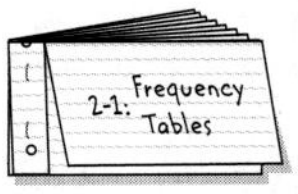

Reading and Writing As you read and study the chapter, fill the booklet with notes and examples of statistical graphs.

GETTING STARTED

This section provides a review of the basic concepts needed before beginning Chapter 2. Page references are included for additional student help.

Prerequisite Skills review skills needed in this chapter.

Exercises	Needed for Lesson	Skill
4–7	2-5	Name Place Value
8–11	2-3	Order Decimals, Least to Greatest
12–15	2-3	Order Decimals, Greatest to Least
16–20	2-4	Add and Divide Decimals

For more prerequisite skills, see:

- Getting Ready for the Next Lesson, at the end of each lesson
- Prerequisite Skills in the Student Handbook on pp. 554–563
- *Math Skills Maintenance Masters*

FOLDABLES™ Study Organizer

For more information about Foldables, see *Teaching Mathematics with Foldables.*

After students make their eight-page Foldables, have them label the front tab of each section to correspond to the eight lessons in this chapter. Under the tabs of their Foldables, students take notes, define terms, record concepts, and write examples. On the front of each tab, ask students to design a visual (graph, diagram, picture, chart, or concept map) that presents the information introduced in the lesson in a concise, easy-to-study format. Encourage students to label their visuals clearly.

2-1 Lesson Notes

Lesson Resources

Chapter 2 Resource Masters

- Study Guide and Intervention, p. 67
- Practice: Skills, p. 68
- Practice: Word Problems, p. 69
- Reading to Learn Mathematics, p. 70
- Enrichment, p. 71

Key Teaching Transparencies, 5

Parent and Student Study Guide Workbook, 2-1

Real-Life Career Masters, p. 16

FAST FILE

1 Focus

5-Minute Check 2-1

INTERACTIVE CHALKBOARD
PowerPoint® Presentations

Use as a quiz or review of Chapter 1. (Also available as a transparency)

Bellringer

Problem Solving Have students randomly select one sentence from any book and copy it onto a sheet of paper. Students should then try to find a way to show what letters are used in the sentence. Some students may tally the letters. Others may rewrite the sentence placing like letters together. Show and discuss the table below with students. Explain that in this lesson, they will learn to organize data in a similar way.
(Continued in bottom margin.)

Reading to Learn Mathematics, p. 70 ELL

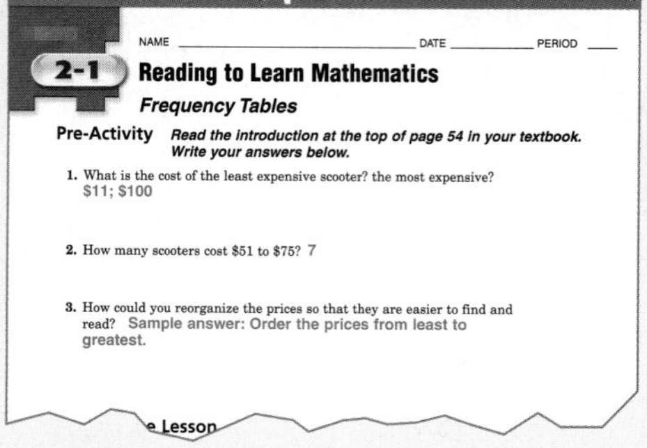
NAME ______________ DATE ________ PERIOD ____

2-1 **Reading to Learn Mathematics**

Frequency Tables

Pre-Activity *Read the introduction at the top of page 54 in your textbook. Write your answers below.*

1. What is the cost of the least expensive scooter? the most expensive? $11; $100
2. How many scooters cost $51 to $75? 7
3. How could you reorganize the prices so that they are easier to find and read? Sample answer: Order the prices from least to greatest.

...e Lesson

2-1 Frequency Tables

What You'll Learn
Organize and interpret data in a frequency table.

NEW Vocabulary
statistics
data
frequency table
scale
interval

WHEN am I ever going to use this?

SCOOTERS Riding a scooter is a fun way to get exercise. The table shows prices for 35 types of nonmotorized scooters found at an online store.

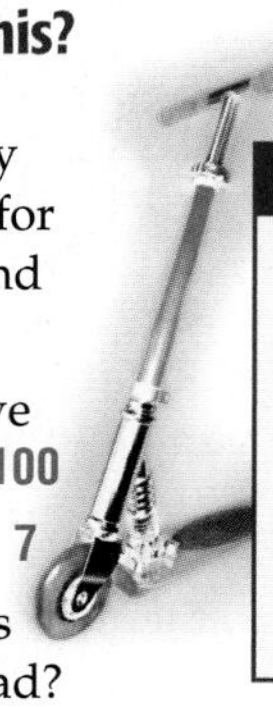

Prices of Scooters			
60	25	29	25
35	50	80	30
80	90	80	55
40	100	52	45
99	60	100	89
99	100	49	35
92	20	49	80

1. What is the cost of the least expensive scooter? the most expensive? **$11; $100**
2. How many scooters cost $51 to $75? **7**
3. How could you reorganize the prices so that they are easier to find and read? **Sample answer: Order the prices from least to greatest.**

Statistics deals with collecting, organizing, and interpreting data. **Data** are pieces of information, which are often numerical. Large amounts of data can be organized in a frequency table. A **frequency table** shows the number of pieces of data that fall within given intervals.

Price ($)	Tally	Frequency
1–25	||||	4
26–50	~~||||~~ ~~||||~~ |	11
51–75	~~||||~~ ||	7
76–100	~~||||~~ ~~||||~~ |||	13

The **scale** allows you to record all of the data. It includes the least value, 11, and the greatest value, 100. The scale is 1 to 100.

The **interval** separates the scale into equal parts. The interval is 25.

Tally marks are counters used to record items in a group.

From the frequency table, you can see that just over half the scooters cost between $51 and $100.

EXAMPLE Make a Frequency Table

1 RACING ***Cyclo-cross*** **is cycle racing that takes place on short off-road courses. The table shows the total points earned by the top 20 racers in the cyclo-cross World Cup events. Make a frequency table of the data.**

Cyclo-Cross Points			
310	179	116	91
225	173	107	76
223	150	105	55
218	149	100	52
190	123	100	50

Source: www.cyclo-cross.com

Step 1 Choose an appropriate interval and scale for the data. The scale should include the least value, 50, and the greatest value, 310.

interval: 50
scale: 50 to 349

The scale includes all of the data, and the interval separates it into equal parts.

TEACHING TIP
There is more than one way to choose the interval and the scale for a set of data. Students should be sure that all the data fall in the given intervals. When the data are not numerical, as in Example 2, there is no interval or scale.

Sentence: Riding a scooter is a fun way to get exercise.							
Tally							
A	|||	F	||	O	|||	U	|
C	||	G	||	R	|||	W	|
D	|	I	||||	S	|||	X	|
E	~~||||~~	N	|	T	|||	Y	|

TEACHING TIP
ere should be at least e number in the data t in the highest tegory and one in the west category.

Step 2 Draw a table with three columns and label the columns *Points*, *Tally*, and *Frequency*.

Step 3 Complete the table.

Points	Tally	Frequency
50–99	ᚎ	5
100–149	ᚎ II	7
150–199	IIII	4
200–249	III	3
250–299		0
300–349	I	1

Frequency tables can also have categories without scales and intervals.

EXAMPLE Make and Use a Frequency Table

2 OLYMPICS Enrique asked his classmates to name their favorite winter Olympic event. The results are shown at the right. Make a frequency table of the data. Then determine the overall favorite event.

Favorite Winter Olympic Events					
S	F	F	S	B	F
F	S	H	J	B	S
J	F	J	F	H	F
F	B	S	F	S	H
S	B	H	F	J	J

B = bobsleigh, F = figure skating, H = hockey, J = ski jumping, S = snowboarding

Draw a table with three columns. In the first column, list the events. Then complete the rest of the table.

Event	Tally	Frequency
bobsleigh	IIII	4
figure skating	ᚎ ᚎ	10
hockey	IIII	4
ski jumping	ᚎ	5
snowboarding	ᚎ II	7

Figure skating, with 10 tallies, was the favorite event. Snowboarding was the second favorite, with 7 tallies.

EAL-LIFE MATH

YMPICS In 2002, ıetta Flowers became the t African-American to win old medal in the Winter mpics. She and Jill ken won the first ever woman bobsleigh event h a two-mile time of ninute 37.76 seconds.

rce: www.olympic-usa.org

Frequency tables are useful for interpreting data.

EXAMPLE Interpret Data

3 HISTORY The frequency table shows the reigns of the last forty rulers of England. How many of them ruled England for more than 30 years?

There are three categories with more than 30 years.

31–45 years = 6 rulers

46–60 years = 4 rulers

61–75 years = 1 ruler

So, 6 + 4 + 1 or 11 of the last forty rulers each ruled England for more than 30 years.

Rulers of England		
Reign (years)	Tally	Frequency
1–15	ᚎ ᚎ ᚎ III	18
16–30	ᚎ ᚎ I	11
31–45	ᚎ I	6
46–60	IIII	4
61–75	I	1

Source: www.scotlandroyalty.org/kings.html

TEACHING TIP
ote that the frequency ble at the right shows *summary* of the data, ot the individual pieces data. The frequency ble in Example 2 shows ach data value.

msmath2.net/extra_examples

2 Teach

In-Class Examples

INTERACTIVE CHALKBOARD PowerPoint® Presentations

1 FOOTBALL Winning Super Bowl scores from 1983 to 2002 are listed below. Make a frequency table of the data.

Winning Scores			
20	34	23	34
31	35	27	49
30	52	37	20
55	20	42	39
46	38	38	27

Source: superbowl.com

Scores	Tally	Frequency
20–28	ᚎ I	6
29–37	ᚎ I	6
38–46	ᚎ	5
47–55	III	3

2 MUSIC Kaley asked her classmates about their favorite types of music. The results are shown in the table. Make a frequency table of the data. Then determine the favorite and least favorite types of music.

Favorite Types of Music				
P	R	F	P	F
F	R	F	P	P
F	P	F	C	P
C	J	R	R	F
J	R	P	P	F

R = rock, J = jazz, C = country, F = top 40, P = rap

Music	Tally	Frequency
rock	ᚎ	5
jazz	II	2
country	II	2
top 40	ᚎ III	8
rap	ᚎ III	8

The two favorite types were top 40 and rap. The least favorite types were jazz and country.

DAILY INTERVENTION **Differentiated Instruction** **ELL**

Verbal/Linguistic Have students use recent editions of *The World Almanac* or the Internet to select frequency data. Once students pick a topic, have them work in pairs to create frequency tables of the data. Students can take turns calling out each data point and tallying the points. Hearing the data presented verbally, then discussing the results, can help students process cumbersome data sets. Have students share their results with the class.

In-Class Example

3 TEMPERATURE INTERACTIVE CHALKBOARD PowerPoint® Presentations

The frequency table below shows the record high temperatures reported by each state of the United States. How many states have reported temperatures above 111°? **33**

Temp (°F)	Tally	Frequency
100–105	卌	5
106–111	卌 卌 II	12
112–117	卌 卌 卌 I	16
118–123	卌 卌 IIII	14
124–129	II	2
130–135	I	1

Source: *The World Almanac*

3 Practice/Apply

DATA SENSE
For Exercise 3, remind students that intervals are created to include all data. Often the intervals span values less than the least data value or greater than the greatest data value.

About the Exercises...

Guided Practice Key

Exercises	Examples
4–6	1, 2
7	3

Odd/Even Assignments
Exercises 8–11 and 13–18 are structured so that students practice the same concepts whether they are assigned odd or even problems.

Alert! Exercise 15 requires research on the Internet or another source.

Skill and Concept Check

Writing Math
Exercises 1 & 3

1. **Describe** an advantage and a disadvantage of using a frequency table to display data. **1–2. See margin.**
2. **OPEN ENDED** Make a frequency table that displays thirty data items and has an interval of 5.
3. **DATA SENSE** Explain why the frequency table in Example 1 does not have 300–310 as its last category when the greatest data value is 310. **The interval in each category must be the same, 50.**

GUIDED PRACTICE

For Exercises 4 and 5, use the table below.

Number of Movie Rentals During the Month

10	9	2	17	12	9	11	10
4	14	3	15	0	6	5	9
20	10	11	8	3	1	13	4

4. Choose an appropriate interval and scale to make a frequency table. **Sample answer: interval 3; scale 0 to 20**
5. Make a frequency table of the data.

For Exercises 6 and 7, use the table below.

Favorite Breakfast Foods

e	c	e	p	c	c
c	p	p	e	c	e
c	e	c	p	p	e

c = cereal, e = eggs, p = pancakes

6. Make a frequency table of the data.
7. What is the favorite breakfast food? **cere**

5–6. See Ch. 2 Answer Appendix.

★ **indicates increased difficulty**

Practice and Applications

HOMEWORK HELP

For Exercises	See Examples
8–12	1
13–18	2, 3

Extra Practice See pages 567, 597.

Make a frequency table of each set of data. **8–11. See Ch. 2 Answer Appendix.**

8. **Ages of Students in Ms. Warren's Class**

12	13	13	14	14
14	14	13	13	14
13	14	13	13	12
14	13	14	12	13

9. **Years of Major Hurricanes**

1900	1975	1985	1995	1996
1938	1944	1957	1980	1992
1961	1935	1989	1972	1928
1955	1969	1979	1915	1964
1970	1960	1954	1926	1967
1999	1998			

Source: *The World Almanac*

10. **How Do You Listen to the Most Music?**

I	C	M	I	R	M
R	M	I	M	C	R
R	C	C	R	I	C
M	C	C	R	C	M
R	I	R	R	R	C

C = CDs, R = radio, I = Internet, M = music videos

11. **Which Chore Do Students Like Least?**

D	L	Y	L	V	R	L
D	L	R	D	R	D	Y
V	D	V	R	R	Y	R
R	R	Y	L	D	V	R
D	R	V	R	R	D	L

D = wash dishes, V = vacuum, R = clean room, L = do laundry, Y = do yard work

12. **Determine** whether the categories 0–100, 100–200, and 200–300 could be used in a frequency table displaying data. Explain. **No; categories cannot overlap.**

Answers

1. An advantage of using a frequency table is that it organizes data to provide a quick summary. It makes it easier to see quickly the number of times each item of data appears. A disadvantage is that a frequency table may not show the individual pieces of data.

2. Sample answer:

Number	Tally	Frequency
1–5	III	3
6–10	卌 卌	10
11–15	卌 III	8
16–20	卌 II	7
21–25	II	2

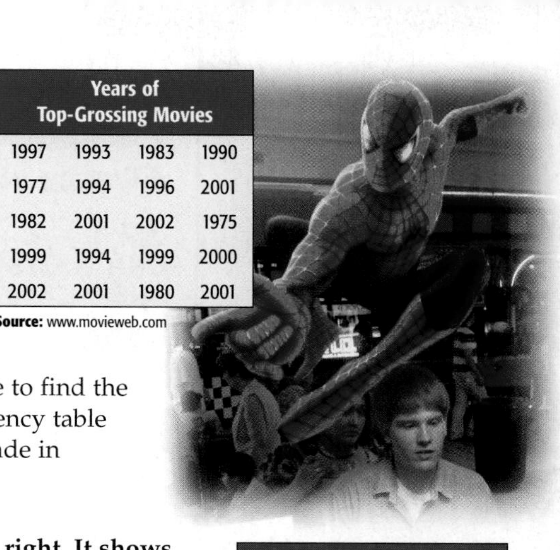

MOVIES **For Exercises 13–15, use the table at the right. It shows when the 20 top-grossing movies of all time were released as of 2002.**

Years of Top-Grossing Movies			
1997	1993	1983	1990
1977	1994	1996	2001
1982	2001	2002	1975
1999	1994	1999	2000
2002	2001	1980	2001

Source: www.movieweb.com

13. Make a frequency table of the data. Use the intervals 1970–1979, 1980–1989, 1990–1999, and 2000–2009. **See Ch. 2 Answer Appendix.**
14. In which years were the greatest number of top-grossing movies released? **1990–1999**

★ 15. **RESEARCH** Use the Internet or another source to find the current 20 top-grossing movies. Make a frequency table of the data and compare it to the table you made in Exercise 13. **See students' work.**

CEREAL **For Exercises 16–18, use the table at the right. It shows the number of Calories contained in one serving of students' favorite cereals.** **17. Sample answer: 91–110 Calories**

Number of Calories per Serving of Dry Cereal				
70	120	100	100	140
70	120	100	100	100
110	110	110	120	100
110	100	100	110	150
90	110	120	110	100
140	130	100	80	100
100	110	80	90	110

16. Make a frequency table of the data. **See Ch. 2 Answer Appendix.**
17. How many Calories are in one serving of most of the cereals?

★ 18. If a student eats two bowls of his favorite cereal, what is the greatest and least number of Calories possible that he consumes? **300; 140**

19. **CRITICAL THINKING** Make a frequency table of the scooter data at the beginning of this lesson using a different scale and interval. Analyze the data in the new table and describe an advantage and a disadvantage of using this scale and interval. **See Ch. 2 Answer Appendix.**

Standardized Test Practice and Mixed Review

20. **MULTIPLE CHOICE** The following are scores of the Duke University women's basketball team in the 2001–2002 conference games: 107, 76, 90, 91, 81, 73, 86, 102, 102, 86, 66, 83, 77, 81, 88, 90. Choose the best scale for making a frequency table of the scores. **C**

    Ⓐ 66–105 Ⓑ 68–102 Ⓒ 66–110 Ⓓ 70–108

21. **MULTIPLE CHOICE** If there are 24 students in math class, how many received a score on the exam in the interval 81–90? **F**

    Ⓕ 7 Ⓖ 6 Ⓗ 5 Ⓘ 4

Score	Frequency
71–80	8
81–90	?
91–100	9

Write each number in standard form. (Lesson 1-9)

22. 1.8×10^3 **1,800**
23. 5.33×10^5 **533,000**
24. 2.4×10^4 **24,000**
25. 6.0×10^7 **60,000,000**

26. **MEASUREMENT** How many centimeters are in 8.05 meters? (Lesson 1-8) **805 cm**

GETTING READY FOR THE NEXT LESSON

PREREQUISITE SKILL

27. At 5:00 P.M. the outside temperature was 81°F; at 6:00 P.M. it was 80°F; at 7:00 P.M. it was 79°F. If the pattern continues, predict the outside temperature at 8:00 P.M. (Lesson 1-7) **78°F**

msmath2.net/self_check_quiz

4 Assess

Open-Ended Assessment

Writing Provide students with a frequency table as shown. Have each student write a scenario or situation from which the table may have been created. Students should include the data in their answers.

Value	Tally	Frequency
10–19	||	2
20–29	||||	4
30–39	𝍸 |||	8
40–49	||||	4
50–59	||	2

Study Guide and Intervention, p. 67

NAME ________ DATE ______ PERIOD ___

2-1 Study Guide and Intervention

Frequency Tables

A **frequency table** uses tally marks to show how many times each piece of **data** appears. If the data is numerical, the table should have a **scale** which includes the least and the greatest numbers. Also, each table should have an **interval** which separates the scale into equal parts.

EXAMPLE 1 ATHLETIC SHOES **The table shows prices of 20 types of athletic shoes at a recent sidewalk sale. Make a frequency table and then determine how many types are available for less than $80.**

Prices of Athletic Shoes ($)				
60	45	120	75	50
70	95	135	65	47
43	110	84	70	53
100	75	70	85	130

Step 1 Choose an appropriate interval and scale for the data. The scale should include the least price, $43, and the greatest price, $135.

Step 2 Draw a table with three columns and label the columns *Price*, *Tally*, and *Frequency*.

Step 3 Complete the table.

Step 4 Two categories include prices less than $80.
$40–$59 = 5 types
$60–$79 = 7 types
So, 5 + 7 or 12 types of shoes cost less than $80.

Price($)	Tally	Frequency
40–59	𝍸	5
60–79	𝍸 ||	7
80–99	|||	3
100–119	||	2
120–139	|||	3

EXERCISES

For Exercises 1 and 2, use the table below.

Hours Spent Studying for Math Exam				
3	7	10	0	2
12	18	3	1	15
10	11	8	5	9
8	12	6	8	12

1. Make a frequency table of the data.
   Sample answer:

Hours	Tally	Frequency
0–4	𝍸	5
5–9	𝍸 ||	7
10–14	𝍸 |	6
15–19	||	2

2. Use your frequency table to determine how many students studied 10 hours or more. 8 students

Practice: Skills, p. 68

NAME ________ DATE ______ PERIOD ___

2-1 Practice: Skills

Frequency Tables

Make a frequency table of each set of data.

1.

Students' Choices of Toothpaste				
R	S	R	W	T
T	T	R	T	W
R	W	T	S	T
R	W	W	T	T

R = regular, W = whitening, T = tartar control, S = sensitive

Toothpaste	Tally	Frequency
R	𝍸	5
W	𝍸	5
T	𝍸 |||	8
S	||	2

2.

Ages of Students in a Class				
11	10	11	12	12
13	11	10	12	10
11	11	12	11	13

Age	Tally	Frequency
10	|||	3
11	𝍸 |	6
12	||||	4
13	||	2

3. Use your frequency table from Exercise 1. Which type of toothpaste is preferred by more students than any other type? tartar control
4. Use your frequency table from Exercise 2. Which age is least common among the students in the class? 13

ALBUMS **For Exercises 5–7, use the data at the right. It shows the number of albums, sold to the nearest million, of the top-selling albums of all time.**

Number of Albums Sold (millions)				
26	19	18	16	21
16	16	27	15	14
22	17	15	19	16
16	15	14	23	15
18	15	16	15	15

5. Make a frequency table of the data. Use the intervals 14–16, 17–19, 20–22, 23–25, and 26–28.

Albums Sold (millions)	Tally	Frequency
14–16	𝍸 𝍸 𝍸	15
17–19	𝍸	5
20–22	||	2
23–25	|	1
26–28	||	2

6. How many albums sold more than 25 million copies? 2 albums
7. How many albums sold less than 20 million copies? 20 albums

Practice: Word Problems, p. 69

NAME ________ DATE ______ PERIOD ___

2-1 Practice: Word Problems

Frequency Tables

FAVORITE COLORS **For Questions 1–3, use the table below. It shows the favorite colors of the students in Mr. Swatzky's class.**

Favorite Colors of Mr. Swatzky's Students									
B	R	R	O	B	Y	G	G	P	B
Y	B	B	Y	R	O	B	R	B	Y
G	B	O	Y	B	Y	G	G	G	G
P	Y	R	R	G					

B = blue, R = red, G = green, Y = yellow, O = orange, P = purple

HOLIDAYS **For Questions 4–6, use the table below. It shows the number of holidays in each month of 2003.**

2003 Holidays			
3	5	5	5
4	4	1	0
2	6	5	2

1. Make a frequency table of the data.

Favorite Color	Tally	Frequency
blue	𝍸 ||||	9
red	𝍸 |	6
green	𝍸 |||	8
yellow	𝍸 ||	7
orange	|||	3
purple	||	2

2. If one student changed his or her vote from blue to yellow, what would be the favorite color of most students? blue, green, and yellow
3. If one student changed his or her vote from red to purple, what would be the favorite color of the fewest students? orange and purple
4. What is wrong with using the intervals 1–2, 3–4, and 5–6 to represent the data in a frequency table? The intervals do not include 0.
5. Make a frequency table of the data.
   Sample answer:

Holidays	Tally	Frequency
0–1	||	2
2–3	|||	3
4–5	𝍸 |	6
6–7	|	1

6. What is the interval and scale of your frequency table from Question 5? Sample answer: interval, 2; scale, 0 to 7

2-2a Lesson Notes

1 Focus

Bellringer

Communication Ask students to describe a situation in which a value increases or decreases over time. For example, as you drive to work, your distance from home increases over time whereas driving from work to home, distance from home decreases over time. Ask students to discuss the appearance of a graph depicting the drive in each case.

2 Teach

In-Class Example

INTERACTIVE CHALKBOARD
PowerPoint® Presentations

Solve. Use the use a graph strategy.

CASSETTE SALES Based on the information in the table, how many music cassettes are expected to be sold at a large media chain in 2004?

Cassette Sales (millions)	
Year	**Cassettes Sold**
1995	273
1996	225
1997	173
1998	159
1999	124
2000	76

If the trend continues, the graph shows sales of cassettes in 2004 to be nonexistent. However, even though sales are declining rapidly, it is not realistic that sales will have completely stopped.

(Continued in bottom margin.)

2-2a Problem-Solving Strategy

A Preview of Lesson 2-2

What You'll Learn
Solve problems by using graphs.

Use a Graph

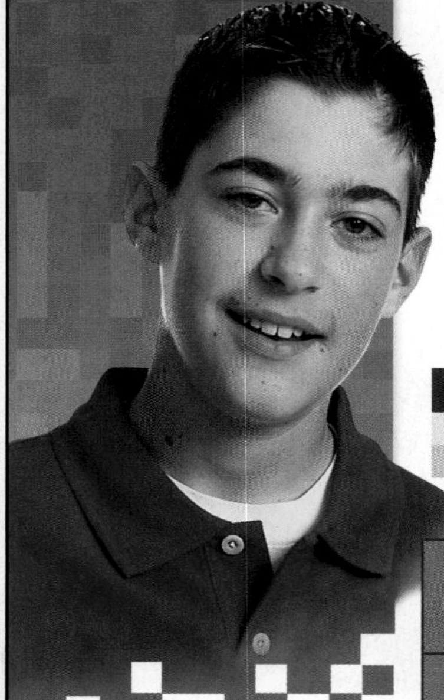

Pittsburgh Pirates Home Runs

Year	'91	'92	'93	'94	'95	'96	'97	'98	'99	'00	'01	'02
Home Runs	126	106	110	80	125	138	129	107	171	168	161	142

Source: www.baseball-almanac.com

Explore	We know the number of home runs from 1991 to 2002. We need to predict how many there will be in 2004.
Plan	If we organize the information in a graph, we may be able to see if there are any trends in the number of home runs.
Solve	We can see from the graph that generally, as the years progress, the number of home runs increases. We can predict that the number of home runs in 2004 will be about 190.
Examine	Draw a line through the points that is close to as many of the points as possible, as shown above. The estimate is close to the line, so the prediction is reasonable.

Pittsburgh Pirates Home Runs

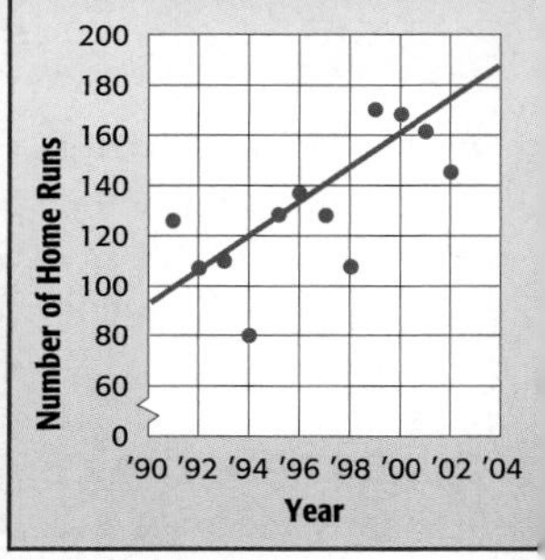

TEACHING TIP

Point out that a line can be drawn through the points in more than one way. The line at the right is just one example.

Analyze the Strategy

1. **Explain** why analyzing a graph is a useful way to quickly make conclusions about a set of data. **See margin.**
2. The last step of the four-step plan asks you to *examine* the solution. **Write** a problem in which using a graph would be a useful way to examine the solution. **See students' work.**

Cassette Sales (millions)

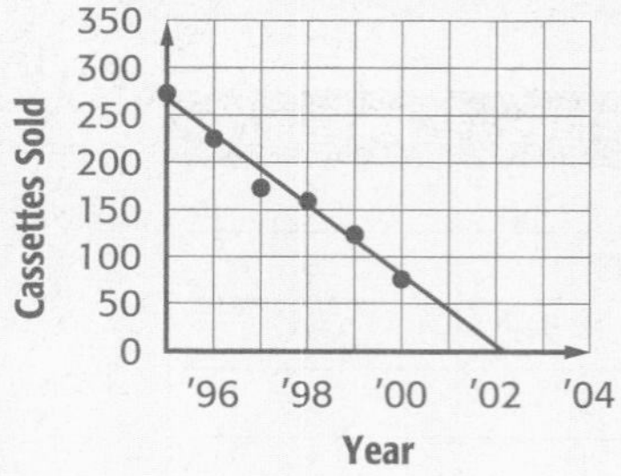

Answer

1. **Sample answer: Graphs provide a visual representation of a situation involving comparisons. A graphical model can sometimes show conclusively what is often difficult to interpret from looking at lists of numbers alone.**

pply the Strategy

olve. Use a graph.

ARTH SCIENCE For Exercises 3 and 4, refer the graph at the right. It shows the average igh and low temperatures each month from nuary to June in Memphis, Tennessee. uppose the trends continue through July.

Predict the average high temperature for the month of July. **Sample answer: 95°F**

Predict the average low temperature for the month of July. **Sample answer: 72°F**

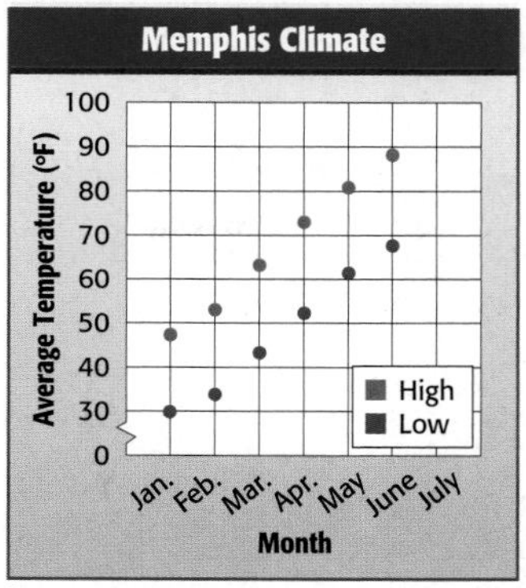

Source: www.weather.com

ixed Problem Solving

olve. Use any strategy.

READING Maya read 10 pages of a 150-page book on Monday and plans to read twice as many pages each day than she did the previous day. On what day will she finish the book? **Thursday**

SPORTS The graph shows the percent of students ages 9–13 who say they know "a lot" about the rules for playing sports.

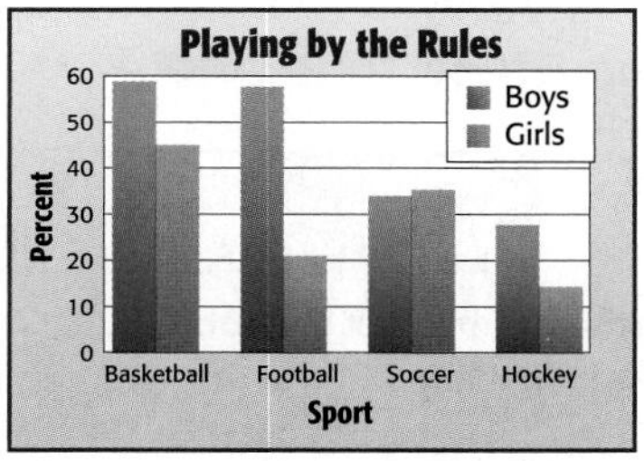

Source: *Sports Illustrated for Kids Omnibus*

According to the graph, do a greater percent of girls understand the rules of basketball or soccer? **basketball**

SAFETY An elevator sign reads *Do not exceed 2,500 pounds.* How many people, each weighing about 150 pounds, can be in the elevator at the same time? **up to 16 people**

8. **NUMBER THEORY** Find two numbers whose sum is 56 and whose product is 783. **27, 29**

9. **BOWLING** Tariq and three of his friends are going bowling, and they have a total of $70 to spend. Suppose they buy a large pizza, four drinks, and each rent bowling shoes. How many games can they bowl if they all bowl the same number of games? **2 games each**

Item	Price
large pizza	$13.75
beverage	$1.50
shoe rental	$3.50
game	$4.00

10. **STANDARDIZED TEST PRACTICE**
The following sequence forms a pattern.

4, 8, 12, 16, …

If this pattern continues, which expression can be used to find the *sixth* term? **D**

(A) $6 + 4$ (B) $6 - 4$
(C) $6 \div 4$ (D) 6×4

You will use a graph to make predictions in the next lesson.

3 Practice/Apply

About the Exercises...

Problem-Solving Strategies
Exercises 5–10 are structured so that students have the opportunity to practice many different problem-solving strategies. You may wish to remind them of some of the strategies they have studied.

- use the four-step plan (p. 6)
- guess and check (p. 22)
- use a graph ✓

4 Assess

Open-Ended Assessment

Modeling Provide students with (or ask them to find) the population of your own state or city for the last 50 years, perhaps at five- or ten-year intervals. Ask students to graph the data and predict the population 10 years from now.

DAILY INTERVENTION

Unlocking Misconceptions

Making Predictions Make sure students understand that predicting from a graph is an approximation process. They must understand that the trend in the data and the line drawn through the data are guides to their predictions. They can use the graph to justify their ideas, but real-life circumstances may alter the actual outcomes.

2-2 Lesson Notes

Lesson Resources

Chapter 2 Resource Masters
- Study Guide and Intervention, p. 72
- Practice: Skills, p. 73
- Practice: Word Problems, p. 74
- Reading to Learn Mathematics, p. 75
- Enrichment, p. 76

Key Teaching Transparencies, 6

Parent and Student Study Guide Workbook, 2-2

Science and Math Lab Masters, pp. 29–32

Teaching Mathematics with Manipulatives, p. 37

FAST FILE

1 Focus

5-Minute Check 2-2

INTERACTIVE CHALKBOARD
PowerPoint® Presentations

Use as a quiz or review of Lesson 2-1. (Also available as a transparency)

Bellringer

Hands-On Activity Have students work with partners. Give each pair of students a tennis ball and a meterstick. Ask students to guess how high the ball will bounce when dropped from a height of 1 meter and test their predictions. Pairs should then drop the ball from several different heights, predicting the bounce height each time. Have students discuss their findings.

Reading to Learn Mathematics, p. 75 **ELL**

NAME ______________ DATE ________ PERIOD ____

2-2 Reading to Learn Mathematics

Making Predictions

Pre-Activity ***Complete the Mini Lab at the top of page 60 in your textbook. Write your answers below.***

1. By how much did the height of the water change after each addition of five marbles? See students' work.
2. Predict the height of the water when thirty marbles are in the drinking glass. Explain how you made your prediction. See students' work.
3. Test your prediction by placing ten more marbles in the glass. See students' work.
4. Draw a graph of the data that you recorded in the table. See students' work.

2-2 Making Predictions

What You'll Learn
Make predictions from graphs.

NEW Vocabulary
line graph
scatter plot

HANDS-ON Mini Lab

Materials
- measuring cup
- drinking glass
- metric ruler
- marbles
- grid paper

Work with a partner.

- Pour 1 cup of water into the drinking glass.
- Measure the height of the water and record it in a table like the one below.
- Place five marbles in the glass. Measure the height of the water and record it in your table.
- Continue adding marbles, five at a time, until there are twenty marbles in the glass. After each time, measure and record the height of the water.

Number of Marbles	Height of Water (cm)
0	
5	
10	
15	
20	

1–4. See students' work.

1. By how much did the height of the water change after each addition of five marbles?
2. Predict the height of the water when thirty marbles are in the drinking glass. Explain how you made your prediction.
3. Test your prediction by placing ten more marbles in the glass.
4. Draw a graph of the data that you recorded in the table.

You created a line graph in the Mini Lab. **Line graphs** can be useful in predicting future events when they show trends over time.

EXAMPLE Use a Line Graph to Predict

1 STATUE OF LIBERTY The table shows the time it takes Kyle to climb the Statue of Liberty. Make a line graph and predict the total time it will take him to climb 354 steps to the top.

Time (min)	0	2	4	6	8	10
Number of Steps	0	51	114	163	210	27

Climbing the Statue of Liberty

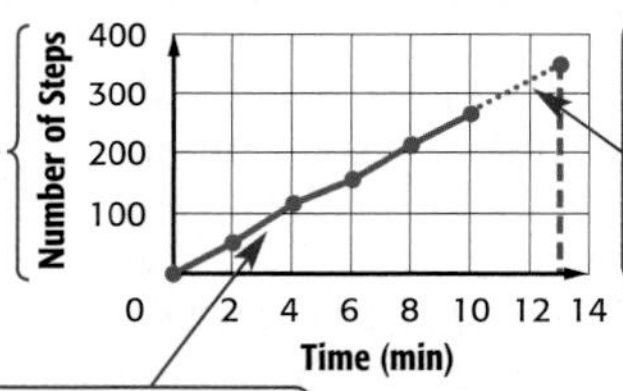

It will take Kyle about 13 minutes to climb 354 steps.

TEACHING TIP
For the Mini Lab, students should use plastic glasses that are cylinder-shaped, or beakers.

DAILY INTERVENTION **Differentiated Instruction**

Kinesthetic Have students work through Examples 1 and 2. As students are extending line graphs or experimenting with scatter plots, provide each of them with a piece of dry spaghetti. They can place the spaghetti on their papers and reposition it to see the line being used for predicting. After visualizing the line, they can draw it in with their pencils.

60 Chapter 2 Statistics: Analyzing Data

READING Math

Scatter Plots Another name for scatter plot is *scattergram*.

A **scatter plot** displays two sets of data on the same graph. Like line graphs, scatter plots are useful for making predictions because they show trends in data. If the points on a scatter plot come close to lying on a straight line, the two sets of data are related.

TEACHING TIP

Data that lie on a curve show a relationship that is not linear. Predictions may be limited to certain intervals. In Example 2, the line would eventually fall below the *x*-axis, which does not make sense in real life.

EXAMPLE Use a Scatter Plot to Predict

2 SWIMMING The scatter plot shows the winning times for the women's 100-meter butterfly event at the Summer Olympics from 1968 to 2000. Use it to predict a winning time for this event at the 2012 Olympics.

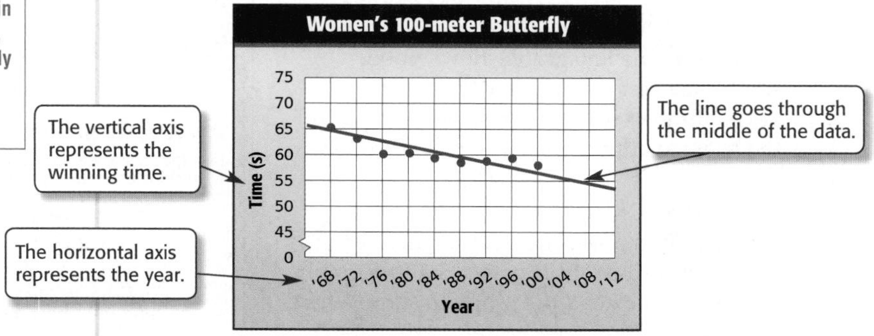

Source: *The World Almanac*

By looking at the pattern in the graph, we can predict that the winning time at the 2012 Olympics will be about 54 seconds.

Skill and Concept Check

Writing Math Exercises 1 & 3

1. **Explain** how a graph can be used to make predictions. **See margin.**
2. **OPEN ENDED** Name two sets of data that can be graphed on a scatter plot. **Sample answer: weight of an animal, speed of an animal**
3. **Which One Doesn't Belong?** Identify the term that does not have the same characteristic as the other three. Explain your reasoning.

line graph	tally mark	frequency table	scatter plot

Tally mark; the other three are ways to display data.

GUIDED PRACTICE

4. **SNACKS** The graph shows the sales, in billions of dollars, of salty snacks like potato chips and pretzels for several years. If the trend continues, about what will the sales be in 2006? **Sample answer: $22 billion**

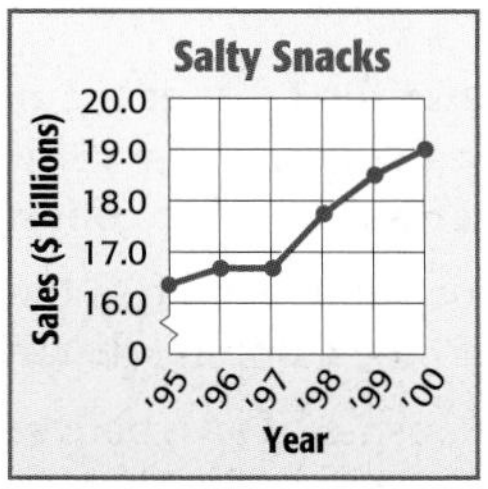

Source: *American Demographics*

2 Teach

In-Class Examples

INTERACTIVE CHALKBOARD PowerPoint® Presentations

1 TYPING Enrique is writing a 600-word paper for class. The table shows the time it has taken Enrique to type the paper so far. Make a line graph and predict the total time it will take him to type his paper. **14 min**

Time (min)	Words Typed
0	0
1	40
2	85
3	128
4	169
5	214
6	258

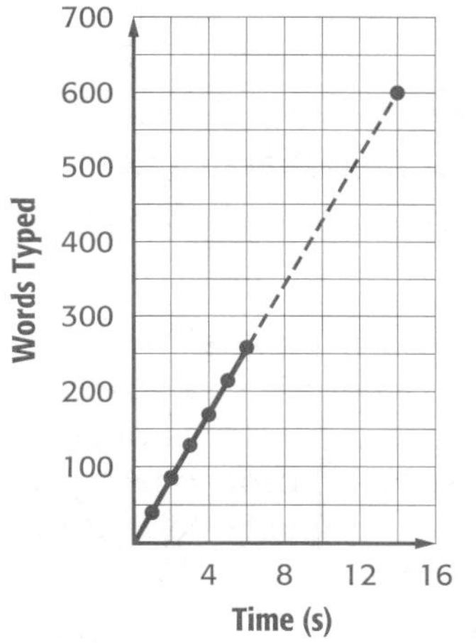

2 POLLUTION The scatter plot shows the number of days that San Bernardino, California, failed to meet air quality standards from 1990 to 1998. Use it to predict the number of days of bad air quality in 2004. **48 days**

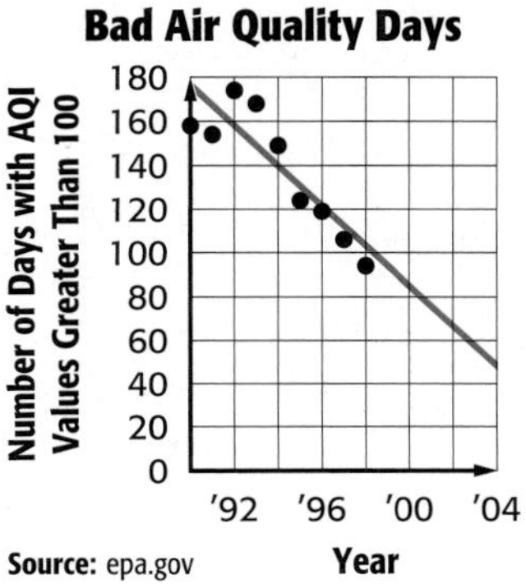

Source: epa.gov

Answer

1. Graphs often show trends over time. If you continue the pattern, you can use it to make a prediction.

Tips for New Teachers

Teaching Strategies As students create graphs and plots, it is often helpful to have them work in pairs, especially if data are not presented with whole number values. Students can share ideas for setting up and labeling the graphs.

3 Practice/Apply

FOLDABLES™ Follow-Up Remind students to use their Foldables to record what they learn about making predictions. Under the second tab, they might explain how they can use information imparted in a graph to make predictions. Encourage students to find examples of graphs in newspapers or magazines, glue or tape these examples into their Foldables, and use them to make predictions.

Which One Doesn't Belong?
In Exercise 3, remind students that tally marks are used in the creation of a frequency table, but the table itself displays the data.

About the Exercises...

Guided Practice Key

Exercise	Example
4	1

Odd/Even Assignments
Exercises 5–12 are structured so that students practice the same concepts whether they are assigned odd or even problems.

Alert! Exercise 7 requires research on the Internet or another source.

★ indicates increased difficulty

Practice and Applications

HOMEWORK HELP

For Exercises	See Examples
5–9	1
10–12	2

Extra Practice
See pages 567, 597.

5. **RUNNING** Mahkah has been training for several weeks for cross-country tryouts. His progress is shown in the graph at the right. To make the team, he must be able to run 1 mile in less than 8 minutes. If the trend continues, will Mahkah make the team? Explain. **Yes; in the 7th week he will be able to run 1 mile in about 7 minutes.**

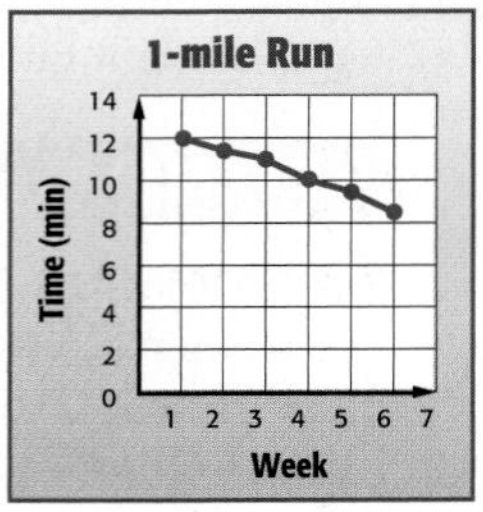

6. **SCHOOLS** The line graph shows public school teachers' salaries for the past few years. If the trend continues, about what will the average annual salary be in 2005? **about $50,000**

7. **RESEARCH** Use the Internet or another source to find a real-life example of a line graph. Write a description of what the graph displays and extend the graph to show where the data will be in the future. **See students' work.**

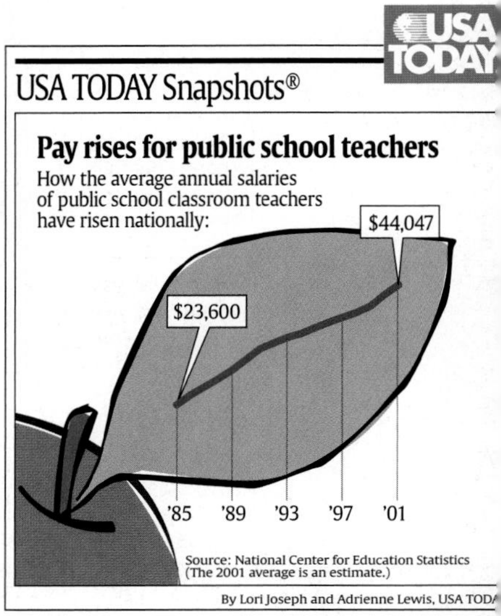

OLYMPICS For Exercises 8 and 9, use the table below. It shows the number of nations that participated in the Summer Olympics from 1948 to 2000.

Year	Number of Nations	Year	Number of Nations
1948	59	1976	92
1952	69	1980	80
1956	72	1984	140
1960	83	1988	159
1964	93	1992	169
1968	112	1996	197
1972	121	2000	199

Source: www.factmonster.com

★ 8. Make a line graph of the data. **See margin.**

★ 9. Predict the number of nations that will participate in the 2008 Summer Olympics. **Sample answer: 210 nations**

For Exercises 10–12, use the table at the right. It shows the relationship between the speed of a vehicle and the distance required to stop in certain weather conditions. **10–11. See margin.**

Speed (mph)	Stopping Distance (ft)
55	273
65	355
75	447

★ 10. Make a scatter plot of the data. Use the speed on the horizontal axis and the stopping distance on the vertical axis.

★ 11. Describe the relationship, if any, between the two sets of data.

★ 12. Predict the stopping distance for a car traveling 45 miles per hour. **Sample answer: about 201 ft**

Answers

8.

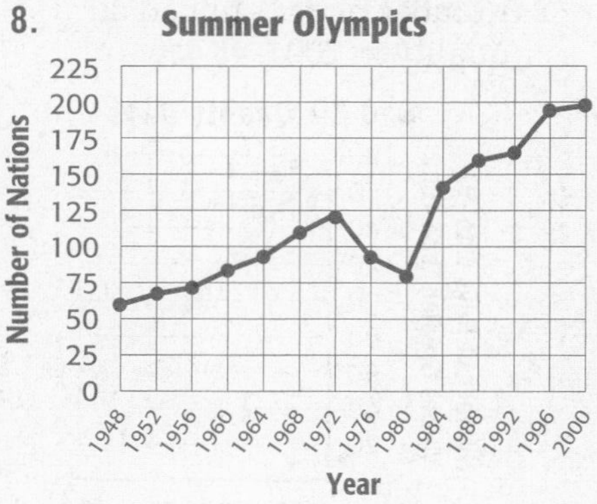

10.

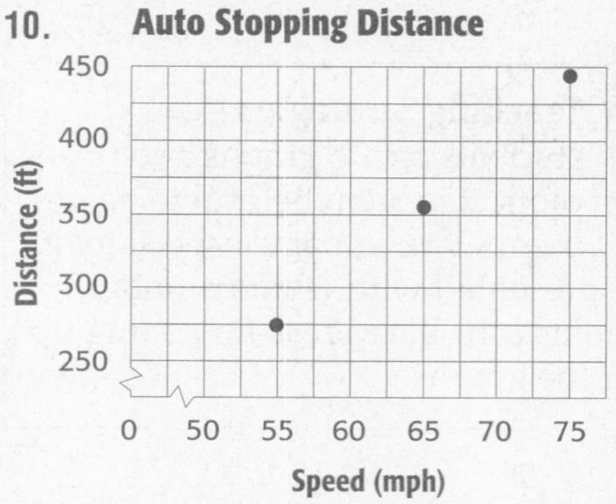

11. **Sample answer: As the speed increases, the distance required to come to a complete stop increases.**

13. **CRITICAL THINKING** What can you conclude about the relationship between students' birth month and height shown in the scatter plot at the right? **The data do not appear to fall along a straight line. So, birth month and height do not seem to be related.**

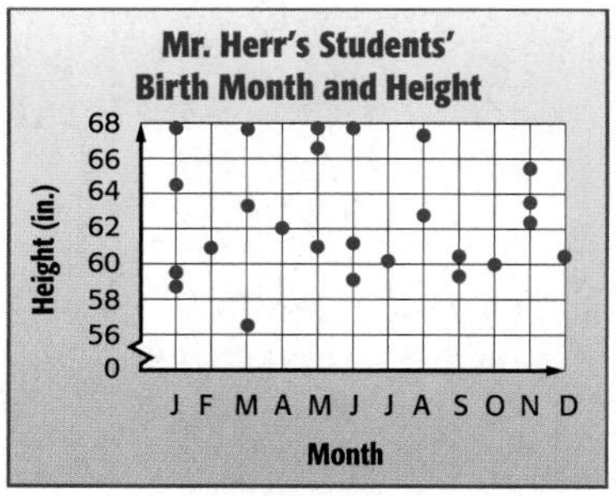

14. **EXTENDING THE LESSON** The graph at the right is called a *multiple-line graph* because it contains more than one line graph.

The multiple-line graph shows the population of Miami-Dade and Broward Counties in Florida from 1950 to 2000. Do you think that the population of Broward County will catch up to the population of Miami-Dade County in the next census of 2010? Explain. **See margin.**

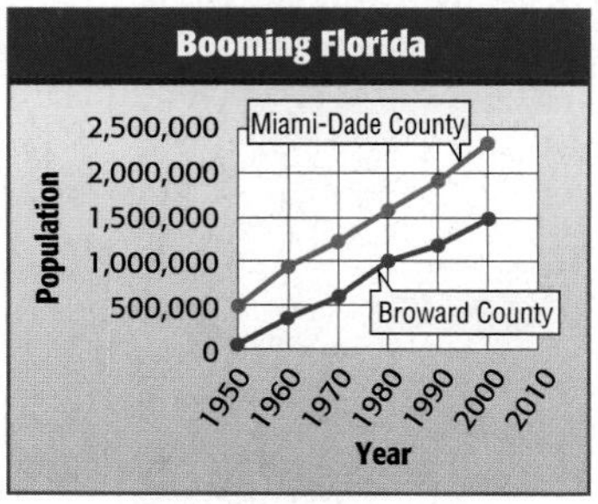

Source: U.S. Census Bureau

Standardized Test Practice and Mixed Review

For Exercises 15 and 16, refer to the graph at the right that shows the number of customers at Clothes Palace at different times during the day.

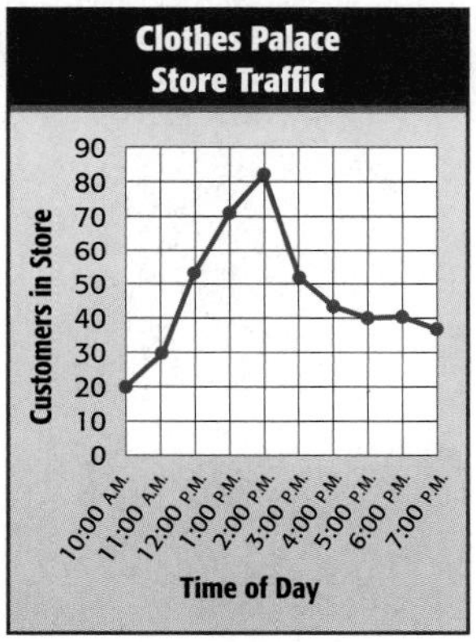

15. **MULTIPLE CHOICE** Extra workers are needed when the number of customers in the store exceeds 50. Between which hours is extra help needed? **A**

Ⓐ 12:00 P.M.–3:00 P.M. Ⓑ 11:00 A.M.–4:00 P.M.

Ⓒ 1:00 P.M.–4:00 P.M. Ⓓ 1:00 P.M.–3:00 P.M.

16. **SHORT RESPONSE** Predict about how many customers are in the store at 8:00 P.M. **about 36**

17. Sample answer: 2; 18–47

17. **STATISTICS** Choose an appropriate interval and scale for the following set of data: 35, 42, 18, 25, 32, 47, 34. (Lesson 2-1)

18. Write 72,000 in scientific notation. (Lesson 1-9) **7.2×10^4**

Write each product in exponential form. (Lesson 1-2)

19. $9 \cdot 9$ **9^2** 20. $14 \cdot 14 \cdot 14 \cdot 14$ **14^4** 21. $7 \cdot 7 \cdot 7$ **7^3** 22. $5 \cdot 5 \cdot 5 \cdot 5 \cdot 5$ **5^5**

GETTING READY FOR THE NEXT LESSON

PREREQUISITE SKILL The following data are the heights of the players, in feet, on the Miami Heat basketball team: 6.8, 7.0, 6.2, 6.7, 6.8, 6.4, 6.7, 6.1, 6.5, 6.2, 6.5, 6.8, 6.8, 7.0, 6.2. (Page 556)

23. Order the heights from least to greatest. **6.1, 6.2, 6.2, 6.2, 6.4, 6.5, 6.5, 6.7, 6.7, 6.8, 6.8, 6.8, 6.8, 7.0, 7.0**

msmath2.net/self_check_quiz

4 Assess

Higher-Level Thinking

Have students find a scatter plot for which predicting would be difficult. Ask them to explain why and give an example of possible data (i.e. a person's age and the number of times he or she sneezes in a day).

Answer

14. No; if the trend continues for both counties, then the population in Miami-Dade County will still be greater than the population in Broward County.

Study Guide and Intervention, p. 72

NAME ______ DATE ______ PERIOD ____

2-2 Study Guide and Intervention

Making Predictions

A **scatter plot** displays two sets of data on the same graph and shows trends in data. A **line graph** is a graph made by connecting the data points from a set of data. A line graph can be useful in predicting future events.

EXAMPLE 1 BASKETBALL **The table shows how many successful free throws Brett made. Make a line graph and predict how many successful free throws he will make in 8 minutes.**

Time (min)	Number of Successful Free Throws
0	0
1	5
2	9
3	12
4	15

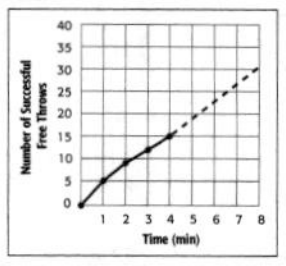

Brett will make about 31 free throws in 8 minutes.

EXERCISES

GROWTH RATE **For Exercises 1 and 2, use the table that shows a puppy's weight by month.**

Month	Weight (lb)
0	7
1	9
2	10
3	13

1. Make a line graph of the data.

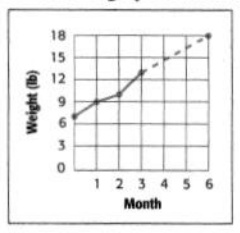

2. Predict her weight at six months. Sample answer: about 18 lb

Practice: Skills, p. 73

NAME ______ DATE ______ PERIOD ____

2-2 Practice: Skills

Making Predictions

1. GOLD The graph shows the gold reserves in the U.S. from 1980 to 2000. If the trend continues, about how many millions of dollars will be in gold reserve in the U.S. in 2010? Sample answer: about 261 million

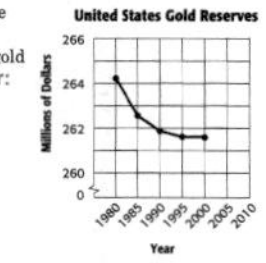

2. VEHICLES The scatter plot shows the number of sport utility vehicles sold in the U.S. from 1995 to 2000. If the trend continues, about how many sport utility vehicles will be sold in the U.S. in 2005? Sample answer: about 4,200,000

3. COMPUTERS The table below shows the number of students per computer in U.S. public schools. Make a line graph of the data.

School Year	Students per Computer
'92–'93	16
'93–'94	14
'94–'95	10.5
'95–'96	10
'96–'97	7.8
'97–'98	6.1
'98–'99	5.7
'99–'00	5.4

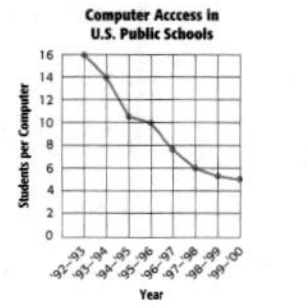

4. Use your line graph from Exercise 3 to predict the number of students per computer in '04–'05, if the trend continues. Sample answer: about 2

Practice: Word Problems, p. 74

NAME ______ DATE ______

2-2 Practice: Word Problems

Making Predictions

IDITAROD RACE **For Exercises 1 and 2, use the data in the table. It shows the number of days it took the winner to complete the Iditarod race.**

Iditarod Race Winning Times	
Year	**Winning Time (rounded to the nearest day)**
1976	19
1982	16
1988	11
1994	11
2000	9

ENTERTAINMENT **For Exercises 3 and 4, use the table. It shows the gross amount collected during several Broadway seasons.**

Broadway Season Earnings	
Year	**Earnings ($ million)**
1997–1998	558
1998–1999	588
1999–2000	603
2000–2001	666

1. Make a line graph of the data. Sample answer:

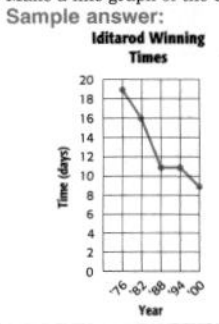

2. Use your line graph from Exercise 2 to predict the winning time in 2006, if the trend continues. about 6 days

3. Make a scatter plot of the data. Sample answer:

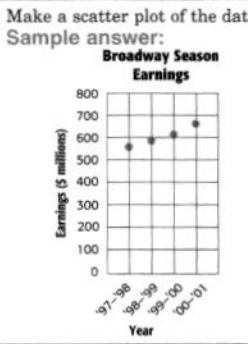

4. Use your scatter plot to predict about how much money a Broadway season will gross in 2005–2006, if the trend continues. Sample answer: about $878,000,000

2-3 Lesson Notes

Lesson Resources

Chapter 2 Resource Masters
- Study Guide and Intervention, p. 77
- Practice: Skills, p. 78
- Practice: Word Problems, p. 79
- Reading to Learn Mathematics, p. 80
- Enrichment, p. 81

Calculator and Spreadsheet Masters, p. 17
Key Teaching Transparencies, 7
Parent and Student Study Guide Workbook, 2-3

1 Focus

5-Minute Check 2-3

INTERACTIVE CHALKBOARD
PowerPoint® Presentations

Use as a quiz or review of Lesson 2-2. (Also available as a transparency)

Bellringer

Problem Solving Tell your students you want to collect data about the number of siblings class members have. Ask students to write how many siblings they have on a sticky note. Take suggestions from the class about ways to organize the notes in order to obtain a quick overview of the data.

Reading to Learn Mathematics, p. 80 **ELL**

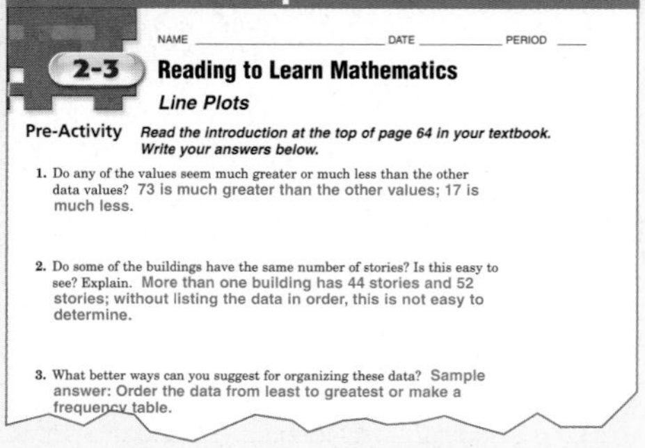
NAME ______ DATE ______ PERIOD ____

2-3 **Reading to Learn Mathematics**
Line Plots

Pre-Activity *Read the introduction at the top of page 64 in your textbook. Write your answers below.*

1. Do any of the values seem much greater or much less than the other data values? 73 is much greater than the other values; 17 is much less.

2. Do some of the buildings have the same number of stories? Is this easy to see? Explain. More than one building has 44 stories and 52 stories; without listing the data in order, this is not easy to determine.

3. What better ways can you suggest for organizing these data? Sample answer: Order the data from least to greatest or make a frequency table.

2-3 Line Plots

What You'll Learn
Construct and interpret line plots.

NEW Vocabulary
- line plot
- cluster
- outlier
- range

LINK To Reading
Everyday Meaning of Range: in music, all the notes between the high and low notes, as in a singer with a wide range

WHEN am I ever going to use this?

BUILDINGS The table shows the number of stories in the 32 tallest buildings in Los Angeles.

Number of Stories			
28	52	44	39
17	44	36	73
32	44	42	53
40	33	33	37
42	62	36	52
34	55	34	28
31	48	41	52
52	54	52	34

Source: *The World Almanac*

1. Do any of the values seem much greater or much less than the other data values?
2. Do some of the buildings have the same number of stories? Is this easy to see? Explain.
3. What better ways can you suggest for organizing these data?

1. 73 is much greater than the other values; 17 is much less.
2. More than one building has 44 stories and 52 stories; without listing the data in order, this is not easy to determine.

3. Sample answer: Order the data from least to greatest or make a frequency table.

One way to show how data are spread out is to use a line plot. A **line plot** is a diagram that shows the frequency of data on a number line. An "×" is placed above a number on a number line each time that dat value occurs.

TEACHING TIP
Point out that a line plot does not need to start at 0. Also, remind students that they cannot leave out numbers on the number line when there are no ×'s above them.

EXAMPLE Make a Line Plot

1 BUILDINGS Make a line plot of the data shown above.

Step 1 Draw a number line. Because the shortest building in the table has 17 stories, and the tallest has 73, you can use a scale of 15 to 75 and an interval of 5. Other scales and intervals could also be used.

Step 2 Put an × above the number that represents the number of stories in each building.

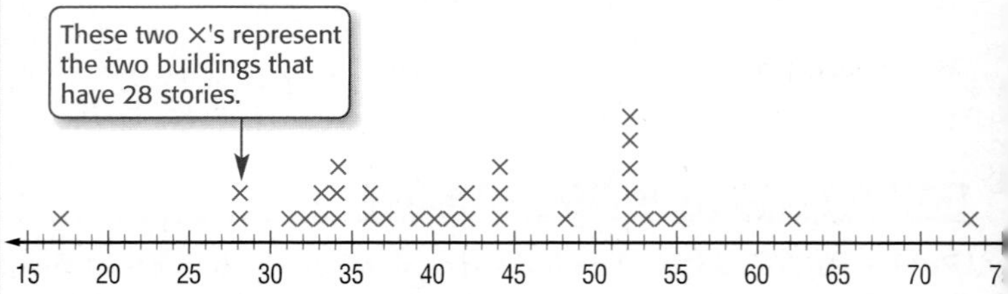

You can see on the line plot how many buildings have the same numbe of stories. This information is not easily found using the table above.

You can make some observations about the *distribution* of data, or how data are grouped together or spread out, using the line plot in Example 1.

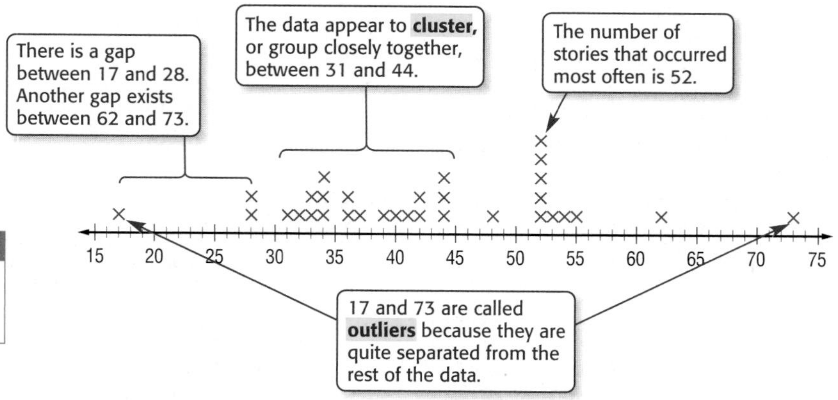

TEACHING TIP
Students will learn the formal definition of *outlier* in Lesson 2-6.

In a line plot, you can also easily find the **range** of the data, or the difference between the greatest and least numbers in the data set. This value is helpful in seeing how spread out the data are.

EXAMPLES Use a Line Plot to Analyze Data

2 SHOES The line plot below shows prices for different basketball shoes. What is the range of the data?

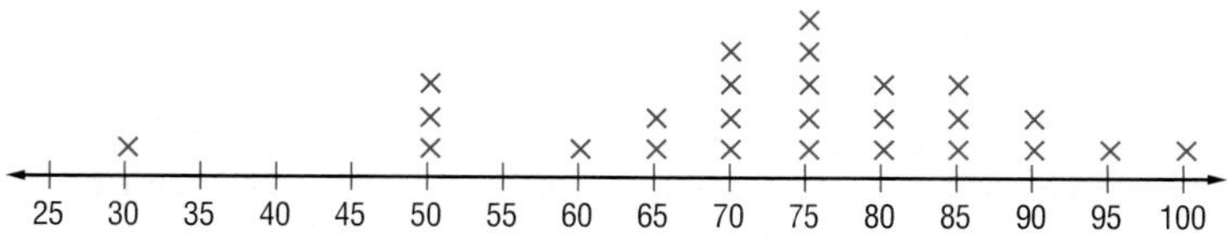

The greatest price is $100, and the lowest price is $30.

range: 100 − 30 or 70

The range of the prices is $70.

STUDY TIP

Clusters You can describe a cluster by stating a range of values or by giving a single value around which the data appear to be grouped.

3 Identify any clusters, gaps, and outliers in the data in Example 2 and explain what they mean.

Many of the data cluster around $75. You could say that most of the shoes cost from $70 to $85.

There is a gap from $30 to $50, so there were no shoes in this price range.

The number 30 appears removed from the rest of the data, so it could be considered an outlier. This means that $30 is an extremely low price and not representative of the whole data set.

Your Turn

a. Identify the range of the data in Example 1. **56**

msmath2.net/extra_examples

2 Teach

In-Class Examples

1 PRESIDENTS INTERACTIVE CHALKBOARD PowerPoint® Presentations The table below shows the ages of the U.S. presidents at the time of their inaugurations. Make a line plot of the data.

Age at Inauguration				
57	51	54	56	61
61	49	49	55	52
57	64	50	51	69
57	50	47	54	64
58	48	55	51	46
57	65	55	60	54
61	52	54	62	68
54	56	42	43	
46	51	55	56	

Source: Factmonster.com

See bottom margin.

2 CLIMATE The second line plot in the bottom margin shows the number of inches of precipitation that fell in several cities west of the Mississippi River during a recent year. What is the range of the data? **45 in.**

3 Identify clusters, gaps, and outliers if any exist in the data in In-Class Example 2 and explain what they mean. **There are data clusters between 11 and 13 inches and between 16 and 18 inches. At least half of the data fall below 25, so most of the selected cities west of the Mississippi River had fewer than 25 inches of precipitation.**

1

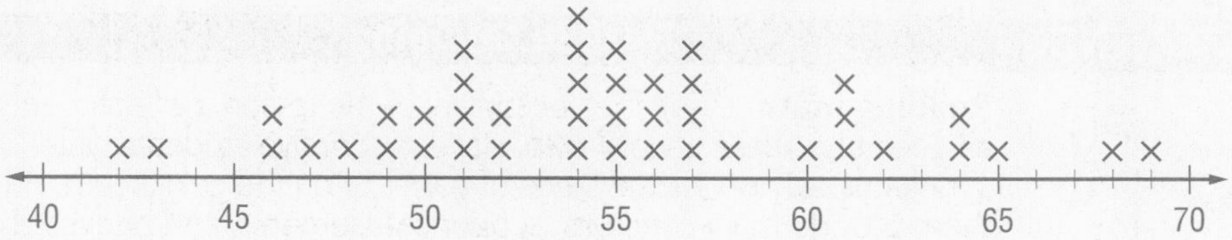

2

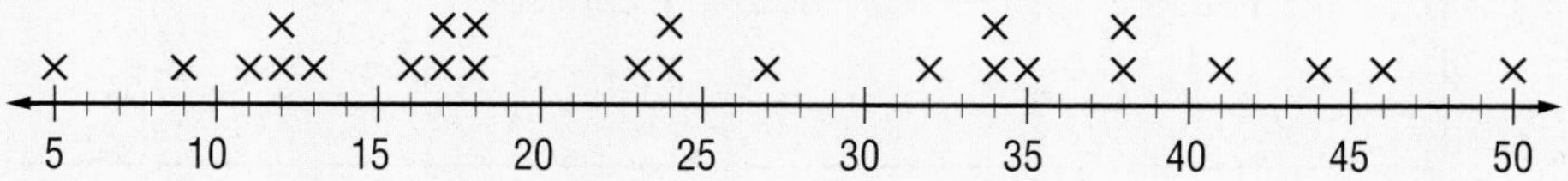

3 Practice/Apply

DAILY INTERVENTION **FIND THE ERROR**

For Exercise 3, emphasize the difference between the value of the data and the frequency of the data. You may need to give meaning to the data, such as assuming that the data represent quiz scores—two students earned scores of 20 and 1 student earned a score of 10.

DATA SENSE

For Exercise 4, ask students which data value is most frequent. 16 Ask whether this value is near the middle of the data. yes Now have students exclude any outliers and find the new range. 5 Within this range, where do most of the data lie? toward the lower end of the range of data

Answers

1. Sample answer: the greatest and least values, clusters of data, gaps in data, how the data are spread out, the range, and the pieces of data that appear most often
6. Sample answer: cluster 6–12; gaps 12–20 and 0–4; outliers 0 and 20
7. Sample answer: Most of the students in the class own 6–12 music CDs.
10. Sample answer: cluster 101–109, indicates that most of the temperatures are between 101°F and 109°F; gaps 85–94, 94–101, and 114–122, indicate where no temperatures fall; outliers 85, 94, and 122, indicate temperatures that are significantly lower and significantly higher than the other temperatures.

Skill and Concept Check

Writing Math Exercises 1 & 3

1. **List** the information you can identify from a line plot that is not obvious in an unorganized list of data. See margin.
2. **OPEN ENDED** Draw a line plot that displays fifteen pieces of data. Arrange the data so that there is a cluster and one outlier. See Ch. 2 Answer Appendix.
3. **FIND THE ERROR** Ryan and Darnell are analyzing the data shown in the line plot at the right. Who is correct? Explain.

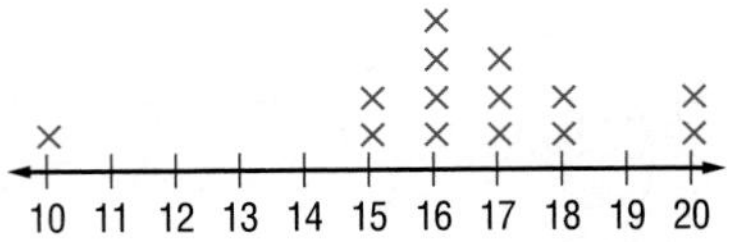

Ryan	Darnell
greatest data value: 16 least data value: 10	greatest data value: 20 least data value: 10

Darnell; 16 is the mode and 20 is the greatest value.

4. **DATA SENSE** Refer to the line plot in Exercise 3. Find the range and identify any clusters, gaps, or outliers. Sample answer: range 10; cluster 15–18; gap 10–15; outlier 10

TEACHING TIP

Be sure students recognize that individual data can be lost in frequency tables, but not in line plots.

GUIDED PRACTICE

For Exercises 5–7, use the table at the right. It shows the number of music CDs owned by students in a classroom.

5. Make a line plot of the data. See Ch. 2 Answer Appendix.
6. Identify any clusters, gaps, or outliers. 6–7. See margin.
7. Summarize the data in a sentence or two.

Number of Music CDs				
12	9	11	8	12
0	7	10	20	11
6	4	10	10	8
10	9	8	11	6

WEATHER For Exercises 8–11, use the line plot below. It shows the record high temperatures recorded by weather stations in each of the fifty states.

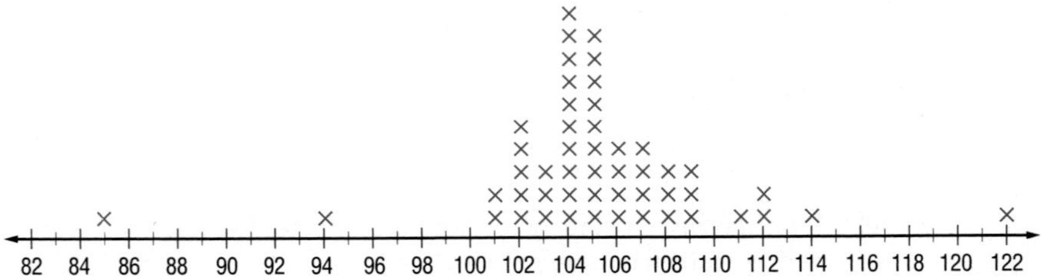

Source: *The World Almanac*

8. What is the range of the data? 37°F
9. What temperature occurred most often? 104°F
10. Identify any clusters, gaps, or outliers and analyze the data by describing what these values represent. See margin.
11. Describe how the range of the data would change if 122 were not part of the data set. The range would be 29 rather than 37.

Data Update What are the record high temperatures for some of the cities in your state? Visit msmath2.net/data_update to learn more.

DAILY INTERVENTION **Flexible Grouping**

Reading Math Using the vocabulary of the lesson, have students work as a team to make a word web. Have four or five students write simultaneously on a large piece of paper or on the chalkboard. Have them provide main concepts, supporting elements, and bridges showing relationships between ideas in a concept.

★ indicates increased difficulty

Practice and Applications

HOMEWORK HELP

For Exercises	See Examples
12–13, 19–22	1–3
14–18	2, 3

Extra Practice
See pages 567, 597.

Make a line plot for each set of data. Identify any clusters, gaps, or outliers. 12–13. See margin.

12.

Heights of Desert Cacti (ft)				
30	10	1	15	10
10	10	10	2	10
20	3	2	15	5

Source: ag.arizona.edu/pubs/garden/mg/cacti

13.

Test Scores					
98	90	97	85	86	92
92	93	95	79	91	92
91	94	88	90	93	92

Determine whether each statement is *sometimes*, *always*, or *never* true. Explain. 14. Always; each data item is plotted.

14. You can determine the number of items in a data set from a line plot.

15. If there is a cluster, it will appear in the center of the line plot. Sometimes; clusters may appear anywhere on the line plot.

SURVEY For Exercises 16–18, use the line plot at the right and the information below.

Jamie asked her classmates how many cans of soda they drink on a typical day. The results are shown on the line plot.

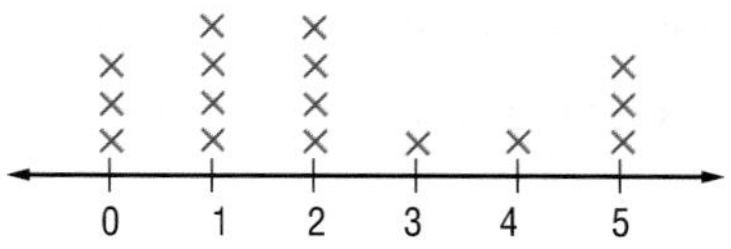

16. What was the most frequent response? 1 or 2 cans of soda per day

17. What is the range? 5

18. Summarize the data in a sentence or two. Sample answer: There are no outliers or gaps in the data. The data do not have a large spread.

ANIMALS For Exercises 19–22, use the table below. 20–21. See margin.

Average Life Spans					
Animal	Span (yr)	Animal	Span (yr)	Animal	Span (yr)
black bear	18	giraffe	10	lion	15
camel	12	gorilla	20	pig	10
cat	12	grizzly bear	25	polar bear	20
chipmunk	6	horse	20	rhesus monkey	15
dog	12	kangaroo	7	squirrel	10
elephant	40	leopard	12	white rhinoceros	20

Source: *The World Almanac for Kids*

TEACHING TIP

Remind students that a line plot is made using a number line. So, only numbers should appear under the line, not, for example, names of animals.

19. Make a line plot of the data. 19, 22. See Ch. 2 Answer Appendix.

20. Find the range and determine any clusters or outliers.

21. Summarize the data on the line plot in a sentence or two.

★ 22. The following data are the *maximum* life spans, in order, of the animals in the table: 36, 50, 28, 10, 20, 77, 36, 54, 50, 50, 24, 23, 30, 27, 45, 37, 24, 50. Make a line plot of the data and compare it to the line plot you made in Exercise 19. Include a discussion about clusters, outliers, range, and gaps in data.

msmath2.net/self_check_quiz

About the Exercises...

Guided Practice Key

Exercises	Examples
5	1
8, 11	2
6, 7, 9, 10	3

Odd/Even Assignments

Exercises 12–23 are structured so that students practice the same concepts whether they are assigned odd or even problems.

Alert! The Data Update following Exercises 8–11 requires access to the Internet.

Answers (pp. 67–68)

20. Sample answer: range 34; cluster 10–12 or 10–20; outlier 40

21. Sample answer: Most of the animals represented on the line plot have a life span of 10–20 years.

23. Sample answer: Bowler B is more consistent because excluding the outlier of 90, the range of scores is 4 points, compared with a range of 10 for Bowler A. Bowler A has scores evenly spread out from 90 to 100.

24. Sample answer: A frequency table shows the number of times data occur by using tally marks. A line plot shows the number of times data occur by using ×s. A frequency table usually shows intervals of data and is useful when a summary is needed. A line plot shows individual data points and is useful when you need to see how all the data points are spread out.

Answers

12.

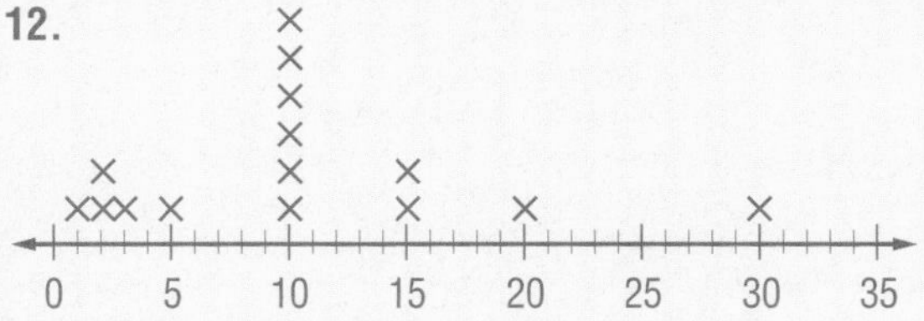

Sample answer: cluster 0–5; gap 20–30; outlier 30

13.

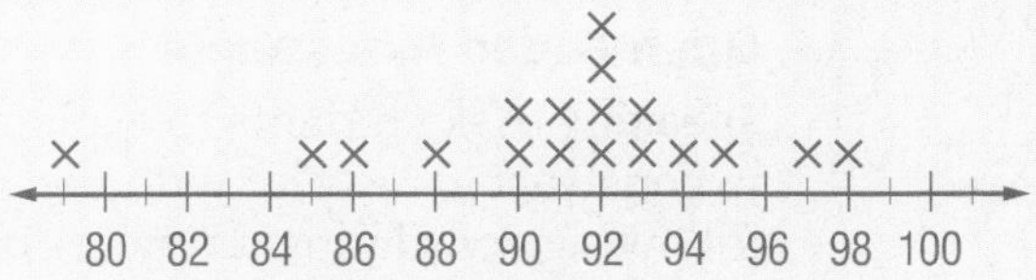

Sample answer: cluster 90–93; gap 79–85; outlier 79

Study Guide and Intervention, p. 77

NAME ______________________ DATE ____________ PERIOD ____

2-3 Study Guide and Intervention

Line Plots

A **line plot** is a diagram that shows the frequency of data on a number line.

EXAMPLE 1 SHOE SIZE **The table shows the shoe size of students in Mr. Kowa's classroom. Make a line plot of the data.**

Shoe Sizes			
10	6	4	6
5	11	10	10
6	9	6	8
7	11	7	14
5	10	6	10

Step 1 Draw a number line. Because the smallest size is 4 and the largest size is 14, you can use a scale of 4 to 14 and an interval of 2.

Step 2 Put an "×" above the number that represents the shoe size of each student.

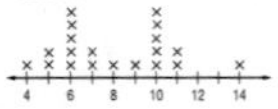

EXAMPLE 2 **Use the line plot in Example 1. Identify any clusters, gaps, or outliers and analyze the data by using these values. What is the range of data?**

Many of the data cluster around 6 and 10. You could say that most of the shoe sizes are 6 or 10. There is a gap between 11 and 14, so there are no shoe sizes in this range. The number 14 appears removed from the rest of the data, so it would be considered an outlier. This means that the shoe size of 14 is very large and is not representative of the whole data set.

The greatest shoe size is 14, and the smallest is 4. The range is 14 – 4 or 10.

EXERCISES

PETS **For Exercises 1–3 use the table at the right that shows the number of pets owned by different families.**

Number of Pets			
2	1	2	0
3	1	1	2
8	3	1	4

1. Make a line plot of the data.

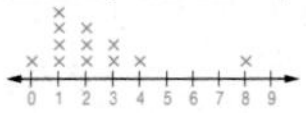

2. Identify any clusters, gaps, or outliers. Sample answer: Many of the data cluster around 1 and 2. There is a gap between 4 and 8. 8 is an outlier.

3. What is the range of the data? 8

Practice: Skills, p. 78

NAME ______________________ DATE ____________ PERIOD ____

2-3 Practice: Skills

Line Plots

For Exercises 1–3, use the data at the right that shows the number of fish each person caught on a fishing trip.

Number of Fish				
3	1	0	1	0
1	2	3	1	4
2	1	2	3	0
1	2	3	2	7

1. Make a line plot of the data.

2. What is the range of the data? 7

3. Identify any clusters, gaps, or outliers and analyze the data by describing what these values represent. There is a cluster around 1, 2, and 3. Although 1 was most common, most people caught 1, 2, or 3 fish. There is a gap between 4 and 7. That means that no one caught 5 or 6 fish. The number 7 appears to be an outlier, which means the person that caught 7 fish is not representative of the whole data set.

Make a line plot for each set of data. Identify any clusters, gaps, or outliers.

4.

Test Scores			
83	84	92	91
82	81	80	94
85	95	96	84
94	98	93	90

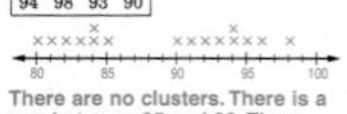

There are no clusters. There is a gap between 85 and 90. There are no outliers.

5.

Rainfall (in.)			
3	2	4	3
1	8	7	3
2	9	4	0

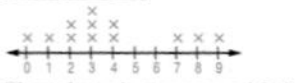

There is a cluster around 2, 3, and 4. There is a gap between 4 and 7. There are no outliers.

For Exercises 6–8, use the line plot at the right.

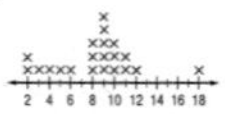

6. What is the range of the data? 16

7. What number occurred most often? 9

8. Identify any clusters, gaps, or outliers. There is a cluster around 8, 9, and 10. There is a gap between 12 and 18. 18 is an outlier.

Practice: Word Problems, p. 79

NAME ______________________ DATE ____________ PERIOD ____

2-3 Practice: Word Problems

Line Plots

TELEVISION SETS **For Exercises 1–6, use the table below. It shows the number of television sets owned by 30 different families.**

Number of TVs					
2	1	2	4	3	0
2	3	2	3	4	2
1	2	2	3	4	0
3	1	3	2	1	2
5	3	4	3	0	0

1. Make a line plot for the data.

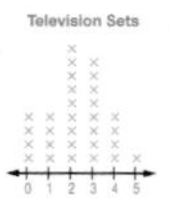

2. How many televisions do most families own? 2 televisions

3. What is the greatest number of televisions owned by a family? 5 televisions

4. What is the range of the data? 5

5. Identify any clusters, gaps, or outliers, if any exist, and explain what they mean. Sample answer: The data clusters around 2 and 3. This means that most of the families owned 2 or 3 televisions. There are no gaps or outliers.

6. Describe how the range of the data would change if 5 were not part of the data set. The range would be 4 instead of 5.

★ 23. **DATA SENSE** The last fourteen bowling scores of two people are shown on the line plots below. Describe which person is more consistent and explain how you know. **See margin.**

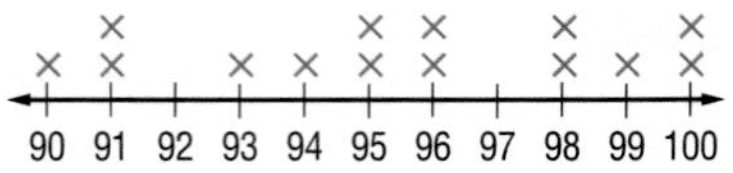

Bowler A

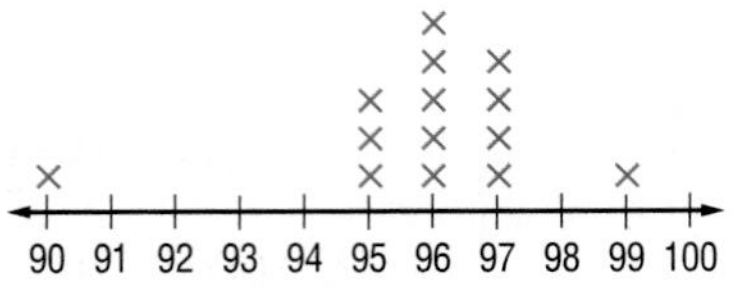

Bowler B

24. **CRITICAL THINKING** Compare and contrast line plots and frequency tables. Include a discussion about what they show and when it is better to use each one. **See margin.**

Standardized Test Practice and Mixed Review

For Exercises 25 and 26, refer to the line plot shown below.

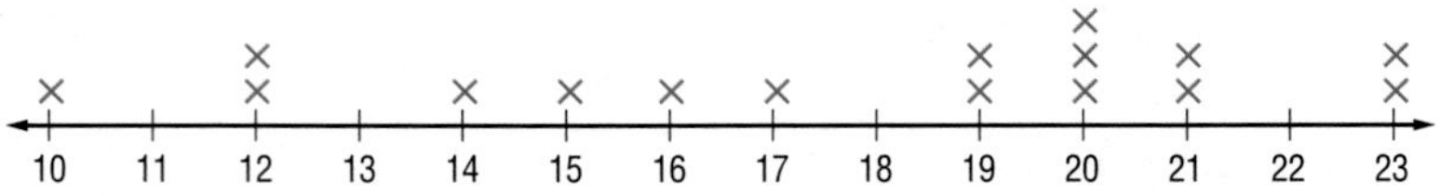

25. **MULTIPLE CHOICE** Which of the following describes a cluster of data? **D**

   Ⓐ 20–23 Ⓑ 14–17 Ⓒ 12–14 Ⓓ 19–21

26. **MULTIPLE CHOICE** Determine which sentence is *not* true. **G**

   Ⓕ The number that occurs most often is 20.

   Ⓖ There are 23 numbers in the data set.

   Ⓗ The majority of the data is greater than 18.

   Ⓘ The range is 13.

27. **RAIN** The scatter plot shows the height of grass on several lawns and the amount of rain that has fallen on each lawn. Write a conclusion based on the graph. (Lesson 2-2) **Sample answer: The more it rains, the taller the grass becomes.**

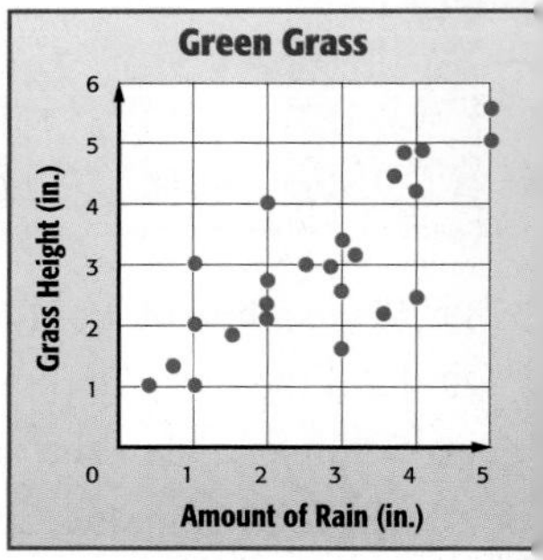

28. *True* or *False*? A frequency table is a good way to organize large amounts of data. (Lesson 2-1) **true**

Solve each equation mentally. (Lesson 1-5)

29. $\frac{t}{5} = 15$ **75**

30. $m + 18 = 33$ **15**

31. $25 = a - 4$ **29**

32. $12x = 120$ **10**

GETTING READY FOR THE NEXT LESSON

BASIC SKILL Add or divide. Round to the nearest tenth if necessary.

33. $16 + 14 + 17$ **47**

34. $32 + 34 + 36$ **102**

35. $4.6 + 2.5 + 9$ **16.1**

36. $77 + 71 + 74$ **222**

37. $\frac{202}{16}$ **12.6**

38. $\frac{56}{3}$ **18.7**

39. $\frac{255}{7}$ **36.4**

40. $\frac{155.3}{30}$ **5.2**

4 Assess

Open-Ended Assessment

Speaking Ask students to explain how to construct a line plot with the following conditions: 25 pieces of data, a range of 18, highest value 30, a cluster around 20, and one outlier.

2-4 Mean, Median, and Mode

/hat You'll Learn
nd the mean,
edian, and mode
a set of data.

EW Vocabulary
easures of
central tendency
ean
edian
ode

LINK To Reading
eryday Meaning
Median: the
iddle paved or
anted section of
highway, as in
edian strip

HANDS-ON Mini Lab

Materials
- centimeter cubes

Work with a partner.

In five days, it snowed 4 inches, 3 inches, 5 inches, 1 inch, and 2 inches.

- Make a stack of cubes to represent the snowfall for each day, as shown at the right.
- Move the cubes from stack to stack until each stack has the same number of cubes.

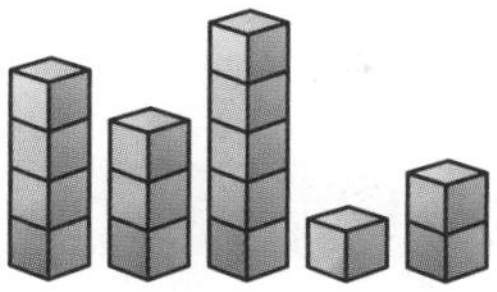

4 in. 3 in. 5 in. 1 in. 2 in.

1. How many cubes are in each stack? **3 cubes**
2. In the past five days, it snowed an average of ___?___ inches per day. **3**
3. Suppose on the sixth day it snowed 9 inches. If you moved the cubes again, how many cubes would be in each stack? **4 cubes**

If you had to choose one number to represent the data set 1, 2, 3, 4, 5, a good choice would be the middle number, 3. When analyzing data, it is often helpful to use a single number to describe the whole set.

Numbers called **measures of central tendency** can be used to describe the *center* of data. The most common of these measures are mean, median, and mode.

Key Concept — Mean

Words The **mean** of a set of data is the sum of the data divided by the number of items in the data set.

Example data set: 1, 1, 5, 2, 2, 4, 2, 5

mean: $\frac{1+1+5+2+2+4+2+5}{8}$ or 2.75

EXAMPLE Find the Mean

1 NUTRITION The table shows the grams of sugar in twenty different breakfast bars. Find the mean.

Grams of Sugar in Breakfast Bars

14	11	7	14	15
22	18	18	21	10
10	10	11	9	11
11	9	12	8	12

Source: www.cspinet.org

$\text{mean} = \frac{14 + 11 + \ldots + 12}{20}$ ← sum of data / ← number of data items

$= \frac{253}{20}$ or 12.65

The mean amount of sugar for the breakfast bars is 12.65 grams.

TEACHING TIP
e mean, also called
ithmetic average, is
hat most people think of
hen using the word
erage.

msmath2.net/extra_examples

DAILY INTERVENTION — Differentiated Instruction

Visual/Spatial Many students benefit from the use of a line plot when data are presented. For Examples 1–3, ask groups of students to create a line plot on the chalkboard or overhead transparency. Then refer to the plot as mean, median, and mode are discussed.

2-4 Lesson Notes

Lesson Resources

Chapter 2 Resource Masters
- Study Guide and Intervention, p. 82
- Practice: Skills, p. 83
- Practice: Word Problems, p. 84
- Reading to Learn Mathematics, p. 85
- Enrichment, p. 86

Calculator and Spreadsheet Masters, p. 18

Classroom Math Games, pp. 5–6

Parent and Student Study Guide Workbook, 2-4

FAST FILE

1 Focus

5-Minute Check 2-4

INTERACTIVE CHALKBOARD PowerPoint® Presentations

Use as a quiz or review of Lesson 2-3. (Also available as a transparency)

Bellringer

Communication Ask students to describe what *average* means in their own words. Then have each student make a list of several times they have used or observed an average for something. Discuss the lists. Do they have something in common?

Reading to Learn Mathematics, p. 85 ELL

NAME ______ DATE ______ PERIOD ____

2-4 Reading to Learn Mathematics
Mean, Median, and Mode

Pre-Activity *Complete the Mini Lab at the top of page 69 in your textbook. Write your answers below.*

1. How many cubes are in each stack? 3 cubes
2. In the past five days, it snowed an average of ___?___ inches per day. 3
3. Suppose on the sixth day it snowed 9 inches. If you moved the cubes again, how many cubes would be in each stack? 4 cubes

Reading the Lesson

4. Look at the data set 2, 5, 5, 6, 8, 11, 12. What is the mean? the median? the mode? 7; 6; 5
5. Match the measure of central tendency with the description of when it would be most useful.

2 Teach

In-Class Examples

INTERACTIVE CHALKBOARD
PowerPoint® Presentations

1 ANIMALS The table below shows the number of species of animals found at 30 major zoos across the United States. Find the mean.

Number of Species in Major U.S. Zoos				
300	400	283	400	175
617	700	700	715	280
800	290	350	133	400
195	347	488	435	640
232	350	300	300	400
705	400	800	300	659

Source: *The World Almanac*

$436.4\overline{6}$ species

2 OLYMPICS The table below shows the number of gold medals won by each country participating in the 2002 Winter Olympic games. Find the mean, median, and mode of the data.

2002 Winter Olympics: Gold Medals Won				
12	6	4	3	0
10	6	4	2	3
11	2	3	4	2
1	1	0	2	2
1	0	0	0	0

Source: CBSSportsline.com

mean = 3.16, median = 2, mode = 0

3 FIRST FAMILIES The line plot in the bottom margin shows the number of children of United States presidents. Would the mean, median, or mode best represent the number of children? **The mean, median, and mode are close with values of 3.1, 2.5, and 2 respectively. Any of the three could be used to represent the data.**

Key Concept — Median

Words The **median** of a set of data is the middle number of the ordered data, or the mean of the middle two numbers.

Example data set: 7, 11, 15, 17, 20, 20

median: $\frac{15 + 17}{2}$ or 16 — The median divides the data in half.

Mode

Words The **mode** or modes of a set of data is the number or numbers that occur most often.

Example data set: 50, 45, 45, 52, 49, 56, 56

modes: 45 and 56

REAL-LIFE MATH

SPACE The International Space Station measures 356 feet by 290 feet, and contains almost an acre of solar panels.

Source: *The World Almanac*

EXAMPLE Find the Mean, Median, and Mode

2 SPACE Twenty-seven countries have sent people into space. The table shows the number of individuals from each country. Find the mean, median, and mode of the data.

People in Space								
267	1	9	8	1	1	1	1	1
97	1	1	1	3	1	1	2	1
11	2	1	1	5	1	1	1	1

Source: *The World Almanac*

mean: sum of data divided by 27, or 15.6
median: 14th number of the ordered data, or 1
mode: number appearing most often, or 1

The mean of the data in Example 2 does not accurately describe the data set because it is greater than most of the data items. This happens when one or more of the data is an outlier.

EXAMPLE Analyze Data

3 TESTS The line plot shows the test scores of the students in Mrs. Hiroshi's math class. Would the mean, median, or mode best represent the test scores?

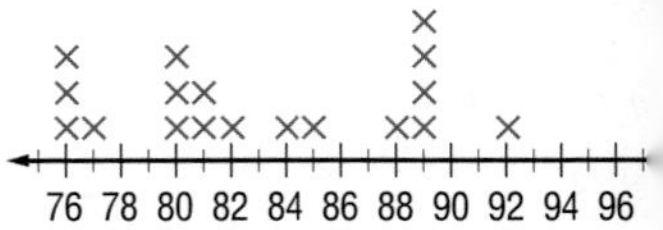

mean: $\frac{76 + 76 + 76 + \ldots + 92}{18}$ or 83

median: $\frac{\text{9th term} + \text{10th term}}{2} = \frac{81 + 82}{2}$ or 81.5

mode: 89

The mode of 89 misrepresents the scores as being higher than they really were. Either the mean of 83 or the median of 81.5 could be used to represent the scores.

STUDY TIP

Median When there is an odd number of data, the median is the middle number of the ordered set. When there is an even number of data, the median is the mean of the two middle numbers.

3

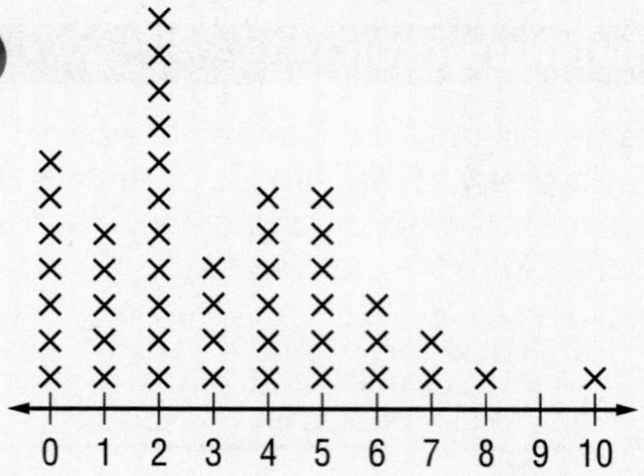

Answers

1. Sometimes; if there is an odd number of items, the median is the middle number. If there is an even number of items, the median is the mean of the two middle numbers.
2. Sample answer: 4, 5, 2, 2, 3, 3, 1, 0, 1, 2, 4, 68, 5; the mean, 7.7, is greater than all the data items except one.

ou can find measures of central tendency from charts, tables, graphs, and ne plots. Below are some general guidelines for using these measures.

TEACHING TIP

Range is a *measure of dispersion*; it describes the spread of data. Mean, median, and mode are *measures of central tendency*; they describe the center of data.

Concept Summary — Mean, Median, and Mode

Measure	Most Useful When . . .
Mean	• data set has no outliers
Median	• data set has outliers • there are no big gaps in the middle of the data
Mode	• data set has many identical numbers

Skill and Concept Check

Writing Math Exercises 1–3

1. **Determine** whether the median is *sometimes*, *always*, or *never* part of the data set. Explain. 1–2. See margin.

2. **OPEN ENDED** Write a problem in which the mean is not the best representation of the data set. Explain why not.

3. **FIND THE ERROR** Jared and Marissa are analyzing the data in the frequency table at the right. Who is correct? Explain.

Time (min)	Tally	Frequency
1	𝍸	5
2	𝍸 I	6
3	𝍸 IIII	9
4	I	1

Jared
median: 2 min
mode: 3 min

Marissa
median: 2.5 min
mode: 9 min

Jared; the median is the 11th term of the ordered data, or 2. The mode is the number that appears most often, or 3.

GUIDED PRACTICE

Find the mean, median, and mode for each set of data. Round to the nearest tenth if necessary. 4. 9; 7; 7 and 14 5. 52.3; 57; 59

4. 14, 5, 16, 7, 4, 15, 9, 7, 2, 14, 6

5. 29, 14, 80, 59, 78, 30, 59, 69, 55, 50

6. 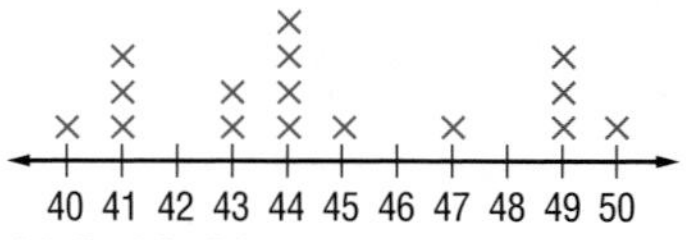

44.6; 44; 44

7.

Number of Wins	Tally	Frequency
7	𝍸 𝍸 II	12
8	IIII	4
9	II	2
10	𝍸 II	7

8.2; 8; 7

SCHOOL For Exercises 8 and 9, use the line plot. It shows the number of times per day that students drink from the water fountain at school.

8. Find the mean, median, and mode. 3.1; 2.5; 2

9. Which measure best describes the data? Explain. See margin.

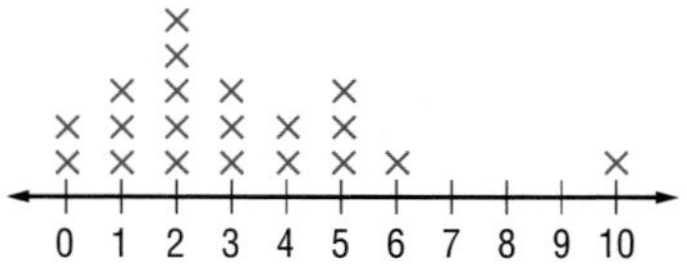

3 Practice/Apply

FOLDABLES™ Follow-Up Remind students to use their Foldables to record what they learn about finding the mean, median, and mode of a set of data. Under the fourth tab, they might define and differentiate between mean, median, and mode.

DAILY INTERVENTION FIND THE ERROR

For Exercise 3, point out that mode refers to the actual number with the greatest frequency (in this case 3), and not the frequency of its occurrence (9). Marissa also has confused median with mean. Suggest that students give themselves mental quizzes of these math terms and their meanings frequently. Time can be made for such study while walking to class or riding to school.

About the Exercises...

Guided Practice Key

Exercises	Examples
4–8	1, 2
9	3

Odd/Even Assignments

Exercises 10–15 are structured so that students practice the same concepts whether they are assigned odd or even problems.

Alert! The Data Update following Exercise 16 requires access to the Internet.

Answers (pp. 71–72)

9. Sample answer: Any of the three measures could be used to represent the data. The mean is slightly greater than most of the data items, and therefore is a less accurate description of the data.

17. Sample answer: The mean or median best represents the data because half of the items are less than these values and half are greater. Since there are four modes, the mode does not best represent the data.

18. Mean; a mode must be a member of the data set, and it is impossible to have 3.28 family members.

19. Sample answer: The mean would be most affected; the mode would be least affected, since it would not change.

Study Guide and Intervention, p. 82

NAME ______________ DATE ________ PERIOD ____

2-4 Study Guide and Intervention

Mean, Median, and Mode

The **mean** is the sum of the data divided by the number of data items. The **median** is the middle number of the ordered data, or the mean of the middle two numbers. The **mode** is the number (or numbers) that occur most often. The mean, median, and mode are each **measures of central tendency**.

EXAMPLE 1 **The table shows the number of hours students spent practicing for a music recital. Find the mean, median, and mode of the data.**

Numbers of Hours Spent Practicing				
3	12	10	8	7
18	11	12	10	3
8	6	0	1	5
8	2	15	9	12

mean $= \frac{3 + 12 + 10 + \ldots + 12}{20} = \frac{160}{20}$ or 8.

To find the median, the data must be ordered.

0, 1, 2, 3, 3, 5, 6, 7, 8, 8, 8, 9, 10, 10, 11, 12, 12, 12, 15, 18

$$\frac{8 + 8}{2} = 8$$

To find the mode, look for the number that occurs most often. Since 8 and 12 each occur 3 times, the modes are 8 and 12.

EXERCISES

Find the mean, median, and mode for each set of data. Round to the nearest tenth if necessary.

1. 27, 56, 34, 19, 41, 56, 27, 25, 34, 56 mean, 37.5; median, 34; mode, 56
2. 7, 3, 12, 4, 6, 3, 4, 8, 7, 3, 20 mean, 7; median, 6; mode, 3
3. 1, 23, 4, 6, 7, 20, 7, 5, 3, 4, 6, 7, 11, 6 mean, 7.9 median, 6; mode, 7 and 6
4. 3, 3, 3, 3, 3, 3, 3 mean, 3; median, 3; mode, 3
5. 2, 4, 1, 3, 5, 6, 1, 1, 3, 4, 3, 1 mean, 2.8; median, 3; mode, 1
6. 4, 0, 12, 10, 0, 5, 7, 16, 12, 10, 12, 12 mean, 8.3, median, 10; mode, 12

Practice: Skills, p. 83

NAME ______________ DATE ________ PERIOD ____

2-4 Practice: Skills

Mean, Median, and Mode

Find the mean, median, and mode for each set of data. Round to the nearest tenth if necessary.

1. 5, 9, 6, 6, 11, 8, 4 7; 6; 6
2. 1, 3, 5, 2, 4, 8, 4, 7, 2 4; 4; 2 and 4
3. 1, 9, 4, 7, 5, 3, 16, 11 7; 6; none
4. 3, 4, 4, 4, 4, 3, 6 4; 4; 4
5. 3, 7, 2, 5, 5, 6, 5, 10, 11, 5 5.9; 5; 5
6. 19, 17, 24, 11, 19, 25, 15, 15, 19, 16, 16 17.8; 17; 19

7.
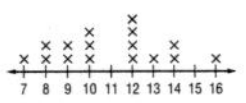

11; 11; 12

8.
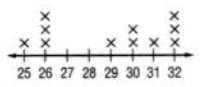

29; 30; 26 and 32

9.
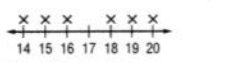

17; 17; none

10.
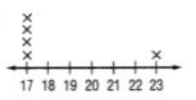

18.2; 17; 17

11.

Prize ($)	Tally	Frequency
6.00	𝍸 IIII	9
7.00	𝍸 𝍸	10
7.50	𝍸	5
8.00	I	1

$6.78; $7.00; $7.00

12.

Size	Tally	Frequency
6	𝍸 𝍸 III	13
7	𝍸 II	7
8	𝍸 II	7
9	III	3

7; 7; 6

13.

Time (min)	Tally	Frequency
8	𝍸 𝍸	10
9	𝍸 𝍸	10
10	𝍸	5

8.8; 9; 8 and 9

14.

Hits	Tally	Frequency
0	𝍸	5
1	𝍸 I	6
2	III	3
3	I	1

1; 1; 1

Practice: Word Problems, p. 84

NAME ______________ DATE ________ PERIOD ____

2-4 Practice: Word Problems

Mean, Median, and Mode

SCHOOL **For Exercises 1–6, use the table below. It shows the number of times per day that students go to their lockers.**

Student Locker Visits							
2	2	0	1	2	2	3	4
0	5	2	5	2	5	2	4
2	4	6	4	5	6	5	6
2	2	0	1	4	6	10	2

1. Make a frequency table of the data. Sample answer:

Number of Visits	Tally	Frequency
0–2	𝍸 𝍸 𝍸 I	16
3–5	𝍸 𝍸 I	11
6–8	IIII	4
9–11	I	1

2. What is the range of the data? 10
3. Find the mean, median, and mode of the data. Round to the nearest tenth if necessary. mean: 3.3; median: 2.5; mode: 2
4. Would the mean, median, or mode best represent the data? Explain. Mode; the data has many identical numbers.
5. Explain why the mean does not best represent the data. Sample answer: The mean should not be used when the data has outliers, and 10 is an outlier.
6. If the value 10 were dropped from the data, find the median and the mode of the remaining data. median: 2; mode: 2

★indicates increased difficulty

Practice and Applications

HOMEWORK HELP	
For Exercises	See Examples
10–16	1, 2
17, 18	3
Extra Practice See pages 568, 597.	

Find the mean, median, and mode for each set of data. Round to the nearest tenth if necessary.

10. 65, 56, 57, 75, 76, 66, 64 65.6; 65; none
11. 9.3, 8.4, 8.0, 9.1, 9.4, 9.0, 7.8, 9.3, 8.0 8.7; 9.0; 8.0 and 9.3
12. 95, 90, 92, 94, 91, 90, 98, 88, 89, 100 92.7; 91.5; 90
13. $10, $18, $11, $6, $6, $5, $10, $11, $46, $7, $6, $8 $12; $9; $6

14.
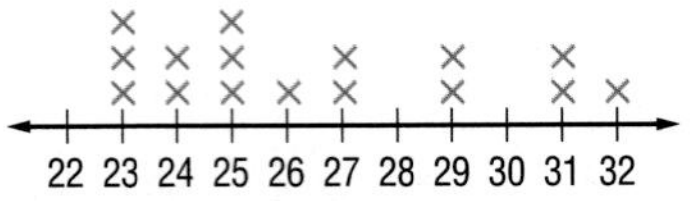

26.5; 25.5; 23 and 25

15.

Price	Tally	Frequency
$7.50	𝍸 III	8
$8.00	III	3
$8.50	𝍸 I	6

$7.90; $8; $7.50

MUSIC For Exercises 16 and 17, use the data at the right. It shows the number of weeks that songs have been on the *Top 20 Hot Country Singles & Tracks* chart. 16. 18.55; 18.5; 12, 15, 18, and 21

Number of Weeks in Top 20				
8	15	14	24	21
28	12	20	19	18
12	7	11	30	25
15	21	22	18	31

Source: www.billboard.com

16. Find the mean, median, and mode of the data.
17. Determine which measure best represents the data. Explain. See p. 71 margin.

Data Update What is the mean, median, and mode of the number of weeks of the current songs that are on the *Top 20 Hot Country Singles & Tracks* chart? Visit msmath2.net/data_update to learn more.

★ 18. **DATA SENSE** According to the U.S. Census Bureau, the average number of family members per household in Texas is 3.28. State whether this average is a mean or mode. Explain how you know. See p. 71 margin.

19. **CRITICAL THINKING** Without calculating, would the mean, median, or mode be most affected by eliminating 1,000 from the list below? Which would be the least affected? Explain. See p. 71 margin.

50, 100, 75, 60, 75, 1,000, 90, 100

Standardized Test Practice and Mixed Review

20. **MULTIPLE CHOICE** Which measure describes the center of a set of data? D

Ⓐ mean Ⓑ median Ⓒ mode Ⓓ all of the above

21. **GRID IN** The Cardinals baseball team played 8 games and scored a total of 96 runs. What was the mean number of runs scored per game? 12

22. **NUTRITION** The grams of fiber in one serving of fifteen different cereals are 5, 5, 4, 3, 3, 3, 1, 1, 1, 2, 1, 1, 1, 1, and 0. Make a line plot of the data. (Lesson 2-3) See margin.

23. What type of graph often shows trends over time? (Lesson 2-2) line graph

GETTING READY FOR THE NEXT LESSON

BASIC SKILL Name the place value of the highlighted digit.

24. 581 hundreds
25. 6,295 ones
26. 4,369 thousands
27. 284 tens

msmath2.net/self_check_quiz

4 Assess

Higher-Level Thinking

Have students draw three line plots, each showing a different measure of central tendency. Have students label the plots and include them in their portfolios.

Answer

22.
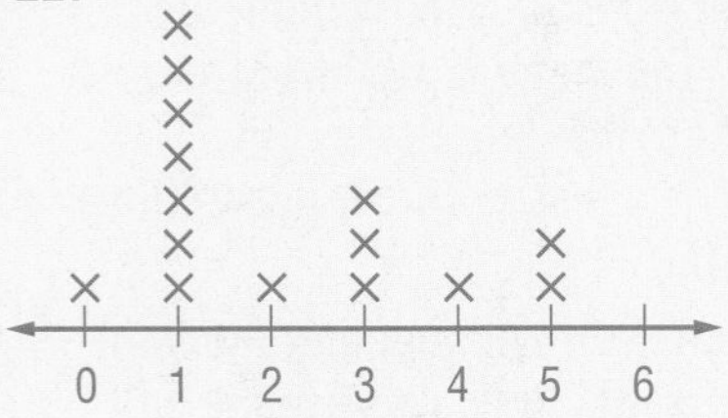

2-4b HANDS-ON LAB

A Follow-Up of Lesson 2-4

What You'll Learn
se mean, median, node, and range to escribe a set of data.

Materials
- markers
- ruler

Are You Average?

INVESTIGATE *Work in groups of four.*

What do you think of when you hear the word "average"? How would you describe an "average" student in your school? In this lab, you will investigate what an average student in your math class is like.

STEP 1 Your group will create a survey of ten appropriate questions to ask your classmates that will help you describe the "average" student in your class. For example:

- What is your height in inches?
- How many hours per week do you spend doing homework?
- What is your favorite subject in school?
- What is your favorite extracurricular activity?

STEP 2 After your teacher approves your list of questions, have your classmates complete your survey.

STEP 3 Compile the data in a frequency table. Find the mean, median, mode, and range of the data for each question if appropriate. Which measure best describes each set of data? Justify your answer.

STEP 4 Create appropriate graphs to display your data from all ten questions. Each group will present their graphs with a description of what the "average" student in your classroom is like.

Writing Math

1. **Explain** why results may vary if you survey another class in your school.
2. **Explain** why your group selected the graph that you did to display your data. Could you have used another type of graph? Explain.
3. The word *bias* means to influence. **Describe** any factors that could have unfairly influenced the responses given by your classmates. Is there a way to limit bias in a survey? See margin.

Sample answer: ırvey results could ıry if another class ıd older or younger udents.

Sample answer: The aphs should be ıpropriate presentations of ıtegorical data, such s line plots.

2-4b HANDS-ON LAB
Lesson Notes

Lesson Resources

Teaching Mathematics with Manipulatives
- student recording sheet 2-4b

GETTING STARTED

Teaching Tip Plan enough class time to carry out the group activities. One day the groups can devise their questions and obtain teacher approval, then bring the surveys to class the next day to begin the data collection process.

TEACH

- The **Investigate** section demonstrates that different questions build a different picture of an "average" person. Note that not all questions will have numerical answers. In this case, frequency tables and mode will best display and describe the data.
- After each group presents its findings, discuss with the class which questions students would use again and what new questions they might ask if they were to repeat this lab.

Answer

3. Sample answer: Oral responses could influence outcomes. Limit the survey to written responses. Popularity could influence responses. Try to remove the pressures of responding like the group. Fads could influence responses, therefore the same survey given six months later could have different outcomes.

ASSESS

Observe students as they respond to the **Writing Math** portion of the lab. Students should:

- demonstrate an understanding of the concept of average, and
- understand when a particular type of graph is appropriate.

Discuss Exercise 3. Students may mention that peer pressure or the presence of the teacher may cause students to respond to a survey in a particular way. They could limit this bias by making the surveys anonymous.

Chapter 2 Mid-Chapter Practice Test

Assessment Option

Mid-Chapter Test (Lessons 2-1 through 2-4) is available on p. 123 of the *Chapter 2 Resource Masters*.

ExamView® Pro Test Bank CD-ROM

Use the networkable ExamView® Pro Test Bank CD-ROM to:

- Create **Multiple versions** of tests.
- Create **Modified** tests for **Inclusion** students with one mouse click.
- **Edit** existing questions and **Add** your own questions.
- Build tests aligned with state standards using built-in **State Curriculum Correlations.**
- Change **English** tests to **Spanish** with one mouse click and vice versa.

FOLDABLES™ Follow-Up

Before students comple the Mid-Chapter Practice Test, encourage them to review the notes, definitions, concepts, and main ideas they recorded under the first four tabs of their Foldables.

Answers

1. Sample answer: the number of pieces of data that fall within given intervals
2. Sample answer: predicting profit earned by a company for the next year
3. Sample answer: the greatest and least values, clusters of data, gaps in data, how the data are spread out, the range, and the pieces of data that appear most often

Chapter 2 Mid-Chapter Practice Test

Vocabulary and Concepts

1. **Explain** what a frequency table shows. (Lesson 2-1) **1–3. See margin.**
2. **Describe** a situation in which a line graph would be useful. (Lesson 2-2)
3. **List** the information that you can identify from a line plot that is not obvious in an unorganized list of data. (Lesson 2-3)
4. **Define** *mean*. (Lesson 2-4) **sum of data divided by the number of items in the data set**

Skills and Applications

For Exercises 5–10, use the data below that shows the ages of twenty people when they got their driver's licenses. (Lessons 2-1, 2-3, and 2-4) **5–6. See margin.**

16 17 16 16 18 21 16 16 18 18 17 25 16 17 17 17 17 16 20 16

5. Make a frequency table of the data.
6. Make a line plot of the data.
7. Identify any clusters or gaps. **Sample answer: cluster 16–18; gap 21–25**
8. Identify any outliers. **25**
9. Find the mean, median, and mode. **17.5; 17; 16**
10. What is the range of the data? **9**
11. **HEATING** Use the scatter plot to make a conclusion about the average monthly outside temperature and heating costs of a home. (Lesson 2-2) **Sample answer: As the temperature goes down, the heating costs rise.**

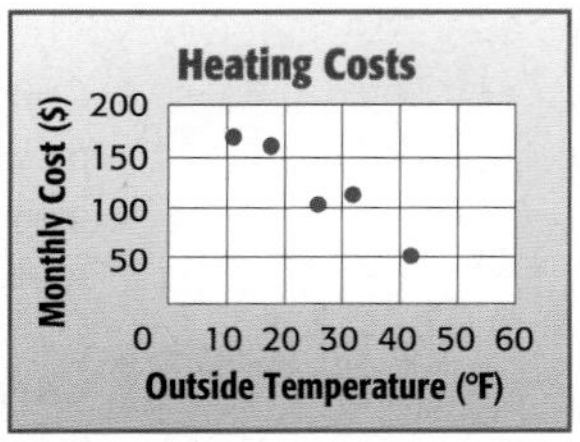

Find the mean, median, and mode for each set of data. (Lesson 2-4)

12. 27, 32, 18, 24, 32, 39, 41 **30.4; 32; 32**
13. 7.6, 6.8, 6.5, 7, 7.9, 6.8, 7, 7, 6.8, 8.1 **7.15; 7; 6.8 and 7**

Standardized Test Practice

14. **GRID IN** Below are the quiz scores of a science class.

    89 92 67 75 95 89 82 92 89
    89 92 96 85 94 90 79 94 87

    What is the median quiz score? (Lesson 2-4) **89**

15. **MULTIPLE CHOICE** Which may *not* show individual pieces of data? (Lessons 2-1, 2-2, and 2-3) **B**

    (A) scatter plot
    (B) frequency table
    (C) line plot
    (D) an ordered list

5.

Age (yr)	Tally	Frequency
16–17	~~IIII~~ ~~IIII~~ IIII	14
18–19	III	3
20–21	II	2
22–23		
24–25	I	1

6.

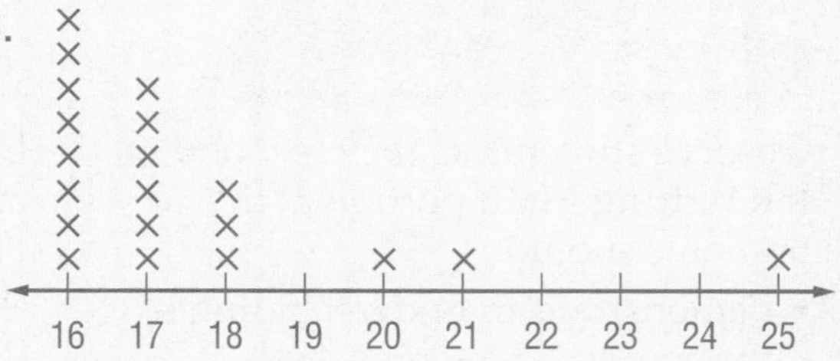

The Game Zone

A Place To Practice Your Math Skills

Can You Guess?

- **GET READY!**

Players: the entire class divided into three-person teams
Materials: cup, coins, water dropper, self-adhesive notes

- **GET SET!**

- Draw a number line on the chalkboard with a scale from 10 to 25.
- Each team should fill a cup with water.

- **GO!**

- Use the water dropper and find how many drops of water your team can get to stay on the head of a dime. Write the results on a sticky note. Post it on the chalkboard above the corresponding number on the number line. Find the mean of the numbers.

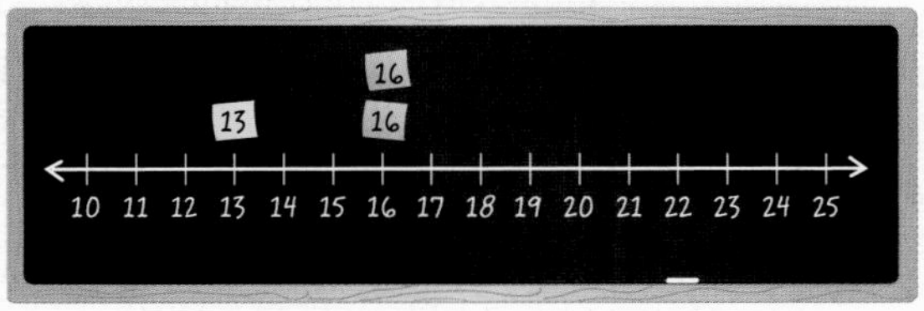

- Next, use the mean to predict how many drops of water will fit on the head of a penny. Record your team's prediction on a piece of paper.
- Find how many drops of water will stay on the head of a penny. Record the result on a sticky note and post it on the chalkboard. Find the mean of all of the results for the pennies. If your prediction is within three drops of the mean, your team gets to stay in the game.
- Continue predicting and testing using a nickel and a quarter.
- **Who Wins?** The teams left at the end of the game are the winners.

The Game Zone

Glencoe Mathematics Classroom Manipulative Kit
- cups

GETTING READY

Objective Use a number line to find the mean.

MANAGEMENT TIPS

Recommended Time
20 minutes

Tips
- Have students stay at their seats while you deliver the cups of water.
- You may want to have students try each coin three times and take the average of their results.
- Note that depending on the size of the droppers, you may need to adjust the scale of the number line to include 40 or more.
- You may wish to use the game as an assessment tool for the concepts the students have been learning in Lessons 2-1 through 2-4.

For an additional game for this chapter, visit: msmath2.net/gamezone.

2-5 Lesson Notes

Lesson Resources

Chapter 2 Resource Masters
- Study Guide and Intervention, p. 87
- Practice: Skills, p. 88
- Practice: Word Problems, p. 89
- Reading to Learn Mathematics, p. 90
- Enrichment, p. 91

Key Teaching Transparencies, 8
Parent and Student Study Guide Workbook, 2-5

1 Focus

5-Minute Check 2-5
Use as a quiz or review of Lesson 2-4. (Also available as a transparency)

INTERACTIVE CHALKBOARD PowerPoint® Presentations

Bellringer
Hands-On Activity Provide students with a list of data with some values in the tens and some values in the hundreds. Have students search for the least value and the greatest value and circle them. Then have them highlight the least place value of each. **the ones place** Tell students that this lesson presents a tool for organizing lists of data visually.

Reading to Learn Mathematics, p. 90 **ELL**

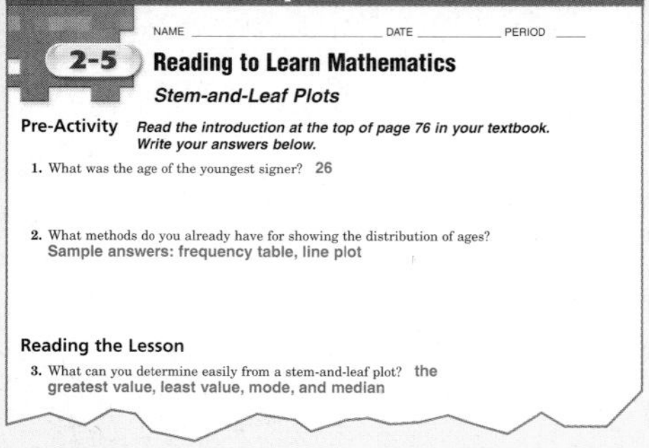
NAME ______ DATE ______ PERIOD ____
2-5 Reading to Learn Mathematics
Stem-and-Leaf Plots
Pre-Activity *Read the introduction at the top of page 76 in your textbook. Write your answers below.*
1. What was the age of the youngest signer? 26
2. What methods do you already have for showing the distribution of ages? Sample answers: frequency table, line plot
Reading the Lesson
3. What can you determine easily from a stem-and-leaf plot? the greatest value, least value, mode, and median

2-5 Stem-and-Leaf Plots

What You'll Learn
Construct and interpret stem-and-leaf plots.

NEW Vocabulary
- stem-and-leaf plot
- leaf
- stem

WHEN am I ever going to use this?

HISTORY Benjamin Franklin was the oldest signer of the Declaration of Independence. The table shows the ages in years of all the people who signed the Declaration of Independence.

Ages of Signers of Declaration of Independence

40	70	69	34	46	53	31
38	50	30	45	46	39	38
35	50	37	48	41	41	52
39	50	65	46	29	34	45
33	41	44	63	60	26	42
42	52	37	35	45	35	42
47	26	55	57	45	33	60
62	35	33	53	49	50	46

Source: *The World Almanac*

1. What was the age of the youngest signer? **26**
2. What methods do you already have for showing the distribution of ages?
Sample answers: frequency table, line plot

A useful way to organize data as you collect it is a stem-and-leaf plot. In a **stem-and-leaf plot**, the data are organized from least to greatest. The digits of the least place value usually form the **leaves**, and the next place value digits form the **stems**.

TEACHING TIP
There should be at least one leaf for the first stem and one for the last stem. There may be gaps in between where some stems have no leaves, but do not leave out these stems.

EXAMPLE Construct a Stem-and-Leaf Plot

1 HISTORY Make a stem-and-leaf plot of the data above.

Step 1 The digits in the least place value will form the leaves and the remaining digits will form the stems. In this data, 26 is the least value, and 70 is the greatest. So, the ones digits will form the leaves and the tens digits will form the stems.

Step 2 List the stems 2 to 7 in order from least to greatest in the *Stem* column. Write the leaves, the ones digits of the ages, to the right of the corresponding stems. The ages in the first column of the chart are shown at the right.

Stem	Leaf
2	
3	8 5 9
4	0 2 7
5	
6	2
7	

Step 3 Order the leaves and write a *key* that explains how to read the stems and leaves.

Stem	Leaf
2	6 6 9
3	0 1 3 3 3 4 4 5 5 5 5 7 7 8 8 9 9
4	0 1 1 1 2 2 2 4 5 5 5 5 6 6 6 6 7 8 9
5	0 0 0 0 2 2 3 3 5 7
6	0 0 2 3 5 9
7	0

3|1 = 31 yr

The tens digits of the data form the stems.

A key shows how the digits are related.

The ones digits of the data form the leaves.

TEACHING TIP

enerally, diagrams that re called *plots* show dividual data points ather than bars or lines.

Stem-and-leaf plots are useful in analyzing data because you can see all the data values, including the greatest, least, mode, and median value.

EXAMPLE Analyze Data

2 PETS The stem-and-leaf plot shows the average chick weight in grams of sixteen different species of pet birds. Find the range, median, and mode of the data.

Stem	Leaf
0	5 6 7
1	0 0 1 1 2 2 2 3 8 9
2	0 1 1 5 *1|2 = 12 g*

Source: www.upatsix.com

range: greatest weight − least weight = 25 − 5 or 20 grams

median: middle value, or 12 grams

mode: most frequent value, or 12 grams

EAL-IFE CAREERS

ow Does a Nutritionist se Math?

utritionists use statistical aphs to summarize and mpare the nutritional lues of various foods.

Online Research

r information about a career a nutritionist, visit: math2.net/careers

Stem-and-leaf plots are also useful for making conclusions about data. Most of the data in the plot above are clustered in the 10–19 interval. So, you could say that most of those birds weigh 10 to 19 grams as chicks.

EXAMPLE Make Conclusions About Data

3 NUTRITION The table shows the average amounts of pasta that people in different countries consume each year. Make a stem-and-leaf plot of the data. Then use it to describe how the data are spread out.

Country	Pasta per Person (lb)	Country	Pasta per Person (lb)
Argentina	15	Netherlands	9
Australia	5	Portugal	15
Canada	14	Russia	15
Egypt	15	Spain	9
France	15	Sweden	8
Germany	10	Switzerland	20
Greece	19	Turkey	11
Italy	59	United Kingdom	4
Japan	3	United States	20
Mexico	8	Venezuela	28

Source: National Pasta & National Restaurant Associations

The least value is 3, and the greatest value is 59. So, the tens digits form the stems, and the ones digits form the leaves.

Stem	Leaf
0	3 4 5 8 8 9 9
1	0 1 4 5 5 5 5 5 9
2	0 0 8
3	
4	
5	9 *2|8 = 28 lb*

The data that do not have a digit in the tens place have 0 for a stem.

It is easy to see in the stem-and-leaf plot that people in most of the countries consume less than 20 pounds of pasta per year. You can also see that the mode is 15 and that 59 is an outlier, with a significant gap between that value and the next smaller value.

TEACHING TIP

o find how many ems there are in a tem-and-leaf plot, ask tudents whether they hould count the stems, e leaves, or both. nswer: the leaves

msmath2.net/extra_examples

2 Teach

In-Class Examples

INTERACTIVE CHALKBOARD PowerPoint® Presentations

1 BASEBALL The table below shows the number of home runs that Babe Ruth hit during his career from 1914 to 1935. Make a stem-and-leaf plot of the data.

Home Runs			
0	54	25	46
4	59	47	41
3	35	60	34
2	41	54	6
11	22	46	
29	46	49	

Source: baberuth.com

Stem	Leaf
0	0 2 3 4 6
1	1
2	2 5 9
3	4 5
4	1 1 6 6 6 7 9
5	4 4 9
6	0

2\|5 = 25 home runs

2 FITNESS The stem-and-leaf plot below shows the number of miles that Megan biked each day during July. Find the range, median, and mode of the data. **25; 12; 10**

Stem	Leaf
0	5 5 5 6
1	0 0 0 0 1 2 2 5 8 8 9
2	1 2 5 8
3	0

2\|5 = 25 miles

DAILY INTERVENTION Differentiated Instruction

Interpersonal For Examples 1 and 2, pair students together to create the stem-and-leaf plots. They should discuss how to set up each plot. Many students find it beneficial to have one partner read out the data points while the other partner writes them down in the plot.

In-Class Example

3 ANIMALS The table in the bottom margin shows the average life span of several animals. Make a stem-and-leaf plot of the data. Then use it to describe how the data are spread out.

Stem	Leaf
0	3 4 6 8
1	0 0 2 2 2 5 5 6 8
2	0 0 0 0
3	
4	0

1|0 = 10 years

Most animals live less than 20 years. The mode is 20. Forty is an outlier with a significant gap between that value and the next smaller value.

3 Practice/Apply

DAILY INTERVENTION

FIND THE ERROR For Exercise 4, remind students that a 0 must be recorded in the leaf column for any multiple of 10.

About the Exercises...

Guided Practice Key

Exercises	Examples
5, 6	1
7–9	2
10	3

Odd/Even Assignments
Exercises 11–18 are structured so that students practice the same concepts whether they are assigned odd or even problems.

Alert! Exercise 22 requires research on the Internet or another source.

Skill and Concept Check

Writing Math
Exercises 1 & 4

1. **Compare and contrast** a stem-and-leaf plot and a frequency table. **1–2. See Ch. 2 Answer Appendix.**
2. **OPEN ENDED** Make a stem-and-leaf plot in which the median of the data is 25.
3. **Write** the stems that would be used in a stem-and-leaf plot for the following set of data: 21, 8, 12, 7, 18, 9, 22, 31, 1, 14, 19, 11, 2. **0, 1, 2, 3**
4. **FIND THE ERROR** Selena and Lauren are analyzing the data in the stem-and-leaf plot at the right. Who is correct? Explain.

Stem	Leaf
2	6 6 9
3	
4	6
5	3 6

2|6 = 26 in.

Selena
None of the data have 3 in the tens place.

Lauren
Thirty is one of the data values.

Selena; 3|0 = 30, but the stem 3 has no leaves.

GUIDED PRACTICE

Make a stem-and-leaf plot for each set of data. **5–6. See Ch. 2 Answer Appendix.**

5. 42, 47, 19, 16, 21, 23, 25, 25, 29, 31, 33, 34, 35, 39, 48
6. 225, 227, 230, 229, 246, 243, 269, 269, 267, 278, 278, 278

CHESS For Exercises 7–10, use the stem-and-leaf plot at the right. It shows the number of matches won by each member of the Avery Middle School Chess Team.

Stem	Leaf
0	8 8 9
1	9
2	0 0 2 4 4 8 9
3	1 1 2 4 5 5 6 6 7 7
4	0 0 0 3 8 9
5	2 4
6	1

3|2 = 32 wi

7. How many members are on the chess team? **31**
8. What is the range of the number of matches won? **53**
9. Find the mean, median, and mode for these data. **32.9; 35; 40**
10. Describe how the data are spread out. **See Ch. 2 Answer Appendix.**

★ indicates increased difficulty

Practice and Applications

HOMEWORK HELP

For Exercises	See Examples
11–14, 19	1
15–17, 20	2
18, 21	3

Extra Practice
See pages 568, 597.

Make a stem-and-leaf plot for each set of data. **11–14. See Ch. 2 Answer Appendix.**

11. 15, 13, 28, 32, 38, 30, 31, 13, 36, 35, 38, 32, 38, 24
12. 56, 70, 96, 62, 80, 80, 69, 93, 66, 55, 95, 63, 90, 93, 60, 91, 67, 60
13. 451, 458, 439, 403, 411, 434, 433, 454, 457, 442, 449, 459, 454
14. ★ 8.2, 9.3, 10, 9.9, 8, 9.2, 8.7, 8, 8.2, 9, 9.9, 8.7, 8.5, 8.1, 8.8, 9.3

SHOPPING For Exercises 15–18, use the stem-and-leaf plot at the right. It shows the prices of portable CD players found on an Internet shopping site.

Stem	Leaf
7	0
8	2 5 5
9	9 9
10	0 0 2 5 6 8
11	0 0 5 5 5 9 9
12	5 7 7
13	
14	
15	0 0 0 5

11|5 = $115

15. How many CD players are represented on the stem-and-leaf plot? **26**
16. What is the range of the prices? **$85**
17. Find the median and mode for these data. **$110; $115 and $150**
18. Write a sentence or two to describe how the prices of the portable CD players are spread out. **See Ch. 2 Answer Appendix.**

3

Animal	Years	Animal	Years	Animal	Years
Baboon	20	Chipmunk	6	Guinea Pig	4
Black Bear	18	Cow	15	Horse	20
Polar Bear	20	Deer	8	Mouse	3
Camel	12	Dog	12	Squirrel	10
Cat	12	Elephant	40	Tiger	16
Chimpanzee	20	Giraffe	10	Zebra	15

Source: *The World Almanac*

BASKETBALL For Exercises 19–21, use the table at the right. It shows the points scored in each post-season tournament game by the Indiana Hoosiers men's basketball teams from 1990 to 2002.

Points Scored in Post-Season Games					
63	79	82	65	94	52
106	78	97	73	82	
67	68	60	51	62	
85	84	68	61	89	
108	77	94	57	73	
75	76	74	81	73	

Source: www.iuhoosiers.com

19. Make a stem-and-leaf-plot of the data.
20. Find the mean, median, mode, and range of the data. 75.9; 75; 73; 57
21. Write a sentence or two to describe how the points shown in the stem-and-leaf plot are spread out.

19, 21. See Ch. 2 Answer Appendix.

22. **RESEARCH** Use the Internet or another source to find an interesting set of data that can be displayed using a stem-and-leaf plot. Then make a stem-and-leaf plot of the data. See students' work.
23. **CRITICAL THINKING** Make a frequency table, a line plot, and a stem-and-leaf plot of the data at the right. Describe the similarities and differences among the representations and state which of the three you prefer to make and use. Explain your reasoning. See Ch. 2 Answer Appendix.

Price of Jeans ($)			
40	45	38	30
35	32	33	24
26	36	56	36
26	38	49	34
28	40	40	35

Standardized Test Practice and Mixed Review

24. **MULTIPLE CHOICE** The table shows how many minutes the students in Mr. Davis' class spent doing their homework one evening. Choose the correct stems for a stem-and-leaf plot of the data. B

Homework Time (min)				
42	5	75	30	55
45	47	0	24	75
45	51	56	23	45
39	30	49	58	35

Ⓐ 1, 2, 3, 4, 5, 6, 7 Ⓑ 0, 1, 2, 3, 4, 5, 6, 7
Ⓒ 0, 2, 3, 4, 5, 7 Ⓓ 2, 3, 4, 5, 6, 7

25. **GRID IN** Determine the mode of the data shown in the stem-and-leaf plot at the right. 97

Stem	Leaf
7	2 3 5 5
8	0 0 0 1 2
9	6 7 7 7 7

7|2 = 72

Find the mean, median, and mode for each set of data. (Lesson 2-4)

26. 80, 23, 55, 58, 45, 32, 40, 55, 50 48.7; 50; 55
27. 3.6, 2.4, 3.0, 7.9, 7.8, 2.4, 3.6, 3.9 4.3; 3.6; 3.6 and 2.4
28. Make a line plot of the following test scores. (Lesson 2-3)
83, 94, 78, 78, 85, 86, 88, 83, 82, 92, 90, 77, 83, 81, 89, 90, 88, 87, 88, 85, 84, 81, 83, 85, 91 See Ch. 2 Answer Appendix.

Evaluate each expression. (Lesson 1-3)

29. 18 + 9 ÷ 3 21
30. 24 ÷ (6 + 2) 3
31. 3(7 − 3) + 1 13
32. (3 + 5) ÷ 4 − 2 0

GETTING READY FOR THE NEXT LESSON

PREREQUISITE SKILL Order each set of decimals from least to greatest. (Page 556)

33. 6, 8.4, 8, 7.9 6, 7.9, 8, 8.4
34. 4.0, 2.3, 4.5, 3.0, 3.8 2.3, 3.0, 3.8, 4.0, 4.5
35. 0.14, 0.33, 0.7, 0.09 0.09, 0.14, 0.33, 0.7
36. 5.01, 5.8, 4.99, 5.64, 4.25 4.25, 4.99, 5.01, 5.64, 5.8

4 Assess

Open-Ended Assessment

Writing Have students write a question that can be answered using a stem-and-leaf plot. Students should write out an answer to the question explaining how the plot is used.

Study Guide and Intervention, p. 87

NAME ______ DATE ______ PERIOD ___

2-5 Study Guide and Intervention

Stem-and-Leaf Plots

In a **stem-and-leaf plot**, the data are organized from least to greatest. The digits of the least place value usually form the **leaves**, and the next place value digits form the **stems**.

EXAMPLE 1 **Make a stem-and-leaf plot of the data below. Then find the range, median, and mode of the data.**
42, 45, 37, 46, 35, 49, 47, 35, 45, 63, 45

Order the data from least to greatest.

35, 35, 37, 42, 45, 45, 45, 46, 47, 49, 63

The least value is 35, and the greatest value is 63. So, the tens digits form the stems, and the ones digits form the leaves.

Stem	Leaf
3	5 5 7
4	2 5 5 5 6 7 9
5	
6	3

6|3 = 63

range: greatest value − least value = 63 − 35 or 28
median: middle value, or 45
mode: most frequent value, or 45

EXERCISES

Make a stem-and-leaf plot for each set of data. Then find the range, median, and mode of the data.

1. 15, 25, 16, 28, 1, 27, 16, 19, 28

Stem	Leaf
0	1
1	5 6 6 9
2	5 7 8 8

2|5 = 25

The range is 27. The median is 19. The mode is 16 and 28.

2. 1, 2, 3, 2, 3, 1, 4, 2, 5, 7, 12, 11, 11, 3, 10

Stem	Leaf
0	1 1 2 2 2 3 3 3 4 5 7
1	0 1 1 2

1|0 = 10

The range is 11. The median is 3. The mode is 2 and 3.

3. 3, 5, 1, 17, 11, 45, 17

Stem	Leaf
0	1 3 5
1	1 7 7
2	
3	
4	5

4|5 = 45

The range is 44. The median is 11. The mode is 17.

4. 4, 7, 10, 5, 8, 12, 7, 6

Stem	Leaf
0	4 5 6 7 7 8
1	0 2

1|0 = 10

The range is 8. The median is 7. The mode is 7.

Practice: Skills, p. 88

NAME ______ DATE ______ PERIOD ___

2-5 Practice: Skills

Stem-and-Leaf Plots

Make a stem-and-leaf plot for each set of data.

1. 23, 36, 25, 13, 24, 25, 32, 33, 17, 26, 24

Stem	Leaf
1	3 7
2	3 4 4 5 5 6
3	2 3 6

3|2 = 32

2. 3, 4, 6, 17, 12, 5, 17, 4, 26, 17, 18, 21, 16, 15, 20

Stem	Leaf
0	3 4 4 5 6
1	2 5 6 7 7 7 8
2	0 1 6

2|0 = 20

3. 26, 27, 23, 23, 24, 26, 31, 45, 33, 32, 41 40, 21, 20

Stem	Leaf
2	0 1 3 3 4 6 6 7
3	1 2 3
4	0 1 5

4|0 = 40

4. 347, 334, 346, 330, 348, 347, 359, 344, 357

Stem	Leaf
33	0 4
34	4 6 7 7 8
35	7 9

35|7 = 357

HOT DOGS **For Exercises 5–7, use the stem-and-leaf plot at the right that shows the number of hot dogs eaten during a contest.**

Stem	Leaf
0	8 8 9
1	1 2 2 4 7 7 7
2	1 1 2

2|1 = 21

5. How many hot dogs are represented on the stem-and-leaf plot? 189
6. What is the range of the number of hot dogs eaten? 14
7. Find the median and mode of the data. 14; 17

Determine the mean, median, and mode of the data shown in each stem-and-leaf plot.

8.

Stem	Leaf
0	1 2 2 3
1	3 4 5 5
2	0 0 0 1 3

2|0 = 20

13; 15; 20

9.

Stem	Leaf
2	0 0 0 2 3 5 7
3	1 2
4	0

4|0 = 40

26; 24; 20

10.

Stem	Leaf
22	1 1 2 7
23	3 3 9
24	0 6 8

24|0 = 240

233; 233; 233

11.

Stem	Leaf
0	1 3 3 4 7
1	2 2 2 4 5 6
2	0 0 0 1

2|0 = 20

12; 12; 12 and 20

Practice: Word Problems, p. 89

NAME ______ DATE ______ PERIOD ___

2-5 Practice: Word Problems

Stem-and-Leaf Plots

ENDANGERED SPECIES **For Exercises 1–6, use the table below. It shows the number of endangered species in the U.S.**

Endangered Species in U.S.			
Group	Number of Species	Group	Number of Species
mammals	63	clams	61
birds	78	snails	20
reptiles	14	insects	33
amphibians	10	arachnids	12
fishes	70	crustaceans	18

1. Make a stem-and-leaf plot of the data.

Stem	Leaf
1	0 2 4 8
2	0
3	3
4	
5	
6	1 3
7	0 8

6|1 = 61

2. What group has the greatest number of endangered species in the U.S.? birds
3. What group has the least number of endangered species in the U.S.? amphibians
4. What is the range of the data? 68
5. Use your stem-and-leaf plot to determine the median and mode. median: 26.5; mode: none
6. How many groups have less than 30 endangered species in the U.S.? 5

2-6 Lesson Notes

Lesson Resources

Chapter 2 Resource Masters
- Study Guide and Intervention, p. 92
- Practice: Skills, p. 93
- Practice: Word Problems, p. 94
- Reading to Learn Mathematics, p. 95
- Enrichment, p. 96

FAST FILE

Calculator and Spreadsheet Masters, p. 19
MathPASS Tutorial CD-ROM, 58
Parent and Student Study Guide Workbook, 2-6
Teaching Mathematics with Manipulatives, pp. 39–40

1 Focus

5-Minute Check 2-6

INTERACTIVE CHALKBOARD
PowerPoint® Presentations

Use as a quiz or review of Lesson 2-5. (Also available as a transparency)

Bellringer

Reading Mathematics Write the following words on the chalkboard: *quart, quarter, quarter note, quarterly, quarterfinal.* If students are not familiar with any of these words, discuss them in class, then have students decide how these words are related. Have them use a dictionary as needed. Is there a mathematics connection to each word?

Reading to Learn Mathematics, p. 95 **ELL**

NAME ______________ DATE ________ PERIOD ____

2-6 Reading to Learn Mathematics
Box-and-Whisker Plots

Pre-Activity *Read the introduction at the top of page 80 in your textbook. Write your answers below.*

1. What is the median of the data? 175

2. Into how many parts does the median divide the data? 2 parts

Reading the Lesson

3. What do the terms *lower quartile, upper quartile,* and *interquartile range* have to do with the word *quarter?* Sample answer: Each quartile contains one-fourth (a quarter) of the data.

2-6 Box-and-Whisker Plots

What You'll Learn
Construct and interpret box-and-whisker plots.

NEW Vocabulary
- lower quartile
- upper quartile
- box-and-whisker plot
- lower extreme
- upper extreme
- interquartile range

REVIEW Vocabulary
outlier: a number separated from the rest of the data (Lesson 2-3)

WHEN am I ever going to use this?

NUTRITION The Calories per serving of items from the meat, poultry, and fish food group are shown in the table.

1. What is the median of the data? **175**
2. Into how many parts does the median divide the data? **2 parts**

Nutrition Facts

Item	Calories	Item	Calories
bacon	110	ham	205
beef steak	240	pork chop	275
bologna	180	roast beef	135
crabmeat	135	salmon	120
fish sticks	70	sardines	175
fried shrimp	200	trout	175
ground beef	245	tuna	165

Source: *The World Almanac*

A median divides a data set into two parts. To divide the data into four parts, find the median of the lower half, called the **lower quartile (LQ)** and the median of the upper half, called the **upper quartile (UQ)**. A **box-and-whisker plot** is a diagram that summarizes data by dividing it into four parts.

EXAMPLE Construct a Box-and-Whisker Plot

1 **NUTRITION** **Make a box-and-whisker plot of the data in the table above.**

Step 1 Order the data from least to greatest.

Step 2 Find the median and the quartiles.

70, 110, 120, (135), 135, 165, 175, 175, 180, 200, (205), 240, 245, 275

median: 175
lower quartile: median of lower half = 135
upper quartile: median of upper half = 205

Step 3 Draw a number line. The scale should include the median, the quartiles, and the least and greatest values, which are called the **lower extreme** and the **upper extreme**, respectively. Graph the values as points above the line.

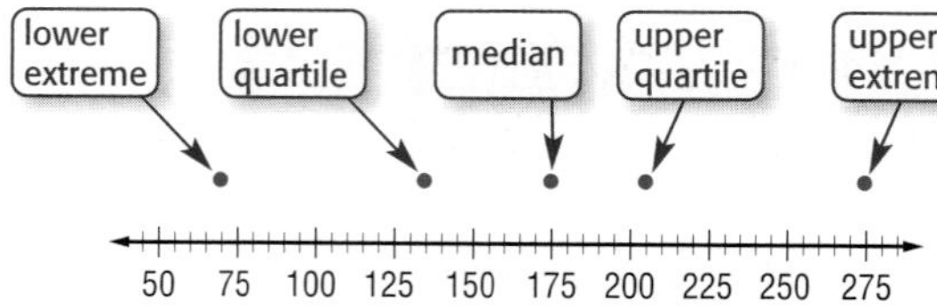

TEACHING TIP
Point out that a box-and-whisker plot does not display all individual data values.

DAILY INTERVENTION **Unlocking Misconceptions**

Outliers Make sure students understand that upper and lower extremes are not necessarily outliers. These values (and others) are only considered outliers if they fall beyond the upper and lower limits. In a box-and-whisker plot, this means occurring more than 1.5 times the interquartile range away from the quartiles.

Step 4 Draw the box and whiskers.

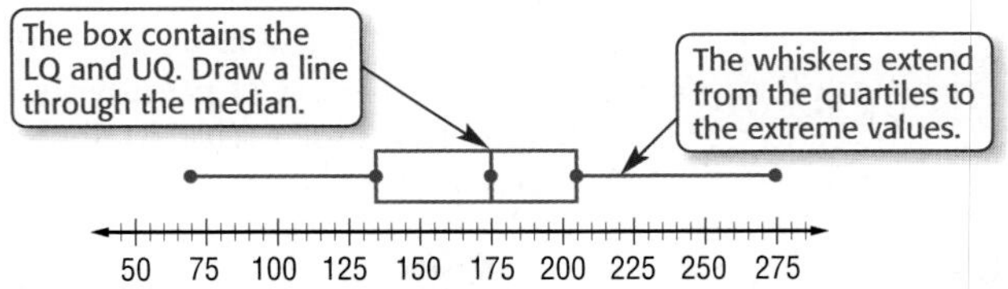

Each of the four parts of a box-and-whisker plot contain one fourth of the data.

STUDY TIP

Length of Parts The parts of a box-and-whisker plot are different lengths because of how the data are clustered or spread out.

- short part: data are clustered together
- long part: data are spread out

TEACHING TIP

emind students that the rst step in making a ox-and-whisker plot is lways to order the data.

EXAMPLE Analyze Data

2 ARCHITECTURE The table shows the heights of the ten tallest buildings in St. Louis, Missouri. Make a box-and-whisker plot of the data. Then use it to describe how the data are spread out.

Building	Height (ft)
Metropolitan Square Tower	593
One Bell Center	588
Thomas F. Eagleton Federal Courthouse	557
Mercantile Center Tower	540
Firstar Center	485
Boatmen's Plaza	420
Laclede Gas Building	400
SW Bell Telephone Building	398
Civil Courts Building	390
One City Center	375

Source: *The World Almanac*

Find the median, the quartiles, and the extremes. Then construct the plot.

median = $\frac{485 + 420}{2}$ or 452.5

LQ = 398 UQ = 557

lower extreme = 375

upper extreme = 593

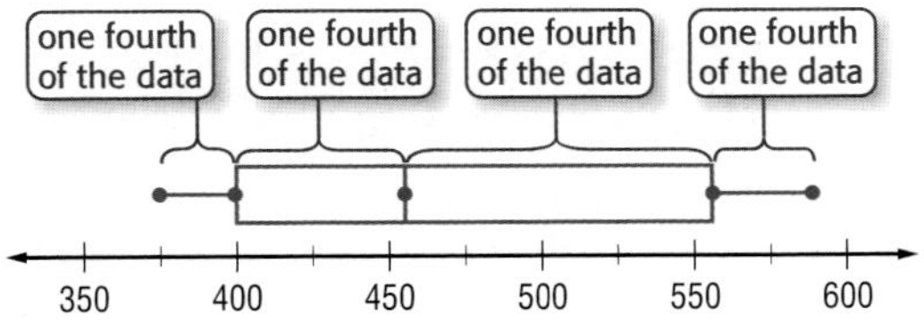

The graph shows that half of the tallest buildings are between 398 and 557 feet high. The largest range of the four quartiles is from 452.5 to 557 feet. One-fourth of the tallest buildings are within these heights.

READING Math

Quartiles A *quartile* can also refer to any of the four parts of a box-and-whisker plot. So, each quartile contains one-fourth, or 25%, of the data.

Another way to describe how data are spread out is to use the **interquartile range**, or the difference between the upper quartile and the lower quartile.

In Lesson 2-3, you learned that outliers are values separated from the rest of the data. In a box-and-whisker plot, outliers are data that are more than 1.5 times the interquartile range from the quartiles. They are not included in the whiskers of a box-and-whisker plot.

msmath2.net/extra_examples

2

NHL Leading Scorers			
Player	**Goals**	**Player**	**Goals**
Wayne Gretzky	894	Steve Yzerman	645
Gordie Howe	801	Phil Esposito	717
Marcel Dionne	731	Ray Bourque	410
Mark Messler	627	Mario Lemieux	648
Ron Francis	487	Paul Coffey	396

Source: *The World Almanac*

2 Teach

In-Class Examples

INTERACTIVE CHALKBOARD PowerPoint® Presentations

1 NUTRITION The grams of fat per serving of items from the meat, poultry, and fish food group are shown in the table. Make a box-and-whisker plot of the data.

Nutrition Facts	
Item	**Fat (gm)**
Bacon	9
Beefsteak	15
Bologna	16
Crabmeat	3
Fish sticks	3
Fried shrimp	10
Ground beef	18
Ham	14
Pork chop	19
Roast beef	5
Salmon	5
Sardines	9
Trout	9
Tuna	7

Source: *The World Almanac*

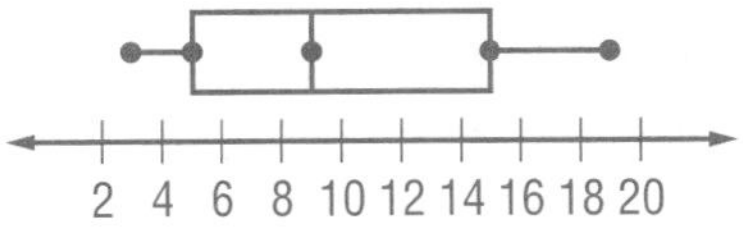

2 HOCKEY The table in the bottom margin shows the ten all-time leading scorers in the National Hockey League through a recent season. Make a box-and-whisker plot of the data. Then use it to describe how the data are spread. **Half of the players scored between 487 and 731 points. The largest range of the quartiles is from 731 to 894. One-fourth of the players scored within this range.**

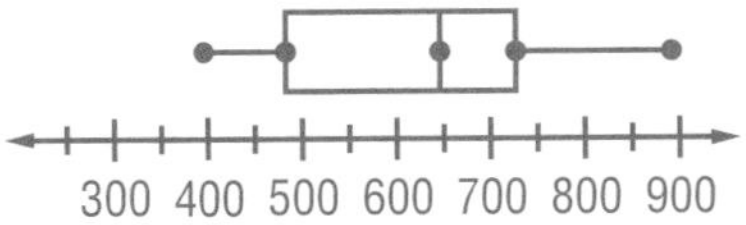

In-Class Example

3 **CANDY SALES** INTERACTIVE CHALKBOARD PowerPoint® Presentations

Twelve members of the music club sold candy bars as a fund-raiser. The table shows the number of candy bars sold by each person. Make a box-and-whisker plot of the data.

Candy Sold per Student			
23	69	27	60
51	46	81	53
46	54	39	55

See bottom margin.

3 Practice/Apply

About the Exercises...

Guided Practice Key

Exercises	Examples
5	1
3, 6	2
4	3

Odd/Even Assignments
Exercises 7–13 are structured so that students practice the same concepts whether they are assigned odd or even problems.

4 Assess

Higher-Level Thinking
Ask students to create a data set and draw a box-and-whisker plot for the following conditions: there are 6 data values with a median value of 20, a lower extreme at 8, and an outlier at 18.

REAL-LIFE MATH

PRESIDENTS The state where the greatest number of presidents was born is Virginia, with 8 presidents.

Source: www.learningnetwork.com

George Washington — Woodrow Wilson

TEACHING TIP
When there is an odd number of data, as in Example 3, students should not use the original median to determine the LQ and UQ.

EXAMPLE Identify and Plot Outliers

3 **PRESIDENTS** **The table shows the ages of the last 21 presidents when they first took office. Make a box-and-whisker plot of the data.**

Ages of Presidents						
47	55	54	42	51	56	55
51	54	51	60	62	43	55
56	61	52	69	64	46	54

Find the median and the quartiles. Plot the points and draw a box.

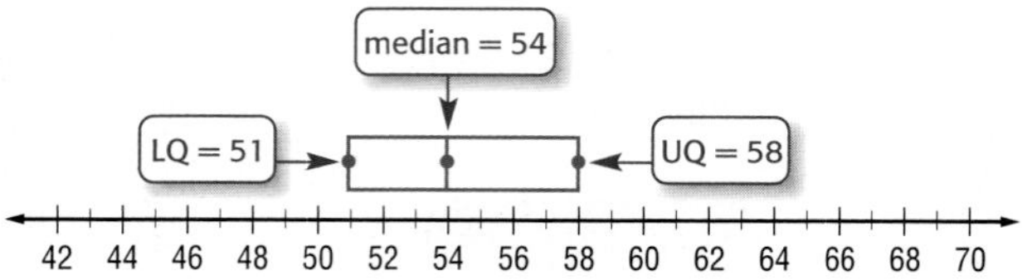

Next, determine whether there are any outliers.

interquartile range: UQ − LQ = 58 − 51 or 7

So, outliers are data more than 1.5(7) or 10.5 from the quartile.

lower limit: LQ − 10.5 = 51 − 10.5 or 40.5
upper limit: UQ + 10.5 = 58 + 10.5 or 68.5

← 40.5 and 68.5 are limits for the outliers

Any data point that is less than 40.5 or greater than 68.5 is an outlier. So, 69 is an outlier. Plot the outlier using a dot. Then draw the lower whisker to the lower extreme, 42, and the upper whisker to the greatest value that is not an outlier, 64.

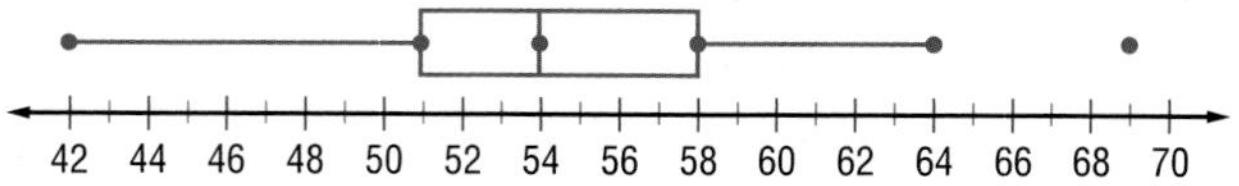

Skill and Concept Check

Writing Math Exercises 1 & 2

1. **Describe** advantages of using a stem-and-leaf plot and advantages of using a box-and-whisker plot to display data. **1–2. See Ch. 2 Answer Appendix.**
2. **OPEN ENDED** Draw a box-and-whisker plot in which the data between the lower quartile and the median are clustered and the data between the median and the upper quartile are spread out.

GUIDED PRACTICE

For Exercises 3–6, refer to the data at the right. It gives the number of Calories in common snacks you can buy at the movies. **3. 270; 350; 380; 588; 1,221**

3. Find the lower extreme, LQ, median, UQ, and upper extreme.
4. Are there any outliers? If so, identify them. **yes; 1,221**
5. Draw a box-and-whisker plot of the data. **See Ch. 2 Answer Appendix.**
6. What is the largest range of the four quartiles? Explain what this means. **From 588 to 901; the data in this range are very spread out.**

Movie Snack Calories		
380	286	588
350	1,221	901
492	360	500
363	270	

3

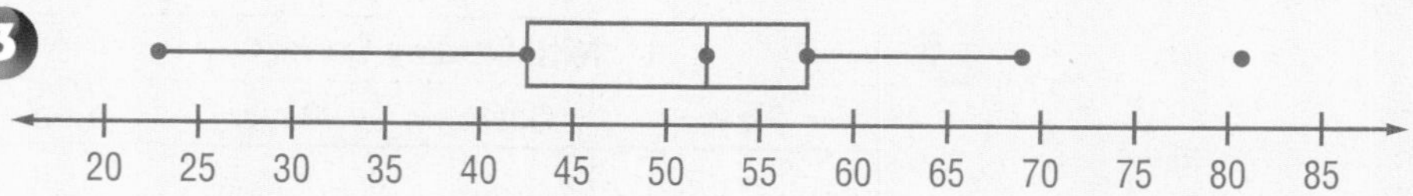

82 Chapter 2 Statistics: Analyzing Data

★ indicates increased difficulty

Practice and Applications

HOMEWORK HELP	
For Exercises	See Examples
7, 8, 12	1
9, 13	2
10, 11	3

Extra Practice
See pages 568, 597.

SPORTS For Exercises 7–9, refer to the table at the right. It shows the regular season games won by each men's professional basketball team in a recent season. 7. 15; 30.5; 45; 51.5; 58

Number of Wins					
25	36	46	15	30	53
40	32	17	45	41	31
56	50	52	47	26	48
43	56	51	50	55	58
44	47	53	23	19	

7. Find the lower extreme, LQ, median, UQ, and upper extreme.
8. Draw a box-and-whisker plot of the data. See Ch. 2 Answer Appendix.
9. What fraction of the data are between 45 and 51.5? one fourth

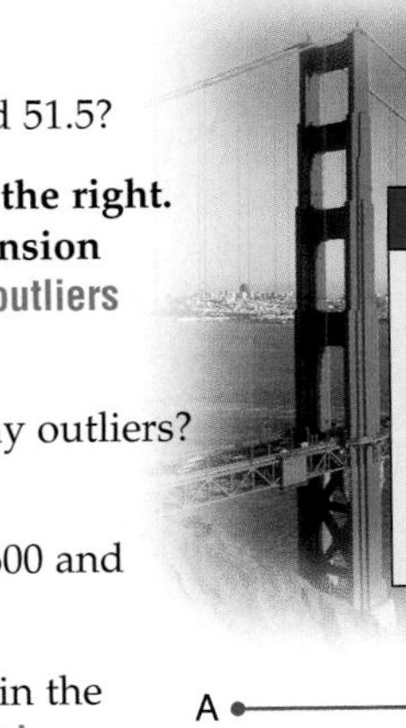

BRIDGES For Exercises 10–13, refer to table at the right. It shows the lengths of 22 of the longest suspension bridges in North America. 11. 0 and 4,600; no outliers

Length of Bridges (ft)			
4,260	2,310	1,800	1,595
4,200	2,300	1,750	1,550
3,800	2,190	1,632	1,500
3,500	2,150	1,600	1,500
2,800	2,000	1,600	
2,800	1,850	1,600	

10. Determine the interquartile range. 1,200
11. What are the limits on outliers? Are there any outliers?
12. Draw a box-and-whisker plot of the data. See Ch. 2 Answer Appendix.
13. What fraction of the bridges are between 1,600 and 2,800 feet long? one half

★ 14. Compare and contrast the data represented in the box-and-whisker plots at the right. See margin.

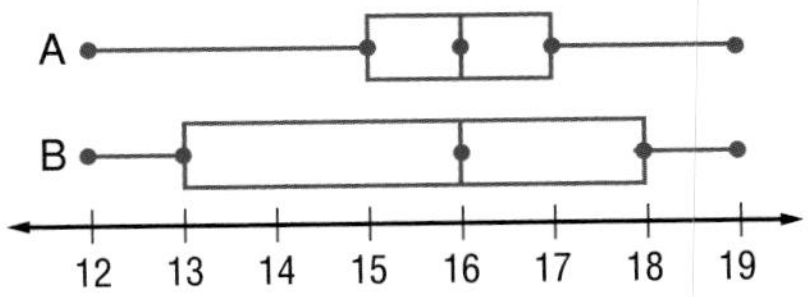

15. **CRITICAL THINKING** Describe a set of data in which there is only one whisker in its box-and-whisker plot. A set in which an extreme value and a quartile are the same has no whisker on that side of the box-and-whisker plot.

Standardized Test Practice and Mixed Review

16. **MULTIPLE CHOICE** What is the lower quartile of the data in the box-and-whisker plot at the right? C

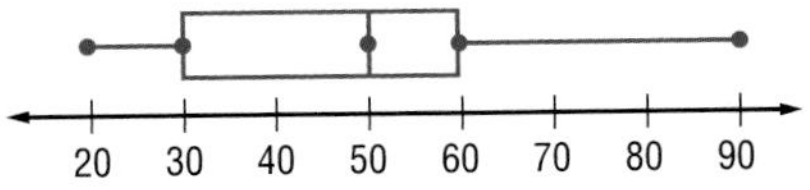

Ⓐ 60 Ⓑ 50 Ⓒ 30 Ⓓ 20

17. **GRID IN** Find the interquartile range of the following set of data.
25, 36, 46, 15, 30, 53, 40, 32, 17, 45, 41, 31, 56, 50, 52, 47, 26 20.5

RAINFALL For Exercises 18 and 19, refer to the table at the right.

Days of Rain per Year		
17	26	29
30	32	37
42	36	35

18. Make a stem-and-leaf plot of these data. (Lesson 2-5) See margin.
19. Find the mean, median, and mode. Round to the nearest tenth if necessary. (Lesson 2-4) 31.6; 32; none

GETTING READY FOR THE NEXT LESSON

PREREQUISITE SKILL Choose an appropriate interval and scale to make a frequency table for each set of data. (Lesson 2-1) 20–21. Sample answers are given.

20. 9, 0, 18, 19, 2, 9, 8, 13, 4 5; 0–19
21. 30, 20, 60, 80, 90, 110, 40 20; 20–119

msmath2.net/self_check_quiz

Answers

14. Sample answer: They have the same median and extreme values. However, the interquartile range of the data in A is 2; the data are more clustered around the median. The interquartile range of the data in B is 5; the data are more widely dispersed.

18.

Stem	Leaf
1	7
2	6 9
3	0 2 5 6 7
4	2

2|6 = 26 days

Study Guide and Intervention, p. 92

NAME ______ DATE ______ PERIOD ____

2-6 Study Guide and Intervention
Box-and-Whisker Plots

A **box-and-whisker plot** is a diagram that divides data into four equal parts. To do this, first find the median of the data, and then find the median of the lower half, called the **lower quartile**, and the median of the upper half, called the **upper quartile**.

EXAMPLE 1 **Make a box-and-whisker plot of the data below.**

12, 23, 6, 17, 9, 10, 19, 7, 11, 15, 7, 12, 13, 20

Step 1 Order the data from least to greatest. 6, 7, 7, (9), 10, 11, 12, 12, 13, 15, (17), 19, 20, 23
Step 2 Find the median and the quartiles.
Step 3 Draw a number line and graph the values you found in Step 2 as points above the line. Also graph the least value (lower extreme) and the greatest value (upper extreme).
Step 4 Draw the box and whiskers.

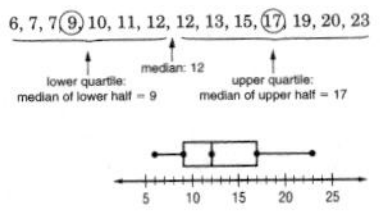

EXERCISES

Make a box-and-whisker plot for each set of data.

1. 15, 16, 7, 8, 5, 5, 3, 4, 8, 12, 10, 9, 6, 13

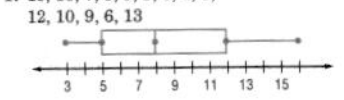

2. 4, 6, 3, 7, 10, 11, 4, 5, 6, 2, 7

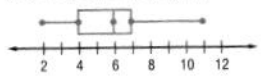

3. 1, 5, 2, 2, 6, 3, 7

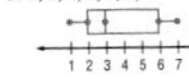

4. 8, 2, 7, 4, 12, 8, 11

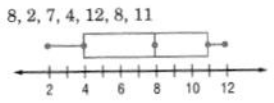

5. 17, 22, 11, 11, 11, 10, 19, 18, 16, 11, 18

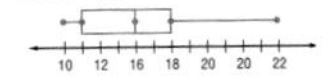

6. 3, 5, 1, 4, 2, 4, 3, 5, 2, 1

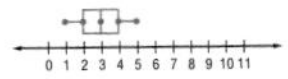

Practice: Skills, p. 93

NAME ______ DATE ______ PERIOD ____

2-6 Practice: Skills
Box-and-Whisker Plots

SPORTS For Exercises 1–6, refer to the table at the right. It shows the regular season games lost by each professional baseball team in the National League in 2001.

Number of Losses			
74	76	80	86
69	69	74	94
100	70	72	76
89	94	96	83

1. Find the lower extreme, LQ, median, UQ, and upper extreme. lower extreme, 69; LQ, 73; median, 78; UQ, 91.5; upper extreme, 100
2. Draw a box-and-whisker plot of the data.

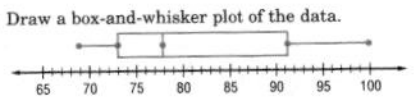

3. What fraction of the data is between 73 and 78? one-fourth
4. Between what two numbers is the largest range of the four quartiles? 78 and 91.5
5. Find the interquartile range. 18.5
6. Are there any outliers? If so, identify them. There are no outliers.

LIFE SCIENCE For Exercises 7–12, refer to the table at the right. It shows average life spans of 21 mammals.

Life Span (yr)				
5	12	4	3	12
12	6	5	8	35
7	8	12	10	12
10	3	7	1	12
10				

7. Find the lower extreme, LQ, median, UQ, and upper extreme. lower extreme, 1; LQ, 5; median, 8; UQ, 12; upper extreme; 35
8. Draw a box-and-whisker plot of the data.

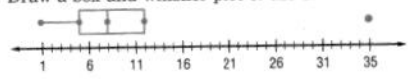

9. What fraction of the data is between 5 and 12? three-quarters
10. Find the interquartile range. 7
11. What are the limits on outliers? 0 and 22.5
12. Are there any outliers? If so, identify them. 35

Practice: Word Problems, p. 94

NAME ______ DATE ______ PERIOD ____

2-6 Practice: Word Problems
Box-and-Whisker Plots

SOCCER For Exercises 1–6, use the table below. It shows the number of wins in a recent major league soccer season

Major League Soccer Wins					
16	13	7	8	10	4
14	13	11	5	16	13

1. Find the lower extreme, LQ, median, UQ, and upper extreme. lower extreme: 4; LQ: 7.5; median: 12; UQ: 13.5; upper extreme: 16
2. Construct a box-and-whisker plot of the data in the table.

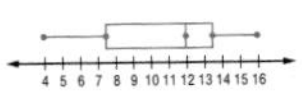

3. What fraction of the data is greater than 7.5? three-fourths
4. What fraction of the data is between 7.5 and 13.5? one-half
5. Determine the interquartile range. 6
6. Use the interquartile range to determine the limits for the outliers. Are there any outliers? 0 and 22.5 are the limits. There are no outliers.

2-6b Graphing Calculator Investigation

GETTING STARTED

Teaching Tip Before students begin entering data, remind students how to edit entries on the calculator in case they press a wrong button. While a list is open, use the up and down arrow keys to highlight the incorrect entry. Type the new entry over the old one. To edit a list later, press STAT ENTER to open L1 to edit it.

TEACH

- After students enter the data, have them verify that the number of entries in L1 equals the number of grades on Miss Romero's test.
- Before students create the box-and-whisker plot, ask students what they expect the calculator to do. **Sample answer: The calculator should find the median and quartiles.**
- Ask students to explain why it is important to choose the display window before viewing the plot. **In choosing the window, you are setting up the number line.**
- You may want to have students experiment with changing the settings on the x-axis. What happens? **The appearance of the plot is elongated or condensed.**

2-6b Graphing Calculator Investigation

A Follow-Up of Lesson 2-6

What You'll Learn
Use a graphing calculator to make box-and-whisker plots.

Box-and-Whisker Plots

You can create a box-and-whisker plot using a TI-83 Plus graphing calculator.

ACTIVITY

Make a box-and-whisker plot of the data at the right. It shows the grades on Miss Romero's last Math 7 test.

Miss Romero's Math 7 Test							
78	94	85	92	72	56	89	92
90	84	98	82	75	100	94	87
92	85	94	70	78	95	70	80

STEP 1 Enter the data.
Begin by clearing any existing data in the first list, L1.
Press STAT ENTER ▲ CLEAR ENTER.

Then enter the data into L1. Input each number and press ENTER.

STEP 2 Choose the type of graph.
Press 2nd [STAT PLOT] ENTER to choose the first plot. Highlight On, the modified box-and-whisker plot for the type, L1 for the Xlist, and 1 as the frequency.

STEP 3 Choose the display window.
Press WINDOW and choose appropriate range settings for the x values. The window 50 to 110 with a scale of 5 includes all of this data.

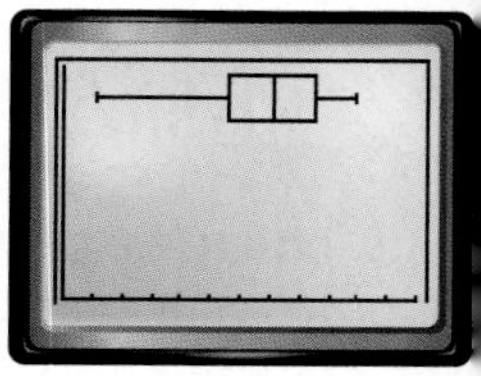

STEP 4 Display the graph.
Press GRAPH. Press TRACE and the arrow keys to determine the five key data points of your graph.

TEACHING TIP
A modified box-and-whisker plot does not graph outliers. This type is the second graph in the second row of the graphing calculator plot screen.

EXERCISES

1. What are the values of the five key data points of the graph? What do they represent? **See margin.**
2. What percent of the test scores are below 78? **25%**
3. What percent of the test scores are above the median? What percent of the test scores are below the median? **50%; 50%**
4. What percent of the scores are between 56 and 86? **50%**
5. Suppose you earned a grade of 80. Describe what percent of students scored higher and what percent scored lower than you.

5. Sample answer: About 75% of the class scored higher, about 25% scored lower.

msmath2.net/other_calculator_keystrokes

ASSESS

Students should:

- identify the lower extreme, upper extreme, median, upper quartile, and lower quartile, and
- recognize the percent of the data below, above, or between the key data points.

Answer

1. minX = 56, **the lowest test score;** Q1 = 78, **the lower quartile;** Med = 86, **the median;** Q3 = 93, **the upper quartile;** maxX = 100, **the highest test score**

2-7 Bar Graphs and Histograms

/hat You'll Learn
onstruct and interpret
ar graphs and
stograms.

EW Vocabulary
ar graph
stogram

EVIEW Vocabulary
equency table: a
ble that shows
ow often an item
ccurs (Lesson 2-1)

WHEN am I ever going to use this?

ANIMALS The cheetah is the fastest known land animal. The table shows its fastest speed and the top speeds of four other land animals.

Animal	Speed (mph)
cheetah	70
wildebeest	50
lion	50
elk	45
zebra	40

Source: *The World Almanac*

1. What are the fastest and slowest speeds recorded in the table? **70 mph; 40 mph**
2. How can you create a visual representation to summarize the data?
3. Do any of these representations show both the animal name and its speed? **no**

Sample answer: Use a
equency table, a line
ot, a stem-and-leaf
ot, or a box-and-
hisker plot.

A **bar graph** is one method of comparing data by using solid bars to represent quantities.

EXAMPLE Construct a Bar Graph

1 **Make a bar graph to display the data in the table above.**

Step 1 Draw a horizontal axis and a vertical axis. Label the axes as shown. In this case, the scale on the vertical axis is chosen so that it includes all the speeds. Add a title.

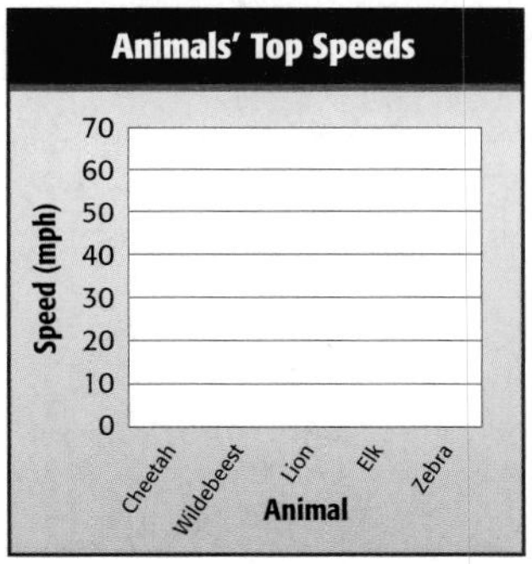

Step 2 Draw a bar to represent each category. In this case, a bar is used to represent the speed of each animal.

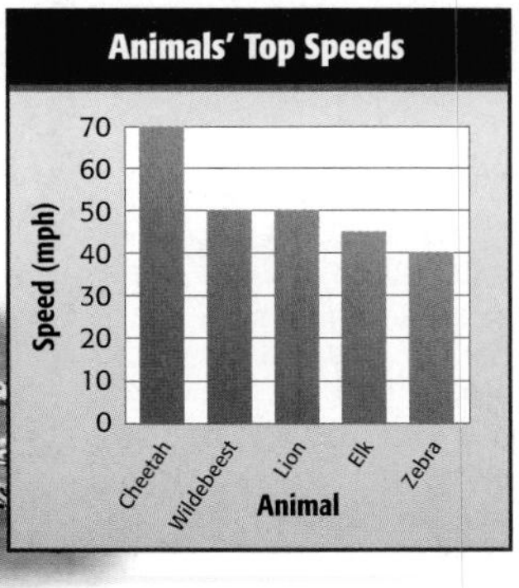

2-7 Lesson Notes

Lesson Resources

Chapter 2 Resource Masters
- Study Guide and Intervention, p. 97
- Practice: Skills, p. 98
- Practice: Word Problems, p. 99
- Reading to Learn Mathematics, p. 100
- Enrichment, p. 101

Parent and Student Study Guide Workbook, 2-7

Science and Math Lab Masters, pp. 81–88

FAST FILE

1 Focus

5-Minute Check 2-7
Use as a quiz or review of Lesson 2-6. (Also available as a transparency)

INTERACTIVE CHALKBOARD
PowerPoint® Presentations

Bellringer

Hands-On Activity On the chalkboard, write the costs of a large 1-item pizza, rounded to the nearest dollar, from 5 different pizza shops. Ask students to use stacking cubes to represent the 5 costs. How could this model be made into a graph?

DAILY INTERVENTION

Differentiated Instruction

Auditory/Musical Relate the visual aspects of bar graphs and histograms to listening to music. Play a short song using 3 or 4 notes. Count the number of times each note is played and compare this to a bar graph. Play the song again and listen to each note played. Record the duration of each note. Compare this to a histogram with time (beats) as intervals.

Reading to Learn Mathematics, p. 100 **ELL**

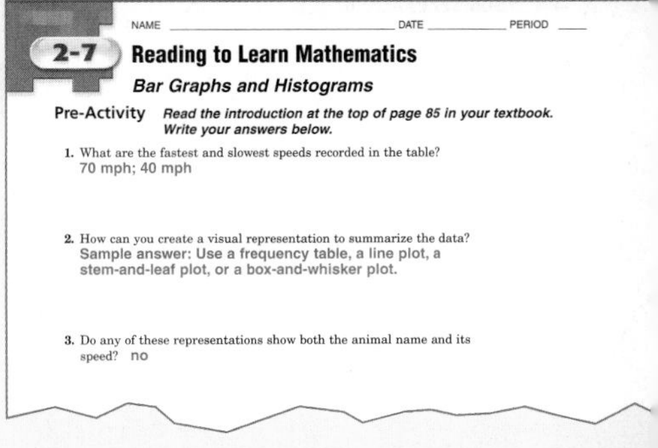
NAME ______ DATE ______ PERIOD ____

2-7 Reading to Learn Mathematics
Bar Graphs and Histograms

Pre-Activity *Read the introduction at the top of page 85 in your textbook. Write your answers below.*

1. What are the fastest and slowest speeds recorded in the table?
70 mph; 40 mph
2. How can you create a visual representation to summarize the data?
Sample answer: Use a frequency table, a line plot, a stem-and-leaf plot, or a box-and-whisker plot.
3. Do any of these representations show both the animal name and its speed? no

2 Teach

In-Class Examples

1 **TOURISM** Make a bar graph to display the data in the table below.

INTERACTIVE CHALKBOARD
PowerPoint® Presentations

Country	Vacation Days per Year
Italy	42
France	37
Germany	35
Brazil	34
United Kingdom	28
Canada	26
Korea	25
Japan	25
United States	13

Source: *The World Almanac*

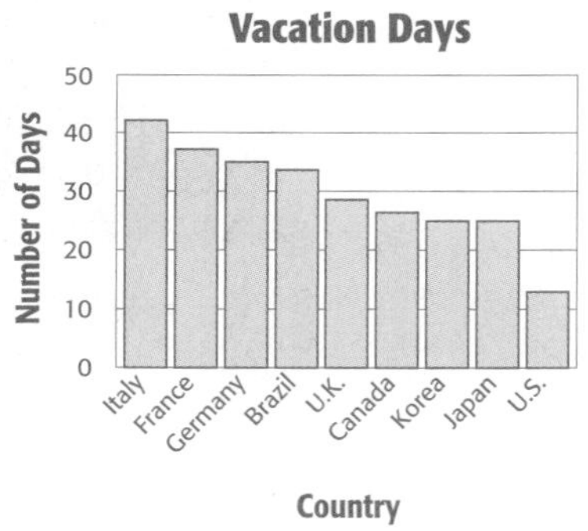

2 **BASKETBALL** The number of wins for the 29 teams of the NBA for the 2000–2001 season have been organized into a frequency table. Make a histogram of the data.

Number of Wins	Frequency
11–20	3
21–30	4
31–40	4
41–50	10
51–60	8

Source: espn.com

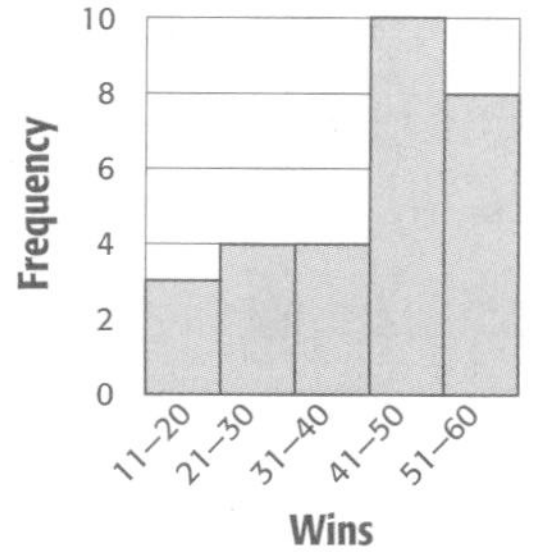

TEACHING TIP

Point out that the data values in Example 2 are whole numbers that fall within the given intervals. If a value like 20.5 had been given, different intervals would be needed to include that value.

STUDY TIP

Histograms
Because the intervals are equal, all of the bars have the same width, with no space between them. The space at 51–60 indicates that there are no data values on that interval.

TEACHING TIP

Use a bar graph to show *what is the value* of different items.
Use a histogram to show *how many items* are contained in each interval.

A special kind of bar graph, called a **histogram**, uses bars to represent the frequency of numerical data that have been organized in intervals.

EXAMPLE Construct a Histogram

2 **ANIMALS** **The top speeds of thirty different land animals have been organized into a frequency table. Make a histogram of the data.**

Maximum Speed (mph)	Frequency
11–20	5
21–30	5
31–40	10
41–50	8
51–60	0
61–70	2

Source: *The World Almanac*

Step 1 Draw and label horizontal and vertical axes. Add a title.

Step 2 Draw a bar to represent the frequency of each interval.

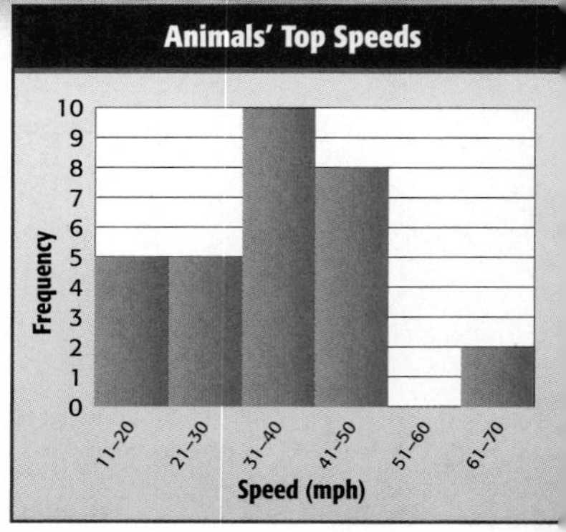

The two highest bars represent a majority of the data. From the graph, you can easily see that most of the animals have a top speed of 31–50 miles per hour.

EXAMPLE Compare Bar Graphs and Histograms

3 **TEXTBOOKS** **Refer to the graphs below.**

Graph A

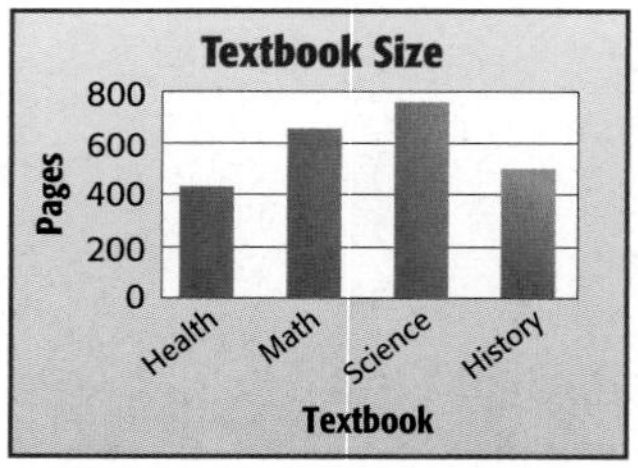

Graph B

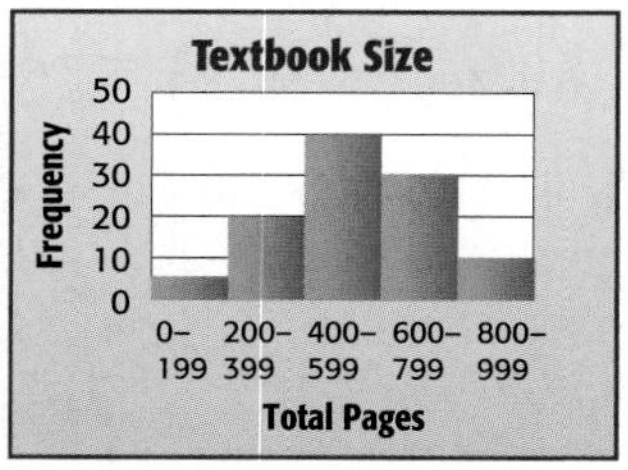

Which graph would you use to determine how many textbooks have less than 600 pages?

Graph B

Which graph would you use to compare the size of a science book and a history book?

Graph A

EXAMPLE Use Graphs to Solve a Problem

Standardized Test Practice

4 MULTIPLE-CHOICE TEST ITEM Which conclusion *cannot* be made about the data in Graph B in Example 3?

(A) There are 105 books in the data set.

(B) Twenty books have between 200 and 399 pages.

(C) Health books have the fewest pages.

(D) The interval with the fewest books is 0–199.

Read the Test Item

Determine which of the four statements is *not* correct.

Solve the Test Item

A is correct; 5 books have 0–199 pages, 20 books have 200–399 pages, and so on, for a total of 105 books.

B is correct; the bar representing 200–399 has a value of 20.

C is incorrect; the graph does not describe individual books.

D is correct; only 5 books have 0 to 199 pages, which is less than the other intervals. The answer is C.

est-Taking Tip The Princeton Review

ead a Graph
hen analyzing data in a stogram, ask yourself uestions such as, "What es each bar represent?"

Skill and Concept Check

Writing Math
Exercise 2

1. **Describe** how to determine the number of values in a data set using a histogram.
   1–2. See margin.
2. **OPEN ENDED** Can any data set be displayed using a histogram? If yes, explain why. If no, give a counterexample and explain why not.

3. World Cup Soccer titles of seven countries

GUIDED PRACTICE 4–5. See Ch.2 Answer Appendix.

For Exercises 3–6, use the table at the right.

3. Describe the data in the table.
4. Make a bar graph of the data.
5. What do the horizontal and vertical axes represent?
6. Which country won the most titles? the least titles? Brazil; England and France

Men's World Cup Soccer Titles, 1930–2002

Country	Titles Won
Uruguay	2
Italy	3
Germany	3
Brazil	5
England	1
Argentina	2
France	1

Source: fifaworldcup.yahoo.com/en

msmath2.net/extra_examples

In-Class Examples

INTERACTIVE CHALKBOARD PowerPoint® Presentations

3 AUTOMOBILES Refer to the graphs below.

Car Prices, Graph A

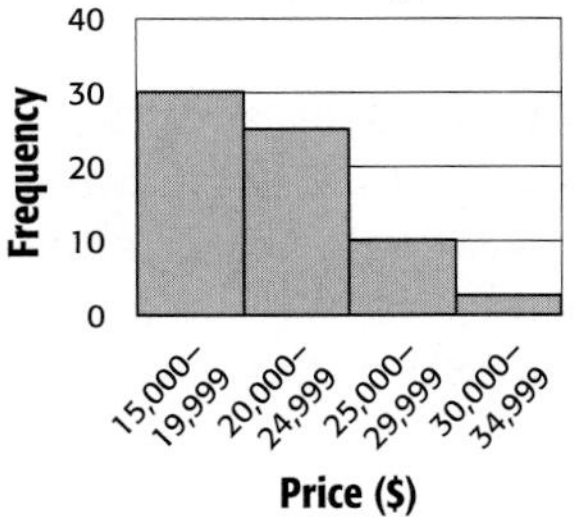

Car Prices, Graph B

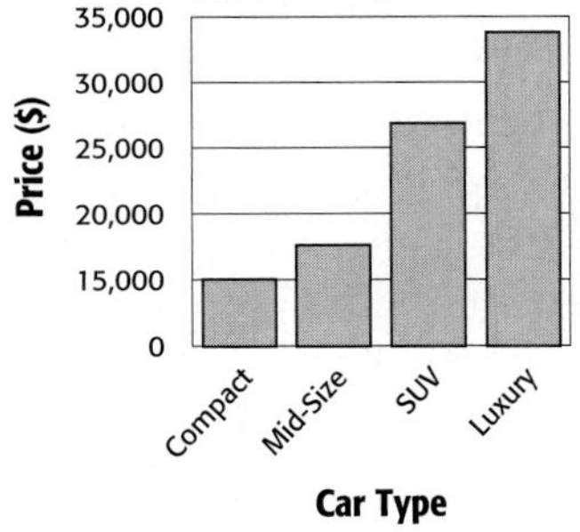

Which graph would you use to tell how many cars under $30,000 were sold? **Graph A**
Which graph would you use to compare the prices of a mid-size car and an SUV? **Graph B**

4 MULTIPLE-CHOICE TEST ITEM Which conclusion *cannot* be made about the data in Graph A in In-Class Example 3? **D**

A There are 67 cars in the data set.

B Two cars are priced between $30,000 and $34,999.

C Most of the cars are priced between $15,000 and $19,999.

D Mid-size cars sell the best.

Answers

1. Sample answer: Each interval represents a portion of the data set. The number of items in each interval is indicated by the frequency, typically shown along the vertical scale. By adding the frequencies for each interval, you can determine the number of values in the data set.
2. No; counterexample: to compare the heights of five different trees, a histogram would not be appropriate because these data are measurements of different items.

3 Practice/Apply

About the Exercises...

Guided Practice Key

Exercises	Example
3–6	1

Odd/Even Assignments
Exercises 7–16 are structured so that students practice the same concepts whether they are assigned odd or even problems.

4 Assess

Open-Ended Assessment

Modeling Present students with 2 or 3 data sets and ask them to display the data with an appropriate bar graph or histogram. They should explain their choice of graph. Include one large data set.

★ indicates increased diffficulty

Practice and Applications

HOMEWORK HELP

For Exercises	See Examples
7–9	1, 2
10–16	3

Extra Practice See pages 569, 597.

For Exercises 7–10, use the table below. 7–10. See margin.

7. Describe the data in the table.
8. What does each interval represent?
9. Make a histogram of the data.
10. Describe general patterns in the histogram.

Candies in Snack-Size Packages	
Number of Candies	Frequency
9–11	8
12–14	15
15–17	24
18–20	3

TECHNOLOGY For Exercises 11–13, use the bar graph below.

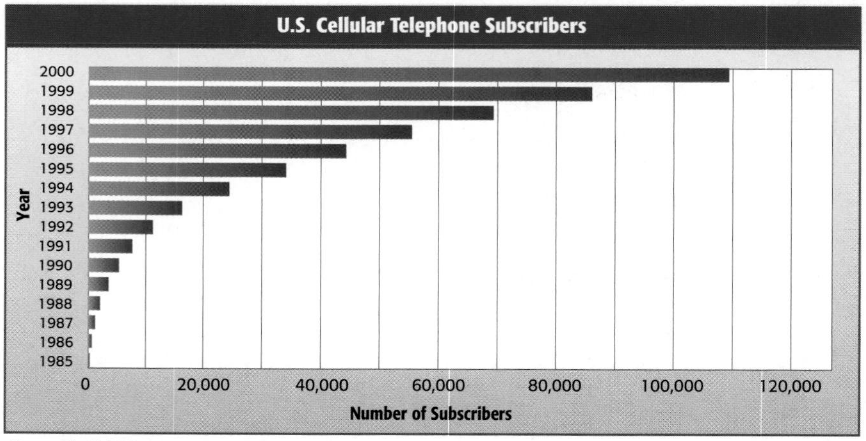

Source: *The World Almanac*

11. How is this bar graph different from the other bar graphs in the lesson?
12. Describe how the number of cell phone subscribers grew over the 15-year period. 11–12. See margin.
13. Use the graph to predict the number of cell phone subscribers in 2006. Sample answer: 200,000

CIVICS For Exercises 14–16, use the histogram at the right and the information below.
Congress is made up of senators and representatives. Each state has 2 senators, and the number of representatives per state is determined by the state's population. 14–15. See margin.

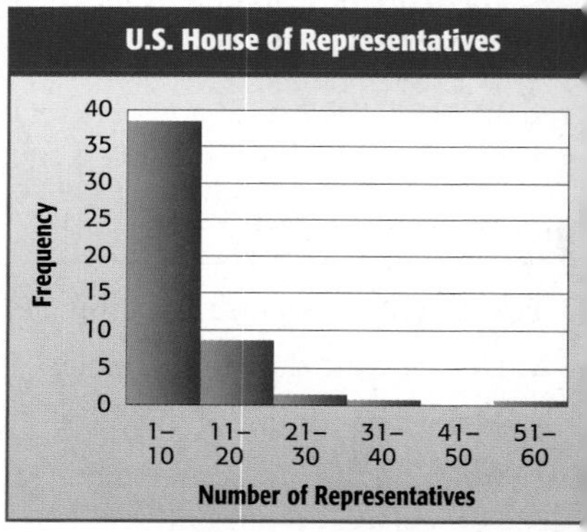

14. What does each interval represent?

★ 15. Describe general patterns in the histogram.

★ 16. There are 435 members in the House of Representatives. Determine the mean number of representatives per state. Compare this value to the value of the median number, according to the histogram. Discuss the significance of 38 states having 10 or fewer representatives. See Ch. 2 Answer Appendix.

Answers

7. the number of candies found in 50 snack-size packages
8. a specific count for the number of candies in a pack
9. **Candies in Snack-Size Packages**

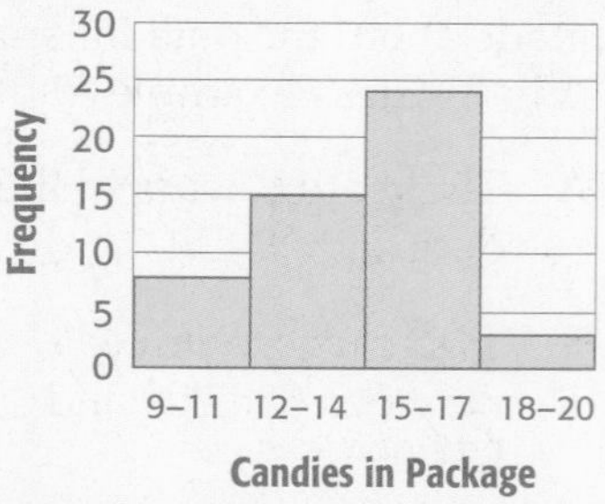

10. Sample answer: The distribution is centered around the middle of the range of possible values. There are lots of packs with a moderate number of candies (from 11 to 18 candies per pack, there were 39 packs) and even a few packs approaching the extremes of 9 candies and 20 candies per pack.
11. Sample answer: The bars are horizontal; the categories are on the vertical axis and the scale is on the horizontal axis.
12. Sample answer: Cellular phone subscription in the U.S. increased dramatically over this time period. Every year, the amount of increase was larger than for the previous year's increase.
14. the number of representatives per state
15. Sample answer: The histogram is strongly skewed toward the lower end, with a high concentration in the number of states with 10 or fewer representatives. The distribution spreads out from there, with fewer and fewer states as the number per state increases.

17. **CRITICAL THINKING** The histograms show players' salaries to the nearest million dollars for the two major league baseball teams that played in the 2001 World Series. Compare the salary distributions of the two teams. See Ch. 2 Answer Appendix.

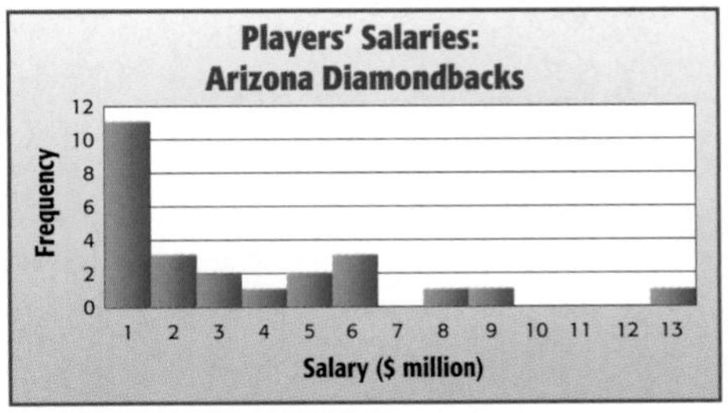

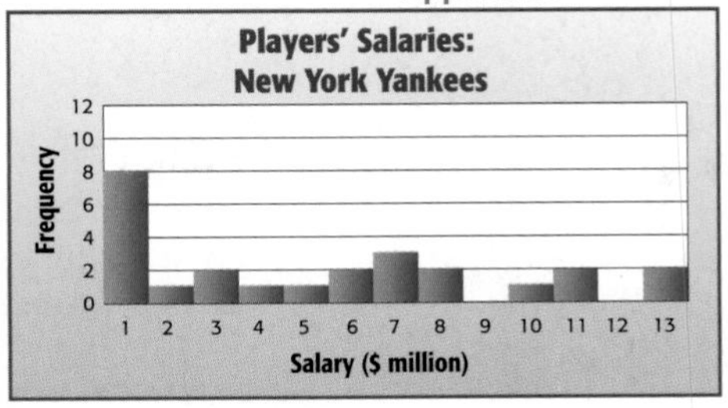

EXTENDING THE LESSON

A *multiple-bar graph* shows two or more sets of data on the same graph. The multiple-bar graph compares the annual costs of using old model appliances and using equivalent new model appliances.

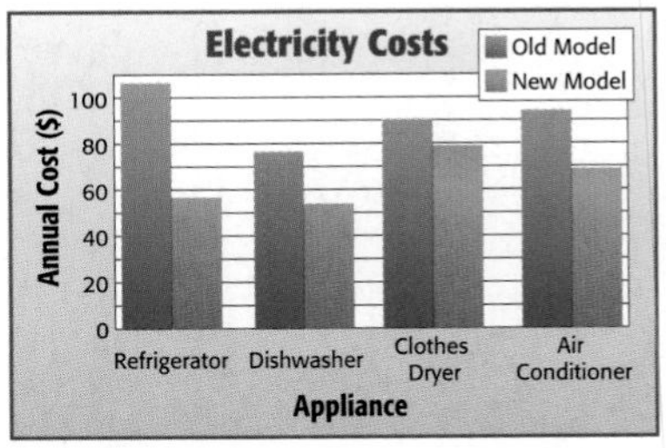

Source: Association of Home Appliance Manufacturers

18. For which appliance is the difference in electricity costs between the old and new model the greatest? Explain. See margin.

19. Describe an advantage of using a multiple-bar graph rather than two separate graphs to compare data. Sample answer: It is easier to compare two sets of data.

Standardized Test Practice and Mixed Review

20. **MULTIPLE CHOICE** The histogram at the right shows the number of bordering states for each of the fifty states. Which is *not* a true statement? D

    A The mode of the data is 4.

    B The greatest number of bordering states is 8.

    C Ten states have five bordering states.

    D No states have only 1 bordering state.

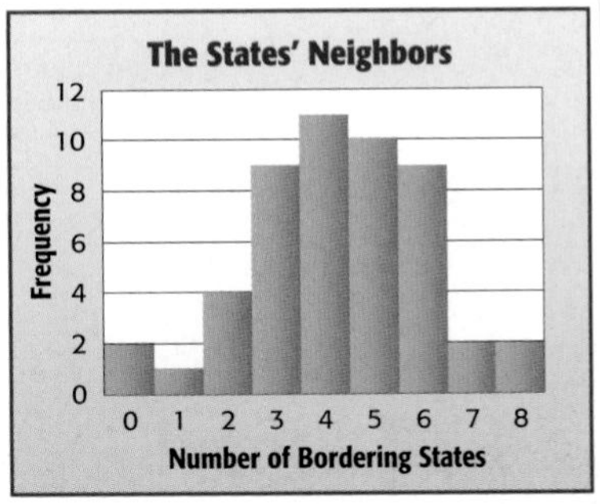

21. **SHORT RESPONSE** Which type of graph can be drawn using the data in a frequency table? histogram

For Exercises 22 and 23, use the data 12, 17, 23, 5, 9, 25, 13, 16, 2, 25, 31, and 10.

22. Make a box-and-whisker plot of the data. (Lesson 2-6) See Ch. 2 Answer Appendix.

23. Make a stem-and-leaf plot of the data. (Lesson 2-5) See margin.

GETTING READY FOR THE NEXT LESSON

PREREQUISITE SKILL Determine whether each statement is *true* or *false*.
(Lesson 2-2)

24. The vertical scale on a line graph must have equal intervals. true

25. You do not need to label the axes of a line graph. false

Answers

18. Refrigerator; the difference in annual costs between an old model and a new model is $106 – $56 or $50, which is greater than the differences for the other appliances.

23.

Stem	Leaf
0	2 5 9
1	0 2 3 6 7
2	3 5 5
3	1

$2|3 = 23$ days

Study Guide and Intervention, p. 97

NAME ______ DATE ______ PERIOD ____

2-7 Study Guide and Intervention
Bar Graphs and Histograms

A **bar graph** is one method of comparing data by using solid bars to represent quantities. A **histogram** is a special kind of bar graph. It uses bars to represent the frequency of numerical data that have been organized into intervals.

EXAMPLE 1 SIBLINGS **Make a bar graph to display the data in the table below.**

Student	Number of Siblings
Sue	1
Isfu	6
Margarita	3
Akira	2

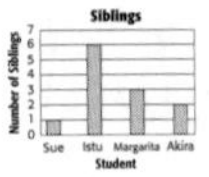

Step 1 Draw a horizontal and a vertical axis. Label the axes as shown. Add a title.

Step 2 Draw a bar to represent each student. In this case, a bar is used to represent the number of siblings for each student.

EXAMPLE 2 SIBLINGS **The number of siblings of 17 students have been organized into a table. Make a histogram of the data.**

Number of Siblings	Frequency
0–1	4
2–3	10
4–5	2
6–7	1

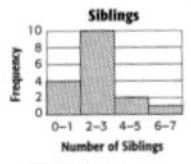

Step 1 Draw and label horizontal and vertical axes. Add a title.

Step 2 Draw a bar to represent the frequency of each interval.

EXERCISES

1. Make a bar graph for the data in the table.

Student	Number of Free Throws
Luis	6
Laura	10
Opal	4
Gad	14

See students' work.

2. Make a histogram for the data in the table.

Number of Free Throws	Frequency
0–1	1
2–3	5
4–5	10
6–7	4

See students' work.

Practice: Skills, p. 98

NAME ______ DATE ______ PERIOD ____

2-7 Practice: Skills
Bar Graphs and Histograms

ZOOS **For Exercises 1 and 2, use the table. It shows the number of species at several zoological parks.**

Zoo	Species
Los Angeles	350
Lincoln Park	290
Cincinnati	700
Bronx	530
Oklahoma City	600

1. Make a bar graph of the data.
Sample answer:

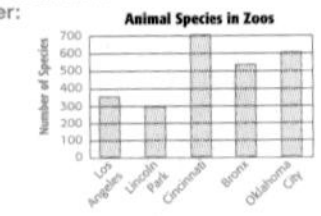

2. Which zoological park has the most species? Cincinnati

ZOOS **For Exercises 3 and 4, use the table at the right. It shows the number of species at 37 major U.S. public zoological parks.**

Number of Species				
200	700	290	600	681
300	643	350	794	400
360	600	134	200	800
305	384	500	330	250
530	715	303	200	475
465	340	347	300	708
184	800	375	350	450
337	221			

3. Make a histogram of the data. Use intervals of 101–200, 201–300, 301–400, 401–500, 501–600, 601–700, and 701–800 for the horizontal axis. Sample answer:

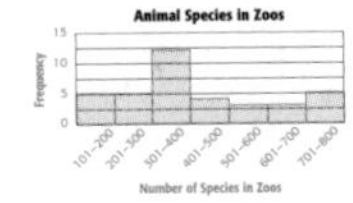

4. Which interval has the largest frequency? 301–400

HEALTH **For Exercises 5 and 6, use the graph at the right.**

5. What does each bar represent? The number of Calories consumed per day by one person.

6. Determine whether the graph is a bar graph or a histogram. Explain how you know. bar graph; the data is not in intervals

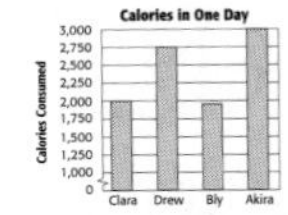

Practice: Word Problems, p. 99

NAME ______ DATE ______ PERIOD ____

2-7 Practice: Word Problems
Bar Graphs and Histograms

PUPPIES **For Exercises 1 and 2, use the table below. It shows the results of a survey in which students were asked what name they would most like to give a new pet puppy.**

Name	Votes
Max	15
Tiger	5
Lady	13
Shadow	10
Molly	9
Buster	2

EARTH SCIENCE **In Exercises 3–6, use the table below. It shows the highest wind speeds in 30 U.S. cities.**

Highest Wind Speeds (mph)									
52	75	60	80	55	54	91	60	81	58
53	73	46	76	53	46	73	46	51	49
57	58	56	47	65	49	56	51	54	51

1. Make a bar graph to display the data.
Sample answer:

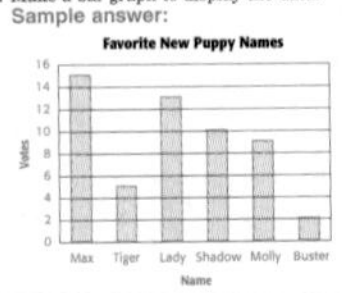

2. Use your bar graph from Exercise 1. Compare the number of votes the name Shadow received to the number of votes the name Tiger received. Shadow received twice as many votes as Tiger.

3. Make a histogram of the data.

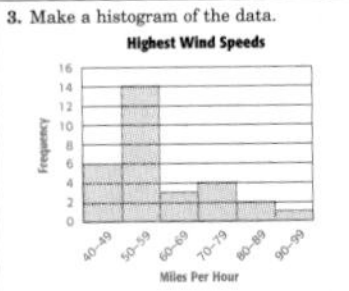

4. What is the top wind speed of most of the cities? 50–59 mph

5. How many cities recorded wind speeds of 80 miles per hour or more? 3

6. How many cities recorded their highest wind speeds at 60 miles per hour or more? 10

2-7b Spreadsheet Investigation

GETTING STARTED

Teaching Tip Have students review the multiple-line graph on page 63 and the multiple-bar graph on page 89. Ask students to describe the scale of each graph and tell what is needed for data to be plotted on a multiple-line or multiple-bar graph. **Sample answer: Each graph has one scale for both sets of data. Two or more sets of data must be measured in the same units or convertible to the same units in order to be plotted on the same graph.**

Teaching Tip Before starting this activity, you may want to familiarize students with the layout of a Microsoft Excel® spreadsheet. Photocopy a portion of a blank sheet from an Excel workbook for students. Ask them to locate specific cells using the column letters and row numbers, such as A4 and C3.

TEACH

Activity 1

- Before students set up their spreadsheets, briefly demonstrate to students how to maneuver around a spreadsheet using the arrow, tab, and enter keys and how to highlight cells using the mouse or keyboard shortcuts.
- If students have hand drawn a multiple-line graph of the data, have them compare their spreadsheet line charts to the hand-drawn graphs after completing Step 11.

2-7b Spreadsheet Investigation

A Follow-Up of Lesson 2-2

What You'll Learn
Use a spreadsheet to make a multiple-line graph and a multiple-bar graph.

Multiple-Line and Multiple-Bar Graphs

In Lessons 2-2 and 2-7, you interpreted data in a multiple-line graph and in a multiple-bar graph, respectively. You can use a Microsoft Excel® spreadsheet to make these two types of graphs.

ACTIVITY 1

The stopping distances on dry pavement and on wet pavement are shown in the table at the right.

Speed (mph)	Stopping Distance (ft)	
	Dry Pavement	Wet Pavement
50	200	250
60	271	333
70	342	430
80	422	532

Source: Continental Teves

Set up a spreadsheet like the one shown below.

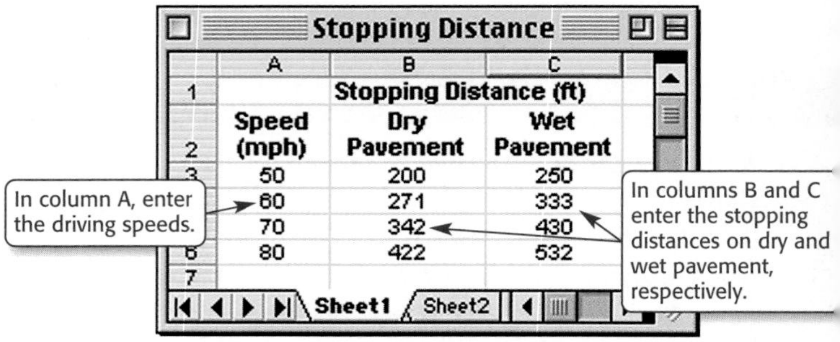

TEACHING TIP
Have students draw a multiple-line graph and a multiple-bar graph for this data by hand first. Then use the spreadsheet.

TEACHING TIP
The icon in steps 4 and 6 is called the *Collapse Dialog* button.

The next step is to "tell" the spreadsheet to make a double-line graph for the data.

1. Highlight the data in columns B and C, from B2 through C6.
2. Click on the Chart Wizard icon.
3. Choose the line graph and click Next.
4. To set the x-axis, choose the Series tab and press the icon next to the Category (X) axis labels.
5. On the spreadsheet, highlight the data in column A, from A3 through A6.
6. Press the icon on the bottom of the Chart Wizard box to automatically paste the information.
7. Click Next and enter the chart title and labels for the x- and y-axe
8. Click Next and then Finish.

This tells the spreadsheet to read the data in colum B and C.

Multiple-Line Graph

	A	B	C
1		Stopping Distance (ft)	
2	Speed (mph)	Dry Pavement	Wet Pavement
3	50	200	250
4	60	271	333
5	70	342	430
6	80	422	532

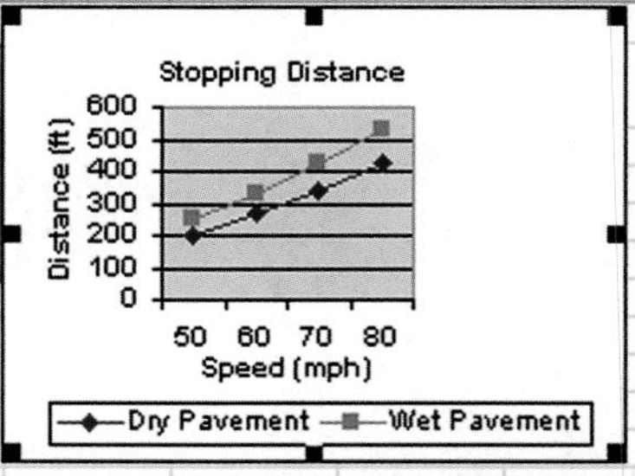

Sheet1 / Sheet2 / Sheet3

TEACHING TIP

tudents can enlarge eir graph by clicking nd dragging one of the boxes on the frame.

ACTIVITY 2

Use the same data to make a multiple-bar graph. Set up the spreadsheet shown in Activity 1. The next step is to "tell" the spreadsheet to make a bar graph for the data.

- Highlight the data in columns B and C, from B2 through C6.
- Click on the Chart Wizard icon.
- Click on Column and Next to choose the vertical bar graph.
- Complete steps 4–8 from Activity 1.

Multiple-Bar Graph

	A	B	C
1		Stopping Distance (ft)	
2	Speed (mph)	Dry Pavement	Wet Pavement
3	50	200	250
4	60	271	333
5	70	342	430
6	80	422	532

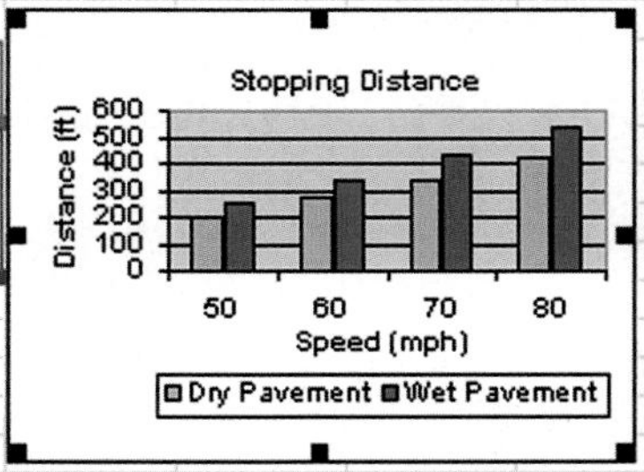

Sheet1 / Sheet2 / Sheet3

EXERCISES

1. **Explain** the steps you would take to make a multiple-line graph of the stopping distances that include the speeds 55, 65, and 75.

Insert the numbers 5, 65, and 75 in the ppropriate places in olumn A, along with the opping distances for y and wet pavement.

2. **OPEN ENDED** Collect some data that can be displayed in a multiple-line graph or a multiple-bar graph. Record the data in a spreadsheet. Then make a graph to display the data.

See students' work.

3. **Discuss** the advantages and disadvantages of using a spreadsheet to make a multiple-bar graph. See margin.

Activity 2

- Have students compare their completed spreadsheet column charts to any hand-drawn bar graphs they have made of the data.
- Students may wish to print their final Excel worksheet for their portfolios. If this option is available, show students how to use the mouse to drag their charts to suitable positions on the page for printing and have them click outside the charts before hitting the print key. Alternatively, students can click on an individual graph to print just that chart.

ASSESS

Observe as students solve the exercises. They should:

- correctly explain how to insert additional rows of data into the spreadsheet,
- be able to choose appropriate data for a multiple-line or multiple-bar graph and create a graph to display it, and
- demonstrate an understanding of the benefits of using spreadsheet software to create multiple-line and multiple-bar graphs.

Answer

3. Sample answer: One advantage of using spreadsheet software is the ease of changing data or graph type. One disadvantage is that if you enter data incorrectly in the spreadsheet, the graph will be wrong.

Lesson 2-7b Spreadsheet Investigation: Multiple-Line and Multiple-Bar Graphs **91**

2-8 Lesson Notes

Lesson Resources

Chapter 2 Resource Masters
- Study Guide and Intervention, p. 102
- Practice: Skills, p. 103
- Practice: Word Problems, p. 104
- Reading to Learn Mathematics, p. 105
- Enrichment, p. 106

Calculator and Spreadsheet Masters, p. 20
MathPASS Tutorial CD-ROM, 57
Parent and Student Study Guide Workbook, 2-8
Teaching Mathematics with Manipulatives, p. 41

FAST FILE

1 Focus

5-Minute Check 2-8
Use as a quiz or review of Lesson 2-7. (Also available as a transparency)

INTERACTIVE CHALKBOARD
PowerPoint® Presentations

Bellringer
Reading Mathematics Give students examples of graphs from newspapers, magazines, or the Internet. Ask whether the graphs are straightforward and easy to understand. How could they be modified to be harder or easier to read?

Reading to Learn Mathematics, p. 105 **ELL**

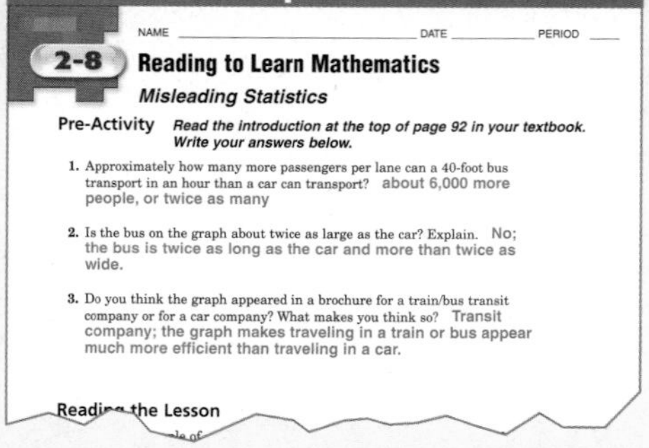
NAME ______ DATE ______ PERIOD ____

2-8 Reading to Learn Mathematics
Misleading Statistics

Pre-Activity ***Read the introduction at the top of page 92 in your textbook. Write your answers below.***

1. Approximately how many more passengers per lane can a 40-foot bus transport in an hour than a car can transport? about 6,000 more people, or twice as many
2. Is the bus on the graph about twice as large as the car? Explain. No; the bus is twice as long as the car and more than twice as wide.
3. Do you think the graph appeared in a brochure for a train/bus transit company or for a car company? What makes you think so? Transit company; the graph makes traveling in a train or bus appear much more efficient than traveling in a car.

Reading the Lesson

2-8 Misleading Statistics

What You'll Learn
Recognize when statistics and graphs are misleading.

WHEN am I ever going to use this?

TRANSPORTATION A graph like the one at the right appeared in a brochure describing various modes of transportation. **1–3. See margin.**

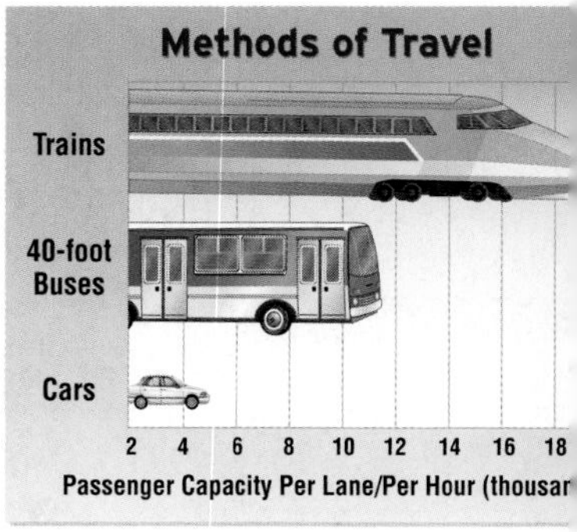

1. Approximately how many more passengers per lane can a 40-foot bus transport in an hour than a car can transport?
2. Is the bus on the graph about twice as large as the car? Explain.
3. Do you think the graph appeared in a brochure for a train/bus transit company or for a car company? What makes you think so?

The graph above shows that a car can transport about 5,000 people per lane, per hour. A train can transport about 4 times that many. However the train on the graph is much more than 4 times larger than the car. For this reason, the graph could be considered misleading.

Other ways that graphs could be misleading are if there is no title, there are no labels on the axes or scales, or the intervals on a scale are not equal.

TEACHING TIP
Figures that represent quantities should be drawn to scale. Otherwise, the representations are not accurate, as in the graph at the beginning of the lesson.

EXAMPLE Misleading Graphs

1 BUSINESS The line graphs below show monthly profits of a company from October to March. Which graph is misleading? Explain.

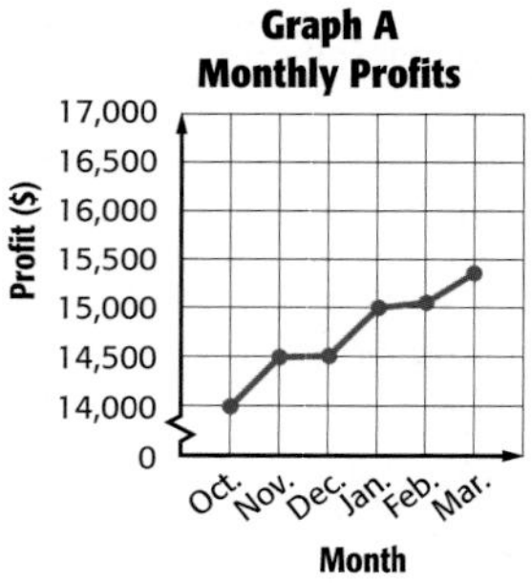

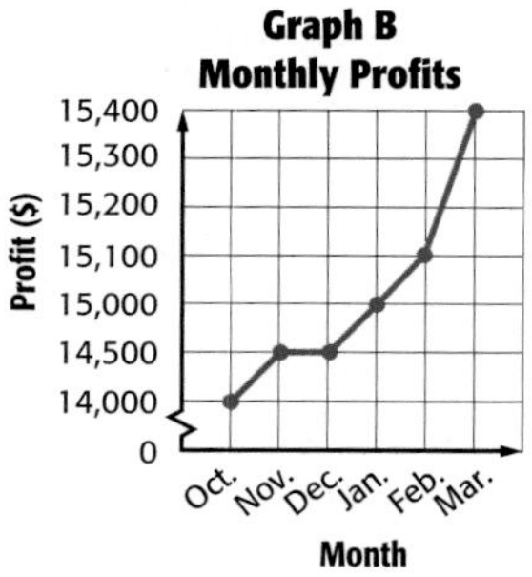

Graph B is misleading because the vertical scale has intervals of both \$500 and \$100. The increase in profits are exaggerated.

DAILY INTERVENTION **Flexible Grouping**

Misleading Graphs Using the numbered-heads-together strategy, divide the class into small groups of four and have students in each group count off from 1 to 4. Provide the class with a graph that has been exaggerated and give each group 30 seconds to discuss how this graph could be used to persuade people. Then randomly call a number from 1 to 4. Students with that number should be polled for their groups' conclusions.

Statistical averages used to describe data may also be misleading. For example, whenever there are outliers, the mean may not be a good way to describe a data set because it is affected the most by outliers.

EXAMPLE Misleading Statistics

2 SALARY ABC Corporation claims the average salary for its employees is more than $60,000, while the average salary at XYZ Incorporated is only $25,000. Use the table to explain their reasoning and determine where you would prefer to work.

Position	Salary ($)	
	ABC Corp.	XYZ Inc.
President	500,000	120,000
1st Vice President	400,000	85,000
2nd Vice President	240,000	75,000
Sales Staff (5)	20,000	40,000
Supporting Staff (8)	15,000	25,000
Catalog Staff (7)	9,000	22,500

	ABC Corp.	XYZ Inc.
mean salary	$61,870	$36,413
median salary	$15,000	$25,000

ABC Corporation used the mean to represent its average salary. They used the median to represent the average salary at XYZ Inc.

Unless you could be a president or vice president, it would be better to work for XYZ Inc. because its 20 lowest-paid employees are all better paid than any of the 20 lowest-paid employees at ABC Corporation.

Skill and Concept Check

Writing Math
Exercises 1 & 2

1. Describe two ways in which the presentation of data can be misleading. See margin.
2. **OPEN ENDED** Find an example of a graph in a newspaper or magazine that could be considered misleading. See students' work.

GUIDED PRACTICE

For Exercises 3–6, use the graph at the right. It shows the prices for natural gas charged by an Illinois natural gas supplier.

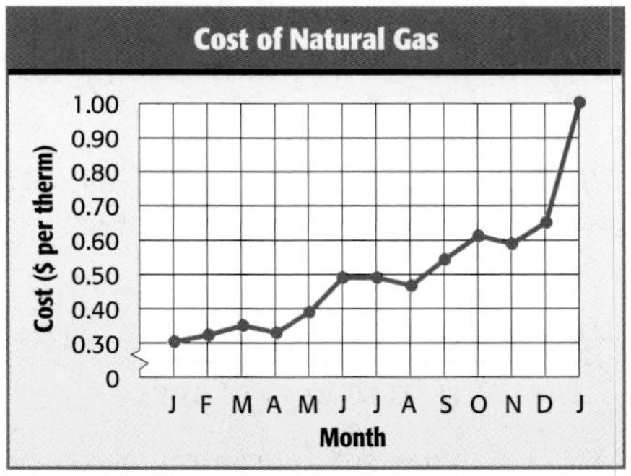

3. What trend seems to be revealed in the line graph? 3–6. See Ch. 2 Answer Appendix.
4. What problems might there be in using this graph to predict future prices of natural gas?
5. Which measure of central tendency might the gas supplier use to advertise its prices? Explain.
6. Which measure of central tendency could competitors use to discourage customers from choosing that natural gas supplier? Explain.

Answers (When am I ever going to use this?)

1. about 6,000 more people, or twice as many
2. No; the bus is twice as long as the car and more than twice as wide.
3. Transit company; the graph makes traveling in a train or bus appear much more efficient than traveling in a car.

Answer

1. Sample answer: Outliers may distort measures of central tendency; data shown in graphs may be exaggerated or minimized by manipulating scales and intervals.

2 Teach

In-Class Examples

INTERACTIVE CHALKBOARD PowerPoint® Presentations

1 BUSINESS The line graphs below show the last 10 weeks of sales for the Crumby Cookie Bakery.

Sales, Graph A

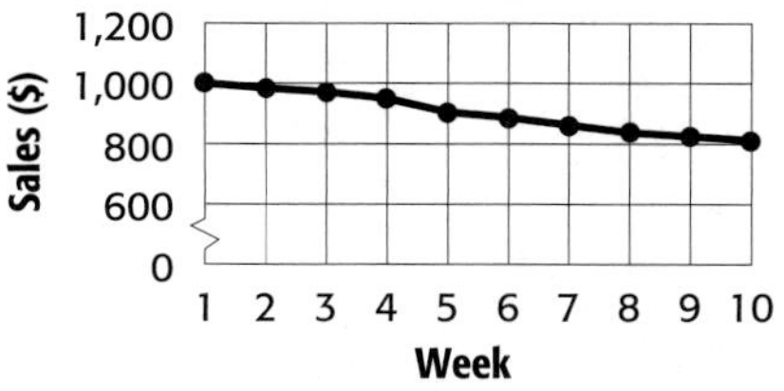

Sales, Graph B

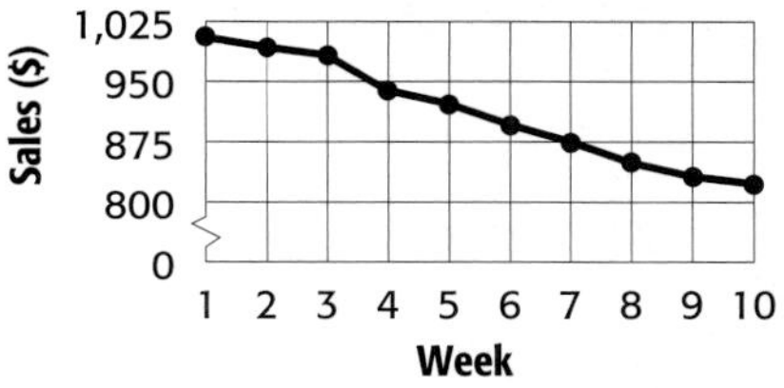

Which graph would be better to help convince a bank loan officer to open a $20,000 loan to remodel the kitchen? Graph A Why might this graph be considered misleading? It does not show the consistent decline in sales.

2 GRADES Michael and Melissa both claim to be earning a C average, 70% to 79%, in their Latin class. Use the table below to explain their reasoning and determine which student is earning a C average.

Test	Grade (%)	
	Michael	Melissa
1	80	88
2	76	83
3	73	75
4	70	70
5	40	60
6	25	65
7	10	62

	Michael	Melissa
mean	53.4%	71.9%
median	70%	70%

Michael is using the median to describe his grade rather than the mean. Only Melissa's mean or average is 70% or better.

3 Practice/Apply

FOLDABLES™ Follow-Up Remind students to use their Foldables to record what they learn about misleading statistics. Under the eighth tab, they might explain how to recognize misleading graphs and statistics.

About the Exercises...

Guided Practice Key

Exercises	Examples
3, 4	1
5, 6	2

Odd/Even Assignments
Exercises 7–8 are structured so that students practice the same concepts whether they are assigned odd or even problems.

Alert! Exercise 2 requires a newspaper or a magazine.

Answers

7. Sample answer: Although both graphs show a decrease in network viewers, in Graph B, a change in the vertical scale makes it appear that the viewers are decreasing more rapidly.
10. The mean could be misleading, since the areas of all but one of the islands are less than that value.
13. Yes; all the other test scores were lower than 96.
14. No; for example, the median of 9, 10, 11, 12, 100 is the same as the median of 9, 10, 11, 12, 13.
18. 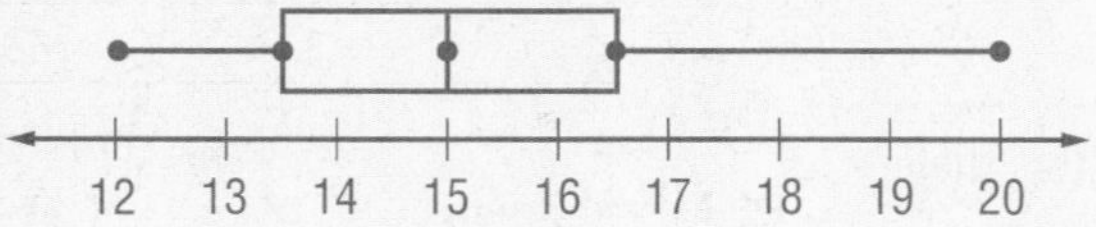

Practice and Applications

HOMEWORK HELP

For Exercises	See Examples
7–8	1
9–13	2

Extra Practice
See pages 569, 597.

7. **ENTERTAINMENT** Both bar graphs show the percent of viewers that watch network television. Which graph makes it appear that the number of viewers is decreasing rapidly? Explain. **See margin.**

Graph A

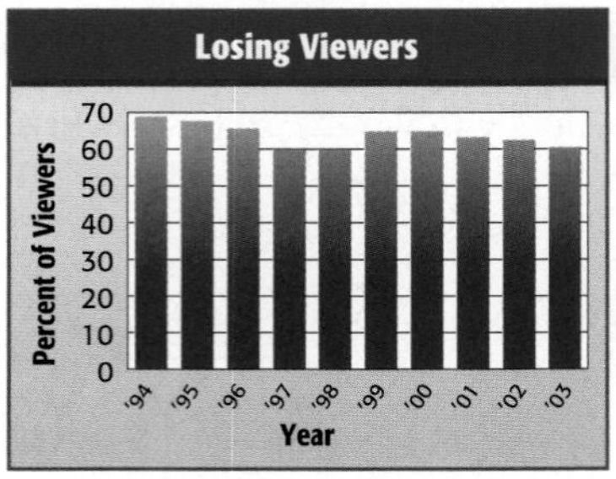

Graph B

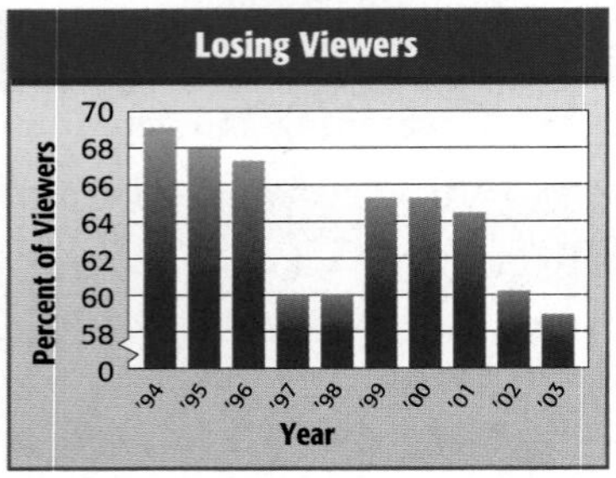

8. **UTILITIES** The line graph shows the monthly electric bill for the condominium that Toshiko is interested in renting. Why is the graph misleading? **Sample answer: The graph does not include the very hot summer months or the very cold winter months, which would have higher electricity bills.**

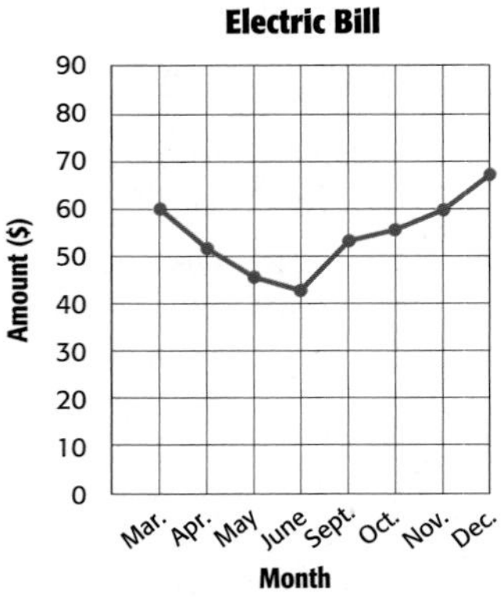

GEOGRAPHY For Exercises 9–11, use the table at the right. **9. 488.3; none; 150**

9. Find the mean, mode, and median of the data.
10. Which measure of central tendency would be misleading in describing the size of the islands? Explain. **See margin.**
11. Which measure would most accurately describe the data? **median**

Caribbean Islands

Island	Area (sq mi)	Island	Area (sq mi
Antigua	108	Martinique	425
Aruba	75	Puerto Rico	3,339
Barbados	166	Tobago	116
Curacao	171	Virgin Islands, UL	59
Dominica	290	Virgin Islands U.S.	134

Source: U.S. Census Bureau

TESTS For Exercises 12 and 13, use Miguel's test scores shown in the line plot.

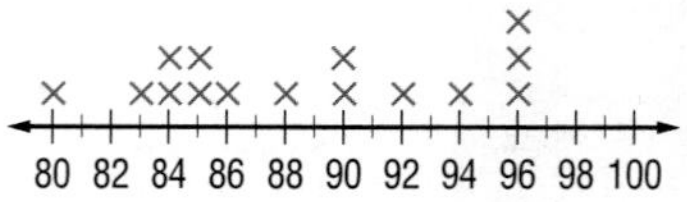

12. Which measure of central tendency might he use to emphasize the high scores? **mode**
13. Would this number be misleading? Explain. **See margin.**

14. **CRITICAL THINKING** Do values that are much greater or much less than the other values affect the median of a set of data? Give an example to support your answer. **See margin.**

Standardized Test Practice and Mixed Review

15. **MULTIPLE CHOICE** The table below shows the population of ten counties in Arizona from the 2000 Census. Which measure of central tendency, if any, is misleading in representing the data? **B**

County	Population	County	Population
Apache	69,423	La Paz	19,715
Cochise	117,755	Maricopa	3,072,149
Coconino	116,320	Mohave	155,032
Gila	51,335	Navajo	97,470
Graham	33,489	Santa Cruz	38,381

Source: U.S. Census Bureau

Ⓐ mode Ⓑ mean Ⓒ median Ⓓ none of them

16. **SHORT RESPONSE** Describe one way that a graph could be misleading. **Sample answer: size of art is not representative of data**

17. **SKATING** Describe what the height of the fourth bar in the graph represents. (Lesson 2-7)

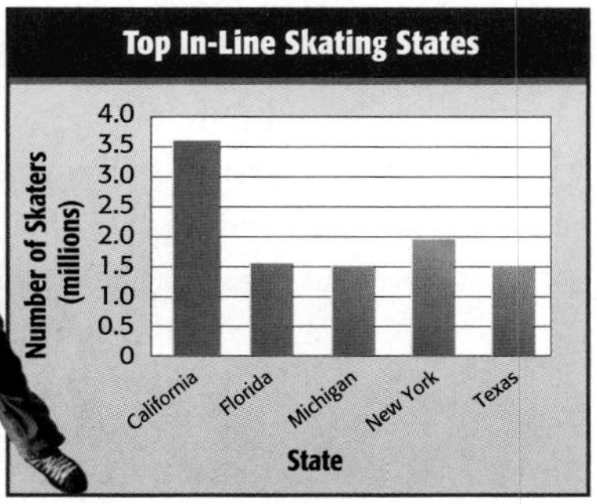

Source: National Sporting Goods Association

18. Make a box-and-whisker plot of the following data. 12, 15, 20, 12, 13, 14, 15, 14, 17, 16, 16, 18, 15 (Lesson 2-6) **See margin.**

17. the number of in-line skaters in New York, almost 2 million

MOVIES For Exercises 19 and 20, use the movie ticket prices below.

$7.50, $6.50, $7.50, $8.00, $8.50, $8.00, $7.00

19. Find the mean, median, and mode. Round to the nearest tenth if necessary. (Lesson 2-4) **$7.57; $7.50; $7.50 and $8.00**

20. Find the range of the data. (Lesson 2-3) **$2.00**

Complete. (Lesson 1-8)

21. 8.5 g = ___?___ mg **8,500**

22. 19 cm = ___?___ mm **190**

23. 643 mg = ___?___ g **0.643**

24. 5,375 mL = ___?___ L **5.375**

WebQuest Interdisciplinary Project

Lions, Tigers, and Bears

It's time to complete your project. Use the information and data you have gathered about the animals you selected to prepare a video or poster. Be sure to include a spreadsheet and all the necessary graphs with your project.

msmath2.net/webquest

msmath2.net/self_check_quiz

4 Assess

Open-Ended Assessment

Speaking Ask students to write an advertisement for a hotel to persuade vacationing visitors to stay with them. Have students present their ad to the class and explain how the ad may mislead the public. Rates for the hotel are: Sun.–Thurs. $49 per night; Fri. and Sat. $89 per night.

Study Guide and Intervention, p. 102

NAME ______________ DATE ________ PERIOD ____

2-8 Study Guide and Intervention

Misleading Statistics

Graphs can be misleading for many reasons: there is no title, the scale does not include 0; there are no labels on either axis; the intervals on a scale are not equal; or the size of the graphics misrepresents the data.

EXAMPLE 1 WEEKLY CHORES **The line graphs below show the total hours Salomon spent doing his chores one month. Which graph would be best to use to convince his parents he deserves a raise in his allowance? Explain.**

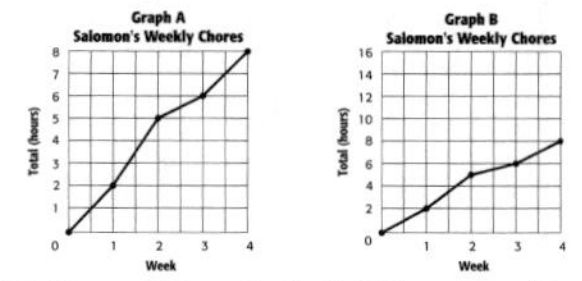

He should use graph A because it makes the total hours seem much larger.

EXERCISES

PROFITS **For Exercises 1 and 2, use the graphs below. It shows a company's profits over a four-month period.**

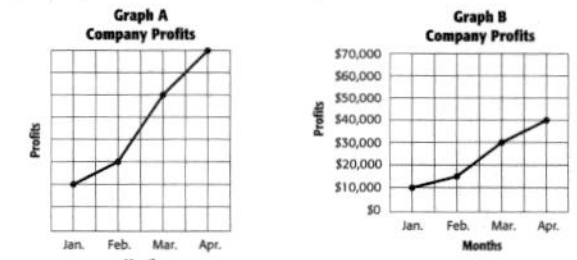

1. Which graph would be best to use to convince potential investors to invest in this company? Graph A

2. Why might the graph be misleading? There is no vertical scale.

Practice: Skills, p. 103

NAME ______________ DATE ________ PERIOD ____

2-8 Practice: Skills

Misleading Statistics

1. INCOME The bar graphs below show the total U.S. national income (nonfarm). Which graph could be misleading? Explain.

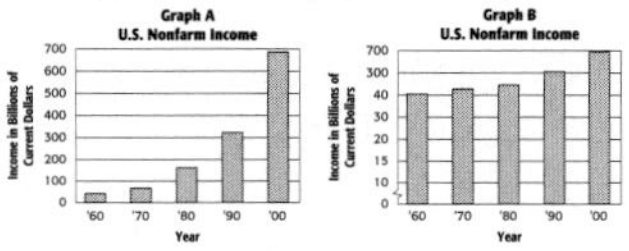

Graph B; Sample answer: The scale on the vertical axis does not have equal intervals. It makes the income growth appear to be slower.

GEOGRAPHY **For Exercises 2–4, use the table that shows the miles of shoreline for five states.**

Miles of Shoreline	
State	Length of Shoreline (mi)
Virginia	3,315
Maryland	3,190
Washington	3,026
North Carolina	3,375
Pennsylvania	89

2. Find the mean, median, and mode of the data. 2,599; 3,190; none

3. Which measure of central tendency is misleading in describing the miles of shoreline for the states? Explain. The mean is misleading. Sample answer: All states but one have over 3,000 miles of shoreline. The outlier of 89 causes the mean to be a poor choice to describe the data.

4. Which measure of central tendency most accurately describes the data? median

Practice: Word Problems, p. 104

NAME ______________ DATE ________ PERIOD ____

2-8 Practice: Word Problems

Misleading Statistics

QUIZ SCORES **For Exercises 1 and 2, use the data shown in the table below. The table shows the quiz grades for Ms. Andrey's and Mr. Luna's classes.**

Quiz Scores	
Ms. Andrey's Class	Mr. Luna's Class
10	20
15	20
25	25
25	29
12	26

BOOK SALES **For Exercises 3 and 4, use the table below. It shows the number of books sold each day for 20 days.**

Book Sales Per Day			
23	18	23	15
24	16	0	11
19	10	13	17
12	23	11	16
36	24	12	27

1. Ms. Andrey claims the average score on a quiz in her class was 25. Mr. Luna claims the average score on a quiz in his class is 25. Explain how they arrived at these figures. Ms. Andrey used the mode, while Mr. Luna used the median.

2. What additional information could be useful in analyzing the data? Sample answer: There is no information concerning total possible points on each quiz. Without this information, the data is meaningless.

3. Find the mean, median, and mode of the data. Which measure of central tendency would be misleading in describing the book sales? Explain. Mean: 17.5; median: 16.5; mode: 23; the mode would be misleading. Sample answer: Only on 6 days out of 20 did sales reach 23 or more.

4. Which value would most accurately describe the data? Explain. Sample answer: The median; the data has outliers, so the mean would not be accurate. Also, there are no big gaps in the middle of the data.

CHAPTER 2 Study Guide and Review

Vocabulary Assessment

- This alphabetical list of vocabulary terms in Chapter 2 includes a page reference where each term was introduced.
- A vocabulary review/test for Chapter 2 is available on p. 120 of the *Chapter 2 Resource Masters*.

Vocabulary PuzzleMaker

ELL The online Vocabulary PuzzleMaker improves students' mathematics vocabulary using four puzzle formats – crossword, scramble, word search using a word list, and word search using clues. Students can work on a computer screen or from a printed handout.

MindJogger Videoquizzes

ELL MindJogger Videoquizzes provide an alternative review of concepts presented in this chapter. Students work in teams in a game show format to gain points for correct answers. The questions are presented in three rounds.

Round 1 Concepts (5 questions)
Round 2 Skills (4 questions)
Round 3 Problem Solving (4 questions)

ExamView® Pro Test Bank CD-ROM

Use the networkable ExamView® Pro Test Bank CD-ROM to:

- Create **Multiple versions** of tests.
- Create **Modified** tests for **Inclusion** students with one mouse click.
- **Edit** existing questions and **Add** your own questions.
- Build tests aligned with state standards using built-in **State Curriculum Correlations.**
- Change **English** tests to **Spanish** with one mouse click and vice versa.

CHAPTER 2 Study Guide and Review

Vocabulary and Concept Check

bar graph (p. 85)	line graph (p. 60)	outlier (p. 65)
box-and-whisker plot (p. 80)	line plot (p. 64)	range (p. 65)
cluster (p. 65)	lower extreme (p. 80)	scale (p. 54)
data (p. 54)	lower quartile (p. 80)	scatter plot (p. 61)
frequency table (p. 54)	mean (p. 69)	statistics (p. 54)
histogram (p. 86)	measures of central tendency (p. 69)	stem (p. 76)
interquartile range (p. 81)	median (p. 70)	stem-and-leaf plot (p. 76)
interval (p. 54)	mode (p. 70)	upper extreme (p. 80)
leaf (p. 76)		upper quartile (p. 80)

State whether each sentence is *true* or *false*. If *false*, replace the underlined word to make a true sentence. 5. false; outlier

1. The range is the difference between the greatest and the least values in a set of data. true
2. The mode divides a set of data in half. false; median
3. A graph that uses bars to make comparisons is a bar graph. true
4. A scatter plot shows two sets of related data on the same graph. true
5. The median is a data value that is quite separated from the rest of the data.
6. The mean is the arithmetic average of a set of data. true
7. The number or item that appears most often in a set of data is the mode. true
8. The lower quartile is the middle number in a set of data when the data are arranged in numerical order. false; median

Lesson-by-Lesson Exercises and Examples

2-1 Frequency Tables (pp. 54–57)

Make a frequency table of each set of data.

9. 9–10. See margin.

Hours of Sleep Per Night					
9	6	7	8	9	5	8	7
6	7	6	7	7	8	9	8

10.

Favorite Sport									
B	F	H	S	K	F	S	B	B	F
B	F	K	S	F	F	B	K	F	S

B = baseball, F = football, H = hockey, S = soccer, K = basketball

Example 1 Make a frequency table of the data at the right.

Age of Pet (yr)					
4	2	2	3	8	9
5	9	8	7	10	4

interval: 4
scale: 0–11

Age (yr)	Tally	Frequency
0–3	III	3
4–7	IIII	4
8–11	~~IIII~~	5

msmath2.net/vocabulary_review

FOLDABLES™ Study Organizer

For more information about Foldables, see *Teaching Mathematics with Foldables.*

Have students look through the chapter to make sure they have included examples in their Foldables for each lesson or topic in the chapter.

Encourage students to refer to their Foldables while completing the Study Guide and Review and to use them in preparing for the Chapter Test.

2-2 Making Predictions (pp. 60–63)

11. **ENTERTAINMENT** The graph shows the attendance at a theme park. If the trend continues, predict the attendence in 2005. **Sample answer: 17 million**

That's Entertainment

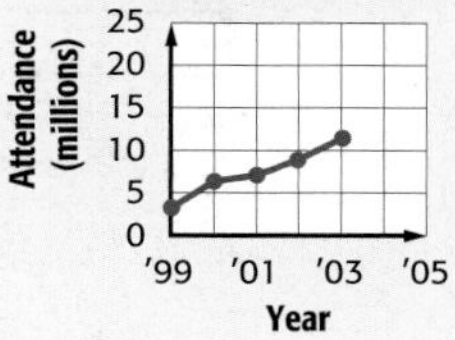

Example 2

SPORTS The graph shows the 400-meter winning times in the Summer Olympics. Predict the winning time in 2004.

By extending the graph, it appears that the winning time in 2004 will be about 42.1 seconds.

400-meter Run

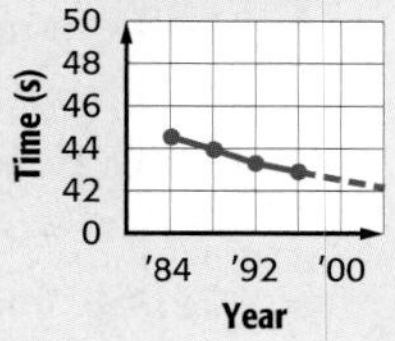

2-3 Line Plots (pp. 64–68)

Make a line plot for each set of data. Identify any clusters, gaps, or outliers.

12. 10, 12, 10, 8, 13, 10, 8, 12
13. 7, 8, 8, 9, 14, 9, 8, 7
14. 43, 41, 42, 45, 43, 42, 43, 46, 44, 44
15. 21, 23, 19, 21, 22, 28, 20, 22, 23, 19

12–15. See margin.

Example 3 **Make a line plot for the data set 72, 75, 73, 72, 74, 68, 73, 74, 74, 75, and 73. Identify any clusters, gaps, or outliers.**

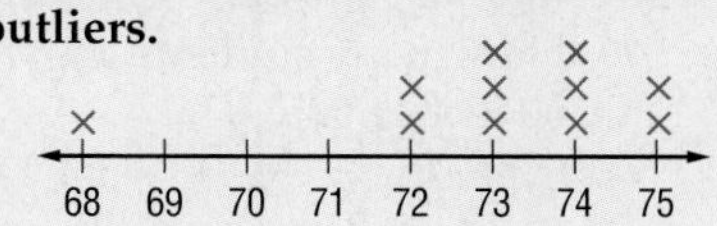

cluster: 72–75, gap: 68–72, outlier: 68

2-4 Mean, Median, and Mode (pp. 69–72)

Find the mean, median, and mode for each set of data. **17. 24.4; 24; 18 and 31**

16. 2, 3, 4, 3, 4, 3, 8, 7, 2 **4; 3; 3**
17. 31, 24, 26, 18, 23, 31, 18
18. 89, 76, 93, 100, 72, 86, 74 **84.3; 86; none**
19. 125, 108, 172, 136, 121, 112, 148, 136 **132.3, 130.5, 136**

Example 4 **Find the mean, median, and mode for the data set 23, 22, 19, 19, and 20.**

mean: $\frac{23 + 22 + 19 + 19 + 20}{5}$ or 20.6

median: 20 middle value of the ordered set

mode: 19 value that occurs most often

2-5 Stem-and-Leaf Plots (pp. 76–79)

Make a stem-and-leaf plot for each set of data. **20–23. See margin.**

20. 29, 54, 31, 26, 38, 46, 23, 21, 32, 37
21. 75, 83, 78, 85, 87, 92, 78, 53, 87, 89, 91
22. 9, 5, 12, 21, 18, 7, 16, 24, 11, 10, 3, 14
23. 234, 218, 229, 204, 221, 219, 201, 225

Example 5 **Make a stem-and-leaf plot for the data set 12, 15, 17, 20, 22, 22, 23, 25, 27, and 45.**

The tens digits form the stems, and the ones digits form the leaves.

Stem	Leaf
1	2 5 7
2	0 2 2 3 5 7
3	
4	5

$2|3 = 23$

Answers

9. Sample answer:

Hours of Sleep	Tally	Frequency
5	I	1
6	III	3
7	~~IIII~~	5
8	IIII	4
9	III	3

10. Sample answer:

Favorite Sport	Tally	Frequency
baseball	~~IIII~~	5
football	~~IIII~~ II	7
hockey	I	1
soccer	IIII	4
basketball	III	3

Answers

12.

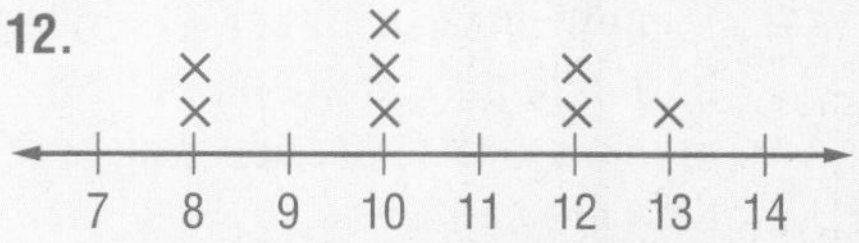

Sample answer: no significant clusters, gaps, or outliers

13.

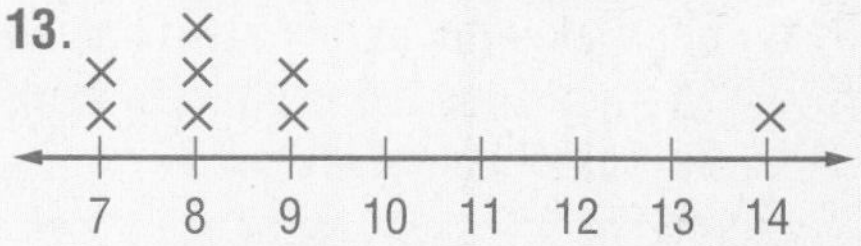

Sample answer: cluster 7–9; gap 9–14; outlier 14

14.

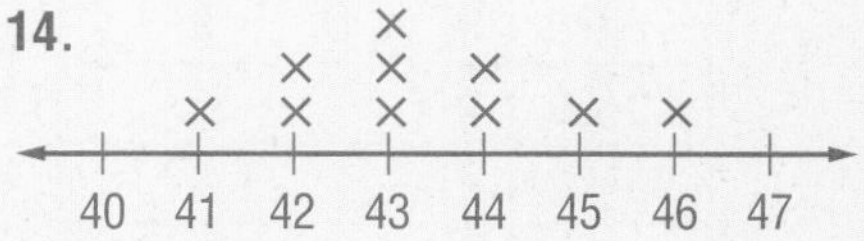

Sample answer: cluster 42–44; no gaps or outliers

15.

Sample answer: cluster 19–23; gap 23–28; outlier 28

20.

Stem	Leaf
2	1 3 6 9
3	1 2 7 8
4	6
5	4

$3|2 = 32$

21.

Stem	Leaf
5	3
6	
7	5 8 8
8	3 5 7 7 9
9	1 2

$7|5 = 75$

22.

Stem	Leaf
0	3 5 7 9
1	0 1 2 4 6 8
2	1 4

$1|2 = 12$

23.

Stem	Leaf
20	1 4
21	8 9
22	1 5 9
23	4

$23|4 = 234$

2-6 Box-and-Whisker Plots (pp. 80–83)

GRADES For Exercises 24–27, use the following math test scores.

83, 92, 75, 59, 94, 82, 88, 79, 92, 90, 85, 96

24. Find the lower extreme, LQ, median, UQ, and upper extreme.
25. Determine the interquartile range. 11.5
26. Identify any outliers. 59
27. Draw a box-and-whisker plot of the data. See Ch. 2 Answer Appendix.

24. 59; 80.5; 86.5; 92; 96

Example 6 **Make a box-and-whisker plot of the data set 12, 12, 13, 14, 14, 15, 15, 15, 16, 16, 17, 18, and 20.**

lower extreme = 12 upper extreme = 20
median = 15 LQ = 13.5 UQ = 16.5

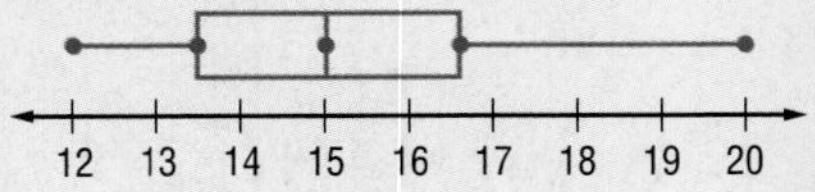

2-7 Bar Graphs and Histograms (pp. 85–89)

For Exercises 28 and 29, refer to the graph at the right.

28. How many students are 50 to 59 inches tall? 13
29. How many students are at least 60 inches tall? 8

Example 7 **The graph shows the heights to the nearest inch of students in a classroom. Is this a bar graph or a histogram? Explain.**

This is a histogram because it shows the number of students having the heights described by each interval.

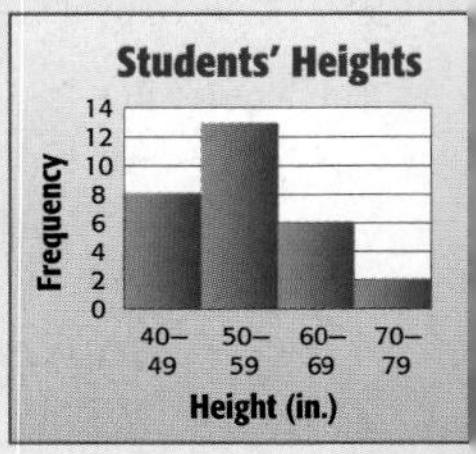

2-8 Misleading Statistics (pp. 92–95)

30. **SALES** The graph below shows the monthly CD sales at the Music Barn. Why is the graph misleading?

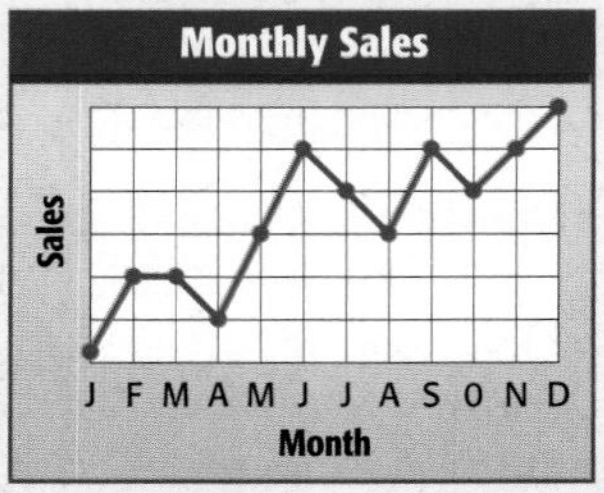

There are no labels on the vertical scale.

Example 8 **The graph shows the pounds of aluminum cans that students recycled. Why is this graph misleading?**

The vertical scale does not have equal intervals. It has intervals of 200 and 50.

Answers (Practice Test)

1. Sample answer: Range describes how data are spread out; mean, median, and mode describe how data are centered.
2. Sample answer: When the data are ordered, the median separates the set into two halves. The quartiles then separate each half into two halves.
3.

Grade	Tally	Frequency
60–69	I	1
70–79	III	3
80–89	~~IIII~~	5
90–99	IIII	4

4. See bottom margin.

9.

Stem	Leaf
2	6
3	1 7
4	2
5	3 7 8 9

4|2 = 42

10.

Stem	Leaf
42	2 5 5 9
43	0
44	
45	
46	1 7

43|1 = 431

12. See bottom margin.
13. Sample answer: The vertical axis has no labels. The sales could be in dollars, hundreds of dollars, or thousands of dollars.
14. Sample answer:

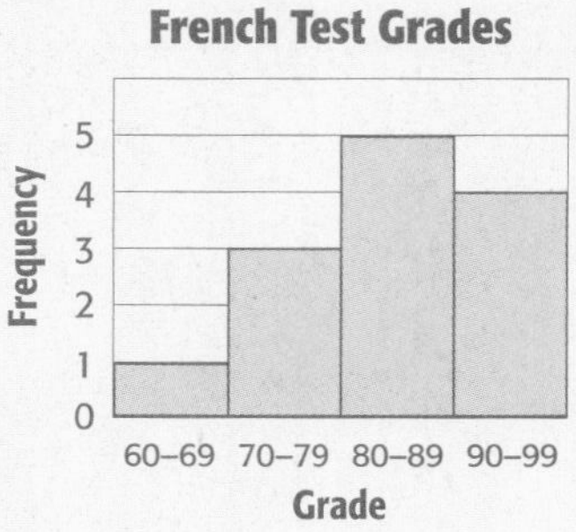

Answers (Practice Test)

4.

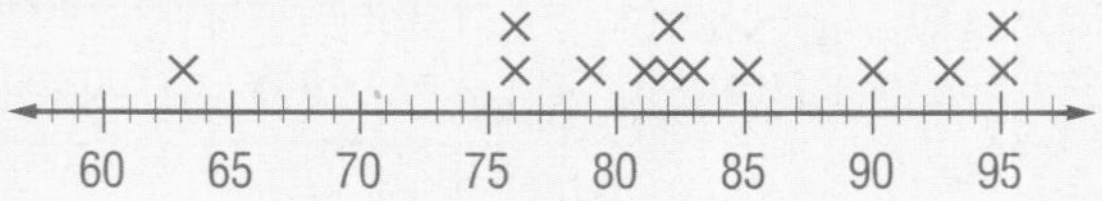

12.

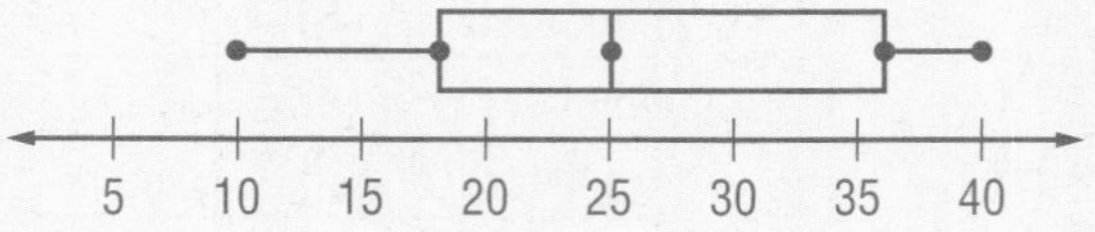

Chapter 2 Practice Test

Vocabulary and Concepts

1. **Explain** how the range is different from the mean, median, and mode. **1–2. See margin.**
2. **Describe** how quartiles separate a set of data.

Skills and Applications

TESTS **For Exercises 3–5, use the French test grades shown below.** **3–4. See margin.**

95, 76, 82, 90, 83, 76, 79, 82, 95, 85, 93, 81, 63

3. Make a frequency table of the data.
4. Make a line plot of the data.
5. Identify any clusters, gaps, or outliers. **Sample answer: cluster 76–85; gap 63–76; outlier 63**
6. **EMPLOYMENT** The line graph shows the percent of women who had jobs outside the home from 1975 to 2000. Use the graph to predict the percent of women who will have jobs outside the home in 2005. **Sample answer: 46%**

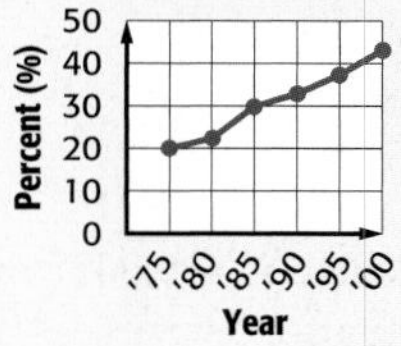

Find the mean, median, and mode for each set of data. Round to the nearest tenth if necessary.

7. 4, 6, 11, 7, 4, 11, 4 **6.7; 6; 4**
8. 12.4, 17.9, 16.5, 10.2 **14.3; 14.5; none**

Make a stem-and-leaf plot for each set of data. **9–10. See margin.**

9. 37, 59, 26, 42, 57, 53, 31, 58
10. 461, 422, 430, 425, 425, 467, 429

ALLOWANCES **For Exercises 11 and 12, use the monthly allowances below.**

$25, $32, $40, $10, $18, $15, $30, $40, $25, $36, $20

11. Find the lower extreme, LQ, median, UQ, and upper extreme. **10; 18; 25; 36; 40**
12. Make a box-and-whisker plot of the data. **See margin.**
13. **SALES** The line graph shows ship sales at Marvin's Marina in thousands of dollars. Why is the graph misleading? **See margin.**

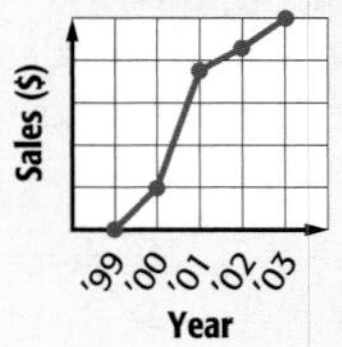

14. Make a histogram for the French test grades given for Exercises 3–5. **See margin.**

Standardized Test Practice

15. **MULTIPLE CHOICE** Choose the statement that is *not* correct. **B**
    - (A) The bars on a histogram represent the frequency of data in intervals.
    - (B) The bars on a bar graph represent the number of data items in each interval.
    - (C) The bars on a bar graph have space between them.
    - (D) There is no space between the bars on a histogram.

msmath2.net/chapter_test

Chapter 2 Practice Test

Assessment Options

Vocabulary Test A vocabulary review/test for Chapter 2 can be found on p. 120 of the *Chapter 2 Resource Masters.*

Chapter Tests There are six Chapter 2 Tests and an Extended Response Assessment task available in the *Chapter 2 Resource Masters.*

Chapter 2 Tests			
Form	**Type**	**Level**	**Pages**
1	MC	basic	107–108
2A	MC	average	109–110
2B	MC	average	111–112
2C	FR	average	113–114
2D	FR	average	115–116
3	FR	advanced	117–118

MC = multiple-choice questions
FR = free-response questions

Extended Response Assessment

Extended Response tasks for Chapter 2 can be found on p. 119 of the *Chapter 2 Resource Masters.* A sample scoring rubric for these tasks appears on p. A28.

Unit 1 Test

A unit test/review can be found on pp. 127–128 of the *Chapter 2 Resource Masters.*

Portfolio Suggestion

Introduction Have students write a paragraph about the different methods they studied for presenting data. Have them include an example of each.

Ask students to write about how statistics can be presented to mislead the intended audience. Have them include a few reasons why this is a concern.

Chapter 2 Standardized Test Practice

These two pages contain practice questions in the various formats that can be found on the most frequently given standardized tests.

A student recording sheet for these two pages can be found on page A1 of the *Chapter 2 Resource Masters.*

Standardized Test Practice Student Recording Sheet, p. A1

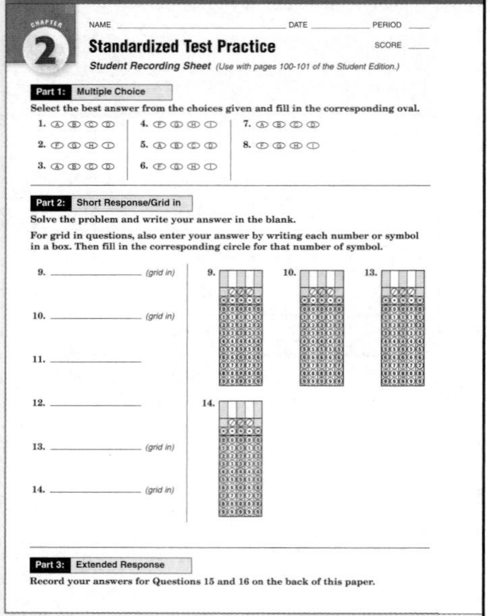
NAME ______ DATE ______ PERIOD ______

Standardized Test Practice SCORE ______

Student Recording Sheet (Use with pages 100-101 of the Student Edition.)

Part 1: Multiple Choice

Select the best answer from the choices given and fill in the corresponding oval.

Part 2: Short Response/Grid in

Solve the problem and write your answer in the blank.

For grid in questions, also enter your answer by writing each number or symbol in a box. Then fill in the corresponding circle for that number of symbol.

Part 3: Extended Response

Record your answers for Questions 15 and 16 on the back of this paper.

Additional Practice

See pp. 125–126 of the *Chapter 2 Resource Masters* for additional standardized test practice.

Answer (p. 101)

16a. Swimming Practice

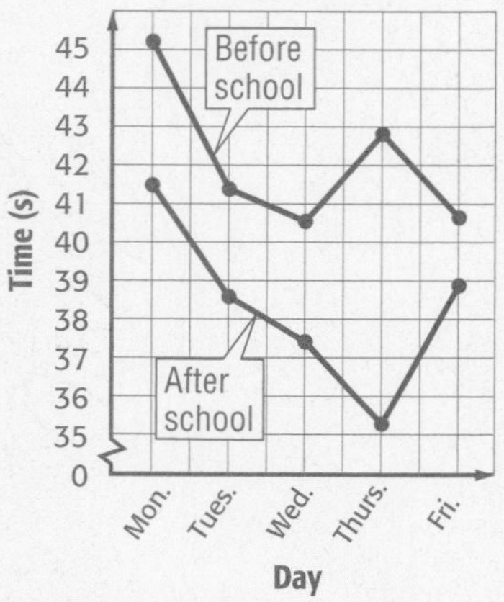

Chapter 2 Standardized Test Practice

PART 1 Multiple Choice

Record your answers on the answer sheet provided by your teacher or on a sheet of paper.

1. Which is greater than 6.8? (Prerequisite Skill, p. 556) **D**

   Ⓐ 6.08 Ⓑ 6.79
   Ⓒ 6.80 Ⓓ 6.9

2. What is the value of the expression $5 + 2 \times 4^2 - 9$? (Lesson 1-3) **H**

   Ⓕ 4 Ⓖ 19
   Ⓗ 28 Ⓘ 103

3. Which is 430,000 written in scientific notation? (Lesson 1-9) **A**

   Ⓐ 4.3×10^5
   Ⓑ 4.3×10^6
   Ⓒ 43×10^5
   Ⓓ 43×10^6

4. The line graph shows a baby's length during the first 8 months. If the pattern of growth continues, predict the baby's length at 9 months old. (Lesson 2-2) **G**

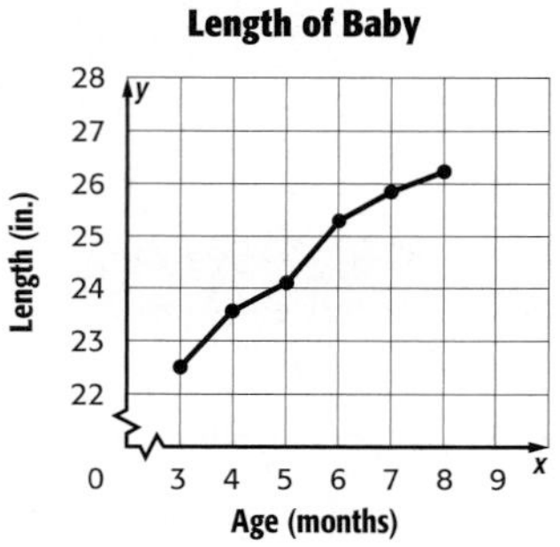

   Ⓕ 26 in. Ⓖ 27 in.
   Ⓗ 28 in. Ⓘ 29 in.

5. The line plot shows the number of minutes ten different students spent on homework last night. What is the range of the data? (Lesson 2-3) **B**

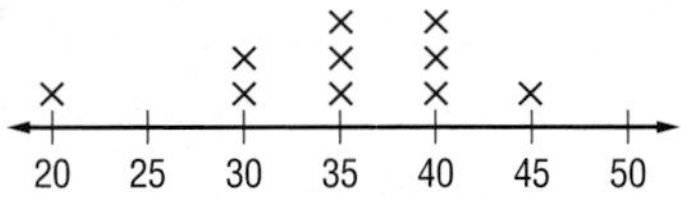

   Ⓐ 20 Ⓑ 25 Ⓒ 30 Ⓓ 50

6. The table shows the length of the phone calls made in one household on a given day. What is the median? (Lesson 2-4) **H**

   Ⓕ 2 min
   Ⓖ 6 min
   Ⓗ 7 min
   Ⓘ 8 min

Length of Phone Calls (min)				
2	3	10	2	8	1
6	7	25	7	6	

7. The stem-and-leaf plot shows the average monthly temperature in one city for a year. What is the mode? (Lesson 2-5) **A**

Stem	Leaf
4	2 2 2 5
5	5 6
6	4
7	0 8 8
8	2 5 5|6 = 56

   Ⓐ 42 Ⓑ 55 Ⓒ 65 Ⓓ 78

8. The box-and-whisker plot below summarizes the ages of Leila's cousins. Which statement is *not* true? (Lesson 2-6) **G**

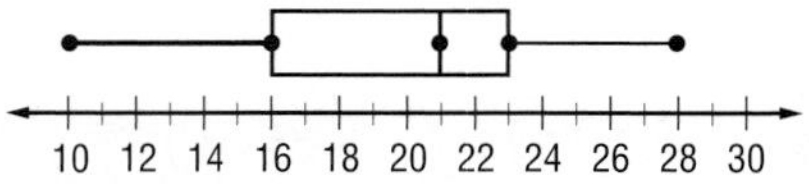

   Ⓕ One fourth of Leila's cousins are between 10 and 16 years old.
   Ⓖ Half of Leila's cousins are between 16 and 21 years old.
   Ⓗ The range of ages is 18 years.
   Ⓘ Half of Leila's cousins are older than 2

The Princeton Review — ***Log On*** **for Test Practice**

The Princeton Review offers additional test-taking tips and practice problems at their web site. Visit **princetonreview.com** or **review.com**

ExamView® Pro Test Bank CD-ROM

Special banks of standardized test questions similar to those on the TIMSS and NAEP grade 8 tests can be found on this CD-ROM.

Aligned and verified by The Princeton Review

ART 2 Short Response/Grid In

ecord your answers on the answer sheet ovided by your teacher or on a sheet of per.

. What is the value of $14 + x^2$ if $x = 3$? (Lesson 1-4) **23**

. Find the missing number in the sequence below. (Lesson 1-7) **25.3**

24.7, 25.0, __?__, 25.6, 25.9, …

. The frequency table summarizes the number of miles Dino ran daily to train for a race. What interval shows the number of miles that he ran most frequently? (Lesson 2-1) **5–9**

Number of Miles	Tally	Frequency
0–4	𝍸 IIII	9
5–9	𝍸 𝍸 III	13
10–14	IIII	4
15–19	I	1

. If the number 18 was taken out of the number data set 7, 18, 46, 24, 2, 44, 31, 11, would the median or the mean change the most? (Lesson 2-4) **median**

. What is the upper extreme of the box-and-whisker plot below? (Lesson 2-6) **40**

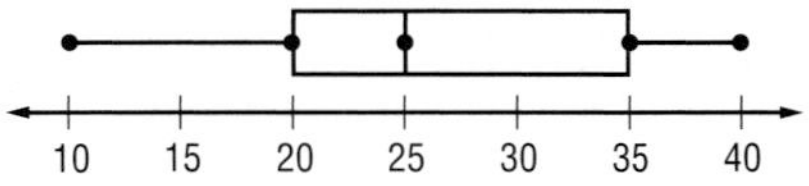

. The graph below shows the sales of compact discs during a five-day sale. How many more compact discs were sold on Saturday than on Wednesday? (Lesson 2-7) **40**

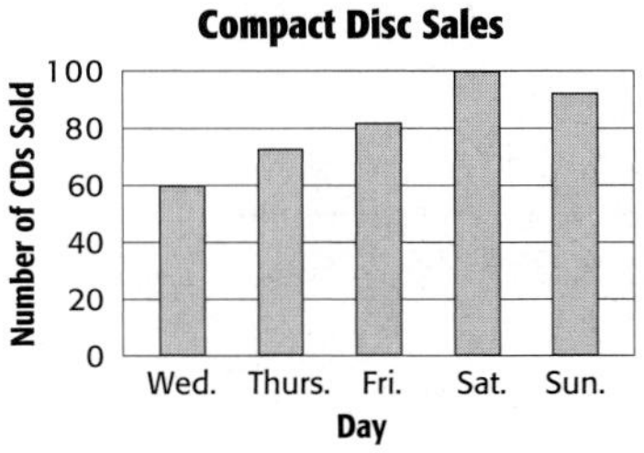

TEST-TAKING TIP

Question 15 Make sure you have answered every part of an open-ended question. If you find that you cannot answer the question entirely, do as much as you can, because you may earn partial credit.

PART 3 Extended Response

Record your answers on a sheet of paper. Show your work.

15. The table shows how the value of a painting increased over ten years. (Lesson 2-2)

Year	Value	Year	Value
1993	$350	1998	$1,851
1994	$650	1999	$2,151
1995	$950	2000	$2,451
1996	$1,250	2001	$2,752
1997	$1,551	2002	$3,052

a. Make a line graph of the data. **See margin.**

b. Use the graph to predict what the value of the painting will be in 2006. **Sample answer: $4,000**

16. Dominique had swimming practice before school one week and after school the following week. Her timed results are in the table below. (Lesson 2-7) **a–b. See margin.**

Day	Time Before School (s)	Time After School (s)
Monday	45.2	41.5
Tuesday	41.3	38.7
Wednesday	40.6	37.4
Thursday	42.9	35.2
Friday	40.7	38.9

a. Make a double-line graph of the data. Let the x-axis represent the days and let the y-axis represent the times. Title the graph and label each part.

b. Write a sentence comparing the data.

msmath2.net/standardized_test

Standardized Test Practice

Rubrics The following rubrics are sample scoring devices for short response and extended response questions.

Short Response

Points	Description
2	The student demonstrates a <u>thorough understanding</u> of the mathematics of the task. The response may contain minor flaws that do not detract from the demonstration of a thorough understanding.
1	The student has provided a response that is only <u>partially correct</u>.
0	The student has provided a <u>completely incorrect</u> solution or no response at all.

Extended Response

Points	Description
4	The student demonstrates a <u>thorough understanding</u> of the mathematics of the task. The response may contain minor flaws that do not detract from the demonstration of a thorough understanding.
3	The student demonstrates an <u>understanding</u> of the mathematics of the task. The response is essentially correct and demonstrates an essential but less than thorough understanding of the mathematics.
2	The student demonstrates only a <u>partial understanding</u> of the mathematics of the task. Although the student may have used the correct approach to a solution or may have provided a correct solution, the work lacks an essential understanding of the underlying mathematical concepts.
1	The student demonstrates a <u>very limited understanding</u> of the mathematics of the task. The response is incomplete and exhibits many flaws.
0	The student provides a <u>completely incorrect</u> solution or no response at all.

A more detailed 4-point rubric is provided on page A2 of the *Chapter 2 Resource Masters*.

Answers

15a.

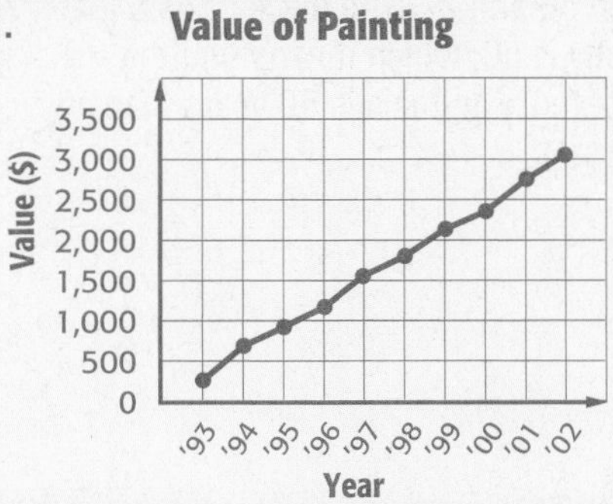

16a. See p. 100 side margin.

16b. Sample answer: The times after school were all less than the times before school, so Dominique swam faster after school.

Pages 56–57, Lesson 2-1

5. Sample answer:

Number of Movie Rentals	Tally	Frequency
0–2	𝍷𝍷𝍷	3
3–5	𝍸	5
6–8	𝍷𝍷	2
9–11	𝍸 𝍷𝍷𝍷	8
12–14	𝍷𝍷𝍷	3
15–17	𝍷𝍷	2
18–20	𝍷	1

6. Sample answer:

Favorite Breakfast Food	Tally	Frequency
cereal	𝍸 𝍷𝍷	7
eggs	𝍸 𝍷	6
pancakes	𝍸	5

8. Sample answer:

Age of Students	Tally	Frequency
12	𝍷𝍷𝍷	3
13	𝍸 𝍷𝍷𝍷𝍷	9
14	𝍸 𝍷𝍷𝍷	8

9. Sample answer:

Year	Tally	Frequency
1900–1919	𝍷𝍷	2
1920–1939	𝍷𝍷𝍷𝍷	4
1940–1959	𝍷𝍷𝍷𝍷	4
1960–1979	𝍸 𝍷𝍷𝍷𝍷	9
1980–1999	𝍸 𝍷𝍷𝍷	8

10. Sample answer:

Music Source	Tally	Frequency
CDs	𝍸 𝍷𝍷𝍷𝍷	9
radio	𝍸 𝍸	10
Internet	𝍸	5
music videos	𝍸 𝍷	6

11. Sample answer:

Chore	Tally	Frequency
wash dishes	𝍸 𝍷𝍷𝍷	8
vacuum	𝍸	5
clean room	𝍸 𝍸 𝍷𝍷	12
do laundry	𝍸 𝍷	6
do yard work	𝍷𝍷𝍷𝍷	4

13. Sample answer:

Years	Tally	Frequency
1970–1979	𝍷𝍷	2
1980–1989	𝍷𝍷𝍷	3
1990–1999	𝍸 𝍷𝍷𝍷	8
2000–2009	𝍸 𝍷𝍷	7

16. Sample answer:

Number of Calories	Tally	Frequency
51–70	𝍷𝍷	2
71–90	𝍷𝍷𝍷𝍷	4
91–110	𝍸 𝍸 𝍸 𝍸 𝍷	21
111–130	𝍸	5
131–150	𝍷𝍷𝍷	3

19. Sample answer: Most of the scooter prices are spread out evenly among the intervals, with the \$91–\$100 category containing the greatest number. An advantage of using this scale and interval is that you can get a more detailed distribution. A disadvantage is that it is difficult to see clusters or trends.

Price ($)	Tally	Frequency
11–20	𝍷𝍷	2
21–30	𝍷𝍷𝍷𝍷	4
31–40	𝍷𝍷𝍷𝍷	4
41–50	𝍸	5
51–60	𝍸	5
61–70	𝍷𝍷	2
71–80	𝍷𝍷𝍷𝍷	4
81–90	𝍷𝍷𝍷	3
91–100	𝍸 𝍷	6

Pages 66–67, Lesson 2-3

2. Sample answer:

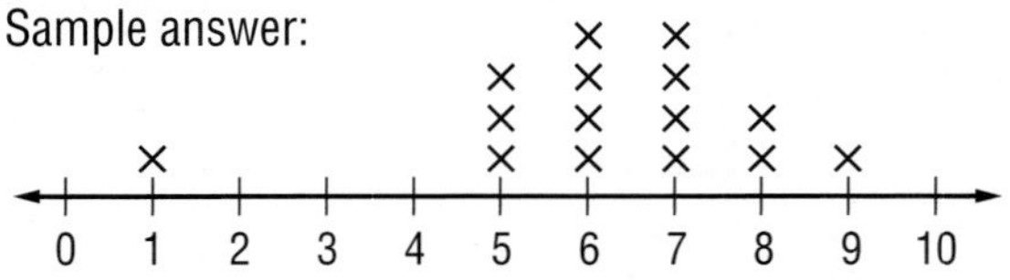

5.

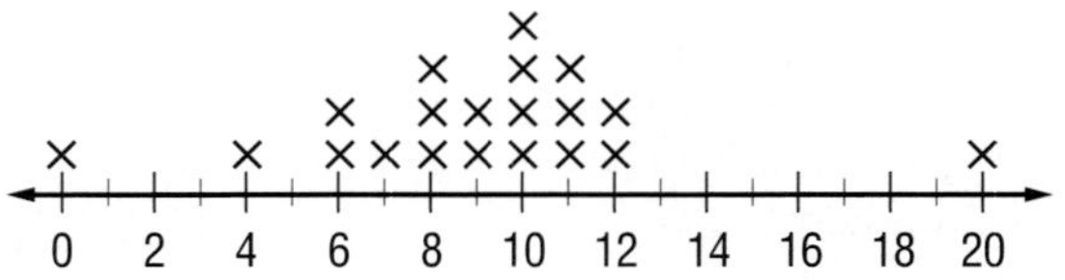

19.

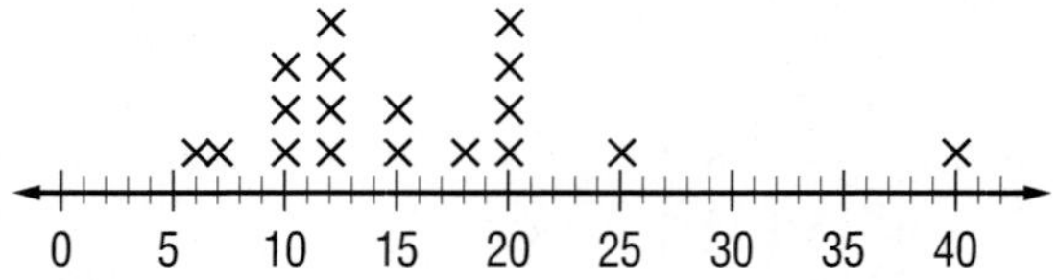

22. Sample answer:

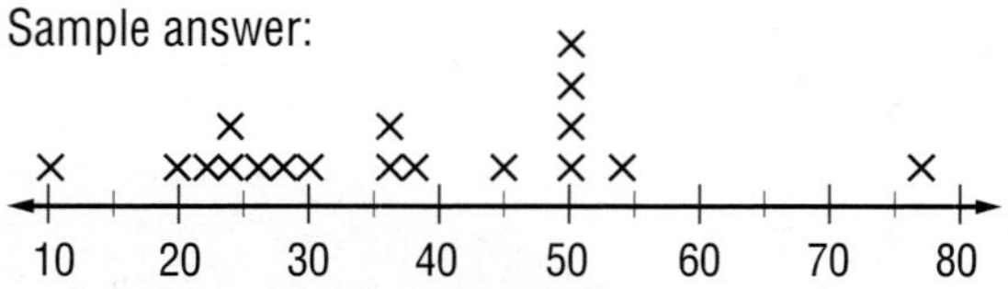

The data are more spread out, with a range of 67 compared with the previous range of 34. There is a cluster from 20 to 30 and a gap between 54 and 77. The piece of data that occurs most frequently is 50, which means that the maximum life span of more animals is 50 years than any other number of years.

Pages 78–79 Lesson 2-5

1. Both a stem-and-leaf plot and frequency table show the frequency of data occurring. However, a stem-and-leaf plot shows individual data values and a frequency table shows only the number of data items occurring in each interval.

2. Sample answer:

Stem	Leaf
1	1 2 3 4
2	5 5 6 7
3	
4	7 8

2|5 = 25

5.

Stem	Leaf
1	6 9
2	1 3 5 5 9
3	1 3 4 5 9
4	2 7 8

3|4 = 34

6.

Stem	Leaf
22	5 7 9
23	0
24	3 6
25	
26	7 9 9
27	8 8 8

26|7 = 267

10. Sample answer: The mean and median number of wins are close, indicating the data are fairly well centered in the middle of the range of possible wins. About half the data values cluster from 28 to 40. There is a small gap from 54 to the largest value, 61.

11.

Stem	Leaf
1	3 3 5
2	4 8
3	0 1 2 2 5 6 8 8 8

1|3 = 13

12.

Stem	Leaf
5	5 6
6	0 0 2 3 6 7 9
7	0
8	0 0
9	0 1 3 3 5 6

8|0 = 80

13.

Stem	Leaf
40	3
41	1
42	
43	3 4 9
44	2 9
45	1 4 4 7 8 9

43|9 = 439

14.

Stem	Leaf
8	0 0 1 2 2 5 7 7 8
9	0 2 3 3 9 9
10	0

8|0 = 8.0

18. Sample answer: About half the prices are from $100 to $119. There is a gap from $127 to $150, in which no prices are listed.

19.

Stem	Leaf
5	1 2 7
6	0 1 2 3 5 7 8 8
7	3 3 3 4 5 6 7 8 9
8	1 2 2 4 5 9
9	4 4 7
10	6 8

9|4 = 94

21. Sample answer: Most of the game point totals are clustered from 60 to 89 points. There do not appear to be any gaps or outliers.

23. Sample answer:

Price ($)	Tally	Frequency
20–29	||||	4
30–39	𝍸 𝍸	10
40–49	𝍸	5
50–59	|	1

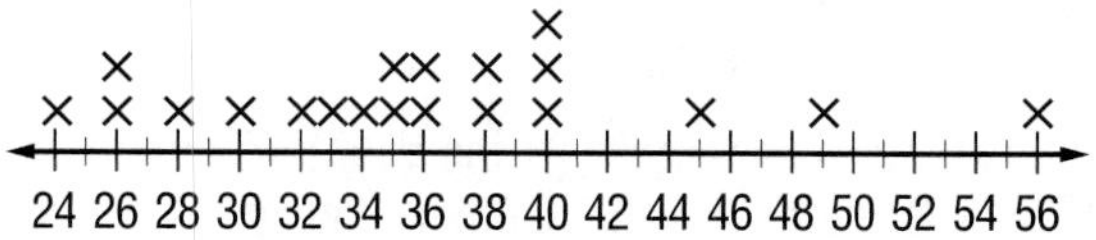

Stem	Leaf
2	4 6 6 8
3	0 2 3 4 5 5 6 6 8 8
4	0 0 0 5 9
5	6

3|5 = $35

All three representations show the frequency of data occurring. The frequency table shows intervals of data and is useful in comparing price ranges. The line plot gives a good picture of the spread of the data. The stem-and-leaf plot shows individual prices as in the line plot, as well as intervals, such as $20s, $30s, and so on. See students' favorites and reasons.

28.

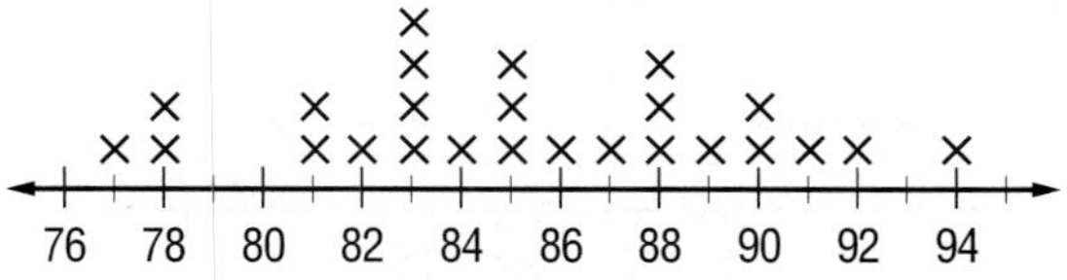

Pages 82–83, Lesson 2-6

1. Sample answer: In a stem-and-leaf plot, you can see each data item and how the data are clustered. In a box-and-whisker plot, you can immediately see the median and how the data are dispersed.

2. Sample answer:

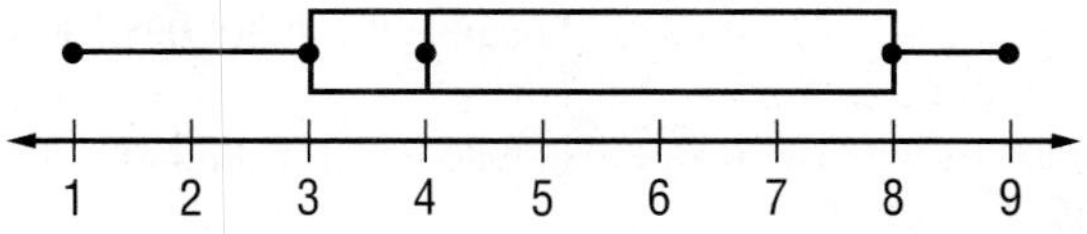

5.

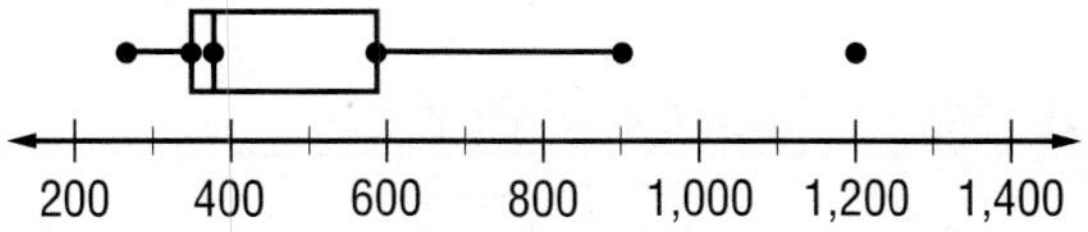

8.

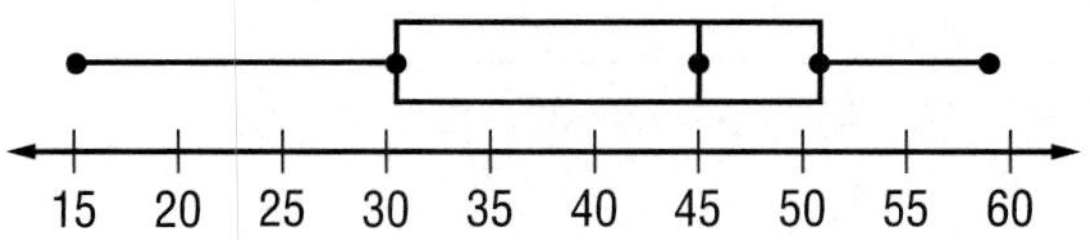

12.

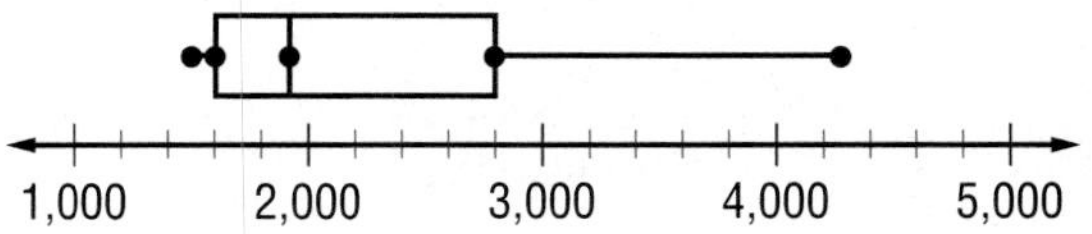

Pages 87–89, Lesson 2-7

4.

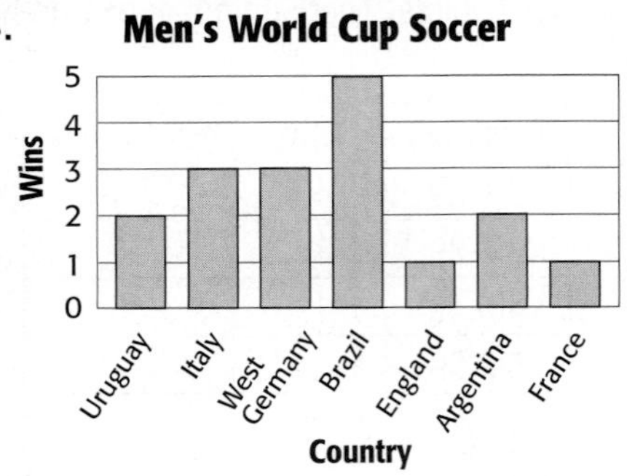

5. The horizontal axis represents the countries and the vertical axis represents the number of wins.

16. Sample answer: The mean is 8.7 representatives per state. The median is somewhere in the interval from 1 to 10 representatives. The 53 for California is an outlier from the rest of the data. The fact that 38 states have 10 or fewer members in the House of Representatives is significant because it means that just a few states, including California, have large numbers of representatives that could carry influence in the House.

17. Sample answer: Both histograms show that many players have salaries in the lowest interval of the histogram, less than $1 million. Both histograms show one, two, or three players in several of the other intervals. The Diamondbacks' salaries show only three players' salaries of more than $7 million, while the Yankees have seven players in that same range.

22.

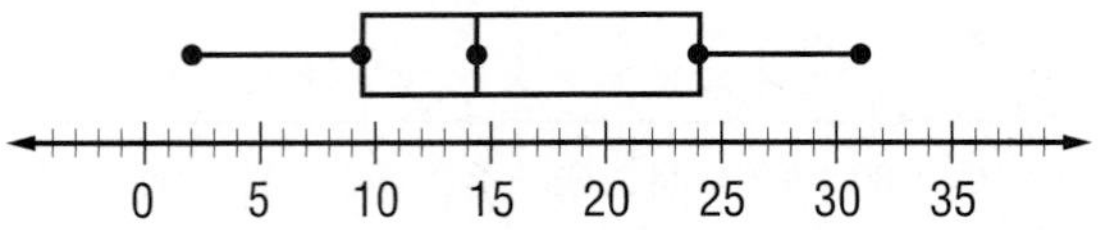

Page 93, Lesson 2-8

3. The cost is increasing over time. In about a year's time, the cost has more than tripled.

4. Sample answer: The pattern of the cost tripling each year is not likely to continue.

5. Sample answer: The median, because it will not be affected by the high prices.

6. Sample answer: The mean, because it will be higher due to the increase in prices.

Page 98, Study Guide and Review

27.

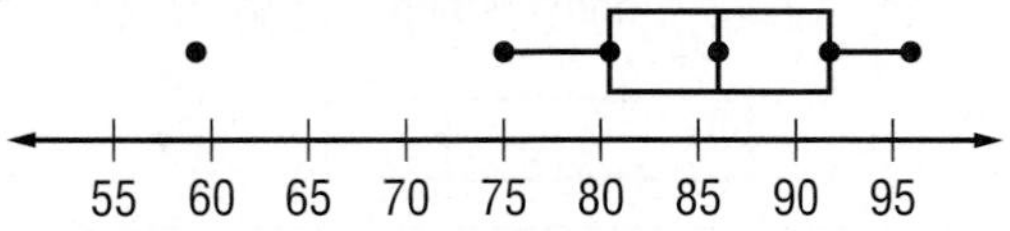

NOTES

UNIT 2 Notes

Introduction

In this unit, students will explore integers. They will find absolute values, compare and order integers, and add, subtract, multiply, and divide integers. Students will use a coordinate plane to locate and graph ordered pairs as points and represent algebraic relationships.

Students will apply their knowledge of operations with integers to solve one- and two-step equations. They also will extend their graphing skills to graph linear equations and inequalities. Students will also learn to find the slope of a line.

Assessment Options

Unit 2 Test Pages 241–242 of the *Chapter 4 Resource Masters* may be used as a test or a review for Unit 2. This assessment contains both multiple-choice and short-response items.

UNIT 2

Integers and Algebra

Chapter 3

Algebra: Integers

Chapter 4

Algebra: Linear Equations and Functions

Your study of math includes more than just whole numbers and decimals. In this unit, you will use negative numbers to describe many real-life situations and you will solve and graph equations that represent them.

What's Math Got to Do With It? Real-Life Videos

What's Math Got to Do With It? Real-Life Videos engage students, showing them how math is used in everyday situations. Use Video 2 to discuss how linear equations and slope are used in soccer, baseball, and horseback riding.

WebQuest INTERDISCIPLINARY PROJECT — **MATH and GEOGRAPHY**

THE WIDE WORLD OF SOCCER

Soccer fans, get up on your feet! You've been selected by an elite committee to join us on a world-wide soccer adventure. Along the way, you'll be gathering data about the geography of countries where soccer is the favorite sport. You'll also make some predictions about the future of soccer in the United States. We will be leaving on our adventure very shortly, so pack your math tools and your thinking cap. This is one adventure you don't want to miss.

Log on to msmath2.net/webquest to begin your WebQuest.

WebQuest INTERDISCIPLINARY PROJECT

The Wide World of Soccer

A WebQuest is an online project in which students do research on the Internet, gather data, and make presentations using word processing, graphing, page-making, or presentation software. In each chapter, students advance to the next step in their WebQuest. At the end of Chapter 4, the project culminates with a presentation of their findings.

Teaching suggestions and sample answers are available in the WebQuests, Projects, and Interdisciplinary Investigations.

Math and Geography

You can use this WebQuest Interdisciplinary Project with your students' social studies teacher to make the connection from mathematics to the geography your students are studying.

More Interdisciplinary Connections

You may wish to share these suggestions with your students' other teachers.

Math and Language Arts

Research the game of soccer. Write an essay about some aspect of the game such as the origins of the game, the World Cup competition, or the career of a famous soccer player.

Math and Science

Research the physics of soccer. Have students kick a soccer ball at different angles and investigate how far the ball travels.

CHAPTER 3

Chapter Overview

Algebra: Integers

	LESSON OBJECTIVES	NCTM Standards	NAEP 2005	State/Local Objectives
3-1	**Integers and Absolute Value** *(pp. 106–108)* • Read and write integers, and find the absolute value of an integer.	1, 2, 6–10	N1b, N1d, N1g, N1i	
3-2	**Comparing and Ordering Integers** *(pp. 109–111)* • Compare and order integers.	1, 2, 6–10	D1a, N1i, N1j	
3-3	**The Coordinate Plane** *(pp. 112–115)* • Graph points on a coordinate plane.	1, 2, 3, 6–10	A2c, A2d	
3-4	**Adding Integers** *(pp. 118–124)* **3-4a Hands-On Lab:** Use counters to model the addition of integers. • Add integers.	1, 2, 6–10	N1b, N3a, N3g, A3b, N5e	
3-5	**Subtracting Integers** *(pp. 126–131)* **3-5a Hands-On Lab:** Use counters to model the subtraction of integers. • Subtract integers.	1, 2, 6–10	N1b, N3f, A3b, N3a, N3g	
3-6	**Multiplying Integers** *(pp. 132–137)* **3-6a Problem-Solving Strategy:** Solve problems using the look for a pattern strategy. • Multiply integers.	1, 2, 6–10	A1a, A3b, N3g, A2c, N1b, N3f, N3a	
3-7	**Dividing Integers** *(pp. 138–141)* • Divide integers.	1, 2, 6–10	N1b, N3f, N3a, N3g, A4e, A3b	

Key to NCTM Standards:

1=Number & Operations, 2=Algebra, 3=Geometry, 4=Measurement, 5=Data Analysis & Probability, 6=Problem Solving, 7=Reasoning & Proof, 8=Communication, 9=Connections, 10=Representation

NAEP 2005 Standards: See math.glencoe.com

Pacing suggestions for the entire year can be found on page T20.

Chapter Resource Manager

Time-Saving Tool
TeacherWorks™
All-In-One Planner and Resource Center
See page T5.

FAST FILE
Chapter 3 Resource Masters

Study Guide and Intervention	Practice (Skills and Word Problems)	Reading to Learn Mathematics	Enrichment	Assessment	Applications*	Parent and Student Study Guide Workbook	5-Minute Check Transparencies	Key Teaching Transparencies	MathPASS: Tutorial Plus (lessons)	Teaching Mathematics with Manipulatives	Optional Materials
129	130–131	132	133		CS 21	3-1	3-1				
134	135–136	137	138	179	RC 17	3-2	3-2				
139	140–141	142	143		CS 22	3-3	3-3		37	44	
144	145–146	147	148	179, 182	CS 23	3-4	3-4	9		45	
149	150–151	152	153		CM 7–9	3-5	3-5	10		46	
154	155–156	157	158	180		3-6	3-6	11		47–48	
159	160–161	162	163	165–178, 180, 182–184	CS 24	3-7	3-7	12	27		

*** Key to Abbreviations:** *CS = Calculator and Spreadsheet Masters, RC = Real-Life Career Masters, CM = Classroom Math Games (online), SM = Science and Math Lab Manual (online)*

INTERACTIVE CHALKBOARD

PowerPoint® Presentations

This CD-ROM is a customizable Microsoft® PowerPoint® presentation that includes:

- In-Class Examples
- Links to Online Study Tools
- 5-Minute Check Transparencies
- Bellringers

Math Skills Maintenance

Use the Math Skills Maintenance Masters to keep your students' basic math skills fresh. Students can review skills one at a time or in a combined format. The Masters are also available online at msmath2.net/maintenance.

CHAPTER 3

Mathematical Content and Teaching Strategies

Continuity of Instruction

Building on Prior Knowledge

Prior to this course, students gained a basic understanding of nonnegative number systems and how to use counters and number lines to visualize operations on nonnegative numbers. Students learned how to add, subtract, multiply, and divide nonnegative numbers. In Chapter 1, students studied the order of operations and properties.

Learning in This Chapter

Students will be introduced to the integer number system and will learn how to determine the absolute value of a number and how to compare and order integers. They also will learn how to add, subtract, multiply, and divide integers and determine the sign of sums, differences, products, and quotients. Students also will be introduced to the Cartesian coordinate system and will learn how to plot points on the grid using their coordinates.

Making Future Connections

Students will use their knowledge of integers and mathematical operations throughout this course and in all subsequent math courses, including algebra and statistics courses. Positive and negative integers are used in many real-life applications, including temperature, acceleration, and altitude. Coordinate planes are used as reference grids in many disciplines including geography.

Mathematical Background

3-1 Integers and Absolute Value

An integer is defined as a real whole number. The number may be positive, negative, or equal to zero. Integers are whole numbers that describe opposites. The set of integers is made up of the natural numbers, their opposites, and zero. Integers form a countably infinite set. Even if a fraction will simplify to an integer, the fraction form is not an integer. The absolute value of an integer is its distance from zero on the number line. A distance can only be positive which is why the absolute value of a number is always positive.

Bulletin Board Starters

Post a number line at the top of a bulletin board under the title "Integers." Make the number line as long as the bulletin board is wide. Put a zero in the middle and extend positive and negative number labels along the full length of the board. Refer to the number line as you define integers and absolute value, and again as you discuss the ordering of integers. Have students use velcro dots or sticky notes to plot integers on the number line. As students work through Lessons 3-4 to 3-7, have them add examples of the rules they are learning to the board.

3-2 Comparing and Ordering Integers

Beginning with zero, numbers increase in value to the right and decrease in value to the left. The order that an integer is placed on the number line will determine whether it is greater than or less than other integers. Being able to compare integers is a fundamental understanding in math. It provides a basis for working with decimals and fractions.

3-3 The Coordinate Plane

A coordinate plane is formed by two perpendicular number lines called axes. Every point on the plane can be named by an ordered pair of numbers. René Descartes, a French philosopher and mathematician, founded analytical geometry and originated the Cartesian coordinates. Descartes lived from 1596 to 1650 and coined the phrase: *Cogito, ergo sum*—"I think, therefore I am." The coordinate plane is often called the Cartesian plane in honor of Descartes. The x-coordinate is called the abscissa and the y-coordinate is called the ordinate.

In-Class Speaker

Invite a member of the American Congress on Surveying and Mapping to come to your class and describe how the coordinate system is used in making maps. Information on your nearest local affiliate can be found through the organization's Web site at www.acsm.net.

3-4 Adding Integers

Adding integers is similar to the addition that students already know, except that integers may be negative. Students need to have a very thorough understanding of the number line and what a negative integer means. Beginning with manipulatives will allow students to build their understanding.

Teacher to Teacher

Mollie Van Veckhoven-Boeving
White Hall Junior High
White Hall, AR

"I teach students that if two integers are being added with opposite signs, it is like a tug of war. The positive one tries to go one way, and the negative one tries to go the other. So, you take the sign of the larger one since it is the winner. If there is a tie, there is no winner, and the solution is zero."

Mollie Van Veckhoven-Boeving

3-5 Subtracting Integers

Subtraction means to add the opposite. Therefore, students can rewrite any subtraction problem as an addition problem.

3-6 Multiplying Integers

The main key to multiplying integers is remembering the sign rules. For example, positive times positive equals positive. Negative times negative equals positive. Positive times negative equals negative. Students also need to remember that multiplication is associative so the order of the numbers does not matter. A negative integer can be written as the number times negative one, if that helps students. For example, $-5 = (-1)(5)$.

3-7 Dividing Integers

The sign rules for dividing integers are the same as for multiplying. Students may find this easier to understand when the division problem is written as a fraction and two of the same sign cancel out.

McGraw-Hill Professional Development

Mathematics, YES! is a self-paced, web-based professional development program. Units are based on major mathematical concepts and are standards-based. Actual classroom videos provide a springboard for discussion and activities. Units appropriate for this chapter:

- Unit 3, Integers
- Unit 5, Coordinate Pairs and Graphing

CHAPTER 3

DAILY INTERVENTION and Assessment

	Type	Student Edition	Teacher Resources	Technology/Internet
INTERVENTION	Ongoing	Prerequisite Skills, pp. 105, 108, 111, 124 Mid-Chapter Practice Test, p. 116	5-Minute Check Transparencies Quizzes 1–4, *CRM*, pp. 179, 180 Mid-Chapter Test, *CRM*, p. 181 Study Guide and Intervention, *CRM*, pp. 129, 134, 139, 144, 149, 154, 159	MathPASS: Tutorial Plus msmath2.net/self_check_quiz msmath2.net/extra_examples
	Mixed Review	pp. 108, 111, 115, 124, 131, 137, 141	Cumulative Review, *CRM*, p. 182	
	Error Analysis	Find the Error, pp. 122, 130	Find the Error, *TWE*, pp. 122, 130 Unlocking Misconceptions, *TWE*, pp. 106, 135 Tips for New Teachers, *TWE*, p. 129	
	Standardized Test Practice	pp. 108, 110, 111, 115, 116, 124, 131, 133, 137, 141, 145, 146–147	Standardized Test Practice, *CRM*, pp. 183–184	Standardized Test Practice CD-ROM msmath2.net/standardized_test
	Open-Ended Assessment	Writing Math, pp. 107, 110, 114, 119, 122, 127, 130, 136, 140 Open Ended, pp. 107, 110, 114, 122, 130, 136, 140 Standardized Test, p. 147	Modeling: *TWE*, pp. 108, 137 Speaking: *TWE*, pp. 131, 141 Writing: *TWE*, pp. 111, 125, 127, 133 Extended Response Assessment, *CRM*, p. 177	
ASSESSMENT	Chapter Assessment	Study Guide, pp. 142–144 Practice Test, p. 145	Multiple-Choice Tests (Forms 1, 2A, 2B), *CRM*, pp. 165–170 Free-Response Tests (Forms 2C, 2D, 3), *CRM*, pp. 171–176 Vocabulary Test/Review, *CRM*, p. 178	ExamView® Pro Test Bank (see below) MindJogger Videoquizzes msmath2.net/vocabulary_review msmath2.net/chapter_test

* **Key to Abbreviations:** *TWE* = Teacher Wraparound Edition, *CRM* = Chapter Resource Masters

Additional Intervention Resources

Skills Intervention for Middle School Mathematics workbook and CD-ROM includes 88 skills with pretests, tutorials, guided practice, and posttests for diagnosis and remediation.

Guide to Daily Intervention booklet details each intervention resource available and guides teachers in using them as intervention tools everyday to assist students who may be having difficulty.

ExamView® Pro

Use ExamView® Pro Test Bank CD-ROM to:

- **Create Multiple versions of tests.**
- **Create Modified tests for Inclusion students with one mouse click.**
- **Edit existing questions and Add your own questions.**
- **Build tests aligned with state standards using built-in State Curriculum Correlations.**
- **Change English tests to Spanish with one mouse click and vice versa.**

Reading and Writing Mathematics

• Student Edition

Foldables Study Organizer, p. 105

Study Skill: How to Take Good Notes, p. 125

Concept Check questions at the beginning of each exercise set require students to verbalize and write about what they have learned in the lesson.

- **Open Ended,** pp. 107, 110, 114, 122, 130, 136, 140
- **Find the Error,** pp. 122, 130
- **Which One Doesn't Belong?,** pp. 107, 140

Reading Math Tip, pp. 107, 113, 118, 127, 139

• Teacher Wraparound Edition

Foldables Follow-Up, pp. 114, 116, 122, 130, 136, 140

Writing Activities, pp. 111, 125, 127, 133

ELL Resources, *see page T9 for a complete listing.*

• Additional Resources

Vocabulary Builder, *Chapter 3 Resource Masters*

Reading to Learn Mathematics, *Chapter 3 Resource Masters*

Vocabulary PuzzleMaker (online)

Teaching Mathematics with Foldables

Reading and Writing in the Mathematics Classroom

WebQuests, Projects, and Interdisciplinary Investigations

For detailed information on these resources, see pp. T6–T7.

• Recommended Outside Reading for Students

Nonfiction

- Reimer, Luetta. *Mathematicians Are People, Too: Stories from the Lives of Great Mathematicians.* Pearson Learning, ©1990.

  This book relates stories of mathematical discovery through the ages and includes biographies of great mathematicians. The historical approach to math provides an enjoyable tool for interdisciplinary study.

Fiction

- Tahan, Malba. *The Man Who Counted: A Collection of Mathematical Adventures.* W. W. Norton & Co., ©1993.

  This set of stories and puzzles follows the adventures of a mathematically talented man who solves conflicts and gives advice in old Arabia.

PROJECT CRISS℠ Study Skill

A pre-reading organizer can help students examine their background knowledge before they begin to study a new topic. The pre-reading organizer at the right shows the background a student might have for the lesson on integers and absolute value. Have students skim through a lesson and note any familiar concepts. Then have them complete a pre-reading organizer based on their individual experiences. Students can use their organizers as the basis for a discussion of the topics presented in the lesson.

Topic	What I Already Know
integers	• positive and negative numbers and zero • graphing points on number lines
absolute value	• always a nonnegative number • distances on number lines

CReating Independence through Student-owned Strategies

For more information on Project CRISS, see page T9.

CHAPTER 3 Making Connections

School-To-Home

Parent and Student Study Guide

Parents and students may work together to reinforce the concepts and skills of this chapter. These workbooks are available in English and Spanish. A sample lesson is shown below. Also available at msmath2.net/parent_student.

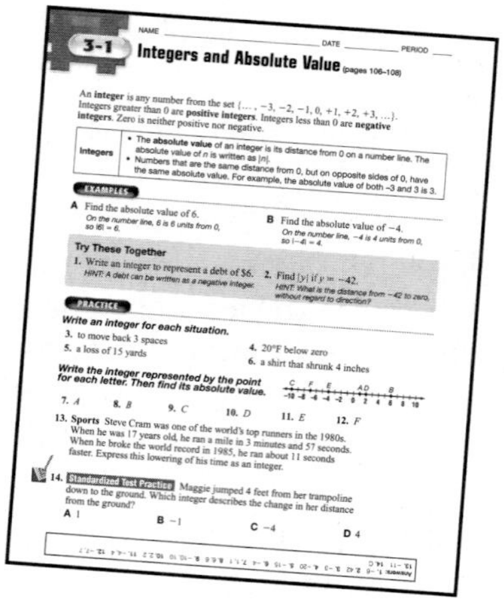

NAME ______ DATE ______ PERIOD ______

3-1 Integers and Absolute Value (pages 106–108)

An **integer** is any number from the set {..., −3, −2, −1, 0, +1, +2, +3, ...}. Integers greater than 0 are **positive integers**. Integers less than 0 are **negative integers**. Zero is neither positive nor negative.

Integers	• The **absolute value** of an integer is its distance from 0 on a number line. The absolute value of n is written as $\|n\|$. • Numbers that are the same distance from 0, but on opposite sides of 0, have the same absolute value. For example, the absolute value of both −3 and 3 is 3.

EXAMPLES

A Find the absolute value of 6.
On the number line, 6 is 6 units from 0, so $|6| = 6$.

B Find the absolute value of −4.
On the number line, −4 is 4 units from 0, so $|-4| = 4$.

Try These Together

1. Write an integer to represent a debt of $6.
HINT: A debt can be written as a negative integer.

2. Find $|y|$ if $y = -42$.
HINT: What is the distance from −42 to zero, without regard to direction?

PRACTICE

Write an integer for each situation.

3. to move back 3 spaces
4. 20°F below zero
5. a loss of 15 yards
6. a shirt that shrunk 4 inches

Write the integer represented by the point for each letter. Then find its absolute value.

7. *A* 8. *B* 9. *C* 10. *D* 11. *E* 12. *F*

13. **Sports** Steve Cram was one of the world's top runners in the 1980s. When he was 17 years old, he ran a mile in 3 minutes and 57 seconds. When he broke the world record in 1985, he ran about 11 seconds faster. Express this lowering of his time as an integer.

14. **Standardized Test Practice** Maggie jumped 4 feet from her trampoline down to the ground. Which integer describes the change in her distance from the ground?
A 1 B −1 C −4 D 4

Family Letters and Activities

Send a letter and activity for this chapter home to encourage family members to become active participants in their students' learning. (English: *Chapter 3 Resource Masters*, pp. ix–x; Spanish: log on to msmath2.net/family_letters)

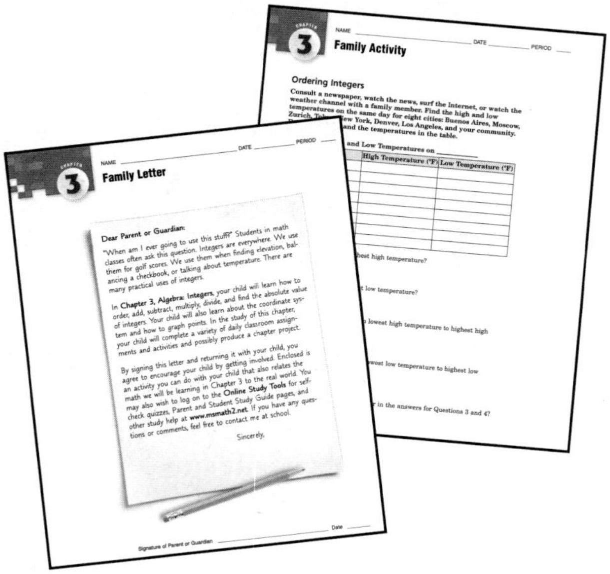

NAME ______ DATE ______ PERIOD ______

Chapter 3 Family Activity

Ordering Integers

Consult a newspaper, watch the news, surf the Internet, or watch the weather channel with a family member. Find the high and low temperatures on the same day for eight cities: Buenos Aires, Moscow, Zurich, ... New York, Denver, Los Angeles, and your community. ...and the temperatures in the table.

NAME ______ DATE ______ PERIOD ______

Chapter 3 Family Letter

Dear Parent or Guardian:

"When am I ever going to use this stuff?" Students in math classes often ask this question. Integers are everywhere. We use them for golf scores. We use them when finding elevation, balancing a checkbook, or talking about temperature. There are many practical uses of integers.

In **Chapter 3, Algebra: Integers**, your child will learn how to order, add, subtract, multiply, divide, and find the absolute value of integers. Your child will also learn about the coordinate system and how to graph points. In the study of this chapter, your child will complete a variety of daily classroom assignments and activities and possibly produce a chapter project.

By signing this letter and returning it with your child, you agree to encourage your child by getting involved. Enclosed is an activity you can do with your child that also relates the math we will be learning in Chapter 3 to the real world. You may also wish to log on to the **Online Study Tools** for self-check quizzes, Parent and Student Study Guide pages, and other study help at **www.msmath2.net**. If you have any questions or comments, feel free to contact me at school.

Sincerely,

Signature of Parent or Guardian ______ Date ______

English Language Learner ELL

Lesson 3-1

Using Manipulatives

Have students work in groups with chart paper, rulers, and markers. Have them make a number line with tick marks equally spaced so that they understand that each one represents 1 unit. Have students label the number line from −10 to 10. Explain that zero (0) is neither positive nor negative. Discuss absolute value by asking them the distance that 5 is from 0 and the distance that −5 is from 0. Reinforce absolute values by discussing the following question: If a student lives 2 miles from school, how many miles did he travel to get there and return?

Lesson 3-2

Flexible Groups

Give groups of six students a set of six cards with an integer on each card. Have the students order the numbers from least to greatest. On a wall of the room, mark 0, positive, and negative to denote a number line. Have the students come to the number line by group and position themselves in order on the number line. The students can check each other for accuracy and assist the ELL students.

Lesson 3-3

Using Models

Have students draw a coordinate plane on grid paper, labeling the x- and y-axes, the origin, and all four quadrants. Discuss with students what an ordered pair is and how to locate one on the coordinate plane. Then, use a shower curtain to represent a coordinate plane. Mark each part with masking tape, including the grid. Have students stand on the positions noted by the ordered pairs you have written on index cards. Have the rest of the class help lead those students to the correct points. Also, give every student the opportunity to graph a point.

MathScape™

This project was supported, in part, by the **National Science Foundation**

MathScape is a mathematics curriculum for grades 6–8 developed by the Seeing and Thinking Mathematically Project at the Education Development Center.

Generalizing About Numbers

Making Mathematical Arguments

Summary Students will understand signed numbers and will add, subtract, multiply, and divide with integers. The order of operations agreement is used to evaluate expressions with integers.

USA TODAY Education Online Lesson Planner

USA TODAY® Education's Online site offers resources and interactive features connected to each day's newspaper. *Experience Today,* USA TODAY's® daily lesson plan, is available on the site and delivered daily to subscribers. This plan provides instruction for integrating USA TODAY® graphics and key editorial features into your mathematics classroom. Log on to **education.usatoday.com**.

Quick Review Math Handbook

is Glencoe's mathematical handbook for students. The Hot Words section includes a glossary of terms while the Hot Topics section consists of explanations of key mathematical concepts. An exercise set is included to check students' understanding of the concepts. This valuable resource can be used as a reference in the classroom or for home study.

Lesson	Hot Topics Section	Lesson	Hot Topics Section
3-1	1-5	**3-4, 3-5**	1-5
3-2	1-1, 1-5	**3-6**	1-5
3-3	6-7	**3-7**	1-5

Chapter 3 Notes

Key Chapter Objectives

Lesson 3-1 Read and write integers, and find the absolute value of an integer.

Lesson 3-3 Graph points on a coordinate plane.

Lessons 3-4 through 3-7 Add, subtract, multiply, and divide integers.

"What does lightning have to do with math?"

Point out to students that this is only one of many applications of the math that they are studying. Others are provided in the introduction to each lesson.

Vocabulary Builder, pp. vii–viii ELL

Chapter 3 NAME ______ DATE ______ PERIOD ____

Reading to Learn Mathematics

Vocabulary Builder

This is an alphabetical list of new vocabulary terms you will learn in Chapter 3. As you study the chapter, complete each term's definition or description. Remember to add the page number where you found the term. Add this page to your math study notebook to review vocabulary at the end of the chapter.

Vocabulary Term	Found on Page	Definition/Description/Example
absolute value		
additive inverse		
coordinate grid		
coordinate plane		
graph		
integer [IHN-tih-juhr]		
negative integer		
opposite		

Chapter 3 Algebra: Integers

"What does lightning have to do with math?"

Have you heard of the phrase "opposites attract"? During thunderstorms, negatively-charged electrons in the clouds are attracted to positively-charged protons on the ground. This opposite attraction causes lightning. In mathematics, you can use opposites to help you add and subtract positive and negative integers.

You will solve problems about thunderstorms in Lesson 3-4.

Vocabulary Builder

For a thorough vocabulary list with pronunciations of new words, give students the Vocabulary Builder worksheets from the *Chapter 3 Resource Masters.* Encourage them to complete the definition of each term as they progress through the chapter.

GETTING STARTED

Take this quiz to see whether you are ready to begin Chapter 3. Refer to the lesson or page number in parentheses if you need more review.

Vocabulary Review

State whether each sentence is *true* or *false*. If *false*, replace the underlined word to make a true sentence.

1. The <u>mean</u> of 1, 3, and 6 is 3. (Lesson 2-4) **false; median**
2. The difference between the greatest number and the least number in a set of data is called the <u>range</u>. (Lesson 2-3) **true**

Prerequisite Skills

Replace each ● with < or > to make a true sentence. (Page 556)

3. 1,458 ● 1,548 **<**
4. 36 ● 34 **>**
5. 1.02 ● 1.20 **<**
6. 76.7 ● 77.6 **<**

Add.

7. 84 + 39 **123**
8. 198 + 289 **487**
9. 826 + 904 **1,730**
10. 3,068 + 5,294 **8,362**

Multiply.

11. $2 \cdot 5 \cdot 3$ **30**
12. $18 \cdot 9$ **162**
13. $15 \cdot 6$ **90**
14. $10 \cdot 4 \cdot 7$ **280**

Divide.

15. $63 \div 9$ **7**
16. $96 \div 12$ **8**
17. $125 \div 5$ **25**
18. $187 \div 17$ **11**

Find the mean and range for each set of data. (Lessons 2-3 and 2-4) **19. 14; 17 21. 100; 34**

19. 12, 8, 25, 16, 9
20. 34, 57, 60, 45 **49; 26**
21. 93, 88, 122, 97
22. 7, 10, 2, 6, 13 **7.6, 11**

FOLDABLES™ Study Organizer

Integers Make this Foldable to help you organize information about integers. Begin with two sheets of $8\frac{1}{2}'' \times 11''$ paper.

STEP 1 Fold and Cut One Sheet
Fold in half from top to bottom. Cut along fold from edges to margin.

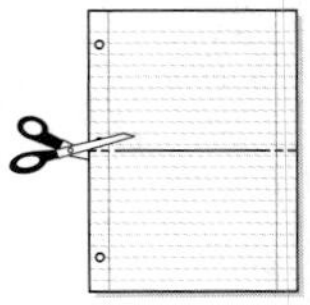

STEP 2 Fold and Cut the Other Sheet
Fold in half from top to bottom. Cut along fold between margins.

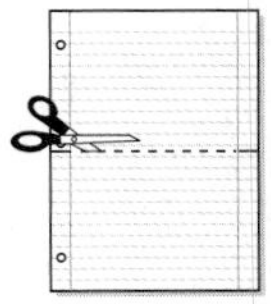

STEP 3 Fold
Insert first sheet through second sheet and align folds.

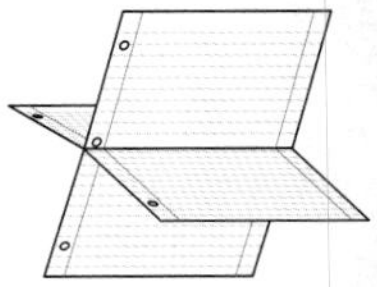

STEP 4 Label
Label each page with a lesson number and title.

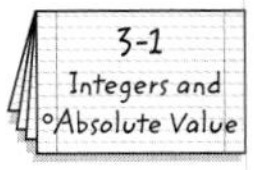

Reading and Writing As you read and study the chapter, fill the journal with notes, diagrams, and examples of integers.

GETTING STARTED

This section provides a review of the basic concepts needed before beginning Chapter 3. Page references are included for additional student help.

Prerequisite Skills review skills needed in this chapter.

Exercises	Needed for Lesson	Skill
3–6	3-2	Compare and Order Integers
7–10	3-4	Add Integers
11–14	3-6	Multiply Integers
15–18	3-7	Divide Integers
19–22	3-5, 3-7	Find the Mean and Range

For more prerequisite skills, see:

- Getting Ready for the Next Lesson, at the end of each lesson
- Prerequisite Skills in the Student Handbook on pp. 554–563
- *Math Skills Maintenance Masters*

FOLDABLES™ Study Organizer

For more information about Foldables, see *Teaching Mathematics with Foldables.*

After students make their Foldable journals, have them label the six inside pages to corrrespond to the first six lessons in the chapter, and label the back of the book with the title of lesson seven. Students use their Foldables to take notes, define terms, record concepts, and write examples. Foldables can also be used to record their progress, to describe positive and negative experiences during learning, and to list ways the subject matter relates to daily life.

3-1 Lesson Notes

Lesson Resources

Chapter 3 Resource Masters

- Study Guide and Intervention, p. 129
- Practice: Skills, p. 130
- Practice: Word Problems, p. 131
- Reading to Learn Mathematics, p. 132
- Enrichment, p. 133

Calculator and Spreadsheet Masters, p. 21

Parent and Student Study Guide Workbook, 3-1

FAST FILE

1 Focus

5-Minute Check 3-1

INTERACTIVE CHALKBOARD
PowerPoint® Presentations

Use as a quiz or review of Chapter 2. (Also available as a transparency)

Bellringer

Reading Mathematics Have students look up elevations of various landmarks. Ask students why these elevations include the words *above* or *below sea level*. **Sea level is considered the zero point for measuring elevation.** Can they find any landmarks that are *below sea level*? **Dead Sea, Israel-Jordan; Death Valley, California** Discuss ways students can easily tell whether measurements are above or below sea level. **positive or negative sign**

Reading to Learn Mathematics, p. 132 ELL

NAME ______ DATE ______ PERIOD ____

3-1 Reading to Learn Mathematics
Integers and Absolute Value

Pre-Activity *Read the introduction at the top of page 106 in your textbook. Write your answers below.*

1. What does a value of −2 represent? a 2-yard loss
2. On which down did they lose the most yards? 3rd down
3. How can you represent a gain of 9 yards? +9 or 9

Reading the Lesson

4. Express each of the following in words.

3-1 Integers and Absolute Value

What You'll Learn
Read and write integers, and find the absolute value of an integer.

NEW Vocabulary

- integer
- graph
- positive integer
- negative integer
- absolute value

***MATH* Symbols**

- $+3$ positive three
- -3 negative three
- $|3|$ absolute value of three

WHEN am I ever going to use this?

FOOTBALL The graph shows the number of yards the Bears gained or lost on the first four downs. A value of −3 represents a 3-yard loss.

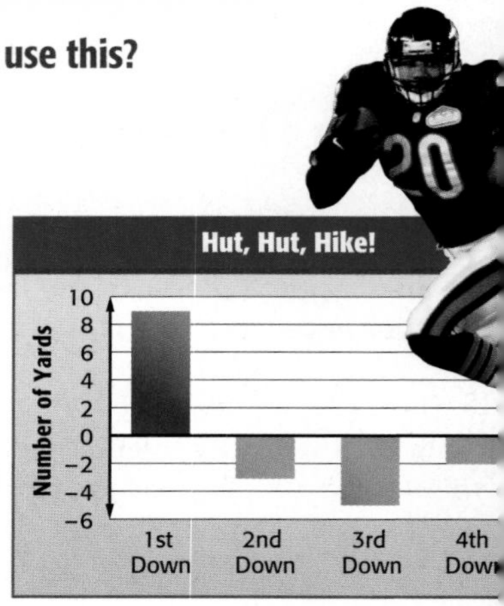

1. What does a value of −2 represent? **a 2-yard loss**
2. On which down did they lose the most yards? **3rd down**
3. How can you represent a gain of 9 yards? **+9 or 9**

Numbers like 9 and −2 are called integers. An **integer** is any number from the set {…, −4, −3, −2, −1, 0, 1, 2, 3, 4, …}. Integers can be graphed on a number line. To **graph** a point on the number line, draw a point on the line at its location.

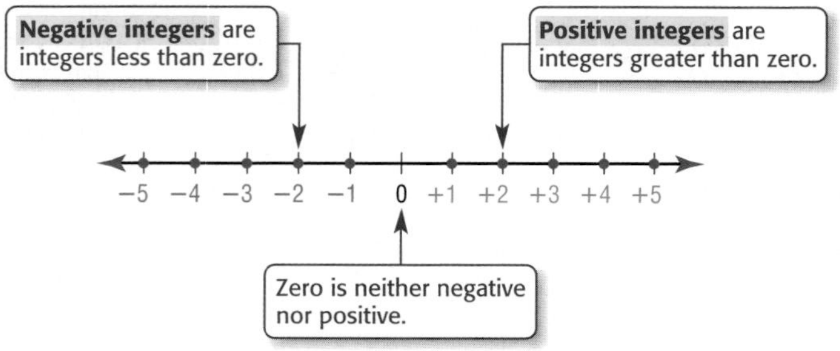

STUDY TIP

Integers Positive integers are usually written without the + sign.

EXAMPLES Write Integers for Real-Life Situations

WEATHER Write an integer for each situation.

1 The average temperature in Tennessee for May was 5 degrees below normal.

Because it represents *below* normal, the integer is −5.

2 The average rainfall in Virginia for November was 5 inches above normal.

Because it represents *above* normal, the integer is +5 or 5.

Your Turn **Write an integer for each situation.**

a. 6 degrees above normal **6**   b. 2 inches below normal **−2**

DAILY INTERVENTION

Unlocking Misconceptions

Absolute Value It is not unusual for students to think that the absolute value of a number is the opposite of that number. Practice having students find the absolute value of a number by making a number line of masking tape on the floor and having the students take steps from various positive and negative integers toward zero. Make sure the units for the number line are a comfortable stride length. Then students will find the absolute value of a number by seeing how many steps or units it is from zero.

EADING Math

Set Theory The number 5 is an *element*, or member, of the set of integers.

The set {−5, 5} is a *subset* of the set of integers.

You can represent the integers from Examples 1 and 2 on a number line.

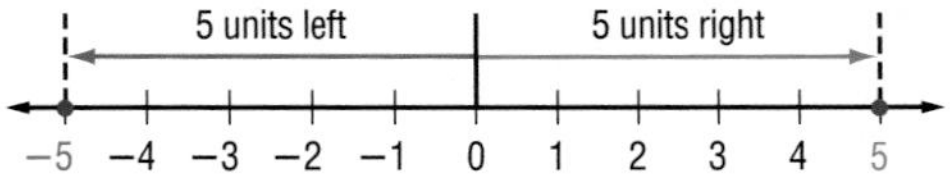

The numbers −5 and 5 are the same distance from 0, but on opposite sides of 0. So, −5 and 5 have the same **absolute value**.

Key Concept — **Absolute Value**

Words The absolute value of an integer is the distance the number is from zero on a number line.

Examples $|6| = 6$ $|-6| = 6$

EXAMPLES Evaluate Expressions

Evaluate each expression.

3 $|-4|$

On the number line, the graph of −4 is 4 units from 0.

So, $|-4| = 4$.

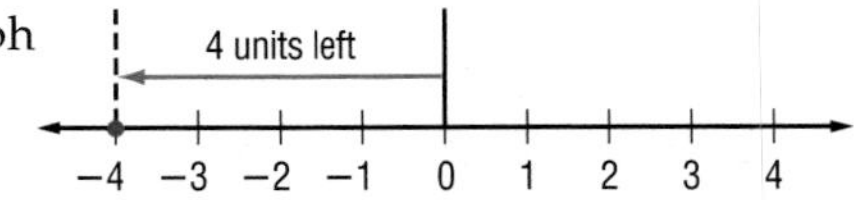

4 $|-5| - |2|$

$|-5| - |2| = 5 - 2$ $|-5| = 5, |2| = 2$

$= 3$ Subtract.

Skill and Concept Check

Writing Math
Exercises 1 & 3

1. **Describe** a situation in everyday life where negative numbers are used.
2. **OPEN ENDED** On a number line, graph two different points that have the same absolute value. **1–2. See margin.**
3. **Which One Doesn't Belong?** Identify the expression that does not have the same value as the other three. Explain your reasoning. **−3; The other expressions equal 3.**

$\lvert -3 \rvert$	-3	$\lvert 3 \rvert$	3

GUIDED PRACTICE

Write an integer for each situation.

4. 6°F below 0 **−6**
5. a loss of 11 yards **−11**
6. a deposit of $16 **16**

Evaluate each expression.

7. $|7|$ **7**
8. $|-4|$ **4**
9. $|-7| - |1|$ **6**
10. **STOCK MARKET** The price of a company's stock fell 21 points in two days. Write an integer to represent the amount the stock price fell. **−21**

2 Teach

In-Class Examples

INTERACTIVE CHALKBOARD
PowerPoint® Presentations

WEATHER Write an integer for each situation.

1 The seasonal snowfall for Barrow, Alaska, was 3 inches above normal. **+3**

2 The total rainfall this month is 2 inches below normal. **−2**

Evaluate each expression.

3 $|-5|$ **5**

4 $|-4| - |-3|$ **1**

3 Practice/Apply

About the Exercises...

Guided Practice Key

Exercises	Examples
4–6, 10	1, 2
7–9	3, 4

Odd/Even Assignments
Exercises 11–28 are structured so that students practice the same concepts whether they are assigned odd or even problems.

Which One Doesn't Belong?
In Exercise 3, without evaluating, students should recognize that $|-3|$ and $|3|$ are positive. Only −3 is negative.

Answers

1. **Sample answers: temperature, golf scores, elevation, loss of money**
2. 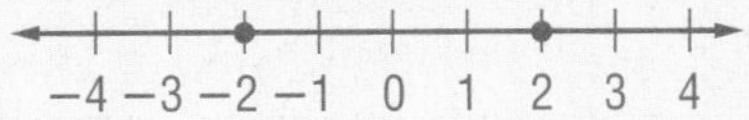

Tips for New Teachers

Using Applications Sometimes numbers are modified to make a problem easier. For example, Exercise 29 uses integers less than 10. In reality, the number of protons and electrons is in the billions.

Study Guide and Intervention, p. 129

NAME ______________ DATE ________ PERIOD ____

3-1 Study Guide and Intervention

Integers and Absolute Value

Integers less than zero are **negative integers**. Integers greater than zero are **positive integers**.

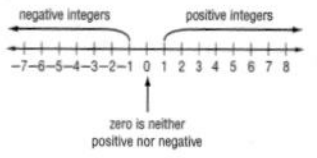

The absolute value of an integer is the distance the number is from zero on a number line. Two vertical bars are used to represent absolute value. The symbol for absolute value of 3 is $|3|$.

EXAMPLE 1 **Write an integer that represents 160 feet below sea level.**

Because it represents *below* sea level, the integer is −160.

EXAMPLE 2 **Evaluate $|-2|$.**

On the number line, the graph of −2 is 2 units away from 0. So, $|-2| = 2$.

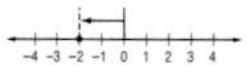

EXERCISES

Write an integer for each situation.

1. 12°C above 0 12
2. a loss of $24 −24
3. a gain of 20 pounds 20
4. falling 6 feet −6

Evaluate each expression.

5. $|12|$ 12
6. $|-150|$ 150
7. $|-8|$ 8
8. $|75|$ 75
9. $|-19|$ 19
10. $|84|$ 84

Practice: Skills, p. 130

NAME ______________ DATE ________ PERIOD ____

3-1 Practice: Skills

Integers and Absolute Value

Write an integer for each situation.

1. 15°C below 0 −15
2. a profit of $27 27
3. 2010 A.D. 2010
4. average attendance is down 38 people −38
5. 376 feet above sea level 376
6. a withdrawal of $200 −200
7. 3 points lost −3
8. a bonus of $150 150
9. a deposit of $41 41
10. 240 B.C. −240
11. a wage increase of $120 120
12. 60 feet below sea level −60

Evaluate each expression.

13. $|-1|$ 1
14. $|9|$ 9
15. $|23|$ 23
16. $|-107|$ 107
17. $|-45|$ 45
18. $|19|$ 19
19. $|0|$ 0
20. $|6| - |-2|$ 4
21. $|-8| + |4|$ 12
22. $|-12| - |12|$ 0

Graph each set of integers on a number line.

23. {0, 2, −3}

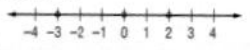

24. {−4, −1, 3}

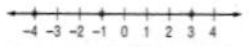

Practice: Word Problems, p. 131

NAME ______________ DATE ________ PERIOD ____

3-1 Practice: Word Problems

Integers and Absolute Value

1. **DEATH VALLEY** The lowest point in the United States is Death Valley in California. Its altitude is 282 feet below sea level. Write an integer to represent the altitude of Death Valley. −282

2. **RAIN** A meteorologist reported that in the month of April there were 3 inches more rainfall than normal. Write an integer to represent the amount of rainfall above normal in April. 3

3. **ARCHIMEDES** A famous mathematician and physicist named Archimedes was born in 287 B.C. Write an integer to express the year of his birth. −287

4. **TEMPERATURE** In our world's tropical rain forests, the average temperature of every month is 64 degrees above zero or higher. Write an integer to express this temperature. 64

5. **STOCK MARKET** A certain stock gained 5 points in one day and lost 4 points the next day. Write integers to represent the stock's gains and losses for the two days. 5; −4

6. **ALTITUDE** An airplane pilot changed his altitude by 100 meters. Describe what this could mean. Sample answer: The airplane could be flying 100 meters higher or 100 meters lower, 100 or −100.

★ indicates increased difficulty

Practice and Applications

HOMEWORK HELP	
For Exercises	See Examples
11–18, 29	1, 2
19–22, 27, 28	3
23–26	4

Extra Practice See pages 569, 598.

Write an integer for each situation.

11. a profit of $9 **9**
12. 53°C below 0 **−53**
13. no gain on first down **0**
14. an elevator goes up 12 floors **12**
15. 2008 A.D. **2008**
16. 160 feet above sea level **160**
17. a bank withdrawal of $50 **−50**
18. 1000 B.C. **−1000**

Evaluate each expression.

19. $|6|$ **6**
20. $|-12|$ **12**
21. $|-9|$ **9**
22. $|21|$ **21**
23. $|12| - |-8|$ **4**
24. $|-10| - 5$ **5**
25. $|-9| + |5|$ **14**
26. $|26| + |-4|$ **30**

27. What is the absolute value of 0? **0**

★ 28. Find $|x|$ if $x = -6$. **6**

29. **STATIC ELECTRICITY** Electrical charges are made up of positively-charged protons and negatively-charged electrons. Suppose you rub a balloon through your hair to make the balloon stick to a wall. There are 2 protons on the wall and 5 electrons on the balloon. Write an integer for each charge. **2 protons: 2; 5 electrons: −5**

Graph each set of integers on a number line. **30–33. See Ch. 3 Answer Appendix.**

30. {0, 1, −3}
31. {−4, 5, 4}
32. {−5, −1, 10, −9}
33. {−2, −4, −6, −8}

★ 34. **WEATHER** A meteorologist reports a 20° change in the temperature from yesterday to today. Describe what this could mean. **The temperature could be 20° warmer or 20° cooler, +20 or −20.**

CRITICAL THINKING **Determine whether each statement is *true* or *false*. If *false*, give a counterexample.**

35. Every integer has an absolute value. **true**

36. The absolute value of every integer is positive. **false; 0**

Standardized Test Practice and Mixed Review

37. **MULTIPLE CHOICE** Identify the point that represents −5. **B**

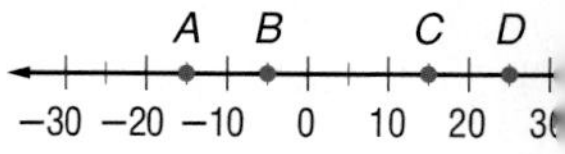

Ⓐ A Ⓑ B Ⓒ C Ⓓ D

38. **SHORT RESPONSE** Write an integer for 23°F below 0. **−23**

STATISTICS **For Exercises 39 and 40, use the following information.**
The mean income for a group of accountants is $36,266.67. Their incomes are $27,500, $36,100, $29,800, $33,400, $31,300, and $59,500.

39. In what way is the mean misleading? (Lesson 2-8) **The very high salary increases the mean.**

40. Draw a bar graph of the data. (Lesson 2-7) **See Ch. 3 Answer Appendix.**

GETTING READY FOR THE NEXT LESSON

PREREQUISITE SKILL **Replace each ● with < or > to make a true sentence.** (Page 556)

41. 16 ● 6 **>**
42. 2.3 ● 3.2 **<**
43. 101 ● 111 **<**
44. 87.3 ● 83.7 **>**
45. 1,051 ● 1,015 **>**

msmath2.net/self_check_quiz

4 Assess

Open-Ended Assessment

Modeling Have students each find three examples of positive and negative integers in a newspaper, on the Internet, or in a current magazine. Ask them to plot the integers on a large number line on the chalkboard.

3-2 Comparing and Ordering Integers

/hat You'll Learn
ompare and order
tegers.

***EVIEW* Vocabulary**
edian: the middle
ımber in an
'dered data set
esson 2-4)

***ATH* Symbols**
< is less than
> is greater than

WHEN am I ever going to use this?

WEATHER The Wind Chill Temperature Index table shows how cold air feels on human skin.

WIND CHILL					
Wind (mph)	Temperature (°F)				
	15	10	5	0	25
5	7	1	−5	−11	−16
10	3	−4	−10	−16	−22
15	0	−7	−13	−19	−26
20	−2	−9	−15	−22	−29

1. What is the wind chill if there is a wind at 20 miles per hour and the temperature is 5°? **−15**
2. Which is colder, a temperature of 15° with a 20 mile-per-hour wind or a temperature of 10° with a 10 mile-per-hour wind? **10° with 10 mph wind**
3. Graph both wind chills found in Exercise 2 on a number line. **See margin.**

When two numbers are graphed on a number line, the number to the left is always less than the number to the right. The number to the right is always greater than the number to the left.

Key Concept — **Comparing Integers**

Model (number line: −5 −4 −3 −2 −1 0 1, with points at −4 and −2)

Words −4 is less than −2. −2 is greater than −4.

Symbols $-4 < -2$ $-2 > -4$

The symbol points to the lesser number.

EXAMPLE Compare Integers

1 **Replace the ● with < or > to make −5 ● −3 a true sentence.**

Graph each integer on a number line.

(number line: −6 −5 −4 −3 −2 −1 0 1 2 3 4 5 6, with points at −5 and −3)

Since −5 is to the left of −3, $-5 < -3$.

Your Turn **Replace each ● with < or > to make a true sentence.**

a. −8 ● −4 **<** b. 5 ● −1 **>** c. −10 ● −13 **>**

msmath2.net/extra_examples

3-2 Lesson Notes

Lesson Resources

Chapter 3 Resource Masters
- Study Guide and Intervention, p. 134
- Practice: Skills, p. 135
- Practice: Word Problems, p. 136
- Reading to Learn Mathematics, p. 137
- Enrichment, p. 138

Parent and Student Study Guide Workbook, 3-2

Real-Life Career Masters, p. 17

1 Focus

5-Minute Check 3-2

INTERACTIVE CHALKBOARD
PowerPoint® Presentations

Use as a quiz or review of Lesson 3-1. (Also available as a transparency)

Bellringer

Problem Solving Using the table in the student edition, find the wind chill if the temperature is 0° and there is a wind at 10 miles per hour. **−16** Which is colder, a temperature of 5° with a 20 mile-per-hour wind or a temperature of 0° with a 5 mile-per-hour wind? **5° with a 20 mile-per-hour wind**

Answer (When am I ever going to use this?)

3. (number line: −5 −4 −3 −2 −1 0 1, with points at −4 and −2)

Reading to Learn Mathematics, p. 137 **ELL**

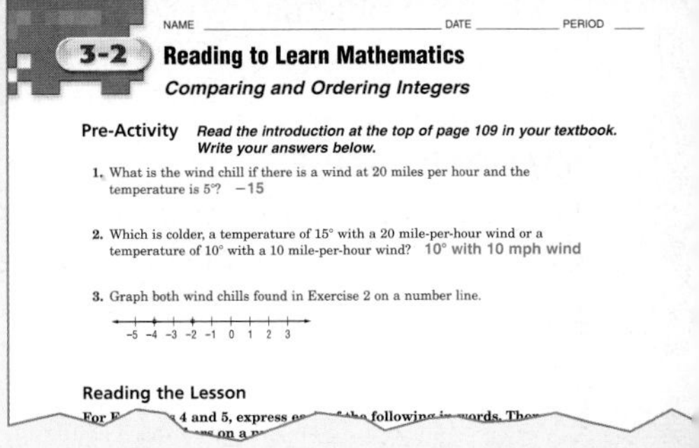

NAME ______ DATE ______ PERIOD ______

3-2 Reading to Learn Mathematics
Comparing and Ordering Integers

Pre-Activity *Read the introduction at the top of page 109 in your textbook. Write your answers below.*

1. What is the wind chill if there is a wind at 20 miles per hour and the temperature is 5°? −15
2. Which is colder, a temperature of 15° with a 20 mile-per-hour wind or a temperature of 10° with a 10 mile-per-hour wind? 10° with 10 mph wind
3. Graph both wind chills found in Exercise 2 on a number line.

Reading the Lesson

DAILY INTERVENTION — **Flexible Grouping**

Researching Integers Have students work in groups of four to research how integers are used in journalism. Have each group member research a different section of a print or online newspaper. Sections should include Sports, Weather, News, and Business. When the students in each group have gathered as much information as possible, have them combine their knowledge to write a comprehensive report to present to the class.

2 Teach

In-Class Examples

INTERACTIVE CHALKBOARD PowerPoint® Presentations

Replace the ● with < or > to make a true sentence.

1 −9 ● −5 <

2 MULTIPLE-CHOICE TEST ITEM The lowest temperatures in Europe, Greenland, Oceania, and Antarctica are listed in the table.

Continent	Record Low Temperature (°F)
Europe	−67
Greenland	−87
Oceania	14
Antarctica	−129

Source: *The World Almanac*

Order the temperatures from least to greatest. **C**

A −67, −87, 14, −129
B 14, −67, −87, −129
C −129, −87, −67, 14
D −67, −87, −129, 14

3 Practice/Apply

About the Exercises...

Guided Practice Key

Exercises	Examples
5–7	1
8–10	2

Odd/Even Assignments
Exercises 11–22 and 29–32 are structured so that students practice the same concepts whether they are assigned odd or even problems.

NUMBER SENSE
For Exercise 3, have students draw a number line and plot the points to check their answers.

Integers are used to compare numbers in many real-life situations.

Standardized Test Practice

EXAMPLE 2 Order Integers

MULTIPLE-CHOICE TEST ITEM The lowest temperatures in Alaska, Florida, Hawaii, and Montana are listed in the table. Order the temperatures from least to greatest.

State	Record Low Temperature (°F)
Alaska	−80
Florida	−2
Hawaii	12
Montana	−70

Source: *The World Almanac and Book of Facts*

(A) −80, −70, 12, −2
(B) −80, −70, −2, 12
(C) 12, −2, −70, −80
(D) −2, 12, −70, −80

Test-Taking Tip *The Princeton Review*

Eliminating Answer Choices
If you are unsure of the correct answer, eliminate the choices you know are incorrect. Then consider the remaining choices. You can eliminate choice C since the list begins with a positive number.

Read the Test Item To order the integers, graph them on a number line.

Solve the Test Item

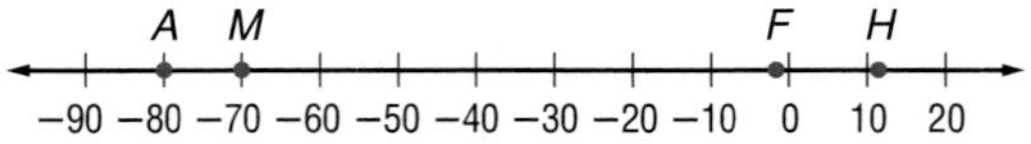

Order the integers from least to greatest by reading from left to right: −80, −70, −2, 12. So, the answer is B.

Skill and Concept Check

Writing Math Exercise 2

1. **Draw** a number line to show that −5 is less than −1. **See margin.**
2. **OPEN ENDED** Write an integer that is less than −9. Explain. **Sample answer: −12 < −9 because −12 is to the left of −9 on a number line.**
3. **NUMBER SENSE** Complete the sentence: −7 is greater than −12 because −7 lies to the __?__ of −12 on a number line. **right**
4. **Name** the greatest negative integer. **−1**

GUIDED PRACTICE

Replace each ● with < or > to make a true sentence.

5. −5 ● −6 **>**
6. −2 ● 8 **<**
7. 0 ● −10 **>**
8. Order 51, −63, 49, −24, −38, and 38 from least to greatest. **−63, −38, −24, 38, 49, 51**

MONEY For Exercises 9 and 10, use the information below and at the right.
Marva is saving money for a new bike and has already saved \$21. She begins a log to keep track of her money.

9. Write each entry as an integer. **15, −13, 10, −4, 13, −5**
10. Order the integers from least to greatest. **−13, −5, −4, 10, 13, 15**

Tips for New Teachers

Using Applications You may want to choose a few exercises for students to expand upon. For example, for Exercise 34, students could find additional inventions and put the events in order.

Answer

1.
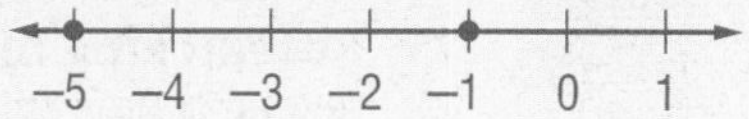

★ indicates increased difficulty

Practice and Applications

HOMEWORK HELP	
For Exercises	See Examples
11–28	1
29, 31–32	2
Extra Practice See pages 570, 598.	

Replace each ● with < or > to make a true sentence.

11. −17 ● −20 >
12. −21 ● −12 <
13. 3 ● −10 >
14. −5 ● 17 <
15. 4 ● −4 >
16. −25 ● −20 <
17. −52 ● −72 >
18. 100 ● −10 >
★19. |−8| ● 0 >
★20. −13 ● |−14| <
★21. |36| ● −37 >
★22. |−29| ● |92| <

Determine whether each sentence is *true* or *false*. If *false*, change one number to make the sentence true. 23, 26, 28. Sample answers are given.

23. −8 > 5 false; 8 > 5
24. −7 < 0 true
25. |−9| = 9 true
26. |5| < −6 false; |5| < 6
27. 10 > |−8| true
28. 7 > |−7| false; 8 > |−7|

WEATHER For Exercises 29 and 30, use the information in the table. It shows the record low temperatures in Indianapolis, Indiana, for March 1–7 of a recent year. 29. See margin.

Day	Temperature (°F)
Mar. 1	−4
Mar. 2	−7
Mar. 3	1
Mar. 4	3
Mar. 5	−1
Mar. 6	−6
Mar. 7	6

Source: www.weather.com

29. Arrange the dates from the coldest temperature to the warmest.
30. Find the median temperature. −1°F
31. Order −7, 5, −6, −4, 1, and 3 from least to greatest. See margin.

★32. Order |51|, −53, |−52|, 55, −56, and −57 from greatest to least. 55, |−52|, |51|, −53, −56, −57

33. **CRITICAL THINKING** If 0 is the greatest integer in a set of five integers, what can you conclude about the other four integers? They are negative integers.

Standardized Test Practice and Mixed Review

34. **MULTIPLE CHOICE** The table shows the inventions of several toys. Order the inventions from earliest to most recent. B

Toy	Year
Yo-Yo	1000 B.C.
Teddy Bear	1902 A.D.
Chess	600 A.D.
Checkers	2000 B.C.

Ⓐ chess, yo-yo, teddy bear, checkers
Ⓑ checkers, yo-yo, chess, teddy bear
Ⓒ yo-yo, teddy bear, checkers, chess
Ⓓ chess, teddy bear, yo-yo, checkers

35. **GRID IN** Which is greater, −12 or 7? 7

Write an integer for each situation. (Lesson 3-1)

36. 9°C below 0 −9
37. a gain of 20 feet 20

38. **PROFITS** The daily profits of T-shirts sold last week were \$55, \$35, \$25, \$30, and \$55. Which average might be misleading: the mode, the median, or the mean? Explain. (Lesson 2-8) See margin.

GETTING READY FOR THE NEXT LESSON 39–42. See Ch. 3 Answer Appendix.

PREREQUISITE SKILL Graph the solution of each equation on a number line. (Lesson 1-5)

39. $x + 3 = 5$
40. $x - 4 = 8$
41. $3x = 9$
42. $5x = 30$

msmath2.net/self_check_quiz

4 Assess

Open-Ended Assessment

Writing Have students write an explanation of how they order positive and negative integers. Have them include a number line to illustrate the process.

Answers

29. Mar. 2, Mar. 6, Mar. 1, Mar. 5, Mar. 3, Mar. 4, Mar. 7
31. −7, −6, −4, 1, 3, 5
38. mode; Sample answer: The mode is \$55, which does not represent the true average of the data.

Study Guide and Intervention, p. 134

NAME ______ DATE ______ PERIOD ____

3-2 Study Guide and Intervention
Comparing and Ordering Integers

When two numbers are graphed on a number line, the number to the left is always less than (<) the number to the right. The number to the right is always greater than (>) the number to the left.

Model

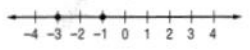

Words −3 is less than −1. −1 is greater than −1.
Symbols −3 < −1 −1 > −3

The symbol points to the lesser number.

EXAMPLE 1 **Replace the ● with < or > to make −1 ● −6 a true sentence.**

Graph each integer on a number line.

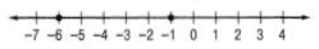

Since −1 is to the right of −6, −1 > −6.

EXAMPLE 2 **Order the integers 2, −3, 0, −5 from least to greatest.**

To order the integers, graph them on a number line.

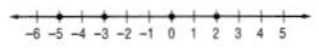

Order the integers by reading from left to right: −5, −3, 0, 2.

EXERCISES

1. Replace the ● with < or > to make −5 ● −10 a true sentence. >
2. Order −1, 5, −3, and 2 from least to greatest. −3, −1, 2, 5
3. Order 0, −4, −2, and 7 from greatest to least. 7, 0, −2, −4
4. Order −3, |−2|, 4, 0, and −5 from greatest to least. 4, |−2|, 0, −3, −5

Practice: Skills, p. 135

NAME ______ DATE ______ PERIOD ____

3-2 Practice: Skills
Comparing and Ordering Integers

Replace each ● with < or > to make a true sentence.

1. −15 ● −16 >
2. −8 ● −7 <
3. 0 ● −2 >
4. −2 ● −5 >
5. −25 ● 3 <
6. −14 ● |−20| <
7. |−4| ● 3 >
8. |−6| ● |−7| <
9. |−7| ● |2| >
10. −8 ● |−9| <

Determine whether each sentence is *true* or *false*. If *false*, change one number to make the sentence true.

11. −7 < 3 true
12. 2 > 0 true
13. −20 < −22 false; Sample answer: −20 < 22
14. 12 < 15 true
15. 3 > |−5| false; Sample answer: 3 > |−1|
16. |−2| < −3 false; Sample answer: |−2| < 3
17. |8| < |−10| true
18. |−11| = 11 true
19. −4 < 4 true
20. |−9| < |−10| true

Order the integers from least to greatest.

21. 12, −6, 20, −47, −11 −47, −11, −6, 12, 20
22. 9, −6, 0, −4, 17, −11 −11, −6, −4, 0, 9, 17

Order the integers from greatest to least.

23. −40, 65, −7, 24, −6, 15 65, 24, 15, −6, −7, −40
24. |−13|, 0, 7, −8, −5, |2| |−13|, 7, |2|, 0, −5, −8

Practice: Word Problems, p. 136

NAME ______ DATE ______ PERIOD ____

3-2 Practice: Word Problems
Comparing and Ordering Integers

HISTORY OF WRITING For Exercises 1 and 2, use the table below. It shows important events in the history of writing.

Event	Aprox. Year
The *Iliad* and the *Odyssey* are composed by Homer.	700 BC
T'sai Lun invents paper.	105 AD
Date of oldest existing papyrus	2200 BC
Ovid wrote *Metamorphosis*.	5 AD
Torah is compiled.	450 BC
Metal type developed in Korea	1241 AD

EXTREME TEMPERATURES For Exercises 3–5, use the table below. It shows the extreme temperatures for four states. Temperatures are in degrees Fahrenheit.

Extreme Temperatures (°F)

State	Highest	Lowest
Alabama	104	3
Nebraska	118	−47
Maine	101	−30
Florida	109	−2

1. Write each year as an integer. −700, 105, −2200, 5, −450, 1241
2. Order the integers from Exercise 1 from least to greatest. Write a sentence describing the earliest and most recent events in the table. −2200, −700, −450, 5, 105, 1241; See students' work.
3. Arrange the highest temperatures from greatest to least. 118, 109, 104, 101
4. What is the median low temperature for these four states? −16
5. Nebraska's lowest temperature was −47°F, and Maine's lowest temperature was −30°F. Write a true statement using the two temperatures with the symbol > or <. Sample answer: −47 < −30
6. **MONEY** Mr. Firewalks pays close attention to how much money is in his checking account. One week he deposited \$230, spent \$15 on a lunch, and loaned \$25 to a friend. Write each transaction as an integer, and list them from least to greatest. −25, −15, 230

3-3 Lesson Notes

Lesson Resources

Chapter 3 Resource Masters
- Study Guide and Intervention, p. 139
- Practice: Skills, p. 140
- Practice: Word Problems, p. 141
- Reading to Learn Mathematics, p. 142
- Enrichment, p. 143

Calculator and Spreadsheet Masters, p. 22
MathPASS Tutorial CD-ROM, 37
Parent and Student Study Guide Workbook, 3-3
Teaching Mathematics with Manipulatives, p. 44

FAST FILE

1 Focus

5-Minute Check 3-3

INTERACTIVE CHALKBOARD PowerPoint® Presentations

Use as a quiz or review of Lesson 3-2. (Also available as a transparency)

Bellringer
Hands-On Activity Have students fill in their schedules for a typical week in a table such as the one below. Then have them identify 3 or 4 class periods as follows:

(T, 8:00) = math
(W, 10:00) = English

Show students they can label each point on the schedule with two pieces of information: the time and the day of the week.
(Continued in bottom margin.)

Reading to Learn Mathematics, p. 142 **ELL**

NAME ______________ DATE ________ PERIOD ____

3-3 Reading to Learn Mathematics
The Coordinate Plane

Pre-Activity *Read the introduction at the top of page 112 in your textbook. Write your answers below.*

1. Suppose Terrell starts at the corner of Russel and Main and walks 1 block north and 2 blocks east. Name the intersection of his location.
Highland and Mobile

2. Using the words *north, south, west,* and *east,* write directions to go from the corner of School and Highland to the corner of Main and Oak.
Sample answer: Walk 6 blocks west, 1 block south, and 1 block west.

Reading the Lesson

3. The word *coordinate* comes from two Latin words that mean "to arrange together." How are coordinates used together to locate a point in a coordinate plane? Sample answer: The location of a point in a ...

3-3 The Coordinate Plane

What You'll Learn
Graph points on a coordinate plane.

NEW Vocabulary
- coordinate plane
- coordinate grid
- *x*-axis
- *y*-axis
- origin
- ordered pair
- *x*-coordinate
- *y*-coordinate
- quadrant

WHEN am I ever going to use this?

MAPS A map of Terrell's neighborhood is shown.

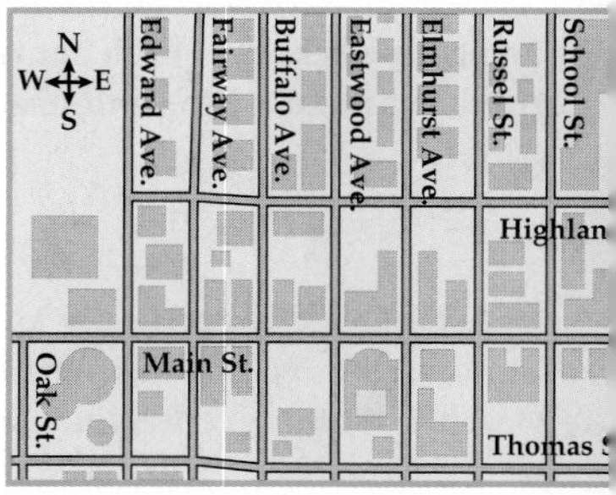

1. Suppose Terrell starts at the corner of Russel and Main and walks 1 block north and 2 blocks east. Name the intersection of his location. **Highland and Mobile**
2. Using the words *north, south, west,* and *east,* write directions to go from the corner of School and Highland to the corner of Main and Oak.

Sample answer: walk 6 blocks west, 1 block south, and 1 block west

A **coordinate plane** is used to locate points. It is a plane in which a horizontal number line and a vertical number line intersect at their zero points. A coordinate plane is also called a **coordinate grid**.

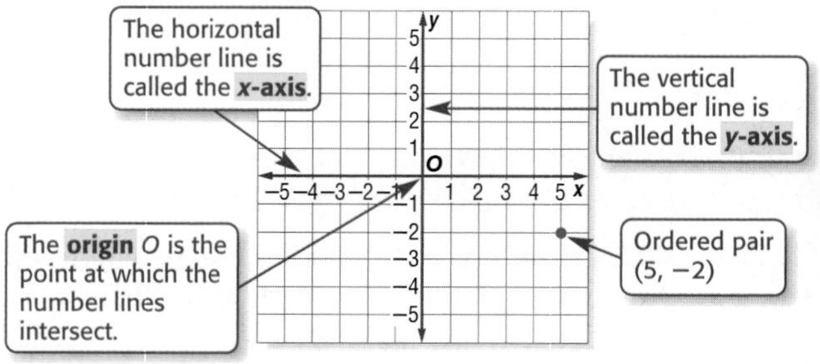

An **ordered pair** is a pair of numbers such as $(5, -2)$ used to locate a point in the coordinate plane.

The **x-coordinate** is the first number, and it corresponds to a number on the *x*-axis. $(5, -2)$ The **y-coordinate** is the second number, and it corresponds to a number on the *y*-axis.

STUDY TIP

Graphing Moving right or up on a coordinate plane is in the *positive direction*. Moving left or down is in the *negative direction*.

EXAMPLE Name an Ordered Pair

1 Name the ordered pair for point *P*.

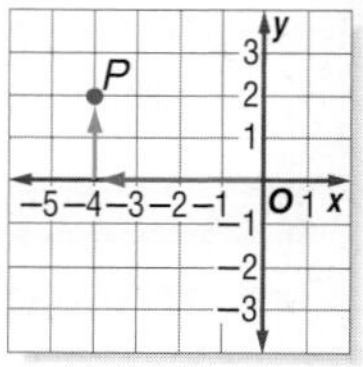

- Start at the origin.
- Move left to find the *x*-coordinate of point *P*, which is -4.
- Move up to find the *y*-coordinate, which is 2.

So, the ordered pair for point *P* is $(-4, 2)$.

	M	T	W	Th	F
8:00					
9:00					
10:00					
11:00					
12:00					
1:00					
2:00					
3:00					

READING Math

Scale When no numbers are shown on the *x*- or *y*-axis, you can assume that each grid is 1 unit long on each side.

EXAMPLE Graph an Ordered Pair

2 Graph and label the point $Q(2, -5)$.

- Draw a coordinate plane.
- Move 2 units to the right. Then move 5 units down.
- Draw a dot and label it $Q(2, -5)$.

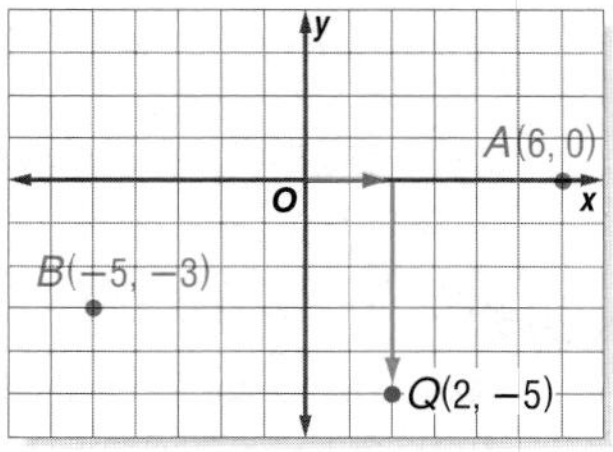

Your Turn **Graph each point.**

a. $A(6, 0)$ b. $B(-5, -3)$

The coordinate plane is separated into four sections called **quadrants**.

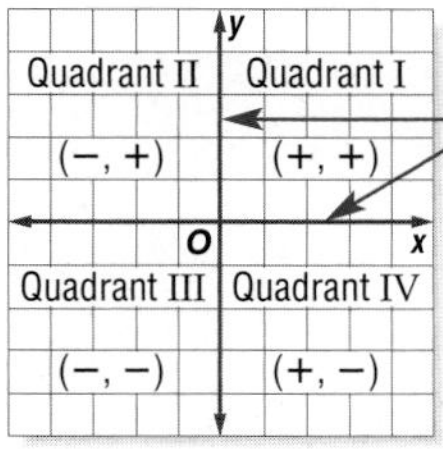

Axes is the plural form of axis. The axes are not located in any of the quadrants.

REAL-LIFE CAREERS

How Does a Cartographer Use Math?
Cartographers use coordinates to prepare geographic, political, and cultural maps of large areas.

Online Research
For more information about a career as a cartographer, visit: msmath2.net/careers

EXAMPLES Identify Quadrants

3 GEOGRAPHY The world map can be divided into a coordinate grid where (x, y) represents (degrees longitude, degrees latitude). In which quadrant is the United States located?

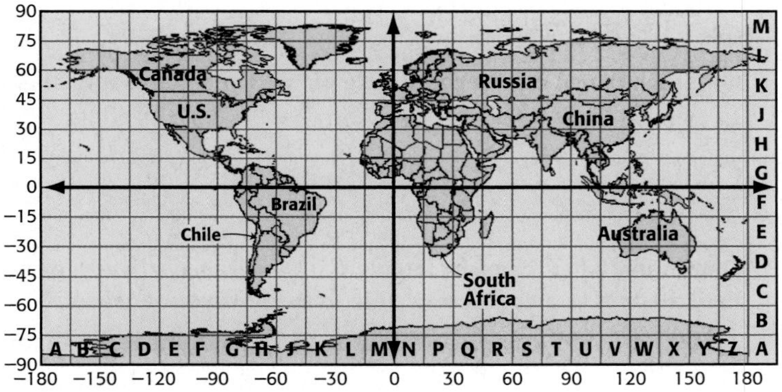

Source: www.colorado.edu

The United States is located in the upper left quadrant, quadrant II.

4 Name a country from the map that is located in quadrant III.

Quadrant III is the bottom-left quadrant. So, Chile is in quadrant III.

2 Teach

In-Class Examples

INTERACTIVE CHALKBOARD PowerPoint® Presentations

1 Name the ordered pair for point *R* graphed below. **(−2, 4)**

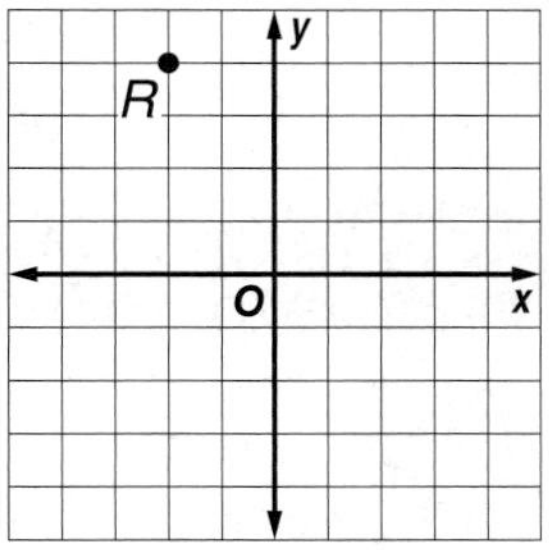

2 Graph and label the point $M(3, 5)$.

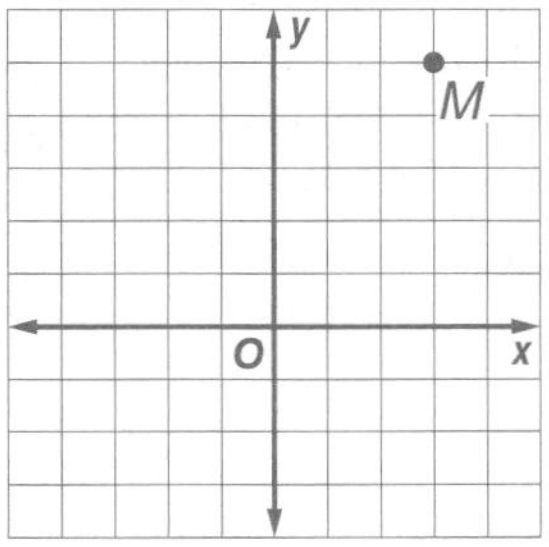

3 **GEOGRAPHY** Use the map of Utah shown below. In which quadrant is Vernal located? **Quadrant I**

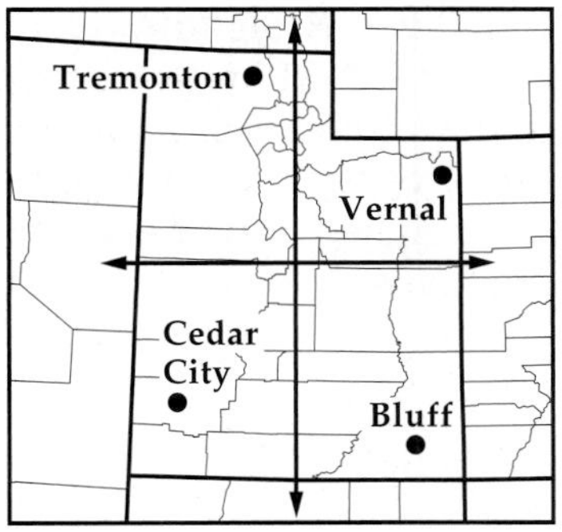

4 Name a city from the map that is located in quadrant IV. **Bluff**

DAILY INTERVENTION — Differentiated Instruction

Visual/Spatial If possible, acquire an aerial photo or schematic of the school and school grounds to photocopy. Have students draw a grid on the photo and label where different features are, such as ball fields, parking lots, tennis courts, and outbuildings. Then take students outside and have them walk between landmarks to get a sense of the distance shown on the map. Have them plot their paths on their grids.

Skill and Concept Check

Writing Math Exercise 1

1. **Explain** why point $A(1, -2)$ is different from point $B(-2, 1)$. **1–2. See margin.**

2. **OPEN ENDED** Name and graph a point in quadrant IV.

GUIDED PRACTICE

Name the ordered pair for each point graphed at the right. Then identify the quadrant in which each point lies.

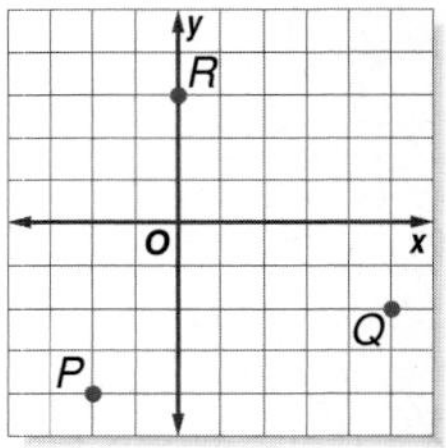

3. P **(−2, −4), III**
4. Q **(5, −2), IV**
5. R **(0, 3), y-axis**

On graph paper, draw a coordinate plane. Then graph and label each point. 6–8. See margin.

6. $S(2, 3)$
7. $T(-4, 6)$
8. $U(-5, 0)$

★ indicates increased difficulty

Practice and Applications

HOMEWORK HELP

For Examples	See Examples
9–18	1, 3
19–31	2
35–42	3, 4

Extra Practice See pages 570, 598.

Name the ordered pair for each point graphed at the right. Then identify the quadrant in which each point lies. 11. (−5, 0), x-axis

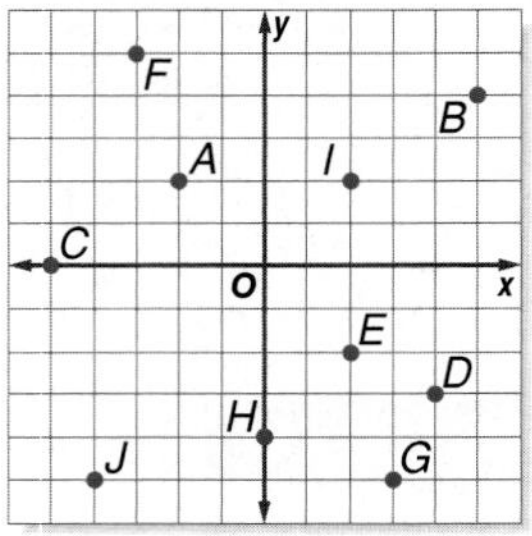

9. A **(−2, 2), II**
10. B **(5, 4), I**
11. C
12. D **(4, −3), IV**
13. E **(2, −2), IV**
14. F **(−3, 5), II**
15. G **(3, −5), IV**
16. H **(0, −4), y-axis**
17. I **(2, 2), I**
18. J **(−4, −5), III**

19. Write the ordered pair for the point that lies on the y-axis and is 32 units down from the origin. **(0, −32)**

20–31. See Ch. 3 Answer Appendix.

On graph paper, draw a coordinate plane. Then graph and label each point.

20. $M(5, 6)$
21. $N(-2, 10)$
22. $P(7, -8)$
23. $Q(3, 0)$
24. $R(-1, -7)$
25. $S(0, 6)$
26. $T(-3, 7)$
27. $U(5, -2)$
28. $V(8, 1)$
29. $W(-5, -7)$
★30. $X(1.5, -3)$
★31. $Y(-6.5, 6.5)$

Determine whether each statement is *sometimes*, *always*, or *never* true. Explain or give a counterexample to support your answer. 32–34. See margin.

32. Both x- and y-coordinates of a point in quadrant I are negative.
33. The x-coordinate of a point that lies on the x-axis is negative.
34. The y-coordinate of a point in quadrant IV is negative.

GEOGRAPHY For Exercises 35 and 36, use the map in Example 3.

35. In what country is the point (105°, 30°) located? **China**
36. Find an ordered pair that can represent the location of California. **Sample answer: (−122°, 40°)**

Tips for New Teachers

Classroom Management You often can obtain maps for class use from your local AAA, Department of Transportation, or Bureau of Motor Vehicles. Keep a supply on hand.

3 Practice/Apply

FOLDABLES™ Follow-Up Remind students to use their Foldable journals to record what they learn about the coordinate system. For example, they might record and define key terms presented in this lesson and give examples of each.

About the Exercises...

Guided Practice Key

Exercises	Examples
3–5	1
6–8	2

Odd/Even Assignments
Exercises 9–31 are structured so that students practice the same concepts whether they are assigned odd or even problems.

Answers

1. Sample answer: Point A is 1 unit to the right and 2 units down from the origin, in quadrant IV. Point B is 2 units to the left and 1 unit up from the original, in quadrant II.

2. Sample answer: $G(3, -1)$

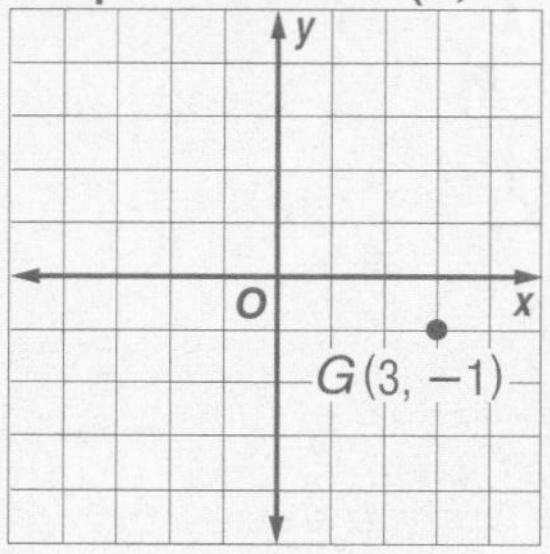

6–8.

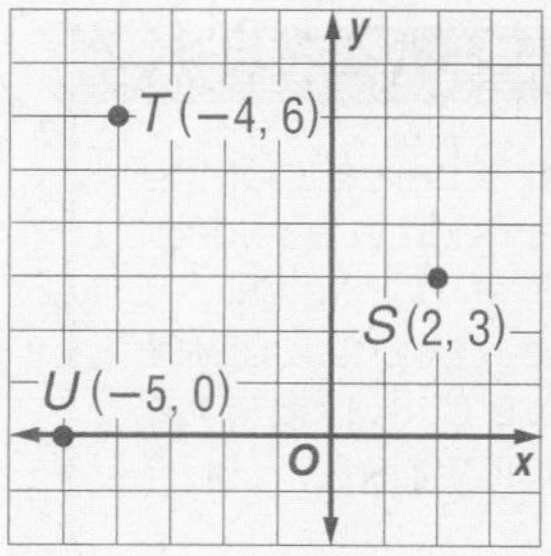

32. Never; both coordinates are positive.
33. Sometimes; both (−2, 0) and (2, 0) lie on the x-axis.
34. Always; every point in quadrant IV has a positive x-coordinate and a negative y-coordinate.

42.

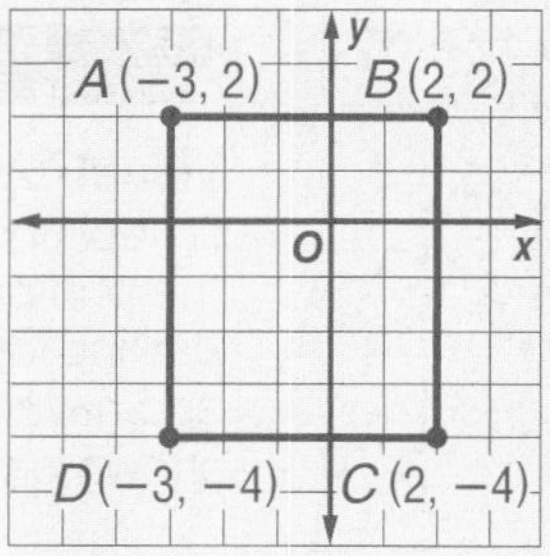

For Exercises 37–41, use the map of the Brookfield Zoo.

Brookfield Zoo, Brookfield, IL

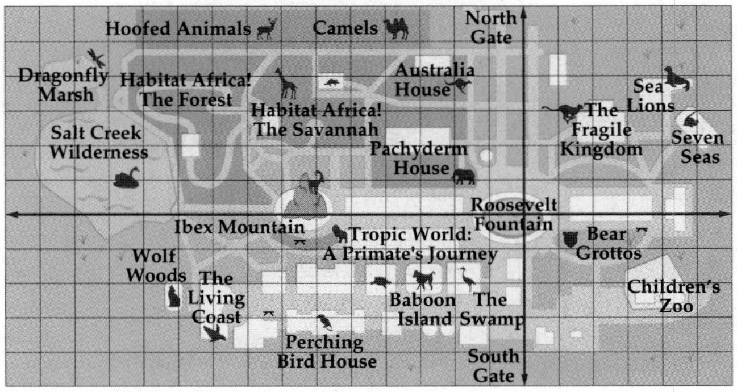

Source: www.brookfieldzoo.org

37. What exhibit is located at $(4, -2)$? **Children's Zoo**

38. In which quadrant is the Dragonfly Marsh exhibit located? **II**

39. Find the ordered pair that represents the location of Baboon Island. **$(-3, -2)$**

40. What is located at the origin? **Roosevelt Fountain**

★ 41. Describe how you would walk from the entrance of the Pachyderm House at $(-2, 2)$ to the entrance of The Swamp at $(-1, -2)$. **Sample answer: In the coordinate plane, I would walk 1 unit right, then 4 units down.**

★ 42. **GEOMETRY** Graph the points $A(-3, 2)$, $B(2, 2)$, $C(2, -4)$, and $D(-3, -4)$ on the same coordinate plane. Connect the points from A to B, B to C, C to D, and D to A. Name the figure. **42–43. See margin.**

43. **CRITICAL THINKING** Find the possible locations for any ordered pair whose x- and y-coordinates are always the same integer. Explain.

Standardized Test Practice and Mixed Review

A B C D

For Exercises 44–46, use the coordinate plane at the right.

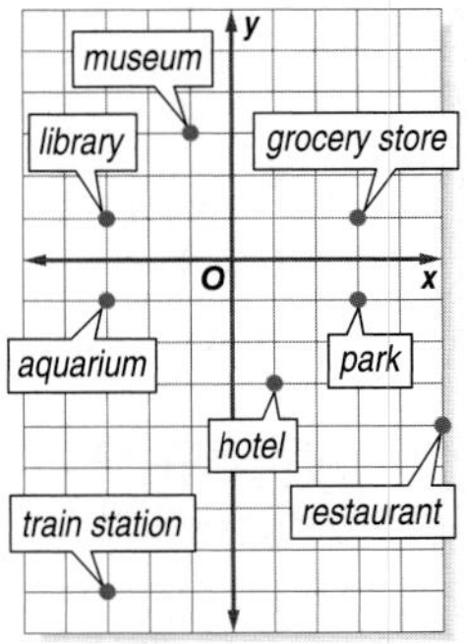

44. **MULTIPLE CHOICE** Which building has the coordinates of $(-3, -1)$? **D**

Ⓐ museum Ⓑ park Ⓒ library Ⓓ aquarium

45. **MULTIPLE CHOICE** What are the coordinates of the point that shows the location of the hotel? **G**

Ⓕ $(-1, -3)$ Ⓖ $(1, -3)$ Ⓗ $(-1, 3)$ Ⓘ $(1, 3)$

46. **SHORT RESPONSE** In which quadrant is the grocery store located? **quadrant I**

Replace each ● with < or > to make a true sentence. (Lesson 3-2)

47. $14 \bullet |-15|$ **<**
48. $-8 \bullet -3$ **<**
49. $26 \bullet -30$ **>**
50. $-40 \bullet |40|$ **<**

51. Find the absolute value of -101. (Lesson 3-1) **101**

52. **SPORTS** A triathlon competition consists of swimming 3 miles, running 10 miles, and bicycling 35 miles. How many miles does an athlete travel during the competition? (Lesson 1-1) **48 mi**

GETTING READY FOR THE NEXT LESSON

BASIC SKILL Add.

53. $138 + 246$ **384**
54. $814 + 512$ **1,326**
55. $2,653 + 4,817$ **7,470**
56. $6,003 + 5,734$ **11,737**

msmath2.net/self_check_quiz

4 Assess

Higher-Level Thinking

What is true of any ordered pair whose x- or y-coordinate is 0?
Its plotted point lies on either the x- or y-axis.

Answer

43. Quadrant I when both x- and y-coordinates are positive; Quadrant III when both x- and y-coordinates are negative; the origin when both x- and y-coordinates are 0.

Study Guide and Intervention, p. 139

NAME ________ DATE ______ PERIOD ____

3-3 Study Guide and Intervention

The Coordinate Plane

The **coordinate plane** is used to locate points. The horizontal number line is the **x-axis**. The vertical number line is the **y-axis**. Their intersection is the **origin**.

Points are located using **ordered pairs**. The first number in an ordered pair is the **x-coordinate**; the second number is the **y-coordinate**.

The coordinate plane is separated into four sections called **quadrants**.

EXAMPLE 1 Name the ordered pair for point P. Then identify the quadrant in which P lies.

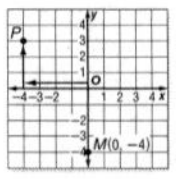

- Start at the origin.
- Move 4 units left along the x-axis.
- Move 3 units up on the y-axis.
  The ordered pair for point P is $(-4, 3)$.
  P is in the upper left quadrant or quadrant II.

EXAMPLE 2 Graph and label the point $M(0, -4)$.

- Start at the origin.
- Move 0 units along the x-axis.
- Move 4 units down on the y-axis.
- Draw a dot and label it $M(0, -4)$.

EXERCISES

Name the ordered pair for each point graphed at the right. Then identify the quadrant in which each point lies.

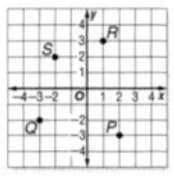

1. P (2, −3), IV
2. Q (−3, −2), III
3. R (1, 3), I
4. S (−2, 2), II

Graph and label each point on the coordinate plane.

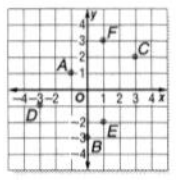

5. $A(-1, 1)$
6. $B(0, -3)$
7. $C(3, 2)$
8. $D(-3, -1)$
9. $E(1, -2)$
10. $F(1, 3)$

Practice: Skills, p. 140

NAME ________ DATE ______ PERIOD ____

3-3 Practice: Skills

The Coordinate Plane

Name the ordered pair for each point graphed at the right. Then identify the quadrant in which each point lies.

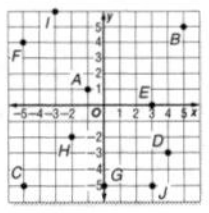

1. A (−1, 1), II
2. B (5, 5), I
3. C (−5, −5), III
4. D (4, −3), IV
5. E (3, 0), x-axis
6. F (−5, 4), II
7. G (0, −5), y-axis
8. H (−2, −2), III
9. I (−3, 6), II
10. J (3, −5), IV

Graph and label each point on the coordinate plane.

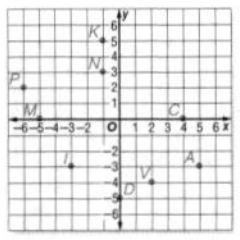

11. $N(-1, 3)$
12. $V(2, -4)$
13. $C(4, 0)$
14. $P(-6, 2)$
15. $M(-5, 0)$
16. $K(-1, 5)$
17. $I(-3, -3)$
18. $A(5, -3)$
19. $D(0, -5)$

Name the ordered pair for each point on the city map at the right.

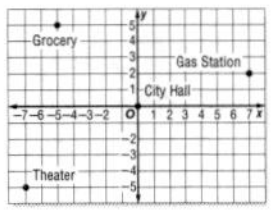

20. City Hall (0, 0)
21. Theater (−7, −5)
22. Gas Station (7, 2)
23. Grocery (−5, 5)

Practice: Word Problems, p. 141

NAME ________ DATE ______ PERIOD ____

3-3 Practice: Word Problems

The Coordinate Plane

SCHOOL For Exercises 1–4, use the coordinate plane at the right. It shows a map of the rooms in a junior high school.

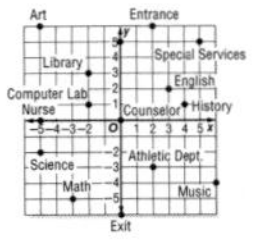

1. Thalia is in the room located at $(-2, 1)$. What room is she in? Describe in words how to get from the origin to this point. Computer Lab; move 2 units to the left and one unit up.	2. Thalia's next class is 8 units to the right and 5 units down on the map from where she is now. In what room is Thalia's next class? Find the ordered pair that represents the location of that room. Music room; (6, −4)
3. Tyrone is in the Art room, but his next class is in the History room. Give Tyrone directions on how to get to the History room. Sample answer: Walk 9 units to the right, and then 5 units down.	4. On the map, which classrooms are located in the third quadrant? Describe the coordinates of all points in the third quadrant. Science and Math rooms; the x- and y-coordinates are all negative.
5. NEIGHBORHOOD Delsin made a map of his neighborhood in such a way that each intersection is a point on a coordinate plane. Right now, Delsin stands at point $(-4, -3)$. Give the ordered pair of where he will be if moves 5 units to the right and 7 units up on the map. (1, 4)	6. NEIGHBORHOOD Refer to Exercise 5. In which quadrant is Delsin when he is done walking? Describe this quadrant. Quadrant I; Sample answer: Quadrant I is the upper-right quadrant.

Chapter 3 Mid-Chapter Practice Test

Assessment Option

Mid-Chapter Test (Lessons 3-1 through 3-3) is available on p. 181 of the *Chapter 3 Resource Masters*.

ExamView® Pro Test Bank CD-ROM

Use the networkable ExamView® Pro Test Bank CD-ROM to:

- Create **Multiple versions** of tests.
- Create **Modified** tests for **Inclusion** students with one mouse click.
- **Edit** existing questions and **Add** your own questions.
- Build tests aligned with state standards using built-in **State Curriculum Correlations.**
- Change **English** tests to **Spanish** with one mouse click and vice versa.

FOLDABLES™ Follow-Up

Before students complete the Mid-Chapter Practice Test, encourage them to review the notes, definitions, concepts, and main ideas they recorded on the first pages of their Foldable journals.

Answers

1. **Sample answer: The absolute value of an integer is its distance from 0 on a number line.**

3\.

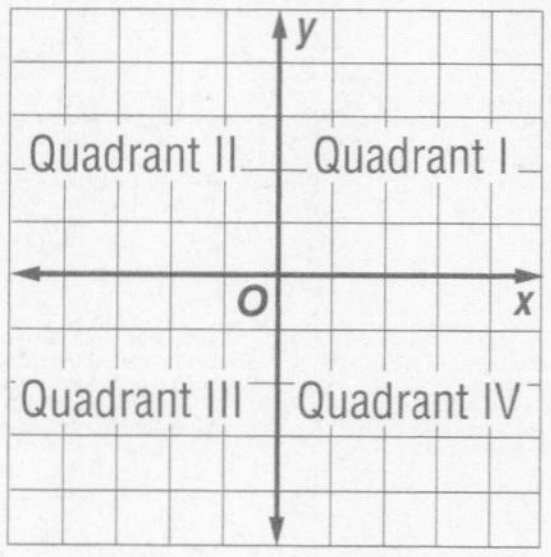

Chapter 3 Mid-Chapter Practice Test

Vocabulary and Concepts

1. **Define** *absolute value.* (Lesson 3-1) **See margin.**
2. **Write** the ordered pair which identifies a point 4 units to the left of the *y*-axis and three units above the *x*-axis. (Lesson 3-3) **(−4, 3)**
3. **Draw** a coordinate plane, and label the quadrants. (Lesson 3-3) **See margin.**

Skills and Applications

Write an integer for each situation. (Lesson 3-1)

4. 45 feet below sea level **−45**
5. a deposit of $100 **100**
6. a gain of 8 yards **8**
7. lost a $5 bill **−5**

Replace each ● with < or > to make a true sentence. (Lesson 3-2)

8. −12 ● −9 **<**
9. −4 ● 4 **<**
10. $|-14|$ ● $|3|$ **>**
11. **FOOTBALL** The Tigers have recorded the following yardage on the past six plays: 9, −2, 5, 0, 12, and −7. Order the integers from least to greatest. (Lesson 3-2) **−7, −2, 0, 5, 9, 12**

On graph paper, draw a coordinate plane. Then graph and label each point. (Lesson 3-3) **12–14. See margin.**

12. *D*(4, −3)
13. *E*(1, 3)
14. *F*(0, −5)

Standardized Test Practice

15. **MULTIPLE CHOICE** Which of the following points represents a number and its absolute value? (Lesson 3-1) **A**

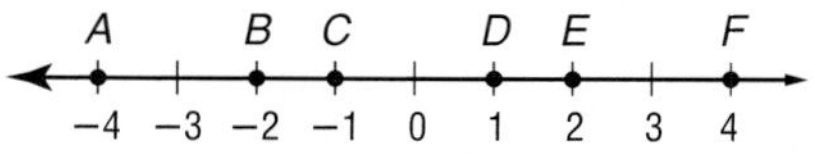

Ⓐ *B* and *E* Ⓑ *C* and *F* Ⓒ *B* and *D* Ⓓ *A* and *E*

16. **SHORT RESPONSE** The table shows the number of inches of monthly precipitation above or below normal for a midwestern city in a recent year. Find the median monthly precipitation above or below normal. (Lesson 3-2) **−1 in.**

Monthly Precipitation Above or Below Normal

J	F	M	A	M	J
4	−1	6	−2	−3	1
J	**A**	**S**	**O**	**N**	**D**
−2	−1	2	−3	1	−3

12–14.

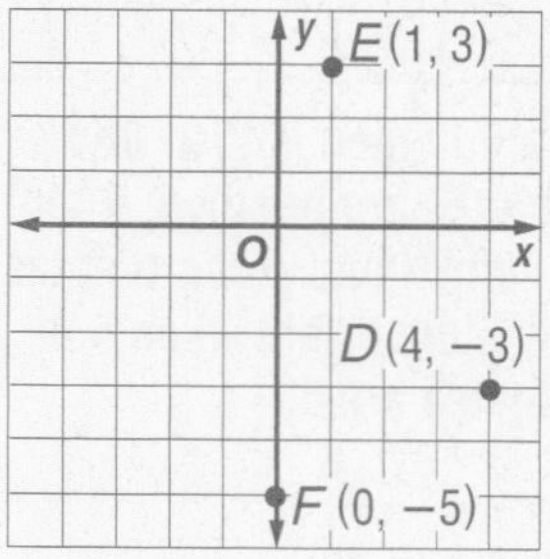

The Game Zone

A Place To Practice Your Math Skills

Tic-Tac-Toe

GET READY!

Players: two
Materials: grid paper

GET SET!

- Draw a coordinate plane on grid paper.
- This game is similar to tic-tac-toe, except players must get four Xs or four Os in a row.

GO!

- Player 1 chooses two numbers: the first number is the *x*-coordinate of an ordered pair, and the second number is the *y*-coordinate. Each number must be between -5 and 5. Then Player 1 announces the ordered pair and plots the X or O on the coordinate plane.
- Player 2 then chooses his or her numbers, announces them, and plots the points.
- An ordered pair cannot be changed after it has been announced.
- If a player announces an ordered pair that has already been used or graphs an ordered pair incorrectly, the player loses a turn.
- **Who Wins?** The first player to get four Xs or Os in a row is the winner.

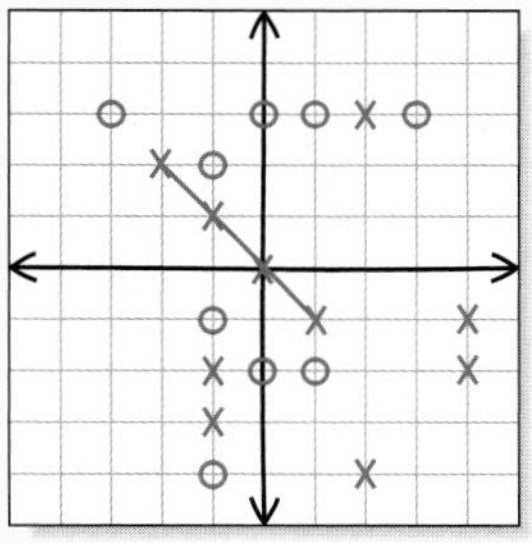

The Game Zone

Resources

For an easy-to-make manipulative for this game, visit: msmath2.net/gamezone.

Glencoe Mathematics Classroom Manipulative Kit

- coordinate grid stamp

GETTING READY

Objective Graph points on a coordinate plane.

MANAGEMENT TIPS

Recommended Time
15 minutes

Tips

- You may want to have 10×10 grids already drawn with the axes marked to save time.
- Have one player plot all points as X and the other player plot all points as O. You may also wish to suggest that students use different colored markers to help make winning rows stand out.
- You may wish to use the game as an assessment tool for the concepts the students have been learning in Lessons 3-1 through 3-3.

For an additional game for this chapter, visit: msmath2.net/gamezone.

The Game Zone: Graphing Points on a Coordinate Plane **117**

3-4a HANDS-ON LAB Lesson Notes

Lesson Resources

Teaching Mathematics with Manipulatives

- integer counters, p. 6
- integer mat, p. 8
- student recording sheet 3-4a

Glencoe Mathematics Classroom Manipulative Kit

- counters
- equation mat

GETTING STARTED

Teaching Tip You may want to let students practice writing simple addition sentences with positive integers and using the counters to support their solutions.

TEACH

- You may want to expand upon the opening activity by adding pairs of positive and negative counters to the counters on the mat. Continue until students see that as long as you add one counter of each sign, the result is still −6.
- **Activity 1** shows the addition of integers with the same signs. Ask students how adding two negatives differs from adding two positives. The only difference is that the answer is negative.

3-4a HANDS-ON LAB

A Preview of Lesson 3-4

What You'll Learn

Use counters to model the addition of integers.

Materials

- counters
- integer mat

Adding Integers

You can use positive and negative counters to model the addition of integers. The counter ⊕ represents 1, and the counter ⊖ represents −1. Remember that addition means *combining* two sets.

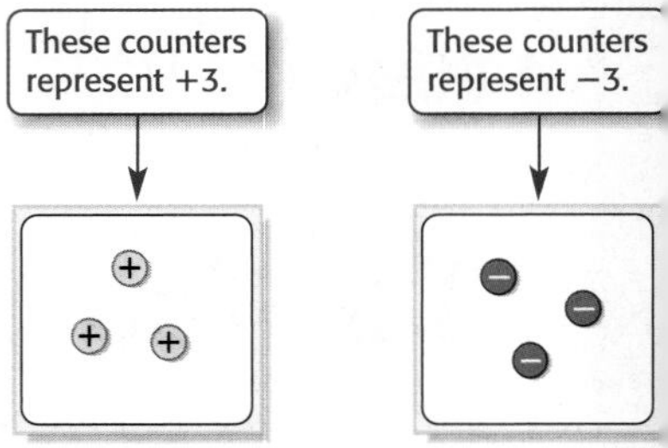

ACTIVITY *Work with a partner.*

1 **Use counters to find −2 + (−4).**

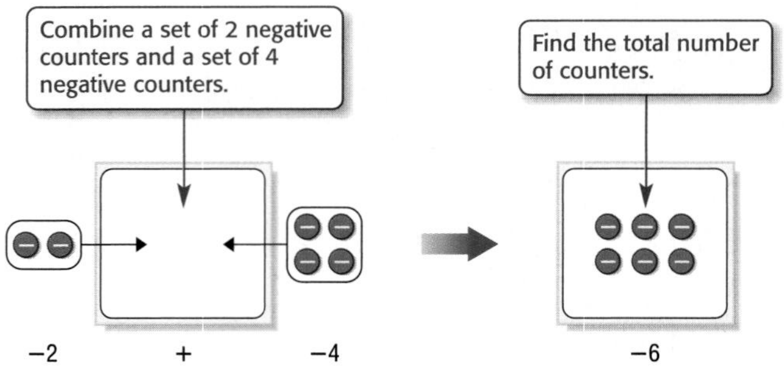

So, −2 + (−4) = −6.

Your Turn **Use counters to find each sum.** a–f. See margin for models.

a. 5 + 6 11
b. −3 + (−5) −8
c. −5 + (−4) −9
d. 7 + 3 10
e. −2 + (−5) −7
f. −8 + (−6) −14

READING Math

Addends and Sums The numbers you add are called *addends*. The result is called the *sum*.

The following two properties are important when modeling operations with integers.

- When one positive counter is paired with one negative counter, the result is called a **zero pair**. The value of a zero pair is 0.
- You can add or remove zero pairs from a mat because adding or removing zero does not change the value of the counters on the mat.

You will use zero pairs in Activity 2 and Activity 3.

Answers (Your Turn)

a.

5 + 6 = 11

b.

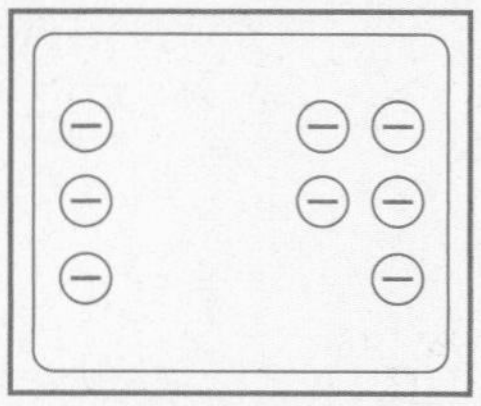

−3 + (−5) = −8

c.

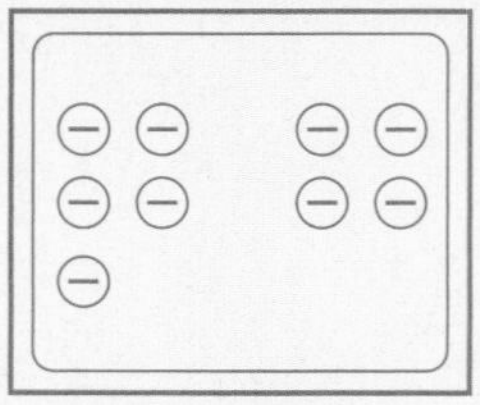

−5 + (−4) = −9

d.

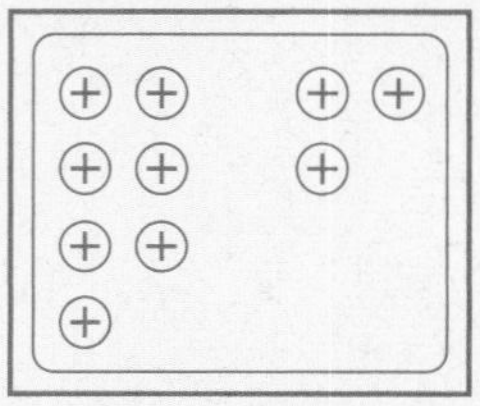

7 + 3 = 10

- **Activities 2** and **3** demonstrate the addition of integers with different signs. Have students predict the value of each expression before removing the zero pairs.

ASSESS

Observe as students solve the **Your Turn** exercises. They should:

- correctly add two positive integers,
- correctly add two negative integers, and
- correctly add a positive and a negative integer.

Discuss the **Writing Math** exercises. Have students check the rules they made in Exercise 3 to see whether they work for all the sentences they wrote for Exercises 1 and 2.

Answer (Writing Math)

3. **Sample answer: If two integers have the same sign, add and keep the sign. If two integers have different signs, subtract and keep the sign of the integer with the greater absolute value.**

ACTIVITIES *Work with a partner.*

Use counters to find each sum.

2 $-5 + 3$

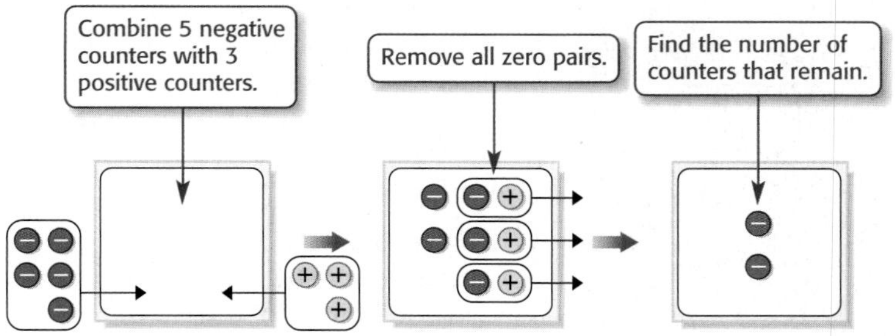

So, $-5 + 3 = -2$.

3 $4 + (-1)$

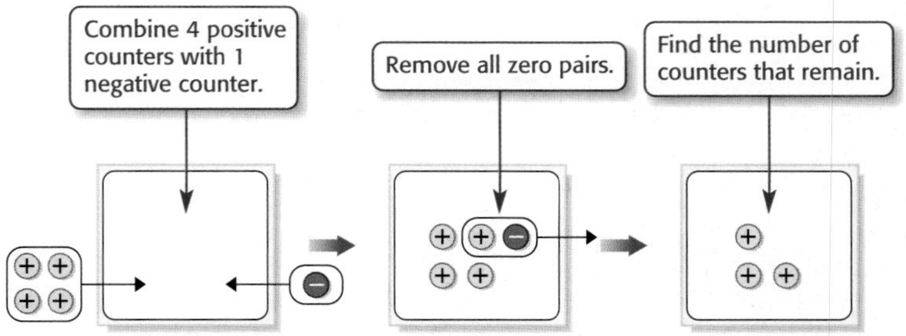

So, $4 + (-1) = 3$.

Your Turn **Use counters to find each sum.** g–l. See Ch. 3 Answer Appendix for models.

g. $-6 + 5$ **−1** h. $3 + (-6)$ **−3** i. $-2 + 7$ **5**

j. $8 + (-3)$ **5** k. $-9 + 1$ **−8** l. $-4 + 10$ **6**

Writing Math

1. **Write** two addition sentences where the sum is positive. In each sentence, one addend should be positive and the other negative.
2. **Write** two addition sentences where the sum is negative. In each sentence, one addend should be positive and the other negative.
3. **MAKE A CONJECTURE** Write a rule that will help you determine the sign when finding the sum of integers. **See margin.**

. **Sample answer:** ·1 + 2 = 1; + (−4) = 1

. **Sample answer:** + (−7) = −5; ·10 + 5 = −5

e.

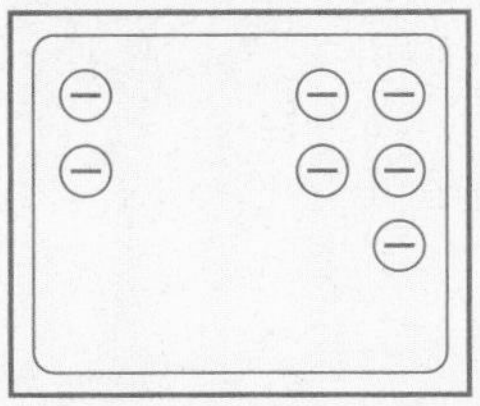

$-2 + (-5) = -7$

f.

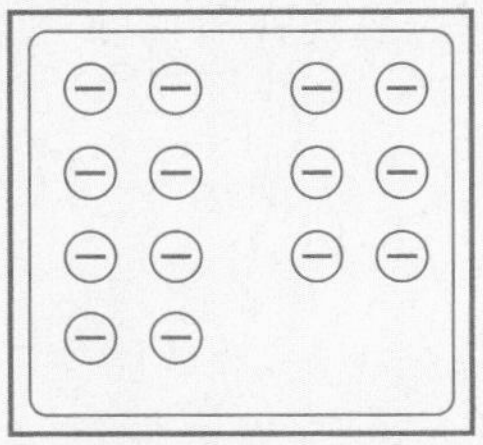

$-8 + (-6) = -14$

3-4 Lesson Notes

Lesson Resources

Chapter 3 Resource Masters

- Study Guide and Intervention, p. 144
- Practice: Skills, p. 145
- Practice: Word Problems, p. 146
- Reading to Learn Mathematics, p. 147
- Enrichment, p. 148

Calculator and Spreadsheet Masters, p. 23

Key Teaching Transparencies, 9

Parent and Student Study Guide Workbook, 3-4

1 Focus

5-Minute Check 3-4

INTERACTIVE CHALKBOARD PowerPoint® Presentations

Use as a quiz or review of Lesson 3-3. (Also available as a transparency)

Bellringer

Communication Have students brainstorm examples of opposites, such as night and day, up and down, or north and south. Then have them consider cases in which opposites may cancel each other out. For example, walking 3 paces forward and then 3 paces backward would get you back to the place you started.

Reading to Learn Mathematics, p. 147 ELL

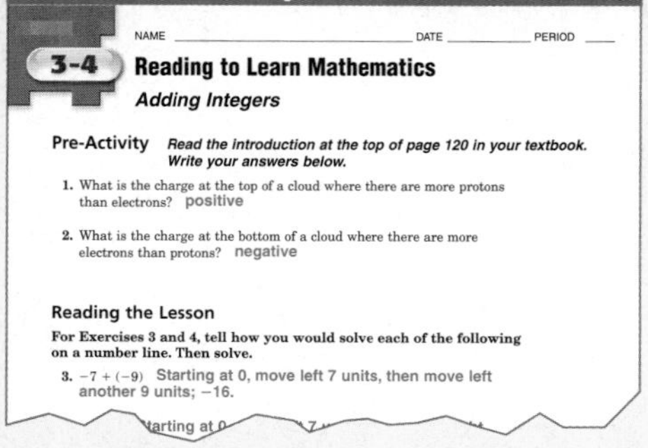

NAME ______ DATE ______ PERIOD ____

3-4 Reading to Learn Mathematics

Adding Integers

Pre-Activity *Read the introduction at the top of page 120 in your textbook. Write your answers below.*

1. What is the charge at the top of a cloud where there are more protons than electrons? positive
2. What is the charge at the bottom of a cloud where there are more electrons than protons? negative

Reading the Lesson

For Exercises 3 and 4, tell how you would solve each of the following on a number line. Then solve.

3. $-7 + (-9)$ Starting at 0, move left 7 units, then move left another 9 units; -16.

3-4 Adding Integers

What You'll Learn
Add integers.

NEW Vocabulary
opposites
additive inverse

LINK To Reading
Everyday Meaning of Opposite: something that is across from or is facing the other way, as in running the opposite way

WHEN am I ever going to use this?

EARTH SCIENCE Thunderstorms are made of both positive and negative electrical charges. The negative charges (electrons) are at the bottom of a thundercloud and positive charges (protons) are at the top.

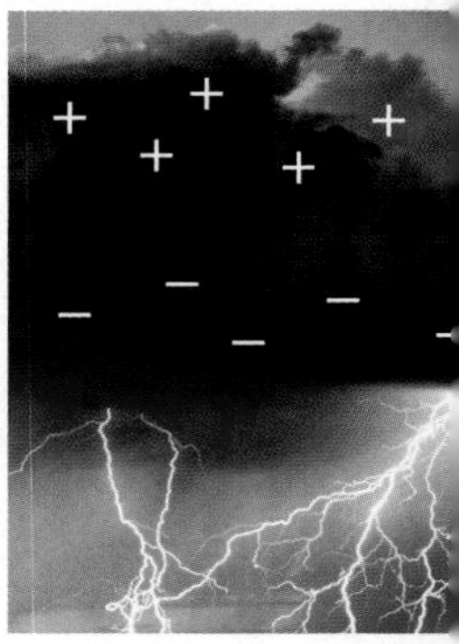

1. What is the charge at the top of a cloud where there are more protons than electrons? **positive**
2. What is the charge at the bottom of a cloud where there are more electrons than protons? **negative**

Combining positive and negative electrical charges in a thunderstorm is similar to adding integers.

EXAMPLE Add Integers with the Same Sign

1 Find $-3 + (-2)$.

Use a number line.

- Start at 0.
- Move 3 units left to show -3.
- From there, move 2 units left to show -2.

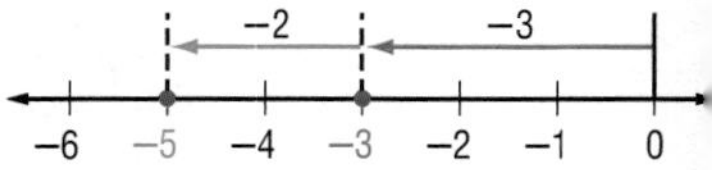

So, $-3 + (-2) = -5$.

Key Concept **Adding Integers with the Same Sign**

Words	The sum of two positive integers is positive. The sum of two negative integers is negative.
Examples	$7 + 4 = 11$ $\quad -7 + (-4) = -11$

EXAMPLE Add Integers with the Same Sign

2 Find $-26 + (-17)$.

$-26 + (-17) = -43$ The sum of two negative integers is negative.

Your Turn Add.

a. $-14 + (-16)$ **−30** b. $23 + 38$ **61** c. $-35 + (-49)$ **−84**

STUDY TIP

Technology To enter a negative integer on a calculator, use the (−) key. For example, to enter −5, press (−) 5.

The integers 43 and −43 are called **opposites** of each other because they are the same distance from 0, but on opposite sides of 0. Two integers that are opposites are also called **additive inverses**.

TEACHING TIP
'hen students use ɔunters to find sums of tegers, tell them that sing zero pairs is the ame as applying the dditive Inverse Property.

Key Concept — **Additive Inverse Property**

Words The sum of any number and its additive inverse is 0.
Example $5 + (-5) = 0$

EXAMPLES Add Integers with Different Signs

3 Find $5 + (-3)$.

Use counters.

Remove all zero pairs.

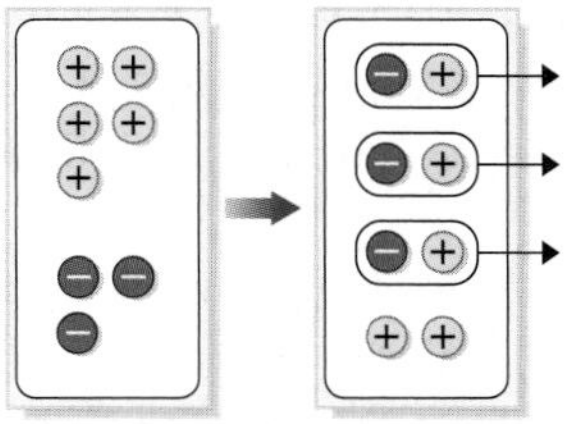

$5 + (-3)$ $5 + (-3) = 2$

So, $5 + (-3) = 2$.

4 Find $-3 + 2$.

Use a number line.

- Start at 0.
- Move 3 units left.
- Then move 2 units right.

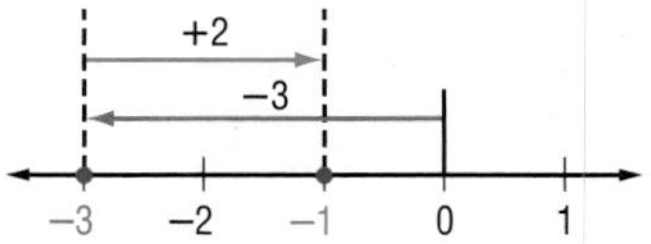

So, $-3 + 2 = -1$.

Key Concept — **Adding Integers with Different Signs**

Words To add integers with different signs, subtract their absolute values. The sum is:

- positive if the positive integer has the greater absolute value.
- negative if the negative integer has the greater absolute value.

Examples $9 + (-4) = 5$ $-9 + 4 = -5$

STUDY TIP

Look Back You can review **absolute value** in Lesson 3-1.

EXAMPLES Add Integers with Different Signs

5 Find $7 + (-1)$.

$7 + (-1) = 6$ Subtract. Keep the sign of the integer with the greater absolute value.

6 Find $-8 + 4$.

$-8 + 4 = -4$ Subtract. Keep the sign of the integer with the greater absolute value.

Your Turn **Add.**

d. $10 + (-12)$ **−2** e. $-16 + 9$ **−7** f. $-13 + 18$ **5**

msmath2.net/extra_examples

2 Teach

In-Class Examples

INTERACTIVE CHALKBOARD
PowerPoint® Presentations

1. Find $-6 + (-3)$. **−9**
2. Find $-34 + (-21)$. **−55**
3. Find $8 + (-7)$. **1**
4. Find $-5 + 4$. **−1**
5. Find $2 + (-7)$. **−5**
6. Find $-9 + 6$. **−3**

Tips for New Teachers

Using Manipulatives You may wish to set up a small table in a corner of the room with counters and integer mats. Provide time for students to use them if they are struggling with operations of integers. Students also could use other models, such as number lines.

DAILY INTERVENTION — **Differentiated Instruction**

Kinesthetic Pair students and give each pair 10 lima beans that have been painted red on one side and white on the other. Determine which side will represent negative integers and which will represent positive integers. Have one student in each pair toss the beans onto a desk. The other student adds up the negative and positive beans, removes the zero pairs, and writes down the sum or score for that toss. After a set playing time, have students find the total score for each player by adding all the scores.

In-Class Examples

7 ALGEBRA Simplify $(-9) + t + 8$. **$-1 + t$**

INTERACTIVE CHALKBOARD
PowerPoint® Presentations

8 OCEANOGRAPHY Oceanographers divide the ocean into three light zones. The deeper the water, the less light shines through. The middle zone is called the Twilight Zone. The lowest part of this zone is 1,000 meters below the surface of the water. The top of the zone lies 800 meters above the lowest zone. What is the depth of the top of the zone? **200 m**

3 Practice/Apply

FOLDABLES™ Follow-Up Remind students to use their Foldable journals to record what they learn about adding integers. For example, they might outline the steps necessary to add integers using an original example.

DAILY INTERVENTION

FIND THE ERROR For Exercise 3, have students plot both Brooke's and Javier's problems on a number line using arrows to see why the sum has the same sign as the addend with the greater absolute value.

NUMBER SENSE
For Exercise 4, remind students to compare the absolute values of the addends when determining the signs of the sums.

EXAMPLE Simplify an Expression with Integers

7 ALGEBRA Simplify $12 + x + (-20)$.

$12 + x + (-20) = 12 + (-20) + x$ Commutative Property of Addition

$= -8 + x$ Add.

Your Turn **Simplify each expression.**

g. $1 + y + (-5)$ **$-4 + y$** h. $z + (-2) + 9$ **$z + 7$**

REAL-LIFE MATH

OCEANOGRAPHY The Great Barrier Reef off the coast of Australia is home to 1,500 species of fish and 215 species of birds.

Source: www.ozramp.net.au

EXAMPLE Use Integers to Solve a Problem

8 OCEANOGRAPHY Anna was scuba diving near the Great Barrier Reef 16 meters below the surface of the water. She saw a dolphin swim by 7 meters above her. What was the depth of the dolphin?

Anna is 16 meters underwater, and the dolphin is 7 meters above her. So, the depth of the dolphin can be represented by the expression $-16 + 7$, or -9.

The dolphin is 9 meters below the surface of the water.

Skill and Concept Check

Writing Math
Exercises 2 & 3

1. **Draw** a model to show $2 + (-7)$. **See margin.**
2. **OPEN ENDED** Give an example of integers that are additive inverses. **Sample answer: 2, −2**
3. **FIND THE ERROR** Brooke and Javier are finding $-12 + 13$. Who is correct? Explain. **Brooke; Javier added correctly but did not apply the sign of the greater absolute value.**

Brooke	Javier
$-12 + 13 = 1$	$-12 + 13 = -1$

4. **NUMBER SENSE** **Tell whether each sum is *positive*, *negative*, or *zero* without adding.**

a. $-6 + (-7)$ **negative** b. $-8 + 10$ **positive** c. $-14 + 14$ **zero**

GUIDED PRACTICE

Add.

5. $-6 + (-8)$ **−14** 6. $-3 + 10$ **7** 7. $7 + (-11)$ **−4** 8. $9 + (-9)$ **0**

9. **MONEY** You pay your brother \$42 that you owe him. The same week you earn \$35 dog-sitting for the neighbors. Do you have more or less money than at the beginning of the week? **less since −\$42 + \$35 = −\$7**

10. **ALGEBRA** Simplify $12 + y + (-8)$. **$4 + y$**

Answer

1.

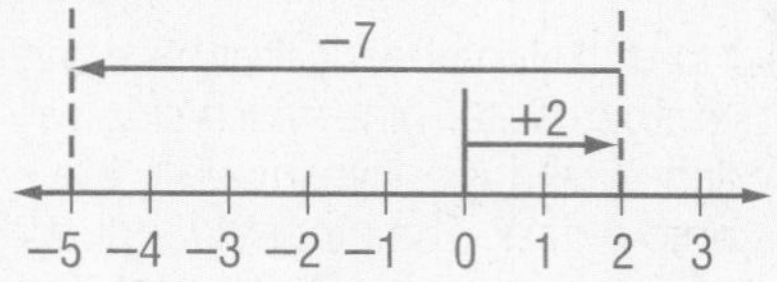

★ indicates increased difficulty

Practice and Applications

HOMEWORK HELP

For Exercises	See Examples
11–42	1–6
46–57	7
43–45, 58	8

Extra Practice See pages 570, 598.

Add. 13. −6 15. −25 17. −14 20. −2 21. −38

11. $-8 + 8$ 0
12. $-9 + 11$ 2
13. $13 + (-19)$
14. $6 + 10$ 16
15. $-10 + (-15)$
16. $-12 + 10$ −2
17. $-30 + 16$
18. $18 + (-5)$ 13
19. $21 + (-21)$ 0
20. $18 + (-20)$
21. $-22 + (-16)$
22. $-24 + 19$ −5

★ 23. $-11 + 13 + 6$ 8
★ 24. $-16 + (-21) + 15$ −22
★ 25. $12 + (-17) + (-25)$ −30
★ 26. $20 + (-30) + (-40)$ −50

Write an addition expression to describe each situation. Then find each sum.

27. **WEATHER** The temperature outside is −3°F. The temperature drops 6°. −3 + (−6); −9° F

28. **SUBMARINE** A submarine dives 106 feet below the water. Then, it rises 63 feet. −106 + 63; −43 ft

29. **SKATEBOARDING** Hakeem starts at the bottom of a half pipe 6 feet below street level. He rises 14 feet at the top of his kickturn. −6 + 14; 8 ft

30. **MONEY MATTERS** Stephanie has $43 in the bank. She withdraws $35. 43 + (−35); $8

ALGEBRA Evaluate each expression if $x = -10$, $y = 7$, and $z = -8$.

31. $x + 14$ 4
32. $6 + y$ 13
33. $z + (-5)$ −13
34. $-17 + y$ −10
35. $20 + z$ 12
36. $-10 + x$ −20
37. $z + 8$ 0
38. $15 + x$ 5
39. $x + y$ −3
40. $y + z$ −1
41. $x + z$ −18
42. $x + y + z$ −11

GOLF For Exercises 43–45, use the information below.

Scores over par in a golf tournament are recorded as positive integers. Scores under par are recorded as negative integers. Even par is recorded as 0. The person with the lowest total score wins. The table shows the top two finishers in the 2001 LPGA Championship.

	Round 1	Round 2	Round 3	Round 4
Laura Diaz	−4	0	−5	−3
Karrie Webb	−4	−7	−1	−2

Source: www.lpga.com

43. Find Laura Diaz's final score. −12
44. Find Karrie Webb's final score. −14
45. Who had the better score? Explain. Webb; $-14 < -12$

Data Update What were the four-round scores of the latest winners of the LPGA Championship? Visit msmath2.net/data_update to learn more.

ALGEBRA Simplify.

46. $x + (-5) + 1$ $x + (-4)$
47. $4 + y + (-2)$ $y + 2$
48. $-9 + m + (-6)$ $m + (-15)$
49. $8 + (-8) + n$ n
50. $-1 + a + 7$ $a + 6$
51. $f + (-19) + 11$ $f + (-8)$

msmath2.net/self_check_quiz

About the Exercises...

Guided Practice Key

Exercises	Examples
5	1, 2
6–8	3–6
10	7
9	8

Odd/Even Assignments
Exercises 11–42 and 46–57 are structured so that students practice the same concepts whether they are assigned odd or even problems.

Alert! Exercises 43–45 involve research on the Internet or other reference materials for the Data Update.

Study Guide and Intervention, p. 144

NAME ______________ DATE ________ PERIOD ____

3-4 Study Guide and Intervention

Adding Integers

For integers with the same sign:
- the sum of two positive integers is positive.
- the sum of two negative integers is negative.

For integers with different signs, subtract their absolute values. The sum is:
- positive if the positive integer has the greater absolute value.
- negative if the negative integer has the greater absolute value.

To add integers, it is helpful to use counters or a number line.

EXAMPLE 1 **Find 4 + (−6).**

Method 1 Use counters.
Combine a set of 4 positive counters and a set of 6 negative counters on a mat.

Method 2 Use a number line.
- Start at 0.
- Move 4 units right.
- Then move 6 units left.

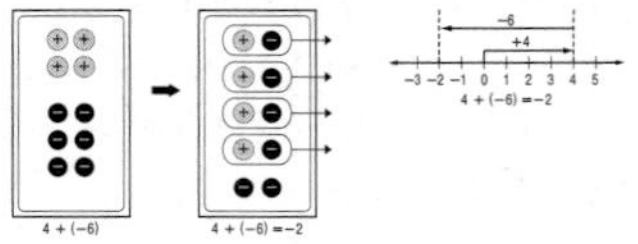

EXERCISES

Add.

1. −5 + (−2) −7
2. 8 + 1 9
3. −7 + 10 3
4. 16 + (−11) 5
5. −22 + (−7) −29
6. −50 + 50 0
7. −10 + (−10) −20
8. 100 + (−25) 75
9. −35 + −20 −15

Evaluate each expression if $a = 8$, $b = -8$, and $c = 4$.

10. $a + 15$ 23
11. $b + (-9)$ −17
12. $a + b$ 0
13. $b + c$ −4
14. $-10 + c$ −6
15. $12 + b$ 4

Practice: Skills, p. 145

NAME ______________ DATE ________ PERIOD ____

3-4 Practice: Skills

Adding Integers

Tell whether each sum is positive, negative, or zero without adding.
Add.

1. 5 + (−8) −3
2. −3 + 3 0
3. −3 + (−8) −11
4. −7 + (−7) −14
5. −8 + 10 2
6. −7 + 13 6
7. 15 + (−10) 5
8. −11 + (−12) −23
9. 25 + (−12) 13
10. −14 + (−13) −27
11. 14 + (−27) −13
12. −28 + 16 −12

Evaluate each expression if $a = -8$, $b = 12$, and $c = -4$.

13. $5 + a$ −3
14. $b + (-9)$ 3
15. $c + (-5)$ −9
16. $a + b$ 4
17. $a + 0$ −8
18. $b + c$ 8
19. $-12 + b$ 0
20. $a + (-7)$ −15
21. $21 + c$ 17
22. $a + c$ −12

Practice: Word Problems, p. 146

NAME ______________ DATE ________ PERIOD ____

3-4 Practice: Word Problems

Adding Integers

Write an addition expression to describe each situation. Then find each sum.

1. **FOOTBALL** A team gains 20 yards. Then they lose 7 yards. 20 + (−7); 13	2. **MONEY** Roger owes his mom \$5. He borrows another \$6 from her. (−5) + (−6); −11
3. **GOLF** Juanita's score was 5 over par on the first 9 holes. Her score was 4 under par on the second 9 holes. 5 + (−4); 1	4. **HOT AIR BALLOON** A balloon rises 340 feet into the air. Then it descends 130 feet. 340 + (−130); 210
5. **CYCLING** A cyclist travels downhill for 125 feet. Then she travels up a hill 50 feet. −125 + 50; −75	6. **AIRPLANE** A plane descends 1,200 feet. Then it descends another 500 feet. −1,200 + (−500); −1,700

Explain how the Commutative and Associative Properties of Addition can help you find each sum mentally. Then find each sum. 52–57. See Ch. 3 Answer Appendix.

52. 7 + (−2) + (−7)
53. −6 + 9 + (−4)
54. −5 + (−6) + (−3)
55. 8 + 10 + (−8)
56. −5 + (−7) + (−10)
57. 8 + (−9) + 9

58. **STOCK MARKET** The members of the Investment Club purchased a stock for \$50. The next day the price of the stock dropped \$18. On the second and third days, the price dropped another \$16 and then rose \$21. How much was the stock worth at the end of the third day? **\$37**

59. **WRITE A PROBLEM** Write about a real-life problem using the addition sentence $-8 + 11 = t$. Then solve the equation and explain what the solution represents. **See students' work.**

CRITICAL THINKING **For Exercises 60 and 61, use a number line to find each sum. Does the order of the addends make a difference? Explain.**

60. 3 + (−8) and −8 + 3 **−5; −5; No, addition is commutative.**

61. [7 + (−3)] + (−6) and 7 + [−3 + (−6)] **−2; −2; No, addition is associative.**

Standardized Test Practice and Mixed Review

62. **MULTIPLE CHOICE** In a game with a standard deck of cards and the scoring system at the right, three cards are dealt and added together to get a final score. Dylan is dealt the 4 of hearts, the king of spades, and the 3 of diamonds. What is his final score? **C**

Ⓐ −11 Ⓑ −3 Ⓒ 3 Ⓓ 9

Card	Points
Face cards	10
Aces	1
2–10	equal their value
Spades and clubs (black)	positive value
Hearts and diamonds (red)	negative value

63. **SHORT RESPONSE** Jeremy owes his sister \$5. Then he borrows \$6 more from her. Write the total amount he owes as an integer. **−11**

Name the ordered pair for each point graphed at the right. Then identify the quadrant in which each point lies. (Lesson 3-3)

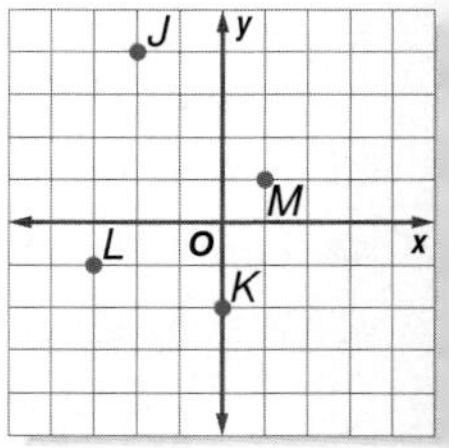

64. J **(−2, 4), II**
65. K **(0, −2), y-axis**
66. L **(−3, −1), III**
67. M **(1, 1), I**

68. Order 6, −3, 0, 4, −8, 1, and −4 from least to greatest. (Lesson 3-2) **−8, −4, −3, 0, 1, 4, 6**

69. **STATISTICS** Construct a line plot for the following test scores: 81, 83, 75, 81, 82, 81, 75, 82, 82, 86, 83, 81, and 79. (Lesson 2-3) **See margin.**

GETTING READY FOR THE NEXT LESSON

PREREQUISITE SKILL **Find the range for each set of data.** (Lesson 2-3)

70. 13, 7, 6, 22, 21 **16**
71. 54, 32, 43, 49, 30 **24**
72. 62, 59, 85, 74, 82 **26**

4 Assess

Higher-Level Thinking

A boat is passing over a reef, the top of which comes up to 3 feet below the surface of the water. If the bottom of the boat is 4 feet below water, how far from the reef is the boat? Is the boat safe? **−1 ft; No, the boat will hit the reef.**

Answer

69.

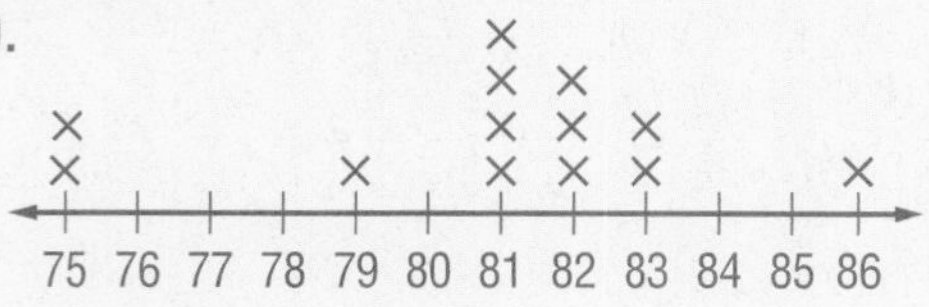

tudy Skill

HOW TO...

Take Good Notes

FLOWCHARTS

ave you ever tried
 solve a math
roblem and then
alized you left out
n important step?
y using a flowchart
hen you take notes
 map out the steps
u should follow.

A flowchart is like a map that tells you how to get from the beginning of a problem to the end.

	Flowchart Symbols
	A diamond contains a question. You need to stop and make a decision.
	A rectangle tells you what to do.
	An oval indicates the beginning or end.

Here's a flowchart for adding two integers. Just follow the arrows.

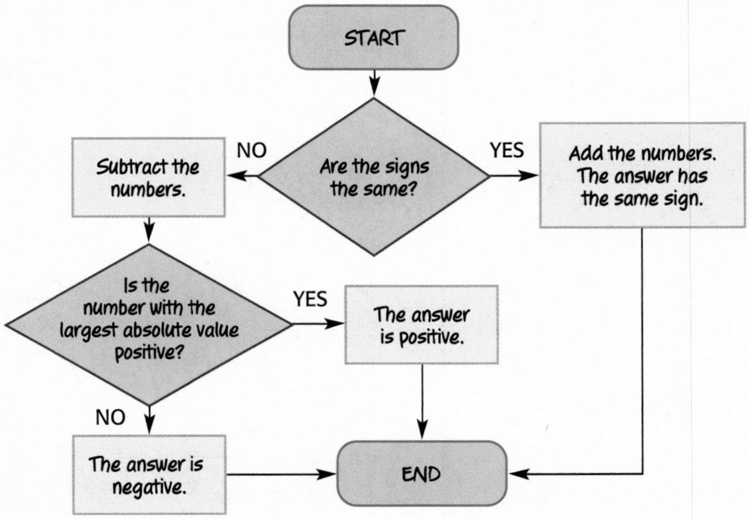

SKILL PRACTICE 1–2. See students' work.

Make a flowchart for each kind of problem.

1. rounding a decimal to a given place (See page 557.)
2. evaluating an expression using order of operations (See Lesson 1-3.)

Study Skill

Objective Use a flowchart to help take good notes.

GETTING STARTED

Flowchart Uses Flowcharts are very useful for keeping track of the steps of a math problem and taking good notes. Alternate uses for flowcharts include robotics, outlining college coursework, tracking computer programming and software development, showing the hierarchical structure of a company, following data flow, tracking product development, and displaying an electric circuit.

TEACH

Flowchart suggestions:

- Write a couple of examples for each kind of problem for which you will be making a flowchart. Flowcharts are easier to make from concrete examples rather than abstract thought processes.
- Start the flowchart while doing an example problem. Write every thought process down in a box.
- Think about all possible outcomes for any decision-making diamond. Some questions cannot be answered by a simple yes or no.

ASSESS

Open-Ended Assessment

Writing Have students summarize a flowchart in paragraph form. Have them evaluate using a flowchart versus solving a problem without a guide.

3-5a HANDS-ON LAB

Lesson Notes

Lesson Resources

Teaching Mathematics with Manipulatives
- integer counters, p. 6
- integer mat, p. 8
- student recording sheet 3-5a

Glencoe Mathematics Classroom Manipulative Kit
- counters

GETTING STARTED

Teaching Tip Ask volunteers to describe briefly how to use counters to add integers. Have students try to predict how they will use counters to subtract integers.

TEACH

- Explain to students that subtracting is the same as adding the opposite. For example, in **Activity 1**, students can rewrite $6 - 4$ as $6 + (-4)$.
- In **Activities 1** and **2** students are subtracting from positive integers.
- Remind students that zero pairs can be added without changing the total value on the mat.
- After students complete **Activity 2**, you may wish to have them model the addition of 3 and 2. Ask them to compare the result with the result of Activity 2. **The results are the same.**

3-5a HANDS-ON LAB

A Preview of Lesson 3-5

What You'll Learn
Use counters to model the subtraction of integers.

Materials
- counters
- integer mat

Subtracting Integers

You can also use counters to model subtraction of integers. Remember one meaning of subtraction is to *take away*.

ACTIVITIES *Work with a partner.*

Use counters to find each difference.

1 $6 - 4$

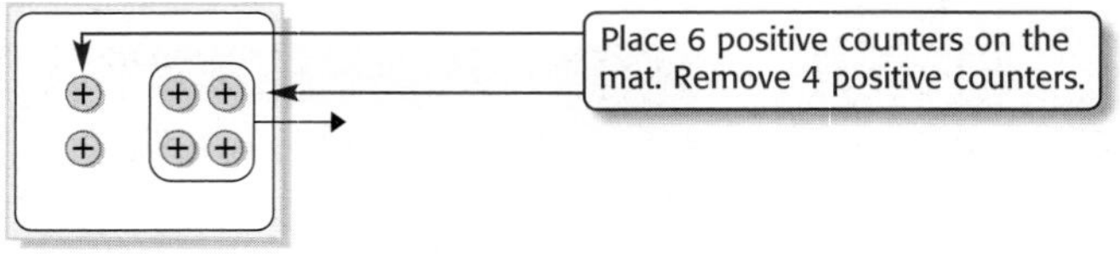

So, $6 - 4 = 2$.

2 $3 - (-2)$

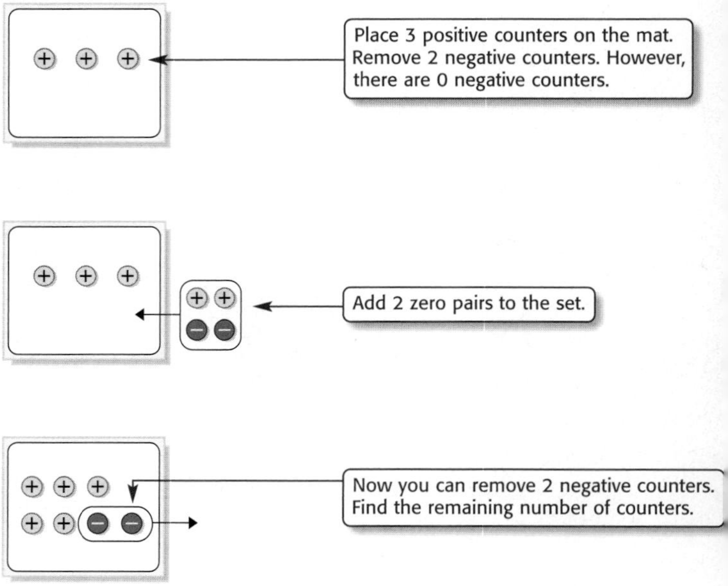

So, $3 - (-2) = 5$.

Your Turn **Use counters to find each difference.** **a–d. See margin for models.**

a. $7 - 6$ **1** b. $5 - (-3)$ **8** c. $6 - (-3)$ **9** d. $5 - 8$ **−3**

Answers (Your Turn)

a.

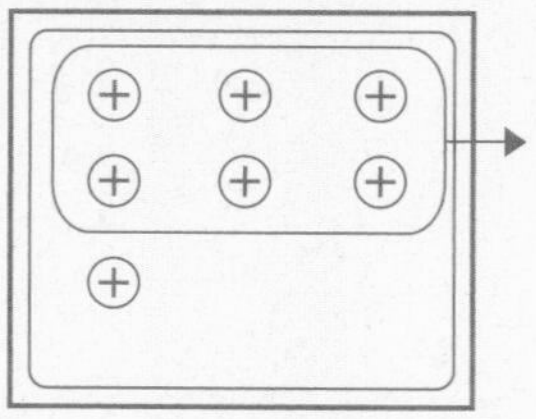

$7 - 6 = 1$

b.

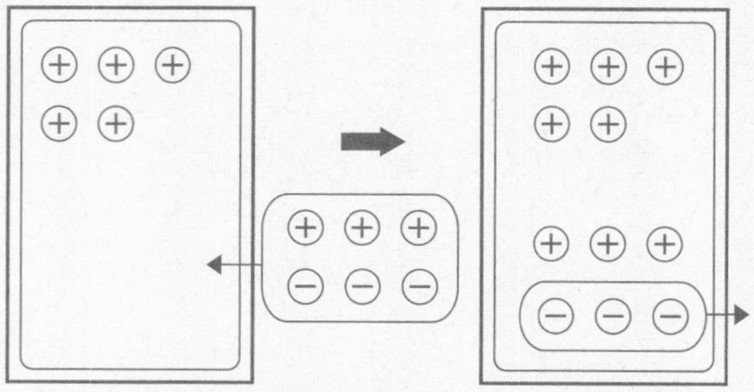

$5 - (-3) = 8$

ACTIVITIES *Work with a partner.*

Use counters to find each difference.

3 $-5 - (-2)$

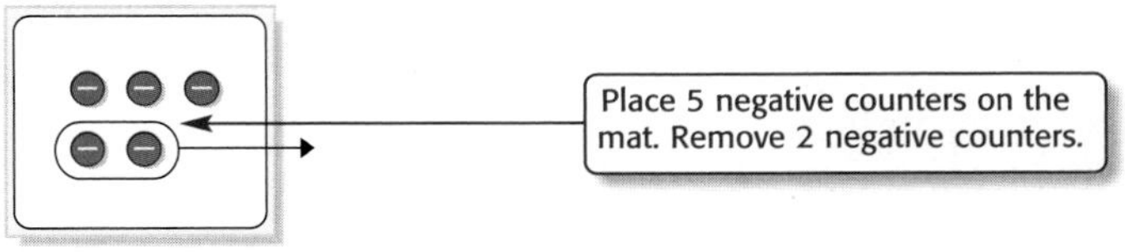

So, $-5 - (-2) = -3$.

4 $-4 - 3$

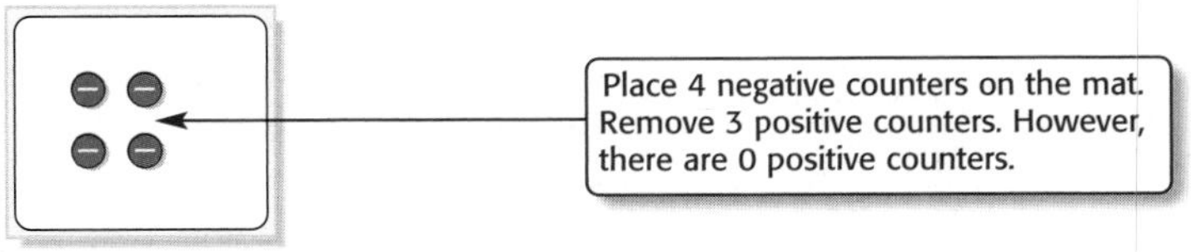

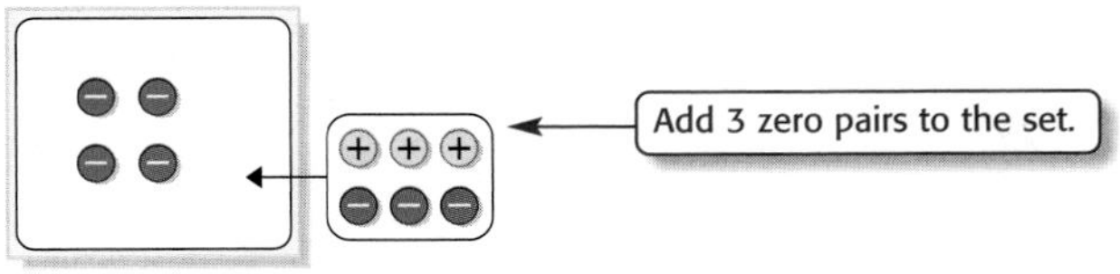

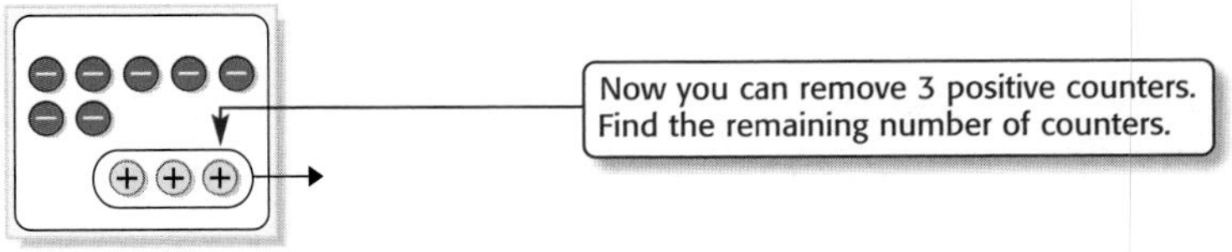

So, $-4 - 3 = -7$.

Your Turn **Use counters to find each difference.**

e. $-6 - (-3)$ **-3** f. $-7 - 3$ **-10** g. $-5 - (-7)$ **2**

e. See margin for models. f–g. See Ch. 3 Answer Appendix for models.

Writing Math

Work with a partner.

1. **Write** two subtraction sentences where the difference is positive. Make sure you use a combination of positive and negative integers.
2. **Write** two subtraction sentences where the difference is negative. Make sure you use a combination of positive and negative integers.
3. **MAKE A CONJECTURE** Write a rule that will help you determine the sign of the difference of two integers. See margin.

EADING Math

Minuends, Subtrahends, and Differences In the subtraction sentence $-4 - 3 = -7$, -4 is the *minuend,* 3 is the *subtrahend,* and -7 is the *difference.*

. Sample answers:
$- 2 = 6$;
$- (-6) = 13$;
$3 - (-6) = 3$

. Sample answers:
$5 - (-2) = -3$;
$1 - 4 = -5$;
$- 3 = -1$

- In **Activities 3** and **4** students are subtracting from negative integers.
- After students complete **Activity 4**, you may wish to have them model the addition of -4 and -3. Ask them to compare the result with the result of **Activity 4**. The results are the same.

ASSESS

Observe as students solve the **Your Turn** exercises. They should:

- correctly find positive differences of integers,
- correctly find negative differences of integers, and
- correctly use zero pairs when necessary when subtracting integers with counters.

Discuss the **Writing Math** exercises. Use Exercise 3 to determine whether students have discovered the rules for subtracting integers. You may want to have students include their answers in their Foldables for Chapter 3.

Answer (Writing Math)

3. Sample answer: The difference of two integers is positive when the minuend is greater than the subtrahend. The difference of two integers is negative when the minuend is less than the subtrahend.

c.

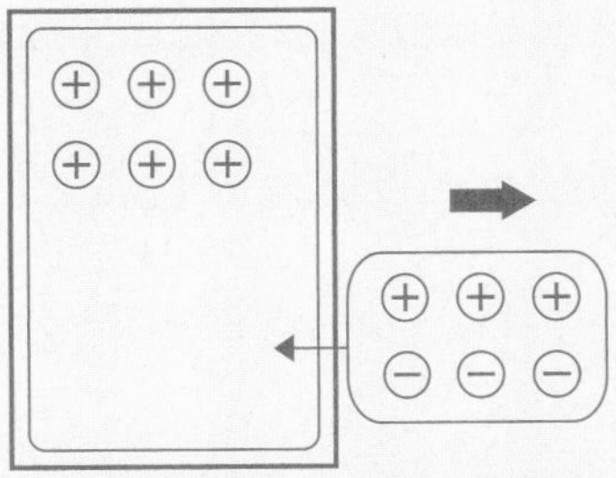

$6 - (-3) = 9$

d.

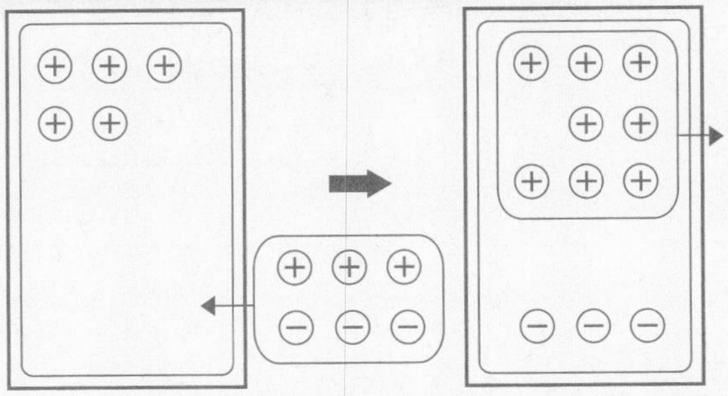

$5 - 8 = -3$

e.

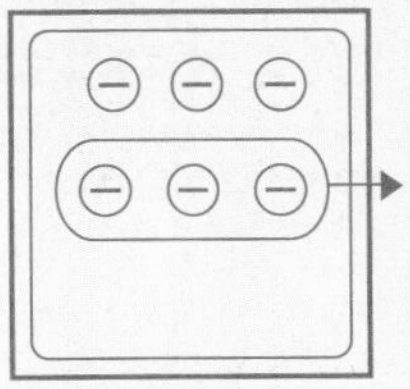

$-6 - (-3) = -3$

Lesson 3-5a Hands-On Lab: Subtracting Integers **127**

3-5 Lesson Notes

Lesson Resources

Chapter 3 Resource Masters
- Study Guide and Intervention, p. 149
- Practice: Skills, p. 150
- Practice: Word Problems, p. 151
- Reading to Learn Mathematics, p. 152
- Enrichment, p. 153

Classroom Math Games, pp. 7–9
Key Teaching Transparencies, 10
Parent and Student Study Guide Workbook, 3-5

FAST FILE

1 Focus

5-Minute Check 3-5

INTERACTIVE CHALKBOARD
PowerPoint® Presentations

Use as a quiz or review of Lesson 3-4. (Also available as a transparency)

Bellringer
Communication Have students compare the use of counters with the use of a number line to subtract integers. How do zero pairs of counters translate to a number line? **The place on a number line where the arrows overlap can be likened to zero pairs—they cancel each other out.**

Reading to Learn Mathematics, p. 152 **ELL**

NAME ______________ DATE ________ PERIOD ____

3-5 Reading to Learn Mathematics
Subtracting Integers

Pre-Activity *Complete the Mini Lab at the top of page 128 in your textbook. Write your answers below.*

1. Write a related addition sentence for each subtraction sentence. Sample answers: 3 + (−5) = −2; −1 + (−4) = −5

Use a number line to find each difference. Write an equivalent addition sentence for each.

2. 1 − 5 1 − 5 = −4; 1 + (−5) = −4
3. −2 − 1 −2 − 1 = −3; −2 + (−1) = −3
4. −3 − 4 −3 − 4 = −7; −3 + (−4) = −7
5. 0 − 5 0 − 5 = −5; 0 + (−5) = −5
6. **Compare and contrast** subtraction sentences with their related addition sentences. Sample answer: The subtraction problems giv... ame answer as ... related addition probl... positi...

3-5 Subtracting Integers

What You'll Learn
Subtract integers.

REVIEW Vocabulary
range: the difference between the greatest number and the least number in a set of data (Lesson 2-3)

HANDS-ON Mini Lab

Materials
- graph paper

Work with a partner.

The subtraction problems below are modeled on number lines.

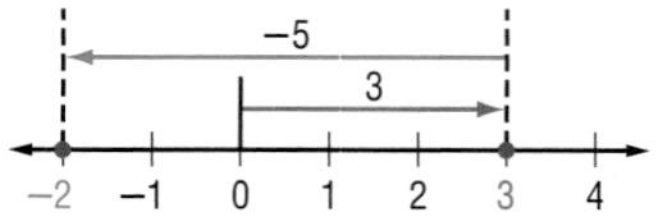

3 − 5 = −2

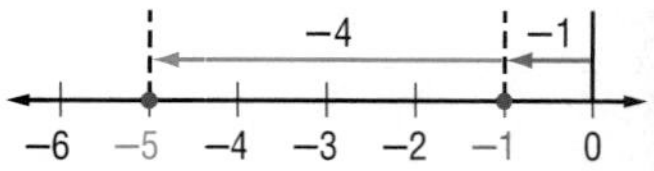

−1 − 4 = −5

1. Write a related addition sentence for each subtraction sentence. **Sample answers: 3 + (−5) = −2; −1 + (−4) = −5**

Use a number line to find each difference. Write an equivalent addition sentence for each. **2–6. See margin.**

2. 1 − 5
3. −2 − 1
4. −3 − 4
5. 0 − 5
6. **Compare and contrast** subtraction sentences with their related addition sentences.

When you subtract 5, as shown in the Mini Lab, the result is the same a[s] adding −5. When you subtract 4, the result is the same as adding −4.

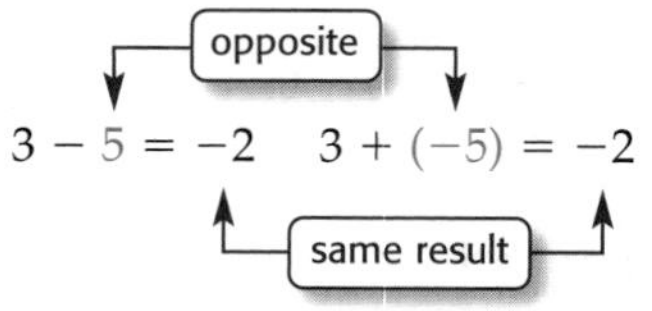

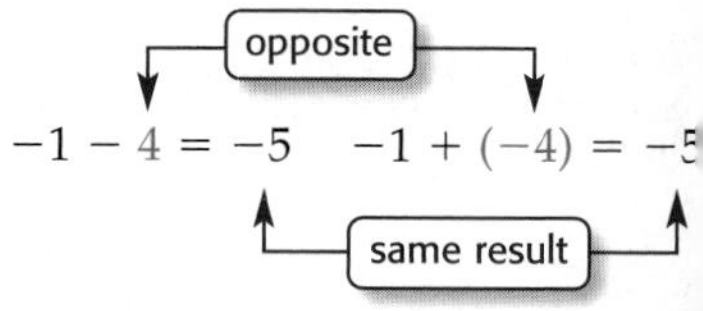

Key Concept **Subtracting Integers**

Words	To subtract an integer, add its opposite.
Example	4 − 9 = 4 + (−9) = −5

EXAMPLES Subtract Positive Integers

Subtract.

1 **8 − 13**

$8 - 13 = 8 + (-13)$ To subtract 13, add −13.
$= -5$ Simplify.

2 **−10 − 7**

$-10 - 7 = -10 + (-7)$ To subtract 7, add −7.
$= -17$ Simplify.

Answers (Hands-On Mini Lab)

2.

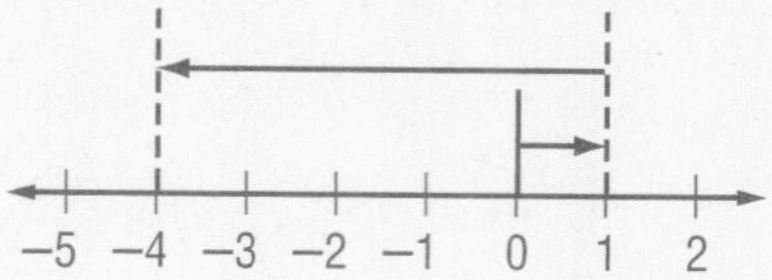

1 − 5 = −4

3.

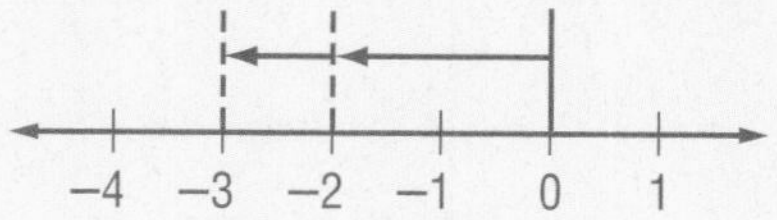

−2 − 1 = −3

TEACHING TIP

tudents can use a umber line to check eir results. First, ave them change the ubtraction sentence to n addition sentence. hen, they can draw a umber line.

STUDY TIP

Technology Use the [−] key to subtract a number. Use the [(−)] key to enter a negative number.

EXAMPLES Subtract Negative Integers

3 **Find $1 - (-2)$.**

$1 - (-2) = 1 + 2$ To subtract -2, add 2.

$= 3$ Simplify.

4 **Find $-10 - (-7)$.**

$-10 - (-7) = -10 + 7$ To subtract -7, add 7.

$= -3$ Simplify.

Your Turn **Subtract.**

a. $4 - (-12)$ **16** b. $-15 - (-5)$ **−10** c. $18 - (-6)$ **24**

EXAMPLE Evaluate an Expression

5 **ALGEBRA** **Evaluate $x - y$ if $x = -6$ and $y = 5$.**

$x - y = -6 - 5$ Replace x with -6 and y with 5.

$= -6 + (-5)$ To subtract 5, add -5.

$= -11$ Simplify.

Your Turn **Evaluate each expression if $a = 5$, $b = -8$, and $c = -9$.**

d. $b - 10$ **−18** e. $a - b$ **13** f. $c - a$ **−14**

EXAMPLE Use Integers to Solve a Problem

REAL-LIFE MATH

ARTH SCIENCE Yellowfin na can be found in waters at are between 17.8°C d 26.7°C.

urce: www.sstcharts.com

6 **EARTH SCIENCE** **The legend on the sea-surface temperature map shows the minimum temperature at −2°C and the maximum temperature at 31°C. What is the range of temperatures on the map?**

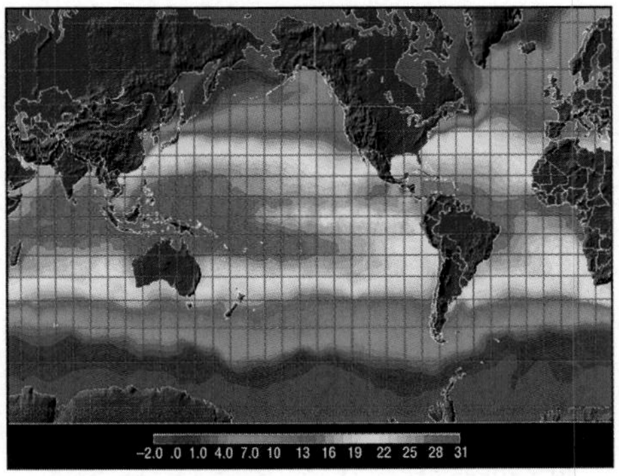

To find the range, or difference in temperatures, subtract the lowest temperature from the highest temperature.

$31 - (-2) = 31 + 2$ To subtract -2, add 2.

$= 33$ Simplify.

So, the range of temperatures is 33°C.

msmath2.net/extra_examples

2 Teach

In-Class Examples

INTERACTIVE CHALKBOARD PowerPoint® Presentations

Subtract.

1 $2 - 15$ **−13**

2 $-13 - 8$ **−21**

3 Find $12 - (-6)$. **18**

4 Find $21 - (-8)$. **29**

5 **ALGEBRA** Evaluate $g - h$ if $g = -2$ and $h = -7$. **5**

6 **GEOGRAPHY** In Mongolia, the temperature can dip down to −45°C in January. The temperature in July may reach 40°C. What is the temperature range in Mongolia? **85°C**

Tips for New Teachers

Intervention Many students find it helpful to have a large number line available for reference, perhaps at the front of the classroom above the chalkboard. Use masking tape and 3 × 5 cards, and extend the line into the twenties.

Answers (Hands-On Mini Lab)

4. 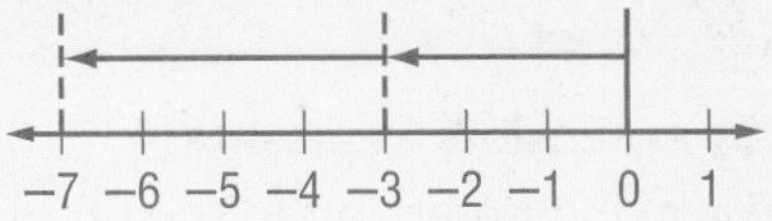

$-3 - 4 = -7$

5. 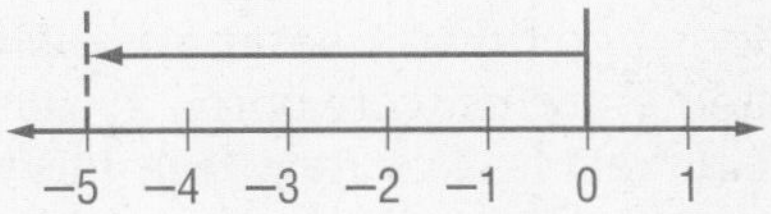

$0 - 5 = -5$

6. **Sample answer: The subtraction problems give the same answer as their related addition problems. Subtracting a positive integer is the same as adding a negative integer.**

DAILY INTERVENTION

Differentiated Instruction

Interpersonal Have teams of students use the Internet or atlases to research the elevations of cities and/or landmarks around the world. Have the teams write problems that can be solved by writing and solving equations with integers. Then have the teams trade problems and solve.

3 Practice/Apply

FOLDABLES™ Follow-Up Remind students to use their Foldable journals to record what they learn about subtracting integers. For example, they might describe the sequence of steps they use to subtract integers and give an original example.

DAILY INTERVENTION **FIND THE ERROR** For Exercise 3, have students work through the problem themselves before viewing Bradley's and Mitsu's work. Then discuss what was done incorrectly.

About the Exercises...

Guided Practice Key

Exercises	Examples
4, 5	1, 2
6, 7	3, 4
8–10	5
11	6

Odd/Even Assignments
Exercises 12–48 are structured so that students practice the same concepts whether they are assigned odd or even problems.

4 Assess

Open-Ended Assessment

Speaking Referring to the timeline for Exercises 46–48, have students explain how to determine how long ago Julius Caesar, Augustus, and Tiberius were born. Add the birth year to the current year. This will work because all three men were born in years B.C.

Skill and Concept Check

Writing Math Exercises 1 & 3

1. **Explain** how additive inverses are used in subtraction.
Sample answer: To subtract an integer, add its additive inverse.
2. **OPEN ENDED** Write a subtraction sentence using integers. Then, write the equivalent addition sentence and find the sum.
Sample answer: $-5 - 11 = -5 + (-11) = -16$
3. **FIND THE ERROR** Bradley and Mitsu are finding $-16 - (-19)$. Who is correct? Explain. Mitsu; Bradley did not add the additive inverse of -19.

Bradley	Mitsu
$-16 - (-19) = -16 + (-19)$ $= -35$	$-16 - (-19) = -16 + (19)$ $= 3$

GUIDED PRACTICE

Subtract.

4. $-4 - 8$ -12
5. $14 - 17$ -3
6. $14 - (-10)$ 24
7. $-3 - (-1)$ -2

ALGEBRA Evaluate each expression if $p = 8$, $q = -14$, and $r = -6$.

8. $p - q$ 22
9. $q - r$ -8
10. $r - p$ -14

11. **METEOROLOGY** The highest temperature ever recorded on Earth was 136°F in Libya. The lowest temperature was −129°F in Antarctica. What is the range of the highest and lowest temperatures on Earth? 265°F

★ indicates increased difficulty

Practice and Applications

HOMEWORK HELP

For Exercises	See Examples
12–31	1–4
32–44	5
45–48	6

Extra Practice See pages 571, 598.

Subtract.

12. $-9 - 5$ -14
13. $0 - 10$ -10
14. $-8 - 9$ -17
15. $17 - 13$ 4
16. $27 - (-8)$ 35
17. $-25 - (-5)$ -20
18. $12 - 26$ -14
19. $4 - (-19)$ 23
20. $-11 - 42$ -53
21. $15 - (-14)$ 29
22. $-27 - (-19)$ -8
23. $-18 - (-20)$ 2
24. $31 - 48$ -17
25. $-33 - (-27)$ -6
26. $52 - (-52)$ 104
27. $-44 - (-41)$ -3
★28. $-2 - 9 + 7$ -4
★29. $6 + (-1) - 4$ 1
30. What is -3 minus 4? -7
31. Find $-23 - (-19)$. -4

ALGEBRA Evaluate each expression if $f = -6$, $g = 7$, and $h = 9$.

32. $5 - f$ 11
33. $h - (-9)$ 18
34. $f - g$ -13
35. $g - 7$ 0
36. $h - f$ 15
37. $f - 6$ -12
38. $g - h$ -2
39. $4 - (-g)$ 11
40. $-h - 10$ -19
★41. $-f - h$ -3
★42. $f - g - h$ -22
★43. $h - g - f$ 8

★44. **ALGEBRA** Find $|a - b|$ when $a = -7$ and $b = 11$. 18

45. **GEOGRAPHY** The Dead Sea's deepest part is 799 meters below sea level. A plateau to the east of the Dead Sea rises to about 1,340 meters above sea level. What is the difference between the deepest part of the Dead Sea and the top of the plateau? 2,139 m

130 Chapter 3 Algebra: Integers

HISTORY For Exercises 46–48, use the timeline that shows the lives of three rulers of Rome.

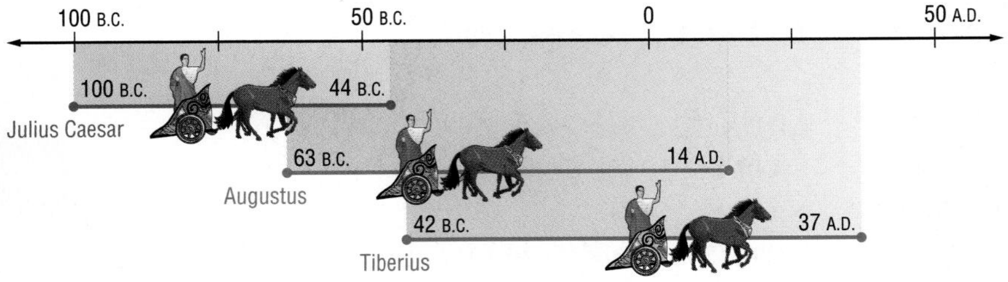

46. How old was Augustus when he died? 77 yr

47. Who lived the longest? How old was he when he died? Tiberius; 79 yr

48. How many years were there between when Julius Caesar was born and when Tiberius died? 137 yr

Determine whether each statement is *sometimes*, *always*, or *never* true. Give an example or counterexample for each answer. 49–52. See margin.

★ 49. negative − positive = negative ★ 50. negative − negative = positive

★ 51. positive − positive = positive ★ 52. positive − negative = negative

53. **CRITICAL THINKING** *True* or *False*? When n is a negative integer, $n - n = 0$. true

Standardized Test Practice and Mixed Review

54. **MULTIPLE CHOICE** Find the correct subtraction sentence shown in the model. D

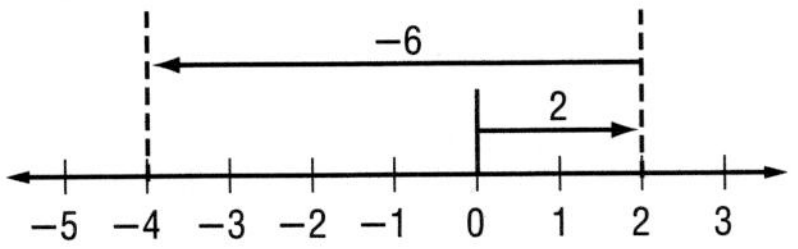

Ⓐ $2 - (-6) = -4$ Ⓑ $-6 - 2 = -4$

Ⓒ $-2 - 4 = -6$ Ⓓ $2 - 6 = -4$

55. **SHORT RESPONSE** The temperatures on the moon vary from −173°C to 127°C. Find the range of temperatures. 300°C

Add. (Lesson 3-4)

56. $10 + (-3)$ 7 57. $-2 + (-9)$ −11 58. $-7 + (-6)$ −13 59. $-18 + 4$ −14

60. In which quadrant do ordered pairs with a positive x-coordinate and a negative y-coordinate lie? (Lesson 3-3) quadrant IV

GETTING READY FOR THE NEXT LESSON

BASIC SKILL Multiply.

61. $14 \cdot 5$ 70 62. $9 \cdot 16$ 144 63. $6 \cdot 8 \cdot 4$ 192 64. $11 \cdot 7 \cdot 7$ 539

msmath2.net/self_check_quiz

Answers

49. always; $-2 - 3 = -5$
50. sometimes; $-1 - (-2) = 1$, $-2 - (-1) = -1$
51. sometimes; $5 - 2 = 3$, $2 - 5 = -3$
52. never; $1 - (-3) = 4$

Study Guide and Intervention, p. 149

NAME ______ DATE ______ PERIOD ____

3-5 Study Guide and Intervention

Subtracting Integers

To subtract an integer, add its opposite.

EXAMPLE 1 Find 6 − 9.

$6 - 9 = 6 + (-9)$ To subtract 9, add −9.
$= -3$ Simplify.

EXAMPLE 2 Find −10 − (−12).

$-10 - (-12) = -10 + 12$ To subtract −12, add 12.
$= 2$ Simplify.

EXAMPLE 3 Evaluate $a - b$ if $a = -3$ and $b = 7$.

$a - b = -3 - 7$ Replace a with −3 and b with 7.
$= -3 + (-7)$ To subtract 7, add −7.
$= -10$ Simplify.

EXERCISES

Subtract.

1. $7 - 9$ −2 2. $20 - (-6)$ 26
3. $-10 - 4$ −14 4. $0 - 12$ −12
5. $-7 - 8$ −15 6. $13 - 18$ −5
7. $-20 - (-5)$ −15 8. $-8 - (-6)$ −2
9. $25 - (-14)$ 39 10. $-75 - 50$ −125
11. $15 - 65$ −50 12. $19 - (-10)$ 29

Evaluate each expression if $m = -2$, $n = 10$, and $p = 5$.

13. $m - 6$ −8 14. $9 - n$ −1
15. $p - (-8)$ 13 16. $p - m$ 7
17. $m - n$ −12 18. $-25 - p$ −30

Practice: Skills, p. 150

NAME ______ DATE ______ PERIOD ____

3-5 Practice: Skills

Subtracting Integers

Subtract.

1. $5 - 2$ 3 2. $6 - (-7)$ 13
3. $-3 - 2$ −5 4. $8 - 13$ −5
5. $-7 - (-7)$ 0 6. $6 - 12$ −6
7. $15 - (-7)$ 22 8. $-15 - 6$ −21
9. $-3 - 8$ −11 10. $-10 - 12$ −22
11. $13 - (-12)$ 25 12. $14 - (-22)$ 36
13. $10 - (-20)$ 30 14. $-16 - 14$ −30
15. $-25 - 25$ −50 16. $6 - (-31)$ 37
17. $-18 - (-40)$ 22 18. $15 - (-61)$ 76

Evaluate each expression if $r = -4$, $s = 10$, and $t = -7$.

19. $r - 7$ −11 20. $t - s$ −17
21. $s - (-8)$ 18 22. $t - r$ −3
23. $s - t$ 17 24. $r - s$ −14

Practice: Word Problems, p. 151

NAME ______ DATE ______ PERIOD ____

3-5 Practice: Word Problems

Subtracting Integers

Subtract.

1. **FOOTBALL** A team gained 5 yards on their first play of the game. Then they lost 6 yards. Find the total change in yardage. −1 yd

2. **CHECKING** Your checking account is overdrawn by \$50. You write a check for \$20. What is the balance in your account? −\$70

3. **TEMPERATURE** The average temperature in Calgary, Canada, is 22°C in July and −11°C in January. Find the range of the highest and lowest temperatures in Calgary. 33°C

4. **ROLLER COASTER** A roller coaster begins at 90 feet above ground level. Then it descends 105 feet. Find the height of the coaster after the first descent. −15 ft

5. **SAVINGS** Sonia has \$235 in her savings account. She withdraws \$45. What is left in her savings account? \$190

6. **BEACH** Wai and Kuri were digging in the sand at the beach. Wai dug a hole that was 15 inches below the surface, and Kuri dug a hole that was 9 inches below the surface. Find the difference in the depths of their holes. 6 in.

3-6a Lesson Notes

1 Focus

Bellringer
Hands-On Activity Have groups of students look through the weekly advertisements from a Sunday newspaper. Have them choose one or two larger items, such as a television or bicycle, that they would like to own. Give each group a time limit to earn the money. Have students determine how much they would have to save each month to be able to buy their item.

2 Teach

In-Class Example

INTERACTIVE CHALKBOARD PowerPoint® Presentations

Solve. Use the look for a pattern strategy.

DONATION Leilani wants to grow an 11-inch ponytail to cut off and donate to a program that makes wigs for children with cancer. She has a 3-inch ponytail right now and her hair grows about an inch every two months. How long will it take for her hair to be long enough? **16 months**

3-6a Problem-Solving Strategy

A Preview of Lesson 3-6

What You'll Learn
Solve problems using the look for a pattern strategy.

Look for a Pattern

We've already saved $155 in four months. If we keep saving our money at the same rate, how long do you think it will take to save enough money to buy a DVD player that costs $330?

I found the table where we listed our savings each month. Let's **look for a pattern** to figure it out.

Month	Balance
1	$50
2	$85
3	$120
4	$155

Explore	We began with $50 and added more money to our savings every month. We need to find the number of months when we will have $330 to buy the DVD player.
Plan	Let's look for a pattern or rule that increases the balance each month. Then use the rule to extend the pattern and find the solution.
Solve	After the initial $50, we saved $35 per month. To extend the pattern, add $35 to each monthly balance until the balance equals $330. We will have enough money saved after 9 months.
Examine	We saved about 2 × $155, or $310 in 8 months. So, 9 months is a reasonable answer.

Month	Balance
1	$50
2	$85
3	$120
4	$155
5	$190
6	$225
7	$260
8	$295
9	$330

Analyze the Strategy

1. **Explain** when you would use the look for a pattern strategy to solve a problem. **1–2. See margin.**
2. **Describe** how to solve a problem using the look for a pattern method as a problem-solving strategy.
3. **Write** a problem that could be solved by looking for a pattern. Explain your answer. **See students' work.**

Answers

1. Sample answer: You would use the look for a pattern strategy when there is a data table, a series of numbers, or a geometric pattern as part of the problem.
2. Sample answer: Usually there is a data table or geometric pattern involved in the problem. You would look for a rule or pattern in the data, use the rule to extend the pattern, and then find the solution to the problem.

pply the Strategy

lve. Use the look for a pattern strategy.

. **LIFE SCIENCE** The table shows about how many times a firefly flashes at different temperatures. Estimate how many times a firefly will flash when the temperature is 36°C. **23 times**

Outside Temperature (°C)	Flashes per Minute
16	8
20	9
24	11
28	14

5. **CEREAL** A display of cereal boxes is stacked in the shape of a pyramid. There are 4 boxes in the top row, 6 boxes in the second row, 8 boxes in the next row, and so on. The display contains 7 rows of boxes. How many boxes are in the display? **70 boxes**

ixed Problem Solving

lve. Use any strategy.

FE SCIENCE For Exercises 6–8, use the formation and the graph.

. What does 0 on this graph represent?

. Write an integer to represent the rainfall for each month. **−3, −2, −4, 2**

. Write a sentence that summarizes the message this graph conveys about this summer's rainfall. **6, 8. See margin.**

. **EARTH SCIENCE** Hydrothermal vents are similar to geysers, but are found on the ocean floor. A hydrothermal vent chimney can grow at an average rate of 9 meters in 18 months. What is the average rate of growth per month? **0.5 m**

). **MULTI STEP** Francisco is on vacation and is planning to send postcards and letters to his friends. He has $3.04 to spend on postage. A stamp for a letter costs 37¢, and a stamp for a postcard costs 23¢. If he is going to spend the entire $3.04 on postage, how many postcards and letters can he send? **Sample answer: 2 letters and 10 postcards**

11. **BASKETBALL** Laura makes 3 free throws out of every 5 she attempts. Find the number of free throws she will make after 15, 20, and 30 attempts. **9, 12, 18**

12. **COINS** Olivia has seven coins that total $1.32. What are the coins? **See margin.**

13. **FOOD** The school cafeteria added a breakfast special to their menu. The table shows the foods that are part of the special and the number of Calories. Estimate how many Calories there are in the special. **Sample answer. 500 Calories**

Food	Calories
whole-wheat bagel	156
skim milk	90
nonfat strawberry yogurt	183
fresh fruit salad	68

14. **STANDARDIZED TEST PRACTICE** The total land area of Alabama is about 50,744 square miles. In 2000, about 87.6 persons per square mile were living in Alabama. What was the approximate population of Alabama in 2000? **C**

Ⓐ 45,000 Ⓑ 450,000

Ⓒ 4,500,000 Ⓓ 45,000,000

You will use the look for a pattern strategy in the next lesson.

3 Practice/Apply

About the Exercises...

Problem-Solving Assignments
Exercises 6–14 are structured so that students have the opportunity to practice many different problem-solving strategies. You may wish to remind them of some of the strategies they have studied.

- use the four-step plan (p. 6)
- guess and check (p. 22)
- use a graph (p. 58)
- look for a pattern ✓

4 Assess

Open-Ended Assessment

Writing Have each student write a problem involving a pattern and describe the rule for that pattern.

Answers

6. normal rainfall
8. Sample answer: There was a drought during the summer.
12. Sample answer: 1-half dollar, 3-quarters, 1-nickel, and 2-pennies

DAILY INTERVENTION **Flexible Grouping**

Look for a Pattern Use the think-pair-share technique of cooperative learning. Assign one of the exercises. Let students read over it individually. Then pair them up and have them look for the pattern to solve the problem. Repeat with pairs rearranged.

3-6 Lesson Notes

Lesson Resources

Chapter 3 Resource Masters
- Study Guide and Intervention, p. 154
- Practice: Skills, p. 155
- Practice: Word Problems, p. 156
- Reading to Learn Mathematics, p. 157
- Enrichment, p. 158

Key Teaching Transparencies, 11
Parent and Student Study Guide Workbook, 3-6
Teaching Mathematics with Manipulatives, pp. 47–48

1 Focus

5-Minute Check 3-6

INTERACTIVE CHALKBOARD
PowerPoint® Presentations

Use as a quiz or review of Lesson 3-5. (Also available as a transparency)

Bellringer
Problem Solving Remind students that *to multiply* means: To add any given number or quantity to itself a certain number of times. Show them an expression such as 3 + 3 + 3 + 3. How would you write this as a multiplication problem? **3 · 4** How would you write the expression −3 + −3 + −3 + −3 as a multiplication problem? **−3 · 4**

Reading to Learn Mathematics, p. 157 ELL

NAME ______ DATE ______ PERIOD ____

3-6 Reading to Learn Mathematics
Multiplying Integers

Pre-Activity *Complete the Mini Lab at the top of page 134 in your textbook. Write your answers below.*

1. Write a multiplication sentence that describes the model. 4(−2) = −8

Find each product using counters.

2. 3(−2) −6
3. 4(−3) −12
4. 1(−7) −7
5. 5(−2) −10

6. Write a rule for finding the sign of the product of a positive and negative integer. A positive integer times a negative integer equals a negative integer.

Reading the Lesson

7. Give an example that shows how multiplication is the same as repeated addition. In your example, tell what the addend is. Sample answer: 3(... −9) + (−9) + (... = −27 The addend is ...

3-6 Multiplying Integers

What You'll Learn
Multiply integers.

HANDS-ON Mini Lab

Materials
- counters
- integer mat

Work with a partner.

Counters can be used to multiply positive and negative integers.

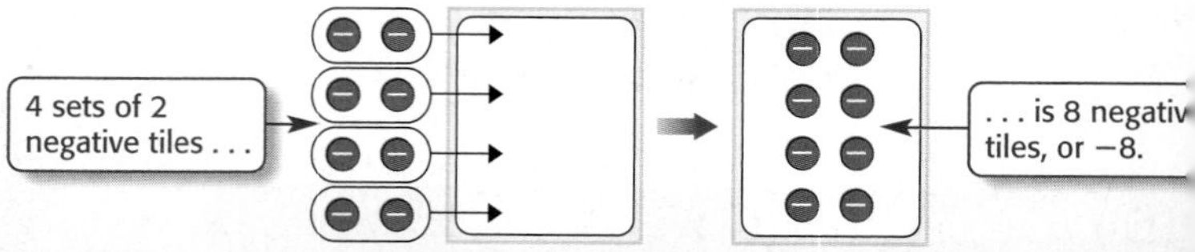

1. Write a multiplication sentence that describes the model above. **4(−2) = −8**

Find each product using counters. **2–5. See margin for models.**

2. 3(−2) **−6**
3. 4(−3) **−12**
4. 1(−7) **−7**
5. 5(−2) **−10**
6. Write a rule for finding the sign of the product of a positive and negative integer. **A positive integer times a negative integer equals a negative integer.**

Remember that multiplication is the same as repeated addition. The multiplication expression 4(−2) in the Mini Lab means that −2 is used as an addend four times.

$$4(-2) = (-2) + (-2) + (-2) + (-2)$$
$$= -8$$

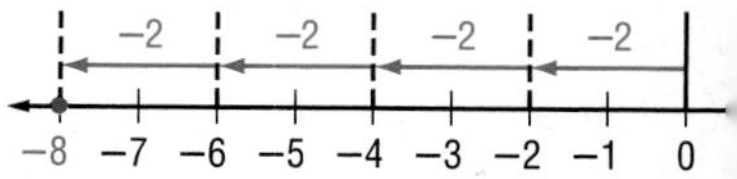

By the Commutative Property of Multiplication, 4(−2) = −2(4). When two integers have different signs, the following rule applies.

Key Concept — **Multiplying Integers with Different Signs**

Words The product of two integers with different signs is negative.
Examples 6(−4) = −24 −5(7) = −35

EXAMPLES Multiply Integers with Different Signs

Multiply.

1 **3(−5)**

3(−5) = −15 The integers have different signs. The product is negative.

2 **−6(8)**

−6(8) = −48 The integers have different signs. The product is negative.

Answers (Hands-On Mini Lab)

2.

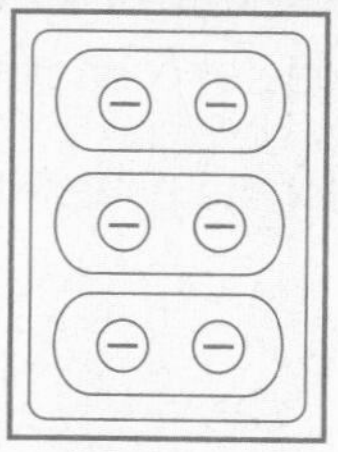

3(−2) = −6

3.

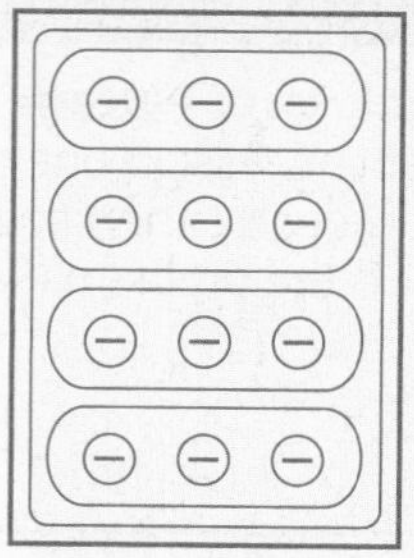

4(−3) = −12

STUDY TIP

Zero The Multiplicative Property of Zero states that when any number is multiplied by zero, the product is zero.

The product of two positive integers is positive. You can use a pattern to find the sign of the product of two negative integers.

negative × positive = negative	$(-6)(2)$	=	-12	
	$(-6)(1)$	=	-6	+6
	$(-6)(0)$	=	0	+6
negative × negative = positive	$(-6)(-1)$	=	6	+6
	$(-6)(-2)$	=	12	+6

Each product is 6 more than the previous product.

When two integers have the same sign, the following rule applies.

Key Concept **Multiplying Integers with the Same Sign**

Words The product of two integers with the same sign is positive.

Examples $2(6) = 12$ $\quad -10(-6) = 60$

EXAMPLES Multiply Integers with the Same Sign

Multiply.

3 $-11(-9)$

$-11(-9) = 99$ The integers have the same sign. The product is positive.

STUDY TIP

Look Back You can review **exponents** in Lesson 1-2.

4 $(-4)^2$

$(-4)^2 = (-4)(-4)$ There are two factors of -4.

$= 16$ The product is positive.

Your Turn Multiply.

a. $-12(-4)$ **48** b. $(-5)^2$ **25** c. $(-2)^3$ **−8**

EXAMPLES Simplify and Evaluate Expressions

5 **ALGEBRA** **Simplify $-2(3x)$.**

$-2(3x) = (-2 \cdot 3)x$ Associative Property of Multiplication

$= -6x$ Simplify.

6 **ALGEBRA** **Evaluate pqr if $p = -3$, $q = 4$, and $r = -1$.**

$pqr = (-3)(4)(-1)$ Replace p with -3, q with 4, and r with -1.

$= (-12)(-1)$ Multiply -3 and 4.

$= 12$ Multiply -12 and -1.

Your Turn

d. Simplify $-5(2y)$. **$-10y$**

e. Evaluate xyz if $x = -7$, $y = -4$, and $z = 2$. **56**

msmath2.net/extra_examples

2 Teach

In-Class Examples

INTERACTIVE CHALKBOARD PowerPoint® Presentations

Multiply.

1. $5(-4)$ **−20**
2. $-3(9)$ **−27**
3. $-6(-8)$ **48**
4. $(-8)^2$ **64**
5. **ALGEBRA** Simplify $-5(-7x)$. **35x**
6. **ALGEBRA** Evaluate abc if $a = -3$, $b = 5$, and $c = -8$. **120**

Answers (Hands-On Mini Lab)

4. 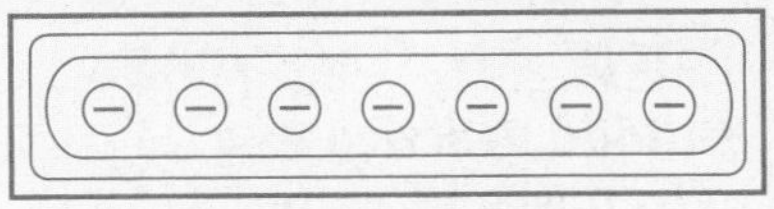

$1(-7) = -7$

5. 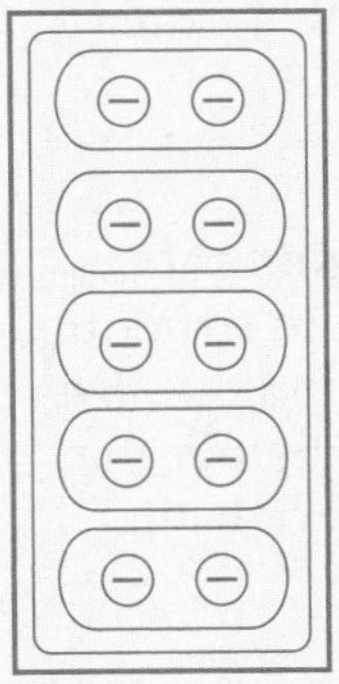

$5(-2) = -10$

DAILY INTERVENTION **Unlocking Misconceptions**

Multiplying Integers Some students may mistakenly assign the wrong sign to the products of one or more negative integers. Suggest that students multiply first, disregarding the signs, and then use the rules provided to affix the correct sign to the product.

3 Practice/Apply

FOLDABLES™ Follow-Up Remind students to use their Foldable journals to record what they learn about subtracting integers. For example, they might describe the sequence of steps they use to subtract integers and give an original example.

NUMBER SENSE
For Exercise 3, have students try two or three sets of negative integers until they are sure of the outcome. Ask students whether the same pattern would exist for any odd number of negative multipliers.

About the Exercises...

Guided Practice Key

Exercises	Examples
4, 13	1, 2
5, 6	3, 4
7–9	5
10–12	6

Odd/Even Assignments
Exercises 14–48 are structured so that students practice the same concepts whether they are assigned odd or even problems.

Skill and Concept Check

Writing Math
Exercise 3

1. **Model** the product of 2 and -3 using counters. Then write the multiplication sentence. **See margin.**
2. **OPEN ENDED** Name two integers whose product is negative. **Sample answer: $4(-9) = -36$**
3. **NUMBER SENSE** What is the sign of the product of three negative integers? Give an example. **negative; Sample answer: $-2(-3)(-4) = -24$**

GUIDED PRACTICE

Multiply.

4. $6(-10)$ **−60**
5. $-15(-3)$ **45**
6. $(-2)^2$ **4**

ALGEBRA Simplify each expression.

7. $-5(2a)$ **$-10a$**
8. $3(-6b)$ **$-18b$**
9. $-5(-9c)$ **$45c$**

ALGEBRA Evaluate each expression if $f = -1$, $g = 7$, and $h = -10$.

10. $5f$ **−5**
11. fgh **70**
12. $-h^2$ **−100**

13. **SUBMARINES** A submarine is diving from the surface of the water at a rate of 125 feet per minute. What is the depth of the submarine after 7 minutes? **−875, or 875 feet below sea level**

★ **indicates increased difficulty**

Practice and Applications

HOMEWORK HELP

For Exercises	See Examples
14–27, 44	1–4
28–35	5
36–43	6

Extra Practice
See pages 571, 598.

Multiply. **17. 170 21. −450**

14. $8(-13)$ **−104**
15. $-16(-5)$ **80**
16. $(-9)^2$ **81**
17. $-10(-17)$
18. $-7(16)$ **−112**
19. $(-6)^2$ **36**
20. $-20(-8)$ **160**
21. $-15(30)$
22. $-31(-5)$ **155**
23. $11(-20)$ **−220**
★24. $-(7^2)$ **−49**
★25. $(-4)^3$ **−64**

26. Find the product of -13 and 13. **−169**
27. Find -7 squared. **49**

ALGEBRA Simplify each expression.

28. $-3(6c)$ **$-18c$**
29. $-7(10d)$ **$-70d$**
30. $5(-4e)$ **$-20e$**
31. $9(-8f)$ **$-72f$**
★32. $-2(-3g)$ **$6g$**
★33. $-6(-4h)$ **$24h$**
★34. $(2x)(-3y)$ **$-6xy$**
★35. $(-5r)(2s)$ **$-10rs$**

ALGEBRA Evaluate each expression if $w = 7$, $x = -8$, $y = 5$, and $z = 10$.

36. $-4w$ **−28**
37. xy **−40**
38. $-2xz$ **160**
39. xyz **−400**
40. $-7wy$ **−245**
41. $-3z^2$ **−300**
42. $12x^2$ **768**
★43. $-wz^2$ **−700**

44. **VOLUNTEERING** The Volunteer Club raked leaves at several senior citizens' homes in the neighborhood. If each group of three students could remove 8 cubic meters of leaves in one hour, find an integer to represent the number of cubic meters of leaves 12 students could remove in five hours. **160**

★45. **PATTERNS** Find the next two numbers in the pattern $1, -2, 4, -8, 16, \ldots$. Then describe the pattern. **−32, 64; geometric: multiply the previous term by −2.**

Answers

1.
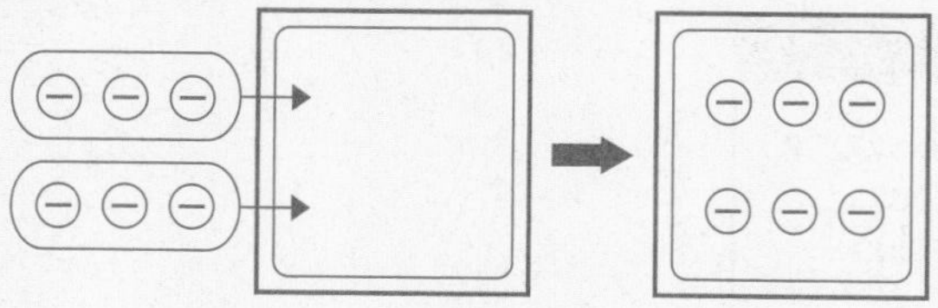

$2(-3) = -6$

46. $P(2, 5)$ is now $(-2, -5)$; $Q(3, 1)$ is now $(-3, -1)$; $R(5, 4)$ is now $(-5, -4)$.

47. Sample answer: New triangle $P'Q'R'$ (at the right) is on the other side of the origin (in quadrant III) from original triangle PQR.

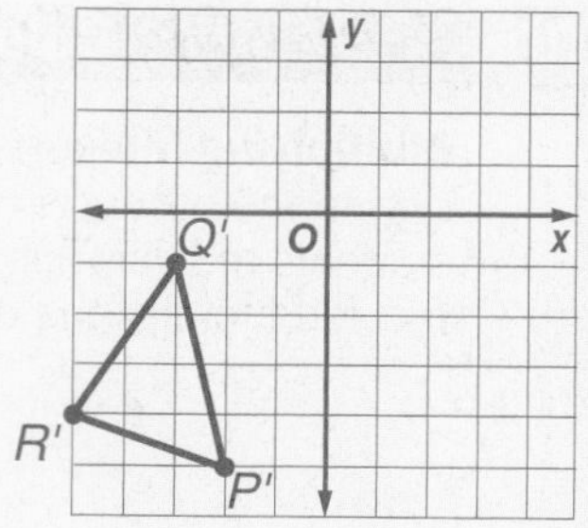

GEOMETRY For Exercises 46–48, use the graph at the right.

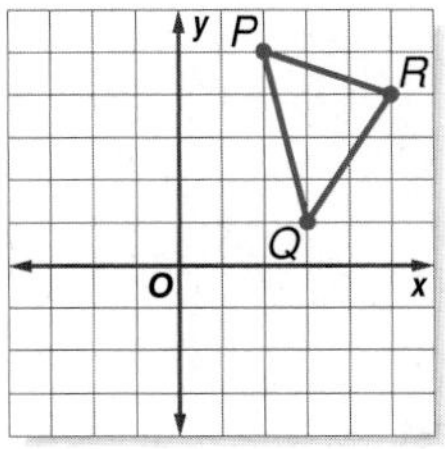

46. Name the ordered pairs for P, Q, and R. Multiply each x- and y-coordinate by -1 to get three new ordered pairs. **46–47. See margin**

47. Graph the ordered pairs and connect them to form a new triangle. Describe its position with respect to the original triangle.

★ 48. In which quadrant does a triangle lie if only the y-coordinates of the original triangle are multiplied by -2? **quadrant IV**

49. **CRITICAL THINKING** For what values of n is $(-2)^n$ positive? **when n is an even number**

Standardized Test Practice and Mixed Review

50. **MULTIPLE CHOICE** An oil rig is drilling into the ground at a rate of 7 feet per minute. What integer represents the position of the oil rig after 42 minutes? **A**

Ⓐ −294 ft Ⓑ −35 ft Ⓒ 35 ft Ⓓ 294 ft

51. **MULTIPLE CHOICE** Monifa has 100 shares of stock each worth $15. If the price drops to $8, what integer represents the change in Monifa's current investment? **F**

Ⓕ −$700 Ⓖ −$7 Ⓗ $7 Ⓘ $700

52. **GRID IN** Evaluate $-6[-2(3) + 0(-5)] + (-4)^2$. **52**

53. Find $-25 - (-33)$. (Lesson 3-5) **8**

ALGEBRA **Evaluate each expression if $x = -4$, $y = 6$, and $z = 1$.** (Lesson 3-4)

54. $x + (-2)$ **−6** 55. $-1 + z$ **0** 56. $-15 + y$ **−9** 57. $x + y$ **2**

58. **EARTH SCIENCE** The low temperatures in degrees Fahrenheit for ten cities on January 23 were −3, 27, 13, −6, −14, 36, 47, 52, −2, and 0. Order these temperatures from greatest to least. (Lesson 3-2)
52, 47, 36, 27, 13, 0, −2, −3, −6, −14

MILITARY For Exercises 59 and 60, use the double-bar graph at the right. (Lesson 2-7)

59. In which age groups are there more members in the Navy than members in the Air Force?

60. About how many 36–40 year olds are in the Air Force?

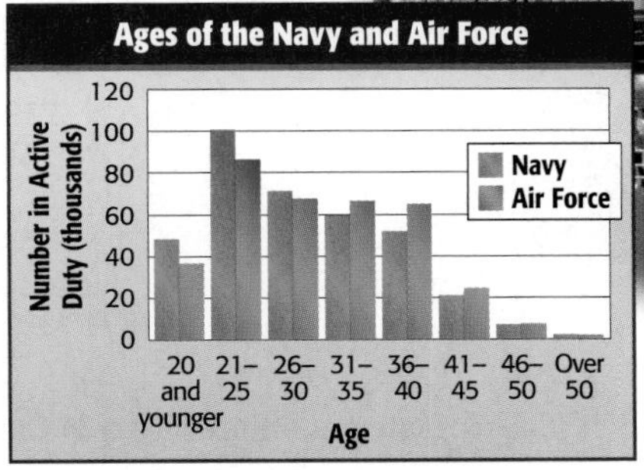

59. 20 and younger, 21–25, 26–30 **60. Sample answer: 65,000**

GETTING READY FOR THE NEXT LESSON

BASIC SKILL Divide.

61. $72 \div 9$ **8** 62. $120 \div 6$ **20** 63. $84 \div 21$ **4** 64. $215 \div 43$ **5**

msmath2.net/self_check_quiz

4 Assess

Open-Ended Assessment

Modeling Have students use number lines to model the multiplication of negative integers by positive integers.

Study Guide and Intervention, p. 154

NAME ________ DATE ________ PERIOD ____

3-6 Study Guide and Intervention

Multiplying Integers

The product of two integers with **different** signs is **negative**.
The product of two integers with the **same** sign is **positive**.

EXAMPLE 1 **Multiply 5(−2).**

$5(-2) = -10$ The integers have different signs. The product is negative.

EXAMPLE 2 **Multiply −3(7).**

$-3(7) = -21$ The integers have different signs. The product is negative.

EXAMPLE 3 **Multiply −6(−9).**

$-6(-9) = 54$ The integers have the same sign. The product is positive.

EXAMPLE 4 **Multiply $(-7)^2$.**

$(-7)^2 = (-7)(-7)$ There are 2 factors of −7.
$= 49$ The product is positive.

EXAMPLE 5 **Simplify −2(6c).**

$-2(6c) = (-2 \cdot 6)c$ Associative Property of Multiplication.
$= -12c$ Simplify.

EXAMPLE 6 **Simplify 2(5x).**

$2(5x) = (2 \cdot 5)x$ Associative Propery of Multiplication.
$= 10x$ Simplify.

EXERCISES

Multiply.

1. $-5(8)$ −40 2. $-3(-7)$ 21 3. $10(-8)$ −80
4. $-8(3)$ −24 5. $-12(-12)$ 144 6. $(-8)^2$ 64

ALGEBRA Simplify each expression.

7. $-5(7a)$ $-35a$ 8. $3(-2x)$ $-6x$ 9. $4(6f)$ $24f$
10. $7(6b)$ $42b$ 11. $-6(-3y)$ $18y$ 12. $7(-8g)$ $-56g$

ALGEBRA Evaluate each expression if $a = -3$, $b = -4$, and $c = 5$.

13. $-2a$ 6 14. $9b$ −36 15. ab 12
16. $-3ac$ 45 17. $-2c^2$ −50 18. abc 60

Practice: Skills, p. 155

NAME ________ DATE ________ PERIOD ____

3-6 Practice: Skills

Multiplying Integers

Multiply.

1. $-4(6)$ −24 2. $-2(-8)$ 16
3. $12(-4)$ −48 4. $-6(5)$ −30
5. $-10(-9)$ 90 6. $-(5)^2$ −25
7. $(-5)^2$ 25 8. $-30(5)$ −150
9. $20(-6)$ −120 10. $-14(-6)$ 84
11. $(-13)^2$ 169 12. $-7(15)$ −105

ALGEBRA Simplify each expression.

13. $-3(4y)$ $-12y$ 14. $7(-3x)$ $-21x$
15. $7(5g)$ $35g$ 16. $7(7w)$ $49w$
17. $3(-3y)$ $-9y$ 18. $-2(-10h)$ $20h$

ALGEBRA Evaluate each expression if $g = -5$, $h = -3$, and $k = 4$.

19. $-3g$ 15 20. $5h$ −15
21. $7gk$ −140 22. $-2gh$ −30
23. $-10h$ 30 24. $-2h^2$ −18

Practice: Word Problems, p. 156

NAME ________ DATE ________ PERIOD ____

3-6 Practice: Word Problems

Multiplying Integers

Multiply.

1. TEMPERATURE Suppose the temperature outside is dropping 3 degrees each hour. How much will the temperature drop in 8 hours? −24°

2. DIVING A deep-sea diver descends below the surface of the water at a rate of 60 feet each minute. What is the depth of the diver after 10 minutes? −600 ft

3. STOCK A computer stock lost 2 points each hour for 6 hours. Find the total points the stock fell. −12 points

4. DROUGHT A drought can cause the level of the local water supply to drop by a few inches each week. Suppose the level of the water supply drops 2 inches each week. How far will it have dropped in 4 weeks? −8 in.

5. MONEY Mrs. Rockwell lost money on an investment at a rate of $4 per day. How much did she lose after two weeks? −$56

6. TENNIS BALLS Josh purchased 8 cans of tennis balls. The cans came with 3 balls in each can. How many balls did Josh purchase? 24 tennis balls

3-7 Lesson Notes

Lesson Resources

Chapter 3 Resource Masters
- Study Guide and Intervention, p. 159
- Practice: Skills, p. 160
- Practice: Word Problems, p. 161
- Reading to Learn Mathematics, p. 162
- Enrichment, p. 163

Calculator and Spreadsheet Masters, p. 24
Key Teaching Transparencies, 12
MathPASS Tutorial CD-ROM, 27
Parent and Student Study Guide Workbook, 3-7

FAST FILE

1 Focus

5-Minute Check 3-7

INTERACTIVE CHALKBOARD
PowerPoint® Presentations

Use as a quiz or review of Lesson 3-6. (Also available as a transparency)

Bellringer

Reading Mathematics The word *division* came into popular usage in the 14th century. It comes from the Latin *dis-* + *-videre* and means "to separate into parts, sections, groups, or branches." A related word, *divisible*, means "able to be divided evenly, without a remainder." Have students consider these meanings when using counters to divide.

Reading to Learn Mathematics, p. 162 ELL

NAME ________________ DATE ________ PERIOD ____

3-7 Reading to Learn Mathematics
Dividing Integers

Pre-Activity *Complete the Mini Lab at the top of page 138 in your textbook. Write your answers below.*

Find each quotient using counters.

1. $-6 \div 2$ -3
2. $-12 \div 3$ -4

Reading the Lesson

Write two division sentences related to each of the following multiplication sentences.

3. $-6(-3) = 18$
   $18 \div (-3) = -6$;
   $18 \div (-6) = -3$
4. $-21(-2) = 42$
   $42 \div (-21) = -2$;
   $42 \div (-2) = -21$

3-7 Dividing Integers

What You'll Learn
Divide integers.

HANDS-ON Mini Lab

Materials
- counters
- integer mat

Work with a partner.

You can use counters to model division of integers. Follow these steps to find $-10 \div 5$.

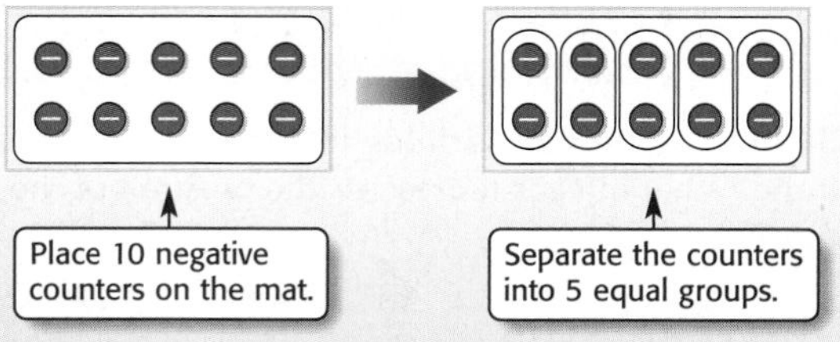

There are 2 negative counters in each group. So, $-10 \div 5 = -2$.

Find each quotient using counters. **1–2. See margin for models.**

1. $-6 \div 2$ **-3**
2. $-12 \div 3$ **-4**

Division of integers is related to multiplication. When finding the quotient of two integers, you can use a related multiplication sentence.

The factor in the multiplication sentence . . .

$2(-6) = -12 \rightarrow -12 \div 2 = -6$

$4(-5) = -20 \rightarrow -20 \div 4 = -5$

. . . is the quotient in the division sentence.

Since multiplication and division sentences are related, you can use them to find the quotient of integers with different signs.

$8(-9) = -72 \rightarrow -72 \div 8 = -9$ (different signs)

$-8(-9) = 72 \rightarrow 72 \div (-8) = -9$ (different signs)

negative quotient

These related sentences lead to the following rule.

Key Concept **Dividing Integers with Different Signs**

Words	The quotient of two integers with different signs is negative.
Examples	$33 \div (-11) = -3$ $-64 \div 8 = -8$

Answers (Hands-On Mini Lab)

1.

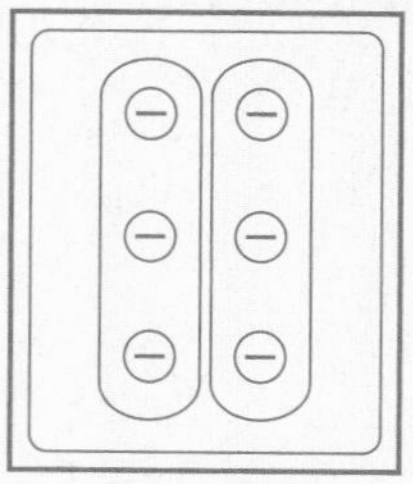

$-6 \div 2 = -3$

2.

$-12 \div 3 = -4$

EADING Math

Division Sentence
In the division sentence $80 \div (-10) = -8$, 80 is called the *dividend*, -10 is called the *divisor*, and -8 is called the *quotient*.

EXAMPLES Dividing Integers with Different Signs

Divide.

1 $\mathbf{80 \div (-10)}$ — The integers have different signs.

$80 \div (-10) = -8$ — The quotient is negative.

2 $\frac{-55}{11}$ — The integers have different signs.

$\frac{-55}{11} = -5$ — The quotient is negative.

Your Turn Divide.

a. $20 \div (-4)$ -5 b. $\frac{-81}{9}$ -9 c. $-45 \div 9$ -5

The quotient of two integers with the same sign is positive.

Key Concept — **Dividing Integers with the Same Sign**

Words	The quotient of two integers with the same sign is positive.
Examples	$15 \div 5 = 3$ $-21 \div (-3) = 7$

EXAMPLE Dividing Integers with the Same Sign

Divide.

3 $\mathbf{-14 \div (-7)}$ — The integers have the same sign.

$-14 \div (-7) = 2$ — The quotient is positive.

Your Turn Divide.

d. $-24 \div (-4)$ 6 e. $-9 \div (-3)$ 3 f. $\frac{-28}{-7}$ 4

EAL-LIFE MATH

TRONOMY The acecraft *Mars Odyssey* as launched in April 2001 study mineralogy and diation of the surface of ars.

urce: *The World Almanac for Kids*

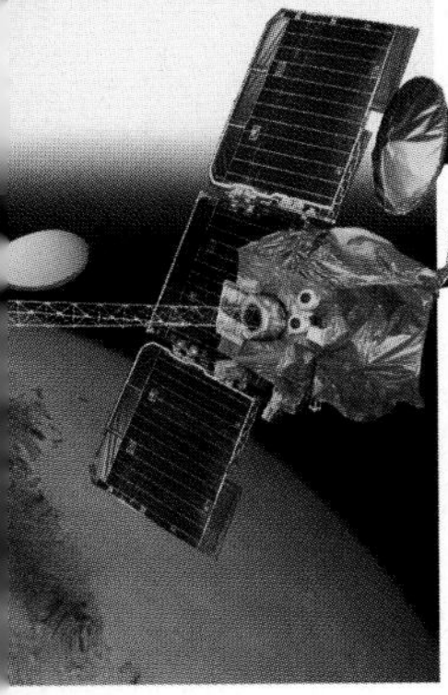

EXAMPLE Evaluate an Expression

4 **ASTRONOMY The average surface temperature on Mars is -81°F. Use the expression $\frac{5(F - 32)}{9}$, where F represents the number of degrees Fahrenheit, to find the temperature on Mars in degrees Celsius.**

$\frac{5(F - 32)}{9} = \frac{5(-81 - 32)}{9}$ — Replace F with -81.

$= \frac{5(-113)}{9}$ — Subtract 32 from -81.

$= \frac{-565}{9}$ — Multiply 5 and -113.

≈ -62.8 — Divide.

The average temperature on the surface of Mars is about -63°C.

msmath2.net/extra_examples

2 Teach

In-Class Examples

INTERACTIVE CHALKBOARD PowerPoint® Presentations

Divide.

1 $51 \div (-3)$ -17

2 $\frac{-81}{9}$ -9

3 $-12 \div (-2)$ 6

4 **PHYSICS** An object's acceleration is determined by the difference between the final and starting speeds divided by the total time. A car is traveling at 80 feet per second and, after 10 seconds, is traveling at 40 feet per second. What is the car's acceleration? -4 ft/sec

DAILY INTERVENTION — **Differentiated Instruction**

Naturalist Assign each student a country and have them find the highest and lowest temperatures in that country. Make a class table that shows these temperatures in both Fahrenheit and Celsius. Use the expression in Example 4 to convert between the temperature scales.

3 Practice/Apply

FOLDABLES™ Follow-Up Remind students to use their Foldable journals to record what they learn about dividing integers. For example, they might describe the process they use to divide integers and then use these notes to present an example to the class as a mini-lesson review.

Which One Doesn't Belong? For Exercise 3, have students find each quotient. Review the rules for dividing integers as they apply to the sign of the answer.

About the Exercises...

Guided Practice Key

Exercises	Examples
4, 5, 7	1, 2
6	3
8–10	4

Odd/Even Assignments
Exercises 11–32 are structured so that students practice the same concepts whether they are assigned odd or even problems.

Answer

2. Sample answer: $-8 \div 2 = -4$

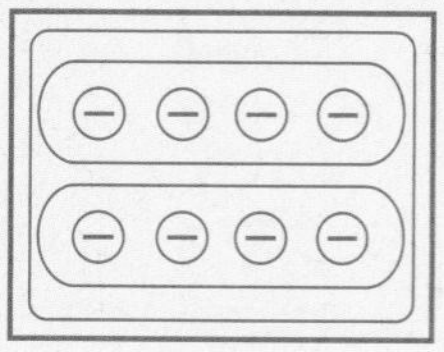

Skill and Concept Check

Writing Math Exercise 3

1. **Write** two division sentences related to the multiplication sentence $-6 \cdot 7 = -42$. **$-42 \div 7 = -6$, $-42 \div (-6) = 7$**
2. **OPEN ENDED** Write a division sentence. Then draw a model to show how the quotient can be found. **See margin.**
3. **Which One Doesn't Belong?** Identify the division expression whose quotient does not have the same sign as the other three. Explain your reasoning. **$-18 \div (-9)$ has a positive quotient; the others have negative quotients.**

-24 ÷ 6	-18 ÷ (-9)	28 ÷ (-7)	-22 ÷ 11

GUIDED PRACTICE

Divide.

4. $32 \div (-8)$ **−4**
5. $-16 \div 2$ **−8**
6. $-60 \div (-5)$ **12**
7. $\frac{-6}{6}$ **−1**

ALGEBRA Evaluate each expression if $d = -9$, $e = 36$, and $f = -6$.

8. $-108 \div f$ **18**
9. $e \div d$ **−4**
10. $\frac{e - f}{f}$ **−7**

★ indicates increased difficulty

Practice and Applications

HOMEWORK HELP	
For Exercises	See Examples
11–24	1–3
25–37	4

Extra Practice See pages 571, 598.

Divide.

11. $-18 \div 9$ **−2**
12. $50 \div (-5)$ **−10**
13. $-15 \div (-3)$ **5**
14. $\frac{21}{-7}$ **−3**
15. $56 \div (-8)$ **−7**
16. $\frac{0}{-5}$ **0**
17. $-52 \div (-13)$ **4**
18. $-34 \div 2$ **−17**
19. $\frac{90}{6}$ **15**
20. $-300 \div 25$ **−12**
21. $99 \div (-99)$ **−1**
22. $-184 \div (-23)$ **8**
23. Find the quotient of −65 and 13. **−5**
24. Divide 200 by −100. **−2**

ALGEBRA Evaluate each expression if $r = 12$, $s = -4$, and $t = -6$.

25. $-12 \div r$ **−1**
26. $72 \div t$ **−12**
27. $r \div s$ **−3**
28. $rs \div 16$ **−3**
29. $\frac{-r}{t}$ **2**
30. $\frac{16 - (-r)}{-s}$ **7**
31. $t^2 \div r$ **3**
32. ★ $\frac{r^2}{s^2}$ **9**
33. **FOOTBALL** During the fourth quarter, the Colts were penalized 3 times for the same amount for a total of 45 yards. Write a division sentence to represent this situation. Then find the number of yards for each penalty. **$p = -45 \div 3$; $p = -15$ or 15-yd penalty**
34. **EARTH SCIENCE** Use the expression $\frac{5(F - 32)}{9}$, where F represents the number of degrees Fahrenheit, to convert 5°F to degrees Celsius. **−15°C**
35. ★ **PATTERNS** Find the next two numbers in the pattern 729, −243, 81, −27, 9, … . Explain your reasoning. **−3, 1; Divide the previous term by −3.**

36. **SALES** The graph shows five magazines that had losses in a recent year. The numbers represent the profit the magazines made in 2002 compared to 2001. What is the mean of the losses for these five magazines? **−$18 million**

No Profit at the Newsstand

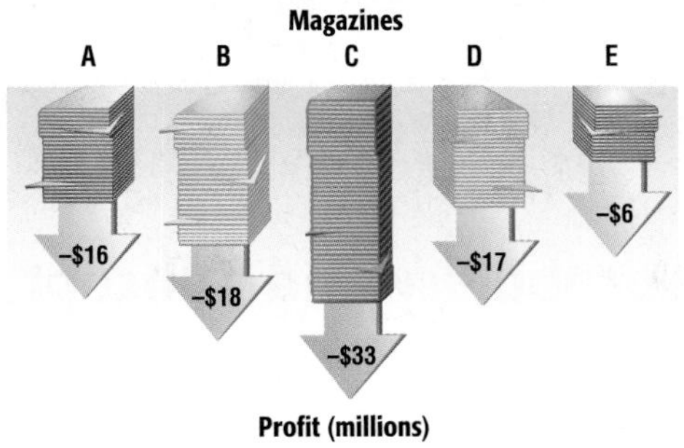

★ 37. **MULTI STEP** The sea otter population is decreasing. There were 2,377 sea otters in 1995. The population dropped to 2,161 in 2001. Find the average rate of change for the sea otter population between 1995 and 2001. **−36 sea otters per year**

38. **WRITE A PROBLEM** Write about a situation in your life in which you used positive and negative integers. Create a problem and solve it using any of the four operations. **See students' work.**

39. **CRITICAL THINKING** List all of the numbers by which −20 is divisible. **1, 2, 4, 5, 10, 20, −1, −2, −4, −5, −10, −20**

Standardized Test Practice and Mixed Review

40. **MULTIPLE CHOICE** Shenequa is driving cross-country. If she expects to drive between 350 and 450 miles per day, which number of days is reasonable for her to drive 3,800 miles? **C**

- (A) fewer than 6 days
- (B) between 6 and 8 days
- (C) between 9 and 11 days
- (D) more than 11 days

41. **MULTIPLE CHOICE** The width of a beach at 8:00 P.M. is 107 feet. At 3:00 A.M., the width of the beach narrows to 23 feet due to the high tide. What is the average rate of change per hour of the beach width? **H**

- (F) −15 ft/h
- (G) −13 ft/h
- (H) −12 ft/h
- (I) −10 ft/h

42. **SHORT RESPONSE** Find the mean of the following low temperatures for a 5-day period: −2°F, −3°F, 1°F, 4°F, and −5°F. **−1°F**

Multiply. (Lesson 3-6)

43. 14(−2) **−28**
44. $(-9)^2$ **81**
45. −20(−3) **60**
46. −5(7) **−35**

47. Find 6 − (−12). (Lesson 3-5) **18**

48. **DIVING** Mark jumped into 12 feet of water and touched the bottom of the pool before he surfaced. Which integer describes where Mark was in relation to the surface of the water? (Lesson 3-1) **−12**

49. **PHYSICAL SCIENCE** A chemistry experiment requires 3 milligrams of potassium chloride. How many grams of potassium chloride are needed? (Lesson 1-8) **0.003 g**

msmath2.net/self_check_quiz

4 Assess

Open-Ended Assessment

Speaking Have students explain how to determine the sign of the quotient based on the signs of the dividend and divisor.

Study Guide and Intervention, p. 159

NAME ________ DATE ______ PERIOD ____

3-7 Study Guide and Intervention

Dividing Integers

The quotient of two integers with different signs is negative.
The quotient of two integers with the same sign is positive.

EXAMPLE 1 **Divide 30 ÷ (−5).**

30 ÷ (−5) The integers have different signs.
30 ÷ (−5) = −6 The quotient is negative.

EXAMPLE 2 **Divide −100 ÷ (−5).**

−100 ÷ (−5) The integers have the same sign.
−100 ÷ (−5) = 20 The quotient is positive.

EXERCISES

Divide.

1. −12 ÷ 4 −3
2. −14 ÷ (−7) 2
3. $\frac{18}{-2}$ −9
4. −6 ÷ (−3) 2
5. −10 ÷ 10 −1
6. $\frac{-80}{-20}$ 4
7. 350 ÷ (−25) −14
8. −420 ÷ (−3) 140
9. $\frac{540}{45}$ 12
10. $\frac{-256}{16}$ −16

ALGEBRA **Evaluate each expression if $d = -24$, $e = -4$, and $f = 8$.**

11. $12 \div e$ −3
12. $40 \div f$ 5
13. $d \div 6$ −4
14. $d \div e$ 6
15. $f \div e$ −2
16. $e^2 \div f$ 2
17. $\frac{-d}{e}$ −6
18. $ef \div 2$ −16
19. $\frac{f^2}{e^2}$ 4
20. $\frac{de}{f}$ 12

Practice: Skills, p. 160

NAME ________ DATE ______ PERIOD ____

3-7 Practice: Skills

Dividing Integers

Divide.

1. −15 ÷ 3 −5
2. −24 ÷ (−8) 3
3. 22 ÷ (−2) −11
4. −49 ÷ (−7) 7
5. −8 ÷ (−8) 1
6. $\frac{36}{-4}$ −9
7. 225 ÷ (−15) −15
8. $\frac{0}{-9}$ 0
9. −38 ÷ 2 −19
10. $\frac{64}{4}$ 16
11. −500 ÷ (−50) 10
12. −189 ÷ (−21) 9

ALGEBRA **Evaluate each expression if $m = -32$, $n = 2$, and $p = -8$.**

13. $m \div n$ −16
14. $p \div 4$ −2
15. $p^2 \div m$ −2
16. $m \div p$ 4
17. $\frac{-p}{n}$ 4
18. $p \div n^2$ −2
19. $\frac{p^2}{n^2}$ 16
20. $\frac{18 - n}{p}$ −2
21. $m \div (np)$ 2
22. $\frac{m}{p} + n$ 6

Practice: Word Problems, p. 161

NAME ________ DATE ______ PERIOD ____

3-7 Practice: Word Problems

Dividing Integers

Divide.

1. **STOCK MARKET** During a 5-day workweek, the stock market decreased by 65 points. Find the average daily change in the market for the week. −13 points

2. **MOTION** Mr. Diaz decreased the speed of his car by 30 miles per hour over a period of 10 seconds. Find the average change in speed each second. −3 mph per s

3. **WEATHER** Over the past seven days, Mrs. Cho found that the temperature outside had dropped a total of 35 degrees. Find the average drop in temperature each day. −5 degrees per day

4. **BASKETBALL** The basketball team lost their last 6 games. They lost by a total of 48 points. Find the average number of points by which each game was lost. −8 points per game

5. **POPULATION** The enrollment at Davis Middle School dropped by 60 students over a 5-year period. What is the average yearly drop in enrollment? −12 students annually

6. **SUBMARINE** A submarine descends at a rate of 60 feet each minute. How long will it take it to descend to a depth of 660 feet below the surface? 11 min

Chapter 3 Study Guide and Review

Vocabulary Assessment

- This alphabetical list of vocabulary terms in Chapter 3 includes a page reference where each term was introduced.
- A vocabulary review/test for Chapter 3 is available on p. 178 of the *Chapter 3 Resource Masters*.

Vocabulary PuzzleMaker

ELL The online Vocabulary PuzzleMaker improves students' mathematics vocabulary using four puzzle formats — crossword, scramble, word search using a word list, and word search using clues. Students can work on a computer screen or from a printed handout.

MindJogger Videoquizzes

ELL MindJogger Videoquizzes provide an alternative review of concepts presented in this chapter. Students work in teams in a game show format to gain points for correct answers. The questions are presented in three rounds.

Round 1 Concepts (5 questions)
Round 2 Skills (4 questions)
Round 3 Problem Solving (4 questions)

ExamView® Pro Test Bank CD-ROM

Use the networkable ExamView® Pro Test Bank CD-ROM to:

- Create **Multiple versions** of tests.
- Create **Modified** tests for **Inclusion** students with one mouse click.
- **Edit** existing questions and **Add** your own questions.
- Build tests aligned with state standards using built-in **State Curriculum Correlations.**
- Change **English** tests to **Spanish** with one mouse click and vice versa.

Chapter 3 Study Guide and Review

Vocabulary and Concept Check

absolute value (p. 107)	negative integer (p. 106)	*x*-axis (p. 112)
additive inverse (p. 121)	opposite (p. 121)	*x*-coordinate (p. 112)
coordinate grid (p. 112)	ordered pair (p. 112)	*y*-axis (p. 112)
coordinate plane (p. 112)	origin (p. 112)	*y*-coordinate (p. 112)
graph (p. 106)	positive integer (p. 106)	
integer (p. 106)	quadrant (p. 113)	

Choose the correct term or number to complete each sentence.

1. Integers less than zero are (positive, negative) integers.
2. Two numbers represented by points that are the same distance from 0 are (opposites, integers).
3. The absolute value of 7 is (7, −7).
4. The opposite of (−12, 12) is −12.
5. The (coordinate plane, origin) is the point where the horizontal and vertical number lines intersect.
6. The *x*-axis and the *y*-axis separate the plane into four (quadrants, coordinates).
7. The first number in an ordered pair is the (*x*-coordinate, *y*-coordinate).
8. The sum of two (positive, negative) integers is negative.
9. The product of a positive and a negative integer is (positive, negative).
10. The quotient of a negative integer and a (positive, negative) integer is negative.

Lesson-by-Lesson Exercises and Examples

3-1 Integers and Absolute Value (pp. 106–108)

Write an integer for each situation.

11. a loss of $150 **−150**
12. 350 feet above sea level **350**
13. a gain of 8 yards **8**
14. 12°F below 0 **−12**

Evaluate each expression.

15. $|-11|$ **11**
16. $|100|$ **100**
17. $|5|$ **5**
18. $|-32|$ **32**
19. $|-16| + |9|$ **25**

Example 1 **Write an integer for 8 feet below sea level.**

Since this situation represents an elevation *below* sea level, −8 represents the situation.

Example 2 **Evaluate $|-10|$.**

On the number line, the graph of −10 is 10 units from 0.

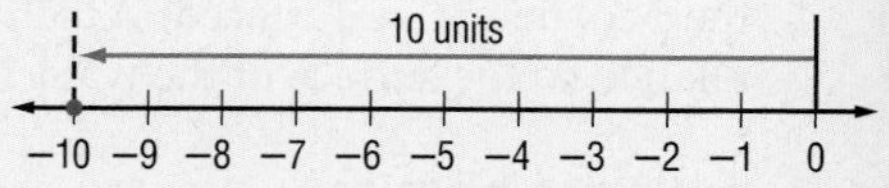

So, $|-10| = 10$.

msmath2.net/vocabulary_review

FOLDABLES™ Study Organizer

For more information about Foldables, see *Teaching Mathematics with Foldables.*

Have students look through the chapter to make sure they have included examples in their Foldables for each lesson or topic in the chapter.

Encourage students to refer to their Foldables while completing the Study Guide and Review and to use them in preparing for the Chapter Test.

3-2 Comparing and Ordering Integers (pp. 109–111)

Replace each ● with < or > to make a true sentence.

20. $-3 ● -9$ >
21. $8 ● -12$ >
22. $-3 ● 3$ <
23. $|-10| ● |-13|$ <
24. $25 ● |8|$ >
25. $0 ● |-4|$ <

Order each set of integers from least to greatest. 26–28. See margin.

26. $-3, 8, -10, 0, 5, -12, 9$
27. $-21, 19, -23, 14, -32, 25$
28. $|-17|, -18, 18, |15|, -16, |16|$

29. **EARTH SCIENCE** The predicted low temperatures for Monday through Friday are 3°, −1°, −2°, 0°, and 1°. Order the temperatures from greatest to least.
    3°, 1°, 0°, −1°, −2°

Example 3 **Replace ● with < or > to make $-4 ● -7$ a true sentence.**

Graph each integer on a number line.

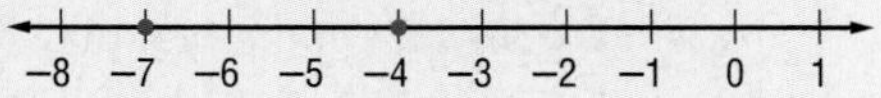

Since −4 is to the right of −7, $-4 > -7$.

Example 4 **Order the integers −4, −3, 5, 3, 0, −2 from least to greatest.**

Graph the integers on a number line.

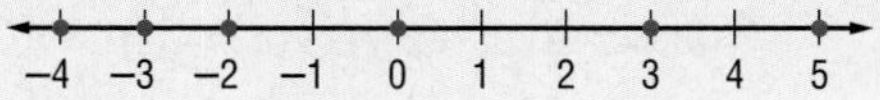

Order the integers by reading from left to right: −4, −3, −2, 0, 3, 5.

3-3 The Coordinate Plane (pp. 112–115)

Name the ordered pair for each point graphed at the right. Then identify the quadrant in which each point lies.

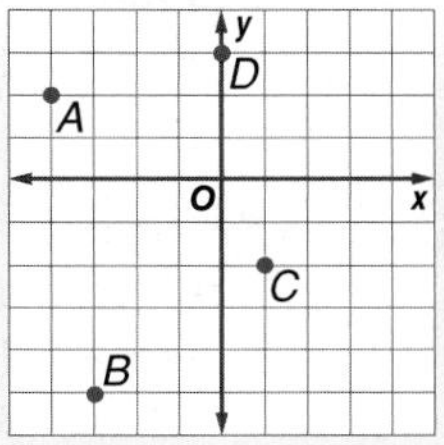

30. *A* (−4, 2), II
31. *B* (−3, −5), III
32. *C* (1, −2), IV
33. *D* (0, 3), *y*-axis

On graph paper, draw a coordinate plane. Then graph and label each point.

34. *E*(1, −4)
35. *F*(−5, 2)
36. *G*(−2, −3)
37. *H*(4, 0)

34–37. See margin.

Example 5 **Name the ordered pair for point *W* graphed at the right. Then identify the quadrant in which point *W* lies.**

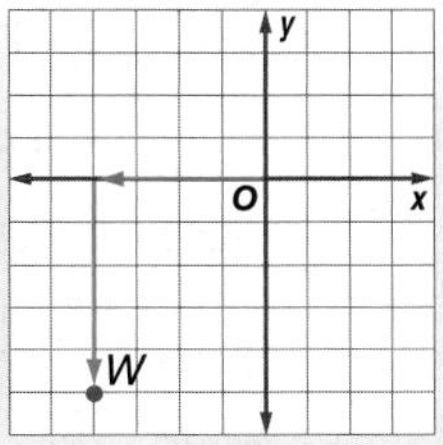

The ordered pair is (−4, −5). Point *W* is in quadrant III.

Example 6 **Graph and label the point *S*(3, −1).**

Draw a coordinate plane. Move 3 units to the right. Then move 1 unit down. Draw a dot and label it *S*(3, −1).

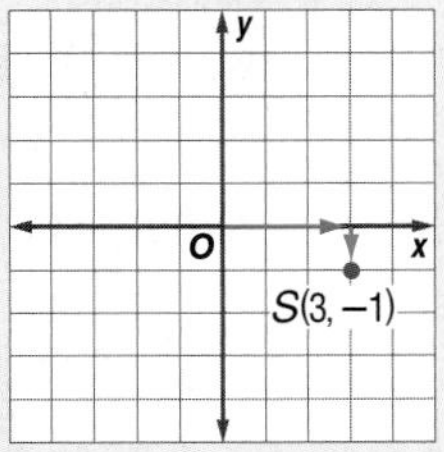

Answers

26. −12, −10, −3, 0, 5, 8, 9
27. −32, −23, −21, 14, 19, 25
28. −18, −16, |15|, |16|, |−17|, 18

34–37.

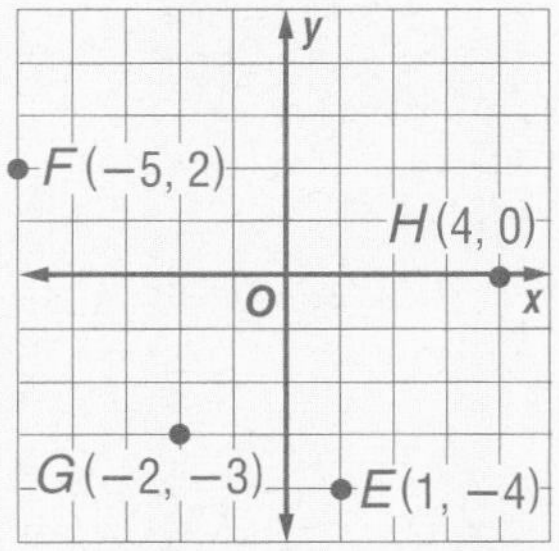

Chapter 3 Study Guide and Review 143

3-4 Adding Integers (pp. 120–124)

Add.

38. $-6 + 8$ **2**
39. $-4 + (-9)$ **−13**
40. $7 + (-12)$ **−5**
41. $-18 + 18$ **0**
42. **FOOTBALL** On the first play of the game, the Bulldogs lost 8 yards. On the second and third plays, they gained 5 yards and then lost 2 yards. Find the result of the first three plays. **−5**

Example 7 Find $-4 + 3$.

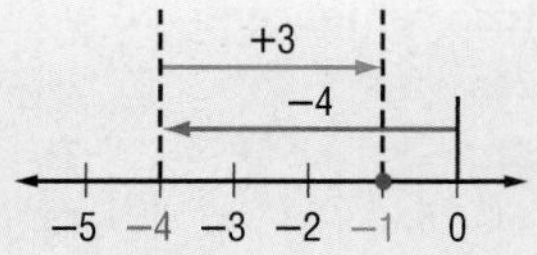

So, $-4 + 3 = -1$.

3-5 Subtracting Integers (pp. 128–131)

Subtract.

43. $-5 - 8$ **−13**
44. $3 - 6$ **−3**
45. $5 - (-2)$ **7**
46. $-4 - (-8)$ **4**

Example 8 Find $-3 - 9$.

$-3 - 9 = -3 + (-9)$ To subtract 9, add −9.
$= -12$ Simplify.

3-6 Multiplying Integers (pp. 134–137)

Multiply.

47. $-4(3)$ **−12**
48. $8(-6)$ **−48**
49. $-5(-7)$ **35**
50. $-2(40)$ **−80**

ALGEBRA Evaluate each expression if $a = -4$, $b = -7$, and $c = 5$.

51. ab **28**
52. $-3c$ **−15**
53. bc **−35**
54. abc **140**

Example 9 Find $-4(3)$.

$-4(3) = -12$ The integers have different signs. The product is negative.

Example 10 Evaluate xyz if $x = -6$, $y = 11$, and $z = -10$.

$xyz = (-6)(11)(-10)$ $x = -6, y = 11, z = -10$
$= (-66)(-10)$ Multiply −6 and 11.
$= 660$ Multiply −66 and −10.

3-7 Dividing Integers (pp. 138–141)

Divide.

55. $-45 \div (-9)$ **5**
56. $36 \div (-12)$ **−3**
57. $-12 \div 6$ **−2**
58. $-81 \div (-9)$ **9**
59. **HIKING** Marta started a hike at sea level and ended the hike 6 hours later at 300 feet below sea level. If Marta hiked at the same pace during the trip, how far did she travel each hour? **−50 ft per h**

Example 11 Find $-72 \div (-9)$.

$-72 \div (-9) = 8$ The integers have the same sign. The quotient is positive.

144 Chapter 3 Algebra: Integers

CHAPTER 3 Practice Test

Vocabulary and Concepts

1. **Explain** what it means for two numbers to be opposites. See margin.
2. **Name** the rule for dividing integers with different signs. The quotient of two integers with different signs is negative.

Skills and Applications

Write an integer for each situation.

3. a stock increased by $5 **5**
4. 1000 B.C. **−1000**
5. an elevator goes down 11 floors **−11**

Replace each ● with < or > to make a true sentence.

6. $-3 \bullet -9$ >
7. $|9| \bullet |-12|$ <
8. $|-7| \bullet 9$ <

9. **WEATHER** The local weather service records the following changes in temperature during the last week: 4, −7, −3, 2, 9, −8, 1. Order these temperature changes from greatest to least. **9, 4, 2, 1, −3, −7, −8**

Name the ordered pair for each point graphed at the right. Then identify the quadrant in which each point lies.

10. P (−3, −3), III
11. Q (−1, 4), II
12. R (5, −2), IV

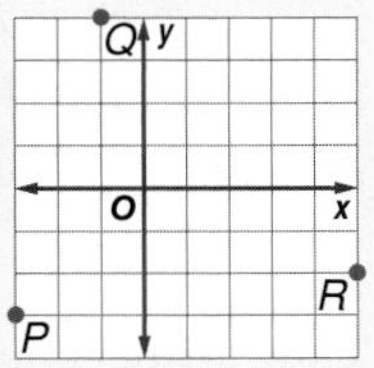

Add, subtract, multiply, or divide.

13. $-3 + 8$ **5**
14. $12 + (-19)$ **−7**
15. $-3 - 8$ **−11**
16. $-7 - (-20)$ **13**
17. $-7(-3)$ **21**
18. $5(-11)$ **−55**
19. $-24 \div 8$ **−3**
20. $-36 \div (-9)$ **4**

ALGEBRA Evaluate each expression if $a = -5$, $b = 4$, and $c = -12$.

21. $c - a$ **−7**
22. ab **−20**
23. $ac \div b$ **15**

24. **STOCK MARKET** The value of a stock went down $3 each week for a period of seven weeks. Describe the change in the value of the stock at the end of the seven week period. **−$21**

Standardized Test Practice

25. **MULTIPLE CHOICE** Choose the graph that shows the ordered pair (2, −1). **D**

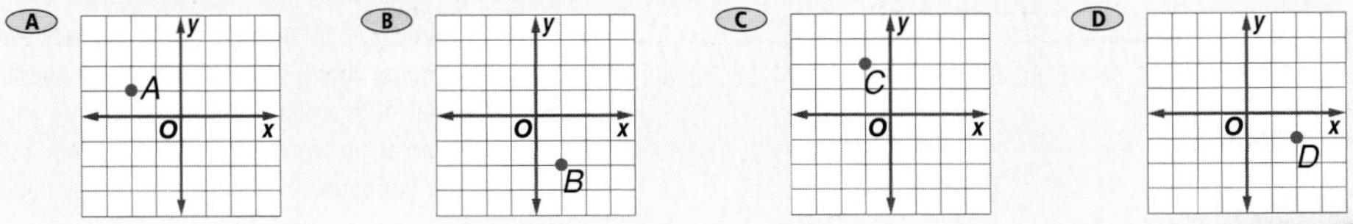

CHAPTER 3 Practice Test

Assessment Options

Vocabulary Test A vocabulary review/test for Chapter 3 can be found on p. 178 of the *Chapter 3 Resource Masters.*

Chapter Tests There are six Chapter 3 Tests and an Extended Response Assessment task available in the *Chapter 3 Resource Masters.*

Chapter 3 Tests			
Form	**Type**	**Level**	**Pages**
1	MC	basic	165–166
2A	MC	average	167–168
2B	MC	average	169–170
2C	FR	average	171–172
2D	FR	average	173–174
3	FR	advanced	175–176

MC = multiple-choice questions
FR = free-response questions

Extended Response Assessment

Extended Response tasks for Chapter 3 can be found on p. 177 of the *Chapter 3 Resource Masters.* A sample scoring rubric for these tasks appears on p. A26.

Answer

1. Sample answer: Two numbers are opposites of one another if they are the same distance from 0, but on opposite sides of 0 on a number line.

Portfolio Suggestion

Introduction Students frequently have problems remembering how to determine the sign of the answer in an integer problem.

Ask Students to write out examples for each possible combination of signs for each operation. Then have students create a chart that presents a summary of the rules for determining signs.

CHAPTER 3 Standardized Test Practice

These two pages contain practice questions in the various formats that can be found on the most frequently given standardized tests.

A student recording sheet for these two pages can be found on page A1 of the *Chapter 3 Resource Masters*.

Standardized Test Practice Student Recording Sheet, p. A1

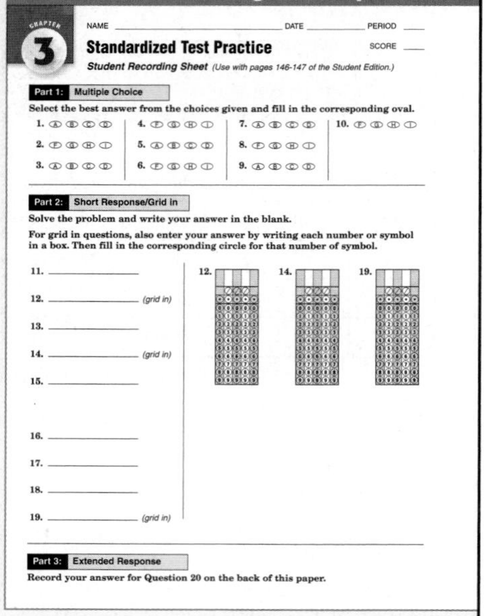

Additional Practice

See pp. 183–184 of the *Chapter 3 Resource Masters* for additional standardized test practice.

CHAPTER 3 Standardized Test Practice

PART 1 Multiple Choice

Record your answers on the answer sheet provided by your teacher or on a sheet of paper.

1. The table shows the five most common languages spoken in the United States other than English. How many more people speak Chinese than French? (Lesson 1-1) **A**

Language	Number of People
Spanish	26,745,067
Chinese	1,976,564
French	1,914,918
German	1,224,213
Tagalog	1,184,939

Source: U.S. Census Bureau

Ⓐ 61,646　Ⓑ 1,914,918
Ⓒ 1,976,564　Ⓓ 3,891,482

2. Which is equivalent to 5^4? (Lesson 1-2) **H**

Ⓕ 20　Ⓖ 125
Ⓗ $5 \cdot 5 \cdot 5 \cdot 5$　Ⓘ $4 \cdot 4 \cdot 4 \cdot 4 \cdot 4$

3. How many millimeters are in 13 centimeters? (Lesson 1-8) **D**

Ⓐ 0.13　Ⓑ 1.3
Ⓒ 13　Ⓓ 130

4. Salvador recorded the number of minutes it took him to drive to work each day for a week. Find the mean for the following times: 12, 23, 10, 14, and 11. (Lesson 2-4) **H**

Ⓕ 12 min　Ⓖ 13 min
Ⓗ 14 min　Ⓘ 15 min

5. Find the interquartile range of the data in the box-and-whisker plot. (Lesson 2-6) **A**

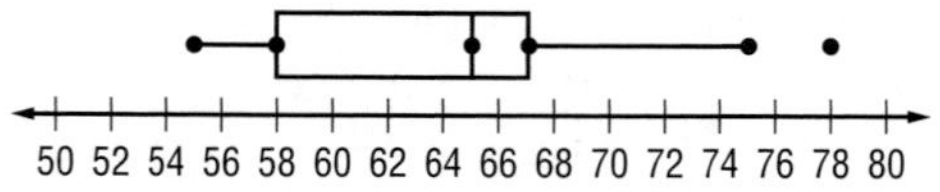

Ⓐ 9　Ⓑ 23
Ⓒ 58　Ⓓ 65

6. What is the value of $|-2|$? (Lesson 3-1) **I**

Ⓕ -2　Ⓖ -1
Ⓗ 0　Ⓘ 2

7. Which of these is the correct order of the integers from greatest to least? (Lesson 3-2) **B**

Ⓐ $0, 1, -2, -5$　Ⓑ $1, 0, -2, -5$
Ⓒ $-5, -2, 0, 1$　Ⓓ $1, -2, -5, 0$

8. Which of these are the coordinates of point Z? (Lesson 3-3) **G**

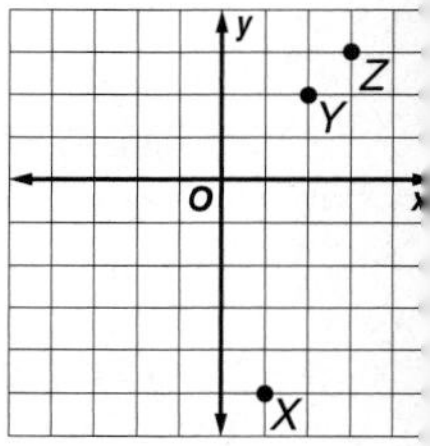

Ⓕ (2, 2)
Ⓖ (3, 3)
Ⓗ $(-2, -2)$
Ⓘ $(-3, -3)$

9. If you graph and connect the following points on a coordinate plane, what shape would you make? (Lesson 3-3) **A**

$(0, -3), (0, 4), (4, 4), (4, -3)$

Ⓐ rectangle　Ⓑ triangle
Ⓒ circle　Ⓓ parallelogram

10. Which of the following points on the number line represents opposites? (Lesson 3-4) **I**

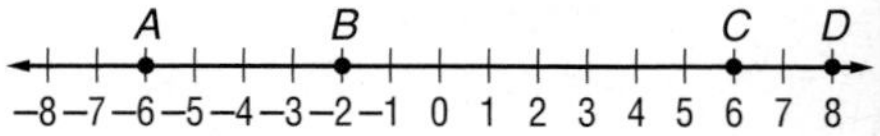

Ⓕ *C* and *D*　Ⓖ *B* and *D*
Ⓗ *A* and *D*　Ⓘ *A* and *C*

TEST-TAKING TIP

Question 10 Always be sure to check every answer choice of a multiple-choice question. Start with answer choice F. Each time you find an incorrect answer choice, cross it off so you remember that you've eliminated it.

The Princeton Review ***Log On*** **for Test Practice**
The Princeton Review offers additional test-taking tips and practice problems at their web site. Visit princetonreview.com or review.com

ExamView® Pro Test Bank CD-ROM

Special banks of standardized test questions similar to those on the TIMSS and NAEP grade 8 tests can be found on this CD-ROM.

Aligned and verified by The Princeton Review

ART 2 Short Response/Grid In

cord your answers on the answer sheet ovided by your teacher or on a sheet of per.

. Wallpaper costs \$16 per roll, and border costs \$9 per roll. If 12 rolls of wallpaper and 6 rolls of border are needed for one room, find the total cost of the wallpaper and border. (Lesson 1-3) **\$246**

. Each triangle in the figure below is made from toothpicks. Extend the pattern. Find the number of toothpicks in the fifth figure. (Lesson 1-7) **27**

. Write 63.5 million in scientific notation. (Lesson 1-9) **6.35×10^7**

r Questions 14 and 15, use the information low.

e graph shows the number of orders taken ch hour one day at a fast food restaurant. sson 2-2)

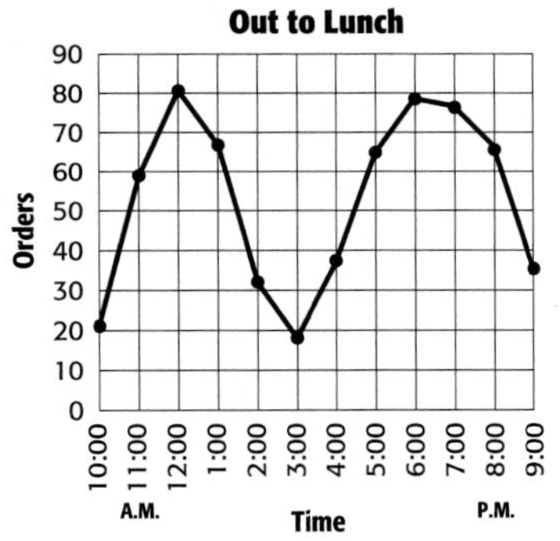

. About how many orders were taken at 11:00 A.M.? **59**

. What time of the day appears to be the least busy at this restaurant? **3 P.M.**

For Questions 16 and 17, use the information below.

The temperature of the liquid in Connor's beaker changed drastically. The temperatures he recorded were −28°F, 59°F, 1°F, −16°F, 24°F, and 8°F. (Lesson 3-2)

16. Order the temperatures from least to greatest. **−28°F, −16°F, 1°F, 8°F, 24°F, 59°F**

17. Find the mean temperature. **8°F**

18. When Elena went hiking at 8 A.M., she started at an elevation of 16 meters below sea level. But at 2 P.M., she was 12 meters higher than when she started. What was Elena's elevation at 2 P.M.? (Lesson 3-5) **−4, or 4 meters below sea level**

19. Find $\frac{-12 + 8(-3)}{-2 - 7}$. (Lesson 3-7) **4**

PART 3 Extended Response

Record your answers on a sheet of paper. Show your work.

20. Use the table of ordered pairs at the right to answer the questions below. (Lesson 3-3)

x	y
−2	1
3	2
−1	−4

a. Plot the ordered pairs in the table on a coordinate plane. **See margin.**

b. In which quadrant is a point on the graph not represented? **quadrant IV**

c. Connect the points on the graph. What shape do they form? **triangle**

d. How would you double the size of the figure you drew above? **See margin.**

e. Add four more columns to the table. Label the columns $x + y$, $x - y$, xy, and $x \div y$. Complete the table. **See margin.**

msmath2.net/standardized_test

Standardized Test Practice

Rubrics The following rubrics are sample scoring devices for short response and extended response questions.

Short Response

Points	Description
2	The student demonstrates a **thorough understanding** of the mathematics of the task. The response may contain minor flaws that do not detract from the demonstration of a thorough understanding.
1	The student has provided a response that is only **partially correct**.
0	The student has provided a **completely incorrect** solution or no response at all.

Extended Response

Points	Description
4	The student demonstrates a **thorough understanding** of the mathematics of the task. The response may contain minor flaws that do not detract from the demonstration of a thorough understanding.
3	The student demonstrates an **understanding** of the mathematics of the task. The response is essentially correct and demonstrates an essential but less than thorough understanding of the mathematics.
2	The student demonstrates only a **partial understanding** of the mathematics of the task. Although the student may have used the correct approach to a solution or may have provided a correct solution, the work lacks an essential understanding of the underlying mathematical concepts.
1	The student demonstrates a **very limited understanding** of the mathematics of the task. The response is incomplete and exhibits many flaws.
0	The student provides a **completely incorrect** solution or no response at all.

A more detailed 4-point rubric is provided on page A2 of the *Chapter 3 Resource Masters*.

Answers

20a.

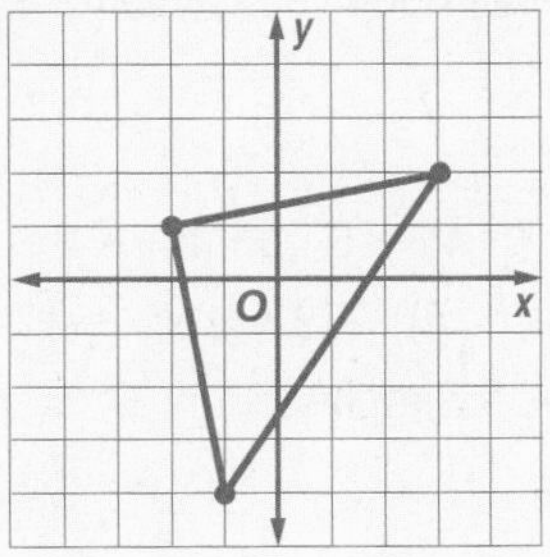

20d. Sample answer: Multiply each ordered pair by 2.

20e.

$x + y$	$x - y$	xy	$x \div y$
−1	−3	−2	−2
5	1	6	1.5
−5	3	4	0.25

Page 108, Lesson 3-1

30.
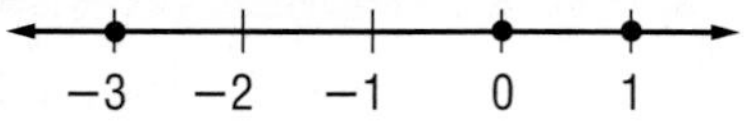

31.
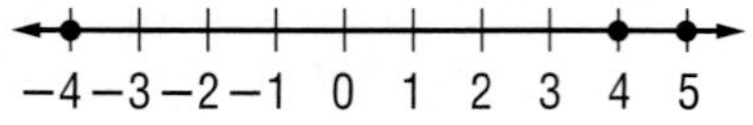

32.
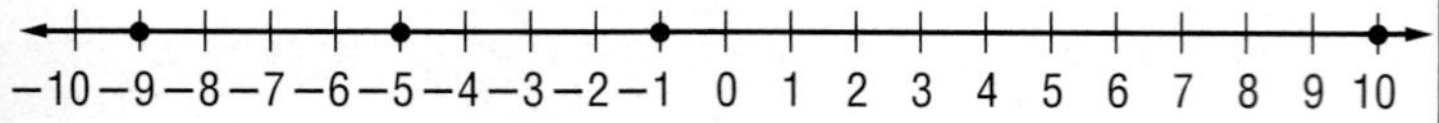

33.
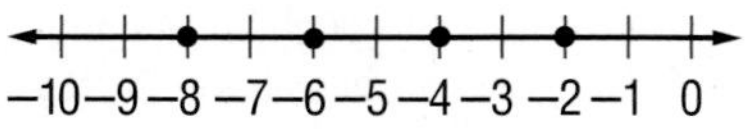

40.
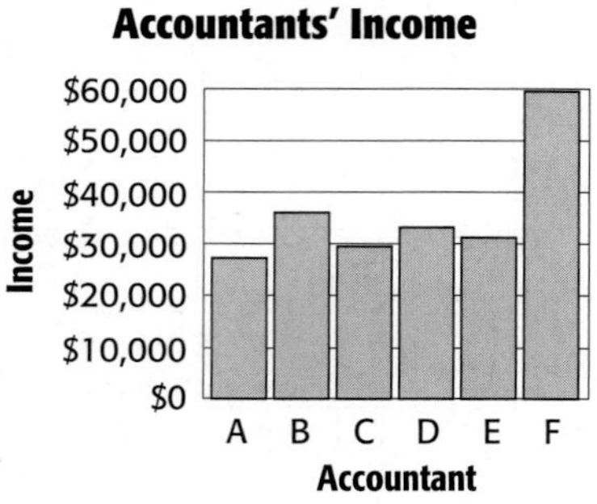

Page 111, Lesson 3-2

39.
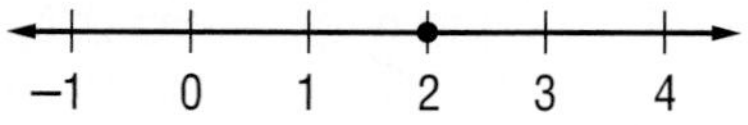

40.

41.
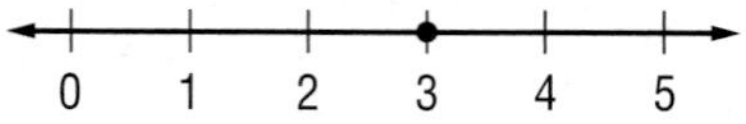

42.

Page 114, Lesson 3-3

20–31.
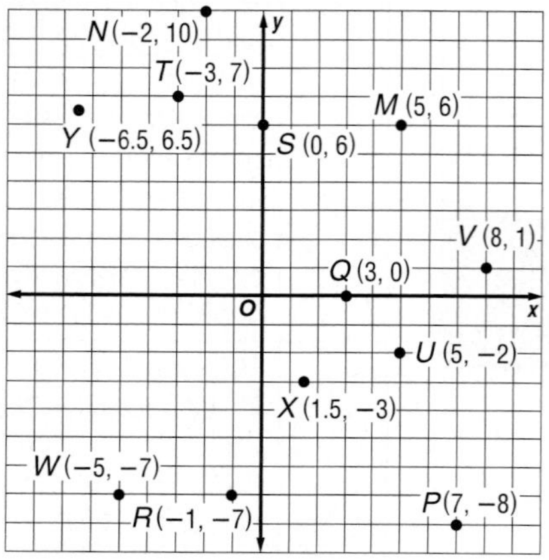

Page 119, Lesson 3-4a (Your Turn)

g.
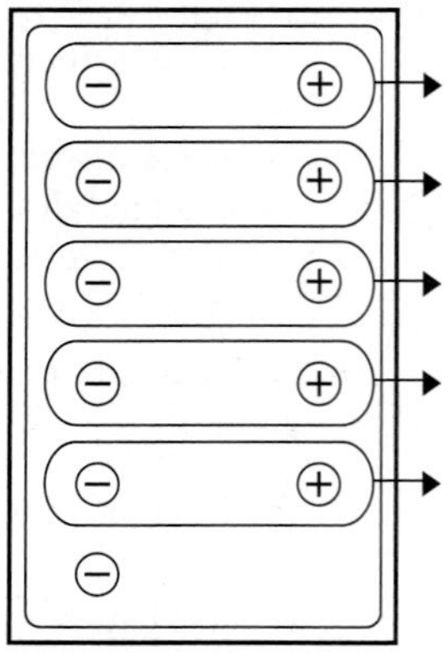

−6 + 5 = −1

h.
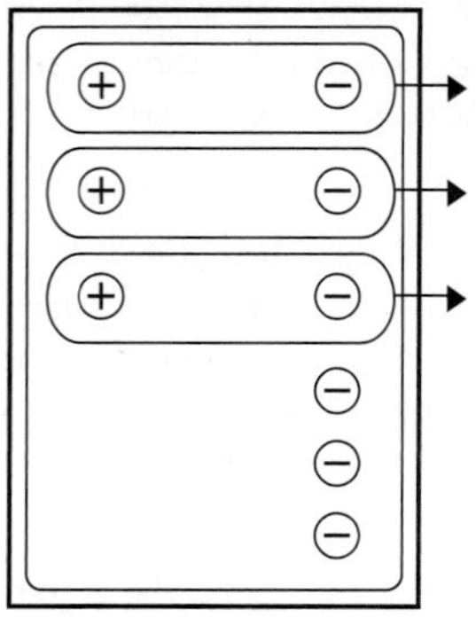

3 + (−6) = −3

i.
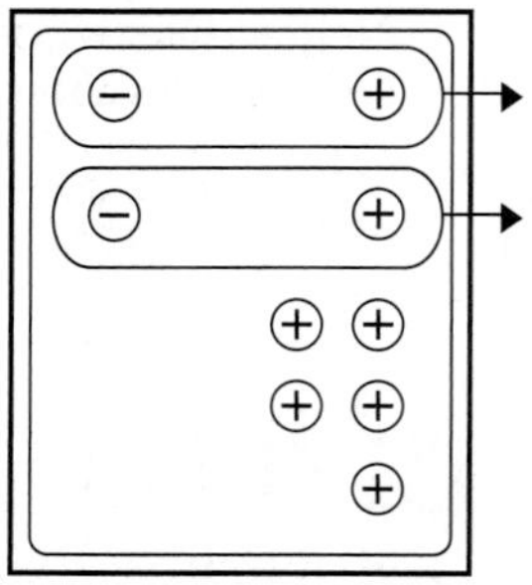

−2 + 7 = 5

j.
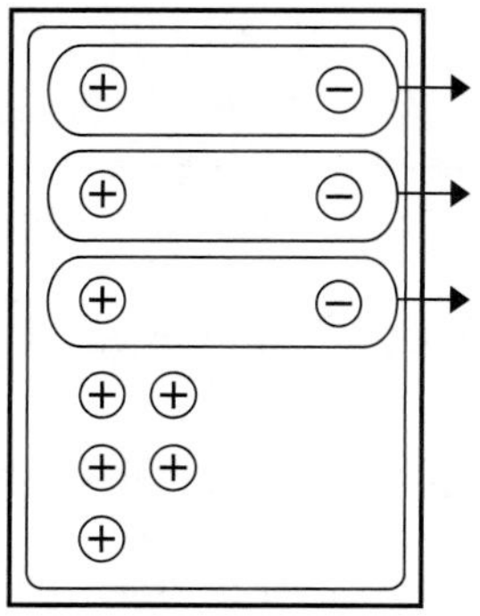

8 + (−3) = 5

k.
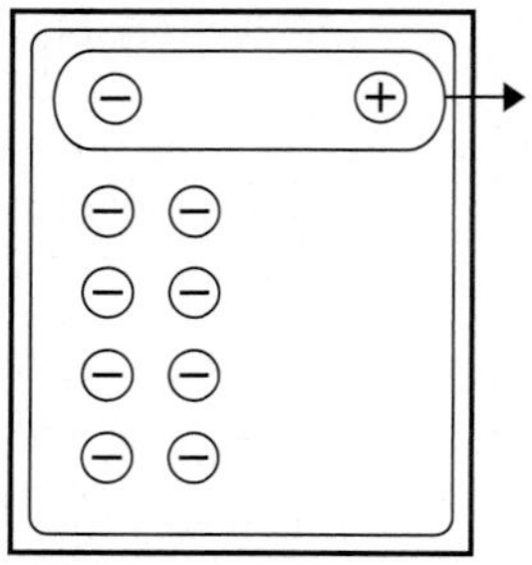

−9 + 1 = −8

l.
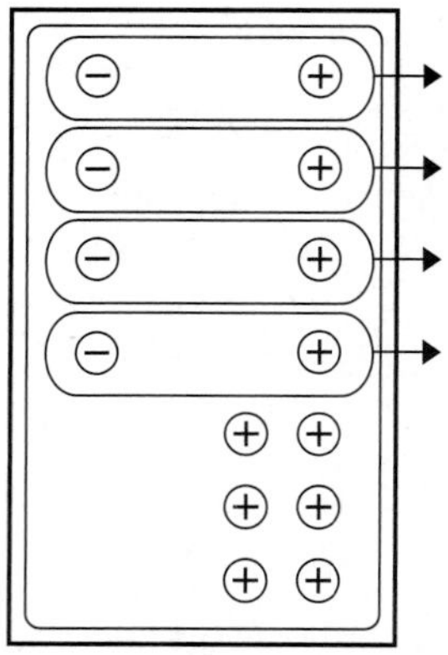

−4 + 10 = 6

Page 124, Lesson 3-4

52. Sample answer:

$7 + (-2) + (-7) = 7 + (-7) + (-2)$ Comm. (+)
$= (7 + (-7)) + (-2)$ Assoc. (+)
$= 0 + (-2)$ Additive Inverse
$= -2$

53. Sample answer:

$-6 + 9 + (-4) = -6 + (-4) + 9$ Comm. (+)
$= (-6 + (-4)) + 9$ Assoc. (+)
$= -10 + 9$ Simplify.
$= -1$

54. Sample answer:

$$-5 + (-6) + -3 = (-5 + (-6)) + -3 \text{ Assoc. } (+)$$
$$= -11 + (-3) \text{ Simplify.}$$
$$= -14$$

55. Sample answer:

$$8 + 10 + (-8) = 8 + (-8) + 10 \text{ Comm. } (+)$$
$$= (8 + (-8)) + 10 \text{ Assoc. } (+)$$
$$= 0 + 10 \text{ Additive Inverse}$$
$$= 10$$

56. Sample answer:

$$-5 + (-7) + (-10) = (-5 + (-7)) + (-10) \text{ Assoc. } (+)$$
$$= -12 + (-10) \text{ Simplify.}$$
$$= -22$$

57. Sample answer:

$$8 + (-9) + 9 = 8 + ((-9) + 9) \text{ Assoc. } (+)$$
$$= 8 + 0 \text{ Additive Inverse}$$
$$= 8$$

Page 127, Lesson 3-5a (Your Turn)

f.

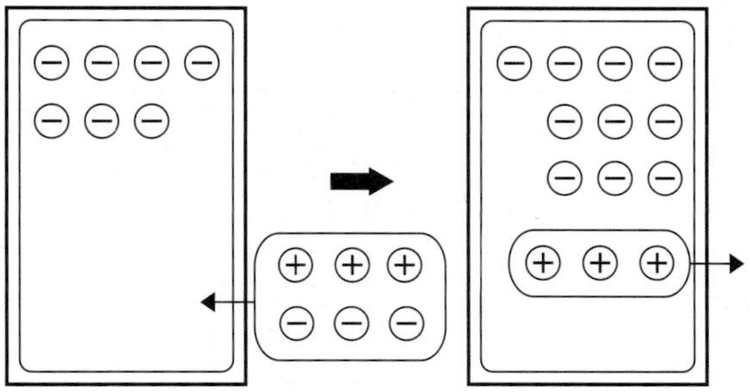

$-7 - 3 = -10$

g.

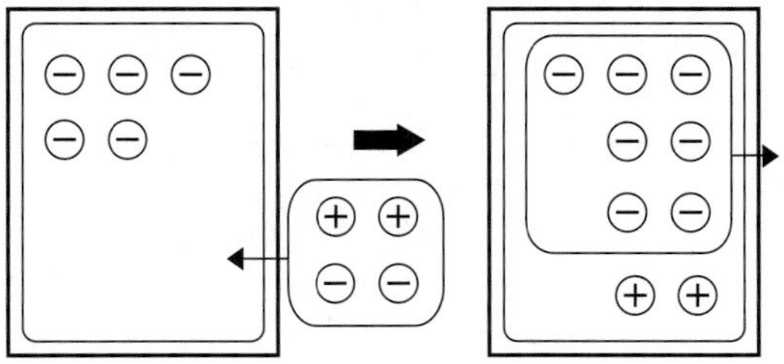

$-5 - (-7) = 2$

CHAPTER 4

Chapter Overview

Algebra: Linear Equations and Functions

	LESSON OBJECTIVES	NCTM Standards	NAEP 2005	State/Local Objectives
4-1	**Writing Expressions and Equations** *(pp. 150–152)* • Write verbal phrases and sentences as simple algebraic expressions and equations.	1, 2, 6–10	A3a, N5a	
4-2	**Solving Addition and Subtraction Equations** *(pp. 154–159)* **4-2a Hands-On Lab:** Solve equations using models. • Solve addition and subtraction equations.	1, 2, 6–10	N1b, A4a, A4b, N5e, N3f, A4c, A3a	
4-3	**Solving Multiplication Equations** *(pp. 160–163)* • Solve multiplication equations.	1, 2, 6–10	N1b, N3f, A3a, A4a, A4b, A4c, N5e	
4-4	**Solving Two-Step Equations** *(pp. 164–169)* **4-4a Problem-Solving Strategy:** Solve problems using the work backward strategy. • Solve two-step equations.	1, 2, 6–10	N1b, N3f, N5e, A3a, A4a, A4b, A4c, A4e, N5f	
4-5	**Inequalities** *(pp. 172–175)* • Solve inequalities.	1, 6–10	N1b, A4a,A3a, A4c	
4-6	**Functions and Linear Equations** *(pp. 176–181)* **4-6a Hands-On Lab:** Graph a function on a scatter plot. • Graph linear equations.	1–3, 6–10	A1a, A1c, A2a, A2c, A1b, A1e, A2g, D1a, A2e, D2e	
4-7	**Lines and Slope** *(pp. 182–185)* • Find the slope of a line.	1–3, 6–10	A4d, A1f, A1e, A2b, A2g, A2c	

Key to NCTM Standards:

1=Number & Operations, 2=Algebra, 3=Geometry, 4=Measurement, 5=Data Analysis & Probability, 6=Problem Solving, 7=Reasoning & Proof, 8=Communication, 9=Connections, 10=Representation

NAEP 2005 Standards: See math.glencoe.com

Pacing suggestions for the entire year can be found on page T20.

Chapter Resource Manager

Time-Saving Tool
TeacherWorks™
All-In-One Planner and Resource Center
See page T5.

Chapter 4 Resource Masters (FAST FILE)											
Study Guide and Intervention	Practice (Skills and Word Problems)	Reading to Learn Mathematics	Enrichment	Assessment	Applications*	Parent and Student Study Guide Workbook	5-Minute Check Transparencies	Key Teaching Transparencies	MathPASS: Tutorial Plus (lessons)	Teaching Mathematics with Manipulatives	Optional Materials
185	186–187	188	189			4-1	4-1		36		
190	191–192	193	194	235	CS 25	4-2	4-2	13		51	algebra tiles
195	196–197	198	199			4-3	4-3	14		52	algebra tiles
200	201–202	203	204	235, 237	RC 18	4-4	4-4	15	39	53–54	
205	206–207	208	209		CS 26	4-5	4-5	16	40		
210	211–212	213	214	236	CS 27 CM 11–14	4-6	4-6		41	55, 56–57	stopwatch, uncooked spaghetti
215	216–217	218	219	221–234, 236, 238–242	CS 28	4-7	4-7				

*** Key to Abbreviations:** *CS = Calculator and Spreadsheet Masters, RC = Real-Life Career Masters, CM = Classroom Math Games (online), SM = Science and Math Lab Manual (online)*

INTERACTIVE CHALKBOARD

PowerPoint® Presentations

This CD-ROM is a customizable Microsoft® PowerPoint® presentation that includes:

- In-Class Examples
- Links to Online Study Tools
- 5-Minute Check Transparencies
- Bellringers

Math Skills Maintenance

Use the Math Skills Maintenance Masters to keep your students' basic math skills fresh. Students can review skills one at a time or in a combined format. The Masters are also available online at msmath2.net/maintenance.

CHAPTER 4

Mathematical Content and Teaching Strategies

Continuity of Instruction

Building on Prior Knowledge

In Chapter 1, students learned the properties of equality. In Chapter 3, students performed operations with integers, including addition, subtraction, multiplication, and division. Students intuitively translated verbal expressions to mathematical expressions each time they solved a simple word problem. Students have interpreted graphs and charts in previous courses and in real life.

Learning in This Chapter

Students will discover the relationship between mathematical expressions and verbal expressions. They will apply their knowledge of basic operations with integers to expressions and equations that include variables. They will use the properties of equality to solve one- and two-step equations and inequalities. Students will investigate graphs and equations of functions and analyze the shapes of graphs to interpret what is happening. Finally, they will learn to find the slope of a line.

Making Future Connections

Writing mathematical expressions and equations for verbal expressions and sentences shows students how math is used in everyday life. Using algebraic properties is essential to solving equations and formulas throughout all mathematics from algebra on. The concepts of function and slope are fundamental to mathematical study, especially in the study of advanced algebra and calculus.

Mathematical Background

4-1 Writing Expressions and Equations

Expressions and equations containing variables are the basis of algebra. Algebraic expressions consist of one or more numbers and variables along with one or more arithmetic operations. In this lesson, students learn to interpret verbal expressions and sentences to write algebraic expressions and equations.

4-2 Solving Addition and Subtraction Equations

In this lesson, students apply their knowledge of the Addition and Subtraction Properties of Equality to solving addition and subtraction equations. An addition equation is solved using the Subtraction Property of Equality. A subtraction equation is solved using the Addition Property of Equality. These concepts can be modeled using cups and counters.

Teacher to Teacher

Harvey Almarode
Mathematics Supervisor
Augusta County Schools
Fisherville, VA

"I use a shoebox to help students move from the concrete to the abstract when solving equations. One side of the box is drawn to represent a balance scale and the other shows an equals sign. Use manipulatives to show the equation on a balance and as an equation."

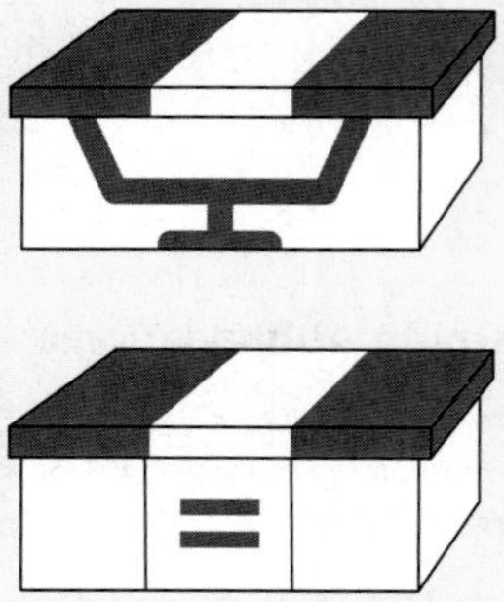

4-3 Solving Multiplication Equations

A multiplication equation involves a variable multiplied by a constant. For example, $4y = 24$ is a multiplication equation. To solve a multiplication equation, students apply the Division Subtraction Property of Equality. Students will study division equations after operations with fractions in Lesson 6-5.

4-4 Solving Two-Step Equations

Students apply the work backward problem-solving strategy in this lesson. A two-step equation contains two operations. In the equation $4x + 1 = 9$, x is multiplied by 4 and then 1 is added. In this lesson, students learn to solve two-step equations by using inverse operations to undo each operation in reverse order. The equation $4x + 1 = 9$ can be solved in two steps. First, subtract 1 from each side. Second, divide each side by 4.

4-5 Inequalities

Just as an equation shows two equal mathematical expressions on opposite sides of an equals sign, an inequality shows two unequal mathematical expressions on opposite sides of the following symbols: $<$ (less than), $\leq$ (less than or equal to), $>$ (greater than), or $\geq$ (greater than or equal to). An inequality compares numbers or quantities. In this lesson, students apply what they have learned in the previous four lessons to writing and solving inequalities.

McGraw-Hill Professional Development

Mathematics, YES! is a self-paced, web-based professional development program. Units are based on major mathematical concepts and are standards-based. Actual classroom videos provide a springboard for discussion and activities. Units appropriate for this chapter:

- Unit 2, Patterns and Functions
- Unit 5, Coordinate Pairs and Graphing
- Unit 7, Solving Equations and Inequalities

4-6 Functions and Linear Equations

A function is a relation that matches one quantity with exactly one other quantity. A function matches one member of the domain with exactly one member of the range. Functions can be represented using a function table, a graph, ordered pairs, and an equation. The equations that they have studied in this chapter are *linear*, meaning that their graphs are lines.

Building a Word Wall

The term *function* may not be clear to all students. Develop a word wall to clarify the concept. Have students volunteer uses of the word *function* outside of mathematics. Record the real-life and mathematical meanings in the word wall.

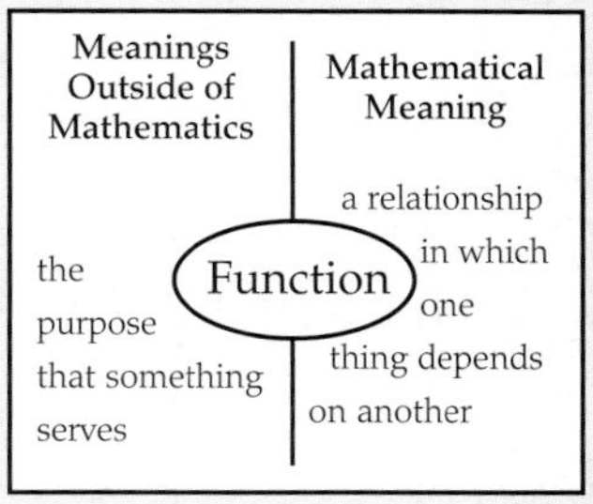

4-7 Lines and Slope

In this lesson, students learn to calculate the slope or incline of a straight line. The formal definition of slope is $\frac{y_1 - y_2}{x_1 - x_2}$. They learn to use slope to graph a line easily. In future courses, particularly calculus, students will apply this basic knowledge of slope to the use of derivatives to find rates of change and the slope of a curve. These applications are useful in economics and physics in real life.

Key Concepts

Additional mathematical information and teaching notes are available at

msmath2.com/key_concepts.

The sections appropriate for the chapter are:

- Solving One- and Two-Step Equations
- Inequalities
- Functions and Graphs

CHAPTER 4

DAILY INTERVENTION and Assessment

	Type	Student Edition	Teacher Resources	Technology/Internet
INTERVENTION	Ongoing	Prerequisite Skills, pp. 149, 152, 159, 163, 169, 175, 181 Mid-Chapter Practice Test, p. 170	5-Minute Check Transparencies Quizzes 1–4, *CRM*, pp. 235–236 Mid-Chapter Test, *CRM*, p. 237 Study Guide and Intervention, *CRM*, pp. 185, 190, 195, 200, 205, 210, 215	MathPASS: Tutorial Plus msmath2.net/self_check_quiz msmath2.net/extra_examples
	Mixed Review	pp. 152, 159, 163, 169, 175, 181, 185	Cumulative Review, *CRM*, p. 238	
	Error Analysis	Find the Error, pp. 151, 162, 168, 174	Find the Error, *TWE*, pp. 151, 162, 168, 174 Unlocking Misconceptions, *TWE*, pp. 164, 173	
ASSESSMENT	Standardized Test Practice	pp. 152, 159, 163, 165, 169, 170, 175, 181, 183, 185, 189, 190–191	Standardized Test Practice, *CRM*, pp. 239–240	Standardized Test Practice CD-ROM msmath2.net/standardized_test
	Open-Ended Assessment	Writing Math, pp. 151, 155, 158, 162, 168, 174, 176, 179, 184 Open-Ended, pp. 151, 158, 162, 168, 174, 179, 184 Standardized Test, p. 191	Modeling: *TWE*, p. 169 Speaking: *TWE*, pp. 163, 181 Writing: *TWE*, pp. 153, 159, 165 Extended Response Assessment, *CRM*, p. 233	
	Chapter Assessment	Study Guide, pp. 186–188 Practice Test, p. 189	Multiple-Choice Tests (Forms 1, 2A, 2B), *CRM*, pp. 221–226 Free-Response Tests (Forms 2C, 2D, 3), *CRM*, pp. 227–232 Vocabulary Test/Review, *CRM*, p. 234	ExamView® Pro Test Bank (see below) MindJogger Videoquizzes msmath2.net/vocabulary_review msmath2.net/chapter_test

* **Key to Abbreviations:** *TWE* = Teacher Wraparound Edition, *CRM* = Chapter Resource Masters

Additional Intervention

Skills Intervention for Middle School Mathematics workbook and CD-ROM includes 88 skills with pretests, tutorials, guided practice, and posttests for diagnosis and remediation.

Guide to Daily Intervention booklet details each intervention resource available and guides teachers in using them as intervention tools everyday to assist students who may be having difficulty.

ExamView® Pro

Use ExamView® Pro Test Bank CD-ROM to:

- **Create Multiple versions of tests.**
- **Create Modified tests for Inclusion students with one mouse click.**
- **Edit existing questions and Add your own questions.**
- **Build tests aligned with state standards using built-in State Curriculum Correlations.**
- **Change English tests to Spanish with one mouse click and vice versa.**

Reading and Writing Mathematics

• Student Edition

Foldables Study Organizer, p. 149

Study Skill: How to Read Math Problems, p. 153

Concept Check questions at the beginning of each exercise set require students to verbalize and write about what they have learned in the lesson.

- **Open Ended,** pp. 151, 158, 162, 168, 174, 179, 184
- **Find the Error,** pp. 151, 162, 168, 174
- **Which One Doesn't Belong?,** p. 158

WebQuest, p. 185

• Teacher Wraparound Edition

Foldables Follow-Up, pp. 151, 158, 162, 168, 170, 184

Writing Activities, pp. 153, 159, 165

ELL Resources, *see page T9 for a complete listing.*

• Additional Resources

Vocabulary Builder, *Chapter 4 Resource Masters*

Reading to Learn Mathematics, *Chapter 4 Resource Masters*

Vocabulary PuzzleMaker (online)

Teaching Mathematics with Foldables

Reading and Writing in the Mathematics Classroom

WebQuests, Projects and Interdisciplinary Investigations

For detailed information on these resources, see pp. T6–T7.

• Recommended Outside Reading for Students

Nonfiction

- Gillings, Richard. *Mathematics in the Time of the Pharaohs.* Dover, ©1992.

  This book examines the development of Egyptian mathematics from its origins in practical and commercial computations to its applications involving proportions and linear equations. Students will learn to write, solve, and graph linear equations such as those described in Chapter 4.

Fiction

- Merrill, Jean. *The Toothpaste Millionaire.* Houghton Mifflin, ©1974.

  The Toothpaste Millionaire is a funny story of how two friends solve real-world mathematics problems involved in starting a toothpaste business and making a profit. They use functions, which students will study in Lessons 4-6 and 4-7.

PROJECT CRISS℠ Study Skill

A story frame can help students organize information contained in word problems. They can use story frames either to solve word problems or to help them develop word problems of their own. The sample story frame at the right describes the following word problem: *A car travels 74 miles less than a truck travels in one day. If the truck travels 252 miles, how far has the car traveled?* Have students use story frames to write equation and inequality word problems as they work through Chapter 4.

Things involved: truck, car

Setting: truck travels 252 miles; car travels 74 miles less

Goal: distance traveled by car

Steps:
1. Write an equation relating the distance traveled by the car to the distance traveled by the truck. ($t - 74 = c$)
2. Insert t, the distance traveled by the truck, 252 miles. ($252 - 74 = c$)
3. Solve for c, the distance traveled by the car.

Solution: $c = 178$ miles

Check your solution. (Is it reasonable? Recheck calculations.)

CReating Independence through Student-owned Strategies

For more information on Project CRISS, see page T9.

Chapter 4 Making Connections

School-To-Home

Parent and Student Study Guide

Parents and students may work together to reinforce the concepts and skills of this chapter. These workbooks are available in English and Spanish. A sample lesson is shown below. Also available at msmath2.net/parent_student.

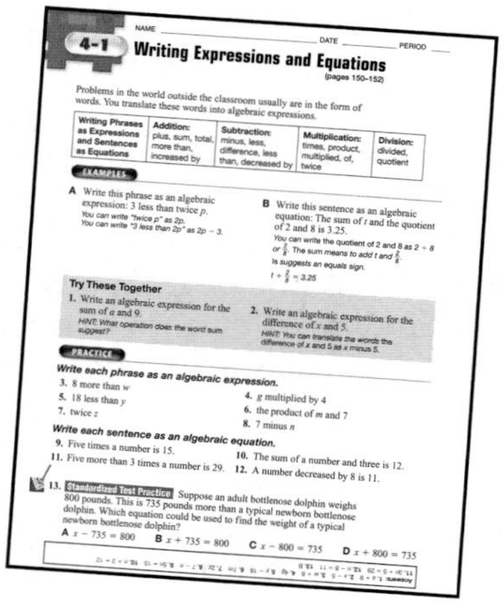

Family Letters and Activities

Send a letter and activity for this chapter home to encourage family members to become active participants in their students' learning. (English: *Chapter 4 Resource Masters*, pp. ix–x; Spanish: log on to msmath2.net/family_letters)

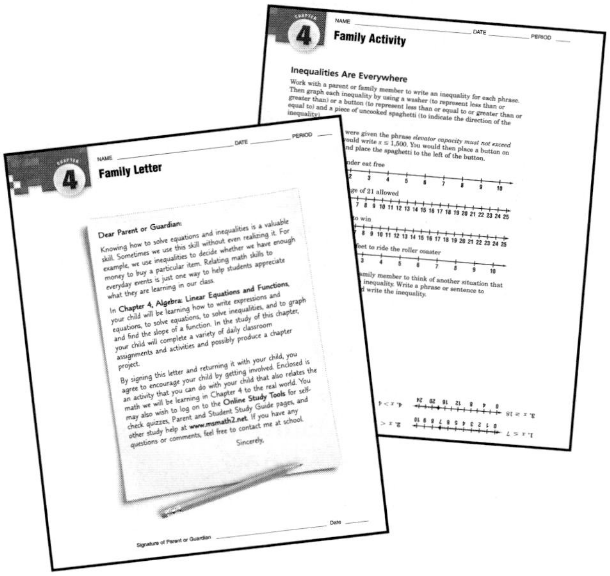

English Language Learner ELL

Lesson 4-1

Language Experience Approach to Illustrations

Write the words *addition, subtraction, multiplication,* and *division* on the chalkboard. Have students come up and write words that signify each operation. Allow for and correct spelling and grammatical errors. Help students with writing mathematical sentences from word sentences by writing one on the chalkboard. Underline phrase by phrase to highlight and translate the text until they have formed the expression or equation. Have the students then practice with a partner for peer tutoring.

Lesson 4-2

Using Authentic Assessment

Once you have modeled how to solve addition and subtraction equations for the students, have them explain the process for each step and why they must perform the operation on both sides of the equation.

Lesson 4-4

Using Modeling

Model solving a two-step equation. For example, to solve $2x + 3 = 11$, first cover the coefficient to show the students the equation $x + 3 = 11$. Ask students how they would solve this equation. Subtract 3 from each side. Now the equation is $2x = 8$. Ask students to solve this equation. Then solve for x. This method allows students to see that two steps are involved and that the addition or subtraction occurs first followed by the multiplication or division. You may wish to do several examples the same way until students see a pattern and make a conjecture about how to solve two-step equations.

MathScape™

This project was supported, in part, by the **National Science Foundation**

MathScape is a mathematics curriculum for grades 6–8 developed by the Seeing and Thinking Mathematically Project at the Education Development Center.

Equations, Tables, and Graphs

The Language of Algebra

Summary Writing expressions and equations will be explored using variables and constants. Students will make a table of values to describe a graphical situation. Slope is investigated using horizontal and vertical lines. Solutions are found for one- and two-step equations and inequalities.

USA TODAY Education — Online Lesson Planner

USA TODAY® Education's Online site offers resources and interactive features connected to each day's newspaper. *Experience Today,* USA TODAY's® daily lesson plan, is available on the site and delivered daily to subscribers. This plan provides instruction for integrating USA TODAY® graphics and key editorial features into your mathematics classroom. Log on to **education.usatoday.com**.

USA TODAY Snapshots® in Chapter 1	
Lesson	**Snapshot**
4-5	Where children get hurt
4-6	Health research spending small

Quick Review Math Handbook

is Glencoe's mathematical handbook for students. The Hot Words section includes a glossary of terms while the Hot Topics section consists of explanations of key mathematical concepts. An exercise set is included to check students' understanding of the concepts. This valuable resource can be used as a reference in the classroom or for home study.

Lesson	Hot Topics Section	Lesson	Hot Topics Section
4-1	6-1	**4-5**	6-6
4-2	6-1, 6-4	**4-6**	6-7, 6-8
4-3, 4-4	6-4	**4-7**	6-7, 6-8

Chapter 4 Notes

Key Chapter Objectives

Lesson 4-1 Write expressions and equations.

Lessons 4-2 through 4-4 Solve equations using the Properties of Equality.

Lesson 4-5 Solve inequalities.

Lessons 4-6 and 4-7 Represent functions with words, equations, and graphs.

"What do horses have to do with math?"

Point out to students that this is only one of many applications of the math that they are studying. Others are provided in the introduction to each lesson.

Vocabulary Builder, pp. vii–viii ELL

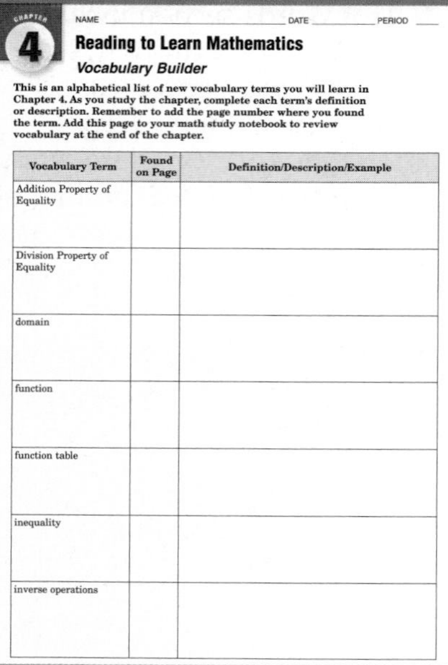

NAME ______ DATE ______ PERIOD ____

Chapter 4

Reading to Learn Mathematics

Vocabulary Builder

This is an alphabetical list of new vocabulary terms you will learn in Chapter 4. As you study the chapter, complete each term's definition or description. Remember to add the page number where you found the term. Add this page to your math study notebook to review vocabulary at the end of the chapter.

Vocabulary Term	Found on Page	Definition/Description/Example
Addition Property of Equality		
Division Property of Equality		
domain		
function		
function table		
inequality		
inverse operations		

Chapter 4 Algebra: Linear Equations and Functions

"What do horses have to do with math?"

A horse can gallop at a speed of 43 miles per hour. You can use the linear equation $d = 43t$ to find the **distance** d a horse gallops in a certain **time** t. In algebra, you will use variables and equations to describe many real-life situations.

You will solve problems about distance, rate, and time in Lesson 4-3.

Vocabulary Builder

For a thorough vocabulary list with pronunciations of new words, give students the Vocabulary Builder worksheets from the *Chapter 4 Resource Masters*. Encourage them to complete the definition of each term as they progress through the chapter.

GETTING STARTED

Take this quiz to see whether you are ready to begin Chapter 4. Refer to the lesson or page number in parentheses if you need more review.

Vocabulary Review

Complete each sentence. **1. solved**

1. When you replace the variable with a number that makes an equation true, you have __?__ the equation. (Lesson 1-5)
2. The first number in an ordered pair is the __?__, and the second number is the __?__. (Lesson 3-3) ***x*-coordinate; *y*-coordinate**

Prerequisite Skills

Name the number that is the solution of the given equation. (Lesson 1-5)

3. $a + 15 = 19$; 4, 5, 6 **4**
4. $11a = 77$; 6, 7, 8 **7**
5. $x + 9 = -2$; 7, −11, 11 **−11**
6. $2x - 8 = 10$; 1, 9, 18 **9**

Graph each point on a coordinate plane. (Lesson 3-3) **7–8. See margin.**

7. $(-4, 3)$
8. $(-2, -1)$

Add. (Lesson 3-4)

9. $-3 + (-5)$ **−8**
10. $-8 + 3$ **−5**
11. $9 + (-5)$ **4**
12. $-10 + 15$ **5**

Subtract. (Lesson 3-5)

13. $-5 - 6$ **−11**
14. $8 - 10$ **−2**
15. $8 - (-6)$ **14**
16. $-3 - (-1)$ **−2**

Divide. (Lesson 3-7)

17. $-6 \div (-3)$ **2**
18. $-12 \div 3$ **−4**
19. $10 \div (-5)$ **−2**
20. $-24 \div (-4)$ **6**

FOLDABLES™ Study Organizer

Solving Equations Make this Foldable to help you organize your notes. Begin with a sheet of $8\frac{1}{2}'' \times 11''$ paper.

STEP 1 Fold Fold the short sides toward the middle.

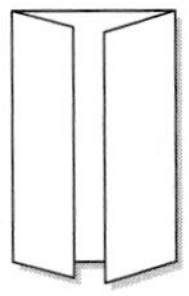

STEP 2 Fold Again Fold the top to the bottom.

STEP 3 Cut Open. Cut along the second fold to make four tabs.

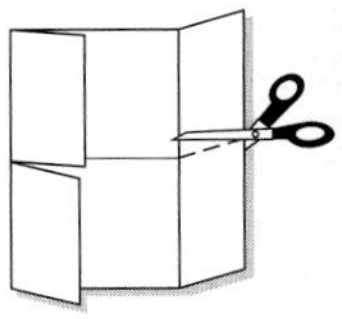

STEP 4 Label Label each of the tabs as shown.

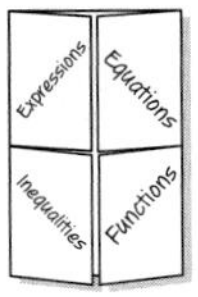

Reading and Writing As you read and study the chapter, write examples under each tab.

GETTING STARTED

This section provides a review of the basic concepts needed before beginning Chapter 4. Page references are included for additional student help.

Prerequisite Skills review skills needed in this chapter.

Exercises	Needed for Lesson	Skill
3–6	4-1	Solve Open Sentences
7–8	4-6	Graph Points on a Coordinate Plane
9–12	4-2	Add Integers
13–16	4-2	Subtract Integers
17–20	4-3	Divide Integers

For more prerequisite skills, see:

- Getting Ready for the Next Lesson, at the end of each lesson
- Prerequisite Skills in the Student Handbook on pp. 554–563
- *Math Skills Maintenance Masters*

Answers

7.

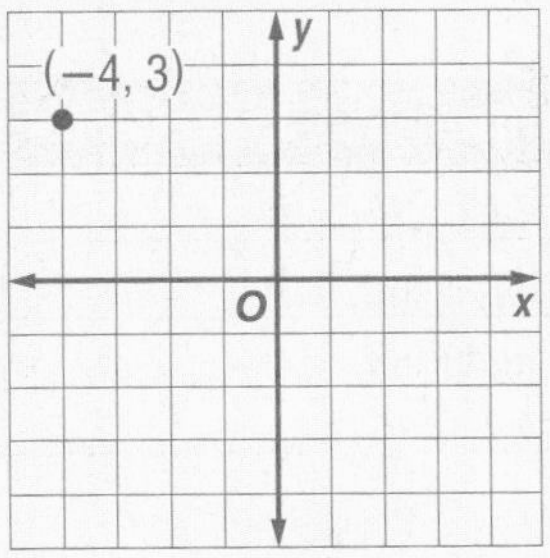

8.

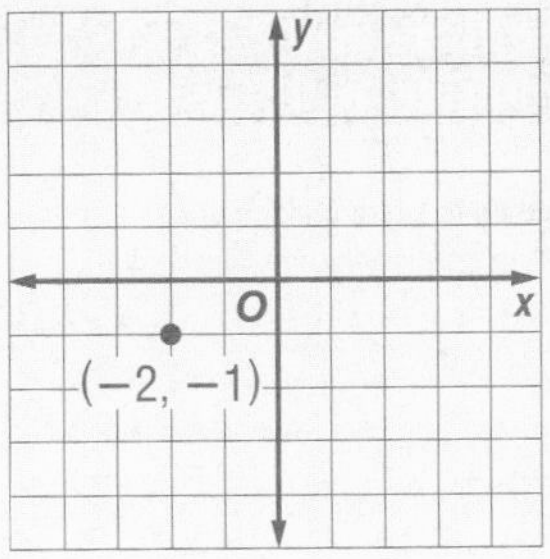

FOLDABLES™ Study Organizer

For more information about Foldables, see *Teaching Mathematics with Foldables.*

Use this Foldable study guide for student writing about linear equations and functions. Have students make and label the Foldable as illustrated. Review the fact that note taking is a skill based upon listening or reading for main ideas and then recording those ideas in a simplified form for future reference. Under the tabs of their Foldables, have students take notes about what they learn about expressions, equations, inequalities, and functions. Encourage students to apply what they have learned by writing examples of each.

4-1 Lesson Notes

Lesson Resources

Chapter 4 Resource Masters

- Study Guide and Intervention, p. 185
- Practice: Skills, p. 186
- Practice: Word Problems, p. 187
- Reading to Learn Mathematics, p. 188
- Enrichment, p. 189

MathPASS Tutorial CD-ROM, 36

Parent and Student Study Guide Workbook, 4-1

FAST FILE

1 Focus

5-Minute Check 4-1

INTERACTIVE CHALKBOARD
PowerPoint® Presentations

Use as a quiz or review of Chapter 3. (Also available as a transparency)

Bellringer

Communication Write the following phrases and expressions on the chalkboard. When students enter the classroom, have them write each and match the phrases with expressions they believe represent them. **1. c 2. a 3. b**

Phrases	Expressions
1. seven plus two	**a.** 7 – 2
2. seven less two	**b.** 7 · 2
3. the product of seven and two	**c.** 7 + 2

Reading to Learn Mathematics, p. 188 **ELL**

NAME ______________ DATE ________ PERIOD ____

4-1 Reading to Learn Mathematics
Writing Expressions and Equations

Pre-Activity *Read the introduction at the top of page 150 in your textbook. Write your answers below.*

1. Suppose the daughter is 12 years old. How old is the son? 15 yr
2. What operation did you use to find the son's age? Explain. addition; the daughter's age plus three years
3. Suppose the comic said that *the son is twice as old as the daughter*. If the daughter is 12 years old, how old is the son? 24 yr
4. What operation did you use to find the son's age? Explain. Multiplication; "twice" means to multiply by 2.

Reading the Lesson

5. Write the symbol that each word or phrase represents.

Symbol	Words

4-1 Writing Expressions and Equations

What You'll Learn
Write verbal phrases and sentences as simple algebraic expressions and equations.

REVIEW Vocabulary
expression: a combination of variables, numbers, and at least one operation (Lesson 1-4)

WHEN am I ever going to use this?

COMICS Even these characters from *Peanuts* are using algebra.

1. Suppose the daughter is 12 years old. How old is the son? **15 yr**
2. What operation did you use to find the son's age? Explain.
3. Suppose the comic said that *the son is twice as old as the daughter.* If the daughter is 12 years old, how old is the son? **24 yr**
4. What operation did you use to find the son's age? Explain.

2. addition; the daughter's age plus three years
4. multiplication; "twice" means to multiply by 2

Words and phrases in problems often suggest addition, subtraction, multiplication, and division. Here are some examples.

Addition and Subtraction		Multiplication and Division	
plus	minus	times	divided
sum	difference	product	quotient
more than	less than	multiplied	per
increased by	less	twice	separate
in all	decreased by	each	

EXAMPLE Write a Phrase as an Expression

1 **Write the phrase *five dollars less than Jennifer earned* as an algebraic expression.**

Words	five dollars less than Jennifer earned
Variable	Let d represent the number of dollars Jennifer earned.
Expression	$d - 5$

TEACHING TIP
Point out that the expression for *5 less than a number* is $n - 5$. But *5 less a number* is $5 - n$.

Your Turn **Write each phrase as an algebraic expression.**

a. twice as many tomatoes as last year **$2t$**

b. 3 more runs than the Pirates scored **$r + 3$**

DAILY INTERVENTION

Flexible Grouping

Writing Expressions and Equations Have students share what they have read after they read each paragraph of the lesson, or after they complete the entire lesson, depending on the overall ability level of the class.

Remember, an equation is a sentence in mathematics that contains an equals sign. When you write a verbal sentence as an equation, you can use the equals sign (=) for the words *equals* or *is*.

REAL-LIFE MATH

FOOD In addition to potato chips and tortilla chips, other snacks, in pounds, consumed during the 2002 Super Bowl were:
Popcorn: 4.2 million
Pretzels: 3.8 million
Nuts: 3.1 million
Source: Snack Food Association

EXAMPLES Write Sentences as Equations

Write each sentence as an algebraic equation.

Sentence	Equation
2 Five more than a number is 20.	$n + 5 = 20$
3 Three times Jack's age equals 12.	$3a = 12$

4 **FOOD It is estimated that 12.4 million pounds of potato chips were consumed during the 2002 Super Bowl. This was 3.1 million pounds more than the number of pounds of tortilla chips consumed. Write an equation that models this situation.**

Words	Potato chips were 3.1 million more than tortilla chips.
Variable	Let t = number of millions of pounds of tortilla chips.
Equation	$12.4 = 3.1 + t$

The equation is $12.4 = 3.1 + t$.

Skill and Concept Check

Writing Math
Exercise 2

1. **OPEN ENDED** Write a verbal sentence that translates into the equation $n + 5 = 8$. **Sample answer: 5 dollars more than my allowance is 8 dollars.**
2. **FIND THE ERROR** Antonio and Julia are writing an algebraic expression for the phrase *2 less than a number.* Who is correct? Explain. **Julia; "two less than a number" indicates that 2 is subtracted from the variable.**

Antonio	Julia
$2 - n$	$n - 2$

GUIDED PRACTICE

Write each phrase as an algebraic expression.

3. eight more than x $x + 8$
4. nine less than t $t - 9$
5. ten times as many hours $10h$
6. -15 divided by some number $\frac{-15}{x}$

Write each sentence as an algebraic equation.

7. The sum of a number and four is -9. $n + 4 = -9$
8. Twice a number equals -10. $2y = -10$
9. **POPULATION** The median age of people living in the United States was 35.3 years in 2000. This is 2.4 years older than the median age in 1990. Write an equation that models this situation. $a + 2.4 = 35.3$

msmath2.net/extra_examples

FOLDABLES™ Follow-Up

Remind students to use their Foldables to take notes on what they learn about writing expressions and equations. Students might write phrases that also can be written as algebraic expressions under the "Expressions" tab and phrases that can be written as equations under the "Equations" tab of their Foldables.

2 Teach

In-Class Examples

INTERACTIVE CHALKBOARD
PowerPoint® Presentations

1 Write the phrase *twenty dollars less the price of a movie ticket* as an algebraic expression. $20 - m$

Write each sentence as an algebraic equation.

2 A number less 4 is 12.
$n - 4 = 12$

3 Twice a number is 18. $2n = 18$

4 **FOOD** An average American adult drinks more soft drinks than any other beverage each year. Three times the number of gallons of soft drinks plus 27 is equal to the total 183 gallons of beverages consumed. Write an equation that models this situation.
$3s + 27 = 183$

3 Practice/Apply

DAILY INTERVENTION **FIND THE ERROR** In Exercise 2, tell students that numbers and variables in a verbal phrase are not always given in the same order that they should be included in an algebraic expression. You must interpret the phrase to determine the correct expression.

About the Exercises...

Guided Practice Key

Exercises	Examples
3–6	1
7–8	2, 3
9	4

Odd/Even Assignments
Exercises 10–25 are structured so that students practice the same concepts whether they are assigned odd or even problems.

Study Guide and Intervention, p. 185

NAME ______________ DATE ________ PERIOD ____

4-1 Study Guide and Intervention

Writing Expressions and Equations

The table below shows phrases written as mathematical expressions.

Phrases	Expression	Phrases	Expression
9 more than a number the sum of 9 and a number a number plus 9 a number increased by 9 the total of x and 9	$x + 9$	4 subtracted from a number a number minus 4 4 less than a number a number decreased by 4 the difference of h and 4	$h - 4$
Phrases	**Expression**	**Phrases**	**Expression**
6 multiplied by g 6 times a number the product of g and 6	$6g$	a number divided by 5 the quotient of t and 5 divide a number by 5	$\frac{t}{5}$

The table below shows sentences written as an equation.

Sentences	Equation
Sixty less than three times the amount is \$59. Three times the amount less 60 is equal to 59. 59 is equal to 60 subtracted from three times a number. A number times three minus 60 equals 59.	$3n - 60 = 59$

EXERCISES

Write each phrase as an algebraic expression.

1. 7 less than m $m - 7$
2. the quotient of 3 and y $\frac{3}{y}$
3. the total of 5 and c $5 + c$
4. the difference of 6 and r $6 - r$
5. n divided by 2 $\frac{n}{2}$
6. the product of k and 9 $9k$

Write each sentence as an algebraic equation.

7. A number increased by 7 is 11. $n + 7 = 11$
8. The price decreased by \$4 is \$29. $p - 4 = 29$
9. Twice as many points as Bob would be 18 points. $2b = 18$
10. After dividing the money 5 ways, each person got \$67. $\frac{m}{5} = 67$
11. Three more than 8 times as many trees is 75 trees. $8t + 3 = 75$
12. Seven less than a number is 15. $n - 7 = 15$

Practice: Skills, p. 186

NAME ______________ DATE ________ PERIOD ____

4-1 Practice: Skills

Writing Expressions and Equations

Write each phrase as an algebraic expression.

1. b plus 1 $b + 1$
2. three more than x $x + 3$
3. twelve minus y $12 - y$
4. seven less than n $n - 7$
5. five years younger than Jessica $j - 5$
6. a number less eleven $n - 11$
7. four increased by a $4 + a$
8. eight dollars more than m $m + 8$
9. the product of c and 10 $10c$
10. twice as many days $2d$
11. three times as many soft drinks $3s$
12. t multiplied by 14 $14t$
13. Emily's age divided by 3 $\frac{e}{3}$
14. 24 divided by some number $\frac{24}{n}$
15. a number divided by 2 $\frac{n}{2}$
16. the quotient of -15 and w $\frac{-15}{w}$

Write each sentence as an algebraic equation.

17. A number plus three is 9. $n + 3 = 9$
18. The sum of x and 2 is 10. $x + 2 = 10$
19. Four cents more than the price is 93¢. $p + 4 = 93$
20. Fifteen minus y is 7. $15 - y = 7$
21. A number decreased by 5 is 12. $n - 5 = 12$
22. Five dollars less than Yumi's pay is \$124. $y - 5 = 124$
23. A number times four is 20. $4n = 20$
24. Twice the number of cars is 40. $2c = 40$
25. The product of z and 6 is 54. $6z = 54$
26. A number divided by 6 is 12. $\frac{n}{6} = 12$
27. 72 divided by y is -9. $\frac{72}{y} = -9$
28. 175 students separated into n classes is 25. $\frac{175}{n} = 25$
29. One more than twice as many CDs is 17. $2c + 1 = 17$
30. Four less than three times a number is 14. $3n - 4 = 14$

Practice: Word Problems, p. 187

NAME ______________ DATE ________ PERIOD ____

4-1 Practice: Word Problems

Writing Expressions and Equations

OLYMPICS **For Exercises 1–4, use the table that shows the number of medals won by each country in the 2002 Winter Olympics.**

2002 Winter Olympic Medals					
Country	**Medals**	**Country**	**Medals**	**Country**	**Medals**
Germany	35	Austria	16	China	8
USA	34	Italy	12	Finland	7
Norway	24	France	11	Sweden	6
Canada	17	Switzerland	11	Croatia	4
Russia	16	Netherlands	8	South Korea	4

Let x represent the number of medals won by Italy.

1. Write an expression using x to represent the number of medals won by Russia. $x + 4$
2. Write an expression using x to represent the number of medals won by Croatia. $x - 8$
3. Which country's number of medals can be represented by $2x$? Norway
4. Which country's number of medals can be represented by $2x + 10$? USA
5. GEOGRAPHY The Virgin Islands were acquired by the United States in 1927. This is 29 years after Puerto Rico was acquired. Write an equation to model this situation. $y + 29 = 1927$
6. POPULATION According to the Census Bureau, the U.S. population grew from 281.4 million in April 2000 to 284.8 million in July 2001. Write an equation to model this situation. $281.4 + x = 284.8$

★ indicates increased difficulty

Practice and Applications

HOMEWORK HELP

For Exercises	See Examples
10–17, 26–27	1
18–25, 28	2–4

Extra Practice See pages 572, 599.

Write each phrase as an algebraic expression.

10. fifteen increased by t $15 + t$
11. five years older than Luis $y + 5$
12. nine dollars less than j $j - 9$
13. a number less six $n - 6$
14. the product of r and 8 $8r$
15. twice as many oranges $2f$
16. Emily's age divided by 3 $\frac{a}{3}$
17. a number divided by -12 $\frac{z}{-12}$

Write each sentence as an algebraic equation. 18–25. See margin.

18. The sum of a number and four is -8.
19. Two more than the number of cookies is 4.
20. The product of a number and five is -20.
21. Ten times the number of students is 280.
22. Ten inches less than her height equals 26.
23. Five less than a number equals 31.
24. Seven more than twice his age is 51.
25. Three more than twice a number is 15.

MONEY For Exercises 26 and 27, use the table. The table shows the average lifespan of several kinds of paper currency in the United States. Let y represent the average lifespan of a \$5 bill.

U.S. Currency	
Kind	**Lifespan (years)**
\$1	1.5
\$5	2
\$10	3
\$20	4
\$50	9
\$100	9

Source: Federal Reserve System

★ 26. Which lifespan can be represented by $2y$? \$20 bill

★ 27. Write an expression to represent the lifespan of a \$50 bill. $y + 7$ or $4.5y$

28. **TOURISM** The Washington Monument is 555 feet tall. It is 75 feet shorter than the Gateway to the West Arch. Write an equation that models this situation. $g - 75 = 555$

29. **CRITICAL THINKING** If x is an odd number, how would you represent the odd number immediately following it? preceding it? $x + 2$; $x - 2$

Standardized Test Practice and Mixed Review

30. **MULTIPLE CHOICE** A mechanic charges a \$35 initial fee and \$32.50 for each hour he works. Which equation could be used to find the cost c of a repair job that lasts h hours? B

    Ⓐ $c = 32.5 + 35h$ Ⓑ $c = 35 + 32.5h$
    Ⓒ $c = 32.5 - 35h$ Ⓓ $c = 32.5(35 - h)$

31. **MULTIPLE CHOICE** Translate *12 more than d* into an algebraic expression. H

    Ⓕ $12d$ Ⓖ $d - 12$ Ⓗ $d + 12$ Ⓘ $12 - d$

Divide. (Lesson 3-7)

32. $-18 \div 3$ -6
33. $25 \div (-5)$ -5
34. $-14 \div (-7)$ 2
35. $72 \div (-9)$ -8

GETTING READY FOR THE NEXT LESSON

PREREQUISITE SKILL Add. (Lesson 3-4)

36. $-8 + (-3)$ -11
37. $-10 + 9$ -1
38. $12 + (-20)$ -8
39. $-15 + 15$ 0

msmath2.net/self_check_quiz

4 Assess

Higher-Level Thinking

Must $2x$ represent an even number? Explain your reasoning. No; $2x$ is even if x is an integer, but if x is a rational number such as $\frac{1}{2}$, then $2x$ is not even.

Answers

18. $n + 4 = -8$
19. $c + 2 = 4$
20. $5d = -20$
21. $10k = 280$
22. $h - 10 = 26$
23. $n - 5 = 31$
24. $2a + 7 = 51$
25. $2z + 3 = 15$

tudy Skill

HOW TO...

Read Math Problems

MPLIFY THE PROBLEM

ave you ever tried
solve a long word
oblem and didn't
now where to start?
y to rewrite the
oblem using fewer
nd fewer words.
hen translate the
oblem into an
quation.

►**READ** the problem.

> Shopping networks on television are a popular way to shop. In addition to the cost of the items, you usually pay a shipping fee. Kylie wants to order several pairs of running shorts that cost \$12 each. The total shipping fee is \$7. How many shorts can she order with \$55?

►**REWRITE** the problem to make it simpler. Keep all of the important information but use fewer words.

Kylie has \$55 to spend on some shorts that cost \$12 each plus a shipping fee of \$7. How many can she buy?

►**REWRITE** the problem using even fewer words. Write a variable for the unknown.

The total cost of x shorts at \$12 each plus \$7 is \$55.

►**TRANSLATE** the words into an equation.

$12x + 7 = 55$

SKILL PRACTICE

Use the method above to write an equation for each word problem.

1. **FLYING** Orville and Wilbur Wright flew their airplane called Flyer I in Kitty Hawk, North Carolina, on December 17, 1903. Wilbur's flight was 364 feet. This was 120 feet longer than Orville's flight. How far was Orville's flight? $f + 120 = 364$

2. **ANIMALS** The cougars that are found in the colder parts of North and South America are about 75 inches long. They are about 1.5 times longer than the cougars that are found in the tropical jungles of Central America. Find the length of the tropical cougar. $1.5c = 75$

3. **MONEY** Akira is saving money to buy a scooter that costs \$125. He has already saved \$80 and plans to save an additional \$5 each week. In how many weeks will he have enough money for the scooter? $80 + 5x = 125$

4. **TRAVEL** A taxi company charges \$1.50 per mile plus a \$10 fee. Suppose Olivia can afford to spend \$19 for a taxi ride from her apartment to the mall. How far can she travel by taxi with \$19? $10 + 1.50m = 19$

Objective Read, simplify, and rewrite word problems in order to solve them.

GETTING STARTED

Problem Solving There are many possible approaches to solving a problem. You can compare solving a word problem to a skydiver approaching jumping out of an airplane. For those who are unprepared, skydiving seems dangerous. You may jump out of the airplane and realize that you do not know how to open the parachute!

In problem solving, being unprepared may mean getting stuck or panicking in problems. However, for those who are thoroughly prepared, doing math problems is not at all scary.

TEACH

Problem solving suggestions to students:

- Read each problem slowly, twice. Make sure that you understand what is being asked.
- Identify what you are solving for. Note the key words and numbers that you will be working with. Assign variables and make an estimate.
- Determine your problem-solving strategy. Create your equation or translate what you are solving for.
- Read the problem to yourself or out loud. Talk yourself through the solution.

ASSESS

Open-Ended Assessment

Writing Have each student write a word problem. Ask students to trade their problems with a partner and use the *simplify the problem* method to write equations to solve them.

4-2a Hands-On Lab Lesson Notes

Lesson Resources

Teaching Mathematics with Manipulatives
- counters, p. 5
- pattern for cup, p. 7
- equation mat, p. 9
- student recording sheet 4-2a

Glencoe Mathematics Classroom Manipulative Kit
- cups and counters
- equation mat

GETTING STARTED

Teaching Tip Model **Activity 1** on an overhead projector with an equation mat and overhead cups and counters. Have students follow along using their cups and counters.

TEACH

Activity 1
- Point out that in **Activity 1**, the equation is solved by subtracting the same quantity from each side to isolate the variable on one side.

Answers (Your Turn)

a.
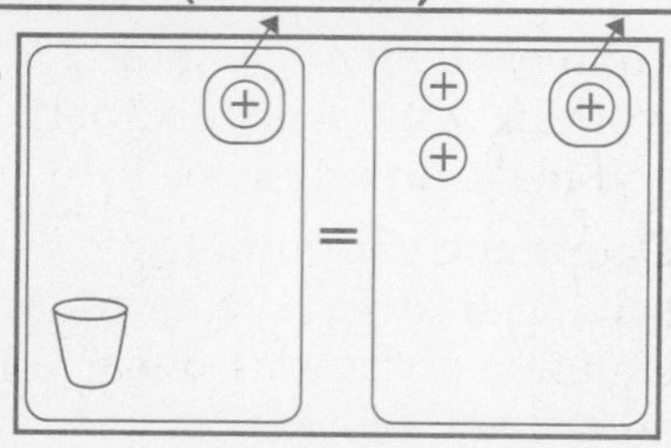

b.
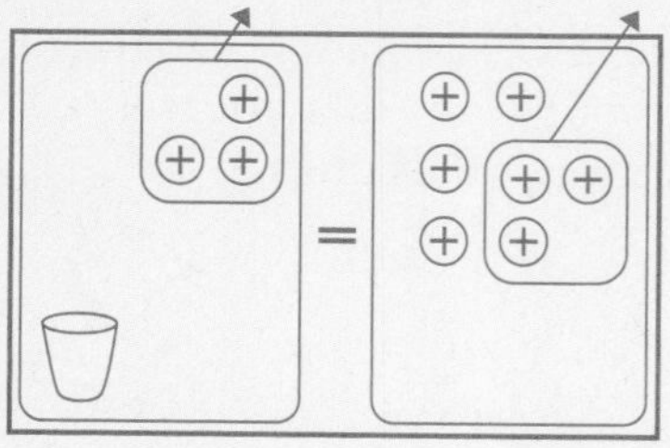

c.
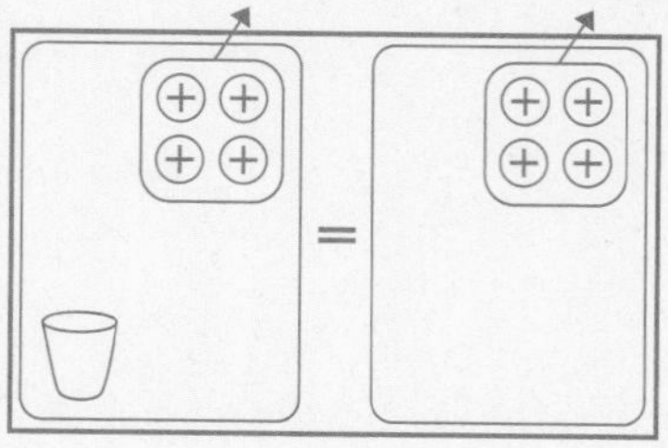

d.
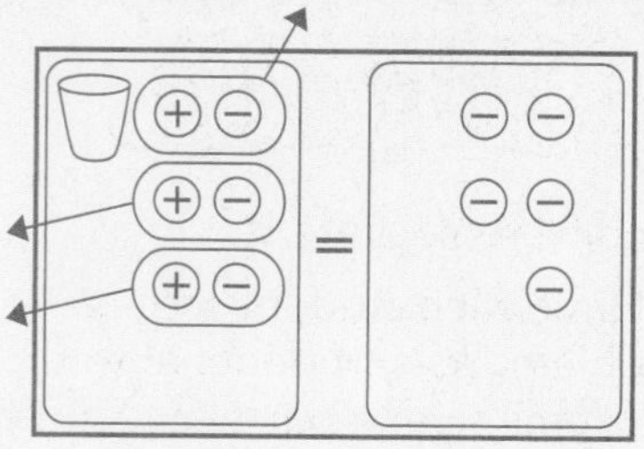

4-2a Hands-On Lab

A Preview of Lesson 4-2

What You'll Learn
Solve equations using models.

Materials
- cups and counters
- equation mat

***REVIEW* Vocabulary**

equation: a mathematical sentence that contains an equals sign **(Lesson 1-5)**

Solving Equations Using Models

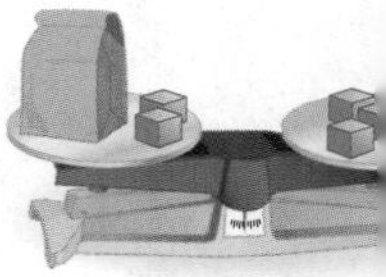

INVESTIGATE The scale at the right is balanced, and the bag contains a certain number of blocks.

1. Suppose you cannot look in the bag. How can you find the number of blocks in the bag?
2. In what way is a balanced scale like an equation? **Both sides are eq**
3. What does it mean to *solve an equation*? **Find a value for the variable that makes the equation a true statement.**

To solve an equation using models, you can use these steps.

- You can add or subtract the same number of counters from each side of the mat.
- You can add or subtract zero from each side of the mat.

1. Find the number, that when added to 2, equals the number of blocks on the other side.

ACTIVITY *Work with a partner.*

1 **Solve $x + 2 = 5$ using models.**

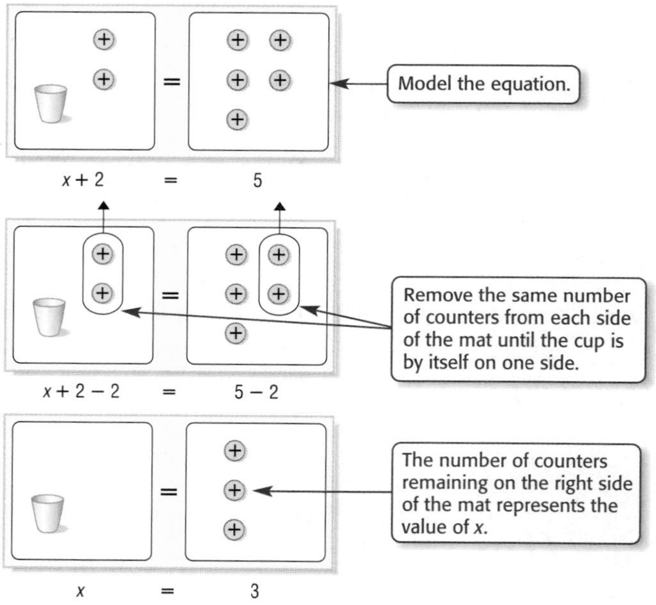

Therefore, $x = 3$. Since $3 + 2 = 5$, the solution is correct.

Your Turn **a–c. See margin for models.**
Solve each equation using models.

a. $x + 1 = 3$ **2** b. $x + 3 = 7$ **4** c. $x + 4 = 4$ **0**

STUDY TIP

Look Back You can review **zero pairs** in Lesson 3-4.

Some equations are solved by using **zero pairs**. You can add or subtract a zero pair from either side of an equation without changing its value, because the value of a zero pair is zero.

ACTIVITY *Work with a partner.*

2 **Solve $x + 2 = -1$ using models.**

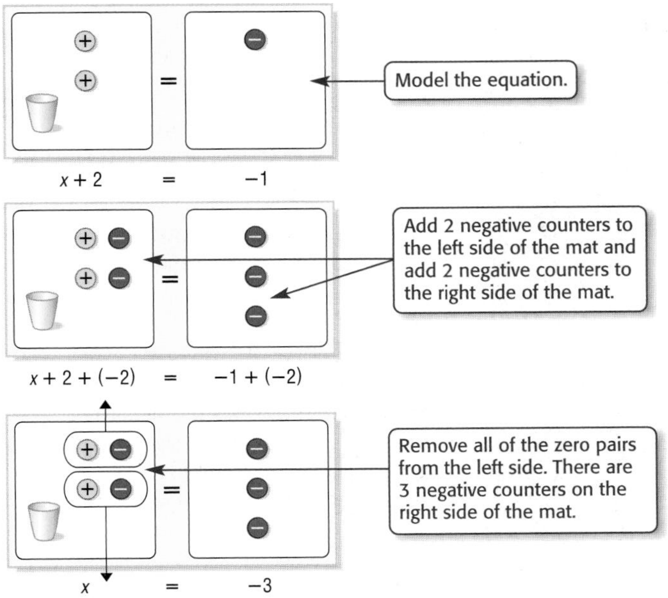

Therefore, $x = -3$. Since $-3 + 2 = -1$, the solution is correct.

Your Turn **Solve each equation using models.**

d. $x + 3 = -2$ **−5** e. $x + 4 = 1$ **−3** f. $-2 = x + 1$ **−3**

g. $x - 3 = -2$ **1** h. $x - 1 = -3$ **−2** i. $4 = x - 2$ **6**

d–i. See margin for models.

Writing Math

1. How is solving an equation similar to keeping a scale in balance?
2. For any equation, how can you determine how many counters to add or subtract from each side? **Add or subtract the number that is on the same side as the variable.**
3. **Identify** the property of numbers that is illustrated by a zero pair. **Additive Inverse**
4. **Identify** the property of numbers that allows you to add or subtract zero without changing the value of a number. **Additive Identity**
5. **MAKE A CONJECTURE** Write a rule that you can use to solve an equation like $x + 3 = 2$ without using models.

1. Sample answer: Whatever is done to one side of the equation must be done to the other side to keep both sides equal.

5. If a number is added to the variable, subtract that number from both sides of the equation. If a number is subtracted from the variable, add that number to both sides of the equation.

Activity 2

- In **Activity 2**, students solve an equation that cannot be solved by subtracting the same quantity from each side. They learn to add the appropriate amount to each side to create zero pairs that can be removed to isolate the variable.

ASSESS

Observe as students solve the **Your Turn** exercises. They should:

- correctly model the equations using cups and counters,
- manipulate cups and counters to solve the equations, and
- correctly use zero pairs when solving equations.

Discuss the **Writing Math** exercises. Use Exercise 5 to assess whether students have intuitively grasped how to solve addition and subtraction equations.

g.

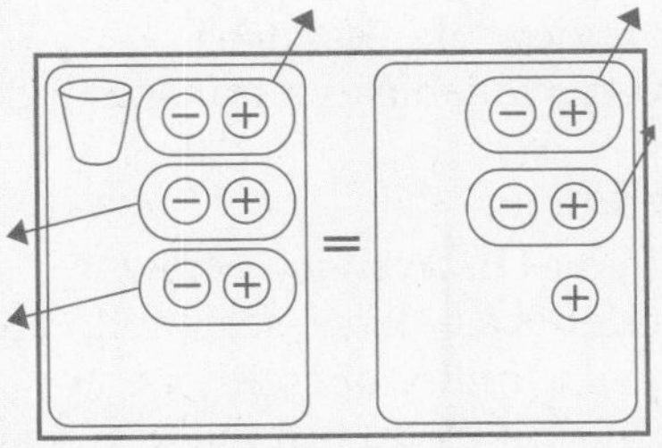

h.

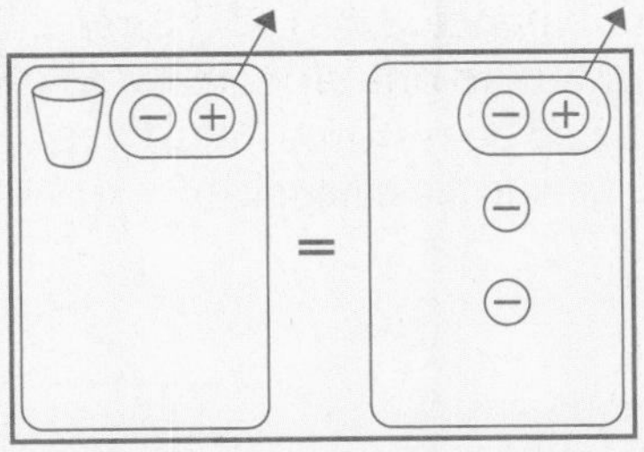

i.

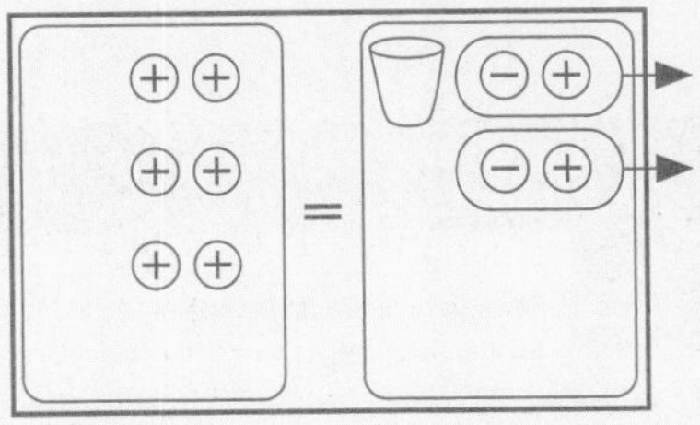

Answers (Your Turn)

e.

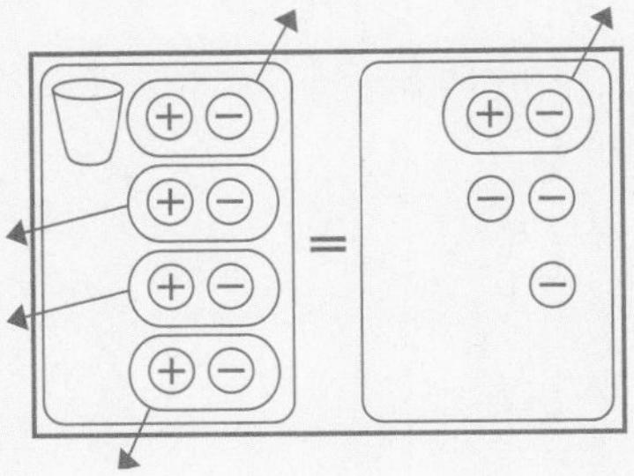

f.

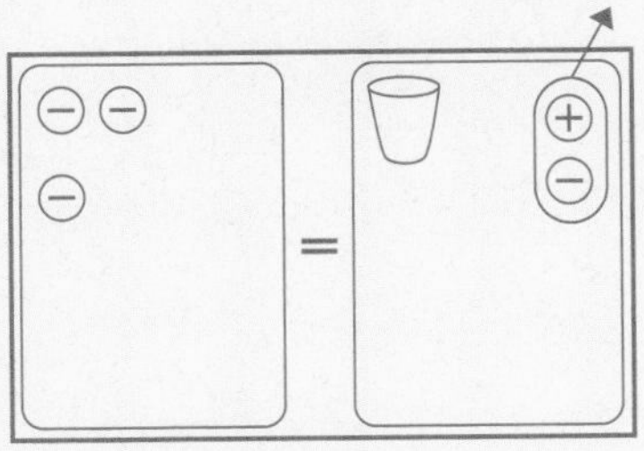

4-2 Lesson Notes

Lesson Resources

FAST FILE

Chapter 4 Resource Masters
- Study Guide and Intervention, p. 190
- Practice: Skills, p. 191
- Practice: Word Problems, p. 192
- Reading to Learn Mathematics, p. 193
- Enrichment, p. 194

Calculator and Spreadsheet Masters, p. 25

Key Teaching Transparencies, 13

Parent and Student Study Guide Workbook, 4-2

1 Focus

INTERACTIVE CHALKBOARD
PowerPoint® Presentations

5-Minute Check 4-2
Use as a quiz or review of Lesson 4-1. (Also available as a transparency)

Bellringer
Hands-On Activity Show students a balance scale with an equal number of counters on each side. Point out that the scale is in balance. Then add three counters to one side. How can balance be restored? **Add three counters to the other side.**

Reading to Learn Mathematics, p. 193 **ELL**

NAME ______________ DATE ________ PERIOD ____

4-2 Reading to Learn Mathematics
Solving Addition and Subtraction Equations

Pre-Activity ***Read the introduction at the top of page 156 in your textbook. Write your answers below.***
1. What does x represent in the figure?
   the length of a dolphin
2. Write an expression to represent *4 meters longer than a dolphin.*
   $x + 4$
3. Write an addition equation you could use to find the length of a dolphin.
   $x + 4 = 9$

Reading the Lesson

4. Match the method of solving with the appropriate equation.

$x + 5 = 9$ ___c___ a. add 2 to each side

$-2 + y = 1$ ___a___ b. add 5 to each side

d

4-2 Solving Addition and Subtraction Equations

What You'll Learn
Solve addition and subtraction equations.

NEW Vocabulary
inverse operations

REVIEW Vocabulary
solve: finding a number that makes the equation true **(Lesson 1-5)**

WHEN am I ever going to use this?

SCIENCE Bottle-nosed dolphins and killer whales are the best-known species of the dolphin family. A killer whale, which can grow to a length of 9 meters, is 4 meters longer than a bottle-nosed dolphin.

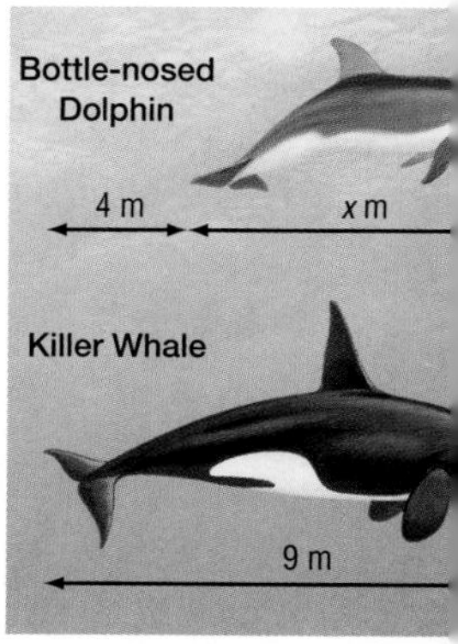

1. What does x represent in the figure?
2. Write an expression to represent *4 meters longer than a dolphin.* **$x + 4$**
3. Write an addition equation you could use to find the length of a dolphin. **$x + 4 = 9$**

1. the length of a dolphin

You can use inverse operations to solve the equation $x + 4 = 9$. **Inverse operations** "undo" each other. To undo the addition of 4 in the equation $x + 4 = 9$, you would subtract 4 from each side of the equation.

EXAMPLE Solve an Addition Equation

1 **Solve $x + 4 = 9$.**

Method 1 Use symbols.

$x + 4 = 9$ Write the equation.

$$\begin{array}{rcr} x + 4 &=& 9 \\ -4 &=& -4 \end{array}$$ Subtract 4 from each side.

$$\begin{array}{rcr} x + 4 &=& 9 \\ -4 &=& -4 \\ \hline x &=& 5 \end{array}$$ Simplify.

The solution is 5.

Method 2 Use models.

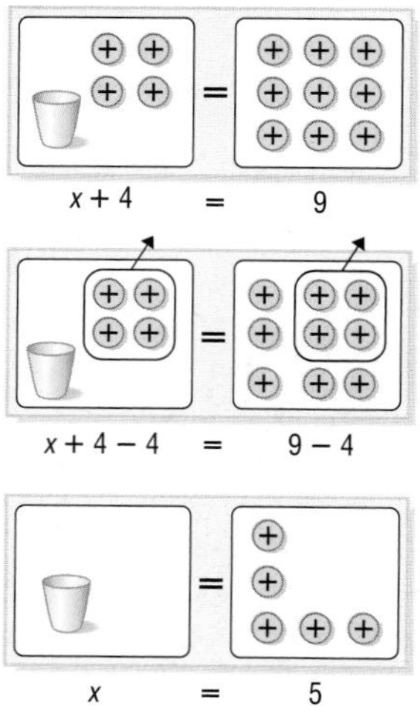

Your Turn **Solve each equation.**

a. $y + 6 = 9$ **3**
b. $x + 3 = 1$ **−2**
c. $-3 = a + 4$ **−7**

You can use the **Subtraction Property of Equality** and the **Addition Property of Equality** to solve equations like $x + 4 = 9$ and $x - 2 = 1$.

Key Concept — Subtraction Property of Equality

Words If you subtract the same number from each side of an equation, the two sides remain equal.

Symbols

Arithmetic	Algebra
$\begin{array}{rcr} 5 &=& 5 \\ -3 &=& -3 \\ \hline 2 &=& 2 \end{array}$	$\begin{array}{rcr} x + 2 &=& 3 \\ -2 &=& -2 \\ \hline x &=& 1 \end{array}$

Key Concept — Addition Property of Equality

Words If you add the same number to each side of an equation, the two sides remain equal.

Symbols

Arithmetic	Algebra
$\begin{array}{rcr} 5 &=& 5 \\ +3 &=& +3 \\ \hline 8 &=& 8 \end{array}$	$\begin{array}{rcr} x - 2 &=& 4 \\ +2 &=& +2 \\ \hline x &=& 6 \end{array}$

EXAMPLE Solve a Subtraction Equation

2 Solve $x - 2 = 1$. Check your solution.

Method 1 Use symbols.

$x - 2 = 1$ Write the equation.

$$\begin{array}{rcr} x - 2 &=& 1 \\ +2 &=& +2 \\ \hline \end{array}$$
Add 2 to each side.

$$\begin{array}{rcr} x - 2 &=& 1 \\ +2 &=& +2 \\ \hline x &=& 3 \end{array}$$
Simplify.

Check $x - 2 = 1$ Write the original equation.

$3 - 2 \stackrel{?}{=} 1$ Replace x with 3.

$1 = 1$ ✔ This sentence is true.

The solution is 3.

Method 2 Use models.

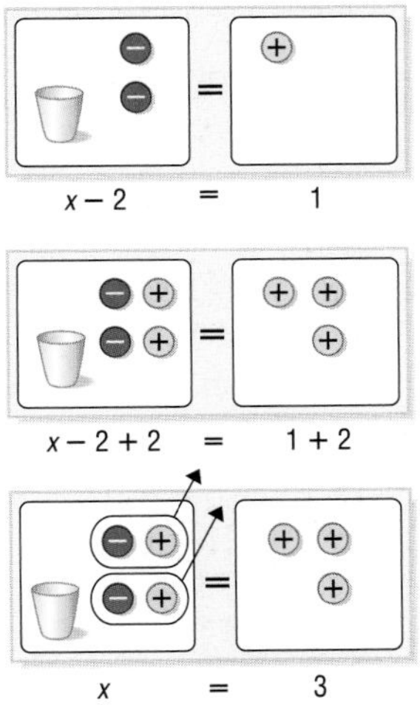

Your Turn Solve each equation.

d. $y - 3 = 4$ 7 e. $r - 4 = -2$ 2 f. $-9 = q - 8$ -1

STUDY TIP

Checking Solutions It is always wise to check your solution. You can often use arithmetic facts to check the solutions of simple equations.

TEACHING TIP

When using cups and counters, remind students that $x - 2 = x + (-2)$.

msmath2.net/extra_examples

2 Teach

In-Class Examples

INTERACTIVE CHALKBOARD PowerPoint® Presentations

1. Solve $14 + y = 20$. 6
2. Solve $z - 8 = 12$. Check your solution. 20

DAILY INTERVENTION — Differentiated Instruction

Kinesthetic Borrow a balance scale from a science teacher. Place cups with equal numbers of counters on each side of the balance. Ask:

- What would happen if you removed two counters from the cup on the right side? **The right side of the balance would be higher than the left.**
- What can you do to the left side of the scale to make the scale be in balance again? **Remove two counters.**

In-Class Example

INTERACTIVE CHALKBOARD
PowerPoint® Presentations

3 **SPORTS** If Tiger Woods had scores of −1, −4, and −3 on his first three rounds in another tournament, what would his fourth round score need to be if his final score was −18? **−10**

3 Practice/Apply

FOLDABLES™ Follow-Up Remind students to use their Foldables to take notes on what they learn about solving addition and subtraction equations. Students might record the Subtraction and Addition Properties of Equality in their own words and symbols under the "Equations" tab.

Which One Doesn't Belong?
For Exercise 3, encourage students to examine each equation, looking for similarities and differences. Then have them solve each equation.

About the Exercises...

Guided Practice Key

Exercises	Examples
4–6	1
7–9	2
10	3

Odd/Even Assignments
Exercises 11–28 and 31–36 are structured so that students practice the same concepts whether they are assigned odd or even problems.

EXAMPLE Use an Equation to Solve a Problem

3 **SPORTS Tiger Woods won the 2002 Masters Championship with a final score of 12 under par, or −12. His scores for the first three rounds were −2, −3, and −6. What was his score for the fourth round?**

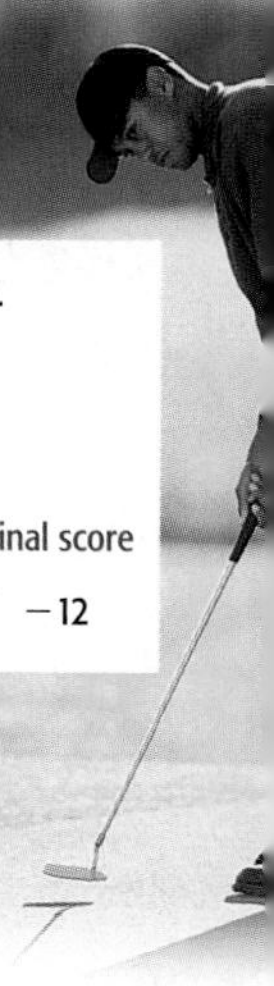

STUDY TIP

Look Back To review **adding integers**, see Lesson 3-4.

Words	The sum of the scores for all of the rounds was −12.				
Variable	Let s represent the score for the fourth round.				
Equation	scores for the first three rounds		score for the fourth round		final score
	$-2 + (-3) + (-6)$	$+$	s	$=$	-12

$-2 + (-3) + (-6) + s = -12$ Write the equation.

$-11 + s = -12$ $-2 + (-3) + (-6) = -11$

$\underline{+11 \quad\quad = +11}$ Add 11 to each side.

$s = -1$ Simplify.

Check You can check the solution by adding.
$-2 + (-3) + (-6) + (-1) = -12$ ✓

Skill and Concept Check

Writing Math
Exercise 3

1. **Tell** what property you would use to solve $x - 4 = -2$.
   Addition Property of Equality
2. **OPEN ENDED** Write two different equations that have −2 as a solution.
   Sample answer: $x + 5 = 3$ and $x - 5 = -7$
3. **Which One Doesn't Belong?** Identify the equation that does not have the same solution as the other three. Explain your reasoning.

$x + 3 = 4$	$b + 5 = 4$	$7 + y = 8$	$8 + a = 9$

$b + 5 = 4$; All of the other equations have solutions equal to 1. This one equals −1.

GUIDED PRACTICE

Solve each equation. Check your solution.

4. $n + 6 = 9$ **3**
5. $m + 5 = 3$ **−2**
6. $-2 = a + 6$ **−8**
7. $x - 5 = 6$ **11**
8. $-1 = c - 6$ **5**
9. $s - 4 = -7$ **−3**

10. **GEOMETRY** The sum of the measures of the angles of a triangle is 180°. Find the missing measure. **100°**

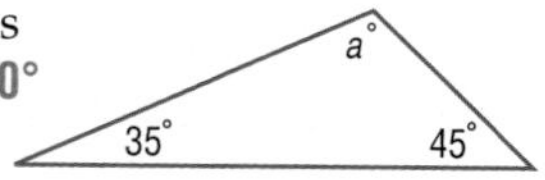

158 Chapter 4 Algebra: Linear Equations and Functions

Practice and Applications

HOMEWORK HELP	
For Exercises	See Examples
11–16	1
17–22	2
29–30, 37–38	3

Extra Practice
See pages 572, 599.

Solve each equation. Check your solution.

11. $a + 3 = 10$ **7** 12. $y + 5 = 11$ **6** 13. $x + 8 = 5$ **−3**
14. $y + 15 = 11$ **−4** 15. $r + 6 = -3$ **−9** 16. $k + 3 = -9$ **−12**
17. $s - 8 = 9$ **17** 18. $w - 7 = 11$ **18** 19. $q - 8 = -1$ **7**
20. $p - 13 = -2$ **11** 21. $x - 5 = -9$ **−4** 22. $w - 9 = -12$ **−3**
23. $34 + r = 95$ **61** 24. $64 + y = 84$ **20** 25. $-23 = x - 18$ **−5**
26. $-59 = m - 11$ **−48** 27. $-18 + c = -30$ **−12** 28. $-34 = t + 9$ **−43**

29. The sum of a number and 3 is −2. Find the number. **−5**
30. If you decrease a number by 4, the result is −5. Find the number. **−1**

Solve each equation. Check your solution.

31. $a - 3.5 = 14.9$ **18.4** 32. $x - 2.8 = 9.5$ **12.3** 33. $r - 8.5 = -2.1$ **6.4**
34. $z - 9.4 = -3.6$ **5.8** 35. $n + 1.4 = 0.72$ **−0.68** 36. $b + 2.25 = 1$ **−1.25**

ROLLER COASTERS For Exercises 37 and 38, use the table.

37. *Superman The Escape* is 105 feet taller than *Millennium Force*. Write and solve an addition equation that you could use to find the height of *Millennium Force*.
38. The difference in the speeds of *Superman The Escape* and *Millennium Force* is 7 miles per hour. If *Superman The Escape* has the greater speed, write and solve a subtraction equation to find its speed.
39. **CRITICAL THINKING** Suppose $x + y = 10$ and the value of x increases by 3. What must happen to the value of y so that $x + y = 10$ is still a true sentence?

Tallest Steel Roller Coasters in North America	Height (feet)	Speed (mph)
Superman The Escape	415	?
Millenium Force	?	93
Titan	245	85
Goliath	235	85
Nitro	230	80

Source: www.rcdb.com

37. $415 = 105 + x$; **310 ft** **38.** $x - 93 = 7$; **100 mph** **39. The value of y must decrease by 3.**

Standardized Test Practice and Mixed Review

40. **MULTIPLE CHOICE** The Sears Tower in Chicago is 1,454 feet tall. It is 204 feet taller than the Empire State Building in New York City. Use the equation $e + 204 = 1{,}454$ to find the height of the Empire State Building. **A**
    (A) 1,250 ft (B) 1,350 ft (C) 1,450 ft (D) 1,650 ft
41. **GRID IN** If $x + 3 = 2$, what is the value of $2x + 5$? **3**
42. **ALGEBRA** Write an expression for 6 inches less than w. (Lesson 4-1) **$w - 6$**

Divide. (Lesson 3-7)

43. $-15 \div 3$ **−5** 44. $36 \div (-9)$ **−4** 45. $-63 \div (-7)$ **9** 46. $-27 \div 9$ **−3**

GETTING READY FOR THE NEXT LESSON

PREREQUISITE SKILL Divide. (Page 562)

47. $15.6 \div 13$ **1.2** 48. $8.84 \div 3.4$ **2.6** 49. $75.25 \div 0.25$ **301** 50. $0.76 \div 0.5$ **1.52**

msmath2.net/self_check_quiz

4 Assess

Open-Ended Assessment

Writing Have students write two equations—one that can be solved by using the Addition Property of Equality and one that can be solved by using the Subtraction Property of Equality.

Study Guide and Intervention, p. 190

NAME ______ DATE ______ PERIOD ____

4-2 Study Guide and Intervention
Solving Addition and Subtraction Equations

Remember, equations must always remain balanced. If you subtract the same number from each side of an equation, the two sides remain equal. Also, if you add the same number to each side of an equation, the two sides remain equal.

EXAMPLE 1 Solve $x + 5 = 11$. Check your solution.

$x + 5 = 11$ Write the equation.
$-5 = -5$ Subtract 5 from each side.
$x = 6$ Simplify.

Check $x + 5 = 11$ Write the equation.
$6 + 5 \stackrel{?}{=} 11$ Replace x with 6.
$11 = 11$ ✓ This sentence is true.

The solution is 6.

EXAMPLE 2 Solve $15 = t - 12$. Check your solution.

$15 = t - 12$ Write the equation.
$+12 = +12$ Add 12 to each side.
$27 = t$ Simplify.

Check $15 = t - 12$ Write the equation.
$15 \stackrel{?}{=} 27 - 12$ Replace t with 27.
$15 = 15$ ✓ This sentence is true.

The solution is 27.

EXERCISES

Solve each equation. Check your solution.

1. $h + 3 = 14$ 11 2. $m + 8 = 22$ 14 3. $p + 5 = 15$ 10 4. $17 = y + 8$ 9
5. $w + 4 = -1$ −5 6. $k + 5 = -3$ −8 7. $25 = 14 + r$ 11 8. $57 + z = 97$ 40
9. $b - 3 = 6$ 9 10. $7 = c - 5$ 12 11. $j - 12 = 18$ 30 12. $v - 4 = 18$ 22
13. $-9 = w - 12$ 3 14. $y - 8 = -12$ −4 15. $14 = f - 2$ 16 16. $23 = n - 12$ 35

Practice: Skills, p. 191

NAME ______ DATE ______ PERIOD ____

4-2 Practice: Skills
Solving Addition and Subtraction Equations

Solve each equation. Check your solution.

1. $x + 2 = 8$ 6 2. $y + 7 = 9$ 2 3. $a + 5 = 12$ 7
4. $16 = n + 6$ 10 5. $q + 10 = 22$ 12 6. $m + 9 = 17$ 8
7. $b - 4 = 9$ 13 8. $8 = c - 4$ 12 9. $11 = t - 7$ 18
10. $d - 10 = 8$ 18 11. $x - 11 = 9$ 20 12. $2 = z - 14$ 16
13. $72 = 24 + w$ 48 14. $86 + y = 99$ 13 15. $6 + y = -8$ −14
16. $-5 = m + 11$ −16 17. $n + 3.5 = 6.7$ 3.2 18. $x + 1.6 = 0.8$ −0.8
19. $98 = t - 18$ 116 20. $12 = g - 56$ 68 21. $x - 18 = -2$ 16
22. $p - 11 = -5$ 6 23. $a - 1.5 = 4.2$ 5.7 24. $7.4 = n - 2.6$ 10

Practice: Word Problems, p. 192

NAME ______ DATE ______ PERIOD ____

4-2 Practice: Word Problems
Solving Addition and Subtraction Equations

ANIMALS For Exercises 1–4, use the table.

The average lifespans of several different types of animals are shown in the table.

Average Lifespans of Animals			
Animal	Lifespan (yr)	Animal	Lifespan (yr)
Black Bear	18	Guinea Pig	4
Dog	12	Puma	?
Giraffe	10	Tiger	16
Gray Squirrel	10	Zebra	?

1. The lifespan of a black bear is 3 years longer than the lifespan of a zebra. Write an addition equation that you could use to find the lifespan of a zebra. $z + 3 = 18$
2. Solve the equation you wrote in Exercise 1. What is the lifespan of a zebra? 15 yr
3. The lifespan of a guinea pig is 8 years shorter than the lifespan of a puma. Write a subtraction equation that you could use to find the lifespan of a puma. $p - 8 = 4$
4. Solve the equation you wrote in Exercise 3. What is the lifespan of a puma? 12 yr
5. TECHNOLOGY A survey of teens showed that teens in Pittsburgh aged 12-17 spend 15.8 hours per week online. Teens in Miami/Ft. Lauderdale spend 14.2 hours per week online. Write and solve an addition equation to find the difference in time spent online by teens in these cities. $14.2 + d = 15.8$; 1.6 hours per week
6. SPORTS Annika Sorenstam won the 2002 LPGA Takefuji Classic with a final score of 14 under par, or −14. Her scores for the first two of the three rounds were −6 and −4. What was Ms. Sorenstam's score for the third round? −4

4-3 Lesson Notes

Lesson Resources

Chapter 4 Resource Masters

- Study Guide and Intervention, p. 195
- Practice: Skills, p. 196
- Practice: Word Problems, p. 197
- Reading to Learn Mathematics, p. 198
- Enrichment, p. 199

Key Teaching Transparencies, 14

Parent and Student Study Guide Workbook, 4-3

Teaching Mathematics with Manipulatives, p. 52

FAST FILE

1 Focus

5-Minute Check 4-3

INTERACTIVE CHALKBOARD
PowerPoint® Presentations

Use as a quiz or review of Lesson 4-2. (Also available as a transparency)

Bellringer

Problem Solving Point out to students that in the previous lesson they solved addition and subtraction equations by using the inverse operation. What is the inverse operation for multiplication? **division** Ask students to make a conjecture about how to solve an equation involving multiplication like the one in the Hands-On Mini Lab. **Sample answer: Divide both sides by the coefficient of the variable.**

Reading to Learn Mathematics, p. 198 **ELL**

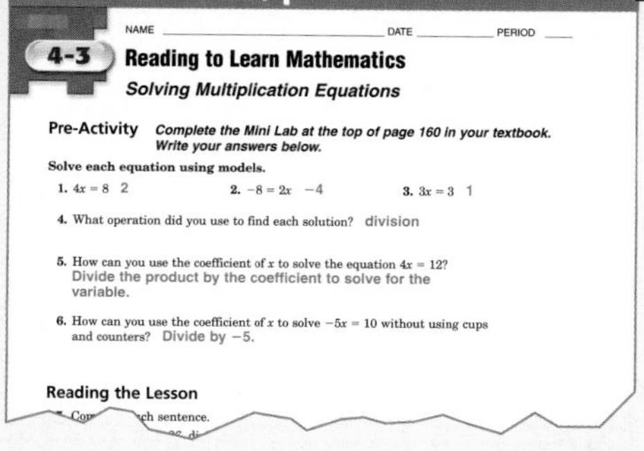

NAME ____________ DATE ________ PERIOD ____

4-3 Reading to Learn Mathematics

Solving Multiplication Equations

Pre-Activity *Complete the Mini Lab at the top of page 160 in your textbook. Write your answers below.*

Solve each equation using models.

1. $4x = 8$ 2
2. $-8 = 2x$ −4
3. $3x = 3$ 1

4. What operation did you use to find each solution? division

5. How can you use the coefficient of x to solve the equation $4x = 12$? Divide the product by the coefficient to solve for the variable.

6. How can you use the coefficient of x to solve $-5x = 10$ without using cups and counters? Divide by −5.

Reading the Lesson

4-3 Solving Multiplication Equations

What You'll Learn
Solve multiplication equations.

REVIEW Vocabulary

coefficient: the numerical factor of a multiplication expression **(Lesson 1-4)**

HANDS-ON Mini Lab

Materials
- cups and counters
- equation mat

Work with a partner.

Equations like $2x = -6$ are called multiplication equations because the expression $2x$ means *2 times the value of x.* Follow these steps to solve $2x = -6$.

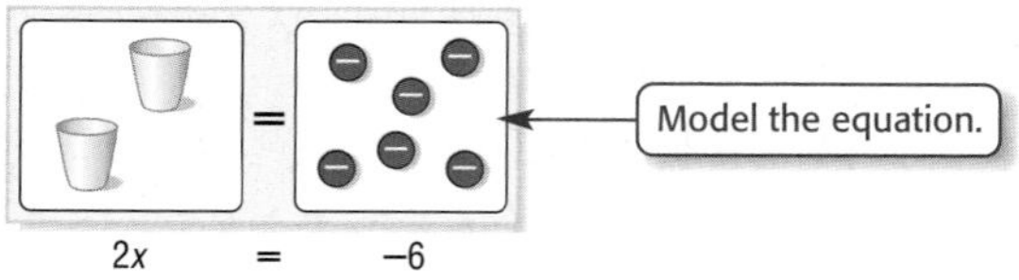

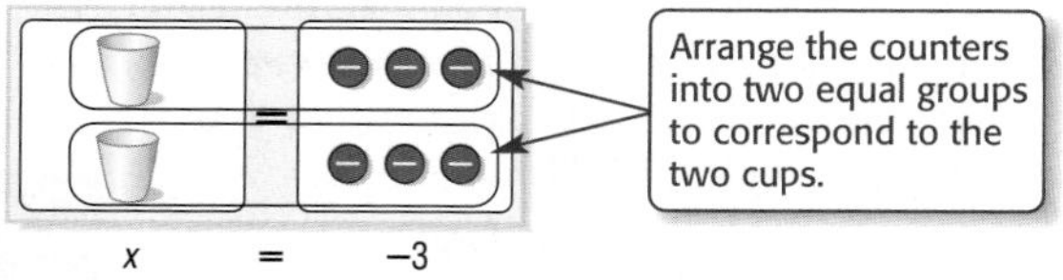

Each cup is matched with 3 negative counters. So, $x = -3$.

Solve each equation using models. **1–3. See margin for models.**

1. $4x = 8$ **2**
2. $-8 = 2x$ **−4**
3. $3x = 3$ **1**
4. What operation did you use to find each solution? **division**
5. How can you use the coefficient of x to solve $4x = 12$?
6. How can you use the coefficient of x to solve $-5x = 10$ without using cups and counters? **Divide both sides by −5.**

5. Divide the product by the coefficient to solve for the variable.

In the Mini Lab above, you matched each cup with an equal number of counters. This suggests the operation of division. You can use the **Division Property of Equality** to solve equations like $2x = -6$.

Key Concept — **Division Property of Equality**

Words If you divide each side of an equation by the same nonzero number, the two sides remain equal.

Symbols

Arithmetic	Algebra
$8 = 8$	$2x = -6$
$\frac{8}{2} = \frac{8}{2}$	$\frac{2x}{2} = \frac{-6}{2}$
$4 = 4$	$x = -3$

Notice that the division expression $8 \div 2$ can be written as the fraction $\frac{8}{2}$.

Answers (Hands-On Mini Lab)

1.

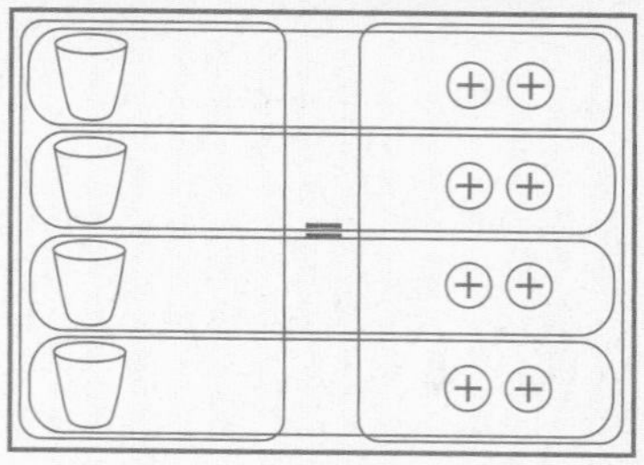

2.

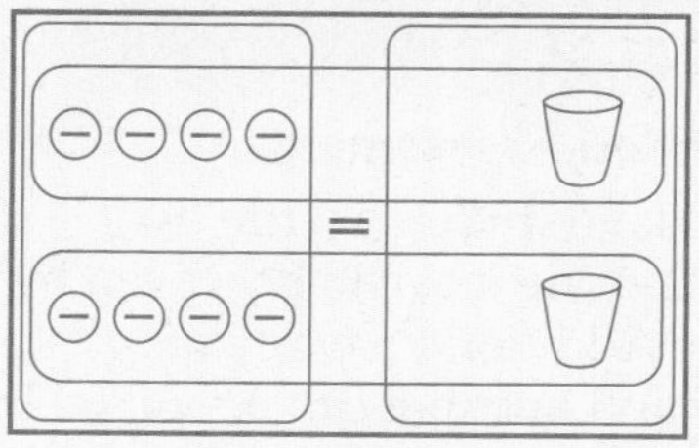

TEACHING TIP

…mind students of the …portance of checking …eir solutions.

EXAMPLES Solve Multiplication Equations

1 Solve $20 = 4x$. Check your solution.

$20 = 4x$ Write the equation.

$\frac{20}{4} = \frac{4x}{4}$ Divide each side of the equation by 4.

$5 = x$ $20 \div 4 = 5$

Check $20 = 4x$ Write the original equation.

$20 \stackrel{?}{=} 4(5)$ Replace x with 5. Is this sentence true?

$20 = 20$ ✓

The solution is 5.

2 Solve $-8y = 24$. Check your solution.

$-8y = 24$ Write the equation.

$\frac{-8y}{-8} = \frac{24}{-8}$ Divide each side by -8.

$y = -3$ $24 \div (-8) = -3$

Check $-8y = 24$ Write the original equation.

$-8(-3) \stackrel{?}{=} 24$ Replace y with -3. Is this sentence true?

$24 = 24$ ✓

The solution is -3.

Your Turn **Solve each equation.**

a. $30 = 6x$ **5** b. $-6a = 36$ **-6** c. $-9d = -72$ **8**

…EAL-LIFE MATH

…IMALS The tortoise is …e of the slowest land …mals, reaching a top …eed of about 0.25 mile …hour. Dragonflies, the …test insect, can fly at a …eed of about 18 miles …hour.

…rce: *Incredible Comparisons*

The equation $d = rt$ shows the relationship between the variables d (distance), r (rate or speed), and t (time).

EXAMPLE Use an Equation to Solve a Problem

3 ANIMALS Refer to the information at the left. At this speed, how long will it take a tortoise to travel 1.5 miles?

Words	Distance is equal to the rate times the time.				
Variables	d	$=$	r	$\cdot$	t
Equation	1.5	$=$	$0.25t$		

$1.5 = 0.25t$ Write the equation.

$\frac{1.5}{0.25} = \frac{0.25t}{0.25}$ Divide each side by 0.25.

$6 = t$ $1.5 \div 0.25 = 6$

At this speed, it would take a tortoise 6 hours to travel 1.5 miles. Check this solution.

msmath2.net/extra_examples

2 Teach

In-Class Examples

1 Solve $39 = 3y$. Check your solution. **13**

INTERACTIVE CHALKBOARD PowerPoint® Presentations

2 Solve $-4z = 60$. Check your solution. **-15**

3 **ROLLER COASTERS** With a lengthy 6,072 feet, *California Screamin'* is the longest steel roller coaster in the world. The ride takes 2 minutes 30 seconds to complete. Find the speed of the roller coaster in feet per second. **40.48 ft/s**

Tips for New Teachers

Using Manipulatives A cupcake liner serves as an inexpensive, portable cup for cups and counters activities. You may wish to give each student a set of counters and cupcake liners in a zippered sandwich bag that you have 3-hole punched. They can keep this set of cups and counters in their math notebooks for homework help or for future activities.

Answer (Hands-On Mini Lab)

3.

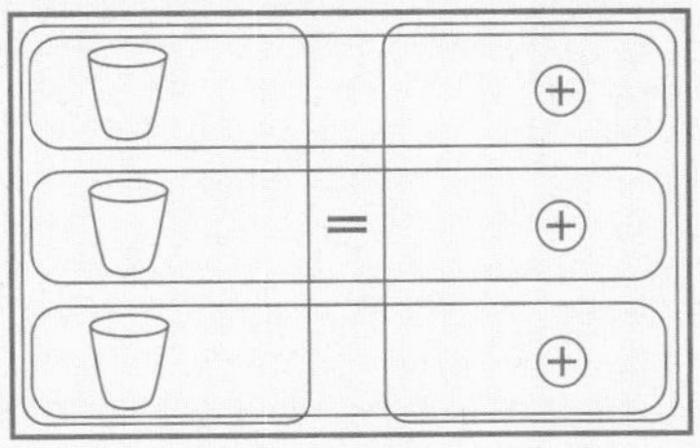

DAILY INTERVENTION

Differentiated Instruction

Intrapersonal Have students write the equation in the Hands-On Mini Lab at the beginning of the lesson, $2x = -6$.

- Before beginning the activity, have students write down what they do not yet know about solving an equation like this one. For example, they may not know how to find the value of x.
- After completing the activity, have students look over what they wrote. Are there things they still need to learn?

3 Practice/Apply

FOLDABLES™ Follow-Up Remind students to use their Foldables to take notes on what they learn about solving multiplication equations. Students might record the Division Property of Equality in their own words and symbols under the "Equations" tab.

DAILY INTERVENTION **FIND THE ERROR** In Exercise 3, ask students to think about the rules for operations with integers to determine which answer is reasonable. Then they should look back at the solutions to determine where the error was made.

About the Exercises...

Guided Practice Key

Exercises	Examples
4–5, 10–11	1
6–9	2
12–13	3

Odd/Even Assignments
Exercises 14–25 and 32–37 are structured so that students practice the same concepts whether they are assigned odd or even problems.

Skill and Concept Check

Writing Math Exercises 1 & 3

1. **Tell** whether -4 is a solution of $-3x = -12$. Explain. **No; $-3(-4)$ equals 12.**
2. **OPEN ENDED** Write two different multiplication equations that have a negative integer as a solution. **Sample answer: $2x = -2$; $4x = -8$**
3. **FIND THE ERROR** Jesse and Haley are solving $-5x = 30$. Who is correct? Explain. **Jesse; the variable is multiplied by -5. To solve for x, you need to divide each side of the equation by the entire coefficient, -5.**

Jesse	Haley
$-5x = 30$	$-5x = 30$
$\frac{-5x}{-5} = \frac{30}{-5}$	$\frac{-5x}{5} = \frac{30}{5}$
$x = -6$	$x = -6$

GUIDED PRACTICE

Solve each equation. Check your solution.

4. $6c = 18$ **3**
5. $10y = 20$ **2**
6. $-6s = 24$ **-4**
7. $-9r = 36$ **-4**
8. $-8z = -40$ **5**
9. $-11r = -77$ **7**
10. $15 = 5z$ **3**
11. $72 = 12r$ **6**
12. The product of a number and -4 is 64. Find the number. **-16**
13. If you multiply a number by 3, the result is -21. What is the number? **-7**

★ indicates increased difficulty

Practice and Applications

HOMEWORK HELP

For Exercises	See Examples
14–19	1
20–25	2
26–31, 38–39	3

Extra Practice See pages 572, 599.

Solve each equation. Check your solution.

14. $7a = 49$ **7**
15. $9e = 27$ **3**
16. $35 = 5v$ **7**
17. $112 = 8p$ **14**
18. $4j = -36$ **-9**
19. $12y = -60$ **-5**
20. $48 = -6r$ **-8**
21. $266 = -2t$ **-133**
22. $-3w = -36$ **12**
23. $-10g = -100$ **10**
24. $-28 = -7f$ **4**
25. $-275 = -5s$ **55**
26. When a number is multiplied by -12, the result is -168. Find the number. **14**
27. The product of a number and 25 is 1,000. What is the number? **40**
28. **BABY-SITTING** Gracia earns $5 per hour when she baby-sits. How many hours does she need to work to earn $75? **15 h**
29. **TRAVEL** A Boeing 747 aircraft has a cruising speed of about 600 miles per hour. At that speed, how long will it take to travel 1,500 miles? Use the formula $d = rt$. **2.5 h**

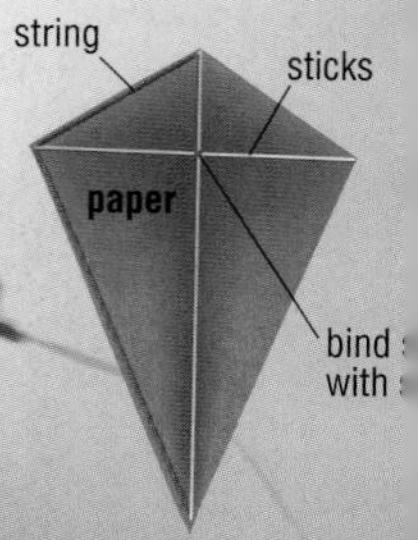

KITES For Exercises 30 and 31, use the following information.
In a simple kite, the length of the longer stick should be 1.5 times the length of the shorter stick. **30. $1.5d = 36$**

30. Suppose the length of the longer stick is 36 inches. Write a multiplication equation to find the length of the shorter stick.
31. Solve the equation. **24 in.**

162 Chapter 4 Algebra: Linear Equations and Functions

Solve each equation. Check your solution.

32. $4x = 9.2$ **2.3** 33. $9y = 13.5$ **1.5** 34. $5.4 = 0.3p$ **18**

35. $9.72 = 1.8a$ **5.4** 36. $3.9y = 18.33$ **4.7** 37. $2.6b = 2.08$ **0.8**

EARTHQUAKES For Exercises 38 and 39, use the following information. Scientists determine the epicenter of an earthquake by measuring the time it takes for surface waves to travel between two places. Surface waves travel about 6 kilometers per second through Earth's crust.

38. The distance from Los Angeles, California, to Phoenix, Arizona, is 600 kilometers. Write a multiplication equation to find how long it would take surface waves to travel from Los Angeles to Phoenix. **$6k = 600$**

39. Solve the equation. **100 s**

ROLLER COASTERS For Exercises 40 and 41 use the table.

★ 40. Without calculating, explain whether the *Blue Streak* or *Magnum* has the greater speed. **See margin.**

★ 41. Find the speed of each roller coaster in feet per second.

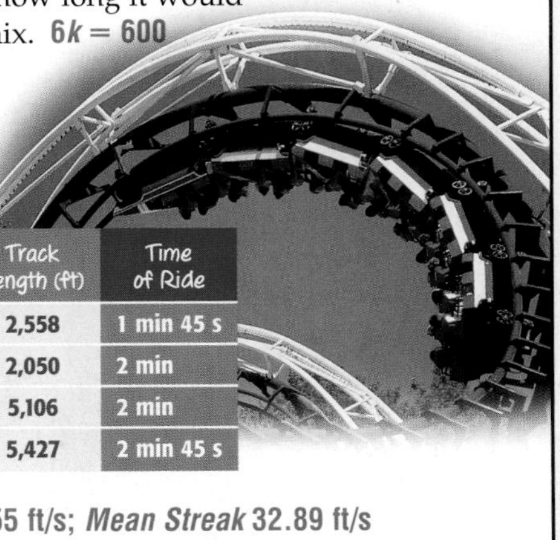

Name	Track Length (ft)	Time of Ride
Blue Streak	2,558	1 min 45 s
Corkscrew	2,050	2 min
Magnum	5,106	2 min
Mean Streak	5,427	2 min 45 s

Source: www.rcdb.com

42. **CRITICAL THINKING** Solve $3|x| = 12$. **$x = 4$ or $x = -4$**

41. ***Blue Streak* 24.36 ft/s; *Corkscrew* 17.08 ft/s; *Magnum* 42.55 ft/s; *Mean Streak* 32.89 ft/s**

Standardized Test Practice and Mixed Review

43. **MULTIPLE CHOICE** A car is traveling at a speed of 75 feet per second. How many yards will the car travel in 90 seconds if it maintains the same speed? **C**

Ⓐ 75 yd Ⓑ 270 yd Ⓒ 2,250 yd Ⓓ 6,750 yd

44. **MULTIPLE CHOICE** Use the formula $A = bh$ to find the height of a parallelogram with a base of 34 millimeters and an area of 612 square millimeters. **I**

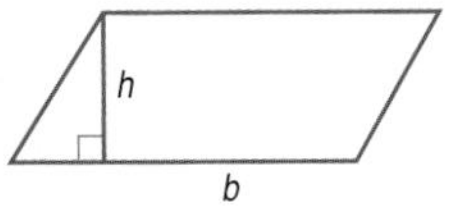

Ⓕ 20,800 mm Ⓖ 646 mm

Ⓗ 578 mm Ⓘ 18 mm

ALGEBRA Solve each equation. (Lesson 4-2)

45. $y + 8 = -2$ **−10** 46. $x - 7 = -2$ **5** 47. $20 = z + 23$ **−3**

ALGEBRA Write an algebraic expression for each phrase. (Lesson 4-1)

48. the product of −3 and y **$-3y$** 49. 5 years older than Rafael **$a + 5$**

50. 10 fewer students than last year **$y - 10$** 51. twice as many runs as the Marlins scored **$2r$**

GETTING READY FOR THE NEXT LESSON

PREREQUISITE SKILL Subtract. (Lesson 3-5)

52. $8 - (-2)$ **10** 53. $-7 - 7$ **−14** 54. $-3 - (-9)$ **6** 55. $-3 - 18$ **−21**

4 Assess

Open-Ended Assessment

Speaking Have students describe how to solve a multiplication equation. Have one person give the first step, another give the second step, and so on.

Answer

40. Sample answer: The *Magnum* has greater speed. The track is almost twice as long, and yet the ride only takes 15 more seconds than the *Blue Streak*.

Study Guide and Intervention, p. 195

NAME ______ DATE ______ PERIOD ____

4-3 Study Guide and Intervention

Solving Multiplication Equations

If each side of an equation is multiplied by the same number, the resulting equation is equivalent to the given one. You can use this property to solve equations involving multiplication and division.

EXAMPLE 1 **Solve 45 = 5x. Check your solution.**

$45 = 5x$ Write the equation.

$\frac{45}{5} = \frac{5x}{5}$ Divide each side of the equation by 5.

$9 = x$ $45 \div 5 = 9$

Check $45 = 5x$ Write the original equation.

$45 \stackrel{?}{=} 5(9)$ Replace x with 9. Is this sentence true?

$45 = 45$ ✓

The solution is 9.

EXAMPLE 2 **Solve −21 = −3y. Check your solution.**

$-21 = -3y$ Write the equation.

$\frac{-21}{-3} = \frac{-3y}{-3}$ Divide each side by −3.

$7 = y$ $-21 \div (-3) = 7$

Check $-21 = -3y$ Write the original equation.

$-21 \stackrel{?}{=} -3(7)$ Replace y with 7. Is this sentence true?

$-21 = -21$ ✓

The solution is 7.

EXERCISES

Solve each equation. Then check your solution.

1. $8q = 56$ 7 2. $4p = 32$ 8 3. $42 = 6m$ 7 4. $104 = 13h$ 8

5. $-6n = 30$ −5 6. $-18x = 36$ −2 7. $48 = -8y$ −6 8. $72 = -3b$ −24

9. $-9a = -45$ 5 10. $-12m = -120$ 10 11. $-66 = -11t$ 6 12. $-144 = -9r$ 16

13. $3a = 4.5$ 1.5 14. $2h = 3.8$ 1.9 15. $4.9 = 0.7k$ 7 16. $9.75 = 2.5z$ 3.9

Practice: Skills, p. 196

NAME ______ DATE ______ PERIOD ____

4-3 Practice: Skills

Solving Multiplication Equations

Solve each equation. Check your solution.

1. $4c = 16$ 4 2. $10x = 50$ 5 3. $42 = 6s$ 7

4. $9c = 45$ 5 5. $49 = 7y$ 7 6. $11t = 44$ 4

7. $15a = 60$ 4 8. $72 = 12c$ 6 9. $18x = 162$ 9

10. $14d = 154$ 11 11. $24z = 288$ 12 12. $16v = 256$ 16

13. $-5b = 40$ −8 14. $32 = -2f$ −16 15. $-9x = -63$ 7

16. $4g = -52$ −13 17. $-5x = -85$ 17 18. $-63 = 7a$ −9

19. $0.6m = 1.8$ 3 20. $1.5z = 6$ 4 21. $0.6q = 3.6$ 6

22. $1.8a = 0.9$ 0.5 23. $1.2r = 4.8$ 4 24. $2.4 = 0.2t$ 12

Practice: Word Problems, p. 197

NAME ______ DATE ______ PERIOD ____

4-3 Practice: Word Problems

Solving Multiplication Equations

1. **TRAVEL** The speed limit on an Arizona highway is 75 miles per hour. Suppose a truck traveling at the speed limit drives 225 miles before the driver stops for a break. Write a multiplication equation to find the length of time the truck has traveled. $75t = 225$

2. **TRAVEL** Solve the equation you wrote in Exercise 1. How long did the truck travel? 3 h

3. **FLOWERS** A gardening expert recommends that flower bulbs be planted to a depth of three times their height. Suppose Jenna determines that a certain bulb should be planted at a depth of 4.5 inches. Write a multiplication equation to find the height of the bulb. $3h = 4.5$

4. **FLOWERS** Solve the equation you wrote in Exercise 3. What is the height of the bulb? 1.5 in.

5. **EXERCISE** A 125-pound person uses 4.4 Calories per minute when walking. Write a multiplication equation to find the number of minutes of walking it will take for a 125-pound person to use 198 Calories. $4.4m = 198$

6. **EXERCISE** Solve the equation you wrote in Exercise 5. How many minutes of walking it will take for a 125-pound person to use 198 Calories? 45 min

7. **ELECTRICITY** The electric company charges $0.06 per kilowatt hour of electricity used. Write a multiplication equation to find the number of kilowatt hours of electricity for which the Estevez family was charged if their electric bill was $45.84. $0.06k = 45.84$

8. **ELECTRICITY** Solve the equation you wrote in Exercise 7. For how many kilowatt hours of electricity was the Estevez family charged? 764 kilowatt h

4-4a Lesson Notes

1 Focus

Bellringer

Problem Solving Ask students to describe how to do and undo an everyday task. For example, to put on your shoes and socks you would put on your socks, put on your shoes, and tie the shoes. To take the socks and shoes off, you would untie the shoes, take the shoes off, and then take the socks off. Point out that to undo the process, you reverse the steps or work backward.

2 Teach

In-Class Example

INTERACTIVE CHALKBOARD PowerPoint® Presentations

Solve. Use the work backward strategy.

SHOPPING Lucy and Elena went shopping. Each bought a CD that cost $16.50. If they each bought a $3.50 popcorn and a $2.50 drink and had $5.00 left, how much did they have to spend? **$50**

Answer

3. Sample answer: In the first four games, Hannah scored a total of 83 points. In the fourth game she scored 19 points. In game three, she scored 27 points and in the second game she scored 22 points. How many points did she score in the first game? To solve, first subtract 19 from 83, which is 64. Then subtract 27 from 64 to get 37. Finally, subtract 22 from 37. So, Hannah scored 15 points in her first game.

4-4a Problem-Solving Strategy

A Preview of Lesson 4-4

What You'll Learn
Solve problems using the work backward strategy.

Work Backward

Alicia, how much money did we take to the movies today?

Well, we spent $9 on movie tickets. Then we spent $5 on candy, and one half of what was left on popcorn. We have $3 left. Let's **work backward** to figure it out.

Explore	We know we have $3 left and the amounts spent. We need to find the initial amount we had.	
Plan	Let's start with the end result and work backward.	
Solve	We have $3 left. **Undo** the half spent on popcorn.	3 × 2
	Undo the $5 spent on candy.	6 + 5
	Undo the $9 spent on movie tickets. So, we initially had $20.	11 + 9 20
Examine	Assume that we started with $20. After buying movie tickets, we had $20 − $9 or $11. We spent $5 on candy, so we had $11 − $5 or $6. Then we spent one half of the remaining money on popcorn, so we had $6 ÷ 2 or $3. So, our answer of $20 is correct.	

1. When you are given the final total and you are asked to find an earlier amount.
2. Begin by taking the last value in the problem and perform opposite operations with each subsequent value until you arrive at the initial value.

Analyze the Strategy

1. **Explain** when you would use the work backward strategy to solve a problem.
2. **Describe** how to solve a problem by working backward.
3. **Write** a problem that can be solved by working backward. Then write the steps you would take to find the solution to your problem. **See margin.**

DAILY INTERVENTION — Unlocking Misconceptions

Work Backward Make sure that students correctly identify the order in which events occurred in order to solve problems using the work backward strategy. Tell them to read the problems carefully, as steps are not always presented in order.

pply the Strategy

lve. Use the work backward strategy.

MONEY Mia spent $4.50 at the bakery and then spent four times that amount at the grocery. She had $12.80 left. How much money did she have initially? **$35.30**

5. **NUMBER THEORY** A number is multiplied by 4. Then 6 is subtracted from the product. After adding 5, the result is 15. What is the number? **4**

ixed Problem Solving

lve. Use any strategy.

. **PATTERNS** Draw the sixth figure in the pattern shown. **See margin.**

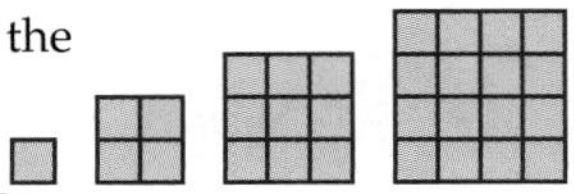

. **FOOD** Mateo goes to the grocery store and buys a ham for $24.98 and a vegetable tray for $17.49. There is no tax on food. He gives the cashier one bill and receives less than $10 in change. What was the denomination of the bill Mateo gave the cashier? **$50 bill**

. **TRAVEL** A shuttle bus schedule is shown. What is the earliest time after noon when the bus departs? **12:07 P.M.**

Departs	Arrives
8:55 A.M.	9:20 A.M.
9:43 A.M.	10:08 A.M.
10:31 A.M.	10:56 A.M.
11:19 A.M.	11:44 A.M.

. **GEOGRAPHY** The land area of Texas is 267,277 square miles. This is about 5 times the land area of Arkansas. Estimate the land area of Arkansas. **about 53,000 sq mi**

. **NUMBER THEORY** How many different two-digit numbers can you make using the digits 2, 4, and 9 if no number is repeated? **6**

. **AGE** Maya is two years older than her sister Jenna. Jenna is 5 years older than their brother Trevor, who is 9 years younger than their brother Trent. Trent is 17 years old. How old is Maya? **15**

FOOD For Exercises 12 and 13, use the graph below.

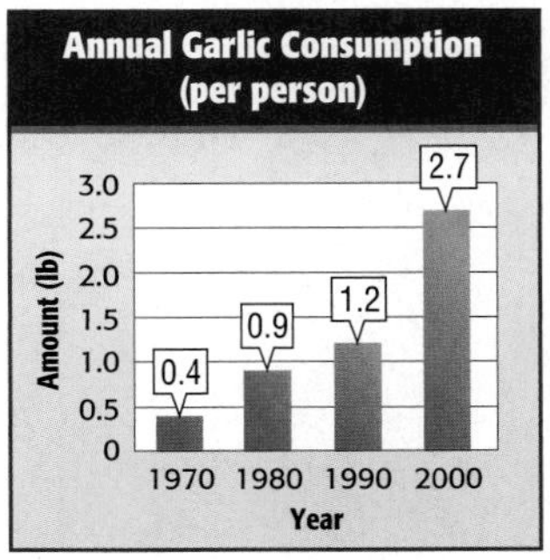

12. How much more garlic was eaten in 2000 than in 1990? **about 1.5 lb**

13. Find the average monthly consumption of garlic in 2000. **about 0.225 lb/mo**

14. **VIDEOS** Carlos rented 2 times as many DVDs as Ashley last month. Ashley rented 4 fewer than Greg, but four more than Grace. Greg rented 9 DVDs. How many DVDs did each person rent? **Greg 9; Ashley 5; Grace 1; Carlos 10**

15. **STANDARDIZED TEST PRACTICE** Which of the following is the most reasonable total amount for the items purchased? **C**

Pens	$ 2.09
Ruler	$ 0.99
Paper	$ 1.49
Book	$14.99
Candy	$ 0.49
Glue	$ 0.89
Folder	$ 1.19
Erasers	$ 1.99
Pencils	$ 1.87

Ⓐ $17 Ⓑ $20

Ⓒ $26 Ⓓ $30

You will use the work backward strategy in the next lesson.

3 Practice/Apply

About the Exercises...

Problem-Solving Strategies
Exercises 6–15 are structured so that students have the opportunity to practice many different problem-solving strategies. You may wish to remind them of some of the strategies they have studied.

- guess and check (p. 22)
- use a graph (p. 58)
- look for a pattern (p. 132)
- work backward ✓

Answer

6.

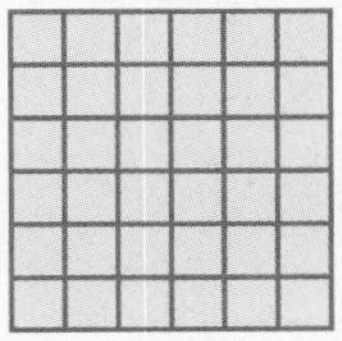

4 Assess

Open-Ended Assessment

Writing Ask students to imagine that they have received a gift of $100. Have them write a budget, working backward from $100, and include the costs of at least three items they would like to buy. Will they allot any money to savings?

4-4 Lesson Notes

Lesson Resources

Chapter 4 Resource Masters

- Study Guide and Intervention, p. 200
- Practice: Skills, p. 201
- Practice: Word Problems, p. 202
- Reading to Learn Mathematics, p. 203
- Enrichment, p. 204

Key Teaching Transparencies, 15

MathPASS Tutorial CD-ROM, 39

Parent and Student Study Guide Workbook, 4-4

Real-Life Career Masters, p. 18

Teaching Mathematics with Manipulatives, pp. 53–54

FAST FILE

1 Focus

5-Minute Check 4-4

INTERACTIVE CHALKBOARD PowerPoint® Presentations

Use as a quiz or review of Lesson 4-3. (Also available as a transparency)

Bellringer

Hands-On Activity Before students complete the Hands-On Mini Lab, model the solution of $3x + 4 = 7$ using cups and counters. Point out that you must use two inverse operations—subtraction and division—to solve the equation. First subtract to get $3x = 3$, then divide to solve for $x = 1$. Adding 3 cups to each side in order to solve for x leads you to the equation $6x = 3x + 3$, taking you *further* from the solution, $x = 1$.

Reading to Learn Mathematics, p. 203 **ELL**

NAME ______ DATE ______ PERIOD ____

4-4 Reading to Learn Mathematics

Solving Two-Step Equations

Pre-Activity *Complete the Mini Lab at the top of page 166 in your textbook. Write your answers below.*

Solve each equation using models.

1. $3x + 1 = 7$ 2   2. $2x - 4 = 2$ 3   3. $2x + 3 = -3$ −3

Reading the Lesson

4. Describe in words each step shown for solving the equation.

$12 + 7s = -9$

$12 + 7s = -9$ Write the equation.

$-12 \quad = -12$ Subtract 12 from each side.

$7s = -21$ Simplify.

$\frac{7s}{7} = \frac{-21}{7}$ Divide each side by 7.

$s = -3$ Simplify.

4-4 Solving Two-Step Equations

What You'll Learn
Solve two-step equations.

NEW Vocabulary
two-step equation

HANDS-ON Mini Lab

Materials
- cups and counters
- equation mat

Work with a partner.

A **two-step equation** has two different operations. Follow these steps to solve $2x - 3 = 1$.

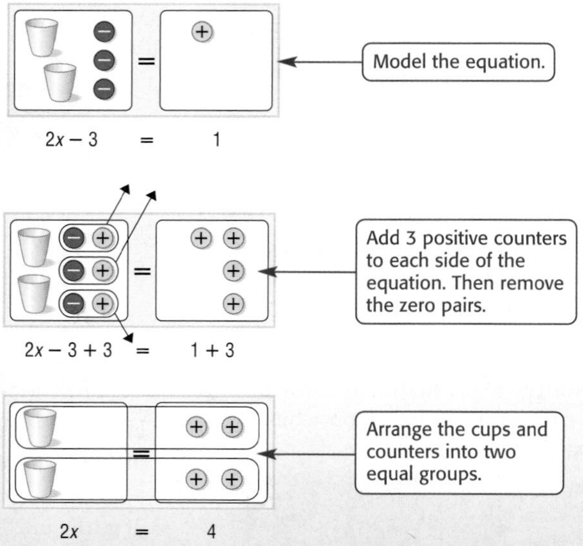

Each cup is matched with 2 positive counters. So, $x = 2$.

Solve each equation using models.

1. $3x + 1 = 7$ **2**   2. $2x - 4 = 2$ **3**   3. $2x + 3 = -3$ **−3**

1–3. See Ch. 4 Answer Appendix for models.

To solve two-step equations, "undo" the operations in reverse order of the order of operations. You are using the *work backward* strategy.

TEACHING TIP

Tell students that in $3x + 2 = 23$, x is first multiplied by 3. Then 2 is added. So, they should reverse the operations to solve.

EXAMPLE Solve a Two-Step Equation

1 Solve $3x + 2 = 23$. Check your solution.

$3x + 2 = 23$ Write the equation.

$\underline{\quad -2 = -2}$ Subtract 2 from each side.

$3x = 21$ Simplify.

$\frac{3x}{3} = \frac{21}{3}$ Divide each side by 3.

$x = 7$ Simplify.

The solution is 7. Check the solution.

EXAMPLES Solve Two-Step Equations

TUDY TIP

Solving Equations When you solve a simple equation like $-2y = 10$, you can mentally divide each side by -2.

Solve each equation. Check your solution.

2 $-2y - 7 = 3$

$-2y - 7 = 3$	Write the equation.
$+7 = +7$	Add 7 to each side.
$-2y = 10$	Simplify.
$\frac{-2y}{-2} = \frac{10}{-2}$	Divide each side by -2.
$y = -5$	Simplify.

The solution is -5. Check the solution.

3 $4 + 5r = -11$

$4 + 5r = -11$	Write the equation.
$-4 = -4$	Subtract 4 from each side.
$5r = -15$	Simplify.
$\frac{5r}{5} = \frac{-15}{5}$	Divide each side by 5.
$r = -3$	Simplify.

The solution is -3. Check the solution.

Your Turn **Solve each equation.**

a. $4x + 5 = 13$ **2** b. $-3n - 8 = 7$ **−5** c. $1 + 2y = -3$ **−2**

EAL-LIFE MATH

WLING *Handicapping* ws bowlers with varying s to score about the ne. Your handicap h is led to your score s, so r final score is $h + s$.

ce: www.bowl.com

Some problems start with a given amount and increase at a certain rate.

EXAMPLE Use an Equation to Solve a Problem

4 **BOWLING Suppose you spend \$6 to rent bowling shoes at The Bowling Alley. Each game costs \$3.50. How many games can you bowl if you have \$20 to spend?**

Variable	Let x = the number of games.
Words	Cost of the shoes plus the cost of the games equals \$20.
Equation	$6 + 3.50x = 20$

$6 + 3.50x = 20$	Write the equation.
$-6 = -6$	Subtract 6 from each side.
$3.50x = 14$	Simplify.
$\frac{3.50x}{3.50} = \frac{14}{3.50}$	Divide each side by 3.50.
$x = 4$	$14 \div 3.50 = 4$

You can bowl 4 games. Is this answer reasonable?

msmath2.net/extra_examples

2 Teach

In-Class Examples

INTERACTIVE CHALKBOARD PowerPoint® Presentations

1 Solve $4x + 3 = 19$. Check your solution. **4**

Solve each equation. Check your solution.

2 $-3c + 9 = 3$ **2**

3 $6 + 3t = 0$ **−2**

4 **PARKS** There are 76 thousand acres of state parkland in Georgia. This is 4 thousand acres more than three times the number of acres of state parkland in Mississippi. How many acres of state parkland are there in Mississippi? **24,000 acres**

Tips for New Teachers

Classroom Management If you have students who finish an in-class assignment or activity more quickly than the rest of the group, consider having them work together as a small group to check their work. They can compare solutions and discuss any differences they find.

DAILY INTERVENTION — **Differentiated Instruction**

Interpersonal Have students work in pairs to write and solve two-step equations. Write an equation on a piece of paper. For example, $x = 3$. Have the first person multiply both sides of the equation by an integer of his or her choice, such as 6. Then the second partner adds or subtracts a number of his or her choice, such as adding 2. This gives $6x + 2 = 20$. Then have each pair reverse the steps to solve the equation. Provide several two-step equations for the pairs to solve by working backward.

3 Practice/Apply

FOLDABLES™ Follow-Up Remind students to use their Foldables to take notes on what they learn about solving two-step equations. Under the "Equations" tab, students might explain a two-step equation in their own words and give examples illustrating methods for solving the equation.

DAILY INTERVENTION **FIND THE ERROR** In Exercise 2, point out that since the answers are different, students can substitute each answer for x in order to determine which is correct. Then they can look back at the solutions to determine where the error was made.

About the Exercises...

Guided Practice Key

Exercises	Examples
3, 5–6	1, 3
4	2
7	4

Odd/Even Assignments
Exercises 8–37 are structured so that students practice the same concepts whether they are assigned odd or even problems.

Alert! The Data Update following Exercise 41 requires access to the Internet.

Skill and Concept Check

Writing Math
Exercise 2

1. **OPEN ENDED** Write a two-step equation that has -2 as the solution. **Sample answer: $3x + 10 = 4$**
2. **FIND THE ERROR** Jackson and Michele are solving $4 + 2x = -8$. Who is correct? Explain. **Michele; Jackson incorrectly divided by the coefficient before subtracting 4 from each side of the equation.**

Jackson
$$4 + 2x = -8$$
$$4 + \frac{2x}{2} = \frac{-8}{2}$$
$$4 + x = -4$$
$$x = -8$$

Michele
$$4 + 2x = -8$$
$$\underline{-4 \qquad = -4}$$
$$2x = -12$$
$$x = -6$$

GUIDED PRACTICE

Solve each equation. Check your solution.

3. $3x + 1 = 7$ **2**
4. $4h - 6 = 22$ **7**
5. $-6r + 1 = -17$ **3**
6. $13 = 1 + 4s$ **3**
7. Five more than three times a number is 23. Find the number. **6**

★ indicates increased difficulty

Practice and Applications

HOMEWORK HELP

For Exercises	See Examples
8–11	1
12–15	2
16–19	3
26–29, 38–41	4

Extra Practice
See pages 573, 599.

Solve each equation. Check your solution.

8. $3x + 1 = 10$ **3**
9. $5x + 4 = 19$ **3**
10. $2t + 7 = -1$ **−4**
11. $6m + 1 = -23$ **−4**
12. $-4w - 4 = 8$ **−3**
13. $-7y + 3 = -25$ **4**
14. $-8s + 1 = 33$ **−4**
15. $-2x + 5 = -13$ **9**
16. $3 + 8n = -5$ **−1**
17. $5 + 4d = 37$ **8**
18. $14 + 2p = 8$ **−3**
19. $25 + 2y = 47$ **11**
20. $2 = 3t - 13$ **5**
21. $57 = -8x - 7$ **−8**
22. $18 = 9d - 18$ **4**
23. $4 = 4 + 7f$ **0**
24. $21 + 11x = -1$ **−2**
25. $15x + 4 = 49$ **3**
26. Three more than the product of a number and 4 is 15. Find the number. **3**
27. Five less than three times a number is 1. Find the number. **2**
28. The product of 2 and a number is increased by 9. The result is -17. Find the number. **−13**
29. If you subtract 3 from twice a number, the result is 25. Find the number. **14**

Solve each equation. Check your solution.

30. $2r - 3.1 = 1.7$ **2.4**
31. $4t + 3.5 = 12.5$ **2.25**
32. $16b - 6.5 = 9.5$ **1**
33. $5w + 9.2 = 19.7$ **2.**
34. $16 = 0.5r - 8$ **48**
35. $0.2n + 3 = 8.6$ **28**
36. $7.5s - 2 = 28$ **4**
37. $1.5v - 16 = 8$ **16**
38. **MONEY MATTERS** Joshua has saved \$74 toward a new sound system that costs \$149. He plans on saving an additional \$15 each week. How many weeks will it take Joshua to save enough money to buy the sound system? **5 wk**
39. **WEATHER** The temperature is 20°F. It is expected to rise at a rate of 4° each hour for the next several hours. In how many hours will the temperature be 32°? **3 h**

168 Chapter 4 Algebra: Linear Equations and Functions

TEMPERATURE For Exercises 40 and 41, use the following information and the graph.

Temperature is usually measured on the Fahrenheit scale (°F) or the Celsius scale (°C). Use the formula $F = 1.8C + 32$ to convert from one scale to the other.

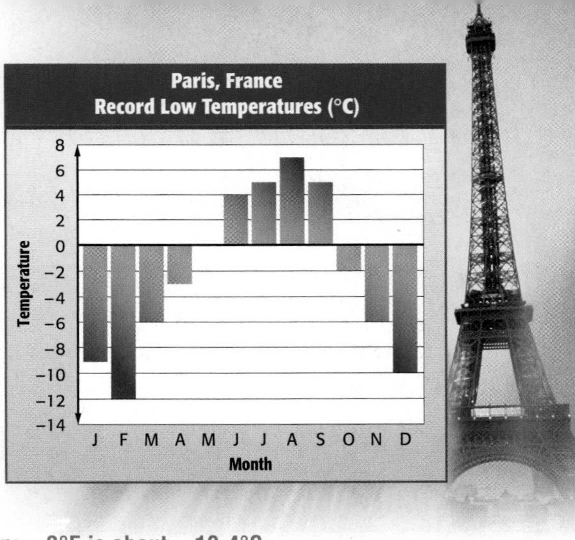

40. The highest temperature ever recorded in Virginia Beach, Virginia, was 104°F. Find this temperature in degrees Celsius. **40°C**

★ 41. **MULTI STEP** The lowest temperature ever recorded in Virginia Beach was −3°F. Is this temperature greater or less than the lowest temperature ever recorded in Paris, France? **Less than; −3°F is about −19.4°C.**

Data Update What are the record high and low temperatures for a large city near you? Visit msmath2.net/data_update to learn more.

42. **CRITICAL THINKING** Is there a temperature at which the number of Celsius degrees is the same as the number of Fahrenheit degrees? If so, find it. If not, explain why not. **yes; −40°C = −40°F**

Standardized Test Practice and Mixed Review

43. **MULTIPLE CHOICE** A taxi driver charges $2.00 plus $0.80 for each mile traveled. Which expression could be used to find the cost of a taxi ride if m miles are traveled? **C**

Ⓐ $2.80m$ Ⓑ $2m + 0.80$ Ⓒ $2 + 0.80m$ Ⓓ $2m + 0.80m$

44. **SHORT RESPONSE** In 2002, the first-class postage rates were $0.37 for the first ounce plus an additional cost for each ounce.

Weight (oz)	1	2	3	4	5	6
Postage (dollars)	0.37	0.60	0.83	1.06	1.29	1.52

What is the cost, in dollars, for a 12-ounce letter? **$2.90**

ALGEBRA Solve each equation. Check your solution. (Lessons 4-2 and 4-3)

45. $4f = 28$ **7**
46. $6p = -72$ **−12**
47. $p - 14 = 27$ **41**
48. $26 = n + 2$ **24**

49. **DIVING** Find the distance between two divers if one diver is 27 feet below sea level and the other diver is 13 feet below sea level. (Lesson 3-5) **14 ft**

GETTING READY FOR THE NEXT LESSON

PREREQUISITE SKILL Replace each ● with < or > to make a true sentence. (Lesson 3-2)

50. −3 ● −12 >
51. 1 ● −1 >
52. 7 ● −18 >
53. −120 ● −30 <

msmath2.net/self_check_quiz

4 Assess

Open-Ended Assessment

Modeling Have students model the solution to $2x + 4 = -2$ using cups and counters. −3

Study Guide and Intervention, p. 200

NAME ____ DATE ____ PERIOD ____

4-4 Study Guide and Intervention

Solving Two-Step Equations

To solve two-step equations, you need to add or subtract first. Then divide to solve the equation.

EXAMPLE 1 Solve $7v - 3 = 25$. Check your solution.

$7v - 3 = 25$ Write the equation.
$+3 = +3$ Add 3 to each side.
$7v = 28$ Simplify.
$\frac{7v}{7} = \frac{28}{7}$ Divide each side by 7.
$v = 4$ Simplify.

Check $7v - 3 = 25$ Write the original equation.
$7(4) - 3 \stackrel{?}{=} 25$ Replace v with 4.
$28 - 3 \stackrel{?}{=} 25$ Multiply.
$25 = 25$ ✓ The solution checks.

The solution is 4.

EXAMPLE 2 Solve $-10 = 8 + 3x$. Check your solution.

$-10 = 8 + 3x$ Write the equation.
$-8 = -8$ Subtract 8 from each side.
$-18 = 3x$ Simplify.
$\frac{-18}{3} = \frac{3x}{3}$ Divide each side by 3.
$-6 = x$ Simplify.

Check $-10 = 8 + 3x$ Write the original equation.
$-10 \stackrel{?}{=} 8 + 3(-6)$ Replace x with −6.
$-10 \stackrel{?}{=} 8 + (-18)$ Multiply.
$-10 = -10$ ✓ The solution checks.

The solution is −6.

EXERCISES

Solve each equation. Check your solution.

1. $4y + 1 = 13$ 3
2. $6x + 2 = 26$ 4
3. $-3 = 5k + 7$ −2
4. $6n + 4 = -26$ −5
5. $7 = -3c - 2$ −3
6. $-8p + 3 = -29$ 4
7. $-5 = -5t - 5$ 0
8. $-9r + 12 = -24$ 4
9. $11 + 7n = 4$ −1
10. $35 = 7 + 4b$ 7
11. $15 + 2p = 9$ −3
12. $49 = 16 + 3y$ 11
13. $2 = 4t - 14$ 4
14. $-9x - 10 = 62$ −8
15. $30 = 12z - 18$ 4
16. $7 + 4g = 7$ 0
17. $24 + 9x = -3$ −3
18. $50 = 16q + 2$ 3
19. $3c - 2.5 = 4.1$ 2.2
20. $9y + 4.8 = 17.4$ 1.4

Practice: Skills, p. 201

NAME ____ DATE ____ PERIOD ____

4-4 Practice: Skills

Solving Two-Step Equations

Solve each equation. Check your solution.

1. $2x + 1 = 9$ 4
2. $5b + 2 = 17$ 3
3. $3w + 5 = 23$ 6
4. $8n + 1 = 25$ 3
5. $4t - 2 = 14$ 4
6. $7k - 3 = 32$ 5
7. $8x - 1 = 63$ 8
8. $2x - 5 = 15$ 10
9. $3 + 6v = 45$ 7
10. $9 + 4b = 17$ 2
11. $2p + 14 = 0$ −7
12. $3y + 10 = -2$ −4
13. $3w + 5 = 2$ −1
14. $8x + 7 = -9$ −2
15. $5d - 1 = -11$ −2
16. $4d - 35 = -3$ 8
17. $11x - 24 = -2$ 2
18. $15a - 54 = -9$ 3
19. $3g - 49 = -7$ 14
20. $-2x - 4 = 8$ −6
21. $-9d - 1 = 17$ −2
22. $-4f + 1 = 13$ −3
23. $-5b + 24 = -1$ 5
24. $-6x + 4 = -2$ 1

Practice: Word Problems, p. 202

NAME ____ DATE ____ PERIOD ____

4-4 Practice: Word Problems

Solving Two-Step Equations

1. **GOLF** It costs $12 to attend a golf clinic with a local pro. Buckets of balls for practice during the clinic cost $3 each. How many buckets can you buy at the clinic if you have $30 to spend? 6 buckets

2. **MONEY** Paulo has $145 in his savings account. He earns $36 a week mowing lawns. If Paulo saves all of his earnings, after how many weeks will he have $433 saved? 8 wk

3. **RETAIL** An online retailer charges $6.99 plus $0.55 per pound to ship electronics purchases. How many pounds is a DVD player for which the shipping charge is $11.94? 9 lb

4. **MONEY** Caitlin has a $10 gift certificate to the music store. She has chosen a number of CDs from the $7 bargain bin. If the cost of the CDs is $32 after the gift certificate is credited, how many CDs did Caitlin buy? 6 CDs

5. **EMPLOYMENT** Mrs. Jackson earned a $500 bonus for signing a one-year contract to work as a nurse. Her salary is $22 per hour. If her first week's check including the bonus is $1,204, how many hours did Mrs. Jackson work? 32 h

6. **PHOTOGRAPHY** Morgan subscribes to a website for processing her digital pictures. The subscription is $5.95 per month and 4 by 6 inch prints are $0.19 each. How many prints did Morgan purchase if the charge for January was $15.83? 52 prints

Chapter 4 Mid-Chapter Practice Test

Assessment Option

Mid-Chapter Test (Lessons 4-1 through 4-4) is available on p. 237 of the *Chapter 4 Resource Masters*.

ExamView® Pro Test Bank CD-ROM

Use the networkabe ExamView® Pro Test Bank CD-ROM to:

- Create **Multiple versions** of tests.
- Create **Modified** tests for **Inclusion** students with one mouse click.
- **Edit** existing questions and **Add** your own questions.
- Build tests aligned with state standards using built-in **State Curriculum Correlations.**
- Change **English** tests to **Spanish** with one mouse click and vice versa.

FOLDABLES™ Follow-Up

Before students complete the Mid-Chapter Practice Test, have them review the notes, definitions, concepts, and main ideas they recorded under the four tabs of their Foldables.

Chapter 4 Mid-Chapter Practice Test

1. Addition Property of Equality 2. Sample answer: $13a - 10 = -114$; $a = -8$

Vocabulary and Concepts

1. **State** the property of equality used to solve $a - 7 = -2$. (Lesson 4-2)
2. **Write** a two-step equation. Then solve your equation. (Lesson 4-4)
3. **Define** *inverse operation* and give an example. (Lesson 4-2)
   operations that undo each other; Sample answer: addition and subtraction

Skills and Applications

Write each sentence as an algebraic equation. (Lesson 4-1)

4. The product of a number and 3 is -16. $3a = -16$
5. 10 less than a number is 45. $z - 10 = 45$

Solve each equation. Check your solution. (Lessons 4-2, 4-3, and 4-4)

6. $21 + m = 33$ **12**
7. $a - 5 = -12$ **−7**
8. $7y = 63$ **9**
9. $5f = -75$ **−15**
10. $-28 = -2d$ **14**
11. $-1.6w = 4.8$ **−3**
12. $3z - 7 = 17$ **8**
13. $2g - 9 = -5$ **2**
14. $-4c - 1 = 11$ **−3**

15. **FLYING** An airplane is flying at an altitude of t feet before it increases its altitude by 1,000 feet. Write an expression for its new altitude. (Lesson 4-1)
    $t + 1{,}000$
16. **GEOMETRY** The sum of the measures of the angles of a triangle is 180°. Find the missing measure. (Lesson 4-2) **115°**

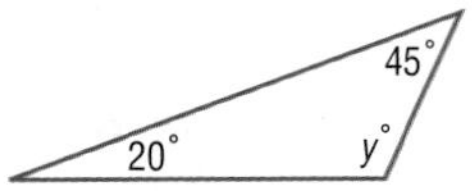

LAWN SERVICE For Exercises 17 and 18, use the following information.
Alex earned $326 this summer mowing lawns. The total was 4 times more than what he earned last summer. (Lesson 4-3)

17. Write a multiplication equation to find how much Alex earned last summer. $4d = 326$
18. Solve the equation. **$81.50**

Standardized Test Practice

19. **MULTIPLE CHOICE** Kim's time for the 5K race was four minutes less than Tanya's time. If Tanya's time is t, which expression represents Kim's time? (Lesson 4-1) **B**

    Ⓐ $4 - t$ Ⓑ $t - 4$
    Ⓒ $t + 4$ Ⓓ $4t$

20. **GRID IN** Last baseball season, Ryan had four less than twice the number of hits that Marcus had. Ryan had 48 hits. How many hits did Marcus have last season? (Lesson 4-4) **26**

170 Chapter 4 Algebra: Linear Equations and Functions

The Game Zone

A Place To Practice Your Math Skills

Math-O

GET READY!

Players: two, three, or four
Materials: scissors, 26 index cards, and 4 different colored markers

GET SET!

- Cut each index card in half, making 52 cards.
- To make a set of four cards, use the markers to put a different-colored stripe at the top of each card.
- Then write a different equation on each card. The solution of each equation should be 1.
- Continue to make sets of four cards having equations with solutions of 2, 3, 4, 5, 6, 0, −1, −2, −3, −4, and −5.
- Mark the remaining set of four cards "Wild."

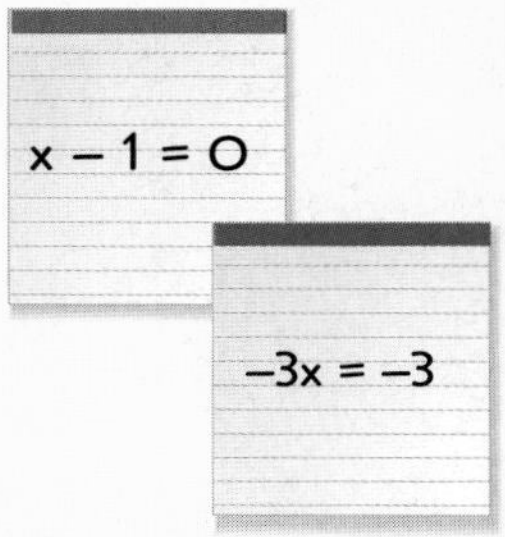

GO!

- The dealer shuffles the cards and deals five to each person. The remaining cards are placed in a pile facedown in the middle of the table. The dealer turns the top card faceup.
- The player to the left of the dealer plays a card with the same color or solution as the faceup card. Wild cards can be played any time. If the player cannot play a card, he or she takes a card from the pile and plays it, if possible. If it is not possible to play, the player places the card in his or her hand, and it is the next player's turn.
- **Who Wins?** The first person to play all cards in his or her hand is the winner.

The Game Zone

Glencoe Mathematics Classroom Manipulatives
- scissors

GETTING READY

Objective Write and solve one- and two-step algebraic equations.

MANAGEMENT TIPS

Recommended Time
30 minutes

Tips
- Consider having the index cards precut, colored, and packaged in sets to give to students at the beginning of the activity in order to save time and confusion in these steps.
- If the students will play more than one game, have groups exchange decks of cards so that they will solve different equations during the game.
- You may wish to use the game as an assessment tool for the concepts the students have been learning in Lessons 4-1 through 4-4.

For an additional game for this chapter, visit: msmath2.net/gamezone.

4-5 Lesson Notes

Lesson Resources

Chapter 4 Resource Masters

- Study Guide and Intervention, p. 205
- Practice: Skills, p. 206
- Practice: Word Problems, p. 207
- Reading to Learn Mathematics, p 208
- Enrichment, p. 209

Calculator and Spreadsheet Masters, p. 26

Key Teaching Transparencies, 16

MathPASS Tutorial CD-ROM, 40

Parent and Student Study Guide Workbook, 4-5

1 Focus

5-Minute Check 4-5

INTERACTIVE CHALKBOARD PowerPoint® Presentations

Use as a quiz or review of Lesson 4-4. (Also available as a transparency)

Bellringer

Communication Make a table like the one below. Starting with the words in the table in the lesson opener, work with students to make lists of words that might be represented by the symbols <, >, ≤, and ≥. Post the lists and add to them while working on this lesson.

Symbol	Words
<	• less than
>	• more than
≤	• at most
≥	• at least

Reading to Learn Mathematics, p. 208 ELL

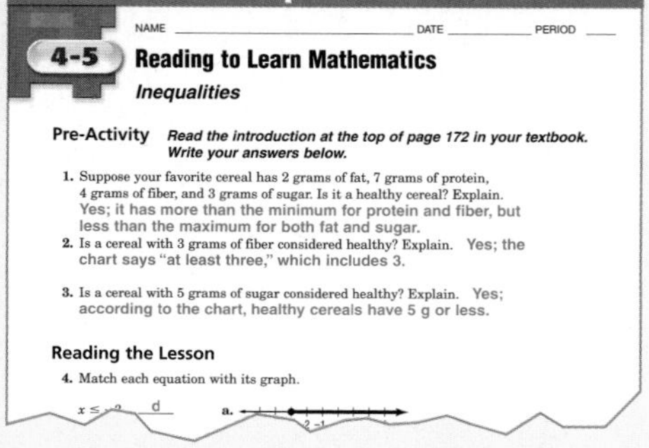
NAME ______ DATE ______ PERIOD ____

4-5 **Reading to Learn Mathematics**
Inequalities

Pre-Activity *Read the introduction at the top of page 172 in your textbook. Write your answers below.*

1. Suppose your favorite cereal has 2 grams of fat, 7 grams of protein, 4 grams of fiber, and 3 grams of sugar. Is it a healthy cereal? Explain. Yes; it has more than the minimum for protein and fiber, but less than the maximum for both fat and sugar.
2. Is a cereal with 3 grams of fiber considered healthy? Explain. Yes; the chart says "at least three," which includes 3.
3. Is a cereal with 5 grams of sugar considered healthy? Explain. Yes; according to the chart, healthy cereals have 5 g or less.

Reading the Lesson

4. Match each equation with its graph.

4-5 Inequalities

What You'll Learn
Solve inequalities.

NEW Vocabulary
inequality

MATH Symbols
< is less than
> is greater than
≤ is less than or equal to
≥ is greater than or equal to

WHEN am I ever going to use this?

BREAKFAST The table shows the nutrition requirements for a healthy breakfast cereal with milk.

Healthy Breafast Cereals (per serving)	
Fat	less than 3 g
Protein	more than 5 g
Fiber	at least 3 g
Sugar	at most 5 g

1. Suppose your favorite cereal has 2 grams of fat, 7 grams of protein, 4 grams of fiber, and 3 grams of sugar. Is it a healthy cereal? Explain.
2. Is a cereal with 3 grams of fiber considered healthy? Explain.
3. Is a cereal with 5 grams of sugar considered healthy? Explain.

1. Yes; it has more than the minimum for protein and fiber, but less than the maximum for both fat and sugar.
2. Yes; the chart says "at least three," which includes 3.
3. Yes; according to the chart, healthy cereals have 5 g or less.

An **inequality** is a mathematical sentence that contains the symbols <, >, ≤, or ≥.

Inequalities				
Words	is less than	is greater than	is less than or equal to is at most	is greater than or equal to is at least
Symbols	<	>	≤	≥

Any number that makes the inequality true is a solution of the inequality. Inequalities may have many solutions. The solutions are shown by shading a number line.

EXAMPLES Graph Solutions of Inequalities

Graph each inequality on a number line.

1 $x < 3$

2 $x \geq 3$

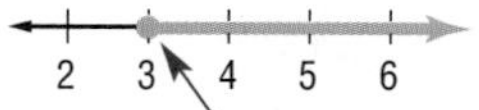

The open circle means the number is *not* included in the solution.

The closed circle means the number *is* included in the solution.

3 $x > 5$

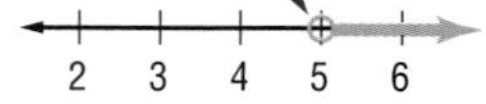

4 $x \leq 5$

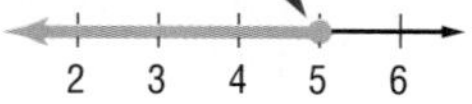

Your Turn **Graph each inequality on a number line.**

a. $x < -1$ b. $x \geq -2$ c. $x > 0$ d. $x \leq 3$

a–d. See margin.

Answers (Your Turn)

a.

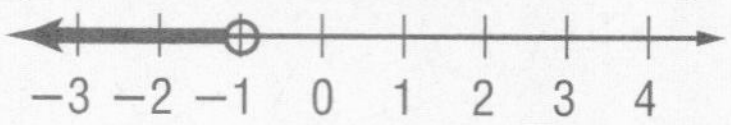

b.

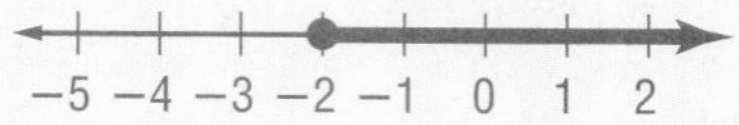

c.

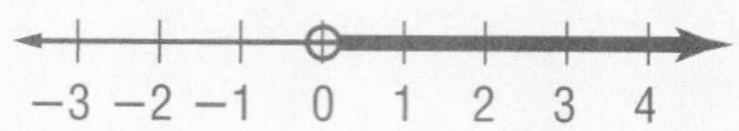

d.

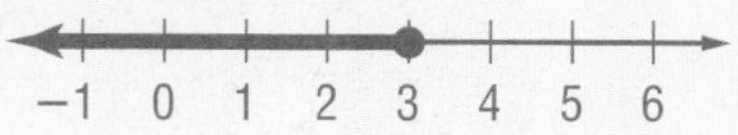

TEACHING TIP
iscuss that $x > 5$
cludes every number
eater than 5.

STUDY TIP

Checking Solutions You may also want to check $x = 5$. The solution of the equation is the *boundary* between numbers that are solutions and numbers that are not solutions.

To solve an inequality, follow the same steps you use to solve an equation.

EXAMPLES Solve One-Step Inequalities

5 **Solve $x + 4 > 9$. Check your solution. Then graph the solution.**

$x + 4 > 9$	Write the inequality.
$-4 = -4$	Subtract 4 from each side.
$x > 5$	Simplify.

Check Try 6, a number greater than 5.

$x + 4 > 9$	Write the inequality.
$6 + 4 \overset{?}{>} 9$	Replace x with 6. Is this sentence true?
$10 > 9$ ✓	

The solution is all numbers greater than 5.

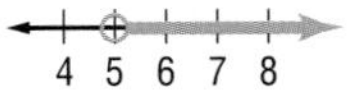

6 **Solve $4y \leq 8$. Graph the solution.**

$4y \leq 8$	Write the inequality.
$\frac{4y}{4} \leq \frac{8}{4}$	Divide each side by 4.
$y \leq 2$	Check this solution.

The solution is all numbers less than or equal to 2.

Your Turn **Solve each inequality. Graph the solution.**

e. $x + 6 > 8$ **$x > 2$** f. $x - 4 \geq -7$ **$x \geq -3$** g. $5x < 25$ **$x < 5$**

e–g. See margin for graphs.

REAL-LIFE CAREERS

ow Does a Nurse se Math?
irses use inequalities ien they record test sults like a person's blood essure.

Online Research
r information about a career a nurse, visit:
math2.net/careers

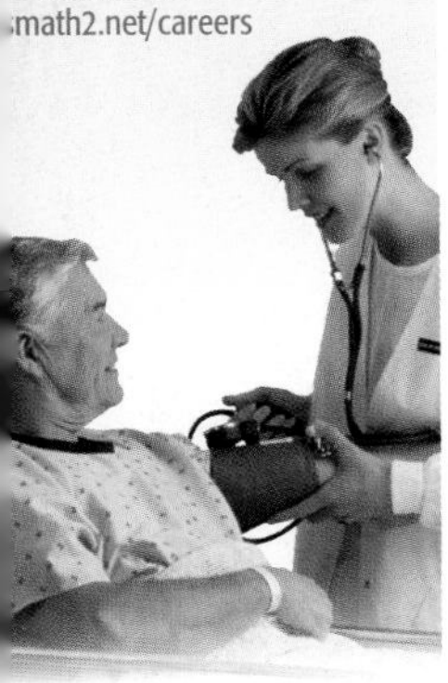

EXAMPLE Use an Inequality to Solve a Problem

7 **HEALTH The formula $110 + 0.5A < P$ is used to estimate whether a person has high blood pressure. In the formula, A is the person's age and P is the blood pressure. For which ages is a blood pressure of 120 considered high?**

$110 + 0.5A < P$	Write the formula.
$110 + 0.5A < 120$	Replace P with 120.
$-110 = -110$	Subtract 110 from each side.
$0.5A < 10$	Simplify.
$\frac{0.5A}{0.5} < \frac{10}{0.5}$	Divide each side by 0.5.
$A < 20$	$10 \div 0.5 = 20$

A blood pressure of 120 is considered high for a person who is younger than 20 years old.

msmath2.net/extra_examples

2 Teach

In-Class Examples

INTERACTIVE CHALKBOARD PowerPoint® Presentations

Graph each inequality on a number line.

1 $x < -2$

2 $x \geq -1$

3 $x > -3$

4 $x \leq 0$

5 Solve $x - 7 < 2$. Check your solution. Then graph the solution. **$x < 9$**

6 Solve $6x \geq 24$. Check your solution. Then graph the solution. **$x \geq 4$**

7 **BASEBALL CARDS** Jacob is buying uncirculated baseball cards online. The cards he has chosen are $6.70 each and the Web site charges a $1.50 service charge for each sale. If Jacob has no more than $35 to spend, how many cards can he buy? **no more than 5**

Answers (Your Turn)

e.

f.

g.

DAILY INTERVENTION

Unlocking Misconceptions

Inequalities Students often have difficulty interpreting the phrases "no more than" and "no less than." Discuss examples using these phrases by having students determine whether specific numbers are part of the solution. For example, given the statement, "The number of hours Arturo works in a week is no more than 15," ask students whether Arturo could have worked 4, 8, 12, 15, or 20 hours in a week.

3 Practice/Apply

About the Exercises...

Guided Practice Key

Exercises	Examples
5–7	1–4
8–10	5, 6

Odd/Even Assignments
Exercises 11–35 are structured so that students practice the same concepts whether they are assigned odd or even problems.

Alert! The Data Update following Exercise 35 requires access to the Internet.

Answers

1. 

2. The first inequality indicates all values greater than 3, while the second inequality indicates 3 and all values greater as solutions.
3. Sample answer: $2x < 4$ and $x + 1 < 3$
4. Diego; "At least two hours" indicates 2 hours or more. Diego's answer shows that 2 and all values greater than 2 are solutions.

5. 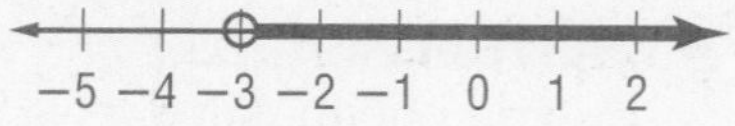

6. 

7. 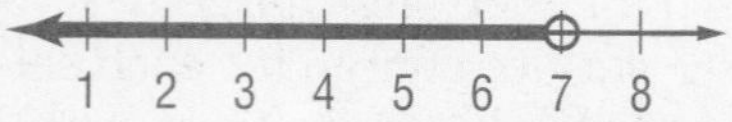

11. 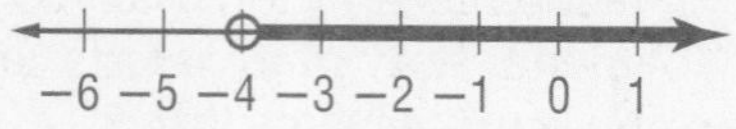

12. 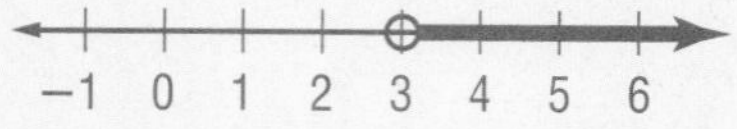

13. 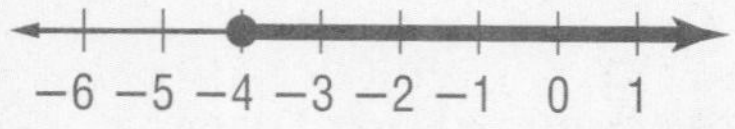

14. 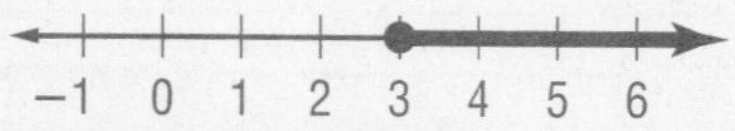

15. 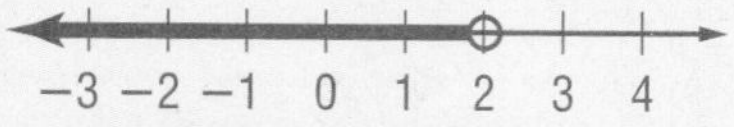

16. 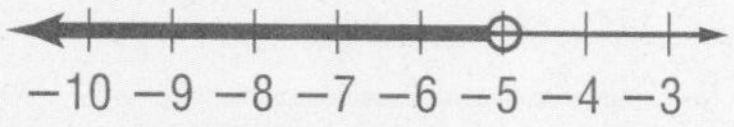

17. 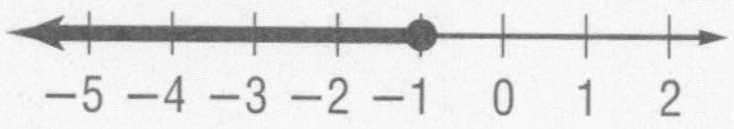

Skill and Concept Check

Writing Math Exercises 2 & 4

1. **Draw** a number line that shows all numbers less than or equal to −4. **1–4. See margin.**
2. Explain the difference between $x > 3$ and $x \geq 3$.
3. **OPEN ENDED** Write two different inequalities whose solution is $x < 2$.
4. **FIND THE ERROR** Courtney and Diego are writing an inequality for the expression *at least 2 hours of homework.* Who is correct? Explain.

Courtney	Diego
$h \leq 2$	$h \geq 2$

GUIDED PRACTICE

Graph each inequality on a number line. **5–7. See margin.**

5. $y > -3$ 6. $x \geq 0$ 7. $d < 7$

Solve each inequality.

8. $x + 3 > -4$ $x > -7$ 9. $6d \geq 24$ $d \geq 4$ 10. $3x + 8 < 15$ $x < \frac{7}{3}$

★ indicates increased difficulty

Practice and Applications

HOMEWORK HELP

For Exercises	See Examples
11–19	1–4
20–25, 32–33	5–6
26–27, 40–41	7

Extra Practice See pages 573, 599.

Graph each inequality on a number line. **11–19. See margin.**

11. $s > -4$ 12. $y > 3$ 13. $t \geq -4$
14. $a \geq 3$ 15. $h < 2$ 16. $g < -5$
17. $z \leq -1$ 18. $p \leq 0$ 19. $w \leq -6$

Solve each inequality.

20. $y + 5 < 14$ $y < 9$ 21. $x + 6 < 0$ $x < -6$ 22. $a - 3 \geq -5$ $a \geq -2$ 23. $g - 5 > 2$ $g > 7$
24. $3r \leq 18$ $r \leq 6$ 25. $6u \geq 36$ $u \geq 6$ 26. $2x + 8 < 24$ $x < 8$ 27. $3y + 1 \leq 5$ $y \leq \frac{4}{3}$

Write the inequality for each graph.

28. 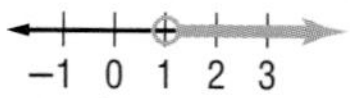

$z > 1$

29. 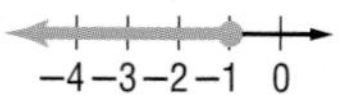

$p \leq -1$

30. 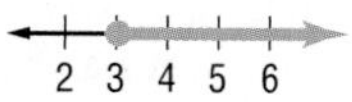

$w \geq 3$

31. 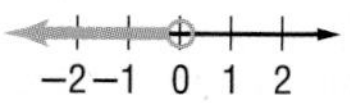

$d < 0$

Write an inequality for each sentence. Then solve the inequality.

32. Five times a number is greater than 25. $5x > 25$; $x > 5$
33. The sum of a number and 1 is at least 5. $w + 1 \geq 5$; $w \geq 4$
34. **WEATHER** A tropical depression has maximum sustained winds of less than 39 miles per hour. Write an inequality showing the wind speeds. $w < 39$ mph
35. **DRIVING** In Ohio, you can get a driver's license if you are at least 16 years old. Write an inequality showing the age of all drivers in Ohio. $a \geq 16$

Data Update What are the minimum ages for getting a driver's license in other states? Visit msmath2.net/data_update to learn more.

18. 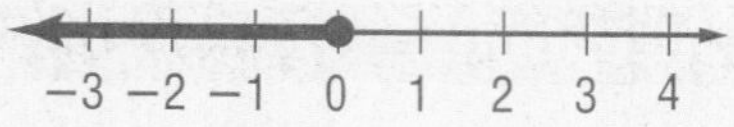

19. 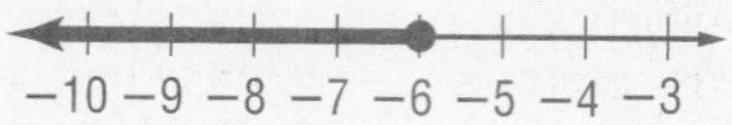

42a. intersection
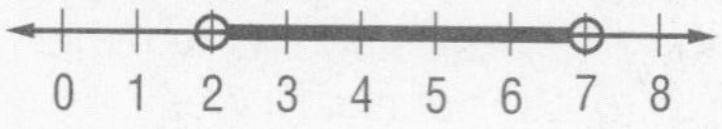

42b. union
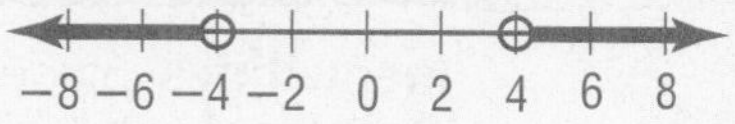

SPORTS For Exercises 36–39, use the graphic.

36. In which sport(s) are more than 400,000 children hurt? **bicycling, basketball**

37. In which sport(s) are at least 185,000 children hurt? **all except baseball-softball**

38. Of the sports listed, which have less than 300,000 injuries? **roller sports, playgrounds, soccer, baseball-softball**

39. Write an inequality that describes the number of children hurt each year playing sports. **$h > 2{,}000{,}000$**

SHOPPING For Exercises 40 and 41, use the following information. **40. $29j + 8 \leq 70$**

Suppose a pair of jeans costs \$29 and a necklace costs \$8. You have \$70 to spend on both.

★ 40. Write an inequality to find how many pairs of jeans you can buy along with one necklace.

★ 41. Solve the inequality. **$j \leq 2.14$, so 1 or 2 pairs of jeans**

42. **CRITICAL THINKING** A *compound inequality* is formed by two inequalities connected by the words *and* or *or*.

Type	Example	Solution	Graph
Intersection	$x > -3$ *and* $x < 4$	Both inequalities are true. $\{x \mid -3 < x < 4\}$	−5 −4 −3 −2 −1 0 1 2 3 4 5
Union	$x < 2$ *or* $x > 5$	Either inequality is true. $\{x \mid x < 2$ *or* $x > 5\}$	−5 −4 −3 −2 −1 0 1 2 3 4 5

Identify each compound inequality as an intersection or union. Then find and graph the solution. **a–d. See margin.**

a. $x > 2$ and $x < 7$ b. $x < -4$ or $x > 4$ c. $x \geq -3$ or $x < -6$ d. $x \leq 10$ and $x > 0$

Standardized Test Practice and Mixed Review

43. **MULTIPLE CHOICE** Which inequality is graphed on the number line? **B**

Ⓐ $x < 3{,}000$ Ⓑ $x \leq 3{,}000$ Ⓒ $x > 3{,}000$ Ⓓ $x \geq 3{,}000$

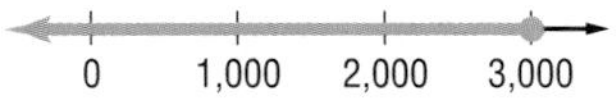

44. **MULTIPLE CHOICE** Ayano has \$26 to spend. A hat costs \$8. Which inequality shows how much he can spend for a T-shirt? **H**

Ⓕ $s + 8 < 26$ Ⓖ $s + 8 > 26$ Ⓗ $s + 8 \leq 26$ Ⓘ $s + 8 \geq 26$

ALGEBRA Solve each equation. Check your solution. (Lessons 4-2, 4-3, and 4-4)

45. $13 = s + 5$ **8** 46. $18x = -54$ **−3** 47. $2q + 6 = -20$ **−13** 48. $-7q + 4.6 = -0.3$ **0.7**

GETTING READY FOR THE NEXT LESSON

PREREQUISITE SKILL Graph each point on a coordinate plane. (Lesson 3-3) **49–52. See Ch. 4 Answer Appendix.**

49. $(-4, 2)$ 50. $(3, -1)$ 51. $(-3, -4)$ 52. $(2, 0)$

msmath2.net/self_check_quiz

4 Assess

Higher-Level Thinking

The school cafeteria sells cookies for \$0.60 each. If Ashley has \$2.70 to spend, how many cookies can she buy? (*Hint:* Ashley cannot buy a fraction of a cookie.) **up to 4 cookies**

Answers

42c. **union**

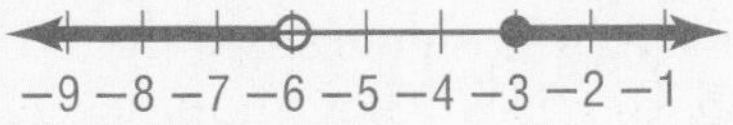

42d. **intersection**

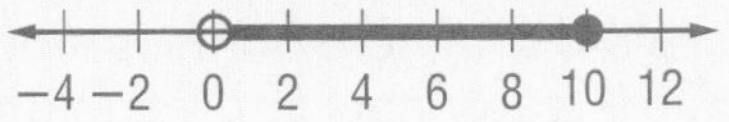

Study Guide and Intervention, p. 205

NAME ______ DATE ______ PERIOD ____

4-5 Study Guide and Intervention

Inequalities

An **inequality** is a mathematical sentence that contains the symbols <, >, ≤, or ≥.

Words	Symbols
m is greater than 7.	$m > 7$
r is less than −4.	$r < -4$
t is greater than or equal to 6.	$t \geq 6$
y is less than or equal to 1.	$y \leq 1$

EXAMPLE 1 **Solve $v + 3 < 5$. Then graph the solution.**

$v + 3 < 5$ Write the inequality.
$-3 = -3$ Subtract 3 from each side.
$v < 2$ Simplify.

Check Try 1, a number less than 2.
$v + 3 < 5$ Write the inequality.
$1 + 3 \overset{?}{<} 5$ Replace v with 1.
$4 < 5$ ✓ The solution checks.

The solution is all numbers less than 2. (−4 −3 −2 −1 0 1 2 3 4)

EXAMPLE 2 **Solve $3y + 2 \geq 26$. Then graph the solution.**

$3y + 2 \geq 26$ Write the inequality.
$-2 \geq -2$ Subtract 2 from each side.
$3y \geq 24$ Simplify.
$\frac{3y}{3} \geq \frac{24}{3}$ Divide each side by 4.
$y \geq 8$ $24 \div 3 = 8$

The solution is all numbers greater than or equal to 8. (5 6 7 8 9 10 11 12 13 14)

EXERCISES

Graph each inequality on the number line.

1. $c < 5$ (−2 −1 0 1 2 3 4 5 6 7 8)
2. $y > -5$ (−7 −6 −5 −4 −3 −2 −1 0 1 2 3)
3. $x \geq 10$ (7 8 9 10 11 12 13 14 15 16 17)
4. $n \leq -1$ (−7 −6 −5 −4 −3 −2 −1 0 1 2 3)

Solve each inequality. Graph the solution.

5. $y + 6 < 11$ $y < 5$
6. $a - 6 > 0$ $a > 6$
7. $4q \leq 32$ $q \leq 8$
8. $9w > 45$ $w > 5$
9. $2r + 5 \leq 19$ $r \leq 7$
10. $5x - 7 \geq -2$ $x \geq 1$

Practice: Skills, p. 206

NAME ______ DATE ______ PERIOD ____

4-5 Practice: Skills

Inequalities

Graph each inequality on a number line.

1. $x > 2$
2. $y > -3$
3. $b \geq 1$
4. $c \geq -5$
5. $z < 3$
6. $q < -2$
7. $a \leq 6$
8. $r \leq 0$

Solve each inequality.

9. $a + 4 < 10$ $a < 6$
10. $c + 5 < 9$ $c < 4$
11. $d - 1 \geq 8$ $d \geq 9$
12. $g - 11 > 2$ $g > 13$
13. $t + 4 \geq -6$ $t \geq -10$
14. $a + 12 < 8$ $a < -4$
15. $x - 7 \leq -8$ $x \leq -1$
16. $3t \leq 15$ $t \leq 5$
17. $3w \geq 30$ $w \geq 10$
18. $4n + 8 < 24$ $n < 4$
19. $6y + 1 \leq 19$ $y \leq 3$
20. $2r - 8 > 6$ $r > 7$
21. $b - 5 > -2$ $b > 3$
22. $2y + 1 \leq -5$ $y \leq -3$
23. $4x - 6 > -10$ $x > -1$

Practice: Word Problems, p. 207

NAME ______ DATE ______ PERIOD ____

4-5 Practice: Word Problems

Inequalities

1. **HURRICANES** The National Hurricane Center issues a hurricane warning if sustained winds of 74 miles per hour or higher associated with a hurricane are expected in a coastal area. Write an inequality showing the wind speeds for which a hurricane warning is issued. $w \geq 74$

2. **BUSINESS** An Internet company reported that its earnings will be less than the 24 cents per share that was predicted. Write an inequality showing the possible earnings per share. $e < 24$

3. **EXERCISE** Jumping rope can burn 600 Calories per hour. Write and solve an inequality to find the number of hours of jumping rope that it would take for you to burn at least 450 Calories. $600x \geq 450$; at least 0.75 h

4. **TRAVEL** The speed limit on highways in Florida is 70 miles per hour. Write and solve an inequality to find how long will it take you to travel the 105 miles from Orlando to St. Augustine if you travel at or below the speed limit. $70x \geq 105$; 1.5 h or more

5. **SHOPPING** Suppose a DVD costs \$19 and a CD costs \$14. Write an inequality to find how many CDs you can buy along with one DVD if you have \$65 to spend. Solve the inequality. $14c + 19 \leq 65$; up to 3 CDs

6. **TRAVEL** Jeanine has driven 35 miles of a 230-mile trip. If she is traveling at or below the 65-mile-per-hour speed limit, much longer can Jeanine expect to spend completing the trip? 3 h or more

Lesson Notes

Lesson Resources

Teaching Mathematics with Manipulatives
- centimeter grid, p. 11
- student recording sheet 4-6a

Glencoe Mathematics Classroom Manipulative Kit
- stopwatch

GETTING STARTED

Teaching Tip You may wish to graph the data on an overhead transparency as they are collected.

TEACH

- If time allows, you may wish to repeat the experiment for different numbers of students a few times and average the results. This may provide a more accurate function.
- Point out that students probably will be able to choose several different places for the line. Each line may pass through or close to a number of points. Each is a reasonable estimate and will give a different prediction for the time needed to complete the wave.

What You'll Learn
Graph a function on a scatter plot.

Materials
- stopwatch
- grid paper
- uncooked spaghetti

Functions and Graphs

INVESTIGATE *Work as a class.*

Have you ever been at a sporting event when the crowd does the "wave"? In this Lab, you will investigate how long it would take the students at your school to complete the "wave."

STEP 1 Begin with five students sitting in a row.

STEP 2 At the timer's signal, the first student stands up, waves his or her arms overhead, and sits down. Each student repeats the wave in order.

STEP 3 When the last student sits down, the timer records the time in seconds.

STEP 4 Repeat for 6, 7, 8, and so on, up to 25 students.

Writing Math

Work with a partner. **1–5. See students' work.**

1. **Graph** the ordered pairs (number of students, time) on a coordinate grid like the one at the right.

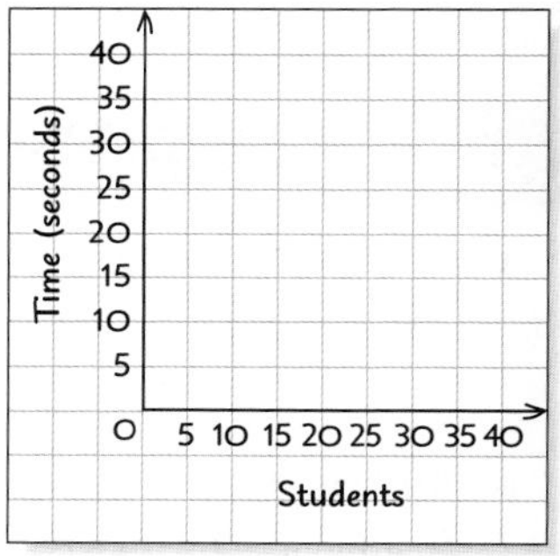

2. **Describe** how the points appear on your graph.
3. Place one piece of uncooked spaghetti on your graph so that it covers as many of the points as possible. **Predict** how long it would take 30 students to complete the "wave." Make a prediction for 50 students.
4. **Find a pattern** in the data and use the pattern to predict how long it would take the students in your school to complete the "wave." Explain your reasoning.
5. A *function* describes the relationship between two quantities. In a function, one quantity *depends on* the other. **Complete the sentence:** The time it takes to do the "wave" depends on __?__.

ASSESS

Observe as students solve the **Writing Math** exercises. They should:
- correctly graph the data,
- place the spaghetti on the graph to make a reasonable approximation of the function,
- and use the graph to make a reasonable prediction for the time it would take for the whole school to do the wave.

Use **Exercise 5** to assess whether students have grasped the concept of a *function*.

4-6 Functions and Linear Equations

hat You'll Learn
aph linear equations.

:W Vocabulary
nction
nction table
›main
nge
ear equation

:VIEW Vocabulary
dered pair: a pair numbers used to :ate a point in a ordinate system
sson 3-3)

WHEN am I ever going to use this?

FAST FOOD Suppose you can buy hamburgers for $2 each.

1. Copy and complete the table to find the cost of 2, 3, and 4 hamburgers.
2. On grid paper, graph the ordered pairs (number, cost). Describe how the points appear on the grid.
3. What happens to the cost as the number of hamburgers increases?

Hamburgers		
Number	Multiply by 2	Cost ($)
1	2 × 1	2
2	2 × 2	4
3	2 × 3	6
4	2 × 4	8

1–2. See Ch. 4 Answer ›pendix for graph; they e in a straight line.
3. The cost increases 2 for each time the ımber of hamburgers creases by 1.

The total cost of the hamburgers depends on the number of hamburgers. A relationship where one thing depends on another is called a **function**. In a function, you start with an *input* number, perform one or more operations on it, and get an *output* number.

You can organize the input numbers, output numbers, and the function rule in a **function table**.

EXAMPLE Make a Function Table

1 MONEY MATTERS Suppose you earn $5 each week. Make a function table that shows your total earnings after 1, 2, 3, and 4 weeks.

Input	Function Rule	Output
Number of Weeks	Multiply by 5	Total Earnings ($)
1	5 × 1	5
2	5 × 2	10
3	5 × 3	15
4	5 × 4	20

The set of input values is called the **domain**, and the set of output values is called the **range**. In Example 1, the domain is {1, 2, 3, 4}, and the range is {5, 10, 15, 20}.

4-6 Lesson Notes

Lesson Resources

Chapter 4 Resource Masters
- Study Guide and Intervention, p. 210
- Practice: Skills, p. 211
- Practice: Word Problems, p. 212
- Reading to Learn Mathematics, p. 213
- Enrichment, p. 214

FAST FILE

Calculator and Spreadsheet Masters, p. 27
Classroom Math Games, pp. 11–14
MathPASS Tutorial CD-ROM, 41
Parent and Student Study Guide Workbook, 4-6
Teaching Mathematics with Manipulatives, pp. 56–57

1 Focus

5-Minute Check 4-6

INTERACTIVE CHALKBOARD
PowerPoint® Presentations

Use as a quiz or review of Lesson 4-5. (Also available as a transparency)

Bellringer
Reading Mathematics The term *function* is one that will be important in students' future studies in mathematics. Have students discuss the term as it relates to its everyday uses. Compare these uses to its mathematical meaning.

D A I L Y INTERVENTION — Differentiated Instruction

Auditory/Musical "Think aloud" as you demonstrate the steps of completing a function table and graphing a function. Tell students exactly what you are thinking and why as you complete each step. Verbalize explicitly what you would normally perform automatically.

Reading to Learn Mathematics, p. 213 ELL

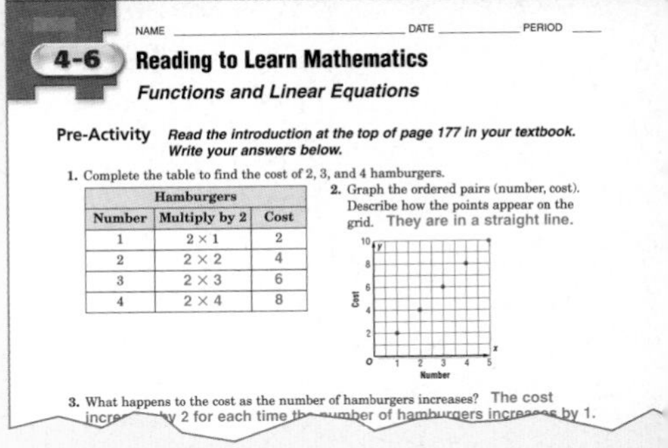
NAME ________ DATE ________ PERIOD ____

4-6 **Reading to Learn Mathematics**
Functions and Linear Equations

Pre-Activity *Read the introduction at the top of page 177 in your textbook. Write your answers below.*

1. Complete the table to find the cost of 2, 3, and 4 hamburgers.

Hamburgers		
Number	Multiply by 2	Cost
1	2 × 1	2
2	2 × 2	4
3	2 × 3	6
4	2 × 4	8

2. Graph the ordered pairs (number, cost). Describe how the points appear on the grid. They are in a straight line.

3. What happens to the cost as the number of hamburgers increases? The cost incre[illegible] by 2 for each time th[illegible] number of hamburgers increa[illegible] by 1.

2 Teach

In-Class Examples

INTERACTIVE CHALKBOARD PowerPoint® Presentations

1 **WORK** Asha makes \$6.00 an hour working at a grocery store. Make a function table that shows Asha's total earnings for working 1, 2, 3, and 4 hours.

Number of Hours	Multipy by 6	Total Earnings (\$)
1	6×1	6
2	6×2	12
3	6×3	18
4	6×4	24

2 Graph $y = x + 3$.

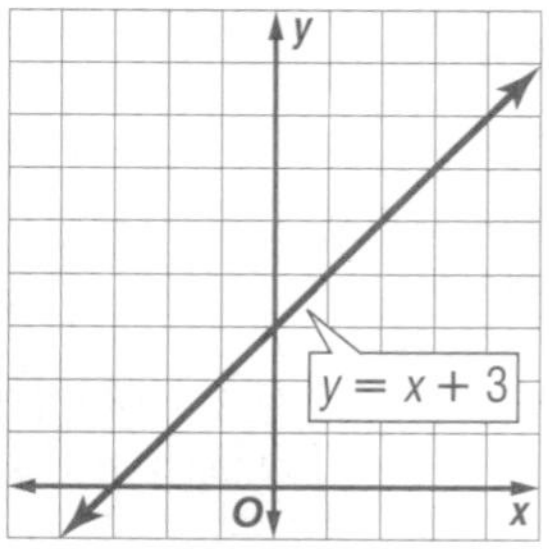

Functions are often written as equations with two variables—one to represent the input and one to represent the output. Here's an equation for the situation in Example 1.

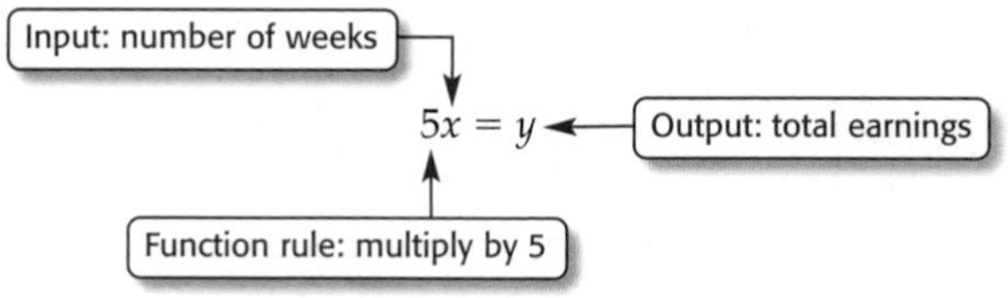

STUDY TIP

Input and Output When *x* and *y* are used in an equation, *x* usually represents the input, and *y* usually represents the output.

The solution of an equation with two variables consists of two numbers, one for each variable, that make the equation true. The solution is usually written as an ordered pair (x, y), which can be graphed.

TEACHING TIP

Point out that in science, *y* is called the *dependent variable* because it depends on the input, or *independent variable*, *x*.

EXAMPLE Graph Solutions of Linear Equations

2 **Graph $y = 2x + 1$.**

Select any four values for the input x. We chose 2, 1, 0, and -1. Substitute these values for x to find the output y.

x	$2x + 1$	y	(x, y)
2	$2(2) + 1$	5	(2, 5)
1	$2(1) + 1$	3	(1, 3)
0	$2(0) + 1$	1	(0, 1)
-1	$2(-1) + 1$	-1	$(-1, -1)$

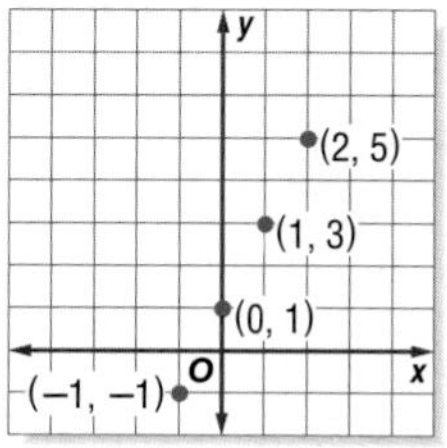

Four solutions are (2, 5), (1, 3), (0, 1), and $(-1, -1)$. The graph is shown above at the right.

Your Turn **Graph each equation.** a–c. See margin.

a. $y = x - 3$ b. $y = -3x$ c. $y = -3x + 2$

STUDY TIP

Graphing Equations Only two points are needed to graph the line. However, graph more points to check accuracy.

Notice that all four points in the graph lie on a line. Draw a line through the points to graph *all* solutions of the equation $y = 2x + 1$. The graph of (3, 7) is also on the line.

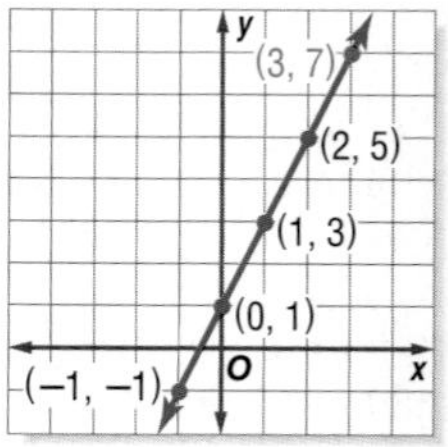

$y = 2x + 1$	Write the equation.
$7 \stackrel{?}{=} 2(3) + 1$	Replace *x* with 3 and *y* with 7.
$7 = 7$ ✓	This sentence is true.

So, (3, 7) is also a solution of $y = 2x + 1$.

An equation like $y = 2x + 1$ is called a **linear equation** because its graph is a straight line.

Answers (Your Turn)

a.

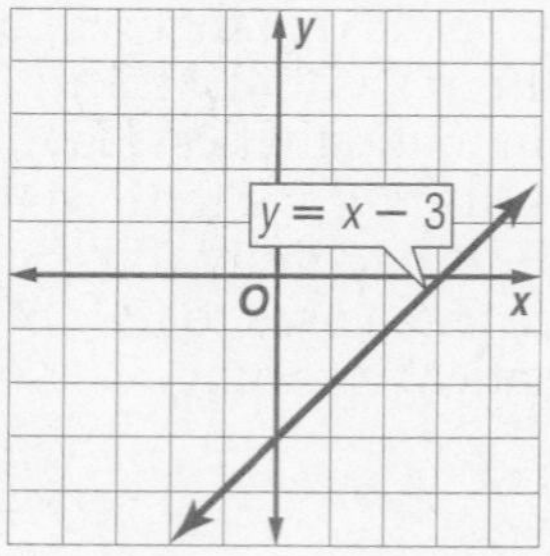

b.

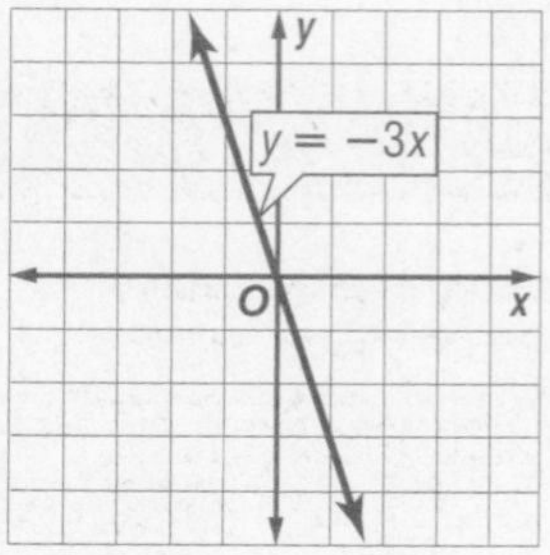

c.

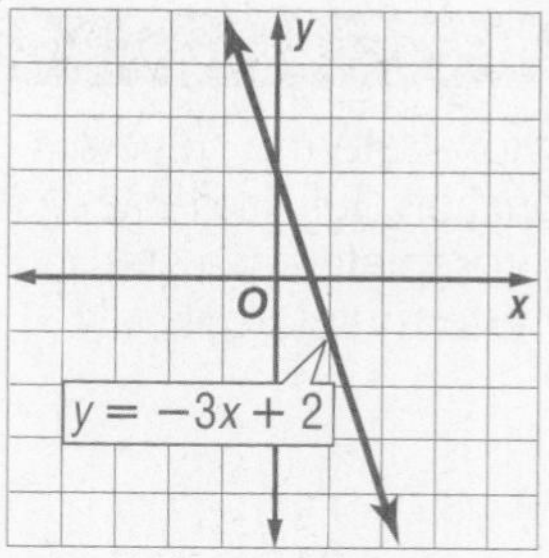

EXAMPLE Represent Real-World Functions

3 SKATEBOARDING The top speed reached by a standing skateboarder is 80 feet per second. The equation $d = 80t$ describes the distance d that a skateboarder can travel in time t. Represent the function with a graph.

Step 1 Select any four values for t. Select only positive numbers because t represents time. Make a function table.

t	$80t$	d	(t, d)
1	80(1)	80	(1, 80)
2	80(2)	160	(2, 160)
3	80(3)	240	(3, 240)
4	80(4)	320	(4, 320)

Step 2 Graph the ordered pairs and draw a line through the points.

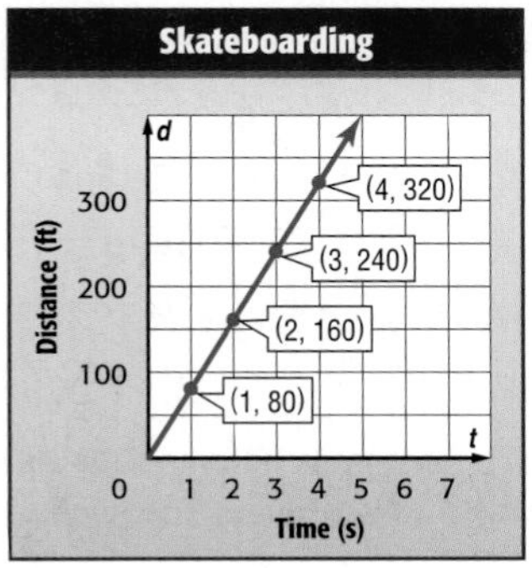

EAL-LIFE MATH

ATEBOARDING out one fourth of all teboarding purchases made by teenagers m 14 to 17 years old.

rce: *Statistical Abstract of the United es*

In-Class Example

3 ANIMALS Blue whales can reach a speed of 30 miles per hour in bursts when in danger. The equation $d = 30t$ describes the distance d that a whale traveling at that speed can travel in time t. Represent the function with a graph.

INTERACTIVE CHALKBOARD PowerPoint® Presentations

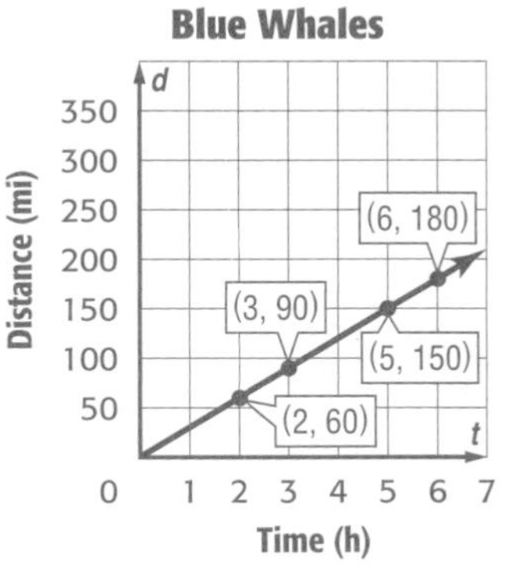

Skill and Concept Check

Writing Math Exercise 2

1. **OPEN ENDED** Write an equation that has (1, 2) as a solution. Sample answer: $y = 2x$
2. **Explain** the relationship among *input*, *output*, and *function rule*. See margin.

GUIDED PRACTICE

Copy and complete each function table. Identify the domain and range.

3. $y = x - 2$

domain: {1, 2, 3, 4}; range: {−1, 0, 1, 2}

x	$x - 2$	y
1	1 − 2	−1
2	2 − 2	0
3	3 − 2	1
4	4 − 2	2

4. $y = 4x$

domain: {−1, 0, 1, 2}; range: {−4, 0, 4, 8}

x	$4x$	y
−1	4(−1)	−4
0	4(0)	0
1	4(1)	4
2	4(2)	8

Graph each equation. 5–7. See margin.

5. $y = x - 1$
6. $y = -1x$
7. $y = -2x + 3$

msmath2.net/extra_examples

3 Practice/Apply

About the Exercises...

Guided Practice Key

Exercises	Examples
3–4	1
5–7	2

Odd/Even Assignments

Exercises 8–34 are structured so that students practice the same concepts whether they are assigned odd or even problems.

Answers

2. You start with an input number, perform the operations in the function rule, and the result is the output number.

5.

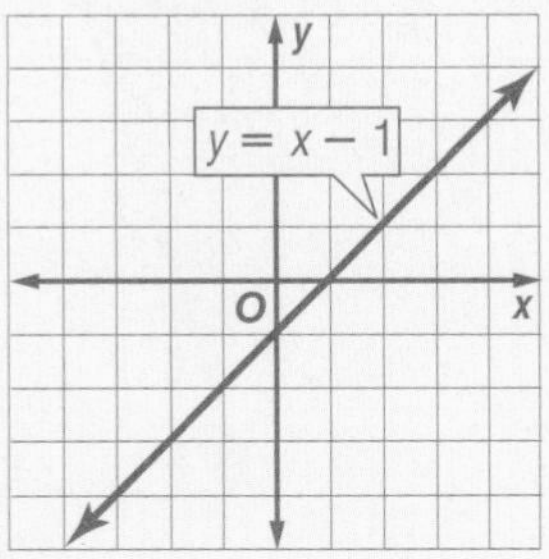

6.

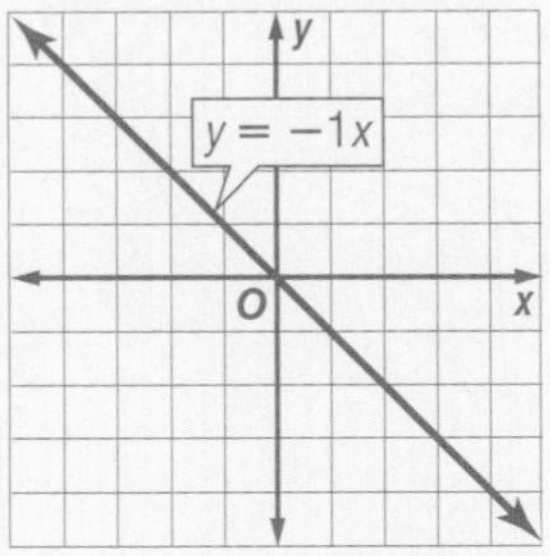

7.

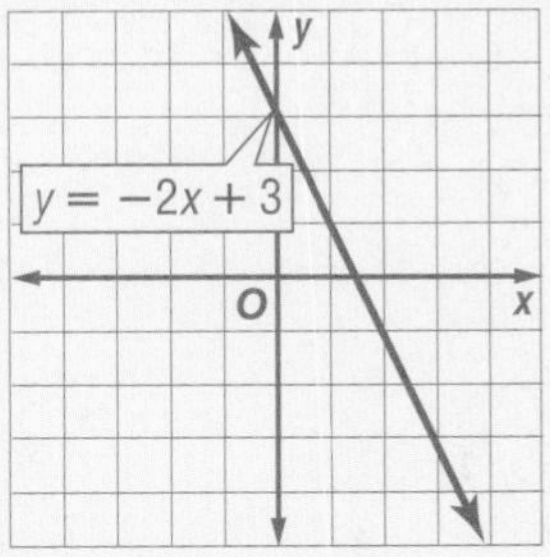

Answers

8. domain: {1, 2, 3, 4}; range: {−3, −2, −1, 0}
9. domain: {1, 2, 3, 4}; range: {6, 7, 8, 9}
10. domain: {−1, 0, 1, 2}; range: {−2, 0, 2, 4}
11. domain: {−1, 0, 1, 2}; range: {6, 0, −6, −12}
12. domain: {1, 2, 3, 4}; range: {1, 3, 5, 7}
13. domain: {−1, 0, 1, 2}; range: {0, −2, −4, −6}

14.
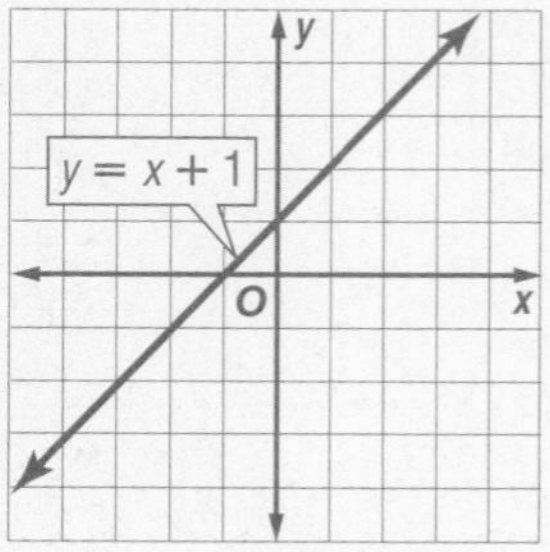

15.
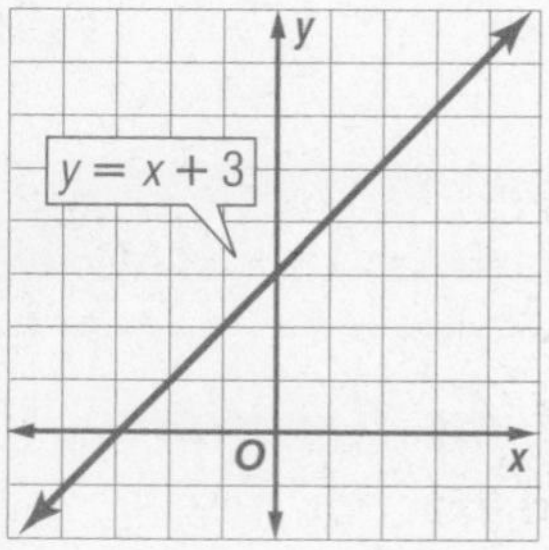

16.
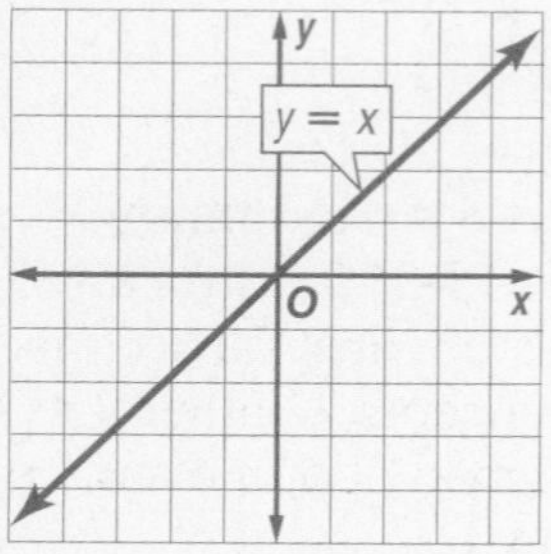

17.
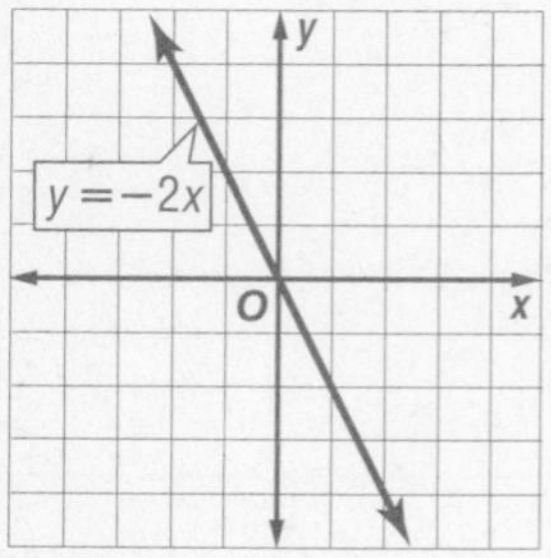

★ indicates increased difficulty

Practice and Applications

HOMEWORK HELP

For Example	See Examples
8–13	1
14–25	2
29–35	3

Extra Practice See pages 573, 599.

Copy and complete each function table. Identify the domain and range. 8–13. See margin for domain and range.

8. $y = x - 4$

x	$x - 4$	y
1	1 − 4	−3
2	2 − 4	−2
3	3 − 4	−1
4	4 − 4	0

9. $y = x + 5$

x	$x + 5$	y
1	1 + 5	6
2	2 + 5	7
3	3 + 5	8
4	4 + 5	9

10. $y = 2x$

x	$2x$	y
−1		−2
0	2(0)	0
1	2(1)	2
2	2(2)	4

2(−1)

11. $y = -6x$

x	$-6x$	y
−1		
0		
1		
2		

11. −6(−1) 6; −6(0) 0; −6(1) −6; −6(2) −12

12. $y = 2x - 1$

x	$2x - 1$	y
1	2(1)−1	1
2	2(2)−1	3
3	2(3)−1	5
4	2(4)−1	7

13. $y = -2x - 2$

x	$-2x - 2$	y
−1		0
0	−2(0)−2	−2
1	−2(1)−2	−4
2	−2(2)−2	−6

−2(−1) − 2

Graph each equation. 14–20. See margin. 21–25. See Ch. 4 Answer Appendix.

14. $y = x + 1$
15. $y = x + 3$
16. $y = x$
17. $y = -2x$
18. $y = 2x + 3$
19. $y = 3x - 1$
20. $y = 4x - 2$
21. $y = 2x + 5$
22. $y = x + 0.5$
23. $y = 0.25x$
24. $y = 0.5x - 1$
25. $y = 2x - 1.5$

Make a function table for each sentence. Then write an equation using x to represent the first number and y to represent the second number.

26. The second number is three more than the first number. 26–28. See Ch. 4 Answer Appendix.
27. The second number is five less than the first number.
28. The second number is ten times the first number.

SPENDING For Exercises 29–31, use the graph.

29. Make a function table that shows the total average defense spending per person for 1, 2, 3, and 4 days.
30. Write an equation in which x represents the days and y represents the total spending. $y = 2.67x$
31. Graph the equation.

29, 31. See Ch. 4 Answer Appendix.

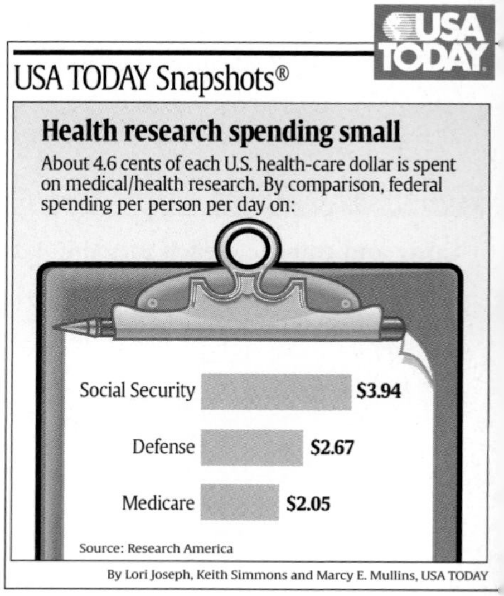

INTERNET For Exercises 32–34, use the following information. 33. $y = 20x$

An Internet provider charges $20 each month.

32. Make a function table that shows the total charge for 1, 2, 3, and 4 months of service.
33. Write an equation in which x represents months and y represents the total charge.
34. Graph the equation. 32, 34. See Ch. 4 Answer Appendix.

18.
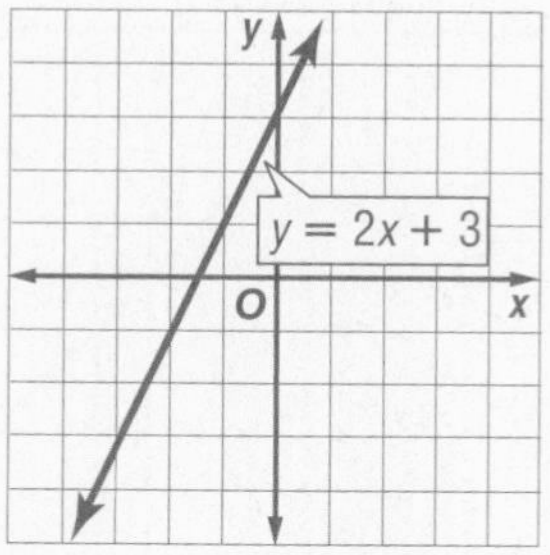

19.
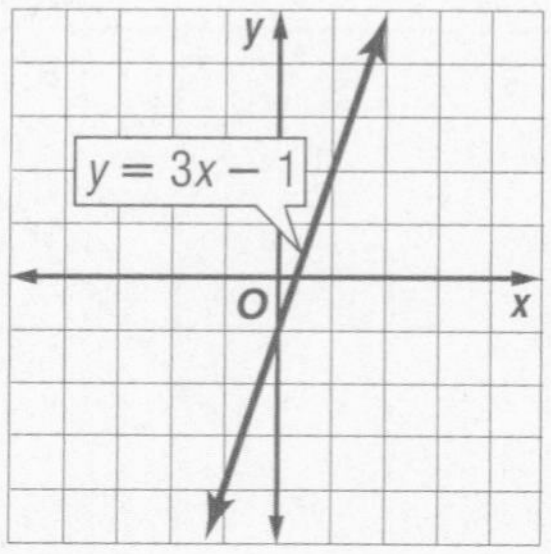

20.
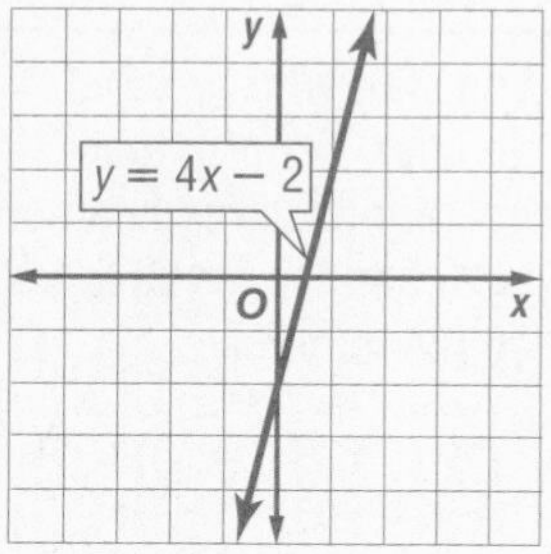

35. **GEOMETRY** The formula for the area A of a rectangle whose length is 5 units is $A = 5w$, where w is the width. Graph the function. **See Ch. 4 Answer Appendix.**

36. **WRITE A PROBLEM** Write about a real-life situation that can be represented by the equation $y = 3x$. **See margin.**

CRITICAL THINKING For Exercises 37–40, write an equation for the function shown in each function table.

37.

x	y
1	3
2	4
3	5
4	6

$y = x + 2$

38.

x	y
2	6
4	12
6	18
8	24

$y = 3x$

39.

x	y
−1	4
0	0
1	−4
2	−8

$y = -4x$

40.

x	y
1	3
2	5
3	7
4	9

$y = 2x + 1$

EXTENDING THE LESSON Not all equations have graphs that are straight lines. The graphs at the right show two *nonlinear equations.* **41–46. See Ch. 4 Answer Appendix.**

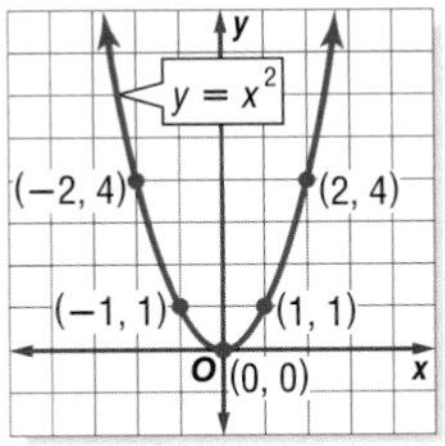

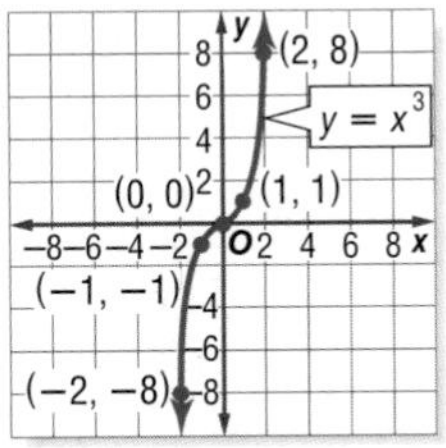

Graph each equation. Determine whether it is *linear* or *nonlinear*.

★ 41. $y = x^2 + 1$ ★ 42. $y = x + 1$

★ 43. $y = 2x$ ★ 44. $xy = 12$

★ 45. $y = 2x^2$ ★ 46. $y = x^3 - 1$

Standardized Test Practice and Mixed Review

47. **MULTIPLE CHOICE** Choose the equation that is graphed at the right. **D**

Ⓐ $y = 2x$ Ⓑ $y = x$

Ⓒ $y = x - 1$ Ⓓ $y = 2x - 2$

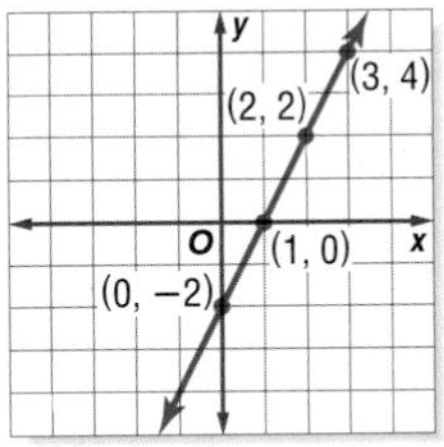

48. **MULTIPLE CHOICE** Which ordered pairs are solutions of the equation $y = 2x - 4$? **H**

Ⓕ (−2, −3), (0, 2) Ⓖ (−2, −1), (2, − 3)

Ⓗ (1, −2), (3, 2) Ⓘ (−3, 2), (0, − 4)

49. **CIVICS** To serve as a U.S. Representative, a person must be at least 25 years old and a citizen of the United States for at least 7 years. Write an inequality showing the age of a person who may be a U.S. Representative. (Lesson 4-5) $x \geq 25$

ALGEBRA Solve each equation. Check your solution. (Lesson 4-4)

50. $8 = 3h - 1$ **3** 51. $2q + 6 = -20$ **−13** 52. $32 = -4 + 9m$ **4**

GETTING READY FOR THE NEXT LESSON

PREREQUISITE SKILL Divide. (Lesson 3-7)

53. $-4 \div 2$ **−2** 54. $10 \div (-5)$ **−2** 55. $-12 \div (-4)$ **3** 56. $-16 \div 16$ **−1**

4 Assess

Open-Ended Assessment

Speaking Have students explain how a graph of a linear equation shows solutions of that equation. **Sample answer: It is a line connecting all points whose ordered pairs satisfy the equation.**

Answer

36. **Sample answer: Sam charges $3 for each dog that he walks. In the equation $y = 3x$, x represents the number of dogs and y represents the total amount of money earned.**

Study Guide and Intervention, p. 210

NAME ________ DATE ________ PERIOD ____

4-6 Study Guide and Intervention

Functions and Linear Equations

The solution of an equation with two variables consists of two numbers, one for each variable, that make the equation true. The solution is usually written as an ordered pair (x, y), which can be graphed. If the graph for an equation is a straight line, then the equation is a linear equation.

EXAMPLE 1 Graph $y = 3x - 2$.

Select any four values for the input x. We chose 3, 2, 0, and −1. Substitute these values for x to find the output y.

x	$3x - 2$	y	(x, y)
2	3(2) − 2	4	(2, 4)
1	3(1) − 2	1	(1, 1)
0	3(0) − 2	−2	(0, −2)
−1	3(−1) − 2	−5	(−1, −5)

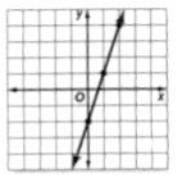

Four solutions are (2, 4), (1, 1), (0, −2), and (−1, −5). The graph is shown at the right.

EXERCISES

Graph each equation.

1. $y = x - 1$ 2. $y = x + 2$ 3. $y = -x$

4. $y = 4x$ 5. $y = 2x + 4$ 6. $y = 3x - 1$

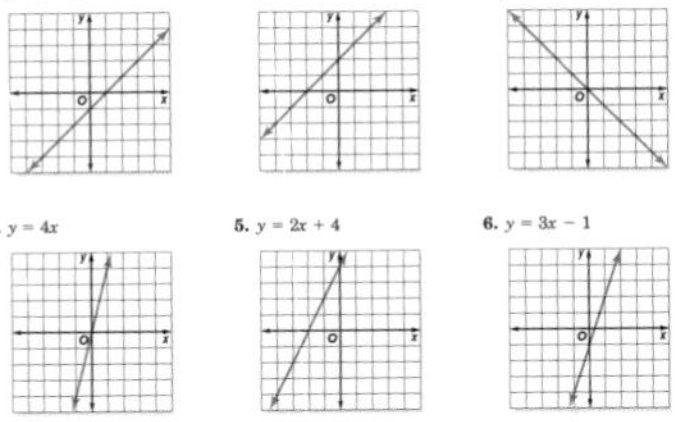

Practice: Skills, p. 211

NAME ________ DATE ________ PERIOD ____

4-6 Practice: Skills

Functions and Linear Equations

Copy and complete each function table. Identify the domain and range.

1. $y = x - 1$

x	$x - 1$	y
1	1 − 1	0
2	2 − 1	1
3	3 − 1	2
4	4 − 1	3

2. $y = x + 7$

x	$x + 7$	y
1	1 + 7	8
2	2 + 7	9
3	3 + 7	10
4	4 + 7	11

3. $y = 3x$

x	$3x$	y
1	3(1)	3
2	3(2)	6
3	3(3)	9
4	3(4)	12

4. $y = -4x$

x	$-4x$	y
−1	−4(−1)	4
0	−4(0)	0
1	−4(1)	−4
2	−4(2)	−8

5. $y = 3x + 1$

x	$3x + 1$	y
−1	3(−1) + 1	−2
0	3(0) + 1	1
1	3(1) + 1	4
2	3(2) + 1	7

6. $y = -2x + 3$

x	$-2x + 3$	y
−1	−2(−1) + 3	5
0	−2(0) + 3	3
1	−2(1) + 3	1
2	−2(2) + 3	−1

Graph each equation.

7. $y = x - 2$ 8. $y = x + 4$ 9. $y = -3x$

10. $y = 2x$ 11. $y = 2x + 2$ 12. $y = 3x - 2$

13. $y = 0.75x$ 14. $y = 0.5x + 1$ 15. $y = 2x - 0.5$

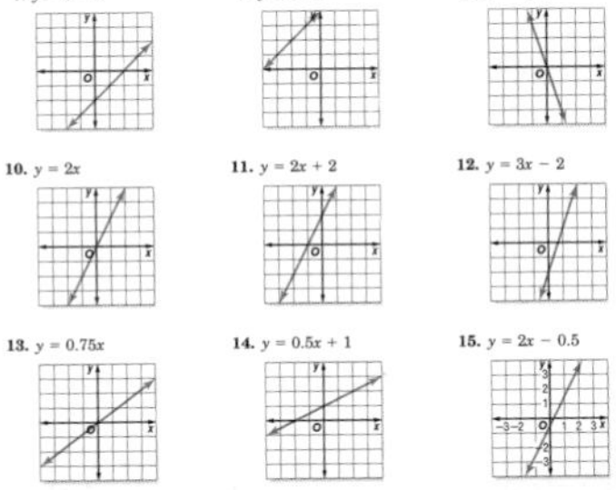

Practice: Word Problems, p. 212

NAME ________ DATE ________ PERIOD ____

4-6 Practice: Word Problems

Functions and Linear Equations

1. **TECHNOLOGY** The fee for your pager service is $22 per month. Make a function table that shows your total charge for 1, 2, 3, and 4 months of service.

x, Months	y, Total Charge
1	22
2	44
3	66
4	88

2. **TECHNOLOGY** Use the information in Exercise 1 to write an equation in which x represents the number of months and y represents the total charge. Then graph the equation. $y = 22x$

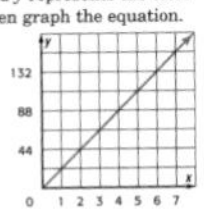

3. **TRAINS** Between Hiroshima and Kokura, Japan, the bullet train averages a speed of 164 miles per hour, which is the fastest scheduled train service in the world. Make a function table that shows the distance traveled at that speed in 1, 2, 3, and 4 hours.

x, Hours	y, Distance
1	164
2	328
3	492
4	656

4. **TRAINS** Use the information in Exercise 3 to write an equation in which x represents the number of hours and y represents the distance. Then graph the equation. $y = 164x$

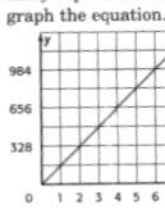

5. **GEOMETRY** The formula for the volume of a rectangular prism whose base has an area of 8 square units is $V = 8h$, where V is the volume and h is the height. Graph the function.

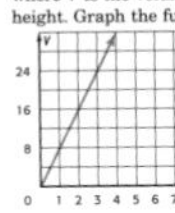

6. **ANIMALS** The fastest insect in the world is the dragonfly with a top speed of 36 miles per hour. Write an equation using x to represent hours and y to represent distance. Then graph the equation. $y = 36x$

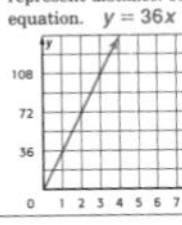

4-7 Lesson Notes

Lesson Resources

Chapter 4 Resource Masters
- Study Guide and Intervention, p. 215
- Practice: Skills, p. 216
- Practice: Word Problems, p. 217
- Reading to Learn Mathematics, p. 218
- Enrichment, p. 219

Calculator and Spreadsheet Masters, p. 28

Parent and Student Study Guide Workbook, 4-7

1 Focus

5-Minute Check 4-7
Use as a quiz or review of Lesson 4-6. (Also available as a transparency)

INTERACTIVE CHALKBOARD
PowerPoint® Presentations

Bellringer
Hands-On Activity After discussing the questions in the lesson opener, have students graph the equations $y = 0.5x$ and $y = 4x$. Compare to the graphs of $y = x$ and $y = 3x$. As the coefficient of x becomes closer to 0, does the line become closer to vertical or horizontal? **horizontal**

Reading to Learn Mathematics, p. 218 **ELL**

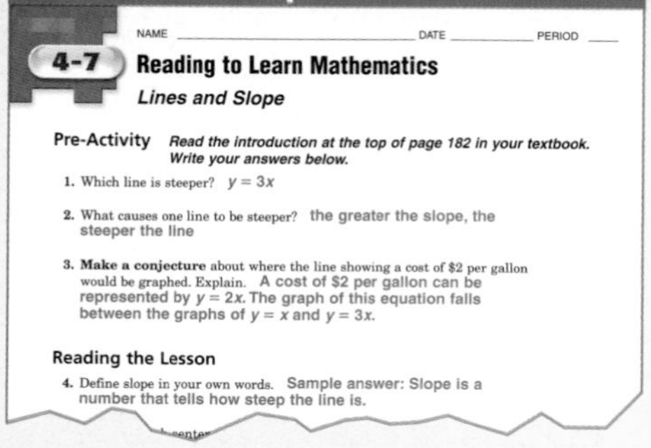
NAME ________________ DATE ________ PERIOD ____

4-7 Reading to Learn Mathematics
Lines and Slope

Pre-Activity *Read the introduction at the top of page 182 in your textbook. Write your answers below.*

1. Which line is steeper? $y = 3x$
2. What causes one line to be steeper? the greater the slope, the steeper the line
3. **Make a conjecture** about where the line showing a cost of \$2 per gallon would be graphed. Explain. A cost of \$2 per gallon can be represented by $y = 2x$. The graph of this equation falls between the graphs of $y = x$ and $y = 3x$.

Reading the Lesson

4. Define slope in your own words. Sample answer: Slope is a number that tells how steep the line is.

4-7 Lines and Slope

What You'll Learn
Find the slope of a line.

NEW Vocabulary
slope

LINK To Reading
Everyday Meaning of Slope: ground that forms an incline, as in a ski slope

WHEN am I ever going to use this?

COST OF GASOLINE In recent years, the cost of one gallon of gasoline has varied from a low of about \$1 per gallon to a high of about \$3 per gallon. The equations $y = 1x$ and $y = 3x$ are graphed.

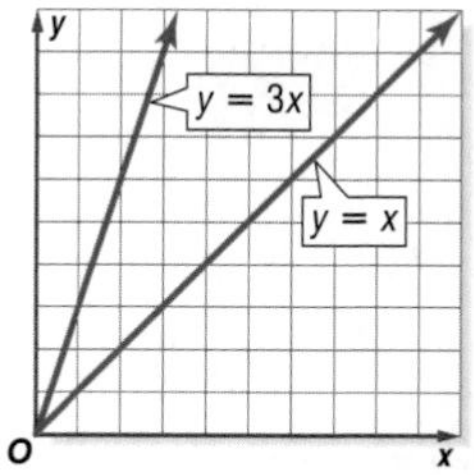

1. Which line is steeper? **$y = 3x$**
2. What causes one line to be steeper?
3. **Make a conjecture** about where the line showing a cost of \$2 per gallon would be graphed. Explain.

2. the greater the slope, the steeper the line
3. A cost of \$2 per gallon can be represented by $y = 2x$. The graph of this equation falls between the graphs of $y = x$ and $y = 3x$.

The function table shows the total cost y of x gallons of gasoline at \$2 per gallon. The equation $y = 2x$ is graphed below.

x	$2x$	y
1	2(1)	2
2	2(2)	4
3	2(3)	6
4	2(4)	8

(+1 for each step in x; +2 for each step in y)

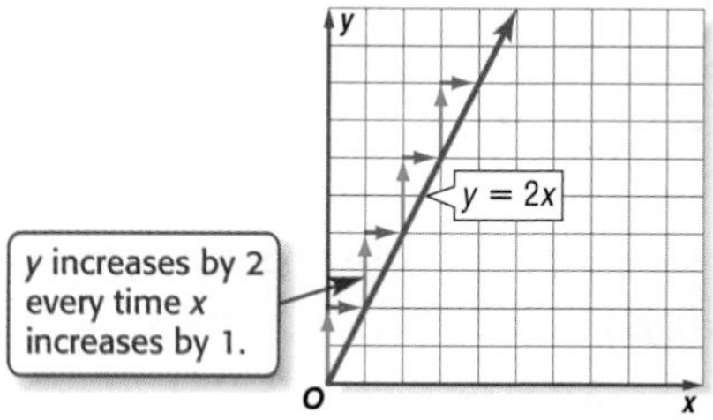

The change in y with respect to the change in x is called the **slope** of a line. Slope is a number that tells how steep the line is.

$$\text{slope} = \frac{\text{change in } y}{\text{change in } x} \quad \begin{matrix} \leftarrow \text{vertical change} \\ \leftarrow \text{horizontal change} \end{matrix}$$

$$= \frac{2}{1} \text{ or } 2$$

The slope is the same for any two points on a straight line.

EXAMPLE Positive Slope

1 **Find the slope of the line.**

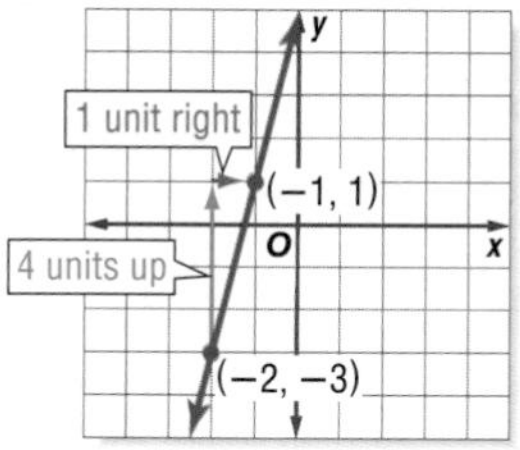

$$\text{slope} = \frac{\text{change in } y}{\text{change in } x}$$

$$= \frac{4}{1} \text{ or } 4$$

The slope of the line is 4.

When the value of y decreases as the value of x increases, the slope is a negative number. The slope of a line can also be a fraction.

EXAMPLES Negative Slope

STUDY TIP

Slope A line with a *positive slope* rises to the right. A line with a *negative slope* falls to the right.

2 **Find the slope of each line.**

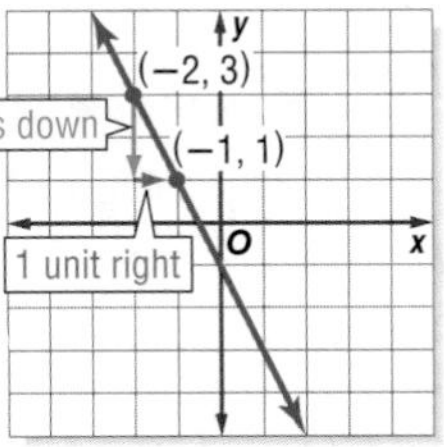

$$\text{slope} = \frac{\text{change in } y}{\text{change in } x} = \frac{-2}{1} \text{ or } -2$$

The slope of the line is −2.

3

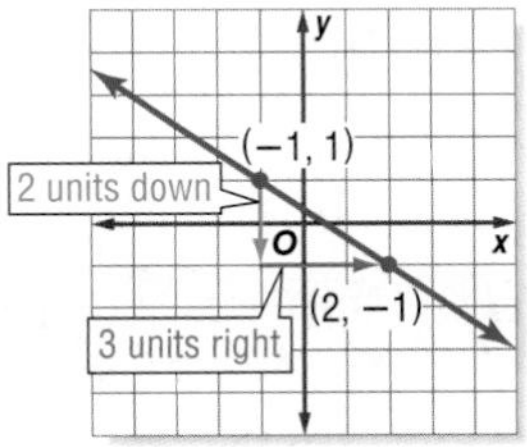

$$\text{slope} = \frac{\text{change in } y}{\text{change in } x} = \frac{-2}{3} \text{ or } -\frac{2}{3}$$

The slope of the line is $-\frac{2}{3}$.

TEACHING TIP

Point out that $\frac{-2}{3} = -\frac{2}{3}$.

Your Turn **Find the slope of the line that passes through each pair of points.**

a. (−2, −4), (1, 5) **3**   b. (0, 3), (4, −1) **−1**   c. (2, 2), (5, 3) $\frac{1}{3}$

EXAMPLE Compare Slopes

Standardized Test Practice

4 **MULTIPLE-CHOICE TEST ITEM** The table shows information about the rise and run of three ski slopes in Pennsylvania. Which has the steepest slope?

Ⓐ Giant Boulder   Ⓑ Giant Steps

Ⓒ Gunner   Ⓓ They have the same slope.

Ski Slope	Rise (ft)	Run (ft)
Giant Boulder	750	4,000
Giant Steps	750	3,200
Gunner	750	3,360

Test-Taking Tip *The Princeton Review*

Make a Drawing Whenever possible, make a drawing of the problem. Then use the drawing to estimate the answer.

Read the Test Item

The *rise* corresponds to the vertical change, or change in y. The *run* corresponds to the horizontal change, or change in x.

Solve the Test Item

Giant Boulder: $\frac{750}{4{,}000}$ Giant Steps: $\frac{750}{3{,}200}$ Gunner: $\frac{750}{3{,}360}$

All three fractions have the same rise. But Giant Steps has the shortest run. Therefore, its slope is greater. The answer is B.

msmath2.net/extra_examples

2 Teach

In-Class Examples

INTERACTIVE CHALKBOARD PowerPoint® Presentations

Find the slope of each line.

1

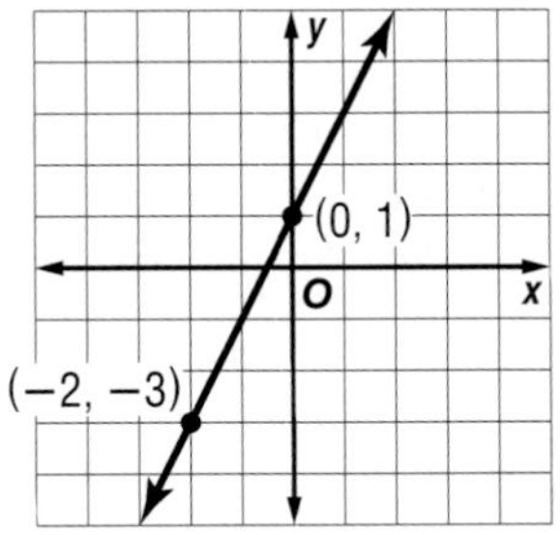

2

2

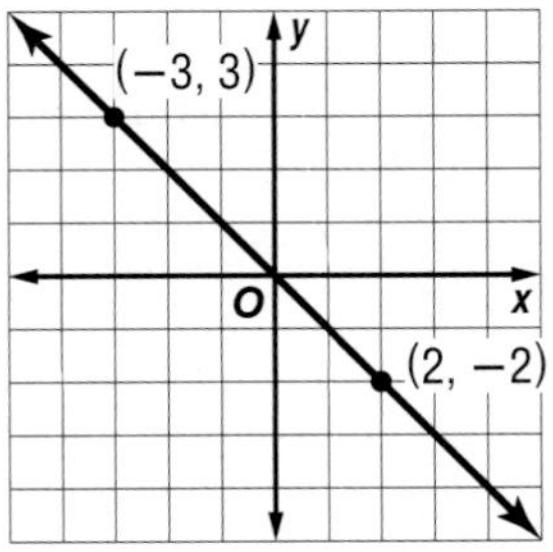

−1

3

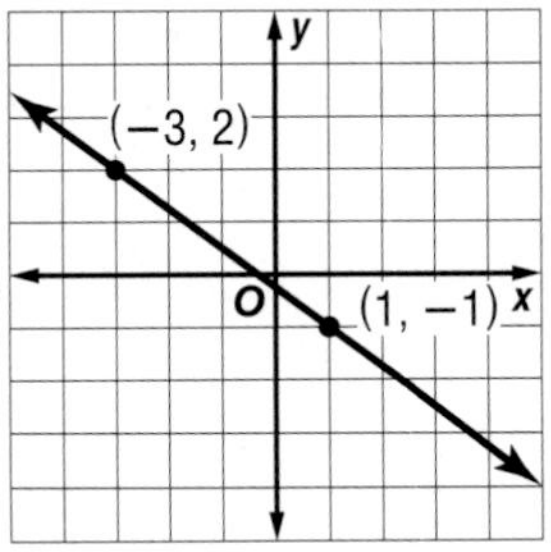

$-\frac{3}{4}$

4 **MULTIPLE-CHOICE TEST ITEM** The Americans with Disabilities Act states that the maximum slope of a ramp in new construction shall be $\frac{1}{12}$. Which ramp does *not* meet this requirement? **A**

A rise: 30 in.; run: 300 in.

B rise: 30 in.; run: 360 in.

C rise: 30 in.; run: 380 in.

D rise: 30 in.; run: 400 in.

DAILY INTERVENTION **Flexible Grouping**

Lines and Slope Have each student in a group graph a different line with positive slope. Compare the slopes and the steepness of the lines. Is a line with slope 7 steeper or less steep than one with slope 4? **steeper** Repeat the activity with negative slopes and compare.

3 Practice/Apply

FOLDABLES™ Follow-Up Remind students to use their Foldables to take notes on what they learn about lines and slope. Under the "Functions" tab, students might explain how they would use a verbal description, a table of values, an equation, or a graph to find the slope of a line.

About the Exercises...

Guided Practice Key

Exercises	Examples
4, 6	1
5	2, 3

Odd/Even Assignments
Exercises 7–16 are structured so that students practice the same concepts whether they are assigned odd or even problems.

Answers

3. Sample answer:

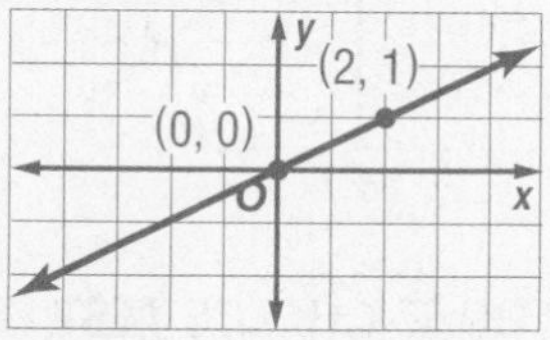

17. Monica; her earnings are increasing faster.
18. Greg: 4; Monica: 5
21. 1994–1995; The slope is steepest.
22. Between 1995–1996 the sales decreased; the slope is negative.
23. The sales remained constant; there is no change in the slope of the line. It is horizontal and the slope is equal to zero.

Skill and Concept Check

Writing Math Exercises 1 & 2

1. **Define** *slope*. the steepness of a line from left to right
2. **Explain** how a line can have a negative slope.
3. **OPEN ENDED** On a coordinate plane, draw a line with a slope of $\frac{1}{2}$. See margin.

2. Negative slope occurs when the y value decreases but the x value increases.

GUIDED PRACTICE

Find the slope of the line that passes through each pair of points.

4.

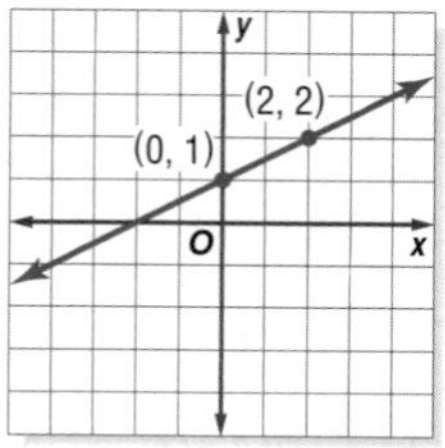

$\frac{1}{2}$

5.

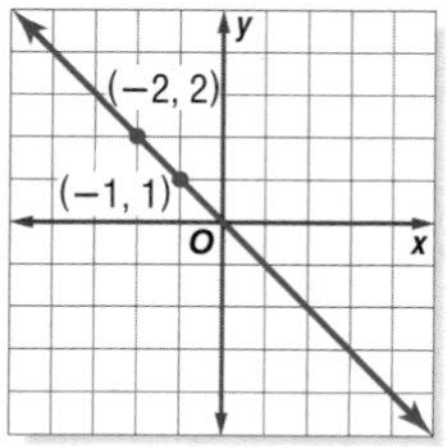

−1

6. (0, 0), (1, 3) 3

★ indicates increased difficulty

Practice and Applications

HOMEWORK HELP

For Exercises	See Examples
7–16	1–3
17–20	4

Extra Practice See pages 574, 599.

Find the slope of the line that passes through each pair of points.

7.

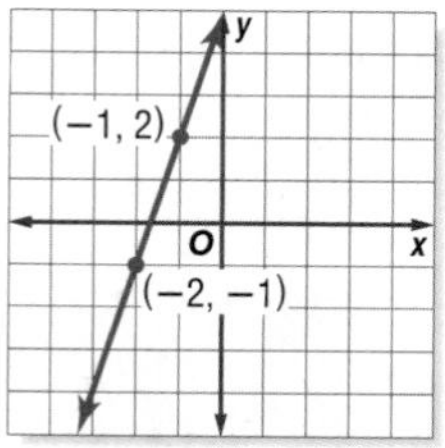

3

8.

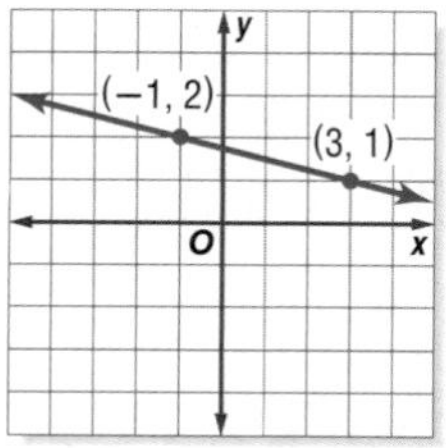

$-\frac{1}{4}$

9. (−2, 0), (1, 3) 1
10. (3, 4), (4, 6) 2
11. (1, 0), (2, −3) −3
12. (−2, 2), (−1, −2) −4
13. (1, 1), (4, 2) $\frac{1}{3}$
14. (−2, 1), (3, 3) $\frac{2}{5}$
15. (1, −1), (3, −2) $-\frac{1}{2}$
16. (0, 0), (3, −2) $-\frac{2}{3}$

EARNINGS For Exercises 17–19, use the table at the right. 17–18. See margin.

Hours Worked	Earnings ($)	
	Greg	Monica
1	4	5
2	8	10
3	12	15
4	16	20

17. Suppose each of the functions in the table was graphed on a coordinate plane. Which line is steeper? Explain.
18. Find the slope of each line.
19. What does the slope of each line represent? The slope represents the amount of money earned with respect to the number of hours worked.

20. **SKIING** Aerial skiers launch themselves into the air from a ramp like the one shown at the right. Is the slope of the ramp greater than one or less than one? less than one

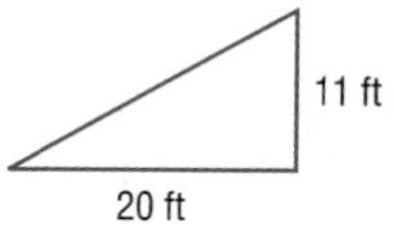

27.

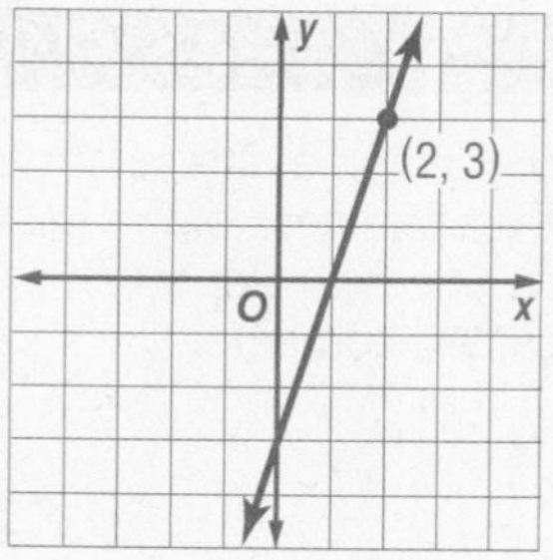

28.

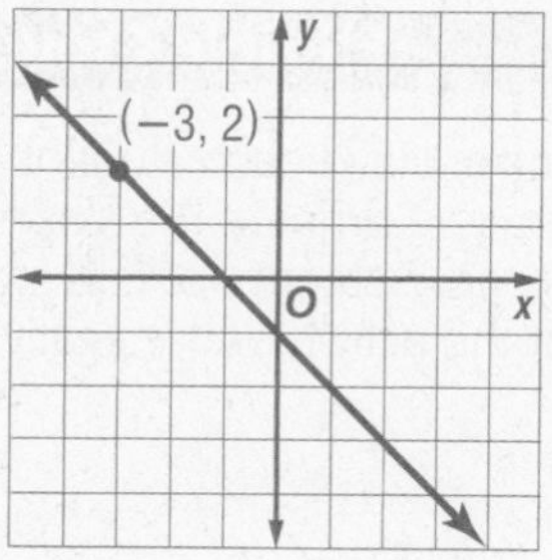

29.

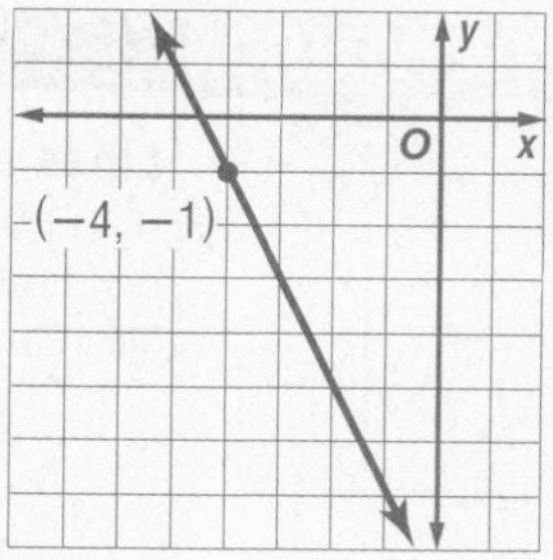

CAMPING For Exercises 21–23, use the graph.

★ **21.** Which section of the graph shows the greatest increase in sales of camping gear? Describe the slope of this part of the graph. 21–23. See margin.

★ **22.** What happened to sales between 1995 and 1996? Describe the slope of this part of the graph.

★ **23.** What happened to sales between 1997 and 1998? Describe the slope of this part of the graph.

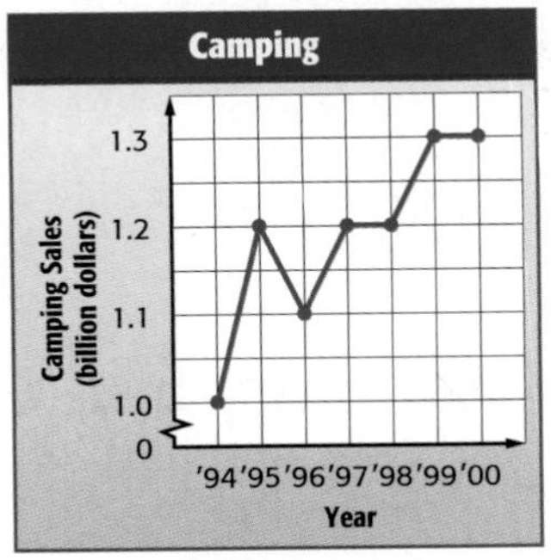

Source: National Sporting Goods Association

CRITICAL THINKING A linear function has a constant slope. Determine whether each function is *linear* or *nonlinear*.

24.

x	y
1	5
2	6
3	7
4	8

linear

25.

x	y
1	3
2	4
3	6
4	9

nonlinear

26.

x	y
−1	1
0	0
1	1
2	4

nonlinear

EXTENDING THE LESSON You can graph a line if you know the slope and the coordinates of a point on the line. The figure at the right shows how to graph a line with slope 2 that passes through $(1, -1)$.

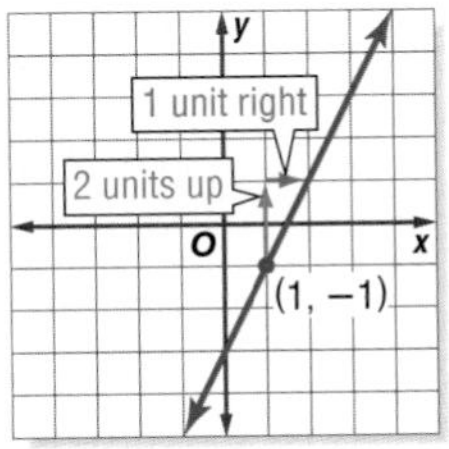

Graph each line with the given slope that passes through the given point. 27–30. See margin.

★ **27.** slope = 3; (2, 3)

★ **28.** slope = −1; (−3, 2)

★ **29.** slope = −2; (−4, −1)

★ **30.** slope = 5; (0, −4)

Standardized Test Practice and Mixed Review

31. MULTIPLE CHOICE What is the slope of the line in the graph? C

Ⓐ 3   Ⓑ −3   Ⓒ $\frac{1}{3}$ Ⓓ $-\frac{1}{3}$

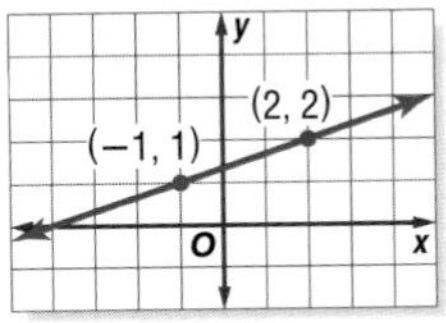

32. GRID IN The graph of a straight line contains the points $(0, -3)$, $(1, 2)$, and $(2, y)$. What is the value of y? 7

Graph each equation. (Lesson 4-6) 33–35. See Ch. 4 Answer Appendix.

33. $y = 2x - 1$ **34.** $y = -3x$ **35.** $y = x + 4$

36. Solve $2x + 6 < 8$. (Lesson 4-5) $x < 1$

WebQuest Interdisciplinary Project

The Wide World of Soccer

It is time to complete your project. Use the information and data you have gathered about countries where soccer is a favorite sport to prepare a Web page or poster. Be sure to include a graph with your project.

msmath2.net/webquest

msmath2.net/self_check_quiz

4 Assess

Higher-Level Thinking

What type of line has a slope of 0? (*Hint:* A line that contains points with the same y-coordinate has a slope of 0.) a horizontal line

Answer

30.

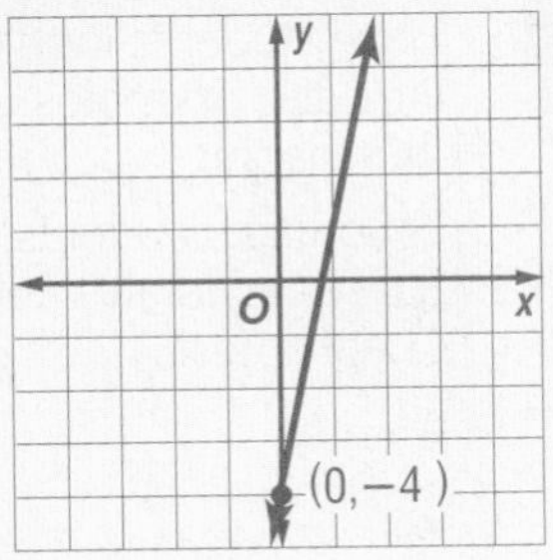

Study Guide and Intervention, p. 215

NAME ________ DATE ______ PERIOD ___

4-7 Study Guide and Intervention

Lines and Slope

Slope is a number that tells how steep the line is.

$\text{slope} = \frac{\text{change in } y}{\text{change in } x}$ ← vertical change ← horizontal change

EXAMPLE 1 Find the slope of the line.

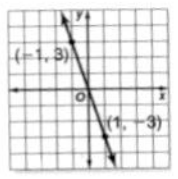

$\text{slope} = \frac{\text{change in } y}{\text{change in } x}$

$= \frac{-6}{2}$ or -3

The slope of the line is −3.

EXERCISES

Find the slope of the line that passes through each pair of points.

1. 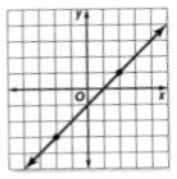 1

2. 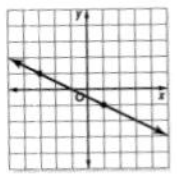 $-\frac{1}{2}$

3. 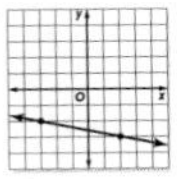 $-\frac{1}{5}$

4. (5, 3), (3, 1) 1
5. (5, 3), (6, 5) 2
6. (3, 4), (−1, −4) 2
7. (2, −6), (3, −3) 3
8. (4, 6), (5, 2) −4
9. (4, 4), (5, 2) −2
10. (0, 3), (1, −4) −7
11. (−3, 2), (−1, −4) −3
12. (2, 0), (6, 1) $\frac{1}{4}$
13. (11, 4), (7, 1) $\frac{3}{4}$
14. (2, 7), (−3, 4) $\frac{3}{5}$
15. (−1, −3), (7, 3) $\frac{3}{4}$
16. (7, −4), (1, 0) $-\frac{2}{3}$
17. (5, −2), (7, −3) $-\frac{1}{2}$
18. (0, 0), (6, −1) $-\frac{1}{6}$

Practice: Skills, p. 216

NAME ________ DATE ______ PERIOD ___

4-7 Practice: Skills

Lines and Slope

Find the slope of the line that passes through each pair of points.

1. 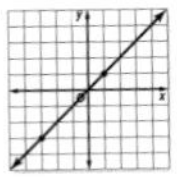 1

2. 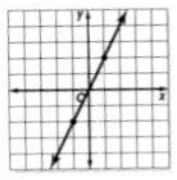 2

3. 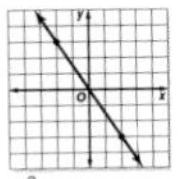 $-\frac{3}{2}$

4. 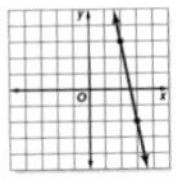 −5

5. 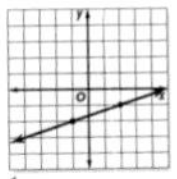 $\frac{1}{3}$

6. 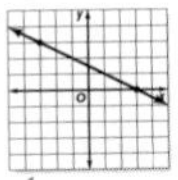 $-\frac{1}{2}$

7. (3, 4), (2, 2) 2
8. (3, 6), (2, 1) 5
9. (8, 4), (12, 6) $\frac{1}{2}$
10. (1, 6), (15, 8) $\frac{1}{7}$
11. (2, −4), (1, −5) 1
12. (3, −3), (2, −5) 2
13. (9, 5), (8, 10) −5
14. (5, 0), (6, 1) 1
15. (2, 7), (5, 6) $-\frac{1}{3}$
16. (2, 0), (−10, 2) $-\frac{1}{6}$
17. (12, −4), (7, 1) −1
18. (1, −2), (2, −5) −3
19. (7, −5), (1, 7) −2
20. (4, 6), (8, 4) $-\frac{1}{2}$
21. (0, 2), (6, 0) $-\frac{1}{3}$
22. (−8, −1), (−4, 2) $\frac{3}{4}$
23. (9, 0), (−3, −3) $\frac{1}{4}$
24. (13, −4), (7, 1) $-\frac{5}{6}$

Practice: Word Problems, p. 217

NAME ________ DATE ______ PERIOD ___

4-7 Practice: Word Problems

Lines and Slope

1. **TRAVEL** Suppose each of these functions was graphed on a coordinate plane. Which line is steeper? Explain.

Hours Traveled	Distance (mi)	
	Scott	Amparo
1	55	60
2	110	120
3	165	180
4	220	240

Amparo's line; for each x-value, the corresponding y-value is greater.

2. **TRAVEL** Use the data in Exercise 1. Find the slope of each line. Scott's line: 55; Amparo's line: 60

3. **TRAVEL** What does the slope of each line that you found in Exercise 2 represent? the rate in miles per hour

4. **NATIONAL PARKS** Which section of the graph shows the least increase in recreational visits? Describe the slope of this part of the graph.

1997 to 1998; positive and close to 0

5. **NATIONAL PARKS** Use the graph in Exercise 4. What happened to recreational visits between 1998 and 1999? Describe the slope of this part of the graph. increased; positive

6. **NATIONAL PARKS** Use the graph in Exercise 4. What happened to the sales between 2000 and 2001? Describe the slope of this part of the graph. decreased; negative

CHAPTER 4 Study Guide and Review

Vocabulary Assessment

- This alphabetical list of vocabulary terms in Chapter 4 includes a page reference where each term was introduced.
- A vocabulary review/test for Chapter 4 is available on p. 234 of the *Chapter 4 Resource Masters*.

Vocabulary PuzzleMaker

ELL The online Vocabulary PuzzleMaker improves students' mathematics vocabulary using four puzzle formats — crossword, scramble, word search using a word list, and word search using clues. Students can work on a computer screen or from a printed handout.

MindJogger Videoquizzes

ELL MindJogger Videoquizzes provide an alternative review of concepts presented in this chapter. Students work in teams in a game show format to gain points for correct answers. The questions are presented in three rounds.

Round 1 Concepts (5 questions)
Round 2 Skills (4 questions)
Round 3 Problem Solving (4 questions)

ExamView® Pro Test Bank CD-ROM

Use the networkable ExamView® Pro Test Bank CD-ROM to:

- Create **Multiple versions** of tests.
- Create **Modified** tests for **Inclusion** students with one mouse click.
- **Edit** existing questions and **Add** your own questions.
- Build tests aligned with state standards using built-in **State Curriculum Correlations.**
- Change **English** tests to **Spanish** with one mouse click and vice versa.

CHAPTER 4 Study Guide and Review

Vocabulary and Concept Check

domain (p. 177)	inverse operations (p. 156)	slope (p. 182)
function (p. 177)	linear equation (p. 178)	two-step equation (p. 166)
function table (p. 177)	range (p. 177)	work backward strategy (p. 164)
inequality (p. 172)		

State whether each sentence is *true* or *false*. If *false*, replace the underlined word or number to make a true sentence. 2. false; addition 3. true

1. Inverse operations "undo" each other. true
2. The words "more than" sometimes suggest the operation of multiplication.
3. An inequality is a mathematical sentence that contains the symbols $<, >, \leq$, or $\geq$.
4. When graphing $t < 2$ on a number line, an open circle should be used to show that 2 is not included in the solution. true
5. Slope is a number that tells how steep a line is. true
6. An equation is called a linear equation if its graph is a point. false; straight line
7. The solution of $m + 5 = 12$ is 17. false; 7
8. The solution of $g - 4 = 18$ is 22. true
9. Six is the solution of $-8w = 48$. false; -6
10. The solution of $2y \leq 18$ is $y \leq 16$. false; $y \leq 9$

Lesson-by-Lesson Exercises and Examples

4-1 Writing Expressions and Equations (pp. 150–152)

Write each phrase as an algebraic expression.

11. the sum of a number and 5 $n + 5$
12. six inches less than her height $h - 6$
13. twice as many apples $2a$
14. four times the number of dollars $4d$

Write each sentence as an algebraic equation.

15. Ten years older than Theresa's age is 23.
16. Four less than a number is 19.
17. The quotient of 56 and a number is 14.
18. Eight times the number of students is 64. $8k = 64$

Example 1 **Write the phrase as an algebraic expression.**

four times the price

Let p represent the price.

The algebraic expression is $4p$.

Example 2 **Write the sentence as an algebraic equation.**

Six less than the number of cookies is 24.

Let c represent the number of cookies.

The equation is $c - 6 = 24$.

15. $a + 10 = 23$ 16. $x - 4 = 19$
17. $\frac{56}{z} = 14$

msmath2.net/vocabulary_review

FOLDABLES™ Study Organizer

For more information about Foldables, see *Teaching Mathematics with Foldables.*

Have students look through the chapter to make sure they have included examples in their Foldables for each lesson or topic in the chapter.

Encourage students to refer to their Foldables while completing the Study Guide and Review and to use them in preparing for the Chapter Test.

4-2 Solving Addition and Subtraction Equations (pp. 156–159)

Solve each equation. Check your solution.

19. $x + 5 = 8$ **3**
20. $y + 4 = 12$ **8**
21. $a + 6 = 5$ **−1**
22. $r + 8 = 2$ **−6**
23. $p + 9 = -4$ **−13**
24. $d + 14 = 23$ **9**
25. $s - 8 = 15$ **23**
26. $t - 6 = 7$ **13**
27. $n - 1 = -3$ **−2**
28. $w - 9 = 28$ **37**

Example 3 Solve $x + 6 = 4$.

$x + 6 = 4$
$\underline{-6 = -6}$ Subtract 6 from each side.
$x = -2$

Example 4 Solve $y - 3 = -2$.

$y - 3 = -2$
$\underline{+3 = +3}$ Add 3 to each side.
$y = 1$

4-3 Solving Multiplication Equations (pp. 160–163)

Solve each equation. Check your solution.

29. $7c = 28$ **4**
30. $9y = 45$ **5**
31. $-2h = 24$ **−12**
32. $-8w = 72$ **−9**
33. $10y = -90$ **−9**
34. $6q = -18$ **−3**
35. $-11f = -121$ **11**
36. $-12r = -36$ **3**

Example 5 Solve $-4b = 32$.

$-4b = 32$
$\frac{-4b}{-4} = \frac{32}{-4}$ Divide each side by −4.
$b = -8$

4-4 Solving Two-Step Equations (pp. 166–169)

Solve each equation. Check your solution.

37. $3y - 12 = 6$ **6**
38. $6x - 4 = 20$ **4**
39. $2x + 5 = 3$ **−1**
40. $5m + 6 = -4$ **−2**
41. $10c - 8 = 90$ **9.8**
42. $3r - 20 = -5$ **5**
43. Ten more than five times a number is 25. Find the number. **3**

Example 6 Solve $3p - 4 = 8$.

$3p - 4 = 8$
$\underline{+4 = +4}$ Add 4 to each side.
$3p = 12$
$\frac{3p}{3} = \frac{12}{3}$ Divide each side by 3.
$p = 4$

4-5 Inequalities (pp. 172–175) **44–53. See margin for graphs.**

Solve each inequality. Graph the solution. **50. $y < -5$ 51. $m > -3$**

44. $x + 3 < 8$ **$x < 5$**
45. $y + 2 > 5$ **$y > 3$**
46. $a + 4 \geq 10$ **$a \geq 6$**
47. $d + 1 \leq 6$ **$d \leq 5$**
48. $h - 5 \geq 7$ **$h \geq 12$**
49. $s - 2 \leq 9$ **$s \leq 11$**
50. $y + 2 < -3$
51. $m - 7 > -10$
52. $b + 9 \leq -11$ **$b \leq -20$**
53. $t - 10 \geq -8$ **$t \geq 2$**

Example 7 Solve $g + 8 \leq 10$. Graph your solution.

$g + 8 \leq 10$
$\underline{-8 = -8}$ Subtract 8 from each side.
$g \leq 2$

Answers

44.
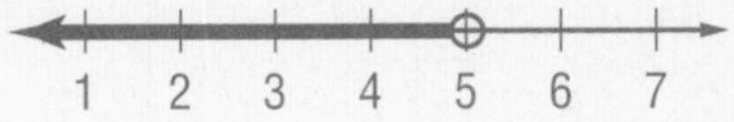

45.
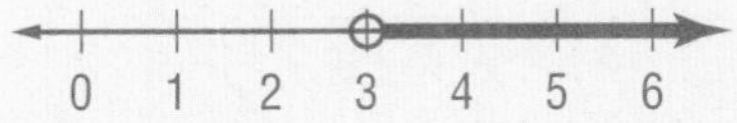

46.
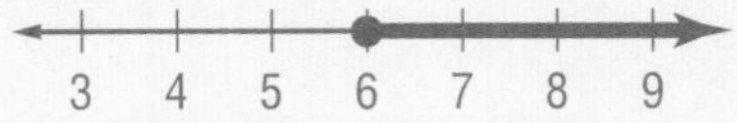

47.
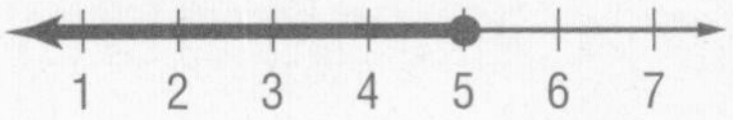

48.
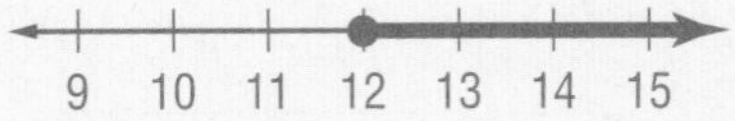

49.
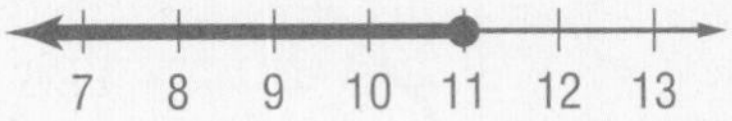

50.
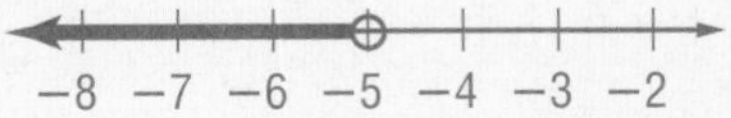

51.
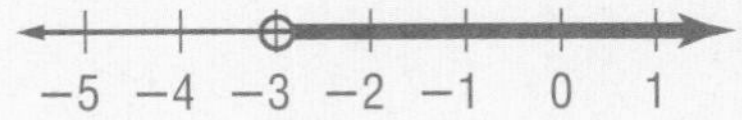

52.
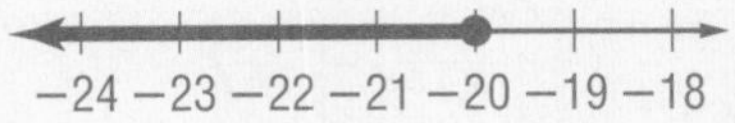

53.

Answers

54.

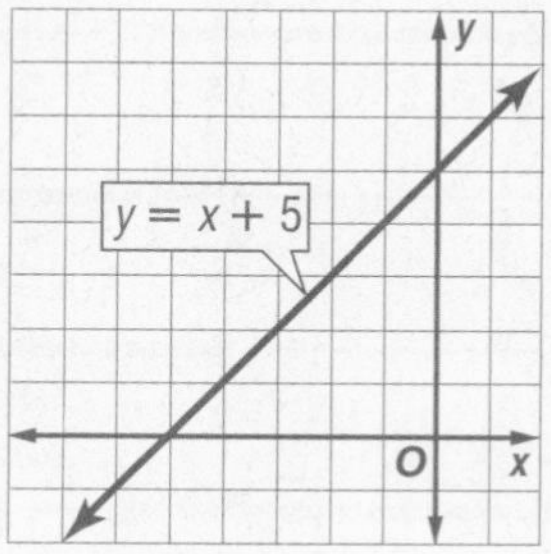

55.

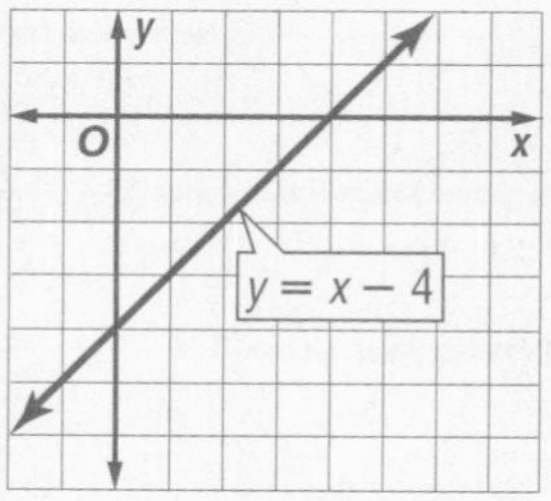

56.

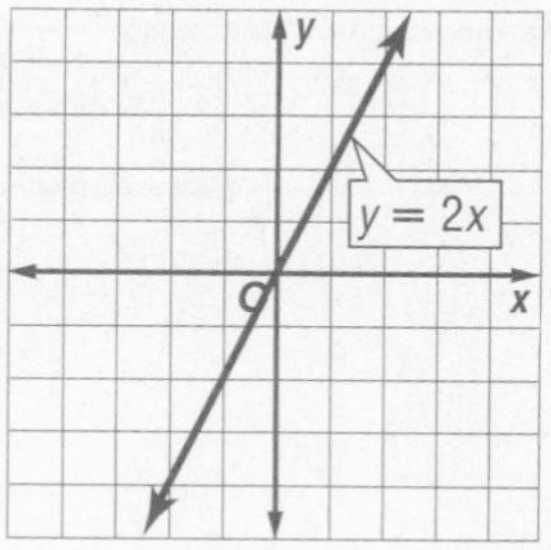

57.

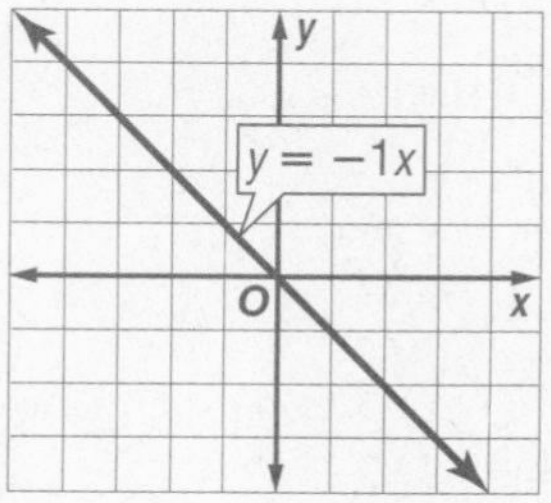

58.

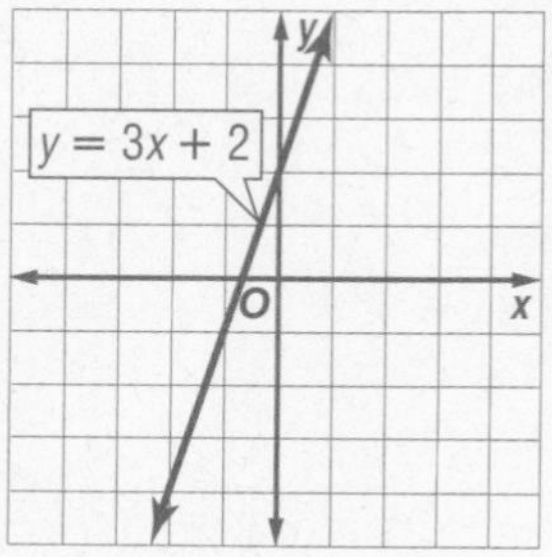

4-6 Functions and Linear Equations (pp. 177–181)

Graph each equation. **54–59. See margin.**

54. $y = x + 5$

55. $y = x - 4$

56. $y = 2x$

57. $y = -1x$

58. $y = 3x + 2$

59. $y = -2x + 3$

MONEY MATTERS **For Exercises 60–62, use the following infomation.**

Angel earns $6 per hour working at the Ice Cream Shop.

60. Make a table that shows her total earnings for working 3, 5, 7, and 9 hours. **See margin.**

61. Write an equation in which x represents the number of hours and y represents Angel's total earnings. $y = 6x$

62. Graph the equation. **See margin.**

Example 8 **Graph $y = x + 3$.**

Select four values for x. Substitute these values for x to find values for y.

x	$x + 3$	y
−1	−1 + 3	2
0	0 + 3	3
1	1 + 3	4
2	2 + 3	5

Four solutions are (−1, 2), (0, 3), (1, 4), and (2, 5). The graph is shown below.

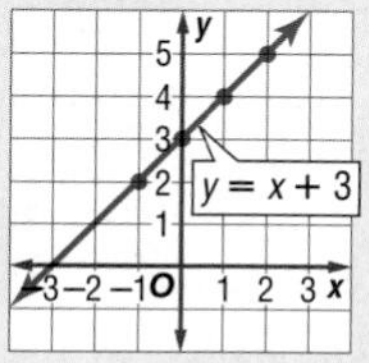

4-7 Lines and Slope (pp. 182–185)

Find the slope of the line that passes through each pair of points.

63.

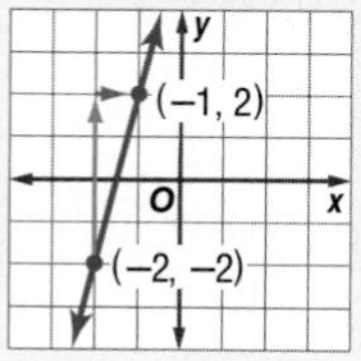

4

64.

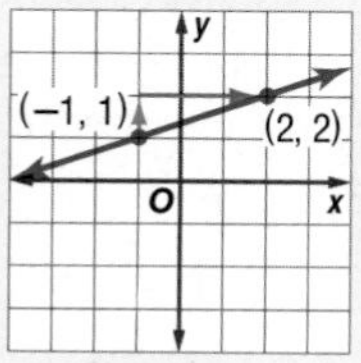

$\frac{1}{3}$

65. (1, 1), (2, 0) **−1**

66. (2, 3), (4, −1) **−2**

Example 9 **Find the slope of the line.**

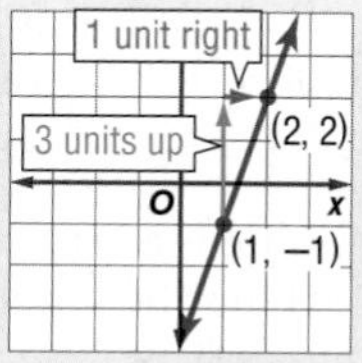

$$\text{slope} = \frac{\text{change in } y}{\text{change in } x}$$

$$= \frac{3}{1} \text{ or } 3$$

The slope of the line is 3.

59.

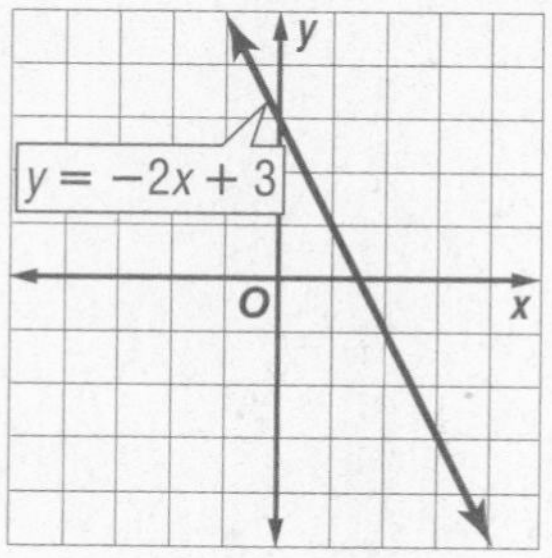

60.

Hours Worked	Dollars Earned
3	$18
5	$30
7	$42
9	$54

62.

Angel's Earnings

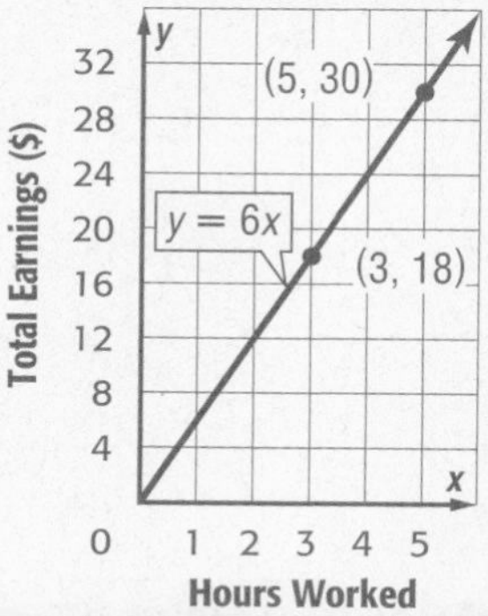

CHAPTER 4 Practice Test

Vocabulary and Concepts

1. **Explain** the difference between an equation and an inequality. An equation always has an equals sign. An inequality contains $<, \leq, >,$ or $\geq$.
2. **Define** *function* and give an example. See margin.

Skills and Applications

Write each phrase as an algebraic expression.

3. \$5 less than Matt has $a - 5$
4. 4 years older than Hana $y + 4$

Solve each equation. Check your solution.

5. $x + 5 = -8$ -13
6. $y - 11 = 15$ 26
7. $9z = -81$ -9
8. $-6k - 4 = 38$ -7

Solve each inequality. Graph the solution. 9–10. See margin for graphs.

9. $p - 4 \geq -3$ $p \geq 1$
10. $j + 5 > 2$ $j > -3$

SHOPPING For Exercises 11 and 12, use the following information.
Suppose you want to buy 3 CDs and a new CD case that costs \$7. Each CD costs the same amount.

11. If you spend \$46, write an equation to find the cost of each CD. $3c + 7 = 46$
12. Solve the equation. \$13

Graph each equation. 13–15. See Ch. 4 Answer Appendix.

13. $y = 3x - 2$
14. $y = -2x + 4$
15. $y = 0.5x$

MOVIES For Exercises 16–18, use the following information.
A student ticket to the movies costs \$3.

16. Make a table that shows the total cost of 2, 4, and 6 tickets. See Ch. 4 Answer Appendix.
17. Write an equation in which x represents the number of tickets, and y represents the total cost. $y = 3x$
18. Graph the equation. See Ch. 4 Answer Appendix.
19. Find the slope of the line that passes through $(-2, 3)$ and $(-1, 2)$. -1

Standardized Test Practice

20. **MULTIPLE CHOICE** Which line has a slope of 2? C

A 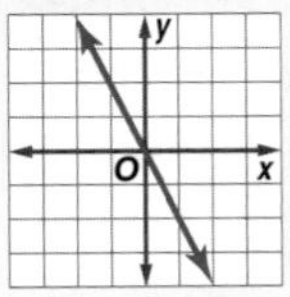

B 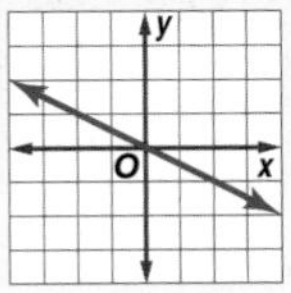

C 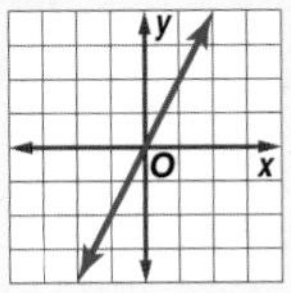

D 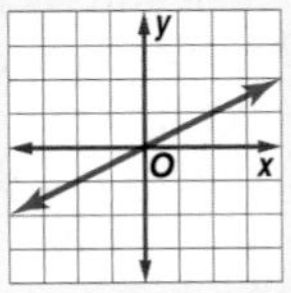

msmath2.net/chapter_test

Portfolio Suggestion

Introduction After working through a chapter of material, students may still feel uncertain about a problem that they cannot solve or concerned that that type of problem may appear on a test.

Ask Students to choose a problem they still cannot solve and write as much of the solution as they can. Ask them to analyze what about the problem gives them difficulty and then find help from the teacher or another student. They should include this information in their portfolios.

CHAPTER 4 Practice Test

Assessment Options

Vocabulary Test A vocabulary review/test for Chapter 4 can be found on p. 234 of the *Chapter 4 Resource Masters.*

Chapter Tests There are six Chapter 4 Tests and an Extended Response Assessment task available in the *Chapter 4 Resource Masters.*

Chapter 4 Tests			
Form	**Type**	**Level**	**Pages**
1	MC	basic	221–222
2A	MC	average	223–224
2B	MC	average	225–226
2C	FR	average	227–228
2D	FR	average	229–230
3	FR	advanced	231–232

MC = multiple-choice questions
FR = free-response questions

Extended Response Assessment

Extended Response tasks for Chapter 4 can be found on p. 233 of the *Chapter 4 Resource Masters.* A sample scoring rubric for these tasks appears on p. A26.

Unit 2 Test

A unit test/review can be found on pp. 241–242 of the *Chapter 4 Resource Masters.*

Answers

2. A function describes the relationship between quantities; sample answer: $y = x$.

9. 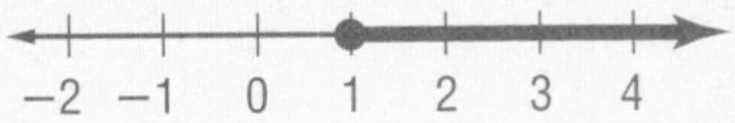

10. 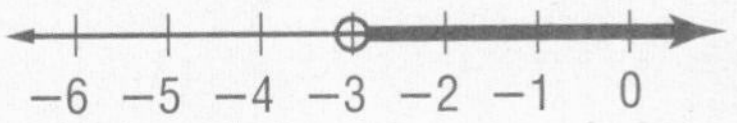

Chapter 4 Standardized Test Practice

These two pages contain practice questions in the various formats that can be found on the most frequently given standardized tests.

A student recording sheet for these two pages can be found on page A1 of the *Chapter 4 Resource Masters*.

Standardized Test Practice Student Recording Sheet, p. A1

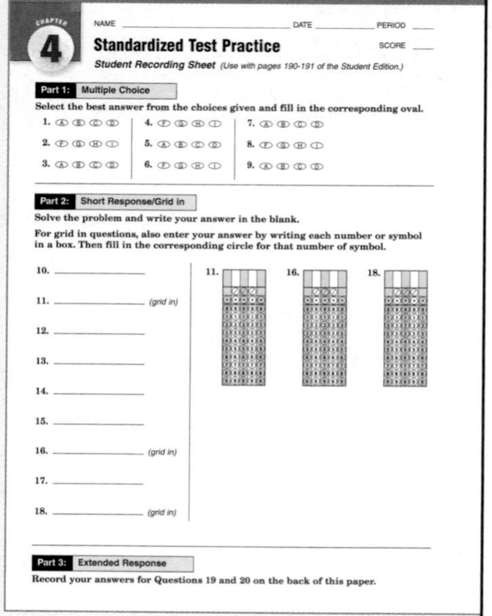
Chapter 4 NAME ______ DATE ______ PERIOD ______

Standardized Test Practice SCORE ______

Student Recording Sheet (Use with pages 190–191 of the Student Edition.)

Part 1: Multiple Choice

Select the best answer from the choices given and fill in the corresponding oval.

1. Ⓐ Ⓑ Ⓒ Ⓓ 4. Ⓕ Ⓖ Ⓗ Ⓘ 7. Ⓐ Ⓑ Ⓒ Ⓓ
2. Ⓕ Ⓖ Ⓗ Ⓘ 5. Ⓐ Ⓑ Ⓒ Ⓓ 8. Ⓕ Ⓖ Ⓗ Ⓘ
3. Ⓐ Ⓑ Ⓒ Ⓓ 6. Ⓕ Ⓖ Ⓗ Ⓘ 9. Ⓐ Ⓑ Ⓒ Ⓓ

Part 2: Short Response/Grid in

Solve the problem and write your answer in the blank.

For grid in questions, also enter your answer by writing each number or symbol in a box. Then fill in the corresponding circle for that number of symbol.

10. ______
11. ______ (grid in)
12. ______
13. ______
14. ______
15. ______
16. ______ (grid in)
17. ______
18. ______ (grid in)

11. 16. 18.

Part 3: Extended Response

Record your answers for Questions 19 and 20 on the back of this paper.

Additional Practice

See pp. 239–240 of the *Chapter 4 Resource Masters* for additional standardized test practice.

Chapter 4 Standardized Test Practice

PART 1 Multiple Choice

Record your answers on the answer sheet provided by your teacher or on a sheet of paper.

1. Felicia's family wants to buy her a gift for \$220. Her parents will pay half. Her older sister will pay \$50. If her three other siblings split the remaining cost, how much will each pay? (Lesson 1-1) **A**

   Ⓐ \$20 Ⓑ \$60 Ⓒ \$130 Ⓓ \$170

2. Which is equivalent to 3^6? (Lesson 1-2) **I**

   Ⓕ 18 Ⓖ 36
   Ⓗ $6 \cdot 6 \cdot 6$ Ⓘ $3 \cdot 3 \cdot 3 \cdot 3 \cdot 3 \cdot 3$

3. Suppose you need 0.65 liter of water for a science experiment, but the container is measured in milliliters. How many milliliters of water do you need? (Lesson 1-5) **D**

   Ⓐ 0.0065 mL Ⓑ 0.65 mL
   Ⓒ 6.5 mL Ⓓ 650 mL

4. If 18 is added to the data set below, which statement is true? (Lesson 2-4) **H**

   16, 14, 22, 16, 16, 18, 15, 25

   Ⓕ The mode increases.
   Ⓖ The mean decreases.
   Ⓗ The mean increases.
   Ⓘ The median increases.

5. The stem-and-leaf plot shows the number of points scored by the Bears in each of their basketball games this season. In how many games did they score at least 30 points? (Lesson 2-5) **C**

Stem	Leaf
1	8 9
2	0 2 3 3 6 8 8 9
3	0 1 4 4 5 6 8 9
4	0 1 2

1 | 8 = 18 points

   Ⓐ 8 Ⓑ 9 Ⓒ 11 Ⓓ 20

6. Suppose points given by (x, y) in the table are graphed. Which statement is true about the graphs? (Lesson 3-3) **G**

x	y
−4	5
−5	10
−2	6

   Ⓕ The graphs of the points are located in Quadrant I.
   Ⓖ The graphs of the points are located in Quadrant II.
   Ⓗ The graphs of the points are located in Quadrant III.
   Ⓘ The graphs of the points are located in Quadrant IV.

7. The temperature at 6:00 A.M. was −5°F. What was the temperature at 8:00 A.M. if it had risen 7 degrees? (Lesson 3-4) **C**

   Ⓐ −12°F Ⓑ −2°F
   Ⓒ 2°F Ⓓ 12°F

8. The Tigers scored four more runs than the Giants scored. Which expression represents the number of runs the Giants scored if the Tigers scored n runs? (Lesson 4-1) **G**

   Ⓕ $n + 4$ Ⓖ $n - 4$
   Ⓗ $4 - n$ Ⓘ $4n$

9. Which is the graph of the equation $y = 3x - 2$? (Lesson 4-6) **B**

   Ⓐ
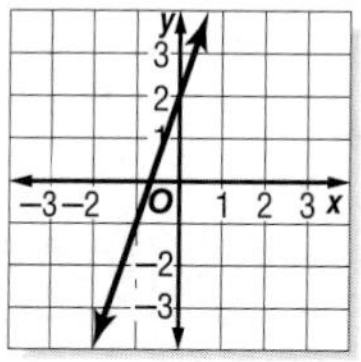

   Ⓑ
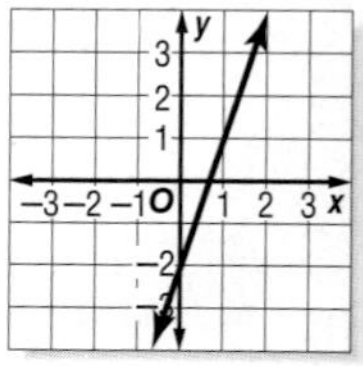

   Ⓒ
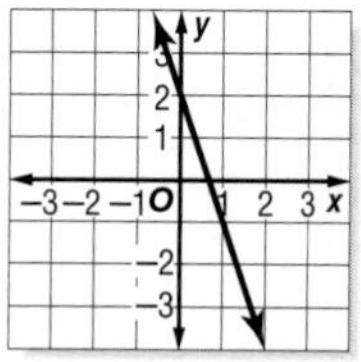

   Ⓓ
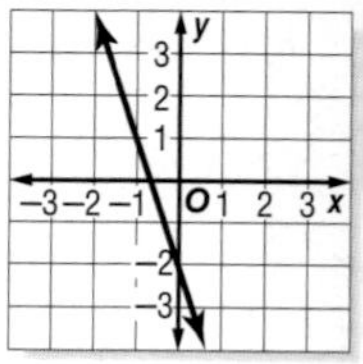

The Princeton Review **Log On** for Test Practice

The Princeton Review offers additional test-taking tips and practice problems at their web site. Visit princetonreview.com or review.com

ExamView® Pro Test Bank CD-ROM

Special banks of standardized test questions similar to those on the TIMSS and NAEP grade 8 tests can be found on this CD-ROM.

Aligned and verified by The Princeton Review

ART 2 Short Response/Grid In

ecord your answers on the answer sheet ovided by your teacher or on a sheet paper.

. The charge to enter a park is a flat amount per vehicle plus a fee for each person in the vehicle. The table shows the charge for vehicles holding up to 4 people.

Number of people	Charge (dollars)
1	2.00
2	2.50
3	3.00
4	3.50

What is the charge, in dollars, for a vehicle holding 8 people? (Lesson 1-1) **\$5.50**

. Evaluate $2(8 + 5^2)$. (Lesson 1-3) **66**

. The line plot shows how far in kilometers some students live from the school. How many students are represented in the plot? (Lesson 2-3) **9 students**

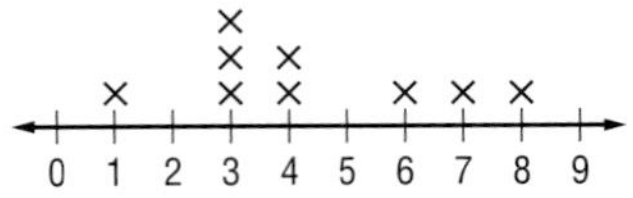

. The ordered pairs (1, 2), (6, 2), and (1, −5) are coordinates of three of the vertices of a rectangle. What is the y-coordinate of the ordered pair that represents the fourth vertex? (Lesson 3-3) **−5**

. What number should replace y in the table? (Lesson 3-6) **−1**

x	$2x - 5$
1	−3
2	y
4	3
8	11
16	27

TEST-TAKING TIP

Question 13 This question does not include a drawing. Make one. It can help you quickly see how to solve the problem.

15. What value of x makes $x - 2 = -4$ a true statement? (Lesson 4-2) **−2**

16. Twice a number plus 7 is 35. Find the number. (Lesson 4-4) **14**

17. Winston earns \$7 an hour landscaping. He wants to buy a DVD player that costs \$140. Write an inequality for the least number of hours h he needs to work to reach his goal. (Lesson 4-5) $h \geq 20$

18. The graph of a line contains (1, 2) and (3, y). What is the value of y if the slope of the line is 2? (Lesson 4-7) **6**

PART 3 Extended Response

Record your answers on a sheet of paper. Show your work.

19. Pete's Paints charges a \$20 fee plus \$7.50 per hour to rent a paint sprayer.
    a. Write an equation that could be used to determine the cost c of renting a paint sprayer for h hours. $7.50h + 20 = c$
    b. Suppose you want to spend no more than \$50 to rent the paint sprayer. Write an inequality for this situation. $7.50h + 20 \leq 50$
    c. Solve the inequality in part **b**. Explain the meaning of the solution. $h \leq 4$; **You can rent the sprayer for 4 hours or less.**

20. The distances traveled by a bicycle rider are given in the table. (Lesson 4-6) **b.** $d = 10t$
    a. Graph the ordered pairs. **See margin.**
    b. Write an equation that relates the time t to the distance d.
    c. Use your equation to predict the distance traveled in 3.5 hours. **35 mi**

Time (hours)	Distance (miles)
1	10
2	20
3	30
4	40
5	50

msmath2.net/standardized_test

Answer

20a. **Bicycle Ride**

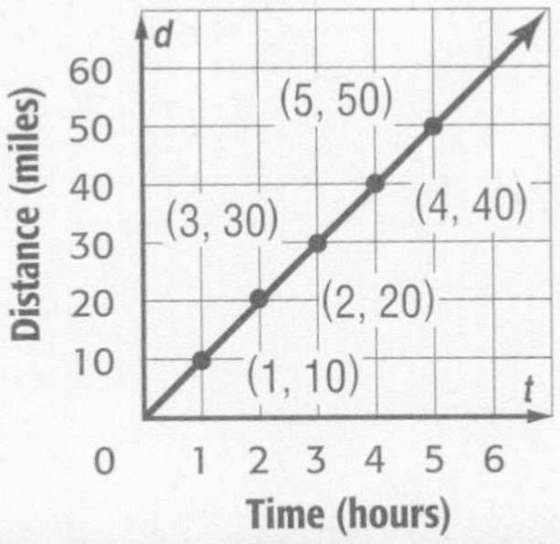

Standardized Test Practice

Rubrics The following rubrics are sample scoring devices for short response and extended response questions.

Short Response

Points	Description
2	The student demonstrates a thorough understanding of the mathematics of the task. The response may contain minor flaws that do not detract from the demonstration of a thorough understanding.
1	The student has provided a response that is only partially correct.
0	The student has provided a completely incorrect solution or no response at all.

Extended Response

Points	Description
4	The student demonstrates a thorough understanding of the mathematics of the task. The response may contain minor flaws that do not detract from the demonstration of a thorough understanding.
3	The student demonstrates an understanding of the mathematics of the task. The response is essentially correct and demonstrates an essential but less than thorough understanding of the mathematics.
2	The student demonstrates only a partial understanding of the mathematics of the task. Although the student may have used the correct approach to a solution or may have provided a correct solution, the work lacks an essential understanding of the underlying mathematical concepts.
1	The student demonstrates a very limited understanding of the mathematics of the task. The response is incomplete and exhibits many flaws.
0	The student provides a completely incorrect solution or no response at all.

A more detailed 4-point rubric is provided on page A2 of the *Chapter 4 Resource Masters*.

Page 166, Lesson 4-4 (Hands-On Mini Lab)

1.

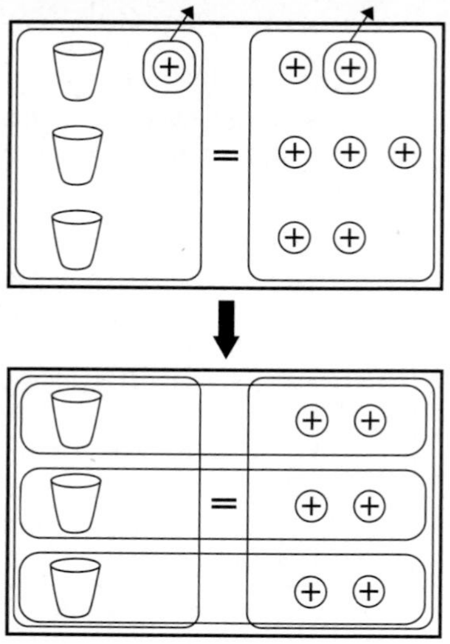

2.

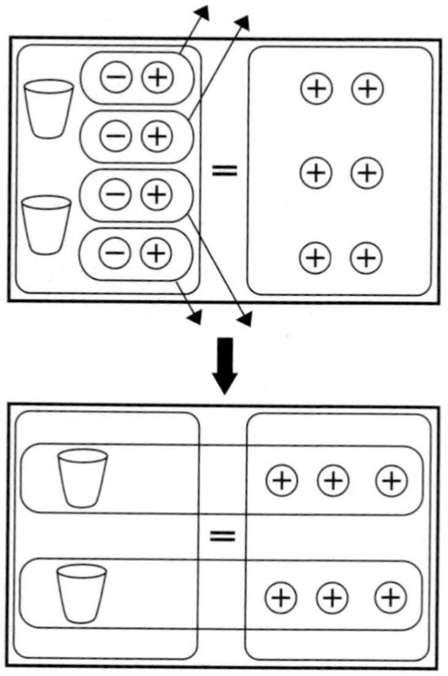

3.

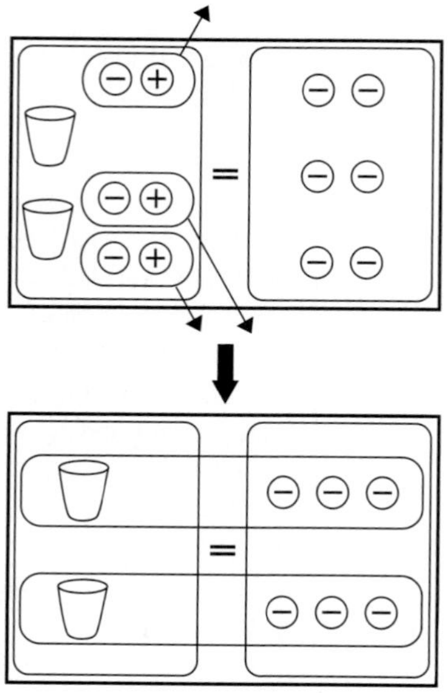

Page 175, Lesson 4-5

49.

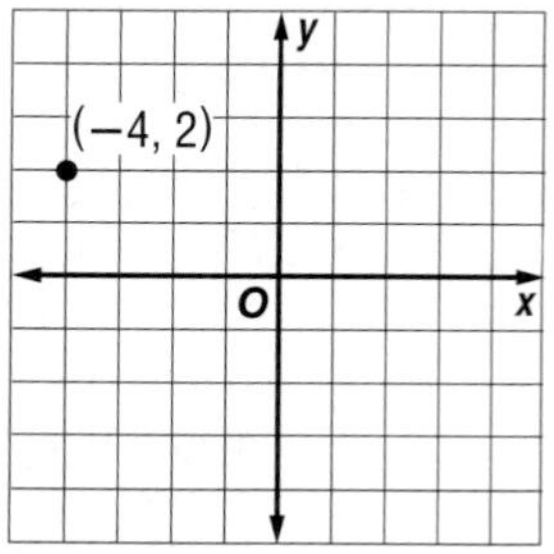

50.

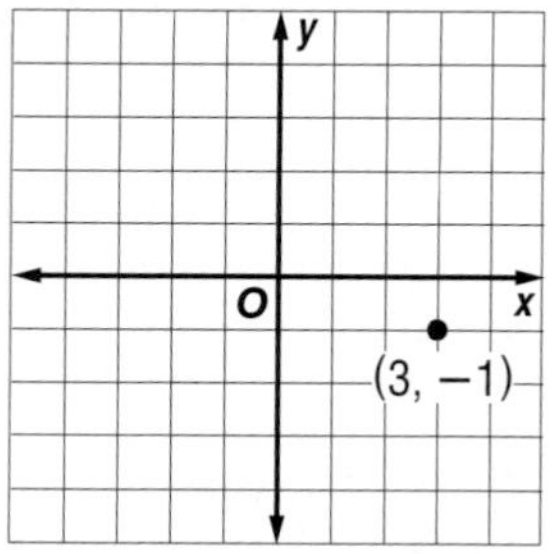

51.

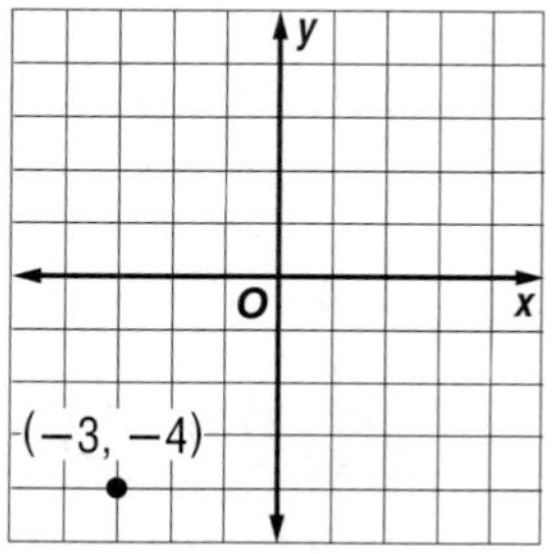

52.

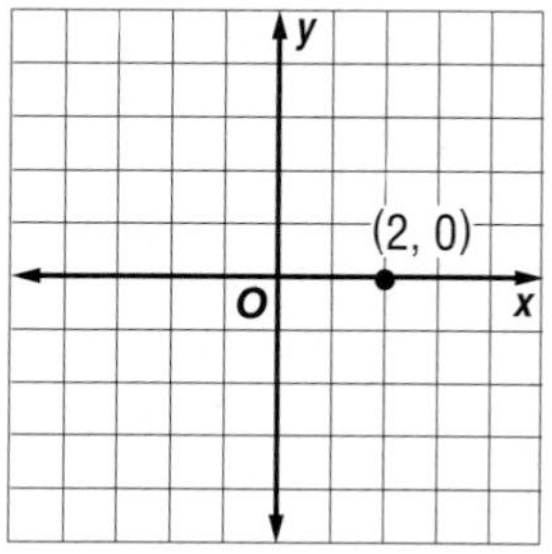

Page 177, Lesson 4-6 (When am I ever going to use this?)

2.

Hamburgers

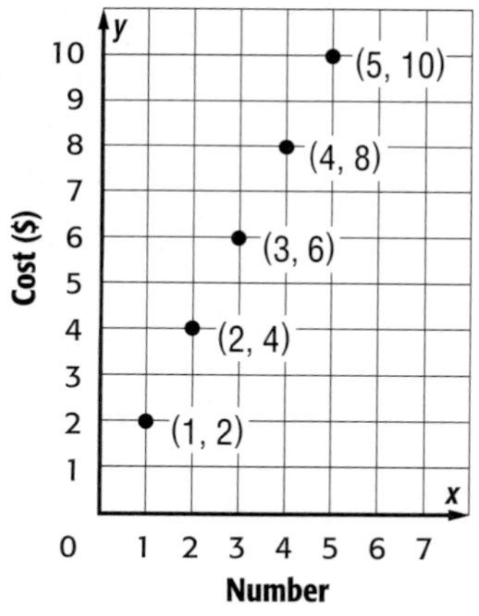

Pages 180–181, Lesson 4-6

21.

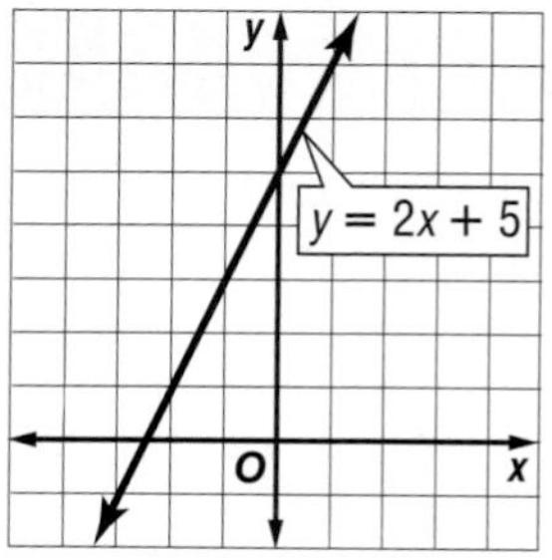

22.

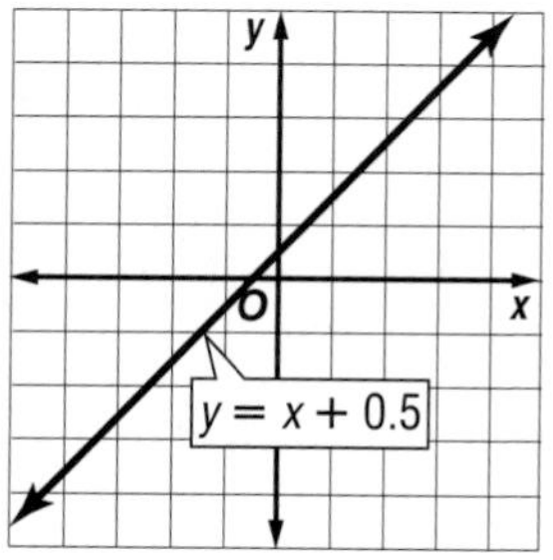

23.

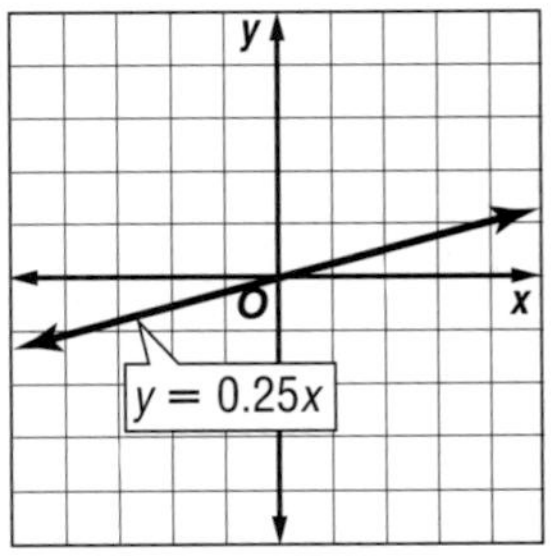

24.

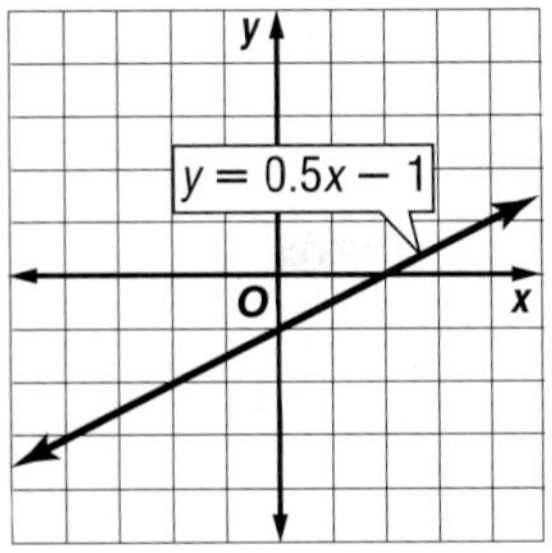

25.

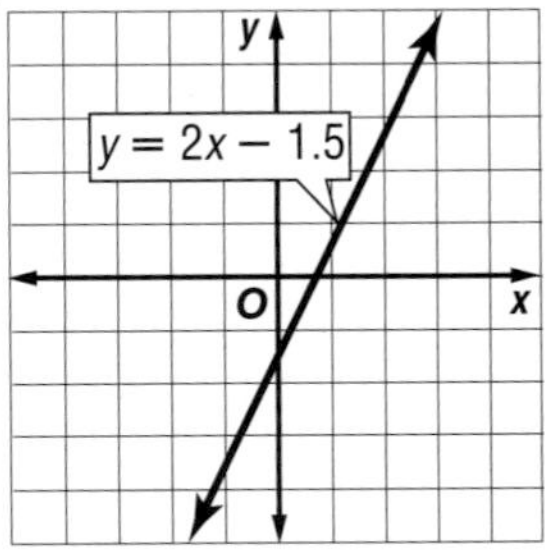

26. $y = x + 3$

x	$x + 3$	y
−1	−1 + 3	2
0	0 + 3	3
1	1 + 3	4
2	2 + 3	5

27. $y = x - 5$

x	$x - 5$	y
−1	−1 − 5	−6
0	0 − 5	−5
1	1 − 5	−4
2	2 − 5	−3

28. $y = 10x$

x	$10x$	y
1	10(1)	10
2	10(2)	20
3	10(3)	30
4	10(4)	40

29.

x	$2.67x$	y
1	2.67(1)	2.67
2	2.67(2)	5.34
3	2.67(3)	8.01
4	2.67(4)	10.68

31. **Average Defense Spending (per person)**

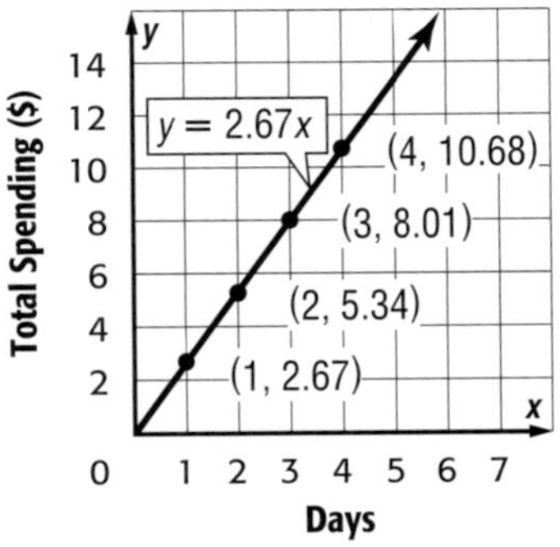

32.

x	$20x$	y
1	20(1)	20
2	20(2)	40
3	20(3)	60
4	20(4)	80

34. **Internet Charges**

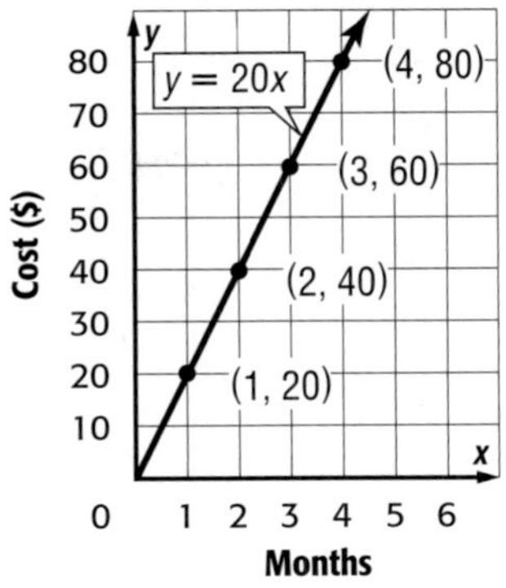

35. **Rectangles**

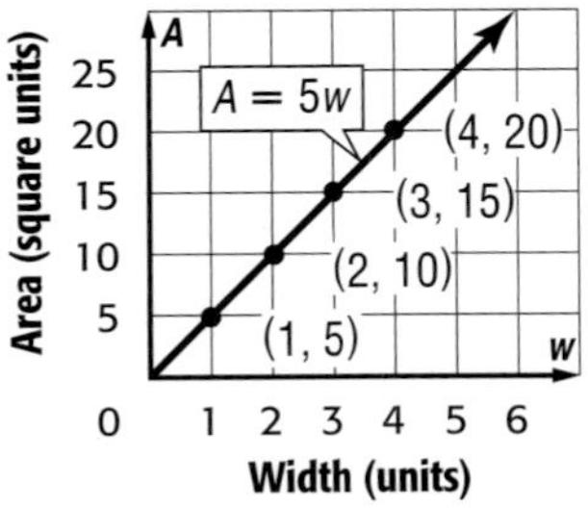

41. nonlinear

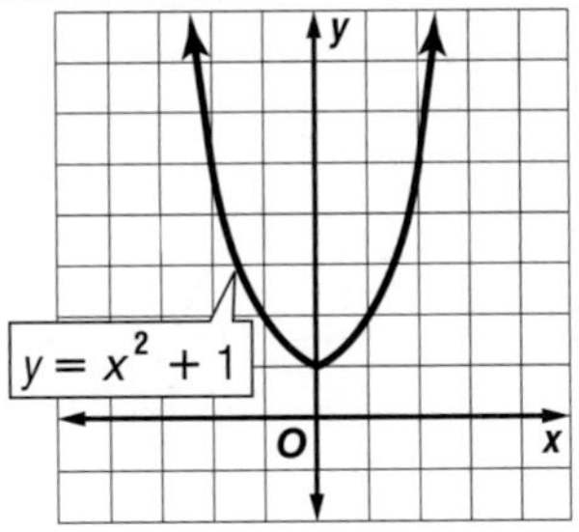

42. linear

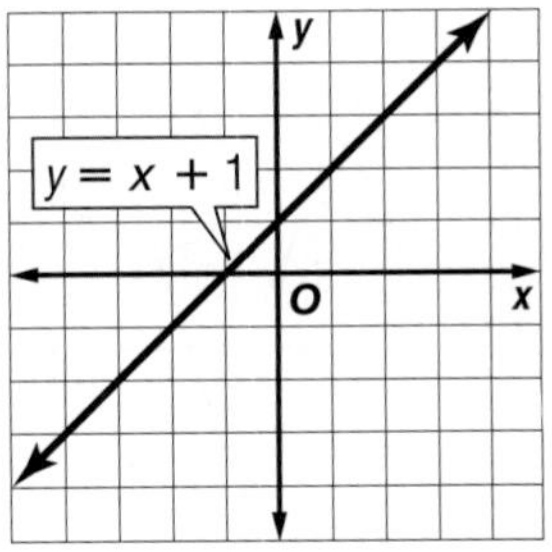

43. linear

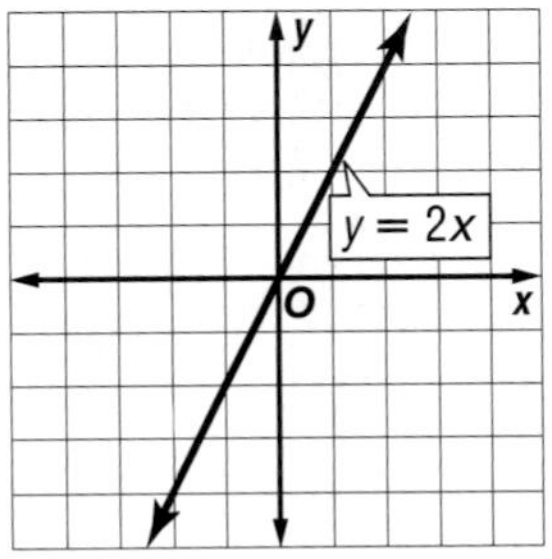

44. nonlinear

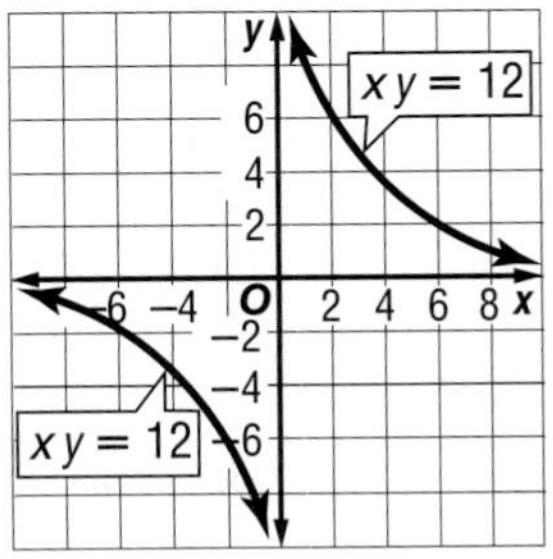

45. nonlinear

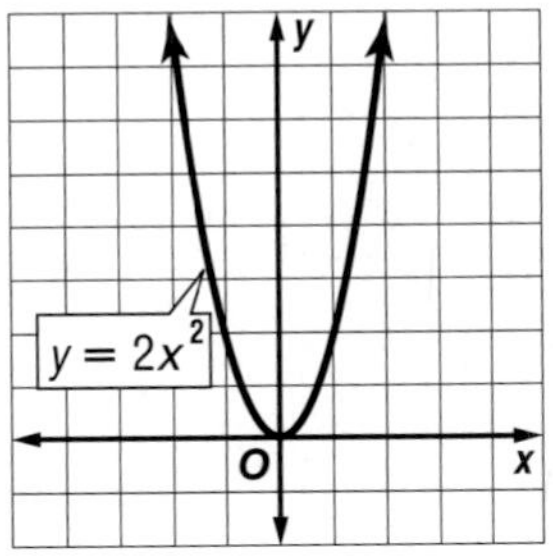

46. nonlinear

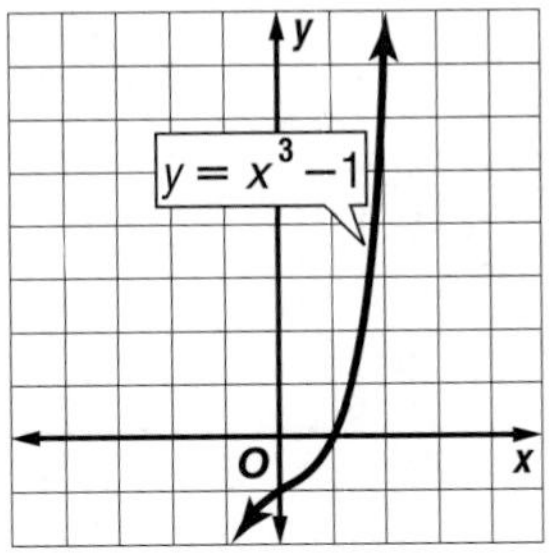

Page 185, Lesson 4-7

33.

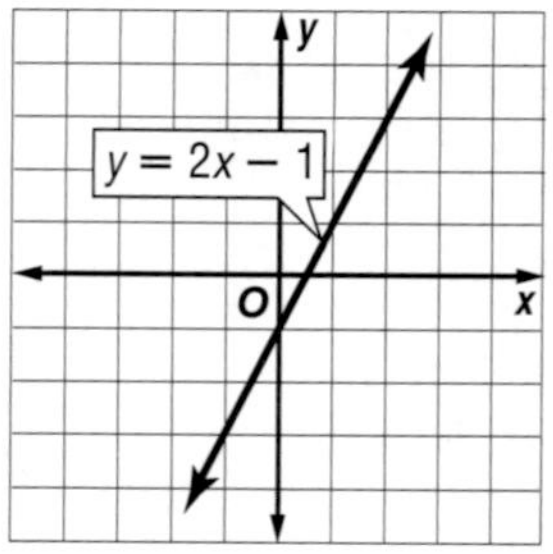

34.

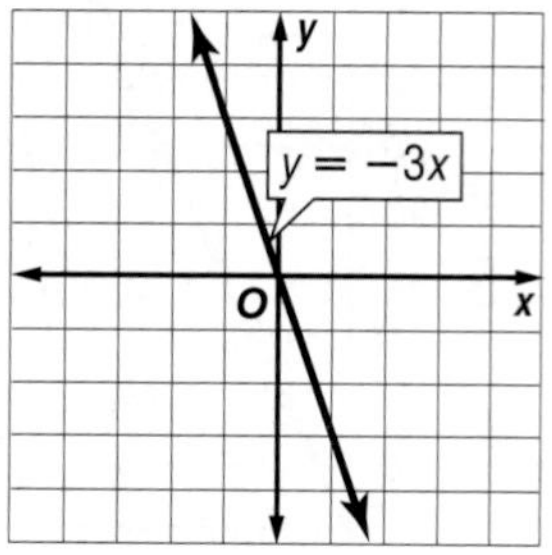

35.

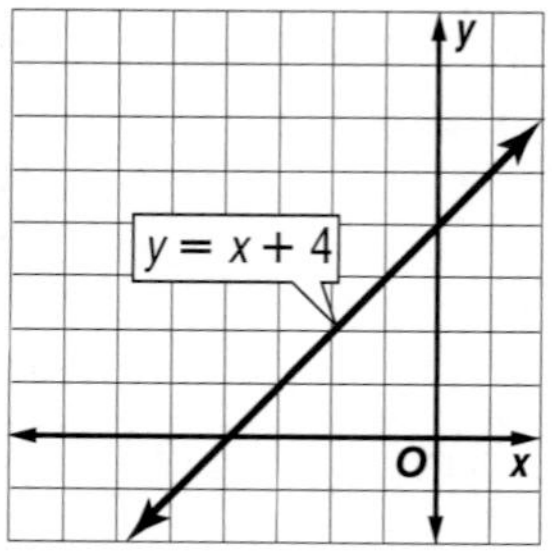

Page 189, Practice Test

13.

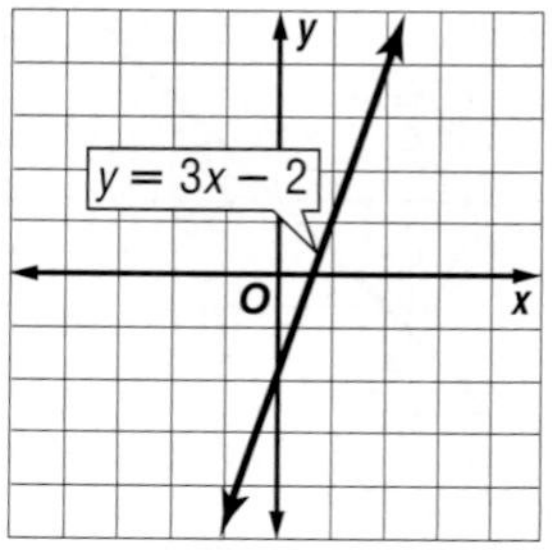

14.

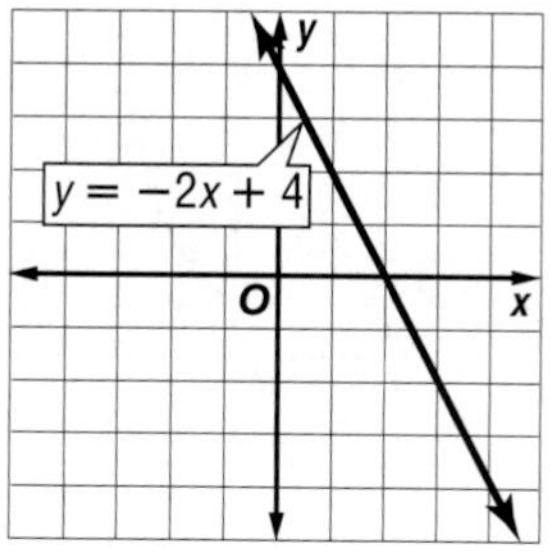

15.

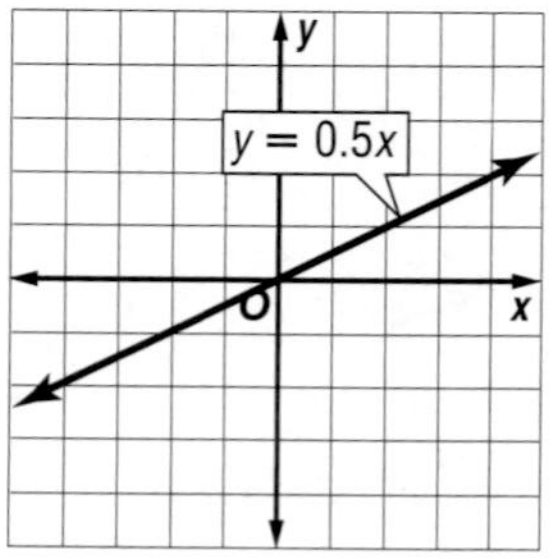

16.

Tickets Purchased	Total Cost
2	\$6
4	\$12
6	\$18

18.

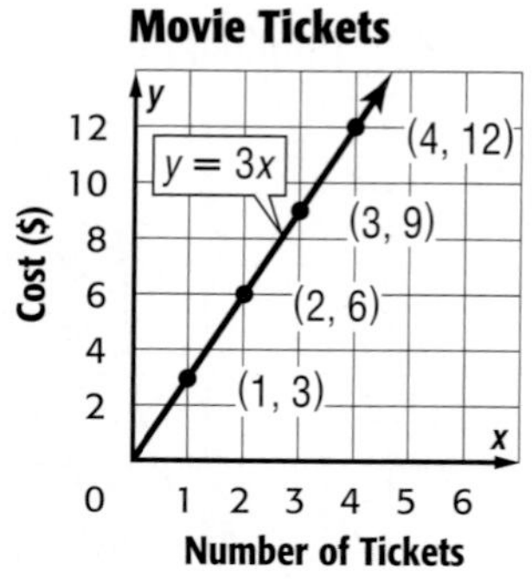

NOTES

UNIT 3 Notes

Introduction

In this unit, students will learn to simplify fractions using the greatest common factor (GCF) of the numerator and denominator and by using prime factorization. They will also learn to express fractions as decimals that either terminate or repeat. In addition, students will learn to convert fractions to percents and percents to simplified fractions. They then will use the least common multiple (LCM) of a set of fractions to compare and order fractions and other rational numbers.

Students will apply their knowledge of fractions to the adding, subtracting, multiplying, and dividing of fractions and mixed numbers. They also will learn how to change customary units. Finally, they will explore the geometric measurements of perimeter, area, and circumference.

Assessment Options

Unit 3 Test Pages 369–370 of the *Chapter 6 Resource Masters* may be used as a test or a review for Unit 3. This assessment contains both multiple-choice and short-response items.

UNIT 3 Fractions

Chapter 5

Fractions, Decimals, and Percents

Chapter 6

Applying Fractions

Fractions, decimals, and percents are different ways of representing the same values. In this unit, you will describe real-life situations, perform basic operations, and solve equations with fractions.

What's Math Got to Do With It? Real-Life Videos

What's Math Got to Do With It? Real-Life Videos engage students, showing them how math is used in everyday situations. Use Video 3 to discuss how fractions, percents, and circumference are used in amusement parks and in movie concessions.

WebQuest INTERDISCIPLINARY PROJECT — MATH and HEALTH

A WELL-BALANCED DIET

You are what you eat! So are you ice cream or broccoli? You're on a mission to find out! Along the way, you'll collect and analyze data about what you eat over a period of five days. You'll take into account serving sizes, Calories, and fat grams. You'll also take on the roll of a nutritionist, researching the Food Pyramid and creating a healthy meal plan. So bring a hearty appetite and your math tool kit. This adventure will tantalize your taste buds!

Log on to msmath2.net/webquest to begin your WebQuest.

WebQuest INTERDISCIPLINARY PROJECT

A Well-Balanced Diet

A WebQuest is an online project in which students do research on the Internet, gather data, and make presentations using word processing, graphing, page-making, or presentation software. In each chapter, students advance to the next step in their WebQuest. At the end of Chapter 6, the project culminates with a presentation of their findings.

Teaching suggestions and sample answers are available in the WebQuests, Projects, and Interdisciplinary Investigations.

Math and Health

You can use this WebQuest Interdisciplinary Project with your students' health teacher to make the connection from mathematics to the health topics your students are studying.

More Interdisciplinary Connections

You may wish to share these suggestions with your students' other teachers.

Math and Life Skills

Imagine that your task is to prepare a well-balanced lunch for 100 middle-school students. Suggest a menu and find recipes for the dishes. Calculate the amount of ingredients needed to make 100 servings as well as the amount of preparation and cooking time.

Math and Language Arts

Write an anthropological account of the food eaten by U.S. teenagers from the perspective of a visitor from another galaxy. Describe the colors, textures, smells, and sounds of various teen favorites.

CHAPTER 5

Chapter Overview

Fractions, Decimals, and Percents

	LESSON OBJECTIVES	NCTM Standards	NAEP 2005	State/Local Objectives
5-1	**Prime Factorization** *(pp. 196–200)* **5-1a Hands-On Lab:** Discover factors of whole numbers. • Find the prime factorization of a composite number.	1, 2, 6, 9, 10	A1a, A1b, A3b, N5b, N5c, N5d, N5f	
5-2	**Greatest Common Factor** *(pp. 201–206)* **5-2a Problem-Solving Strategy:** Solve problems by making an organized list. • Find the greatest common factor of two or more numbers.	1, 2, 6, 9, 10	D4e, N5b, N5c, N5d, N5f	
5-3	**Simplifying Fractions** *(pp. 207–209)* • Write fractions in simplest form.	1, 2, 6, 9, 10	N1b, N1d	
5-4	**Fractions and Decimals** *(pp. 210–213)* • Write fractions as terminating or repeating decimals and write decimals as fractions.	1, 2, 6, 9, 10	A1a, N1d, N1e, N3a, N5f	
5-5	**Fractions and Percents** *(pp. 216–219)* • Write fractions as percents and percents as fractions.	1, 2, 6, 9, 10	N1b, N1e, N4a, N4b, N4c	
5-6	**Percents and Decimals** *(pp. 220–223)* • Write percents as decimals and vice versa.	1, 2, 6, 9, 10	N1e, N1j	
5-7	**Least Common Multiple** *(pp. 224–226)* • Find the least common multiple of two or more numbers.	1, 2, 6, 9, 10	N5b, N5c	
5-8	**Comparing and Ordering Rational Numbers** *(pp. 227–231)* • Compare and order fractions, decimals, and percents.	1, 2, 6, 9, 10	N1d, N1e, N1i, N1j, N2c, N4a, N4b, N4c	

Key to NCTM Standards:
1=Number & Operations, 2=Algebra, 3=Geometry, 4=Measurement, 5=Data Analysis & Probability, 6=Problem Solving, 7=Reasoning & Proof, 8=Communication, 9=Connections, 10=Representation

NAEP 2005 Standards: See math.glencoe.com

Pacing suggestions for the entire year can be found on page T20.

Chapter Resource Manager

Time-Saving Tool
TeacherWorks™
All-In-One Planner and Resource Center
See page T5.

Chapter 5 Resource Masters

Study Guide and Intervention	Practice (Skills and Word Problems)	Reading to Learn Mathematics	Enrichment	Assessment	Applications*	Parent and Student Study Guide Workbook	5-Minute Check Transparencies	Key Teaching Transparencies	MathPASS: Tutorial Plus (lessons)	Teaching Mathematics with Manipulatives	Optional Materials
243	244–245	246	247		CM 15–16	5-1	5-1			59	index cards, markers
248	249–250	251	252	297	CS 29	5-2	5-2	17	2		
253	254–255	256	257			5-3	5-3		12		colored pencils
258	259–260	261	262	297, 299	RC 19	5-4	5-4	18			
263	264–265	266	267		CS 30, 31	5-5	5-5			60	
268	269–270	271	272	298		5-6	5-6	19	31		
273	274–275	276	277			5-7	5-7	20			centimeter cubes
278	279–280	281	282	283–296, 298, 300–302	CS 32	5-8	5-8		15		

*** Key to Abbreviations:** *CS = Calculator and Spreadsheet Masters, RC = Real-Life Career Masters, CM = Classroom Math Games (online), SM = Science and Math Lab Manual (online)*

Interactive Chalkboard

PowerPoint® Presentations

This CD-ROM is a customizable Microsoft® PowerPoint® presentation that includes:

- In-Class Examples
- Links to Online Study Tools
- 5-Minute Check Transparencies
- Bellringers

Math Skills Maintenance

Use the Math Skills Maintenance Masters to keep your students' basic math skills fresh. Students can review skills one at a time or in a combined format. The Masters are also available online at msmath2.net/maintenance.

CHAPTER 5

Mathematical Content and Teaching Strategies

Continuity of Instruction

Building on Prior Knowledge

In previous courses, students have worked with decimals and fractions. Operations with decimals as well as comparing and ordering decimals have been addressed. They have also performed operations with fractions. Models have been used to represent both fractions and decimals, and the concept of percent has been introduced.

Learning in This Chapter

Fractions are further explored utilizing greatest common factor, or GCF, to simplify fractions to simplest form. Least common multiple, or LCM, is introduced and used to find common denominators to compare rational numbers. These ideas help students work with fractions, decimals, and percents interchangeably. Converting from one form to another and comparing values is stressed throughout the chapter.

Making Future Connections

Students will build upon the concepts in this chapter when studying proportional reasoning and ratios. They will apply these skills to real-life situations involving percents. Ultimately the ideas will be generalized in algebra when working with rational expressions.

Mathematical Background

5-1 Prime Factorization

Prime numbers have exactly two factors, one and the number itself. Composite numbers have more than two factors. You may want to review with your students why 0 and 1 are neither prime nor composite. Students learn this difference and two factoring techniques to use on composite numbers: factor trees and division by prime numbers. They apply these techniques to algebraic monomials. Students need to remember that multiplication is commutative, so the order of the factors does not matter.

5-2 Greatest Common Factor

Knowing how to find the greatest common factor, or GCF, will help students simplify fractions. They begin by visualizing the GCF using a Venn diagram. They then use prime factorization to find the GCF of two and three numbers and of algebraic expressions.

5-3 Simplifying Fractions

Fractions are most easily manipulated when they are in their simplest form. Students visualize the fractions using models and find equivalent fractions. They then simplify the numerator and the denominator by dividing each by the GCF of the denominator.

Teacher to Teacher

M. J. Richards
Davis Middle School
Dublin, OH

"I give the students a list of fractions that are not simplified, which they use to fill in their blank bingo cards. I call out fractions in simplest form, and they place markers on all fractions which are equivalent. I often let students play in pairs and switch cards for each game."

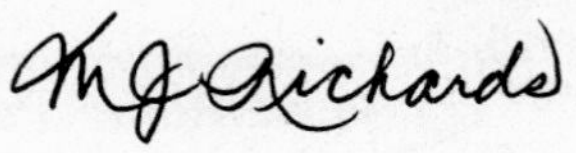

5-4 Fractions and Decimals

Because fractions can be inconvenient to use in some situations, students learn to convert fractions into decimals and then back to fractions. Fractions convert into either terminating decimals or repeating decimals. Repeating decimals are expressed using bar notation. The group of numbers under the bar is called the repetend.

5-5 Fractions and Percents

Percent is a ratio that compares a number to 100. Students will see percents used everyday, in situations ranging from sale prices at stores to news stories. Students learn to write a percent as a fraction and a fraction as a percent. Fractions with denominators other than 100 are rewritten as equivalent fractions with denominators of 100.

Using the Media

USA Today Snapshots® provide many examples of bar graphs or circle graphs using percents. Present students with a graph and ask them to write a paragraph explaining the information presented in the graph. Be sure to have them emphasize the use of percents and have them write the percents as fractions.

5-6 Percents and Decimals

Another method that students can use to evaluate percents is to convert them into decimals. Baseball statistics are a common example of percents written as decimals. Students learn how to multiply and divide by 100 to convert between decimals and percents.

McGraw-Hill Professional Development

Mathematics, YES! is a self-paced, web-based professional development program. Units are based on major mathematical concepts and are standards-based. Actual classroom videos provide a springboard for discussion and activities. Unit appropriate for this chapter:

- **Unit 4, Fractions, Decimals, and Percentages**

5-7 Least Common Multiple

Another common method for simplifying fractions is using the least common multiple, or LCM. Students initially visualize this concept using a Venn diagram. They then use prime factorization to find the LCM for two and three numbers.

5-8 Comparing and Ordering Rational Numbers

In order for students to understand and utilize fractions, decimals, and percents, they must be able to compare and order them. Using the symbols $>$, $<$, and $=$, students write numbers from least to greatest and from greatest to least. Ratios written in any form—fraction, decimal, or percent—are compared using consistent representations. This lesson is a basic introduction to rational numbers. Point out that the word rational contains the word ratio which is considered a form of a fraction. Students will further their study of rational numbers in Chapter 11.

In-Class Speaker

Invite a loan officer or other bank employee to visit your classroom. Ask the visitor to talk about how percents are used in his or her field of work. The guest could discuss with the class how to compare interest rates for loans to obtain the best deal. For example, is the interest rate for a new car the same as the interest rate for a used car? Or, is an interest rate of $5\frac{3}{4}\%$ better or worse than a rate of $5\frac{7}{8}\%$?

Key Concepts

Additional mathematical information and teaching notes are available at

msmath2.com/key_concepts.

The sections appropriate for the chapter are:

- **Simplifying Fractions and Ratios**
- **Fractions and Decimals**

Chapter 5

Daily Intervention and Assessment

	Type	Student Edition	Teacher Resources	Technology/Internet
INTERVENTION	Ongoing	Prerequisite Skills, pp. 195, 200, 209, 223, 226 Mid-Chapter Practice Test, p. 214	5-Minute Check Transparencies Quizzes 1–4, *CRM*, pp. 297–298 Mid-Chapter Test, *CRM*, p. 299 Study Guide and Intervention, *CRM*, pp. 243, 248, 253, 258, 263, 268, 273, 278	MathPASS: Tutorial Plus msmath2.net/self_check_quiz msmath2.net/extra_examples
	Mixed Review	pp. 200, 206, 209, 213, 219, 223, 226, 231	Cumulative Review, *CRM*, p. 300	
	Error Analysis	Find the Error, pp. 205, 208, 222	Find the Error, *TWE*, pp. 205, 208, 222 Unlocking Misconceptions, *TWE*, pp. 208, 221 Tips for New Teachers, *TWE*, p. 198	
ASSESSMENT	Standardized Test Practice	pp. 200, 202, 206, 209, 213, 214, 219, 223, 226, 229, 231, 235, 236–237	Standardized Test Practice, *CRM*, pp. 301–302	Standardized Test Practice CD-ROM msmath2.net/standardized_test
	Open-Ended Assessment	Writing Math, pp. 196, 199, 205, 208, 212, 218, 222, 225, 229 Open-Ended, pp. 199, 205, 208, 212, 218, 222, 225, 229 Standardized Test, p. 237	Modeling: *TWE*, pp. 219, 231 Speaking: *TWE*, pp. 209, 223 Writing: *TWE*, pp. 200, 202, 206 Extended Response Assessment, *CRM*, p. 295	
	Chapter Assessment	Study Guide, pp. 232–234 Practice Test, p. 235	Multiple-Choice Tests (Forms 1, 2A, 2B), *CRM*, pp. 283–288 Free-Response Tests (Forms 2C, 2D, 3), *CRM*, pp. 289–294 Vocabulary Test/Review, *CRM*, p. 296	ExamView® Pro Test Bank (see below) MindJogger Videoquizzes msmath2.net/vocabulary_review msmath2.net/chapter_test

* **Key to Abbreviations:** *TWE* = Teacher Wraparound Edition, *CRM* = Chapter Resource Masters

Additional Intervention

Skills Intervention for Middle School Mathematics workbook and CD-ROM includes 88 skills with pretests, tutorials, guided practice, and posttests for diagnosis and remediation.

Guide to Daily Intervention booklet details each intervention resource available and guides teachers in using them as intervention tools everyday to assist students who may be having difficulty.

ExamView® Pro

Use ExamView® Pro Test Bank CD-ROM to:

- **Create Multiple versions of tests.**
- **Create Modified tests for Inclusion students with one mouse click.**
- **Edit existing questions and Add your own questions.**
- **Build tests aligned with state standards using built-in State Curriculum Correlations.**
- **Change English tests to Spanish with one mouse click and vice versa.**

Reading and Writing Mathematics

• Student Edition

Foldables Study Organizer, p. 195

Concept Check questions at the beginning of each exercise set require students to verbalize and write about what they have learned in the lesson.

- **Open Ended,** pp. 199, 205, 208, 212, 218, 222, 225, 229
- **Find the Error,** pp. 205, 208, 222
- **Which One Doesn't Belong?,** pp. 199, 218, 229

Reading Study Tip, p. 217

• Teacher Wraparound Edition

Foldables Follow-Up, pp. 199, 205, 212, 214, 218, 222, 225, 229

Writing Activities, pp. 200, 202, 206

Differentiated Instruction, (Verbal/Linguistic), p. 217

ELL Resources, *see page T9 for a complete listing.*

• Additional Resources

Vocabulary Builder, *Chapter 5 Resource Masters*

Reading to Learn Mathematics, *Chapter 5 Resource Masters*

Vocabulary PuzzleMaker (online)

Teaching Mathematics with Foldables

Reading and Writing in the Mathematics Classroom

WebQuests, Projects, and Interdisciplinary Investigations

For detailed information on these resources, see pp. T6–T7.

• Recommended Outside Reading for Students

Nonfiction

- Wright, Joseph & Salvadori, Mario. *Math Games for Middle School: Challenges and Skill Builders for Students at Every Level.* Chicago Review Press, © 1998.

  This book uses activities and games to reinforce middle school mathematics concepts. Fractions and decimals are among the topics addressed in the book.

Fiction

- Greenburg, Dan. *Funny and Fabulous Fraction Stories.* Scholastic, © 1996.

  This book is a series of stories written to reinforce skills with fractions, decimals, and ratios. The stories are appropriate for students in Grades 4–8.

PROJECT CRISS℠ Study Skill

A comparison map can help students understand the differences in two or more similar concepts. Show students the sample map at the right that compares the way prime factorization can be used to find the greatest common factor (GCF) and least common multiple (LCM) of two numbers. While studying Chapter 5, have students work in small cooperative groups to design comparison maps showing how to write fractions as percents, percents as fractions, percents as decimals, and decimals as percents.

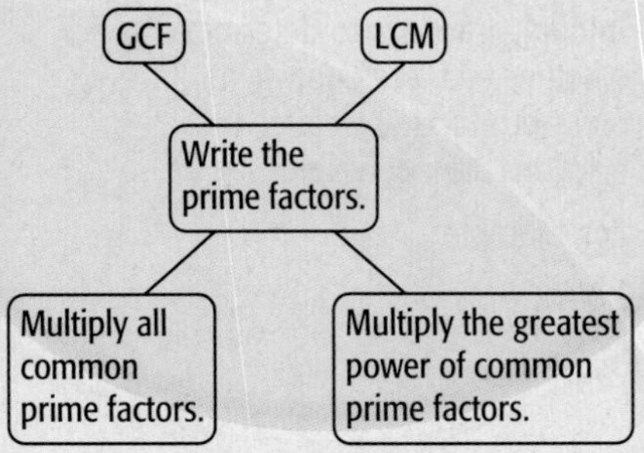

CReating Independence through Student-owned Strategies

For more information on Project CRISS, see page T9.

CHAPTER 5 Making Connections

School-To-Home

Parent and Student Study Guide

Parents and students may work together to reinforce the concepts and skills of this chapter. These workbooks are available in English and Spanish. A sample lesson is shown below. Also available at msmath2.net/parent_student.

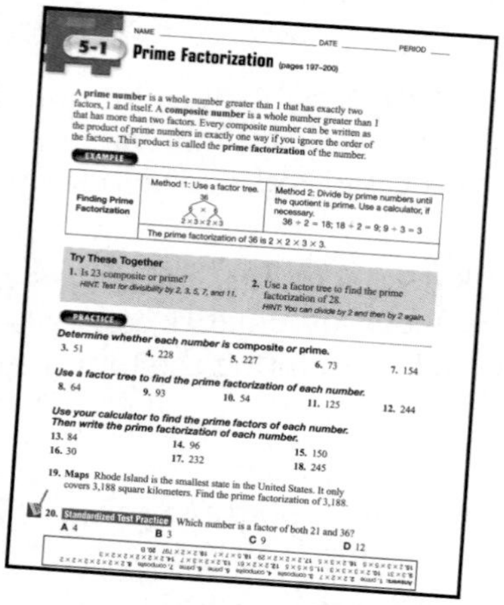

NAME ______ DATE ______ PERIOD ____

5-1 Prime Factorization (pages 197–200)

A **prime number** is a whole number greater than 1 that has exactly two factors, 1 and itself. A **composite number** is a whole number greater than 1 that has more than two factors. Every composite number can be written as the product of prime numbers in exactly one way if you ignore the order of the factors. This product is called the **prime factorization** of the number.

EXAMPLE

Finding Prime Factorization	Method 1: Use a factor tree. 36; $2 \times 3 \times 2 \times 3$	Method 2: Divide by prime numbers until the quotient is prime. Use a calculator, if necessary. $36 \div 2 = 18$; $18 \div 2 = 9$; $9 \div 3 = 3$
	The prime factorization of 36 is $2 \times 2 \times 3 \times 3$.	

Try These Together

1. Is 23 composite or prime? *HINT: Test for divisibility by 2, 3, 5, 7, and 11.*
2. Use a factor tree to find the prime factorization of 28. *HINT: You can divide by 2 and then by 2 again.*

PRACTICE

Determine whether each number is composite or prime.

3. 51 4. 228 5. 227 6. 73 7. 154

Use a factor tree to find the prime factorization of each number.

8. 64 9. 93 10. 54 11. 125 12. 244

Use your calculator to find the prime factors of each number. Then write the prime factorization of each number.

13. 84 14. 96 15. 150 16. 30 17. 232 18. 245

19. **Maps** Rhode Island is the smallest state in the United States. It only covers 3,188 square kilometers. Find the prime factorization of 3,188.

20. **Standardized Test Practice** Which number is a factor of both 21 and 36?
A 4 B 3 C 9 D 12

Family Letters and Activities

Send a letter and activity for this chapter home to encourage family members to become active participants in their students' learning. (English: *Chapter 5 Resource Masters*, pp. ix–x; Spanish: log on to msmath2.net/family_letters)

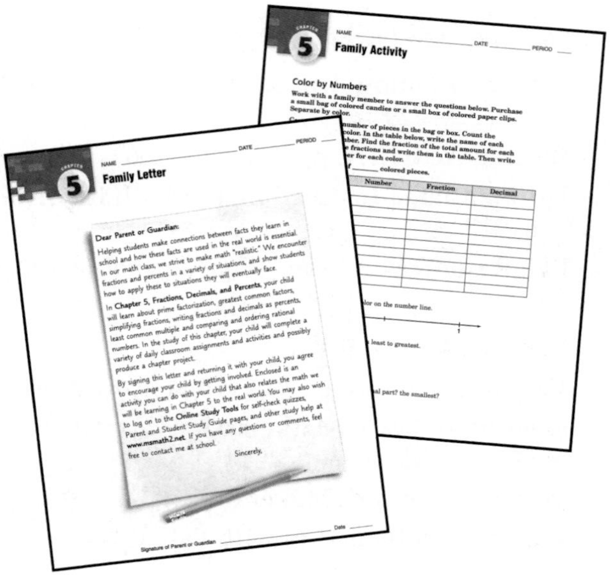

NAME ______ DATE ______ PERIOD ____

Family Activity

Color by Numbers

Work with a family member to answer the questions below. Purchase a small bag of colored candies or a small box of colored paper clips. Separate by color.

...number of pieces in the bag or box. Count the ...color. In the table below, write the name of each ...ber. Find the fraction of the total amount for each ...fractions and write them in the table. Then write ...er for each color.

...f ______ colored pieces.

Number	Fraction	Decimal

...lor on the number line.

...least to greatest.

...al part? the smallest?

NAME ______ DATE ______ PERIOD ____

Family Letter

Dear Parent or Guardian:

Helping students make connections between facts they learn in school and how these facts are used in the real world is essential. In our math class, we strive to make math "realistic." We encounter fractions and percents in a variety of situations, and show students how to apply these to situations they will eventually face.

In **Chapter 5, Fractions, Decimals, and Percents**, your child will learn about prime factorization, greatest common factors, simplifying fractions, writing fractions and decimals as percents, least common multiple and comparing and ordering rational numbers. In the study of this chapter, your child will complete a variety of daily classroom assignments and activities and possibly produce a chapter project.

By signing this letter and returning it with your child, you agree to encourage your child by getting involved. Enclosed is an activity you can do with your child that also relates the math we will be learning in Chapter 5 to the real world. You may also wish to log on to the **Online Study Tools** for self-check quizzes, Parent and Student Study Guide pages, and other study help at **www.msmath2.net**. If you have any questions or comments, feel free to contact me at school.

Sincerely,

Signature of Parent or Guardian ______ Date ______

English Language Learner ELL

Lesson 5-3

Using Models

Show the students that you can cross multiply fractions to determine whether they are equivalent. If the cross products are equal, the fractions are equivalent.

For example:

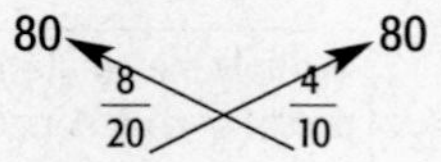

Lesson 5-4

Using Manipulatives

Give groups of students a bag with fraction cards inside. Have one student draw a card and find a decimal equivalent for the fraction using a calculator. Make sure students know how to input the numbers into the calculator. Have students write down the answer from the calculator and ask them what they notice. Have the students categorize the quotients as terminating or repeating decimals.

Lesson 5-8

Flexible Groups

Give groups of six students a set of six cards with a rational number, such as 1, -1, $\frac{1}{2}$, -3.75, $4\frac{3}{4}$, and so on, on each card. Have the students order the numbers from least to greatest. On a wall of the room, mark *0, positive,* and *negative* to denote a number line. Have the students come to the number line by group and position themselves in order on the number line. The students can check each other for accuracy.

MathScape™

This project was supported, in part, by the **National Science Foundation**

MathScape is a mathematics curriculum for grades 6–8 developed by the Seeing and Thinking Mathematically Project at the Education Development Center.

Generalizing About Numbers

Making Mathematical Arguments

Summary Students will understand multiples, factors, primes, and divisibility. Generalizing patterns of multiples is used as a problem-solving strategy to find factors.

USA TODAY Education Online Lesson Planner

USA TODAY® Education's Online site offers resources and interactive features connected to each day's newspaper. *Experience Today,* USA TODAY's® daily lesson plan, is available on the site and delivered daily to subscribers. This plan provides instruction for integrating USA TODAY® graphics and key editorial features into your mathematics classroom. Log on to **education.usatoday.com**.

USA TODAY Snapshots® in Chapter 5

Lesson	Snapshot
5-3	Where parking costs most
5-6	Seles soars at Australian Open

Quick Review Math Handbook

is Glencoe's mathematical handbook for students. The Hot Words section includes a glossary of terms while the Hot Topics section consists of explanations of key mathematical concepts. An exercise set is included to check students' understanding of the concepts. This valuable resource can be used as a reference in the classroom or for home study.

Lesson	Hot Topics Section	Lesson	Hot Topics Section
5-1, 5-2	1-4	5-5, 5-6	2-9
5-3	2-1	5-7	1-4
5-4	2-5, 2-9	5-8	2-2, 2-5

CHAPTER 5 Notes

Key Chapter Objectives

Lesson 5-1 Find the prime factorization of a composite number.

Lesson 5-2 Find the greatest common factor of two or more numbers.

Lesson 5-3 Write fractions in simplest form.

Lesson 5-4 Write fractions as terminating or repeating decimals and write decimals as fractions.

Lessons 5-5 and 5-6 Write fractions and decimals as percents and percents as fractions and decimals.

Lesson 5-7 Find the least common multiple of two or more numbers.

Lesson 5-8 Compare and order fractions, decimals, and percents.

"What does music have to do with math?"

Point out to students that this is only one of many applications of the math that they are studying. Others are provided in the introduction to each lesson.

Vocabulary Builder, pp. vii–viii ELL

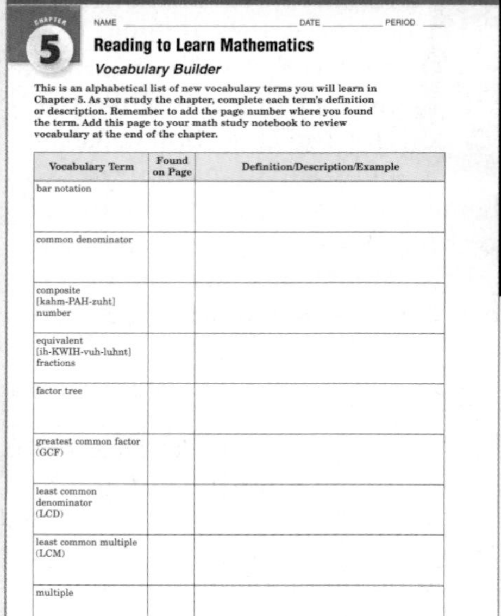

CHAPTER 5 NAME ______ DATE ______ PERIOD ____

Reading to Learn Mathematics

Vocabulary Builder

This is an alphabetical list of new vocabulary terms you will learn in Chapter 5. As you study the chapter, complete each term's definition or description. Remember to add the page number where you found the term. Add this page to your math study notebook to review vocabulary at the end of the chapter.

Vocabulary Term	Found on Page	Definition/Description/Example
bar notation		
common denominator		
composite [kahm-PAH-zuht] number		
equivalent [ih-KWIH-vuh-luhnt] fractions		
factor tree		
greatest common factor (GCF)		
least common denominator (LCD)		
least common multiple (LCM)		
multiple		

CHAPTER 5 Fractions, Decimals, and Percents

"What does music have to do with math?"

The *frequencies* of two musical notes can be compared using a fraction. If the fraction can be simplified, the two notes are harmonious. In mathematics, you will use fractions to describe many real-life situations.

You will solve problems about music in Lesson 5-3.

194

Vocabulary Builder

For a thorough vocabulary list with pronunciations of new words, give students the Vocabulary Builder worksheets from the *Chapter 5 Resource Masters.* Encourage them to complete the definition of each term as they progress through the chapter.

GETTING STARTED

Take this quiz to see whether you are ready to begin Chapter 5. Refer to the lesson or page number in parentheses if you need more review.

Vocabulary Review

Complete each sentence.

1. After comparing the decimals 0.41 and 0.04, you find that 0.41 is __?__ than 0.04. (Page 556) **greater**
2. Two or more numbers that are multiplied together to form a product are called __?__. (Lesson 1-2) **factors**

Prerequisite Skills

State which decimal is greater. (Page 556)

3. 0.6, 0.61 **0.61**
4. 1.25, 1.52 **1.52**
5. 0.33, 0.13 **0.33**
6. 1.08, 10.8 **10.8**

Use divisibility rules to determine whether each number is divisible by 2, 3, 5, 6, or 10. (Page 554)

7. 125 **5**
8. 78 **2, 3, 6**
9. 37 **none**
10. 214 **2**

Divide. (Page 562)

11. $5\overline{)2.0}$ **0.4**
12. $4\overline{)3.00}$ **0.75**

Write each power as a product of the same factor. (Lesson 1-2) **14. $5 \times 5 \times 5 \times 5 \times 5$**

13. 2^3 **$2 \times 2 \times 2$**
14. 5^5
15. 7^2 **7×7**
16. 6^4 **$6 \times 6 \times 6 \times 6$**

Write each product in exponential form. (Lesson 1-2) **19. 1^5**

17. $4 \times 4 \times 4$ **4^3**
18. 8×8 **8^2**
19. $1 \times 1 \times 1 \times 1 \times 1$
20. $7 \times 7 \times 7 \times 7$ **7^4**

FOLDABLES™ Study Organizer

Fractions, Decimals, and Percents Begin with a sheet of $8\frac{1}{2}''$ by 11″ construction paper and two sheets of notebook paper.

STEP 1 **Fold and Label** Fold the construction paper in half lengthwise. Label the chapter title on the outside.

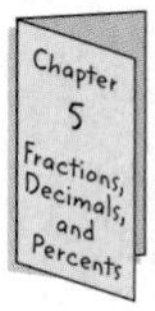

STEP 2 **Fold** Fold the sheets of notebook paper in half lengthwise. Then fold top to bottom twice.

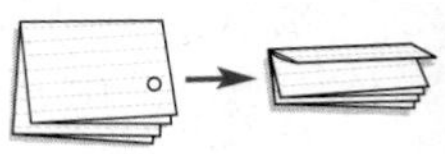

STEP 3 **Cut** Open the notebook paper. Cut along the second folds to make four tabs.

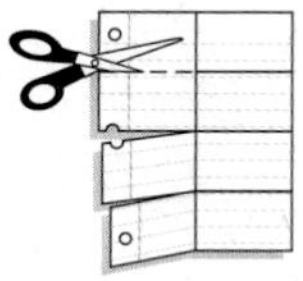

STEP 4 **Glue and Label** Glue the uncut halves of notebook paper side by side onto the construction paper. Label each tab with the lesson number.

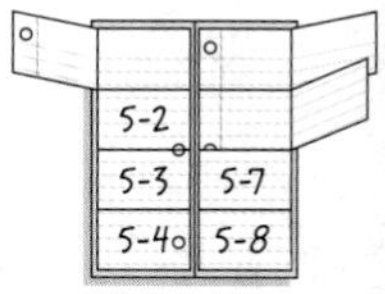

Reading and Writing As you read and study the chapter, write notes and examples for each lesson under the appropriate tab.

195

GETTING STARTED

This section provides a review of the basic concepts needed before beginning Chapter 5. Page references are included for additional student help.

Prerequisite Skills review skills needed in this chapter.

Exercises	Needed for Lesson	Skill
3–6	5-8	Compare Decimals
7–10	5-1, 5-2, 5-3	Use Divisibility Rules
11–12	5-4	Divide Decimals
13–16	5-1	Write Powers as Products
17–20	5-1	Write Products in Exponential Form

For more prerequisite skills, see:

- Getting Ready for the Next Lesson, at the end of each lesson
- Prerequisite Skills in the Student Handbook on pp. 554–563
- *Math Skills Maintenance Masters*

FOLDABLES™ Study Organizer

For more information about Foldables, see *Teaching Mathematics with Foldables.*

Use this Foldable for student writing about fractions, decimals, and percents. Before beginning each lesson, ask students to skim through the lesson and write one question that comes to mind during their preview. Write the question on the front of the corresponding lesson tab. As students read and work through the lesson, ask them to record the answer to their question under the tab. Self-questioning is a strategy that helps students stay focused during reading, writing, and skills practice.

5-1a HANDS-ON LAB
Lesson Notes

Lesson Resources

Teaching Mathematics with Manipulatives

- student recording sheet 5-1a

GETTING STARTED

Teaching Tip Use the largest index cards available so they can be seen around the room. Explain that students will be moving in and out of seats so they need to arrange belongings accordingly.

TEACH

- You may want to record student comments as the activity progresses. They will likely make observations you will want to discuss later.
- If needed, model the movement with a transparency on the overhead projector.
- Discuss which card(s) have the most numbers.

ASSESS

Discuss the **Writing Math** exercises. Students should:

- realize the numbers on the back of the card are the factors for the number on the front,
- correctly identify the prime numbers as having exactly two factors and square numbers as having an odd number of factors, and
- accurately predict the factors for numbers greater than 30.

5-1a HANDS-ON LAB

A Preview of Lesson 5-1

What You'll Learn
Discover factors of whole numbers.

Materials
- 15 index cards cut in half
- markers

Exploring Factors

INVESTIGATE *Work as a class.*

STEP 1 Number each index card consecutively with the numbers 1 through 30.

STEP 2 In order around the classroom, give each of thirty students one index card. Have each student stand up and write the number 1 on the back of his or her card.

STEP 3 Begin with the student holding the "2" card. Have this student and every second student sit down and write the number 2 on the back of his or her card. The other students remain standing.

STEP 4 Next, start with the student holding the "3" card. Have this student and every third student stand up or sit down (depending on whether the student is already sitting or standing) and write the number 3 on the back of his or her card.

STEP 5 Continue this process for each of the remaining numbers up to 30. The thirtieth student ends the activity by standing or sitting and writing the number 30 on the back of his or her card.

Writing Math

Work as a class.

1. How many students are standing at the end of the activity? Which cards are they holding? **5; 1, 4, 9, 16, and 25**
2. **LOOK FOR A PATTERN** Suppose there were 100 students holding index cards. Extend the pattern in Exercise 1 to predict the numbers that would be held by students standing at the end of the activity.
3. **Explain** the relationship between the numbers on the front and the back of the cards.
4. Separate the cards into two groups: one group with exactly two numbers on the back of the card and one group with more than two numbers. **Describe** any special characteristics of each group.

2. 1, 4, 9, 16, 25, 36, 49, 64, 81, 100
3. Sample answer: The numbers on the back are factors of the numbers on the front.
4. See margin.

Answer

4. **Sample answer: The group with exactly two numbers is made up of prime numbers and the number 1. The group with more than two numbers is made up of composite numbers.**

5-1 Prime Factorization

/hat You'll Learn
nd the prime
ctorization of a
omposite number.

EW Vocabulary

rime number
omposite number
rime factorization
ctor tree

EVIEW Vocabulary

ctors: two or more
umbers that are
ultiplied together
form a product
esson 1-2)

HANDS-ON Mini Lab

Materials
- grid paper

Work with a partner.

There is only one way that 2 can be expressed as the product of whole numbers. The figure shows that there is only one way that two squares can form a rectangle.

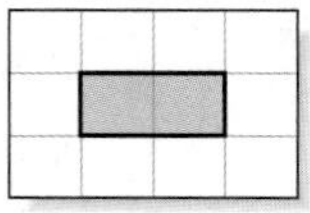

1. Using your grid paper, draw as many different rectangles as possible containing 3, 4, 5, 6, 7, 8, 9, and 10 squares.
2. Which numbers of squares can be drawn in only one rectangle? In more than one rectangle? **3, 5, 7; 4, 6, 8, 9, 10**

1. See Ch. 5 Answer Appendix.

The numbers of squares that can only be drawn in exactly one rectangle are prime numbers. The numbers that can be drawn in more than one rectangle are composite numbers.

A **prime number** is a whole number greater than 1 that has exactly two factors, 1 and itself.

A **composite number** is a whole number greater than 1 that has more than two factors.

The numbers 0 and 1 are neither prime nor composite.

Whole Numbers	Factors
2	1, 2
3	1, 3
5	1, 5
7	1, 7
4	1, 2, 4
6	1, 2, 3, 6
8	1, 2, 4, 8
9	1, 3, 9
10	1, 2, 5, 10
0	many
1	1

EXAMPLES Identify Numbers as Prime or Composite

Determine whether each number is *prime* or *composite*.

1 17

The number 17 has only two factors, 1 and 17, so it is prime.

2 12

The number 12 has six factors: 1, 2, 3, 4, 6, and 12. So, it is composite.

Your Turn **Determine whether each number is *prime* or *composite*.**

a. 11 **prime** b. 15 **composite** c. 24 **composite**

msmath2.net/extra_examples

5-1 Lesson Notes

Lesson Resources

Chapter 5 Resource Masters
- Study Guide and Intervention, p. 243
- Practice: Skills, p. 244
- Practice: Word Problems, p. 245
- Reading to Learn Mathematics, p. 246
- Enrichment, p. 247

Classroom Math Games, pp. 15–16
Parent and Student Study Guide Workbook, 5-1

FAST FILE

1 Focus

5-Minute Check 5-1
Use as a quiz or review of Chapter 4. (Also available as a transparency)

INTERACTIVE CHALKBOARD
PowerPoint® Presentations

Bellringer
Hands-On Activity Give pairs of students 18 small items, such as dried beans or paper clips. Have them separate the items into equal groups and record their findings. Challenge them to find as many equal groups as possible. **Groupings: 1 of 18, 18 of 1, 2 of 9, 9 of 2, 3 of 6, 6 of 3**

Reading to Learn Mathematics, p. 246 **ELL**

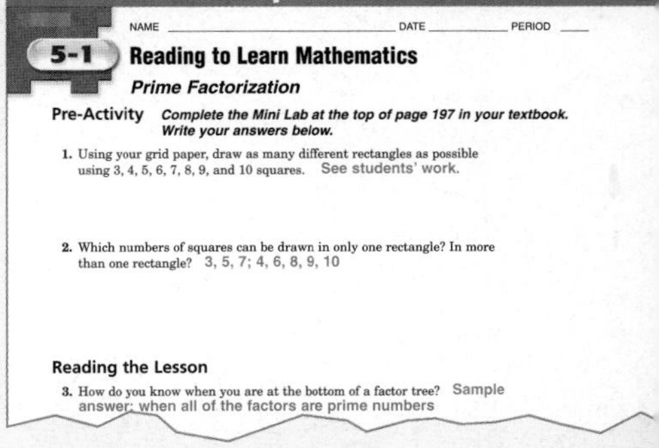
NAME ____________ DATE ________ PERIOD ____

5-1 **Reading to Learn Mathematics**
Prime Factorization

Pre-Activity *Complete the Mini Lab at the top of page 197 in your textbook. Write your answers below.*

1. Using your grid paper, draw as many different rectangles as possible using 3, 4, 5, 6, 7, 8, 9, and 10 squares. See students' work.

2. Which numbers of squares can be drawn in only one rectangle? In more than one rectangle? 3, 5, 7; 4, 6, 8, 9, 10

Reading the Lesson

3. How do you know when you are at the bottom of a factor tree? Sample answer: when all of the factors are prime numbers

2 Teach

In-Class Examples

INTERACTIVE CHALKBOARD PowerPoint® Presentations

Determine whether each number is *prime* or *composite*.

1. 63 **composite**
2. 29 **prime**
3. Find the prime factorization of 100. $2^2 \cdot 5^2$
4. **ALGEBRA** Factor $21m^2n$. $3 \cdot 7 \cdot m \cdot m \cdot n$

Tips for New Teachers

Intervention Explain to students that the dimensions of each rectangle are the factors of the number representing the area. Make sure students understand that a 1-by-2 rectangle and a 2-by-1 rectangle are considered the same rectangle. They are just drawn differently.

STUDY TIP

Commutative Property Multiplication is commutative, so the order of the factors does not matter.

Every composite number can be written as a product of prime numbers in exactly one way. The product is called the **prime factorization** of the number. You can use a **factor tree** to find the prime factorization.

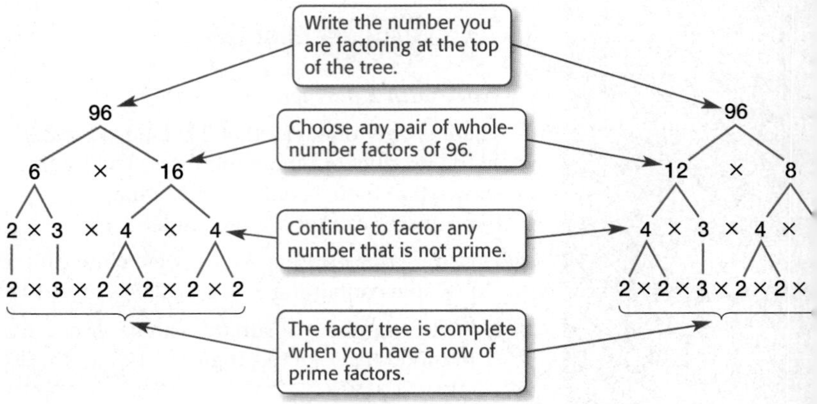

STUDY TIP

Mental Math You can use divisibility rules to find prime factors. See page 554.

Both trees give the same prime factorization of 96, but in different orders. The prime factorization of 96 is $2 \times 2 \times 2 \times 2 \times 2 \times 3$, or $2^5 \times 3$ using exponents.

EXAMPLE Find the Prime Factorization

3 Find the prime factorization of 24.

Method 1 Use a factor tree.

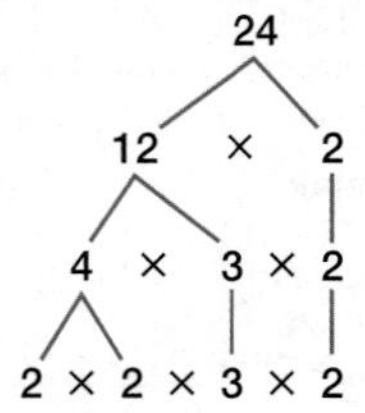

Method 2 Divide by prime numbers.

$$\begin{array}{r} 2 \\ 2\overline{)4} \\ 3\overline{)12} \\ 2\overline{)24} \end{array}$$

← Start here.

The divisors are 2, 2, 3, and 2.

The prime factorization of 24 is $2 \times 2 \times 3 \times 2$ or $2^3 \times 3$.

Your Turn **Find the prime factorization of each number.**

d. 18 2×3^2 e. 28 $2^2 \times 7$ f. 16 2^4

TEACHING TIP

Tell students that finding the prime factorization and factoring mean the same thing.

You can factor an expression like $6ab$ as the product of prime numbers and variables.

EXAMPLE Factor an Algebraic Expression

4 ALGEBRA Factor $6ab$.

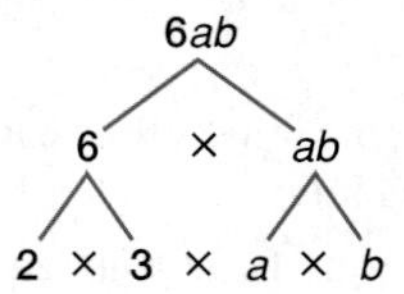

$6ab = 2 \cdot 3 \cdot a \cdot b$ Recall that a raised dot means multiplication.

DAILY INTERVENTION **Differentiated Instruction**

Logical Give pairs of students a 3- or 4-digit number. One student uses pencil and paper and the other uses a calculator to find the prime factorization. After a few minutes, ask students if the calculator helped in determining the factorization. **Possible answer: the calculator made the factoring quicker.**

Skill and Concept Check

Writing Math Exercises 1 & 3

1. **Compare and contrast** prime and composite numbers. **1–2. See margin.**
2. **OPEN ENDED** Draw two different factor trees to find the prime factorization of 36.
3. **Explain** why $2^3 \times 9$ is *not* the prime factorization of 72. **Sample answer: 9 is not prime.**
4. **Which One Doesn't Belong?** Identify the number that is not prime. **51**

11	37	17	51

GUIDED PRACTICE

Determine whether each number is *prime* or *composite*.

5. 7 **prime**
6. 50 **composite**
7. 67 **prime**

Find the prime factorization of each number.

8. 34 **2×17**
9. 30 **$2 \times 3 \times 5$**
10. 12 **$2^2 \times 3$**

Factor each expression.

11. $10ac$ **$2 \cdot 5 \cdot a \cdot c$**
12. $16x^2$ **$2 \cdot 2 \cdot 2 \cdot 2 \cdot x \cdot x$**

★ indicates increased difficulty

Practice and Applications

HOMEWORK HELP	
For Exercises	See Examples
13–20, 29	1, 2
21–28, 36–38, 45–47	3
30–35	4

Extra Practice See pages 574, 600.

Determine whether each number is *prime* or *composite*.

13. 22 **composite**
14. 44 **composite**
15. 13 **prime**
16. 39 **composite**
17. 81 **composite**
18. 31 **prime**
19. 97 **prime**
20. 43 **prime**

Find the prime factorization of each number.

21. 36 **$2^2 \times 3^2$**
22. 42 **$2 \times 3 \times 7$**
23. 99 **$3^2 \times 11$**
24. 64 **2^6**
25. 210 **$2 \times 3 \times 5 \times 7$**
26. 180 **$2^2 \times 3^2 \times 5$**
27. 126 **$2 \times 3^2 \times 7$**
28. 375 **3×5^3**
29. **ALGEBRA** Is the value of $3a + 6b$ *prime* or *composite* if $a = 1$ and $b = 5$? **composite**

Factor each expression.

30. $15mn$ **$3 \cdot 5 \cdot m \cdot n$**
31. $20pq$ **$2 \cdot 2 \cdot 5 \cdot p \cdot q$**
32. $34jkl$ **$2 \cdot 17 \cdot j \cdot k \cdot l$**
33. $49y^2$ **$7 \cdot 7 \cdot y \cdot y$**
34. $52gh^2$ **$2 \cdot 2 \cdot 13 \cdot g \cdot h \cdot h$**
35. $48a^2b^2$ **$2 \cdot 2 \cdot 2 \cdot 2 \cdot 3 \cdot a \cdot a \cdot b \cdot b$**

PACKAGING For Exercises 36–38, use the following information.
A juice company is designing a package that holds 36 juice cans. For example, a package could be 1 can high, 6 cans wide, and 6 cans long. Such a package would contain $1 \times 6 \times 6$, or 36 cans.

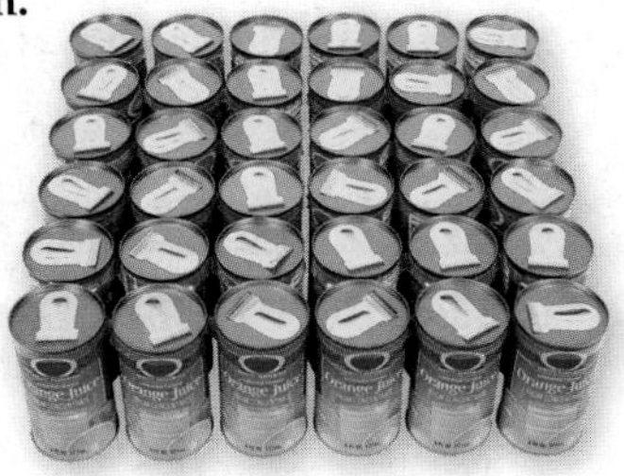

36. List two other ways to arrange 36 cans. **See margin.**
37. How many different ways can you arrange 36 cans? **8**
38. ★ In which arrangement should the 36 cans be packaged? Explain your reasoning. **See margin.**

msmath2.net/self_check_quiz

36. Sample answer: $1 \times 1 \times 36$, $1 \times 2 \times 18$, $1 \times 3 \times 12$, $1 \times 4 \times 9$, $1 \times 6 \times 6$, $2 \times 2 \times 9$, $2 \times 3 \times 6$, or $3 \times 3 \times 4$

38. Sample answer: The $3 \times 3 \times 4$ arrangement takes the least amount of space.

3 Practice/Apply

FOLDABLES™ Follow-Up Remind students to write a question about prime factorization on the front of the first tab of their Foldables before they begin the lesson. As students progress through the lesson, encourage them to take notes and record information that answers their question under the tab.

Which One Doesn't Belong?
In Exercise 4, if students are not sure if a number is prime, remind them to use the divisibility rules to get started.

About the Exercises...

Guided Practice Key

Exercises	Examples
5–7	1, 2
8–10	3
11–12	4

Odd/Even Assignments
Exercises 13–28 and 30–35 are structured so that students practice the same concepts whether they are assigned odd or even problems.

Answers

1. Sample answer: Both prime and composite numbers are whole numbers greater than 1. A prime number has exactly two factors, 1 and itself. Composite numbers have more than two factors.
2. Sample answer:

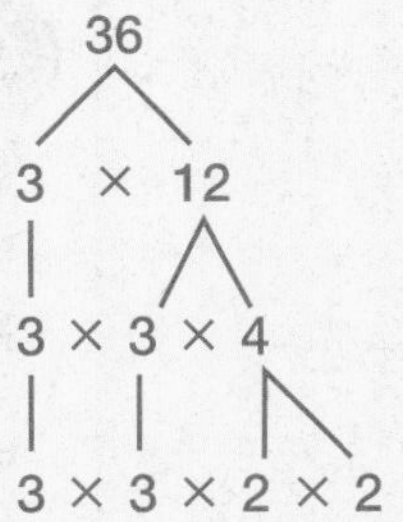

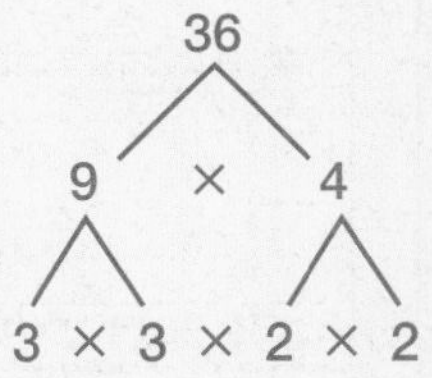

Replace each ■ with prime factors to make a true sentence.

★ 39. $2^3 \times ■ \times 11 = 616$ **7** ★ 40. $2 \times ■ \times 5^2 = 450$ **3^2** ★ 41. $3 \times 2^4 \times ■ = 1{,}200$ **5^2**

42. **NUMBER THEORY** *True* or *False*? All prime numbers are odd. If false, give a counterexample. **False; 2 is an even prime number.**

43. **NUMBER THEORY** Primes that differ by two are called *twin primes*. For example, 59 and 61 are twin primes. Give three examples of twin primes that are less than 50. **Sample answer: 5 and 7, 11 and 13, and 17 and 19**

LANDSCAPING For Exercises 44–47, use the following information.
Mrs. Franks is building a patio that covers an area measuring 8 feet by 15 feet. She will use rectangular concrete tiles for the patio.

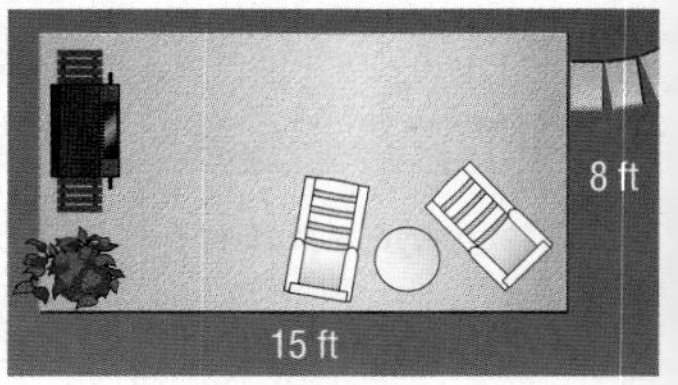

44. What is the area of the patio? **120 ft²**

45. Find the prime factorization of the area. **$2^3 \times 3 \times 5$**

46. If she can only buy tiles with dimensions that are prime numbers, name the dimensions of the concrete tiles available for her patio. **Sample answer: 2 ft by 3 ft or 2 ft by 5 ft**

47. How many tiles will she need to cover the patio? **Sample answer: twenty 2-ft-by-3-ft tiles, or twelve 2-ft-by-5-ft tiles**

★ 48. **NUMBER THEORY** Suppose n represents a whole number. Is $2n$ prime or composite? Explain. **If $n = 1$, $2n$ is a prime number. If $n > 1$, $2n$ is a composite number with at least 3 factors.**

49. **CRITICAL THINKING** Find the mystery number from the following clues. **36**
- This whole number is between 30 and 40.
- It has only two prime factors.
- The sum of its prime factors is 5.

Standardized Test Practice and Mixed Review

50. **MULTIPLE CHOICE** Which expression represents the prime factorization of 126? **B**

Ⓐ $2^2 \times 3 \times 7$ Ⓑ $2 \times 3^2 \times 7$ Ⓒ $2 \times 3 \times 7^2$ Ⓓ $2^2 \times 3^2 \times 7^2$

51. **SHORT RESPONSE** Determine whether the value of the expression $x^2 - 3x + 11$ is prime or composite when $x = 4$. **composite**

52. **ALGEBRA** What is the slope of the line that passes through points at $(-2, 5)$ and $(3, -1)$? (Lesson 4-7) **$-\frac{6}{5}$**

53. **ALGEBRA** Graph $y = 3x + 1$. (Lesson 4-6) **See margin.**

Tell whether each sum is *positive, negative,* or *zero*. (Lesson 3-4)

54. $6 + (-4)$ **positive** 55. $-13 + 9$ **negative** 56. $25 + (-26)$ **negative** 57. $-5 + 5$ **zero**

GETTING READY FOR THE NEXT LESSON

PREREQUISITE SKILL State whether each number is divisible by 2, 3, 5, 6, 9, or 10. (Page 554)

58. 24 **2, 3, 6** 59. 70 **2, 5, 10** 60. 120 **2, 3, 5, 6, 10** 61. 99 **3, 9** 62. 125 **5**

4 Assess

Open-Ended Assessment

Writing Ask students to find the prime factorization of 60 in two different ways. Have them write a short paragraph summarizing their methods and results.

Answer

53.

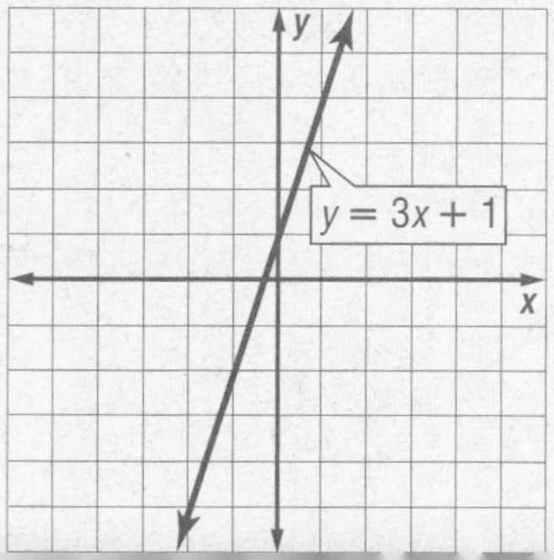

Study Guide and Intervention, p. 243

NAME ______ DATE ______ PERIOD ____

5-1 Study Guide and Intervention
Prime Factorization

A **prime number** is a whole number greater than 1 that has exactly two factors, 1 and itself. A whole number greater than 1 that has more than two factors is a **composite number**. Every composite number can be written as the product of prime numbers in exactly one way. This is called the **prime factorization** of the number.

EXAMPLE 1 **Determine whether 21 is *prime* or *composite*.**

The number 21 has four factors: 1, 3, 7, and 21. So, it is composite.

EXAMPLE 2 **Find the prime factorization of 420.**

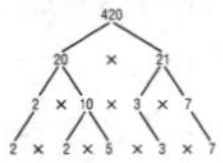

Write 420 as the product of two factors.

Keep factoring until all of the factors are prime numbers.

The prime factorization of 420 is $2 \times 2 \times 5 \times 3 \times 7$, or $2^2 \times 3 \times 5 \times 7$.

EXERCISES

Determine whether each number is *prime* or *composite*.

1. 97 prime 2. 91 composite 3. 327 composite

Find the prime factorization of each number.

4. 60 $2^2 \times 3 \times 5$ 5. 441 $3^2 \times 7^2$ 6. 140 $2^2 \times 5 \times 7$

7. 450 $2 \times 3^2 \times 5^2$ 8. 156 $2^2 \times 3 \times 13$ 9. 216 $2^3 \times 3^3$

Practice: Skills, p. 244

NAME ______ DATE ______ PERIOD ____

5-1 Practice: Skills
Prime Factorization

Determine whether each number is *prime* or *composite*.

1. 34 composite 2. 77 composite 3. 37 prime

4. 89 prime 5. 69 composite 6. 67 prime

7. 123 composite 8. 71 prime 9. 2 prime

10. 45 composite 11. 29 prime 12. 90 composite

Find the prime factorization of each number.

13. 48 $2^4 \times 3$ 14. 54 2×3^3 15. 108 $2^2 \times 3^3$

16. 80 $2^4 \times 5$ 17. 125 5^3 18. 66 $2 \times 3 \times 11$

19. 250 2×5^3 20. 187 11×17 21. 242 2×11^2

Write each expression as a product of its factors.

22. $56ab$ $2^3 \cdot 7 \cdot a \cdot b$ 23. $24bc$ $2^3 \cdot 3 \cdot b \cdot c$ 24. $147abc$ $3 \cdot 7^2 \cdot a \cdot b \cdot c$

Practice: Word Problems, p. 245

NAME ______ DATE ______ PERIOD ____

5-1 Practice: Word Problems
Prime Factorization

1. **FLAG** When the United States had 48 states, the stars were arranged in a 6 × 8 rectangular arrangement. What other rectangular arrangements of 48 stars are possible? 1 × 48; 2 × 24; 3 × 16; 4 × 12

2. **MARCHING BAND** A marching band has 72 members. If they are to march with an equal number of people in each row, state all possible numbers of rows and numbers of people in each row. 1 × 72; 2 × 36; 3 × 24; 4 × 18; 6 × 12; 8 × 9

3. **BIOLOGY** The human face uses 14 muscles to smile and 43 to frown. Which number is prime and which is composite? Explain. 43 is prime; 14 is composite; The only factors of 43 are 1 and 43 but 14 has several factors.

4. **BASEBALL CARDS** Jack is arranging his prized baseball cards in a frame. If he has 24 cards, in how many different numbers of rows and columns can he display them if each row has the same number of cards? 1 × 24; 2 × 12; 3 × 8; 4 × 6

5. **HISTORY** It is estimated that Sophocles, an ancient Greek dramatist, died in 406 B.C. Find the prime factorization of 406. 2 × 7 × 29

6. **ANATOMY** There are 230 joints in the human body. Find the prime factorization of 230. 2 × 5 × 23

7. **PHOTOS** Bill is arranging 108 photos to display on a poster board for a presentation. If he arranges an equal number of photos in each row, in how many different numbers of rows and columns can he arrange the photos? 1 × 108; 2 × 54; 3 × 36; 4 × 27; 6 × 18; 9 × 12

8. **ART** The supreme example of Renaissance genius, Leonardo da Vinci, passed away in 1519 A.D. Is 1519 a composite or prime number? Explain. Composite; 1519 = 7 × 217

5-2a Problem-Solving Strategy

A Preview of Lesson 5-2

hat You'll Learn
lve problems by
aking an organized
t.

Make an Organized List

I wonder how many different ways a woodwind trio can be made if either a bass clarinet or a bassoon fills the first position and either a clarinet, oboe, or flute fills the other two positions?

Let's **make an organized list** of all of the possibilities.

Explore We know that either a bass clarinet or a bassoon must fill the first position. The other two positions are to be filled with two out of the three remaining instruments. We have to find all possible ways the trio can be made.

Plan Let's make an organized list.

Solve

	Trio 1	Trio 2	Trio 3	Trio 4	Trio 5	Trio 6
Position 1	bass clarinet	bass clarinet	bass clarinet	bassoon	bassoon	bassoon
Position 2	clarinet	oboe	flute	clarinet	oboe	flute
Position 3	oboe	flute	clarinet	oboe	flute	clarinet

There are six possibilities.

Examine We can draw a tree diagram to check our result.

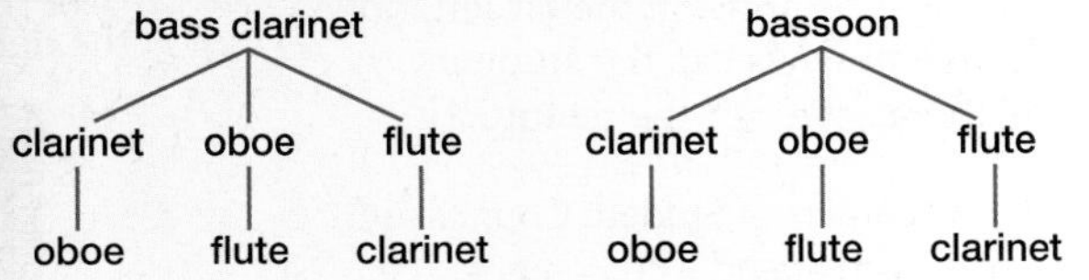

. Sample answer:
aking an organized
st helps you see and
ount all of the possible
ays a woodwind trio
an be made.

nalyze the Strategy

Explain why making an organized list was a useful strategy in solving this problem.

Explain why a tree diagram is another method for making an organized list. 2–3. See margin.

Write a problem that could be solved by making an organized list. Explain.

DAILY INTERVENTION — **Differentiated Instruction**

Interpersonal Have groups of students solve a problem using the organized list strategy by making their lists on posterboard. Then have each group present its list to the rest of the class. Compare the lists. Did any group forget a step? Did that affect the answer?

5-2a Lesson Notes

1 Focus

Bellringer

Problem Solving Present students with a menu from a local pizza shop. Ask them to predict how many different 2-item pizzas they could order based on type of crusts and toppings available.

2 Teach

In-Class Example

INTERACTIVE CHALKBOARD PowerPoint® Presentations

Solve. Use the make an organized list strategy.

PASSWORD In order to log on to the computer at school, Miranda must use a password. The password is 2 characters. The first character is the letter A or B followed by a single numeric digit. How many passwords does Miranda have to choose from? 20

Answers

2. Sample answer: A tree diagram is another method of showing organization that accounts for all combinations. Reading down a path is like reading a list.
3. Sample answer: How many different ways can a two-topping pizza be made with the following five toppings: pepperoni, mushroom, green pepper, sausage, and onion? Using an organized list, there will be 10 different possibilities.

3 Practice/Apply

About the Exercises...

Problem-Solving Strategies
Exercises 6–12 are structured so that students have the opportunity to practice many different problem-solving strategies. You may wish to remind them of some of the strategies they have studied.

- use the four-step plan (p. 6)
- guess and check (p. 22)
- use a graph (p. 58)
- look for a pattern (p. 132)
- work backward (p. 164)
- make an organized list ✓

4 Assess

Open-Ended Assessment

Writing Have students go back and solve the pizza problem presented in the bellringer. They should determine the number of single-item pizzas and write an explanation of their work using the organized list problem solving process.

Answer

5. 15; ABC, ABD, ABE, ABF, ADG, ADH, ADE, ADB, AEB, AEC, AED, AEF, AEG, AEH, AEI

Apply the Strategy

Solve. Use the make an organized list strategy.

4. **FOOD** Daniel is making a peanut butter and jelly sandwich. His choices are creamy or crunchy peanut butter, white or wheat bread, and grape, apple, or strawberry jelly. How many different types of sandwiches can Daniel make? **12**

5. **GAMES** On the game board, you plan to move two spaces away from square A. You can either move horizontally, vertically, or diagonally. How many different moves can you make from square A? List them. **See margin.**

A	B	C
D	E	F
G	H	I

Mixed Problem Solving

Solve. Use any strategy.

6. **SAVINGS** The graph below shows deposits and withdrawals. During which week was the difference between deposits and withdrawals the greatest? **Week 3 shows a $450 difference.**

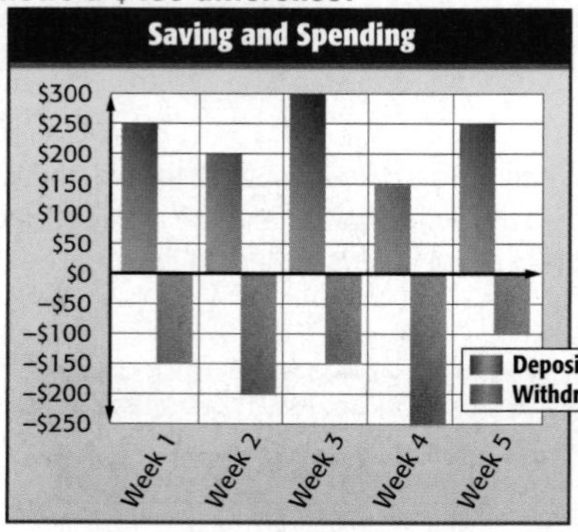

7. **SCIENCE FAIR** Mrs. Cassidy is taking a picture with the first, second, and third place winners of the science fair. If Mrs. Cassidy always stands on the far left, how many different ways can the students arrange themselves for the picture? **6**

8. **SCHOOL** Members of Student Council are raising money to attend a conference. The total costs of the conference are shown in the table. If 24 students are attending the conference, how much does each student have to raise? **$77**

Transportation	$1,200
Registration	$288
Luncheon	$360

9. **NUMBER THEORY** Find two consecutive odd numbers that when added equal 56, and when multiplied equal 783. **27, 29**

10. **MULTI STEP** Terrez took a bag of cookies to play rehearsal. Half were given to the musicians and five to the director of the play. Terrez was left with 15 cookies. How many cookies did he take to rehearsal? **40**

11. **EARTH SCIENCE** Giant kelp seaweed is found in the Pacific Ocean. One plant grows 3 feet the first two days. If it continues to grow at the same rate, what would be the length of the seaweed at the end of 80 days? **120 ft**

12. **STANDARDIZED TEST PRACTICE**
Mary has to make deliveries to three neighbors. She lives at house b on the map. Which is the shortest route to make the deliveries and return home? **A**

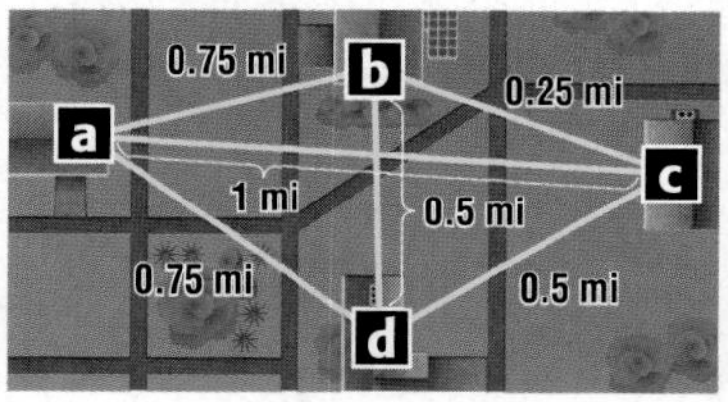

Ⓐ b-a-d-c-b Ⓑ b-a-c-d-b
Ⓒ b-c-a-d-b Ⓓ b-d-a-c-b

You will use the make an organized list strategy in the next lesson.

5-2 Greatest Common Factor

Vhat You'll Learn
ind the greatest ommon factor of two r more numbers.

'EW Vocabulary
- enn diagram
- reatest common factor (GCF)

WHEN am I ever going to use this?

INTERNET A group of friends spent time in two Internet chat rooms. The diagram shows the chat rooms Angel, Sydney, Ian, Candace, and Christine visited. The friends were able to stay in one chat room or go to the other one.

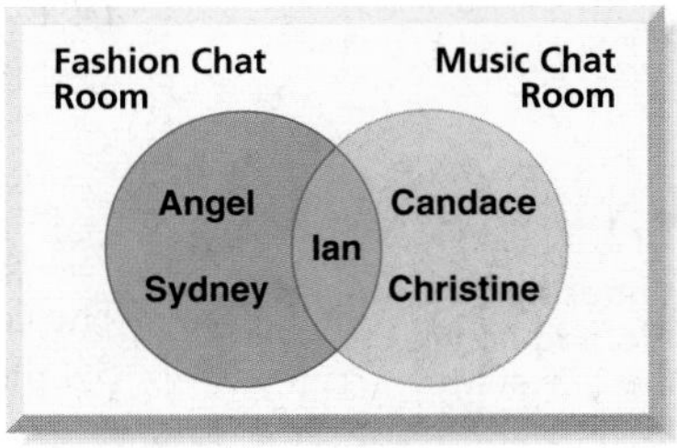

1. Who visited the Fashion Chat Room? **Angel, Sydney, and Ian**
2. Who visited the Music Chat Room? **Candace, Christine, and Ian**
3. Who visited both chat rooms? **Ian**

The diagram above is called a **Venn diagram**. It uses circles to show how elements among sets of numbers or objects are related. The region where circles overlap represents items that are common to two or more sets. It shows that Ian visited both chat rooms.

Venn diagrams can also show factors that are common to two or more numbers. The greatest of these common factors is called the **greatest common factor (GCF)**. The GCF of prime numbers is 1.

EXAMPLE Find the GCF by Listing Factors

1 Find the GCF of 20 and 24.

First, list the factors of 20 and 24.
factors of 20: 1, 2, 4, 5, 10, 20
factors of 24: 1, 2, 3, 4, 6, 8, 12, 24
} common factors: 1, 2, 4

Notice that 1, 2, and 4 are common factors of 20 and 24. So, the GCF is 4.

Check You can draw a Venn diagram to check your answer.

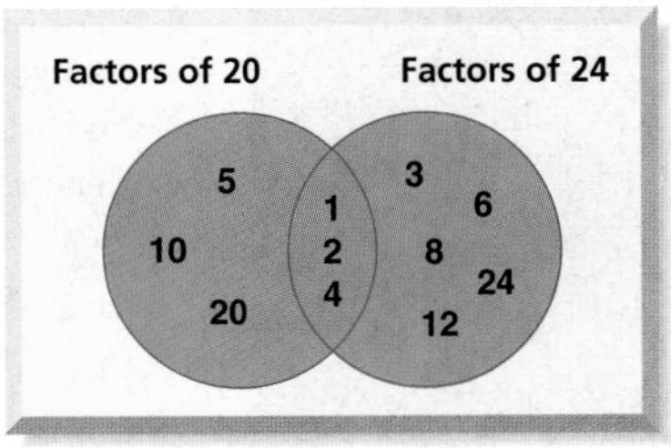

Your Turn **Find the GCF of each pair of numbers.**

a. 8 and 10 **2** b. 6 and 12 **6** c. 10, 17 **1**

msmath2.net/extra_examples

5-2 Lesson Notes

Lesson Resources

Chapter 5 Resource Masters
- Study Guide and Intervention, p. 248
- Practice: Skills, p. 249
- Practice: Word Problems, p. 250
- Reading to Learn Mathematics, p. 251
- Enrichment, p. 252

Calculator and Spreadsheet Masters, p. 29
Key Teaching Transparencies, 17
MathPASS Tutorial CD-ROM, 2
Parent and Student Study Guide Workbook, 5-2

FAST FILE

1 Focus

5-Minute Check 5-2
INTERACTIVE CHALKBOARD PowerPoint® Presentations
Use as a quiz or review of Lesson 5-1. (Also available as a transparency)

Bellringer
Hands-On Activity Place the Venn diagram in the bottom margin on the chalkboard. Have students take turns coming to the board and writing their initials in the Venn diagram based on their activities of the previous evening. Have students keep track in a Venn diagram at their seats. Discuss how many students did each activity and how many did both.

(Continued in bottom margin.)

Reading to Learn Mathematics, p. 251 ELL

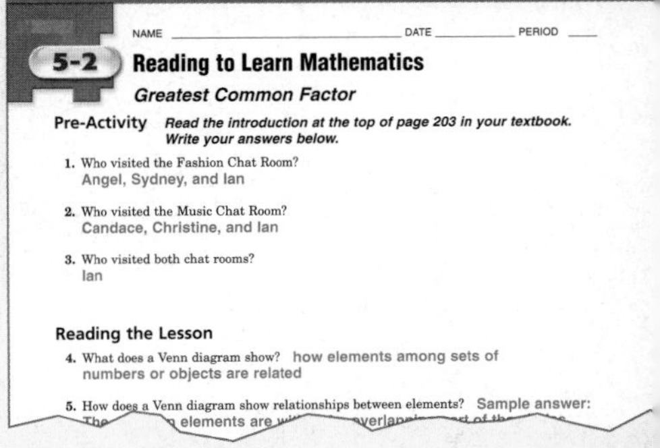
NAME ______ DATE ______ PERIOD ____

5-2 Reading to Learn Mathematics
Greatest Common Factor

Pre-Activity *Read the introduction at the top of page 203 in your textbook. Write your answers below.*

1. Who visited the Fashion Chat Room?
Angel, Sydney, and Ian
2. Who visited the Music Chat Room?
Candace, Christine, and Ian
3. Who visited both chat rooms?
Ian

Reading the Lesson

4. What does a Venn diagram show? how elements among sets of numbers or objects are related
5. How does a Venn diagram show relationships between elements? Sample answer:

2 Teach

In-Class Examples

INTERACTIVE CHALKBOARD
PowerPoint® Presentations

1. Find the GCF of 28 and 42. **14**

Find the GCF of each set of numbers.

2. 20, 32 **4**

3. 21, 42, 63 **21**

4. **ALGEBRA** Find the GCF of $12p^2$ and $30p^3$. $6p^2$

5. **ART** Searra wants to cut a 15-centimeter by 25-centimeter piece of tag board into squares for an art project. She does not want to waste any of the tag board and she wants the largest squares possible. What is the length of the side of the squares she should use? **5 cm**

Tips for New Teachers

Classroom Management You may want to use small groups or partners for assignments related to this section. Make calculators available as needed. Some students will find factors more quickly than others. The dialogue between students can help the more able students clarify their own thinking while helping the other students understand the concepts.

TEACHING TIP

You can use the TI-83 Plus graphing calculator to find the GCF. For 18 and 48, press MATH, select NUM, select 9:gcd(ENTER 18,48) ENTER.

STUDY TIP

Choosing a Method To find the GCF of two numbers, it is easier to:
- list factors for small numbers, and
- find the prime factorization or use division for large numbers.

EXAMPLES Find the GCF Using Prime Factors

Find the GCF of each set of numbers.

2 **18, 48**

Method 1 Write the prime factorization.

$18 = 2 \times 3 \times 3$
$48 = 2 \times 2 \times 2 \times 2 \times 3$

Method 2 Divide by prime numbers.

Divide both 18 and 48 by 2. Then divide the quotients by 3.

```
   3    8
3)9    24
2)18   48  ← Start here.
```

The common prime factors are 2 and 3.
So, the GCF of 18 and 48 is 2×3, or 6.

3 **12, 24, 60**

$12 = 2 \times 2 \times 3$
$24 = 2 \times 2 \times 2 \times 3$
$60 = 2 \times 2 \times 3 \times 5$

Circle the common factors.

The common prime factors are 2, 2, and 3.
So, the GCF is $2 \times 2 \times 3$, or 12.

Your Turn **Find the GCF of each set of numbers.**

d. 36, 48 **12** e. 10, 35 **5** f. 30, 45, 75 **15**

REAL-LIFE MATH

PARADES Forty pounds of dye is used to turn the Chicago River green for the annual St. Patrick's Day Parade.

Source: www.chicagostpatsparade.com

EXAMPLE Find the GCF of an Algebraic Expression

4 **ALGEBRA** **Find the GCF of $10a$ and $15a^2$.**

Factor each expression.

$10a = 2 \cdot 5 \cdot a$
$15a^2 = 3 \cdot 5 \cdot a \cdot a$

Circle the common factors.

The GCF is $5 \cdot a$, or $5a$.

EXAMPLE Use the GCF to Solve a Problem

5 **PARADES** **In a parade, 64 eighth-graders are to march in front of 88 seventh-graders. Both groups should have the same number of students in each row. Find the greatest number of students in each row.**

The greatest number of students in each row is the GCF of the number of students in each group.

$64 = 2 \times 2 \times 2 \times 2 \times 2 \times 2 = 2^6$
$88 = 2 \times 2 \times 2 \times 11 = 2^3 \times 11$

The GCF of 64 and 88 is 2^3, or 8.

So, there should be 8 students in each row.

DAILY INTERVENTION **Differentiated Instruction**

Visual/Spatial Have students solve the examples using a hundreds chart and colored chips to cover factors of the numbers. Have students mark all factors of the first number with one color and all factors of the second number in another color. What factors do they have in common? What is the GCF?

Skill and Concept Check

Writing Math
Exercise 3

1. **OPEN ENDED** Find two numbers whose GCF is 16. Sample answer: 32 and 48
2. **Draw** a Venn diagram showing the factors of 30 and 42. Name the GCF. See margin.
3. **FIND THE ERROR** Charles and Tiffany both found the GCF of 12, 18, and 27. Who is correct? Explain. Tiffany is correct; Charles included extra factors of 2 and 3.

Charles	Tiffany
$12 = 2 \cdot 2 \cdot 3$	$12 = 2 \cdot 2 \cdot 3$
$18 = 2 \cdot 3 \cdot 3$	$18 = 2 \cdot 3 \cdot 3$
$27 = 3 \cdot 3 \cdot 3$	$27 = 3 \cdot 3 \cdot 3$
GCF: $2 \cdot 3 \cdot 3 = 18$	GCF: 3

GUIDED PRACTICE

Find the GCF of each set of numbers.

4. 18, 30 6
5. 45, 60 15
6. 6, 8, 12 2
7. 18, 42, 60 6
8. Find the GCF of $14xy$ and $7x^2$. $7x$

★ indicates increased difficulty

Practice and Applications

HOMEWORK HELP	
For Exercises	See Examples
9–16, 21–22	1, 2
17–20	3
26–29	4
30–33	5

Extra Practice
See pages 574, 600.

Find the GCF of each set of numbers.

9. 12, 78 6
10. 40, 50 10
11. 20, 45 5
12. 32, 48 16
13. 24, 48 24
14. 45, 75 15
15. 56, 96 8
16. 40, 125 5
17. 18, 24, 30 6
18. 36, 60, 84 12
19. 35, 49, 84 7
20. 36, 50, 130 2
21. What is the GCF of $2^4 \times 5$ and $5^2 \cdot 7$? 5
22. Find the GCF of $2^3 \times 3^2 \times 7$ and $2^2 \times 3 \times 11$. $2^2 \times 3$ or 12

Determine whether each statement is *sometimes*, *always*, or *never* true.

23. The GCF of two numbers is greater than both numbers. never
24. If two numbers have no common prime factors, the GCF is 1. always
25. The GCF of two numbers is one of the numbers. sometimes

Find the GCF of each set of algebraic expressions.

26. $24a$, $6a$ $6a$
27. $27b^2$, $36b$ $9b$
28. $16r^3$, $28r$ $4r$
★29. $20x^2$, $50xy^2$ $10x$

INDUSTRIAL TECHNOLOGY For Exercises 30 and 31, use the following information.
Kibbe is building scale models of high-rise buildings for his class project. He designs the models using one-inch cubes. He is planning to build three buildings, the first with 108 blue cubes, the second with 270 red cubes, and the third with 225 yellow cubes. All of the buildings must be the same height, but not necessarily the same length and width.

30. What is the maximum height of each high-rise Kibbe can build? 9 in.
31. What are the dimensions of all three buildings? See margin.

msmath2.net/self_check_quiz

3 Practice/Apply

FOLDABLES™ Follow-Up Remind students to write a question about greatest common factors on the front of the second tab of their Foldables before they begin the lesson. As students progress through the lesson, encourage them to take notes and record information that helps them answer their question under the tab.

DAILY INTERVENTION FIND THE ERROR In Exercise 3, emphasize the different approaches used by each person. Remind students that they should only use factors common of all three numbers to find the GCF.

About the Exercises...

Guided Practice Key

Exercises	Examples
4–7	1–3
8	4

Odd/Even Assignments
Exercises 9–24 and 26–29 are structured so that students practice the same concepts whether they are assigned odd or even problems.

Alert! Exercise 34 uses the Internet or another source for research.

Answers

2.
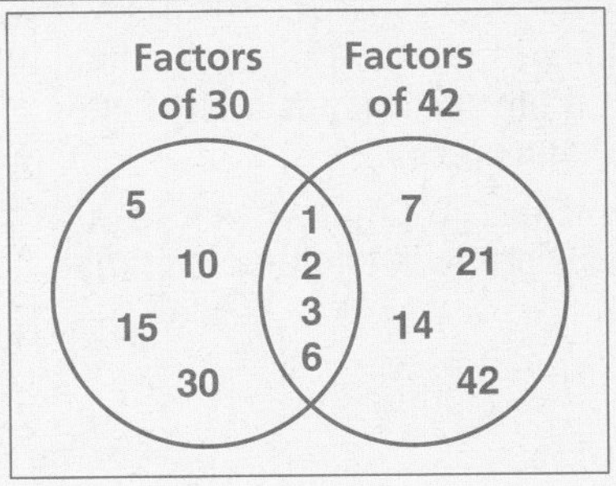

The GCF is 6.

31. Sample answer: The first building is 9 in. high, 3 in. long, and 4 in. wide. The second building is 9 in. high, 6 in. long, and 5 in. wide. The third building is 9 in. high, 5 in. long, and 5 in. wide.

FOOD **For Exercises 32 and 33, use the following information.** The track-and-field coaches threw a party at the end of the season and bought a 32-inch, a 20-inch, and a 12-inch submarine sandwich. Suppose the sandwiches are cut into equal-sized pieces.

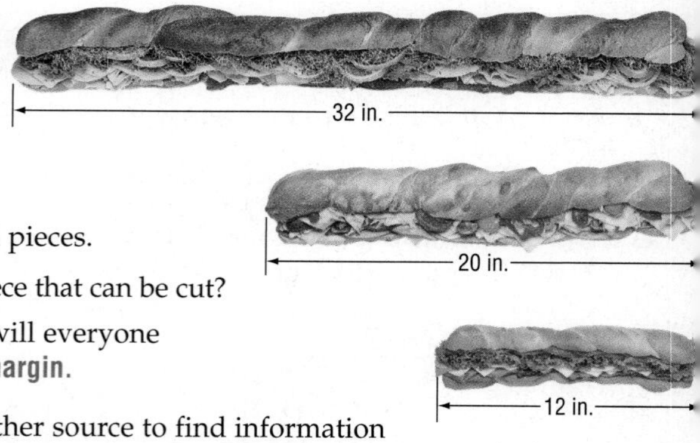

★32. What is the length of the longest piece that can be cut? **4 in.**

33. If there are 18 people on the team, will everyone receive a sandwich? Explain. **See margin.**

34. **RESEARCH** Use the Internet or another source to find information about the Sieve of Eratosthenes. Explain how you can find the GCF of 18 and 54 using the Sieve. **See students' work.**

CRITICAL THINKING **For Exercises 35–37, use the following information.** Numbers that have a GCF of 1 are *relatively prime*. Use this definition to determine whether each statement is true or false. Explain your answer.

35. Any two prime numbers are relatively prime. **True; their only common factor is 1.**

36. If two numbers are relatively prime, one of them must be prime. **See margin.**

37. Two consecutive numbers are always relatively prime. **True; their only common factor is 1.**

Standardized Test Practice and Mixed Review

38. **MULTIPLE CHOICE** Felisa is making shelves to store her books. How many shelves measuring 8 inches by 12 inches can be cut from a 36-inch-by-56-inch piece of plywood so that there is no waste? **D**

Ⓐ 17 Ⓑ 18 Ⓒ 20 Ⓓ 21

39. **MULTIPLE CHOICE** Which set of numbers does *not* have a GCF of 6? **G**

Ⓕ 42, 150 Ⓖ 54, 18 Ⓗ 24, 6 Ⓘ 18, 30

40. **SHORT RESPONSE** Find the GCF of $20st^2$ and $50s^2$. **$10s$**

Determine whether each number is *prime* or *composite*. (Lesson 5-1)

41. 21 **composite** 42. 31 **prime** 43. 65 **composite** 44. 129 **composite**

45. **ALGEBRA** Find the slope of the line that passes through the pair of points at the right. (Lesson 4-7) $\frac{1}{2}$

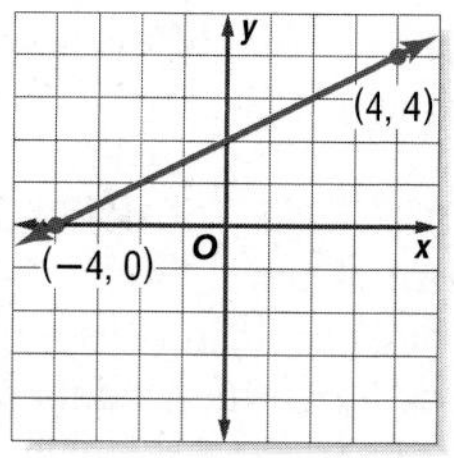

46. **GEOMETRY** Write the ordered pair that is 16 units up from and 11 units to the left of the origin. (Lesson 3-3) **(−11, 16)**

47. **MEASUREMENT** Twenty milliliters is equal to how many liters? (Lesson 1-8) **0.02 L**

GETTING READY FOR THE NEXT LESSON

BASIC SKILL **Divide.**

48. 27 ÷ 3 **9** 49. 48 ÷ 2 **24** 50. 160 ÷ 40 **4** 51. 96 ÷ 6 **16**

4 Assess

Open-Ended Assessment

Writing Have each student write three questions about determining greatest common factor. They should also include fully explained solutions to all their questions.

Answers

33. No; Sample answer: There are a total of 16 pieces if the sandwiches are cut into 4-inch long pieces.

36. False; counterexample: 4 and 9 are relatively prime but neither is prime.

Study Guide and Intervention, p. 248

NAME ______________ DATE ________ PERIOD ____

5-2 Study Guide and Intervention

Greatest Common Factor

The **greatest common factor (GCF)** of two or more numbers is the largest number that is a factor of

EXAMPLE 1 **Find the GCF of 72 and 108 by listing factors.**

factors of 72: 1, 2, 3, 4, 6, 8, 9, 12, 18, 24, 36, 72

factors of 108: 1, 2, 3, 4, 6, 9, 12, 18, 27, 36, 54, 108

common factors: 1, 2, 3, 4, 6, 9, 12, 18, 36

The GCF of 72 and 108 is 36.

EXAMPLE 2 **Find the GCF of 42 and 60 using prime factors.**

Method 1 Write the prime factorization.

$60 = 2 \times 2 \times 3 \times 5$
$42 = 2 \times 3 \times 7$

Method 2 Divide by prime numbers. Divide both 42 and 60 by 2. Then divide the quotients by 3.

```
   7   10
3)21   30
2)42   60   ← Start here.
```

The common prime factors are 2 and 3. The GCF of 42 and 60 is 2 × 3, or 6.

EXERCISES

Find the GCF of each set of numbers.

1. 18, 30 6 2. 60, 45 15 3. 24, 72 24
4. 32, 48 16 5. 100, 30 10 6. 54, 36 18
7. 3, 97, 5 1 8. 4, 20, 24 4 9. 36, 9, 45 9

Practice: Skills, p. 249

NAME ______________ DATE ________ PERIOD ____

5-2 Practice: Skills

Greatest Common Factor

Find the GCF of each set of numbers.

1. 14, 20 2 2. 16, 42 2
3. 8, 18 2 4. 24, 36 12
5. 72, 22 2 6. 77, 15 1
7. 32, 80 16 8. 90, 120 30
9. 45, 30 15 10. 12, 62 2
11. 15, 27 3 12. 21, 28 7
13. 12, 20, 26 2 14. 15, 20, 25 5
15. 60, 72, 36 12 16. 32, 48, 64 16
17. 36, 48, 30 6 18. 28, 56, 42 14
19. 80, 110, 90 10 20. 9, 25, 49 1

Find the GCF of each set of algebraic expressions.

21. $21ab$, $14b$ $7b$ 22. $20a^2$, $36a$ $4a$
23. $15ab$, $5b^2$ $5b$ 24. $35a^2$, $85ab$ $5a$

25. Find the GCF of $2^3 \times 3^2 \times 5$ and $2^2 \times 3 \times 5^2$. $2^2 \times 3 \times 5$ or 60

Practice: Word Problems, p. 250

NAME ______________ DATE ________ PERIOD ____

5-2 Practice: Word Problems

Greatest Common Factor

1. **TABLE TENNIS** Rebecca has 20 table tennis balls and 16 table tennis paddles. She wants to sell packages of balls and paddles bundled together. What is the greatest number of packages she can sell with no leftover balls or paddles? 4

2. **TUMBLING** Mr. Nicolet wants to organize equal-sized groups of boys and girls for tumbling exercises. If there are 12 boys and 18 girls and each group is all boys or all girls, what is the largest size group he can organize? 6

3. **BAKE SALE** Volunteers at a bake sale want to sell slices of banana nut bread and raisin bread packaged together. They have 63 slices of banana nut bread and 45 slices of raisin bread, and they plan to use all the bread. What is the greatest number of packages they can put together? How many slices of each type of bread are in a package? 9; 7 slices of banana nut bread and 5 slices of raisin bread

4. **DOG TREATS** Krista wants to give her dog a special treat. She has 81 dog bones and 54 pieces of beef jerky. If she wants to give her dog the same number of treats every day, what is the greatest number of days she can feed the dog these treats? How many of each type should she give the dog? 27; 3 dog bones and 2 pieces of beef jerky

5. **FRUIT TREES** Mr. Farber has 84 pear trees and 180 apple trees. He wants to plant the trees in rows of equal width. Find the most trees that can be planted in a row if each row has only one type of tree. 12

6. **BOARDS** A scouting troop has three boards of lengths 14 feet, 28 feet, and 21 feet. If the boards must be cut to produce equal-sized pieces, what is the longest piece that can be cut with no waste? 7 ft

5-3 Simplifying Fractions

What You'll Learn
Write fractions in simplest form.

NEW Vocabulary
equivalent fractions
simplest form

HANDS-ON Mini Lab

Materials
- grid paper
- colored pencils

Work with a partner.

On grid paper, draw the two figures at the right. Both figures should be the same size. Shade 8 out of the 20 squares in one figure. Shade 2 out of the 5 rectangles in the other.

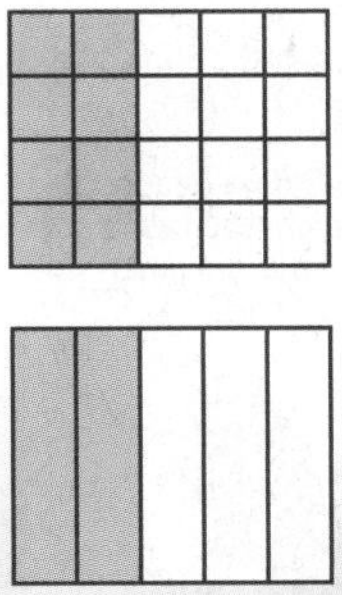

1. Write a fraction to describe each figure: $\frac{\text{number of shaded parts}}{\text{total number of parts}}$. **$\frac{8}{20}$, $\frac{2}{5}$**
2. Which figure has a greater portion of its parts shaded?
3. What can you conclude about the fractions you wrote above? **They are equal.**

2. Neither, they both have an equal portion shaded.

The fractions $\frac{8}{20}$ and $\frac{2}{5}$ are **equivalent fractions** because they have the same value. The fraction $\frac{2}{5}$ is in **simplest form** because the GCF of the numerator and denominator is 1.

EXAMPLES Write Fractions in Simplest Form

Write each fraction in simplest form.

1 $\frac{8}{20}$

First, find the GCF of the numerator and denominator.

factors of 8: 1, 2, 4, 8
factors of 20: 1, 2, 4, 5, 10, 20

The GCF of 8 and 20 is 4.

Then, divide the numerator and the denominator by the GCF.

$\frac{8}{20} = \frac{8 \div 4}{20 \div 4} = \frac{2}{5}$ So, $\frac{8}{20}$ written in simplest form is $\frac{2}{5}$.

Check Multiply the numerator and denominator of the answer by the GCF. The result should be the original fraction.

2 $\frac{18}{30}$

$18 = 2 \cdot 3 \cdot 3$

$30 = 2 \cdot 3 \cdot 5$

GCF: $2 \cdot 3 = 6$

$\frac{18}{30} = \frac{18 \div 6}{30 \div 6} = \frac{3}{5}$

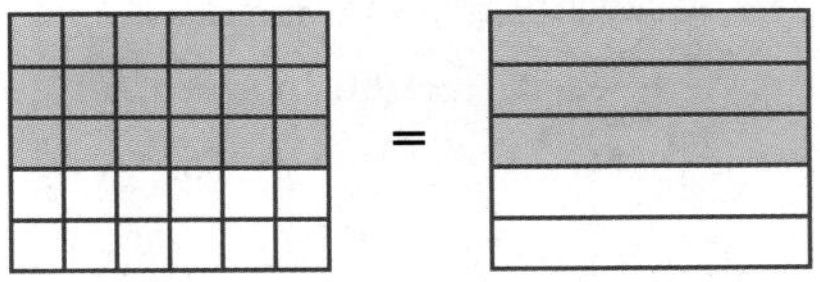

$\frac{18}{30} = \frac{3}{5}$

So, $\frac{18}{30}$ written in simplest form is $\frac{3}{5}$.

STUDY TIP

Simplest Form A fraction is in simplest form when the GCF of the numerator and denominator is 1.

msmath2.net/extra_examples

5-3 Lesson Notes

Lesson Resources

Chapter 5 Resource Masters FAST FILE
- Study Guide and Intervention, p. 253
- Practice: Skills, p. 254
- Practice: Word Problems, p. 255
- Reading to Learn Mathematics, p. 256
- Enrichment, p. 257

MathPASS Tutorial CD-ROM, 12
Parent and Student Study Guide Workbook, 5-3

1 Focus

5-Minute Check 5-3 INTERACTIVE CHALKBOARD PowerPoint® Presentations

Use as a quiz or review of Lesson 5-2. (Also available as a transparency)

Bellringer
Communication Give groups of students 12 blocks: 6 blue and 6 red. Have them represent $\frac{1}{2}$ in several ways using the blocks.

Reading to Learn Mathematics, p. 256 ELL

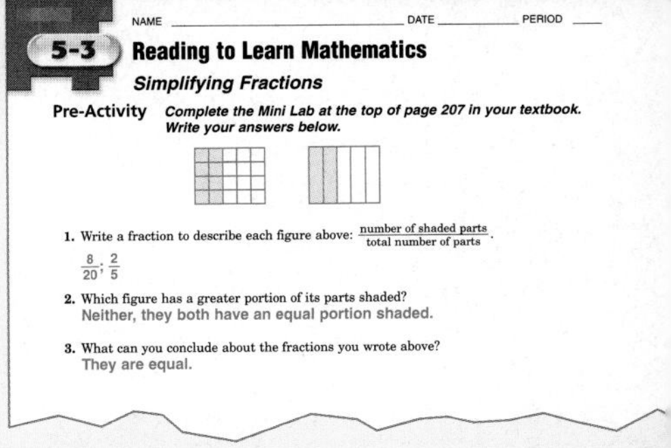
NAME ______ DATE ______ PERIOD ____

5-3 Reading to Learn Mathematics
Simplifying Fractions

Pre-Activity *Complete the Mini Lab at the top of page 207 in your textbook. Write your answers below.*

1. Write a fraction to describe each figure above: $\frac{\text{number of shaded parts}}{\text{total number of parts}}$.
$\frac{8}{20}$, $\frac{2}{5}$
2. Which figure has a greater portion of its parts shaded?
Neither, they both have an equal portion shaded.
3. What can you conclude about the fractions you wrote above?
They are equal.

2 Teach

In-Class Examples

Write each fraction in simplest form.

INTERACTIVE CHALKBOARD
PowerPoint® Presentations

1. $\frac{12}{45}$ **$\frac{4}{15}$**
2. $\frac{40}{64}$ **$\frac{5}{8}$**
3. **MUSIC** Two notes form a *perfect fifth* if the simplified fraction of the frequencies of the notes equals $\frac{3}{4}$. If note D = 294 Hertz and note G = 392 Hertz, do they form a *perfect fifth?* **yes**

3 Practice/Apply

DAILY INTERVENTION **FIND THE ERROR** In Exercise 3, have students correct Luther's work using one more step. **Divide 8 and 18 by 2.** Discuss why this extra step led to the correct answer.

About the Exercises...

Guided Practice Key

Exercises	Examples
4–6	1, 2
7	3

Odd/Even Assignments

Exercises 8–23 are structured so that students practice the same concepts whether they are assigned odd or even problems.

Answers

1. **Sample answer: A fraction is in simplest form if the GCF of the numerator and denominator is 1.**

31. **Sample answer:**

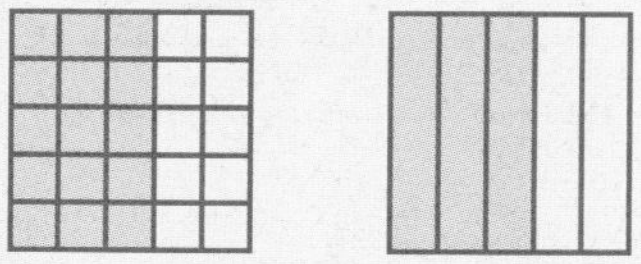

REAL-LIFE CAREERS

How Does a Music Composer Use Math? Music composers write notes that can vary in length. There are whole notes, half notes, quarter notes, eighth notes, and sixteenth notes.

Online Reasearch For information about a career as a music composer, visit: msmath2.net/careers

EXAMPLE Use Fractions to Solve a Problem

3. **MUSIC If the fraction of the frequencies of two notes can be simplified, the two notes are harmonious. Use the graphic below to find the simplified fraction of the frequency of notes C and E.**

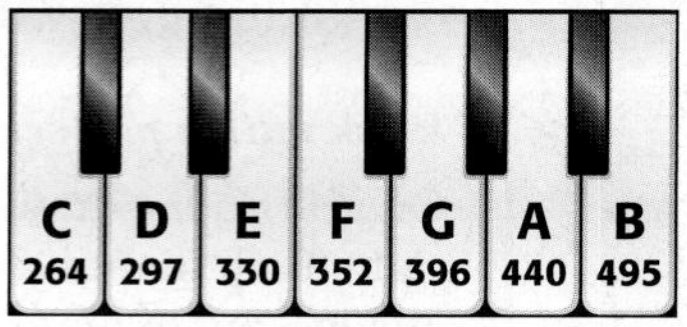

Frequency Chart (Hertz)

$$\frac{\text{frequency of note C}}{\text{frequency of note E}} = \frac{264}{330}$$

$$= \frac{\overset{1}{\cancel{2}} \times 2 \times 2 \times \overset{1}{\cancel{3}} \times \overset{1}{\cancel{11}}}{\underset{1}{\cancel{2}} \times \underset{1}{\cancel{3}} \times 5 \times \underset{1}{\cancel{11}}}$$

$$= \frac{4}{5}$$

The slashes mean that part of the numerator and part of the denominator are both divided by the same number. For example, $3 \div 3 = 1$.

The fraction of the frequency of notes C and E is $\frac{4}{5}$.

Your Turn **Find the simplified fraction of the frequency of each pair of notes.**

a. D and G **$\frac{3}{4}$**   b. E and F **$\frac{15}{16}$**

Skill and Concept Check

Writing Math Exercises 1 & 3

1. **Explain** how to determine whether a fraction is in simplest form. **See margin.**
2. **OPEN ENDED** Write a fraction that is *not* in simplest form. Then, simplify it. **Sample answer: $\frac{6}{20} = \frac{3}{10}$**
3. **FIND THE ERROR** Seki and Luther both wrote $\frac{16}{36}$ in simplest form. Who is correct? Explain.

Seki: $\frac{16}{36} = \frac{16 \div 4}{36 \div 4} = \frac{4}{9}$

Luther: $\frac{16}{36} = \frac{16 \div 2}{36 \div 2} = \frac{8}{18}$

Seki; Luther did not divide the numerator and denominator by the GCF.

GUIDED PRACTICE

Write each fraction in simplest form.

4. $\frac{3}{9}$ **$\frac{1}{3}$**   5. $\frac{18}{30}$ **$\frac{3}{5}$**   6. $\frac{81}{90}$ **$\frac{9}{10}$**

7. **AGRICULTURE** George Washington Carver produced about 100 products from sweet potatoes and about 325 from peanuts. Write the number of sweet potato products as a fraction of the number of peanut products in simplest form. **$\frac{4}{13}$**

208 Chapter 5 Fractions, Decimals, and Percents

DAILY INTERVENTION **Unlocking Misconceptions**

Simplest Form Make sure students understand that to obtain simplest form, they must divide both the numerator and denominator by the GCF. Using any other common factor will not give simplest form. Also, make sure they understand that fractions can be simplified in steps, one factor at a time, to simplest form.

★ indicates increased difficulty

Practice and Applications

For Exercises	See Examples
8–23	1, 2
24–28	3

HOMEWORK HELP — Extra Practice See pages 575, 600.

Write each fraction in simplest form.

8. $\frac{9}{12}$ $\frac{3}{4}$ 9. $\frac{25}{35}$ $\frac{5}{7}$ 10. $\frac{16}{32}$ $\frac{1}{2}$ 11. $\frac{14}{20}$ $\frac{7}{10}$

12. $\frac{10}{20}$ $\frac{1}{2}$ 13. $\frac{12}{21}$ $\frac{4}{7}$ 14. $\frac{15}{25}$ $\frac{3}{5}$ 15. $\frac{24}{28}$ $\frac{6}{7}$

16. $\frac{48}{64}$ $\frac{3}{4}$ 17. $\frac{32}{32}$ 1 18. $\frac{50}{300}$ $\frac{1}{6}$ 19. $\frac{80}{96}$ $\frac{5}{6}$

Write two fractions that are equivalent to each fraction. 20–23. Sample answers are given.

20. $\frac{1}{2}$ $\frac{2}{4}, \frac{5}{10}$ 21. $\frac{3}{5}$ $\frac{6}{10}, \frac{15}{25}$ 22. $\frac{6}{7}$ $\frac{12}{14}, \frac{18}{21}$ 23. $\frac{5}{8}$ $\frac{10}{16}, \frac{20}{32}$

24. **PARKING** New York holds the number one spot when it comes to the cost for parking your car. Use the graphic to express the cost of parking a car in Chicago as a fraction of the cost of parking a car in Boston in simplest form. $\frac{4}{5}$

25. **MUSIC** Find the simplified fraction of the frequency of notes C and A in Example 3 on page 208. $\frac{3}{5}$

Write each fraction in simplest form.

★ 26. Fifteen minutes is what part of one hour? $\frac{1}{4}$

★ 27. Nine inches is what part of one foot? $\frac{3}{4}$

★ 28. Four days is what part of the month of April? $\frac{2}{15}$

29. **CRITICAL THINKING** Both the numerator and denominator of a fraction are even. Is the fraction in simplest form? Explain.
No, because both the numerator and denominator can be divided by 2.

Standardized Test Practice and Mixed Review

30. **MULTIPLE CHOICE** The drive to the football game took 56 minutes. Twenty-four minutes of the time was spent stopped in traffic. What fraction of the drive was spent stopped in traffic? A

Ⓐ $\frac{3}{7}$ Ⓑ $\frac{6}{14}$ Ⓒ $\frac{15}{35}$ Ⓓ $\frac{1}{3}$

31. **SHORT RESPONSE** Draw a model that shows $\frac{15}{25} = \frac{3}{5}$. See margin.

32. Find the GCF of 345, 253, and 115. (Lesson 5-2) 23

33. Find the prime factorization of 630. (Lesson 5-1) $2 \times 3^2 \times 5 \times 7$

GETTING READY FOR THE NEXT LESSON

PREREQUISITE SKILL **Divide.** (Page 562)

34. $2\overline{)1.0}$ 0.5 35. $4\overline{)1.00}$ 0.25 36. $10\overline{)7.0}$ 0.7 37. $8\overline{)3.000}$ 0.375

msmath2.net/self_check_quiz

4 Assess

Open-Ended Assessment

Speaking Give students a fraction with 8 in the numerator and an open denominator. Have them suggest a denominator that would give a fraction in simplest form. Then repeat the question for a fraction that is not in simplest form. Have students explain their answers.

Study Guide and Intervention, p. 253

NAME ______ DATE ______ PERIOD ____

5-3 Study Guide and Intervention
Simplifying Fractions

Fractions that have the same value are called **equivalent fractions**. A fraction is in **simplest form** when the GCF of the numerator and denominator is 1.

EXAMPLE 1 **Write $\frac{36}{54}$ in simplest form.**

First, find the GCF of the numerator and denominator.
factors of 36: 1, 2, 3, 4, 6, 9, 12, 18, 36
factors of 54: 1, 2, 3, 6, 9, 18, 27, 54
The GCF of 36 and 54 is 18.

Then, divide the numerator and the denominator by the GCF.
$\frac{36}{54} = \frac{36 \div 18}{54 \div 18} = \frac{2}{3}$ So, $\frac{36}{54}$ written in simplest form is $\frac{2}{3}$.

EXAMPLE 2 **Write $\frac{8}{12}$ in simplest form.**

$8 = (2) \cdot (2) \cdot 2$
$12 = (2) \cdot (2) \cdot 3$
GCF: $2 \cdot 2 = 4$
$\frac{8}{12} = \frac{8 \div 4}{12 \div 4} = \frac{2}{3}$
So, $\frac{8}{12}$ written in simplest form is $\frac{2}{3}$.

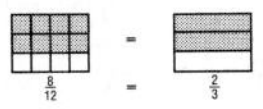

$\frac{8}{12} = \frac{2}{3}$

EXERCISES

Write each fraction in simplest form.

1. $\frac{42}{72}$ $\frac{7}{12}$ 2. $\frac{40}{64}$ $\frac{5}{8}$ 3. $\frac{21}{35}$ $\frac{3}{5}$

4. $\frac{25}{100}$ $\frac{1}{4}$ 5. $\frac{99}{132}$ $\frac{3}{4}$ 6. $\frac{17}{85}$ $\frac{1}{5}$

Practice: Skills, p. 254

NAME ______ DATE ______ PERIOD ____

5-3 Practice: Skills
Simplifying Fractions

Write each fraction in simplest form.

1. $\frac{49}{70}$ $\frac{7}{10}$ 2. $\frac{5}{30}$ $\frac{1}{6}$ 3. $\frac{6}{14}$ $\frac{3}{7}$

4. $\frac{14}{28}$ $\frac{1}{2}$ 5. $\frac{72}{72}$ 1 6. $\frac{18}{21}$ $\frac{6}{7}$

7. $\frac{45}{75}$ $\frac{3}{5}$ 8. $\frac{50}{200}$ $\frac{1}{4}$ 9. $\frac{32}{50}$ $\frac{16}{25}$

10. $\frac{56}{64}$ $\frac{7}{8}$ 11. $\frac{14}{35}$ $\frac{2}{5}$ 12. $\frac{39}{45}$ $\frac{13}{15}$

13. $\frac{48}{66}$ $\frac{8}{11}$ 14. $\frac{42}{45}$ $\frac{14}{15}$ 15. $\frac{78}{130}$ $\frac{3}{5}$

Write two fractions that are equivalent to each fraction.
16–21 Sample answers given.

16. $\frac{3}{4}$ $\frac{6}{8}, \frac{9}{12}$ 17. $\frac{7}{9}$ $\frac{14}{18}, \frac{21}{27}$ 18. $\frac{7}{11}$ $\frac{14}{22}, \frac{21}{33}$

19. $\frac{14}{17}$ $\frac{28}{34}, \frac{42}{51}$ 20. $\frac{21}{23}$ $\frac{42}{46}, \frac{63}{69}$ 21. $\frac{11}{17}$ $\frac{22}{34}, \frac{33}{51}$

Practice: Word Problems, p. 255

NAME ______ DATE ______ PERIOD ____

5-3 Practice: Word Problems
Simplifying Fractions

1. **EXAM** Mr. Bonilla gave an exam and 20 out of 25 students passed the exam. What fraction of the students passed the exam? Write the answer in simplest form. $\frac{4}{5}$

2. **GASOLINE** Aisha filled her car's 24-gallon gas tank. She took a short trip and used 8 gallons of gas. What fraction of the full gas tank was used on the trip? Write the answer in simplest form. $\frac{1}{3}$

3. **BICYCLES** A local community college has 860 students. Of these 860 students, 220 ride bicycles. Write the number of bike riders as a fraction of the number of students at the college in simplest form. $\frac{11}{43}$

4. **PRESIDENTS** Of the first 22 presidents, 8 were from New York. Write the number of presidents from New York as a fraction of the first 22 presidents in simplest form. $\frac{4}{11}$

5. **TIME** Ten hours is what part of a day? Write the fraction in simplest form. $\frac{5}{12}$

6. **MEASUREMENT** Eighteen inches is what part of a yard? Write the fraction in simplest form. $\frac{1}{2}$

5-4 Lesson Notes

Lesson Resources

Chapter 5 Resource Masters
- Study Guide and Intervention, p. 258
- Practice: Skills, p. 259
- Practice: Word Problems, p. 260
- Reading to Learn Mathematics, p. 261
- Enrichment, p. 262

Key Teaching Transparencies, 18
Parent and Student Study Guide Workbook, 5-4
Real-Life Career Masters, p.19

FAST FILE

1 Focus

5-Minute Check 5-4
Use as a quiz or review of Lesson 5-3. (Also available as a transparency)

INTERACTIVE CHALKBOARD
PowerPoint® Presentations

Bellringer
Reading Mathematics Put the following phrases on the chalkboard.

a. four tenths
b. sixteen hundredths
c. one hundred twenty-five ten-thousandths

Ask students to write each phrase as a numeral. Discuss their ideas. Did anyone write decimal numbers? Did anyone write fractions?

Reading to Learn Mathematics, p. 261 **ELL**

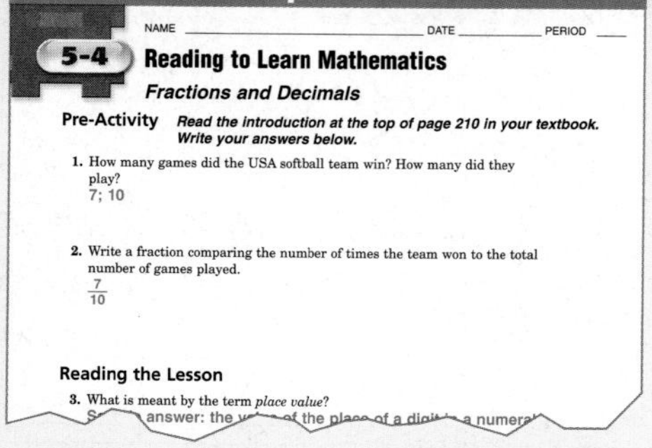
NAME ________________ DATE ________ PERIOD ____
5-4 Reading to Learn Mathematics
Fractions and Decimals
Pre-Activity *Read the introduction at the top of page 210 in your textbook. Write your answers below.*
1. How many games did the USA softball team win? How many did they play?
7; 10
2. Write a fraction comparing the number of times the team won to the total number of games played.
$\frac{7}{10}$
Reading the Lesson
3. What is meant by the term *place value*?
Sample answer: the value of the place of a digit in a numeral

5-4 Fractions and Decimals

What You'll Learn
Write fractions as terminating or repeating decimals and write decimals as fractions.

***NEW* Vocabulary**
terminating decimals
repeating decimals
bar notation

***LINK* To Reading**
Everyday Meaning of terminate: coming to an end, as in terminate a game

WHEN am I ever going to use this?

SOFTBALL The United States Women's softball team won the gold medal in the 2000 Summer Olympic Games. The table shows the results of the games they played in Sydney, Australia. The number of runs scored by the United States is listed first.

Opponent	Score
Japan	2–1
Australia	1–0
China	3–0
Italy	6–0
New Zealand	2–0
Australia	1–2
China	0–2
Japan	1–2
Cuba	3–0
Canada	6–0

Source: www.usasoftball.com

1. How many games did the USA softball team win? How many did they play? **7; 10**
2. Write a fraction comparing the number of times the team won to the total number of games played. $\frac{7}{10}$

The softball team won $\frac{7}{10}$ of the games played. When the denominator is a power of 10, you can use place value to write the fraction as a decimal.

Words	Model	Fraction	Decimal
seven tenths	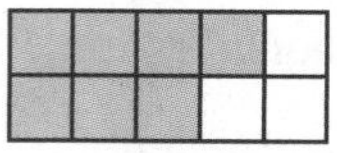	$\frac{7}{10}$	0.7

If the denominator is *not* a power of 10, you can write the fraction as a decimal using division.

EXAMPLES Write Fractions as Decimals

Write each fraction or mixed number as a decimal.

1 $\frac{1}{4}$

The fraction $\frac{1}{4}$ indicates $1 \div 4$.

Method 1 Use paper and pencil.

$$\begin{array}{r} 0.25 \\ 4\overline{)1.00} \\ \underline{8} \\ 20 \\ \underline{20} \\ 0 \end{array}$$

Division ends when the remainder is 0.

Method 2 Use a calculator.

1 [÷] 4 [ENTER] **0.25**

So, $\frac{1}{4} = 0.25$.

2 $5\frac{3}{4}$

The mixed number $5\frac{3}{4}$ is $5 + \frac{3}{4}$.

Method 1 Use paper and pencil.

$5\frac{3}{4} = 5 + \frac{3}{4}$ Write as a sum.
$= 5 + 0.75$ Write $\frac{3}{4}$ as 0.75.
$= 5.75$ Add.

Method 2 Use a calculator.

3 [÷] 4 [+] 5 [ENTER] **5.75**

So, $5\frac{3}{4} = 5.75$.

In Example 1, the division ends, or terminates, when the remainder is zero. So, 0.25 is called a **terminating decimal**.

Repeating decimals have a pattern in the digits that repeats forever.

Consider $\frac{1}{3}$.

$$\begin{array}{r} 0.333\ldots \\ 3\overline{)1.000} \\ \underline{9} \\ 10 \\ \underline{9} \\ 10 \\ \underline{9} \\ 1 \end{array}$$

The number 3 repeats. It is represented by three dots.

The remainder after each step is 1.

You can use **bar notation** to indicate that a number repeats forever. A bar is written over the digits that repeat.

$0.33333\ldots = 0.\overline{3}$ $\quad 0.121212\ldots = 0.\overline{12}$ $\quad 11.3858585\ldots = 11.3\overline{85}$

TEACHING TIP

ou can use an example f a fraction with 7 in the enominator to show that ny number of digits can epeat.

EXAMPLES Write Fractions as Repeating Decimals

Write each fraction or mixed number as a decimal.

3 $\frac{7}{9}$

Method 1 Use paper and pencil.

$$\begin{array}{r} 0.77\ldots \\ 9\overline{)7.00} \\ \underline{6\,3} \\ 70 \\ \underline{63} \\ 7 \end{array}$$

Method 2 Use a calculator.

7 [÷] 9 [ENTER] 0.7777...

So, $\frac{7}{9} = 0.\overline{7}$.

4 $8\frac{1}{3}$

Method 1 Use paper and pencil.

$8\frac{1}{3} = 8 + \frac{1}{3}$ Write as a sum.

$= 8 + 0.\overline{3}$ Write $\frac{1}{3}$ as $0.\overline{3}$.

$= 8.\overline{3}$ Add.

Method 2 Use a calculator.

1 [÷] 3 [+] 8 [ENTER] 8.3333...

So, $8\frac{1}{3} = 8.\overline{3}$.

STUDY TIP

Mental Math It will be helpful to memorize the following fraction-decimal equivalencies.

$\frac{1}{2} = 0.5$

$\frac{1}{3} = 0.\overline{3}$ $\quad \frac{2}{3} = 0.\overline{6}$

$\frac{1}{4} = 0.25$ $\quad \frac{3}{4} = 0.75$

$\frac{1}{5} = 0.2$ $\quad \frac{1}{10} = 0.1$

$\frac{1}{8} = 0.125$

Your Turn **Write each fraction or mixed number as a decimal. Use bar notation if the decimal is a repeating decimal.**

a. $\frac{5}{8}$ **0.625** b. $\frac{2}{3}$ **$0.\overline{6}$** c. $1\frac{3}{11}$ **$1.\overline{27}$** d. $6\frac{2}{5}$ **6.4** e. $4\frac{5}{6}$ **$4.8\overline{3}$**

You can use a power of 10 to write a decimal as a fraction. Use the place value of the final digit as the denominator.

EXAMPLE Write Decimals as Fractions

5 **Write 0.48 as a fraction in simplest form.**

$0.48 = \frac{48}{100}$ The 8 is in the hundredths place.

$= \frac{12}{25}$ Simplify.

Your Turn **Write each number as a fraction in simplest form.**

f. 0.56 **$\frac{14}{25}$** g. 0.3 **$\frac{3}{10}$** h. 2.8 **$2\frac{4}{5}$**

msmath2.net/extra_examples

2 Teach

In-Class Examples

INTERACTIVE CHALKBOARD PowerPoint® Presentations

Write each fraction or mixed number as a decimal.

1 $\frac{1}{8}$ **0.125**

2 $7\frac{3}{5}$ **7.6**

3 $\frac{1}{11}$ **$0.\overline{09}$**

4 $6\frac{4}{9}$ **$6.\overline{4}$**

5 Write 0.72 as a fraction in simplest form. **$\frac{18}{25}$**

Tips for New Teachers

Using Manipulatives When students are unfamiliar with a topic, encourage them to use models when solving their homework exercises. When working with decimals, they could use grid paper or play money to help them.

DAILY INTERVENTION **Flexible Grouping**

Solving Method Have students work through a number of problems in pairs. Have one student solve the example with the calculator and the other with paper and pencil. Partners should switch methods with every problem. Have students compare answers, aiding one another in finding any mistakes in calculations.

3 Practice/Apply

FOLDABLES™ Follow-Up Remind students to write a question about decimals and fractions on the front of the fourth tab of their Foldables before they begin the lesson. As students progress through the lesson, encourage them to take notes and record information that helps them answer their question under the tab.

NUMBER SENSE

In Exercise 2, remind students that both values rounded to the nearest tenth are 0.3. However, the fractions are not equal since $\frac{1}{3} = 0.\overline{3} = 0.333\ldots$ and $\frac{3}{10} = 0.3$.

About the Exercises...

Guided Practice Key

Exercises	Examples
6, 13	1
7	2
3, 8	3
4–5, 9	4
10–12	5

Odd/Even Assignments

Exercises 14–31 and 33–42 are structured so that students practice the same concepts whether they are assigned odd or even problems.

Answers

2. No; Sample answer: 1 out of 3 produces the repeating decimal $0.\overline{3}$, which is greater than the terminating decimal 0.3 (3 out of 10).
46. Sample answer: The first set of numbers can be factored into primes of 2s. The second set of numbers can be factored into primes of 2s and 3s; the factor of 3 produces the repeating decimal.

Skill and Concept Check

Writing Math Exercise 2

1. **OPEN ENDED** Write a fraction that is equivalent to a terminating decimal and one that is equivalent to a repeating decimal. **Sample answer:** $\frac{1}{2} = 0.5$; $\frac{2}{9} = 0.\overline{2}$
2. **NUMBER SENSE** Compare 1 out of 3 and 3 out of 10. Are these values equal? Explain your reasoning. **See margin.**

GUIDED PRACTICE

Write each repeating decimal using bar notation.

3. 0.6333... $0.6\overline{3}$
4. 5.313131... $5.\overline{31}$
5. 12.470470470... $12.\overline{470}$

Write each fraction or mixed number as a decimal. Use bar notation if the decimal is a repeating decimal.

6. $\frac{2}{5}$ 0.4
7. $3\frac{5}{8}$ 3.625
8. $\frac{5}{9}$ $0.\overline{5}$
9. $1\frac{5}{6}$ $1.8\overline{3}$

Write each decimal as a fraction in simplest form.

10. 0.22 $\frac{11}{50}$
11. 0.1 $\frac{1}{10}$
12. 4.6 $4\frac{3}{5}$

13. **SOCCER** Cirilo surveyed his physical education class and found that 12 out of 23 students chose soccer as their favorite sport. Write the fraction of students who chose soccer as a decimal to the nearest thousandth. **0.522**

★ indicates increased difficulty

Practice and Applications

HOMEWORK HELP

For Exercises	See Examples
20–21, 24–27, 32, 43	1, 2
14–19, 22–23, 28–31, 44	3, 4
33–42	5

Extra Practice See pages 575, 600.

Write each repeating decimal using bar notation.

14. 0.999... $0.\overline{9}$
15. 5.92111... $5.92\overline{1}$
16. 2.010101... $2.\overline{01}$
17. 13.1464646... $13.1\overline{46}$
18. 26.993993... $26.\overline{993}$
19. 30.6841841841... $30.6\overline{841}$

Write each fraction or mixed number as a decimal. Use bar notation if the decimal is a repeating decimal.

20. $\frac{3}{8}$ 0.375
21. $\frac{4}{5}$ 0.8
22. $\frac{4}{9}$ $0.\overline{4}$
23. $\frac{8}{11}$ $0.\overline{72}$
24. $\frac{5}{32}$ 0.15625
25. $\frac{3}{20}$ 0.15
26. $1\frac{7}{20}$ 1.35
27. $6\frac{17}{40}$ 6.425
28. ★ $\frac{13}{24}$ $0.541\overline{6}$
29. ★ $\frac{5}{7}$ $0.\overline{714285}$
30. ★ $8\frac{1}{33}$ $8.\overline{03}$
31. ★ $4\frac{203}{999}$ $4.\overline{203}$

32. **DOGS** There are an estimated 53 billion dogs in the U.S. The table shows the results of a survey taken recently of dog owners. Write the fraction of people who pet and hug their dogs daily as a decimal. **0.95**

Activity	Fraction of People
Play with your dog daily	$\frac{23}{25}$
Take your dog on vacation	$\frac{9}{20}$
Pet and hug your dog daily	$\frac{19}{20}$
Celebrate your dog's birthday	$\frac{43}{100}$

Source: Ralston Purina

Write each decimal as a fraction in simplest form.

33. 0.75 $\frac{3}{4}$
34. 0.34 $\frac{17}{50}$
35. 0.2 $\frac{1}{5}$
36. 0.9 $\frac{9}{10}$
37. 5.96 $5\frac{24}{25}$
38. 2.66 $2\frac{33}{50}$

BASKETBALL Use the graph to write the number of rebounds and blocks per game for each player as a mixed number in simplest form.

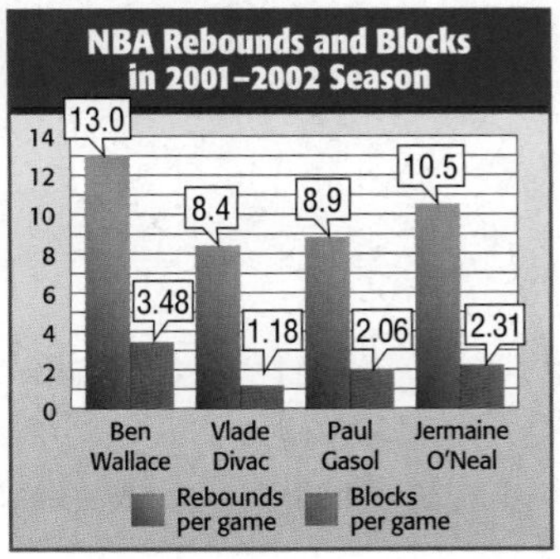

Source: www.nba.com

39. Wallace 13; $3\frac{12}{25}$
40. Divac $8\frac{2}{5}$; $1\frac{9}{50}$
41. Gasol $8\frac{9}{10}$; $2\frac{3}{50}$
42. O'Neal $10\frac{1}{2}$; $2\frac{31}{100}$

43. **MULTI STEP** Kelsey practiced playing the cello for 2 hours and 18 minutes. Write the time Kelsey spent practicing in hours as a decimal. **2.3 h**

MATH HISTORY For Exercises 44 and 45, use the following information.
The value of pi (π) is 3.1415927… . Pi is a nonrepeating, nonterminating decimal. Mathematicians have used many methods to find the value of π.

44. Archimedes believed that π was between $3\frac{1}{7}$ and $3\frac{10}{71}$. Write each fraction as a decimal rounded to the nearest thousandth. Was Archimedes correct? **3.143, 3.141; yes**

45. The Rhind Papyrus states that the Egyptians used $\frac{256}{81}$ for π. Write the fraction as a decimal rounded to the nearest thousandth. Which is closer to the actual value of π, Archimedes' value or the Egyptians' value? **3.160; Archimedes'**

46. **CRITICAL THINKING** Fractions with denominators of 2, 4, 8, 16, and 32 produce terminating decimals. Fractions with denominators of 6, 12, 18, and 24 produce repeating decimals. What do you think causes the difference? Explain. **See margin.**

Standardized Test Practice and Mixed Review

47. **MULTIPLE CHOICE** Express $\frac{9}{25}$ as a decimal. **B**

Ⓐ 0.036 Ⓑ 0.36 Ⓒ $0.\overline{36}$ Ⓓ 3.6

48. **SHORT REPONSE** In 2001, Ichiro Suzuki had 242 hits in 692 at-bats. The batting average of a baseball player is the number of hits divided by the number of at-bats. Find Suzuki's batting average to the nearest thousandth. **0.350**

Write each fraction in simplest form. (Lesson 5-3)

49. $\frac{10}{24}$ $\frac{5}{12}$
50. $\frac{39}{81}$ $\frac{13}{27}$
51. $\frac{28}{98}$ $\frac{2}{7}$
52. $\frac{51}{68}$ $\frac{3}{4}$

53. Find the GCF of 36 and 48. (Lesson 5-2) **12**

GETTING READY FOR THE NEXT LESSON

BASIC SKILL Write a fraction for the number of squares shaded to the total number of squares.

54. 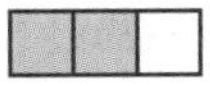 $\frac{2}{3}$

55. 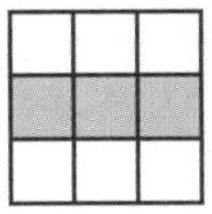 $\frac{3}{9}$

56. 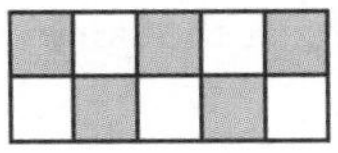 $\frac{5}{10}$

57. 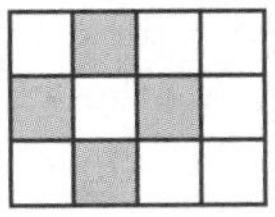 $\frac{4}{12}$

msmath2.net/self_check_quiz

4 Assess

Higher-Level Thinking

Have students use calculators to find the decimal equivalents of $\frac{1}{9}, \frac{2}{9}, \frac{3}{9}, \frac{4}{9},$ and $\frac{5}{9}$. What do they observe? **$0.\overline{1}$, $0.\overline{2}$, $0.\overline{3}$, $0.\overline{4}$, $0.\overline{5}$; They are the numerator as a repeating decimal.** Make and test a conjecture about $\frac{6}{9}, \frac{7}{9},$ and $\frac{8}{9}$. Then look for a pattern in $\frac{1}{99}, \frac{5}{99}, \frac{12}{99},$ and other fractions with a denominator of 99.

Study Guide and Intervention, p. 258

NAME ____________ DATE ________ PERIOD ____

5-4 Study Guide and Intervention
Fractions and Decimals

To write a decimal as a fraction, divide the numerator of the fraction by the denominator. Use a power of ten to change a decimal to a fraction.

EXAMPLE 1 **Write $\frac{5}{9}$ as a decimal.**

Method 1 Use pencil and paper.

0.555…
9)5.000
4 5
50
45
50
45
5

The remainder after each step is 5.

Method 2 Use a calculator.

5 ÷ 9 = 0.55555556

You can use bar notation $0.\overline{5}$ to indicate that 5 repeats forever. So, $\frac{5}{9} = 0.\overline{5}$.

EXAMPLE 2 **Write 0.32 as a fraction in simplest form.**

$0.32 = \frac{32}{100}$ The 2 is in the hundredths place.

$= \frac{8}{25}$ Simplify.

EXERCISES

Write each fraction or mixed number as a decimal. Use bar notation if the decimal is a repeating decimal.

1. $\frac{8}{10}$ 0.8
2. $\frac{3}{5}$ 0.6
3. $\frac{7}{11}$ $0.\overline{63}$
4. $4\frac{7}{8}$ 4.875
5. $\frac{13}{15}$ $0.8\overline{6}$
6. $3\frac{47}{99}$ $3.\overline{47}$

Write each decimal as a fraction in simplest form.

7. 0.14 $\frac{7}{50}$
8. 0.3 $\frac{3}{10}$
9. 0.94 $\frac{47}{50}$

Practice: Skills, p. 259

NAME ____________ DATE ________ PERIOD ____

5-4 Practice: Skills
Fractions and Decimals

Write each repeating decimal using bar notation.

1. 0.7353535… $0.7\overline{35}$
2. 0.424242… $0.\overline{42}$
3. 5.126126126… $5.\overline{126}$

Write each fraction or mixed number as a decimal. Use bar notation if the decimal is a repeating decimal.

4. $\frac{3}{5}$ 0.6
5. $\frac{19}{20}$ 0.95
6. $3\frac{4}{5}$ 3.8
7. $\frac{23}{50}$ 0.46
8. $1\frac{5}{8}$ 1.625
9. $\frac{19}{25}$ 0.76
10. $4\frac{17}{37}$ $4.\overline{459}$
11. $5\frac{3}{11}$ $5.\overline{27}$
12. $\frac{17}{24}$ $0.708\overline{3}$
13. $6\frac{7}{32}$ 6.21875
14. $7\frac{9}{22}$ $7.4\overline{09}$
15. $1\frac{17}{48}$ $1.3541\overline{6}$

Write each decimal as a fraction in simplest form.

16. 0.8 $\frac{4}{5}$
17. 0.52 $\frac{13}{25}$
18. 0.92 $\frac{23}{25}$
19. 0.48 $\frac{12}{25}$
20. 0.86 $\frac{43}{50}$
21. 0.76 $\frac{19}{25}$

Practice: Word Problems, p. 260

NAME ____________ DATE ________ PERIOD ____

5-4 Practice: Word Problems
Fractions and Decimals

1. **BOYS AND GIRLS** There were 6 girls and 18 boys in Mrs. Johnson's math class. Write the number of girls as a fraction of the number of boys. Then write the fraction as a repeating decimal.
$\frac{6}{18}$; $0.\overline{3}$

2. **CATS** In a neighborhood of 72 families, 18 families own one or more cats. Write the number of families who own one or more cats as a fraction. Then write the fraction as a decimal.
$\frac{18}{72}$; 0.25

3. **CELLULAR PHONES** In Italy, about 74 of every 100 people use cellular telephones. Write the fraction of cellular phone users in Italy. Then write the fraction as a decimal.
$\frac{74}{100}$; 0.74

4. **FRUITS** Ms. Rockwell surveyed her class and found that 12 out of the 30 students chose peaches as their favorite fruit. Write the number of students who chose peaches as a fraction in simplest form. Then write the fraction as a decimal.
$\frac{2}{5}$; 0.4

5. **TRAVEL** Tora took a short trip of 320 miles. He stopped to have lunch after he had driven 120 miles. Write the fraction of the trip he had completed by lunch in simplest form. Then write the fraction as a decimal.
$\frac{3}{8}$; 0.375

6. **VOTING** In a recent school election, 208 of the 325 freshmen voted in their class election. Write the fraction of freshmen who voted. Then write the fraction as a decimal.
$\frac{208}{325}$; 0.64

Lesson 5-4 Fractions and Decimals **213**

CHAPTER 5 Mid-Chapter Practice Test

Assessment Option

Mid-Chapter Test (Lessons 5-1 through 5-4) is available on p. 299 of the *Chapter 5 Resource Masters*.

ExamView® Pro Test Bank CD-ROM

Use the networkable ExamView® Pro Test Bank CD-ROM to:

- Create **Multiple versions** of tests.
- Create **Modified** tests for **Inclusion** students with one mouse click.
- **Edit** existing questions and **Add** your own questions.
- Build tests aligned with state standards using built-in **State Curriculum Correlations.**
- Change **English** tests to **Spanish** with one mouse click and vice versa.

FOLDABLES™ Follow-Up Before completing the Mid-Chapter Practice Test, encourage students to review the notes, definitions, concepts, and main ideas they recorded on the first four tabs of their Foldables.

Answer

1. **Sample answer: a whole number greater than 1 that has exactly two factors, 1 and itself; 3**

CHAPTER 5 Mid-Chapter Practice Test

Vocabulary and Concepts

1. **Define** *prime number* and give an example. (Lesson 5-1) **See margin.**
2. **Describe** how you know when a fraction is in simplest form. (Lesson 5-3) **The GCF of the numerator and denominator is 1.**

Skills and Applications

3. Determine whether 24 is *prime* or *composite*. (Lesson 5-1) **composite**

Factor each expression. (Lesson 5-1)

4. 30 $2 \times 3 \times 5$
5. 120 $2^3 \times 3 \times 5$
6. $14x^2y$ $2 \cdot 7 \cdot x \cdot x \cdot y$
7. $50mn$ $2 \cdot 5 \cdot 5 \cdot m \cdot n$

Find the GCF of each set of numbers or algebraic expressions. (Lesson 5-2)

8. 16, 40 **8**
9. 65, 100 **5**
10. $12x$, $20x^3$ $4x$
11. $45a^2b^2$, $81ab^3$ $9ab^2$

12. **CYCLING** The *gear ratio* of a bicycle is the comparison of the number of teeth on a chainwheel to the number on a freewheel. If the gear ratio for Alexis' 10-speed bike is $\frac{52}{16}$, write this fraction in simplest form. (Lesson 5-3) $\frac{13}{4}$

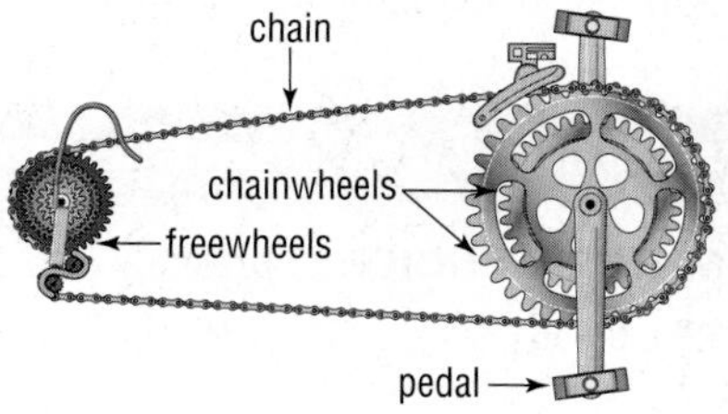

Write each fraction or mixed number as a decimal. Use bar notation if the decimal is a repeating decimal. (Lesson 5-4)

13. $\frac{7}{8}$ **0.875**
14. $\frac{2}{9}$ $0.\overline{2}$
15. $3\frac{13}{20}$ **3.65**

Write each decimal as a fraction in simplest form. (Lesson 5-4)

16. 0.6 $\frac{3}{5}$
17. 0.48 $\frac{12}{25}$
18. 7.02 $7\frac{1}{50}$

Standardized Test Practice

19. **MULTIPLE CHOICE** What is the greatest number of crayons in each row of an 8-, a 64-, and a 96-crayon box if all rows have the same number? (Lesson 5-2) **C**

    Ⓐ 2 Ⓑ 4 Ⓒ 8 Ⓓ 12

20. **GRID IN** Amy spent 35 minutes in the afternoon practicing the piano. What part of one hour did she spend practicing? Write as a fraction in simplest form. (Lesson 5-3) **7/12**

The Game Zone

A Place To Practice Your Math Skills

The Factor Fair

GET READY!

Players: two teams
Materials: 36 index cards, tape

GET SET!

- Number the index cards 1 through 36.
- Tape the index cards in order on the chalkboard.
- Divide the class into two teams.

GO!

- Team A chooses one of the index cards, such as the 8 card, and takes it off the chalkboard. Team A gets 8 points. Team B gets all the cards that are factors of 8 that have not yet been taken. The sum of the factors are the points they receive. For example, Team B would receive 1 + 2 + 4, or 7 points.

Team A	Team B
8	1
	2
	+ 4
	7

- Team B then chooses a card and gets those points.
- A team loses a turn if it selects an illegal number. A number is considered illegal if it does not have at least one factor available.
- Teams continue to take turns until there are no legal plays remaining.
- **Who Wins?** The team with the most points is the winner.

The Game Zone

Resources

For an easy-to-make manipulative for this game, visit: msmath2.net/gamezone.

GETTING READY

Objective Find all the factors of a number.

MANAGEMENT TIPS

Recommended Time
30 minutes

Tips

- Record the team scores on the chalkboard. Update the score after each play.
- After playing the game with the whole class, you may want to separate students into smaller groups and repeat the game.
- After playing the game, discuss any strategies the students discovered for winning. What is a good first choice for Team A?
- You may wish to use the game as an assessment tool for the concepts the students have been learning in Lessons 5-1 through 5-4.

For an additional game for this chapter, visit: msmath2.net/gamezone.

5-5 Lesson Notes

Lesson Resources

Chapter 5 Resource Masters
- Study Guide and Intervention, p. 263
- Practice: Skills, p. 264
- Practice: Word Problems, p. 265
- Reading to Learn Mathematics, p. 266
- Enrichment, p. 267

Calculator and Spreadsheet Masters, pp. 30, 31
Parent and Student Study Guide Workbook, 5-5
Teaching Mathematics with Manipulatives, p. 60

FAST FILE

1 Focus

5-Minute Check 5-5

INTERACTIVE CHALKBOARD
PowerPoint® Presentations

Use as a quiz or review of Lesson 5-4. (Also available as a transparency)

Bellringer
Communication Tell students that a pupil scored 16 out of 20 points on a quiz and 80 out of 100 points on a test. Ask them what the student's grades were on the two assessments.

Reading to Learn Mathematics, p. 266 ELL

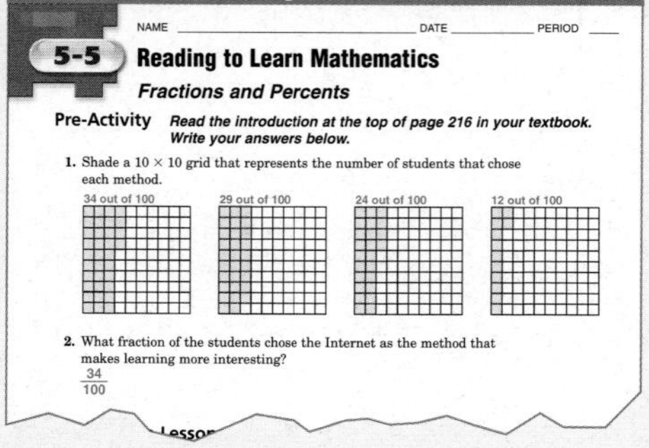
NAME ______ DATE ______ PERIOD ____

5-5 **Reading to Learn Mathematics**
Fractions and Percents

Pre-Activity *Read the introduction at the top of page 216 in your textbook. Write your answers below.*

1. Shade a 10 × 10 grid that represents the number of students that chose each method.

34 out of 100 | 29 out of 100 | 24 out of 100 | 12 out of 100

2. What fraction of the students chose the Internet as the method that makes learning more interesting? $\frac{34}{100}$

5-5 Fractions and Percents

What You'll Learn
Write fractions as percents and percents as fractions.

NEW Vocabulary
ratio
percent

REVIEW Vocabulary
simplest form: a fraction whose numerator and denominator have a GCF of 1 **(Lesson 5-3)**

WHEN am I ever going to use this?

EDUCATION The table shows the results of a survey in which students were asked to choose methods that make learning new subjects more interesting.

Method	Number of Students
Internet	34 out of 100
Teacher	29 out of 100
TV Program	24 out of 100
Textbook	12 out of 100

Source: Opinion Research Corporation

1. Shade a 10 × 10 grid that represents the number of students that chose each method. **See margin.**
2. What fraction of the students chose the Internet as the method that makes learning more interesting? $\frac{34}{100}$

A **ratio** is a comparison of two numbers by division. Ratios, like the ones above, can be written in several different ways.

34 out of 100 34:100 $\frac{34}{100}$

When a ratio compares a number to 100, it can be written as a **percent**.

Key Concept **Percent**

Words A percent is a ratio that compares a number to 100.

Symbols $\frac{n}{100} = n\%$

Ratio 34 to 100

Model

EXAMPLES **Write Ratios as Percents**

Write each ratio as a percent.

1 **Annie answered 90 out of 100 questions correctly.**

You can represent 90 out of 100 with a model.

$\frac{90}{100} = 90\%$

2 **On average, 50.5 out of 100 students own a pet.**

$\frac{50.5}{100} = 50.5\%$

TEACHING TIP
Show students that they can shade half of a square in the model to represent 0.5%.

Your Turn **Write each ratio as a percent.**

a. $\frac{45}{100}$ **45%** b. 19.2 out of 100 **19.2%** c. \$3.30:\$100 **3.3%**

Answer (When am I ever going to use this?)

1. 34 out of 100

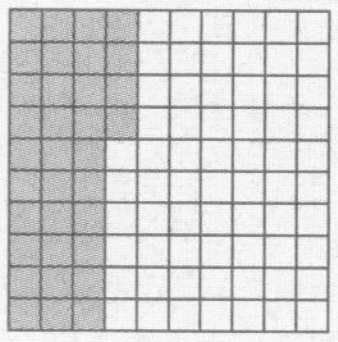

29 out of 100

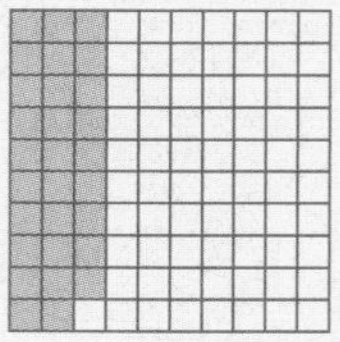

24 out of 100

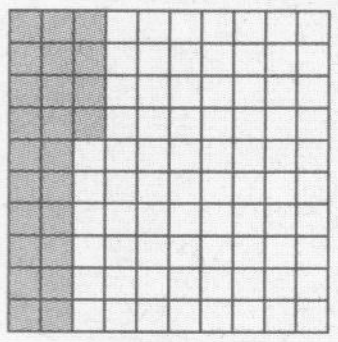

12 out of 100

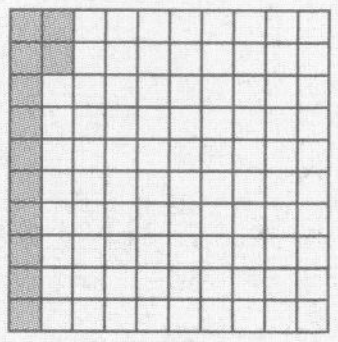

READING Math

Percent *Percent* means *per hundred* or *hundredths*. The symbol % means percent.

Fractions and percents are ratios that can represent the same number. You can write a fraction as a percent by finding an equivalent fraction with a denominator of 100.

EXAMPLE Write a Fraction as a Percent

3 **Write $\frac{3}{20}$ as a percent.**

Since 100 ÷ 20 = 5, . . . → $\frac{3}{20} = \frac{15}{100}$ (×5 numerator, ×5 denominator) ← . . . multiply the numerator and denominator by 5.

$\frac{15}{100} = 15\%$

So, $\frac{3}{20} = 15\%$.

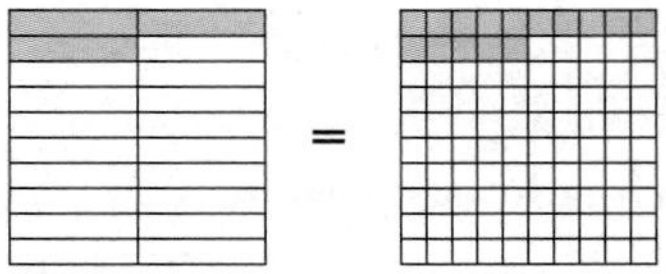

$\frac{3}{20} = \frac{15}{100}$

Your Turn **Write each fraction as a percent.**

d. $\frac{17}{20}$ **85%** e. $\frac{3}{5}$ **60%** f. $\frac{2}{25}$ **8%**

You can also use the meaning of percent to write percents as fractions.

EXAMPLE Write a Percent as a Fraction

4 **Write 48% as a fraction in simplest form.**

$48\% = \frac{48}{100}$ Definition of percent

$= \frac{12}{25}$ Simplify.

Your Turn **Write each percent as a fraction.**

g. 40% $\frac{2}{5}$ h. 6% $\frac{3}{50}$ i. 24% $\frac{6}{25}$

REAL-LIFE MATH

COINS Each new quarter has a diameter of 24.26 millimeters, a thickness of 1.75 millimeters, and weighs 5.670 grams.

Source: www.usmint.gov

EXAMPLE Use Percent to Solve a Problem

5 **COINS The 50 State Quarters® Program allows the United States Mint to release a new quarter every ten weeks from 1999 through 2008 commemorating the 50 states. By the end of 2004, 30 state coins will have been released. What percent of the coins is this?**

By the end of 2004, 30 out of 50 state coins will have been released.

$\frac{30}{50} = \frac{60}{100}$ Write an equivalent fraction with a denominator of 100.

$= 60\%$ $\frac{60}{100} = 60\%$

So, 60% of the state coins will have been released by the end of 2004.

msmath2.net/extra_examples

DAILY INTERVENTION

Differentiated Instruction ELL

Verbal/Linguistic Have students choose a paragraph in a literature book and record the percent that each vowel is of the total number of vowels. Have students share their findings and try to determine the most common vowel. **The most common vowel is e.**

2 Teach

In-Class Examples

INTERACTIVE CHALKBOARD PowerPoint® Presentations

Write each ratio as a percent.

1. Diana scored 63 goals out of 100 attempts. **63%**
2. 31.9 out of 100 people bought crunchy peanut butter. **31.9%**
3. Write $\frac{16}{25}$ as a percent. **64%**
4. Write 22% as a fraction in simplest form. $\frac{11}{50}$
5. **RECREATION** The graphic shows the most popular outdoor activities according to parents with children ages 4–14. What fraction of parents prefer hiking as a favorite outdoor activity? $\frac{15}{74}$

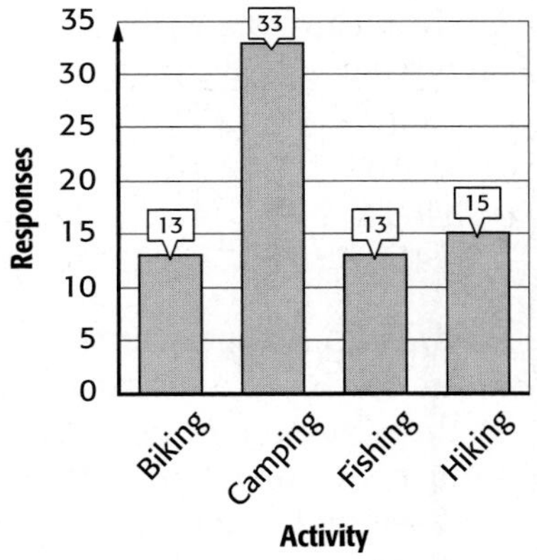

Tips for New Teachers

Teaching Strategies Students who consistently use the four-step plan will find word problems to be much easier to understand. Encourage students to write the four-step plan on a note card that they can have nearby while doing homework. They can refer to the card to make sure they do not skip a step.

3 Practice/Apply

FOLDABLES™ Follow-Up Remind students to write a question about fractions and percents on the front of the fifth tab of their Foldables before they begin the lesson. As students progress through the lesson, encourage them to take notes and record information that helps them answer their question under the tab.

Which One Doesn't Belong? For Exercise 3, encourage students to write each value as a fraction to help them make their selection.

About the Exercises...

Guided Practice Key

Exercises	Examples
4–6	1, 2
7–10	3
11–14	4
15	5

Odd/Even Assignments
Exercises 16–35 are structured so that students practice the same concepts whether they are assigned odd or even problems.

Answer

2. Sample answer:

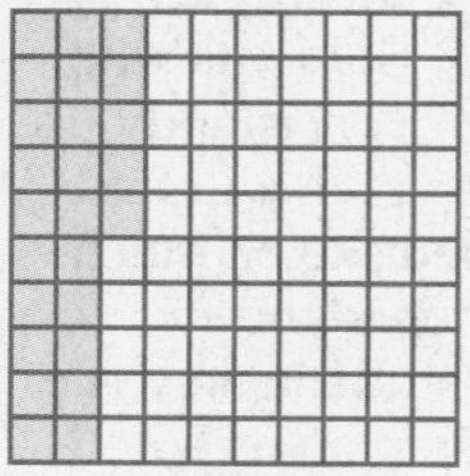

Skill and Concept Check

Writing Math Exercise 3

1. **Write** a fraction in simplest form and a percent to represent the shaded area at the right. $\frac{9}{25}$; 36%

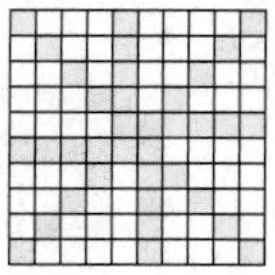

2. **OPEN ENDED** Draw a model that shows 25%. See margin.
3. **Which One Doesn't Belong?** Identify the number that does not have the same value as the other three. Explain your reasoning.

$\frac{10}{100}$	1%	$\frac{1}{10}$	100 out of 1,000

1%; the other ratios equal 10%

GUIDED PRACTICE

Write each ratio as a percent.

4. $\frac{57}{100}$ 57%
5. \$29.20 per \$100 29.2%
6. 11 teams out of 100 11%

Write each fraction as a percent.

7. $\frac{1}{4}$ 25%
8. $\frac{6}{10}$ 60%
9. $\frac{4}{5}$ 80%
10. $\frac{17}{20}$ 85%

Write each percent as a fraction in simplest form.

11. 90% $\frac{9}{10}$
12. 80% $\frac{4}{5}$
13. 75% $\frac{3}{4}$
14. 22% $\frac{11}{50}$
15. **ELECTRICITY** Electrical power is measured in a unit called the watt. Write a percent that compares the amount of power given off by a 75-watt bulb to the power given off by a 100-watt bulb. 75%

★ indicates increased difficulty

Practice and Applications

HOMEWORK HELP

For Exercises	See Examples
16–23	1, 2
24–35	3
39–50	4
36–38	5

Extra Practice See pages 575, 600.

Write each ratio as a percent. 19. $11\frac{2}{7}$%

16. $\frac{87}{100}$ 87%
17. $\frac{12.2}{100}$ 12.2%
18. 42 per 100 42%
19. $11\frac{2}{7}$ out of 100
20. $66\frac{2}{3}$:100 $66\frac{2}{3}$%
21. 99.9:100 99.9%
22. Fifty-one out of every 100 households have at least one computer. 51%
23. Twenty out of every 100 adults use the Internet to check on news, weather, and sports. 20%

Write each fraction as a percent.

24. $\frac{7}{10}$ 70%
25. $\frac{16}{20}$ 80%
26. $\frac{15}{25}$ 60%
27. $\frac{37}{50}$ 74%
28. $\frac{73}{100}$ 73%
29. $\frac{13}{50}$ 26%
30. $\frac{1}{5}$ 20%
31. $\frac{19}{20}$ 95%
32. $\frac{3}{5}$ 60%
33. $\frac{3}{25}$ 12%
34. $\frac{10}{10}$ 100%
35. $\frac{19}{25}$ 76%
36. **SPECIAL OLYMPICS** Mila and Randi are volunteers for their school district's Special Olympics competition. They are expecting 100 participants, but only 67 have registered so far. What percent of the Special Olympics participants still need to register before the competition begins? 33%

CHORES For Exercises 37 and 38, use the graphic. It shows the percent of chores that parents have to remind their children to do the most.

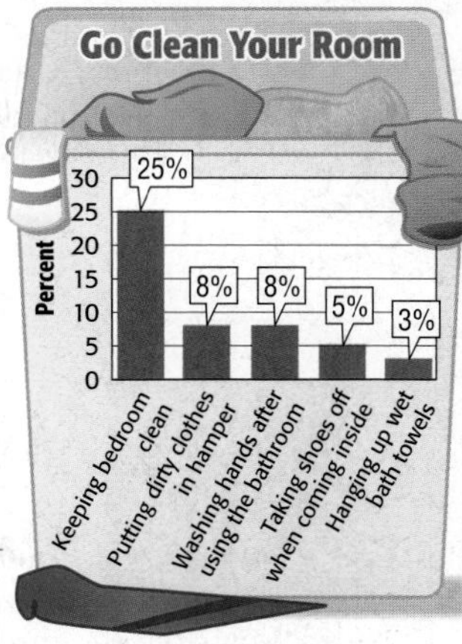

Source: Opinion Research Corp. for The Soap and Detergent Association

37. What fraction of parents remind their children to clean their rooms? $\frac{1}{4}$

★ 38. If 1,500 parents were surveyed, how many have to remind their children to put their dirty clothes in the hamper? 120

Write each percent as a fraction in simplest form.

39. 45% $\frac{9}{20}$	40. 30% $\frac{3}{10}$	41. 50% $\frac{1}{2}$
42. 20% $\frac{1}{5}$	43. 62% $\frac{31}{50}$	44. 88% $\frac{22}{25}$
45. 68% $\frac{17}{25}$	46. 13% $\frac{13}{100}$	47. 44% $\frac{11}{25}$
48. 52% $\frac{13}{25}$	49. 2% $\frac{1}{50}$	50. 14% $\frac{7}{50}$

★ 51. **GEOMETRY** What percent of the larger rectangle at the right is shaded? 36%

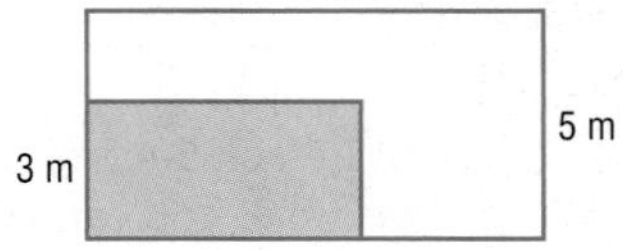

★ 52. What percent of 100 is 1? 1%

★ 53. 17% of 100 is what number? 17

54. **CRITICAL THINKING** Kenneth and Rico both finished a 1-minute typing test on the computer. Kenneth misspelled one word and received a score of 96% correct. Rico also misspelled one word but received a score of 90% correct. Explain why both students received different scores with only one misspelled word. See margin.

Standardized Test Practice and Mixed Review

55. **MULTIPLE CHOICE** Find the ratio that represents 20%. D

Ⓐ $\frac{20}{100}$ Ⓑ 2:10

Ⓒ 200 per 1,000 Ⓓ all of the above

56. **SHORT RESPONSE** Write a percent for a $7.50 donation from a $100 check. 7.5%

Write each repeating decimal using bar notation. (Lesson 5-4)

57. 0.6555… $0.6\overline{5}$ 58. 4.232323… $4.\overline{23}$ 59. 0.414141… $0.\overline{41}$

60. **MUSIC** In a survey, 12 of the 78 people preferred classical music. Write this ratio as a fraction in simplest form. (Lesson 5-3) $\frac{2}{13}$

GETTING READY FOR THE NEXT LESSON

BASIC SKILL **Multiply or divide.**

61. 16.2×10 162 62. 0.71×100 71 63. $14.4 \div 100$ 0.144 64. $791 \div 1{,}000$ 0.791

msmath2.net/self_check_quiz

4 Assess

Open-Ended Assessment

Modeling Present students with a variety of percents and fractions. Ask students to represent each value as a percent, fraction, and model.

Answer

54. Sample answer: For Rico to receive a 90% with only one misspelled word means he correctly typed 9 words out of 10, or $\frac{9}{10}$ correct. Kenneth correctly typed 24 words out of 25, or $\frac{24}{25}$ correct for a score of 96%.

Study Guide and Intervention, p. 263

NAME ______ DATE ______ PERIOD ____

5-5 Study Guide and Intervention

Fractions and Percents

A **ratio** is a comparison of two numbers by division. When a ratio compares a number to 100, it can be written as a **percent**. To write a ratio or fraction as a percent, find an equivalent fraction with a denominator of 100. You can also use the meaning of percent to change percents to fractions.

EXAMPLE 1 Write $\frac{19}{20}$ as a percent.

$\frac{19}{20} \xrightarrow{\times 5} \frac{95}{100} = 95\%$ Since 100 ÷ 20 = 5, multiply the numerator and denominator by 5.

EXAMPLE 2 Write 92% as a fraction in simplest form.

$92\% = \frac{92}{100}$ Definition of percent

$= \frac{23}{25}$ Simplify.

EXERCISES

Write each ratio as a percent.

1. $\frac{14}{100}$ 14% 2. $\frac{27}{100}$ 27% 3. 34.5 per 100 34.5%

4. 18 per 100 18% 5. 21:100 21% 6. 96:100 96%

Write each fraction as a percent.

7. $\frac{3}{100}$ 3% 8. $\frac{14}{100}$ 14% 9. $\frac{2}{5}$ 40%

10. $\frac{1}{20}$ 5% 11. $\frac{13}{25}$ 52% 12. $\frac{4}{10}$ 40%

Write each percent as a fraction in simplest form.

13. 35% $\frac{7}{20}$ 14. 18% $\frac{9}{100}$ 15. 75% $\frac{3}{4}$

16. 80% $\frac{4}{5}$ 17. 16% $\frac{4}{25}$ 18. 15% $\frac{3}{20}$

Practice: Skills, p. 264

NAME ______ DATE ______ PERIOD ____

5-5 Practice: Skills

Fractions and Percents

Write each ratio as a percent.

1. 26 out of 100 26% 2. 5 per 100 5% 3. 13:100 13%

4. $\frac{39}{100}$ 39% 5. 12.5 per 100 12.5% 6. 51 out of 100 51%

Write each fraction as a percent.

7. $\frac{7}{10}$ 70% 8. $\frac{6}{50}$ 12% 9. $\frac{13}{20}$ 65%

10. $\frac{30}{50}$ 60% 11. $\frac{7}{20}$ 35% 12. $\frac{12}{20}$ 60%

13. $\frac{23}{25}$ 92% 14. $\frac{3}{10}$ 30% 15. $\frac{17}{50}$ 34%

Write each percent as a fraction in simplest form.

16. 15% $\frac{3}{20}$ 17. 85% $\frac{17}{20}$ 18. 1% $\frac{1}{100}$

19. 70% $\frac{7}{10}$ 20. 25% $\frac{1}{4}$ 21. 19% $\frac{19}{100}$

22. 33% $\frac{33}{100}$ 23. 22% $\frac{11}{50}$ 24. 95% $\frac{19}{20}$

Practice: Word Problems, p. 265

NAME ______ DATE ______ PERIOD ____

5-5 Practice: Word Problems

Fractions and Percents

1. **LUNCHES** Three out of every 10 students in Mr. Chan's class bring their lunch to school. Write this ratio as a percent. 30%

2. **COMPUTERS** In 2000, 57 out of every 100 school age children (ages 6 to 17 years) had access to a computer both at home and at school. Write this ratio as a percent. 57%

3. **SALES TAX** In one town, the sales tax is 8%. Write this percent as a fraction in simplest form. $\frac{2}{25}$

4. **HYGEINE** Ms. Agosto surveyed her class and found that 15 out of 30 students brushed their teeth more than twice a day. What percent of students brushed more than twice a day? 50%

5. **DISCOUNT** A local retail store was having a sale and offered all their merchandise at a 25% discount. Write this percent as a fraction in simplest form. $\frac{1}{4}$

6. **SPACE FLIGHT** About 64% of all individuals who have flown in space from 1961 to 2001 are from the United States. Write this percent as a fraction in simplest form. $\frac{16}{25}$

5-6 Lesson Notes

Lesson Resources

Chapter 5 Resource Masters

- Study Guide and Intervention, p. 268
- Practice: Skills, p. 269
- Practice: Word Problems, p. 270
- Reading to Learn Mathematics, p. 271
- Enrichment, p. 272

Key Teaching Transparencies, 19
MathPASS Tutorial CD-ROM, 31
Parent and Student Study Guide Workbook, 5-6

FAST FILE

1 Focus

5-Minute Check 5-6

INTERACTIVE CHALKBOARD PowerPoint® Presentations

Use as a quiz or review of Lesson 5-5. (Also available as a transparency)

Bellringer

Communication Have students find examples of percents in newspapers and magazines. Using what they learned in Lesson 5-5, have students write the percents as fractions. Using what they learned in Lesson 5-4, have them write the fractions as decimals. Ask students if there might be an easier way to convert from percents into decimals.

Reading to Learn Mathematics, p. 271 **ELL**

NAME ________________ DATE ________ PERIOD ____

5-6 Reading to Learn Mathematics

Percents and Decimals

Pre-Activity ***Read the introduction at the top of page 220 in your textbook. Write your answers below.***

1. Write the percent of students who read for fun as a fraction. $\frac{42}{100}$ or $\frac{21}{50}$

2. Write the fraction as a decimal. 0.42

Reading the Lesson

3. Describe each step in changing a percent to a fraction. Sample answer: 1) Write the percent as a fraction with the denominator equal to 100. 2) Write the fraction as a decimal, moving the decimal point two places to the left.

5-6 Percents and Decimals

What You'll Learn
Write percents as decimals and decimals as percents.

WHEN am I ever going to use this?

READING The graphic shows the reasons that students in 6th through 12th grades read.

1. Write the percent of students who read for fun as a fraction.
2. Write the fraction as a decimal. **0.42**

1. $\frac{42}{100}$ or $\frac{21}{50}$

Source: National Education Association, *American Demographics*

In Lesson 5-4, you learned that any fraction can be written as a decimal. You can use this fact to write percents as decimals.

EXAMPLES Write Percents as Decimals

1 GEOGRAPHY Alaska is the largest state, making up more than 16% of the land area of the United States. Write 16% as a decimal.

$16\% = \frac{16}{100}$ Write the percent as a fraction.

$= 0.16$ Write the fraction as a decimal.

So, 16% = 0.16.

2 Write 85.3% as a decimal.

$85.3\% = \frac{85.3}{100}$ Write the percent as a fraction.

$= \frac{85.3 \times 10}{100 \times 10}$ Multiply by 10 to remove the decimal in the numerator.

$= \frac{853}{1{,}000}$ Simplify.

$= 0.853$ Write the fraction as a decimal.

So, 85.3% = 0.853.

Study the pattern in the percents and the equivalent decimals in Examples 1 and 2. Notice that you can write the percent as a decimal by dividing the number by 100 and removing the percent symbol.

Key Concept **Writing Percents as Decimals**

Words To write a percent as a decimal, divide the percent by 100 and remove the percent symbol.

Symbols 25% = 25 = 0.25

TEACHING TIP

oint out that $3\frac{1}{4}\%$ is not ıual to 3.25. It is equal 3.25%.

EXAMPLES Write Percents as Decimals

3 Write 23% as a decimal.

$23\% = 23$ Divide by 100.

$= 0.23$ Remove the %.

So, 23% = 0.23.

4 Write $3\frac{1}{4}\%$ as a decimal.

$3\frac{1}{4}\% = 3.25\%$ Write $\frac{1}{4}$ as 0.25.

$= 03.25$ Divide by 100.

$= 0.0325$ Remove the %.

So, $3\frac{1}{4}\% = 0.0325$.

Your Turn **Write each percent as a decimal.**

a. 76% **0.76** b. 8.5% **0.085** c. $92\frac{1}{2}\%$ **0.925**

EAL-LIFE MATH

OGRAPHY A large tarctic iceberg can igh 400 million tons, 10 stories above the face of the water, and pply water for 3 million ople for a year.

rce: www.chennaionline.com

You can also use fractions to help you write decimals as percents.

EXAMPLES Write Decimals as Percents

5 GEOGRAPHY Nearly 0.02 of the world's fresh water comes from the Antarctic Icecap. Write 0.02 as a percent.

$0.02 = \frac{2}{100}$ Definition of decimal

$= 2\%$ Definition of percent

So, 0.02 = 2%.

6 Write 0.347 as a percent.

$0.347 = \frac{347}{1,000}$ Definition of decimal

$= \frac{34.7}{100}$ Divide both numerator and denominator by 10.

$= 34.7\%$ Definition of percent

So, 0.347 = 34.7%.

Study the pattern in the decimals and the equivalent percents in Examples 5 and 6. Notice that you can write the decimal as a percent by multiplying the number by 100 and adding the percent symbol.

Key Concept **Writing Decimals as Percents**

Words To write a decimal as a percent, multiply the percent by 100 and add a percent symbol.

Symbols 0.58 = 0.58 = 58%

STUDY TIP

Mental Math
- To write a percent as a decimal, move the decimal point two places to the left.
- To write a decimal as a percent, move the decimal point two places to the right.

EXAMPLES Write Decimals as Percents

7 Write 0.64 as a percent.

$0.64 = 0.64$ Multiply by 100.

$= 64\%$ Add the %.

So, 0.64 = 64%.

8 Write 0.875 as a percent.

$0.875 = 0.875$ Multiply by 100.

$= 87.5\%$ Add the %.

So, 0.875 = 87.5%.

msmath2.net/extra_examples

2 Teach

In-Class Examples

INTERACTIVE CHALKBOARD PowerPoint® Presentations

1 POPULATION According to the Administration on Aging, about 28% of the population of the United States is 19 years of age or younger. Write 28% as a decimal. **0.28**

2 Write 47.8% as a decimal. **0.478**

3 Write 95.3% as a decimal. **0.953**

4 Write $8\frac{1}{5}\%$ as a decimal. **0.082**

5 POPULATION In 1790, about 0.05 of the population of the United States lived in an urban setting. Write 0.05 as a percent. **5%**

6 Write 0.121 as a percent. **12.1%**

7 Write 0.33 as a percent. **33%**

8 Write 0.419 as a percent. **41.9%**

Tips for New Teachers

Teaching Strategies Ask students to discuss why the mental math tool in the Study Tip works.

DAILY INTERVENTION **Unlocking Misconceptions**

Converting Percents to Decimals Students may become confused when percents have a decimal component, such as 55.8%. Remind students that such a percent can be converted to a decimal in the standard way: dividing by 100 and moving the decimal two places to the left (0.558).

3 Practice/Apply

FOLDABLES™ Follow-Up Remind students to write a question about fractions, decimals, and percents on the front of the sixth tab of their Foldables before they begin the lesson. As students progress through the lesson, encourage them to take notes and record information that helps them answer their question under the tab.

DAILY INTERVENTION **FIND THE ERROR** In Exercise 2, explain to students that the closer a decimal is to 1.0, the closer the percent form will be to 100%.

About the Exercises...

Guided Practice Key

Exercises	Examples
11	1
5	2
3–4	3
6	4
7–10	5–8

Odd/Even Assignments
Exercises 12–31 are structured so that students practice the same concepts whether they are assigned odd or even problems.

Skill and Concept Check

Writing Math Exercise 2

1. **OPEN ENDED** Choose a decimal. Then write it as a fraction in simplest form and as a percent. **Sample answer: 0.25, $\frac{1}{4}$, 25%**
2. **FIND THE ERROR** Jessica and Gregorio both wrote 0.881 as a percent. Who is correct? Explain. **Jessica; Gregorio divided by 100 instead of multiplying by 100.**

Jessica
0.881 = 0.881
= 88.1%

Gregorio
0.881 = 000.881
= 0.00881%

GUIDED PRACTICE

Write each percent as a decimal.

3. 68% **0.68** 4. 3% **0.03** 5. 27.6% **0.276** 6. $45\frac{1}{2}$% **0.455**

Write each decimal as a percent.

7. 0.56 **56%** 8. 0.12 **12%** 9. 0.08 **8%** 10. 0.399 **39.9%**

11. **GEOGRAPHY** Pennsylvania makes up 1.2% of the landmass of the United States. Write this percent as a decimal. **0.012**

★ indicates increased difficulty

Practice and Applications

HOMEWORK HELP

For Exercises	See Examples
12–19, 34–37	1, 2, 3, 4
20–33	5, 6, 7, 8

Extra Practice
See pages 576, 600.

Write each percent as a decimal.

12. 27% **0.27** 13. 70% **0.7** 14. 6% **0.06** 15. 1% **0.01**
16. 18.5% **0.185** 17. 2.2% **0.022** 18. $15\frac{1}{2}$% **0.155** 19. $30\frac{1}{4}$% **0.3025**

Write each decimal as a percent.

20. 0.95 **95%** 21. 0.08 **8%**
22. 0.17 **17%** 23. 0.6 **60%**
24. 0.675 **67.5%** 25. 0.145 **14.5%**
26. 0.012 **1.2%** 27. 0.7025 **70.25%**

TENNIS For Exercises 28–32, use the graphic at the right. Write each decimal as a percent for the following players.

28. Monica Seles **94.9%**
29. Martina Hingis **90.5%**
30. Venus Williams **81.3%**
31. Jennifer Capriati **80.8%**
★ 32. Based on her winning percentage, which player wins 413 out of every 500 games played? **Lindsay Davenport**

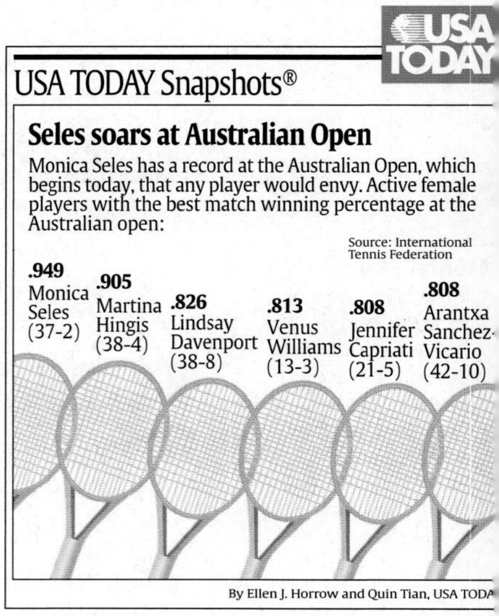

222 Chapter 5 Fractions, Decimals, and Percents

33. **BIRDS** Birds shown in a field guide are 0.10 to 0.67 their actual size. What percent of their actual size are they? **10% to 67%**

MODEL TRAINS For Exercises 34–37, use the table at the right.

34. The five scales at the right can be used to build a model train based on the life-size original. For example, a G-scale train is about 4% of the size of the original. So, the model of a 75-foot-long locomotive measures 40 inches. Find the decimal equivalents for each scale. **See margin.**

35. Which scale is the smallest? **Z**

★ 36. Find the approximate length of a 75-foot train in the N scale. **5.625 i**

★ 37. Name the scale used when a model of a 75-foot train measures approximately 18.75 inches. **O**

Scale	Approximate Percent Equivalent
G	4.4%
O	2.083%
HO	1.15%
N	0.625%
Z	0.45%

Source: www.internettrains.com

CRITICAL THINKING For Exercises 38–43, choose the greater number in each pair.

38. $1\frac{1}{4}$%, 0.125 **0.125**
39. 0.76, 76.5% **76.5%**
40. 42%, 4.2 **4.2**
41. 99%, 0.985 **99%**
42. $\frac{13}{40}$, 0.30 **$\frac{13}{40}$**
43. 0.56, $\frac{45}{80}$ **$\frac{45}{80}$**

Standardized Test Practice and Mixed Review

44. **MULTIPLE CHOICE** Which is 0.08 written as a percent? **C**

Ⓐ 0.08% Ⓑ 0.8% Ⓒ 8% Ⓓ 80%

45. **MULTIPLE CHOICE** Which is 96.3% written as a decimal? **G**

Ⓕ 0.0963 Ⓖ 0.963 Ⓗ 9.63 Ⓘ 963

46. **GRID IN** Write $78\frac{1}{2}$% as a decimal. **0.785**

Write each ratio as a percent. (Lesson 5-5)

47. $\frac{72}{100}$ **72%**
48. 90.9:100 **90.9%**
49. 18 per 100 **18%**
50. 23.1 out of 100 **23.1%**

51. Write $9\frac{3}{8}$ as a decimal. (Lesson 5-4) **9.375**

52. **BALLOONS** Write an integer to represent a hot-air balloon descending 83 feet. (Lesson 3-1) **−83**

TESTS For Exercises 53 and 54, use the stem-and-leaf plot of Chapter 5 test scores at the right.

53. How many students took the test? (Lesson 2-5) **23 students**

54. What was the median score? (Lesson 2-4) **82 points**

Stem	Leaf
10	0
9	2 3 3 3 4 5 9
8	0 1 1 2 3 5 9 9
7	0 4 6 8 8 9
6	4

8|1 = 81 points

GETTING READY FOR THE NEXT LESSON

PREREQUISITE SKILL Use exponents to write the prime factorization of each number. (Lesson 5-1)

55. 50 **2×5^2**
56. 32 **2^5**
57. 76 **$2^2 \times 19$**
58. 105 **$3 \times 5 \times 7$**

msmath2.net/self_check_quiz

4 Assess

Open-Ended Assessment

Speaking Ask students to describe how to compare 0.8, $\frac{8}{1,000}$, and 8%.

Answer

34. 0.044; 0.02083; 0.0115; 0.00625; 0.0045

Study Guide and Intervention, p. 268

NAME ____________ DATE ________ PERIOD ____

5-6 Study Guide and Intervention

Percents and Decimals

To write a percent as a decimal, divide the percent by 100 and remove the percent symbol. To write a decimal as a percent, multiply the decimal by 100 and add the percent symbol.

EXAMPLE 1 **Write 42.5% as a decimal.**

$42.5\% = \frac{42.5}{100}$ Write the percent as a fraction.

$= \frac{42.5 \times 10}{100 \times 10}$ Multiply by 10 to remove the decimal in the numerator.

$= \frac{425}{1,000}$ Simplify.

$= 0.425$ Write the fraction as a decimal.

EXAMPLE 2 **Write 0.625 as a percent.**

$0.625 = 062.5$ Multiply by 100.

$= 62.5\%$ Add the % symbol.

EXERCISES

Write each percent as a decimal.

1. 6% 0.06
2. 28% 0.28
3. 81% 0.81
4. 84% 0.84
5. 35.5% 0.355
6. 12.5% 0.125
7. 14.2% 0.142
8. 11.1% 0.111

Write each decimal as a percent.

9. 0.47 47%
10. 0.03 3%
11. 0.075 7.5%
12. 0.914 91.4%

Practice: Skills, p. 269

NAME ____________ DATE ________ PERIOD ____

5-6 Practice: Skills

Percents and Decimals

Write each percent as a decimal.

1. 5% 0.05
2. 20% 0.2
3. 21% 0.21
4. 83% 0.83
5. 7% 0.07
6. 56% 0.56
7. 16% 0.16
8. 45% 0.45
9. 27.3% 0.273
10. 14.9% 0.149
11. 91.5% 0.915
12. 29.3% 0.293
13. 14.4% 0.144
14. 80% 0.8
15. 7.5% 0.075
16. $10\frac{1}{2}$% 0.105

Write each decimal as a percent.

17. 0.06 6%
18. 0.13 13%
19. 0.5 50%
20. 0.74 74%
21. 0.14 14%
22. 0.92 92%
23. 0.54 54%
24. 0.66 66%
25. 0.192 19.2%
26. 0.295 29.5%
27. 0.911 91.1%
28. 0.247 24.7%
29. 0.4165 41.65%
30. 0.2199 21.99%
31. 0.7601 76.01%
32. 0.4833 48.33%

Practice: Word Problems, p. 270

NAME ____________ DATE ________ PERIOD ____

5-6 Practice: Word Problems

Percents and Decimals

1. **AREA** New Mexico's land area is about 0.03 of the total area of the United States. What percent is New Mexico's land area of the total area of the United States? 3%

2. **SCALE MODEL** A scale model of a building is 0.25 the actual size. What percent of the actual size of the building is the model? 25%

3. **NFL COACHES** Don Shula ranks among the most successful coaches in the National Football League. In his career, he won 0.665 of his games. Write the decimal as a percent. 66.5%

4. **SOFTBALL** Jenny's batting average is 0.346. Write the decimal as a percent. 34.6%

5. **VITAMINS** A multiple vitamin contains 450 milligrams of calcium. This is 45% of the recommended daily allowance. Write the percent as a decimal. 0.45

6. **BASKETBALL** Tao makes 74% of his free throws. Write the percent as a decimal. 0.74

7. **SALES TAX** The sales tax in a town is 7.25%. Write the percent as a decimal. 0.0725

8. **FIELD TRIP** In Ms. Silver's English class, $20\frac{1}{4}$% of the students signed up to visit a local museum. Write the percent as a decimal. 0.2025

Lesson 5-6 Percents and Decimals **223**

5-7 Lesson Notes

Lesson Resources

Chapter 5 Resource Masters
- Study Guide and Intervention, p. 273
- Practice: Skills, p. 274
- Practice: Word Problems, p. 275
- Reading to Learn Mathematics, p. 276
- Enrichment, p. 277

Key Teaching Transparencies, 20
Parent and Student Study Guide Workbook, 5-7

FAST FILE

1 Focus

5-Minute Check 5-7
Use as a quiz or review of Lesson 5-6. (Also available as a transparency)

INTERACTIVE CHALKBOARD
PowerPoint® Presentations

Bellringer
Problem Solving Present the following problem. Maria has a large collection of dimes and quarters. In order to count the coins, she wants to make several piles of each coin that are equal in value. What could be the value of each pile? **$0.50, $1, $1.50, . . .** Is there more than one possible answer? **yes**

Reading to Learn Mathematics, p. 276 **ELL**

NAME ________ DATE ______ PERIOD ____

5-7 Reading to Learn Mathematics
Least Common Multiple

Pre-Activity ***Complete the Mini Lab at the top of page 224 in your textbook. Write your answers below.***

1. Add a second floor to each building. Record the total number of cubes used in a table like the one shown below.

Number of Floors	1	2	3	4	5
Number of Cubes in Building 1	9	18	27	36	45
Number of Cubes in Building 2	12	24	36	48	60

2. Continue adding floors until each building has five floors. Record your results.
3. Describe two buildings that have the same number of cubes. Both buildings have the same number of cubes when Building 1 has 4 floors and Building 2 has 3 floors.
4. If you keep adding floors, will the two buildings have the same number of cubes again? Explain. Yes; Sample answer: Every fourth floor in Building 1 will have the same number of cubes as every ...

5-7 Least Common Multiple

What You'll Learn
Find the least common multiple of two or more numbers.

NEW Vocabulary
multiple
least common multiple (LCM)

HANDS-ON Mini Lab

Materials
- centimeter cubes

Work with a partner.

Use 9 centimeter cubes to model the first floor of Building 1 and 12 centimeter cubes to model the first floor of Building 2, as shown.

Building 1

Building 2

1. Add a second floor to each building. Record the total number of cubes used in a table like the one shown below.

18 27 36 45

Number of Floors	1	2	3	4	5
Number of Cubes in Building 1	9	?	?	?	?
Number of Cubes in Building 2	12	?	?	?	?

24 36 48 60

2. Continue adding floors until each building has five floors. Record your results.
3. Describe two buildings that have the same number of cubes.
4. If you keep adding floors, will the two buildings have the same number of cubes again? Explain. **3–4. See margin.**

In the Mini Lab, you listed multiples of 9 and 12. A **multiple** is the product of a number and any whole number. The **least common multiple**, or **(LCM)**, of two or more numbers is the least of their common multiples, excluding zero. So, the LCM of 9 and 12 is 36.

You can use several strategies to find the LCM of two or more numbers.

EXAMPLE Find the LCM by Listing Multiples

1 Find the LCM of 6 and 10.

First, list the multiples of 6 and 10.

multiples of 6: 6, 12, 18, 24, 30, 36, 42, 48, 54, 60, …
multiples of 10: 10, 20, 30, 40, 50, 60, 70, 80, …

Notice that 30, 60, …, are common multiples. So, the LCM of 6 and 10 is 30.

Your Turn **Find the LCM of each pair of numbers.**

a. 3, 12 **12** b. 10, 12 **60** c. 4, 5 **20**

TEACHING TIP
Point out that 0 is a multiple of every number but is never the LCM.

You can also use prime factorization to find the LCM. The LCM is the smallest product that contains the prime factors of each number.

Answers (Hands-On Mini Lab)

3. Both buildings have the same number of cubes when Building 1 has 4 floors and Building 2 has 3 floors.
4. Yes; Sample answer: Every fourth floor in Building 1 will have the same number of cubes as every third floor in Building 2.

EXAMPLES Find the LCM Using Prime Factors

2 **Find the LCM of 24 and 36.**

Method 1 Write the prime factorization.

$24 = 2 \times 2 \times 2 \times 3 = 2^3 \times 3$

$36 = 2 \times 2 \times 3 \times 3 = 2^2 \times 3^2$

The prime factors of 24 and 36 are 2 and 3. Multiply the greatest power of both 2 and 3.

Method 2 Divide by prime numbers.

```
   2    3
3)6     9
2)12   18
2)24   36  <---- Start here.
```

Start dividing by prime factors until both numbers cannot be divided by the same divisor. Then multiply the divisors and quotients to find the LCM.

The LCM of 24 and 36 is $2^3 \times 3^2$, or 72.

STUDY TIP

Choosing a Method To find the LCM of two or more numbers, it is easier to:
- list factors for small numbers, and
- use prime factorization or division for large numbers.

3 **Find the LCM of 4, 10, and 15.**

$4 = 2^2$

$10 = 2 \times 5$

$15 = 3 \times 5$

LCM: $2^2 \times 3 \times 5 = 60$

So, the LCM of 4, 10, and 15 is 60.

Your Turn **Find the LCM of each set of numbers.**

d. 9, 15 **45** e. 10, 20, 30 **60** f. 6, 17, 34 **102**

Skill and Concept Check

Writing Math Exercises 1 & 3

1. **Describe** the relationship between 12, 4, and 3 using the words *multiple* and *factor*. **3 and 4 are factors of 12; 12 is a multiple (and LCM) of 3 and 4.**
2. **OPEN ENDED** Find three numbers that each have a multiple of 30. **Sample answer: 2, 10, and 15**
3. **NUMBER SENSE** Determine whether the LCM for all pairs of odd numbers is *sometimes*, *always*, or *never* their product. Explain. **Sometimes; the LCM of 3 and 9 is 9; the LCM of 3 and 5 is 15.**

GUIDED PRACTICE

Find the LCM of each set of numbers.

4. 4, 10 **20**
5. 6, 7 **42**
6. 3, 5, 12 **60**
7. **PICNIC** Flavio is having a cookout for his class and serving hot dogs. Hot dogs come in packages of 10. Hot dog buns come in packages of 8. If Flavio wants to have the same number of hot dogs and buns, what is the least number of each that he will have to buy? **4 packages of hot dogs and 5 packages of buns**

msmath2.net/extra_examples

DAILY INTERVENTION Differentiated Instruction

Kinesthetic Write the numbers 3, 5, 10, and 15 on index cards, one number per card. Have students in groups of four each take a card. Looking at a clock with a second hand, have students tap their pencils at 3, 5, 10, or 15 seconds depending on their card. Stop when all four members tap their pencils at the same time. How many seconds did this take? What is the least common multiple of 3, 5, 10, and 15? **30**

2 Teach

In-Class Examples

INTERACTIVE CHALKBOARD PowerPoint® Presentations

1. Find the LCM of 4 and 6. **12**
2. Find the LCM of 4 and 15. **60**
3. Find the LCM of 18, 24, and 48. **144**

3 Practice/Apply

FOLDABLES™ Follow-Up Remind students to write a question about least common multiples on the front of the seventh tab of their Foldables before they begin the lesson. As students progress through the lesson, encourage them to take notes and record information that helps them answer their question under the tab.

NUMBER SENSE
Encourage students to investigate several examples for Exercise 3. They should find that the LCM is the product of the numbers when the numbers have no common factors.

About the Exercises...

Guided Practice Key

Exercises	Examples
4–5, 7	1, 2
6	3

Odd/Even Assignments
Exercises 8–19 are structured so that students practice the same concepts whether they are assigned odd or even problems.

Study Guide and Intervention, p. 273

NAME ______________ DATE ________ PERIOD ____

5-7 Study Guide and Intervention

Least Common Multiple

A **multiple** of a number is the product of that number and any whole number. The least nonzero multiple of two or more numbers is the **least common multiple (LCM)** of the numbers.

EXAMPLE 1 **Find the LCM of 15 and 20 by listing multiples.**

List the multiples.

multiples of 15: 15, 30, 45, **60**, 75, 90, 105, **120**, ...

multiples of 20: 20, 40, **60**, 80, 100, **120**, 140, ...

Notice that 60, 120, ..., are common multiples. So, the LCM of 15 and 20 is 60.

EXAMPLE 2 **Find the LCM of 8 and 12 using prime factors.**

Method 1 Write the prime factorization.

$8 = 2 \times 2 \times 2 = 2^3$

$12 = 2 \times 2 \times 3 = 2^2 \times 3$

The prime factors of 8 and 12 are 2 and 3. Multiply the greatest power of both 2 and 3.

The LCM of 8 and 12 is $2^3 \times 3$, or 24.

Method 2 Divide by prime numbers.

2 3
2)4 6
2)8 12 ← Start here.

Start dividing by prime factors until both numbers cannot be divided by the same divisor. Then multiply the divisors and quotients to get the LCM.

EXERCISES

Find the LCM of each set of numbers.

1. 4, 6 12
2. 6, 9 18
3. 5, 9 45
4. 8, 10 40
5. 12, 15 60
6. 15, 21 105
7. 4, 15 60
8. 8, 20 40
9. 8, 16 16
10. 6, 14 42
11. 12, 20 60
12. 9, 12 36
13. 14, 21 42
14. 6, 15 30
15. 4, 6, 8 24
16. 3, 5, 6 30

Practice: Skills, p. 274

NAME ______________ DATE ________ PERIOD ____

5-7 Practice: Skills

Least Common Multiple

Find the LCM of each set of numbers.

1. 2, 8 8
2. 6, 10 30
3. 10, 11 110
4. 10, 12 60
5. 9, 18 18
6. 4, 22 44
7. 12, 30 60
8. 4, 13 52
9. 25, 30 150
10. 250, 30 750
11. 200, 18 1,800
12. 70, 90 630
13. 18, 54 54
14. 30, 65 390
15. 180, 252 1,260
16. 20, 55 220
17. 21, 60 420
18. 3, 5, 10 30
19. 3, 4, 13 156
20. 4, 10, 12 60
21. 6, 15, 20 60
22. 48, 16, 3 48
23. 66, 55, 44 660
24. 29, 58, 4 116

Practice: Word Problems, p. 275

NAME ______________ DATE ________ PERIOD ____

5-7 Practice: Word Problems

Least Common Multiple

1. **PROMOTION** In a promotion for a local delicatessen, every eighth customer will get a free sandwich and every sixth customer will get a free drink. Which customer will be first to get both a free sandwich and a free drink? the 24th customer

2. **WORK** Alano and Abey both work at night. Alano has every fourth night off and Abey has every sixth night off. If they are both off tonight, how long will it be before they are both off again? 12 nights

3. **RADIO** A radio station is giving away a discount coupon to every twelfth caller and a free concert ticket to every twentieth caller. Which caller will be first to win both the coupon and the ticket? the 60th caller

4. **MUSIC** Faith spent the same amount of money on cassette tapes as she did on CDs. If tapes cost \$12 and CDs cost \$16, what is the least amount of money she could have spent on each? \$48

5. **BIKES** Three bike riders ride around a circular path. The first rider circles the path in 12 minutes, the second in 18 minutes, and the third in 24 minutes. If they all start at the same place, at the same time, and go in the same direction, after how many minutes will they meet at the starting point? 72 min

6. **PAPER GOODS** At a party store, paper cups come in packages of 15, paper plates come in packages of 30, and napkins come in packages of 20. In order to have the same number of cups, plates, and napkins, what is the least number of each that must be purchased? 60

★ indicates increased difficulty

Practice and Applications

HOMEWORK HELP

For Exercises	See Examples
8–13, 20, 24	1, 2
14–19, 22–23	3

Extra Practice See pages 576, 600.

Find the LCM of each set of numbers.

8. 6, 8 **24**
9. 12, 16 **48**
10. 30, 45 **90**
11. 8, 18 **72**
12. 11, 12 **132**
13. 45, 63 **315**
14. 2, 3, 5 **30**
15. 6, 8, 9 **72**
16. 8, 12, 16 **48**
17. 12, 15, 28 **420**
18. 22, 33, 44 **132**
19. 12, 16, 36 **144**

20. Find the LCM of 2×3^2 and $2^3 \times 3$. **72**

★ 21. **PATTERNS** Write a rule that describes the common multiples of 9 and 15 that are greater than 180. **45*n*, where *n* = 5, 6, 7, ...**

PLANETS For Exercises 22 and 23, use the following information and the table at the right.

The nine planets in our solar system rotate at different orbital speeds.

22. When will the number of miles rotated by Earth, Mercury, and Pluto be the same? **when they all rotate 90 miles**

23. In how many seconds does Earth rotate the number of miles in Exercise 22? Mercury? Pluto? **5 s; 3 s; 30 s**

Planet	Orbital Speed (mi/s)
Earth	18
Jupiter	30
Pluto	3

Source: infoplease.com

★ 24. **NUMBER THEORY** The LCM of two consecutive numbers is greater than 200 and is a multiple of 7. Name the numbers. **14 and 15**

25. **CRITICAL THINKING** Two numbers have an LCM of $2^2 \times 3 \times 5^2$. Their GCF is 2×5. If one of the numbers is $2 \times 3 \times 5$, what is the other number? **$2^2 \times 5^2$, or 100**

Standardized Test Practice and Mixed Review

26. **MULTIPLE CHOICE** Find the LCM of 27 and 30. **C**

Ⓐ 810 Ⓑ 300 Ⓒ 270 Ⓓ 3

27. **SHORT RESPONSE** Presidents are elected every four years. Senators are elected every six years. If a senator was elected in the presidential election year 2000, in what year will she campaign again during a presidential election year? **2012**

28. Write 55.7% as a decimal to the nearest hundredth. (Lesson 5-6) **0.56**

DIAMONDS The graph shows the percent of diamond engagement rings that are different shapes. Write each percent as a fraction in simplest form. (Lesson 5-5)

29. round $\frac{17}{25}$
30. marquise $\frac{1}{5}$
31. others $\frac{3}{25}$

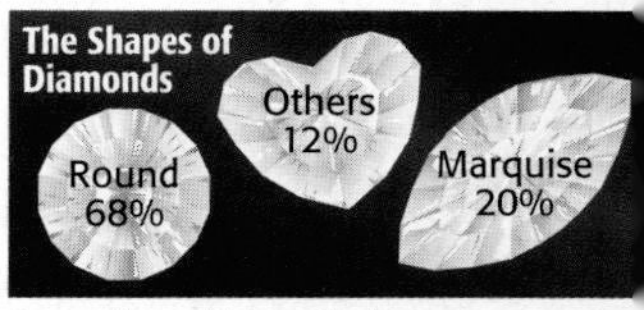

Source: Diamond Information Center

GETTING READY FOR THE NEXT LESSON

PREREQUISITE SKILL Replace each ● with < or > to make a true sentence. (Page 556)

32. 6.85 ● 5.68 $>$
33. 2.34 ● 2.43 $<$
34. 6.9 ● 5.99 $>$
35. 1.001 ● 1.1 $<$

msmath2.net/self_check_quiz

4 Assess

Higher-Level Thinking

The product of two numbers is 324 and the GCF is 6. Find the numbers and the LCM. **54, 6; 54**

5-8 Comparing and Ordering Rational Numbers

hat You'll Learn
ompare and order
actions, decimals,
nd percents.

EW Vocabulary
- mmon denominator
- ast common denominator (LCD)
- tional numbers

WHEN am I ever going to use this?

SOFTBALL The table shows how Latanya and Patrick did during the first month of the softball season.

Player	At-bats	Hits
Latanya	12	7
Patrick	18	8

1. A batting average is the ratio of hits to at-bats. Write each player's batting average as a fraction.
2. Estimate which fraction is greater than $\frac{1}{2}$. Which is less than $\frac{1}{2}$? $\frac{7}{12}$; $\frac{8}{18}$
3. Which player has the better batting average? Latanya

1. Latanya: $\frac{7}{12}$; Patrick: $\frac{8}{18}$

To compare and order fractions like the ones above, you can rewrite each fraction with a common denominator and then compare the numerators.

A **common denominator** is a common multiple of the denominators of two or more fractions. The **least common denominator (LCD)** is the LCM of the denominators.

EXAMPLE Compare Fractions

1 Who has the better batting average, Latanya or Patrick?

Method 1 Rename using the LCD. Then compare numerators.

Latanya: $\frac{7}{12} = \frac{7 \times 3}{12 \times 3} = \frac{21}{36}$

Patrick: $\frac{8}{18} = \frac{8 \times 2}{18 \times 2} = \frac{16}{36}$

The LCD is 36.

Since $\frac{21}{36} > \frac{16}{36}$, then $\frac{7}{12} > \frac{8}{18}$. Latanya has the better average.

Method 2 Write each fraction as a decimal. Then compare decimals.

Latanya: $\frac{7}{12} = 0.58\overline{3}$

Patrick: $\frac{8}{18} = 0.\overline{4}$

Since $0.58\overline{3} > 0.\overline{4}$, then $\frac{7}{12} > \frac{8}{18}$. Latanya has the better average.

Your Turn **Replace each ● with <, >, or = to make a true sentence.**

a. $\frac{5}{6}$ ● $\frac{7}{9}$ > b. $\frac{1}{5}$ ● $\frac{7}{50}$ > c. $\frac{9}{16}$ ● $\frac{7}{10}$ <

5-8 Lesson Notes

Lesson Resources

Chapter 5 Resource Masters FAST FILE
- Study Guide and Intervention, p. 278
- Practice: Skills, p. 279
- Practice: Word Problems, p. 280
- Reading to Learn Mathematics, p. 281
- Enrichment, p. 282

Calculator and Spreadsheet Masters, p. 32
MathPASS Tutorial CD-ROM, 15
Parent and Student Study Guide Workbook, 5-8

1 Focus

5-Minute Check 5-8

INTERACTIVE CHALKBOARD PowerPoint® Presentations

Use as a quiz or review of Lesson 5-7. (Also available as a transparency)

Bellringer

Reading Mathematics Post the following information on the chalkboard.

George's Market	Fred's Foodmart
Double Coupons!	Triple Coupons!

Ask students at which store they would rather shop and why.

Reading to Learn Mathematics, p. 281 ELL

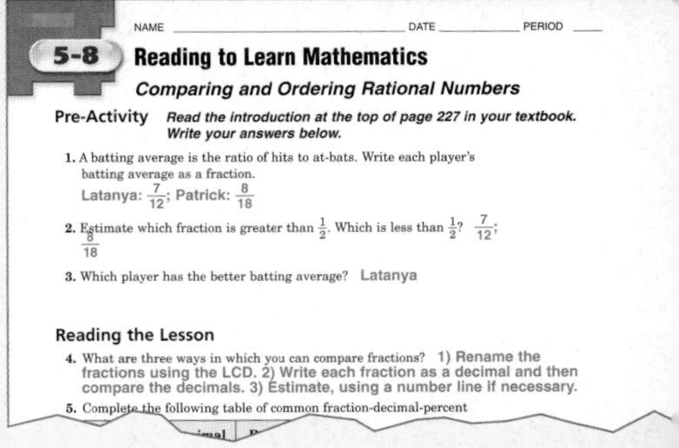
NAME ______ DATE ______ PERIOD ____

5-8 **Reading to Learn Mathematics**
Comparing and Ordering Rational Numbers

Pre-Activity *Read the introduction at the top of page 227 in your textbook. Write your answers below.*

1. A batting average is the ratio of hits to at-bats. Write each player's batting average as a fraction.
Latanya: $\frac{7}{12}$; Patrick: $\frac{8}{18}$
2. Estimate which fraction is greater than $\frac{1}{2}$. Which is less than $\frac{1}{2}$? $\frac{7}{12}$; $\frac{8}{18}$
3. Which player has the better batting average? Latanya

Reading the Lesson

4. What are three ways in which you can compare fractions? 1) Rename the fractions using the LCD. 2) Write each fraction as a decimal and then compare the decimals. 3) Estimate, using a number line if necessary.
5. Complete the following table of common fraction-decimal-percent

2 Teach

In-Class Examples

INTERACTIVE CHALKBOARD
PowerPoint® Presentations

1 **GRADES** Enrique and his younger brother both had a math test last Friday. Enrique scored 48 points out of 60 and his brother scored 55 points out of 75. Who got the better score, Enrique or his brother? **Enrique**

2 **DOGS** According to the Pet Food Manufacturer's Association, 11 out of 25 people own large dogs and 13 out of 50 own medium dogs. Do more people own large or medium dogs? **large**

3 Order $\frac{7}{10}$, 0.6, and 72% from least to greatest. **0.6, $\frac{7}{10}$, 72%**

REAL-LIFE MATH

MOUNTAIN BIKING Approximately 6.3 million people participated in off-road mountain biking in 2001.

Source: National Sporting Goods Association

TEACHING TIP

Point out the pattern in each column of the table.

d. $\frac{2}{10}$, 22%, 0.3
e. 2%, 0.18, $\frac{1}{5}$
f. 70%, 0.74, $\frac{3}{4}$

EXAMPLE Compare Ratios

2 **MOUNTAIN BIKING In Coach Ito's first period class, 19 out of 32 students owned a mountain bike. In his seventh period class, 16 out of 28 students owned a mountain bike. Did a greater fraction of students own a mountain bike in the first period class or the seventh period class?**

Estimate Both fractions are slightly greater than $\frac{1}{2}$.

Since the denominators are large, write $\frac{19}{32}$ and $\frac{16}{28}$ as decimals and then compare.

$19 \div 32 \approx 0.5938$ $16 \div 28 \approx 0.5714$ Use a calculator.

Since $0.5938 > 0.5714$, then $\frac{19}{32} > \frac{16}{28}$. So, a greater fraction of student in the first period class owned a mountain bike.

Some fractions, decimals, and percents are used more frequently than others. So, it is a good idea to be familiar with these common fraction-decimal-percent equivalents.

Concept Summary **Fraction-Decimal-Percent Equivalents**

$\frac{1}{5} = 0.2 = 20\%$	$\frac{1}{10} = 0.1 = 10\%$	$\frac{1}{4} = 0.25 = 25\%$
$\frac{2}{5} = 0.4 = 40\%$	$\frac{3}{10} = 0.3 = 30\%$	$\frac{1}{2} = 0.5 = 50\%$
$\frac{3}{5} = 0.6 = 60\%$	$\frac{7}{10} = 0.7 = 70\%$	$\frac{3}{4} = 0.75 = 75\%$
$\frac{4}{5} = 0.8 = 80\%$	$\frac{9}{10} = 0.9 = 90\%$	

EXAMPLE Order Ratios

3 **Order 0.6, 48%, and $\frac{1}{2}$ from least to greatest.**

Use estimation to find that 48% is the least. Then, compare $\frac{1}{2}$ and 0.6 using decimals.

$\frac{1}{2} = 0.5$ and $0.5 < 0.6$ So, $48\% < \frac{1}{2} < 0.6$.

Check You can change 48% and 0.6 to fractions, then compare all three fractions using the LCD.

$0.6 = \frac{60}{100}$ $48\% = \frac{48}{100}$ $\frac{1}{2} = \frac{50}{100}$

Since $\frac{48}{100} < \frac{50}{100} < \frac{60}{100}$, $48\% < \frac{1}{2} < 0.6$.

Your Turn **Order each set of ratios from least to greatest.**

d. 22%, 0.3, $\frac{2}{10}$ **e.** $\frac{1}{5}$, 2%, 0.18 **f.** 0.74, $\frac{3}{4}$, 70%

DAILY INTERVENTION **Flexible Grouping**

Ordering Rational Numbers Place students in groups of four. Give each group 12 cards, each of which will show a fraction, decimal, or percent. Each member of the group draws one card. After the cards are drawn, have students order the cards from least to greatest. Repeat the activity several times.

228 Chapter 5 Fractions, Decimals, and Percents

TEACHING TIP

his is a basic introduction to rational numbers. tudents will further their tudy of rational numbers n Chapter 11.

Fractions, terminating and repeating decimals, and integers are called **rational numbers** because they can be written as fractions.

Key Concept — Rational Numbers

Words Rational numbers are numbers that can be written as fractions.

Symbols $\frac{a}{b}$, where a and b are integers and $b \neq 0$

Model

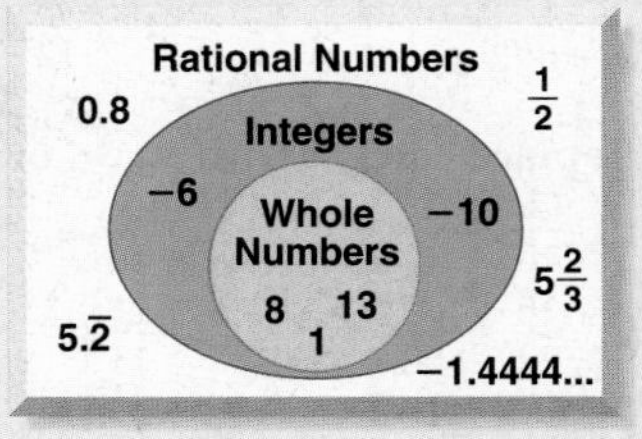

EXAMPLE Identify Rational Numbers

Standardized Test Practice

Test-Taking Tip (The Princeton Review)

Eliminate Possibilities Eliminate the possibilities that you know are incorrect. Then consider the choices that are left.

4 MULTIPLE-CHOICE TEST ITEM

Find the number that is *not* rational.

Ⓐ -8 Ⓑ $0.0\overline{9}$ Ⓒ $\frac{3}{2}$ Ⓓ $2.010010001\ldots$

Read the Test Item To find the number that is not rational, identify three rational numbers.

Solve the Test Item -8, $0.0\overline{9}$, and $\frac{3}{2}$ can be expressed as fractions. So, 2.010010001… is not rational, and the answer is D.

Skill and Concept Check

Writing Math — Exercise 2

1. **OPEN ENDED** Write two fractions whose LCD is 30. **See margin.**

2. **Which One Doesn't Belong?** Identify the ratio that does not have the same value as the other three. Explain your reasoning.

9 out of 15	0.06	60%	$\frac{3}{5}$

0.06; 0.06 equals 6% and the other ratios equal 60%.

GUIDED PRACTICE

Find the LCD for each pair of fractions.

3. $\frac{4}{5}, \frac{3}{9}$ **45**
4. $\frac{3}{8}, \frac{11}{32}$ **32**

Replace each ● with <, >, or = to make a true sentence.

5. $\frac{6}{15}$ ● $\frac{3}{8}$ >
6. $\frac{1}{6}$ ● $\frac{3}{18}$ =
7. $\frac{5}{4}$ ● $\frac{8}{7}$ >
8. $\frac{1}{8}$ ● 0.2 <

Order each set of ratios from least to greatest.

9. 60%, $\frac{31}{50}$, 0.59 **0.59, 60%, $\frac{31}{50}$**
10. $\frac{3}{20}$, 0.02, 16% **0.02, $\frac{3}{20}$, 16%**

Determine whether each number is rational. Write *yes* or *no*.

11. $\frac{1}{11}$ **yes**
12. 1.121121112… **no**

msmath2.net/extra_examples

In-Class Example

INTERACTIVE CHALKBOARD PowerPoint® Presentations

4 MULTIPLE-CHOICE TEST ITEM Find the number that is rational. **B**

A 0.12345 . . .
B 0.123123 . . .
C π
D 0.102030 . . .

3 Practice/Apply

FOLDABLES™ Follow-Up Remind students to write a question about comparing and ordering rational numbers on the front of the eighth tab of their Foldables before they begin the lesson. As students progress through the lesson, encourage them to take notes and record information that helps them answer their question under the tab.

Which One Doesn't Belong? For Exercise 2, encourage students to write all the ratios as decimals or as fractions with common denominators to identify the one that doesn't belong.

About the Exercises…

Guided Practice Key

Exercises	Examples
3–8	1
9–10	3
11–12	4

Odd/Even Assignments Exercises 13–34 are structured so that students practice the same concepts whether they are assigned odd or even problems.

Answer

1. Sample answer: $\frac{2}{15}, \frac{5}{6}$

Answers

51. Sample answer: $2\frac{3}{4}$ is the same as 2.75. Fill the cup to halfway between 2.7 and 2.8.

54. McGwire: $\frac{70}{155} = 0.452$; Sosa: $\frac{66}{159} = 0.415$; McGwire

55. McGwire: $\frac{70}{509} = 0.138$; Sosa: $\frac{66}{643} = 0.103$; McGwire

56. Sample answer: Determine whether Mark McGwire in 1998 or Barry Bonds in 2001 had a greater home run-to-game ratio; Barry Bonds, with a ratio of about 0.477

★ indicates increased difficulty

Practice and Applications

HOMEWORK HELP

For Exercises	See Examples
13–30, 36, 51–53	1
31–35, 43, 50, 54–55	2
37–42	3
44–49	4

Extra Practice
See pages 576, 600.

Find the LCD for each pair of fractions.

13. $\frac{1}{4}, \frac{3}{10}$ **20**
14. $\frac{5}{6}, \frac{7}{15}$ **30**
15. $\frac{9}{14}, \frac{2}{5}$ **70**
16. $\frac{1}{6}, \frac{3}{8}$ **24**
17. $\frac{7}{12}, \frac{13}{36}$ **36**
18. $\frac{13}{17}, \frac{3}{4}$ **68**

Replace each ● with <, >, or = to make a true sentence.

19. $\frac{7}{10}$ ● $\frac{2}{3}$ **>**
20. $\frac{5}{14}$ ● $\frac{3}{7}$ **<**
21. $\frac{4}{7}$ ● $\frac{5}{8}$ **<**
22. $\frac{2}{3}$ ● $\frac{10}{15}$ **=**
23. $\frac{16}{20}$ ● $\frac{40}{50}$ **=**
24. $\frac{11}{15}$ ● $\frac{5}{9}$ **>**
25. $\frac{9}{13}$ ● $\frac{14}{20}$ **<**
26. $\frac{17}{24}$ ● $\frac{11}{12}$ **<**
27. $\frac{7}{5}$ ● $\frac{14}{11}$ **>**
28. $2\frac{3}{4}$ ● $2\frac{2}{3}$ **>**
29. $\frac{11}{16}$ ● $\frac{7}{8}$ **<**
30. $\frac{3}{2}$ ● $1\frac{1}{2}$ **=**
31. 0.82 ● $\frac{5}{6}$ **<**
32. $\frac{9}{20}$ ● 0.45 **=**
33. $\frac{3}{5}$ ● 59% **>**
34. 40% ● $\frac{11}{25}$ **<**

35. Which average is better for a soccer goalie: 3 saves out of 4 or 7 saves out of 11? Explain. **3 saves out of 4; 3 out of 4 has an average of 0.75; 7 out of 11 has an average of about 0.64.**

36. Order the fractions from greatest to least: $\frac{7}{8}, \frac{4}{5}, \frac{22}{25}$. **$\frac{22}{25}, \frac{7}{8}, \frac{4}{5}$**

Order each set of ratios from least to greatest.

37. 0.23, 19%, $\frac{1}{5}$ **19%, $\frac{1}{5}$, 0.23**
38. $\frac{8}{10}$, 81%, 0.805 **$\frac{8}{10}$, 0.805, 81%**
39. 0.615, $\frac{5}{8}$, 62% **0.615, 62%, $\frac{5}{8}$**
40. 0.4, $\frac{1}{25}$, 0.25 **$\frac{1}{25}$, 0.25, 0.4**
41. 0.49, $\frac{49}{50}$, 0.5 **0.49, 0.5, $\frac{49}{50}$**
42. $\frac{4}{7}$, 44%, 0.47 **44%, 0.47, $\frac{4}{7}$**

43. **DARTS** Bianca and Christopher were playing darts. Bianca hit the bull's-eye 5 out of 18 times. Christopher missed the bull's-eye 4 out of 15 times. Who hit the bull's-eye a greater fraction of the time? **Christopher**

Determine whether each number is rational. Write *yes* or *no*.

44. $\frac{1}{9}$ **yes**
45. 1.141141114... **no**
46. 1.2345... **no**
47. π **no**
48. $-3\frac{4}{5}$ **yes**
49. $5.\overline{23}$ **yes**

50. **WEATHER** Refer to the table at the right that shows about how much rain falls in Albuquerque and Denver. Which city has the greater fraction of inches of rain per day? Explain. **Denver, because $\frac{15}{90} > \frac{9}{60}$.**

City	Amount of Rain (in.)	Number of Days
Albuquerque, NM	9	60
Denver, CO	15	90

Source: www.weather.com

★ 51. **BAKING** Ofelia needs $2\frac{3}{4}$ cups of water for a cake recipe. If her measuring cup only shows decimals in increments of tenths, how will Ofelia measure out the water for the recipe? **See margin.**

52. **MULTI STEP** Find one terminating decimal and one repeating decimal between $\frac{2}{3}$ and $\frac{3}{4}$. **Sample answer: 0.7, $0.\overline{71}$**

230 Chapter 5 Fractions, Decimals, and Percents

★ 53. **NUMBER SENSE** Is $1\frac{15}{16}$, $\frac{17}{8}$, or $\frac{63}{32}$ nearest to 2? Explain. **$\frac{63}{32}$; the difference of $\frac{63}{32}$ and 2 is the least.**

BASEBALL For Exercises 54–56, use the following information and the table below.
Mark McGwire of the St. Louis Cardinals and Sammy Sosa of the Chicago Cubs competed in a home run race during the 1998 baseball season. **54–55. See margin.**

54. Write the ratio of home runs to games as a decimal for both players. Who had a better chance of hitting a home run during a game?

55. Write the ratio of home runs to at-bats as a decimal for both players. Who had a better chance of hitting a home run during an at-bat?

56. **WRITE A PROBLEM** In 2001, Barry Bonds of the San Francisco Giants set the single-season home run record. He had 476 at-bats and hit 73 home runs in 153 games. Use this information and the information in Exercises 54 and 55 to write and then solve a problem involving ratios. **See margin.**

Player	Games	At-bats	Home Runs
McGwire	155	509	70
Sosa	159	643	66

Data Update Has anyone surpassed Bonds' single-season mark of 73 home runs? Visit msmath2.net/data_update to learn more.

57. **CRITICAL THINKING** How do you order fractions that have the same numerator? Explain. **Sample answer: When the numerators are the same, the fraction with the largest denominator is the least fraction.**

Standardized Test Practice and Mixed Review

58. **MULTIPLE CHOICE** Find the least fraction. **C**

Ⓐ $\frac{3}{4}$ Ⓑ $\frac{7}{9}$ Ⓒ $\frac{13}{18}$ Ⓓ $\frac{11}{15}$

59. **SHORT RESPONSE** Order the ratios from least to greatest:
$47.\overline{4}$, $\frac{4}{7}$, $47\frac{2}{5}$, 47.41, 47%. **47%, $\frac{4}{7}$, $47\frac{2}{5}$, 47.41, $47.\overline{4}$**

60. Find the LCM of 14 and 21. (Lesson 5-7) **42**

FOOD The graph shows what percent of Americans chose the given pizza toppings as their favorite. Write the percent of people who chose each topping as a decimal. (Lesson 5-6)

61. onions **0.04**

62. vegetables **0.13**

63. mushrooms **0.14**

What's On Top?

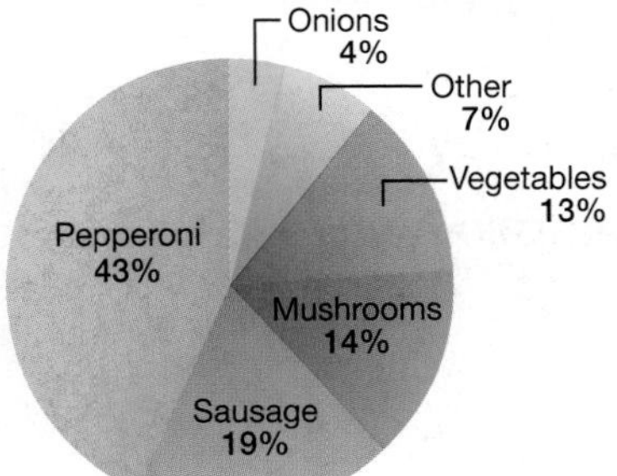

Source: Market Facts for Bolla

64. **ALGEBRA** Translate the following sentence into an algebraic equation. *Four less than two times a number is 30.* (Lesson 4-1) **$2n - 4 = 30$**

msmath2.net/self_check_quiz

4 Assess

Open-Ended Assessment

Modeling For Example 3, have students model each number using grid paper. Have them cut the models out and compare them. Have them tape the models on a piece of paper in order from least to greatest.

Study Guide and Intervention, p. 278

NAME ________ DATE ______ PERIOD ____

5-8 Study Guide and Intervention

Comparing and Ordering Rational Numbers

To compare fractions, rewrite them so they have the same denominator. The **least common denominator (LCD)** of two fractions is the LCM of their denominators.
Another way to compare fractions is to express them as decimals. Then compare the decimals.

EXAMPLE 1 Which fraction is greater, $\frac{3}{4}$ or $\frac{4}{5}$?

Method 1 Rename using the LCD.

$\frac{3}{4} = \frac{3 \times 5}{4 \times 5} = \frac{15}{20}$

$\frac{4}{5} = \frac{4 \times 4}{5 \times 4} = \frac{16}{20}$

The LCD is 20.

Because the denominators are the same, compare numerators.
Since $\frac{16}{20} > \frac{15}{20}$, then $\frac{4}{5} > \frac{3}{4}$.

Method 2 Write each fraction as a decimal. Then compare decimals.

$\frac{3}{4} = 0.75$

$\frac{4}{5} = 0.8$

Since $0.8 > 0.75$, then $\frac{4}{5} > \frac{3}{4}$.

EXERCISES

Find the LCD of each pair of fractions.

1. $\frac{1}{2}, \frac{1}{8}$ 8
2. $\frac{1}{3}, \frac{3}{4}$ 12
3. $\frac{3}{4}, \frac{7}{10}$ 20

Replace each ● with <, >, or = to make a true sentence.

4. $\frac{1}{2}$ ● $\frac{4}{9}$ >
5. $\frac{4}{5}$ ● $\frac{8}{10}$ =
6. $\frac{3}{4}$ ● $\frac{7}{8}$ <
7. $\frac{1}{2}$ ● $\frac{5}{9}$ <
8. $\frac{9}{14}$ ● $\frac{10}{17}$ >
9. $\frac{5}{7}$ ● $\frac{6}{11}$ >
10. $\frac{8}{17}$ ● $\frac{1}{2}$ <
11. $\frac{9}{10}$ ● $\frac{17}{19}$ >

Practice: Skills, p. 279

NAME ________ DATE ______ PERIOD ____

5-8 Practice: Skills

Comparing and Ordering Rational Numbers

Find the LCD of each pair of fractions.

1. $\frac{4}{7}, \frac{3}{5}$ 35
2. $\frac{5}{12}, \frac{7}{24}$ 24
3. $\frac{6}{28}, \frac{3}{7}$ 28
4. $\frac{7}{15}, \frac{1}{4}$ 60
5. $\frac{7}{11}, \frac{3}{5}$ 55
6. $\frac{5}{17}, \frac{7}{8}$ 136
7. $\frac{5}{12}, \frac{7}{10}$ 60
8. $\frac{15}{16}, \frac{1}{4}$ 16
9. $\frac{5}{8}, \frac{3}{5}$ 40

Replace each ● with <, >, or = to make a true sentence.

10. $\frac{3}{10}$ ● $\frac{2}{9}$ >
11. $\frac{3}{7}$ ● $\frac{5}{11}$ <
12. $\frac{9}{12}$ ● $\frac{3}{4}$ =
13. $\frac{12}{13}$ ● $\frac{14}{15}$ <
14. $\frac{4}{5}$ ● $\frac{5}{4}$ <
15. $\frac{17}{30}$ ● $\frac{13}{20}$ <
16. $\frac{35}{60}$ ● $\frac{49}{84}$ =
17. $3\frac{4}{11}$ ● $3\frac{7}{20}$ >
18. $1\frac{2}{3}$ ● $\frac{9}{7}$ >

Order each set of ratios from least to greatest.

19. 0.48, 0.46, $\frac{9}{20}$ — $\frac{9}{20}$, 0.46, 0.48
20. 0.99, 0.89, $\frac{7}{8}$ — $\frac{7}{8}$, 0.89, 0.99
21. $\frac{1}{4}$, 0.2, 0.4 — 0.2, $\frac{1}{4}$, 0.4

Determine whether each number is rational. Write *yes* or *no*.

22. 2.323323332... no
23. $\frac{7}{19}$ yes
24. $4.\overline{3}$ yes

Practice: Word Problems, p. 280

NAME ________ DATE ______ PERIOD ____

5-8 Practice: Word Problems

Comparing and Ordering Rational Numbers

1. **RAIN** The amount of rainfall was measured after a recent storm. The north side of town received $\frac{7}{8}$ inch of rain, and the south side received $\frac{13}{15}$ inch of rain. Which side of town received more rain from the storm? north

2. **MOVIES** Because he sees movies at his local theater so often, Delmar is being offered a discount. He can have either $\frac{1}{3}$ off his next ticket or 30% off his next ticket. Which discount should Delmar choose? Explain. $\frac{1}{3}$ off; $\frac{1}{3} > 30\%$

3. **TRACK** Willie runs the 110-meter hurdles in $17\frac{3}{5}$ seconds, and Anier runs it in $17\frac{6}{11}$ seconds. Which runner is faster? Anier

4. **FARMING** Cassie successfully harvested $\frac{7}{12}$ of her crop, and Robert successfully harvested 58% of his crop. Which person successfully harvested the larger portion of his or her crop? Cassie

5. **TRANSPORTATION** My-Lien has enough room in her truck to move 3.385 tons of gravel. Her father has asked her to move $3\frac{5}{16}$ tons. Will My-Lien be able to move all of the gravel in only one trip? Explain. Yes; $3\frac{5}{16} < 3.385$

6. **WOOD WORKING** Kishi has a bolt that is $\frac{5}{8}$ inch wide, and she drilled a hole 0.6 inch wide. Is the hole large enough to fit the bolt? Explain. No; $\frac{5}{8} > 0.6$

7. **PIZZA** In a recent pizza-eating contest, Alfonso ate $1\frac{3}{8}$ pizzas, Della ate $1\frac{3}{10}$ pizzas, and Delsin ate $1\frac{4}{9}$ pizzas. Which person won the contest? Delsin

8. **STUDYING** For a recent algebra exam, Pat studied $1\frac{8}{15}$ hours, Toni studied $1\frac{11}{20}$ hours, and Morgan studied $1\frac{9}{16}$ hours. List the students in order by who studied the most. Morgan, Toni, Pat

Vocabulary Assessment

- This alphabetical list of vocabulary terms in Chapter 5 includes a page reference where each term was introduced.
- A vocabulary review/test for Chapter 5 is available on p. 296 of the *Chapter 5 Resource Masters*.

Vocabulary PuzzleMaker

ELL The online Vocabulary PuzzleMaker improves students' mathematics vocabulary using four puzzle formats – crossword, scramble, word search using a word list, and word search using clues. Students can work on a computer screen or from a printed handout.

MindJogger Videoquizzes

ELL MindJogger Videoquizzes provide an alternative review of concepts presented in this chapter. Students work in teams in a game show format to gain points for correct answers. The questions are presented in three rounds.

Round 1 Concepts (5 questions)
Round 2 Skills (4 questions)
Round 3 Problem Solving (4 questions)

ExamView® Pro Test Bank CD-ROM

Use the networkable ExamView® Pro Test Bank CD-ROM to:

- Create **Multiple versions** of tests.
- Create **Modified** tests for **Inclusion** students with one mouse click.
- **Edit** existing questions and **Add** your own questions.
- Build tests aligned with state standards using built-in **State Curriculum Correlations.**
- Change **English** tests to **Spanish** with one mouse click and vice versa.

CHAPTER 5 Study Guide and Review

Vocabulary and Concept Check

bar notation (p. 211)	least common denominator (LCD) (p. 227)	prime number (p. 197)
common denominator (p. 227)	least common multiple (LCM) (p. 224)	ratio (p. 216)
composite number (p. 197)	multiple (p. 224)	rational numbers (p. 229)
equivalent fractions (p. 207)	percent (p. 216)	repeating decimals (p. 211)
factor tree (p. 198)	prime factorization (p. 198)	simplest form (p. 207)
greatest common factor (GCF) (p. 203)		terminating decimals (p. 211)
		Venn diagram (p. 203)

Choose the letter of the term or number that best matches each phrase.

1. a common multiple of the denominators of two or more fractions **c**
2. a fraction in which the GCF of the numerator and denominator is 1 **b**
3. a ratio that compares a number to 100 **d**
4. a whole number greater than 1 that has exactly two factors, 1 and itself **e**
5. a comparison of two numbers by division **a**

a. ratio
b. simplest form
c. common denominator
d. percent
e. prime number
f. multiple
g. bar notation

Lesson-by-Lesson Exercises and Examples

5-1 Prime Factorization (pp. 197–200)

Determine whether each number is *prime* or *composite*.

6. 45 **composite**
7. 37 **prime**
8. 91 **composite**
9. 118 **composite**

Find the prime factorization of each number.

10. 54 2×3^3
11. 128 2^7
12. 68 $2^2 \times 17$
13. 95 5×19
14. **ALGEBRA** Factor the expression $36x^2yz^3$. $2 \cdot 2 \cdot 3 \cdot 3 \cdot x \cdot x \cdot y \cdot z \cdot z \cdot z$

Example 1 **Determine whether 21 is *prime* or *composite*.**

The number 21 has factors 1, 3, 7, and 21. So, 21 is a composite number.

Example 2 **Find the prime factorization of 18.**

Use a factor tree.

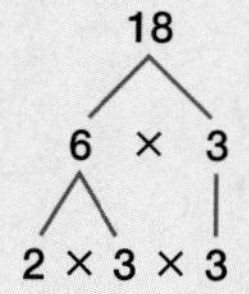

The prime factorization of 18 is 2×3^2.

msmath2.net/vocabulary_review

FOLDABLES™ Study Organizer

For more information about Foldables, see *Teaching Mathematics with Foldables.*

Have students look through the chapter to make sure they have included examples in their Foldables for each lesson or topic in the chapter.

Encourage students to refer to their Foldables while completing the Study Guide and Review and to use them in preparing for the Chapter Test.

5-2 Greatest Common Factor (pp. 203–206)

Find the GCF of each set of numbers.

15. 18, 27 **9**
16. 30, 72 **6**
17. 28, 70, 98 **14**
18. 42, 63, 105 **21**

Find the GCF of each set of algebraic expressions.

19. $18w, 54w^2y$ **$18w$**
20. $21de^2, 35d^3e^3$ **$7de^2$**

Example 3 **Find the GCF of 24 and 56.**

First, make a list of all of the factors of 24 and 56.

factors of 24: 1, 2, 3, 4, 6, 8, 12, 24
factors of 56: 1, 2, 4, 7, 8, 14, 28, 56
common factors: 1, 2, 4, 8

The GCF of 24 and 56 is 8.

5-3 Simplifying Fractions (pp. 207–209)

Write each fraction in simplest form.

21. $\frac{12}{15}$ **$\frac{4}{5}$**
22. $\frac{35}{60}$ **$\frac{7}{12}$**
23. $\frac{11}{121}$ **$\frac{1}{11}$**
24. $\frac{14}{63}$ **$\frac{2}{9}$**
25. $\frac{37}{45}$ **$\frac{37}{45}$**
26. $\frac{55}{110}$ **$\frac{1}{2}$**
27. **LIFE SCIENCE** The human body has 60,000 miles of blood vessels. The distance across the United States from east to west is 3,000 miles. Write a fraction in simplest form that compares the distance across the United States to the miles of blood vessels. **$\frac{1}{20}$**

Example 4 **Write $\frac{24}{32}$ in simplest form.**

Find the GCF of the numerator and denominator.

$24 = 2 \times 2 \times 2 \times 3$
$32 = 2 \times 2 \times 2 \times 2 \times 2$

GCF: $2 \times 2 \times 2 = 8$

$\frac{24}{32} = \frac{24 \div 8}{32 \div 8} = \frac{3}{4}$ Divide the numerator and denominator by the GCF.

So, $\frac{24}{32}$ in simplest form is $\frac{3}{4}$.

5-4 Fractions and Decimals (pp. 210–213)

Write each fraction or mixed number as a decimal. Use bar notation if the decimal is a repeating decimal.

28. $\frac{3}{4}$ **0.75**
29. $\frac{7}{8}$ **0.875**
30. $\frac{5}{9}$ **$0.\overline{5}$**
31. $4\frac{1}{3}$ **$4.\overline{3}$**
32. $6\frac{2}{5}$ **6.4**
33. $1\frac{6}{7}$ **$1.\overline{857142}$**

Write each decimal as a fraction in simplest form.

34. 0.7 **$\frac{7}{10}$**
35. 0.44 **$\frac{11}{25}$**
36. 0.05 **$\frac{1}{20}$**
37. 0.18 **$\frac{9}{50}$**
38. 0.54 **$\frac{27}{50}$**
39. 0.08 **$\frac{2}{25}$**

Example 5 **Write $\frac{3}{8}$ as a decimal.**

$$\begin{array}{r} 0.375 \\ 8\overline{)3.000} \\ \underline{24} \\ 60 \\ \underline{56} \\ 40 \\ \underline{40} \\ 0 \end{array}$$

So, $\frac{3}{8} = 0.375$.

Example 6 **Write 0.64 as a fraction in simplest form.**

$0.64 = \frac{64}{100}$ Write as a fraction with a denominator of 100.

$= \frac{16}{25}$ Simplify.

Chapter 5 Study Guide and Review 233

5-5 Fractions and Percents (pp. 216–219)

Write each fraction as a percent.

40. $\frac{32}{100}$ 32% 41. $\frac{11}{25}$ 44% 42. $\frac{47}{50}$ 94% 43. $\frac{8}{20}$ 40%

Write each percent as a fraction in simplest form.

44. 68% $\frac{17}{25}$ 45. 95% $\frac{19}{20}$ 46. 42% $\frac{21}{50}$ 47. 16% $\frac{4}{25}$

48. **FOOD** A recent survey showed that 58% of school children chose peanut butter and jelly sandwiches as their favorite lunch food. Write this percent as a fraction in simplest form. $\frac{29}{50}$

Example 7 Write $\frac{27}{50}$ as a percent.

$\frac{27}{50} = \frac{54}{100}$ Write an equivalent fraction with a denominator of 100.

$= 54\%$ Definition of percent

Example 8 Write 96% as a fraction in simplest form.

$96\% = \frac{96}{100}$ Definition of percent

$= \frac{24}{25}$ Simplify.

5-6 Percents and Decimals (pp. 220–223)

Write each percent as a decimal.

49. 48% 0.48 50. 7% 0.07 51. 12.5% 0.125 52. $75\frac{1}{4}\%$ 0.7525

Write each decimal as a percent.

53. 0.61 61% 54. 0.055 5.5% 55. 0.19 19% 56. 0.999 99.9%

57. **WEATHER** Twenty-three percent of the days last month were rainy. Write this percent as a decimal. 0.23

Example 9 Write 35% as a decimal.

$35\% = \frac{35}{100}$ Write the percent as a fraction.

$= 0.35$ Write the fraction as a decimal.

Example 10 Write 0.625 as a percent.

$0.625 = 0.625$ Multiply by 100.

$= 62.5\%$ Add the %.

5-7 Least Common Multiple (pp. 224–226)

Find the LCM of each set of numbers.

58. 9, 15 45 59. 4, 8 8

60. 16, 24 48 61. 3, 8, 12 24

62. 4, 9, 12 36 63. 15, 24, 30 120

Example 11 Find the LCM of 8 and 36.

Write each prime factorization.

$8 = 2^3$ $36 = 2^2 \times 3^2$

LCM: $2^3 \times 3^2 = 72$

The LCM of 8 and 36 is 72.

5-8 Comparing and Ordering Rational Numbers (pp. 227–231)

Replace each ● with <, >, or = to make a true sentence.

64. $\frac{3}{8}$ ● $\frac{2}{3}$ < 65. 0.45 ● $\frac{9}{20}$ =

66. $\frac{8}{9}$ ● 85% > 67. $3\frac{3}{4}$ ● $3\frac{5}{8}$ >

68. 72% ● $\frac{8}{11}$ < 69. $\frac{5}{7}$ ● $\frac{60}{84}$ =

Example 12 Replace ● with <, >, or = to make $\frac{3}{5}$ ● $\frac{5}{8}$ a true sentence.

Find equivalent fractions. The LCD is 40.

$\frac{3}{5} = \frac{3 \times 8}{5 \times 8} = \frac{24}{40}$ $\frac{5}{8} = \frac{5 \times 5}{8 \times 5} = \frac{25}{40}$

Since $\frac{24}{40} < \frac{25}{40}$, then $\frac{3}{5} < \frac{5}{8}$.

234 Chapter 5 Fractions, Decimals, and Percents

CHAPTER 5 Practice Test

Vocabulary and Concepts

1. Which of the fractions modeled at the right is greater? Explain. **See margin.**

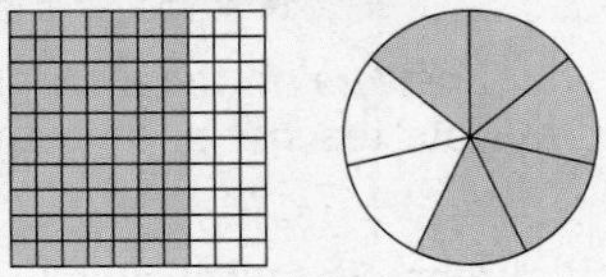

2. **Explain** how to write a percent as a decimal. **Sample answer: Divide the percent by 100 and remove the % symbol.**

Skills and Applications

3. Find the prime factorization of 72. $2^3 \times 3^2$

4. Find the GCF of 24 and 40. **8**

Write each fraction in simplest form.

5. $\frac{24}{60}$ $\frac{2}{5}$
6. $\frac{64}{72}$ $\frac{8}{9}$

Write each fraction, mixed number, or percent as a decimal. Use bar notation if the decimal is a repeating decimal.

7. $\frac{7}{9}$ $0.\overline{7}$
8. $4\frac{5}{8}$ **4.625**
9. 91% **0.91**

Write each decimal or percent as a fraction in simplest form.

10. 0.84 $\frac{21}{25}$
11. 0.006 $\frac{3}{500}$
12. 34% $\frac{17}{50}$

Write each fraction or decimal as a percent.

13. $\frac{15}{25}$ **60%**
14. 0.26 **26%**
15. 0.135 **13.5%**

16. **PAINTING** Suppose you are painting a wall that measures 10 feet by 10 feet. You have already painted 46 square feet. Write a percent for the portion of the wall that remains unpainted. **54%**

Replace each ● with $<$, $>$, or $=$ to make a true sentence.

17. $\frac{3}{5}$ ● $\frac{5}{9}$ $>$
18. $\frac{7}{12}$ ● $\frac{6}{8}$ $<$
19. $\frac{13}{20}$ ● 65% $=$

Standardized Test Practice

20. **MULTIPLE CHOICE** One type of cicada emerges from hibernation every 17 years. Another type emerges every 13 years. If both types come out of hibernation one year, in how many years would this happen again? **D**

Ⓐ 30 yr Ⓑ 68 yr Ⓒ 120 yr Ⓓ 221 yr

msmath2.net/chapter_test

CHAPTER 5 Practice Test

Assessment Options

Vocabulary Test A vocabulary review/test for Chapter 5 can be found on p. 296 of the *Chapter 5 Resource Masters.*

Chapter Tests There are six Chapter 5 Tests and an Extended Response Assessment task available in the *Chapter 5 Resource Masters.*

Chapter 5 Tests			
Form	**Type**	**Level**	**Pages**
1	MC	basic	283–284
2A	MC	average	285–286
2B	MC	average	287–288
2C	FR	average	289–290
2D	FR	average	291–292
3	FR	advanced	293–294

MC = multiple-choice questions
FR = free-response questions

Extended Response Assessment

Extended Response tasks for Chapter 5 can be found on p. 295 of the *Chapter 5 Resource Masters.* A sample scoring rubric for these tasks appears on p. A28.

Answer

1. **Sample answer: The circular model represents $\frac{5}{7}$. The 10-by-10 grid represents $\frac{7}{10}$. Since $\frac{5}{7} > \frac{7}{10}$, the circular model represents the greater fraction.**

Portfolio Suggestion

Introduction Have students look at recent newspapers and magazines. Have them take notice of the use of percentages, decimals, and fractions.

Ask Students to make a list in their portfolio describing instances in which percentages, decimals, and fractions are used in current media.

Chapter 5 Standardized Test Practice

These two pages contain practice questions in the various formats that can be found on the most frequently given standardized tests.

A student recording sheet for these two pages can be found on page A1 of the *Chapter 5 Resource Masters*.

Standardized Test Practice Student Recording Sheet, p. A1

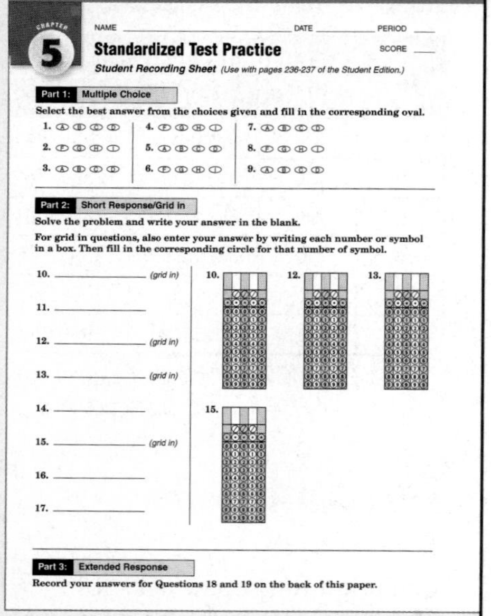
Chapter 5 Standardized Test Practice
NAME ____ DATE ____ PERIOD ____
SCORE ____
Student Recording Sheet (Use with pages 236-237 of the Student Edition.)

Part 1: Multiple Choice
Select the best answer from the choices given and fill in the corresponding oval.
1. A B C D 4. F G H I 7. A B C D
2. F G H I 5. A B C D 8. F G H I
3. A B C D 6. F G H I 9. A B C D

Part 2: Short Response/Grid In
Solve the problem and write your answer in the blank.
For grid in questions, also enter your answer by writing each number or symbol in a box. Then fill in the corresponding circle for that number of symbol.
10. ____ (grid in)
11. ____
12. ____ (grid in)
13. ____ (grid in)
14. ____
15. ____ (grid in)
16. ____
17. ____

Part 3: Extended Response
Record your answers for Questions 18 and 19 on the back of this paper.

Additional Practice

See pp. 301–302 of the *Chapter 5 Resource Masters* for additional standardized test practice.

Chapter 5 Standardized Test Practice

PART 1 Multiple Choice

Record your answers on the answer sheet provided by your teacher or on a sheet of paper.

1. Which number is divisible by 2, 3, 5, 6, 9, and 10? (Prerequisite Skill, p. 554) **A**
   - (A) 90
   - (B) 120
   - (C) 150
   - (D) 200

2. Students at Lincoln High School have collected money this year for charity. According to the graph, how much did all four classes collect this year? (Lesson 1-1) **I**

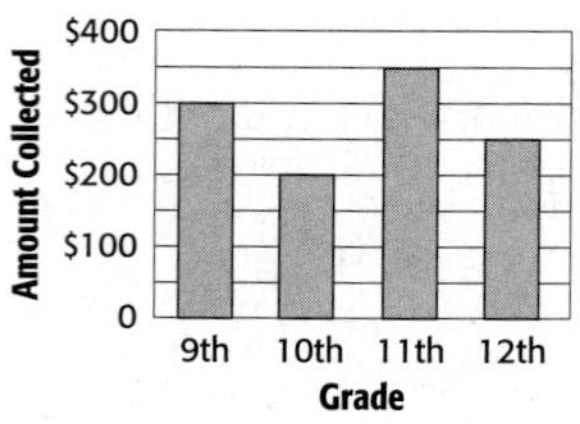

   - (F) \$950
   - (G) \$1,000
   - (H) \$1,050
   - (I) \$1,100

3. Which expression is the same as k^6? (Lesson 1-2) **D**
   - (A) $k + k + k + k + k + k$
   - (B) $6k$
   - (C) $k^3 + k^3$
   - (D) $k \times k \times k \times k \times k \times k$

TEST-TAKING TIP

Question 2 When an item includes a graph, scan the graph to see what kind of information it includes and how the information is organized. Read each answer choice and compare it with the graph to see if the information in the answer choice is correct. Eliminate any wrong answer choices.

4. Which sentence shows the Commutative Property of Multiplication? (Lesson 1-6) **H**
   - (F) $f \cdot (g \cdot h) = (f \cdot g) \cdot h$
   - (G) $q \cdot (r + s) = q \cdot r + q \cdot s$
   - (H) $x \cdot y = y \cdot x$
   - (I) $a + b = b + a$

5. In which quadrant is $(-5, -3)$ located? (Lesson 3-3) **C**
   - (A) quadrant I
   - (B) quadrant II
   - (C) quadrant III
   - (D) quadrant IV

6. Write an addition sentence modeled by the number line. (Lesson 3-4) **F**

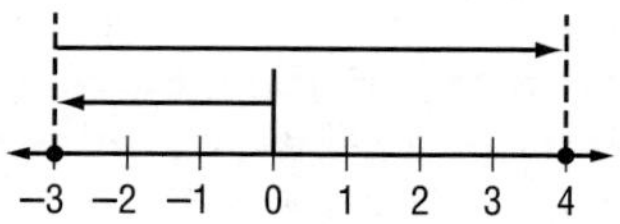

   - (F) $-3 + 7 = 4$
   - (G) $3 + 4 = 7$
   - (H) $-7 + 3 = -4$
   - (I) $3 + (-7) = -4$

7. Which of the ordered pairs is *not* a solution of $y = -3x + 7$? (Lesson 4-6) **D**
   - (A) $(-1, 10)$
   - (B) $(1, 4)$
   - (C) $(2, 1)$
   - (D) $(3, 2)$

8. Find the percent that represents the shaded area. (Lesson 5-5) **G**

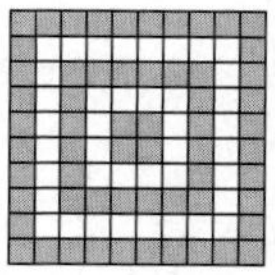

   - (F) 75%
   - (G) 60%
   - (H) 50%
   - (I) 40%

9. Mrs. Benitez hired a contractor for a kitchen renovation project. The contractor renovated 60% of the kitchen the first week. What part of the project is left to do? (Lesson 5-6) **B**
   - (A) 0.04
   - (B) 0.4
   - (C) 0.44
   - (D) 4.00

The Princeton Review

Log On **for Test Practice**

The Princeton Review offers additional test-taking tips and practice problems at their web site. Visit princetonreview.com or review.com

ExamView® Pro Test Bank CD-ROM

Special banks of standardized test questions similar to those on the TIMSS and NAEP grade 8 tests can be found on this CD-ROM.

Aligned and verified by The Princeton Review

RT 2 Short Response/Grid In

:cord your answers on the answer sheet ovided by your teacher or on a sheet of per.

. Find the eighth term in the following sequence. 1, 3, 7, 13, 21, ... (Lesson 1-7) **57**

r Questions 11 and 12, use the information low.

ie table gives weekly water-level readings of ail Creek Lake over an 8-week period. The adings indicate the number of feet above or low flood level.

Week	1	2	3	4	5	6	7	8
Water Level	−8	−8	−6	−5	−2	−5	0	3

. Find the median of the readings. (Lesson 3-2) **−5**

. Find the range of readings. (Lesson 3-5) **11**

. The equation $4.5h + 75 = 165$, where h is the number of hours, describes how long Hannah must work to earn $165. How many hours must she work to earn $165? (Lesson 4-4) **20**

. Write an equation for the function shown in the table. (Lesson 4-6) **$y = -2x + 1$**

x	−1	0	1	2
y	3	1	−1	−3

. Aaron invited 30 friends to a party, but only 22 of them were able to attend. What fraction of the friends could not attend? (Lesson 5-5) **4/15**

. A string of red, green, and blue lights blink at different rates. The red lights blink once every 4 seconds, the green lights blink once every 7 seconds, and the blue lights blink once every 9 seconds. How many seconds will go by between times when all three colors will blink at the same time? (Lesson 5-7) **252 s**

17. Tino painted a mural on the side of his shed. He used $\frac{1}{7}$ can of yellow paint, $\frac{1}{4}$ can of red paint, and $\frac{2}{9}$ can of white paint. Of which color did he use the least? (Lesson 5-8) **yellow**

PART 3 Extended Response

Record your answers on a sheet of paper. Show your work.

18. A soft drink company sells its beverages in packages of 12 and 24 plastic bottles. It added a package with 30 bottles. (Lessons 5-2 and 5-7)

    a. Suppose the company designed boxes so that all three packages have the same number of bottles in each row. What is the greatest number of bottles in each row? **6**

    b. The time it takes to seal a package is equal to 1 second per bottle. If three assembly lines containing the 12, 24, and 30 packages begin at the same time, when is the first time that all three packages are sealed at the same time? **120 s**

19. a, d. See margin.

19. The table shows the number of students who volunteered at the food bank. (Lesson 5-8)

Classroom	Number of Students Volunteering	Total Number of Students
A	7	24
B	3	8
C	6	18

    a. Model each ratio of the number of students volunteering to the total number of students on grid paper.

    b. Which classroom had the greatest number of students volunteering? **A**

    c. Which classroom had the greatest fraction of students volunteering? **B**

    d. Write each ratio as a fraction in simplest form, as a decimal, and as a percent.

Answers

19a. Sample answers: 7:24

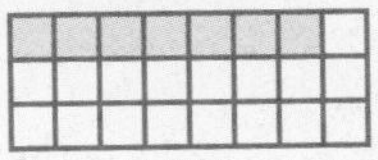

6:18

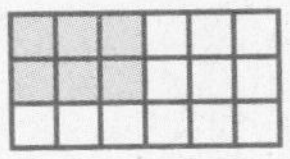

3:8

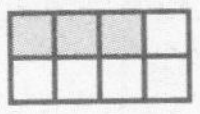

19d. $\frac{7}{24} = 0.291\overline{6} = 29.2\%$; $\frac{3}{8} = 0.375 = 37.5\%$; $\frac{6}{18} = \frac{1}{3} = 0.\overline{3} = 33.3\%$

Standardized Test Practice

Rubrics The following rubrics are sample scoring devices for short response and extended response questions.

Short Response

Points	Description
2	The student demonstrates a **thorough understanding** of the mathematics of the task. The response may contain minor flaws that do not detract from the demonstration of a thorough understanding.
1	The student has provided a response that is only **partially correct**.
0	The student has provided a **completely incorrect** solution or no response at all.

Extended Response

Points	Description
4	The student demonstrates a **thorough understanding** of the mathematics of the task. The response may contain minor flaws that do not detract from the demonstration of a thorough understanding.
3	The student demonstrates an **understanding** of the mathematics of the task. The response is essentially correct and demonstrates an essential but less than thorough understanding of the mathematics.
2	The student demonstrates only a **partial understanding** of the mathematics of the task. Although the student may have used the correct approach to a solution or may have provided a correct solution, the work lacks an essential understanding of the underlying mathematical concepts.
1	The student demonstrates a **very limited understanding** of the mathematics of the task. The response is incomplete and exhibits many flaws.
0	The student provides a **completely incorrect** solution or no response at all.

A more detailed 4-point rubric is provided on page A2 of the *Chapter 5 Resource Masters*.

Page 197, Lesson 5-1 (Hands-On Mini Lab)

1. 3 squares

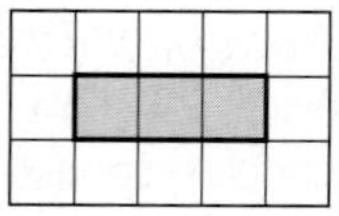

4 squares

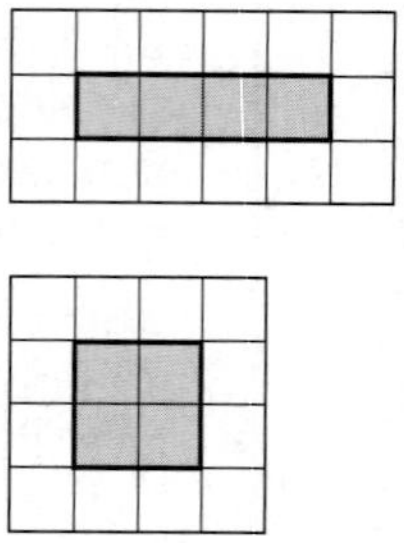

5 squares

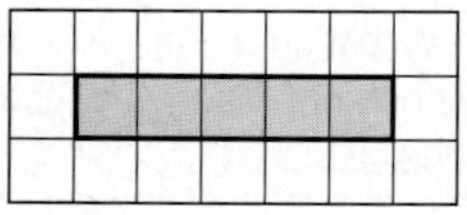

6 squares

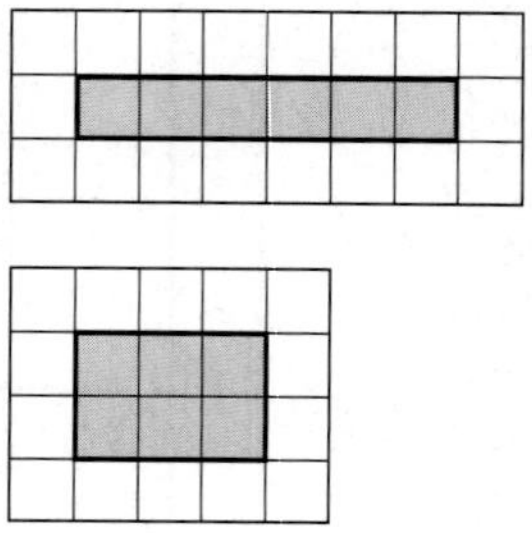

7 squares

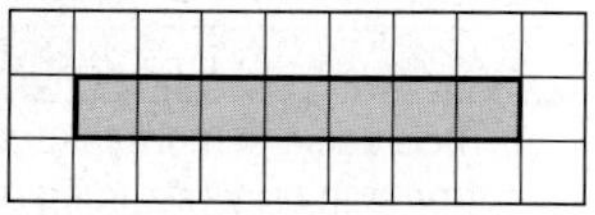

8 squares

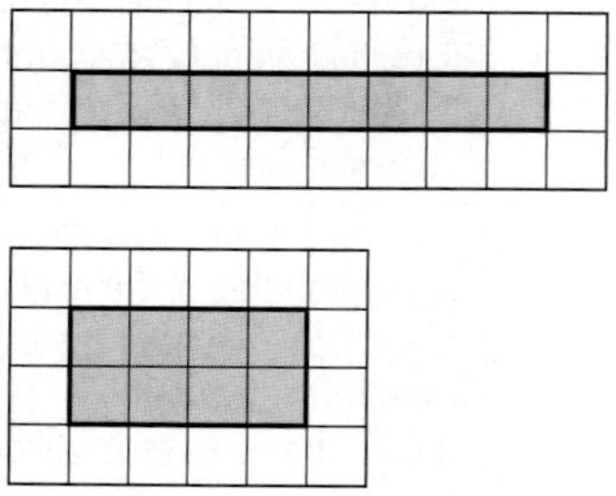

9 squares

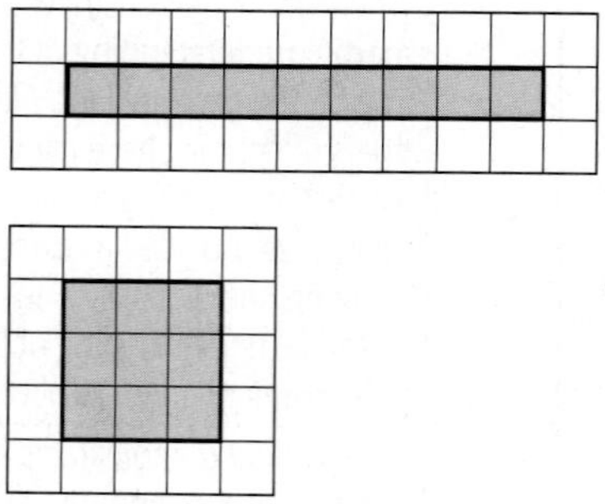

10 squares

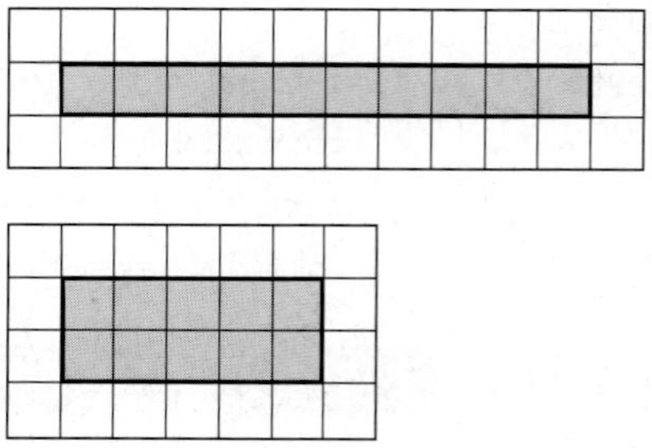

NOTES

CHAPTER 6

Chapter Overview

Applying Fractions

	LESSON OBJECTIVES	NCTM Standards	NAEP 2005	State/Local Objectives
6-1	**Estimating with Fractions** *(pp. 240–243)* • Estimate sums, differences, products, and quotients of fractions and mixed numbers.	1, 6–10	N2a, N2b, N1b, N3a, N3g, N3d	
6-2	**Adding and Subtracting Fractions** *(pp. 244–247)* • Add and subtract fractions.	1, 4, 6–10	N1b, N1d, N2c, A3b, N3a, N3g	
6-3	**Adding and Subtracting Mixed Numbers** *(pp. 248–253)* • Add and subtract mixed numbers. **6-3b Problem-Solving Strategy:** Solve problems by eliminating possibilities.	1, 4, 6–10	N1b, N1d, N2c, N3a, N3g	
6-4	**Multiplying Fractions and Mixed Numbers** *(pp. 254–257)* • Multiply fractions and mixed numbers.	1, 4, 6–10	N1b, N1d, N2c, N3a, N3g, N3d	
6-5	**Algebra: Solving Equations** *(pp. 258–261)* • Solve equations with rational number solutions.	1, 2, 6–10	N1b, N1d, N3f, N5e, A4a, A4b, N3a, N3g	
6-6	**Dividing Fractions and Mixed Numbers** *(pp. 264–266)* • Divide fractions and mixed numbers.	1, 2, 6–10	N1b, N1d, N3f, N2c, N3a, N3g, N3d	
6-7	**Measurement: Changing Customary Units** *(pp. 267–269)* • Change units in the customary system.	1, 4, 6–10	N3a, M2b	
6-8	**Geometry: Perimeter and Area** *(pp. 270–273)* • Find the perimeters and areas of figures.	1, 2, 3, 4, 6–10	M1h, M2a, A4e	
6-9	**Geometry: Circles and Circumference** *(pp. 274–277)* **6-9a Hands-On Lab:** Find a relationship between circumference and diameter. • Find the circumference of circles.	1, 2, 3, 4, 6–10	M1h, A3a, A4e, N3a, N3g, N2a, M1g, A2c	

Key to NCTM Standards:

1=Number & Operations, 2=Algebra, 3=Geometry, 4=Measurement, 5=Data Analysis & Probability, 6=Problem Solving, 7=Reasoning & Proof, 8=Communication, 9=Connections, 10=Representation

NAEP 2005 Standards: See math.glencoe.com

Pacing suggestions for the entire year can be found on page T20.

Chapter Resource Manager

Time-Saving Tool
TeacherWorks™
All-In-One Planner and Resource Center
See page T5.

FAST FILE Chapter 6 Resource Masters

Study Guide and Intervention	Practice (Skills and Word Problems)	Reading to Learn Mathematics	Enrichment	Assessment	Applications*	Parent and Student Study Guide Workbook	5-Minute Check Transparencies	Key Teaching Transparencies	MathPASS: Tutorial Plus (lessons)	Teaching Mathematics with Manipulatives	Optional Materials
303	304–305	306	307			6-1	6-1				
308	309–310	311	312			6-2	6-2	21	16, 17		ruler
313	314–315	316	317	363		6-3	6-3		18, 19		
318	319–320	321	322		CS 33	6-4	6-4	22	20, 21	62–63	
323	324–325	326	327	363, 365		6-5	6-5				colored pencils
328	329–330	331	332		CM 17–18	6-6	6-6		22, 23, 24		
333	334–335	336	337	364		6-7	6-7	23	55		
338	339–340	341	342		CS 34 RC 20	6-8	6-8			64	
343	344–345	346	347	349–362, 364, 366–374	CS 35, 36	6-9	6-9	24	51	65	ruler, measuring tape, circular objects

*** Key to Abbreviations:** *CS = Calculator and Spreadsheet Masters, RC = Real-Life Career Masters, CM = Classroom Math Games (online), SM = Science and Math Lab Manual (online)*

Interactive Chalkboard

PowerPoint® Presentations

This CD-ROM is a customizable Microsoft® PowerPoint® presentation that includes:

- In-Class Examples
- Links to Online Study Tools
- 5-Minute Check Transparencies
- Bellringers

Math Skills Maintenance

Use the Math Skills Maintenance Masters to keep your students' basic math skills fresh. Students can review skills one at a time or in a combined format. The Masters are also available online at msmath2.net/maintenance.

CHAPTER 6

Mathematical Content and Teaching Strategies

Continuity of Instruction

Building on Prior Knowledge

In Chapter 3, students added, subtracted, multiplied, and divided integers. Students solved equations involving integers in Chapter 4. In the previous chapter, they developed skills in working with fractions, including simplifying, finding the greatest common factor and least common multiple, comparing fractions, and relating fractions to decimals and percents. Throughout the previous chapters, students have learned problem-solving strategies that can be applied to situations involving various types of numbers, including fractions.

Learning in This Chapter

Students will estimate fractional quantities, sums, differences, products, and quotients based on their understanding of fractions. They will use their knowledge of integers and working with fractions to add, subtract, multiply, and divide fractions and mixed numbers. Students will learn about customary units of measure and how to convert among these units. Finally, they will learn about area, perimeter, and circumference, and to apply their knowledge to calculate these quantities.

Making Future Connections

Fractions are numbers that are all around us in everyday life. Acquiring competency in adding, subtracting, multiplying, and dividing fractions and solving equations with fractions is essential for problem solving and making accurate calculations in all areas of mathematics. Changing units in the customary system is fundamental for everyday problem solving. Knowledge of area, perimeter, and circumference will be increasingly emphasized as students move through geometry, trigonometry, and calculus courses.

Mathematical Background

6-1 Estimating with Fractions

Estimating with fractions is a fundamental skill that allows students to determine the accuracy of calculated answers. Estimation is also very useful in situations where a timely and reasonable, yet inexact, answer is required. In this lesson, students learn to estimate sums, differences, products, and quotients of fractions and mixed numbers. These concepts can be modeled using a number line.

6-2 Adding and Subtracting Fractions

In this lesson, students build on their knowledge of least common denominators and addition and subtraction of integers to add and subtract fractions. The lesson begins with fractions with common denominators, and then continues with those involving finding the LCD. Students use prior knowledge to reduce answers to simplest form. Concepts in this lesson can be modeled using rulers and algebra tiles.

6-3 Adding and Subtracting Mixed Numbers

In this lesson, students add and subtract mixed numbers. They use their skills in renaming fractions to enable subtraction when the second number's fraction is larger than the first number's fraction.

6-4 Multiplying Fractions and Mixed Numbers

When multiplying fractions and mixed numbers, students apply their knowledge of the multiplication of integers to multiply, respectively, the integer and fractional parts of a number. They carry out the same task by renaming mixed numbers as fractions. Students write answers in simplest form by simplifying before multiplying or by reducing the final answers.

Team Teaching

Inform the other teachers on your team that your classes are studying multiplication of fractions and mixed numbers. Suggestions for curriculum integration are:

Art: multiplication of fractions when determining scale
Home Economics: halving, doubling, tripling, and so on, of measures of ingredients in recipes to provide an appropriate number of servings
Social Studies: fractional representations of ethnic heritage

Teacher to Teacher

Joyce Felton
Jackson Middle School
Grand Prairie, TX

"I help students remember how to divide fractions by using this rhyme. *Dividing fractions is easy as pie. Flip the second number and multiply!*"

Joyce Felton

6-5 Algebra: Solving Equations

In this lesson, students combine knowledge of operations on fractions with skills in solving equations. *Multiplicative inverses,* also known as *reciprocals,* are two numbers whose product is 1. Students learn to find multiplicative inverses, which multiply both sides of an equation to eliminate fractional coefficients. This method applies the Multiplication Property of Equality.

6-6 Dividing Fractions and Mixed Numbers

In this lesson, students find multiplicative inverses and use the relationship between division and multiplication to divide fractions and mixed numbers. Mixed numbers are renamed as fractions before dividing. Then, division of fractions is carried out by multiplying the dividend by the reciprocal of the divisor. Modeling of division can be done using algebra tiles or rulers.

McGraw-Hill Professional Development

Mathematics, YES! is a self-paced, web-based professional development program. Units are based on major mathematical concepts and are standards-based. Actual classroom videos provide a springboard for discussion and activities. Units appropriate for this chapter:

- Unit 4, Fractions, Decimals, and Percentages
- Unit 7, Solving Equations and Inequalities
- Unit 8, Geometry and Measurement of Plane Figures

6-7 Measurement: Changing Customary Units

Customary units are standard U.S. units of measurement based on British Imperial units. They include the pound, ounce, ton, cup, pint, quart, gallon, inch, foot, yard, and mile. Changing units in the customary system is a skill that is necessary for solving problems involving different units for the same system and category of measure.

6-8 Geometry: Perimeter and Area

Perimeter is the distance around a geometric figure, and *area* is the measure of the surface enclosed by a figure. In this lesson, students learn to find both area and perimeter by using formulas. The formula for perimeter, $P = 2\ell + 2w$, and for area, $A = \ell w$, are used to solve problems in geometry as well as in real-life applications.

6-9 Geometry: Circles and Circumference

In this lesson, students learn that the relationship between the circumference and the diameter, which is the distance across a circle through its center, is given in the following way: $\frac{C}{d} = \pi = 3.1415926\ldots$. This lesson also relates circumference to the radius, which is the distance from the center to any point on a circle. The formulas for circumference are given as $C = \pi d$ and $C = 2\pi r$.

Key Concepts

Additional mathematical information and teaching notes are available at msmath2.com/key_concepts.

The section appropriate for the chapter is:

- Adding and Subtracting Fractions

CHAPTER 6

DAILY INTERVENTION and Assessment

	Type	Student Edition	Teacher Resources	Technology/Internet
INTERVENTION	Ongoing	Prerequisite Skills, pp. 239, 243, 251, 257, 261, 266, 269, 273 Mid-Chapter Practice Test, p. 262	5-Minute Check Transparencies Quizzes 1–4, *CRM*, pp. 263–264 Mid-Chapter Test, *CRM*, p. 265 Study Guide and Intervention, *CRM*, pp. 303, 308, 313, 318, 323, 328, 333, 338, 343	MathPASS: Tutorial Plus msmath2.net/self_check_quiz msmath2.net/extra_examples
	Mixed Review	pp. 243, 247, 251, 257, 261, 266, 269, 273, 277	Cumulative Review, *CRM*, p. 366	
	Error Analysis	Find the Error, pp. 246, 250, 276	Find the Error, *TWE*, pp. 246, 250, 276 Unlocking Misconceptions, *TWE*, pp. 259, 275 Tips for New Teachers, *TWE*, p. 253	
	Standardized Test Practice	pp. 243, 247, 251, 253, 257, 259, 261, 262, 266, 269, 273, 277, 281, 282–283	Standardized Test Practice, *CRM*, pp. 367–368	Standardized Test Practice CD-ROM msmath2.net/standardized_test
ASSESSMENT	Open-Ended Assessment	Writing Math, pp. 242, 246, 250, 256, 260, 266, 268, 272, 274, 276 Open-Ended, pp. 242, 246, 250, 256, 260, 266, 268, 272, 276 Standardized Test, p. 283	Modeling: *TWE*, pp. 253, 266 Speaking: *TWE*, pp. 251, 269 Writing: *TWE*, pp. 243, 273 Extended Response Assessment, *CRM*, p. 361	
	Chapter Assessment	Study Guide, pp. 278–280 Practice Test, p. 281	Multiple-Choice Tests (Forms 1, 2A, 2B), *CRM*, pp. 349–354 Free-Response Tests (Forms 2C, 2D, 3), *CRM*, pp. 355–360 Vocabulary Test/Review, *CRM*, p. 362	ExamView® Pro Test Bank (see below) MindJogger Videoquizzes msmath2.net/vocabulary_review msmath2.net/chapter_test

*** Key to Abbreviations:** *TWE* = Teacher Wraparound Edition, *CRM* = Chapter Resource Masters

Additional Intervention

Skills Intervention for Middle School Mathematics workbook and CD-ROM includes 88 skills with pretests, tutorials, guided practice, and posttests for diagnosis and remediation.

Guide to Daily Intervention booklet details each intervention resource available and guides teachers in using them as intervention tools everyday to assist students who may be having difficulty.

ExamView® Pro

Use ExamView® Pro Test Bank CD-ROM to:

- **Create Multiple versions of tests.**
- **Create Modified tests for Inclusion students with one mouse click.**
- **Edit existing questions and Add your own questions.**
- **Build tests aligned with state standards using built-in State Curriculum Correlations.**
- **Change English tests to Spanish with one mouse click and vice versa.**

Reading and Writing Mathematics

- **Student Edition**

Foldables Study Organizer, p. 239

Concept Check questions at the beginning of each exercise set require students to verbalize and write about what they have learned in the lesson.

- **Open Ended,** pp. 242, 246, 250, 256, 260, 266, 268, 272, 276
- **Find the Error,** pp. 246, 250, 276
- **Which One Doesn't Belong?,** pp. 256, 260, 268

WebQuest, p. 277

- **Teacher Wraparound Edition**

Foldables Follow-Up, pp. 246, 250, 256, 262

Writing Activities, pp. 243, 273

Differentiated Instruction, (Verbal/Linguistic), p. 265

ELL Resources, *see page T9 for a complete listing.*

- **Additional Resources**

Vocabulary Builder, *Chapter 6 Resource Masters*

Reading to Learn Mathematics, *Chapter 6 Resource Masters*

Vocabulary PuzzleMaker (online)

Teaching Mathematics with Foldables

Reading and Writing in the Mathematics Classroom

WebQuests, Projects, and Interdisciplinary Investigations

For detailed information on these resources, see pp. T6–T7.

- **Recommended Outside Reading for Students**

Nonfiction

- Swetz, F. *Learning Activities from the History of Mathematics.* J Weston Walch, © 1993.

  Biographies of 23 mathematicians from many centuries and cultures are featured in this book, along with study questions. The book investigates a variety of historical problems, including the geometry concepts studied in this chapter—perimeter, area, and circumference.

Fiction

- Neuschwander, Cindy. *Sir Cumference and the Dragon of Pi.* Charlesbridge Publishing, © 1999.

  This book takes readers on an adventure as young boy, Radius, gives his father, Sir Cumference, the wrong potion to cure a bellyache. Radius must solve the riddle of the ratio of a circle's circumference to its diameter to get the dosage to cure his father. Written as a fantasy, with a dragon as a main character, the book introduces the concept of pi.

PROJECT CRISS[SM] Study Skill

Taking good notes will help students become actively involved in the learning process. For each lesson, have students read and then write notes about the topic. You may wish to show them the sample notes for Lesson 6-1 at the right to use as a guide. Afterward, allow class time for students to discuss their notes. Encourage students to talk about the procedures used in the lesson for solving the problems and how they addressed those procedures in their notes.

Lesson 6-1

Estimate with mixed numbers

- round to nearest whole number
- Example: $7\frac{1}{5} + 5\frac{7}{10} \approx 7 + 6$ or 13

Estimate with fractions

- round to 0, $\frac{1}{2}$, or 1
- Example: $\frac{2}{3} + \frac{1}{5} \approx 1 + 0$ or 1

CReating Independence through Student-owned Strategies

For more information on Project CRISS, see page T9.

CHAPTER 6 Making Connections

School-To-Home

Parent and Student Study Guide

Parents and students may work together to reinforce the concepts and skills of this chapter. These workbooks are available in English and Spanish. A sample lesson is shown below. Also available at msmath2.net/parent_student.

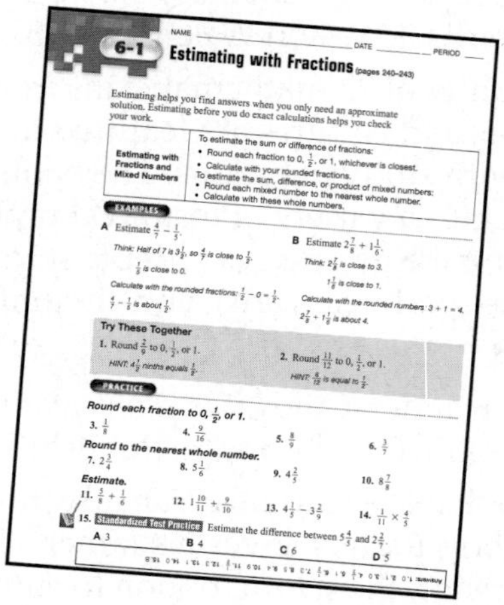

Family Letters and Activities

Send a letter and activity for this chapter home to encourage family members to become active participants in their students' learning. (English: *Chapter 6 Resource Masters*, pp. ix–x; Spanish: log on to msmath2.net/family_letters)

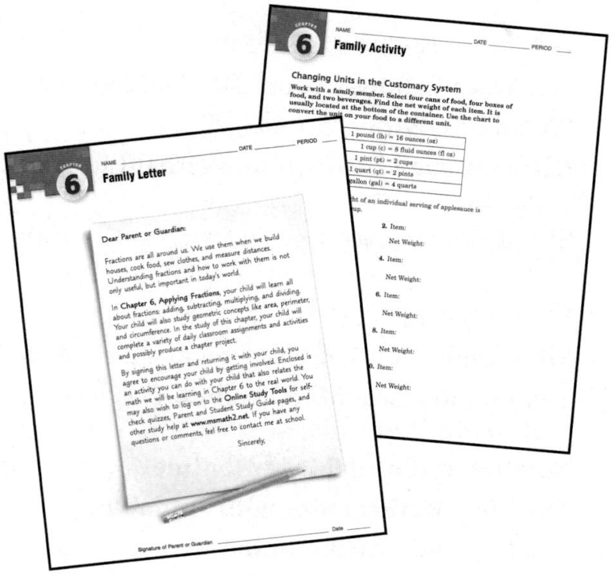

English Language Learner ELL

Lesson 6-5

Language Experience Approach to Illustrations

Show students several examples of taking the reciprocal of a number. Have them come up with their own definition for multiplicative inverse or reciprocal using the proper terminology, such as *numerator, denominator,* and *inverse.*

Lesson 6-7

Using Multisensory Activities

Bring a scale and measuring cup to class and have students measure objects. For liquid measures, have students pour liquid into a measuring cup and tell the amount of liquid. Also, ask students what unit of measure would best be used to measure different quantities.

Lesson 6-9a

Using Manipulatives

From the activity, students will discover that the circumference divided by the diameter is equivalent to pi. Ask or show students how to solve the equation for the circumference. Ask students to show each algebraic step involved. It is also important to point out the relationship between the radius and diameter and how $C = \pi d$ is equivalent to $C = 2\pi r$.

MathScape™

This project was supported, in part, by the **National Science Foundation**

MathScape is a mathematics curriculum for grades 6–8 developed by the Seeing and Thinking Mathematically Project at the Education Development Center.

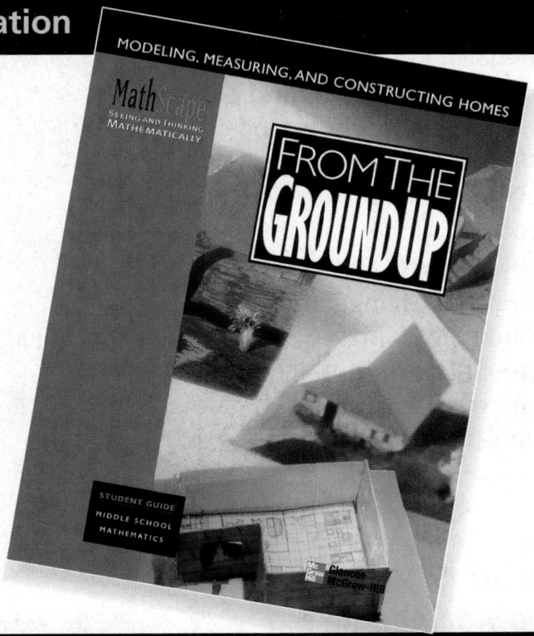

Modeling, Measuring, and Constructing Homes

From the Ground Up

Summary Students will explore area and perimeter of triangles, squares, rectangles, trapezoids, and irregular shapes.

USA TODAY Education Online Lesson Planner

USA TODAY® Education's Online site offers resources and interactive features connected to each day's newspaper. *Experience Today,* USA TODAY's® daily lesson plan, is available on the site and delivered daily to subscribers. This plan provides instruction for integrating USA TODAY® graphics and key editorial features into your mathematics classroom. Log on to **education.usatoday.com**.

USA TODAY Snapshots® in Chapter 6	
Lesson	**Snapshot**
6-5	Villanova finds indoor track success

Quick Review Math Handbook

is Glencoe's mathematical handbook for students. The Hot Words section includes a glossary of terms while the Hot Topics section consists of explanations of key mathematical concepts. An exercise set is included to check students' understanding of the concepts. This valuable resource can be used as a reference in the classroom or for home study.

Lesson	Hot Topics Section	Lesson	Hot Topics Section
6-1	2-3	**6-5**	2-4, 6-4
6-2	1-4, 2-3	**6-7**	8-1
6-3	1-4, 2-3, 2-4	**6-8**	7-4, 7-5, 8-3
6-4,6-6	2-4	**6-9**	7-8

CHAPTER 6 Notes

Key Chapter Objectives

Lesson 6-1 Estimate with fractions and mixed numbers.

Lessons 6-2, 6-3, 6-4, and 6-6 Add, subtract, multiply, and divide fractions and mixed numbers.

Lesson 6-5 Solve equations with rational number solutions.

Lesson 6-7 Change units in the customary system.

Lessons 6-8 and 6-9 Find perimeters, areas, and circumferences.

"What do Ferris wheels have to do with math?"

Point out to students that this is only one of many applications of the math that they are studying. Others are provided in the introduction to each lesson.

Vocabulary Builder, pp. vii–viii ELL

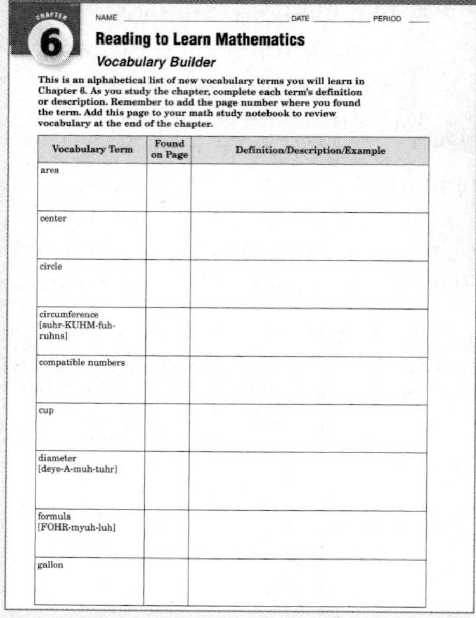
CHAPTER 6 NAME ________ DATE ________ PERIOD ____

Reading to Learn Mathematics

Vocabulary Builder

This is an alphabetical list of new vocabulary terms you will learn in Chapter 6. As you study the chapter, complete each term's definition or description. Remember to add the page number where you found the term. Add this page to your math study notebook to review vocabulary at the end of the chapter.

Vocabulary Term	Found on Page	Definition/Description/Example
area		
center		
circle		
circumference [suhr-KUHM-fuh-ruhns]		
compatible numbers		
cup		
diameter [deye-A-muh-tuhr]		
formula [FOHR-myuh-luh]		
gallon		

CHAPTER 6 Applying Fractions

"What do Ferris wheels have to do with math?"

A Ferris wheel follows a circular path. To find the distance that you travel when you go one time around a Ferris wheel, you can use the formula $C = \pi d$, where π is approximately $\frac{22}{7}$, or 3.14. In geometry, you will use formulas to solve many real-life problems.

You will solve problems about Ferris wheels in Lesson 6-9.

238 Chapter 6 Applying Fractions

Vocabulary Builder

For a thorough vocabulary list with pronunciations of new words, give students the Vocabulary Builder worksheets from the *Chapter 6 Resource Masters.* Encourage them to complete the definition of each term as they progress through the chapter.

GETTING STARTED

Take this quiz to see whether you are ready to begin Chapter 6. Refer to the lesson or page number in parentheses if you need more review.

Vocabulary Review

Choose the correct term to complete each sentence.

1. The Division (Identity, Property) of Equality states that if you divide each side of an equation by the same nonzero number, the two sides remain equal. (Lesson 4-3)
2. The (GCF, LCD) of 12 and 16 is 4. (Lessons 5-2 and 5-7)

Prerequisite Skills

Find the LCD of each pair of fractions. (Lesson 5-8)

3. $\frac{5}{7}, \frac{3}{5}$ **35**
4. $\frac{1}{2}, \frac{4}{9}$ **18**
5. $\frac{8}{15}, \frac{1}{6}$ **30**
6. $\frac{3}{4}, \frac{7}{10}$ **20**

Multiply or divide. (Pages 560, 562)

7. 1.8×12 **21.6**
8. $99 \div 12$ **8.25**
9. $83 \div 100$ **0.83**
10. 4.6×0.3 **1.38**
11. 15×0.5 **7.5**
12. $216 \div 16$ **13.5**

Complete to show equivalent mixed numbers. (Page 563)

13. $3\frac{1}{5} = 2\frac{\blacksquare}{5}$ $2\frac{6}{5}$
14. $9\frac{2}{3} = \blacksquare\frac{5}{3}$ $8\frac{5}{3}$
15. $6\frac{1}{4} = 5\frac{\blacksquare}{4}$ $5\frac{5}{4}$
16. $8\frac{6}{7} = 7\frac{\blacksquare}{7}$ $7\frac{13}{7}$

Write each mixed number as an improper fraction. (Page 563)

17. $10\frac{3}{4}$ $\frac{43}{4}$
18. $1\frac{7}{8}$ $\frac{15}{8}$
19. $4\frac{2}{5}$ $\frac{22}{5}$
20. $7\frac{2}{9}$ $\frac{65}{9}$

FOLDABLES™ Study Organizer

Fractions Make this Foldable to help you organize your notes. Begin with a sheet of $8\frac{1}{2}''$ by 11″ paper, four index cards, and glue.

STEP 1 **Fold**
Fold the paper in half widthwise.

STEP 2 **Open and Fold Again**
Open and fold along the length about $2\frac{1}{2}''$ from the bottom.

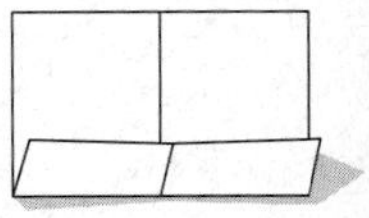

STEP 3 **Glue**
Glue the edges on either side to form two pockets.

STEP 4 **Label**
Label the pockets *Fractions* and *Mixed Numbers*, respectively. Place two index cards in each pocket.

Reading and Writing As you read and study the chapter, write the rules for performing operations on fractions and mixed numbers. Show the rules by giving real-life examples.

GETTING STARTED

This section provides a review of the basic concepts needed before beginning Chapter 6. Page references are included for additional student help.

Prerequisite Skills review skills needed in this chapter.

Exercises	Needed for Lesson	Skill
3–6	6-2	Find the Least Common Denominator
7–12	6-5	Multiply or Divide Decimals
13–16	6-3	Show Equivalent Mixed Numbers
17–20	6-4	Write Mixed Numbers as Improper Fractions

For more prerequisite skills, see:

- Getting Ready for the Next Lesson, at the end of each lesson
- Prerequisite Skills in the Student Handbook on pp. 554–563
- *Math Skills Maintenance Masters*

FOLDABLES™ Study Organizer

For more information about Foldables, see *Teaching Mathematics with Foldables.*

Use this Foldable study guide for student writing about fractions and mixed numbers. Have students make a Foldable and label the two pockets as illustrated. As students work their way through the nine lessons in this chapter, have them record what they learn about fractions and mixed numbers on study cards and sort them into the appropriate pocket of their Foldables. Encourage students to review what they have learned by reading their study cards.

6-1 Lesson Notes

Lesson Resources

Chapter 6 Resource Masters
- Study Guide and Intervention, p. 303
- Practice: Skills, p. 304
- Practice: Word Problems, p. 305
- Reading to Learn Mathematics, p. 306
- Enrichment, p. 307

Parent and Student Study Guide Workbook, 6-1

FAST FILE

1 Focus

INTERACTIVE CHALKBOARD
PowerPoint® Presentations

5-Minute Check 6-1
Use as a quiz or review of Chapter 5. (Also available as a transparency)

Bellringer
Problem Solving Ask students to suggest ways to solve the following problem. *Chloe celebrated her thirteenth birthday with a big pizza party. Afterward, there were $1\frac{1}{3}$ pepperoni pizzas, $\frac{1}{2}$ of a mushroom pizza, $\frac{3}{4}$ of a cheese pizza, and $\frac{2}{3}$ of a veggie pizza left. Assuming that all of the original pizzas were the same size, about how many rearranged "whole pizzas" are left over for Chloe's family?*
(Continued in bottom margin.)

Reading to Learn Mathematics, p. 306 **ELL**

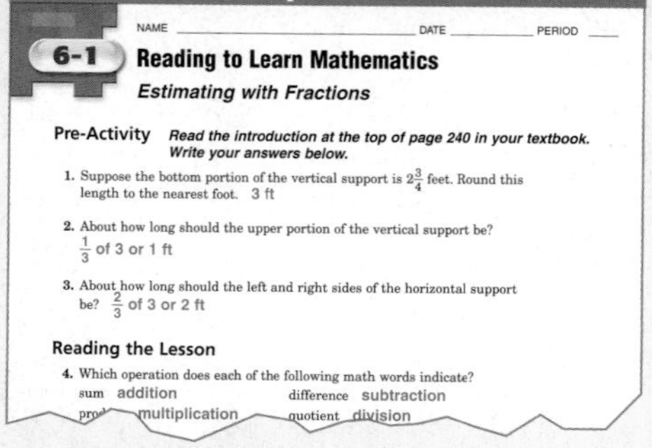
NAME ______ DATE ______ PERIOD ______

6-1 Reading to Learn Mathematics
Estimating with Fractions

Pre-Activity *Read the introduction at the top of page 240 in your textbook. Write your answers below.*

1. Suppose the bottom portion of the vertical support is $2\frac{3}{4}$ feet. Round this length to the nearest foot. 3 ft
2. About how long should the upper portion of the vertical support be? $\frac{1}{3}$ of 3 or 1 ft
3. About how long should the left and right sides of the horizontal support be? $\frac{2}{3}$ of 3 or 2 ft

Reading the Lesson

4. Which operation does each of the following math words indicate?
sum addition   difference subtraction
pro... multiplication   quotient division

6-1 Estimating with Fractions

What You'll Learn
Estimate sums, differences, products, and quotients of fractions and mixed numbers.

NEW Vocabulary
compatible numbers

MATH Symbols
$\approx$ is approximately equal to

WHEN am I ever going to use this?

KITES For a kite to have balance while flying, the left and right sides of the horizontal support must each be $\frac{2}{3}$ as long as the bottom of the vertical support. Also, the top must be $\frac{1}{3}$ as long as the bottom portion.

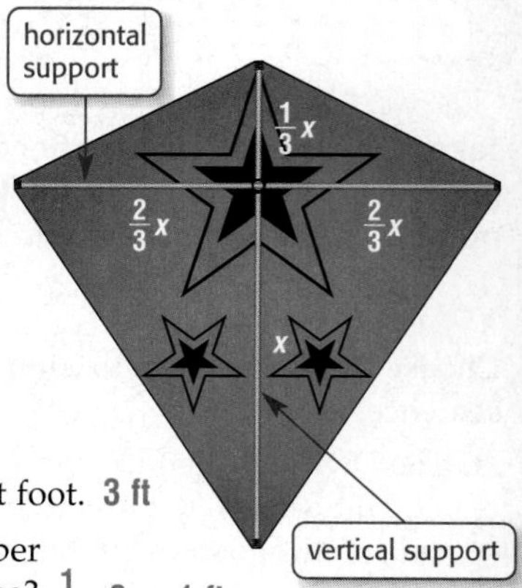

1. Suppose the bottom portion of the vertical support is $2\frac{3}{4}$ feet. Round this length to the nearest foot. **3 ft**
2. About how long should the upper portion of the vertical support be? **$\frac{1}{3}\cdot 3$ or 1 ft**
3. About how long should the left and right sides of the horizontal support be? **$\frac{2}{3}\cdot 3$ or 2 ft**

To estimate the sum, difference, product, or quotient of mixed numbers, round the mixed numbers to the nearest whole number.

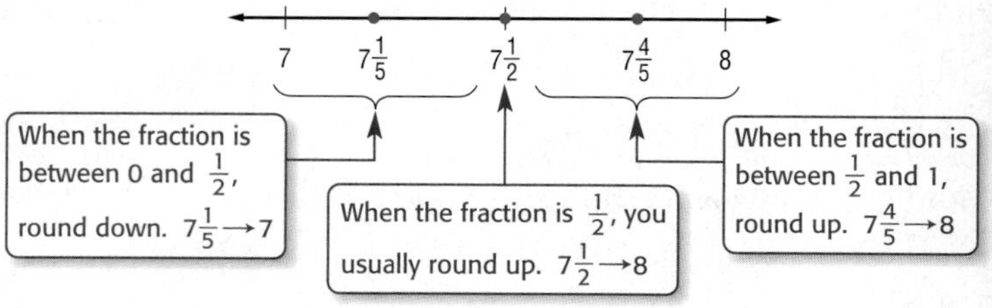

STUDY TIP

Front-End Estimation
Add the integers, estimate the sum of the fractions, and find the total sum.
$3\frac{2}{3} + 5\frac{1}{6}$
$\approx 8 + 1$
≈ 9

EXAMPLES Estimate with Mixed Numbers

Estimate.

1 $3\frac{2}{3} + 5\frac{1}{6}$

$3\frac{2}{3} + 5\frac{1}{6} \rightarrow 4 + 5 = 9$

The sum is *about* 9.

2 $6\frac{2}{5} \times 1\frac{7}{8}$

$6\frac{2}{5} \times 1\frac{7}{8} \rightarrow 6 \times 2 = 12$

The product is *about* 12.

Your Turn **Estimate.** **a–c. Sample answers are given.**

a. $2\frac{1}{5} + 3\frac{1}{2}$ **$2 + 4 = 6$**

b. $4\frac{3}{8} \times 5\frac{1}{4}$ **$4 \times 5 = 20$**

c. $8\frac{7}{9} \div 2\frac{3}{4}$ **$9 \div 3 = 3$**

about 3; Some students may suggest converting the fractions to decimals, then adding. Others may suggest using fraction strips or number lines to add the fractions. Lead the students to see that $1\frac{1}{3}$ can be added mentally to $\frac{2}{3}$ to make 2, and $\frac{1}{2} + \frac{3}{4}$ is a little more than 1.

To estimate the sum, difference, product, or quotient of fractions, round each fraction to 0, $\frac{1}{2}$, or 1, whichever is closest. Number lines and fraction models can help you decide how to round.

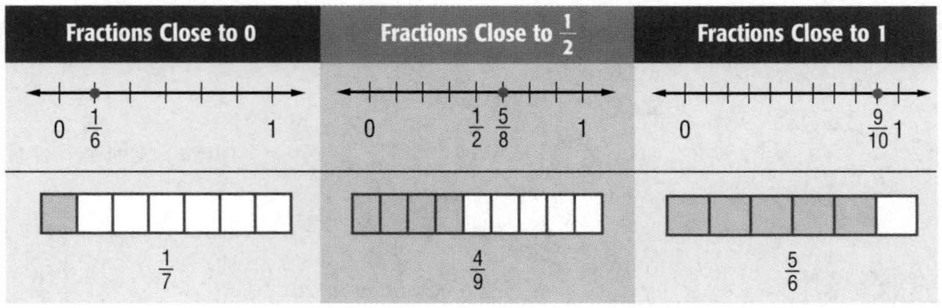

EXAMPLES Estimate with Fractions

Estimate.

3 $\frac{1}{8} + \frac{2}{3}$

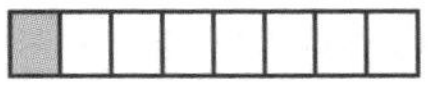

$\frac{1}{8}$ is about 0.

$\frac{2}{3}$ is about $\frac{1}{2}$.

$\frac{1}{8} + \frac{2}{3} \rightarrow 0 + \frac{1}{2} = \frac{1}{2}$ The sum is *about* $\frac{1}{2}$.

4 $\frac{6}{7} - \frac{7}{10}$

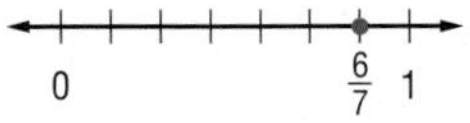

$\frac{6}{7}$ is about 1.

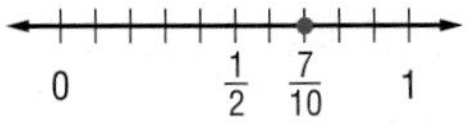

$\frac{7}{10}$ is about $\frac{1}{2}$.

$\frac{6}{7} - \frac{7}{10} \rightarrow 1 - \frac{1}{2} = \frac{1}{2}$ The difference is *about* $\frac{1}{2}$.

5 $\frac{8}{9} \div \frac{5}{6}$

$\frac{8}{9} \div \frac{5}{6} \rightarrow 1 \div 1 = 1$ $\frac{8}{9}$ is about 1, and $\frac{5}{6}$ is about 1.

The quotient is *about* 1.

Your Turn Estimate. **d–f. Sample answers are given.**

d. $\frac{4}{5} + \frac{2}{7}$ $1 + 0 = 1$ e. $\frac{5}{8} - \frac{3}{7}$ $\frac{1}{2} - \frac{1}{2} = 0$ f. $\frac{3}{5} \times \frac{11}{12}$ $\frac{1}{2} \times 1 = \frac{1}{2}$

TEACHING TIP

/hen rounding fractions, and $\frac{3}{4}$ may be rounded $\frac{1}{2}$. When rounding ixed numbers that ontain $\frac{2}{3}$ or $\frac{3}{4}$, the umbers are usually ounded to the nearest teger.

STUDY TIP

Fractions When the numerator and the denominator of a fraction are very close in value, such as $\frac{6}{7}$ or $\frac{8}{9}$, the fraction is close to 1.

msmath2.net/extra_examples

2 Teach

In-Class Examples

INTERACTIVE CHALKBOARD PowerPoint® Presentations

Estimate.
1–5. Sample answers are given.

1 $5\frac{1}{4} + 3\frac{5}{8}$ $5 + 4 = 9$

2 $7\frac{3}{4} \times 1\frac{7}{8}$ $8 \times 2 = 16$

3 $\frac{1}{3} + \frac{4}{7}$ $\frac{1}{2} + \frac{1}{2} = 1$

4 $\frac{5}{8} - \frac{1}{4}$ $\frac{1}{2} - 0 = \frac{1}{2}$

5 $\frac{5}{6} \div \frac{3}{5}$ $1 \div \frac{1}{2} = 2$

Tips for New Teachers

Teaching Strategies For problems involving fractions, suggest that students use models to help them better visualize the sizes of the fractions. Useful tools include number lines, fraction strips, and fraction models. You may wish to provide students with fraction strips made from scrap paper. Show them how to shade or fold the strips to represent fractional portions.

DAILY INTERVENTION **Differentiated Instruction**

Logical Have students draw two columns on a sheet of paper. They should title the chart "Estimation," and label the columns "OK" and "Not OK." Have students list situations when estimating mixed numbers makes sense and situations when it does not make sense. For example, when making a loaf of bread, estimating how much flour to buy makes sense; estimating how much flour to mix in the bowl does not.

In-Class Examples

INTERACTIVE CHALKBOARD PowerPoint® Presentations

Estimate. Sample answers are given.

6 $\frac{3}{4} \times 21$ $\frac{3}{4} \times 20 = 15$

7 $15\frac{8}{9} \div 3\frac{2}{5}$ $16 \div 4 = 4$

3 Practice/Apply

FOLDABLES™ Follow-Up Remind students to record main ideas, definitions of vocabulary words, and other notes on study cards as they progress through the lesson. Students store their study cards from this lesson in the "Fractions" pocket of their Foldables.

NUMBER SENSE
For Exercise 3, have students draw fraction models and then compare the "missing" part (the unshaded part) of the larger fraction with the shaded part of the smaller fraction. If the smaller fraction fits into the blank space with room to spare, the sum is less than one.

About the Exercises...

Guided Practice Key

Exercises	Examples
4	1
5	2
6	3
7	4
8	6
9–10	7

Odd/Even Assignments
Exercises 11–38 are structured so that students practice the same concepts whether they are assigned odd or even problems.

Sometimes it makes sense to round fractions to the nearest $\frac{1}{2}$, or mixed numbers to the nearest integer. Other times, it is useful to use **compatible numbers**, or numbers that are easy to compute mentally.

EXAMPLES Use Compatible Numbers

Estimate.

6 $\frac{1}{3} \times 14$ **THINK** What is $\frac{1}{3}$ of 14?

$\frac{1}{3} \times 14 \rightarrow \frac{1}{3} \times 15 = 5$ Round 14 to 15, since 15 is divisible by 3.

The product is *about* 5.

7 $9\frac{7}{8} \div 4\frac{1}{5}$

$9\frac{7}{8} \div 4\frac{1}{5} \rightarrow 10 \div 4\frac{1}{5}$ Round $9\frac{7}{8}$ to 10.

$\rightarrow 10 \div 5 = 2$ Round $4\frac{1}{5}$ to 5, since 10 is divisible by 5.

The quotient is *about* 2.

STUDY TIP
Compatible Numbers When dividing mixed numbers, round so that the dividend is a multiple of the divisor.

Your Turn Estimate. g–i. Sample answers are given.

g. $\frac{1}{4} \cdot 21$ $\frac{1}{4} \cdot 20 = 5$ h. $\frac{1}{2} \times 17$ $\frac{1}{2} \times 18 = 9$ i. $12 \div 6\frac{2}{3}$ $12 \div 6 = 2$

Skill and Concept Check

Writing Math Exercises 1 & 3

1. **Explain** how models are useful when estimating with fractions.
1–3. See margin.
2. **OPEN ENDED** Describe when estimation is a better method for solving a problem rather than using pencil and paper, a calculator, or a computer. Then give a real-life example.
3. **NUMBER SENSE** Determine which of the following has a sum that is greater than 1. Write *yes* or *no* and explain.

a. $\frac{1}{2} + \frac{4}{7}$ b. $\frac{3}{4} + \frac{5}{8}$ c. $\frac{2}{5} + \frac{1}{6}$

GUIDED PRACTICE

Estimate. 4–10. Sample answers are given.

4. $8\frac{3}{8} + 1\frac{4}{5}$ $8 + 2 = 10$
5. $5\frac{5}{7} \times 2\frac{7}{8}$ $6 \times 3 = 18$
6. $\frac{1}{6} + \frac{2}{5}$ $0 + \frac{1}{2} = \frac{1}{2}$
7. $\frac{6}{7} - \frac{1}{5}$ $1 - 0 = 1$
8. $\frac{1}{4} \cdot 15$ $\frac{1}{4} \cdot 16 = 4$
9. $21\frac{5}{6} \div 9\frac{3}{4}$ $22 \div 11 = 2$
10. **CONSTRUCTION** About how many bookcase shelves shown at the right can a carpenter cut from a board that is 1 foot wide and 12 feet long? about $12 \div 4$ or 3 shelves

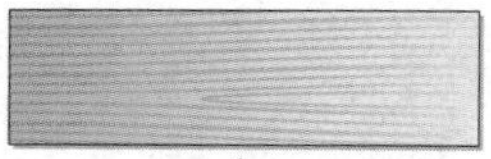

TEACHING TIP
Rounding mixed numbe[rs] is similar to rounding [dec]imals. If the decimal p[art] of a positive number is
- < 0.5, round down,
- ≥ 0.5, round up.

Answers

1. Sample answer: Models are useful to determine whether fractions are closer to 0, $\frac{1}{2}$, or 1, and to check estimates.
2. Sample answer: Using estimation is acceptable when no advantage is gained by getting an exact answer. An example is determining the amount of tip to leave. It is not necessary to know the amount to the exact cent.

3a. Yes; the sum of $\frac{1}{2}$ and a fraction greater than $\frac{1}{2}$ is greater than 1.

3b. Yes; both fractions are greater than $\frac{1}{2}$, so the sum is greater than $\frac{1}{2} + \frac{1}{2}$ or 1.

3c. No; both fractions are less than $\frac{1}{2}$, so the sum is less than 1.

242 Chapter 6 Applying Fractions

★ indicates increased difficulty

Practice and Applications

HOMEWORK HELP	
For Exercises	See Examples
11–16, 37–38	1, 2
17–24	3–5
25–28	6–7

Extra Practice
See pages 577, 601.

Estimate. 11–30. See Ch. 6 Answer Appendix.

11. $3\frac{3}{4} + 4\frac{5}{6}$ 12. $1\frac{1}{8} + 5\frac{11}{12}$ 13. $5\frac{1}{3} - 3\frac{1}{6}$ 14. $4\frac{2}{5} - 1\frac{1}{2}$

15. $2\frac{2}{3} \cdot 6\frac{1}{3}$ 16. $6\frac{1}{8} \div 1\frac{2}{3}$ 17. $\frac{3}{4} + \frac{3}{8}$ 18. $\frac{5}{8} + \frac{3}{7}$

19. $\frac{5}{9} - \frac{1}{6}$ 20. $\frac{3}{4} - \frac{3}{5}$ 21. $\frac{1}{8} \times \frac{3}{4}$ 22. $\frac{4}{9} \cdot \frac{11}{12}$

23. $\frac{4}{5} \div \frac{7}{8}$ 24. $\frac{1}{10} \div \frac{5}{6}$ 25. $\frac{1}{2} \times 13$ 26. $\frac{1}{4} \times 39$

27. $25\frac{3}{10} \div 5\frac{2}{3}$ 28. $27\frac{5}{8} \div 6\frac{1}{5}$ 29. $12\frac{2}{7} \div 4\frac{1}{3}$ 30. $5\frac{9}{10} \cdot 4\frac{1}{6}$

31–36. Sample answers are given.

31. Estimate $23\frac{2}{9}$ divided by 3. $24 \div 3 = 8$

★ 32. Estimate the sum of $4\frac{1}{8}$, $1\frac{5}{6}$, and $\frac{7}{9}$. $4 + 2 + 1 = 7$

Estimate. 34. $1\left(-\frac{1}{2}\right) = -\frac{1}{2}$ 35. $-\frac{1}{6} \times (-66) = 11$

★ 33. $-3\frac{2}{7} \times 8\frac{3}{4}$ $-3 \times 9 = -27$ ★ 34. $\frac{4}{5}\left(-\frac{5}{8}\right)$ ★ 35. $-\frac{1}{6} \times (-65)$ ★ 36. $12\frac{1}{4} \cdot 2\frac{7}{9}$ $12 \cdot 3 = 36$

37. **BAKING** Kayla wants to make the bread recipe shown at the right, but she has only $1\frac{1}{3}$ cups of flour. About how much more flour does she need? $3 - 1$ or 2 c

38. **SEWING** A skirt that is $15\frac{7}{8}$ inches long has a hem of $1\frac{1}{8}$ inches. Approximately how long will the skirt be if the hem is let down? $16 + 1$ or 17 in.

39. **CRITICAL THINKING** If a number being divided is rounded up and the divisor is rounded down, what is the effect on the quotient? It will be greater than the actual quotient.

Standardized Test Practice and Mixed Review

40. **MULTIPLE CHOICE** Choose the best estimate for $2\frac{1}{5} + 3\frac{3}{4}$. A

Ⓐ 6 Ⓑ 5 Ⓒ 4 Ⓓ 2

41. **MULTIPLE CHOICE** If Lucas's car gets $23\frac{1}{3}$ miles per gallon, about how many miles can he drive on $1\frac{3}{4}$ gallons? H

Ⓕ 0.46 mi Ⓖ 4.6 mi Ⓗ 46 mi Ⓘ 460 mi

Replace each ● with <, >, or = to make a true sentence. (Lesson 5-8)

42. $\frac{7}{8}$ ● 0.75 > 43. $\frac{4}{5}$ ● $\frac{5}{7}$ > 44. $2\frac{1}{3}$ ● $\frac{7}{3}$ = 45. $\frac{6}{11}$ ● $\frac{9}{14}$ <

46. Find the LCM of 9 and 12. (Lesson 5-7) 36

GETTING READY FOR THE NEXT LESSON

PREREQUISITE SKILL Find the LCD of each pair of fractions. (Lesson 5-8)

47. $\frac{3}{4}, \frac{5}{12}$ 12 48. $\frac{1}{2}, \frac{7}{10}$ 10 49. $\frac{1}{6}, \frac{1}{8}$ 24 50. $\frac{4}{5}, \frac{2}{3}$ 15

msmath2.net/self_check_quiz

4 Assess

Open-Ended Assessment

Writing Have students write a word problem that uses the estimated sum or difference of two or more mixed numbers. Students can share their problems with the class or trade with a classmate.

Study Guide and Intervention, p. 303

NAME ______ DATE ______ PERIOD ____

6-1 Study Guide and Intervention

Estimating with Fractions

Use rounding to estimate with fractions.

Estimating: For mixed numbers, round to the nearest whole number. $4\frac{1}{6} + 3\frac{7}{8} \rightarrow 4 + 4 = 8$; $4\frac{1}{6} + 3\frac{7}{8}$ is *about* 8.

For fractions, round to 0, $\frac{1}{2}$, or 1. $\frac{11}{12} - \frac{4}{9} \rightarrow 1 - \frac{1}{2} = \frac{1}{2}$; $\frac{11}{12} - \frac{4}{9}$ is *about* $\frac{1}{2}$.

EXAMPLE 1 Estimate $2\frac{2}{3} \times 4\frac{1}{4}$.

$2\frac{2}{3} \times 4\frac{1}{4} \rightarrow 3 \times 4 = 12$

The product is *about* 12.

EXAMPLE 2 Estimate $\frac{6}{7} - \frac{3}{5}$.

$\frac{6}{7}$ is about 1.

$\frac{3}{5}$ is about $\frac{1}{2}$.

$\frac{6}{7} - \frac{3}{5} \rightarrow 1 - \frac{1}{2} = \frac{1}{2}$ The difference is *about* $\frac{1}{2}$.

EXERCISES

Estimate.

1. $4\frac{1}{3} + 3\frac{4}{5}$ $4 + 4 = 8$
2. $2\frac{1}{6} \times 3\frac{2}{3}$ $2 \times 4 = 8$
3. $\frac{7}{12} - \frac{1}{10}$ $\frac{1}{2} - 0 = \frac{1}{2}$
4. $5\frac{1}{4} - 1\frac{1}{2}$ $5 - 2 = 3$
5. $4\frac{3}{4} + 1\frac{1}{5}$ $5 + 1 = 6$
6. $\frac{5}{9} \times \frac{13}{14}$ $\frac{1}{2} \times 1 = \frac{1}{2}$
7. $\frac{1}{6} \div \frac{8}{9}$ $0 \div 1 = 0$
8. $\frac{6}{7} \div \frac{9}{10}$ $1 \div 1 = 1$
9. $13\frac{4}{5} \div 1\frac{7}{8}$ $14 \div 2 = 7$
10. $12\frac{1}{4} \div 5\frac{7}{8}$ $12 \div 6 = 2$

Practice: Skills, p. 304

NAME ______ DATE ______ PERIOD ____

6-1 Practice: Skills

Estimating with Fractions

Estimate.

1. $\frac{4}{5} + \frac{2}{11}$ $1 + 0 = 1$
2. $\frac{4}{7} + \frac{1}{5}$ $\frac{1}{2} + 0 = \frac{1}{2}$
3. $\frac{7}{9} - \frac{1}{5}$ $1 - 0 = 1$
4. $\frac{9}{10} - \frac{1}{23}$ $1 - 0 = 1$
5. $\frac{3}{5} + \frac{9}{11}$ $\frac{1}{2} + 1 = 1\frac{1}{2}$
6. $\frac{4}{5} - \frac{4}{9}$ $1 - \frac{1}{2} = \frac{1}{2}$
7. $5\frac{1}{7} + 7\frac{9}{11}$ $5 + 8 = 13$
8. $3\frac{10}{11} - 2\frac{1}{6}$ $4 - 2 = 2$
9. $5\frac{1}{4} - \frac{1}{7}$ $5 - 0 = 5$
10. $8\frac{3}{7} - 2\frac{1}{2}$ $8 - 3 = 5$
11. $2\frac{1}{8} + 6\frac{9}{10}$ $2 + 7 = 9$
12. $10\frac{1}{8} - 3\frac{1}{4}$ $10 - 3 = 7$
13. $\frac{4}{5} \times \frac{8}{9}$ $1 \times 1 = 1$
14. $\frac{6}{7} \div \frac{10}{11}$ $1 \div 1 = 1$
15. $3\frac{6}{7} \times 2\frac{1}{10}$ $4 \times 2 = 8$
16. $16\frac{1}{3} \div 3\frac{8}{9}$ $16 \div 4 = 4$
17. $31\frac{3}{4} \div 2\frac{1}{8}$ $32 \div 2 = 16$
18. $3\frac{4}{5} \cdot 1\frac{1}{4}$ $4 \cdot 1 = 4$
19. $12 \div 2\frac{6}{7}$ $12 \div 3 = 4$
20. $44\frac{1}{5} \div 3\frac{7}{8}$ $44 \div 4 = 11$
21. $10\frac{1}{7} \cdot 4\frac{1}{3}$ $10 \cdot 4 = 40$
22. $5\frac{1}{8} \cdot 6\frac{9}{11}$ $5 \cdot 7 = 35$
23. $\frac{3}{12} \div 4\frac{4}{5}$ $0 \div 5 = 0$
24. $2\frac{1}{2} \div 3\frac{1}{3}$ $3 \div 3 = 1$
25. Estimate $36\frac{1}{4}$ divided by 6. $36 \div 6 = 6$
26. Estimate the sum of $7\frac{9}{10}$, $2\frac{1}{5}$, and $3\frac{2}{3}$. $8 + 2 + 4 = 14$

Practice: Word Problems, p. 305

NAME ______ DATE ______ PERIOD ____

6-1 Practice: Word Problems

Estimating with Fractions

COOKING For Exercises 1–4, use the recipe shown below.

Lightning Creamed Potatoes
$\frac{1}{3}$ cup water
$1\frac{1}{2}$ teaspoon salt
$3\frac{3}{4}$ cups pared potatoes, cut in bite-size pieces
$\frac{1}{3}$ cup finely chopped onion
$\frac{1}{2}$ cup light cream

serves 6

1. Daniel wants to serve twelve people the Lightning Creamed Potatoes. Estimate how much salt he will need if he doubles the recipe. 2×2 or 4 tsp
2. Rosita wants to triple the recipe above. Estimate how many cups of pared potatoes she will need. 3×4 or 12 c
3. Alvin is going to serve six people. He only has $1\frac{1}{4}$ cups of pared potatoes. About how many cups of potatoes will he have to borrow? $4 - 1$ or 3 c
4. Katrina wants to make half of the recipe. About how many cups of potatoes will she need? $\frac{1}{2} \times 4$ or 2 c
5. CARPENTRY A board is $17\frac{3}{4}$ inches long. Carmen wants to shorten the length by about $1\frac{7}{8}$ inches. Estimate the length of the board after the board has been shortened. $18 - 2$ or 16 in.
6. TRACK Akira ran two miles. He ran the first mile in $7\frac{3}{4}$ minutes and the second mile in $8\frac{3}{4}$ minutes. Estimate how long it took Akira to run two miles. $8 + 9$ or 17 min

6-2 Lesson Notes

Lesson Resources

Chapter 6 Resource Masters
- Study Guide and Intervention, p. 308
- Practice: Skills, p. 309
- Practice: Word Problems, p. 310
- Reading to Learn Mathematics, p. 311
- Enrichment, p. 312

Key Teaching Transparencies, 21
MathPASS Tutorial CD-ROM, 16–17
Parent and Student Study Guide Workbook, 6-2

FAST FILE

1 Focus

5-Minute Check 6-2

INTERACTIVE CHALKBOARD
PowerPoint® Presentations

Use as a quiz or review of Lesson 6-1. (Also available as a transparency)

Bellringer

Problem Solving Draw the following on the chalkboard.

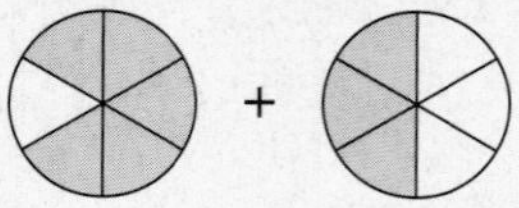

Have students write out the problem in numbers and symbols, then draw the model for the answer.

$\frac{5}{6} + \frac{1}{2} = 1\frac{2}{6}$ or $1\frac{1}{3}$

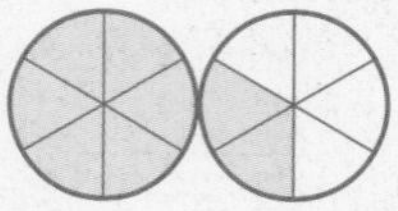

Reading to Learn Mathematics, p. 311 **ELL**

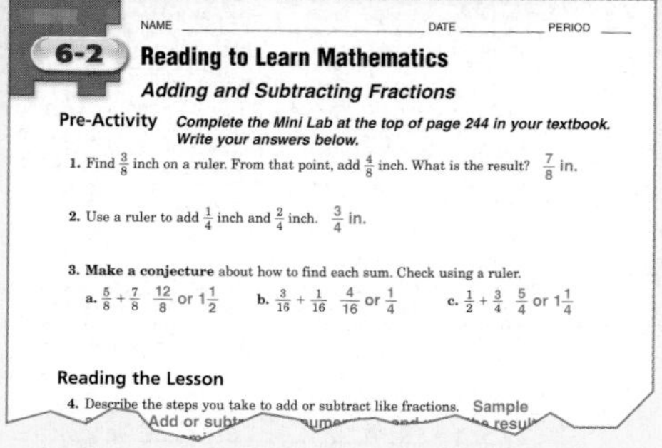
NAME ______ DATE ______ PERIOD ____

6-2 Reading to Learn Mathematics
Adding and Subtracting Fractions

Pre-Activity ***Complete the Mini Lab at the top of page 244 in your textbook. Write your answers below.***

1. Find $\frac{3}{8}$ inch on a ruler. From that point, add $\frac{4}{8}$ inch. What is the result? $\frac{7}{8}$ in.
2. Use a ruler to add $\frac{1}{4}$ inch and $\frac{2}{4}$ inch. $\frac{3}{4}$ in.
3. **Make a conjecture** about how to find each sum. Check using a ruler.
   a. $\frac{5}{8} + \frac{7}{8}$ $\frac{12}{8}$ or $1\frac{1}{2}$ b. $\frac{3}{16} + \frac{1}{16}$ $\frac{4}{16}$ or $\frac{1}{4}$ c. $\frac{1}{2} + \frac{3}{4}$ $\frac{5}{4}$ or $1\frac{1}{4}$

Reading the Lesson

4. Describe the steps you take to add or subtract like fractions. Sample ... Add or subt... ume... the resul...

6-2 Adding and Subtracting Fractions

What You'll Learn
Add and subtract fractions.

***REVIEW* Vocabulary**

LCD: the least common multiple of the denominators of two or more fractions **(Lesson 5-8)**

HANDS-ON Mini Lab

Materials
- ruler

Work with a partner.

1. Find $\frac{3}{8}$ inch on a ruler. From that point, add $\frac{4}{8}$ inch. What is the result? $\frac{7}{8}$ in.
2. Use a ruler to add $\frac{1}{4}$ inch and $\frac{2}{4}$ inch. $\frac{3}{4}$ in.
3. **Make a conjecture** about how to find each sum. Check using a ruler.
   a. $\frac{5}{8} + \frac{7}{8}$ $\frac{12}{8}$ or $1\frac{1}{2}$ b. $\frac{3}{16} + \frac{1}{16}$ $\frac{4}{16}$ or $\frac{1}{4}$ c. $\frac{1}{2} + \frac{3}{4}$ $\frac{5}{4}$ or $1\frac{1}{4}$

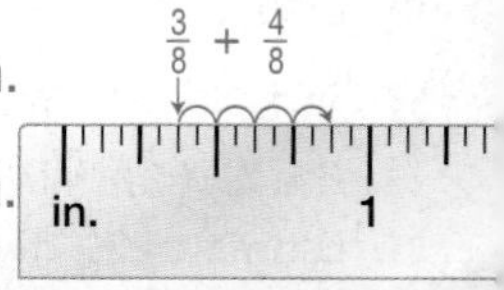

Fractions that have the same denominators are called *like fractions.*

Key Concept **Adding and Subtracting Like Fractions**

Words To add or subtract like fractions, add or subtract the numerators and write the result over the denominator. Simplify if necessary.

Symbols

	Arithmetic	Algebra
	$\frac{1}{5} + \frac{3}{5} = \frac{4}{5}$	$\frac{a}{c} + \frac{b}{c} = \frac{a+b}{c}$, where $c \neq 0$
	$\frac{11}{12} - \frac{7}{12} = \frac{4}{12}$ or $\frac{1}{3}$	$\frac{a}{c} - \frac{b}{c} = \frac{a-b}{c}$, where $c \neq 0$

EXAMPLES Add and Subtract Like Fractions

Add or subtract. Write in simplest form.

1 $\frac{5}{9} + \frac{2}{9}$

$\frac{5}{9} + \frac{2}{9} = \frac{5+2}{9}$ Add the numerators.

$= \frac{7}{9}$ Write the sum over the denominator.

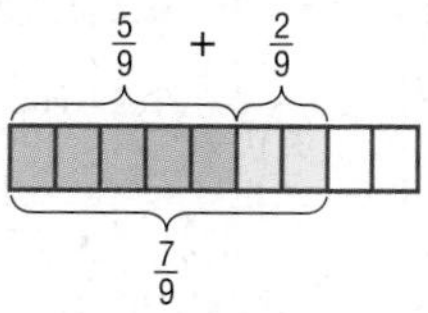

2 $\frac{9}{10} - \frac{1}{10}$

$\frac{9}{10} - \frac{1}{10} = \frac{9-1}{10}$ Subtract the numerators.

$= \frac{8}{10}$ Write the difference over the denominator.

$= \frac{4}{5}$ Simplify.

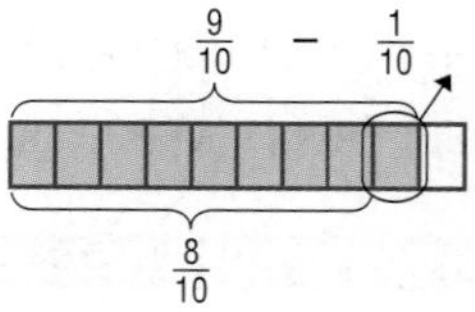

To add or subtract *unlike fractions*, or fractions with different denominators, rename the fractions using the LCD. Then add or subtract as with like fractions.

EXAMPLES Add and Subtract Unlike Fractions

Add or subtract. Write in simplest form.

3 $\frac{1}{2} + \frac{1}{6}$ **Estimate** $\frac{1}{2} + 0 = \frac{1}{2}$

The least common denominator of 2 and 6 is 6.

$\frac{1}{2} = \frac{1 \times 3}{2 \times 3} = \frac{3}{6}$ Rename $\frac{1}{2}$ using the LCD.

$$\begin{array}{rcr} \frac{1}{2} & \rightarrow & \frac{3}{6} \\ +\frac{1}{6} & \rightarrow & +\frac{1}{6} \\ \hline & & \frac{4}{6} \text{ or } \frac{2}{3} \end{array}$$

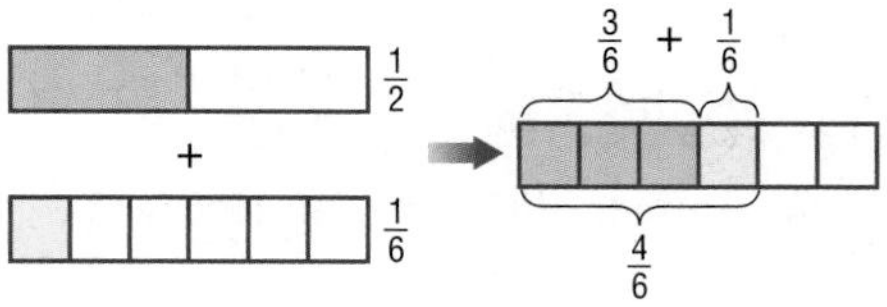

$\frac{2}{3}$ is close to the estimate. So, $\frac{1}{2} + \frac{1}{6} = \frac{2}{3}$.

4 $\frac{7}{8} - \frac{5}{6}$ **Estimate** $1 - 1 = 0$

The LCD of 8 and 6 is 24.

$$\begin{array}{rcrcr} \frac{7}{8} & \rightarrow & \frac{7 \times 3}{8 \times 3} & \rightarrow & \frac{21}{24} \\ -\frac{5}{6} & \rightarrow & \frac{5 \times 4}{6 \times 4} & \rightarrow & -\frac{20}{24} \\ \hline & & & & \frac{1}{24} \end{array}$$

So, $\frac{7}{8} - \frac{5}{6} = \frac{1}{24}$. Compare to the estimate.

REAL-LIFE MATH

WEATHER The average precipitation for November and December in Seattle, Washington, is $5\frac{7}{10}$ inches and 6 inches, respectively.

Source: *The World Almanac*

5 **WEATHER The average precipitation for November and December in Grand Junction, Colorado, is $\frac{7}{10}$ inch and $\frac{3}{5}$ inch, respectively. What is the total of the average precipitation for these two months?**

Estimate $\frac{1}{2} + \frac{1}{2} = 1$

$\frac{7}{10} + \frac{3}{5} = \frac{7}{10} + \frac{3 \times 2}{5 \times 2}$ The LCD of 10 and 5 is 10.

$= \frac{7}{10} + \frac{6}{10}$ Rename the fractions with the LCD.

$= \frac{13}{10}$ Add the numerators.

$= 1\frac{3}{10}$ Rename $\frac{13}{10}$ as a mixed number.

The total amount of precipitation is $1\frac{3}{10}$ inches.

Your Turn **Add or subtract. Write in simplest form.**

a. $\frac{8}{9} - \frac{2}{9}$ $\frac{2}{3}$ b. $\frac{5}{6} - \frac{3}{8}$ $\frac{11}{24}$ c. $\frac{7}{8} + \frac{3}{4}$ $1\frac{5}{8}$ d. $\frac{2}{3} - \frac{1}{6}$ $\frac{1}{2}$

2 Teach

In-Class Examples

INTERACTIVE CHALKBOARD PowerPoint® Presentations

Add or subtract. Write in simplest form.

1. $\frac{7}{12} + \frac{4}{12}$ $\frac{11}{12}$
2. $\frac{5}{6} - \frac{1}{6}$ $\frac{2}{3}$
3. $\frac{1}{3} + \frac{1}{9}$ $\frac{4}{9}$
4. $\frac{3}{4} - \frac{1}{6}$ $\frac{7}{12}$
5. **ART** A picture is mounted on art board that is $\frac{1}{8}$ inch thick. The frame selected for the picture is $\frac{1}{2}$ inch thick. The thickness of the picture is so small that it can be ignored. What thickness of cardboard is needed to hold the picture in place? $\frac{3}{8}$ in.

Tips for New Teachers

Teaching Strategies Use timed competitions to keep students engaged in activities such as finding the LCD and estimating, subtracting, and adding fractions. Give out small prizes, extra-credit points, or in-class privileges.

DAILY INTERVENTION

Differentiated Instruction

Kinesthetic Have pairs of students measure each other's vertical jumps in fractions of inches. Have them place yardsticks, tip-down, on the ground, and mark the highest point reached by the jumper's feet. Have students compare each other's vertical jumps by finding the differences in height. Students can jump multiple times and find their amount of improvement. Anyone who chooses not to jump should help tabulate classwide data.

3 Practice/Apply

DAILY INTERVENTION **FIND THE ERROR** For Exercise 3, reinforce adding only the numerators of fractions by having students think of concrete items. For example, a raspberry pie is divided into 6 slices, and a peach pie is divided into 8 slices. Three raspberry slices and 2 peach slices are $\frac{3}{6} + \frac{2}{8}$ or $\frac{1}{2} + \frac{1}{4}$.

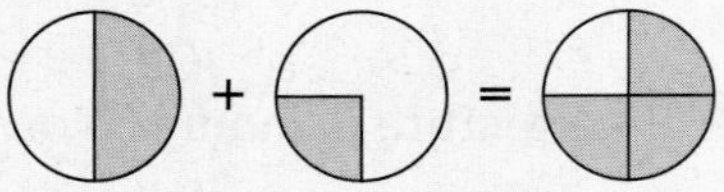

$\frac{1}{2} + \frac{1}{4} = \frac{3}{4}$

About the Exercises...

Guided Practice Key

Exercises	Examples
4, 5, 8, 9	1, 2
6, 7, 10–13	3, 4
14	5

Odd/Even Assignments
Exercises 15–38 are structured so that students practice the same concepts whether they are assigned odd or even problems.

Answer

1. Sample answer:

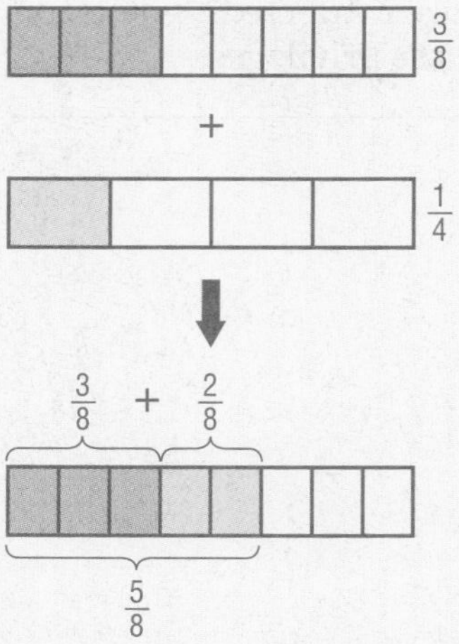

Skill and Concept Check

Writing Math Exercise 3

1. **Draw** a model to show $\frac{3}{8} + \frac{1}{4}$. **See margin.**
2. **OPEN ENDED** Write a subtraction problem with fractions in which the difference is $\frac{2}{7}$. **Sample answer: $\frac{6}{7} - \frac{4}{7}$**
3. **FIND THE ERROR** Marissa and Jacinta are finding $\frac{5}{9} + \frac{1}{6}$. Who is correct? Explain. **Jacinta; Marissa did not rename the fractions using the LCD.**

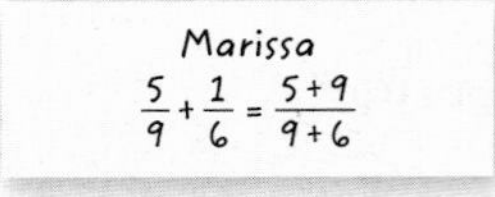

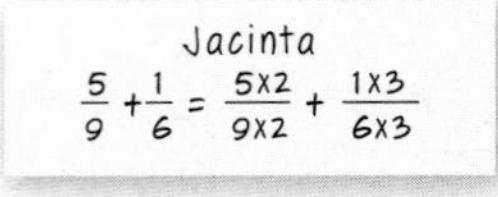

GUIDED PRACTICE

Add or subtract. Write in simplest form.

4. $\frac{4}{5} - \frac{2}{5}$ **$\frac{2}{5}$**
5. $\frac{6}{7} + \frac{3}{7}$ **$1\frac{2}{7}$**
6. $\frac{1}{6} + \frac{3}{8}$ **$\frac{13}{24}$**
7. $\frac{5}{6} - \frac{7}{12}$ **$\frac{1}{4}$**
8. $\frac{4}{9} + \frac{2}{9}$ **$\frac{2}{3}$**
9. $\frac{3}{8} - \frac{1}{8}$ **$\frac{1}{4}$**
10. $\frac{2}{3} + \frac{5}{6}$ **$1\frac{1}{2}$**
11. $\frac{5}{6} + \frac{4}{9}$ **$1\frac{5}{18}$**
12. $\frac{3}{4} - \frac{1}{3}$ **$\frac{5}{12}$**
13. $\frac{7}{10} - \frac{1}{6}$ **$\frac{8}{15}$**
14. **PRESIDENTS** As of 2002, $\frac{5}{14}$ of the United States presidents were born in either Virginia or Ohio, and $\frac{1}{6}$ were born in either Massachusetts or New York. What fraction were born in one of these four states? **$\frac{11}{21}$**

TEACHING TIP
Remind students that the denominators of fractions are not added or subtracted.

★ indicates increased difficulty

Practice and Applications

HOMEWORK HELP

For Exercises	See Examples
15–18, 23–26	1, 2
19–22, 27–41	3–5

Extra Practice
See pages 577, 601.

Add or subtract. Write in simplest form.

15. $\frac{3}{7} + \frac{1}{7}$ **$\frac{4}{7}$**
16. $\frac{5}{8} + \frac{7}{8}$ **$1\frac{1}{2}$**
17. $\frac{5}{6} - \frac{1}{6}$ **$\frac{2}{3}$**
18. $\frac{7}{10} - \frac{3}{10}$ **$\frac{2}{5}$**
19. $\frac{1}{15} + \frac{3}{5}$ **$\frac{2}{3}$**
20. $\frac{7}{9} - \frac{1}{3}$ **$\frac{4}{9}$**
21. $\frac{4}{5} - \frac{1}{6}$ **$\frac{19}{30}$**
22. $\frac{7}{12} + \frac{7}{10}$ **$1\frac{17}{60}$**
23. $\frac{4}{5} - \frac{3}{5}$ **$\frac{1}{5}$**
24. $\frac{8}{9} - \frac{5}{9}$ **$\frac{1}{3}$**
25. $\frac{3}{8} + \frac{7}{8}$ **$1\frac{1}{4}$**
26. $\frac{5}{6} + \frac{5}{6}$ **$1\frac{2}{3}$**
27. $\frac{5}{8} - \frac{7}{12}$ **$\frac{1}{24}$**
28. $\frac{2}{15} + \frac{4}{9}$ **$\frac{26}{45}$**
29. $\frac{5}{8} + \frac{11}{12}$ **$1\frac{13}{24}$**
30. $\frac{3}{8} - \frac{1}{12}$ **$\frac{7}{24}$**
31. $\frac{3}{10} - \frac{1}{4}$ **$\frac{1}{20}$**
32. $\frac{9}{10} + \frac{4}{15}$ **$1\frac{1}{6}$**
33. $\frac{9}{10} - \frac{1}{6}$ **$\frac{11}{15}$**
34. $\frac{7}{9} + \frac{5}{6}$ **$1\frac{11}{18}$**

ALGEBRA **Evaluate each expression if $a = \frac{3}{4}$ and $b = \frac{5}{6}$.**

35. $\frac{1}{12} + a$ **$\frac{5}{6}$**
36. $b - \frac{7}{10}$ **$\frac{2}{15}$**
37. $b - a$ **$\frac{1}{12}$**
38. $a + b$ **$1\frac{7}{12}$**

39. **MULTI STEP** After 1 hour, Jon had finished $\frac{5}{6}$ of a long-distance race, and Ling had finished $\frac{7}{9}$ of it. At that time, who had finished a greater fraction of the race, and by how much? **Jon; $\frac{1}{18}$ of the race**

MONEY For Exercises 40 and 41, use the following information and the table at the right.
Sierra and Jacob each receive an equal allowance. The table shows the fraction of their allowance that they each deposit into their savings account and the fraction they each spend at the mall.

Where Money Goes	Fraction of Allowance Sierra	Jacob
savings account	$\frac{1}{2}$	$\frac{1}{3}$
spend at mall	$\frac{1}{4}$	$\frac{3}{5}$
left over	?	?

★ 40. What fraction of Jacob's allowance goes into his savings account or is spent at the mall? **$\frac{14}{15}$**

★ 41. Who has more money left over? Explain. **See margin.**

42. **CRITICAL THINKING** Does $\frac{1}{3} + \frac{5}{9} - \frac{5}{12} = \frac{5}{9} + \frac{5}{12} - \frac{1}{3}$? Explain.
No; the Commutative Property does not hold for subtraction.

Standardized Test Practice and Mixed Review

43. **MULTIPLE CHOICE** Makayla uses $\frac{1}{5}$ pound of ham and $\frac{1}{8}$ pound of turkey for her sandwich. How much meat does she use in all? **C**

Ⓐ $\frac{1}{13}$ lb Ⓑ $\frac{2}{13}$ lb Ⓒ $\frac{13}{40}$ lb Ⓓ $\frac{7}{20}$ lb

44. **MULTIPLE CHOICE** Jamal used a bucket that was $\frac{7}{9}$ full with soapy water to wash his mother's car. After washing the car, the bucket was only $\frac{1}{6}$ full. What part of the bucket of soapy water did Jamal use? **H**

Ⓕ $\frac{1}{9}$ Ⓖ $\frac{8}{15}$ Ⓗ $\frac{11}{18}$ Ⓘ $\frac{17}{18}$

Estimate. (Lesson 6-1) **45–48. Sample answers are given.**

45. $\frac{6}{7} - \frac{5}{12}$ **$1 - \frac{1}{2} = \frac{1}{2}$**
46. $4\frac{1}{9} + 3\frac{3}{4}$ **$4 + 4 = 8$**
47. $16\frac{2}{3} \div 8\frac{1}{5}$ **$16 \div 8 = 2$**
48. $5\frac{4}{5} \cdot 3\frac{1}{3}$ **$6 \cdot 3 = 18$**

PETS For Exercises 49 and 50, refer to the table at the right. It shows where pet owners get their pets. (Lesson 5-8)

Pet Source	Portion of Pet Owners
animal shelter	$\frac{3}{20}$
friend/family	$\frac{21}{50}$
pet store	0.07
find as stray	0.14

Source: Yankelovich Partners

49. Where do the greatest number of people get their pets? **friend/family**

50. Of the sources listed, where do the fewest people get their pets? **pet store**

GETTING READY FOR THE NEXT LESSON

BASIC SKILL Complete.

Example: $8\frac{1}{2} = 7\frac{3}{2}$

51. $5\frac{2}{3} = 4\frac{■}{3}$ **$4\frac{5}{3}$**
52. $7\frac{8}{9} = 6\frac{■}{9}$ **$6\frac{17}{9}$**
53. $12\frac{1}{5} = ■\frac{6}{5}$ **$11\frac{6}{5}$**
54. $4\frac{3}{8} = ■\frac{11}{8}$ **$3\frac{11}{8}$**

msmath2.net/self_check_quiz

4 Assess

Higher-Level Thinking

Have students evaluate the following expression if $a = \frac{1}{2}$, $b = \frac{2}{3}$, and $c = \frac{5}{8}$.

$(a + b) - c$ **$\frac{13}{24}$**

Answer

41. Sierra; she has $\frac{1}{4}$ of her allowance left over; Jacob has $\frac{1}{15}$ left over.

Study Guide and Intervention, p. 308

NAME ____________ DATE ________ PERIOD ____

6-2 Study Guide and Intervention
Adding and Subtracting Fractions

Like fractions are fractions that have the same denominator. To add or subtract like fractions, add or subtract the numerators and write the result over the denominator.
Simplify if necessary.
To add or subtract *unlike fractions*, rename the fractions with a least common denominator. Then add or subtract as with like fractions.

EXAMPLE 1 **Subtract $\frac{3}{4} - \frac{1}{4}$. Write in simplest form.**

$\frac{3}{4} - \frac{1}{4} = \frac{3-1}{4}$ Subtract the numerators.
$= \frac{2}{4}$ Write the difference over the denominator.
$= \frac{1}{2}$ Simplify.

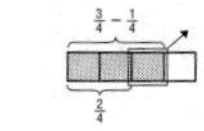

EXAMPLE 2 **Add $\frac{2}{3} + \frac{1}{12}$. Write in simplest form.**

The least common denominator of 3 and 12 is 12.

$\frac{2}{3} = \frac{2 \times 4}{3 \times 4} = \frac{8}{12}$ Rename $\frac{2}{3}$ using the LCD.

$\frac{2}{3} \rightarrow \frac{8}{12}$
$+\frac{1}{12} \rightarrow +\frac{1}{12}$
$\frac{9}{12}$ or $\frac{3}{4}$ Add the numerators and simplify.

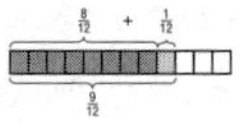

EXERCISES

Add or subtract. Write in simplest form.

1. $\frac{5}{8} + \frac{1}{8}$ $\frac{3}{4}$
2. $\frac{7}{9} - \frac{2}{9}$ $\frac{5}{9}$
3. $\frac{1}{2} + \frac{3}{4}$ $1\frac{1}{4}$
4. $\frac{7}{8} - \frac{5}{6}$ $\frac{1}{24}$
5. $\frac{5}{9} + \frac{5}{6}$ $1\frac{7}{18}$
6. $\frac{3}{8} - \frac{1}{12}$ $\frac{7}{24}$
7. $\frac{3}{10} + \frac{7}{12}$ $\frac{53}{60}$
8. $\frac{2}{5} - \frac{1}{3}$ $\frac{1}{15}$
9. $\frac{7}{15} + \frac{5}{6}$ $1\frac{3}{10}$
10. $\frac{7}{9} - \frac{1}{2}$ $\frac{5}{18}$

Practice: Skills, p. 309

NAME ____________ DATE ________ PERIOD ____

6-2 Practice: Skills
Adding and Subtracting Fractions

Add or subtract. Write in simplest form.

1. $\frac{3}{8} + \frac{3}{8}$ $\frac{3}{4}$
2. $\frac{7}{10} - \frac{5}{10}$ $\frac{1}{5}$
3. $\frac{9}{10} + \frac{3}{10}$ $1\frac{1}{5}$
4. $\frac{4}{7} - \frac{2}{7}$ $\frac{2}{7}$
5. $\frac{2}{3} + \frac{2}{3}$ $1\frac{1}{3}$
6. $\frac{5}{9} - \frac{2}{9}$ $\frac{1}{3}$
7. $\frac{8}{15} - \frac{1}{5}$ $\frac{1}{3}$
8. $\frac{5}{6} + \frac{5}{12}$ $1\frac{1}{4}$
9. $\frac{3}{5} - \frac{3}{10}$ $\frac{3}{10}$
10. $\frac{7}{16} + \frac{3}{8}$ $\frac{13}{16}$
11. $\frac{19}{20} - \frac{3}{10}$ $\frac{13}{20}$
12. $\frac{5}{9} + \frac{7}{9}$ $1\frac{1}{3}$
13. $\frac{4}{9} - \frac{1}{12}$ $\frac{13}{36}$
14. $\frac{2}{3} + \frac{3}{7}$ $1\frac{2}{21}$
15. $\frac{3}{4} + \frac{1}{7}$ $\frac{25}{28}$
16. $\frac{7}{8} - \frac{2}{3}$ $\frac{5}{24}$
17. $\frac{8}{9} - \frac{5}{6}$ $\frac{1}{18}$
18. $\frac{5}{12} - \frac{3}{10}$ $\frac{7}{60}$
19. $\frac{7}{9} + \frac{2}{3}$ $1\frac{4}{9}$
20. $\frac{3}{5} + \frac{4}{11}$ $\frac{53}{55}$
21. $\frac{11}{12} - \frac{1}{4}$ $\frac{2}{3}$

ALGEBRA Evaluate each expression if $a = \frac{5}{6}$ and $b = \frac{3}{8}$.

22. $a + b$ $1\frac{5}{24}$
23. $a - b$ $\frac{11}{24}$
24. $\frac{9}{10} - a$ $\frac{1}{15}$

Practice: Word Problems, p. 310

NAME ____________ DATE ________ PERIOD ____

6-2 Practice: Word Problems
Adding and Subtracting Fractions

RETAIL STORES For Exercises 1–4 use the table below. It shows what fraction of the stores at a mall fall into seven categories.

Type of Store	Fraction of Stores in Mall
jewelry	$\frac{1}{30}$
clothing	$\frac{8}{15}$
gifts	$\frac{3}{20}$
electronics	$\frac{1}{20}$
department	$\frac{1}{15}$
shoes	$\frac{1}{15}$
athletic	$\frac{1}{10}$

1. What fraction of the stores are jewelry or gift stores? $\frac{11}{60}$

2. What fraction of the stores are clothing or electronics stores? $\frac{7}{12}$

3. Which type of store has the greatest number of stores? clothing stores

4. How many more clothing stores are there than athletic stores? Write as a fraction. $\frac{13}{30}$

5. **SEWING** Jin wants to make a scarf and matching hat for his sister. The patterns call for $\frac{7}{8}$ yard of fabric for the scarf and $\frac{1}{2}$ yard of fabric for the hat. How much fabric should Jin buy? Jin should buy $1\frac{3}{8}$ yards.

6. **RESTAURANT** Ms. Malle owns a restaurant. Typically, $\frac{1}{5}$ of the customers order fish, while $\frac{1}{4}$ of the customers order poultry. What fraction of her customers order either fish or poultry? $\frac{9}{20}$

Lesson 6-2 Adding and Subtracting Fractions **247**

6-3 Lesson Notes

Lesson Resources

Chapter 6 Resource Masters
- Study Guide and Intervention, p. 313
- Practice: Skills, p. 314
- Practice: Word Problems, p. 315
- Reading to Learn Mathematics, p. 316
- Enrichment, p. 317

MathPASS Tutorial CD-ROM, 18–19

Parent and Student Study Guide Workbook, 6-3

FAST FILE

1 Focus

INTERACTIVE CHALKBOARD
PowerPoint® Presentations

5-Minute Check 6-3
Use as a quiz or review of Lesson 6-2. (Also available as a transparency)

Bellringer

Hands-On Activity Direct students to fold four pieces of scrap paper into eight equal parts as shown.

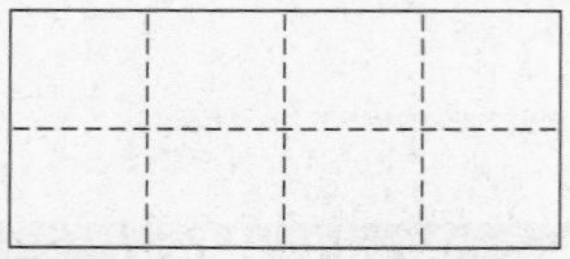

Have students remove $\frac{1}{8}$ sheet by cutting off one section of one piece of paper. Ask, "With all four sheets, what mixed number do we have?" $3\frac{7}{8}$

(Continued in bottom margin.)

Reading to Learn Mathematics, p. 316 ELL

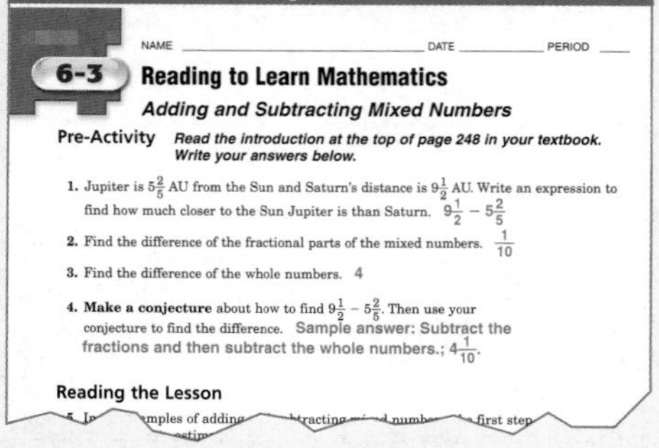
NAME ______________ DATE ________ PERIOD ____

6-3 Reading to Learn Mathematics
Adding and Subtracting Mixed Numbers

Pre-Activity *Read the introduction at the top of page 248 in your textbook. Write your answers below.*

1. Jupiter is $5\frac{2}{5}$ AU from the Sun and Saturn's distance is $9\frac{1}{2}$ AU. Write an expression to find how much closer to the Sun Jupiter is than Saturn. $9\frac{1}{2} - 5\frac{2}{5}$
2. Find the difference of the fractional parts of the mixed numbers. $\frac{1}{10}$
3. Find the difference of the whole numbers. 4
4. **Make a conjecture** about how to find $9\frac{1}{2} - 5\frac{2}{5}$. Then use your conjecture to find the difference. Sample answer: Subtract the fractions and then subtract the whole numbers.; $4\frac{1}{10}$.

Reading the Lesson

6-3 Adding and Subtracting Mixed Numbers

What You'll Learn
Add and subtract mixed numbers.

WHEN am I ever going to use this?

ASTRONOMY Astronomers use *astronomical units* (AU) to represent large distances in space. One AU is the average distance from Earth to the Sun. Mercury is about $\frac{2}{5}$ AU from the Sun.

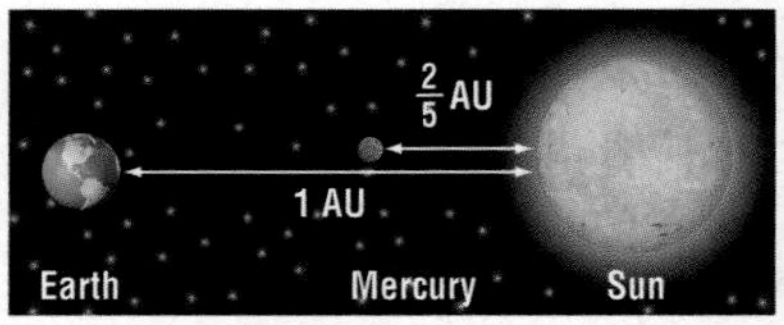

Diagram is not drawn to scale.

1. Jupiter is $5\frac{2}{5}$ AU from the Sun and Saturn's distance is $9\frac{1}{2}$ AU. Write an expression to find how much closer to the Sun Jupiter is than Saturn. $9\frac{1}{2} - 5\frac{2}{5}$
2. Find the difference of the fractional parts of the mixed numbers. $\frac{1}{10}$
3. Find the difference of the whole numbers. 4
4. **Make a conjecture** about how to find $9\frac{1}{2} - 5\frac{2}{5}$. Then use your conjecture to find the difference. **See margin.**

To add or subtract mixed numbers, first add or subtract the fractions. If necessary, rename them using the LCD. Then add or subtract the whole numbers and simplify if necessary.

EXAMPLES Add and Subtract Mixed Numbers

Add or subtract. Write in simplest form.

1 $7\frac{4}{9} + 10\frac{2}{9}$

Estimate $7 + 10 = 17$

$$\begin{array}{r} 7\frac{4}{9} \\ +\ 10\frac{2}{9} \\ \hline 17\frac{6}{9} \text{ or } 17\frac{2}{3} \end{array}$$

Add the whole numbers and fractions separately.

Simplify.

Compare each sum to its estimate.

2 $8\frac{5}{6} - 2\frac{1}{3}$

Estimate $9 - 2 = 7$

$$\begin{array}{rcr} 8\frac{5}{6} & \rightarrow & 8\frac{5}{6} \\ -\ 2\frac{1}{3} & \rightarrow & -\ 2\frac{2}{6} \\ \hline & & 6\frac{3}{6} \text{ or } 6\frac{1}{2} \end{array}$$

Rename the fraction using the LCD.

Simplify.

TEACHING TIP
When adding mixed numbers, students can also add the whole numbers first and then add the fractions.

Your Turn **Add or subtract. Write in simplest form.**

a. $6\frac{1}{8} + 2\frac{5}{8}$ $8\frac{3}{4}$
b. $13\frac{7}{8} - 9\frac{3}{4}$ $4\frac{1}{8}$
c. $1\frac{5}{9} + 4\frac{1}{6}$ $5\frac{13}{18}$

Then have students tear off $1\frac{3}{8}$ from the remaining $3\frac{7}{8}$. Have them state the remaining pieces of paper as a mixed number. $2\frac{4}{8}$ or $2\frac{1}{2}$ Finally, write the following on the chalkboard.

$3\frac{7}{8} - 1\frac{3}{8} = 2\frac{1}{2}$

Tell students they have just solved this problem.

EXAMPLE Use Mixed Numbers to Solve a Problem

3 **BUILDING** **There should be $2\frac{2}{3}$ feet of clearance space in front of a bathtub. What is the total width of the bathtub and clearance space shown at the right?**

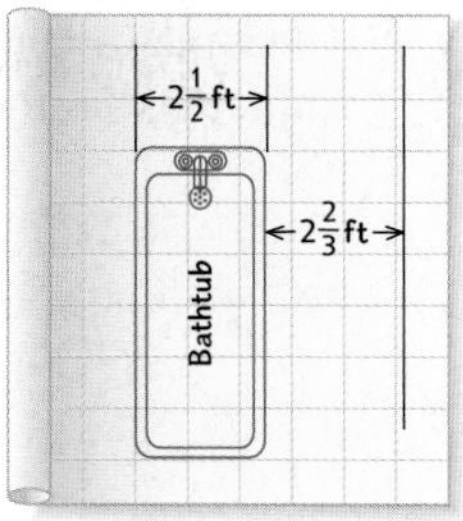

$2\frac{1}{2} + 2\frac{2}{3} = 2\frac{3}{6} + 2\frac{4}{6}$ Rename the fractions.

$= 4 + \frac{7}{6}$ Add the whole numbers and add the fractions.

$= 4 + 1\frac{1}{6}$ Rename $\frac{7}{6}$ as $1\frac{1}{6}$.

$= 5\frac{1}{6}$ Simplify.

The total width is $5\frac{1}{6}$ feet.

EAL-IFE CAREERS

w Does a Plumber e Math?

mbers add and subtract xed numbers when culating the dimensions installing house fixtures ch as shower stalls and ks.

Online Research
r information about a eer as a plumber, visit: math2.net/careers

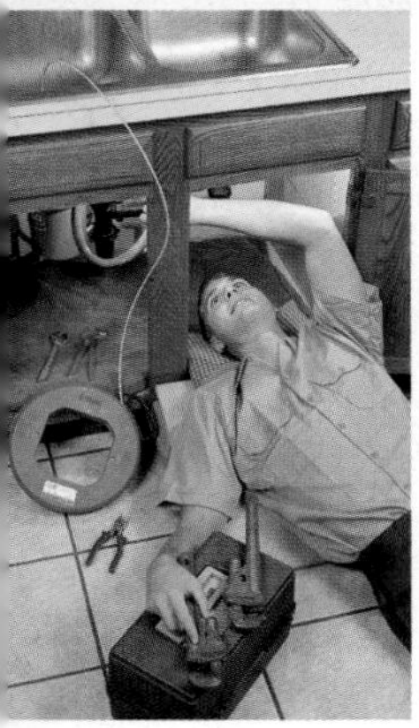

TEACHING TIP

oint out that both odels in Example 4 rep- sent the same quantity.

Sometimes when you subtract mixed numbers, the fraction in the first mixed number is less than the fraction in the second.

EXAMPLES Rename Mixed Numbers to Subtract

4 **Find $2\frac{1}{3} - 1\frac{2}{3}$.**

Rename $2\frac{1}{3}$ before subtracting.

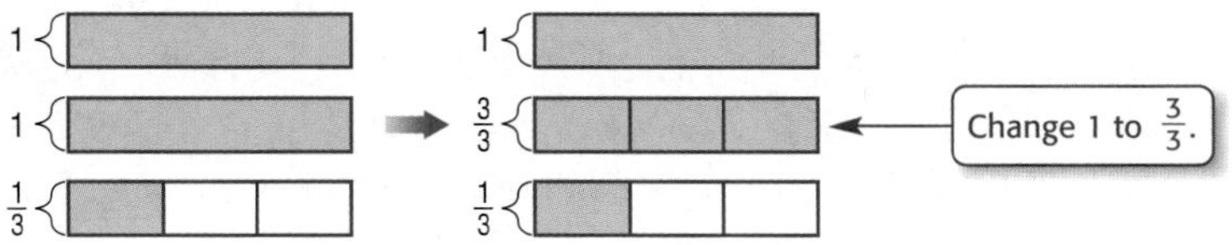

$2\frac{1}{3} = 1\frac{3}{3} + \frac{1}{3}$ or $1\frac{4}{3}$

$$\begin{array}{rcrl} 2\frac{1}{3} & \rightarrow & 1\frac{4}{3} & \text{Rename } 2\frac{1}{3} \text{ as } 1\frac{4}{3}. \\ -\,1\frac{2}{3} & \rightarrow & -\,1\frac{2}{3} & \text{First subtract the whole numbers and then the fractions.} \\ & & \frac{2}{3} & \end{array}$$

So, $2\frac{1}{3} - 1\frac{2}{3} = \frac{2}{3}$.

5 **Find $8\frac{7}{12} - 3\frac{3}{4}$.** **Estimate** $9 - 4 = 5$

$$\begin{array}{rcrcr} 8\frac{7}{12} & \rightarrow & 8\frac{7}{12} & \rightarrow & 7\frac{19}{12} \\ -\,3\frac{3}{4} & \rightarrow & -\,3\frac{9}{12} & \rightarrow & -\,3\frac{9}{12} \\ & & & & 4\frac{10}{12} \text{ or } 4\frac{5}{6} \end{array}$$

$\frac{7}{12} < \frac{9}{12}$, so rename $8\frac{7}{12}$ as $7\frac{19}{12}$.

Subtract.

The difference is close to the estimate.

Your Turn **Subtract. Write in simplest form.**

d. $11\frac{2}{5} - 2\frac{3}{5}$ $8\frac{4}{5}$ e. $5\frac{3}{8} - 4\frac{11}{12}$ $\frac{11}{24}$ f. $7 - 1\frac{1}{2}$ $5\frac{1}{2}$

msmath2.net/extra_examples

2 Teach

In-Class Examples

INTERACTIVE CHALKBOARD PowerPoint® Presentations

Add or subtract. Write in simplest form.

1 $3\frac{1}{12} + 14\frac{7}{12}$ $17\frac{2}{3}$

2 $9\frac{7}{10} - 4\frac{3}{5}$ $5\frac{1}{10}$

3 **COOKING** A quiche recipe calls for $2\frac{3}{4}$ cups of grated cheese. A recipe for quesadillas requires $1\frac{1}{3}$ cups of grated cheese. What is the total amount of grated cheese needed for both recipes? $4\frac{1}{12}$ c

4 Find $8\frac{1}{5} - 3\frac{3}{5}$. $4\frac{3}{5}$

5 Find $11\frac{5}{9} - 8\frac{2}{3}$. $2\frac{8}{9}$

Answer (When am I ever going to use this?)

4. Sample answer: Subtract the fractions and then subtract the whole numbers; $4\frac{1}{10}$.

DAILY INTERVENTION **Flexible Grouping**

Adding Mixed Numbers Prepare a set of index cards, each with a mixed number or fraction written on it. Give each student one card, then call out a sum. Students should mingle, trying to find a student whose number, when added to theirs, makes the sum. Plan well, using lists of exercises such as Exercises 15–34, to be sure that each student finds a match after three or four rounds. You may also wish to extend the activity to subtraction.

3 Practice/Apply

FOLDABLES™ Follow-Up Remind students to record main ideas, definitions of vocabulary words, and other notes on study cards as they progress through the lesson. Students store their study cards on adding and subtracting mixed numbers in the "Mixed Numbers" pocket of their Foldables.

DAILY INTERVENTION

FIND THE ERROR For students answering "Evan" to Exercise 2, write $3\frac{6}{8} - 1\frac{7}{8}$ on the chalkboard or an overhead transparency. Then circle or highlight the 6 and the 7. Ask students what is needed to subtract 7 from 6 in this case. **Borrow $\frac{8}{8}$ from the 3 and add them to $\frac{6}{8}$. This way, you now can subtract $\frac{14}{8} - \frac{7}{8}$.**

NUMBER SENSE
When solving Exercise 3, students can estimate and mark off the sums on two different rulers, then compare estimated distances on the rulers.

About the Exercises...

Guided Practice Key

Exercises	Examples
4, 6, 7, 9, 10, 11	1
5, 8	2
14	3
12, 13	4, 5

Odd/Even Assignments
Exercises 15–34 and 39–42 are structured so that students practice the same concepts whether they are assigned odd or even problems.

Skill and Concept Check

Writing Math Exercises 2 & 3

1. **OPEN ENDED** Write a problem that can be solved by finding $8\frac{1}{2} + 2\frac{1}{3}$. Explain what the numbers represent and find the sum. **See margin.**

2. **FIND THE ERROR** Evan and José are finding $3\frac{3}{4} - 1\frac{7}{8}$. Who is correct? Explain. **José; Evan incorrectly subtracted the first fraction from the second.**

Evan

$3\frac{3}{4} - 1\frac{7}{8} = 3\frac{6}{8} - 1\frac{7}{8}$
$= 2\frac{1}{8}$

José

$3\frac{3}{4} - 1\frac{7}{8} = 3\frac{6}{8} - 1\frac{7}{8}$
$= 2\frac{14}{8} - 1\frac{7}{8}$ or $1\frac{7}{8}$

3. **NUMBER SENSE** Use estimation to determine whether $6\frac{3}{4} + \frac{4}{5}$ is *greater than, less than,* or *equal to* $2\frac{1}{9} + 6\frac{7}{8}$. Explain. **less than; $7 + 1 < 2 + 7$**

GUIDED PRACTICE

Add or subtract. Write in simplest form.

4. $1\frac{5}{7} + 8\frac{1}{7}$ **$9\frac{6}{7}$**
5. $7\frac{5}{6} - 3\frac{1}{6}$ **$4\frac{2}{3}$**
6. $8\frac{1}{2} + 3\frac{4}{5}$ **$12\frac{3}{10}$**
7. $6\frac{3}{4} + 2\frac{9}{10}$ **$9\frac{13}{20}$**
8. $9\frac{4}{5} - 2\frac{3}{5}$ **$7\frac{1}{5}$**
9. $2\frac{3}{8} + 5\frac{7}{8}$ **$8\frac{1}{4}$**
10. $4\frac{3}{10} + 4\frac{9}{10}$ **$9\frac{1}{5}$**
11. $7\frac{5}{6} + 9\frac{3}{8}$ **$17\frac{5}{24}$**
12. $3\frac{1}{4} - 1\frac{3}{4}$ **$1\frac{1}{2}$**
13. $11 - 6\frac{3}{8}$ **$4\frac{5}{8}$**

14. **AQUARIUMS** A fish tank that holds $18\frac{2}{3}$ gallons of water has $10\frac{1}{2}$ gallons so far. How much more water can be added? **$8\frac{1}{6}$ gal**

★ indicates increased difficulty

Practice and Applications

HOMEWORK HELP

For Exercises	See Examples
15–25, 28–30	1, 2
35–38	3
26–27, 31–34	4, 5

Extra Practice
See pages 577, 601.

Add or subtract. Write in simplest form.

15. $2\frac{1}{9} + 7\frac{4}{9}$ **$9\frac{5}{9}$**
16. $10\frac{4}{5} - 2\frac{1}{5}$ **$8\frac{3}{5}$**
17. $3\frac{1}{6} + 5\frac{1}{6}$ **$8\frac{1}{3}$**
18. $6\frac{5}{7} + 8\frac{6}{7}$ **$15\frac{4}{7}$**
19. $11\frac{3}{4} - 4\frac{1}{3}$ **$7\frac{5}{12}$**
20. $4\frac{3}{8} + 10\frac{5}{12}$ **$14\frac{19}{24}$**
21. $9\frac{4}{5} - 2\frac{3}{10}$ **$7\frac{1}{2}$**
22. $6\frac{5}{6} - 2\frac{1}{3}$ **$4\frac{1}{2}$**
23. $2\frac{3}{8} + 5\frac{7}{8}$ **$8\frac{1}{4}$**
24. $8\frac{3}{7} + 1\frac{4}{7}$ **10**
25. $6\frac{2}{3} - 1\frac{3}{5}$ **$5\frac{1}{15}$**
26. $4\frac{3}{10} - 1\frac{3}{4}$ **$2\frac{11}{20}$**
27. $14\frac{1}{6} - 7\frac{1}{3}$ **$6\frac{5}{6}$**
28. $3\frac{7}{9} + 3\frac{5}{9}$ **$7\frac{1}{3}$**
29. $8\frac{5}{12} + 11\frac{1}{4}$ **$19\frac{2}{3}$**
30. $7\frac{7}{8} + 10\frac{5}{6}$ **$18\frac{17}{24}$**
31. $9\frac{1}{5} - 2\frac{3}{5}$ **$6\frac{3}{5}$**
32. $12\frac{1}{2} - 6\frac{5}{8}$ **$5\frac{7}{8}$**
33. $8 - 3\frac{2}{3}$ **$4\frac{1}{3}$**
34. $13 - 5\frac{5}{6}$ **$7\frac{1}{6}$**

Answer

1. Sample answer: A table is $8\frac{1}{2}$ feet long. A tablecloth is $2\frac{1}{3}$ feet longer. How long is the tablecloth? Answer: $8\frac{1}{2} + 2\frac{1}{3}$ or $10\frac{5}{6}$ ft

STOCK MARKET **For Exercises 35–37, use the following information.**
Until several years ago, stock prices were listed as mixed numbers. Find the difference between the high and low price of each restaurant chain stock shown in the table.

35. Restaurant A $31\frac{3}{16}$
36. Restaurant B $11\frac{1}{2}$
37. Restaurant C $38\frac{5}{8}$

Stock Prices		
Company	High Price	Low Price
Restaurant A	$52\frac{5}{16}$	$21\frac{1}{8}$
Restaurant B	$42\frac{1}{4}$	$30\frac{3}{4}$
Restaurant C	$68\frac{3}{8}$	$29\frac{3}{4}$

★ 38. **MONUMENTS** The Washington Monument is 555 feet $5\frac{1}{8}$ inches tall. The San Jacinto Monument near Houston, Texas, is 14 feet $6\frac{7}{8}$ inches taller. How tall is the San Jacinto Monument? **570 ft**

Add or subtract. Write in simplest form. **41. $13\frac{7}{18}$**

★ 39. $10 - 3\frac{5}{11}$ $6\frac{6}{11}$ ★ 40. $24 - 8\frac{3}{4}$ $15\frac{1}{4}$ ★ 41. $6\frac{1}{6} + 1\frac{2}{3} + 5\frac{5}{9}$ ★ 42. $3\frac{1}{4} + 2\frac{5}{6} - 4\frac{1}{3}$ $1\frac{3}{4}$

43. **CRITICAL THINKING** A string is cut in half, and one of the halves is used to bundle newspapers. Then one fifth of the remaining string is cut off. The piece left is 8 feet long. How long was the string originally? **20 ft**

Standardized Test Practice and Mixed Review

44. **MULTIPLE CHOICE** What is the sum of $7\frac{1}{2}$ and $2\frac{1}{6}$? **C**

Ⓐ $5\frac{1}{3}$ Ⓑ $9\frac{1}{2}$ Ⓒ $9\frac{2}{3}$ Ⓓ $10\frac{2}{3}$

45. **MULTIPLE CHOICE** Melanie had $4\frac{2}{3}$ pounds of chopped walnuts. She used $1\frac{1}{4}$ pounds in a recipe. How many pounds of chopped walnuts did she have left? **H**

Ⓕ $2\frac{1}{3}$ lb Ⓖ $2\frac{5}{12}$ lb Ⓗ $3\frac{5}{12}$ lb Ⓘ $3\frac{1}{2}$ lb

46. Find $\frac{7}{10} - \frac{1}{3}$. Write in simplest form. (Lesson 6-2) $\frac{11}{30}$

Estimate. (Lesson 6-1) **47–50. Sample answers are given.**

47. $\frac{8}{9} \div \frac{9}{10}$ $1 \div 1 = 1$
48. $3\frac{1}{2} + 6\frac{2}{3}$ $4 + 7 = 11$
49. $8\frac{4}{5} \times 7\frac{1}{9}$ $9 \times 7 = 63$
50. $4\frac{2}{9} - 1\frac{1}{4}$ $4 - 1 = 3$

Replace each ● with <, >, or = to make a true sentence. (Lesson 5-8)

51. $\frac{4}{5}$ ● $\frac{7}{9}$ > 52. $\frac{2}{3}$ ● $\frac{5}{6}$ < 53. $\frac{1}{8}$ ● 0.15 < 54. $\frac{3}{7}$ ● 0.4 >

GETTING READY FOR THE NEXT LESSON

PREREQUISITE SKILL **Write each mixed number as an improper fraction.** (Page 563)

55. $2\frac{3}{8}$ $\frac{19}{8}$ 56. $1\frac{2}{7}$ $\frac{9}{7}$ 57. $5\frac{1}{10}$ $\frac{51}{10}$ 58. $6\frac{4}{5}$ $\frac{34}{5}$

msmath2.net/self_check_quiz

4 Assess

Open-Ended Assessment

Speaking Ask, "How do we subtract $2\frac{3}{5}$ from 4?" Have a student or several students (in turn) describe the steps for solving the problem as you write the suggested steps on the chalkboard.

Study Guide and Intervention, p. 313

NAME ________ DATE ________ PERIOD ____

6-3 Study Guide and Intervention
Adding and Subtracting Mixed Numbers

To add or subtract mixed numbers:
1. Add or subtract the fractions. Rename using the LCD if necessary.
2. Add or subtract the whole numbers.
3. Simplify if necessary.

EXAMPLE 1 Find $14\frac{1}{2} + 18\frac{2}{3}$.

$14\frac{1}{2} \rightarrow 14\frac{3}{6}$ Rename the fractions.
$+18\frac{2}{3} \rightarrow +18\frac{4}{6}$ Add the whole numbers and add the fractions.
$32\frac{7}{6}$ or $33\frac{1}{6}$ Simplify.

EXAMPLE 2 Find $21 - 12\frac{5}{8}$.

$21 \rightarrow 20\frac{8}{8}$ Rename 21 as $20\frac{8}{8}$.
$-12\frac{5}{8} \rightarrow -12\frac{5}{8}$ First subtract the whole numbers and then the fractions.
$8\frac{3}{8}$

EXERCISES

Add or subtract. Write in simplest form.

1. $7\frac{3}{4} + 2\frac{3}{4}$ $10\frac{1}{2}$ 2. $14\frac{2}{9} - 6\frac{1}{9}$ $8\frac{1}{9}$ 3. $9\frac{1}{5} - 4\frac{3}{4}$ $4\frac{9}{20}$
4. $7\frac{1}{8} + 5\frac{3}{8}$ $12\frac{1}{2}$ 5. $7\frac{3}{4} + 2\frac{2}{3}$ $10\frac{5}{12}$ 6. $5\frac{1}{2} - 5\frac{1}{3}$ $\frac{1}{6}$
7. $5\frac{1}{2} - 3\frac{1}{4}$ $2\frac{1}{4}$ 8. $6\frac{1}{3} + 2\frac{1}{6}$ $8\frac{1}{2}$ 9. $9 - 3\frac{2}{5}$ $5\frac{3}{5}$
10. $2\frac{2}{3} + 7\frac{1}{2}$ $10\frac{1}{6}$ 11. $6\frac{1}{2} - 6\frac{1}{3}$ $\frac{1}{6}$ 12. $18\frac{1}{2} + 5\frac{5}{8}$ $24\frac{1}{8}$

Practice: Skills, p. 314

NAME ________ DATE ________ PERIOD ____

6-3 Practice: Skills
Adding and Subtracting Mixed Numbers

Add or subtract. Write in simplest form.

1. $3\frac{2}{7} + 2\frac{1}{7}$ $5\frac{3}{7}$ 2. $7\frac{1}{3} + 7\frac{1}{3}$ $14\frac{2}{3}$ 3. $9\frac{3}{5} - 2\frac{1}{5}$ $7\frac{2}{5}$
4. $7\frac{3}{4} - 5\frac{1}{4}$ $2\frac{1}{2}$ 5. $3\frac{1}{4} + 5\frac{1}{4}$ $8\frac{1}{2}$ 6. $6\frac{3}{4} - 5\frac{3}{4}$ 1
7. $12\frac{1}{8} + 9\frac{3}{8}$ $21\frac{1}{2}$ 8. $5\frac{2}{3} - 2\frac{1}{3}$ $3\frac{1}{3}$ 9. $14\frac{3}{5} - 9\frac{2}{5}$ $5\frac{1}{5}$
10. $5\frac{1}{2} + 3\frac{1}{4}$ $8\frac{3}{4}$ 11. $2\frac{1}{3} + 6\frac{1}{6}$ $8\frac{1}{2}$ 12. $6\frac{1}{3} - 6\frac{1}{4}$ $\frac{1}{12}$
13. $7\frac{5}{6} - 2\frac{2}{3}$ $5\frac{1}{6}$ 14. $6\frac{7}{10} + 5\frac{1}{4}$ $11\frac{19}{20}$ 15. $12\frac{3}{8} - 9\frac{1}{3}$ $3\frac{1}{24}$
16. $12\frac{13}{15} + 4\frac{1}{9}$ $16\frac{44}{45}$ 17. $15\frac{2}{3} - 7\frac{1}{5}$ $8\frac{7}{15}$ 18. $4\frac{7}{12} - 2\frac{3}{16}$ $2\frac{19}{48}$
19. $8\frac{3}{4} + 3\frac{2}{5}$ $12\frac{3}{20}$ 20. $12\frac{1}{3} - 8\frac{5}{9}$ $3\frac{7}{9}$ 21. $8 - 3\frac{2}{5}$ $4\frac{3}{5}$
22. $7\frac{7}{9} + 6\frac{7}{8}$ $14\frac{47}{72}$ 23. $7\frac{7}{9} - 6\frac{7}{8}$ $\frac{65}{72}$ 24. $10\frac{3}{8} + 7\frac{11}{12}$ $18\frac{7}{24}$

Practice: Word Problems, p. 315

NAME ________ DATE ________ PERIOD ____

6-3 Practice: Word Problems
Adding and Subtracting Mixed Numbers

HEIGHT **For Exercises 1–6, use the table below. It shows the heights of five students. Write all answers in simplest form.**

Student	Height (ft)
Karen	$5\frac{1}{6}$
Arturo	6
Felisa	$4\frac{11}{12}$
Max	$5\frac{3}{4}$
Silvia	$5\frac{1}{4}$

1. How much taller is Max than Felisa? $\frac{5}{6}$ ft
2. How much shorter is Karen than Arturo? $\frac{5}{6}$ ft
3. If Silvia stood on a box that was $1\frac{5}{6}$ feet high, how far would the top of her head be from the floor? $7\frac{1}{12}$ ft
4. What is the sum of Felisa and Silvia's heights? $10\frac{1}{6}$ ft
5. The distance from the floor to Karen's knee is $1\frac{1}{4}$ feet. What is the distance from her knee to the top of her head? $3\frac{11}{12}$ ft
6. Max grew $\frac{1}{3}$ foot last year. How tall was he last year? $5\frac{5}{12}$ ft

6-3b Problem-Solving Strategy

A Follow-Up of Lesson 6-3

What You'll Learn
Solve problems by eliminating possibilities.

Eliminate Possibilities

I recorded $3\frac{1}{4}$ hours of a miniseries on a videotape that can record 6 hours of programming. What is the most that I can record on the rest of the same tape—2 hours, $2\frac{1}{2}$ hours, or 3 hours?

Well, we can **eliminate** some **possibilities** by estimating.

Explore	We know the combined hours of programming must be less than or equal to 6 hours.
Plan	Let's eliminate answers that are not reasonable.
Solve	You couldn't record 3 more hours on the tape because $3\frac{1}{4} + 3 = 6\frac{1}{4}$. So, we can eliminate that choice. Now let's check the choice of $2\frac{1}{2}$ hours. $3\frac{1}{4} + 2\frac{1}{2} = 5\frac{3}{4}$ Since this is less than 6 hours, this choice is correct. You could record $2\frac{1}{2}$ more hours on the tape.
Examine	Recording 2 more hours would give $3\frac{1}{4} + 2$ or $5\frac{1}{4}$ hours. This is less than the 6-hour maximum, but not the most that you could record.

Analyze the Strategy

1. **Describe** different ways that you can eliminate possibilities when solving problems. **Sample answer: Use estimation, look for a pattern, work backward.**
2. **Explain** how the strategy of eliminating possibilities is useful for taking multiple choice tests. **2–3. See Ch. 6 Answer Appendix.**
3. **Write** a problem that could be solved by eliminating the possibilities. Explain your answer.

6-3b Lesson Notes

1 Focus

Bellringer

Communication Write the following problem on the chalkboard. Ask students to explain how they could find the answer without calculating.

$3\frac{3}{7} - 2\frac{1}{2}$

A 0 B $\frac{7}{8}$

C $\frac{13}{14}$ D 1

Sample answer: We can estimate the answer to be about 1. So, it cannot be A. Eight is not a common denominator of 7 and 2. So, the answer is not B. To get an answer of 1, $\frac{3}{7} - \frac{1}{2}$ would have to equal 0, which it does not. So, the answer is not D. So, C must be the answer.

2 Teach

In-Class Example

INTERACTIVE CHALKBOARD PowerPoint® Presentations

Solve. Use the eliminate possibilities strategy.

GAMES On a television game show, the winning contestant must answer three questions correctly to take home the grand prize. Each question is worth twice as many points as the question before it. The third question is worth 1,000 points. How much is the first question worth—250 points, 500 points, or 2,000 points?
250 points

DAILY INTERVENTION — Flexible Grouping

Eliminate Possibilities Organize students into groups of four or five. Provide each student with a worksheet of multiple-choice problems. Tell students, that for each problem, they first are to discuss the problem, eliminate possibilities, and find the solution. Only then they are to record the answer on their worksheets. Inform students that you will be collecting only one sheet from each group to determine the group members' scores.

pply the Strategy

olve. Use the eliminate possibilities strategy.

JUICE Lauren has a 3-gallon cooler with $1\frac{3}{4}$ gallons of juice in it. If she wants the cooler full for her soccer game, how much juice does she need to add? **C**

Ⓐ 4 gal Ⓑ $3\frac{1}{4}$ gal

Ⓒ $1\frac{1}{4}$ gal Ⓓ $\frac{1}{4}$ gal

5. **ELEPHANTS** An elephant in a zoo eats 58 cabbages in a week. About how many cabbages does an elephant eat in one year? **I**

Ⓕ 7 Ⓖ 700

Ⓗ 1,500 Ⓘ 3,000

ixed Problem Solving

olve. Use any strategy. **7. Sample answer: The grade scale does not have equal intervals.**

RAIN FOREST In some areas of the rain forest, 325 inches of rain may fall in a year. Which is the *best* estimate for the average rainfall per day in such an area? **B**

Ⓐ $\frac{1}{3}$ in. Ⓑ 1 in.

Ⓒ 5 in. Ⓓ 33 in.

GRADES Explain why the graph showing a student's science grades is misleading.

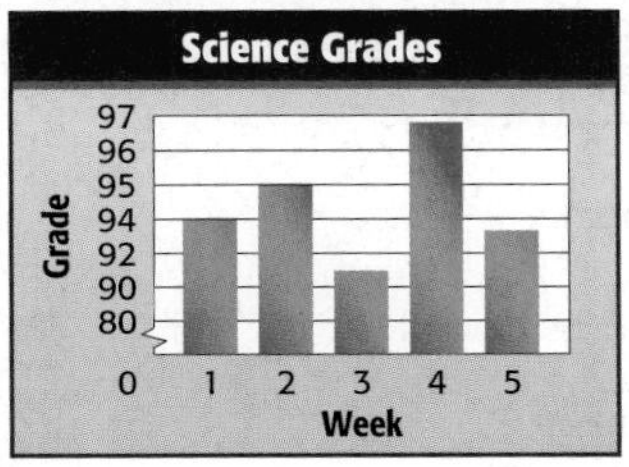

ELEVATORS An elevator can hold a maximum weight of 3,500 pounds. Which is the *best* estimate for the number of adults that the elevator can hold? **G**

Ⓕ 10 adults Ⓖ 20 adults

Ⓗ 35 adults Ⓘ 80 adults

SUPPLIES Vanessa has $55 to buy school supplies. She bought a backpack that costs $23.50, a combination lock that costs $6.25, and 4 binders that are $3.99 each. If mechanical pencils are $2.50 per pack, how many packs can she buy? **3 packs**

10. **SHOPPING** Abby bought the items at the right for a party. Which is the *best* estimate of the total cost of the items, not including tax? **B**

Item	Price
balloons	$2.95
cups and plates	$9.30
streamers	$4.50

Ⓐ less than $15

Ⓑ between $15 and $25

Ⓒ between $25 and $35

Ⓓ more than $35

11. **TRAVEL** Mr. Rollins drove 780 miles on a five-day trip. He rented a car for $23 per day plus $0.15 per mile after 500 free miles. About how much did the rental car cost? **H**

Ⓕ $100 Ⓖ $130 Ⓗ $160 Ⓘ $180

12. **STANDARDIZED TEST PRACTICE**
If the trend in the graph continues, which is the best estimate for the number of acres of wheat harvested in the United States in 2004?

Ⓐ 60 million

Ⓑ 55 million

Ⓒ 45 million

Ⓓ 40 million

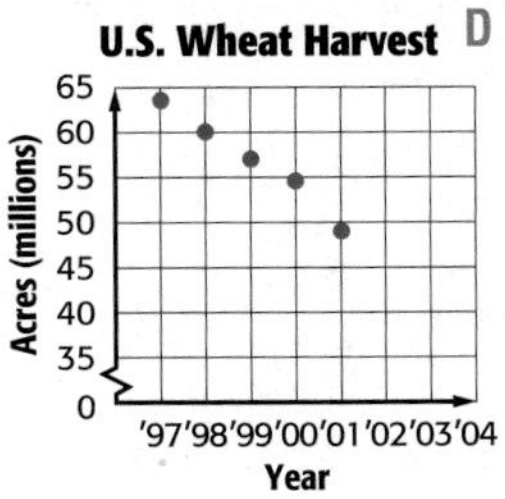

Tips for New Teachers

Intervention You may wish to refer to this lesson whenever preparing students for standardized tests. Such tests commonly have at least one answer choice that can be eliminated quickly before any calculations are made. Other choices often are erroneous solutions found by making only one error in calculation. Have students write some multiple-choice questions, using these guidelines for the answer choices, to reinforce this concept.

3 Practice/Apply

About the Exercises...

Problem-Solving Strategies
Exercises 6–12 are structured so that students have the opportunity to practice many different problem-solving strategies. You may wish to remind them of some of the strategies they have studied.

- use the four-step plan (p. 4)
- guess and check (p. 22)
- use a graph (p. 58)
- look for a pattern (p. 132)
- work backward (p. 164)
- make an organized list (p. 201)
- eliminate possibilities ✓

4 Assess

Open-Ended Assessment

Modeling Have students write a brief paragraph explaining how to use the eliminate possibilities strategy to solve problems, as well as how this method can help when taking tests.

6-4 Lesson Notes

Lesson Resources

Chapter 6 Resource Masters
- Study Guide and Intervention, p. 318
- Practice: Skills, p. 319
- Practice: Word Problems, p. 320
- Reading to Learn Mathematics, p. 321
- Enrichment, p. 322

Calculator and Spreadsheet Masters, p. 33
Key Teaching Transparencies, 22
MathPASS Tutorial CD-ROM, 20–21
Parent and Student Study Guide Workbook, 6-4
Teaching Mathematics with Manipulatives, pp. 62–63

FAST FILE

1 Focus

5-Minute Check 6-4

INTERACTIVE CHALKBOARD PowerPoint® Presentations

Use as a quiz or review of Lesson 6-3. (Also available as a transparency)

Bellringer
Problem Solving Write the questions in the bottom margin on the chalkboard. Direct students to solve, using sketches and manipulatives if they want. Students should simplify answers.

(Continued in bottom margin.)

Reading to Learn Mathematics, p. 321 ELL

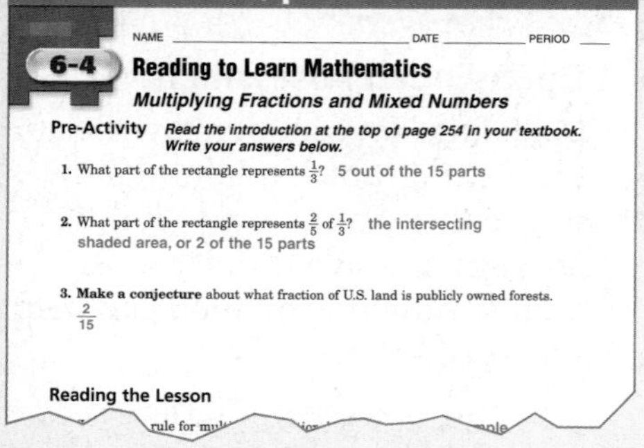
NAME ______ DATE ______ PERIOD ____

6-4 Reading to Learn Mathematics
Multiplying Fractions and Mixed Numbers

Pre-Activity *Read the introduction at the top of page 254 in your textbook. Write your answers below.*

1. What part of the rectangle represents $\frac{1}{3}$? 5 out of the 15 parts

2. What part of the rectangle represents $\frac{2}{5}$ of $\frac{1}{3}$? the intersecting shaded area, or 2 of the 15 parts

3. **Make a conjecture** about what fraction of U.S. land is publicly owned forests. $\frac{2}{15}$

Reading the Lesson

6-4 Multiplying Fractions and Mixed Numbers

What You'll Learn
Multiply fractions and mixed numbers.

REVIEW Vocabulary
GCF: the greatest of the common factors of two or more numbers **(Lesson 5-2)**

WHEN am I ever going to use this?

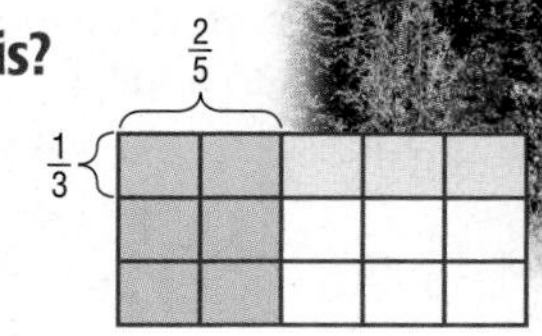

EARTH SCIENCE About $\frac{1}{3}$ of the land in the United States is forests. About $\frac{2}{5}$ of U.S. forests are publicly owned.

1. What part of the rectangle represents $\frac{1}{3}$? **5 out of the 15 parts**
2. What part of the rectangle represents $\frac{2}{5}$ of $\frac{1}{3}$?
3. **Make a conjecture** about what fraction of U.S. land is publicly owned forests. $\frac{2}{15}$

2. the intersecting shaded area, or 2 of the 15 parts

You can multiply fractions by using the rule below.

Key Concept — **Multiplying Fractions**

Words To multiply fractions, multiply the numerators and multiply the denominators.

Symbols

Arithmetic $\frac{1}{3} \times \frac{2}{5} = \frac{1 \times 2}{3 \times 5} = \frac{2}{15}$

Algebra $\frac{a}{b} \times \frac{c}{d} = \frac{a \times c}{b \times d} = \frac{ac}{bd}$

EXAMPLES Multiply Fractions

Multiply. Write in simplest form.

1 $\frac{1}{2} \times \frac{1}{3}$

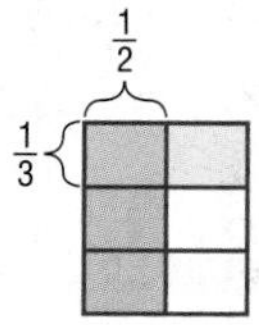

$\frac{1}{2} \times \frac{1}{3} = \frac{1 \times 1}{2 \times 3}$ ← Multiply the numerators. ← Multiply the denominators.

$= \frac{1}{6}$ Simplify.

2 $2 \times \frac{3}{4}$

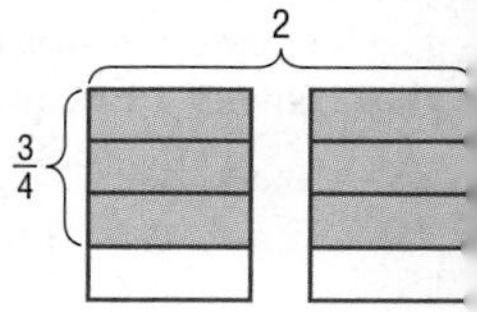

$2 \times \frac{3}{4} = \frac{2}{1} \times \frac{3}{4}$ Write 2 as $\frac{2}{1}$.

$= \frac{2 \times 3}{1 \times 4}$ Multiply the numerators and multiply the denominators.

$= \frac{6}{4}$ or $1\frac{1}{2}$ Simplify.

TEACHING TIP
Just as 3 × 2 means 3 sets of 2 units, $\frac{1}{2} \times \frac{1}{3}$ means $\frac{1}{2}$ set of $\frac{1}{3}$ unit.

Your Turn **Multiply. Write in simplest form.**

a. $\frac{3}{5} \times \frac{1}{2}$ $\frac{3}{10}$
b. $\frac{1}{3} \times \frac{3}{4}$ $\frac{1}{4}$
c. $\frac{2}{3} \times 4$ $2\frac{2}{3}$

What is half of $\frac{1}{4}$? $\frac{1}{8}$

What is three times $\frac{2}{5}$? $1\frac{1}{5}$

What is three fourths of 7? $5\frac{1}{4}$

If the numerator and denominator of either fraction have common factors, you can simplify before multiplying.

STUDY TIP

Simplifying You can simplify by dividing a numerator and a denominator by the same factor. Doing this step before you multiply is easier than doing it after you multiply.

EXAMPLE Simplify Before Multiplying

3 **Find $\frac{2}{7} \times \frac{3}{8}$. Write in simplest form.**

$\frac{2}{7} \times \frac{3}{8} = \frac{\overset{1}{\cancel{2}}}{7} \times \frac{3}{\underset{4}{\cancel{8}}}$ Divide 2 and 8 by their GCF, 2.

$= \frac{1 \times 3}{7 \times 4}$ Multiply the numerators and multiply the denominators.

$= \frac{3}{28}$ Simplify.

Your Turn **Multiply. Write in simplest form.**

d. $\frac{1}{3} \times \frac{3}{7}$ $\frac{1}{7}$ e. $\frac{4}{9} \times \frac{1}{8}$ $\frac{1}{18}$ f. $\frac{5}{6} \times \frac{3}{5}$ $\frac{1}{2}$

EXAMPLE Multiply Mixed Numbers

4 **Find $\frac{1}{2} \times 4\frac{2}{5}$. Write in simplest form.** **Estimate** $\frac{1}{2} \times 4 = 2$

Method 1 Rename the mixed number.

$\frac{1}{2} \times 4\frac{2}{5} = \frac{1}{\underset{1}{\cancel{2}}} \times \frac{\overset{11}{\cancel{22}}}{5}$ Rename $4\frac{2}{5}$ as an improper fraction, $\frac{22}{5}$.

$= \frac{1 \times 11}{1 \times 5}$ Multiply.

$= \frac{11}{5}$ or $2\frac{1}{5}$ Simplify.

The product is close to the estimate.

STUDY TIP

Mental Math When you see a problem like $\frac{1}{2} \times 4\frac{2}{5}$, you can use the Distributive Property. Think, "What is $\frac{1}{2}$ of 4 and what is $\frac{1}{2}$ of $\frac{2}{5}$?" This is equal to $\frac{1}{2}\left(4 + \frac{2}{5}\right)$.

Method 2 Use mental math.

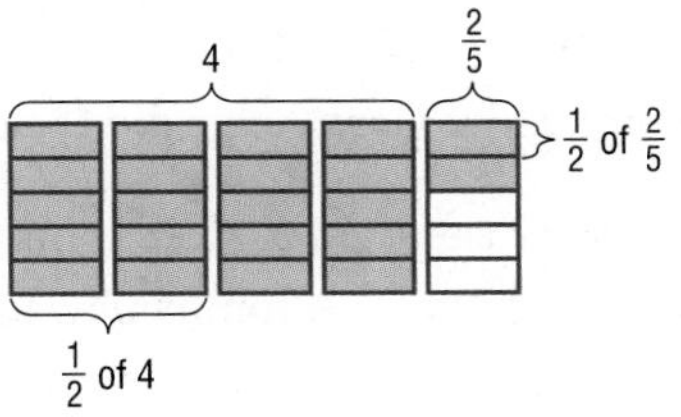

$\frac{1}{2} \times 4\frac{2}{5} = \frac{1}{2} \times \left(4 + \frac{2}{5}\right)$ Write $4\frac{2}{5}$ as a sum of its parts.

$= \left(\frac{1}{2} \times 4\right) + \left(\frac{1}{2} \times \frac{2}{5}\right)$ Distributive Property

$= 2 + \frac{1}{5}$ or $2\frac{1}{5}$ Multiply.

Compare this product to the model shown above.

Your Turn **Multiply. Write in simplest form.**

g. $\frac{1}{4} \times 8\frac{4}{9}$ $2\frac{1}{9}$ h. $3 \times 5\frac{1}{3}$ 16   i. $1\frac{7}{8} \times 2\frac{2}{5}$ $4\frac{1}{2}$

msmath2.net/extra_examples

2 Teach

In-Class Examples

INTERACTIVE CHALKBOARD PowerPoint® Presentations

Multiply. Write in simplest form.

1. $\frac{1}{8} \times \frac{1}{9}$ $\frac{1}{72}$
2. $6 \times \frac{1}{3}$ 2
3. $\frac{3}{12} \times \frac{4}{5}$ $\frac{1}{5}$
4. $\frac{1}{3} \times 6\frac{6}{7}$ $2\frac{2}{7}$

DAILY INTERVENTION

Differentiated Instruction

Auditory/Musical Bring in sheet music to a popular song or show tune. Have the class sing a bar of the music, keeping the beat by clapping. Then have them double all of the notes, then cut all the notes in half, singing each version. Ask them, "If you double the notes, what does a quarter note become?" **a half note** Discuss how the new versions of the music relate to the pace of the music. **Doubling the notes makes the pace twice as slow; halving the notes makes the pace twice as fast.**

3 Practice/Apply

FOLDABLES™ Follow-Up Remind students to reword main ideas, definitions of vocabulary words, and other notes on study cards as they progress through the lesson. Students store their study cards on multiplying fractions in the "Fractions" pocket and their notes on multiplying numbers in the "Mixed Numbers" pocket of their Foldables.

Which One Doesn't Belong?
For Exercise 2, have students model each expression with counters as you read the expressions aloud. As they begin modeling the third choice, students should quickly see that it is different. Ask them how to rewrite the expression so it matches the others. $25 \div 5$

NUMBER SENSE
For Exercise 3, write these expressions on the chalkboard.

$4 \times \frac{5}{6}$ $\quad 1 \times \frac{2}{3}$

$4 \times \frac{6}{5}$ $\quad 1 \times \frac{3}{2}$

Direct students to come up with the following rule about multiplying by fractions. *Multiplying a number by a fraction less than 1 yields a smaller number; multiplying by a fraction greater than 1 yields a larger number.*

About the Exercises...

Guided Practice Key

Exercises	Examples
4	1
7, 10	2
5, 6	3
8, 9	4

Odd/Even Assignments
Exercises 11–36 are structured so that students practice the same concepts whether they are assigned odd or even problems.

Skill and Concept Check

Writing Math Exercises 2 & 3

1. **OPEN ENDED** Write a pair of fractions whose product is $\frac{8}{15}$. 1–2. See margin.
2. **Which One Doesn't Belong?** Identify the expression that does not have the same value as the other three. Explain your reasoning.

$\frac{1}{5}$ of 25	$25 \times \frac{1}{5}$	$5 \div 25$	$\frac{1}{5} \cdot \frac{25}{1}$

3. **NUMBER SENSE** Is $18 \times \frac{4}{5}$ *greater than, less than,* or *equal to* 18? Explain. Less than; $\frac{4}{5}$ of 18 is a fraction of 18.

GUIDED PRACTICE

Multiply. Write in simplest form.

4. $\frac{2}{3} \times \frac{1}{3}$ $\frac{2}{9}$
5. $\frac{1}{2} \times \frac{2}{5}$ $\frac{1}{5}$
6. $\frac{2}{3} \times \frac{3}{8}$ $\frac{1}{4}$
7. $\frac{1}{6} \times 4$ $\frac{2}{3}$
8. $2\frac{1}{4} \times \frac{2}{3}$ $1\frac{1}{2}$
9. $1\frac{5}{6} \times 3\frac{3}{5}$ $6\frac{3}{5}$
10. **FOOD** An average slice of American cheese is about $\frac{1}{8}$ inch thick. What is the height of a package containing 20 slices? $2\frac{1}{2}$ in.

★ indicates increased difficulty

Practice and Applications

HOMEWORK HELP

For Exercises	See Examples
11–12	1
21–24, 34	2
13–20, 33	3
25–32, 35–36	4

Extra Practice
See pages 578, 601.

Multiply. Write in simplest form.

11. $\frac{3}{4} \times \frac{1}{8}$ $\frac{3}{32}$
12. $\frac{2}{5} \times \frac{2}{3}$ $\frac{4}{15}$
13. $\frac{1}{5} \times \frac{5}{6}$ $\frac{1}{6}$
14. $\frac{4}{9} \times \frac{1}{4}$ $\frac{1}{9}$
15. $\frac{2}{3} \times \frac{1}{4}$ $\frac{1}{6}$
16. $\frac{1}{12} \times \frac{3}{5}$ $\frac{1}{20}$
17. $\frac{4}{7} \times \frac{7}{8}$ $\frac{1}{2}$
18. $\frac{2}{5} \times \frac{15}{16}$ $\frac{3}{8}$
19. $\frac{3}{8} \times \frac{10}{27}$ $\frac{5}{36}$
20. $\frac{9}{10} \times \frac{5}{6}$ $\frac{3}{4}$
21. $9 \times \frac{1}{2}$ $4\frac{1}{2}$
22. $\frac{4}{5} \times 6$ $4\frac{4}{5}$
23. **ELECTIONS** In an election in which 4,500 votes were cast, one candidate received $\frac{3}{5}$ of the votes. How many votes did the candidate receive? 2,700 votes
24. **PACKAGING** The plastic cases used to store compact disks and DVDs are about $\frac{1}{5}$-inch thick. A company wants to sell 10 of these cases in plastic wrapping. What is the height of 10 cases? 2 in.

Multiply. Write in simplest form.

25. $4\frac{2}{3} \times \frac{4}{7}$ $2\frac{2}{3}$
26. $\frac{5}{8} \times 2\frac{1}{2}$ $1\frac{9}{16}$
27. $14 \times 1\frac{1}{7}$ 16
28. $3\frac{3}{4} \times 8$ 30
29. $9 \times 4\frac{2}{3}$ 42
30. $4 \times 7\frac{5}{6}$ $31\frac{1}{3}$
31. $3\frac{1}{4} \times 2\frac{2}{3}$ $8\frac{2}{3}$
32. $5\frac{1}{3} \times 3\frac{3}{4}$ 20
33. **TELEVISION** A media research survey showed that one evening, $\frac{2}{3}$ of all U.S. households had their TVs on, and $\frac{3}{8}$ of them were watching a World Series baseball game. What fraction of U.S. households was watching the game? $\frac{1}{4}$

Answers

1. Sample answer: $\frac{2}{3}$ and $\frac{4}{5}$
2. $5 \div 25$, because it equals $\frac{1}{5}$. The other expressions equal 5.

34. **ANIMALS** Komodo dragons are the largest lizards in the world. A 250-pound komodo dragon can eat enough at one time to increase its weight by $\frac{3}{4}$. Find $\frac{3}{4} \times 250$ to determine how much weight a 250-pound komodo dragon could gain after eating. **$187\frac{1}{2}$ lb**

★ 35. **TURTLES** A giant tortoise can travel about one tenth of a kilometer in an hour. At this rate, how far can it travel in $1\frac{3}{4}$ hours? **$\frac{7}{40}$ or 0.175 km**

★ 36. **FLAGS** By law, the length of an official United States flag must be $1\frac{9}{10}$ times its width. What is the length of the flag shown at the right? **$6\frac{13}{20}$ ft**

37. **CRITICAL THINKING** Two positive fractions which are *not* improper are multiplied. Is the product *sometimes*, *always*, or *never* less than 1? Explain. **See margin.**

Standardized Test Practice and Mixed Review

38. **MULTIPLE CHOICE** A box of books weighs $8\frac{2}{3}$ pounds. How much do $4\frac{1}{2}$ boxes weigh? **D**

Ⓐ 19 lb Ⓑ $27\frac{2}{3}$ lb Ⓒ $32\frac{1}{2}$ lb Ⓓ 39 lb

39. **GRID IN** Jeanette and Vanesa are each taking half of the leftover pizza shown at the right. What fraction of the whole pizza does each person take? **7/16**

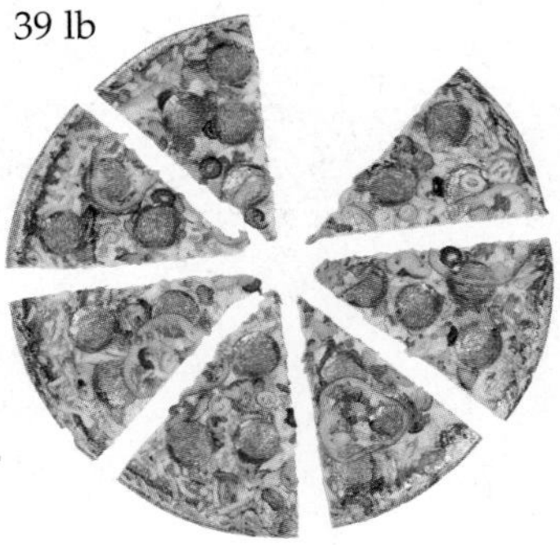

40. **LIFE SCIENCE** A female anglerfish is approximately $2\frac{1}{2}$ inches long, and a male is about $\frac{2}{5}$ inch long. How much longer is the female than the male? (Lesson 6-3) **$2\frac{1}{10}$ in.**

Add or subtract. Write in simplest form. (Lesson 6-2)

41. $\frac{2}{7} + \frac{4}{7}$ **$\frac{6}{7}$**
42. $\frac{1}{2} - \frac{1}{8}$ **$\frac{3}{8}$**
43. $\frac{5}{9} + \frac{5}{6}$ **$1\frac{7}{18}$**
44. $\frac{3}{4} - \frac{1}{6}$ **$\frac{7}{12}$**

Find the greatest common factor of each set of numbers. (Lesson 5-2)

45. 44, 60 **4**
46. 25, 75 **25**
47. 6, 15, 27 **3**
48. 12, 30, 48 **6**

49. **ALGEBRA** On a January day in Buffalo, New York, the temperature dropped to −20.8°C. Find this temperature in degrees Fahrenheit by using the formula $C = 0.6F - 17.8$. (Lesson 4-4) **−5°F**

GETTING READY FOR THE NEXT LESSON

PREREQUISITE SKILL **Multiply.** (Page 560)

50. $2.8 \cdot 5$ **14**
51. $1.9 \cdot 33$ **62.7**
52. $7 \cdot 12.5$ **87.5**
53. $3.6 \cdot 0.8$ **2.88**

msmath2.net/self_check_quiz

4 Assess

Higher-Level Thinking

Have students write a problem using multiplication of fractions whose solution is $\frac{5}{6}$. **Sample answer: $10 \times \frac{1}{12} = \frac{10}{12}$ or $\frac{5}{6}$** You may also wish to have them make this a word problem.

Answer

37. **Always; sample answer: the numerators of the fractions are less than the denominators, so the product of the numerators will be less than the product of the denominators, resulting in a fraction that is less than 1.**

Study Guide and Intervention, p. 318

NAME ______ DATE ______ PERIOD ____

6-4 Study Guide and Intervention

Multiplying Fractions and Mixed Numbers

To multiply fractions, multiply the numerators and then multiply the denominators.

$$\frac{5}{6} \times \frac{3}{5} = \frac{5 \times 3}{6 \times 5} = \frac{15}{30} = \frac{1}{2}$$

To multiply mixed numbers, rename each mixed number as a fraction. Then multiply the fractions.

$$2\frac{2}{3} \times 1\frac{1}{4} = \frac{8}{3} \times \frac{5}{4} = \frac{40}{12} = 3\frac{1}{3}$$

EXAMPLE 1 **Find $\frac{2}{3} \times \frac{4}{5}$. Write in simplest form.**

$\frac{2}{3} \times \frac{4}{5} = \frac{2 \times 4}{3 \times 5}$ ← Multiply the numerators. ← Multiply the denominators.

$= \frac{8}{15}$ Simplify.

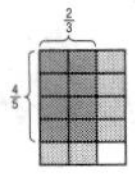

EXAMPLE 2 **Find $\frac{1}{3} \times 2\frac{1}{2}$. Write in simplest form.**

$\frac{1}{3} \times 2\frac{1}{2} = \frac{1}{3} \times \frac{5}{2}$ Rename $2\frac{1}{2}$ as an improper fraction, $\frac{5}{2}$.

$= \frac{1 \times 5}{3 \times 2}$ Multiply.

$= \frac{5}{3}$ Simplify.

EXERCISES

Multiply. Write in simplest form.

1. $\frac{2}{3} \times \frac{2}{3}$ $\frac{4}{9}$
2. $\frac{1}{2} \times \frac{7}{8}$ $\frac{7}{16}$
3. $\frac{1}{3} \times \frac{3}{5}$ $\frac{1}{5}$
4. $\frac{5}{9} \times 4$ $2\frac{2}{9}$
5. $1\frac{2}{3} \times \frac{3}{5}$ 1
6. $3\frac{3}{4} \times 1\frac{1}{6}$ $4\frac{3}{8}$
7. $\frac{3}{4} \times 1\frac{2}{3}$ $1\frac{1}{4}$
8. $3\frac{1}{3} \times 2\frac{1}{2}$ $8\frac{1}{3}$
9. $4\frac{1}{5} \times \frac{1}{7}$ $\frac{3}{5}$
10. $\frac{7}{5} \times 8$ $11\frac{1}{5}$
11. $2\frac{1}{3} \times \frac{4}{6}$ $1\frac{5}{9}$
12. $\frac{1}{8} \times 2\frac{3}{4}$ $\frac{11}{32}$

Practice: Skills, p. 319

NAME ______ DATE ______ PERIOD ____

6-4 Practice: Skills

Multiplying Fractions and Mixed Numbers

Multiply. Write in simplest form.

1. $\frac{1}{2} \times \frac{4}{5}$ $\frac{2}{5}$
2. $\frac{1}{9} \times \frac{3}{5}$ $\frac{1}{15}$
3. $\frac{15}{24} \times \frac{3}{20}$ $\frac{3}{32}$
4. $\frac{1}{7} \times \frac{1}{5}$ $\frac{1}{35}$
5. $\frac{5}{7} \times \frac{14}{15}$ $\frac{2}{3}$
6. $\frac{9}{10} \times \frac{5}{9}$ $\frac{1}{2}$
7. $\frac{4}{11} \times \frac{3}{8}$ $\frac{3}{22}$
8. $\frac{2}{3} \times \frac{7}{9}$ $\frac{14}{27}$
9. $\frac{9}{13} \times \frac{26}{27}$ $\frac{2}{3}$
10. $\frac{4}{9} \times 5$ $2\frac{2}{9}$
11. $7 \times \frac{2}{7}$ 2
12. $2\frac{4}{5} \times \frac{1}{3}$ $\frac{14}{15}$
13. $4\frac{1}{2} \times \frac{1}{3}$ $1\frac{1}{2}$
14. $5\frac{3}{4} \times 12$ 69
15. $14 \times 2\frac{3}{7}$ 34
16. $2\frac{3}{5} \times 1\frac{3}{7}$ $3\frac{5}{7}$
17. $1\frac{4}{9} \times 2\frac{4}{7}$ $3\frac{5}{7}$
18. $5\frac{5}{6} \times 6\frac{3}{8}$ $37\frac{3}{16}$
19. $10\frac{7}{9} \times 4\frac{1}{4}$ $45\frac{29}{36}$
20. $9\frac{7}{9} \times 7\frac{3}{4}$ $75\frac{7}{9}$
21. $3\frac{3}{4} \times 2\frac{4}{7}$ $9\frac{9}{14}$

Practice: Word Problems, p. 320

NAME ______ DATE ______ PERIOD ____

6-4 Practice: Word Problems

Multiplying Fractions and Mixed Numbers

1. **POPULATION** If $\frac{4}{5}$ of the population of a certain town is considered to be middle class and the population of the town is 2,000, how many people are considered middle class? 1,600 people

2. **READING** Robin has read $\frac{3}{4}$ of a book. Mark said he had read $\frac{1}{2}$ as much as Robin. What fraction of the book has Mark read? $\frac{3}{8}$

3. **RADIO** A radio station spends $\frac{1}{40}$ of each 24 hours on public service announcements. How much time is spent on public service announcements each day? $\frac{3}{5}$ h

4. **SALE** A bicycle is on sale for $\frac{2}{3}$ of its original price. If the original price is $354, what is the sale price? $236

5. **STUDENT POPULATION** One sixth of the students at a local college are seniors. The number of freshmen students is $2\frac{1}{2}$ times that amount. What fraction of the students are freshmen? $\frac{5}{12}$

6. **SEWING** Anna wants to make 4 sets of curtains. Each set requires $5\frac{1}{8}$ yards of fabric. How much fabric does she need? $20\frac{1}{2}$ yd

7. **RUNNING** It takes Awan $8\frac{1}{3}$ minutes to run one mile. It takes Kate $1\frac{1}{5}$ times longer. How long does it take Kate to run one mile? 10 min

8. **STOCK** Carl bought some stock at $25 a share. The stock increased to $1\frac{1}{2}$ times its value. How much is the stock per share? $37.50 per share

6-5 Lesson Notes

Lesson Resources

Chapter 6 Resource Masters
- Study Guide and Intervention, p. 323
- Practice: Skills, p. 324
- Practice: Word Problems, p. 325
- Reading to Learn Mathematics, p. 326
- Enrichment, p. 327

Parent and Student Study Guide Workbook, 6-5

FAST FILE

1 Focus

5-Minute Check 6-5
Use as a quiz or review of Lesson 6-4. (Also available as a transparency)

INTERACTIVE CHALKBOARD
PowerPoint® Presentations

Bellringer
Communication Write the following pairs of numbers on the chalkboard: 4 and $\frac{1}{4}$, $\frac{2}{3}$ and $\frac{3}{2}$, $\frac{5}{4}$ and $\frac{4}{5}$, $\frac{11}{19}$ and $\frac{19}{11}$, 6 and $\frac{1}{6}$. Have students compare the pairs in writing. **Sample answer: The numbers in each pair, when multiplied together, equal 1.**

Reading to Learn Mathematics, p. 326 **ELL**

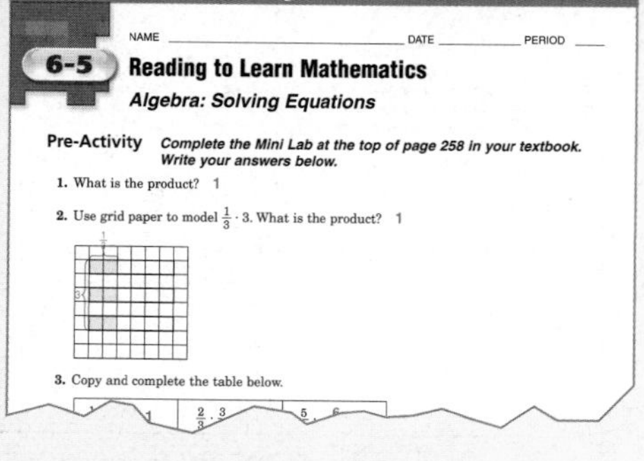
NAME ________ DATE ________ PERIOD ____
6-5 **Reading to Learn Mathematics**
Algebra: Solving Equations
Pre-Activity *Complete the Mini Lab at the top of page 258 in your textbook. Write your answers below.*
1. What is the product? 1
2. Use grid paper to model $\frac{1}{3} \cdot 3$. What is the product? 1
3. Copy and complete the table below.

6-5 Algebra: Solving Equations

What You'll Learn
Solve equations with rational number solutions.

***NEW* Vocabulary**
- multiplicative inverse
- reciprocal

***LINK* To Reading**
Everyday Meaning of Inverse: opposite in order, as in an inverse statement in logic

HANDS-ON Mini Lab

Materials
- grid paper
- colored pencils

Work with a partner.

The model below shows $\frac{1}{2} \cdot 2$.

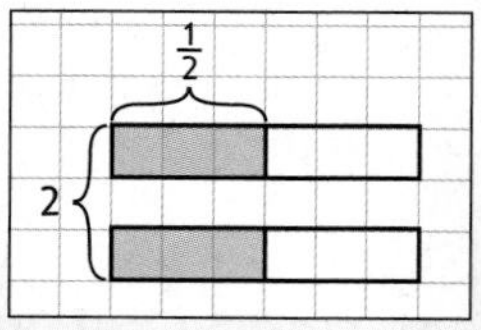

1. What is the product? **1**
2. Use grid paper to model $\frac{1}{3} \cdot 3$. What is the product? **1; See margin for model.**
3. Copy and complete the table below.

$\frac{1}{2} \cdot 2 = \underline{\ ?\ }$ **1**	$\frac{2}{3} \cdot \frac{3}{2} = \underline{\ ?\ }$ **1**	$\frac{5}{6} \cdot \underline{\ ?\ } = 1$ **$\frac{6}{5}$**
$\frac{1}{3} \cdot 3 = \underline{\ ?\ }$ **1**	$\frac{3}{5} \cdot \frac{5}{3} = \underline{\ ?\ }$ **1**	$\frac{9}{20} \cdot \underline{\ ?\ } = 1$ **$\frac{20}{9}$**
$\frac{1}{4} \cdot \frac{4}{1} = \underline{\ ?\ }$ **1**	$\frac{11}{12} \cdot \frac{12}{11} = \underline{\ ?\ }$ **1**	$\underline{\ ?\ } \cdot \frac{17}{8} = 1$ **$\frac{8}{17}$**

Two numbers whose product is 1 are called **multiplicative inverses**, or **reciprocals**.

Key Concept — **Multiplicative Inverse Property**

Words The product of a number and its multiplicative inverse is 1.

Symbols
Arithmetic: $\frac{3}{4} \times \frac{4}{3} = 1$
Algebra: $\frac{a}{b} \times \frac{b}{a} = 1$, for $a, b \neq 0$

EXAMPLES Find Multiplicative Inverses

Find the multiplicative inverse of each number.

1 $\frac{2}{5}$

$\frac{2}{5} \cdot \frac{5}{2} = 1$ Multiply $\frac{2}{5}$ by $\frac{5}{2}$ to get the product 1.

The multiplicative inverse of $\frac{2}{5}$ is $\frac{5}{2}$, or $2\frac{1}{2}$.

2 $2\frac{1}{3}$

$2\frac{1}{3} = \frac{7}{3}$ Rename the mixed number as an improper fraction.

$\frac{7}{3} \cdot \frac{3}{7} = 1$ Multiply $\frac{7}{3}$ by $\frac{3}{7}$ to get the product 1.

The multiplicative inverse of $2\frac{1}{3}$ is $\frac{3}{7}$.

Your Turn **Find the multiplicative inverse of each number.**

a. $\frac{5}{6}$ **$\frac{6}{5}$ or $1\frac{1}{5}$**   b. $1\frac{1}{2}$ **$\frac{2}{3}$**   c. 8 **$\frac{1}{8}$**   d. $-\frac{4}{3}$ **$-\frac{3}{4}$**

Answer (Hands-On Mini Lab)

2. Sample answer:

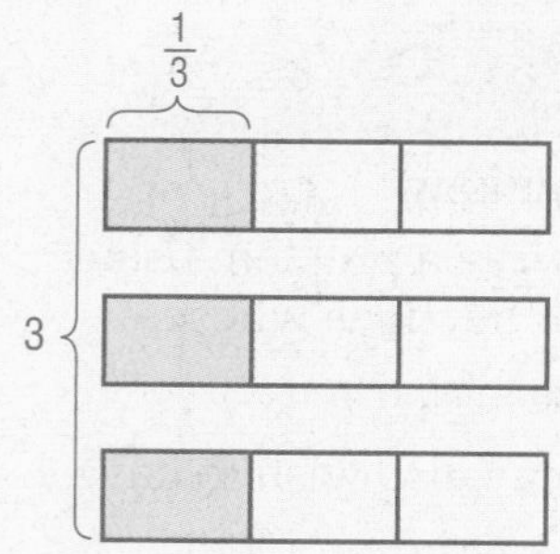

STUDY TIP

Look Back You can review the other **properties of equality** in Lessons 4-2 and 4-3.

In Chapter 4, you learned to solve equations using the Addition, Subtraction, and Division Properties of Equality. You can also solve equations by multiplying each side by the same number. This is called the **Multiplication Property of Equality**.

Key Concept — **Multiplication Property of Equality**

Words If you multiply each side of an equation by the same nonzero number, the two sides remain equal.

Symbols

Arithmetic	Algebra	
$5 = 5$	$\frac{x}{2} = -3$	$\frac{2}{3}x = 4$
$5 \cdot 2 = 5 \cdot 2$	$\frac{x}{2}(2) = -3(2)$	$\frac{3}{2} \cdot \frac{2}{3}x = \frac{3}{2} \cdot 4$
$10 = 10$	$x = -6$	$x = 6$

TEACHING TIP
[R]emind students that $\frac{n}{4}$ [m]eans $n \div 4$ or $n \cdot \frac{1}{4}$. [W]hen they multiply each [s]ide by 4, they are using [i]nverse operations.

EXAMPLE Solve a Division Equation

3 **Solve $\frac{n}{4} = 7$. Check your solution.**

$\frac{n}{4} = 7$ Write the equation.

$\frac{n}{4} \cdot 4 = 7 \cdot 4$ Multiply each side of the equation by 4.

$n = 28$ Simplify.

The solution is 28.

When the coefficient of x is a fraction, multiply each side of the equation by its reciprocal.

EXAMPLE Use a Reciprocal to Solve an Equation

Standardized Test Practice

4 **MULTIPLE-CHOICE TEST ITEM**

What value of x makes $\frac{2}{3}x = -9$ a true sentence?

(A) $\frac{3}{2}$ (B) $-\frac{18}{27}$ (C) $-\frac{9}{1}$ (D) $-13\frac{1}{2}$

Read the Test Item To find the value of x, solve the equation.

Solve the Test Item

$\frac{2}{3}x = -9$ Write the equation.

$\left(\frac{3}{2}\right)\frac{2}{3}x = \left(\frac{3}{2}\right)(-9)$ Multiply each side by the reciprocal of $\frac{2}{3}$, $\frac{3}{2}$.

$x = -\frac{27}{2}$ or $-13\frac{1}{2}$ Simplify.

The answer is D.

[T]est-Taking Tip (The Princeton Review)

Backsolving
Sometimes it may be easier to substitute each of the choices into the given equation until you get a true sentence.

Your Turn **Solve each equation. Check your solution.**

e. $\frac{a}{6} = 2$ 12 f. $6 = \frac{3}{8}d$ 16 g. $\frac{x}{2.5} = -3$ −7.5

msmath2.net/extra_examples

2 Teach

In-Class Examples (INTERACTIVE CHALKBOARD PowerPoint® Presentations)

Find the multiplicative inverse of each number.

1 $\frac{4}{7}$ $\frac{7}{4}$ or $1\frac{3}{4}$

2 $6\frac{1}{4}$ $\frac{4}{25}$

3 Solve $\frac{p}{6} = 11$. Check your solution. $p = 66$

4 **MULTIPLE-CHOICE TEST ITEM**
What value of y makes $\frac{4}{5}y = -8$ a true sentence? D

A $\frac{5}{4}$

B $-\frac{1}{8}$

C $-6\frac{2}{5}$

D -10

DAILY INTERVENTION — **Unlocking Misconceptions**

Solving Equations with Fractions Write the following problem on the chalkboard: $-\frac{5}{6}q = 8$. Students may mistakenly multiply $-5q \cdot 8 = 6 \cdot 8$ instead of $5q \cdot 1$. Likewise, they may multiply $-5 \cdot 1 = 6q \cdot 8$, assuming that the q is in the denominator. Point out these errors, as well as the correct method to solve.
$-\frac{6}{5} \cdot -\frac{5}{6}q = -\frac{6}{5} \cdot 8$; $q = -\frac{48}{5}$ or $-9\frac{3}{5}$

3 Practice/Apply

Which One Doesn't Belong? For Exercise 3, point out that whole numbers can always be written as fractions by writing the number in the numerator and a "1" in the denominator.

While $\frac{5}{1}$ is the reciprocal of $\frac{1}{5}$, $\frac{3}{1}$ is *not* the reciprocal of $\frac{2}{3}$.

Encourage students to use this approach when finding reciprocals.

About the Exercises...

Guided Practice Key

Exercises	Examples
4–7	1, 2
8, 9, 13, 14	3
10–12	4

Odd/Even Assignments
Exercises 15–36 are structured so that students practice the same concepts whether they are assigned odd or even problems.

Alert! The Data Update following Exercise 36 requires access to the Internet.

Skill and Concept Check

Writing Math
Exercise 3

1. **Tell** whether 8 is a solution of $\frac{n}{3} = 24$. Explain. **no; $\frac{8}{3} \neq 24$**
2. **OPEN ENDED** Write a division equation that can be solved by multiplying each side by $\frac{9}{4}$. **Sample answer: $\frac{4}{9}x = 2$**
3. **Which One Doesn't Belong?** Identify the pair of numbers that does not have the same relationship as the other three. Explain your reasoning. **$\frac{2}{3}$, 3; The other pairs of numbers are reciprocals, but these are not.**

$\frac{7}{8}, \frac{8}{7}$	$5, \frac{1}{5}$	$\frac{2}{3}, 3$	$\frac{10}{3}, \frac{3}{10}$

GUIDED PRACTICE

Find the multiplicative inverse of each number.

4. $\frac{8}{5}$ **$\frac{5}{8}$**
5. $\frac{2}{9}$ **$\frac{9}{2}$ or $4\frac{1}{2}$**
6. -9 **$-\frac{1}{9}$**
7. $5\frac{4}{5}$ **$\frac{5}{29}$**

Solve each equation. Check your solution.

8. $\frac{k}{16} = 2$ **32**
9. $-4 = \frac{y}{3}$ **−12**
10. $6 = \frac{4}{7}u$ **$10\frac{1}{2}$**
11. $\frac{1}{4}t = \frac{3}{8}$ **$1\frac{1}{2}$**
12. $\frac{5}{7}y = -1.5$ **−2.1**
13. $\frac{b}{8.2} = 2.5$ **20.5**

14. **MEASUREMENT** The weight in pounds p of an object with a mass m of 25 kilograms is given by the equation $\frac{p}{m} = 2.2$. How many pounds does the object weigh? **55 lb**

Practice and Applications

Find the multiplicative inverse of each number.

15. $\frac{11}{2}$ **$\frac{2}{11}$**
16. $-\frac{9}{5}$ **$-\frac{5}{9}$**
17. $-\frac{3}{8}$ **$-\frac{8}{3}$ or $-2\frac{2}{3}$**
18. $\frac{1}{6}$ **$\frac{6}{1}$ or 6**
19. 3 **$\frac{1}{3}$**
20. -14 **$-\frac{1}{14}$**
21. $4\frac{2}{5}$ **$\frac{5}{22}$**
22. $6\frac{2}{3}$ **$\frac{3}{20}$**

HOMEWORK HELP

For Exercises	See Examples
15–22	1, 2
23–36	3, 4

Extra Practice
See pages 578, 601.

Solve each equation. Check your solution.

23. $\frac{x}{12} = 3$ **36**
24. $\frac{d}{4} = 28$ **112**
25. $-\frac{2}{5}t = -12$ **30**
26. $-24 = \frac{3}{4}a$ **−32**
27. $\frac{7}{8}k = -21$ **−24**
28. $14 = \frac{8}{3}b$ **$5\frac{1}{4}$**
29. $\frac{1}{2}z = -\frac{2}{5}$ **$-\frac{4}{5}$**
30. $\frac{3}{5} = \frac{3}{7}r$ **$1\frac{2}{5}$**
31. $35.1 = \frac{5}{6}m$ **42.12**
32. $-\frac{a}{3.2} = 5$ **−16**
33. $0.8 = \frac{h}{3.6}$ **2.88**
34. $\frac{m}{4.6} = 2.8$ **12.88**

35. **VACATION** The distance Katie travels in her car while driving 55 miles per hour for 2.5 hours is given by the equation $\frac{d}{2.5} = 55$. How far did she travel? **137.5 mi**

260 Chapter 6 Applying Fractions

36. **MONEY** Based on recent exchange rates, the equation $d = \frac{31}{50}c$ shows the value in U.S. dollars d for an amount of Canadian dollars c. To the nearest cent, find the value in Canadian currency for \$250 in U.S. dollars. **\$403.23**

Data Update What is the value in Canadian currency for \$250 in U.S. dollars today? Visit msmath2.net/data_update to learn more.

37. **CRITICAL THINKING** In Lesson 11-5, you will learn that the area of a triangle A is given by the equation $A = \frac{1}{2}bh$, where b is the base of the triangle and h is the height. Explain how you can use the properties of equality to find the value of b in terms of A and h. Then solve for b. **See margin.**

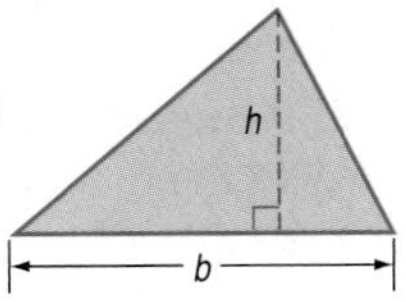

Standardized Test Practice and Mixed Review

38. **MULTIPLE CHOICE** What is the reciprocal of $2\frac{1}{5}$? **C**

(A) $2\frac{5}{1}$ (B) $\frac{11}{5}$ (C) $\frac{5}{11}$ (D) $\frac{1}{10}$

39. **GRID IN** Aaron is serving a 12-pound turkey at a dinner party. As a rule, you should allow about $\frac{3}{4}$ of a pound of meat per person. Use $\frac{3}{4}p = 12$ to find the number of people p that can be served at the dinner party. **16**

Multiply. Write in simplest form. (Lesson 6-4)

40. $\frac{3}{8} \times \frac{4}{9}$ **$\frac{1}{6}$**
41. $1\frac{1}{2} \times 6$ **9**
42. $2\frac{2}{5} \times \frac{1}{6}$ **$\frac{2}{5}$**
43. $1\frac{1}{2} \times 1\frac{7}{9}$ **$2\frac{2}{3}$**

44. Find $7\frac{1}{3} - 3\frac{5}{9}$. (Lesson 6-3) **$3\frac{7}{9}$**

For Exercises 45–47, use the graphic at the right. Write a fraction that compares the number of women champions to the total number of champions for each college. Write in simplest form. (Lesson 5-3)

45. Villanova **$\frac{6}{17}$**
46. Texas–El Paso **$\frac{1}{20}$**
47. Texas **$\frac{5}{6}$**

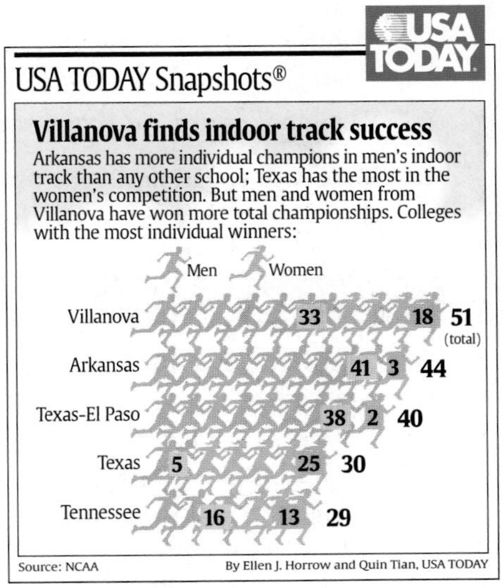

GETTING READY FOR THE NEXT LESSON

PREREQUISITE SKILL **Estimate.** (Lesson 6-1) **48–51. Sample answers are given.**

48. $18\frac{1}{6} \div 3$ **$18 \div 3 = 6$**
49. $24\frac{3}{8} \div 11\frac{7}{9}$ **$24 \div 12 = 2$**
50. $\frac{2}{11} \div \frac{11}{12}$ **$0 \div 1 = 0$**
51. $\frac{9}{10} \div \frac{6}{7}$ **$1 \div 1 = 1$**

msmath2.net/self_check_quiz

Study Guide and Intervention, p. 323

NAME ______ DATE ______ PERIOD ____

6-5 Study Guide and Intervention

Algebra: Solving Equations

Multiplicative inverses, or **reciprocals**, are two numbers whose product is 1. To solve an equation in which the coefficient is a fraction, multiply each side of the equation by the reciprocal of the coefficient.

EXAMPLE 1 **Find the multiplicative inverse of $3\frac{1}{4}$.**

$3\frac{1}{4} = \frac{13}{4}$ Rename the mixed number as an improper fraction.

$\frac{13}{4} \cdot \frac{4}{13} = 1$ Multiply $\frac{13}{4}$ by $\frac{4}{13}$ to get the product 1.

The multiplicative inverse of $3\frac{1}{4}$ is $\frac{4}{13}$.

EXAMPLE 2 **Solve $\frac{4}{5}x = 8$. Check your solution.**

$\frac{4}{5}x = 8$ Write the equation.

$\left(\frac{5}{4}\right)\frac{4}{5}x = \left(\frac{5}{4}\right)8$ Multiply each side by the reciprocal of $\frac{4}{5}$, $\frac{5}{4}$.

$x = 10$ Simplify.

The solution is 10.

EXERCISES

Find the multiplicative inverse of each number.

1. $\frac{4}{9}$ $\frac{9}{4}$ or $2\frac{1}{4}$
2. $\frac{12}{13}$ $\frac{13}{12}$ or $1\frac{1}{12}$
3. $-\frac{15}{4}$ $-\frac{4}{15}$
4. $6\frac{1}{7}$ $\frac{7}{43}$

Solve each equation. Check your solution.

5. $\frac{3}{5}x = 12$ 20
6. $16 = \frac{10}{3}a$ $4\frac{4}{5}$
7. $\frac{c}{2} = 7$ 14
8. $\frac{15}{7}y = 3$ $1\frac{2}{5}$
9. $\frac{m}{6} = -4$ −24
10. $14 = -\frac{7}{9}b$ −18

Practice: Skills, p. 324

NAME ______ DATE ______ PERIOD ____

6-5 Practice: Skills

Algebra: Solving Equations

Find the multiplicative inverse of each number.

1. $\frac{3}{7}$ $\frac{7}{3}$
2. $-\frac{4}{11}$ $-\frac{11}{4}$ or $-2\frac{3}{4}$
3. $\frac{7}{2}$ $\frac{2}{7}$
4. $\frac{9}{5}$ $\frac{5}{9}$
5. −5 $-\frac{1}{5}$
6. $6\frac{1}{3}$ $\frac{3}{19}$
7. $4\frac{1}{9}$ $\frac{9}{37}$
8. $17\frac{1}{2}$ $\frac{2}{35}$
9. $-15\frac{2}{3}$ $-\frac{3}{47}$

Solve each equation. Check your solution.

10. $\frac{x}{10} = 3$ 30
11. $7 = \frac{m}{4}$ 28
12. $\frac{a}{5} = -11$ −55
13. $-8 = \frac{t}{5}$ −40
14. $\frac{3}{4}c = 12$ 16
15. $-7 = -\frac{w}{6}$ 42
16. $\frac{3}{5}y = 6$ 10
17. $15 = \frac{3}{7}b$ 35
18. $\frac{6}{7}c = 18$ 21
19. $\frac{7}{3}x = \frac{2}{3}$ $\frac{2}{7}$
20. $\frac{11}{12} = \frac{3}{4}h$ $1\frac{2}{9}$
21. $\frac{9}{14}y = \frac{3}{7}$ $\frac{2}{3}$
22. $\frac{m}{2.6} = -5$ −13
23. $0.6 = \frac{n}{5}$ 3
24. $\frac{r}{2.6} = -1.3$ −3.38

Practice: Word Problems, p. 325

NAME ______ DATE ______ PERIOD ____

6-5 Practice: Word Problems

Algebra: Solving Equations

1. **BIKING** The speed s that Brandon can ride his bike if he rides $\frac{3}{5}$ of an hour and travels 4 miles is given by the equation $4 = \frac{3}{5}s$. What is Brandon's speed? $6\frac{2}{3}$ mph

2. **BAND** The woodwind section of the middle school band makes up $\frac{1}{4}$ of the band. There are 9 members in the woodwind section. Use the equation $\frac{1}{4}m = 9$ to find the number of members m in the band. 36 members

3. **SALE** A coat is selling for $\frac{3}{4}$ of the original price. The sale price is \$180. The original price p can be found using the equation $\frac{3}{4}p = 180$. Find the original price. \$240

4. **SALARIES** Aaron's annual salary is $\frac{2}{3}$ as much as Juanita's salary. Aaron makes \$46,000. Find Juanita's salary x using the equation $46{,}000 = \frac{2}{3}x$. \$69,000

5. **ENDANGERED SPECIES** In the U. S., there are $\frac{14}{29}$ as many endagered species of birds as endangered species of reptiles. The number of endangered species of birds b can be compared to the 14 endangered species of reptiles using $\frac{14}{29}b = 14$. Find the number of endangered species of birds. 29

6. **SALES TAX** The sticker price p of a purchase with $\frac{1}{10}$ sales tax and a total price (including tax) of \$5.28 can be found using the equation $\frac{11}{10}p = 5.28$. What is the sticker price? \$4.80

7. **SEWING** Each costume uses $\frac{3}{4}$ yard of fabric. The number of costumes c that can be made using $11\frac{1}{4}$ yards of fabric can be found using the equation $\frac{3}{4}c = 11\frac{1}{4}$. Find the number of costumes that can be made. 15 costumes

8. **SAVINGS** Jasmine saves \$46 each month from her part-time job. She saves $\frac{2}{5}$ of her earnings. Her earnings a can be found by using the equation equation $\frac{2}{5}a = 46$. Find her earnings. \$115

4 Assess

Higher-Level Thinking

Can the Multiplicative Inverse Property be used to solve equations such as $12x = 72$? Why or why not? **Yes; In this case, you would multiply both sides by $\frac{1}{12}$, which is the same as dividing both sides by 12.**

Answer

37. **Sample answer: Use the Multiplicative Property to multiply each side by 2. Then use the Division Property to divide each side by h. So, $b = \frac{2A}{h}$.**

Chapter 6 Mid-Chapter Practice Test

Assessment Option

Mid-Chapter Test (Lessons 6-1 through 6-5) is available on p. 365 of the *Chapter 6 Resource Masters*.

ExamView® Pro Test Bank CD-ROM

Use the networkable ExamView® Pro Test Bank CD-ROM to:

- Create **Multiple versions** of tests.
- Create **Modified** tests for **Inclusion** students with one mouse click.
- **Edit** existing questions and **Add** your own questions.
- Build tests aligned with state standards using built-in **State Curriculum Correlations.**
- Change **English** tests to **Spanish** with one mouse click and vice versa.

FOLDABLES™ Follow-Up

Before completing the Mid-Chapter Practice Test, encourage students to review the notes, definitions, concepts, and main ideas they recorded in their Foldables.

Chapter 6 Mid-Chapter Practice Test

Vocabulary and Concepts

1. **Write** an addition expression involving fractions shown by the model at the right. Then find the sum. Write in simplest form. (Lesson 6-2) $\frac{2}{3} + \frac{5}{6} = 1\frac{1}{2}$

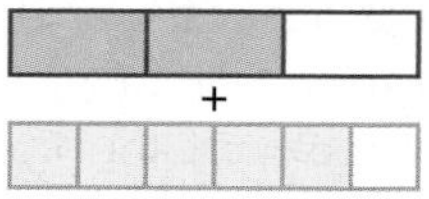

2. **Define** reciprocals. (Lesson 6-5) **two numbers whose product is 1**

Skills and Applications

Estimate. (Lesson 6-1) **3–5. Sample answers are given.**

3. $7\frac{1}{9} + 1\frac{1}{6}$ **$7 + 1 = 8$**
4. $13\frac{1}{2} \div 7\frac{2}{9}$ **$14 \div 7 = 2$**
5. $\frac{11}{20} - \frac{5}{8}$ **$\frac{1}{2} - \frac{1}{2} = 0$**

6. **SAVINGS** Jessica saves $\frac{1}{3}$ of the money she earns baby-sitting. If she earns \$25 one evening, estimate the amount she saves. (Lesson 6-1) **Sample answer: $\frac{1}{3} \cdot 24 = \$8$**

Add, subtract, or multiply. Write in simplest form. (Lessons 6-2, 6-3, and 6-4)

7. $\frac{11}{15} - \frac{1}{15}$ **$\frac{2}{3}$**
8. $\frac{4}{7} - \frac{3}{14}$ **$\frac{5}{14}$**
9. $\frac{5}{8} + \frac{3}{4}$ **$1\frac{3}{8}$**
10. $5\frac{1}{6} - 1\frac{1}{3}$ **$3\frac{5}{6}$**
11. $\frac{7}{12} \times \frac{4}{9}$ **$\frac{7}{27}$**
12. $2\frac{3}{5} + 6\frac{13}{15}$ **$9\frac{7}{15}$**
13. $2\frac{3}{4} \times 12$ **33**
14. $4\frac{2}{7} \times 5\frac{5}{6}$ **25**

15. **AIRPLANES** The aircraft *Voyager* weighed 2,000 pounds. In 1986, it carried about $3\frac{1}{2}$ times its weight in fuel to fly nonstop around the world. How many pounds of fuel did *Voyager* carry? (Lesson 6-4) **about 7,000 lb**

Solve each equation. Check your solution. (Lesson 6-5)

16. $\frac{t}{5} = -11$ **-55**
17. $2 = \frac{3}{8}y$ **$5\frac{1}{3}$**
18. $16.2 = \frac{3}{4}k$ **21.6**

Standardized Test Practice

19. **MULTIPLE CHOICE** One batch of cookies uses $2\frac{1}{2}$ cups of flour and $1\frac{2}{3}$ cups of sugar. Which is the best estimate of the total amount of flour and sugar used in eight batches of cookies? (Lesson 6-1) **B**
    - (A) less than 30 c
    - (B) between 30 c and 45 c
    - (C) between 45 c and 55 c
    - (D) more than 55 c

20. **MULTIPLE CHOICE** How much does a $12\frac{3}{4}$-pound package weigh after a $3\frac{5}{8}$-pound book is taken out of it? (Lesson 6-3) **H**
    - (F) $8\frac{1}{8}$ lb
    - (G) 9 lb
    - (H) $9\frac{1}{8}$ lb
    - (I) 15 lb

262 Chapter 6 Applying Fractions

The Game Zone

A Place To Practice Your Math Skills

Totally Mental

GET READY!

Players: two
Materials: 2 index cards, spinner with the digits 1 through 9

GET SET!

- Each player should make a game sheet on an index card like the one shown at the right.

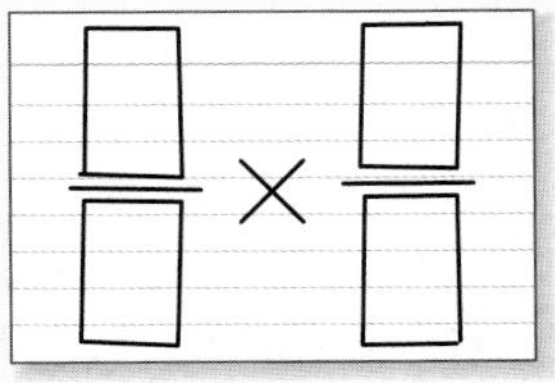

GO!

- One player spins the spinner. The number that is spun should be written in one of the four boxes on his or her game sheet.

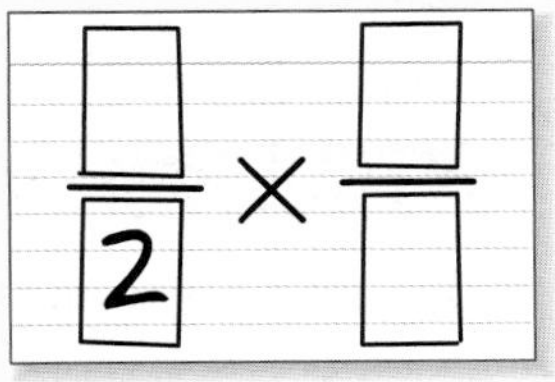

- The second player spins the spinner and writes the number from the spinner in one of the boxes on his or her game sheet.
- Continue until each person has spun the spinner four times.
- Find the product of the fractions on your game sheet.
- **Who Wins?** After four spins, the player with the greatest product is the winner.

The Game Zone

Resources

For an easy-to-make manipulative for this game, visit: msmath2.net/gamezone.

Glencoe Mathematics Classroom Manipulative Kit

- spinner

GETTING READY

Objective Practice skills in multiplying fractions.

MANAGEMENT TIPS

Recommended Time
5 minutes per round of play

Tips

- You may wish to save the game cards for reuse as flashcards. If so, have students make a new game sheet for each round of play.
- You may also wish to use playing cards numbers A–9 to generate the numbers instead of a spinner.
- Ask students what strategies they developed as they played additional rounds of the game.
- This game can be used to practice other operations with fractions by changing the operation symbol on the game sheet. Practice operations on mixed numbers by amending the game sheet.

(Continued in bottom margin.)

For an additional game for this chapter, visit: msmath2.net/gamezone.

- As a variation of the game, have students roll a number cube before the game to determine a target number. The player whose product is closest to the target number is the winner.
- You may wish to use the game as an assessment tool for the concepts the students have been learning in Lessons 6-1 through 6-5.

6-6 Lesson Notes

Lesson Resources

Chapter 6 Resource Masters
- Study Guide and Intervention, p. 328
- Practice: Skills, p. 329
- Practice: Word Problems, p. 330
- Reading to Learn Mathematics, p. 331
- Enrichment, p. 332

Classroom Math Games, pp. 17–18
MathPASS Tutorial CD-ROM, 22–24
Parent and Student Study Guide Workbook, 6-6

FAST FILE

1 Focus

5-Minute Check 6-6
Use as a quiz or review of Lesson 6-5. (Also available as a transparency)

INTERACTIVE CHALKBOARD
PowerPoint® Presentations

Bellringer
Communication Write the following questions on the chalkboard: What is one half of sixteen? **8** What is sixteen *divided by* one half? **32** Ask students, "What is the difference between those two questions?" Discuss that the first question requires multiplying by $\frac{1}{2}$. The second question involves dividing by $\frac{1}{2}$, which is the same as finding out how many $\frac{1}{2}$s are in 16.

Reading to Learn Mathematics, p. 331 **ELL**

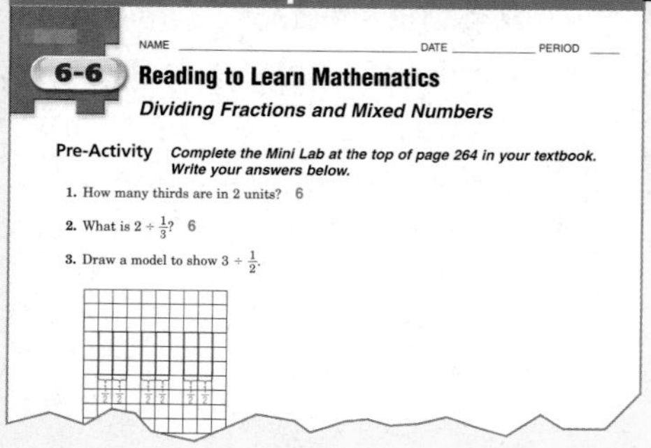
NAME ______________ DATE __________ PERIOD _____
6-6 Reading to Learn Mathematics
Dividing Fractions and Mixed Numbers

Pre-Activity *Complete the Mini Lab at the top of page 264 in your textbook. Write your answers below.*

1. How many thirds are in 2 units? 6
2. What is $2 \div \frac{1}{3}$? 6
3. Draw a model to show $3 \div \frac{1}{2}$.

6-6 Dividing Fractions and Mixed Numbers

What You'll Learn
Divide fractions and mixed numbers.

HANDS-ON Mini Lab

Materials
- grid paper

Work with a partner.

The model at the right shows 2 units divided into thirds, or $2 \div \frac{1}{3}$.

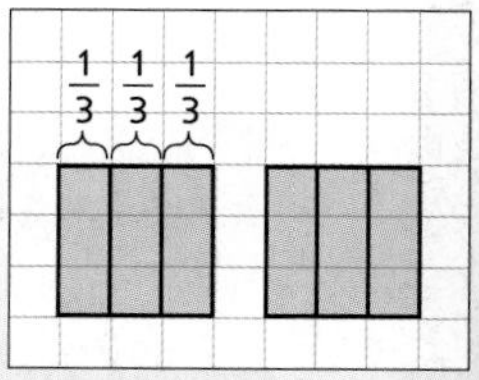

1. How many thirds are in 2 units? **6**
2. What is $2 \div \frac{1}{3}$? **6**
3. Draw a model to show $3 \div \frac{1}{2}$. **See margin.**
4. What is $3 \div \frac{1}{2}$? **6**

Dividing 8 by 2 gives the same result as multiplying 8 by the reciprocal of 2, or $\frac{1}{2}$.

reciprocals

$8 \div 2 = 4 \qquad 8 \cdot \frac{1}{2} = 4$

same result

In the same way, dividing 4 by $\frac{1}{3}$ is the same as multiplying 4 by the reciprocal of $\frac{1}{3}$, or 3. This pattern is true for any rational number.

Key Concept — Division by Fractions

Words To divide by a fraction, multiply by its multiplicative inverse, or reciprocal.

Symbols
Arithmetic: $\frac{7}{8} \div \frac{3}{4} = \frac{7}{8} \cdot \frac{4}{3}$
Algebra: $\frac{a}{b} \div \frac{c}{d} = \frac{a}{b} \cdot \frac{d}{c}$, where $b, c, d \neq 0$

EXAMPLE Divide by a Fraction

1 **Find $\frac{3}{4} \div \frac{1}{2}$. Write in simplest form.**

$\frac{3}{4} \div \frac{1}{2} = \frac{3}{4} \cdot \frac{2}{1}$ Multiply by the reciprocal of $\frac{1}{2}$, which is $\frac{2}{1}$.

$= \frac{3}{\cancel{4}_2} \cdot \frac{\cancel{2}^1}{1}$ Divide by the GCF, 2.

$= \frac{3}{2}$ or $1\frac{1}{2}$ Multiply and simplify.

Your Turn **Divide. Write in simplest form.**

a. $\frac{3}{4} \div \frac{1}{4}$ **3** b. $\frac{4}{5} \div \frac{8}{9}$ $\mathbf{\frac{9}{10}}$ c. $\frac{5}{6} \div \frac{2}{3}$ $\mathbf{1\frac{1}{4}}$

Answer (Hands-On Mini Lab)

3.
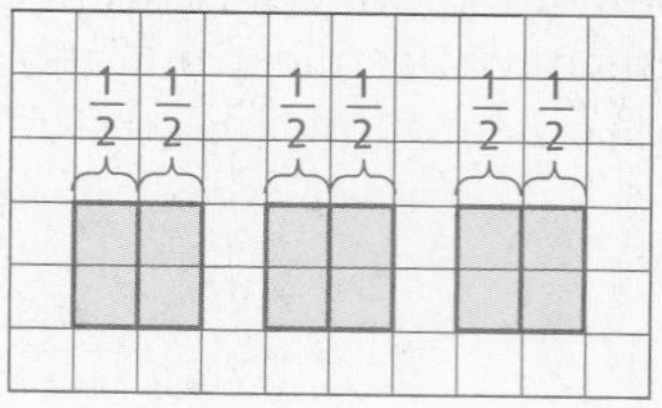

To divide by a mixed number, rename the mixed number as an improper fraction.

EXAMPLES Divide by Mixed Numbers

2 SURVEYING Fifteen acres of land are to be divided into $1\frac{1}{2}$-acre lots. How many lots will there be?

Estimate $16 \div 2 = 8$

$15 \div 1\frac{1}{2} = 15 \div \frac{3}{2}$ Rename $1\frac{1}{2}$ as an improper fraction.

$= 15 \cdot \frac{2}{3}$ Multiply by the reciprocal of $\frac{3}{2}$, which is $\frac{2}{3}$.

$= \frac{\overset{5}{\cancel{15}}}{1} \cdot \frac{2}{\underset{1}{\cancel{3}}}$ Divide out common factors.

$= 10$ Multiply.

There will be 10 lots.

EAL-LIFE MATH

RVEYING A surveyor n survey a line for nine iles before having to use herical geometry and gonometry to correct for rth's curved surface.

urce: Kansas Society of Land veyors

3 Find $\frac{2}{3} \div 3\frac{1}{3}$. Write in simplest form.

Estimate $\frac{1}{2} \div 3 = \frac{1}{2} \times \frac{1}{3}$ or $\frac{1}{6}$

$\frac{2}{3} \div 3\frac{1}{3} = \frac{2}{3} \div \frac{10}{3}$ Rename $3\frac{1}{3}$ as an improper fraction.

$= \frac{2}{3} \cdot \frac{3}{10}$ Multiply by the reciprocal of $\frac{10}{3}$, which is $\frac{3}{10}$.

$= \frac{\overset{1}{\cancel{2}}}{\underset{1}{\cancel{3}}} \cdot \frac{\overset{1}{\cancel{3}}}{\underset{5}{\cancel{10}}}$ Divide out common factors.

$= \frac{1}{5}$ Multiply.

The quotient is close to the estimate.

4 Find $6\frac{1}{2} \div 3\frac{5}{7}$.

Estimate $6 \div 3 = 2 \leftarrow$ compatible numbers

$6\frac{1}{2} \div 3\frac{5}{7} = \frac{13}{2} \div \frac{26}{7}$ Rename the mixed numbers as improper fractions.

$= \frac{13}{2} \cdot \frac{7}{26}$ Multiply by the reciprocal of $\frac{26}{7}$, which is $\frac{7}{26}$.

$= \frac{\overset{1}{\cancel{13}}}{2} \cdot \frac{7}{\underset{2}{\cancel{26}}}$ Divide out common factors.

$= \frac{7}{4}$ Multiply.

$= 1\frac{3}{4}$ Simplify.

The quotient, $1\frac{3}{4}$, is close to the estimate.

Your Turn **Divide. Write in simplest form.**

d. $5 \div 1\frac{1}{3}$ $3\frac{3}{4}$ e. $\frac{3}{4} \div 1\frac{1}{2}$ $\frac{1}{2}$ f. $2\frac{1}{3} \div 5\frac{5}{6}$ $\frac{2}{5}$

msmath2.net/extra_examples

2 Teach

In-Class Examples

INTERACTIVE CHALKBOARD PowerPoint® Presentations

1 Find $\frac{2}{3} \div \frac{4}{9}$. Write in simplest form. $1\frac{1}{2}$

2 **JETS** An Air Force unit has 34 million dollars to spend on new jets. Each jet costs $4\frac{1}{4}$ million dollars. How many jets can be bought? **8 jets**

3 Find $\frac{5}{6} \div 2\frac{1}{2}$. Write in simplest form. $\frac{1}{3}$

4 Find $4\frac{3}{4} \div 6\frac{1}{3}$. $\frac{3}{4}$

3 Practice/Apply

About the Exercises...

Guided Practice Key

Exercises	Examples
3, 4	1
7	2
5	3
6	4

Odd/Even Assignments
Exercises 8–20 are structured so that students practice the same concepts whether they are assigned odd or even problems.

Alert! Exercise 22 illustrates the Closure Property of Rational Numbers.

Answers (p. 266)

1. **Sample answer: How many $\frac{1}{4}$-pound hamburgers can be made with 10 pounds of ground beef?**
2. **Sample answer: Multiply 6 by the multiplicative inverse of $\frac{3}{4}$, which is $\frac{4}{3}$. Then simplify to get 8.**

DAILY INTERVENTION **Differentiated Instruction** **ELL**

Verbal/Linguistic Have students write a simple poem or rhyme that incorporates at least one method or rule for the division of fractions. Rules and methods for other operations on fractions can be included, as well. You may wish to have students work in groups and put together larger works of poetry or verse.

Study Guide and Intervention, p. 328

NAME ______________ DATE ________ PERIOD ____

6-6 Study Guide and Intervention

Dividing Fractions and Mixed Numbers

To divide by a fraction, multiply by its multiplicative inverse or reciprocal.
To divide by a mixed number, rename the mixed number as an improper fraction.

EXAMPLE 1 **Find $3\frac{1}{3} \div \frac{2}{9}$. Write in simplest form.**

$3\frac{1}{3} \div \frac{2}{9} = \frac{10}{3} \div \frac{2}{9}$ Rename $3\frac{1}{3}$ as an improper fraction.

$= \frac{10}{3} \cdot \frac{9}{2}$ Multiply by the reciprocal of $\frac{2}{9}$, which is $\frac{9}{2}$.

$= \frac{\overset{5}{\cancel{10}}}{\underset{1}{\cancel{3}}} \cdot \frac{\overset{3}{\cancel{9}}}{\underset{1}{\cancel{2}}}$ Divide out common factors.

$= 15$ Multiply.

EXERCISES

Divide. Write in simplest form.

1. $\frac{2}{3} \div \frac{1}{4}$ $2\frac{2}{3}$
2. $\frac{2}{5} \div \frac{5}{6}$ $\frac{12}{25}$
3. $\frac{1}{2} \div \frac{1}{5}$ $2\frac{1}{2}$
4. $5 \div \frac{1}{2}$ 10
5. $\frac{5}{8} \div 10$ $\frac{1}{16}$
6. $7\frac{1}{3} \div 2$ $3\frac{2}{3}$
7. $\frac{5}{6} \div 3\frac{1}{2}$ $\frac{5}{21}$
8. $36 \div 1\frac{1}{2}$ 24
9. $2\frac{1}{2} \div 10$ $\frac{1}{4}$
10. $5\frac{2}{5} \div 1\frac{4}{5}$ 3
11. $6\frac{2}{3} \div 3\frac{1}{9}$ $2\frac{1}{7}$
12. $4\frac{1}{4} \div \frac{3}{8}$ $11\frac{1}{3}$
13. $4\frac{6}{7} \div 2\frac{3}{7}$ 2
14. $12 \div 2\frac{1}{2}$ $4\frac{4}{5}$
15. $4\frac{1}{6} \div 3\frac{1}{6}$ $1\frac{6}{19}$

Practice: Skills, p. 329

NAME ______________ DATE ________ PERIOD ____

6-6 Practice: Skills

Dividing Fractions and Mixed Numbers

Divide. Write in simplest form.

1. $\frac{1}{6} \div \frac{1}{5}$ $\frac{5}{6}$
2. $5 \div \frac{3}{5}$ $8\frac{1}{3}$
3. $\frac{6}{7} \div \frac{1}{7}$ 6
4. $\frac{3}{4} \div \frac{1}{2}$ $1\frac{1}{2}$
5. $8 \div \frac{1}{3}$ 24
6. $\frac{1}{5} \div \frac{1}{4}$ $\frac{4}{5}$
7. $7 \div \frac{3}{7}$ $16\frac{1}{3}$
8. $\frac{4}{7} \div \frac{8}{9}$ $\frac{9}{14}$
9. $8\frac{1}{3} \div 5$ $1\frac{2}{3}$
10. $\frac{9}{7} \div \frac{3}{14}$ 6
11. $\frac{12}{5} \div \frac{3}{10}$ 8
12. $5 \div 3\frac{3}{4}$ $1\frac{1}{3}$
13. $6\frac{4}{5} \div 17$ $\frac{2}{5}$
14. $7\frac{1}{3} \div 4$ $1\frac{5}{6}$
15. $\frac{3}{4} \div 5\frac{1}{2}$ $\frac{3}{22}$
16. $\frac{2}{7} \div 1\frac{13}{14}$ $\frac{4}{27}$
17. $\frac{3}{8} \div 6\frac{1}{4}$ $\frac{3}{50}$
18. $7\frac{1}{2} \div 2\frac{5}{6}$ $2\frac{11}{17}$
19. $3\frac{4}{9} \div 2\frac{1}{3}$ $1\frac{10}{21}$
20. $2\frac{2}{3} \div 1\frac{1}{6}$ $2\frac{2}{7}$
21. $4\frac{3}{4} \div 2\frac{1}{2}$ $1\frac{9}{10}$

Practice: Word Problems, p. 330

NAME ______________ DATE ________ PERIOD ____

6-6 Practice: Word Problems

Dividing Fractions and Mixed Numbers

1. PUPPETS If a puppet requires $\frac{3}{4}$ yards of material, how many puppets can be made from 9 yards of material? 12 puppets	2. COOKING A batch of cookies requires $1\frac{1}{2}$ cups of sugar. How many batches can Ty make with $7\frac{1}{2}$ cups of sugar? 5 batches
3. FOOD Julia has $3\frac{1}{2}$ pounds of dog food. She plans to split it equally among her 7 dogs. How much will dog food will each dog receive? $\frac{1}{2}$ lb	4. SNOW CONES Roger has a 28-pound block of ice for his snow cone stand. If each snow cone requires $\frac{2}{3}$ pound of ice, how many snow cones can Roger make? 42 snow cones
5. APPLES Juan took 6 apples and cut each into one-eighths. How many pieces of apple did he have? 48 pieces	6. VACATION The Torres family drove 1,375 miles during their $6\frac{1}{4}$-day vacation. Find the average number of miles they traveled each day. 220 mi
7. RUNNING Hugo just joined the cross-country team and can run at a rate of $\frac{1}{7}$ mile each minute. How long will it take him to run a 5-mile race? 35 min	8. LUMBER Mrs. Shin has a piece of lumber that is $11\frac{5}{8}$ inches wide. She plans to split the width of lumber into 3 equal pieces. How wide will each piece be? $3\frac{7}{8}$ in.

Skill and Concept Check

Writing Math Exercises 1 & 2

1. **OPEN ENDED** Write a problem that is solved by finding $10 \div \frac{1}{4}$.
2. **Describe** the steps you would take to find *six divided by three-fourths.*
1–2. See margin p. 265.

GUIDED PRACTICE

Divide. Write in simplest form.

3. $\frac{3}{5} \div \frac{1}{4}$ $2\frac{2}{5}$
4. $\frac{3}{4} \div 6$ $\frac{1}{8}$
5. $\frac{1}{2} \div 7\frac{1}{2}$ $\frac{1}{15}$
6. $5\frac{3}{5} \div 4\frac{2}{3}$ $1\frac{1}{5}$
7. **FOOD** How many $\frac{1}{8}$-pound boxes of mints can be made with 3 pounds? **24 boxes**

★ indicates increased difficulty

Practice and Applications

HOMEWORK HELP

For Exercises	See Examples
8–14	1
15–17, 20	2, 3
18–19	4

Extra Practice See pags 579, 601.

Divide. Write in simplest form.

8. $\frac{3}{8} \div \frac{6}{7}$ $\frac{7}{16}$
9. $\frac{5}{9} \div \frac{5}{6}$ $\frac{2}{3}$
10. $\frac{2}{3} \div \frac{1}{2}$ $1\frac{1}{3}$
11. $\frac{7}{8} \div \frac{3}{4}$ $1\frac{1}{6}$
12. $6 \div \frac{1}{2}$ 12
13. $\frac{4}{9} \div 2$ $\frac{2}{9}$
14. $2\frac{2}{3} \div 4$ $\frac{2}{3}$
15. $5 \div 1\frac{1}{3}$ $3\frac{3}{4}$
16. $\frac{2}{3} \div 2\frac{1}{2}$ $\frac{4}{15}$
17. $\frac{8}{9} \div 5\frac{1}{3}$ $\frac{1}{6}$
18. $4\frac{1}{2} \div 6\frac{3}{4}$ $\frac{2}{3}$
19. $5\frac{2}{7} \div 2\frac{1}{7}$ $2\frac{7}{15}$

★ 20. **CRAFTS** Jared is making bookmarks like the one shown at the right. How many bookmarks can he make from a 15-yard spool of ribbon? **51 bookmarks**

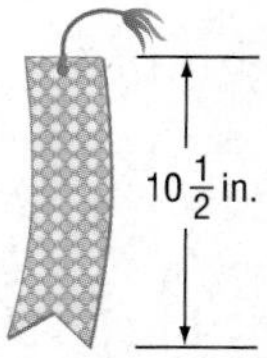

21. **CRITICAL THINKING** Will the quotient $7\frac{1}{6} \div 3\frac{2}{3}$ be a fraction less than 1 or greater than 1? Explain. **greater than 1 because $7\frac{1}{6} > 3\frac{2}{3}$**
22. **EXTENDING THE LESSON** If you add any two rational numbers, the sum is always a rational number. So, the set of rational numbers is *closed* under addition. Is the set of rational numbers closed under subtraction, multiplication, and division also? Explain. **See Ch. 6 Answer Appendix.**

Standardized Test Practice and Mixed Review

23. **SHORT RESPONSE** Leticia is dividing $2\frac{3}{4}$ pounds of trail mix equally among each of her four friends. How much does each receive? **$\frac{11}{16}$ lb**
24. **GRID IN** What is $\frac{8}{9}$ divided by $\frac{2}{9}$? **4**

Find the multiplicative inverse of each number. (Lesson 6-5)

25. $\frac{6}{7}$ $\frac{7}{6}$ or $1\frac{1}{6}$
26. $\frac{4}{13}$ $\frac{13}{4}$ or $3\frac{1}{4}$
27. 8 $\frac{1}{8}$
28. $5\frac{1}{4}$ $\frac{4}{21}$
29. Find $\frac{1}{10} \times \frac{5}{8}$. Write in simplest form. (Lesson 6-4) $\frac{1}{16}$

GETTING READY FOR THE NEXT LESSON

PREREQUISITE SKILL Multiply or divide. (Pages 560, 562)

30. 2.5×20 **50**
31. 3.5×4 **14**
32. $4{,}200 \div 2.1$ **2,000**
33. $104 \div 6.5$ **16**

msmath2.net/self_check_quiz

4 Assess

Open-Ended Assessment

Modeling Write the following problem on the chalkboard: $7\frac{1}{2} \div 1\frac{1}{2}$. Direct students to draw a model of this problem to find the answer. 5

Sample answer:

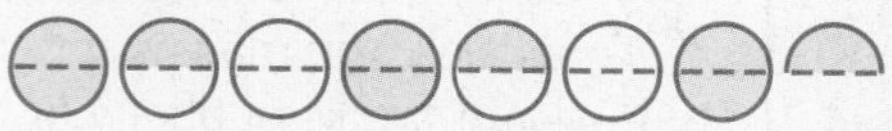

6-7 Measurement: Changing Customary Units

What You'll Learn
Change units in the customary system.

NEW Vocabulary
pound
ounce
ton
cup
pint
quart
gallon

WHEN am I ever going to use this?

ANIMALS The largest creature that has ever lived on Earth is still alive today. This mighty creature is the blue whale. Some blue whales have been estimated to be as large as 150 tons.

1. There are 2,000 pounds in 1 ton. How many pounds are in 150 tons? **300,000 lb**
2. What operation did you use to find the weight in pounds? Explain. **Multiplication; 1 T = 2,000 lb, so 150 T = 150 × 2,000 lb or 300,000 lb.**

The relationships among customary units of length, weight, and capacity are shown in the table at the right.

To change units, use the following rules.

- To convert from larger units to smaller units, multiply.
- To convert from smaller units to larger units, divide.

Customary Units of Length
1 foot (ft) = 12 inches (in.)
1 yard (yd) = 3 feet
1 mile (mi) = 5,280 feet
Customary Units of Weight
1 **pound** (lb) = 16 **ounces** (oz)
1 **ton** (T) = 2,000 pounds
Customary Units of Capacity
1 **cup** (c) = 8 fluid ounces (fl oz)
1 **pint** (pt) = 2 cups
1 **quart** (qt) = 2 pints
1 **gallon** (gal) = 4 quarts

larger units → ← smaller units

EXAMPLES Convert Larger Units to Smaller Units

Complete.

1 20 ft = ___?___ in.

Since 1 foot = 12 inches, multiply by 12.

$20 \times 12 = 240$

20 feet = 240 inches

2 $3\frac{1}{2}$ lb = ___?___ oz

Since 1 pound = 16 ounces, multiply by 16.

$3\frac{1}{2} \times 16 = 56$

$3\frac{1}{2}$ pounds = 56 ounces

Your Turn Complete. **b. 5,500**

a. 36 yd = ___?___ ft **108** b. $2\frac{3}{4}$ T = ___?___ lb c. $1\frac{1}{2}$ c = ___?___ fl oz **12**

TEACHING TIP
Remind students that they multiply to change from larger units to smaller units because it takes more smaller units than larger units to measure an object.

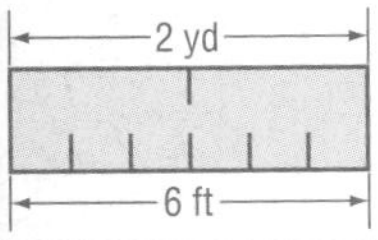

msmath2.net/extra_examples

6-7 Lesson Notes

Lesson Resources

Chapter 6 Resource Masters
- Study Guide and Intervention, p. 333
- Practice: Skills, p. 334
- Practice: Word Problems, p. 335
- Reading to Learn Mathematics, p. 336
- Enrichment, p. 337

Key Teaching Transparencies, 23
MathPASS Tutorial CD-ROM, 55
Parent and Student Study Guide Workbook, 6-7

1 Focus

5-Minute Check 6-7
INTERACTIVE CHALKBOARD PowerPoint® Presentations
Use as a quiz or review of Lesson 6-6. (Also available as a transparency)

Bellringer
Hands-On Activity Group students in small groups of two or three and give each group a yardstick. Direct students to answer the following questions by counting on the yardstick.
How many inches are in $2\frac{1}{2}$ feet? **30 in.** *How many feet are in $1\frac{2}{3}$ yards?* **5 ft**

Reading to Learn Mathematics, p. 336 **ELL**

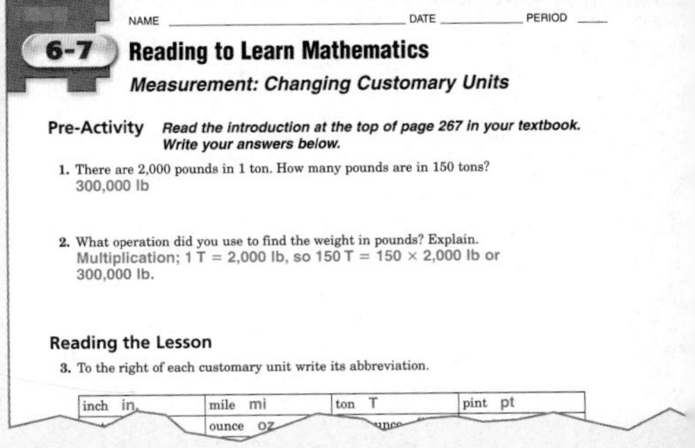
NAME ______ DATE ______ PERIOD ______

6-7 **Reading to Learn Mathematics**
Measurement: Changing Customary Units

Pre-Activity *Read the introduction at the top of page 267 in your textbook. Write your answers below.*

1. There are 2,000 pounds in 1 ton. How many pounds are in 150 tons? 300,000 lb
2. What operation did you use to find the weight in pounds? Explain. Multiplication; 1 T = 2,000 lb, so 150 T = 150 × 2,000 lb or 300,000 lb.

Reading the Lesson
3. To the right of each customary unit write its abbreviation.

inch in.	mile mi	ton T	pint pt
	ounce oz		

DAILY INTERVENTION **Flexible Grouping**

Investigating Customary Units Have students work in groups to research other customary units of measure, for example, bales of cotton, barrels of oil, bushels, drams, gills, pecks, and tablespoons. Students should present their findings accompanied by a conversion table similar to the one on this page. Assign individual tasks based on students' strengths, such as who will do the research, who will write the report, who will make the table, and who will make a presentation to the class.

2 Teach

In-Class Examples

INTERACTIVE CHALKBOARD
PowerPoint® Presentations

Complete.

1. 2 mi = __?__ ft
10,560

2. 5 pt = __?__ c **10**

3. **ELEVATOR** The elevator in an office building has a weight limit posted of one and a half tons. How many pounds can the elevator safely hold?
3,000 lb

Complete.

4. 1,000 in. = __?__ ft **$83\frac{1}{3}$**

5. 6 qt = __?__ gal **$1\frac{1}{2}$**

3 Practice/Apply

Which One Doesn't Belong?
For Exercise 2, have students quickly scan Lesson 1-8 to review the three categories of units in the metric system—length, mass, and capacity.

About the Exercises...

Guided Practice Key

Exercises	Examples
4, 7	1, 2
9–10	3
3, 5, 6, 8	4, 5

Odd/Even Assignments
Exercises 11–32 are structured so that students practice the same concepts whether they are assigned odd or even problems.

REAL-LIFE MATH

SKIING Ross Anderson is an American Indian speed skier. He has been clocked on skis going as fast as 137 miles per hour.
Source: *Native Peoples*

EXAMPLE Convert Units to Solve a Problem

3. **SKIING Speed skiing takes place on a course that is about two thirds of a mile long. How many feet long is the course?**

$\frac{2}{3} \times 5{,}280 = 3{,}520$ Multiply by 5,280 since there are 5,280 feet in 1 mile.

So, the course is about 3,520 feet long.

To convert from smaller units to larger units, divide.

EXAMPLES Convert Smaller Units to Larger Units

Complete.

4. **750 lb = __?__ T**

Since 2,000 pounds are in 1 ton, divide by 2,000.

$750 \div 2{,}000 = \frac{3}{8}$

750 pounds $= \frac{3}{8}$ ton

5. **10 c = __?__ pt**

Since 2 cups are in 1 pint, divide by 2.

$10 \div 2 = 5$

10 cups = 5 pints

Your Turn **Complete.** **d. $\frac{1}{2}$**

d. 2,640 ft = __?__ mi e. 100 oz = __?__ lb **$6\frac{1}{4}$** f. 14 pt = __?__ qt **7**

Skill and Concept Check

Writing Math
Exercises 1 & 2

1. **OPEN ENDED** Write a problem in which you would need to convert pints to cups. **Sample answer: A recipe calls for 2 pints of strawberries. How many cups is this? Answer: 4 c**

2. **Which One Doesn't Belong?** Identify the unit of measure that does not have the same characteristic as the other three. Explain your reasoning.

gallon	pint	fluid ounce	pound

Pound; it is a measure of weight, while the others are measures of liquid capacity.

GUIDED PRACTICE

Complete.

3. 48 oz = __?__ lb **3**
4. $5\frac{1}{3}$ yd = __?__ ft **16**
5. 12 qt = __?__ gal **3**
6. 28 in. = __?__ ft **$2\frac{1}{3}$**
7. $\frac{1}{4}$ T = __?__ lb **500**
8. 15 pt = __?__ qt **$7\frac{1}{2}$**

9. **DINOSAURS** The average weight of the dinosaur *Argentinosaurus* was estimated to be 200,000 pounds. How many tons did it weigh? **100 T**

10. **HISTORY** Liquid products such as oil and vinegar were once shipped in huge containers called *hogsheads*. A hogshead contained 63 gallons of liquid. How many quarts did it contain? **252 qt**

Answers

32. No; the ammonia and vinegar make $2\frac{5}{6}$ cups. A $\frac{1}{2}$-quart pan holds only 2 cups.

33.

Pounds	Ounces
1	16
2	32
3	48
4	64

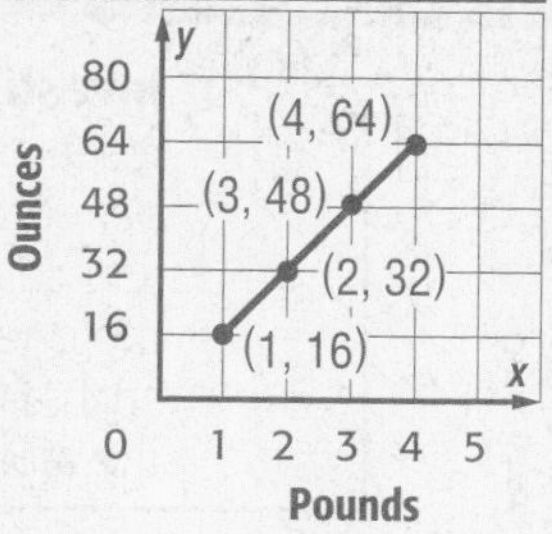

★ indicates increased difficulty

Practice and Applications

HOMEWORK HELP	
For Exercises	See Examples
11–30	1, 2, 4, 5
31–32	3
Extra Practice See pages 579, 601.	

Complete. 14. $2\frac{1}{2}$ 18. 6,600 23. $1\frac{1}{3}$

11. 18 ft = __?__ yd **6** 12. 2 lb = __?__ oz **32** 13. 4 gal = __?__ qt **16**

14. 5,000 lb = __?__ T 15. $4\frac{1}{2}$ pt = __?__ c **9** 16. 72 oz = __?__ lb **$4\frac{1}{2}$**

17. 2 mi = __?__ ft **10,560** 18. $1\frac{1}{4}$ mi = __?__ ft 19. 9 c = __?__ pt **$4\frac{1}{2}$**

20. 3 c = __?__ fl oz **24** 21. $2\frac{3}{4}$ qt = __?__ pt **$5\frac{1}{2}$** 22. 120 ft = __?__ yd **40**

23. 7,040 ft = __?__ mi 24. $3\frac{3}{8}$ T = __?__ lb **6,750** 25. 172 oz = __?__ lb **$10\frac{3}{4}$**

26. If 4 cups = 1 quart, then 9 cups = __?__ quarts. **$2\frac{1}{4}$**

27. If 36 inches = 1 yard, then 2.3 yards = __?__ inches. **82.8**

Complete.

★ 28. $1\frac{1}{4}$ gal = __?__ c **20** ★ 29. 880 yd = __?__ mi **$\frac{1}{2}$** ★ 30. 24 fl oz = __?__ qt **$\frac{3}{4}$**

★ 31. **MULTI STEP** Suppose a car repair company changes the oil of 50 cars and they recover an average of $3\frac{1}{2}$ quarts of oil from each car. How many gallons of oil did they recover? **$43\frac{3}{4}$ gal**

★ 32. **MULTI STEP** A window-washing solution can be made by mixing $1\frac{1}{3}$ cups of ammonia and $1\frac{1}{2}$ cups of vinegar with baking soda and water. Will the solution fit in a $\frac{1}{2}$-quart pan? Explain. **See margin.**

33. **CRITICAL THINKING** Make a table that shows the number of ounces in 1, 2, 3, and 4 pounds. Graph the ordered pairs (pounds, ounces) on a coordinate plane and connect the points. Describe the graph. **See margin for graph; The graph is a straight line. For each *x*-value increase of 1, the *y*-value increases by 16.**

Standardized Test Practice and Mixed Review

34. **SHORT RESPONSE** How many cups of milk are shown at the right? **$1\frac{1}{4}$ c**

35. **MULTIPLE CHOICE** A can of orange juice concentrate makes 48 fluid ounces of orange juice. How many pints is this? **A**

Ⓐ 3 pt Ⓑ 4 pt Ⓒ 6 pt Ⓓ 12 pt

36. Find $1\frac{4}{7} \div 1\frac{5}{6}$. Write in simplest form. (Lesson 6-6) **$\frac{6}{7}$**

Solve each equation. Check your solution. (Lesson 6-5)

37. $\frac{y}{4} = 7$ **28** 38. $\frac{1}{3}x = \frac{5}{9}$ **$1\frac{2}{3}$** 39. $-4 = \frac{p}{2.7}$ **−10.8** 40. $6n = -15$ **−2.5**

GETTING READY FOR THE NEXT LESSON

PREREQUISITE SKILL **Evaluate each expression.** (Lesson 1-3)

41. $2 \cdot 8 + 2 \cdot 9$ **34** 42. 3(7) + 4(2) **29** 43. 2(6.5 + 3) **19** 44. $5 \cdot 2 + 5 \cdot 8.4$ **52**

4 Assess

Open-Ended Assessment

Speaking Write the following on the chalkboard.

56 in. $\stackrel{?}{=}$ (56 · 12) ft

Discuss with students whether this is a true sentence. Then have them explain what would make it true if it is not. **no; 56 in. = (56 ÷ 12)ft**

Study Guide and Intervention, p. 333

NAME ______ DATE ______ PERIOD ____

6-7 Study Guide and Intervention

Measurement: Changing Customary Units

Customary Units		
Length	Weight	Capacity
1 foot (ft) = 12 inches (in.)	1 pound (lb) = 16 ounces (oz)	1 cup (c) = 8 fluid ounces (fl oz)
1 yard (yd) = 3 feet	1 ton (T) = 2,000 pounds	1 pint (pt) = 2 cups
1 mile (mi) = 5,280 feet		1 quart (qt) = 2 pints
		1 gallon (gal) = 4 quarts

EXAMPLE 1 $5\frac{1}{2}$ lb = __?__ oz

To change from larger units to smaller units, multiply.

$5\frac{1}{2} \times 16 = 88$ Since 1 pound is 16 ounces, multiply by 16.

$5\frac{1}{2}$ pounds = 88 ounces

EXAMPLE 2 28 fl oz = __?__ c

To change from smaller units to larger units, divide.

$28 \div 8 = 3\frac{1}{2}$ Since 8 fluid ounces are in 1 cup, divide by 8.

28 fluid ounces = $3\frac{1}{2}$ cups

EXERCISES

Complete.

1. 5 lb = ____ oz 80 2. 48 in. = ____ ft 4 3. 6 yd = ____ ft 18
4. 7 qt = ____ pt 14 5. 8,000 lb = ____ T 4 6. $3\frac{1}{4}$ mi = ____ ft 17,160
7. 4 c = ____ fl oz 32 8. 6 c = ____ pt 3 9. $\frac{1}{2}$ gal = ____ qt 2
10. 3 ft = ____ in. 36 11. 9 qt = ____ gal $2\frac{1}{4}$ 12. 30 fl oz = ____ c $3\frac{3}{4}$
13. 6,864 ft = ____ mi $1\frac{3}{10}$ 14. 40 oz = ____ lb $2\frac{1}{2}$ 15. 9 pt = ____ c 18
16. 18 ft = ____ yd 6 17. 11 pt = ____ qt $5\frac{1}{2}$ 18. $2\frac{3}{4}$ T = ____ lb 5,500

Practice: Skills, p. 334

NAME ______ DATE ______ PERIOD ____

6-7 Practice: Skills

Measurement: Changing Customary Units

Complete.

1. 3 lb = ____ oz 48 2. 16 qt = ____ gal 4
3. 24 in. = ____ ft 2 4. 12 ft = ____ yd 4
5. 3 mi = ____ ft 15,840 6. 12,000 lb = ____ T 6
7. 64 oz = ____ lb 4 8. 6 pt = ____ qt 3
9. 3 pt = ____ c 6 10. $5\frac{1}{2}$ ft = ____ in. 66
11. 22 yd = ____ ft 66 12. $\frac{1}{4}$ mi = ____ ft 1,320
13. 15 T = ____ lb 30,000 14. 7 lb = ____ oz 112
15. $8\frac{1}{2}$ qt = ____ pt 17 16. 5 gal = ____ qt 20
17. 8 c = ____ pt 4 18. 16 in = ____ ft $1\frac{1}{3}$
19. 24 fl oz = ____ c 3 20. 60 ft = ____ yd 20
21. 6,600 ft = ____ mi $1\frac{1}{4}$ 22. 7.5 T = ____ lb 15,000
23. 88 oz = ____ lb $5\frac{1}{2}$ 24. 70 qt = ____ gal $17\frac{1}{2}$

Practice: Word Problems, p. 335

NAME ______ DATE ______ PERIOD ____

6-7 Practice: Word Problems

Measurement: Changing Customary Units

1. **WEIGHT** The average weight of a baby at birth is 7 pounds. How many ounces is this? 112 oz

2. **WATERFALLS** The height of Niagara Falls is 182 feet. How many yards high is it? $60\frac{2}{3}$ yd

3. **GASOLINE** The gasoline tank of a minivan holds 18 gallons. How many quarts is this? 72 qt

4. **TELEPHONES** Portable telephones can weigh as little as 8 ounces. How many pounds is this? $\frac{1}{2}$ lb

5. **RECIPE** A recipe for ice cream calls for 56 fluid ounces of milk. How many cups of milk is this? 7 c

6. **STATUE** The Statue of Liberty weighs 450,000 pounds. How many tons is this? 225 T

7. **TUNNEL** The Ted Williams Tunnel under Boston Harbor is 8,448 feet long. How many yards is this? 2,816 yd

8. **COAL** The United States exports over 200 billion pounds of coal. How many tons is this? 100 million T

6-8 Lesson Notes

Lesson Resources

Chapter 6 Resource Masters
- Study Guide and Intervention, p. 338
- Practice: Skills, p. 339
- Practice: Word Problems, p. 340
- Reading to Learn Mathematics, p. 341
- Enrichment, p. 342

Calculator and Spreadsheet Masters, p. 34
Parent and Student Study Guide Workbook, 6-8
Real-Life Career Masters, p. 20
Teaching Mathematics with Manipulatives, p. 64

FAST FILE

1 Focus

5-Minute Check 6-8

INTERACTIVE CHALKBOARD
PowerPoint® Presentations

Use as a quiz or review of Lesson 6-7. (Also available as a transparency)

Bellringer

Reading Mathematics Tell students that the word *perimeter* comes from the Greek words *peri*, meaning around or near, and *metron*, meaning measure. So, perimeter means to measure around a figure. Ask students whether they know any everyday uses of the word perimeter. **Sample answers: perimeter search, walking the perimeter**

Reading to Learn Mathematics, p. 341 **ELL**

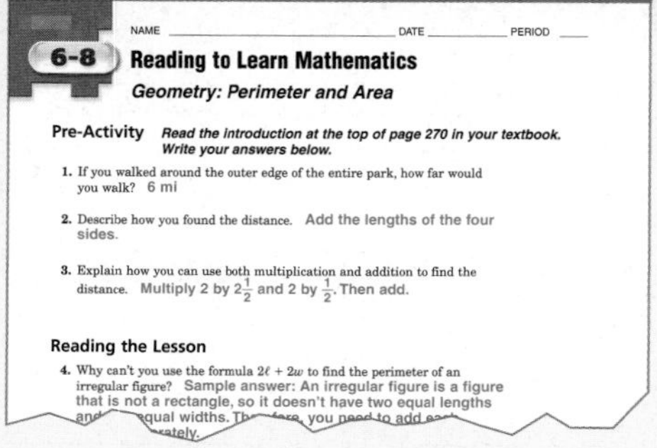
NAME ______________ DATE ________ PERIOD ____

6-8 Reading to Learn Mathematics
Geometry: Perimeter and Area

Pre-Activity *Read the introduction at the top of page 270 in your textbook. Write your answers below.*

1. If you walked around the outer edge of the entire park, how far would you walk? 6 mi
2. Describe how you found the distance. Add the lengths of the four sides.
3. Explain how you can use both multiplication and addition to find the distance. Multiply 2 by $2\frac{1}{2}$ and 2 by $\frac{1}{2}$. Then add.

Reading the Lesson

4. Why can't you use the formula $2\ell + 2w$ to find the perimeter of an irregular figure? Sample answer: An irregular figure is a figure that is not a rectangle, so it doesn't have two equal lengths and equal widths. Therefore, you need to add each ... separately.

6-8 Geometry: Perimeter and Area

What You'll Learn
Find the perimeters and areas of figures.

NEW Vocabulary
perimeter
formula
area

WHEN am I ever going to use this?

PARKS Central Park in New York City contains a running track, walking paths, playgrounds, and even a carousel.

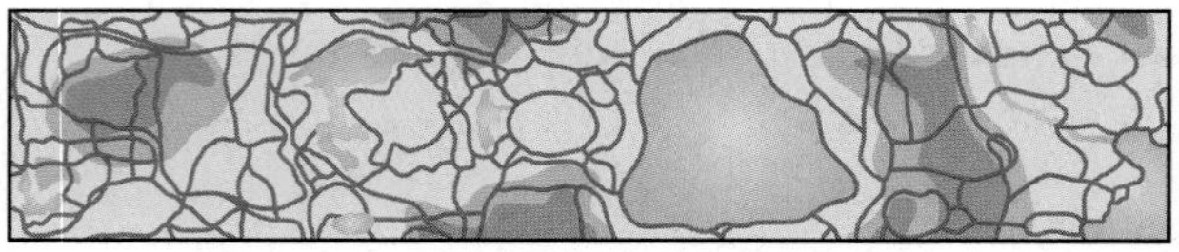

1. If you walked around the outer edge of the entire park, how far would you walk? **6 mi**
2. Describe how you found the distance. **Add the lengths of the four**
3. Explain how you can use both multiplication and addition to find the distance. **Multiply 2 by $2\frac{1}{2}$ and 2 by $\frac{1}{2}$. Then add.**

The distance around a geometric figure is called the **perimeter**. To fin the perimeter P of a rectangle, add the measures of the four sides.

Key Concept — **Perimeter of a Rectangle**

Words The perimeter P of a rectangle is twice the sum of the length ℓ and width w.

Symbols $P = \ell + \ell + w + w$
$P = 2\ell + 2w$ or $2(\ell + w)$

Model (rectangle with sides ℓ and w)

TEACHING TIP
Explain to students that the length of a rectangle is usually understood to be the longest side, but they can choose either side to be the length.

EXAMPLE Find the Perimeter of a Rectangle

1 Find the perimeter of the rectangle shown at the right.

(rectangle: 4 cm by 15 cm)

$P = 2\ell + 2w$	Perimeter of a rectangle
$P = 2(15) + 2(4)$	Replace ℓ with 15 and w with 4.
$P = 30 + 8$	Multiply.
$P = 38$	Add.

The perimeter is 38 centimeters.

You can find the perimeter of irregular figures by adding the lengths of the sides.

EXAMPLE Find the Perimeter of an Irregular Figure

2 **Find the perimeter of the figure.**

Estimate $3 + 5 + 7 + 3 = 18$ in.

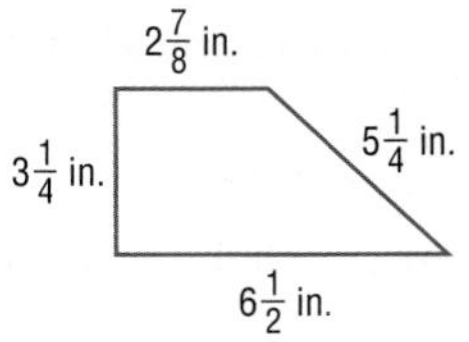

$P = 2\frac{7}{8} + 5\frac{1}{4} + 6\frac{1}{2} + 3\frac{1}{4}$

$P = 2\frac{7}{8} + 5\frac{2}{8} + 6\frac{4}{8} + 3\frac{2}{8}$

$P = 16\frac{15}{8}$ or $17\frac{7}{8}$

The perimeter is $17\frac{7}{8}$ inches. This is close to the estimate.

Your Turn

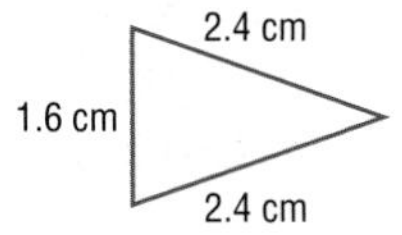

a. Find the perimeter of the figure. **6.4 cm**

The equation $P = 2\ell + 2w$ is called a **formula** because it shows a relationship among quantities. The formula for the **area** of a rectangle, or the measure of the surface enclosed by a figure, is shown below.

Key Concept — Area of a Rectangle

Words The area A of a rectangle is the product of the length ℓ and width w.

Symbol $A = \ell \cdot w$

Model

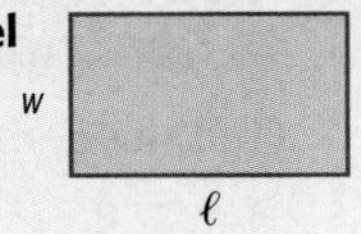

EXAMPLE Find the Area of a Rectangle

3 **VOLLEYBALL Find the area of the volleyball court shown at the right.**

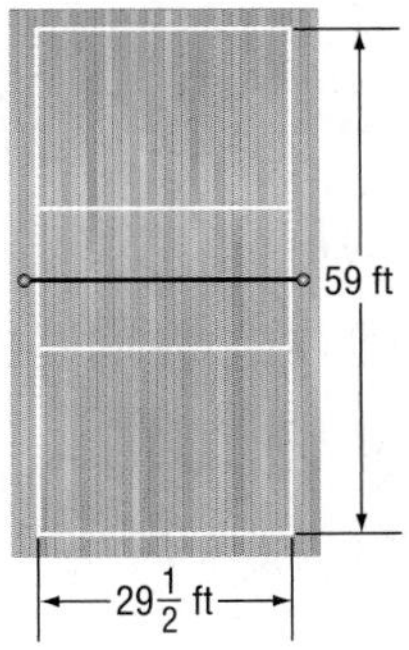

$A = \ell \cdot w$ — Area of a rectangle

$A = 59 \cdot 29\frac{1}{2}$ — Replace ℓ with 59 and w with $29\frac{1}{2}$.

$A = \frac{59}{1} \cdot \frac{59}{2}$ — Rename 59 and $29\frac{1}{2}$.

$A = 1{,}740\frac{1}{2}$ — Multiply and simplify.

The area is $1{,}740\frac{1}{2}$ square feet.

Your Turn Find the perimeter and area of each rectangle.

b.

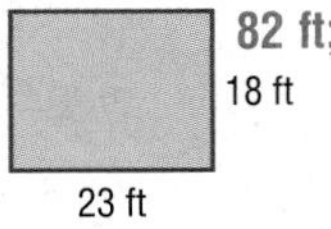

82 ft; 414 ft²

c. length = 13.2 mm
width = 8 mm
42.4 mm; 105.6 mm²

STUDY TIP

Area Units When finding area, the units are also multiplied. So, area is given in *square units*. Consider a rectangle 2 ft by 3 ft.

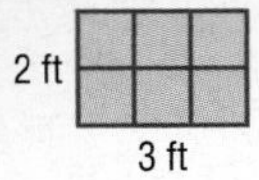

$A = 2 \text{ ft} \times 3 \text{ ft}$
$A = (2 \times 3)(\text{ft} \times \text{ft})$
$A = 6 \text{ ft}^2$

2 Teach

In-Class Examples

INTERACTIVE CHALKBOARD PowerPoint® Presentations

1 Find the perimeter of the rectangle. **40 ft**

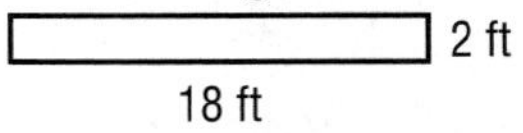

2 Find the perimeter of the figure. **$7\frac{1}{4}$ in.**

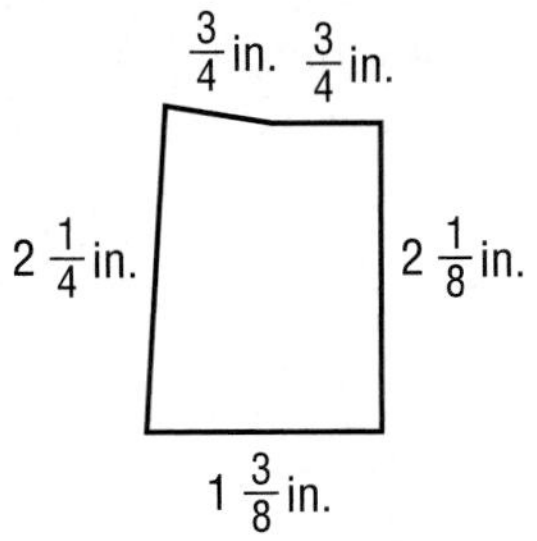

3 **FRESHWATER** Find the area of the surface of the reservoir shown below. **$2\frac{1}{2}$ mi²**

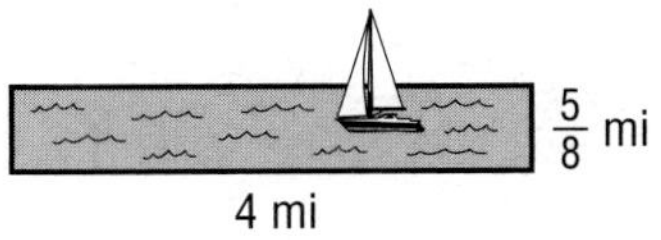

DAILY INTERVENTION

Differentiated Instruction

Naturalist Have students explore the school and grounds, measuring lengths. Suggestions include exterior walls, sidewalks, and sides of rectangular regions bounded by trees, rocks, and so forth. They should chart their data, then calculate perimeter and area for the buildings and regions measured.

3 Practice/Apply

About the Exercises...

Guided Practice Key

Exercises	Examples
3–6	1, 3
7	2

Odd/Even Assignments
Exercises 8–19 and 23–24 are structured so that students practice the same concepts whether they are assigned odd or even problems.

Answers

1. Perimeter is the sum of measures, so the units are the same. Area is the product of measures, so the units are squared.
2. Sample answer:

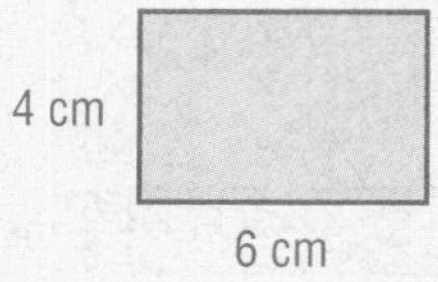

The perimeter is 20 cm.

Skill and Concept Check

Writing Math
Exercise 1

1. **Explain** why perimeter is given in units and area is given in square units. **1–2. See margin.**

2. **OPEN ENDED** Draw and label a rectangle that has an area of 24 square centimeters. What is the perimeter of your rectangle?

GUIDED PRACTICE

Find the perimeter and area of each rectangle.

3. 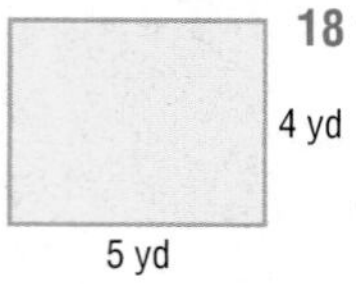

18 yd; 20 yd^2

4. **9.6 m; 3.8 m^2**
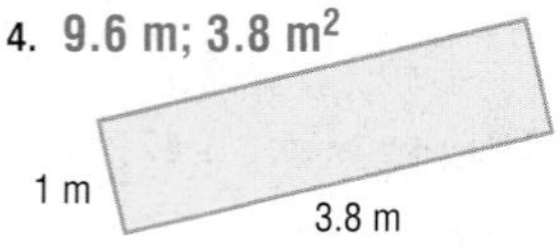

5. $\ell = 7$ cm, $w = 6$ cm **26 cm; 42 cm^2**

6. $\ell = 5\frac{1}{2}$ in., $w = 3$ in. **17 in.; $16\frac{1}{2}$ in^2**

7. Find the perimeter of the figure at the right. **$23\frac{1}{3}$ft**

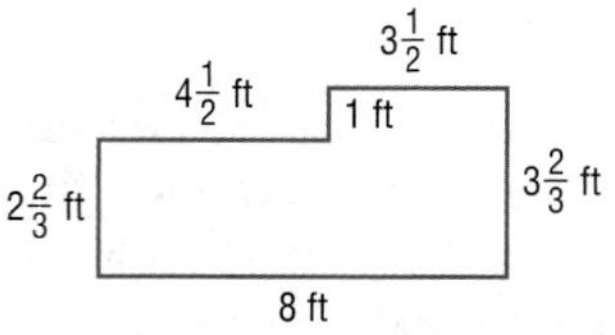

★ indicates increased difficulty

Practice and Applications

HOMEWORK HELP

For Exercises	See Examples
8–22, 25–28	1, 3
23–24	2

Extra Practice
See pages 579, 601.

Find the perimeter and area of each rectangle.

8. 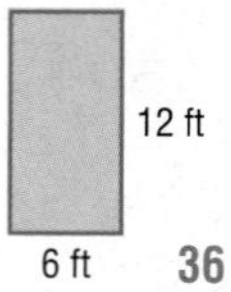

36 ft; 72 ft^2

9. 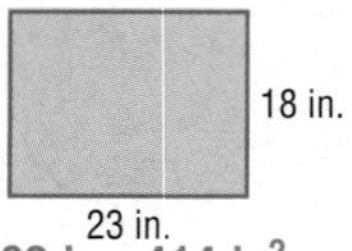

82 in.; 414 in^2

10. 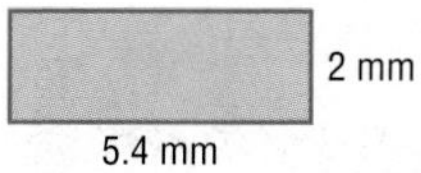

14.8 mm; 10.8 mm^2

11. 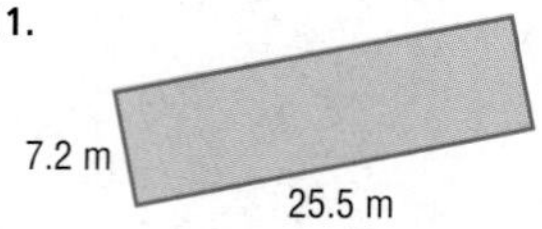

65.4 m; 183.6 m^2

12. 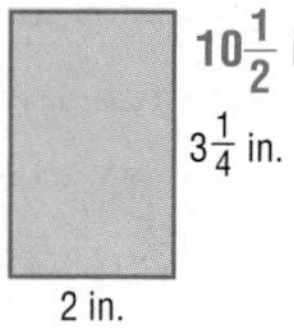

$10\frac{1}{2}$ in.; $6\frac{1}{2}$ in^2

13. 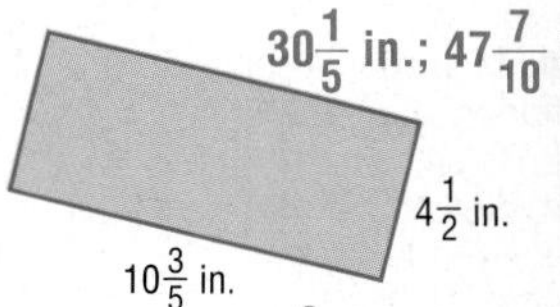

$30\frac{1}{5}$ in.; $47\frac{7}{10}$ i

14. $\ell = 5$ ft, $w = 1$ ft **12 ft; 5 ft^2**
15. $\ell = 18$ cm, $w = 12$ cm
16. $\ell = 6.5$ m, $w = 4$ m
17. $\ell = 4\frac{3}{8}$ in., $w = 3\frac{1}{5}$ in.
★ 18. $\ell = 2$ ft, $w = 18$ in.
★ 19. $\ell = 35$ ft, $w = 7$ yd

15. 60 cm; 216 cm^2 16. 21 m; 26 m^2 17. $15\frac{3}{20}$ in.; 14 in^2 18. 7 ft or 84 in.; 3 ft^2 or 432 in^2
19. $37\frac{1}{3}$ yd or 112 ft; $81\frac{2}{3}$ yd^2 or 735 ft

For Exercises 20 and 21, use the square at the right.

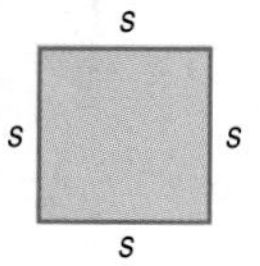

★ 20. Write formulas for the perimeter P and area A of the square. **$P = 4s$; $A = s^2$**

★ 21. If the side length is doubled, what happens to the perimeter and area? **The perimeter is doubled, the area is quadrupled.**

★ 22. Find the width of a rectangle with an area of 30 square inches and a length of 5 inches. **6 in.**

272 Chapter 6 Applying Fractions

Find the perimeter of each figure.

23. 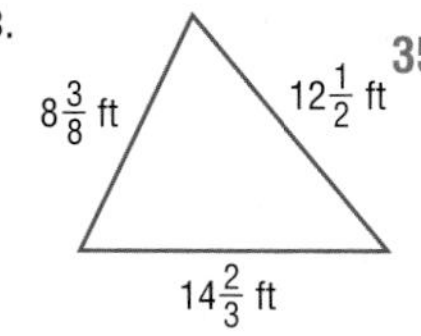

$35\frac{13}{24}$ ft

24. 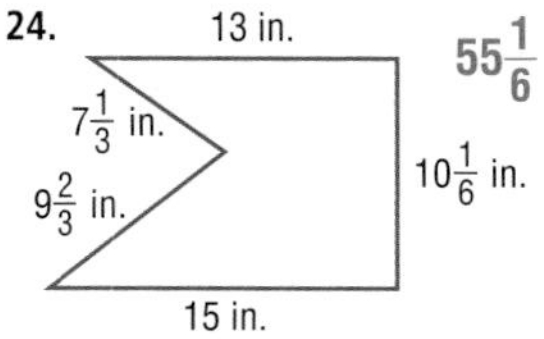

$55\frac{1}{6}$ in.

25. **FLOORING** Jasmine plans to use 1-foot square tiles to tile her kitchen floor, which measures 18 feet by 14 feet. If there are 40 tiles per box, how many boxes must she buy? 7 boxes

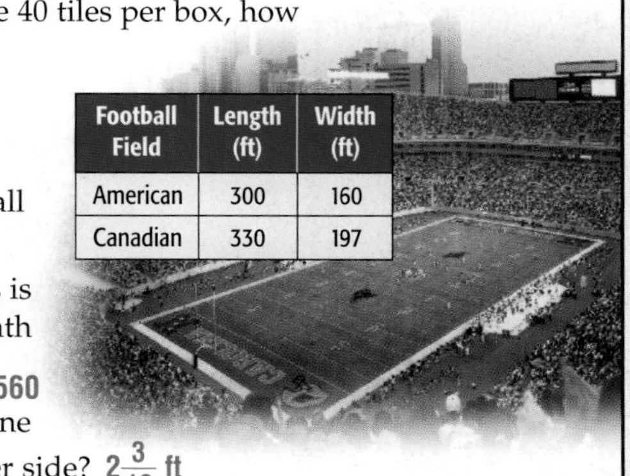

FOOTBALL For Exercises 26 and 27, use the table at the right.

Football Field	Length (ft)	Width (ft)
American	300	160
Canadian	330	197

26. How much greater is the area of a Canadian football field than an American football field? 17,010 ft^2

27. An *acre* equals 43,560 square feet. How many acres is a Canadian football field? Round to the nearest tenth of an acre and explain your method. 1.5 acres; 65,010 ÷ 43,560

★ 28. A rectangle is made with exactly 9 feet of string. One side is $2\frac{5}{16}$ feet long. What is the length of the other side? $2\frac{3}{16}$ ft

★ 29. **CRITICAL THINKING** Compare and contrast the perimeters and areas of rectangles that have the following dimensions: 1 by 9, 2 by 8, 3 by 7, and 4 by 6. See margin.

EXTENDING THE LESSON For Exercises 30–32, refer to the figures at the right.

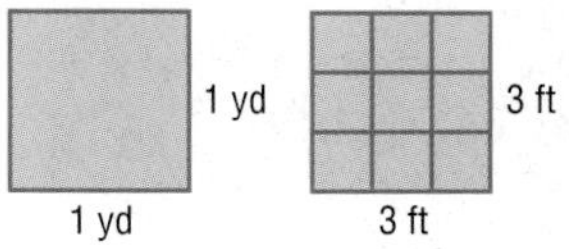

30. 1 yd^2 = __?__ ft^2 9
31. 4 yd^2 = __?__ ft^2 36
32. 1 ft^2 = __?__ in^2 144

Standardized Test Practice and Mixed Review

33. **MULTIPLE CHOICE** The perimeter of the rectangle shown at the right is $41\frac{1}{2}$ feet. What is the value of x? A

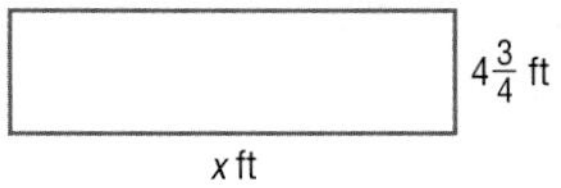

(A) 16 (B) 32 (C) $36\frac{3}{4}$ (D) 64

34. **SHORT RESPONSE** Determine the area of a rectangle that is 3 centimeters wide and 7 centimeters long. 21 cm^2

Complete. (Lesson 6-7)

35. $5\frac{1}{4}$ T = __?__ lb 10,500
36. 8 yd = __?__ ft 24
37. 15 pt = __?__ qt $7\frac{1}{2}$
38. 72 in. = __?__ ft 6

39. Find $22 \div \frac{2}{3}$. Write in simplest form. (Lesson 6-6) 33

GETTING READY FOR THE NEXT LESSON

PREREQUISITE SKILL Multiply. Write in simplest form. (Lesson 6-4)

40. $\frac{9}{8} \cdot 16$ 18
41. $\frac{22}{7} \cdot 14$ 44
42. $2 \cdot \frac{3}{7} \cdot 35$ 30
43. $\frac{22}{7} \cdot 1\frac{1}{2}$ $4\frac{5}{7}$

msmath2.net/self_check_quiz

Study Guide and Intervention, p. 338

NAME ______ DATE ______ PERIOD ____

6-8 Study Guide and Intervention

Geometry: Perimeter and Area

The distance around a geometric figure is called the **perimeter**.
To find the perimeter of any geometric figure, add the measures of its sides.
The perimeter of a rectangle is twice the length ℓ plus twice the width w.

$$P = 2\ell + 2w$$

EXAMPLE 1 Find the perimeter of the figure at the right.

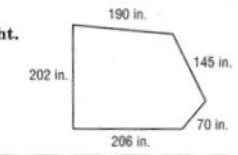

$P = 145 + 70 + 206 + 202 + 190$
$= 813$

The perimeter is 813 inches.

The measure of the surface enclosed by a geometric figure is called the **area**.
The area of a rectangle is the product of the length ℓ and width w.

$$A = \ell \cdot w.$$

EXAMPLE 2 Find the area of the rectangle.

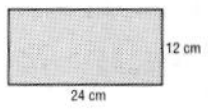

$A = \ell \cdot w$
$= 24 \cdot 12$
$= 288$

The area is 288 square centimeters.

EXERCISES

Find the perimeter of each figure.

1. 7 cm, 9 cm, 15 cm 31 cm
2. $7\frac{7}{8}$ ft, $3\frac{3}{4}$ ft, $12\frac{1}{8}$ ft, 10 ft, $6\frac{1}{4}$ ft, 20 ft 60 ft

Find the perimeter and area of each rectangle.

3. 9 ft by 4 ft 26 ft; 36 ft^2
4. 3 in. by 11 in. 28 in.; 33 in^2
5. $\ell = 8$ ft, $w = 5$ ft 26 ft; 40 ft^2
6. $\ell = 3.5$ m, $w = 2$ m 11 m; 7 m^2
7. $\ell = 8$ yd, $w = 4\frac{1}{3}$ yd $24\frac{2}{3}$ yd; $34\frac{2}{3}$ yd^2
8. $\ell = 29$ cm, $w = 7.3$ cm 72.6 cm; 211.7 cm^2

Practice: Skills, p. 339

NAME ______ DATE ______ PERIOD ____

6-8 Practice: Skills

Geometry: Perimeter and Area

Find the perimeter of each figure.

1. 25 ft, 20 ft, 20 ft, 15 ft 80 ft
2. 12 ft, 12 ft, 13 ft, 13 ft, 15 ft 65 ft
3. 10 yd, 9 yd, 9 yd, 10 yd 38 yd
4. 25 cm, 28 cm, 26 cm, 20 cm, 16 cm 115 cm

Find the perimeter and area of each rectangle.

5. 15 yd by 5 yd 40 yd; 75 yd^2
6. 20 cm by 20 cm 80 cm; 400 cm^2
7. 30 m by 6 m 72 m; 180 m^2
8. 15 cm by 8 cm 46 cm; 120 cm^2
9. $\ell = 6$ yd, $w = 4$ yd 20 yd; 24 yd^2
10. $\ell = 8.2$ m, $w = 7.1$ m 30.6 m; 58.22 m^2
11. $\ell = 50$ in., $w = 10$ in. 120 in.; 500 in^2
12. $\ell = 10$ cm, $w = 4\frac{1}{2}$ cm 29 cm; 45 cm^2
13. $\ell = 4.5$ ft, $w = 3$ ft 15 ft; 13.5 ft^2
14. $\ell = 7\frac{1}{2}$ mm, $w = 6\frac{3}{8}$ mm $27\frac{3}{4}$ mm; $47\frac{13}{16}$ mm^2

Practice: Word Problems, p. 340

NAME ______ DATE ______ PERIOD ____

6-8 Practice: Word Problems

Geometry: Perimeter and Area

1. **PYRAMID** The Great Pyramid in Egypt has a square base with sides 230 meters long. Find the perimeter and area of the base of the Great Pyramid. 920 m; 52,900 m^2

2. **WINDOWS** Mrs. Johnson was planning to caulk around the frame of her patio doors that measure 5 feet by $6\frac{1}{2}$ feet. In order to help her to know how much caulk to buy, find the perimeter of the doors. 23 ft

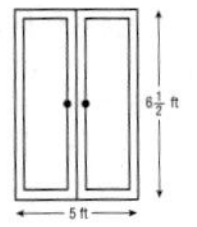

3. **PENTAGON** The Pentagon is a five-sided building that houses the U.S. Department of Defense in Arlington, Virginia. If each side of the Pentagon measures 921 feet, what is the walking distance around its exterior? 4,605 ft

4. **FENCING** Mr. Lao is planning to build a rectangular cattle pen that measures 50 feet by 75 feet. Find the total length of fencing that he will need to purchase. 250 ft

5. **CARPET** Mr. Yuji plans on buying carpet for his bedroom that measures 12 feet by 12 feet. So he will know how much carpet to buy, find the area of his bedroom. 144 ft^2

6. **BORDER** Mrs. Jackson is going to put up a wallpaper border along the top of the walls in her dining room. If the dining room measures 16 meters by 12 meters, how much border should she buy? 56 m

7. **LOBBY** A hotel lobby measures 40 yards by 60 yards. Find the area and perimeter of the lobby's floor. 2,400 yd^2; 200 yd

8. **MURAL** An artist painted a mural measuring 9 feet by $20\frac{1}{2}$ feet. Find the area and perimeter of the mural. $184\frac{1}{2}$ ft^2; 59 ft

4 Assess

Open-Ended Assessment

Writing Have students write an area word problem and a perimeter word problem and exchange problems with a classmate. The classmate should write, in words and symbols, each step of the solution to each problem.

Answer

29. The rectangles have the same perimeters; as the lengths and widths get closer, the areas increase.

6-9a HANDS-ON LAB Lesson Notes

Lesson Resources

Teaching Mathematics with Manipulatives

- student recording sheet 6-9a

Glencoe Mathematics Classroom Manipulative Kit

- rulers
- tape measures

GETTING STARTED

Teaching Tips

- Provide pairs of students with different sets of circular objects to expand the data sample.
- You may wish to record and graph students' sets of data on an overhead transparency as they are collected.

TEACH

- For irregular objects, make sure that students measure the circumference corresponding to the diameter that they measure. If they measure the diameter of the bottom of a can, for example, they should measure the circumference at the bottom as well.
- If students finish early, have them graph the ordered pairs from neighbors' measurements on the same graph used before in a contrasting ink.

6-9a HANDS-ON LAB

A Preview of Lesson 6-9

What You'll Learn
Find a relationship between circumference and diameter.

Materials

- ruler
- measuring tape
- circular objects

Circumference

INVESTIGATE *Work with a partner.*

In this lab, you will investigate how *circumference*, or the distance around a circle, is related to its *diameter*, or the distance across a circle.

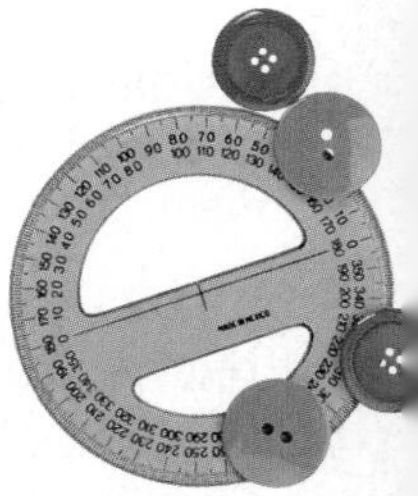

STEP 1 Use a ruler to measure the diameter of a circular object. Record the measure in a table like the one shown below.

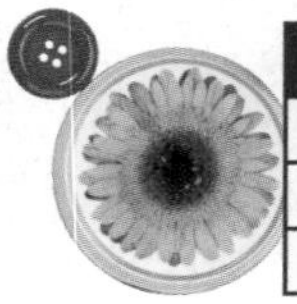

Object	Diameter (cm)	Circumference (cm)

STEP 2 Make a small mark at the edge of the circular object. The measuring tape should be on a flat surface. Place the mark at the beginning of the measuring tape. Roll the object along the tape for one revolution, until you reach the mark again.

STEP 3 Record the length in the table. This is the circumference.

STEP 4 Repeat this activity with circular objects of various sizes.

Writing Math

Work with a partner.

1. For each object, divide the circumference by the diameter. Add another column to your table and record the results. Round to the nearest tenth if necessary. **See students' work.**
2. What do you notice about the ratios? **They are slightly greater than 3.**
3. **Graph** the ordered pair (diameter, circumference) on a coordinate plane for each object. What do you find?
4. Select two points on the graph and find the slope between them. Select two different points and find the slope. What do you observe about the slopes? **They are both slightly greater than 3.**
5. Use the graph to predict the circumference of a circular object that has a diameter of 18 centimeters. **Sample answer: about 56.5 cm**
6. **Write** a rule describing how you would find the circumference C of a circle if you know the diameter d. **Sample answer: $C \approx 3.1d$**

STUDY TIP

Look Back You can review **slope** in Lesson 4-7.

3. See students' graphs. All points lie on a diagonal line.

ASSESS

Observe as students work the **Writing Math** exercises. They should:

- correctly divide and record the circumference/diameter ratios, and
- correctly graph the data.

Use **Exercises 5** and **6** to assess whether students have grasped the connection between circumference and diameter.

6-9 Geometry: Circles and Circumference

/hat You'll Learn
nd the circumference
f circles.

EW Vocabulary

rcle
enter
ameter
dius
rcumference

IATH Symbols

(pi)

WHEN am I ever going to use this?

FERRIS WHEELS The London Eye Ferris wheel measures 450 feet across.

1. Which point appears to be the center of the Ferris wheel? *G*
2. Is the distance from *G* to *F* greater than, less than, or equal to the distance from *G* to *J*? equal to

A **circle** is the set of all points in a plane that are the same distance from a given point, called the **center**.

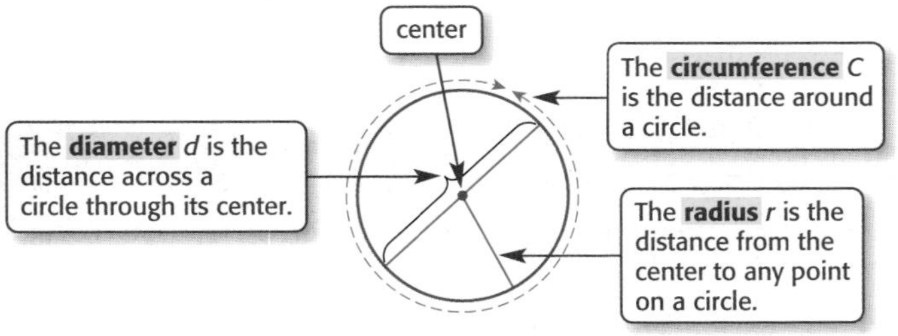

The diameter of a circle is 2 times the radius, or $d = 2r$. Another relationship that is true of all circles is $\frac{C}{d} = 3.1415926\ldots$. This nonterminating and nonrepeating number is represented by the Greek letter **π (pi)**. An approximation often used for π is 3.14.

Key Concept — **Circumference of a Circle**

Words The circumference *C* of a circle is equal to its diameter *d* times π, or 2 times its radius *r* times π.

Symbols $C = \pi d$ or $C = 2\pi r$

TEACHING TIP
'oint out that all
ircumferences are
stimates since 3.14 is
n estimated value of pi.
'he *exact* circumference
n Example 1 is 450π.

EXAMPLE Find Circumference

1 FERRIS WHEELS Find the circumference of the Ferris wheel above.

$C = \pi d$ — Circumference of a circle

$C \approx 3.14(450)$ — Replace π with 3.14 and *d* with 450.

$C \approx 1{,}413$ — Multiply.

So, the distance around the Ferris wheel is about 1,413 feet.

msmath2.net/extra_examples

6-9 Lesson Notes

Lesson Resources

Chapter 6 Resource Masters
- Study Guide and Intervention, p. 343
- Practice: Skills, p. 344
- Practice: Word Problems, p. 345
- Reading to Learn Mathematics, p. 346
- Enrichment, p. 347

Calculator and Spreadsheet Masters, pp. 35–36
Key Teaching Transparencies, 24
MathPASS Tutorial CD-ROM, 51
Parent and Student Study Guide Workbook, 6-9

FAST FILE

1 Focus

5-Minute Check 6-9

INTERACTIVE CHALKBOARD
PowerPoint® Presentations

Use as a quiz or review of Lesson 6-8. (Also available as a transparency)

Bellringer

Reading Mathematics Write the words *circle* and *circumference* on the chalkboard. Point out that *circle* and *circum-* are related to the Greek word for ring, as is *circus*. Then point out that *-ference* comes from the Latin word meaning to carry. The combined form, *circumferre,* means to carry around. Ask students to suggest a way to remember the meaning of circumference.

Reading to Learn Mathematics, p. 346 ELL

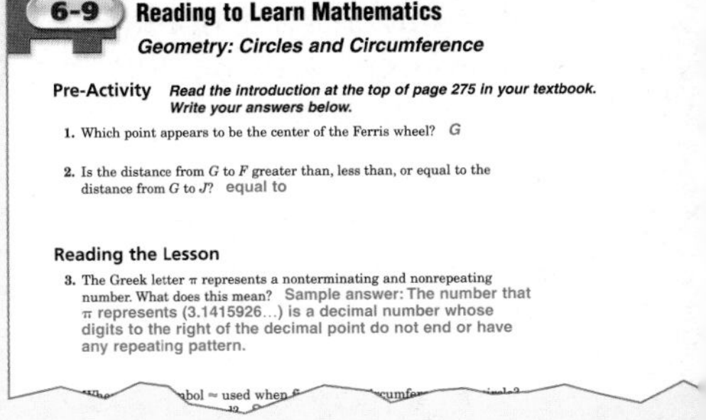
NAME ________ DATE ________ PERIOD ____

6-9 Reading to Learn Mathematics
Geometry: Circles and Circumference

Pre-Activity ***Read the introduction at the top of page 275 in your textbook. Write your answers below.***

1. Which point appears to be the center of the Ferris wheel? G
2. Is the distance from *G* to *F* greater than, less than, or equal to the distance from *G* to *J*? equal to

Reading the Lesson

3. The Greek letter π represents a nonterminating and nonrepeating number. What does this mean? Sample answer: The number that π represents (3.1415926...) is a decimal number whose digits to the right of the decimal point do not end or have any repeating pattern.

DAILY INTERVENTION — **Unlocking Misconceptions**

Pi Students sometimes mistakenly believe that pi (π) is a variable since measurements are often written as 2π, 3.46π, and so on. Write the following on the chalkboard and ask what the difference is between the two measurements: 12π cm and 12 cm. **The first is about 3.14 times the second.** Also point out the difference between the following: $6x$ and 6π; $3.4y$ and 3.4π. **The *x* and *y* are variables and can represent any number. In contrast, π is a constant; its value does not change.**

2 Teach

In-Class Examples

1 **PETS** Find the circumference around the hamster's running wheel shown below. Round to the nearest tenth. **18.8 in.**

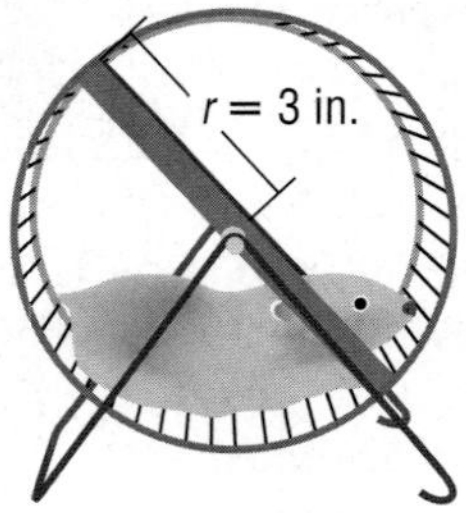

2 Find the circumference of a circle with a diameter of 49 centimeters. **about 154 cm**

3 Practice/Apply

DAILY INTERVENTION

FIND THE ERROR For Exercise 3, have students recite both formulas several times to help commit them to memory. To help them remember the difference between radius and diameter, have students make a reference drawing of a circle with the parts clearly marked.

About the Exercises...

Guided Practice Key

Exercises	Examples
4, 7, 9, 10	1
5, 6, 8	2

Odd/Even Assignments

Exercises 11–22 are structured so that students practice the same concepts whether they are assigned odd or even problems.

Alert! Exercise 24 requires research on the Internet or another source.

Another approximation for π is $\frac{22}{7}$. Use this value when the radius or diameter of a circle is a multiple of 7 or has a multiple of 7 in its numerator.

STUDY TIP

Technology You can use a calculator to find the circumference. To find $2\pi(21)$, press 2 [×] [π] [×] 21 [ENTER]. The circumference is about 131.9468915.

EXAMPLE 2 Find Circumference

Find the circumference of a circle with a radius of 21 inches.

Since 21 is a multiple of 7, use $\frac{22}{7}$ for π.

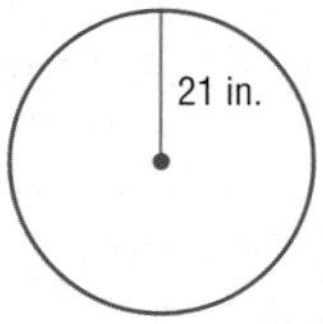

$C = 2\pi r$ Circumference of a circle

$C \approx 2 \cdot \frac{22}{7} \cdot 21$ Replace π with $\frac{22}{7}$ and r with 21.

$C \approx 2 \cdot \frac{22}{\cancel{7}_1} \cdot \frac{\cancel{21}^3}{1}$ Divide by the GCF, 7.

$C \approx 132$ Simplify.

The circumference of the circle is about 132 inches.

Your Turn **Find the circumference of each circle. Use 3.14 or $\frac{22}{7}$ for π. Round to the nearest tenth if necessary.**

a. diameter = 4.5 cm **14.1 cm**
b. radius = $\frac{7}{8}$ ft **$5\frac{1}{2}$ ft**
c. radius = 35 in. **220**

Skill and Concept Check

Writing Math Exercises 1 & 3

1. **Explain** how circumference is affected by an increase in radius. **The circumference increases by 2π for every 1-unit increase in r.**
2. **OPEN ENDED** Describe a real-life situation in which finding the circumference of a circle would be useful. **Sample answer: sewing fringe on a circular tablecloth**
3. **FIND THE ERROR** Aidan and Mya are finding the circumference of a circle with a radius of 5 inches. Who is correct? Explain.

Aidan	Mya
$C = \pi \cdot 5 = 5\pi$	$C = 2 \cdot \pi \cdot 5 = 10\pi$

Mya; Aidan incorrectly applied the formula that uses the diameter.

GUIDED PRACTICE

Find the circumference of each circle. Use 3.14 or $\frac{22}{7}$ for π. Round to the nearest tenth if necessary.

4.
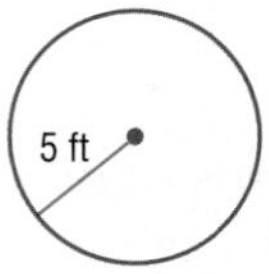

31.4 ft

5.
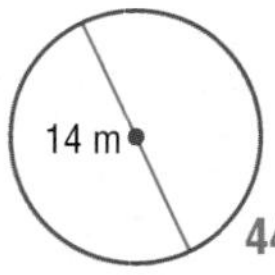

44 m

6.
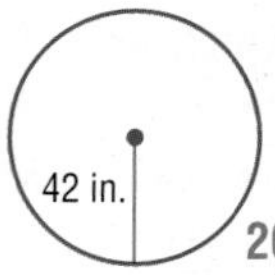

264 in.

7. radius = 11.7 cm **73.5 cm**
8. radius = 28 ft **176 ft**
9. diameter = $3\frac{1}{2}$ yd **11 yd**
10. **MUSIC** Purdue University's marching band has a drum with a diameter of 8 feet. What is its circumference to the nearest tenth? **25.1 ft**

TEACHING TIP

Exercises 5, 6, 8, and 9 were solved using $\frac{22}{7}$ for π because the measures given are multiples of 7 or have multiples of 7 in the numerator. This rule also applies to the exercises on page 277.

4 Assess

Higher-Level Thinking

A swing is attached to a crossbar as shown to the right. The swing's ropes each measure 8 feet long. If Mae Lin made an arc by starting forward at the height of the bar and swinging back to the height of the bar, how far did she swing? $\left(\frac{1}{2}\right)2 \cdot \pi \cdot 8 = 8\pi$ **ft**

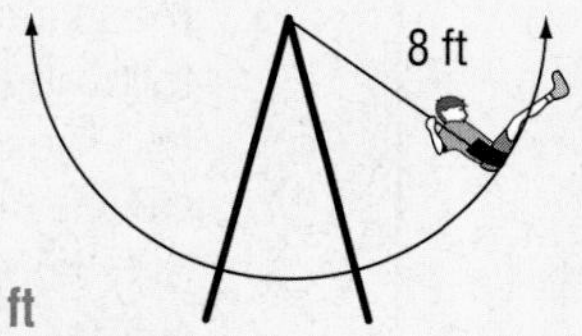

Practice and Applications

HOMEWORK HELP	
For Exercises	See Examples
11–23	1, 2
Extra Practice See pages 579, 601.	

Find the circumference of each circle. Use 3.14 or $\frac{22}{7}$ for π. Round to the nearest tenth if necessary.

11. 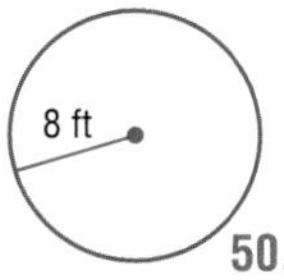

50.2 ft

12. 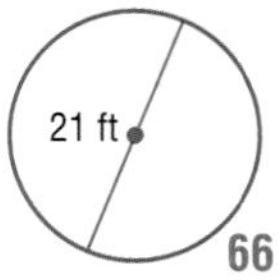

66 ft

13. 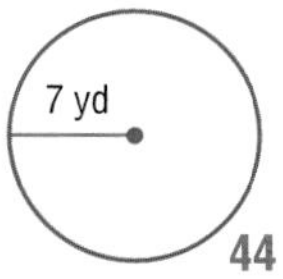

44 yd

14. radius = 38.4 cm **241.2 cm**
15. diameter = 15.1 m **47.4 m**
16. radius = $1\frac{3}{4}$ in. **11 in.**
17. radius = 2 km **12.6 km**
18. diameter = 10 ft **31.4 ft**
19. radius = 45.5 m **285.7 m**
20. radius = 56 cm **352 cm**
21. radius = $2\frac{5}{8}$ in. **$16\frac{1}{2}$ in.**
22. diameter = $10\frac{1}{2}$ in. **33 in.**
23. **CROPS** The mysterious crop circle shown at the right was created in England in a single night by an unknown source. What is the circumference of the circle? **785 ft**

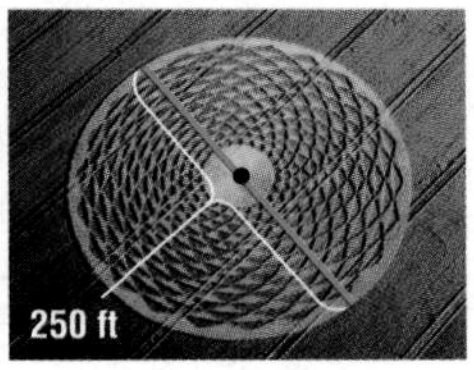

24. **RESEARCH** Use the Internet or another source to find three other parts of circles: *arcs*, *central angles*, and *inscribed angles*. Draw a circle and label these parts. **See margin.**

CRITICAL THINKING For Exercises 25 and 26, refer to the figure at the right.

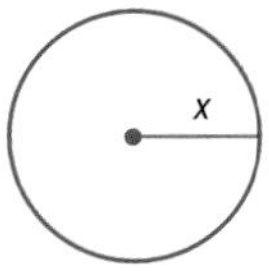

25. How many lengths x will fit on the circumference of the circle? **2π or 6.28**
26. If the value of x is doubled, what effect will this have on the diameter? on the circumference? **Both will be doubled.**

Standardized Test Practice and Mixed Review

27. **SHORT RESPONSE** What is the circumference of the tree trunk whose cross section is shown at the right? Use 3.14 for π. **69.1 in.**

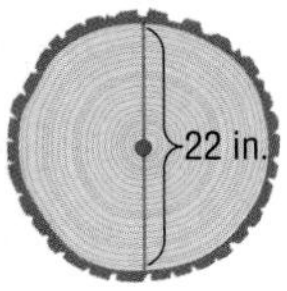

28. **MULTIPLE CHOICE** About how far does a bicycle wheel travel in 150 revolutions of the wheel if its diameter is 11 inches? Use 3.14 for π. **A**

Ⓐ 431.75 ft Ⓑ 863.5 ft Ⓒ 5,181 ft Ⓓ 10,362 ft

Find the perimeter and area of each rectangle. (Lesson 6-8)

29. ℓ = 5 cm, w = 3.8 cm **17.6 cm; 19 cm^2**
30. $\ell = 2\frac{1}{4}$ ft, w = 11 ft **$26\frac{1}{2}$ ft; $24\frac{3}{4}$ ft^2**

31. Twenty-four feet equals how many yards? (Lesson 6-7) **8 yd**

WebQuest Interdisciplinary Project

A Well-Balanced Diet

It's time to complete your project. Use the data you have gathered about the food you eat and about the Food Pyramid to prepare a Web page or poster. Be sure to include a chart and calculations with your project.

msmath2.net/webquest

msmath2.net/self_check_quiz

Answer

24. Sample answer:

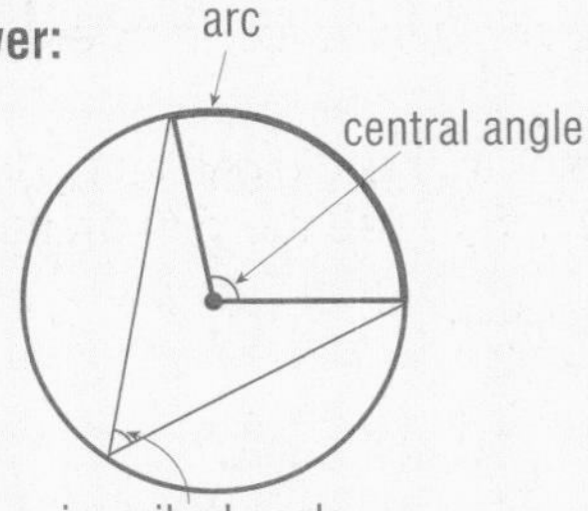

Study Guide and Intervention, p. 343

NAME ______________ DATE ________ PERIOD ____

6-9 Study Guide and Intervention

Geometry: Circles and Circumference

A **circle** is the set of all points in a plane that are the same distance from a given point, called the **center**. The **diameter** d is the distance across the circle through its center. The **radius** r is the distance from the center to any point on the circle. The **circumference** C is the distance around the circle. The circumference of a circle is equal to its diameter d times π, or 2 times its radius r times π.

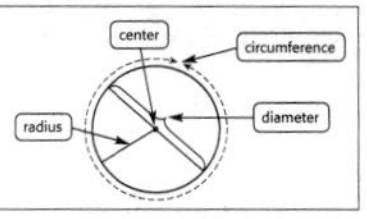

EXAMPLE 1 **Find the circumference of a circle with a diameter of 7.5 centimeters.**

$C = \pi d$

$C \approx 3.14 \times 7.5$ Use 3.14 for π.

$C \approx 23.55$ The circumference of the circle is about 23.55 centimeters.

EXAMPLE 2 **If the radius of a circle is 14 inches, what is its circumference?**

$C = 2\pi r$

$C \approx 2 \times \frac{22}{7} \times 14$ Use $\frac{22}{7}$ for π.

$C \approx 88$ The circumference of the circle is about 88 inches.

EXERCISES

Find the circumference of each circle. Use 3.14 or $\frac{22}{7}$ for π. Round to the nearest tenth if necessary.

1. 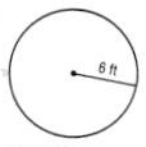

37.7 ft

2. 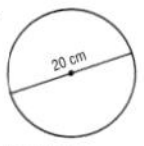

62.8 cm

3. 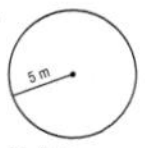

31.4 m

4. 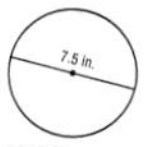

23.6 in.

5. diameter = 15 km 47.1 km
6. radius = 21 mi 131.9 mi
7. radius = 50 m 314 m
8. diameter = 600 ft 1,884 ft
9. radius = 62 mm 389.4 mm
10. diameter = 7 km 22.0 km
11. radius = 95 in. 596.6 in.
12. diameter = 6.3 m 19.8 m
13. diameter = $5\frac{1}{4}$ cm 16.5 cm

Practice: Skills, p. 344

NAME ______________ DATE ________ PERIOD ____

6-9 Practice: Skills

Geometry: Circles and Circumference

Find the circumference of each circle. Use 3.14 or $\frac{22}{7}$ for π. Round to the nearest tenth if necessary.

1. 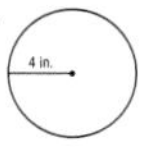

25.1 in.

2. 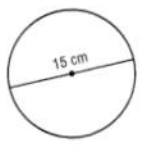

47.1 cm

3. 

50.2 ft

4. 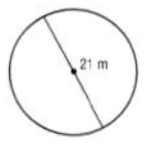

66.0 m

5. 

100.5 km

6. 

116.2 mm

7. radius = 3 km 18.8 km
8. radius = 46 cm 288.9 cm
9. diameter = 30 in. 94.2 in.
10. diameter = 25 m 78.5 m
11. radius = 5 ft 31.4 ft
12. diameter = $9\frac{1}{2}$ in. 29.8 in.
13. radius = $3\frac{1}{2}$ ft 22.0 ft
14. diameter = 9.7 mm 30.5 mm
15. radius = 5.2 km 32.7 km
16. diameter = 25 m 78.5 m
17. radius = 22 ft 138.2 ft
18. diameter = 9.4 in. 29.5 in.
19. radius = 100 m 628 m
20. radius = 65 mi 408.2 mi
21. diameter = $10\frac{1}{2}$ in. 33.0 in.
22. diameter = 8.5 cm 26.7 cm

Practice: Word Problems, p. 345

NAME ______________ DATE ________ PERIOD ____

6-9 Practice: Word Problems

Geometry: Circles and Circumference

1. **PLATES** A manufacturing company is producing dinner plates with a diameter of 12 inches. They plan to put a gold edge on each plate. Determine how much gold edging they need for each plate by finding the circumference of each plate. Round to the nearest tenth. 37.7 in.	2. **MONEY** A dime has a radius of $8\frac{1}{2}$ millimeters. Find the circumference of a dime to the nearest tenth. 53.4 mm
3. **MERRY-GO-ROUND** Mr. Osterhout is putting trim around the edge of a circular merry-go-round that has a diameter of 15 feet. How much trim does he need to buy to the nearest tenth? 47.1 ft	4. **PIZZA** Find the circumference of a pizza with a diameter of 10 inches. Round to the nearest tenth. 31.4 in.
5. **RACING** A circular racetrack has a diameter of $\frac{1}{2}$ mile. How far does a car travel in one lap around the track? Round to the nearest tenth. 1.6 mi	6. **TIRE** A bicycle tire has a radius of 15 inches. What is the circumference of the tire? Round to the nearest tenth. 94.2 in.
7. **EQUATOR** Earth's diameter at the equator is 7,926 miles. Find the distance around Earth at its equator to the nearest tenth. 24,887.6 mi	8. **SATURN** The ring system around Saturn has a diameter of 170,000 miles. Find the circumference of the ring system. 533,800 mi

Chapter 6 Study Guide and Review

Vocabulary Assessment Check

- This alphabetical list of vocabulary terms in Chapter 6 includes a page reference where each term was introduced.
- A vocabulary review/test for Chapter 6 is available on p. 362 of the *Chapter 6 Resource Masters*.

Vocabulary PuzzleMaker

ELL The online Vocabulary PuzzleMaker improves students' mathematics vocabulary using four puzzle formats — crossword, scramble, word search using a word list, and word search using clues. Students can work on a computer screen or from a printed handout.

MindJogger Videoquizzes

ELL MindJogger Videoquizzes provide an alternative review of concepts presented in this chapter. Students work in teams in a game show format to gain points for correct answers. The questions are presented in three rounds.

Round 1 Concepts (5 questions)
Round 2 Skills (4 questions)
Round 3 Problem Solving (4 questions)

ExamView® Pro Test Bank CD-ROM

Use the networkable ExamView® Pro Test Bank CD-ROM to:

- Create **Multiple versions** of tests.
- Create **Modified** tests for **Inclusion** students with one mouse click.
- **Edit** existing questions and **Add** your own questions.
- Build tests aligned with state standards using built-in **State Curriculum Correlations.**
- Change **English** tests to **Spanish** with one mouse click and vice versa.

Chapter 6 Study Guide and Review

Vocabulary and Concept Check

area (p. 271)	diameter (p. 275)	pint (p. 267)
center (p. 275)	formula (p. 271)	pound (p. 267)
circle (p. 275)	gallon (p. 267)	quart (p. 267)
circumference (p. 275)	multiplicative inverse (p. 258)	radius (p. 275)
compatible numbers (p. 242)	ounce (p. 267)	reciprocal (p. 258)
cup (p. 267)	perimeter (p. 270)	ton (p. 267)

Choose the correct term or number to complete each sentence.

1. The (radius, diameter) is the distance across a circle through its center. diameter
2. To add like fractions, add the (numerators, denominators). numerators
3. (Perimeter, Area) is the measure of the surface enclosed by a figure. Area
4. Fractions with different denominators are called (like, unlike) fractions. unlike
5. When dividing by a fraction, multiply by its (value, reciprocal). reciprocal
6. One cup is equivalent to (8, 12) fluid ounces. 8

Lesson-by-Lesson Exercises and Examples

6-1 Estimating with Fractions (pp. 240–243) 7. $3 \div 1 = 3$

Estimate. 7–12. Sample answers are given.

7. $2\frac{9}{10} \div 1\frac{1}{8}$
8. $6\frac{2}{9} - 5\frac{1}{7}$ $6 - 5 = 1$
9. $\frac{13}{15} \times \frac{1}{5}$ $1 \times 0 = 0$
10. $\frac{1}{2} + \frac{3}{8}$ $\frac{1}{2} + \frac{1}{2} = 1$
11. $\frac{1}{2} \cdot 25$ $\frac{1}{2} \cdot 26 = 13$
12. $15\frac{6}{7} \div 7\frac{1}{3}$ $16 \div 8 = 2$

Example 1 Estimate $5\frac{1}{12} + 2\frac{5}{6}$.

$5\frac{1}{12} + 2\frac{5}{6} \rightarrow 5 + 3 = 8$

Example 2 Estimate $\frac{7}{8} - \frac{4}{7}$.

$\frac{7}{8} - \frac{4}{7} \rightarrow 1 - \frac{1}{2} = \frac{1}{2}$

6-2 Adding and Subtracting Fractions (pp. 244–247)

Add or subtract. Write in simplest form.

13. $\frac{2}{6} - \frac{1}{6}$ $\frac{1}{6}$
14. $\frac{3}{7} + \frac{9}{14}$ $1\frac{1}{14}$
15. $\frac{5}{6} - \frac{3}{4}$ $\frac{1}{12}$
16. $\frac{1}{9} + \frac{5}{9}$ $\frac{2}{3}$
17. $\frac{4}{5} + \frac{4}{5}$ $1\frac{3}{5}$
18. $\frac{9}{10} - \frac{3}{10}$ $\frac{3}{5}$
19. $\frac{5}{8} - \frac{5}{12}$ $\frac{5}{24}$
20. $\frac{11}{12} - \frac{1}{6}$ $\frac{3}{4}$
21. $\frac{3}{4} + \frac{7}{20}$ $1\frac{1}{10}$

Example 3 Find $\frac{1}{8} + \frac{3}{8}$.

$\frac{1}{8} + \frac{3}{8} = \frac{1+3}{8}$ Add the numerators.

$= \frac{4}{8}$ or $\frac{1}{2}$ Simplify.

Example 4 Find $\frac{3}{10} - \frac{1}{4}$.

$\frac{3}{10} - \frac{1}{4} = \frac{6}{20} - \frac{5}{20}$ or $\frac{1}{20}$

msmath2.net/vocabulary_review

FOLDABLES™ Study Organizer

For more information about Foldables, see *Teaching Mathematics with Foldables.*

Have students look through the chapter to make sure they have included examples in their Foldables for each lesson or topic in the chapter.

Encourage students to refer to their Foldables while completing the Study Guide and Review and to use them in preparing for the Chapter Test.

6-3 Adding and Subtracting Mixed Numbers (pp. 248–251)

Add or subtract. Write in simplest form.

22. $3\frac{2}{15} + 6\frac{9}{15}$ $9\frac{11}{15}$
23. $9\frac{4}{5} - 2\frac{1}{5}$ $7\frac{3}{5}$
24. $4\frac{1}{3} - 2\frac{2}{3}$ $1\frac{2}{3}$
25. $8\frac{2}{7} + 1\frac{6}{7}$ $10\frac{1}{7}$
26. $7\frac{11}{12} - 4\frac{3}{12}$ $3\frac{2}{3}$
27. $7\frac{3}{5} - 5\frac{1}{3}$ $2\frac{4}{15}$
28. $5\frac{3}{4} + 1\frac{1}{6}$ $6\frac{11}{12}$
29. $3\frac{5}{8} + 11\frac{1}{2}$ $15\frac{1}{8}$
30. $4\frac{3}{10} - 2\frac{4}{5}$ $1\frac{1}{2}$

Example 5 Find $5\frac{2}{3} + 3\frac{1}{2}$.

$5\frac{2}{3} + 3\frac{1}{2} = 5\frac{4}{6} + 3\frac{3}{6}$ Rename the fractions.

$= 8\frac{7}{6}$ or $9\frac{1}{6}$ Add the whole numbers and add the fractions.

Example 6 Find $4\frac{1}{5} - 2\frac{3}{5}$.

$4\frac{1}{5} - 2\frac{3}{5} = 3\frac{6}{5} - 2\frac{3}{5}$ Rename $4\frac{1}{5}$ as $3\frac{6}{5}$.

$= 1\frac{3}{5}$ Subtract the whole numbers and subtract the fractions.

6-4 Multiplying Fractions and Mixed Numbers (pp. 254–257)

Multiply. Write in simplest form.

31. $\frac{3}{5} \times \frac{2}{7}$ $\frac{6}{35}$
32. $\frac{5}{12} \times \frac{4}{9}$ $\frac{5}{27}$
33. $\frac{3}{5} \times \frac{10}{21}$ $\frac{2}{7}$
34. $4 \times \frac{13}{20}$ $2\frac{3}{5}$
35. $2\frac{1}{3} \times \frac{3}{4}$ $1\frac{3}{4}$
36. $4\frac{1}{2} \times 2\frac{1}{12}$ $9\frac{3}{8}$
37. **TRACK AND FIELD** One lap around the high school track is $\frac{3}{8}$ of a mile. If Matthew runs $4\frac{1}{2}$ laps, how far does he run? $1\frac{11}{16}$ mi

Example 7 Find $\frac{5}{9} \times \frac{2}{3}$.

$\frac{5}{9} \times \frac{2}{3} = \frac{5 \times 2}{9 \times 3}$ Multiply the numerators and multiply the denominators.

$= \frac{10}{27}$ Simplify.

Example 8 Find $3\frac{1}{2} \times 2\frac{3}{4}$.

$3\frac{1}{2} \times 2\frac{3}{4} = \frac{7}{2} \times \frac{11}{4}$ Rename $3\frac{1}{2}$ and $2\frac{3}{4}$.

$= \frac{7 \times 11}{2 \times 4}$ Multiply the numerators and multiply the denominators.

$= \frac{77}{8}$ or $9\frac{5}{8}$ Simplify.

6-5 Algebra: Solving Equations (pp. 258–261)

Find the multiplicative inverse of each number. 42. 15 43. −22.8

38. $\frac{7}{12}$ $\frac{12}{7}$ or $1\frac{5}{7}$
39. 5 $\frac{1}{5}$
40. $3\frac{1}{3}$ $\frac{3}{10}$

Solve each equation. Check your solution.

41. $8 = \frac{w}{2}$ 16
42. $\frac{4}{5}b = 12$
43. $-7.6 = \frac{n}{3}$
44. **EARTH SCIENCE** In 1996, a new planet was discovered. Earth's diameter, 7,970 miles, is only $\frac{5}{86}$ the size of this planet's diameter. Solve $\frac{5}{86}d = 7{,}970$ to find d, the diameter of the new planet in miles. 137,084 mi

Example 9 Find the multiplicative inverse of $\frac{9}{5}$.

$\frac{9}{5} \cdot \frac{5}{9} = 1$ The product of $\frac{9}{5}$ and $\frac{5}{9}$ is 1.

The multiplicative inverse of $\frac{9}{5}$ is $\frac{5}{9}$.

Example 10 Solve $\frac{3}{4}g = 2$.

$\frac{3}{4}g = 2$ Write the equation.

$\frac{4}{3} \cdot \frac{3}{4}g = \frac{4}{3} \cdot 2$ Multiply each side by the reciprocal of $\frac{3}{4}$.

$g = \frac{8}{3}$ or $2\frac{2}{3}$ Simplify.

Chapter 6 Study Guide and Review 279

Answers

60. 38 cm; 78 cm^2
61. $6\frac{1}{4}$ ft; $1\frac{5}{16}$ ft^2
66. 26.4 cm
67. 25.1 yd
68. 4.0 ft
69. 26.4 ft

6-6 Dividing Fractions and Mixed Numbers (pp. 264–266) 49. $3\frac{3}{10}$

Divide. Write in simplest form.

45. $\frac{3}{5} \div \frac{6}{7}$ $\frac{7}{10}$ 46. $\frac{1}{2} \div \frac{1}{3}$ $1\frac{1}{2}$ 47. $5 \div \frac{10}{13}$ $6\frac{1}{2}$

48. $4 \div \frac{2}{3}$ 6 49. $2\frac{3}{4} \div \frac{5}{6}$ 50. $\frac{2}{5} \div 3$ $\frac{2}{15}$

51. $\frac{6}{11} \div 4$ $\frac{3}{22}$ 52. $4\frac{3}{10} \div 2\frac{1}{5}$ $1\frac{21}{22}$ 53. $\frac{2}{7} \div \frac{8}{21}$ $\frac{3}{4}$

Example 11 Find $2\frac{4}{5} \div \frac{7}{10}$.

$2\frac{4}{5} \div \frac{7}{10} = \frac{14}{5} \div \frac{7}{10}$ Rename $2\frac{4}{5}$.

$= \frac{\overset{2}{\cancel{14}}}{\underset{1}{\cancel{5}}} \cdot \frac{\overset{2}{\cancel{10}}}{\underset{1}{\cancel{7}}}$ Multiply by the reciprocal of $\frac{7}{10}$.

$= \frac{4}{1}$ or 4 Simplify.

6-7 Measurement: Changing Customary Units (pp. 267–269)

Complete. 57. 4

54. 4 qt = __?__ pt 8 55. 6 gal = __?__ qt 24

56. 48 oz = __?__ lb 3 57. 8,000 lb = __?__ T

58. 9 c = __?__ pt $4\frac{1}{2}$ 59. 36 in. = __?__ ft 3

Example 12 Complete: 32 qt = __?__ gal

Since 4 quarts are in 1 gallon, divide by 4.

$32 \div 4 = 8$

32 quarts = 8 gallons

6-8 Geometry: Area and Perimeter (pp. 270–273)

Find the perimeter and area of each rectangle. 60–61. See margin.

60.

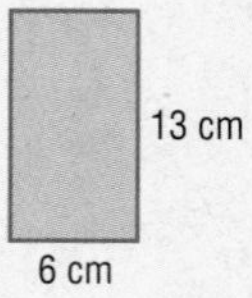

61.

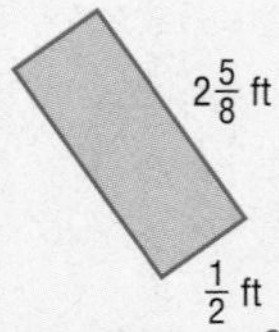

62. $\ell = 9$ cm, $w = 4$ cm 26 cm; 36 cm^2

63. $\ell = 5$ in., $w = \frac{1}{2}$ in. 11 in.; $2\frac{1}{2}$ in^2

64. $\ell = 3.2$ m, $w = 6$ m 18.4 m; 19.2 m^2

65. $\ell = 4\frac{1}{2}$ ft, $w = 2\frac{1}{3}$ ft $13\frac{2}{3}$ ft; $10\frac{1}{2}$ ft^2

Example 13 Find the perimeter and area of the rectangle.

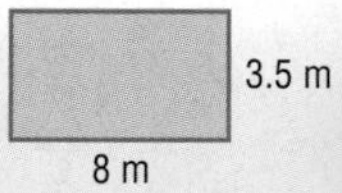

$P = 2\ell + 2w$ Perimeter of a rectangle

$P = 2(8) + 2(3.5)$ Substitution

$P = 23$ Simplify.

The perimeter is 23 meters.

$A = \ell \cdot w$ Area of a rectangle

$A = 8 \cdot 3.5$ Replace ℓ with 8 and w with 3.5.

$A = 28$ Multiply.

The area is 28 square meters.

6-9 Geometry: Circles and Circumference (pp. 275–277)

Find the circumference of each circle. Use 3.14 or $\frac{22}{7}$ for π. Round to the nearest tenth if necessary. 66–69. See margin.

66. $r = 4.2$ cm 67. $d = 8$ yd

68. $r = \frac{7}{11}$ ft 69. $d = 8\frac{2}{5}$ ft

Example 14 Find the circumference of a circle with a diameter of 12.2 meters. Round to the nearest tenth.

$C = \pi d$ Circumference of a circle

$C \approx 3.14(12.2)$ $\pi \approx 3.14$ and $d = 12.2$

$C \approx 38.308$ Multiply.

The circumference is about 38.3 meters.

Chapter 6 Practice Test

Vocabulary and Concepts

1. **Explain** how to add unlike fractions. **See margin.**
2. **Define** perimeter. **the distance around a geometric figure**

Skills and Applications

Estimate. **3–5. Sample answers are given.**

3. $5\frac{7}{9} - 1\frac{2}{13}$ **$6 - 1 = 5$**
4. $3\frac{1}{12} + 6\frac{5}{7}$ **$3 + 7 = 10$**
5. $\frac{3}{7} \times \frac{13}{15}$ **$\frac{1}{2} \times 1 = \frac{1}{2}$**

Add, subtract, multiply, or divide. Write in simplest form.

6. $\frac{4}{15} + \frac{8}{15}$ **$\frac{4}{5}$**
7. $\frac{7}{10} - \frac{1}{6}$ **$\frac{8}{15}$**
8. $\frac{5}{8} \times \frac{2}{5}$ **$\frac{1}{4}$**
9. $6 \times \frac{8}{21}$ **$2\frac{2}{7}$**
10. $4\frac{5}{12} - 2\frac{1}{12}$ **$2\frac{1}{3}$**
11. $6\frac{7}{9} + 3\frac{5}{12}$ **$10\frac{7}{36}$**
12. $8\frac{2}{7} - 1\frac{5}{14}$ **$6\frac{13}{14}$**
13. $\frac{5}{6} \div \frac{2}{3}$ **$1\frac{1}{4}$**
14. $\frac{8}{9} \div 5\frac{1}{3}$ **$\frac{1}{6}$**

15. **COOKING** Taylor wants to make $2\frac{1}{2}$ times the quantity given in a recipe. The recipe calls for $1\frac{3}{4}$ cups of flour. How much flour will Taylor need? **$4\frac{3}{8}$ c**

16. **FLAG DAY** A giant cake decorated as an American flag measured 60 feet by 90 feet. What was the perimeter of the cake? **300 ft**

Solve each equation. Check your solution.

17. $\frac{y}{3} = 8$ **24**
18. $-6 = \frac{2}{5}m$ **−15**
19. $\frac{3}{4} = \frac{5}{8}x$ **$1\frac{1}{5}$**

Complete.

20. 42 ft = __?__ yd **14**
21. 9 qt = __?__ pt **18**
22. 7,600 lb = __?__ T **3.8**

23. Find the perimeter and area of the rectangle. **13 in.; 9 in^2**

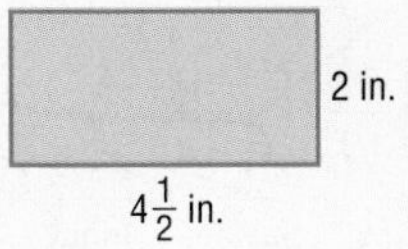

24. Find the circumference of a circle with a radius of 5 meters. Round to the nearest tenth. **31.4 m**

Standardized Test Practice

25. **MULTIPLE CHOICE** In the 1999–2000 school year, the average backpack weighed $7\frac{1}{2}$ pounds. In the 2001–2002 school year, the average backpack weighed $7\frac{1}{5}$ pounds. By how much did the average backpack weight decrease? **B**

(A) $\frac{1}{5}$ lb (B) $\frac{3}{10}$ lb (C) $\frac{1}{2}$ lb (D) $\frac{7}{10}$ lb

msmath2.net/chapter_test

Portfolio Suggestion

Introduction Have students search around at home for examples of items that they can measure for area, perimeter, and circumference.

Ask Students how knowing the measurements of furniture might help an interior decorator. Have students include the answer and supporting sketches in their portfolios.

Chapter 6 Practice Test

Assessment Options

Vocabulary Test A vocabulary review/test for Chapter 6 can be found on p. 362 of the *Chapter 6 Resource Masters.*

Chapter Tests There are six Chapter 6 Tests and an Extended Response Assessment task available in the *Chapter 6 Resource Masters.*

Chapter 6 Tests

Form	Type	Level	Pages
1	MC	basic	349–350
2A	MC	average	351–352
2B	MC	average	353–354
2C	FR	average	355–356
2D	FR	average	357–358
3	FR	advanced	359–360

MC = multiple-choice questions
FR = free-response questions

Extended Response Assessment

Extended Response tasks for Chapter 6 can be found on p. 361 of the *Chapter 6 Resource Masters.* A sample scoring rubric for these tasks appears on p. A31.

Unit 3 Test

A unit test/review can be found on pp. 369–370 of the *Chapter 6 Resource Masters.*

Answer

1. **Rename the fractions with the LCD. Then add the numerators.**

CHAPTER 6 Standardized Test Practice

These two pages contain practice questions in the various formats that can be found on the most frequently given standardized tests.

A student recording sheet for these two pages can be found on page A1 of the *Chapter 6 Resource Masters.*

Standardized Test Practice
Student Recording Sheet, p. A1

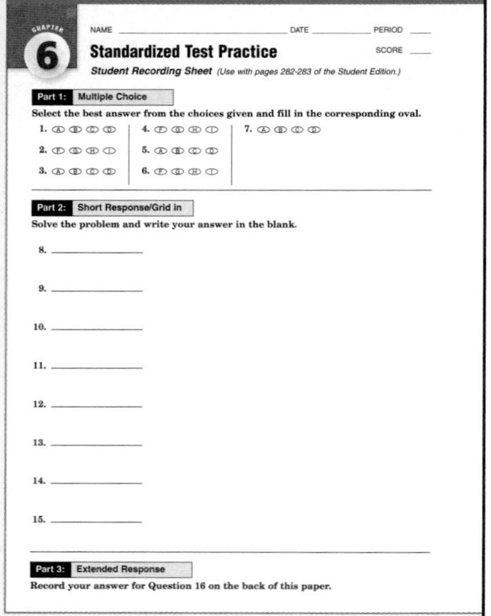
CHAPTER 6 NAME ______ DATE ______ PERIOD ______

Standardized Test Practice SCORE ______

Student Recording Sheet (Use with pages 282-283 of the Student Edition.)

Part 1: Multiple Choice

Select the best answer from the choices given and fill in the corresponding oval.

1. Ⓐ Ⓑ Ⓒ Ⓓ
2. Ⓕ Ⓖ Ⓗ Ⓘ
3. Ⓐ Ⓑ Ⓒ Ⓓ
4. Ⓕ Ⓖ Ⓗ Ⓘ
5. Ⓐ Ⓑ Ⓒ Ⓓ
6. Ⓕ Ⓖ Ⓗ Ⓘ
7. Ⓐ Ⓑ Ⓒ Ⓓ

Part 2: Short Response/Grid In

Solve the problem and write your answer in the blank.

8. ______
9. ______
10. ______
11. ______
12. ______
13. ______
14. ______
15. ______

Part 3: Extended Response

Record your answer for Question 16 on the back of this paper.

Additional Practice

See pp. 367–368 of the *Chapter 6 Resource Masters* for additional standardized test practice.

Answer

16a.

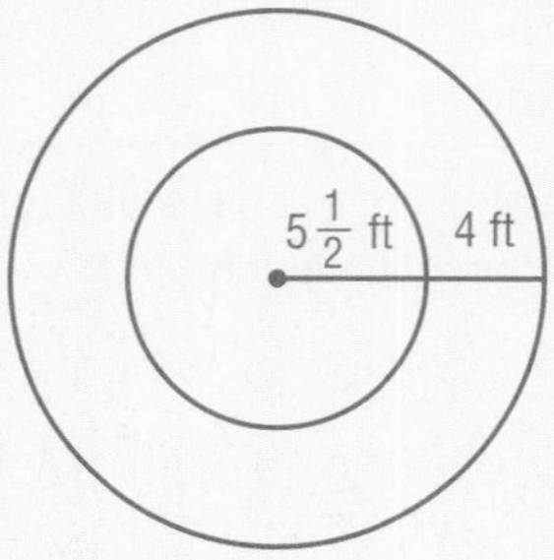

CHAPTER 6 Standardized Test Practice

PART 1 Multiple Choice

Record your answers on the answer sheet provided by your teacher or on a sheet of paper.

1. The table shows four major rivers that run through Texas. Which is the most appropriate way to display this information? (Lessons 2-2 and 2-7) **A**

River	Length (mi)
Brazos	950
Sabine	380
Trinity	360
Washita	500

Source: *The World Almanac*

Ⓐ

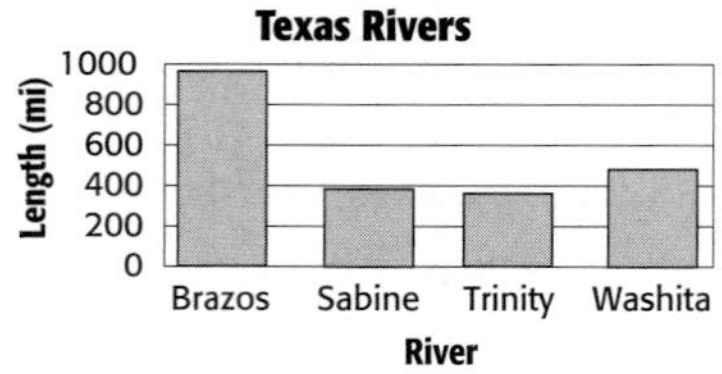

Ⓑ

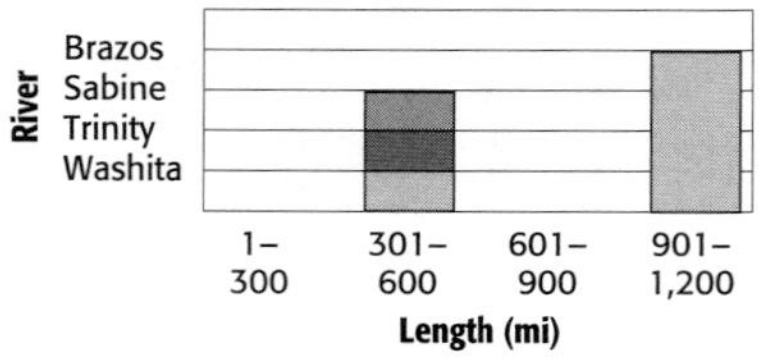

Ⓒ

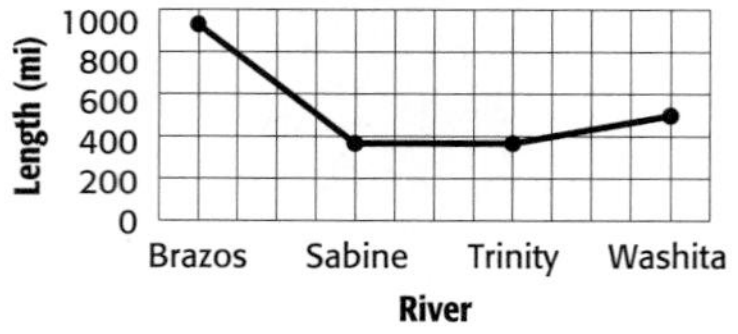

Ⓓ

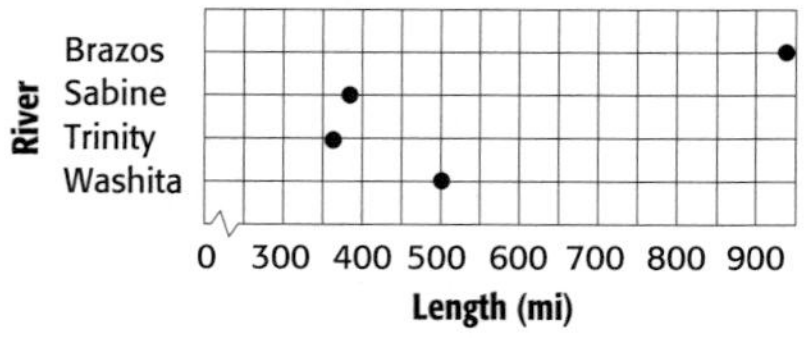

2. Which is the solution of $2x - 5 > 13$? (Lesson 4-5) **I**

Ⓕ

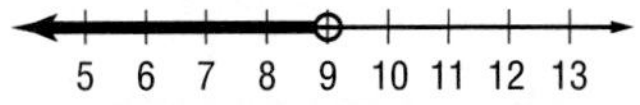

Ⓖ

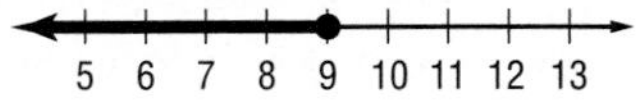

Ⓗ

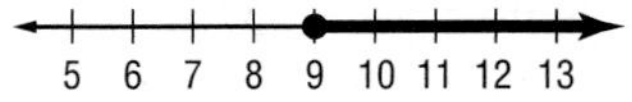

Ⓘ

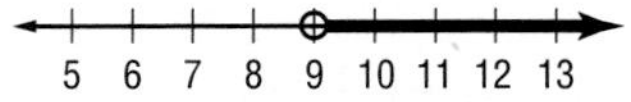

3. A jellyfish's body is made up of 95% water. What is 95% written as a decimal? (Lesson 5-6) **B**

Ⓐ 0.095 Ⓑ 0.95
Ⓒ 9.5 Ⓓ 95

4. Dimitri has $6\frac{1}{8}$ ounces of orange juice and $10\frac{2}{3}$ ounces of cranberry juice. What is the best estimate of the total ounces of juice that he has? (Lesson 6-1) **H**

Ⓕ 15 oz Ⓖ 16 oz
Ⓗ 17 oz Ⓘ 18 oz

5. Cole had $\frac{7}{8}$ of a tank of gas in the lawn mower. After mowing the grass, he had $\frac{1}{4}$ of a tank. What fraction of a tank did Cole use mowing the lawn? (Lesson 6-2) **C**

Ⓐ $\frac{1}{8}$ Ⓑ $\frac{3}{8}$ Ⓒ $\frac{5}{8}$ Ⓓ $\frac{3}{4}$

6. What is the solution of $27 = \frac{3}{4}t$? (Lesson 6-5) **G**

Ⓕ 108 Ⓖ 36
Ⓗ 20.25 Ⓘ 9

7. Kaylee bought 16 gallons of fruit juices and soda for the school's graduation picnic. How many quarts of drinks did Kaylee buy? (Lesson 6-7) **D**

Ⓐ 4 qt Ⓑ 16 qt Ⓒ 18 qt Ⓓ 64 qt

Log On for Test Practice

The Princeton Review

The Princeton Review offers additional test-taking tips and practice problems at their web site. Visit princetonreview.com or review.com

ExamView® Pro Test Bank CD-ROM

Special banks of standardized test questions similar to those on the TIMSS and NAEP grade 8 tests can be found on this CD-ROM.

ART 2 Short Response/Grid In

ecord your answers on the answer sheet rovided by your teacher or on a sheet of aper.

8. If 4 computers are needed for every 7 students in a grade, how many computers are needed for 280 students? (Lesson 1-1) **160 computers**

9. What are the coordinates of point K? (Lesson 3-3) **(−3, 1)**

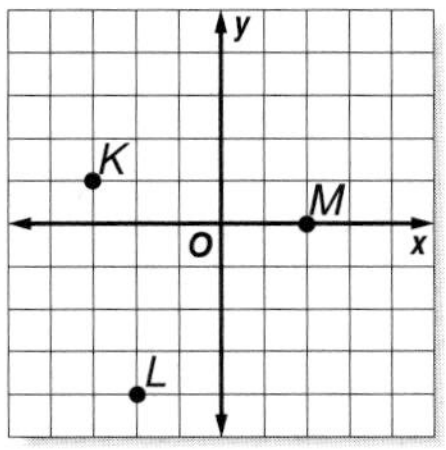

10. The graph shows the average number of greeting cards purchased yearly by the average person in the United States.

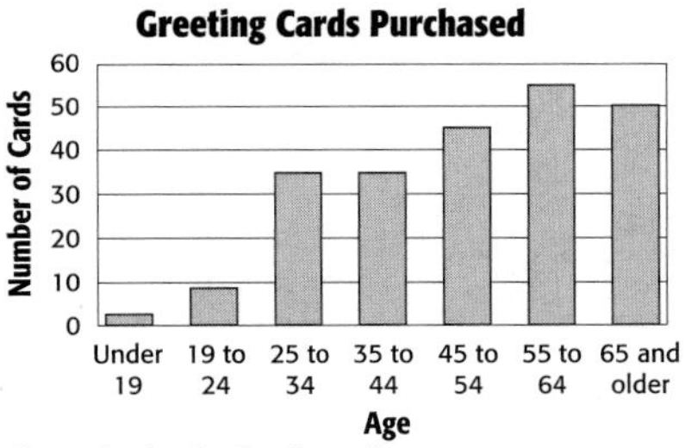

Source: American Greetings Corporation

Let c represent the number of cards purchased by Americans ages 35 to 44. Write an expression using c to represent the number of cards purchased by Americans ages 45 to 54. (Lesson 4-1) **$c + 10$**

11. Write a percent to represent the shaded area. (Lesson 5-5) **36%**

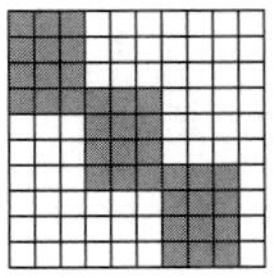

12. Find $7\frac{2}{7} + 3\frac{1}{4}$. (Lesson 6-3) **$10\frac{15}{28}$**

13. To make one batch of cookies, you need $\frac{3}{4}$ cup of butter. How much butter would you need to make $2\frac{1}{2}$ batches of cookies? (Lesson 6-4) **$1\frac{7}{8}$ c**

14. A box of laundry detergent contains 35 cups. If you use $1\frac{1}{4}$ cups per load of laundry, how many loads can you wash with 1 box? (Lesson 6-6) **28 loads**

15. What is the area of the rectangle? (Lesson 6-8) **40 in²**

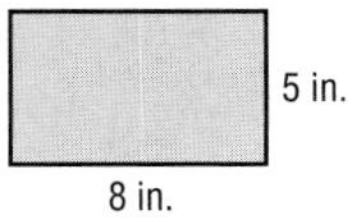

PART 3 Extended Response

Record your answers on a sheet of paper. Show your work.

16. Suppose you want to build a 4-foot wide deck around a circular swimming pool that has a radius of 66 inches. You also want to put a fence around the deck. (Lesson 6-9) **a–c. See margin.**
   a. Make a drawing of the problem. Include labels.
   b. About how much fencing will you need to the nearest foot?
   c. Suppose if you put a fence around just the pool instead of the deck, you save \$375. If this is the case, about how much does one foot of fencing cost? Explain.

TEST-TAKING TIP

Question 16 Many standardized tests include any necessary formulas in the test booklet. It helps to be familiar with formulas such as the area of a rectangle and the circumference of a circle, but use any formulas that are given to you with the test.

Standardized Test Practice

Rubrics The following rubrics are sample scoring devices for short response and extended response questions.

Short Response

Points	Description
2	The student demonstrates a **thorough understanding** of the mathematics of the task. The response may contain minor flaws that do not detract from the demonstration of a thorough understanding.
1	The student has provided a response that is only **partially correct**.
0	The student has provided a **completely incorrect** solution or no response at all.

Extended Response

Points	Description
4	The student demonstrates a **thorough understanding** of the mathematics of the task. The response may contain minor flaws that do not detract from the demonstration of a thorough understanding.
3	The student demonstrates an **understanding** of the mathematics of the task. The response is essentially correct and demonstrates an essential but less than thorough understanding of the mathematics.
2	The student demonstrates only a **partial understanding** of the mathematics of the task. Although the student may have used the correct approach to a solution or may have provided a correct solution, the work lacks an essential understanding of the underlying mathematical concepts.
1	The student demonstrates a **very limited understanding** of the mathematics of the task. The response is incomplete and exhibits many flaws.
0	The student provides a **completely incorrect** solution or no response at all.

A more detailed 4-point rubric is provided on page A2 of the *Chapter 6 Resource Masters*.

Answers

16b. Since 66 inches equals $5\frac{1}{2}$ feet and the deck is 4 feet wide, the total radius of the area to be fenced is $9\frac{1}{2}$ feet. So, the circumference is $2 \cdot \pi \cdot 9\frac{1}{2}$ or about 59.7 feet. You need 60 feet of fencing.

16c. \$15; The circumference of the pool is 11π or about 35 feet. The circumference of the deck and pool minus the circumference of the pool is 60 − 35 or 25 feet. So, the cost of one foot of fencing is \$375 ÷ 25 feet or \$15.

Page 243, Lesson 6-1

11–30. Sample answers are given.

11. $4 + 5 = 9$

12. $1 + 6 = 7$

13. $5 - 3 = 2$

14. $4 - 2 = 2$

15. $3 \cdot 6 = 18$

16. $6 \div 2 = 3$

17. $1 + \frac{1}{2} = 1\frac{1}{2}$

18. $\frac{1}{2} + \frac{1}{2} = 1$

19. $\frac{1}{2} - 0 = \frac{1}{2}$

20. $1 - \frac{1}{2} = \frac{1}{2}$

21. $0 \times 1 = 0$

22. $\frac{1}{2} \cdot 1 = \frac{1}{2}$

23. $1 \div 1 = 1$

24. $0 \div 1 = 0$

25. $\frac{1}{2} \times 14 = 7$

26. $\frac{1}{4} \times 40 = 10$

27. $25 \div 5 = 5$

28. $28 \div 7 = 4$

29. $12 \div 4 = 3$

30. $6 \cdot 4 = 24$

Page 252, Lesson 6-3b

2. Sample answer: By eliminating possibilities, you can reduce the time it takes to solve the problem since there are fewer answers to consider.

3. Sample answer: A fishbowl holds $1\frac{1}{2}$ gallons of water. If there is $\frac{1}{3}$ gallon of water in the bowl, how many more gallons are needed to fill the bowl; $\frac{2}{3}$ gal, $1\frac{1}{6}$ gal, $1\frac{5}{6}$ gal, or $2\frac{1}{6}$ gal? Answer: $\frac{2}{3}$ gal is not enough because that would make exactly 1 gallon of water in the tank. $2\frac{1}{6}$ gal is too much, since the tank only holds $1\frac{1}{2}$ gallons. The answer is $1\frac{1}{6}$ gal.

Page 266, Lesson 6-6

22. Rational numbers are closed under subtraction and multiplication because the difference or product of any two rational numbers is a rational number. The set is not closed under division because a number divided by 0 is undefined, not a rational number.

NOTES

UNIT 4 Notes

Introduction

In this unit, students will examine ratios and rates and solve proportions. They will use this knowledge to make and interpret scale drawings. Students will convert among fractions, decimals, and percents and will use their knowledge of fractions and decimals to understand percents greater than 100% and less than 1%. In addition, they will learn how to use a percent proportion to calculate percent of change, sales tax and discounts, and simple interest. Students will also study how to estimate percents and use statistics to make predictions.

Students will apply their skills to the study of probability and combinatorics. They will use tree diagrams and the Fundamental Counting Principle to find the number of possible outcomes of simple events. Students will explore the connection between possible outcomes and permutations and combinations of items. Finally, they will investigate the differences between theoretical and experimental probability and between independent and dependent events.

Assessment Options

Unit 4 Test Pages 541–542 of the *Chapter 9 Resource Masters* may be used as a test or a review for Unit 4. This assessment contains both multiple-choice and short-response items.

UNIT 4
Proportional Reasoning

Chapter 7
Ratios and Proportions

Chapter 8
Applying Percent

Chapter 9
Probability

In Unit 3, you learned about percents. In this unit, you will solve real-life problems involving proportions and percents. You will also discover how percents are used to describe probabilities.

284 Unit 4 Proportional Reasoning

What's Math Got to Do With It? Real-Life Videos

What's Math Got to Do With It? Real-Life Videos engage students, showing them how math is used in everyday situations. Use Video 4 to discuss how percents and proportions are used in wildlife management and shopping.

![WebQuest] **INTERDISCIPLINARY PROJECT** — **MATH and RECREATION**

STEP RIGHT UP AND WIN A PRIZE

What are my chances of winning? You probably ask yourself that question any time you play a game that offers a prize for winning. You're about to embark on a gaming adventure. You'll investigate the mathematical probabilities of winning various carnival games. You'll also research and design a game of your own. So, come on and take a chance! Sharpen up that hand-eye coordination and grab your probability tool kit. This adventure is a win-win situation!

Log on to msmath2.net/webquest to begin your WebQuest.

WebQuest **INTERDISCIPLINARY PROJECT**

Step Right Up and Win a Prize

A WebQuest is an online project in which students do research on the Internet, gather data, and make presentations using word processing, graphing, page-making, or presentation software. In each chapter, students advance to the next step in their WebQuest. At the end of Chapter 9, the project culminates with a presentation of their findings.

Teaching suggestions and sample answers are available in the WebQuests, Projects, and Interdisciplinary Investigations.

Math and Recreation

You can use this WebQuest Interdisciplinary Project with your students' physical education teacher to make the connection from mathematics to the recreational activities your students are studying.

More Interdisciplinary Connections

You may wish to share these suggestions with your students' other teachers.

Math and Social Studies

Research the history of carnivals around the world. Prepare a report including the significance of games, masks, and fatty foods in carnivals around the world. You may wish to focus on a particular place such as Venice, Italy; Rio de Janeiro, Brazil; or New Orleans in your report.

Math and Art

Design a poster for a carnival that is sure to attract customers. Include drawings of several carnival games and prizes, as well as foods, beverages, and entertainment. Use catchy names, attractive colors, and bold-sized graphics.

Unit 4 Proportional Reasoning 285

CHAPTER 7

Chapter Overview

Ratios and Proportions

	LESSON OBJECTIVES	NCTM Standards	NAEP 2005	State/Local Objectives
7-1	**Ratios** *(pp. 288–291)* • Write ratios as fractions and determine whether two ratios are equivalent.	1, 4, 9, 10	N4a, N4b, N1b, M2b, N1j	
7-2	**Rates** *(pp. 292–296)* • Determine unit rates. **7-2b Hands-On Lab:** Investigate rate of change.	1, 4, 6, 9, 10	N4a, N4b, N4c, N3a, M1l, A1a, A2c, A1f	
7-3	**Solving Proportions** *(pp. 297–301)* • Solve proportions. **7-3b Hands-On Lab:** Use proportions to estimate.	1, 2, 4, 6, 9, 10	N4a, N4b, N4c, D3b, D2a	
7-4	**Scale Drawings** *(pp. 302–309)* **7-4a Problem-Solving Strategy:** Solve problems by drawing a diagram. • Solve problems involving scale drawings. **7-4b Spreadsheet Investigation:** Use a spreadsheet to calculate measurements for scale drawings.	1, 2, 4, 6, 9, 10	N3a, N4b, N4c, M2f, N1b	
7-5	**Fractions, Decimals, and Percents** *(pp. 312–315)* • Write percents as fractions, and vice versa.	1, 2, 6, 9, 10	N1d, N1e, N2c, N4c	
7-6	**Percents Greater Than 100% and Percents Less Than 1%** *(pp. 316–318)* • Write percents greater than 100% and percents less than 1% as fractions and as decimals, and vice versa.	1, 9, 10	N1b, N1d, N1e	
7-7	**Percent of a Number** *(pp. 319–321)* • Find the percent of a number.	1, 2, 6, 9, 10	N4a, N4b, N4d, N5f	
7-8	**The Percent Proportion** *(pp. 322–325)* **7-8a Hands-On Lab:** Use a percent model to find a part. • Solve problems using the percent proportion.	1, 2, 6, 9, 10	N4a, N4b, N4d, N5f, N1b	

Key to NCTM Standards:

1=Number & Operations, 2=Algebra, 3=Geometry, 4=Measurement, 5=Data Analysis & Probability, 6=Problem Solving, 7=Reasoning & Proof, 8=Communication, 9=Connections, 10=Representation

NAEP 2005 Standards: See math.glencoe.com

Pacing suggestions for the entire year can be found on page T20.

Chapter Resource Manager

Time-Saving Tool
TeacherWorks™
All-In-One Planner and Resource Center
See page T5.

Chapter 7 Resource Masters

FAST FILE

Study Guide and Intervention	Practice (Skills and Word Problems)	Reading to Learn Mathematics	Enrichment	Assessment	Applications*	Parent and Student Study Guide Workbook	5-Minute Check Transparencies	Key Teaching Transparencies	MathPASS: Tutorial Plus (lessons)	Teaching Mathematics with Manipulatives	Optional Materials
375	376–377	378	379			7-1	7-1	25		68–69	textbook, computer, stopwatch
380	381–382	383	384	429		7-2	7-2		28	70	square tiles
385	386–387	388	389		CS 37 SM 33–36	7-3	7-3	26	29	71	dried beans, paper cup, markers, small bowl
390	391–392	393	394	429, 431	RC 21	7-4	7-4			72	measuring tape, straightedge
395	396–397	398	399		CS 38	7-5	7-5		30		
400	401–402	403	404	430	CM 19–21	7-6	7-6	27			colored pencils
405	406–407	408	409		CS 39	7-7	7-7		32		
410	411–412	413	414	415–428, 430, 432–434	CS 40	7-8	7-8	28		73	

*** Key to Abbreviations:** *CS = Calculator and Spreadsheet Masters, RC = Real-Life Career Masters, CM = Classroom Math Games (online), SM = Science and Math Lab Manual (online)*

INTERACTIVE CHALKBOARD

PowerPoint® Presentations

This CD-ROM is a customizable Microsoft® PowerPoint® presentation that includes:

- In-Class Examples
- Links to Online Study Tools
- 5-Minute Check Transparencies
- Bellringers

Math Skills Maintenance

Use the Math Skills Maintenance Masters to keep your students' basic math skills fresh. Students can review skills one at a time or in a combined format. The Masters are also available online at msmath2.net/maintenance.

CHAPTER 7

Mathematical Content and Teaching Strategies

Continuity of Instruction

Building on Prior Knowledge

In Chapter 5 and in previous courses, students worked with decimals and equivalent fractions. In the previous chapter, they performed multiplication and division on fractions and mixed numbers. In earlier courses, students used various estimation techniques. Finally, students solved algebraic equations involving multiplication and division in Chapter 4.

Learning in This Chapter

Students write ratios, and learn how to convert ratios to unit rates. They apply their knowledge of equivalent fractions as they set up and solve proportions, and work with scale drawings and models. They explore the relationship between fractions, decimals, and percents, and learn how to solve basic percent problems using proportions and decimal multiplication.

Making Future Connections

Ratios, rates, and proportions are fundamental concepts that students will use throughout middle school and high school mathematics and science courses, as well as in everyday life. Working with percents is also an essential everyday life skill, and one that will help students understand information that they read and see reported in the news media.

Mathematical Background

7-1 Ratios

Ratios are a comparison of two numbers and the basis for solving proportion and percent problems. Ratios are used in photography, architecture, economics, politics, marketing, retail shopping, and so on. Students will use ratios at some point in any class that uses numbers.

Bulletin Board Starters

Newspapers and magazines use ratios and percents every day to report news and statistical data. Have students collect clippings of columns, articles, and graphs that use ratios and percents to describe a wide variety of situations. Ask them to cut out and contribute their findings to a class bulletin board.

7-2 Rates

Unit rates are ratios with a denominator of 1. Unit rates give an equal basis for comparison, and have many real-life applications, such as miles per hour, price per unit of weight, and Calories per serving. In this lesson, students convert ratios to unit rates. They compare rates in applications such as determining the better buy between two products.

7-3 Solving Proportions

A proportion is an equation stating that two ratios are equivalent. Two ratios form a proportion if their cross products are equal. Students learn to set up and solve proportions for unknown values. The cross products can be used to solve for a missing number in the proportion. Point out that comparing cross products to determine whether two ratios are equivalent is valid because it is the same as renaming the ratios with the same denominator and comparing the numerators. The denominator used would be the product of the denominators of the original ratios.

7-4 Scale Drawings

When actual distances or dimensions are very large or small, they may be represented in more conveniently-sized scale drawings or models. This process involves a scale factor, a ratio which is used to form a proportion. Students set up and solve proportions to find actual or scale measurements.

7-5 Fractions, Decimals, and Percents

Being able to convert percents into fractions and decimals enhances understanding of the meaning of percents. In the media, most numbers are in percent or decimal form. Students need to be able to convert between fractions, decimals, and percents easily.

Teacher to Teacher

Barbara Larson
Anderson Middle School
Omaha, NE

"I make decks of cards, each card showing a decimal, a fraction, or a percent. Students use the cards to play war where the highest card wins. They practice changing between decimals, fractions, and percents and also review comparing rational numbers with this one game."

Barbara B Larson

McGraw-Hill Professional Development

Mathematics, YES! is a self-paced, web-based professional development program. Units are based on major mathematical concepts and are standards-based. Actual classroom videos provide a springboard for discussion and activities. Units appropriate for this chapter:

- Unit 4, Fractions, Decimals, and Percentages
- Unit 6, Ratio and Proportional Reasoning

7-6 Percents Greater Than 100% and Percents Less Than 1%

In this lesson, students discover the conceptual meaning of percents greater than 100%, and of those less than 1%. Although these percents are not common, students need to understand what they mean. They gain practice converting between decimals and percents by moving the decimal point.

7-7 Percent of a Number

When an item is on sale, it is usually listed as a percent off the price. Students can use estimation, proportions, or decimal multiplication to determine the sale price. Other applications include surveys, test scores, salary bonuses, and taxes.

7-8 The Percent Proportion

Percent problems typically leave one value unknown. It may be one of the numbers in a ratio when the percent is known, or it may be the percent itself. In this lesson, students learn to solve all of these basic types of percent problems by writing and solving proportions.

In-Class Speaker

Statisticians analyze data, convert ratios to percents, and design surveys that are mathematically meaningful. The applications they work with range from sports statistics to opinion polls. Invite a statistician to speak to the class about the many uses of percents in his or her career.

Key Concepts

Additional mathematical information and teaching notes are available at

msmath2.com/key_concepts.

The sections appropriate for the chapter are:

- Percents and Fractions
- Percents and Decimals
- Percent of a Number

CHAPTER 7

DAILY INTERVENTION and Assessment

	Type	Student Edition	Teacher Resources	Technology/Internet
INTERVENTION	Ongoing	Prerequisite Skills, pp. 287, 291, 295, 300, 308, 315, 318, 321 Mid-Chapter Practice Test, p. 310	5-Minute Check Transparencies Quizzes 1–4, *CRM*, pp. 429–430 Mid-Chapter Test, *CRM*, p. 431 Study Guide and Intervention, *CRM*, pp. 375, 380, 385, 390, 395, 400, 405, 410	MathPASS: Tutorial Plus msmath2.net/self_check_quiz msmath2.net/extra_examples
INTERVENTION	Mixed Review	pp. 291, 295, 300, 308, 315, 318, 321, 325	Cumulative Review, *CRM*, p. 432	
INTERVENTION	Error Analysis	Find the Error, pp. 290, 294	Find the Error, *TWE*, pp. 290, 294 Unlocking Misconceptions, *TWE*, pp. 298, 313	
ASSESSMENT	Standardized Test Practice	pp. 291, 293, 295, 300, 303, 308, 310, 315, 318, 321, 325, 329, 330–331	Standardized Test Practice, *CRM*, pp. 433–434	Standardized Test Practice CD-ROM msmath2.net/standardized_test
ASSESSMENT	Open-Ended Assessment	Writing Math, pp. 290, 294, 299, 301, 306, 314, 317, 320, 322, 325 Open-Ended, pp. 290, 294, 299, 306, 314, 317, 320, 325 Standardized Test, p. 331	Modeling: *TWE*, pp. 291, 325 Speaking: *TWE*, pp. 303, 318 Writing: *TWE*, pp. 300, 308 Extended Response Assessment, *CRM*, p. 427	
ASSESSMENT	Chapter Assessment	Study Guide, pp. 326–328 Practice Test, p. 329	Multiple-Choice Tests (Forms 1, 2A, 2B), *CRM*, pp. 415–420 Free-Response Tests (Forms 2C, 2D, 3), *CRM*, pp. 421–426 Vocabulary Test/Review, *CRM*, p. 428	ExamView® Pro Test Bank (see below) MindJogger Videoquizzes msmath2.net/vocabulary_review msmath2.net/chapter_test

* **Key to Abbreviations:** *TWE* = Teacher Wraparound Edition, *CRM* = Chapter Resource Masters

Additional Intervention

Skills Intervention for Middle School Mathematics workbook and CD-ROM includes 88 skills with pretests, tutorials, guided practice, and posttests for diagnosis and remediation.

Guide to Daily Intervention booklet details each intervention resource available and guides teachers in using them as intervention tools everyday to assist students who may be having difficulty.

ExamView® Pro

Use ExamView® Pro Test Bank CD-ROM to:

- **Create Multiple versions of tests.**
- **Create Modified tests for Inclusion students with one mouse click.**
- **Edit existing questions and Add your own questions.**
- **Build tests aligned with state standards using built-in State Curriculum Correlations.**
- **Change English tests to Spanish with one mouse click and vice versa.**

Reading and Writing Mathematics

• Student Edition

Foldables Study Organizer, p. 287

Concept Check questions at the beginning of each exercise set require students to verbalize and write about what they have learned in the lesson.

- **Open Ended,** pp. 290, 294, 299, 306, 314, 317, 320, 325
- **Find the Error,** pp. 290, 294, 294
- **Which One Doesn't Belong?,** pp. 290, 299, 317

Reading Math Tip, p. 317

• Teacher Wraparound Edition

Foldables Follow-Up, pp. 290, 299, 306, 310, 314, 324

Writing Activities, pp. 300, 308

ELL Resources, *see page T9 for a complete listing.*

• Additional Resources

Vocabulary Builder, *Chapter 7 Resource Masters*

Reading to Learn Mathematics, *Chapter 7 Resource Masters*

Vocabulary PuzzleMaker (online)

Teaching Mathematics with Foldables

Reading and Writing in the Mathematics Classroom

WebQuests, Projects, and Interdisciplinary Investigations

For detailed information on these resources, see pp. T6–T7.

• Recommended Outside Reading for Students

Nonfiction

- Elam, Kimberly. *Geometry of Design: Studies in Proportion and Composition.* Princeton Architectural Press, ©2001.

  This book examines the relationship between mathematics and beauty by showing how artists and designers have used proportions in their work. This book includes common ratios and proportions.

Fiction

- Sundby, Scott. *Cut Down to Size at High Noon.* Charlesbridge Publishing, ©2000.

  This is a funny, Wild West adventure about two barbers who sculpture haircuts to scale in a showdown. This book gives a different look at scale drawings.

PROJECT CRISS℠ Study Skill

A power map can help students organize and display their knowledge of a topic. Show students the sample power map at the right that deals with scale. Explain that the levels of detail in the map are represented by different geometric shapes. As students work through Chapter 7, ask them to think about some ways ratios can be used. Have them design a power map that illustrates what they learn.

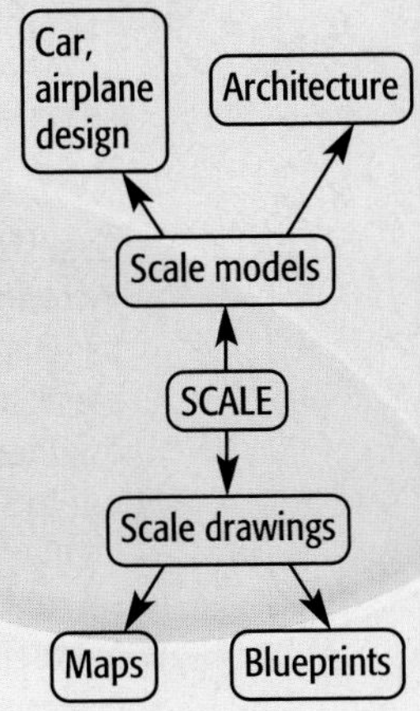

CReating Independence through Student-owned Strategies

For more information on Project CRISS, see page T9.

Chapter 7 Making Connections

School-To-Home

Parent and Student Study Guide

Parents and students may work together to reinforce the concepts and skills of this chapter. These workbooks are available in English and Spanish. A sample lesson is shown below. Also available at msmath2.net/parent_student.

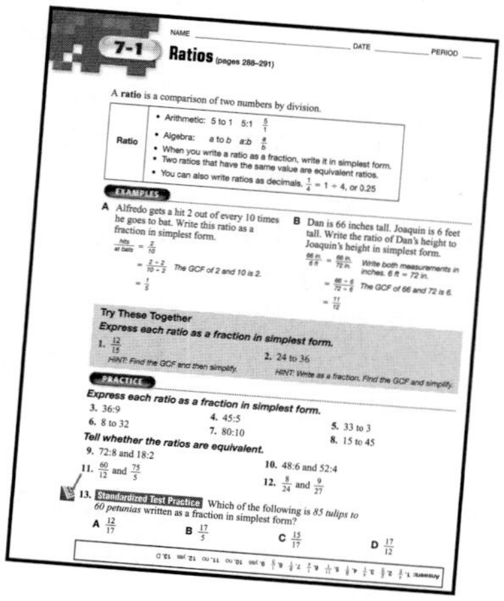

NAME ______ DATE ______ PERIOD ____

7-1 Ratios (pages 288–291)

A **ratio** is a comparison of two numbers by division.

Ratio	• Arithmetic: 5 to 1 5:1 $\frac{5}{1}$ • Algebra: a to b $a:b$ $\frac{a}{b}$ • When you write a ratio as a fraction, write it in simplest form. • Two ratios that have the same value are equivalent ratios. • You can also write ratios as decimals. $\frac{1}{4} = 1 \div 4$, or 0.25

EXAMPLES

A Alfredo gets a hit 2 out of every 10 times he goes to bat. Write this ratio as a fraction in simplest form.

$\frac{\text{hits}}{\text{at bats}} = \frac{2}{10}$

$= \frac{2 \div 2}{10 \div 2}$ *The GCF of 2 and 10 is 2.*

$= \frac{1}{5}$

B Dan is 66 inches tall. Joaquin is 6 feet tall. Write the ratio of Dan's height to Joaquin's height in simplest form.

$\frac{66 \text{ in.}}{6 \text{ ft}} = \frac{66 \text{ in.}}{72 \text{ in.}}$ *Write both measurements in inches. 6 ft = 72 in.*

$= \frac{66 \div 6}{72 \div 6}$ *The GCF of 66 and 72 is 6.*

$= \frac{11}{12}$

Try These Together

Express each ratio as a fraction in simplest form.

1. $\frac{12}{15}$
   HINT: Find the GCF and then simplify.
2. 24 to 36
   HINT: Write as a fraction. Find the GCF and simplify.

PRACTICE

Express each ratio as a fraction in simplest form.

3. 36:9
4. 45:5
5. 33 to 3
6. 8 to 32
7. 80:10
8. 15 to 45

Tell whether the ratios are equivalent.

9. 72:8 and 18:2
10. 48:6 and 52:4
11. $\frac{60}{12}$ and $\frac{75}{5}$
12. $\frac{8}{24}$ and $\frac{9}{27}$
13. **Standardized Test Practice** Which of the following is *85 tulips to 60 petunias* written as a fraction in simplest form?
    A $\frac{12}{17}$ **B** $\frac{17}{5}$ **C** $\frac{15}{17}$ **D** $\frac{17}{12}$

Family Letters and Activities

Send a letter and activity for this chapter home to encourage familiar members to become active participants in their students' learning. (English: *Chapter 7 Resource Masters*, pp. ix–x; Spanish: log on to msmath2.net/family_letters)

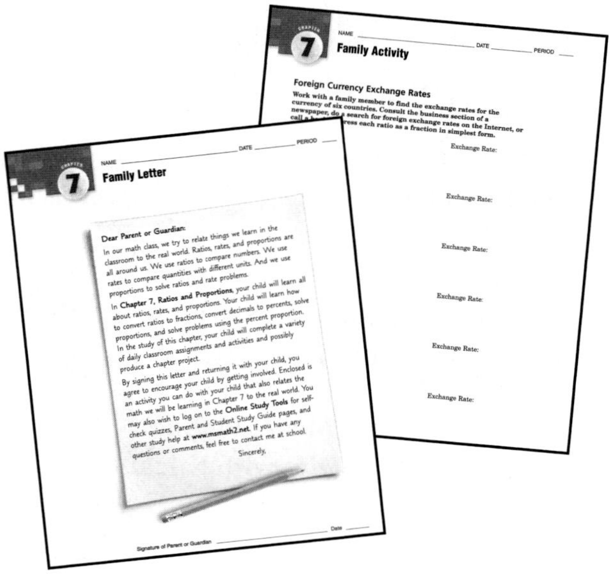

NAME ______ DATE ______ PERIOD ____

Chapter 7 Family Activity

Foreign Currency Exchange Rates

Work with a family member to find the exchange rates for the currency of six countries. Consult the business section of a newspaper, do a search for foreign exchange rates on the Internet, or call a b... ress each ratio as a fraction in simplest form.

Exchange Rate:

Exchange Rate:

Exchange Rate:

Exchange Rate:

Exchange Rate:

Exchange Rate:

NAME ______ DATE ______ PERIOD ____

Chapter 7 Family Letter

Dear Parent or Guardian:

In our math class, we try to relate things we learn in the classroom to the real world. Ratios, rates, and proportions are all around us. We use ratios to compare numbers. We use rates to compare quantities with different units. And we use proportions to solve ratios and rate problems.

In **Chapter 7, Ratios and Proportions**, your child will learn all about ratios, rates, and proportions. Your child will learn how to convert ratios to fractions, convert decimals to percents, solve proportions, and solve problems using the percent proportion. In the study of this chapter, your child will complete a variety of daily classroom assignments and activities and possibly produce a chapter project.

By signing this letter and returning it with your child, you agree to encourage your child by getting involved. Enclosed is an activity you can do with your child that also relates the math we will be learning in Chapter 7 to the real world. You may also wish to log on to the **Online Study Tools** for self-check quizzes, Parent and Student Study Guide pages, and other study help at **www.msmath2.net**. If you have any questions or comments, feel free to contact me at school.

Sincerely,

Signature of Parent or Guardian ______ Date ______

English Language Learners ELL

Lesson 7-3

Peer Tutoring

Give students several word problems involving proportions. Pair English Language Learners with English-speaking or bilingual students and ask them to set up a proportion, but not to solve them. Students should identify what is being asked and then set up both ratios in the same form to create a proportion. Have students label each part of the ratios to express the similarity among numerators and denominators.

Lesson 7-6

Using Applications

Give students real-life examples of how a stock can go up in value more than 100%. Ask students what would it look like if a stock that was originally worth \$25 increased its value by 100%. Then, ask students to calculate how much that stock would be worth if its value had increased by 120%.

MathScape™

This project was supported, in part, by the **National Science Foundation**

MathScape is a mathematics curriculum for grades 6–8 developed by the Seeing and Thinking Mathematically Project at the Education Development Center.

Rates, Ratios, Percents, and Proportions

Buyer Beware

Summary Students will find unit prices and compare prices to find a better buy. Ratios are used to compare data. Cross products are used to compare ratios and to solve proportions. Students will use proportions to evaluate and analyze circle graphs.

USA TODAY Education Online Lesson Planner

USA TODAY® Education's Online site offers resources and interactive features connected to each day's newspaper. *Experience Today,* USA TODAY's® daily lesson plan, is available on the site and delivered daily to subscribers. This plan provides instruction for integrating USA TODAY® graphics and key editorial features into your mathematics classroom. Log on to **education.usatoday.com**.

USA TODAY Snapshots® in Chapter 7	
Lesson	**Snapshot**
7-7	Low-tech tools rule at meetings

Quick Review Math Handbook

is Glencoe's mathematical handbook for students. The Hot Words section includes a glossary of terms while the Hot Topics section consists of explanations of key mathematical concepts. An exercise set is included to check students' understanding of the concepts. This valuable resource can be used as a reference in the classroom or for home study.

Lesson	Hot Topics Section	Lesson	Hot Topics Section
7-1	6-5	7-5	2-7, 2-9
7-2	6-5	7-6	2-7
7-3	6-5	7-7	2-8, 6-5
7-4	6-5, 8-6	7-8	2-8, 6-5

Chapter 7 Notes

Key Chapter Objectives

Lesson 7-1 Write ratios as fractions and determine whether two ratios are equivalent.

Lesson 7-3 Solve proportions.

Lesson 7-4 Solve problems involving scale drawings.

Lessons 7-5 and 7-6 Write percents as fractions, and vice versa.

Lesson 7-8 Solve problems using the percent proportion.

"How was math used to build Mount Rushmore?"

Point out to students that this is only one of many applications of the math that they are studying. Others are provided in the introduction to each lesson.

Vocabulary Builder, pp. vii–viii — ELL

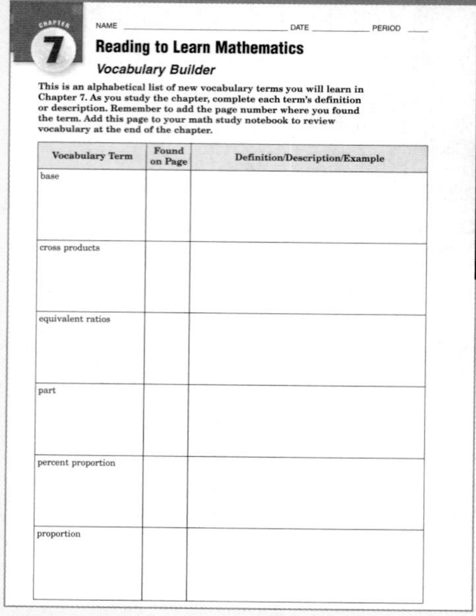
Chapter 7 — NAME ____ DATE ____ PERIOD ____

Reading to Learn Mathematics

Vocabulary Builder

This is an alphabetical list of new vocabulary terms you will learn in Chapter 7. As you study the chapter, complete each term's definition or description. Remember to add the page number where you found the term. Add this page to your math study notebook to review vocabulary at the end of the chapter.

Vocabulary Term	Found on Page	Definition/Description/Example
base		
cross products		
equivalent ratios		
part		
percent proportion		
proportion		

Chapter 7 Ratios and Proportions

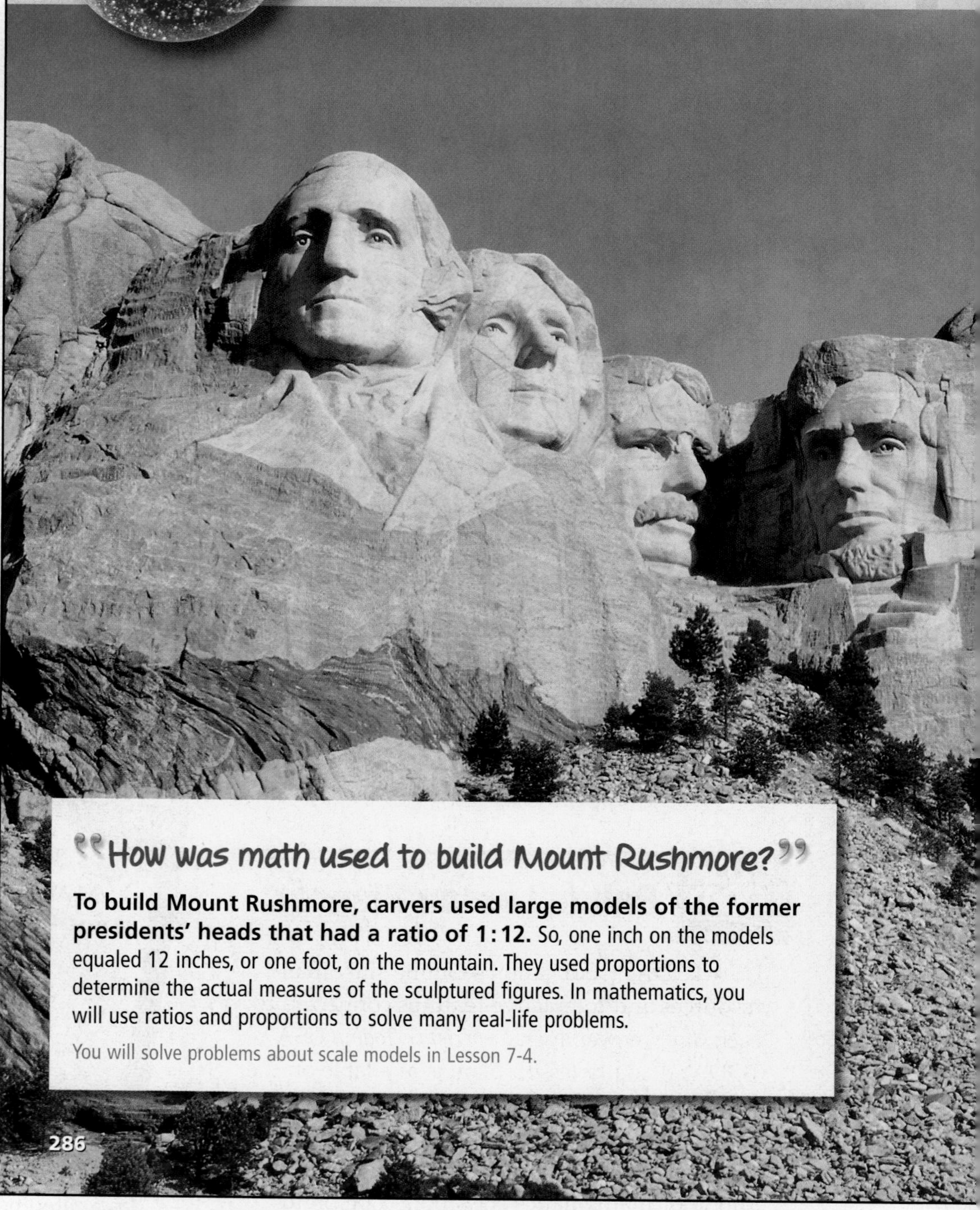

"How was math used to build Mount Rushmore?"

To build Mount Rushmore, carvers used large models of the former presidents' heads that had a ratio of 1 : 12. So, one inch on the models equaled 12 inches, or one foot, on the mountain. They used proportions to determine the actual measures of the sculptured figures. In mathematics, you will use ratios and proportions to solve many real-life problems.

You will solve problems about scale models in Lesson 7-4.

286

Vocabulary Builder

For a thorough vocabulary list with pronunciations of new words, give students the Vocabulary Builder worksheets from the *Chapter 7 Resource Masters.* Encourage them to complete the definition of each term as they progress through the chapter.

GETTING STARTED

Take this quiz to see whether you are ready to begin Chapter 7. Refer to the lesson or page number in parentheses if you need more review.

Vocabulary Review

Complete each sentence.

1. A comparison of two numbers by division is a ____. (Lesson 5-5) **ratio**
2. A ratio that compares a number to 100 is called a ____. (Lesson 5-5) **percent**

Prerequisite Skills

Evaluate each expression. Round to the nearest tenth if necessary. (Lesson 1-3)

3. $100 \times 25 \div 52$ **48.1**
4. $10 \div 4 \times 31$ **77.5**
5. $\frac{63 \times 4}{34}$ **7.4**
6. $\frac{2 \times 100}{68}$ **2.9**

Write each fraction in simplest form. (Lesson 5-3)

7. $\frac{9}{45}$ $\frac{1}{5}$
8. $\frac{16}{24}$ $\frac{2}{3}$
9. $\frac{38}{46}$ $\frac{19}{23}$
10. $\frac{21}{35}$ $\frac{3}{5}$

Write each decimal as a fraction in simplest form. (Lesson 5-4)

11. 0.78 $\frac{39}{50}$
12. 0.320 $\frac{8}{25}$
13. 0.06 $\frac{3}{50}$
14. 0.90 $\frac{9}{10}$

Solve each equation. (Lesson 6-5)

15. $\frac{1}{4}x = 6$ **24**
16. $\frac{a}{5} = 8$ **40**
17. $4r = 22$ $5\frac{1}{2}$
18. $2 = \frac{3}{5}z$ $3\frac{1}{3}$

Complete. (Lesson 6-7)

19. 2 yd = __?__ ft **6**
20. 48 oz = __?__ lb **3**
21. $4\frac{1}{2}$ ft = __?__ yd $1\frac{1}{2}$
22. $3\frac{1}{4}$ h = __?__ min **195**

FOLDABLES™ Study Organizer

Ratios and Proportions Make this Foldable to help you organize your notes. Begin with a sheet of notebook paper.

STEP 1 Fold Fold lengthwise to the holes.

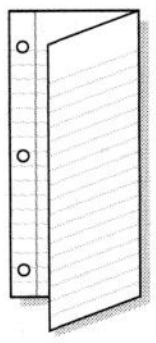

STEP 2 Cut Cut along the top line and then make equal cuts to form 10 tabs.

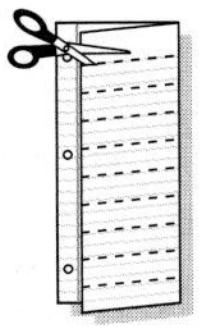

STEP 3 Label Label the major topics as shown.

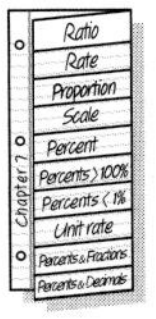

Reading and Writing As you read and study the lessons in this chapter, you can write notes under the corresponding tabs.

GETTING STARTED

This section provides a review of the basic concepts needed before beginning Chapter 7. Page references are included for additional student help.

Prerequisite Skills review skills needed in this chapter.

Exercises	Needed for Lesson	Skill
3–6	7-3	Evaluate Expressions
7–10	7-1	Simplify Fractions
11–14	7-6	Write Decimals as Fractions in Simplest Form
15–18	7-3	Solve Equations
19–22	7-1	Convert Units

For more prerequisite skills, see:

- Getting Ready for the Next Lesson, at the end of each lesson
- Prerequisite Skills in the Student Handbook on pp. 554–563
- *Math Skills Maintenance Masters*

FOLDABLES™ Study Organizer

For more information about Foldables, see *Teaching Mathematics with Foldables.*

Use this Foldable study guide to help students organize information they learn about ratios and proportions and to give them practice writing concise definitions in their own words. While working through the chapter, have students select one word or concept per lesson that they need to learn, and write definitions for each on the back of the appropriate tabs. To the right of each definition, ask students to produce their own examples or illustration of the concept or vocabulary word listed.

7-1 Lesson Notes

Lesson Resources

Chapter 7 Resource Masters

- Study Guide and Intervention, p. 375
- Practice: Skills, p. 376
- Practice: Word Problems, p. 377
- Reading to Learn Mathematics, p. 378
- Enrichment, p. 379

Key Teaching Transparencies, 25

Parent and Student Study Guide Workbook, 7-1

Teaching Mathematics with Manipulatives, pp. 68–69

1 Focus

5-Minute Check 7-1

INTERACTIVE CHALKBOARD PowerPoint® Presentations

Use as a quiz or review of Chapter 6. (Also available as a transparency)

Bellringer

Problem Solving Have students count the number of girls and the number of boys in their classroom. Ask them to write the comparison of these numbers as a fraction, and then to simplify the fraction if possible. They should also write a fraction comparing the number of girls to the total number of students, and a fraction comparing any other pairs of values they can think of.

Reading to Learn Mathematics, p. 378 ELL

NAME ______ DATE ______ PERIOD ____

7-1 Reading to Learn Mathematics

Ratios

Pre-Activity *Read the introduction at the top of page 288 in your textbook. Write your answers below.*

1. Express the gear ratio as a fraction. Then write it as a fraction with a denominator of 1. $\frac{80}{20} = \frac{4}{1}$
2. How many times does the smaller gear turn for every turn of the larger gear? 4 times
3. Describe some of the possible sizes of two gears that have a gear ratio of 3 to 1. Sample answers: 30 teeth and 10 teeth; 60 teeth and 20 teeth

Reading the Lesson

For Exercises 4 and 5, review the introduction to this lesson.

4. What two things are being compared? Sample answer: two gears ...nt sizes, ... teeth ... teeth

7-1 Ratios

What You'll Learn
Write ratios as fractions and determine whether two ratios are equivalent.

NEW Vocabulary
equivalent ratios

REVIEW Vocabulary
ratio: a comparison of two numbers by division **(Lesson 5-5)**

WHEN am I ever going to use this?

GEARS Gears are used in objects that have spinning parts, such as bicycles and pendulum clocks. Suppose you have a larger gear that has 80 teeth being turned by a smaller gear that has 20 teeth. The comparison of the size of the larger gear to the size of the smaller gear is called a *gear ratio*.

1. Express the gear ratio as a fraction. Then write it as a fraction with a denominator of 1. $\frac{80}{20} = \frac{4}{1}$
2. How many times does the smaller gear turn for every turn of the larger gear? **4 times**
3. Describe some of the possible sizes of two gears that have a gear ratio of 3 to 1. **Sample answers: 30 teeth and 10 teeth; 60 teeth and 20 teeth**

In Lesson 5-5, you learned that ratios can be written in several different ways. Ratios can also be simplified as shown below.

$$80 \text{ teeth} : 20 \text{ teeth} = \frac{\overset{4}{\cancel{80 \text{ teeth}}}}{\underset{1}{\cancel{20 \text{ teeth}}}} = \frac{4}{1}$$

EXAMPLES Write Ratios in Simplest Form

Write each ratio as a fraction in simplest form.

1 25 to 10

$25 \text{ to } 10 = \frac{25}{10}$ Write the ratio as a fraction.

$= \frac{5}{2}$ Simplify.

Written as a fraction in simplest form, the ratio 25 to 10 is $\frac{5}{2}$.

For ratios, leave the fraction as an improper fraction.

2 6 : 18

$6:18 = \frac{6}{18}$ Write the ratio as a fraction.

$= \frac{1}{3}$ Simplify.

The ratio 6 : 18 is $\frac{1}{3}$ in simplest form.

TEACHING TIP
The ratio 6 : 18 could also be written in simplest form as 1 : 3.

Your Turn **Write each ratio as a fraction in simplest form.**

a. 9 to 12 $\frac{3}{4}$ **b.** 27 : 15 $\frac{9}{5}$ **c.** 8 to 56 $\frac{1}{7}$

When writing ratios comparing units of length, units of time, units of weight, and so on, both measures should have the same unit.

EXAMPLE Write a Ratio by Converting Units

3 **Write the ratio 2 feet to 10 inches as a fraction in simplest form.**

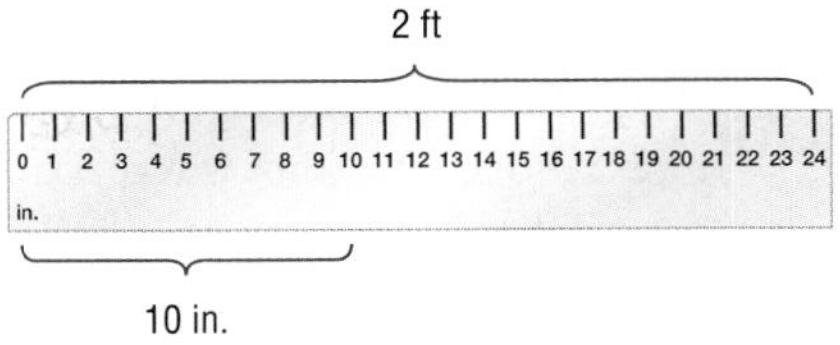

$\frac{2 \text{ feet}}{10 \text{ inches}} = \frac{24 \text{ inches}}{10 \text{ inches}}$ Convert 2 feet to inches.

$= \frac{\overset{12}{\cancel{24} \text{ inches}}}{\underset{5}{\cancel{10} \text{ inches}}}$ Divide by the GCF, 2 inches.

$= \frac{12}{5}$ Simplify.

So, the ratio 2 feet to 10 inches can be written as $\frac{12}{5}$.

Your Turn **Write each ratio as a fraction in simplest form.**

d. 4 feet : 4 yards **$\frac{1}{3}$**   **e.** 15 ounces to 3 pounds **$\frac{5}{16}$**

STUDY TIP

Canceling Units You can cancel units that are the same in the numerator and in the denominator.

TEACHING TIP

Ask students why $\frac{2}{10}$ does not represent the ratio 2 feet to 10 inches. Explain that $\frac{2}{10}$ represents 2 feet to 10 feet or 2 inches to 10 inches.

Two ratios that have the same value are **equivalent ratios**.

EXAMPLES Compare Ratios

4 **Determine whether 6 : 8 and 36 : 48 are equivalent.**

Write each ratio as a fraction in simplest form.

$6:8 = \frac{6 \div 2}{8 \div 2}$ or $\frac{3}{4}$ The GCF of 6 and 8 is 2. $36:48 = \frac{36 \div 12}{48 \div 12}$ or $\frac{3}{4}$ The GCF of 36 and 48 is 12.

The ratios in simplest form both equal $\frac{3}{4}$. So, 6 : 8 and 36 : 48 are equivalent ratios.

5 **PONDS It is recommended that for every 9 square feet of surface, a pond should have 2 fish. A pond that has a surface of 45 square feet contains 6 fish. Is this the correct number of fish based on the above recommendation?**

9 to 2 or $\frac{9}{2}$ Recommended ratio

45 to 6 or $\frac{45}{6} = \frac{15}{2}$ Actual ratio

Since $\frac{9}{2} \neq \frac{15}{2}$, the ratios are not equivalent.

So, the number of fish is not correct for the pond.

Your Turn **Determine whether the ratios are equivalent.**

f. $\frac{3}{8}$ and $\frac{6}{12}$ **g.** 35 students : 5 adults and 14 students : 2 adults

REAL-LIFE MATH

PONDS The recommended number of water lilies for a square-foot pond is 1.

no; $\frac{6}{12} = \frac{1}{2}$

yes; $35:5 = \frac{7}{1}$ and $14:2 = \frac{7}{1}$

msmath2.net/extra_examples

2 Teach

In-Class Examples

INTERACTIVE CHALKBOARD
PowerPoint® Presentations

Write each ratio as a fraction in simplest form.

1 30 to 9 **$\frac{10}{3}$**

2 4:24 **$\frac{1}{6}$**

3 Write the ratio 3 feet to 8 inches as a fraction in simplest form. **$\frac{9}{2}$**

4 Determine whether 12:15 and 32:40 are equivalent. **yes**

5 **POOLS** It is recommended that no more than one person be allowed into the shallow end of an outdoor public pool for every 15 square feet of surface area. If a local pool's shallow end has a surface area of 1,800 square feet, are the lifeguards correct to allow 120 people into that part of the pool? **yes**

DAILY INTERVENTION

Differentiated Instruction

Interpersonal Arrange students in small groups and have them make lists of how ratios are used in their daily lives. If students have trouble getting started, suggest applications such as paper or photograph sizes, playing field or room dimensions, or any news polls they have recently seen. Have them discuss the specific ratios involved, if they know them, as well as the importance of the ratios they mention.

3 Practice/Apply

FOLDABLES™ Follow-Up Remind students to select a term or key concept that they need to know for each lesson and write definitions in words and symbols under the tabs of their Foldables. As students progress through the chapter, encourage them to use their Foldables to self-check their understanding and mastery of these terms.

Which One Doesn't Belong?
In Exercise 3, encourage students to write all of these ratios as fractions in simplest form. They can then compare the ratios more easily.

DAILY INTERVENTION FIND THE ERROR In Exercise 4, ask students to look at the units, not just at the numbers. You may wish to suggest that students include the units when writing fractions. Students should make sure the units match.

About the Exercises...

Guided Practice Key

Exercises	Examples
5–7	1, 2
8–10	3
11–12	4
13	5

Odd/Even Assignments
Exercises 14–31 are structured so that students practice the same concepts whether they are assigned odd or even problems.

Skill and Concept Check

Writing Math Exercises 3 & 4

1. **Write** the ratio 8 feet out of 15 feet in three different ways. $\frac{8}{15}$, 8:15, 8 to 15
2. **OPEN ENDED** Write two different ratios that are equivalent to 3:2. Sample answer: 6:4, 9:6
3. **Which One Doesn't Belong?** Identify the item that does not have the same ratio as the other three. Explain your reasoning. 3 to 2; The other ratios equal $\frac{2}{3}$.

$\frac{30}{45}$	4 to 6	26:39	3 to 2

4. **FIND THE ERROR** Marcus and Nicole are writing the ratio 2 days to 18 hours as a fraction in simplest form. Who is correct? Explain.

Marcus	Nicole
$2:18 = \frac{2}{18}$	$48:18 = \frac{48}{18}$
$= \frac{1}{9}$	$= \frac{8}{3}$

Nicole; Marcus did not convert days to hours to make both measures of time have the same units

GUIDED PRACTICE

Write each ratio as a fraction in simplest form.

5. 8:30 $\frac{4}{15}$
6. 27 to 36 $\frac{3}{4}$
7. 11 to 4 $\frac{11}{4}$
8. 5 meters:1 meter $\frac{5}{1}$
9. 28 feet:7 yards $\frac{4}{3}$
10. 9 hours to 3 days $\frac{1}{8}$

Determine whether the ratios are equivalent. Explain.

11. $\frac{3}{5}$ and $\frac{9}{15}$ yes; $\frac{9}{15} = \frac{3}{5}$
12. 2:3 and 3:9 no; $2:3 = \frac{2}{3}$, $3:9 = \frac{1}{3}$, and $\frac{2}{3} \neq \frac{1}{3}$

13. **HEIGHTS** Megan is 5 feet 3 inches tall, and her brother Troy is 5 feet 9 inches tall. Write a ratio in simplest form that compares Megan's height to Troy's height in inches. $\frac{21}{23}$

★ indicates increased difficulty

Practice and Applications

HOMEWORK HELP

For Exercises	See Examples
14–19	1, 2
20–25	3
26–31, 33–37	4, 5

Extra Practice See pages 580, 602.

Write each ratio as a fraction in simplest form.

14. 24 to 9 $\frac{8}{3}$
15. 14 to 70 $\frac{1}{5}$
16. 18:19 $\frac{18}{19}$
17. 45:21 $\frac{15}{7}$
18. 120:30 $\frac{4}{1}$
19. 33 to 90 $\frac{11}{30}$
20. 66 inches:72 inches $\frac{11}{12}$
21. 12 pounds to 64 pounds $\frac{3}{16}$
22. 4 pounds to 12 ounces $\frac{16}{3}$
23. $8\frac{1}{3}$ yards:5 feet $\frac{5}{1}$
24. 18 weeks:1 year $\frac{9}{26}$
25. 45 minutes to 2 hours $\frac{3}{8}$

Determine whether the ratios are equivalent. Explain. 26–31. See margin.

26. $\frac{6}{9}$ and $\frac{2}{3}$
27. $\frac{9}{15}$ and $\frac{6}{10}$
28. 4:7 and 16:49
29. 8:21 and 16:42
★30. 14 in.:9 ft and 16 ft:12 yd
★31. 6 lb:72 oz and 2 lb:24 oz

Answers

26. yes; $\frac{6}{9} = \frac{2}{3}$
27. yes; $\frac{9}{15} = \frac{3}{5}$ and $\frac{6}{10} = \frac{3}{5}$
28. no; $4:7 = \frac{4}{7}$, $16:49 = \frac{16}{49}$, and $\frac{4}{7} \neq \frac{16}{49}$
29. yes; $8:21 = \frac{8}{21}$ and $16:42 = \frac{8}{21}$
30. no; 14 in.:9 ft $= \frac{7}{54}$, 16 ft:12 yd $= \frac{4}{9}$, and $\frac{7}{54} \neq \frac{4}{9}$
31. yes; 6 lb:72 oz $= \frac{4}{3}$ and 2 lb:24 oz $= \frac{4}{3}$
37. $\frac{2}{5}$; $\frac{7}{3}$; No, the ratio is greater for the dollar bill.
38. $\frac{1}{4}, \frac{4}{9}, \frac{9}{16}, \frac{16}{25}, \frac{25}{36}, \frac{36}{49}, \frac{49}{64}, \frac{64}{81}$; The values of the ratios approach 1, but never equal 1.

32. **WRITE A PROBLEM** Write a problem that could be represented by using a ratio. **Sample answer: The ratio of length to width of a rectangle is 8 in. to 5 in. or 8:5.**

33. **MOVIES** One week in 2002, a movie about the mathematician John Nash earned 6 million dollars at the box office. Total revenues for all movies that week were 78 million dollars. Write a ratio to compare the money earned by that movie to the money earned by all the movies. $\frac{1}{13}$

★ **PHOTOGRAPHS The *aspect ratio* of a photograph is a ratio comparing the length and width. A 35 mm negative has an aspect ratio of 1.5:1. Photo sizes with the same aspect ratio can be printed full frame without cropping. Determine which size photos can be printed full frame from a 35 mm negative.**

34. 10″ × 15″ **yes**
35. 8″ × 10″ **no**
36. 5″ × 7.5″ **yes**

★ 37. **MONEY** In simplest form, write the ratio of the cost in cents to time in years for each type of currency shown in the table. Are they equivalent? Explain. **37–38. See margin.**

Currency	Cost to Make	Time in Use
Sacagawea dollar coin	12¢	30 years
Dollar bill	3.5¢	18 months

Source: www.sciencenews.org

★ 38. **FIND A PATTERN** In decimal form, write the ratios of the first eight consecutive square numbers $1^2:2^2$, $2^2:3^2$, $3^2:4^2$, … . What do you notice about the ratios?

39. **CRITICAL THINKING** Find the next number in the pattern 480, 240, 80, 20, __?__. Explain your reasoning. (*Hint*: Look at the ratios of successive numbers.) **4; The ratios of successive terms increase by 1.**

Standardized Test Practice and Mixed Review

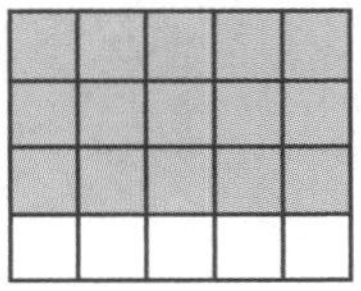

40. **MULTIPLE CHOICE** Express the ratio of shaded squares to total squares as a fraction in simplest form. **A**

Ⓐ $\frac{3}{4}$ Ⓑ $\frac{15}{20}$ Ⓒ $\frac{7}{10}$ Ⓓ $\frac{3}{5}$

41. **MULTIPLE CHOICE** In the seventh grade, 16 of the 120 students are left-handed. Write the ratio of left-handed students to right-handed students as a fraction in simplest form. **G**

Ⓕ $\frac{2}{15}$ Ⓖ $\frac{2}{13}$ Ⓗ $\frac{13}{2}$ Ⓘ $\frac{15}{2}$

42. **TREES** A giant sequoia tree has a diameter of about 30 feet. What is the circumference? Round to the nearest foot. (Lesson 6-9) **94 ft**

Find the perimeter and area of each rectangle. (Lesson 6-8)

43. $\ell = 9$ ft, $w = 4$ ft **26 ft; 36 ft²**
44. $\ell = 12$ cm, $w = 10.6$ cm **45.2 cm; 127.2 cm²**
45. $\ell = 7.5$ cm, $w = 2$ cm **19 cm; 15 cm²**
46. $\ell = 1\frac{2}{5}$ in., $w = \frac{1}{2}$ in. **$3\frac{4}{5}$ in.; $\frac{7}{10}$ in²**

GETTING READY FOR THE NEXT LESSON

PREREQUISITE SKILL Divide. Round to the nearest hundredth if necessary. (Page 562)

47. 9.8 ÷ 2 **4.9**
48. \$4.30 ÷ 5 **\$0.86**
49. \$12.49 ÷ 40 **\$0.31**
50. 27.36 ÷ 3.2 **8.55**

msmath2.net/self_check_quiz

4 Assess

Open-Ended Assessment

Modeling Have students use counters of different colors to represent the ratio 18 to 30 in two simpler ways, including simplest form. By grouping the counters, students can reduce the ratio easily.

Study Guide and Intervention, p. 375

NAME ______ DATE ______ PERIOD ____

7-1 Study Guide and Intervention

Ratios

Any ratio can be written as a fraction. To write a ratio comparing measurements, such as units of length or units of time, both quantities must have the same unit of measure. Two ratios that have the same value are **equivalent ratios**.

EXAMPLE 1 Write the ratio 15 to 9 as a fraction in simplest form.

$15 \text{ to } 9 = \frac{15}{9}$ Write the ratio as a fraction.

$= \frac{5}{3}$ Simplify.

Written as a fraction in simplest form, the ratio 15 to 9 is $\frac{5}{3}$.

EXAMPLE 2 Write 40 centimeters to 2 meters as a fraction in simplest form.

$\frac{40 \text{ centimeters}}{2 \text{ meters}} = \frac{40 \text{ centimeters}}{200 \text{ centimeters}}$ Convert 2 meters to centimeters.

$= \frac{\overset{1}{\cancel{40}} \cancel{\text{centimeters}}}{\underset{5}{\cancel{200}} \cancel{\text{centimeters}}}$ Divide by the GCF, 40 centimeters.

$= \frac{1}{5}$ Simplify.

EXERCISES

Write each ratio as a fraction in simplest form.

1. 30 to 12 $\frac{5}{2}$
2. 5:20 $\frac{1}{4}$
3. 49:42 $\frac{7}{6}$
4. 15 to 13 $\frac{15}{13}$
5. 28 feet:35 feet $\frac{4}{5}$
6. 24 minutes to 18 minutes $\frac{4}{3}$
7. 75 seconds:2 minutes $\frac{5}{8}$
8. 12 feet:10 yards $\frac{2}{5}$

Determine whether the ratios are equivalent. Explain.

9. $\frac{3}{4}$ and $\frac{12}{16}$ yes; $\frac{3}{4} = \frac{12}{16}$
10. 12:17 and 10:15 no; $12:17 = \frac{12}{17}$, $10:15 = \frac{10}{15}$, and $\frac{12}{17} \neq \frac{10}{15}$
11. $\frac{25}{35}$ and $\frac{10}{14}$ yes; $\frac{25}{35} = \frac{5}{7}$ and $\frac{10}{14} = \frac{5}{7}$
12. 2 lb:36 oz and 3 lb:44 oz no; 2 lb:36 oz $= \frac{8}{9}$, 3 lb:44 oz $= \frac{12}{11}$, and $\frac{8}{9} \neq \frac{12}{11}$
13. 3 ft:12 in. and 2 yd:2 ft yes; 3 ft:12 in. $= \frac{3}{1}$ and 2 yd:2 ft $= \frac{3}{1}$

Practice: Skills, p. 376

NAME ______ DATE ______ PERIOD ____

7-1 Practice: Skills

Ratios

Write each ratio as a fraction in simplest form.

1. 14 to 6 $\frac{7}{3}$
2. 18:3 $\frac{6}{1}$
3. 4:22 $\frac{2}{11}$
4. 7:21 $\frac{1}{3}$
5. 18:12 $\frac{3}{2}$
6. 20 to 9 $\frac{20}{9}$
7. 25 to 20 $\frac{5}{4}$
8. 4:10 $\frac{2}{5}$
9. 18:21 $\frac{6}{7}$
10. 84 to 16 $\frac{21}{4}$
11. 33 ounces to 11 ounces $\frac{3}{1}$
12. 45 minutes:25 minutes $\frac{9}{5}$
13. 77 cups:49 cups $\frac{11}{7}$
14. 15 pounds to 39 pounds $\frac{5}{13}$
15. 40 seconds to 6 minutes $\frac{1}{9}$
16. 140 centimeters to 3 meters $\frac{7}{15}$
17. 9 weeks:9 days $\frac{7}{1}$
18. 1 yard to 11 feet $\frac{3}{11}$

Determine whether the ratios are equivalent. Explain.

19. $\frac{3}{16}$ and $\frac{9}{48}$ yes; $\frac{3}{16} = \frac{9}{48}$
20. $\frac{7}{10}$ and $\frac{8}{11}$ no; $\frac{7}{10} \neq \frac{8}{11}$
21. 18 in.:3 ft and 12 in.:2 ft yes; 18 in.:3 ft $= \frac{1}{2}$ and 12 in.:2 ft $= \frac{1}{2}$
22. 6 mos:2 yr and 8 mos:3 yr no; 6 mos:2 yr $= \frac{1}{4}$, 8 mos:3 yr $= \frac{2}{9}$, and $\frac{1}{4} \neq \frac{2}{9}$

Practice: Word Problems, p. 377

NAME ______ DATE ______ PERIOD ____

7-1 Practice: Word Problems

Ratios

1. **ELECTIONS** In an election for sheriff, 210 people voted. If there were 1,260 possible voters, write a ratio to compare the number of people who voted to the number of possible voters. $\frac{1}{6}$	2. **DENTAL CARE** Taru surveyed 60 dentists and found that 48 favored the use of fluoride toothpaste. Write a ratio to compare the number of dentists favoring the use of a flouride toothpaste to all dentists surveyed. $\frac{4}{5}$
3. **E-MAIL** One morning, Mirna counted 15 junk E-mails out of 21 E-mails in her inbox. Write a ratio comparing the number of junk E-mails to the total number of E-mails. $\frac{5}{7}$	4. **SURFING** One evening at his local surf spot, Jeff counted 28 surfers in the water. Among those, he counted 21 that had hoods on their wetsuits. Write a ratio comparing the number of surfers with hoods to the total number of surfers. $\frac{3}{4}$
5. **MUSIC** A music company signed 12 new artists to its label in 2002. Out of the 12, 10 artists have hit songs. Write a ratio to compare the number of artists with hit songs to the total number of artists signed in 2002. $\frac{5}{6}$	6. **COOKING** Charlene poured 18 fluid ounces of juice into a 1-quart pitcher. Write a ratio comparing the amount of juice in the pitcher to the total capacity of the pitcher. $\frac{9}{16}$
7. **BASEBALL** In baseball, David has 10 hits out of 14 at bats. Adam has 15 hits out of 21 at bats. For each player, write a ratio that represents his total number of hits out of times at bat. Are these ratios equivalent? 10:14 and 15:21; yes	8. **DRIVING** Sarah can drive 198 miles on 11 gallons of gasoline. On 6 gallons of gasoline, Rachel can travel 138 miles. Write a ratio that compares miles traveled per gallon of gasoline for each car. Do the cars get the same mileage? 198:11 and 138:6; no

7-2 Lesson Notes

Lesson Resources

Chapter 7 Resource Masters
- Study Guide and Intervention, p. 380
- Practice: Skills, p. 381
- Practice: Word Problems, p. 382
- Reading to Learn Mathematics, p. 383
- Enrichment, p. 384

MathPASS Tutorial CD-ROM, 28

Parent and Student Study Guide Workbook, 7-2

FAST FILE

1 Focus

5-Minute Check 7-2

INTERACTIVE CHALKBOARD
PowerPoint® Presentations

Use as a quiz or review of Lesson 7-1. (Also available as a transparency)

Bellringer

Hands-On Activity Ask students how they would measure the thickness of a sheet of paper. Have them measure the thickness of a stack of papers and count how many sheets are in the stack. Then have them write a ratio of the thickness of the stack to the number of sheets in the stack. Show the class how to determine the thickness of a single sheet by dividing both the numerator and denominator by the total number of sheets of paper in the stack.

Reading to Learn Mathematics, p. 383 ELL

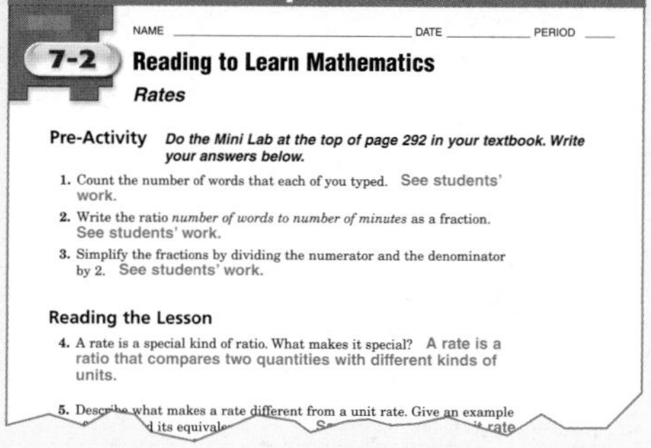
NAME ______________ DATE ________ PERIOD ____

7-2 Reading to Learn Mathematics
Rates

Pre-Activity ***Do the Mini Lab at the top of page 292 in your textbook. Write your answers below.***

1. Count the number of words that each of you typed. See students' work.
2. Write the ratio *number of words to number of minutes* as a fraction. See students' work.
3. Simplify the fractions by dividing the numerator and the denominator by 2. See students' work.

Reading the Lesson

4. A rate is a special kind of ratio. What makes it special? A rate is a ratio that compares two quantities with different kinds of units.
5. Describe what makes a rate different from a unit rate. Give an example

7-2 Rates

What You'll Learn
Determine unit rates.

NEW Vocabulary
- rate
- unit rate

HANDS-ON Mini Lab

Materials
- textbook
- computer
- stopwatch

Work with a partner. 1–3. See students' work.

Choose a page in a textbook and take turns typing as much as possible in 2 minutes.

1. Count the number of words that each of you typed.
2. Write the ratio *number of words to number of minutes* as a fraction.
3. Simplify the fractions by dividing the numerator and the denominator by 2.

A ratio that compares two quantities with different kinds of units is called a **rate**.

$\frac{100 \text{ words}}{2 \text{ minutes}}$ — The units *words* and *minutes* are different.

When a rate is simplified so that it has a denominator of 1 unit, it is called a **unit rate**.

$\frac{50 \text{ words}}{1 \text{ minute}}$ — The denominator is 1 unit.

The unit rate $\frac{50 \text{ words}}{1 \text{ minute}}$ can be read as *50 words per minute*. The table below shows some common unit rates used in everyday life.

Ratio	Unit Rate	Abbreviation	Name
$\frac{\text{number of miles}}{1 \text{ hour}}$	miles per hour	mi/h or mph	speed
$\frac{\text{number of miles}}{1 \text{ gallon}}$	miles per gallon	mi/gal or mpg	gas mileage
$\frac{\text{number of dollars}}{1 \text{ pound}}$	price per pound	dollars/lb	unit price
$\frac{\text{number of dollars}}{1 \text{ hour}}$	dollars per hour	dollars/h	hourly wage

TEACHING TIP
- Point out that the word *rate* is often understood to mean unit rate.
- In Example 1, *average* means the typical speed, not the exact speed at every moment.

EXAMPLE Find a Unit Rate

1 RUNNING Alethia ran 24 miles in 3 hours. What is her average speed in miles per hour?

Write the rate as a fraction. Then find an equivalent rate with a denominator of 1.

$$24 \text{ miles in } 3 \text{ hours} = \frac{24 \text{ mi}}{3 \text{ h}} \quad \text{Write the rate as a fraction.}$$

$$= \frac{24 \text{ mi} \div 3}{3 \text{ h} \div 3} \quad \text{Divide the numerator and the denominator by 3.}$$

$$= \frac{8 \text{ mi}}{1 \text{ h}} \quad \text{Simplify.}$$

The average speed, or unit rate, is 8 miles per hour.

292 Chapter 7 Ratios and Proportions

Standardized Test Practice

EXAMPLES Find Unit Rates

2 GRID-IN TEST ITEM Write *228 feet in 24 seconds* as a unit rate in feet per second.

Read the Test Item

Write the ratio as a fraction. Then divide to get a denominator of 1.

Solve the Test Item

228 feet in 24 seconds

$= \frac{228 \text{ ft}}{24 \text{ s}}$ Write the rate as a fraction.

$= \frac{228 \text{ ft} \div 24}{24 \text{ s} \div 24}$ Divide the numerator and the denominator by 24.

$= \frac{9.5 \text{ ft}}{1 \text{ s}}$ Simplify.

The unit rate is 9.5 feet per second.

Fill in the Grid

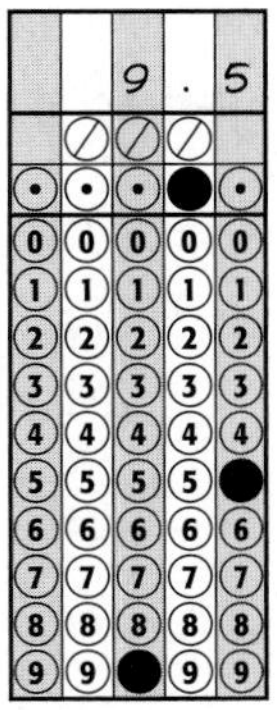

est-Taking Tip The Princeton Review

rid-In Answers ɔ fill in the grid, first ɑade the appropriate ubbles. Then write the nswer in the spaces ɔove. You can align at the ght as shown, or start in e left column.

3 Find the unit price per orange if it costs $2 for six oranges. Round to the nearest hundredth if necessary.

$2 for six oranges $= \frac{\$2}{6 \text{ oranges}}$ Write the rate as a fraction.

$= \frac{\$2 \div 6}{6 \text{ oranges} \div 6}$ Divide the numerator and the denominator by 6.

$\approx \frac{\$0.33}{1 \text{ orange}}$ Simplify.

The unit price is about $0.33 per orange.

TEACHING TIP
ɔte that $\frac{6 \text{ oranges}}{\$2}$ gives e unit rate 3 oranges r $1.

Your Turn **Find each unit rate. Round to the nearest hundredth if necessary.**

a. 300 tickets in 6 days b. 220 miles in 8 gallons

a. 50 tickets per day b. 27.5 mi per gal

In Example 3, you found a special kind of unit rate, called the *unit price*. This is the price per unit and is useful when you want to compare the cost of an item that comes in different sizes.

EXAMPLE Choose the Best Buy

4 The costs of different sizes of peanut butter are shown at the right. Which jar costs the least per ounce?

Size	Price
12 oz	$2.49
40 oz	$5.30
80 oz	$10.89

Find the unit price, or the cost per ounce, of each jar. Divide the price by the number of ounces.

12-ounce jar $2.49 ÷ 12 ounces ≈ $0.21 per ounce

40-ounce jar $5.30 ÷ 40 ounces ≈ $0.13 per ounce

80-ounce jar $10.89 ÷ 80 ounces ≈ $0.14 per ounce

The 40-ounce jar costs the least per ounce.

STUDY TIP

Estimation You can use compatible numbers to estimate the unit prices. For example, $2.49 ÷ 12 oz ≈ $2.40 ÷ 12 oz, or $0.20/oz.

msmath2.net/extra_examples

2 Teach

In-Class Examples

INTERACTIVE CHALKBOARD PowerPoint® Presentations

1 **READING** Julia read 52 pages in 2 hours. What is the average number of pages she read per hour? **26 pages per hour**

2 **GRID-IN TEST ITEM** Write *440 miles in 8 hours* as a unit rate in miles per hour. **55 mph**

3 Find the unit price per can if it costs $3 for 6 cans of soda. Round to the nearest hundredth if necessary. **$0.50 per can**

4 The costs of different sizes of orange juice are shown in the table. Which container costs the least per ounce? **96 oz**

Amount	Total Cost
16 oz	$1.28
32 oz	$1.92
64 oz	$2.56
96 oz	$3.36

Tips for New Teachers

Using Applications Encourage students to be on the lookout for real-life examples of their math lessons. For example, the next time their families go shopping, they can help find the best buys by computing unit rates with their calculators.

DAILY INTERVENTION **Differentiated Instruction**

Logical Many grocery store prices are in multiple units, such as $2.50 per six-pack of cola. Have students find multiple unit prices in grocery ads for competing stores and determine which store has the better price. Items could include potatoes, cans of soda, or detergent. Students will need to determine the unit price for each store and compare them. **See students' work.**

3 Practice/Apply

DAILY INTERVENTION **FIND THE ERROR**
In Exercise 2, remind students that when they write a rate as a fraction, the word following *per* is the unit that is written in the denominator.

NUMBER SENSE
In Exercise 3, students may find it helpful to replace the variables with numbers that are easily calculated.

About the Exercises...

Guided Practice Key

Exercises	Examples
4–7	1, 2
8	3
9	4

Odd/Even Assignments
Exercises 10–23 and 25–28 are structured so that students practice the same concepts whether they are assigned odd or even problems.

Alert! Exercise 24 uses the Internet or another source for research.

Skill and Concept Check

Writing Math
Exercises 1–3

1. **OPEN ENDED** Write a rate and then convert it to a unit rate.
Sample answer: $\frac{30 \text{ mi}}{2 \text{ h}} = \frac{15 \text{ mi}}{\text{h}}$
2. **FIND THE ERROR** Mikasi and Julie are determining which size sports drink is the better buy per ounce: a 16-ounce bottle for \$1.95 or a 36-ounce bottle for \$3.05. Who is correct? Explain.

Mikasi	Julie
16-oz bottle: 12.2¢ per ounce	16-oz bottle: 8.2¢ per ounce
36-oz bottle: 8.5¢ per ounce	36-oz bottle: 11.8¢ per ounce

Mikasi; Julie incorrectly divided the ounces by the cost.
3. **NUMBER SENSE** In which situation will the rate $\frac{x \text{ ft}}{y \text{ min}}$ increase? Give an example to explain your reasoning. See margin.
a. x increases, y is unchanged
b. x is unchanged, y increases

GUIDED PRACTICE

Find each unit rate. Round to the nearest hundredth if necessary. 4. 25.63 mi per gal

4. 410 miles in 16 gallons
5. 1,500 words in 25 minutes 60 words per min
6. 5 pounds for \$2.49 \$0.50 per lb
7. 3 cans of juice for \$2.95 \$0.98 per can
8. Which has the better unit price: a 6-pack of soda for \$2.99 or a 12-pack for \$4.50? 12-pack for \$4.50
9. **MUSIC** Ethan can buy 4 CDs for \$71.96 at CD Express or 9 CDs for \$134.55 at Music Rox. Which has the better unit cost? Explain.
Music Rox; \$14.95/CD < \$17.99/CD

★ indicates increased difficulty

Practice and Applications

HOMEWORK HELP

For Exercises	See Examples
10–17, 21–23	1–3
18–20	4

Extra Practice
See pages 580, 602.

Find each unit rate. Round to the nearest hundredth if necessary.

10. 360 miles in 6 hours
11. 6,840 customers in 45 days
12. 150 people for 5 classes
13. \$7.40 for 5 pounds
14. \$1.12 for 8.2 ounces
15. 810 Calories in 3 servings
16. 40 meters in 13 seconds
17. 144 miles in 4.5 gallons

10–17. See margin.

Choose the best unit price.

18. \$3.99 for a 16-ounce bag of candy or \$2.99 for a 12-ounce bag \$2.99 for a 12-oz bag
★ 19. aspirin sold in bottles of 50 for \$5.49, 100 for \$8.29, or 150 for \$11.99 150 for \$11.99
20. **NUTRITION** Which soft drink described in the table has a lower amount of sodium per ounce? Explain. B; it has 4.38 mg/oz compared to 5.83 mg/oz in A.

Soft Drink	Serving Size (oz)	Sodium (mg)
A	12	70
B	8	35

21. **FOUNTAINS** The Prometheus Fountain in New York City pumps 60,000 gallons of water every 15 minutes. What is the unit rate? 4,000 gal per min

Answers

3. a; Sample answer: $\frac{20 \text{ ft}}{1 \text{ min}} \rightarrow \frac{40 \text{ ft}}{1 \text{ min}}$ represents an increase in speed since 40 ft/min > 20 ft/min. If the same distance is covered in a longer period of time, then the rate is decreased. Consider $\frac{20 \text{ ft}}{1 \text{ min}} \rightarrow \frac{20 \text{ ft}}{2 \text{ min}}$. Since $\frac{20 \text{ ft}}{2 \text{ min}} = 10$ ft/min, the rate is decreased.
10. 60 mi per h
11. 152 customers per day
12. 30 people per class
13. \$1.48 per lb
14. \$0.14 per oz
15. 270 C per serving
16. 3.08 m per s
17. 32 mi per gal
25. Sometimes; a ratio that compares two measurements with different units is a rate, such as $\frac{2 \text{ mi}}{10 \text{ min}}$. A ratio that compares two numbers or two measurements with like units is not a rate, such as $\frac{2 \text{ c}}{3 \text{ c}}$.

22. **WHALES** A humpback whale can migrate 3,000 miles in 30 days. Find a unit rate to describe the average speed per day during migration. **100 mi per day**

23. **POPULATION** The population of Virginia is approximately 7.1 million people, and its land area is approximately 39,600 square miles. Find the *population density*, or the population per square mile. **about 179 people per sq mi**

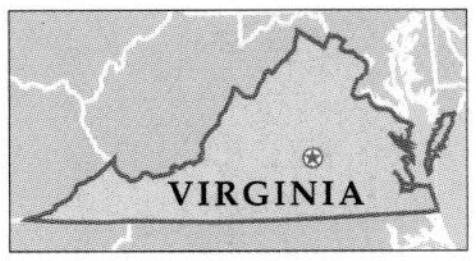

24. **RESEARCH** Use the Internet or another source to find the population and land area of your state. Determine the population density. **See students' work.**

Determine whether the following statements are *sometimes*, *always*, or *never* true. Explain by giving an example or a counterexample.

25. A ratio is a rate.

26. A rate is a ratio. **25–26. See margin.**

LIFE SCIENCE For Exercises 27 and 28, use the graph. It shows the average number of heartbeats for an adult elephant and an adult human.

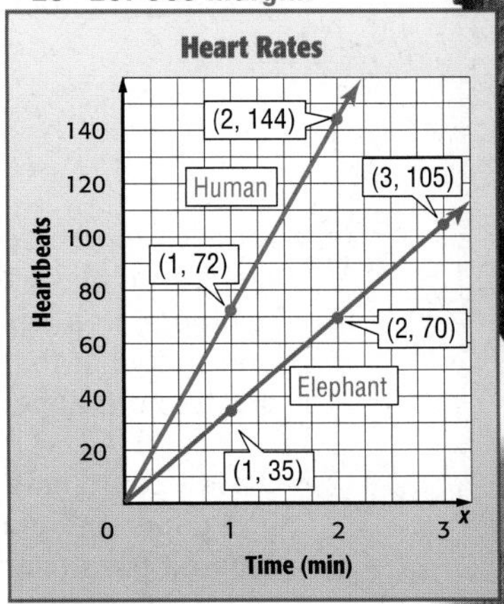

27. Whose heart rate is greater? Explain. **See margin.**

28. Whose heart beats more times in one hour? **human**

29. **CRITICAL THINKING** If 8 tickets are sold in 15 minutes, find the rate in tickets per hour. **32 tickets per h**

Standardized Test Practice and Mixed Review

30. **MULTIPLE CHOICE** Which of the following has the lowest unit price? **B**

(A) 18 oz: \$5.40 (B) 16 oz: \$4.64 (C) 12 oz: \$3.72 (D) 10 oz: \$3.30

31. **MULTIPLE CHOICE** Which rate is the fastest? **I**

(F) 589 miles in 11 hours (G) 360 miles in 7 hours

(H) 111 miles in 2 hours (I) 283 miles in 5 hours

Write each ratio as a fraction in simplest form. (Lesson 7-1)

32. 11:12 $\frac{11}{12}$ 33. 24:4 $\frac{6}{1}$ 34. 15 to 25 $\frac{3}{5}$ 35. 20 to 14 $\frac{10}{7}$

36. A serving plate has a radius of 6 inches. Find the circumference to the nearest tenth. (Lesson 6-9) **37.7 in.**

GETTING READY FOR THE NEXT LESSON

PREREQUISITE SKILL Solve each equation. (Lesson 6-5)

37. $\frac{n}{8} = 7$ **56** 38. $\frac{2}{3}x = 5$ **$7\frac{1}{2}$** 39. $9t = 12$ **$1\frac{1}{3}$** 40. $\frac{r}{5.2} = 4$ **20.8**

msmath2.net/self_check_quiz

Study Guide and Intervention, p. 380

NAME ______ DATE ______ PERIOD ______

7-2 Study Guide and Intervention
Rates

A ratio that compares two quantities with different kinds of units is called a **rate**. When a rate is simplified so that it has a denominator of 1 unit, it is called a **unit rate**.

EXAMPLE 1 **DRIVING Alita drove her car 78 miles and used 3 gallons of gas. What is the car's gas mileage in miles per gallon?**

Write the rate as a fraction. Then find an equivalent rate with a denominator of 1.

78 miles using 3 gallons $= \frac{78 \text{ mi}}{3 \text{ gal}}$ Write the rate as a fraction.

$= \frac{78 \text{ mi} \div 3}{3 \text{ gal} \div 3}$ Divide the numerator and the denominator by 3.

$= \frac{26 \text{ mi}}{1 \text{ gal}}$ Simplify.

The car's gas mileage, or unit rate, is 26 miles per gallon.

EXAMPLE 2 **SHOPPING Joe has two different sizes of boxes of cereal from which to choose. The 12-ounce box costs \$2.54, and the 18-ounce box costs \$3.50. Which box costs less per ounce?**

Find the unit price, or the cost per ounce, of each box. Divide the price by the number of ounces.

12-ounce box \$2.54 ÷ 12 ounces ≈ \$0.21 per ounce
18-ounce box \$3.50 ÷ 18 ounces ≈ \$0.19 per ounce

The 18-ounce box costs less per ounce.

EXERCISES

Find each unit rate. Round to the nearest hundredth if necessary.

1. 18 people in 3 vans **6 people per van**
2. \$156 for 3 books **\$52 per book**
3. 115 miles in 2 hours **57.5 mi per h**
4. 8 hits in 22 games **0.36 hits per game**
5. 65 miles in 2.7 gallons **24.07 mi per gal**
6. 2,500 Calories in 24 hours **104.17 C per h**

Choose the best unit price.

7. \$12.95 for 3 pounds of nuts or \$21.45 for 5 pounds of nuts **\$21.45 for 5 lb**
8. A 32-ounce bottle of apple juice for \$2.50 or a 48-ounce bottle for \$3.84. **\$2.50 for a 32-oz bottle**

Practice: Skills, p. 381

NAME ______ DATE ______ PERIOD ______

7-2 Practice: Skills
Rates

Find each unit rate. Round to the nearest hundredth if necessary.

1. \$112 in 8 hours **\$14 per h**
2. 150 miles in 6 gallons **25 mi per gal**
3. 49 points in 7 games **7 points per game**
4. 105 students in 3 classes **35 students per class**
5. 120 problems in 5 hours **24 problems per h**
6. 3 accidents in 12 months **0.25 accident per mo**
7. 6 eggs in 7 days **0.86 egg per day**
8. 8 batteries in 3 months **2.67 batteries per mo**
9. 122 patients in 4 weeks **30.5 patients per wk**
10. 51 gallons in 14 minutes **3.64 gal per min**
11. \$8.43 for 3 pounds **\$2.81 per lb**
12. 357 miles in 6.3 hours **56.67 mi per h**
13. 25 letters in 4 days **6.25 letters per day**
14. \$99 for 12 CDs **\$8.25 per CD**
15. 5 breaks in 8 hours **0.63 break per h**
16. 3 trips in 14 months **0.21 trip per mo**
17. 2 pay raises in 3 years **0.67 raise per yr**
18. 7 errors in 60 minutes **0.12 error per min**
19. 15 pounds in 6 weeks **2.5 lb per wk**
20. 8 commercials in 15 minutes **0.53 commercial per min**
21. 8 glasses every 24 hours **0.33 glass per h**
22. 13 feet in 5 steps **2.6 ft per step**

Choose the best unit price.

23. \$4.99 for 6 cans or \$7.99 for 10 cans **\$7.99 for 10 cans**
24. \$21.50 for 4 pounds of lunch meat or \$15.10 for 3 pounds of lunch meat **\$15.10 for 3 pounds**

Practice: Word Problems, p. 382

NAME ______ DATE ______ F

7-2 Practice: Word Problems
Rates

1. **TRAVEL** During Sonia's trip across the country, she traveled 2,884 miles. Her trip took 7 days. Find a unit rate to represent the average miles she traveled per day during the trip. **412 mi per day**

2. **BUDGET** Steve was offered \$5,025 per year for a weekend lifeguarding job at a local pool. He wants to know how much his monthly income will be at this salary level. What is his rate of pay in dollars per month? **\$418.75 per mo**

3. **MUSIC** Randall recorded 8 songs on his most recent CD. The total length of the CD is 49 minutes. Find a unit rate to represent the average length per song on the CD. **6.125 min per song**

4. **CARPETING** Hana paid \$1,200 for the carpet in her living room. The room has an area of 251.2 square feet. What was her unit cost of carpeting in dollars per square foot? Round to the nearest cent. **\$4.78 per ft²**

5. **SHOPPING** An 8-ounce can of tomatoes costs \$1.14. A 12-ounce can costs \$1.75. Which can of tomatoes has the better unit price? **8-oz can**

6. **PETS** Last month, Hao's dog ate 40 cans of dog food in 31 days. How many cans should Hao buy to feed his dog for the next 6 days? **8 cans**

4 Assess

Higher-Level Thinking

If three pens cost \$2.70, how much do ten pens cost? How do you use the unit rate to solve this problem? **\$9; Find the rate for one pen and multiply by 10.**

Answers

26. **Always; every rate is a ratio, because it is a comparison of two quantities by division.**

27. **human, 72 beats/min; elephant, 35 beats/min**

7-2b Hands-On Lab
Lesson Notes

Lesson Resources

Teaching Mathematics with Manipulatives

- centimeter grid, p. 11
- student recording sheet 7-2b

GETTING STARTED

Teaching Tip You may want to draw the table and record the data on the chalkboard or an overhead transparency. You may also want to create the graph as an overhead transparency.

Teaching Tip If square tiles are not available, students can shade squares on grid paper.

TEACH

- Challenge students to determine why the perimeter increases by 4 each time two new tiles are added.
- Remind students that in graphing, the x-value is the independent variable and the y-value is the dependent variable. The number of tiles affects the perimeter, therefore the number of tiles is represented by x, while the perimeter is represented by y.

ASSESS

Observe as students solve the **Writing Math** exercises. They should:

- find the rate of change between two sets of data, and
- compare the rate of change and the slope of a graph.

7-2b Hands-On Lab

A Follow-Up of Lesson 7

What You'll Learn
Investigate rate of change.

Materials
- square tiles
- grid paper

Rate of Change

INVESTIGATE *Work in groups of three.*

A *rate of change* is a ratio that shows a change in one quantity with respect to a change in another quantity. In this lab, you will use tables and graphs to represent *constant* rates of change.

STEP 1 Use tiles to build the models shown below. Then continue the pattern to build the fourth and fifth models.

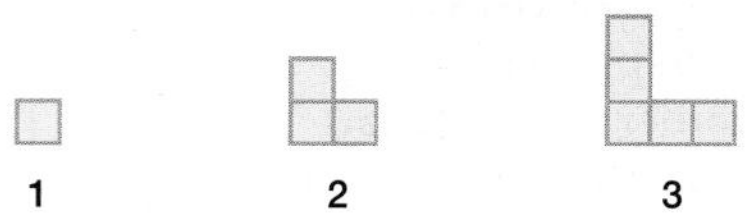

STEP 2 For each model, record the number of tiles and the perimeter in a table like the one shown at the right. The first two are done for you.

Model	Number of Tiles (x)	Perimeter (y)
1	1	4
2	3	8
3	5	12
4	7	16
5	9	20

STEP 3 Draw a coordinate plane on grid paper and graph the ordered pairs (x, y).

Writing Math

2. $\frac{4}{2}$ or $\frac{2}{1}$; They are all equal.

1. What do you notice about the points? **They lie on a straight line.**
2. Find the ratio $\frac{\text{change in perimeter}}{\text{change in tiles}}$ between the second and third points, the third and fourth points, and the fourth and fifth points. Each ratio is a rate of change. **Describe** what you observe.
3. **Complete:** As the number of tiles increases by 2 units, the perimeter of the models increases by __?__ units. **4**
4. Refer to the table at the right that appeared in Lesson 4-7. Find the ratio $\frac{\text{change in earnings}}{\text{change in hours worked}}$ for Greg and Monica. How do these values compare to the slopes that you found in Lesson 4-7? **$\frac{4}{1}$; $\frac{5}{1}$; They equal the slopes.**

Hours Worked	Earnings ($)	
	Greg	Monica
1	4	5
2	8	10
3	12	15
4	16	20

5. **Make a conjecture** about the relationship between rate of change and a graph of the two quantities. **Sample answer: If the rate of change is constant, then the graph is a straight line with slope equal to the rate of change.**

TEACHING TIP
Point out that a straight line represents a constant rate of change, or a linear function.

STUDY TIP
Look Back You can review **slope** in Lesson 4–7.

Answer

Step 3

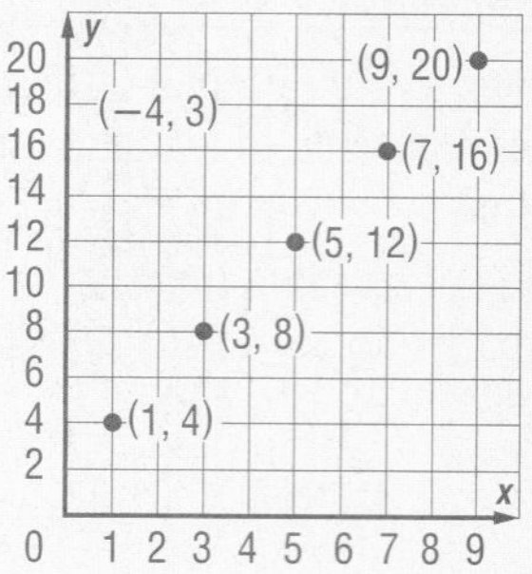

296 Chapter 7 Ratios and Proportions

7-3 Solving Proportions

What You'll Learn
Solve proportions.

NEW Vocabulary
proportion
cross product

WHEN am I ever going to use this?

NUTRITION As part of a healthy diet, teens need 60 milligrams of vitamin C each day.

Vitamin C (mg)	Amount of Cereal (c)
15	0.5
60	2

1. Write the ratio $\frac{\text{vitamin C}}{\text{amount of cereal}}$ for a half-cup serving of cereal. **$\frac{15 \text{ mg}}{0.5 \text{ c}}$**
2. Rewrite the ratio in Exercise 1 to find the unit rate in milligrams per cup. **30 mg/c**
3. Simplify $\frac{60 \text{ mg}}{2 \text{ c}}$ to find the unit rate. **$\frac{30 \text{ mg}}{1 \text{ c}}$**
4. **Make a conjecture** about the rates $\frac{15 \text{ mg}}{0.5 \text{ c}}$ and $\frac{60 \text{ mg}}{2 \text{ c}}$. **They are equivalent.**

When two ratios are equivalent, they form a **proportion**. Since rates are types of ratios, they can also form proportions.

Key Concept — **Proportion**

Words A proportion is an equation stating that two ratios are equivalent.

Symbols
Arithmetic: $\frac{1}{2} = \frac{3}{6}, \frac{8 \text{ ft}}{10 \text{ s}} = \frac{4 \text{ ft}}{5 \text{ s}}$
Algebra: $\frac{a}{b} = \frac{c}{d}$, where $b, d \neq 0$

In a proportion, a **cross product** is the product of the numerator of one ratio and the denominator of the other ratio.

$\frac{1}{2} = \frac{3}{6}$

1 × 6 is a cross product. 2 × 3 is a cross product.

$1 \times 6 = 2 \times 3$

$6 = 6$ ← The cross products of a proportion are equal.

EXAMPLE Identify a Proportion

1 **Determine whether $\frac{6}{10}$ and $\frac{3}{5}$ form a proportion.**

$\frac{6}{10} \stackrel{?}{=} \frac{3}{5}$ Write a proportion.

$6 \times 5 \stackrel{?}{=} 10 \times 3$ Find the cross products.

$30 = 30$ ✓ Multiply.

The cross products are equal, so the ratios form a proportion.

TEACHING TIP
$\frac{6}{10} = \frac{3}{5}$ is sometimes read *6 is to 10 as 3 is to 5.*

msmath2.net/extra_examples

7-3 Lesson Notes

Lesson Resources

Chapter 7 Resource Masters
- Study Guide and Intervention, p. 385
- Practice: Skills, p. 386
- Practice: Word Problems, p. 387
- Reading to Learn Mathematics, p. 388
- Enrichment, p. 389

Calculator and Spreadsheet Masters, p. 37
Key Teaching Transparencies, 26
MathPASS Tutorial CD-ROM, 29
Parent and Student Study Guide Workbook, 7-3
Science and Math Lab Masters, pp. 33–36

1 Focus

5-Minute Check 7-3

INTERACTIVE CHALKBOARD
PowerPoint® Presentations

Use as a quiz or review of Lesson 7-2. (Also available as a transparency)

Bellringer

Communication When students enter the classroom, ask them to write each phrase below as a fraction. Then have them write each fraction as a unit rate and make a conjecture about the relationship between the phrases.

(Continued in bottom margin.)

15 miles on a half-gallon of gas **$\frac{15}{0.5}$**

90 miles on 3 gallons of gas **$\frac{90}{3}$**

300 miles on 10 gallons of gas **$\frac{300}{10}$**

$\frac{30}{1}$; All phrases are equivalent to 30 miles per gallon.

Reading to Learn Mathematics, p. 388 **ELL**

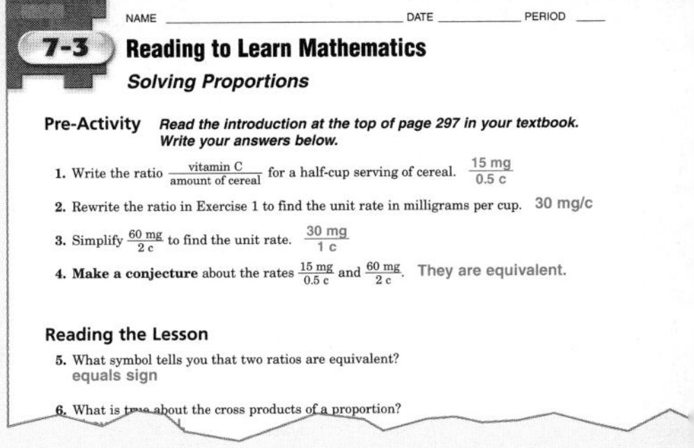
NAME ______ DATE ______ PERIOD ____

7-3 Reading to Learn Mathematics
Solving Proportions

Pre-Activity *Read the introduction at the top of page 297 in your textbook. Write your answers below.*

1. Write the ratio $\frac{\text{vitamin C}}{\text{amount of cereal}}$ for a half-cup serving of cereal. $\frac{15 \text{ mg}}{0.5 \text{ c}}$
2. Rewrite the ratio in Exercise 1 to find the unit rate in milligrams per cup. 30 mg/c
3. Simplify $\frac{60 \text{ mg}}{2 \text{ c}}$ to find the unit rate. $\frac{30 \text{ mg}}{1 \text{ c}}$
4. **Make a conjecture** about the rates $\frac{15 \text{ mg}}{0.5 \text{ c}}$ and $\frac{60 \text{ mg}}{2 \text{ c}}$. They are equivalent.

Reading the Lesson
5. What symbol tells you that two ratios are equivalent? equals sign
6. What is true about the cross products of a proportion?

2 Teach

In-Class Examples

INTERACTIVE CHALKBOARD
PowerPoint® Presentations

1 Determine whether $\frac{8}{12}$ and $\frac{2}{3}$ form a proportion. yes

2 Determine whether $\frac{\$5}{8 \text{ oz}}$ and $\frac{\$18}{32 \text{ oz}}$ form a proportion. no

3 Solve $\frac{3.5}{14} = \frac{6}{n}$. 24

4 **FLAGS** According to specifications, the ratio of the length of the U.S. flag to its width must be 1.9 to 1. How long must a U.S. flag be if it is designed to have a width of 2.5 feet? 4.75 ft

Tips for New Teachers

Teaching Strategies Often students will think a process is a math "trick" because they do not understand how the process works. For example, when determining the cross products of a proportion, students are really getting a common denominator and comparing numerators. To understand the full process, have students multiply the numerator and denominator of each ratio by the denominator of the other ratio in the proportion. They will then see that the denominators are equal and, if the pair of ratios form a proportion, the numerators are also equal.

EXAMPLE Identify a Proportion

2 **Determine whether $\frac{\$4}{9 \text{ oz}}$ and $\frac{\$21}{45 \text{ oz}}$ form a proportion.**

$\frac{4}{9} \stackrel{?}{=} \frac{21}{45}$ Write a proportion.

$4 \times 45 \stackrel{?}{=} 9 \times 21$ Find the cross products.

$180 \neq 189$ Multiply.

The cross products are not equal, so the ratios do not form a proportion.

Your Turn **Determine whether each pair of ratios forms a proportion.**

a. $\frac{4}{9}$ and $\frac{2}{3}$ no b. $\frac{15}{9}$ and $\frac{10}{6}$ yes c. $\frac{8 \text{ in.}}{21 \text{ s}}$ and $\frac{12 \text{ in.}}{31.5 \text{ s}}$ yes

You can use cross products to find a missing term in a proportion. This is known as *solving the proportion*. Solving a proportion is similar to solving an equation.

EXAMPLES Solve Proportions

3 **Solve $\frac{2.6}{13} = \frac{8}{n}$.**

$\frac{2.6}{13} = \frac{8}{n}$ Write the proportion.

$2.6 \cdot n = 13 \cdot 8$ Find the cross products.

$2.6n = 104$ Multiply.

$\frac{2.6n}{2.6} = \frac{104}{2.6}$ Divide each side by 2.6.

$n = 40$ Simplify.

The solution is 40.

REAL-LIFE CAREERS

How Does a Paleontologist Use Math?
In 2001, paleontologists discovered a 6-foot skull of a prehistoric crocodile called *SuperCroc*. By using proportions, they estimated its total length as 40 feet.

Source: *National Geographic*

Online Research
For information about a career as a paleontologist, visit: msmath2.net/careers

4 **LIFE SCIENCE The skull length of a crocodile is proportional to its body length. In one species, a 6-foot crocodile has a 2-foot skull. Estimate the length of a crocodile of that species that has a 3.5-foot skull.**

body length → / skull length → $\frac{6 \text{ ft}}{2 \text{ ft}} = \frac{c \text{ ft}}{3.5 \text{ ft}}$ Write a proportion.

$6 \cdot 3.5 = 2 \cdot c$ Find the cross products.

$21 = 2c$ Multiply.

$\frac{21}{2} = \frac{2c}{2}$ Divide each side by 2.

$10.5 = c$ Simplify.

So, a crocodile with a 3.5-foot skull is about 10.5 feet long.

Your Turn **Solve each proportion.**

d. $\frac{1}{5} = \frac{x}{60}$ 12 e. $\frac{10}{k} = \frac{2.5}{4}$ 16 f. $\frac{2 \text{ ft}}{6 \text{ min}} = \frac{5 \text{ ft}}{t \text{ min}}$ 15

DAILY INTERVENTION — Unlocking Misconceptions

Using Other Methods Remind students that they can solve some proportions without using cross products. Encourage them to try using techniques they already know for some problems, such as simplifying fractions or multiplying or dividing both the numerator and the denominator by the same number to find equivalent fractions.

298 Chapter 7 Ratios and Proportions

Skill and Concept Check

Writing Math
Exercises 1 & 3

1. **Explain** how you can determine whether two ratios are equivalent.
1–2. See margin.
2. **OPEN ENDED** Write a proportion that has $\frac{12}{15}$ as one of the ratios.
3. **Which One Doesn't Belong?** Identify the ratio that is not equivalent to the other three. Explain your reasoning. **$\frac{2}{9}$; The other ratios equal $\frac{2}{3}$.**

$\frac{2}{3}$ $\frac{6}{9}$ $\frac{4}{6}$ $\frac{2}{9}$

GUIDED PRACTICE

Determine whether each pair of ratios forms a proportion.

4. $\frac{2}{3}$ and $\frac{6}{9}$ **yes**
5. $\frac{16}{30}$ and $\frac{8}{10.5}$ **no**
6. $\frac{5 \text{ ft}}{7 \text{ s}}$ and $\frac{25 \text{ ft}}{49 \text{ s}}$ **no**

Solve each proportion.

7. $\frac{24}{6} = \frac{8}{x}$ **2**
8. $\frac{n}{5} = \frac{20}{25}$ **4**
9. $\frac{7}{t} = \frac{3}{9}$ **21**
10. $\frac{8}{12} = \frac{n}{3}$ **2**
11. $\frac{a}{6} = \frac{12}{8}$ **9**
12. $\frac{13}{25} = \frac{39}{r}$ **75**
13. $\frac{21}{5} = \frac{c}{7}$ **29.4**
14. $\frac{7}{x} = \frac{3.5}{9}$ **18**
15. **NUTRITION** An 8-ounce glass of orange juice contains 72 milligrams of vitamin C. How much juice contains 60 milligrams of vitamin C? **$6\frac{2}{3}$ or about 6.7 oz**

★ indicates increased difficulty

Practice and Applications

HOMEWORK HELP

For Exercises	See Examples
16–21	1, 2
22–46	3, 4

Extra Practice
See pages 580, 602.

Determine whether each pair of ratios forms a proportion.

16. $\frac{1}{6}$ and $\frac{4}{24}$ **yes**
17. $\frac{3}{9}$ and $\frac{2}{15}$ **no**
18. $\frac{27}{6}$ and $\frac{9}{2}$ **yes**
19. $\frac{1}{4}$ and $\frac{4}{1}$ **no**
20. $\frac{\$8}{20 \text{ lb}}$ and $\frac{\$24}{60 \text{ lb}}$ **yes**
21. $\frac{21 \text{ mi}}{6 \text{ h}}$ and $\frac{14.3 \text{ mi}}{3 \text{ h}}$ **no**
22. If there are 12 wheels, how many bicycles are there? **6 bicycles**
23. If there are 32 paws, how many kittens are there? **8 kittens**

Solve each proportion.

24. $\frac{5}{9} = \frac{10}{x}$ **18**
25. $\frac{2}{12} = \frac{a}{36}$ **6**
26. $\frac{16}{t} = \frac{2}{3}$ **24**
27. $\frac{n}{8} = \frac{3}{4}$ **6**
28. $\frac{c}{7} = \frac{18}{42}$ **3**
29. $\frac{30}{w} = \frac{8}{20}$ **75**
30. $\frac{9}{15} = \frac{b}{10}$ **6**
31. $\frac{3}{n} = \frac{27}{18}$ **2**
32. $\frac{x}{12} = \frac{12}{4}$ **36**
33. $\frac{45}{5} = \frac{t}{7}$ **63**
34. $\frac{6}{3} = \frac{5}{w}$ **2.5**
35. $\frac{y}{36} = \frac{15}{24}$ **22.5**
36. $\frac{350}{a} = \frac{2}{10}$ **1,750**
37. $\frac{380}{520} = \frac{760}{n}$ **1,040**
★38. $\frac{3}{5} = \frac{0.2}{d}$ **$\frac{1}{3}$**
★39. $\frac{2.5}{4.5} = \frac{7.5}{x}$ **13.5**
40. **MODELS** The toy car shown at the right is modeled after a real car. If the real car is 4.79 meters long, how wide is it? Write the width to the nearest hundredth of a meter. **2.05 m**

41. **TESTS** Suppose you received a score of 13 out of 15 on a quiz. Write a proportion to find the percent. Then solve. Round to the nearest percent. **$\frac{13}{15} = \frac{x}{100}$; 87%**

msmath2.net/self_check_quiz

3 Practice/Apply

FOLDABLES™ Follow-Up Remind students to select a term or key concept that they need to know for each lesson and write definitions in words and symbols under the tabs of their Foldables. As students progress through the chapter, encourage them to use their Foldables to self-check their understanding and mastery of these terms.

Which One Doesn't Belong?
In Exercise 3, encourage students to write proportions using various combinations of the given fractions. Students can also write each fraction in simplest form.

About the Exercises...

Guided Practice Key

Exercises	Examples
4–6	1, 2
7–14	3
15	4

Odd/Even Assignments
Exercises 16–43 are structured so that students practice the same concepts whether they are assigned odd or even problems.

Alert! The Data Update following Exercise 46 requires access to the Internet.

Answers

1. **Compare the cross products. If they are equal, the ratios are equivalent.**
2. **Sample answer: $\frac{12}{15} = \frac{4}{5}$**

Study Guide and Intervention, p. 385

NAME ______________________ DATE ____________ PERIOD _____

7-3 Study Guide and Intervention

Solving Proportions

A **proportion** is an equation stating that two ratios are equivalent. Since rates are types of ratios, they can also form proportions. In a proportion, a **cross product** is the product of the numerator of one ratio and the denominator of the other ratio.

EXAMPLE 1 **Determine whether $\frac{2}{3}$ and $\frac{10}{15}$ form a proportion.**

$\frac{2}{3} \stackrel{?}{=} \frac{10}{15}$	Write a proportion.
$2 \times 15 \stackrel{?}{=} 3 \times 10$	Find the cross products.
$30 = 30$ ✓	Multiply.

The cross products are equal, so the ratios form a proportion.

EXAMPLE 2 **Solve $\frac{8}{a} = \frac{10}{15}$.**

$\frac{8}{a} = \frac{10}{15}$	Write the proportion.
$8 \times 15 = a \times 10$	Find the cross products.
$120 = 10a$	Multiply.
$\frac{120}{10} = \frac{10a}{10}$	Divide each side by 10.
$12 = a$	Simplify.

The solution is 12.

EXERCISES

Determine whether each pair of ratios forms a proportion.

1. $\frac{8}{10}$ and $\frac{4}{5}$ yes
2. $\frac{9}{4}$ and $\frac{11}{6}$ no
3. $\frac{6}{14}$ and $\frac{9}{21}$ yes
4. $\frac{15}{12}$ and $\frac{9}{6}$ no
5. $\frac{\$2.48}{4 \text{ oz}}$ and $\frac{\$3.72}{6 \text{ oz}}$ yes
6. $\frac{125 \text{ mi}}{5.7 \text{ gal}}$ and $\frac{120 \text{ mi}}{5.6 \text{ gal}}$ no

Solve each proportion.

7. $\frac{y}{7} = \frac{16}{28}$ 4
8. $\frac{5}{15} = \frac{15}{w}$ 45
9. $\frac{20}{b} = \frac{70}{28}$ 8
10. $\frac{52}{8} = \frac{m}{9}$ 58.5

Practice: Skills, p. 386

NAME ______________________ DATE ____________ PERIOD _____

7-3 Practice: Skills

Solving Proportions

Determine whether each pair of ratios forms a proportion.

1. $\frac{9}{5}$ and $\frac{27}{15}$ yes
2. $\frac{16}{10}$ and $\frac{24}{15}$ yes
3. $\frac{6}{18}$ and $\frac{9}{25}$ no
4. $\frac{42}{63}$ and $\frac{28}{42}$ yes
5. $\frac{11}{8}$ and $\frac{13}{10}$ no
6. $\frac{22}{33}$ and $\frac{12}{18}$ yes
7. $\frac{14}{17}$ and $\frac{29}{35}$ no
8. $\frac{36}{22}$ and $\frac{30}{19}$ no
9. $\frac{32}{48}$ and $\frac{10}{15}$ yes
10. $\frac{320 \text{ mi}}{6 \text{ h}}$ and $\frac{420 \text{ mi}}{8 \text{ h}}$ no
11. $\frac{\$4.96}{8 \text{ oz}}$ and $\frac{\$3.72}{6 \text{ oz}}$ yes
12. $\frac{25 \text{ mg}}{1.5 \text{ c}}$ and $\frac{100 \text{ mg}}{6 \text{ c}}$ yes

Solve each proportion.

13. $\frac{24}{13} = \frac{a}{26}$ 48
14. $\frac{18}{x} = \frac{3}{36}$ 216
15. $\frac{3}{u} = \frac{5}{15}$ 9
16. $\frac{650}{6.5} = \frac{z}{5}$ 500
17. $\frac{2.8}{4} = \frac{7}{q}$ 10
18. $\frac{c}{17} = \frac{0.01}{8.5}$ 0.02
19. $\frac{0.1}{8.2} = \frac{1.8}{b}$ 147.6
20. $\frac{300}{24} = \frac{18}{j}$ 1.44
21. $\frac{4.2}{t} = \frac{8}{5}$ 2.625
22. $\frac{120}{75} = \frac{8}{m}$ 5

Practice: Word Problems, p. 387

NAME ______________________ DATE ____________ PERIOD _____

7-3 Practice: Word Problems

Solving Proportions

1. **COOKING** Theo wants to use a cookie recipe that makes 36 cookies but he wants to reduce the number of cookies to 24. If the recipe specifies using 2 cups of sugar, how much sugar should he use? $1\frac{1}{3}$ c
2. **MEDICINE** In order to determine her pulse rate, June's nurse counted 18 beats in her pulse in 15 seconds. At this rate, how many beats would she have in 60 seconds? 72 beats
3. **LABOR** Ed earned $112 for 8 hours of work. At this rate, how much will he earn for 40 hours of work? $560
4. **TRAVEL** Rita traveled 1,250 miles in the first 3 days of her trip. At this rate, how long will it take her to travel 1,875 miles? $4\frac{1}{2}$ days
5. **MODELS** An architect built a model of a 220-foot tall building he is designing. The model is 25 inches tall and 10 inches wide. How wide is the actual building? 88 ft
6. **TESTING** Mary is preparing for her college entrance exams. In a practice test, she answered 12 problems in 30 minutes. At this rate, how many questions can she expect to answer in 150 minutes? 60 questions

COOKING **For Exercises 42–44, refer to the table.**

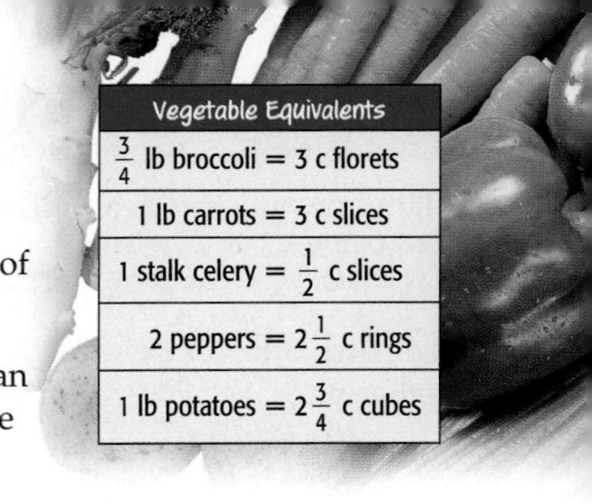

Vegetable Equivalents
$\frac{3}{4}$ lb broccoli = 3 c florets
1 lb carrots = 3 c slices
1 stalk celery = $\frac{1}{2}$ c slices
2 peppers = $2\frac{1}{2}$ c rings
1 lb potatoes = $2\frac{3}{4}$ c cubes

42. How many stalks of celery would you need to get $3\frac{1}{2}$ cups of celery slices? **7 stalks**

43. How many cups of broccoli florets would 1 pound of broccoli produce? **4 c**

44. **WRITE A PROBLEM** Write a real-life problem that can be solved using a proportion and one of the vegetable equivalents. Then write a proportion and solve. **See margin.**

45. **MULTI STEP** Amanda can word process 7 words in 6 seconds. At this rate, how many words can she word process in 3 minutes? **210 words**

★ 46. **MONEY** In 2002, France adopted the *Euro* as its currency, replacing the French franc. At that time, one Euro was worth $0.90 in American currency. In Paris, France, a quart of milk cost 3 Euros. In Los Angeles, a quart of milk cost $1.65. In which city was the quart of milk more expensive? **Paris**

Data Update What is the current U.S. value of one Euro? Visit msmath2.net/data_update to learn more.

47. **CRITICAL THINKING** In the proportion $\frac{3}{6} = \frac{6}{12}$, the number 6 appears in two of the diagonal positions. This repeated number is called the *geometric mean* of the other two numbers. Find a pair of numbers other than 3 and 12 for which 6 is the geometric mean. **Sample answers: 1 and 36; 2 and 18; 4 and 9**

Standardized Test Practice and Mixed Review

48. **MULTIPLE CHOICE** Ming is paid an hourly rate. One week he earned $157.50 by working 30 hours. If he works 35 hours the next week, how much will he earn? **C**

    Ⓐ $135.00 Ⓑ $160.50 Ⓒ $183.75 Ⓓ $210.25

49. **GRID IN** The ratio of boys to girls in the seventh grade is 7 to 8. If there are 120 students in the seventh grade, how many of them are girls? **64**

50. Choose the better unit price: 5 pounds of onions for $2.99 or 3 pounds of onions for $1.29. (Lesson 7-2) **3 lb for $1.29**

Write each ratio as a fraction in simplest form. (Lesson 7-1)

51. 45:81 $\frac{5}{9}$
52. 49 to 14 $\frac{7}{2}$
53. 14 ounces to 5 pounds $\frac{7}{40}$

MEASUREMENT **Complete.** (Lesson 6-7)

54. 5 qt = __?__ pt **10**
55. $3\frac{1}{2}$ lb = __?__ oz **56**
56. 28 c = __?__ qt **7**

GETTING READY FOR THE NEXT LESSON

PREREQUISITE SKILL **Multiply. Write in simplest form.** (Lesson 6-4)

57. $\frac{1}{5} \cdot 8$ $1\frac{3}{5}$
58. $1\frac{3}{4} \cdot 5$ $8\frac{3}{4}$
59. $16 \cdot 3\frac{1}{4}$ **52**
60. $24 \cdot 1\frac{3}{8}$ **33**

4 Assess

Open-Ended Assessment

Writing Have students write two pairs of ratios, one that forms a proportion and one that does not. Have them include an explanation of why the pair is or is not a proportion.

Answer

44. Sample answer: How many cups of sliced carrots can you get from a 2-pound bag? $\frac{1 \text{ lb}}{3 \text{ c}} = \frac{2 \text{ lb}}{x}$; Answer: 6 c

300 Chapter 7 Ratios and Proportions

7-3b HANDS-ON LAB

A Follow-Up of Lesson 7-3

/hat You'll Learn
se proportions to stimate.

laterials
- small bowl
- dried beans
- paper cup
- markers

Wildlife Sampling

INVESTIGATE *Work in groups of three.*

Naturalists can estimate the population in a wildlife preserve by using the *capture-recapture* technique. In this lab, you will model this technique using dried beans in a bowl to represent deer in a forest.

STEP 1 Fill a small bowl with dried beans.

STEP 2 Use the paper cup to scoop out some of the beans. These represent the original *captured* deer. Record the number in a table like the one shown at the right. Mark each bean with an × on both sides. Then return these beans to the bowl and mix well.

Original Number Captured ____			
Trial	Sample	Recaptured	P
A			
B			
C			
⋮			
J			
Total			

STEP 3 Scoop another cup of beans from the bowl and count them. This is the *sample* for Trial A. Count the beans with the ×'s. These are the *recaptured* deer. Record both numbers.

STEP 4 Use the proportion below to estimate the total number of beans in the bowl. This represents the total population P. Record the value of P in the table.

$$\frac{\text{captured}}{\text{total population } (P)} = \frac{\text{recaptured}}{\text{sample}}$$

STEP 5 Return all of the beans to the bowl.

STEP 6 Repeat Steps 3–5 nine times.

Writing Math

Work in groups of three.

1. Find the average of the estimates in column P. Is this a good estimate of the number of beans in the bowl? **Explain** your reasoning.
2. **Count** the actual number of beans in the bowl. How does this number compare to your estimate? **1–2. See students' work.**

7-3b HANDS-ON LAB Lesson Notes

Lesson Resources

Teaching Mathematics with Manipulatives
- pattern for cup, p. 7
- student recording sheet 7-3b

Glencoe Mathematics Classroom Manipulative Kit
- cups

GETTING STARTED

Teaching Tip Consider asking one group of students to demonstrate Steps 1–5 in front of the class before the rest of the groups begin the lab.

TEACH

- Have each group record the average of its estimates on the chalkboard, along with the actual number of beans in the bowl. Discuss why the estimates may vary.
- Walk around the room to make sure that each group is performing the task and recording its data properly.

ASSESS

Observe as students solve the **Writing Math** exercises. They should:

- find the average of a group of numbers, and
- compare their estimates to the actual value.

7-4a Lesson Notes

1 Focus

Bellringer
Problem Solving Copy the instructional diagram for a computer or DVD player and pass it out to the class. Why do instruction writers include diagrams with the written instructions? **Many people understand things they can see more easily than written instructions.** How could a diagram help you solve a problem? **It helps you think logically and visualize what is being asked.**

2 Teach

In-Class Example

Solve. Draw a diagram.

INTERACTIVE CHALKBOARD
PowerPoint® Presentations

ROCK CLIMBING
A rock climber stops to rest at a ledge 90 feet above the ground. If this represents 75% of the total climb, how high above the ground is the top of the rock? **120 ft**

Answers

1. **Sample answer: It is often easier to solve a problem by drawing a diagram than by writing an equation or solving mentally.**
2. **Sample answer:**

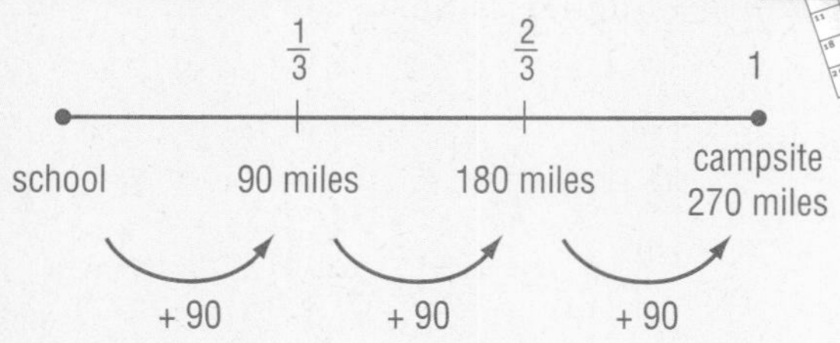

7-4a Problem-Solving Strategy

A Preview of Lesson 7-4

What You'll Learn
Solve problems by drawing a diagram.

Draw a Diagram

Mrs. Dixon said that we've gone about 90 miles, or $\frac{2}{3}$ of the way to the campsite.

So, how much farther do we have to go? I bet if we **draw a diagram**, we could figure it out.

Explore	We know that 90 miles is about $\frac{2}{3}$ of the total distance.
Plan	Let's draw a diagram showing the distance that we've already gone and the fractional part that it represents.
Solve	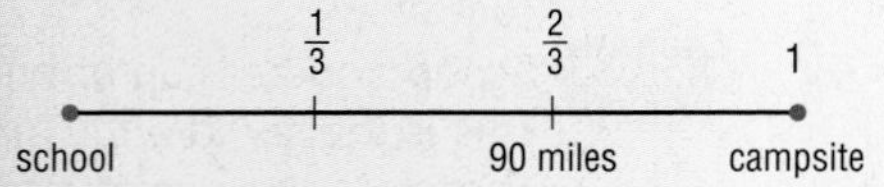 If $\frac{2}{3}$ of the distance is 90 miles, then $\frac{1}{3}$ of the distance would be 45 miles. So, the missing third must be another 45 miles. 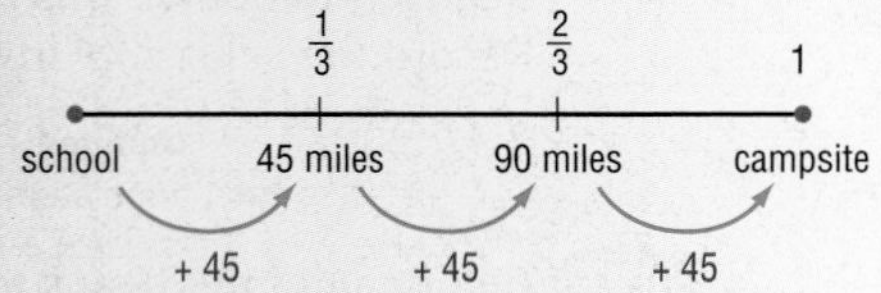 The total distance is 90 + 45 or 135 miles.
Examine	Since $\frac{2}{3}$ of the total distance equals 90 miles, the equation $\frac{2}{3}x = 90$ represents this problem. Solving, we get $x = 135$ miles. So, the solution checks.

Analyze the Strategy

1. **Explain** why drawing a diagram can be a useful problem-solving strategy. **See margin.**
2. **Determine** how far the trip would have been if 90 miles was only $\frac{1}{3}$ of the total distance. Draw a new diagram for this situation. **270 mi; See margin for diagram.**
3. **Write** a problem that could be solved by drawing a diagram. Exchange your problem with a classmate and solve. **See students' work.**

DAILY INTERVENTION **Differentiated Instruction**

Kinesthetic Have students work in pairs with number tiles to solve the following problem. *If tiles measuring 2 feet by 2 feet are used to cover a 16-foot by 12-foot floor, how many tiles are needed?* **48** Then have students draw a diagram for the problem based on the model they used.

pply the Strategy

lve. Use the draw a diagram strategy.

PHYSICAL SCIENCE A ball is dropped from 10 feet above the ground. It hits the ground and bounces up half as high as it fell. This is true for each successive bounce. What height does the ball reach on the fourth bounce? **$\frac{5}{8}$ ft**

5. **FAMILY** At Nate's family reunion, 80% of the people are 18 years of age or older. Half of the remaining people are under 12 years old. If 20 children are under 12 years old, how many people are at the reunion? **200 people**

ixed Problem Solving

lve. Use any strategy.

. **TESTS** The scores on a social studies test are found by adding or subtracting points as shown at the right.

Answer	Points
Correct	+8
Incorrect	−4
No answer	−2

If Mario's score on a 15-question test was 86 points, how many of his answers were correct, incorrect, and blank? **12 correct, 2 incorrect, 1 blank**

. **GAMES** Six members of a video game club are having a tournament. In the first round, every player will play a video game against every other player. How many games will be in the first round of the tournament? **15 games**

. **BUSINESS** The graph shows the annual spending by five industries.

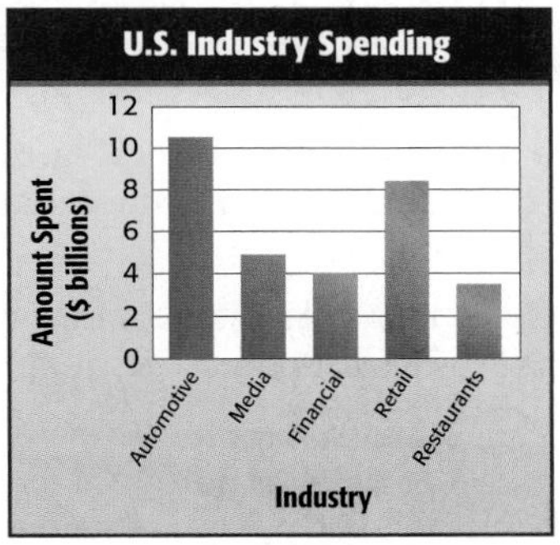

Source: *The Top Ten of Everything*

Estimate how much more the automotive industry spends per year than the retail industry. Write in scientific notation. **Sample answer: $\$2 \times 10^9$**

9. **ENVIRONMENT** What fraction of landfills is paper and plastic packaging? Write in simplest form. **$\frac{4}{25}$**

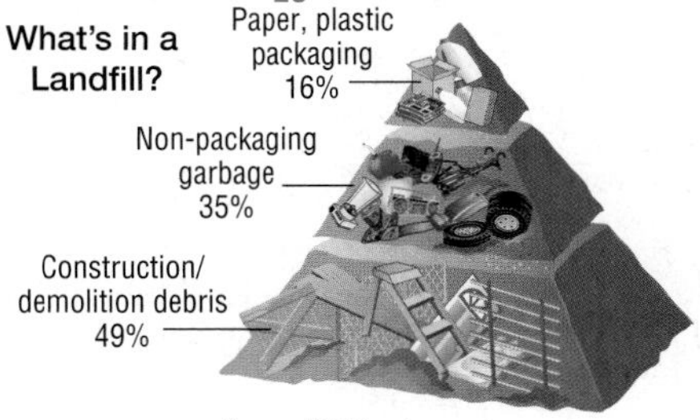

Source: *ULS Report*

10. **MONEY** Mr. Li has $240 in his checking account after writing checks for $15.70, $43.20, and $18. What was his balance before he wrote the three checks? **$316.90**

11. **STANDARDIZED TEST PRACTICE** Danielle is adding 3 feet to the length and width of her rectangular garden, as shown at the right. Which expression represents the area of land that will be added to the garden? **B**

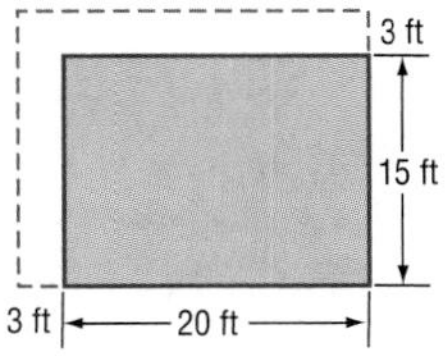

Ⓐ $(15 + 3)(20 + 3)$

Ⓑ $(15 + 3)(20 + 3) - (15)(20)$

Ⓒ $(15 - 3)(20 - 3)$

Ⓓ $(15 - 3)(20 - 3) - (15)(20)$

You will use the draw a diagram strategy in the next lesson.

3 Practice/Apply

About the Exercises...

Problem-Solving Strategies
Exercises 6–11 are structured so that students have the opportunity to practice many different problem-solving strategies. You may wish to remind them of some of the strategies they have studied.

- use a graph (p. 58)
- work backward (p. 164)
- make an organized list (p. 201)
- draw a diagram ✓

4 Assess

Open-Ended Assessment

Speaking Have students work in pairs to discuss methods for solving problems that ask for a total distance or amount when a certain portion of the total is given. One student should describe how to solve this type of problem when a fraction is given, and the other student should describe how it would be solved if a percent is given. Have them refer to a specific example or problem in the lesson to aid their explanations.

7-4 Lesson Notes

Lesson Resources

Chapter 7 Resource Masters

- Study Guide and Intervention, p. 390
- Practice: Skills, p. 391
- Practice: Word Problems, p. 392
- Reading to Learn Mathematics, p. 393
- Enrichment, p. 394

Parent and Student Study Guide Workbook, 7-4

Real-Life Career Masters, p. 21

Teaching Mathematics with Manipulatives, p. 72

1 Focus

5-Minute Check 7-4

INTERACTIVE CHALKBOARD PowerPoint® Presentations

Use as a quiz or review of Lesson 7-3. (Also available as a transparency)

Bellringer

Reading Mathematics The term *scale* is used in mathematics and in everyday life. Have students discuss the ways in which they have seen this term used. Relate these applications to its mathematical meaning. Look up *scale* in the dictionary. Write all of the definitions on the chalkboard. Discuss the different uses of the word.

Reading to Learn Mathematics, p. 393 ELL

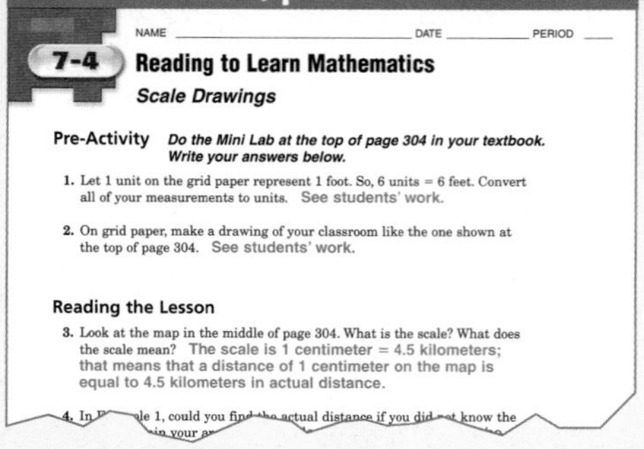
NAME ______________ DATE ________ PERIOD ____

7-4 Reading to Learn Mathematics

Scale Drawings

Pre-Activity *Do the Mini Lab at the top of page 304 in your textbook. Write your answers below.*

1. Let 1 unit on the grid paper represent 1 foot. So, 6 units = 6 feet. Convert all of your measurements to units. See students' work.
2. On grid paper, make a drawing of your classroom like the one shown at the top of page 304. See students' work.

Reading the Lesson

3. Look at the map in the middle of page 304. What is the scale? What does the scale mean? The scale is 1 centimeter = 4.5 kilometers; that means that a distance of 1 centimeter on the map is equal to 4.5 kilometers in actual distance.
4. In Example 1, could you find the actual distance if you did not know the ...

7-4 Scale Drawings

What You'll Learn
Solve problems involving scale drawings.

NEW Vocabulary
- scale drawing
- scale
- scale factor
- scale model

HANDS-ON Mini Lab

Materials
- measuring tape
- straightedge
- $\frac{1}{4}$-inch grid paper

Work with a partner.

- Measure the length of each wall, door, window, and chalkboard in your classroom.
- Record each length to the nearest $\frac{1}{2}$ foot.

1–2. See students' work.

1. Let 1 unit on the grid paper represent 1 foot. So, 6 units = 6 feet. Convert all of your measurements to units.
2. On grid paper, make a drawing of your classroom like the one shown at the right.

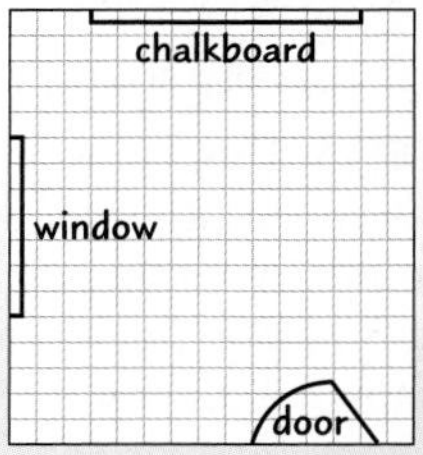

A **scale drawing** represents something that is too large or too small to be drawn at actual size.

The map is a scale drawing. Its **scale** gives the relationship between the distance on the map and the actual distance.

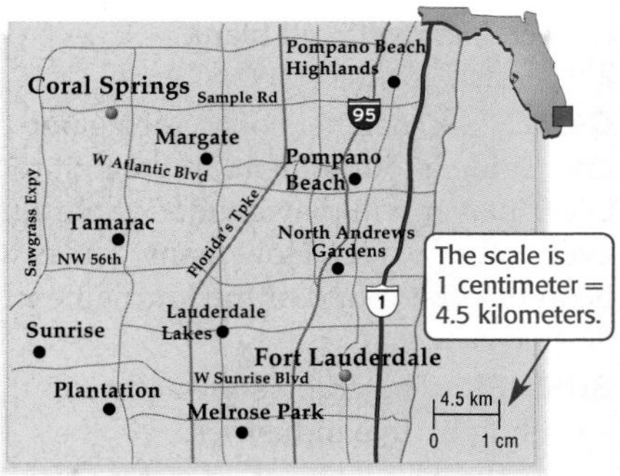

TEACHING TIP
When solving proportions, remind students to give the correct units in the answer. In Example 1, the answer is 18.5 kilometers, not 18.5 centimeters.

EXAMPLE Use a Scale Drawing

1 MAPS On the map above, the distance between Coral Springs and Fort Lauderdale is about 4.1 centimeters. What is the actual distance? Round to the nearest tenth.

Let d = the actual distance between the cities. Write and solve a proportion. Use the scale written as a fraction.

	Scale		Coral Springs to Ft. Lauderdale	
map →	1 centimeter	=	4.1 centimeters	← map
actual →	4.5 kilometers		d kilometers	← actual

$$\frac{1 \text{ centimeter}}{4.5 \text{ kilometers}} = \frac{4.1 \text{ centimeters}}{d \text{ kilometers}}$$

$1 \cdot d = 4.5 \cdot 4.1$ Cross products

$d = 18.45$ Simplify.

The distance between Coral Springs and Fort Lauderdale is about 18.5 kilometers.

DAILY INTERVENTION Differentiated Instruction

Visual/Spatial Have students make a scale drawing of a room in their home. The drawing should be an overhead view, including furniture that is rectangular in shape. Each student should come up with a scale factor that is reasonable for the task.

A blueprint is another example of a scale drawing.

EXAMPLE Read a Scale Drawing

2 POOLS On the blueprint of the pool, each square has a side length of $\frac{1}{4}$ inch. What is the actual width of the pool?

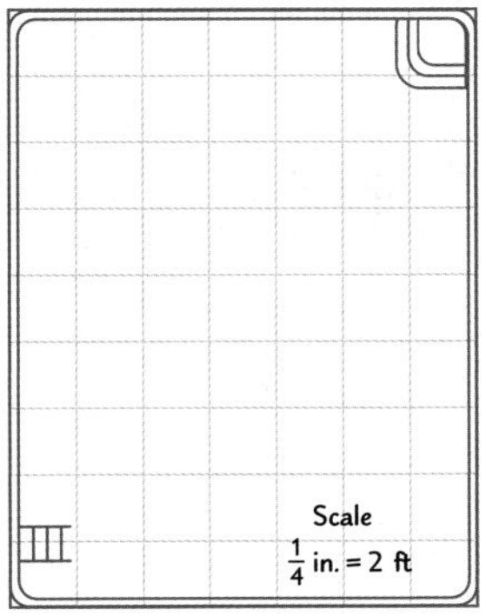

The pool on the blueprint is $1\frac{3}{4}$ inches wide. Let w = the actual width of the pool. Write and solve a proportion using the scale.

	Scale		Width of Pool	
blueprint → / actual →	$\frac{\frac{1}{4}\text{ inch}}{2\text{ feet}}$	=	$\frac{1\frac{3}{4}\text{ inches}}{w\text{ feet}}$	← blueprint / ← actual

$\frac{1}{4} \cdot w = 2 \cdot 1\frac{3}{4}$ Cross products

$\frac{1}{4}w = \frac{14}{4}$ Multiply.

$w = 14$ Simplify. Multiply each side by 4.

The width of the pool is 14 feet.

In Lesson 7-1, you simplified ratios by converting units. You can use the same method to simplify the scale in Example 2. A scale written as a ratio in simplest form is called the **scale factor**.

scale → $\frac{\frac{1}{4}\text{ inch}}{2\text{ feet}} = \frac{\frac{1}{4}\text{ inch}}{24\text{ inches}}$ Convert 2 feet to inches.

$= \frac{4}{4} \cdot \frac{\frac{1}{4}\text{ inch}}{24\text{ inches}}$ Multiply by $\frac{4}{4}$ to eliminate the fraction in the numerator.

$= \frac{1}{96}$ ← scale factor

TEACHING TIP
The scale factor at the right can be written as $\frac{1}{96}$ or 1:96. Scale factors do not have units.

EXAMPLE Find the Scale Factor

3 Find the scale factor of a blueprint if the scale is 1 inch = 8 feet.

Write the ratio of 1 inch to 8 feet in simplest form.

$\frac{1\text{ inch}}{8\text{ feet}} = \frac{1\text{ inch}}{96\text{ inches}}$ Convert 8 feet to inches.

$= \frac{1}{96}$ Cancel the units.

The scale factor is $\frac{1}{96}$. That is, each measure on the blueprint is $\frac{1}{96}$ the actual measure.

STUDY TIP

Equivalent Scales The scales below are equivalent because their scale factors are equal, $\frac{1}{96}$.
- 1 inch = 8 feet
- $\frac{1}{4}$ inch = 2 feet

Your Turn

a. On a scale drawing of a new classroom, the scale is 1 centimeter = 2.5 meters. What is the scale factor? $\frac{1}{250}$

msmath2.net/extra_examples

2 Teach

In-Class Examples

INTERACTIVE CHALKBOARD PowerPoint® Presentations

1 MAPS On the map below, the distance between Portland and Olympia is about $1\frac{7}{8}$ inches. What is the actual distance? **about 125 mi**

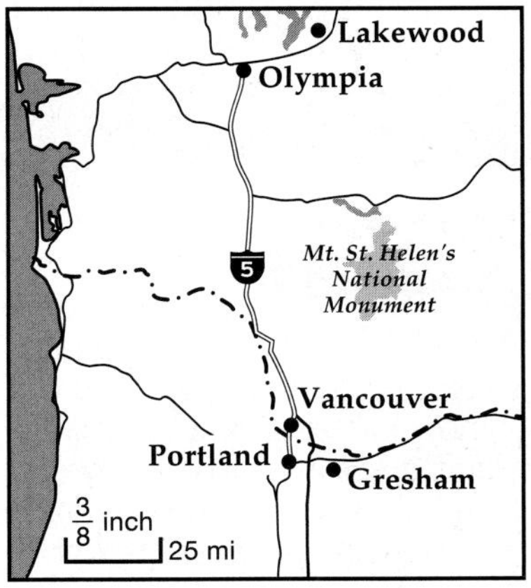

2 ARCHITECTURE On the blueprint of a new house, one-quarter inch represents $2\frac{1}{2}$ feet. If the length of a bedroom on the blueprint is $1\frac{1}{2}$ inches, what is the actual length of the room? **15 ft**

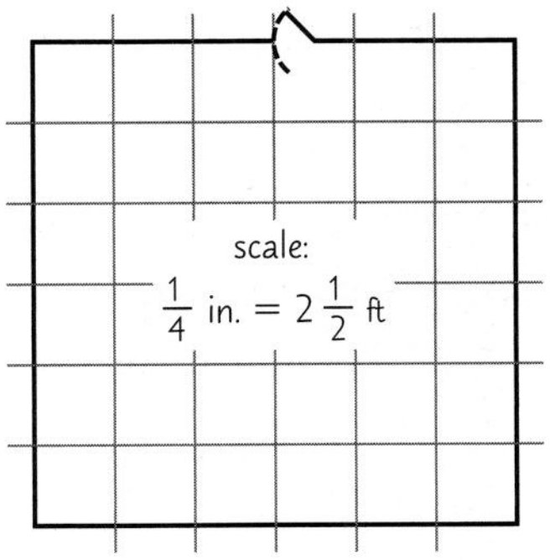

3 Find the scale factor of a blueprint if the scale is $\frac{1}{2}$ inch = 3 feet. $\frac{1}{72}$

Tips for New Teachers

Teaching Strategies You may want to periodically review previous topics by tying them in with the current lesson. For example, if students are creating scale drawings, you might have them find the area and perimeter of those drawings. Students learned about area and perimeter in Chapter 6.

In-Class Example

4 PHOTOGRAPHY INTERACTIVE CHALKBOARD PowerPoint® Presentations

A model is being created from a picture frame which has a length of $4\frac{3}{4}$ inches. If the scale to be used is 8 inches = 1 inch, what is the length of the model? **38 in.**

3 Practice/Apply

FOLDABLES™ Follow-Up Remind students to select a term or key concept that they need to know for each lesson and write definitions in words and symbols under the tabs of their Foldables. As students progress through the chapter, encourage them to use their Foldables to self-check their understanding and mastery of these terms.

NUMBER SENSE
In Exercise 2, encourage students to set up a proportion and to solve for the actual length of the racecar.

About the Exercises...

Guided Practice Key

Exercises	Examples
3–4	1
5–6	2
7	3
8	4

Odd/Even Assignments
Exercises 9–20 and 22–25 are structured so that students practice the same concepts whether they are assigned odd or even problems.

If you know the actual length of an object and the scale, you can build a scale model. A **scale model** can be used to represent something that is too large or too small for an actual-size model.

STUDY TIP

Scale The scale is the ratio of the drawing/model measure to the actual measure. It is not always the ratio of a smaller measure to a larger measure.

EXAMPLE Make a Scale Model

4 COMPUTERS Designers are creating a larger model of the computer memory board to use in design work. If they use a scale of 20 inches = 1 inch, what is the length of the model?

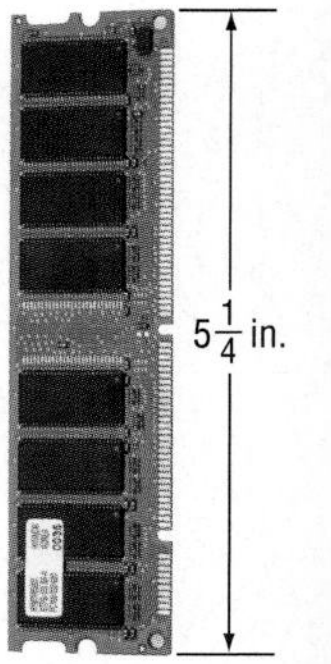

Write a proportion using the scale.

	Scale		Length of Model	
model →	20 inches	=	m inches	← model
actual →	1 inch		$5\frac{1}{4}$ inches	← actual

$$\frac{20 \text{ inches}}{1 \text{ inch}} = \frac{m \text{ inches}}{5\frac{1}{4} \text{ inches}}$$

$20 \cdot 5\frac{1}{4} = 1 \cdot m$ Find the cross products.

$105 = m$ Multiply.

The scale model is 105 inches long.

Skill and Concept Check

Writing Math Exercise 2

1. **OPEN ENDED** On grid paper, make a scale drawing of a room in your home. Include the scale that you used. **See students' work.**

2. **NUMBER SENSE** Roberto built a model racecar that is 3 inches long and has a scale of 0.5 inch = 2 feet. He determined that the actual racecar must be $3 \cdot 2$ or 6 feet long. Is this correct? Explain. **No; he assumed that 1 inch = 2 feet, which is not the correct scale.**

GUIDED PRACTICE

Find the actual distance between each pair of cities. Round to the nearest tenth if necessary.

	Cities	Map Distance	Scale	
3.	Kokomo, Indiana, and Chicago, Illinois	8 cm	1 cm = 25 km	**200 km**
4.	Raleigh and Charlotte, North Carolina	$3\frac{1}{4}$ in.	1 in. = 40 mi	**130 mi**

BLUEPRINTS For Exercises 5–7, use the blueprint at the right. Each square has a side length of 0.5 centimeter.

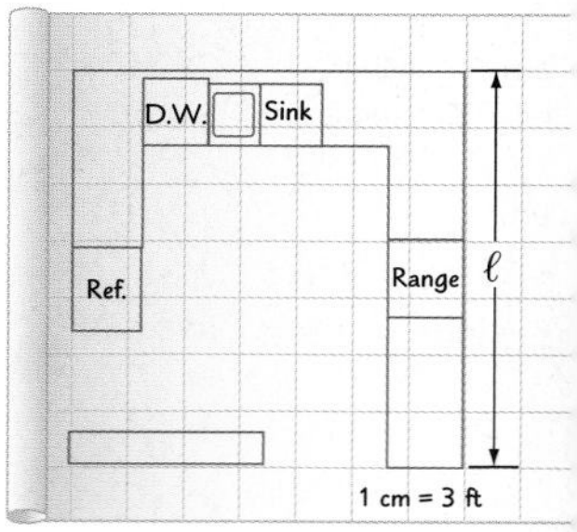

5. What is the length ℓ of the kitchen on the blueprint? **3.5 cm**
6. Find the actual length of the kitchen. **10.5 ft**
7. Find the scale factor. (*Hint*: 1 ft ≈ 30 cm) $\frac{1}{90}$
8. **BRIDGES** A bridge is 28 meters long. Find the length of a scale model if the scale is 1 centimeter = 5.5 meters. Round to the nearest tenth. **5.1 cm**

★ indicates increased difficulty

Practice and Applications

HOMEWORK HELP

For Exercises	See Examples
9–12, 23	1
17–20	2
13–16, 21, 25	3
22, 24, 26	4

Extra Practice
See pages 581, 602.

Find the actual distance between each pair of cities. Round to the nearest tenth if necessary.

	Cities	Map Distance	Scale	
9.	Virginia Beach, Virginia, and Washington, D.C.	8.5 cm	1 cm = 30 km	255 km
10.	Mobile and Huntsville, Alabama	$14\frac{1}{2}$ in.	1 in. = 20 mi	290 mi
11.	Knoxville, Tennessee, and Choctaw, Oklahoma	19.28 cm	2 cm = 125 km	1,205 km
12.	Quebec, Canada, and Paris, France	$4\frac{5}{8}$ in.	$\frac{1}{2}$ in. = 355 mi	3,283.8 mi

Suppose you are making a scale drawing. Find the length of each object on the scale drawing with the given scale. Then find the scale factor.

13. a back yard 120 feet deep; 1 inch : 20 feet 6 in.; $\frac{1}{240}$
14. an airplane 87 feet long; 2 inches = 15 feet $11\frac{3}{5}$ in.; $\frac{1}{90}$
15. an amusement park ride 36 meters high; 0.5 centimeter = 1.5 meters 12 cm; $\frac{1}{300}$
16. a surgical instrument $5\frac{7}{8}$ inches long; 1 inch : $\frac{1}{2}$ inch $11\frac{3}{4}$ in.; $\frac{2}{1}$

For Exercises 17–21, refer to the blueprint of the garden shown below. Each square has a side length of $\frac{1}{4}$ inch.

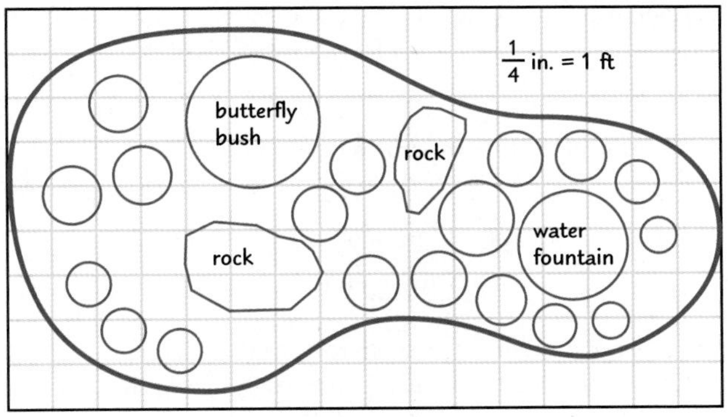

17. What is the length of the garden on the blueprint? 4 in.
18. Find the actual length of the garden. 16 ft
19. What is the diameter of the butterfly bush on the blueprint? $\frac{3}{4}$ in.
★ 20. What is the actual circumference of the butterfly bush? Use 3.14 for π and round to the nearest tenth. 9.4 ft
21. Find the scale factor. $\frac{1}{48}$
22. **BRIDGES** The Natchez Trace Bridge in Franklin, Tennessee, is 1,500 feet long. Suppose you build a model of the bridge using a scale of 1 inch : 25 feet. How long is the model? 60 in. or 5 ft

Lesson 7-4 Scale Drawings **307**

23. **GEOGRAPHY** A map of Brasilia has a scale of 1 inch to 5 miles. If the city is $2\frac{7}{16}$ inches across on the map, what is the distance across the actual city? Use estimation to check your answer. **$12\frac{3}{16}$ mi**

ART For Exercises 24 and 25, refer to the figure at the right and the information below.

Mount Rushmore is a sculpture that was carved using a model with a scale of 1 inch : 1 foot, as shown at the right.

24. How high is George Washington's face on the actual sculpture of Mount Rushmore? **60 ft**

25. What is the scale factor? **$\frac{1}{12}$**

★ 26. **MODEL** The diameter of an igloo is 10 feet. If you were to make a model of the igloo with sugar cubes, which might be an appropriate scale: 1 inch = 1 foot or 1 inch = 50 feet? Explain. **See margin.**

27. **CRITICAL THINKING** The distance between Atlanta and Savannah, Georgia, is 245 miles. The distance on a map is $3\frac{1}{2}$ inches. What is the scale of the map? **1 in. = 70 mi**

Standardized Test Practice and Mixed Review

28. **SHORT RESPONSE** A scale model of a new office building is shown at the right. If the scale is 5 inches = 18 feet, how tall is the actual building? **72 ft**

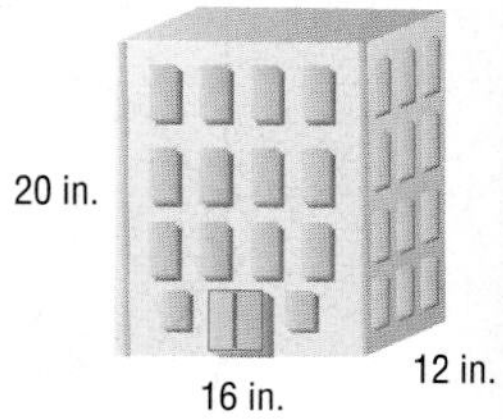

29. **MULTIPLE CHOICE** Which of the following scales does *not* have a scale factor of $\frac{1}{60}$? **D**

   Ⓐ 1 in. = 5 ft Ⓑ 1 cm = 60 cm

   Ⓒ 1 ft = 20 yd Ⓓ 1 in. = 60 ft

30. **BIRDS** A binocular has a *magnification factor* of 35 to 1; that is, objects that are actually 35 feet away appear to be only 1 foot away using the binoculars. If a goldfinch is 368 feet away, what will the distance appear to be in the binoculars? Round to the nearest tenth. (Lesson 7-3) **10.5 ft**

Find each unit rate. (Lesson 7-2)

31. 200 miles in 5 hours **40 mi/h**
32. 99¢ for 30 ounces **3.3¢ per oz**
33. $6.20 for 4 pounds **$1.55 per lb**
34. 150 meters in 12 seconds **12.5 m per s**

GETTING READY FOR THE NEXT LESSON

PREREQUISITE SKILL Divide. Write in simplest form. (Lesson 6-6)

35. $2\frac{3}{4} \div 10$ **$\frac{11}{40}$**
36. $4\frac{1}{3} \div 10$ **$\frac{13}{30}$**
37. $30\frac{2}{3} \div 100$ **$\frac{23}{75}$**
38. $87\frac{1}{2} \div 100$ **$\frac{7}{8}$**

4 Assess

Open-Ended Assessment

Writing Have students find the actual distance between any two points on a map. Then have them write a paragraph explaining how they calculated their answer.

Answer

26. 1 in. = 1 ft because it would make the model 10 inches in diameter; the second scale is more appropriate for modeling a much larger object. It would make the igloo model only $\frac{1}{5}$ inch in diameter.

Study Guide and Intervention, p. 390

NAME ______ DATE ______ PERIOD ____

7-4 Study Guide and Intervention

Scale Drawings

A **scale drawing** represents something that is too large or too small to be drawn at actual size. Similarly, a **scale model** can be used to represent something that is too large or too small for an actual-size model. The **scale** gives the relationship between the drawing/model measure and the actual measure.

EXAMPLE 1 **On this map, each grid unit represents 50 yards. Find the distance from Patrick's Point to Agate Beach.**

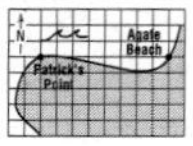

	Scale		Patrick's Point to Agate Beach		
map → / actual →	$\frac{1 \text{ unit}}{50 \text{ yards}}$	=	$\frac{8 \text{ units}}{x \text{ yards}}$	← map / ← actual	
	$1 \times x =$		50×8		Cross products
	$x =$		400		Simplify.

It is 400 yards from Patrick's Point to Agate Beach.

EXERCISES

Find the actual distance between each pair of cities. Round to the nearest tenth if necessary.

	Cities	Map Distance	Scale	Actual Distance
1.	Los Angeles and San Diego, California	6.35 cm	1 cm = 20 mi	127 mi
2.	Lexington and Louisville, Kentucky	15.6 cm	1 cm = 5 mi	78 mi
3.	Des Moines and Cedar Rapids, Iowa	16.27 cm	2 cm = 15 mi	122.0 mi
4.	Miami and Jacksonville, Florida	11.73 cm	$\frac{1}{2}$ cm = 20 mi	469.2 mi

Suppose you are making a scale drawing. Find the length of each object on the scale drawing with the given scale. Then find the scale factor.

5. an automobile 16 feet long; 1 inch:6 inches 32 in.; $\frac{1}{6}$
6. a lake 85 feet across; 1 inch = 4 feet $21\frac{1}{4}$ in.; $\frac{1}{48}$
7. a parking lot 200 meters wide; 1 centimeter:25 meters 8 cm; $\frac{1}{2,500}$
8. a flag 5 feet wide; 2 inches = 1 foot 10 in.; $\frac{1}{6}$

Practice: Skills, p. 391

NAME ______ DATE ______ PERIOD ____

7-4 Practice: Skills

Scale Drawings

ARCHITECTURE The scale on a set of architectural drawings for a house is $\frac{1}{2}$ inch = $1\frac{1}{2}$ feet. Find the length of each part of the house.

	Room	Drawing Length	Actual Length
1.	Living Room	5 inches	15 ft
2.	Dining Room	4 inches	12 ft
3.	Kitchen	$5\frac{1}{2}$ inches	$16\frac{1}{2}$ ft
4.	Laundry Room	$3\frac{1}{4}$ inches	$9\frac{3}{4}$ ft
5.	Basement	10 inches	30 ft
6.	Garage	$8\frac{1}{3}$ inches	25 ft

ARCHITECTURE As part of a city building refurbishment project, architects have constructed a scale model of several city buildings to present to the city commission for approval. The scale of the model is 1 inch = 9 feet.

7. The courthouse is the tallest building in the city. If it is $7\frac{1}{2}$ inches tall in the model, how tall is the actual building? $67\frac{1}{2}$ ft

8. The city commission would like to install new flagpoles that are each 45 feet tall. How tall are the flagpoles in the model? 5 in.

9. In the model, two of the flagpoles are 4 inches apart. How far apart will they be when they are installed? 36 ft

10. The model includes a new park in the center of the city. If the dimensions of the park in the model are 9 inches by 17 inches, what are the actual dimensions of the park? 81 ft by 153 ft

11. Find the scale factor. $\frac{1}{108}$

Practice: Word Problems, p. 392

NAME ______ DATE ______ PERIOD ____

7-4 Practice: Word Problems

Scale Drawings

1. **CARS** A scale drawing of an automobile has a scale of 1 inch = $\frac{1}{2}$ foot. The actual width of the car is 8 feet. What is the width on the scale drawing? 16 in.

2. **MODELS** A model ship is built to a scale of 1 centimeter:5 meters. The length of the model is 30 centimeters. What is the length of the actual ship? 150 m

3. **BUILDING** Jose wants to build a model of a 180-meter tall building. He will be using a scale of 1.5 centimeters = 3.5 meters. How tall will the model be? Round your answer to the nearest tenth. 77.1 cm

4. **TRAVEL** Susan is driving to Mount Shasta. On her map, she is a distance of $7\frac{3}{4}$ inches away. The scale of the map is $\frac{1}{2}$ inch = 50 miles. How far must Susan travel to reach her destination? 775 mi

5. **MAPS** A map of Levi's property is being made with a scale of 2 centimeters: 3 meters. What is the scale factor? $\frac{1}{150}$

6. **LANDSCAPING** A pond is being dug according to plans that have a scale of 1 inch = 6.5 feet. The maximum distance across the pond is 9.75 inches on the plans. What will be the actual maximum distance across the pond? 63.375 ft

7-4b Spreadsheet Investigation

A Follow-Up of Lesson 7-4

/hat You'll Learn
se a spreadsheet
calculate
easurements for
ale drawings.

Scale Drawings

A computer spreadsheet is a useful tool for calculating measures for scale drawings. You can change the scale factors and the dimensions, and the spreadsheet will automatically calculate the new values.

ACTIVITY

Suppose you want to make a scale drawing of your school. Set up a spreadsheet like the one shown below. In this spreadsheet, the actual measures are in feet, and the scale drawing measures are in inches.

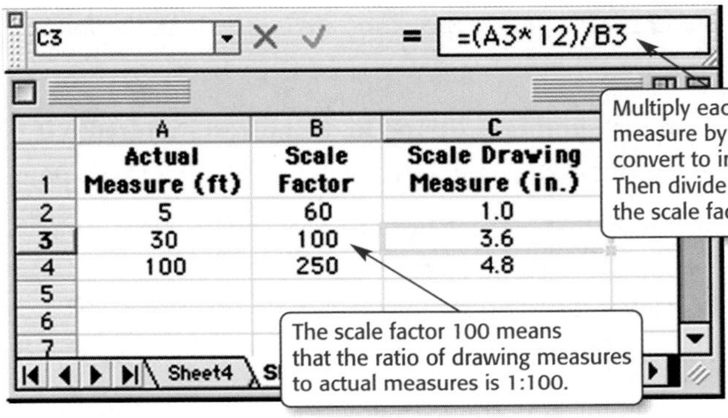

TEACHING TIP
'ou may want to use
lrawing software to help
tudents visualize the
cale drawings.

EXERCISES

1. The length of one side of the school building is 100 feet. If you use a scale factor of 1:250, what is the length on your scale drawing? **4.8 in.**
2. The length of a classroom is 30 feet. What is the scale factor if the length of the classroom on a scale drawing is 3.6 inches? **1:100**
3. Calculate the length of a 30-foot classroom on a scale drawing if the scale factor is 1:10. **36 in. or 3 ft**
4. Suppose the actual measures of your school are given in meters. Describe how you could use a spreadsheet to calculate the scale drawing measures in centimeters using a scale factor of 1:50.
5. Choose three rooms in your home and use a spreadsheet to make scale drawings. First, choose an appropriate scale and calculate the scale factor. Include a sketch of the furniture in each room. **See students' work.**

. Sample answer: Enter
0 in B2 and enter the
›rmula (A2*100)/B2 in
2. Then enter the actual
imensions in A2.

7-4b Spreadsheet Investigation

GETTING STARTED

Teaching Tip If you take your students to a computer lab, remind them to bring their textbooks with them so that they will have access to the data and the exercises.

TEACH

- Ask students to try out different scale factors to see how they affect their drawings.

ASSESS

Observe as students solve the **Exercises**. They should:

- correctly line up the titles for the columns and include the units of measure,
- change the spreadsheet to make the calculations using metric measures, and
- make a scale drawing of three rooms in their houses.

CHAPTER 7 Mid-Chapter Practice Test

Assessment Option

Mid-Chapter Test (Lessons 7-1 through 7-4) is available on p. 431 of the *Chapter 7 Resource Masters.*

ExamView® Pro Test Bank CD-ROM

Use the networkable ExamView® Pro Test Bank CD-ROM to:

- Create **Multiple versions** of tests.
- Create **Modified** tests for **Inclusion** students with one mouse click.
- **Edit** existing questions and **Add** your own questions.
- Build tests aligned with state standards using built-in **State Curriculum Correlations.**
- Change **English** tests to **Spanish** with one mouse click and vice versa.

FOLDABLES™ Follow-Up

Before students complete the Mid-Chapter Practice Test, encourage them to review the notes, definitions, concepts, and main ideas they recorded on the first four tabs of their Foldables.

Answers

1. A rate is a ratio that compares quantities with different units, such as $\frac{16\text{ ft}}{8\text{ s}}$. A unit rate is a rate that is simplified so that it has a denominator of 1 unit, such as $\frac{2\text{ ft}}{1\text{ s}}$ or 2 ft/s.
2. In a proportion, the cross product is the product of the numerator of one ratio and the denominator of the other ratio.

CHAPTER 7 Mid-Chapter Practice Test

Vocabulary and Concepts

1. **Explain** the difference between a rate and a unit rate and give an example of each. (Lesson 7-2) **1–2. See margin.**
2. **Define** cross product. (Lesson 7-3)

Skills and Applications

Write each ratio as a fraction in simplest form. (Lesson 7-1)

3. 18 to 12 $\frac{3}{2}$
4. 7 to 49 $\frac{1}{7}$
5. 20 meters : 44 meters $\frac{5}{11}$

Determine whether the ratios are equivalent. Explain. (Lesson 7-1)

6. $\frac{6}{9}$ and $\frac{2}{3}$ **yes; $\frac{6}{9} = \frac{2}{3}$**
7. 150 : 15 and 3 : 1 **no; 150 : 15 = 10 : 1**
8. 24 : 4 and 72 : 8 **no; 24 : 4 = 6 : 1 and 72 : 8 = 9 : 1**

Find each unit rate. Round to the nearest hundredth if necessary. (Lesson 7-2)

9. 200 miles in 5 hours **40 mph**
10. \$6.20 for 5 pounds **\$1.24 per lb**

11. **SHOPPING** Which box of cereal shown in the table costs the least per ounce? (Lesson 7-2) **18-oz box**

Weight (oz)	Cost ($)
12	2.50
18	3.69
24	4.95

12. Determine whether the ratios $\frac{5}{8}$ and $\frac{25}{40}$ form a proportion. (Lesson 7-3) **yes**

Solve each proportion. (Lesson 7-3)

13. $\frac{3}{d} = \frac{12}{20}$ **5**
14. $\frac{7}{8} = \frac{m}{48}$ **42**
15. $\frac{w}{8} = \frac{1}{3}$ **$2\frac{2}{3}$**

16. **TRUCKS** A truck is 26 feet long. Find the length on a scale model if 1 inch represents $5\frac{1}{2}$ feet. Round to the nearest tenth if necessary. (Lesson 7-4) **4.7 in.**

Suppose you are making a scale drawing. Find the length of each object on the scale drawing with the given scale. Then find the scale factor. (Lesson 7-4)

17. a bedroom 12 feet long; 1 in. = 3 ft **4 in.; $\frac{1}{36}$**
18. a park area 38 meters wide; 0.5 cm = 1 m **19 cm; $\frac{1}{200}$**

Standardized Test Practice

19. **GRID IN** A train travels 146 miles in 2 hours. At this rate, how many miles will it travel in 3.5 hours? (Lesson 7-3) **255.5**
20. **SHORT RESPONSE** On a scale drawing, a dog pen is $4\frac{1}{2}$ inches long. If 1 inch = 3 feet, what is the actual length of the dog pen? (Lesson 7-4) **13.5 ft**

Math Skill

Solving Proportions

The Game Zone

A Place To Practice Your Math Skills

Racing with Proportions

GET READY!

Players: two
Materials: centimeter grid paper, spinner, centimeter cubes

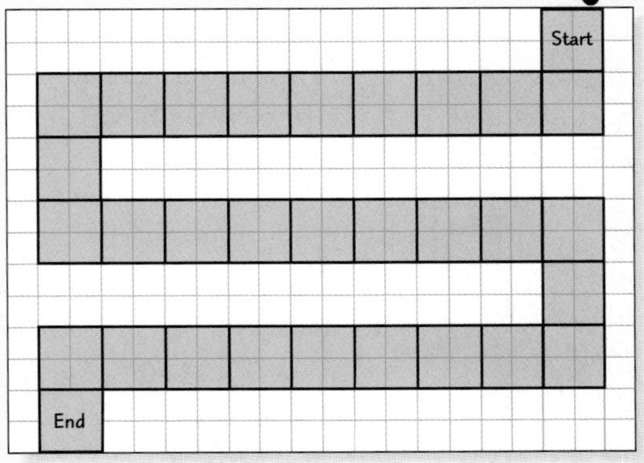

GET SET!

- Copy the game board at the right onto grid paper.
- Label equal sections of a spinner with 10, 12, 15, 20, 24, 30, 40, and 60.

GO!

- Each player places a cube on the *Start* square. (Note that both cubes will not fit on the same square. Place them side by side.)
- Each player spins the spinner and substitutes the number on the spinner for x in the proportion $\frac{y}{15} = \frac{8}{x}$. The player solves the proportion and moves his or her centimeter cube y spaces.
- Then each player spins the spinner and substitutes the number on the spinner for y in the given proportion. He or she solves the proportion and moves his or her centimeter cube x spaces.
- Continue to spin the spinner and solve the proportion for y and x.
- **Who Wins?** The first person to reach the *End* square wins the round. It is not necessary to land on *End* with an exact roll.

The Game Zone

Resources

For an easy-to-make manipulative for this game, visit: msmath2.net/gamezone.

Glencoe Mathematics Classroom Manipulative Kit

- spinners
- centimeter cubes

GETTING READY

Objective Solve proportions.

MANAGEMENT TIPS

Recommended Time
20 minutes

Tips

- Have the game boards and spinners made prior to the activity.
- Encourage students to assist each other and to check each other's work by solving all proportions.
- As an extension, after each player has a turn, provide a new proportion such as $\frac{y}{5} = \frac{24}{x}$ or $\frac{y}{3} = \frac{40}{x}$.
- You may wish to use the game as an assessment tool for the concepts the students have been learning in Lessons 7-1 through 7-4.

For an additional game for this chapter, visit: msmath2.net/gamezone.

7-5 Lesson Notes

Lesson Resources

Chapter 7 Resource Masters
- Study Guide and Intervention, p. 395
- Practice: Skills, p. 396
- Practice: Word Problems, p. 397
- Reading to Learn Mathematics, p. 398
- Enrichment, p. 399

Calculator and Spreadsheet Masters, p. 38

MathPASS Tutorial CD-ROM, 30

Parent and Student Study Guide Workbook, 7-5

FAST FILE

1 Focus

5-Minute Check 7-5

INTERACTIVE CHALKBOARD
PowerPoint® Presentations

Use as a quiz or review of Lesson 7-4. (Also available as a transparency)

Bellringer

Communication Write the following percents and ratios on the chalkboard. When students enter the classroom, have them write each percent. Then ask them to match the percent with the ratio they believe is equivalent. **1. b 2. c 3. a**

Percents	Ratios
1. 15%	**a.** $\frac{4}{5}$
2. 24%	**b.** $\frac{3}{20}$
3. 80%	**c.** $\frac{6}{25}$

Reading to Learn Mathematics, p. 398 ELL

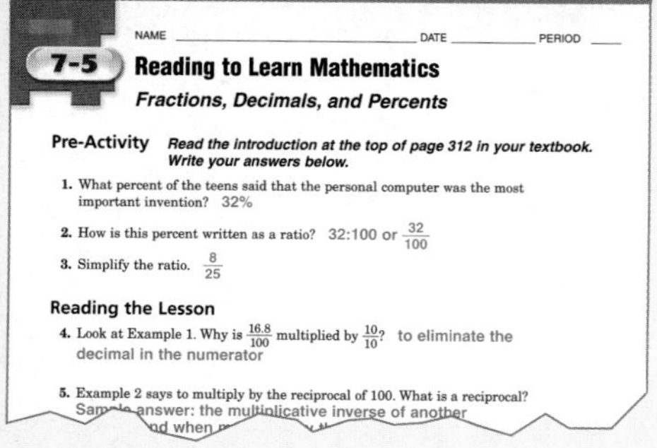
NAME ______ DATE ______ PERIOD ____

7-5 Reading to Learn Mathematics
Fractions, Decimals, and Percents

Pre-Activity *Read the introduction at the top of page 312 in your textbook. Write your answers below.*

1. What percent of the teens said that the personal computer was the most important invention? 32%
2. How is this percent written as a ratio? 32:100 or $\frac{32}{100}$
3. Simplify the ratio. $\frac{8}{25}$

Reading the Lesson

4. Look at Example 1. Why is $\frac{16.8}{100}$ multiplied by $\frac{10}{10}$? to eliminate the decimal in the numerator
5. Example 2 says to multiply by the reciprocal of 100. What is a reciprocal? Sample answer: the multiplicative inverse of another

7-5 Fractions, Decimals, and Percents

What You'll Learn
Write percents as fractions, and vice versa.

REVIEW Vocabulary
percent: a ratio that compares a number to 100 **(Lesson 5-5)**

WHEN am I ever going to use this?

SURVEYS The graph shows the results of a survey in which teens were asked to name the most important invention of the 20th century.

1. What percent of the teens said that the personal computer was the most important invention? **32%**
2. How is this percent written as a ratio? **32 : 100 or $\frac{32}{100}$**
3. Simplify the ratio. **$\frac{8}{25}$**

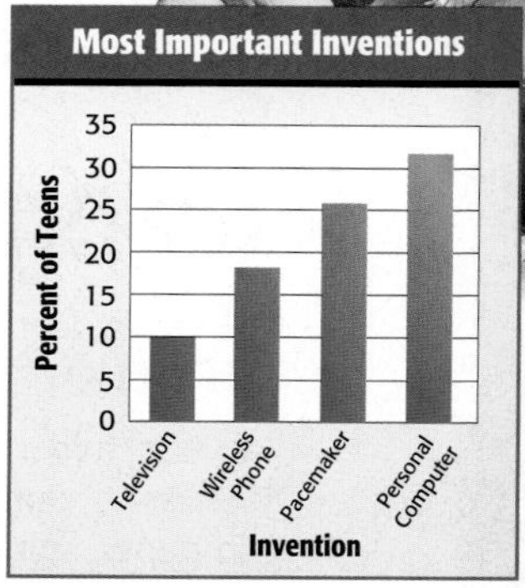

Source: Lemelson-MIT Program

In Lesson 5-5, you wrote percents like 32% as fractions by writing fractions with denominators of 100 and then simplifying. You can use the same method to write percents like 16.8% and $8\frac{1}{3}$% as fractions.

EXAMPLES Percents as Fractions

1 HOCKEY In men's college hockey, 16.8% of the players are from Ontario, Canada. What fraction is this? Write in simplest form.

$16.8\% = \frac{16.8}{100}$ Write a fraction with a denominator of 100.

$= \frac{16.8}{100} \cdot \frac{10}{10}$ Multiply by $\frac{10}{10}$ to eliminate the decimal in the numerator.

$= \frac{168}{1{,}000}$ or $\frac{21}{125}$ Simplify.

So, $\frac{21}{125}$ of the players are from Ontario, Canada.

2 Write $8\frac{1}{3}$% as a fraction in simplest form.

$8\frac{1}{3}\% = \frac{8\frac{1}{3}}{100}$ Write a fraction.

$= 8\frac{1}{3} \div 100$ Divide.

$= \frac{25}{3} \div 100$ Write $8\frac{1}{3}$ as an improper fraction.

$= \frac{25}{3} \times \frac{1}{100}$ Multiply by the reciprocal of 100, which is $\frac{1}{100}$.

$= \frac{25}{300}$ or $\frac{1}{12}$ Simplify.

To write a fraction like $\frac{8}{25}$ as a percent, multiply the numerator and the denominator by a number so that the denominator is 100. If the denominator is not a factor of 100, you can write fractions as percents by using a proportion.

EXAMPLES Fractions as Percents

3 TESTS On a math test, Mary got 7 questions correct out of 8. Find her grade as a percent.

To find her grade, write $\frac{7}{8}$ as a percent. **Estimate** $\frac{7}{8}$ is greater than $\frac{6}{8} = \frac{3}{4}$, or greater than 75%.

$\frac{7}{8} = \frac{n}{100}$ Write a proportion using $\frac{n}{100}$.

$700 = 8n$ Find the cross products.

$\frac{700}{8} = \frac{8n}{8}$ Divide each side by 8.

$87\frac{1}{2} = n$ Simplify.

So, $\frac{7}{8} = 87\frac{1}{2}\%$ or 87.5%.

This is greater than 75%, which was the estimate.

4 Write $\frac{4}{15}$ as a percent. Round to the nearest hundredth.

Estimate $\frac{4}{15}$ is about $\frac{4}{16}$, which equals $\frac{1}{4}$ or 25%.

$\frac{4}{15} = \frac{n}{100}$ Write a proportion using $\frac{n}{100}$.

$400 = 15n$ Find the cross products.

400 [÷] 15 [ENTER] **26.66666667** Use a calculator to simplify.

So, $\frac{4}{15}$ is about 26.67%. This result is close to the estimate.

Your Turn **Write each fraction as a percent. Round to the nearest hundredth if necessary.**

a. $\frac{2}{15}$ **13.33%** b. $\frac{7}{16}$ **43.75%** c. $\frac{17}{24}$ **70.83%**

STUDY TIP

Choose the Method To write a fraction as a percent,
- use *multiplication* when a fraction has a denominator that is a factor of 100,
- use a *proportion* for any type of fraction.

In this lesson, you have written percents as fractions and fractions as percents. In Chapter 5, you wrote percents and fractions as decimals. You can also write a fraction as a percent by first writing the fraction as a decimal and then writing the decimal as a percent.

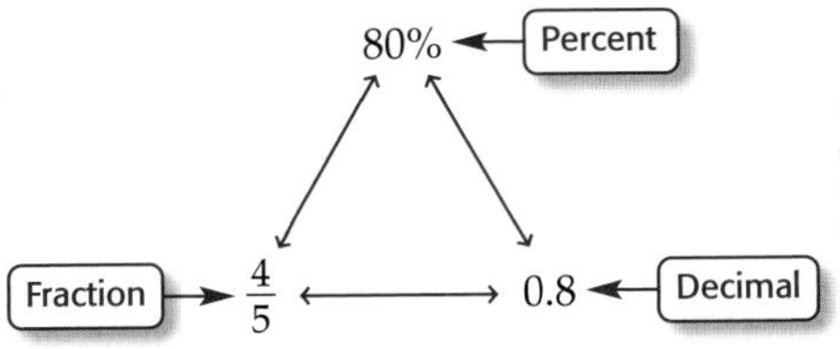

Percents, fractions, and decimals are all different names that represent the same number.

2 Teach

In-Class Examples

INTERACTIVE CHALKBOARD PowerPoint® Presentations

1 NUTRITION In a recent consumer poll, 41.8% of the people surveyed said they gained nutrition knowledge from family and friends. What fraction is this? Write in simplest form. $\frac{209}{500}$

2 Write $12\frac{1}{2}\%$ as a fraction in simplest form. $\frac{1}{8}$

3 PRODUCE In one shipment of fruit to a grocery store, 5 out of 8 bananas were still green. Find this amount as a percent. **62.5%**

4 Write $\frac{5}{12}$ as a percent. Round to the nearest hundredth. **41.67%**

Tips for New Teachers

Teaching Strategies Encourage students to make concept maps, flowcharts, or other graphic organizers to help them remember procedures. For example, they could create a flowchart containing the words *fraction, decimal,* and *percent* as main entries. To help them remember how to convert from one form to another, students can write the rules for changing those forms on the arrows connecting the main entries.

DAILY INTERVENTION **Unlocking Misconceptions**

Repeating Decimals Make sure that students realize that a number such as $\frac{2}{3}$, which is represented as 0.6666667 on a calculator screen, is a repeating decimal that has been rounded by the calculator because of its space limitations.

In-Class Examples

INTERACTIVE CHALKBOARD
PowerPoint® Presentations

5 Write $\frac{9}{20}$ as a percent. **45%**

6 Write $\frac{3}{7}$ as a percent. Round to the nearest hundredth. **42.86%**

3 Practice/Apply

FOLDABLES™ Follow-Up Remind students to select a term or key concept that they need to know for each lesson and write definitions in words and symbols under the tabs of their Foldables. As students progress through the chapter, encourage them to use their Foldables to self-check their understanding and mastery of these terms.

About the Exercises...

Guided Practice Key

Exercises	Examples
4–7	1, 2
8–12	3–6

Odd/Even Assignments
Exercises 13–36 are structured so that students practice the same concepts whether they are assigned odd or even problems.

Answer

1. Sample answer: Write a proportion using the given fraction and $\frac{n}{100}$. Then solve for n and add a %. Or, write the fraction as a decimal. Then multiply by 100 and add a %.

STUDY TIP

Look Back You can review **writing fractions as decimals** in Lesson 5-4.

TEACHING TIP

In this lesson, the percents are less than 100% because the fractions are between 0 and 1. Point out that it is possible to have percents greater than 100%, as students will learn in the next lesson.

EXAMPLES Fractions as Percents

5 **Write $\frac{1}{8}$ as a percent.**

$\frac{1}{8} = 0.125$ Write $\frac{1}{8}$ as a decimal.

$= 12.5\%$ Multiply by 100 and add the %.

6 **Write $\frac{5}{6}$ as a percent. Round to the nearest hundredth.**

$\frac{5}{6} = 0.833333333\ldots$ Write $\frac{5}{6}$ as a decimal.

$\approx 83.33\%$ Multiply by 100 and add the %.

Your Turn **Write each fraction as a percent. Round to the nearest hundredth if necessary.**

d. $\frac{5}{16}$ **31.25%** e. $\frac{7}{12}$ **58.33%** f. $\frac{2}{9}$ **22.22%**

Some fractions whose denominators are not factors of 100 are used often in everyday situations. It is helpful to be familiar with these fractions and their equivalent decimals and percents.

Key Concept **Common Fraction/Decimal/Percent Equivalents**

Fraction	Decimal	Percent	Fraction	Decimal	Percent
$\frac{1}{3}$	$0.\overline{3}$	$33\frac{1}{3}\%$	$\frac{3}{8}$	0.375	$37\frac{1}{2}\%$
$\frac{2}{3}$	$0.\overline{6}$	$66\frac{2}{3}\%$	$\frac{5}{8}$	0.625	$62\frac{1}{2}\%$
$\frac{1}{8}$	0.125	$12\frac{1}{2}\%$	$\frac{7}{8}$	0.875	$87\frac{1}{2}\%$

Skill and Concept Check

Writing Math
Exercises 1 & 3

1. **Describe** two methods for writing a fraction as a percent. **See margin.**
2. **OPEN ENDED** Write a percent that is between 15.2% and 15.8%. Then write it as a fraction in simplest form. **Sample answer: 15.6% = $\frac{39}{250}$**
3. **Explain** why percents are used in statistical graphs rather than fractions. **Sample answer: It is easier to compare percents than it is to compare fractions.**

GUIDED PRACTICE

Write each percent as a fraction in simplest form.

4. 13.5% $\frac{27}{200}$ 5. 18.75% $\frac{3}{16}$ 6. $66\frac{2}{3}\%$ $\frac{2}{3}$ 7. $57\frac{1}{2}\%$ $\frac{23}{40}$

Write each fraction as a percent. Round to the nearest hundredth if necessary.

8. $\frac{4}{25}$ **16%** 9. $\frac{5}{8}$ **62.5%** 10. $\frac{9}{40}$ **22.5%** 11. $\frac{4}{11}$ **36.36%**

12. **LANGUAGES** In Virginia, 1 person out of every 40 speaks an Asian language at home. What percent is this? **2.5%**

Practice and Applications

HOMEWORK HELP	
For Exercises	See Examples
13–24, 38	1, 2
25–37, 39	3–6

Extra Practice
See pages 581, 602.

Write each percent as a fraction in simplest form.

13. 17.5% $\frac{7}{40}$ 14. 62.5% $\frac{5}{8}$ 15. 6.2% $\frac{31}{500}$ 16. 34.5% $\frac{69}{200}$

17. 28.75% $\frac{23}{80}$ 18. 56.25% $\frac{9}{16}$ 19. $33\frac{1}{3}\%$ $\frac{1}{3}$ 20. $16\frac{2}{3}\%$ $\frac{1}{6}$

21. $81\frac{1}{4}\%$ $\frac{13}{16}$ 22. $93\frac{3}{4}\%$ $\frac{15}{16}$ 23. $78\frac{3}{4}\%$ $\frac{63}{80}$ 24. $23\frac{1}{3}\%$ $\frac{7}{30}$

Write each fraction as a percent. Round to the nearest hundredth if necessary.

25. $\frac{11}{20}$ 55% 26. $\frac{18}{25}$ 72% 27. $\frac{3}{8}$ 37.5% 28. $\frac{21}{40}$ 52.5% 29. $\frac{29}{30}$ 96.67% 30. $\frac{8}{9}$ 88.89%

31. $\frac{5}{7}$ 71.43% 32. $\frac{1}{16}$ 6.25% 33. $\frac{1}{80}$ 1.25% 34. $\frac{57}{200}$ 28.5% 35. $\frac{5}{12}$ 41.67% 36. $\frac{7}{15}$ 46.67%

37. What percent is equivalent to $\frac{60}{125}$? 48%

CARS For Exercises 38 and 39, use the table at the right. It shows the percent of people who keep the listed items in their car.

38. What fraction of people keep a first-aid kit in their car? $\frac{191}{500}$

39. Approximately 26 out of 125 people surveyed keep a hairbrush in their car. Is this greater or less than the percent who keep sports equipment? Explain. less than; $\frac{26}{125} = 20.8\%$

40. **CRITICAL THINKING** For what value of x is $\frac{1}{x} = x\%$ a true sentence? 10

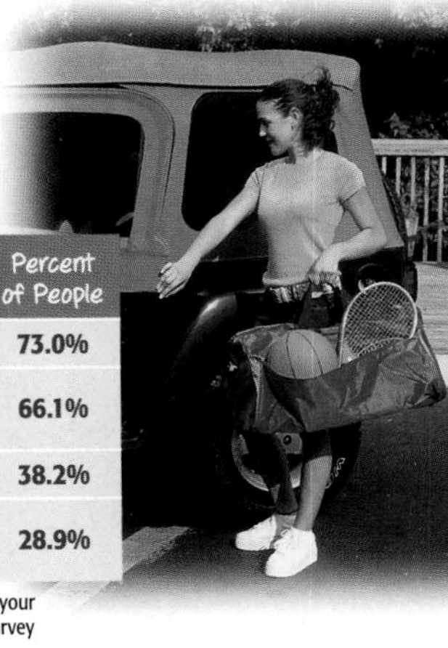

Items in Car	Percent of People
pen/pencil	73.0%
cassette tapes/CDs	66.1%
first-aid kit	38.2%
sports equipment	28.9%

Source: Penzoil "How's your Automotive Attitude?" survey

Standardized Test Practice and Mixed Review

41. **MULTIPLE CHOICE** Two-thirds of Kodiak Island's 2 million acres has been set aside as a wildlife preserve for the Kodiak brown bear. What percent of the island is a preserve? D

(A) $2\frac{2}{3}\%$ (B) 23% (C) $33\frac{1}{3}\%$ (D) $66\frac{2}{3}\%$

42. **SHORT RESPONSE** What is $\frac{7}{16}$ written as a percent? 43.75%

43. **MODELS** On a scale model of a building, 2 inches represents 15 feet. If the model is 9 inches tall, how tall is the actual building? (Lesson 7-4) 67.5 ft

Solve each proportion. (Lesson 7-3)

44. $\frac{6}{9} = \frac{4}{x}$ 6 45. $\frac{3}{t} = \frac{9}{5}$ $1\frac{2}{3}$ 46. $\frac{n}{5} = \frac{2.8}{7}$ 2 47. $\frac{7}{12} = \frac{r}{30}$ 17.5

GETTING READY FOR THE NEXT LESSON

PREREQUISITE SKILL Write each decimal as a fraction with a denominator of 100. (Lesson 5-4)

48. 0.23 $\frac{23}{100}$ 49. 0.09 $\frac{9}{100}$ 50. 0.368 $\frac{36.8}{100}$ 51. 0.425 $\frac{42.5}{100}$

msmath2.net/self_check_quiz

4 Assess

Higher-Level Thinking

Have students compare the use of percents and the use of fractions in advertising. How is the choice of form important for the ad? What things might the writers of the ad think about when using percents or fractions?

Ad writers want consumers to buy their products. They will use whatever mathematical form makes their product seem like the best bargain for the money, thereby making the product more attractive to consumers.

Study Guide and Intervention, p. 395

NAME ______ DATE ______ PERIOD ____

7-5 Study Guide and Intervention

Fractions, Decimals, and Percents

EXAMPLE 1 Write $4\frac{3}{8}\%$ as a fraction in simplest form.

$4\frac{3}{8}\% = \frac{4\frac{3}{8}}{100}$ Write a fraction.

$= 4\frac{3}{8} \div 100$ Divide.

$= \frac{35}{8} \div 100$ Write $4\frac{3}{8}$ as an improper fraction.

$= \frac{35}{8} \times \frac{1}{100}$ Multiply by the reciprocal of 100, which is $\frac{1}{100}$.

$= \frac{35}{800}$ or $\frac{7}{160}$ Simplify.

EXAMPLE 2 Write $\frac{5}{16}$ as a percent.

$\frac{5}{16} = \frac{n}{100}$ Write a proportion using $\frac{n}{100}$.

$500 = 16n$ Find the cross products.

$\frac{500}{16} = \frac{16n}{16}$ Divide each side by 16.

$31\frac{1}{4} = n$ Simplify.

So, $\frac{5}{16} = 31\frac{1}{4}\%$ or 31.25%.

EXERCISES

Write each percent as a fraction in simplest form.

1. 60% $\frac{3}{5}$ 2. $68\frac{3}{4}\%$ $\frac{11}{16}$

3. $27\frac{1}{2}\%$ $\frac{11}{40}$ 4. 37.5% $\frac{3}{8}$

Write each fraction as a percent. Round to the nearest hundredth if necessary.

5. $\frac{2}{5}$ 40% 6. $\frac{5}{8}$ 62.5%

7. $\frac{9}{16}$ 56.25% 8. $\frac{2}{3}$ 66.67%

Practice: Skills, p. 396

NAME ______ DATE ______ PERIOD ____

7-5 Practice: Skills

Fractions, Decimals, and Percents

Write each percent as a fraction in simplest form.

1. 18% $\frac{9}{50}$ 2. 67.5% $\frac{27}{40}$ 3. 21.25% $\frac{17}{80}$

4. 87.5% $\frac{7}{8}$ 5. $31\frac{1}{4}\%$ $\frac{5}{16}$ 6. 17.5% $\frac{7}{40}$

7. $18\frac{3}{4}\%$ $\frac{3}{16}$ 8. $68\frac{3}{4}\%$ $\frac{11}{16}$ 9. 7.5% $\frac{3}{40}$

10. 12.5% $\frac{1}{8}$ 11. 36.75% $\frac{147}{400}$ 12. $5\frac{1}{2}\%$ $\frac{11}{200}$

Write each fraction as a percent. Round to the nearest hundredth if necessary.

13. $\frac{3}{5}$ 60% 14. $\frac{3}{8}$ 37.5% 15. $\frac{2}{18}$ 11.11%

16. $\frac{3}{16}$ 18.75% 17. $\frac{7}{9}$ 77.78% 18. $\frac{21}{50}$ 42%

19. $\frac{1}{3}$ 33.33% 20. $\frac{40}{42}$ 95.24% 21. $\frac{7}{16}$ 43.75%

22. $\frac{64}{125}$ 51.2% 23. $\frac{11}{12}$ 91.67% 24. $\frac{11}{15}$ 73.33%

Practice: Word Problems, p. 397

NAME ______ DATE ______ PERIOD ____

7-5 Practice: Word Problems

Fractions, Decimals and Percents

INTERNET For Exercises 1–4, use the table. It shows the percents of online shopping purchases made by all Internet users and the percents made by Internet users over age 55.

Most Popular Online Purchases		
	Internet Users Over 55	All Internet Users
computer software	43%	19%
books	43%	21%
computer hardware	24%	13%
music CDs	29%	22%
clothing	19%	8%

1. What fraction of Internet users over 55 bought clothing online? $\frac{19}{100}$

2. What fraction of all Internet users bought clothing online? $\frac{2}{25}$

3. What fraction of all Internet users bought music CDs online? $\frac{11}{50}$

4. Is the fraction of Internet users over 55 who bought books online greater or less than $\frac{22}{50}$? Explain. Less than; $\frac{22}{50} = 44\%$, and 43% is less than 44%.

5. **FOOTBALL** In 2001 San Francisco quarterback Jeff Garcia completed 316 out of 504 passes. What was his pass completion percentage to the nearest tenth? 62.7%

6. **COMPUTERS** In Joan's math class, there are 20 computers and 32 students. What percent of students will be able to use a computer without sharing? 62.5%

7. **VEHICLES** In the town of Orick, 5 out of 13 vehicles are trucks. What percent of the vehicles are trucks? Round to the nearest tenth. 38.5%

8. **DENTISTRY** Dana has fillings in 4 of her 32 teeth. What percent of her teeth have fillings? 12.5%

7-6 Lesson Notes

Lesson Resources

Chapter 7 Resource Masters

FAST FILE

- Study Guide and Intervention, p. 400
- Practice: Skills, p. 401
- Practice: Word Problems, p. 402
- Reading to Learn Mathematics, p. 403
- Enrichment, p. 404

Classroom Math Games, pp. 19–21
Key Teaching Transparencies, 27
Parent and Student Study Guide Workbook, 7-6

1 Focus

5-Minute Check 7-6

INTERACTIVE CHALKBOARD PowerPoint® Presentations

Use as a quiz or review of Lesson 7-5. (Also available as a transparency)

Bellringer
Hands-On Activity Have students research to find one application of a percent less than 1% and one of a percent greater than 100%. Discuss what such percents mean.

Answers (Hands-On Mini Lab)

3a.

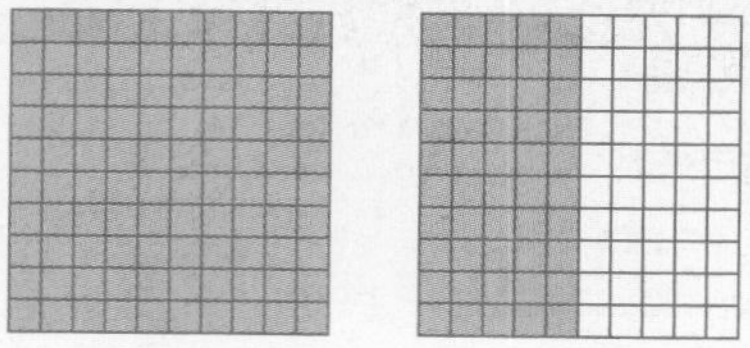

Reading to Learn Mathematics, p. 403 ELL

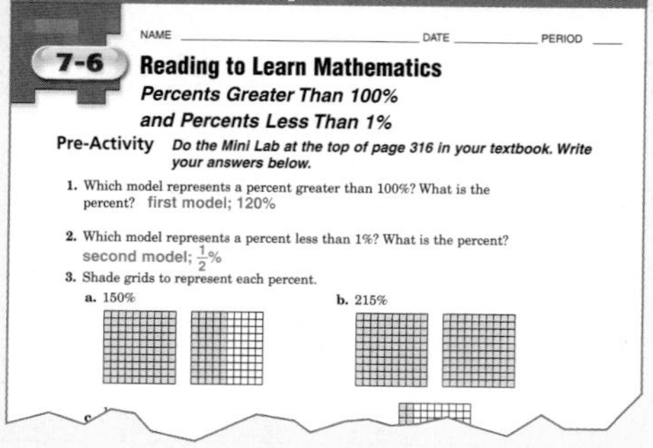
NAME ______ DATE ______ PERIOD ______

7-6 Reading to Learn Mathematics
Percents Greater Than 100% and Percents Less Than 1%

Pre-Activity *Do the Mini Lab at the top of page 316 in your textbook. Write your answers below.*

1. Which model represents a percent greater than 100%? What is the percent? first model; 120%
2. Which model represents a percent less than 1%? What is the percent? second model; $\frac{1}{2}$%
3. Shade grids to represent each percent.
   a. 150%
   b. 215%

7-6 Percents Greater Than 100% and Percents Less Than 1%

What You'll Learn
Write percents greater than 100% and percents less than 1% as fractions and as decimals, and vice versa.

HANDS-ON Mini Lab

Materials
- grid paper
- colored pencils

Work with a partner.

- Draw three 10 × 10 squares on a piece of grid paper. Each large square represents 100%, and each small square represents 1%.
- For the first model, use two grids to shade 120 small squares.
- For the second model, shade half of one small square on a grid. Use a different color than on the first model.

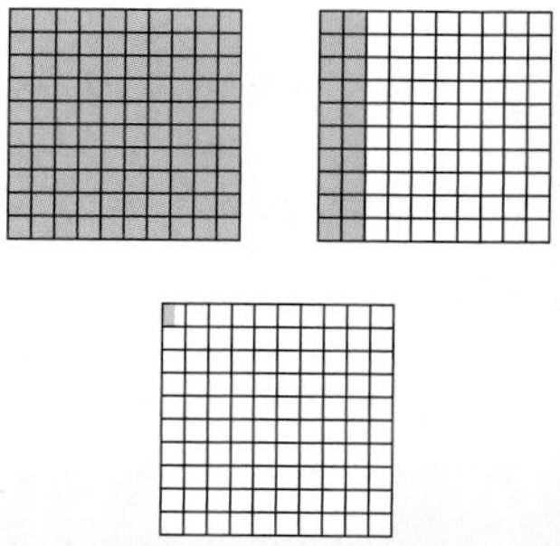

1. Which model represents a percent greater than 100%? What is the percent? **first model; 120%**
2. Which model represents a percent less than 1%? What is the percent? **second model; $\frac{1}{2}$%**
3. Shade grids to represent each percent. **a–c. See margin.**
   a. 150% b. 215% c. $\frac{1}{4}$%

Percents greater than 100% or less than 1% can be written as decimals using the methods you learned in Chapter 5. They can also be written as mixed numbers or fractions.

$$250\% = \frac{250}{100} = 2.5 \text{ or } 2\frac{1}{2}$$

A percent greater than 100% equals a number greater than 1.

$$0.5\% = \frac{0.5}{100} = 0.005 \text{ or } \frac{1}{200}$$

A percent less than 1% equals a number less than 0.01 or $\frac{1}{100}$.

STUDY TIP

Alternate Method
To divide by 100, move the decimal point 2 places to the left.
410% = 4.10 or 4.1
0.2% = 00.2 or 0.002

EXAMPLES Percents as Decimals or Fractions

Write each percent as a decimal and as a mixed number or fraction in simplest form.

1 410%

$410\% = \frac{410}{100}$ Definition of percent

$= 4.1 \text{ or } 4\frac{1}{10}$

2 0.2%

$0.2\% = \frac{0.2}{100}$ Definition of percent

$= 0.002 \text{ or } \frac{1}{500}$

msmath2.net/extra_examples

3b.

3c.

EXAMPLE Real-Life Percents as Decimals

(3) BEARS Before bears hibernate, their need for Calories increases 300%. Write 300% as a decimal.

300% = 300 Divide by 100.

= 3.0 or 3

READING Math

Percents In the example at the right, the solution 3 means *3 times as much.* So, before hibernation bears need 3 times as many Calories as usual.

TEACHING TIP

Point out that percents like 300% or 500% are written as whole numbers.

You can write decimals less than 0.01 and decimals greater than 1 as percents by multiplying by 100 and adding a %.

EXAMPLES Decimals as Percents

Write each decimal as a percent.

(4) 1.68

1.68 = 1.68 Multiply by 100.

= 168%

(5) 0.0075

0.0075 = 0.0075 Multiply by 100.

= 0.75%

Your Turn **Write each decimal as a percent.**

a. 4.5 **450%** b. 0.002 **0.2%** c. 0.0016 **0.16%**

Skill and Concept Check

Writing Math Exercise 3

1. **Write** the fraction of squares that are shaded as a percent and as a decimal. **170%; 1.7**

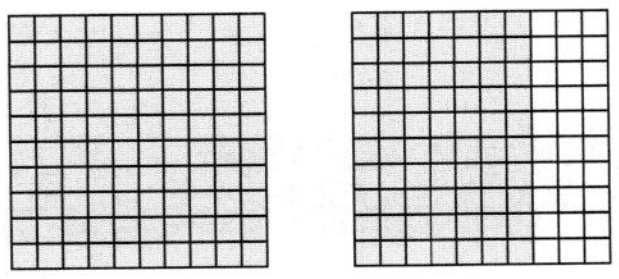

2. **OPEN ENDED** Write a number between 6 and 7 as a percent. **Sample answer: 6.5 = 650%**
3. **Which One Doesn't Belong?** Identify the number that does not have the same value as the other three. Explain your reasoning.

63.5%	0.635	$63\frac{1}{2}$	$63\frac{1}{2}\%$

$63\frac{1}{2}$; It is the only number not equivalent to $63\frac{1}{2}\%$.

GUIDED PRACTICE

Write each percent as a decimal and as a mixed number or fraction in simplest form.

4. 325% **3.25; $3\frac{1}{4}$**
5. 200% **2; 2**
6. 0.7% **0.007; $\frac{7}{1,000}$**
7. 0.15% **0.0015; $\frac{3}{2,000}$**

Write each decimal as a percent.

8. 1.8 **180%**
9. 0.0015 **0.15%**
10. 2.75 **275%**
11. 0.0005 **0.05%**

12. **PHOTOGRAPHY** A photograph developed from 35 mm film is 428% larger than the negative. Express this percent as a decimal. **4.28**

TEACHING TIP

Some students may have been taught the misconception that percents cannot be greater than 100%. Ask them how percents in real life can be greater than 100%. Sample response: extra credit

2 Teach

In-Class Examples

INTERACTIVE CHALKBOARD PowerPoint® Presentations

Write each percent as a decimal and as a mixed number or fraction in simplest form.

1. 220% **2.2 or $2\frac{1}{5}$**
2. 0.6% **0.006 or $\frac{3}{500}$**
3. **STOCKS** During a stock market rally, a company's stock increased in value by 200%. Write 200% as a decimal. **2**

Write each decimal as a percent.

4. 5.12 **512%**
5. 0.0015 **0.15%**

3 Practice/Apply

Which One Doesn't Belong?
For Exercise 3, suggest that students change all numbers to either decimals or percents before comparing.

About the Exercises...

Guided Practice Key

Exercises	Examples
4–7	1, 2
12	3
8–11	4, 5

Odd/Even Assignments
Exercises 13–38 are structured so that students practice the same concepts whether they are assigned odd or even problems.

Alert! Exercise 40 uses the Internet or another source for research.

DAILY INTERVENTION Flexible Grouping

Converting Percents Have students work in pairs and take turns asking their partners to change various large and small percents to decimals, or decimals to percents. Have a large number of real-life examples of percents and decimals available for students to use as a data source.

Study Guide and Intervention, p. 400

NAME ______________ DATE ________ PERIOD ____

7-6 **Study Guide and Intervention**
Percents Greater Than 100% and Percents Less Than 1%

A percent greater than 100% equals a number greater than 1. A percent less than 1% equals a number less than 0.01 or $\frac{1}{100}$.

EXAMPLES Write each percent as a decimal and as a mixed number or fraction in simplest form.

1 280%

$280\% = \frac{280}{100}$ Definition of percent
$= 2.8$ or $2\frac{4}{5}$

2 0.12%

$0.12\% = \frac{0.12}{100}$ Definition of percent
$= 0.0012$ or $\frac{3}{2,500}$

EXAMPLES Write each decimal as a percent.

3 2.17

2.17 = 217. Multiply by 100.
= 217%

4 0.0034

0.0034 = 000.34 Multiply by 100.
= 0.34%

EXERCISES

Write each percent as a decimal and as a mixed number or fraction in simplest form.

1. 200% 2
2. 750% 7.5; $7\frac{1}{2}$
3. 325% 3.25; $3\frac{1}{4}$
4. 0.3% 0.003; $\frac{3}{1,000}$
5. 0.8% 0.008; $\frac{1}{125}$
6. 0.48% 0.0048; $\frac{3}{625}$

Write each decimal as a percent.

7. 2.6 260%
8. 19 1,900%
9. 5.14 514%
10. 0.008 0.8%
11. 0.0014 0.14%
12. 0.0067 0.67%

Practice: Skills, p. 401

NAME ______________ DATE ________ PERIOD ____

7-6 **Practice: Skills**
Percents Greater Than 100% and Percents Less Than 1%

Write each percent as a decimal and as a mixed number or fraction in simplest form.

1. 900% 9
2. 150% 1.5; $1\frac{1}{2}$
3. 675% 6.75; $6\frac{3}{4}$
4. 245% 2.45; $2\frac{9}{20}$
5. 120% 1.2; $1\frac{1}{5}$
6. 0.2% 0.002; $\frac{1}{500}$
7. 0.08% 0.0008; $\frac{1}{1,250}$
8. 0.12% 0.0012; $\frac{3}{2,500}$
9. 0.35% 0.0035; $\frac{7}{2,000}$

Write each decimal as a percent.

10. 3.9 390%
11. 81 8,100%
12. 25 2,500%
13. 6.75 675%
14. 2.81 281%
15. 0.001 0.1%
16. 0.0046 0.46%
17. 0.0069 0.69%
18. 0.0083 0.83%

Write each number as a percent.

19. $6\frac{1}{2}$ 650%
20. $2\frac{1}{2}$ 250%
21. $5\frac{1}{4}$ 525%
22. $\frac{1}{200}$ 0.5%
23. $\frac{2}{250}$ 0.8%
24. $\frac{3}{500}$ 0.6%

Practice: Word Problems, p. 402

NAME ______________ DATE ________

7-6 **Practice: Word Problems**
Percents Greater Than 100% and Percents Less Than 1%

1. **INTERNET** Kilroy found that 0.70% of the documents from the Internet had names that ended in ".txt" or ".text". Write this percent as a decimal and as a fraction. 0.007; $\frac{7}{1,000}$
2. **BUSINESS** Jocelyn expects her new software company to increase its sales next year $2\frac{3}{8}$ times their present value. Write this increase as a percent. 237.5%
3. **UTILITIES** City records showed that 0.8% of new homes had no access to electricity. Write this percent as a decimal and as a fraction in simplest form. 0.008; $\frac{1}{125}$
4. **PETS** Berto got a puppy 8 weeks ago. In this time, the puppy's weight increased 215%. Write this percent as a decimal and as a fraction. 2.15; $2\frac{3}{20}$
5. **MANUFACTURING** The Quality Assurance department at a gear company found that there were 3 defective gears for every 675 produced. Write this as a fraction and as a percent. Round to the nearest hundredth. $\frac{1}{225}$; 0.44%
6. **MEDICINE** Estelle's pharmacist said that the medicine Estelle was taking contained no more than $\frac{1}{500}$ gram of impurities per gram of total weight. Write this fraction as a percent. 0.2%

★ indicates increased difficulty

Practice and Applications

HOMEWORK HELP	
For Exercises	See Examples
13–24, 33–37	1, 2
39–40	3
25–32, 38	4, 5

Extra Practice
See pages 581, 602.

Write each percent as a decimal and as a mixed number or fraction in simplest form.

13. 350% 3.5; $3\frac{1}{2}$
14. 475% 4.75; $4\frac{3}{4}$
15. 600% 6; 6
16. 400% 4; 4
17. 0.6% 0.006; $\frac{3}{500}$
18. 0.1% 0.001; $\frac{1}{1,000}$
19. 0.55% 0.0055; $\frac{11}{2,000}$
20. 0.48% 0.0048; $\frac{3}{625}$
21. 260% 2.6; $2\frac{3}{5}$
22. 195% 1.95; $1\frac{19}{20}$
23. 0.05% 0.0005; $\frac{1}{2,000}$
24. 0.04% 0.0004; $\frac{1}{2,500}$

Write each decimal as a percent.

25. 8.5 850%
26. 35 3,500%
27. 0.009 0.9%
28. 0.003 0.3%
29. 2.64 264%
30. 1.07 107%
31. 0.0034 0.34%
32. 0.0077 0.77%

Write each number as a percent.

33. $3\frac{1}{2}$ 350%
34. $9\frac{3}{4}$ 975%
35. $\frac{2}{500}$ 0.4%
36. $\frac{1}{400}$ 0.25%

★ 37. Write $1\frac{5}{8}$ as a percent. 162.5%

★ 38. Write $\frac{1}{8}$% as a decimal. 0.00125

39. **POPULATION** In 2000, about 0.3% of the population in the United States was Japanese. Write this percent as a decimal and as a fraction. 0.003; $\frac{3}{1,000}$

40. **RESEARCH** Use the Internet or another source to find the percent of people in the United States that come from three other countries. Write the percents as decimals. See students' work.

★ 41. **MULTI STEP** One hour is what percent of one week? about 0.595%

42. **CRITICAL THINKING** Some North African ostriches are 150% of the height of a 6-foot human. What is the height of some ostriches? 9 ft

Standardized Test Practice and Mixed Review

43. **MULTIPLE CHOICE** A certain stock increased its value by 467% over 10 years. What is this percent written as a decimal? A

Ⓐ 4.67 Ⓑ 46.7 Ⓒ 0.467 Ⓓ 0.00467

44. **MULTIPLE CHOICE** Choose the best estimate for 100.5%. G

Ⓕ 0 Ⓖ 1 Ⓗ 100 Ⓘ 1,000

Write each percent as a fraction in simplest form. (Lesson 7-5)

45. 7.5% $\frac{3}{40}$
46. 1.2% $\frac{3}{250}$
47. $6\frac{1}{4}$% $\frac{1}{16}$
48. $92\frac{1}{2}$% $\frac{37}{40}$

49. **SCALE DRAWING** A garage is 18 feet wide. In a scale drawing, 1 inch = 3 feet. What is the width of the garage on the drawing? (Lesson 7-4) 6 in.

GETTING READY FOR THE NEXT LESSON

PREREQUISITE SKILL **Write each percent as a decimal.** (Lesson 5-6)

50. 85% 0.85
51. 6.5% 0.065
52. 36.9% 0.369
53. 12.3% 0.123

4 Assess

Open-Ended Assessment

Speaking Given a percent greater than 100% or less than 1%, have students write a real-life problem using the percent. Have students take turns reading their problem to the class.

7-7 Percent of a Number

What You'll Learn
Find the percent of a number.

WHEN am I ever going to use this?

SURVEYS The graph at the right shows the results of a survey in which 1,016 people were asked to name the things they feared.

1. About how many of them said they were afraid of snakes?
2. How did you estimate?
3. Suppose you pass 50 people at a mall. Based on the results of the survey, do you think more or less than 25 of them are afraid of mice? Explain. See margin.

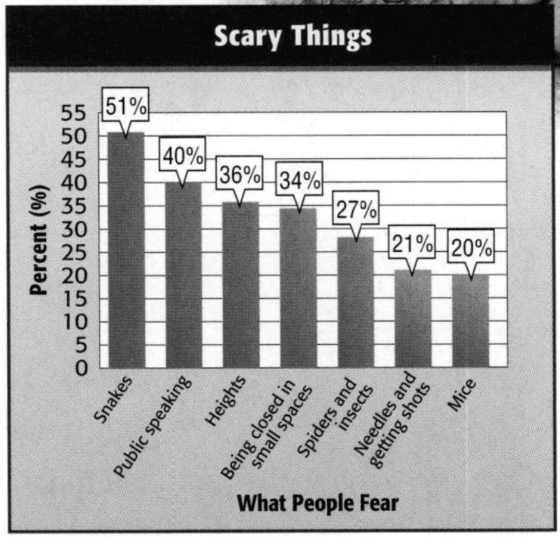

Source: Gallup Poll

1. Sample answer: about 500
2. 51% or about one half of 1,000 is 500.

One way to find the percent of a number is to use a proportion.

EXAMPLE Use a Proportion to Find a Percent

1 SURVEYS The graphic above shows that 36% of 1,016 adults surveyed were afraid of heights. How many of the people surveyed were afraid of heights?

36% means that 36 out of 100 people were afraid of heights. Find an equivalent ratio x out of 1,016 and write a proportion.

number afraid of heights → / total number in survey → $\frac{x}{1{,}016} = \frac{36}{100}$ } percent of people afraid of heights

Now solve the proportion.

$\frac{x}{1{,}016} = \frac{36}{100}$ Write the proportion.

$x \cdot 100 = 1{,}016(36)$ Find the cross products.

$100x = 36{,}576$ Multiply.

$\frac{100x}{100} = \frac{36{,}576}{100}$ Divide each side by 100.

$x = 365.76$ Simplify.

So, about 366 of the 1,016 people surveyed were afraid of heights.

TEACHING TIP
Point out that the percents in the graphic are rounded, which is why the solution in Example 1 is not a whole number.

msmath2.net/extra_examples

Answer (When am I ever going to use this?)

3. Less; Sample answer: 25 is half the people, or 50%. However, only about 20% of them would be afraid of mice according to the survey.

7-7 Lesson Notes

Lesson Resources

Chapter 7 Resource Masters FAST FILE
- Study Guide and Intervention, p. 405
- Practice: Skills, p. 406
- Practice: Word Problems, p. 407
- Reading to Learn Mathematics, p. 408
- Enrichment, p. 409

Calculator and Spreadsheet Masters, p. 39

MathPASS Tutorial CD-ROM, 32

Parent and Student Study Guide Workbook, 7-7

1 Focus

5-Minute Check 7-7
INTERACTIVE CHALKBOARD PowerPoint® Presentations
Use as a quiz or review of Lesson 7-6. (Also available as a transparency)

Bellringer
Reading Mathematics Give students the weekly ads for local retail stores. Ask them to find examples of percents. How are these percents used? They usually denote a sale. How do students figure out the sale price? Can they get any clues from the ads? Some ads will list the sale price.

Reading to Learn Mathematics, p. 408 ELL

NAME ______ DATE ______ PERIOD ____

7-7 Reading to Learn Mathematics
Percent of a Number

Pre-Activity *Read the introduction at the top of page 319 in your textbook. Write your answers below.*

1. About how many of them said they were afraid of snakes? Sample answer: about 500
2. How did you estimate? 51% or about one half of 1,000 is 500.
3. Suppose you pass 50 people at a mall. Based on the results of the survey, do you think more or less than 25 of them are afraid of mice? Explain. Less; Sample answer: 25 is half the people, or 50%. However, only about 20% of them would be afraid of mice according to the survey.

Reading the Lesson

4. What are two methods for finding the percent of a number? Use a proportion or use multiplication.

2 Teach

In-Class Examples

INTERACTIVE CHALKBOARD
PowerPoint® Presentations

1. **SURVEYS** Out of 1,423 adults surveyed, 30% said they knew the name of their mail carrier. How many of the people surveyed knew their mail carrier's name? **about 427**

2. What number is 120% of 24? **28.8**

3. Find 25% of $600. **$150**

3 Practice/Apply

NUMBER SENSE
In Exercise 2, ask students to first use their knowledge of percents to determine whether the statement makes sense, without using proportions or decimal multiplication.

About the Exercises...

Guided Practice Key

Exercises	Examples
8	1
5–6	2
4, 7	3

Odd/Even Assignments
Exercises 9–22 are structured so that students practice the same concepts whether they are assigned odd or even problems.

Answers

2. No; Sample answers: 50% of $150 is $75, so 30% of $150 is less than $75. Or, $\frac{1}{3}$ of 150 is $50, and 30% is less than $\frac{1}{3}$, so 30% is less than $50.
3. Sample answer: Find the number of people surveyed who use a white erasable markerboard to record information during a meeting; 525 people.

Another method for finding the percent of a number is to use multiplication.

TEACHING TIP
In these examples, the word *of* means to multiply. So, 140% of 32 equals 140% *times* 32.

EXAMPLES Use Multiplication to Find a Percent

2 What number is 140% of 32?

140% of 32 = 140% × 32   Write a multiplication expression.
= 1.4 × 32   Write 140% as a decimal.
= 44.8   Multiply.

So, 140% of 32 is 44.8.

3 Find 75% of $800.

75% of $800 = 75% × $800   Write a multiplication expression.
= 0.75 × 800   Write 75% as a decimal.
= 600   Multiply.

So, 75% of $800 is $600.

STUDY TIP
Mental Math You can find 75% of 800 mentally by using the fraction $\frac{3}{4}$. $\frac{3}{4}$ of 800 is 600.

Your Turn **Find each number.**

a. Fifty percent of 140 is what number? **70**
b. Find 37.5% of 64. **24**

Skill and Concept Check

Writing Math
Exercises 2 & 3

1. **Write** a proportion that can be used to find 0.5% of 22. $\frac{x}{22} = \frac{0.5}{100}$
2. **NUMBER SENSE** Use mental math to determine whether 30% of $150 is $80. Explain. **See margin.**
3. **OPEN ENDED** Write a problem that can be solved using the graphic at the right. Then solve the problem. **See margin.**

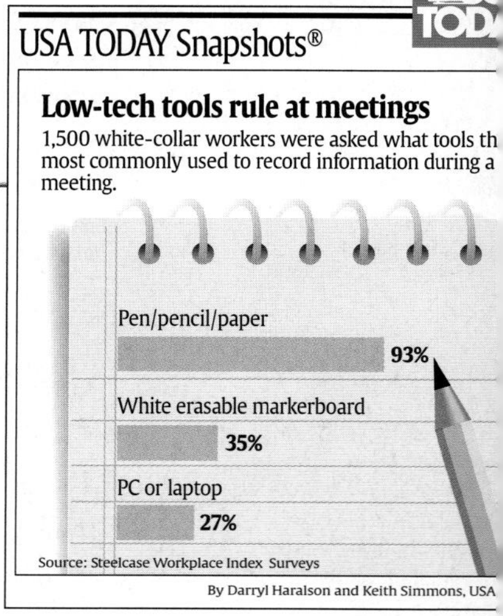

GUIDED PRACTICE

Find each number. Round to the nearest tenth if necessary.

4. Find 95% of $40. **$38**
5. Forty-two percent of 263 is what number? **110.5**
6. What number is 200% of 75? **150**
7. Find 0.25% of 40. **0.1**

8. **COMMISSION** Ms. Sierra earns a 2% *commission*, or fee paid based on a percent of her sales, on every vacation package that she sells. One day, she sold vacation packages worth $6,500. What was her commission? **$130**

DAILY INTERVENTION

Differentiated Instruction

Intrapersonal Have students write one advantage of using proportions and one advantage of multiplying decimals when finding percents of numbers. Ask them to choose the technique they prefer, and to explain, in writing, why they chose this technique.

★ indicates increased difficulty

Practice and Applications

HOMEWORK HELP	
For Exercises	See Examples
9–23	1–3
Extra Practice See pages 582, 602.	

Find each number. Round to the nearest tenth if necessary. 12. 92.5

9. What is 65% of 100? 65
10. Find 45% of $400. $180
11. Find 25% of $640. $160
12. What number is 18.5% of 500?
13. 67.5% of 76 is what? 51.3
14. 64% of 88 is what number? 56.3
15. What is 125% of 10? 12.5
16. Find 300% of 14. 42
17. Find 0.15% of 250 0.4
18. What number is 0.5% of 80? 0.4

★ 19. Find $20\frac{1}{4}$% of 3. 0.6

★ 20. 250.5% of 25 is what number? 62.6

SNACKING For Exercises 21 and 22, use the graphic at the right. It shows the results of a poll of 1,746 college students.

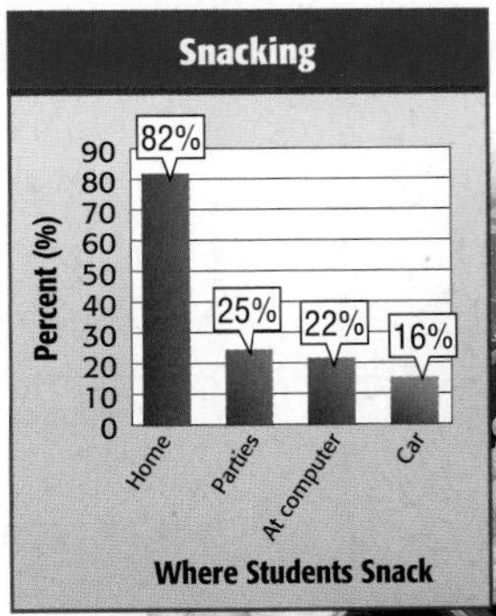

Source: Snack Food Association

21. Write a proportion that can be used to find how many students like to snack at their computer. Then solve. See margin.
22. How many students surveyed like to snack at home? about 1,432 students
23. **DELIVERY** A mountain bike costs $288 plus a 4.5% delivery charge. What is the cost of the bike including the delivery charge? $300.96
24. **CRITICAL THINKING** Suppose you add 10% of a number to the number. Then you subtract 10% of the total. Is the result *greater than*, *less than*, or *equal to* the original number? Explain your reasoning. See margin.

Standardized Test Practice and Mixed Review

25. **MULTIPLE CHOICE** A football player has made 80% of the field goals he has attempted in his career. If he attempts 5 field goals in a game, how many would he be expected to make? B

Ⓐ 5 Ⓑ 4 Ⓒ 3 Ⓓ 2

26. **SHORT RESPONSE** What number is 12% of 16.5? 1.98

Write each decimal as a percent. (Lesson 7-6)

27. 7.5 750%
28. 9 900%
29. 0.0004 0.04%
30. 0.0018 0.18%
31. Write $\frac{11}{15}$ as a percent. Round to the nearest hundredth. (Lesson 7-5) 73.33%

GETTING READY FOR THE NEXT LESSON

PREREQUISITE SKILL Evaluate each expression. Round to the nearest tenth if necessary. (Lesson 1-3)

32. $\frac{36 \cdot 2}{80}$ 0.9
33. $\frac{14(5)}{3.5}$ 20
34. $\frac{9 \times 100}{56}$ 16.1
35. $22 \cdot 100 \div 490$ 4.5

msmath2.net/self_check_quiz

4 Assess

Higher-Level Thinking

Bill bought a jacket for 75% of its regular price of $48. Ted wants to buy the jacket. If Bill charges Ted 10% more than what he paid, how much does Bill charge Ted? $39.60

Answers

21. $\frac{x}{1{,}746} = \frac{22}{100}$; about 384 students

24. Less than; Sample answer: you are subtracting 10% of a greater number, so you are subtracting more than you added.

Study Guide and Intervention, p. 405

NAME ______ DATE ______ PERIOD ____

7-7 Study Guide and Intervention

Percent of a Number

You can use a proportion or multiplication to find the percent of a number.

EXAMPLE 1 **SURVEY A survey asked 2,415 people whether they would buy the restored version of The Beatles's *A Hard Day's Night*. 74.95% of the people said they would *not* buy it. How many people would not buy the restored version of this movie?**

number that would not buy → $\frac{x}{2{,}415} = \frac{74.95}{100}$ ← percent of people who would not buy the restored movie; total number in survey →

$\frac{x}{2{,}415} = \frac{74.95}{100}$ Write the proportion.

$x \cdot 100 = 2{,}415(74.95)$ Find the cross products.

$100x = 181{,}004.25$ Multiply.

$\frac{100x}{100} = \frac{181{,}004.25}{100}$ Divide each side by 100.

$x = 1{,}810.0425$ Simplify.

So, about 1,810 of the 2,415 people surveyed would not buy the restored version of *A Hard Day's Night*.

EXAMPLE 2 **What number is 15% of 200?**

15% of $200 = 15\% \times 200$ Write a multiplication expression.
$= 0.15 \times 200$ Write 15% as a decimal.
$= 30$ Multiply.

So, 15% of 200 is 30.

EXERCISES

Find each number.

1. Find 20% of 50. 10
2. What is 55% of $400? $220
3. 5% of 1,500 is what number? 75
4. Find 190% of 20. 38
5. What is 24% of $500? $120
6. 8% of $300 is how much? $24
7. What is 12.5% of 60? 7.5
8. Find 0.2% of 40. 0.08
9. Find 3% of $800. $24
10. What is 0.5% of 180? 0.9
11. 0.25% of 42 is what number? 0.105
12. What is 0.02% of 280? 0.056

Practice: Skills, p. 406

NAME ______ DATE ______ PERIOD ____

7-7 Practice: Skills

Percent of a Number

Find each number.

1. Find 80% of 80. 64
2. What is 95% of 600? 570
3. 35% of 20 is what number? 7
4. Find 60% of $150. $90
5. What is 75% of 240? 180
6. 380% of 30 is what number? 114
7. Find 40% of 80. 32
8. What is 30% of $320? $96
9. 12% of 150 is what number? 18
10. Find 58% of 200. 116
11. What is 18% of $450? $81
12. What is 70% of 1,760? 1,232
13. Find 92% of 120. 110.4
14. 45% of 156 is what number? 70.2
15. What is 12% of 12? 1.44
16. Find 60% of 264. 158.4
17. 37.5% of 16 is what number? 6
18. What is 82.5% of 400? 330
19. What is 0.25% of 900? 2.25
20. Find 1.5% of 220. 3.3

Practice: Word Problems, p. 407

NAME ______ DATE ______ PERIOD ____

7-7 Practice: Word Problems

Percent of a Number

SPORTS For Exercises 1 and 2, use the graph below. It shows the results of a poll of 440 ninth grade students. Round answers to the nearest whole number.

PETS For Exercises 3 and 4, use the table below. It shows the pet ownership in Los Angeles, California. Assume that the same percents apply to a town of 1,650 households. Round answers to the nearest whole number.

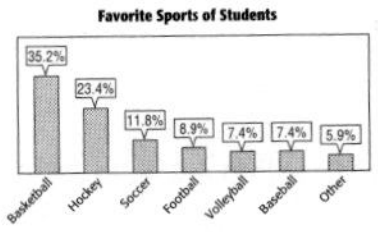

Pets in Household	Percent
at least one dog or cat	26.7
at least one dog	19.9
at least one cat	13
at least one dog and one cat	6.19

1. Write a proportion that can be used to find how many students surveyed chose hockey as their their favorite sport. Then solve. $\frac{23.4}{100} = \frac{n}{440}$; 103 students
2. How many students surveyed chose basketball as their favorite sport? 155 students
3. Write a proportion that can be used to find how many households have at least one dog. Then solve. $\frac{19.9}{100} = \frac{n}{1{,}650}$; 328 households
4. How many households have at least one dog or cat? 441 households
5. **VOTING** Going into a recent election, only about 62% of people old enough to vote were registered. In a community of about 55,200 eligible voters, how many people are registered? 34,224 people
6. **COLLEGE** A local college recently reported that enrollment increased to 108% percent of last year. If enrollment last year was at 17,113, about how many students enrolled this year? Round to the nearest whole number. 18,482 students

7-8a HANDS-ON LAB Lesson Notes

Lesson Resources

Teaching Mathematics with Manipulatives
- centimeter grid, p. 11
- student recording sheet 7-8a

GETTING STARTED

Teaching Tip Students should exercise care when writing numbers along the rectangle. If they do not have room to place the numbers clearly and evenly, they should draw a larger rectangle.

TEACH

- Remind students that there are often several ways to label units along the top of the rectangle depending on the size of the number.
- Ask students to come up with an example where this method only yields an approximate answer.

ASSESS

Observe as students solve the **Your Turn** exercises. They should:
- draw their models neatly and accurately,
- examine the percent before deciding how to divide the given number into parts, and
- use the model to estimate a percent.

Discuss the **Writing Math** exercises. You may wish to have students include an example of this model in their Foldables for Chapter 7.

7-8a HANDS-ON LAB

A Preview of Lesson 7-8

What You'll Learn
Use a percent model to find a part.

Materials
- grid paper

Using a Percent Model

You can find the percent of a number or the *part* of the whole by using a model.

ACTIVITY *Work with a partner.*

Suppose you and a friend eat dinner at the Pizza Palace and your total bill is about $25. If you want to leave a 20% tip, how much should you leave? Use a model to find the amount.

Draw a 10-unit by 1-unit rectangle on grid paper.

Label equal units from 0 to 25 to represent parts of the whole quantity, 25.

Label equal units from 0 to 100 to represent percents.

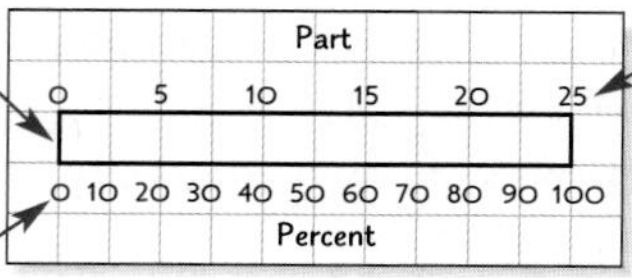

Draw a vertical line at 20%.

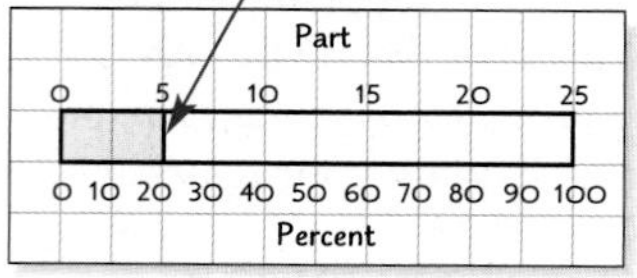

So, you should leave $5 as a tip.

Your Turn **Draw a model to find each part.**

a. 70% of 50 **35** b. 45% of 20 **9** c. 20% of 75 **15**

a–c. See Ch. 7 Answer Appendix for models.

Writing Math

1. Suppose your whole family had dinner with you and the total bill was $50. How could you change your model to find the tip?
2. **Write** a sentence that describes what is represented by the model at the right.

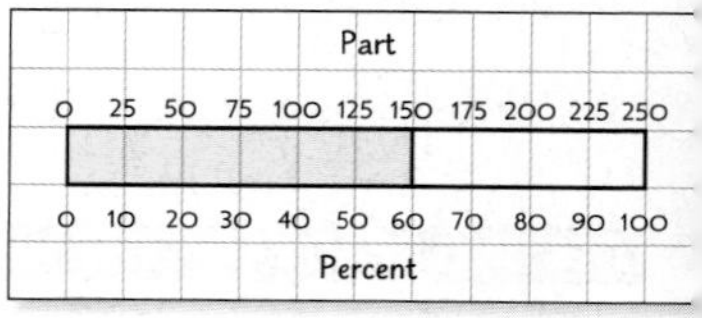

3. **Write** a percent problem that can be represented and solved using a model. **See students' work.**
4. **Make a conjecture** about how you could use a model to estimate the percent represented by the ratio 8 out of 50. **See margin.**

1. Mark equal units from 0 to 50 on the top.
2. 60% of 250 is 150.

Answer (Writing Math)

4. **Sample answer: Draw a model and label the bottom units 0 to 100 for the percent and the top units 0 to 50 for the part. Draw a vertical line from 8 on the top. Then estimate the corresponding percent on the bottom.**

7-8 The Percent Proportion

hat You'll Learn
lve problems using
e percent proportion.

W Vocabulary
rcent proportion
rt
se

WHEN am I ever going to use this?

SPACE The engine on the space shuttle weighs approximately 19,700 pounds. The entire space shuttle weighs 178,000 pounds.

1. Write the ratio of engine weight to total weight as a fraction. $\frac{19{,}700}{178{,}000}$
2. Use a calculator to write the fraction as a decimal to the nearest hundredth. **0.11**
3. What percent of the space shuttle's weight is the engine? **11%**

A **percent proportion** compares **part** of a quantity to the whole quantity, called the **base**, using a percent. You can use a percent proportion to find what percent of a space shuttle's weight is the engine.

part: engine weight → / base: total weight → $\frac{19{,}700}{178{,}000} = \frac{p}{100}$ ← The percent is written as a fraction with a denominator of 100.

Key Concept — **Percent Proportion**

Words The percent proportion is $\frac{\text{part}}{\text{base}} = \frac{\text{percent}}{100}$.

Symbols $\frac{a}{b} = \frac{p}{100}$, where a is the part, b is the base, and p is the percent.

To find the percent, write the percent proportion and solve for p.

EXAMPLE Find the Percent

1 What percent of \$15 is \$9?

9 is the part, and 15 is the base. You need to find the percent p.

$\frac{a}{b} = \frac{p}{100}$	Percent proportion
$\frac{9}{15} = \frac{p}{100}$	Replace a with 9 and b with 15.
$9 \cdot 100 = 15 \cdot p$	Find the cross products.
$900 = 15p$	Simplify.
$\frac{900}{15} = \frac{15p}{15}$	Divide each side by 15.
$60 = p$	So, 60% of \$15 is \$9. Compare the answer to the model.

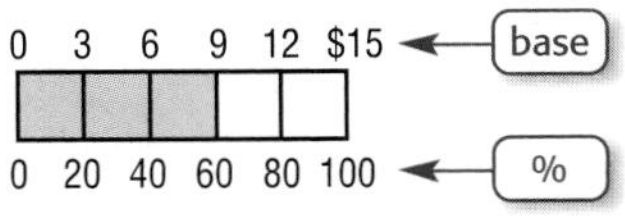

TEACHING TIP
may be helpful for
udents to use $\frac{\text{is}}{\text{of}} = \frac{\%}{100}$
help them remember
e percent proportion.

msmath2.net/extra_examples

7-8 Lesson Notes

Lesson Resources

Chapter 7 Resource Masters
- Study Guide and Intervention, p. 410
- Practice: Skills, p. 411
- Practice: Word Problems, p. 412
- Reading to Learn Mathematics, p. 413
- Enrichment, p. 414

Calculator and Spreadsheet Masters, p. 40

Key Teaching Transparencies, 28

Parent and Student Study Guide Workbook, 7-8

FAST FILE

1 Focus

5-Minute Check 7-8
Use as a quiz or review of Lesson 7-7. (Also available as a transparency)

INTERACTIVE CHALKBOARD
PowerPoint® Presentations

Bellringer
Problem Solving Review with students how they have used proportions. **comparing ratios**
Ask students how they can use a proportion to solve a percent problem. For example, Ellie ate 2 of the 8 slices of pizza. What percent of the pizza did Ellie eat?
$\frac{2}{8} = \frac{x}{100}$; **25%**

Reading to Learn Mathematics, p. 413 **ELL**

NAME ______ DATE ______ PERIOD ____

7-8 Reading to Learn Mathematics
The Percent Proportion

Pre-Activity ***Read the introduction at the top of page 323 in your textbook. Write your answers below.***

1. Write the ratio of engine weight to total weight as a fraction. $\frac{19{,}700}{178{,}000}$
2. Use a calculator to write the fraction as a decimal to the nearest hundredth. 0.11
3. What percent of the space shuttle's weight is the engine? 11%

Reading the Lesson

4. What is a percent proportion? a proportion that compares part of a quantity to the whole quantity using a percent
5. In the formula $\frac{a}{b} = \frac{p}{100}$, what does each letter stand for?
   a is the part, b is the base, and p is the percent.
6. Select the information that can be found by solving each percent problem.
   ... number is 30% of ... Find the ba...

2 Teach

In-Class Examples

INTERACTIVE CHALKBOARD
PowerPoint® Presentations

1. What percent of 24 is 18? **75%**
2. What number is 30% of 150? **45**
3. 12 is 80% of what number? **15**

3 Practice/Apply

FOLDABLES™ Follow-Up Remind students to select a term or key concept that they need to know for each lesson and write definitions in words and symbols under the tabs of their Foldables. As students progress through the chapter, encourage them to use their Foldables to self-check their understanding and mastery of these terms.

About the Exercises...

Guided Practice Key

Exercises	Examples
3	1
4–5	2
6	3

Odd/Even Assignments
Exercises 7–22 are structured so that students practice the same concepts whether they are assigned odd or even problems.

Answer

23. **20% of 500, 20% of 100, 5% of 100; Sample answer: If the percent is the same but the base is greater, then the part is greater. If the base is the same but the percent is greater, then the part is greater.**

STUDY TIP

Finding the Base In a percent problem, the base follows the word *of*.

TEACHING TIP

This is the method that students used to solve the problem in Example 1, Lesson 7-7. Only the variables are different.

You can also use the percent proportion to find a missing part or base

Concept Summary — **Types of Percent Problems**

Type	Example	Proportion
Find the Percent	What percent of 6 is 3?	$\frac{3}{6} = \frac{p}{100}$
Find the Part	What number is 50% of 6?	$\frac{a}{6} = \frac{50}{100}$
Find the Base	3 is 50% of what number?	$\frac{3}{b} = \frac{50}{100}$

To find the part, write the percent proportion and solve for *a*. To find the base, write the percent proportion and solve for *b*.

EXAMPLE Find the Part

2 What number is 40% of 120?

40 is the percent and 120 is the base. You need to find the part.

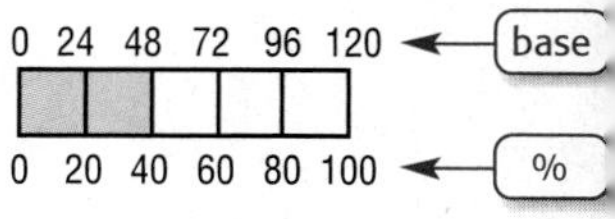

$\frac{a}{b} = \frac{p}{100}$ Percent proportion

$\frac{a}{120} = \frac{40}{100}$ Replace *b* with 120 and *p* with 40.

$a \cdot 100 = 120 \cdot 40$ Find the cross products.

$100a = 4{,}800$ Simplify.

$\frac{100a}{100} = \frac{4{,}800}{100}$ Divide each side by 100.

$a = 48$ So, 40% of 120 is 48. Compare to the model.

EXAMPLE Find the Base

3 18 is 25% of what number?

18 is the part and 25 is the percent. You need to find the base, or the whole quantity.

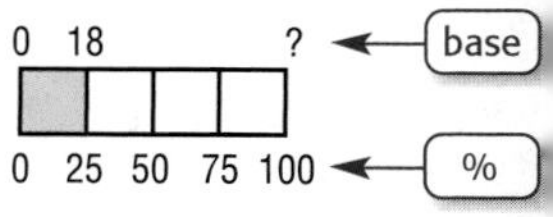

$\frac{a}{b} = \frac{p}{100}$ Percent proportion

$\frac{18}{b} = \frac{25}{100}$ Replace *a* with 18 and *p* with 25.

$18 \cdot 100 = b \cdot 25$ Find the cross products.

$1{,}800 = 25b$ Simplify.

$\frac{1{,}800}{25} = \frac{25b}{25}$ Divide each side by 25.

$72 = b$ So, 18 is 25% of 72.

Check In the model above, the whole quantity is $18 \cdot 4$ or 72.

Your Turn **Find each number. Round to the nearest tenth if necessary.**

a. What number is 5% of 60? **3**   b. 80 is 75% of what number? **106**

DAILY INTERVENTION — **Differentiated Instruction**

Naturalist Take students outside and have them choose an object that they can see, such as a tree, squirrel, leaf, branch, building, or car. Then list categories such as green, alive, or soft. Name a category aloud, and have students whose objects fit that category stand up. After each category is given, have students write an estimate of the percent of the class standing, and then have them use a proportion to more accurately determine what percent of the class is standing. Repeat until every student has been in at least one category.

Skill and Concept Check

Writing Math
Exercise 2

1. **Choose** the problem that can be solved by using $\frac{8}{b} = \frac{16}{100}$. **b**
   a. What number is 16% of 8? b. 8 is 16% of what number?

2. **OPEN ENDED** Write a proportion that can be used to find the part if you know the base and the percent. **Sample answer:** $\frac{a}{20} = \frac{25}{100}$

GUIDED PRACTICE

Find each number. Round to the nearest tenth if necessary.

3. What percent of \$50 is \$18? **36%**
4. What number is 25% of 180? **45**
5. What number is 2% of 35? **0.7**
6. 62 is 90.5% of what number? **68.5**

★ indicates increased difficulty

Practice and Applications

HOMEWORK HELP

For Exercises	See Examples
7–22	1–3

Extra Practice
See pages 582, 602.

Find each number. Round to the nearest tenth if necessary.

7. What percent of 60 is 15? **25%**
8. 9 is 45% of what number? **20**
9. What number is 12% of 72? **8.6**
10. What percent of \$12 is \$9? **75%**
11. 75 is 20% of what number? **375**
12. What number is 40% of 80? **32**
13. \$3 is what percent of \$40? **7.5%**
14. 12.5% of what number is 24? **192**
15. 8.2% of 50 is what number? **4.1**
16. What percent of 300 is 0.6? **0.2%**
17. 45 is 150% of what number? **30**
18. What percent of 25 is 30? **120%**
★ 19. What number is 0.5% of 8? **0.04**
★ 20. 6% of what number is $10\frac{1}{2}$? **175**

21. **SHOPPING** A paperback book originally priced at \$12.50 is on sale for \$7.50. What percent of the original cost is the sale price? **60%**

22. **MOVIES** Sarah and Monique spent \$14, or 35% of their money, on movie tickets. How much money did they have to start with? **\$40**

23. **CRITICAL THINKING** Without calculating, order the following from greatest to least value. Explain your reasoning. **See margin.**
    20% of 100, 20% of 500, 5% of 100

Standardized Test Practice and Mixed Review

24. **MULTIPLE CHOICE** There are about 53,500 hotels in the United States. Use the table to find how many are located along highways. **A**
    (A) 22,577 (B) 22,470 (C) 17,976 (D) 5,457

U.S. Hotels

Location	Percent
highways	42.2%
suburbs	33.6%
city	10.2%

25. **SHORT RESPONSE** 95 of 273 students volunteered. About what percent of the students did *not* volunteer? **65%**

Find each number. Round to the nearest tenth if necessary. (Lesson 7-7)

26. What is 25% of 120? **30**
27. Find 45% of 70. **31.5**

28. Write 1.2 as a percent. (Lesson 7-6) **120%**

msmath2.net/self_check_quiz

4 Assess

Open-Ended Assessment

Modeling Have students work in pairs. One student should draw a rectangle on grid paper and ask his or her partner to shade a specific percent of the squares. The other student should then shade a number of squares in another rectangle and ask his or her partner to determine the percent of squares that are shaded. Have students alternate tasks.

Study Guide and Intervention, p. 410

NAME ______ DATE ______ PERIOD ____

7-8 Study Guide and Intervention
The Percent Proportion

A **percent proportion** compares **part** of a quantity to the whole quantity, called the **base**, using a percent. In symbols, the percent proportion can be written as $\frac{a}{b} = \frac{p}{100}$, where a is the part, b is the base, and p is the percent.

EXAMPLE 1 **What percent of 24 is 18?**

18 is the part, and 24 is the base. You need to find the percent p.

$\frac{a}{b} = \frac{p}{100}$	Percent proportion
$\frac{18}{24} = \frac{p}{100}$	Replace a with 18 and b with 24.
$18 \cdot 100 = 24 \cdot p$	Find the cross products.
$1{,}800 = 24p$	Simplify.
$\frac{1{,}800}{24} = \frac{24p}{24}$	Divide each side by 24.
$75 = p$	Simplify.

So, 75% of 24 is 18.

EXAMPLE 2 **What number is 60% of 150?**

60 is the percent, and 150 is the base. You need to find the part.

$\frac{a}{b} = \frac{p}{100}$	Percent proportion
$\frac{a}{150} = \frac{60}{100}$	Replace b with 150 and p with 60.
$a \cdot 100 = 150 \cdot 60$	Find the cross products.
$100a = 9000$	Simplify.
$\frac{100a}{100} = \frac{9{,}000}{100}$	Divide each side by 100.
$a = 90$	Simplify.

So, 60% of 150 is 90.

EXERCISES

Find each number. Round to the nearest tenth if necessary.

1. What number is 25% of 20? 5
2. What percent of 50 is 20? 40%
3. 30 is 75% of what number? 40
4. 40% of what number is 36? 90
5. What number is 20% of 625? 125
6. 12 is what percent of 30? 40%

Practice: Skills, p. 411

NAME ______ DATE ______ PERIOD ____

7-8 Practice: Skills
The Percent Proportion

Find each number. Round to the nearest tenth if necessary.

1. 50 is 20% of what number? 250
2. What percent of 20 is 4? 20%
3. What number is 70% of 250? 175
4. 10 is 5% of what number? 200
5. What number is 45% of 180? 81
6. 40% of what number is 82? 205
7. What percent of 90 is 36? 40%
8. 60 is 25% of what number? 240
9. What number is 32% of 1,000? 320
10. What percent of 125 is 5? 4%
11. 73 is 20% of what number? 365
12. 57% of 109 is what number? 62.1
13. What percent of 185 is 35? 18.9%
14. 25 is what percent of 365? 6.8%
15. 85% of 190 is what number? 161.5
16. 12.5 is 25% of what number? 50
17. What percent of 128 is 24? 18.8%
18. 5.25% of 170 is what number? 8.9
19. What is 82% of 230? 188.6
20. What percent of 49 is 7? 14.3%

Practice: Word Problems, p. 412

NAME ______ DATE ______

7-8 Practice: Word Problems
The Percent Proportion

1. **DRIVING** David installed a device on his car that guaranteed to increase his gas mileage by 15%. He currently gets 22 miles per gallon. How much will the gas mileage increase after installing the device? 3.3 mi per gal

2. **POPULATION** The number of students at Marita's school decreased to 98% of last year's number. Currently, there are 1,170 students. How many students were there last year? Round to the nearest whole number. 1,194 students

3. **VOTING** Yolanda's club has 35 members. Its rules require that 60% of them must be present for any vote. At least how many members must be present to have a vote? 21 members

4. **GARBAGE** This month, Chun's office produced 690 pounds of garbage. Chun wants to reduce the weight of garbage produced to 85% of the weight produced this month. What is the target weight for the garbage produced next month? 586.5 lb

5. **SALARIES** Alma just received a 6% raise in salary. Before the raise, she was making \$52,000 per year. How much more will Alma earn next year? \$3,120

6. **SPORTS** Sally's soccer team played 25 games and won 17 of them. What percent did the team win? 68%

CHAPTER 7 Study Guide and Review

Vocabulary Assessment

- This alphabetical list of vocabulary terms in Chapter 7 includes a page reference where each term was introduced.
- A vocabulary review/test for Chapter 7 is available on p. 428 of the *Chapter 7 Resource Masters*.

Vocabulary PuzzleMaker

ELL The online Vocabulary PuzzleMaker improves students' mathematics vocabulary using four puzzle formats — crossword, scramble, word search using a word list, and word search using clues. Students can work on a computer screen or from a printed handout.

MindJogger Videoquizzes

ELL MindJogger Videoquizzes provide an alternative review of concepts presented in this chapter. Students work in teams in a game show format to gain points for correct answers. The questions are presented in three rounds.

Round 1 Concepts (5 questions)
Round 2 Skills (4 questions)
Round 3 Problem Solving (4 questions)

ExamView® Pro Test Bank CD-ROM

Use the networkable ExamView® Pro Test Bank CD-ROM to:

- Create **Multiple versions** of tests.
- Create **Modified** tests for **Inclusion** students with one mouse click.
- **Edit** existing questions and **Add** your own questions.
- Build tests aligned with state standards using built-in **State Curriculum Correlations.**
- Change **English** tests to **Spanish** with one mouse click and vice versa.

CHAPTER 7 Study Guide and Review

Vocabulary and Concept Check

base (p. 323)	percent proportion (p. 323)	scale drawing (p. 304)
cross product (p. 297)	proportion (p. 297)	scale factor (p. 305)
equivalent ratios (p. 289)	rate (p. 292)	scale model (p. 306)
part (p. 323)	scale (p. 304)	unit rate (p. 292)

Choose the letter of the term that best matches each phrase.

1. a comparison of two numbers by division e
2. two ratios that have the same value i
3. a ratio of two measurements with different units a
4. an equation that shows that two ratios are equivalent h
5. used to represent something that is too large or too small for an actual-size drawing c
6. the ratio of the distance on a map to the actual distance f
7. the number that is compared to the whole quantity in a percent proportion g
8. product of the numerator of one ratio in a proportion and the denominator of the other ratio j
9. a rate that is simplified so that it has a denominator of 1 d
10. the number that represents the whole quantity in a percent proportion b

a. rate
b. base
c. scale drawing
d. unit rate
e. ratio
f. scale
g. part
h. proportion
i. equivalent ratios
j. cross products

Lesson-by-Lesson Exercises and Examples

7-1 Ratios (pp. 288–291)

Write each ratio as a fraction in simplest form.

11. 16:12 $\frac{4}{3}$
12. 5 to 25 $\frac{1}{5}$
13. 50 cm to 75 cm $\frac{2}{3}$
14. 6 ft:12 in. $\frac{6}{1}$

Determine whether the ratios are equivalent. Explain. 15–18. See margin.

15. $\frac{3}{5}$ and $\frac{21}{35}$
16. $\frac{18}{24}$ and $\frac{5}{20}$
17. 27:15 and 9:5
18. 4:21 and 2:7

Example 1 **Write the ratio 32 to 18 as a fraction in simplest form.**

$32 \text{ to } 18 = \frac{32}{18}$ Write the ratio as a fraction.

$= \frac{16}{9}$ Simplify.

Example 2 **Determine whether 5:6 and 15:18 are equivalent.**

$5:6 = \frac{5}{6}$ $\quad 15:18 = \frac{15}{18}$ or $\frac{5}{6}$

The ratios in simplest form both equal $\frac{5}{6}$. So, 5:6 and 15:18 are equivalent.

msmath2.net/vocabulary_review

FOLDABLES™ Study Organizer

For more information about Foldables, see *Teaching Mathematics with Foldables.*

Have students look through the chapter to make sure they have included examples in their Foldables for each lesson or topic in the chapter.

Encourage students to refer to their Foldables while completing the Study Guide and Review and to use them in preparing for the Chapter Test.

7-2 Rates (pp. 292–295)

Find each unit rate. Round to the nearest hundredth if necessary.

19. \$23.75 for 5 pounds **\$4.75 per lb**
20. 810 miles in 9 days **90 mi per day**
21. \$38 in 4 hours **\$9.50 per h**
22. 24 gerbils in 3 cages **8 gerbils per cage**
23. 14 laps in 4 minutes **$3\frac{1}{2}$ laps per min**
24. **SHOPPING** Which bottle of laundry detergent shown at the right costs the least per ounce? **64-oz bottle**

Bottle	Price
12 oz	\$2.16
32 oz	\$5.60
64 oz	\$10.24

Example 3 **Find the unit price per ounce of a 16-ounce box of pasta that is on sale for 96¢.**

To find the cost per ounce, write the rate as a fraction and then simplify.

$$\text{16-oz box for 96¢} = \frac{96¢}{16 \text{ ounces}} = \frac{96¢ \div 16}{16 \text{ ounces} \div 16} = \frac{6¢}{1 \text{ ounce}}$$

The unit price is 6¢ per ounce.

7-3 Solving Proportions (pp. 297–300)

Solve each proportion.

25. $\frac{x}{10} = \frac{3}{5}$ **6**
26. $\frac{4}{9} = \frac{24}{m}$ **54**
27. $\frac{2}{t} = \frac{8}{50}$ **12.5**
28. $\frac{15}{w} = \frac{35}{21}$ **9**
29. $\frac{12}{8} = \frac{a}{6}$ **9**
30. $\frac{7}{18} = \frac{d}{6}$ **$2\frac{1}{3}$**
31. **MODELS** A Boeing 747 jet is 70.5 meters long and has a wingspan of 60 meters. A model of the 747 has a wingspan of 80 centimeters. What is its length? **94 cm**

Example 4 **Solve $\frac{6}{9} = \frac{n}{12}$.**

$\frac{6}{9} = \frac{n}{12}$	Write the proportion.
$6 \times 12 = 9 \times n$	Find the cross products.
$72 = 9n$	Multiply.
$\frac{72}{9} = \frac{9n}{9}$	Divide each side by 9.
$8 = n$	Simplify.

The solution is 8.

7-4 Scale Drawings (pp. 304–308)

Suppose you are making a scale drawing. Find the length of each object on the drawing with the given scale.

32. a minivan 10 feet long; 2 inches = 1 foot **20 in.**
33. a rose garden 9 meters wide; 3 centimeters = 1 meter **27 cm**
34. a Ferris wheel with a diameter of 42 feet; 2 inches = 3 feet **28 in.**
35. a library 10.4 meters wide; 1 centimeter = 2 meters **5.2 cm**

Example 5 **On a map, the distance between Chicago, Illinois, and Mexico City, Mexico, is 10.9 centimeters. If the scale is 1 centimeter = 250 kilometers, what is the actual distance?**

Scale = Chicago to Mexico City	
$\frac{1 \text{ cm}}{250 \text{ km}} = \frac{10.9 \text{ cm}}{n \text{ km}}$	Write a proportion.
$1 \times n = 250 \times 10.9$	Find the cross products.
$n = 2{,}725$	Simplify.

The actual distance is 2,725 kilometers.

Answers

15. yes; $\frac{3}{5} = \frac{21}{35}$
16. no; $\frac{18}{24} = \frac{3}{4}, \frac{5}{20} = \frac{1}{4}$, and $\frac{3}{4} \neq \frac{1}{4}$
17. yes; $27{:}15 = \frac{9}{5}$ and $9{:}5 = \frac{9}{5}$
18. no; $4{:}21 = \frac{4}{21}$, $2{:}7 = \frac{2}{7}$, and $\frac{4}{21} \neq \frac{2}{7}$

7-5 Fractions, Decimals, and Percents (pp. 312–315)

Write each percent as a fraction in simplest form.

36. 27.5% $\frac{11}{40}$ 37. 5.4% $\frac{27}{500}$ 38. $45\frac{1}{4}\%$ $\frac{181}{400}$

Write each fraction as a percent. Round to the nearest hundredth if necessary.

39. $\frac{1}{8}$ 12.5% 40. $\frac{5}{6}$ 83.33% 41. $\frac{7}{40}$ 17.5%

Example 6 **Write 82.5% as a fraction in simplest form.**

$82.5\% = \frac{82.5}{100}$ Write a fraction with a denominator of 100.

$= \frac{825}{1{,}000}$ Multiply 82.5 and 100 by 10 to eliminate the decimal.

$= \frac{33}{40}$ Simplify.

7-6 Percents Greater Than 100% and Percents Less Than 1% (pp. 316–318)

Write each percent as a decimal and as a mixed number or fraction in simplest form.

42. 125% 1.25; $1\frac{1}{4}$ 43. 0.75% 0.0075; $\frac{3}{400}$

44. 563% 5.63; $5\frac{63}{100}$ 45. 0.5% 0.005; $\frac{1}{200}$

Write each decimal as a percent.

46. 0.002 0.2% 47. 4.75 475%

48. 7.5 750% 49. 0.0095 0.95%

Example 7 **Write 235% as a decimal and as a mixed number in simplest form.**

235% = 2.35 Definition of percent

$= 2.35$ or $2\frac{7}{20}$

Example 8 **Write 0.008 as a percent.**

0.008 = 0.008 Multiply by 100.

= 0.8%

7-7 Percent of a Number (pp. 319–321)

Find each number. Round to the nearest tenth if necessary.

50. Find 78% of 50. 39
51. 45.5% of 75 is what number? 34.1
52. What is 225% of 60? 135
53. 0.75% of 80 is what number? 0.6

Example 9 **Find 24% of 200.**

24% of 200

$= 24\% \times 200$ Write a multiplication expression.

$= 0.24 \times 200$ Write 24% as a decimal.

$= 48$ Multiply.

So, 24% of 200 is 48.

7-8 The Percent Proportion (pp. 323–325)

Find each number. Round to the nearest tenth if necessary.

54. 6 is what percent of 120? 5%
55. Find 0.8% of 35. 0.3
56. What percent of 375 is 40? 10.7%
57. Find 310% of 42. 130.2

Example 10 **What percent of 90 is 18?**

$\frac{a}{b} = \frac{p}{100}$ Percent proportion

$\frac{18}{90} = \frac{p}{100}$ Replace *a* with 18 and *b* with 90.

$18 \cdot 100 = 90 \cdot p$ Find the cross products.

$20 = p$ So, 18 is 20% of 90.

328 Chapter 7 Ratios and Proportions

CHAPTER 7 Practice Test

Vocabulary and Concepts

1. **Define** *equivalent ratios.* two ratios that have the same value
2. **Write** a proportion that can be used to find 76% of 512. $\frac{x}{512} = \frac{76}{100}$

Skills and Applications

Write each ratio as a fraction in simplest form.

3. 45 to 18 $\frac{5}{2}$
4. 24:88 $\frac{3}{11}$
5. 15 minutes to 2 hours $\frac{1}{8}$

Find each unit rate. Round to the nearest hundredth if necessary.

6. 24 cards for $4.80 $0.20 per card
7. 330 miles on 15 gallons of gasoline 22 mi per gal

Solve each proportion.

8. $\frac{2}{3} = \frac{x}{42}$ 28
9. $\frac{t}{21} = \frac{15}{14}$ 22.5
10. $\frac{9}{m} = \frac{12}{36}$ 27

11. **BLUEPRINTS** The dimensions of a rectangular room in a new home are shown in the blueprint at the right. If the scale is 1 centimeter = 2.5 meters, find the length ℓ of the room on the scale drawing. Then find the scale factor. 1.84 cm; $\frac{1}{250}$

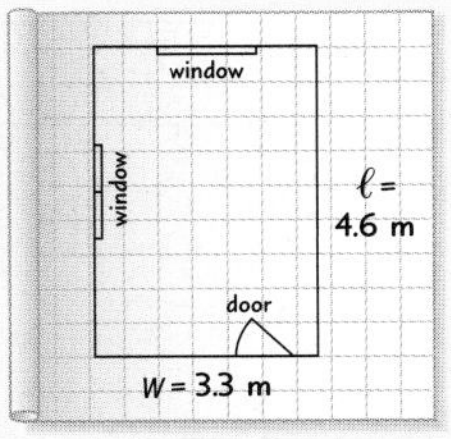

12. **SWIMMING** Alyssa swims 3 laps in 12 minutes. At this same rate, how many laps will she swim in 30 minutes? $7\frac{1}{2}$ laps

Write each percent as a fraction in simplest form.

13. $56\frac{1}{4}\%$ $\frac{9}{16}$
14. $82\frac{1}{4}\%$ $\frac{329}{400}$
15. $83\frac{1}{3}\%$ $\frac{5}{6}$

Write each fraction as a percent. Round to the nearest hundredth if necessary.

16. $\frac{5}{8}$ 62.5%
17. $\frac{7}{15}$ 46.67%
18. $\frac{33}{40}$ 82.5%
19. Write 0.45% as a decimal and as a fraction in simplest form. 0.0045; $\frac{9}{2{,}000}$
20. Write 8.25 as a percent. 825%

Find each number. Round to the nearest tenth if necessary.

21. What is 18% of 30? 5.4
22. What number is 162.2% of 50? 81.1
23. What percent of 64 is 12.8? 20%
24. 458 is 105% of what number? 436.2

Standardized Test Practice

25. **MULTIPLE CHOICE** The purchase price of a bicycle is $140. The state tax rate is $6\frac{1}{2}\%$ of the purchase price. What is the tax rate expressed as a decimal? C

Ⓐ 6.5 Ⓑ 0.65 Ⓒ 0.065 Ⓓ 0.0065

msmath2.net/chapter_test

CHAPTER 7 Practice Test

Assessment Options

Vocabulary Test A vocabulary review/test for Chapter 7 can be found on p. 428 of the *Chapter 7 Resource Masters.*

Chapter Tests There are six Chapter 7 Tests and an Extended Response Assessment task available in the *Chapter 7 Resource Masters.*

Chapter 7 Tests			
Form	Type	Level	Pages
1	MC	basic	415–416
2A	MC	average	417–418
2B	MC	average	419–420
2C	FR	average	421–422
2D	FR	average	423–424
3	FR	advanced	425–426

MC = multiple-choice questions
FR = free-response questions

Extended Response Assessment

Extended Response tasks for Chapter 7 can be found on p. 427 of the *Chapter 7 Resource Masters.* A sample scoring rubric for these tasks appears on p. A28.

Portfolio Suggestion

Introduction Ratios and proportions are found in the daily newspaper and magazines, and on the Internet. They are common methods for comparison.

Ask Students to include examples of real-life uses of ratios and proportions in their portfolios.

CHAPTER 7 Standardized Test Practice

These two pages contain practice questions in the various formats that can be found on the most frequently given standardized tests.

A student recording sheet for these two pages can be found on page A1 of the *Chapter 7 Resource Masters*.

Standardized Test Practice Student Recording Sheet, p. A1

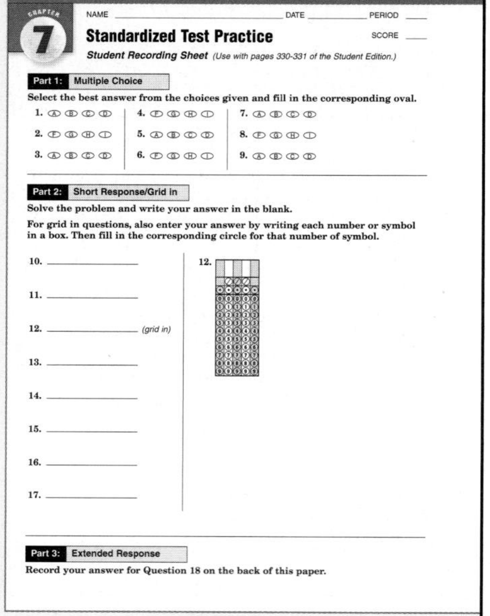

Additional Practice

See pp. 433–434 of the *Chapter 7 Resource Masters* for additional standardized test practice.

CHAPTER 7 Standardized Test Practice

PART 1 Multiple Choice

Record your answers on the answer sheet provided by your teacher or on a sheet of paper.

1. Andre charges $4 per day for taking care of the neighbor's dog. Which expression represents the amount of money he earns in n days? (Lesson 1-4) **A**

   (A) $4n$ (B) $4 + n$
   (C) $4 - n$ (D) n^4

2. The box-and-whisker plot shows the prices of single-scoop ice cream cones. What is the median price? (Lesson 2-6) **G**

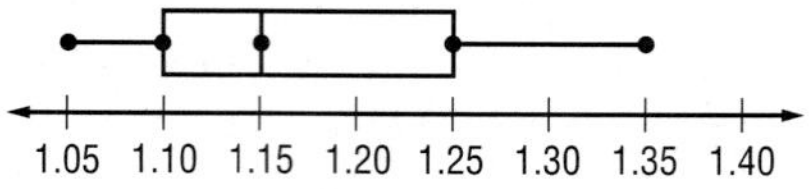

   (F) $1.05 (G) $1.15
   (H) $1.25 (I) $1.35

3. The graph shows the relationship between which two units of measure? (Lessons 4-6 and 6-7) **D**

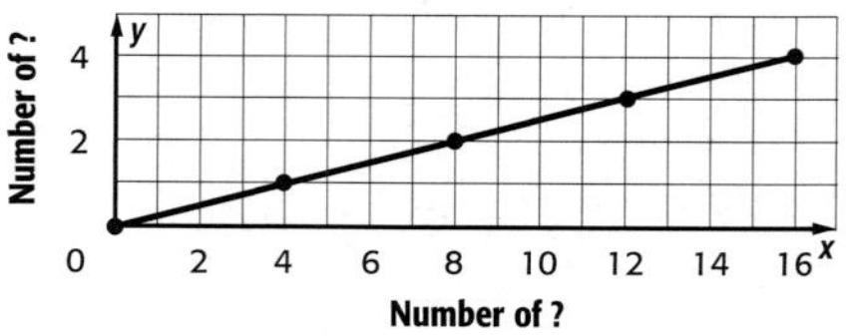

   (A) feet and inches
   (B) yards and feet
   (C) pounds and ounces
   (D) quarts and gallons

4. At Igoe's Campsite, 5 of the 15 sites were rented by campers. Which decimal equals the fraction of sites that were rented? (Lesson 5-4) **F**

   (F) $0.\overline{33}$ (G) 0.30 (H) 3.30 (I) 33.30

5. A survey showed that the ratio of tourists at the Grand Canyon who had cameras to those who did not was 8 : 10. A similar survey was conducted at the Statue of Liberty. If the ratios are equivalent, what could be the ratio at the Statue of Liberty? (Lesson 7-1) **D**

   (A) 32 : 30 (B) 30 : 24
   (C) 30 : 32 (D) 24 : 30

6. Tre can play 18 holes of golf in 180 minutes. What is his average rate in number of minutes per hole? (Lesson 7-2) **I**

   (F) 1.8 (G) 3
   (H) 6 (I) 10

7. Which of the following is the value of p in $\frac{p}{7} = \frac{21}{49}$? (Lesson 7-3) **A**

   (A) 3 (B) 6
   (C) 11 (D) 16

8. A scale model of a computer circuit board is 1 centimeter = 0.2 centimeter. The actual circuit board is 4 centimeters wide. What is the width of the model? (Lesson 7-4) **H**

   (F) 0.2 cm (G) 2 cm
   (H) 20 cm (I) 200 cm

9. The table shows the percent of Internet users in each region of the United States who are 3–17 years old. What fraction of Internet users in the South region are 3–17 years old? (Lesson 7-5) **B**

Internet Users 3–17 Years Old

Region	Percent
Midwest	32.1
Northeast	35.5
South	27.6
West	29.0

Source: U.S. Census Bureau

   (A) $\frac{69}{500}$ (B) $\frac{69}{250}$
   (C) $\frac{7}{25}$ (D) $\frac{7}{50}$

ExamView® Pro Test Bank CD-ROM

Special banks of standardized test questions similar to those on the TIMSS and NAEP grade 8 tests can be found on this CD-ROM.

RT 2 Short Response/Grid In

cord your answers on the answer sheet
ovided by your teacher or on a sheet of
per.

The graph shows the number of school days per year in four countries. How many more days per year do Japanese students attend school than students in the United States? (Lesson 2-7) **63 days**

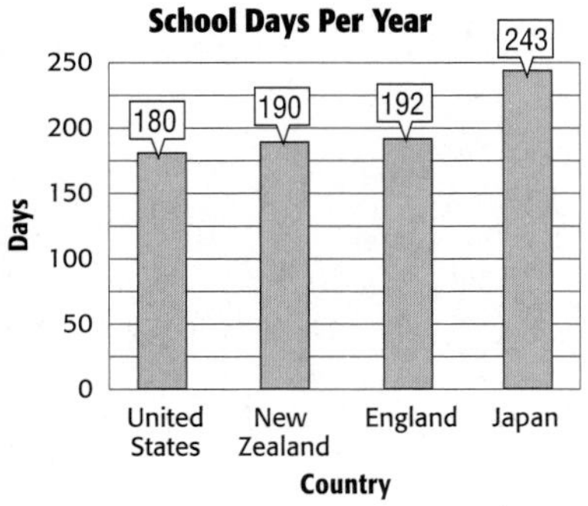

A train goes 65 miles per hour and travels 320 miles. How many hours will it take for the train to reach its destination? Round to the nearest tenth. (Lesson 3-7) **4.9 h**

The table at the right represents a function. What is the missing x-coordinate? (Lesson 4-6) **9**

x	y
3	9
4	12
7	21
?	27

Mellia has written $\frac{3}{5}$ of a research paper on fine art and French literature. What percent of the paper has she written? (Lesson 5-5) **60%**

The area of the rectangle is 18 square inches. What is the width w? (Lesson 6-8) **$2\frac{2}{3}$ in.**

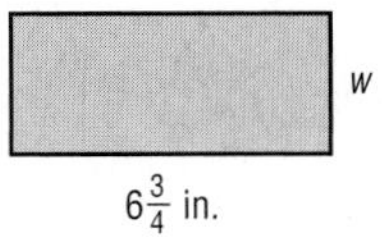

Ron's camera has a shutter speed of $\frac{1}{1{,}000}$. What is this shutter speed expressed as a percent? (Lesson 7-6) **0.1%**

16. Jean practices her headstand in gymnastics class every day. She has been able to stay in the headstand pose without tipping for 45% of the last 60 classes. How many classes is this? (Lesson 7-7) **27 classes**

17. The table shows the nutrition amounts in one serving of pretzels and the percent of the recommended daily amounts.

Nutrition Facts of Pretzels		
Item	**Amount**	**Percent***
fat	1 g	1.5%
sodium	486 mg	20%
carbohydrate	22 g	7.5%

Source: www.diet-data.com
*based on a 2,000-Calorie diet

How many daily milligrams of sodium are recommended? (Lesson 7-8) **2,430 mg**

PART 3 Extended Response

Record your answers on a sheet of paper. Show your work.

18. Jessie built a model of her town using items from her family's recycling bin.

a. One centimeter in Jessie's model equals 10 kilometers in the real town. If Lansdowne Road is 12 centimeters long in the model, how long is the real Lansdowne Road? (Lesson 7-4) **120 km**

b. Jessie has cut up milk cartons to construct $\frac{4}{5}$ of her model town. What percent is this? (Lesson 7-5) **80%**

c. Jessie used a total of 70 recycled items to build her model town. Explain how to find the number of items that were milk cartons. Then solve. (Lesson 7-8) **See margin.**

TEST-TAKING TIP

Question 18 When a question involves an answer from a previous question, make sure to check that answer before you move on.

msmath2.net/standardized_test

Answer

18c. Sample answer: Write and solve the proportion $\frac{4}{5} = \frac{x}{70}$. So, 56 of the items were milk cartons.

Standardized Test Practice

Rubrics The following rubrics are sample scoring devices for short response and extended response questions.

Short Response

Points	Description
2	The student demonstrates a thorough understanding of the mathematics of the task. The response may contain minor flaws that do not detract from the demonstration of a thorough understanding.
1	The student has provided a response that is only partially correct.
0	The student has provided a completely incorrect solution or no response at all.

Extended Response

Points	Description
4	The student demonstrates a thorough understanding of the mathematics of the task. The response may contain minor flaws that do not detract from the demonstration of a thorough understanding.
3	The student demonstrates an understanding of the mathematics of the task. The response is essentially correct and demonstrates an essential but less than thorough understanding of the mathematics.
2	The student demonstrates only a partial understanding of the mathematics of the task. Although the student may have used the correct approach to a solution or may have provided a correct solution, the work lacks an essential understanding of the underlying mathematical concepts.
1	The student demonstrates a very limited understanding of the mathematics of the task. The response is incomplete and exhibits many flaws.
0	The student provides a completely incorrect solution or no response at all.

A more detailed 4-point rubric is provided on page A2 of the *Chapter 7 Resource Masters*.

Page 322, Lesson 7-8a (Your Turn)

a.

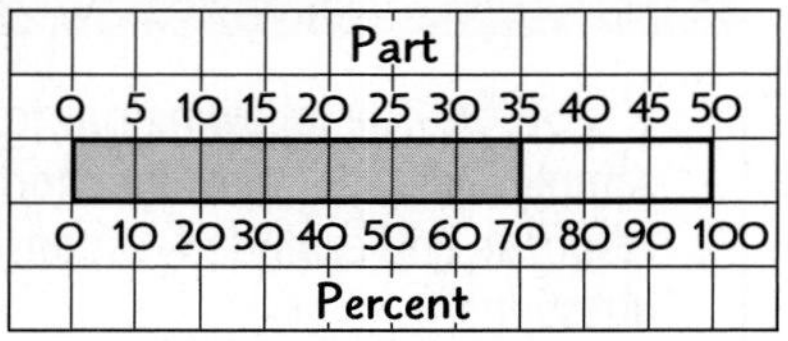

b.

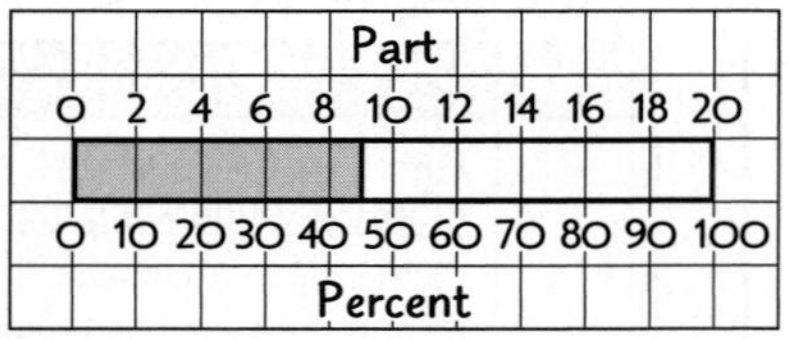

c.

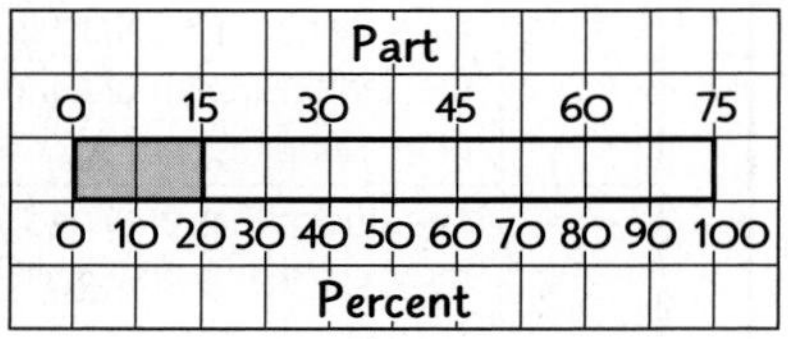

NOTES

CHAPTER 8

Chapter Overview

Applying Percent

LESSON OBJECTIVES		NCTM Standards	NAEP 2005	State/Local Objectives
8-1	**Percent and Estimation** *(pp. 334–339)* • Estimate percents by using fractions and decimals. **8-1b Problem-Solving Strategy:** Solve problems by determining reasonable answers.	1, 6–10	N2a, N2b, N2c, N3a, N3g	
8-2	**The Percent Equation** *(pp. 340–343)* • Solve problems by using the percent equation.	1, 2, 6–10	N2c, N3a, N4d, N4f, A4a	
8-3	**Using Statistics to Predict** *(pp. 344–347)* **8-3a Hands-On Lab:** Use sampling to collect data. • Predict actions of a larger group by using a sample.	1, 2, 5, 6–10	D3b, N4c, N4d, A4a, D1a, D3a	
8-4	**Percent of Change** *(pp. 350–353)* • Find the percent of increase or decrease.	1, 2, 6–10	N1b, N4b, N4d, N3a, N3g, N5f	
8-5	**Sales Tax and Discount** *(pp. 354–357)* • Solve problems involving sales tax and discount.	1, 2, 6–10	N4d, N3a, N3g, A4a, N5f	
8-6	**Simple Interest** *(pp. 358–361)* • Solve problems involving simple interest. **8-6b Spreadsheet Investigation:** Use a spreadsheet to calculate simple interest.	1, 2, 6–10	N4d, A4e, N3a, N3g	

Key to NCTM Standards:

1=Number & Operations, 2=Algebra, 3=Geometry, 4=Measurement, 5=Data Analysis & Probability, 6=Problem Solving, 7=Reasoning & Proof, 8=Communication, 9=Connections, 10=Representation

NAEP 2005 Standards: See math.glencoe.com

Pacing suggestions for the entire year can be found on page T20.

Chapter Resource Manager

Time-Saving Tool
TeacherWorks™
All-In-One Planner and Resource Center
See page T5.

FAST FILE Chapter 8 Resource Masters

Study Guide and Intervention	Practice (Skills and Word Problems)	Reading to Learn Mathematics	Enrichment	Assessment	Applications*	Parent and Student Study Guide Workbook	5-Minute Check Transparencies	Key Teaching Transparencies	MathPASS: Tutorial Plus (lessons)	Teaching Mathematics with Manipulatives	Optional Materials
435	436–437	438	439		SM 89–92	8-1	8-1			75	
440	441–442	443	444	479	RC 22	8-2	8-2	29	34		
445	446–447	448	449	479, 481	SM 37–40	8-3	8-3	30		76	
450	451–452	453	454			8-4	8-4			77–78	paper strips, scissors, tape
455	456–457	458	459	480	CS 41 CM 23–24	8-5	8-5	31			
460	461–462	463	464	465–478, 480, 482–484	CS 42	8-6	8-6	32	35		

* **Key to Abbreviations:** *CS = Calculator and Spreadsheet Masters, RC = Real-Life Career Masters, CM = Classroom Math Games (online), SM = Science and Math Lab Manual (online)*

Interactive Chalkboard

PowerPoint® Presentations

This CD-ROM is a customizable Microsoft® PowerPoint® presentation that includes:

- In-Class Examples
- Links to Online Study Tools
- 5-Minute Check Transparencies
- Bellringers

Math Skills Maintenance

Use the Math Skills Maintenance Masters to keep your students' basic math skills fresh. Students can review skills one at a time or in a combined format. The Masters are also available online at msmath2.net/maintenance.

CHAPTER 8 Mathematical Content and Teaching Strategies

Continuity of Instruction

Building on Prior Knowledge

In previous courses, students have explored the concept of percent. In addition, they have gained experience using fractions, decimals, and percents in Chapters 5, 6, and 7. They have worked with visual models to understand these concepts and practiced computations. Students know how to convert among fractions, decimals, and percents and can use the percent proportion to solve problems.

Learning in This Chapter

In this chapter, students continue to build on their experiences with decimals and percents. They use benchmarks to estimate percents and rewrite the percent proportion to establish the percent equation. Students explore applications including percent of change, sales tax, discount, and simple interest.

Making Future Connections

Skills students gained rewriting the percent equation to find an unknown form the basis for solving linear equations in algebra. Additionally, the applications in this chapter have a strong link to students' everyday lives. Percents are presented as statistics in news stories and discounts in advertisements, so the ability to understand percents will help students evaluate data critically and arrive at good decisions.

Mathematical Background

8-1 Percent and Estimation

In this lesson, students use fractions and decimals to estimate percents of whole numbers. While students may be familiar with estimating 50% of a number, in this lesson they also learn to estimate percents greater than 100 or less than 1. The ability to estimate percents quickly and accurately will help students make predictions and determine whether an answer is reasonable. Real-life applications include finding tips at restaurants.

8-2 The Percent Equation

In Chapter 7, students used the percent proportion to solve problems. In this lesson, they use the percent proportion to derive the percent equation. Both the percent proportion and the percent equation can be used to find the missing part, percent, or base in a percent problem. Manipulating the three parts of the percent equation will help students learn to rearrange terms in equations and solve for an unknown, a basic algebra skill.

Team Teaching

Inform the other teachers on your team that you are studying percents. Coordinate with the others to discuss at least one example in each class, then have students report what they discussed in each setting. They should explain how the mathematics they are learning helps them understand their other subjects.

Physical Education: sports statistics, percent of improvement

Physical Science: percent change in speed while accelerating

Social Studies: population growth, consumer math, economics

Teacher to Teacher

Carolyn Catto
Harney Middle School
Las Vegas, NV

"For an interdisciplinary activity, I have students find their target heart rates for exercise. First find the resting heart rate. Then use the formula 220 – age – resting heart rate = base. Finally, multiply the base by 60%, 70%, or 80% to find target heart rates for different levels of intensity. You can also expand the activity to review measures of central tendency by finding the class average target heart rate."

Carolyn Catto

8-3 Using Statistics to Predict

In this lesson, students use the percent proportion and the percent equation to make predictions based on survey data. The ability to use statistics to predict is one of the core skills in statistical reasoning. This type of reasoning will help students design statistical studies, form predictions about outcomes, and understand whether the results are significant. Real-life applications include analyzing data presented by the news media.

8-4 Percent of Change

Students will encounter percent of change frequently, including in reports of fluctuation in the stock market, property values, and levels of public support for political leaders or legislation. This lesson will give them the skills needed to make sense of such information. Tell students to watch for phrases such as "more than doubled" or "cut in half," which indicate that percent of change is being discussed.

8-5 Sales Tax and Discount

Consumers must make choices everyday on what to buy, and making a decision based on price is made more complex with the addition of sales tax and promotional discounts. This lesson will give students the ability to use percents to calculate the final cost of an item. Point out that students are building on their percent of change skills—for example, a 6% sales tax represents a 6% increase in cost.

8-6 Simple Interest

Whether financing a large purchase, setting up a savings plan, or leaving even a small balance on a credit card, understanding simple interest will help students evaluate the benefits and consequences of their economic decisions. Students will learn the formula for simple interest, which can be manipulated in a fashion similar to their previous manipulation of the percent equation.

In-Class Speaker

Consider inviting a representative of a bank to visit your mathematics class. He or she could discuss the services provided by the bank as they relate to percents as presented in this chapter, particularly Lesson 8-6. The visitor could address savings account options, CDs, simple interest loans, and credit cards.

McGraw-Hill Professional Development

Mathematics, YES! is a self-paced, web-based professional development program. Units are based on major mathematical concepts and are standards-based. Actual classroom videos provide a springboard for discussion and activities. Units appropriate for this chapter:

- **Unit 4, Fractions, Decimals, and Percentages**
- **Unit 10, Statistics, Data Analysis, and Probability**

Key Concepts

Additional mathematical information and teaching notes are available at

msmath2.com/key_concepts.

The sections appropriate for the chapter are:

- **Percent of Change**
- **Simple Interest**

CHAPTER 8 DAILY INTERVENTION and Assessment

	Type	Student Edition	Teacher Resources	Technology/Internet
INTERVENTION	Ongoing	Prerequisite Skills, pp. 333, 337, 343, 347, 353, 357 Mid-Chapter Practice Test, p. 348	5-Minute Check Transparencies Quizzes 1–4, *CRM*, pp. 479–480 Mid-Chapter Test, *CRM*, p. 481 Study Guide and Intervention, *CRM*, pp. 435, 440, 445, 450, 455, 460	MathPASS: Tutorial Plus msmath2.net/self_check_quiz msmath2.net/extra_examples
	Mixed Review	pp. 337, 343, 347, 353, 357, 360	Cumulative Review, *CRM*, p. 482	
	Error Analysis	Find the Error, pp. 336, 352	Find the Error, *TWE*, pp. 336, 352 Unlocking Misconceptions, *TWE*, p. 351	
	Standardized Test Practice	pp. 337, 339, 343, 347, 348, 353, 355, 357, 360, 365, 366–367	Standardized Test Practice, *CRM*, pp. 483–484	Standardized Test Practice CD-ROM msmath2.net/standardized_test
ASSESSMENT	Open-Ended Assessment	Writing Math, pp. 336, 342, 344, 346, 352, 356, 359 Open-Ended, pp. 336, 342, 346, 352, 356, 359 Standardized Test, p. 367	Modeling: *TWE*, p. 343 Speaking: *TWE*, p. 347 Writing: *TWE*, pp. 337, 339 Extended Response Assessment, *CRM*, p. 477	
	Chapter Assessment	Study Guide, pp. 362–364 Practice Test, p. 365	Multiple-Choice Tests (Forms 1, 2A, 2B), *CRM*, pp. 465–470 Free-Response Tests (Forms 2C, 2D, 3), *CRM*, pp. 471–476 Vocabulary Test/Review, *CRM*, p. 478	ExamView Pro® Test Bank (see below) MindJogger Videoquizzes msmath2.net/vocabulary_review msmath2.net/chapter_test

* **Key to Abbreviations:** *TWE* = Teacher Wraparound Edition, *CRM* = Chapter Resource Masters

Additional Intervention

Skills Intervention for Middle School Mathematics workbook and CD-ROM includes 88 skills with pretests, tutorials, guided practice, and posttests for diagnosis and remediation.

Guide to Daily Intervention booklet details each intervention resource available and guides teachers in using them as intervention tools everyday to assist students who may be having difficulty.

ExamView® Pro

Use ExamView® Pro Test Bank CD-ROM to:

- **Create Multiple versions of tests.**
- **Create Modified tests for Inclusion students with one mouse click.**
- **Edit existing questions and Add your own questions.**
- **Build tests aligned with state standards using built-in State Curriculum Correlations.**
- **Change English tests to Spanish with one mouse click and vice versa.**

Reading and Writing Mathematics

- **Student Edition**

Foldables Study Organizer, p. 333

Concept Check questions at the beginning of each exercise set require students to verbalize and write about what they have learned in the lesson.

- **Open Ended,** pp. 336, 342, 346, 352, 356, 359
- **Find the Error,** pp. 336, 352

Reading Math Tip, p. 359

- **Teacher Wraparound Edition**

Foldables Follow-Up, pp. 336, 342, 348, 352, 356

Writing Activities, pp. 337, 339

Differentiated Instruction, (Verbal/Linguistic), p. 335

ELL Resources, *see page T9 for a complete listing.*

- **Additional Resources**

Vocabulary Builder, *Chapter 8 Resource Masters*

Reading to Learn Mathematics, *Chapter 8 Resource Masters*

Vocabulary PuzzleMaker (online)

Teaching Mathematics with Foldables

Reading and Writing in the Mathematics Classroom

WebQuests, Projects, and Interdisciplinary Investigations

For detailed information on these resources, see pp. T6–T7.

- **Recommended Outside Reading for Students**

Nonfiction

- Brumbaugh, Doug & Brumbaugh, Linda. *Scratch Your Brain Where It Itches.* Critical Thinking Books and Software, ©1994.

  Book C in this series uses concepts of percents, fractions, and decimals in a collection of engaging brain teasers and puzzles. Written for Grades 7–9.

Fiction

- Haver, D., Kozoil, A., Haven, E., & Mulligan, D. *57 Great Math Stories and the Problems They Present.* Instructional Fair, ©1998.

  Appropriate for Grades 5–8. Each story is a humorous yet realistic account of everyday life and the problems life presents. Many stories throughout the book utilize percents. Often the reader must make calculations using the percent equation or proportion to understand the outcome of the story.

PROJECT CRISS℠ Study Skill

Writing out the steps in a mathematical process using a point-by-point format can help students better understand that process. In order to explain a process, they must not only understand how to perform each step, but they must also understand the reasoning behind each step. Provide students with the description at the right as an example of using a point-by-point format to explain how to multiply a fraction and a whole number. After reading Lessons 8-5 and 8-6, have students write a description of how to solve sales tax and simple interest problems. Students may also find two-column notes to be a useful tool in this chapter.

Multiplying a Fraction and a Whole Number

1. Write the whole number as a fraction with a denominator of 1. For example, 8 would be written as $\frac{8}{1}$.
2. Multiply the fractions. The numerators should be multiplied and the denominators should be multiplied.
3. If the numerator is greater than the denominator, write the answer as a mixed number.
4. Be sure the answer is written in simplest form. If not, rename.

CReating Independence through Student-owned Strategies

For more information on Project CRISS, see page T9.

CHAPTER 8

Making Connections

School-To-Home

Parent and Student Study Guide

Parents and students may work together to reinforce the concepts and skills of this chapter. These workbooks are available in English and Spanish. A sample lesson is shown below. Also available at msmath2.net/parent_student.

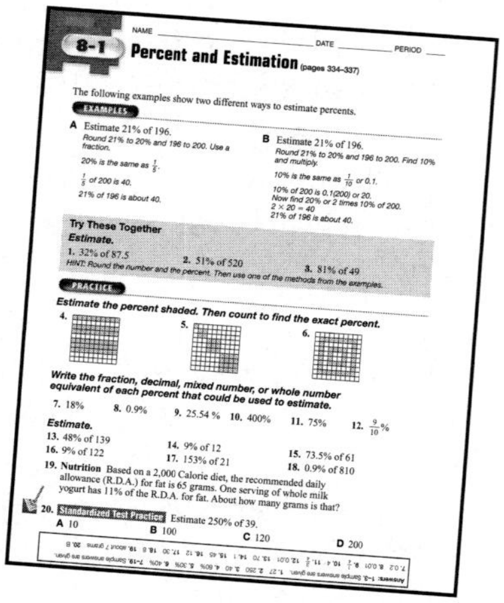

NAME ______ DATE ______ PERIOD ______

8-1 Percent and Estimation (pages 334–337)

The following examples show two different ways to estimate percents.

EXAMPLES

A Estimate 21% of 196.
Round 21% to 20% and 196 to 200. Use a fraction.
20% is the same as $\frac{1}{5}$.
$\frac{1}{5}$ of 200 is 40.
21% of 196 is about 40.

B Estimate 21% of 196.
Round 21% to 20% and 196 to 200. Find 10% and multiply.
10% is the same as $\frac{1}{10}$ or 0.1.
10% of 200 is 0.1(200) or 20.
Now find 20% or 2 times 10% of 200.
$2 \times 20 = 40$
21% of 196 is about 40.

Try These Together
Estimate.
1. 32% of 87.5
2. 51% of 520
3. 81% of 49

HINT: Round the number and the percent. Then use one of the methods from the examples.

PRACTICE

Estimate the percent shaded. Then count to find the exact percent.
4. 5. 6.

Write the fraction, decimal, mixed number, or whole number equivalent of each percent that could be used to estimate.
7. 18% 8. 0.9% 9. 25.54 % 10. 400% 11. 75% 12. $\frac{9}{10}$%

Estimate.
13. 48% of 139 14. 9% of 12 15. 73.5% of 61
16. 9% of 122 17. 153% of 21 18. 0.9% of 810

19. **Nutrition** Based on a 2,000 Calorie diet, the recommended daily allowance (R.D.A.) for fat is 65 grams. One serving of whole milk yogurt has 11% of the R.D.A. for fat. About how many grams is that?

20. **Standardized Test Practice** Estimate 250% of 39.
A 10 B 100 C 120 D 200

Answers: 1–3. Sample answers are given. 1. 27 2. 250 3. 40 4. 60% 5. 30% 6. 40% 7–19. Sample answers are given. 7. 0.2 8. 0.01 9. $\frac{1}{4}$ 10. 4 11. $\frac{3}{4}$ 12. 0.001 13. 70 14. 1 15. 45 16. 12 17. 30 18. 8 19. about 7 grams 20. B

Family Letters and Activities

Send a letter and activity for this chapter home to encourage family members to become active participants in their students' learning. (English: *Chapter 8 Resource Masters*, pp. ix–x; Spanish: log on to msmath2.net/family_letters)

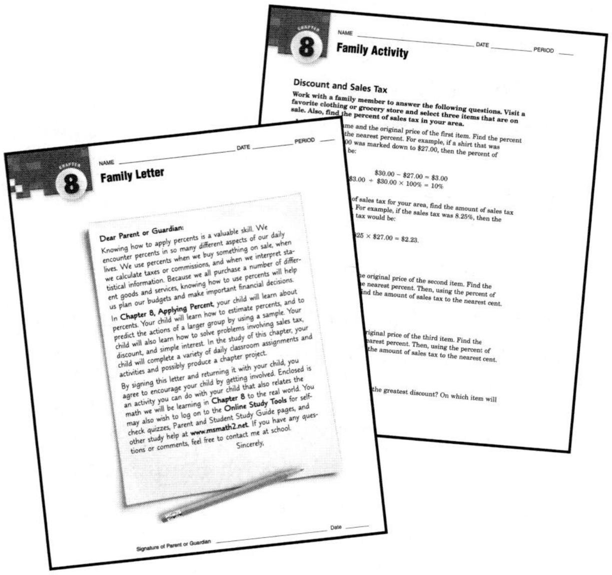

CHAPTER 8 **Family Activity**

NAME ______ DATE ______ PERIOD ______

Discount and Sales Tax

Work with a family member to answer the following questions. Visit a favorite clothing or grocery store and select three items that are on sale. Also, find the percent of sales tax in your area.

...me and the original price of the first item. Find the percent ...the nearest percent. For example, if a shirt that was ...00 was marked down to $27.00, then the percent of ...be:

$30.00 − $27.00 = $3.00
$3.00 ÷ $30.00 × 100% = 10%

...of sales tax for your area, find the amount of sales tax ... For example, if the sales tax was 8.25%, then the ...tax would be:

...25 × $27.00 = $2.23.

...e original price of the second item. Find the ...e nearest percent. Then, using the percent of ...ind the amount of sales tax to the nearest cent.

...riginal price of the third item. Find the ...earest percent. Then, using the percent of ...the amount of sales tax to the nearest cent.

...the greatest discount? On which item will

CHAPTER 8 **Family Letter**

NAME ______ DATE ______ PERIOD ______

Dear Parent or Guardian:

Knowing how to apply percents is a valuable skill. We encounter percents in so many different aspects of our daily lives. We use percents when we buy something on sale, when we calculate taxes or commissions, and when we interpret statistical information. Because we all purchase a number of different goods and services, knowing how to use percents will help us plan our budgets and make important financial decisions.

In **Chapter 8, Applying Percent**, your child will learn about percents. Your child will learn how to estimate percents, and to predict the actions of a larger group by using a sample. Your child will also learn how to solve problems involving sales tax, discount, and simple interest. In the study of this chapter, your child will complete a variety of daily classroom assignments and activities and possibly produce a chapter project.

By signing this letter and returning it with your child, you agree to encourage your child by getting involved. Enclosed is an activity you can do with your child that also relates the math we will be learning in **Chapter 8** to the real world. You may also wish to log on to the **Online Study Tools** for self-check quizzes, Parent and Student Study Guide pages, and other study help at **www.msmath2.net**. If you have any questions or comments, feel free to contact me at school.

Sincerely,

Signature of Parent or Guardian ______ Date ______

English Language Learner ELL

Lesson 8-4

Peer Tutoring

Give students a set of percent of change problems. Have them work with a partner to read each problem and state whether the change is an increase or decrease. Ask students to also state the amount of the increase or decrease. Students can then form a fraction with the amount of change as the numerator and the original amount as the denominator.

Lesson 8-5

Using Manipulatives

Bring several items such as CDs, T-shirts, and hair accessories. Have a price tag on each item and categorize the items as having different percent-off sale prices. Have students come up in groups and write at least 3 items they would be interested in purchasing and the percent off for each one. Tell the students they have $20 to spend and that they need to take sales tax of 6% into consideration. In groups, have the students calculate how much they will spend once they have calculated the sale prices and included the sales tax.

Lesson 8-6

Using Manipulatives

Bring in a section of the newspaper which lists interest rates for local banks. Have the students look through the paper in their groups and find the bank that gives the most interest for a deposit of $2,500. Using the formula, have students show their calculations for several banks. Then have students explain which bank they would choose to invest their money for 18 months and why they would choose that particular one.

MathScape™

This project was supported, in part, by the **National Science Foundation**

MathScape is a mathematics curriculum for grades 6–8 developed by the Seeing and Thinking Mathematically Project at the Education Development Center.

Rates, Ratios, Percents, and Proportions

Buyer Beware

Summary Students will understand estimating, counting, and rounding with percents. Calculating discounts by a variety of methods is explored. To find percents in a circle graph, percent proportions are used.

USA TODAY Education — Online Lesson Planner

USA TODAY® Education's Online site offers resources and interactive features connected to each day's newspaper. *Experience Today,* USA TODAY's® daily lesson plan, is available on the site and delivered daily to subscribers. This plan provides instruction for integrating USA TODAY® graphics and key editorial features into your mathematics classroom. Log on to **education.usatoday.com**.

USA TODAY Snapshots® in Chapter 8	
Lesson	**Snapshot**
8-2	Texas is the king of cotton
8-3	Many kids watch TV unrestricted

Quick Review Math Handbook

is Glencoe's mathematical handbook for students. The Hot Words section includes a glossary of terms while the Hot Topics section consists of explanations of key mathematical concepts. An exercise set is included to check students' understanding of the concepts. This valuable resource can be used as a reference in the classroom or for home study.

Lesson	Hot Topics Section	Lesson	Hot Topics Section
8-1	2-8	**8-4**	2-8, 6-5
8-2	2-8, 6-5	**8-5**	2-8, 6-5
8-3	4-1, 4-3	**8-6**	2-8, 9-4

CHAPTER 8 Notes

Key Chapter Objectives

Lesson 8-2 Solve problems by using the percent equation.

Lesson 8-3 Predict actions of a larger group by using a sample.

Lesson 8-4 Find the percent of increase or decrease.

Lesson 8-6 Solve problems involving simple interest.

"What do snowboards have to do with math?"

Point out to students that this is only one of many applications of the math that they are studying. Others are provided in the introduction to each lesson.

Vocabulary Builder, pp. vii–viii **ELL**

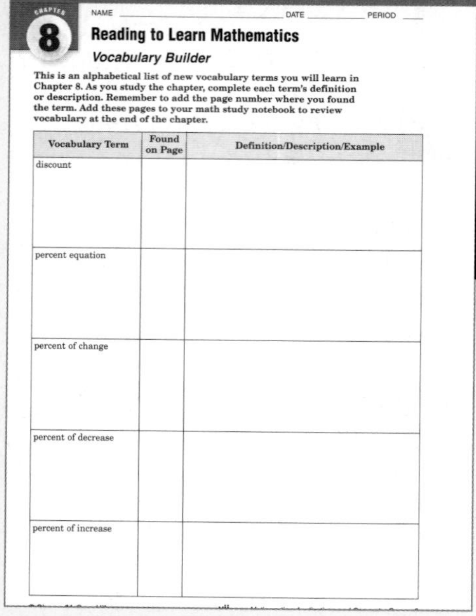

NAME ________ DATE ________ PERIOD ____

CHAPTER 8 **Reading to Learn Mathematics**

Vocabulary Builder

This is an alphabetical list of new vocabulary terms you will learn in Chapter 8. As you study the chapter, complete each term's definition or description. Remember to add the page number where you found the term. Add these pages to your math study notebook to review vocabulary at the end of the chapter.

Vocabulary Term	Found on Page	Definition/Description/Example
discount		
percent equation		
percent of change		
percent of decrease		
percent of increase		

CHAPTER 8 Applying Percent

"What do snowboards have to do with math?"

How much money would you save buying a snowboard that is 35% off its original price of $169?** You could use the equation **$d = 0.35 \cdot 169$ to find the **discount d**. Equations are useful in finding the total cost of an item including the discount and the sales tax.

You will solve problems involving discount and sales tax in Lesson 8-5.

Vocabulary Builder

For a thorough vocabulary list with pronunciations of new words, give students the Vocabulary Builder worksheets from the *Chapter 8 Resource Masters.* Encourage them to complete the definition of each term as they progress through the chapter.

GETTING STARTED

Take this quiz to see whether you are ready to begin Chapter 8. Refer to the lesson number in parentheses if you need more review.

Vocabulary Review

State whether each sentence is *true* or *false*. If *false*, replace the underlined word to make a true sentence.

1. Data refers to pieces of information that are often numerical. (Lesson 2-1) **true**
2. The percent equation is written as $\frac{\text{part}}{\text{base}} = \frac{\text{percent}}{100}$. (Lesson 7-8) **false; proportion**

Prerequisite Skills

Multiply. (Page 560)

3. $300 \times 0.02 \times 8$ **48**
4. $85 \times 0.25 \times 3$ **63.75**
5. $560 \times 0.6 \times 4.5$ **1,512**
6. $154 \times 0.12 \times 5$ **92.4**

Simplify. Write as a decimal. (Lesson 1-3)

7. $\frac{22 - 8}{8}$ **1.75**
8. $\frac{50 - 33}{50}$ **0.34**
9. $\frac{35 - 7}{35}$ **0.8**
10. $\frac{63 - 14}{14}$ **3.5**

Solve. Round to the nearest tenth if necessary. (Lesson 4-3)

11. $0.4m = 52$ **130**
12. $21 = 0.28a$ **75**
13. $13 = 0.06s$ **216.7**
14. $0.95z = 37$ **38.9**

Write each percent as a decimal. (Lesson 7-6)

15. 40% **0.4**
16. 3.25% **0.0325**
17. 7.5% **0.075**
18. 2% **0.02**

Use the percent proportion to find each number. Round to the nearest tenth if necessary. (Lesson 7-8)

19. What percent of 86 is 34? **39.5%**
20. 20% of what number is 55? **275**

FOLDABLES Study Organizer

Percents Make this Foldable to help you organize information about percents. Begin with a piece of 11″ by 17″ paper.

STEP 1 Fold
Fold a 2″ tab along the long side of the paper. Then fold the rest in half.

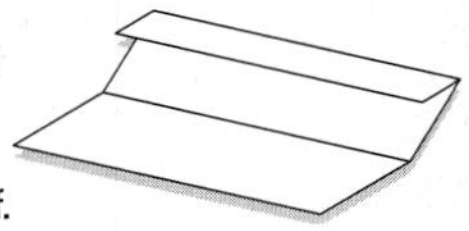

STEP 2 Open and Fold
Open the paper and fold in half widthwise 3 times to make 8 columns.

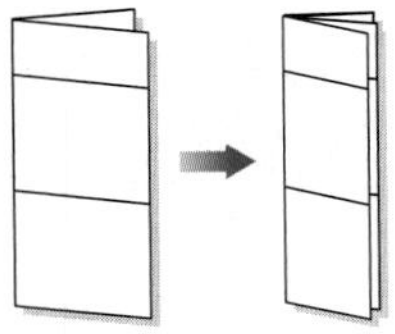

STEP 3 Open and Label
Draw lines along the folds and label as shown.

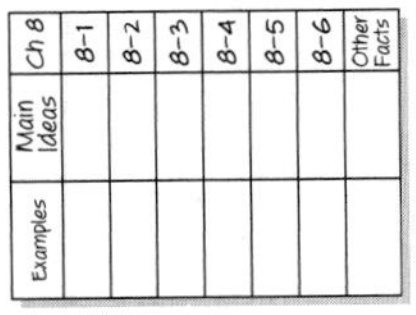

Reading and Writing As you read and study each lesson, write the main ideas and some examples in the corresponding columns. In the last column, write other important facts that may apply to multiple lessons.

GETTING STARTED

This section provides a review of the basic concepts needed before beginning Chapter 8. Page references are included for additional student help.

Prerequisite Skills review skills needed in this chapter.

Exercises	Needed for Lesson	Skill
3–6	8-5 and 8-6	Multiply Decimals
7–10	8-4	Write Fractions as Decimals
11–14	8-2 and 8-5	Solve Equations
15–18	8-1 and 8-6	Write Percents as Decimals
19–20	8-3	Use the Percent Proportion

For more prerequisite skills, see:

- Getting Ready for the Next Lesson, at the end of each lesson
- Prerequisite Skills in the Student Handbook on pp. 554–563
- *Math Skills Maintenance Masters*

FOLDABLES Study Organizer

For more information about Foldables, see *Teaching Mathematics with Foldables.*

Use this Foldable table for students to write about percent. After students make their Foldable tables, have them label the columns and rows as illustrated. Students use their tables to take notes, define terms, record concepts, and write examples of what they learned in each lesson. Encourage students to use the data recorded to pose questions, to present their own ideas or alternative methods, and to reflect on how what they have learned applies to their daily lives.

8-1 Lesson Notes

Lesson Resources

Chapter 8 Resource Masters

FAST FILE

- Study Guide and Intervention, p. 435
- Practice: Skills, p. 436
- Practice: Word Problems, p. 437
- Reading to Learn Mathematics, p. 438
- Enrichment, p. 439

Parent and Student Study Guide Workbook, 8-1

Science and Math Lab Masters, pp. 89–92

Teaching Mathematics with Manipulatives, p. 75

1 Focus

5-Minute Check 8-1

INTERACTIVE CHALKBOARD
PowerPoint® Presentations

Use as a quiz or review of Chapter 7. (Also available as a transparency)

Bellringer

Communication Make a list of fractions and a corresponding list of decimals in scrambled order. Have students match each fraction with its equivalent decimal.

Reading to Learn Mathematics, p. 438 ELL

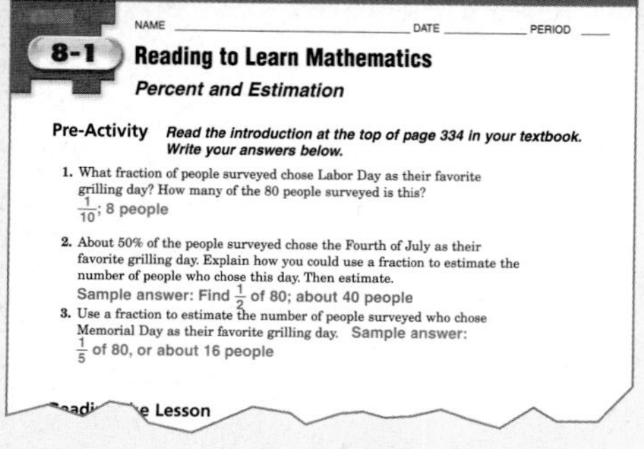
NAME ______ DATE ______ PERIOD ____

8-1 Reading to Learn Mathematics

Percent and Estimation

Pre-Activity ***Read the introduction at the top of page 334 in your textbook. Write your answers below.***

1. What fraction of people surveyed chose Labor Day as their favorite grilling day? How many of the 80 people surveyed is this?
$\frac{1}{10}$; 8 people

2. About 50% of the people surveyed chose the Fourth of July as their favorite grilling day. Explain how you could use a fraction to estimate the number of people who chose this day. Then estimate.
Sample answer: Find $\frac{1}{2}$ of 80; about 40 people

3. Use a fraction to estimate the number of people surveyed who chose Memorial Day as their favorite grilling day. Sample answer: $\frac{1}{5}$ of 80, or about 16 people

Readi... e Lesson

8-1 Percent and Estimation

What You'll Learn
Estimate percents by using fractions and decimals.

WHEN am I ever going to use this?

GRILLING The graph shows the results of a survey in which people were asked which holiday was their favorite for grilling outdoors. Suppose 80 people were surveyed.

Source: Market Facts for Butterball Turkey

1. What fraction of people surveyed chose Labor Day as their favorite grilling day? How many of the 80 people surveyed is this? **$\frac{1}{10}$; 8 people**
2. About 50% of the people surveyed chose the Fourth of July as their favorite grilling day. Explain how you could use a fraction to estimate the number of people who chose this day. Then estimate. **Sample answer: Find $\frac{1}{2}$ of 80; about 40 people**
3. Use a fraction to estimate the number of people surveyed who chose Memorial Day as their favorite grilling day.

Sample answer: $\frac{1}{5}$ of 80, or about 16 people

Sometimes an exact answer is not needed when using percents. In these cases, you can estimate. One way to estimate the percent of a number is to use a fraction.

TEACHING TIP
To estimate the percent of a number, round the percent, round the number, or round both.

EXAMPLES Use Fractions to Estimate

1 Estimate 48% of 60.

48% is about 50% or $\frac{1}{2}$.

48% of 60 $\approx \frac{1}{2} \cdot 60$ Use $\frac{1}{2}$ to estimate.

≈ 30 Multiply.

So, 48% of 60 is about 30.

2 Estimate 82% of 195.

82% is about 80%, which is $\frac{8}{10}$ or $\frac{4}{5}$.

82% of 195 $\approx \frac{4}{5} \cdot 200$ Use $\frac{4}{5}$ to estimate and round 195 to 200.

≈ 160 Multiply.

So, 82% of 195 is about 160.

Your Turn **Estimate by using a fraction.**

a. 26% of 80 b. 75% of 23 c. 62% of 507

a–c. Sample answers are given.

a. $\frac{1}{4} \cdot 80 = 20$

b. $\frac{3}{4} \cdot 24 = 18$

c. $\frac{3}{5} \cdot 500 = 300$

Another method for estimating the percent of a number is to first find 10% of the number and then multiply. For example, 70% = 7 · 10%. So, 70% of a number equals 7 times 10% of the number.

STUDY TIP

Percents To use decimals in estimating, first round percents to the nearest 10%. To find 10% of a number, move the decimal point one place to the left.

EXAMPLE Estimate by Using 10%

3 Estimate 71% of 300.

Step 1 Find 10% of the number.

10% of $300 = 0.1 \cdot 300$ To multiply by 10%, move the decimal point one place to the left.

$= 30$

Step 2 Multiply.

71% is about 70%.

70% of 300 is 7 times 10% of 300.

$7 \cdot 30 = 210$

So, 71% of 300 is about 210.

Your Turn **Estimate by using 10%.**

d. 19% of 40 e. 30% of 217 f. 63% of 91

-f. Sample answers are ven.
0.1 · 40 = 4 and · 4 = 8
0.1 · 210 = 21 and · 21 = 63
0.1 · 90 = 9 and · 9 = 54

You can also estimate percents of numbers when the percent is greater than 100 or the percent is less than 1.

EXAMPLES Percents Greater Than 100 or Less Than 1

4 Estimate 122% of 50.

122% is more than 100%, so 122% of 50 is greater than 50.

122% is about 120%.

120% of $50 = (100\%$ of $50) + (20\%$ of $50)$ 120% = 100% + 20%

$= (1 \cdot 50) + \left(\frac{1}{5} \cdot 50\right)$ 100% = 1 and 20% = $\frac{1}{5}$

$= 50 + 10$ or 60 Simplify.

So, 122% of 50 is about 60.

TEACHING TIP
oint out that Example 4 s a use of the istributive Property.

5 Estimate $\frac{1}{4}$% of 589.

$\frac{1}{4}$% is one fourth of 1%. 589 is about 600.

1% of $600 = 0.01 \cdot 600$ To multiply by 1%, move the decimal point two places to the left.

$= 6$

One fourth of 1% is $\frac{1}{4} \cdot 6$ or 1.5.

So, $\frac{1}{4}$% of 589 is about 1.5.

TEACHING TIP
In this example, students find 1% first, then find $\frac{1}{4}$ of that.

Your Turn **Estimate.**

g. 174% of 200 h. 298% of 45 i. 0.25% of 789

–i. Sample answers are iven.
. $(1 \cdot 200) + \left(\frac{3}{4} \cdot 200\right) = 350$
. $3 \cdot 45 = 135$
. $0.01 \cdot 800 = 8$ and
$\cdot 8$ is 2

msmath2.net/extra_examples

2 Teach

In-Class Examples

INTERACTIVE CHALKBOARD PowerPoint® Presentations

1. Estimate 61% of 407. **about 240**
2. Estimate 18% of 296. **about 60**
3. Estimate 59% of 500. **about 300**
4. Estimate 173% of 60. **about 105**
5. Estimate $\frac{1}{3}$% of 898. **about 3**

Tips for New Teachers

Teaching Strategies When studying estimation, you may want to have students work in pairs. Encourage each student to estimate separately and then compare answers. Ask students whether it is possible to have two different estimates for the same problem.

DAILY INTERVENTION **Differentiated Instruction** **ELL**

Verbal/Linguistic Have students write a paragraph describing a situation where estimating with percents would be appropriate. Their paragraphs should both describe the situation and include the numbers used as estimates. They may look through newspapers for ideas. Have them present their paragraphs to the class.

In-Class Example

INTERACTIVE CHALKBOARD PowerPoint® Presentations

6 GAME ATTENDANCE Last season 96,081 people were in attendance at a college football game. About 0.25% of those were reporters from newspapers and television and radio stations. About how many media people were present? **about 240 media people**

3 Practice/Apply

DAILY INTERVENTION FIND THE ERROR In Exercise 3, remind students that 1.5% means 1.5 per hundred. So $1.5 \times 4 = 6$ is reasonable.

NUMBER SENSE
In Exercise 4, encourage students to think about the relationship between estimating and rounding. The direction in which numbers are rounded will determine whether the estimate is greater than or less than the actual percent.

About the Exercises...

Guided Practice Key

Exercises	Examples
5–7	1, 2
8–10	3
11–13	4, 5
14	6

Odd/Even Assignments
Exercises 15–34 are structured so that students practice the same concepts whether they are assigned odd or even problems.

TEACHING TIP
Ask students to explain whether the estimate in Example 6 is greater or less than the actual number.

EXAMPLE Estimate Percent to Solve a Problem

6 POPULATION In 2000, about 0.5% of the people in Montana were of Asian descent. If Montana had a population of 902,195, estimate the number of people who were of Asian descent.

0.5% is half of 1%.

1% of 900,000 $= 0.01 \cdot 900{,}000$ 902,195 is about 900,000.

$= 9{,}000$

So, 0.5% of 902,195 is about $\frac{1}{2}$ of 9,000 or 4,500.

So, about 4,500 people in Montana were of Asian descent.

Skill and Concept Check

Writing Math — Exercises 3 & 4

1. **Describe** two different ways to estimate 22% of 136. **See margin.**
2. **OPEN ENDED** Write a problem in which the answer can be found by estimating 12% of 50. **See margin.**
3. **FIND THE ERROR** Ian and Mandy are estimating 1.5% of 420. Who is correct? Explain. **Ian; Mandy incorrectly changed 1.5% to 1.5, which is 150%.**

Ian
1.5% of 420
$\approx$ 1% of 400 + 0.5% of 400
$= 0.01 \cdot 400 + \frac{1}{2}(0.01 \cdot 400)$
$= 4 + 2$ or 6

Mandy
1.5% of 420
$= 1 \cdot 400 + 0.5 \cdot 400$
$= 400 + 200$
$= 600$

4. **NUMBER SENSE** Explain whether an estimate for the percent of a number is *always*, *sometimes*, or *never* greater than the actual percent of the number. Give an example or a counterexample to support your answer. **See margin.**

GUIDED PRACTICE

Estimate by using fractions. **5. $\frac{1}{2} \cdot 160 = 80$ 5–7. Sample answers are given.**

5. 52% of 160
6. 30% of 79 **$\frac{3}{10} \cdot 80 = 24$**
7. 77% of 22 **$\frac{3}{4} \cdot 20 = 15$**

Estimate by using 10%. **8–13. See margin.**

8. 40% of 62
9. 23% of 400
10. 89% of 98

Estimate.

11. 151% of 70
12. 305% of 6
13. $\frac{1}{2}$% of 82

14. **LIFE SCIENCE** The 639 muscles in your body make up about 40% of your total weight. If a person weighs 120 pounds, about how much of the weight is muscle? **Sample answer: $\frac{2}{5} \cdot 120 = 48$ lb**

Answers

1. Sample answer: Find $\frac{1}{5} \cdot 140$ or find 10% of 140 and multiply by 2; both equal 28.
2. Sample answer: 12% of the students surveyed said tacos were their favorite meal. If 50 students were surveyed, about how many chose tacos as their favorite meal? Answer: $0.1 \cdot 50$ or about 5 students
4. Sometimes; sample answer: 37% of 60 is about $\frac{2}{5} \cdot 60 = 24$. This is greater than the actual answer because $\frac{2}{5}$ is greater than 37%. Another estimate is $\frac{1}{3} \cdot 60 = 20$. This is less than the actual answer because $\frac{1}{3}$ is less than 37%.
8. $0.1 \cdot 60 = 6$ and $4 \cdot 6 = 24$
9. $0.1 \cdot 400 = 40$ and $2 \cdot 40 = 80$
10. $0.1 \cdot 100 = 10$ and $9 \cdot 10 = 90$
11. $(1 \cdot 70) + \left(\frac{1}{2} \cdot 70\right) = 105$
12. $3 \cdot 6 = 18$
13. $0.01 \cdot 80 = 0.8$ and $\frac{1}{2} \cdot 0.8 = 0.4$

Practice and Applications

Estimate by using fractions. 15–32. Sample answers are given.

15. 25% of 408 **100**
16. 80% of 37 **32**
17. 76% of 280 **210**
18. 39% of 20 **8**
19. 67% of 15.2 **10**
20. 10.5% of 238 **24**

15-20. See margin for equations.

Estimate by using 10%.

21. 60% of 39 **24**
22. 20% of 132 **26**
23. 76% of 80 **64**
24. 37% of 250 **100**
25. 28% of 121 **36**
26. 88% of 207 **180**

Estimate.

27. 132% of 54 **72**
28. 224% of 320 **720**
29. 410% of 12 **48**
30. 198% of 33 **66**
31. 0.4% of 400 **2**
32. 0.9% of 74 **0.7**

HOMEWORK HELP

For Exercises	See Examples
15–20	1, 2
21–26	3
27–34	4–6

Extra Practice
See pages 582, 603.

21–32. See Ch. 8 Answer Appendix for equations.

GEOLOGY For Exercises 33 and 34, use the following information.
Granite, a stone found in New Hampshire and Vermont, is 0.8% water.

33. About how many pounds of water are there in 3,000 pounds of granite? Sample answer: $0.01 \cdot 3{,}000 = 30$ lb
34. About how much water is contained in a 15-pound piece of granite? Sample answer: $0.01 \cdot 15 = 0.15$ lb
35. **CRITICAL THINKING** Explain how you could find $\frac{3}{8}$% of a number. Sample answer: Find 1% of a number, then multiply by $\frac{3}{8}$.

Standardized Test Practice and Mixed Review

36. **MULTIPLE CHOICE** Estimate 15% of 61. **D**
   - (A) 30
   - (B) 18
   - (C) 15
   - (D) 9

37. **MULTIPLE CHOICE** In a survey, 1,031 people were asked to choose the greatest athlete of the 20th century. The top five choices are shown at the right. About how many more people chose Michael Jordan than Muhammad Ali? Choose the best estimate. **H**

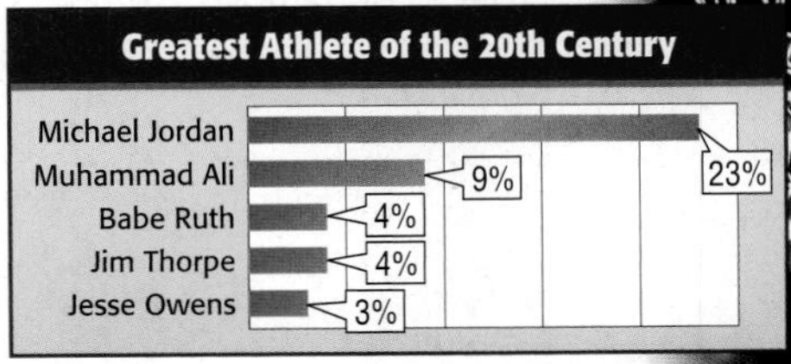

Source: Gallup Poll

   - (F) 90 people
   - (G) 100 people
   - (H) 150 people
   - (I) 200 people

Find each number. Round to the nearest tenth if necessary. (Lesson 7-8)

38. 6 is what percent of 15? **40%**
39. Find 72% of 90. **64.8**
40. What number is 120% of 60? **72**
41. 35% of what number is 55? **157.1**
42. Find 22% of 85. (Lesson 7-7) **18.7**

GETTING READY FOR THE NEXT LESSON

PREREQUISITE SKILL Solve. Round to the nearest tenth if necessary. (Lesson 4-3)

43. $40 = 0.8x$ **50**
44. $10r = 61$ **6.1**
45. $25 = 0.07t$ **357.1**
46. $56 = 0.32n$ **175**

msmath2.net/self_check_quiz

4 Assess

Open-Ended Assessment

Writing Have students write an explanation of how they estimate a percent of a number.

Answers

15. $\frac{1}{4} \cdot 400 = 100$
16. $\frac{4}{5} \cdot 40 = 32$
17. $\frac{3}{4} \cdot 280 = 210$
18. $\frac{2}{5} \cdot 20 = 8$
19. $\frac{2}{3} \cdot 15 = 10$
20. $\frac{1}{10} \cdot 240 = 24$

Study Guide and Intervention, p. 435

NAME ______________ DATE ________ PERIOD ____

8-1 Study Guide and Intervention

Percent and Estimation

To estimate the percent of a number, you can use a fraction or a multiple of 10% or 1%.

EXAMPLE 1 **Estimate 77% of 800.**

77% is about 75% or $\frac{3}{4}$.

$77\% \text{ of } 800 \approx \frac{3}{4} \cdot 800$ Use $\frac{3}{4}$ to estimate.
≈ 600 Multiply.

So, 77% of 800 is about 600.

EXAMPLE 2 **Estimate 137% of 50.**

137% is more than 100%, so 137% of 50 is greater than 50.
137% is about 140%.

$140\% \text{ of } 50 = (100\% \text{ of } 50) + (40\% \text{ of } 50)$ 140% = 100% + 40%
$= (1 \cdot 50) + \left(\frac{2}{5} \cdot 50\right)$ 100% = 1 and 40% = $\frac{2}{5}$
$= 50 + 20 \text{ or } 70$ Simplify.

So, 137% of 50 is about 70.

EXAMPLE 3 **Estimate 0.5% of 692.**

0.5% is half of 1%. 692 is about 700.

$1\% \text{ of } 700 = 0.01 \cdot 700$ To mutiply by 1%, move the decimal point two places to the left.
$= 7$

One half of 7 is $\frac{1}{2} \cdot 7$ or 3.5.
So, 0.5% of 697 is about 3.5.

EXERCISES

Estimate. 1–12. Sample answers are given.

1. 24% of 36 — $\frac{1}{4} \cdot 36 = 9$
2. 81% of 25 — $\frac{4}{5} \cdot 25 = 20$
3. 11% of 67 — $0.1 \cdot 70 = 7$
4. 150% of 179 — $(1 \cdot 180) + \left(\frac{1}{2} \cdot 180\right) = 270$
5. 67% of 450 — $\frac{2}{3} \cdot 450 = 300$
6. 79% of 590 — $\frac{3}{4} \cdot 600 = 450$
7. 0.4% of 200 — $0.01 \cdot 200 = 2$ and $\frac{1}{2} \cdot 2 = 1$
8. 42% of 61 — $0.1 \cdot 60 = 6$ and $4 \cdot 6 = 24$
9. 19% of 41 — $0.1 \cdot 40 = 4$ and $2 \cdot 4 = 8$
10. 129% of 54 — $(1 \cdot 54) + \left(\frac{1}{3} \cdot 54\right) = 72$
11. 32% of 66 — $\frac{1}{3} \cdot 66 = 22$
12. 0.2% of 150 — $0.01 \cdot 150 = 1.5$ and $\frac{1}{5} \cdot 1.5 = 0.3$

Practice: Skills, p. 436

NAME ______________ DATE ________ PERIOD ____

8-1 Practice: Skills

Percent and Estimation

Estimate by using fractions. 1–6. Sample answers are given.

1. 51% of 128 — $\frac{1}{2} \cdot 128 = 64$
2. 76% of 200 — $\frac{3}{4} \cdot 200 = 150$
3. 32.9% of 90 — $\frac{1}{3} \cdot 90 = 30$
4. 23% of 8 — $\frac{1}{4} \cdot 8 = 2$
5. 19% of 45 — $\frac{1}{5} \cdot 45 = 9$
6. 81% of 16 — $\frac{4}{5}$ of $15 = 12$

Estimate by using 10%. 7–12. Sample answers are given.

7. 12% of 98 — $0.1 \cdot 100 = 10$
8. 89% of 300 — $0.1 \cdot 300 = 30$ and $9 \cdot 30 = 270$
9. 31% of 80 — $0.1 \cdot 80 = 8$ and $3 \cdot 8 = 24$
10. 28% of 49 — $0.1 \cdot 50 = 5$ and $3 \cdot 5 = 15$
11. 62% of 13 — $0.1 \cdot 10 = 1$ and $6 \cdot 1 = 6$
12. 77% of 28 — $0.1 \cdot 30 = 3$ and $8 \cdot 3 = 24$

Estimate. 13–21. Sample answers are given.

13. 308% of 500 — $3 \cdot 500 = 1{,}500$
14. 0.5% of 87 — $0.01 \cdot 90 = 0.9$ and $\frac{1}{2} \cdot 0.9 = 0.45$
15. 153% of 20 — $(1 \cdot 20) + \left(\frac{1}{2} \cdot 20\right) = 30$
16. 0.6% of 41 — $0.01 \cdot 40 = 0.4$ and $\frac{1}{2} \cdot 0.4 = 0.2$
17. 231% of 54 — $(2 \cdot 54) + \left(\frac{1}{3} \cdot 54\right) = 126$
18. 0.9% of 116 — $0.01 \cdot 116 = 1.16$
19. 0.26% of 36 — $0.01 \cdot 36 = 0.36$ and $\frac{1}{4} \cdot 0.36 = 0.09$
20. 425% of 119 — $(4 \cdot 120) + \left(\frac{1}{4} \cdot 120\right) = 510$

Practice: Word Problems, p. 437

NAME ______________ DATE ________ PERIOD ____

8-1 Practice: Word Problems

Percent and Estimation

1. **ORCHESTRA** The orchestra at Millard Middle School has 120 members. Of these, 17% are eighth-grade students. Estimate the number of eighth-grade students in the orchestra. Sample answer: $\frac{1}{5} \cdot 120 = 24$ students
2. **RESTAURANTS** In one west coast city, 34% of the restaurants are on the river. The city has 178 restaurants. Estimate the number of restaurants that are on the river. Sample answer: $\frac{1}{3} \cdot 180 = 60$ restaurants
3. **FARMING** Rhonda planted green beans on 67% of her farm. Rhonda's farm has 598 acres of land. Estimate the number of acres of green beans on Rhonda's farm. Sample answer: $\frac{2}{3} \cdot 600 = 400$ acres
4. **HOTELS** At the Eastward Inn hotel, 47% of the rooms face the pool. The hotel has 92 rooms. Estimate the number of rooms that face the pool. Sample answer: $\frac{1}{2} \cdot 90 = 45$ rooms
5. **TREES** The students in Leon's seventh grade science class determined that 42% of the trees at a local park are pine trees. If there are 632 trees in the park, about how many of them are pine trees? Sample answer: $0.1 \cdot 600 = 60$ and $4 \cdot 60 = 240$ trees
6. **BOOKS** Jenna has read 0.7% of a book. If the book has 431 pages, estimate the number of pages Jenna has read. Sample answer: $0.01 \cdot 420 = 4.2$ and $\frac{2}{3} \cdot 4.2 = 2.8$ pages
7. **FITNESS** At the office where Mika works, 40% of the 18 employees exercise at least three times a week. Estimate the number of people who exercise at least three times a week. Sample answer: $0.1 \cdot 20 = 2$ and $4 \cdot 2 = 8$ people
8. **PETS** Of all seventh grade students at Hart Middle School, 0.3% of the students own a pet iguana. If there are 610 seventh grade students at Hart, about how many own pet iguanas? Sample answer: $0.01 \cdot 600 = 6$ and $\frac{1}{3} \cdot 6 = 2$ students

8-1b Lesson Notes

1 Focus

Bellringer
Reading Mathematics Have students discuss the meanings of the words *estimate, round,* and *reasonable*. Encourage them to reference a dictionary. Ask them to explain how rounding and estimation can help generate a reasonable answer.

2 Teach

In-Class Example

INTERACTIVE CHALKBOARD
PowerPoint® Presentations

Solve. Use the reasonable answer strategy.
During a candy sale for a fundraiser, the soccer team gets to keep 25% of the sales. The candy bars sell for $1.50 and the team has sold 510 bars so far. Can the team now buy new shirts costing a total of $175? Explain. **Yes, they have sold 500 · 1.50 or about $750 worth of candy bars. So, they have earned $\frac{1}{4}$ · 750 or about $188, which is more than $175.**

8-1b Problem-Solving Strategy

A Follow-Up of Lesson 8-1

What You'll Learn
Solve problems by determining reasonable answers.

Reasonable Answers

The meals for our group cost a total of $38.95. I think we should leave a 15% tip, which would be about $4.

I think that the tip should be more than that. Let's estimate to find a **reasonable answer**.

Explore	We know that the total bill is $38.95 and we want to leave a 15% tip.
Plan	We can round $38.95 to $40 and then use mental math to find 15% of 40.
Solve	10% of 40 = 0.1 · 40 or 4 10% = 0.1 5% of 40 = $\frac{1}{2}$ · 4 or 2 15% of 40 = (10% of 40) + (5% of 40) = 4 + 2 or 6 So, $6 would be a better amount to leave for a tip.
Examine	Use a calculator to check. .15 [×] 38.95 [ENTER] **5.8425** 5.8425 is close to 6, so the answer is reasonable.

Analyze the Strategy

1. The last step of the four-step plan for problem solving asks you to examine your solution. **Explain** how you can use estimation with decimals to help you examine a solution. **Sample answer: Estimate before you calculate.**
2. **Write** a problem that has an unreasonable answer and ask a classmate to explain why they think the answer is unreasonable. **See students' work.**
3. **Describe** other problem-solving strategies that you could use to determine whether answers are reasonable. **Sample answer: use estimation, guess and check, look for a pattern**

DAILY INTERVENTION **Differentiated Instruction**

Logical Have students develop a strategy for determining reasonable answers in another area of study. For example, if students are taking an Earth Science course, they may need to understand radiocarbon dating. This process is based on the half-life of carbon-14, or 5,730 years. Rounding the half-life of carbon-14 to 6,000 years might help students determine a reasonable, but not exact, answer to a radiocarbon dating problem.

pply the Strategy

lve. Use the reasonable answer strategy.

COMMUNICATION Sandra makes a long distance phone call to her grandparents and talks for 45 minutes. The phone company charges a rate of \$0.20 per minute. How much does the call cost? **\$9**

5. **SHOPPING** Suppose you are buying an entertainment system for \$1,301.90 and the speakers are 57.6% of the total cost. What is a reasonable estimate for the cost of the speakers? Explain. **Sample answer: 60% or $\frac{3}{5} \cdot 1,300 = \$780$**

xed Problem Solving

lve. Use any strategy.

CHORES Cameron is using a 2.5-liter container to fill a tank that holds 24 liters of water. How many times will he need to fill the container? **10 times**

MUSIC A survey showed that 73% of teens who use computers listen to music at the same time. Suppose there are 410 teens in your school who use computers. Estimate how many of them listen to music while on the computer. **See margin.**

VACATION The graph below shows how the Mason family spent their time during their summer vacation. What percent of the time shown was spent touring historic sites? **15%**

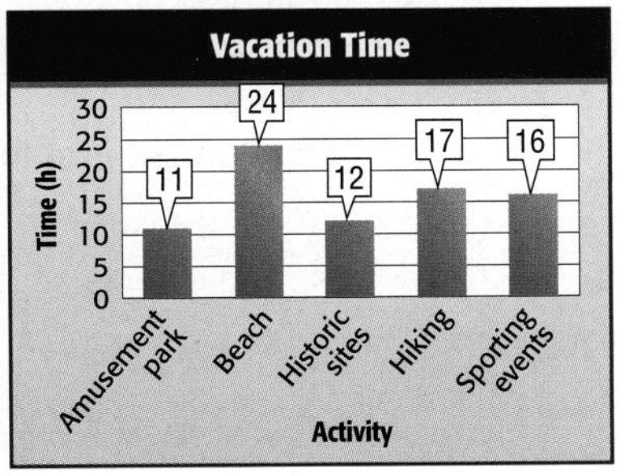

. **BASKETBALL** Keisha made 18 points in one basketball game. How many possible shot combinations of 2- and 3-pointers could she have made? List the combinations in a table. **See margin.**

. **POPULATION** About 9.4% of the people in Texas live in Houston. If the population of Texas is about 20,852,000, estimate the population of Houston. **Sample answer: about $0.1 \cdot 21,000,000$ or 2,100,000 people**

11. **SEAT BELTS** The graph shows the percent of drivers that wore seat belts. Predict the percent of drivers that will wear seat belts in 2006. Explain why your answer is reasonable. **See margin.**

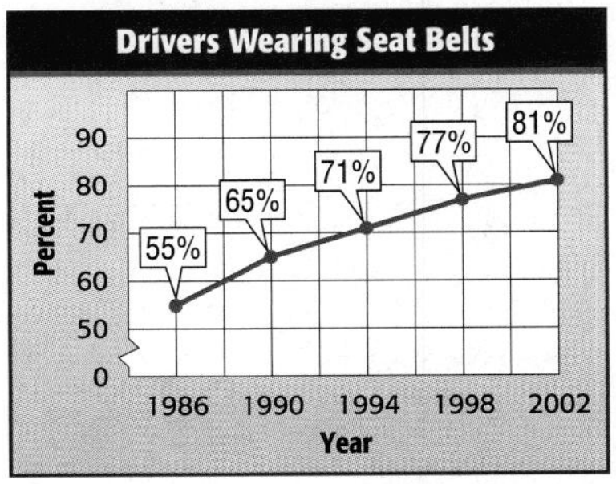

Source: Harris Interactive

12. **GEOMETRY** A rectangle has a length of $4\frac{1}{4}$ inches and a width of $3\frac{2}{5}$ inches. Is between 7 and 8 inches a reasonable estimate for the perimeter? Explain. **See margin.**

13. **STANDARDIZED TEST PRACTICE** Mr. Camacho is purchasing carpet for the two rooms described in the table. Which expression shows the number of square yards of carpet that he needs? **A**

Room	Dimensions
living room	15 ft by 18 ft
TV room	18 ft by 20 ft

Ⓐ $[(15 \times 18) + (18 \times 20)] \div (3 \times 3)$

Ⓑ $[(15 \times 18(2) \times 20)] \div 3$

Ⓒ $[(15 \times 18) + (18 \times 20)] \div 3$

Ⓓ $(15 \times 18 \times 20) \div (3 \times 3)$

3 Practice/Apply

About the Exercises...

Problem-Solving Strategies

Exercises 6–13 are structured so that students have the opportunity to practice many different problem-solving strategies. You may wish to remind them of some of the strategies they have studied.

- look for a pattern (p. 132)
- work backward (p. 164)
- make an organized list (p. 201)
- eliminate possibilities (p. 252)
- reasonable answers ✓

4 Assess

Open-Ended Assessment

Writing Have students write a strategy to determine a reasonable amount that they would save on a \$78 sweater that is 25% off during a sale. They should check their totals with a calculator.

Answers

7. Sample answer: $\frac{3}{4} \cdot 400 =$ 300 teens

9. 4 combinations;

2-pointers	3-pointers	Total
9	0	18
6	2	18
3	4	18
0	6	18

11. Sample answer: 85%; the percent of people who wear seatbelts has been increasing, but at a slower rate.

12. No; the perimeter is twice the sum of the length and the width, so a reasonable estimate would be 2(4) + 2(3) or 14.

8-2 Lesson Notes

Lesson Resources

Chapter 8 Resource Masters
- Study Guide and Intervention, p. 440
- Practice: Skills, p. 441
- Practice: Word Problems, p. 442
- Reading to Learn Mathematics, p. 443
- Enrichment, p. 444

Key Teaching Transparencies, 29
MathPASS Tutorial CD-ROM, 34
Parent and Student Study Guide Workbook, 8-2
Real-Life Career Masters, p. 22

FAST FILE

1 Focus

5-Minute Check 8-2

INTERACTIVE CHALKBOARD
PowerPoint® Presentations

Use as a quiz or review of Lesson 8-1. (Also available as a transparency)

Bellringer

Problem Solving Tell students you have a quiz paper with a grade of 90%. Ask students to write down different ways to obtain that score. For example, the student may score 9 points out of a possible 10 or 18 out of 20.

Reading to Learn Mathematics, p. 443 ELL

NAME ______________ DATE ________ PERIOD ____

8-2 Reading to Learn Mathematics
The Percent Equation

Pre-Activity *Read the introduction at the top of page 340 in your textbook. Write your answers below.*

1. Suppose a person weighs 120 pounds. Use the percent proportion to find the number of pounds of oxygen, carbon, and hydrogen. 78 lb; 21.6 lb; 12 lb
2. Express the percent of each element as a decimal. 0.65; 0.18; 0.1
3. Multiply each decimal by 120. Record your results. 78; 21.6; 12
4. Compare the answers to Exercises 1 and 3. They are the same.

Reading the Lesson

5. The word *percent* is used in both the percent proportion and the percent equation. There is one major difference in the way percent is represented i... hat is the diffe... Sample ...wer: In th... ercent p... rtion...

8-2 The Percent Equation

What You'll Learn
Solve problems by using the percent equation.

NEW Vocabulary
percent equation

LINK To Reading
Everyday Meaning of Base: the bottom of something considered as its support, as in the base of a column

WHEN am I ever going to use this?

PHYSICAL SCIENCE The graph shows three main elements that make up most of the human body.

Elements of the Body

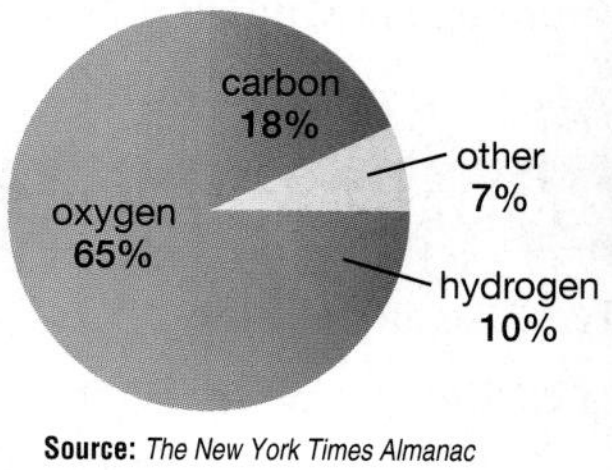

Source: *The New York Times Almanac*

1. Suppose a person weighs 120 pounds. Use the percent proportion to find the number of pounds of oxygen, carbon, and hydrogen. **78 lb; 21.6 lb; 12 lb**
2. Express the percent of each element as a decimal. **0.65; 0.18; 0.1**
3. Multiply each decimal by 120. Record your results. **78; 21.6; 12**
4. Compare the answers to Exercises 1 and 3. **They are the same.**

In Lesson 7-8, you used the percent proportion $\frac{\text{part}}{\text{base}} = \frac{\text{percent}}{100}$ to find the missing part, percent, or base. You can also use an equation.

$\frac{\text{part}}{\text{base}} = \text{percent}$ — The percent is written as a decimal.

$\frac{\text{part}}{\text{base}} \cdot \text{base} = \text{percent} \cdot \text{base}$ — Multiply each side by the base.

$\text{part} = \text{percent} \cdot \text{base}$ ← This form is called the **percent equation**.

Concept Summary — **Types of Percent Problems**

Type	Example	Equation
Find the Part	What number is 50% of 6?	$n = 0.5 \cdot 6$
Find the Percent	3 is what percent of 6?	$3 = n \cdot 6$
Find the Base	3 is 50% of what number?	$3 = 0.5 \cdot n$

STUDY TIP

Percents When using the percent equation, always write the percent as a decimal or fraction before computing.

EXAMPLE Find the Part

1 What number is 12% of 350? **Estimate** $0.1 \cdot 350 = 35$

12% or 0.12 is the percent and 350 is the base. Let n represent the part.

$\underbrace{\text{part}} = \underbrace{\text{percent}} \cdot \underbrace{\text{base}}$

$n = 0.12 \cdot 350$ — Write an equation.

$n = 42$ — Multiply. The part is 42.

So, 42 is 12% of 350. — This is close to the estimate.

EXAMPLE Find the Percent

2 **21 is what percent of 40?** **Estimate** $\frac{21}{40} \approx \frac{1}{2}$ or 50%

Let n represent the percent.

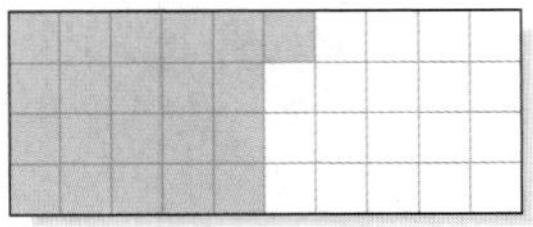

$$\underbrace{\text{part}} = \underbrace{\text{percent}} \cdot \underbrace{\text{base}}$$

$21 = n \cdot 40$	Write an equation.
$\frac{21}{40} = \frac{40n}{40}$	Divide each side by 40.
$0.525 = n$	Simplify.
$52.5\% = n$	Write 0.525 as a percent.
So, 21 is 52.5% of 40.	This is close to the estimate.

STUDY TIP

Percent Remember to write the decimal as a percent in your final answer.

EXAMPLE Find the Base

3 **13 is 26% of what number?** **Estimate** 13 is 25% or $\frac{1}{4}$ of 52.

Let n represent the base.

$$\underbrace{\text{part}} = \underbrace{\text{percent}} \cdot \underbrace{\text{base}}$$

$13 = 0.26 \cdot n$	Write an equation.
$\frac{13}{0.26} = \frac{0.26n}{0.26}$	Divide each side by 0.26.
$50 = n$	The base is 50.
So, 13 is 26% of 50.	Compare to the estimate.

TEACHING TIP

ncourage students to stimate the unknown alue in the percent quation first, then use a alculator to find the olution.

Your Turn **Write an equation for each problem. Then solve. Round to the nearest tenth if necessary.**

a. What percent of 125 is 75? **$75 = n \cdot 125$; 60%**

b. 39 is 84% of what number? **$39 = 0.84 \cdot n$; 46.4**

EAL-LIFE MATH

CHNOLOGY The following ies in the United States ve the highest percent of ll phone users.

Anchorage, AK	56%
Chicago, IL	55%
Honolulu, HI	54%
Atlanta, GA	52%
Detroit, MI	52%

urce: Polk's Research

EXAMPLE Apply the Percent Equation

4 **TECHNOLOGY Finland has the highest percent of people who have cell phones, 67.8%. If there are about 3,499,000 people with cell phones, what is the population of Finland?**

Words	3,499,000 is 67.8% of what number?
Symbols	Let n represent the base.
Equation	$3{,}499{,}000 = 0.678 \cdot n$

67.8% is written as the decimal 0.678.

$3{,}499{,}000 = 0.678 \cdot n$	Write the equation.
$\frac{3{,}499{,}000}{0.678} = \frac{0.678n}{0.678}$	Divide each side by 0.678. Use a calculator.
$5{,}160{,}767 \approx n$	Simplify.

The population of Finland is about 5,160,767.

msmath2.net/extra_examples

2 Teach

In-Class Examples

INTERACTIVE CHALKBOARD PowerPoint® Presentations

1. What number is 46% of 200? **92**
2. 26 is what percent of 32? **81.25%**
3. 12 is 40% of what number? **30**
4. **TELEVISIONS** In the year 2000, 102.2 million homes in the United States had at least one television. Of these, 86% also had VCRs. How many homes had VCRs? **about 87.9 million**

Tips for New Teachers

Using Manipulatives Keep a supply of grid paper or number cubes on hand for students who need to sketch diagrams or use models to better understand the concepts.

DAILY INTERVENTION

Differentiated Instruction

Visual/Spatial Ask students to draw an 8×12 rectangle on grid paper. Have them fold the rectangle into quarters and shade one of the quarters. Ask students: How many squares are shaded? What is 25% of 96? **24** Repeat using different percents or different rectangles.

3 Practice/Apply

FOLDABLES™ Follow-Up Remind students to record main ideas and to give examples of what they learn about the percent equation in the Lesson 8-2 section of their Foldable tables.

About the Exercises...

Guided Practice Key

Exercises	Examples
3	1
4	2
5–6	3
7	4

Odd/Even Assignments
Exercises 8–23 and 25–28 are structured so that students practice the same concepts whether they are assigned odd or even problems.

Alert! The Data Update following Exercise 7 requires access to the Internet.

Answers

2. Sample answer: A jacket is on sale for 35% off the regular price of $52. What is the sale price? To solve, find the part and subtract from $52; $33.80.
8. $n = 0.65 \cdot 98$; 63.7
9. $n = 0.53 \cdot 470$; 249.1
10. $9 = n \cdot 45$; 20%
11. $26 = n \cdot 96$; 27.1%
12. $84 = 0.75 \cdot n$; 112
13. $17 = 0.40 \cdot n$; 42.5
14. $n = 0.24 \cdot 25$; 6
15. $30 = n \cdot 64$; 46.9%
16. $98 = n \cdot 392$; 25%
17. $1.45 = 0.33 \cdot n$; 4.4
18. $n = 0.135 \cdot 520$; 70.2
19. $n = 0.752 \cdot 600$; 451.2
20. $n = 0.04 \cdot 82.1$; 3.3
21. $14 = 0.028 \cdot n$; 500
22. $n = 1.35 \cdot 64$; 86.4
23. $230 = n \cdot 200$; 115%

Skill and Concept Check

Writing Math Exercise 2

1. **State** whether the following problem represents a missing part, a missing percent, or a missing base. **missing base**
17 is 10% of what number?

2. **OPEN ENDED** Write a real-life problem that can be solved by using the percent equation. State whether you need to find the part, the percent, or the base, and then solve. **See margin.**

GUIDED PRACTICE

Write an equation for each problem. Then solve. Round to the nearest tenth if necessary. 3. $n = 0.88 \cdot 300$; 264 4. $75 = n \cdot 150$; 50%

3. What number is 88% of 300?
4. 75 is what percent of 150?
5. 3 is 12% of what number? $3 = 0.12 \cdot n$; 25
6. 84 is 60% of what number? $84 = 0.6 \cdot n$; 140
7. **BASEBALL** In 2002, Derek Jeter had 191 hits in 644 times at bat. What was his *batting average*, or the percent of times at bat that were hits? Round the percent to the nearest tenth. Then write as a decimal. **29.7%; 0.297**

Data Update How do current batting averages of other baseball players compare to Derek Jeter's average in 2002? Visit msmath2.net/data_update to learn more.

★ indicates increased difficulty

Practice and Applications

HOMEWORK HELP

For Exercises	See Examples
8–23	1–3
24–28	4

Extra Practice See pages 583, 603.

Write an equation for each problem. Then solve. Round to the nearest tenth if necessary. 8–23. See margin.

8. What number is 65% of 98?
9. Find 53% of 470.
10. 9 is what percent of 45?
11. 26 is what percent of 96?
12. 84 is 75% of what number?
13. 17 is 40% of what number?
14. Find 24% of 25.
15. What percent of 64 is 30?
16. 98 is what percent of 392?
17. 33% of what number is 1.45?
18. Find 13.5% of 520.
19. What number is 75.2% of 600?
20. What number is 4% of 82.1?
21. 14 is 2.8% of what number?
22. ★ Find 135% of 64.
23. ★ What percent of 200 is 230?
24. **SALES** Ms. Allon received a $325 *commission*, which is a fee paid based on a percent of her sales. If her sales totaled $8,125, what is the percent that she earns? **4%**

★ **NUMBER THEORY For Exercises 25 and 26, consider the whole numbers 1 through 100.**

25. For what percent of these numbers will the digits add to 5? **6%**
26. For what percent of these numbers will the digits add to an even number? an odd number? **49%; 51%**

CROPS For Exercises 27–29, use the graphic at the right. It shows the acres of cotton planted in the top five states. Round to the nearest percent.

27. About what percent of the cotton planted in the United States is planted in Texas? **42%**
28. About what percent of U.S. cotton is planted in North Carolina? **6%**
★ 29. What percent of U.S. cotton is planted in states other than those listed on the graphic? **28%**

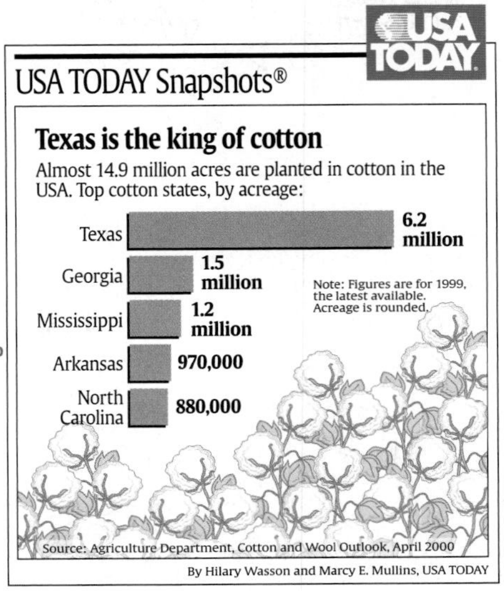

★ 30. **MULTI STEP** A cargo of 18,000 tons of grain is 14% water due to moisture. Another cargo of 26,000 tons of grain is 12.2% water. Which contains more water? How much more? **26,000-ton cargo; 652 tons**
31. **CRITICAL THINKING** If you need to find the percent of a number, explain how you can predict whether the part will be less than, greater than, or equal to the number. **See margin.**

Standardized Test Practice and Mixed Review

32. **MULTIPLE CHOICE** 23 is what percent of 64? Round to the nearest tenth. **D**

Ⓐ 2.8% Ⓑ 33.4% Ⓒ 34.0% Ⓓ 35.9%

33. **MULTIPLE CHOICE** 56 is 16% of what number? Round to the nearest tenth. **I**

Ⓕ 300.5 Ⓖ 305.0 Ⓗ 305.5 Ⓘ 350.0

Estimate by using fractions. (Lesson 8-1) **34–36. Sample answers are given.**

34. 11% of 79 $\frac{1}{10} \cdot 80 = 8$
35. 30.5% of 50 $\frac{3}{10} \cdot 50 = 15$
36. 48% of 311 $\frac{1}{2} \cdot 300 = 150$

37. **NUTRITION** A cup of fruit yogurt has approximately 315 milligrams of calcium. This is about 24% of the recommended daily allowance for people 9 to 18 years old. How many milligrams of calcium is recommended per day for people 9 to 18 years old? (Lesson 7-8) **1,312.5 mg**

38. **GEOMETRY** Find the circumference of the circle. Use $\frac{22}{7}$ for π. (Lesson 6-7) **176 mm**

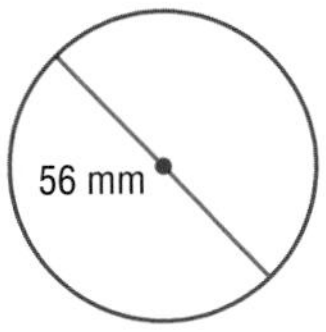

39. Find the greatest common factor of 72 and 270. (Lesson 5-2) **18**

GETTING READY FOR THE NEXT LESSON

PREREQUISITE SKILL Use the percent proportion to find each number. Round to the nearest tenth if necessary. (Lesson 7-9)

40. What number is 45% of 60? **27**
41. What percent of 180 is 30? **16.7%**
42. 20% of what number is 18? **90**
43. 50% of what number is 15.8? **31.6**

Study Guide and Intervention, p. 440

NAME ______ DATE ______ PERIOD ____

8-2 Study Guide and Intervention

The Percent Equation

To solve any type of percent problem, you can use the percent equation, part = percent · base, where the percent is written as a decimal.

EXAMPLE 1 **600 is what percent of 750?**

600 is the part and 750 is the base. Let n represent the percent.

part = percent · base

$600 = n \cdot 750$ Write an equation.

$\frac{600}{750} = \frac{750n}{750}$ Divide each side by 750.

$0.8 = n$ Simplify.

$80\% = n$ Write 0.8 as a percent.

So, 600 is 80% of 750.

EXAMPLE 2 **45 is 90% of what number?**

45 is the part and 90% or 0.9 is the percent. Let n represent the base.

part = percent · base

$45 = 0.9 \cdot n$ Write an equation.

$\frac{45}{0.9} = \frac{0.9n}{0.9}$ Divide each side by 0.9.

$50 = n$ The base is 50.

So, 45 is 90% of 50.

EXERCISES

Write an equation for each problem. Then solve. Round to the nearest tenth if necessary.

1. What percent of 56 is 14? $14 = n \cdot 56$; 25%
2. 36 is what percent of 40? $36 = n \cdot 40$; 90%
3. 80 is 40% of what number? $80 = 0.4 \cdot n$; 200
4. 65% of what number is 78? $78 = 0.65 \cdot n$; 120
5. What percent of 2,000 is 8? $8 = n \cdot 2{,}000$; 0.4%
6. What is 110% of 80? $n = 1.1 \cdot 80$; 88
7. 85 is what percent of 170? $85 = n \cdot 170$; 50%
8. Find 30% of 70. $n = 0.3 \cdot 70$; 21

Practice: Skills, p. 441

NAME ______ DATE ______ PERIOD ____

8-2 Practice: Skills

The Percent Equation

Write an equation for each problem. Then solve. Round to the nearest tenth if necessary.

1. 25% of 176 is what number? $n = 0.25 \cdot 176$; 44
2. What is 90% of 20? $n = 0.9 \cdot 20$; 18
3. 24 is what percent of 30? $24 = n \cdot 30$; 80%
4. 80% of what number is 94? $94 = 0.8 \cdot n$; 117.5
5. What is 60% of 45? $n = 0.6 \cdot 45$; 27
6. 9 is what percent of 30? $9 = n \cdot 30$; 30%
7. What percent of 125 is 25? $25 = n \cdot 125$; 20%
8. What is 120% of 20? $n = 1.2 \cdot 20$; 24
9. 2% of what number is 5? $5 = 0.02 \cdot n$; 250
10. 15% of 290 is what number? $n = 0.15 \cdot 290$; 43.5
11. 16 is what percent of 4,000? $16 = n \cdot 4{,}000$; 0.4%
12. What is 140% of 60? $n = 1.4 \cdot 60$; 84
13. 344.8 is what percent of 862? $344.8 = n \cdot 862$; 40%
14. 6% of what number is 21? $21 = 0.06 \cdot n$; 350
15. What number is 60% of 605? $n = 0.6 \cdot 605$; 363
16. 32% of 250 is what number? $n = 0.32 \cdot 250$; 80
17. Find 30% of 70. $n = 0.3 \cdot 70$; 21
18. What is 80% of 65? $n = 0.8 \cdot 65$; 52

Practice: Word Problems, p. 442

NAME ______ DATE ______

8-2 Practice: Word Problems

The Percent Equation

1. **DINING** Jonas and Linda's restaurant bill comes to $23.40. They are planning to tip the waiter 15% of their bill. How much money should they leave for a tip? $3.51
2. **CHESS** The Briarwood Middle School chess club has 55 members. 22 of the members are in seventh grade. What percent of the members of the chess club are in seventh grade? 40%
3. **TENNIS** In the city of Springfield, 75% of the parks have tennis courts. If 15 parks have tennis courts, how many parks does Springfield have altogether? 20 parks
4. **COLLEGE** There are 225 students in eighth grade at Jefferson Middle School. A survey shows that 64% of them are planning to attend college. How many Jefferson eighth grade students are planning to attend college? 144 students
5. **BASEBALL** In the 2001 season, the Chicago Cubs won 88 out of 162 games. What percent of games did the Cubs win? Round to the nearest tenth if necessary. 54.3%
6. **HOUSING** In the Stoneridge apartment complex, 35% of the apartments have one bedroom. If there are 49 one-bedroom apartments, what is the total number of apartments at Stoneridge? 140 apartments
7. **SPACE** On Mars, an object weighs 38% as much as on Earth. How much would a person who weighs 165 pounds on Earth weigh on Mars? 62.7 pounds
8. **FOOTBALL** In the 2001 season, quarterback Matt Hasselbeck of the Seattle Seahawks had 8 passes intercepted out of 321 attempts. What percent of Matt Hasselbeck's passes were intercepted? Round to the nearest tenth if necessary. 2.5%

4 Assess

Open-Ended Assessment

Modeling Ask students to use manipulatives or decimal pieces to model the solutions to questions such as, "What number is 40% of 3?" Have them demonstrate that "40% of 3" and "0.4×3" are equivalent expressions.

Answer

31. Sample answer: If the percent is less than 100%, then the part is less than the number; if the percent equals 100%, then the part equals the number; if the percent is greater than 100%, then the part is greater than the number.

8-3a HANDS-ON LAB Lesson Notes

Lesson Resources

Teaching Mathematics with Manipulatives
- student recording sheet 8-3a

GETTING STARTED

Teaching Tip Before starting this lab, collect several examples of survey results from newspapers or magazines to show students.

TEACH

- Be sure to allow enough time to carry out the tasks. Students will need time to develop and conduct the survey. Organizing and displaying results will likely be done over multiple days.
- You may want to discuss giving an oral survey as an alternative to a written one.
- Discuss appropriate times for collecting the data. Students may wish to survey classmates at lunch. You may also want to make arrangements to conduct the surveys with other classes.

ASSESS

Observe as students solve the **Writing Math** exercises. They should:
- design and conduct a survey using a random sample,
- analyze any bias in the survey, and
- evaluate the hypothesis based on the results of the survey.

8-3a HANDS-ON LAB

What You'll Learn
Use sampling to collect data.

Sampling

INVESTIGATE *Work as a class.*

To determine the favorite lunches of students at your school, you could ask each student. A more practical method is to survey a representative group of students and use that data to make a conclusion. This method is called **sampling**.

Term	Definition
population	total group of people or items in which the survey is interested
sample	part of the population
sample size	number of people or items in the survey

It is important to obtain a sample that is **unbiased**, that is, not limited or not favoring a particular outcome. Suppose you put the ID number of all the students in a box and randomly choose 65 numbers. This sample is unbiased because it is:

- a *random sample*. Each member of the population, or students at your school, has an equal chance of being selected.
- large enough to provide accurate data.

STUDY TIP
Unbiased An unbiased sample is representative of the population.

Writing Math

Work in groups of three.

State whether each sample is random. Explain. 1–2. See margin.

1. To determine the favorite spectator sport for women over 25 years old, 350 women over 25 are surveyed at a professional basketball game.
2. To collect data about the study habits of middle school students in the Franklin School District, the name of every middle school student in the district is placed in a bag and 250 names are randomly selected.

For Exercises 3–7, refer to the information below.
Suppose you are to survey students in your school. 3–7. See students' work.

3. Formulate a hypothesis about students' activities.
4. Design and conduct a survey. Describe the technique that you used to get a random sample.
5. Organize and display the results of your survey in a table or graph.
6. Analyze ways in which the wording of your questions might have changed the outcome of the survey.
7. Use the results of your survey to evaluate your hypothesis.

Answers

1. No; the sample is biased since women at a professional basketball game may prefer basketball as their favorite sport.
2. Yes; each middle school student in the school district is equally likely to be selected.

344 **Chapter 8** Applying Percent

8-3 Using Statistics to Predict

hat You'll Learn
edict actions of a
ger group by using
sample.

W Vocabulary
rvey
pulation
ndom sample

VIEW Vocabulary
ta: pieces of
ormation that are
en numerical
sson 2-1)

WHEN am I ever going to use this?

TELEVISION The graphic shows the results of a survey in which children were asked whether they had rules about what they watched on television.

1. What ages of children are represented? **8–17**
2. Can you tell how many were surveyed? Explain.
3. Describe how you could use the graphic to predict how many students in your school have no restrictions on their television viewing.

2. No; the graphic only gives percents of the group. 3. See margin.

A **survey** is a question or set of questions designed to collect data about a specific group of people, called the **population**. If a survey uses a **random sample** of a population, or a sample chosen without preference, you can assume that the sample represents the population. Then you can use the results to make predictions about the population.

EXAMPLE Predict Using Percent Proportion

1 TELEVISION Refer to the survey results in the graphic above. Predict how many of the 1,250 Gallatin Middle School students have some restrictions on the television that they watch.

You can use the percent proportion and the survey results to predict what part of the population of students have TV restrictions.

(part of the population → a; entire population → b) $\frac{a}{b} = \frac{p}{100}$	Percent proportion
$\frac{a}{1{,}250} = \frac{54}{100}$	Survey results: $54\% = \frac{54}{100}$
$100a = 1{,}250(54)$	Cross products
$a = 675$	Simplify.

So, about 675 of the Gallatin Middle School students have some television restrictions.

TEACHING TIP
sk students to list
amples of random
mples and examples
samples that are not
ndom.

msmath2.net/extra_examples

Answer (When am I ever going to use this?)

3. Sample answer: Assume the students in your school are similar to those represented in the graph. Then find 46% of the number of students in your school.

8-3 Lesson Notes

Lesson Resources

Chapter 8 Resource Masters FAST FILE
- Study Guide and Intervention, p. 445
- Practice: Skills, p. 446
- Practice: Word Problems, p. 447
- Reading to Learn Mathematics, p. 448
- Enrichment, p. 449

Key Teaching Transparencies, 30
Parent and Student Study Guide Workbook, 8-3
Science and Math Lab Masters, pp. 37–40

1 Focus

5-Minute Check 8-3
INTERACTIVE CHALKBOARD PowerPoint® Presentations
Use as a quiz or review of Lesson 8-2. (Also available as a transparency)

Bellringer
Hands-On Using grid paper, have students design a game board of any shape. It should have four point regions. Have students drop paper clips onto the board from one yard above the board. Have students determine the percent of paper clips that landed in each region. Ask students if they think they can predict where the next paper clip will drop based on the previous calculations.

Reading to Learn Mathematics, p. 448 ELL

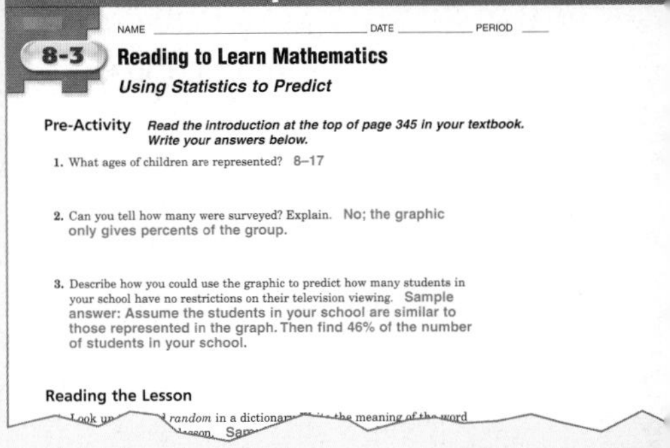
NAME ______ DATE ______ PERIOD ____

8-3 Reading to Learn Mathematics
Using Statistics to Predict

Pre-Activity *Read the introduction at the top of page 345 in your textbook. Write your answers below.*

1. What ages of children are represented? 8–17
2. Can you tell how many were surveyed? Explain. No; the graphic only gives percents of the group.
3. Describe how you could use the graphic to predict how many students in your school have no restrictions on their television viewing. Sample answer: Assume the students in your school are similar to those represented in the graph. Then find 46% of the number of students in your school.

Reading the Lesson

2 Teach

In-Class Examples

INTERACTIVE CHALKBOARD
PowerPoint® Presentations

1 **PETS** The table shows the results of a survey in which people were asked whether their house pets watch television.

Does your pet watch television?	
Response	**Percent**
yes	38%
no	60%
don't know	2%

Predict how many of the 540 people surveyed said their pets watch TV. **about 205**

2 **SUMMER JOBS** According to one survey, 25% of high school students reported they would not get a summer job. Predict how many of the 948 students at Mohawk High School will not get a summer job. **237 students**

3 Practice/Apply

About the Exercises...

Guided Practice Key

Exercises	Examples
3–4	1, 2

Odd/Even Assignments
Exercises 5–8 are structured so that students practice the same concepts whether they are assigned odd or even problems.

Alert! For Exercise 2, students will need access to newspapers or magazines.

Answer

1. **Sample answer: Randomly select a part of the group to get a sample. Find their preferences, and use the results to find the percent of the total group.**

REAL-LIFE MATH

COMMUNICATION The survey at the right also found that 81% of people 12 to 17 years old use the Internet to send E-mails.

You can also use the percent equation to make predictions using survey results.

EXAMPLE **Predict Using Percent Equation**

2 **COMMUNICATION A survey showed that 70% of people 12 to 17 years old use the Internet to send instant messages. Predict how many of the 2,640 students at Washington Middle School send instant messages.**

You need to predict how many of the 2,640 students send instant messages.

Words	What number is 70% of 2,640?
Variable	Let n represent the number.
Equation	$n = 0.7 \cdot 2{,}640$

$n = 0.7 \cdot 2{,}640$ Write the equation.

$n = 1{,}848$ Multiply.

So, you could predict that about 1,848 students at Washington Middle School use the Internet to send instant messages.

Skill and Concept Check

Writing Math Exercises 1 & 2

1. **Explain** how to use a sample to predict what a group of people prefer. **See margin.**
2. **OPEN ENDED** Select a newspaper or magazine article that contains a table or graph. Identify the population and explain how you think the results were found. **See students' work.**

GUIDED PRACTICE

3. **SPENDING** The circle graph shows the results of a poll to which 60,000 teens responded. Predict how many of the approximately 28 million teens in the United States would buy a music CD if they were given $20. **8.96 million**

How Would You Spend a Gift of $20?

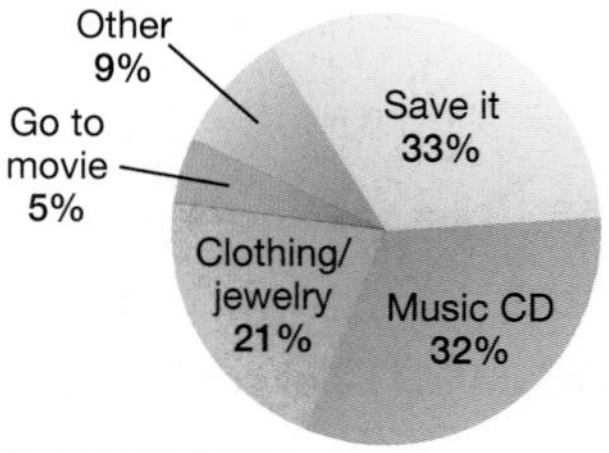

Source: *USA WEEKEND*

4. **TRANSPORTATION** The graph shows the results of a survey in which working adults in America were asked how they get to work. Use the data to predict how many working adults in America out of 143 millio[n] walk or ride a bicycle to work. **5.72 million**

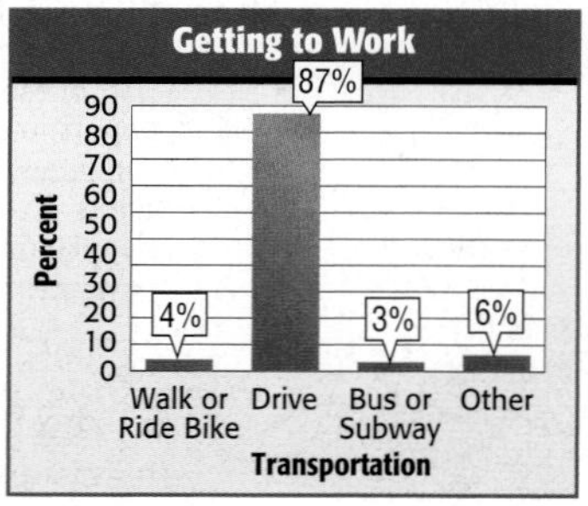

Source: Gallup Poll

346 Chapter 8 Applying Percent

DAILY INTERVENTION **Flexible Grouping**

Using Statistics to Predict Have students work in small groups. Each student should read an example problem to the group, then discuss the information presented. Have part of the group solve the problem using the percent proportion and the other part using the percent equation. Students should share their results within their groups.

Practice and Applications

HOMEWORK HELP	
For Exercises	See Examples
5–8	1, 2
Extra Practice See pages 583, 603.	

5. **COMPUTERS** The table shows the results of a survey in which students were asked how they use a personal computer at home. Use the results to predict how many of the 1,745 Allegheny Valley Middle School students use a PC for homework. **1,396 students**

PC Use	Percent
educational programs	93%
homework	80%
games	70%

Source: U.S. Census Bureau

6. **CAMERAS** In a survey, 14% of teens said they own a digital camera. Predict how many of the 420,000 teens in Arizona own digital cameras. **58,800 teens**

CATS For Exercises 7 and 8, use the graphic at the right. It shows the percent of cat owners who train their cats to prevent each problem.

7. Out of 255 cat owners, how many would you predict have trained their cat not to climb on furniture? **about 110 owners**
8. Predict how many cat owners in a group of 316 have trained their cat not to claw furniture. **about 92 owners**

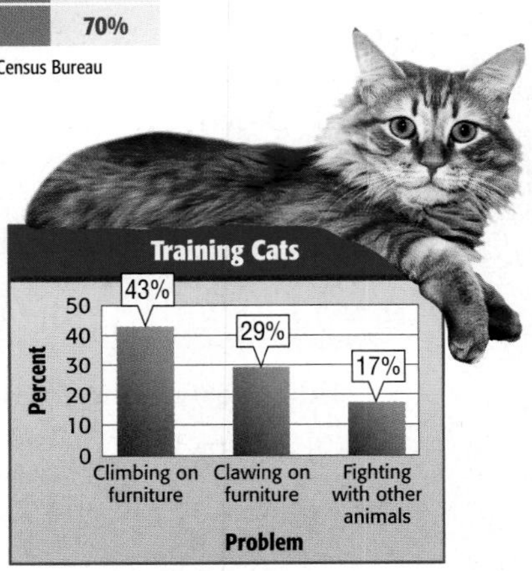

Source: Purina Cat Chow

9. **CRITICAL THINKING** A survey found that 80% of teens enjoy going to the movies in their free time. This was the response of 5,200 teens that were surveyed. What was the total number of teens surveyed? **6,500 teens**

Standardized Test Practice and Mixed Review

10. **MULTIPLE CHOICE** A survey of 80 seventh graders at Lincoln Middle School was taken to find how they get to school each day. The results are shown in the table. Of the 423 seventh graders in the school, predict how many walk to school. **C**

Ⓐ 23 Ⓑ 64 Ⓒ 123 Ⓓ 394

Getting to School	Percent
take a bus	33%
walk	29%
adult drives	18%
other	20%

11. **SHORT RESPONSE** A survey showed that 16% of the people in Tennessee over the age of 16 belong to a fitness center. Predict how many of the 5 million people in Tennessee over the age of 16 belong to a fitness center. **800,000 people**

Write an equation for each problem. Then solve. Round to the nearest tenth if necessary. (Lesson 8-2)

12. What number is 12% of 60? **$n = 0.12 \cdot 60$; 7.2**
13. 54 is 72% of what number? **$54 = 0.72 \cdot n$; 75**
14. Estimate 30% of 149. (Lesson 8-1) **Sample answer: $0.1 \cdot 150 = 15$ and $3 \cdot 15 = 45$**

GETTING READY FOR THE NEXT LESSON

PREREQUISITE SKILL Simplify. Write as a decimal. (Lesson 1-3)

15. $\frac{10-7}{10}$ **0.3**
16. $\frac{50-18}{50}$ **0.64**
17. $\frac{22-4}{4}$ **4.5**
18. $\frac{39-15}{15}$ **1.6**

msmath2.net/self_check_quiz

4 Assess

Open-Ended Assessment

Speaking Discuss with students how accurate predictions seem to be. Ask them how the number of people surveyed can affect predictions.

Sample answer: If only a small population is surveyed, the results will not be as accurate as if a large population were used.

Study Guide and Intervention, p. 445

NAME ____________ DATE ________ PERIOD ____

8-3 Study Guide and Intervention

Using Statistics to Predict

Data gathered by surveying a random sample of the population may be used to make predictions about the entire population.

EXAMPLE 1 **In a survey, 200 people from a town were asked if they thought the town needed more bicycle paths. The results are shown in the table. Predict how many of the 28,000 people in the town think more bicycle paths are needed.**

More Bicycle Paths Needed?	
Response	**Percent**
yes	39%
no	42%
undecided	19%

Use the percent proportion.

$\frac{a}{b} = \frac{p}{100}$ Percent proportion

part of the population → / entire population → $\frac{a}{28{,}000} = \frac{39}{100}$ Survey results: $39\% = \frac{39}{100}$

$100a = 28{,}000(39)$ Cross products

$a = 10{,}920$ Simplify.

So, about 10,920 people in the town think more bicycle paths are needed.

EXERCISES

1. **VOTES** In a survey of voters in Binghamton, 55% of those surveyed said they would vote for Armas for city council. If 24,000 people vote in the election, about how many will vote for Armas? 13,200 people
2. **LUNCH** A survey shows that 43% of high school and middle school students buy school lunches. If a school district has 2,900 high school and middle school students, about how many buy school lunches? 1,247 students
3. **CLASS TRIP** Students of a seventh grade class were surveyed to find out how much they would be willing to pay to go on a class trip. 24% of the students surveyed said they would pay \$21 to \$30. If there are 360 students in the seventh grade class, about how many would be willing to pay for a trip that cost \$21 to \$30? about 86 students

Practice: Skills, p. 446

NAME ____________ DATE ________ PERIOD ____

8-3 Practice: Skills

Using Statistics to Predict

CELL PHONES For Exercises 1–3, use the table at the right. It shows the results of a survey in which students 12 to 17 years old were asked how often they use a cell phone.

Frequency of Use	Percent
more than twice a week	32%
once or twice a week	16%
once or twice a month	23%
less than once a month	12%
never used one	17%

1. Out of 215 students 12 to 17 years old, how many would you predict use a cell phone once or twice a week? about 34 students
2. Predict how many students 12 to 17 years old in a group of 375 have never used a cell phone. about 64 students
3. How many students 12 to 17 years old out of 1,200 would you expect use a cell phone at least once or twice a week? 576 students

PIZZA For Exercises 4–6, use the table at the right. It shows the results of a survey in which a random sample of seventh graders at Kiewit Middle School were asked to name their favorite pizza topping.

Pizza Topping	Percent
pepperoni	46%
peppers	28%
olives	8%
onions	2%
pineapple	4%
mushrooms	12%

4. There are 32 students in Mrs. Chen's seventh grade class. Predict how many would choose olives as their favorite topping. about 3 students
5. There are 210 seventh grade students eating lunch in the cafeteria. How many of them would choose peppers as their favorite topping? about 59 students
6. Predict how many of the 524 seventh graders at Kiewit Middle School would choose pepperoni as their favorite pizza topping. about 241 seventh graders
7. **BACKPACKS** A survey showed that 78% of students who take a bus to school carry a backpack. Predict how many of the 654 students who take a bus also carry a backpack. about 510 students

Practice: Word Problems, p. 447

NAME ____________ DATE ________ PERIOD ____

8-3 Practice: Word Problems

Using Statistics to Predict

1. **SHOES** The table shows the results of a survey in which seventh graders were asked how many pairs of shoes they own. Predict how many of the 632 seventh graders at Seneca West Middle School own more than 7 pairs of shoes.

Shoes	Percent
3 or less	10%
4	20%
5	21%
6	22%
7	19%
more than 7	8%

about 51 seventh graders

2. **ACTIVITIES** Of the students listed as members of a high school academic team, 75% were involved in sports, speech, music or debate. If 111 students were listed as part of the teams, how many were involved in sports, speech, music, or debate? about 83 students
3. **MOVIEGOERS** A research study found that about 63% of people 18 or older who go to the movies at least once a month own a personal computer. Out of 500 people 18 and older who go to the movies once or more a month, how many of them would you expect to own a personal computer? 315 people
4. **HAIR** A survey showed that 37% of people 12 to 17 years old use hair gel. Predict how many of the 30 students in Mr. Avalon's ninth grade class use hair gel. about 11 students
5. **GRADUATION** A survey of first-year students at North Carolina State University showed that about 73% expect to complete their degree in 4 years. If there are 3,333 first-year students, how many of them expect to complete their degree in 4 years? about 2,433 students
6. **INTERNET** A recent survey conducted by the Millard school district showed that 87% of households of students have Internet access at home. If there are 19,000 Millard households, how many have Internet access? 16,530 households

Lesson 8-3 Using Statistics to Predict 347

CHAPTER 8 Mid-Chapter Practice Test

Assessment Option

Mid-Chapter Test (Lessons 8-1 through 8-3) is available on p. 481 of the *Chapter 8 Resource Masters*.

ExamView® Pro Test Bank CD-ROM

Use the networkable ExamView® Pro Test Bank CD-ROM to:

- Create **Multiple versions** of tests.
- Create **Modified** tests for **Inclusion** students with one mouse click.
- **Edit** existing questions and **Add** your own questions.
- Build tests aligned with state standards using built-in **State Curriculum Correlations.**
- Change **English** tests to **Spanish** with one mouse click and vice versa.

FOLDABLES™ Follow-Up

Before students complete the Mid-Chapter Practice Test, encourage them to review the notes, definitions, concepts, and main ideas they recorded in their Foldable tables while working through the first three lessons of the chapter.

Answers

3. $\frac{1}{5} \cdot 400 = 80$
4. $\frac{4}{5} \cdot 110 = 88$
5. $\frac{1}{2} \cdot 300 = 150$
6. $0.1 \cdot 40 = 4$ and $3 \cdot 4 = 12$
7. $0.1 \cdot 90 = 9$ and $8 \cdot 9 = 72$
8. $0.1 \cdot 210 = 21$ and $4 \cdot 21 = 84$
9. $n = 0.35 \cdot 72$; 25.2
10. $16.1 = n \cdot 70$; 23%
11. $27.2 = 0.68 \cdot n$; 40
12. $n = 0.16 \cdot 32$; 5.1

CHAPTER 8 Mid-Chapter Practice Test

Vocabulary and Concepts

1. **Describe** two methods that can be used to estimate the percent of a number. (Lesson 8-1) **Sample answer: Use a fraction or find 10% of a number and multiply.**
2. **Explain** what the *population* is in a survey. (Lesson 8-3) **the total group in which the survey is interested**

Skills and Applications

Estimate by using fractions. (Lesson 8-1) **3–5. See margin.**

3. 20% of 392
4. 78% of 112
5. 52% of 295

Estimate by using 10%. (Lesson 8-1) **6–8. See margin.**

6. 30% of 42
7. 79% of 88
8. 41.5% of 212

Write an equation for each problem. Then solve. Round to the nearest tenth if necessary. (Lesson 8-2) **9–12. See margin.**

9. What number is 35% of 72?
10. 16.1 is what percent of 70?
11. 27.2 is 68% of what number?
12. 16% of 32 is what number?

ENTERTAINMENT For Exercises 13 and 14, refer to the graph at the right. It shows the results of a survey of students' favorite TV programs at Morgan Middle School. (Lesson 8-3)

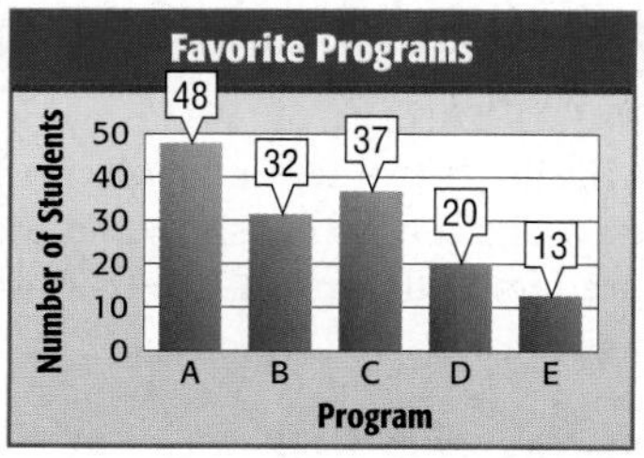

13. What percent of the students surveyed preferred Program A? **32%**
14. How many of the 925 students in the school would you expect to choose Program A as their favorite? **296 students**

Standardized Test Practice

15. **MULTIPLE CHOICE** Miyoki has read 82% of a book that has 214 pages. Which is the best estimate of the number of pages that she has read? (Lesson 8-1) **D**

    Ⓐ 16 pages Ⓑ 50 pages
    Ⓒ 80 pages Ⓓ 160 pages

16. **MULTIPLE CHOICE** A cookie company received 1,600 E-mails in one week. Of those E-mails, 12.5% were people requesting catalogues. How many people requested catalogues that week? (Lesson 8-2) **H**

    Ⓕ 2,000 people Ⓖ 1,280 people
    Ⓗ 200 people Ⓘ 128 people

The Game Zone

A Place To Practice Your Math Skills

Spinning for Percents

GET READY!

Players: two or three
Materials: 6 index cards, 2 spinners

GET SET!

- Copy the numbers 11, 28, 45, 62, 84, and 98 onto index cards, one number per card.
- Shuffle the cards and place them facedown in a pile.
- Label equal sections of two spinners with the digits 0 through 9.
- Decide which spinner will stand for digits in the tens place and which spinner will stand for the ones place. The number formed by spinning both spinners is the percent.

These spinners represent 62%.

GO!

- One player selects the top card from the pile.
- The player spins both spinners and records the percent that is formed. He or she then finds that percent of the number that is on the index card. Round to the nearest whole number if necessary. This is the player's score for this turn.
- Continue in this way, taking turns selecting a card, until no cards remain in the pile.
- **Who Wins?** The player with the greatest total score wins.

The Game Zone

Resources

For an easy-to-make manipulative for this game, visit: msmath2.net/gamezone.

Glencoe Mathematics Classroom Manipulative Kit

- spinners

GETTING READY

Objective Practice calculating the percent of a number.

MANAGEMENT TIPS

Recommended Time
20 minutes

Tips

- You may want to provide students with a recording sheet to keep track of their work and their total points.
- For a simpler game, you may wish to use two number cubes instead of spinners to form the percents.
- Consider adding a third spinner with two sections, 0 and 1, to allow for percents greater than 100.
- You may wish to use the game as an assessment tool for the concepts the students have been learning in Lessons 8-1 through 8-3.

For an additional game for this chapter, visit: msmath2.net/gamezone.

8-4 Lesson Notes

Lesson Resources

Chapter 8 Resource Masters

- Study Guide and Intervention, p. 450
- Practice: Skills, p. 451
- Practice: Word Problems, p. 452
- Reading to Learn Mathematics, p. 453
- Enrichment, p. 454

Parent and Student Study Guide Workbook, 8-4

Teaching Mathematics with Manipulatives, pp. 77–78

FAST FILE

1 Focus

5-Minute Check 8-4

INTERACTIVE CHALKBOARD
PowerPoint® Presentations

Use as a quiz or review of Lesson 8-3. (Also available as a transparency)

Bellringer

Communication Have students describe situations in which they have heard a percent increase or percent decrease discussed. For example, students might mention hearing a report that school revenues have increased by five percent.

Reading to Learn Mathematics, p. 453 **ELL**

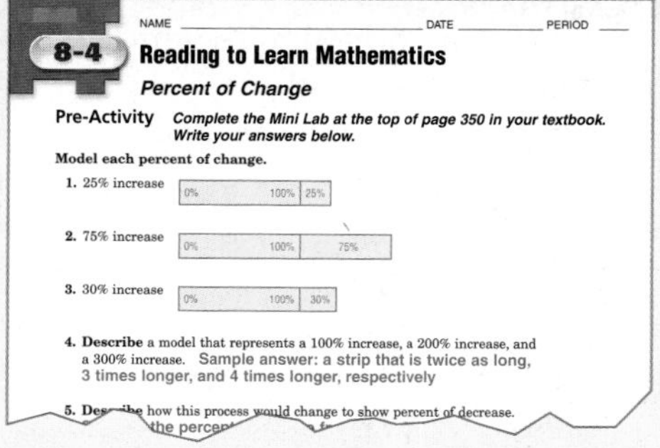
NAME ______ DATE ______ PERIOD ____

8-4 **Reading to Learn Mathematics**

Percent of Change

Pre-Activity *Complete the Mini Lab at the top of page 350 in your textbook. Write your answers below.*

Model each percent of change.

1. 25% increase
2. 75% increase
3. 30% increase
4. **Describe** a model that represents a 100% increase, a 200% increase, and a 300% increase. Sample answer: a strip that is twice as long, 3 times longer, and 4 times longer, respectively
5. **Describe** how this process would change to show percent of decrease.

8-4 Percent of Change

What You'll Learn
Find the percent of increase or decrease.

NEW Vocabulary
percent of change
percent of increase
percent of decrease

HANDS-ON Mini Lab

Materials
- paper strips
- scissors
- tape

Work with a partner.

You can use paper strips to model a 50% increase.

- Begin with two paper strips. On each strip, label 0% on the left side and 100% on the right side.

- Fold one of the paper strips in half. Mark 50% in the center.

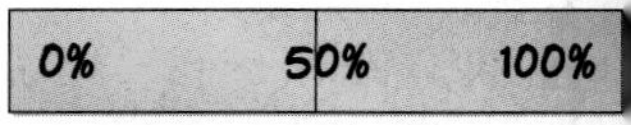

- Cut the second strip at the 50% mark and tape the piece onto the end of the first strip. The new longer strip represents a 50% increase or 150%.

Model each percent of change. 1–4. See margin.

1. 25% increase
2. 75% increase
3. 30% increase
4. **Describe** a model that represents a 100% increase, a 200% increase, and a 300% increase.
5. **Describe** how this process would change to show percent of decrease. **Subtract the percent of change from the original strip.**

One way to describe a change in quantities is to use percent of change.

Key Concept — **Percent of Change**

Words A **percent of change** is a ratio that compares the change in quantity to the original amount.

Equation $\text{percent of change} = \dfrac{\text{amount of change}}{\text{original amount}}$

If the original quantity is increased, then it is called a **percent of increase**. If the original quantity is decreased, then it is called a **percent of decrease**.

$$\text{percent of increase} = \frac{\text{amount of increase}}{\text{original amount}} \leftarrow \text{new amount} - \text{original amount}$$

$$\text{percent of decrease} = \frac{\text{amount of decrease}}{\text{original amount}} \leftarrow \text{original amount} - \text{new amount}$$

TEACHING TIP
Point out that the amount of change is always a positive number.

Answers (Hands-On Mini Lab)

1.

2.
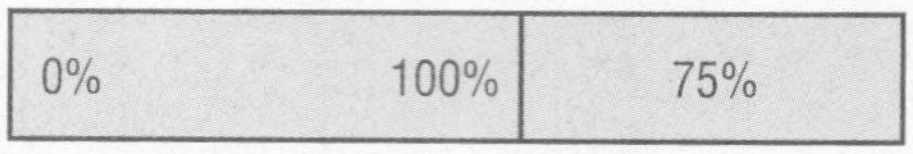

3.

4. Sample answer: a strip that is twice as long, 3 times longer, and 4 times longer, respectively

EXAMPLE Find Percent of Increase

1 TREES Find the percent of change in tree height from year 1 to year 10. Round to the nearest whole percent if necessary.

Since the new height is greater than the original height, this is a percent of increase. The amount of increase is 9.6 − 4 or 5.6 feet.

$$\text{percent of increase} = \frac{\text{amount of increase}}{\text{original amount}}$$

$$= \frac{5.6}{4} \quad \text{Substitution}$$

$$= 1.4 \quad \text{Simplify.}$$

$$= 140\% \quad \text{Write 1.4 as a percent.}$$

The percent of increase in the tree height is 140%.

EXAMPLE Find Percent of Decrease

2 STOCKS Find the percent of change if the original price of a stock was $75 and the new price is $60. Round to the nearest whole percent if necessary.

Stock	Price
original	$75
new	$60

Since the new price is less than the original price, this is a percent of decrease. The amount of decrease is 75 − 60 or $15.

$$\text{percent of decrease} = \frac{\text{amount of decrease}}{\text{original amount}}$$

$$= \frac{15}{75} \quad \text{Substitution}$$

$$= 0.2 \quad \text{Simplify.}$$

$$= 20\% \quad \text{Write 0.2 as a percent.}$$

The percent of decrease of the stock is 20%.

Your Turn Find each percent of change. Round to the nearest whole percent if necessary. State whether the percent of change is an *increase* or *decrease*.

a. original: 10
new: 13 **30% increase**

b. original: 20
new: 15 **25% decrease**

TEACHING TIP

oint out that an increase f 140% means the tree ore than doubled in eight.

TEACHING TIP

emind students to hange all answers to ercents.

EAL-IFE CAREERS

w Does a Stockbroker e Math?

stockbroker must be able compare percents of ange of different stock ces.

Online Research information about a eer as a stockbroker, visit: math2.net/careers

TEACHING TIP

he stockbrokers above re learning signals for uying and selling stocks n a stock market trading oor.

msmath2.net/extra_examples

2 Teach

In-Class Examples

INTERACTIVE CHALKBOARD PowerPoint® Presentations

1 SHOPPING Last year a sweater sold for $56. This year the same sweater sells for $60. Find the percent of change in the cost of the sweater. Round to the nearest whole percent if necessary. **7% increase**

2 ATTENDANCE On the first day of school this year, 435 students reported to Howard Middle School. Last year on the first day, 460 students attended. Find the percent of change for the first day attendance. Round to the nearest whole percent if necessary. **5% decrease**

DAILY INTERVENTION **Unlocking Misconceptions**

Percent of Change Some students may mistakenly want to divide the greater number into the lesser number. Encourage them to take the time to identify the amount of change and the original amount and write them as a ratio before calculating.

3 Practice/Apply

FOLDABLES™ Follow-Up Remind students to record main ideas and to give examples of what they learn about percent of change in the Lesson 8-4 section of their Foldable tables.

NUMBER SENSE
In Exercise 2, encourage students to consider the fractions $\frac{10}{90}$ and $\frac{10}{60}$ as they answer the question. The fraction with the greater value indicates the greater percent of decrease.

DAILY INTERVENTION **FIND THE ERROR** In Exercise 3, remind students to use the original amount in the denominator when finding the percent of change. It might be helpful for students to compare the amount of the change, 84, to the original value, 46. This comparison quickly shows that the percent of change is greater than 100.

About the Exercises...

Guided Practice Key

Exercises	Examples
5–7, 9–10	1
4, 8	2

Odd/Even Assignments
Exercises 11–26 are structured so that students practice the same concepts whether they are assigned odd or even problems.

Skill and Concept Check

Writing Math Exercises 1–3

1. **OPEN ENDED** Write a percent of change problem using 14 and 25. State whether there is a percent of increase or decrease. Then solve.
   1–3. See margin.
2. **NUMBER SENSE** The costs of two different shelf stereo systems are decreased by $10. The original costs were $90 and $60, respectively. Without calculating, which had greater percent of decrease? Explain.
3. **FIND THE ERROR** Jada and Miranda are finding the percent of change from 46 to 130. Who is correct? Explain.

Jada: $\frac{130 - 46}{46} \approx 1.83$ or 183%

Miranda: $\frac{130 - 46}{130} \approx 0.65$ or 65%

GUIDED PRACTICE

Find each percent of change. Round to the nearest whole percent if necessary. State whether the percent of change is an *increase* or *decrease*.

4. original: 30
   new: 24 **20% dec.**
5. original: $126
   new: $150 **19% inc.**
6. original: 624
   new: 702 **13% inc.**
7. original: $75.80
   new: $94.75 **25% inc.**
8. original: 1.6
   new: 0.95 **41% dec.**
9. original: 20.5
   new: 35.5 **73% inc.**
10. **SPORTS** The table shows the number of people ages 7 to 17 who played soccer. What was the percent of increase in soccer players from 1990 to 2000? Round to the nearest whole percent. **18%**

Playing Soccer

Year	Number (millions)
1990	10.9
2000	12.9

Source: National Sporting Goods Association

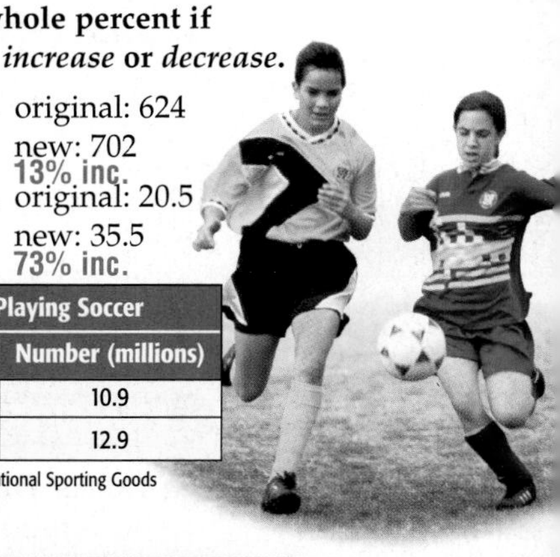

★ **indicates increased difficulty**

Practice and Applications

HOMEWORK HELP

For Exercises	See Examples
11–31	1, 2

Extra Practice See pages 583, 603.

Find each percent of change. Round to the nearest whole percent if necessary. State whether the percent of change is an *increase* or *decrease*.

11. original: 15
    new: 18 **20% inc.**
12. original: 100
    new: 140 **40% inc.**
13. original: $12
    new: $6 **50% dec.**
14. original: $240
    new: $320 **33% inc.**
15. original: 48
    new: 14 **71% dec.**
16. original: 360
    new: 120 **67% dec.**
17. original: 125
    new: 87.5 **30% dec.**
18. original: $15.60
    new: $11.70 **25% dec.**
19. original: $89.50
    new: $105.20 **18% inc.**
20. original: 132
    new: 125.4 **5% dec.**
21. original: 782
    new: 789.82 **1% inc.**
22. original: 12
    new: 60 **400% inc.**

GEOMETRY **For Exercises 23 and 24, refer to the rectangle at the right. Suppose the side lengths are doubled.**

8 cm

3 cm

23. Find the percent of increase in the perimeter. **100%**
24. Find the percent of increase in the area. **300%**

Answers

1. Sample answer: The sale price of a T-shirt is $14. If the shirt originally cost $25, what is the percent discount? Answer: 44% decrease
2. The $60 stereo, because10 is a greater part of 60 than of 90.
3. Jada; Miranda did not write a ratio comparing the change to the original amount.

32. No; after a 10% increase, the quantity is greater than the original quantity. Decreasing the greater quantity by 10% does not equal the original quantity.

25. **SALES** Use the graphic at the right to find the percent of change in ketchup sales from 2000 to 2001. **21% inc.**

Source: Heinz North America

26. **EDUCATION** A person with a high school diploma earns an average of $16,053 per year. A person with some college earns an average of $25,686 per year. What is the percent of increase? **60%**

27. **POPULATION** On July 1, 2001, the U.S. had an estimated 284,796,887 residents. This is 3,374,981 more than on April 1, 2000. To the nearest tenth, what was the percent of change in population? **1.2% increase**

★ 28. **ART** In 2001, the sales of fine art were $40.8 billion. Art sales are projected to increase by 25% from 2001 to 2005. What is the projected amount of art sales in 2005? **$51 billion**

ALLOWANCES For Exercises 29–31, refer to the table at the right.

Age	Average Weekly Allowance
12	$9.58
13	$9.52
14	$13.47
15	$15.57
16	$17.84
17	$30.66

Source: www.kidsmoney.org

29. Find the percent of increase of allowance from age 13 to 14. Round to the nearest whole percent. **41%**

30. State two consecutive ages where the change in allowance is a percent of decrease. Then find the percent of decrease. Round to the nearest tenth. **12 to 13; 0.6%**

★ 31. Between which two consecutive years is the percent of increase the greatest? What is the percent of increase? Round to the nearest whole percent. **16 to 17; 72%**

32. **CRITICAL THINKING** If a quantity increases by 10% and then decreases by 10%, will the result be the original quantity? Explain. **See margin.**

Standardized Test Practice and Mixed Review

33. **MULTIPLE CHOICE** The table shows the average number of hours that Americans worked in 1990 and in 2000. Find the percent of increase to the nearest tenth. **C**

Year	Time (h)
1990	1,943
2000	1,979

Source: International Labor Organization

Ⓐ 19.0% Ⓑ 2.0% Ⓒ 1.9% Ⓓ 0.2%

34. **SHORT RESPONSE** Find the percent of decrease from 85 to 68. **20%**

35. **FOOD** In a survey of 150 students at Kennedy Middle School, 48% said that their favorite type of pizza crust is thick crust. Predict how many of the 1,375 students in the school prefer thick crust pizza. (Lesson 8-3) **660 students**

Write an equation for each problem. Then solve. Round to the nearest tenth if necessary. (Lesson 8-2)

36. 30% of what number is 17? $0.3n = 17$; **56.7**
37. What is 21% of 62? $n = 0.21 \cdot 62$; **13.0**

GETTING READY FOR THE NEXT LESSON

PREREQUISITE SKILL Write each percent as a decimal. (Lesson 7-5)

38. 6.5% **0.065**
39. $5\frac{1}{2}$% **0.055**
40. $8\frac{1}{4}$% **0.0825**
41. $6\frac{3}{4}$% **0.0675**

4 Assess

Higher-Level Thinking

A department store has all blue jeans on sale for 25% off. If you use the store's coupon from the newspaper, the cashier deducts an additional 15% off your purchase. Is the final price of a pair of jeans 40% off the regular price? Explain. **No, each percent decrease is calculated separately, so the additional 15% is not deducted from the original price but from the lower sale price.**

Study Guide and Intervention, p. 450

NAME ________ DATE ________ PERIOD ____

8-4 Study Guide and Intervention

Percent of Change

A **percent of change** is a ratio that compares the change in quantity to the original amount. If the original quantity is increased, it is a **percent of increase**. If the original quantity is decreased, it is a **percent of decrease**.

EXAMPLE 1 **Last year, 2,376 people attended the rodeo. This year, attendance was 2,950. What was the percent of change in attendance to the nearest whole percent?**

Since this year's attendance is greater than last year's attendance, this is a percent of increase.

The amount of increase is 2,950 – 2,376 or 574.

percent of increase = $\frac{\text{amount of increase}}{\text{original amount}}$ ← new amount – original amount

$= \frac{574}{2{,}376}$ Substitution

≈ 0.24 or 24% Simplify.

Rodeo attendance increased by about 24%.

EXAMPLE 2 **John's grade on the first math exam was 94. His grade on the second math exam was 86. What was the percent of change in John's grade to the nearest whole percent?**

Since the second grade is less than the first grade, this is a percent of decrease. The amount of decrease is 94 – 86 or 8.

percent of decrease = $\frac{\text{amount of decrease}}{\text{original amount}}$ ← original amount – new amount

$= \frac{8}{94}$ Substitution

≈ 0.09 or 9% Simplify.

John's math grade decreased by about 9%.

EXERCISES

Find each percent of change. Round to the nearest whole percent if necessary. State whether the percent of change is an *increase* or *decrease*.

1. original: 4, new: 5 — 25%; increase
2. original: 1.0, new: 1.3 — 30%; increase
3. original: 15, new: 12 — 20%; decrease
4. original: $30, new: $18 — 40%; decrease
5. original: 60, new: 63 — 5%; increase
6. original: 160, new: 136 — 15%; decrease
7. original: 7.7, new: 10.5 — 36%; increase
8. original: 9.6, new: 5.9 — 39%; decrease

Practice: Skills, p. 451

NAME ________ DATE ________ PERIOD ____

8-4 Practice: Skills

Percent of Change

Find each percent of change. Round to the nearest whole percent if necessary. State whether the percent of change is an *increase* or *decrease*.

1. original: 35, new: 70 — 100%; inc.
2. original: 8, new: 12 — 50%; inc.
3. original: 45, new: 30 — 33%; dec.
4. original: $350, new: $400 — 14%; inc.
5. original: $75, new: $60 — 20%; dec.
6. original: 250, new: 100 — 60%; dec.
7. original: $30, new: $110 — 267%; inc.
8. original: 35, new: 28 — 20%; dec.
9. original: $12.50, new: $15 — 20%; inc.
10. original: 80, new: 52 — 35%; dec.
11. original: 45, new: 63 — 40%; inc.
12. original: 120, new: 132 — 10%; inc.
13. original: $210, new: $105 — 50%; dec.
14. original: 84, new: 111 — 32%; inc.
15. original: $84, new: $100 — 19%; inc.
16. original: 6.8, new: 8.2 — 21%; inc.
17. original: 1.5, new: 2.5 — 67%; inc.
18. original: 91, new: 77 — 15%; dec.
19. original: $465.50, new: $350 — 25%; dec.
20. original: $87.05, new: $100 — 15%; inc.
21. original: 144, new: 108 — 25%; dec.
22. original: 20.8, new: 12.2 — 41%; dec.
23. original: $75, new: $15 — 80%; dec.
24. original: 8.6, new: 7 — 19%; dec.

Practice: Word Problems, p. 452

NAME ________ DATE ________ F

8-4 Practice: Word Problems

Percent of Change

1. **SHOES** A popular brand of running shoes costs a local store $68 for each pair. If the store sells the shoes for $119, what is the percent of increase in the price? 75%

2. **CLUBS** Last year the backgammon club had 30 members. This year the club has 24 members. Find the percent of decrease in the number of members. 20%

3. **READING** In the seventh grade, Rachel read 15 books. In the eighth grade, she read 18 books. Find the percent of increase in the number of books Rachel read. 20%

4. **VOTES** Last year 762 students voted in the student council election at San Bruno Middle School. This year 721 students voted. To the nearest tenth, what was the percent of change in the number of students that voted? 5.4% decrease

5. **HEIGHT** When Hugo was 9 years old he was 56 inches tall. Hugo is now 12 years old and he is 62 inches tall. Find the percent of increase in Hugo's height to the nearest tenth. 10.7%

6. **PLANTS** Alicia planted 45 tulip bulbs last year. This year she plans to plant 65 bulbs. Find the percent of increase in the number of tulip bulbs to the nearest tenth. 44.4%

7. **PICTURES** The 2002 yearbook at Middleton Middle School had 236 candid pictures of students. The 2001 yearbook had 214 candid pictures of students. To the nearest tenth, what was the percent of change in the number of candid student pictures? 10.3% increase

8. **POPULATION** In 1990, there were 4,298,000 Mexican immigrants living in the United States. In 2000 this number had increased to 7,858,000. Find the percent of increase to the nearest tenth. 82.8%

8-5 Lesson Notes

Lesson Resources

Chapter 8 Resource Masters
- Study Guide and Intervention, p. 455
- Practice: Skills, p. 456
- Practice: Word Problems, p. 457
- Reading to Learn Mathematics, p. 458
- Enrichment, p. 459

Calculator and Spreadsheet Masters, p. 41
Classroom Math Games, pp. 23–24
Key Teaching Transparencies, 31
Parent and Student Study Guide Workbook, 8-5

FAST FILE

1 Focus

5-Minute Check 8-5

INTERACTIVE CHALKBOARD PowerPoint® Presentations

Use as a quiz or review of Lesson 8-4. (Also available as a transparency)

Bellringer

Reading Mathematics Ask students to provide everyday meanings for *sales tax* and *discount*. Sample answer: Sales tax is the extra amount of money that you have to pay for most items in addition to the price tag amount. A discount is an amount of money taken off the price tag amount.

(Continued in bottom margin.)

Reading to Learn Mathematics, p. 458 ELL

NAME ______ DATE ______ PERIOD ____

8-5 Reading to Learn Mathematics
Sales Tax and Discount

Pre-Activity *Read the introduction at the top of page 354 in your textbook. Write your answers below.*

1. Calculate the sales tax by finding 6% of $1,299. $77.94
2. What will be the total cost including the sales tax? $1,376.94
3. Use a calculator to multiply 1.06 and 1,299. How does the result compare to your answer in Exercise 2? 1,376.94; It is the same.

Reading the Lesson

4. In Example 1, the ≈ is used when the sales tax is found. Why is the value of 0.0425 times 90 rounded? Sample answer: For dollar amounts, if the decimal has more than two places the decimal is rounded to the nearest cent.
5. Cal... the total cost of th... item in Example 1 using the ...thod shown ... What ...

8-5 Sales Tax and Discount

What You'll Learn
Solve problems involving sales tax and discount.

NEW Vocabulary
sales tax
discount

WHEN am I ever going to use this?

COMPUTERS Julie Ann plans to buy a new computer with a flat screen that costs $1,299. She lives in Florida where there is a sales tax of 6%.

1. Calculate the sales tax by finding 6% of $1,299. **$77.94**
2. What will be the total cost including the sales tax? **$1,376.94**
3. Use a calculator to multiply 1.06 and 1,299. How does the result compare to your answer in Exercise 2? **1,376.94; It is the same.**

One everyday use of percent is sales tax. **Sales tax** is an additional amount of money charged on items that people buy. The local, state, or federal government receives this money. The total cost of an item is the regular price plus the sales tax.

EXAMPLE Find the Total Cost

1 **TECHNOLOGY A graphing calculator costs $90, and the sales tax is 4.25%. What is the total cost of the calculator?**

First, find the sales tax.

4.25% of $\$90 = 0.0425 \cdot 90$

≈ 3.83 — The sales tax is $3.83.

Next, add the sales tax to the regular price.

$3.83 + 90 = 93.83$

The total cost of the calculator is $93.83.

Another way to find the cost of an item with sales tax is to add the percent of tax to 100%.

EXAMPLE Find the Total Cost

2 **CLOTHES What is the total cost of a sweatshirt if the regular price is $42 and the sales tax is $5\frac{1}{2}\%$?**

$100\% + 5\frac{1}{2}\% = 105\frac{1}{2}\%$ — Add the percent of tax to 100%.

The total cost is $105\frac{1}{2}\%$ of the regular price.

$105\frac{1}{2}\%$ of $\$42 = 1.055 \cdot 42$ — Use a calculator.

$= 44.31$

The total cost of the sweatshirt is $44.31.

TEACHING TIP
Remind students that the cost of an item with sales tax is always greater than the regular price.

Ask students whether a 5 percent sales tax is a percent of increase or a percent of decrease. A 15-percent discount? percent of increase; percent of decrease

Discount is the amount by which the regular price of an item is reduced. The sale price is the regular price minus the discount.

Standardized Test Practice

EXAMPLE Find the Sale Price

3 MULTIPLE-CHOICE TEST ITEM Alan wants to buy a snowboard that has a regular price of \$169. This week, the snowboard is on sale at a 35% discount. What is the sale price of the snowboard?

(A) \$59.15 (B) \$109.85 (C) \$134.00 (D) \$228.15

Read the Test Item

The sale price is \$169 minus the discount.

Solve the Test Item

Method 1 First, find the amount of the discount d.

part = percent · base	
$d = 0.35 \cdot 169$	Use the percent equation.
$d = 59.15$	The discount is \$59.15.

So, the sale price is \$169 − \$59.15 or \$109.85.

Method 2 First, subtract the percent of discount from 100%.

100% − 35% = 65%

So, the sale price s is 65% of the regular price.

$s = 0.65 \cdot 169$	Use the percent equation.
$s = 109.85$	The sale price is \$109.85.

So, the sale price of the snowboard is \$109.85. The answer is B.

Test-Taking Tip (The Princeton Review)

Elimination The sale price of an item is less than the regular price. Eliminate any choices in which the sale price given is greater than the regular price.

EXAMPLE Find the Percent of Discount

4 MUSIC An electric guitar is on sale as shown at the right. What is the percent of discount?

First, find the *amount* of discount.

\$299.95 − \$179.99 = \$119.96

Next, use the percent equation to find the percent discount.

Words	\$119.96 is what percent of \$299.95?
Variable	Let n represent the percent.
Equation	$119.96 = n \cdot 299.95$

$119.96 = n \cdot 299.95$	Write the equation.
$0.40 \approx n$	Divide each side by 299.95 and simplify.

The percent of discount is about 40%.

Your Turn

a. Find the percent of discount if the sale price of the guitar is \$224.96. 25%

TEACHING TIP This example is like finding the percent of decrease.

STUDY TIP

Discount You also could have used the percent proportion to find what percent \$119.96 is of \$299.95.

msmath2.net/extra_examples

2 Teach

In-Class Examples

INTERACTIVE CHALKBOARD PowerPoint® Presentations

1 GOLF A set of golf balls sells for \$20 and the sales tax is 5.75%. What is the total cost of the set? **\$21.15**

2 TENNIS What is the total cost of a tennis racket if the regular price is \$130 and the sales tax is $6\frac{1}{2}$%? **\$138.45**

3 MULTIPLE-CHOICE TEST ITEM Whitney wants to buy a new coat that has a regular price of \$185. This weekend, the coat is on sale at a 33% discount. What is the sale price of the coat? **B**

A \$61.05 B \$123.95

C \$152.00 D \$246.05

4 WATCHES A sports watch with an original price of \$86 is on sale for \$60.20. What is the percent of discount? **30%**

DAILY INTERVENTION

Differentiated Instruction

Intrapersonal Bring in advertisements from your local newspaper showing sale or discounted items. Allot students hypothetical budgets that allow them to purchase some, but not all, of the items. Tell students that they must stay within that budget. Have them select their items, calculate the total cost, and write a paragraph explaining why they made the selections that they did.

3 Practice/Apply

FOLDABLES™ Follow-Up Remind students to record main ideas and to give examples of what they learn about discounts and sales tax in the Lesson 8-5 section of their Foldable tables.

About the Exercises...

Guided Practice Key

Exercises	Examples
4, 7	1, 2
5, 6	3
8–10	4

Odd/Even Assignments
Exercises 11–26 are structured so that students practice the same concepts whether they are assigned odd or even problems.

Alert! Exercise 30 requires research on the Internet or another resource.

Answers

2. One method is to find 30% of the regular price, then subtract that amount from the regular price. Another method is to find 70% of the regular price. Sample answer: The second method is more efficient because it can be done in one step rather than in two.
3. Sample answer: The regular price of a CD is $17.95. The total cost is $18.98.

Skill and Concept Check

Writing Math Exercises 2 & 3

1. **Find** the sales tax of a $98 chair if the tax rate is 7%. **$6.86**
2. **Describe** two methods for finding the sale price of an item that is discounted 30%. Which method do you prefer? Explain. **2–3. See margin.**
3. **OPEN ENDED** Give an example of the regular price of an item and the total cost including sales tax if the tax rate is 5.75%.

GUIDED PRACTICE

Find the total cost or sale price to the nearest cent.

4. $2.95 notebook; 5% tax **$3.10**
5. $1,575 computer; 15% discount **$1,338.75**
6. $119.50 skateboard; 20% off **$95.60**
7. $46 shoes; 2.9% tax **$47.33**

TEACHING TIP
Point out that in real l… tax is always rounded to the next whole cent

Find the percent of discount to the nearest percent.

8. lotion: regular price, $4.50; sale price, $2.25 **50%**
9. in-line skates: regular price, $99; sale price, $90 **9%**

10. **VIDEO GAMES** What is the sales tax of a $178.90 video game system if the tax rate is 3.75%? **$6.71**

★ indicates increased difficulty

Practice and Applications

HOMEWORK HELP

For Exercises	See Examples
11–19, 25–31	1–3
20–24	4

Extra Practice
See pages 584, 603.

Find the total cost or sale price to the nearest cent. **11. $46.40 14. $4.90**

11. $58 ski lift ticket; 20% discount
12. $1,500 computer; 7% tax **$1,605**
13. $99 CD player; 5% tax **$103.95**
14. $12.25 pen set; 60% discount
15. $4.30 notebook; 40% discount
16. $7.50 meal; 6.5% tax **$7.99**
17. $39.60 sweater; 33% discount
18. $89.75 scooter; $7\frac{1}{4}$% sales tax **$96.26**

15. $2.58 17. $26.53

19. **BOOKS** A book about candle making is $24.95. Find the total cost of the book including the 4.5% sales tax. **$26.07**
20. **TICKETS** A local movie theater is selling movie tickets for $5.25 during the first week of May. The regular price of a ticket is $6.75. What is the percent of discount to the nearest percent? **22%**

Find the percent of discount to the nearest percent.

21. calendar: regular price, $9; sale price, $2.25 **75%**
22. telescope: regular price, $180; sale price, $126 **30%**
23. concert tickets: regular price, $44; sale price, $34 **23%**
24. TV: regular price, $625; sale price, $562.50 **10%**
25. **MULTI STEP** A sound system has a regular price of $249. Find the total cost if it is on sale for 50% off and the sales tax is 5.75%. **$131.66**

★ 26. **MULTI STEP** Suppose your restaurant bill comes to $28.35. Find your total cost if the tax is 6.25% and you leave a 20% tip on the amount before tax. **$35.79**

356 Chapter 8 Applying Percent

MOUNTAIN BIKES **For Exercises 27–30, use the information below and at the right.**

State	2002 Sales Tax Rate
Alabama	4%
Florida	6%
Indiana	5%
North Carolina	4.5%

Source: www.salestaxinstitute.com

A mountain bike has a regular price of $575.

27. Suppose Antonio lives in North Carolina. What is the total cost of the bike including tax? **$600.88**

★ 28. If the mountain bike is on sale with a 25% discount, how much will Antonio pay for the bike, including tax? **$450.66**

★ 29. Before 8 A.M., the bike will be discounted an additional 15% off the already discounted price. What will be the sale price, not including tax? **$366.56**

30. **RESEARCH** Use the Internet or another source to find the current tax rates of the states listed in the table or of other states. Find the cost of the bike including tax in one of these states. **See students' work.**

★ 31. **PROFIT** To make a profit, stores sell items for more than they paid. The increase in price is called the *markup*. Suppose Sports Galore purchases tennis racquets for $45 each. Find the markup price if the racquets are sold for 28% over the price paid for them. **$57.60**

32. **CRITICAL THINKING** Find the total percent of change on the price of an item if it is 15% off and the sales tax is 5%. Does it matter in which order the discount and the sales tax are applied? Explain. **10.75% decrease; No, multiplication is commutative.**

Standardized Test Practice and Mixed Review

33. **MULTIPLE CHOICE** A T-shirt at the mall costs $14.95. It is on sale for 30% off. What is the sale price to the nearest cent? **B**

(A) $4.49 (B) $10.47 (C) $10.50 (D) $10.65

34. **SHORT RESPONSE** All of a department store's jackets are 20% off. To the nearest cent, what is the total cost of a jacket if the original price is $74.99 and the sales tax is $8\frac{1}{2}$%? **$65.09**

Find each percent of change. Round to the nearest whole percent if necessary. State whether the percent of change is an *increase* or *decrease*. (Lesson 8-4)

35. original: 4
new: 6 **50% inc.**

36. original: $556
new: $500 **10% dec.**

37. original: 20.5
new: 35.5 **73% inc.**

38. **FOOD** The graph shows the results of a survey in which magazine readers were asked to name their favorite take-out foods. Predict how many out of 5,000 readers would choose Chinese food as one of their favorite take-out foods. (Lesson 8-3) **2,800 readers**

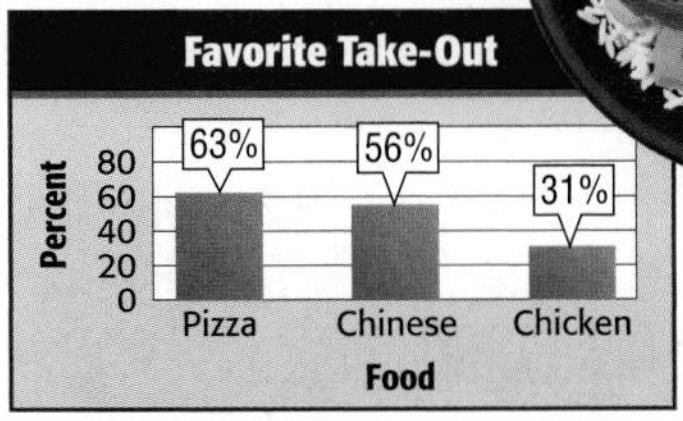

Source: *Bon Appetit*

GETTING READY FOR THE NEXT LESSON

PREREQUISITE SKILL **Multiply.** (Lesson 6-4) **41. 99** **42. 141.75**

39. $790 \cdot 0.185 \cdot 2$ **292.3**

40. $45 \cdot 0.04 \cdot 9$ **16.2**

41. $600 \cdot 0.03 \cdot 5\frac{1}{2}$

42. $162 \cdot 0.25 \cdot 3\frac{1}{2}$

msmath2.net/self_check_quiz

4 Assess

Higher-Level Thinking

Ask students to find the cost of a $100 item after a discount of 30% and sales tax of $5\frac{1}{2}$%. **$73.85** Challenge the class to develop two different orders of operations for this calculation.

You could find the sale price of $70 and then add the sales tax, or you could add the regular price and tax, $105.50, and discount that amount by 30%.

Study Guide and Intervention, p. 455

NAME ______ DATE ______ PERIOD ____

8-5 Study Guide and Intervention

Sales Tax and Discount

Sales tax is a percent of the purchase price and is an amount paid in addition to the purchase price. **Discount** is the amount by which the regular price of an item is reduced.

EXAMPLE 1 SOCCER **Find the total price of a $17.75 soccer ball if the sales tax is 6%.**

Method 1

First, find the sales tax.
6% of $17.75 = $0.06 \cdot 17.75$
≈ 1.07
The sales tax is $1.07.

Next, add the sales tax to the regular price.

$1.07 + 17.75 = 18.82$

The total cost of the soccer ball is $18.82.

Method 2

100% + 6 % = 106% Add the percent of tax to 100%.
The total cost is 106% of the regular price.

106% of $17.75 = $1.06 \cdot 17.75$
≈ 18.82

EXAMPLE 2 TENNIS **Find the price of a $69.50 tennis racket that is on sale for 20% off.**

First, find the amount of the discount d.
part = percent · base

$d = 0.2 \cdot 69.50$ Use the percent equation.
$d = 13.90$ The discount is $13.90.

So, the sale price of the tennis racket is $69.50 − $13.90 or $55.60.

EXERCISES

Find the total cost or sale price to the nearest cent.

1. $22.95 shirt; 7% sales tax $24.56
2. $39.00 jeans; 25% discount $29.25
3. $35 belt; 40% discount $21
4. $115.48 watch; 6% sales tax $122.41
5. $16.99 book; 5% off $16.14
6. $349 television; 6.5% sales tax $371.69

Practice: Skills, p. 456

NAME ______ DATE ______ PERIOD ____

8-5 Practice: Skills

Sales Tax and Discount

Find the total cost or sale price to the nearest cent.

1. $49.95 CD player; 5% discount $47.45
2. $69 shoes; 6% sales tax $73.14
3. $2.99 socks; 5.5% sales tax $3.15
4. $119 coat; 40% discount $71.40
5. $299 DVD player; 7% sales tax $319.93
6. $49 tie; 15% discount $41.65
7. $59 power tool; 5% sales tax $61.95
8. $17.99 CD; 10% discount $16.19
9. $79 cell phone; 20% discount $63.20
10. $65 concert ticket; 7.5% sales tax $69.88
11. $459 television; 30% discount $321.30
12. $19,995 car; 6.5% sales tax $21,294.68

Find the percent of discount to the nearest percent.

13. boots: regular price, $89
sale price, $62.50
30%
14. video game: regular price, $14.99
sale price, $12.64
16%
15. drum set: regular price, $1,240
sale price, $1,099
11%
16. gloves: regular price, $24
sale price, $16.40
32%
17. sweater: regular price, $48
sale price, $34
29%
18. sunglasses: regular price, $80
sale price, $62.95
21%
19. dinner for two: regular price, $75
sale price, $70
7%
20. bicycle: regular price, $189
sale price, $147.85
22%

Practice: Word Problems, p. 457

NAME ______ DATE ______ PERIOD ____

8-5 Practice: Word Problems

Sales Tax and Discount

1. **SKATEBOARDS** Ines wants to buy a skateboard but she does not know if she has enough money. The price of the skateboard is $85 and the sales tax is 6%. What will be the total cost of the skateboard? $90.10

2. **PRETZELS** The Spanish club sold hot pretzels as a fund-raiser. The pretzels normally sold for $1.50, but near the end of the sale they wanted to sell as many as possible, so they reduced the price by 30%. What was the new price for a hot pretzel? $1.05

3. **COMPUTERS** Andrea ordered a computer on the Internet. The computer cost $1,499 plus $7\frac{1}{2}$% sales tax. What was the total amount Andrea paid for her computer? $1,611.43

4. **BOOKS** Nate went shopping at a bookstore. The price of the book he selected was $14.95, but it had a sale sticker on it. When he paid for the book, he was charged $12.71 before sales tax was added. What was the percent of discount to the nearest percent? 15%

5. **CELL PHONES** Justin is buying a cell phone that has a regular price of $149. The cell phone is on sale for 15% off the regular price. What will be the sale price? $126.65

6. **MAGAZINES** Ivan bought two magazines for $4.95 each. If the sales tax was 6.75%, what was the total amount that he paid for the magazines? $10.57

7. **MOVIES** A video store is having a sale in which DVDs are on sale for 20% off. During this sale, what is the cost of three DVDs that regularly cost $16.99? $40.78

8. **MODELS** The original price of a collectible model airplane is $115. The discounted price is $99. What is the percent of discount to the nearest percent? 14%

8-6 Lesson Notes

Lesson Resources

Chapter 8 Resource Masters
- Study Guide and Intervention, p. 460
- Practice: Skills, p. 461
- Practice: Word Problems, p. 462
- Reading to Learn Mathematics, p. 463
- Enrichment, p. 464

Calculator and Spreadsheet Masters, p. 42
Key Teaching Transparencies, 32
MathPASS Tutorial CD-ROM, 35
Parent and Student Study Guide Workbook, 8-6

FAST FILE

1 Focus

5-Minute Check 8-6

INTERACTIVE CHALKBOARD PowerPoint® Presentations

Use as a quiz or review of Lesson 8-5. (Also available as a transparency)

Bellringer

Communication Ask students to discuss the reasons why people deposit money in banks. **Students will probably mention security and interest.** Ask them what they think a savings account and a loan from a bank have in common. **Interest—in one case the bank is paying you for your money, in the other you pay the bank to borrow money.**

Reading to Learn Mathematics, p. 463 ELL

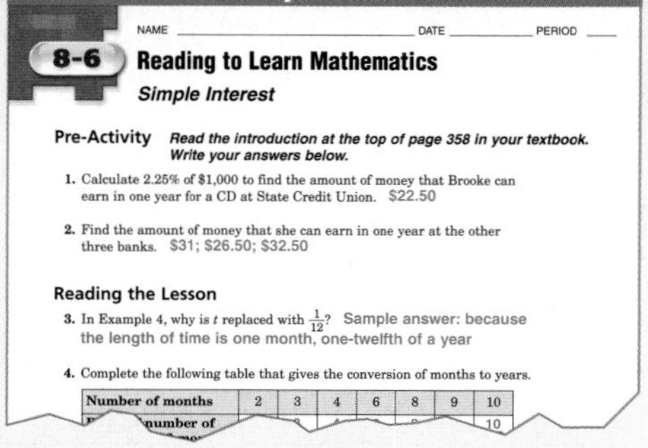

NAME ______ DATE ______ PERIOD ____

8-6 **Reading to Learn Mathematics**
Simple Interest

Pre-Activity ***Read the introduction at the top of page 358 in your textbook. Write your answers below.***

1. Calculate 2.25% of $1,000 to find the amount of money that Brooke can earn in one year for a CD at State Credit Union. $22.50
2. Find the amount of money that she can earn in one year at the other three banks. $31; $26.50; $32.50

Reading the Lesson

3. In Example 4, why is t replaced with $\frac{1}{12}$? Sample answer: because the length of time is one month, one-twelfth of a year
4. Complete the following table that gives the conversion of months to years.

Number of months	2	3	4	6	8	9	10
[illegible] number of [illegible]							10

8-6 Simple Interest

What You'll Learn
Solve problems involving simple interest.

NEW Vocabulary
simple interest
principal

WHEN am I ever going to use this?

INVESTING Brooke plans to invest $1,000 in a certificate of deposit (CD). The graph shows CD rates for one year at various banks.

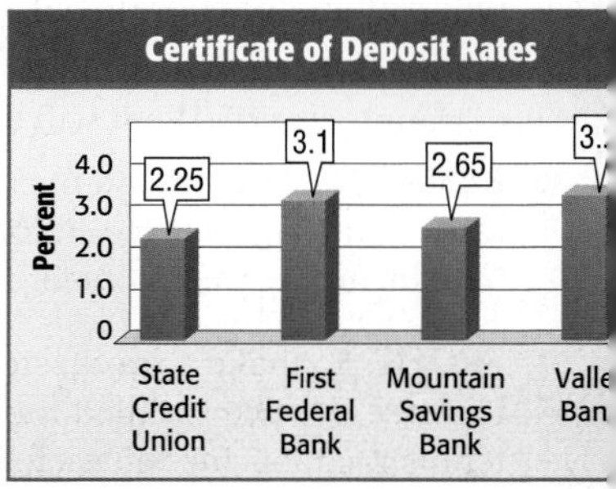

1. Calculate 2.25% of $1,000 to find the amount of money that Brooke can earn in one year for a CD at State Credit Union. **$22.50**
2. Find the amount of money that she can earn in one year at the other three banks. **$31; $26.50; $32.50**

When you deposit money in a CD, the amount that you earn is called interest. **Simple interest** is the amount paid or earned for the use of money. To find simple interest I, use the following formula.

Principal is the amount of money deposited or invested.

$$I = prt$$

annual interest rate written as a decimal

amount of time that money is invested or borrowed (years)

EXAMPLES Find Interest Earned

SAVINGS Raini has $750 in a savings account that pays 3% simple interest. How much interest will he earn in each amount of time?

1 4 years

$I = prt$	Formula for simple interest
$I = 750 \cdot 0.03 \cdot 4$	Replace p with $750, r with 0.03, and t with 4.
$I = 90$	Simplify.

Raini will earn $90 in interest in 4 years.

2 9 months

9 months $= \frac{9}{12}$ or 0.75 year	Write the time as years.
$I = prt$	Formula for simple interest
$I = 750 \cdot 0.03 \cdot 0.75$	$p = \$750, r = 0.03, t = 0.75$
$I \approx 16.88$	Simplify.

Raini will earn $16.88 in interest in 9 months.

TEACHING TIP
Explain to students that interest is the amount of money a bank pays you for loaning them your money.

DAILY INTERVENTION

Flexible Grouping

Banking Group students in pairs so that students with greater math skills work with students with less skill. Have pairs role-play banking scenarios similar to those presented in the lesson. The banker must present options, such as better rates for longer time periods, and provide calculations for interest earned or paid. The customer must verify the calculations and choose an option. Students should take turns playing each role. Each pair should write a summary of the transactions.

READING Math

Formulas Read $I = prt$ as *Interest is equal to principal times rate times time.*

The formula $I = prt$ can also be used to find the interest owed when you borrow money. In this case, p is the amount of money borrowed, and t is the amount of time the money is borrowed.

EXAMPLE Find Interest Paid on a Loan

3 LOANS Emilio's father borrows $1,200 from the bank for a riding lawn mower. The interest rate is 8% per year. How much simple interest will he pay if he takes 2 years to repay the loan?

$I = prt$	Formula for simple interest
$I = 1{,}200 \cdot 0.08 \cdot 2$	Replace p with $1,200, r with 0.08, and t with 2.
$I = 192$	Simplify.

Emilio's father will pay $192 in interest in 2 years.

TEACHING TIP

In Example 5 and Exercise 25, assume that simple interest is owed on the given balance. Usually, credit card interest is not accrued if the entire balance is paid off within a month.

EXAMPLE Find Total Paid on a Credit Card

4 CREDIT CARDS Cory charged a $600 TV on his credit card with an interest rate of 21%. If he has no other charges on the card, how much money will he owe after one month?

$I = prt$	Formula for simple interest
$I = 600 \cdot 0.21 \cdot \frac{1}{12}$	Replace p with $600, r with 0.19, and t with $\frac{1}{12}$.
$I = 10.5$	Simplify.

The interest owed after one month is $10.50. So, the total amount owed would be $600 + $10.50 or $610.50.

Skill and Concept Check

Writing Math Exercise 2

1. **List** the steps you would use to find the simple interest on a $500 loan at 6% interest rate for 18 months. See margin.
2. **OPEN ENDED** Suppose you earn 3% on a $1,200 deposit for 5 years. Investigate how the interest is affected if the rate or the time is increased. See margin.

GUIDED PRACTICE

Find the interest earned to the nearest cent for each principal, interest rate, and time.

3. $640, 3%, 2 years $38.40
4. $1,500, 4.25%, 4 years $255

Find the interest paid to the nearest cent for each loan balance, interest rate, and time.

5. $4,500, 9%, 3.5 years $1,417.50
6. $290, 12.5%, 6 months $18.13

7. **HISTORY** In 2002, a $5 bank note from 1886 was sold to a collector for $103,500. Suppose a person had deposited the $5 in a bank in 1886 with an interest rate of 4%. After 116 years, how much simple interest would have been earned on the account? $23.20

msmath2.net/extra_examples

2 Teach

In-Class Examples

INTERACTIVE CHALKBOARD PowerPoint® Presentations

SAVINGS Brandon found a bank offering a certificate of deposit that pays 4% simple interest. He has $1,500 to invest. How much interest will he earn in each amount of time?

1. 3 years $180
2. 30 months $150
3. **LOANS** Laura borrowed $2,000 from her credit union to buy a computer. The interest rate is 9% per year. How much interest will she pay if it takes 8 months to repay the loan? $120
4. **CREDIT CARDS** Alma charged a $30 handbag on her credit card with an interest rate of 8%. If she has no other charges on the card, how much money would she owe after six months? $31.20

3 Practice/Apply

About the Exercises...

Guided Practice Key

Exercises	Examples
3–4, 7	1, 2
5–6	3

Odd/Even Assignments

Exercises 8–19 are structured so that students practice the same concepts whether they are assigned odd or even problems.

Answers

1. Sample answer: Write the interest rate as a decimal and write the months in years. Then multiply $500 \cdot 0.06 \cdot 1.5$.
2. Sample answer: If the rate is increased by 1%, then the interest earned is $60 more. If the time is increased by 1 year, then the interest earned is $36 more.

★ indicates increased difficulty

Practice and Applications

HOMEWORK HELP	
For Exercises	See Examples
8–13	1, 2
14–19	3
20–22	4

Extra Practice
See pages 584, 603.

Find the interest earned to the nearest cent for each principal, interest rate, and time.

8. \$1,050, 4.6%, 2 years **\$96.60**
9. \$250, 2.85%, 3 years **\$21.38**
10. \$500, 3.75%, 1 year **\$18.75**
11. \$3,000, 5.5%, 2 years **\$330**
12. \$875, 6%, 4 months **\$17.50**
13. \$98.50, $6\frac{1}{2}$%, 16 months **\$8.54**

Find the interest paid to the nearest cent for each loan balance, interest rate, and time.

14. \$1,000, 7%, 2 years **\$140**
15. \$725, 6.25%, 1 year **\$45.31**
16. \$2,700, 8.2%, 3 years **\$664.20**
17. \$175.80, 12%, 1.25 years **\$26.37**
18. \$925, $19\frac{1}{2}$%, 3 months **\$45.09**
19. \$800, 10.5%, 30 months **\$210**

TEACHING TIP
Remind students to account for the fraction of a year's interest.

INVESTING **For Exercises 20–22, use the following information.**
Marcus has \$1,800 from his summer job to invest.

20. If he invests in a CD for 3 years at a rate of 5.25%, how much will the CD be worth after 3 years? **\$2,083.50**

★ 21. Suppose he invests the \$1,800 for 2 years and earns \$144. What was the rate of interest? **4%**

★ 22. Marcus would like to have \$2,340 altogether. If he invests his money at 5% interest, in how many years will he have \$2,340? **6 yr**

23. **CRITICAL THINKING** Mrs. Williams deposits \$600 in an account that pays 4.5% annually. At the end of the year, the interest earned is added to the principal. Find the total amount in her account each year for 3 years. **\$627, \$655.22, \$684.70**

Standardized Test Practice and Mixed Review

24. **MULTIPLE CHOICE** Antonia opened a savings account that pays 6.5% simple interest. How much money will be in Antonia's account after 3 years if she deposited \$250 at the beginning and never made any more deposits? **C**

(A) \$48.75 (B) \$248.75 (C) \$298.75 (D) \$300.00

25. **MULTIPLE CHOICE** Mr. McMahon bought a \$562 freezer using a credit card that charges 18% annual interest. If he does not make any payments or any additional charges, how much will he owe after 1 month? **G**

(F) \$553.57 (G) \$570.43 (H) \$578.86 (I) \$663.16

26. Find the total cost of a \$13.99 music CD if the tax rate is 7%. (Lesson 8-5) **\$14.97**

Find each percent of change. Round to the nearest whole percent if necessary. State whether the percent of change is an *increase* or *decrease*. (Lesson 8-4)

27. original: 35
new: 45 **29% inc.**
28. original: 60
new: 38 **37% dec.**
29. original: \$2.75
new: \$1.80 **35% dec.**

msmath2.net/self_check_quiz

4 Assess

Higher-Level Thinking

Shelby wants to buy a new television for \$900. Is it better to get a bank loan that charges 14% interest and pay it off in 18 months or borrow the money from her mom and pay back the \$900 plus \$125 interest?

It is better to borrow the money from her mother since the interest on the bank loan is \$189.

Study Guide and Intervention, p. 460

NAME ______ DATE ______ PERIOD ____

8-6 Study Guide and Intervention
Simple Interest

Simple interest is the amount of money paid or earned for the use of money. To find simple interest I, use the formula $I = prt$. Principal p is the amount of money deposited or invested. Rate r is the annual interest rate written as a decimal. Time t is the amount of time the money is invested in years.

EXAMPLE 1 **Find the simple interest earned in a savings account where \$136 is deposited for 2 years if the interest rate is 7.5% per year.**

$I = prt$ Formula for simple interest
$I = 136 \cdot 0.075 \cdot 2$ Replace p with \$136, r with 0.075, and t with 2.
$I = 20.40$ Simplify.
The simple interest earned is \$20.40.

EXAMPLE 2 **Find the simple interest for \$600 invested at 8.5% for 6 months.**

6 months $= \frac{6}{12}$ or 0.5 year Write the time as years.
$I = prt$ Formula for simple interest
$I = 600 \cdot 0.085 \cdot 0.5$ $p = \$600$, $r = 0.085$, $t = 0.5$
$I = 25.50$ Simplify.
The simple interest is \$25.50.

EXERCISES

Find the interest earned to the nearest cent for each principal, interest rate, and time.

1. \$300, 5%, 2 years
\$30
2. \$650, 8%, 3 years
\$156
3. \$575, 4.5%, 4 years
\$103.50
4. \$735, 7%, $2\frac{1}{2}$ years
\$128.63
5. \$1,665, 6.75%, 3 years
\$337.16
6. \$2,105, 11%, $1\frac{3}{4}$ years
\$405.21
7. \$903, 8.75%, 18 months
\$118.52
8. \$4,275, 19%, 3 months
\$203.06

Practice: Skills, p. 461

NAME ______ DATE ______ PERIOD ____

8-6 Practice: Skills
Simple Interest

Find the interest earned to the nearest cent for each principal, interest rate, and time.

1. \$500, 4%, 2 years
\$40
2. \$350, 6.2%, 3 years
\$65.10
3. \$740, 3.25%, 2 years
\$48.10
4. \$725, 4.3%, $2\frac{1}{2}$ years
\$77.94
5. \$955, 6.75%, $3\frac{1}{4}$ years
\$209.50
6. \$1,540, 8.25%, 2 years
\$254.10
7. \$3,500, 4.2%, $1\frac{3}{4}$ years
\$257.25
8. \$568, 16%, 8 months
\$60.59

Find the interest paid to the nearest cent for each loan balance, interest rate, and time.

9. \$800, 9%, 4 years
\$288
10. \$280, 5.5%, 4 years
\$61.60
11. \$1,150, 7.6%, 5 years
\$437
12. \$266, 5.2%, 3 years
\$41.50
13. \$450, 22%, 1 year
\$99
14. \$2,180, 7.7%, $2\frac{1}{2}$ years
\$419.65
15. \$2,650, 3.65%, $4\frac{1}{2}$ years
\$435.26
16. \$1,245, 5.4%, 6 months
\$33.62

Practice: Word Problems, p. 462

NAME ______ DATE ______

8-6 Practice: Word Problems
Simple Interest

1. **SAVINGS ACCOUNT** How much interest will Hannah earn in 4 years if she deposits \$630 in a savings account at 6.5% simple interest? \$163.80
2. **INVESTMENTS** Terry invested \$2,200 in the stock market for 2 years. If the investment earned 12% simple interest, how much money did Terry earn in interest in 2 years? \$528
3. **SAVINGS ACCOUNT** Malik deposited \$1,050 in a savings account, and it earned \$241.50 in simple interest after four years. Find the interest rate on Malik's savings account. 5.75%
4. **INHERITANCE** Kelli Rae's inheritance from her great-grandmother was \$220,000 after taxes. If Kelli Rae invests this money in a savings account that earns \$18,260 in simple interest every year, what is the interest rate on her account? 8.3%
5. **RETIREMENT** Mr. Pham has \$410,000 in a retirement account that earns 3.85% simple interest each year. Find the amount earned each year by this investment. \$15,785
6. **COLLEGE FUND** When Melissa was born, her parents put \$8,000 into a college fund account that earned 9% simple interest. Find the total amount in the account after 18 years. \$20,960
7. **LOTTERY** Raj won \$900,000 in a regional lottery. After paying \$350,000 in taxes, he invested the remaining money in a savings account at 4.25% simple interest. How much money is in the account if Raj makes no deposits or withdrawals for two years? \$596,750
8. **SAVINGS** Mona opened a savings account with a \$500 deposit and a simple interest rate of 5.6%. If there were no deposits or withdrawals, how much money is in the account after $8\frac{1}{2}$ years? \$738

360 Chapter 8 Applying Percent

8-6b Spreadsheet Investigation

A Follow-Up of Lesson 8-6

/hat You'll Learn
se a spreadsheet
calculate simple
terest.

Simple Interest

A computer spreadsheet is a useful tool for quickly calculating simple interest for different values of principal, rate, and time.

ACTIVITY

Max plans on opening a "Young Savers" account at his bank. The current rate on the account is 4%. He wants to see how different starting balances, rates, and times will affect his account balance. To find the balance at the end of 2 years for different principal amounts, he enters the values B2 = 4 and C2 = 2 into the spreadsheet below.

Simple Interest.xls

	A	B	C	D	E
1	Principal (*p*)	Rate (*r*)	Time (*t*)	Interest (*I*)	New Balance
2		4	2		
3	500	=B2/100	=C2	=A3*B3*C3	=A3+D3
4	1000	=B2/100	=C2	=A4*B4*C4	=A4+D4
5	1500	=B2/100	=C2	=A5*B5*C5	=A5+D5
6	2000	=B2/100	=C2	=A6*B6*C6	=A6+D6
7	2500	=B2/100	=C2	=A7*B7*C7	=A7+D7
8					

Sheet1 / Sheet2 / Sheet3

Ready

For each principal given in column A, simple interest is calculated for any values of rate and time entered in B2 and C2, respectively.

The spreadsheet adds simple interest to the principal.

EXERCISES

1. Why is the rate in column B divided by 100?
2. What is the balance in Max's account after 2 years if the principal is $1,500 and the simple interest rate is 4%? **$1,620**
3. How much interest does Max earn in 2 years if his account has a principal of $2,000 and an interest rate of 4%? **$160**
4. Suppose you wanted to add a new row to the spreadsheet that represents a principal of $3,000. List each of the cell entries (A8, B8, C8, D8, and E8) that you would enter.
5. What entries for cells B2 and C2 would you use to calculate the simple interest on a principal of $1,500 at a rate of 7% for a 9-month period? What is the balance of this account at the end of the 9 months? **B2: 7, C2: 0.75; $1,578.75**
6. Explain how a spreadsheet is more useful than a pencil and paper when finding simple interest. **Sample answer: When you change *p*, *r*, or *t*, a spreadsheet automatically calculates the new interest.**

. to change the percent
a decimal

. A8: 3000; B8:
=B2/100; C8: =C2;
8: =A8*B8*C8;
8: =A8+D8

TEACHING TIP

For Exercises 2, 3, and 5, students should assume that no additional money is added to the accounts during the times given.

8-6b Spreadsheet Investigation

GETTING STARTED

Teaching Tip This investigation uses the same simple interest formula students used in their pencil and paper calculations ($I = prt$). Interest (column D) equals principal (column A) times rate (column B) multiplied by time (column C). Some students may need to have this pointed out at the start in order to make sense of the investigation. Other students will profit from discovering this for themselves. If you use the latter approach, at the end of the investigation ask students to explain how the spreadsheet uses the simple interest formula.

TEACH

- You may want to provide students with a hard copy of a spreadsheet screen and discuss the location of the variables before you start.
- Have students predict the amounts of interest Max could earn before using the spreadsheet. Then have students compare their estimates to the calculated values.

ASSESS

Observe as students solve the **Exercises**. They should:

- correctly identify interest earnings and account balances from the cells of the spreadsheet,
- correctly input values for interest, rate, and time, and
- correctly explain the purpose of each spreadsheet cell, including those in column B.

Use Exercises 4 and 5 to assess whether students fully understand how simple interest is calculated in the spreadsheet.

Vocabulary Assessment

- This alphabetical list of vocabulary terms in Chapter 8 includes a page reference where each term was introduced.
- A vocabulary review/test for Chapter 8 is available on p. 478 of the *Chapter 8 Resource Masters*.

Vocabulary PuzzleMaker

ELL The online Vocabulary PuzzleMaker improves students' mathematics vocabulary using four puzzle formats — crossword, scramble, word search using a word list, and word search using clues. Students can work on a computer screen or from a printed handout.

MindJogger Videoquizzes

ELL MindJogger Videoquizzes provide an alternative review of concepts presented in this chapter. Students work in teams in a game show format to gain points for correct answers. The questions are presented in three rounds.

Round 1 Concepts (5 questions)
Round 2 Skills (4 questions)
Round 3 Problem Solving (4 questions)

ExamView® Pro Test Bank CD-ROM

Use the networkable ExamView® Pro Test Bank CD-ROM to:

- Create **Multiple versions** of tests.
- Create **Modified** tests for **Inclusion** students with one mouse click.
- **Edit** existing questions and **Add** your own questions.
- Build tests aligned with state standards using built-in **State Curriculum Correlations.**
- Change **English** tests to **Spanish** with one mouse click and vice versa.

Chapter 8 Study Guide and Review

Vocabulary and Concept Check

discount (p. 355)	percent of increase (p. 350)	sales tax (p. 354)
percent equation (p. 340)	population (p. 345)	simple interest (p. 358)
percent of change (p. 350)	principal (p. 358)	survey (p. 345)
percent of decrease (p. 350)	random sample (p. 345)	

State whether each sentence is *true* or *false*. If *false*, replace the underlined word, number, or equation to make a true sentence.

1. The sale price of an item is the regular price <u>minus</u> the discount. **true**
2. When taking a survey, the total group of people that the survey is interested in is called the <u>sample</u>. **false; population**
3. To find a percent of increase, compare the amount of the increase to the <u>new</u> amount. **false; original**
4. The formula for simple interest is $\underline{I = prt}$. **true**
5. A sample is representative of the population if it is <u>random</u>. **true**
6. A method for estimating the percent of a number is to find <u>21%</u> of the number and then multiply. **false; 10%**
7. The percent equation is <u>part = percent · base</u>. **true**
8. A <u>tax</u> is the amount by which the regular price of an item is reduced. **false; discount**
9. The <u>principal</u> is the amount of money deposited or borrowed. **true**
10. A <u>sample</u> is a question or set of questions designed to collect data about a specific group of people. **false; survey**

Lesson-by-Lesson Exercises and Examples

8-1 Percent and Estimation (pp. 334–337) **11–18. See margin.**

Estimate by using fractions.

11. 25% of 81
12. 33% of 122
13. 77% of 38
14. 19.5% of 96

Estimate by using 10%.

15. 12% of 77
16. 88% of 400
17. 52% of 1,000
18. 21% of 53

19. **PETS** About 12% of 291 households in a neighborhood have fish. Estimate how many households have fish.

Sample answer: $\frac{1}{10} \cdot 300 = 30$ **households**

Example 1 **Estimate 52% of 495.**

52% is about 50% or $\frac{1}{2}$, and 495 is about 500.

52% of 495 $\approx \frac{1}{2} \cdot 500$ or 250

So, 52% of 495 is about 250.

Example 2 **Estimate 68% of 80.**

10% of 80 $= 0.1 \cdot 80$ or 8 — Find 10% of 80.

68% is about 70%.

$7 \cdot 8 = 56$ — 70% of 80 ≈ 7 · (10% of 80)

So, 68% of 80 is about 56.

msmath2.net/vocabulary_review

FOLDABLES™ Study Organizer

For more information about Foldables, see *Teaching Mathematics with Foldables.*

Have students look through the chapter to make sure they have included examples in their Foldables for each lesson or topic in the chapter.

Encourage students to refer to their Foldables while completing the Study Guide and Review and to use them in preparing for the Chapter Test.

8-2 The Percent Equation (pp. 340–343)

20. $32 = n \cdot 50$; 64% 21. $39 = 0.65 \cdot n$; 60
23. $n = 0.07 \cdot 92$; 6.4

Write an equation for each problem. Then solve. Round to the nearest tenth if necessary.

20. 32 is what percent of 50?
21. 65% of what number is 39?
22. Find 42% of 300. $n = 0.42 \cdot 300$; 126
23. 7% of 92 is what number?
24. 12% of what number is 108? $108 = 0.12 \cdot n$; 900

Example 3 27 is what percent of 90?

27 is the part and 90 is the base.

Let n represent the percent.

$$\underbrace{\text{part}} = \underbrace{\text{percent}} \cdot \underbrace{\text{base}}$$

$27 = n \cdot 90$ Write an equation.

$\frac{27}{90} = \frac{90n}{90}$ Divide each side by 90.

$0.3 = n$ The percent is 30%.

So, 27 is 30% of 90.

8-3 Using Statistics to Predict (pp. 345–347)

CAREERS In a university survey, 5.5% of the incoming freshmen chose elementary teacher as a career goal and 6.4% chose engineer.

25. Predict how many of the 3,775 freshmen would choose a career as an elementary teacher. about 208
26. How many of the 3,775 freshmen would you expect to choose a career as an engineer? about 242

Example 4 In a survey of 150 students at McAuliffe Middle School, 12% said they have after-school jobs. Predict how many of the 644 students at the school have after-school jobs.

Find 12% of 644.

$n = 0.12 \cdot 644$ Write an equation.

$= 77.28$ Multiply.

So, you could predict that about 77 students at McAuliffe Middle School have after-school jobs.

8-4 Percent of Change (pp. 350–353)

Find each percent of change. Round to the nearest whole percent if necessary. State whether the percent of change is an *increase* or *decrease*. 30. 10% inc.

27. original: 172 new: 254 48% inc.
28. original: \$200 new: \$386 93% inc.
29. original: 75 new: 60 20% dec.
30. original: \$49.95 new: \$54.95
31. **GAMES** A computer game that sold for \$24.95 last year is now priced at \$27.95. Find the percent of change. about 12% increase

Example 5 A magazine that originally cost \$2.75 is now \$3.55. Find the percent of change. Round to the nearest whole percent.

The new price is greater than the original price, so this is a percent of increase.

amount of increase = 3.55 − 2.75 or 0.80

$$\text{percent of increase} = \frac{\text{amount of increase}}{\text{original amount}}$$

$= \frac{0.80}{2.75}$ Substitution

≈ 0.29 Simplify.

The percent of increase is about 29%.

Answers

11–18. Sample answers are given.

11. $\frac{1}{4} \cdot 80 = 20$
12. $\frac{1}{3} \cdot 120 = 40$
13. $\frac{3}{4} \cdot 40 = 30$
14. $\frac{1}{5} \cdot 100 = 20$
15. $0.1 \cdot 80 = 8$
16. $0.1 \cdot 400 = 40$ and $9 \cdot 40 = 360$
17. $0.1 \cdot 1{,}000 = 100$ and $5 \cdot 100 = 500$
18. $0.1 \cdot 50 = 5$ and $2 \cdot 5 = 10$

8-5 Sales Tax and Discount (pp. 354–357)

Find the total cost or sale price to the nearest cent.

32. \$25 backpack; 7% tax **\$26.75**
33. \$210 bicycle; 15% discount **\$178.50**
34. \$8,000 car; $5\frac{1}{2}$% tax **\$8,440**
35. \$40 sweater; 33% discount **\$26.80**
36. \$6.25 address book, 40% discount **\$3.75**

Find the percent of discount to the nearest percent.

37. shirt: regular price: \$42
    sale price: \$36 **14%**
38. snack: regular price, \$2.50
    sale price: \$1 **60%**
39. boots: regular price, \$78
    sale price: \$70 **10%**
40. DVD: regular price, \$24.99
    sale price: \$19.99 **20%**

Example 6 **A new computer system is priced at \$2,499. Find the total cost if the sales tax is 6.5%.**

First, find the sales tax.

$$6.5\% \text{ of } \$2{,}499 = 0.065 \cdot 2{,}499$$
$$\approx 162.44$$

Next, add the sales tax to the original price
The total cost is 162.44 + 2,499 or \$2,661.44

Example 7 **A pass at a water park is \$58. At the end of the season, the same pass costs \$46.40. What is the percent of discount?**

$58 - 46.40 = 11.60$ Find the amount of discount.

Next, find what percent of 58 is 11.60.

$11.60 = n \cdot 58$ Write an equation.

$0.2 = n$ Divide each side by 58.

The percent of discount is 20%.

8-6 Simple Interest (pp. 358–360)

Find the interest earned to the nearest cent for each principal, interest rate, and time.

41. \$475, 5%, 2 years **\$47.50**
42. \$5,000, 10%, 3 years **\$1,500**
43. \$2,500, 11%, $1\frac{1}{2}$ years **\$412.50**

Find the interest paid to the nearest cent for each loan balance, interest rate, and time.

44. \$3,200, 8%, 4 years **\$1,024**
45. \$450, 13.5%, 2 years **\$121.50**
46. \$1,980, 21%, 9 months **\$311.85**
47. **LOANS** Brian has a loan balance of \$1,000. If he pays off the balance over 2 years at an annual simple interest rate of 18%, what is the total amount that he will pay? **\$1,360**

Example 8 **Find the interest earned on \$400 at 9% for 3 years.**

$I = prt$ Formula for simple interest

$I = 400 \cdot 0.09 \cdot 3$ $p = \$400, r = 0.09, t = 3$

$I = 108$ Simplify.

The interest earned is \$108.

Example 9 **Elisa has a loan for \$1,300. The interest rate is 7%. If she pays it off in 6 months, how much interest will she pay?**

$I = prt$ Formula for simple interest

$I = 1{,}300 \cdot 0.07 \cdot 0.5$ $p = \$1{,}300, r = 0.07, t = 0.5$

$I = 45.5$ Simplify.

The interest she will pay after 6 months is \$45.50.

CHAPTER 8 Practice Test

Vocabulary and Concepts

1. **Describe** percent of change. See margin.
2. **State** the formula used to compute simple interest. $I = prt$

Skills and Applications

Estimate. 3–5. Sample answers are given.

3. 18% of 246 $\frac{1}{5} \cdot 250 = 50$
4. 145% of 81 $(1 \cdot 80) + (\frac{1}{2} \cdot 80) = 120$
5. 71% of 324 $\frac{7}{10} \cdot 320 = 224$

Write an equation for each problem. Then solve. Round to the nearest tenth if necessary. 7. $n = 0.36 \cdot 294$; 105.8

6. Find 14% of 65. $n = 0.14 \cdot 65$; 9.1
7. What number is 36% of 294?
8. 82% of what number is 73.8? $73.8 = 0.82 \cdot n$; 90
9. 75 is what percent of 50? $75 = n \cdot 50$; 150%

BOOKS **For Exercises 10 and 11, refer to the table. It shows the results of a survey in which students at Haskell Middle School were asked to name their favorite types of fiction.**

Type of Fiction	Percent
historical fiction	8%
mystery	24%
science fiction	38%
sports	30%

10. Predict how many of the 845 students at Haskell Middle School would choose science fiction as their favorite type.
11. How many of the 845 students would you expect to select mystery as their favorite type of fiction? about 203 students

10. about 321 students

Find each percent of change. Round to the nearest whole percent if necessary. State whether the percent of change is an *increase* or *decrease*.

12. original: \$60 new: \$75 25% inc.
13. original: 145 new: 216 49% inc.
14. original: 48 new: 40 17% dec.

Find the total cost or sale price to the nearest cent.

15. \$1,730 treadmill, $6\frac{1}{2}$% sales tax \$1,842.45
16. \$16 hat, 55% discount \$7.20

Find the interest earned for each principal, interest rate, and time.

17. \$3,000, 5.5%, 5 years \$825
18. \$2,600, 4%, 3 months \$26
19. **LOANS** Leah borrows \$2,200 to buy new furniture. Her loan has an annual interest rate of 16%. Find the simple interest that Leah will owe after 1 year. \$352

Standardized Test Practice

20. **MULTIPLE CHOICE** James earned \$38 last week from mowing lawns in his neighborhood. This week, he earned \$52. What was the percent of change? D

(A) 37% decrease (B) 27% decrease (C) 27% increase (D) 37% increase

Assessment Options

Vocabulary Test A vocabulary review/test for Chapter 8 can be found on p. 478 of the *Chapter 8 Resource Masters.*

Chapter Tests There are six Chapter 8 Tests and an Extended Response Assessment task available in the *Chapter 8 Resource Masters.*

Chapter 8 Tests			
Form	Type	Level	Pages
1	MC	basic	465–466
2A	MC	average	467–468
2B	MC	average	469–470
2C	FR	average	471–472
2D	FR	average	473–474
3	FR	advanced	475–476

MC = multiple-choice questions
FR = free-response questions

Extended Response Assessment

Extended Response tasks for Chapter 8 can be found on p. 477 of the *Chapter 8 Resource Masters.* A sample scoring rubric for these tasks appears on p. A23.

Answer

1. a ratio that compares the change in quantity to the original amount

Portfolio Suggestion

Introduction Percents are used in many ways in this chapter and in real life.

Ask Students to list five real-life uses of percents in their portfolios. Have them include examples of each, along with any formulas used.

Chapter 8 Standardized Test Practice

These two pages contain practice questions in the various formats that can be found on the most frequently given standardized tests.

A student recording sheet for these two pages can be found on page A1 of the *Chapter 8 Resource Masters.*

Standardized Test Practice Student Recording Sheet, p. A1

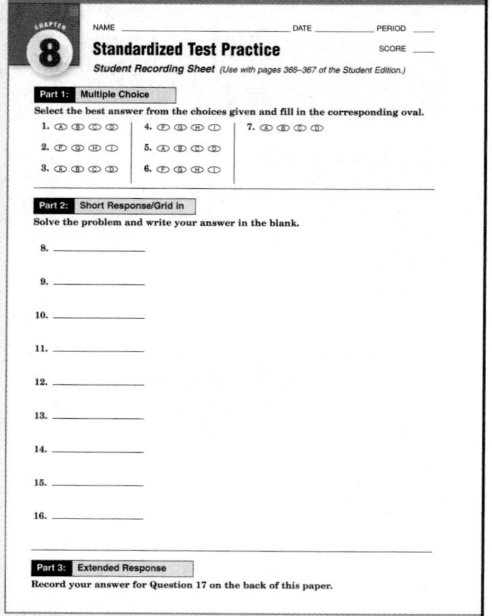
NAME ______ DATE ______ PERIOD ______

Standardized Test Practice SCORE ______

Student Recording Sheet (Use with pages 366–367 of the Student Edition.)

Part 1: Multiple Choice

Select the best answer from the choices given and fill in the corresponding oval.

1. A B C D
2. F G H I
3. A B C D
4. F G H I
5. A B C D
6. F G H I
7. A B C D

Part 2: Short Response/Grid in

Solve the problem and write your answer in the blank.

8. ______
9. ______
10. ______
11. ______
12. ______
13. ______
14. ______
15. ______
16. ______

Part 3: Extended Response

Record your answer for Question 17 on the back of this paper.

Additional Practice

See pp. 483–484 of the *Chapter 8 Resource Masters* for additional standardized test practice.

Standardized Test Practice

PART 1 Multiple Choice

Record your answers on the answer sheet provided by your teacher or on a sheet of paper.

1. Students at Karlon School are collecting canned foods for the local food pantry. The graph shows how many cans students in four of the grades collected in the first week.

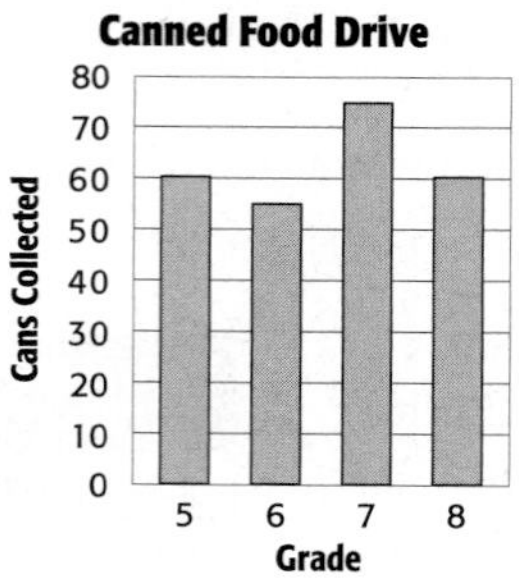

How many total cans did students in the 5th, 6th, 7th, and 8th grades collect during the first week? (Lesson 2-7) **D**

Ⓐ 75 cans Ⓑ 120 cans
Ⓒ 240 cans Ⓓ 250 cans

2. Greenapple Books has 25 copies of this week's national best-selling book. On Friday, 12 copies of the book were sold. On Saturday, 7 copies of the book were returned. Which represents the copies of this book that Greenapple Books now has available? (Lesson 3-5) **I**

Ⓕ -5 Ⓖ 5 Ⓗ 19 Ⓘ 20

3. Find the slope of the line graphed below. (Lesson 4-7) **B**

Ⓐ -2
Ⓑ $-\frac{1}{2}$
Ⓒ $\frac{1}{2}$
Ⓓ 2

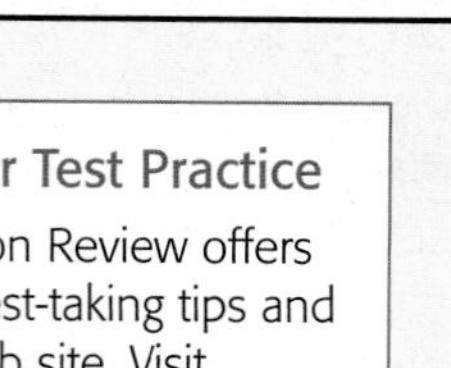

4. How far will you travel after 10 complete turns of the bicycle wheel shown at the right? Use 3.14 for π. (Lesson 6-9) **H**

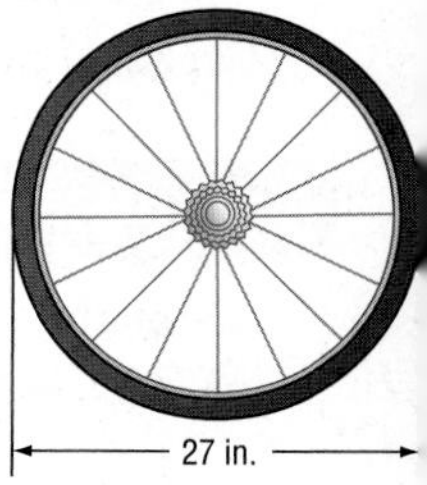

Ⓕ 84.8 in.
Ⓖ 270 in.
Ⓗ 847.8 in.
Ⓘ 1,695.6 in.

5. The graph shows the results of an online poll to which 1,721 high school students responded.

Estimate how many of the students already had a summer job when they responded to the poll. (Lesson 8-1) **C**

Ⓐ 50 Ⓑ 500 Ⓒ 850 Ⓓ 1,850

6. Zoe finished reading 85% of her book. If the book is 280 pages long, how many pages has she read? (Lesson 8-2) **H**

Ⓕ 185 Ⓖ 208 Ⓗ 238 Ⓘ 265

7. A survey showed that 85% of youths who went to summer camp when they were 11 years old also attended summer camp when they were 12 years old. Out of 300 11-year-olds who went to summer camp this year, predict how many will go to summer camp next year. (Lesson 8-3) **B**

Ⓐ 385 Ⓑ 255 Ⓒ 245 Ⓓ 45

The Princeton Review *Log On* for Test Practice

The Princeton Review offers additional test-taking tips and practice problems at their web site. Visit princetonreview.com or review.com

ExamView® Pro Test Bank CD-ROM

Special banks of standardized test questions similar to those on the TIMSS and NAEP grade 8 tests can be found on this CD-ROM.

Aligned and verified by The Princeton Review

Standardized Test Practice

.RT 2 Short Response/Grid In

·cord your answers on the answer sheet ovided by your teacher or on a sheet of ·per.

Jamal bought 2.4 meters of blue ribbon and 110 centimeters of white ribbon. How many centimeters longer is the blue ribbon than the white ribbon? (Lesson 1-8) **130 cm**

. Write a percent to represent the shaded area of the model. (Lesson 5-5) **30%**

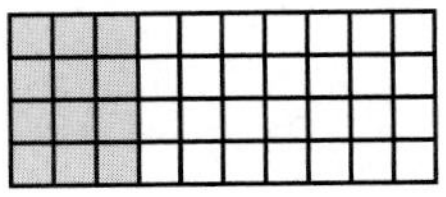

. If a 5-pound bag of potatoes costs $3.75, how much does a 4-pound bag of the same potatoes cost? (Lesson 7-3) **$3**

. A survey showed that 9 out of 10 teens expect to make a charitable contribution or volunteer during the holidays. Based on that survey, how many teenagers in a class of 400 expect to give money or volunteer the next holiday season? (Lesson 8-3) **360 teens**

. What was the percent of increase in the price of first class stamps from 2001 to 2002? Round to the nearest whole percent. (Lesson 8-4) **9%**

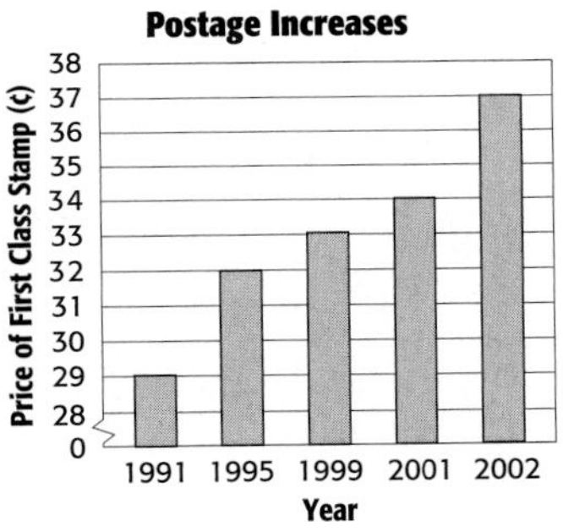

Source: U.S. Postal Service

3. Midtown Veterinarians cared for 125 animals on Monday and 100 animals on Tuesday. What is the percent of decrease in number of animals? (Lesson 8-4) **20%**

msmath2.net/standardized_test

14. Mr. Martinez wants to purchase a sound system that costs $245. If 6.2% sales tax is added, what will be the total cost of the system? (Lesson 8-5) **$260.19**

15. Kaitlyn and Camilia each have money in savings accounts, as shown below.

Money in the Bank		
Name	**Principal**	**Rate**
Kaitlyn	$400	3%
Camilia	$300	6%

Whose savings account will earn more interest after 1 year? (Lesson 8-6) **Camilia's account**

16. If a principal amount of $600 earned $75 in $2\frac{1}{2}$ years, find the interest rate. (Lesson 8-6) **5%**

PART 3 Extended Response

Record your answers on a sheet of paper. Show your work.

17. The Sybil family is working on its budget.

a. Their rent increased from $750 to $907.50 a month. What is the percent of increase? (Lesson 8-4) **21%**

b. Their landlord said that she would decrease the $907.50 monthly rent payment by 6% if the family swept the stairwell of the apartment building each week. If the Sybil family takes the landlord's offer, describe two ways to find how much rent they would pay. (Lesson 8-5) **See margin.**

TEST-TAKING TIP

Question 17 Remember to show all of your work. You may be able to get partial credit for your answers, even if they are not entirely correct.

Answer

17b. Sample answer: Find 6% of $907.50 and then subtract that amount from $907.50 to get $853.05. Another method is to find 100% – 6% or 94% of $907.50 to get $853.05.

Rubrics The following rubrics are sample scoring devices for short response and extended response questions.

Short Response

Points	Description
2	The student demonstrates a **thorough understanding** of the mathematics of the task. The response may contain minor flaws that do not detract from the demonstration of a thorough understanding.
1	The student has provided a response that is only **partially correct**.
0	The student has provided a **completely incorrect** solution or no response at all.

Extended Response

Points	Description
4	The student demonstrates a **thorough understanding** of the mathematics of the task. The response may contain minor flaws that do not detract from the demonstration of a thorough understanding.
3	The student demonstrates an **understanding** of the mathematics of the task. The response is essentially correct and demonstrates an essential but less than thorough understanding of the mathematics.
2	The student demonstrates only a **partial understanding** of the mathematics of the task. Although the student may have used the correct approach to a solution or may have provided a correct solution, the work lacks an essential understanding of the underlying mathematical concepts.
1	The student demonstrates a **very limited understanding** of the mathematics of the task. The response is incomplete and exhibits many flaws.
0	The student provides a **completely incorrect** solution or no response at all.

A more detailed 4-point rubric is provided on page A2 of the *Chapter 8 Resource Masters*.

Page 337, Lesson 8-1

21. $0.1 \cdot 40 = 4$ and $6 \cdot 4 = 24$
22. $0.1 \cdot 130 = 13$ and $2 \cdot 13 = 26$
23. $0.1 \cdot 80 = 8$ and $8 \cdot 8 = 64$
24. $0.1 \cdot 250 = 25$ and $4 \cdot 25 = 100$
25. $0.1 \cdot 120 = 12$ and $3 \cdot 12 = 36$
26. $0.1 \cdot 200 = 20$ and $9 \cdot 20 = 180$
27. $(1 \cdot 54) + \left(\frac{1}{3} \cdot 54\right) = 72$
28. $(2 \cdot 320) + \left(\frac{1}{4} \cdot 320\right) = 720$
29. $4 \cdot 12 = 48$
30. $2 \cdot 33 = 66$
31. $0.01 \cdot 400 = 4$ and $\frac{1}{2} \cdot 4 = 2$
32. $0.01 \cdot 70 = 0.7$

NOTES

Additional Answers for Chapter 8

CHAPTER 9

Chapter Overview

Probability

LESSON OBJECTIVES	NCTM Standards	NAEP 2005	State/Local Objectives
9-1 **Simple Events** *(pp. 370–373)* • Find the probability of a simple event.	1, 4, 9, 10	D4a, D4b, D4g, D4j	
9-2 **Tree Diagrams** *(pp. 374–377)* • Use tree diagrams to count outcomes and find probabilities.	1, 4, 6, 9, 10	D4a, D4b, D4e, D4f, D4g, D4j	
9-3 **The Fundamental Counting Principle** *(pp. 378–380)* • Use multiplication to count outcomes.	1, 2, 4, 6, 9, 10	D4e	
9-4 **Permutations** *(pp. 381–383)* • Find the number of permutations of a set of objects.	1, 2, 4, 6, 9, 10	D4e, N3f	
9-5 **Combinations** *(pp. 386–390)* **9-5a Hands-On Lab:** Find combinations. • Find the number of combinations of a set of objects.	1, 2, 6, 9, 10	D4e, N3f	
9-6 **Theoretical and Experimental Probability** *(pp. 391–397)* **9-6a Problem-Solving Strategy:** Solve problems by acting it out. • Find and compare experimental and theoretical probabilities. **9-6b Hands-On Lab**: Investigate experimental probability.	1, 9, 10	D4a, D4b, D4c, D4d, D4g, N4c	
9-7 **Independent and Dependent Events** *(pp. 398–401)* • Find the probability of independent and dependent events.	1, 2, 6, 9, 10	D4a, D4b, D4g, D4h, D4j	

Key to NCTM Standards:

1=Number & Operations, 2=Algebra, 3=Geometry, 4=Measurement, 5=Data Analysis & Probability, 6=Problem Solving, 7=Reasoning & Proof, 8=Communication, 9=Connections, 10=Representation

NAEP 2005 Standards: See math.glencoe.com

Pacing suggestions for the entire year can be found on page T20.

Chapter Resource Manager

Time-Saving Tool
TeacherWorks™
All-In-One Planner and Resource Center
See page T5.

Chapter 9 Resource Masters

Study Guide and Intervention	Practice (Skills and Word Problems)	Reading to Learn Mathematics	Enrichment	Assessment	Applications*	Parent and Student Study Guide Workbook	5-Minute Check Transparencies	Key Teaching Transparencies	MathPASS: Tutorial Plus (lessons)	Teaching Mathematics with Manipulatives	Optional Materials
485	486–487	488	489			9-1	9-1	33			
490	491–492	493	494	535	RC 23 CS 43	9-2	9-2				markers
495	496–497	498	499			9-3	9-3	34			
500	501–502	503	504	535, 537	CS 44	9-4	9-4			81–82	3 index cards
505	506–507	508	509		CS 45	9-5	9-5	35		83	
510	511–512	513	514	536	CM 25–27	9-6	9-6			84, 85	scissors, paper
515	516–517	518	519	521–534, 536, 538–540	SM 93–96 CS 46	9-7	9-7	36	59		

* **Key to Abbreviations:** *CS = Calculator and Spreadsheet Masters, RC = Real-Life Career Masters, CM = Classroom Math Games (online), SM = Science and Math Lab Manual (online)*

Interactive Chalkboard

PowerPoint® Presentations

This CD-ROM is a customizable Microsoft® PowerPoint® presentation that includes:

- In-Class Examples
- Links to Online Study Tools
- 5-Minute Check Transparencies
- Bellringers

Math Skills Maintenance

Use the Math Skills Maintenance Masters to keep your students' basic math skills fresh. Students can review skills one at a time or in a combined format. The Masters are also available online at msmath2.net/maintenance.

CHAPTER 9 Mathematical Content and Teaching Strategies

Continuity of Instruction

Building on Prior Knowledge

In previous chapters, students worked with ratios. They have learned how to represent ratios as fractions and how to convert between fractions and decimals. Students have also learned different ways of organizing and presenting data, such as tables, bar graphs, and histograms.

Learning in This Chapter

In this chapter, students will learn various ways of representing the sample space, or the set of all possible outcomes. They begin by calculating simple probabilities in situations where the total number of outcomes is given; then they use tree diagrams to count outcomes in problems where that information is not given. From tree diagrams, they learn to use the Fundamental Counting Principle. They are also introduced to permutations and combinations, experimental and theoretical probabilities, and compound events.

Making Future Connections

Knowledge of probabilities is critical for making predictions and arriving at good decisions. For example, a scientist analyzes all the information about a subject, but then needs to make a prediction about the outcome of an experiment before she or he conducts it. A group such as the school board might need to make predictions about the most effective ways to spend a limited amount of money. In each of these cases, the people involved need to understand probable outcomes to make good decisions.

Mathematical Background

9-1 Simple Events

This lesson introduces students to the most basic definition of probability. By using examples involving events like the roll of a number cube or the spin of a spinner, students can easily visualize the possible outcomes for any given example. Encourage students to work carefully, writing down all the possible outcomes and highlighting favorable outcomes.

9-2 Tree Diagrams

Tree diagrams are used to count outcomes. Visual learners will appreciate tree diagrams as a way to keep organized, and all students will benefit from using tree diagrams to more thoroughly understand how arrangements are computed.

9-3 The Fundamental Counting Principle

While tree diagrams are extremely useful, the number of possible arrangements of even a few choices quickly becomes too difficult to display visually. The Fundamental Counting Principle is an easier way to count outcomes. In order to make connections between tree diagrams and the Fundamental Counting Principle, encourage students to apply the Fundamental Counting Principle to problems they have already solved with tree diagrams. As they arrive at the same number of outcomes with tree diagrams and with the Fundamental Counting Principle, they will develop confidence in their ability to count outcomes.

9-4 Permutations

Calculating a permutation means counting outcomes for a set of objects when order matters. Use visual examples, such as arranging four students in different orders at the front of the class, to help students conceptualize the impact of order on the number of outcomes. Visual examples will also help students make the connection between permutations, factorials, and the Fundamental Counting Principle—when order matters, the choices for each position have one less option than those for the preceding position.

9-5 Combinations

In this lesson, students learn to discriminate combinations from permutations and to compute the number of combinations. Realizing that fewer arrangements result when order does not matter will help students understand the reason they must divide when calculating combinations.

Building a Word Wall

The word *combination* has a very specific mathematical meaning. Develop a word wall to clarify the concept. Have students volunteer uses of the word *combination* outside of mathematics. Record the real-world and mathematical meanings in the word wall.

Meanings Outside of Mathematics	Mathematical Meaning
a mixture	a listing of objects in which order is not important
an order of numbers (as in a lock)	
Combination	

9-6 Theoretical and Experimental Probability

At this point, students may perceive *probability* as meaning *theoretical probability,* which is the type of probability involved when a single outcome is compared to all potential outcomes under ideal circumstances (the probability of heads when you toss a fair coin, for example). However, it is not always possible to determine the theoretical probability. In cases when theoretical probability can be calculated and the experimental probability determined, the values will be close when the experiment is repeated many times.

9-7 Independent and Dependent Events

In this section, students advance from simple to compound events. The distinction between independent and dependent events can be difficult for students, so planning hands-on activities to illustrate the differences will be very helpful. A deck of cards or a bingo hopper will be useful for dependent events. Multiple tosses of coins and rolls of number cubes, as well as drawing and replacing cards, can illustrate independent events.

Teacher to Teacher

Pam Berry
Hartselle Junior High School
Hartselle, AL

"To explore experimental probability, we play *SKUNK.* Students write the letters SKUNK each at the top of a column on paper. Then stand while I roll two number cubes for the first column. If I do not roll a 1, students record the sum as points for that column. Then they can choose to sit or stand for the next roll. If I roll a 1 and they are standing, they lose any points in that column and we move to the next column. Students keep their points when sitting, but earn no points then. If I roll double 1s, standing students lose all their points in all columns. The student with the most points at the end is the winner. We discuss the probabilities of rolling 1s and double 1s after the game."

Pam Berry

McGraw-Hill Professional Development

Mathematics, YES! is a self-paced,web-based professional development program. Units are based on major mathematical concepts and are standards-based. Actual classroom videos provide a springboard for discussion and activities. Unit appropriate for this chapter:

- Unit 10, Statistics, Data Analysis, and Probability

Key Concepts

Additional mathematical information and teaching notes are available at

msmath2.com/key_concepts.

The sections appropriate for the chapter are:

- Permutations
- Combinations
- Theoretical and Experimental Probability

DAILY INTERVENTION and Assessment

	Type	Student Edition	Teacher Resources	Technology/Internet
INTERVENTION	Ongoing	Prerequisite Skills, pp. 369, 373, 383, 390, 396 Mid-Chapter Practice Test, p. 384	5-Minute Check Transparencies Quizzes 1–4, *CRM*, pp. 535–536 Mid-Chapter Test, *CRM*, p. 537 Study Guide and Intervention, *CRM*, pp. 485, 490, 495, 500, 505, 510, 515	MathPASS: Tutorial Plus msmath2.net/self_check_quiz msmath2.net/extra_examples
	Mixed Review	pp. 373, 377, 380, 383, 390, 396, 401	Cumulative Review, *CRM*, p. 538	
	Error Analysis	Find the Error, pp. 389, 399	Find the Error, *TWE*, pp. 389, 400 Unlocking Misconceptions, *TWE*, pp. 375, 388	
	Standardized Test Practice	pp. 373, 377, 379, 380, 383, 384, 390, 392, 396, 401, 405, 406–407	Standardized Test Practice, *CRM*, pp. 539–540	Standardized Test Practice CD-ROM msmath2.net/standardized_test
ASSESSMENT	Open-Ended Assessment	Writing Math, pp. 372, 376, 379, 382, 386, 389, 395, 397, 399 Open-Ended, pp. 372, 376, 379, 382, 389, 395, 399 Standardized Test, p. 407	Modeling: *TWE*, pp. 373, 401 Speaking: *TWE*, p. 380 Writing: *TWE*, pp. 390, 395 Extended Response Assessment, *CRM*, p. 533	
	Chapter Assessment	Study Guide, pp. 402–404 Practice Test, p. 405	Multiple-Choice Tests (Forms 1, 2A, 2B), *CRM*, pp. 521–526 Free-Response Tests (Forms 2C, 2D, 3), *CRM*, pp. 527–532 Vocabulary Test/Review, *CRM*, p. 534	ExamView® Pro Test Bank (see below) MindJogger Videoquizzes msmath2.net/vocabulary_review msmath2.net/chapter_test

* **Key to Abbreviations:** *TWE* = Teacher Wraparound Edition, *CRM* = Chapter Resource Masters

Additional Intervention

Skills Intervention for Middle School Mathematics workbook and CD-ROM includes 88 skills with pretests, tutorials, guided practice, and posttests for diagnosis and remediation.

Guide to Daily Intervention booklet details each intervention resource available and guides teachers in using them as intervention tools everyday to assist students who may be having difficulty.

ExamView® Pro

Use ExamView® Pro Test Bank CD-ROM to:

- **Create Multiple versions of tests.**
- **Create Modified tests for Inclusion students with one mouse click.**
- **Edit existing questions and Add your own questions.**
- **Build tests aligned with state standards using built-in State Curriculum Correlations.**
- **Change English tests to Spanish with one mouse click and vice versa.**

Reading and Writing Mathematics

• Student Edition

Foldables Study Organizer, p. 369

Concept Check questions at the beginning of each exercise set require students to verbalize and write about what they have learned in the lesson.

- **Open Ended,** pp. 372, 376, 379, 382, 389, 395, 399
- **Find the Error,** pp. 389, 399
- **Which One Doesn't Belong?,** pp. 372, 379

Reading Math Tip, p. 398

WebQuest, p. 401

• Teacher Wraparound Edition

Foldables Follow-Up, pp. 372, 384, 389, 395, 399

Writing Activities, pp. 390, 395

ELL Resources, *see page T9 for a complete listing.*

• Additional Resources

Vocabulary Builder, *Chapter 9 Resource Masters*

Reading to Learn Mathematics, *Chapter 9 Resource Masters*

Vocabulary PuzzleMaker (online)

Teaching Mathematics with Foldables

Reading and Writing in the Mathematics Classroom

WebQuests, Projects, and Interdisciplinary Investigations

For detailed information on these resources, see pp. T6–T7.

• Recommended Outside Reading for Students

Nonfiction

- Peterson, Ivars and Henderson, Nancy. *Math Trek: Adventures in the Math Zone.* John Wiley and Sons, ©1999.

  This book shows students the math of an amusement park. Games and activities are included. "Trek 13: Luck on the Boredwalk" will be of special interest to students after completing their study of probability.

Fiction

- Bruce, Colin. *Conned Again, Watson! Cautionary Tales of Logic, Math, and Probability.* Perseus Publishing, ©2002.

  Using the stories of Sherlock Holmes, the author creates mysteries that require mathematical skill to be solved. This book provides interesting insight as probability and statistics are used by Sherlock Holmes to outwit con men.

PROJECT CRISS℠ Study Skill

Mapping can help students create useful representations of new subject matter to use as study tools. The sample map at the right shows how to find the probability of rolling an even number on a number cube. Have students work in cooperative groups to design maps for lessons of their choosing from Chapter 9. The maps should be designed after each student has carefully read the lesson. Afterward, have each group present their map to the class.

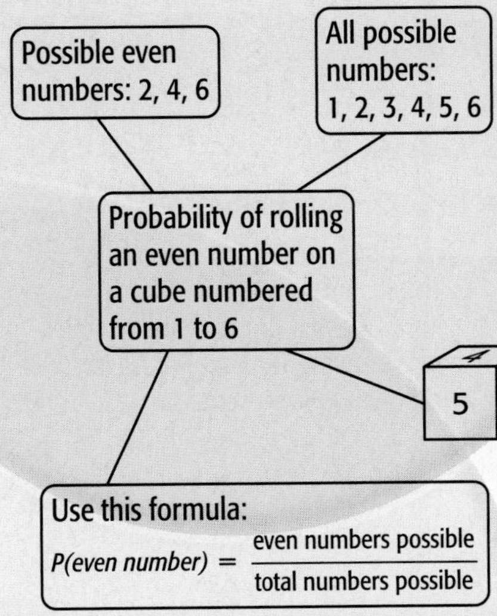

CReating Independence through Student-owned Strategies

For more information on Project CRISS, see page T9.

Making Connections

School-To-Home

Parent and Student Study Guide

Parents and students may work together to reinforce the concepts and skills of this chapter. These workbooks are available in English and Spanish. A sample lesson is shown below. Also available at msmath2.net/parent_student.

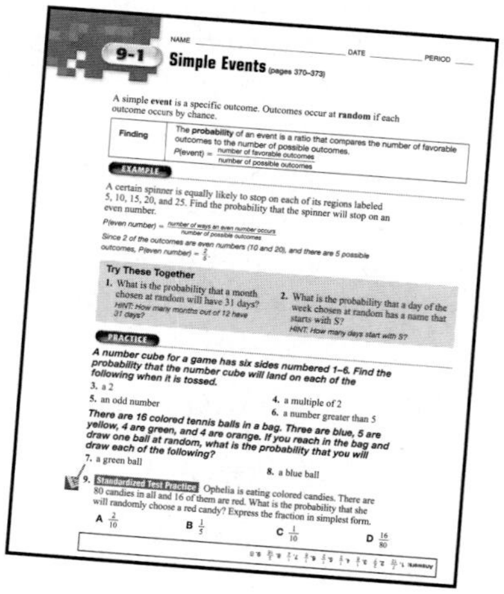

NAME ________ DATE ________ PERIOD ____

9-1 Simple Events (pages 370–373)

A simple **event** is a specific outcome. Outcomes occur at **random** if each outcome occurs by chance.

Finding	The **probability** of an event is a ratio that compares the number of favorable outcomes to the number of possible outcomes. $P(\text{event}) = \frac{\text{number of favorable outcomes}}{\text{number of possible outcomes}}$

EXAMPLE

A certain spinner is equally likely to stop on each of its regions labeled 5, 10, 15, 20, and 25. Find the probability that the spinner will stop on an even number.

$P(\text{even number}) = \frac{\text{number of ways an even number occurs}}{\text{number of possible outcomes}}$

Since 2 of the outcomes are even numbers (10 and 20), and there are 5 possible outcomes, P(even number) = $\frac{2}{5}$.

Try These Together

1. What is the probability that a month chosen at random will have 31 days? *HINT: How many months out of 12 have 31 days?*
2. What is the probability that a day of the week chosen at random has a name that starts with S? *HINT: How many days start with S?*

PRACTICE

A number cube for a game has six sides numbered 1–6. Find the probability that the number cube will land on each of the following when it is tossed.

3. a 2
4. a multiple of 2
5. an odd number
6. a number greater than 5

There are 16 colored tennis balls in a bag. Three are blue, 5 are yellow, 4 are green, and 4 are orange. If you reach in the bag and draw one ball at random, what is the probability that you will draw each of the following?

7. a green ball
8. a blue ball
9. **Standardized Test Practice** Ophelia is eating colored candies. There are 80 candies in all and 16 of them are red. What is the probability that she will randomly choose a red candy? Express the fraction in simplest form.

A $\frac{2}{10}$ B $\frac{1}{5}$ C $\frac{1}{10}$ D $\frac{16}{80}$

Family Letters and Activities

Send a letter and activity for this chapter home to encourage family members to become active participants in their students' learning. (English: *Chapter 9 Resource Masters*, pp. ix–x; Spanish: log on to msmath2.net/family_letters)

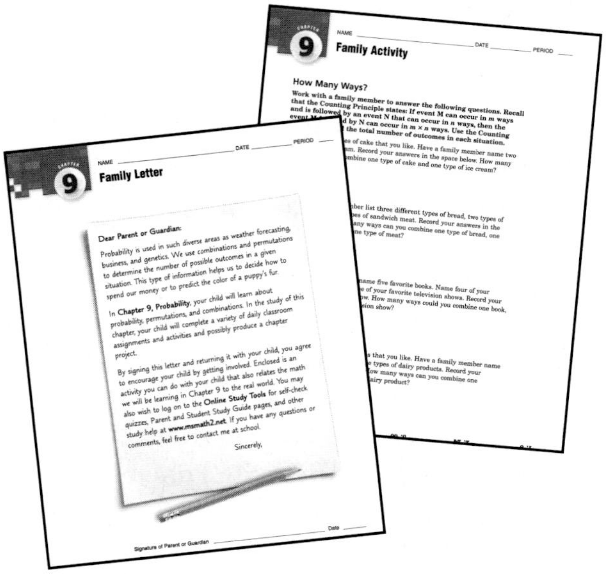

NAME ________ DATE ________ PERIOD ____

Chapter 9 Family Activity

How Many Ways?

Work with a family member to answer the following questions. Recall that the Counting Principle states: If event M can occur in *m* ways and is followed by an event N that can occur in *n* ways, then the event M … d by N can occur in *m* × *n* ways. Use the Counting … the total number of outcomes in each situation.

… es of cake that you like. Have a family member name two … m. Record your answers in the space below. How many … mbine one type of cake and one type of ice cream?

… ber list three different types of bread, two types of … es of sandwich meat. Record your answers in the … any ways can you combine one type of bread, one … e type of meat?

NAME ________ DATE ________ PERIOD ____

Chapter 9 Family Letter

Dear Parent or Guardian:

Probability is used in such diverse areas as weather forecasting, business, and genetics. We use combinations and permutations to determine the number of possible outcomes in a given situation. This type of information helps us to decide how to spend our money or to predict the color of a puppy's fur.

In **Chapter 9, Probability**, your child will learn about probability, permutations, and combinations. In the study of this chapter, your child will complete a variety of daily classroom assignments and activities and possibly produce a chapter project.

By signing this letter and returning it with your child, you agree to encourage your child by getting involved. Enclosed is an activity you can do with your child that also relates the math we will be learning in Chapter 9 to the real world. You may also wish to log on to the **Online Study Tools** for self-check quizzes, Parent and Student Study Guide pages, and other study help at **www.msmath2.net**. If you have any questions or comments, feel free to contact me at school.

Sincerely,

Signature of Parent or Guardian ________ Date ____

English Language Learner ELL

Lesson 9-4

Higher-Level Thinking

After showing students several examples of permutations and factorials, ask them to write in their own words the meaning of *factorial* and how it is represented.

Lesson 9-5

Building on Prior Knowledge

Ask students how they would define *combination.* Have students give real-life examples of how and when they use combinations. For instance cooking, clothing, arranging a wardrobe or packing for a trip. Then relate their applications to the mathematical definition of *combination.*

Lesson 9-6

Higher-Level Thinking

After completing the Hands-On Mini Lab, ask students to find the difference between their experimental probability and the theoretical probability. Ask them to do the same when they combine the results for the class. Write each difference as a percent. Was the class experimental probability or their own closer to the theoretical probability? Why do they think this is true?

MathScape™

This project was supported, in part, by the **National Science Foundation**

MathScape is a mathematics curriculum for grades 6–8 developed by the Seeing and Thinking Mathematically Project at the Education Development Center.

Probability in Games and Simulations
Chance Encounters

Summary Students will understand chance and outcomes. Theoretical and experimental probabilities are investigated. Students will conduct probability experiments. Probabilities are described with decimals, fractions, and percentages. Students will represent probabilities verbally, visually, and numerically.

USA TODAY Education Online Lesson Planner

USA TODAY® Education's Online site offers resources and interactive features connected to each day's newspaper. *Experience Today,* USA TODAY's® daily lesson plan, is available on the site and delivered daily to subscribers. This plan provides instruction for integrating USA TODAY® graphics and key editorial features into your mathematics classroom. Log on to **education.usatoday.com**.

USA TODAY Snapshots® in Chapter 9	
Lesson	**Snapshot**
9-6	Cards, flowers top Mother's Day gifts

Quick Review Math Handbook

is Glencoe's mathematical handbook for students. The Hot Words section includes a glossary of terms while the Hot Topics section consists of explanations of key mathematical concepts. An exercise set is included to check students' understanding of the concepts. This valuable resource can be used as a reference in the classroom or for home study.

Lesson	Hot Topics Section	Lesson	Hot Topics Section
9-1	4-6	**9-4, 9-5**	4-5
9-2	4-5	**9-6**	4-6
9-3	4-6	**9-7**	4-6

Chapter 9 Notes

Key Chapter Objectives

Lesson 9-1 Find the probability of a simple event.

Lesson 9-3 Use multiplication to count outcomes.

Lessons 9-4 and 9-5 Find the number of permutations and combinations of a set of objects.

Lesson 9-7 Find the probability of independent and dependent events.

"What do games have to do with math?"

Point out to students that this is only one of many applications of the math that they are studying. Others are provided in the introduction to each lesson.

Vocabulary Builder, pp. vii–viii ELL

NAME ______________ DATE ________ PERIOD ____

Reading to Learn Mathematics

Vocabulary Builder

This is an alphabetical list of new vocabulary terms you will learn in Chapter 9. As you study the chapter, complete each term's definition or description. Remember to add the page number where you found the term. Add these pages to your math study notebook to review vocabulary at the end of the chapter.

Vocabulary Term	Found on Page	Definition/Description/Example
combination		
complementary [KAHM-pluh-MEHN-tuh-ree] event		
compound event		
dependent event		
experimental [ihk-SPEHR-uh-MEHN-tuhl] probability		
factorial [fak-TAWR-ee-uhl]		
fair game		
Fundamental Counting Principle		

Chapter 9 Probability

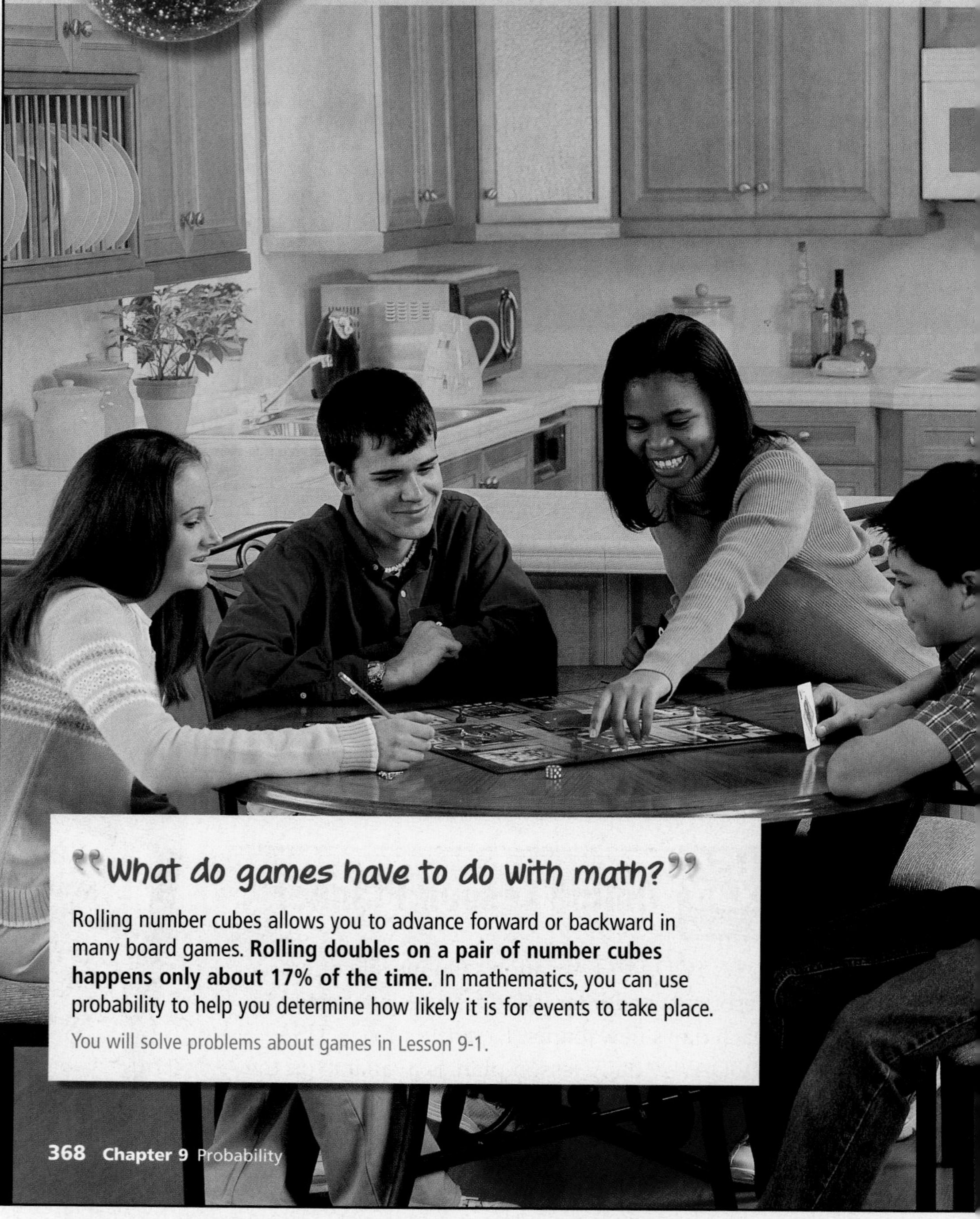

"What do games have to do with math?"

Rolling number cubes allows you to advance forward or backward in many board games. **Rolling doubles on a pair of number cubes happens only about 17% of the time.** In mathematics, you can use probability to help you determine how likely it is for events to take place.

You will solve problems about games in Lesson 9-1.

Vocabulary Builder

For a thorough vocabulary list with pronunciations of new words, give students the Vocabulary Builder worksheets from the *Chapter 9 Resource Masters*. Encourage them to complete the definition of each term as they progress through the chapter.

GETTING STARTED

Take this quiz to see whether you are ready to begin Chapter 9. Refer to the lesson or page number in parentheses if you need more review.

Vocabulary Review

Choose the correct term or number to complete each sentence.

1. A fraction is in (factored, simplest) form when the GCF of the numerator and denominator is 1. (Lesson 5-3) **simplest**
2. The fraction $\frac{12}{38}$ in simplest form is $\left(\frac{6}{19}, \frac{3}{17}\right)$. (Lesson 5-3) $\frac{6}{19}$

Prerequisite Skills

Multiply.

3. 7×15 **105**
4. 24×6 **144**
5. 13×4 **52**
6. 8×21 **168**
7. 5×32 **160**
8. 30×8 **240**
9. $6 \cdot 5 \cdot 4 \cdot 3$ **360**
10. $7 \cdot 6 \cdot 5$ **210**
11. $8 \cdot 7 \cdot 6$ **336**
12. $4 \cdot 3 \cdot 2 \cdot 1$ **24**
13. $10 \cdot 9 \cdot 8 \cdot 7$ **5,040**
14. $11 \cdot 10 \cdot 9$ **990**

Write each fraction in simplest form. Write *simplified* if the fraction is already in simplest form. (Lesson 5-3)

15. $\frac{8}{12}$ $\frac{2}{3}$
16. $\frac{3}{18}$ $\frac{1}{6}$
17. $\frac{4}{9}$ **simplified**
18. $\frac{5}{15}$ $\frac{1}{3}$

Find each value. (Lesson 5-3)

19. $\frac{6 \cdot 5}{3 \cdot 2}$ **5**
20. $\frac{9 \cdot 8 \cdot 7}{5 \cdot 4 \cdot 3}$ **8.4**
21. $\frac{4 \cdot 3 \cdot 2}{3 \cdot 2 \cdot 1}$ **4**
22. $\frac{7 \cdot 6 \cdot 5 \cdot 4}{4 \cdot 3 \cdot 2 \cdot 1}$ **35**

FOLDABLES™ Study Organizer

Probability Make this Foldable to help you organize the topics in this chapter. Begin with four sheets of $8\frac{1}{2}''$ by 11″ paper.

STEP 1 **Stack Pages**
Place 4 sheets of paper $\frac{3}{4}$ inch apart.

STEP 2 **Roll Up Bottom Edges**
All tabs should be the same size.

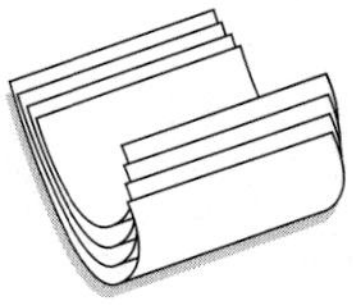

STEP 3 **Crease and Staple**
Staple along fold.

STEP 4 **Label**
Write the chapter title on the front. Label each tab with a lesson number and title.

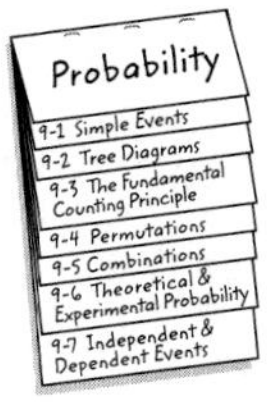

Reading and Writing As you read and study each lesson, write notes and examples under the corresponding tab.

GETTING STARTED

This section provides a review of the basic concepts needed before beginning Chapter 9. Page references are included for additional student help.

Prerequisite Skills review skills needed in this chapter.

Exercises	Needed for Lesson	Skill
3–14	9-3, 9-4, 9-5	Find the Product
15–18	9-1, 9-6	Simplify Fractions
19–22	9-5, 9-6, 9-7	Simplify Expressions

For more prerequisite skills, see:

- Getting Ready for the Next Lesson, at the end of each lesson
- Prerequisite Skills in the Student Handbook on pp. 554–563
- *Math Skills Maintenance Masters*

FOLDABLES™ Study Organizer

For more information about Foldables, see *Teaching Mathematics with Foldables.*

After students make their Foldables, have them label the tabs to correspond to the seven lessons in this chapter. At the end of each lesson, ask students to write a descriptive paragraph sharing their experiences with the concepts, the computational skills, and the graphics presented. For example, students might write about how they felt when they were first asked to use a tree diagram or about how difficult the lesson appeared before they understood the concepts presented and how it appeared after mastery.

9-1 Lesson Notes

Lesson Resources

Chapter 9 Resource Masters
- Study Guide and Intervention, p. 485
- Practice: Skills, p. 486
- Practice: Word Problems, p. 487
- Reading to Learn Mathematics, p. 488
- Enrichment, p. 489

Key Teaching Transparancies, p. 33
Parent and Student Study Guide Workbook, 9-1

FAST FILE

1 Focus

5-Minute Check 9-1
Use as a quiz or review of Chapter 8. (Also available as a transparency)

INTERACTIVE CHALKBOARD
PowerPoint® Presentations

Bellringer
Reading Mathematics Ask students to discuss three everyday situations in which they have heard the word *probability*. You may want to have them scan a newspaper, including the weather report, for examples of how probability is used.

Reading to Learn Mathematics, p. 488 **ELL**

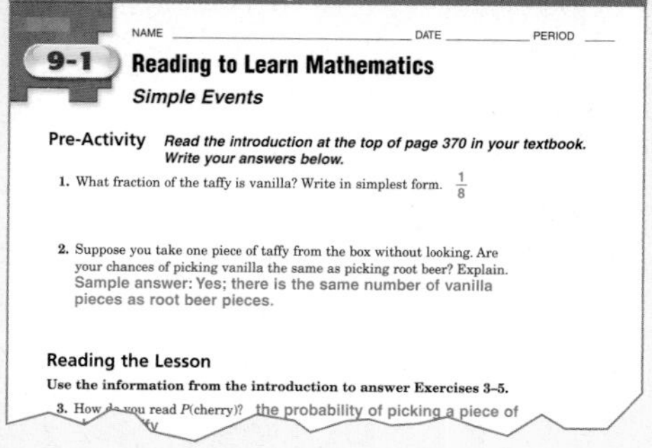
NAME ______ DATE ______ PERIOD ____

9-1 Reading to Learn Mathematics
Simple Events

Pre-Activity ***Read the introduction at the top of page 370 in your textbook. Write your answers below.***

1. What fraction of the taffy is vanilla? Write in simplest form. $\frac{1}{8}$

2. Suppose you take one piece of taffy from the box without looking. Are your chances of picking vanilla the same as picking root beer? Explain. Sample answer: Yes; there is the same number of vanilla pieces as root beer pieces.

Reading the Lesson
Use the information from the introduction to answer Exercises 3–5.

3. How do you read *P*(cherry)? the probability of picking a piece of

9-1 Simple Events

What You'll Learn
Find the probability of a simple event.

***NEW* Vocabulary**
- outcome
- simple event
- probability
- random
- complementary event

WHEN am I ever going to use this?

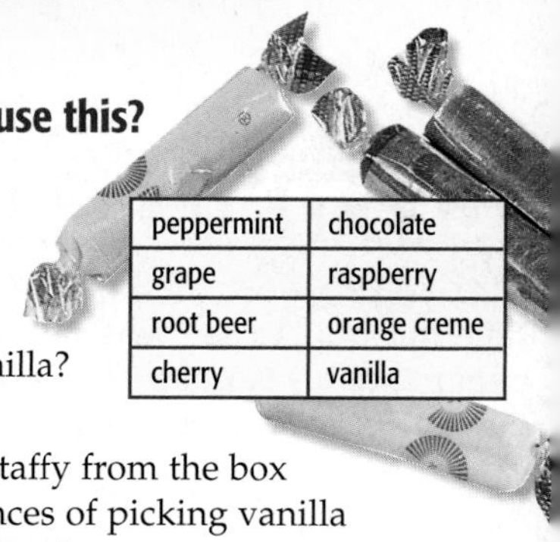

CANDY A box of saltwater taffy contains 48 pieces, six pieces of each flavor shown at the right.

peppermint	chocolate
grape	raspberry
root beer	orange creme
cherry	vanilla

1. What fraction of the taffy is vanilla? Write in simplest form. $\frac{1}{8}$
2. Suppose you take one piece of taffy from the box without looking. Are your chances of picking vanilla the same as picking root beer? Explain.

2. Sample answer: Yes; there is the same number of vanilla pieces as root beer pieces.

The 48 pieces of taffy in the box above are called **outcomes**. A **simple event** is one outcome or a collection of outcomes. For example, picking a piece of vanilla taffy is a simple event. The chance of that event happening is called **probability**.

Key Concept **Probability**

Words The probability of an event is a ratio that compares the number of favorable outcomes to the number of possible outcomes.

Symbols $P(\text{event}) = \frac{\text{number of favorable outcomes}}{\text{number of possible outcomes}}$

Outcomes occur at **random** if each outcome occurs by chance. A piece of vanilla taffy was selected at random in the activity above.

Read *P*(vanilla) as *the probability of picking a piece of vanilla taffy.*

$P(\text{vanilla}) = \frac{6}{48}$ ← number of pieces of vanilla taffy / total number of pieces of taffy

$= \frac{1}{8}$, or 12.5% Simplify.

EXAMPLE **Find Probability**

1 **What is the probability of rolling an even number on a number cube marked with 1, 2, 3, 4, 5, and 6 on its faces?**

$P(\text{even number}) = \frac{\text{even numbers possible}}{\text{total numbers possible}}$

$= \frac{3}{6}$ Three numbers are even: 2, 4, and 6.

$= \frac{1}{2}$ Simplify.

The probability of rolling an even number is $\frac{1}{2}$ or 50%.

TEACHING TIP
Tell students that they can assume all number cubes have six faces, unless it is indicated otherwise.

370 Chapter 9 Probability

EXAMPLE Find Probability

2 PARCHEESI Jewel rolls two number cubes. She can move her game piece if a total of 5 is shown on the number cubes, or if a 5 is shown on at least one number cube. What is the probability that she can move her game piece on one roll of the number cubes?

List all the possible outcomes. Then, find the pairs that total 5, or if a 5 is shown on at least one of the number cubes.

1, 1	1, 2	1, 3	1, 4	1, 5	1, 6
2, 1	2, 2	2, 3	2, 4	2, 5	2, 6
3, 1	3, 2	3, 3	3, 4	3, 5	3, 6
4, 1	4, 2	4, 3	4, 4	4, 5	4, 6
5, 1	5, 2	5, 3	5, 4	5, 5	5, 6
6, 1	6, 2	6, 3	6, 4	6, 5	6, 6

There are 36 possible outcomes and 15 of them are favorable. So, the probability that Jewel moves her game piece in one roll is $\frac{15}{36}$, or $\frac{5}{12}$.

EAL-LIFE MATH

CHEESI Parcheesi is ved from the Indian word is, which means 25.

ce: www.yesterdayland.com

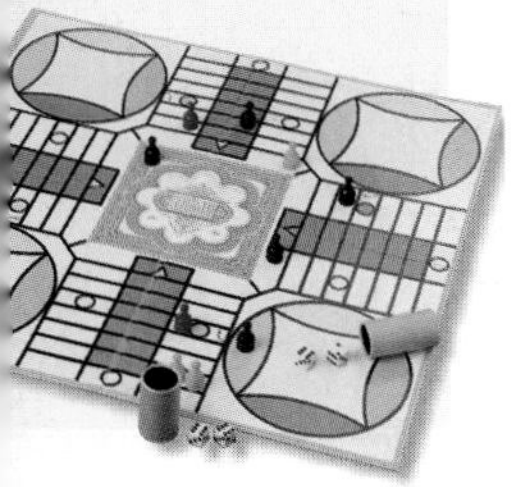

The probability that an event will happen is somewhere between 0 and 1. It can be shown on a number line.

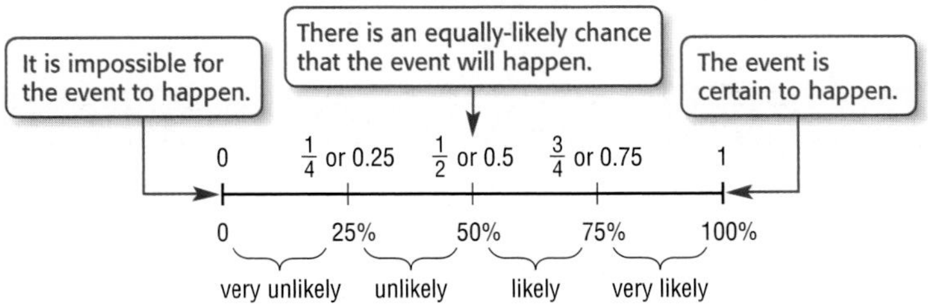

TEACHING TIP

ll students that mplementary events e two events in which ther one or the other ust take place, but they nnot both happen at e same time.

Either Jewel will be able to move her game piece or she will *not* be able to move it. Two events that are the only ones that can possibly happen are examples of **complementary events**.

EXAMPLE Find a Complementary Event

3 PARCHEESI Refer to Example 2. Find the probability that Jewel *cannot* move her game piece on one roll of the number cubes.

$P(A) + P(\text{not } A) = 1$

$\frac{5}{12} + P(\text{not } A) = 1$ Substitute $\frac{5}{12}$ for $P(A)$.

$-\frac{5}{12} \qquad -\frac{5}{12}$ Subtract $\frac{5}{12}$ from each side.

$P(\text{not } A) = \frac{7}{12}$ Simplify.

So, the probability that Jewel *cannot* move her game piece is $\frac{7}{12}$.

TUDY TIP

Check Your Answer You can check the answer by counting the number of outcomes that do not total 5 and that do not have a 5 on at least one number cube.

msmath2.net/extra_examples

2 Teach

In-Class Examples

INTERACTIVE CHALKBOARD PowerPoint® Presentations

1 If the spinner shown below is spun once, what is the probability of it landing on an odd number? $\frac{1}{2}$

2 GAMES A game requires spinning the spinner shown above twice. If the sum of the two spins is 6 or greater, the player wins. What is the probability of winning the game? $\frac{3}{8}$

3 GAMES What is the probability of not winning the game described above? $\frac{5}{8}$

Tips for New Teachers

Using Manipulatives There are many opportunities to use manipulatives in probability lessons. Preparing manipulatives for use before class and predetermining groups will make their use more productive and educational.

DAILY INTERVENTION **Differentiated Instruction**

Kinesthetic Have each student toss a number cube 30 times and record the result of each toss. Have students compare their results with the predicted results in Example 1. Then collect individual results and compile them as one large data set. Save both the individual results and the larger data set for review when students are learning about experimental probability.

3 Practice/Apply

FOLDABLES™ Follow-Up Remind students to use their Foldables to record what they learn about simple events. Point out that they might describe how they found the probability of a simple event.

Which One Doesn't Belong?
In Exercise 3, remind students that both numbers should be in the same form (decimals or fractions with common denominators) in order to work with them. By checking the two pairs of probabilities that are already in the same form, students should quickly notice that 0.65 and 0.55 do not add to 1.

About the Exercises...

Guided Practice Key

Exercises	Examples
4–8	1
9	2, 3

Odd/Even Assignments
Exercises 10–25 are structured so that students practice the same concepts whether they are assigned odd or even problems.

Answers

1. Sample answer: Since 0.7 is greater than 0.5 and less than 0.75, the event is likely to happen.
2. Sample answer: a spinner with 5 equal sections labeled A, B, C, D, and E. Spinning the letter B has a probability of $\frac{1}{5}$.
3. 0.65, 0.55 are probabilities that are not complementary because $0.65 + 0.55 \neq 1$. The other pairs are complementary.

16. Sample answer: Since 0.28 is greater than 0.25 and less than 0.5, the event is unlikely to happen.
17. Sample answer: The complementary event is the chance of no rain. Its probability is 63%.

Skill and Concept Check

Writing Math Exercises 1 & 3

1. **Explain** why an event with a probability of 0.7 is likely to happen. 1–3. See margin.
2. **OPEN ENDED** Describe an event that has a probability of $\frac{1}{5}$.
3. **Which One Doesn't Belong?** Identify the pair of probabilities that do not represent probabilities of complementary events. Explain.

$\frac{5}{8}, \frac{3}{8}$	0.65, 0.55	$\frac{4}{6}, \frac{1}{3}$	0.875, $\frac{1}{8}$

GUIDED PRACTICE

Use the spinner at the right to find each probability. Write as a fraction in simplest form.

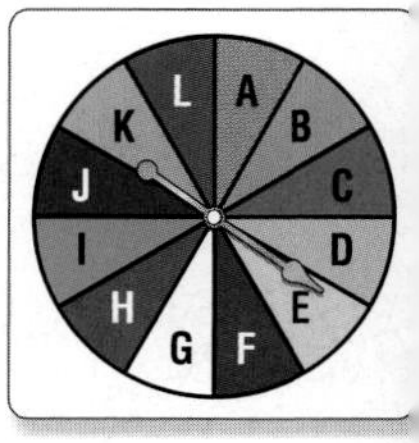

4. $P(\text{J})$ $\frac{1}{12}$
5. $P(\text{vowel})$ $\frac{1}{4}$

A bag contains 7 blue, 5 purple, 12 red, and 6 orange marbles. Find each probability if you draw one marble at random from the bag. Write as a fraction in simplest form.

6. $P(\text{purple})$ $\frac{1}{6}$
7. $P(\text{red or orange})$ $\frac{3}{5}$
8. $P(\text{not blue})$ $\frac{23}{30}$
9. **CANDY** Refer to the activity on the top of page 370. What is the probability of picking a piece of taffy that is *not* peppermint nor root beer? Write the probability as a percent. 75%

★ indicates increased difficulty

Practice and Applications

HOMEWORK HELP

For Exercises	See Examples
10–15	1
18, 20–23, 25–27	2
17, 19, 24, 28–29	3

Extra Practice See pages 584, 604.

A set of 20 cards is numbered 1, 2, 3, ..., 20. Suppose you pick a card at random without looking. Find the probability of each event. Write as a fraction in simplest form.

10. $P(1)$ $\frac{1}{20}$
11. $P(\text{not a factor of 10})$ $\frac{4}{5}$
12. $P(\text{multiple of 3})$ $\frac{3}{10}$
13. $P(\text{even number})$ $\frac{1}{2}$
14. $P(\text{less than or equal to 20})$ 1
15. $P(3 \text{ or } 13)$ $\frac{1}{10}$

16–17. See margin.

16. How likely is it that an event with a probability of 0.28 will occur?
17. The forecast for tomorrow says that there is a 37% chance of rain. Describe the complementary event and its probability.

STUDENT COUNCIL **The table shows the members of the Student Council. Suppose one student is randomly selected as the president. Find the probability of each event. Write as a fraction in simplest form.**

Student Council	
girls	30
boys	20
8th graders	25
7th graders	15
6th graders	10

18. $P(\text{girl})$ $\frac{3}{5}$
19. $P(\text{not 7th grader})$ $\frac{7}{10}$
20. $P(\text{boy})$ $\frac{2}{5}$
21. $P(\text{8th grader})$ $\frac{1}{2}$
22. $P(\text{boy or girl})$ 1
23. $P(\text{6th or 8th grader})$ $\frac{7}{10}$
24. $P(\text{not 6th grader})$ $\frac{4}{5}$
25. $P(\text{5th grader})$ 0
26. Which event has a greater chance of happening: picking a president who is a girl or a president who is not an 8th grader? Explain. See margin.

26. The probability of picking a girl for president is $\frac{3}{5}$. The probability of picking a student who is not an 8th grader is $\frac{1}{2}$. Since $\frac{3}{5} > \frac{1}{2}$, picking a girl for president has a greater chance of happening than picking a student who is not an eighth grader.

30. No; the chance of rolling a 5 $\left(\frac{4}{36}\right)$, 6 $\left(\frac{5}{36}\right)$, 7 $\left(\frac{6}{36}\right)$, or 8 $\left(\frac{5}{36}\right)$ gives Melissa $\frac{20}{36}$ chance of winning a point. The probability of Hakan winning a point is only $\frac{16}{36}$. Since $\frac{20}{36} > \frac{16}{36}$, Melissa has a better chance of winning.

27. **TECHNOLOGY** The graph shows the cost of 22 digital cameras. If one of the 22 cameras is chosen at random, what is the probability that it costs between \$111 and \$160? $\frac{1}{11}$

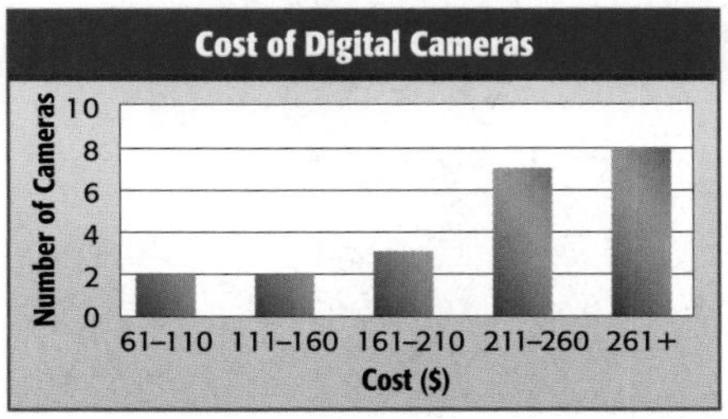

28. **MOVIES** The probability of buying a defective DVD is 0.002. What is the probability of buying a DVD that is *not* defective? **0.998**

★ 29. **MULTI STEP** The Jefferson Middle School Booster Club is selling raffle tickets for a new computer system. They sold 1,000 tickets at \$2 each. Emilia's parents spent \$200 on tickets. What is the probability that they will *not* win? $\frac{9}{10}$

30. **CRITICAL THINKING** Melissa and Hakan are playing a game by rolling two number cubes. Hakan gets a point each time the sum of the number cubes is 2, 3, 4, 9, 10, 11, or 12. Melissa gets a point when the sum is 5, 6, 7, or 8. Does each player have an equal chance to win? Explain. **See margin.**

Standardized Test Practice and Mixed Review

31. **MULTIPLE CHOICE** A bookshelf contains the books at the right. If you randomly choose a book from the shelf, what is the probability that it is science fiction? **A**

Ⓐ $\frac{1}{10}$ Ⓑ $\frac{3}{20}$ Ⓒ $\frac{9}{40}$ Ⓓ $\frac{1}{4}$

Biography	20
Children's	18
Self-help	13
Romance	12
Science Fiction	8
Mystery	5
Horror	4

32. **MULTIPLE CHOICE** In a school raffle, one ticket will be drawn out of a total of 500 tickets. If the Hawkins family has 12 tickets, what is the probability that they will win? **H**

Ⓕ 0.002 Ⓖ 2% Ⓗ $\frac{3}{125}$ Ⓘ $0.08\overline{3}$

Find the interest earned to the nearest cent for each principal, interest rate, and time. (Lesson 8-6)

33. \$300, 10%, 2 years **\$60**

34. \$900, 5.5%, 4.5 years **\$222.75**

35. **FOOD** The United States produced almost 11 billion pounds of apples in 2000. Use the information in the graph to find how many pounds of apples were used to make juice and cider. (Lesson 7-8) **about 2.5 billion**

Uses of Apples in the United States

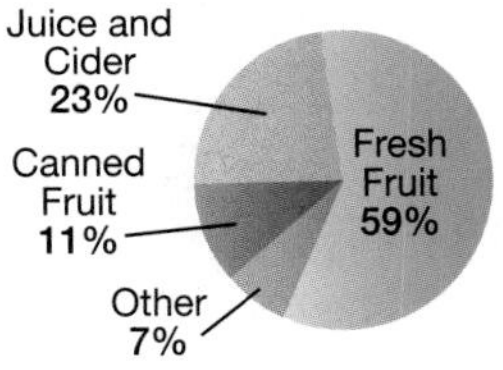

Source: usapple.org

Write each fraction or mixed number as a decimal. Use bar notation if necessary. (Lesson 5-4)

36. $\frac{3}{5}$ **0.6**

37. $\frac{9}{20}$ **0.45**

38. $6\frac{1}{8}$ **6.125**

39. $1\frac{7}{9}$ **$1.\overline{7}$**

GETTING READY FOR THE NEXT LESSON

PREREQUISITE SKILL Write each fraction in simplest form. Write *simplified* if the fraction is already in simplest form. (Lesson 5-3)

40. $\frac{2}{6}$ $\frac{1}{3}$

41. $\frac{3}{8}$ **simplified**

42. $\frac{15}{30}$ $\frac{1}{2}$

43. $\frac{6}{16}$ $\frac{3}{8}$

44. $\frac{18}{32}$ $\frac{9}{16}$

msmath2.net/self_check_quiz

4 Assess

Open-Ended Assessment

Modeling A cafeteria serves 1 side dish each day. They choose at random from 3 side dishes. How can you model the side dish selection with a 6-sided number cube? **Assign each side dish two numbers on the 6-sided number cube. The side dishes each will have a $\frac{1}{3}$ chance of being selected on a roll.**

Study Guide and Intervention, p. 485

NAME ________ DATE ________ PERIOD ____

9-1 Study Guide and Intervention

Simple Events

The **probability** of an event is a ratio that compares the number of favorable outcomes to the number of possible outcomes. Outcomes occur at **random** if each outcome occurs by chance.

Two events that are the only ones that can possibly happen are **complementary events**. The sum of the probabilities of complementary events is 1.

EXAMPLE 1 **What is the probability of rolling a multiple of 3 on a number cube marked with 1, 2, 3, 4, 5, and 6 on its faces?**

$P(\text{multiple of } 3) = \frac{\text{multiples of 3 possible}}{\text{total numbers possible}}$

$= \frac{2}{6}$ Two numbers are multiples of 3: 3 and 6.

$= \frac{1}{3}$ Simplify.

The probability of rolling a multiple of 3 is $\frac{1}{3}$ or about 33.3%.

EXAMPLE 2 **What is the probability of *not* rolling a multiple of 3 on a number cube marked with 1, 2, 3, 4, 5, and 6 on its faces?**

$P(A) + P(\text{not } A) = 1$

$\frac{1}{3} + P(\text{not } A) = 1$ Substitute $\frac{1}{3}$ for $P(A)$.

$-\frac{1}{3} \qquad -\frac{1}{3}$ Subtract $\frac{1}{3}$ from each side

$P(\text{not } A) = \frac{2}{3}$ Simplify.

The probability of *not* rolling a multiple of 3 is $\frac{2}{3}$ or about 66.7%.

EXERCISES

A set of 30 cards is numbered 1, 2, 3, ..., 30. Suppose you pick a card at random without looking. Find the probability of each event. Write as a fraction in simplest form.

1. $P(12)$ $\frac{1}{30}$

2. $P(2 \text{ or } 3)$ $\frac{1}{15}$

3. $P(\text{odd number})$ $\frac{1}{2}$

4. $P(\text{a multiple of } 5)$ $\frac{1}{5}$

5. $P(\text{not a multiple of } 5)$ $\frac{4}{5}$

6. $P(\text{less than or equal to } 10)$ $\frac{1}{3}$

Practice: Skills, p. 486

NAME ________ DATE ________ PERIOD ____

9-1 Practice: Skills

Simple Events

A set of 12 cards is numbered 1, 2, 3, ...12. Suppose you pick a card at random without looking. Find the probability of each event. Write as a fraction in simplest form.

1. $P(5)$ $\frac{1}{12}$

2. $P(6 \text{ or } 8)$ $\frac{1}{6}$

3. $P(\text{a multiple of } 3)$ $\frac{1}{3}$

4. $P(\text{an even number})$ $\frac{1}{2}$

5. $P(\text{a multiple of } 4)$ $\frac{1}{4}$

6. $P(\text{less than or equal to } 8)$ $\frac{2}{3}$

7. $P(\text{a factor of } 12)$ $\frac{1}{2}$

8. $P(\text{not a multiple of } 4)$ $\frac{3}{4}$

9. $P(1, 3, \text{ or } 11)$ $\frac{1}{4}$

10. $P(\text{a multiple a } 5)$ $\frac{1}{6}$

The students at Job's high school were surveyed to determine their favorite foods. The results are shown in the table at the right. Suppose students were randomly selected and asked what their favorite food is. Find the probability of each event. Write as a fraction in simplest form.

Favorite Food	Responses
pizza	19
steak	8
chow mein	5
seafood	4
spaghetti	3
cereal	1

11. $P(\text{steak})$ $\frac{1}{5}$

12. $P(\text{spaghetti})$ $\frac{3}{40}$

13. $P(\text{cereal or seafood})$ $\frac{1}{8}$

14. $P(\text{not chow mein})$ $\frac{7}{8}$

15. $P(\text{pizza})$ $\frac{19}{40}$

16. $P(\text{cereal or steak})$ $\frac{9}{40}$

17. $P(\text{not steak})$ $\frac{4}{5}$

18. $P(\text{not cereal or seafood})$ $\frac{7}{8}$

19. $P(\text{chicken})$ 0

20. $P(\text{chow mein or spaghetti})$ $\frac{1}{5}$

Practice: Word Problems, p. 487

NAME ________ DATE ________ PERIOD ____

9-1 Practice: Word Problems

Simple Events

COINS Susan opened her piggy bank and counted the number of each coin. The table at the right shows the results. For Exercises 1–3, assume that the coins are put in a bag and one is chosen at random.

Coin	Number
quarters	15
dimes	21
nickels	22
pennies	32

1. What is the probability that a quarter is chosen? $\frac{1}{6}$

2. What is the probability that a nickel or a dime is chosen? $\frac{43}{90}$

3. What is the probability that the chosen coin is worth more than 5 cents? $\frac{2}{5}$

4. **NUMBER CUBES** Juan has two number cubes, each with faces numbered 1, 2, ...6. What is the probability that he can roll the cubes so that the sum of the faces showing equals 11? $\frac{1}{18}$

5. **SKATEBOARDS** Carlotta bought a new skateboard for which the probability of having a defective wheel is 0.015. What is the probability of not having a defective wheel? 0.985

6. **CALCULATORS** Jake's teacher had 6 calculators for 28 students to use. If the first students to use the calculators are chosen at random, what is the probability that Jake will get one? $\frac{3}{14}$

7. **VEHICLES** The rental car company had 14 sedans and 8 minivans available to rent. If the next customer picks a vehicle at random, what is the probability that a minivan is chosen? $\frac{4}{11}$

8. **MUSIC** Tina has 16 pop CDs, 6 classical, and 2 rock. Tina chooses a CD at random. What is the probability she does not choose a classical CD? $\frac{3}{4}$

9-2 Lesson Notes

Lesson Resources

Chapter 9 Resource Masters
- Study Guide and Intervention, p. 490
- Practice: Skills, p. 491
- Practice: Word Problems, p. 492
- Reading to Learn Mathematics, p. 493
- Enrichment, p. 494

Calculator and Spreadsheet Masters, p. 43

Parent and Student Study Guide Workbook, 9-2

Real-Life Career Masters, p. 23

FAST FILE

1 Focus

5-Minute Check 9-2

INTERACTIVE CHALKBOARD
PowerPoint® Presentations

Use as a quiz or review of Lesson 9-1. (Also available as a transparency)

Bellringer

Problem Solving Ask students to make a table listing all the possible outcomes of three coin tosses. Have students explain their table's format and how it displays the possible outcomes.

9-2 Tree Diagrams

What You'll Learn
Use tree diagrams to count outcomes and find probabilities.

NEW Vocabulary
- fair game
- tree diagram
- sample space

HANDS-ON Mini Lab

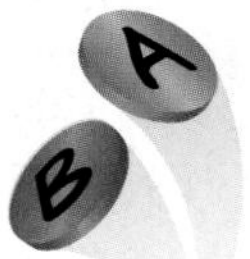

Materials
- 2 counters
- marker

Work with a partner.

Here is a probability game that you can play with two counters.

- Mark one side of the first counter A. Mark the other side B. Mark both sides of the second counter A.
- Player 1 tosses the counters. If both sides shown are the same, Player 1 wins a point. If the sides are different, Player 2 wins a point. Record your results.
- Player 2 then tosses the counters and the results are recorded. Continue alternating the tosses until each player has tossed the counters ten times. The player with the most points wins.

1. Before you play, make a conjecture. Do you think that each player has an equal chance of winning? Explain. **See students' work**
2. Now, play the game. Who won? What was the final score?
3. Collect the data from the entire class. What is the combined score for Player 1 versus Player 2?
4. Do you want to change the conjecture you made in Exercise 1? Explain. **See students' work.**

2. Note: After 20 tosses, the games may be tied.
3. Note: The number of wins for each player will be about the same.

A game in which players of equal skill have an equal chance of winning is a **fair game**. One way you can analyze whether games are fair is by drawing a **tree diagram**. A tree diagram is used to show all of the possible outcomes, or **sample space**, in a probability experiment.

TEACHING TIP
Tell students that the outcome BA means the first counter came up B and the second counter came up A.

EXAMPLE Draw a Tree Diagram

1 GAMES Refer to the Mini Lab above. Draw a tree diagram to show the sample space. Then determine whether the game is fair.

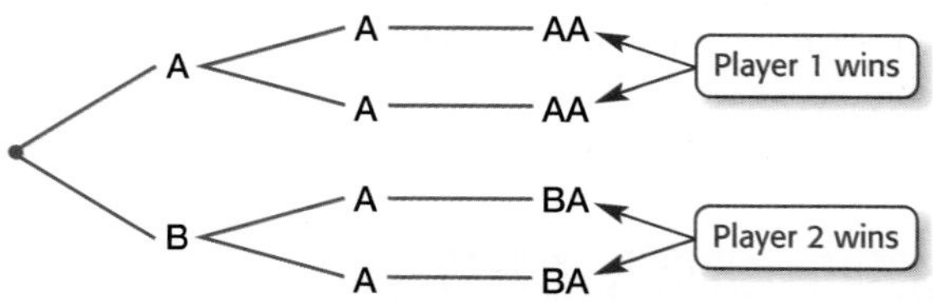

There are four equally-likely outcomes with two favoring each player. So, the probability that each player can win is $\frac{1}{2}$. Thus, the game is fair.

Reading to Learn Mathematics, p. 493 ELL

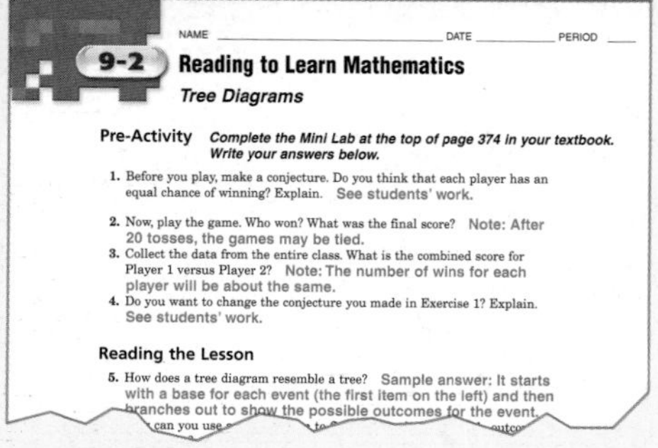
NAME ____________ DATE ________ PERIOD ____

9-2 **Reading to Learn Mathematics**
Tree Diagrams

Pre-Activity *Complete the Mini Lab at the top of page 374 in your textbook. Write your answers below.*

1. Before you play, make a conjecture. Do you think that each player has an equal chance of winning? Explain. See students' work.
2. Now, play the game. Who won? What was the final score? Note: After 20 tosses, the games may be tied.
3. Collect the data from the entire class. What is the combined score for Player 1 versus Player 2? Note: The number of wins for each player will be about the same.
4. Do you want to change the conjecture you made in Exercise 1? Explain. See students' work.

Reading the Lesson

5. How does a tree diagram resemble a tree? Sample answer: It starts with a base for each event (the first item on the left) and then branches out to show the possible outcomes for the event.

EXAMPLE Find the Number of Outcomes

EAL-LIFE MATH

OOTERS Kickboard oters can travel roximately 4 miles per ir.

ce: www.factmonster.com

2 SCOOTERS A certain type of kickboard scooter comes in silver, red, or purple with wheel sizes of 125 millimeters or 180 millimeters. Find the total number of color-wheel size combinations.

Make a tree diagram to show the sample space.

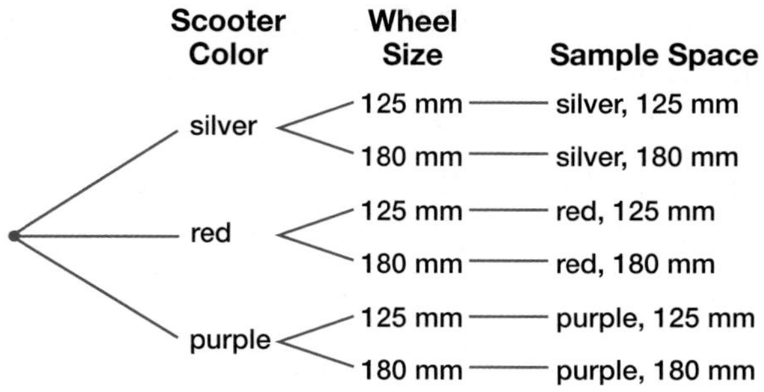

There are six different color and wheel size combinations.

Your Turn

a. How many outcomes are there if the manufacturer added an orange color and a 150-millimeter wheel size? 12

You can also use tree diagrams to help you find the probability of events.

EXAMPLE Find Probability Using Tree Diagrams

3 COINS Suppose Pablo tosses three pennies. Find the probability that all three will show heads.

Make a tree diagram to show the sample space. Then, find the probability of the three pennies showing heads.

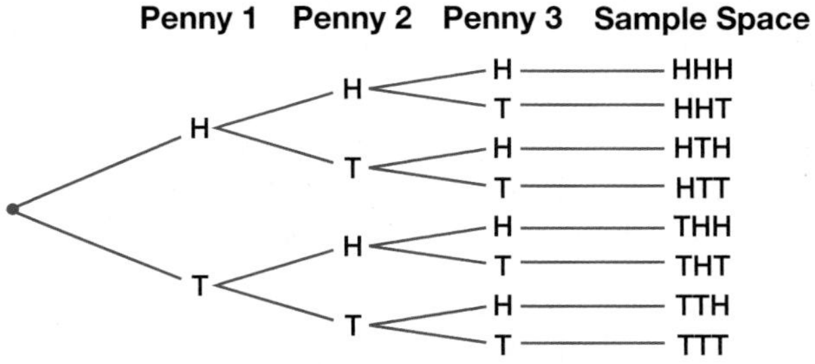

The sample space contains 8 possible outcomes. Only 1 outcome has all pennies showing heads. So, the probability of three pennies showing heads is $\frac{1}{8}$, or 12.5%.

Your Turn **Find each probability.**

b. P(3 tails) $\frac{1}{8}$ c. P(exactly 2 heads) $\frac{3}{8}$ d. P(at least 1 tail) $\frac{7}{8}$

2 Teach

In-Class Examples

INTERACTIVE CHALKBOARD PowerPoint® Presentations

1 CHILDREN A family has two children. Draw a tree diagram to show the sample space of the children's genders. Then determine the probability of the family having two girls. $\frac{1}{4}$

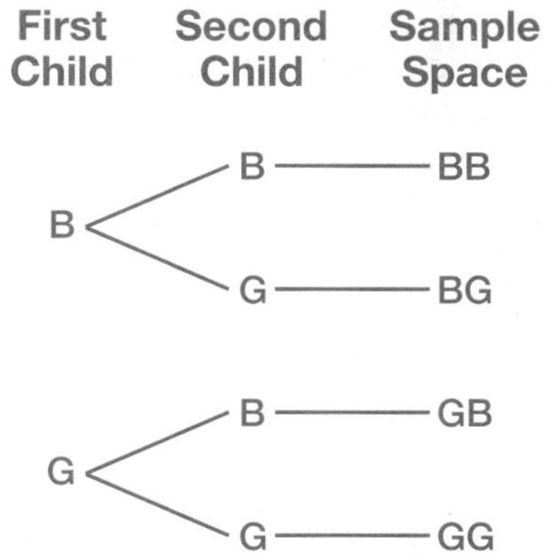

2 ICE CREAM An ice cream sundae at the Ice Cream Shoppe is made from one flavor of ice cream and one topping. For ice cream flavors, you can choose from chocolate, vanilla, and strawberry. For toppings you can have hot fudge, butterscotch, or marshmallow. Find the number of different sundaes that are possible. 9

3 ICE CREAM If you are given a sundae at random from the shop above, what is the probability that it has vanilla ice cream? $\frac{1}{3}$

DAILY INTERVENTION Unlocking Misconceptions

Repeated Events When dealing with repeated events, having students think of one event as first and one as second helps them conceptualize the sample space. However, they should understand that the events could be simultaneous without affecting the sample space. For example, the tree diagram for one number cube rolled twice is the same as the tree diagram for two number cubes rolled at the same time.

3 Practice/Apply

About the Exercises...

Guided Practice Key

Exercises	Examples
3, 6	1
4	2
5	3

Odd/Even Assignments
Exercises 7–14 are structured so that students practice the same concepts whether they are assigned odd or even problems.

Alert! Exercise 20 requires the use of the Internet or other reference materials.

4 Assess

Higher-Level Thinking
Tell students to create tree diagrams showing the possible outcomes of one, two, and three coin tosses. Then challenge them to write a general rule describing the possible results for n coin tosses. 2^n

Answers

1. Sample game: Each player tosses a coin 10 times. If it comes up heads, player 1 receives 1 point. If it comes up tails, player 2 receives 1 point. The player with the most points at the end of 10 tosses wins.
2. Sample answer: Lunch at a fast-food restaurant comes in a choice of a hamburger, chicken sandwich, fish sandwich, or hot dog with French fries or potato chips.

Skill and Concept Check

Writing Math
Exercises 1 & 2

1. **Describe** a fair game between two players using one coin.
1–2. See margin.
2. **OPEN ENDED** Give an example of a situation that has 8 outcomes.

GUIDED PRACTICE

The spinner is spun twice.

3. Draw a tree diagram to represent the situation. See Ch. 9 Answer Appendix.
4. How many outcomes are possible? 16
5. What is the probability of spinning two Cs? $\frac{1}{16}$
6. **DELICATESSEN** A neighborhood deli sells sandwiches that can be made with ham, turkey, roast beef, salami, or bologna on rye, white, pumpernickel, or sourdough breads. Make a tree diagram to show all of the possible meat-bread choices. See Ch. 9 Answer Appendix.

★ indicates increased difficulty

Practice and Applications

HOMEWORK HELP	
For Exercises	See Examples
7–15, 18, 21	1, 2
16–17, 19	3

Extra Practice
See pages 585, 604.

For each situation, make a tree diagram to show the sample space. Then give the total number of outcomes. 7–14. See Ch. 9 Answer Appendix for tree diagrams.

7. tossing a coin and rolling a number cube 12
8. choosing black, blue, or brown socks with boots, gym shoes, or dress shoes 9
9. picking a number from 1 to 5 and choosing the color red, white, or blue 15
10. choosing a card with a shape and spinning the spinner from the choices at the right 10

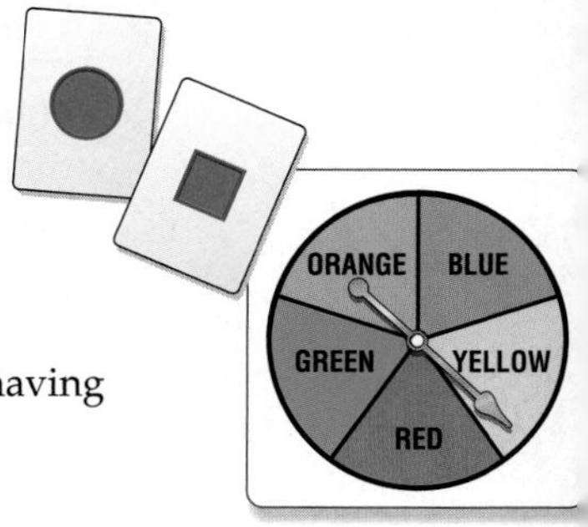

11. choosing a letter from the word SPACE and choosing a consonant from the word MATH 15
12. choosing a purple, green, black, or silver mountain bike having 10, 18, 21, or 24 speeds 16
13. ★ tossing a quarter, a dime, and a nickel 8
14. ★ rolling a number cube, tossing a coin, and choosing a card from the cards marked A and B 24

FAMILY For Exercises 15–17, use the information below.
Mr. and Mrs. Chen have three children. Suppose the chance of having either a boy or a girl is 50%.

15. Draw a tree diagram showing the possible arrangements of gender of the three children. See Ch. 9 Answer Appendix.
16. What is the probability of having 2 boys and 1 girl? $\frac{3}{8}$
17. What is the probability of having the first two children boys and the last child a girl? $\frac{1}{8}$

UNIFORMS For Exercises 18 and 19, use the information below and at the right.

Hunter is a big fan of the Houston Astros baseball team and wears a different jersey and cap every time he goes to a game. The table shows the number of different jerseys and caps Hunter owns.

	Home Jersey	Road Jersey	Practice Jersey	Cap
White	2	0	2	0
Black	1	0	0	1
Gray	0	1	0	0
Orange	1	1	0	2

18. How many jersey/cap combinations can Hunter wear when he goes to the baseball game to cheer on the Astros? 24

★ 19. If Hunter picks a jersey/cap combination at random, what is the probability that he will wear a home jersey with a black cap? $\frac{4}{24}$ or $\frac{1}{6}$

20. **RESEARCH** Use the Internet or another resource to find the number of jerseys and caps your favorite major league baseball team has as part of its uniform. How many jersey/cap combinations are there for the team you picked? See students' work.

★ 21. **GAMES** Determine whether the following game for two players is fair. Explain. See margin.

	Side 1	Side 2
Counter 1	A	B
Counter 2	A	C
Counter 3	B	C

- Three counters are labeled according to the table at the right.
- Toss the three counters.
- If exactly 2 counters match, Player 1 scores a point. Otherwise, Player 2 scores a point.

22. **CRITICAL THINKING** Refer to Exercise 21. Adjust the scoring of the game so that it is fair. See margin.

Standardized Test Practice and Mixed Review

23. **MULTIPLE CHOICE** How many three-course dinners are possible if Yolanda chooses from an appetizer of soup or salad; an entree of steak, chicken, or fish; and a dessert of cake or pie? C

   (A) 7 (B) 10 (C) 12 (D) 15

24. **SHORT RESPONSE** Find the sample space of the tree diagram. short-sleeve gray, short-sleeve white, long-sleeve gray, long-sleeve white

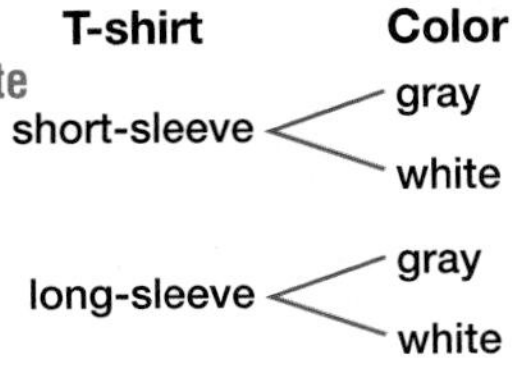

PROBABILITY A spinner is equally likely to stop on each of its regions numbered 1 to 20. Find each probability as a fraction in simplest form. (Lesson 9-1)

25. a prime number $\frac{2}{5}$ 26. GCF(12, 18) $\frac{1}{20}$ 27. multiple of 2 or 3 $\frac{13}{20}$

28. **INTEREST** If Carlota invests \$2,100 in a CD for 5 years at a simple interest rate of 4.75%, how much will the CD be worth after 5 years? (Lesson 8-7) \$2,598.75

GETTING READY FOR THE NEXT LESSON

BASIC SKILL Multiply.

29. 7 · 22 154 30. 5 · 36 180 31. 11 · 16 176 32. 23 · 20 460 33. 131 · 4 524

msmath2.net/self_check_quiz

Answers

21. $P(\text{Player 1}) = \frac{6}{8}$ or $\frac{3}{4}$; $P(\text{Player 2}) = \frac{2}{8}$ or $\frac{1}{4}$; Player 2 does not have an equal chance of winning.

22. Sample answer: New scoring for a fair game—if exactly 2 counters match, Player 1 scores one point. Otherwise, Player 2 scores three points.

Study Guide and Intervention, p. 490

NAME ______ DATE ______ PERIOD ____

9-2 Study Guide and Intervention

Tree Diagrams

A game in which players of equal skill have an equal chance of winning is a **fair game**. A **tree diagram** is used to show all of the possible outcomes, or **sample space**, in a probability experiment.

EXAMPLE 1 WATCHES **A certain type of watch comes in brown or black and in a small or large size. Find the number of color-size combinations that are possible.**

Make a tree diagram to show the sample space. Then give the total number of outcomes.

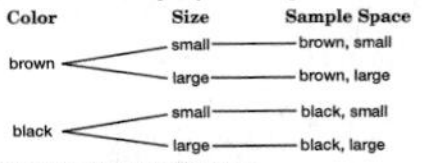

There are four different color and size combinations.

EXAMPLE 2 CHILDREN **The chance of having either a boy or a girl is 50%. What is the probability of the Smith's having two girls?**

Make a tree diagram to show the sample space. Then find the probability of having two girls.

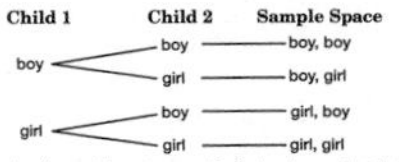

The sample space contains 4 possible outcomes. Only 1 outcome has both children being girls. So, the probability of having two girls is $\frac{1}{4}$.

EXERCISES

For each situation, make a tree diagram to show the sample space. Then give the total number of outcomes.

1. choosing an outfit from a green shirt, blue shirt, or a red shirt, and black pants or blue pants See students' work. There are 6 outcomes.
2. choosing a vowel from the word COUNTING and a consonant from the word PRIME See students' work. There are 9 outcomes.

Practice: Skills, p. 491

NAME ______ DATE ______ PERIOD ____

9-2 Practice: Skills

Tree Diagrams

The spinner at the right is spun twice.

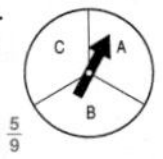

1. Draw a tree diagram to represent the situation.

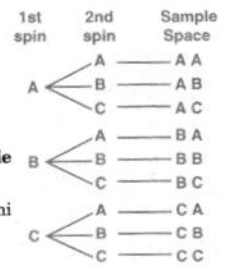

2. What is the probability of getting at least one A? $\frac{5}{9}$

For each situation, make a tree diagram to show the sample space. Then give the total number of outcomes.

3. choosing a hamburger or hot dog and potato salad or macaroni salad

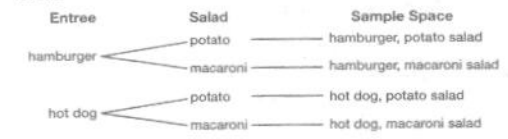

There are 4 possible outcomes.

4. choosing a vowel from the word COMPUTER and a consonant from the word BOOK

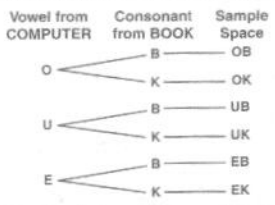

There are 6 possible outcomes.

5. choosing between the numbers 1, 2 or 3, and the colors blue, red, or green

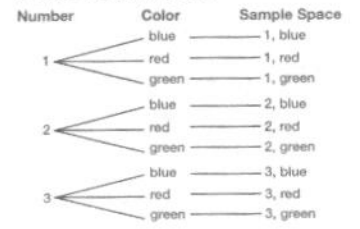

There are 9 possible outcomes.

Practice: Word Problems, p. 492

NAME ______ DATE ______ PERIOD ____

9-2 Practice: Word Problems

Tree Diagrams

1. **GASOLINE** Craig stops at a gas station to fill his gas tank. He must choose between full-service or self-service and between regular, midgrade, and premium gasoline. Draw a tree diagram showing the possible combinations of service and gasoline type. How many possible combinations are there?

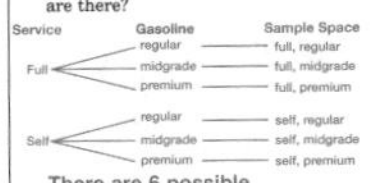

There are 6 possible combinations.

2. **COINS** Judy tosses a coin 4 times. Draw a tree diagram showing the possible outcomes. What is the probability of getting at least 2 tails? $\frac{11}{16}$

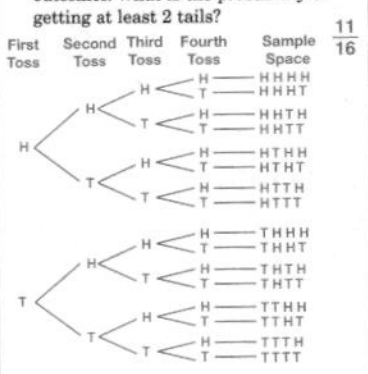

3. **COINS** In Exercise 2, what is the probability of getting 2 heads, then 2 tails? $\frac{1}{16}$

4. **EQUIPMENT** The computer accessory that Joanne is considering selling at her store comes in white, beige, gray, or black and as an optical mouse, mechanical mouse, or trackball. How many combinations of color and model must she stock in order to have at least one of every possible combination? 12

9-3 Lesson Notes

Lesson Resources

Chapter 9 Resource Masters
- Study Guide and Intervention, p. 495
- Practice: Skills, p. 496
- Practice: Word Problems, p. 497
- Reading to Learn Mathematics, p. 498
- Enrichment, p. 499

Key Teaching Transparencies, 34

Parent and Student Study Guide Workbook, 9-3

FAST FILE

1 Focus

5-Minute Check 9-3
Use as a quiz or review of Lesson 9-2. (Also available as a transparency)

INTERACTIVE CHALKBOARD
PowerPoint® Presentations

Bellringer

Hands-On Activity Tell students that counters for a game consist of squares, circles, or triangles in the colors of red, green, blue, or yellow. Have students draw all possible counters. Ask them to draw a tree diagram to verify their work.

Reading to Learn Mathematics, p. 498 ELL

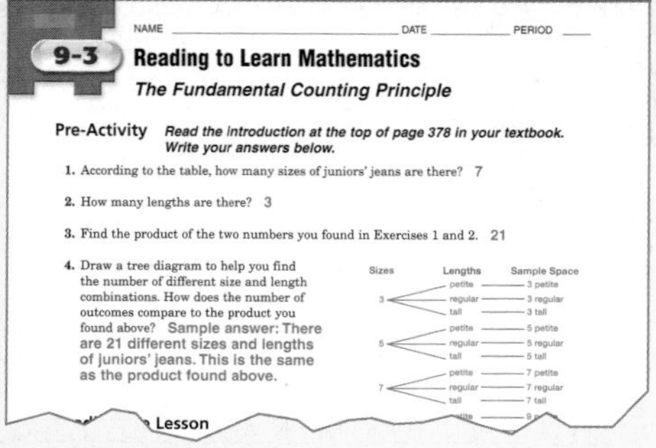
NAME ______ DATE ______ PERIOD ____

9-3 **Reading to Learn Mathematics**
The Fundamental Counting Principle

Pre-Activity *Read the introduction at the top of page 378 in your textbook. Write your answers below.*

1. According to the table, how many sizes of juniors' jeans are there? 7
2. How many lengths are there? 3
3. Find the product of the two numbers you found in Exercises 1 and 2. 21
4. Draw a tree diagram to help you find the number of different size and length combinations. How does the number of outcomes compare to the product you found above? Sample answer: There are 21 different sizes and lengths of juniors' jeans. This is the same as the product found above.

Lesson

9-3 The Fundamental Counting Principle

What You'll Learn
Use multiplication to count outcomes.

NEW Vocabulary
- Fundamental Counting Principle

WHEN am I ever going to use this?

RETAIL SALES The Jean Factory sells juniors' jeans in different sizes and lengths. The table shows what they have available.

Sizes	Lengths
3	petite
5	regular
7	tall
9	
11	
13	
15	

1. According to the table, how many sizes of juniors' jeans are there? 7
2. How many lengths are there? 3
3. Find the product of the two numbers you found in Exercises 1 and 2. 21
4. Draw a tree diagram to help you find the number of different size and length combinations. How does the number of outcomes compare to the product you found above? See Ch. 9 Answer Appendix.

In the activity above, you discovered that multiplication, instead of a tree diagram, can be used to find the number of possible outcomes in a sample space. This is called the **Fundamental Counting Principle**.

Key Concept **Fundamental Counting Principle**

If event M can occur in m ways and is followed by event N that can occur in n ways, then the event M followed by N can occur in $m \times n$ ways.

EXAMPLE **Use the Fundamental Counting Principle**

1 FOOD A famous steak house allows customers to create their own steak dinners. The choices are shown at the right. How many different steak dinners are possible?

451 Steakhouse

Steak	How Steaks Are Cooked	Potatoes
New York Strip	rare	mashed
Ribeye	medium	baked
Filet	well	twice baked
Porterhouse		au gratin
T-Bone		

$$\underbrace{5}_{\text{types of steaks}} \cdot \underbrace{3}_{\text{number of ways steaks can be cooked}} \cdot \underbrace{4}_{\text{types of potatoes}} = \underbrace{60}_{\text{total number of steak dinners}}$$

There are 60 different ways of choosing a steak dinner.

Check You can check your work by drawing a tree diagram and listing the 60 outcomes.

DAILY INTERVENTION **Flexible Grouping**

Fundamental Counting Principle Some students may be comfortable with tree diagrams and small numbers but may have difficulty applying the Fundamental Counting Principle to problems with larger numbers. Have them work with stronger students to solve problems with more difficult numbers.

378 Chapter 9 Probability

Standardized Test Practice

EXAMPLE Find Outcomes

2 MULTIPLE-CHOICE TEST ITEM There are 2 roads connecting Eastland and Harping, 3 roads connecting Harping and Sinclair, and 2 roads connecting Sinclair and Johnstown. Find the number of ways to drive from Eastland to Johnstown.

Ⓐ 8 Ⓑ 12 Ⓒ 16 Ⓓ 20

Read the Test Item
To find the number of ways to drive from Eastland to Johnstown, multiply the number of roads from Eastland to Harping, from Harping to Sinclair, and from Sinclair to Johnstown.

Solve the Test Item
There are 2 roads from Eastland to Harping, 3 roads from Harping to Sinclair, and 2 roads from Sinclair to Johnstown. So, there are $2 \cdot 3 \cdot 2$, or 12 ways to drive from Eastland to Johnstown. So the answer is B.

est-Taking Tip The Princeton Review

ducated Guess Find out there is a penalty for correct answers. If there no penalty, making an ducated guess can only crease your score, or worst, leave your score he same.

Skill and Concept Check

Writing Math Exercises 1–4

1. **Explain** how to use the Fundamental Counting Principle to count outcomes. **See margin.**
2. **OPEN ENDED** Give an example of a situation that has 48 outcomes using three events. **Sample answer: a sweatshirt that comes in 4 colors, 4 sizes, and 3 designs**
3. **NUMBER SENSE** Can a pair of events have a number of outcomes that is prime? Explain. **See margin.**
4. **Which One Doesn't Belong?** Identify the pair of events that do not have the same number of outcomes as the other three. Explain your reasoning.

12 meats, 8 breads	16 T-shirts, 6 colors
18 hats, 8 sizes	24 teams, 4 sports

18 hats, 8 sizes have 144 outcomes; the others have 96 outcomes

GUIDED PRACTICE

Use the Fundamental Counting Principle to find the total number of outcomes in each situation.

5. tossing a quarter, a dime, and a nickel **8**
6. choosing scrambled, sunny-side up, or poached eggs with bacon or sausage and milk, orange juice, or apple juice **18**
7. choosing a card from twenty different cards and spinning the spinner at the right **60**

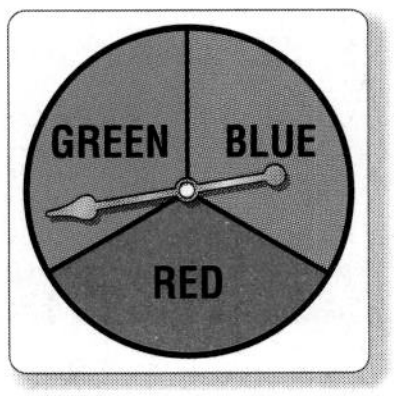

8. **STUDENT GOVERNMENT** If Opal, Jacob, Luanda, Tyrone, and Erica have an equal chance of being an officer of student council, what is the probability that Luanda will be an officer? $\frac{1}{5}$

Answers

1. Sample answer: Multiply the number of ways each event can occur to find the total number of outcomes.
3. Sample answer: Yes, a pair of events can have a number of outcomes that is prime. For example, 11 different pairs of shoes can come in 1 color— $11 \cdot 1$, or 11 outcomes.

2 Teach

In-Class Examples

INTERACTIVE CHALKBOARD PowerPoint® Presentations

1 CLOTHING The table below shows the shirts, shorts, and shoes in Gerry's wardrobe. How many possible outfits can he choose consisting of one shirt, one pair of shorts, and one pair of shoes? **24**

Shirts	Shorts	Shoes
red	beige	black
blue	green	brown
white	blue	
yellow		

2 MULTIPLE-CHOICE TEST ITEM An orchestra has one opening for a violinist, one opening for a cellist, and one opening for an oboist. Three musicians are trying out for violin, five for cello, and three for oboe. Find the number of ways the openings can be filled. **D**

A 9 **B** 11
C 15 **D** 45

3 Practice/Apply

Which One Doesn't Belong?
In Exercise 4, point out to students that they are looking for the pair of numbers that gives a unique answer when multiplied together. Two of the pairs involve multiplying by 8 (8 breads, 8 sizes), so the answer must be one of these pairs.

About the Exercises...

Guided Practice Key

Exercises	Examples
5–8	1, 2

Odd/Even Assignments
Exercises 9–16 are structured so that students practice the same concepts whether they are assigned odd or even problems.

Study Guide and Intervention, p. 495

NAME ______________ DATE ________ PERIOD ____

9-3 Study Guide and Intervention

The Fundamental Counting Principle

If event M can occur in m ways and is followed by event N that can occur in n ways, then the event M followed by N can occur in $m \times n$ ways. This is called the **Fundamental Counting Principle.**

EXAMPLE 1 CLOTHING **Andy has 5 shirts, 3 pairs of pants, and 6 pairs of socks. How many different outfits can Andy choose with a shirt, pair of pants, and pair of socks?**

number of shirts		number of pants		number of socks		total number of outfits
5	·	3	·	6	=	90

Andy can choose 90 different outfits.

EXERCISES

Use the Fundamental Counting Principle to find the total number of outcomes in each situation.

1. rolling two number cubes 36
2. tossing 3 coins 8
3. picking one consonant and one vowel 105
4. choosing one of 3 processor speeds, 2 sizes of memory, and 4 sizes of hard drive 24
5. choosing a 4-, 6-, or 8-cylinder engine and 2- or 4-wheel drive 6
6. rolling 2 number cubes and tossing 2 coins 144
7. choosing a color from 4 colors and a number from 4 to 10 28

Practice: Skills, p. 496

NAME ______________ DATE ________ PERIOD ____

9-3 Practice: Skills

The Fundamental Counting Principle

Use the Fundamental Counting Principle to find the total number of outcomes in each situation.

1. rolling two number cubes and tossing one coin 72
2. choosing rye or Bermuda grass and 3 different mixtures of fertilizer 6
3. making a sandwich with ham, turkey, or roast beef; Swiss or provolone cheese; and mustard or mayonaisse 12
4. tossing 4 coins 16
5. choosing from 3 sizes of distilled, filtered, or spring water 9
6. choosing from 3 flavors of juice and 3 sizes 9
7. choosing from 35 flavors of ice cream; one, two, or three scoops; and sugar or waffle cone 210
8. picking a day of the week and a date in the month of April 210
9. rolling 3 number cubes and tossing 2 coins 864
10. choosing a 4-letter password using only vowels 625
11. choosing a bicycle with or without shock absorbers; with or without lights; and 5 color choices 20
12. a license plate that has 3 numbers from 0 to 9 and 2 letters 676,000

Practice: Word Problems, p. 497

NAME ______________ DATE ________ PERIOD ____

9-3 Practice: Word Problems

The Fundamental Counting Principle

1. SURFBOARD Jay owns 3 surfboards and 2 wetsuits. If he takes one surfboard and one wetsuit to the beach, how many different combinations can he choose? 6	2. SHOPPING John is trying to decide which bag of dog food to buy. The brand he wants comes in 4 flavors and 3 sizes. How many choices are there? 12
3. LOTTERY To purchase a lottery ticket, you must select 4 numbers from 0 to 9. How many possible lottery tickets can be chosen? 10,000	4. RESTAURANTS Miriam's favorite restaurant has 3 specials every day. Each special has 2 choices of vegetable and 3 choices of dessert. How many different meals could Miriam have? 18
5. ROUTES When Sunil goes to the building where he works, he can go through 4 different doors into the lobby. Then he can go to the seventh floor by taking 2 different elevators or 2 different stairways. How many different ways can Sunil get from outside the building to the seventh floor? 16	6. STEREOS Jailin went to her local stereo store. Given her budget and the available selection, she can choose between 2 CD players, 5 amplifiers, and 3 pairs of speakers. How many different stereos can Jailin purchase? 30
7. DESSERT For dessert you can choose apple, cherry, blueberry, or peach pie to eat, and milk or juice to drink. How many different combinations can you choose from? 8	8. TESTS John is taking a true or false quiz. There are six questions on the quiz. How many ways can the quiz be answered? 64

★ indicates increased difficulty

Practice and Applications

HOMEWORK HELP	
For Exercises	See Examples
9–14	1
15–16	2

Extra Practice
See pages 585, 604.

Use the Fundamental Counting Principle to find the total number of outcomes in each situation.

9. rolling a number cube and tossing two coins 24
10. choosing a number from 1 to 20 and a color from 7 colors 140
11. choosing a plain, blueberry, garlic, or cinnamon-and-raisin bagel, with plain, chive, or sun-dried tomato cream cheese 12
12. choosing iced tea in regular, raspberry, lemon, or peach flavors; sweetened or unsweetened; and in a glass or in a plastic container 16
13. picking a date in the month of May and a day of the week 217
14. choosing a 4-digit Personal Identification Number (PIN) if the digits cannot be repeated 5,040
15. **ADVERTISING** The Wake-Up Restaurant advertises that you can have a different pancake breakfast every day of the year. It offers 25 different kinds of pancakes and 14 flavored syrups. If the restaurant is open every day of the year, is its claim valid? Explain.
No; the number of selections is 350, which is less than 365.

★ 16. **GAMES** A player rolls five number cubes to score the maximum number of points. Find the number of outcomes possible in one roll. 7,776

CRITICAL THINKING How many outcomes are possible if you toss each of the following?

17. one coin 2
18. two coins 4
19. three coins 8
20. n coins 2^n

Standardized Test Practice and Mixed Review

21. **MULTIPLE CHOICE** Elizabeth has 3 sweaters, 5 blouses, and 6 skirts that coordinate. How many different outfits can Elizabeth make? B

(A) 105 (B) 90 (C) 45 (D) 30

22. **GRID IN** WritePen makes 8 different styles of pens in several colors with 2 types of grips. If the company makes 112 kinds of pens, how many different colors do they make? 7

PROBABILITY What is the probability that the spinner shown at the right will stop on each of the following numbers? Write as a fraction in simplest form. (Lesson 9-1)

23. an even number $\frac{1}{2}$
24. a multiple of 4 $\frac{1}{4}$
25. a number less than 10 $\frac{3}{4}$
26. **SCHOOL** Gustavo must choose from two geography classes, three history classes, and two statistics classes for next year. Make a tree diagram to show all of the possible schedules he can arrange. (Lesson 9-2)
See Ch. 9 Answer Appendix.

GETTING READY FOR THE NEXT LESSON

BASIC SKILL Multiply.

27. $3 \cdot 2 \cdot 1$ 6
28. $9 \cdot 8 \cdot 7$ 504
29. $5 \cdot 4 \cdot 3 \cdot 2$ 120
30. $7 \cdot 6 \cdot 5 \cdot 4$ 840

msmath2.net/self_check_quiz

4 Assess

Open-Ended Assessment

Speaking Solve a problem on the chalkboard using a tree diagram and the Fundamental Counting Principle. Ask students to explain each procedure and to explain why tree diagrams are not practical for many problems.

9-4 Permutations

What You'll Learn
nd the number of ermutations of a set objects.

EW Vocabulary
ermutation
ctorial

ATH Symbols
four factorial

HANDS-ON Mini Lab

Materials
- 3 index cards

Work with a partner.

How many different ways are there to arrange your first three classes if they are math, science, and language arts?

STEP 1 Write math, science, and language arts on the index cards.

STEP 2 Choose one of the subjects as the first class of the day. Choose one of the remaining two subjects for the second class. The third class is the card that remains.

MATH	SCIENCE	LANGUAGE ARTS

STEP 3 Record this arrangement of classes.

STEP 4 Change the order of the classes. Record this arrangement. Continue rearranging the cards until you have found all of the possible arrangements.

1. When you first started to make your list, how many choices did you have for your first class? **3**
2. Once your first class was selected, how many choices did you have for the second class? Then, the third class? **2; 1**
3. Explain how you can use the Fundamental Counting Principle to find the number of arrangements. **$3 \times 2 \times 1 = 6$**

A **permutation** is an arrangement, or listing, of objects in which order is important. You can use the Fundamental Counting Principle to find the number of possible arrangements.

There are **3** choices for the 1st class.
There are **2** choices that remain for the 2nd class.
There is **1** choice that remains for the 3rd class.

$3 \cdot 2 \cdot 1 = 6$ ← The number of arrangements of 3 classes.

The expression $3 \cdot 2 \cdot 1$ can be written as 3!, which is read *three factorial.*

Key Concept — **Factorial**

The expression *n* **factorial** (*n*!) is the product of all counting numbers beginning with *n* and counting backward to 1.

TEACHING TIP
xplain to students hat they are using the undamental Counting rinciple when finding ermutations.

DAILY INTERVENTION — **Differentiated Instruction**

Visual/Spatial Encourage students to model the simpler permutations presented in this lesson. This can involve standing in different orders and counting those orders. It could also involve using different colored counters to represent different elements. As students see more and more of these patterns represented visually, they will develop a better conceptual understanding of permutations.

9-4 Lesson Notes

Lesson Resources

Chapter 9 Resource Masters
- Study Guide and Intervention, p. 500
- Practice: Skills, p. 501
- Practice: Word Problems, p. 502
- Reading to Learn Mathematics, p. 503
- Enrichment, p. 504

Calculator and Spreadsheet Masters, p. 44

Parent and Student Study Guide Workbook, 9-4

Teaching Mathematics with Manipulatives, pp. 81–82

FAST FILE

1 Focus

5-Minute Check 9-4

INTERACTIVE CHALKBOARD
PowerPoint® Presentations

Use as a quiz or review of Lesson 9-3. (Also available as a transparency)

Bellringer

Communication Have students explain how many ways a president and a vice-president can be chosen from a group of four club members. Students can explain using words, pictures, or a list. **There are 12 ways to choose a president and vice-president from a group of 4.**

Reading to Learn Mathematics, p. 503 **ELL**

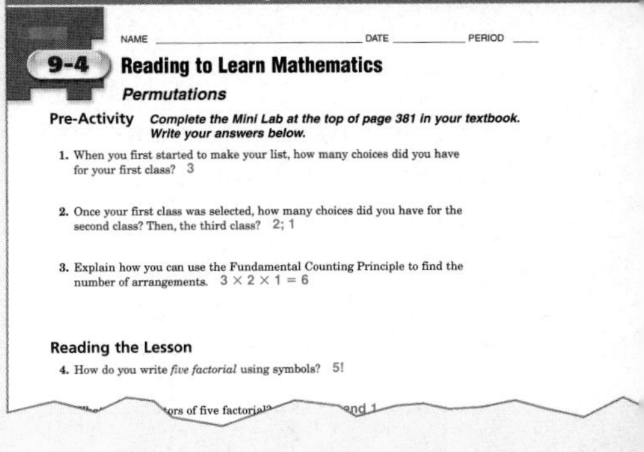
NAME ________ DATE ______ PERIOD ____

9-4 Reading to Learn Mathematics
Permutations

Pre-Activity *Complete the Mini Lab at the top of page 381 in your textbook. Write your answers below.*

1. When you first started to make your list, how many choices did you have for your first class? 3
2. Once your first class was selected, how many choices did you have for the second class? Then, the third class? 2; 1
3. Explain how you can use the Fundamental Counting Principle to find the number of arrangements. $3 \times 2 \times 1 = 6$

Reading the Lesson

4. How do you write *five factorial* using symbols? 5!

2 Teach

In-Class Examples

INTERACTIVE CHALKBOARD
PowerPoint® Presentations

Find the value of each expression.

1. 4! **24**
2. 3! · 5! **720**
3. **BOWLING** A team of bowlers has five members who bowl one at a time. In how many orders can they bowl? **120**
4. **RAFFLE** A school fair holds a raffle with 1st, 2nd, and 3rd prizes. Seven people enter the raffle. How many ways can the three prizes be awarded? **210**

Tips for New Teachers

Using Applications After students understand the concepts in this lesson, introduce them to the factorial and permutation functions on graphing calculators.

3 Practice/Apply

About the Exercises...

Guided Practice Key

Exercises	Examples
3, 4	1, 2
5, 6	3, 4

Odd/Even Assignments
Exercises 7–16 are structured so that students practice the same concepts whether they are assigned odd or even problems.

Alert! For Exercises 7 through 20, students will need a scientific calculator.

Alert! Students will need to access the Internet for the Data Update after Exercise 19.

STUDY TIP

Technology You can find the factorial of a number on a calculator. To find 3!, press 3 [PRB] [▶] [▶] [ENTER] [ENTER].

EXAMPLES Evaluate Factorials

Find the value of each expression.

1 **7!**

$7! = 7 \cdot 6 \cdot 5 \cdot 4 \cdot 3 \cdot 2 \cdot 1$ Definition of factorial

$= 5{,}040$ Simplify.

2 **2! · 3!**

$2! \cdot 3! = 2 \cdot 1 \cdot 3 \cdot 2 \cdot 1$ Definition of factorial

$= 12$ Simplify.

EXAMPLE Find a Permutation

3 **VOLLEYBALL In how many ways can the starting six players of a volleyball team stand in a row for a picture?**

This is a permutation that can be written as 6!.

$6! = 6 \cdot 5 \cdot 4 \cdot 3 \cdot 2 \cdot 1$ Definition of factorial

$= 720$ Simplify.

So, there are 720 ways the six starting players can stand in a row.

REAL-LIFE MATH

SWIMMING In 1996, swimmer Amy Van Dyken was the first American woman to win 4 gold medals at one Olympic Games.

Source: sportsillustrated.cnn.com

Arrangements can include all or only part of a group.

EXAMPLE Find a Permutation

4 **SWIMMING The finals of the Middle School Appalachian League features 8 swimmers. In how many ways can the swimmers finish in first or second place?**

There are 8 choices for first place and 7 choices that remain for second place. So, there are 8 · 7 or 56 choices for first and second place.

Skill and Concept Check

Writing Math
Exercise 1

1. **Explain** the difference between 5! and 5 · 4 · 3. **1–2. See margin.**
2. **OPEN ENDED** Describe a real-life situation that has 6 permutations.

GUIDED PRACTICE

Find the value of each expression.

3. 3! **6**
4. 6! · 2! **1,440**
5. In how many ways can you arrange the letters in the word *equal*? **120**
6. **TRANSPORTATION** There are 7 students waiting at the bus stop. In how many ways can the students board the bus when it arrives? **5,040**

382 Chapter 9 Probability

msmath2.net/extra_examples

Answers

1. Sample answer: Five factorial has extra factors of 2 and 1, 5! = 5 · 4 · 3 · 2 · 1.
2. Sample answer: the number of ways you can order 3 books on a shelf

★ indicates increased difficulty

Practice and Applications

HOMEWORK HELP	
For Exercises	See Examples
7–14	1, 2
16	3
15, 17–18	4

Extra Practice See pages 585, 604.

Find the value of each expression.

7. 5! **120**
8. 9! **362,880**
9. 4! · 3! **144**
10. four factorial **24**
11. 3! · 6! **4,320**
12. 10 · 9 · 8 **720**
13. 5! · 4! **2,880**
14. 8! · 2! **80,640**

15. In how many ways can a softball manager arrange the first four batters in a lineup of nine players? **3,024**

16. How many different 5-digit zip codes are there if no digit is repeated? **30,240**

17. **MUSIC** The chromatic scale has 12 notes. In how many ways can a song start with 4 different notes from that scale? **11,880**

DOGS For Exercises 18 and 19, use the information below and at the right.

During the annual Westminster Dog Show, the best dog in each breed competes to win one of four top ribbons in the group.

2002 Westminster Dog Show

Group	Number of Breeds
Herding	19
Hounds	25
Non-sporting	18
Sporting	27
Terriers	27
Toy	22
Working	21

Source: westminsterkennelclub.org

18. In how many ways can a ribbon be awarded to a breed of dog in the Working group? **143,640**

★ 19. The top dog in each group competes against the other six group winners for Best of Show. If each dog has an equally-likely chance of winning Best of Show, what is the probability that a terrier will win? $\frac{1}{7}$

Data Update Which breed of dog has won the most Best of Shows? Visit msmath2.net/data_update to learn more.

20. **CRITICAL THINKING** There are 1,320 ways for three students to win first, second, and third place during a debate match. How many students are there on the debate team? **12**

Standardized Test Practice and Mixed Review

21. **SHORT RESPONSE** In a Battle of the Bands contest, how many ways can the four participating bands be ordered? **24**

22. **GRID IN** How many different three-digit security codes can be made from the digits 1, 2, 3, 4, and 5 if no digit is repeated in a code? **60**

23. **BREAKFAST** Find the total number of outcomes if you can choose from 8 kinds of muffins, 3 sizes, and 4 beverages. (Lesson 9-3) **96**

24. **LUNCH** Make a tree diagram showing different ways to make a sandwich with turkey, ham, or salami and either cheddar or Swiss cheese. (Lesson 9-2) **See Ch. 9 Answer Appendix.**

GETTING READY FOR THE NEXT LESSON

PREREQUISITE SKILL Find each value. (Lesson 5-3)

25. $\frac{5 \cdot 4}{2 \cdot 1}$ **10**
26. $\frac{8 \cdot 7 \cdot 6}{3 \cdot 2 \cdot 1}$ **56**
27. $\frac{5 \cdot 4 \cdot 3}{4 \cdot 3 \cdot 2}$ **2.5**
28. $\frac{10 \cdot 9 \cdot 8 \cdot 7}{8 \cdot 7 \cdot 6 \cdot 5}$ **3**

msmath2.net/self_check_quiz

4 Assess

Higher-Level Thinking

When 3-digit area codes for telephone numbers were first developed, the following rules applied.

1. The first digit could NOT be 0 or 1.
2. The second digit must be 0 or 1.
3. The third digit could be any digit, 0 to 9.

How many possible area codes were there under this system? **160**

Study Guide and Intervention, p. 500

NAME ______ DATE ______ PERIOD ____

9-4 Study Guide and Intervention
Permutations

The expression *n* **factorial** (*n*!) is the product of all counting numbers beginning with *n* and counting backward to 1. A **permutation** is an arrangement, or listing, of objects in which order is important. You can use the Fundamental Counting Principle to find the number of possible arrangements.

EXAMPLE 1 Find the value of 5!.

5! = 5 · 4 · 3 · 2 · 1 Definition of factorial
= 120 Simplify.

EXAMPLE 2 Find the value of 4! · 2!.

4! · 2! = 4 · 3 · 2 · 1 · 2 · 1 Definition of factorial
= 48 Simplify.

EXAMPLE 3 BOOKS How many ways can 4 different books be arranged on a bookshelf?

This is a permutation that can be written as 4!. Suppose the books are placed on the shelf from left to right.

There are 4 choices for the first book.
There are 3 choices that remain for the second book.
There are 2 choices that remain for the third book.
There is 1 choice that remains for the fourth book.

4! = 4 · 3 · 2 · 1 Definition of factorial
= 24 Simplify.

So, there are 24 ways to arrange 4 different books on a bookshelf.

EXERCISES

Find the value of each expression.

1. 3! 6
2. seven factorial 5,040
3. 6! · 3! 4,320
4. 9 · 8 · 7 504
5. How many ways can you arrange the letters in the word GROUP? 120
6. How many different 4-digit numbers can be created if no digit can be repeated? Remember, a number cannot begin with 0. 4,536

Practice: Skills, p. 501

NAME ______ DATE ______ PERIOD ____

9-4 Practice: Skills
Permutations

Find the value of each expression.

1. 2! 2
2. 4! 24
3. 3! · 5! 720
4. nine factorial 362,880
5. 2! · 8! 80,640
6. 3! · 2! 12
7. 11 · 10 · 9 990
8. 10! 3,628,800
9. 5! · 2! 240
10. 5 · 4 · 3 · 2 120
11. 8! · 6! 29,030,400
12. 6! · 4! 17,280
13. How many ways can you arrange the letters in the word PRIME? 120
14. How many ways can you arrange 8 different crates on a shelf if they are placed from left to right? 40,320

Practice: Word Problems, p. 502

NAME ______ DATE ______

9-4 Practice: Word Problems
Permutations

1. AREA CODES How many different 3-digit area codes can be created if no digit can be repeated? 720	2. CARDS Jason is dealt five playing cards. In how many different orders could Jason have been dealt the same hand? 120
3. PASSWORDS How many different 3-letter passwords are possible if no letter may be repeated? 15,600	4. RACING All 22 students in Amy's class are going to run the 100-meter dash. In how many ways can the students finish in first, second, and third place? 9,240
5. LETTERS How many ways can you arrange the letters in the word HISTORY? 5,040	6. PARKING The parking lot for a company has three parking spaces for compact cars. The company has 8 employees with compact cars. How many ways can the compact parking spaces be filled? 336
7. SERIAL NUMBERS How many different 6-digit serial numbers are available if no digit can be repeated? 151,200	8. WINNERS There are 156 ways for 2 cars to win first and second place in a race. How many cars are in the race? 13

Chapter 9 Mid-Chapter Practice Test

Assessment Option

Mid-Chapter Test (Lessons 9-1 through 9-4) is available on p. 537 of the *Chapter 9 Resource Masters.*

ExamView® Pro Test Bank CD-ROM

Use the networkable ExamView® Pro Test Bank CD-ROM to:

- Create **Multiple versions** of tests.
- Create **Modified** tests for **Inclusion** students with one mouse click.
- **Edit** existing questions and **Add** your own questions.
- Build tests aligned with state standards using built-in **State Curriculum Correlations.**
- Change **English** tests to **Spanish** with one mouse click and vice versa.

FOLDABLES™ Follow-Up Before completing the Mid-Chapter Practice Test, encourage students to review the notes, definitions, concepts, and main ideas they recorded on the first four tabs of their Foldables.

Answers

1. Sample answer: Each outcome is equally likely to occur.
2. A method for finding the number of ways that two or more events can occur by multiplying the number of ways that each event can occur.

Chapter 9 Mid-Chapter Practice Test

Vocabulary and Concepts

1. **Explain** what it means for an event to occur at random. (Lesson 9-1)
2. **State** the Fundamental Counting Principle. (Lesson 9-3) **1–2. See margin.**
3. **Define** *permutation.* (Lesson 9-4) **Sample answer: An arrangement in which order is important.**

Skills and Applications

Two number cubes are rolled. Find each probability. (Lesson 9-1)

4. $P(\text{sum of } 6)$ $\frac{5}{36}$
5. $P(\text{sum less than } 7)$ $\frac{5}{12}$
6. $P(\text{sum of } 7 \text{ or } 11)$ $\frac{2}{9}$

For each situation, make a tree diagram to show the sample space. Then give the total number of outcomes. (Lesson 9-2) **7–8. See Ch. 9 Answer Appendix for tree diagrams.**

7. tossing a penny and tossing a dime **4**
8. choosing cereal, French toast, or pancakes and choosing orange, apple, cranberry or grapefruit juice **12**

Use the Fundamental Counting Principle to find the total number of outcomes in each situation. (Lesson 9-3)

Entrée	Salad	Dessert
ham	potato	ice cream
beef	tossed	pie
turkey	cole slaw	cake

9. choosing a dinner with one entrée, one salad, and one dessert **27**
10. rolling a number cube and tossing three coins **48**

Find the value of each expression. (Lesson 9-4)

11. 5! **120**
12. 8! **40,320**
13. $2! \cdot 7!$ **10,080**

14. **STUDENT COUNCIL** In how many ways can a president, treasurer, and a secretary be chosen from among 8 candidates? (Lesson 9-4) **336 ways**

Standardized Test Practice

15. **GRID IN** Jeffrey has 2 action, 3 comedy, and 4 drama DVDs. If he randomly picks one DVD to watch, what is the probability that it will be a comedy? (Lesson 9-1) **1/3**
16. **SHORT RESPONSE** Molly has three windows to display three of five best-selling books. How many different displays can she make if she puts one book in each window? (Lesson 9-4) **60**

384 Chapter 9 Probability

The Game Zone

A Place To Practice Your Math Skills

Cherokee Butterbean Game

GET READY!

Players: two, three, or four
Materials: 6 dry lima beans, bowl, marker

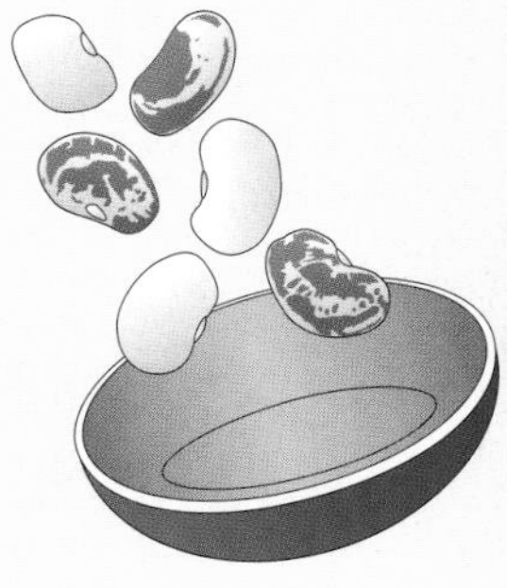

GET SET!

- This is a variation of a traditional Cherokee game.
- Color one side of each bean with the marker.

GO!

- The first player places the beans in the bowl, gently tosses the beans into the air, and catches them in the bowl. Points are scored as follows.
  - If all of the beans land with the marked or unmarked sides up, score 6 points.
  - If exactly one bean lands with the marked or unmarked side up, score 3 points.
  - If three beans are marked and three beans are unmarked, score 1 point.
- If a toss scores points, the player takes another turn. If a toss does not score any points, it is the next player's turn.
- **Who Wins?** The person with the most points after a given number of rounds is the winner.

The Game Zone

GETTING READY

Objective Use experimental probability.

MANAGEMENT TIPS

Recommended Time
20 minutes

Tips

- Have students use the Fundamental Counting Principle to determine the probability of all the beans landing with the marked sides up and the probability of all beans landing with the unmarked sides up. **Both probabilities are $\frac{1}{64}$.**
- Point out that, unlike a fair coin, beans are irregularly shaped and may not have an equal chance of landing on each side. Have students calculate the probability for each outcome of a toss of two beans. **both marked = $\frac{1}{4}$; both unmarked = $\frac{1}{4}$; one marked, one unmarked = $\frac{1}{2}$** Then have them perform fifty repetitions of a two-bean toss, recording the result of each repetition. Compare students' results with the theoretical calculation and discuss whether the beans are fair beans. Consider giving extra points to students who are willing to calculate the total number of scoring outcomes in the six-bean game. **At this point, students must use a tree diagram to total the scoring outcomes. Thirty-four tosses score, while thirty tosses do not score.**

(Continued in bottom margin.)

- Suggest that students use a table to keep track of their scores for each round. Be sure to tell how many rounds to play before play begins.
- You may wish to use the game as an assessment tool for the concepts the students have been learning in Lessons 9-1 through 9-4.

For an additional game for this chapter, visit: msmath2.net/gamezone.

9-5a HANDS-ON LAB
Lesson Notes

Lesson Resources

Teaching Mathematics with Manipulatives
- student recording sheet 9-5a

GETTING STARTED

Teaching Tip You may want to pair students of weaker math ability with students of stronger math ability. This will help weaker students understand the concept while giving the stronger students a teaching opportunity.

TEACH

- As students make their lists of topping combinations, encourage them to list any duplicate combinations (same toppings, different order) on a separate sheet of paper. This strategy will form a permutations list, and it will give students a visual reference for comparing and contrasting permutations and combinations in Exercise 4.
- As an extension, have students compute the number of combinations and permutations possible if a sundae has three toppings.

ASSESS

Observe as students solve the **Writing Math** exercises. They should:
- use the index cards to answer each question correctly, and
- be able to distinguish between permutations and combinations.

Discuss Exercise 5 in the **Writing Math** section. Students should be able to use their previous answers to prove their conclusion.

9-5a HANDS-ON LAB

A Preview of Lesson 9-5

What You'll Learn
Find combinations.

Materials
- 5 index cards

Exploring Combinations

INVESTIGATE *Work with a partner.*

The student safety club at Mahomet Middle School is planning to sell sundaes with two different toppings at the summer carnival. The choices are shown at the right.

Mahomet Middle School Safety Club Sundae Toppings:	
hot fudge	nuts
cherry	caramel

STEP 1 Write the name of the four sundae toppings on the index cards.

STEP 2 To make a sundae, select any pair of cards. Make a list of all the different combinations that are possible. Note that the order of the toppings is not important.

Cherry and Nuts

is considered the same as

Nuts and Cherry

Writing Math

Work with a partner.

1. How many different combinations are possible with two toppings?
2. How many different sundaes could be made with two toppings if the order of the toppings *was* important? What is this type of arrangement called? **12; permutation**
3. **Write** a fifth topping on an index card. Now find all two-topping sundae combinations. How many are there? How many sundaes can be made with two toppings if the order of the toppings was important? **10; 20**
4. **Compare and contrast** the way you can find a permutation with the way you find a combination. **See margin.**
5. **Make a conjecture** about how to find a combination given a permutation.

5. Sample answer: To find a combination, divide the permutation by the number of ways to arrange the items.

Answer

4. Sample answer: Both permutations and combinations are an arrangement of items in a group. In a permutation, you find all the different arrangements of a set of objects where order of the objects is important. In a combination, you find all the different arrangements of a set of objects where order of the objects is not important.

9-5 Combinations

/hat You'll Learn
nd the number of
›mbinations of a set
objects.

EW Vocabulary
›mbination

LINK To Reading
/eryday Meaning
combination: the
sult of putting
gether objects, as
a combination of
gredients

WHEN am I ever going to use this?

BASKETBALL Coach Chávez wants to select co-captains for her basketball team. She will select two girls from the four oldest members on the team: Alita, Bailey, Charmaine, and Danielle.

1. Use the first letter of each name to list all of the permutations of co-captains. How many are there?
2. Cross out any arrangement that contains the same letters as another one in the list. How many are there now? **6**
3. Explain the difference between the two lists above. **See margin.**

1. AB, AC, AD, BA, BC, BD, CA, CB, CD, DA, DB, DC; 12

An arrangement, or listing, of objects in which order is *not* important is called a **combination**. For example, in the activity above, choosing Alita and Bailey is the same as choosing Bailey and Alita.

Permutations and combinations are related. You can find the number of combinations of objects by dividing the number of permutations of the entire set by the number of ways each smaller set can be arranged.

A permutation of 4 players, taken 2 at a time.

$$\frac{4 \cdot 3}{2!} = \frac{4 \cdot 3}{2 \cdot 1} = \frac{12}{2} = 6$$

There are 2! ways to arrange 2 players.

EXAMPLE Find the Number of Combinations

1 FOOD Paul's Pizza Parlor is offering a large two-topping pizza for \$14.99. There are five toppings from which to choose. How many different two-topping pizzas are possible?

Method 1 Make a list.

The five toppings are labeled pepperoni (p), sausage (s), onions (o), mushrooms (m), and green pepper (g).

p, s	p, m	p, o	p, g	s, m
s, o	s, g	m, o	m, g	o, g

Method 2 Use a permutation.

There are $5 \cdot 4$ permutations of two toppings chosen from five.

There are 2! ways to arrange the two toppings.

$$\frac{5 \cdot 4}{2!} = \frac{20}{2} = 10$$

So, there are 10 different two-topping pizzas.

TEACHING TIP
:xplain to students that
o find a combination,
hey have to divide the
ermutation by the
umber of ways they
an arrange the items.

msmath2.net/extra_examples

Answer (When am I ever going to use this?)

3. The list in Exercise 1 is a permutation where order is important. The list in Exercise 2 does not take order into account.

9-5 Lesson Notes

Lesson Resources

Chapter 9 Resource Masters
- Study Guide and Intervention, p. 505
- Practice: Skills, p. 506
- Practice: Word Problems, p. 507
- Reading to Learn Mathematics, p. 508
- Enrichment, p. 509

Calculator and Spreadsheet Masters, p. 45
Key Teaching Transparencies, 35
Parent and Student Study Guide Workbook, 9-5

FAST FILE

1 Focus

5-Minute Check 9-5
Use as a quiz or review of Lesson 9-4. (Also available as a transparency)

INTERACTIVE CHALKBOARD
PowerPoint® Presentations

Bellringer
Communication Discuss the difference between ordered and nonordered arrangements. Have students describe two situations in which order of selection is important and two situations in which it is not. **Sample answer: Order matters when you are looking at the order of finish in a contest and when you are assigning actors to play roles; order does not matter when you are selecting pizza toppings or when you are selecting flowers for a floral arrangement.**

Reading to Learn Mathematics, p. 508 ELL

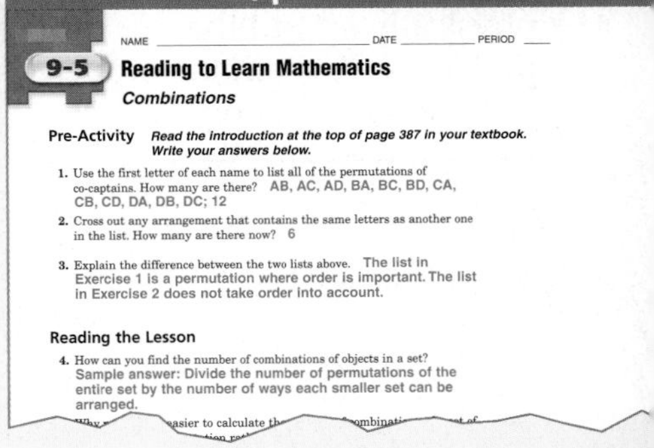
NAME ______ DATE ______ PERIOD ____

9-5 **Reading to Learn Mathematics**
Combinations

Pre-Activity *Read the introduction at the top of page 387 in your textbook. Write your answers below.*

1. Use the first letter of each name to list all of the permutations of co-captains. How many are there? AB, AC, AD, BA, BC, BD, CA, CB, CD, DA, DB, DC; 12
2. Cross out any arrangement that contains the same letters as another one in the list. How many are there now? 6
3. Explain the difference between the two lists above. The list in Exercise 1 is a permutation where order is important. The list in Exercise 2 does not take order into account.

Reading the Lesson

4. How can you find the number of combinations of objects in a set? Sample answer: Divide the number of permutations of the entire set by the number of ways each smaller set can be arranged.

2 Teach

In-Class Examples

INTERACTIVE CHALKBOARD
PowerPoint® Presentations

1 DECORATING Ada can select from seven paint colors for her room. She wants to choose two colors. How many different pairs of colors can she choose? **21 pairs**

2 INTRODUCTIONS Ten managers attend a business meeting. Each person exchanges names with each other person once. How many introductions will there be? **45 introductions**

Tell whether each situation represents a *permutation* or *combination*. Then solve the problem.

3 TRACK From an eight-member track team, three members will be selected to represent the team at the state meet. How many ways can these three members be selected? **combination; 56 ways**

4 In how many ways can you choose the first, second, and third runners in a relay race from the eight members of the track team above? **permutation; 336 ways**

Tips for New Teachers

Teaching Strategies To help students understand complex subject matter such as combinations and permutations, have students physically act out the problems. To maintain order, you direct the action.

EXAMPLE Use a Combination to Solve a Problem

2 CHECKERS A checkers tournament features each of the top 8 regional players playing every opponent one time. The 2 players with the best records will then play in a final round to determine the champion. How many matches will be played if there are no ties?

Find the number of ways 2 players can be chosen from a group of 8.

There are 8 · 7 ways to choose 2 people. →
There are 2! ways to arrange 2 people. →

$$\frac{8 \cdot 7}{2!} = \frac{56}{2} = 28$$

There are 28 matches plus 1 final match to determine the champion. So, there will be 29 matches played.

Check Make a diagram in which each person is represented by a point. Draw line segments between two points to represent the games. There are 28 line segments. Then add the final-round match to make a total of 29 matches.

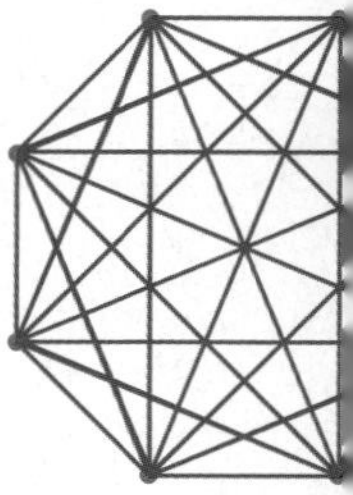

Your Turn

a. How many matches will be played if the top 16 players were invited to play? **121**

REAL-LIFE MATH

CHECKERS Checkers is played on an 8-by-8 board in the U.S., England, Australia, and Ireland. People in eastern Europe play on a 10-by-10 board.

Source: www.about.com

The difference between permutations and combinations is that order is important in permutations, while order is *not* important in combinations.

EXAMPLES Identify Permutations and Combinations

Tell whether each situation represents a *permutation* or *combination*. Then solve the problem.

3 STUDENT GOVERNMENT **The six students listed at the right are members of Student Council. How many ways can you choose a president, vice president, and treasurer from this group?**

STUDENT COUNCIL **BALLOT**

- Marissa ☐
- Santos ☐
- Paige ☐
- Travis ☐
- Sareeta ☐
- Kenji ☐

This is a permutation because the order of president, vice president, and treasurer is important. So, the number of ways you can choose the three officers is 6 · 5 · 4, or 120 ways.

4 In how many ways can you choose a committee of three students from the six members in student council shown above?

This is a combination because the order of the students in the committee is not important.

There are 6 · 5 · 4 ways to choose 3 people. →
There are 3! ways to arrange 3 people. →

$$\frac{6 \cdot 5 \cdot 4}{3!} = \frac{120}{6} = 20$$

So, there are 20 ways to choose the committee.

DAILY INTERVENTION — Unlocking Misconceptions

Combinations and Permutations Some students might misconstrue a *combination* as indicating a situation where order matters, such as a combination lock. Tell them that the word *permutation* comes from Latin words meaning thoroughly changed. The word *combination* means a collection of items that have been brought together. Discuss with students ways to use these definitions to help them distinguish permutations from combinations.

Skill and Concept Check

Writing Math
Exercises 1 & 3

1. **Explain** why a combination lock should be called a permutation lock. **Sample answer: The order of the numbers is important in a lock.**
2. **OPEN ENDED** Give an example of a permutation and combination. **2–3. See margin.**
3. **FIND THE ERROR** Allison and Francisca are calculating the number of ways that a 3-member committee can be chosen from a 7-member club. Who is correct? Explain.

Allison
$7 \cdot 6 \cdot 5 = 210$ ways

Francisca
$\frac{7 \cdot 6 \cdot 5}{3!} = \frac{210}{6} = 35$ ways

GUIDED PRACTICE

4. **VOLLEYBALL** Coach Malone has an 8-member volleyball team. He told his team that he would start six different players every game. How many games would it take to do this? **28**

Tell whether each problem represents a *permutation* or *combination*. Then solve the problem.

5. How many ways can 10 students finish first, second, or third at the science fair? **permutation; 720 ways**
6. How many ways can you pick 2 puppies from a litter of 7 puppies? **combination; 21 ways**

Practice and Applications

HOMEWORK HELP

For Exercises	See Examples
7–10	1, 2
11–16	3, 4

Extra Practice
See pages 586, 604.

7. **FOOD** The International Club is selling hot dogs at the Spring Carnival. Customers can select three toppings from among chili, onions, cheese, mustard, or relish. How many combinations of three-topping hot dogs are there? **10**

CIVICS For Exercises 8 and 9, use the information below and at the right.
If five of the nine Justices on the 2002 United States Supreme Court agree on a decision, they can issue a majority opinion.

8. How many different combinations of five Supreme Court Justices are there? **126**
9. Before they take the Bench each day, the Justices engage in the "Conference handshake." Each Justice shakes hands with each of the other eight. How many handshakes take place? **36**

U.S. Supreme Court, 2002	
Chief Justice, William H. Rehnquist	
Stephan G. Breyer	Antonin Scalia
Ruth Bader Ginsburg	David H. Souter
Anthony M. Kennedy	John Paul Stevens
Sandra Day O'Connor	Clarence Thomas

Source: www.washingtonpost.com

10. **SOCCER** There are 21 players trying out for 15 spots on the soccer team. How many ways does the coach have to create her team? **54,264**

msmath2.net/self_check_quiz

3 Practice/Apply

FOLDABLES™ Follow-Up Remind students to use their Foldables to record what they learn about combinations. Point out that they might describe how they found the number of possible combinations of something in their own life. Students might compare and contrast combinations and permutations.

DAILY INTERVENTION FIND THE ERROR Ask students to determine whether order is important in Exercise 3. By determining that order is not important, they should identify which calculation represents a combination.

About the Exercises...

Guided Practice Key

Exercises	Examples
4	2
5	3
6	4

Odd/Even Assignments
Exercises 7–16 are structured so that students practice the same concepts whether they are assigned odd or even problems.

Answers

2. Sample answer: permutation: the number of ways the first three runners can finish a race with eight runners; combination: the number of ways three runners can be chosen as team captains on a team of eight runners
3. Francisca; Allison found the number of permutations of three people chosen from a group of seven.

Tell whether each problem represents a *permutation* or *combination*. Then solve the problem.

11. How many ways can you select four essay questions out of a total of 10 on the exam? **combination; 210**
12. Six children remain in a game of musical chairs. If two chairs are removed, how many different groups of four students can remain? **combination; 15**
13. How many ways can three flute players be seated in the first, second, or third seats in the orchestra? **permutation; 6 ways**
14. In how many ways can four paintings be displayed from a collection of 15? **combination; 1,365**
15. How many ways can seven students line up to buy concert tickets? **permutation; 5,040**
16. Given 12 Web sites, how many ways can you visit half of them? **combination; 924**
17. **WRITE A PROBLEM** Write about a real-life situation that can be solved using a permutation and one that can be solved using a combination. Then solve both problems. **See students' work.**
18. **CRITICAL THINKING** At a party, there were 105 handshakes. If each person shook hands exactly once with every other person, how many people were at the party? **15**

Standardized Test Practice and Mixed Review

19. **MULTIPLE CHOICE** In how many ways can three gymnasts from a team of 10 be chosen to compete in a meet? **A**

    (A) 120 (B) 180 (C) 240 (D) 720

20. **SHORT RESPONSE** Find $\frac{(n+6)(n+5)(n+4)}{n!}$ if $n = 3$. **84**
21. **SAILING** When Mr. Elms purchased his sailboat, it came with six different-colored flags to be used for sending signals. A specific signal depended on the order of the flags. How many different three-flag signals can he send? (Lesson 9-4) **120 signals**
22. **CARS** A certain brand of car has a choice of a 2.5, 3.1, or 4.0 liter engine; a radio with a cassette-player, CD-player, or CD-changer; cloth or leather seats; and silver, green, red, or white. How many different cars are possible? (Lesson 9-3) **72**

Estimate. (Lesson 6-1) **23–26. Sample answers are given.**

23. $\frac{1}{10} + \frac{7}{8}$ **$0 + 1 = 1$** 24. $\frac{5}{12} - \frac{1}{9}$ **$\frac{1}{2} - 0 = \frac{1}{2}$** 25. $\frac{4}{9} \cdot 20$ **$\frac{1}{2} \cdot 20 = 10$** 26. $15\frac{7}{9} \div 3\frac{3}{5}$ **$16 \div 4 =$**

GETTING READY FOR THE NEXT LESSON

PREREQUISITE SKILL Suppose you choose a card from a set of eight cards labeled 1–8. Write the probability of each event as a fraction in simplest form. (Lesson 9-1)

27. P(even number) **$\frac{1}{2}$**
28. P(number greater than 6) **$\frac{1}{4}$**
29. P(multiple of 8) **$\frac{1}{8}$**
30. P(number less than 9) **1**
31. P(prime number) **$\frac{1}{2}$**
32. P(*not* 2 or 5) **$\frac{3}{4}$**

4 Assess

Open-Ended Assessment

Writing Ask students to write a paragraph describing the differences between combinations and permutations. Have them include one specific example of each in their descriptions.

Study Guide and Intervention, p. 505

NAME ______________ DATE ________ PERIOD ____

9-5 Study Guide and Intervention

Combinations

An arrangement, or listing, of objects in which order is *not* important is called a **combination**. You can find the number of combinations of objects by dividing the number of permutations of the entire set by the number of ways each smaller set can be arranged.

EXAMPLE 1 **Jill was asked by her teacher to choose 3 topics from the 8 topics given to her. How many different three-topic groups could she choose?**

There are $8 \cdot 7 \cdot 6$ permutations of three-topic groups chosen from eight. There are 3! ways to arrange the groups.

$\frac{8 \cdot 7 \cdot 6}{3!} = \frac{336}{6} = 56$

So, there are 56 different three-topic groups.

EXAMPLES **Tell whether each situation represents a *permutation* or *combination*. Then solve the problem.**

2 **On a quiz, you are allowed to answer any 4 out of the 6 questions. How many ways can you choose the questions?**

This is a combination because the order of the 4 questions is not important. So, there are $6 \cdot 5 \cdot 4 \cdot 3$ permutations of four questions chosen from six. There are 4! or $4 \cdot 3 \cdot 2 \cdot 1$ orders in which these questions can be chosen.

$\frac{6 \cdot 5 \cdot 4 \cdot 3}{4!} = \frac{360}{24} = 15$

So, there are 15 ways to choose the questions.

3 **Five different cars enter a parking lot with only 3 empty spaces. How many ways can these spaces be filled?**

This is a permutation because each arrangement of the same 3 cars counts as a distinct arrrangement. So, there are $5 \cdot 4 \cdot 3$ or 60 ways the spaces can be filled.

EXERCISES

Tell whether each situation represents a *permutation* or *combination*. Then solve the problem.

1. How many ways can 4 people be chosen from a group of 11? combination; 330
2. How many ways can 3 people sit in 4 chairs? permutation; 24
3. How many ways can 2 goldfish be chosen from a tank containing 15 goldfish? combination; 105

Practice: Skills, p. 506

NAME ______________ DATE ________ PERIOD ____

9-5 Practice: Skills

Combinations

Tell whether each situation represents a *permutation* or *combination*. Then solve the problem.

1. You are allowed to omit two out of 12 questions on a quiz. How many ways can you select the questions to omit? combination; 66
2. Six students are to be chosen from a class of 18 to represent the class at a math contest. How many ways can the six students be chosen? combination; 18,564
3. How many different 5-digit zip codes are possible if no digits are repeated? permutation; 30,240
4. In a race with six runners, how many ways can the runners finish first, second, or third? permutation; 120
5. How many ways can two names be chosen from 76 in a raffle if only one entry per person is allowed? combination; 2,850
6. How many ways can six students be arranged in a lunch line? permutation; 720
7. A family has a bike rack that fits seven bikes but they only have five bikes. How many ways can the bikes fit in the bike rack? permutation; 2,520
8. How many ways can you select three sheriff deputies from eight candidates? combination; 56
9. How many ways can four finalists be selected from 50 contestants? combination; 230,300
10. How many 4-digit pin numbers are available if no number is repeated? permutation; 5,040
11. How many handshakes can occur between five people if everyone shakes hands? combination; 10

Practice: Word Problems, p. 507

NAME ______________ DATE ________ PERIOD ____

9-5 Practice: Word Problems

Combinations

1. **SNACKS** A vending machine can display six snacks. If there are eight different kinds of snacks available, how many different groups of six different snacks can be displayed? 28
2. **MUSIC** Each month, Jose purchases two CDs from a selection of 20 bestselling CDs. How many different pairs of CDs can Jose choose if he chooses two different CDs? 190
3. **TESTS** On a math test, you can choose any 20 out of 23 questions. How many different groups of 20 questions can you choose? 1,771
4. **RESTAURANTS** The dinner special at a local pizza parlor gives you the choice of two toppings from a selection of six toppings. How many different choices are possible if two different toppings are chosen? 15
5. **TESTING** In a science fair experiment, two units are selected for testing from every 500 units produced. How many ways can these two units be selected? 124,750
6. **MEETINGS** Linda's teacher divided the class into groups of five and required each member of a group to meet with every other member of that group. How many meetings will each group have? 10
7. **BASEBALL** A baseball coach has 13 players to fill nine positions. How many different teams could he put together? 715
8. **GEOMETRY** Ten points are marked on a circle. How many different triangles can be drawn between any three points? 120

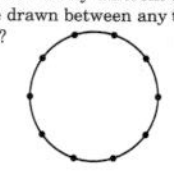

390 Chapter 9 Probability

9-6a Problem-Solving Strategy

A Preview of Lesson 9-6

hat You'll Learn
lve problems by
ting them out.

Act It Out

I wonder if tossing a coin would be a good way to answer a 5-question true-false quiz?

I'm not so sure, Whitney. Let's do an experiment with a coin and **act it out**!

Explore We know that there are five true-false questions on the quiz. We can carry out an experiment to test whether tossing a coin would be a good way to answer the questions and get a good grade.

Plan Let's toss a coin five times. If the coin shows tails, we will answer T. If the coin shows heads, we will answer F. Let's do four trials.

Solve Suppose the correct answers are T, F, F, T, F. Let's circle them.

Answers	T	F	F	T	F	Number Correct
Trial 1	Ⓣ	T	Ⓕ	F	T	2
Trial 2	F	Ⓕ	T	Ⓣ	Ⓕ	3
Trial 3	Ⓣ	Ⓕ	T	F	T	2
Trial 4	F	Ⓕ	T	F	Ⓕ	2

The experiment produces 2 or 3 correct answers.

Examine There is an equally likely chance that tossing a coin will produce a correct answer or a wrong answer for each question. Since the experiment produced about 2–3 correct answers on a 5-question quiz, it shows that tossing a coin to answer a true-false quiz is not the way to get a good grade.

nalyze the Strategy

Explain whether the results of the experiment would be the same if it were repeated. **Sample answer: Results would vary slightly.**

Explain an advantage of using the act it out strategy to solve a problem. **See margin.**

9-6a Lesson Notes

1 Focus

Bellringer

Hands-On Activity Have pairs of students place 50 play coins in a shoe box with a cover on top. Ask them what the probability is that a coin will be faceup after shaking the box. $\frac{1}{2}$ Have them shake their boxes and predict how many coins will land faceup. **Students should predict 25 coins.** Then have them remove the lids and count the coins landing faceup. Record all of the actual results on the chalkboard and find the median result. Ask students whether their predictions came close to the median.

2 Teach

In-Class Example

INTERACTIVE CHALKBOARD
PowerPoint® Presentations

Solve. Use the act it out strategy.

Salvador is looking for his lunch money, which he put in one of the pockets of his backpack this morning. If the backpack has six pockets, what is the probability that he will find the money in the first pocket that he checks? $\frac{1}{6}$; **Check students' experiments to see whether they accurately represented the situation.**

DAILY INTERVENTION **Flexible Grouping**

Act It Out When you group students together to discuss problems to solve, try placing students with different interests in each group. Students will naturally choose act it out strategies based on their own experiences. Mixing students interested in sports, music, computers, and photography, for example, will enrich students' visions of possible act it out strategies.

Answer

2. **Sample answer: You can make a prediction about what will actually happen in the problem.**

3 Practice/Apply

About the Exercises...

Problem-Solving Strategies
Exercises 5–13 are structured so that students have the opportunity to practice many different problem-solving strategies. You may wish to remind them of some of the strategies they have studied.

- use a graph (p. 58)
- look for a pattern (p. 132)
- draw a diagram (p. 302)
- act it out ✓

4 Assess

Open-Ended Assessment

Speaking Have students describe a situation in which they could use the act it out strategy to solve a probability problem. They should outline both the problem and the method they would use to test the problem.

Answers

3. No; Sample answer: the experiment produces about 1–2 correct answers, so using a spinner with 4 sections is not a good way to answer a 5-question multiple-choice quiz.
9. Sample answer: Since a number cube has six sides, roll a number cube until all six numbers are rolled.
10. 6th row: 1, 5, 10, 10, 5, 1; 7th row: 1, 6, 15, 20, 15, 6, 1

Apply the Strategy

Solve. Use the act it out strategy.

3. **POP QUIZ** Determine whether using a spinner with four equal sections is a good way to answer a 5-question multiple-choice quiz. Each question has choices A, B, C, and D. Explain. See margin.

4. **PHOTOGRAPHS** Samuel is taking a picture of the Spanish Club's five officers. The club president will always stand on the left, and the vice president will always stand on the right. How many different ways can he arrange the officers for the picture? 6

Mixed Problem Solving

Solve. Use any strategy.

5. **MONEY MATTERS** Lola purchased a $35 book bag at a sale price of $27.50. What was the percent of decrease from the original price to the sales price? about 21%

6. **PICNIC** Examine the bar graph.

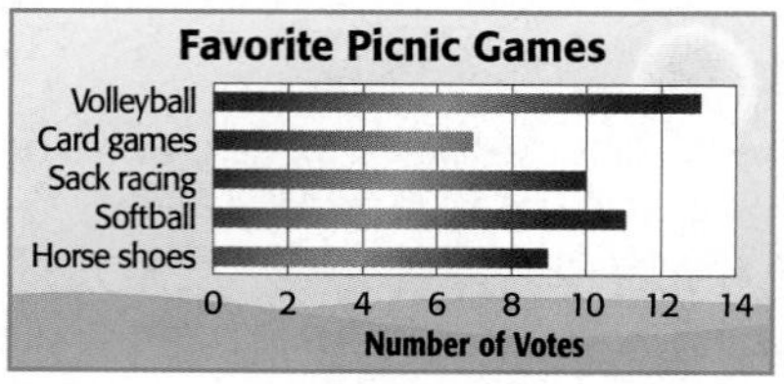

What is the probability that a student's favorite picnic game is sack racing? $\frac{1}{5}$

7. **CROQUET** Sixteen teams are playing in a croquet tournament. If a team loses one game, it is eliminated. How many total games will be played in the tournament? 15

8. **EARTH SCIENCE** The Hawaiian mountain Mauna Kea is about 1.38×10^4 feet tall. Mt. Everest is about 29,000 feet. Which mountain is taller? Mt. Everest

9. **BASEBALL** Each package of baseball cards contains a puzzle piece. If you collect all 6 different pieces, you win two tickets to a major league game. There is an equally-likely chance of getting a different puzzle piece each time. Explain how you would model this problem to find the answer. See margin.

10. **PATTERNS** The pattern below is known as Pascal's Triangle. Find the pattern and complete the 6th and 7th rows. See margin.

1st Row					1				
2nd Row				1		1			
3rd Row			1		2		1		
4th Row		1		3		3		1	
5th Row	1		4		6		4		1

11. **SPACE SCIENCE** A 21-kilogram sample of rocks from the moon is composed of about 40% oxygen and about 19.2% silicon. How much more is the oxygen mass of the rocks than the silicon mass? about 4.368 kg more

12. **TRANSPORTATION** A taxi charges $2.35 for the first 0.4 mile and $0.75 for each additional 0.4 mile. Find the cost of a 4-mile taxi ride. $9.10

13. **STANDARDIZED TEST PRACTICE**
In a middle school of 900 students, about how many prefer the season of autumn? C

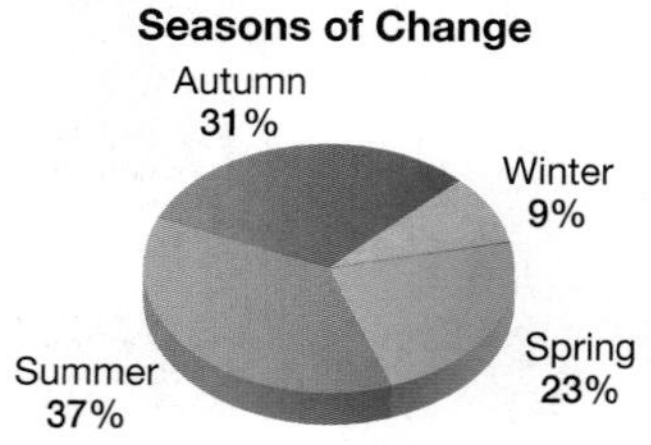

Ⓐ 81 Ⓑ 207 Ⓒ 279 Ⓓ 333

You will use the act it out strategy in the next lesson.

9-6 Theoretical and Experimental Probability

hat You'll Learn
nd and compare
perimental and
eoretical probabilities.

:W Vocabulary
perimental probability
eoretical probability

:VIEW Vocabulary
mple space: a set all possible itcomes **(Lesson 9-2)**

HANDS-ON Mini Lab

Materials
- 2 number cubes

Work with a partner.

STEP 1 Roll two number cubes 36 times. Record the number of times the sum of the number cubes is 7.

STEP 2 Use an addition table from 1 to 6 to help you find the expected number of times the sum of 7 should come up after rolling two number cubes 36 times. The top row represents one number cube, and the left column represents the other number cube. The table above has been started for you.

+	1	2	3
1	2	3	4
2	3	4	5

1. How many times did you roll a sum of 7? What is the probability of rolling a sum of 7? **1, 2, 4. See students' work.**
2. How does your result compare to the results of other groups? Explain.
3. What is the expected probability of rolling a sum of 7? $\frac{6}{36}$ **or** $\frac{1}{6}$
4. How does your result compare to the expected probability of rolling a sum of 7? Explain any differences.

In activity above, you found the experimental probability of rolling a sum of 7 on two number cubes. **Experimental probability** is found using frequencies obtained in an experiment or game.

The expected probability of an event occurring is called **theoretical probability**. This is the probability that you have been using since Lesson 9-1. The theoretical probability of rolling a sum of 7 on two number cubes is $\frac{6}{36}$, or $\frac{1}{6}$.

EXAMPLE Experimental Probability

1 **Two number cubes are rolled seventy-five times and a sum of 9 is rolled ten times.**

What is the experimental probability of rolling a sum of 9?

$$P(9) = \frac{\text{number of times a sum of 9 occurs}}{\text{number of possible outcomes}}$$

$$= \frac{10}{75} \text{ or } \frac{2}{15}$$

The experimental probability of rolling a sum of 9 is $\frac{2}{15}$.

msmath2.net/extra_examples

9-6 Lesson Notes

Lesson Resources

Chapter 9 Resource Masters FAST FILE
- Study Guide and Intervention, p. 510
- Practice: Skills, p. 511
- Practice: Word Problems, p. 512
- Reading to Learn Mathematics, p. 513
- Enrichment, p. 514

Classroom Math Games pp. 25–27

Parent and Student Study Guide Workbook, 9-6

Teaching Mathematics with Manipulatives, p. 84

1 Focus

5-Minute Check 9-6 INTERACTIVE CHALKBOARD PowerPoint® Presentations

Use as a quiz or review of Lesson 9-5. (Also available as a transparency)

Bellringer

Reading Mathematics Have students look up the words *theoretical* and *experimental*. Encourage them to discuss the differences between the two words and speculate about how their meanings relate to probability.

Reading to Learn Mathematics, p. 513 **ELL**

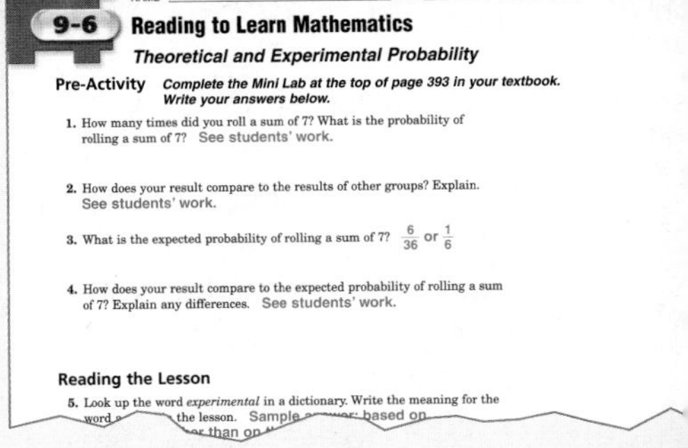
NAME ______ DATE ______ PERIOD ____

9-6 Reading to Learn Mathematics

Theoretical and Experimental Probability

Pre-Activity *Complete the Mini Lab at the top of page 393 in your textbook. Write your answers below.*

1. How many times did you roll a sum of 7? What is the probability of rolling a sum of 7? See students' work.
2. How does your result compare to the results of other groups? Explain. See students' work.
3. What is the expected probability of rolling a sum of 7? $\frac{6}{36}$ or $\frac{1}{6}$
4. How does your result compare to the expected probability of rolling a sum of 7? Explain any differences. See students' work.

Reading the Lesson

5. Look up the word *experimental* in a dictionary. Write the meaning for the word ... the lesson. Sample ... based on ...

2 Teach

In-Class Examples

INTERACTIVE CHALKBOARD
PowerPoint® Presentations

1. A spinner is spun 50 times, and it lands on the color blue 15 times. What is the experimental probability of spinning blue? $P(\text{blue}) = \frac{15}{50} = \frac{3}{10}$

2. The graph below shows the results of an experiment in which a number cube is rolled 30 times. Find the experimental probability of rolling a 5. $\frac{2}{15}$

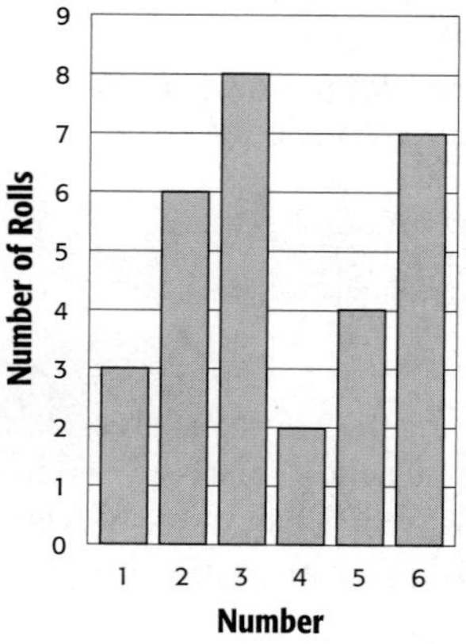

3. Compare the experimental probability you found in In-Class Example 2 to its theoretical probability. **The theoretical probability $\left(\frac{1}{6}\right)$ is close to the experimental probability.**

4. **MOVIES** In a survey, 50 people were asked to pick which movie they would see this weekend. Twenty chose *Horror Story*, 15 chose *The Ink Well*, 10 chose *The Monkey House*, and 5 chose *Little Rabbit*. What is the experimental probability of someone wanting to see *The Monkey House*? $\frac{1}{5}$

5. Suppose 300 people are expected to attend a movie theater this weekend to see one of the four movies mentioned above. How many can be expected to see *The Monkey House*? **60**

TEACHING TIP

Sometimes the experiments produce skewed results indicating inequality in the trials. Assure students that events in this chapter have outcomes that are equally likely to occur, unless it is indicated otherwise.

EXAMPLES Experimental and Theoretical Probabilit

2 The graph shows the results of an experiment in which a coin was tossed thirty times. Find the experimental probability of tossing tails for this experiment.

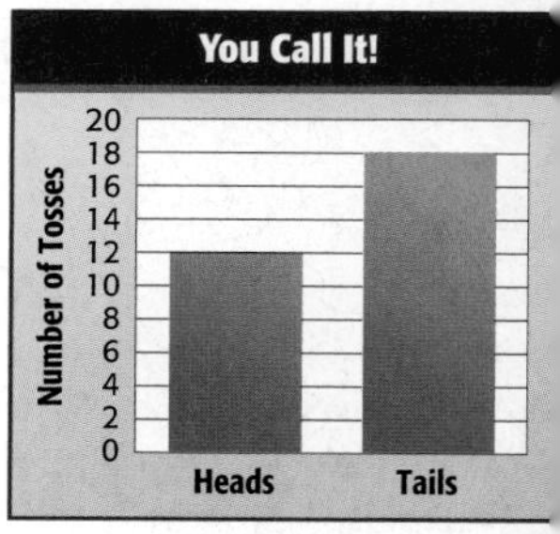

$P(\text{tails}) = \frac{\text{number of times tails occurs}}{\text{number of possible outcomes}}$

$= \frac{18}{30}$ or $\frac{3}{5}$

The experimental probability of tossing tails is $\frac{3}{5}$.

3 Compare the experimental probability you found in Example 2 to its theoretical probability.

The theoretical probability of tossing tails on a coin is $\frac{1}{2}$. So, the experimental probability is close to the theoretical probability.

Experimental probability can also be based on past performances and be used to make predictions on future events.

EXAMPLES Predict Future Events

4 FOOD In a survey, 100 people were asked to name their favorite Independence Day side dishes. What is the experimental probability of macaroni salad being someone's favorite dish?

Side Dish	Number of People
potato salad	55
green salad or vegetables	25
macaroni salad	12
coleslaw	8

There were 100 people surveyed and 12 chose macaroni salad. So, the experimental probability is $\frac{12}{100}$, or $\frac{3}{25}$.

5 Suppose 250 people attend the city's Independence Day barbecue. How many can be expected to choose macaroni salad as their favorite side dish?

$\frac{3}{25} = \frac{x}{250}$ Write a proportion.

$3 \cdot 250 = 25x$ Find the cross products.

$30 = x$ About 30 will choose macaroni salad.

Your Turn

a. What is the experimental probability of potato salad being someone's favorite dish? $\frac{11}{20}$

b. About how many people can be expected to choose potato salad as their favorite dish if 250 attend the barbecue? **138**

REAL-LIFE CAREERS

How Does a Chef Use Math?

In addition to cooking, chefs order enough food and supplies based on customers' preferences and recent sales.

Online Research
For more information about a career as a chef, visit: msmath2.net/careers

DAILY INTERVENTION — Differentiated Instruction

Interpersonal Have students work in small groups to create a survey. Questions should have either yes/no or multiple-choice answers. Each group should conduct its survey with the other students in your class. Students should use their experimental results to predict how many students from the entire grade or school can be expected to have the same answers. Check the surveys for content and format before they are handed out.

Skill and Concept Check

Writing Math
Exercises 1 & 2

1. **Compare and contrast** experimental probability and theoretical probability. **See margin.**

2. **OPEN ENDED** Give an example of a situation that has a theoretical probability of $\frac{1}{10}$. **Sample answer: ten cards numbered 1–10; randomly choosing 8 has a probability of $\frac{1}{10}$**

GUIDED PRACTICE

For Exercises 3 and 4, a coin is tossed 50 times, and it lands heads 28 times.

3. Find the experimental probability of the coin landing heads. $\frac{14}{25}$
4. Find the theoretical probability of the coin landing heads. $\frac{1}{2}$

5. **TRAFFIC** Between 3:00 P.M. and 4:00 P.M., 11 sedans, 15 trucks, and 6 sports cars drove through an intersection. Based on this information, what is the probability that a vehicle that drives through the intersection is a sports car? $\frac{3}{16}$

★ indicates increased difficulty

Practice and Applications

HOMEWORK HELP

For Exercises	See Examples
6–9	1–3
10, 15	4
11–14, 16–17	5

Extra Practice
See pages 586, 604.

For Exercises 6–9, a number cube is tossed 20 times and lands on 1 two times and on 5 four times.

6. Find the experimental probability of landing on 5. $\frac{1}{5}$
7. Find the theoretical probability of *not* landing on 5. $\frac{5}{6}$
8. Find the theoretical probability of landing on 1. $\frac{1}{6}$
9. Find the experimental probability of *not* landing on 1. $\frac{9}{10}$

X GAMES For Exercises 10–12, use the graph of a survey of 50 students asked to name their favorite X Game sport.

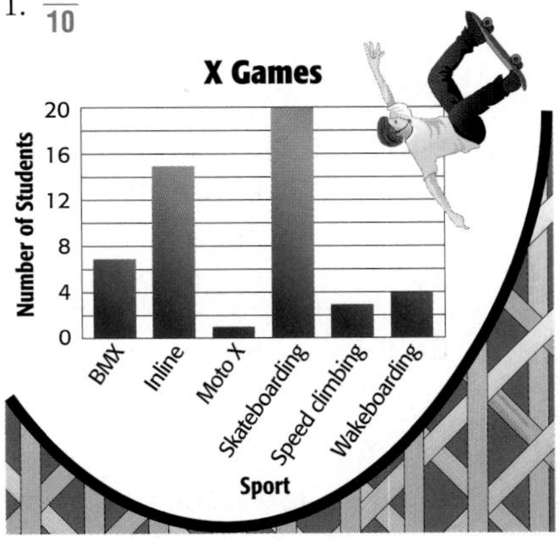

10. What is the probability of inline being someone's favorite sport? $\frac{3}{10}$
11. Suppose 500 people attend the X Games. How many can be expected to choose inline as their favorite sport? **150**
12. Suppose 500 people attend the X Games. How many can be expected to choose speed climbing as their favorite sport? **30**

13. **REFRESHMENTS** In a survey taken at the beach, 47 people preferred cola, 28 preferred root beer, and 25 preferred ginger ale. If the manager of the Beach Hut is going to buy 50 cases of soda for the next day, about how many cases should be root beer? **14**

★ 14. **SPINNERS** A spinner marked with three sections A, B, and C was spun 100 times. The results are shown in the table. Make a drawing of the spinner based on its experimental probabilities. **See margin.**

Section	Frequency
A	24
B	50
C	26

msmath2.net/self_check_quiz

3 Practice/Apply

FOLDABLES™ Follow-Up Remind students to use their Foldables to record what they learn about theoretical and experimental probability. Point out that they might describe an instance in which they have used either theoretical or experimental probability.

About the Exercises...

Guided Practice Key

Exercises	Examples
3	1, 2
4	3
5	4

Odd/Even Assignments
Exercises 6–13 are structured so that students practice the same concepts whether they are assigned odd or even problems.

Answers

1. Sample answer: Both probabilities are ratios that compare the number of favorable outcomes to the total number of possible outcomes. Experimental probability is the result of an experiment. Theoretical probability is what is expected to happen.

14.

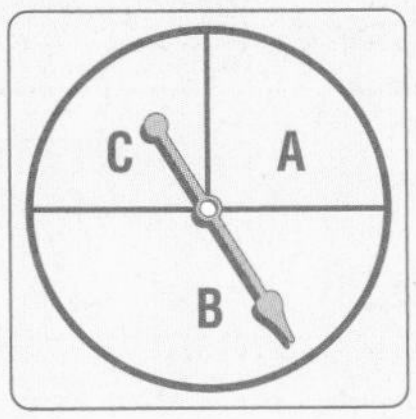

4 Assess

Open-Ended Assessment

Writing Ask students to write a paragraph explaining the difference between theoretical and experimental probabilities. In their paragraphs, they should use one concrete example, such as the roll of a number cube, to illustrate why the experimental probabilities for the example can change but the theoretical probability remains the same.

Study Guide and Intervention, p. 510

NAME ______________ DATE __________ PERIOD ____

9-6 Study Guide and Intervention

Theoretical and Experimental Probability

Experimental probability is found using frequencies obtained in an experiment or game. **Theoretical probability** is the expected probability of an event occurring.

EXAMPLE 1 The graph shows the results of an experiment in which a number cube was rolled 100 times. Find the experimental probability of rolling a 3 for this experiment.

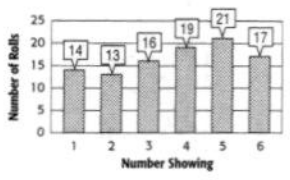

$P(3) = \frac{\text{number of times 3 occurs}}{\text{number of possible outcomes}}$

$= \frac{16}{100}$ or $\frac{4}{25}$

The experimental probability of rolling a 3 is $\frac{4}{25}$, which is close to its theoretical probability of $\frac{1}{6}$.

EXAMPLES

2 In a telephone poll, 225 people were asked for whom they planned to vote in the race for mayor. What is the experimental probability of Juarez being elected?

Candidate	Number of People
Juarez	75
Davis	67
Abramson	83

Of the 225 people polled, 75 planned to vote for Juarez.

So, the experimental probability is $\frac{75}{225}$ or $\frac{1}{3}$.

3 Suppose 5,700 people vote in the election. How many can be expected to vote for Juarez?

$\frac{1}{3} \cdot 5{,}700 = 1{,}900$

About 1,900 will vote for Juarez.

EXERCISES

For Exercises 1–3, use the graph of a survey of 150 students asked whether they prefer cats or dogs.

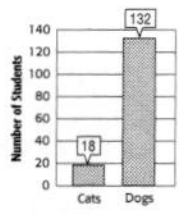

1. What is the probability of a student preferring dogs? $\frac{22}{25}$
2. Suppose 100 students were surveyed. How many can be expected to prefer dogs? 88
3. Suppose 300 students were surveyed. How many can be expected to prefer cats? 36

Practice: Skills, p. 511

NAME ______________ DATE __________ PERIOD ____

9-6 Practice: Skills

Theoretical and Experimental Probability

For Exercises 1–5, a number cube is rolled 50 times and the results are shown in the graph below.

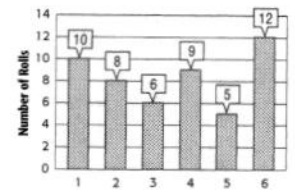

1. Find the experimental probability of rolling a 2. $\frac{4}{25}$
2. What is the theoretical probability of rolling a 2? $\frac{1}{6}$
3. Find the experimental probability of *not* rolling a 2. $\frac{21}{25}$
4. What is the theoretical probability of *not* rolling a 2? $\frac{5}{6}$
5. Find the experimental probability of rolling a 1. $\frac{1}{5}$

For Exercises 6–9, use the results of the survey at the right.

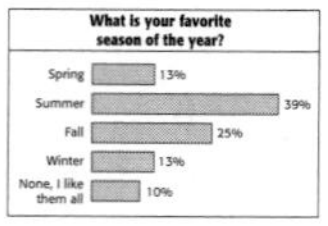

6. What is the probability that a person's favorite season is fall? Write the probability as a fraction. $\frac{1}{4}$
7. Out of 300 people, how many would you expect to say that fall is their favorite season? 75
8. Out of 20 people, how many would you expect to say that they like all the seasons? 2
9. Out of 650 people, how many more would you expect to say that they like summer than say that they like winter? 169

Practice: Word Problems, p. 512

NAME ______________ DATE __________

9-6 Practice: Word Problems

Theoretical and Experimental Probability

HOBBIES For Exercises 1–3, use the graph of a survey of 24 seventh grade students asked to name their favorite hobby.

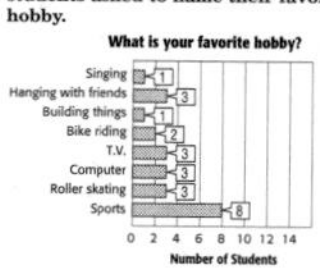

TELEVISION For Exercises 5 and 6, use the graph of a survey with 104 responses in which respondents were asked about their favorite winter activities.

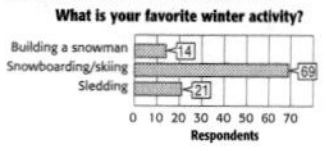

1. What is the probability that a student's favorite hobby is roller skating? $\frac{1}{8}$
2. Suppose 200 seventh grade students were surveyed. How many can be expected to say that roller skating is their favorite hobby? 25
3. Suppose 60 seventh grade students were surveyed. How many can be expected to say that bike riding is their favorite hobby? 5
4. MARBLES A bag contains 5 blue, 4 red, 9 white, and 6 green marbles. If a marble is drawn at random and replaced 100 times, how many times would you expect to draw a green marble? 25
5. What is the probability that someone's favorite winter activity is building a snowman? Write the probability as a fraction. $\frac{7}{52}$
6. If 500 people had responded, how many would have been expected to list sledding as their favorite winter activity? Round to the nearest whole person. 101

GIFTS For Exercises 15–17, use the results of the survey at the right.

15. What is the probability that a mother will receive a gift of flowers? Write the probability as a fraction. $\frac{9}{25}$
16. Out of 750 mothers that receive gifts, how many would you expect to receive flowers? 270
17. Out of 750 mothers that receive gifts, how many would you expect to receive jewelry? 120

18. **CRITICAL THINKING** *Hot numbers* in a lottery are numbers that keep coming up. *Cold numbers* are numbers that have not "hit" in awhile. Some people select hot numbers or cold numbers when buying a ticket. Using probability, explain whether this makes sense. See margin.

EXTENDING THE LESSON

The *odds* of an event occurring is the ratio that compares the number of ways an event can occur (success) to the number of ways it cannot occur (failure).

Example Find the odds of rolling a number less than 5 on a number cube.

ways to occur: 4 ways *not* to occur: 2

sample space: 1, 2, 3, 4, 5, 6 → 4:2, or 2:1

Find the odds of each outcome if a number cube is rolled.

19. 6 1:5
20. odd number 3:3 or 1:1
21. number greater than 4 2:4 or 1:2

Standardized Test Practice and Mixed Review

22. **MULTIPLE CHOICE** Myron has four dimes in his wallet with dates 1998, 1995, 2002, and 2000. If he randomly picks one dime from his wallet, what is the probability that it will have a date in the 1990s? B

    (A) $\frac{1}{4}$ (B) $\frac{1}{2}$ (C) $\frac{2}{3}$ (D) $\frac{3}{4}$

23. **GRID IN** Rachel spins the spinner at the right 50 times, and it lands on red 15 times. What is the theoretical probability of spinning red? 1/5

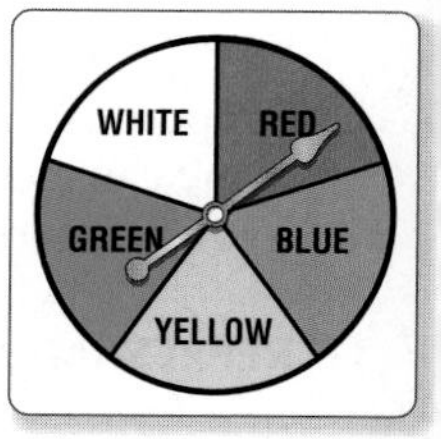

24. **BASEBALL** How many ways can a baseball coach select four starting pitchers from a pitching staff of eight? (Lesson 9-5) 70
25. **POLE VAULT** A pole vault competition has 10 people in it. In how many ways can first-, second- and third-place ribbons be awarded? (Lesson 9-4) 720

GETTING READY FOR THE NEXT LESSON

PREREQUISITE SKILL Three coins are tossed. Find each probability. (Lesson 9-1) 28. $\frac{1}{2}$

26. P(3 tails) $\frac{1}{8}$
27. P(2 heads) $\frac{3}{8}$
28. P(at least 2 heads)
29. P(at least 1 tail) $\frac{7}{8}$

Answer

18. Sample answer: If a lottery is fair, it doesn't matter whether a number is hot or cold. Each number has an equally-likely chance of hitting during any given drawing.

hat You'll Learn
vestigate
perimental
obability.

aterials
aper
cissors
pinner arrows

Experimental Probability

INVESTIGATE *Work with a partner.*

Shannon is on the basketball team. She makes 75% of her free throws. In this lab, you will model, or *simulate* Shannon shooting free throws with a spinner and investigate the experimental probabilities of her being able to make a free throw.

- Create and cut out the spinner at the right.
- Spin the spinner 25 times to model Shannon shooting 25 free throws. Record your results in the table below. If the spinner lands on a line, spin again.

Outcome	Tally	Total
Made Free Throw		
Missed Free Throw		

- Combine your data with four other groups so you have 100 data points.

Writing Math

Work with a partner.

1. **Explain** why you combined your data with four other groups.
   Sample answer: his replicates the xperiment (Shannon ooting free throws) 0 times. It is easier convert fractions with enominators of 100 to their decimal and ercent equivalents.
2. From your table, what is the experimental probability of making a free throw? **Probabilities should be close to 75%.**
3. How does the experimental probability compare with Shannon's past performance? **Probabilities should be close to 75%.**
4. What is the experimental probability of making two free throws in a row? **Explain** how you found your answer. **4–6. See margin.**
5. Suppose Shannon did shoot 100 free throws in a row. What factors would influence her making *or* not making 75 free throws? **Explain** your reasoning.
6. **Describe** the size of both sections of the spinner.
7. **Describe** another way to simulate Shannon shooting free throws.
8. **Draw** a spinner that simulates a free-throw shooter making 60% of her free throws. **7–8. See Ch. 9 Answer Appendix.**

Answers

4. 56.25%; Sample answer: The probability of Shannon making one free throw is 75% or 0.75. The probability of Shannon making the second free throw is also 0.75. So, $P(\text{2 free throws}) = P(\text{first free throw}) \times P(\text{second free throw}) = 0.75 \times 0.75 = 0.5625$, or 56.25%.
5. Sample answer: Shannon would probably make less than 75 free throws because her arms would get tired from shooting the basketball.
6. Sample answer: The Made Free Throws section is three-quarters, or 75% of the whole spinner. The Missed Free Throws section is one-quarter, or 25% of the whole spinner.

9-6b HANDS-ON LAB Lesson Notes

Lesson Resources

Teaching Mathematics with Manipulatives
- spinners, p. 19
- student recording sheet 9-6b

Glencoe Mathematics Classroom Manipulative Kit
- scissors
- spinners

GETTING STARTED

Teaching Tip If spinner arrows are unavailable, show students how to use a pencil and a paper clip to form a pointer: Open the paper clip part-way and lay it flat over the spinner. Hold the pencil vertically and place its tip on the center of the circle, through the paper clip. Flick the end of the paper clip like a spinner arrow.

TEACH

- Determine which groups will combine their data in Step 3. Have them write their results on the chalkboard.

ASSESS

Observe as students solve the **Writing Math** exercises. They should:
- manipulate the experimental data accurately and
- observe how experimental data differed from theoretical probability.

Discuss Exercise 5 in the **Writing Math** exercises. Ask students how skill and practice might affect Shannon's results.

9-7 Lesson Notes

Lesson Resources

Chapter 9 Resource Masters

- Study Guide and Intervention, p. 515
- Practice: Skills, p. 516
- Practice: Word Problems, p. 517
- Reading to Learn Mathematics, p. 518
- Enrichment, p. 519

Calculator and Spreadsheet Masters, p. 46

Key Teaching Transparencies, 36

MathPASS Tutorial CD-ROM, 59

Parent and Student Study Guide Workbook, 9-7

Science and Math Lab Masters, pp. 93–96

1 Focus

5-Minute Check 9-7

INTERACTIVE CHALKBOARD PowerPoint® Presentations

Use as a quiz or review of Lesson 9-6. (Also available as a transparency)

Bellringer

Problem Solving Tell students that they have a bag containing three blue marbles and two red marbles. They pick one marble, and it is blue.

1. If the blue marble is replaced and another marble drawn, what is the probability the second marble will be blue? $\frac{3}{5}$

(Continued in bottom margin.)

Reading to Learn Mathematics, p. 518 ELL

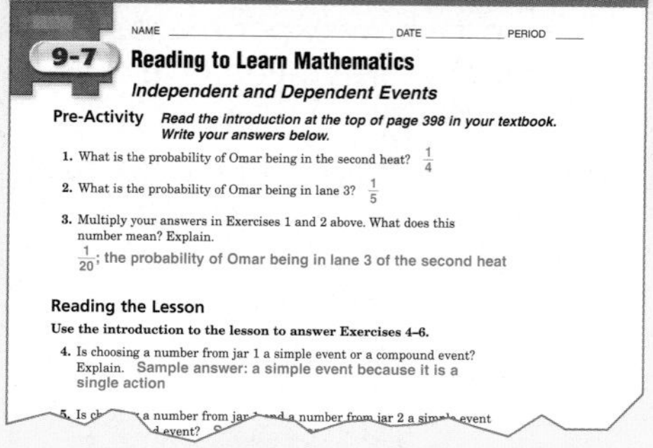
NAME ______________ DATE ________ PERIOD ____

9-7 Reading to Learn Mathematics

Independent and Dependent Events

Pre-Activity ***Read the introduction at the top of page 398 in your textbook. Write your answers below.***

1. What is the probability of Omar being in the second heat? $\frac{1}{4}$
2. What is the probability of Omar being in lane 3? $\frac{1}{5}$
3. Multiply your answers in Exercises 1 and 2 above. What does this number mean? Explain. $\frac{1}{20}$; the probability of Omar being in lane 3 of the second heat

Reading the Lesson

Use the introduction to the lesson to answer Exercises 4–6.

4. Is choosing a number from jar 1 a simple event or a compound event? Explain. Sample answer: a simple event because it is a single action
5. Is ch... a number from jar ... a number from jar 2 a sim... event ...d event? ...

9-7 Independent and Dependent Events

What You'll Learn
Find the probability of independent and dependent events.

NEW Vocabulary
- compound event
- independent event
- dependent event

READING Math

Probability Notation *P(A and B) is read the probability of A followed by B.*

WHEN am I ever going to use this?

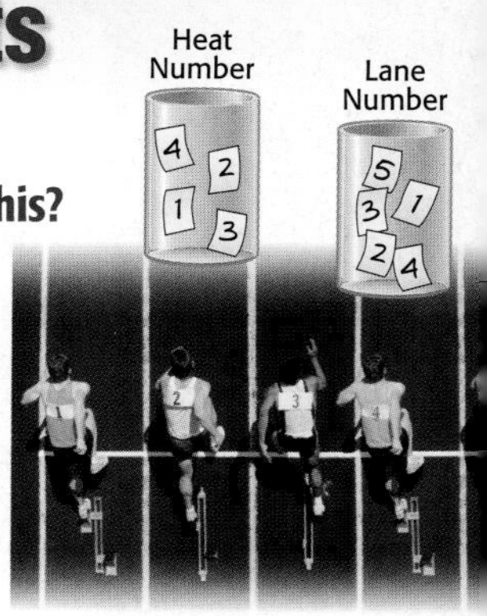

TRACK AND FIELD The 100-meter dash features 20 runners competing in a preliminary round of 4 heats. The winner of each heat advances to the final race. Before the race, each runner chooses a number from jar 1 to determine the heat in which he runs and a number from jar 2 to determine one of five lanes he occupies. Omar is the first runner to choose from the jars.

1. What is the probability of Omar being in the second heat? $\frac{1}{4}$
2. What is the probability of Omar being in lane 3? $\frac{1}{5}$
3. Multiply your answers in Exercises 1 and 2 above. What does this number mean? Explain. $\frac{1}{20}$; **The probability of Omar being in lane 3 of the second heat.**

In the Mini Lab, choosing the heat and the lane is a compound event. A **compound event** consists of two or more simple events. Since choosing the heat number does not affect choosing the lane number, both events are called **independent events**.

Key Concept — Probability of Two Independent Events

Words The probability of two independent events can be found by multiplying the probability of the first event by the probability of the second event.

Symbols $P(A \text{ and } B) = P(A) \cdot P(B)$

EXAMPLE Independent Events

1 **A number cube is rolled, and the spinner at the right is spun. Find the probability of rolling a 2 and spinning a vowel.**

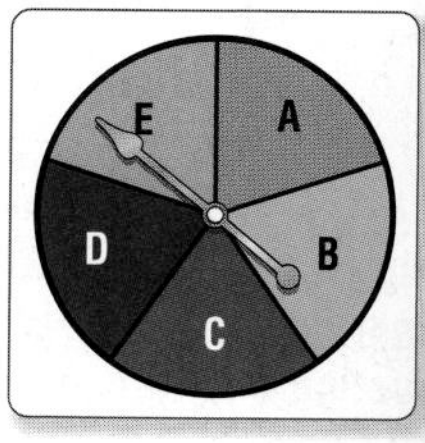

$P(2) = \frac{1}{6}$ $\quad P(\text{vowel}) = \frac{2}{5}$

$P(2 \text{ and vowel}) = \frac{1}{\cancel{6}_3} \cdot \frac{\cancel{2}^1}{5} \text{ or } \frac{1}{15}$

So, the probability of rolling a 2 and spinning a vowel is $\frac{1}{15}$.

Check You can make a tree diagram to check your answer.

2. If the blue marble is not replaced and another marble drawn, what is the probability the second marble will be blue? $\frac{1}{2}$

If the outcome of one event affects the outcome of a second event, the events are called **dependent events**.

TEACHING TIP

...udents can assume the ...st event is a success ...ring dependent events.

EXAMPLE Dependent Events

2 SNACK BARS A box contains 2 oatmeal, 3 strawberry, and 6 cinnamon snack bars. Ruby reaches in the box and randomly takes two snack bars, one after the other. Find the probability that she will choose a cinnamon bar and then a strawberry bar.

$P(\text{cinnamon}) = \frac{6}{11}$ ← 11 snack bars, 6 are cinnamon

$P(\text{strawberry}) = \frac{3}{10}$ ← 10 snack bars after 1 cinnamon snack bar has been removed, 3 are strawberry

$P(\text{cinnamon, then strawberry}) = \frac{\overset{3}{\cancel{6}}}{11} \cdot \frac{3}{\underset{5}{\cancel{10}}} \text{ or } \frac{9}{55}$

So, the probability that Ruby will choose a cinnamon snack bar and then a strawberry snack bar is $\frac{9}{55}$, or about 16%.

STUDY TIP

Dependent Events The probability of *B* following *A* is the same as the probability of *A* following *B*.

Key Concept — Probability of Two Dependent Events

Words	The probability of two dependent events is the probability of the first event times the probability that the second event occurs after the first.
Symbols	$P(A \text{ and } B) = P(A) \cdot P(B \text{ following } A)$

Skill and Concept Check

Writing Math — Exercises 1 & 3

1. **Explain** the difference between independent and dependent events.
2. **OPEN ENDED** Describe two dependent events. **1–2. See margin.**
3. **FIND THE ERROR** Kimi and Shane are finding the probability of spinning an even number on two consecutive spins. Who is correct? Explain. **Shane; Kimi treated spinning the spinner twice as dependent events.**

Kimi: $\frac{2}{5} \cdot \frac{1}{4} = \frac{2}{20} = \frac{1}{10}$

Shane: $\frac{2}{5} \cdot \frac{2}{5} = \frac{4}{25}$

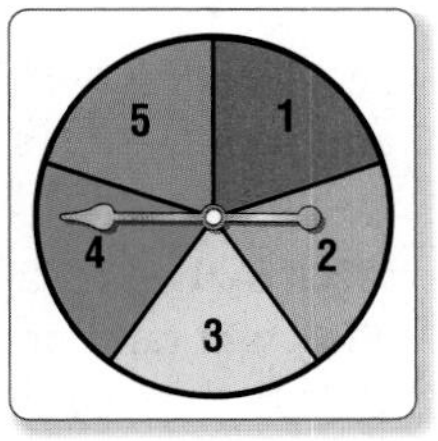

GUIDED PRACTICE

4. A coin is tossed, and a number cube is rolled. What is the probability of tossing heads and rolling a 3 or a 5? $\frac{1}{6}$
5. A bag contains 5 red apples and 3 yellow apples. What is the probability of picking 2 red apples without the first being replaced? $\frac{5}{14}$

2 Teach

In-Class Examples

INTERACTIVE CHALKBOARD — PowerPoint® Presentations

1. **LUNCH** For lunch, Jessica may choose from a turkey sandwich, a tuna sandwich, a salad, or a soup. For a drink, she can choose juice, milk, or water. If she chooses a lunch at random, what is the probability that she chooses a sandwich (of either kind) and juice? $\frac{1}{6}$

2. **COMMITTEE SELECTION** Mrs. Tierney will select two students from her class to be on the principal's committee. She places the name of each student in a bag and selects one at a time. The class contains 15 girls and 12 boys. What is the probability she selects a girl's name first, then a boy's name? $\frac{10}{39}$

3 Practice/Apply

FOLDABLES™ Follow-Up Remind students to use their Foldables to record what they learn about independent and dependent events. Point out that they might describe how they found the probability of either an independent or dependent event.

Answers

1. Sample answer: Events are independent when neither outcome affects the other. Events are dependent when the outcome of the first event affects the outcome of the second event.
2. Sample answer: choosing a card from a standard deck of cards, then choosing a second card without replacing the first

DAILY INTERVENTION — Differentiated Instruction

Logical Place the names of all your students in a hat. Tell students you are going to divide the class into two groups by drawing names from the hat. Ask students to predict the gender composition of each group. Before each draw, tell students to compute the probability that a boy will be chosen and the probability that a girl will be chosen. (Remind them that the two probabilities must add to one.) Ask them to notice how the probabilities change as names are removed from the hat.

DAILY INTERVENTION **FIND THE ERROR** In Exercise 3, point out to students that spinning a spinner is similar to rolling a number cube or tossing a coin. Since one spin, roll, or toss does not affect the next in any way, these are independent events. Since the sample space does not change, the probability of an even number does not change, so Shane is the only possible correct choice.

About the Exercises...

Guided Practice Key

Exercises	Examples
4	1
5	2

Odd/Even Assignments
Exercises 6–9 and 12–18 are structured so that students practice the same concepts whether they are assigned odd or even problems.

Alert! For Exercises 12–14, students need to understand the composition of a standard deck of cards.

Answer

10.

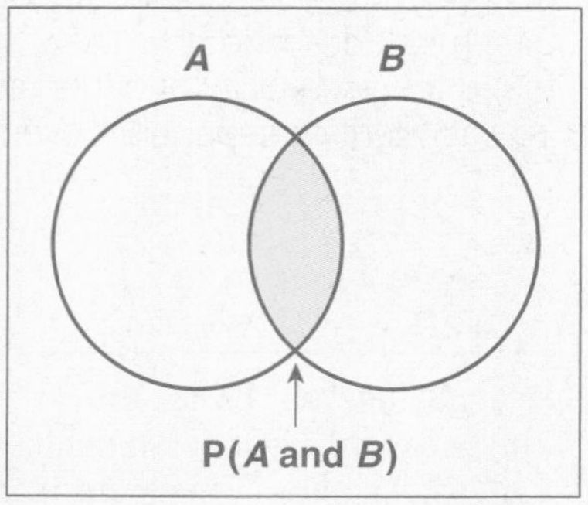

★ indicates increased difficulty

Practice and Applications

HOMEWORK HELP

For Exercises	See Examples
6–7, 17–19	1
8–9, 11–14	2

Extra Practice
See pages 586, 604.

TEACHING TIP
Explain to students that any time an item is *not* replaced into a group of items, the probability of the second event is dependent on the first.

6. A red and a blue number cube are rolled. Find the probability that an odd number is rolled on the red cube and a number greater than 1 is rolled on the blue cube. $\frac{5}{12}$

7. Find the probability of heads on three consecutive tosses of a coin. $\frac{1}{8}$

8. A cooler is filled with 12 colas and 9 diet colas. If Victor randomly chooses two without replacing the first, what is the probability that he will choose a cola and then a diet cola? $\frac{9}{35}$

9. A deck of 30 cards is made up of the numbers 1–10 in three colors: red, purple, and green. Two cards are selected without either being replaced. Find the probability of choosing a purple 5 and then a red or green card. $\frac{2}{87}$

10. Draw a Venn diagram to show the probability of two independent events *A* and *B*. **See margin.**

11. **CIVICS** In the 108th Congress, Tennessee had 4 Republicans and 5 Democrats serving in the House of Representatives. If a subcommittee of 2 representatives was formed to study Internet usage among middle school students, what is the probability that both would be Republicans? $\frac{1}{6}$

CARDS For Exercises 12–14, use the information below.
A standard deck of playing cards contains 52 cards in four suits of 13 cards each. Two suits are red and two suits are black. Two cards are chosen from the deck one after another. Find each probability.

12. P(2 hearts) $\frac{1}{17}$

13. P(red, black) $\frac{13}{51}$

14. P(Ace, King) $\frac{4}{663}$

Determine whether each event is *independent* or *dependent*.

15. choosing a student in the 7th grade and a student from the 8th grade **independent**

16. choosing a pair of shoes to try on, then choosing a smaller pair to try on **dependent**

FASHION For Exercises 17 and 18, use the graphic at the right. Write your answers as decimals to the nearest hundredth.

★ 17. What is the probability that a teen chosen at random said that friends and sports figures influenced her fashion choices? **0.03**

★ 18. What is the probability that a teen chosen at random said that personal style and celebrities influenced his fashion choices? **0.04**

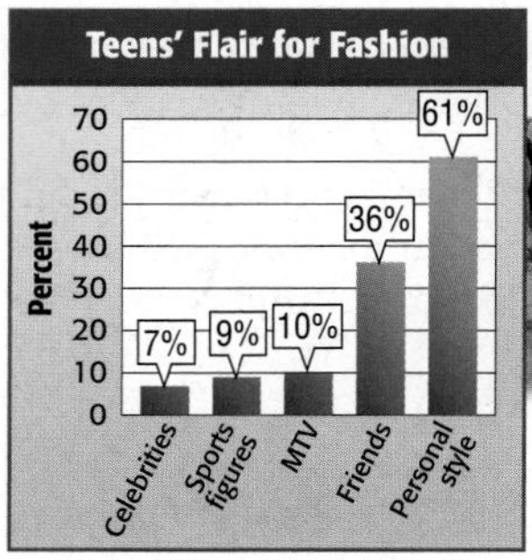

Source: American Express Retail Index

★ 19. **TRAFFIC SIGNALS** There are three consecutive traffic signals on a street that operate independently of each other. The first is green 45% of the time, the second is green 60% of the time, and the third is green 50% of the time. What is the probability of a person driving down the street making all three green lights? Write as a percent to the nearest tenth. **13.5%**

400 Chapter 9 Probability

CRITICAL THINKING For Exercises 20 and 21, use the spinner.

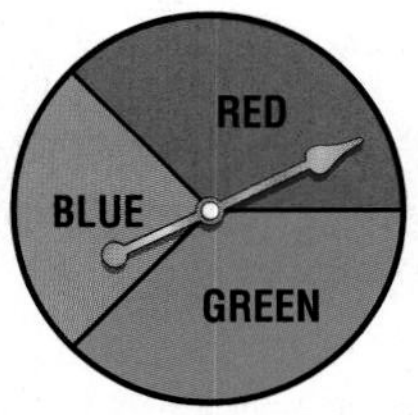

20. Make a tree diagram of all the possible outcomes of three successive spins of the spinner shown. How many paths in the tree diagram represent two red spins and one blue?

21. Suppose the spinner is designed so that for each spin there is a 40% probability of spinning red and a 20% chance of spinning blue. What is the probability of spinning two reds and one blue? 0.096

20. See Ch. 9 Answer Appendix.

EXTENDING THE LESSON Events that cannot happen at the same time are *mutually exclusive.* The two events are connected by the word *or* and the probability is found by *adding* their individual probabilities.

Example Suppose a number cube is rolled. Find P(even or 5).

$$P(\text{even or }5) = \frac{3}{6} + \frac{1}{6} = \frac{4}{6} \text{ or } \frac{2}{3}$$

Find each probability as a fraction in simplest form.

22. P(1 or 3) $\frac{1}{3}$

23. P(5 or multiple of 3) $\frac{1}{2}$

24. P(6 or less than 5) $\frac{5}{6}$

Standardized Test Practice and Mixed Review

25. **MULTIPLE CHOICE** You guess the answers on a two-question multiple-choice quiz with five choices: a, b, c, d, and e. What is the probability that you will get both correct? B

Ⓐ $\frac{1}{50}$ Ⓑ $\frac{1}{25}$ Ⓒ $\frac{2}{25}$ Ⓓ $\frac{1}{5}$

26. **GRID IN** Find the probability of rolling an even number on a number cube, tossing heads on a coin, and choosing an ace in a standard deck of cards. 1/52

27. **PROBABILITY** Paz performed a probability experiment by spinning a spinner 20 times. The results are shown in the table. If the spinner is divided into four equal sections, how many sections would you expect to be blue? (Lesson 9-6) 1

Color	Frequency
red	𝍸
green	𝍸 𝍸
blue	𝍸

28. **LOTTERY** Balls numbered 1 to 51 are dropped into a machine and mixed together. Six balls are then selected in any order to make up the winning numbers. How many different ways can the six numbers be chosen? (Lesson 9-5) 18,009,460 ways

Evaluate each expression if $a = 6$, $b = -4$, and $c = -3$. (Lesson 3-6)

29. $9c$ -27

30. $-8a$ -48

31. $2bc$ 24

32. $5b^2$ 80

WebQuest Interdisciplinary Project

Step Right Up and Win a Prize

It's time to complete your project. Use the information and data you have gathered about carnival games to prepare a Web page or poster. Be sure to include a scale drawing of the game you design with your project.

msmath2.net/webquest

msmath2.net/self_check_quiz

4 Assess

Open-Ended Assessment

Modeling Ask students to create a tree diagram to show all possible outcomes for Example 1 on page 398. Confirm that the calculated probability for rolling a 2 and spinning a vowel is the same as the probability indicated by the tree diagram.

Study Guide and Intervention, p. 515

NAME ______ DATE ______ PERIOD ____

9-7 Study Guide and Intervention

Independent and Dependent Events

A **compound event** consists of two or more simple events. If the outcome of one event does not affect the outcome of a second event, the events are called **independent events**. The probability of two independent events can be found by multiplying the probability of the first event by the probability of the second event. If the outcome of one event affects the outcome of a second event, the events are called **dependent events**.

EXAMPLE 1 **A coin is tossed and a number cube is rolled. Find the probability of tossing tails and rolling a 5.**

$P(\text{tails}) = \frac{1}{2}$ $P(5) = \frac{1}{6}$

$P(\text{tails and }5) = \frac{1}{2} \cdot \frac{1}{6}$ or $\frac{1}{12}$

So, the probability of tossing tails and rolling a 5 is $\frac{1}{12}$.

EXAMPLE 2 MARBLES **A bag contains 7 blue, 3 green, and 3 red marbles. If Agnes randomly draws two marbles from the bag, one after another, what is the probability of drawing a green and then a blue marble?**

$P(\text{green}) = \frac{3}{13}$ 13 marbles, 3 are green

$P(\text{blue}) = \frac{7}{12}$ 12 marbles after 1 green marble is removed, 7 are blue

$P(\text{green, then blue}) = \frac{3}{13} \cdot \frac{7}{12} = \frac{7}{52}$

So, the probability that Agnes will draw a green, then a blue marble is $\frac{7}{52}$.

EXERCISES

1. Find the probability of rolling a 2 and then an even number on two consecutive rolls of a number cube. $\frac{1}{12}$

2. A penny and a dime are tossed. What is the probability that the penny lands on heads and the dime lands on tails? $\frac{1}{4}$

3. Lazlo's sock drawer contains 8 blue and 5 black socks. If he randomly pulls out two socks, what is the probability that he picks two blue socks? $\frac{14}{39}$

Practice: Skills, p. 516

NAME ______ DATE ______ PERIOD ____

9-7 Practice: Skills

Independent and Dependent Events

1. Four coins are tossed. What is the probability of tossing all heads? $\frac{1}{16}$

2. One letter is randomly selected from the word PRIME and one letter is randomly selected from the word MATH. What is the probability that both letters selected are vowels? $\frac{1}{10}$

3. A card is chosen at random from a deck of 52 cards. It is then replaced and a second card is chosen. What is the probability of getting a jack and then an eight? $\frac{1}{169}$

For Exercises 4–6, use the information below.

A standard deck of playing cards contains 52 cards in four suits of 13 cards each. Two suits are red and two suits are black. Find each probability. Assume the first card is replaced before the second card is drawn.

4. P(black, queen) $\frac{1}{26}$

5. P(black, diamond) $\frac{1}{8}$

6. P(jack, queen) $\frac{1}{169}$

7. What is the probability of spinning a number greater than 5 on a spinner numbered 1 to 8 and tossing a tail on a coin? $\frac{3}{16}$

8. Two cards are chosen at random from a standard deck of cards with replacement. What is the probability of getting 2 aces? $\frac{1}{169}$

9. A CD rack has 8 classical CDs, 5 pop CDs, and 3 rock CDs. Two CDs are chosen without replacement. What is the probability of choosing a rock CD then a classical CD? $\frac{1}{10}$

10. A jar holds 15 red pencils and 10 blue pencils. What is the probability of drawing two red pencils from the jar? $\frac{7}{20}$

Practice: Word Problems, p. 517

NAME ______ DATE ______ PERIOD ____

9-7 Practice: Word Problems

Independent and Dependent Events

1. **SAFETY** Eighty percent of all California drivers wear seat belts. If three drivers are pulled over, what is the probability that all would be wearing their seat belts? Write as a percent to the nearest tenth. 51.2%

2. **VEGETABLES** A nationwide survey showed that 65% of all children in the United States dislike eating vegetables. If three children are chosen at random, what is the probability that all three dislike eating vegetables? Write as a percent to the nearest tenth. 27.5%

3. **QUALITY** In a shipment of 50 calculators, 4 are defective. Two calculators are randomly selected and tested. What is the probability that both are defective if the first one is not replaced after being tested? $\frac{6}{1,225}$

4. **MARBLES** A bag contains 6 green marbles, 2 blue marbles, and 3 white marbles. Gwen draws one marble from the jar, and then Jeff draws one marble from the remaining marbles. What is the probability that Gwen draws a blue marble and Jeff draws a white marble? $\frac{3}{55}$

5. **DEMONSTRATION** Ms. Morris needs two students to help her with a demonstration for her class of 12 girls and 14 boys. She randomly chooses two students. What is the probability that she chooses two girls? $\frac{66}{325}$

6. **SURVEY** Ruben surveyed his class and found that 4 out of 22 students walk to school. If two of the 22 students are selected at random, what is the probability that both walk to school? $\frac{2}{77}$

CHAPTER 9 Study Guide and Review

Vocabulary Assessment

- This alphabetical list of vocabulary terms in Chapter 9 includes a page reference where each term was introduced.
- A vocabulary review/test for Chapter 9 is available on p. 534 of the *Chapter 9 Resource Masters*.

Vocabulary PuzzleMaker

ELL The online Vocabulary PuzzleMaker improves students' mathematics vocabulary using four puzzle formats — crossword, scramble, word search using a word list, and word search using clues. Students can work on a computer screen or from a printed handout.

MindJogger Videoquizzes

ELL MindJogger Videoquizzes provide an alternative review of concepts presented in this chapter. Students work in teams in a game show format to gain points for correct answers. The questions are presented in three rounds.

Round 1 Concepts (5 questions)
Round 2 Skills (4 questions)
Round 3 Problem Solving (4 questions)

ExamView® Pro Test Bank CD-ROM

Use the networkable ExamView® Pro Test Bank CD-ROM to:

- Create **Multiple versions** of tests.
- Create **Modified** tests for **Inclusion** students with one mouse click.
- **Edit** existing questions and **Add** your own questions.
- Build tests aligned with state standards using built-in **State Curriculum Correlations.**
- Change **English** tests to **Spanish** with one mouse click and vice versa.

CHAPTER 9 Study Guide and Review

Vocabulary and Concept Check

combination (p. 387)	fair game (p. 374)	random (p. 370)
complementary events (p. 371)	Fundamental Counting Principle (p. 378)	sample space (p. 374)
compound event (p. 398)	independent events (p. 398)	simple event (p. 370)
dependent events (p. 399)	outcome (p. 370)	theoretical probability (p. 393)
experimental probability (p. 393)	permutation (p. 381)	tree diagram (p. 374)
factorial (p. 381)	probability (p. 370)	

Choose the correct term to complete the sentence.

1. The set of all possible outcomes for an experiment is called the (, probability).
2. The Fundamental Counting Principle counts the number of possible outcomes using the operation of (addition,).
3. The ratio of the number of times an event occurs to the number of trials completed is called the (theoretical,) probability.
4. When the outcome of one event influences the outcome of a second event, the events are called (independent,).
5. A (, combination) is a listing of objects in which order is important.
6. A (complementary,) event consists of two or more simple events.
7. A (simple,) event occurs by chance.
8. The probability of an event is the (product,) of the number of ways an event can occur to the number of possible outcomes.
9. When using a combination, the order of the arrangement (is,) important.
10. The expression $n!$ is the (sum,) of all counting numbers beginning with n and counting backward to 1.

Lesson-by-Lesson Exercises and Examples

9-1 Simple Events (pp. 370–373)

A bag contains 6 red, 3 pink, and 3 white bows. Suppose you draw a bow at random. Find the probability of each event. Write each fraction in simplest form. 15. $\frac{1}{2}$

11. $P(\text{red})$ $\frac{1}{2}$
12. $P(\text{pink})$ $\frac{1}{4}$
13. $P(\text{white})$ $\frac{1}{4}$
14. $P(\text{red or white})$ $\frac{3}{4}$
15. $P(\text{pink or white})$
16. $P(\textit{not}\text{ white})$ $\frac{3}{4}$

Example 1 **What is the probability of rolling an odd number on a number cube?**

$P(\text{odd}) = \frac{\text{odd numbers possible}}{\text{total numbers possible}}$

$= \frac{3}{6}$ Three numbers are odd: 1, 3, and 5.

$= \frac{1}{2}$ Simplify.

Therefore, $P(\text{odd}) = \frac{1}{2}$.

msmath2.net/vocabulary_reviev

FOLDABLES™ Study Organizer

For more information about Foldables, see *Teaching Mathematics with Foldables.*

Have students look through the chapter to make sure they have included examples in their Foldables for each lesson or topic in the chapter.

Encourage students to refer to their Foldables while completing the Study Guide and Review and to use them in preparing for the Chapter Test.

402 Chapter 9 Probability

9-2 Tree Diagrams (pp. 374–377) 17–20. See Ch. 9 Answer Appendix for tree diagrams.

For each situation, make a tree diagram to show the sample space. Then give the total number of outcomes.

17. rolling a number cube and then tossing a coin 12
18. choosing a red, blue, or white shirt with either black or gray lettering 6
19. tossing a coin and choosing a card from seven cards numbered from 1 to 7 14
20. choosing from white, wheat, or rye bread and turkey, ham, or salami to make a sandwich 9

Example 2 **Make a tree diagram to describe the possible outcomes for a family with two children.**

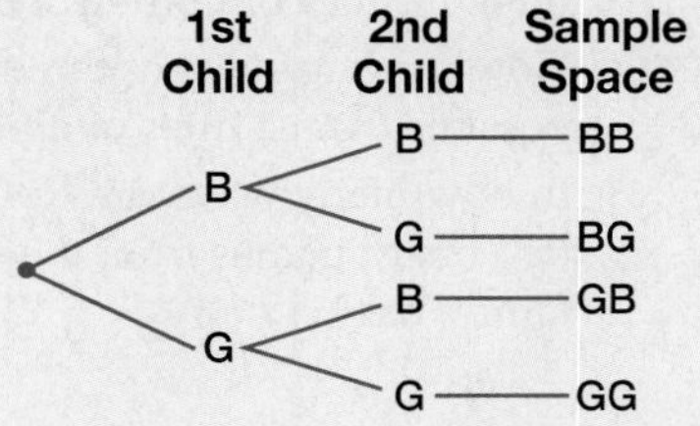

If a family has two children, there are 4 possible outcomes.

9-3 The Fundamental Counting Principle (pp. 378–380)

Use the Fundamental Counting Principle to find the total number of outcomes in each situation.

21. rolling two number cubes 36
22. selecting a car from 3 styles, 3 interior colors, and 3 exterior colors 27
23. making an ice cream sundae selecting from 5 flavors of ice cream and 4 different toppings 20
24. **SHOPPING** One catalog offers a jogging suit in two colors, gray and black. It comes in sizes S, M, L, XL, and XXL. How many possible jogging suits can be ordered? 10 jogging suits

Example 3 **Use the Fundamental Counting Principle to find the total number of outcomes for a family that has three children.**

There are 2 possible outcomes each time a family has a child, a boy or a girl. If a family has 3 children, there are $2 \cdot 2 \cdot 2$, or 8 outcomes.

9-4 Permutations (pp. 381–383)

Find the value of each expression.

25. $4!$ 24
26. $9!$ 362,880
27. $8 \cdot 7 \cdot 6 \cdot 5$ 1,680
28. $3! \cdot 4!$ 144
29. $5! \cdot 3!$ 720
30. $7! \cdot 4!$ 120,960
31. **BASKETBALL** In how many ways can five basketball players be placed in three positions? 60

Example 4 **In how many ways can a director and assistant director be chosen from among 4 candidates?**

There are 4 choices for director and then 3 choices remain for assistant director. So, there are $4 \cdot 3$, or 12, ways the director and assistant director can be chosen.

Chapter 9 Study Guide and Review 403

9-5 Combinations (pp. 387–390) 32. 56 pizzas 33. 126 groups

32. How many three-topping pizzas are possible given eight different toppings?
33. How many groups of five people are there from a committee of nine?
34. How many groups of three kittens are possible from a litter of nine? 84 groups
35. In how many ways can Rondell select two board games from the ten that his family has? 45 ways

Example 5 In how many ways can you choose two items from a menu with four items?

There are $4 \cdot 3$ permutations of two menu items chosen from four.

There are 2! ways to arrange the two menu items.

$\frac{4 \cdot 3}{2!} = \frac{12}{2}$ or 6

So, there are 6 ways to choose two items.

9-6 Theoretical and Experimental Probability (pp. 393–396)

The results of spinning a spinner labeled A–E fifty times are given. Find the experimental probability of each event.

Letter	Frequency
A	8
B	17
C	9
D	6
E	10

36. $P(A)$ $\frac{4}{25}$ 37. $P(D)$ $\frac{3}{25}$ 38. $P(E)$ $\frac{1}{5}$
39. If the spinner is equally likely to land on each section, what is the theoretical probability of landing on B? $\frac{1}{5}$

Example 6 A coin is tossed 75 times, and it lands on tails 55 times. What is the experimental probability of the coin landing on heads?

The coin landed on heads 20 times.

$$P(\text{heads}) = \frac{\text{number of times heads occurs}}{\text{number of possible outcomes}}$$

$$= \frac{20}{75} \text{ or } \frac{4}{15}$$

So, the experimental probability of the coin landing on heads is $\frac{4}{15}$.

9-7 Independent and Dependent Events (pp. 398–401)

For Exercises 40–43, a bag contains 6 green, 8 white, and 2 blue counters. Two counters are randomly drawn. Find each probability if the first counter is replaced before the second counter is drawn.

40. $P(\text{green, blue})$ $\frac{3}{64}$ 41. $P(2 \text{ white})$ $\frac{1}{4}$

Find each probability if the first counter is *not* replaced before the second counter is drawn.

42. $P(2 \text{ green})$ $\frac{1}{8}$ 43. $P(\text{blue, white})$ $\frac{1}{15}$

Example 7 A box contains 12 solid, 14 striped, and 10 spotted marbles. Suppose you reach in and grab two marbles. Find the probability of choosing a striped marble, replacing it, and then choosing a spotted marble.

$P(\text{striped}) = \frac{14}{36}$

$P(\text{spotted, after replacing the striped}) = \frac{10}{36}$

$\frac{14}{36} \cdot \frac{10}{36} = \frac{140}{1{,}296}$ or $\frac{35}{324}$

So, the probability of choosing a striped marble, replacing it, and then choosing a spotted marble is $\frac{35}{324}$, or about 11%.

CHAPTER 9 Practice Test

Vocabulary and Concepts

1. **Explain** the purpose of a tree diagram. See margin.
2. **Define** *combination.* An arrangement, or listing, of objects in which order is not important.

Skills and Applications

A spinner with sections labeled 1–8 has an equal chance of landing on each number. Find each probability.

3. P(odd number) $\frac{1}{2}$
4. P(number greater than 1) $\frac{7}{8}$
5. P(1 or 7) $\frac{1}{4}$

For each situation, make a tree diagram to show the sample space. Then give the total number of outcomes. 6–7. See Ch. 9 Answer Appendix for tree diagrams.

6. tossing a coin three times 8
7. choosing a letter from the word MATH and then a digit from the number 123 12

Use the Fundamental Counting Principle to find the total number of outcomes in each situation.

8. choosing a 3-digit security code 1,000
9. rolling four number cubes 1,296

Find the value of each expression.

10. 8! 40,320
11. $10 \cdot 9 \cdot 8 \cdot 7$ 5,040
12. $5! \cdot 4!$ 2,880

13. **PARADES** If there are 50 floats in a parade, how many ways can a first place and a second place trophy be awarded? 2,450 ways

14. **SURVEY** Two hundred fifty teenagers were asked what type of pet they owned. The results of the survey are in the table. What is the experimental probability that a teenager owns a pet? Write as a percent. 94%

Pet	Number of Teenage Pet Owners
fish	26
cat	65
dog	86
bird	20
other	38
no pet	15

15. **YOGURT** A brand of yogurt has 15 different flavors. In how many ways can you choose three flavors? 455 ways

Standardized Test Practice

16. **GRID IN** A variety bag of popcorn contains 10 pieces of regular popcorn, 6 pieces of cheese popcorn, and 4 pieces of caramel popcorn. Katie pulls out one piece of popcorn and eats it. Then she pulls out a second piece and eats it. What is the probability that her first piece was caramel and her second piece was regular? 2/19

msmath2.net/chapter_test

CHAPTER 9 Practice Test

Assessment Options

Vocabulary Test A vocabulary review/test for Chapter 9 can be found on p. 534 of the *Chapter 9 Resource Masters.*

Chapter Tests There are six Chapter 9 Tests and an Extended Response Assessment task available in the *Chapter 9 Resource Masters.*

Chapter 9 Tests			
Form	**Type**	**Level**	**Pages**
1	MC	basic	521–522
2A	MC	average	523–524
2B	MC	average	525–526
2C	FR	average	527–528
2D	FR	average	529–530
3	FR	advanced	531–532

MC = multiple-choice questions
FR = free-response questions

Extended Response Assessment

Extended Response tasks for Chapter 9 can be found on p. 533 of the *Chapter 9 Resource Masters.* A sample scoring rubric for these tasks appears on p. A26.

Unit 4 Test

A unit test/review can be found on pp. 541–542 of the *Chapter 9 Resource Masters.*

Answer

1. Sample answer: to show the total number of possible outcomes in a probability experiment

Portfolio Suggestion

Introduction After studying this chapter, students might be surprised to note the extent to which probabilities affect their everyday lives.

Ask Students to keep track of situations in their everyday lives that require an understanding of probabilities. They might note games involving number cubes or news reports involving data. They should describe the situations and, if possible, calculate the relevant probabilities.

Chapter 9 Standardized Test Practice

These two pages contain practice questions in the various formats that can be found on the most frequently given standardized tests.

A student recording sheet for these two pages can be found on page A1 of the *Chapter 9 Resource Masters*.

Standardized Test Practice Student Recording Sheet, p. A1

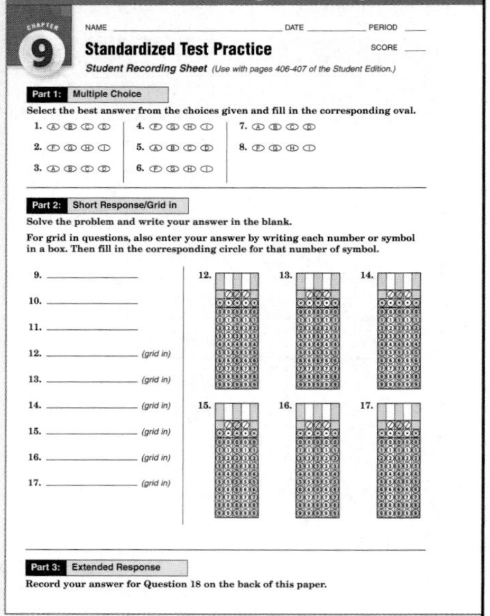

Additional Practice

See pp. 539–540 of the *Chapter 9 Resource Masters* for additional standardized test practice.

Chapter 9 Standardized Test Practice

PART 1 Multiple Choice

Record your answers on the answer sheet provided by your teacher or on a sheet of paper.

1. The ornithologist said there is a 0.6754 chance of the bird returning next year to its nest in a tree in the Glowinski family's backyard. What is the value of the 5 in this number? (Prerequisite Skill, p. 555) **C**

(A) five tenths

(B) five hundredths

(C) five thousandths

(D) five ten thousandths

2. The line plot below shows the weight in grams of a single serving of different brands of cookies. Between which weights is there a cluster of data? (Lesson 2-3) **I**

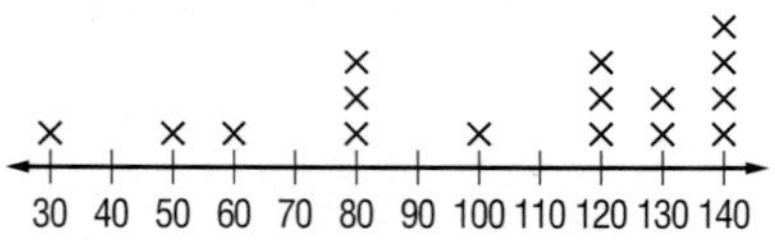

(F) between 30 and 50

(G) between 60 and 80

(H) between 90 and 110

(I) between 120 and 140

3. Park rangers measured the depth of a nearby river. It is 12 inches shallower than it was 4 years ago. Which formula best expresses the yearly change in the river? (Lesson 3-7) **A**

(A) $-12 \div 4$ (B) 12×4

(C) -12×4 (D) $-12 \div -4$

4. Which of the following is the greatest common factor of 84 and 49? (Lesson 5-2) **H**

(F) 2 (G) 6

(H) 7 (I) 12

5. If $792 was made in the sale of pens, what was the total amount of all sales to the nearest dollar? (Lesson 8-2) **B**

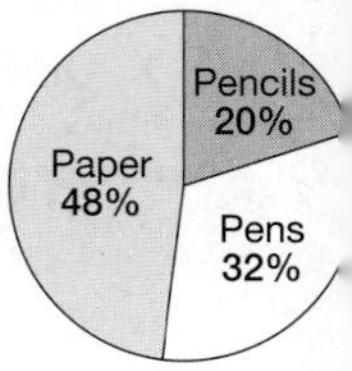

(A) $2,534 (B) $2,475

(C) $253 (D) $248

6. The tree diagram shows the number of ways a customer can order lunch in a diner. If all choices are equally likely, what is the probability that a customer will order meat, rice, and salad? (Lesson 9-2) **G**

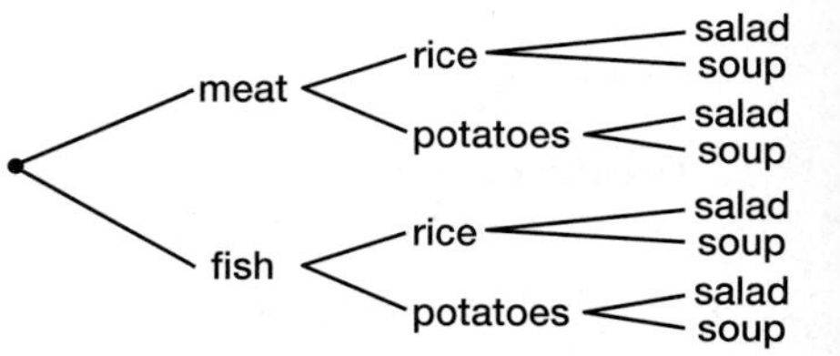

(F) $\frac{1}{16}$ (G) $\frac{1}{8}$ (H) $\frac{1}{4}$ (I) $\frac{3}{5}$

7. Fernando has to wash carrots, celery, onions, and potatoes for dinner. Assuming that he washes each vegetable only once, in how many different orders can Fernando wash all of the vegetables? (Lesson 9-4) **D**

(A) 4 (B) 16 (C) 20 (D) 24

8. What is the theoretical probability of choosing a vowel from the word MATHEMATICS? (Lesson 9-6) **F**

(F) $\frac{4}{11}$ (G) $\frac{4}{9}$ (H) $\frac{4}{7}$ (I) $\frac{7}{11}$

TEST-TAKING TIP

Question 7 Find your own answer before looking at the answer choices. Doing so keeps you from being tempted by wrong answer choices that look correct, but are still wrong.

Log On for Test Practice

The Princeton Review

The Princeton Review offers additional test-taking tips and practice problems at their web site. Visit princetonreview.com or review.com

ExamView® Pro Test Bank CD-ROM

Special banks of standardized test questions similar to those on the TIMSS and NAEP grade 8 tests can be found on this CD-ROM.

Aligned and verified by The Princeton Review

RT 2 Short Response/Grid In

cord your answers on the answer sheet ovided by your teacher or on a sheet of per.

Ebony shot −4, −2, 1, and −3 in four rounds of golf. The final score is the sum of the scores. What was her final score? (Lesson 3-4) **−8**

How much greater is the area of a professional basketball court than a college basketball court? (Lesson 6-8) **500 ft^2**

Basketball Court	Length (ft)	Width (ft)
college	84	50
professional	94	50

What is the percent of increase in two-vehicle households from 1990 to 2000? Write as a percent to the nearest tenth. (Lesson 8-4) **17.7%**

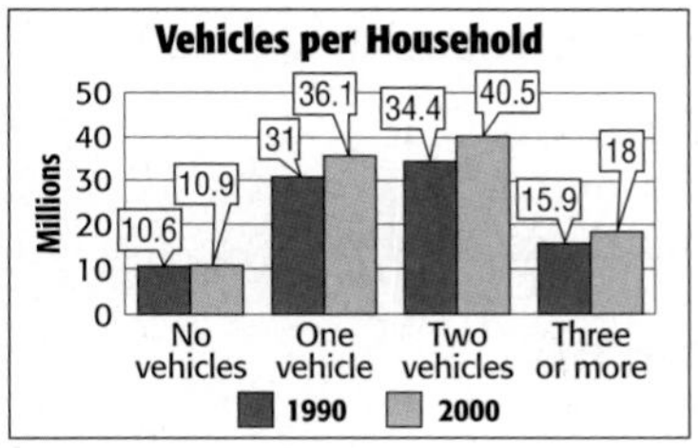

Source: Census Bureau

If two number cubes are rolled, what is the probability that their sum is 5 or 9? (Lesson 9-1) **2/9**

Suppose you want to spend some time on the water. In how many ways can you choose a type of watercraft, the length of time, and the river? (Lesson 9-3) **24**

Type of Watercraft	Time (hours)	River
paddleboat	1	Hamlin
canoe	2	Dustin
kayak	3	
water tricycle		

14. In how many ways can the Geiger family sit in a car if the two parents are in the front seat and the three children are in the back seat? (Lesson 9-4) **12**

15. Twelve points are drawn on a circle. How many different line segments can be drawn between any two points? (Lesson 9-5) **66**

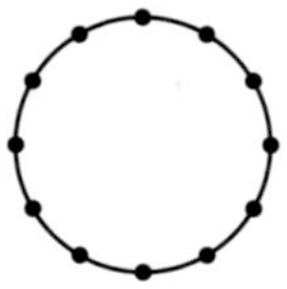

For Questions 16 and 17, use the information below and at the right.

A survey asked 400 students what type of materials they prefer to read. The table shows the results. (Lesson 9-6)

Type of Reading Materials	Number of Students
book	225
newspaper	35
magazine	140

16. What is the probability that a student prefers to read a magazine? **7/20**

17. Suppose 1,000 students attend Franklin Middle School. About how many would prefer to read a book? **563**

PART 3 Extended Response

Record your answers on a sheet of paper. Show your work. **18a. See Ch. 9 Answer Appendix.**

18. Paula has five hobbies: jogging, painting, stamp collecting, quilting, and listening to music. She can enjoy her hobbies on either Friday night or Saturday night.

    a. Make a tree diagram that shows all of her possible outcomes. (Lesson 9-2)

    b. What is the probability that Paula will paint or jog on Friday night? (Lesson 9-2) $\frac{1}{5}$

    c. Paula will not do the same hobby on both nights. What is the probability that Paula will quilt on Friday night and listen to music on Saturday night? (Lesson 9-7) $\frac{1}{20}$

msmath2.net/standardized_test

Standardized Test Practice

Rubrics The following rubrics are sample scoring devices for short response and extended response questions.

Short Response

Points	Description
2	The student demonstrates a <u>thorough understanding</u> of the mathematics of the task. The response may contain minor flaws that do not detract from the demonstration of a thorough understanding.
1	The student has provided a response that is only <u>partially correct</u>.
0	The student has provided a <u>completely incorrect</u> solution or no response at all.

Extended Response

Points	Description
4	The student demonstrates a <u>thorough understanding</u> of the mathematics of the task. The response may contain minor flaws that do not detract from the demonstration of a thorough understanding.
3	The student demonstrates an <u>understanding</u> of the mathematics of the task. The response is essentially correct and demonstrates an essential but less than thorough understanding of the mathematics.
2	The student demonstrates only a <u>partial understanding</u> of the mathematics of the task. Although the student may have used the correct approach to a solution or may have provided a correct solution, the work lacks an essential understanding of the underlying mathematical concepts.
1	The student demonstrates a <u>very limited understanding</u> of the mathematics of the task. The response is incomplete and exhibits many flaws.
0	The student provides a <u>completely incorrect</u> solution or no response at all.

A more detailed 4-point rubric is provided on page A2 of the *Chapter 9 Resource Masters*.

Page 376, Lesson 9-2

3.

1st Spin	2nd Spin	Sample Space
A	A	AA
	B	AB
	C	AC
	D	AD
B	A	BA
	B	BB
	C	BC
	D	BD
C	A	CA
	B	CB
	C	CC
	D	CD
D	A	DA
	B	DB
	C	DC
	D	DD

6. There are 20 choices.

Meat	Bread	Sample Space
ham	rye	ham, rye
	white	ham, white
	pumpernickel	ham, pumpernickel
	sourdough	ham, sourdough
turkey	rye	turkey, rye
	white	turkey, white
	pumpernickel	turkey, pumpernickel
	sourdough	turkey, sourdough
roast beef	rye	roast beef, rye
	white	roast beef, white
	pumpernickel	roast beef, pumpernickel
	sourdough	roast beef, sourdough
salami	rye	salami, rye
	white	salami, white
	pumpernickel	salami, pumpernickel
	sourdough	salami, sourdough
bologna	rye	bologna, rye
	white	bologna, white
	pumpernickel	bologna, pumpernickel
	sourdough	bologna, sourdough

7.

Coin	Number Cube	Sample Space
H	1	H1
	2	H2
	3	H3
	4	H4
	5	H5
	6	H6
T	1	T1
	2	T2
	3	T3
	4	T4
	5	T5
	6	T6

8.

Socks	Shoes	Sample Space
black	boots	black socks, boots
	gym	black socks, gym shoes
	dress	black socks, dress shoes
blue	boots	blue socks, boots
	gym	blue socks, gym shoes
	dress	blue socks, dress shoes
brown	boots	brown socks, boots
	gym	brown socks, gym shoes
	dress	brown socks, dress shoes

9.

Number	Color	Sample Space
1	red	1 red
	white	1 white
	blue	1 blue
2	red	2 red
	white	2 white
	blue	2 blue
3	red	3 red
	white	3 white
	blue	3 blue
4	red	4 red
	white	4 white
	blue	4 blue
5	red	5 red
	white	5 white
	blue	5 blue

10.

Shape	Color	Sample Space
square	orange	square, orange
	blue	square, blue
	yellow	square, yellow
	red	square, red
	green	square, green
circle	orange	circle, orange
	blue	circle, blue
	yellow	circle, yellow
	red	circle, red
	green	circle, green

11.

Space	Math	Sample Space
S	M	SM
	T	ST
	H	SH
P	M	PM
	T	PT
	H	PH
A	M	AM
	T	AT
	H	AH
C	M	CM
	T	CT
	H	CH
E	M	EM
	T	ET
	H	EH

12.

Color	Speeds	Sample Space
purple	10	purple, 10-speed
	18	purple, 18-speed
	21	purple, 21-speed
	24	purple, 24-speed
green	10	green, 10-speed
	18	green, 18-speed
	21	green, 21-speed
	24	green, 24-speed
black	10	black, 10-speed
	18	black, 18-speed
	21	black, 21-speed
	24	black, 24-speed
silver	10	silver, 10-speed
	18	silver, 18-speed
	21	silver, 21-speed
	24	silver, 24-speed

13.

Quarter	Dime	Nickel	Sample Space
H	H	H	HHH
		T	HHT
	T	H	HTH
		T	HTT
T	H	H	THH
		T	THT
	T	H	TTH
		T	TTT

14.

Number Cube	Coin	Card	Sample Space
1	H	A	1HA
		B	1HB
	T	A	1TA
		B	1TB
2	H	A	2HA
		B	2HB
	T	A	2TA
		B	2TB
3	H	A	3HA
		B	3HB
	T	A	3TA
		B	3TB
4	H	A	4HA
		B	4HB
	T	A	4TA
		B	4TB
5	H	A	5HA
		B	5HB
	T	A	5TA
		B	5TB
6	H	A	6HA
		B	6HB
	T	A	6TA
		B	6TB

15. There are 8 outcomes.

Child 1	Child 2	Child 3	Sample Space
B	B	B	BBB
		G	BBG
	G	B	BGB
		G	BGG
G	B	B	GBB
		G	GBG
	G	B	GGB
		G	GGG

Page 378, Lesson 9-3 (When am I ever going to use this?)

4. Sample answer: There are 21 different sizes and lengths of junior's jeans. This is the same as the product found in Exercise 3.

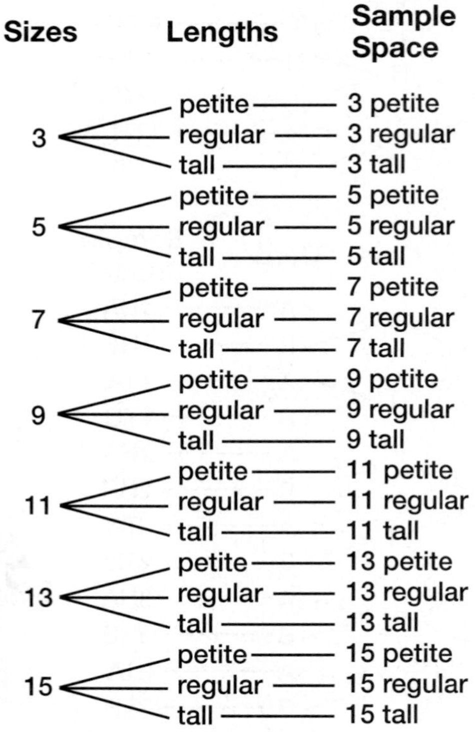

Page 380, Lesson 9-3

26. There are 12 outcomes.

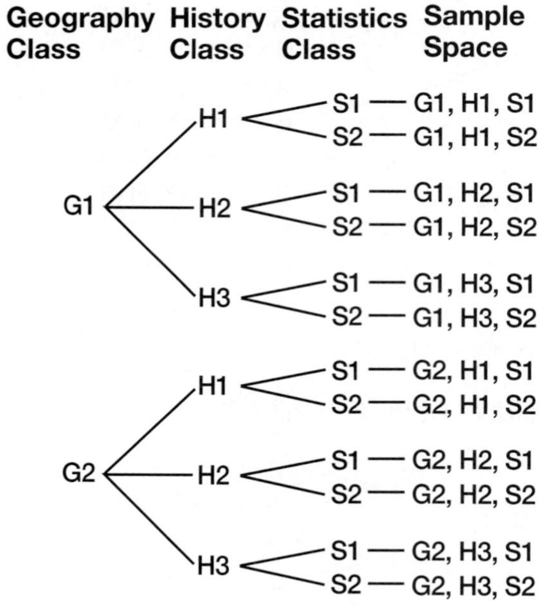

Page 383, Lesson 9-4

24. There are 6 outcomes.

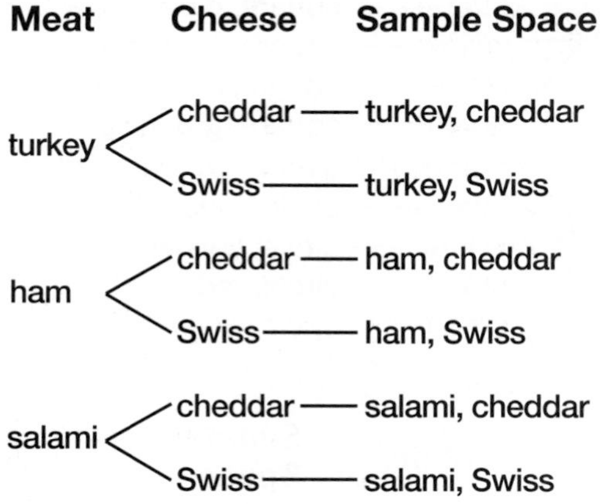

Page 384, Mid-Chapter Practice Test

7.

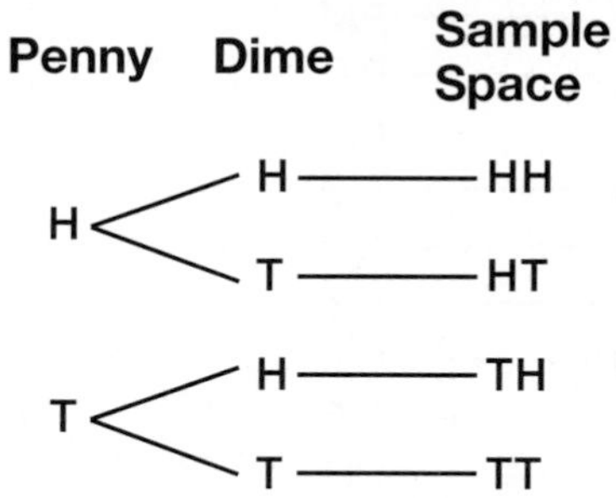

8.

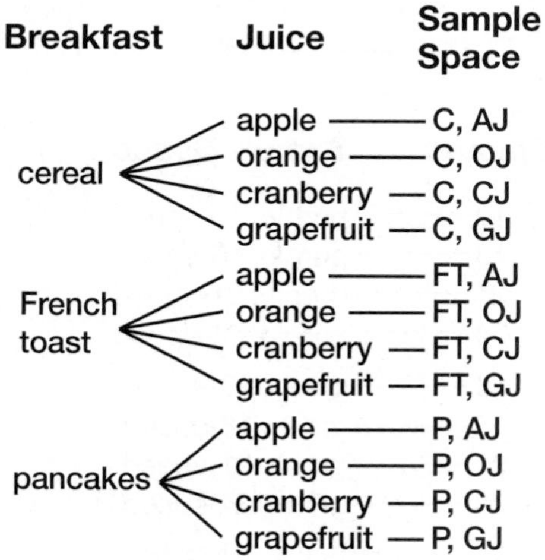

Page 397, Lesson 9-6b

7. Sample answer: Place 3 red marbles and 1 blue marble in a hat. Randomly select one marble. If the marble is red, Shannon made a free throw. If the marble is blue, Shannon missed a free throw.

8.

Page 401, Lesson 9-7

20. 3

1st Spin	2nd Spin	3rd Spin	Sample Space
B	B	B	BBB
		R	BBR
		G	BBG
	R	B	BRB
		R	BRR
		G	BRG
	G	B	BGB
		R	BGR
		G	BGG
R	B	B	RBB
		R	RBR
		G	RBG
	R	B	RRB
		R	RRR
		G	RRG
	G	B	RGB
		R	RGR
		G	RGG
G	B	B	GBB
		R	GBR
		G	GBG
	R	B	GRB
		R	GRR
		G	GRG
	G	B	GGB
		R	GGR
		G	GGG

Page 403, Study Guide and Review

17.

Number Cube	Coin	Sample Space
1	H	1H
	T	1T
2	H	2H
	T	2T
3	H	3H
	T	3T
4	H	4H
	T	4T
5	H	5H
	T	5T
6	H	6H
	T	6T

18.

Shirt	Lettering	Sample Space
red	black	red, black
	gray	red, gray
blue	black	blue, black
	gray	blue, gray
white	black	white, black
	gray	white, gray

19.

Coin	Card	Sample Space
H	1	H1
	2	H2
	3	H3
	4	H4
	5	H5
	6	H6
	7	H7
T	1	T1
	2	T2
	3	T3
	4	T4
	5	T5
	6	T6
	7	T7

20.

Bread	Meat	Sample Space
white	turkey	white,turkey
	ham	white, ham
	salami	white, salami
wheat	turkey	wheat, turkey
	ham	wheat, ham
	salami	wheat, salami
rye	turkey	rye, turkey
	ham	rye, ham
	salami	rye, salami

Page 405, Practice Test

6.

Coin 1	Coin 2	Coin 3	Sample Space
H	H	H	HHH
		T	HHT
	T	H	HTH
		T	HTT
T	H	H	THH
		T	THT
	T	H	TTH
		T	TTT

7.

Letter	Number	Sample Space
M	1	M1
	2	M2
	3	M3
A	1	A1
	2	A2
	3	A3
T	1	T1
	2	T2
	3	T3
H	1	H1
	2	H2
	3	H3

Page 407, Standardized Test Practice

18a.

Hobby	Night	Sample Space
jogging	Fri.	jogging, Fri.
	Sat.	jogging, Sat.
painting	Fri.	painting, Fri.
	Sat.	painting, Sat.
quilting	Fri.	quilting, Fri.
	Sat.	quilting, Sat.
stamp collecting	Fri.	stamp collecting, Fri.
	Sat.	stamp collecting, Sat.
listening to music	Fri.	listening to music, Fri.
	Sat.	listening to music, Sat.

UNIT 5 Notes

Introduction

In this unit, students will explore the geometry and measurement of two-dimensional and three-dimensional figures. They will learn to classify angles based on their measurements and relationships to lines and other angles. Students will construct angles, parallel lines, and triangles and bisect angles using a compass and a straightedge. They will apply their knowledge of angles to similar figures, the classification of polygons, and tessellations. Students also will learn how to transform two-dimensional figures by translating, dilating, reflecting, and rotating them on a coordinate plane.

Students then will investigate squares, square roots, and the Pythagorean Theorem as they learn how to find the area of parallelograms, triangles, and trapezoids. They will see how dividing a circle into many similar triangles can be used to estimate the area of a circle. Finally, students apply their knowledge of area to finding the volume and surface area of three-dimensional figures.

Assessment Options

Unit 5 Test Pages 719–720 of the *Chapter 12 Resource Masters* may be used as a test or a review for Unit 5. This assessment contains both multiple-choice and short-response items.

UNIT 5

Geometry and Measurement

Chapter 10
Geometry

Chapter 11
Measuring Two-Dimensional Figures

Chapter 12
Measuring Three-Dimensional Figures

In this unit, you will explore line and angle relationships, two- and three-dimensional figures, and the areas and volumes of these figures.

408 Unit 5 Geometry and Measurement

What's Math Got to Do With It? Real-Life Videos

What's Math Got to Do With It? Real-Life Videos engage students, showing them how math is used in everyday situations. Use Video 5 to discuss how polygons and area are used in construction and sports.

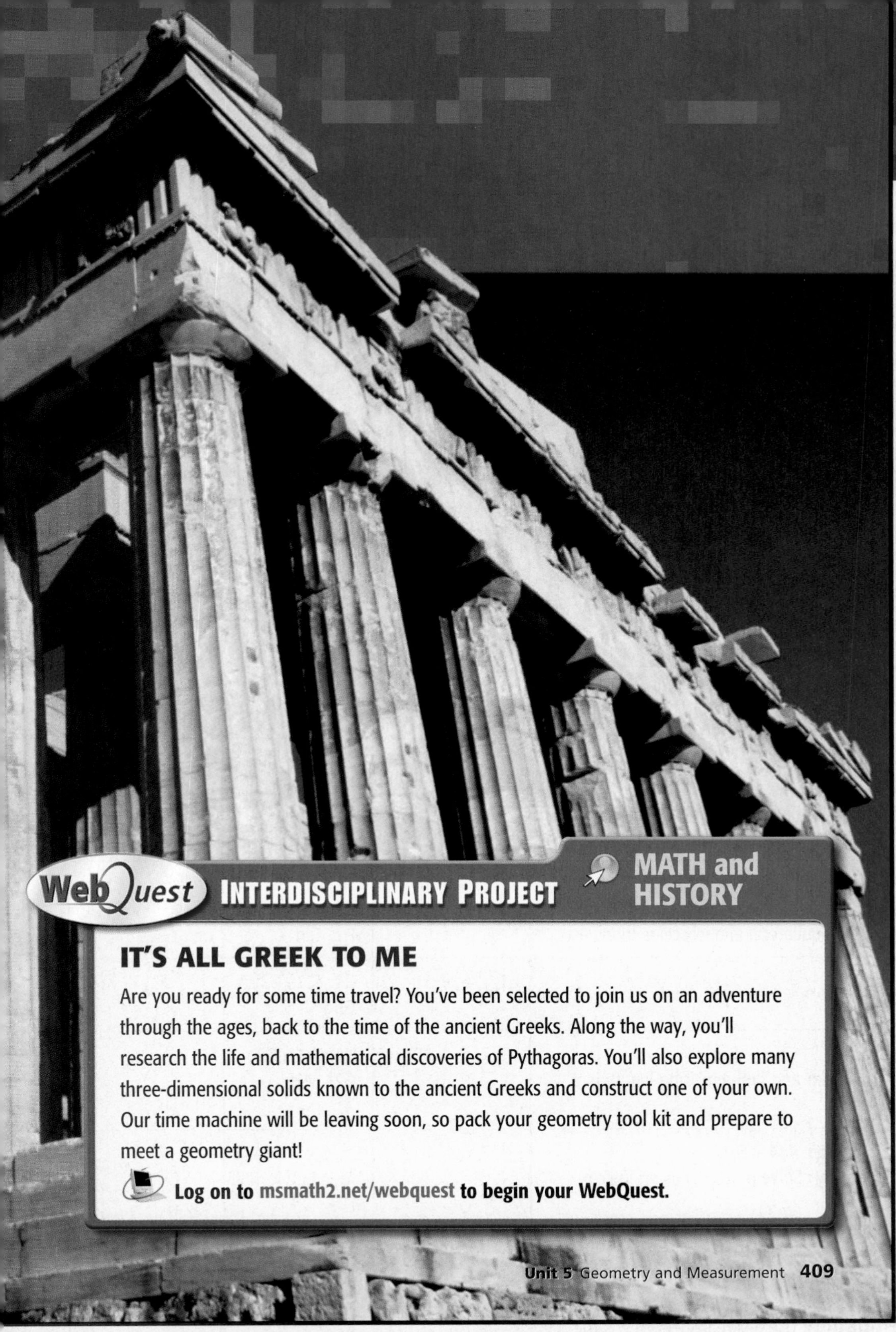

WebQuest INTERDISCIPLINARY PROJECT — MATH and HISTORY

IT'S ALL GREEK TO ME

Are you ready for some time travel? You've been selected to join us on an adventure through the ages, back to the time of the ancient Greeks. Along the way, you'll research the life and mathematical discoveries of Pythagoras. You'll also explore many three-dimensional solids known to the ancient Greeks and construct one of your own. Our time machine will be leaving soon, so pack your geometry tool kit and prepare to meet a geometry giant!

Log on to msmath2.net/webquest to begin your WebQuest.

WebQuest INTERDISCIPLINARY PROJECT

It's All Greek to Me

A WebQuest is an online project in which students do research on the Internet, gather data, and make presentations using word processing, graphing, page-making, or presentation software. In each chapter, students advance to the next step in their WebQuest. At the end of Chapter 12, the project culminates with a presentation of their findings.

Teaching suggestions and sample answers are available in the WebQuests, Projects, and Interdisciplinary Investigations.

Math and History

You can use this WebQuest Interdisciplinary Project with your students' social studies teacher to make the connection from mathematics to the history your students are studying.

More Interdisciplinary Connections

You may wish to share these suggestions with your students' other teachers.

Math and Art

Research the art and architecture of the ancient Greeks. Investigate the two- and three-dimensional shapes favored by the Greeks. Make a poster or diorama depicting examples of shapes you found in art works and buildings.

Math and Language Arts

Read about Plato (427–347 B.C.) and his views on ideal beauty and mathematics. What are the platonic solids, in particular the dodecahedron, and how do they exemplify Plato's vision of the universe?

CHAPTER 10

Chapter Overview

Geometry

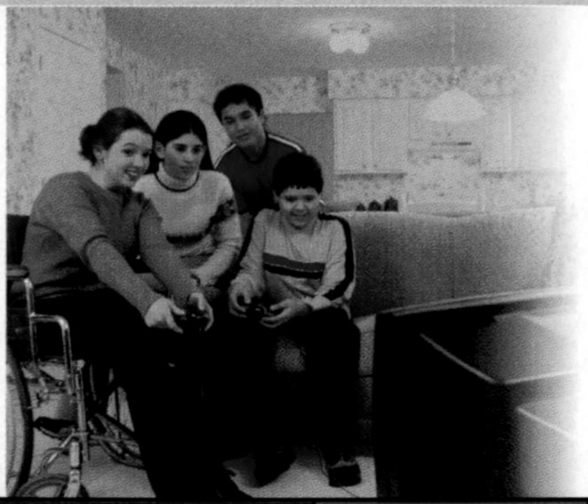

	LESSON OBJECTIVES	NCTM Standards	NAEP 2005	State/Local Objectives
10-1	**Angles** *(pp. 412–417)* **10-1a Hands-On Lab:** Measure angles. • Classify and draw angles. **10-1b Hands-On Lab:** Construct and bisect angles.	1, 2, 6, 9, 10	M1c, M1g, G1c, G1f, G4a	
10-2	**Making Circle Graphs** *(pp. 418–421)* • Construct and interpret circle graphs.	1, 2, 6, 9, 10	D1a, D1b, N1e, N3a, D1e, M1g	
10-3	**Angle Relationships** *(pp. 422–427)* • Identify and apply angle relationships. **10-3b Hands-On Lab:** Construct parallel lines, and discover angle relationships.	1, 2, 6, 9, 10	G1f, A3a, A4a, A4b, A4c, G3b, G3g	
10-4	**Triangles** *(pp. 428–433)* • Identify and classify triangles. **10-4b Hands-On Lab:** Construct equilateral and isosceles triangles.	1, 2, 6, 9, 10	G1c, G1f, G3b, G2f, A3a, A4a, A4b, A4c, M1g, G3f	
10-5	**Quadrilaterals** *(pp. 434–437)* • Identify and classify quadrilaterals.	1, 2, 6, 9, 10	G1c, G1f, G3f, G3g	
10-6	**Similar Figures** *(pp. 440–443)* • Determine whether figures are similar and find a missing length in a pair of similar figures.	1, 2, 6, 9, 10	N4b, N4c, G2e, G2f, M1k	
10-7	**Polygons and Tessellations** *(pp. 444–450)* **10-7a Problem-Solving Strategy:** Solve problems using logical reasoning. • Classify polygons and determine which polygons can form a tesselation.	1, 2, 6, 9, 10	G1c, G1f, G3b, A3a, A4a, A4b, A4c, N5f, M1h	
10-8	**Translations** *(pp. 451–455)* • Graph translations of polygons on a coordinate plane. **10-8b Spreadsheet Investigation:** Use a spreadsheet to enlarge and reduce polygons.	1, 2, 6, 9, 10	G1f, G2d, G2c, A2c, A1d, G4d, A2d	
10-9	**Reflections** *(pp. 456–461)* • Identify figures with line symmetry and graph reflections on a coordinate plane. **10-9b Hands-On Lab:** Graph rotations on a coordinate plane.	1, 2, 6, 9, 10	G2a, A2c, A2d, G2c, G4d	

Key to NCTM Standards:
1=Number & Operations, 2=Algebra, 3=Geometry, 4=Measurement, 5=Data Analysis & Probability, 6=Problem Solving, 7=Reasoning & Proof, 8=Communication, 9=Connections, 10=Representation

NAEP 2005 Standards: See math.glencoe.com

Pacing suggestions for the entire year can be found on page T20.

Chapter Resource Manager

Time-Saving Tool
TeacherWorks™
All-In-One Planner and Resource Center
See page T5.

FAST FILE — CHAPTER 10 RESOURCE MASTERS

	Study Guide and Intervention	Practice (Skills and Word Problems)	Reading to Learn Mathematics	Enrichment	Assessment	Applications*	Parent and Student Study Guide Workbook	5-Minute Check Transparencies	Key Teaching Transparencies	MathPASS: Tutorial Plus (lessons)	Teaching Mathematics with Manipulatives	Optional Materials
	543	544–555	546	547		SM 41–44	10-1	10-1			90, 91	straightedge, compass
	548	549–550	551	552		CS 47	10-2	10-2	37		92	
	553	554–555	556	557	603		10-3	10-3		43	93	straightedge, compass, colored pencils, notebook paper
	558	559–560	561	562			10-4	10-4			94	paper, scissors, straightedge, compass, ruler
	563	564–565	566	567	603, 605		10-5	10-5			95–96	
	568	569–570	571	572		CS 48	10-6	10-6	38	44		
	573	574–575	576	577	604	CM 29–31 CS 49	10-7	10-7	39		97, 98–99, 100–101	
	578	579–580	581	582		CS 50	10-8	10-8			102–103	paper, scissors
	583	584–585	586	587	589–602, 604, 606–608	RC 24	10–9	10-9	40		104	geomirror, paper, ruler, graph paper

*** Key to Abbreviations:** *CS = Calculator and Spreadsheet Masters, RC = Real-Life Career Masters, CM = Classroom Math Games (online), SM = Science and Math Lab Manual (online)*

CHAPTER 10 Mathematical Content and Teaching Strategies

Continuity of Instruction

Building on Prior Knowledge

In previous chapters, students have examined different ways of displaying data. Students used these methods to reach conclusions about the data. Students have also examined basic algebraic equations and relations. Proportions were also used to solve problems.

Learning in This Chapter

In this chapter, students investigate spatial relations. Students discover the relationships between angles. They use their knowledge of angles to create circle graphs. They learn to classify different triangles and quadrilaterals, and how to construct these types of figures. Students will also be able to use similar figures to find an unknown length. They will use their knowledge of polygons to create tessellations. At the end of the chapter, students will investigate four types of transformations: translations, dilations, reflections, and rotations.

Making Future Connections

Understanding basic angle and polygon relationships is crucial to an understanding of geometry. Constructions also help students gain a deeper understanding of spatial relations. The concept of similar figures is used in scale drawings. Tessellations and transformations help students gain an understanding of how geometric figures can be manipulated.

Mathematical Background

10-1 Angles

In this lesson, students learn the basic definitions of four angle types. These definitions will be used to classify different types of triangles and quadrilaterals. Students also practice drawing angles with a protractor, which builds the foundation for constructing geometric figures in later lessons.

10-2 Making Circle Graphs

Circle graphs are a common way to display data. In this lesson, students learn how to construct a circle graph with appropriate angles for different percents. Students will also interpret data displayed in a circle graph. Knowledge of proportions and ratios will be necessary to determine proper angles. Remind students that there are other types of graphs and that circle graphs are the best choice for displaying parts of a whole.

Using the Media

Have students complete a project of their choice on circle graphs by using information from the media. Students might find a number of examples of circle graphs in newspapers and magazines and interpret the information. Or, students could research numerical data in the media, such as survey results, and create their own circle graphs.

10-3 Angle Relationships

In this lesson, students explain relationships between angles. These relationships will be used in numerous future lessons about geometry and include important vocabulary to use in discussing spatial relations. For example, in physics, the law of reflection states the angle of incidence equals the angle of reflection. Students will need to understand angle relationships to understand and apply this law.

Teacher to Teacher

Judy Lecocq
Murphysboro Middle School
Murphysboro, IL

"I teach definitions of complementary and supplementary angles by considering the *p* in each word as a backward 9. Therefore, one *p* in *complementary* represents one 9, or 90 degrees. The two *p*'s in *supplementary* represent two 9's, or 180 degrees."

Judy Lecocq

10-4 Triangles

In this lesson, students use their knowledge of angles to classify triangles and to find missing angle measures. These skills are important in solving algebraic equations and provide a vital foundation for trigonometry, which is based on the properties of triangles.

10-5 Quadrilaterals

In this lesson, students learn to classify quadrilaterals by their sides and by their angles. Students continue to practice their algebra skills by finding missing angle measures.

10-6 Similar Figures

Students will use their knowledge of proportion to determine if figures are similar and to find the lengths of sides in similar figures. The ability to set up ratios and solve for one unknown in equivalent ratios will be used repeatedly in future math and science classes.

McGraw-Hill Professional Development

Mathematics, YES! is a a self-paced, web-based professional development program. Units are based on major mathematical concepts and are standards-based. Actual classroom videos provide a springboard for discussion and activities. Unit appropriate for this chapter:

- Unit 8, Geometry and Measurement of Plane Figures

10-7 Polygons and Tessellations

When students first think of polygons, they will probably think of regular polygons, such as squares and stop-sign-shaped figures (regular hexagons). Remind them that a polygon is any figure enclosed by line segments that meet only at their endpoints. To help students visualize the way in which a regular polygon can be divided into triangles, have them draw several regular polygons on cardboard and then cut out the triangles.

Bulletin Board Starter

Have students create a shape that will tessellate. Then have them create a tessellation and display their artwork as well as the original shape on a bulletin board. Students might include short descriptions of why their shapes tessellate.

10-8 Translations

Students will learn to translate figures on a coordinate grid. Many students, especially visual learners, will benefit from hands-on activities that involve making a tesselation by performing translations.

10-9 Reflections

Students begin their study of reflections by identifying lines of symmetry in different figures. Line symmetry applies to the study of biology; as students are introduced to the scientific classification of animals, they will need to indicate whether an animal has radial symmetry, bilateral symmetry, or no symmetry.

Key Concepts

Additional mathematical information and teaching notes are available at

msmath2.com/key_concepts.

The section appropriate for the chapter is:

- Triangles and Quadrilaterals

CHAPTER 10

DAILY INTERVENTION and Assessment

	Type	Student Edition	Teacher Resources	Technology/Internet
INTERVENTION	Ongoing	Prerequisite Skills, pp. 411, 415, 421, 425, 431, 437, 443, 450 Mid-Chapter Practice Test, p. 438	5-Minute Check Transparencies Quizzes 1–4, *CRM*, pp. 603–604 Mid-Chapter Practice Test, *CRM*, p. 605 Study Guide and Intervention, *CRM*, pp. 543, 548, 553, 558, 563, 568, 573, 578, 583	MathPASS: Tutorial Plus msmath2.net/self_check_quiz msmath2.net/extra_examples
INTERVENTION	Mixed Review	pp. 415, 421, 425, 431, 437, 443, 450, 454, 459	Cumulative Review, *CRM*, p. 606	
INTERVENTION	Error Analysis	Find the Error, p. 435	Find the Error, *TWE*, p. 436 Unlocking Misconceptions, *TWE*, pp. 413, 456 Tips for New Teachers, *TWE*, pp. 422, 441	
ASSESSMENT	Standardized Test Practice	pp. 415, 421, 425, 431, 437, 438, 441, 443, 445, 450, 454, 459, 465, 466–467	Standardized Test Practice, *CRM*, pp. 607–608	Standardized Test Practice CD-ROM msmath2.net/standardized_test
ASSESSMENT	Open-Ended Assessment	Writing Math, pp. 412, 414, 417, 420, 424, 427, 430, 433, 435, 442, 448, 453, 458, 461 Open-Ended, pp. 414, 420, 424, 430, 435, 442, 448, 453, 458 Standardized Test, p. 467	Modeling: *TWE*, pp. 415, 431 Speaking: *TWE*, pp. 443, 445, 459 Writing: *TWE*, p. 437 Extended Response Assessment, *CRM*, p. 601	
ASSESSMENT	Chapter Assessment	Study Guide, pp. 462–464 Practice Test, p. 465	Multiple-Choice Tests (Forms 1, 2A, 2B), *CRM*, pp. 589–594 Free-Response Tests (Forms 2C, 2D, 3), *CRM*, pp. 595–600 Vocabulary Test/Review, *CRM*, p. 602	ExamView® Pro Test Bank (see below) MindJogger Videoquizzes msmath2.net/vocabulary_review msmath2.net/chapter_test

*** Key to Abbreviations:** *TWE* = Teacher Wraparound Edition, *CRM* = Chapter Resource Masters

Additional Intervention

Skills Intervention for Middle School Mathematics workbook and CD-ROM includes 88 skills with pretests, tutorials, guided practice, and posttests for diagnosis and remediation.

Guide to Daily Intervention booklet details each intervention resource available and guides teachers in using them as intervention tools everyday to assist students who may be having difficulty.

ExamView® Pro

Use ExamView® Pro Test Bank CD-ROM to:

- **Create Multiple versions of tests.**
- **Create Modified tests for Inclusion students with one mouse click.**
- **Edit existing questions and Add your own questions.**
- **Build tests aligned with state standards using built-in State Curriculum Correlations.**
- **Change English tests to Spanish with one mouse click and vice versa.**

Reading and Writing Mathematics

• Student Edition

Foldables Study Organizer, p. 411

Concept Check questions at the beginning of each exercise set require students to verbalize and write about what they have learned in the lesson.

- **Open Ended,** pp. 414, 420, 424, 430, 435, 442, 448, 453, 458
- **Find the Error,** p. 435
- **Which One Doesn't Belong?,** pp. 453, 458

Reading Math Tip, pp. 441, 460

• Teacher Wraparound Edition

Foldables Follow-Up, pp. 430, 436, 438, 442, 448

Writing Activities, p. 437

Differentiated Instruction, (Verbal/Linguistic), p. 418, 434

ELL Resources, *see page T9 for a complete listing.*

• Additional Resources

Vocabulary Builder, *Chapter 10 Resource Masters*

Reading to Learn Mathematics, *Chapter 10 Resource Masters*

Vocabulary PuzzleMaker (online)

Teaching Mathematics with Foldables

Reading and Writing in the Mathematics Classroom

WebQuests, Projects, and Interdisciplinary Investigations

For detailed information on these resources, see pp. T6–T7.

• Recommended Outside Reading for Students

Nonfiction

- Arnone, Wendy. *Geometry for Dummies.* John Wiley & Sons, ©2001.

  This book is an easy-to-read introduction to geometry. It contains material on vocabulary and terms, angles, and angle relationships, and career connections with geometry.

Fiction

- Neuschwander, Cindy. *Sir Cumference and the Great Knight of Angleland: A Math Adventure.* Illustrator: Wayne Geehan. Charlesbridge Publishing, ©2001.

  This is the third book in the *Sir Cumference* series. In this adventure, Radius needs to earn his knighthood by completing tasks, examining every angle along the way. Students will be able to apply their knowledge of angles and shapes when reading this book.

PROJECT CRISS℠ Study Skill

Concept maps can be developed as part of a class discussion to help students understand newly introduced mathematical relationships. After students have read Lesson 10-1, write "Types of Angles" on the chalkboard, as shown in the sample map at the right. Allow students to complete the concept map. Encourage them to make drawings to accompany their descriptions. Students can develop similar concept maps as part of class discussions about "Types of Triangles" and "Types of Quadrilaterals."

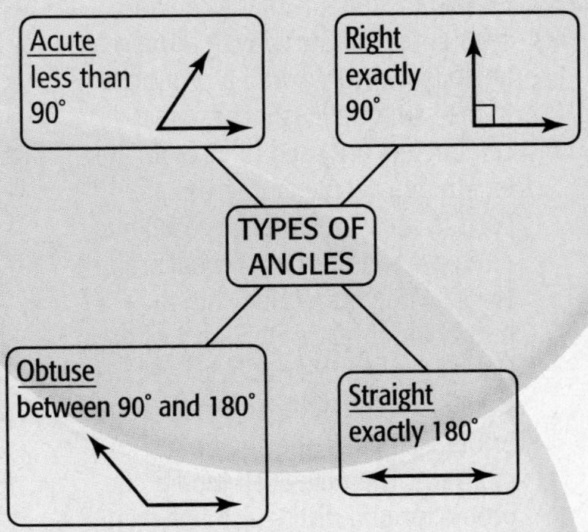

CReating Independence through Student-owned Strategies

For more information on Project CRISS, see page T9.

CHAPTER 10 Making Connections

School-To-Home

Parent and Student Study Guide

Parents and students may work together to reinforce the concepts and skills of this chapter. These workbooks are available in English and Spanish. A sample lesson is shown below. Also available at msmath2.net/parent_student.

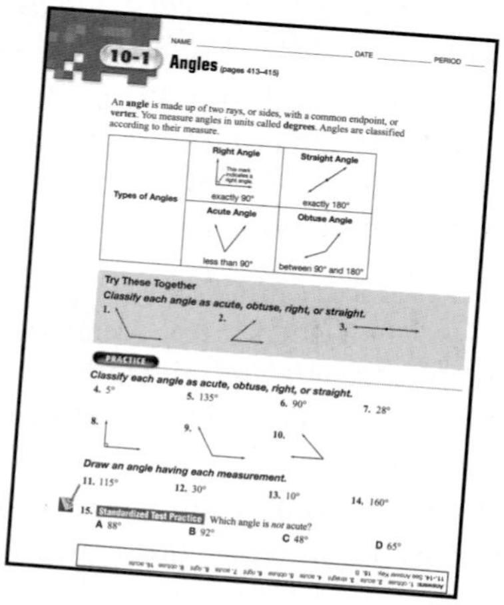

10-1 Angles (pages 413–415)

An **angle** is made up of two rays, or sides, with a common endpoint, or **vertex**. You measure angles in units called **degrees**. Angles are classified according to their measure.

Types of Angles	Right Angle (exactly 90°)	Straight Angle (exactly 180°)
	Acute Angle (less than 90°)	Obtuse Angle (between 90° and 180°)

Try These Together

Classify each angle as acute, obtuse, right, or straight.

PRACTICE

Classify each angle as acute, obtuse, right, or straight.

4. 5° 5. 135° 6. 90° 7. 28°

Draw an angle having each measurement.

11. 115° 12. 30° 13. 10° 14. 160°

15. Standardized Test Practice Which angle is *not* acute?
A 88° B 92° C 48° D 65°

Family Letters and Activities

Send a letter and activity for this chapter home to encourage family members to become active participants in their students' learning. (English: *Chapter 10 Resource Masters*, pp. ix–x; Spanish: log on to msmath2.net/family_letters)

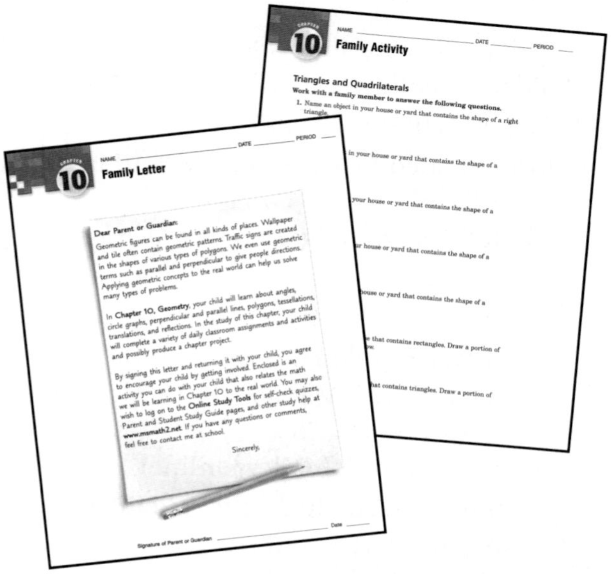

10 Family Activity

Triangles and Quadrilaterals

Work with a family member to answer the following questions.

1. Name an object in your house or yard that contains the shape of a right triangle.

10 Family Letter

Dear Parent or Guardian:

Geometric figures can be found in all kinds of places. Wallpaper and tile often contain geometric patterns. Traffic signs are created in the shapes of various types of polygons. We even use geometric terms such as parallel and perpendicular to give people directions. Applying geometric concepts to the real world can help us solve many types of problems.

In Chapter 10, Geometry, your child will learn about angles, circle graphs, perpendicular and parallel lines, polygons, tessellations, translations, and reflections. In the study of this chapter, your child will complete a variety of daily classroom assignments and activities and possibly produce a chapter project.

By signing this letter and returning it with your child, you agree to encourage your child by getting involved. Enclosed is an activity you can do with your child that also relates the math we will be learning in Chapter 10 to the real world. You may also wish to log on to the Online Study Tools for self-check quizzes, Parent and Student Study Guide pages, and other study help at www.msmath2.net. If you have any questions or comments, feel free to contact me at school.

Sincerely,

Signature of Parent or Guardian Date

English Language Learner ELL

Lesson 10-1a

Building on Prior Knowledge

Explain to students that as a line segment is measured with a ruler, an angle is measured with a protractor. In the same way, as inches or centimeters are used to indicate the length of a segment, degrees are used to measure an angle. Students often get confused as to what scale to use when measuring with a protractor. If following Step 2 of the Hands-On Lab doesn't help, ask students to label the angles as acute or obtuse before they begin to measure. Then they can look at the protractor and distinguish between which measure is reasonable for the angle they are measuring.

Lesson 10-8

Language Experience Approach to Illustrations

Write the words *transformation* and *translation* on the chalkboard. Ask students what these words have in common. Underline the prefix *trans* and have the students look this up in a dictionary or the Internet. They will find that trans means "across". This will help students to remember that they are moving an object or polygon across a plane.

Lesson 10-9

Flexible Groups

Give students two pieces of grid paper with the origin and the *x*- and *y*-axes labeled. Have the students draw a polygon in Quadrant I and label each vertex with its ordered pair. Have them copy the polygon onto the second piece of grid paper. Have students fold the new graph along the *y*-axis so that the graph is on the outside of the paper. Then, flip the paper over and reproduce the original shape in Quadrant II. Point out that they have reflected the polygon over the *y*-axis. Students can work together to reflect the polygon over the *x*-axis.

MathScape™ This project was supported, in part, by the **National Science Foundation**

MathScape is a mathematics curriculum for grades 6–8 developed by the Seeing and Thinking Mathematically Project at the Education Development Center.

Exploring Two-Dimensional Figures

Getting in Shape

Summary Students will measure, classify, and investigate angles. Classification of triangles is determined as well as the exploration of quadrilateral and polygons. Students will investigate tessellations to make designs.

USA TODAY Education Online Lesson Planner

USA TODAY® Education's Online site offers resources and interactive features connected to each day's newspaper. *Experience Today,* USA TODAY's® daily lesson plan, is available on the site and delivered daily to subscribers. This plan provides instruction for integrating USA TODAY® graphics and key editorial features into your mathematics classroom. Log on to **education.usatoday.com**.

USA TODAY Snapshots® in Chapter 10	
Lesson	**Snapshot**
10-2	Teens feel political power
10-3	Where the wild things are

Quick Review Math Handbook

is Glencoe's mathematical handbook for students. The Hot Words section includes a glossary of terms while the Hot Topics section consists of explanations of key mathematical concepts. An exercise set is included to check students' understanding of the concepts. This valuable resource can be used as a reference in the classroom or for home study.

Lesson	Hot Topics Section	Lesson	Hot Topics Section
10-1	7-1, 9-3	**10-6**	6-5
10-2	7-8	**10-7**	6-7, 7-2
10-3, 10-4	7-1, 9-3	**10-8**	6-7, 7-3, 9-4
10-5	7-2	**10-9**	6-7, 7-3

CHAPTER 10 Notes

Key Chapter Objectives

Lesson 10-3 Identify and apply angle relationships.

Lessons 10-4 and 10-5 Identify and classify triangles and quadrilaterals.

Lesson 10-6 Determine whether figures are similar and find a missing length in a pair of similar figures.

"What do video games have to do with math?"

Point out to students that this is only one of many applications of the math that they are studying. Others are provided in the introduction to each lesson.

Vocabulary Builder, pp. vii–viii ELL

CHAPTER 10 NAME ______ DATE ______ PERIOD ____

Reading to Learn Mathematics

Vocabulary Builder

This is an alphabetical list of new vocabulary terms you will learn in Chapter 10. As you study the chapter, complete each term's definition or description. Remember to add the page number where you found the term. Add these pages to your math study notebook to review vocabulary at the end of the chapter.

Vocabulary Term	Found on Page	Definition/Description/Example
acute triangle		
complementary angles		
congruent angles		
congruent segments		
equilateral [EH-kwuh-LA-tuh-rul] triangle		
indirect measurement		
isosceles [y-SAHS-LEEZ] triangle		
line symmetry		
obtuse triangle		
parallelogram		

CHAPTER 10 Geometry

"What do video games have to do with math?"

The first video game programmers used quadrilaterals like rectangles, squares, and trapezoids in designing video games. Now they use these basic figures as a starting point to make the games more elaborate with real-life characters, breath-taking scenery, and silly animations.

You will solve problems about quadrilaterals in Lesson 10-5.

Vocabulary Builder

For a thorough vocabulary list with pronunciations of new words, give students the Vocabulary Builder worksheets from the *Chapter 10 Resource Masters.* Encourage them to complete the definition of each term as they progress through the chapter.

GETTING STARTED

Take this quiz to see whether you are ready to begin Chapter 10. Refer to the lesson or page number in parentheses if you need more review.

Vocabulary Review

Complete each sentence. **1. ratio**

1. A comparison of two numbers by division is called a __?__. (Lesson 7-1)
2. The __?__ is written as part = percent · base. (Lesson 8-2) **percent equation**

Prerequisite Skills

Multiply or divide. Round to the nearest hundredth if necessary. (Pages 560 and 562)

3. $360 \cdot 0.85$ **306**
4. $48 \div 191$ **0.25**
5. $24 \div 156$ **0.15**
6. $0.37 \cdot 360$ **133.2**
7. $33 \div 307$ **0.11**
8. $0.69 \cdot 360$ **248.4**

Solve each equation. (Lessons 4-2 and 4-3)

9. $b + 36 = 89$ **53**
10. $74 = 22 + s$ **52**
11. $15 + r = 146$ **131**
12. $m + 78 = 93$ **15**
13. $153 = d + 61$ **92**
14. $6x = 360$ **60**
15. $180 = 2f$ **90**
16. $120 = 3c$ **40**
17. $5n = 270$ **54**
18. $15z = 90$ **6**

Solve each proportion. (Lesson 7-3)

19. $\frac{4}{a} = \frac{3}{9}$ **12**
20. $\frac{7}{16} = \frac{h}{32}$ **14**
21. $\frac{5}{8} = \frac{15}{y}$ **24**
22. $\frac{t}{42} = \frac{6}{7}$ **36**
23. $\frac{s}{12} = \frac{2}{3}$ **8**
24. $\frac{24}{p} = \frac{2}{3}$ **36**

FOLDABLES™ Study Organizer

Geometry Make this Foldable to help you organize your notes. Begin with a sheet of 11″ × 17″ paper and six index cards.

STEP 1 Fold
Fold lengthwise about 3″ from the bottom.

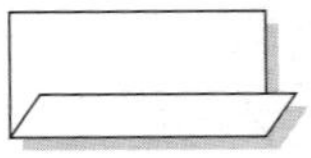

STEP 2 Fold Again
Fold the paper in thirds.

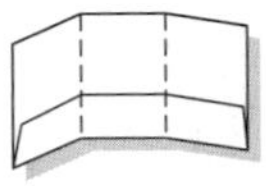

STEP 3 Open and Staple
Staple the edges on either side to form three pockets.

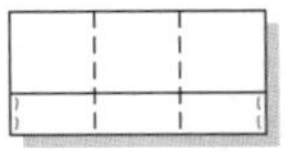

STEP 4 Label
Label the pockets as shown. Place two index cards in each pocket.

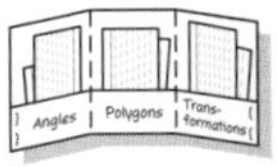

Reading and Writing As you read and study the chapter, write the properties and characteristics for angles, polygons, and transformations on the index cards.

GETTING STARTED

This section provides a review of the basic concepts needed before beginning Chapter 10. Page references are included for additional student help.

Prerequisite Skills review skills needed in this chapter.

Exercises	Needed for Lesson	Skill
3–8	10-2	Multiplying and Dividing Rational Numbers
9–18	10-3, 10-4, 10-5, 10-7	Solving Equations
19–24	10-6	Solving Proportions

For more prerequisite skills, see:

- Getting Ready for the Next Lesson, at the end of each lesson
- Prerequisite Skills in the Student Handbook on pp. 554–563
- *Math Skills Maintenance Masters*

FOLDABLES™ Study Organizer

For more information about Foldables, see *Teaching Mathematics with Foldables.*

Use this Foldable to collect student writing about geometry. Students use 3″ x 5″ index cards or quarter sheets of notebook paper to take notes, define terms, record concepts, and write examples. As students progress through the lessons in this chapter, note the order in which geometry concepts are presented. Ask students to explain in writing why the concepts and computations in this chapter were presented to them in this sequence.

10-1a Hands-On Lab
Lesson Notes

Lesson Resources

Teaching Mathematics with Manipulatives
- protractors, p. 21
- student recording sheet 10-1a

Glencoe Mathematics Classroom Manipulative Kit
- protractors

GETTING STARTED

Teaching Tip Make sure that students have their own protractors, even when working with a partner.

TEACH

- Students do not yet know the vocabulary *acute* and *obtuse*, but you can tell them to use the smaller number on a protractor for small angles and the bigger number for big angles. This tip will help them focus on the correct scale as they measure the angles.
- If the angles are too short to measure easily with a protractor, have students extend the rays with a ruler or straightedge.

ASSESS

Observe as students solve the **Your Turn** exercises. They should:
- take enough time for both partners to measure each angle, and
- be able to explain why they chose the inner or the outer scale for the angles they measured.

10-1a Hands-On Lab

What You'll Learn
Measure angles.

Materials
- protractor

Measuring Angles

To measure lines, you use a ruler. To measure angles, you use a *protractor*. Angles are measured in units called **degrees**. In this lab, you will learn how to measure angles using a protractor.

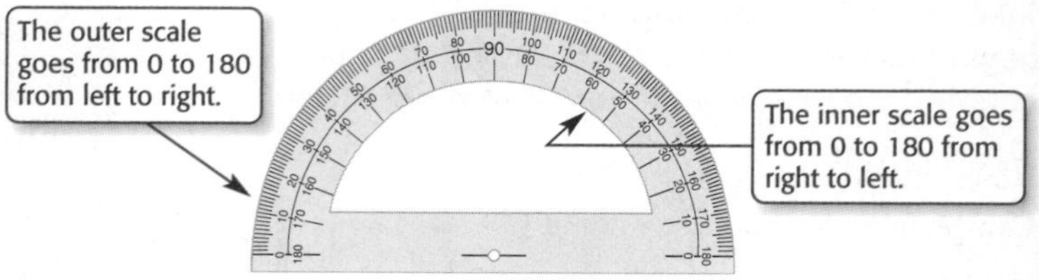

ACTIVITY *Work with a partner.*

Find the measure of the angle below.

STEP 1 Place the protractor on the angle so that the center is on the vertex of the angle and one side goes through 0° on the protractor.

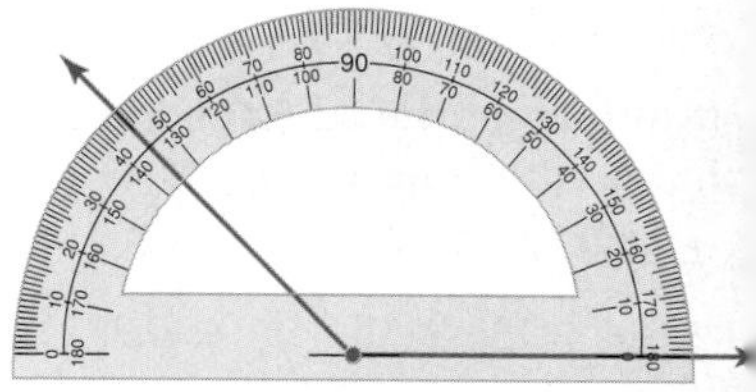

STEP 2 In this case, 0° is on the inner scale. So, follow the inner scale to the point where the other side of the angle meets the protractor. The inner number is the angle's degree measure.

So, the measure of this angle is 135°.

Your Turn **Find the measure of each angle.**

a. 75°

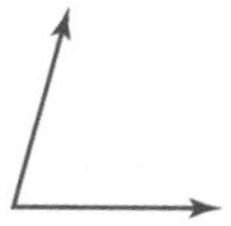

b. 118°

c. 42°

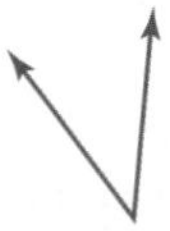

Writing Math

1. **Explain** why there are two scales on the protractor.
2. Can you place either side of an angle through 0° and get the same angle measure? **Explain** your reasoning. **See margin.**
3. What angle has the same measure on both the inner and outer scales?
4. **Explain** how the two scales on the protractor are related.

1. Sample answer: So you can find the measure of an angle from the left or from the right.
3. 90°
4. Sample answer: Both scales go from 0° to 180°.

Discuss the **Writing Math** exercises. Have students demonstrate their answers to Exercise 2.

Answer

2. Yes; Sample answer: Whichever side of the angle you place through 0°, you have to read the correct scale.

10-1 Angles

/hat You'll Learn
assify and draw
ngles.

EW Vocabulary
ngle
egrees
ertex
cute angle
ght angle
btuse angle
raight angle

LINK To Reading
/veryday Meaning
f Acute:
haracterized by
harpness, as in
cute pain

TEACHING TIP
Show students that a ray
starts at a point and goes
without end in one
direction.

WHEN am I ever going to use this?

CLOCKS The hour and minute hands of a clock form angles of different sizes.

3:10	3:00	3:40
less than 90°	90°	greater than 90°

1. Name other times in which the hands of a clock form an angle less than 90°, equal to 90°, and greater than 90°.
2. How many degrees is the angle that is formed by clock hands at 6:00? **180°**

1. Sample answer: 7:45; 9:00; 2:50

An **angle** is made up of two rays with a common endpoint and is measured in units called **degrees**. If a circle were divided into 360 equal-sized parts, each part would have an angle measure of 1 degree.

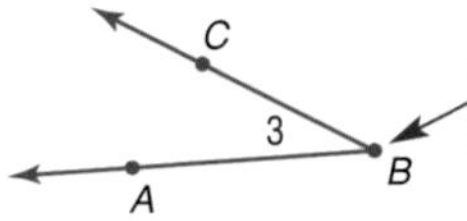

The point where the sides meet is called the **vertex**.

You can use the symbol for angle, ∠, to name an angle.

Use the vertex as the middle letter and a point from each side.	∠*ABC* or ∠*CBA*
Use the vertex only.	∠*B*
Use a number.	∠3

Angles are classified according to their measure.

Key Concept — **Types of Angles**

Acute Angle

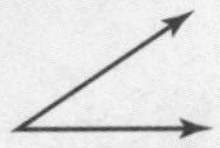

less than 90°

Right Angle

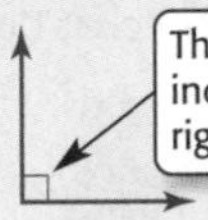

This symbol indicates a right angle.

exactly 90°

Obtuse Angle

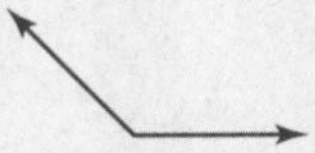

between 90° and 180°

Straight Angle

exactly 180°

10-1 Lesson Notes

Lesson Resources

Chapter 10 Resource Masters FAST FILE
- Study Guide and Intervention, p. 543
- Practice: Skills, p. 544
- Practice: Word Problems, p. 545
- Reading to Learn Mathematics, p. 546
- Enrichment, p. 547

Parent and Student Study Guide Workbook, 10-1

Science and Math Lab Masters, pp. 41–44

1 Focus

5-Minute Check 10-1 INTERACTIVE CHALKBOARD PowerPoint® Presentations

Use as a quiz or review of Chapter 9. (Also available as a transparency)

Bellringer

Hands-On Activity Have students find examples of lines meeting to form angles in the classroom. Ask them to describe the angles and to suggest different ways to categorize them. For example, a student might suggest that the edges of books and the edges of desks meet to form angles that are the same.

Reading to Learn Mathematics, p. 546 ELL

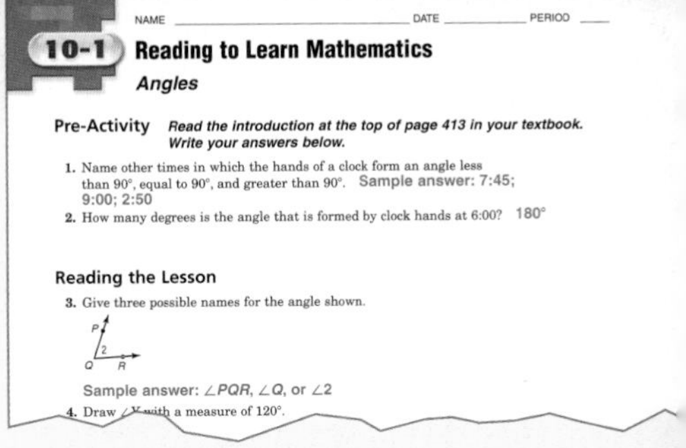
NAME ______ DATE ______ PERIOD ____

10-1 Reading to Learn Mathematics
Angles

Pre-Activity *Read the introduction at the top of page 413 in your textbook. Write your answers below.*

1. Name other times in which the hands of a clock form an angle less than 90°, equal to 90°, and greater than 90°. Sample answer: 7:45; 9:00; 2:50
2. How many degrees is the angle that is formed by clock hands at 6:00? 180°

Reading the Lesson

3. Give three possible names for the angle shown.

Sample answer: ∠*PQR*, ∠*Q*, or ∠2

4. Draw ∠... with a measure of 120°.

DAILY INTERVENTION

Unlocking Misconceptions

Using a Protractor Many students have difficulty using a protractor. In particular, they have difficulty determining which set of numbers to use. By encouraging students to determine whether an angle is acute or obtuse before measuring, they should be able to determine which number on the protractor is correct for that angle.

2 Teach

In-Class Examples

INTERACTIVE CHALKBOARD
PowerPoint® Presentations

Classify each angle as *acute, obtuse, right,* or *straight.*

1  straight

2 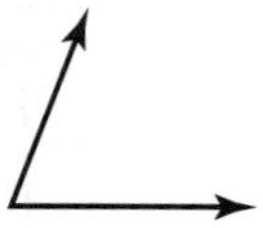 acute

3 Draw an angle with a measure of 140°. Classify the angle as *acute, obtuse, right,* or *straight.* obtuse

Tips for New Teachers

Using Manipulatives Insist that students use a straightedge when drawing lines. This will ensure that their figures are accurate.

3 Practice/Apply

About the Exercises...

Guided Practice Key

Exercises	Examples
3–4, 8	1, 2
5–7	3

Odd/Even Assignments
Exercises 12–17 are structured so that students practice the same concepts whether they are assigned odd or even problems.

TEACHING TIP

Students can use the corner of a piece of paper to help them determine if angles are acute, right, or obtuse. They can also use the edge of the paper to help them determine if an angle is straight.

EXAMPLES Classify Angles

Classify each angle as *acute, obtuse, right,* or *straight.*

1 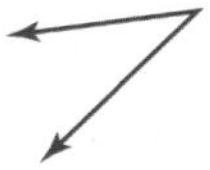

The angle is less than 90°, so it is an acute angle.

2 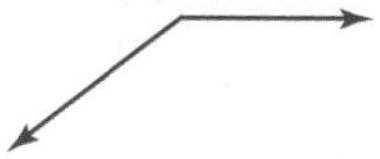

The angle is greater than 90°, so it is an obtuse angle.

EXAMPLE Draw an Angle

3 **Draw ∠*A* with a measure of 75°.**

- Draw a ray with endpoint *A*.
- Place the center point of a protractor on *A*. Align the ray with 0°.
- Use the scale that begins with 0°. Locate the mark labeled 75. Then draw the other side of the angle.

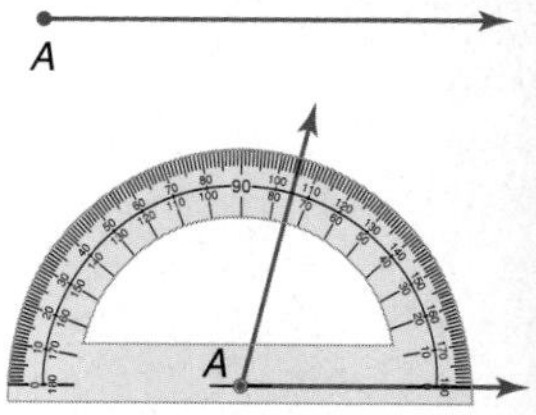

Skill and Concept Check

Writing Math
Exercise 1

1. **Describe** a right angle. 1–2. See margin.
2. **OPEN ENDED** Draw and label acute angle *XYZ*. Then find its measure.

GUIDED PRACTICE

Classify each angle as *acute, obtuse, right,* or *straight.*

3.  obtuse
4. 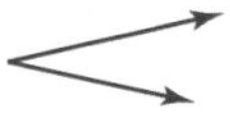 acute

Draw an angle having each measurement. Then classify the angle as *acute, obtuse, right,* or *straight.* 5–7. See margin for angles.

5. 90° right
6. 170° obtuse
7. 11° acute
8. **BICYCLE SAFETY** A cyclist should be familiar with arm and hand signals indicating his or her movements. Describe the angles formed by the arm of the cyclist below. See margin.

msmath2.net/extra_examples

Answers

1. Sample answer: An angle with a measure of exactly 90°, which is indicated by a right angle symbol.
2. Sample answer:

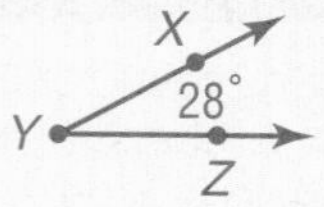

5. 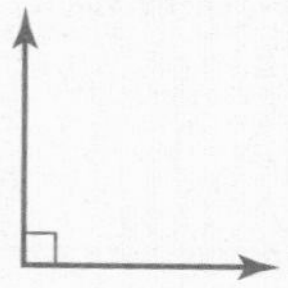

6. 

7. 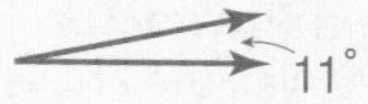

8. signal left turn: straight angle; signal right turn and signal slow or stop: right angle

★ indicates increased difficulty

Practice and Applications

HOMEWORK HELP	
For Exercises	See Examples
9–11, 18, 21	1, 2
12–17	3
Extra Practice See pages 587, 605.	

Classify each angle as *acute, obtuse, right,* or *straight*.

9. acute 10. right 11. obtuse

Draw an angle having each measurement. Then classify each angle as *acute, obtuse, right,* or *straight*. 12–17. See Ch. 10 Answer Appendix.

12. 56° acute 13. 147° obtuse 14. 180° straight 15. 99° obtuse 16. 8° acute 17. 90° right

For Exercises 18–20, use the figure at the right.

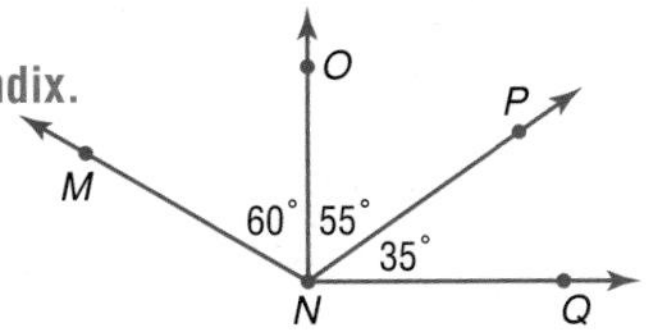

18. Name the angles that are obtuse. See Ch. 10 Answer Appendix.
19. The symbol $m\angle MNQ$ means the *measure of angle MNQ*. Find $m\angle MNQ$. 150°
20. Angles are said to be *adjacent* if they have a common vertex and a common side between them. So, $\angle MNO$ is adjacent to $\angle ONP$. Name another angle adjacent to $\angle ONP$. $\angle PNQ$

21. **SKI JUMPING** When skiers jump off a hill, they want to make the angle between their bodies and the front of their skis as small as possible. Describe where the ski jumper's legs and skis form acute and obtuse angles in the photograph. See Ch. 10 Answer Appendix.

★ 22. **EARTH SCIENCE** Earth rotates 360° degrees in one day. Through how many degrees does it rotate in one hour? 15°

23. **CRITICAL THINKING** How many times do the hands of a clock make a right angle in a 24-hour time period? 48 times

Standardized Test Practice and Mixed Review

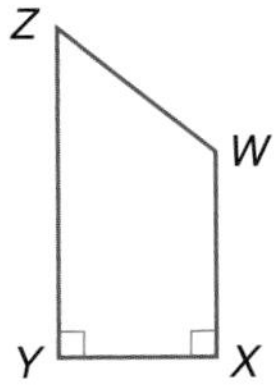

24. **SHORT RESPONSE** Name the obtuse angle in the figure at the right. $\angle ZWX$
25. **MULTIPLE CHOICE** Which is the measure of an acute angle? A

 (A) 87° (B) 95° (C) 120° (D) 180°

PROBABILITY A number cube is rolled, and a coin is tossed. Find each probability. (Lesson 9-7)

26. P(5 and heads) $\frac{1}{12}$ 27. P(odd and tails) $\frac{1}{4}$ 28. P(7 and heads or tails) 0

29. Anica spins a spinner fifty times, and it lands on 3 fifteen times. What is the experimental probability of *not* landing on 3? (Lesson 9-6) $\frac{7}{10}$

GETTING READY FOR THE NEXT LESSON

PREREQUISITE SKILL Multiply or divide. Round to the nearest hundredth if necessary. (Pages 560 and 562)

30. $0.62 \cdot 360$ 223.2 31. $360 \cdot 0.25$ 90 32. $17 \div 146$ 0.12 33. $63 \div 199$ 0.32

msmath2.net/self_check_quiz

4 Assess

Open-Ended Assessment

Modeling Have students make sketches of real-life examples of acute, obtuse, and right angles. For example, music students might sketch the triangle from music class, or students interested in buildings might sketch triangular supports incorporated into some designs.

Study Guide and Intervention, p. 543

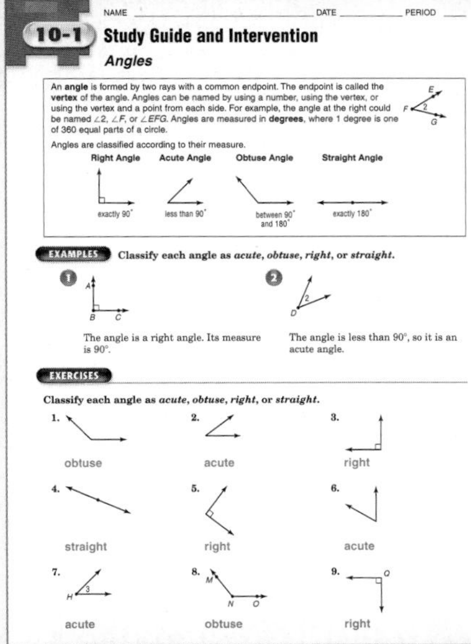

NAME ____ DATE ____ PERIOD ____

10-1 Study Guide and Intervention

Angles

An **angle** is formed by two rays with a common endpoint. The endpoint is called the **vertex** of the angle. Angles can be named by using a number, using the vertex, or using the vertex and a point from each side. For example, the angle at the right could be named ∠2, ∠F, or ∠EFG. Angles are measured in **degrees**, where 1 degree is one of 360 equal parts of a circle.

Angles are classified according to their measure.

Right Angle: exactly 90°; Acute Angle: less than 90°; Obtuse Angle: between 90° and 180°; Straight Angle: exactly 180°

EXAMPLES Classify each angle as *acute, obtuse, right,* or *straight*.

1. The angle is a right angle. Its measure is 90°.
2. The angle is less than 90°, so it is an acute angle.

EXERCISES

Classify each angle as *acute, obtuse, right,* or *straight*.

1. obtuse 2. acute 3. right
4. straight 5. right 6. acute
7. acute 8. obtuse 9. right

Practice: Skills, p. 544

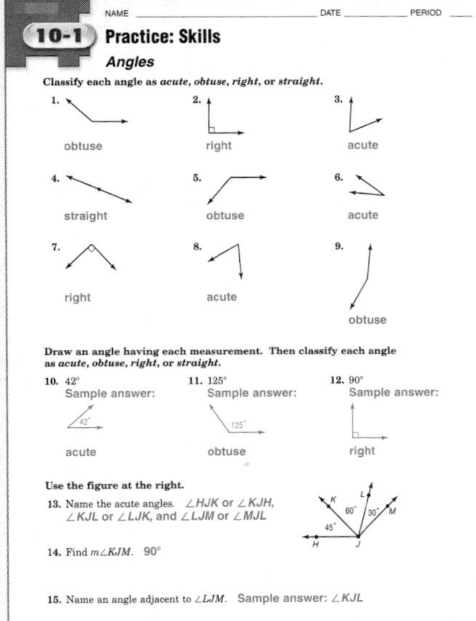

NAME ____ DATE ____ PERIOD ____

10-1 Practice: Skills

Angles

Classify each angle as *acute, obtuse, right,* or *straight*.

1. obtuse 2. right 3. acute
4. straight 5. obtuse 6. acute
7. right 8. acute 9. obtuse

Draw an angle having each measurement. Then classify each angle as *acute, obtuse, right,* or *straight*.

10. 42° Sample answer: acute 11. 125° Sample answer: obtuse 12. 90° Sample answer: right

Use the figure at the right.

13. Name the acute angles. ∠HJK or ∠KJH, ∠KJL or ∠LJK, and ∠LJM or ∠MJL
14. Find m∠KJM. 90°
15. Name an angle adjacent to ∠LJM. Sample answer: ∠KJL

Practice: Word Problems, p. 545

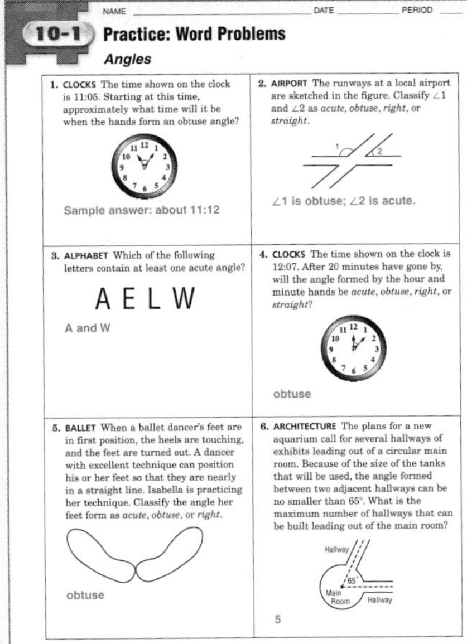

NAME ____ DATE ____ PERIOD ____

10-1 Practice: Word Problems

Angles

1. **CLOCKS** The time shown on the clock is 11:05. Starting at this time, approximately what time will it be when the hands form an obtuse angle? Sample answer: about 11:12
2. **AIRPORT** The runways at a local airport are sketched in the figure. Classify ∠1 and ∠2 as *acute, obtuse, right,* or *straight*. ∠1 is obtuse; ∠2 is acute.
3. **ALPHABET** Which of the following letters contain at least one acute angle? A E L W — A and W
4. **CLOCKS** The time shown on the clock is 12:07. After 20 minutes have gone by, will the angle formed by the hour and minute hands be *acute, obtuse, right,* or *straight*? obtuse
5. **BALLET** When a ballet dancer's feet are in first position, the heels are touching, and the feet are turned out. A dancer with excellent technique can position his or her feet so that they are nearly in a straight line. Isabella is practicing her technique. Classify the angle her feet form as *acute, obtuse,* or *right*. obtuse
6. **ARCHITECTURE** The plans for a new aquarium call for several hallways of exhibits leading out of a circular main room. Because of the size of the tanks that will be used, the angle formed between two adjacent hallways can be no smaller than 65°. What is the maximum number of hallways that can be built leading out of the main room? 5

10-1b Hands-On Lab
Lesson Notes

Lesson Resources

Teaching Mathematics with Manipulatives
- student recording sheet 10-1b

Glencoe Mathematics Classroom Manipulative Kit
- compasses
- rulers (straightedges)

GETTING STARTED

Teaching Tip Make sure that each student has a straightedge and compass.

TEACH

Activity 1
- Remind students to use a straightedge and compass only. They will not be using a ruler to measure length or a protractor to measure angles for this activity.
- Encourage students to read all of the steps before starting their constructions. Doing so will give them a chance to ask questions and make it easier to follow each step once they begin.

10-1b Hands-On Lab

A Follow-Up of Lesson 1

What You'll Learn
Construct and bisect angles.

Materials
- straightedge
- compass

Constructing and Bisecting Angles

Two angles that have the same measure are **congruent angles**. To **bisect** an angle means to divide it into two congruent angles. In this lab, you will learn how to construct congruent angles and bisect angle

ACTIVITY *Work with a partner.*

1 Construct an angle congruent to $\angle ABC$.

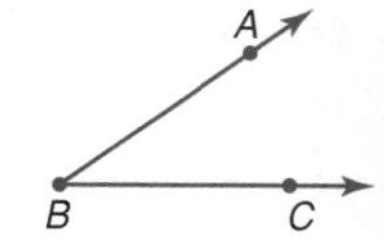

STEP 1 Use a straightedge to draw $\overrightarrow{LK}$.

STEP 2 With the compass at point B, draw an arc that intersects both sides of $\angle ABC$. Label the two points of intersection as X and Y.

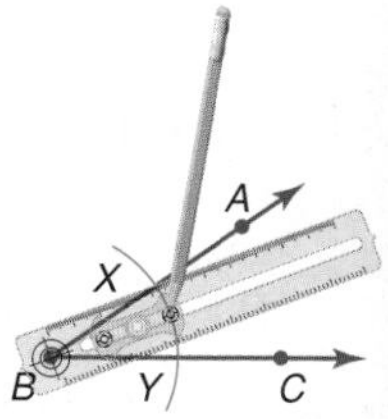

STEP 3 With the same setting on your compass, place your compass at point L. Draw another arc. Label the intersection R.

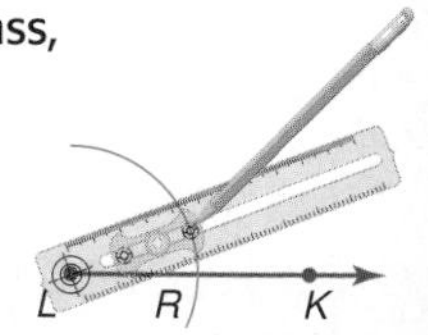

STEP 4 Open your compass to the same width as the distance between points X and Y. Then place the compass at point R. Draw an arc that intersects the arc you drew in Step 3. Label this point of intersection S.

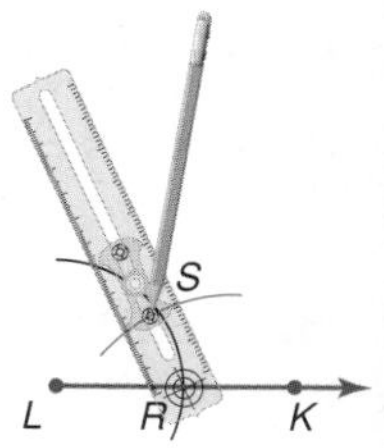

STEP 5 Draw $\overrightarrow{LM}$ through point S. Angle MLK is congruent to $\angle ABC$.

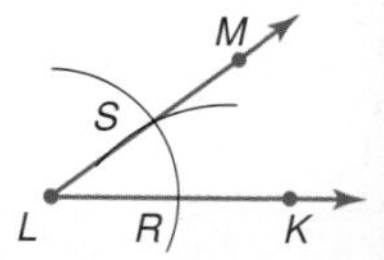

Your Turn **a. See margin.**

a. Draw an obtuse angle. Then construct an angle congruent to it.

Answer (Your Turn)

a.

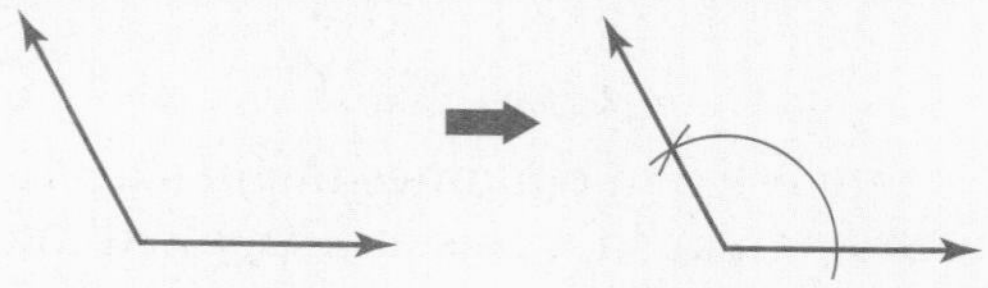

ACTIVITY *Work with a partner.*

2 **Bisect ∠*DEF*.**

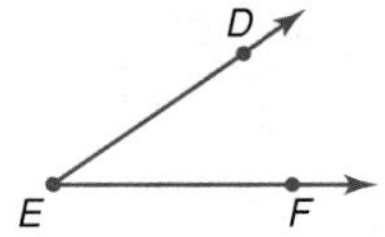

STEP 1 Place the compass at point *E* and draw an arc that intersects both sides of ∠*DEF*. Mark the two points of intersection as *H* and *I*.

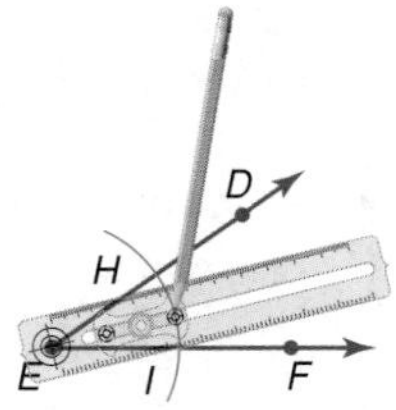

STEP 2 Place your compass at point *H*. Draw an arc inside ∠*DEF*. Using the same setting on your compass, place your compass at point *I*. Draw another arc inside ∠*DEF* intersecting the first arc. Label the intersection point *G*.

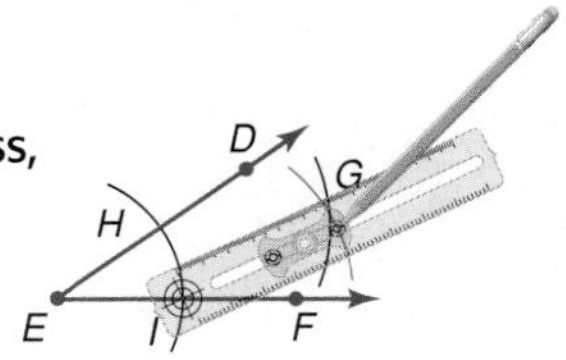

STEP 3 Draw $\overrightarrow{EG}$. $\overrightarrow{EG}$ is the bisector of ∠*DEF*.

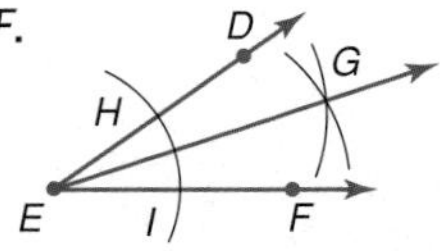

Your Turn

b. Draw a right angle. Then bisect it. **See margin.**

Writing Math

Work with a partner.

1. **Explain** how you could verify that ∠*ABC* and ∠*MLK* in Activity 1 are congruent.
2. **Explain** how you could verify that ∠*DEF* in Activity 2 is bisected.
3. A straight angle has a measure of 180°. What kind of angle is formed if you bisect a straight angle? **right**
4. **Perpendicular lines** are lines that meet to form right angles. In the figure at the right, $\overleftrightarrow{PS}$ is perpendicular to $\overleftrightarrow{QR}$. Use a compass and a straightedge to construct perpendicular lines. (*Hint*: Use the procedure for bisecting angles.) **See margin.**

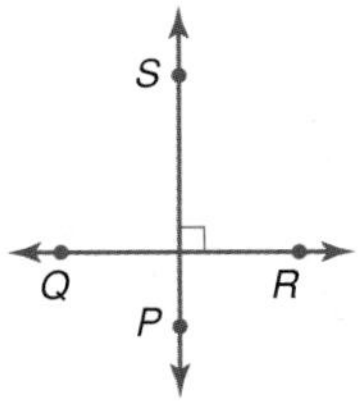

. Sample answer:
easure both angles to
eck that the measures
e the same.

. Sample answer:
easure each smaller
gle and take the sum.
e sum must be the
easure of the original
gle, and each smaller
gle must have the
me measure.

Activity 2

- Write the following statement on the chalkboard:

  $m\angle DEG = m\angle GEF = \frac{1}{2}(m\angle DEF)$.

  After students complete Step 3, have them explain this statement, pointing to each angle as they do so.

ASSESS

Observe as students solve the **Your Turn** exercises. They should:

- begin with an obtuse or a right angle, and
- follow the same steps as in the activities to complete their constructions.

Discuss the **Writing Math** exercises. After students have answered Exercise 4, discuss the angles that would be formed if students were to bisect the right angles they made. Be sure students can see this bisector would form a 45° angle with $\overline{QR}$ and a 45° angle with $\overline{PS}$. You might also wish to point out that the bisector forms an obtuse angle with $\overline{QR}$, and the obtuse angle has a measure of 135°.

Answer (Writing Math)

4. Sample answer:

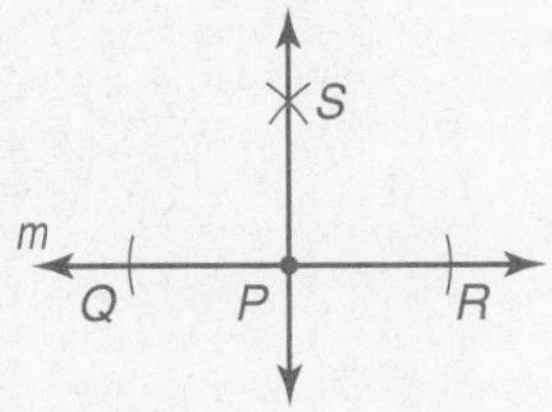

Answer (Your Turn)

b.

10-2 Lesson Notes

Lesson Resources

Chapter 10 Resource Masters
- Study Guide and Intervention, p. 548
- Practice: Skills, p. 549
- Practice: Word Problems, p. 550
- Reading to Learn Mathematics, p. 551
- Enrichment, p. 552

Calculator and Spreadsheet Masters, p. 47
Key Teaching Transparencies, 37
Parent and Student Study Guide Workbook, 10-2
Teaching Mathematics with Manipulatives, p. 92

FAST FILE

1 Focus

5-Minute Check 10-2

INTERACTIVE CHALKBOARD
PowerPoint® Presentations

Use as a quiz or review of Lesson 10-1. (Also available as a transparency)

Bellringer

Reading Mathematics Have students locate examples of circle graphs in newspapers or magazines. Ask students to describe what information they can learn from each graph.

Reading to Learn Mathematics, p. 551 ELL

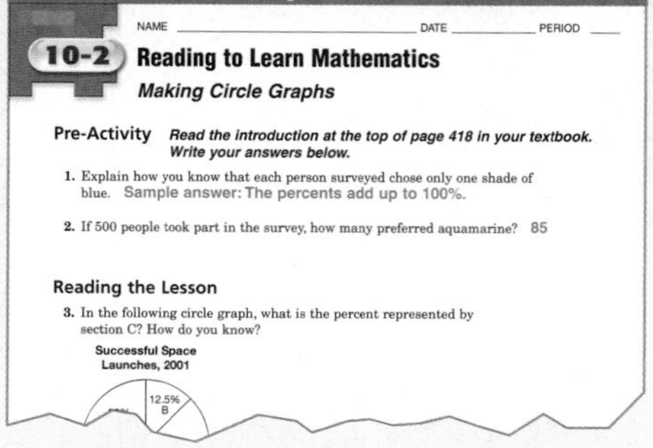
NAME ________ DATE ________ PERIOD ____

10-2 Reading to Learn Mathematics
Making Circle Graphs

Pre-Activity *Read the introduction at the top of page 418 in your textbook. Write your answers below.*

1. Explain how you know that each person surveyed chose only one shade of blue. Sample answer: The percents add up to 100%.
2. If 500 people took part in the survey, how many preferred aquamarine? 85

Reading the Lesson

3. In the following circle graph, what is the percent represented by section C? How do you know?

10-2 Making Circle Graphs

What You'll Learn
Construct and interpret circle graphs.

NEW Vocabulary
circle graph

REVIEW Vocabulary
ratio: a comparison of two numbers by division (Lesson 7-1)
percent equation: part = percent · base (Lesson 8-2)

WHEN am I ever going to use this?

COLORS In a recent survey, people ages 13–20 were asked to choose their favorite shade of blue. The results are shown in the table.

1. Explain how you know that each person surveyed chose only one shade of blue.
2. If 500 people took part in the survey, how many preferred aquamarine? **85**

Favorite Shades of Blue for People Ages 13–20

Shade	Percent
navy	35%
sky/light blue	30%
aquamarine	17%
other	18%

Source: *American Demographics*

1. Sample answer: The percents add up to 100%.

A graph used to compare parts of a whole is called a **circle graph**. In a circle graph, the percents add up to 100.

EXAMPLE Construct a Circle Graph

1 COLORS Make a circle graph of the data in the table above.

- Find the degrees for each part. Round to the nearest whole degree.

  35% of 360° = 0.35 · 360° or 126°

  30% of 360° = 0.30 · 360° or 108°

  17% of 360° = 0.17 · 360° or about 61°

  18% of 360° = 0.18 · 360° or about 65°

- Use a compass to draw a circle with a radius as shown. Then use a protractor to draw the first angle, in this case 126°. Repeat this step for each section.

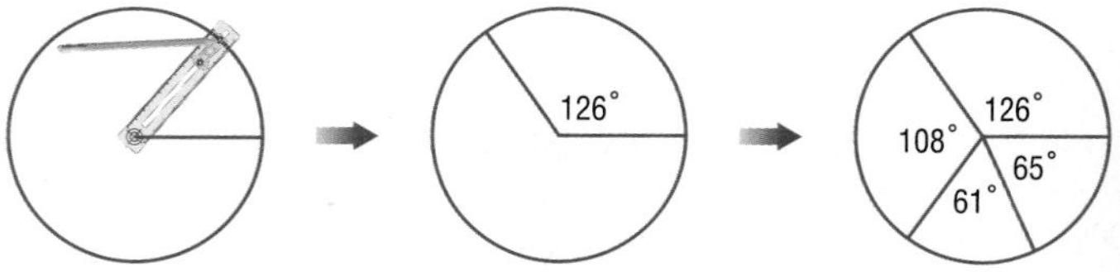

- Label each section of the graph with the category and percent. Give the graph a title.

Check To draw an accurate circle graph, make sure the sum of the angle measures equals 360°.

Favorite Shades of Blue for People Ages 13–20

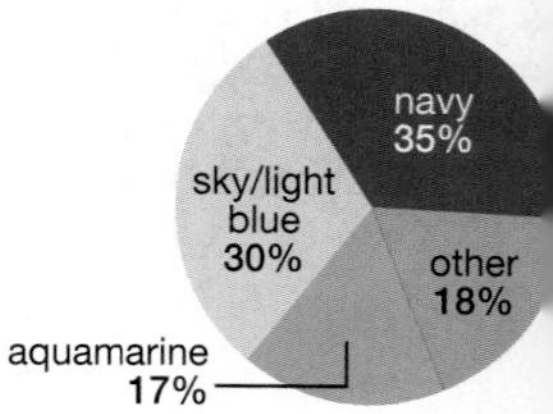

TEACHING TIP
Students may find it easier to draw the angle with the least number of degrees first, in this case 61°.

DAILY INTERVENTION

Differentiated Instruction ELL

Verbal/Linguistic Have students write a survey question and pass it out to their classmates. They should use a multiple-choice format with four or five responses. For example, a question might read: What is your favorite subject? A) Science B) Math C) History D) English E) Other. Be sure to review the questions for content and format before they are distributed. After the questions are answered, have students create a circle graph to display their results.

EAL-LIFE MATH

YMPICS The United tes has won 2,117 mmer Olympic medals ough 2000, the most by country.

rce: infoplease.com

When constructing a circle graph, it is sometimes necessary to first convert the data to percents or decimals, and then to degrees.

EXAMPLE Construct a Circle Graph

2 **OLYMPICS The table shows the number of each type of medal won by the United States during the Summer Olympics from 1896 to 2000. Make a circle graph of the data.**

U.S. Summer Olympic Medals

Type	Number
gold	872
silver	658
bronze	587

Source: infoplease.com

- Find the total number of medals: 872 + 658 + 587 or 2,117.
- Find the ratio that compares each number with the total. Write the ratio as a decimal rounded to the nearest hundredth.

  gold: $\frac{872}{2{,}117} \approx 0.41$ silver: $\frac{658}{2{,}117} \approx 0.31$ bronze: $\frac{587}{2{,}117} \approx 0.28$

- Find the number of degrees for each section of the graph.

  gold: $0.41 \cdot 360° \approx 148°$

  silver: $0.31 \cdot 360° \approx 112°$

  bronze: $0.28 \cdot 360° \approx 101°$

  Because of rounding, the sum of the degrees is 361°.

- Draw the circle graph.

U.S. Summer Olympic Medals

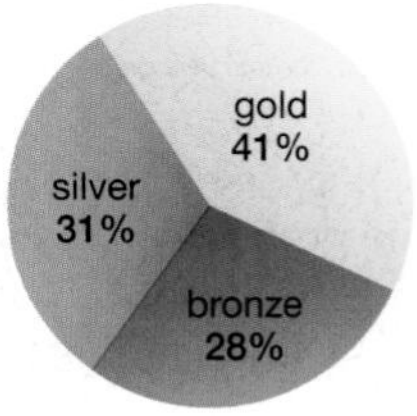

Check After drawing the first two sections, you can measure the last section of a circle graph to verify that the angles have the correct measures.

TEACHING TIP

udents can also use oportions to make rcle graphs. To find the egrees in Example 2, e following proportions uld be used.

$\frac{1}{00} = \frac{x}{360}$

$\frac{1}{00} = \frac{x}{360}$

$\frac{8}{00} = \frac{x}{360}$

EXAMPLE Interpret a Circle Graph

3 **MONEY The circle graph shows the percent of Americans who favor, oppose, or don't know how they feel about a common currency for North America. Use the graph to describe the opinion of most Americans.**

Do Americans Favor Common North American Currency?

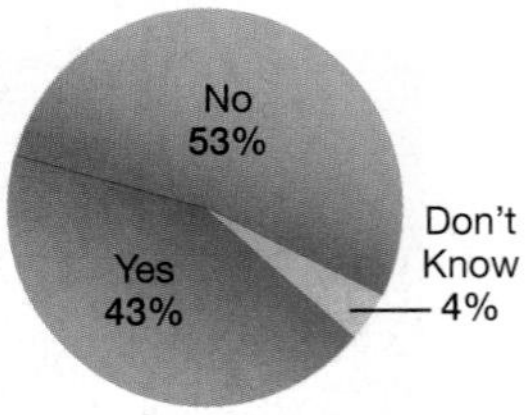

Source: Coinstar

The greatest percent of the circle graph is the section representing the "No" response.

So, most Americans do not favor a common North American currency.

msmath2.net/extra_examples

2 Teach

In-Class Examples

INTERACTIVE CHALKBOARD PowerPoint® Presentations

1 **SPORTS** In a survey, a group of middle school students were asked to name their favorite sport. The results are shown in the table. Make a circle graph of the data.

Sport	Percent
football	30%
basketball	25%
baseball	22%
tennis	8%
other	15%

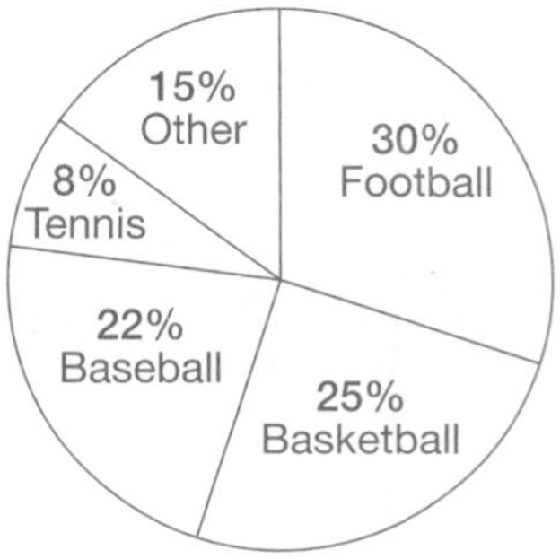

2 **MOVIES** Gina has the following types of movies in her DVD collection. Make a circle graph of the data.

Type of Movie	Number
action	24
comedy	15
science fiction	7

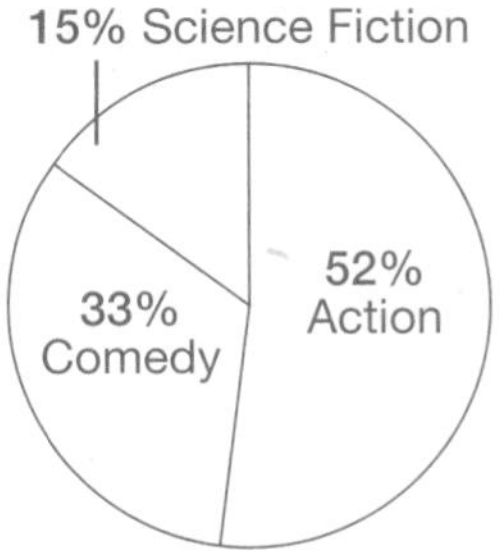

Tips for New Teachers

Using Applications Point out to students that, even with a compass and protractor, it is very easy to make mistakes when constructing circle graphs. Colorful, accurate circle graphs can also be constructed using computer programs such as spreadsheets, databases, and statistics software. Show students how to use one of these applications to make a circle graph.

In-Class Example

INTERACTIVE CHALKBOARD
PowerPoint® Presentations

3 **VOTING** The circle graph below shows the percent of voters in a town who are registered with a political party. Use the graph to describe the voters in this town.

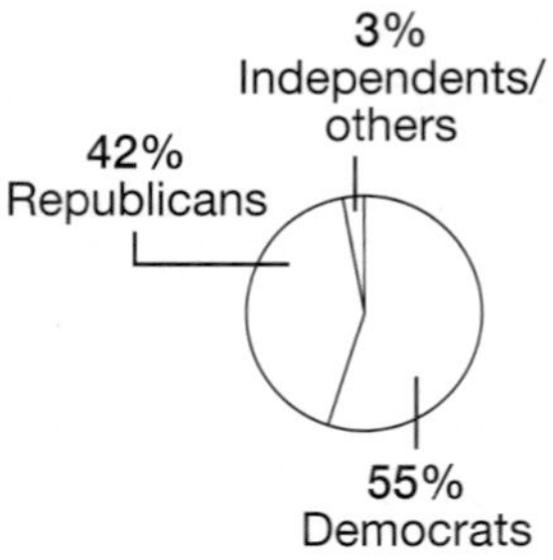

The majority of voters are Democrats. Only a very small portion are Independents.

3 Practice/Apply

About the Exercises...

Guided Practice Key

Exercises	Examples
3–4	1, 2
5–6	3

Odd/Even Assignments
Exercises 7–8 and 10–11 are structured so that students practice the same concepts whether they are assigned odd or even problems.

Alert! Exercise 15 requires research on the Internet or another source.

Skill and Concept Check

Writing Math
Exercises 1 & 2

1. **OPEN ENDED** Draw a circle graph that has three parts. Label the parts and describe what they represent. **See students' work.**

2. **DATA SENSE** The table shows the percent of people that use the sweeteners listed. Can the data be represented in a circle graph? Explain. **Sample answer: No; the sum of the percents is greater than 100. The people surveyed must have been able to choose more than one sweetener.**

Type of Sweetener	Percent
sugar	74%
honey, molasses, syrup	46%
low calorie sweeteners	36%

Source: Yankelovich Partners

GUIDED PRACTICE

Make a circle graph of the data in each table. **3–4. See margin.**

3.

Favorite Shades of Blue for People Ages 21–34	
Shade	**Percent**
navy	48%
sky/light blue	23%
aquamarine	12%
other	18%

Source: *American Demographics*

4.

Speed Limit (mph)	Number of States
55	1
65	20
70	18
75	11

Source: *The World Almanac*

TEACHING TIP
Tell students that the percents in Exercise 3 add up to 101%, due to rounding.

5. **EDUCATION** The circle graph shows the percent of students by grade level in U.S. schools. In which grades are most students? **grades 1–8**

Grade Level of U.S. Students

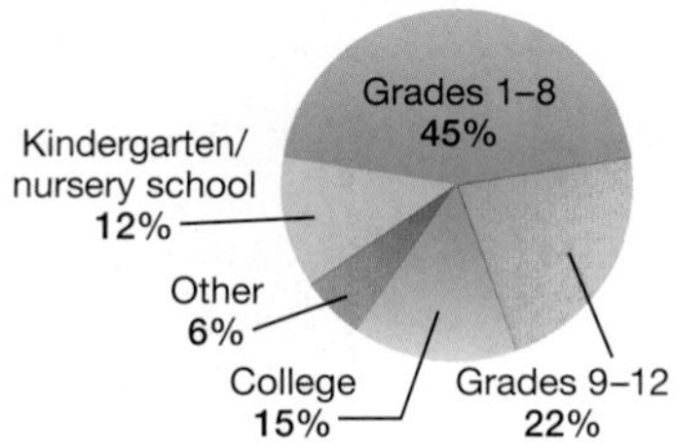

Source: U.S. Census Bureau

6. Refer to the circle graph in Example 1 and the graph you drew in Exercise 3. What can you conclude about the favorite shades of blue for people in different age groups? **See margin.**

★ indicates increased difficulty

Practice and Applications

HOMEWORK HELP

For Exercises	See Examples
7–8, 12–13	1
10–11	2
9, 14	3

Extra Practice
See pages 587, 605.

Make a circle graph of the data in each table. **7–9. See Ch. 10 Answer Appendix.**

7.

New York City Commuters	
Transportation	Percent
driving alone	24%
carpool	9%
public transit	53%
other	14%

Source: *Time Almanac*

8.

Los Angeles Commuters	
Transportation	Percent
driving alone	65%
carpool	15%
public transit	11%
other	9%

Source: *Time Almanac*

9. Compare and contrast the data from the circle graphs you drew in Exercises 7 and 8.

Answers

3. **Favorite Shades of Blue**

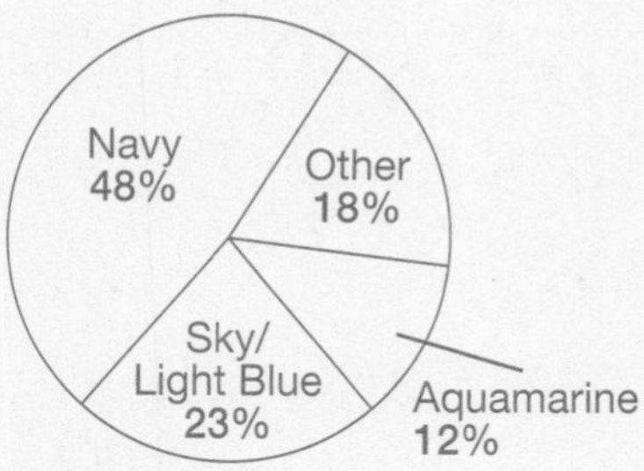

4. **Speed Limit**

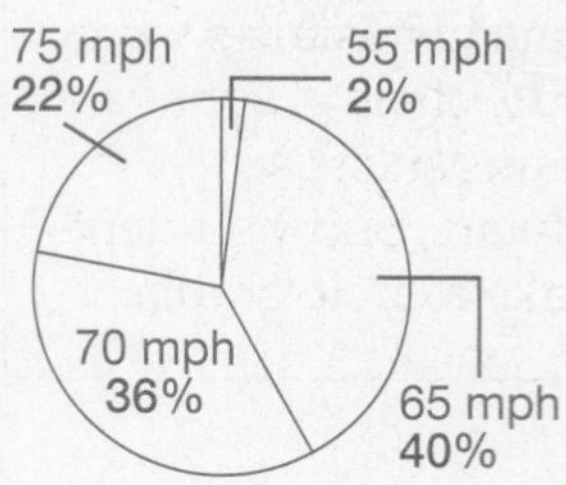

6. **Sample answer: A greater percent of people 21–34 years old preferred navy blue than people 13–20 years old. A greater percent of people in the younger age group preferred light blue.**

14. **Sample answer: The measured angles are different from the angle measures in Exercise 14 because the graph is not exactly a circle. Since the graph is not a circle, the measured angles will not be accurate.**

Make a circle graph of the data in each table. **10–11. See Ch. 10 Answer Appendix.**

10.

Endangered Species in U.S.	Number of Species
mammals	63
birds	78
reptiles	14
amphibians	10

Source: U.S. Fish and Wildlife Service

11.

U.S. Regions	Population (millions)
Northeast	54
Midwest	64
South	100
West	63

Source: *Time Almanac*

POLITICS For Exercises 12–14, refer to the graphic at the right.

12. How many 13- to 15-year olds participated in the survey? **502**

13. Calculate the angle measure for each section.

14. Measure the angles for each section on the graph. Explain any differences. **See margin.**

13. yes: 262.8°; no: 90°; not sure: 7.2°

15. **RESEARCH** Use the Internet or another source to find data that add up to 100%. Make a circle graph of the data. **See students' work.**

16. **CRITICAL THINKING** Line graphs are usually best for data that show change over time. When might it be more appropriate to display data in a circle graph? Give an example. **See margin.**

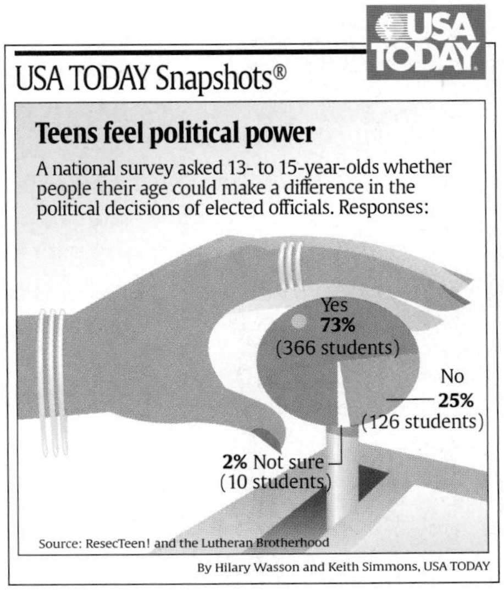

Standardized Test Practice and Mixed Review

17. **MULTIPLE CHOICE** On a circle graph of the data in the table, which section would have an angle measure of about 38°? **A**

 Ⓐ Lake Erie Ⓑ Lake Huron
 Ⓒ Lake Michigan Ⓓ Lake Ontario

Lake	Area (sq mi)
Erie	9,930
Huron	23,010
Michigan	22,400
Ontario	7,520
Superior	31,820

Source: www.infoplease.com

18. **SHORT RESPONSE** Find the angle measure of the Lake Michigan section in the circle graph referred to in Exercise 17. Round to the nearest whole degree. **85°**

Classify each angle as *acute, obtuse, right,* or *straight.* (Lesson 10-1)

19. 65° **acute** 20. 102° **obtuse** 21. 90° **right**

22. **PROBABILITY** Corey has an after-school activity 15% of the time. Li-Cheng has one 20% of the time. Find the probability that they will both have an after-school activity on the same day. (Lesson 9-7) **3%**

GETTING READY FOR THE NEXT LESSON

PREREQUISITE SKILL Solve each equation. (Lesson 4-2)

23. $a + 18 = 90$ **72** 24. $44 + x = 90$ **46** 25. $180 = 39 + n$ **141** 26. $123 + t = 180$ **57**

4 Assess

Higher-Level Thinking

Give students examples of a circle graph and a line graph. Ask them to identify situations when each is most useful. **Circle graphs show parts of a whole. Line graphs show change over time.**

Answer

16. **Sample answer: when you want to show how parts are related to the whole, such as a budget**

Study Guide and Intervention, p. 548

NAME ________ DATE ______ PERIOD ____

10-2 Study Guide and Intervention

Making Circle Graphs

A graph used to compare parts of a whole is called a **circle graph**. In a circle graph, the percents add up to 100. When percents are not given, you must first determine what part of the whole each item represents.

EXAMPLE 1 ENERGY **Make a circle graph of the data in the table.**

Nuclear Reactors in Operation, 2001	
Country	Number of Reactors
United States	104
France	59
Japan	54
Other Countries	222

Step 1 Find the total number of reactors: $104 + 59 + 54 + 222 = 439$.

Step 2 Find the ratio that compares each number with the total. Write the ratio as a decimal rounded to the nearest hundredth.

United States: $\frac{104}{439} \approx 0.24$ Japan: $\frac{54}{439} \approx 0.12$

France: $\frac{59}{439} \approx 0.13$ Other: $\frac{222}{439} \approx 0.51$

Step 3 Find the number of degrees for each section of the graph.

United States: $0.24 \cdot 360° \approx 86°$ Japan: $0.12 \cdot 360° \approx 43°$

France: $0.13 \cdot 360° \approx 47°$ Other: $0.51 \cdot 360° \approx 183°$

Step 4 Use a compass to draw a circle with a radius. Then use a protractor to draw an 86° angle. This represents the percent of nuclear reactors in the United States.

Step 5 From the new radius, draw a 47° angle for France. Repeat this step for the other two sections. Label each section and give the graph a title.

Nuclear Reactors in Operation, 2001

EXERCISES

1. SWIMMING The table shows the number of members of the swim team who competed at the swim meet. Each competed in only one event. Make a circle graph of the data.

Swim Team Member Participation	
Event	Number
Freestyle	18
Breaststroke	7
Backstroke	5
Butterfly	2

Swim Team Member Participation

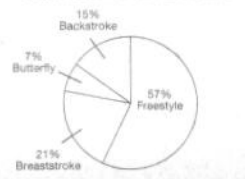

Practice: Skills, p. 549

NAME ________ DATE ______ PERIOD ____

10-2 Practice: Skills

Making Circle Graphs

For each table, find the number of degrees in each section of a circle graph. Then make a circle graph of the data.

1.

United States Energy Usage, 2001	
Category	Percent
Commercial and Industrial	52%
Residential	20%
Transportation	27%
Other	1%

Commercial and Industrial: 187°
Residential: 72°
Transportation: 97°; Other: 4°

United States Energy Usage, 2001

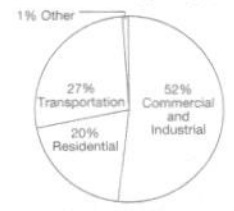

2.

Family Members Students Confide In	
Family Member	Percent
Mom	52%
Dad	17%
Brother/Sister	16%
Grandparent/Other	15%

Mom: 187°; Dad: 61°
Brother/Sister: 58°
Grandparent/Other: 54°

Family Members Students Confide In

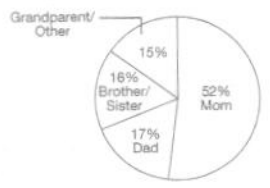

3.

Successful Space Launches, 2001	
Country	Number
India	2
United States	23
European Space Agency	7
China	1

India: 22°; United States: 251°
European Space Agency: 76°
China: 11°

Successful Space Launches, 2001

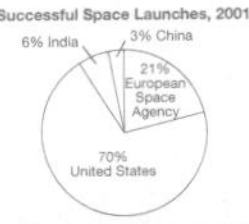

4.

United States Coastline	
Coast	Length (mi)
Atlantic	2,100
Pacific	7,600
Gulf	1,600
Arctic	1,100

Atlantic: 61°; Pacific: 221°
Gulf: 47°
Arctic: 32°

United States Coastline

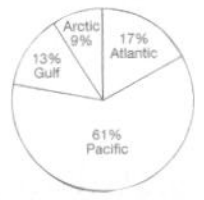

Practice: Word Problems, p. 550

NAME ________ DATE ______ PERIOD ____

10-2 Practice: Word Problems

Making Circle Graphs

LANGUAGES **For Exercises 1 and 2, use the table that shows the number of people that speak the five languages that are spoken by the most people in the world.**

Languages Spoken by the Most People	
Language	Speakers (millions)
Chinese, Mandarin	874
Hindi	366
English	341
Spanish	322
Bengali	207

1. Find the degrees for each part of a circle graph that shows the data.
Chinese, Mandarin: 149°
Hindi: 62°
English: 58°
Spanish: 55°
Bengali: 35°

2. Make a circle graph of the data. Which three languages account for 41% of the total?
Bengali, Spanish, and English

Languages Spoken by the Most People

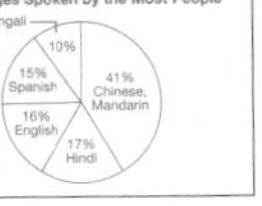

MILITARY **For Exercises 3 and 4, use the table that shows the number of people active in the United States military in 2002.**

United States Military, Active Duty, 2002	
Branch	Personnel (thousands)
Army	486
Navy	385
Marine Corps	174
Air Force	368
Coast Guard	38

3. Make a circle graph of the data.

United States Military Personnel Active Duty, 2002

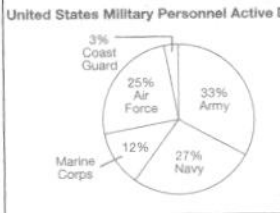

4. Which two branches taken together account for almost half of the total?
Air Force and Navy

10-3 Lesson Notes

Lesson Resources

Chapter 10 Resource Masters
- Study Guide and Intervention, p. 553
- Practice: Skills, p. 554
- Practice: Word Problems, p. 555
- Reading to Learn Mathematics, p. 556
- Enrichment, p. 557

MathPASS Tutorial CD-ROM, 43

Parent and Student Study Guide Workbook, 10-3

FAST FILE

1 Focus

5-Minute Check 10-3 Use as a quiz or review of Lesson 10-2. (Also available as a transparency)

INTERACTIVE CHALKBOARD
PowerPoint® Presentations

Bellringer

Problem Solving Draw a straight angle on the chalkboard and ask students to give the angle's measure. 180° Then draw a ray extending from the vertex of the straight angle. Have students use a protractor to measure one of the angles formed. Challenge them to determine the measure of the other angle without using a protractor.

Reading to Learn Mathematics, p. 556 **ELL**

NAME ______________ DATE ________ PERIOD ____

10-3 Reading to Learn Mathematics

Angle Relationships

Pre-Activity ***Complete the Mini Lab at the top of page 422 in your textbook. Write your answers below.***

1. Which angles have the same measure? ∠1 and ∠4; ∠2 and ∠3
2. Draw two other pairs of intersecting lines. Measure their angles. **Make a conjecture** involving four angles created by intersecting lines. Opposite angles have the same measure.
3. What is the relationship between the measures of ∠1 and ∠2? ∠3 and ∠4? $m\angle 1 + m\angle 2 = 180°$; $m\angle 3 + m\angle 4 = 180°$
4. Do other pairs of angles share the same relationship? Explain. Yes; ∠2 and ∠4; ∠1 and ∠3; the sum of each pair of adjacent angles is 180°.

10-3 Angle Relationships

What You'll Learn
Identify and apply angle relationships.

NEW Vocabulary
- vertical angles
- congruent angles
- supplementary angles
- complementary angles

MATH Symbol
≅ is congruent to

HANDS-ON Mini Lab

Materials
- protractor

Work with a partner.

Draw two intersecting lines and label the angles as shown. Then measure each angle with your protractor and record the measurements.

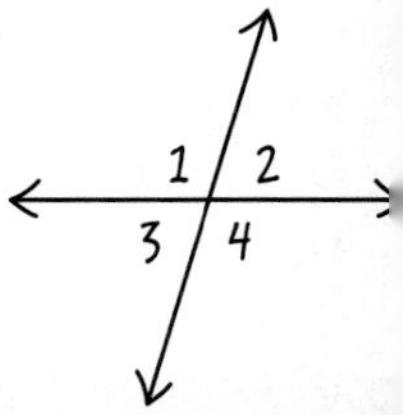

1. Which angles have the same measure?
2. Draw two other pairs of intersecting lines. Measure their angles. **Make a conjecture** involving four angles created by intersecting lines. **Opposite angles have the same measure.**
3. What is the relationship between the measures of ∠1 and ∠2? ∠3 and ∠4? $m\angle 1 + m\angle 2 = 180°$; $m\angle 3 + m\angle 4 = 180°$
4. Do other pairs of angles share the same relationship? Explain.

1. ∠1 and ∠4; ∠2 and ∠3

4. Yes; ∠2 and ∠4; ∠1 and ∠3; the sum of each pair of adjacent angles is 180°.

When two lines intersect, they form two pairs of opposite angles called **vertical angles**. In the Mini Lab, you found that vertical angles have the same measure. Angles with the same measure are **congruent angles**.

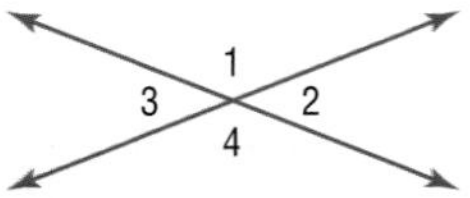

∠1 ≅ ∠4
∠2 ≅ ∠3

The symbol ≅ is used to show that the angles are congruent.

Pairs of angles can also have other relationships. In the Mini Lab, you found pairs of angles whose sum is 180°. Two angles are **supplementary** if the sum of their measures is 180°.

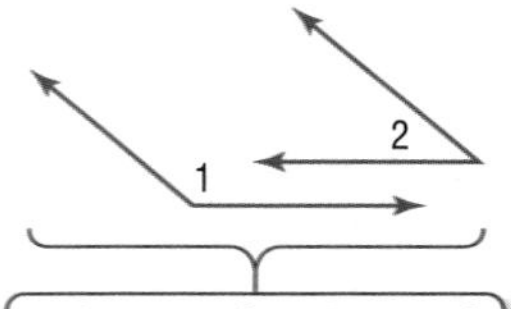

$m\angle 1 = 140°$, $m\angle 2 = 40°$,
$m\angle 1 + m\angle 2 = 180°$

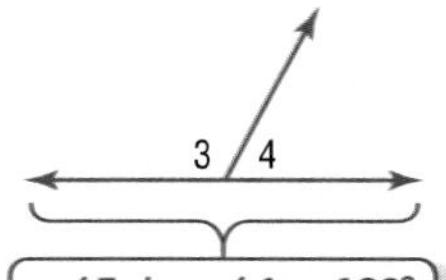

$m\angle 3 + m\angle 4 = 180°$

Two angles are **complementary** if the sum of their measures is 90°.

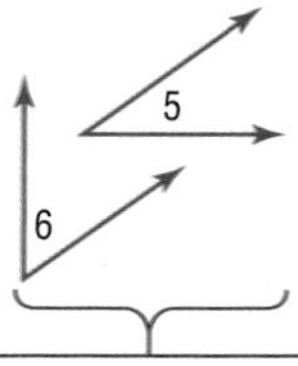

$m\angle 5 = 35°$, $m\angle 6 = 55°$,
$m\angle 5 + m\angle 6 = 90°$

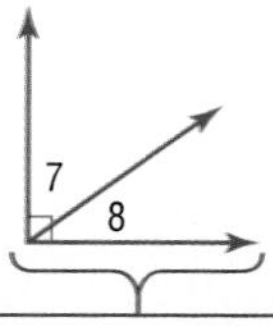

$m\angle 7 + m\angle 8 = 90°$

TEACHING TIP
Help students understand that when two angles are *congruent*, the measures of the angles are *equal*. If $\angle A \cong \angle B$, then $m\angle A = m\angle B$.

Tips for New Teachers

Intervention To help students learn new geometry vocabulary, have them create flash cards with the word on one side and a drawing on the other.

EXAMPLES Classify Angles

Classify each pair of angles as *complementary, supplementary,* or *neither.*

1

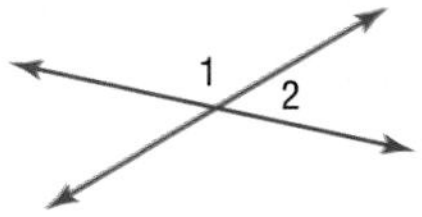

$\angle 1$ and $\angle 2$ form a straight line. So, the angles are supplementary.

2

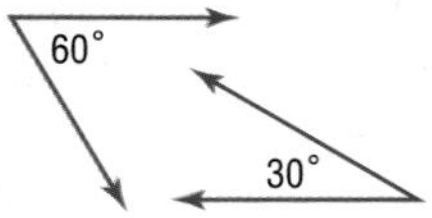

$60° + 30° = 90°$
The angles are complementary.

You can use angle relationships to find missing measures.

EXAMPLE Find a Missing Angle Measure

3 **Angles *ABC* and *CBD* are complementary. If $m\angle ABC = 28°$, find $m\angle CBD$.**

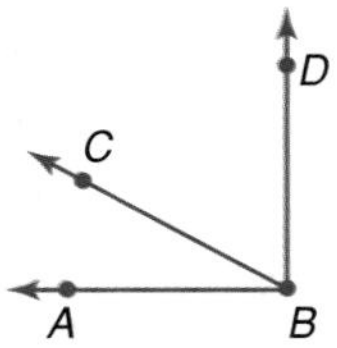

Since $\angle ABC$ and $\angle CBD$ are complementary, $m\angle ABC + m\angle CBD = 90°$.

$m\angle ABC + m\angle CBD = 90$	Write the equation.
$28 + m\angle CBD = 90$	Replace $m\angle ABC$ with 28.
$-28 \qquad\qquad -28$	Subtract 28 from each side.
$m\angle CBD = 62$	$90 - 28 = 62$

The measure of $\angle CBD$ is 62°.

Your Turn **Find the value of x in each figure.**

a. 134

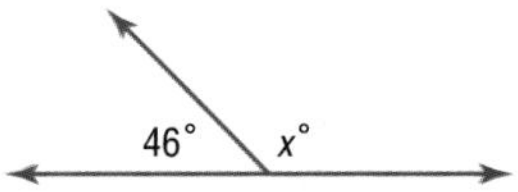

b. 17

REAL-LIFE MATH

BASKETBALL Canadian James Naismith invented the game of basketball in 1891 using a ball the size of a soccer ball and two peach baskets.

Source: www.allsands.com

EXAMPLE Use Angles to Solve A Problem

4 **BASKETBALL Erin wants to make a bounce pass to Mackenzie. Find the value of x so that Erin's pass hits Mackenzie in the hands.**

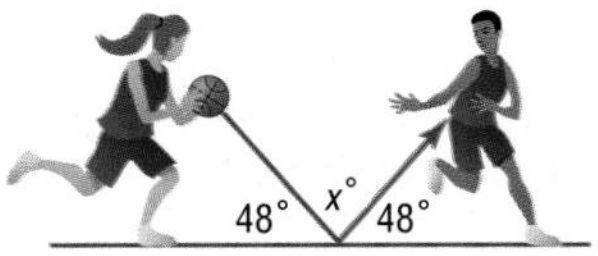

Since the sum of the three angles is 180°, $48° + x° + 48° = 180°$.

$48 + x + 48 = 180$	Write the equation.
$x + 96 = 180$	Simplify.
$-96 \qquad -96$	Subtract 96 from each side.
$x = 84$	$180 - 96 = 84$

So, the angle must be 84° for the ball to hit Mackenzie in the hands.

2 Teach

In-Class Examples

INTERACTIVE CHALKBOARD PowerPoint® Presentations

Classify each pair of angles as *complementary, supplementary,* or *neither*.

1

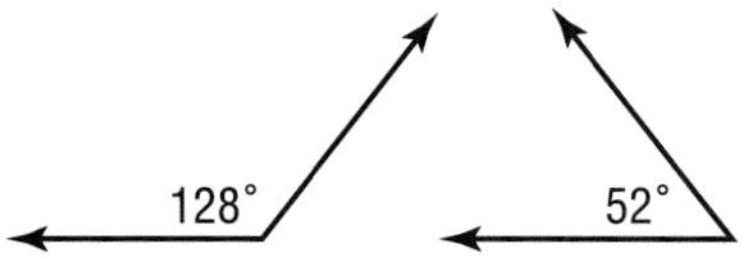

supplementary

2

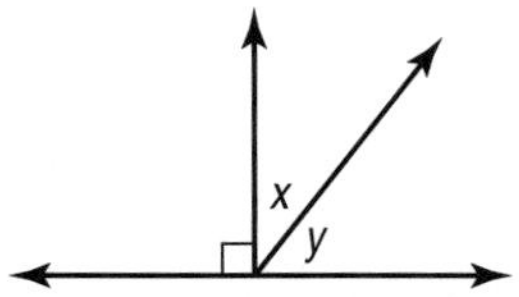

complementary

3 Angles *PQS* and *RQS* are supplementary. If $m\angle PQS = 56°$, find $m\angle RQS$. **124°**

4 **GEOMETRY** The rectangle shown is divided by a diagonal. Find the value of x. **20**

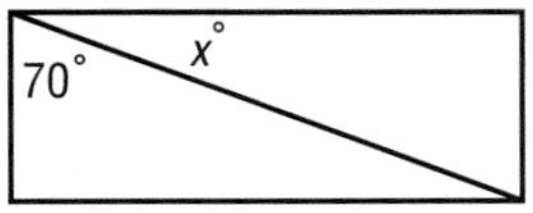

Tips for New Teachers

Teaching Strategies Students sometimes think that angles must be adjacent in order to be complementary or supplementary. Use angles such as those in Example 2 to prove that this is not true.

DAILY INTERVENTION **Differentiated Instruction**

Kinesthetic Open the classroom door so that it forms a right angle with the wall. Use masking tape on the floor to mark the door's position. Then have pairs of students move the door to a new position and mark this new position with masking tape. Have each pair of students measure one of the new angles with a large classroom protractor and then determine the missing angle measure.

3 Practice/Apply

NUMBER SENSE
Exercise 2 can be used to practice simple algebra concepts. Call the measure of one angle $x°$. Since the pair are vertical, the other angle must also be $x°$. Set up the equation $x + x = 180$ and solve.

About the Exercises...

Guided Practice Key

Exercises	Examples
3–5	1, 2
6–8	3
9	4

Odd/Even Assignments
Exercises 10–21 and 24–27 are structured so that students practice the same concepts whether they are assigned odd or even problems.

Answers

1. Sample answer: $m\angle 1 + m\angle 2 = 90°$

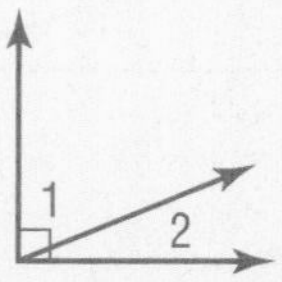

2. yes; $\angle 1$ and $\angle 2$ are vertical angles and $m\angle 1 + m\angle 2 = 180°$

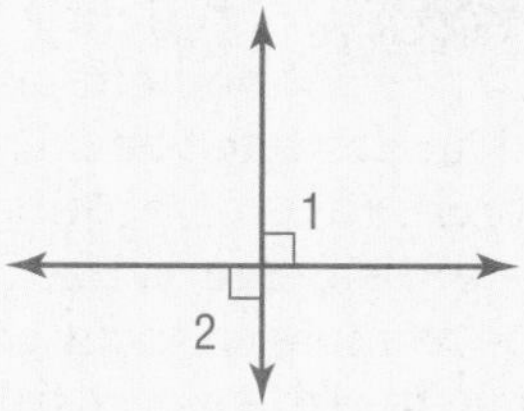

Skill and Concept Check

Writing Math
Exercise 2

1. **OPEN ENDED** Draw two angles that are complementary.
2. **NUMBER SENSE** Can a pair of angles be vertical and supplementary? Give an example or nonexample, with angle measurements, to support your answer. **1–2. See margin.**

GUIDED PRACTICE

Classify each pair of angles as *complementary, supplementary,* or *neither.*

3. 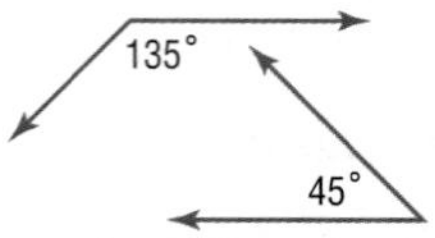

supplementary

4. 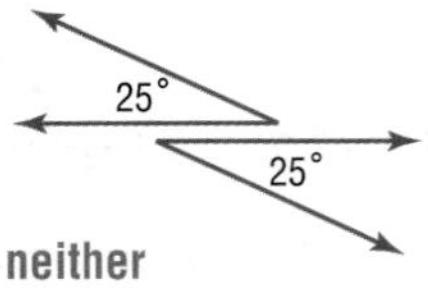

neither

5. 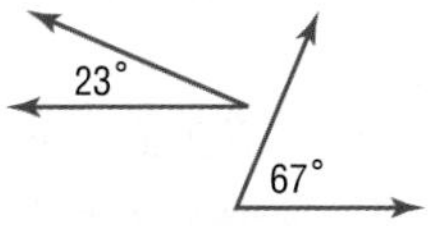

complementary

Find the value of *x* in each figure.

6. 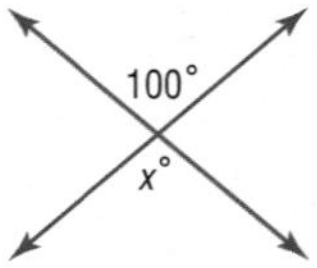

100

7. 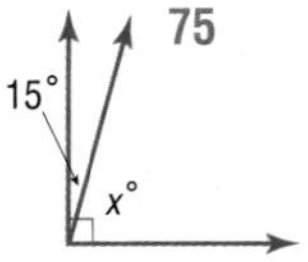

75

8. 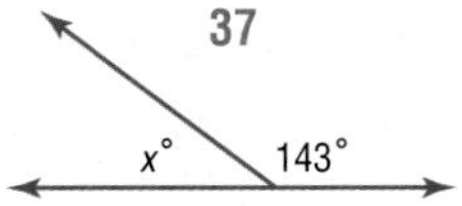

37

9. Suppose $\angle 1$ and $\angle 2$ are supplementary. If $m\angle 1 = 84°$, find $m\angle 2$. **96°**

★ indicates increased difficulty

Practice and Applications

HOMEWORK HELP

For Exercises	See Examples
10–15, 22–23	1, 2
16–21	3, 4

Extra Practice
See pages 587, 605.

Classify each pair of angles as *complementary, supplementary,* or *neither.*

10. 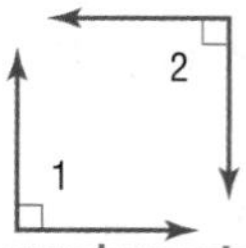

supplementary

11. 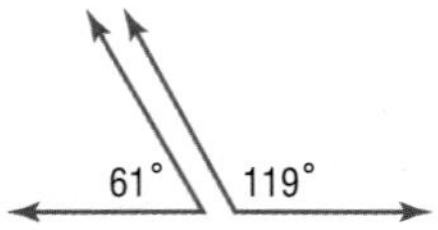

supplementary

12. 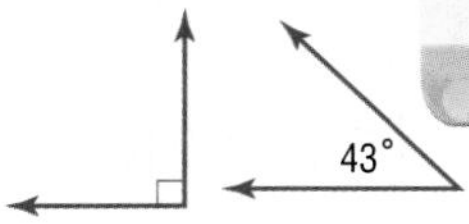

neither

13. 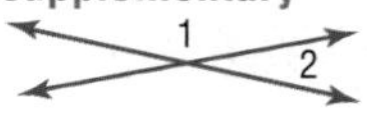

supplementary

14. 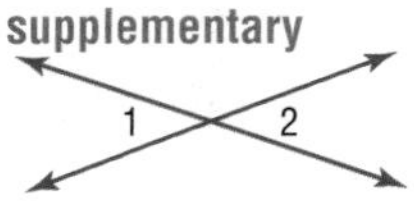

neither

15. 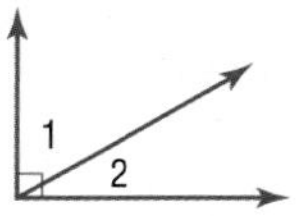

complementary

Find the value of *x* in each figure.

16. 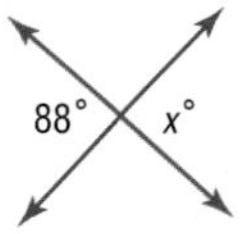

88

17. 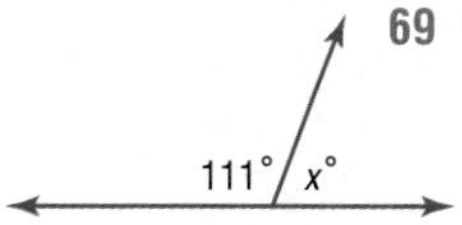

69

18. 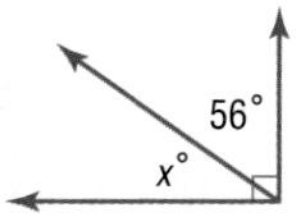

34

19. 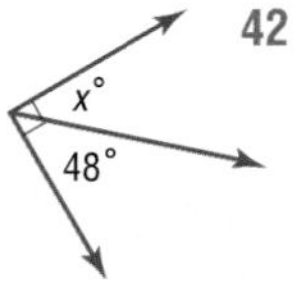

42

20. 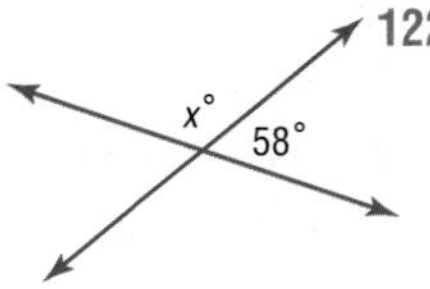

122

21. 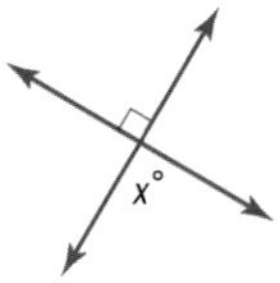

90

22. Restaurant and Movie Theater Seat, or Airplane and Don't know/none of these
24. sometimes; Sample answer: Adjacent angles can add up to 90° or any other degree measure.
25. never; Sample answer: One straight angle measures 180°.
26. always; Sample answer: Vertical angles are congruent.
27. sometimes; Sample answer: Adjacent angles can have the same or different angle measure.
32. 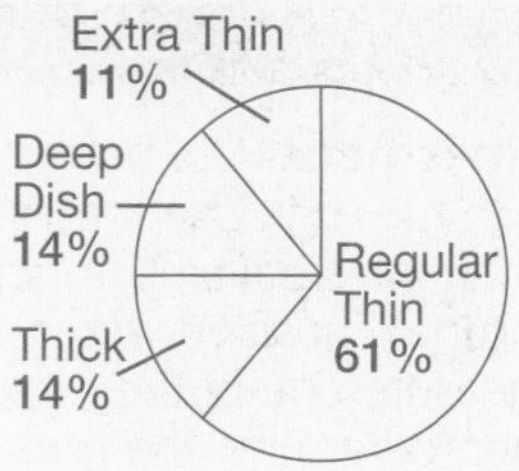

HEALTH For Exercises 22 and 23, use the graphic at the right.

★ 22. Which two sections have angles that are complementary? **See margin.**

★ 23. Which two sections have angles that are supplementary? **Public Restroom and Subway/train**

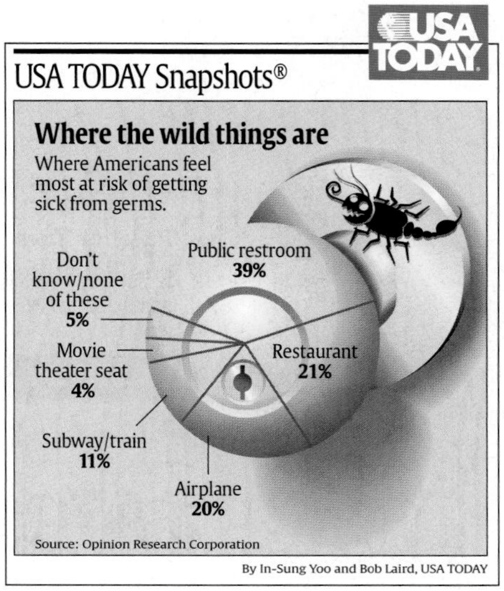

Determine whether each statement is *sometimes*, *always*, or *never* true. Explain or give an example to support your answer. **24–27. See margin.**

24. Adjacent angles are complementary.
25. Two straight angles are supplementary.
26. Vertical angles have the same angle measure.
27. Adjacent angles are congruent.

28. **WRITE A PROBLEM** Write about a real-life object that has angles. Draw a diagram of it, measure all angles, and identify any complementary, supplementary, and vertical angles. **See students' work.**

29. **CRITICAL THINKING** Angles E and F are complementary. If $m\angle E = x - 10$ and $m\angle F = x + 2$, find the measure of each angle. $m\angle E = 39°$, $m\angle F = 51°$

Standardized Test Practice and Mixed Review

30. **SHORT RESPONSE** Find the measure of two angles that are vertical and complementary. **Both angles have measures of 45°.**

31. **MULTIPLE CHOICE** Find the value of a in the figure. **C**

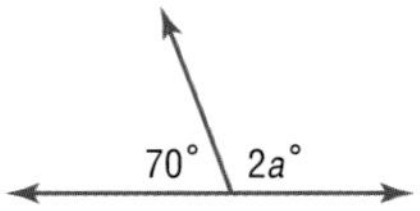

Ⓐ 90 Ⓑ 75 Ⓒ 55 Ⓓ 45

32. **STATISTICS** A company surveyed people about the type of crust they preferred on their pizza. Make a circle graph of the results shown at the right. (Lesson 10-2) **See margin.**

Type of Crust	Percent
regular thin	61%
thick	14%
deep dish	14%
extra thin	11%

Source: CREST

33. **TIME** Classify the angle the hands of a clock make at 3:30. (Lesson 10-1) **acute**

34. **REAL ESTATE** A house for sale has a rectangular lot with a length of 250 feet and a width of 120 feet. What is the area of the lot? (Lesson 6-8) **30,000 ft²**

Write each percent as a decimal. (Lessons 5-5 and 7-6)

35. 53% **0.53** 36. 78.5% **0.785** 37. 431% **4.31** 38. 0.23% **0.0023**

GETTING READY FOR THE NEXT LESSON

PREREQUISITE SKILL Solve each equation. (Lesson 4-2)

39. $x + 112 = 180$ **68** 40. $50 + t = 180$ **130** 41. $180 = 79 + y$ **101** 42. $180 = h + 125$ **55**

4 Assess

Higher-Level Thinking

Angles 1 and 2 are supplementary. Angles 2 and 3 are complementary. Angles 3 and 4 are vertical angles, and the measure of angle 4 is 50°. What is the measure of ∠1? **140°**

Study Guide and Intervention, p. 553

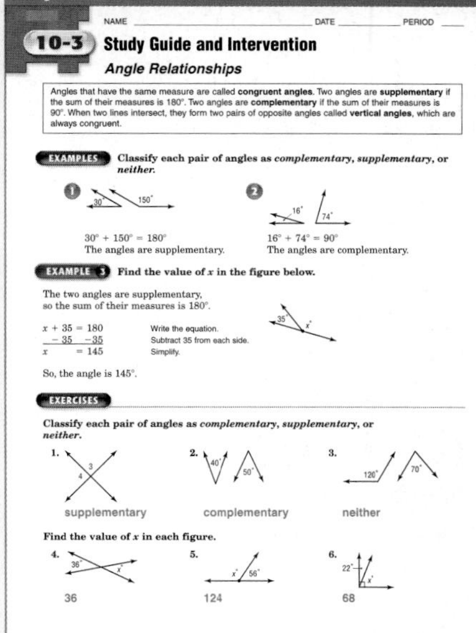
NAME ______ DATE ______ PERIOD ____

10-3 Study Guide and Intervention

Angle Relationships

Angles that have the same measure are called **congruent angles**. Two angles are **supplementary** if the sum of their measures is 180°. Two angles are **complementary** if the sum of their measures is 90°. When two lines intersect, they form two pairs of opposite angles called **vertical angles**, which are always congruent.

EXAMPLES Classify each pair of angles as *complementary*, *supplementary*, or *neither*.

1. 30° + 150° = 180° The angles are supplementary.
2. 16° + 74° = 90° The angles are complementary.

EXAMPLE 3 Find the value of x in the figure below.

The two angles are supplementary, so the sum of their measures is 180°.

$x + 35 = 180$ Write the equation.
$-35 \quad -35$ Subtract 35 from each side.
$x = 145$ Simplify.

So, the angle is 145°.

EXERCISES

Classify each pair of angles as *complementary*, *supplementary*, or *neither*.

1. supplementary 2. complementary 3. neither

Find the value of x in each figure.

4. 36 5. 124 6. 68

Practice: Skills, p. 554

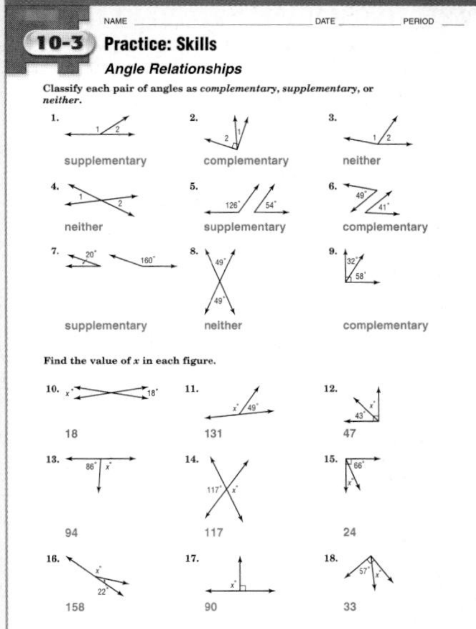
NAME ______ DATE ______ PERIOD ____

10-3 Practice: Skills

Angle Relationships

Classify each pair of angles as *complementary*, *supplementary*, or *neither*.

1. supplementary 2. complementary 3. neither
4. neither 5. supplementary 6. complementary
7. supplementary 8. neither 9. complementary

Find the value of x in each figure.

10. 18 11. 131 12. 47
13. 94 14. 117 15. 24
16. 158 17. 90 18. 33

Practice: Word Problems, p. 555

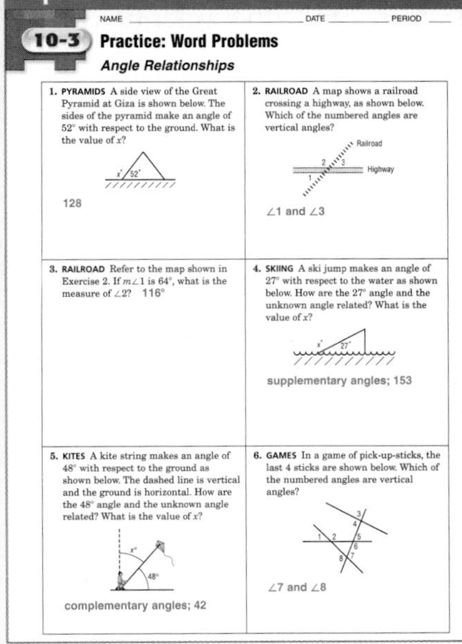
NAME ______ DATE ______ PERIOD ____

10-3 Practice: Word Problems

Angle Relationships

1. **PYRAMIDS** A side view of the Great Pyramid at Giza is shown below. The sides of the pyramid make an angle of 52° with respect to the ground. What is the value of x? 128

2. **RAILROAD** A map shows a railroad crossing a highway, as shown below. Which of the numbered angles are vertical angles? ∠1 and ∠3

3. **RAILROAD** Refer to the map shown in Exercise 2. If $m\angle 1$ is 64°, what is the measure of ∠2? 116°

4. **SKIING** A ski jump makes an angle of 27° with respect to the water as shown below. How are the 27° angle and the unknown angle related? What is the value of x? supplementary angles; 153

5. **KITES** A kite string makes an angle of 48° with respect to the ground as shown below. The dashed line is vertical and the ground is horizontal. How are the 48° angle and the unknown angle related? What is the value of x? complementary angles; 42

6. **GAMES** In a game of pick-up-sticks, the last 4 sticks are shown below. Which of the numbered angles are vertical angles? ∠7 and ∠8

10-3b Hands-On Lab
Lesson Notes

Lesson Resources

Teaching Mathematics with Manipulatives
- protractors, p. 21
- student recording sheet 10-3b

Glencoe Mathematics Classroom Manipulative Kit
- rulers (straightedges)
- compasses
- protractors

GETTING STARTED

Teaching Tip Encourage students to leave plenty of space between line ℓ and point *A*. This will make it easier to draw the necessary arcs. A larger diagram will also make it easier to measure the angles in the second part of the lab.

TEACH

Activity 1

- Remind students that to complete constructions they will use only a straightedge and compass. They will not measure with a ruler or protractor.

10-3b Hands-On Lab

A Follow-Up of Lesson 10

What You'll Learn
Construct parallel lines, and discover angle relationships.

Materials
- straightedge
- compass
- protractor
- colored pencils
- notebook paper

Constructing Parallel Lines

Parallel lines are lines that do not intersect. The symbol ∥ means parallel. In this lab, you will learn how to construct parallel lines and discover angle relationships that are created from parallel lines.

ACTIVITY *Work with a partner.*

1 **Construct a line parallel to line ℓ.**

STEP 1 Draw line ℓ. Choose a point *A* not on the line. Then draw a line through point *A* that intersects ℓ. Label the point of intersection *B*.

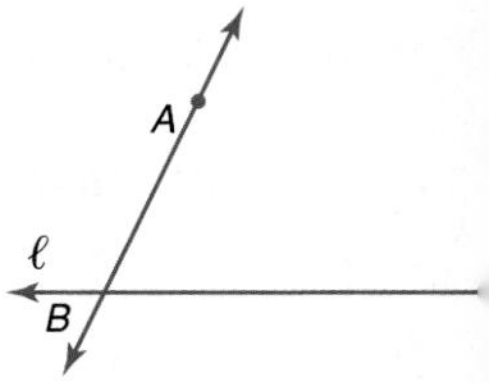

STEP 2 Place the compass at vertex *B* and draw an arc. Label the points of intersection *C* and *D*. With the same compass setting, place the compass at point *A* and draw an arc. Label the point of intersection *E*.

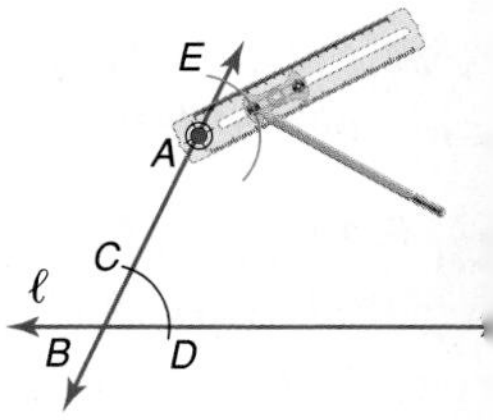

STEP 3 Open your compass to the same width as the distance between *C* and *D*. Then place the compass at point *E*. Draw an arc that intersects the arc you drew in Step 2. Label this point of intersection *F*.

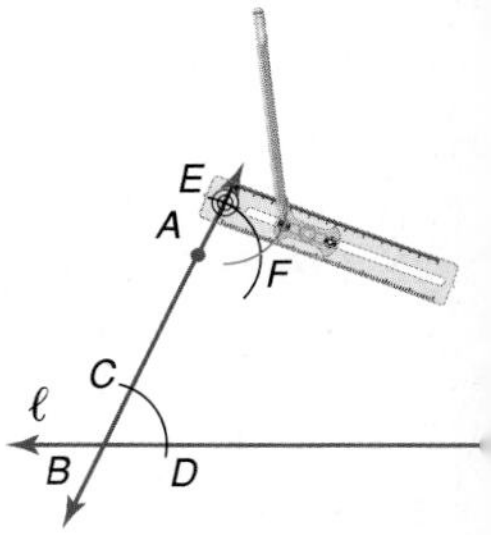

STEP 4 Draw a line through points *A* and *F*. Label the line *m*. Line *m* is parallel to line ℓ.

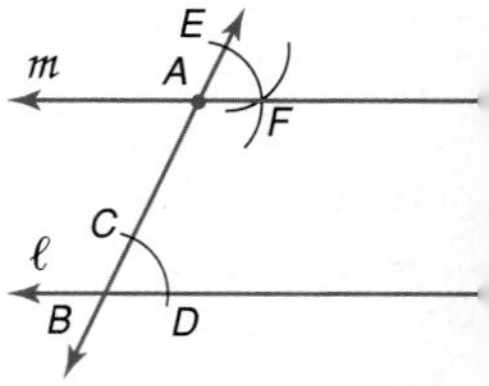

Your Turn

a. Draw a line. Then construct a line parallel to it. **See students' work**

A line that intersects parallel lines is called a **transversal**. Line t is a transversal for parallel lines ℓ and m. When a transversal cuts two parallel lines, there are several angle relationships formed.

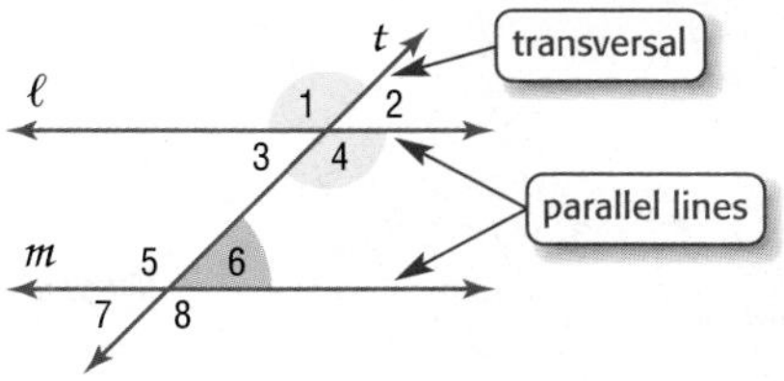

TEACHING TIP

lake sure students use
e lines on the notebook
aper to help them draw
arallel lines.

ACTIVITY *Work with a partner.*

2 **Find angle relationships in the figure above.**

STEP 1 Draw parallel lines and a transversal on notebook paper. Label the angles as shown above.

STEP 2 Measure each angle with your protractor. Record each measure.

STEP 3 Using colored pencils, shade any angles with the same measure one color. If there are angles with a different measure, shade them a second color. For example, ∠1 and ∠4 shown above are both shaded yellow because they have the same measure. Angle 6 is shaded blue because it has a different measure.

In the figure above, certain pairs of angles have special names. The angles that make up each pair have the same measure.

alternate interior angles	∠3 and ∠6, ∠4 and ∠5
alternate exterior angles	∠1 and∠8, ∠2 and ∠7
corresponding angles	∠1 and ∠5, ∠2 and ∠6, ∠3 and ∠7, ∠4 and ∠8
vertical angles	∠1 and ∠4, ∠2 and ∠3, ∠5 and ∠8, ∠6 and ∠7

Writing Math

1. **Explain** why you think ∠1, ∠2, ∠7, and ∠8 are called exterior angles.
2. **Explain** why you think ∠3, ∠4, ∠5, and ∠6 are called interior angles.
3. What are ∠2 and ∠4 called? **supplementary**
4. If you know only one angle measure in the figure, **explain** how you can find the measures of the other angles without measuring.
5. Predict the measure of the other angles in the figure at the right using the 45° angle. Then copy the figure onto notebook paper and check by using a protractor.

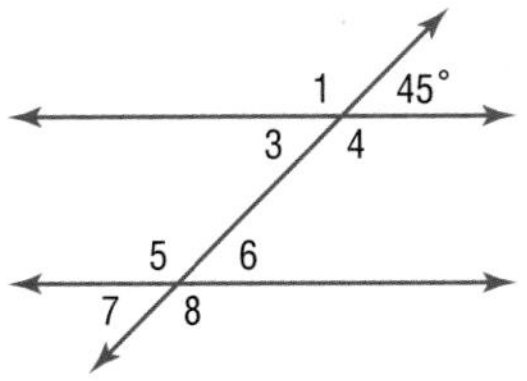

Sample answer: Those
ngles are to the exterior
the parallel lines.

Sample answer: Those
ngles are to the interior
the parallel lines.

Sample answer:
u can use the angle
lationships to find
e measures of the
her angles without
easuring.

m∠1 = m∠4 =
∠5 = m∠8 = 135°,
∠3 = m∠6 = m∠7 =
5°

Activity 2

- Before measuring the angles in Activity 2, have students observe which angles are obtuse and which are acute. From there, have them make predictions about which are congruent.
- Have students write an outline of the procedure from this lab into their math journals or Foldables for Chapter 10.

ASSESS

Observe as students solve the **Your Turn** exercise. They should:

- begin with a horizontal line, and
- carefully follow the directions for the activity.

Discuss the **Writing Math** exercises. For Exercise 5, encourage students to carefully read the table showing the angle relationships.

Lesson 10-3b Hands-On Lab: Constructing Parallel Lines **427**

10-4 Lesson Notes

Lesson Resources

Chapter 10 Resource Masters

FAST FILE

- Study Guide and Intervention, p. 558
- Practice: Skills, p. 559
- Practice: Word Problems, p. 560
- Reading to Learn Mathematics, p. 561
- Enrichment, p. 562

Parent and Student Study Guide Workbook, 10-4

1 Focus

5-Minute Check 10-4

INTERACTIVE CHALKBOARD
PowerPoint® Presentations

Use as a quiz or review of Lesson 10-3. (Also available as a transparency)

Bellringer

Hands-On Activity Tell students to take out a piece of paper and make three dots anywhere on the page. Have students use a straightedge to connect the dots and make a triangle; then have them measure the angles with their protractors. They should find the sum of the angles and record this result on the chalkboard. Help students recognize that any three points on the page can be connected to make a triangle, and discuss the table showing the sum of the angle measures.

Reading to Learn Mathematics, p. 561 ELL

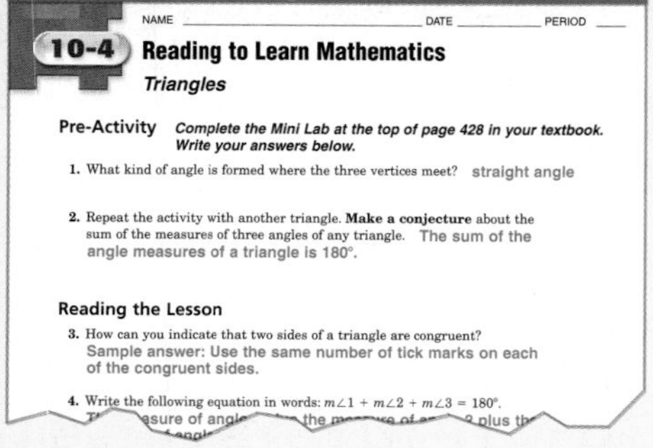
NAME ______ DATE ______ PERIOD ____

10-4 Reading to Learn Mathematics

Triangles

Pre-Activity ***Complete the Mini Lab at the top of page 428 in your textbook. Write your answers below.***

1. What kind of angle is formed where the three vertices meet? straight angle
2. Repeat the activity with another triangle. **Make a conjecture** about the sum of the measures of three angles of any triangle. The sum of the angle measures of a triangle is 180°.

Reading the Lesson

3. How can you indicate that two sides of a triangle are congruent? Sample answer: Use the same number of tick marks on each of the congruent sides.
4. Write the following equation in words: $m\angle 1 + m\angle 2 + m\angle 3 = 180°$.

10-4 Triangles

What You'll Learn
Identify and classify triangles.

NEW Vocabulary

- triangle
- acute triangle
- right triangle
- obtuse triangle
- congruent segments
- scalene triangle
- isosceles triangle
- equilateral triangle

MATH Symbols

$m\angle 1$ measure of angle 1

HANDS-ON Mini Lab

Materials
- paper
- straightedge
- scissors

Work with a partner.

STEP 1 Use a straightedge to draw a triangle with three acute angles. Label the angles *A*, *B*, and *C*. Cut out the triangle.

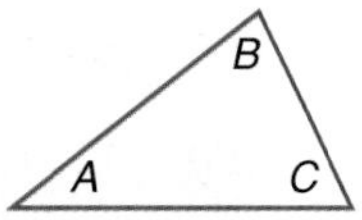

STEP 2 Fold $\angle B$ so that its vertex touches the line between angles *A* and *C*. The fold should be parallel to the base of the triangle.

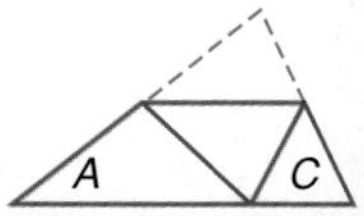

STEP 3 Fold $\angle A$ and $\angle C$ so the vertices meet.

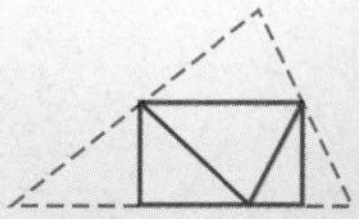

1. straight angle

1. What kind of angle is formed where the three vertices meet?
2. Repeat the activity with another triangle. **Make a conjecture** about the sum of the measures of three angles of any triangle.

The sum of the angle measures of a triangle is 180°.

READING Math

Triangle The word *triangle* means *three angles.*

A **triangle** is a figure with three sides and three angles. In the Mini Lab, you discovered a relationship among the three angles in a triangle.

Key Concept — **Angles of a Triangle**

Words The sum of the measures of the angles of a triangle is 180°.

Symbols $m\angle 1 + m\angle 2 + m\angle 3 = 180°$

Model

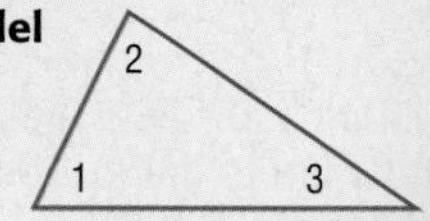

EXAMPLE **Find Angle Measures of Triangles**

1 KITES A kite is constructed with two triangles. Find the missing measure.

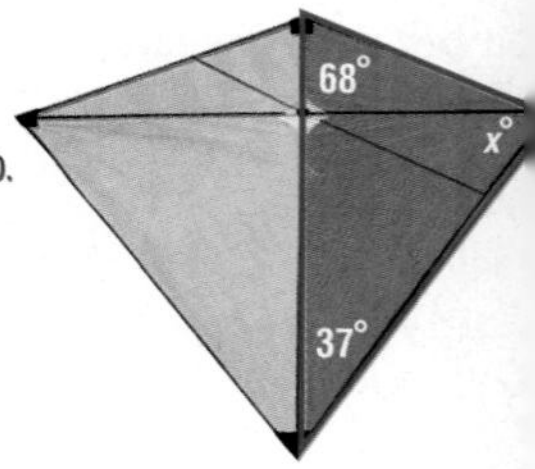

$x + 68 + 37 = 180$		The sum of the measures is 180.
$x + 105 = 180$		Simplify.
$-105 \quad -105$		Subtract 105 from each side.
$x = 75$		

The missing measure is 75°.

STUDY TIP

Check the Reasonableness of Results You can check whether you found the correct measure by adding 43 + 119 + 18. The sum should be 180.

EXAMPLE Find a Missing Measure

2 **ALGEBRA Find $m\angle Z$ in $\triangle XYZ$ if $m\angle X = 43°$ and $m\angle Y = 119°$.**

Draw the triangle. Then write and solve an equation to find $m\angle Z$.

$m\angle X + m\angle Y + m\angle Z =$	180	The sum of the measures is 180.
$43 + 119 + m\angle Z =$	180	Substitute.
$162 + m\angle Z =$	180	Simplify.
-162	-162	Subtract 162 from each side.
$m\angle Z =$	18	

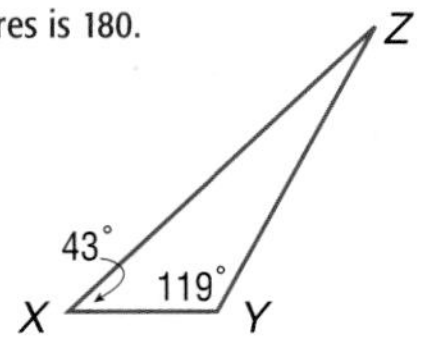

So, the measure of $\angle Z$ is 18°.

Since every triangle has at least two acute angles, one way you can classify a triangle is by using the third angle.

Key Concept — Classify Triangles Using Angles

Acute Triangle

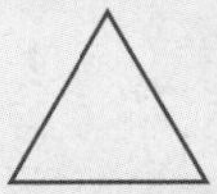

all acute angles

Right Triangle

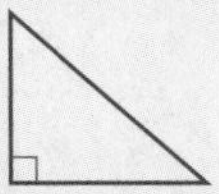

1 right angle

Obtuse Triangle

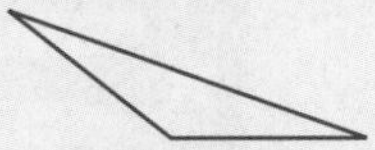

1 obtuse angle

Another way to classify triangles is by their sides. Sides with the same length are **congruent segments**.

STUDY TIP

Congruent Segments The marks on the sides of the triangles indicate that those sides are congruent.

Key Concept — Classify Triangles Using Sides

Scalene Triangle

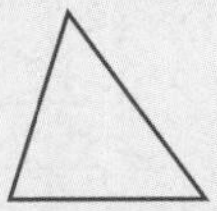

no congruent sides

Isosceles Triangle

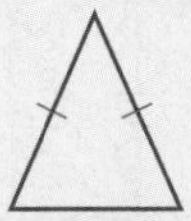

at least 2 congruent sides

Equilateral Triangle

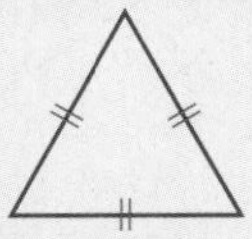

3 congruent sides

TEACHING TIP

ave students use the :k marks to help them assify triangles.

EXAMPLES Classify Triangles

Classify each triangle by its angles and by its sides.

3

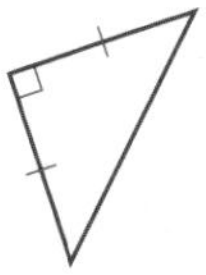

The triangle has a right angle and two congruent sides. So, it is a right, isosceles triangle.

4

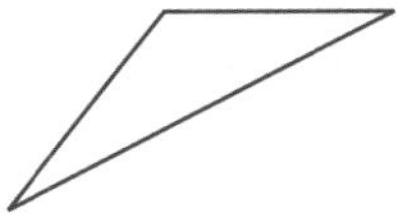

The triangle has one obtuse angle and no congruent sides. So, it is an obtuse, scalene triangle.

msmath2.net/extra_examples

2 Teach

In-Class Examples

INTERACTIVE CHALKBOARD PowerPoint® Presentations

1 **PLANES** An airplane has wings that are shaped like triangles. Find the missing measure.

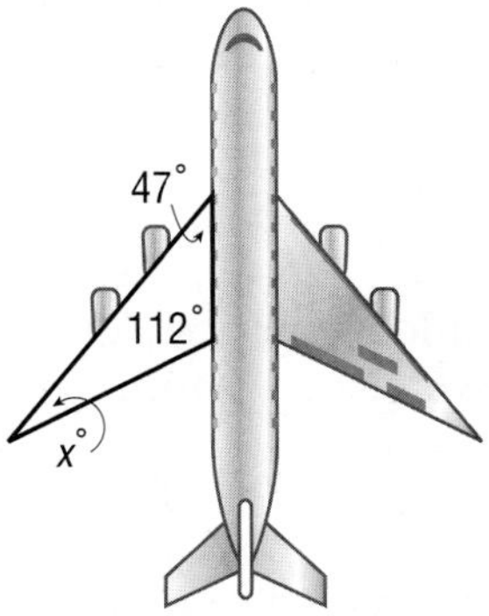

21

2 **ALGEBRA** Find $m\angle A$ in $\triangle ABC$ if $m\angle A = m\angle B$ and $m\angle C = 80°$. **50°**

Classify each triangle by its angles and by its sides.

3

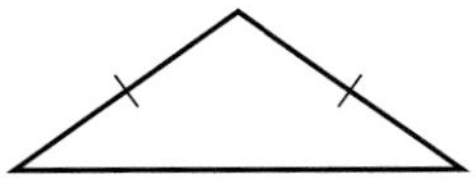

obtuse, isosceles

4

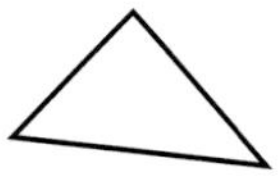

acute, scalene

DAILY INTERVENTION — **Flexible Grouping**

Angles in Triangles Try to group students who understand the quantitative relationships among angles with those who are good at modeling and drawing angles. Encourage each student to teach one aspect of an activity to the other group members. For example, one student might demonstrate the algebraic way to find the missing angle of a triangle, and another might demonstrate drawing an angle with the missing measure.

3 Practice/Apply

FOLDABLES™ Follow-Up Remind students to use study cards to record what they learn about the classification of triangles. Encourage students to illustrate and explain how triangles can be classified by their sides and angles.

About the Exercises...

Guided Practice Key

Exercises	Examples
3–4	1, 2
5–7	3, 4

Odd/Even Assignments
Exercises 8–13 and 18–23 are structured so that students practice the same concepts whether they are assigned odd or even problems.

Answer

2. Sample answer:

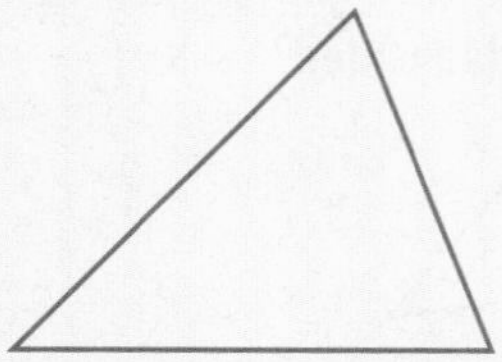

Skill and Concept Check

Writing Math Exercise 1

1. **Describe** the angles in an obtuse triangle.
   An obtuse triangle has one obtuse angle and two acute angles.
2. **OPEN ENDED** Draw an acute scalene triangle. See margin.

GUIDED PRACTICE

Find the missing measure in each triangle. Then classify the triangle as *acute, right,* or *obtuse.*

TEACHING TIP
Have students research the location of the countries represented by the flags in Exercises 5–7.
5. Caribbean Sea
6. Indian Ocean
7. Europe

3. 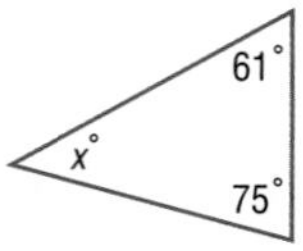

44°; acute

4. 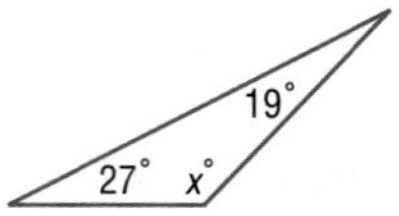

134°; obtuse

Classify the marked triangle in each flag by its angles and by its sides.

5. Puerto Rico

acute, equilateral

6. Seychelles Islands
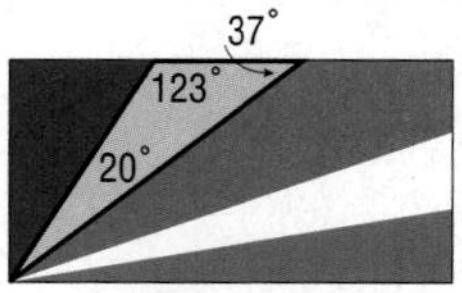

obtuse, scalene

7. Bosnia-Herzegovina
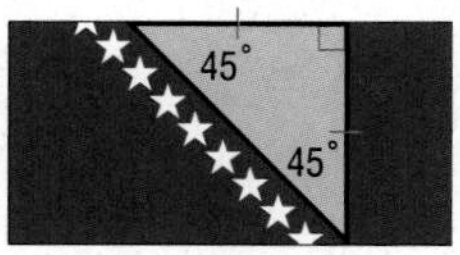

right, isosceles

Practice and Applications

HOMEWORK HELP

For Exercises	See Examples
8–14, 17, 24	1, 2
8–13, 15–16, 18–23	3, 4

Extra Practice
See pages 588, 605.

Find the missing measure in each triangle. Then classify the triangle as *acute, right,* or *obtuse.*

8. 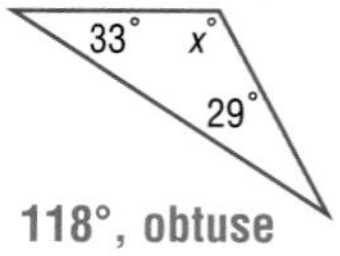

118°, obtuse

9. 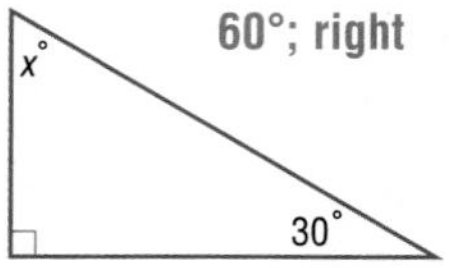

60°; right

10. 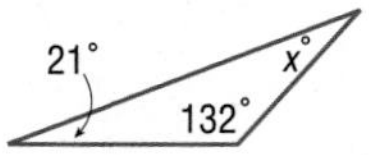

27°; obtuse

11. 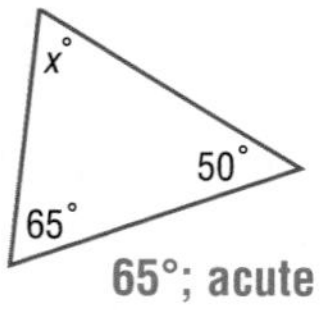

65°; acute

12. 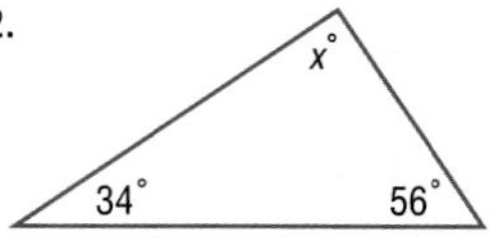

90°; right

13. 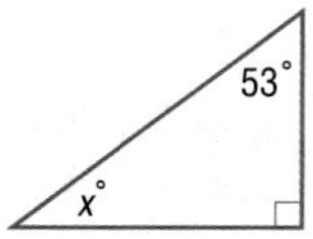

37°; right

14. Find the third angle of a right triangle if the measure of one of the angles is 10°. 80°

15. Three angles of a triangle measure 30°, 60°, and 90°. Classify the triangle by its angles. right

16. Three sides of a triangle measure 5 meters, 8 meters, and 8 meters. Classify the triangle by its sides. isosceles

17. **ALGEBRA** Find $m\angle T$ in $\triangle RST$ if $m\angle R = 88°$ and $m\angle S = 75°$. 17°

Classify each triangle by its angles and by its sides.

18. 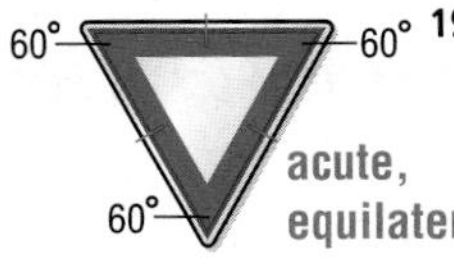

acute, equilateral

19. 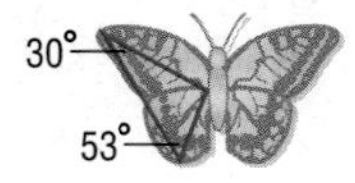

obtuse, scalene

20. 

right, isosceles

21. 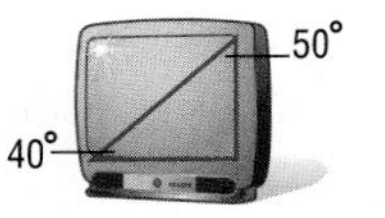

right, scalene

22. 

acute, scalene

23. 

acute, isosceles

24. **GEOGRAPHY** A triangular area in the southern Atlantic Ocean where airplanes and boats have disappeared is known as the *Bermuda Triangle*. Find the missing angle measurement at Bermuda in the figure at the right. **49°**

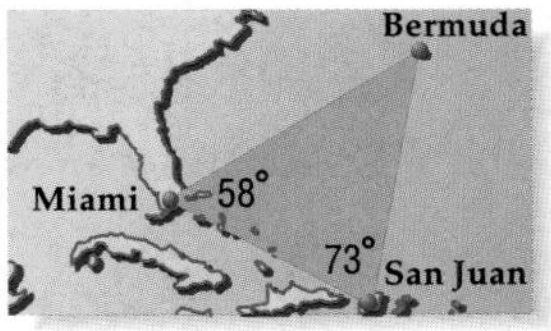

25. **WRITE A PROBLEM** Write about a real-life situation or object involving a triangle and its measures. Classify the triangle by its sides and its angles. **See students' work.**

26. **CRITICAL THINKING** Find the missing angle measures in the figure. **$a = 55°$; $b = 65°$; $c = 60°$; $d = 30°$**

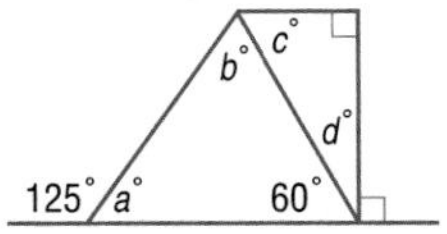

Standardized Test Practice and Mixed Review

27. **MULTIPLE CHOICE** How would you find $m\angle R$? **B**

(A) Add 30° to 180°.
(B) Subtract 60° from 180°.
(C) Subtract 30° from 90°.
(D) Subtract 180° from 60°.

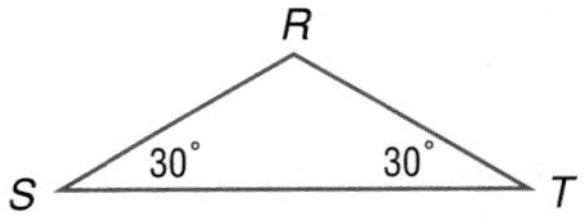

28. **SHORT RESPONSE** Find the measure of two congruent angles of a triangle if the third angle measure is 54°. **63°**

Classify each pair of angles as *complementary*, *supplementary*, or *neither*.
(Lesson 10-3)

29. 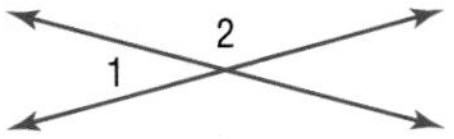

supplementary

30. 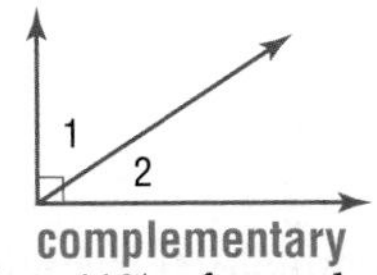

complementary

31. 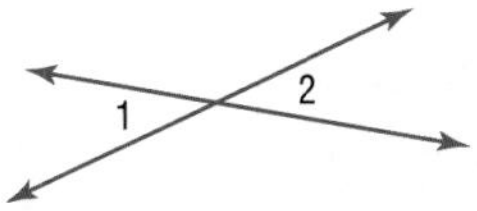

neither

32. **BIRDS** A circle graph shows that 41% of people who bird watch live in the Northeast region of the United States. What is the measure of the angle of the Northeast section of the graph? (Lesson 10-2) **147.6°**

GETTING READY FOR THE NEXT LESSON

PREREQUISITE SKILL Solve each equation. (Lesson 4-2)

33. $x + 120 + 120 + 60 = 360$ **60**
34. $73 + 119 + x + 50 = 360$ **118**

msmath2.net/self_check_quiz

4 Assess

Open-Ended Assessment

Modeling Create a Venn diagram that shows how different types of triangles relate to each other. For example, students might draw three overlapping circles labeled *acute triangle, right triangle,* and *obtuse triangle*. They would place *m∠1 + m∠2 + m∠3 = 180°* in the common area, and *1 angle > 90°* in the obtuse triangle's circle.

Study Guide and Intervention, p. 558

NAME ______ DATE ______ PERIOD ____

10-4 Study Guide and Intervention
Triangles

A **triangle** is a figure with three sides and three angles. The sum of the measures of the angles of a triangle is 180°. You can use this to find a missing angle measure in a triangle.

EXAMPLE 1 Find the value of x in $\triangle ABC$.

$x + 66 + 52 = 180$ The sum of the measures is 180.
$x + 118 = 180$ Simplify.
$-118 - 118$ Subtract 118 from each side.
$x = 62$

The missing angle is 62°.

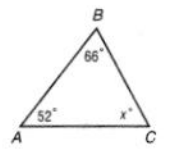

Triangles can be classified by the measures of their angles. An **acute triangle** has three acute angles. An **obtuse triangle** has one obtuse angle. A **right triangle** has one right angle.

Triangles can also be classified by the lengths of their sides. Sides that are the same length are **congruent segments** and are often marked by tick marks. In a **scalene triangle**, all sides have different lengths. An **isosceles** triangle has at least two congruent sides. An **equilateral triangle** has all three sides congruent.

EXAMPLE 2 Classify the triangle by its angles and by its sides.

The triangle has one obtuse angle and two sides the same length. So, it is an obtuse, isosceles triangle.

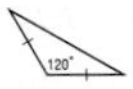

EXERCISES

Find the missing measure in each triangle. Then classify the triangle as *acute, right,* or *obtuse*.

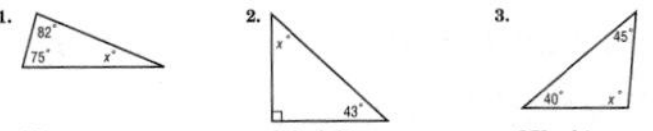

1. 23°; acute
2. 47°; right
3. 95°; obtuse

Classify each triangle by its angles and by its sides.

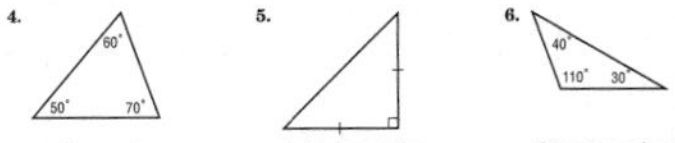

4. acute, scalene
5. right, isosceles
6. obtuse, scalene

Practice: Skills, p. 559

NAME ______ DATE ______ PERIOD ____

10-4 Practice: Skills
Triangles

Find the missing measure in each triangle. Then classify the triangle as *acute, right,* or *obtuse*.

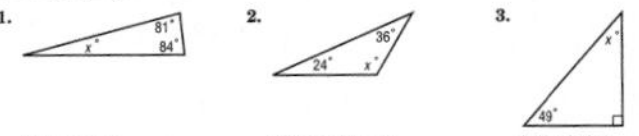

1. 15°; acute
2. 120°; obtuse
3. 41°; right

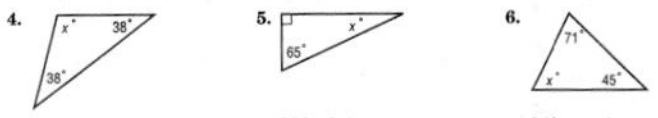

4. 104°; obtuse
5. 25°; right
6. 64°; acute

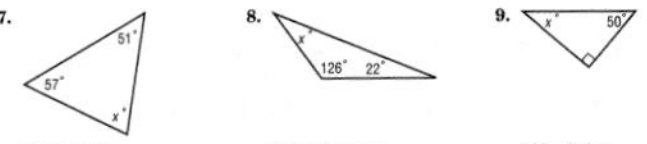

7. 72°; acute
8. 32°; obtuse
9. 40°; right

Classify each triangle by its angles and by its sides.

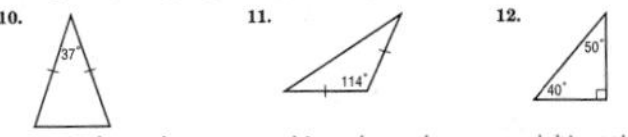

10. acute, isosceles
11. obtuse, isosceles
12. right, scalene

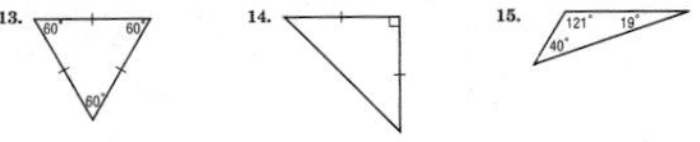

13. acute, equilateral
14. right, isosceles
15. obtuse, scalene

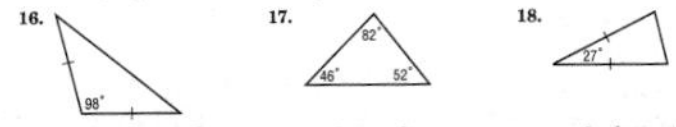

16. obtuse, isosceles
17. acute, scalene
18. acute, isosceles

Practice: Word Problems, p. 560

NAME ______ DATE ______ PERIOD ____

10-4 Practice: Word Problems
Triangles

1. **TAILORING** Each lapel on a suit jacket is in the shape of a triangle. The three angles of each triangle measure 47°, 68°, and 65°. Classify the triangle by its angles. acute

2. **FLAGS** A naval distress signal flag is in the shape of a triangle. The three sides of the triangle measure 5 feet, 9 feet, and 9 feet. Classify the triangle by its sides. isosceles

3. **CARPENTRY** The supports of a wood table are in the shape of a right triangle. Find the third angle of the triangle if the measure of one of the angles is 23°. 67°

4. **MAPS** The three towns of Ripon, Sparta, and Walker form a triangle as shown below. Classify the triangle by its angles and by its sides. What is the value of x in the triangle?

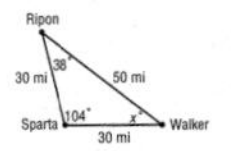

obtuse, isosceles; 38

5. **HIKING** The figure shows the Oak Creek trail, which is shaped like a triangle. Classify the triangle by its angles and by its sides. What is the value of x in the figure?

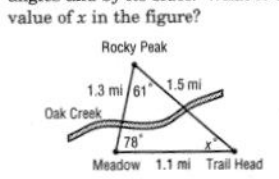

acute, scalene; 41

6. **LADDER** The figure shows a ladder leaning against a wall, forming a triangle. Classify the triangle by its angles and by its sides. What is the value of x in the figure?

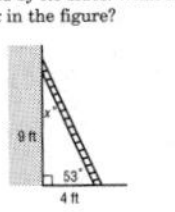

right, scalene; 27

10-4b HANDS-ON LAB Lesson Notes

Lesson Resources

Teaching Mathematics with Manipulatives
- protractors, p. 21
- student recording sheet 10-4b

Glencoe Mathematics Classroom Manipulative Kit
- rulers (straightedges)
- compasses
- protractors

GETTING STARTED

Teaching Tip Before starting the lab, prepare a table on the chalkboard for students to record their measurements for **Writing Math** Exercises 1 and 3. This table will make it easier for students to compare their measurements with those of other groups.

TEACH

Activity 1

- Remind students that constructions are made with a straightedge and compass.
- Before beginning the constructions, remind students of the definitions of equilateral and isosceles triangles.
- Ask students why it would be very difficult to construct an equilateral triangle using only a ruler. **It would be very difficult to get the angles just right.** As they make their constructions, encourage them to discuss how the compass solves this problem.

10-4b HANDS-ON LAB

A Follow-Up of Lesson 1

What You'll Learn
Construct equilateral and isosceles triangles.

Materials
- straightedge
- compass
- protractor
- ruler

Constructing Triangles

Equilateral triangles are triangles with three congruent sides. Isosceles triangles are triangles with at least two congruent sides. In this lab, you will learn how to construct both types of triangles.

ACTIVITY *Work with a partner.*

1 Construct an equilateral triangle.

STEP 1 Use your straightedge to draw line segment *AB*.

STEP 2 Open the compass to the same length as $\overline{AB}$. With the compass at point *A*, draw an arc above the line.

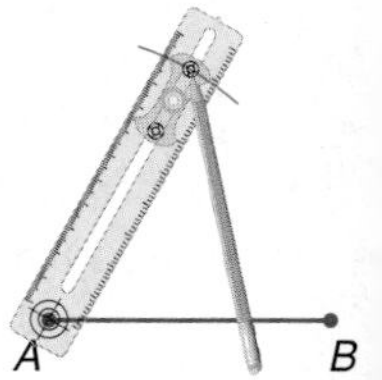

STEP 3 With the same compass setting, place the compass at point *B*. Draw an arc that intersects the arc you drew in Step 2. Label the intersection *C*.

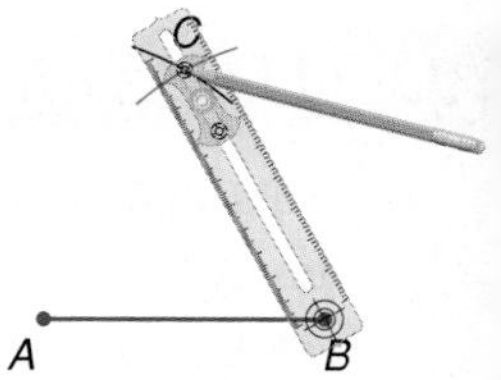

STEP 4 Connect the points to complete equilateral △*ABC*.

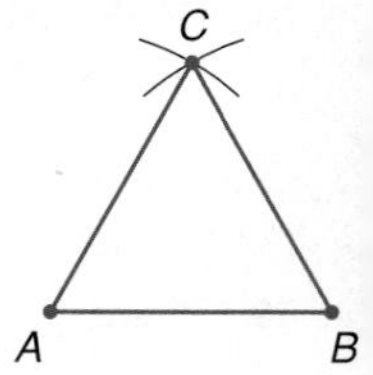

STUDY TIP

Check You can measure each side of the triangle to see if the sides are congruent.

Your Turn

a. Construct equilateral triangle *TUV* with sides measuring 3 inches. **See students' work.**

In order to construct equilateral triangles in the activity above, you copied line segments using a compass. You constructed line segments with the same length by copying $\overline{AB}$ and making $\overline{AC}$ and $\overline{BC}$ congruent. So, a compass is also a useful tool to check whether line segments have the same length.

ACTIVITY *Work with a partner.*

2 **Construct an isosceles triangle.**

STEP 1 Use your straightedge to draw line segment *DE*.

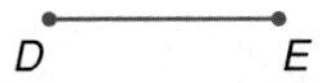

STEP 2 Open the compass to a length greater than $\overline{DE}$. With the compass at point *D*, draw an arc above the line.

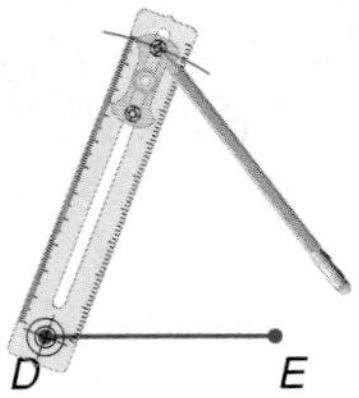

STEP 3 With the same compass setting, place the compass at point *E*. Draw an arc that intersects the arc you drew in Step 2. Label the intersection *F*.

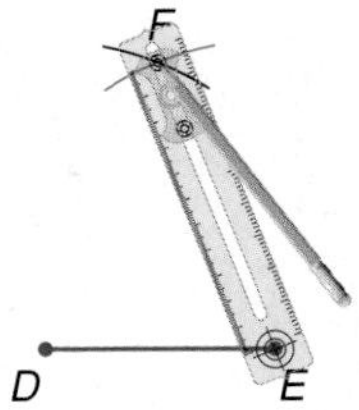

STEP 4 Connect the points to complete isosceles $\triangle DEF$.

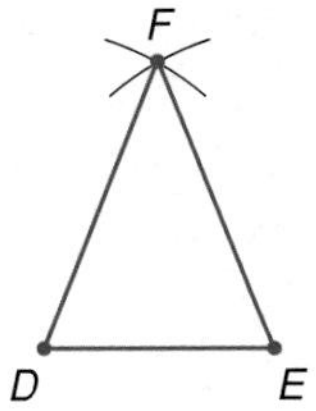

STUDY TIP

Check You can measure $\overline{DF}$ and $\overline{EF}$ to check if they are congruent.

Your Turn

b. Construct isosceles triangle *XYZ* with sides measuring 3 centimeters, 5 centimeters, and 5 centimeters. **See students' work.**

Writing Math

Work with a partner. **1. All angle measures are approximately 60°.**

1. **Measure** the angles in the equilateral triangle you constructed with a protractor. Compare your measurements with other groups.
2. **Make a conjecture** about the measure of the angles of an equilateral triangle.
3. **Measure** the angles in the isosceles triangle you constructed with a protractor. Compare your measurements with other groups.
4. **Make a conjecture** about the measure of the angles of an isosceles triangle.

. Sample answer: The ngle measures of an quilateral triangle are 0°.

. Angles *D* and *E* are ongruent.

. Sample answer: The ase angles of an sosceles triangle are ongruent.

Activity 2

- Ask students to outline a procedure for constructing $\angle DEF$ such that it measures 80° and $\triangle DEF$ is an isosceles triangle. **Students will have to first construct an 80° angle; then they can use their protractors or rulers to make the congruent sides.**

ASSESS

Observe as students solve the **Your Turn** exercises. They should:

- use a ruler to draw the base of each triangle,
- complete the remainder of the construction without the assistance of a ruler or protractor, and
- follow the directions shown in the activity.

Discuss the **Writing Math** exercises. Include a discussion of the reasons measurements for Exercises 1 and 3 might vary.

Lesson 10-4b Hands-On Lab: Constructing Triangles **433**

10-5 Lesson Notes

Lesson Resources

Chapter 10 Resource Masters
- Study Guide and Intervention, p. 563
- Practice: Skills, p. 564
- Practice: Word Problems, p. 565
- Reading to Learn Mathematics, p. 566
- Enrichment, p. 567

Parent and Student Study Guide Workbook, 10-5

Teaching Mathematics with Manipulatives, pp. 95–96

FAST FILE

1 Focus

5-Minute Check 10-5

INTERACTIVE CHALKBOARD
PowerPoint® Presentations

Use as a quiz or review of Lesson 10-4. (Also available as a transparency)

Bellringer

Communication Show students a cardboard box with both ends open. Ask them to describe the shape of the bottom of the box. square or rectangular Then compress the box slightly so that there are no 90° angles. Ask students to discuss this new shape.

Answer (When am I ever going to use this?)

1. Sample answer: The angles on the left side of the lane are acute and the ones on the right side are obtuse.

Reading to Learn Mathematics, p. 566 ELL

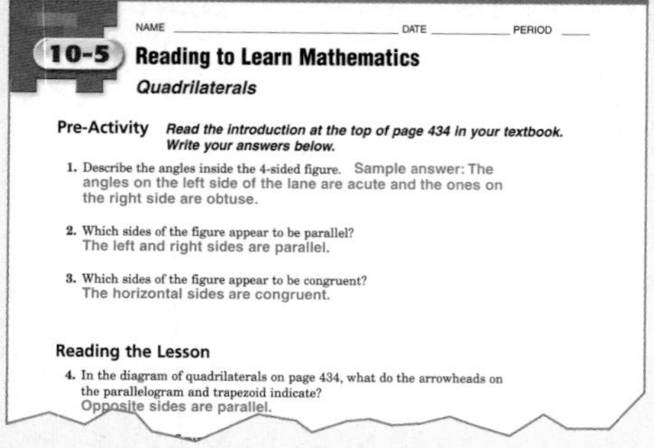
NAME ______ DATE ______ PERIOD ______

10-5 Reading to Learn Mathematics
Quadrilaterals

Pre-Activity *Read the introduction at the top of page 434 in your textbook. Write your answers below.*

1. Describe the angles inside the 4-sided figure. Sample answer: The angles on the left side of the lane are acute and the ones on the right side are obtuse.
2. Which sides of the figure appear to be parallel? The left and right sides are parallel.
3. Which sides of the figure appear to be congruent? The horizontal sides are congruent.

Reading the Lesson

4. In the diagram of quadrilaterals on page 434, what do the arrowheads on the parallelogram and trapezoid indicate? Opposite sides are parallel.

10-5 Quadrilaterals

What You'll Learn
Identify and classify quadrilaterals.

NEW Vocabulary
- quadrilateral
- parallelogram
- trapezoid
- rhombus

REVIEW Vocabulary
parallel: lines that do not intersect (Lesson 10-3b)

WHEN am I ever going to use this?

BASKETBALL The photograph shows the free throw lane used in international basketball.

1. Describe the angles inside the 4-sided figure. See margin.
2. Which sides of the figure appear to be parallel? 2. The left and right sides are parallel.
3. Which sides of the figure appear to be congruent? The horizontal sides are congruent.

The shape of the free-throw lane above is called a trapezoid. Squares, rectangles, and trapezoids are examples of quadrilaterals. A **quadrilateral** is a closed figure with four sides and four angles. Quadrilaterals are named based on their sides and angles.

The diagram shows how quadrilaterals are related. Notice how it goes from the most general to the most specific.

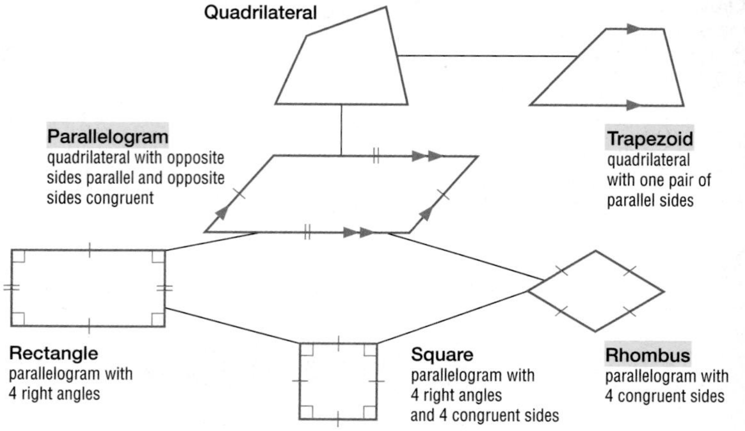

READING Math

Parallel Lines The sides with matching arrows are parallel.

EXAMPLES Classify Quadrilaterals

Classify the quadrilateral using the name that *best* describes it.

1 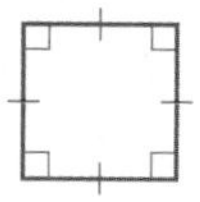
The quadrilateral has 4 right angles and 4 congruent sides. It is a square.

2 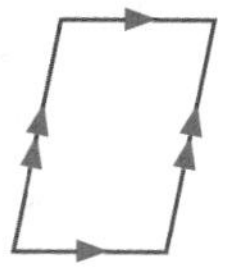
The quadrilateral has opposite sides parallel. It is a parallelogram.

DAILY INTERVENTION

Differentiated Instruction ELL

Verbal/Linguistic Have students write detailed descriptions of each quadrilateral studied in this lesson. Other students should be able to read the descriptions and draw the quadrilaterals.

A quadrilateral can be separated into two triangles, *A* and *B*. Since the sum of the angle measures of each triangle is 180°, the sum of the angle measures of both triangles is $2 \cdot 180$, or 360°.

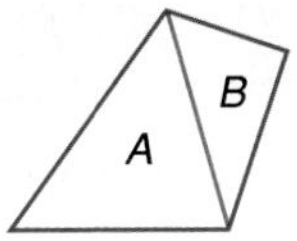

Key Concept — **Angles of a Quadrilateral**

Words The sum of the measures of the angles of a quadrilateral is 360°.

Symbols $m\angle 1 + m\angle 2 + m\angle 3 + m\angle 4 = 360°$

Model

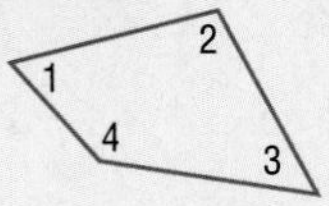

EXAMPLE Find a Missing Measure

3 **Find the missing angle measure in the quadrilateral.**

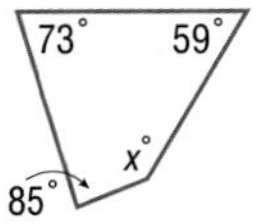

$$85 + 73 + 59 + x = 360 \quad \text{The sum of the measures is 360°.}$$
$$217 + x = 360 \quad \text{Simplify.}$$
$$-217 \quad -217 \quad \text{Subtract 217 from each side.}$$
$$x = 143$$

So, the missing angle measure is 143°.

Skill and Concept Check

Writing Math
Exercises 1 & 3

1. **Explain** why all squares are rectangles but not all rectangles are squares.
   See margin.
2. **OPEN ENDED** Describe a real-life example of a rhombus.
   Sample answer: the diamond suit on the face of a playing card
3. **FIND THE ERROR** Venus and Justin are describing a square. Who is correct? Explain.

Venus	Justin
a parallelogram with 4 right angles	a rhombus with 4 right angles

Justin; Venus did not mention the 4 congruent sides of a square.

GUIDED PRACTICE

Classify the quadrilateral using the name that *best* describes it.

4. 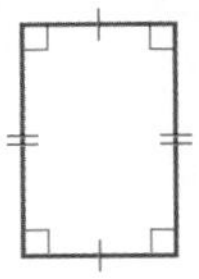 rectangle

5. 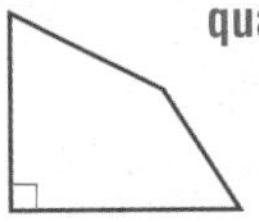 quadrilateral

6. 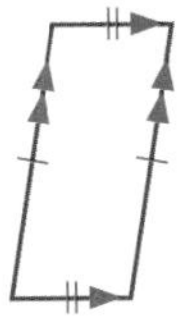 parallelogram

Find the missing angle measure in each quadrilateral.

7. 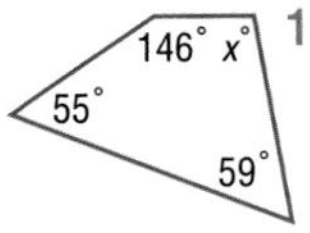

100°

8. 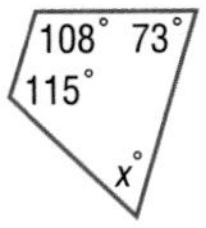

64°

9. 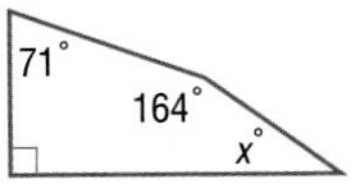

3

msmath2.net/extra_examples

2 Teach

In-Class Examples

INTERACTIVE CHALKBOARD
PowerPoint® Presentations

Classify the quadrilateral using the name that *best* describes it.

1 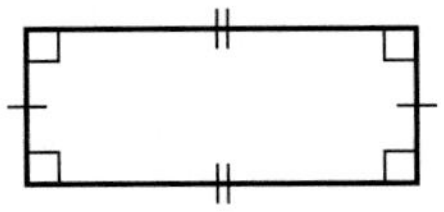

rectangle

2 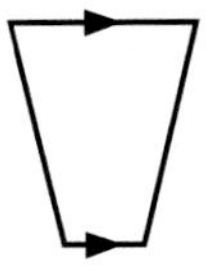

trapezoid

3 Find the missing angle measure in the quadrilateral.
120°

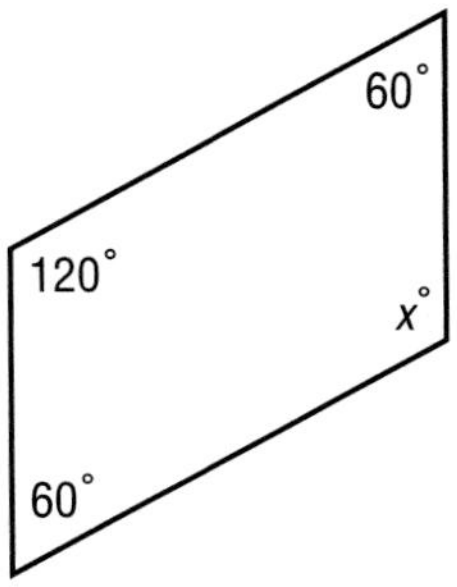

3 Practice/Apply

DAILY INTERVENTION **FIND THE ERROR** In Exercise 3, refer students to the quadrilaterals on page 434 and encourage them to draw pictures of the figures described by Justin and Venus. If they draw the figures accurately, they will see Venus described a rectangle and Justin described a square.

Answer

1. Sample answer: A square has all the properties of a rectangle. However, a rectangle does not have four congruent sides, which is a property of a square.

FOLDABLES™ Follow-Up Remind students to use study cards to record what they learn about quadrilaterals. Encourage students to illustrate and describe the five types of quadrilaterals discussed in this chapter—square, rectangle, parallelogram, rhombus, and trapezoid.

About the Exercises...

Guided Practice Key

Exercises	Examples
4–6	1, 2
7–9	3

Odd/Even Assignments
Exercises 10–24 are structured so that students practice the same concepts whether they are assigned odd or even problems.

Alert! Excercise 26 requires research on the Internet or another source.

Answers

16. sometimes; Sample answer: A quadrilateral can also be a rectangle.
17. sometimes; Sample answer: A parallelogram can also have angles with different degree measures.
18. never; Sample answer: A trapezoid has only one pair of parallel sides. A parallelogram has 2 pairs of parallel sides.
29. Yes; A square is a rhombus and a rectangle.
30. No; A quadrilateral with three right angles will have both pairs of opposite sides parallel. So, it cannot be a trapezoid.
31. Yes; One pair of opposite sides are parallel, while the other pair of opposite sides are congruent. This figure is called an isosceles trapezoid.

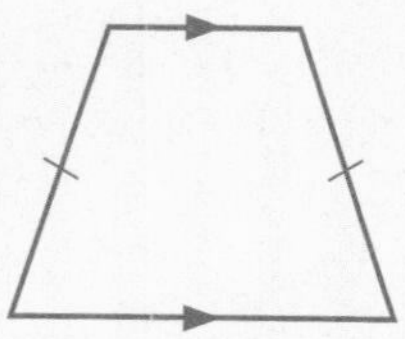

★ indicates increased difficulty

Practice and Applications

HOMEWORK HELP

For Exercises	See Examples
10–15, 25, 27, 29–31	1, 2
19–24, 28	3

Extra Practice
See pages 588, 605.

Classify the quadrilateral using the name that *best* describes it.

10. 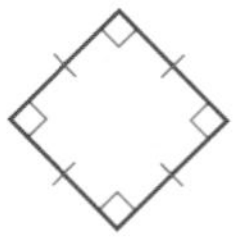 square

11. 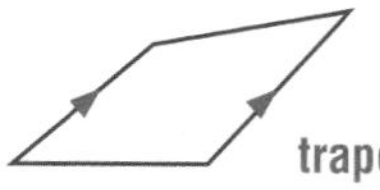 trapezoid

12. 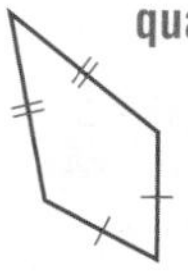 quadrilateral

13. 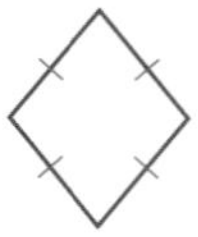 rhombus

14.  trapezoid

15. 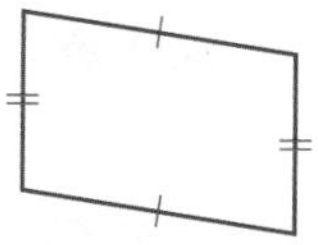 parallelogram

Determine whether each statement is *sometimes, always,* or *never* true. Explain. 16–18. See margin.

16. A quadrilateral is a trapezoid.
17. A parallelogram is a rectangle.
18. A trapezoid is a parallelogram.

Find the missing angle measure in each quadrilateral.

19. 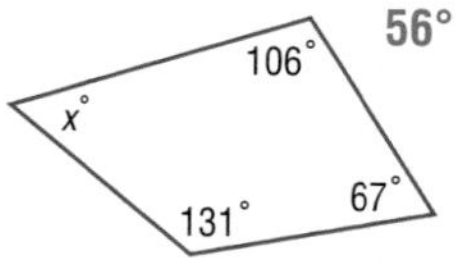

56°

20. 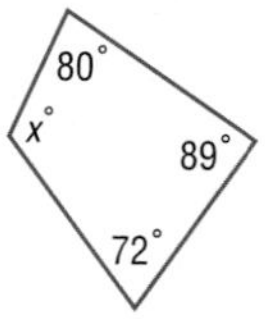

119°

21. 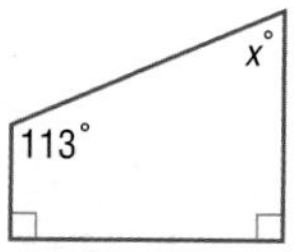

67°

22. 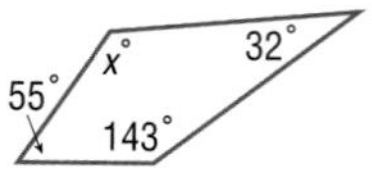

130°

23. 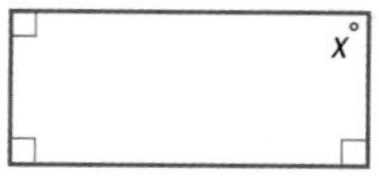

90°

★ 24. 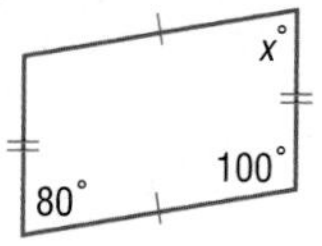

80°

25. **VIDEO GAMES** The first video games used basic shapes like rectangles, squares, and circles in their interface. Name the quadrilaterals used in the video game shown at the right. rhombus, rectangle

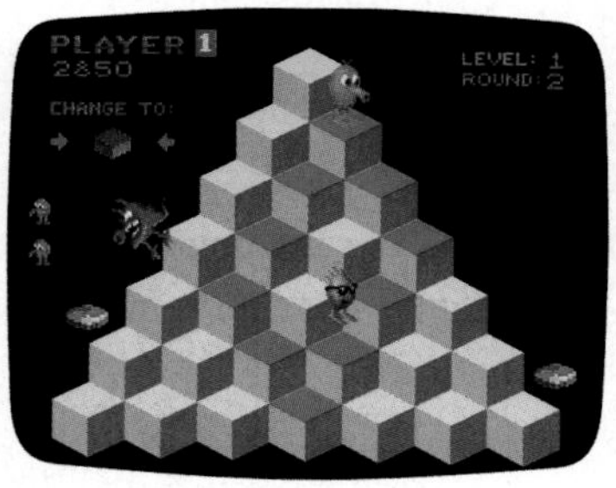

26. **RESEARCH** Use the Internet or another source to find other early video games that used different types of quadrilaterals as part of their interface. Classify the quadrilaterals in the games you found. See students' work.

27. **ART** Design and draw a stained glass window that contains the following types of quadrilaterals: trapezoid, parallelogram, and rhombus. See students' work.

28. **ALGEBRA** Find $m\angle B$ in quadrilateral $ABCD$ if $m\angle A = 87°$, $m\angle C = 135°$, and $m\angle D = 22°$. 116°

Determine whether each figure described below can be drawn. Explain. 29–31. See margin.

★ **29.** a quadrilateral that is both a rhombus and a rectangle

★ **30.** a trapezoid with 3 right angles

★ **31.** a trapezoid with two congruent sides

32. CRITICAL THINKING Find the value of x in the quadrilateral. Then find the measure of each angle. $x = 36$; 36°, 72°, 108°, 144°

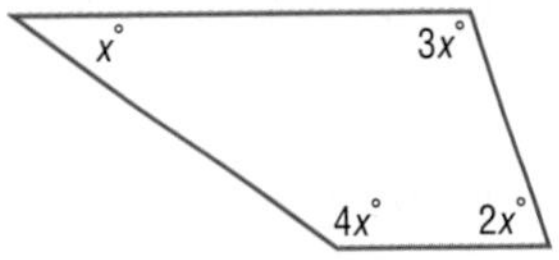

TEACHING TIP

Have students solve Exercise 32 algebraically, not by using a protractor.

Standardized Test Practice and Mixed Review

33. MULTIPLE CHOICE Which property is *not* characteristic of a rhombus? B

Ⓐ opposite sides parallel　　Ⓑ 4 right angles

Ⓒ 4 congruent sides　　Ⓓ opposite sides congruent

34. MULTIPLE CHOICE Which quadrilateral does *not* have opposite sides congruent? H

Ⓕ parallelogram　Ⓖ square　Ⓗ trapezoid　Ⓘ rectangle

Classify each triangle by its angles and by its sides. (Lesson 10-4)

35. 
obtuse, isosceles

36. 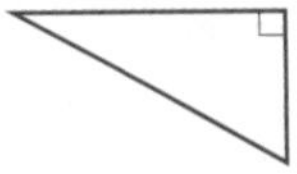
right, scalene

37. 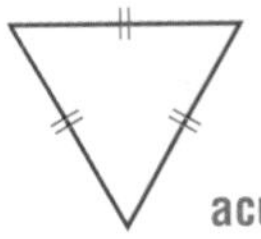
acute, equilateral

38. Suppose $\angle RST$ and $\angle TSU$ are supplementary angles. Find $m\angle TSU$ if $m\angle RST$ is 76°. (Lesson 10-3) 104°

Find the sales tax or discount to the nearest cent. (Lesson 8-5)

39. \$54 jacket; 7% sales tax \$3.78

40. \$23 hat; 15% discount \$3.45

Find each number. Round to the nearest tenth if necessary. (Lesson 7-7)

41. 20% of what number is 17? 85

42. What number is 45% of 160? 72

43. 5 is 12% of what number? 41.7

44. 15 is what percent of 24? 62.5%

45. TRAVEL On his summer vacation, Timothy drove 250 miles in 5 hours on the first day. He continued driving at the same rate the second day and drove for 8 hours. How many miles did Timothy drive the second day of his vacation? (Lesson 7-2) 400 mi

GETTING READY FOR THE NEXT LESSON

PREREQUISITE SKILL **Solve each proportion.** (Lesson 7-3)

46. $\frac{3}{5} = \frac{x}{75}$ 45　**47.** $\frac{a}{7} = \frac{18}{42}$ 3　**48.** $\frac{7}{9} = \frac{28}{m}$ 36　**49.** $\frac{3.5}{t} = \frac{16}{32}$ 7　**50.** $\frac{3}{6} = \frac{c}{5}$ 2.5

msmath2.net/self_check_quiz

4 Assess

Open-Ended Assessment

Writing The words *equilateral* and *equiangular* are used to classify triangles. Ask students to write a paragraph describing how these words can be applied to quadrilaterals. Ask them to identify the quadrilaterals that are equilateral and the ones that are equiangular. Rhombuses and squares are equilateral; squares and rectangles are equiangular.

Study Guide and Intervention, p. 563

NAME ______ DATE ______ PERIOD ____

10-5 Study Guide and Intervention

Quadrilaterals

Quadrilaterals can be classified using their angles and sides. The best description of a quadrilateral is the one that is the most specific.

Trapezoid	Parallelogram	Rectangle	Rhombus	Square
quadrilateral with one pair of parallel sides	quadrilateral with opposite sides parallel and opposite sides congruent	parallelogram with 4 right angles	parallelogram with 4 congruent sides	parallelogram with 4 right angles and 4 congruent sides

EXAMPLES **Classify the quadrilateral using the name that *best* describes it.**

1 The quadrilateral is a parallelogram with 4 congruent sides. It is a rhombus.

2 The quadrilateral has one pair of parallel sides. It is a trapezoid.

3 The quadrilateral is a parallelogram with 4 right angles. It is a rectangle.

EXERCISES

Classify the quadrilateral using the name that *best* describes it.

1. quadrilateral　**2.** rhombus　**3.** parallelogram

Find the missing measure in each quadrilateral.

4. 135°, x°, 45°, 88° — 92°　**5.** x°, 72°, 82°, 95° — 111°　**6.** 54°, 125°, x°, 96° — 85°

Practice: Skills, p. 564

NAME ______ DATE ______ PERIOD ____

10-5 Practice: Skills

Quadrilaterals

Classify the quadrilateral using the name that *best* describes it.

1. parallelogram　**2.** rectangle　**3.** square

4. trapezoid　**5.** quadrilateral　**6.** rhombus

7. parallelogram　**8.** rhombus　**9.** trapezoid

Find the missing angle measure of each quadrilateral.

10. 52°, x° — 128°　**11.** 135°, 110°, x°, 75° — 40°　**12.** 89°, x°, 92°, 64° — 115°

13. 121°, 72°, x° — 77°　**14.** 99°, 109°, 77°, x° — 55°　**15.** x°, 64° — 116°

Practice: Word Problems, p. 565

NAME ______ DATE ______ PERIOD ____

10-5 Practice: Word Problems

Quadrilaterals

1. KITES A kite is shown below. What is the best name to classify the shape of the kite? Explain.

quadrilateral; Sample answer: The kite is a four-sided figure. It does not fit the description of any other quadrilateral in the lesson.

2. MAPS A map showing the road connecting the towns of Pike, Hudson, Placid, and Alton is shown. The road connecting Pike and Hudson is parallel to the road connecting Alton and Placid. What is the best name to classify the shape of the roads connecting the four towns? Explain.

N　Pike　Hudson　Alton　Placid

trapezoid; Sample answer: Two of the roads are going East to West, so they are parallel. So, the quadrilateral is a trapezoid.

3. ART A picture frame is shown below. What is the *best* name to classify the shape of the frame?

1 ft　1 ft

square

4. SCHOOL SUPPLIES The side view of an eraser is shown below. What is the best name to classify the shape of the eraser?

parallelogram

5. PARTY The front of a birthday party invitation is shown below. Find the measure of the missing angle.

120°　x°　PARTY!　60°　120°

60°

6. TABLE The top of Mr. Bautista's new coffee table is shown below. Find the measure of the missing angle.

60°　x°　120°　100°

80°

Chapter 10 Mid-Chapter Practice Test

Assessment Option

Mid-Chapter Test (Lessons 10-1 through 10-5) is available on p. 605 of the *Chapter 10 Resource Masters*.

ExamView® Pro Test Bank CD-ROM

Use the networkable ExamView® Pro Test Bank CD-ROM to:

- Create **Multiple versions** of tests.
- Create **Modified** tests for **Inclusion** students with one mouse click.
- **Edit** existing questions and **Add** your own questions.
- Build tests aligned with state standards using built-in **State Curriculum Correlations.**
- Change **English** tests to **Spanish** with one mouse click and vice versa.

FOLDABLES™ Follow-Up

Before completing the Mid-Chapter Practice Test, encourage students to review the notes, definitions, concepts, and main ideas they recorded on the study cards they collected in the three pockets of their Foldables.

Answers

1. Sample answer: Two angles whose measures sum to 180°.

3.

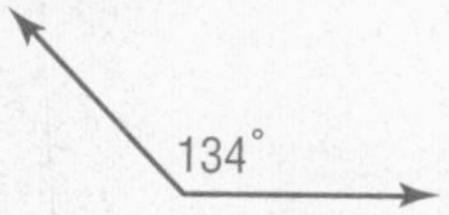

4.

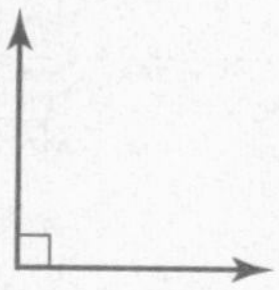

5.

6.

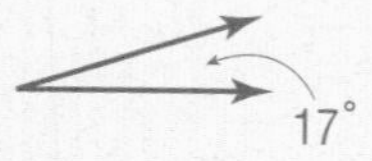

CHAPTER 10 Mid-Chapter Practice Test

2. Sample answer: Both have four congruent sides, but a square has 4 right angles and a rhombus may not.

Vocabulary and Concepts

1. **Define** *supplementary angles.* (Lesson 10-3) See margin.
2. **Describe** the difference between a square and a rhombus. (Lesson 10-5)

Skills and Applications

Draw an angle having each measurement. Then classify each angle as *acute, obtuse, right,* or *straight*. (Lesson 10-1) 3–6. See margin for angles.

3. 134° obtuse
4. 90° right
5. 180° straight
6. 17° acute

7. **SOCCER** Make a circle graph of total injuries of high school girls soccer players by position. (Lesson 10-2) See margin.

Injuries of High School Girls Soccer Players	
Position	**Percent**
halfbacks	37%
fullbacks	23%
forward line	28%
goalkeepers	12%

Source: National Athletic Trainers' Association

Find the value of x in each figure. (Lesson 10-3)

8. 137

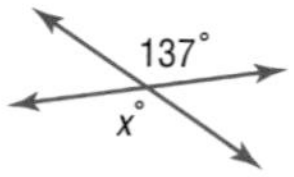

9. 71

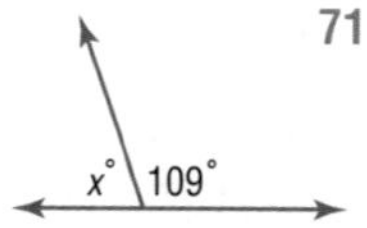

10. 6

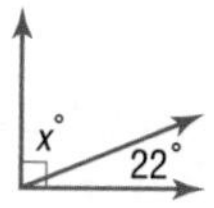

11. **ALGEBRA** Find $m\angle B$ in $\triangle ABC$ if $m\angle A = 62°$ and $m\angle C = 44°$. (Lesson 10-4) 74°

Classify the quadrilateral using the name that *best* describes it. (Lesson 10-5)

12.

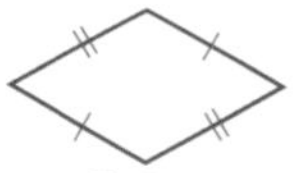

parallelogram

13.

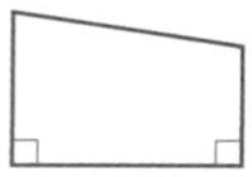

trapezoid

14.

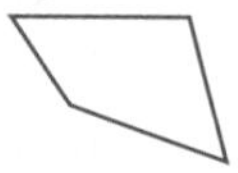

quadrilateral

Standardized Test Practice

15. **MULTIPLE CHOICE** Which angle is complementary to $\angle CBD$? (Lesson 10-1) A

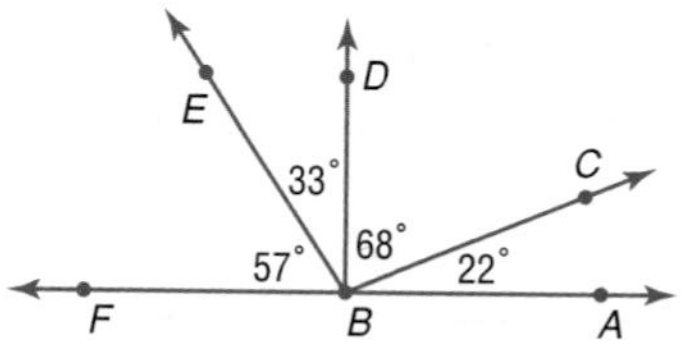

Ⓐ $\angle ABC$ Ⓑ $\angle DBE$

Ⓒ $\angle FBC$ Ⓓ $\angle EBF$

16. **SHORT RESPONSE** Which section of the circle graph has an angle measure of 63°? (Lesson 10-2) milk

What Do You Drink With Dinner?

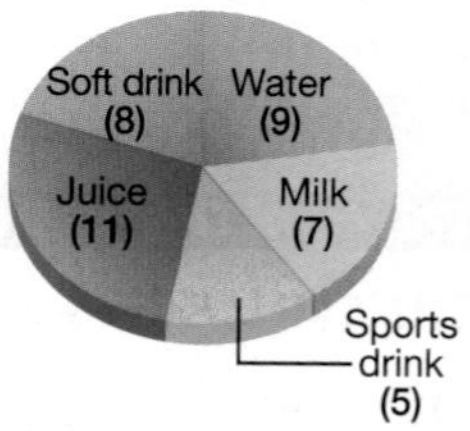

7. Sample answer:

Soccer Injuries

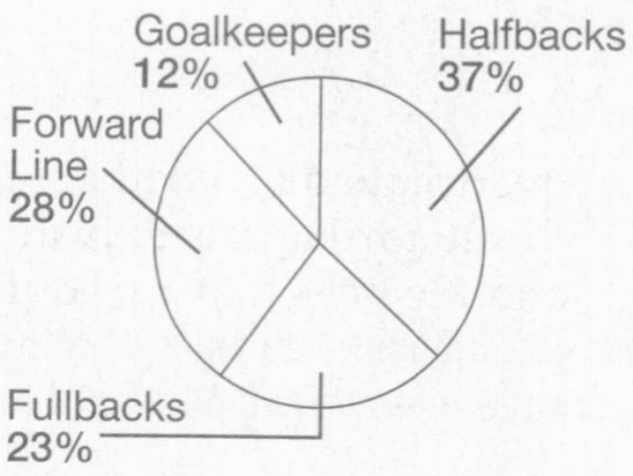

The Game Zone

A Place To Practice Your Math Skills

Squares Everywhere!

GET READY!

Players: two

Materials: 8 red counters, 8 yellow counters, dot paper, straight edge

GET SET!

- Copy the game board onto dot paper.

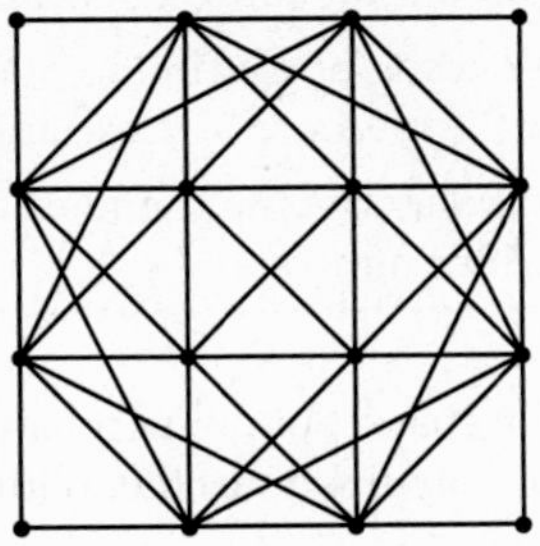

GO!

- The first player covers any dot with a counter. Then, players alternate turns.
- The object of the game is to cover the four vertices of a square with one set of counters. Note that the sides of a square don't have to be vertical or horizontal. The square can be "tilted" to one side.
- **Who Wins?** The first player to cover the four vertices of a square with his or her counters wins.
- Once you have played a game covering the four vertices of a square, change the game so you cover the four vertices of a rectangle or a parallelogram.

The Game Zone

Resources

For an easy-to-make manipulative for this game, visit: msmath2.net/gamezone.

Glencoe Mathematics Classroom Manipulative Kit

- counters

GETTING READY

Objective Find squares.

MANAGEMENT TIPS

Recommended Time
25 minutes

Tips

- Review the different types of quadrilaterals, especially squares, before beginning the game.
- The game can also be played with students covering other quadrilaterals, such as rectangles and trapezoids.
- You may wish to use the game as an assessment tool for the concepts the students have been learning in Lessons 10-1 through 10-5.

For an additional game for this chapter, visit: msmath2.net/gamezone.

10-6 Lesson Notes

Lesson Resources

Chapter 10 Resource Masters

- Study Guide and Intervention, p. 568
- Practice: Skills, p. 569
- Practice: Word Problems, p. 570
- Reading to Learn Mathematics, p. 571
- Enrichment, p. 572

Calculator and Spreadsheet Masters, p. 48

Key Teaching Transparencies, 38

MathPASS Tutorial CD-ROM, 44

Parent and Student Study Guide Workbook, 10-6

1 Focus

5-Minute Check 10-6

Use as a quiz or review of Lesson 10-5. (Also available as a transparency)

INTERACTIVE CHALKBOARD PowerPoint® Presentations

Bellringer

Communication Have students discuss everyday situations in which they might want a smaller or larger model of an object. For example, a model of an atom might be useful in chemistry class, and a model of a house might communicate the design before building the full-sized house. Ask students how the parts of these models relate to the actual objects.

Reading to Learn Mathematics, p. 571 **ELL**

NAME ______ DATE ______ PERIOD ______

10-6 Reading to Learn Mathematics

Similar Figures

Pre-Activity *Complete the Mini Lab at the top of page 440 in your textbook. Write your answers below.*

1. Write each fraction in simplest form.
 a. $\frac{AB}{EF}; \frac{BC}{FG}; \frac{DC}{HG}; \frac{AD}{EH}$ $\frac{1}{3}; \frac{1}{3}; \frac{1}{3}; \frac{1}{3}$ b. $\frac{RS}{XY}; \frac{ST}{YZ}; \frac{RT}{XZ}$ 2; 2; 2
2. What do you notice about the ratios of corresponding sides? They are equal.
3. Measure the corresponding angles in the figures above. What do you notice about the measure of these angles? They are equal.
4. The rectangles are similar, and the triangles are similar. **Make a conjecture** about similar figures. Sample answer: Similar figures have corresponding sides that form equal ratios and their corresponding angles are congruent.

10-6 Similar Figures

What You'll Learn
Determine whether figures are similar and find a missing length in a pair of similar figures.

NEW Vocabulary
- similar figures
- indirect measurement

REVIEW Vocabulary
proportion: an equation that shows that two ratios are equivalent (Lesson 7-3)

MATH Symbol
~ is similar to

HANDS-ON Mini Lab

Materials
- dot paper
- protractor

Work with a partner.

Copy each pair of rectangles and triangles onto dot paper. Then find the measure of each angle and the length of each side. Letters such as *AB* refer to the measure of the segment with those endpoints.

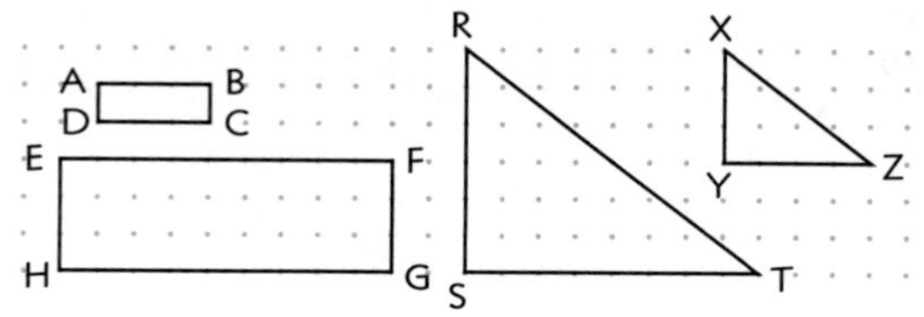

- Each pair of figures has the same shape but different sizes.
- Side *AB* "goes with" side *EF*. So, they are *corresponding sides*.

1. Write each fraction in simplest form.
 a. $\frac{AB}{EF}; \frac{BC}{FG}; \frac{DC}{HG}; \frac{AD}{EH}$ $\frac{1}{3}; \frac{1}{3}; \frac{1}{3}; \frac{1}{3}$ b. $\frac{RS}{XY}; \frac{ST}{YZ}; \frac{RT}{XZ}$ 2; 2; 2
2. What do you notice about the ratios of corresponding sides?
3. Measure the corresponding angles in the figures above. What do you notice about the measure of these angles?
4. The rectangles are similar, and the triangles are similar. **Make a conjecture** about similar figures.

2–3. They are equal.
4. Sample answer: Similar figures have corresponding sides that form equal ratios and their corresponding angles are congruent.

Figures that have the same shape but not necessarily the same size are **similar figures**. The Mini Lab illustrates the following definition.

Key Concept **Similar Figures**

Words If two figures are similar, then
- the corresponding sides are proportional, and
- the corresponding angles are congruent.

Models

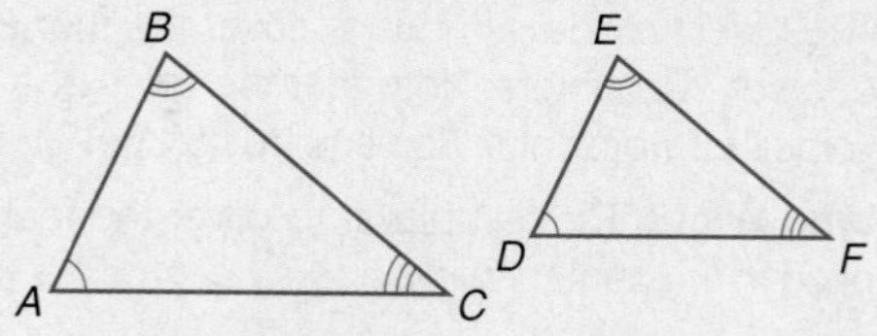

Symbols $\triangle ABC \sim \triangle DEF$ The symbol ~ means *is similar to.*
corresponding sides: $\frac{AB}{DE} = \frac{BC}{EF} = \frac{AC}{DF}$
corresponding angles: $\angle A \cong \angle D$; $\angle B \cong \angle E$; $\angle C \cong \angle F$

Tips for New Teachers

Teaching Strategies
Make sure that students understand the relationship between congruent and similar figures. Congruent figures are also similar, but similar figures are not necessarily congruent.

READING Math

Line Segments The symbol $\overline{XY}$ means the line segment XY. XY means the measure of $\overline{XY}$.

EXAMPLE Find Side Measures of Similar Triangles

1 **If $\triangle RST \sim \triangle XYZ$, find the length of $\overline{XY}$.**

Since the two triangles are similar, the ratios of their corresponding sides are equal. So, you can write and solve a proportion to find XY.

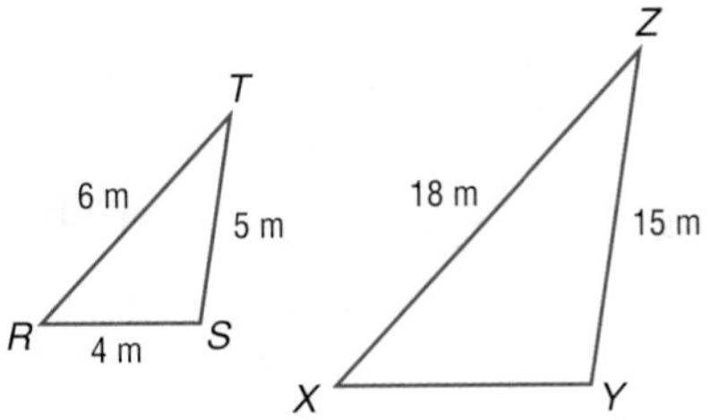

$\frac{RT}{XZ} = \frac{RS}{XY}$ Write a proportion.

$\frac{6}{18} = \frac{4}{n}$ Let n represent the length of $\overline{XY}$. Then substitute.

$6n = 18(4)$ Find the cross products.

$6n = 72$ Simplify.

$n = 12$ Divide each side by 6.

The length of $\overline{XY}$ is 12 meters.

STUDY TIP

Look Back You can review **solving proportions** in Lesson 7-3.

You can use similar figures to find the length, width, or height of objects that are too difficult to measure directly. This type of measurement is called **indirect measurement**.

EXAMPLE Use Indirect Measurement

Standardized Test Practice

2 **GRID-IN TEST ITEM** Natalie wants to resize a 4-inch wide by 5-inch long photograph for the school newspaper. It is to fit in a space that is 2 inches wide. What is the length of the resized photograph in inches?

Read the Test Item

To find the length of the resized photograph, draw a picture and write a proportion.

Solve the Test Item

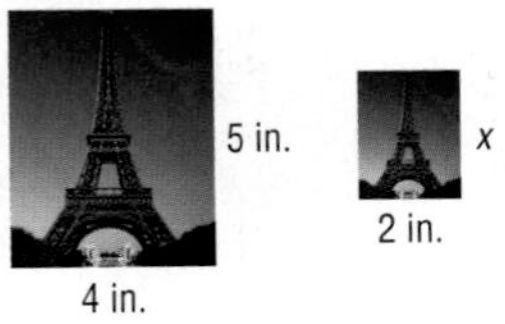

$\frac{4}{2} = \frac{5}{x}$ Write a proportion.

$4x = 10$ Find the cross products.

$x = 2.5$ Divide each side by 4.

So, the length of the photograph is 2.5 inches.

Fill in the Grid

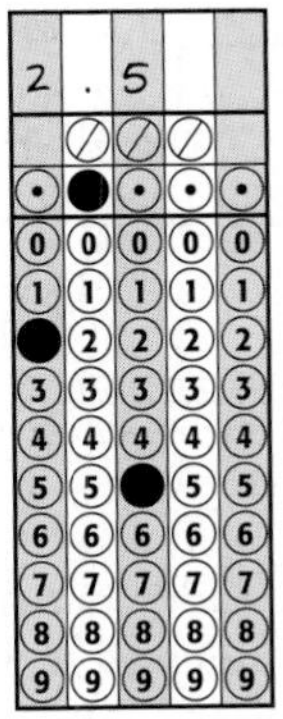

Test-Taking Tip The Princeton Review

Grid In When filling in the response grid, you can print your answer with the first digit in the left answer box, or with the last digit in the right answer box.

msmath2.net/extra_examples

2 Teach

In-Class Examples

INTERACTIVE CHALKBOARD PowerPoint® Presentations

1 If $\triangle ABC \sim \triangle DEF$, find the length of $\overline{DF}$. **16.5 cm**

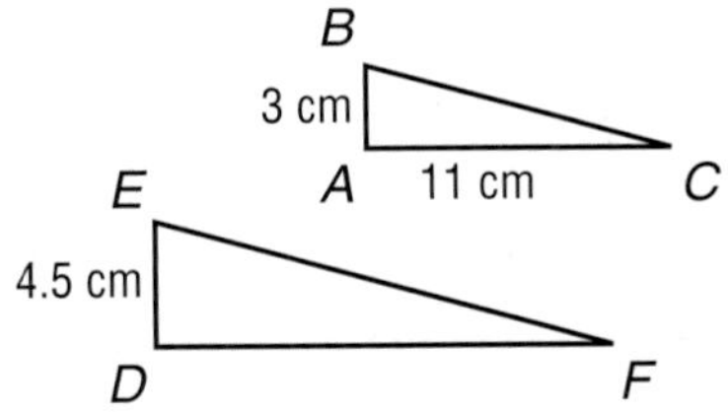

2 **GRID-IN TEST ITEM** A rectangular picture window 12-feet long and 6-feet wide needs to be shortened to 9 feet in length to fit a redesigned wall. If the architect wants the new window to be similar to the old window, how wide will the new window be? **4.5 ft**

Tips for New Teachers

Intervention As students are faced with word problems, remind them to sketch a diagram of the problem and to write down all the given information. Doing so will make the problem much more manageable.

DAILY INTERVENTION **Differentiated Instruction**

Naturalist Have students determine the heights of trees or other natural objects by using similar triangles. Sketch one example problem for them on the chalkboard. (Refer them to Exercise 5 on page 442 for another example.) Then tell students to follow this procedure for two other natural objects. They should sketch the problem, indicate the measurements, and calculate the unknown height.

3 Practice/Apply

FOLDABLES™ Follow-Up Remind students to use study cards to record what they learn about similar figures. Encourage students to describe similar figures using their own words.

About the Exercises...

Guided Practice Key

Exercises	Examples
3–4	1
5	2

Odd/Even Assignments
Exercises 6–15 are structured so that students practice the same concepts whether they are assigned odd or even problems.

Skill and Concept Check

Writing Math
Exercise 1

1. **Describe** characteristics common to two similar triangles.
1–2. See margin.
2. **OPEN ENDED** Draw two similar quadrilaterals and label the vertices. Then write equivalent ratios comparing all corresponding sides.

GUIDED PRACTICE

Find the value of x in each pair of similar figures.

3. 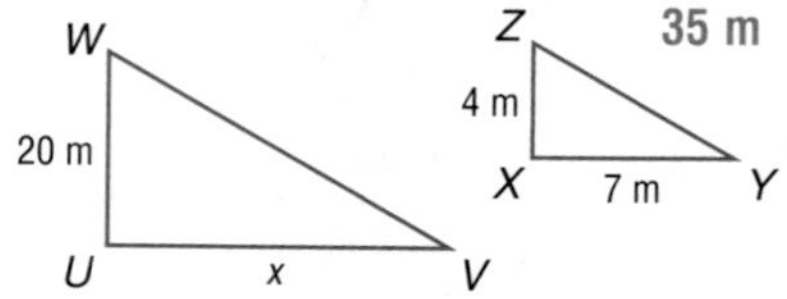

4. 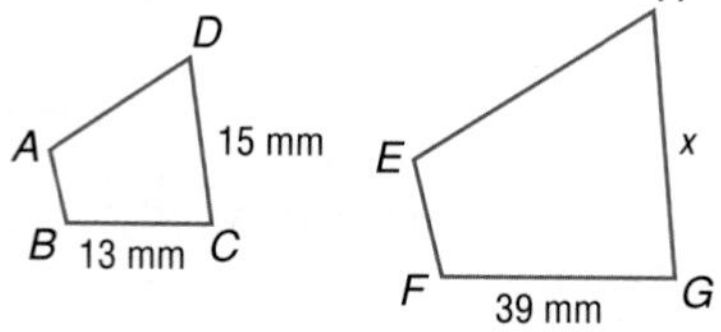

45 mm

5. **MEASUREMENT** The height of an object and its shadow is proportional to the height of another object and its shadow. Suppose you are 6 feet tall and you cast a shadow 5 feet long. Find the height of the tree if it casts a shadow 40 feet long. **48 ft**

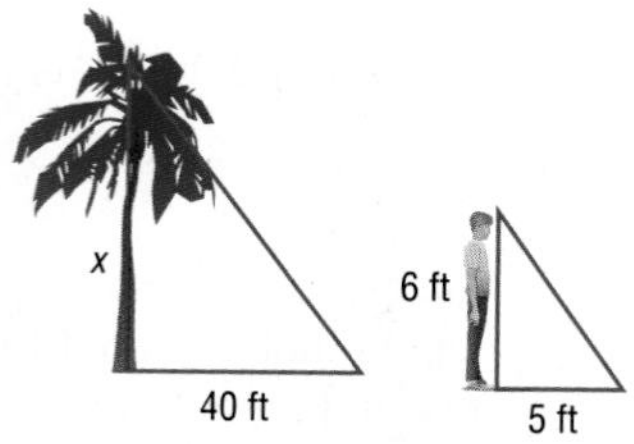

★ **indicates increased difficulty**

Practice and Applications

HOMEWORK HELP

For Exercises	See Examples
6–9, 11–12	1
10, 13–15	2

Extra Practice
See pages 588, 605.

Find the value of x in each pair of similar figures.

6. 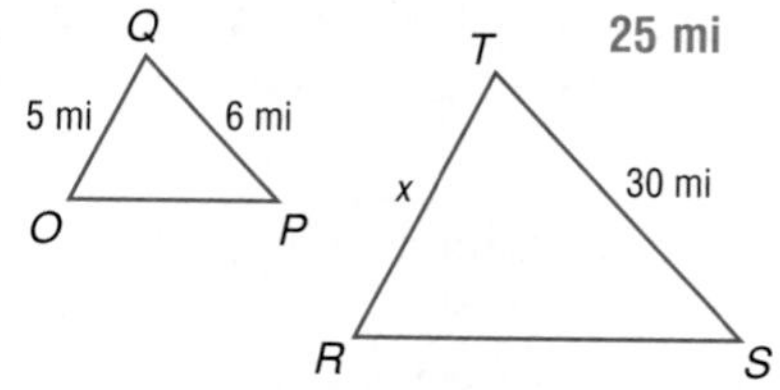

25 mi

7. 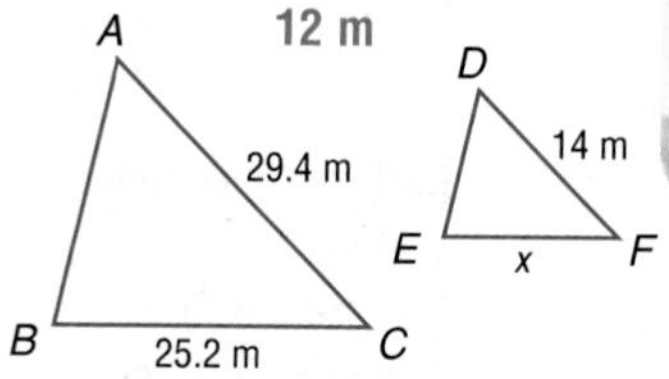

12 m

8. 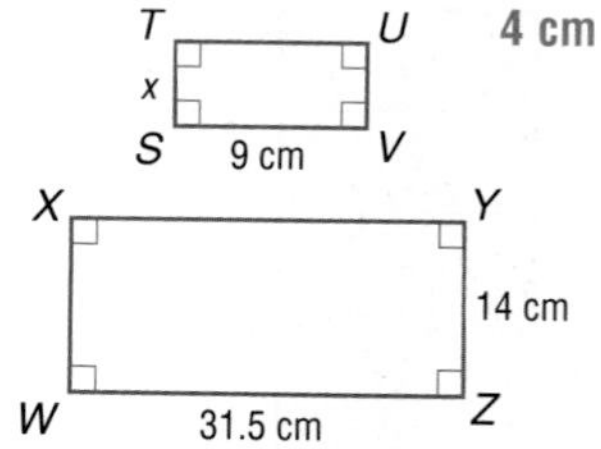

4 cm

★9. 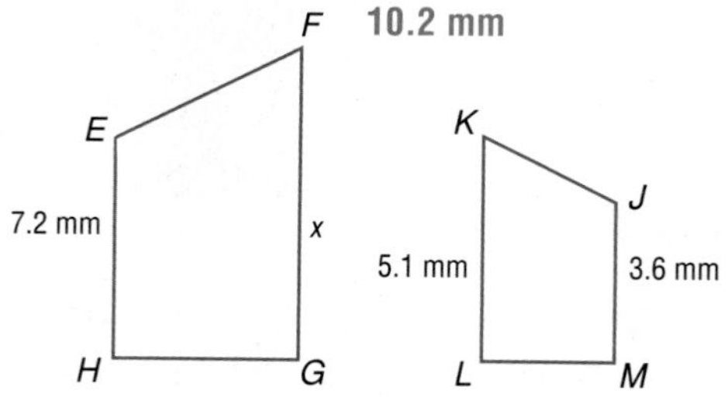

10.2 mm

10. **FURNITURE** Kidco Furniture produces children's furniture that is similar in shape to full-sized furniture. The top of a full-sized desk measures 54 inches long by 36 inches wide. If the top of a child's desk is 24 inches wide, what is the length? **36 in.**

Answers

1. Corresponding sides are proportional and corresponding angles are congruent.

2. Sample answer: $\frac{AD}{EH} = \frac{DC}{HG} = \frac{CB}{GF} = \frac{BA}{FE} = 3$

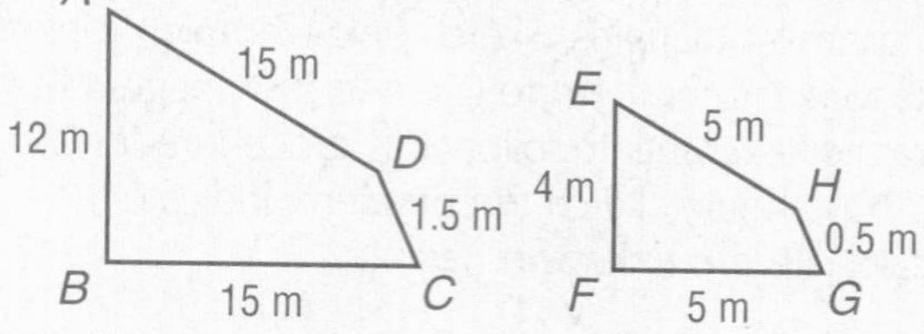

Determine whether each pair of figures is similar. Justify your answer. 11–12. See margin.

11. 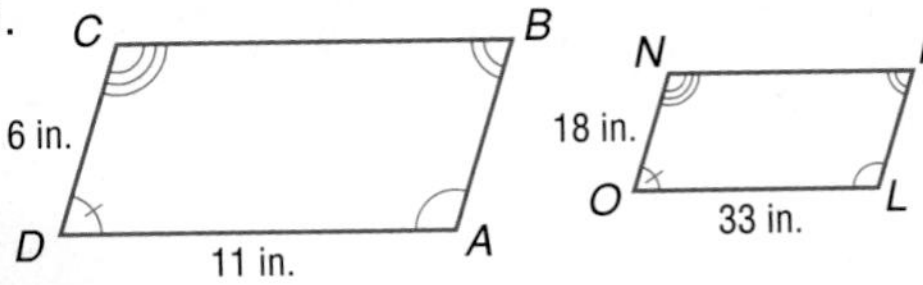

12. 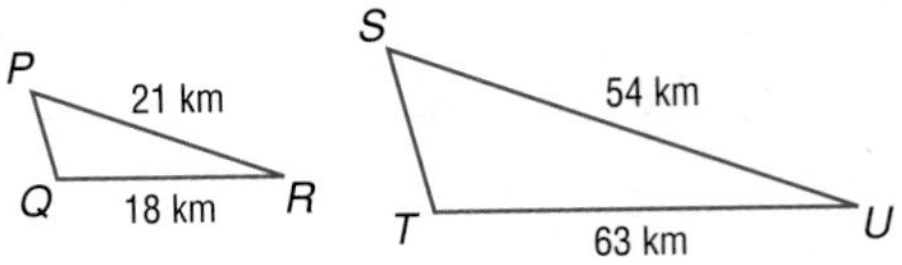

STATUES For Exercises 13 and 14, use the information below and at the right. Alyssa bought a miniature replica of the Statue of Liberty. The replica is 9 inches tall, and the length of the statue's right arm holding the torch is $1\frac{1}{4}$ inches. 13–14. See margin.

13. **MULTI STEP** About how long is the Statue of Liberty's right arm?

14. **MULTI STEP** Alyssa's friend Garcia brought home a smaller replica. The length of the statue's right arm is $\frac{3}{4}$ inch. How tall is Garcia's statue?

★ 15. **GEOMETRY** The ratio of square H's length to square I's length is 3:5. If the length of square H is 18 meters, what is the perimeter of square I? 120 m

CRITICAL THINKING Two rectangles are similar. The ratio of their corresponding sides is 1:4.

16. Find the ratio of their perimeters. 1:4

17. Find the ratio of their areas. 1:16

Standardized Test Practice and Mixed Review

18. **MULTIPLE CHOICE** Which of the following is *not* a true statement? B

- (A) All squares are similar.
- (B) All rhombi are similar.
- (C) Some trapezoids are similar.
- (D) All equilateral triangles are similar.

19. **SHORT RESPONSE** Old Faithful in Yellowstone National Park shoots water 60 feet into the air and casts a shadow of 42 feet. What is the height of a nearby tree if it casts a shadow 63 feet long? 90 ft

Classify the quadrilateral using the name that *best* describes it. (Lesson 10-5)

20. 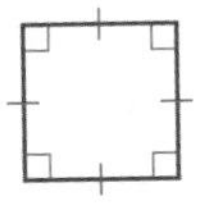 square

21. 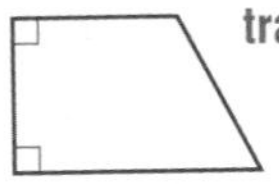 trapezoid

22. 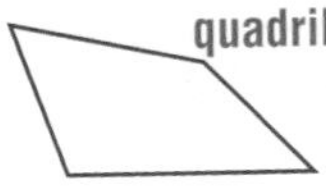 quadrilateral

23. **SAILING** A triangular-shaped sail has angle measures of 44° and 67°. Find the measure of the third angle. (Lesson 10-4) 69°

GETTING READY FOR THE NEXT LESSON

PREREQUISITE SKILL Solve each equation. (Lesson 4-3)

24. $5a = 120$ 24
25. $360 = 4a$ 90
26. $940 = 8n$ 117.5
27. $6t = 720$ 120

msmath2.net/self_check_quiz

4 Assess

Open-Ended Assessment

Speaking Have students explain how to use proportions to make a scale model of a very large or very small object.

Answers

11. yes; $\frac{6}{18} = \frac{11}{33}$
12. no; $\frac{21}{54} \neq \frac{18}{63}$
13. about 42 ft
14. about 5.4 in.

Study Guide and Intervention, p. 568

NAME ________ DATE ________ PERIOD ____

10-6 Study Guide and Intervention

Similar Figures

Two figures are **similar** if corresponding angles are congruent and corresponding sides are proportional. The symbol ~ means *is similar to*. You can use proportions to find the missing length of a side in a pair of similar figures.

For example $\triangle ABC \sim \triangle DEF$.

Corresponding angles: $\angle A \cong \angle D$, $\angle B \cong \angle E$, $\angle C \cong \angle F$

Corresponding sides: $\frac{5}{10} = \frac{4}{8} = \frac{3}{6}$

EXAMPLE 1 If $MNOP \sim RSTU$, find the length of $\overline{ST}$.

Since the two figures are similar, the ratios of their corresponding sides are equal. You can write and solve a proportion to find $\overline{ST}$.

$\frac{PO}{UT} = \frac{NO}{ST}$ Write a proportion.

$\frac{7}{28} = \frac{5}{n}$ Let n represent the length of $\overline{ST}$. Then substitute.

$7n = 28(5)$ Find the cross products.

$7n = 140$ Simplify.

$n = 20$ Divide each side by 7.

The length of $\overline{ST}$ is 20 feet.

EXERCISES

Find the value of x in each pair of similar figures.

1. 15 m
2. 6 ft
3. 18 yd
4. 15 in.

Practice: Skills, p. 569

NAME ________ DATE ________ PERIOD ____

10-6 Practice: Skills

Similar Figures

Find the value of x in each pair of similar figures.

1. 8 ft
2. 21 m
3. 21 ft
4. 30 cm
5. 6 mm
6. 16 in.

Determine whether each pair of figures is similar. Justify your answer.

7. yes; Sample answer: $\frac{9}{18} = \frac{7}{14}$
8. no; Sample answer: $\frac{8}{24} \neq \frac{15}{40}$

Practice: Word Problems, p. 570

NAME ________ DATE ________ PERIOD ____

10-6 Practice: Word Problems

Similar Figures

MODEL CARS For Exercises 1 and 2, use the following information. A scale model racing car is 11 inches long, 3 inches wide, and 2 inches tall. The actual racing car is shown at the right.

1. How wide is the actual racing car? 6 ft
2. How tall is the actual racing car? 4 ft

PHOTOGRAPHY For Exercises 3–4, use the given information. James wants to enlarge a photograph that is 6 inches wide and 4 inches tall so that it fits into the frame shown.

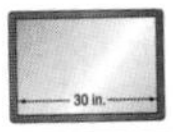

3. How tall must the frame be for the picture to fit? 20 in.
4. Suppose James cuts 1 inch from the width of the photo, so that it is 5 inches wide, before he makes the enlargement. How tall will the frame have to be for the picture to fit? 24 in.
5. **MAPS** A map below shows the towns of Dover, Butler, and Lodi. If the actual distance between Dover and Butler is 24 miles, how far is it from Dover to Lodi? 32 mi

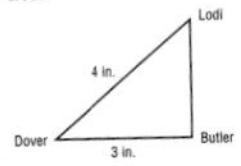

6. **BLUEPRINTS** A blueprint for a house is shown below. If the front of the house is actually 30 feet wide, how tall is the house? 35 ft

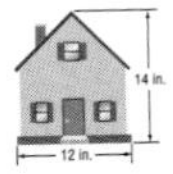

10-7a Lesson Notes

1 Focus

Bellringer

Problem Solving Give students several general statements and ask them to give one bit of evidence that would disprove the generalization. Possible generalizations include:

all triangles are right triangles **one nonright triangle**

all motor vehicles have four wheels **one motor vehicle that does not have four wheels, such as a motorcycle or semi-truck**

all swans are white **one nonwhite swan**

After students answer three or four such problems, have them discuss the meaning of the word *logical*.

2 Teach

In-Class Example

INTERACTIVE CHALKBOARD PowerPoint® Presentations

Solve. Use logical reasoning.

Draw an isosceles triangle. How can you confirm that it is isosceles? **Drawings will vary. The triangle will have two sides of equal length and two angles of equal measure.**

10-7a Problem-Solving Strategy

A Preview of Lesson 10-7

What You'll Learn
Solve problems using logical reasoning.

Use Logical Reasoning

Dion, how can we check that the garden we dug is in the shape of an equilateral triangle?

There is a relationship between the angles of an equilateral triangle that can help us. Nathaniel, let's **use logical reasoning** to find out!

Explore	Equilateral triangles have sides that are congruent. We need to find a relationship between the angles.
Plan	Let's draw several equilateral triangles and measure the angles.
Solve	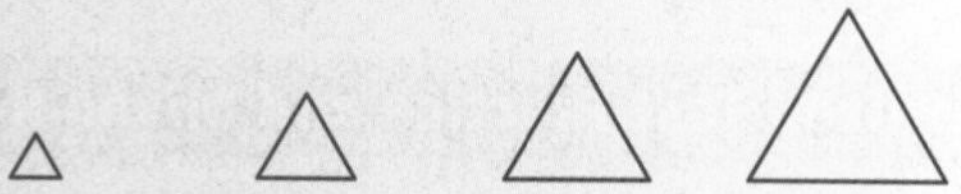 sides = 1 cm, sides = 2 cm, sides = 3 cm, sides = 4 cm. Each angle of the triangles is 60°. So, if the angles of our triangular garden are 60°, then the garden is in the shape of an equilateral triangle.
Examine	Any triangle with angle measures of 60° is equilateral. If we try to draw a triangle with angle measures of 60° and different side lengths, the drawing would not be in the shape of a triangle. 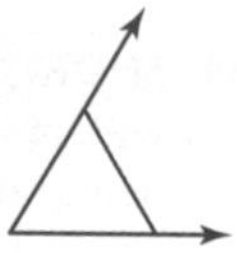

Analyze the Strategy

1. When you use *inductive reasoning*, you make a rule after seeing several examples. When you use *deductive reasoning*, you use a rule to make a decision. What type of reasoning did Nathaniel and Dion use to solve the problem? **Explain** your reasoning. **1–2. See margin.**
2. **Explain** how Nathaniel and Dion could have solved the problem using deductive reasoning.
3. **Explain** how the *look for a pattern* problem-solving strategy is similar to inductive reasoning. **Sample answer: both strategies base a conclusion on a pattern.**

Answers

1. Sample answer: They used inductive reasoning because they made a rule after seeing four examples.
2. Sample answer: To solve the problem using deductive reasoning, they could have measured the lengths of the sides of the triangular garden. They dug out an equilateral triangle if the sides are congruent.

pply the Strategy

lve. Use logical reasoning.

GEOMETRY Draw several isosceles triangles and measure their angles. What can you conclude about the measures of the angles of an isosceles triangle? See Exercise 1. Did you use inductive or deductive reasoning? **Two of the angles are equal; inductive.**

5. **TRAVEL** Use $d = rt$ where d is distance, r is rate, and t is time to find how far Mrs. Petricca drives if she drives 55 miles per hour for seven hours. See Exercise 1. Did you use inductive or deductive reasoning? **385 mi; deductive**

xed Problem Solving

lve. Use any strategy.

PHOTOCOPYING Suppose you enlarge a drawing to 120% of its original size on the copy machine. If the drawing is 2 inches long and 3 inches wide, what are the dimensions of the copy? **2.4 in. by 3.6 in.**

TOYS Hannah was finding the relationship between the time it took a yo-yo to swing back and forth and its length. Predict the length of a yo-yo if it takes 5 seconds to swing back and forth. **25 units**

Time (seconds)	Length (units)
1	1
2	4
3	9
4	16

EOMETRY For Exercises 8 and 9, draw several ctangles and measure their diagonals.

. Find a relationship between the diagonals of a rectangle. **They are equal.**

. See Exercise 1. Did you use inductive or deductive reasoning in Exercise 8? **inductive**

. **BASKETBALL** Placido, Dexter, and Scott play guard, forward, and center on a team, but not necessarily in that order. Placido and the center drove Scott to practice on Saturday. Placido does not play guard. Who is the guard? **Scott**

. **GARDENING** Shelby wants to fence in her garden with the dimensions shown. How much fencing will she need? **$119\frac{1}{2}$ ft**

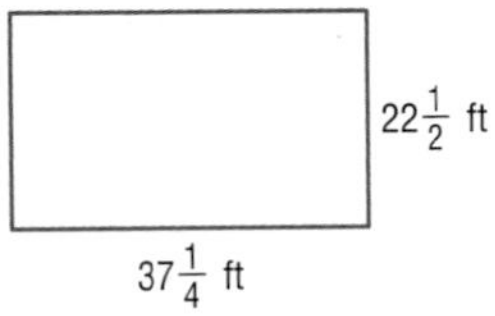

12. **CLOTHES** Careta put a \$5 bill into one of seven pockets of her jacket. If she randomly reaches into one pocket, what is the probability that she will find the \$5 bill? $\frac{1}{7}$

13. **GEOMETRY** The large square has been divided into 9 squares. The lengths of the sides of two squares are given. Find the area of the entire square. **196 sq units**

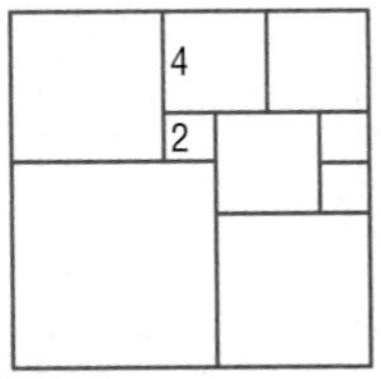

14. **STANDARDIZED TEST PRACTICE**
What can you conclude from the survey about favorite athletic shoes? **B**

Ⓐ Brand A is the most popular among both boys and girls.

Ⓑ Brand B is two times more popular among boys than girls.

Ⓒ Brand C cost more than Brand D.

Ⓓ Brand A is less popular than Brand C.

You will use logical reasoning in the next lesson.

3 Practice/Apply

About the Exercises...

Problem-Solving Strategies
Exercises 6–14 are structured so that students have the opportunity to practice many different problem-solving strategies. You may wish to remind them of some of the strategies they have studied.

- use the four-step plan
- use a graph (p. 58)
- look for a pattern (p. 132)
- eliminate possibilities (p. 252)
- act it out (p. 391)
- use logical reasoning ✓

4 Assess

Open-Ended Assessment

Speaking Have each student create a number pattern. Students should be able to direct the class or a partner in finding a rule for their pattern. After students discuss the pattern and understand the rule, have them determine the next few numbers in the sequence.

DAILY INTERVENTION **Flexible Grouping**

Logical Reasoning After completing the problem-solving exercises, have students work in pairs to determine which problems they solved by inductive reasoning and which by deductive reasoning.

10-7 Lesson Notes

Lesson Resources

Chapter 10 Resource Masters
- Study Guide and Intervention, p. 573
- Practice: Skills, p. 574
- Practice: Word Problems, p. 575
- Reading to Learn Mathematics, p. 576
- Enrichment, p. 577

Calculator and Spreadsheet Masters, p. 49
Classroom Math Games, pp. 29–31
Key Teaching Transparencies, 39
Parent and Student Study Guide Workbook, 10-7
Teaching Mathematics with Manipulatives, pp. 97–101

FAST FILE

1 Focus

5-Minute Check 10-7

INTERACTIVE CHALKBOARD
PowerPoint® Presentations

Use as a quiz or review of Lesson 10-6. (Also available as a transparency)

Bellringer
Reading Mathematics Have students research the meaning of the root *poly-*. Then have them find three words using the same root. Discuss students' findings with the class. Also discuss how *poly-* might relate to geometry.

Reading to Learn Mathematics, p. 576 ELL

NAME ______________ DATE ________ PERIOD ____

10-7 Reading to Learn Mathematics
Polygons and Tessellations

Pre-Activity ***Read the introduction at the top of page 446 in your textbook. Write your answers below.***

1. Find the difference between the shapes of the states in Group 1 and the shapes of the states in Group 2. Sample answer: The borders of states in Group 1 are straight. The borders of states in Group 2 are not straight.

2. Why do most states have boundaries that are not straight line segments? Sample answer: They use natural boundaries like rivers and lakes.

Reading the Lesson

3. How many straight lines can a polygon have? three or more

4. What is a simple figure? Sample answer: a figure that does not have lines that cross each other

10-7 Polygons and Tessellations

What You'll Learn
Classify polygons and determine which polygons can form a tessellation.

NEW Vocabulary
polygon
pentagon
hexagon
heptagon
octagon
nonagon
decagon
regular polygon
tessellation

WHEN am I ever going to use this?

GEOGRAPHY The size and shape of each state in the United States is different. Analyze the shapes of the states in both groups at the right.

Group 1	Group 2
Utah	Florida
Wyoming	Texas
New Mexico	Indiana
Colorado	North Carolina

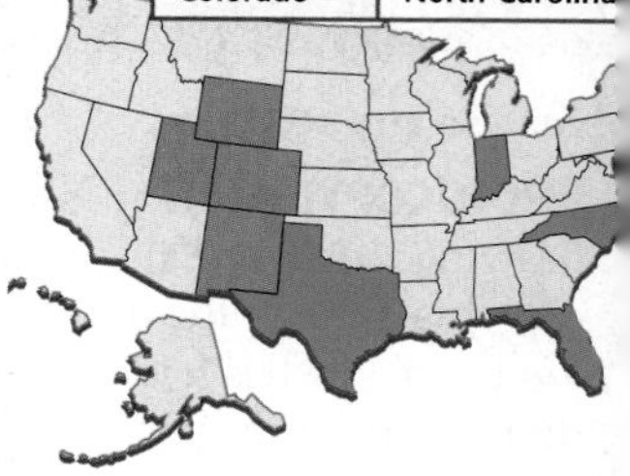

1. Find the difference between the shapes of the states in Group 1 and the shapes of the states in Group 2. **See margin.**
2. Why do most states have boundaries that are not straight line segments?

2. **Sample answer: They use natural boundaries like rivers and lakes.**

The states listed in Group 1 above are examples of polygons. A **polygon** is a simple, closed figure formed by three or more straight lines. A simple figure does not have lines that cross each other. You have drawn a closed figure when your pencil ends up where it started.

Polygons	Not Polygons

• Line segments are called sides. • Sides meet only at their endpoints. • Points of intersection are called vertices.	• Figures whose sides overlap. • Figures that are open. • Figures that have curved sides.

READING Math
Polygons You can name polygons with more than 10 sides *n-gons*, with *n* representing the number of sides. For example, an 11-gon is a polygon with 11 sides.

Just as a triangle has three sides and a quadrilateral has four sides, other polygons can be classified by the number of sides they have.

Words	pentagon	hexagon	heptagon	octagon	nonagon	decagon
Number of Sides	5	6	7	8	9	10
Models	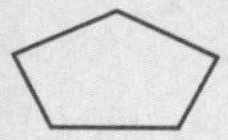		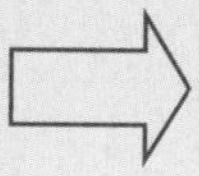		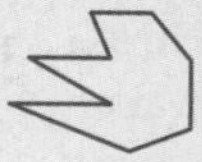	

A polygon that has all sides congruent and all angles congruent is called a **regular polygon**. Equilateral triangles and squares are regular polygons.

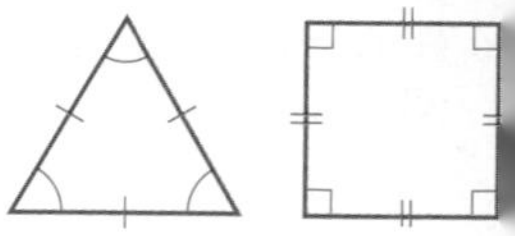

Answer (When am I ever going to use this?)

1. **Sample answer: The borders of states in Group 1 are straight. The borders of states in Group 2 are not straight.**

EADING Math

Regular Polygons Since regular polygons have *equal-sized angles*, they are also called *equiangular*.

EXAMPLES Classify Polygons

Determine whether each figure is a polygon. If it is, classify the polygon and state whether it is regular. If it is *not* a polygon, explain why.

1

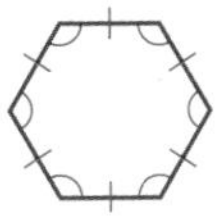

The figure has 6 congruent sides and 6 congruent angles. It is a regular hexagon.

2

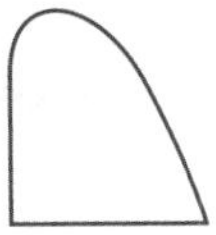

The figure is not a polygon since it has a curved side.

Your Turn **Determine whether each figure is a polygon. If it is, classify the polygon and state whether it is regular. If it is *not* a polygon, explain why.**

a. 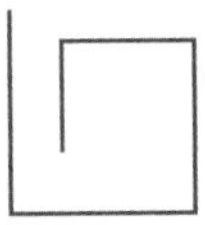 not a polygon; it is not a closed figure

b. 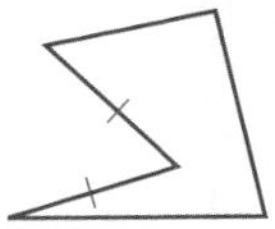 pentagon

The sum of the measures of the angles of a triangle is 180°. You can use this relationship to find the measures of the angles of regular polygons.

TUDY TIP

Diagonal A diagonal of a polygon is a line segment that connects two vertices that are not next to each other.

EXAMPLE Angle Measures of a Polygon

3 **ALGEBRA Find the measure of each angle of a regular pentagon.**

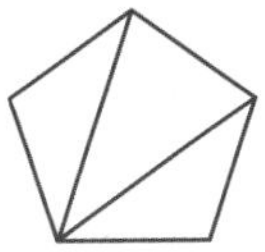

- Draw all of the diagonals from one vertex as shown and count the number of triangles formed.
- Find the sum of the angle measures in the polygon.

number of triangles formed × 180° = sum of angle measures in polygon

$$3 \times 180° = 540°$$

- Find the measure of each angle of the polygon. Let n represent the measure of one angle in the pentagon.

$5n = 540$ There are five congruent angles.

$n = 108$ Divide each side by 5.

The measure of each angle in a regular pentagon is 108°.

TEACHING TIP

emind students that the ım of the angles in a iangle is 180° and in a ıadrilateral, 360°.

Your Turn **Find the measure of an angle in each polygon.**

c. regular octagon 135°

d. equilateral triangle 60°

A repetitive pattern of polygons that fit together with no overlaps or holes is called a **tessellation**. The surface of a chessboard is an example of a tessellation of squares.

msmath2.net/extra_examples

2 Teach

In-Class Examples

INTERACTIVE CHALKBOARD PowerPoint® Presentations

Determine whether each figure is a polygon. If it is, classify the polygon and state whether it is regular. If it is *not* a polygon, explain why.

1

not a polygon, curved side

2

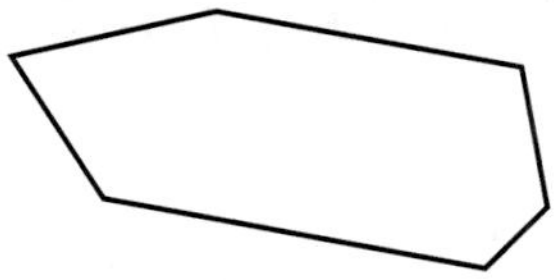

hexagon, not regular

3 **ALGEBRA** Find the measure of each angle of a regular heptagon. Round to the nearest tenth of a degree. 128.6°

DAILY INTERVENTION

Differentiated Instruction

Visual/Spatial Have each student pick a shape that can tessellate and cut it out of cardboard. Then have them trace it multiple times to make a tessellation. Have them color their design. As an additional challenge, encourage students to try to create a tessellation of two or three different shapes.

In-Class Example

INTERACTIVE CHALKBOARD
PowerPoint® Presentations

4 **PATTERNS** Ms. Pena is creating a pattern on her wall. She wants to use triangles with angles 120°, 30°, and 30°. Can Ms. Pena tessellate with these triangles? **Yes, they can be arranged in a way that the angles where the vertices meet make 360°.**

3 Practice/Apply

FOLDABLES™ Follow-Up Remind students to use study cards to record what they learn about tessellations. Encourage students to explain how a tessellation can be made with several kinds of polygons.

About the Exercises...

Guided Practice Key

Exercises	Examples
4–5	1, 2
6	3
7	4

Odd/Even Assignments
Exercises 8–17 are structured so that students practice the same concepts whether they are assigned odd or even problems.

Alert! For Exercise 29, students will use the Internet or another source.

Answers

1. the point where vertices meet: $45° + 45° + 135° + 135° = 360°$
2. Sample answer: square

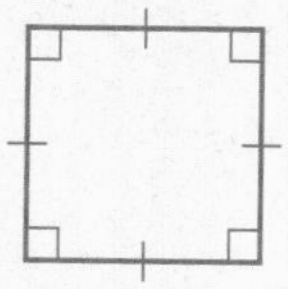

REAL-LIFE CAREERS

How Does a Landscape Architect Use Math?
Landscape architects multiply the length times the width of a garden to find the amount of mulch needed to cover the garden.

Online Research
For information about a career as a landscape architect, visit: msmath2.net/careers

The sum of the measures of the angles where the vertices meet in a tessellation is 360°. The diagrams below show tessellations of equilateral triangles and squares.

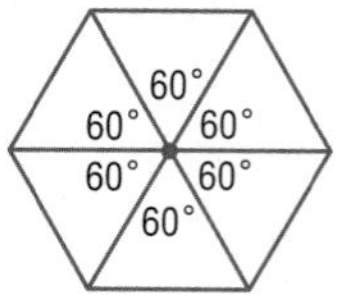

$6 \times 60° = 360°$

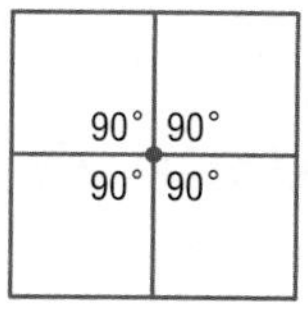

$4 \times 90° = 360°$

EXAMPLE Tessellations

4 **LANDSCAPING Mr. Brooks bought hexagonal-shaped stones to pave his patio. The stones are regular hexagons. Can Mr. Brooks tessellate his patio with the stones?**

The measure of each angle in a regular hexagon is 120°.

The sum of the measures of the angles where the vertices meet mus be 360°. So, solve $120n = 360$.

Since the solution, $n = 3$, is a whole number, a regular hexagon makes a tessellation.

Check You can check if your answer is correct by drawing a tessellation of regular hexagons.

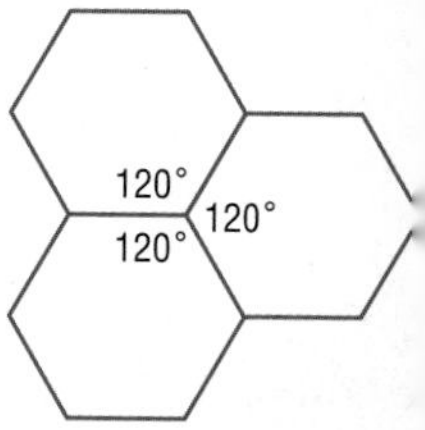

Skill and Concept Check

Writing Math
Exercises 1 & 3

1. **Explain** how you know the parallelogram at the right can be used by itself to make a tessellation. **1–2. See margin.**

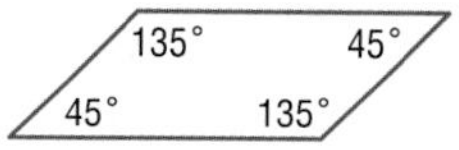

2. **OPEN ENDED** Draw and name a regular polygon.
3. **Explain** why a rhombus is not a regular polygon. **All of the angles of a rhombus do not necessarily have the same measure.**

GUIDED PRACTICE

Determine whether each figure is a polygon. If it is, classify the polygon and state whether it is regular. If it is *not* a polygon, explain why.

4.  **decagon; not regular**

5. 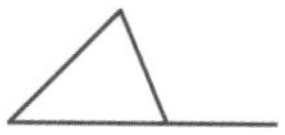 **not a polygon, it is not closed**

6. Find the measure of each angle of a regular hexagon. **120°**
7. Can a regular polygon with an angle measure of 140° be used by itself to make a tessellation? Explain. **No; 140 does not go into 360 evenly.**

★ indicates increased difficulty

Practice and Applications

HOMEWORK HELP	
For Exercises	See Examples
8–13, 18, 20, 24–25, 27	1, 2
14–17, 19, 26	3
21–23, 28	4

Extra Practice
See pages 589, 605.

Determine whether each figure is a polygon. If it is, classify the polygon and state whether it is regular. If it is *not* a polygon, explain why. 8–13. See margin.

8. 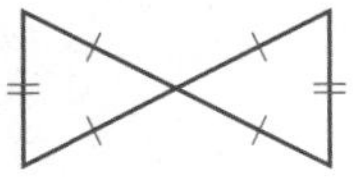

9. 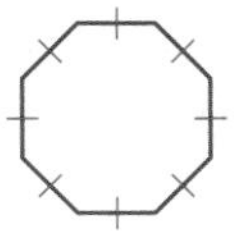

10. 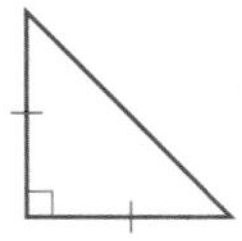

11. 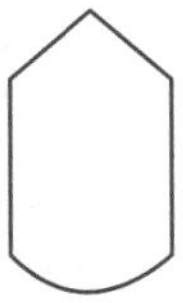

12. 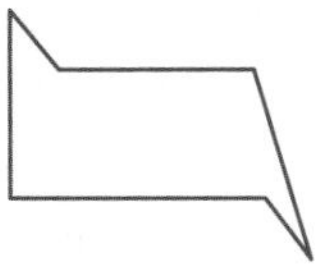

13. 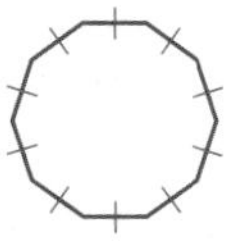

Find the measure of an angle in each polygon if the polygon is regular. Round to the nearest tenth of a degree if necessary.

14. decagon 144°
15. nonagon 140°
16. 13-gon 152.3°
17. 20-gon 162°

MANUFACTURING **For Exercises 18–20, use the information below and the figure at the right.**
A company designs cafeteria trays so that four people can place their trays around a square table without bumping corners. The trays are similar to the one at the right. The top and bottom sides of the tray are parallel.

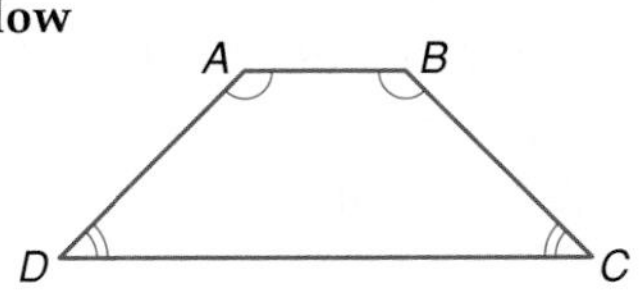

18. Classify the shape of the tray. trapezoid
19. If $\angle A \cong \angle B$, $\angle C \cong \angle D$, and $m\angle A = 135°$, find $m\angle B$, $m\angle C$, and $m\angle D$. 135°, 45°, 45°
★ 20. Name the polygon formed by the outside of four trays when they are placed around the table with their sides touching. square

Identify the polygons that are used to create each tessellation.

21. 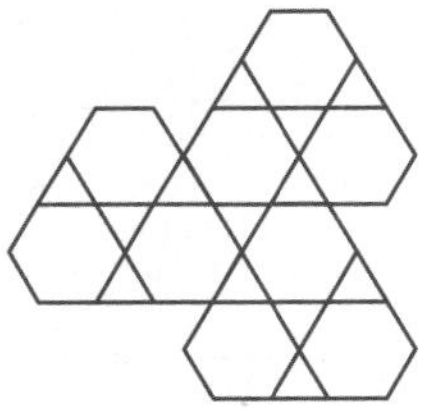
hexagon and triangle

22. 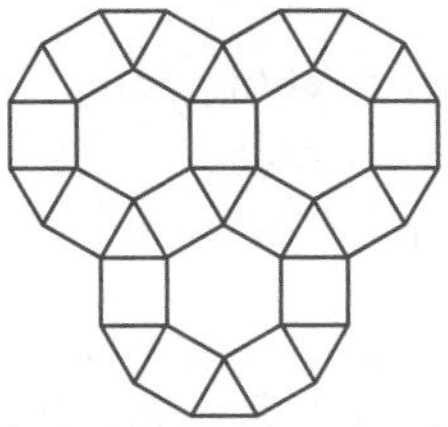
hexagon, square, and triangle

23. 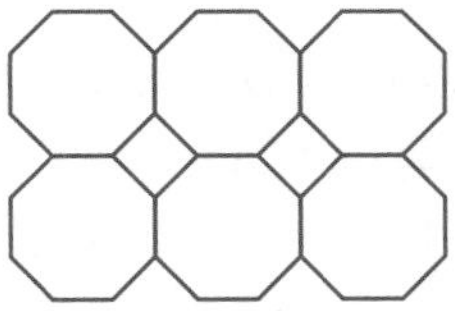
octagons and squares

24. **ART** The mosaic at the right is decorated with handmade tiles from Pakistan. Name the polygons used in the tessellation. triangle, trapezoid, and hexagon

25. What is the perimeter of a nonagon with sides 4.8 centimeters long? 43.2 cm

★ 26. Name the regular polygon if the sum of the measures of the polygon is 2,880° and the measure of one angle is 160°. 18-gon

msmath2.net/self_check_quiz

Answers (pp. 449–450)

8. not a polygon; the figure is not simple
9. regular octagon
10. isosceles triangle
11. not a polygon; the figure has a curved side
12. hexagon
13. regular decagon
28. No; the stop sign is an octagon in shape. An octagon cannot make a tessellation. So, there will be some steel that is wasted after the nine signs are cut from the sheet.
31. Yes; the sum of the measures of angles of any triangle is 180°.

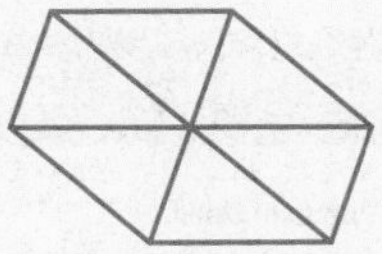

42–45. 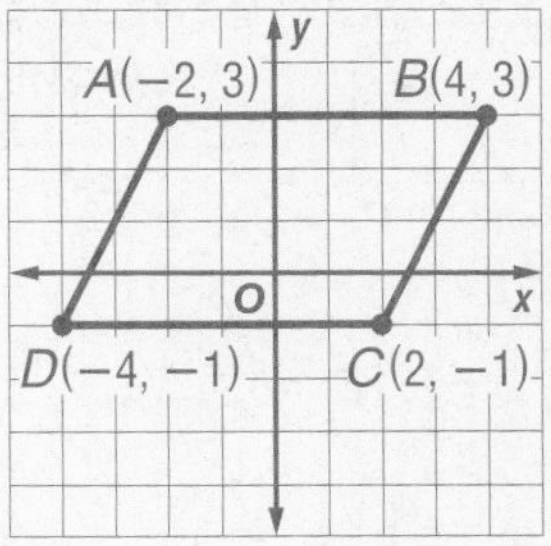

Study Guide and Intervention, p. 573

NAME ________________ DATE ________ PERIOD ____

10-7 Study Guide and Intervention

Polygons and Tessellations

A **polygon** is a simple, closed figure formed by three or more straight lines. A simple figure does not have lines that cross each other. You have drawn a closed figure when your pencil ends up where it started. Polygons can be classified by the number of sides they have.

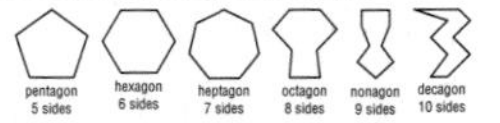

A polygon that has all sides congruent and all angles congruent is called a **regular polygon**.

EXAMPLES Determine whether each figure is a polygon. If it is, classify the polygon and state whether it is regular. If it is *not* a polygon, explain why.

1. The figure has 5 congruent sides and 5 congruent angles. It is a regular pentagon.

2. The figure is not a polygon because it has sides that overlap.

EXERCISES

Determine whether each figure is a polygon. If it is, classify the polygon and state whether it is regular. If it is *not* a polygon, explain why.

1. triangle
2. Not a polygon; it has a curved side.
3. regular pentagon
4. Not a polygon; it is not closed.
5. regular quadrilateral (square)
6. hexagon

Practice: Skills, p. 574

NAME ________________ DATE ________ PERIOD ____

10-7 Practice: Skills

Polygons and Tessellations

Determine whether each figure is a polygon. If it is, classify the polygon and state whether it is regular. If it is *not* a polygon, explain why.

1. quadrilateral
2. Not a polygon; the figure is circular.
3. regular hexagon
4. Not a polygon; the figure is not closed.
5. equilateral (regular) triangle
6. pentagon
7. regular heptagon
8. octagon
9. Not a polygon; the figure is not closed.

Find the measure of an angle in each polygon.

10. regular 15-gon 156°
11. regular 18-gon 160°
12. regular 24-gon 165°

Identify the polygons that are used to create each tessellation.

13. square and triangle
14. hexagon and triangle

15. What is the perimeter of a regular pentagon with sides 8.4 inches long? 42 in.

Practice: Word Problems, p. 575

NAME ________________ DATE ________ PERIOD ____

10-7 Practice: Word Problems

Polygons and Tessellations

1. ROYALTY The outline of a crown worn by a king is shown below. Is the figure a polygon? If it is, classify the polygon and state whether it is regular. If it is *not* a polygon, explain why. yes; nonagon; not regular

2. ALCHEMY The symbol shown is one of the signs for *salt alkali* used in 17th-century chemistry. Is the symbol a polygon? If it is, classify the polygon and state whether it is regular. If it is *not* a polygon, explain why. yes; octagon; not regular

3. JEWELRY The symbol shown is often used to represent gems. Is the symbol a polygon? If it is, classify the polygon and state whether it is regular. If it is *not* a polygon, explain why. No; the figure has overlapping sides.

4. SYMBOLS The 5-pointed star shown has sides of equal length. Is the symbol a polygon? If it is, classify the polygon and state whether it is regular. If it is *not* a polygon, explain why. yes; decagon; not regular

5. STAIRS The figure shows a side view of a set of stairs. Is the figure a polygon? If it is, classify the polygon and state whether it is regular. If it is *not* a polygon, explain why. yes; 14-gon; not regular

6. TESSELLATIONS Identify the polygons that are used to create the tessellation shown in the figure. triangle and hexagon

27. Find the perimeter of a pentagon having sides $7\frac{1}{2}$ yards long. $37\frac{1}{2}$ yd

★ 28. **SIGNS** Stop signs are made from large sheets of steel. Suppose nine stop signs can be cut from one sheet of steel. Can all nine signs be arranged on the sheet so that none of the steel goes to waste? Explain your reasoning. See margin.

29. **RESEARCH** Use the Internet or another source to find the shape of other traffic signs. Name the type of sign, its shape, and whether it is regular. See students' work.

30. **WRITE A PROBLEM** Write about a real-life object or situation that contains a tessellation. Name the polygons used and draw a picture of the tessellation. See students' work.

31. **CRITICAL THINKING** You can make a tessellation with equilateral triangles. Can you make a tessellation with any isosceles or scalene triangles? If so, explain your reasoning and make a drawing. See margin.

Standardized Test Practice and Mixed Review

32. **MULTIPLE CHOICE** Which of the following flowchart symbols is *not* a polygon? C

Ⓐ process box Ⓑ decision box Ⓒ stop box Ⓓ preparation box

33. **MULTIPLE CHOICE** Which of the following regular shapes *cannot* be used by itself to make a tessellation? I

Ⓕ hexagon Ⓖ triangle Ⓗ square Ⓘ pentagon

34. **SHORT RESPONSE** What is the sum of the measures of the angles of a quadrilateral? 360°

For Exercises 35 and 36, use the figures at the right.

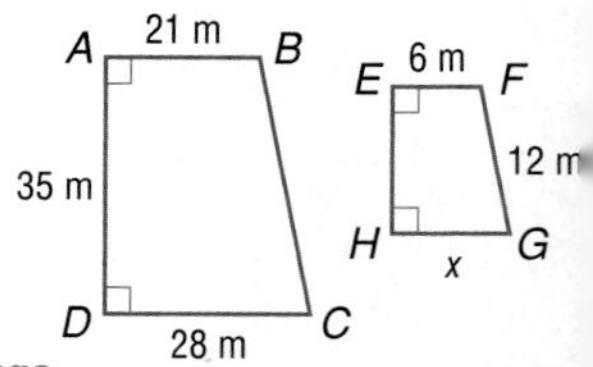

35. Classify figure *ABCD*. (Lesson 10-5) trapezoid

36. The quadrilaterals are similar. Find the value of x. (Lesson 10-6) 8 m

37. **MEASUREMENT** How many $\frac{1}{2}$-cup servings of ice cream are there in a gallon of chocolate ice cream? (Lesson 6-5) 32 servings

Add or subtract. Write each sum or difference in simplest form. (Lesson 6-3)

38. $3\frac{2}{9} + 5\frac{4}{9}$ $8\frac{2}{3}$
39. $5\frac{1}{3} - 2\frac{1}{6}$ $3\frac{1}{6}$
40. $1\frac{3}{7} + 6\frac{1}{4}$ $7\frac{19}{28}$
41. $9\frac{4}{5} - 4\frac{7}{8}$ $4\frac{37}{40}$

GETTING READY FOR THE NEXT LESSON

PREREQUISITE SKILL On graph paper, draw a coordinate plane. Then graph and label each point. Connect the points in order. (Lesson 3-3) 42–45. See margin.

42. $A(-2, 3)$
43. $B(4, 3)$
44. $C(2, -1)$
45. $D(-4, -1)$

4 Assess

Higher-Level Thinking

Draw all possible diagonals in regular polygons with four, five, six, seven, and eight sides. Create a table with the number of sides and number of diagonals. From your table, determine the number of diagonals that can be drawn in a decagon. Explain how you found your answer. 35 diagonals; To find this answer, students must extend the pattern they observe for the first five figures to include nonagons and decagons.

10-8 Translations

What You'll Learn
Graph translations of polygons on a coordinate plane.

NEW Vocabulary
transformation
translation

LINK To Reading
Everyday Meaning
of Translation: a change from one language to another, as in an English-to-Spanish translation

HANDS-ON Mini Lab

Materials
- paper
- scissors

Work with a partner.

You can make changes in the polygons that tessellate to create new shapes that also tessellate.

STEP 1 Draw a square and cut a shape from it as shown.

STEP 2 Slide the shape to the opposite side without turning it.

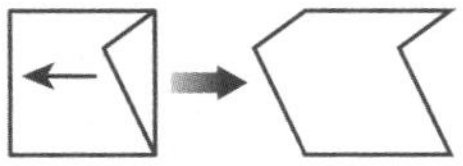

STEP 3 The new shape will make a tessellation.

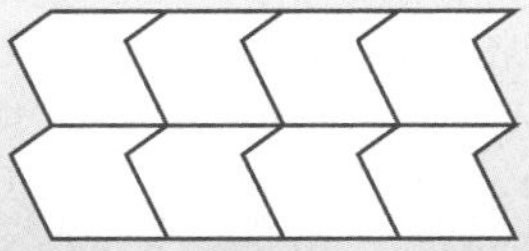

1. Make your own tessellation. Use different colors and shapes to make an interesting design. **See students' work.**

Anytime you move a geometric figure, it is called a **transformation**. In the Mini Lab, you slid the polygon to a new position without turning it. This sliding motion is called a **translation**.

When translating a figure, every point of the original figure is moved the same distance and in the same direction.

EXAMPLE Graph a Translation

1 Translate trapezoid *HIJK* 3 units left and 5 units up.

- Move each vertex of the figure 3 units left and 5 units up. Label the new vertices H', I', J', and K'.
- Connect the vertices to draw the trapezoid. The coordinates of the vertices of the new figure are $H'(-2, 1)$, $I'(1, 1)$, $J'(1, 5)$, and $K'(-2, 4)$.

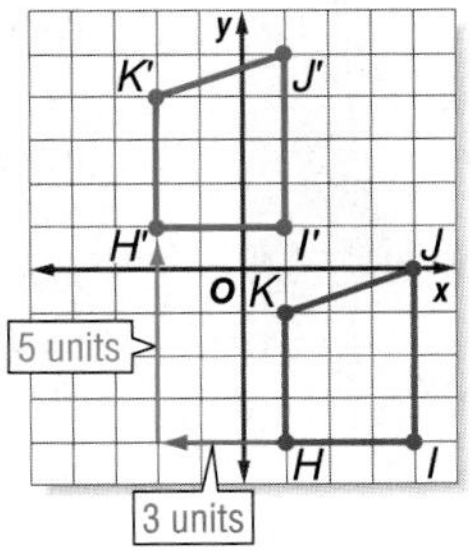

STUDY TIP

Transformations Whenever a figure is translated, use prime symbols for the vertices in the transformed image.
$A \rightarrow A'$
$B \rightarrow B'$
$C \rightarrow C'$

msmath2.net/extra_examples

10-8 Lesson Notes

Lesson Resources

Chapter 10 Resource Masters FAST FILE
- Study Guide and Intervention, p. 578
- Practice: Skills, p. 579
- Practice: Word Problems, p. 580
- Reading to Learn Mathematics, p. 581
- Enrichment, p. 582

Calculator and Spreadsheet Masters, p. 50

Parent and Student Study Guide Workbook, 10-8

1 Focus

5-Minute Check 10-8
Use as a quiz or review of Lesson 10-7. (Also available as a transparency)

INTERACTIVE CHALKBOARD PowerPoint® Presentations

Bellringer
Hands-On Activity Have students draw a coordinate plane on grid paper. Tell them to start with their pencils at the origin and to follow your directions. Describe movements for them to make such as, "Three units up and four units left." After each movement, have students give their location on the coordinate plane as an ordered pair.

Reading to Learn Mathematics, p. 581 ELL

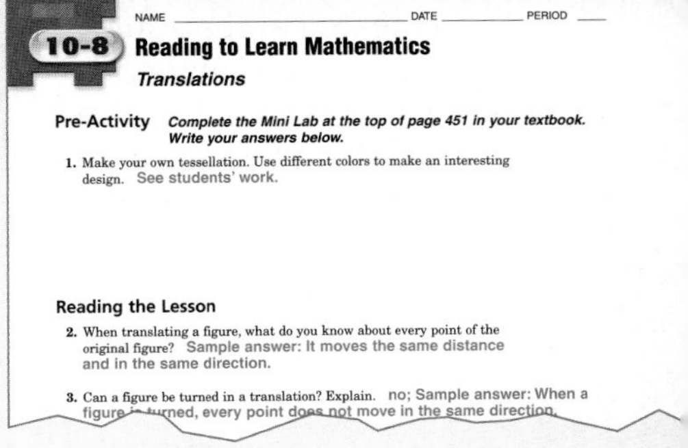
NAME ______ DATE ______ PERIOD ____

10-8 Reading to Learn Mathematics
Translations

Pre-Activity *Complete the Mini Lab at the top of page 451 in your textbook. Write your answers below.*

1. Make your own tessellation. Use different colors to make an interesting design. See students' work.

Reading the Lesson

2. When translating a figure, what do you know about every point of the original figure? Sample answer: It moves the same distance and in the same direction.
3. Can a figure be turned in a translation? Explain. no; Sample answer: When a figure is turned, every point does not move in the same direction.

DAILY INTERVENTION — Flexible Grouping

Translations Have students work in groups of three or four. One student should describe the translation of a figure while other students draw the translation. The students can then compare results and assist each other in completing the translation correctly.

2 Teach

In-Class Examples

INTERACTIVE CHALKBOARD
PowerPoint® Presentations

1 Translate $\triangle ABC$ 5 units left and 1 unit up.

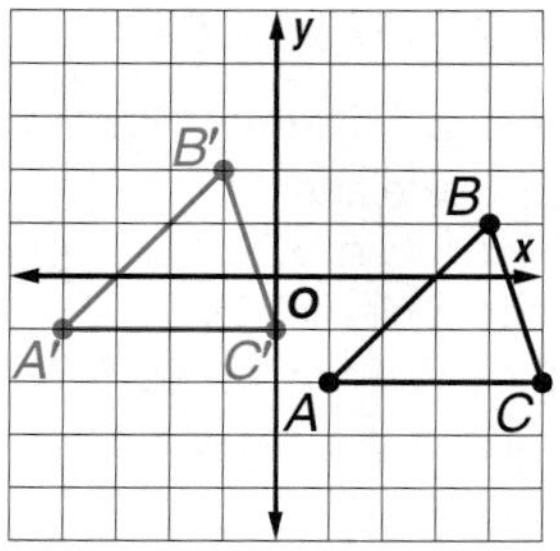

2 Trapezoid *GHIJ* has vertices $G(-4, 1)$, $H(-4, 3)$, $I(-2, 3)$, and $J(-1, 1)$. Find the vertices of trapezoid $G'H'I'J'$ after a translation of 5 units right and 3 units down. Then graph the figure and its translated image. **$G'(1, -2)$, $H'(1, 0)$, $I'(3, 0)$, $J'(4, -2)$**

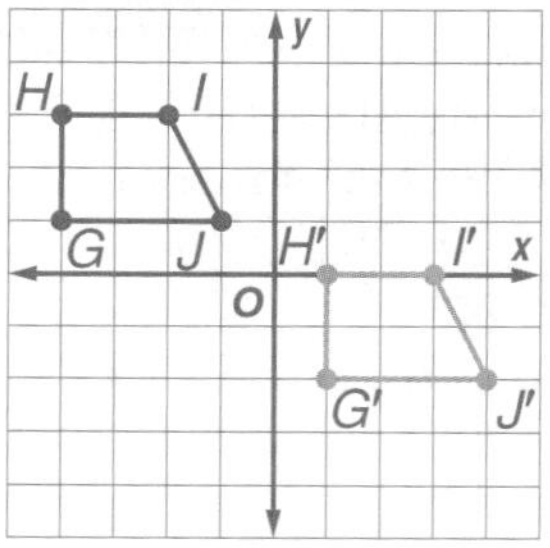

3 **CLASSROOMS** The squares below represent desks in a classroom.

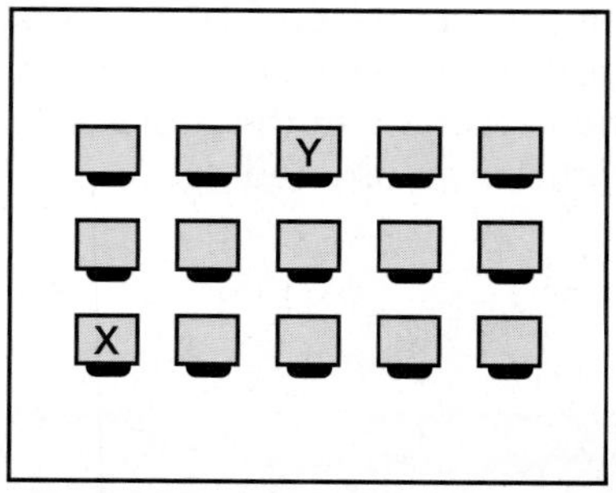

Ana is seated at the square marked X. She is moved to the seat marked Y. Describe this translation as an ordered pair **She moved 2 places right and 2 places up. So, the translation can be written (2, 2).**

TEACHING TIP

Tell students that the order of translating a figure does not matter. For example, moving a figure 3 units left and then 6 units up is the same as moving the figure 6 units up and then 3 units left.

STUDY TIP

Translation
A *positive* integer describes a translation right or up on a coordinate plane. A *negative* integer describes a translation left or down.

When a figure has been translated, the original figure and the translated figure, or *image*, are congruent. In Example 1, trapezoid *HIJK* is congruent to trapezoid $H'I'J'K'$.

You can also add to or subtract from the coordinates of the vertices of a figure to find the translated vertices.

EXAMPLE Find Coordinates of a Translation

2 **Triangle *LMN* has vertices $L(-1, -2)$, $M(6, -3)$, and $N(2, -5)$. Find the vertices of $\triangle L'M'N'$ after a translation of 6 units left and 4 units up. Then graph the figure and its translated image.**

Add −6 to each *x*-coordinate.

Add 4 to each *y*-coordinate.

Vertices of $\triangle LMN$	$(x + (-6), y + 4)$	Vertices of $\triangle L'M'N'$
$L(-1, -2)$	$(-1 + (-6), -2 + 4)$	$L'(-7, 2)$
$M(6, -3)$	$(6 + (-6), -3 + 4)$	$M'(0, 1)$
$N(2, -5)$	$(2 + (-6), -5 + 4)$	$N'(-4, -1)$

The coordinates of the vertices of $\triangle L'M'N'$ are $L'(-7, 2)$, $M'(0, 1)$, and $N'(-4, -1)$.

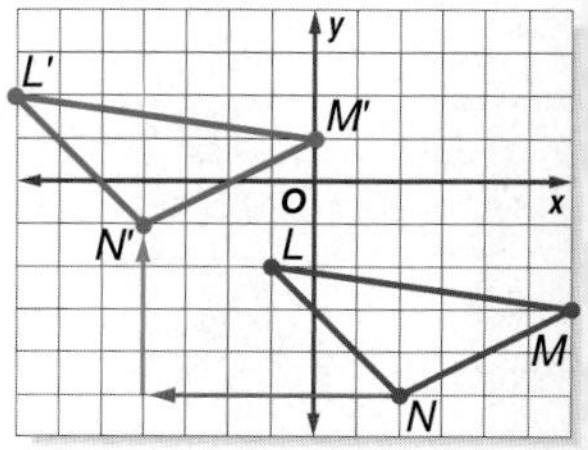

Your Turn **Triangle *TUV* has vertices $T(6, -3)$, $U(-2, 0)$, and $V(-1, 2)$. Find the vertices of $\triangle T'U'V'$ after each translation. Then graph the figure and its translated image.** **a–b. See Ch. 10 Answer Appendix.**

a. 3 units right and 4 units down
b. 1 unit left and 6 units up

In Example 2, $\triangle LMN$ was translated 6 units left and 4 units up. This translation can be described using the ordered pair $(-6, 4)$.

REAL-LIFE MATH

CHESS In 1977, Vlastimil Hort of Czechoslovakia played 550 opponents, 201 simultaneously, and lost only 10 games in just over 30 hours.

Source: www.research.ibm.com

EXAMPLE Naming Translations with Ordered Pairs

3 **GAMES When playing chess, the rook can only move vertically or horizontally across a chessboard. The chessboard at the right shows the movement of a rook after two turns. Describe this translation as an ordered pair.**

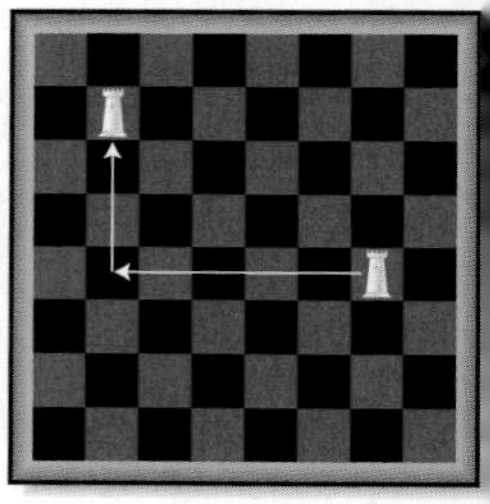

The rook moved 5 places left and 3 places up. So, the translation can be written as $(-5, 3)$.

Tips for New Teachers

Using Manipulatives
Patty paper is an inexpensive manipulative for geometry activities. You may be able to obtain some from your school cafeteria or from a food service supply store.

452 Chapter 10 Geometry

Skill and Concept Check

Writing Math
Exercise 2

1. **OPEN ENDED** On a coordinate plane, draw a triangle and its translation 1 unit right and 4 units down. **See margin.**

2. **Which One Doesn't Belong?** Identify the translation that is not the same as the other three. Explain your reasoning.

(-5, -4) (1, -3) (6, -6) (4, -2)

(−5, −4), the translation is to the left and down; the others show a translation to the right and down.

GUIDED PRACTICE

3. Translate $\triangle ABC$ 3 units left and 3 units down. Graph $\triangle A'B'C'$. **See margin.**

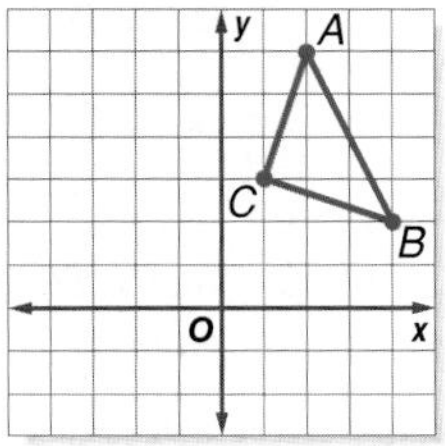

Quadrilateral *DEFG* has vertices *D*(1, 0), *E*(−2, −2), *F*(2, 4), and *G*(6, −3). Find the vertices of *D'E'F'G'* after each translation. Then graph the figure and its translated image. **4–5. See Ch. 10 Answer Appendix.**

4. 4 units right, 5 units down
5. 6 units right

★ indicates increased difficulty

Practice and Applications

HOMEWORK HELP	
For Exercises	See Examples
6–7	1
8–11, 15	2
12, 16	3

Extra Practice
See pages 589, 605.

6. Translate $\triangle HIJ$ 2 units right and 6 units down. Graph $\triangle H'I'J'$.

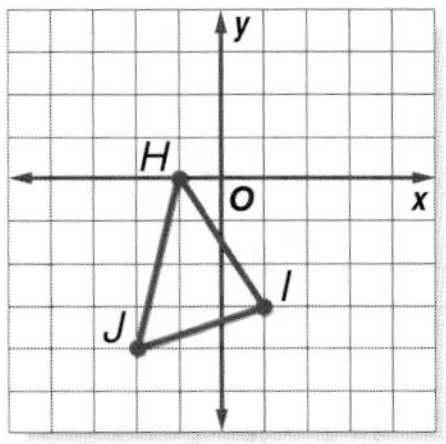

7. Translate rectangle *KLMN* 1 unit left and 3 units up. Graph rectangle *K'L'M'N'*.

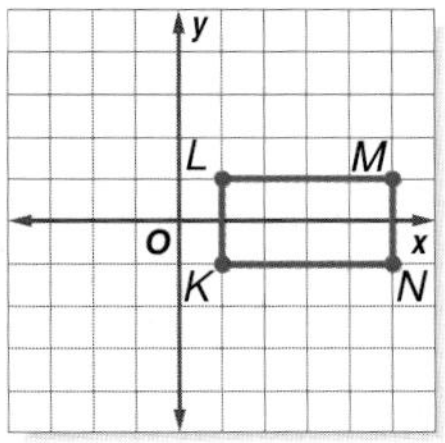

6–7. See Ch. 10 Answer Appendix.

Triangle *PQR* has vertices *P*(0, 0), *Q*(5, −2), and *R*(−3, 6). Find the vertices of *P'Q'R'* after each translation. Then graph the figure and its translated image. **8–11. See Ch. 10 Answer Appendix.**

8. 6 units right, 5 units up
9. 8 units left, 1 unit down
10. 3 units left
11. 9 units down

12. **MAPS** Payat lives at the corner of Wabash and Ohio. The school he attends is located at Huron and Dearborn. Describe Payat's walk from school to home as an ordered pair of the number of blocks. **(2, −3)**

msmath2.net/self_check_quiz

3 Practice/Apply

Which One Doesn't Belong?
In Exercise 2, students can draw arrows on a coordinate plane to illustrate each translation.

About the Exercises...

Guided Practice Key

Exercises	Examples
3	1
4–5	2

Odd/Even Assignments
Exercises 6–11 are structured so that students practice the same concepts whether they are assigned odd or even problems.

Alert! Have students use grid paper for Exercises 6–11.

Alert! Exercise 14 requires students to do research on the Internet or to use another source.

Alert! Exercise 17 requires patty paper.

Answers

1. **Sample answer:**

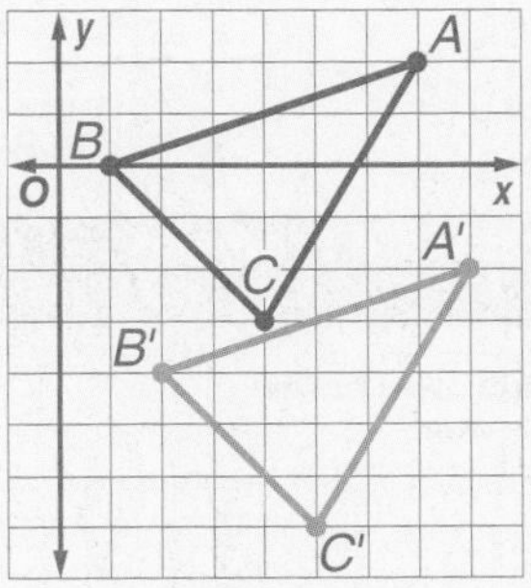

3.

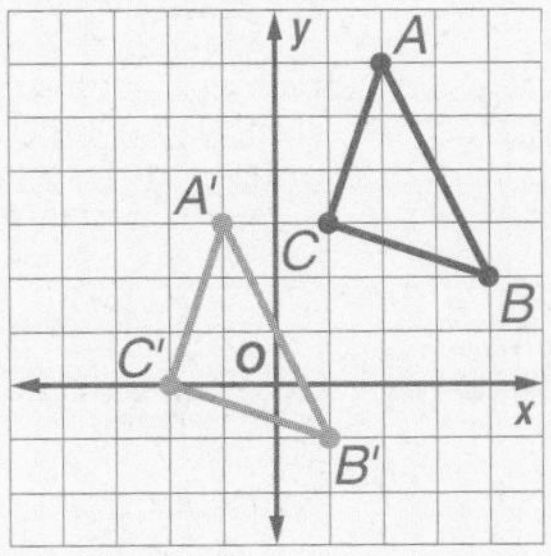

Study Guide and Intervention, p. 573

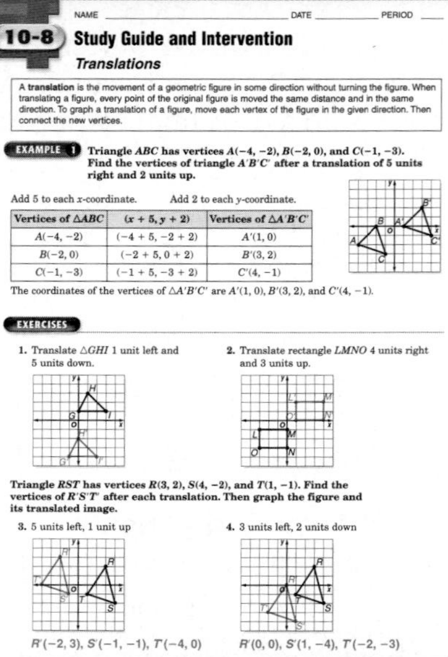
NAME ______ DATE ______ PERIOD ____

10-8 Study Guide and Intervention
Translations

A **translation** is the movement of a geometric figure in some direction without turning the figure. When translating a figure, every point of the original figure is moved the same distance and in the same direction. To graph a translation of a figure, move each vertex of the figure in the given direction. Then connect the new vertices.

EXAMPLE 1 Triangle *ABC* has vertices *A*(−4, −2), *B*(−2, 0), and *C*(−1, −3). Find the vertices of triangle *A′B′C′* after a translation of 5 units right and 2 units up.

Add 5 to each *x*-coordinate. Add 2 to each *y*-coordinate.

Vertices of △ABC	(x + 5, y + 2)	Vertices of △A′B′C′
A(−4, −2)	(−4 + 5, −2 + 2)	A′(1, 0)
B(−2, 0)	(−2 + 5, 0 + 2)	B′(3, 2)
C(−1, −3)	(−1 + 5, −3 + 2)	C′(4, −1)

The coordinates of the vertices of △A′B′C′ are A′(1, 0), B′(3, 2), and C′(4, −1).

EXERCISES

1. Translate △GHI 1 unit left and 5 units down.
2. Translate rectangle LMNO 4 units right and 3 units up.

Triangle *RST* has vertices *R*(3, 2), *S*(4, −2), and *T*(1, −1). Find the vertices of *R′S′T′* after each translation. Then graph the figure and its translated image.

3. 5 units left, 1 unit up
4. 3 units left, 2 units down

R′(−2, 3), S′(−1, −1), T′(−4, 0) R′(0, 0), S′(1, −4), T′(−2, −3)

Practice: Skills, p. 574

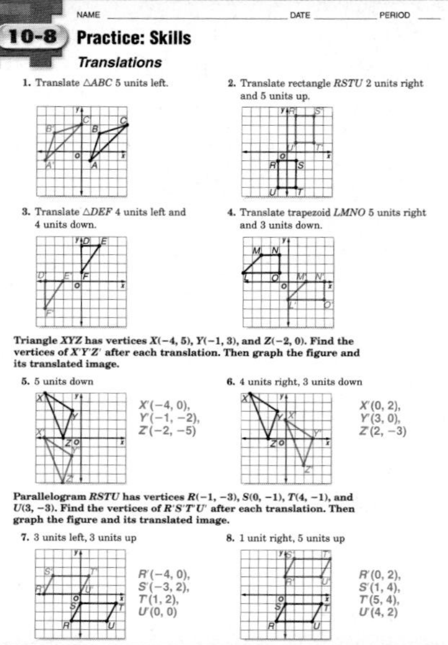
NAME ______ DATE ______ PERIOD ____

10-8 Practice: Skills
Translations

1. Translate △ABC 5 units left.
2. Translate rectangle RSTU 2 units right and 5 units up.
3. Translate △DEF 4 units left and 4 units down.
4. Translate trapezoid LMNO 5 units right and 3 units down.

Triangle *XYZ* has vertices *X*(−4, 5), *Y*(−1, 3), and *Z*(−2, 0). Find the vertices of *X′Y′Z′* after each translation. Then graph the figure and its translated image.

5. 5 units down X′(−4, 0), Y′(−1, −2), Z′(−2, −5)
6. 4 units right, 3 units down X′(0, 2), Y′(3, 0), Z′(2, −3)

Parallelogram *RSTU* has vertices *R*(−1, −3), *S*(0, −1), *T*(4, −1), and *U*(3, −3). Find the vertices of *R′S′T′U′* after each translation. Then graph the figure and its translated image.

7. 3 units left, 3 units up R′(−4, 0), S′(−3, 2), T′(1, 2), U′(0, 0)
8. 1 unit right, 5 units up R′(0, 2), S′(1, 4), T′(5, 4), U′(4, 2)

Practice: Word Problems, p. 575

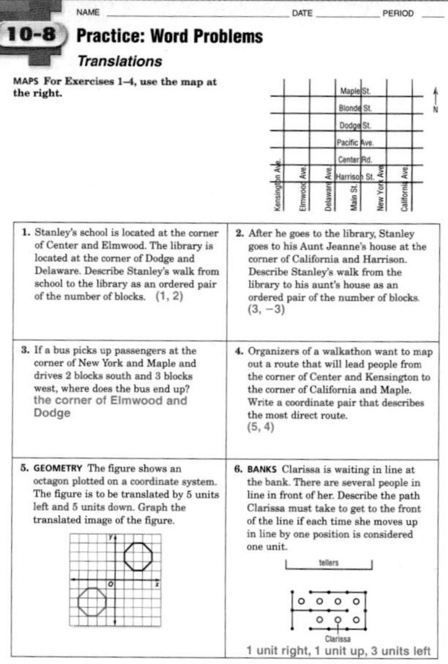
NAME ______ DATE ______ PERIOD ____

10-8 Practice: Word Problems
Translations

MAPS For Exercises 1–4, use the map at the right.

1. Stanley's school is located at the corner of Center and Elmwood. The library is located at the corner of Dodge and Delaware. Describe Stanley's walk from school to the library as an ordered pair of the number of blocks. (1, 2)
2. After he goes to the library, Stanley goes to his Aunt Jeanne's house at the corner of California and Harrison. Describe Stanley's walk from the library to his aunt's house as an ordered pair of the number of blocks. (3, −3)
3. If a bus picks up passengers at the corner of New York and Maple and drives 2 blocks south and 3 blocks west, where does the bus end up? the corner of Elmwood and Dodge
4. Organizers of a walkathon want to map out a route that will lead people from the corner of Center and Kensington to the corner of California and Maple. Write a coordinate pair that describes the most direct route. (5, 4)
5. GEOMETRY The figure shows an octagon plotted on a coordinate system. The figure is to be translated by 5 units left and 5 units down. Graph the translated image of the figure.
6. BANKS Clarissa is waiting in line at the bank. There are several people in line in front of her. Describe the path Clarissa must take to get to the front of the line if each time she moves up in line by one position is considered one unit. 1 unit right, 1 unit up, 3 units left

13. **ART** Explain how translations and tessellations were used in *Horsemen*, created by M.C. Escher at the right. **See margin.**

Horsemen by M.C. Escher.

14. **RESEARCH** Use the Internet or another source to find other pieces of art that contain tessellations of translations. Describe how the artists incorporated both ideas into their work. **See students' work.**

MULTI STEP For Exercises 15 and 16, use the following information. **15. Sample answer: △X′Y′Z′ was translated 5 units left and 2 units down.**

$\triangle XYZ$ has vertices $X(7, 6)$, $Y(-3, 4)$, and $Z(1, -5)$. The translated figure $\triangle X'Y'Z'$ has vertices $X'(2, 4)$, $Y'(-8, 2)$, and $Z'(-4, -7)$.

★ 15. Describe the translation.

★ 16. Write the translation as an ordered pair. **(−5, −2)**

17. **TESSELLATIONS** Use the figure at the right and patty paper to make a tessellation using translations. **See Ch. 10 Answer Appendix.**

18. **CRITICAL THINKING** Is it possible to make a tessellation with translations of equilateral triangles? Explain your reasoning. **No; The triangle has to be flipped in order to make a tessellation.**

Standardized Test Practice and Mixed Review

19. **MULTIPLE CHOICE** Which graph shows a translation of the letter Z? **C**

(A)
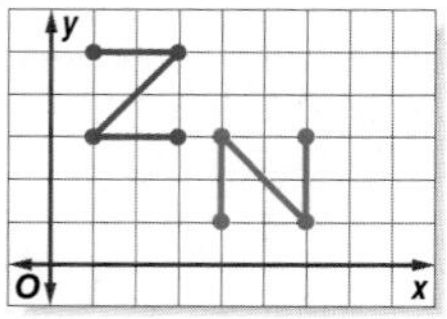
(B)
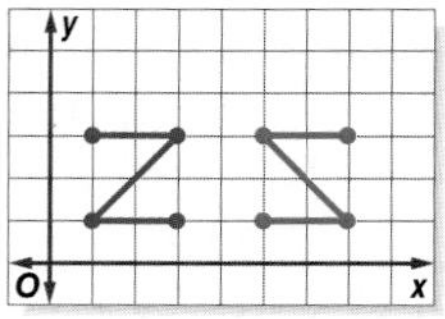
(C)
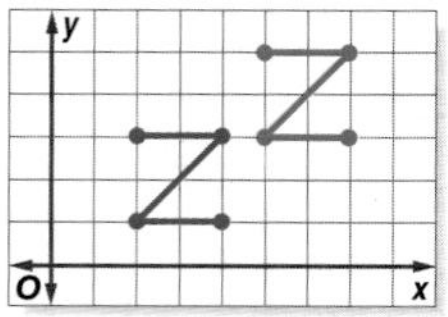
(D)
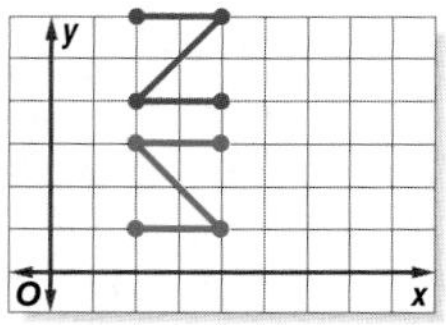

20. **SHORT RESPONSE** Triangle *ABC* with vertices $A(-5, 10)$, $B(5, 10)$, and $C(-10, 5)$ is translated by $(-10, 5)$. What are the coordinates of A'? **(−15, 15)**

21. **GEOMETRY** What is the name of a polygon with eight sides? (Lesson 10-7) **octagon**

22. **TOWERS** Horatio is 167 centimeters tall and casts a shadow 93 centimeters long. What is the approximate height of a cellphone tower if it casts a shadow 763 centimeters long? (Lesson 10-6) **1,370 cm**

GETTING READY FOR THE NEXT LESSON

BASIC SKILL Determine whether each figure can be folded in half so that one side matches the other. Write *yes* or *no*.

23. 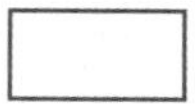 **yes**
24.  **no**
25. 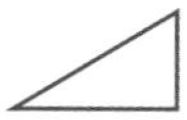 **no**
26. 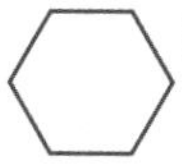 **yes**

4 Assess

Higher-Level Thinking

A point (x, y) is translated 5 units right and 2 units down. How would you write the coordinates of the translated point? **$(x + 5, y - 2)$**

Answer

13. **Sample answer: there are two main images, the brown horseman and the yellow horseman. Both main horsemen are translated to different parts of the picture. These translations allow for the tessellation of both horsemen.**

10-8b Spreadsheet Investigation

A Follow-Up of Lesson 10-8

hat You'll Learn
e a spreadsheet to
large and reduce
lygons.

Dilations

A computer spreadsheet is a useful tool for calculating the vertices of polygons. You can enlarge or reduce the polygons by using a spreadsheet to automatically calculate the new coordinates of the vertices. Enlarging or reducing a figure is a transformation called a **dilation**.

ACTIVITY

Emma has plotted a pentagon on graph paper. The coordinates of the vertices of the pentagon are (2, 2), (4, 2), (5, 4), (3, 6), and (1, 4). She wants to multiply the coordinates by 3 to enlarge the pentagon. She enters the coordinates on a spreadsheet as shown below.

Set up the spreadsheet like the one shown below.

Pentagon Dilation

	A	B	C	D
1	original		dilation	
2	x-coordinate	y-coordinate	x-coordinate	y-coordinate
3	2	2	=A3*3	=B3*3
4	4	2	=A4*3	=B4*3
5	5	4		
6	3	6		
7	1	4		
8				

Sheet1 / Sheet2 / Sheet3

Continue entering the formulas in columns C and D to complete the dilation.

EXERCISE

1. How will the formulas in columns C and D change the original pentagon? How do you know?
2. Graph the original pentagon and its dilation on graph paper.
3. What is the percent of increase of the original pentagon to its dilation? **300%**
4. Find the coordinates of the pentagon enlarged 5 times.
5. Find the coordinates of the pentagon reduced by one-half.
6. What type of dilation had occurred if the new coordinates of the pentagon are (5, 5), (10, 5), (12.5, 10), (7.5, 15), and (2.5, 10)? What is the scale factor? **enlargement; 2.5**
7. Select another geometric figure and plot its points on graph paper. Set up a spreadsheet to find two dilations, one enlargement and one reduction of the same figure. **See students' work.**

Sample answer:
large it by a scale
:tor of 3. The new
ordinates are three
nes greater than the
ginal coordinates.

See Ch. 10 Answer
pendix.

(10, 10), (20, 10),
5, 20), (15, 30),
, 20)

(1, 1), (2, 1), (2.5, 2),
.5, 3), (0.5, 2)

10-8b Spreadsheet Investigation

Lesson Resources

Teaching Mathematics with Manipulatives, pp. 102–103

GETTING STARTED

Teaching Tip If students will need to share computers, assign groups before going to the computer lab.

TEACH

- As you discuss dilations, show students examples of enlargements and reductions done on a photocopy machine.
- Remind students to be careful as they enter the formulas for the dilation in columns C and D. They are entering the cell to be multiplied, not the value of the *x*- or *y*-coordinate to be multiplied.
- You may wish to have students graph the polygons using the spreadsheet software. To do this, have them enter the ordered pairs in the order they should be connected, repeating the first ordered pair at the end of the list. Then choose the *xy* connected graph type.

ASSESS

Observe as students solve the **Exercises**. They should:

- create the spreadsheet and graph the results,
- use their work in Exercises 1 and 2 to answer the other questions, and
- be able to describe their dilations in Exercise 7 so that other students could reproduce the figures.

10-9 Lesson Notes

Lesson Resources

Chapter 10 Resource Masters

FAST FILE

- Study Guide and Intervention, p. 583
- Practice: Skills, p. 584
- Practice: Word Problems, p. 585
- Reading to Learn Mathematics, p. 586
- Enrichment, p. 587

Key Teaching Transparencies, 40

Parent and Student Study Guide Workbook, 10-9

Real-Life Career Masters, p. 24

1 Focus

5-Minute Check 10-9

INTERACTIVE CHALKBOARD
PowerPoint® Presentations

Use as a quiz or review of Lesson 10-8. (Also available as a transparency)

Bellringer

Communication Have students describe the reflections of some everyday objects in the mirror. Ask them what is different between an object and its reflection. Consider showing students several objects, some with a line of symmetry and some without, to get the discussion rolling.

Reading to Learn Mathematics, p. 586 **ELL**

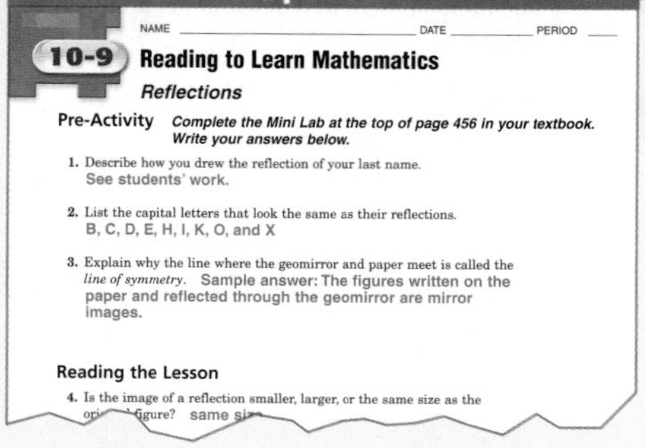
NAME ______ DATE ______ PERIOD ____

10-9 Reading to Learn Mathematics

Reflections

Pre-Activity *Complete the Mini Lab at the top of page 456 in your textbook. Write your answers below.*

1. Describe how you drew the reflection of your last name.
   See students' work.
2. List the capital letters that look the same as their reflections.
   B, C, D, E, H, I, K, O, and X
3. Explain why the line where the geomirror and paper meet is called the *line of symmetry*. Sample answer: The figures written on the paper and reflected through the geomirror are mirror images.

Reading the Lesson

4. Is the image of a reflection smaller, larger, or the same size as the original figure? same size

10-9 Reflections

What You'll Learn
Identify figures with line symmetry and graph reflections on a coordinate plane.

***NEW* Vocabulary**
line symmetry
line of symmetry
reflection

HANDS-ON Mini Lab

Materials
- geomirror
- paper

Work with a partner.

- Write your first name in capital letters on a sheet of paper.
- Use the geomirror to trace the reflection of the letters in your name.
- Write your last name. Draw the reflection of the letters without using the geomirror.

2. B, C, D, E, H, I, K, O, and X

1. Describe how you drew the reflection of your last name. **See students' work.**
2. List the capital letters that look the same as their reflections.
3. Explain why the line where the geomirror and paper meet is called the *line of symmetry*. **Sample answer: The figures written on the paper and reflected through the geomirror are mirror images.**

Figures that match exactly when folded in half have **line symmetry**. The figures below have line symmetry.

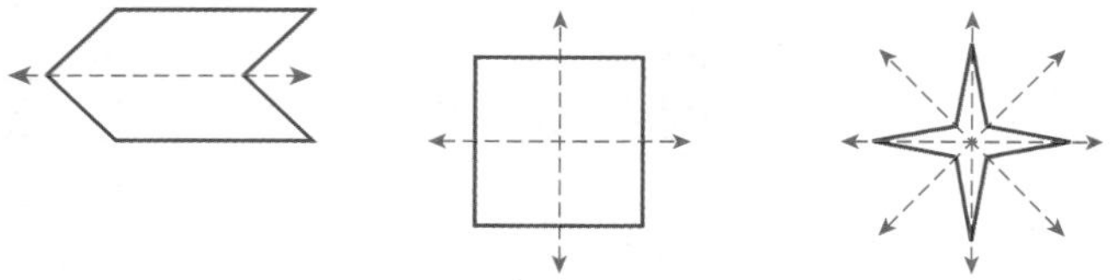

Each fold line is called a **line of symmetry**. Some figures can be folded in more than one way to show symmetry, so they have more than one line of symmetry.

TEACHING TIP
You can have students trace the figures on patty paper, cut them out, and fold them to check for lines of symmetry.

EXAMPLES Identify Lines of Symmetry

Determine whether each figure has line symmetry. If so, copy the figure and draw all lines of symmetry.

1 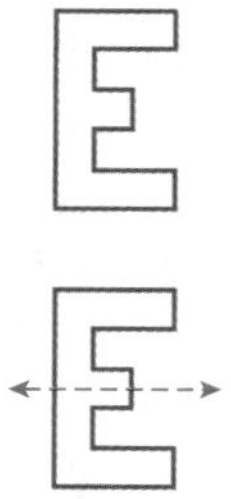

2 

no symmetry

3 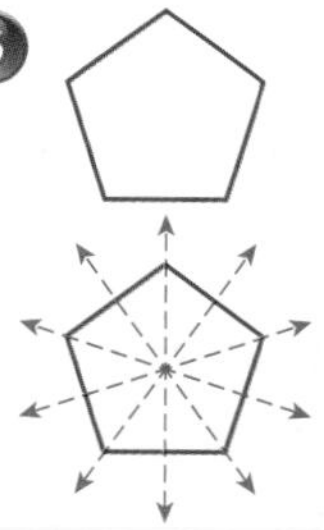

DAILY INTERVENTION **Unlocking Misconceptions**

Reflections Students may be confused as to which coordinate changes when reflecting a figure over an axis. Have them draw a triangle in Quadrant I and label the vertices. First have them determine the location of the triangle after a reflection over the *x*-axis. Point out that the move to Quadrant IV requires a change in the *y*-coordinate, with the *x*-coordinate remaining constant. Repeat this procedure with a reflection of the original figure over the *y*-axis.

A type of transformation where a figure is flipped over a line of symmetry is a **reflection**. As with translations, the original figure and the reflected image are congruent.

TUDY TIP

x-axis When eflecting over the x-axis, change the y-coordinates to their opposites.

EXAMPLE Reflect a Figure Over the *x*-axis

4 **Triangle *ABC* has vertices *A*(5, 2), *B*(1, 3), and *C*(−1, 1). Find the coordinates of *ABC* after a reflection over the *x*-axis. Then graph the figure and its reflected image.**

Vertices of $\triangle ABC$	Distance from *x*-axis	Vertices of $\triangle A'B'C'$
A(5, 2)	2	A′(5, −2)
B(1, 3)	3	B′(1, −3)
C(−1, 1)	1	C′(−1, −1)

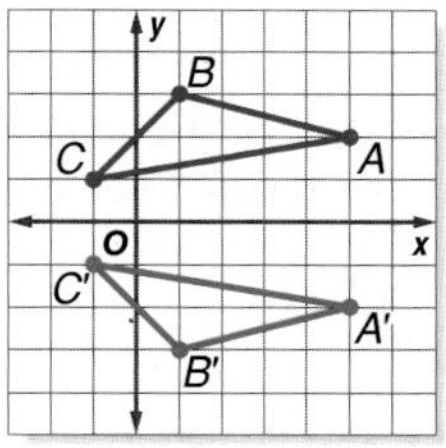

Plot the vertices and connect to form $\triangle ABC$. The *x*-axis is the line of symmetry. So, the distance from each point on $\triangle ABC$ to the line of symmetry is the same as the distance from the line of symmetry to $\triangle A'B'C'$.

TUDY TIP

y-axis When eflecting over the y-axis, change the x-coordinates to their opposites.

EXAMPLE Reflect a Figure Over the *y*-axis

5 **Quadrilateral *KLMN* has vertices *K*(2, 3), *L*(5, 1), *M*(4, −2), and *N*(1, −1). Find the coordinates of *KLMN* after a reflection over the *y*-axis. Then graph the figure and its reflected image.**

Vertices of quad *KLMN*	Distance from *y*-axis	Vertices of quad *K′L′M′N′*
K(2, 3)	2	K′(−2, 3)
L(5, 1)	5	L′(−5, 1)
M(4, −2)	4	M′(−4, −2)
N(1, −1)	1	N′(−1, −1)

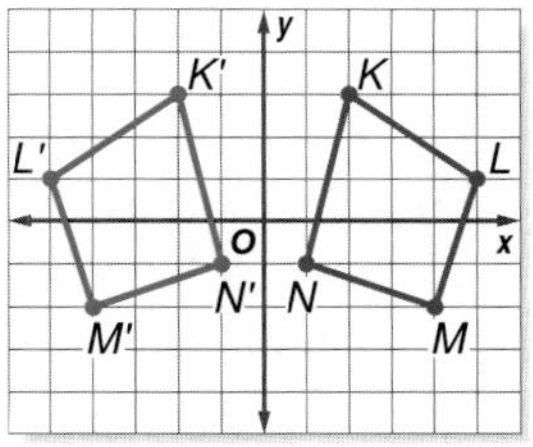

Plot the vertices and connect to form quadrilateral *KLMN*. The *y*-axis is the line of symmetry. So, the distance from each point on quadrilateral *KLMN* to the line of symmetry is the same as the distance from the line of symmetry to quadrilateral *K′L′M′N′*.

Your Turn a–b. See margin for graphs.

a. Find the coordinates of rectangle *GHIJ* with vertices *G*(3, −4), *H*(3, −1), *I*(−2, −1), and *J*(−2, −4) after a reflection over the *x*-axis. Then graph the figure and its reflected image.

b. Find the coordinates of triangle *PQR* with vertices *P*(1, 5), *Q*(3, 7), and *R*(5, −1) after a reflection over the *y*-axis. Then graph the figure and its reflected image.

G′(3, 4), H′(3, 1), −2, 1), and J′(−2, 4)
P′(−1, 5), Q′(−3, 7), d R′(−5, −1)

Answers (Your Turn)

a.

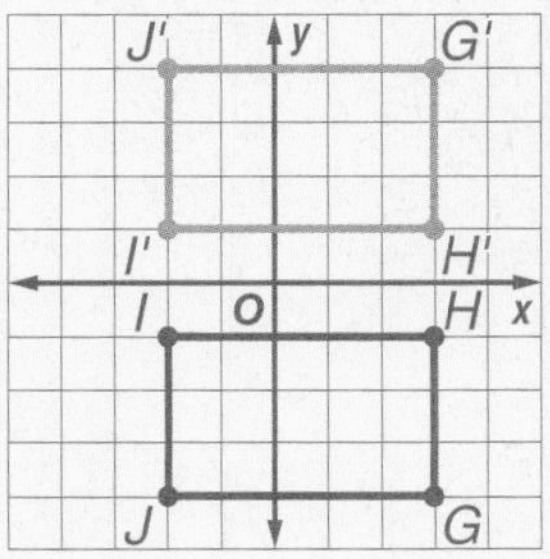

b.

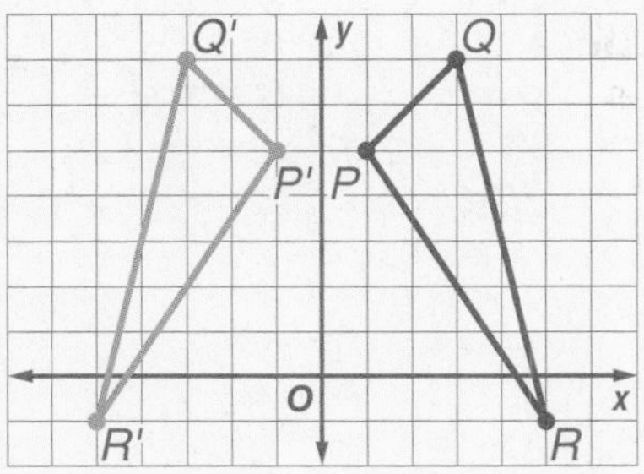

2 Teach

In-Class Examples

INTERACTIVE CHALKBOARD PowerPoint® Presentations

Determine whether each figure has line symmetry. If so, copy the figure and draw all lines of symmetry.

1

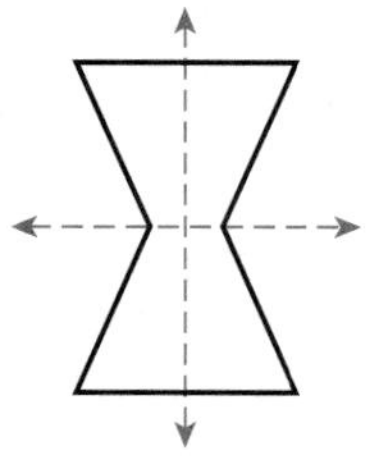

2

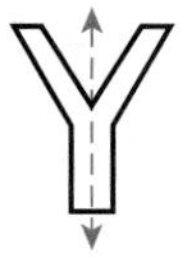

3

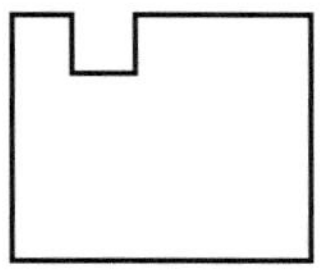

no symmetry

4 Quadrilateral *QRST* has vertices *Q*(−1, 1), *R*(0, 3), *S*(3, 2), and *T*(4, 0). Find the coordinates of *QRST* after a reflection over the *x*-axis. Then graph the figure and its reflected image. **Q′(−1, −1), R′(0, −3), S′(3, −2), T′(4, 0)**

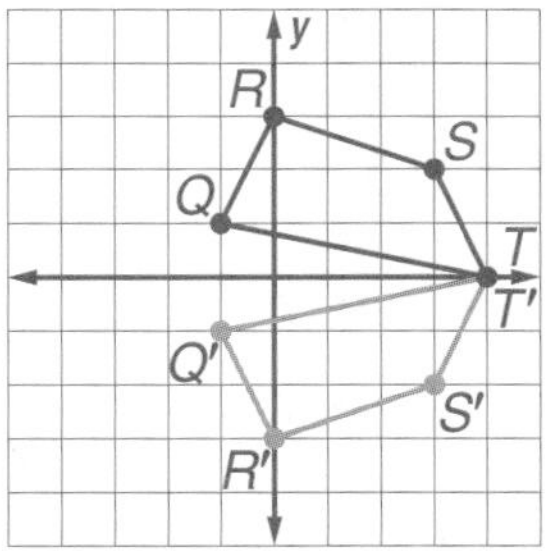

5 Triangle *XYZ* has vertices *X*(1, 2), *Y*(2, 1), and *Z*(1, −2). Find the coordinates of *XYZ* after a reflection over the *y*-axis. Then graph the figure and its reflected image. **X′(−1, 2), Y′(−2, 1), Z′(−1, −2)**

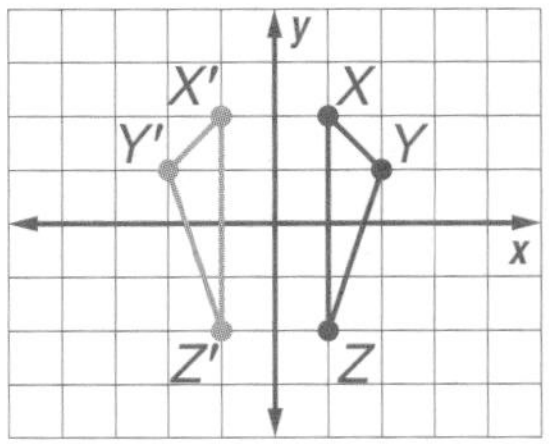

3 Practice/Apply

Which One Doesn't Belong?
In Exercise 2, students need to understand the difference between reflections and translations. Tell students to imagine placing a small mirror along the arrows and perpendicular to the plane of the page. Ask them to indicate which figure would be duplicated by looking into the mirror. (If you have a hand-held mirror, give students the opportunity to do this.)

About the Exercises...

Guided Practice Key

Exercises	Examples
3–4	1–3
5	4
6	5

Odd/Even Assignments
Exercises 7–14 are structured so that students practice the same concepts whether they are assigned odd or even problems.

Alert! The Data Update after Exercise 17 requires access to the Internet.

Answers

1.

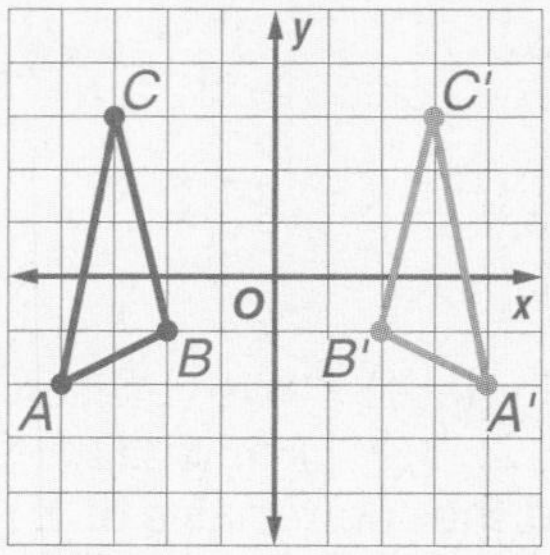

15. vertical line that divides the building in half and a line at the top of the water from the reflection in the water

16. Sample answer: It is written in this way so that drivers looking in their rear-view mirrors can see the word "AMBULANCE" written correctly.

17. All four flags have line symmetry.

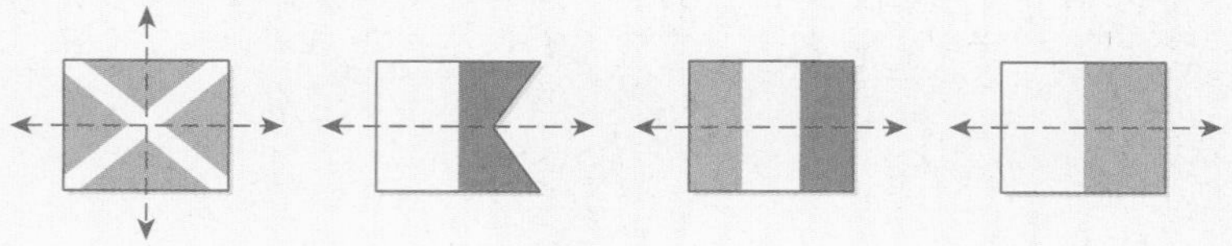

Skill and Concept Check

Writing Math
Exercise 2

1. **OPEN ENDED** Draw a figure on a coordinate plane and its reflection over the *y*-axis. See margin.

2. **Which One Doesn't Belong?** Identify the transformation that is not the same as the other three. Explain your reasoning.

The first transformation is a reflection; the others are translations.

GUIDED PRACTICE

Determine which figures have line symmetry. Write *yes* or *no*. If *yes*, draw all lines of symmetry.

3. 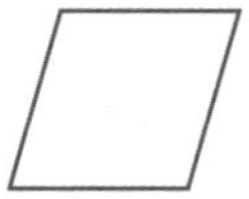 no

4. 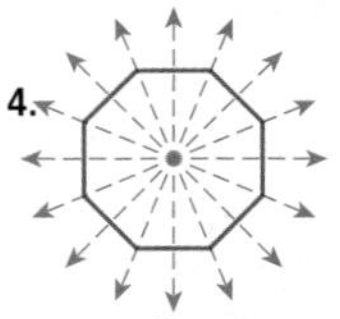 yes

Find the coordinates of each figure after a reflection over the given axis. Then graph the figure and its reflected image.

5. $\triangle ABC$ with vertices $A(5, 8)$, $B(1, 2)$, and $C(6, 4)$; x-axis

6. parallelogram $WXYZ$ with vertices $W(-4, -2)$, $X(-4, 3)$, $Y(-2, 4)$, and $Z(-2, -1)$; y-axis 5–6. See Ch. 10 Answer Appendix.

★ indicates increased difficulty

Practice and Applications

HOMEWORK HELP

For Exercises	See Examples
7–10, 15–17	1–3
11–12	4
13–14	5

Extra Practice
See pages 589, 605.

Determine which figures have line symmetry. Write *yes* or *no*. If *yes*, draw all lines of symmetry.

7. 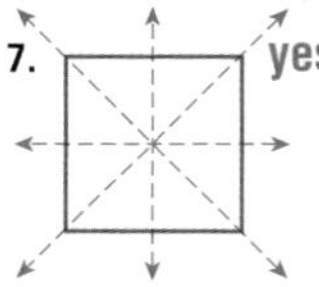 yes

8. 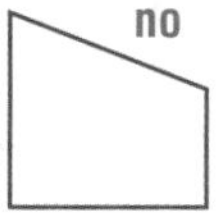 no

9. 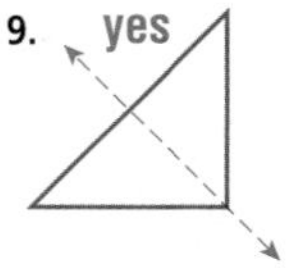 yes

10. 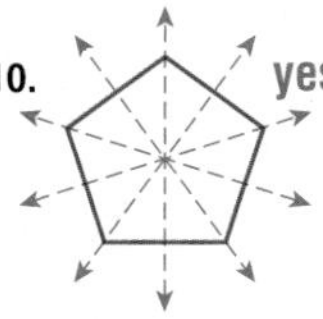 yes

Find the coordinates of each figure after a reflection over the *x*-axis. Then graph the figure and its reflected image. 11–14. See Ch. 10 Answer Appendix.

11. quadrilateral $DEFG$ with vertices $D(-3, 6)$, $E(-2, -3)$, $F(2, 2)$, and $G(4, 9)$

12. $\triangle TUV$ with vertices $T(-6, 1)$, $U(-2, -3)$, and $V(5, -4)$

Find the coordinates of each figure after a reflection over the *y*-axis. Then graph the figure and its reflected image.

13. $\triangle QRS$ with vertices $Q(2, -5)$, $R(4, -5)$, and $S(2, 3)$

14. parallelogram $HIJK$ with vertices $H(-1, 3)$, $I(-1, -1)$, $J(2, -2)$, and $K(2, 2)$

15. **BUILDINGS** Describe the location of the line(s) of symmetry in the photograph of the Taj Mahal. See margin.

16. **AMBULANCE** Explain why the word "AMBULANCE" is written backward and from right to left on the front of the emergency vehicle. **16–17. See margin.**

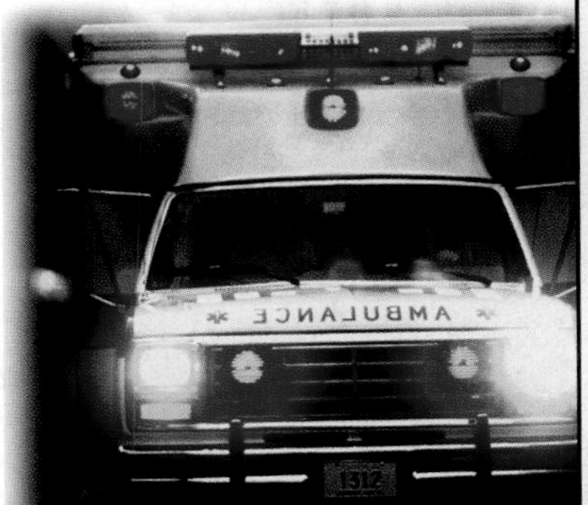

17. **SIGNAL FLAGS** International Marine Signal Flags are used by sailors to send messages at sea. A flag represents each letter of the alphabet. The flags below spell out the word *MATH*. Which flags have line symmetry? Draw all lines of symmetry.

Data Update What other maritime signal flags have line symmetry? Visit msmath2.net/data_update for more information.

★ 18. **TESSELLATIONS** Make a tessellation using a combination of translations and reflections of polygons. Explain how you created your tessellation. **See students' work.**

19. **CRITICAL THINKING** Triangle *JKL* has vertices $J(-7, 4)$, $K(7, 1)$, and $L(2, -2)$. Without graphing, reflect the triangle over the *x*-axis and then over the *y*-axis. What are the new coordinates of the triangle after the double reflection? $J''(7, -4)$, $K''(-7, -1)$, **and** $L''(-2, 2)$

TEACHING TIP

In Exercise 19, tell students that the vertices of a figure receive double prime symbols (″) after they have been transformed twice.

Standardized Test Practice and Mixed Review

20. **MULTIPLE CHOICE** $\triangle ABC$ with vertices $A(-2, -5)$, $B(4, 1)$, and $C(3, -2)$ is reflected over the *y*-axis. Find the coordinates of the new figure. **A**

    Ⓐ $A'(2, -5), B'(-4, 1), C'(-3, -2)$
    Ⓑ $A'(-2, 5), B'(4, -1), C'(3, 2)$
    Ⓒ $A'(2, 5), B'(-4, -1), C'(-3, 2)$
    Ⓓ $A'(2, 5), B'(4, 1), C'(3, 2)$

21. **SHORT RESPONSE** How many lines of symmetry, if any, are in the rhombus at the right? **2**

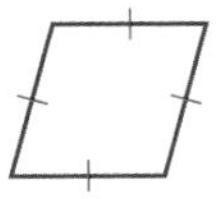

22. **GEOMETRY** Graph $\triangle FGH$ with vertices $F(-3, 7)$, $G(-1, 5)$, and $H(-2, 2)$ and its translation 4 units right and 1 unit down. Write the ordered pairs for the vertices of the new figure. (Lesson 10-8) **See margin.**

23. **ART** Aisha wishes to construct a tessellation for a wall hanging made only from regular decagons. Is this possible? Explain. (Lesson 10-7) **No, the measure of one angle is 144°, which is not a factor of 360°.**

Estimate. (Lesson 6-1) **24–26. Sample answers are given.**

24. $\frac{4}{9} + 8\frac{1}{8}$ $\quad \frac{1}{2} + 8 = 8\frac{1}{2}$

25. $\frac{1}{9} \times \frac{2}{5}$ $\quad 0 \times \frac{1}{2} = 0$

26. $12\frac{1}{4} \div 5\frac{6}{7}$ $\quad 12 \div 6 = 2$

27. Write an inequality for *six times a number is less than or equal to 18*. Then solve the inequality. (Lesson 4-5) $6n \le 18$; $n \le 3$

msmath2.net/self_check_quiz

4 Assess

Open-Ended Assessment

Speaking Ask students to explain how to determine the number of lines of symmetry in a regular octagon.

Answer

22. $F'(1, 6)$, $G'(3, 4)$, and $H'(2, 1)$

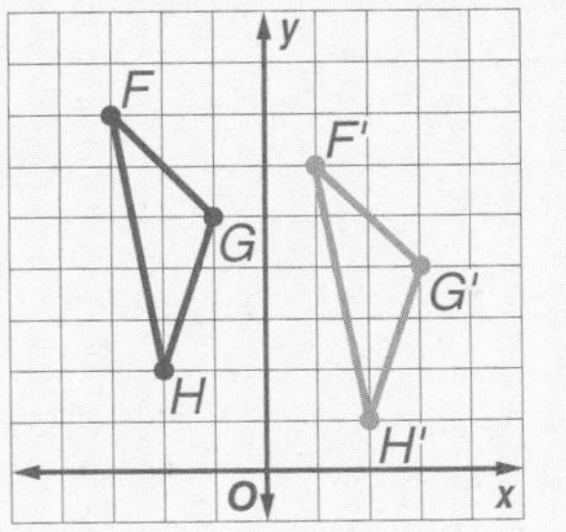

Study Guide and Intervention, p. 583

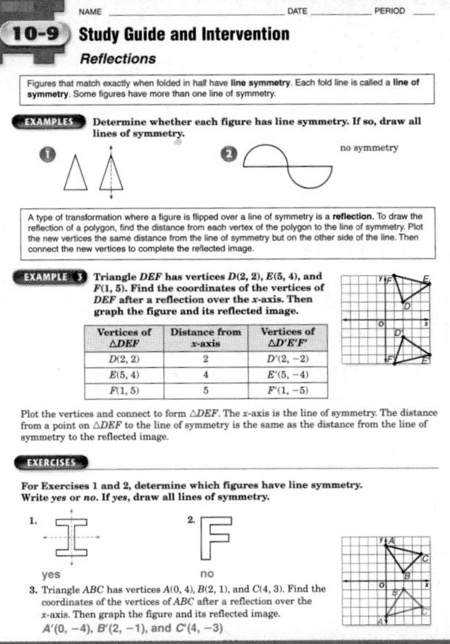

NAME ______ DATE ______ PERIOD ____

10-9 Study Guide and Intervention

Reflections

Figures that match exactly when folded in half have **line symmetry**. Each fold line is called a **line of symmetry**. Some figures have more than one line of symmetry.

EXAMPLES Determine whether each figure has line symmetry. If so, draw all lines of symmetry.

1

2 no symmetry

A type of transformation where a figure is flipped over a line of symmetry is a **reflection**. To draw the reflection of a polygon, find the distance from each vertex of the polygon to the line of symmetry. Plot the new vertices the same distance from the line of symmetry but on the other side of the line. Then connect the new vertices to complete the reflected image.

EXAMPLE 3 Triangle *DEF* has vertices $D(2, 2)$, $E(5, 4)$, and $F(1, 5)$. Find the coordinates of the vertices of *DEF* after a reflection over the *x*-axis. Then graph the figure and its reflected image.

Vertices of $\triangle DEF$	Distance from *x*-axis	Vertices of $\triangle D'E'F'$
$D(2, 2)$	2	$D'(2, -2)$
$E(5, 4)$	4	$E'(5, -4)$
$F(1, 5)$	5	$F'(1, -5)$

Plot the vertices and connect to form $\triangle DEF$. The *x*-axis is the line of symmetry. The distance from a point on $\triangle DEF$ to the line of symmetry is the same as the distance from the line of symmetry to the reflected image.

EXERCISES

For Exercises 1 and 2, determine which figures have line symmetry. Write *yes* or *no*. If *yes*, draw all lines of symmetry.

1. yes
2. no

3. Triangle *ABC* has vertices $A(0, 4)$, $B(2, 1)$, and $C(4, 3)$. Find the coordinates of the vertices of *ABC* after a reflection over the *x*-axis. Then graph the figure and its reflected image. $A'(0, -4)$, $B'(2, -1)$, and $C'(4, -3)$

Practice: Skills, p. 584

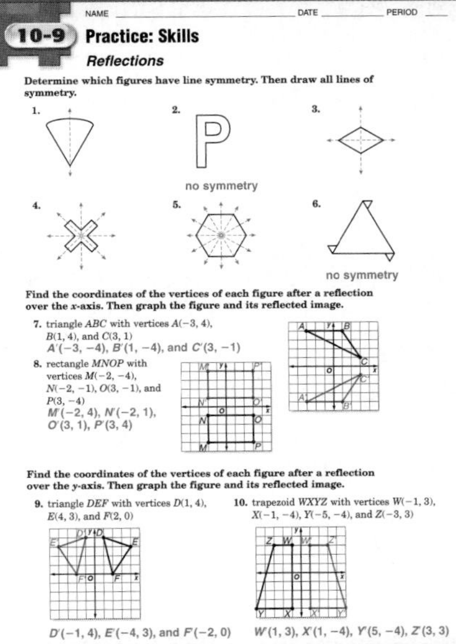

NAME ______ DATE ______ PERIOD ____

10-9 Practice: Skills

Reflections

Determine which figures have line symmetry. Then draw all lines of symmetry.

1.
2. no symmetry
3.
4.
5.
6. no symmetry

Find the coordinates of the vertices of each figure after a reflection over the *x*-axis. Then graph the figure and its reflected image.

7. triangle *ABC* with vertices $A(-3, 4)$, $B(1, 4)$, and $C(3, 1)$ $A'(-3, -4)$, $B'(1, -4)$, and $C'(3, -1)$

8. rectangle *MNOP* with vertices $M(-2, -4)$, $N(-2, -1)$, $O(3, -1)$, and $P(3, -4)$ $M'(-2, 4)$, $N'(-2, 1)$, $O'(3, 1)$, $P'(3, 4)$

Find the coordinates of the vertices of each figure after a reflection over the *y*-axis. Then graph the figure and its reflected image.

9. triangle *DEF* with vertices $D(1, 4)$, $E(4, 3)$, and $F(2, 0)$ $D'(-1, 4)$, $E'(-4, 3)$, and $F'(-2, 0)$

10. trapezoid *WXYZ* with vertices $W(-1, 3)$, $X(-1, -4)$, $Y(-5, -4)$, and $Z(-3, 3)$ $W'(1, 3)$, $X'(1, -4)$, $Y'(5, -4)$, $Z'(3, 3)$

Practice: Word Problems, p. 585

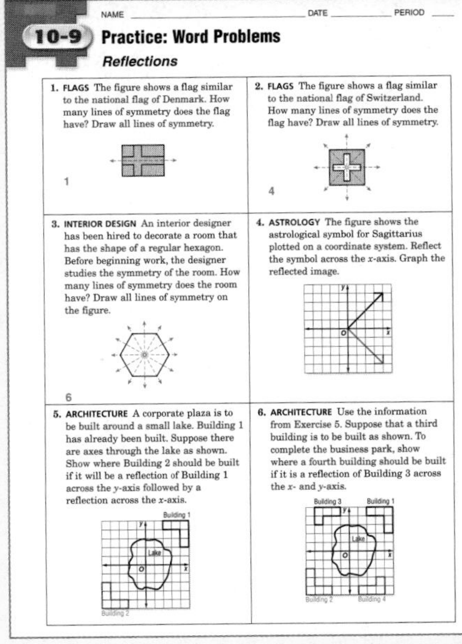

NAME ______ DATE ______ PERIOD ____

10-9 Practice: Word Problems

Reflections

1. **FLAGS** The figure shows a flag similar to the national flag of Denmark. How many lines of symmetry does the flag have? Draw all lines of symmetry. 1

2. **FLAGS** The figure shows a flag similar to the national flag of Switzerland. How many lines of symmetry does the flag have? Draw all lines of symmetry. 4

3. **INTERIOR DESIGN** An interior designer has been hired to decorate a room that has the shape of a regular hexagon. Before beginning work, the designer studies the symmetry of the room. How many lines of symmetry does the room have? Draw all lines of symmetry on the figure. 6

4. **ASTROLOGY** The figure shows the astrological symbol for Sagittarius plotted on a coordinate system. Reflect the symbol across the *x*-axis. Graph the reflected image.

5. **ARCHITECTURE** A corporate plaza is to be built around a small lake. Building 1 has already been built. Suppose there are axes through the lake as shown. Show where Building 2 should be built if it will be a reflection of Building 1 across the *y*-axis followed by a reflection across the *x*-axis.

6. **ARCHITECTURE** Use the information from Exercise 5. Suppose that a third building is to be built as shown. To complete the business park, show where a fourth building should be built if it is a reflection of Building 3 across the *x*- and *y*-axis.

10-9b Hands-On Lab Lesson Notes

Lesson Resources

Teaching Mathematics with Manipulatives

- coordinate planes, p. 17
- protractors, p. 21
- student recording sheet 10-9b

Glencoe Mathematics Classroom Manipulative Kit

- coordinate grid stamp
- stamp pad
- rulers

GETTING STARTED

Teaching Tip Students can use any straightedge as a ruler for this activity.

TEACH

Activity 1

- Students should work with a partner but create their own graphs.

Answers (Your Turn)

a.

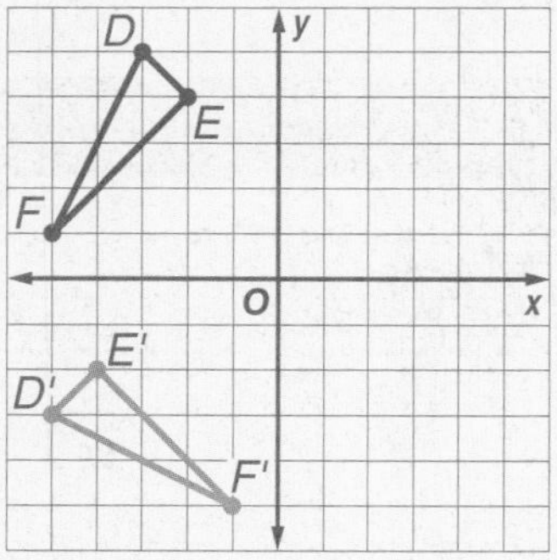

b.

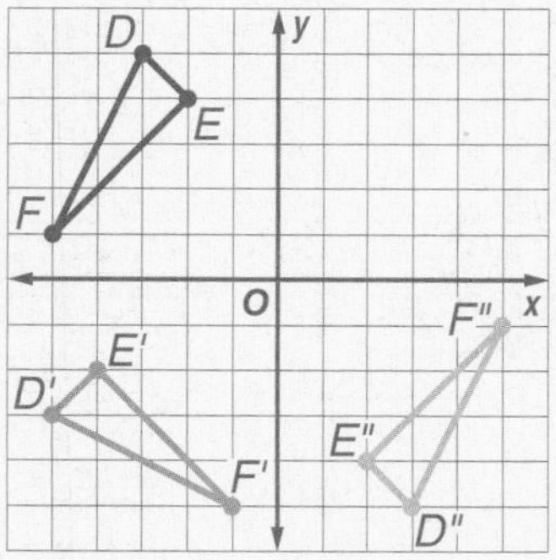

10-9b Hands-On Lab

A Follow-Up of Lesson 1

What You'll Learn
Graph rotations on a coordinate plane.

Materials

- graph paper
- protractor
- ruler

Rotations

Another type of transformation is a rotation. A **rotation** moves a figure around a central point. Another name for a rotation is a *turn*. In this lab, you will learn how to rotate a figure on a coordinate plane.

ACTIVITY *Work with a partner.*

1 **Rotate $\triangle ABC$ 90° counterclockwise about the origin. Then graph $\triangle A'B'C'$.**

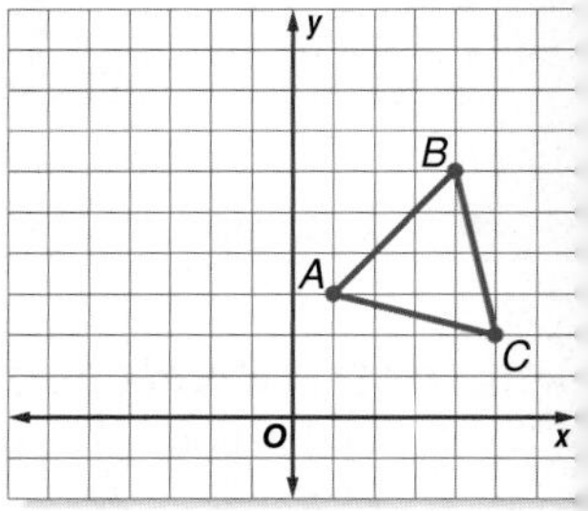

STEP 1 Draw a ray $\overrightarrow{OA}$ from the origin through point A.

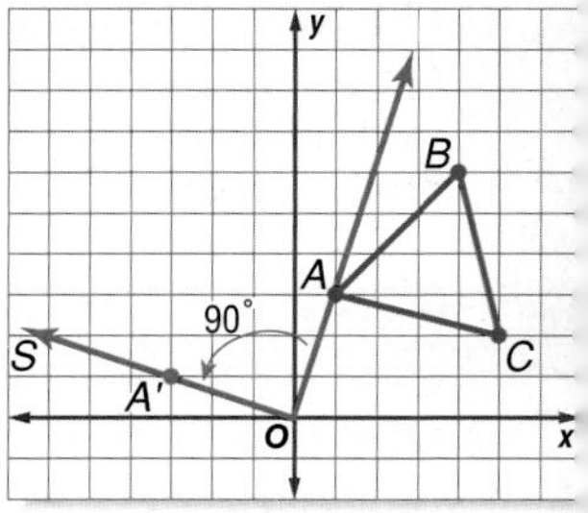

STEP 2 Draw a 90° angle from $\overrightarrow{OA}$ rotated counterclockwise. Label the ray $\overrightarrow{OS}$.

STEP 3 Measure $\overline{OA}$. Then measure the same distance from O on $\overrightarrow{OS}$. This is the location of rotated vertex A'.

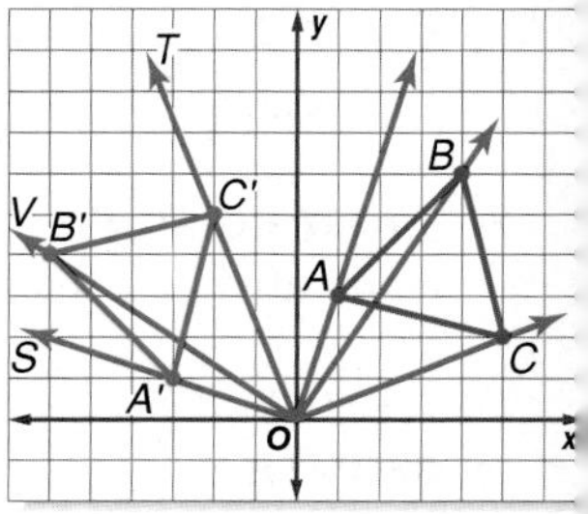

STEP 4 Rotate the other vertices in the same way by drawing rotated rays $\overrightarrow{OV}$ and $\overrightarrow{OT}$ to obtain vertices B' and C'.

STEP 5 Connect the rotated points to make $\triangle A'B'C'$.

Rotated figures are congruent because the side lengths remain the same. So, $\triangle ABC \cong \triangle A'B'C'$.

READING Math

Rotations If you rotate an image once, the vertices contain a prime symbol ('). If you rotate a rotated image, the vertices contain a double prime symbol (").

Your Turn a–b. See margin.

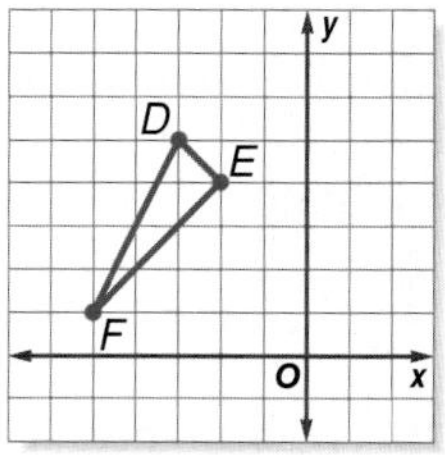

a. Rotate $\triangle DEF$ 90° counterclockwise about the origin. Then graph $\triangle D'E'F'$.

b. Rotate $\triangle D'E'F'$ from Part a above 90° counterclockwise about the origin. Then graph $\triangle D''E''F''$.

ACTIVITY *Work with a partner.*

2 Identify the transformation as a *translation, reflection,* or *rotation.* The original figure is blue and the transformation is green.

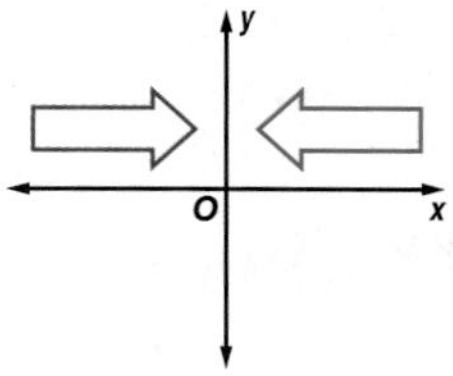

The distance from the point of the blue arrow to the line of symmetry is the same as the distance from the point of the green arrow to the line of symmetry. The y-axis is a line of symmetry. Therefore, the transformation is a reflection.

Your Turn **Identify the transformation as a *translation, reflection,* or *rotation.* The original figure is blue and the transformation is green.**

c. 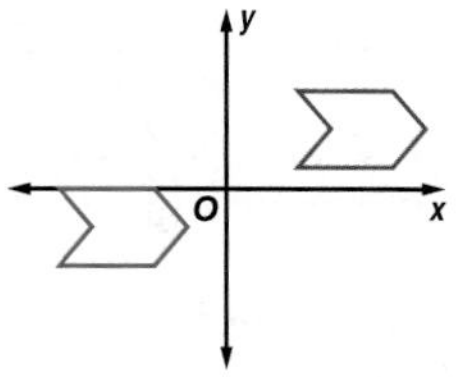

d. 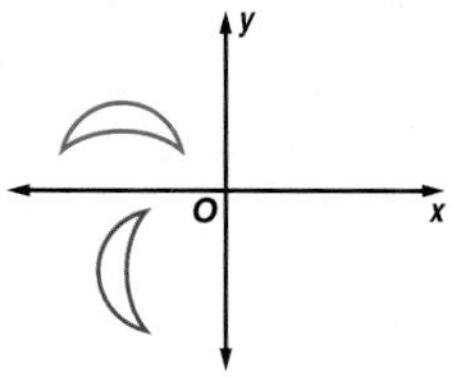

e. 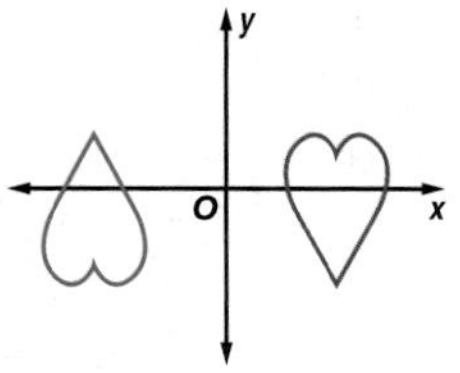

f. 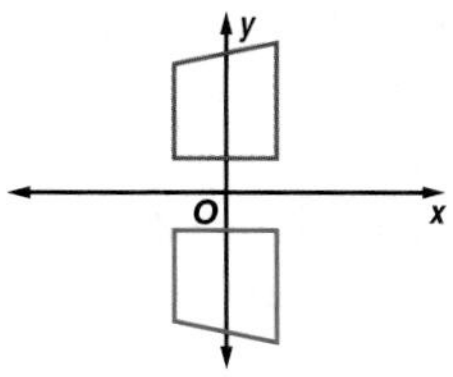

translation
rotation
rotation
eflection

Writing Math

Work with a partner.

1. Rotate $\triangle DEF$ from Your Turn a 180° counterclockwise. Compare your rotated figure with your answer from Your Turn b.
2. **Make a conjecture** about rotating figures 180°.
3. Without rotating $\triangle DEF$ 360°, **make a conjecture** about rotating figures 360°. Describe the figure after a 360° rotation.
4. A 60° *counterclockwise* rotation is the same as a −300° *clockwise* rotation. Find an equivalent rotation of a 90° counterclockwise rotation.
5. Complete the following sentence. Since you can rotate figures in clockwise or counterclockwise directions, a 270° counterclockwise rotation produces the same result as a __?__ clockwise rotation. **90°**

They are the same
age.
Sample answer:
ures that are rotated
0° are the same as
ures rotated 90°
ce.
Sample answer:
ures that are rotated
0° are the same as the
ginal figure.
−270° clockwise rotation

Activity 2

- Contrast the four types of transformations (translations, dilations, reflections, and rotations) for students.
- Many students will have difficulty visualizing whether the transformations in the **Your Turn** exercises are translations, reflections, or rotations. Have them reproduce and cut out the figures. They can then slide the figures through the different transformations until they arrive at the correct answer.

ASSESS

Observe as students solve the **Your Turn** exercises. They should:

- follow the directions from the activity to create their own rotations,
- draw all lines using a straightedge, and
- compare their results with their partner.

Discuss the **Writing Math** exercises. For Exercise 5, have students give at least one other example of equivalent rotations.

CHAPTER 10 Study Guide and Review

Vocabulary Assessment

- This alphabetical list of vocabulary terms in Chapter 10 includes a page reference where each term was introduced.
- A vocabulary review/test for Chapter 10 is available on p. 602 of the *Chapter 10 Resource Masters*.

Vocabulary PuzzleMaker

ELL The online Vocabulary PuzzleMaker improves students' mathematics vocabulary using four puzzle formats — crossword, scramble, word search using a word list, and word search using clues. Students can work on a computer screen or from a printed handout.

MindJogger Videoquizzes

ELL MindJogger Videoquizzes provide an alternative review of concepts presented in this chapter. Students work in teams in a game show format to gain points for correct answers. The questions are presented in three rounds.

Round 1 Concepts (5 questions)
Round 2 Skills (4 questions)
Round 3 Problem Solving (4 questions)

ExamView® Pro Test Bank CD-ROM

Use the networkable ExamView® Pro Test Bank CD-ROM to:

- Create **Multiple versions** of tests.
- Create **Modified** tests for **Inclusion** students with one mouse click.
- **Edit** existing questions and **Add** your own questions.
- Build tests aligned with state standards using built-in **State Curriculum Correlations.**
- Change **English** tests to **Spanish** with one mouse click and vice versa.

CHAPTER 10 Study Guide and Review

Vocabulary and Concept Check

acute angle (p. 413)	line of symmetry (p. 456)	right triangle (p. 429)
acute triangle (p. 429)	line symmetry (p. 456)	scalene triangle (p. 429)
angle (p. 413)	nonagon (p. 446)	similar figures (p. 440)
circle graph (p. 418)	obtuse angle (p. 413)	straight angle (p. 413)
complementary angles (p. 422)	obtuse triangle (p. 429)	supplementary angles (p. 422)
congruent angles (p. 422)	octagon (p. 446)	tessellation (p. 447)
congruent segments (p. 429)	parallelogram (p. 434)	transformation (p. 451)
decagon (p. 446)	pentagon (p. 446)	translation (p. 451)
degrees (p. 413)	polygon (p. 446)	trapezoid (p. 434)
equilateral triangle (p. 429)	quadrilateral (p. 434)	triangle (p. 428)
heptagon (p. 446)	reflection (p. 457)	vertex (p. 413)
hexagon (p. 446)	regular polygon (p. 446)	vertical angles (p. 422)
indirect measurement (p. 441)	rhombus (p. 434)	
isosceles triangle (p. 429)	right angle (p. 413)	

Choose the letter of the term that best matches each phrase.

1. the point where the sides of an angle meet **c**
2. an angle whose measure is less than 90° **e**
3. two angles whose measures add to 180° **f**
4. a polygon with six sides **d**
5. a triangle with no congruent sides **a**

a. scalene triangle
b. complementary angles
c. vertex
d. hexagon
e. acute angle
f. supplementary angles

Lesson-by-Lesson Exercises and Examples

10-1 Angles (pp. 413–415)

Classify each angle as *acute, obtuse, right,* or *straight*.

6. **obtuse**

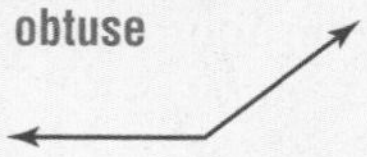

7. **right**

8. **straight**

9. **acute**

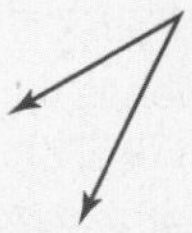

10. Draw $\angle PQR$ with a measure of 128°. **See margin.**

Example 1 **Classify the angle as *acute, obtuse, right,* or *straight*.**

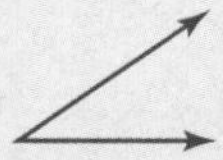

The angle is an acute angle because its measure is less than 90°.

msmath2.net/vocabulary_review

FOLDABLES™ Study Organizer

For more information about Foldables, see *Teaching Mathematics with Foldables.*

Have students look through the chapter to make sure they have included examples in their Foldables for each lesson or topic in the chapter.

Encourage students to refer to their Foldables while completing the Study Guide and Review and to use them in preparing for the Chapter Test.

462 Chapter 10 Geometry

10-2 Making Circle Graphs (pp. 418–421)

11. Make a circle graph of the data. **See margin.**

Favorite Soft Drink	Percent
Cola	36%
Diet Cola	28%
Root Beer	15%
Lemon Lime	7%
Other	14%

Example 2 Make a circle graph of the data of favorite season: 40% spring, 26% summer, 22% fall, and 12% winter.

First find the degrees for each part. Then construct the circle graph.

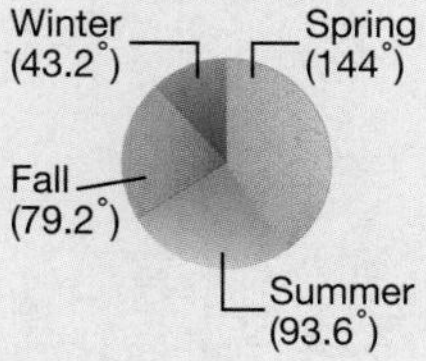

10-3 Angle Relationships (pp. 422–425)

Find the value of x in each figure.

12. **74**

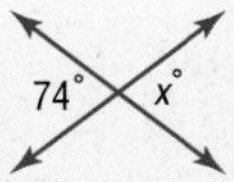

13. **79**

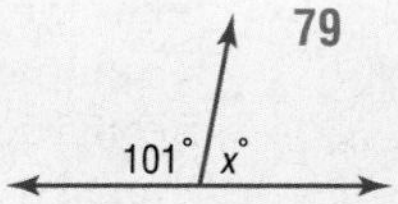

14. If $\angle Y$ and $\angle Z$ are complementary angles and $m\angle Z = 35°$, find $m\angle Y$. **55°**

Example 3 Find the value of x.

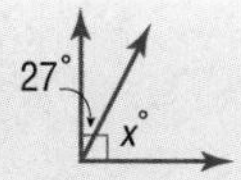

Since the angles are complementary, $x + 27 = 90$.

$x + 27 = 90$ Write the equation.

$-27 \quad -27$ Subtract 27 from each side.

$x = 63$

So, the value of x is 63.

10-4 Triangles (pp. 428–431)

Classify each triangle by its angles and by its sides. **15–16. See margin.**

15.

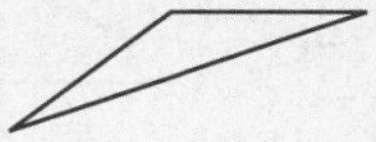

16.

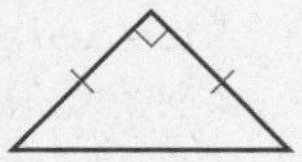

17. Find $m\angle S$ in $\triangle RST$ if $m\angle R = 28°$ and $m\angle T = 13°$. **139°**

Example 4 Classify the triangle by its angles and by its sides.

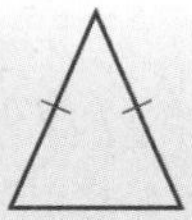

The triangle is acute since all three angles are acute. It is also isosceles because it has two congruent sides.

10-5 Quadrilaterals (pp. 434–437)

Classify the quadrilateral using the name that *best* describes it. **18–19. See margin.**

18.

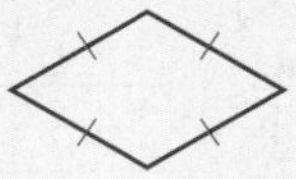

19.

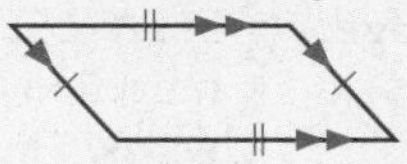

20. What type of quadrilateral may *not* have opposite sides congruent? **trapezoid**

Example 5 Classify the quadrilateral using the name that *best* describes it.

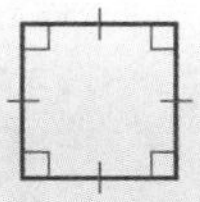

The quadrilateral is a parallelogram with 4 right angles and 4 congruent sides. It is a square.

Answers

10.

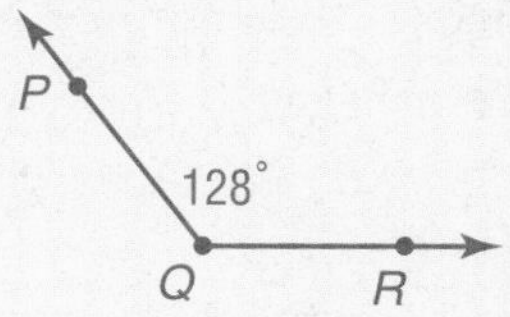

11.

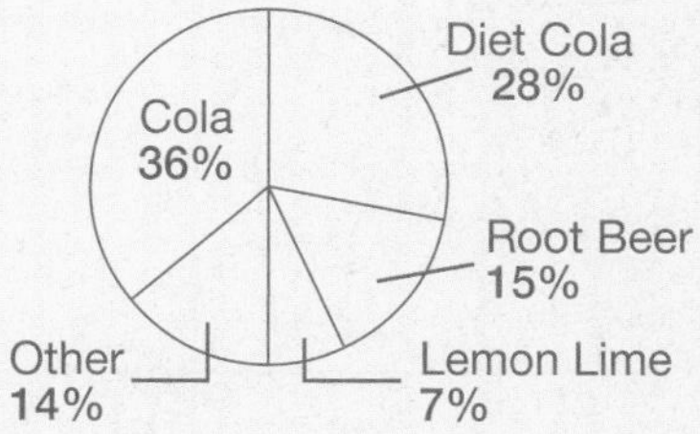

15. **obtuse, scalene**
16. **right, isosceles**
18. **rhombus**
19. **parallelogram**

Answers

23. heptagon
24. regular nonagon
26. $P'(-2, 1)$, $Q'(-8, 0)$, and $R'(-7, 9)$

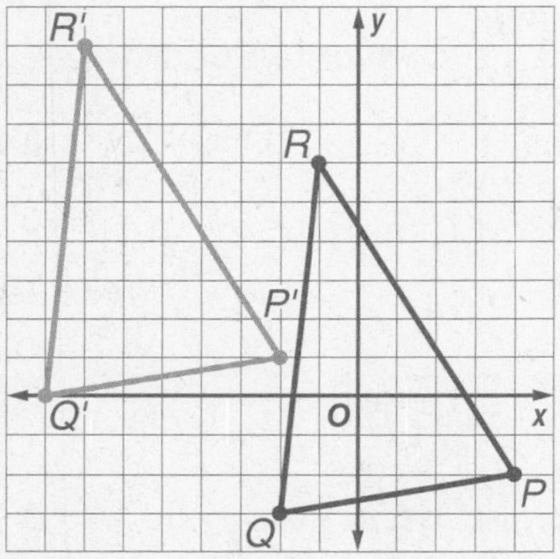

27. $P'(8, -3)$, $Q'(2, -4)$, and $R'(3, 5)$

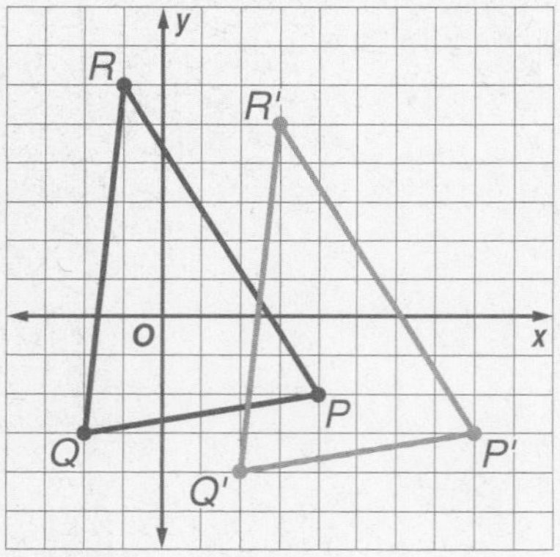

28. $P'(1, -2)$, $Q'(-5, -3)$, and $R'(-4, 6)$

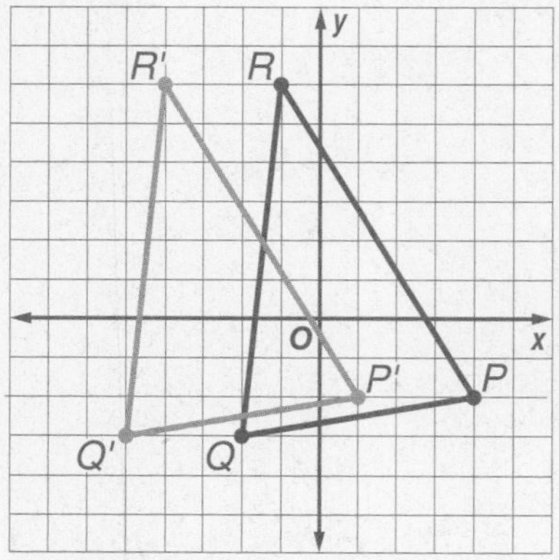

10-6 Similar Figures (pp. 440–443)

Find the value of x in each pair of similar figures.

21.

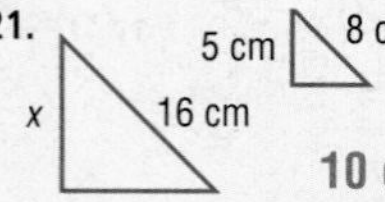

10 cm

22.

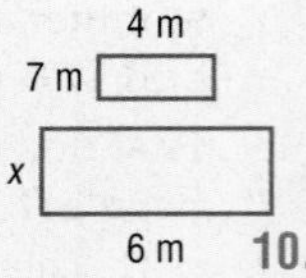

10.5 m

Example 6 **Find the value of x in the pair of similar figures.**

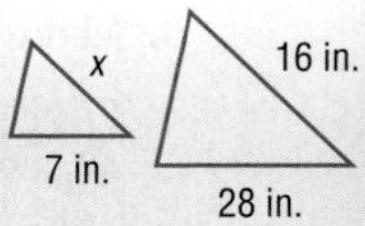

$\frac{7}{28} = \frac{x}{16}$ Write a proportion.

$28x = 112$ Multiply.

$x = 4$ Divide.

10-7 Polygons and Tessellations (pp. 446–450)

Classify each polygon below and state whether it is regular. 23–24. See margin.

23.

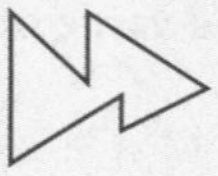

24.

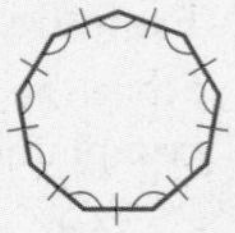

25. **ALGEBRA** Find the measure of each angle of a regular 15-gon. 156°

Example 7 **Classify the polygon and state whether it is regular.**

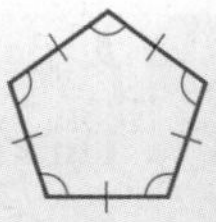

Since the polygon has 5 congruent sides and 5 congruent angles, it is a regular pentagon.

10-8 Translations (pp. 451–454) 26–28. See margin.

Triangle PQR has coordinates $P(4, -2)$, $Q(-2, -3)$, and $R(-1, 6)$. Find the coordinates of $P'Q'R'$ after each translation. Then graph each translation.

26. 6 units left, 3 units up
27. 4 units right, 1 unit down
28. 3 units left
29. 7 units down See Ch. 10 Answer Appendix.

Example 8
Find the coordinates of $\triangle G'H'I'$ after a translation of 2 units left and 4 units up.

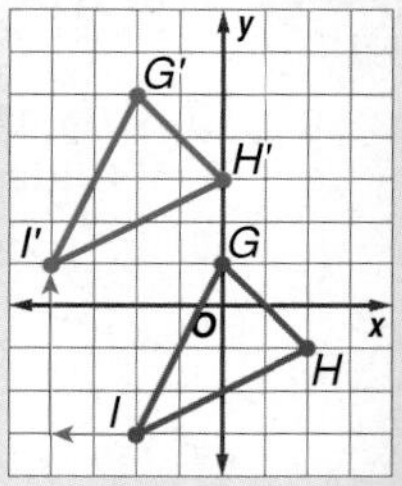

The vertices of $\triangle G'H'I'$ are $G'(-2, 5)$, $H'(0, 3)$, and $I'(-4, 1)$.

10-9 Reflections (pp. 456–459)

Find the coordinates of each figure after a reflection over the given axis. Then graph the figure and its reflected image.

30. $\triangle RST$ with coordinates $R(-1, 3)$, $S(2, 6)$, and $T(6, 1)$; x-axis
31. parallelogram $ABCD$ with coordinates $A(1, 3)$, $B(2, -1)$, $C(5, -1)$, and $D(4, 3)$; y-axis 30–31. See Ch. 10 Answer Appendix.

Example 9
Find the coordinates of $\triangle C'D'E'$ after a reflection over the y-axis. Then graph its reflected image.

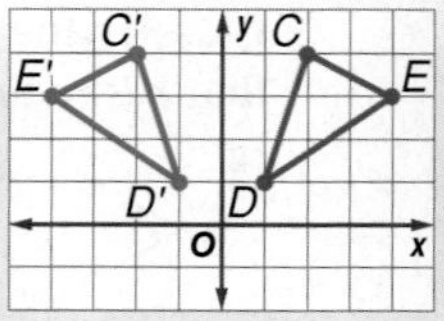

The vertices of $\triangle C'D'E'$ are $C'(-2, 4)$, $D'(-1, 1)$, and $E'(-4, 3)$.

464 Chapter 10 Geometry

CHAPTER 10 Practice Test

Vocabulary and Concepts

1. **List** the number of sides and number of angles in an octagon. **8; 8**
2. **Explain** how to translate a geometric figure. **Sample answer: Every point of the original figure is moved the same distance and in the same direction.**

Skills and Applications

Classify each angle as *acute, obtuse, right,* or *straight.*

3. 53° **acute**
4. 97° **obtuse**
5. 180° **straight**

Classify each pair of angles as *complementary, supplementary,* or *neither.*

6. 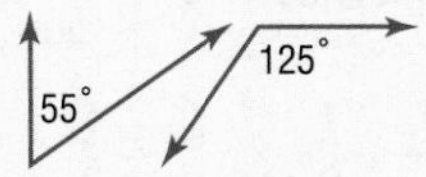

supplementary

7. 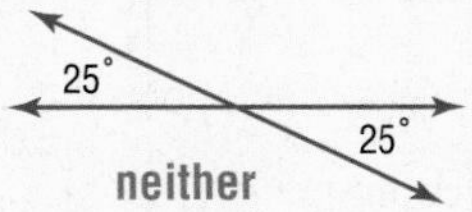

neither

8. 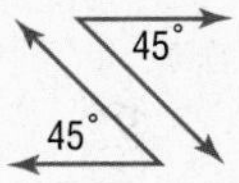

complementary

9. **FOOD** Angelo's Pizza Parlor makes square pizzas. After baking, the pizzas are cut along one diagonal into two triangles. Describe the triangles. **right, isosceles**

Find the missing measure in each quadrilateral.

10. 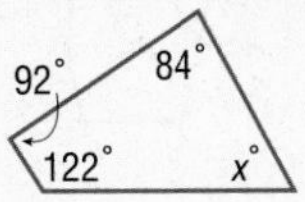

62°

11. 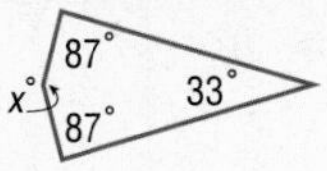

153°

12. 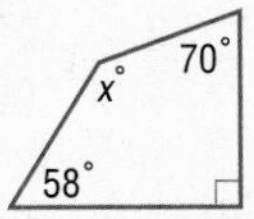

142°

13. **GEOGRAPHY** Luz is drawing a map similar to one found in the atlas. The map in the atlas is 7 inches wide and 10 inches long. If she draws the width 17.5 inches, how long should she draw the map? **25 in.**
14. Can a regular heptagon, whose angle measures total 900°, be used by itself to make a tessellation? Explain your reasoning. **See margin.**
15. Draw a figure with exactly one line of symmetry. **See students' work.**

Standardized Test Practice

16. **MULTIPLE CHOICE** The table shows the results of a student survey of favorite type of book. Choose the statement that is *false*. **D**

Favorite Type of Books	
Mystery	24
Science Fiction	8
Sports	26
Romance	30

(A) The science fiction section on a circle graph of the data at the right has an angle measure of about 33°.

(B) Romance books were the most favorite types of books.

(C) About 30% of students chose sports books as their favorite.

(D) The mystery and sports sections on a circle graph of the data at the right have supplementary angles.

msmath2.net/chapter_test

CHAPTER 10 Practice Test

Assessment Options

Vocabulary Test A vocabulary review/test for Chapter 10 can be found on p. 602 of the *Chapter 10 Resource Masters*.

Chapter Tests here are six Chapter 10 Tests and an Extended Response Assessment task available in the *Chapter 10 Resource Masters*.

Chapter 10 Tests			
Form	**Type**	**Level**	**Pages**
1	MC	basic	589–590
2A	MC	average	591–592
2B	MC	average	593–594
2C	FR	average	595–596
2D	FR	average	597–598
3	FR	advanced	599–600

MC = multiple-choice questions
FR = free-response questions

Extended Response Assessment

Extended Response tasks for Chapter 10 can be found on p. 601 of the *Chapter 10 Resource Masters*. A sample scoring rubric for these tasks appears on p. A31.

Answer

14. **No; since each angle is about 128.6°, 128.6 cannot go into 360 evenly.**

Portfolio Suggestion

Introduction Students have been studying geometry, which is one of the oldest branches of mathematics.

Ask Students to research the origins of geometry. Have them write a brief description of one contribution from Mesopotamian, Egyptian, or Greek culture. Students might also explain why the word *geometry* contains the prefix *geo-*, which means *earth.*

Chapter 10 Standardized Test Practice

These two pages contain practice questions in the various formats that can be found on the most frequently given standardized tests.

A student recording sheet for these two pages can be found on p. A1 of the *Chapter 10 Resource Masters*.

Standardized Test Practice Student Recording Sheet, p. A1

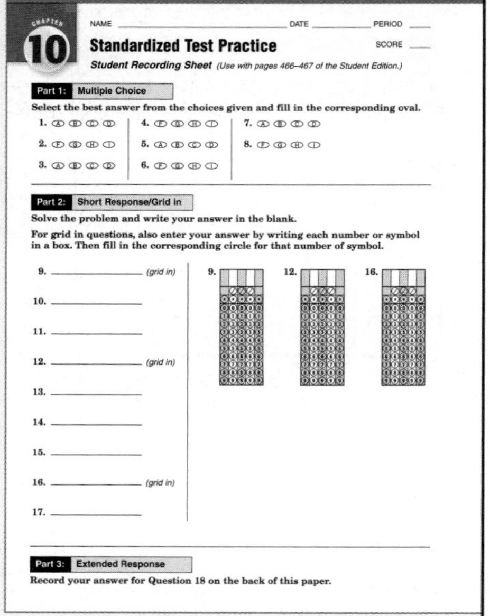
NAME ______ DATE ______ PERIOD ______

Chapter 10 Standardized Test Practice SCORE ______

Student Recording Sheet (Use with pages 466–467 of the Student Edition.)

Part 1: Multiple Choice

Select the best answer from the choices given and fill in the corresponding oval.

1. A B C D 2. F G H I 3. A B C D 4. F G H I 5. A B C D 6. F G H I 7. A B C D 8. F G H I

Part 2: Short Response/Grid in

Solve the problem and write your answer in the blank.

For grid in questions, also enter your answer by writing each number or symbol in a box. Then fill in the corresponding circle for that number of symbol.

9. ______ (grid in) 10. ______ 11. ______ 12. ______ (grid in) 13. ______ 14. ______ 15. ______ 16. ______ (grid in) 17. ______

Part 3: Extended Response

Record your answer for Question 18 on the back of this paper.

Additional Practice

See pp. 607–608 in the *Chapter 10 Resource Masters* for additional standardized test practice.

Chapter 10 Standardized Test Practice

PART 1 Multiple Choice

Record your answers on the answer sheet provided by your teacher or on a sheet of paper.

1. Which inequality is false? (Prerequisite Skill, p. 556) **D**
   - (A) $0.0059 > 0.0005$
   - (B) $6.1530 < 6.1532$
   - (C) $89.13 > 89.10$
   - (D) $5.06 < 5.006$

2. On Spring Street, there is an apartment building that houses four families, each with four members. One of the families moves away. Express this change in residents as an integer. (Lesson 3-1) **H**
   - (F) 12
   - (G) 4
   - (H) -4
   - (I) -1

3. Find the algebraic expression for the following statement. *Blanca has five more than three times as many CDs as Garrett.* (Lesson 4-1) **D**
   - (A) $B = \frac{3g}{5}$
   - (B) $B = 3g - 5$
   - (C) $B = 3(g + 5)$
   - (D) $B = 5 + 3g$

4. Which is the equation for the graphed line? (Lesson 4-7) **F**

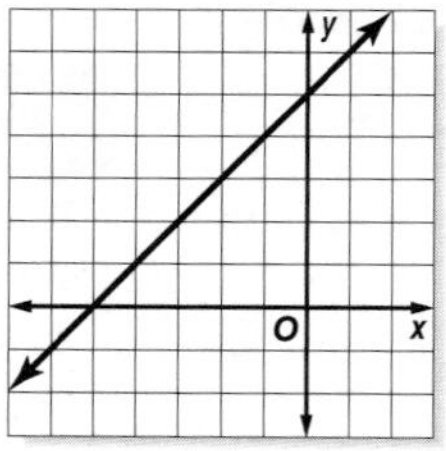

   - (F) $y = x + 5$
   - (G) $y = x - 5$
   - (H) $y = 5x$
   - (I) $y = 5x + 5$

5. The tree diagram shows the types of notebooks Esther can buy for school. If Esther selects one at random, what are the chances that she will select an orange notebook without lines? (Lesson 9-2) **C**

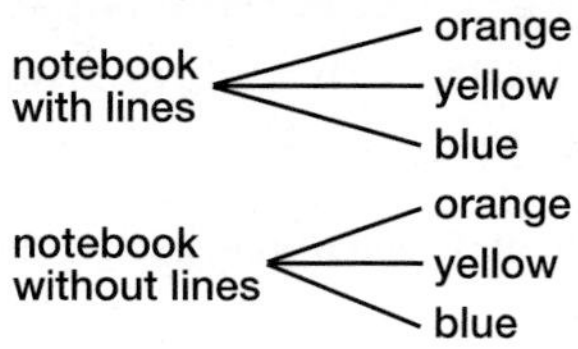

   - (A) $\frac{1}{12}$
   - (B) $\frac{1}{8}$
   - (C) $\frac{1}{6}$
   - (D) $\frac{1}{5}$

6. Which two of the figures below appear to be similar? (Lesson 10-6) **F**

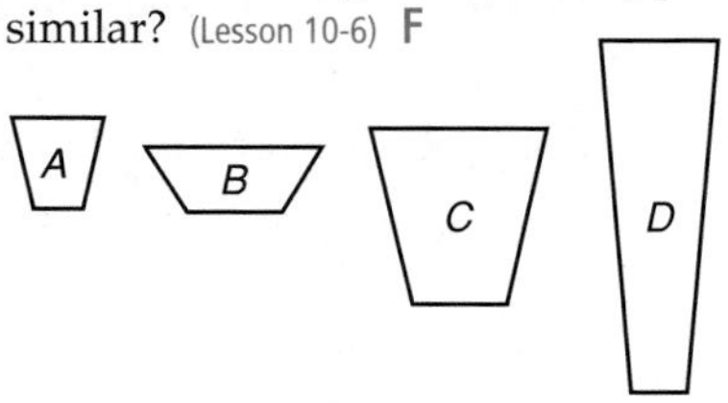

   - (F) figure A and figure C
   - (G) figure B and figure C
   - (H) figure A and figure D
   - (I) figure A and figure B

7. Kendrick chose a frame for a painting that looks like the polygon. Is this a regular polygon? (Lesson 10-7) **B**

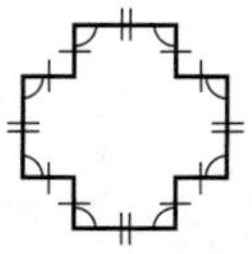

   - (A) Yes, because the angles are congruent.
   - (B) No, because the sides are not congruent.
   - (C) Yes, because the polygon is made up of line segments.
   - (D) Yes, because the angles and sides are congruent.

8. Which letter has two lines of symmetry? (Lesson 10-9) **I**
   - (F) B
   - (G) E
   - (H) L
   - (I) X

466 **Chapter 10** Geometry

The Princeton Review ***Log On*** **for Test Practice** The Princeton Review offers additional test-taking tips and practice problems at their web site. Visit princetonreview.com or review.com

ExamView® Pro Test Bank CD-ROM

Special banks of standardized test questions similar to those on the TIMSS and NAEP grade 8 tests can be found on this CD-ROM.

Aligned and verified by The Princeton Review

RT 2 Short Response/Grid In

cord your answers on the answer sheet ovided by your teacher or on a sheet of per.

According to the box-and-whisker plot, what is the interquartile range? (Lesson 2-6) **7**

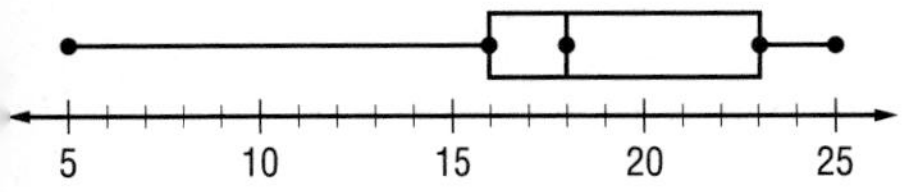

The table shows the values of p and q, where the values of p and q form a proportion. What are the values of Y and Z? (Lesson 7-3) **28, 36**

p	4	16	Z
q	7	Y	63

The veterinary hospital treated 162 animals last week. It treated 204 animals this week. What was the percent of increase? (Lesson 8-4) **about 26%**

Rebecca can choose seven different classes from five different class periods offered during the school day. How many possible ways can Rebecca arrange her class schedule? (Lesson 9-3) **35**

How many degrees do the hands of a clock make at 9:00? (Lesson 10-1) **90°**

How long should the missing sides of the triangle be to make it isosceles with a perimeter of 15 centimeters? (Lesson 10-4) **6 cm each**

Find the sum of the angle measures of a hexagon. (Lesson 10-7) **720°**

How many lines of symmetry does the figure at the right have? (Lesson 10-9) **10**

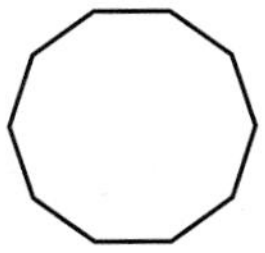

TEST-TAKING TIP

Question 17 Most standardized tests allow you to write in the test booklet. Use it to work out problems. For example, plot a figure and its translation on the coordinate graph given with the question in the booklet. Doing so will help keep you from making careless errors.

17. Triangle ABC with vertices $A(3, -7)$, $B(2, -3)$, and $C(8, -3)$ was translated to $A'B'C'$ with vertices $A'(-3, -3)$, $B'(-4, 1)$, and $C'(2, 1)$. Represent the translation as an ordered pair. (Lesson 10-8) **(−6, 4)**

PART 3 Extended Response

Record your answers on a sheet of paper. Show your work.

18. Edmundo plotted polygon $JKLM$ on the coordinate plane below.

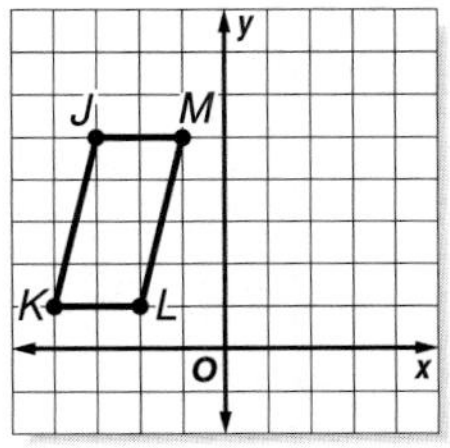

f. $J'(-3, -5)$, $K'(-4, -1)$, $L'(-2, -1)$, and $M'(-1, -5)$

a. Classify $\angle J$. (Lesson 10-1) **obtuse**

b. Classify $\angle M$. (Lesson 10-1) **acute**

c. Classify polygon $JKLM$ using the name that *best* describes it. Be as specific as possible. (Lesson 10-5) **parallelogram**

d. Can polygon $JKLM$ be used by itself to make a tessellation? Explain your reasoning. (Lesson 10-7) **d–e. See margin.**

e. If polygon $JKLM$ is translated 2 units right and 5 units down, what are the coordinates of the new figure? (Lesson 10-8)

f. If polygon $JKLM$ is reflected over the x-axis, what are the coordinates of the new figure? (Lesson 10-9)

msmath2.net/standardized_test

Answers

18d. Yes; the sum of the angles where the vertices meet is 360°.

18e. $J'(-1, 0)$, $K'(-2, -4)$, $L'(0, -4)$, and $M'(1, 0)$

Standardized Test Practice

Rubrics The following rubrics are sample scoring devices for short response and extended response questions.

Short Response

Points	Description
2	The student demonstrates a **thorough understanding** of the mathematics of the task. The response may contain minor flaws that do not detract from the demonstration of a thorough understanding.
1	The student has provided a response that is only **partially correct**.
0	The student has provided a **completely incorrect** solution or no response at all.

Extended Response

Points	Description
4	The student demonstrates a **thorough understanding** of the mathematics of the task. The response may contain minor flaws that do not detract from the demonstration of a thorough understanding.
3	The student demonstrates an **understanding** of the mathematics of the task. The response is essentially correct and demonstrates an essential but less than thorough understanding of the mathematics.
2	The student demonstrates only a **partial understanding** of the mathematics of the task. Although the student may have used the correct approach to a solution or may have provided a correct solution, the work lacks an essential understanding of the underlying mathematical concepts.
1	The student demonstrates a **very limited understanding** of the mathematics of the task. The response is incomplete and exhibits many flaws.
0	The student provides a **completely incorrect** solution or no response at all.

A more detailed 4-point rubric is provided on page A2 of the *Chapter 10 Resource Masters*.

Page 415, Lesson 10-1

12.

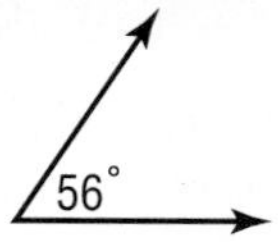

13.

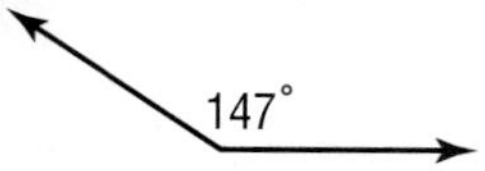

14.

15.

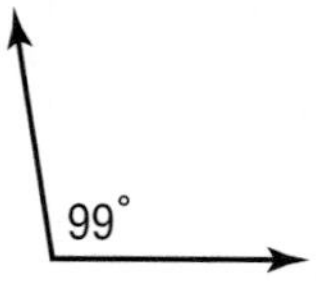

16.

17.

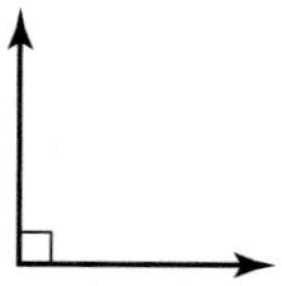

18. $\angle MNQ$ or $\angle QNM$ and $\angle MNP$ or $\angle PNM$

21. Sample answer: acute angles: between their legs and the front of their skis; obtuse angles: between their legs and the back of their skis

Pages 420–421, Lesson 10-2

7. **NYC Commuters**

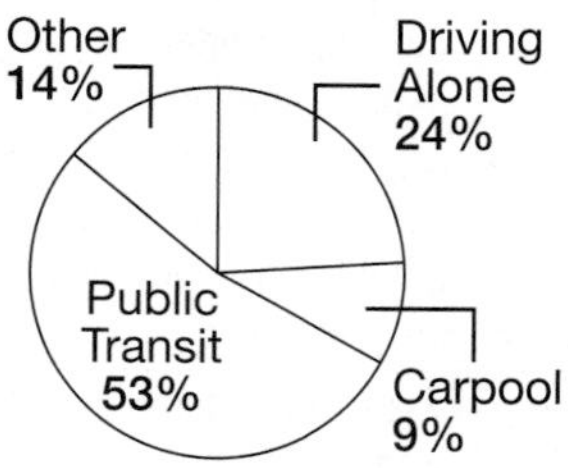

8. **LA Commuters**

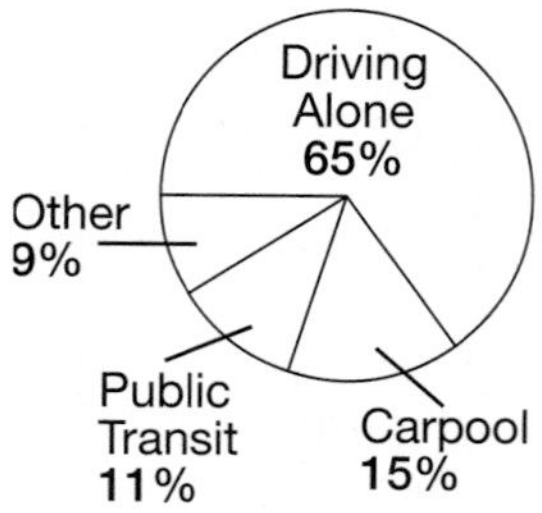

9. Sample answer: A greater percent of commuters in Los Angeles drive alone, while in New York, a greater percent use public transportation. A greater percent of people carpool in Los Angeles than in New York.

10. **Endangered Species in the U.S.**

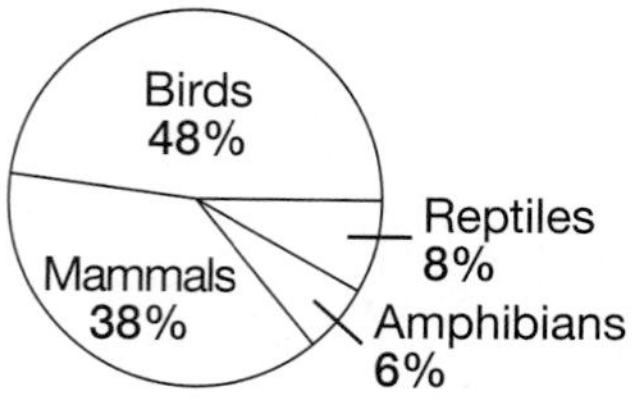

11. **U.S. Regions**

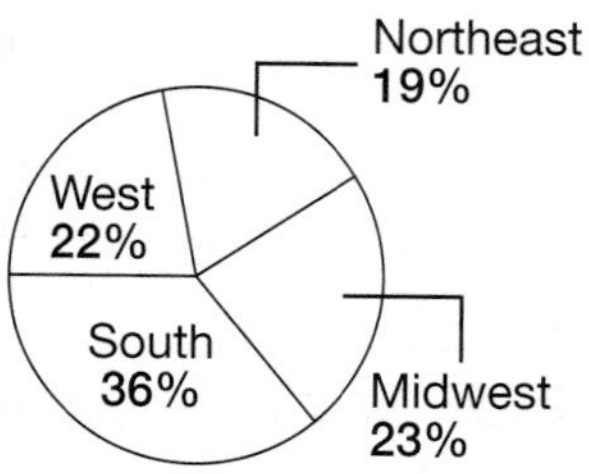

Page 452, Lesson 10-8 (Your Turn)

a. $T'(9, -7)$, $U'(1, -4)$, and $V'(2, -2)$

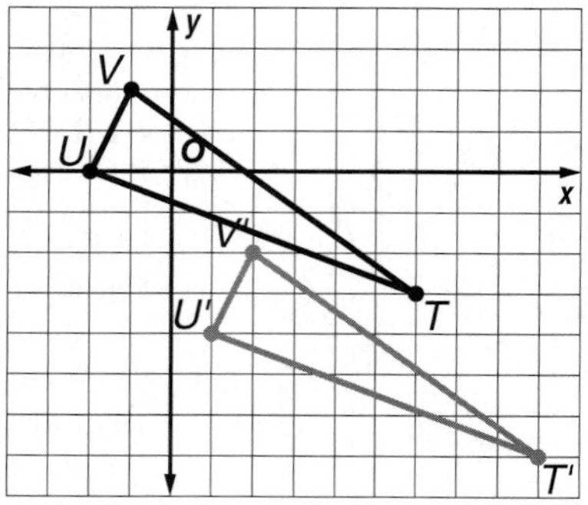

b. $T'(5, 3)$, $U'(-3, 6)$, and $V'(-2, 8)$

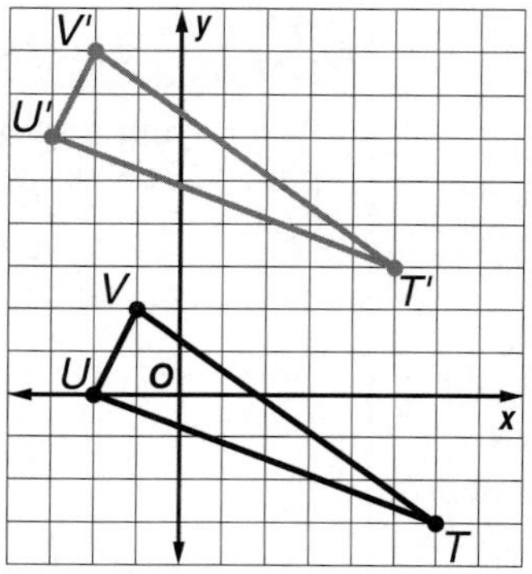

Page 453, Lesson 10-8

4. $D'(5, -5)$, $E'(2, -7)$, $F'(6, -1)$, $G'(10, -8)$

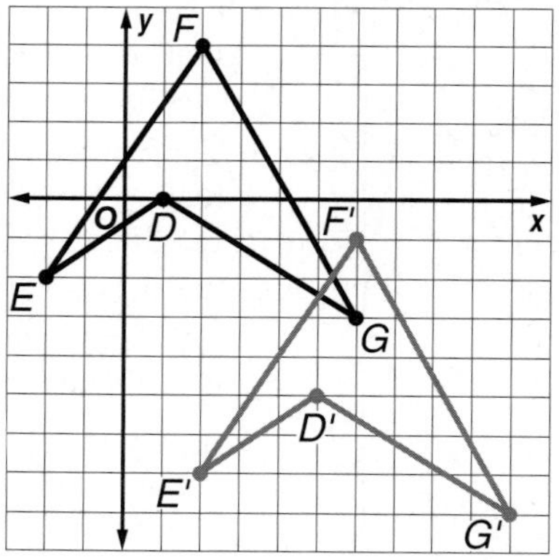

5. $D'(7, 0)$, $E'(4, -2)$, $F'(8, 4)$, $G'(12, -3)$

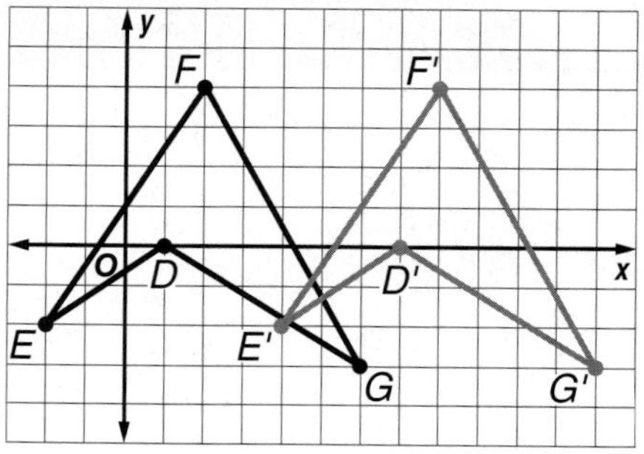

6.

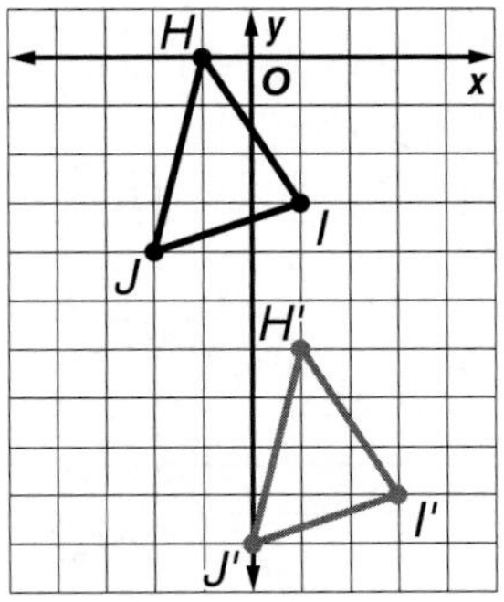

7.

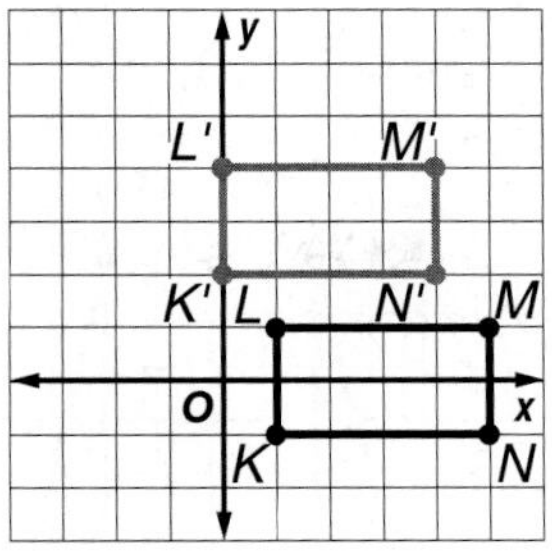

8. $P'(6, 5)$, $Q'(11, 3)$, and $R'(3, 11)$

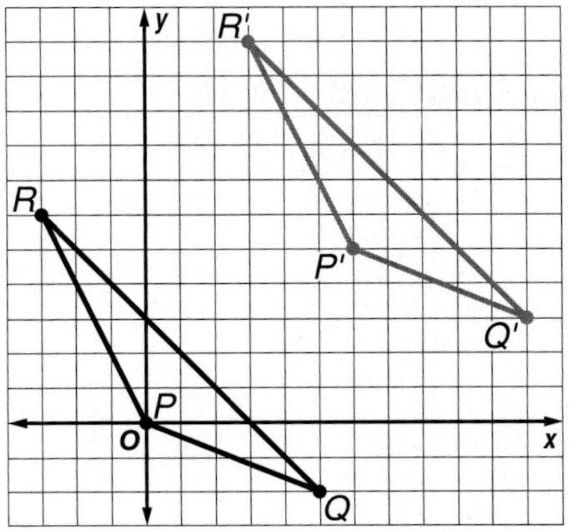

9. $P'(-8, -1)$, $Q'(-3, -3)$, *and* $R'(-11, 5)$

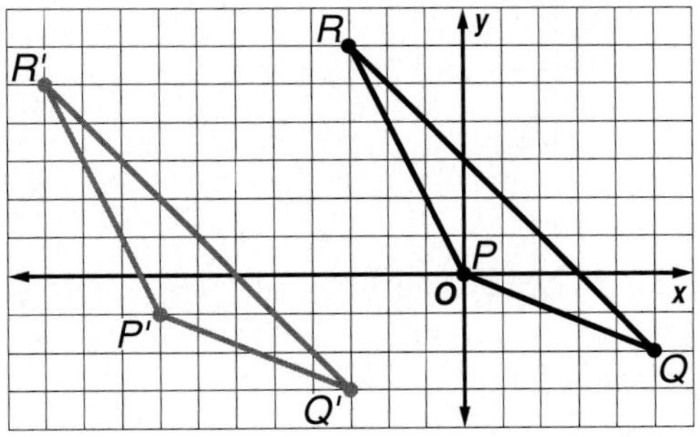

10. $P'(-3, 0)$, $Q'(2, -2)$, and $R'(-6, 6)$

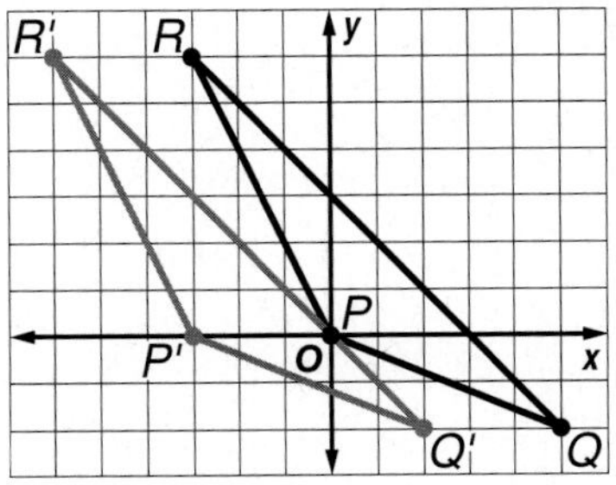

11. $P'(0, -9)$, $Q'(5, -11)$, and $R'(-3, -3)$

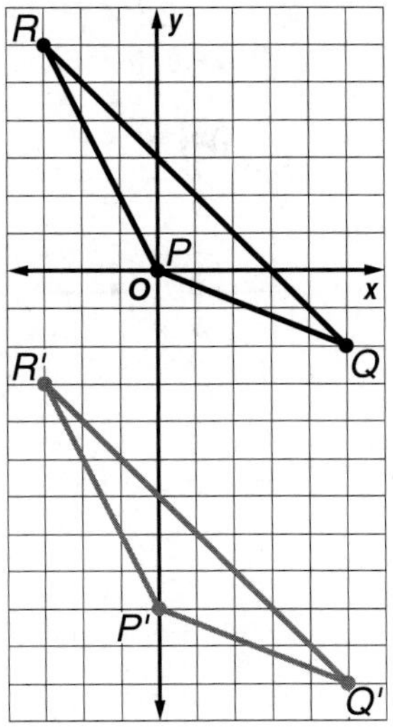

Page 454, Lesson 10-8

17.

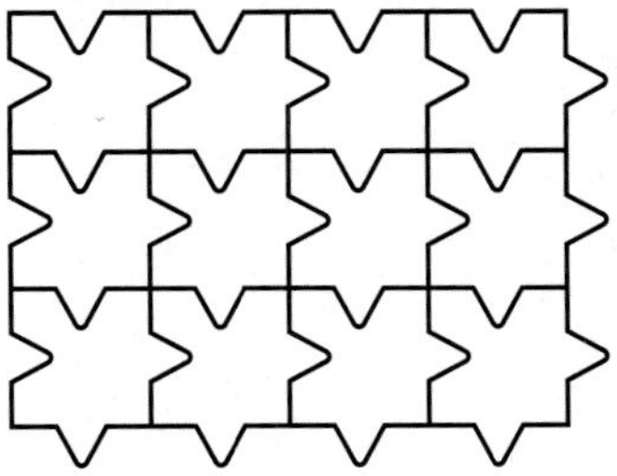

Page 455, Lesson 10-8b (Spreadsheet Investigation)

2.

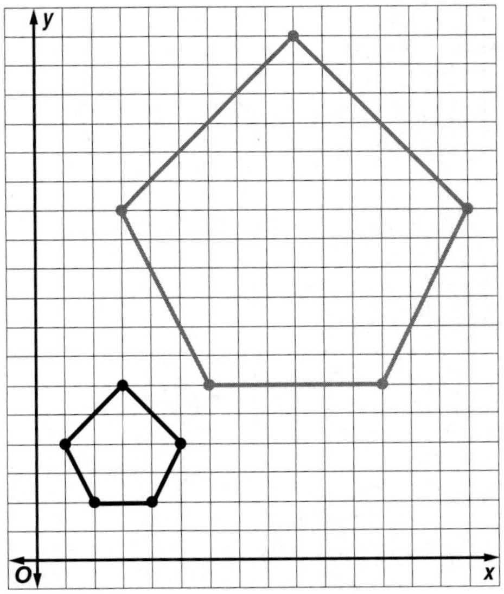

Page 458, Lesson 10-9

5. $A'(5, -8)$, $B'(1, -2)$, and $C'(6, -4)$

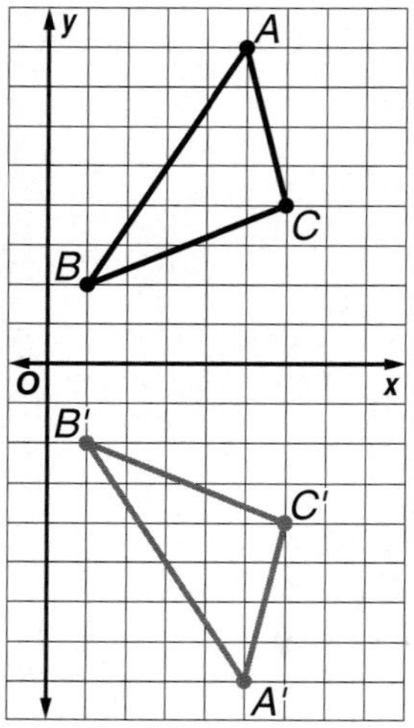

6. $W'(4, -2)$, $X'(4, 3)$, $Y'(2, 4)$, and $Z'(2, -1)$

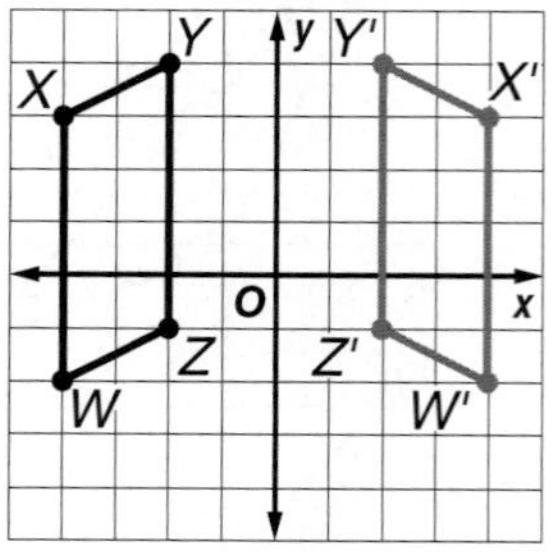

11. $D'(-3, -6)$, $E'(-2, 3)$, $F'(2, -2)$, and $G'(4, -9)$

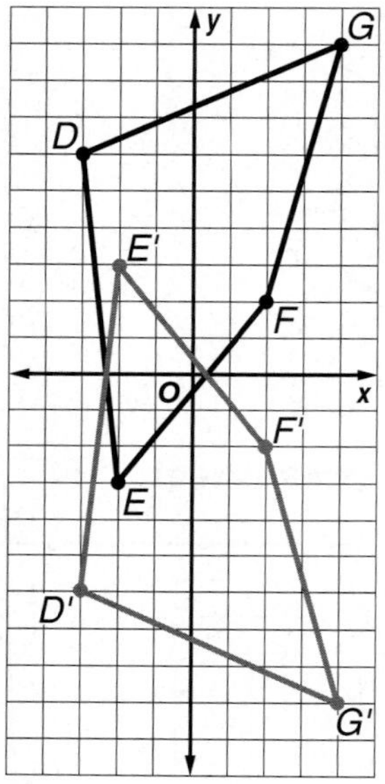

12. $T'(-6, -1)$, $U'(-2, 3)$, and $V'(5, 4)$

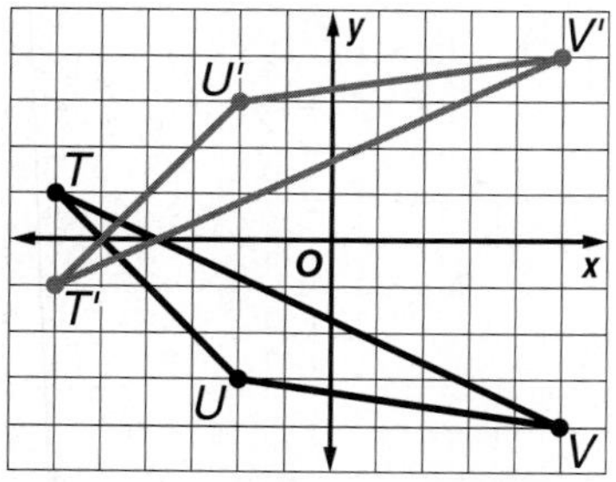

13. $Q'(-2, -5)$, $R'(-4, -5)$, and $S'(-2, 3)$

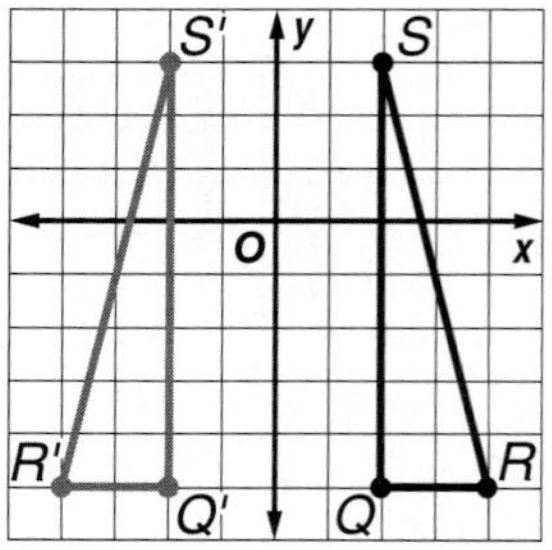

14. $H'(1, 3)$, $I'(1, -1)$, $J'(-2, -2)$, and $K'(-2, 2)$

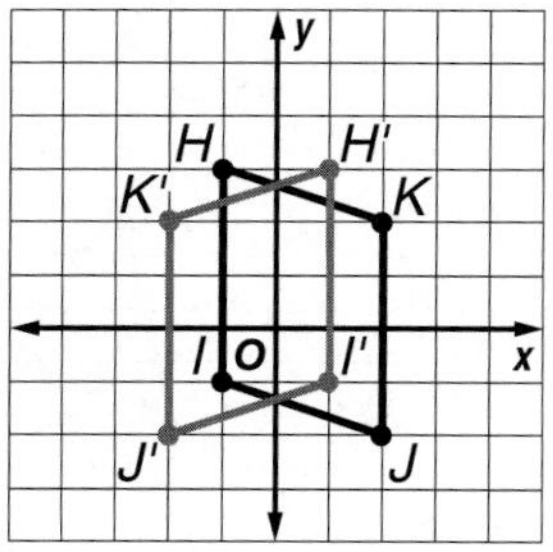

Page 464, Study Guide and Review

29. $P'(4, -9)$, $Q'(-2, -10)$, and $R'(-1, -1)$

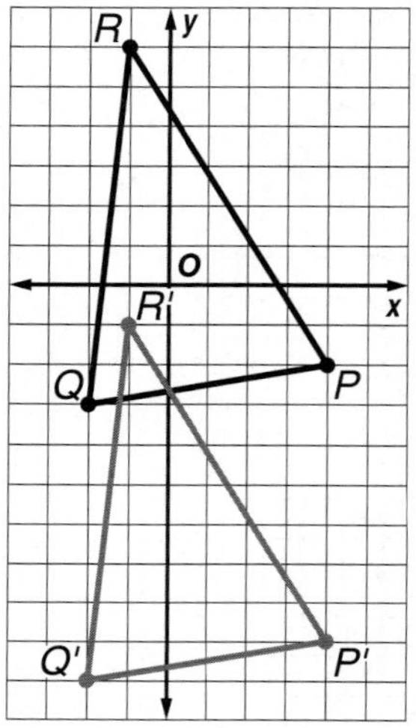

30. $R'(-1, -3)$, $S'(2, -6)$, and $T'(6, -1)$

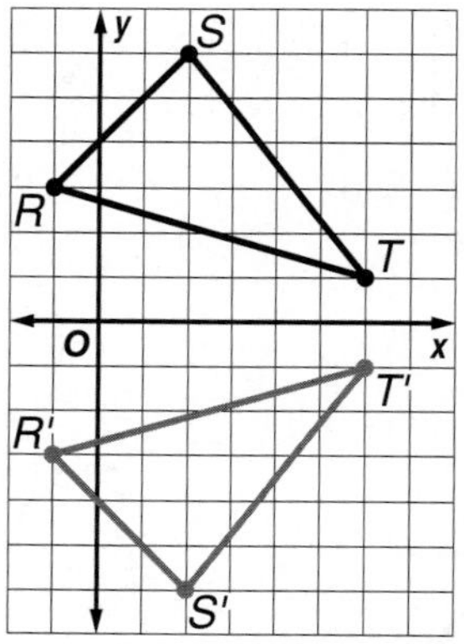

31. $A'(-1, 3)$, $B'(-2, -1)$, $C'(-5, -1)$, $D'(-4, 3)$

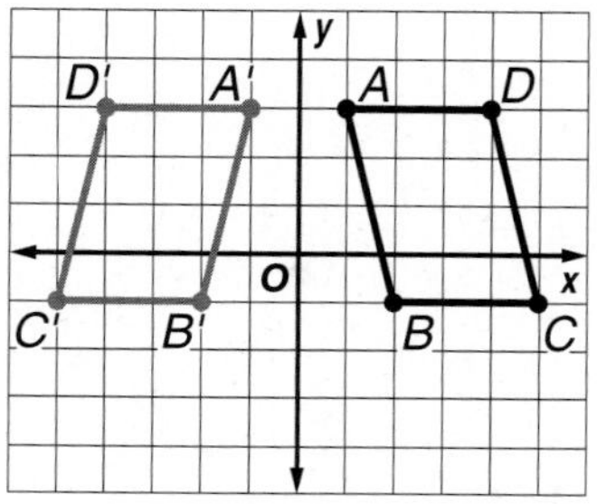

NOTES

CHAPTER

11 Chapter Overview

Geometry: Measuring Two-Dimensional Figures

	LESSON OBJECTIVES	NCTM Standards	NAEP 2005	State/Local Objectives
11-1	**Squares and Square Roots** *(pp. 470–473)* • Find squares of numbers and square roots of perfect squares.	1, 2, 6, 7, 9, 10	N3f, A4e, M1h, A2c, A2g	
11-2	**Estimating Square Roots** *(pp. 475–477)* • Estimate square roots.	1, 2, 6, 7, 9, 10	N2d, N3f	
11-3	**The Pythagorean Theorem** *(pp. 478–482)* **11-3a Hands-On Lab:** Find the relationship among the sides of a right triangle. • Find length using the Pythagorean Theorem.	1, 2, 3, 6, 7, 9, 10	G3d, N3a, N3f, A4e, N1b, M1h	
11-4	**Area of Parallelograms** *(pp. 483–485)* • Find the areas of parallelograms.	1, 2, 3, 4, 6, 7, 9, 10	G2d, N3a, M1h, A4e, M2a	
11-5	**Area of Triangles and Trapezoids** *(pp. 488–492)* **11-5a Hands-On Lab:** Find the areas of triangles and trapezoids using models. •Find the areas of triangles and trapezoids.	1, 2, 3, 4, 6, 7, 9, 10	N3a, M1h, A4e, G1d, M2a, G2d	
11-6	**Area of Circles** *(pp. 493–495)* • Find the areas of circles.	1, 2, 3, 4, 6, 7, 9, 10	G2d, M1h, A4e, N3a, M2a	
11-7	**Area of Complex Figures** *(pp. 496–500)* **11-7a Problem-Solving Strategy:** Solve problems by solving a simpler problem. • Find the areas of complex figures.	1, 2, 3, 4, 6, 7, 9, 10	M1h, A4e, N3a, M2a	
11-8	**Area Models and Probability** *(pp. 501–503)* • Find probability using area models.	1, 2, 3, 4, 6, 7, 9, 10	M1h, D4a, D4b, D4g	

Key to NCTM Standards:

1=Number & Operations, 2=Algebra, 3=Geometry, 4=Measurement, 5=Data Analysis & Probability, 6=Problem Solving, 7=Reasoning & Proof, 8=Communication, 9=Connections, 10=Representation

NAEP 2005 Standards: See math.glencoe.com

Pacing suggestions for the entire year can be found on page T20.

Chapter Resource Manager

Time-Saving Tool
TeacherWorks™
All-In-One Planner and Resource Center
See page T5.

Chapter 11 Resource Masters

FAST FILE

	Study Guide and Intervention	Practice (Skills and Word Problems)	Reading to Learn Mathematics	Enrichment	Assessment	Applications*	Parent and Student Study Guide Workbook	5-Minute Check Transparencies	Key Teaching Transparencies	MathPASS: Tutorial Plus (lessons)	Teaching Mathematics with Manipulatives	Optional Materials
	609	610–611	612	613		CS 51	11-1	11-1	41			algebra tiles
	614	615–616	617	618	663	CS 52	11-2	11-2	42	3, 33		
	619	620–621	622	623		CS 53	11-3	11-3	43	47	107	ruler, scissors
	624	625–626	627	628	663, 665	CS 54	11-4	11-4			108–109	
	629	630–631	632	633		CM 33–34	11-5	11-5	44		110	straightedge, scissors, tape
	634	635–636	637	638	664	RC 25	11-6	11-6		49, 50	111	large paper plate, scissors
	639	640–641	642	643			11-7	11-7				
	644	645–646	647	648	649–662, 664, 666–668		11-8	11-8			112–113	

*** Key to Abbreviations:** *CS = Calculator and Spreadsheet Masters, RC = Real-Life Career Masters, CM = Classroom Math Games (online), SM = Science and Math Lab Manual (online)*

Interactive Chalkboard

PowerPoint® Presentations

This CD-ROM is a customizable Microsoft® PowerPoint® presentation that includes:

- In-Class Examples
- Links to Online Study Tools
- 5-Minute Check Transparencies
- Bellringers

Math Skills Maintenance

Use the Math Skills Maintenance Masters to keep your students' basic math skills fresh. Students can review skills one at a time or in a combined format. The Masters are also available online at msmath2.net/maintenance.

Chapter 11 Mathematical Content and Teaching Strategies

Continuity of Instruction

Building on Prior Knowledge

Students have learned about exponents in previous chapters of this book. They understand, for example, that $x^2 = x \cdot x$. Students have also learned how to solve simple linear equations, such as $6x + 3 = 27$. The concept of ratios has been presented to students as a way to express the probability that an event will occur. For example, if a number cube is rolled, what is the probability that the number is even?

Learning in This Chapter

Students begin the chapter by exploring perfect squares and square roots. For example, 36 is a perfect square because its square root is an integer. Students next solve equations with a squared variable, such as $x^2 + 4 = 20$. They apply this concept to right triangles using the Pythagorean Theorem, $a^2 + b^2 = c^2$. A major focus of the chapter is calculating areas of two-dimensional figures. Students combine this with their knowledge of ratios to determine probabilities in area models. An example is the probability that a dart will strike the center circle of a dartboard.

Making Future Connections

In Chapter 12 of this book, students will learn to calculate the volumes of three-dimensional figures, such as prisms, cones, and spheres. They will extend these skills by calculating the surface areas of these solid objects. Students will apply what they have learned about ratios and probability to explore similar polygons and similar triangles. For example, if two polygons have congruent angles and proportional sides, then the lengths of corresponding sides are related by ratios.

Mathematical Background

11-1 Squares and Square Roots

Calculating squares and square roots is an important skill in many fields, such as science, engineering, and finance. Squared numbers and square roots are used in many mathematical formulas.

Team Teaching

Ask other teachers on your team to think of ways to encourage students to apply what they have learned about square numbers and square roots to science.

Physical Science: square numbers are used when calculating the velocity or acceleration of objects

Earth Science and Life Science: square numbers are used when calculating land area or habitat area

11-2 Estimating Square Roots

With the availability of calculators, estimating square roots is simple. Students learn this skill both with and without calculators. Estimating square roots without using a calculator is accomplished by relating them to perfect squares.

11-3 The Pythagorean Theorem

The Pythagorean Theorem describes the relationship between the legs and hypotenuse of a right triangle. This theorem is probably the most useful mathematical relationship for real-world problems. The Pythagorean Theorem is used in many fields, such as architecture, surveying, and construction. Since the Pythagorean Theorem can be used to find the missing measure of a right triangle, in future classes students will use it as a step in finding the area of a right triangle or of a figure that includes a right triangle.

In-Class Speaker

Ask a building contractor to speak to the class about how the Pythagorean Theorem and formulas for area of geometric figures are used in construction. Contractors need to use these formulas to estimate the cost of materials in a building. Ask the contractor to bring a set of construction plans so students can see the figures in the floor plan of a building.

11-4 Area of Parallelograms

A parallelogram is a four-sided figure with opposite sides parallel and equal in length. Finding the area of a parallelogram is a helpful skill when finding the areas of other, more complex figures.

11-5 Area of Triangles and Trapezoids

The formula for determining the area of a triangle can be found by dividing a parallelogram into two equal triangles. The area formula for a trapezoid can also be found from a parallelogram. These formulas can be applied to determine areas in everyday situations, such as the area of a garden plot or a geographic region.

Barbara Van Den Berg
Woodrow Wilson Middle School
Clifton, NJ

"When moving from the area of a triangle to discovering the area of a trapezoid, have students fold a triangle in half vertically and cut on the fold to form a smaller triangle and a trapezoid. To find the area of the trapezoid, use the formula for the area of a triangle, substituting $(b_1 + b_2)$ for b, since the trapezoid has two bases. This helps students remember the formula for the area of a trapezoid because they know how it developed."

Barbara C. Van Den Berg

11-6 Area of Circles

Determining the area of a circle is sometimes difficult for students because, unlike other areas, it depends on the number π. The formula for the area of a circle is $A = \pi r^2$, where r is the length of the circle's radius.

11-7 Area of Complex Figures

Complex figures can be seen as combinations of simple shapes, such as circles, trapezoids, and triangles. The area of a complex figure is the sum of the areas of these simpler shapes. Calculating the area of a complex figure is especially important in architecture.

11-8 Area Models and Probability

Students have previously learned to find the probability that a certain event occurs. Probability can be related to area models. For example, given a geometric figure such as a circle, the probability of a randomly-dropped counter falling onto a certain portion of that figure can be determined

McGraw-Hill Professional Development

Mathematics, YES! is a self-paced, web-based professional development program. Units are based on major mathematical concepts and are standards-based. Actual classroom videos provide a springboard for discussion and activities. Unit appropriate for this chapter:

- **Unit 8, Geometry and Measurement of Plane Figures**

Key Concepts

Additional mathematical information and teaching notes are available at

msmath2.com/key_concepts.

The section appropriate for the chapter is:

- **The Pythagorean Theorem**

CHAPTER 11 DAILY INTERVENTION and Assessment

	Type	Student Edition	Teacher Resources	Technology/Internet
INTERVENTION	Ongoing	Prerequisite Skills, pp. 469, 473, 477, 482, 485, 500 Mid-Chapter Practice Test, p. 486	5-Minute Check Transparencies Quizzes 1–4, *CRM*, pp. 663–664 Mid-Chapter Test, *CRM*, p. 665 Study Guide and Intervention, *CRM*, pp. 609, 614, 619, 624, 629, 634, 639, 644	MathPASS: Tutorial Plus msmath2.net/self_check_quiz msmath2.net/extra_examples
	Mixed Review	pp. 473, 477, 482, 485, 492, 495, 500, 503	Cumulative Review, *CRM*, p. 666	
	Error Analysis	Find the Error, pp. 481, 494	Find the Error, *TWE*, pp. 481, 494 Unlocking Misconceptions, *TWE*, pp. 485, 501 Tips for New Teachers, *TWE*, p. 480	
	Standardized Test Practice	pp. 473, 477, 482, 485, 486, 492, 495, 497, 499, 500, 503, 507, 508–509	Standardized Test Practice, *CRM*, pp. 667–668	Standardized Test Practice CD-ROM msmath2.net/standardized_test
ASSESSMENT	Open-Ended Assessment	Writing Math, pp. 472, 476, 478, 481, 484, 488, 491, 494, 499, 502 Open-Ended, pp. 472, 476, 481, 484, 491, 494, 499, 502 Standardized Test, p. 509	Modeling: *TWE*, pp. 473, 495, 503 Speaking: *TWE*, pp. 485, 497 Writing: *TWE*, pp. 474, 482 Extended Response Assessment, *CRM*, p. 661	
	Chapter Assessment	Study Guide, pp. 504–506 Practice Test, p. 507	Multiple-Choice Tests (Forms 1, 2A, 2B), *CRM*, pp. 649–654 Free-Response Tests (Forms 2C, 2D, 3), *CRM*, pp. 655–660 Vocabulary Test/Review, *CRM*, p. 662	ExamView® Pro Test Bank (see below) MindJogger Videoquizzes msmath2.net/vocabulary_review msmath2.net/chapter_test

* **Key to Abbreviations:** *TWE* = Teacher Wraparound Edition, *CRM* = Chapter Resource Masters

Additional Intervention

Skills Intervention for Middle School Mathematics workbook and CD-ROM includes 88 skills with pretests, tutorials, guided practice, and posttests for diagnosis and remediation.

Guide to Daily Intervention booklet details each intervention resource available and guides teachers in using them as intervention tools everyday to assist students who may be having difficulty.

ExamView® Pro

Use ExamView® Pro Test Bank CD-ROM to:

- **Create Multiple versions of tests.**
- **Create Modified tests for Inclusion students with one mouse click.**
- **Edit existing questions and Add your own questions.**
- **Build tests aligned with state standards using built-in State Curriculum Correlations.**
- **Change English tests to Spanish with one mouse click and vice versa.**

Reading and Writing Mathematics

- **Student Edition**

  Foldables Study Organizer, p. 469

  Study Skill: How to Study Math Vocabulary, p. 474

  Concept Check questions at the beginning of each exercise set require students to verbalize and write about what they have learned in the lesson.
  - **Open Ended,** pp. 472, 476, 481, 484, 491, 494, 499, 502
  - **Find the Error,** pp. 481, 494
  - **Which One Doesn't Belong?,** p. 472

  Reading Math Tip, pp. 470, 471, 480, 490

- **Teacher Wraparound Edition**

  Foldables Follow-Up, pp. 472, 481, 486, 491

  Writing Activities, pp. 474, 482

  Differentiated Instruction, (Verbal/Linguistic), p. 476

  ELL Resources, *see page T9 for a complete listing.*

- **Additional Resources**

  Vocabulary Builder, *Chapter 11 Resource Masters*

  Reading to Learn Mathematics, *Chapter 11 Resource Masters*

  Vocabulary PuzzleMaker (online)

  Teaching Mathematics with Foldables

  Reading and Writing in the Mathematics Classroom

  WebQuests, Projects, and Interdisciplinary Investigations

  For detailed information on these resources, see pp. T6–T7.

- **Recommended Outside Reading for Students**

  Nonfiction
  - Vancleave, Janice Pratt. *Janice Vancleave's Geometry for Every Kid: Easy Activities That Make Learning Geometry Fun.* John Wiley & Sons, ©1994.

    This book from the Science for Every Kid series presents geometry concepts through fun activities. Students can use the activities to reinforce what they learn in this chapter about areas of parallelograms, triangles, trapezoids, and circles.

  Fiction
  - Neuschwander, Cindy. *Sir Cumference and the Great Knight of Angleland: A Math Adventure.* Charlesbridge Publishing, ©2001.

    In this story, Sir Cumference's son, Radius, must complete a quest. Students will enjoy reading this fanciful tale about the geometric figures they study in this chapter.

PROJECT CRISS[SM] Study Skill

Producing an accurate, detailed written summary can help students' understanding of new material. Demonstrate the summarization process using the example at the right for Lesson 11-1. Begin by having students notice lesson titles, boldfaced vocabulary words, and illustrations. Next, have them read the lesson, then summarize it on the chalkboard, as shown. Finally, students should write a summary, in their own words, based on the notes. Have students work through this process individually or in cooperative groups for the remaining lessons in Chapter 11.

Lesson 11-1

- The square of a number is the product of the number and itself. Example: $4^2 = 16$
- Perfect squares are squares of rational numbers. Example: 9 is a perfect square with factors 3, 3
- Square roots are the factors multiplied to form perfect squares.
- A radical sign indicates the positive square root of a number.

CReating Independence through Student-owned Strategies

For more information on Project CRISS, see page T9.

Making Connections

School-To-Home

Parent and Student Study Guide

Parents and students may work together to reinforce the concepts and skills of this chapter. These workbooks are available in English and Spanish. A sample lesson is shown below. Also available at msmath2.net/parent_student.

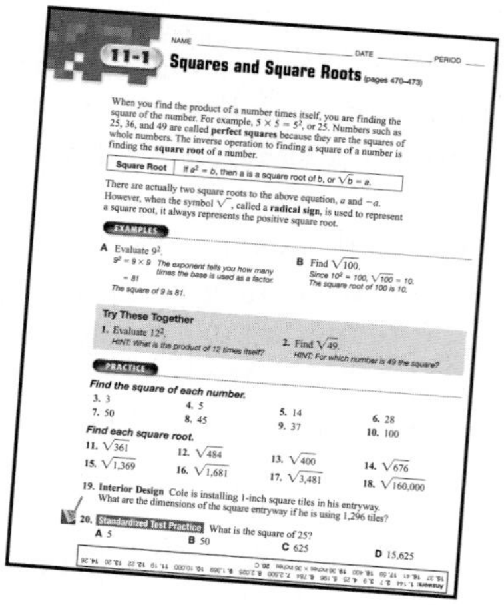
NAME ______ DATE ______ PERIOD ______

11-1 Squares and Square Roots (pages 470–473)

When you find the product of a number times itself, you are finding the square of the number. For example, $5 \times 5 = 5^2$, or 25. Numbers such as 25, 36, and 49 are called **perfect squares** because they are the squares of whole numbers. The inverse operation to finding a square of a number is finding the **square root** of a number.

Square Root	If $a^2 = b$, then a is a square root of b, or $\sqrt{b} = a$.

There are actually two square roots to the above equation, a and $-a$. However, when the symbol $\sqrt{\ }$, called a **radical sign**, is used to represent a square root, it always represents the positive square root.

EXAMPLES

A Evaluate 9^2.
$9^2 = 9 \times 9$ *The exponent tells you how many times the base is used as a factor.*
$= 81$
The square of 9 is 81.

B Find $\sqrt{100}$.
Since $10^2 = 100$, $\sqrt{100} = 10$. The square root of 100 is 10.

Try These Together

1. Evaluate 12^2. *HINT: What is the product of 12 times itself?*
2. Find $\sqrt{49}$. *HINT: For which number is 49 the square?*

PRACTICE

Find the square of each number.

3. 3
4. 5
5. 14
6. 28
7. 50
8. 45
9. 37
10. 100

Find each square root.

11. $\sqrt{361}$
12. $\sqrt{484}$
13. $\sqrt{400}$
14. $\sqrt{676}$
15. $\sqrt{1{,}369}$
16. $\sqrt{1{,}681}$
17. $\sqrt{3{,}481}$
18. $\sqrt{160{,}000}$
19. **Interior Design** Cole is installing 1-inch square tiles in his entryway. What are the dimensions of the square entryway if he is using 1,296 tiles?
20. **Standardized Test Practice** What is the square of 25?
A 5 B 50 C 625 D 15,625

Family Letters and Activities

Send a letter and activity for this chapter home to encourage family members to become active participants in their students' learning. (English: *Chapter 11 Resource Masters*, pp. ix–x; Spanish: log on to msmath2.net/family_letters)

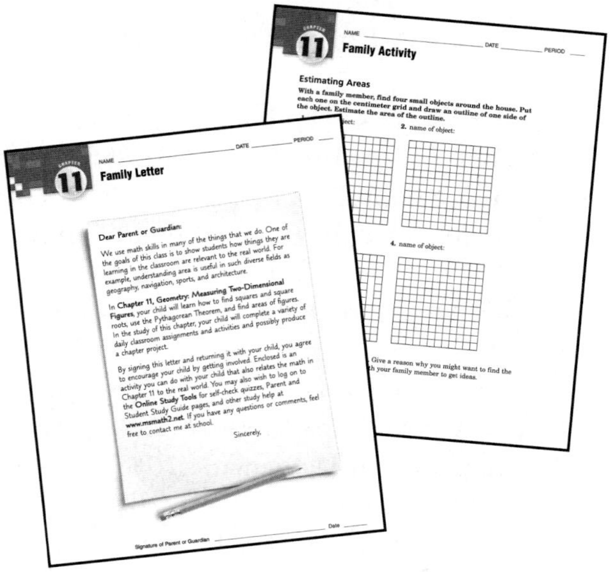
Family Activity

NAME ______ DATE ______ PERIOD ______

Estimating Areas

With a family member, find four small objects around the house. Put each one on the centimeter grid and draw an outline of one side of the object. Estimate the area of the outline.

2. name of object:

4. name of object:

Family Letter

NAME ______ DATE ______ PERIOD ______

Dear Parent or Guardian:

We use math skills in many of the things that we do. One of the goals of this class is to show students how things they are learning in the classroom are relevant to the real world. For example, understanding area is useful in such diverse fields as geography, navigation, sports, and architecture.

In **Chapter 11, Geometry: Measuring Two-Dimensional Figures**, your child will learn how to find squares and square roots, use the Pythagorean Theorem, and find areas of figures. In the study of this chapter, your child will complete a variety of daily classroom assignments and activities and possibly produce a chapter project.

By signing this letter and returning it with your child, you agree to encourage your child by getting involved. Enclosed is an activity you can do with your child that also relates the math in Chapter 11 to the real world. You may also wish to log on to the **Online Study Tools** for self-check quizzes, Parent and Student Study Guide pages, and other study help at **www.msmath2.net**. If you have any questions or comments, feel free to contact me at school.

Sincerely,

Signature of Parent or Guardian ______ Date ______

English Language Learner ELL

Lesson 11-1

Higher-Level Thinking

Have the students list the integers from 1 to 20 and then multiply each number by itself. Have the students write the products in exponential form and then in standard form. Explain to students the concept of a square root or radical and how taking a square root is the inverse operation of squaring a number. Show students one example: $2 \times 2 = 2^2$ or 4. The square root of 4 is 2.

Lesson 11-3

Reading and Writing

If possible, pair each English Language Learner with a bilingual student. Give pairs of students several right triangles cut from posterboard in different sizes, each with the right angle labeled. Have students label the legs and the hypotenuse. Ask them to explain in their own words how they know which side of the triangle is the hypotenuse.

Lesson 11-7

Using Manipulatives

Give groups of students diagrams of different irregular figures. Ask students to list the shapes for which they know area formulas and the formulas. Next, have students cut the irregular figures into the shapes they have listed. Have them use rulers to get exact measurements. Then have students find the area of each piece and ask them how they would find the area of the whole irregular figure. The students should realize that they must find the sum of all of the areas to get the area of the entire figure.

MathScape™

This project was supported, in part, by the **National Science Foundation**

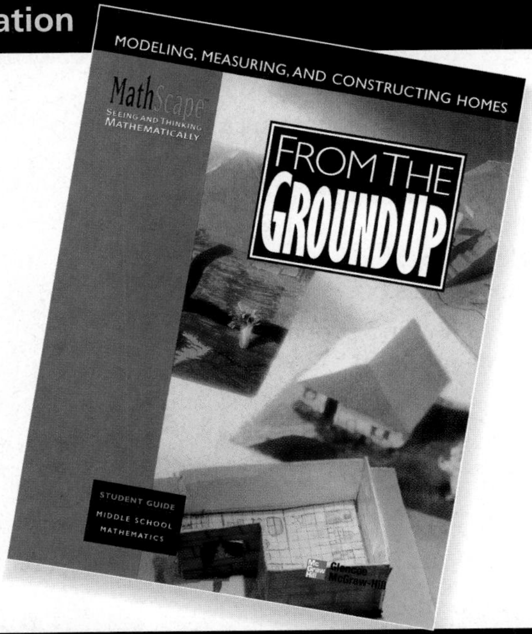

MathScape is a mathematics curriculum for grades 6–8 developed by the Seeing and Thinking Mathematically Project at the Education Development Center.

Modeling, Measuring, and Constructing Homes

From the Ground Up

Summary Students will investigate two-dimensional shape and find their areas and perimeters. Drawings are used to explore triangles, rectangles, and trapezoids.

USA TODAY Education Online Lesson Planner

USA TODAY® Education's Online site offers resources and interactive features connected to each day's newspaper. *Experience Today,* USA TODAY's® daily lesson plan, is available on the site and delivered daily to subscribers. This plan provides instruction for integrating USA TODAY® graphics and key editorial features into your mathematics classroom. Log on to **education.usatoday.com**.

USA TODAY Snapshots® in Chapter 11	
Lesson	**Snapshot**
11-5	Hours on the job

Quick Review Math Handbook

is Glencoe's mathematical handbook for students. The Hot Words section includes a glossary of terms while the Hot Topics section consists of explanations of key mathematical concepts. An exercise set is included to check students' understanding of the concepts. This valuable resource can be used as a reference in the classroom or for home study.

Lesson	Hot Topics Section	Lesson	Hot Topics Section
11-1	3-2	**11-5**	7-5, 8-3
11-2	3-2	**11-6**	7-8, 8-3
11-3	3-2, 7-9	**11-7**	7-5, 8-3
11-4	7-5, 8-3	**11-8**	7-5, 8-3

CHAPTER 11 Notes

Key Chapter Objectives

Lesson 11-1 Find squares of numbers and square roots of perfect squares.

Lesson 11-2 Find length using the Pythagorean Theorem.

Lessons 11-4 through 11-7 Find the areas of parallelograms, triangles, trapezoids, circles, and complex figures.

Lesson 11-8 Find probability using area models.

"What does landscape design have to do with math?"

Point out to students that this is only one of many applications of the math that they are studying. Others are provided in the introduction to each lesson.

Vocabulary Builder, pp. vii–viii ELL

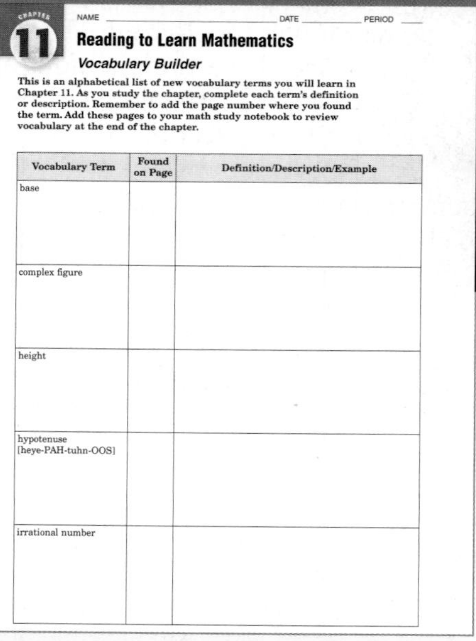

CHAPTER 11 NAME ________ DATE ________ PERIOD ____

Reading to Learn Mathematics

Vocabulary Builder

This is an alphabetical list of new vocabulary terms you will learn in Chapter 11. As you study the chapter, complete each term's definition or description. Remember to add the page number where you found the term. Add these pages to your math study notebook to review vocabulary at the end of the chapter.

Vocabulary Term	Found on Page	Definition/Description/Example
base		
complex figure		
height		
hypotenuse [heye-PAH-tuhn-OOS]		
irrational number		

CHAPTER 11 Geometry: Measuring Two-Dimensional Figures

"What does landscape design have to do with math?"

In designing a circular path, pool, or fountain, landscape architects calculate the area of the region. **They use the formula $A = \pi r^2$, where r is the radius of the circle.** There are many formulas involving two-dimensional figures that are useful in real life.

You will solve problems involving areas of circles in Lesson 11-6.

Vocabulary Builder

For a thorough vocabulary list with pronunciations of new words, give students the Vocabulary Builder worksheets from the *Chapter 11 Resource Masters*. Encourage them to complete the definition of each term as they progress through the chapter.

GETTING STARTED

Take this quiz to see whether you are ready to begin Chapter 11. Refer to the lesson or page number in parentheses if you need more review.

Vocabulary Review

State whether each sentence is *true* or *false*. If *false*, replace the underlined word or number to make a true sentence.

1. 0.5 is a <u>rational</u> number. (Lesson 5-4) **true**
2. π is approximately equal to <u>3.14</u>. (Lesson 6-9) **true**
3. A triangle with angle measures of 60°, 90°, and 30° is an <u>acute</u> triangle. (Lesson 10-4) **false; right**

Prerequisite Skills

Replace each ● with <, >, or = to make a true sentence. (Page 556)

4. $67 \bullet 8.2 \cdot 8.2$ $<$
5. $11.1 \times 11.1 \bullet 123$ $>$
6. $5.9(5.9) \bullet 34.9$ $<$
7. $12.25 \bullet 3.5 \cdot 3.5$ $=$

Evaluate each expression. (Lesson 1-2)

8. 3^2 **9**
9. 8 squared **64**
10. 5 to the third power **125**
11. 6 to the second power **36**

Find each value. (Lesson 1-3)

12. $\frac{1}{2}(5)(6)$ **15**
13. $\frac{1}{2}(4)(12 + 18)$ **60**
14. $9(3 + 3)$ **54**
15. $7(2)(8)$ **112**
16. $2(5)(4 + 1)$ **50**

Find the probability of rolling each number on a number cube. (Lesson 9-1)

17. $P(3)$ $\frac{1}{6}$
18. $P(6 \text{ or } 2)$ $\frac{1}{3}$
19. $P(\text{odd})$ $\frac{1}{2}$
20. $P(\text{greater than } 4)$ $\frac{1}{3}$

FOLDABLES™ Study Organizer

Measuring Figures Make this Foldable to help you organize your notes. Begin with a piece of 11″ by 17″ paper.

STEP 1 Fold Fold a 2″ tab along the long side of the paper.

STEP 2 Open and Fold Unfold the paper and fold in thirds widthwise.

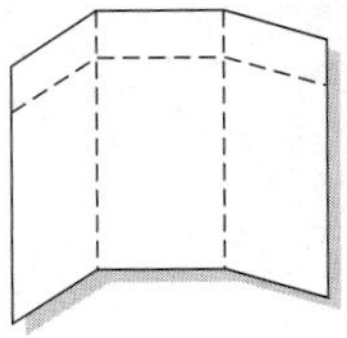

STEP 3 Open and Label Draw lines along the folds and label the head of each column as shown. Label the front of the folded table with the chapter title.

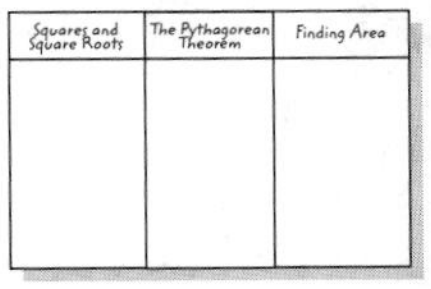

Reading and Writing As you read and study the chapter, write notes and examples in the appropriate columns of the table.

GETTING STARTED

This section provides a review of the basic concepts needed before beginning Chapter 11. Page references are included for additional student help.

Prerequisite Skills review skills needed in this chapter.

Exercises	Needed for Lesson	Skill
4–7	11-1	Solve Inequalities Involving Decimals
8–11	11-1, 11-3	Evaluate Expressions
12–16	11-5, 11-6	Find Values Using the Associative Property
17–20	11-8	Find Probability

For more prerequisite skills, see:

- Getting Ready for the Next Lesson at the end of each lesson
- Prerequisite Skills in the Student Handbook on pp. 554–563
- *Math Skills Maintenance Masters*

FOLDABLES™ Study Organizer

For more information about Foldables, see *Teaching Mathematics with Foldables.*

Use this Foldable table to help students organize information they learn in the chapter. Have students write key vocabulary words, definitions, concepts, or procedures in the appropriate columns of their tables. Ask students to record their own examples of the concepts presented in the chapter. Students can use this Foldable as a study guide to review what they know and what they are learning about finding squares and square roots, the Pythagorean Theorem, and finding the area of two-dimensional figures.

11-1 Lesson Notes

Lesson Resources

Chapter 11 Resource Masters
- Study Guide and Intervention, p. 609
- Practice: Skills, p. 610
- Practice: Word Problems, p. 611
- Reading to Learn Mathematics, p. 612
- Enrichment, p. 613

Calculator and Spreadsheet Masters, p. 51

Key Teaching Transparencies, 41

Parent and Student Study Guide Workbook, 11-1

FAST FILE

1 Focus

5-Minute Check 11-1

INTERACTIVE CHALKBOARD PowerPoint® Presentations

Use as a quiz or review of Chapter 10. (Also available as a transparency)

Bellringer

Communication Have students find the area of the square below.

$1 \text{ ft} \times 1 \text{ ft} = 1 \text{ ft}^2$ Next, ask them for two ways to find the area of a 2-square by 2-square figure. **Add the squares' areas or multiply the sides—$2 \text{ ft} \times 2 \text{ ft} = 4 \text{ ft}^2$.**

Reading to Learn Mathematics, p. 612 **ELL**

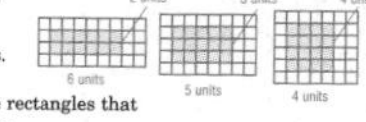

NAME ____________ DATE ________ PERIOD ____

11-1 Reading to Learn Mathematics

Squares and Square Roots

Pre-Activity ***Complete the Mini Lab at the top of page 470 in your textbook. Write your answers below.***

1. On grid paper, draw and label three other rectangles that have a perimeter of 16 units.

2. Summarize the dimensions and areas of the rectangles that you drew in a table like the one shown below.

Drawing	Dimensions (units)	Area (square units)
	1 × 7	7
	2 × 6	12
	3 × 5	15
	4	

11-1 Squares and Square Roots

What You'll Learn
Find squares of numbers and square roots of perfect squares.

NEW Vocabulary
- square
- perfect squares
- square root
- radical sign

HANDS-ON Mini Lab

Materials
- grid paper

Work with a partner.

The rectangle has a perimeter of 16 units and an area of 7 square units.

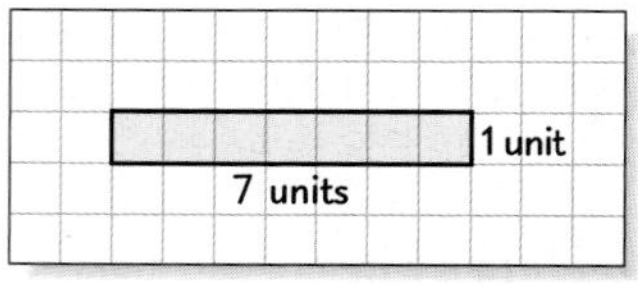

1. On grid paper, draw and label three other rectangles that have a perimeter of 16 units. **1–3. See Ch. 11 Answer Appendix.**
2. Summarize the dimensions and areas of the rectangles that you drew in a table like the one shown below.

Drawing	Dimensions (units)	Area (sq units)
	1 × 7	7

3. Draw three different rectangles that have a perimeter of 12 units and find their areas.
4. What do you notice about the rectangles with the greatest areas? **They are squares.**

The area of the square at the right is $4 \cdot 4$ or 16 square units. The product of a number and itself is the **square** of the number. So, the square of 4 is 16.

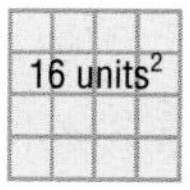

READING Math

Square a Number *To square a number* means to multiply that number by itself.

EXAMPLES Find Squares of Numbers

1 Find the square of 3.

$3 \cdot 3 = 9$

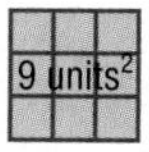

2 Find the square of 15.

15 [x²] [ENTER] 225

Your Turn **Find the square of each number.**

a. 8 **64** b. 12 **144** c. 23 **529**

Tips for New Teachers

Teaching Strategies
You may wish to have students use square tiles to create the rectangles in the Hands-On Mini Lab. As an extension of the lab, you can have students graph data they generate on the length of the side of a square related to its area.

EXAMPLE Find a Square to Solve a Problem

3 PHYSICAL SCIENCE The falling distance of an object in feet d after t seconds is given by the formula $d = \frac{1}{2}(32)t^2$. If you went bungee jumping, how far would you fall 2.5 seconds after being released?

$d = \frac{1}{2}(32)t^2$ Write the formula.

$= \frac{1}{2}(32)(2.5)^2$ Replace t with 2.5.

$= \frac{1}{2}(32)(6.25)$ Use a calculator to square 2.5.

$= 100$ Simplify.

So, after 2.5 seconds, you would fall 100 feet.

Numbers like 9, 16, 225, and 6.25 are called **perfect squares** because they are squares of rational numbers. The factors multiplied to form perfect squares are called **square roots**.

Key Concept — Square Root

Words A square root of a number is one of its two equal factors.

Symbols

Arithmetic	Algebra
$4 \cdot 4 = 16$, so 4 is a square root of 16.	If $x \cdot x$ or $x^2 = y$, then x is a square root of y.

Both $4 \cdot 4$ and $(-4)(-4)$ equal 16. So, 16 has two square roots, 4 and -4. A **radical sign**, $\sqrt{\ }$, is the symbol used to indicate the *positive* square root of a number. So, $\sqrt{16} = 4$.

EXAMPLES Find Square Roots

4 Find $\sqrt{81}$.

$9 \cdot 9 = 81$, so $\sqrt{81} = 9$.

5 Find $\sqrt{196}$.

2nd [$\sqrt{\ }$] 196 ENTER 14

So, $\sqrt{196} = 14$.

6 SPORTS A boxing ring is a square with an area of 400 square feet. What are the dimensions of the ring?

2nd [$\sqrt{\ }$] 400 ENTER 20 Find the square root of 400.

So, a boxing ring measures 20 feet by 20 feet.

Your Turn **Find each square root.**

d. $\sqrt{25}$ 5 e. $\sqrt{64}$ 8 f. $\sqrt{289}$ 17

EAL-LIFE MATH

YSICAL SCIENCE The gee bridge in Squaw ey, California, is at an vation of 8,200 feet, the est in the world.

TEACHING TIP

this course, $\sqrt{\ }$ is ed to indicate a posi- e square root. In future urses, the symbol will so be used to indicate gative square roots.

EADING Math

Square Roots Read $\sqrt{81} = 9$ as *the square root of 81 is 9.*

TEACHING TIP

elate Examples 4 and 5 the square models on ge 470. Ask students hat the length of one de of a square would be the area were 81 units 196 units.

msmath2.net/extra_examples

2 Teach

In-Class Examples

INTERACTIVE CHALKBOARD PowerPoint® Presentations

1. Find the square of 5. **25**
2. Find the square of 19. **361**
3. **PHYSICAL SCIENCE** A 6 kilogram ball is soaring through the air at 5 meters per second. If kinetic energy $= \frac{1}{2}mv^2$, where m is the mass and v is the speed, what is the ball's kinetic energy? **$75 \text{ kg} \cdot \text{m}^2/\text{s}^2$**
4. Find $\sqrt{36}$. **6**
5. Find $\sqrt{676}$. **26**
6. **GAMES** A checkerboard is a square with an area of 1,225 square centimeters. What are the dimensions of the checkerboard? **35 cm × 35 cm**

Tips for New Teachers

Using Applications You may want students to do some research to find additional formulas that utilize the math concept you are studying. For example, $d = \frac{1}{2}(32)t^2$ is a physics formula. Additional areas in which to look for formulas could include compound interest and population growth.

DAILY INTERVENTION — Flexible Grouping

Real-Life Problems Have students work in small groups to write problems that can be solved using squares or square roots. Encourage them to include some real-life problems related to activities that interest them. Have groups trade problems and work through them.

3 Practice/Apply

FOLDABLES™ Follow-Up Remind students to use their Foldables to record what they learned about squares and square roots. Encourage students to explain in words and symbols how they find squares of numbers and square roots of perfect squares.

Which One Doesn't Belong? Encourage students to relate the numbers in Exercise 3 to perfect squares that they know. For example, they may know that $10^2 = 100$ and $11^2 = 121$. Since 116 is between 100 and 121, 116 cannot be a perfect square.

About the Exercises...

Guided Practice Key

Exercises	Examples
4–7	1, 2
12	3
8–11	4, 5

Odd/Even Assignments
Exercises 13–28 are structured so that students practice the same concepts whether they are assigned odd or even problems.

Alert! Students must use grid paper to complete Exercises 40–44.

Skill and Concept Check

Writing Math — Exercises 1 & 3

1. **Explain** how finding the square of a number is similar to finding the area of a square. **Both involve multiplying a number by itself.**
2. **OPEN ENDED** Write a number whose square is between 200 and 300. **Sample answer: 15**
3. **Which One Doesn't Belong?** Identify the number that is not a perfect square. Explain your reasoning. **116; There are not two equal factors whose product is 116.**

256   121   529   116

GUIDED PRACTICE

Find the square of each number.

4. 6 **36**   5. 10 **100**   6. 17 **289**   7. 30 **900**

Find each square root.

8. $\sqrt{9}$ **3**   9. $\sqrt{121}$ **11**   10. $\sqrt{169}$ **13**   11. $\sqrt{529}$ **23**

12. **PHYSICAL SCIENCE** A model rocket is launched straight up into the air at an initial speed of 115 feet per second. The height of the rocket h after t seconds is given by the formula $h = -16t^2 + 115t$. What is the height of the rocket 3.5 seconds after it is launched? **206.5 ft**

★ **indicates increased difficulty**

Practice and Applications

HOMEWORK HELP

For Exercises	See Examples
13–20, 29–30, 34–35	1–3
21–28, 31–33, 36	4–6

Extra Practice See pages 590, 606.

Find the square of each number.

13. 5 **25**   14. 1 **1**   15. 7 **49**   16. 11 **121**
17. 16 **256**   18. 20 **400**   19. 18 **324**   20. 34 **1,156**

Find each square root.

21. $\sqrt{4}$ **2**   22. $\sqrt{49}$ **7**   23. $\sqrt{144}$ **12**   24. $\sqrt{225}$ **15**
25. $\sqrt{729}$ **27**   26. $\sqrt{625}$ **25**   27. $\sqrt{1{,}225}$ **35**   28. $\sqrt{1{,}600}$ **40**

29. What is the square of -22? **484**   30. Square 5.8. **33.64**
31. Find both square roots of 100. **10, −10**   ★ 32. Find $-\sqrt{361}$. **−19**
★ 33. **ALGEBRA** Evaluate $a^2 + \sqrt{b}$ if $a = 36$ and $b = 256$. **1,312**

GEOGRAPHY For Exercises 34–36, refer to the squares in the diagram at the right. They represent the approximate areas of Texas, Michigan, and Florida.

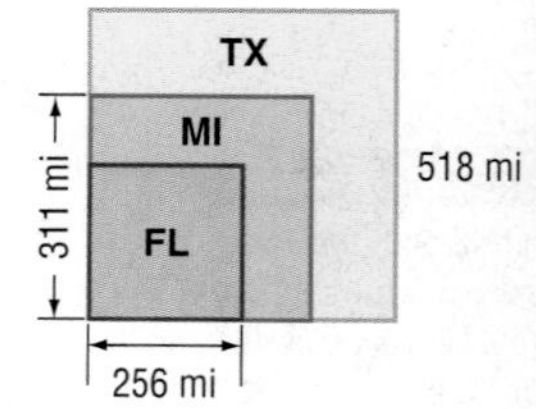

34. What is the area of Michigan? **96,721 mi²**
35. How much larger is Texas than Florida? **202,788 mi²**
★ 36. The water areas of Texas, Michigan, and Florida are about 6,724 square miles, 40,000 square miles, and 11,664 square miles, respectively. Make a similar diagram comparing the water areas of these states. Label the squares. **See margin.**

Answers

36.

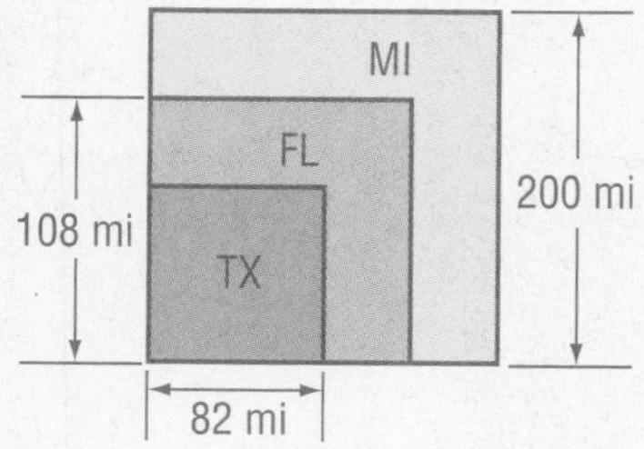

37. Yes; for example, a garden that measures 30 ft by 30 ft has the same perimeter, but its area is 900 ft², which is greater than 500 ft².
38. A square that measures 30 ft on each side; the perimeter is the same and the area is 900 ft², or 400 ft² greater than that of the original garden.
42. Both graphs increase as the lengths of the sides increase. The graph for the area is a curve and the graph for the perimeter is a straight line.
43. when the side length is between 0 and 4; when the side length is 0 or 4

MEASUREMENT **For Exercises 37 and 38, refer to the garden, which is enclosed on all sides by a fence.**

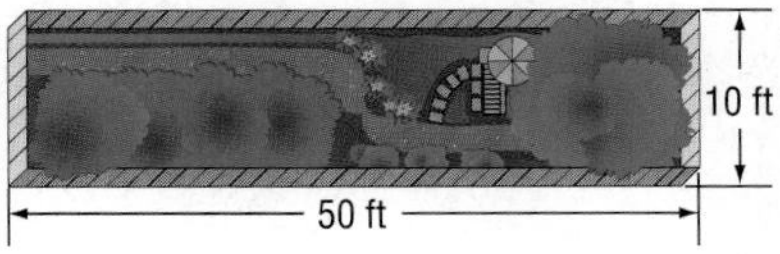

37. Could the garden area be made larger using the same fencing? Explain. **37–38. See margin.**

38. Describe the largest garden area possible using the same amount of fencing. How do the perimeter and area compare to the original garden?

★ 39. **PROBABILITY** A set consists of all the perfect squares from 1 to 100. What is the probability that a number chosen at random from this set is divisible by 4 or 5? $\frac{3}{5}$

ALGEBRA **For Exercises 40–44, let the x-axis of a coordinate plane represent the side length of a square.** **40–41. See Ch. 11 Answer Appendix.**

★ 40. Let the y-axis represent the area. Graph the points for squares with sides 0, 1, 2, 3, 4, and 5 units long. Draw a line or curve that goes through each point.

★ 41. On the same coordinate plane, let the y-axis represent the perimeter of a square. Graph the points for squares with sides 0, 1, 2, 3, 4, and 5 units long. Draw a line or curve that goes through each point.

★ 42. Compare and contrast the two graphs. **See margin.**

★ 43. For what side lengths is the value of the perimeter greater than the value of the area? When are the values equal? **See margin.**

★ 44. Why do these graphs only make sense in the first quadrant? **Sample answer: Side length, area, and perimeter cannot be negative.**

45. **CRITICAL THINKING** The area of a square 8 meters by 8 meters is how much greater than the area of a square containing 9 square meters? Explain. **55 m^2 greater; The area of an 8-by-8-meter square is 64 m^2.**

Standardized Test Practice and Mixed Review

46. **MULTIPLE CHOICE** A square plot of land has an area of 1,156 square feet. What is the perimeter of the plot? **C**

Ⓐ 34 ft Ⓑ 102 ft Ⓒ 136 ft Ⓓ 289 ft

47. **SHORT RESPONSE** The perimeter of a square is 128 centimeters. Find its area. **1,024 cm^2**

For Exercises 48 and 49, refer to $\triangle ABC$ at the right. Find the vertices of $\triangle A'B'C'$ after each transformation. Then graph the triangle and its reflected or translated image. **48–49. See Ch. 11 Answer Appendix for graphs.**

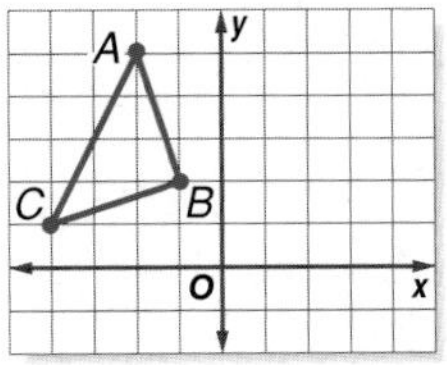

48. $\triangle ABC$ reflected over the y-axis (Lesson 10-9) **$A'(2, 5)$, $B'(1, 2)$, $C'(4, 1)$**

49. $\triangle ABC$ translated 5 units right and 1 unit up (Lesson 10-8) **$A'(3, 6)$, $B'(4, 3)$, $C'(1, 2)$**

GETTING READY FOR THE NEXT LESSON

PREREQUISITE SKILL **Replace each ● with <, >, or = to make a true sentence.** (Lesson 4-5)

50. $7 ● \sqrt{49}$ **=** 51. $\sqrt{25} ● 4$ **>** 52. $7.9 ● \sqrt{64}$ **<** 53. $10.5 ● \sqrt{100}$ **>**

4 Assess

Open-Ended Assessment

Modeling Provide students with various colors of posterboard. Have them work in small groups to create models that show relationships among square numbers, square roots, and perfect squares. For example, they may show that a 16-block square is made from 4 rows of 4 blocks, but a 12-block rectangle is not a perfect square.

Study Guide and Intervention, p. 609

NAME ________ DATE ________ PERIOD ____

11-1 Study Guide and Intervention

Squares and Square Roots

The product of a number and itself is the **square** of the number. Numbers like 4, 25, and 2.25 are called **perfect squares** because they are squares of rational numbers. The factors multiplied to form perfect squares are called **square roots**. Both 5 · 5 and (−5)(−5) equal 25. So, 25 has two square roots, 5 and −5. A **radical sign**, $\sqrt{\ }$, is the symbol used to indicate the *positive* square root of a number. So, $\sqrt{25} = 5$.

EXAMPLES

1 Find the square of 5.

$5 \cdot 5 = 25$

2 Find the square of 16.

16 [x²] [ENTER] 256

3 Find $\sqrt{49}$.

$7 \cdot 7 = 49$, so $\sqrt{49} = 7$.

4 Find $\sqrt{169}$.

[2nd] [√] 169 [ENTER] 13

So, $\sqrt{169} = 13$.

EXAMPLE 5 **A square tile has an area of 144 square inches. What are the dimensions of the tile?**

[2nd] [√] 144 [ENTER] 12 Find the square root of 144.

So, the tile measures 12 inches by 12 inches.

EXERCISES

Find the square of each number.

1. 2 4 2. 9 81 3. 14 196

4. 15 225 5. 21 441 6. 45 2,025

Find each square root.

9. $\sqrt{16}$ 4 10. $\sqrt{36}$ 6 11. $\sqrt{256}$ 16

12. $\sqrt{1{,}024}$ 32 13. $\sqrt{361}$ 19 14. $\sqrt{484}$ 22

Practice: Skills, p. 610

NAME ________ DATE ________ PERIOD ____

11-1 Practice: Skills

Squares and Square Roots

Find the square of each number.

1. 3 9 2. 22 484

3. 25 625 4. 24 576

5. 35 1,225 6. 26 676

7. 37 1,369 8. 50 2,500

Find each square root.

9. $\sqrt{25}$ 5 10. $\sqrt{100}$ 10

11. $\sqrt{441}$ 21 12. $\sqrt{900}$ 30

13. $\sqrt{961}$ 31 14. $\sqrt{784}$ 28

15. $\sqrt{3{,}600}$ 60 16. $\sqrt{1{,}936}$ 44

17. What is the square of −37? 1,369 18. Find both square roots of 4,900. 70, −70

19. Square 7.2. 51.84 20. Square 4.5. 20.25

Practice: Word Problems, p. 611

NAME ________ DATE ________ PERIOD ____

11-1 Practice: Word Problems

Squares and Square Roots

1. **FERTILIZER** John bought a bag of lawn fertilizer that will cover 400 square feet. What are the dimensions of the largest square plot of lawn that the bag of fertilizer will cover? 20 ft by 20 ft

2. **GEOMETRY** The area A of a circle in square feet with a radius r in feet is given approximately by the formula $A \approx 3.14r^2$. What is the approximate area of a circle with a radius of 3 feet? 28.26 ft^2

3. **MOTION** The time t in seconds for an object dropped from a height of h feet to hit the ground is given by the formula $t = \sqrt{\frac{2h}{32}}$. How long will it take an object dropped from a height of 500 feet to hit the ground? Round to the nearest tenth. 5.6 s

4. **PACKAGING** A cardboard envelope for a compact disc is a square with an area of 171.61 square centimeters. What are the dimensions of the envelope? 13.1 cm by 13.1 cm

5. **GEOGRAPHY** Refer to the squares below. They represent the approximate areas of California, Alabama, and Nebraska. Find the area of Alabama.

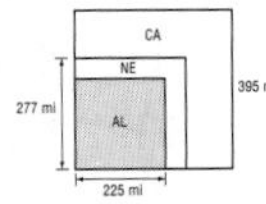

50,625 mi^2

6. Use the figure in Exercise 5. How much larger is California than Nebraska? 79,296 mi^2

Lesson 11-1 Squares and Square Roots **473**

Study Skill

Objective Use webs to study math vocabulary.

GETTING STARTED

Vocabulary Before students start the **Study Skill**, have them discuss their current methods of studying math vocabulary. They will likely mention flash cards, rereading, and other forms of repetition and rehearsal.

TEACH

- Ask students whether they have ever tried to memorize vocabulary terms. Do they think this approach to studying vocabulary is effective? For most people, memorizing terms out of context is very difficult. Point out that webs require students to organize vocabulary terms in relation to one another. This process requires an understanding of the concepts being presented. Since the words have a context, their meanings will be easier to recall.
- Remind students to use webs for the vocabulary in the remaining chapters of this course and in their other courses. Doing so will help them practice using webs and establish effective study habits.

ASSESS

Open-Ended Assessment

Writing Have students write a paragraph comparing webs to other methods for studying vocabulary, such as flash cards. Suggest that they describe how the various methods can be used together to create a more complete approach to studying vocabulary.

Study Skill HOW TO... Study Math Vocabulary

WEBS

If you've surfed the World Wide Web, you know it is a collection of documents linked together to form a huge electronic library. In mathematics, a web helps you understand how concepts are linked together.

A web can help you understand how math concepts are related to each other. To make a web, write the major topic in a box in the center of a sheet of paper. Then, draw "arms" from the center for as many categories as you need. You can label the arms to indicate the type of information that you are listing.

Here is a partial web for the major topic of *polygons*.

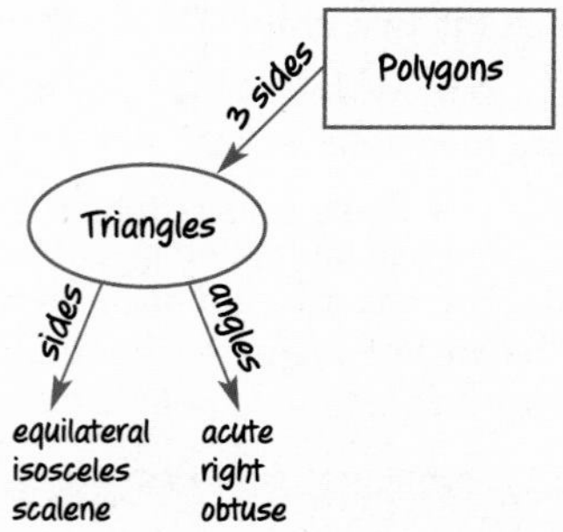

SKILL PRACTICE

1. Continue the web above by adding an arm for *quadrilaterals*. **See students' work.**
2. In Lesson 5-8, you learned that the set of rational numbers contains fractions, terminating and repeating decimals, and integers. You also know that there are proper and improper fractions and that integers are whole numbers and their opposites. Complete the web below for *rational numbers*. **See margin.**

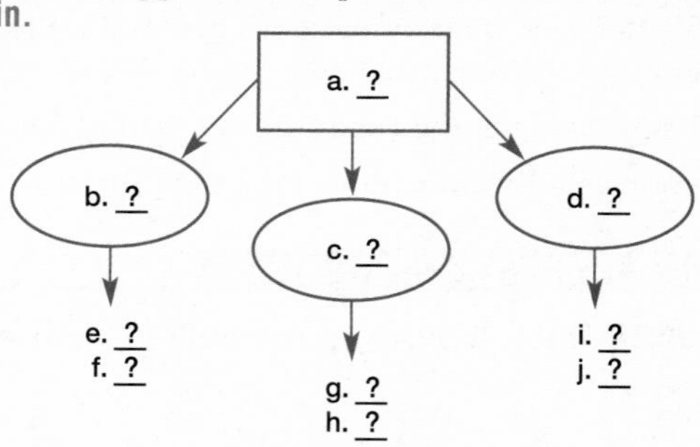

Answers

2. Sample answers are given.
2a. rational numbers
2b. fractions
2c. decimals
2d. integers
2e. proper
2f. improper
2g. terminating decimals
2h. repeating decimals
2i. whole numbers
2j. opposites

11-2 Estimating Square Roots

hat You'll Learn
:imate square roots.

'W Vocabulary
ational number

'VIEW Vocabulary
ional number: a
mber that can be
itten as a fraction
sson 5-4)

HANDS-ON Mini Lab

Materials
- algebra tiles

Work with a partner.

You can use algebra tiles to estimate the square root of 30.

- Arrange 30 tiles into the largest square possible. In this case, the largest possible square has 25 tiles, with 5 left over.

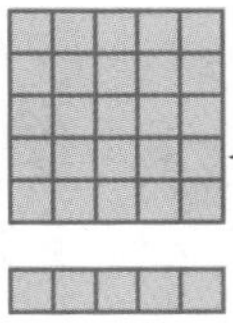

The square root of 25 is 5.

- Add tiles until you have the next larger square. So, add 6 tiles to make a square with 36 tiles.

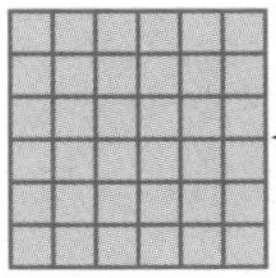

The square root of 36 is 6.

- The square root of 30 is between 5 and 6. $\sqrt{30}$ is closer to 5 because 30 is closer to 25 than to 36.

Use algebra tiles to estimate the square root of each number to the nearest whole number.

1. 40 **6** **2.** 28 **5** **3.** 85 **9** **4.** 62 **8**

5. Describe another method that you could use to estimate the square root of a number.

5. Sample answer: Square numbers using guess and check.

Recall that the square root of a perfect square is a rational number. You can find an estimate for the square root of a number that is *not* a perfect square.

EXAMPLE Estimate the Square Root

1 Estimate $\sqrt{75}$ to the nearest whole number.

List some perfect squares.

1, 4, 9, 16, 25, 36, 49, 64, 81, …

(75 falls between 64 and 81)

$64 < 75 < 81$	75 is between the perfect squares 64 and 81.
$\sqrt{64} < \sqrt{75} < \sqrt{81}$	Find the square root of each number.
$8 < \sqrt{75} < 9$	$\sqrt{64} = 8$ and $\sqrt{81} = 9$

So, $\sqrt{75}$ is between 8 and 9. Since 75 is closer to 81 than to 64, the best whole number estimate is 9. Verify with a calculator.

msmath2.net/extra_examples

11-2 Lesson Notes

Lesson Resources

Chapter 11 Resource Masters FAST FILE
- Study Guide and Intervention, p. 614
- Practice: Skills, p. 615
- Practice: Word Problems, p. 616
- Reading to Learn Mathematics, p. 617
- Enrichment, p. 618

Calculator and Spreadsheet Masters, p. 52
Key Teaching Transparencies, 42
MathPASS Tutorial CD-ROM, 3, 33
Parent and Student Study Guide Workbook, 11-2

1 Focus

5-Minute Check 11-2

INTERACTIVE CHALKBOARD PowerPoint® Presentations

Use as a quiz or review of Lesson 11-1. (Also available as a transparency)

Bellringer

Hands-On Activity Have students work in pairs. Provide each pair of students with 12 index cards. Have students make up flash cards for the first 12 perfect squares. For example, one side would have $\sqrt{144}$, and the other side would have 12. Students should quiz each other using the flash cards. To save class time, you may wish to make the flash cards ahead of time.

Reading to Learn Mathematics, p. 617 ELL

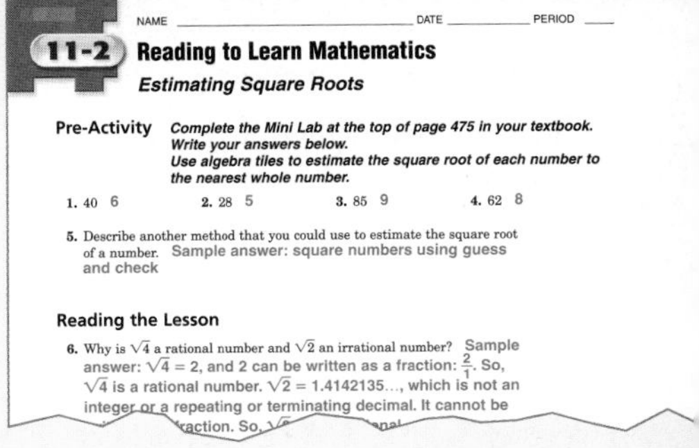

NAME ______ DATE ______ PERIOD ____

11-2 Reading to Learn Mathematics
Estimating Square Roots

Pre-Activity *Complete the Mini Lab at the top of page 475 in your textbook. Write your answers below. Use algebra tiles to estimate the square root of each number to the nearest whole number.*

1. 40 6 2. 28 5 3. 85 9 4. 62 8

5. Describe another method that you could use to estimate the square root of a number. Sample answer: square numbers using guess and check

Reading the Lesson

6. Why is $\sqrt{4}$ a rational number and $\sqrt{2}$ an irrational number? Sample answer: $\sqrt{4} = 2$, and 2 can be written as a fraction: $\frac{2}{1}$. So, $\sqrt{4}$ is a rational number. $\sqrt{2} = 1.4142135...$, which is not an integer or a repeating or terminating decimal. It cannot be ...raction. So, ...

2 Teach

In-Class Examples

INTERACTIVE CHALKBOARD
PowerPoint® Presentations

1. Estimate $\sqrt{96}$ to the nearest whole number. **10**

2. Use a calculator to find the value of $\sqrt{37}$ to the nearest tenth. **6.1**

3 Practice/Apply

NUMBER SENSE
For Exercise 3, remind students that they should consider the perfect squares just below 51 and just above 51 when determining the best estimate. If they have difficulty with this, encourage them to spend time memorizing the perfect squares up to at least 12^2.

About the Exercises...

Guided Practice Key

Exercises	Examples
4–7	1
8–12	2

Odd/Even Assignments
Exercises 13–28 are structured so that students practice the same concepts whether they are assigned odd or even problems.

Alert! Exercise 33 requires use of the Internet or other reference materials.

4 Assess

Higher-Level Thinking
Estimate the following values to the nearest whole number without using a calculator: $5\sqrt{13}$, $3\sqrt{20}$, $8\sqrt{73}$, $6\sqrt{85}$.
20, 12, 72, 54

STUDY TIP

Irrational Numbers The square root of any number that is not a perfect square is an irrational number.

In Lesson 5-8, you learned that any number that can be written as a fraction is a rational number. These include integers as well as terminating and repeating decimals. A number that *cannot* be written as a fraction is an **irrational number**.

Rational Numbers $\sqrt{4}, 3\frac{1}{7}, 0.\overline{63}$

Irrational Numbers $\sqrt{2}, \pi, 0.636336333\ldots$

You can use a calculator to estimate square roots that are irrational numbers.

TEACHING TIP
In this lesson, only real numbers are being considered. In future courses, imaginary numbers will be studied, which also cannot be written as fractions.

Point out that decimals used to represent irrational numbers are estimates, not exact values.

EXAMPLE Use a Calculator to Estimate

2 **Use a calculator to find the value of $\sqrt{42}$ to the nearest tenth.**

2nd [$\sqrt{\ }$] 42 ENTER 6.480740698

$\sqrt{42} \approx 6.5$

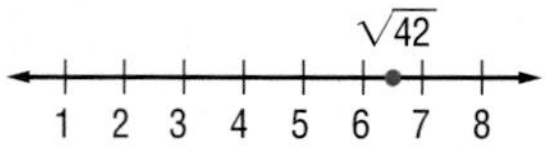

Check $6^2 = 36$ and $7^2 = 49$. Since 42 is between 36 and 49, the answer, 6.5, is reasonable.

Your Turn **Use a calculator to find each square root to the nearest tenth.**

a. $\sqrt{6}$ **2.4** b. $\sqrt{23}$ **4.8** c. $\sqrt{309}$ **17.6**

Skill and Concept Check

Writing Math Exercises 1 & 3

1. **Explain** why $\sqrt{30}$ is an irrational number. **It cannot be written as a fraction.**

2. **OPEN ENDED** List three numbers that have square roots between 4 and 5. **Sample answer: 17, 18, 19**

3. **NUMBER SENSE** Explain why 7 is the best whole number estimate for $\sqrt{51}$. **51 is between the perfect squares 49 and 64. Since 51 is closer to 49 than 64, $\sqrt{51}$ is closer to $\sqrt{49}$, or 7.**

GUIDED PRACTICE

Estimate each square root to the nearest whole number.

4. $\sqrt{39}$ **6** 5. $\sqrt{106}$ **10** 6. $\sqrt{90}$ **9** 7. $\sqrt{140}$ **12**

Use a calculator to find each square root to the nearest tenth.

8. $\sqrt{7}$ **2.6** 9. $\sqrt{51}$ **7.1** 10. $\sqrt{135}$ **11.6** 11. $\sqrt{462}$ **21.5**

12. **GEOMETRY** Use a calculator to find the side length of the square at the right. Round to the nearest tenth. **9.7 cm**

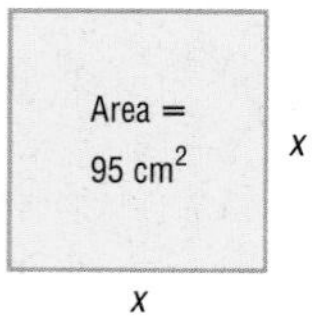

DAILY INTERVENTION **Differentiated Instruction** **ELL**

Verbal/Linguistic Write any two- or three-digit number on the chalkboard. Have a student explain how to estimate the square root of the number to the nearest whole number. You may wish to write some numbers on small cards, place the numbers in a bag, and have each student draw out a number and explain how to estimate its square root.

Practice and Applications

HOMEWORK HELP	
For Exercises	See Examples
13–20	1
21–32	2

Extra Practice
See pages 590, 606.

Estimate each square root to the nearest whole number.

13. $\sqrt{11}$ **3**   14. $\sqrt{20}$ **4**   15. $\sqrt{35}$ **6**   16. $\sqrt{65}$ **8**

17. $\sqrt{89}$ **9**   18. $\sqrt{116}$ **11**   19. $\sqrt{137}$ **12**   20. $\sqrt{409}$ **20**

Use a calculator to find each square root to the nearest tenth.

21. $\sqrt{15}$ **3.9**   22. $\sqrt{8}$ **2.8**   23. $\sqrt{44}$ **6.6**   24. $\sqrt{89}$ **9.4**

25. $\sqrt{160}$ **12.6**   26. $\sqrt{573}$ **23.9**   27. $\sqrt{645}$ **25.4**   28. $\sqrt{2{,}798}$ **52.9**

29. Order $\sqrt{87}$, 10, π, and $\frac{14}{3}$ from least to greatest. **π, $\frac{14}{3}$, $\sqrt{87}$, 10**

30. Graph $\sqrt{34}$ and $\sqrt{92}$ on the same number line. **See margin.**

31. **ALGEBRA** Evaluate $\sqrt{a + b}$ if $a = 8$ and $b = 3.7$. **3.4**

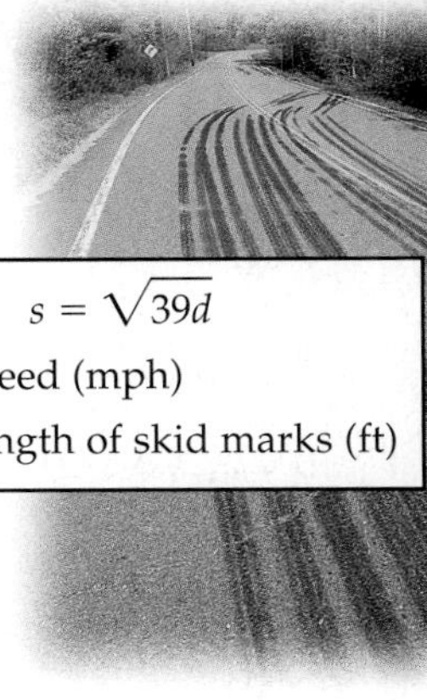

32. **DRIVING** Police officers can use a formula and skid marks to calculate the speed of a car. Use the formula at the right to estimate how fast a car was going if it left skid marks 83 feet long. Round to the nearest tenth. **56.9 mph**

$$s = \sqrt{39d}$$
- s = speed (mph)
- d = length of skid marks (ft)

33. **RESEARCH** In the 1990s, over 50 billion decimal places of pi had been computed. Use the Internet or another source to find the current number of decimal places of pi that have been computed. **See students' work.**

34. **CRITICAL THINKING** You can use *Hero's formula* to find the area A of a triangle if you know the measures of its sides, a, b, and c. The formula is $A = \sqrt{s(s - a)(s - b)(s - c)}$, where s is half of the perimeter. Find the area of the triangle to the nearest tenth. **159.0 in²**

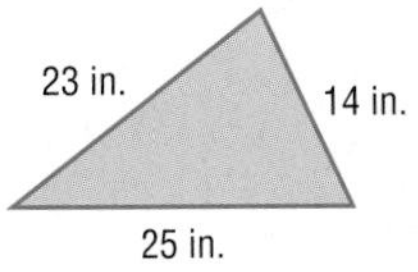

Standardized Test Practice and Mixed Review

35. **MULTIPLE CHOICE** Identify the number that is irrational. **D**

Ⓐ −4.1   Ⓑ 0   Ⓒ $\frac{3}{8}$ Ⓓ $\sqrt{6}$

36. **SHORT RESPONSE** Name the point that best represents the graph of $\sqrt{209}$. **N**

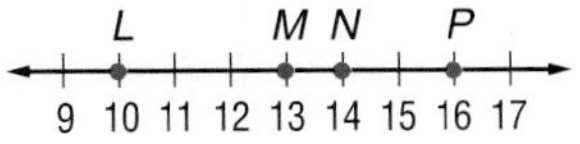

Find each square root. (Lesson 11-1)

37. $\sqrt{169}$ **13**   38. $\sqrt{2{,}025}$ **45**   39. $\sqrt{784}$ **28**

40. Graph $\triangle JLK$ with vertices $J(-1, -4)$, $K(1, 1)$, and $L(3, -2)$ and its reflection over the x-axis. Write the ordered pairs for the vertices of the new figure. (Lesson 10-9) **See margin.**

GETTING READY FOR THE NEXT LESSON

PREREQUISITE SKILL Solve each equation. (Lesson 1-5)

41. $7^2 + 5^2 = c$ **74**   42. $4^2 + b = 36$ **20**   43. $3^2 + a = 25$ **16**   44. $9^2 + 2^2 = c$ **85**

msmath2.net/self_check_quiz

Answers

30.

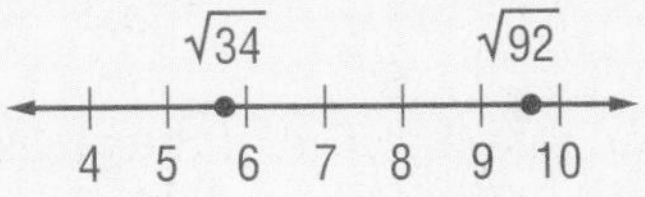

40.

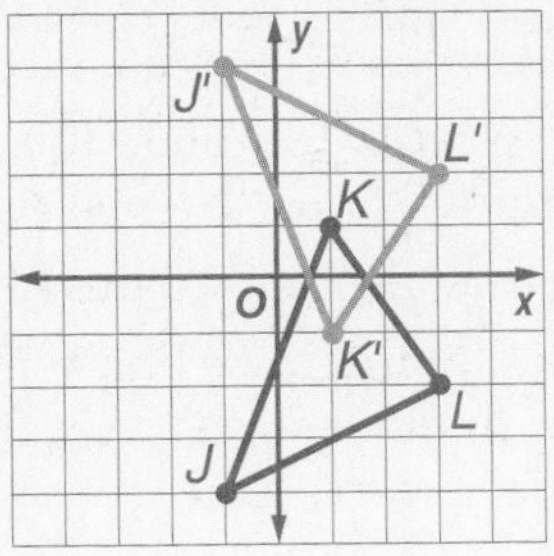

$J'(-1, 4)$, $K'(1, -1)$, $L'(3, 2)$

Study Guide and Intervention, p. 614

NAME ______ DATE ______ PERIOD ____

11-2 Study Guide and Intervention
Estimating Square Roots

Recall that a perfect square is a square of a rational number. In Lesson 5-8, you learned that any number that can be written as a fraction is a rational number. A number that cannot be written as a fraction is an **irrational number**.

EXAMPLE 1 **Estimate $\sqrt{40}$ to the nearest whole number.**

List some perfect squares.

1, 4, 9, 16, 25, 36, 49, …

$36 < 40 < 49$ 40 is between the perfect squares 36 and 49.

$\sqrt{36} < \sqrt{40} < \sqrt{49}$ Find the square root of each number.

$6 < \sqrt{40} < 7$ $\sqrt{36} = 6$ and $\sqrt{49} = 7$

So, $\sqrt{40}$ is between 6 and 7. Since 40 is closer to 36 than to 49, the best whole number estimate is 6.

EXAMPLE 2 **Use a calculator to find the value of $\sqrt{28}$ to the nearest tenth.**

2nd [√] 28 ENTER 5.291502622

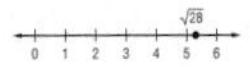

$\sqrt{28} \approx 5.3$

Check Since $5^2 = 25$ and 25 is close to 28, the answer is reasonable.

EXERCISES

Estimate each square root to the nearest whole number.

1. $\sqrt{3}$ 2   2. $\sqrt{8}$ 3
3. $\sqrt{26}$ 5   4. $\sqrt{41}$ 6
5. $\sqrt{61}$ 8   6. $\sqrt{94}$ 10
7. $\sqrt{152}$ 12   8. $\sqrt{850}$ 29

Use a calculator to find each square root to the nearest tenth.

9. $\sqrt{2}$ 1.4   10. $\sqrt{27}$ 5.2
11. $\sqrt{73}$ 8.5   12. $\sqrt{82}$ 9.1
13. $\sqrt{105}$ 10.2   14. $\sqrt{395}$ 19.9
15. $\sqrt{846}$ 29.1   16. $\sqrt{2{,}298}$ 47.9

Practice: Skills, p. 615

NAME ______ DATE ______ PERIOD ____

11-2 Practice: Skills
Estimating Square Roots

Estimate each square root to the nearest whole number.

1. $\sqrt{5}$ 2   2. $\sqrt{10}$ 3   3. $\sqrt{21}$ 5
4. $\sqrt{28}$ 5   5. $\sqrt{78}$ 9   6. $\sqrt{102}$ 10
7. $\sqrt{179}$ 13   8. $\sqrt{274}$ 17   9. $\sqrt{303}$ 17
10. $\sqrt{563}$ 24   11. $\sqrt{592}$ 24   12. $\sqrt{755}$ 27
13. $\sqrt{981}$ 31   14. $\sqrt{1{,}356}$ 37   15. $\sqrt{1{,}688}$ 41
16. $\sqrt{3{,}287}$ 57   17. $\sqrt{3{,}985}$ 63   18. $\sqrt{4{,}125}$ 64

Use a calculator to find each square root to the nearest tenth.

19. $\sqrt{6}$ 2.4   20. $\sqrt{19}$ 4.4   21. $\sqrt{30}$ 5.5
22. $\sqrt{77}$ 8.8   23. $\sqrt{114}$ 10.7   24. $\sqrt{125}$ 11.2
25. $\sqrt{149}$ 12.2   26. $\sqrt{182}$ 13.5   27. $\sqrt{212}$ 14.6
28. $\sqrt{436}$ 20.9   29. $\sqrt{621}$ 24.9   30. $\sqrt{853}$ 29.2
31. $\sqrt{918}$ 30.3   32. $\sqrt{1{,}004}$ 31.7   33. $\sqrt{1{,}270}$ 35.6
34. $\sqrt{5{,}438}$ 73.7   35. $\sqrt{4{,}215}$ 64.9   36. $\sqrt{5{,}786}$ 76.1

37. Order $\frac{25}{7}$, 4.91, and $\sqrt{23}$ from least to greatest. $\frac{25}{7}$, $\sqrt{23}$, 4.91

38. Graph $\sqrt{42}$ and $\sqrt{62}$ on the same number line.

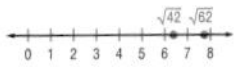

Practice: Word Problems, p. 616

NAME ______ DATE ______ PERIOD ____

11-2 Practice: Word Problems
Estimating Square Roots

1. **GEOMETRY** The diameter d of a circle with area A is given by the formula $d = \sqrt{\frac{4A}{\pi}}$. What is the diameter of a circle with an area of 56 square inches? Use 3.14 for π and round to the nearest tenth. 8.4 in.

2. **FENCING** Carmen wants to buy fencing to enclose a square garden with an area of 500 square feet. How much fencing does Carmen need to buy? Round to the nearest tenth. 89.4 ft

3. **OCEANS** The speed v in feet per second of an ocean wave in shallow water of depth d in feet is given by the formula $v = \sqrt{32d}$. What is the speed of an ocean wave at a depth of 10 feet? Round to the nearest tenth. 17.9 ft

4. **LIGHTING** A new flashlight has a beam whose width w at a distance d from the flashlight is given by the formula $w = 1.2\sqrt{d}$. What is the width of the beam at a distance of 30 feet? Round to the nearest tenth. 6.6 ft

5. **SOUND** The speed of sound in air c in meters per second at a temperature T in degrees Celsius is given approximately by the formula $c = \sqrt{402(T + 273)}$. What is the speed of sound in air at a temperature of 25 degrees Celsius? Round to the nearest tenth. 346.1 m/s

6. **PROJECTILES** The muzzle velocity v in feet per second necessary for a cannon to hit a target x feet away is estimated by the formula $v = \sqrt{32x}$. What muzzle velocity is required to hit a target 3,000 feet away? Round to the nearest tenth. 309.8 ft/s

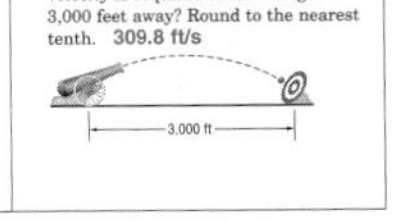

11-3a Hands-On Lab

Lesson Notes

Lesson Resources

Teaching Mathematics with Manipulatives
- centimeter grid, p. 11
- student recording sheet 11-3a

Glencoe Mathematics Classroom Manipulative Kit
- ruler
- scissors

GETTING STARTED

Teaching Tip Prior to class, prepare ropes knotted into 12 equal spaces. Allow students to work with the ropes to determine that a triangle formed with 3 spaces, 4 spaces, and 5 spaces must be a right triangle.

TEACH

- Encourage students to realign the squares to see that it is impossible to form a triangle that does not have a right angle.
- Students should understand that they cannot choose the lengths of the sides for a right triangle randomly. For example, a triangle with sides 4, 5, and 6 can never be a right triangle.

Answer

4. The sum of the squares of the lengths of the two perpendicular sides of a right triangle equals the square of the length of the third side.

11-3a Hands-On Lab

A Preview of Lesson 11-

What You'll Learn
Find the relationship among the sides of a right triangle.

Materials
- centimeter grid paper
- ruler
- scissors

The Pythagorean Theorem

INVESTIGATE *Work as a class.*

Four thousand years ago, the ancient Egyptians used mathematics to lay out their fields with square corners. They took a piece of rope and knotted it into 12 equal spaces. Taking three stakes, they stretched the rope around the stakes to form a right triangle. The sides of the triangle had lengths of 3, 4, and 5 units.

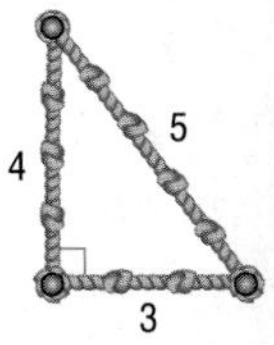

STEP 1 On grid paper, draw a segment that is 3 centimeters long. At one end of this segment, draw a perpendicular segment that is 4 centimeters long. Draw a third segment to form a triangle. Cut out the triangle.

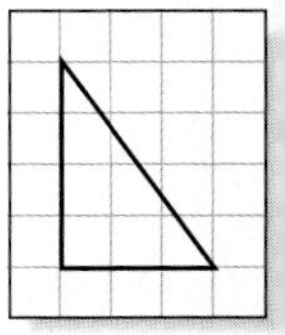

STEP 2 Measure the length of the longest side in centimeters. In this case, it is 5 centimeters.

STEP 3 Cut out three squares: one with 3 centimeters on a side, one with 4 centimeters on a side, and one with 5 centimeters on a side.

STEP 4 Place the edges of the squares against the corresponding sides of the right triangle.

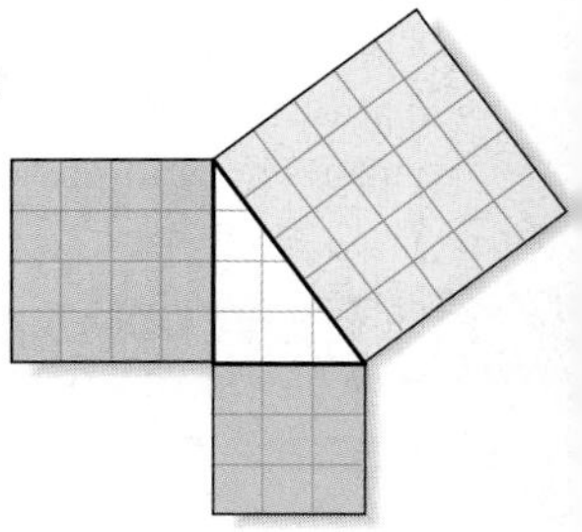

STEP 5 Find the area of each square.

Writing Math

Work with a partner.

1. What relationship exists among the areas of the three squares?

Repeat the activity for each right triangle whose perpendicular sides have the following measures. Write an equation to show your findings.

2. 6 cm, 8 cm $6^2 + 8^2 = 10^2$
3. 5 cm, 12 cm $5^2 + 12^2 = 13^2$
4. **Write** a sentence or two summarizing your findings. **See margin.**
5. **MAKE A CONJECTURE** Determine the length of the third side of a right triangle if the perpendicular sides of the triangle are 9 inches and 12 inches long. **15 in.**

1. The sum of the areas of the two smaller squares equals the area of the largest square.
2–3. See students' drawings.

ASSESS

Observe as students complete the **Writing Math** exercises. They should:
- cut out the triangle and squares neatly,
- measure the hypotenuse and realize that its length determines why the 5 × 5 square is used, and
- conclude that the sum of the areas of the smaller squares equals the area of the square formed by the hypotenuse.

Use Exercise 5 to determine whether students have intuitively grasped the Pythagorean Theorem.

11-3 The Pythagorean Theorem

/hat You'll Learn
nd length using the
/thagorean Theorem.

EW Vocabulary

g
/potenuse
/thagorean
'heorem

EVIEW Vocabulary

ght triangle: a
angle with exactly
ne angle that
easures 90°
ɜsson 10-4)

WHEN am I ever going to use this?

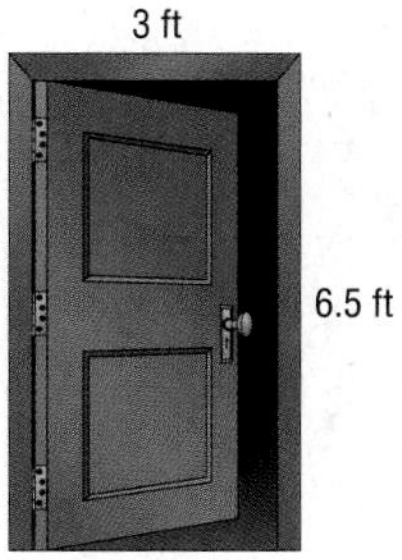

MOVING A square mirror 7 feet on each side must be delivered through the doorway.

1. Can the mirror fit through the doorway? Explain.
2. Make a scale drawing on grid paper to solve the problem. See margin.

It will not fit if the
irror is horizontal or
ertical. It may fit if it is
ted.

The sides of a right triangle have special names, as shown below.

The two sides adjacent to the right angle are the **legs**.

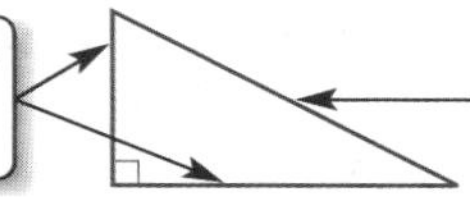

The side opposite the right angle is the **hypotenuse**.

The **Pythagorean Theorem** describes the relationship between the length of the hypotenuse and the lengths of the legs.

Key Concept — **Pythagorean Theorem**

Words In a right triangle, the square of the length of the hypotenuse equals the sum of the squares of the lengths of the legs.

Model

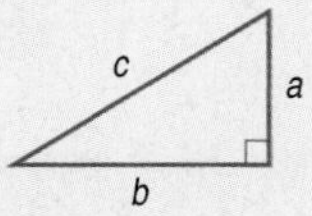

Symbols $c^2 = a^2 + b^2$

You can use the Pythagorean Theorem to find the length of the hypotenuse of a right triangle if the measures of both legs are known.

EXAMPLE Find the Length of the Hypotenuse

1 MOVING Determine whether a 7-foot square mirror will fit diagonally through the doorway shown at the right.

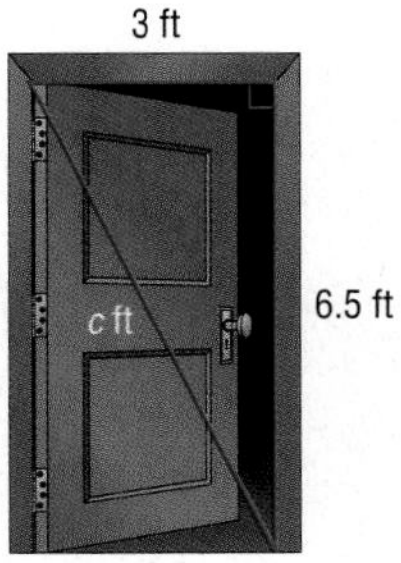

To solve, find the length of the hypotenuse c.

$c^2 = a^2 + b^2$	Pythagorean Theorem
$c^2 = 3^2 + 6.5^2$	Replace a with 3 and b with 6.5.
$c^2 = 9 + 42.25$	Evaluate 3^2 and 6.5^2.
$c^2 = 51.25$	Add.
$\sqrt{c^2} = \sqrt{51.25}$	Take the square root of each side.
$c \approx 7.2$	Simplify.

The length of the diagonal is about 7.2 feet. So, the mirror will fit through the doorway if it is turned diagonally.

TEACHING TIP
oint out that the
ɜlationship described
y the Pythagorean
heorem is true for
ny right triangle.

msmath2.net/extra_examples

11-3 Lesson Notes

Lesson Resources

Chapter 11 Resource Masters FAST FILE
- Study Guide and Intervention, p. 619
- Practice: Skills, p. 620
- Practice: Word Problems, p. 621
- Reading to Learn Mathematics, p. 622
- Enrichment, p. 623

Calculator and Spreadsheet Masters, p. 53
Key Teaching Transparencies, 43
MathPASS Tutorial CD-ROM, 47
Parent and Student Study Guide Workbook, 11-3

1 Focus

5-Minute Check 11-3

INTERACTIVE CHALKBOARD PowerPoint® Presentations

Use as a quiz or review of Lesson 11-2. (Also available as a transparency)

Bellringer

Reading Mathematics On the chalkboard, draw three triangles that are not quite right triangles. Students will probably say the figures are all right triangles. Explain that they should not assume that a figure is a right triangle unless it has a right angle symbol in the corner.

Answer (When am I ever going to use this?)

2. Sample Answer: The mirror will fit if it is turned, since there are more than 7 feet of space between opposite corners of the door.

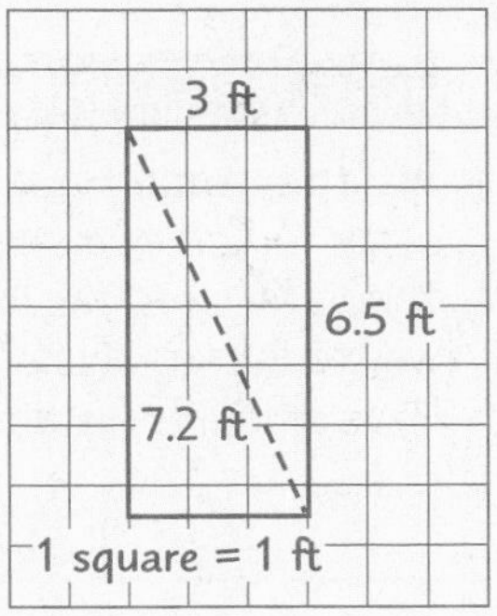

Reading to Learn Mathematics, p. 622 ELL

NAME ______ DATE ______ PERIOD ____

11-3 Reading to Learn Mathematics
The Pythagorean Theorem

Pre-Activity *Read the introduction at the top of page 479 in your textbook. Write your answers below.*

1. Can the mirror fit through the doorway? Explain. It will not fit if the mirror is horizontal or vertical. It may fit if it is tilted.
2. Make a scale drawing on grid paper to solve the problem. Sample answer: The mirror will fit if it is turned, since there is more than 7 feet of space between opposite corners of the door.

Reading the Lesson

3. In the Pythagorean Theorem $c^2 = a^2 + b^2$, which letter represents the length of the hypotenuse? c
4. How do you know that the diagonal of a rectangle is the hypotenuse of two ri... ...les? Sample answer: A rectangle has four right

2 Teach

In-Class Examples

INTERACTIVE CHALKBOARD
PowerPoint® Presentations

1 **GYMNASTICS** A gymnastics tumbling floor is in the shape of a square with sides 12 meters long. If a gymnast flips from one corner to the opposite corner, about how far has he flipped? **about 17 m**

2 Find the missing measure of the triangle below. **12 cm**

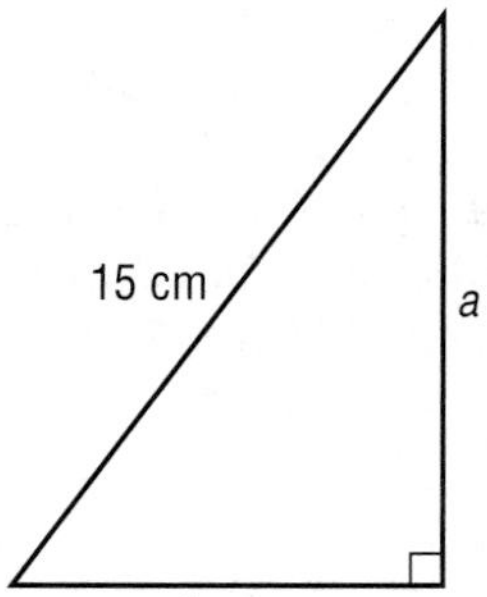

3 **TELEVISION** Televisions are measured according to their diagonal measure. If the diagonal of a television is 36 inches, and its height is 21.6 inches, what is its width? **28.8 in.**

Tips for New Teachers

Intervention Point out to students that in Examples 1 and 3, it doesn't matter which measure is used for a and b. The hypotenuse is always the longest side.

READING Math

Theorem A *theorem* is a statement in mathematics that can be justified by logical reasoning.

You can also use the Pythagorean Theorem to find the measure of a leg if the measure of the other leg and the hypotenuse are known.

EXAMPLE Find the Length of a Leg

2 **Find the missing measure of the triangle at the right.**

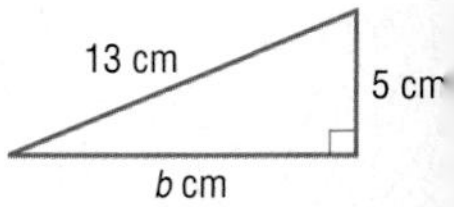

The missing measure is of a leg of the triangle.

$c^2 = a^2 + b^2$	Pythagorean Theorem
$13^2 = 5^2 + b^2$	Replace a with 5 and c with 13.
$169 = 25 + b^2$	Evaluate 13^2 and 5^2.
$169 - 25 = 25 + b^2 - 25$	Subtract 25 from each side.
$144 = b^2$	Simplify.
$\sqrt{144} = \sqrt{b^2}$	Take the square root of each side.
$12 = b$	Simplify.

The length of the leg is 12 centimeters.

Your Turn **Find the missing measure of each right triangle. Round to the nearest tenth if necessary.**

a.

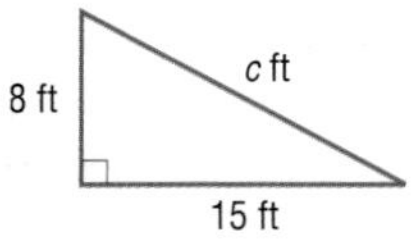

17 ft

b.

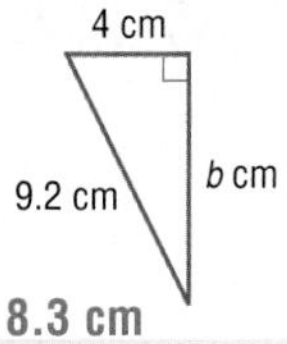

8.3 cm

c. $b = 7$ in., $c = 25$ in. **24 in.**

REAL-LIFE CAREERS

How Does an Archaeologist Use Math? Before digging, archaeologists use the Pythagorean Theorem to calculate the diagonal of an excavation site to be sure that the area is a rectangle.

Online Research For information about a career as an archaeologist, visit: msmath2.net/careers

EXAMPLE Solve a Real-Life Problem

3 **ARCHAEOLOGY** **Archaeologists placed corner stakes to mark a rectangular excavation site, as shown at the right. If their stakes are placed correctly, what is the measure of the diagonal?**

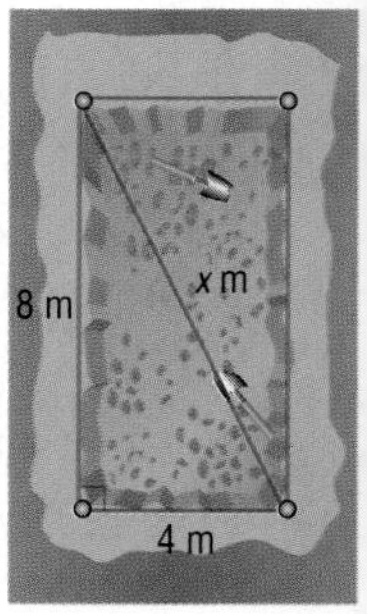

The diagonal of the rectangle is the hypotenuse of a right triangle. Write an equation to solve for x.

$c^2 = a^2 + b^2$	Pythagorean Theorem
$x^2 = 8^2 + 4^2$	Replace a with 8, b with 4, and c with x.
$x^2 = 64 + 16$	Evaluate 8^2 and 4^2.
$x^2 = 80$	Simplify.
$\sqrt{x^2} = \sqrt{80}$	Take the square root of each side.
$x \approx 8.9$	Simplify.

The length of the diagonal is about 8.9 meters.

DAILY INTERVENTION **Differentiated Instruction**

Intrapersonal Have students think about what is meant by *the square of the length of a side* and how this phrase relates to the Pythagorean Theorem. One way to think of this is as multiplying numbers. Another way to think of it is as the area of a square figure having that length. Have students write out their personal interpretations of the phrase.

You can determine whether a triangle is a right triangle by applying the Pythagorean Theorem.

EXAMPLES Identify Right Triangles

Determine whether a triangle with the given lengths is a right triangle.

STUDY TIP

Hypotenuse Remember that the hypotenuse is always the longest side.

4 **1.5 mm, 2 mm, 2.5 mm**

$c^2 = a^2 + b^2$

$2.5^2 \stackrel{?}{=} 1.5^2 + 2^2$

$6.25 \stackrel{?}{=} 2.25 + 4$

$6.25 = 6.25$ ✔

The triangle is a right triangle.

5 **4 ft, 6 ft, 8 ft**

$c^2 = a^2 + b^2$

$8^2 \stackrel{?}{=} 4^2 + 6^2$

$64 \stackrel{?}{=} 16 + 36$

$64 \neq 52$

The triangle is *not* a right triangle.

Your Turn **Determine whether each triangle with the given lengths is a right triangle. Write *yes* or *no*.**

d. 7.5 cm, 8 cm, 12 cm **no**

e. 9 in., 40 in., 41 in. **yes**

Skill and Concept Check

Writing Math Exercises 1 & 3

1. **Describe** the information that you need in order to find the missing measure of a right triangle. **the lengths of both legs or the length of one leg and the hypotenuse**
2. **OPEN ENDED** Draw and label a right triangle that has one side measuring 14 units. Write the length of another side. Then find the length of the third side to the nearest tenth. **See margin.**
3. **FIND THE ERROR** Devin and Jamie are writing an equation to find the missing measure of the triangle at the right. Who is correct? Explain.

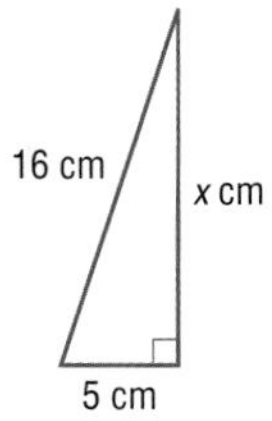

Devin	Jamie
$16^2 = 5^2 + x^2$	$x^2 = 16^2 + 5^2$

Devin; the hypotenuse is 16 cm, not x cm as Jamie assumed.

GUIDED PRACTICE

Find the missing measure of each right triangle. Round to the nearest tenth if necessary.

4. 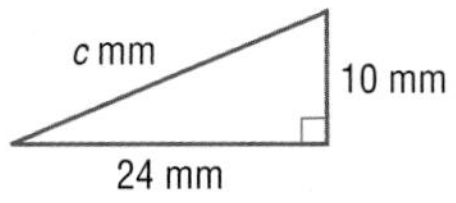

26 mm

5. 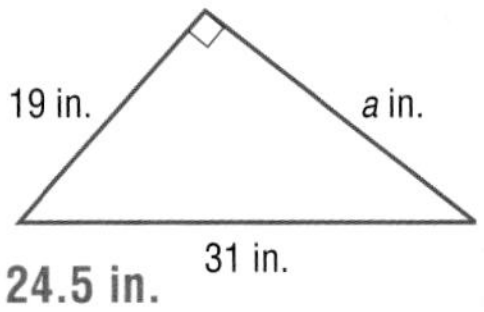

24.5 in.

6. $b = 21$ cm, $c = 28$ cm **18.5 cm**

Determine whether a triangle with the given side lengths is a right triangle. Write *yes* or *no*.

7. 1.4 m, 4.8 m, 5 m **yes**
8. 21 ft, 24 ft, 30 ft **no**

Answer

2. Sample answer:

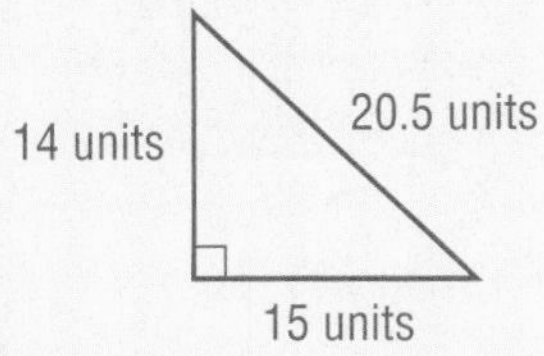

In-Class Examples

INTERACTIVE CHALKBOARD PowerPoint® Presentations

Determine whether a triangle with the given lengths is a right triangle.

4 2.5 cm, 6 cm, 6.5 cm **yes**

5 5 ft, 6 ft, 8 ft **no**

3 Practice/Apply

FOLDABLES™ Follow-Up Remind students to use their Foldables to record what they learned about the Pythagorean Theorem. Encourage them to draw a right triangle, give the length of its two legs, and explain how to find the hypotenuse using the Pythagorean Theorem.

DAILY INTERVENTION **FIND THE ERROR** In Exercise 3, have students label the appropriate sides of the triangle as *a*, *b*, and *c*. Then write out the Pythagorean Theorem and substitute the numbers for the variables. This method will ensure they have the numbers in the correct places.

About the Exercises...

Guided Practice Key

Exercises	Examples
4–6	2
7–8	4, 5

Odd/Even Assignments

Exercises 9–22 are structured so that students practice the same concepts whether they are assigned odd or even problems.

Study Guide and Intervention, p. 619

NAME ______________ DATE ________ PERIOD ____

11-3 Study Guide and Intervention

The Pythagorean Theorem

The sides of a right triangle have special names. The sides adjacent to the right angle are the **legs**. The side opposite the right angle is the **hypotenuse**. The **Pythagorean Theorem** describes the relationship between the length of the hypotenuse and the lengths of the legs. In a right triangle, the square of the length of the hypotenuse equals the sum of the squares of the lengths of the legs.

$c^2 = a^2 + b^2$

EXAMPLE 1 **Find the missing measure of a right triangle if $a = 4$ inches and $b = 3$ inches.**

$c^2 = a^2 + b^2$	Pythagorean Theorem
$c^2 = 4^2 + 3^2$	Replace a with 4 and b with 3.
$c^2 = 16 + 9$	Evaluate 4^2 and 3^2.
$c^2 = 25$	Add.
$\sqrt{c^2} = \sqrt{25}$	Take the square root of each side.
$c = 5$	Simplify.

The length of the hypotenuse is 5 inches.

EXAMPLE 2 **Determine whether a triangle with side lengths of 6 meters, 9 meters, and 12 meters is a right triangle.**

$c^2 = a^2 + b^2$	Pythagorean Theorem
$12^2 \stackrel{?}{=} 6^2 + 9^2$	Replace a with 6, b with 9, and c with 12.
$144 \stackrel{?}{=} 36 + 81$	Simplify.
$144 \neq 117$	Add.

The triangle is *not* a right triangle.

EXERCISES

Find the missing measure of each right triangle. Round to the nearest tenth if necessary.

1. 9 in., 4 in., c in. 9.8 in.
2. 5 m, c m, 7.5 m 9.0 m
3. a cm, 13 cm, 16 cm 9.3 cm

Determine whether each triangle with the given side lengths is a right triangle. Write *yes* or *no*.

4. 15 ft, 8 ft, 17 ft yes
5. 5 in., 13 in., 17 in. no
6. 9 yd, 40 yd, 41 yd yes

Practice: Skills, p. 620

NAME ______________ DATE ________ PERIOD ____

11-3 Practice: Skills

The Pythagorean Theorem

Find the missing measure of each right triangle. Round to the nearest tenth if necessary.

1. b m, 20 m, 7 m 18.7 m
2. 5 in., c in., 5 in. 7.1 in.
3. 12.4 ft, 15 ft, a ft 8.4 ft
4. x cm, 26 cm, 20 cm 32.8 cm
5. c in., 20.3 in., 32 in. 37.9 in.
6. a yd, 2.7 yd, 3 yd 1.3 yd
7. a m, 11.2 m, 6 m 9.5 m
8. 6.7 mm, c mm, 14 mm 15.5 mm
9. $a = 15$ cm, $b = 20$ cm 25 cm
10. $a = 2$ yd, $b = 12$ yd 12.2 yd
11. $a = 13$ in., $c = 16.5$ in. 10.2 in.
12. $b = 8$ mm, $c = 17$ mm 15 mm
13. $a = 1.3$ ft, $b = 4.6$ ft 4.8 ft
14. $a = 14.7$ m, $c = 23$ m 17.7 m

Determine whether each triangle with the given side lengths is a right triangle. Write *yes* or *no*.

15. 10 ft, 24 ft, 26 ft yes
16. 5 in., 8 in., 9 in. no
17. 6 cm, 9 cm, 12 cm no
18. 4.5 mm, 6.0 mm, 7.5 mm yes

Practice: Word Problems, p. 621

NAME ______________ DATE ________ PERIOD ____

11-3 Practice: Word Problems

The Pythagorean Theorem

1. **ORIGAMI** Chee has a piece of paper measuring 8.5 inches by 8.5 inches. If she folds the paper diagonally in half, how long is the folded side? Round to the nearest tenth. 12.0 in.

2. **COMPUTERS** In a computer catalog, a computer monitor is said to be 19 inches. This distance is the diagonal distance across the screen. If the screen is 10 inches high, what is the width of the screen? Round to the nearest tenth. 16.2 in.

3. **ANTENNAS** A wire 10 meters long is supporting a utility pole. The wire is anchored to the ground and is attached to the pole 9 meters above the ground. What is the distance from the bottom of the pole to the point where the wire is attached to the ground? Round to the nearest tenth. 4.4 m

4. **RAMPS** Crystal wants to build a ramp that will rise 4 feet over a horizontal distance of 20 feet. How long will the ramp be? Round to the nearest tenth. 20.4 ft

5. **POOLS** Salomon swims diagonally across his pool every day. If Salomon's pool is 4 meters wide and 16 meters diagonally across, how long is his pool, to the nearest tenth of a meter? 15.5 m

6. **FRAMES** Rosa has a picture frame that measures 12 inches by 18 inches. What is the diagonal distance across the frame? Round to the nearest tenth. 21.6 in.

★ indicates increased difficulty

Practice and Applications

HOMEWORK HELP

For Exercises	See Examples
9–16	1, 2
17–20	4, 5
21–22	3

Extra Practice See pages 590, 606.

Find the missing measure of each right triangle. Round to the nearest tenth if necessary.

9. 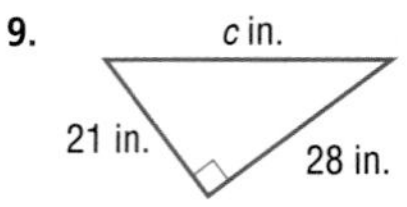

35 in.

10. 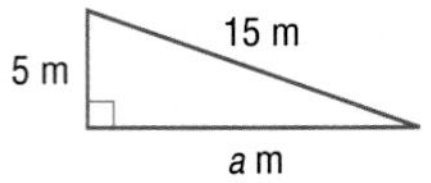

14.1 m

11. 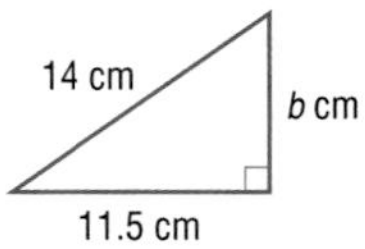

8.0 cm

12. 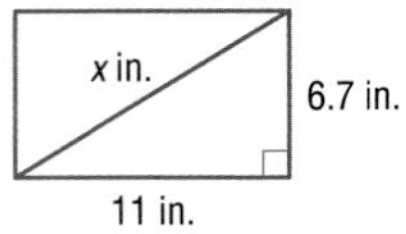

12.9 in.

13. $a = 7$ in., $b = 24$ in. 25 in.
14. $a = 13.5$ mm, $b = 18$ mm 22.5 mm
15. $b = 13$ m, $c = 27$ m 23.7 m
16. $a = 2.4$ yd, $c = 3$ yd 1.8 yd

Determine whether a triangle with the given side lengths is a right triangle. Write *yes* or *no*.

17. 12 cm, 16 cm, 20 cm yes
18. 8 m, 15 m, 17 m yes
19. 11 ft, 14 ft, 17 ft no
20. 18 in., 18 in., 36 in. no

21. **SAFETY** To the nearest tenth of a foot, how far up the wall x does the ladder shown at the right reach? 19.1 ft

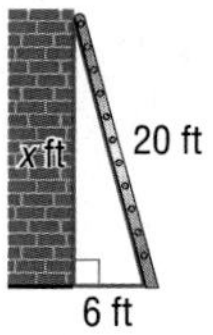

★ 22. **TRAVEL** You drive 80 miles east, then 50 miles north, then 140 miles west, and finally 95 miles south. Make a drawing to find how far you are from your starting point. See margin for drawing; 75 mi.

23. **CRITICAL THINKING** What is the length of the diagonal of the cube shown at the right? about 10.4 in.

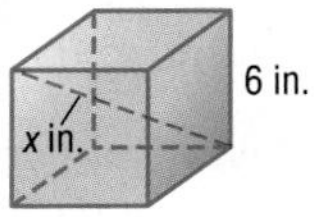

Standardized Test Practice and Mixed Review

24. **MULTIPLE CHOICE** Find the missing measure of a right triangle if $a = 20$ meters and $c = 52$ meters. C
   - Ⓐ 24 m
   - Ⓑ 32 m
   - Ⓒ 48 m
   - Ⓓ 55.7 m

25. **SHORT RESPONSE** An isosceles right triangle has legs that are 8 inches long. Find the length of the hypotenuse to the nearest tenth. 11.3 in.

Estimate each square root to the nearest whole number. (Lesson 11-2)

26. $\sqrt{61}$ 8
27. $\sqrt{147}$ 12
28. $\sqrt{40}$ 6
29. $\sqrt{277}$ 17

30. Find $\sqrt{256}$. (Lesson 11-1) 16

GETTING READY FOR THE NEXT LESSON

PREREQUISITE SKILL **Multiply.** (Lesson 6-4)

31. $17.8 \cdot 12$ 213.6
32. $21.5 \cdot 27.1$ 582.65
33. $3\frac{1}{2} \cdot 8$ 28
34. $15\frac{1}{4} \cdot 18$ $274\frac{1}{2}$

msmath2.net/self_check_quiz

4 Assess

Open-Ended Assessment

Writing Have each student write three word problems that can be solved using the Pythagorean Theorem. Students should then exchange problems and solve them.

Answer

22.

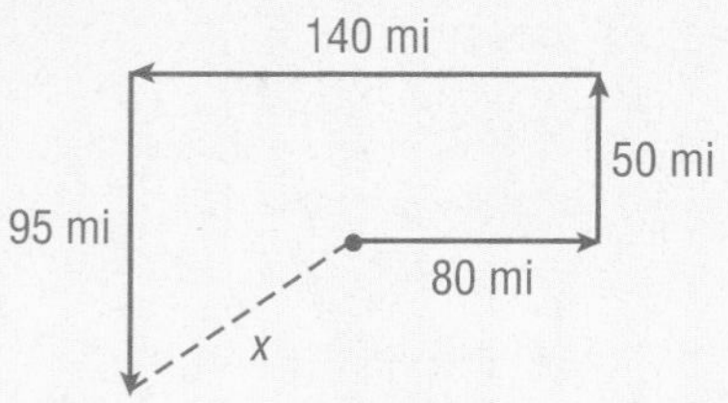

11-4 Area of Parallelograms

What You'll Learn
Find the areas of parallelograms.

NEW Vocabulary
base
height

REVIEW Vocabulary
parallelogram: quadrilateral with opposite sides parallel and opposite sides congruent (Lesson 10-5)

HANDS-ON Mini Lab

Materials
- grid paper
- straightedge

Work with a partner.

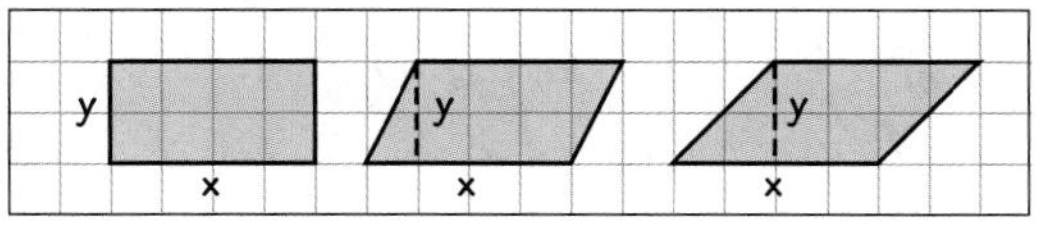

1. What is the value of x and y for each parallelogram?
2. Count the grid squares to find the area of each parallelogram.
3. On grid paper, draw three different parallelograms in which $x = 5$ units and $y = 4$ units. Find the area of each.
4. **Make a conjecture** about how to find the area of a parallelogram if you know the values of x and y. The area equals $x \cdot y$.

1. $x = 4$ units, $y = 2$ units 2. 8 sq units
3. See Ch. 11 Answer Appendix.

In Lesson 10-5, you learned that a parallelogram is a special kind of quadrilateral. You can find the area of a parallelogram by using the values for the base and height, as described below.

The **base** is any side of a parallelogram.

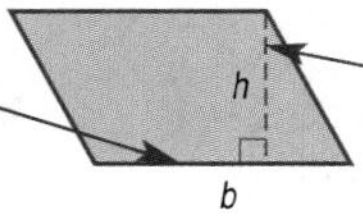

The **height** is the length of the segment perpendicular to the base with endpoints on opposite sides.

Key Concept — Area of a Parallelogram

Words The area A of a parallelogram equals the product of its base b and height h.

Model

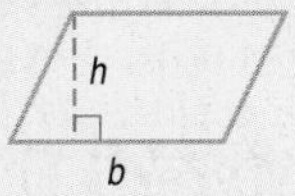

Symbols $A = bh$

EXAMPLE Find the Area of a Parallelogram

1 Find the area of the parallelogram.

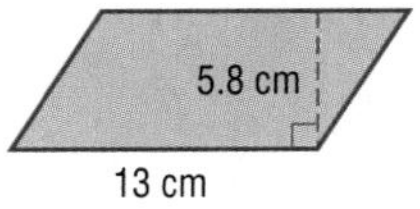

Estimate $A = 13 \cdot 6$ or 78 cm^2

$A = bh$ — Area of a parallelogram

$A = 13 \cdot 5.8$ — Replace b with 13 and h with 5.8.

$A = 75.4$ — Multiply.

The area of the parallelogram is 75.4 square centimeters. This is close to the estimate.

TEACHING TIP
Point out that in the Mini Lab figures, $b = x$ and $h = y$.

TEACHING TIP
More simply, the *height* of a parallelogram is the distance from the base to the opposite side.

msmath2.net/extra_examples

11-4 Lesson Notes

Lesson Resources

Chapter 11 Resource Masters FAST FILE
- Study Guide and Intervention, p. 624
- Practice: Skills, p. 625
- Practice: Word Problems, p. 626
- Reading to Learn Mathematics, p. 627
- Enrichment, p. 628

Calculator and Spreadsheet Masters, p. 54

Parent and Student Study Guide Workbook, 11-4

Teaching Mathematics with Manipulatives, pp. 108–109

1 Focus

5-Minute Check 11-4
INTERACTIVE CHALKBOARD PowerPoint® Presentations

Use as a quiz or review of Lesson 11-3. (Also available as a transparency)

Bellringer
Problem Solving In geometry, students should attempt to visualize the problems they are solving. Draw figures like the ones below on the chalkboard. Have students draw a line on each figure, as shown in the first one, to show how part of the figure could be mentally sliced off and moved to form a rectangle.

(Continued in bottom margin.)

Reading to Learn Mathematics, p. 627 ELL

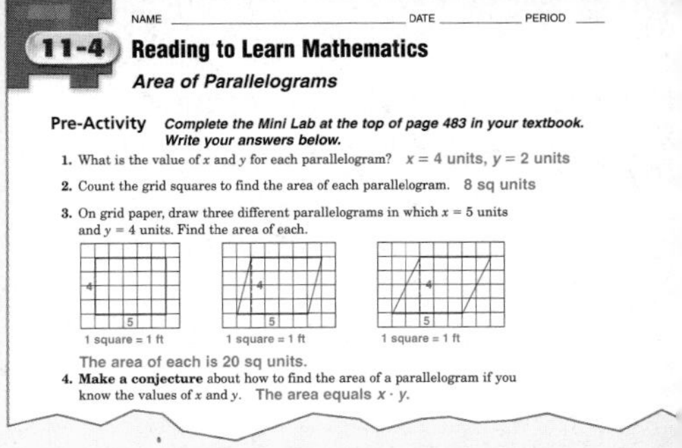
NAME ______ DATE ______ PERIOD ____

11-4 Reading to Learn Mathematics
Area of Parallelograms

Pre-Activity *Complete the Mini Lab at the top of page 483 in your textbook. Write your answers below.*
1. What is the value of x and y for each parallelogram? $x = 4$ units, $y = 2$ units
2. Count the grid squares to find the area of each parallelogram. 8 sq units
3. On grid paper, draw three different parallelograms in which $x = 5$ units and $y = 4$ units. Find the area of each.

1 square = 1 ft 1 square = 1 ft 1 square = 1 ft

The area of each is 20 sq units.
4. **Make a conjecture** about how to find the area of a parallelogram if you know the values of x and y. The area equals $x \cdot y$.

This action should help students relate the formulas for the area of rectangles and the area of parallelograms.

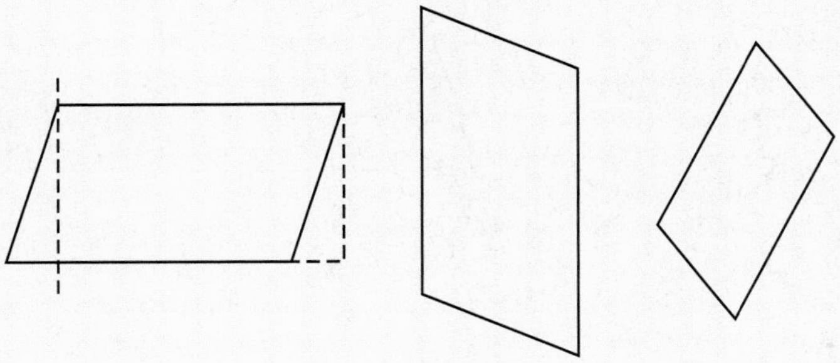

2 Teach

In-Class Examples

INTERACTIVE CHALKBOARD
PowerPoint® Presentations

1 Find the area of the parallelogram. **48 cm²**

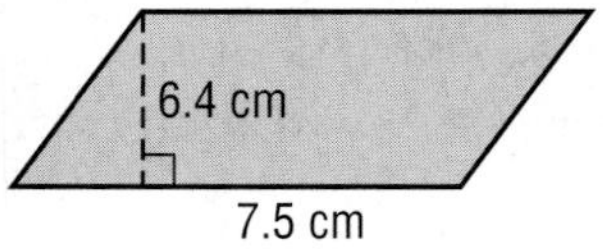

2 Find the area of the parallelogram below. **36 cm²**

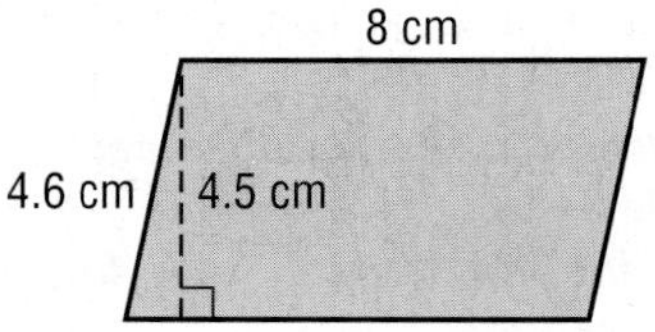

3 Practice/Apply

NUMBER SENSE
To help students understand Exercise 3, have them look at the formula for the area, $A = bh$. If they replace b with $2b$ and h with $2h$, they will see that the new area is *not* $2A$.

About the Exercises...

Guided Practice Key

Exercises	Examples
4–9	1, 2

Odd/Even Assignments
Exercises 10–19 are structured so that students practice the same concepts whether they are assigned odd or even problems.

4 Assess

Open-Ended Assessment

Speaking Have students discuss why the formulas for the area of a rectangle and the area of a parallelogram are the same. Encourage them to suggest how one can be transformed into the other.

STUDY TIP

Common Error
When finding the area of a parallelogram, be sure to use the base and the height, not the base and a side.

EXAMPLE Find the Area of a Parallelogram

2 **Find the area of the parallelogram at the right.**

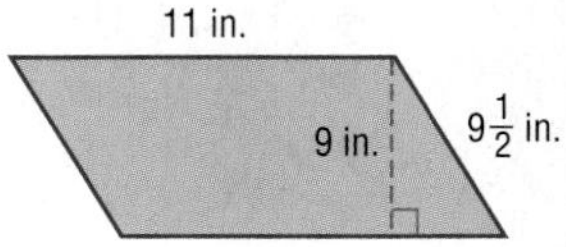

The base is 11 inches, and the height is 9 inches.

Estimate $A = 10 \cdot 10$ or 100 in²

$A = bh$ Area of a parallelogram

$A = 11 \cdot 9$ Replace b with 11 and h with 9.

$A = 99$ Multiply.

The area of the parallelogram is 99 square inches. This is close to the estimate.

Your Turn **Find the area of each parallelogram.**

a. base = 5 cm
height = 8 cm **40 cm²**

b. **71.4 ft²**

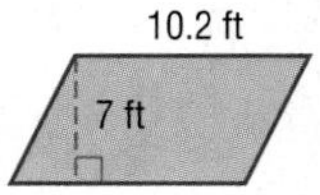

Skill and Concept Check

Writing Math
Exercises 1 & 3

1. **Describe** what values you would substitute in the formula $A = bh$ to find the area of the parallelogram at the right. **$b = 6$, $h = 3$**

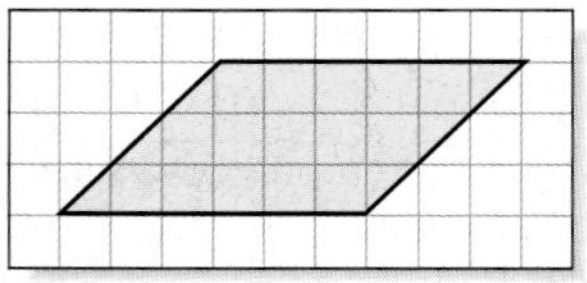

2. **OPEN ENDED** Draw three different parallelograms, each with an area of 24 square units. **See margin.**

3. **NUMBER SENSE** *True* or *False*? The area of a parallelogram doubles if you double the base and the height. Explain or give a counterexample to support your answer. **False; if the base and height are each doubled, then the area is $2b \cdot 2h = 4bh$, or 4 times greater.**

GUIDED PRACTICE

Find the area of each parallelogram. Round to the nearest tenth if necessary.

4. 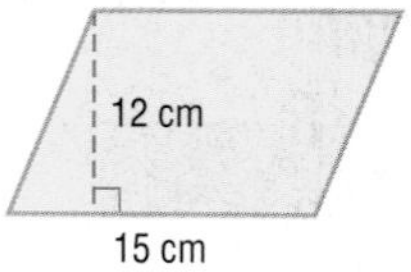

180 cm²

5. 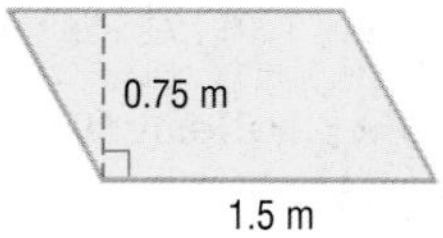

1.1 m²

6. 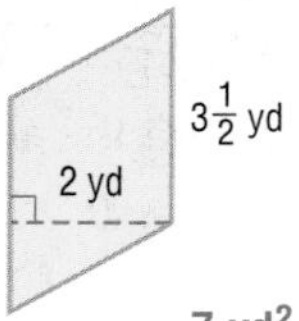

7 yd²

7. base = 16 in.
height = 4 in. **64 in²**

8. base = 3.5 m
height = 5 m **17.5 m²**

9. What is the area of a parallelogram with a base of 25 millimeters and a height that is half the base? **312.5 mm²**

Answer

2. Sample answer:

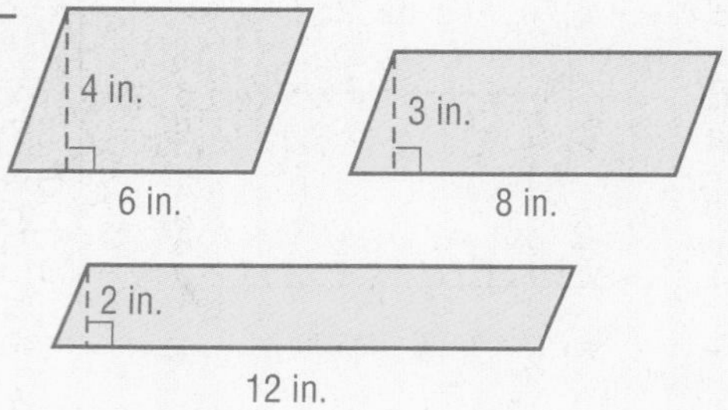

★ indicates increased difficulty

Practice and Applications

HOMEWORK HELP	
For Exercises	See Examples
10–21	1, 2

Extra Practice
See pages 591, 606.

Find the area of each parallelogram. Round to the nearest tenth if necessary.

10. 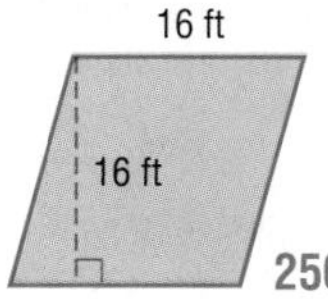

256 ft^2

11. 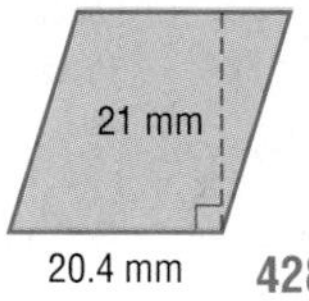

428.4 mm^2

12. 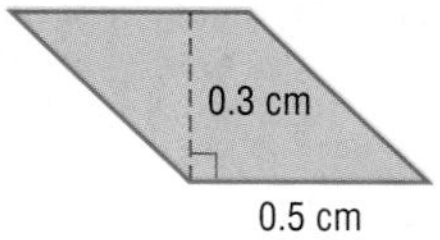

0.2 cm^2

13. 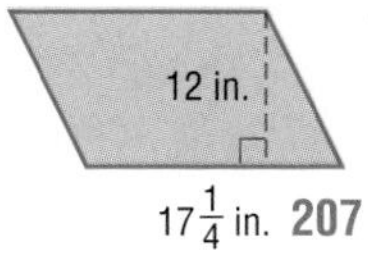

207 in^2

★ 14. 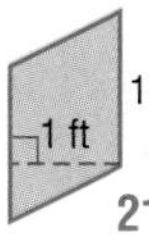

216 in^2 or 1.5 ft^2

★ 15. 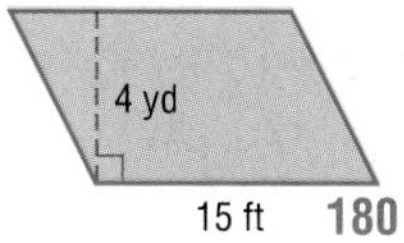

180 ft^2 or 20 yd^2

16. base = 13 mm
height = 6 mm 78 mm^2

17. base = 45 yd
height = 35 yd $1{,}575\text{ yd}^2$

18. base = 8 in.
height = 12.5 in. 100 in^2

19. base = 7.9 cm
height = 7.2 cm 56.9 cm^2

★ 20. **MULTI STEP** A quilted block uses eight parallelogram-shaped pieces of cloth, each with a height of $3\frac{1}{3}$ inches and a base of $3\frac{3}{4}$ inches. How much fabric is needed to make the parallelogram pieces for 24 blocks? Write in square feet. (*Hint*: $144\text{ in}^2 = 1\text{ ft}^2$) $16\frac{2}{3}\text{ ft}^2$

★ 21. What is the height of a parallelogram if the base is 24 inches and the area is 360 square inches? 15 in.

22. **CRITICAL THINKING** Identify two possible measures of base and height for a parallelogram that has an area of 320 square inches.
Sample answer: 32 in. and 10 in., 16 in. and 20 in.

Standardized Test Practice and Mixed Review

23. **MULTIPLE CHOICE** Find the area of the parallelogram. C

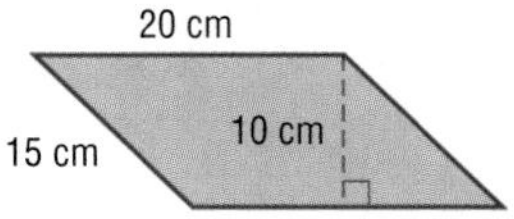

Ⓐ 75 cm^2
Ⓑ 150 cm^2
Ⓒ 200 cm^2
Ⓓ 300 cm^2

24. **SHORT RESPONSE** What is the base of a parallelogram if the height is 18.6 inches and the area is 279 square inches? 15 in.

Determine whether a triangle with the given side lengths is a right triangle. Write *yes* or *no*. (Lesson 11-3)

25. 8 in., 10 in., 12 in. no
26. 12 ft, 16 ft, 20 ft yes
27. 5 cm, 12 cm, 14 cm no

28. Which is closer to $\sqrt{55}$, 7 or 8? (Lesson 11-2) 7

GETTING READY FOR THE NEXT LESSON

PREREQUISITE SKILL Find each value. (Lesson 1-3)

29. $6(4 + 10)$ 84
30. $\frac{1}{2}(8)(8)$ 32
31. $\frac{1}{2}(24 + 15)$ 19.5
32. $\frac{1}{2}(5)(13 + 22)$ 87.5

msmath2.net/self_check_quiz

DAILY INTERVENTION — Unlocking Misconceptions

Parallelograms Be sure students do not think of a parallelogram as a collapsed box that they can mentally straighten to form a rectangle. This may cause them to calculate the area as the length of a base multiplied by the length of a side.

Study Guide and Intervention, p. 624

NAME ______________ DATE ________ PERIOD ____

11-4 Study Guide and Intervention
Area of Parallelograms

The area A of a parallelogram equals the product of its base b and its height h.

$A = bh$

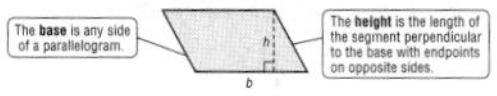

EXAMPLE 1 **Find the area of a parallelogram if the base is 6 inches and the height is 3.7 inches.**

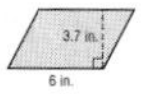

Estimate $A = 6 \cdot 4$ or 24 in^2

$A = bh$ Area of a parallelogram
$A = 6 \cdot 3.7$ Replace b with 6 and h with 3.7.
$A = 22.2$ Multiply.

The area of the parallelogram is 22.2 square inches. This is close to the estimate.

EXAMPLE 2 **Find the area of the parallelogram at the right.**

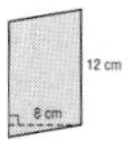

Estimate $A = 10 \cdot 10$ or 100 cm^2

$A = bh$ Area of a parallelogram
$A = 12 \cdot 8$ Replace b with 12 and h with 8.
$A = 96$ Multiply.

The area of the parallelogram is 96 square centimeters. This is close to the estimate.

EXERCISES

Find the area of each parallelogram. Round to the nearest tenth if necessary.

1. 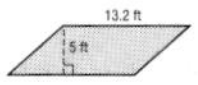

66 ft^2

2. 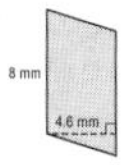

36.8 mm^2

3. 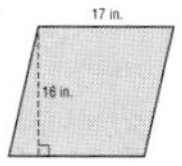

272 in^2

Practice: Skills, p. 625

NAME ______________ DATE ________ PERIOD ____

11-4 Practice: Skills
Area of Parallelograms

Find the area of each parallelogram. Round to the nearest tenth if necessary.

1. base = 5 ft
height = 12 ft 60 ft^2

2. base = 9 in.
height = 2 in. 18 in^2

3. base = 6 cm
height = 5.5 cm 33 cm^2

4. base = $4\frac{2}{5}$ yd
height = 2 yd $8\frac{4}{5}\text{ yd}^2$

5. base = 15.3 mm
height = 8 mm 122.4 mm^2

6. base = 19.6 m
height = 14.5 m 284.2 m^2

7. 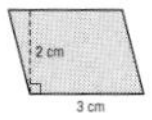

6 cm^2

8. 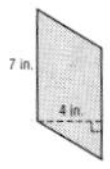

28 in^2

9. 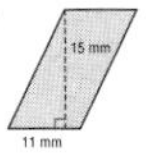

165 mm^2

10. 

108 ft^2

11. 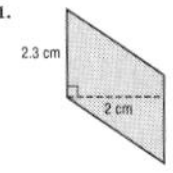

4.6 cm^2

12. 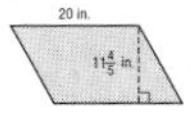

236 in^2

13. 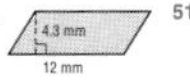

51.6 mm^2

14. 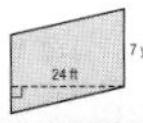

504 ft^2 or 56 yd^2

Practice: Word Problems, p. 626

NAME ______________ DATE ________ PERIOD ____

11-4 Practice: Word Problems
Area of Parallelograms

1. **SAILS** Joyce wants to construct a sail with the dimensions shown. How much material will be used?
350 ft^2

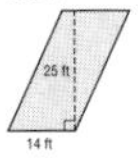

2. **SIGNS** Pedro wants to make the sign in the shape shown and needs to know how much material will be needed. What is the area of the sign?
$1{,}050\text{ in}^2$

3. **SHADING** Alma's engineering firm must determine the area of the largest noontime shadow that a proposed building design will create. What is the area of the shadow? $2{,}240\text{ ft}^2$

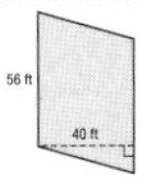

4. **POOLS** Tamika has designed a pool in the shape shown. What is the area of the bottom of the pool if the surface is perfectly flat?
600 m^2

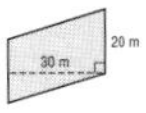

5. **CITY PLANNING** Two parallel streets are cut across by two other parallel streets as shown in the figure, cutting off a parcel of land in the shape of a parallelogram. Find the area of the parcel of land. $85{,}000\text{ ft}^2$

6. **TARPS** Neka wants to cut a tarp in the shape shown. What is the minimum amount of canvas cloth that he will need?
$1{,}440\text{ ft}^2$

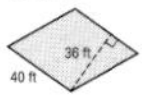

Chapter 11 Mid-Chapter Practice Test

Assessment Option

Mid-Chapter Test (Lessons 11-1 through 11-4) is available on p. 665 of the *Chapter 11 Resource Masters.*

ExamView® Pro Test Bank CD-ROM

Use the networkable ExamView® Pro Test Bank CD-ROM to:

- Create **Multiple versions** of tests.
- Create **Modified** tests for **Inclusion** students with one mouse click.
- **Edit** existing questions and **Add** your own questions.
- Build tests aligned with state standards using built-in **State Curriculum Correlations.**
- Change **English** tests to **Spanish** with one mouse click and vice versa.

FOLDABLES™ Follow-Up

Before students complete the Mid-Chapter Practice Test, encourage them to review the notes, definitions, concepts, and main ideas they recorded in their Foldables as they progressed through the first four lessons of the chapter.

Chapter 11 Mid-Chapter Practice Test

2. In a right triangle, the square of the length of the hypotenuse equals the sum of the squares of the lengths of the legs.

Vocabulary and Concepts

1. **Define** *square root*. (Lesson 11-1) one of two equal factors of a number
2. **State,** in words, the Pythagorean Theorem. (Lesson 11-3)
3. *True* or *False*? The area of any parallelogram equals the length times the width. Explain. (Lesson 11-4) False; the area equals the base times the height. It is true only for parallelograms that are rectangles.

Skills and Applications

Find the square of each number. (Lesson 11-1)

4. 4 16
5. 12 144

Find each square root. (Lesson 11-1)

6. $\sqrt{64}$ 8
7. $\sqrt{289}$ 17

8. **LANDSCAPING** A bag of lawn fertilizer covers 2,500 square feet. Describe the largest square that one bag of fertilizer could cover. (Lesson 11-1) 50 ft by 50 ft

Estimate each square root to the nearest whole number. (Lesson 11-2)

9. $\sqrt{32}$ 6
10. $\sqrt{55}$ 7

Find the missing measure of each right triangle. Round to the nearest tenth if necessary. (Lesson 11-3)

11. 18.0 m

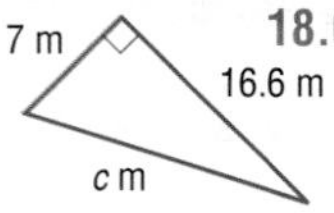

12. $a = 8.2$ m
$b = 15.6$ m 17.6 m

Find the area of each parallelogram. (Lesson 11-4)

13. base = 4.3 in.
height = 9 in. 38.7 in^2

14. 72 mm^2

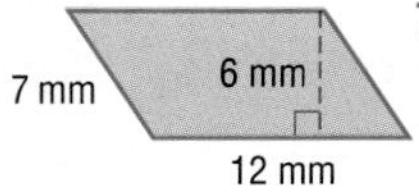

Standardized Test Practice

15. **MULTIPLE CHOICE** Which is the best estimate for $\sqrt{120}$? (Lesson 11-2) B

Ⓐ 10 Ⓑ 11 Ⓒ 12 Ⓓ 15

16. **SHORT RESPONSE** Miranda jogs 5 kilometers north and 5 kilometers west. To the nearest kilometer, how far is she from her starting point? (Lesson 11-3) 7 km

486 Chapter 11 Geometry: Measuring Two-Dimensional Figures

The Game Zone

A Place To Practice Your Math Skills

Tic Tac Root

GET READY!

Players: two to four
Materials: index cards, construction paper

GET SET!

- Use 20 index cards. On each card, write one of the following square roots.

$\sqrt{1}$	$\sqrt{4}$	$\sqrt{9}$	$\sqrt{16}$
$\sqrt{25}$	$\sqrt{36}$	$\sqrt{49}$	$\sqrt{64}$
$\sqrt{81}$	$\sqrt{100}$	$\sqrt{121}$	$\sqrt{144}$
$\sqrt{169}$	$\sqrt{196}$	$\sqrt{225}$	$\sqrt{256}$
$\sqrt{289}$	$\sqrt{324}$	$\sqrt{361}$	$\sqrt{400}$

- Each player should draw a tic-tac-toe board on construction paper. In each square, place a number from 1 to 20, but do not use any number more than once. See the sample board at the right.

4	15	9
11	6	2
20	12	7

GO!

- The dealer shuffles the index cards and places them facedown on the table.
- The player to the left of the dealer chooses the top index card and places it faceup. Any player with the matching square root on his or her board places an X on the appropriate square.
- The next player chooses the top index card and places it faceup on the last card chosen. Players mark their boards accordingly.
- **Who Wins?** The first player to get three Xs in a row wins the game.

The Game Zone

Resources

For an easy-to-make manipulative and recording sheet for this game, visit: msmath2.net/gamezone.

GETTING READY

Objective Find the square roots of numbers.

MANAGEMENT TIPS

Recommended Time
15 minutes

Tips

- Students may need to review the squares of numbers up to 20 before playing the game.
- Some students may need extra time to think of the answers for this game. Try to group students according to their abilities to avoid frustration.
- You may wish to use the game as an assessment tool for the concepts the students have been learning in Lessons 11-1 through 11-4.

For an additional game for this chapter, visit: msmath2.net/gamezone.

11-5a Hands-On Lab

Lesson Notes

Lesson Resources

Teaching Mathematics with Manipulatives

- centimeter grid, p. 11
- student recording sheet 11-5a

Glencoe Mathematics Classroom Manipulative Kit

- ruler (straightedge)
- scissors

GETTING STARTED

Teaching Tip Demonstrate the steps for changing the triangle into a parallelogram on an overhead projector.

TEACH

- Students should understand the relationship between the formulas for a parallelogram and a triangle. It takes two triangles to make the parallelogram.
- Students can cut out the trapezoids by folding the paper in half as they did for the triangle.

ASSESS

Observe as students complete the **Writing Math** exercises. They should:

- compare the area of a parallelogram to the area of a triangle and a trapezoid,
- write a formula for the area of a triangle, and
- derive the formula for the area of a trapezoid from the formula for a parallelogram.

11-5a Hands-On Lab

What You'll Learn
Find the areas of triangles and trapezoids using models.

Materials

- centimeter grid paper
- straightedge
- scissors
- tape

Triangles and Trapezoids

INVESTIGATE *Work as a class.*

STEP 1 On grid paper, draw a triangle with a base of 6 units and a height of 3 units. Label the base b and the height h as shown.

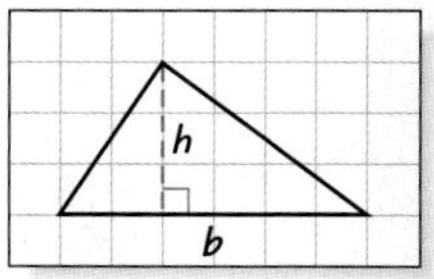

STEP 2 Fold the grid paper in half and cut out the triangle through both sheets so that you have two congruent triangles.

STEP 3 Turn the second triangle upside down and tape it to the first triangle.

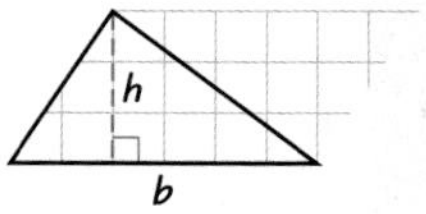

Writing Math

Work with a partner.

1. What figure is formed by the two triangles? **parallelogram**
2. Write the formula for the area of the figure. Then find the area.
3. What is the area of each of the triangles? How do you know?
4. Repeat the activity above, drawing a different triangle in Step 1. Then find the area of each triangle. **See students' work.**
5. Compare the area of a triangle to the area of a parallelogram with the same base and height. **The area of the triangle equals one half the area of the parallelogram.**
6. **MAKE A CONJECTURE** Write a formula for the area of a triangle with base b and height h. $A = \frac{1}{2}bh$

For Exercises 7–9, refer to the information below.

On grid paper, cut out two identical trapezoids. Label the bases b_1 and b_2, respectively, and label the heights h. Then turn one trapezoid upside down and tape it to the other trapezoid as shown.

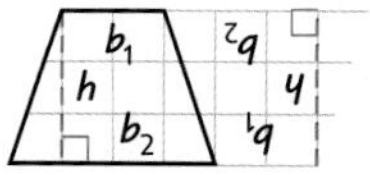

7. Write an expression to represent the base of the parallelogram.
8. Write a formula for the area A of the parallelogram using b_1, b_2, and h.
9. **MAKE A CONJECTURE** Write a formula for the area A of a trapezoid with bases b_1 and b_2, and height h. $A = \frac{1}{2}(b_1 + b_2)h$

2. $A = bh$; 18 units^2
3. 9 units^2; Each triangle makes up half the parallelogram, so it has half the area.
7. $b_1 + b_2$
8. $A = (b_1 + b_2)h$

11-5 Area of Triangles and Trapezoids

/hat You'll Learn
nd the areas of
iangles and
apezoids.

EVIEW Vocabulary
apezoid:
uadrilateral with
ne pair of parallel
des **(Lesson 10-5)**

HANDS-ON Mini Lab

Materials
- grid paper
- straightedge
- scissors

Work with a partner.

- Draw a parallelogram with a base of 6 units and a height of 4 units.
- Draw a diagonal as shown.
- Cut out the parallelogram.

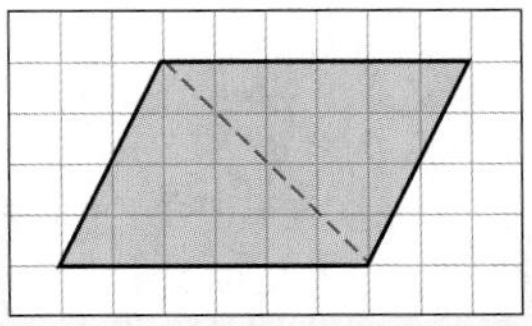

1. What is the area of the parallelogram?
2. Cut along the diagonal. What is true about the triangles formed?
3. What is the area of each triangle? **12 sq units**
4. If the area of a parallelogram is bh, then write an expression for the area A of each of the two congruent triangles that form the parallelogram. $A = \frac{1}{2}bh$

1. 24 sq units 2. They are congruent.

Like parallelograms, you can find the area of a triangle by using the base and height.

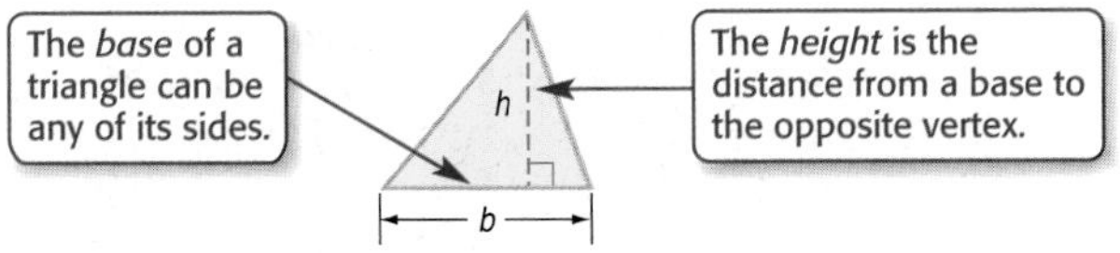

Key Concept **Area of a Triangle**

Words The area A of a triangle equals half the product of its base b and height h.

Symbols $A = \frac{1}{2}bh$

Model

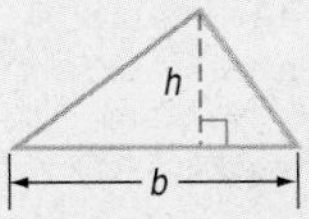

EXAMPLE Find the Area of a Triangle

1 **Find the area of the triangle below.** **Estimate** $\frac{1}{2}(10)(7) = 35$

$A = \frac{1}{2}bh$ Area of a triangle

$A = \frac{1}{2}(10)(6.5)$ Replace b with 10 and h with 6.5.

$A = 32.5$ Multiply.

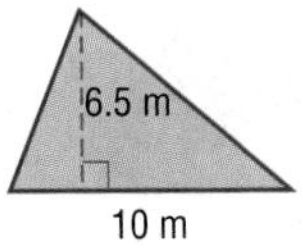

The area of the triangle is 32.5 square meters. This is close to the estimate.

msmath2.net/extra_examples

11-5 Lesson Notes

Lesson Resources

Chapter 11 Resource Masters FAST FILE
- Study Guide and Intervention, p. 629
- Practice: Skills, p. 630
- Practice: Word Problems, p. 631
- Reading to Learn Mathematics, p. 632
- Enrichment, p. 633

Classroom Math Games, pp. 33–34
Key Teaching Transparencies, 44
Parent and Student Study Guide Workbook, 11-5

1 Focus

5-Minute Check 11-5 INTERACTIVE CHALKBOARD PowerPoint® Presentations

Use as a quiz or review of Lesson 11-4. (Also available as a transparency)

Bellringer

Communication Draw a triangle on the chalkboard. Have students identify the vertices and legs. Draw a right triangle. Ask students how they know it is a right triangle. **Sample answer: The square at the 90° vertex shows that it is a right triangle.**

Reading to Learn Mathematics, p. 632 **ELL**

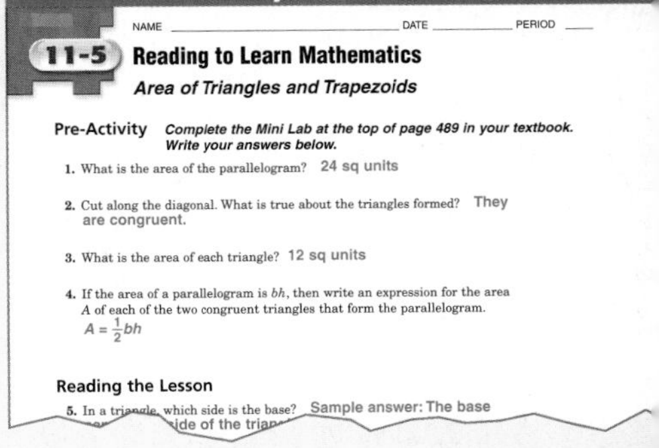
NAME ______ DATE ______ PERIOD ____

11-5 Reading to Learn Mathematics
Area of Triangles and Trapezoids

Pre-Activity *Complete the Mini Lab at the top of page 489 in your textbook. Write your answers below.*

1. What is the area of the parallelogram? 24 sq units
2. Cut along the diagonal. What is true about the triangles formed? They are congruent.
3. What is the area of each triangle? 12 sq units
4. If the area of a parallelogram is bh, then write an expression for the area A of each of the two congruent triangles that form the parallelogram. $A = \frac{1}{2}bh$

Reading the Lesson

5. In a triangle, which side is the base? Sample answer: The base ... side of the trian...

2 Teach

In-Class Examples

INTERACTIVE CHALKBOARD
PowerPoint® Presentations

1 Find the area of the triangle below. **14.4 cm²**

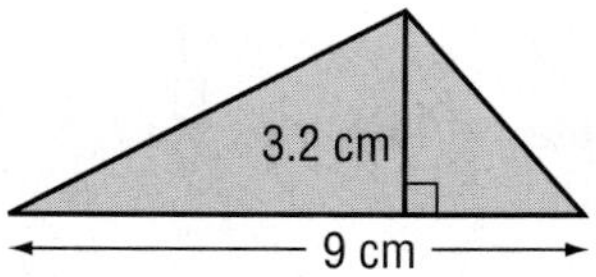

2 Find the area of the trapezoid below. **17.4 m²**

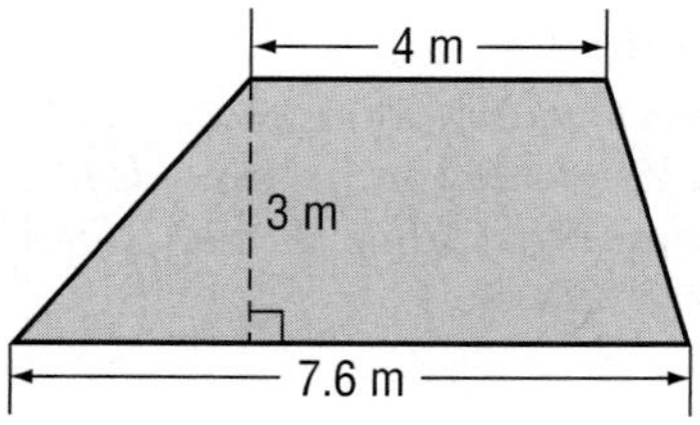

3 **GEOGRAPHY** The shape of the state of Montana resembles a trapezoid. Estimate its area in square miles. **about 145,493 mi²**

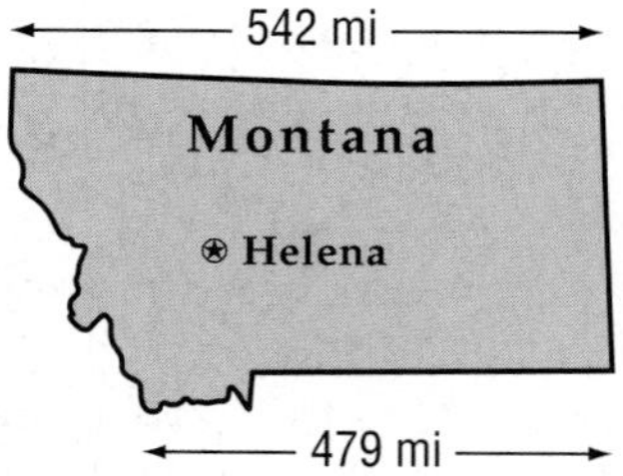

Tips for New Teachers

Teaching Strategies When graphing on an overhead projector, you may have to constantly clean the transparencies. Instead, project the grid onto the chalkboard. You or your students can graph on the chalkboard, then easily erase it.

Answers

2. See students' drawings. Find half of the product of the height and the sum of the bases.
3. The area of a parallelogram is twice the area of a triangle with the same base and height.

READING Math

Subscripts Read b_1 as *b sub 1*. Read b_2 as *b sub 2*. The subscripts mean that b_1 and b_2 represent different variables.

TEACHING TIP

Have students explain why the area of the trapezoid in Exercise c is less than a square foot. Sample answer: The trapezoid has a smaller width than a 1 ft by 1 ft square, which has an area of 1 ft².

REAL-LIFE MATH

GEOGRAPHY The actual area of Tennessee is 42,244 square miles.

Source: *Merriam Webster's Collegiate Dictionary*

A trapezoid has two bases, b_1 and b_2. The height of a trapezoid is the distance between the bases.

Key Concept — Area of a Trapezoid

Words The area A of a trapezoid equals half the product of the height h and the sum of the bases b_1 and b_2.

Symbols $A = \frac{1}{2}h(b_1 + b_2)$

Model

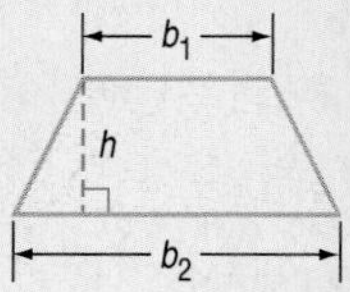

EXAMPLE Find the Area of a Trapezoid

2 **Find the area of the trapezoid at the right.**

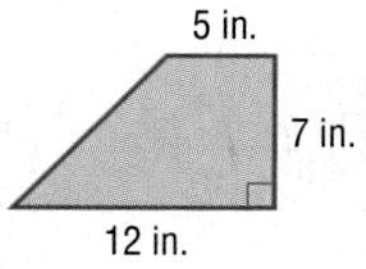

The bases are 5 inches and 12 inches. The height is 7 inches.

$A = \frac{1}{2}h(b_1 + b_2)$ Area of a trapezoid

$A = \frac{1}{2}(7)(5 + 12)$ Replace h with 7, b_1 with 5, and b_2 with 12.

$A = \frac{1}{2}(7)(17)$ Add 5 and 12.

$A = 59.5$ Multiply.

The area of the trapezoid is 59.5 square inches.

Your Turn **Find the area of each triangle or trapezoid. Round to the nearest tenth if necessary.**

a. 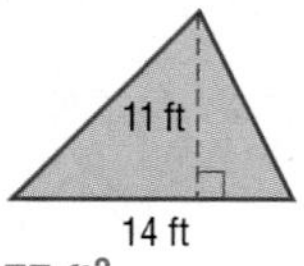

77 ft²

b. 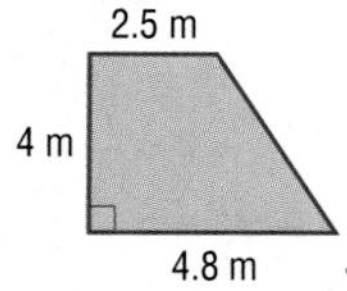

14.6 m²

c. 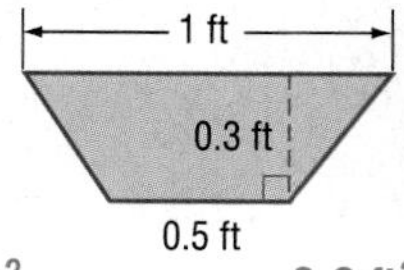

0.2 ft²

EXAMPLE Use a Formula to Estimate Area

3 **GEOGRAPHY The shape of the state of Tennessee resembles a trapezoid. Estimate its area in square miles.**

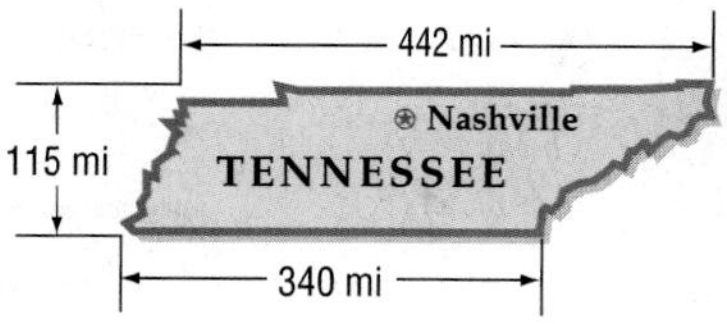

$A = \frac{1}{2}h(b_1 + b_2)$

$A = \frac{1}{2}(115)(442 + 340)$ Replace h with 115, b_1 with 442, and b_2 with 340.

$A = \frac{1}{2}(115)(782)$ Add 442 and 340.

$A = 44{,}965$ Multiply.

The area of Tennessee is about 44,965 square miles.

DAILY INTERVENTION — Differentiated Instruction

Visual/Spatial Give groups of students colored paper, rulers, and scissors. Have each group make a mosaic design using triangles and trapezoids. Have them determine the area of each shape and the total area of the design. You may want to attach the designs to poster board and hang them on the wall.

Skill and Concept Check

Writing Math
Exercises 2 & 3

1. **Estimate** the area of the trapezoid at the right.

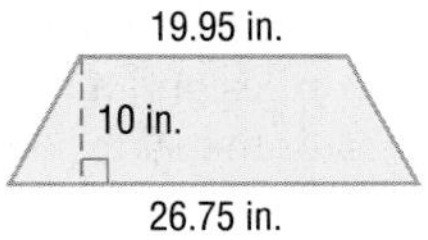

2. **OPEN ENDED** Draw a trapezoid and label the bases and the height. In your own words, explain how to find the area of the trapezoid.

2–3. See margin.

3. **NUMBER SENSE** Describe the relationship between the area of a parallelogram and the area of a triangle with the same height and base.

1. Sample answer: $\frac{1}{2}(10)(20 + 30) = 250$ in^2

GUIDED PRACTICE

Find the area of each figure. Round to the nearest tenth if necessary.

4. **6 in^2**

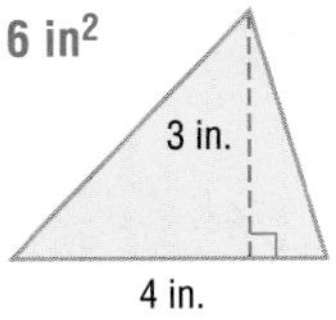

5. **105.6 m^2**

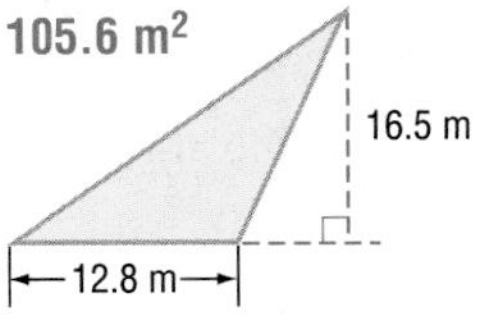

6. **90.4 ft^2**

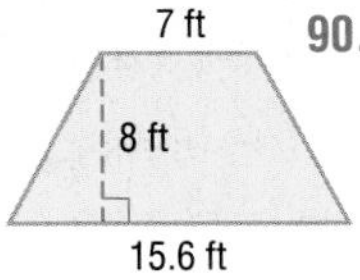

7. **MULTI STEP** The blueprints for a patio are shown at the right. If the cost of the patio is \$4.50 per square foot, what will be the total cost of the patio? **\$1,417.50**

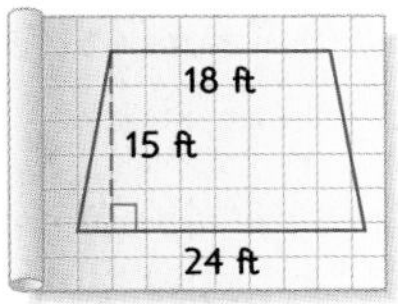

★ **indicates increased difficulty**

Practice and Applications

HOMEWORK HELP

For Exercises	See Examples
8–9, 12, 14	1
10–11, 13, 15	2
20–21	3

Extra Practice
See pages 591, 606.

Find the area of each figure. Round to the nearest tenth if necessary.

8.

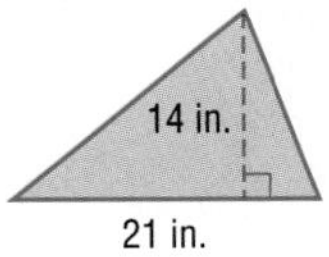

147 in^2

9.

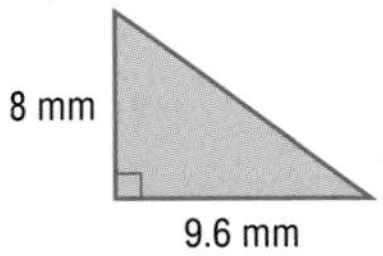

38.4 mm^2

10.

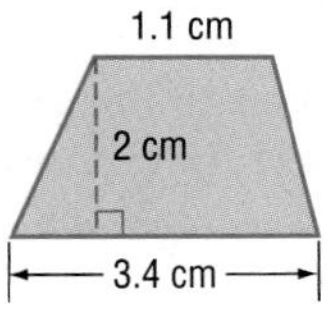

4.5 cm^2

11.

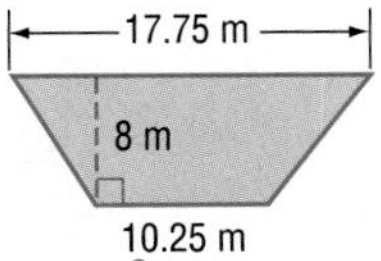

112 m^2

12.

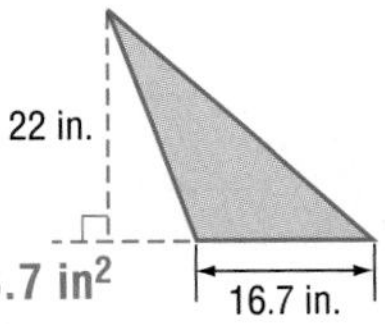

183.7 in^2

★ 13.

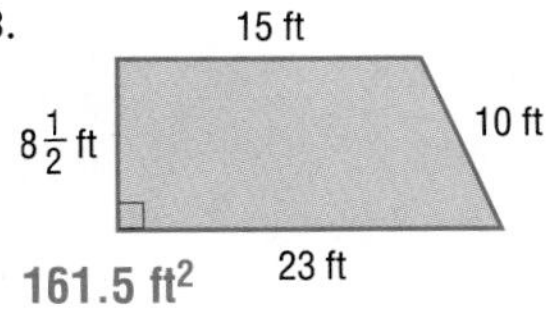

161.5 ft^2

14. triangle: base = 4 cm, height = 7.5 cm **15 cm^2**

★ 15. trapezoid: bases 13 in. and $1\frac{1}{4}$ ft, height 1 ft **168 in^2 or 1.2 ft^2**

16–19. See margin for sample answers.

Draw and label each figure on grid paper. Then find the area.

16. a triangle with no right angles
17. an isosceles triangle with a height greater than 6 units

★ 18. a trapezoid with a right angle and an area of 40 square units

★ 19. a trapezoid with no right angles and an area less than 25 square units

TEACHING TIP

The height of a triangle can be a side of the figure, as shown in Exercise 9, or outside the figure, as shown in Exercises 5 and 12.

3 Practice/Apply

FOLDABLES™ Follow-Up Remind students to use their Foldables to record in words and symbols how they find the area of triangles and trapezoids.

NUMBER SENSE

For Exercise 3, students should compare the respective formulas for the areas of the two figures. They might also approach the question visually. Slicing a parallelogram across a diagonal makes two identical triangles.

About the Exercises...

Guided Practice Key

Exercises	Examples
4–5	1
6–7	2

Odd/Even Assignments

Exercises 8–21 are structured so that students practice the same concepts whether they are assigned odd or even problems.

Alert! Exercises 16–19 require the use of grid paper.

Alert! The Data Update after Exercises 20 and 21 requires students to visit msmath2.net/data_update.

Answers

16. 10 in^2

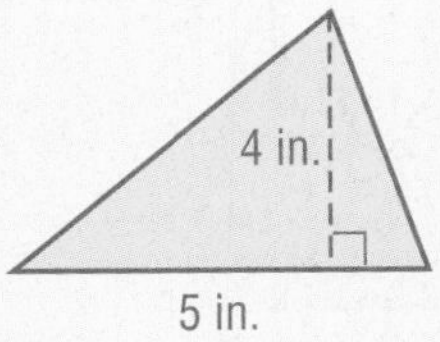

17. 28 in^2

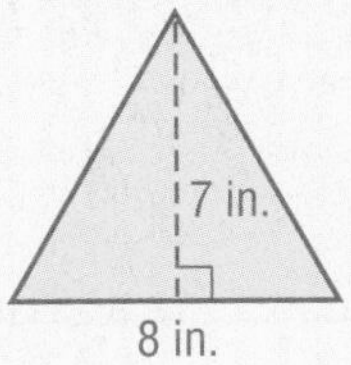

18. 40 m^2

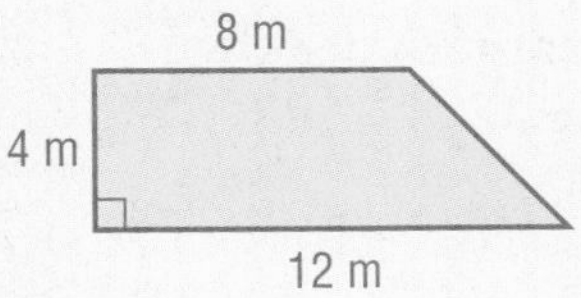

19. 7 ft^2

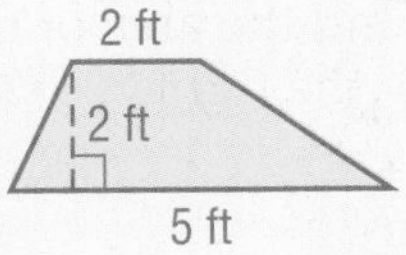

Study Guide and Intervention, p. 629

NAME ______ DATE ______ PERIOD ______

11-5 Study Guide and Intervention

Area of Triangles and Trapezoids

The area A of a triangle equals half the product of its base b and its height h.

$A = \frac{1}{2}bh$ The base of a triangle can be any of its sides. The height is the distance from a base to the opposite vertex.

A trapezoid has two bases, b_1 and b_2. The height of a trapezoid is the distance between the two bases. The area A of a trapezoid equals half the product of the height h and the sum of the bases b_1 and b_2.

$A = \frac{1}{2}h(b_1 + b_2)$

EXAMPLE 1 Find the area of the triangle. Estimate $\frac{1}{2}(6)(5) = 15$

$A = \frac{1}{2}bh$ Area of a triangle
$A = \frac{1}{2}6 \cdot 4.5$ Replace b with 6 and h with 4.5.
$A = 13.5$ Multiply.

The area of the triangle is 13.5 square inches. This is close to the estimate.

EXAMPLE 2 Find the area of the trapezoid.

$A = \frac{1}{2}h(b_1 + b_2)$ Area of a trapezoid
$A = \frac{1}{2}(4)(3 + 6)$ Replace h with 4, b_1 with 3, and b_2 with 6.
$A = 18$ Simplify.

The area of the trapezoid is 18 square centimeters.

EXERCISES

Find the area of each figure. Round to the nearest tenth if necessary.

1. 42 ft²
2. 31.5 mm²
3. 52.5 in²
4. 175.5 cm²

Practice: Skills, p. 630

NAME ______ DATE ______ PERIOD ______

11-5 Practice: Skills

Area of Triangles and Trapezoids

Find the area of each figure. Round to the nearest tenth if necessary.

1. 45 cm²
2. 3 ft²
3. 150 mm²
4. 19 ft²
5. 39.2 cm²
6. 248.4 mm²
7. 270.6 ft²
8. 19.3 in²
9. 45.8 cm²
10. 136.2 mm²
11. triangle: base = 16 cm, height = 9.4 cm 75.2 cm²
12. triangle: base = 13.5 in., height = 6.4 in. 43.2 in²
13. trapezoid: bases 22.8 mm and 19.7 mm, height 36 mm 765 mm²
14. trapezoid: bases 5 ft and $3\frac{1}{2}$ yd, height 7 ft 54.3 ft²

Practice: Word Problems, p. 631

NAME ______ DATE ______ PERIOD ______

11-5 Practice: Word Problems

Area of Triangles and Trapezoids

1. GEOGRAPHY Arkansas has a shape that is similar to a trapezoid with bases of about 182 miles and 267 miles and a height of about 254 miles. Estimate the area of the state. 57,023 mi²
2. PATIOS Greta is making a patio with the dimensions given in the figure. What is the area of the patio? 172.5 ft²
3. FLAGS Malila wants to make the International Marine Signal flag shown which represents the number six. What is the area of the flag? 1,750 in²
4. SIGNS Estimate the area of the yield sign. 390 in²
5. TILING A ceramics company wants to produce tiles in the shape shown. What is the area of the surface of each tile? 36.125 cm²
6. GARDENING Kinu wants to buy topsoil for a section of her garden that has the dimensions shown in the figure. What is the area of this section of Kinu's garden? 7 yd²

20. **GEOGRAPHY** Nevada has a shape that looks like a trapezoid, as shown at the right. Find the approximate area of the state. **about 108,756 mi²**

21. **GEOGRAPHY** Delaware has a shape that is roughly triangular with a base of 39 miles and a height of 96 miles. Find the approximate area of the state. **about 1,872 mi²**

Data Update How do the actual areas of Nevada and Delaware compare to your estimates? Visit msmath2.net/data_update to learn more.

22. **CRITICAL THINKING** A triangle has height h. Its base is 4. Find the area of the triangle. (*Hint*: Express your answer in terms of h.) $A = 2h$

Standardized Test Practice and Mixed Review

23. **SHORT RESPONSE** Find the area of the triangle at the right to the nearest tenth. **382.4 cm²**

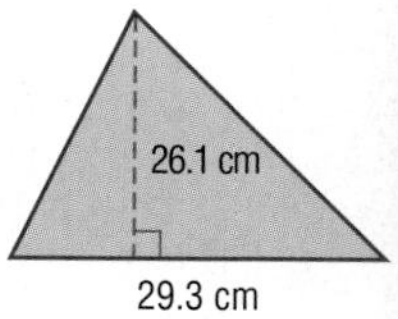

24. **MULTIPLE CHOICE** A trapezoid has bases of 15 meters and 18 meters and a height of 10 meters. What is the area of the trapezoid? **C**

Ⓐ 30 m² Ⓑ 60 m² Ⓒ 165 m² Ⓓ 330 m²

25. **GEOMETRY** Find the area of a parallelogram having a base of 2.3 inches and a height of 1.6 inches. Round to the nearest tenth. (Lesson 11-4) **3.7 in²**

Find the missing measure of each right triangle. Round to the nearest tenth if necessary. (Lesson 11-3)

26. $a = 10$ m, $b = 14$ m **17.2 m**
27. $a = 13$ ft, $c = 18$ ft **12.4 ft**

For Exercises 28–31, refer to the graphic at the right. Classify the angle that represents each category as *acute*, *obtuse*, *right*, or *straight*. (Lesson 10-1)

28. 30–39 hours **acute**
29. 1–29 hours **acute**
30. 40 hours **obtuse**
31. 41–50 hours **obtuse**

32. **MUSIC** Use the Fundamental Counting Principle to find the number of piano instruction books in a series if there are Levels 1, 2, 3, and 4 and each level contains five different books. (Lesson 9-3) **20 books**

GETTING READY FOR THE NEXT LESSON

BASIC SKILL Use a calculator to find each product to the nearest tenth.

33. $\pi \cdot 13$ **40.8**
34. $\pi \cdot 29$ **91.1**
35. $\pi \cdot 16^2$ **804.2**
36. $\pi \cdot 4.8^2$ **72.4**

4 Assess

Higher-Level Thinking

Find the area of the trapezoid shown at the right. **88 cm²**

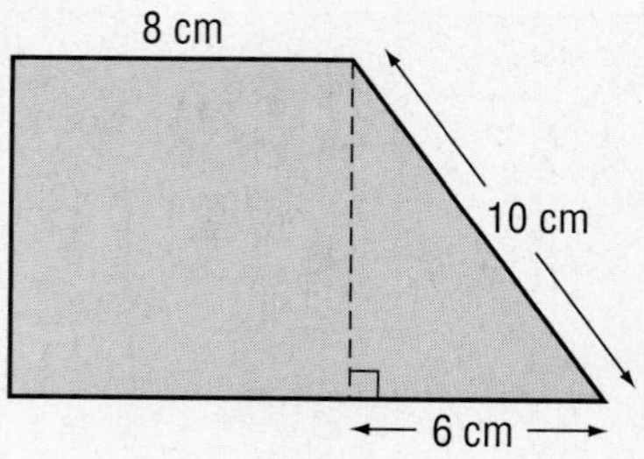

11-6 Area of Circles

at You'll Learn
l the areas of
es.

/IEW Vocabulary

π): the Greek
er that represents
rrational number
roximately equal
.14 **(Lesson 6-9)**

HANDS-ON Mini Lab

Materials

- large paper plate
- scissors

Work with a partner.

- Fold a paper plate in half four times to divide it into 16 equal-sized sections.
- Label the radius r as shown. Let C represent the circumference of the circle.
- Cut out each section; reassemble to form a parallelogram-shaped figure.

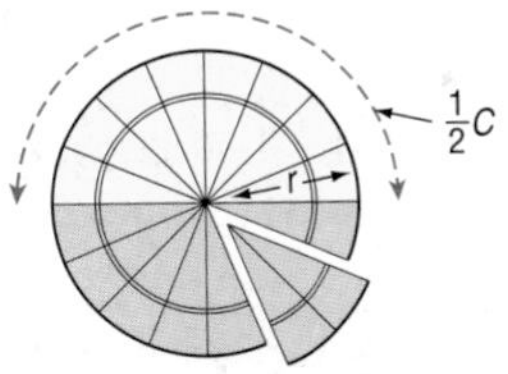

1. What is the measurement of the base and the height? $\frac{1}{2}C$; r

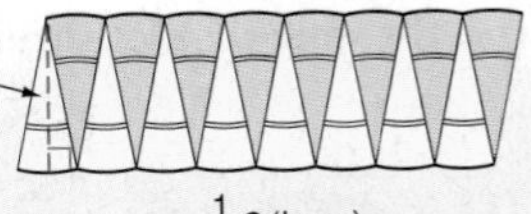

2. Substitute these values into the formula for the area of a parallelogram. $A = \frac{1}{2}C(r)$
3. Replace C with the expression for the circumference of a circle, $2\pi r$. Simplify the equation and describe what it represents.

$A = \frac{1}{2}(2\pi r)(r)$;
$= \pi r^2$; the area of a
cle

In the Mini Lab, the formula for the area of a parallelogram was used to develop a formula for the area of a circle.

Key Concept — **Area of a Circle**

Words The area A of a circle equals the product of pi (π) and the square of its radius r.

Symbols $A = \pi r^2$

Model

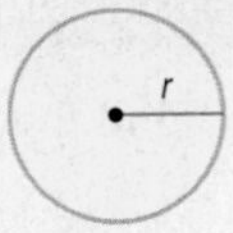

EXAMPLES Find the Areas of Circles

1 Find the area of the circle at the right.

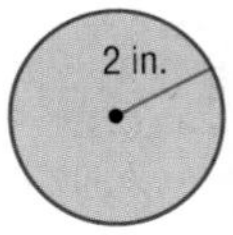

$A = \pi r^2$ Area of a circle

$A = \pi \cdot 2^2$ Replace r with 2.

[π] [×] 2 [x²] [ENTER] 12.56637061

The area of the circle is approximately 12.6 square inches.

2 Find the area of a circle with a diameter of 15.2 centimeters.

$A = \pi r^2$ Area of a circle

$A = \pi \cdot 7.6^2$ Replace r with $15.2 \div 2$ or 7.6.

$A \approx 181.5$ Use a calculator.

The area of the circle is approximately 181.5 square centimeters.

TEACHING TIP

int out that all areas of
rcles are estimates
nce the calculator gives
approximate value of
. The *exact* area in
ample 2 is 57.76π cm^2.

msmath2.net/extra_examples

DAILY INTERVENTION — **Flexible Grouping**

Areas of Circles Divide the class into groups of three. Provide each group with a collection of different-sized circles cut from cardboard. The groups should calculate the area of each circle by first measuring either its radius or its diameter.

11-6 Lesson Notes

Lesson Resources

Chapter 11 Resource Masters

- Study Guide and Intervention, p. 634
- Practice: Skills, p. 635
- Practice: Word Problems, p. 636
- Reading to Learn Mathematics, p. 637
- Enrichment, p. 638

MathPASS Tutorial CD-ROM, 49, 50

Parent and Student Study Guide Workbook, 11-6

Real-Life Career Masters, p. 25

Teaching Mathematics with Manipulatives, p. 111

FAST FILE

1 Focus

5-Minute Check 11-6

INTERACTIVE CHALKBOARD
PowerPoint® Presentations

Use as a quiz or review of Lesson 11-5. (Also available as a transparency)

Bellringer

Reading Mathematics Have students look up the definitions of *circumference, radius,* and *diameter.* Have them draw a circle and label it. Ask them what the difference is between *perimeter* and *circumference.* **Circumference is a circle's perimeter.**

Reading to Learn Mathematics, p. 637 **ELL**

NAME ______ DATE ______ PERIOD ____

11-6 Reading to Learn Mathematics
Area of Circles

Pre-Activity *Complete the Mini Lab at the top of page 493 in your textbook. Write your answers below.*

1. What is the measurement of the base and the height? $\frac{1}{2}C$; r
2. Substitute these values into the formula for the area of a parallelogram. $A = \frac{1}{2}C(r)$
3. Replace C with the expression for the circumference of a circle, $2\pi r$. Simplify the equation and describe what it represents. $A = \frac{1}{2}(\pi r)(r)$; $A = \pi r^2$; the area of a circle

Reading the Lesson

2 Teach

In-Class Examples

INTERACTIVE CHALKBOARD
PowerPoint® Presentations

1 Find the area of the circle shown below. **50.3 cm^2**

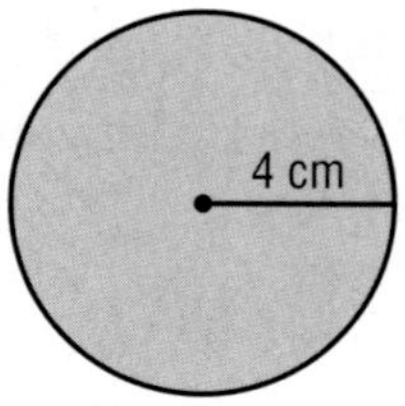

2 Find the area of a circle with a diameter of 11.6 centimeters. **105.7 cm^2**

3 Practice/Apply

DAILY INTERVENTION **FIND THE ERROR** Encourage students to draw a circle for Exercise 2 with the diameter labeled correctly. The drawing should help students recognize Carlos's mistake.

NUMBER SENSE
In Exercise 3, remind students that π stands for the number 3.1415927… .

About the Exercises…

Guided Practice Key

Exercises	Examples
4–6	1
7–10	2

Odd/Even Assignments
Exercises 11–22 are structured so that students practice the same concepts whether they are assigned odd or even problems.

Alert! Exercises 26 and 27 require grid paper.

Skill and Concept Check

Writing Math
Exercises 2 & 3

1. **OPEN ENDED** Draw and label the radius of a circle that has an area less than 10 square units. **See margin.**
2. **FIND THE ERROR** Carlos and Sean are finding the area of a circle that has a diameter of 12 centimeters. Who is correct? Explain. **2. Sean; Carlos incorrectly squared the diameter rather than the radius.**

Carlos	Sean
$A = \pi(12)^2$	$A = \pi(6)^2$
$\approx 452\text{ cm}^2$	$\approx 113\text{ cm}^2$

TEACHING TIP
The answers given in thi[s] lesson were found using the π key on a calculator.

3. **NUMBER SENSE** Without using a calculator, determine which has the greatest value: 2π, $\sqrt{7}$, or 1.5^2. Explain. **2π; It is about $2 \cdot 3$, or 6; $\sqrt{7}$ is less than $\sqrt{9}$, or 3; 1.5^2 is less than 2^2, or 4.**

GUIDED PRACTICE

Find the area of each circle. Round to the nearest tenth.

4. 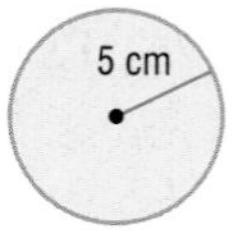

78.5 cm^2

5. 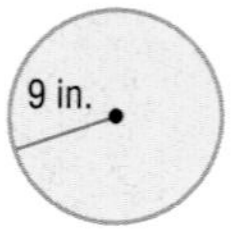

254.5 in^2

6. 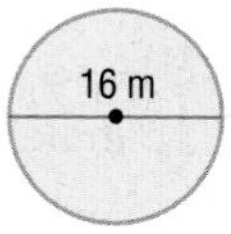

201.1 m^2

7. radius = 4.2 ft **55.4 ft^2**
8. diameter = 13 ft **132.7 ft^2**
9. diameter = 24 mm **452.4 mm^2**
10. **HISTORY** The Roman Pantheon is a circular-shaped structure that was completed about 126 A.D. Find the area of the floor if the diameter is 43 meters. **$1{,}452.2\text{ m}^2$**

★ indicates increased difficulty

Practice and Applications

HOMEWORK HELP

For Exercises	See Examples
11–16, 25	1
17–22, 24	2

Extra Practice
See pages 591, 606.

Find the area of each circle. Round to the nearest tenth.

11. 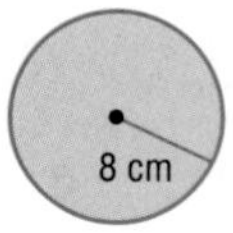

201.1 cm^2

12. 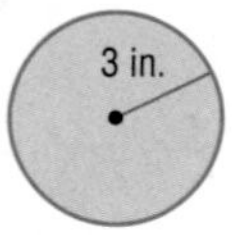

28.3 in^2

13. 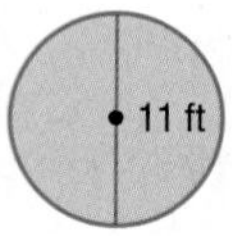

95.0 ft^2

14. 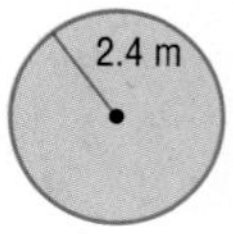

18.1 m^2

15. 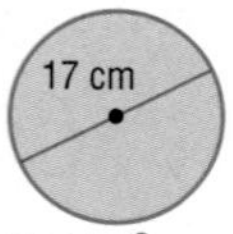

227.0 cm^2

16. 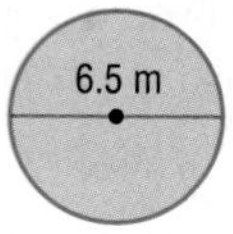

33.2 m^2

17. radius = 6 ft **113.1 ft^2**
18. diameter = 7 ft **18. 38.5 ft^2**
19. diameter = 3 cm **7.1 cm^2**
20. radius = 10.5 mm **346.4 mm^2**
21. radius = $4\frac{1}{2}$ in. **63.6 in^2**
22. diameter = $20\frac{3}{4}$ yd **338.2 yd^2**
23. A *semicircle* is half a circle. Find the area of the semicircle at the right to the nearest tenth. **29.0 m^2**

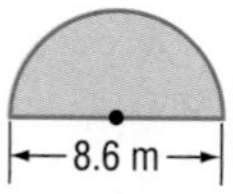

TEACHING TIP
In Exercises 6, 13, 15, and 16, the measures are for the diameters of the circles.

24. **MONEY** Find the area of the face of a Sacagawea \$1 coin if the diameter is 26.5 millimeters. Round to the nearest tenth. **551.5 mm^2**

Answer

1. Sample answer:

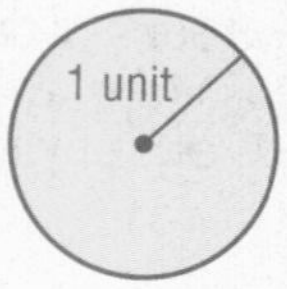

494 Chapter 11 Geometry: Measuring Two-Dimensional Figures

25. **LANDSCAPE DESIGN** A circular stone path is to be installed around a birdbath with radius 1.5 feet, as shown at the right. What is the area of the path? (*Hint*: Find the area of the large circle minus the area of the small circle.) **43.2 ft^2**

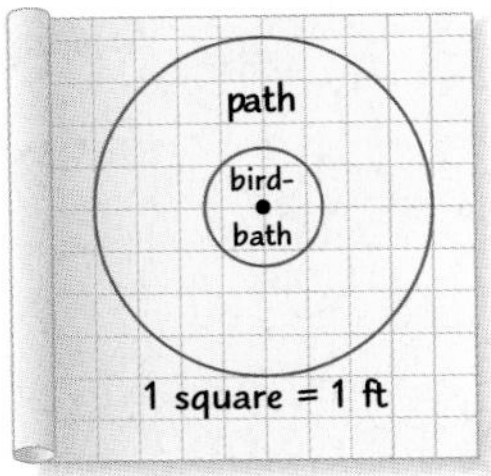

For Exercises 26 and 27, refer to the information below.
Let the x-axis of a coordinate plane represent the radius of a circle and the y-axis represent the area of a circle.

26. Graph the points that represent the circles with radii 0, 1, 2, and 3 units long. Draw a line or curve that goes through each point. **See Ch. 11 Answer Appendix.**

27. Consider a circle with radius of 1 unit and a circle with a radius of 2 units. Write a ratio comparing the radii. Write a ratio comparing the areas. Do these ratios form a proportion? Explain. **1:2; 1:4; No, $\frac{1}{2} \neq \frac{1}{4}$.**

Find the area of the shaded region in each figure. Round to the nearest tenth.

28. **62.8 m^2**

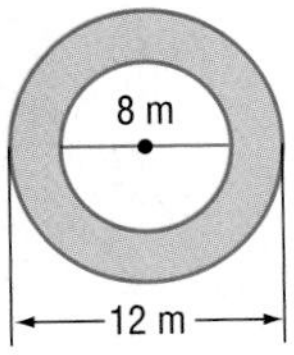

★ 29. **5.9 in^2**

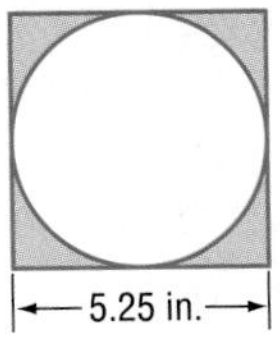

★ 30. **103.5 cm^2**

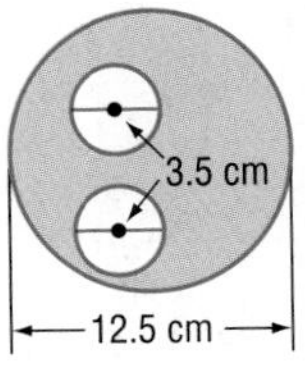

31. **CRITICAL THINKING** Determine whether the area of a circle is *sometimes*, *always*, or *never* doubled when the radius is doubled. Explain. **Never; the area is quadrupled because $A = \pi(2r)^2$, or $A = 4\pi r^2$.**

Standardized Test Practice and Mixed Review

32. **MULTIPLE CHOICE** A CD has a diameter of 12 centimeters. The hole in the middle of the CD has a diameter of 1.5 centimeters. Find the area of one side of the CD to the nearest tenth. Use 3.14 for π. **A**

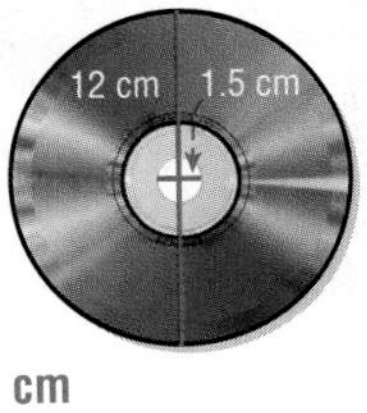

Ⓐ 111.3 cm^2 Ⓑ 113.0 cm^2 Ⓒ 349.4 cm^2 Ⓓ 445.1 cm^2

33. **SHORT RESPONSE** Find the radius of a circle that has an area of 42 square centimeters. Use 3.14 for π and round to the nearest tenth. **3.7 cm**

34. **GEOMETRY** Find the area of a triangle with a base of 21 meters and a height of 27 meters. (Lesson 11-5) **283.5 m^2**

Find the area of each parallelogram. Round to the nearest tenth if necessary. (Lesson 11-4)

35. **120 in^2**

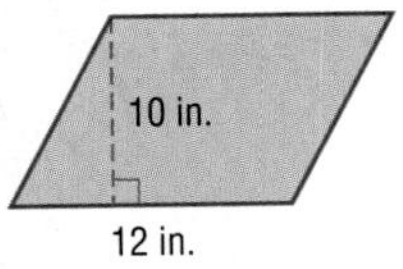

36. **39.5 cm^2**

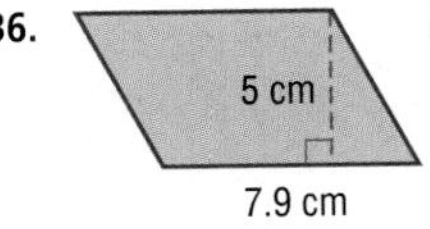

37. **100.1 m^2**

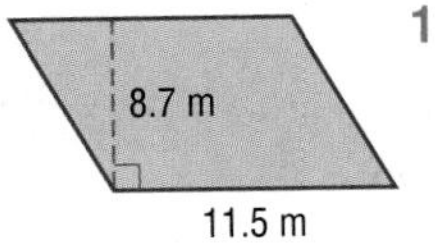

GETTING READY FOR THE NEXT LESSON

BASIC SKILL Simplify each expression. (Lessons 1-2 and 1-3)

38. 8.5^2 **72.25**
39. $3.14 \cdot 6^2$ **113.04**
40. $\frac{1}{2} \cdot 5.4^2 + 11$ **25.58**
41. $\frac{1}{2} \cdot 7^2 + (9)(14)$ **150.5**

msmath2.net/self_check_quiz

4 Assess

Open-Ended Assessment

Modeling Clear a circular space in the classroom and have students stand at the circumference of the circle. Ask one or more students to walk through the center of the circle, and demonstrate how to find the area. Encourage students to use their own feet to measure the radius or diameter.

Study Guide and Intervention, p. 634

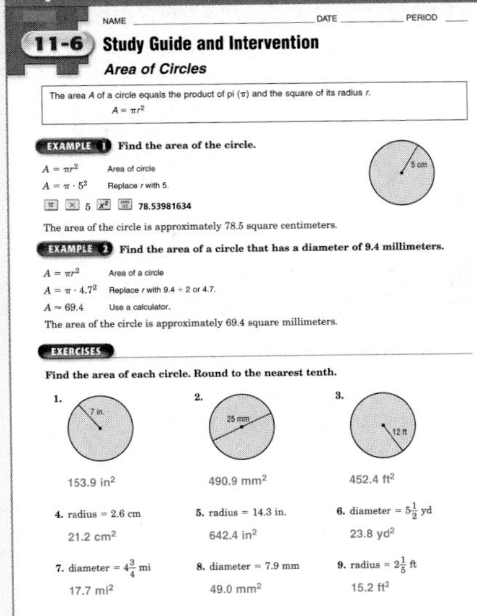

NAME ______ DATE ______ PERIOD ____

11-6 Study Guide and Intervention

Area of Circles

The area A of a circle equals the product of pi (π) and the square of its radius r.
$A = \pi r^2$

EXAMPLE 1 **Find the area of the circle.**

$A = \pi r^2$ Area of circle
$A = \pi \cdot 5^2$ Replace r with 5.
[π] [×] 5 [x²] [ENTER] 78.53981634

The area of the circle is approximately 78.5 square centimeters.

EXAMPLE 2 **Find the area of a circle that has a diameter of 9.4 millimeters.**

$A = \pi r^2$ Area of a circle
$A = \pi \cdot 4.7^2$ Replace r with 9.4 ÷ 2 or 4.7.
$A \approx 69.4$ Use a calculator.

The area of the circle is approximately 69.4 square millimeters.

EXERCISES

Find the area of each circle. Round to the nearest tenth.

1. (7 in.) 153.9 in^2
2. (25 mm) 490.9 mm^2
3. (12 ft) 452.4 ft^2
4. radius = 2.6 cm 21.2 cm^2
5. radius = 14.3 in. 642.4 in^2
6. diameter = $5\frac{1}{2}$ yd 23.8 yd^2
7. diameter = $4\frac{3}{4}$ mi 17.7 mi^2
8. diameter = 7.9 mm 49.0 mm^2
9. radius = $2\frac{1}{5}$ ft 15.2 ft^2

Practice: Skills, p. 635

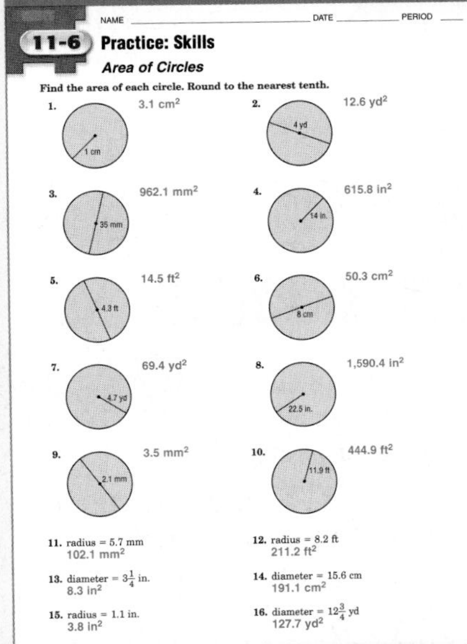

NAME ______ DATE ______ PERIOD ____

11-6 Practice: Skills

Area of Circles

Find the area of each circle. Round to the nearest tenth.

1. (1 cm) 3.1 cm^2
2. (4 yd) 12.6 yd^2
3. (35 mm) 962.1 mm^2
4. (14 in.) 615.8 in^2
5. (4.3 ft) 14.5 ft^2
6. (8 cm) 50.3 cm^2
7. (4.7 yd) 69.4 yd^2
8. (22.5 in.) 1,590.4 in^2
9. (2.1 mm) 3.5 mm^2
10. (11.9 ft) 444.9 ft^2
11. radius = 5.7 mm 102.1 mm^2
12. radius = 8.2 ft 211.2 ft^2
13. diameter = $3\frac{1}{4}$ in. 8.3 in^2
14. diameter = 15.6 cm 191.1 cm^2
15. radius = 1.1 in. 3.8 in^2
16. diameter = $12\frac{3}{4}$ yd 127.7 yd^2

Practice: Word Problems, p. 636

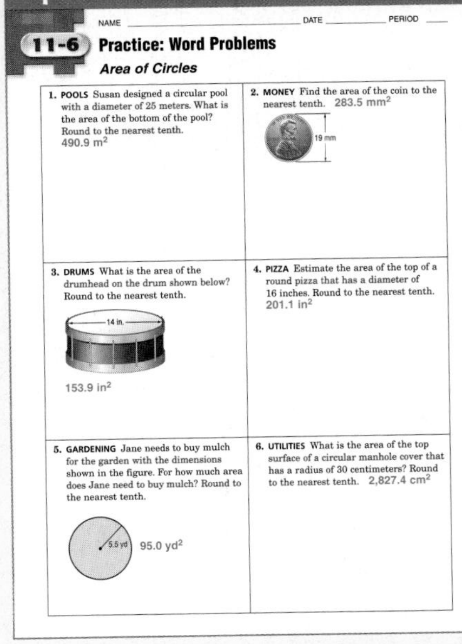

NAME ______ DATE ______ PERIOD ____

11-6 Practice: Word Problems

Area of Circles

1. **POOLS** Susan designed a circular pool with a diameter of 25 meters. What is the area of the bottom of the pool? Round to the nearest tenth. 490.9 m^2

2. **MONEY** Find the area of the coin to the nearest tenth. 283.5 mm^2

3. **DRUMS** What is the area of the drumhead on the drum shown below? Round to the nearest tenth.

153.9 in^2

4. **PIZZA** Estimate the area of the top of a round pizza that has a diameter of 16 inches. Round to the nearest tenth. 201.1 in^2

5. **GARDENING** Jane needs to buy mulch for the garden with the dimensions shown in the figure. For how much area does Jane need to buy mulch? Round to the nearest tenth.

95.0 yd^2

6. **UTILITIES** What is the area of the top surface of a circular manhole cover that has a radius of 30 centimeters? Round to the nearest tenth. 2,827.4 cm^2

11-7a Lesson Notes

1 Focus

Bellringer
Problem Solving Show students the syllabus for the rest of the year. Ask them how they would go about preparing for all of the tests that are scheduled. Suggest that instead of looking at the whole year, they look at one week at a time. Ask students how they prepare for a midterm exam. Do they study everything at once or study one chapter at a time?

2 Teach

In-Class Example

INTERACTIVE CHALKBOARD
PowerPoint® Presentations

Solve. Use the solve a simpler problem strategy.

Ben and Sheila are going to paint the wall of a room as shown. What is the area that will be painted? **67 ft²**

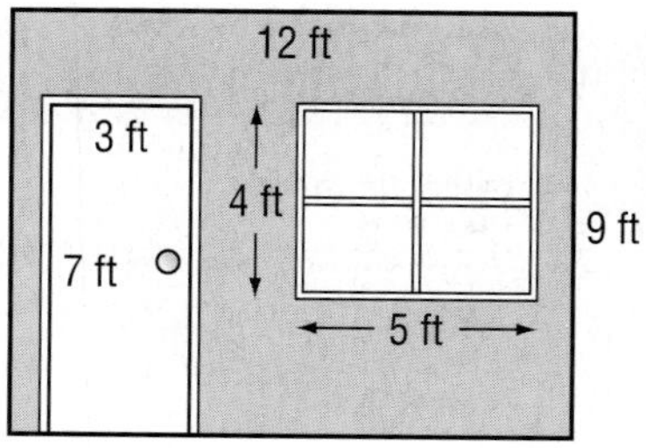

11-7a Problem-Solving Strategy

A Preview of Lesson 11-7

What You'll Learn
Solve problems by solving a simpler problem.

Solve a Simpler Problem

The diagram shows the backdrop for our fall play. How much wallpaper will we need to cover the entire front?

We need to find the total area of the backdrop. Let's **solve a simpler problem** by breaking it down into separate geometric shapes.

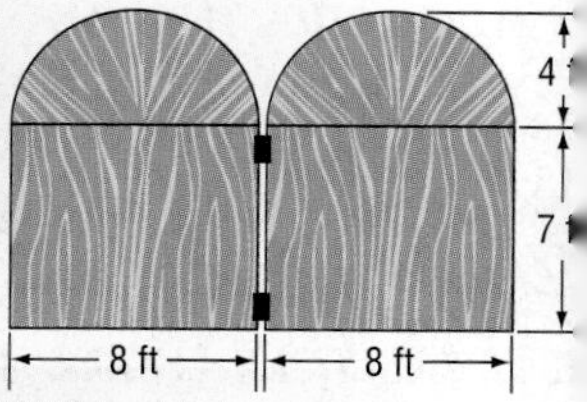

Explore	We know that the backdrop is made of one large rectangle and two semicircles, which equal an entire circle.
Plan	We can find the areas of the rectangle and the circle, and then add.
Solve	area of rectangle: $A = \ell w$ $A = (8 + 8)7$ or 112 area of circle: $A = \pi r^2$ $A = \pi \cdot 4^2$ or about 50.3 total area: 112 + 50.3 or 162.3 square feet So, we need at least 162.3 square feet of wallpaper.
Examine	Use estimation to check. The backdrop is 16 feet long and 11 feet high. However, it is less than a complete rectangle, so the area should be *less than* $16 \cdot 11$ or 176 feet. The area, 162.3 square feet, is less than 176 feet, so the answer is reasonabl

Analyze the Strategy

1. **Explain** why simplifying this problem is a good strategy to solve this problem. **1–3. See margin.**
2. **Describe** another way that the problem could have been solved.
3. **Write** a problem that can be solved by breaking it down into a simpler problem. Solve the problem and explain your answer.

Answers

1. Sample answer: It is easier to find the areas of the separate geometric figures and then add.
2. Sample answer: Find the sum of the areas of the smaller rectangle and semicircle and then multiply by 2.
3. Sample answer: Find the area of the wall at the right to determine how much paint to buy. To solve, find the area of the triangle and the area of the rectangle, then add. The answer is 77 ft².

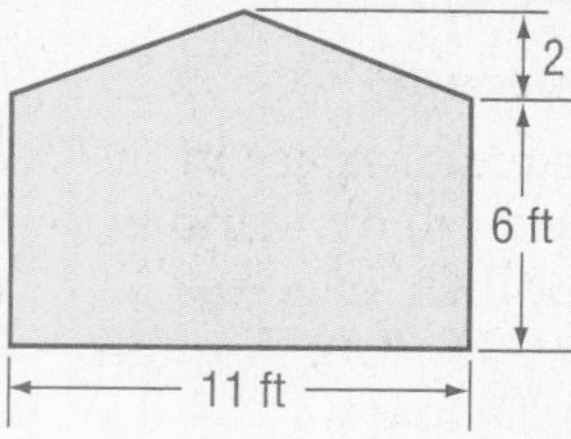

ply the Strategy

ve. Use the solve a simpler problem strategy.

.ANDSCAPING James is helping his father ɔour a circular sidewalk around a flower ɔed, as shown below. What is the area, n square feet, of the sidewalk? Use 3.14 or π. **326.56 ft^2**

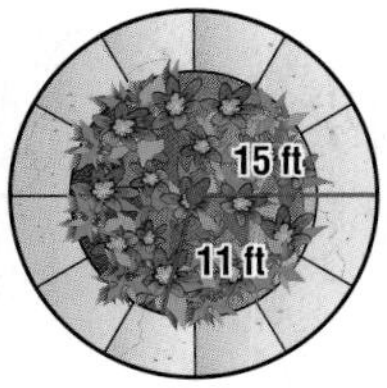

5. **COMMUNICATION** According to a recent report, one city has 2,945,000 phone lines assigned to three different area codes. How many of the phone lines are assigned to each area code?

Area Code	Percent	
888	44.3%	1,304,635
777	23.7%	697,965
555	31.5%	927,675

xed Problem Solving

ve. Use any strategy.

EARTH SCIENCE Earth's atmosphere exerts a pressure of 14.7 pounds per square inch at the ocean's surface. The pressure increases by 12.7 pounds per square inch for every 5 feet that you descend. Find the pressure at 18 feet below the surface. **52.8 lb per sq in.**

SALES Deirdre is trying to sell $3,000 in ads for the school newspaper. The prices of the ads and the number of ads that she has sold are shown in the table. Which is the smallest ad she could sell in order to meet her quota? **one full-page ad**

Ad Size	Cost Per Ad	Number Sold
quarter-page	$75	15
half-page	$125	8
full-page	$175	4

THEATER Mr. Marquez is purchasing fabric for curtains for a theatrical company. The front of the stage is 15 yards wide and 5 yards high. The fabric is sold on bolts that are 60 inches wide and 20 yards long. How many bolts are needed to make the curtains? **3 bolts or 2 whole bolts and one fourth of the fabric from another bolt**

9. Sample answer: $\frac{1}{3} \times 360{,}000$ or 120,000 people

9. **TELEVISION** The graph shows the results of a survey in which 365,750 people were asked to name their favorite television programs. Estimate how many people chose sitcoms as their favorite.

Favorite TV Shows

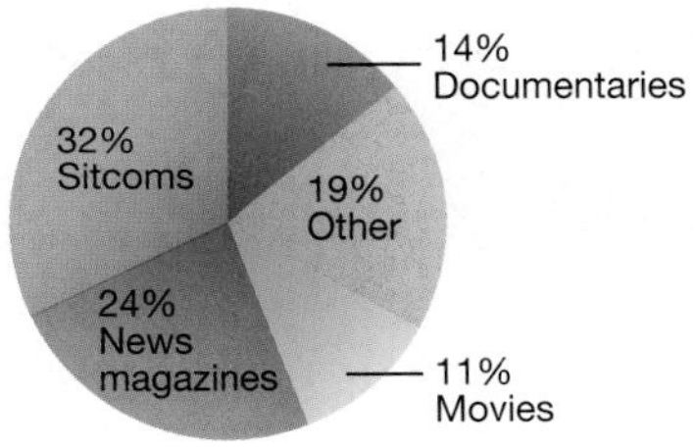

10. **STANDARDIZED TEST PRACTICE**
Kara is painting one wall in her room, as shown by the shaded region below. What is the area that she is painting? **A**

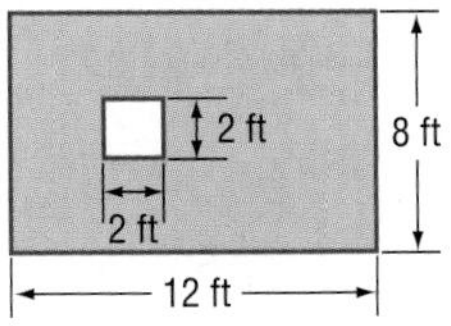

Ⓐ 92 ft^2
Ⓑ 94 ft^2
Ⓒ 96 ft^2
Ⓓ 100 ft^2

You will use the solve a simpler problem strategy in the next lesson.

3 Practice/Apply

About the Exercises...

Problem-Solving Strategies
Exercises 6–10 are structured so that students have the opportunity to practice many different problem-solving strategies. You may wish to remind them of some of the strategies they have studied.

- eliminate possibilities (p. 252)
- reasonable answers (p. 338)
- use logical reasoning (p. 444)
- solve a simpler problem ✓

4 Assess

Open-Ended Assessment

Speaking Have students explain how they can apply the solve a simpler problem strategy to Exercises 6–8. Encourage them to discuss different ideas for simplifying each problem.

DAILY INTERVENTION — **Differentiated Instruction**

Logical Have students write out the steps to solve Exercises 4 and 5 before they actually complete the exercises. Have them draw pictures to help break down the problem into simpler problems. Students could work in pairs and each do a portion of the problem. Have them combine their answers for the total and estimate to check the answer.

11-7 Lesson Notes

Lesson Resources

Chapter 11 Resource Masters

- Study Guide and Intervention, p. 639
- Practice: Skills, p. 640
- Practice: Word Problems, p. 641
- Reading to Learn Mathematics, p. 642
- Enrichment, p. 643

Parent and Student Study Guide Workbook, 11-7

FAST FILE

1 Focus

5-Minute Check 11-7

INTERACTIVE CHALKBOARD
PowerPoint® Presentations

Use as a quiz or review of Lesson 11-6. (Also available as a transparency)

Bellringer

Hands-On Activity Divide the class into small groups. Provide each group with pieces of cardboard cut into the shapes below. Each group should discuss how each shape could be divided into combinations of rectangles, semicircles, triangles, and trapezoids.

(Continued in bottom margin.)

Reading to Learn Mathematics, p. 642 ELL

NAME ______________ DATE ________ PERIOD ____

11-7 Reading to Learn Mathematics

Area of Complex Figures

Pre-Activity *Read the introduction at the top of page 498 in your textbook. Write your answers below.*

1. Describe the shape of the kitchen. rectangle and semicircle
2. How could you determine the area of the kitchen? Sample answer: Find the area of the rectangle and the area of the semicircle, then add.
3. How could you determine the total square footage of a house with rooms shaped like these? Sample answer: Find the area of each room, then add.

Reading the Lesson

4. Look up the term *footage* in a dictionary. Write the meaning that matches the way the term is used in this lesson.
Sample answer: length... quantity expressed in feet ... think ...

11-7 Area of Complex Figures

What You'll Learn
Find the areas of complex figures.

NEW Vocabulary
complex figure

LINK To Reading
Everyday Meaning of Complex: a whole made up of interrelated parts, as in a retail complex that is made up of many different stores

WHEN am I ever going to use this?

ARCHITECTURE Rooms in a house are not always square or rectangular, as shown in the diagram at the right.

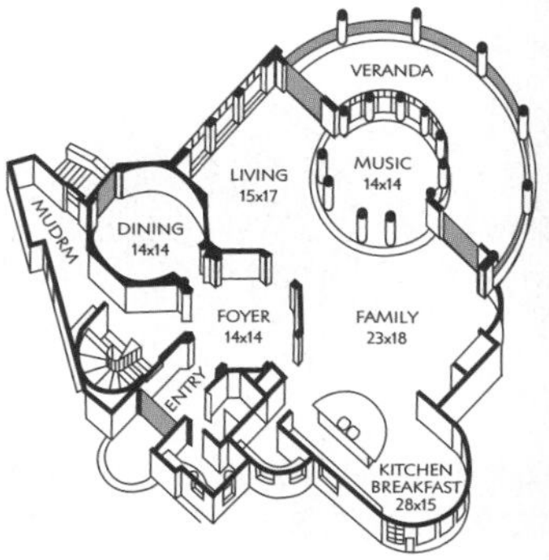

1. Describe the shape of the kitchen. **rectangle and semicircle**
2. How could you determine the area of the kitchen?
3. How could you determine the total square footage of a house with rooms shaped like these?
Sample answer: Find the area of each room, then add.

2. Sample answer: Find the areas of the rectangle and semicircle, then add.

A **complex figure** is made of circles, rectangles, squares, and other two-dimensional figures. To find the area of a complex figure, separate it into figures whose areas you know how to find, and then add the areas.

EXAMPLE Find the Area of an Irregular Room

1 ARCHITECTURE Refer to the diagram of the house above. The kitchen is 28 feet by 15 feet, as shown at the right. Find the area of the kitchen. Round to the nearest tenth.

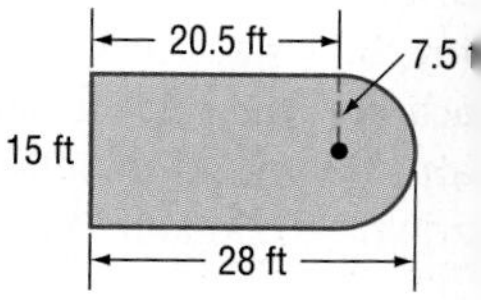

The figure can be separated into a rectangle and a semicircle.

Area of Rectangle

$A = \ell w$ Area of a rectangle

$A = 20.5 \cdot 15$ Replace ℓ with 20.5 and w with 15.

$A = 307.5$ Multiply.

Area of Semicircle

$A = \frac{1}{2}\pi r^2$ Area of a semicircle

$A = \frac{1}{2}\pi(7.5)^2$ Replace r with 7.5.

$A \approx 88.4$ Simplify.

The area of the kitchen is approximately 307.5 + 88.4 or 395.9 square feet.

TEACHING TIP
A bullet signifies the center of a circle. So, in Example 1, students can assume that the round part of the kitchen is a semicircle.

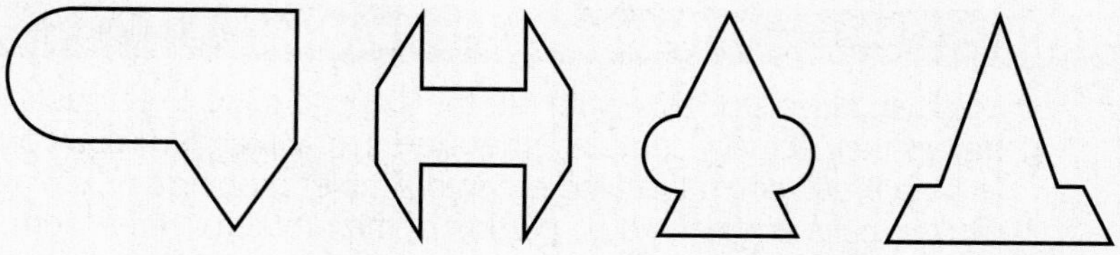

Standardized Test Practice

EXAMPLE Find the Area of a Complex Figure

2 GRID-IN TEST ITEM Find the area of the figure at the right in square inches.

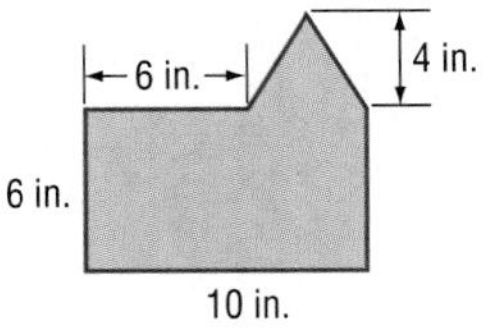

Read the Test Item

The figure can be separated into a rectangle and a triangle. Find the area of each.

Solve the Test Item

Area of Rectangle

$A = \ell w$ Area of a rectangle

$A = 10 \cdot 6$ Replace ℓ with 10 and w with 6.

$A = 60$ Multiply.

Area of Triangle

$A = \frac{1}{2}bh$ Area of a triangle

$A = \frac{1}{2}(4)(4)$ $b = 10 - 6$ or 4, $h = 4$

$A = 8$ Multiply.

The area is 60 + 8 or 68 square inches.

Fill in the Grid

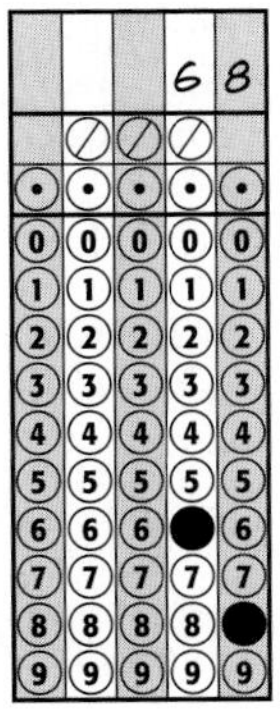

Your Turn Find the area of each figure. Round to the nearest tenth if necessary.

a. 60 ft²

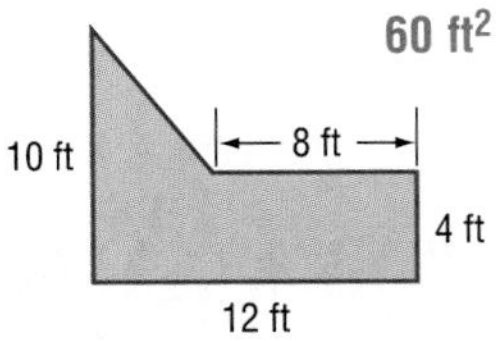

b. 31.4 cm²

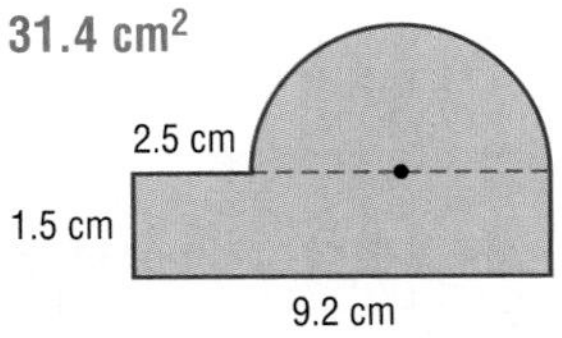

st-Taking Tip The Princeton Review

awings Be sure to add e areas of all the parate figures and not p once you find the ea of part of the figure.

TEACHING TIP

nind students that y can also fill in the d starting in the left umn.

Skill and Concept Check

Writing Math Exercises 1 & 2

1. **Describe** how you would find the area of the figure at the right. Sample answer: Separate it into a rectangle and a triangle, find the area of each, and add.
2. **OPEN ENDED** Sketch a complex figure and describe how you could find the area. See students' work.

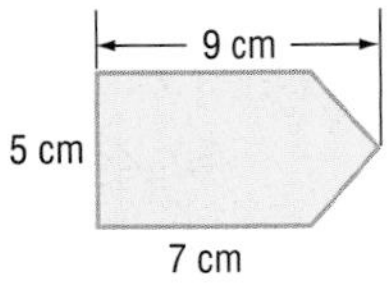

GUIDED PRACTICE

Find the area of each figure. Round to the nearest tenth if necessary.

3.

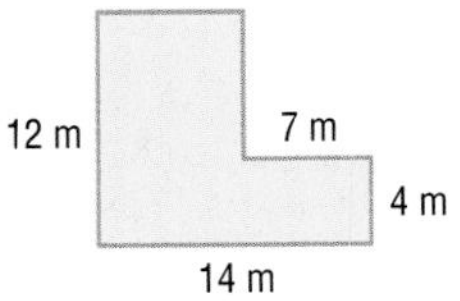

112 m²

4.

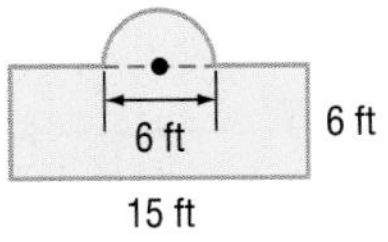

104.1 ft²

5.

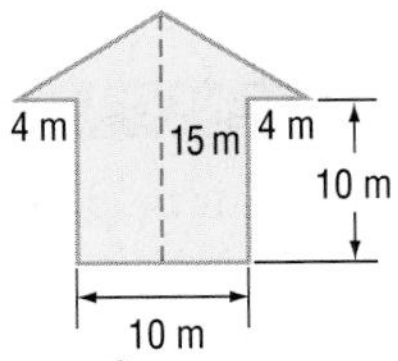

145 m²

TEACHING TIP

There may be several ways to subdivide a figure. In Exercise 3, students could find the area of the whole 12 × 14 rectangle and then subtract the area of the 8 × 7 rectangle.

msmath2.net/extra_examples

DAILY INTERVENTION Differentiated Instruction

Naturalist Have students measure and sketch grassy areas and flowerbeds around the school, or find maps of local, irregularly shaped parks. Then have them calculate the areas by separating the complex figures into simpler figures.

2 Teach

In-Class Examples

INTERACTIVE CHALKBOARD PowerPoint® Presentations

1 WINDOWS The diagram below shows the dimensions of a window that is 3.4 feet by 7.2 feet. Find the area of the window. Round to the nearest tenth. **about 23.2 ft²**

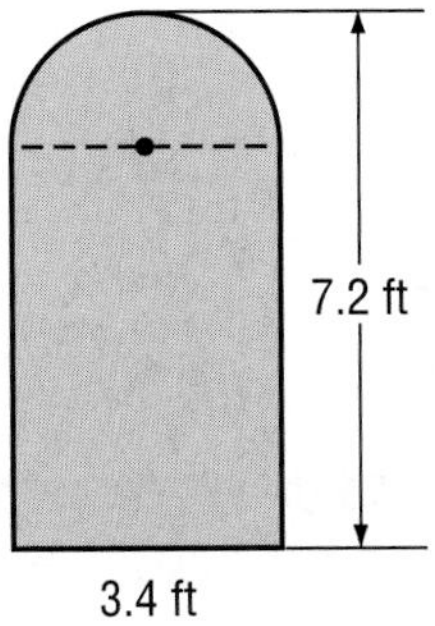

2 GRID-IN TEST ITEM Find the area of the figure below in square centimeters. **160 cm²**

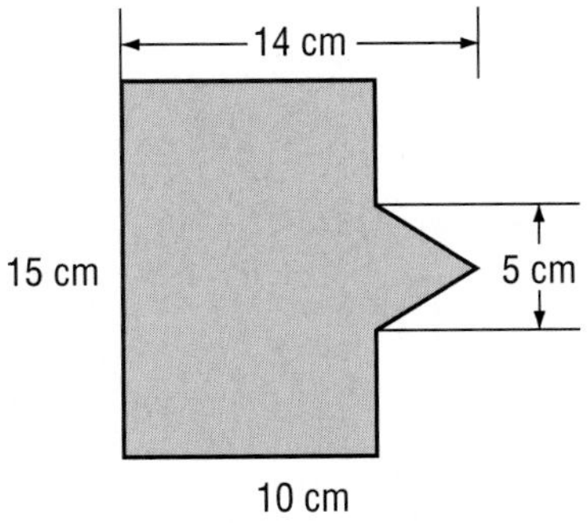

3 Practice/Apply

About the Exercises...

Guided Practice Key

Exercises	Examples
3–5	1, 2

Odd/Even Assignments

Exercises 6–11 are structured so that students practice the same concepts whether they are assigned odd or even problems.

Study Guide and Intervention, p. 639

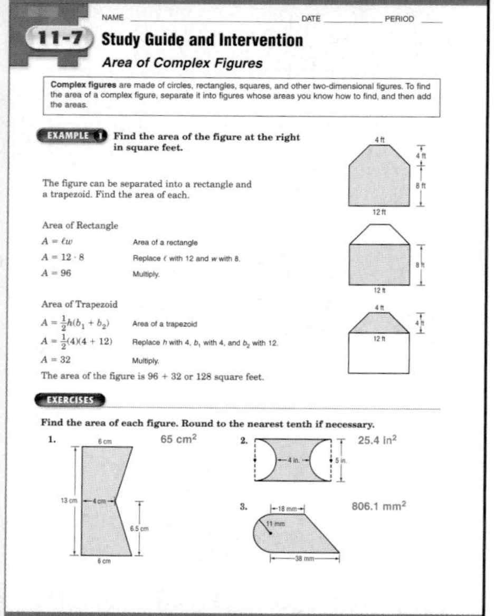
NAME ______ DATE ______ PERIOD ______

11-7 Study Guide and Intervention
Area of Complex Figures

Complex figures are made of circles, rectangles, squares, and other two-dimensional figures. To find the area of a complex figure, separate it into figures whose areas you know how to find, and then add the areas.

EXAMPLE 1 Find the area of the figure at the right in square feet.

The figure can be separated into a rectangle and a trapezoid. Find the area of each.

Area of Rectangle

$A = \ell w$ Area of a rectangle
$A = 12 \cdot 8$ Replace ℓ with 12 and w with 8.
$A = 96$ Multiply.

Area of Trapezoid

$A = \frac{1}{2}h(b_1 + b_2)$ Area of a trapezoid
$A = \frac{1}{2}(4)(4 + 12)$ Replace h with 4, b_1 with 4, and b_2 with 12.
$A = 32$ Multiply.

The area of the figure is 96 + 32 or 128 square feet.

EXERCISES

Find the area of each figure. Round to the nearest tenth if necessary.

1. 65 cm²
2. 25.4 in²
3. 806.1 mm²

Practice: Skills, p. 640

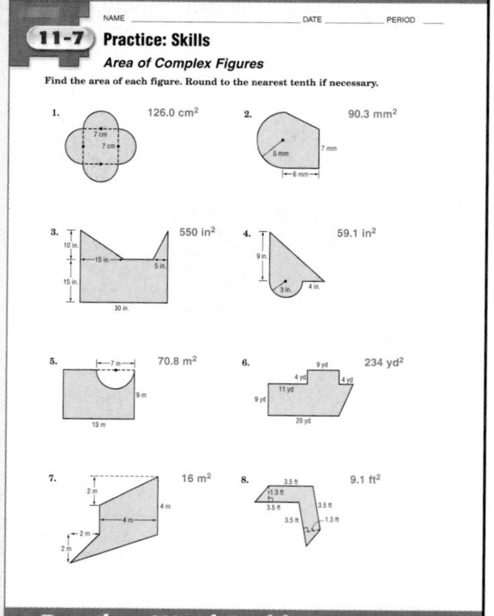
NAME ______ DATE ______ PERIOD ______

11-7 Practice: Skills
Area of Complex Figures

Find the area of each figure. Round to the nearest tenth if necessary.

1. 126.0 cm²
2. 90.3 mm²
3. 550 in²
4. 59.1 in²
5. 70.8 m²
6. 234 yd²
7. 16 m²
8. 9.1 ft²

Practice: Word Problems, p. 641

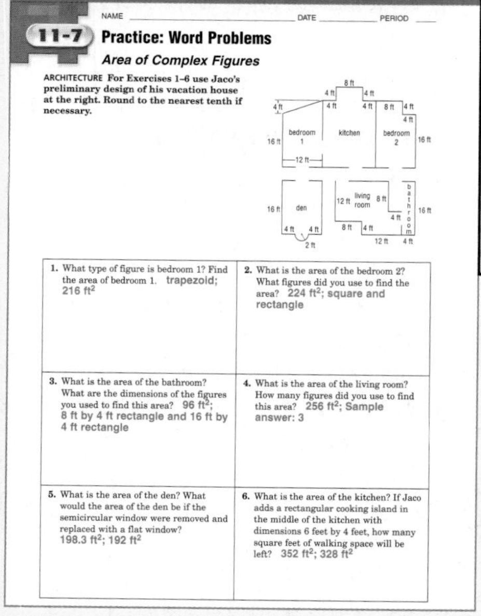
NAME ______ DATE ______ PERIOD ______

11-7 Practice: Word Problems
Area of Complex Figures

ARCHITECTURE For Exercises 1–6 use Jaco's preliminary design of his vacation house at the right. Round to the nearest tenth if necessary.

1. What type of figure is bedroom 1? Find the area of bedroom 1. trapezoid; 216 ft²	2. What is the area of the bedroom 2? What figures did you use to find the area? 224 ft²; square and rectangle
3. What is the area of the bathroom? What are the dimensions of the figures you used to find this area? 96 ft²; 8 ft by 4 ft rectangle and 16 ft by 4 ft rectangle	4. What is the area of the living room? How many figures did you use to find this area? 256 ft²; Sample answer: 3
5. What is the area of the den? What would the area of the den be if the semicircular window were removed and replaced with a flat window? 198.3 ft²; 192 ft²	6. What is the area of the kitchen? If Jaco adds a rectangular cooking island in the middle of the kitchen with dimensions 6 feet by 4 feet, how many square feet of walking space will be left? 352 ft²; 328 ft²

★ indicates increased difficulty

Practice and Applications

HOMEWORK HELP

For Exercises	See Examples
6–12	1, 2

Extra Practice
See pages 592, 606.

Find the area of each figure. Round to the nearest tenth if necessary.

6. 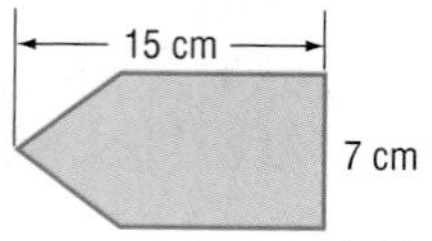

87.5 cm²

7. 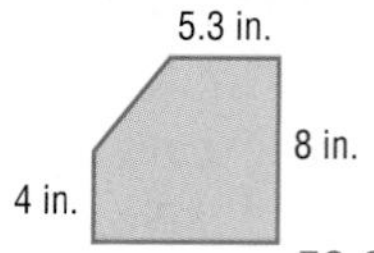

58.6 in²

8. 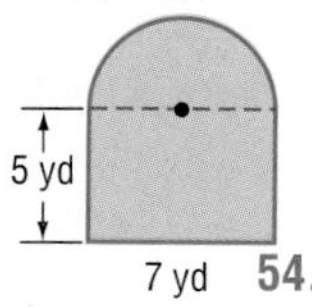

54.2 yd²

9. 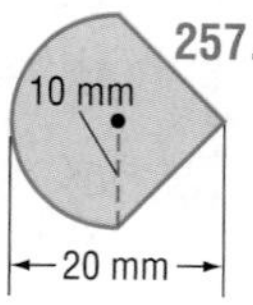

257.1 mm²

★ 10. 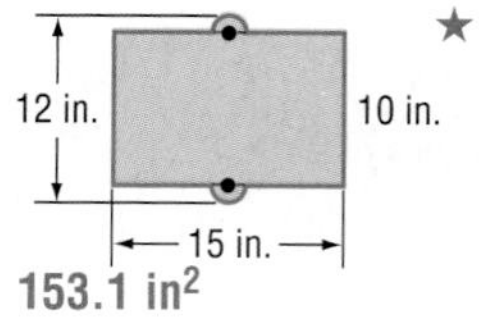

153.1 in²

★ 11. 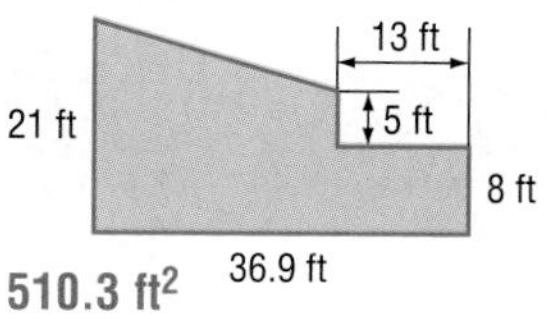

510.3 ft²

12. **INTERIOR DESIGN** The Eppicks' living room, shown at the right, has a bay window. They are planning to have the hardwood floors in the room refinished. What is the total area that needs to be refinished? 195 ft²

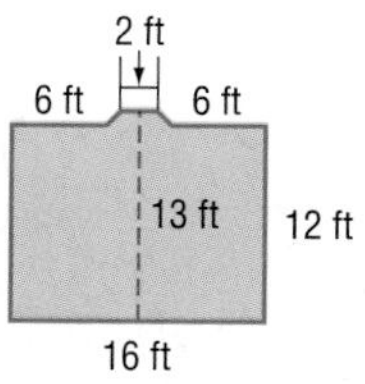

CRITICAL THINKING Describe how you could estimate the area of each state.

13. 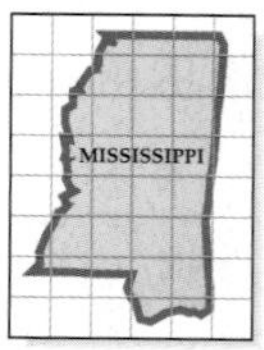

Sample answer: Add the areas of a trapezoid and a small rectangle.

14. 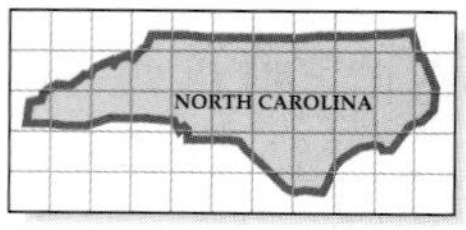

Sample answer: Add the areas of a trapezoid and a triangle.

Standardized Test Practice and Mixed Review

15. **SHORT RESPONSE** Find the area of the figure if each triangle has a height of 3.5 inches and the square has side lengths of 4 inches. 44 in²

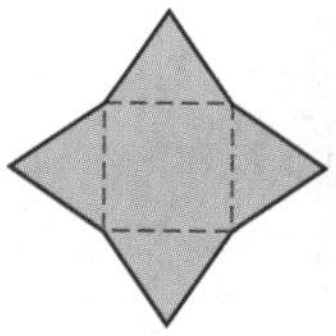

16. **MULTIPLE CHOICE** A rectangular room 14 feet by 12 feet has a semicircular sitting area attached with a diameter of 12 feet. What is the total area of the room and the sitting area? B

Ⓐ 168 ft² Ⓑ 224.5 ft² Ⓒ 281.1 ft² Ⓓ 620.4 ft²

Find the area of each circle. Round to the nearest tenth. (Lesson 11-6)

17. radius = 4.7 cm 69.4 cm²
18. radius = 12 in. 452.4 in²
19. diameter = 15 in. 176.7 in²

20. Find the area of a triangle that has a base of 3.8 meters and a height of 9 meters. (Lesson 11-5) 17.1 m²

GETTING READY FOR THE NEXT LESSON

PREREQUISITE SKILL Find the probability of rolling each number on a number cube. (Lesson 9-1)

21. $P(2)$ $\frac{1}{6}$
22. $P(\text{even})$ $\frac{1}{2}$
23. $P(3 \text{ or } 4)$ $\frac{1}{3}$
24. $P(\text{less than } 5)$ $\frac{2}{3}$

msmath2.net/self_check_quiz

4 Assess

Higher-Level Thinking

Look at the figure in Example 2. If all of the dimensions were tripled, would the area also be tripled? Explain. No; the area would be multiplied by 9. The new area would be 612 in².

11-8 Area Models and Probability

...at You'll Learn
...d probability using ...a models.

***...VIEW* Vocabulary**

...bability: the ratio ...he number of ...s an event can ...ur to the number ...ossible ...comes **(Lesson 9-1)**

HANDS-ON Mini Lab

Materials
- two number cubes

Work with a partner.

- Roll the number cubes and find the product of the numbers rolled.
- Repeat nine more times.
- Collect the data for the entire class. Organize the outcomes in a table.

1. Do certain products occur more often? **See students' work.**
2. Make and complete a table like the one at the right to find all the possible outcomes. **See below.**

×	1	2	3	4	5	6
1	1	2	3			
2	2	4	6			
3						

The grid at the right shows the possible products when two number cubes are rolled. The area of the grid is 36 square units. Notice that 6 and 12 make up $\frac{8}{36}$ of the area. So, the probability of rolling two numbers whose product is 6 or 12 is $\frac{8}{36}$.

1	2	3	4	5	6
2	4	6	8	10	12
3	6	9	12	15	18
4	8	12	16	20	24
5	10	15	20	25	30
6	12	18	24	30	36

The area of geometric shapes can be used to find probabilities.

EXAMPLE Use Area Models to Find Probability

1 PROBABILITY A randomly-dropped counter falls somewhere in the squares. Find the probability that it falls on the shaded squares.

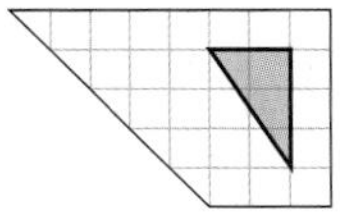

$$\text{probability} = \frac{\text{number of ways to land in shaded squares}}{\text{number of ways to land on squares}}$$

$$= \frac{\text{area of shaded squares}}{\text{area of all squares}}$$

Area of Shaded Squares

$A = \frac{1}{2}bh$ Area of a triangle

$A = \frac{1}{2}(2)(3)$ $b = 2$ and $h = 3$

$A = 3$ Simplify.

Area of All Squares

$A = \frac{1}{2}h(b_1 + b_2)$ Area of a trapezoid

$A = \frac{1}{2}(5)(8 + 3)$ $h = 5, b_1 = 8, b_2 = 3$

$A = 27.5$ Simplify.

So, the probability of a counter falling in the shaded squares is $\frac{3}{27.5}$ or about 10.9%.

TEACHING TIP
...nt out that the total ...nber of squares in ...gure represents the ...nber of possible ...comes. Guide students ...ee that the shaded ...ares represent the ...orable outcomes.

DAILY INTERVENTION — Unlocking Misconceptions

Determining Probability Students may mistakenly think that the probability of an event occurring is the number of ways it could occur divided by the number of ways it could *not* occur. Remind them that they should divide by the total number of possible outcomes.

11-8 Lesson Notes

Lesson Resources

Chapter 11 Resource Masters

FAST FILE

- Study Guide and Intervention, p. 644
- Practice: Skills, p. 645
- Practice: Word Problems, p. 646
- Reading to Learn Mathematics, p. 647
- Enrichment, p. 648

Parent and Student Study Guide Workbook, 11-8

Teaching Mathematics With Manipulatives, pp. 112–113

1 Focus

5-Minute Check 11-8

INTERACTIVE CHALKBOARD PowerPoint® Presentations

Use as a quiz or review of Lesson 11-7. (Also available as a transparency)

Bellringer

Problem Solving Tell students that they will have to think logically to solve probability problems. Write the following on the chalkboard:

(1) What do I know?

(2) What do I need to know?

(3) How can I answer the problem?

Have students use these questions as a guide in determining the ratio of left-handed to right-handed students in the classroom.

Reading to Learn Mathematics, p. 647 ELL

NAME ______ DATE ______ PERIOD ____

11-8 Reading to Learn Mathematics
Area Models and Probability

Pre-Activity *Complete the Mini Lab at the top of page 501 in your textbook. Write your answers below.*

1. Do certain products occur more often? See students' work.
2. Make and complete the table below to find all the possible outcomes.

×	1	2	3	4	5	6
1	1	2	3	4	5	6
2	2	4	6	8	10	12
3	3	6	9	12	15	18
4	4	8	12	16	20	24
5	5	10	15	20	25	30
6	6	12	18	24	30	36

2 Teach

In-Class Examples

INTERACTIVE CHALKBOARD
PowerPoint® Presentations

1 PROBABILITY A randomly dropped counter falls somewhere in the squares below. Find the probability that it falls on the shaded squares. **about 25.6%**

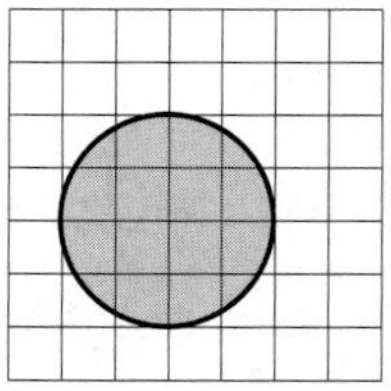

2 GAMES Suppose a dart is equally likely to hit any point on the figure below. What is the probability that it hits in the shaded region? **about 39.3%**

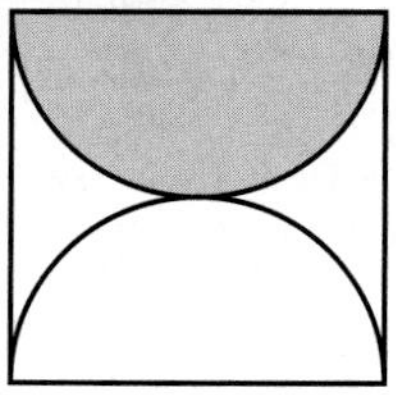

3 Practice/Apply

About the Exercises…

Guided Practice Key

Exercises	Examples
3–4	1
5	2

Odd/Even Assignments
Exercises 6–13 are structured so that students practice the same concepts whether they are assigned odd or even problems.

REAL-LIFE MATH

GAMES The arrangement of numbers around a standard dartboard dates back over 100 years.

Source: mathpages.com

EXAMPLE Find the Probability of Winning a Game

2 GAMES **Suppose a dart is equally likely to hit any point on the board. What is the probability that it hits the white section?**

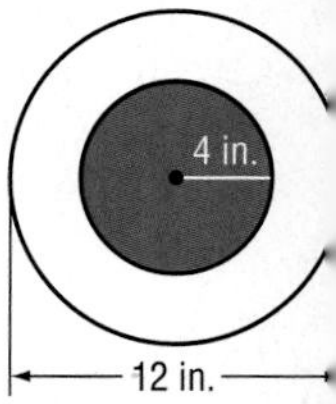

First, find the area of the white section. It equals the area of the large circle minus the area of the small circle.

Area of Large Circle		Area of Small Circle	
$A = \pi r^2$	Area of a circle	$A = \pi r^2$	Area of a circle
$A = \pi(6)^2$	Replace r with 6.	$A = \pi(4)^2$	Replace r with 4.
$A \approx 113.1$	Simplify.	$A \approx 50.3$	Simplify.

Area of White Section $= 113.1 - 50.3$ large circle – small circle
$= 62.8$ Subtract.

$P(\text{white}) = \frac{62.8}{113.1}$ ← area of white section / ← area of entire model

≈ 0.5555 Use a calculator.

So, the probability of hitting the white section is about 55.6%.

Skill and Concept Check

Writing Math
Exercises 1 & 2

1. **Explain** how area models are used to solve probability problems.
2. **OPEN ENDED** Draw a spinner in which the probability of spinning and landing on a blue region is $\frac{1}{6}$. Explain your reasoning. **1–2. See margin.**

GUIDED PRACTICE

A randomly-dropped counter falls in the squares. Find the probability that it falls in the shaded squares. Write as a percent. Round to the nearest tenth if necessary.

3. 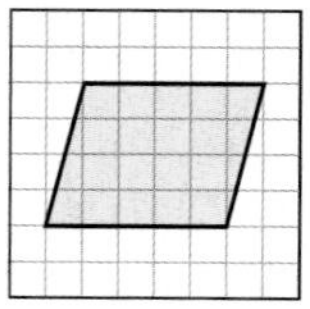
31.3%

4. 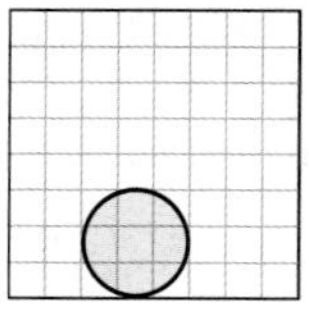
11.0%

5. **GAMES** Suppose a dart is equally likely to hit any point on the dartboard at the right. What is the probability that it hits the red section? **16%**

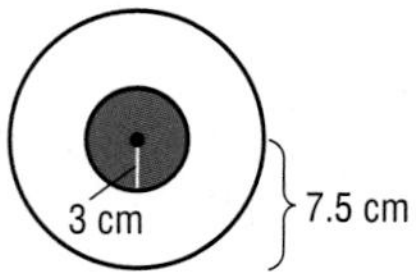

Answers

1. Sample answer: The probability of an event occurring is the ratio of the shaded area (the number of ways an event can occur) to the entire area (the number of possible outcomes).

2. Sample answer:

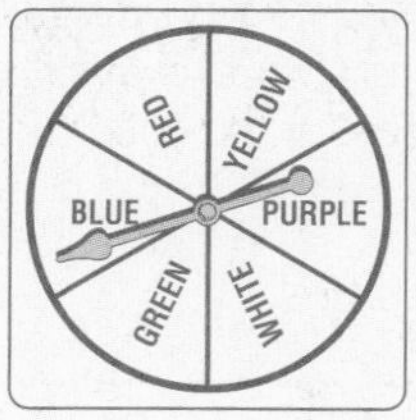

There are 6 possible outcomes, and the number of blue regions is 1. So, the probability of landing on a blue region is $\frac{1}{6}$.

★ indicates increased difficulty

Practice and Applications

HOMEWORK HELP	
For Exercises	See Examples
6–11	1
12–14	2
Extra Practice See pages 592, 602.	

A randomly-dropped counter falls in the squares. Find the probability that it falls in the shaded squares. Write as a percent. Round to the nearest tenth if necessary.

6. 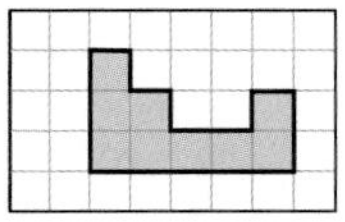 22.5%

7. 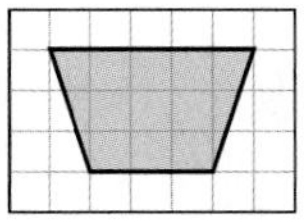 34.3%

8. 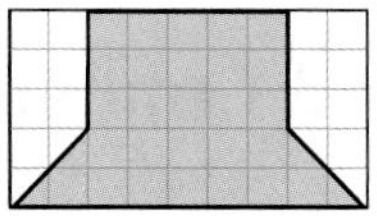 64.4%

9. 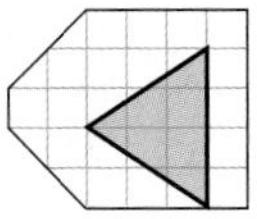 23.1%

10. 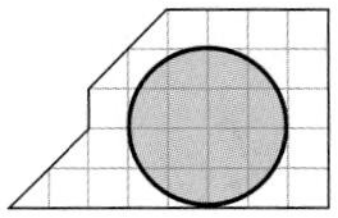 41.9%

★ 11. 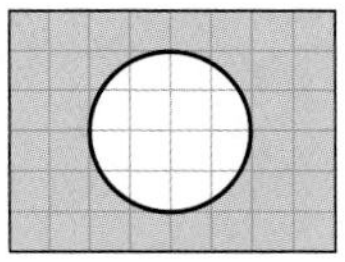 73.8%

GAMES **Each figure represents a dartboard. If it is equally likely that a thrown dart will land anywhere on the dartboard, find the probability that it lands in the shaded region.**

12. 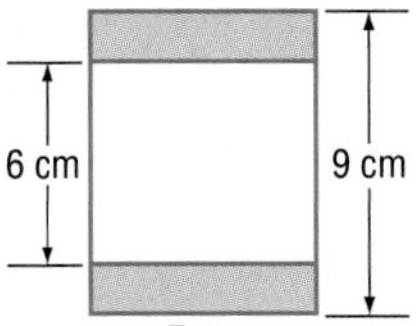

33.3%

13. 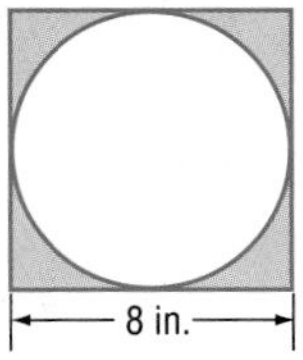

21.5%

14. 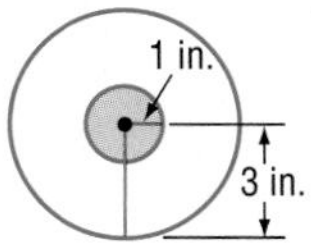

11.1%

15. **CRITICAL THINKING** A quarter is randomly tossed on the grid board at the right. If a quarter has a radius of 12 millimeters, what is the probability that it does *not* touch a line when it lands? (*Hint*: Find the area where the center of the coin could land so that the edges do not touch a line.) $\frac{4}{25}$ or 16%

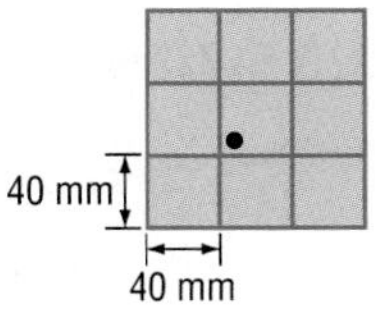

Standardized Test Practice and Mixed Review

16. **MULTIPLE CHOICE** At a carnival, a person wins a prize if their dart pops a balloon on the rectangular wall. If the radius of each circular balloon is 0.4 foot, approximately what is the probability that a person will win? Use 3.14 for π. A

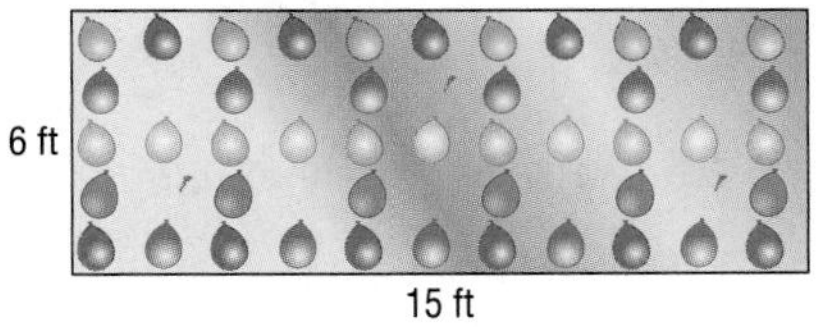

Ⓐ 25% Ⓑ 29% Ⓒ 30% Ⓓ 63%

17. **GRID IN** Find the probability that a randomly dropped counter will fall in the shaded region at the right. Write as a fraction. 3/10

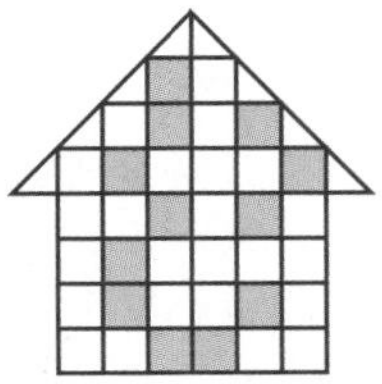

18. Find the area of a figure that is a 6-inch square with a semicircle attached to each side. Each semicircle has a diameter of 6 inches. Round to the nearest tenth. (Lesson 11-7) 92.5 in^2

19. Find the area of a circle with a radius of 5.7 meters. Round to the nearest tenth. (Lesson 11-6) 102.1 m^2

msmath2.net/self_check_quiz

4 Assess

Open-Ended Assessment

Modeling Have students work in small groups to make a cardboard model of the figure shown in Example 2. Place the model on the floor and have a blindfolded student drop a penny onto the model. Repeat 25 times. Students should compare their results with the predicted probability. Repeat with other models if time allows.

Study Guide and Intervention, p. 644

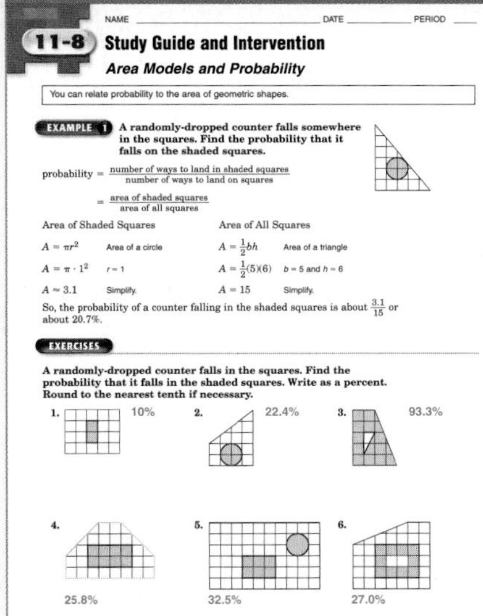
NAME ________ DATE ________ PERIOD ____

11-8 Study Guide and Intervention
Area Models and Probability

You can relate probability to the area of geometric shapes.

EXAMPLE 1 A randomly-dropped counter falls somewhere in the squares. Find the probability that it falls on the shaded squares.

probability = $\frac{\text{number of ways to land in shaded squares}}{\text{number of ways to land on squares}}$ = $\frac{\text{area of shaded squares}}{\text{area of all squares}}$

Area of Shaded Squares: $A = \pi r^2$ Area of a circle; $A = \pi \cdot 1^2$ $r = 1$; $A \approx 3.1$ Simplify.

Area of All Squares: $A = \frac{1}{2}bh$ Area of a triangle; $A = \frac{1}{2}(5)(6)$ $b = 5$ and $h = 6$; $A = 15$ Simplify.

So, the probability of a counter falling in the shaded squares is about $\frac{3.1}{15}$ or about 20.7%.

EXERCISES

A randomly-dropped counter falls in the squares. Find the probability that it falls in the shaded squares. Write as a percent. Round to the nearest tenth if necessary.

1. 10% 2. 22.4% 3. 93.3%
4. 25.8% 5. 32.5% 6. 27.0%

Practice: Skills, p. 645

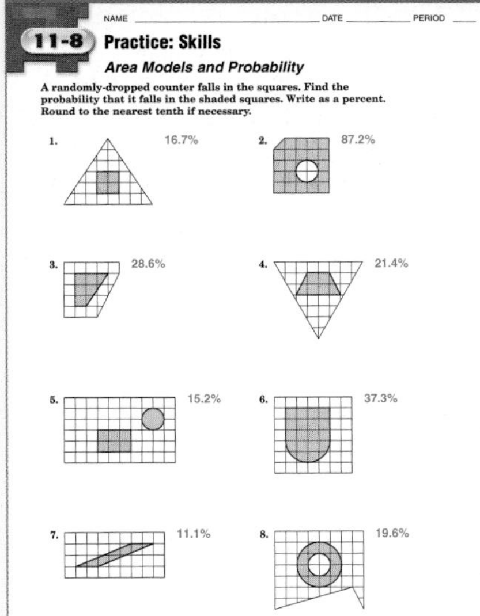
NAME ________ DATE ________ PERIOD ____

11-8 Practice: Skills
Area Models and Probability

A randomly-dropped counter falls in the squares. Find the probability that it falls in the shaded squares. Write as a percent. Round to the nearest tenth if necessary.

1. 16.7% 2. 87.2%
3. 28.6% 4. 21.4%
5. 15.2% 6. 37.3%
7. 11.1% 8. 19.6%

Practice: Word Problems, p. 646

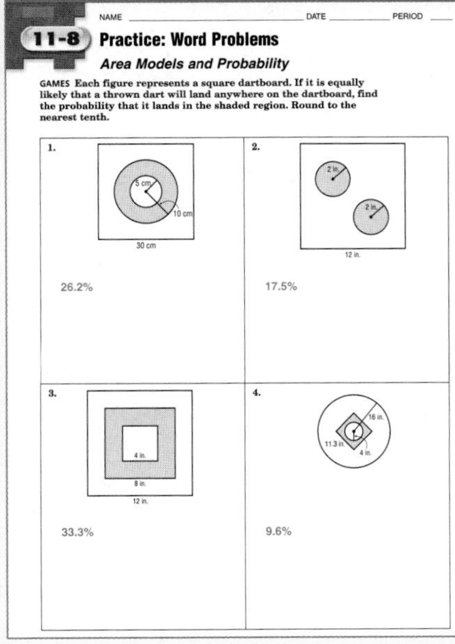
NAME ________ DATE ________ PERIOD ____

11-8 Practice: Word Problems
Area Models and Probability

GAMES Each figure represents a square dartboard. If it is equally likely that a thrown dart will land anywhere on the dartboard, find the probability that it lands in the shaded region. Round to the nearest tenth.

1. 26.2% 2. 17.5%
3. 33.3% 4. 9.6%

Lesson 11-8 Area Models and Probability **503**

Chapter 11 Study Guide and Review

Vocabulary Assessment

- This alphabetical list of vocabulary terms in Chapter 11 includes a page reference where each term was introduced.
- A vocabulary review/test for Chapter 11 is available on p. 662 of the *Chapter 11 Resource Masters*.

Vocabulary PuzzleMaker

ELL The online Vocabulary PuzzleMaker improves students' mathematics vocabulary using four puzzle formats — crossword, scramble, word search using a word list, and word search using clues. Students can work on a computer screen or from a printed handout.

MindJogger Videoquizzes

ELL MindJogger Videoquizzes provide an alternative review of concepts presented in this chapter. Students work in teams in a game show format to gain points for correct answers. The questions are presented in three rounds.

Round 1 Concepts (5 questions)
Round 2 Skills (4 questions)
Round 3 Problem Solving (4 questions)

ExamView® Pro Test Bank CD-ROM

Use the networkable ExamView® Pro Test Bank CD-ROM to:

- Create **Multiple versions** of tests.
- Create **Modified** tests for **Inclusion** students with one mouse click.
- **Edit** existing questions and **Add** your own questions.
- Build tests aligned with state standards using built-in **State Curriculum Correlations.**
- Change **English** tests to **Spanish** with one mouse click and vice versa.

Chapter 11 Study Guide and Review

Vocabulary and Concept Check

base (p. 483)	irrational number (p. 476)	radical sign (p. 471)
complex figure (p. 498)	leg (p. 479)	square (p. 470)
height (p. 483)	perfect square (p. 471)	square roots (p. 471)
hypotenuse (p. 479)	Pythagorean Theorem (p. 479)	

Choose the correct term or number to complete each sentence.

1. In a right triangle, the square of the length of the hypotenuse is (equal to, less than) the sum of the squares of the lengths of the legs.
2. $A = \frac{1}{2}h(a + b)$ is the formula for the area of a (triangle, trapezoid).
3. The square of 25 is (625, 5).
4. A (square, square root) of 49 is 7.
5. The longest side of a right triangle is called the (leg, hypotenuse).
6. The $\sqrt{\ }$ symbol is called a (radical, perfect square) sign.

Lesson-by-Lesson Exercises and Examples

11-1 Squares and Square Roots (pp. 470–473)

Find the square of each number.

7. 6 **36** 8. 14 **196** 9. 23 **529**

Find each square root.

10. $\sqrt{16}$ **4** 11. $\sqrt{256}$ **16** 12. $\sqrt{900}$ **30**

Example 1 Find the square of 9.
$9^2 = 9 \cdot 9$ or 81

Example 2 Find $\sqrt{121}$.
Since $11 \cdot 11 = 121$, $\sqrt{121} = 11$.

11-2 Estimating Square Roots (pp. 475-477)

Estimate each square root to the nearest whole number.

13. $\sqrt{6}$ **2** 14. $\sqrt{99}$ **10** 15. $\sqrt{48}$ **7**
16. $\sqrt{76}$ **9** 17. $\sqrt{19}$ **4** 18. $\sqrt{52}$ **7**

Use a calculator to find each square root to the nearest tenth.

19. $\sqrt{61}$ **7.8** 20. $\sqrt{132}$ **11.5**
21. $\sqrt{444}$ **21.1** 22. $\sqrt{12}$ **3.5**

Example 3 Estimate $\sqrt{29}$ to the neares[t] whole number.

$25 < 29 < 36$ 29 is between the perfect squares 25 and 36.

$\sqrt{25} < \sqrt{29} < \sqrt{36}$ Find the square root of eac[h] number.

$5 < \sqrt{29} < 6$ $\sqrt{25} = 5$ and $\sqrt{36} = 6$

So, $\sqrt{29}$ is between 5 and 6. Since 29 is closer to 25 than to 36, the best whole number estimate is 5.

msmath2.net/vocabulary_revie[w]

FOLDABLES™ Study Organizer

For more information about Foldables, see *Teaching Mathematics with Foldables.*

Have students look through the chapter to make sure they have included examples in their Foldables for each lesson or topic in the chapter.

Encourage students to refer to their Foldables while completing the Study Guide and Review and to use them in preparing for the Chapter Test.

1-3 The Pythagorean Theorem (pp. 479–482)

Find the missing measure of each right triangle. Round to the nearest tenth if necessary.

23. **13 mm**

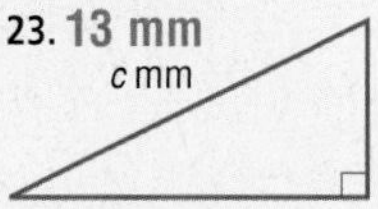

24.

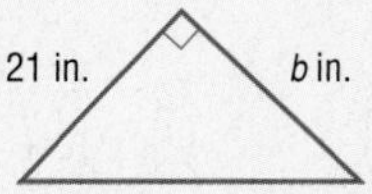

26.7 in.

25. $a = 5$ ft, $b = 6$ ft **7.8 ft**

26. $b = 10$ yd, $c = 12$ yd **6.6 yd**

27. $a = 7$ m, $c = 15$ m **13.3 m**

28. $a = 12$ in., $b = 4$ in. **12.6 in.**

Example 4 **Find the missing measure of the triangle. Round to the nearest tenth if necessary.**

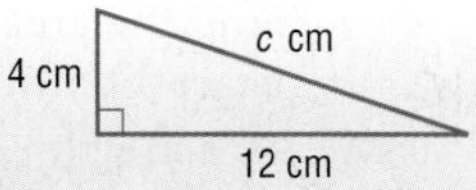

$c^2 = a^2 + b^2$ Pythagorean Theorem

$c^2 = 4^2 + 12^2$ $a = 4, b = 12$

$c^2 = 16 + 144$ Evaluate.

$c^2 = 160$ Add.

$c \approx 12.6$ Take the square root of each side.

The length of the hypotenuse is about 12.6 centimeters.

1-4 Area of Parallelograms (pp. 483–485)

Find the area of each parallelogram. Round to the nearest tenth if necessary.

29. **99 cm²**

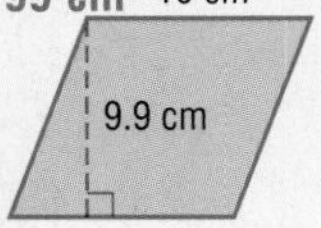

30. **2,520 in²**

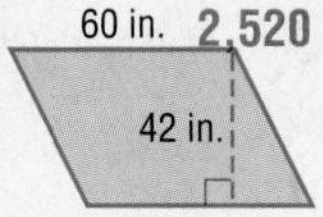

31. base = 9 cm, height = 15 cm **135 cm²**

32. base = 24 m, height = 16.2 m **388.8 m²**

Example 5 **Find the area of a parallelogram if the base is 15 inches and the height is 8 inches.**

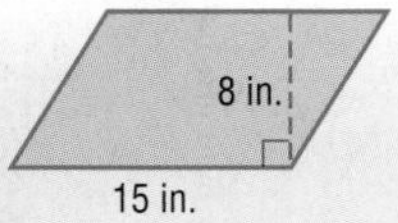

$A = bh$ Area of a parallelogram

$A = 15 \cdot 8$ Replace b with 15 and h with 8.

$A = 120 \text{ in}^2$ Multiply.

1-5 Area of Triangles and Trapezoids (pp. 489–492)

Find the area of each figure. Round to the nearest tenth if necessary.

33. **36 ft²**

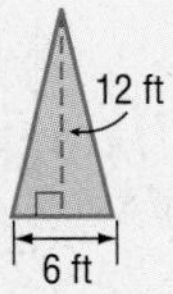

34. **37.5 in²**

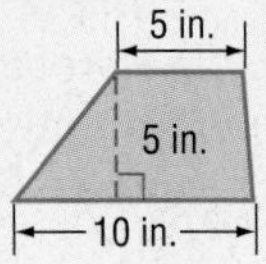

35. triangle: base = 24.7 cm, height = 15.2 cm **187.7 cm²**

36. trapezoid: bases = 22 yd and 35 yd, height = 18.5 yd **527.3 yd²**

Example 6 **Find the area of a triangle with a base of 8 meters and a height of 11.2 meters.**

$A = \frac{1}{2}bh$ Area of a triangle

$A = \frac{1}{2}(8)(11.2)$ or 44.8 m^2 $b = 8, h = 11.2$

Example 7 **Find the area of the trapezoid.**

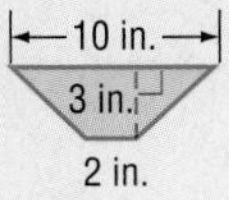

$A = \frac{1}{2}h(b_1 + b_2)$

$A = \frac{1}{2}(3)(2 + 10)$ $h = 3, b_1 = 2, b_2 = 10$

$A = \frac{1}{2}(3)(12)$ or 18 in^2 Simplify.

Chapter 11 Study Guide and Review 505

11-6 Area of Circles (pp. 493–495)

Find the area of each circle. Round to the nearest tenth.

37. radius = 11.4 in. **408.3 in^2**
38. diameter = 44 cm **1,520.5 cm^2**
39. **GARDENING** A lawn sprinkler can water a circular area with a radius of 20 feet. Find the area that can be watered. Round to the nearest tenth. **1,256.6 ft^2**

Example 8 **Find the area of a circle with a radius of 5 inches.**

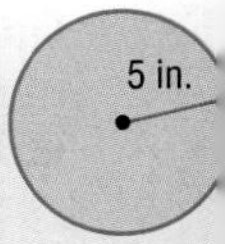

$A = \pi r^2$ Area of a circle

$A = \pi(5)^2$ Replace r with 5.

$A \approx 78.5$ Multiply.

The area of the circle is about 78.5 square inches.

11-7 Area of Complex Figures (pp. 498–500)

Find the area of each figure. Round to the nearest tenth if necessary.

40.

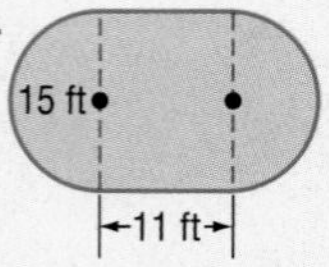

341.7 ft^2

41.

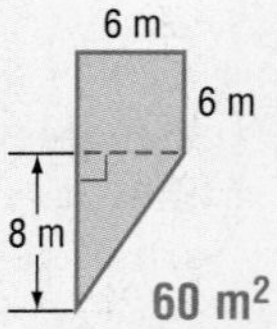

60 m^2

42. **67.8 yd^2**

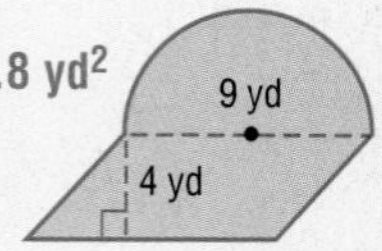

43. **182.8 in^2**

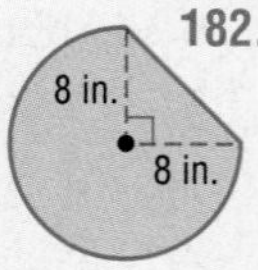

Example 9 **Find the area of the figure.**

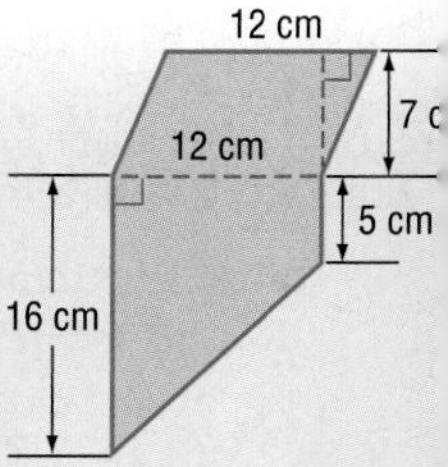

The figure can be separated into a parallelogram and a trapezoid.

parallelogram: $A = bh = (12)(7)$ or 84

trapezoid: $A = \frac{1}{2}h(b_1 + b_2)$

$= \frac{1}{2}(12)(16 + 5)$ or 126

The area of the figure is 84 + 126 or 210 square centimeters.

11-8 Area Models and Probability (pp. 501–503)

A randomly-dropped counter falls in the squares. Find the probability that it falls in the shaded regions. Write as a percent. Round to the nearest tenth if necessary.

44.

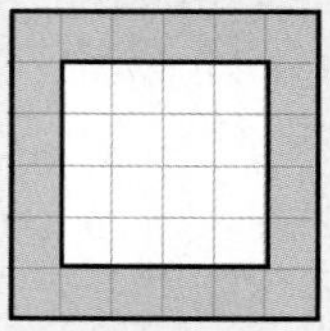

55.6%

45.

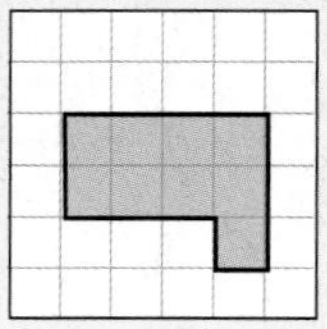

25%

Example 10 **Find the probability that a randomly-dropped counter will land on a white square.**

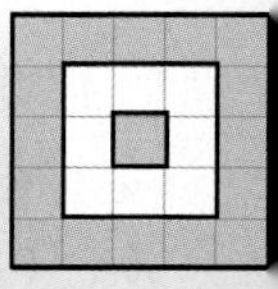

Probability of landing on a white square

$= \frac{\text{number of ways to land on white squares}}{\text{number of ways to land on squares}}$

$= \frac{\text{area of white squares}}{\text{total area}}$

$= \frac{8}{25}$ The probability is $\frac{8}{25}$ or 32%.

Chapter 11 Practice Test

Vocabulary and Concepts

1. **State** the measurements that you need in order to find the area of a parallelogram. base and height
2. **Describe** how to find the area of a complex figure. Separate it into figures whose areas you know how to find, then add the areas.

Skills and Applications

3. Find the square of 9. 81
4. Find $\sqrt{400}$. 20
5. **PHYSICAL FITNESS** This morning, Elisa walked 1 mile north, 0.5 mile west, and then walked straight back to her starting point. How far did Elisa walk? Round to the nearest tenth. 2.6 mi
6. Estimate $\sqrt{23}$ to the nearest whole number. 5
7. Use a calculator to find $\sqrt{133}$ to the nearest tenth. 11.5

Find the missing measure of each right triangle. Round to the nearest tenth if necessary.

8. $a = 5$ m, $b = 4$ m 6.4 m
9. $b = 12$ in., $c = 14$ in. 7.2 in.

Find the area of each figure. Round to the nearest tenth if necessary.

10. 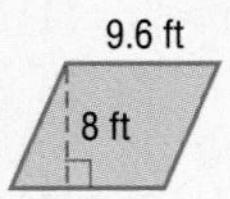

$76.8\ ft^2$

11. 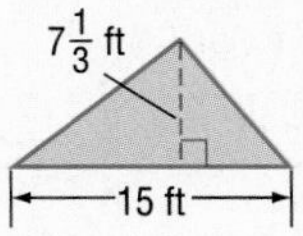

$55\ ft^2$

12. 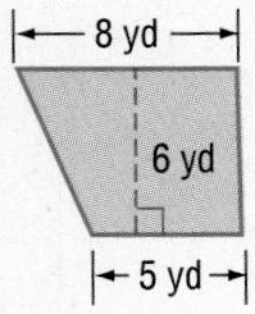

$39\ yd^2$

13. 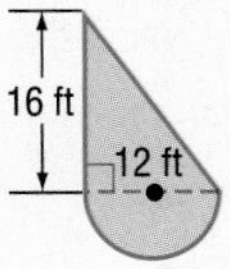

$152.5\ ft^2$

Find the area of each circle. Round to the nearest tenth.

14. radius = 9 ft $254.5\ ft^2$
15. diameter = 5.2 cm $21.2\ cm^2$

Standardized Test Practice

16. **MULTIPLE CHOICE** A randomly dropped counter falls in the squares. Find the probability that it falls in the shaded squares. Write as a percent. Round to the nearest tenth if necessary. C

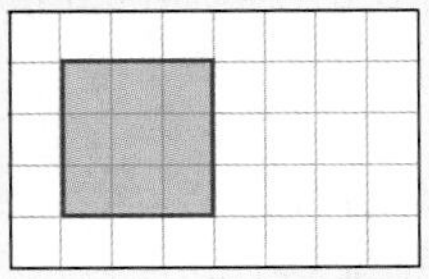

Ⓐ 9% Ⓑ 14.1% Ⓒ 22.5% Ⓓ 40%

msmath2.net/chapter_test

Chapter 11 Practice Test

Assessment Options

Vocabulary Test A vocabulary review/test for Chapter 11 can be found on p. 662 of the *Chapter 11 Resource Masters.*

Chapter Tests There are six Chapter 11 Tests and an Extended Response Assessment task available in the *Chapter 11 Resource Masters.*

Chapter 11 Tests			
Form	**Type**	**Level**	**Pages**
1	MC	basic	649–650
2A	MC	average	651–652
2B	MC	average	653–654
2C	FR	average	655–656
2D	FR	average	657–658
3	FR	advanced	659–660

MC = multiple-choice questions
FR = free-response questions

Extended Response Assessment

Extended Response tasks for Chapter 11 can be found on p. 661 of the *Chapter 11 Resource Masters.* A sample scoring rubric for these tasks appears on p. A28.

Portfolio Suggestion

Introduction Most shapes in real life are complex figures. The basic shapes are combined and modified to make aesthetically pleasing composites.

Ask Students to find examples of complex figures in their school and neighborhoods. Have students draw the complex figures in their portfolios and show the basic shapes that make up the figures. Have them include the formulas they would use to find the area of each figure.

CHAPTER 11

Standardized Test Practice

These two pages contain practice questions in the various formats that can be found on the most frequently given standardized tests.

A student recording sheet for these two pages can be found on page A1 of the *Chapter 11 Resource Masters*.

Standardized Test Practice Student Recording Sheet, p. A1

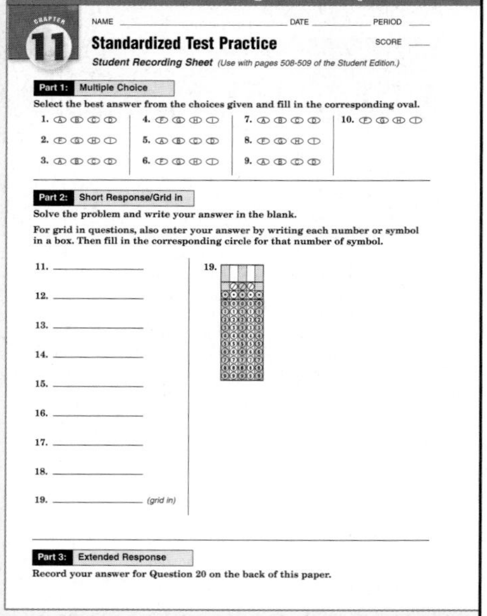
NAME ________ DATE ________ PERIOD ____

CHAPTER 11 **Standardized Test Practice** SCORE ____

Student Recording Sheet (Use with pages 508-509 of the Student Edition.)

Part 1: Multiple Choice

Select the best answer from the choices given and fill in the corresponding oval.

1. Ⓐ Ⓑ Ⓒ Ⓓ 4. Ⓕ Ⓖ Ⓗ Ⓘ 7. Ⓐ Ⓑ Ⓒ Ⓓ 10. Ⓕ Ⓖ Ⓗ Ⓘ
2. Ⓕ Ⓖ Ⓗ Ⓘ 5. Ⓐ Ⓑ Ⓒ Ⓓ 8. Ⓕ Ⓖ Ⓗ Ⓘ
3. Ⓐ Ⓑ Ⓒ Ⓓ 6. Ⓕ Ⓖ Ⓗ Ⓘ 9. Ⓐ Ⓑ Ⓒ Ⓓ

Part 2: Short Response/Grid in

Solve the problem and write your answer in the blank.

For grid in questions, also enter your answer by writing each number or symbol in a box. Then fill in the corresponding circle for that number of symbol.

11. ________
12. ________
13. ________
14. ________
15. ________
16. ________
17. ________
18. ________
19. ________ (grid in)

Part 3: Extended Response

Record your answer for Question 20 on the back of this paper.

Additional Practice

See pp. 667–668 of the *Chapter 11 Resource Masters* for additional standardized test practice.

CHAPTER 11

Standardized Test Practice

PART 1 Multiple Choice

Record your answers on the answer sheet provided by your teacher or on a sheet of paper.

1. Victoria bought 3 notebooks and 2 gel pens. Which expression represents her total cost if n is the cost of each notebook and p is the cost of each gel pen? (Lesson 1-4) **D**

   Ⓐ $n + p$ Ⓑ $5(n + p)$
   Ⓒ $5n \cdot 2p$ Ⓓ $3n + 2p$

2. The table shows the weights of 13 dogs at a dog adoption center. Which measure of central tendency for these data is the least number? (Lesson 2-4) **G**

<table>
<tr><th>Weight (lb)</th><th>Number of Dogs</th></tr>
<tr><td>20</td><td>|||</td></tr>
<tr><td>40</td><td>||</td></tr>
<tr><td>60</td><td>卌 ||</td></tr>
<tr><td>80</td><td>|</td></tr>
</table>

   Ⓕ median Ⓖ mean
   Ⓗ mode Ⓘ range

3. To make muffins, Desiree used $4\frac{1}{8}$ cups of flour, $4\frac{2}{3}$ cups of water, $4\frac{1}{4}$ cups of sugar, and $4\frac{1}{3}$ cups of milk. Desiree used the least amount of which ingredient? (Lesson 5-8) **A**

   Ⓐ flour Ⓑ sugar
   Ⓒ water Ⓓ milk

4. Find $\frac{1}{7} \times \frac{2}{9}$. (Lesson 6-4) **G**

   Ⓕ $\frac{1}{63}$ Ⓖ $\frac{2}{63}$ Ⓗ $\frac{3}{16}$ Ⓘ $\frac{9}{14}$

5. For his job, Marc drives 15,000 miles every 60 days. What is the average number of miles that Marc drives every day? (Lesson 7-2) **A**

   Ⓐ 250 mi Ⓑ 300 mi
   Ⓒ 900 mi Ⓓ 1,500 mi

6. Trevor drives 45 miles per hour. Robin drives 54 miles per hour. What is the percent of increase from 45 miles per hou to 54 miles per hour? (Lesson 8-4) **G**

   Ⓕ 9% Ⓖ 20% Ⓗ 109% Ⓘ 120

7. Chi can take 3 different routes and 4 different modes of transportation to get to school, as shown below.

Route	Transportation
scenic	bike
quick	car
convenient	bus
	walking

   How many possible choices are there for Chi to use to get to school? (Lesson 9-3) **C**

   Ⓐ 7 Ⓑ 9 Ⓒ 12 Ⓓ 16

8. Which is the square root of 441? (Lesson 11-1)

   Ⓕ 21 Ⓖ 22 Ⓗ 23 Ⓘ 24

9. Which is a reasonable *estimate* for the square root of 66? (Lesson 11-2) **C**

   Ⓐ 7.4 Ⓑ 7.8 Ⓒ 8.1 Ⓓ 8.9

10. What is the value of x in the triangle? (Lesson 11-3) **G**

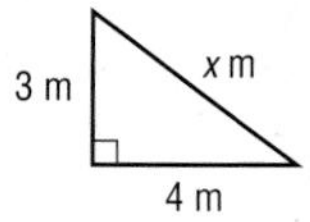

   Ⓕ 2 Ⓖ 5
   Ⓗ 7 Ⓘ 125

TEST-TAKING TIP

Question 8 You can use the answer choices to help you answer a multiple-choice question. For example, if you are asked to find the square root of a number, you can find the correct answer by calculating the square of the number in each answer choice.

ExamView® Pro Test Bank CD-ROM

Special banks of standardized test questions similar to those on the TIMSS and NAEP grade 8 tests can be found on this CD-ROM.

RT 2 Short Response/Grid In

cord your answers on the answer sheet
ovided by your teacher or on a sheet of
per.

What value is represented by $(1 \times 10^5) + (2 \times 10^4) + (1 \times 10^3) + (1 \times 10^0)$? (Prerequisite Skill, page 555) **121,001**

In the coordinate system, coordinates with a positive x value and a negative y value appear in what quadrant? (Lesson 3-3) **quadrant IV**

Write an equation to represent the following statement. (Lesson 4-1)

The Palmas have 8 less than 2 times the number of trees in their yard as the Kandinskis have.

Sample answer: $p = 2k - 8$

If you translate hexagon $LMNOPQ$ 4 units to the right, what are the new coordinates of point Q? (Lesson 10-8) **(0, 7)**

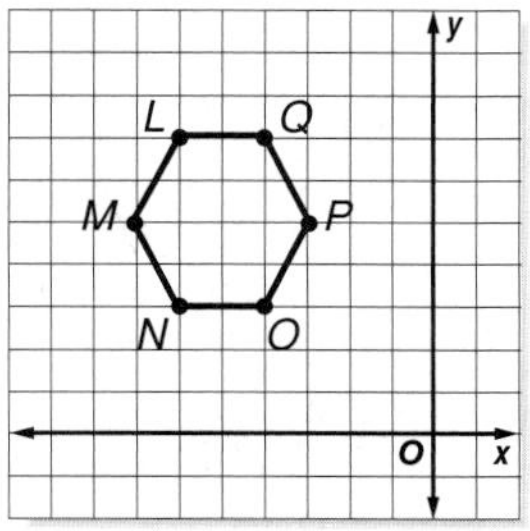

A window shade is being custom made to cover the triangular window shown at the right. What is the minimum area of the shade? (Lesson 11-5) **2,296 in²**

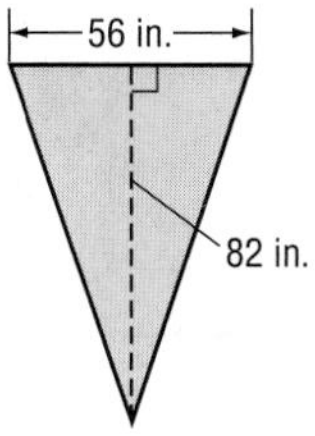

A kitchen chair has a circular seat that measures 14 inches across. What is the area of the seat on the kitchen chair? Use 3.14 for π and round to the nearest tenth. (Lesson 11-6) **153.9 in²**

17. The Connaught Centre building in Hong Kong has 1,748 circular windows. The diameter of each window is 2.4 meters. Find the total area of the glass in the windows. Round to the nearest tenth. (Lesson 11-6) **7,907.8 m²**

18. Find the area of the figure shown at the right. Use 3.14 for π. (Lesson 11-7) **4,672 cm²**

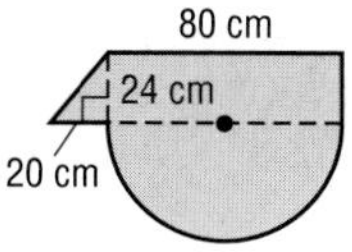

19. Amy plays hopscotch by throwing a small stone onto a numbered triangle. What is the probability that Amy's stone will land on a triangle numbered 5? Write as a fraction. (Lesson 11-8) $\frac{1}{3}$

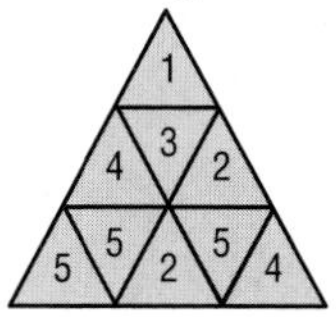

PART 3 Extended Response

Record your answers on a sheet of paper. Show your work.

20. Suppose you bought a new tent with the dimensions shown below. **20b. 1,320 in²**

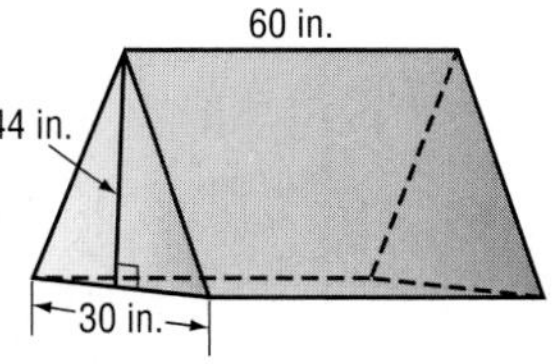

a. Is the area of the parallelogram-shaped side of the tent greater than or less than the area of the floor? Explain. (Lessons 11-3 and 11-4) **See margin.**

b. The front and back triangular regions are covered with screens. What is the total area of the screens? (Lesson 11-5)

msmath2.net/standardized_test

Answer

20a. Greater than; the area of a side is $60 \cdot 46.5$ in² and the area of the floor is $60 \cdot 30$ in².

Standardized Test Practice

Rubrics The following rubrics are sample scoring devices for short response and extended response questions.

Short Response

Points	Description
2	The student demonstrates a <u>thorough understanding</u> of the mathematics of the task. The response may contain minor flaws that do not detract from the demonstration of a thorough understanding.
1	The student has provided a response that is only <u>partially correct</u>.
0	The student has provided a <u>completely incorrect</u> solution or no response at all.

Extended Response

Points	Description
4	The student demonstrates a <u>thorough understanding</u> of the mathematics of the task. The response may contain minor flaws that do not detract from the demonstration of a thorough understanding.
3	The student demonstrates an <u>understanding</u> of the mathematics of the task. The response is essentially correct and demonstrates an essential but less than thorough understanding of the mathematics.
2	The student demonstrates only a <u>partial understanding</u> of the mathematics of the task. Although the student may have used the correct approach to a solution or may have provided a correct solution, the work lacks an essential understanding of the underlying mathematical concepts.
1	The student demonstrates a <u>very limited understanding</u> of the mathematics of the task. The response is incomplete and exhibits many flaws.
0	The student provides a <u>completely incorrect</u> solution or no response at all.

A more detailed 4-point rubric is provided on page A2 of the *Chapter 11 Resource Masters*.

Page 470, Lesson 11-1 (Hands-On Mini Lab)

1.

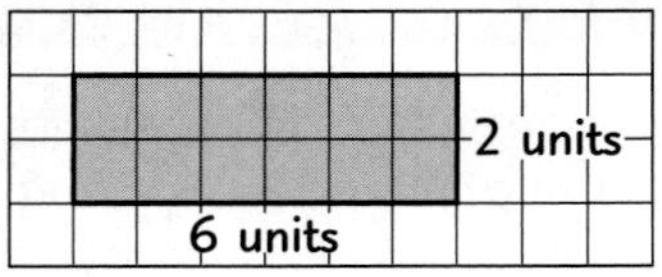

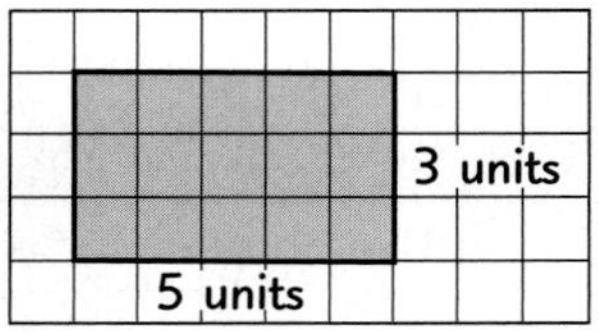

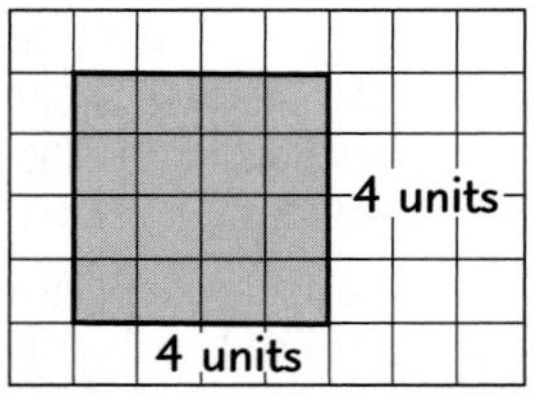

2.

Drawing	Dimensions (units)	Area (sq. units)
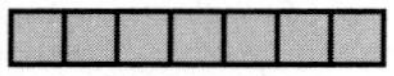	1 × 7	7
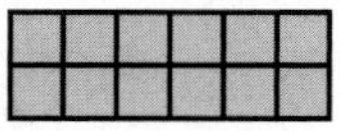	2 × 6	12
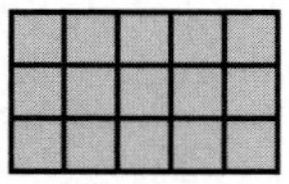	3 × 5	15
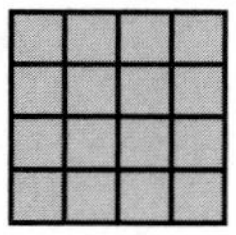	4 × 4	16

3.

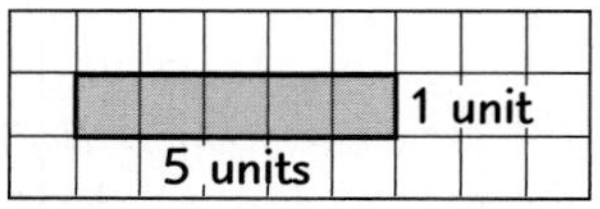

5 sq units

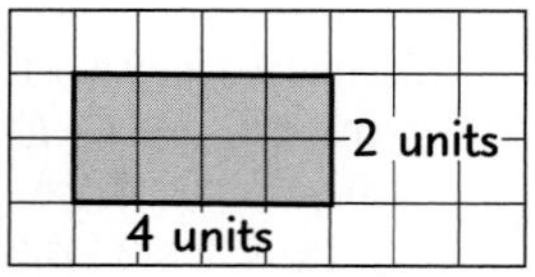

8 sq units

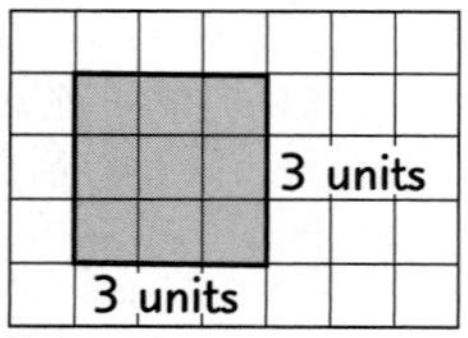

9 sq units

Page 473, Lesson 11-1

40–41.

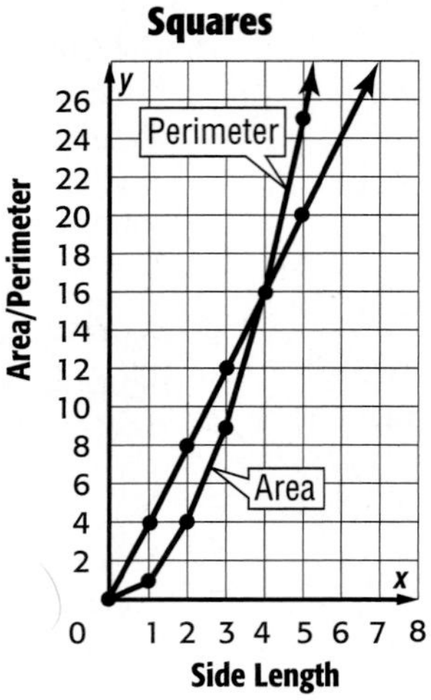

48.

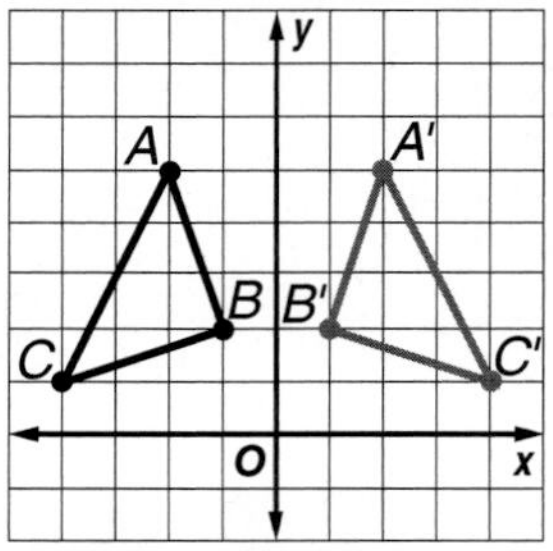

49.

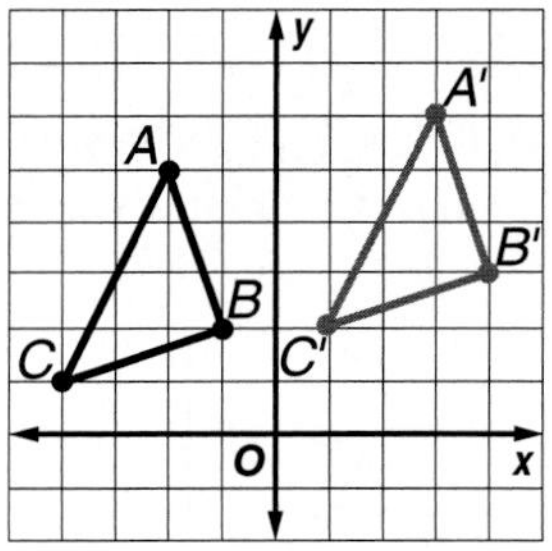

Page 483, Lesson 11-4 (Hands-On Mini Lab)

3. Sample answer:

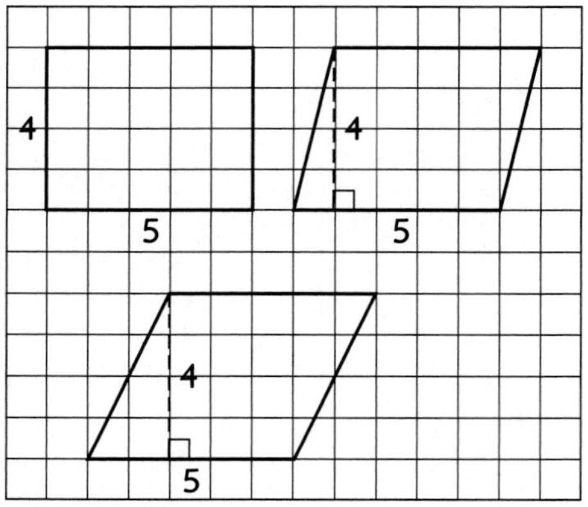

The area of each is 20 square units.

Page 495, Lesson 11-6

26.

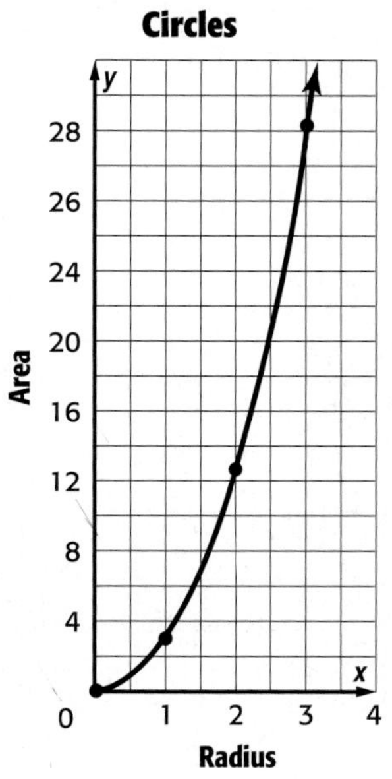

CHAPTER 12

Chapter Overview

Geometry: Measuring Three-Dimensional Figures

	LESSON OBJECTIVES	NCTM Standards	NAEP 2005	State/Local Objectives
12-1	**Drawing Three-Dimensional Figures** *(pp. 512–519)* **12-1a Hands-On Lab:** Build three-dimensional figures given the top, side, and front views. • Draw a three-dimensional figure given the top, side, and front views. **12-1b Problem-Solving Strategy:** Solve problems by making a model.	1, 2, 6–10	G1e, G1f, G3c	
12-2	**Volume of Rectangular Prisms** *(pp. 520–523)* • Find the volumes of rectangular prisms. **12-2b Spreadsheet Investigation:** Use a spreadsheet to investigate the volumes of similar solids.	1, 2, 6–10	G1f, M1j, A4e, N3a, M2a, M2b, A1b	
12-3	**Volume of Cylinders** *(pp. 524–527)* • Find the volume of cylinders.	1, 2, 3, 6–10	G1f, A4e, N3a, M1j, M2a	
12-4	**Surface Area of Rectangular Prisms** *(pp. 530–537)* **12-4a Hands-On Lab:** Use nets to find the surface area of rectangular prisms. • Find the surface areas of rectangular prisms. **12-4b Hands-On Lab:** Investigate changes in volume and surface area.	1, 2, 6–10	G1f, M1j, A4e, M2a, A1a, A1b, N4b, N4C	
12-5	**Surface Area of Cylinders** *(pp. 538–541)* • Find the surface areas of cylinders.	1, 2, 6–10	G1f, M1j, A4e, N3a, N5f	
12-6	**Precision and Measurement** *(pp. 542–545)* • Determine and apply significant digits in a real-life context.	1, 2, 6–10	M2e, M1c	

Key to NCTM Standards:

1=Number & Operations, 2=Algebra, 3=Geometry, 4=Measurement, 5=Data Analysis & Probability, 6=Problem Solving, 7=Reasoning & Proof, 8=Communication, 9=Connections, 10=Representation

NAEP 2005 Standards: See math.glencoe.com

Pacing suggestions for the entire year can be found on page T20.

Chapter Resource Manager

Time-Saving Tool
TeacherWorks™
All-In-One Planner and Resource Center
See page T5.

FAST FILE — CHAPTER 12 RESOURCE MASTERS

Study Guide and Intervention	Practice (Skills and Word Problems)	Reading to Learn Mathematics	Enrichment	Assessment	Applications*	Parent and Student Study Guide Workbook	5-Minute Check Transparencies	Key Teaching Transparencies	MathPASS: Tutorial Plus (lessons)	Teaching Mathematics with Manipulatives	Optional Materials
669	670–671	672	673			12-1	12-1	45	48		
674	675–676	677	678	713	RC 26 CS 55	12-2	12-2		52	117–118	scissors, tape
679	680–681	682	683	713, 715	SM 77–80	12-3	12-3	46		119	soup can, tape, scissors
684	685–686	687	688			12-4	12-4	47		120, 121	soup can, scissors
689	690–691	692	693	714	CM 35–36 CS 56	12-5	12-5	48	54		
694	695–696	697	698	699–712, 714, 716–718		12-6	12-6				

* **Key to Abbreviations:** *CS = Calculator and Spreadsheet Masters, RC = Real-Life Career Masters, CM = Classroom Math Games (online), SM = Science and Math Lab Manual (online)*

INTERACTIVE CHALKBOARD

PowerPoint® Presentations

This CD-ROM is a customizable Microsoft® PowerPoint® presentation that includes:

- In-Class Examples
- Links to Online Study Tools
- 5-Minute Check Transparencies
- Bellringers

Math Skills Maintenance

Use the Math Skills Maintenance Masters to keep your students' basic math skills fresh. Students can review skills one at a time or in a combined format. The Masters are also available online at msmath2.net/maintenance.

CHAPTER 12

Mathematical Content and Teaching Strategies

Continuity of Instruction

Building on Prior Knowledge

Students have learned about three-dimensional figures in previous courses. They know how to multiply fractions and how to multiply decimals. Students have learned to find the areas of rectangles and the areas of circles, and they have learned to find the circumference of a circle. Students have measured objects in previous math courses and in real life.

Learning in This Chapter

Students learn how to classify and draw three-dimensional figures. They apply what they have learned about finding the areas of rectangles to calculate the volumes and surface areas of rectangular prisms. They also apply their knowledge of area and circumference of circles to find volumes and surface areas of cylinders. Finally, they learn to measure objects using significant digits.

Making Future Connections

Three-dimensional figures appear in everyday life, and are studied in mathematics courses throughout middle school and high school. Calculating volumes and surface areas of rectangular prisms and cylinders requires important skills that are essential for understanding concepts studied in high school geometry. Students will encounter significant digits in science classes such as physics and chemistry.

Mathematical Background

12-1 Drawing Three-Dimensional Figures

Drawing three-dimensional figures is an important skill because it allows us to represent what we see around us in our everyday lives. Students already have experience in drawing polygons and circles. As they combine these shapes to draw three-dimensional figures, make sure they use dotted lines appropriately to represent portions of their figures that are not visible from the outside.

Bulletin Board

Have students work in pairs to design a rectangular prism or cylinder that could be used to package a fictitious product. Give students the basic dimensions of the product and ask them to construct their packages using paper or cardboard. They should also find the volume and surface area of the package and record these measurements. As students complete their projects, they should attach them to the bulletin board.

12-2 Volume of Rectangular Prisms

In this lesson, students apply their knowledge of finding the areas of rectangles to calculate the volumes of rectangular prisms. While many students will be inclined to find volume by multiplying the length, width, and height, encourage them to begin by determining the area of the rectangular base and then multiplying by the height. Proceeding in this manner will help them understand future concepts of finding the volumes of cylinders and various right prisms.

In-Class Speaker

Invite an architect to speak to students about applying the concept of volume to the design of buildings and homes. Ask the architect to discuss the procedures for calculating the total amount of space in each room, as well as in the entire structure. Another topic for discussion could be selecting a heating or air conditioning unit that is commensurate with the amount of space in the structure.

12-3 Volume of Cylinders

Understanding how to find the volumes of cylinders can be difficult, since students cannot easily use blocks to build cylinders like they can to build rectangular prisms. Students should build on their knowledge of areas of circles by imagining cylinders with heights of one unit stacked on top of each other. They should then be able to make the connection between the procedure for determining the volume of a rectangular prism and that for finding the volume of a cylinder—multiplying the area of the base by the height.

12-4 Surface Area of Rectangular Prisms

In this lesson, students learn to add the areas of the faces of a rectangular prism to determine its surface area. Students may at first experience difficulty trying to picture all six faces of a rectangular prism and the dimensions of each. Providing them with rectangular blocks and asking them to sum the areas of all six faces may assist them in understanding the concept of surface area.

12-5 Surface Area of Cylinders

The surfaces of a cylinder consist of the top and bottom circles and the curved surface connecting them. Students have already learned how to compute the area of a circle, but may have trouble understanding how to find the area of the curved surface. Wrapping this part of the cylinder in paper and then unwrapping it to reveal a rectangle should help students understand how to find its area.

Teacher to Teacher

Sandy Gavin
Tieton Middle School
Tieton, WA

"I use empty toilet paper rolls to model the surface area of a cylinder. By cutting down the tube, I can open the tube to show that the surface area can be found by multiplying the circumference of the base by the height of the cylinder."

Sandy Gavin

12-6 Precision and Measurement

In this lesson, significant digits and estimated digits are used to explore the concept of precision in measurement. Students might have trouble understanding the mathematical meanings of precision and significant digits. Ask them to imagine a stick, such as a broom handle, that is exactly one meter long but is not marked like a meterstick. Ask them whether they could report the width of a desk in millimeters using this stick. Students should be able to understand that using this stick to report millimeters would be impossible.

McGraw-Hill Professional Development

Mathematics, YES! is a self-paced, web-based professional development program. Units are based on major mathematical concepts and are standards-based. Actual classroom videos provide a springboard for discussion and activities. Units appropriate for this chapter:

- Unit 8, Geometry and Measurement of Plane Figures
- Unit 9, Solid Geometry

Key Concepts

Additional mathematical information and teaching notes are available at

msmath2.com/key_concepts.

The sections appropriate for the chapter are:

- Volume of Rectangular Prisms
- Volume of Cylinders
- Surface Area of Rectangular Prisms
- Surface Area of Cylinders

CHAPTER 12

DAILY INTERVENTION and Assessment

	Type	Student Edition	Teacher Resources	Technology/Internet
INTERVENTION	Ongoing	Prerequisite Skills, pp. 511, 517, 522, 527, 535, 541 Mid-Chapter Practice Test, p. 528	5-Minute Check Transparencies Quizzes 1–4, *CRM*, pp. 713–714 Mid-Chapter Test, *CRM*, p. 715 Study Guide and Intervention, *CRM*, pp. 669, 674, 679, 684, 689, 694	MathPASS: Tutorial Plus msmath2.net/self_check_quiz msmath2.net/extra_examples
	Mixed Review	pp. 517, 522, 527, 535, 541, 545	Cumulative Review, *CRM*, p. 716	
	Error Analysis	Find the Error, pp. 521, 539	Find the Error, *TWE*, pp. 521, 540 Unlocking Misconceptions, *TWE*, pp. 520, 543 Tips for New Teachers, *TWE*, p. 539	
	Standardized Test Practice	pp. 517, 519, 522, 525, 527, 528, 535, 541, 545, 549, 550–551	Standardized Test Practice, *CRM*, pp. 717–718	Standardized Test Practice CD-ROM msmath2.net/standardized_test
ASSESSMENT	Open-Ended Assessment	Writing Math, pp. 513, 515, 521, 525, 531, 533, 537, 539, 543 Open-Ended, pp. 515, 521, 525, 533, 539, 543 Standardized Test, p. 551	Modeling: *TWE*, pp. 517, 541 Speaking: *TWE*, pp. 527, 545 Writing: *TWE*, p. 522 Extended Response Assessment, *CRM*, p. 711	
	Chapter Assessment	Study Guide, pp. 546–548 Practice Test, p. 549	Multiple-Choice Tests (Forms 1, 2A, 2B), *CRM*, pp. 699–704 Free-Response Tests (Forms 2C, 2D, 3), *CRM*, pp. 705–710 Vocabulary Test/Review, *CRM*, p. 712	ExamView® Pro Test Bank (see below) MindJogger Videoquizzes msmath2.net/vocabulary_review msmath2.net/chapter_test

* **Key to Abbreviations:** *TWE* = Teacher Wraparound Edition, *CRM* = Chapter Resource Masters

Additional Intervention

Skills Intervention for Middle School Mathematics workbook and CD-ROM includes 88 skills with pretests, tutorials, guided practice, and posttests for diagnosis and remediation.

Guide to Daily Intervention booklet details each intervention resource available and guides teachers in using them as intervention tools everyday to assist students who may be having difficulty.

ExamView® Pro

Use ExamView® Pro Test Bank CD-ROM to:

- **Create Multiple versions of tests.**
- **Create Modified tests for Inclusion students with one mouse click.**
- **Edit existing questions and Add your own questions.**
- **Build tests aligned with state standards using built-in State Curriculum Correlations.**
- **Change English tests to Spanish with one mouse click and vice versa.**

Reading and Writing Mathematics

- **Student Edition**

Foldables Study Organizer, p. 511

Concept Check questions at the beginning of each exercise set require students to verbalize and write about what they have learned in the lesson.

- **Open Ended,** pp. 515, 521, 525, 533, 539, 543
- **Find the Error,** pp. 521, 539
- **Which One Doesn't Belong?,** pp. 515, 543

WebQuest, p. 545

- **Teacher Wraparound Edition**

Foldables Follow-Up, pp. 516, 526, 528, 534, 540

Writing Activities, p. 522

Differentiated Instruction, (Verbal/Linguistic), p. 539

ELL Resources, *see page T9 for a complete listing.*

- **Additional Resources**

Vocabulary Builder, *Chapter 12 Resource Masters*

Reading to Learn Mathematics, *Chapter 12 Resource Masters*

Vocabulary PuzzleMaker (online)

Teaching Mathematics with Foldables

Reading and Writing in the Mathematics Classroom

WebQuests, Projects, and Interdisciplinary Investigations

For detailed information on these resources, see pp. T6–T7.

- **Recommended Outside Reading for Students**

Nonfiction

- Kremer, Ron. *From Crystals to Kites: Exploring Three Dimensions.* Dale Seymour Publications, ©1997.

In this activity book, students explore the properties of three-dimensional figures by building straw models, making kites with tetrahedrons, and finding the volume of prisms and pyramids. Included with the activities are teacher notes, blackline masters, and patterns.

Fiction

- Abbot, Edwin. *Flatland A Romance of Many Dimensions.* Indypublish.Com, ©2002.

This short, classic fantasy novel, first published in 1880, takes the reader to a two-dimensional world occupied by geometric figures. When one such shape discovers the existence of a third dimension, the reader begins to understand how a fourth dimension, or beyond, might relate to our three-dimensional existence.

PROJECT CRISS[SM] Study Skill

A comparison chart can help students understand mathematical relationships, especially when a number of new formulas are being learned. The chart at the right shows a comparison of the formulas for the volumes of rectangular prisms and the volumes of cylinders. From the sketches, students should note that the volumes could both be written as $V = Bh$, where B is the area of the base. Have students draw a comparison chart or charts for other formulas as they read through Chapter 12. Then, allow time for a class discussion so students can compare charts and formulas.

Figure	Sketch	Volume Formula
rectangular prism	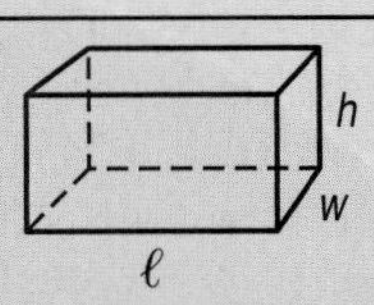	$V = \ell \cdot w \cdot h$
cylinder	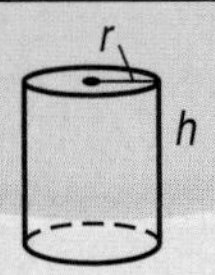	$V = \pi r^2 h$

CReating Independence through Student-owned Strategies

For more information on Project CRISS, see page T9.

CHAPTER

12 Making Connections

School-To-Home

Parent and Student Study Guide

Parents and students may work together to reinforce the concepts and skills of this chapter. These workbooks are available in English and Spanish. A sample lesson is shown below. Also available at msmath2.net/parent_student.

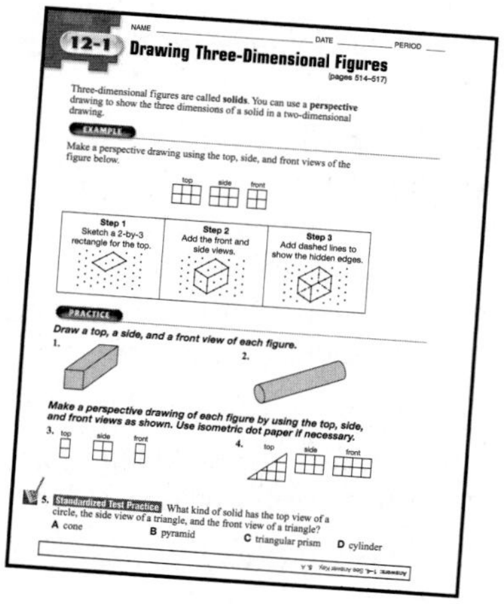

NAME ______ DATE ______ PERIOD ____

12-1 Drawing Three-Dimensional Figures
(pages 514–517)

Three-dimensional figures are called **solids**. You can use a **perspective drawing** to show the three dimensions of a solid in a two-dimensional drawing.

EXAMPLE

Make a perspective drawing using the top, side, and front views of the figure below.

Step 1	Step 2	Step 3
Sketch a 2-by-3 rectangle for the top.	Add the front and side views.	Add dashed lines to show the hidden edges.

PRACTICE

Draw a top, a side, and a front view of each figure.

1.

2.

Make a perspective drawing of each figure by using the top, side, and front views as shown. Use isometric dot paper if necessary.

3. top side front

4. top side front

5. **Standardized Test Practice** What kind of solid has the top view of a circle, the side view of a triangle, and the front view of a triangle?
A cone B pyramid C triangular prism D cylinder

Answers: 1–4. See Answer Key. 5. A

Family Letters and Activities

Send a letter and activity for this chapter home to encourage family members to become active participants in their students' learning. (English: *Chapter 12 Resource Masters*, pp. ix–x; Spanish: log on to msmath2.net/family_letters)

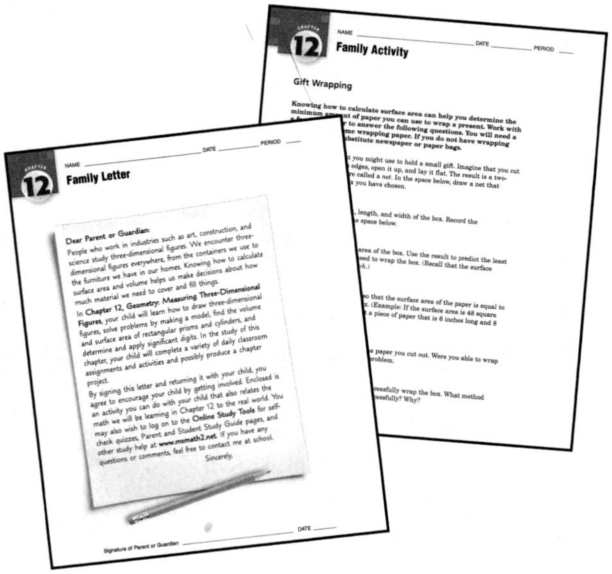

CHAPTER 12 **Family Activity**

NAME ______ DATE ______ PERIOD ____

Gift Wrapping

Knowing how to calculate surface area can help you determine the minimum amount of paper you can use to wrap a present. Work with ... to answer the following questions. You will need a ... wrapping paper. If you do not have wrapping ... substitute newspaper or paper bags.

... you might use to hold a small gift. Imagine that you cut ... edges, open it up, and lay it flat. The result is a two- ... called a *net*. In the space below, draw a net that ... you have chosen.

..., length, and width of the box. Record the ... space below.

... area of the box. Use the result to predict the least ... need to wrap the box. (Recall that the surface ...)

... so that the surface area of the paper is equal to ... (Example: If the surface area is 48 square ... a piece of paper that is 6 inches long and 8 ...)

... paper you cut out. Were you able to wrap ... problem.

... successfully wrap the box. What method ... successfully? Why?

CHAPTER 12 **Family Letter**

NAME ______ DATE ______ PERIOD ____

Dear Parent or Guardian:

People who work in industries such as art, construction, and science study three-dimensional figures. We encounter three-dimensional figures everywhere, from the containers we use to the furniture we have in our homes. Knowing how to calculate surface area and volume helps us make decisions about how much material we need to cover and fill things.

In **Chapter 12, Geometry: Measuring Three-Dimensional Figures**, your child will learn how to draw three-dimensional figures, solve problems by making a model, find the volume and surface area of rectangular prisms and cylinders, and determine and apply significant digits. In the study of this chapter, your child will complete a variety of daily classroom assignments and activities and possibly produce a chapter project.

By signing this letter and returning it with your child, you agree to encourage your child by getting involved. Enclosed is an activity you can do with your child that also relates the math we will be learning in Chapter 12 to the real world. You may also wish to log on to the **Online Study Tools** for self-check quizzes, Parent and Student Study Guide pages, and other study help at **www.msmath2.net**. If you have any questions or comments, feel free to contact me at school.

Sincerely,

Signature of Parent or Guardian ______ DATE ______

English Language Learner ELL

Lesson 12-1
Using Modeling

In order to help students understand three-dimensional figures, bring in soup cans and ice cream cones or newspaper to form a cone. Take a soup can and unwrap the label. Ask students what shape they see. Do the same with a cone. Unwrap a cone, or form a cone with a piece of newspaper, and ask students what shape they see. This activity can help students see how a two-dimensional figure can be transformed into a three-dimensional figure.

Lesson 12-3
Using Manipulatives

Give groups of students scissors, tape, and three pieces of plain paper. Have them roll and tape the first piece of paper lengthwise to make a cylinder. Next, have them roll and tape the second piece of paper widthwise to make a cylinder. Have them make a third cylinder by cutting the last piece of paper in half lengthwise and taping the two pieces end to end. Have them roll and tape this to make the cylinder. Ask students which of these holds the most. After taking predictions, fill the cylinders with rice or unpopped popcorn, comparing the amount it takes to fill each one. Discuss the results.

Lesson 12-4
Higher-Level Thinking

Give students in groups several nets for rectangular prisms. Have the students find the area of each face. Then find the sum of all of the areas to find the surface area. After students have completed several nets in their groups, ask them what they notice about finding the surface area for each one. What commonalities do they see? Allow them time to discover the formula for the surface area of a rectangular prism.

MathScape™

This project was supported, in part, by the **National Science Foundation**

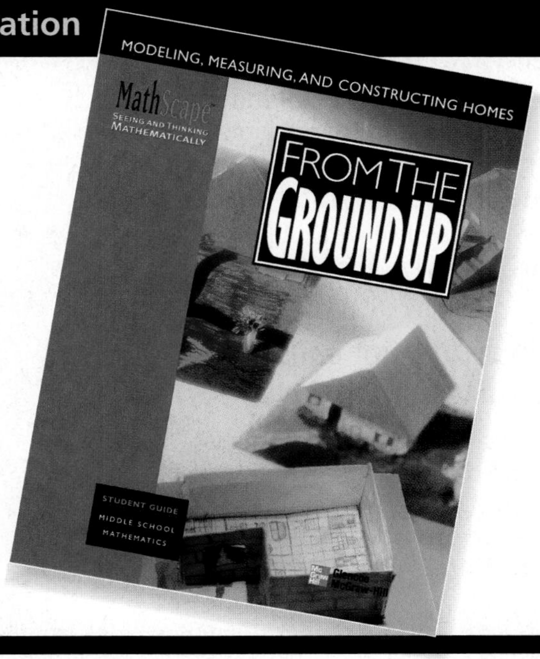

MathScape is a mathematics curriculum for grades 6–8 developed by the Seeing and Thinking Mathematically Project at the Education Development Center.

Modeling, Measuring, and Constructing Homes

From the Ground Up

Summary Students will investigate three-dimensional shapes. Nets, diagrams, and models are used to represent three-dimensional shapes. Students will find surface areas of three-dimensional objects.

USA TODAY Education Online Lesson Planner

USA TODAY® Education's Online site offers resources and interactive features connected to each day's newspaper. *Experience Today,* USA TODAY's® daily lesson plan, is available on the site and delivered daily to subscribers. This plan provides instruction for integrating USA TODAY® graphics and key editorial features into your mathematics classroom. Log on to **education.usatoday.com**.

USA TODAY Snapshots® in Chapter 12	
Lesson	**Snapshot**
12-6	More women attempt triathalon challenge

Quick Review Math Handbook

is Glencoe's mathematical handbook for students. The Hot Words section includes a glossary of terms while the Hot Topics section consists of explanations of key mathematical concepts. An exercise set is included to check students' understanding of the concepts. This valuable resource can be used as a reference in the classroom or for home study.

Lesson	Hot Topics Section	Lesson	Hot Topics Section
12-1	7-2	**12-4**	7-6, 7-7, 9-3
12-2	7-7, 8-6, 9-4	**12-5**	7-6
12-3	7-7	**12-6**	8-1

CHAPTER 12 Notes

Key Chapter Objectives

Lesson 12-1 Draw three-dimensional figures.

Lessons 12-2 and 12-3 Find the volumes of rectangular prisms and cylinders.

Lessons 12-4 and 12-5 Find the surface areas of rectangular prisms and cylinders.

"What do paint cans have to do with math?"

Point out to students that this is only one of many applications of the math that they are studying. Others are provided in the introduction to each lesson.

Vocabulary Builder, pp. vii–viii ELL

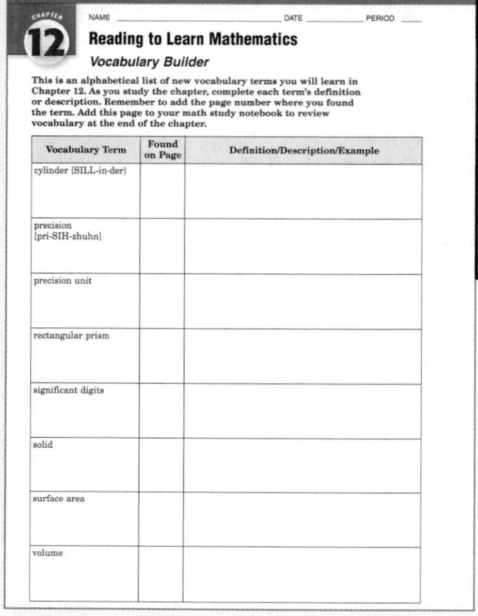

CHAPTER 12 NAME ______ DATE ______ PERIOD ____

Reading to Learn Mathematics

Vocabulary Builder

This is an alphabetical list of new vocabulary terms you will learn in Chapter 12. As you study the chapter, complete each term's definition or description. Remember to add the page number where you found the term. Add this page to your math study notebook to review vocabulary at the end of the chapter.

Vocabulary Term	Found on Page	Definition/Description/Example
cylinder [SILL-in-der]		
precision [pri-SIH-zhuhn]		
precision unit		
rectangular prism		
significant digits		
solid		
surface area		
volume		

CHAPTER 12 Geometry: Measuring Three-Dimensional Figure[s]

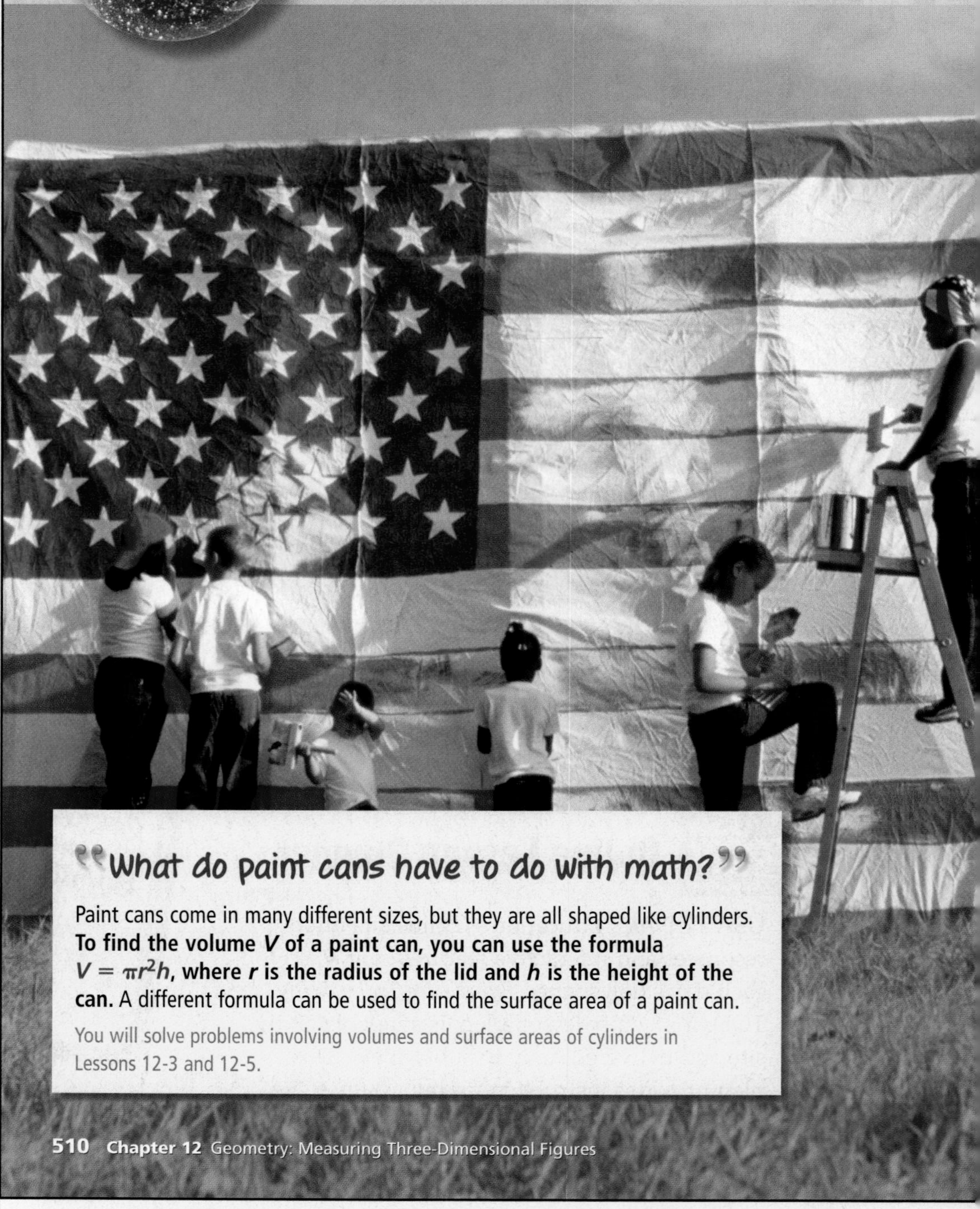

"What do paint cans have to do with math?"

Paint cans come in many different sizes, but they are all shaped like cylinders. **To find the volume V of a paint can, you can use the formula $V = \pi r^2 h$, where r is the radius of the lid and h is the height of the can.** A different formula can be used to find the surface area of a paint can.

You will solve problems involving volumes and surface areas of cylinders in Lessons 12-3 and 12-5.

510 **Chapter 12** Geometry: Measuring Three-Dimensional Figures

Vocabulary Builder

For a thorough vocabulary list with pronunciations of new words, give students the Vocabulary Builder worksheets from the *Chapter 12 Resource Masters*. Encourage them to complete the definition of each term as they progress through the chapter.

GETTING STARTED

Take this quiz to see whether you are ready to begin Chapter 12. Refer to the lesson or page number in parentheses if you need more review.

Vocabulary Review

Choose the correct term to complete each sentence.

1. The distance around a circle is called (perimeter, <u>circumference</u>). (Lesson 6-9)
2. The expression πr^2 is used to find the (circumference, <u>area</u>) of a circle. (Lesson 11-6)

Prerequisite Skills

Estimate each product. (Page 558)

3. $9 \cdot 10.4$ **90**
4. $6.25 \cdot 3.8$ **24**
5. $7.03 \cdot 5.3$ **35**
6. $4.7 \cdot 2.15$ **10**

Evaluate each expression. Round to the nearest tenth if necessary. (Pages 559, 560)

7. $14.45 + 9.62$ **24.1**
8. $8.3 \cdot 6.4$ **53.1**
9. 36×5.2 **187.2**
10. $26.45 - 7.918$ **18.5**

Simplify. (Lesson 1-3) **12. 23.7**

11. $2 \cdot 6 + 4 \cdot 1 \cdot 3$ **24**
12. $1.2 \cdot 4 \cdot 4 + 3 \cdot 1.5$
13. $8 + 1 + 9 \cdot 2 + 5$ **32**
14. $7 + 3 \cdot 5 + 2 \cdot 6$ **34**

Multiply. (Lesson 6-4)

15. $4\frac{1}{2} \cdot 6$ **27**
16. $1\frac{1}{4} \cdot 5\frac{3}{5}$ **7**
17. $\frac{2}{7} \cdot 3\frac{3}{4}$ **$1\frac{1}{14}$**
18. $3 \cdot 2\frac{2}{3}$ **8**

Find the area of each circle. Round to the nearest tenth. (Lesson 11-6) **19. 855.3 cm^2**

19. diameter = 33 cm
20. radius = 3.8 yd **45.4 yd^2**
21. radius = 6 ft **113.1 ft^2**
22. diameter = 18 m **254.5 m^2**

FOLDABLES Study Organizer

Surface Area and Volume Make this Foldable to help you organize information about solids. Begin with a piece of 11″ by 17″ paper.

STEP 1 Fold
Fold the paper in fourths lengthwise.

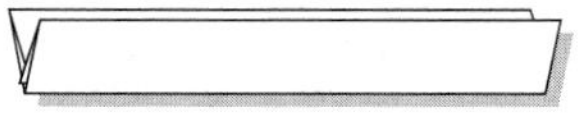

STEP 2 Open and Fold
Fold a 2″ tab along the short side. Then fold the rest in half.

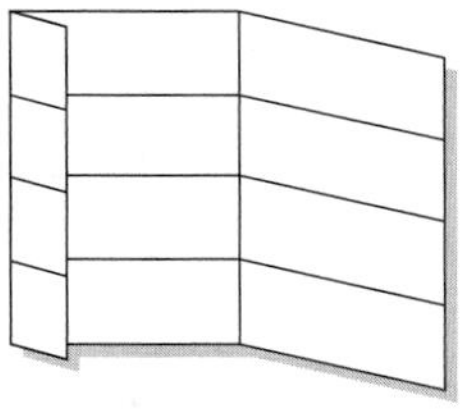

STEP 3 Label
Draw lines along folds and label as shown.

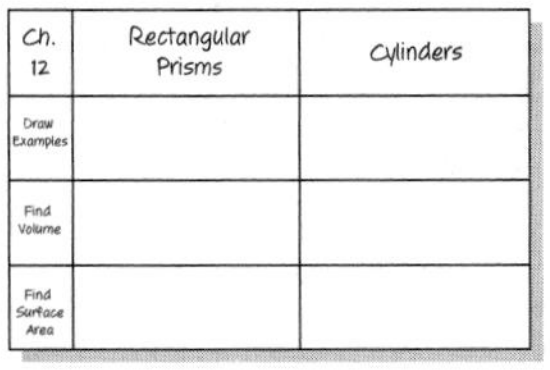

Reading and Writing As you read and study the chapter, write the formulas for volume and surface area of each three-dimensional figure. Draw examples and list the characteristics of the figures.

GETTING STARTED

This section provides a review of the basic concepts needed before beginning Chapter 12. Page references are included for additional student help.

Prerequisite Skills review skills needed in this chapter.

Exercises	Needed for Lesson	Skill
3–6	12-3	Estimate Products
7–10	12-6	Evaluate and Round Expressions
11–14	12-4	Simplify Expressions
15–18	12-2	Multiply Mixed Numbers
19–22	12-5	Find the Area of Circles

For more prerequisite skills, see:
- Getting Ready for the Next Lesson, at the end of each lesson
- Prerequisite Skills in the Student Handbook on pp. 554–563
- *Math Skills Maintenance Masters*

FOLDABLES Study Organizer

For more information about Foldables, see *Teaching Mathematics with Foldables.*

After students make their Foldable tables, have them label the columns and rows as illustrated. Students will use their tables to take notes, draw examples, and explain how to find the volume and surface area of three-dimensional figures. Use the data recorded to compare and contrast rectangular prisms and cylinders. When making comparisons in mathematics, first determine what will be compared, then decide what standards will be used for comparison, and then use what is known to find similarities and differences.

12-1a Hands-On Lab Lesson Notes

Lesson Resources

Teaching Mathematics with Manipulatives
- student recording sheet 12-1a

Glencoe Mathematics Classroom Manipulative Kit
- centimeter cubes

GETTING STARTED

Teaching Tip Bring several shoe boxes or other rectangular boxes to class. Have students in small groups examine the boxes and discuss how they might look if cut at the edges to lie flat. Then have one student per group cut out the top or bottom, one side, and one end. Ask students how these shapes compare to how they predicted they would look.

TEACH

- Model the activity for students by using cubes to build an example figure and showing students the top, side, and front views of the figure.
- Remind students that the figures they are forming with cubes are three-dimensional, but the outside surfaces are only two-dimensional.
- You may wish to have students build their models so that they sit on index cards. They can rotate the cards slowly to view the models from different perspectives without disturbing their structures.

12-1a Hands-On Lab

A Preview of Lesson 12-1

What You'll Learn
Build three-dimensional figures given the top, side, and front views.

Materials
- centimeter cubes

Building Three-Dimensional Figures

Cubes are examples of three-dimensional figures because they have length, width, and depth. In this lab, you will use centimeter cubes to build other three-dimensional figures.

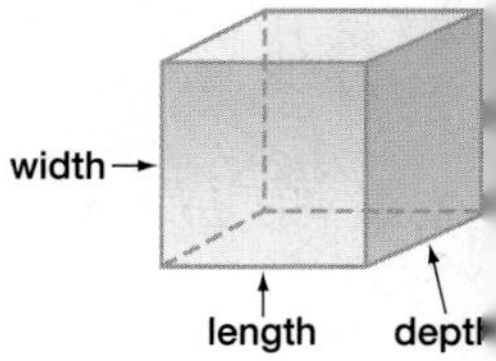

ACTIVITY *Work with a partner.*

The top view, side view, and front view of a three-dimensional figure are shown below. Use centimeter cubes to build the figure.

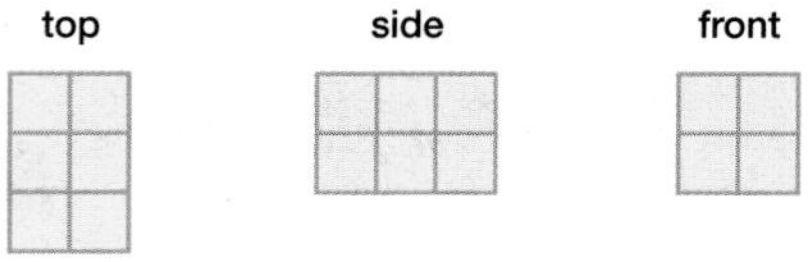

STEP 1 Use the top view to build the base of the figure. It is a 3-by-2 rectangle.

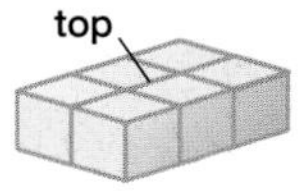

STEP 2 Use the side view to complete the figure. It is a 2-by-3 rectangle.

STEP 3 Use the front view to check the figure. It is a 2-by-2 square. So, the model is correct.

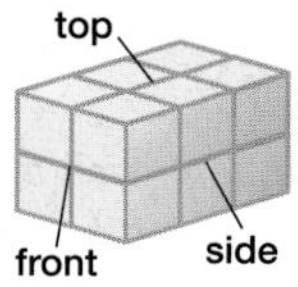

Your Turn **The top view, side view, and front view of each three-dimensional figure are shown. Use centimeter cubes to build the figure. Then make a sketch of the figure.**

a.
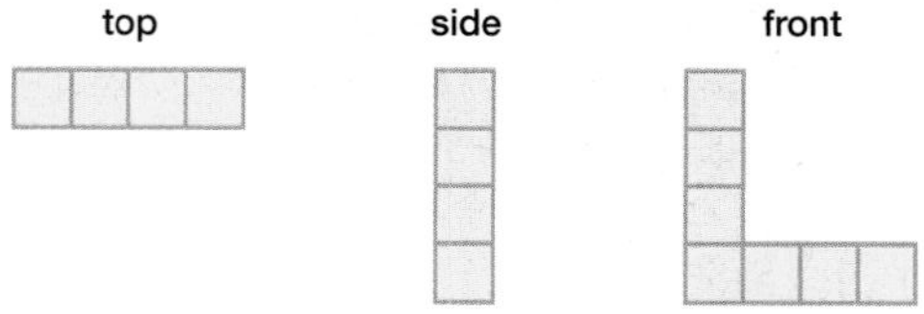

b.
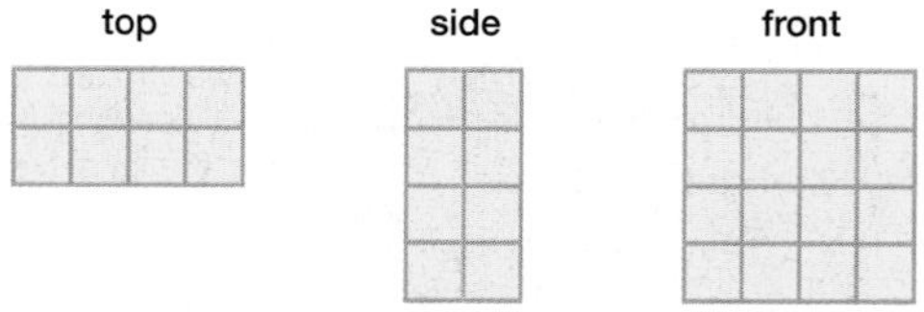

a–b. Sample drawings are given.

a.
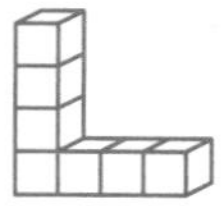

b.
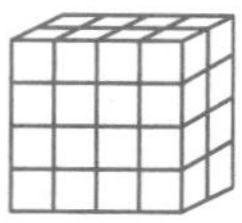

512 **Chapter 12** Geometry: Measuring Three-Dimensional Figures

e. Sample drawings
e given.

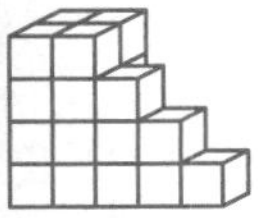

The top view, side view, and front view of each three-dimensional figure are shown. Use centimeter cubes to build the figure. Then make a sketch of the figure.

c.
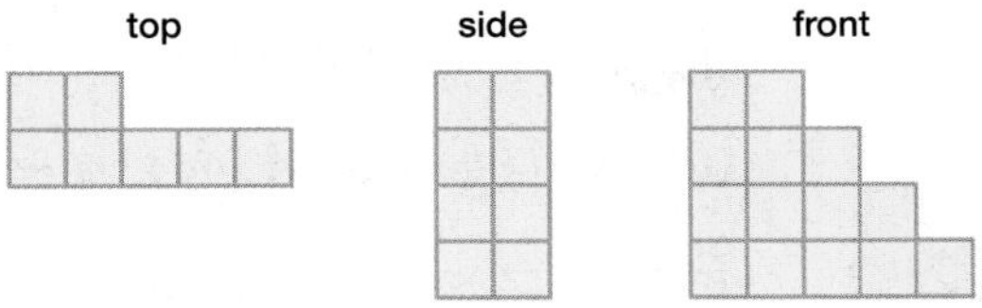

d.
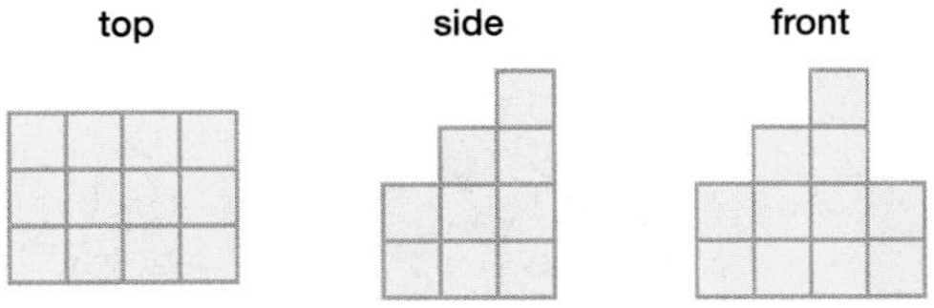

e.
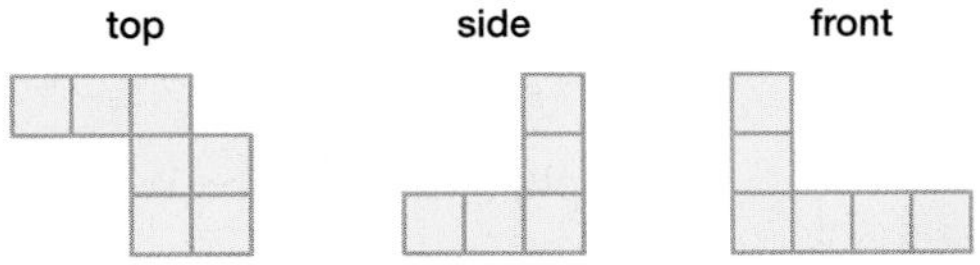

Writing Math

Work with a partner.

1. **Build** a model with cubes and draw the top, side, and front views. Give the drawing of the views to your partner and have him or her build the figure with cubes. Trade roles with your partner and repeat making the drawing and building the figure. See students' work.
2. **Explain** how you began building the figures.
3. **Determine** whether there is more than one way to build each model. Explain your reasoning. See students' work.
4. The figure at the right represents a building with a section that is 15 stories tall and another section that is 20 stories tall. Which view would you use to show the difference in height of each section? side view

5. **Build** two different models that would look the same from two views, but not the third view. Draw a top view, side view, and front view of each model. See margin.
6. **Describe** a real-life situation where it might be necessary to draw a top, side, and front view of a three-dimensional figure.

Sample answer: Start
ith the base.

. Sample answer:
uilding a doghouse

ASSESS

Observe as students solve the **Your Turn** exercises. They should:

- position blocks carefully to achieve desired shapes of figures,
- check their models by examining them from the top, side, and front, and
- sketch each figure before taking it apart.

Discuss the **Writing Math** exercises. Ask students to share their answers to Exercise 6. Ask them how architects and builders might use similar drawings to communicate their designs.

Answer (Writing Math)

5. Sample answer:

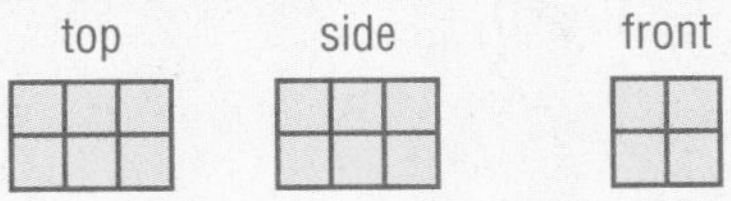

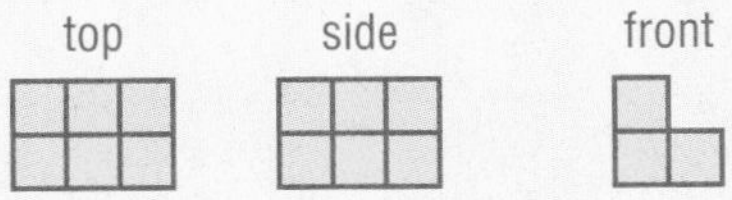

12-1 Lesson Notes

Lesson Resources

Chapter 12 Resource Masters
- Study Guide and Intervention, p. 669
- Practice: Skills, p. 670
- Practice: Word Problems, p. 671
- Reading to Learn Mathematics, p. 672
- Enrichment, p. 673

Key Teaching Transparencies, 45
MathPASS Tutorial CD-ROM, 48
Parent and Student Study Guide Workbook, 12-1

1 Focus

5-Minute Check 12-1 Use as a quiz or review of Chapter 11. (Also available as a transparency)

Bellringer
Communication Have students select any object in the room. Ask them to describe its shape from directly above.

Answer (When am I ever going to use this?)

2. Sample answer:

Reading to Learn Mathematics, p. 672 ELL

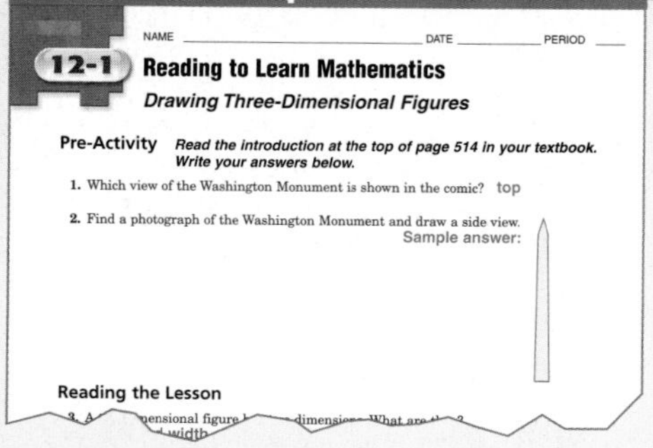
NAME ______ DATE ______ PERIOD ______

12-1 **Reading to Learn Mathematics**
Drawing Three-Dimensional Figures

Pre-Activity *Read the introduction at the top of page 514 in your textbook. Write your answers below.*

1. Which view of the Washington Monument is shown in the comic? top
2. Find a photograph of the Washington Monument and draw a side view. Sample answer:

Reading the Lesson

12-1 Drawing Three-Dimensional Figures

What You'll Learn
Draw a three-dimensional figure given the top, side, and front views.

NEW Vocabulary
solid

WHEN am I ever going to use this?

COMICS For Exercises 1 and 2, refer to the comic below.

SHOE by Jeff MacNelly

1. Which view of the Washington Monument is shown in the comic? top
2. Find a photograph of the Washington Monument and draw a side view. See margin.

A **solid** is a three-dimensional figure because it has length, width, and depth. You can draw different views of solids.

EXAMPLE Draw Different Views of a Solid

1 **Draw a top, a side, and a front view of the figure at the right.**

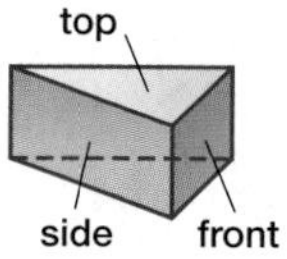

The top view is a triangle.
The side and front views are rectangles.

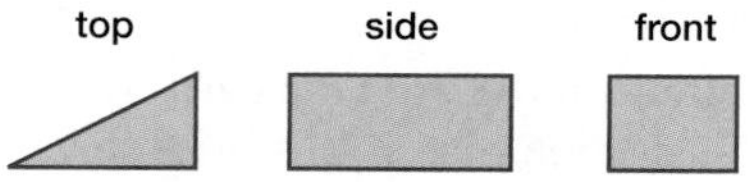

TEACHING TIP
Point out that there is more than one way to draw the different views of a three-dimensional figure. For example, the side and front views in Example 1 could be reversed.

Your Turn **Draw a top, a side, and a front view of each solid.**

a. 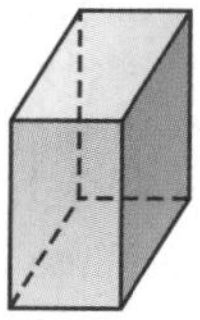

a–b. See margin.

b. 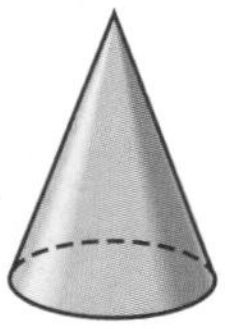

Answers (Your Turn)

a–b. Sample answers given.

a. 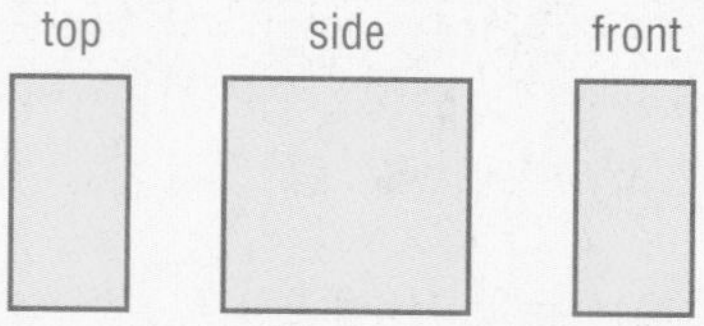

b. 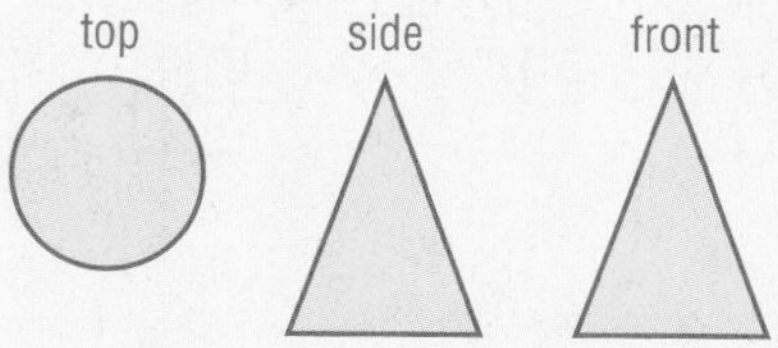

EXAMPLE Draw a Three-Dimensional Figure

2 **Draw the solid using the top, side, and front views shown at the right.**

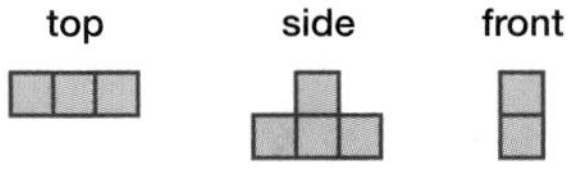

Step 1 Use the top view to draw the base of the figure, a 1-by-3 rectangle.

Step 2 Add edges to make the base a solid figure.

Step 3 Use the side and front views to complete the figure.

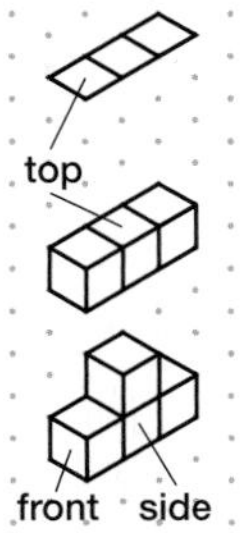

STUDY TIP

Paper Use isometric dot paper for the drawings in this lesson, as shown at the right. This is different from rectangular dot paper.

Your Turn **Draw each solid using the top, side, and front views shown. Use isometric dot paper.**

c. 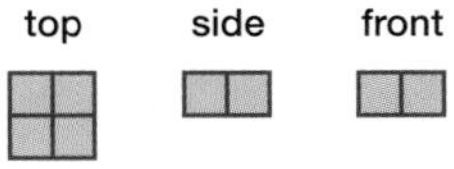

d. 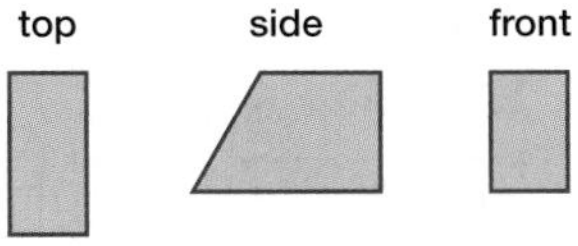

c–d. Sample drawings are given.

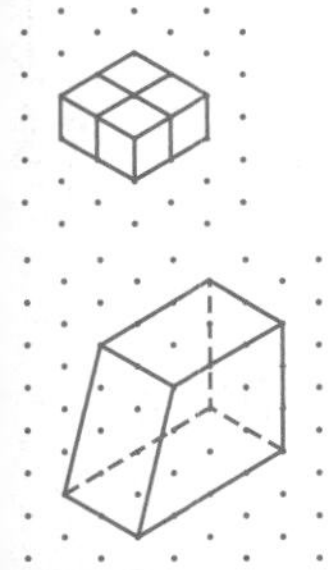

Skill and Concept Check

Writing Math Exercise 2

1. **OPEN ENDED** Draw the top, side, and front view of a solid in your home. **See students' work.**
2. **Which One Doesn't Belong?** Identify the figure that does not have the same characteristic as the other three. Explain your reasoning.

Rectangle, because it is the only two-dimensional figure.

GUIDED PRACTICE

Draw a top, a side, and a front view of each solid. **3–4. See Ch. 12 Answer Appendix.**

3. 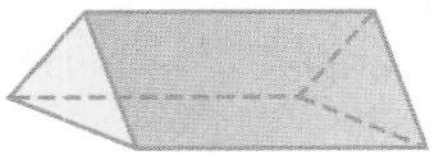

4. 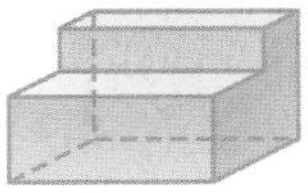

Draw each solid using the top, side, and front views shown. Use isometric dot paper. **5–6. See margin.**

5. 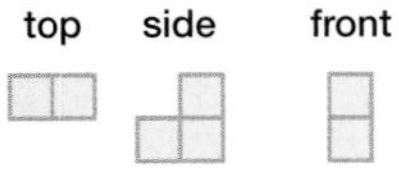

6. 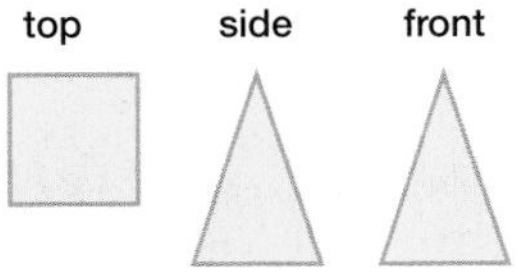

msmath2.net/extra_examples

2 Teach

In-Class Examples

INTERACTIVE CHALKBOARD PowerPoint® Presentations

1 Draw a top, a side, and a front view of the figure below.

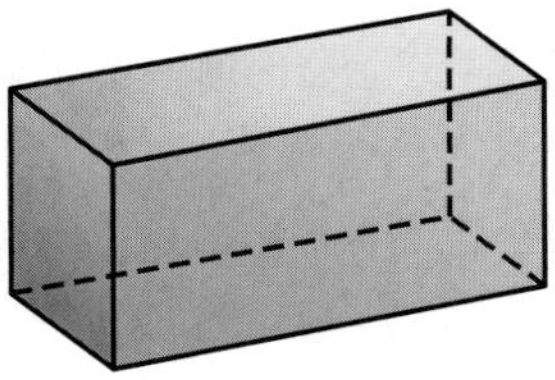

2 Draw the solid using the top, side, and front views shown below. Use isometric dot paper.

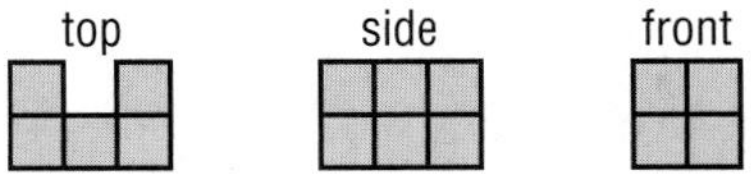

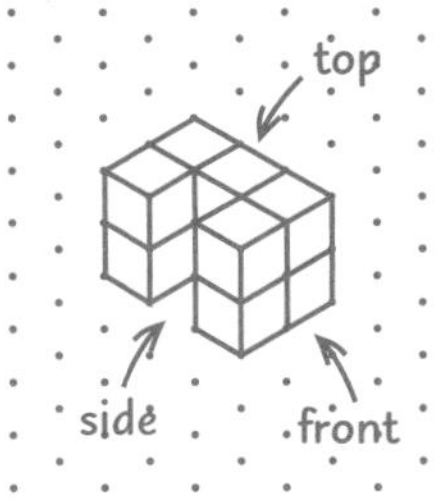

3 Practice/Apply

Which One Doesn't Belong?
In Exercise 2, students might initially think the rectangle is a side or top view of a three-dimensional figure, like the views they have seen in the example problems. Ask them to consider each figure individually and to describe its appearance as if it were sitting on a desk in front of them.

Answers

5. 

6. 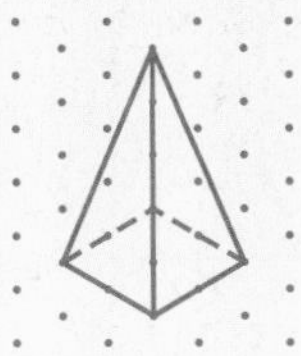

DAILY INTERVENTION **Differentiated Instruction**

Interpersonal Assign students to work in pairs, and have them look around the school grounds or their neighborhoods. They should find a large object, such as a jungle gym or shed, and draw top, side, and front views of the object. Pairs of students can then work together to build a model of the object.

★ indicates increased difficulty

Practice and Applications

HOMEWORK HELP

For Exercises	See Examples
7–12, 18	1
13–17, 19, 20	2

Extra Practice See pages 593, 607.

Draw a top, a side, and a front view of each solid. 7–12. See Ch. 12 Answer Appendix.

7. 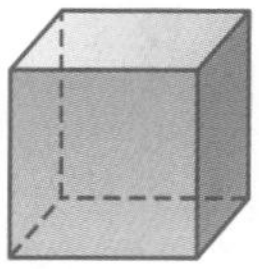

8. 

9. 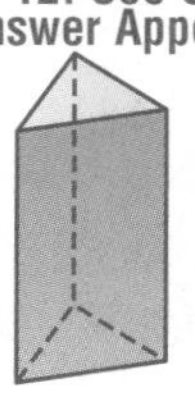

10. 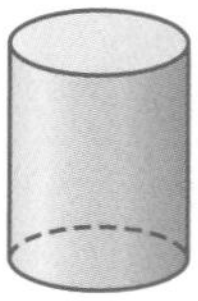

11. 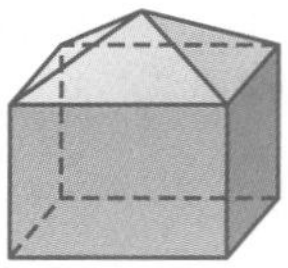

★ 12. 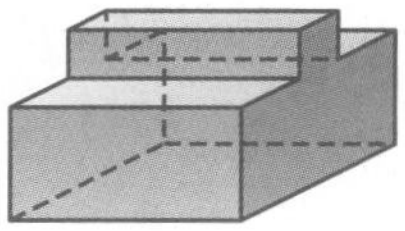

Draw each solid using the top, side, and front views shown. Use isometric dot paper. 13–16. See margin.

13. top side front

14. top side front
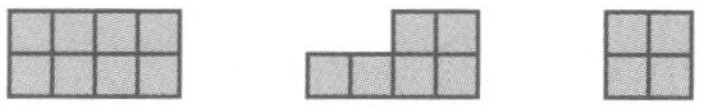

15. top side front

16. top side front
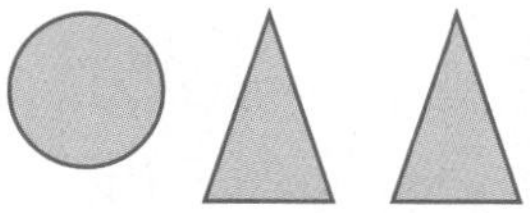

HISTORY For Exercises 17–20, use the information below.

The Lighthouse of Alexandria was the last of the seven Wonders of the Ancient World that disappeared. It contained three different-shaped levels.

top side front
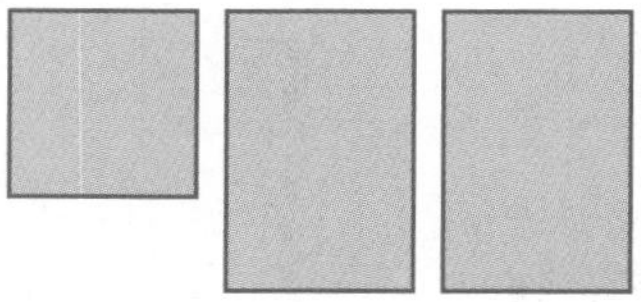

17. The bottom level is the largest. Make a drawing of this level using the top, side, and front views shown at the right. See margin.

★ 18. The middle level is a prism with a base that is a regular octagon. If the height and width of this level is approximately one third the height and width of the bottom level, draw the top, side, and front views. See margin.

19. The top level has the views shown at the right. What kind of solid is this? cylinder

top side front

★ 20. Make a sketch of the Lighthouse of Alexandria that shows all three levels. See margin.

Data Update How does your drawing compare to the actual Lighthouse of Alexandria? Visit msmath2.net/data_update to learn more.

TEACHING TIP

In 1166, an Arab traveler wrote an accurate description of the Lighthouse of Alexandria, which helped modern archaeologists reconstruct the monument. It was estimated to be about 400 feet high.

Source: ce.eng.usf.edu/pharos/

FOLDABLES™ Follow-Up Remind students to record notes about drawing three-dimensional figures in the appropriate sections of their Foldable tables. Encourage students to sketch examples of rectangular prisms and cylinders.

About the Exercises...

Guided Practice Key

Exercises	Examples
3–4	1
5–6	2

Odd/Even Assignments
Exercises 7–16 are structured so that students practice the same concepts whether they are assigned odd or even problems.

Alert! The Data Update after Exercise 20 requires access to the Internet.

Alert! Exercise 21 requires research on the Internet or another source.

Answers

13–18, 20–21. Sample answers given.

13. 

14. 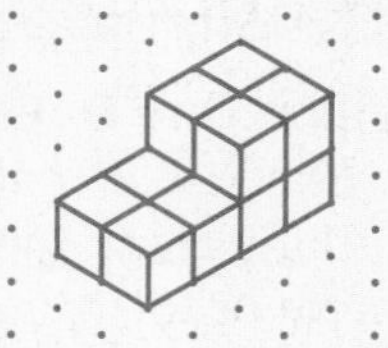

15. 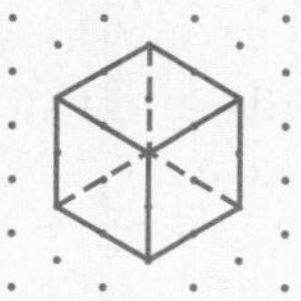

16. 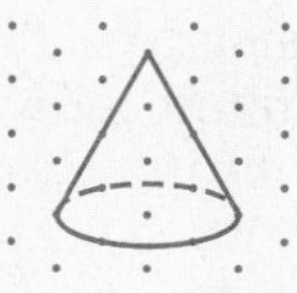

17. 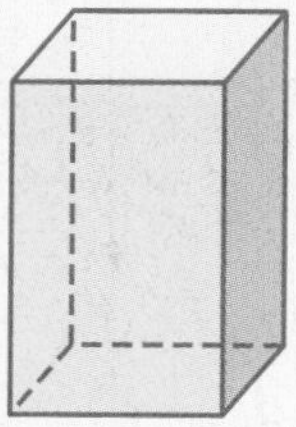

18. top side front

20. 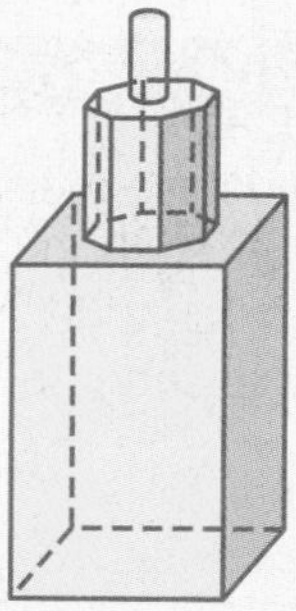

21. top side front
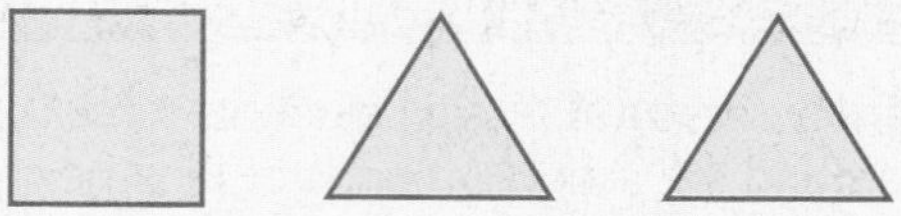

21. **RESEARCH** Use the Internet or another source to find a photograph of the only Wonder of the Ancient World existing today, the Great Pyramid of Giza. Draw a top view, a side view, and a front view of the pyramid. **See margin.**

22. **ARCHITECTURE** When a building is being designed, an architect provides a set of elevation drawings. These drawings show how the building appears from each side. Draw a set of elevation drawings for your home or school. **See students' work.**

23. **CRITICAL THINKING** Draw a three-dimensional figure in which the front and top views each have a line of symmetry but the side view does not. (*Hint*: Refer to Lesson 10-9 to review lines of symmetry.) **See margin.**

Standardized Test Practice and Mixed Review

24. **MULTIPLE CHOICE** Which is the top view of the cylinder at the right? **D**

Ⓐ triangle　Ⓑ rectangle
Ⓒ square　Ⓓ circle

25. **MULTIPLE CHOICE** Which three-dimensional figure has the top, side, and front views shown at the right? **G**

top　side　front

Ⓕ 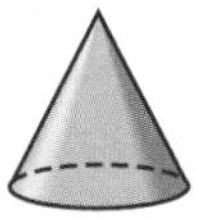

Ⓖ 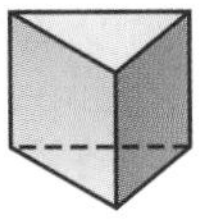

Ⓗ 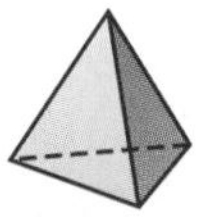

Ⓘ 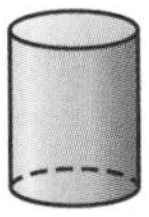

26. Find the probability that a randomly dropped counter will fall in the shaded region. Write as a percent. (Lesson 11-8) **24%**

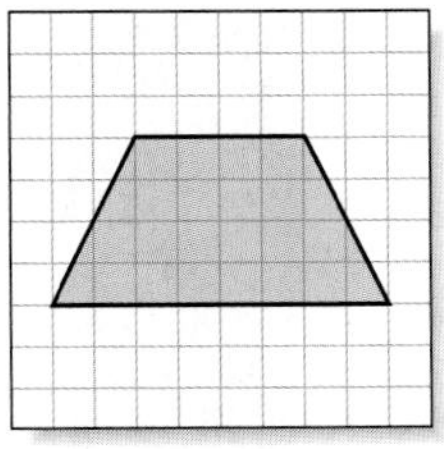

Find the area of each figure. Round to the nearest tenth if necessary.
(Lesson 11-7)

27. 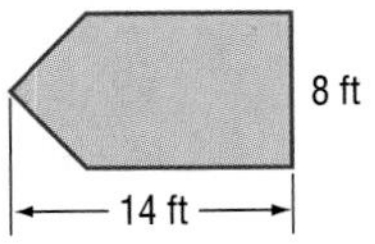

96 ft²

28. 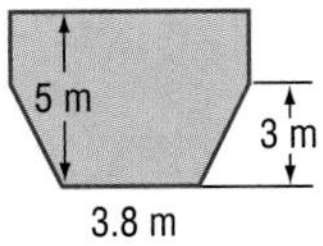

30.2 m²

29. 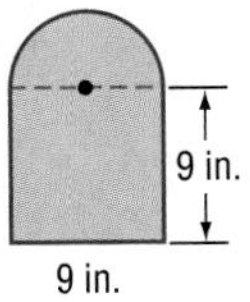

112.8 in²

GETTING READY FOR THE NEXT LESSON

PREREQUISITE SKILL Multiply. (Lesson 6-4)

30. $7\frac{1}{2} \cdot 6$ **45**
31. $8 \cdot 2\frac{3}{4}$ **22**
32. $\frac{5}{6} \cdot 1\frac{4}{5}$ **$1\frac{1}{2}$**
33. $10\frac{1}{5} \cdot 6\frac{2}{3}$ **68**

4 Assess

Open-Ended Assessment

Modeling Have students draw top, side, and front views of a three-dimensional figure. They should then trade drawings with a partner and sketch the figure represented in their partner's drawing.

Answer

23. Sample answer:

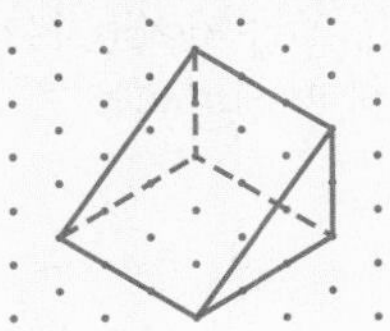

Study Guide and Intervention, p. 669

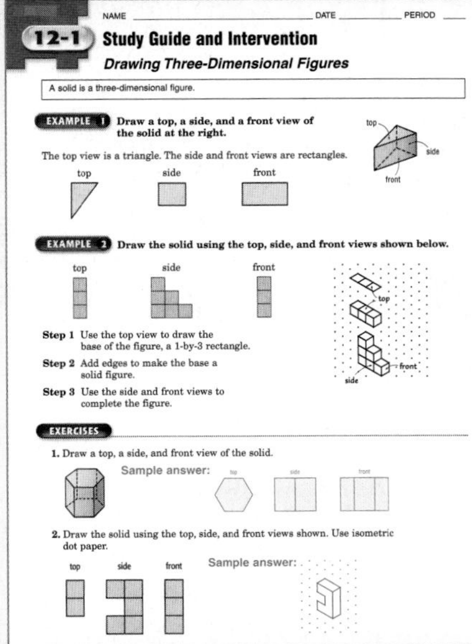

NAME ______ DATE ______ PERIOD ______

12-1 Study Guide and Intervention
Drawing Three-Dimensional Figures

A solid is a three-dimensional figure.

EXAMPLE 1 Draw a top, a side, and a front view of the solid at the right.

The top view is a triangle. The side and front views are rectangles.

EXAMPLE 2 Draw the solid using the top, side, and front views shown below.

Step 1 Use the top view to draw the base of the figure, a 1-by-3 rectangle.
Step 2 Add edges to make the base a solid figure.
Step 3 Use the side and front views to complete the figure.

EXERCISES
1. Draw a top, a side, and front view of the solid. Sample answer:
2. Draw the solid using the top, side, and front views shown. Use isometric dot paper. Sample answer:

Practice: Skills, p. 670

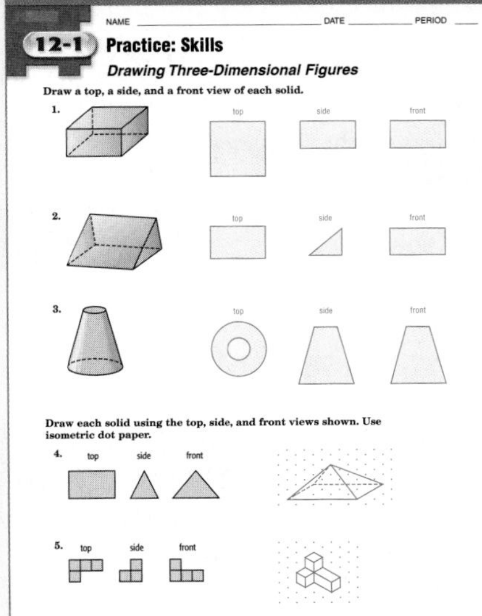

NAME ______ DATE ______ PERIOD ______

12-1 Practice: Skills
Drawing Three-Dimensional Figures

Draw a top, a side, and a front view of each solid.

Draw each solid using the top, side, and front views shown. Use isometric dot paper.

Practice: Word Problems, p. 671

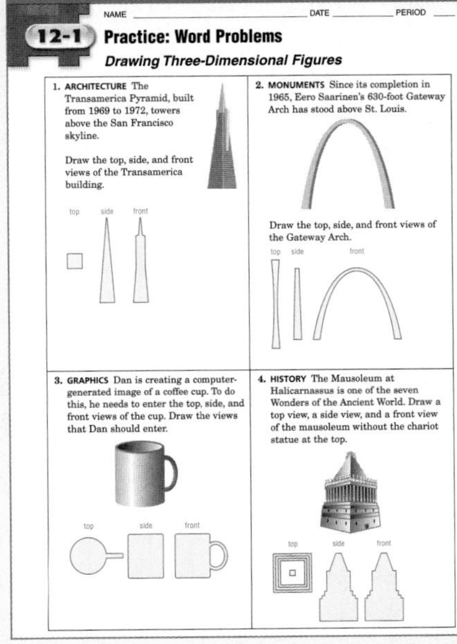

NAME ______ DATE ______ PERIOD ______

12-1 Practice: Word Problems
Drawing Three-Dimensional Figures

1. ARCHITECTURE The Transamerica Pyramid, built from 1969 to 1972, towers above the San Francisco skyline. Draw the top, side, and front views of the Transamerica building.

2. MONUMENTS Since its completion in 1965, Eero Saarinen's 630-foot Gateway Arch has stood above St. Louis. Draw the top, side, and front views of the Gateway Arch.

3. GRAPHICS Dan is creating a computer-generated image of a coffee cup. To do this, he needs to enter the top, side, and front views of the cup. Draw the views that Dan should enter.

4. HISTORY The Mausoleum at Halicarnassus is one of the seven Wonders of the Ancient World. Draw a top view, a side view, and a front view of the mausoleum without the chariot statue at the top.

12-1b Lesson Notes

1 Focus

Bellringer
Hands-On Activity Have students use cubes to form a three-dimensional figure with a length of 4 units, a width of 3 units, and a height of 2 units. Ask them how many squares are on the outside of the figure, including the bottom. 52

2 Teach

In-Class Example

INTERACTIVE CHALKBOARD PowerPoint® Presentations

Solve. Use the make a model strategy.

STORAGE A daycare center plans to make simple wooden storage bins for the 3-inch square alphabet blocks. If each bin will hold 30 blocks, give two possible dimensions for the inside of the bin. **Sample answer: 15 in. by 9 in. by 6 in. or 18 in. by 15 in. by 3 in.**

12-1b Problem-Solving Strategy

A Follow-Up of Lesson 12-1

What You'll Learn
Solve problems by making a model.

Make a Model

I'm going to help my mom make 3-inch soft alphabet blocks for the children at her daycare center. About how much fabric do we need for one cube if there is a $\frac{1}{2}$-inch seam on each side?

We could **make a model** to find out.

Explore	We know that each cube is 3 inches long with $\frac{1}{2}$-inch seams.
Plan	We can make a cardboard model of a cube with sides 3 inches long. We could then cut the model into six squares and add $\frac{1}{2}$-inch paper extensions to each side as seams.
Solve	Make the cardboard model and unfold the cube. 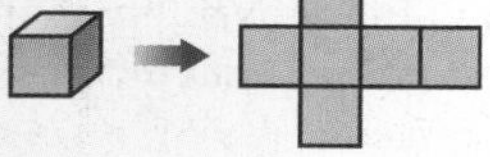 Cut the cube into six squares and tape $\frac{1}{2}$-inch seams on each side. Now each pattern piece is 4 inches by 4 inches or *about* 16 square inches. 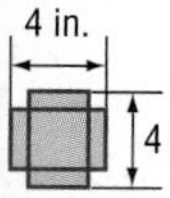 $16 \text{ in}^2 \times 6 \text{ squares} = 96 \text{ in}^2$ So, each cube needs about 96 square inches of fabric.
Examine	If the 4-inch-by-4-inch pattern pieces were laid as two columns and three rows, it would be 8 inches wide and 12 inches long or 96 square inches.

Analyze the Strategy

1. Sample answer: It allows students to see what the shapes look like in three dimensions.

1. **Explain** when making a model is a better strategy than drawing a picture.
2. **Explain** why you think the students started with the three-dimensional model to make their pattern. **See students' work.**
3. **Write** a problem that can be solved by making a model. Then solve the problem. **See students' work.**

518 Chapter 12 Geometry: Measuring Three-Dimensional Figures

DAILY INTERVENTION

Differentiated Instruction

Visual/Spatial Have students draw a simple three-dimensional design, write down its dimensions, and then determine the area of each of its surfaces.

pply the Strategy

lve. Make a model.

ART Dominic is creating a layout of his bedroom for art class. The room measures 15 foot by 12 feet. If he uses a scale of 1 foot = $\frac{3}{4}$ inch, what are the dimensions of his bedroom on the model? **11.25 in. by 9 in.**

5. **BICYCLES** Eight customers lined up outside The Bike Shop with either a bicycle or a tricycle that needed repair. When the owner looked out the window, she counted 21 wheels outside the shop. How many tricycles and bicycles were there? **5 tricycles, 3 bicycles**

ixed Problem Solving

lve. Use any strategy.

COMMUNITY SERVICE There are four drop-off centers for the community food drive. Their total collections are shown in the table.

Center	Number of Cans
A	3,298
B	2,629
C	4,429
D	2,892

A newsletter reported that over 13,000 cans of food were collected. Is this estimate reasonable? Explain. **yes; 3,000 + 3,000 + 4,000 + 3,000 = 13,000**

SWIMMING Yeti can swim one 20-meter lap in 1.25 minutes. How long will it take her to swim 100 meters at the same rate? **6.25 min**

TRAFFIC At the four-way intersection shown below, the traffic lights change every 90 seconds. About 8 cars in one lane travel through the light in this amount of time. Determine the number of cars that travel through the intersection in 3 minutes. **about 64 cars**

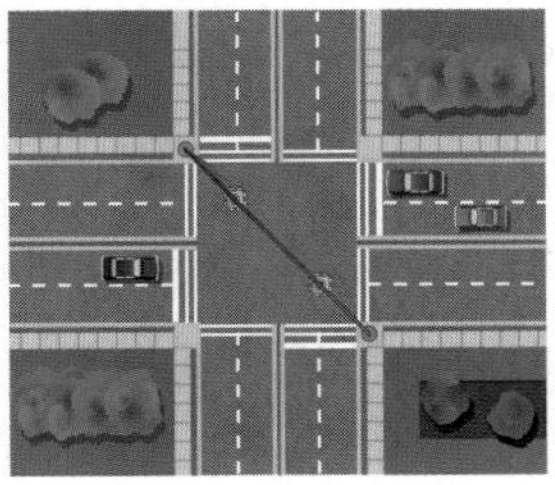

9. **MONEY** A top film actor made 16 films, which grossed over $\$5.09 \times 10^9$. On average, how much did each film make? Write in scientific notation. Round to the nearest tenth. **$\$3.2 \times 10^8$**

10. **DISPLAYS** Identical boxes are stacked in the corner of a store as shown below. How many boxes are *not* visible? **20 boxes**

11. **REMODELING** How many square feet of wallpaper are needed to cover a wall that measures 9 feet by 16 feet? **144 ft^2**

12. **STANDARDIZED TEST PRACTICE**
Yolanda deposited $450 in a new savings account in January, withdrew $175 in February, and then began monthly deposits of $75 from March through December. Which equation shows her balance b after her last deposit in December? **B**

Ⓐ $b = \$275 + 9(\$75)$

Ⓑ $b = \$275 + 10(\$75)$

Ⓒ $b = \$450 + 10(\$75)$

Ⓓ $b = \$275 - 10(\$75)$

3 Practice/Apply

About the Exercises...

Problem-Solving Strategies
Exercises 6–12 are structured so that students have the opportunity to practice many different problem-solving strategies. You may wish to remind them of some of the strategies they have studied.

- guess and check (p. 22)
- look for a pattern (p. 132)
- work backward (p. 164)
- draw a diagram (p. 302)
- act it out (p. 391)
- use logical reasoning (p. 444)
- make a model ✓

4 Assess

Higher-Level Thinking
Ask students to draw a rectangle with a perimeter of 20 units on grid paper so that it has the maximum possible area. Then draw one with the minimum possible area. Have them determine the area of both figures. **maximum area = 25 square units; minimum area = 9 square units**

Lesson 12-1b Problem-Solving Strategy: Make a Model 519

12-2 Lesson Notes

Lesson Resources

Chapter 12 Resource Masters
- Study Guide and Intervention, p. 674
- Practice: Skills, p. 675
- Practice: Word Problems, p. 676
- Reading to Learn Mathematics, p. 677
- Enrichment, p. 678

Calculator and Spreadsheet Masters, p. 55
MathPASS Tutorial CD-ROM, 52
Parent and Student Study Guide Workbook, 12-2
Real-Life Career Masters, p. 26
Teaching Mathematics with Manipulatives, pp. 117–118

1 Focus

5-Minute Check 12-2
Use as a quiz or review of Lesson 12-1. (Also available as a transparency)

INTERACTIVE CHALKBOARD PowerPoint® Presentations

Bellringer
Reading Mathematics Have students describe the relationship between a square and a cube. Ask them to discuss why the area of an object is measured in square units, while the volume of a three-dimensional object is measured in cubic units.

Reading to Learn Mathematics, p. 677 ELL

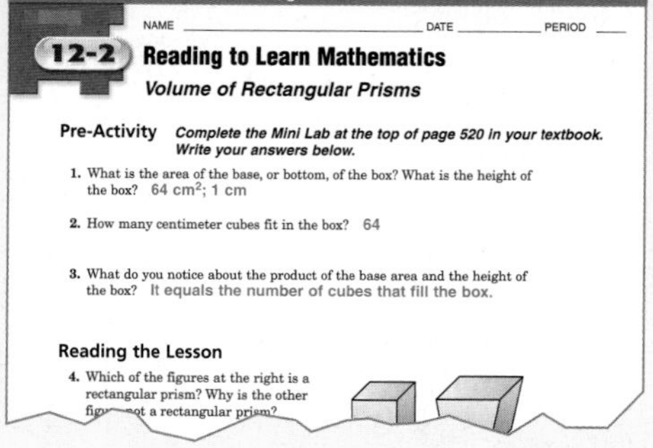

NAME ______ DATE ______ PERIOD ____

12-2 Reading to Learn Mathematics
Volume of Rectangular Prisms

Pre-Activity *Complete the Mini Lab at the top of page 520 in your textbook. Write your answers below.*

1. What is the area of the base, or bottom, of the box? What is the height of the box? 64 cm²; 1 cm
2. How many centimeter cubes fit in the box? 64
3. What do you notice about the product of the base area and the height of the box? It equals the number of cubes that fill the box.

Reading the Lesson

4. Which of the figures at the right is a rectangular prism? Why is the other fig... ot a rectangular prism?

12-2 Volume of Rectangular Prisms

What You'll Learn
Find the volumes of rectangular prisms.

NEW Vocabulary
volume
rectangular prism

LINK To Reading
Everyday Meaning of Volume: bulk or mass, as in shipping a large volume of merchandise

HANDS-ON Mini Lab

Materials
- centimeter grid paper
- scissors
- tape
- centimeter cubes

Work with a partner.

- On a piece of grid paper, cut out a square that is 10 centimeters on each side.
- Cut a 1-centimeter square from each corner. Fold the paper and tape the corners together to make a box.

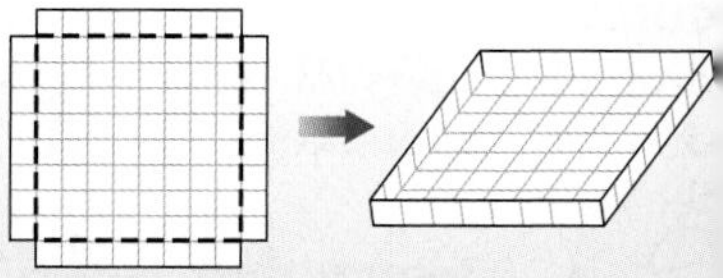

1. What is the area of the base, or bottom, of the box? What is the height of the box? **1. 64 cm²; 1 cm**
2. How many centimeter cubes fit in the box? **64**
3. What do you notice about the product of the base area and the height of the box? **It equals the number of cubes that fill the box.**

The **volume** of a solid is the measure of space occupied by it. It is measured in cubic units such as cubic centimeters (cm^3) or cubic inches (in^3). The volume of the figure at the right can be shown using cubes.

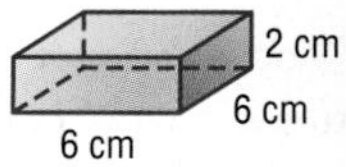

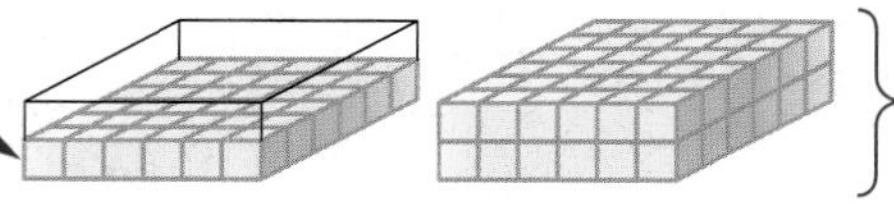

It takes 36 · 2 or 72 cubes to fill the box. So, the volume of the box is 72 cubic centimeters.

A **rectangular prism** is a solid figure that has two parallel and congruent sides, or bases, that are rectangles.

TEACHING TIP
The activity in the Mini Lab can also be done using a spreadsheet.

Key Concept — **Volume of a Rectangular Prism**

Words The volume V of a rectangular prism is the area of the base B times the height h. It is also the product of the length ℓ, the width w, and the height h.

Model

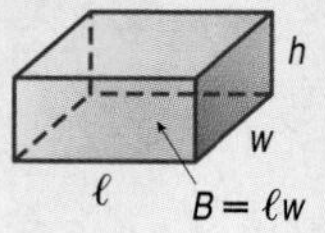

Symbols $V = Bh$ or $V = \ell wh$

DAILY INTERVENTION — **Unlocking Misconceptions**

Volume of Rectangular Prisms Students may have difficulty deciding which side of a rectangular prism to represent as the base. Remind them that there are several ways to look at each figure and that any side of a prism that has a parallel side may be considered the base.

EXAMPLES Find the Volumes of Prisms

1 **Find the volume of the rectangular prism.**

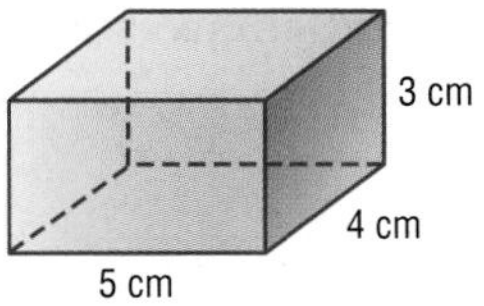

$V = \ell wh$ Volume of a rectangular prism

$V = 5 \cdot 4 \cdot 3$ Replace ℓ with 5, w with 4, and h with 3.

$V = 60$ Multiply.

The volume is 60 cubic centimeters.

2 **CEREAL Find the volume of the cereal box.**

Estimate $V \approx 8 \cdot 3 \cdot 10$ or 240

$V = \ell wh$ Volume of a rectangular prism

$V = 7.5 \cdot 2.5 \cdot 12$ Replace ℓ with 7.5, w with 2.5, and h with 12.

$V = 225$ Multiply.

The volume is 225 cubic inches.

Compare to the estimate.

STUDY TIP

Estimation It is helpful to first estimate the volume of a rectangular prism to determine whether your answer is reasonable.

Skill and Concept Check

Writing Math
Exercises 2 & 3

1. **Write** the abbreviation for cubic yards using an exponent. yd^3
2. **OPEN ENDED** Draw and label a rectangular prism that has a volume between 400 and 600 cubic centimeters. State the volume of the prism. $504\ cm^3$; See margin for drawing.
3. **FIND THE ERROR** Cassandra and Ling are comparing the volumes of the two prisms at the right. Who is correct? Explain.

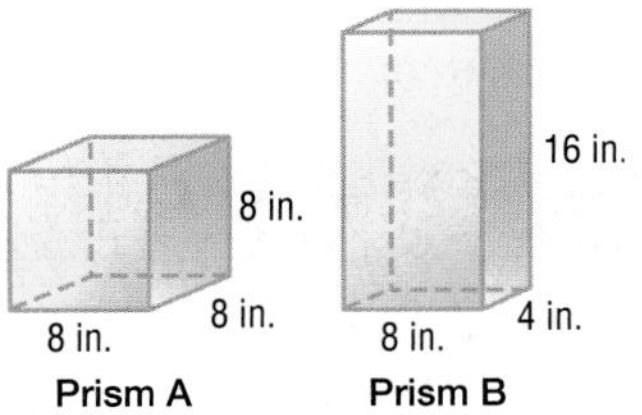

Cassandra
Prism B has a greater volume because it is twice as tall as Prism A.

Ling
Both prisms have the same volume.

3. Ling; the width of Box B is half that of Box A. All three dimensions have to be considered when comparing volumes.

GUIDED PRACTICE

Find the volume of each rectangular prism.

4. 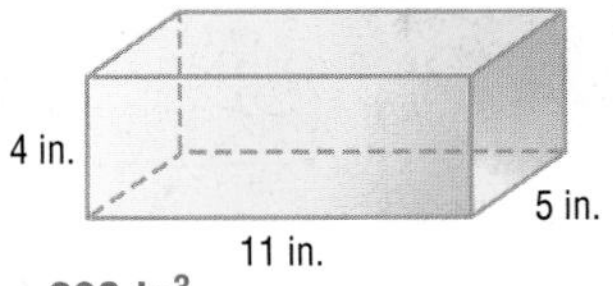

$220\ in^3$

5. 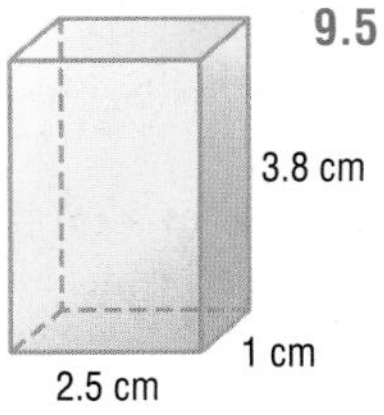

$9.5\ cm^3$

6. 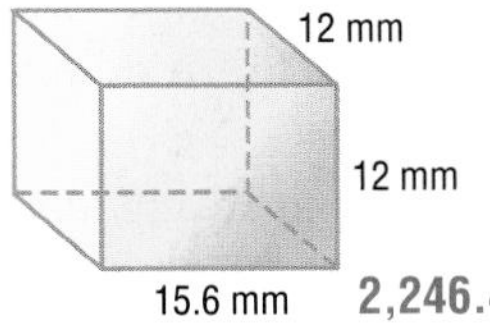

$2{,}246.4\ mm^3$

7. **STORAGE** A cabinet measures 20 inches by 30 inches by 60 inches. What is its volume? $36{,}000\ in^3$

msmath2.net/extra_examples

2 Teach

In-Class Examples

INTERACTIVE CHALKBOARD
PowerPoint® Presentations

1 Find the volume of the rectangular prism. $24\ cm^3$

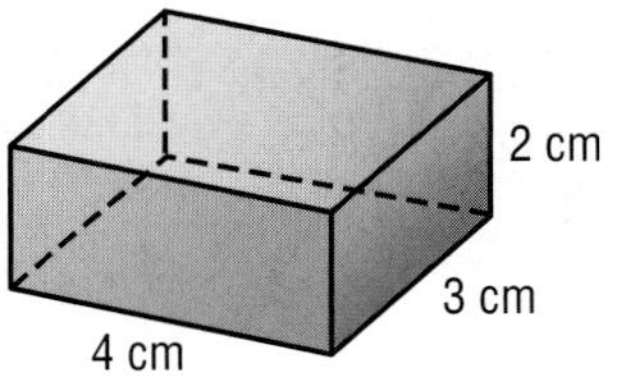

2 **GAMES** Find the volume of the game box. $289\ in^3$

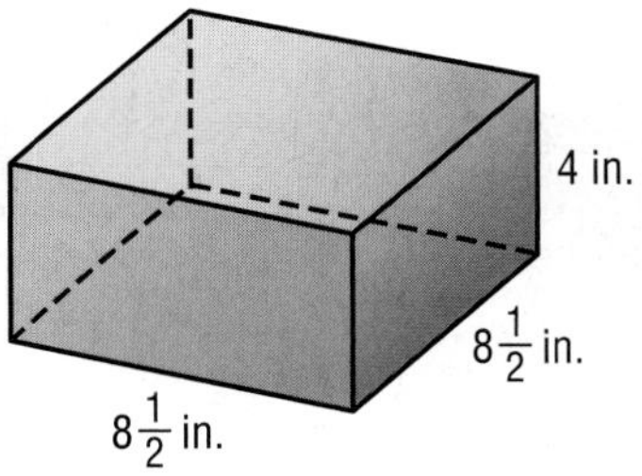

3 Practice/Apply

DAILY INTERVENTION **FIND THE ERROR** In Exercise 3, students might notice that although Prism B's height is twice Prism A's height, B's width is half A's width. Since the two prisms are equal in length, their volumes are equal.

About the Exercises...

Guided Practice Key

Exercises	Examples
4–6	1
7	2

Odd/Even Assignments

Exercises 8–17 are structured so that students practice the same concepts whether they are assigned odd or even problems.

Tips for New Teachers

Using Manipulatives Some students may be more familiar with the volumes of liquids. When working activities with centimeter cubes, consider modeling the liquid equivalents as well. One cubic centimeter equals one milliliter.

Answer

2. 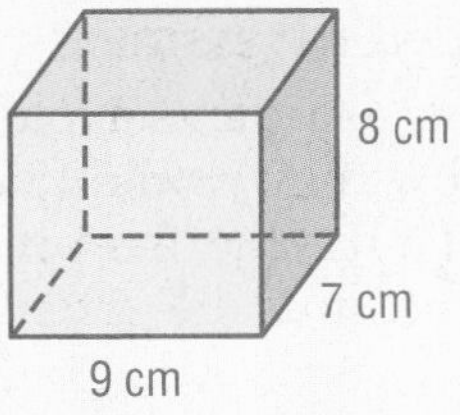

Study Guide and Intervention, p. 674

NAME ________________ DATE ________ PERIOD ____

12-2 Study Guide and Intervention

Volume of Rectangular Prisms

The **volume** of a solid is the measure of space occupied by it. It is measured in cubic units such as cubic centimeters (cm^3) or cubic inches (in^3). The volume of the figure at the right can be shown using cubes.

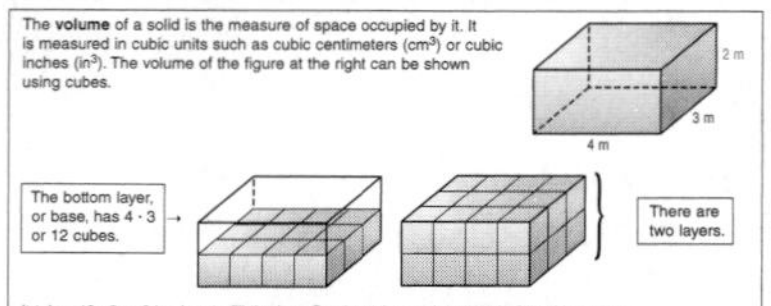

It takes 12 · 2 or 24 cubes to fill the box. So, the volume of the box is 24 cubic meters.

A **rectangular prism** is a solid figure that has two parallel and congruent sides, or bases, that are rectangles. To find the volume of a rectangular prism, multiply the area of the base and the height, or find the product of the length ℓ, the width w, and the height h.

$V = Bh$ or $V = \ell wh$

EXAMPLE 1 **Find the volume of the rectangular prism.**

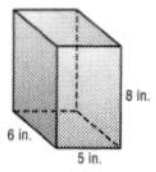

$V = \ell wh$ Volume of a rectangular prism

$V = 5 \cdot 6 \cdot 8$ Replace ℓ with 5, w with 6, and h with 8.

$V = 240$ Multiply.

The volume is 240 cubic inches.

EXERCISES

Find the volume of each rectangular prism. Round to the nearest tenth if necessary.

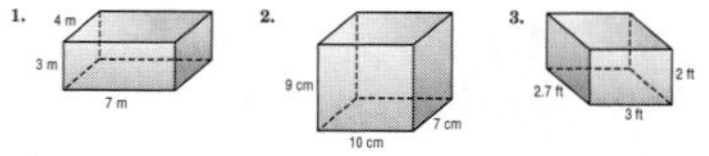

1. 84 m^3 2. 630 cm^3 3. 16.2 ft^3

Practice: Skills, p. 675

NAME ________________ DATE ________ PERIOD ____

12-2 Practice: Skills

Volume of Rectangular Prisms

Find the volume of each rectangular prism. Round to the nearest tenth if necessary.

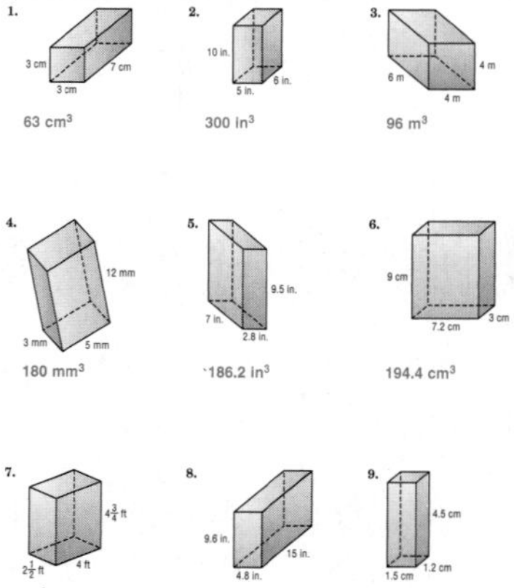

1. 63 cm^3 2. 300 in^3 3. 96 m^3

4. 180 mm^3 5. 186.2 in^3 6. 194.4 cm^3

7. $47\frac{1}{2}$ ft^3 8. 691.2 in^3 9. 8.1 cm^3

Practice: Word Problems, p. 676

NAME ________________ DATE ________

12-2 Practice: Word Problems

Volume of Rectangular Prisms

1. **PACKAGING** A cereal box has a length of 8 inches, a width of $1\frac{3}{4}$ inches, and a height of $12\frac{1}{8}$ inches. What is the volume of the cereal box? $169\frac{3}{4}$ in^3

2. **FOOD STORAGE** Nara wants to determine how much ice it will take to fill her cooler. If the cooler has a length of 22 inches, a width of 12 inches, and a height of $10\frac{1}{2}$ inches, how much ice will her cooler hold? 2,772 in^3

3. **TRANSPORTATION** The cargo-carrying part of Billy's truck has a length of 8.3 meters, a width of 3 meters, and a height of 4.2 meters. What is the maximum volume of sand that Billy's truck can carry? 104.58 m^3

4. **PLUMBING** Alexia's bathroom has a tub in the shape of a rectangular prism with a length of 1.5 meters, a width of 0.5 meter, and a height of 0.4 meter. How many cubic feet of water can it hold? 0.3 m^3

5. **PACKAGING** A box of tissues has a length of 11.2 centimeters, a width of 11.2 centimeters, and a height of 13 centimeters. What is the volume of the tissue box? 1,630.72 cm^3

6. **GEOMETRY** A *pentagonal prism* is a prism that has bases that are pentagons. Use $V = Bh$ where B is the area of the base, to find the volume of the pentagonal prism below.

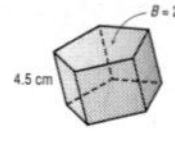

118.35 cm^3

★ indicates increased difficulty

Practice and Applications

Find the volume of each rectangular prism. Round to the nearest tenth if necessary.

HOMEWORK HELP

For Exercises	See Examples
8–15	1, 2

Extra Practice
See pages 593, 607.

8. 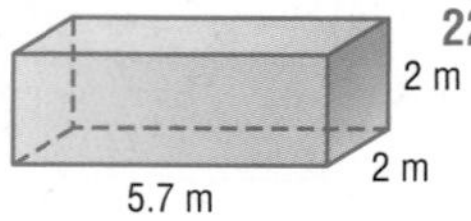

22.8 m^3

9. 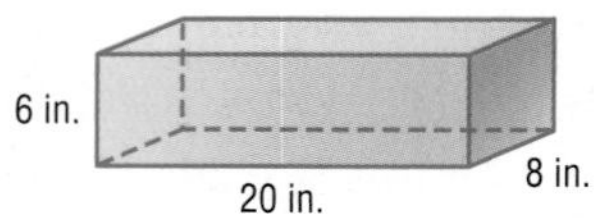

960 in^3

10. 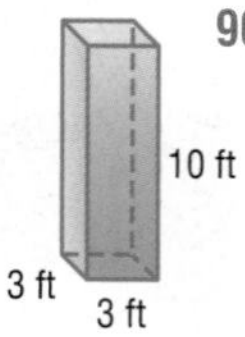

90 ft^3

11. 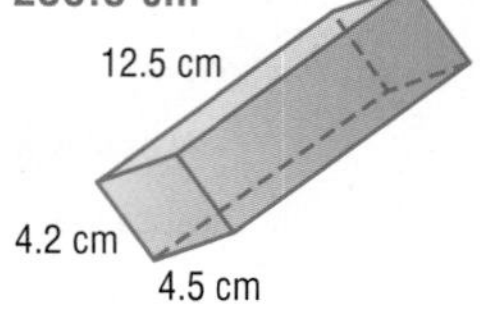

236.3 cm^3

12. 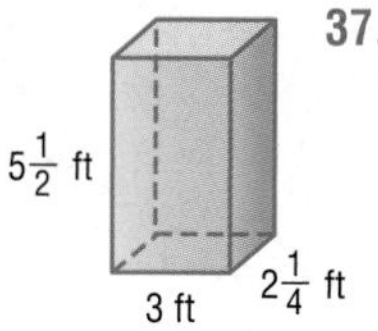

37.1 ft^3

13. A cube has 3-centimeter edges. Find its volume. 27 cm^3

★ 14. The base of a rectangular prism has an area of 19.4 square meters. Find the height if the volume is 306.52 cubic meters. 15.8 m

15. **TRANSPORTATION** Find the approximate volume of the California trolley car shown at the right. 2,425.5 ft^3

★ 16. **MEASUREMENT** How many cubic inches are in a cubic foot? 1,728 in^3

★ 17. **MEASUREMENT** How many cubic centimeters are in a cubic meter? 1,000,000 cm^3

18. **CRITICAL THINKING** A *triangular prism* is a prism that has bases that are triangles. Use $V = Bh$ to find the volume of the triangular prism at the right. 42 cm^3

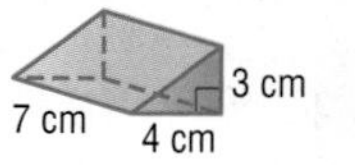

Standardized Test Practice and Mixed Review

19. **MULTIPLE CHOICE** An office is 20 feet long, 15 feet wide, and 12 feet high. It costs about 9¢ per year to air condition one cubic foot of space. On average, how much does it cost to air condition this office for one month? C

Ⓐ $324 Ⓑ $300 Ⓒ $27 Ⓓ $2.25

20. **SHORT RESPONSE** A landscaper wants to cover a 40-foot-by-12-foot rectangular area with small stones. If she uses 120 cubic feet of stones, how deep will they be? 3 in.

21. Draw a top, a side, and a front view of the solid at the right. (Lesson 12-1) See Ch. 12 Answer Appendix.

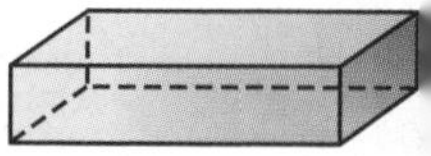

22. Draw an area model so that the probability of a randomly dropped counter falling in a triangular region is 25%. (Lesson 11-8) See margin.

GETTING READY FOR THE NEXT LESSON

PREREQUISITE SKILL **Estimate.** (Page 558) 23–26. Sample answers are given.

23. $3.14 \cdot 6$ 18 24. $5 \cdot 2.7^2$ 45 25. $9.1 \cdot 8.3$ 72 26. $3.1 \cdot 1.75^2 \cdot 2$ 24

msmath2.net/self_check_quiz

4 Assess

Open-Ended Assessment

Writing Have students explain in writing how they would find the volume of a rectangular prism and why volume must be given in cubic units.

Answer

22. Sample answer:

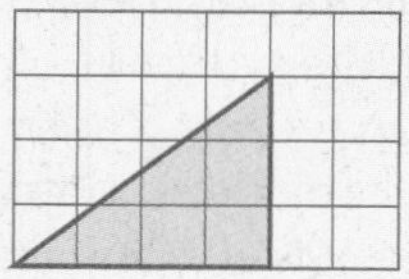

12-2b Spreadsheet Investigation

A Follow-Up of Lesson 12-2

ıat You'll Learn
e a spreadsheet
investigate the
umes of similar
ids.

Similar Solids

A computer spreadsheet can help you calculate the sizes and volumes of similar rectangular prisms. You can enlarge or reduce dimensions of the prisms and the spreadsheet will automatically calculate the new volumes.

ACTIVITY

Mrs. Sanchez owns a box-making factory. She wishes to produce a set of similar boxes so that they will nest inside each other when assembled. She will need to show her customers the dimensions and volume of each type of box she plans to sell. This information can be placed into a spreadsheet.

Set up the spreadsheet like the one shown below.

Similar Boxes

	A	B	C	D	E
1		Dimensions (in.)			
2	Box	L	W	H	Volume (in^3)
3	small	2.5	2.5	2.5	
4	medium	5	5	5	
5	large	7.5	7.5	7.5	
6					

Sheet1 / Sheet2 / Sheet3

1. =B3*C3*D3; =B4*C4*D4; =B5*C5*D5; 15.6 in^3; 125 in^3; 421.9 in^3

TEACHING TIP
mind students how
copy down formulas
her than typing in
ch one.

EXERCISES

1. State the spreadsheet commands you used to find the volumes of the three boxes. Then find the volumes to the nearest tenth.
2. Expand the spreadsheet to calculate the volumes of the boxes when each of the dimensions is doubled. What happens to the volume of the small box when all of the dimensions of the box are doubled? **The volume increases by a factor of 8.**
3. What happens to the volume of the small box when all of the dimensions of the box are tripled?
4. Suppose another box has a volume 216 times greater than the small box. What are the dimensions of this box? Use the spreadsheet to check your answer. **15 in. × 15 in. × 15 in.**
5. Extend the pattern in the volume column. Then express the pattern in exponential form. $2^3, 3^3, 4^3, 5^3, \ldots$

3. The volume increases
a factor of 27.

12-2b Spreadsheet Investigation

GETTING STARTED

Teaching Tip Have students review the concept of similarity as discussed in Chapter 10. Ask them how the sides of similar solids are related. **Sample answer: Each side of one solid would be larger or smaller than its corresponding side of a similar solid. All sides of a solid would have to change by the same factor in order for the solids to be similar.** Ask students to make a conjecture about the volume of similar solids.

TEACH

- Have students make predictions about the change in volume of each new prism before the computation is performed.
- Remind students to compare the computed results with their predictions. Are there any patterns that they can find in the data and the resulting volumes that would facilitate future predictions?

ASSESS

Observe as students solve the **Exercises**. They should:

- correctly use spreadsheet formulas to find the volumes of various sizes of boxes,
- recognize patterns of similarity among the box sizes, and
- correctly express the relationship between the length of a side of a cube and that cube's volume in exponential form.

12-3 Lesson Notes

Lesson Resources

Chapter 12 Resource Masters
- Study Guide and Intervention, p. 679
- Practice: Skills, p. 680
- Practice: Word Problems, p. 681
- Reading to Learn Mathematics, p. 682
- Enrichment, p. 683

Key Teaching Transparencies, 46
Parent and Student Study Guide Workbook, 12-3
Science and Math Lab Masters, pp. 77–80
Teaching Mathematics with Manipulatives, p. 119

FAST FILE

1 Focus

5-Minute Check 12-3

INTERACTIVE CHALKBOARD
PowerPoint® Presentations

Use as a quiz or review of Lesson 12-2. (Also available as a transparency)

Bellringer
Problem Solving Cylindrical objects, such as stacks of round cookies, are often boxed in rectangular prisms. Bring one such item to class and ask students to estimate the volume of the cylinder. **Students should arrive at the conjecture that the volume of the cylinder is close to, but less than, the volume of the rectangular prism.**

Reading to Learn Mathematics, p. 682 ELL

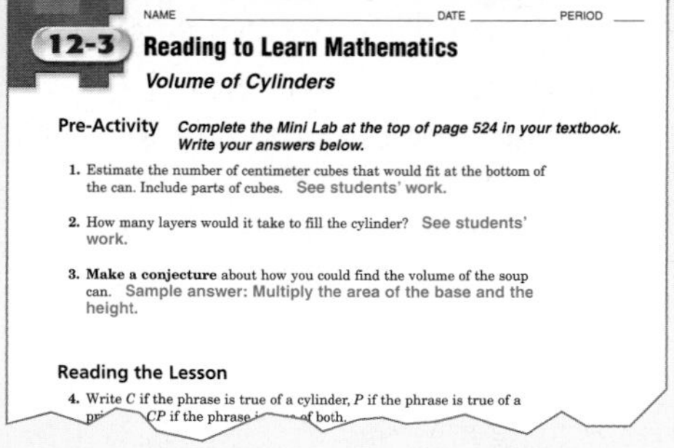
NAME ________________ DATE __________ PERIOD ____

12-3 Reading to Learn Mathematics
Volume of Cylinders

Pre-Activity *Complete the Mini Lab at the top of page 524 in your textbook. Write your answers below.*

1. Estimate the number of centimeter cubes that would fit at the bottom of the can. Include parts of cubes. See students' work.
2. How many layers would it take to fill the cylinder? See students' work.
3. **Make a conjecture** about how you could find the volume of the soup can. Sample answer: Multiply the area of the base and the height.

Reading the Lesson

4. Write *C* if the phrase is true of a cylinder, *P* if the phrase is true of a pri... *CP* if the phrase ... of both.

12-3 Volume of Cylinders

What You'll Learn
Find the volumes of cylinders.

NEW Vocabulary
cylinder

TEACHING TIP
In this chapter, answers involving π were found using the π key on a calculator, unless otherwise specified.

HANDS-ON Mini Lab

Materials
- soup can
- centimeter grid paper
- scissors
- tape

Work with a partner.

Set a soup can on a piece of grid paper and trace around the base, as shown at the right.
1–2. See students' work.

1. Estimate the number of centimeter cubes that would fit at the bottom of the can. Include parts of cubes.
2. How many layers would it take to fill the cylinder?
3. **Make a conjecture** about how you could find the volume of the soup can.
Sample answer: Multiply the area of the base and the height.

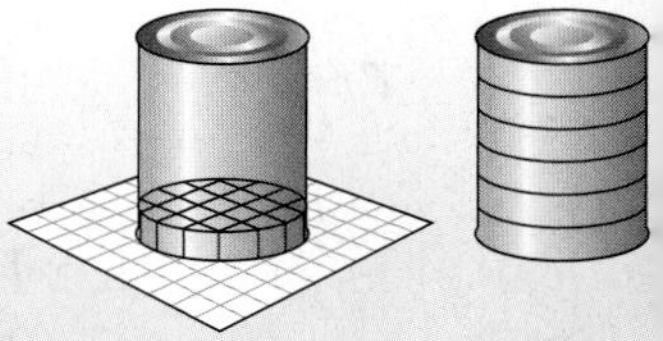

A **cylinder** is a solid figure that has two congruent, parallel circles as its bases. As with prisms, the area of the base tells the number of cubi units in one layer. The height tells how many layers there are.

Key Concept — **Volume of a Cylinde**

Words The volume V of a cylinder with radius r is the area of the base B times the height h.

Symbols $V = Bh$ or $V = \pi r^2 h$, where $B = \pi r^2$

Model

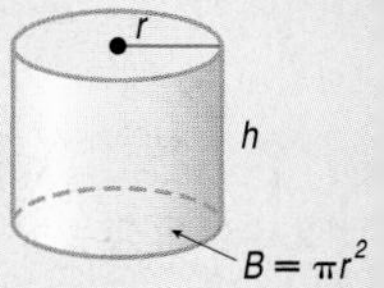

EXAMPLE Find the Volume of a Cylinder

1 **Find the volume of the cylinder. Round to the nearest tenth.**

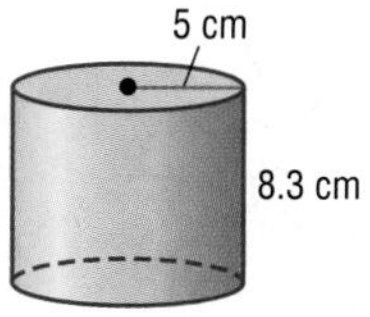

$V = \pi r^2 h$ Volume of a cylinder

$V = \pi(5)^2(8.3)$ Replace r with 5 and h with 8.3.

[π] [×] 5 [x²] [×] 8.3 [ENTER] 651.8804756

The volume is about 651.9 cubic centimeters. Check by using estimation.

524 **Chapter 12** Geometry: Measuring Three-Dimensional Figures

EXAMPLE Find the Volume of a Real-Life Object

Standardized Test Practice

2 **MULTIPLE-CHOICE TEST ITEM** **Find the volume of a cylinder-shaped paint can that has a diameter of 4 inches and a height of 5 inches.**

(A) 20.0 in^3 (B) 62.8 in^3 (C) 125.7 in^3 (D) 251.3 in^3

est-Taking Tip The Princeton Review

eading
ne radius of the can is ÷ 2 or 2 inches. Be sure substitute that value into e formula. If you ıbstitute 4 for *r* into the rmula, you would get the correct answer D.

Read the Test Item

To find the volume, use the formula $V = \pi r^2 h$.

Solve the Test Item

$V = \pi r^2 h$ Volume of a cylinder

$V = \pi(2)^2(5)$ Replace *r* with 2 and *h* with 5.

$V \approx 62.8$ Simplify.

The answer is B.

Skill and Concept Check

Writing Math Exercises 1 & 2

1. **OPEN ENDED** Draw and label a cylinder that has a larger radius, but less volume than the one shown at the right. **See margin.**

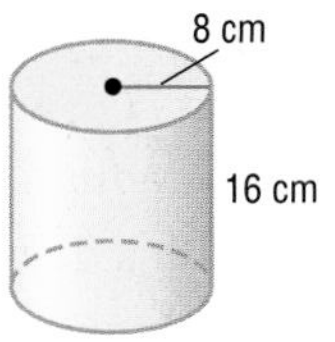

2. **Explain** how the formula for the volume of a cylinder is similar to the formula for the volume of a rectangular prism. **Sample answer: In both, the volume equals the area of the base times the height.**

GUIDED PRACTICE

Find the volume of each cylinder. Round to the nearest tenth.

3. **141.4 in^3**

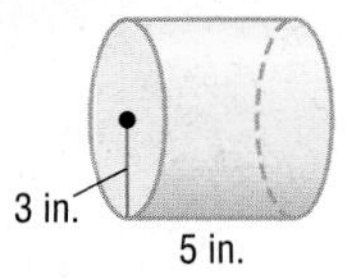

4. **56.5 cm^3**

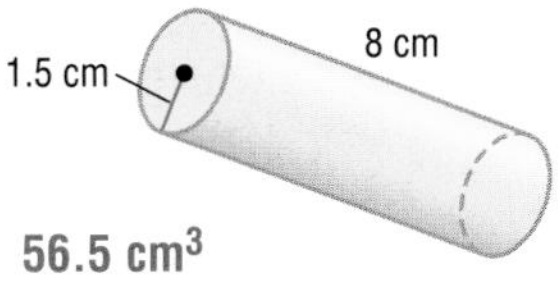

5. **617.7 ft^3**

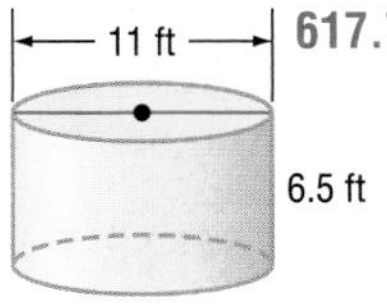

6. radius = 3 in. height = $5\frac{1}{2}$ in. **155.5 in^3**

7. diameter = 12.4 m height = 2 m **241.5 m^3**

8. **BAKING** Which will hold more cake batter, the rectangular pan, or the two round pans? Explain. **Rectangular pan, because its volume is 234 in^3; the total volume of the two round pans is approximately 201.1 in^3.**

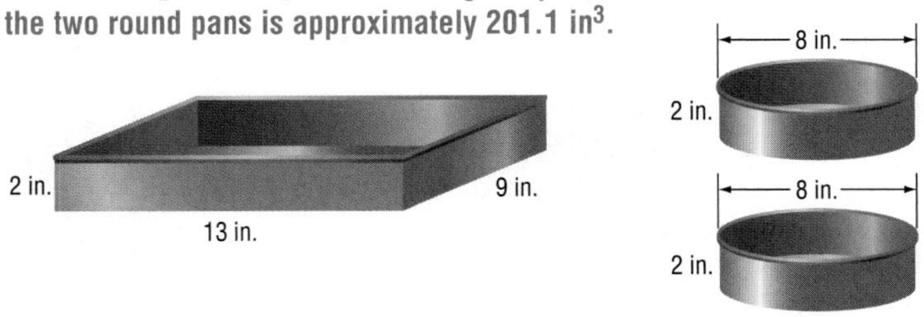

2 Teach

In-Class Examples

INTERACTIVE CHALKBOARD PowerPoint® Presentations

1 Find the volume of the cylinder. Round to the nearest tenth. **855.3 cm^3**

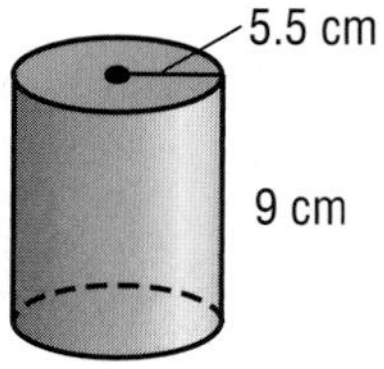

2 **MULTIPLE-CHOICE TEST ITEM** Find the volume of a cylinder-shaped coffee can that has a diameter of 3 inches and a height of 6 inches. **B**

A 28.3 in^3

B 42.4 in^3

C 56.5 in^3

D 169.6 in^3

3 Practice/Apply

About the Exercises...

Guided Practice Key

Exercises	Examples
3–7	1
8	2

Odd/Even Assignments
Exercises 9–22 are structured so that students practice the same concepts whether they are assigned odd or even problems.

DAILY INTERVENTION — Differentiated Instruction

Logical Have students create problems similar to the following: A machinist makes a part by drilling a hole through a block of copper. The copper is in the shape of a rectangular prism with a 6 centimeter by 6 centimeter base and a height of 11 centimeters. If the hole has a radius of 2 centimeters and a height of 11 centimeters, what is the volume of the resulting solid? **257.8 cm^3** After students create their problems, have them trade and solve.

Answer

1. **Sample answer:**

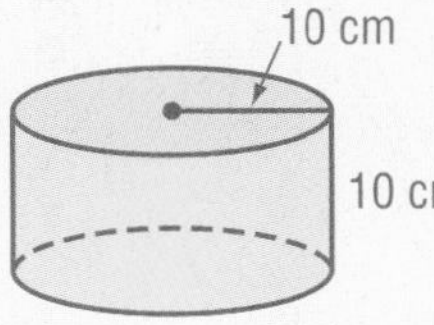

FOLDABLES™ Follow-Up Encourage students to include information in words and symbols on how to find the volume of cylinders in the appropriate section of their tables.

Answers

32. Sample answer:

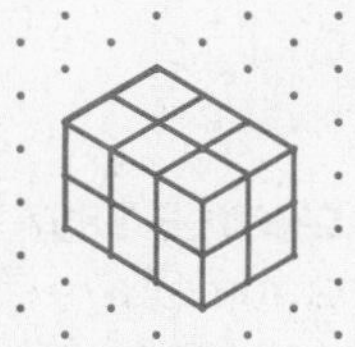

33. Sample answer:

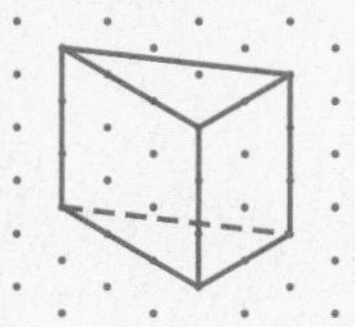

★ indicates increased difficulty

Practice and Applications

HOMEWORK HELP	
For Exercises	See Examples
9–18	1
19–22, 26–27	2

Extra Practice
See pages 594, 607.

Find the volume of each cylinder. Round to the nearest tenth.

9.
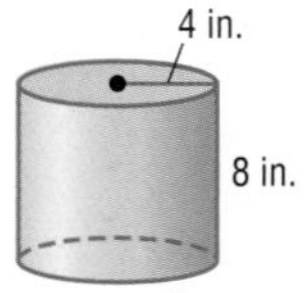

402.1 in^3

10.
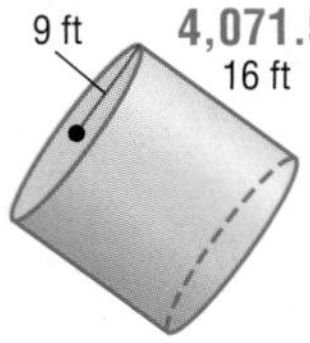

4,071.5 ft^3

11.
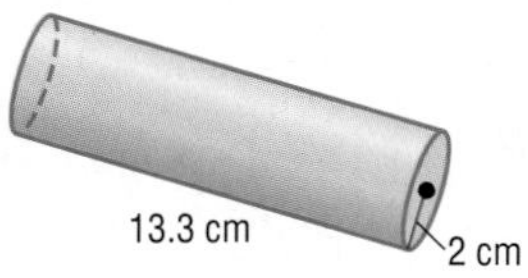

167.1 cm^3

12.
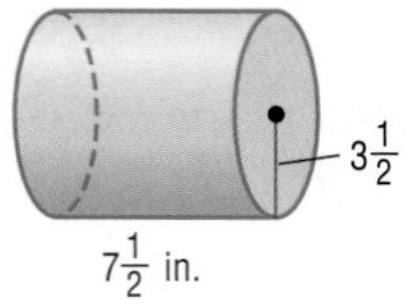

288.6 in^3

13.
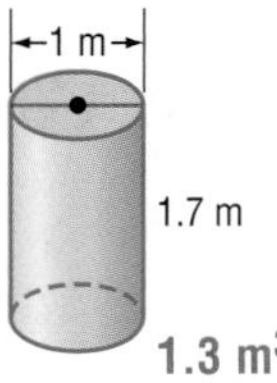

1.3 m^3

14.
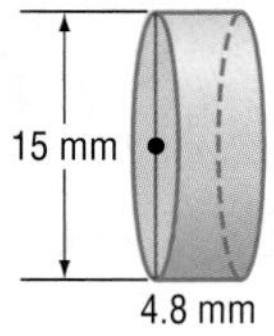

848.2 mm^3

15. diameter = 21 mm
height = 8 mm 2,770.9 mm^3

16. diameter = 4.5 m
height = 6.5 m 103.4 m^3

17. radius = 6 ft
height = $5\frac{1}{3}$ ft 603.2 ft^3

★ 18. radius = 12 ft
height = 11 yd 14,928.8 ft^3 or 552.9 yd^3

19. a soup can with a radius of 4 centimeters and a height of 12 centimeters 603.2 cm^3

20. a hockey puck with a diameter of 3 inches and a height of 1 inch 7.1 in^3

21. a can of potato chips with a radius of $1\frac{1}{2}$ inches and a height of 8 inches 56.5 in^3

22. **SPACE SHUTTLE** The Hubble Space Telescope is cylinder-shaped, approximately the size of a school bus, as shown at the right. What is the volume to the nearest tenth? 163.4 m^3

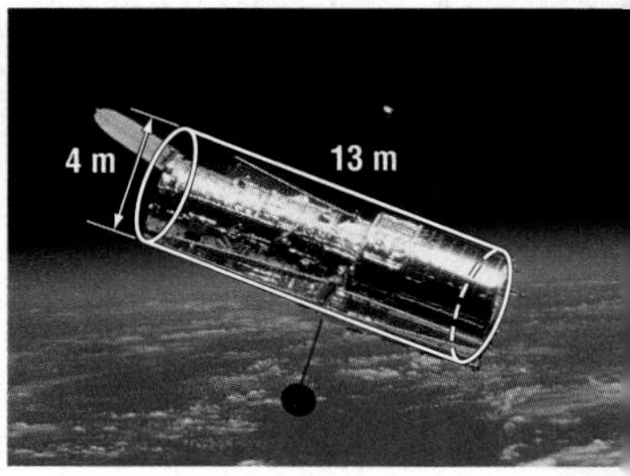

23. **WRITE A PROBLEM** Write a problem in which you find the volume of a cylinder. See students' work.

★ 24. **NUMBER SENSE** What is the ratio of the volume of a cylinder to the volume of a cylinder having twice the height but the same radius? 1 to 2

★ 25. **NUMBER SENSE** What is the ratio of the volume of a cylinder to the volume of a cylinder having the same height but twice the radius? 1 to 4

★ 26. **CONTAINERS** The two cans at the right have the same volume. Find *h*. 9 cm

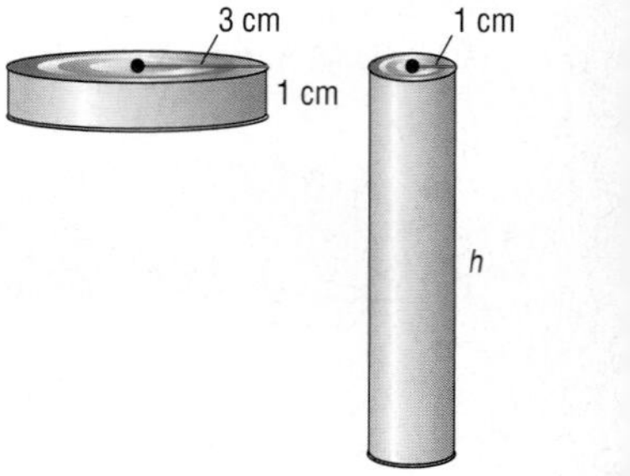

★ 27. **MEASUREMENT** Firewood is usually sold by a unit of measure called a *cord*. A cord is a stack of wood that is 8 feet long, 4 feet wide, and 4 feet high. Suppose a tree has a diameter of 2 feet. Find the height of the tree trunk that would produce about 1 cord of firewood. about 41 ft

526 Chapter 12 Geometry: Measuring Three-Dimensional Figures

28. **CRITICAL THINKING** Two equal-sized sheets of paper are rolled along the length and along the width, as shown at the right. Which cylinder do you think has the greater volume? Explain. **Sample answer: The shorter cylinder, because the radius is larger and that is the squared value in the formula.**

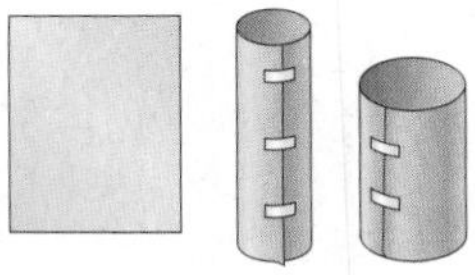

Standardized Test Practice and Mixed Review

29. **MULTIPLE CHOICE** Which statement is true about the volumes of cylinders 1 and 2 shown below? **B**

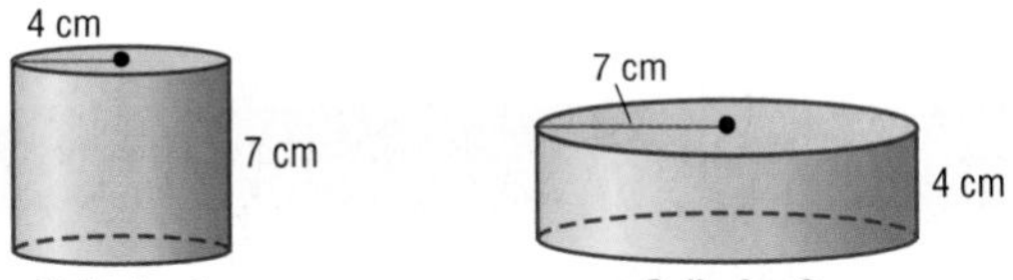

Ⓐ The volume of cylinder 1 is greater.
Ⓑ The volume of cylinder 2 is greater.
Ⓒ The volumes are equal.
Ⓓ cannot tell from the diagrams

30. **MULTIPLE CHOICE** A cylinder-shaped popcorn tin has a height of 1.5 feet and a diameter of 10 inches. Find the volume to the nearest cubic inch. Use 3.14 for π. **H**

Ⓕ 118 in^3 Ⓖ 565 in^3 Ⓗ 1,413 in^3 Ⓘ 5,652 in^3

31. Find the volume of a rectangular prism with a length of 6 meters, a width of 4.9 meters, and a height of 5.2 meters. (Lesson 12-2) **152.88 m^3**

Draw each solid using the top, side, and front views shown. Use isometric dot paper. (Lesson 12-1) **32–33. See margin.**

32. top side front

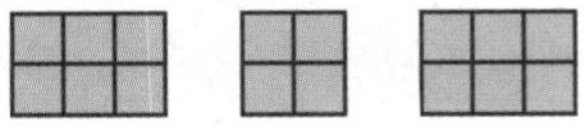

33. top side front

Find the area of each trapezoid. Round to the nearest tenth if necessary. (Lesson 11-5)

34. **56 ft^2**

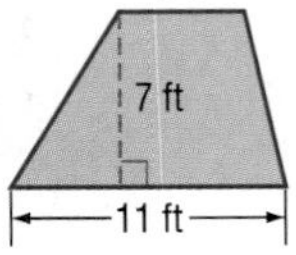

35. **572 m^2**

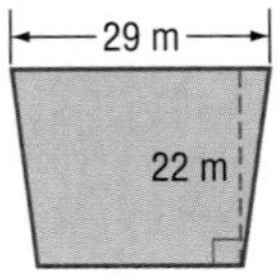

36. **24.5 cm^2**

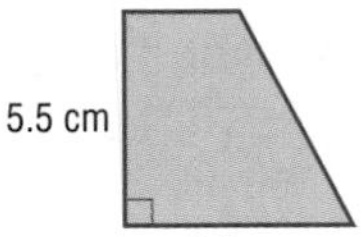

GETTING READY FOR THE NEXT LESSON

PREREQUISITE SKILL **Simplify.** (Lesson 1-3)

37. $3 \cdot 5 \cdot 8 + 2 \cdot 9 \cdot 3$ **174**

38. $7.6 \cdot 11 \cdot 2 + 2 \cdot 7.6 \cdot 3$ **212.8**

39. $2 \cdot 7 \cdot 1.5 + 2 \cdot 4 \cdot 1.5 + 1.5 \cdot 7$ **43.5**

40. $2 \cdot 2\frac{1}{4} \cdot 6 + 2 \cdot 9 \cdot 6 + 2 \cdot 2\frac{1}{4} \cdot 9$ **175.5**

msmath2.net/self_check_quiz

4 Assess

Open-Ended Assessment

Speaking Ask students to compare and contrast finding the volume of a cylinder and finding the volume of a rectangular prism.

Study Guide and Intervention, p. 679

NAME ______ DATE ______ PERIOD ____

12-3 Study Guide and Intervention

Volume of Cylinders

A **cylinder** is a solid figure that has two congruent, parallel circles as its bases. The volume V of a cylinder with radius r is the area of the base B times the height h.

$V = Bh$ or $V = \pi r^2 h$, where $B = \pi r^2$

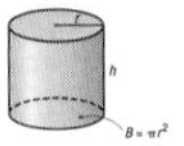

EXAMPLE 1 **Find the volume of the cylinder. Round to the nearest tenth.**

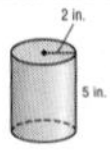

$V = \pi r^2 h$	Volume of a cylinder
$V = \pi(2)^2(5)$	Replace r with 2 and h with 5.
$V \approx 62.8$	Simplify.

The volume is approximately 62.8 cubic inches. Check by using estimation.

EXERCISES

Find the volume of each cylinder. Round to the nearest tenth.

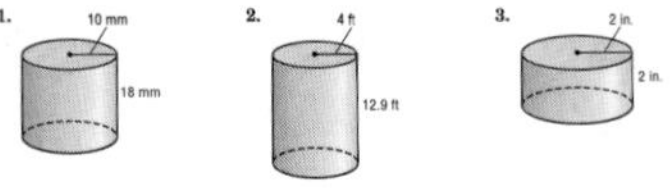

1. 5,654.9 mm^3
2. 648.4 ft^3
3. 25.1 in^3
4. radius = 9.5 yd, height = 2.2 yd 623.8 yd^3
5. diameter = 6 cm, height = 11 cm 311.0 cm^3
6. diameter = $3\frac{2}{5}$ m, height = $1\frac{1}{4}$ m 11.3 m^3

Practice: Skills, p. 680

NAME ______ DATE ______ PERIOD ____

12-3 Practice: Skills

Volume of Cylinders

Find the volume of each cylinder. Round to the nearest tenth.

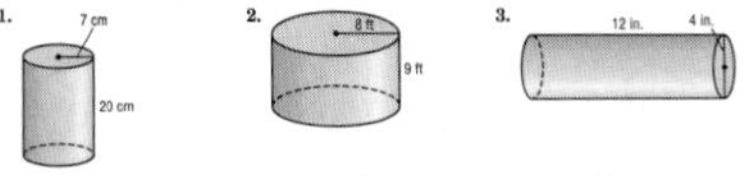

1. 3,078.8 cm^3
2. 1,809.6 ft^3
3. 150.8 in^3

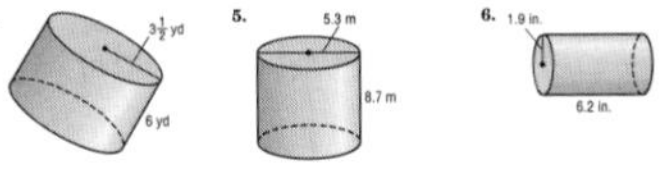

4. 230.9 yd^3
5. 191.9 m^3
6. 70.3 in^3
7. radius = 8.8 cm, height = 4.7 cm 1,143.4 cm^3
8. radius = 4 ft, height = $2\frac{1}{2}$ ft 125.7 ft^3
9. diameter = 10 mm, height = 4 mm 314.2 mm^3
10. diameter = 7.1 in., height = 1 in. 39.6 in^3

Practice: Word Problems, p. 681

NAME ______ DATE ______ PERIOD ____

12-3 Practice: Word Problems

Volume of Cylinders

1. **WATER STORAGE** A cylindrical water tank has a diameter of 5.3 meters and a height of 9 meters. What is the maximum volume that the water tank can hold? Round to the nearest tenth. 198.6 m^3

2. **PACKAGING** A can of corn has a diameter of 6.6 centimeters and a height of 9.9 centimeters. How much corn can the can hold? Round to the nearest tenth. 338.7 cm^3

3. **CONTAINERS** Tionna wants to determine the maximum capacity of a cylindrical bucket that has a radius of 6 inches and a height of 12 inches. What is the capacity of Tionna's bucket? Round to the nearest tenth. 1,357.2 in^3

4. **DESIGN** Rodolfo is designing a new, cylindrical drinking glass. If the glass has a diameter of 8 centimeters and a height of 12.8 centimeters, what is its volume? Round to the nearest tenth. 643.4 cm^3

5. **PAINT** A can of paint is 15 centimeters high and has a diameter of 13.6 cm. What is the volume of the can? Round to the nearest tenth. 2,179 cm^3

6. **SPICES** A spice manufacturer uses a cylindrical dispenser like the one shown. Find the volume of the dispenser to the nearest tenth. 27.2 in^3

Mid-Chapter Practice Test

CHAPTER 12

Assessment Option

Mid-Chapter Test (Lessons 12-1 through 12-3) is available on p. 715 of the *Chapter 12 Resource Masters.*

ExamView® Pro Test Bank CD-ROM

Use the networkable ExamView® Pro Test Bank CD-ROM to:

- Create **Multiple versions** of tests.
- Create **Modified** tests for **Inclusion** students with one mouse click.
- **Edit** existing questions and **Add** your own questions.
- Build tests aligned with state standards using built-in **State Curriculum Correlations.**
- Change **English** tests to **Spanish** with one mouse click and vice versa.

FOLDABLES™ Follow-Up Before students complete the Mid-Chapter Practice Test, encourage them to review the notes, definitions, concepts, and main ideas they recorded in their Foldable tables while working through the first three lessons of this chapter.

Answer

6. Sample answer:

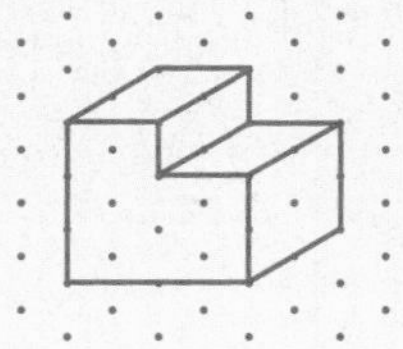

CHAPTER 12 Mid-Chapter Practice Test

Vocabulary and Concepts

1. **Define** *volume.* (Lesson 12-2) **the measure of space occupied by a solid**
2. **State**, in words, the formula for the volume of a cylinder. (Lesson 12-3) **the product of the base area times the height**

Skills and Applications

Draw a top, a side, and a front view of each solid. (Lesson 12-1) **3–5. See Ch. 12 Answer Appendix.**

3. 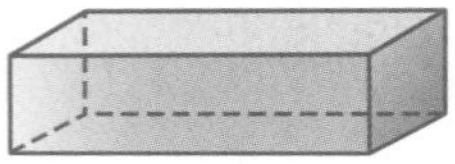

4. 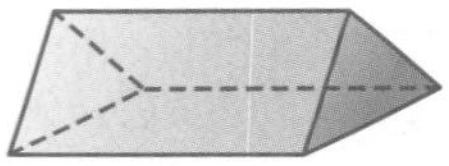

5. 

6. Draw the solid using the top, side, and front views shown at the right. Use isometric dot paper. (Lesson 12-1) **See margin.**

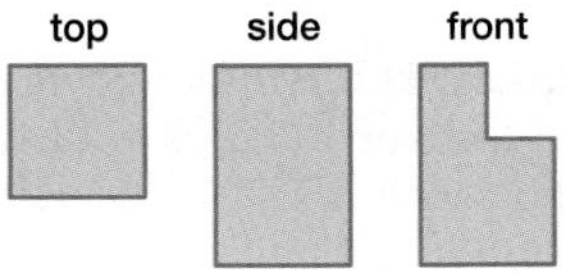

7. **GIFTS** A jewelry box measures 7 centimeters by 12 centimeters by 14 centimeters. What is its volume? (Lesson 12-2) **1,176 cm^3**

8. **TRUCKS** How tall is the trailer of a truck if it is 9 meters long, 7.2 meters wide and has a volume of 226.8 cubic meters? (Lesson 12-2) **3.5 m**

9. **GARDENING** Jocelyn is buying potting soil to fill the window box. If one bag of potting soil contains 576 cubic inches, how many bags should she buy? (Lesson 12-2) **3 bags**

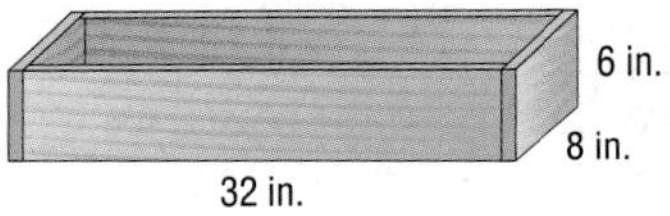

Find the volume of each cylinder. Round to the nearest tenth. (Lesson 12-3)

10. diameter = 25 ft
height = 24 ft
11,781.0 ft^3

11. radius = 2.2 cm
height = 5 cm
76.0 cm^3

12. radius = $1\frac{1}{2}$ in.
height = $4\frac{1}{2}$ in. **31.8 in^3**

Standardized Test Practice

13. **GRID IN** A brick is 4 inches wide, 8 inches long, and 3 inches tall. What is the volume in cubic inches of a stack of 25 bricks? (Lesson 12-2) **2400**

14. **SHORT RESPONSE** A circular pond has a radius of 12 meters and it is 2.5 meters deep. What is its volume? Use 3.14 for π. (Lesson 12-3) **1,130.4 m^3**

The Game Zone

A Place To Practice Your Math Skills

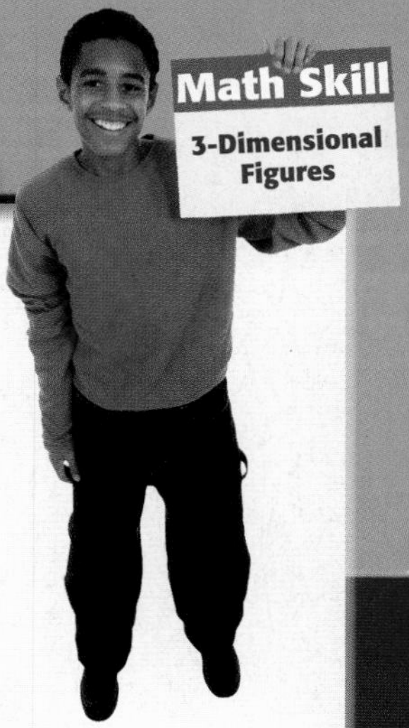

Shape-Tac-Toe

GET READY!

Players: two
Materials: 1 index card, 2 large cubes, 10 two-color counters

GET SET!

One player draws a game board on the index card like the one shown. The second player labels the six faces of each cube with the following words.

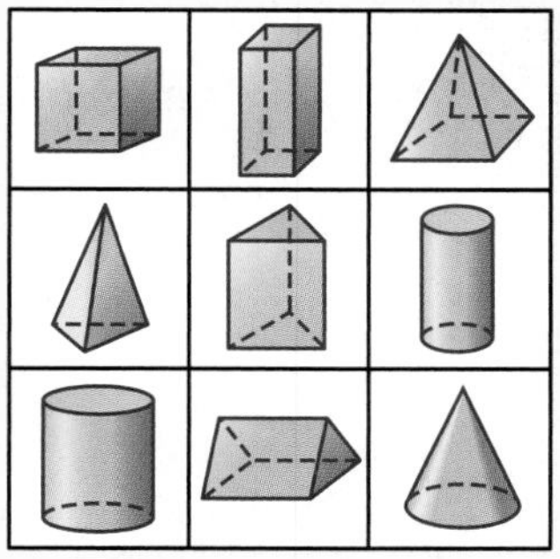

First Cube	**Second Cube**
prism	circular base
pyramid	triangle base
cylinder	square base
cone	2 parallel bases
flat surfaces	1 base
curved surface	congruent faces

GO!

- The first player rolls both cubes and places a counter on any one shape that matches the conditions on the cubes. If it is impossible to cover a shape, the player loses his or her turn.
- Players alternate turns.
- **Who Wins?** The first player to cover three shapes in a row is the winner.

- You may wish to use the game as an assessment tool for the concepts the students have been learning in Lessons 12-1 through 12-3.

For an additional game for this chapter, visit: msmath2.net/gamezone.

The Game Zone

Resources

For an easy-to-make manipulative for this game, visit: msmath2.net/gamezone.

Glencoe Mathematics Classroom Manipulative Kit
- counters

GETTING READY

Objective Identify three-dimensional figures.

MANAGEMENT TIPS

Recommended Time
30 minutes

Tips
- Consider saving time by pre-drawing the game board, and copying it for each pair of students.
- As an alternative to having students label large cubes, have them use two different-colored number cubes and have each number represent one of the items listed. For example, a roll of white-1 and red-3 could represent a roll of prism and square base.
- Inform students that once a counter has been placed on a shape, no other counters may be placed on that shape.
- As an option, you may wish to allow students to win by covering the shapes in the four corners of their cards.
- If some students finish early, have them continue play until they cover the entire board or until time is called.

The Game Zone: Three-Dimensional Figures 529

12-4a Hands-On Lab Lesson Notes

Lesson Resources

Teaching Mathematics with Manipulatives
- rectangular dot paper, p. 13
- student recording sheet 12-4a

Glencoe Mathematics Classroom Manipulative Kit
- scissors

GETTING STARTED

Teaching Tip Provide each student with a small cardboard box such as a box for paper clips, staples, or individual portions of cereal. Have students place their boxes on large pieces of paper and trace around the edges. They should roll the boxes from face to face and mark each box face with an X after it is traced. Have them trace the bottom face first and the top face last. Tell students that the resulting drawings are nets of their boxes. Have students make new nets of the same boxes by changing the order in which they trace their boxes' faces.

TEACH

- Point out to students that when they have drawn the base of the prism in Step 1, they have defined the length and the width of the rectangular prism.
- After students complete Step 2, ask them what the height of the rectangular prism will be. **3 units**
- For Step 3, ask several students to show the class alternate positions for the top of the prism.

12-4a Hands-On Lab

A Preview of Lesson 12-

What You'll Learn
Use nets to find the surface area of rectangular prisms.

Materials
- rectangular dot paper
- scissors

TEACHING TIP
You could have students cut paper prisms into nets. This will show the different nets that will make a rectangular prism.

Nets and Surface Area

Suppose you cut a cardboard box along its edges, open it up, and lay it flat.

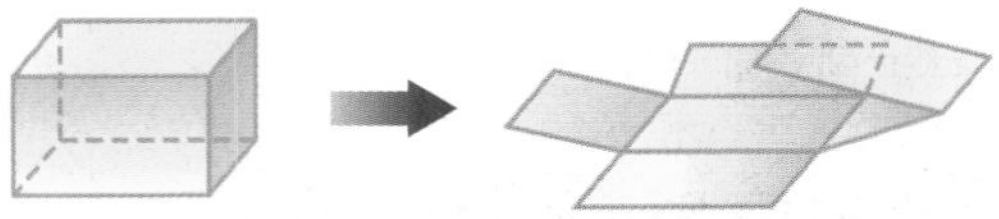

The result is a two-dimensional figure called a **net**. Nets can help you see the regions or faces that make up the surface of the figure.

ACTIVITY *Work with a partner.*

Make a net of the rectangular prism shown at the right.

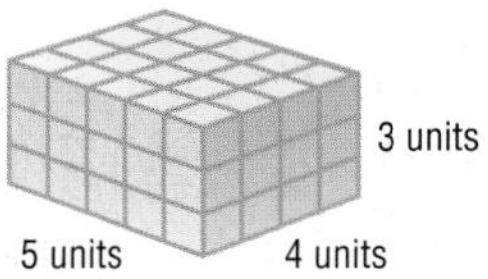

STEP 1 Begin by drawing the base of the prism. On the dot paper, draw a rectangle that is 5 units long and 4 units wide.

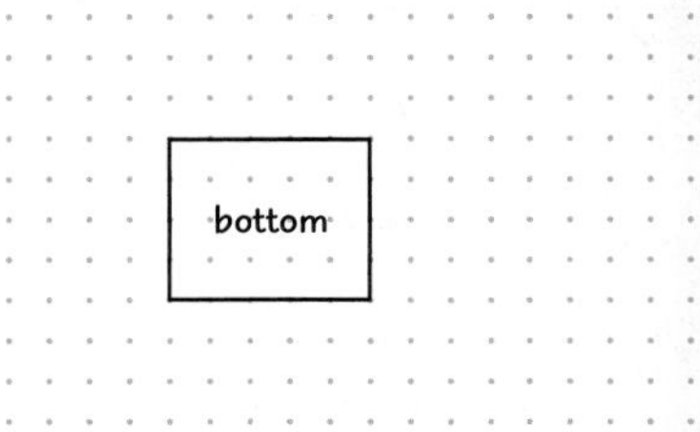

STEP 2 Visualize unfolding the prism along its edges. Draw the rectangles that represent the front, back, and sides of the prism.

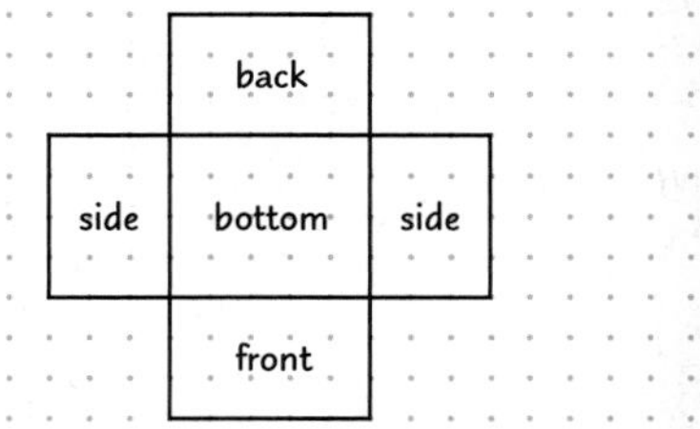

STEP 3 Finally, draw the top of the prism.
This is only one of several possible nets that you could draw.

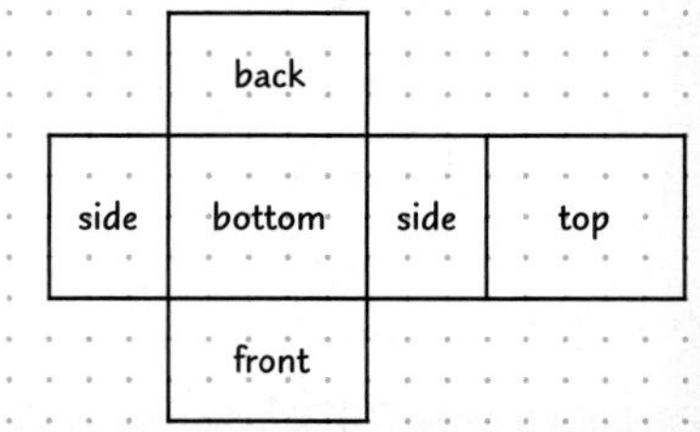

Your Turn

Draw a net for each figure. Find the area of the net. Then cut out the net, fold it, and tape it together to form a three-dimensional figure.

a.

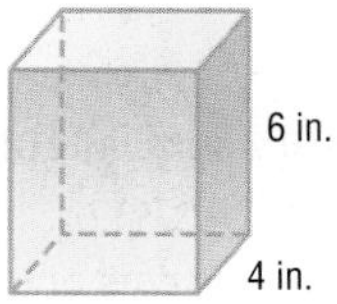

148 in^2

b.

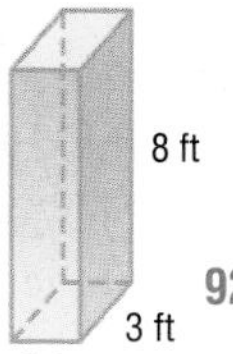

92 ft^2

c.

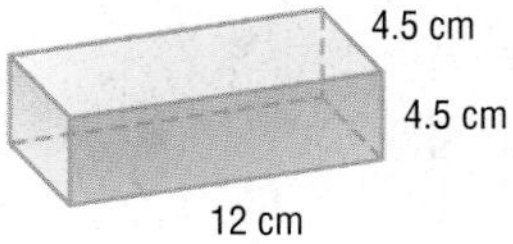

256.5 cm^2

Writing Math

Work in groups of three.

1. The net shown on page 530 is made of rectangles. How many rectangles are in the net? **6**
2. Explain how you can find the total area of the rectangles.
3. The *surface area* of a prism is the total area of its net. Write an equation that shows how to find the surface area of the prism at the right using the length ℓ, width w, and height h.

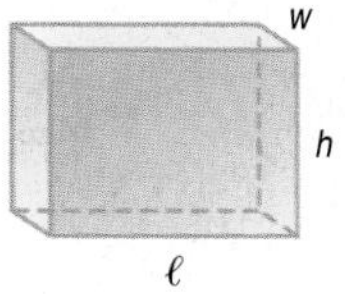

4. Find the surface areas of cubes whose edges are 1 unit, 2 units, and 3 units and graph the ordered pairs (side length, surface area) on a coordinate plane. Describe the graph. **See margin.**
5. **Describe** what happens to the surface area of a cube as its dimensions are doubled? tripled? **increase by 2^2; increase by 3^2**
6. **Describe** how you would find the surface area of a square-based pyramid. **Sample answer: $A = s^2 + 4 \cdot \frac{1}{2}sh$**

Draw a net for each figure. **7–8. See margin.**

7.

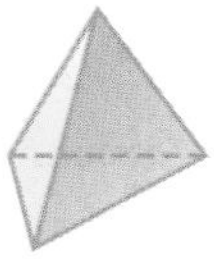

tetrahedron

8.

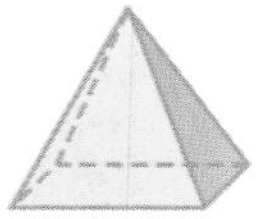

square-based pyramid

9. **Explain** how the formula for the surface area of a tetrahedron differs from the formula for the surface area of a square-based pyramid.

Find the sum of the eas of the rectangles.

surface area = ℓw + w + wh + wh + ℓh +

Sample answer: etrahedron has riangles as sides and area of the base is h. A square-based ramid has 4 triangles sides and the area of base is ℓw.

TEACHING TIP

etrahedron is a angular pyramid; all e faces are triangles. triangular prism is not etrahedron.

ASSESS

Observe as students solve the **Your Turn** exercises. They should:

- correctly draw each net following the steps outlined on page 530,
- correctly find the area of each net, and
- neatly cut out and fold each net into a three-dimensional figure.

Discuss the **Writing Math** exercises. Check that students' answers to Exercises 1–3 are correct before they continue working the exercises.

For Exercise 4, remind students of their work in Lesson 12-2b regarding the relationship of the length of the sides of a cube to its volume. Point out that this relationship could be graphed in the same way they are graphing the relationship between the length of the sides of a cube and its surface area.

Use Exercises 5–9 to determine whether students have grasped the meaning behind surface area formulas.

Answers (Writing Math)

4.

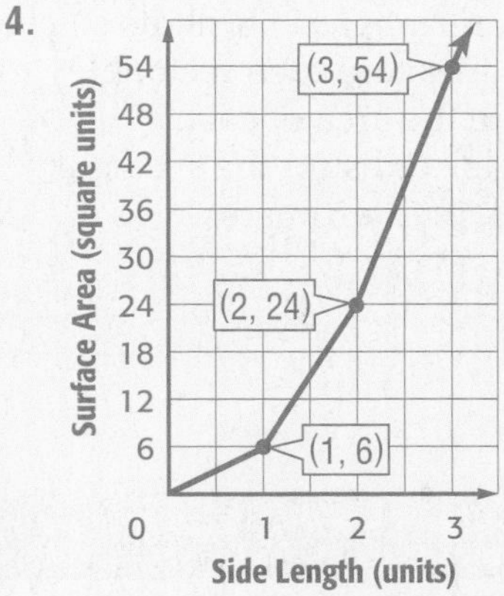

The graph is a curve, sharply increasing as side length increases.

7. Sample answer:

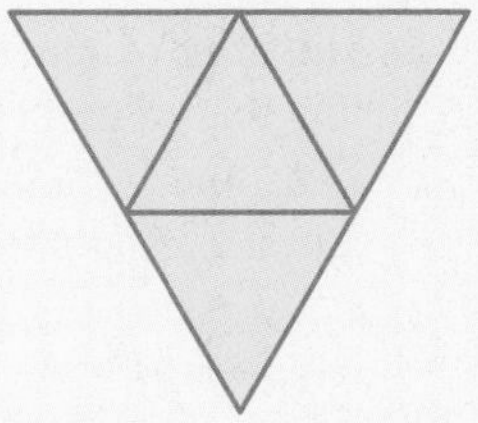

8. Sample answer:

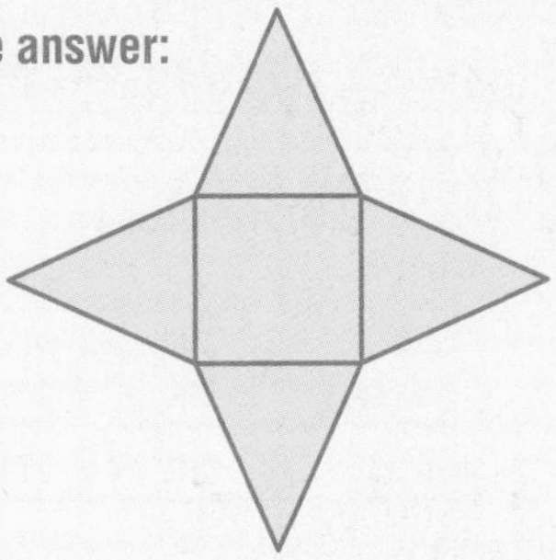

12-4 Lesson Notes

Lesson Resources

Chapter 12 Resource Masters

- Study Guide and Intervention, p. 684
- Practice: Skills, p. 685
- Practice: Word Problems, p. 686
- Reading to Learn Mathematics, p. 687
- Enrichment, p. 688

Key Teaching Transparencies, 47

Parent and Student Study Guide Workbook, 12-4

FAST FILE

1 Focus

5-Minute Check 12-4

INTERACTIVE CHALKBOARD
PowerPoint® Presentations

Use as a quiz or review of Lesson 12-3. (Also available as a transparency)

Bellringer

Hands-On Activity Have students bring a small box, such as a shoe box or an empty box of facial tissues, to class. Ask them to estimate the surface area of the box. Then, have them cut the box into six rectangles. Students should measure each rectangle, calculate the area of each rectangle, and sum the six areas to check their estimates.

Reading to Learn Mathematics, p. 687 ELL

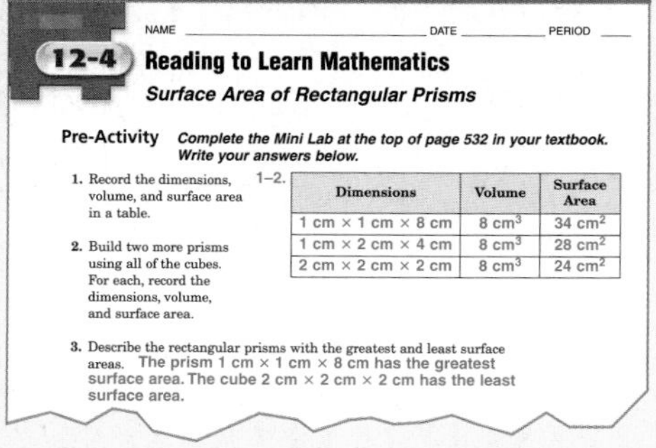
NAME ______ DATE ______ PERIOD ____

12-4 Reading to Learn Mathematics

Surface Area of Rectangular Prisms

Pre-Activity ***Complete the Mini Lab at the top of page 532 in your textbook. Write your answers below.***

1. Record the dimensions, volume, and surface area in a table.
2. Build two more prisms using all of the cubes. For each, record the dimensions, volume, and surface area.

1–2.

Dimensions	Volume	Surface Area
1 cm × 1 cm × 8 cm	8 cm^3	34 cm^2
1 cm × 2 cm × 4 cm	8 cm^3	28 cm^2
2 cm × 2 cm × 2 cm	8 cm^3	24 cm^2

3. Describe the rectangular prisms with the greatest and least surface areas. The prism 1 cm × 1 cm × 8 cm has the greatest surface area. The cube 2 cm × 2 cm × 2 cm has the least surface area.

12-4 Surface Area of Rectangular Prisms

What You'll Learn
Find the surface areas of rectangular prisms.

NEW Vocabulary
surface area

HANDS-ON Mini Lab

Materials
- 8 centimeter cubes

Work with a partner.

- Use the cubes to build a rectangular prism with a length of 8 centimeters.
- Count the number of squares on each face. The sum is the *surface area*. **1–3. See margin.**

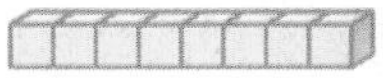

1. Record the dimensions, volume, and surface area in a table.
2. Build two more prisms using all of the cubes. For each, record the dimensions, volume, and surface area.
3. Describe the rectangular prisms with the greatest and least surface areas.

The sum of the areas of all of the surfaces, or faces, of a three-dimensional figure is the **surface area**.

EXAMPLE Use a Net to Find Surface Area

1 Find the surface area of the rectangular prism.

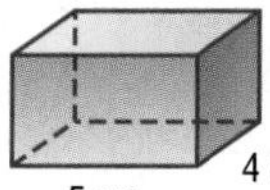

You can use a net of the rectangular prism to find its surface area. There are three pairs of congruent faces.

- top and bottom
- front and back
- two sides

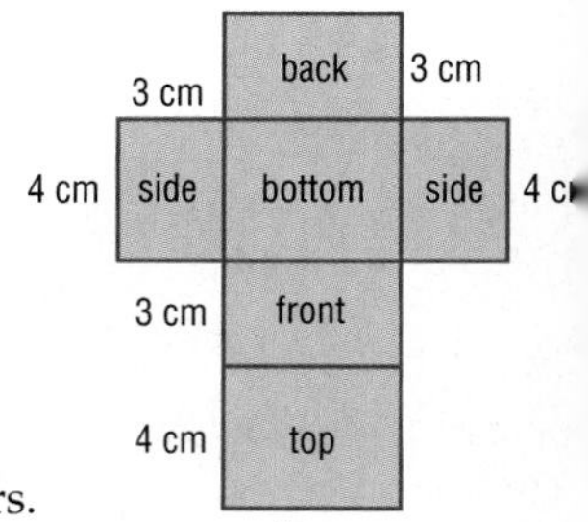

Faces	**Area**
top and bottom	$2(5 \cdot 4) = 40$
front and back	$2(5 \cdot 3) = 30$
two sides	$2(3 \cdot 4) = 24$
Sum of the areas	$40 + 30 + 24 = 94$

The surface area is 94 square centimeters.

TEACHING TIP
A square has the least area compared to other quadrilaterals with the same perimeter; similarly, a cube has the least surface area compared to other rectangular prisms with the same volume.

Key Concept Surface Area of a Rectangular Prism

Words The surface area S of a rectangular prism with length ℓ, width w, and height h is the sum of the areas of the faces.

Model

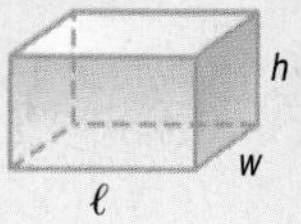

Symbols $S = 2\ell w + 2\ell h + 2wh$

Answers (Hands-On Mini Lab)

1–2.

Dimensions	Volume	Surface Area
1 cm × 1 cm × 8 cm	8 cm^3	34 cm^2
1 cm × 2 cm × 4 cm	8 cm^3	28 cm^2
2 cm × 2 cm × 2 cm	8 cm^3	24 cm^2

3. The prism 1 cm × 1 cm × 8 cm has the greatest surface area. The cube 2 cm × 2 cm × 2 cm has the least surface area.

EXAMPLE Use a Formula to Find Surface Area

2 Find the surface area of the rectangular prism.

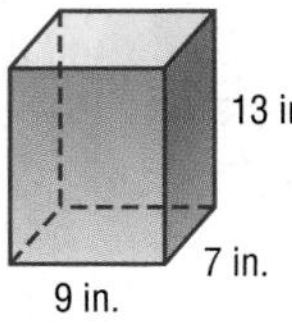

Replace ℓ with 9, w with 7, and h with 13.

$$\begin{aligned}\text{surface area} &= 2\ell w + 2\ell h + 2wh \\ &= 2 \cdot 9 \cdot 7 + 2 \cdot 9 \cdot 13 + 2 \cdot 7 \cdot 13 \\ &= 126 + 234 + 182 \quad \text{Multiply first. Then add.} \\ &= 542\end{aligned}$$

The surface area of the prism is 542 square inches.

EAL-LIFE MATH

FTS Gifts have been
apped since the invention
paper in 105 A.D. in
ina.

rce: www.hallmark.com

EXAMPLE Use Surface Area to Solve a Problem

3 GIFTS Mario is wrapping a package 8 inches long, 2 inches wide, and 11 inches high. He bought a roll of wrapping paper that is 1 foot wide and 2 feet long. Did he buy enough to wrap the package?

Step 1 Find the surface area of the package.

Replace ℓ with 8, w with 2, and h with 11.

$$\begin{aligned}\text{surface area} &= 2 \cdot 8 \cdot 2 + 2 \cdot 8 \cdot 11 + 2 \cdot 2 \cdot 11 \\ &= 252 \text{ square inches}\end{aligned}$$

Step 2 Find the area of the wrapping paper.

area = 12 in. · 24 in. or 288 in^2

Since $288 > 252$, Mario bought enough wrapping paper.

Skill and Concept Check

Writing Math
Exercises 1 & 3

1. **Explain** how to find the surface area of a rectangular prism. **Find the sum of the areas of the faces.**
2. **OPEN ENDED** Draw a rectangular prism and its net with the dimensions labeled on both figures. Then find the surface area. **See margin.**
3. **Explain** why surface area is measured in square units even though the figure is three-dimensional. **Surface area measures the area of the faces, and area is measured in square units.**

GUIDED PRACTICE

Find the surface area of each rectangular prism. Round to the nearest tenth if necessary.

4. 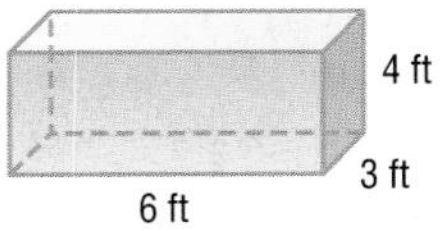

108 ft^2

5. 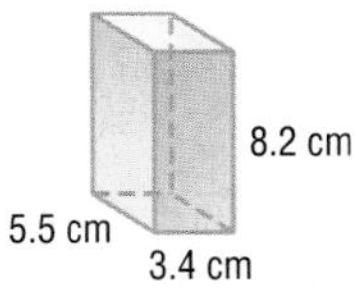

183.4 cm^2

6. Find the surface area of a rectangular prism that has a length of 7 inches, a width of 11 inches, and a height of 9 inches. **478 in^2**

msmath2.net/extra_examples

2 Teach

In-Class Examples

INTERACTIVE CHALKBOARD
PowerPoint® Presentations

1 Find the surface area of the rectangular prism. **72 cm^2**

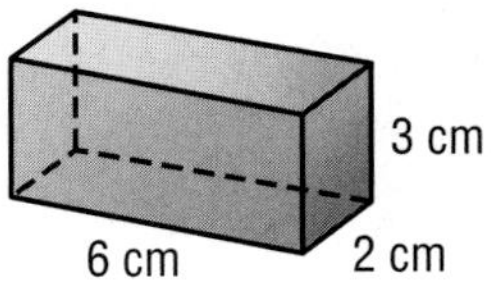

2 Find the surface area of the rectangular prism. **592 in^2**

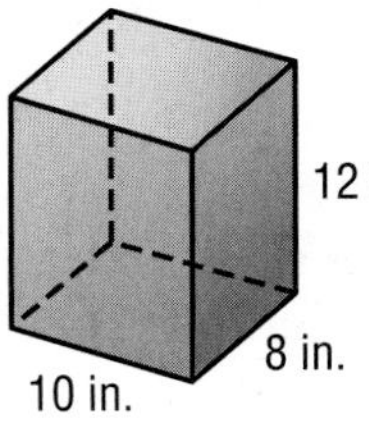

3 **BOXES** Drew is putting together a cardboard box that is 9 inches long, 6 inches wide, and 8 inches high. He bought a roll of wrapping paper that is 1 foot wide and 3 feet long. Did he buy enough to wrap the box? **yes; 348 $\text{in}^2 < 432$ in^2**

Answer

2. Sample answer:

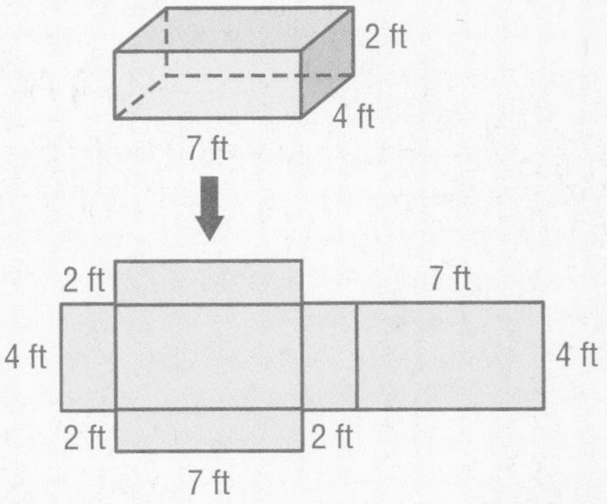

$S = 2 \cdot 7 \cdot 4 + 2 \cdot 7 \cdot 2 + 2 \cdot 4 \cdot 2 = 100 \text{ ft}^2$

DAILY INTERVENTION — Flexible Grouping

Surface Area of Rectangular Prisms Arrange students in small groups, placing students who prefer using nets or visual models with those who prefer to calculate surface area using the formula ($S = 2\ell w + 2\ell h + 2wh$). Give each group the dimensions for several rectangular prisms, and ask them to use two different methods to calculate the surface area of each prism.

3 Practice/Apply

FOLDABLES™ Follow-Up Encourage students to include information in words and symbols on how to find the surface area of rectangular prisms in the appropriate section of their Foldable tables.

About the Exercises...

Guided Practice Key

Exercises	Examples
4–5	1, 2
6	3

Odd/Even Assignments
Exercises 7–14 are structured so that students practice the same concepts whether they are assigned odd or even problems.

★ indicates increased difficulty

Practice and Applications

HOMEWORK HELP

For Exercises	See Examples
7–14	1, 2
15–16, 18–22	3

Extra Practice See pages 594, 607.

Find the surface area of each rectangular prism. Round to the nearest tenth if necessary.

7.
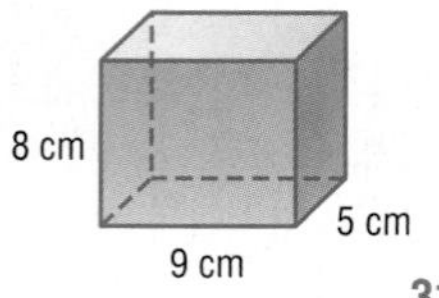

314 cm^2

8.
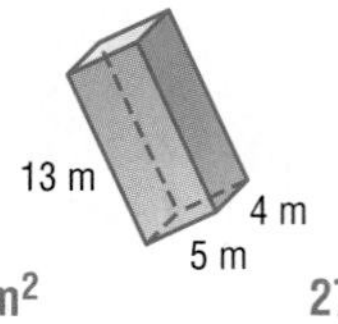

274 m^2

9.
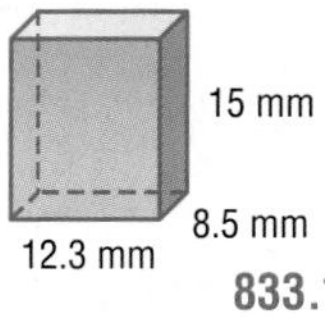

833.1 mm^2

10.
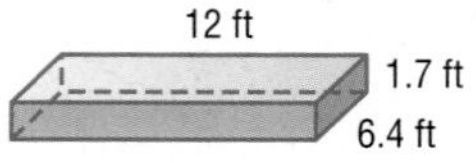

216.2 ft^2

11.
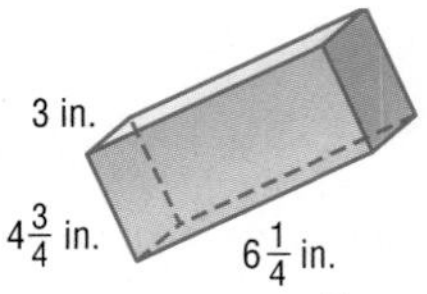

125.4 in^2 or $125\frac{3}{8}$ in^2

12.
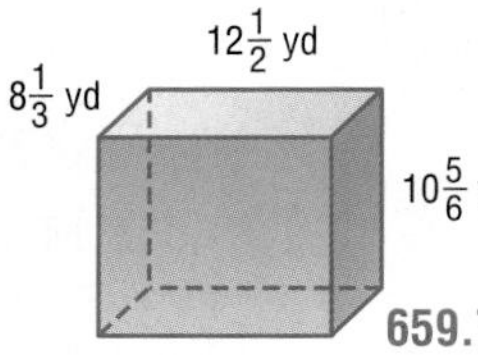

659.7 yd^2 or $659\frac{13}{18}$ yd^2

13. A cube has a surface area of 42 square meters. What is the area of one face? 7 m^2

★ 14. Find the surface area of a rectangular prism that has a length of 5 feet, a width of 3 feet, and a height of 9 inches. 6,048 in^2 or 42 ft^2

BOXES For Exercises 15 and 16, use the information below.
The largest corrugated cardboard box ever constructed was in Helmond, Netherlands. It measured 22.9 feet long, 8.5 feet high, and 7.87 feet wide.

15. Find the volume of the cardboard box to the nearest tenth. 1,531.9 ft^3

16. Find the surface area of the cardboard box to the nearest tenth. 883.5 ft^2

17. **ALGEBRA** Write a formula for the surface area of a cube in which each side measures x units. surface area $= 6x^2$

PHYSICAL SCIENCE For Exercises 18–20, use the information below.
Granulated sugar dissolves faster in water than a sugar cube.

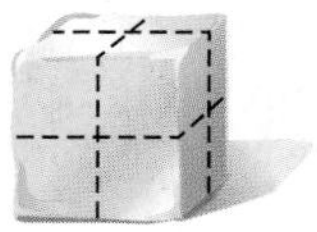

18. Suppose the length of each edge of a sugar cube is 1 centimeter. Find the surface area of the cube. 6 cm^2

19. Imagine cutting the cube once in half horizontally and twice vertically. Find the total surface area of the eight cubes. 12 cm^2

20. Make a conjecture as to why granulated sugar dissolves faster than a sugar cube. Sample answer: The granulated sugar has a greater surface area than the cubes.

CEREAL For the Exercises 21 and 22, use the following information.
Suppose you are designing a trial size cereal box that holds 100 cubic centimeters of cereal.

★ 21. **MULTI STEP** Find the whole number dimensions of the box that would use the least amount of cardboard. 5 cm long, 4 cm wide, 5 cm high

★ 22. **MULTI STEP** If cardboard costs \$0.05 per 100 square centimeters, how much would it cost to make 100 boxes? \$6.50

534 Chapter 12 Geometry: Measuring Three-Dimensional Figures

23. **WRITE A PROBLEM** Write about a real-life situation that involves finding the surface area of a rectangular prism. Then solve the problem. **See students' work.**

24. **CRITICAL THINKING** A model is made by placing a cube with 12-centimeter sides on top of another cube with 15-centimeter sides. Find the surface area. (*Hint*: Do not include the area where the smaller cube covers the larger cube.) **1,926 cm²**

Standardized Test Practice and Mixed Review

25. **MULTIPLE CHOICE** Find the surface area of the rectangular prism to the nearest tenth. **D**

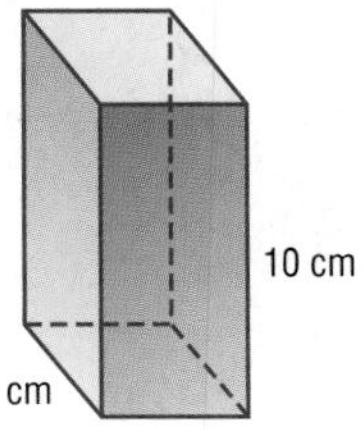

Ⓐ 101.5 cm²   Ⓑ 165.0 cm²
Ⓒ 198.6 cm²   Ⓓ 203.0 cm²

26. **MULTIPLE CHOICE** Find the amount of glass used for an aquarium that is 2.5 feet long, 1.6 feet wide, and 2 feet tall. (*Hint*: The top of the aquarium is open.) **G**

Ⓕ 19.4 ft²   Ⓖ 20.4 ft²   Ⓗ 21.2 ft²   Ⓘ 24.4 ft²

27. **GRID IN** The surface area of a cube is 294 square millimeters. What is the length of one side in millimeters? **7**

Find the volume of each cylinder. Round to the nearest tenth. (Lesson 12-3)

28.
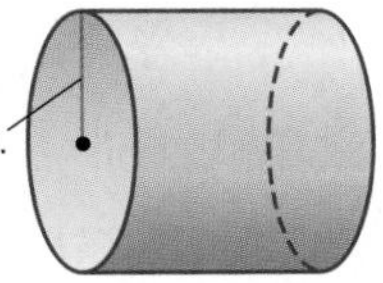

351.9 in³

29.
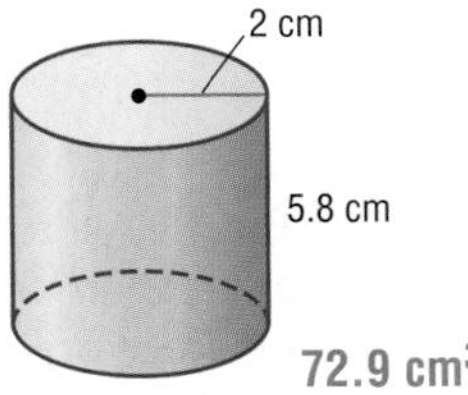

72.9 cm³

30.
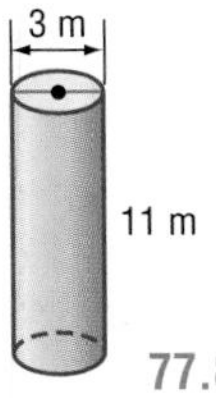

77.8 m³

31. A rectangular prism is 14 inches long, 4.5 inches wide, and 1 inch high. Find its volume. (Lesson 12-2) **63 in³**

32. **CARPENTRY** The deck on a house is $25\frac{3}{4}$ feet long and $12\frac{1}{2}$ feet wide. The longer side of the deck is against the house. How many feet of wood does Jack need to buy to build a railing around the deck? (Lesson 6-6) **$50\frac{3}{4}$ ft**

GETTING READY FOR THE NEXT LESSON

PREREQUISITE SKILL Find the area of each circle. Round to the nearest tenth. (Lesson 11-6)

33.
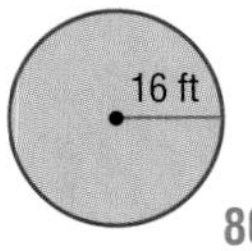

804.2 ft²

34.
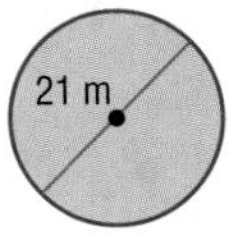

346.4 m²

35. diameter = 13.6 yd **145.3 yd²**

36. radius = 23 km **1,661.9 km²**

msmath2.net/self_check_quiz

Study Guide and Intervention, p. 684

NAME ____________ DATE ________ PERIOD ____

12-4 Study Guide and Intervention

Surface Area of Rectangular Prisms

The sum of the areas of all the surfaces, or faces, of a three-dimensional figure is the **surface area**. The surface area S of a rectangular prism with length ℓ, width w, and height h is found using the following formula.

$S = 2\ell w + 2\ell h + 2wh$

EXAMPLE 1 Find the surface area of the rectangular prism.

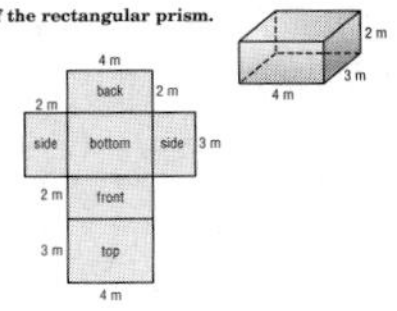

You can use the net of the rectangular prism to find its surface area. There are three pairs of congruent faces in a rectangular prism:

- top and bottom
- front and back
- two sides

Faces	Area
top and bottom	$(4 \cdot 3) + (4 \cdot 3) = 24$
front and back	$(4 \cdot 2) + (4 \cdot 2) = 16$
two sides	$(2 \cdot 3) + (2 \cdot 3) = 12$
Sum of the areas	$24 + 16 + 12 = 52$

Alternatively, replace ℓ with 4, w with 3, and h with 2 in the formula for surface area.

$S = 2\ell w + 2\ell h + 2wh$
$= 2 \cdot 4 \cdot 3 + 2 \cdot 4 \cdot 2 + 2 \cdot 3 \cdot 2$ Follow order of operations.
$= 24 + 16 + 12$
$= 52$

So, the surface area of the rectangular prism is 52 square meters.

EXERCISES

Find the surface area of each rectangular prism.

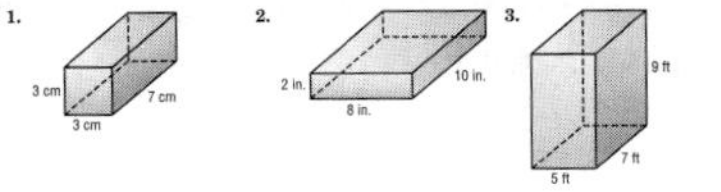

1. 102 cm²   2. 232 in²   3. 286 ft²

Practice: Skills, p. 685

NAME ____________ DATE ________ PERIOD ____

12-4 Practice: Skills

Surface Area of Rectangular Prisms

Find the surface area of each rectangular prism. Round to the nearest tenth if necessary.

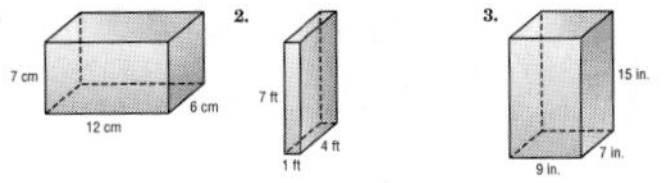

1. 396 cm²   2. 78 ft²   3. 606 in²

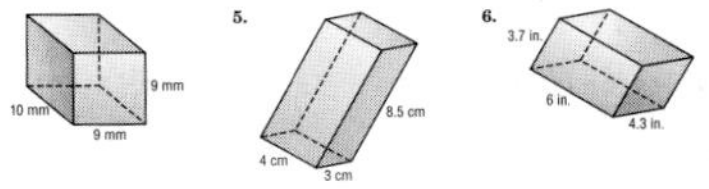

4. 522 mm²   5. 143 cm²   6. 127.8 in²

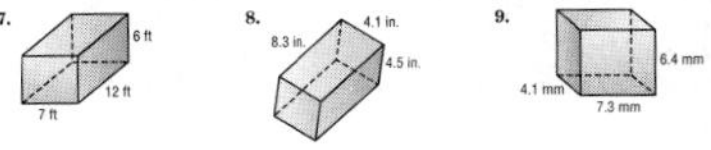

7. 396 ft²   8. 179.7 in²   9. 205.8 mm²

10. A cube has a surface area of 126 square feet. What is the area of one face? 21 ft²

11. Find the surface area of a rectangular prism that has a length of 8 inches, a width of 3 inches, and a height of 6 inches. 180 in²

Practice: Word Problems, p. 686

NAME ____________ DATE ________

12-4 Practice: Word Problems

Surface Area of Rectangular Prisms

1. **PACKAGING** A packaging company needs to know how much cardboard will be required to make boxes 18 inches long, 12 inches wide, and 10 inches high. How much cardboard will be needed for each box if there is no overlap in the construction? 1,032 in²

2. **INSULATION** Jane needs to buy insulation for the inside of a truck container. The container is a rectangular prism 15 feet long, 8 feet wide, and $7\frac{1}{2}$ feet high. How much insulation should Jane buy if all inside surfaces except the floor are to be insulated? 465 ft²

3. **ICE** Suppose the length of each edge of a cube of ice is 4 centimeters. Find the surface area of the cube.

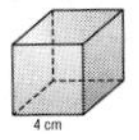

96 cm²

4. **ICE** Suppose you cut the ice cube from Exercise 3 in half horizontally into two smaller rectangular prisms. Find the surface area of one of the two smaller prisms. 64 cm²

5. **CONTAINERS** What is the total surface area of the inside and outside of a container in the shape of a rectangular prism with length of 5 meters, width of 3 meters, and height of 2.2 meters? 130.4 m²

6. **TOYS** Oscar is making a play block for his baby sister by gluing fabric over the entire surface of a foam block. How much fabric will Oscar need?

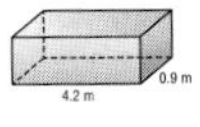

66.7 in²

4 Assess

Higher-Level Thinking

Draw a rectangular prism with a volume of 27 cm³ that has the least possible surface area. Then draw one with the same volume that has the greatest possible surface area.

The rectangular prism with dimensions 3 cm × 3 cm × 3 cm has the least surface area; the one with dimensions 27 cm × 1 cm × 1 cm has the greatest surface area.

12-4b Hands-On Lab Lesson Notes

Lesson Resources

Teaching Mathematics with Manipulatives
- isometric dot paper, p. 13
- student recording sheet 12-4b

Glencoe Mathematics Classroom Manipulative Kit
- isometric dot grid stamp
- stamp pad

GETTING STARTED

Teaching Tip Consider having students use centimeter cubes to build models of the cubes they draw on dot paper. Point out that in cubes with a side length of 3 units or greater, not all of the centimeter cubes used to build the model can be seen at the surface.

TEACH

Activity 1

- After students complete Step 1, check to make sure that they have drawn the cube correctly before proceeding.
- Encourage students to draw each new cube on isometric dot paper to ensure accuracy and to facilitate their understanding of the concept of changes in volume and surface area.

12-4b Hands-On Lab

A Follow-Up of Lesson 1

What You'll Learn
Investigate changes in volume and surface area.

Materials
- isometric dot paper

Changes in Volume and Surface Area

Suppose you have a model of a rectangular prism and you are asked to create a similar model whose dimensions are twice as large. In this lab, you will investigate how changing the dimensions of a three-dimensional figure affects the surface area and volume.

ACTIVITY *Work with a partner.*

1

STEP 1 Draw a cube on dot paper that measures 1 unit on each side. Calculate the volume and the surface area of the cube. Then record the data in a table like the one shown below.

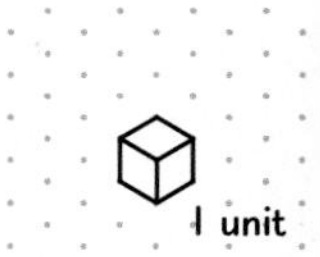

STEP 2 Double the side lengths of the cube. Calculate the volume and the surface area of this cube. Record the data in your table.

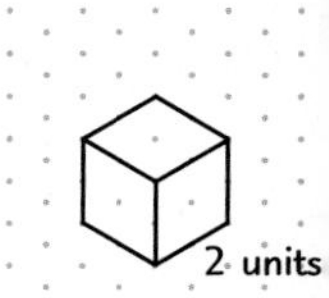

STEP 3 Triple the side lengths of the original cube. Now each side measures 3 units long. Calculate the volume and the surface area of the cube and record the data.

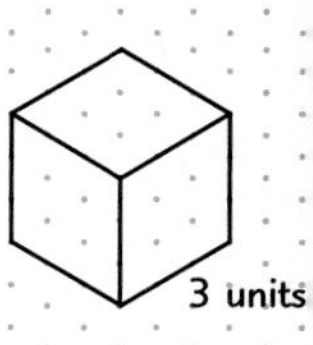

STEP 4 For each cube, write a ratio comparing the side length and the volume. Then write a ratio comparing the side length and the surface area. The first one is done for you.

Side Length (units)	Volume (units^3)	Surface Area (units^2)	Ratio of Side Length to Volume	Ratio of Side Length to Surface Area
1	$1^3 = 1$	$6(1^2) = 6$	1:1	1:6
2				
3				
4				
5				
s				

TEACHING TIP
Remind students that volume is given in cubic units because the measures of three dimensions are multiplied.

Your Turn

a. Complete the table above. **See margin.**

Answer (Your Turn)

a.

Side Length (units)	Volume (units^3)	Surface Area (units^2)	Ratio of Side Length to Volume	Ratio of Side Length to Surface Area
1	$1^3 = 1$	$6(1^2) = 6$	1:1	1:6
2	$2^3 = 8$	$6(2^2) = 24$	2:8	2:24
3	$3^3 = 27$	$6(3^2) = 54$	3:27	3:54
4	$4^3 = 64$	$6(4^2) = 96$	4:64	4:96
5	$5^3 = 125$	$6(5^2) = 150$	5:125	5:150
s	s^3	$6s^2$	$s:s^3$	$s:6s^2$

ACTIVITY *Work with a partner.*

2 **STEP 1** Draw a cube on dot paper that measures 8 units on each side. Calculate the volume and the surface area of the cube. Record the data in a table like the one shown below.

STEP 2 Halve the side lengths of the cube in Step 1. Calculate the volume and the surface area of this cube and record the data.

STEP 3 Halve the side lengths of the cube in Step 2. Calculate the volume and the surface area of the cube and record the data.

STEP 4 For each cube, write a ratio comparing the side length and the volume and a ratio comparing the side length and the surface area. The first one is done for you.

Side Length (units)	Volume (units^3)	Surface Area (units^2)	Ratio of Side Length to Volume	Ratio of Side Length to Surface Area
8	$8^3 = 512$	$6(8^2) = 384$	8:512	8:384
4				
2				
s				

Your Turn

b. Complete the table above. See margin.

Writing Math

Work with a partner.

1. **Write** a formula for the volume V of a cube with side length s. $V = s^3$
2. **Write** a formula for the surface area A of a cube with side length s. $A = 6s^2$

Complete each sentence.

3. If the side length of a cube is doubled, the volume is __?__ times greater. 2^3 or 8
4. If the side length of a cube is doubled, the surface area is __?__ times greater. 2^2 or 4
5. If the side length of a cube is tripled, the volume increases by __?__ times and the surface area increases by __?__ times. 3^3 or 27; 3^2 or 9
6. If the side length of a cube decreases by $\frac{1}{2}$, the surface area decreases by __?__. $\frac{1}{4}$

Activity 2

- Consider having students graph the relationship between side length and volume, as well as the relationship between side length and surface area, for the data generated in each Activity.
- As students complete **Activity 2**, have them compare what is happening as the cubes get smaller to what happened in **Activity 1** as the cubes got larger.

ASSESS

Observe as students solve the **Your Turn** exercises. They should:

- completely fill in each table, and
- correctly calculate the volume, surface area, ratio of side length to volume, and ratio of side length to surface area for all of the cubes.

Discuss the **Writing Math** exercises. Verify the formulas students derived for Exercises 1 and 2.

For Exercises 3–6, be sure that students can explain in their own words the relationships they are investigating.

Answer (Your Turn)

b.

Side Length (units)	Volume (units^3)	Surface Area (units^2)	Ratio of Side Length to Volume	Ratio of Side Length to Surface Area
8	$8^3 = 512$	$6(8^2) = 384$	8:512	8:384
4	$4^3 = 64$	$6(4^2) = 96$	4:64	4:96
2	$2^3 = 8$	$6(2^2) = 24$	2:8	2:24
s	s^3	$6s^2$	$s:s^3$	$s:6s^2$

12-5 Lesson Notes

Lesson Resources

Chapter 12 Resource Masters

- Study Guide and Intervention, p. 689
- Practice: Skills, p. 690
- Practice: Word Problems, p. 691
- Reading to Learn Mathematics, p. 692
- Enrichment, p. 693

Calculator and Spreadsheet Masters, p. 56

Classroom Math Games, pp. 35–36

Key Teaching Transparencies, 48

MathPASS Tutorial CD-ROM, 54

Parent and Student Study Guide Workbook, 12-5

FAST FILE

1 Focus

5-Minute Check 12-5 Use as a quiz or review of Lesson 12-4. (Also available as a transparency)

INTERACTIVE CHALKBOARD PowerPoint® Presentations

Bellringer

Reading Mathematics Have students review the terms *radius, diameter, circumference, surface area,* and *cylinder*. It is essential that they understand the meaning of these words before they begin learning the concepts in this lesson.

Reading to Learn Mathematics, p. 692 **ELL**

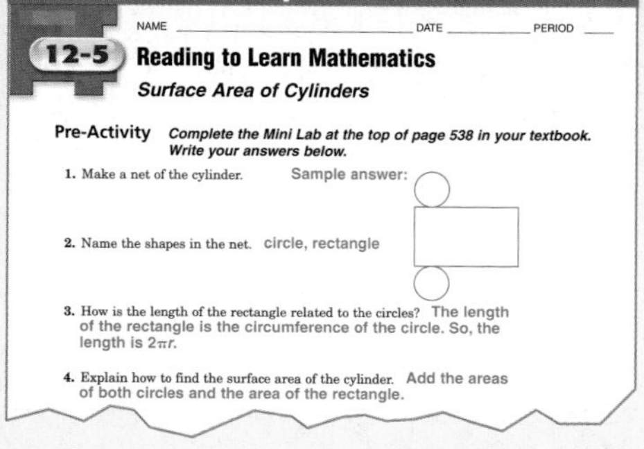
NAME ______ DATE ______ PERIOD ______

12-5 Reading to Learn Mathematics

Surface Area of Cylinders

Pre-Activity *Complete the Mini Lab at the top of page 538 in your textbook. Write your answers below.*

1. Make a net of the cylinder. Sample answer:
2. Name the shapes in the net. circle, rectangle
3. How is the length of the rectangle related to the circles? The length of the rectangle is the circumference of the circle. So, the length is $2\pi r$.
4. Explain how to find the surface area of the cylinder. Add the areas of both circles and the area of the rectangle.

12-5 Surface Area of Cylinders

What You'll Learn
Find the surface areas of cylinders.

REVIEW Vocabulary

circumference: the distance around a circle **(Lesson 6-9)**

HANDS-ON Mini Lab

Materials
- soup can
- grid paper
- scissors

Work with a partner.

- Trace the top and bottom of the can on grid paper. Then cut out the shapes.
- Cut a long rectangle from the grid paper. The width of the rectangle should be the same as the height of the can. Wrap the rectangle around the side of the can. Cut off the excess paper so that the edges just meet.

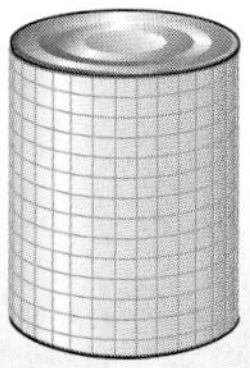

1. Make a net of the cylinder. **See margin.**
2. Name the shapes in the net. **circle, rectangle**
3. How is the length of the rectangle related to the circles?
4. Explain how to find the surface area of the cylinder.

3. See margin. 4. Add the areas of both circles and the area of the rec

The diagram below shows how you can put two circles and a rectangle together to make a cylinder.

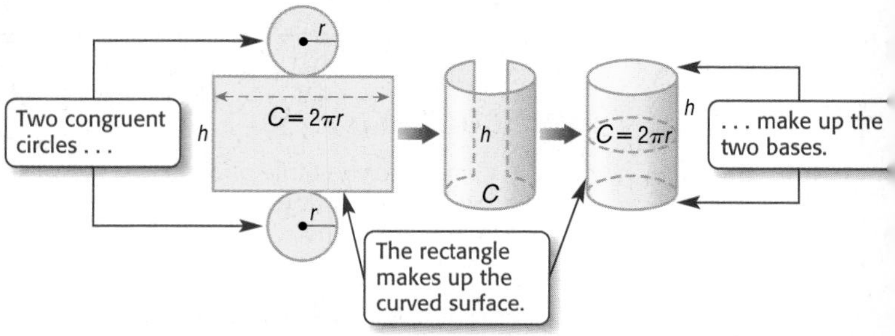

In the diagram above, the length of the rectangle is the same as the circumference of the circle. Also, the width of the rectangle is the same as the height of the cylinder.

The surface area of a cylinder	equals	the area of two bases	plus	the area of the curved surface.
S	$=$	$2(\pi r^2)$	$+$	$(2\pi r)h$

TEACHING TIP
Remind students that the area of each circular base is πr^2.

Key Concept **Surface Area of a Cylinder**

Words The surface area S of a cylinder with height h and radius r is the sum of the areas of the circular bases and the area of the curved surface.

Model

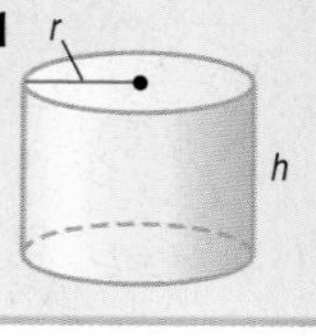

Symbols $S = 2\pi r^2 + 2\pi rh$

Answers (Hands-On Mini Lab)

1. Sample answer:

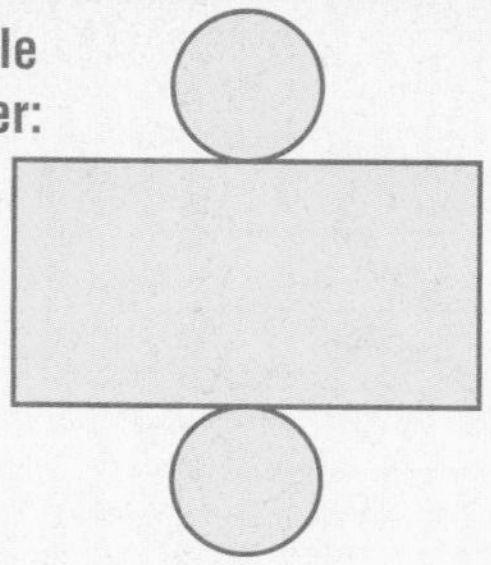

3. The length of the rectangle is the circumference of the circle. So, the length is $2\pi r$.

EXAMPLE Find the Surface Area of a Cylinder

1 Find the surface area of the cylinder. Round to the nearest tenth.

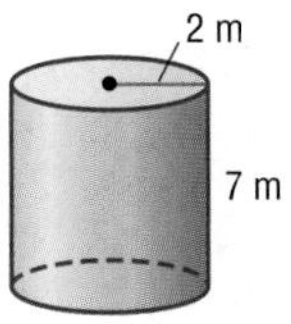

$S = 2\pi r^2 + 2\pi rh$ Surface area of a cylinder

$= 2\pi(2)^2 + 2\pi(2)(7)$ Replace r with 2 and h with 7.

≈ 113.1 Simplify.

The surface area is about 113.1 square meters.

Your Turn

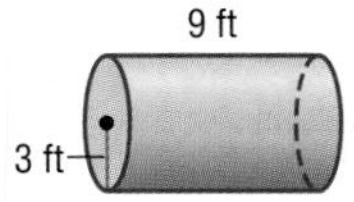

a. Find the surface area of the cylinder. Round to the nearest tenth. **226.2 ft^2**

EAL-IFE CAREERS

w Does a Graphic signer Use Math?

person who designs the ckages of products must ow the size of the ckage before making the sign. This involves culating the surface areas packages that are tangular prisms and inders.

Online Research information about a career a graphic designer, visit: math2.net/careers

EXAMPLE Use Surface Area to Solve a Problem

2 DESIGN A can of soup is 5 inches high, and its base has a diameter of 4 inches. How much paper is needed to make the label on the can?

Since only the curved side of the can has a label, you do not need to include the areas of the top and bottom of the can.

$S = 2\pi rh$ Curved surface of a cylinder

$= 2\pi(2)(5)$ Replace r with 2 and h with 5.

≈ 62.8 Simplify.

So, about 62.8 square inches of paper is needed to make the label.

Skill and Concept Check

Writing Math Exercise 2

1. **OPEN ENDED** Find the surface area of a cylinder found in your home. **See students' work.**
2. **FIND THE ERROR** Matthew and Latonya have each drawn a cylinder with the dimensions shown below and insist that theirs has more surface area. Who is correct? Explain. **See margin.**

Matthew	Latonya
$r = 6$ cm, $h = 3$ cm	$r = 3$ cm, $h = 6$ cm

GUIDED PRACTICE

Find the surface area of each cylinder. Round to the nearest tenth.

3. **88.0 mm^2**

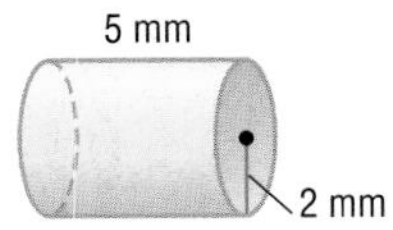

4. **466.5 in^2**

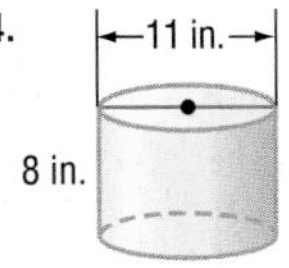

5. The radius of a cylinder is 8 meters, and its height is 5 meters. Find the surface area to the nearest tenth. **653.5 m^2**

msmath2.net/extra_examples

DAILY INTERVENTION Differentiated Instruction ELL

Verbal/Linguistic Have students explain or write a detailed description of the procedure for finding the surface area of a cylinder. In their descriptions they should explain the surface area of a cylinder formula, including the reason there are two parts to this formula.

2 Teach

In-Class Examples

INTERACTIVE CHALKBOARD PowerPoint® Presentations

1 Find the surface area of the cylinder. Round to the nearest tenth. **169.6 cm^2**

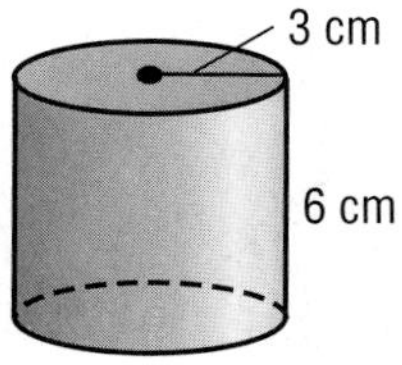

2 **GIFT WRAP** A poster is contained in a cardboard cylinder that is 10 inches high. The cylinder's base has a diameter of 8 inches. How much paper is needed to wrap the cardboard cylinder if the ends are to be left uncovered? **about 251.3 in^2**

Tips for New Teachers

Intervention Some students have difficulty distinguishing circumference from area, radius from diameter, and surface area from volume. Working with the formulas for geometric figures requires a solid understanding of both the terms and the concepts used in the formulas. Use manipulatives, drawings, reference cards, and logical reasoning to help students master the terms and concepts needed for this chapter.

Answer

2. Matthew; Matthew's cylinder has a surface area of 339.3 cm^2; Latonya's cylinder has a surface area of 169.6 cm^2.

3 Practice/Apply

FOLDABLES™ Follow-Up Encourage students to include information in words and symbols on how to find the surface area of a cylinder in the appropriate section of their Foldable tables.

DAILY INTERVENTION **FIND THE ERROR** In Exercise 2, encourage students to attempt to find the answer by examining the formula rather than by carrying out the calculations.

About the Exercises...

Guided Practice Key

Exercises	Examples
3–4	1
5	2

Odd/Even Assignments
Exercises 6–13 are structured so that students practice the same concepts whether they are assigned odd or even problems.

Answers

17. Sample answer: prism: volume = 275 in^3, surface area = 237.5 in^2; cylinder: volume = 274.3 in^3, surface area = 211.1 in^2; Both containers hold about the same amount of popcorn. However, the cylinder uses less cardboard. So, the cylinder would be a more cost-effective container.

19. Sample answer: The areas of both curved surfaces are the same. However, the areas of the top and bottom of cylinder A are greater than the areas of the top and bottom of cylinder B because it has a greater circumference (11 in.).

★ indicates increased difficulty

Practice and Applications

HOMEWORK HELP

For Exercises	See Examples
6–11	1
12–19	2

Extra Practice See pages 595, 607.

Find the surface area of each cylinder. Round to the nearest tenth.

6. 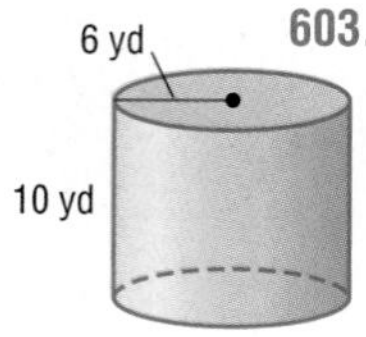

603.2 yd^2

7. 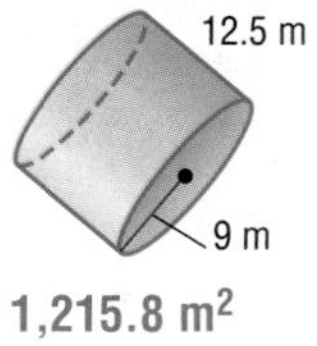

1,215.8 m^2

8. 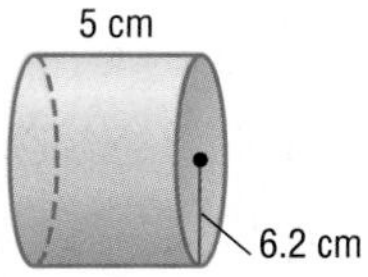

436.3 cm^2

9. 

183.8 ft^2

10. 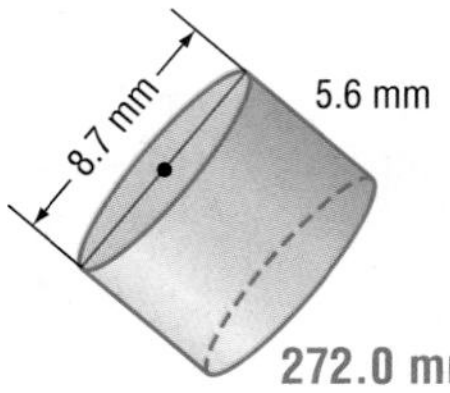

272.0 mm^2

11. 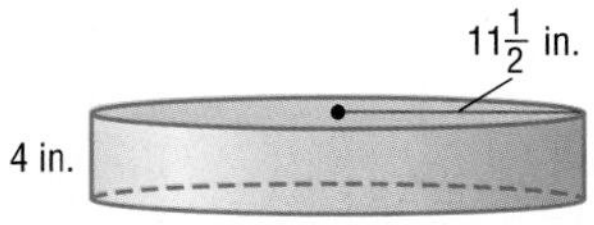

1,120.0 in^2 or $1{,}119\frac{44}{45}$ in^2

12. Find the area of the label on a can of tuna with a radius of 5.1 centimeters and a height of 2.9 centimeters. 92.9 cm^2

13. The height of a water tank is 10 meters and has a diameter of 10 meters. What is the surface area of the tank? 471.2 m^2

★ 14. An underground oil tank has a surface area of 2,915 square feet and a radius of 6 feet. What is the height of the oil tank? 71.3 ft

BAKING **For Exercises 15 and 16, use the following information.**
A *smash cake* is a smaller version of a birthday cake given to a child on his or her first birthday. Mrs. Jones baked a smash cake 3 inches high and 10 inches in diameter for her daughter when she turned 1 year old.

15. Mrs. Jones covers the top and sides of the cake with frosting. Find the area that the frosting covers to the nearest tenth. 172.8 in^2

★ 16. Suppose Mrs. Jones first cuts the cake in two layers, frosts the top and sides of the bottom layer, puts the top layer on, and then frosts the rest of the cake. How much of the cake is covered with frosting now? 251.3 in^2

17. **MOVIES** The Student Council at Southwest Middle School is planning to sell popcorn in one of two open-top containers below at its "Movie Night." The cost depends on the amount of cardboard used to make each container. Which container should Student Council buy? Use volume and surface area measurements to explain your choice. See margin.

21. Sample answer:

18. **CANDY** The largest piece of candy ever created was shaped like a cylinder 4.49 meters long and had a circumference of 1.14 meters. What was the surface area of the candy to the nearest tenth? **5.3 m²**

★ 19. **MODELING** Suppose you took an $8\frac{1}{2}$-×-11-inch piece of paper and made two cylinders; one using the $8\frac{1}{2}$-inch side as the height (cylinder A) and the other using the 11-inch side as the height (cylinder B). Without calculating, which cylinder has the greater surface area? Explain. **See margin.**

20. **CRITICAL THINKING** If you double the height of a cylinder, will its surface area also double? Explain your reasoning. **No; the surface area of the side of the cylinder will double, but the area of the bases will not.**

21. **EXTENDING THE LESSON** In Lesson 12-1, you saw another solid with a curved surface, called a *cone*, as shown at the right. The curved surface is difficult to draw accurately in two dimensions since it is not a polygon. Make a cone with a piece of paper. Then draw the net for the cone using it as a pattern. **See margin.**

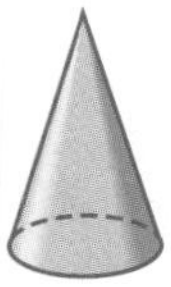

Standardized Test Practice and Mixed Review

22. **MULTIPLE CHOICE** Gilberto completely covers a cylindrical can with construction paper. The can has a height of 15 inches and a radius of 4 inches. About how much construction paper does he use? **A**

   (A) 477.5 in²   (B) 375.2 in²   (B) 104.5 in²   (D) 60.0 in²

23. **MULTIPLE CHOICE** The three metal containers below each hold about 1 liter of liquid. Which container has the greatest surface area? **G**

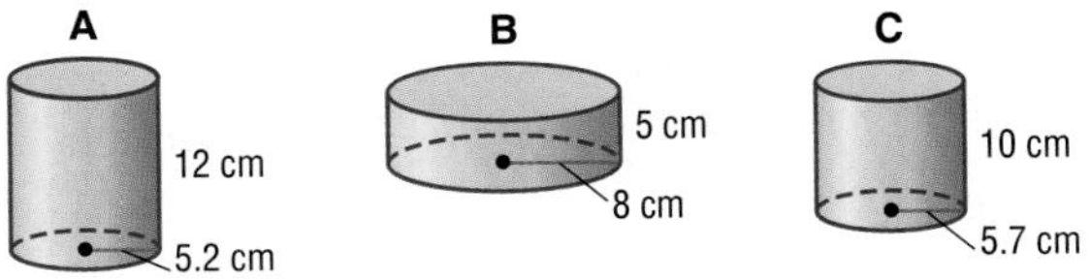

   (F) A   (G) B   (H) C   (I) none of them

Find the surface area of each rectangular prism. (Lesson 12-4)

24. 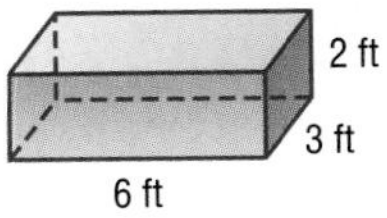

72 ft²

25. 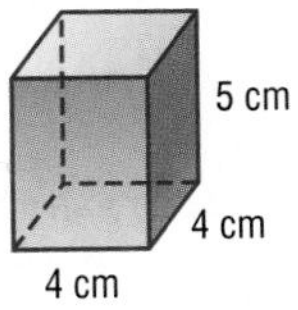

112 cm²

26. 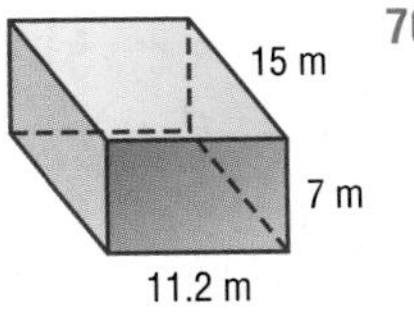

702.8 m²

27. Find the volume of a cylinder having a radius of 4 inches and a height of 6.5 inches. (Lesson 12-3) **326.7 in³**

GETTING READY FOR THE NEXT LESSON

PREREQUISITE SKILL Find the value of each expression. Round to the nearest tenth. (Pages 559, 560)

28. 13.2 + 15.378 **28.6**   29. 23.7 − 9.691 **14.0**   30. 23 · 7.1 **163.3**   31. 1.6(8.5)0.4 **5.4**

msmath2.net/self_check_quiz

4 Assess

Open-Ended Assessment

Modeling Give students tuna cans or soup cans. Have them measure the height and diameter and calculate the areas of the labels. Then have them remove the labels and check.

Study Guide and Intervention, p. 689

NAME ____________ DATE ________ PERIOD ____

12-5 Study Guide and Intervention

Surface Area of Cylinders

The diagram below shows how you can put two circles and a rectangle together to make a cylinder.

The surface area of a cylinder equals the area of two bases plus the area of the curved surface.

$S = 2(\pi r^2) + (2\pi r)h$

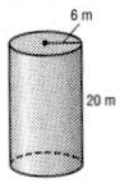

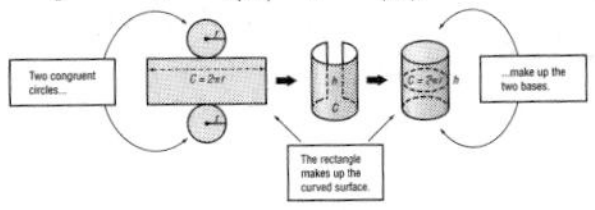

In the diagram above, the length of the rectangle is the same as the circumference of the circle. Also, the width of the rectangle is the same as the height of the cylinder.

EXAMPLE 1 **Find the surface area of the cylinder. Round to the nearest tenth.**

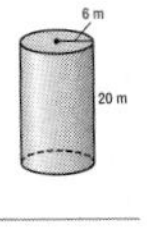

$S = 2\pi r^2 + 2\pi rh$ Surface area of a cylinder.

$S = 2\pi(6)^2 + 2\pi(6)(20)$ Replace r with 6 and h with 20.

≈ 980.2 Simplify.

The surface area is about 980.2 meters.

EXERCISES

Find the surface area of each cylinder. Round to the nearest tenth

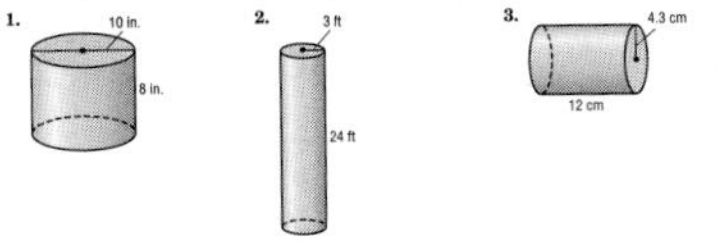

1. 408.4 in²   2. 508.9 ft²   3. 440.4 cm²

Practice: Skills, p. 690

NAME ____________ DATE ________ PERIOD ____

12-5 Practice: Skills

Surface Area of Cylinders

Find the surface area of each cylinder. Round to the nearest tenth.

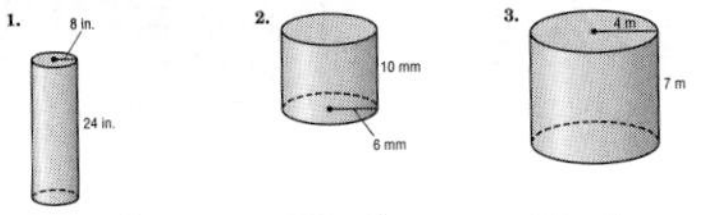

1. 1,608.5 in²   2. 603.2 mm²   3. 276.5 m²

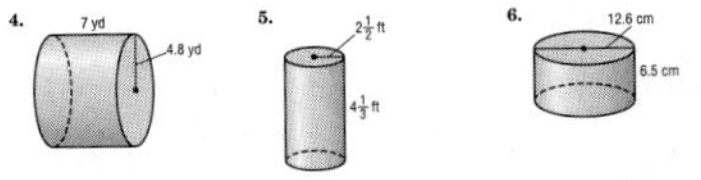

4. 355.9 yd²   5. 107.3 ft²   6. 506.7 cm²

7. Find the surface area of a can with a radius of 4 centimeters and a height of 11 centimeters. 377.0 cm²

8. Find the surface area of the outside of a cylindrical barrel with a diameter of 10 inches and a height of 12 inches. 534.1 in²

9. Find the area of the curved surface of a D battery with a diameter of 3.2 centimeters and a height of 5.6 centimeters. 56.3 cm²

Practice: Word Problems, p. 691

NAME ____________ DATE ________ PERIOD ____

12-5 Practice: Word Problems

Surface Area of Cylinders

1. **PACKAGING** What is the area of the label on a box of oatmeal with a radius of 9.3 centimeters and a height of 16.5 centimeters? Round to the nearest tenth. 964.2 cm²

2. **TIRES** Betty wants to know the total surface area of the tread on one of her tires. If the diameter of the tire is 18 inches and the width of the tire is 5 inches, what is the total surface area of the tire's tread? Round to the nearest tenth. 282.7 in²

3. **CANS** A cylindrical can has a diameter of 6 inches and a height of 7.3 inches. What is the surface area of the can? Round to the nearest tenth. 194.2 in²

4. **CANS** A cylindrical can has a height of 14 centimeters and a radius of 4.2 centimeters. Find the surface area of the can. Round to the nearest tenth. 480.3 cm²

5. **MANUFACTURING** How much sheet metal is required to make a cylindrical trash can with a diameter of 2 feet and height of $4\frac{1}{4}$ feet? Round to the nearest tenth. (*Hint*: Do not include the top.) 29.8 ft²

6. **PLUMBING** How much steel is needed to make a hollow pipe with a radius of 3 inches and a height of 15 inches? Round to the nearest tenth. 282.7 in²

12-6 Lesson Notes

Lesson Resources

Chapter 12 Resource Masters

- Study Guide and Intervention, p. 694
- Practice: Skills, p. 695
- Practice: Word Problems, p. 696
- Reading to Learn Mathematics, p. 697
- Enrichment, p. 698

Parent and Student Study Guide Workbook, 12-6

FAST FILE

1 Focus

5-Minute Check 12-6

INTERACTIVE CHALKBOARD PowerPoint® Presentations

Use as a quiz or review of Lesson 12-5. (Also available as a transparency)

Bellringer

Reading Mathematics Inform students that this lesson discusses how precise a unit of measure is in a given situation. In the mathematical sense, precision is not the same thing as accuracy or correctness. Tell students that the base of precision, *–cise,* comes from the Latin word meaning *to cut*. Have students imagine cutting a unit of measure in half again and again. Each time, the resulting measure is more precise than the one before.

Reading to Learn Mathematics, p. 697 **ELL**

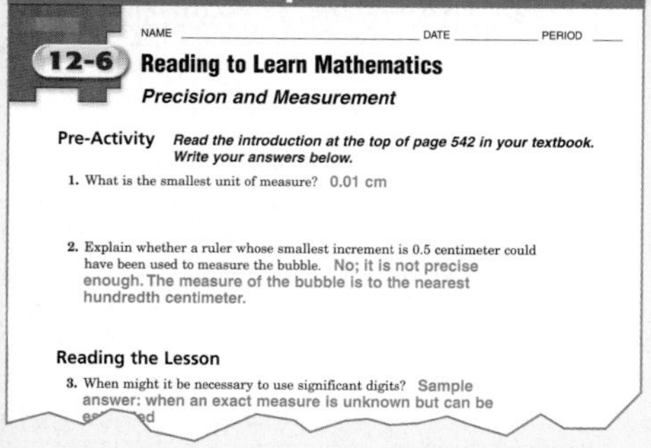
NAME ______________ DATE ______ PERIOD ____

12-6 Reading to Learn Mathematics

Precision and Measurement

Pre-Activity *Read the introduction at the top of page 542 in your textbook. Write your answers below.*

1. What is the smallest unit of measure? 0.01 cm
2. Explain whether a ruler whose smallest increment is 0.5 centimeter could have been used to measure the bubble. No; it is not precise enough. The measure of the bubble is to the nearest hundredth centimeter.

Reading the Lesson

3. When might it be necessary to use significant digits? Sample answer: when an exact measure is unknown but can be es[...]ed

12-6 Precision and Measurement

What You'll Learn
Determine and apply significant digits in a real-life context.

NEW Vocabulary
- precision
- precision unit
- significant digits

WHEN am I ever going to use this?

WORLD RECORDS In 2000, Joyce Samuels blew a bubble gum bubble that had a diameter of 27.94 centimeters.

1. What is the smallest unit of measure? **0.01 cm**
2. Explain whether a ruler whose smallest increment is 0.5 centimeter could have been used to measure the bubble. **No; it is not precise enough. The measure of the bubble is to the nearest hundredth centimeter.**

The **precision** or *exactness* of a measurement depends on the unit of measure. The **precision unit** is the smallest unit on a measuring tool. As the units get smaller, the measurement gets "more precise."

EXAMPLES Identify Precision Units

Identify the precision unit of each measuring tool.

1

The smallest unit is a half centimeter. So, the precision unit is 0.5 centimeter.

2

The smallest unit is an eighth of an inch. So, the precision unit is $\frac{1}{8}$ inch.

Your Turn

a. Identify the precision unit of the scale at the right. **1 kg**

TEACHING TIP
Point out that you can expect a measurement to be correct to the nearest precision unit of the measuring tool.

All measurements are approximate. You could estimate the measure on the scale below as 8 grams. A more precise method is to use significant digits. **Significant digits** include all of the digits of a measurement that you know for sure, plus one estimated digit.

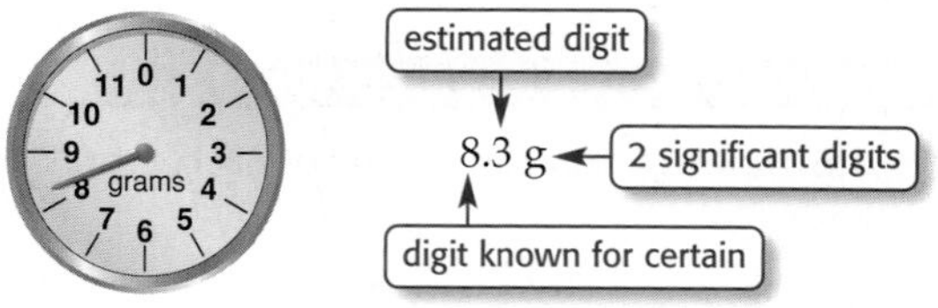

EXAMPLE Apply Significant Digits

3 **State the measure of the pencil using significant digits.**

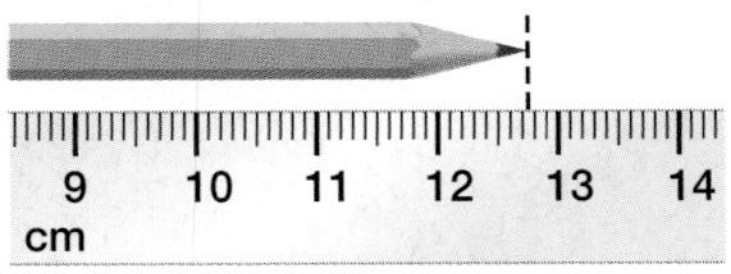

The precision unit is 0.1 centimeter. You know for certain that the length is between 12.7 and 12.8 centimeters. One estimate is 12.75 centimeters.

Your Turn

b. State the measure of the paper clip using significant digits. **Sample answer: 4.7 cm**

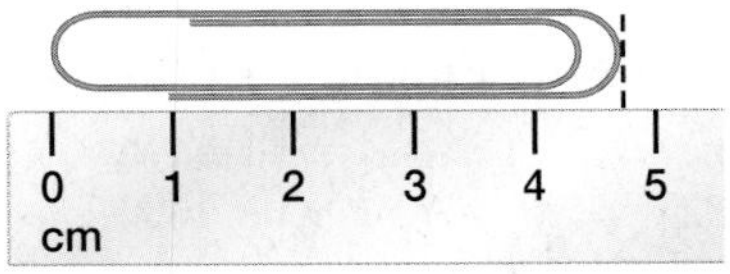

Skill and Concept Check

Writing Math
Exercises 1–3

1. **Choose** the most precise unit of measurement: *inches*, *feet*, or *yards*. Explain. **Inches, because it is the smallest unit of measure.**
2. **OPEN ENDED** Describe some items or situations in which an approximate measure is sufficient. Then describe some items or situations in which exactness in measurement is important and necessary. **See margin.**
3. **Which One Doesn't Belong?** Identify the measurement that does not have the same number of significant digits as the other three. Explain your reasoning.

8.6 mi	14.5 ft	3.22 cm	90.7 yd

8.6 mi, because it has 2 significant digits. The other measures have 3 significant digits.

GUIDED PRACTICE

4. **Identify** the precision unit of the ruler at the right. **$\frac{1}{16}$ in.**

State each measure using significant digits. **5–6. Sample answers are given.**

5. **15.65 cm**

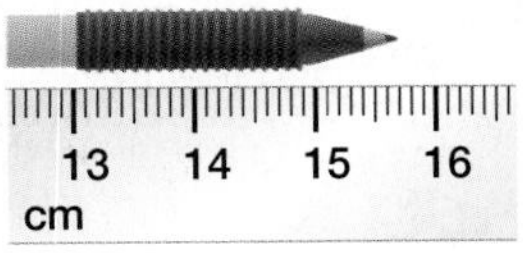

6. **3.6 mL**

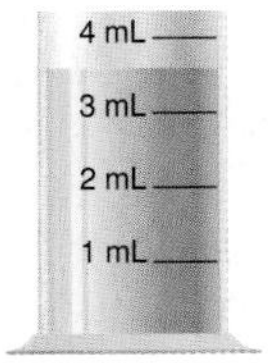

7. **OLYMPICS** Tom Dolan broke the Olympic Swimming World Record in 2000 for the 400-meter race. His time was 4 minutes, 11.76 seconds. Describe the precision of this measure. **Sample answer: The measure is precise to the nearest 0.01 second.**

msmath2.net/extra_examples

DAILY INTERVENTION — Unlocking Misconceptions

Significant Digits In a measurement stated in significant digits, point out that if the estimated digit is zero, it is important to leave it in. Removing a zero that appears as the last decimal place will alter the number of significant digits in the measurement. Since the number of significant digits is related to the precision unit, deleting an estimated zero will cause misunderstandings about the precision of the measurement.

2 Teach

In-Class Examples

INTERACTIVE CHALKBOARD
PowerPoint® Presentations

Identify the precision unit of each measuring tool.

1

1 s

2 **2°F**

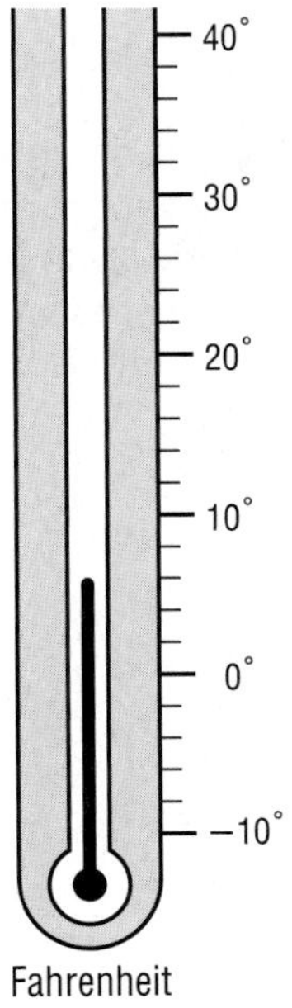

3 State the measure of the piece of chalk using significant digits.

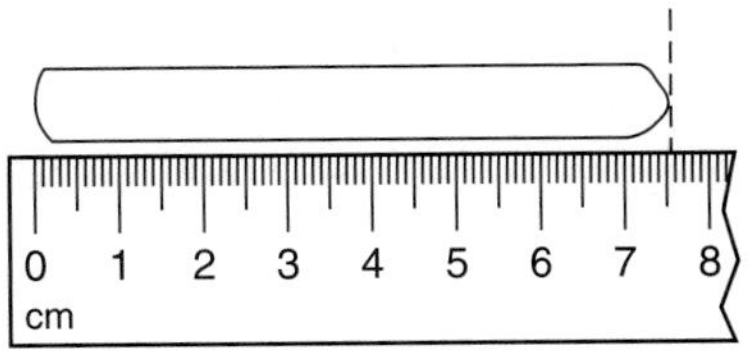

7.55 cm

Answer

2. **Sample answer: The measures of the weight of a baby to buy baby clothes or the distance in miles to visit a friend can be approximate. The measures of the weight of a package to mail or the size of a part to fit a certain tool should be exact.**

3 Practice/Apply

Which One Doesn't Belong? In Exercise 3, students should be able to count the significant digits in each measurement. If they are having difficulty, ask them to indicate which digit in each measurement is the estimated digit.

About the Exercises...

Guided Practice Key

Exercises	Examples
4	1, 2
5–7	3

Odd/Even Assignments Exercises 8–17 are structured so that students practice the same concepts whether they are assigned odd or even problems.

Alert! Exercise 21 requires research on the Internet or another source.

★ indicates increased difficulty

Practice and Applications

HOMEWORK HELP

For Exercises	See Examples
8–11, 19–20	1, 2
12–18	3

Extra Practice
See pages 595, 607.

Identify the precision unit of each measuring tool.

8. 

1 cm

9. 

0.1 cm

10. 

$\frac{1}{4}$ in.

11. 

2 lb

State each measure using significant digits. 12–17. Sample answers are given.

12. 

6.2 oz

13. 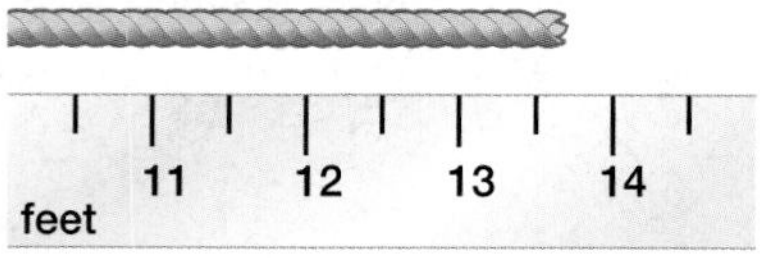

13.6 ft

14. 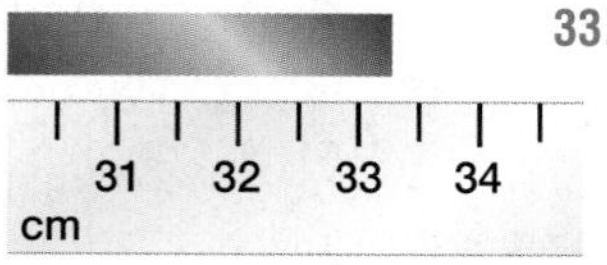

33.3 cm

15. 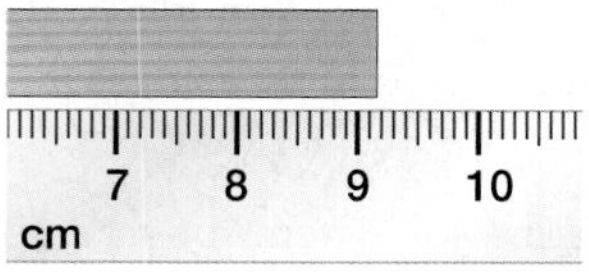

9.15 cm

16. 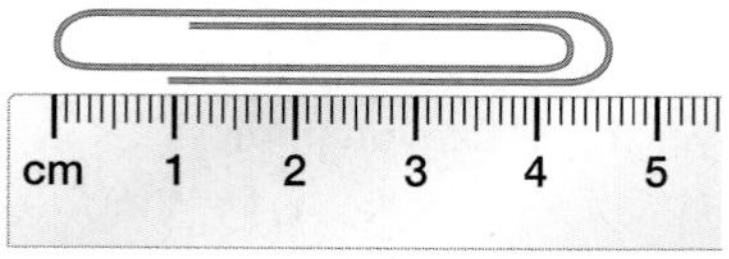

4.61 cm

17. 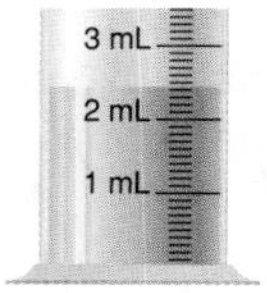

2.45 mL

TRIATHLONS For Exercises 18–20, use the graphic at the right. 19. Neither; they are both given to the nearest tenth of a mile.

18. How many significant digits are in the measurement describing the length of the cycling portion of the triathlon? 3

19. Which is a more precise measure: the number of miles cycled in a triathlon, or the number of miles run? Explain.

★ 20. The running portion of the women's triathlon is equivalent to 5 kilometers. Which unit is more precise: a mile or a kilometer? Explain. Kilometer, because it is a smaller unit of measure.

21. **RESEARCH** The silica tiles on space shuttles are cut accurately to 10,000th of an inch. Use the Internet or another source to investigate the technology that is used to make such precise measurements. See students' work.

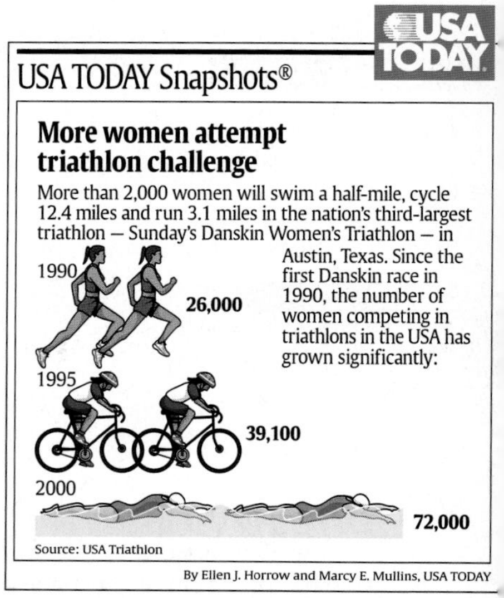

22. If you use a ruler with a precision unit of 1 millimeter, will your measurements *sometimes*, *always*, or *never* be exact? Explain.
Never; all measurements are approximations and could be more precise.

23. **WRITE A PROBLEM** Write a measurement problem in which you use significant digits. **See students' work.**

24. **CRITICAL THINKING** Choose the most precise measures to complete the sentence: The yarn at the right is between __?__ and __?__ inches long. $21\frac{5}{8}$; $21\frac{3}{4}$

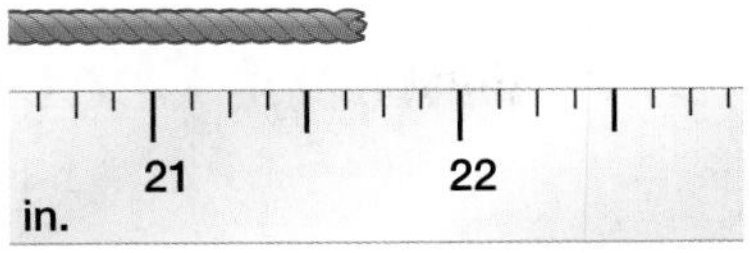

Standardized Test Practice and Mixed Review

25. **MULTIPLE CHOICE** Choose the measuring tool that you would use to estimate the volume of a shoebox. **B**

Ⓐ scale Ⓑ ruler Ⓒ protractor Ⓓ measuring cup

26. **MULTIPLE CHOICE** Choose the best precision unit for estimating the length of a gymnasium. **I**

Ⓕ 0.1 cm Ⓖ 0.5 in. Ⓗ 0.1 mi Ⓘ 0.5 ft

Find the surface area of each cylinder. Round to the nearest tenth. (Lesson 12-5)

27. 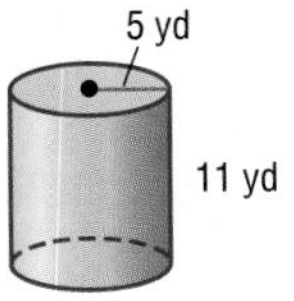

$502.7\ yd^2$

28. 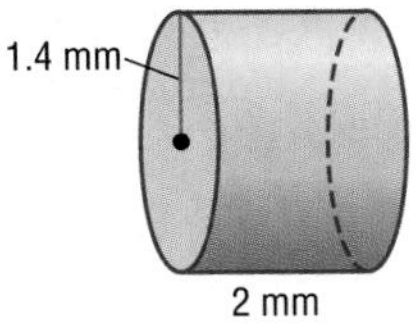

$29.9\ mm^2$

29. 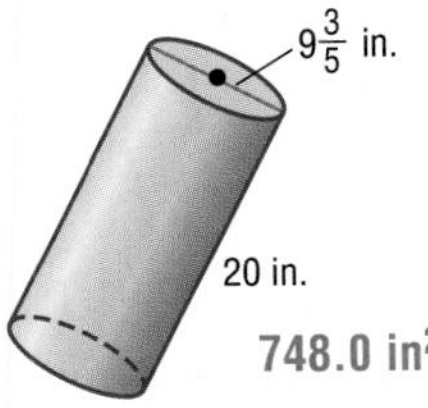

$748.0\ in^2$

30. Find the surface area of a rectangular prism that has a length of 19 centimeters, a width of 8.8 centimeters, and a height of 13 centimeters. (Lesson 12-4) **$1{,}057.2\ cm^2$**

ALGEBRA Solve each equation. Check your solution. (Lesson 6-5)

31. $\frac{n}{15} = 3$ **45**

32. $-26 = \frac{x}{2.4}$ **−62.4**

33. $\frac{3}{8}y = 9$ **24**

WebQuest Interdisciplinary Project

It's All Greek to Me

It's time to complete your project. Use the information and data you have gathered about Pythagoras to prepare a Web page or poster. Be sure to include the three-dimensional solid you created with your project.

msmath2.net/webquest

4 Assess

Open-Ended Assessment

Speaking Have students think of an example of how measurement is used in their everyday lives. Ask them to describe the precision of the measurement in terms of significant and estimated digits.

Study Guide and Intervention, p. 694

NAME ______ DATE ______ PERIOD ____

12-6 Study Guide and Intervention

Precision and Measurement

The **precision** or *exactness* of a measurement depends on the unit of measure. The **precision unit** is the smallest unit on a measuring tool. **Significant digits** include all of the digits of a measurement that you know for sure, plus one estimated digit.

EXAMPLE 1 **Identify the precision unit of the measuring tool.**

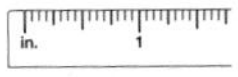

The smallest unit is a sixteenth of an inch. So, the precision unit is $\frac{1}{16}$ inch.

EXAMPLE 2 **State the measurement of the key using significant digits.**

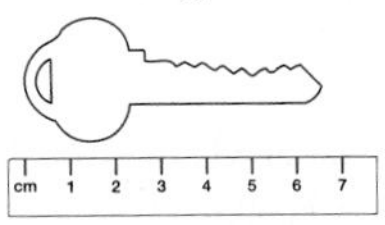

The precision unit is 1 centimeter. You know for certain that the length is between 6 and 7 centimeters. One estimate is 6.6 centimeters.

EXERCISES

Identify the precision unit of each measuring tool.

1. 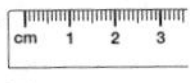

0.1 cm

2. 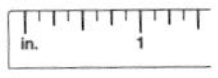

$\frac{1}{8}$ in.

State the measure using significant digits.

3. 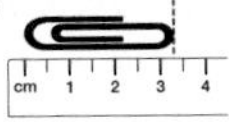

Sample answer: 3.3 cm

4. 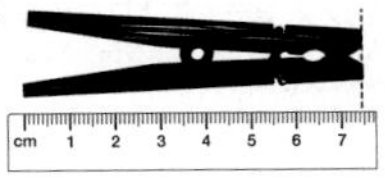

Sample answer: 7.45 cm

Practice: Skills, p. 695

NAME ______ DATE ______ PERIOD ____

12-6 Practice: Skills

Precision and Measurement

Identify the precision unit of each measuring tool.

1. 

1 in.

2. 

0.1 lb

3. 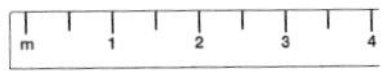

0.5 m

4. 

0.5 oz

State the measure using significant digits. Sample answers given.

5. 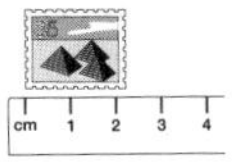

2.2 cm

6. 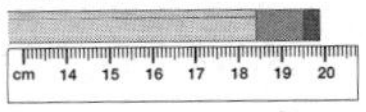

19.89 cm

7. 

51.6 kg

8. 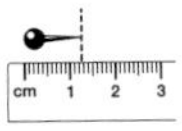

1.25 cm

Practice: Word Problems, p. 696

NAME ______ DATE ______ PERIOD ____

12-6 Practice: Word Problems

Precision and Measurement

FOOTBALL **For Exercises 1–3, use the information about Quentin Griffin.**

University of Oklahoma running back Quentin Griffin ran 248 yards on the ground against No. 2 Texas to earn the USATODAY.com Player of the Week honors. The senior added two touchdowns in the fourth quarter to lift the third-ranked Oklahoma Sooners past the Texas Longhorns 35-24.

1. How many significant digits are in the measurement of the distance that Quentin Griffin ran? 3	2. What is the precision unit on a football field? Would the distance measurement be more precise if the field was measured in meters? Explain. 1 yd; no; Sample answer: Meters are larger than yards, so no additional precision would be gained.
3. How many significant digits are in the score of the game? 2	4. LAND SPEED On October 15, 1997, Andy Green set the land speed record of 763.035 miles per hour. How many significant digits are in the measurement of the land speed record? 6
5. TRACK The record for the men's high jump is held by Javier Sotomayor (Cuba), who became world champion with a distance of 2.45 meters on July 27, 1993, in Salamanca, Spain. How many significant digits are in the measurement of the high jump record? 3	6. OCEANOGRAPHY On March 24, 1995, the Japanese probe *Kaiko* recorded the depth of the Marianas Trench, the deepest point in the ocean, at 10,911 meters or 35,797 feet. Which expression of the measurement is more precise? Explain. 35,797 ft; Sample answer: A foot is a smaller unit of measure than a meter.

Vocabulary Assessment

- This alphabetical list of vocabulary terms in Chapter 12 includes a page reference where each term was introduced.
- A vocabulary review/test for Chapter 12 is available on p. 712 of the *Chapter 12 Resource Masters*.

Vocabulary PuzzleMaker

ELL The online Vocabulary PuzzleMaker improves students' mathematics vocabulary using four puzzle formats — crossword, scramble, word search using a word list, and word search using clues. Students can work on a computer screen or from a printed handout.

MindJogger Videoquizzes

ELL MindJogger Videoquizzes provide an alternative review of concepts presented in this chapter. Students work in teams in a game show format to gain points for correct answers. The questions are presented in three rounds.

Round 1 Concepts (5 questions)
Round 2 Skills (4 questions)
Round 3 Problem Solving (4 questions)

ExamView® Pro Test Bank CD-ROM

Use the networkable ExamView® Pro Test Bank CD-ROM to:

- Create **Multiple versions** of tests.
- Create **Modified** tests for **Inclusion** students with one mouse click.
- **Edit** existing questions and **Add** your own questions.
- Build tests aligned with state standards using built-in **State Curriculum Correlations.**
- Change **English** tests to **Spanish** with one mouse click and vice versa.

CHAPTER 12 Study Guide and Review

Vocabulary and Concept Check

cylinder (p. 524)	precision unit (p. 542)	significant digits (p. 542)	surface area (p. 532)
precision (p. 542)	rectangular prism (p. 520)	solid (p. 514)	volume (p. 520)

Choose the correct term or number to complete each sentence.

1. A (<u>rectangular prism</u>, rectangle) is a solid figure that has three sets of parallel congruent sides.
2. The (<u>volume</u>, surface area) of a solid figure is the measure of the space occupied by it.
3. Volume is measured in (square, <u>cubic</u>) units.
4. The volume of a rectangular prism is found by (adding, <u>multiplying</u>) the length, the width, and the height.
5. A (<u>cylinder</u>, prism) is a solid that has two congruent, parallel circles as its bases.
6. Surface area is the sum of the (<u>areas</u>, volumes) of all of the outside surfaces of a three-dimensional figure.
7. The (<u>smaller</u>, larger) the precision unit, the more precise is the measurement.
8. Significant digits include all of the digits of a measurement that you know for sure, plus (<u>one</u>, two) estimated digit(s).

Lesson-by-Lesson Exercises and Examples

12-1 Drawing Three-Dimensional Figures (pp. 514–517)

Draw a top, a side, and a front view of each solid. 9–10. See margin.

9.
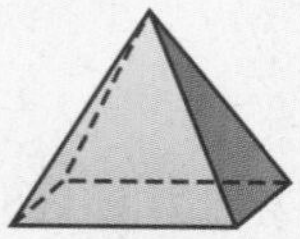

10.
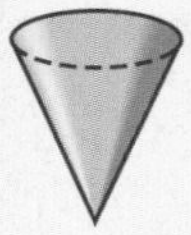

11. Draw a solid using the top, side, and front views shown. Use isometric dot paper. See margin.

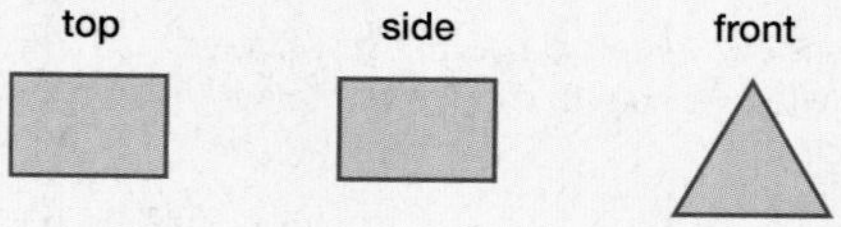

Example 1 **Draw the solid using the top, side, and front views shown.**

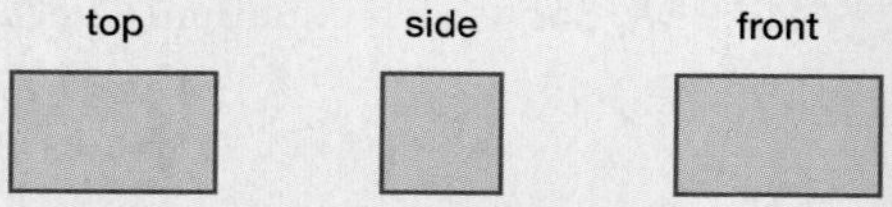

The side view is a square. The top and front views are rectangles.

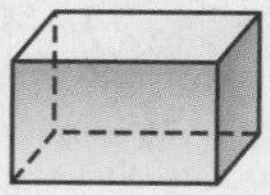

FOLDABLES™ Study Organizer

For more information about Foldables, see *Teaching Mathematics with Foldables.*

Have students look through the chapter to make sure they have included examples in their Foldables for each lesson or topic in the chapter.

Encourage students to refer to their Foldables while completing the Study Guide and Review and to use them in preparing for the Chapter Test.

12-2 Volume of Rectangular Prisms (pp. 520–522)

Find the volume of each rectangular prism. Round to the nearest tenth.

12.

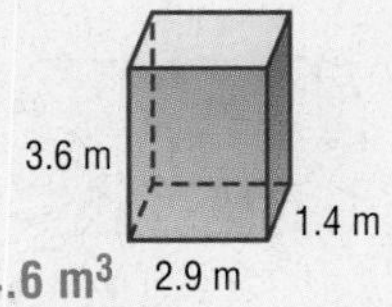

$14.6\ m^3$

13.

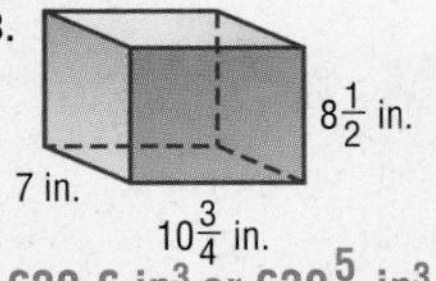

$639.6\ in^3$ or $639\frac{5}{8}\ in^3$

14. **POOLS** A swimming pool 25 yards long has 8 lanes that are each 3 yards wide. The water is 6 feet deep. Find the volume of water in the pool. $32{,}400\ ft^3$ or $1{,}200\ yd^3$

Example 2 **Find the volume of the rectangular prism.**

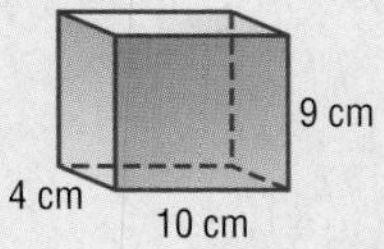

$V = \ell wh$	Volume of a rectangular prism
$V = (10)(4)(9)$	$\ell = 10$, $w = 4$, and $h = 9$.
$V = 360$	Multiply.

The volume is 360 cubic centimeters.

12-3 Volume of Cylinders (pp. 524–527)

Find the volume of each cylinder. Round to the nearest tenth.

15.

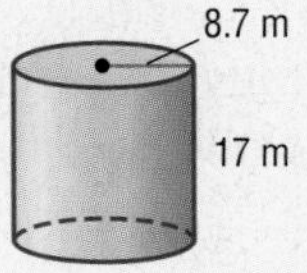

$4{,}042.4\ m^3$

16.

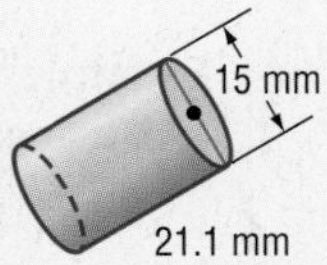

$3{,}728.7\ mm^3$

17. **POTTERY** In art class, Arturo made a vase in the shape of a cylinder. The diameter is 5 inches, and the height is 10 inches. Find the maximum volume of water the vase can hold. $196.3\ in^3$

Example 3 **Find the volume of the cylinder.**

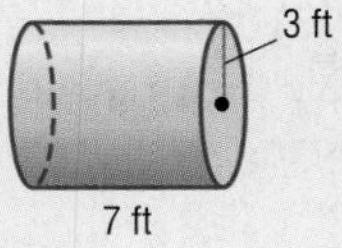

$V = \pi r^2 h$	Volume of a cylinder
$= \pi(3)^2(7)$	Replace r with 3 and h with 7.
≈ 197.9	Multiply.

The volume is about 197.9 cubic feet.

12-4 Surface Area of Rectangular Prisms (pp. 532–535)

Find the surface area of each rectangular prism. Round to the nearest tenth if necessary.

18.

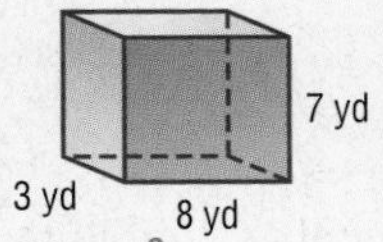

$202\ yd^2$

19.

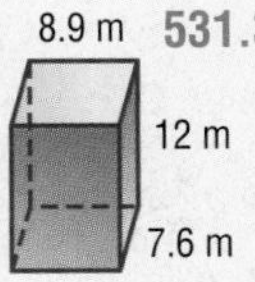

$531.3\ m^2$

20. **PETS** A plastic pet carrier box is 2.5 feet long, 1 foot high, and 1.25 feet wide. How much plastic is used to make this carrier? $13.75\ ft^2$

Example 4 **Find the surface area of the rectangular prism.**

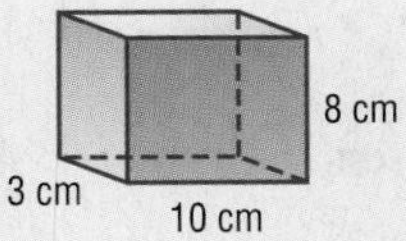

$$\text{surface area} = 2\ell w + 2\ell h + 2wh$$
$$= 2(10)(3) + 2(10)(8) + 2(3)(8)$$
$$= 268$$

The surface area is 268 square centimeters.

Answers

9–11. Sample answers given.

9.

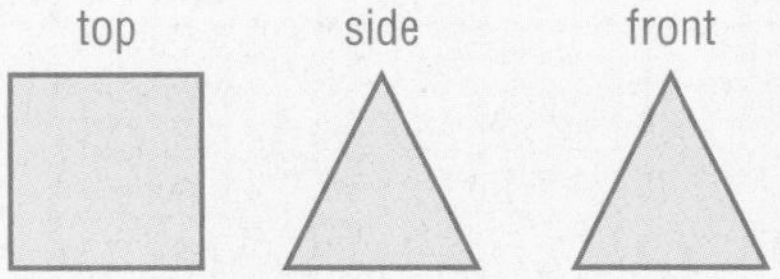

10.

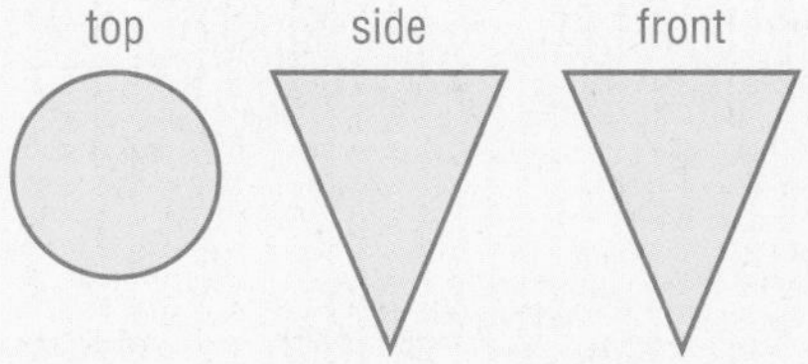

11.

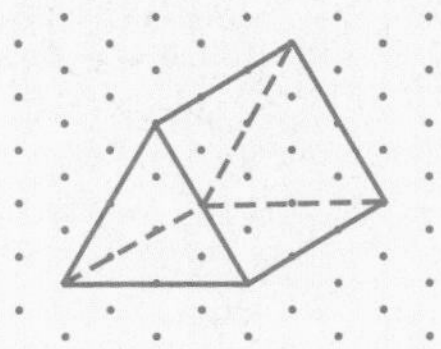

Answers (Practice Test)

3–5. Sample answers given.

3.

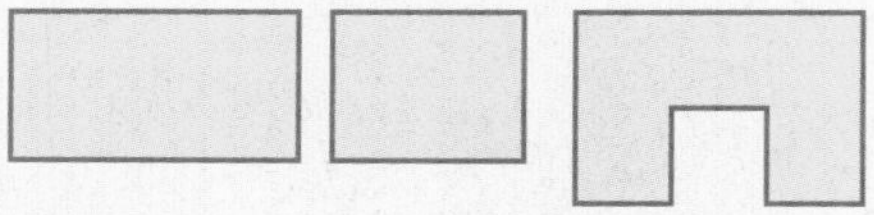

4.

5.

12-5 Surface Area of Cylinders (pp. 538–541)

Find the surface area of each cylinder. Round to the nearest tenth.

21.

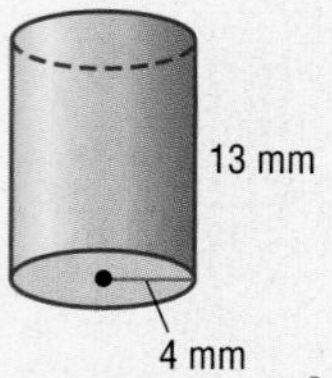

427.3 mm²

22.

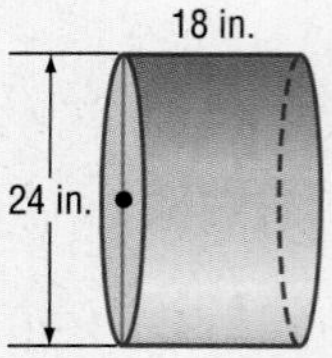

2,261.9 in²

23. **DESIGN** A can of chicken noodle soup is $5\frac{1}{2}$ inches high, and its base has a radius of 2 inches. How much paper is needed to make the label on the can? **69.1 in²**

Example 5 **Find the surface area of the cylinder.**

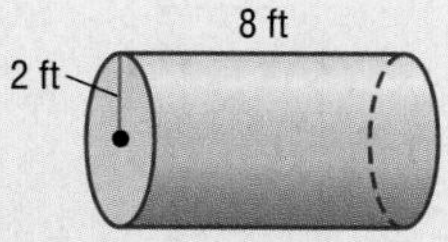

$$\text{surface area} = 2\pi r^2 + 2\pi rh$$
$$= 2\pi(2)^2 + 2\pi(2)(8)$$
$$\approx 125.7 \text{ ft}^2$$

The surface area is about 125.7 square feet.

12-6 Precision and Measurement (pp. 542–545)

Identify the precision unit of each measuring tool.

24.

$\frac{1}{2}$ in.

25.

5 min

26.

1 lb

State each measure using significant digits. **27–28. Sample answers are given.**

27.

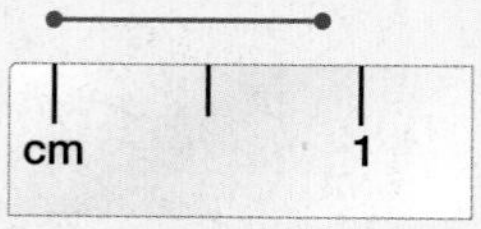

0.8 cm

28.

23 oz

Example 6 **Identify the precision unit of the measuring tool.**

The smallest unit is one eighth of an inch. So, the precision unit is $\frac{1}{8}$ inch.

Example 7 **State the measure of the graduated cylinder using significant digits.**

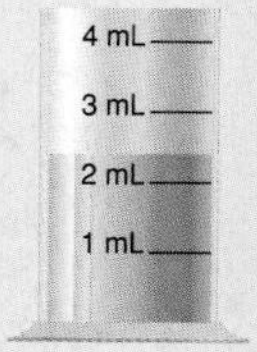

The precision unit is 1 milliliter. You know for certain that the volume is between 2 and 3 milliliters. One estimate is 2.4 milliliters.

Chapter 12 Practice Test

Vocabulary and Concepts

1. **Define** *surface area.* Sample answer: the sum of the areas of all the surfaces, or faces, of a three-dimensional figure.
2. **Explain** what a precision unit is.
Sample answer: the smallest unit on a measuring tool

Skills and Applications

Draw a top, a side, and a front view of each solid. 3–5. See margin.

3. 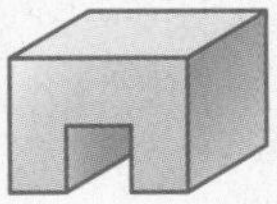

4. 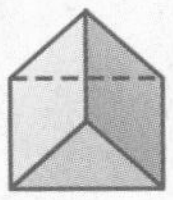

5. 

Find the volume and surface area of each rectangular prism and cylinder. Round to the nearest tenth if necessary.

6. 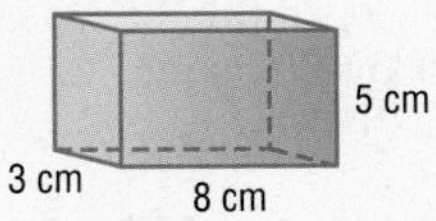

120 cm^3; 158 cm^2

7. rectangular prism;
length = 19.6 m
width = 14 m
height = 26.1 m
7,161.8 m^3; 2,302.7 m^2

8. 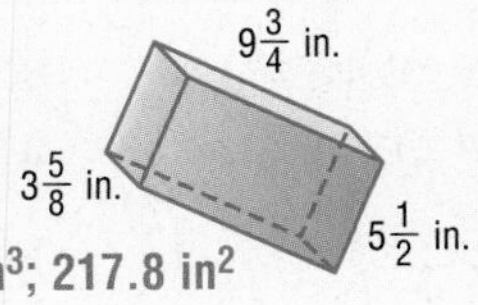

194.4 in^3; 217.8 in^2

9. cylinder;
radius = 6 ft
height = 12 ft
1,357.2 ft^3; 678.6 ft^2

10. 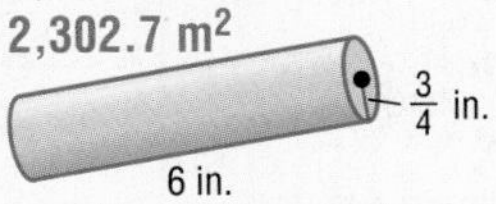

10.6 in^3; 31.8 in^2

11. cylinder;
diameter = 11.5 mm
height = 20.7 mm
2,150.1 mm^3; 955.6 mm^2

12. **DRINKING STRAWS** A drinking straw has a radius of $\frac{1}{8}$ inch and a height of $7\frac{3}{4}$ inches. How much liquid is in a straw that is half full? 0.2 in^3

13. **PACKAGING** What is the least amount of paper needed to wrap a box that is 9 inches by 18 inches by 4 inches? 540 in^2

Identify the precision unit of each measuring tool.

14. 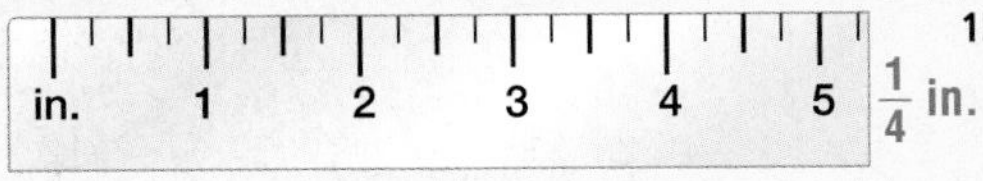

$\frac{1}{4}$ in.

15. 

15 min

Standardized Test Practice

16. **MULTIPLE CHOICE** Find the best estimate of the measure of the paper clip using significant digits. C

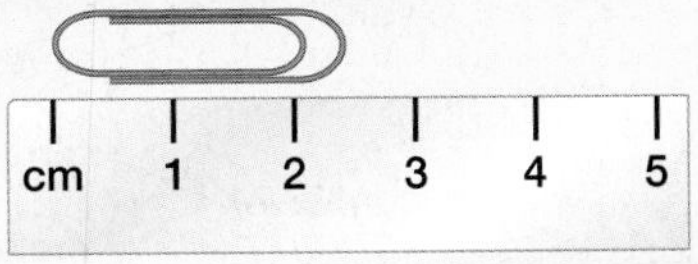

Ⓐ 2 cm
Ⓑ 2.25 cm
Ⓒ 2.4 cm
Ⓓ 2.75 cm

msmath2.net/chapter_test

Portfolio Suggestion

Introduction Students have been studying three-dimensional figures by looking at two-dimensional drawings and nets of solids.

Ask Students to research the ways in which artists use perspective to help them represent three-dimensional figures on a two-dimensional canvas. Students should include a simple illustration of a rectangular prism showing the lines of the prism extending to a vanishing point. This exercise will help them illustrate their figures more accurately.

Chapter 12 Practice Test

Assessment Options

Vocabulary Test A vocabulary review/test for Chapter 12 can be found on p. 712 of the *Chapter 12 Resource Masters.*

Chapter Tests There are six Chapter 12 Tests and an Extended Response Assessment task available in the *Chapter 12 Resource Masters.*

Chapter 12 Tests			
Form	Type	Level	Pages
1	MC	basic	699–700
2A	MC	average	701–702
2B	MC	average	703–704
2C	FR	average	705–706
2D	FR	average	707–708
3	FR	advanced	709–710

MC = multiple-choice questions
FR = free-response questions

Extended Response Assessment

Extended Response tasks for Chapter 12 can be found on p. 711 of the *Chapter 12 Resource Masters.* A sample scoring rubric for these tasks appears on p. A23.

Unit 5 Test

A unit test/review can be found on pp. 719–720 of the *Chapter 12 Resource Masters.*

End-of-Year-Tests

A **Second Semester Test** for Chapters 7–12 and a **Final Test** for Chapters 1–12 can be found on pp. 721–730 of the *Chapter 12 Resource Masters.*

CHAPTER 12 Standardized Test Practice

These two pages contain practice questions in the various formats that can be found on the most frequently given standardized tests.

A student recording sheet for these two pages can be found on p. A1 of the *Chapter 12 Resource Masters*.

Standardized Test Practice Student Recording Sheet, p. A1

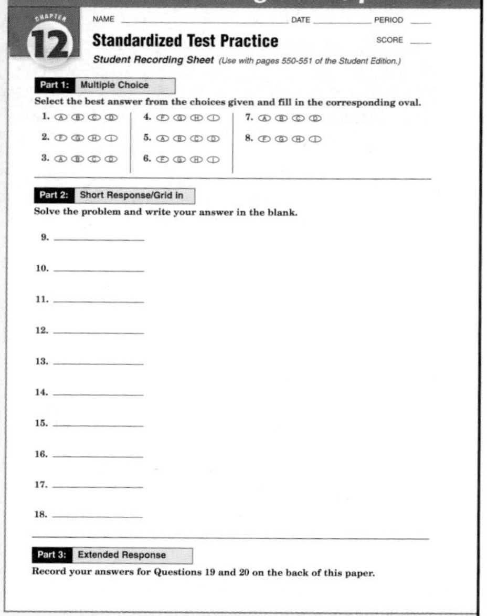
CHAPTER 12 NAME ______ DATE ______ PERIOD ______

Standardized Test Practice SCORE ______

Student Recording Sheet (Use with pages 550–551 of the Student Edition.)

Part 1: Multiple Choice

Select the best answer from the choices given and fill in the corresponding oval.

1. A B C D
2. F G H I
3. A B C D
4. F G H I
5. A B C D
6. F G H I
7. A B C D
8. F G H I

Part 2: Short Response/Grid In

Solve the problem and write your answer in the blank.

9. ______
10. ______
11. ______
12. ______
13. ______
14. ______
15. ______
16. ______
17. ______
18. ______

Part 3: Extended Response

Record your answers for Questions 19 and 20 on the back of this paper.

Additional Practice

See pp. 717–718 in the *Chapter 12 Resource Masters* for additional standardized test practice.

CHAPTER 12 Standardized Test Practice

PART 1 Multiple Choice

Record your answers on the answer sheet provided by your teacher or on a sheet of paper.

1. Which city in the table is the closest to sea level? (Lesson 3-2) **A**

City	Altitude Compared to Sea Level (ft)
Ocean City	−25
Redmond	1,376
Reston	24
West Orange	−13

A West Orange B Reston C Ocean City D Redmond

2. What is the value of w in the equation $6w + 12 = 30$? (Lesson 4-4) **F**

F 3 G 12 H 18 I 30

3. Each inch of Tyrell's train set represents $4\frac{1}{2}$ feet of an actual train system. If the engine is 6 inches long, how long is an actual engine? (Lesson 7-4) **D**

A $1\frac{1}{3}$ ft B 27 in. C $10\frac{1}{2}$ ft D 27 ft

4. The tree diagram shows the choices of flowers and pots sold at Ivitz Gardens. How many different flower-pot choices are possible? (Lesson 9-2) **G**

F 3 G 6 H 9 I 12

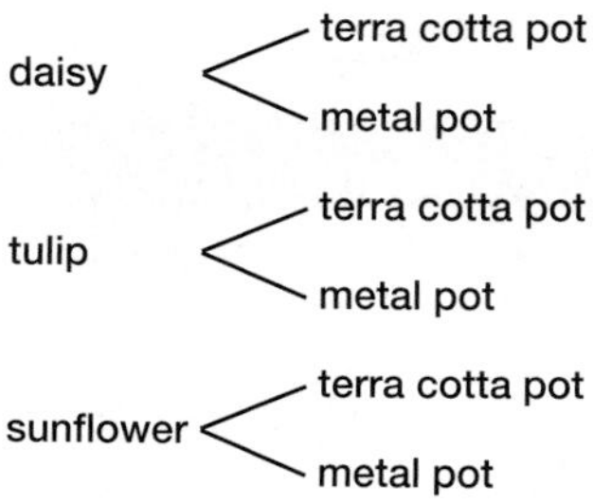

5. Which is the side view of the solid shown at the right? (Lesson 12-1) **C**

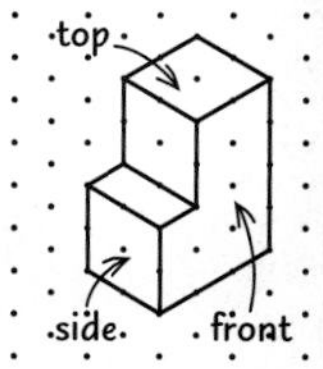

A  B 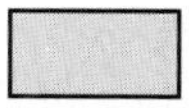

C 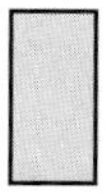 D 

6. Gillian has a window box the shape of a rectangular prism. It is 36 inches wide and 6 inches long. How much soil should she buy in order to fill the flower box with a 4-inch-deep layer of soil? (Lesson 12-2) **H**

F 46 in^3 G 576 in^3 H 864 in^3 I 1,296 in^3

7. Which is the volume of the cylinder? (Lesson 12-3) **B**

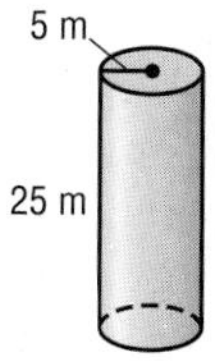

A 94.2 m^3 B 1,963.5 m^3 C 3,125.0 m^3 D 9,812.5 m^3

8. What is the surface area of the rectangular prism? (Lesson 12-4) **G**

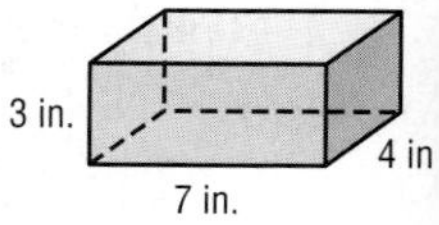

F 84 in^2 G 122 in^2 H 141 in^2 I 588 in^2

TEST-TAKING TIP

Questions 6–8 Most standardized tests include any necessary formulas in the test booklet. It helps to be familiar with formulas such as the volume or surface area of a cylinder, but use any formulas that are given to you.

The Princeton Review **Log On** for Test Practice

The Princeton Review offers additional test-taking tips and practice problems at their web site. Visit princetonreview.com or review.com

ExamView® Pro Test Bank CD-ROM

Special banks of standardized test questions similar to those on the TIMSS and NAEP grade 8 tests can be found on this CD-ROM.

T 2 Short Response/Grid In

ord your answers on the answer sheet vided by your teacher or on a sheet of er.

To find the perimeter of the rectangular tray shown below, Oi Ying wrote the equation $2 \times 8 + 2 \times 14$. What operation should Oi Ying do first? (Lesson 1-3) **Multiply 2 by 8.**

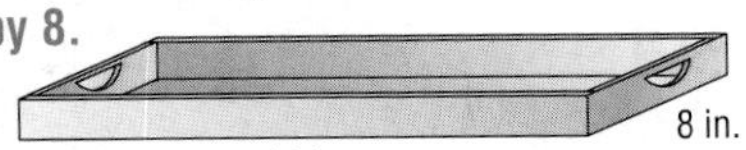

Marcos worked at a car wash during the summer. He worked with other employees to wipe the cars dry. Typically, Marcos wiped down 75% of the cars each day. What are two other ways that 75% can be expressed? (Lesson 5-6) **0.75; $\frac{3}{4}$**

Find $5\frac{1}{2} \div \frac{3}{2}$. (Lesson 6-6) **$3\frac{2}{3}$**

At the Gilmour family reunion, 28 of the 70 family members ate potato salad. What percent of the Gilmours ate potato salad at the reunion? (Lesson 7-5) **40%**

At a sale, Shay finds a $125 coat marked down to $87.50. What percent of decrease is this? (Lesson 8-4) **30%**

Alisa wants a bicycle that usually costs $186. During a sale, the bicycle sells for 20% less than that. What is the price of the bicycle when it is on sale? (Lesson 8-5) **$148.80**

Triangle *TUV* is reflected over the *x*-axis. What are the new coordinates of point *T*? (Lesson 10-9) **(−2, −3)**

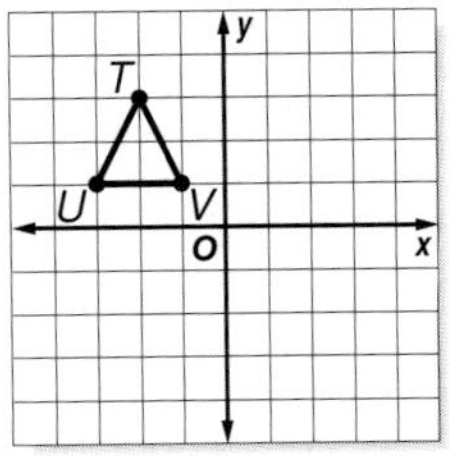

16. Find the total volume of the solid. (Lesson 12-2) **1,209.6 ft^3**

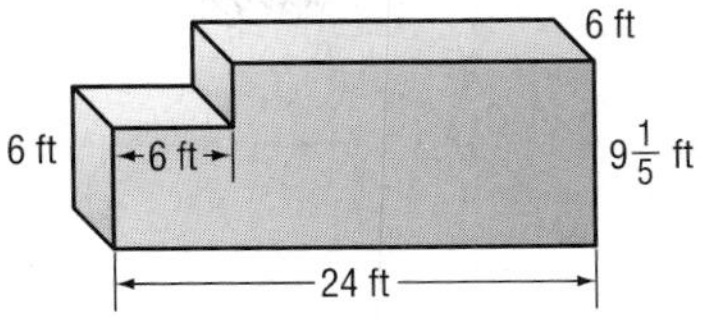

17. What is the surface area of a rectangular prism that is 14.2 centimeters long, 10.5 centimeters wide, and 3 centimeters high? (Lesson 12-4) **446.4 cm^2**

18. The cylindrical gasoline storage tank on Mrs. Simon's farm needs to be painted. The dimensions are shown below.

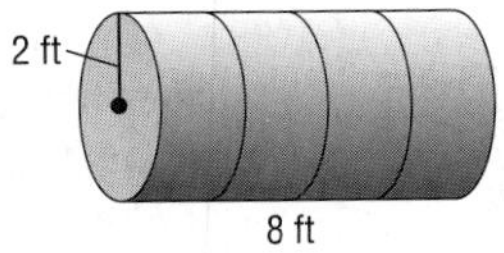

If one quart of paint covers 85 square feet, how many quarts does Mrs. Simon need for this job? (Lesson 12-5) **$1\frac{1}{2}$ qt**

PART 3 Extended Response

Record your answers on a sheet of paper. Show your work.

19. Draw a top view, a front view, and a side view of the *tetrahedron* shown at the right. (Lesson 12-1)

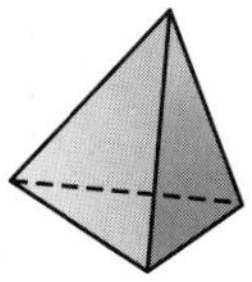

19–20. See margin.

20. Determine whether the given precision unit is appropriate to use in estimating the length of each object. Explain your reasoning. If it is not, describe an appropriate precision unit. (Lesson 12-6)
   a. driveway: 1 inch
   b. compact disc: 1 meter
   c. calculator key: 0.1 centimeter

msmath2.net/standardized_test

Standardized Test Practice

Rubrics The following rubrics are sample scoring devices for short response and extended response questions.

Short Response

Points	Description
2	The student demonstrates a **thorough understanding** of the mathematics of the task. The response may contain minor flaws that do not detract from the demonstration of a thorough understanding.
1	The student has provided a response that is only **partially correct**.
0	The student has provided a **completely incorrect** solution or no response at all.

Extended Response

Points	Description
4	The student demonstrates a **thorough understanding** of the mathematics of the task. The response may contain minor flaws that do not detract from the demonstration of a thorough understanding.
3	The student demonstrates an **understanding** of the mathematics of the task. The response is essentially correct and demonstrates an essential but less than thorough understanding of the mathematics.
2	The student demonstrates only a **partial understanding** of the mathematics of the task. Although the student may have used the correct approach to a solution or may have provided a correct solution, the work lacks an essential understanding of the underlying mathematical concepts.
1	The student demonstrates a **very limited understanding** of the mathematics of the task. The response is incomplete and exhibits many flaws.
0	The student provides a **completely incorrect** solution or no response at all.

A more detailed 4-point rubric is provided on page A2 of the *Chapter 12 Resource Masters*.

ers

mple answer:

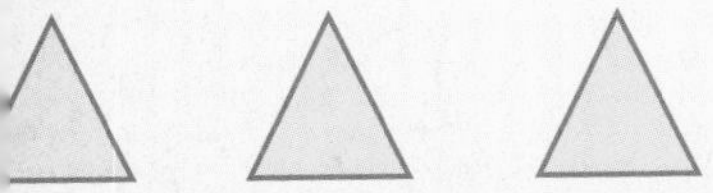

); the precision unit is too small to be useful making an estimate. Sample answer: feet

20b. No; a compact disc is much smaller than 1 meter. So the precision unit is too large. Sample answer: centimeter

20c. Yes; most calculator keys are approximately 1 centimeter long, so a precision unit of 0.1 centimeters would be useful in estimating length.

Pages 515–516, Lesson 12-1

3–4, 7–12. Sample answers given.

3.

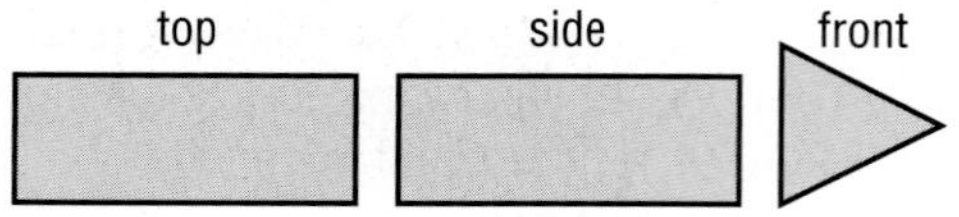

4.

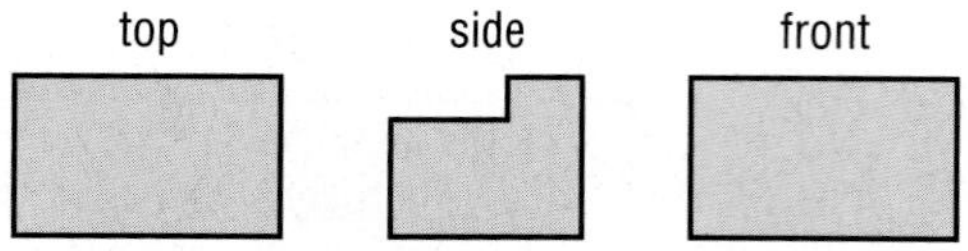

7.

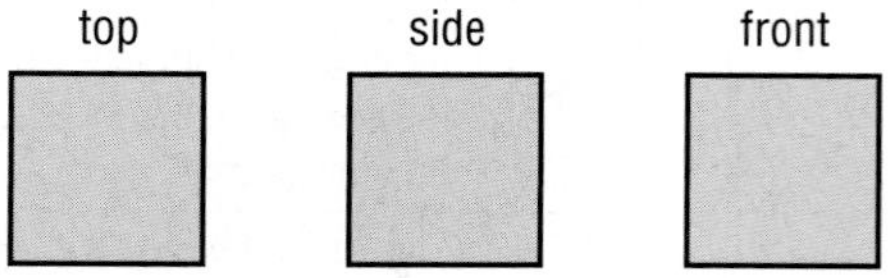

8.

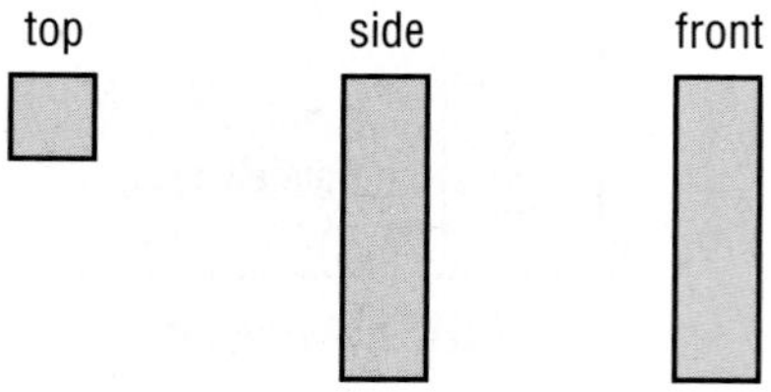

9.

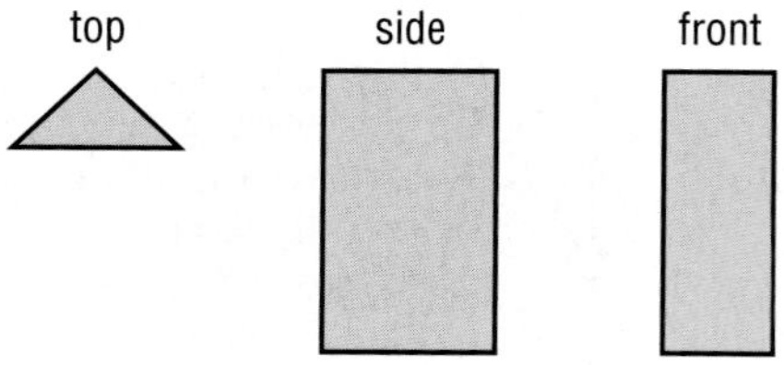

10.

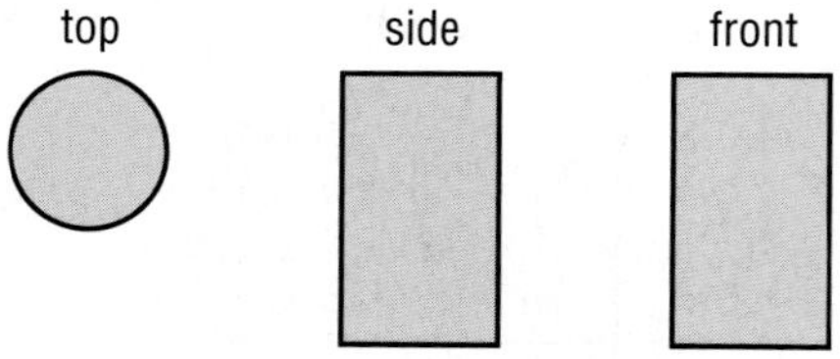

11.

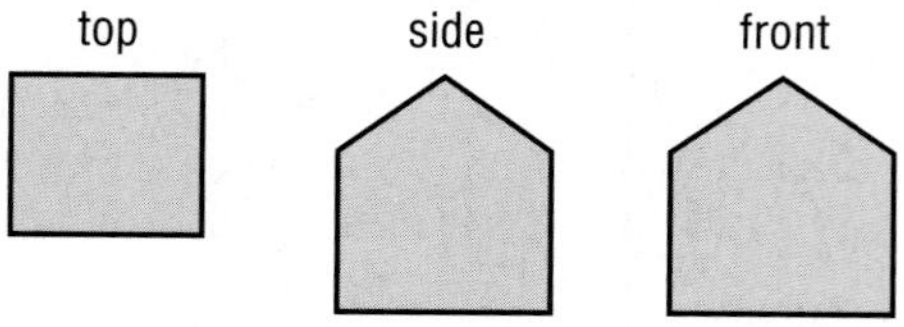

12.

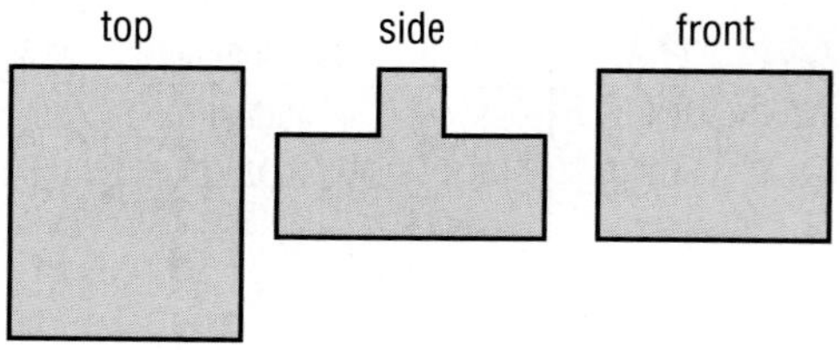

Page 522, Lesson 12-2

21. Sample answer:

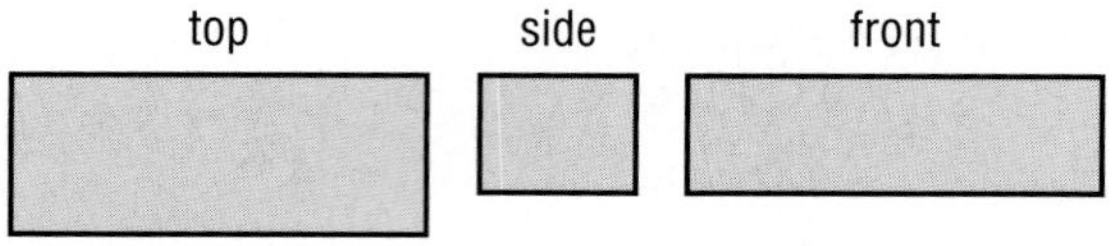

Page 528, Mid-Chapter Practice Test

3–5. Sample answers given.

3.

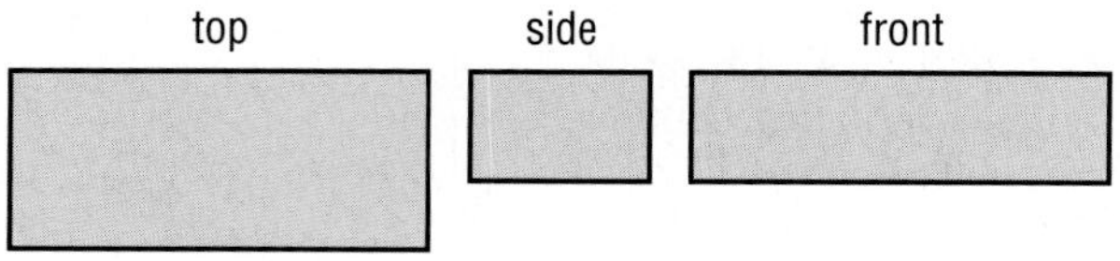

4.

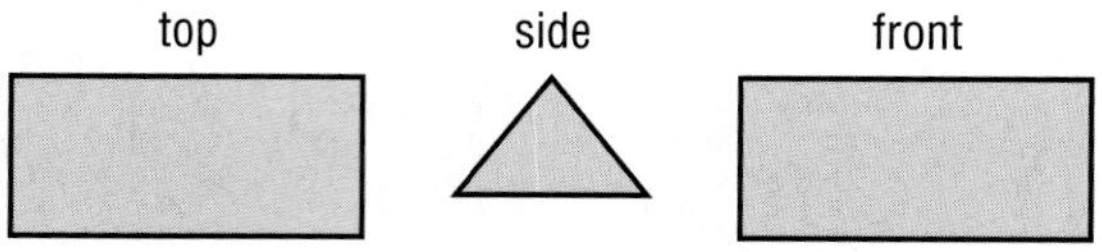

5.

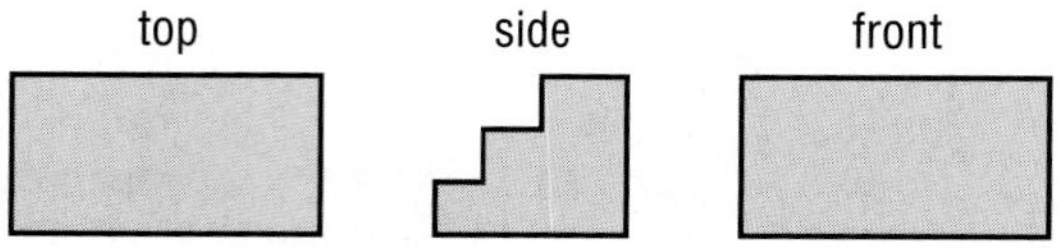

NOTES

STUDENT HANDBOOK

Skills

Prerequisite Skills **554**

Extra Practice **564**

Mixed Problem Solving **596**

Reference

English-Spanish Glossary **608**

Selected Answers **623**

Photo Credits **641**

Index **642**

Formulas and Symbols .... **Inside Back Cover**

How To Cover Your Book ...... **Inside Back Cover**

HOW TO...

USE THE STUDENT HANDBOOK

A Student Handbook is the additional skill and reference material found at the end of books. The Student Handbook can help answer these questions.

What If I Forget What I Learned Last Year?

Use the **Prerequisite Skills** section to refresh your memory about things you have learned in other math classes. Here's a list of the topics covered in your book.

- Divisibility Patterns
- Place Value and Decimals
- Comparing and Ordering Decimals
- Rounding Decimals
- Estimating with Decimals
- Adding and Subtracting Decimals
- Multiplying Decimals
- Powers of Ten
- Dividing Decimals
- Mixed Numbers and Improper Fractions

What If I Need More Practice?

You, or your teacher, may decide that working through some additional problems would be helpful. The **Extra Practice** section provides these problems for each lesson so you have ample opportunity to practice new skills.

What If I Have Trouble with Word Problems?

The **Mixed Problem Solving** portion of the book provides additional word problems that use the skills presented in each chapter. These problems give you real-life situations where the math can be applied.

What If I Forget a Vocabulary Word?

The **English-Spanish Glossary** provides a list of important, or difficult, words used throughout the textbook. It provides a definition in English and Spanish as well as the page number(s) where the word can be found.

What If I Need to Check a Homework Answer?

The answers to the odd-numbered problems are included in **Selected Answers**. Check your answers to make sure you understand how to solve all of the assigned problems.

What If I Need to Find Something Quickly?

The **Index** alphabetically lists the subjects covered throughout the entire textbook and the pages on which each subject can be found.

What If I Forget a Formula?

Inside the back cover of your math book is a list of **Formulas and Symbols** that are used in the book.

Need to Cover Your Book?

Inside the back cover are directions for a Foldable that you can use to cover your math book quickly and easily!

Prerequisite Skills

Prerequisite Skills

Divisibility Patterns

In 54 ÷ 6 = 9, the quotient, 9, is a whole number. So, we say that 54 is **divisible** by 6. You can use the following rules to determine whether a number is divisible by 2, 3, 4, 5, 6, 9, and 10.

A number is divisible by:

- 2 if the ones digit is divisible by 2.
- 3 if the sum of the digits is divisible by 3.
- 4 if the number formed by the last two digits is divisible by 4.
- 5 if the ones digit is 0 or 5.
- 6 if the number is divisible by both 2 and 3.
- 9 if the sum of the digits is divisible by 9.
- 10 if the ones digit is 0.

EXAMPLE Use Divisibility Rules

Determine whether 972 is divisible by 2, 3, 4, 5, 6, 9, or 10.

2: Yes; the ones digit, 2, is divisible by 2.

3: Yes; the sum of the digits, 9 + 7 + 2 = 18, is divisible by 3.

4: Yes; the number formed by the last two digits, 72, is divisible by 4.

5: No; the ones digit is not 0 or 5.

6: Yes; the number is divisible by 2 and 3.

9: Yes; the sum of the digits, 18, is divisible by 9.

10: No; the ones digit is not 0.

Exercises

Use divisibility rules to determine whether the first number is divisible by the second number.

1. 447; 3 yes
2. 135; 6 no
3. 240; 4 yes
4. 419; 3 no
5. 831; 3 yes
6. 4,408; 4 yes
7. 7,110; 5 yes
8. 1,287; 9 yes
9. 2,984; 9 no
10. 7,026; 6 yes
11. 1,260; 10 yes
12. 8,903; 6 no

Determine whether each number is divisible by 2, 3, 4, 5, 6, 9, or 10.

13. 712 2, 4
14. 1,035 3, 5, 9
15. 8,901 3, 9
16. 462 2, 3, 6
17. 270 2, 3, 5, 6, 9, 10
18. 1,005 3, 5
19. 32,221 none
20. 8,340 2, 3, 4, 5, 6, 10
21. 920 2, 4, 5, 10
22. 50,319 3, 9
23. 64,042 2
24. 3,498 2, 3, 6
25. **MEASUREMENT** Jordan has 5,280 feet of rope. Can he cut the rope into 9-foot pieces and use all of the rope? Explain. No; 5,280 is not divisible by 9.

Prerequisite Skills

554 Prerequisite Skills

e Value and Decimals

umber system is based on units of ten.
ers like 1.35, 0.8, and 25.09 are called
ls.

n use a place-value chart like the one at the
 help you read decimals and write them in
 Use the word *and* to represent the decimal

10,000	1,000	100	10	1		0.1	0.01	0.001	0.0001
ten thousands	thousands	hundreds	tens	ones	decimal point	tenths	hundredths	thousandths	ten-thousandths
			2	0	•	0	9		

AMPLES Write Decimals in Words

rite each decimal in words.

.09

...	ones	decimal point	tenths	hundredths
2	0	•	0	9

wenty and nine hundredths

2 6.738

ones	decimal point	tenths	hundredths	thousandths
6	•	7	3	8

six and seven hundred thirty-eight thousandths

n also write decimals in **expanded notation**
lace value and decimals or their fraction
lents as shown at the right.

Decimal	0.1	0.01	0.001	0.0001
Fraction	$\frac{1}{10}$	$\frac{1}{10^2}$	$\frac{1}{10^3}$	$\frac{1}{10^4}$

AMPLE Write a Decimal in Expanded Notation

rite 2.814 in expanded notation using decimals and using fractions.

rite the product of each digit and its place value.

$814 = (2 \times 1) + (8 \times 0.1) + (1 \times 0.01) + (4 \times 0.001)$

$814 = (2 \times 1) + \left(8 \times \frac{1}{10}\right) + \left(1 \times \frac{1}{10^2}\right) + \left(4 \times \frac{1}{10^3}\right)$ Replace the decimals with their fraction equivalents.

ses

ach decimal in words. 1–21. See margin.

7 **2.** 13.7 **3.** 145.66 **4.** 92.03 **5.** 4,280.899

7. 0.69 **8.** 0.0001 **9.** 7.005 **10.** 2000.0002

ach decimal in expanded notation using decimals and using fractions.

12. 5.3 **13.** 6.79 **14.** 9.132 **15.** 65.002

25 **17.** 0.0072 **18.** 100.001 **19.** 548.845 **20.** 4,260.705

TS The northern blossom bat is one of the world's smallest bats. It
ighs just 0.53 ounce. Write its weight in expanded notation using fractions.

12. $(5 \times 1) + (3 \times 0.1)$; $(5 \times 1) + \left(3 \times \frac{1}{10}\right)$
13. $(6 \times 1) + (7 \times 0.1) + (9 \times 0.01)$; $(6 \times 1) + \left(7 \times \frac{1}{10}\right) + \left(9 \times \frac{1}{10^2}\right)$
14. $(9 \times 1) + (1 \times 0.1) + (3 \times 0.01) + (2 \times 0.001)$; $(9 \times 1) + \left(1 \times \frac{1}{10}\right) + \left(3 \times \frac{1}{10^2}\right) + \left(2 \times \frac{1}{10^3}\right)$
15. $(6 \times 10) + (5 \times 1) + (2 \times 0.001)$; $(6 \times 10) + (5 \times 1) + \left(2 \times \frac{1}{10^3}\right)$
16. $(6 \times 0.1) + (2 \times 0.01) + (5 \times 0.001)$; $\left(6 \times \frac{1}{10}\right) + \left(2 \times \frac{1}{10^2}\right) + \left(5 \times \frac{1}{10^3}\right)$
17. $(7 \times 0.001) + (2 \times 0.0001)$; $\left(7 \times \frac{1}{10^3}\right) + \left(2 \times \frac{1}{10^4}\right)$
18. $(1 \times 100) + (1 \times 0.001)$; $(1 \times 100) + \left(1 \times \frac{1}{10^3}\right)$
19. $(5 \times 100) + (4 \times 10) + (8 \times 1) + (8 \times 0.1) + (4 \times 0.01) + (5 \times 0.001)$; $(5 \times 100) + (4 \times 10) + (8 \times 1) + \left(8 \times \frac{1}{10}\right) + \left(4 \times \frac{1}{10^2}\right) + \left(5 \times \frac{1}{10^3}\right)$
20. $(4 \times 1{,}000) + (2 \times 100) + (6 \times 10) + (7 \times 0.1) + (5 \times 0.001)$; $(4 \times 1{,}000) + (2 \times 100) + (6 \times 10) + \left(7 \times \frac{1}{10}\right) + \left(5 \times \frac{1}{10^3}\right)$
21. $\left(5 \times \frac{1}{10}\right) + \left(3 \times \frac{1}{10^2}\right)$

Answers

1. six and thirty-seven hundredths
2. thirteen and seven tenths
3. one hundred forty-five and sixty-six hundredths
4. ninety-two and three hundredths
5. four thousand two hundred eighty and eight hundred ninety-nine thousandths
6. six tenths
7. sixty-nine hundredths
8. one ten-thousandth
9. seven and five thousandths
10. two thousand and two ten-thousandths
11. (8×0.1); $\left(8 \times \frac{1}{10}\right)$

Answers

19. 5.009, 5.07, 5.13
20. 7.088, 7.9, 7.98, 8.02
21. 0.087, 0.901, 1.001, 2
22. 12, 12.008, 12.3, 12.54
23. 60.05, 60.5, 60.505, 60.55
24. 505.09, 505.9, 505.91, 505.99
25. 0.002, 0.02, 0.022, 0.2
26. 0.09, 0.099, 0.90, 0.99, 0.999

Comparing and Ordering Decimals

To determine which of two decimals is greater, you can compare the digits in each place-value position, or you can use a number line.

EXAMPLES Order Decimals

1 Which is greater, 7.4 or 7.63?

Method 1 Use place value.

7.4 Line up the decimal points.

7.63 Starting at the left, compare the digits in each place-value position.
↑

The digits in the tenths place are not the same.

Since 6 tenths > 4 tenths, 7.63 > 7.4.

Method 2 Use a number line.

Graph each decimal on a number line and compare.

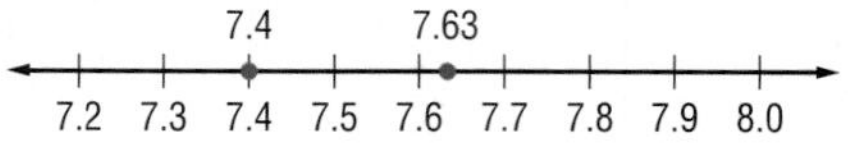

On a number line, numbers to the right are greater than numbers to the left.

7.63 is to the right of 7.4, so 7.63 > 7.4.

2 Order 12.15, 12.08, and 12.103 from least to greatest.

12.15 Line up the decimal points.

12.08 12.08 is less than 12.15 and 12.103 since 0 < 1 in the tenths place.

12.103 12.103 is less than 12.15 since 0 < 5 in the hundredths place.

So, the numbers from least to greatest are 12.08, 12.103, and 12.15.

Exercises

Replace each ● with <, >, or = to make a true sentence.

1. 1.22 ● 1.02 >
2. 0.97 ● 1.06 <
3. 7.90 ● 7.9 =
4. 1.3 ● 1.31 <
5. 4.03 ● 4.01 >
6. 0.77 ● 0.69 >
7. 0.8 ● 0.08 >
8. 0.68 ● 0.680 =
9. 3.28 ● 3.279 >
10. 25.23 ● 25.32 <
11. 77.55 ● 77.65 <
12. 1.29 ● 1.43 <
13. 310.36 ● 310.3600 =
14. 0.0034 ● 0.034 <
15. 9.09 ● 9 >
16. 1.67 ● 0.48 >
17. 738 ● 738.99 <
18. 1,000.01 ● 1,000 >

Order each group of decimals from least to greatest. 19–26. See margin.

19. 5.13, 5.07, 5.009
20. 7.9, 7.088, 8.02, 7.98
21. 0.087, 0.901, 2, 1.001
22. 12.3, 12.008, 12.54, 12
23. 60.5, 60.05, 60.55, 60.505
24. 505.9, 505.09, 505.91, 505.99
25. 0.02, 0.2, 0.022, 0.002
26. 0.99, 0.90, 0.09, 0.999, 0.099

27. Place decimal points in 999, 463, 208, and 175 so that the resulting decimals are in order from least to greatest. Sample answer: 0.999, 4.63, 20.8, 175
28. Order 3, 0.3, 3.33, 0.33, and 3.03 from least to greatest. 0.3, 0.33, 3, 3.03, 3.33

556 Prerequisite Skills

Prerequisite Skills

ıding Decimals

a number has more decimal places than you want or need, you can
it. Use the rules below to round a number to any place value.

ıg ls	Look at the digit to the right of the place being rounded. • If the digit is 4 or less, the digit being rounded remains the same. • If the digit is 5 or greater, the digit being rounded is rounded up.

AMPLES Round Decimals

ɔund 3.92 to the nearest tenth.

ths place

ɔ2 Look at the digit to the right of the tenths place.

Since 2 is less than 5, the digit in the tenths place stays the same.

ɔ2 rounded to the nearest tenth is 3.9.

ɔund 46.297 to the nearest hundredth.

ıdredths place
↓

.297 Look at the digit to the right of the hundredths place.

Since 7 is greater than 5, round 9 up.

.297 rounded to the nearest hundredth is 46.30.

ises

each number to the given place value.

15; tenth **0.3** 2. 0.2456; hundredth **0.25** 3. 17.499; tenth **17.5**

219; hundredth **43.22** 5. 15.522; tenth **15.5** 6. 9.6; ones **10**

375; thousandth **7.038** 8. 16.399; tenth **16.4** 9. 6.95; tenth **7.0**

1283; thousandth **99.128** 11. 1,000.37; tenth **1,000.4** 12. 750.523; ones **751**

45; hundredth **0.45** 14. 490.299; ones **490** 15. 999.99; tenth **1,000.0**

.835; hundredth **476.84** 17. 682.596; tenth **682.6** 18. 1,000.562; ones **1,001**

each number to the underlined place-value position.

89 **0.79** 20. 0.96 **1.0** 21. 1.57246 **1.57**

48 **23** 23. 1.704 **1.70** 24. 0.163 **0.2**

451 **15.5** 26. 4.52988 **4.530** 27. 0.787 **0.8**

56 **39** 29. 59.61 **60** 30. 0.555 **1**

55 **0.6** 32. 0.555 **0.56** 33. 10.929 **10.9**

929 **10.93** 35. 425.599 **426** 36. 0.92929 **0.9293**

RSES A trainer recorded a racehorse running at 47.54 miles per hour. ıund this speed to the nearest mile per hour. **48 mph**

PULATION In 2000, the population of Texas was 20.85 million. Round the nearest million. **21 million**

Prerequisite Skills 557

Answers

16–23. Sample answers are given.

16. $3 \times 20 = 60$
17. $4 \times 60 = 240$
18. $4 \times 40 = 160$
19. $4 \times 80 = 320$
20. $4 \times 240 = 960$
21. $6 \times 10 = 60$
22. $5 \times 50 = 250$
23. $5 \times 100 = 500$

Estimating with Decimals

Estimation can be used to provide quick answers when an exact answer is not necessary. It is also an excellent way to check whether your answer is reasonable. One method of estimating is to use rounding. Round numbers to any place value that makes estimation easier.

EXAMPLES Estimate by Rounding

Estimate by rounding.

1 **23.485 − 9.757**

$$\begin{array}{rcr} 23.485 & \rightarrow & 23 \\ \underline{-9.757} & \rightarrow & \underline{-\ 10} \\ & & 13 \end{array}$$

Round to the nearest whole numbers.

The difference is about 13.

2 **43.9 × 37.5**

$$\begin{array}{rcr} 43.9 & \rightarrow & 40 \\ \underline{\times\ 37.5} & \rightarrow & \underline{\times\ 40} \\ & & 1{,}600 \end{array}$$

Round to the nearest ten.

The product is about 1,600.

3 **6.43 + 2.17 + 9.1 + 4.87**

$$\begin{array}{rcr} 6.43 & \rightarrow & 6 \\ 2.17 & \rightarrow & 2 \\ 9.1 & \rightarrow & 9 \\ \underline{+\ 4.87} & \rightarrow & \underline{+\ 5} \\ & & 22 \end{array}$$

Round to the nearest whole numbers.

The sum is about 22.

4 **432.87 ÷ 8.9**

$8.9 \rightarrow 9$ Round the divisor.

$432.87 \rightarrow 450$ Round the dividend to a multiple of 9.

$432.87 \div 8.9 \rightarrow 450 \div 9 = 50$

The quotient is about 50.

Another way to estimate sums is to use **clustering**. This strategy is used when all the numbers are close to a common value.

EXAMPLE Estimate by Clustering

5 **Estimate 9.775 + 9.862 + 9.475 + 9.724 by clustering.**

All of the numbers are clustered around 10. There are four numbers. So, the sum is about 4×10 or 40.

Exercises

Estimate by rounding. 1–15. Sample answers are given. 14. $122 + 488 = 610$

1. $8.56 + 5.34$ $9 + 5 = 14$
2. $34.84 - 17.69$ $35 - 18 = 17$
3. 6.8×2.4 $7 \times 2 = 14$
4. $40.79 \div 6.8$ $42 \div 7 = 6$
5. $6.9 + 5.2$ $7 + 5 = 12$
6. $23.84 + 12.13$ $24 + 12 = 36$
7. $34.3 - 18.9$ $34 - 19 = 15$
8. 7.5×8.4 $8 \times 8 = 64$
9. $65.48 \div 9.3$ $63 \div 9 = 7$
10. 26.3×9.7 $26 \times 10 = 260$
11. $33.21 - 8.23$ $33 - 8 = 25$
12. $67.86 - 24.35$ $68 - 24 = 44$
13. $8.99 \div 2.6$ $9 \div 3 = 3$
14. $121.5 + 487.8$
15. 32.5×81.4 $30 \times 80 = 2{,}400$

Estimate by clustering. 16–23. See margin.

16. $18.4 + 22.5 + 20.7$
17. $56.9 + 63.2 + 59.3 + 61.1$
18. $42.3 + 41.5 + 39.8 + 40.4$
19. $77.8 + 75.6 + 81.2 + 79.9$
20. $239.8 + 242.43 + 236.20 + 240.77$
21. $9.9 + 10.0 + 10.3 + 11.1 + 9.8 + 11.2$
22. $50.4 + 51.1 + 48.9 + 49.5 + 50.8$
23. $100.5 + 97.8 + 101.6 + 100.2 + 99.3$

Prerequisite Skills

Prerequisite Skills

ing and Subtracting Decimals

. or subtract decimals, write the numbers in a column and line up the l points. In some cases, you may want to *annex*, or place zeros at the the decimals, to help align the columns. Then add or subtract as with numbers and bring down the decimal point. Always estimate first to ether your answer is reasonable.

AMPLES Add and Subtract Decimals

nd each sum or difference.

.6 + 4.5 **Estimate** 13 + 5 = 18

2.6 Line up the decimal points.
4.5 Add.
7.1 The sum is close to the estimate.

2 42.17 − 15.85 **Estimate** 42 − 16 = 26

42.17 Line up the decimal points.
− 15.85 Subtract.
26.32 The difference is close to the estimate.

.125 + 7.63 **Estimate** 57 + 8 = 65

7.125
7.630 Annex a zero to align the columns.
4.755 Compare to the estimate.

4 25 − 15.25 **Estimate** 25 − 15 = 10

25.00 Annex two zeros to align the columns.
− 15.25
9.75 Compare to the estimate.

ises

ach sum or difference.

).132 /).021 / **).111**

2. 3.78 + 0.21 **3.99**

3. 13.2 + 12.8 **26.0**

4. 5.86 − 1.51 **4.35**

5. 42.07 − 38.78 **3.29**

14.7 / 351.82 / **66.52**

7. 42.3 + 0.81 **43.11**

8. 12.3 − 0.847 **11.453**

9. 342.9 − 0.18 **342.72**

10. 282.45 − 111.3 **171.15**

100 / 0.48 / **.52**

12. 82.23 + 0.88 **83.11**

13. 128.01 − 39.117 **88.893**

14. 80.05 − 79.06 **0.99**

15. 104.98 − 0.12 **104.86**

2 + 0.68 **1.10**
17. 0.48 + 2.901 **3.381**
18. 5.8 + 3.92 **9.72**

63 + 38.63 **77.26**
20. 8 − 2.54 **5.46**
21. 16.354 − 0.2 **16.154**

25 + 0.78 **0.905**
23. 8.2 − 6.9 **1.3**
24. 1.245 + 3.842 **5.087**

+ 1.23 **4.43**
26. 0.889 − 0.3 **0.589**
27. 22.22 + 1.475 **23.695**

− 0.25 **9.75**
29. 33.16 − 0.08 **33.08**
30. 1.254 + 0.5 **1.754**

698 − 14.903 **29.795**
32. 10 − 0.005 **9.995**
33. 722.86 + 0.024 **722.884**

).211 + 8.004 **108.215**
35. 86.124 + 32.822 **118.946**
36. 6.9 + 1.1 **8.0**

− 0.24 **74.76**
38. 13 − 0.324 **12.676**
39. 0.8 + 1.2 **2.0**

ıat is the sum of 35.009 and 3.6? **38.609**

ıd 9.1 minus 5.625. **3.475**

R TRAVEL The numbers of passengers, in millions, passing through ır airports in a recent year were 80.2, 72.1, 68.5, and 60.7. Find the total ımber of passengers for the four airports. **281.5 million passengers**

Prerequisite Skills 559

Prerequisite Skills

Multiplying Decimals

To multiply decimals, multiply as with whole numbers. The product has the same number of decimal places as the sum of the decimal places of the factors. Use estimation to determine whether your answers are reasonable.

EXAMPLES Multiply Decimals

Multiply.

1 **1.3 × 0.9** **Estimate** 1 × 1 = 1

1.3	←	1 decimal place
× 0.9	←	1 decimal place
1.17	←	2 decimal places

The product is reasonable.

2 **0.054 × 1.6** **Estimate** 0 × 2 = 0

0.054	←	3 decimal places
× 1.6	←	1 decimal place
324		
540		
0.0864		Annex a zero on the left so the answer has four decimal places. Compare to the estimate.

Exercises

Place the decimal point in each product. Add zeros if necessary.

1. 1.32 × 4 = 528 **5.28**
2. 0.07 × 1.1 = 77 **0.077**
3. 0.4 × 0.7 = 28 **0.28**
4. 1.9 × 0.6 = 114 **1.14**
5. 1.4 × 0.09 = 126 **0.126**
6. 5.48 × 3.6 = 19728 **19.728**
7. 4.5 × 0.34 = 153 **1.53**
8. 0.45 × 0.02 = 9 **0.009**
9. 150.2 × 32.75 = 4919050 **4,**

Multiply.

10. 0.2 × 6 **1.2**
11. 0.3 × 0.9 **0.27**
12. 0.45 × 0.12 **0.054**
13. 0.0023 × 32 **0.0736**
14. 1.5 × 2.7 **4.05**
15. 10.1 × 9 **90.9**
16. 2 × 0.3 **0.6**
17. 6.78 × 1.3 **8.814**
18. 200 × 0.004 **0.8**
19. 0.0023 × 0.35 **0.000805**
20. 15.8 × 11 **173.8**
21. 88 × 2.5 **220**
22. 33 × 0.03 **0.99**
23. 36 × 0.46 **16.56**
24. 0.003 × 482 **1.446**
25. 1.88 × 1.11 **2.0868**
26. 0.6 × 2 **1.2**
27. 38.3 × 29.1 **1,114.53**
28. 0.7 × 18 **12.6**
29. 8 × 0.3 **2.4**
30. 12.2 × 12.4 **151.28**
31. 380 × 1.25 **475**
32. 42 × 0.17 **7.14**
33. 0.4 × 16 **6.4**
34. 0.23 × 0.2 **0.046**
35. 0.44 × 0.5 **0.22**
36. 0.44 × 55 **24.2**
37. 44 × 0.55 **24.2**
38. **JOBS** Antonia earns $10.75 per hour. What are her total weekly earnings if she works 34.5 hours? Round to the nearest cent. **$370.88**

560 Prerequisite Skills

Prerequisite Skills

ers of Ten

use a pattern to mentally find the product of any : and a power of 10. Count the number of zeros in ver of 10 or use the exponent. Then move the point that number of places *to the right*.

Decimal	Power of 10	Product
19.7 × 10^1 (or 10)		= 197
19.7 × 10^2 (or 100)		= 1,970
19.7 × 10^3 (or 1,000)		= 19,700
19.7 × 10^4 (or 10,000)		= 197,000

MPLES Use Mental Math to Multiply

ltiply mentally.

562 × 100

562 × 100 = 12.562 Move the decimal point two places to the right, since 100 has two zeros.

= 1,256.2

9 × 10^4

9 × 10^4 = 0.5900 Move the decimal point four places to the right, since the exponent is 4.

= 5,900

tally multiply by a power of ten that is less count the number of decimal places. Or, if ver is written as a fraction, use the exponent lenominator. Then move the decimal point mber of places *to the left*.

Decimal	Power of 10	Product
19.7 × 0.1 $\left(\text{or } \frac{1}{10^1}\right)$		= 1.97
19.7 × 0.01 $\left(\text{or } \frac{1}{10^2}\right)$		= 0.197
19.7 × 0.001 $\left(\text{or } \frac{1}{10^3}\right)$		= 0.0197

AMPLES Use Mental Math to Multiply

ultiply mentally.

.5 × 0.01

.5 × 0.01 = 10.5 Move the decimal point two places to the left.

= 0.105

284 × 0.00001

284 × 0.00001 = 05284 Move the decimal point five places to the left.

= 0.05284

ises

ly mentally.

53 × 10 **125.3**
2. 4.6 × 10^3 **4,600**
3. 78.4 × 0.01 **0.784**
5 × 100 **5**
5. 4.527 × 10^0 **4.527**
6. 2.78 × 1,000 **2,780**
58 × 0.01 **0.1358**
8. 5.49 × 10^3 **5,490**
9. 0.1 × 0.8 **0.08**
25 × 10 **9.25**
11. 99.44 × 10^2 **9,944**
12. 0.01 × 16 **0.16**
2 × 10^3 **1,320**
14. 0.56 × 10,000 **5,600**
15. 1.4 × 0.001 **0.0014**
23 × 10^5 **1,123,000**
17. 68.94 × 0.01 **0.6894**
18. 0.8 × 10^4 **8,000**
1 × 0.01 **0.281**
20. 9.3 × 10^7 **93,000,000**
21. 625,799 × 0.0001 **62.5799**

Prerequisite Skills 561

Dividing Decimals

To divide two decimals, use the following steps.

- If necessary, change the divisor to a whole number by moving the decimal point to the right. You are multiplying the divisor by a power of ten.
- Move the decimal point in the dividend the same number of places to the right. You are multiplying the dividend by the same power of ten.
- Divide as with whole numbers.

EXAMPLES Divide Decimals

Divide.

1 **25.8 ÷ 2** **Estimate** 26 ÷ 2 = 13

```
   12.9
 2)25.8    The divisor, 2, is already a whole number, so
 − 2       you do not need to move the decimal point.
    5      Divide as with whole numbers. Then place the decimal
  − 4      point directly above the decimal point in the dividend.
    18
  − 18
     0     Compared to the estimate, the quotient, 12.9, is reasonable.
```

2 **199.68 ÷ 9.6** **Estimate** 200 ÷ 10 = 20

```
      20.8
 9.6)199.68   Move each decimal point one place to the right.
   − 192
       7 68
     − 7 68
          0   Compare the answer to the estimate.
```

Exercises

Divide.

1. 0.3)9.81 **32.7**
2. 12)0.12 **0.01**
3. 3.2)5.76 **1.8**
4. 0.22)0.0132 **0.06**
5. 0.04)0.008 **0.2**
6. 3.18)0.636 **0.2**
7. 0.2)8.24 **41.2**
8. 82.3)823 **10**
9. 12.02)24.04 **2**
10. 0.5)85 **170**
11. 74.9)5.992 **0.08**
12. 19.2)4.416 **0.23**
13. 1.9)38.57 **20.3**
14. 13.8)131.1 **9.5**
15. 6.48)259.2 **40**
16. 812 ÷ 0.4 **2,030**
17. 0.34 ÷ 0.2 **1.7**
18. 14.4 ÷ 0.12 **120**
19. 90.175 ÷ 2.5 **36.07**
20. 39.95 ÷ 799 **0.05**
21. 88.8 ÷ 444 **0.2**
22. 613.8 ÷ 66 **9.3**
23. 2,445.3 ÷ 33 **74.1**
24. 20.24 ÷ 2.3 **8.8**
25. 45 ÷ 0.09 **500**
26. 2.475 ÷ 0.03 **82.5**
27. 4.6848 ÷ 0.366 **12.8**
28. 180 ÷ 0.36 **500**
29. 97.812 ÷ 1.1 **88.92**
30. 23 ÷ 0.023 **1,000**
31. 1,680.042 ÷ 44.2 **38.**
32. **OLYMPICS** In the 2000 Olympics, Michael Johnson of the U.S. ran the 400-meter run in 43.84 seconds. To the nearest hundredth, find his speed in meters per second. **9.12 m/s**
33. **PLANETS** It takes Pluto 247.69 Earth years to revolve once around the Sun. It takes Jupiter 11.86 Earth years to revolve once around the Sun. About how many times longer does it take Pluto than Jupiter to revolve once around the Sun? **about 21 times**

Prerequisite Skills
562 Prerequisite Skills

Prerequisite Skills

ed Numbers and Improper Fractions

ed number is the sum of a whole number and a fraction. An
per fraction is a fraction with a numerator that is greater than or
to the denominator.

mixed number: $8\frac{1}{2}$ improper fraction: $\frac{17}{2}$

te a mixed number as an improper fraction, use the following steps.

tiply the whole number by the denominator.

d the numerator.

te the sum as the numerator of the improper fraction.

AMPLES Write Mixed Numbers as Improper Fractions

rite each mixed number as an improper fraction.

$\frac{1}{2}$

$\frac{1}{2} = \frac{(6 \cdot 2) + 1}{2}$ ← Multiply 6 by 2 and add 1. ← The denominator is 2.

$= \frac{12 + 1}{2}$ or $\frac{13}{2}$ Simplify.

2 $3\frac{4}{5}$

$3\frac{4}{5} = \frac{(3 \cdot 5) + 4}{5}$ ← Multiply 3 by 5 and add 4. ← The denominator is 5.

$= \frac{15 + 4}{5}$ or $\frac{19}{5}$ Simplify.

te an improper fraction as a mixed number, divide the numerator by
nominator. Write the remainder as the numerator of the fraction.

AMPLE Write Improper Fractions as Mixed Numbers

rite $\frac{7}{4}$ as a mixed number.

$\div 4 = 1 \text{ R } 3$ Divide the numerator by the denominator.

$= 1\frac{3}{4}$ Write the remainder as the numerator of the fraction.

, $\frac{7}{4} = 1\frac{3}{4}$.

ises

each mixed number as an improper fraction.

$\frac{7}{3}$	2. $1\frac{5}{8}$ $\frac{13}{8}$	3. $4\frac{5}{7}$ $\frac{33}{7}$	4. $3\frac{3}{4}$ $\frac{15}{4}$	5. $9\frac{1}{3}$ $\frac{28}{3}$
$\frac{11}{10}$	7. $10\frac{2}{3}$ $\frac{32}{3}$	8. $12\frac{1}{4}$ $\frac{49}{4}$	9. $6\frac{3}{8}$ $\frac{51}{8}$	10. $11\frac{1}{2}$ $\frac{23}{2}$
$\frac{19}{8}$	12. $20\frac{3}{5}$ $\frac{103}{5}$	13. $5\frac{7}{8}$ $\frac{47}{8}$	14. $27\frac{1}{3}$ $\frac{82}{3}$	15. $5\frac{2}{3}$ $\frac{17}{3}$

each improper fraction as a mixed number.

$7\frac{1}{2}$	17. $\frac{9}{7}$ $1\frac{2}{7}$	18. $\frac{15}{8}$ $1\frac{7}{8}$	19. $\frac{100}{3}$ $33\frac{1}{3}$	20. $\frac{28}{5}$ $5\frac{3}{5}$
1	22. $\frac{19}{8}$ $2\frac{3}{8}$	23. $\frac{17}{10}$ $1\frac{7}{10}$	24. $\frac{51}{3}$ 17	25. $\frac{17}{6}$ $2\frac{5}{6}$
$1\frac{2}{3}$	27. $\frac{99}{3}$ 33	28. $\frac{25}{7}$ $3\frac{4}{7}$	29. $\frac{46}{9}$ $5\frac{1}{9}$	30. $\frac{27}{4}$ $6\frac{3}{4}$

Prerequisite Skills

Prerequisite Skills 563

Extra Practice

Lesson 1-1 (Pages 6–9)

Use the four-step plan to solve each problem.

1. The Reyes family rode their bicycles for 9 miles to the park. The ride back was along a different route for 14 miles. How many miles did they ride in all? **23 mi**
2. A farmer planted 389 acres of land with 78,967 corn plants. How many plants were planted per acre? **203 plants**
3. A group of 251 people is eating dinner at a school fund-raiser. If each person pays $8.00 for their meal, how much money is raised? **$2,008**
4. When Tamika calls home from college, she talks ten minutes per call for 3 calls per week. How many minutes does she call in a 15-week semester? **450 min**
5. Darren runs at 6 feet per second and Kim runs at 7 feet per second. If they both start a race at the same time, how far apart are they after one minute? **60 ft**

Lesson 1-2 (Pages 10–13)

Write each power as a product of the same factor.

1. 13^4 **$13 \cdot 13 \cdot 13 \cdot 13$**
2. 9^6 **$9 \cdot 9 \cdot 9 \cdot 9 \cdot 9 \cdot 9$**
3. 1^7 **$1 \cdot 1 \cdot 1 \cdot 1 \cdot 1 \cdot 1 \cdot 1$**
4. 12^2 **$12 \cdot 12$**
5. 5^8 **$5 \cdot 5 \cdot 5 \cdot 5 \cdot 5 \cdot 5 \cdot 5 \cdot 5$**
6. 15^4 **$15 \cdot 15 \cdot 15 \cdot 15$**

Evaluate each expression.

7. 5^6 **15,625**
8. 17^3 **4,913**
9. 2^{12} **4,096**
10. 3^5 **243**
11. 1^4 **1**
12. 5^3 **125**
13. 10^2 **100**
14. 2^8 **256**
15. 8^2 **64**
16. 7^4 **2,401**
17. 20^3 **8,000**
18. 42^3 **74,088**

Write each product in exponential form.

19. $2 \cdot 2 \cdot 2 \cdot 2 \cdot 2$ **2^5**
20. $3 \cdot 3$ **3^2**
21. $1 \cdot 1 \cdot 1 \cdot 1 \cdot 1 \cdot 1$ **1^6**
22. $18 \cdot 18 \cdot 18 \cdot 18$ **18^4**
23. $9 \cdot 9 \cdot 9 \cdot 9 \cdot 9 \cdot 9 \cdot 9 \cdot 9$ **9^8**
24. $10 \cdot 10 \cdot 10 \cdot 10 \cdot 10 \cdot 10$ **1**

Lesson 1-3 (Pages 14–17)

Evaluate each expression.

1. $14 - (5 + 7)$ **2**
2. $(32 + 10) - 5 \times 6$ **12**
3. $(50 - 6) + (12 + 4)$ **60**
4. $12 - 2 \cdot 3$ **6**
5. $16 + 4 \times 5$ **36**
6. $(5 + 3) \times 4 - 7$ **25**
7. $2 \times 3 + 9 \times 2$ **24**
8. $6 \cdot (8 + 4) \div 2$ **36**
9. $7 \times 6 - 14$ **28**
10. $8 + (12 \times 4) \div 8$ **14**
11. $13 - 6 \cdot 2 + 1$ **2**
12. $(80 \div 10) \times 8$ **64**
13. $14 - 2 \cdot 7 + 0$ **0**
14. $156 - 6 \times 0$ **156**
15. $30 - 14 \cdot 2 + 8$ **10**
16. $3 \times 4 - 3^2$ **3**
17. $10^2 - 5$ **95**
18. $3 + (10 - 5 + 1)^2$ **39**
19. $(4 + 3)^2 \div 7$ **7**
20. 8×10^3 **8,000**
21. $10^4 \times 6$ **60,000**
22. 4.5×10^3 **4,500**
23. 1.8×10^2 **180**
24. $3 + 5(1.7 + 2.3)$ **23**
25. $4(3.6 + 5.4) - 9$ **27**
26. $10 + 3(6.1 + 3.7)$ **39.4**
27. $6(7.5 + 2.1) - 2.3$ **55.3**

564 Extra Practice

n 1-4 (Pages 18–21)

ate each expression if $a = 3$, $b = 4$, $c = 12$, and $d = 1$.

b **7** | 2. $c - d$ **11** | 3. $a + b + c$ **19** | 4. $b - a$ **1**

ab **0** | 6. $a + 2d$ **5** | 7. $b + 2c$ **28** | 8. ab **12**

$3b$ **15** | 10. $6a + c$ **30** | 11. $\frac{c}{d}$ **12** | 12. abc **144**

$+ b)$ **14** | 14. $\frac{2c}{b}$ **6** | 15. $144 - abc$ **0** | 16. $2ab$ **24**

2 | 18. a^2 **9** | 19. $c^2 - 100$ **44** | 20. $a^3 + 3$ **30**

32 | 22. $b^3 + c$ **76** | 23. $\frac{a^2}{d}$ **9** | 24. $5a^2 + 2d^2$ **47**

1 | 26. $\frac{15}{a}$ **5** | 27. $3a^2$ **27** | 28. $10d^3$ **10**

1 | 30. $\frac{(a + b)}{d}$ **7** | 31. $2.5b + c$ **22** | 32. $\frac{10}{d}$ **10**

$\frac{+ b)}{b}$ **7** | 34. $\frac{(b^2 + 2d)}{a}$ **6** | 35. $\frac{(2c + ab)}{c}$ **3** | 36. $\frac{(3.5c + 2)}{11}$ **4**

n 1-5 (Pages 24–27)

each equation mentally.

$- 7 = 12$ **5** | 2. $a + 3 = 15$ **12** | 3. $s + 10 = 23$ **13** | 4. $9 + n = 13$ **4**

$= 24 - n$ **4** | 6. $4x = 36$ **9** | 7. $2y = 10$ **5** | 8. $15 = 5h$ **3**

$3 = 2$ **6** | 10. $14 = w - 4$ **18** | 11. $24 \div k = 6$ **4** | 12. $b - 3 = 12$ **15**

$- 10 = 8$ **80** | 14. $6 = t \div 5$ **30** | 15. $14 + m = 24$ **10** | 16. $3y = 39$ **13**

$= 12$ **24** | 18. $16 = 4v$ **4** | 19. $81 = 80 + a$ **1** | 20. $9 = \frac{72}{x}$ **8**

$= 22m$ **3** | 22. $77 - 12 = a$ **65** | 23. $9k = 81$ **9** | 24. $95 + d = 100$ **5**

$= \frac{72}{6}$ **12** | 26. $z = 15 + 22$ **37** | 27. $15b = 225$ **15** | 28. $43 + s = 57$ **14**

$= 52$ **13** | 30. $e - 10 = 0$ **10** | 31. $62 - d = 12$ **50** | 32. $14f = 14$ **1**

$\div n = 8$ **6** | 34. $a - 82 = 95$ **177** | 35. $\frac{x}{2} = 36$ **72** | 36. $99 = c \div 2$ **198**

00) − 5(50); 750 9. 15(1,000) − 15(200); 12,000 10. 3(2,000) + 3(400); 7,200

n 1-6 **11. 12(1,000) + 12(10); 12,120 12. 7(1,000) − 7(50); 6,650** (Pages 30–33)

e Distributive Property to write each expression as an equivalent ssion. Then evaluate the expression.

$+ 5)$ **3(4) + 3(5); 27** | 2. $(2 + 8)6$ **2(6) + 8(6); 60** | 3. $4(9 - 6)$ **4(9) − 4(6); 12**

$- 3)$ **8(6) − 8(3); 24** | 5. $5(200 - 50)$ | 6. $20(3 + 6)$ **20(3) + 20(6); 180**

$) - 5)8$ **20(8) − 5(8); 120** | 8. $50(8 + 2)$ **50(8) + 50(2); 500** | 9. $15(1{,}000 - 200)$

$2{,}000 + 400)$ | 11. $12(1{,}000 + 10)$ | 12. $7(1{,}000 - 50)$

the property shown by each statement. **13–24. See margin.**

$2 + 3) = 7(2) + 7(3)$ | 14. $7 \times 8 = 8 \times 7$ | 15. $(4 + 2) + 8 = 4 + (2 + 8)$

$+ 3 = 3 + 92$ | 17. $4(12) + 4(3) = 4(12 + 3)$ | 18. $z \cdot 1 = z$

$\times 5) \times 2 = 1 \times (5 \times 2)$ | 20. $72 + (8 + 6) = (72 + 8) + 6$ | 21. $4 \cdot (x \cdot y) = (4 \cdot x) \cdot y$

$+ 0 = a$ | 23. $a \times b = b \times a$ | 24. $43 + z = z + 43$

Extra Practice

Lesson 1-6

13. Distributive
14. Commutative (×)
15. Associative (+)
16. Commutative (+)
17. Distributive
18. Identity (×)
19. Associative (×)
20. Associative (+)
21. Associative (×)
22. Identity (+)
23. Commutative (×)
24. Commutative (+)

Lesson 1-7

1. +4; arithmetic
2. ×2; geometric
3. neither
4. neither
5. ×3; geometric
6. +0.9; arithmetic
7. +0.1; arithmetic
8. +1.1; arithmetic
9. neither
10. 35, 43, 51
11. 150, 100, 50
12. 100, 50, 25
13. 162, 486, 1,458
14. 38, 45, 52
15. 28, 35, 42
16. 625, 3,125; 15,625
17. 89, 83, 77
18. 59,049; 531,441; 4,782,969
19. 108, 36, 12
20. 15,625; 39,062.5; 97,656.25
21. 5.5, 4.4, 3.3
22. 18, 16.5, 15
23. 250, 125, 62.5
24. 1,000; 10,000; 100,000

Extra Practice

Lesson 1-7

(Pages 34–36)

Describe the pattern in each sequence and identify the sequence as *arithmetic, geometric,* or *neither.* 1–9. See margin.

1. 5, 9, 13, 17, …
2. 3, 6, 12, 24, …
3. 10, 15, 25, 40, …
4. 90, 91, 94, 99, …
5. 8, 24, 72, 216, …
6. 4.5, 5.4, 6.3, 7.2, …
7. 0.3, 0.4, 0.5, …
8. 2.3, 3.4, 4.5, 5.6, …
9. 9.1, 8.4, 7.7, 6.0, …

Write the next three terms of each sequence. 10–24. See margin.

10. 3, 11, 19, 27, …
11. 350, 300, 250, 200, …
12. 1,600, 800, 400, 200, …
13. 2, 6, 18, 54, …
14. 10, 17, 24, 31, …
15. 0, 7, 14, 21, …
16. 1, 5, 25, 125, …
17. 113, 107, 101, 95, …
18. 9, 81, 729, 6,561, …
19. 8,748, 2,916, 972, 324, …
20. 400, 1,000, 2,500, 6,250, …
21. 9.9, 8.8, 7.7, 6.6, …
22. 24, 22.5, 21, 19.5, …
23. 4,000, 2,000, 1,000, 500, …
24. 0.1, 1, 10, 100, …

Lesson 1-8

(Pages 38–41)

Complete.

1. 400 mm = __?__ cm 40
2. 4 km = __?__ m 4,000
3. 660 cm = __?__ m 6.6
4. 0.3 km = __?__ m 300
5. 30 mm = __?__ cm 3
6. 84.5 m = __?__ km 0.0845
7. __?__ m = 54 cm 0.54
8. 18 km = __?__ cm 1,800,000
9. __?__ mm = 45 cm 450
10. 4 kg = __?__ g 4,000
11. 632 mg = __?__ g 0.632
12. 4,497 g = __?__ kg 4.497
13. __?__ mg = 21 g 21,000
14. 61.2 mg = __?__ g 0.0612
15. 61 g = __?__ mg 61,000
16. __?__ mg = 0.51 kg 510,000
17. 0.63 kg = __?__ g 630
18. __?__ kg = 563 g 0.563
19. 662 m = __?__ km 0.662
20. 5,283 mL = __?__ L 5.283
21. 0.24 cm = __?__ mm 2.4
22. 380 kL = __?__ L 380,000
23. 10.8 g = __?__ mg 10,800
24. 83,000 mL = __?__ L 83
25. 17.8 m = __?__ cm 1,780
26. 2 km = __?__ m 2,000
27. 0.75 m = __?__ mm 750
28. 125 L = __?__ kL 0.125
29. 3 km = __?__ cm 300,000
30. 99 g = __?__ mg 99,000

Lesson 1-9

(Pages 43–45)

Write each number in standard form.

1. 7.6×10^2 760
2. 2.6×10^3 2,600
3. 4.7×10^5 470,000
4. 1.0×10^2 100
5. 6.4×10^6 6,400,000
6. 8.3×10^4 83,000
7. 2.76×10^3 2,760
8. 6.17×10^5 617,000
9. 8.70×10^5 870,000
10. 1.28×10^7 12,800,000
11. 4.32×10^9 4,320,000,000
12. 5.01×10^2 501
13. 1.89×10^6 1,890,000
14. 7.43×10^4 74,300
15. 3.03×10^9 3,030,000,000

Write each number in scientific notation.

16. 720 7.2×10^2
17. 7,560 7.56×10^3
18. 892 8.92×10^2
19. 1,400 1.4×10^3
20. 91,200 9.12×10^4
21. 51,000 5.1×10^4
22. 145,000 1.45×10^5
23. 90,100 9.01×10^4
24. 123,000,000,000 1.23×10
25. 4,500 4.5×10^3
26. 1,700,000 1.7×10^6
27. 10,000,000 1.0×10^7
28. 820,000,000 8.2×10^8
29. 680,000 6.8×10^5
30. 52,000 5.2×10^4
31. 60,600,000 6.06×10^7
32. 3,970,000 3.97×10^6
33. 574,000,000 5.74×10^8

Lesson 2-1

1.

Time	Tally	Frequency
15	IIII	4
30	卌	5
45	IIII	4
60	III	3

2.

Number	Tally	Frequency
40	III	3
41	卌 I	6
42	I	1
43	I	1
45	III	3
49	II	2

n 2-1

(Pages 54–57)

a frequency table of each set of data. **1–4. See margin.**

ength of Time Walking (min)			
15	30	15	45
45	30	30	60
30	60	15	30
45	45	60	15

2.

Number of Raisins Eaten in Preschool			
40	49	45	49
42	41	45	41
45	41	41	40
41	43	40	41

vorite Type of TV Show				
M	S	N	S	M
M	S	S	S	S
G	M	N	S	M
S	G	G	G	G
N	S	M	G	G
S	S	N	N	N

movie, S = sitcom, N = news show, G = game show

4.

What Type of Transportation Did You Use on Your Last Family Vacation?					
C	A	C	T	C	A
C	C	C	C	S	A
B	B	C	T	A	A
T	S	A	C	A	A
A	A	A	A	A	S

C = car, T = train, S = ship, A = airplane, B = bus

n 2-2

(Pages 60–63)

hel's quiz scores cience have been dily going up e her parents d a tutor for her. e trend continues, dict Rachel's re on the next z. **7**

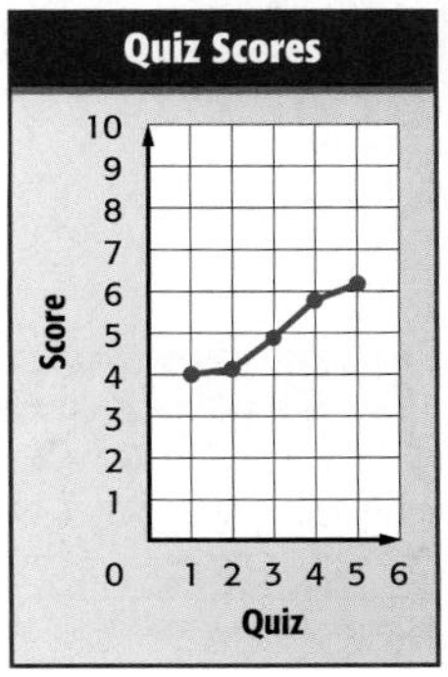

2. The table shows the average price paid to farmers per 100 pounds of sheep they sold.

Year	Price Per 100 Pounds ($)
1940	4
1950	12
1960	6
1970	8
1980	21
1990	23
2000	34

Source: *The World Almanac and Book of Facts*

a. Make a scatter plot of the data. Use the year on the horizontal axis and the price on the vertical axis.

b. Describe the relationship, if any, between the two sets of data. **a–c. See margin.**

c. Predict the price per 100 pounds for 2010. Explain.

n 2-3

(Pages 64–68)

a line plot for each set of data. Identify any clusters, gaps, or rs. **1–4. See Extra Practice Answer Appendix.**

umber of Pets in the Home				
0	1	3	4	0
2	1	0	1	1
10	0	1	5	2

2.

High Temperatures for 18 Days (°F)					
75	81	75	65	76	81
77	80	65	65	80	80
76	85	66	75	80	75

lumber of Stories for Buildings in Denver				
56	43	36	42	29
54	42	32	34	
52	40	32	32	

rce: *The World Almanac and Book of Facts*

4.

Ages of Children at Sunny Day Care (years)					
4	1	6	4	5	3
4	5	1	2	5	4
3	2	4	1	3	3

Lesson 2-1

3.

Favorite Type of TV Show	Tally	Frequency
movie	卌 I	6
sitcom	卌 卌 I	11
news show	卌 I	6
game show	卌 II	7

4.

Type of Transportation	Tally	Frequency
car	卌 IIII	9
train	III	3
ship	III	3
airplane	卌 卌 III	13
bus	II	2

Lesson 2-2

2a. **Price for Sheep**

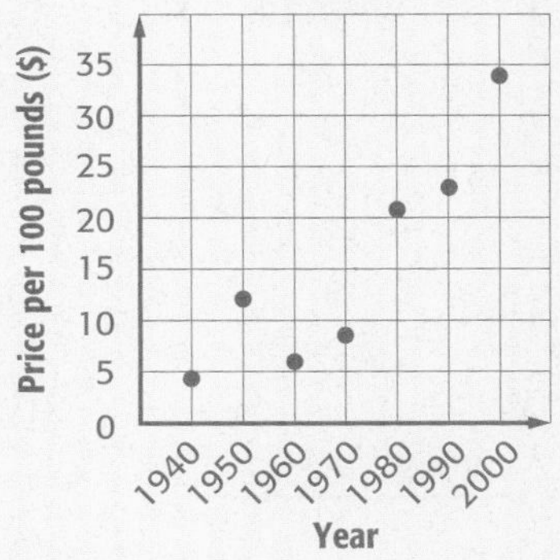

2b. Sample: Except for 1950, the price increases during each ten-year period.

2c. Sample: about $40; if you draw a line lying close to the points from 1960 to 2000, 40 lies at 2010.

Extra Practice

Lesson 2-5

1.

Stem	Leaf
1	5 8
2	3 9
3	9
4	1 2
5	1 2 7
6	8

1|5 = 15

2.

Stem	Leaf
0	5 6 7 9
1	3 4 4 4 8
2	8 8
3	9

0|5 = 5

3.

Stem	Leaf
18	2 3 4 4 9
19	4 6 6 7

18|9 = 189

4.

Stem	Leaf
7	1 4 5 6
8	1 2 3 4
9	2 5 6

7|1 = 71

5.

Stem	Leaf
2	1 4 7
3	4
4	0 6
5	0 8
6	1 7
7	0 2

3|4 = 34

6.

Stem	Leaf
2	0 0 0 3 7
3	0 1 4 4 5 7 8 9
4	2 6 9
5	2 5

2|0 = 0

Extra Practice

Lesson 2-4

(Pages 69–72)

Find the mean, median, and mode for each set of data.

1. 1, 5, 9, 1, 2, 6, 8, 2 **4.25, 3.5, 1 and 2**
2. 2, 5, 8, 9, 7, 6, 3, 5, 1, 4 **5, 5, 5**
3. 1, 2, 1, 2, 2, 1, 2, 1 **1.5, 1.5, 1 and 2**
4. 12, 13, 15, 12, 12, 11 **12.5, 12, 12**
5. 82, 79, 93, 91, 95, 95, 81 **88, 91, 95**
6. 117, 103, 108, 120 **112, 112.5, no mode**
7. 256, 265, 247, 256 **256, 256, 256**
8. 957, 562, 462, 847, 721 **709.8, 721, no mo**
9. 47, 54, 66, 54, 46, 66 **55.5, 54, 54 and 66**
10. 81, 82, 83, 84, 85, 86, 87 **84, 84, no mode**
11.

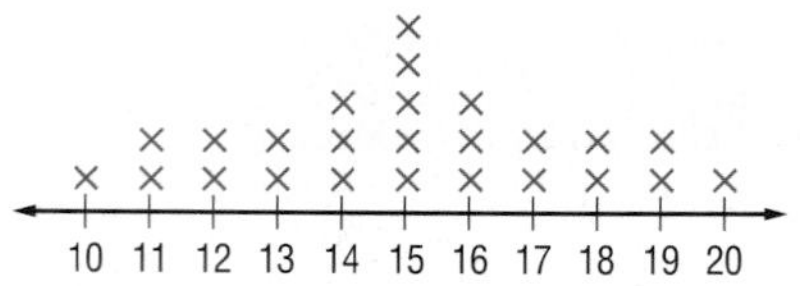

15, 15, 15

12.

Number of Absences	Tally	Frequency
0	||||	4
1	卌 ||||	9
2	卌 |	6
3	卌	5

1.5, 1, 1

Lesson 2-5

(Pages 76–79)

Make a stem-and-leaf plot for each set of data. **1–6. See margin.**

1. 23, 15, 39, 68, 57, 42, 51, 52, 41, 18, 29
2. 5, 14, 39, 28, 14, 6, 7, 18, 13, 28, 9, 14
3. 189, 182, 196, 184, 197, 183, 196, 194, 184
4. 71, 82, 84, 95, 76, 92, 83, 74, 81, 75, 96
5.

Average Monthly High Temperatures in Albany, NY (°F)			
21	46	72	50
24	58	70	40
34	67	61	27

Source: *The World Almanac and Book of Facts*

6.

Super Bowl Winning Scores 1985–2002					
38	42	20	30	35	23
46	20	37	49	31	34
39	55	52	27	34	20

Source: *The World Almanac and Book of Facts*

Lesson 2-6

(Pages 80–83)

For each set of data:

a. **Find the lower extreme, LQ, median, UQ, and upper extreme.**

b. **Draw a box-and-whisker plot of the data.** **1–4. See Extra Practice Answer Appendix.**

1.

Price of Chicken Sandwiches at Various Restaurants			
$1.00	$3.70	$3.50	$3.90
$2.50	$1.70	$2.20	$2.10
$1.20	$1.90	$4.20	

2.

Lengths of Principal Rivers in South America (mi)					
4,000	1,750	1,400	2,485	1,000	910
1,100	2,013	1,600	1,000	1,988	1,000
808	956	1,584	2,100	1,677	1,300

Source: *The World Almanac and Book of Facts*

3.

Height of World's Tallest Free-Standing Towers (ft)			
1,831	1,772	1,535	1,379
1,214	1,230	1,411	
1,815	1,362	1,369	

Source: *The World Almanac and Book of Facts*

4.

Mathematics Test Scores in Ms. Ruff's Class					
65	100	84	80	68	69
92	69	75	80	99	90
78	87	100	72	85	

568 Extra Practice

n 2-7

(Pages 85–89)

ch set of data, make a bar graph. **1–2. See Extra Practice Answer Appendix.**

Longest Snakes	
Snake Name	Length (ft)
yal python	35
aconda	28
dian python	25
amond python	21
ng cobra	19
a constrictor	16

ce: *The Top 10 of Everything*

2.

Least Densely Populated States	
State	People per Square Mile
Alaska	1
Wyoming	5
Montana	6
North Dakota	9
South Dakota	10
New Mexico	15

Source: *The Top 10 of Everything*

ch set of data, make a histogram. **3–4. See margin.**

ost of a Movie Ticket at Selected Theaters			
\$5.25	\$6.50	\$3.50	\$3.75
\$7.50	\$9.25	\$10.40	\$4.75
\$10.00	\$4.50	\$8.75	\$7.25
\$3.50	\$6.75	\$4.20	\$7.50

4.

Highest Recorded Wind Speeds for Selected U.S. Cities (mph)					
52	55	81	46	73	57
75	54	58	76	46	58
60	91	53	53	51	56
80	60	73	46	49	47

Source: *The World Almanac and Book of Facts*

on 2-8

(Pages 92–95)

h graph could be misleading? Why?

th graphs show pounds of grapes sold Westview School in one week.

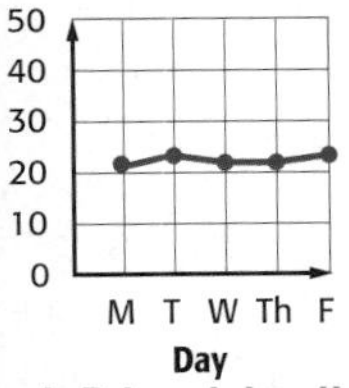

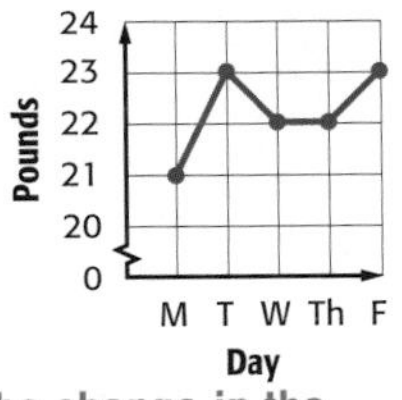

aph B is misleading. The change in the rtical scale is misleading.

2. Both graphs show commissions made by Mr. Turner for a four-week pay period.

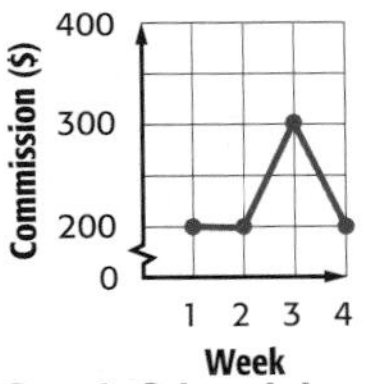

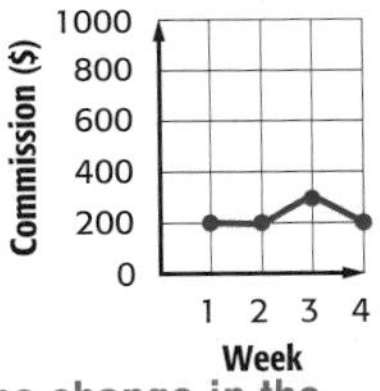

Graph C is misleading. The change in the vertical scale makes it seem as if more money is made than really is.

on 3-1

(Pages 106–108)

an integer for each situation.

ven degrees below zero **−7**
2. a loss of 3 pounds **−3**
3. a loss of 20 yards **−20**
profit of \$25 **25**
5. 112°F above 0 **112**
6. 2,830 feet above sea level **2,830**

ate each expression.

1| **1**
8. |−8| **8**
9. |0| **0**
10. |−82| **82**
64| **64**
12. |−128| **128**
13. |−22| + 5 **27**
14. |−40| − 8 **32**
−18| + |10| **28**
16. |−7| + |−1| **8**
17. |98| − |−5| **93**
18. |−49| − |−10| **39**

h each set of integers on a number line. **19–22. See margin.**

−2, 0, 2}
20. {1, 3, 5}
21. {−2, −5, 3}
22. {7, −1, 4}

Lesson 3-1

19.

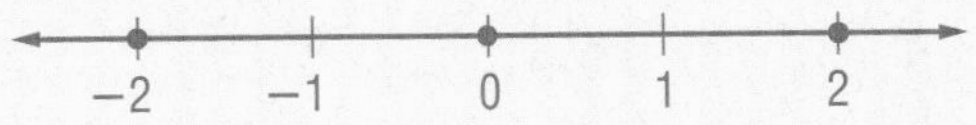

21.

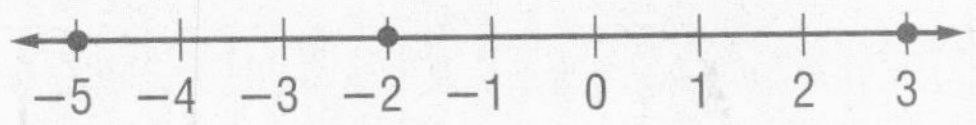

20.

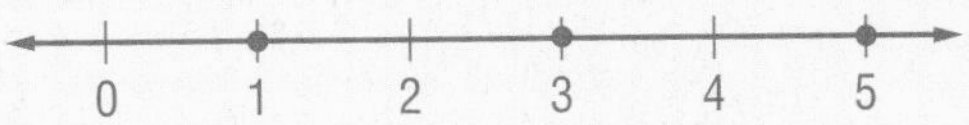

22.

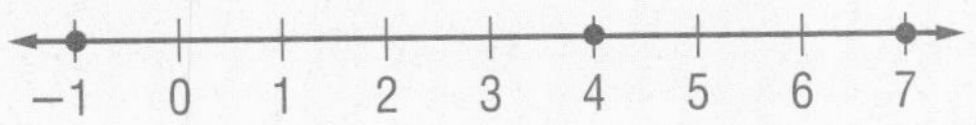

Lesson 2-7

3. Sample answer:

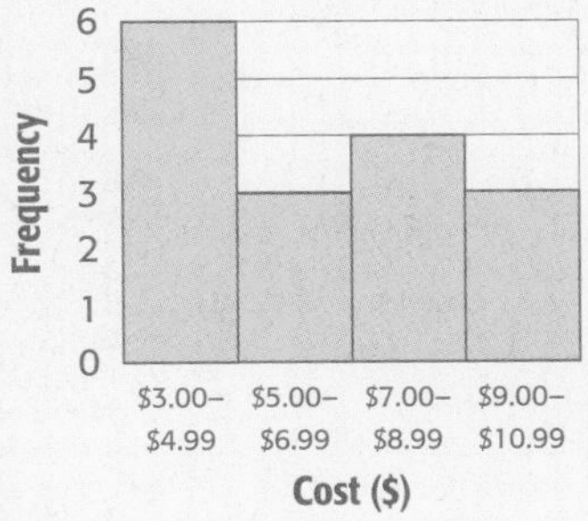

4. Sample answer:

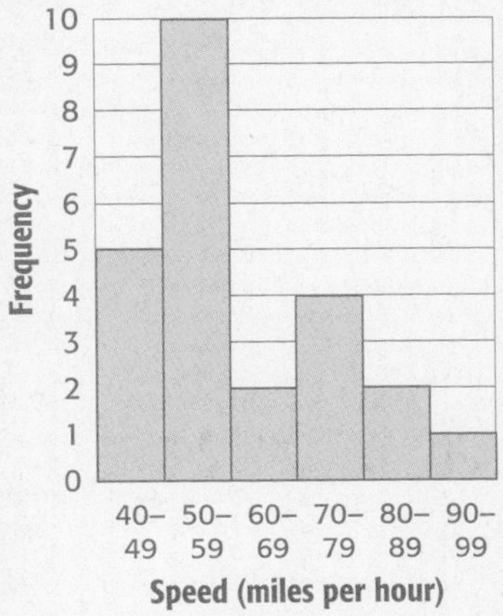

Extra Practice

Lesson 3-3

13–24.

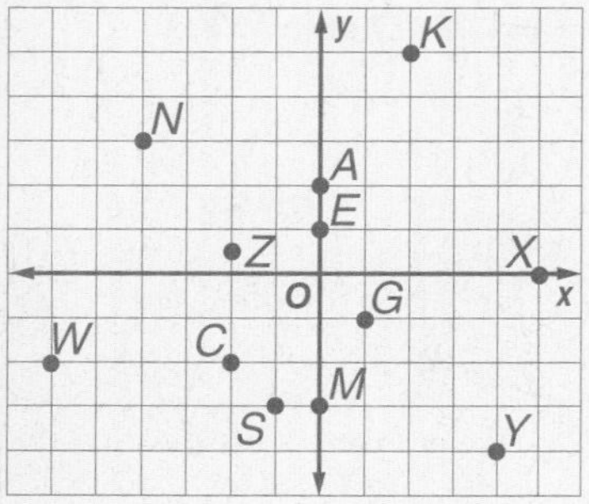

Extra Practice

Lesson 3-2

(Pages 109–111)

Replace each ● with < or > to make a true sentence.

1. 7 ● −7 **>**	2. −8 ● 4 **<**	3. −4 ● −9 **>**	4. −3 ● 0 **<**
5. 8 ● 10 **<**	6. −5 ● −4 **<**	7. 6 ● −7 **>**	8. −12 ● −13 **>**
9. 3 ● 1 **>**	10. −2 ● 2 **<**	11. 7 ● −1 **>**	12. −15 ● −20 **>**
13. −40 ● 30 **<**	14. 0 ● −3 **>**	15. −5 ● 0 **<**	16. 85 ● −17 **>**

Order the integers from least to greatest.

17. −2, −8, 4, 10, −6, −12
18. 19, −19, −21, 32, −14, 18
19. 18, 23, 95, −95, −18, −23, 2
20. 46, −48, −47, −52, −18, 12
21. 0, −10, −6, −8, 12 **−10, −8, −6, 0, 12**
22. −15, 18, −1, 0, 14, −20 **−20, −15, −1, 0,**

17. −12, −8, −6, −2, 4, 10
18. −21, −19, −14, 18, 19, 32
19. −95, −23, −18, 2, 18, 23, 95
20. −52, −48, −47, −18, 12, 46

Lesson 3-3

(Pages 112–115)

Name the ordered pair for each point graphed at the right. Then identify the quadrant in which each point lies.

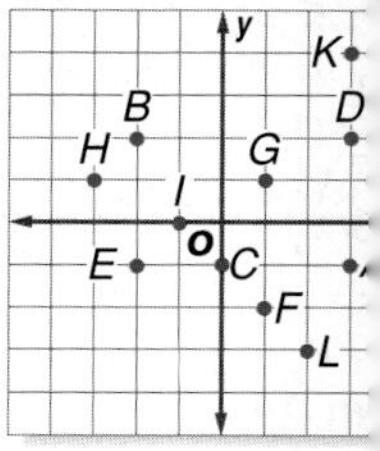

1. *A* **(3, −1), IV**	2. *B* **(−2, 2), II**	3. *C* **(0, −1), *y*-axis**
4. *D* **(3, 2), I**	5. *E* **(−2, −1), III**	6. *F* **(1, −2), IV**
7. *G* **(1, 1), I**	8. *H* **(−3, 1), II**	9. *I* **(−1, 0), *x*-axis**
10. *J* **(4, 3), I**	11. *K* **(3, 4), I**	12. *L* **(2, −3), IV**

On graph paper, draw a coordinate plane. Then graph and label each point.

13. *N*(−4, 3)	14. *K*(2, 5)	15. *W*(−6, −2)	16. *X*(5, 0)
17. *Y*(4, −4)	18. *M*(0, −3)	19. *Z*(−2, 0.5)	20. *S*(−1, −3)
21. *A*(0, 2)	22. *C*(−2, −2)	23. *E*(0, 1)	24. *G*(1, −1)

13–24. See margin.

Lesson 3-4

(Pages 120–124)

Add.

1. −4 + 8 **4**	2. 14 + 16 **30**	3. −7 + (−7) **−14**
4. −9 + (−6) **−15**	5. −18 + 11 **−7**	6. −36 + 40 **4**
7. 42 + (−18) **24**	8. −42 + 29 **−13**	9. 18 + (−32) **−14**
10. 12 + (−9) **3**	11. −24 + 9 **−15**	12. −7 + (−1) **−8**

Evaluate each expression if $a = 6$, $b = -2$, $c = -6$, and $d = 3$.

13. −96 + *a* **−90**	14. *b* + (−5) **−7**	15. *c* + (−32) **−38**
16. *d* + 98 **101**	17. −120 + *b* **−122**	18. −120 + *c* **−126**
19. 5 + *b* **3**	20. *a* + *d* **9**	21. *c* + *a* **0**
22. *d* + (−9) **−6**	23. *b* + *c* **−8**	24. *d* + *c* **−3**

Simplify.

25. *x* + 3 + (−8) ***x* + (−5)**	26. −7 + *b* + 3 ***b* + (−4)**	27. 9 + (−7) + *f* ***f* + 2**
28. 10 + *y* + (−6) ***y* + 4**	29. *m* + (−11) + 11 ***m***	30. −14 + *a* + (−19) ***a* + (−3**

n 3-5

(Pages 128–131)

ict.

- 7 −4 | 2. −5 − 4 −9 | 3. −6 − 2 −8

- 13 −5 | 5. 6 − (−4) 10 | 6. 12 − 9 3

: − 23 −25 | 8. 63 − 78 −15 | 9. 0 − (−14) 14

− 6 9 | 11. 18 − 20 −2 | 12. −5 − 8 −13

− (−37) 58 | 14. −60 − 32 −92 | 15. 57 − 63 −6

ate each expression if $k = -3$, $p = 6$, $n = 1$, and $d = -8$.

− k 58 | 17. $p - 7$ −1 | 18. $d - 15$ −23

- 12 −11 | 20. $-51 - d$ −43 | 21. $k - 21$ −24

- k 4 | 23. $-99 - k$ −96 | 24. $p - k$ 9

- (−1) −7 | 26. $k - d$ 5 | 27. $n - d$ 9

Extra Practice

n 3-6

(Pages 134–137)

ply.

−2) −10 | 2. 6(−4) −24 | 3. 4(21) 84 | 4. −11(−5) 55

5(5) −30 | 6. −50(0) 0 | 7. −5(−5) 25 | 8. −4(8) −32

−13) −39 | 10. 12(−5) −60 | 11. −9(−12) 108 | 12. 15(−8) −120

ate each expression if $a = -5$, $b = 2$, $c = -3$, and $d = 4$.

$2d$ −8 | 14. $6a$ −30 | 15. $3ab$ −30 | 16. $-12d$ −48

$4b^2$ −16 | 18. $-5cd$ 60 | 19. a^2 25 | 20. $13ab$ −130

lify each expression.

$7b)$ $42b$ | 22. $-8(2a)$ $-16a$ | 23. $-3(-5n)$ $15n$ | 24. $5(-6x)$ $-30x$

$9(8c)$ $-72c$ | 26. $14(-2m)$ $-28m$ | 27. $25(3d)$ $75d$ | 28. $-10(17z)$ $-170z$

n 3-7

(Pages 138–141)

le.

÷ (−2) −2 | 2. 16 ÷ (−8) −2 | 3. −14 ÷ (−2) 7 | 4. $\frac{32}{8}$ 4

÷ (−3) −6 | 6. −18 ÷ 3 −6 | 7. 8 ÷ (−8) −1 | 8. 0 ÷ (−1) 0

25 ÷ 5 −5 | 10. $\frac{-14}{-7}$ 2 | 11. −32 ÷ 8 −4 | 12. −56 ÷ (−8) 7

81 ÷ 9 −9 | 14. −42 ÷ (−7) 6 | 15. 121 ÷ (−11) −11 | 16. −81 ÷ (−9) 9

÷ (−2) −9 | 18. $\frac{-55}{11}$ −5 | 19. $\frac{25}{-5}$ −5 | 20. −21 ÷ 3 −7

ate each expression if $a = -2$, $b = -7$, $x = 8$, and $y = -4$.

$64 \div x$ −8 | 22. $\frac{16}{y}$ −4 | 23. $x \div 2$ 4 | 24. $\frac{a}{2}$ −1

$\div y$ 4 | 26. $\frac{bx}{y}$ 14 | 27. $2y \div 1$ −8 | 28. $\frac{x}{ay}$ 1

$y \div a$ −2 | 30. $x^2 \div y$ −16 | 31. $\frac{ab}{1}$ 14 | 32. $\frac{xy}{a}$ 16

Extra Practice

Extra Practice 571

Extra Practice

Lesson 4-1

1. $p - 6$
2. $20 + c$
3. $\frac{a}{b}$
4. $a + 6$
5. $x + 12$
6. $\frac{\$1,000}{z}$
7. $\frac{y}{3}$
8. $7m$
9. $f - 9$
10. $26 - q$
11. $19 - z$
12. $x - 2$
13. $35 + a$
14. $\frac{n}{400}$
15. $h - 19$
16. 2ℓ
17. $8 + x$
18. mn
19. $3x - 4 = 17$
20. $n + 6 = 5$
21. $20 + 2n = -30$
22. $\frac{n}{-2} = -42$
23. $4 + 3n = 18$
24. $5n - 15 = 92$
25. $8n + 12 = 36$
26. $n - 24 = -30$
27. $42 + 5n = 156$
28. $13 + n = 62$
29. $\frac{n}{-4} = -324$
30. $n + 11 = 41$

Lesson 4-5

1.
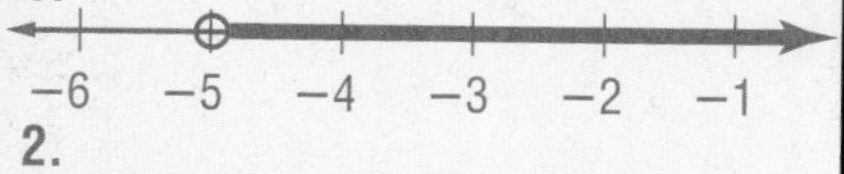

2.
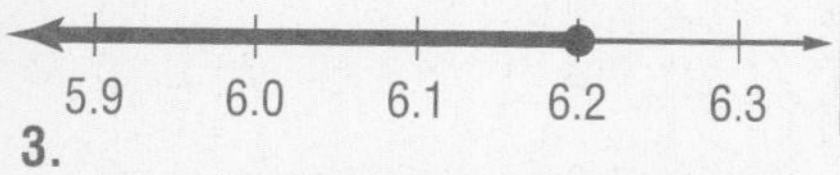

3.
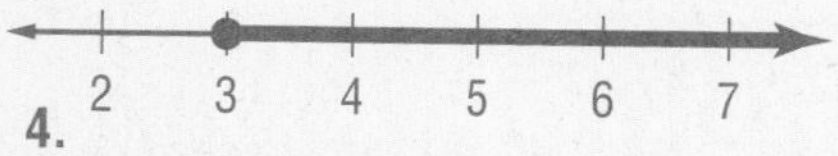

4.
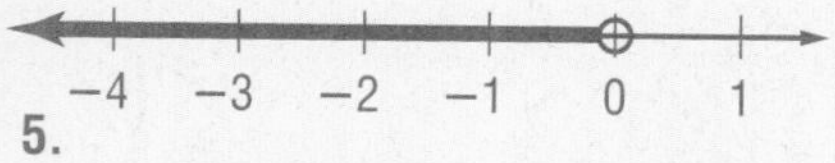

5.
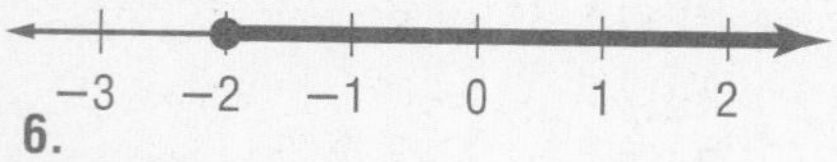

6.
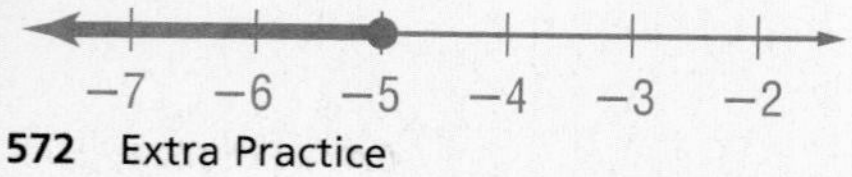

Extra Practice

Lesson 4-1

(Pages 150–152)

Write each phrase as an algebraic expression. **1–18. See margin.**

1. six less than p
2. twenty more than c
3. the quotient of a and b
4. Juana's age plus 6
5. x increased by twelve
6. \$1,000 divided by z
7. 3 divided into y
8. the product of 7 and m
9. the difference of f and 9
10. twenty-six less q
11. 19 decreased by z
12. two less than x
13. thirty-five increased by a
14. four hundred divided into n
15. Joe's height minus ninete
16. twice as many lemons
17. eight more than x
18. the product of m and n

Write each sentence as an algebraic equation. **19–30. See margin.**

19. Three times a number less four is 17.
20. The sum of a number and 6 is 5.
21. Twenty more than twice a number is −30.
22. The quotient of a number and −2 is −42.
23. Four plus three times a number is 18.
24. Five times a number minus 15 is 92.
25. Eight times a number plus twelve is 36.
26. The difference of a number and 24 is −30
27. 42 more than five times a number is 156.
28. Thirteen and a number is 62.
29. The quotient of a number and −4 is −324.
30. The sum of a number and eleven is 41.

Lesson 4-2

(Pages 156–159)

Solve each equation. Check your solution.

1. $r - 3 = 14$ **17**
2. $t + 3 = 21$ **18**
3. $s + 10 = 23$ **13**
4. $7 + a = -10$ **−17**
5. $14 + m = 24$ **10**
6. $-9 + n = 13$ **22**
7. $s - 2 = -6$ **−4**
8. $6 + f = 71$ **65**
9. $x + 27 = 30$ **3**
10. $a - 7 = 23$ **30**
11. $-4 + b = -5$ **−1**
12. $w + 18 = -4$ **−22**
13. $k - 9 = -3$ **6**
14. $j + 12 = 11$ **−1**
15. $-42 + v = -42$ **0**
16. $s + 1.3 = 18$ **16.7**
17. $t + 3.43 = 7.4$ **3.97**
18. $x + 7.4 = 23.5$ **16.1**
19. $p + 3.1 = 18$ **14.9**
20. $q - 2.17 = 21$ **23.17**
21. $w - 3.7 = 4.63$ **8.33**
22. $m - 4.8 = 7.4$ **12.2**
23. $x - 1.3 = 12$ **13.3**
24. $y + 3.4 = 18$ **14.6**
25. $0.013 + h = 4.0$ **3.987**
26. $7.2 + g = 9.1$ **1.9**
27. $z - 12.1 = 14$ **26.1**
28. $v - 18 = 13.7$ **31.7**
29. $w - 0.1 = 0.32$ **0.42**
30. $r + 6.7 = 1.2$ **−5.5**

Lesson 4-3

(Pages 160–163)

Solve each equation. Check your solution.

1. $2m = 18$ **9**
2. $-42 = 6n$ **−7**
3. $72 = 8k$ **9**
4. $-20r = 20$ **−1**
5. $420 = 5s$ **84**
6. $325 = 25t$ **13**
7. $-14 = -2p$ **7**
8. $18q = 36$ **2**
9. $40 = 10a$ **4**
10. $100 = 20b$ **5**
11. $416 = 4c$ **104**
12. $45 = 9d$ **5**
13. $0.5m = 3.5$ **7**
14. $1.8 = 0.6x$ **3**
15. $0.4y = 2$ **5**
16. $1.86 = 6.2z$ **0.3**
17. $-8x = 24$ **−3**
18. $8.34 = 2r$ **4.17**
19. $1.67t = 10.02$ **6**
20. $243 = 27a$ **9**
21. $0.9x = 4.5$ **5**
22. $4.08 = 1.2y$ **3.4**
23. $8d = 112$ **14**
24. $5f = 180.5$ **36.1**
25. $59.66 = 3.14m$ **19**
26. $98.4 = 8p$ **12.3**
27. $208 = 26k$ **8**

7.
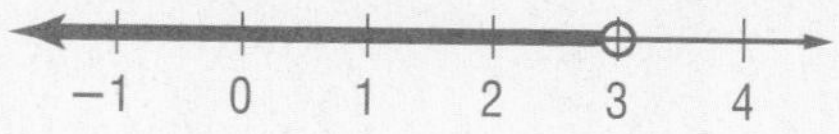

8.
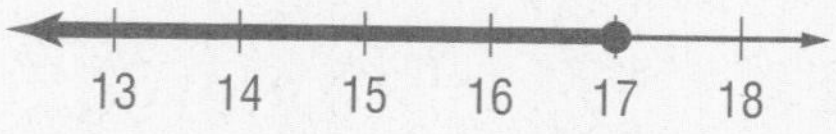

9.
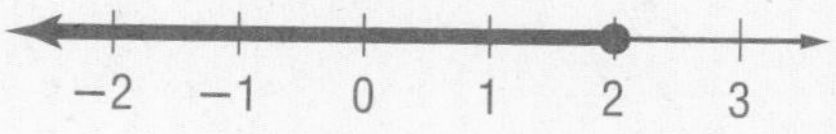

10.

11.
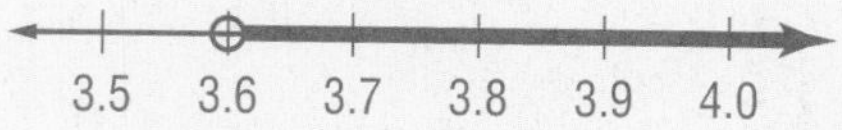

12.
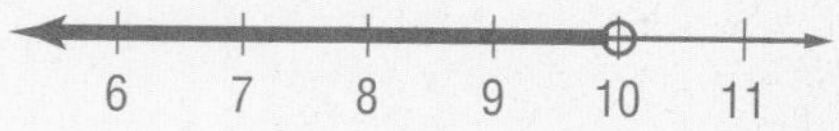

n 4-4

(Pages 166–169)

each equation. Check your solution.

+ 6 = 6 **0** | 2. $2r - 7 = -1$ **3** | 3. $-10 + 2d = 8$ **9**

+ 4 = −8 **−6** | 5. $5w - 12 = 3$ **3** | 6. $5t - 4 = 6$ **2**

− 6 = 4 **5** | 8. $2g - 3 = -9$ **−3** | 9. $15 = 6y + 3$ **2**

− 4 = 8 **4** | 11. $18 - 7f = 4$ **2** | 12. $13 + 3p = 7$ **−2**

r + 2 = −28 **−4** | 14. $4.2 + 7z = 2.8$ **−0.2** | 15. $-9m - 9 = 9$ **−2**

+ 0.2c = 1 **−155** | 17. $5t - 14 = -14$ **0** | 18. $-0.25x + 0.5 = 4$ **−14**

− 4 = 8 **2.4** | 20. $4d - 3 = 9$ **3** | 21. $2g - 16 = -9$ **3.5**

+ 13 = 20 **1.75** | 23. $7 = 5 - 2x$ **−1** | 24. $8z + 15 = -1$ **−2**

− 16b = 12 **5** | 26. $14e + 14 = 28$ **1** | 27. $1.1j + 2 = 7.5$ **5**

n 4-5

(Pages 172–175)

each inequality. Graph the solution on a number line. See margin for graphs.

+ 2 > −3 $x > -5$ | 2. $x + 2.9 \leq 9.1$ $x \leq 6.2$ | 3. $8t \geq 24$ $t \geq 3$

− 3 < −3 $v < 0$ | 5. $6y \geq -12$ $y \geq -2$ | 6. $a + 3 \leq -2$ $a \leq -5$

− 5 < −2 $k < 3$ | 8. $q - 3 \leq 14$ $q \leq 17$ | 9. $c - 4 \leq -2$ $c \leq 2$

+ 2 > −5 $n > -7$ | 11. $j + 1.2 > 4.8$ $j > 3.6$ | 12. $4x < 40$ $x < 10$

≤ 10 $y \leq 5$ | 14. $g + 8 < 10$ $g < 2$ | 15. $2 + b < 4$ $b < 2$

an inequality for each sentence. Then solve the inequality.

ne product of a number and four is at least 32. $4m \geq 32$; $m \geq 8$

ne sum of a number and −6 is at most 27. $t + (-6) \leq 27$; $t \leq 33$

x times a number is greater than 48. $6x > 48$; $x > 8$

ne difference between a number and 7 is less than 18. $a - 7 < 18$; $a < 25$

n 4-6

(Pages 177–181)

and complete each function table. Identify the domain and range.

x	2x	y	
2		4	2(2)
1		2	2(1)
0		0	2(0)
-1		−2	2(−1)

d: {2, 1, 0, −1}
r: {4, 2, 0, −2}

2.

x	3x + 1	y	
1		4	3(1) + 1
0		1	3(0) + 1
−1		−2	3(−1) + 1
−2		−5	3(−2) + 1

d: {1, 0, −1, −2} r: {4, 1, −2, −5}

h each equation. 3–14. See margin.

= 3x | 4. $y = 2x + 3$ | 5. $y = -x$

= 4x + 2 | 7. $y = 0.5x + 2$ | 8. $y = -x + 3$

= 0.25x + 6 | 10. $y = -3x + 6$ | 11. $y = 2x + 7$

= −5x + 1 | 13. $y = 13 + x$ | 14. $y = 5 - 0.5x$

e a function table for each sentence. Then write an equation using epresent the first number and y to represent the second number. 15–17. See margin.

he second number is four less than the first number.

he second number is twelve times the first number.

he second number is seven more than the first number.

Lesson 4-5

13. 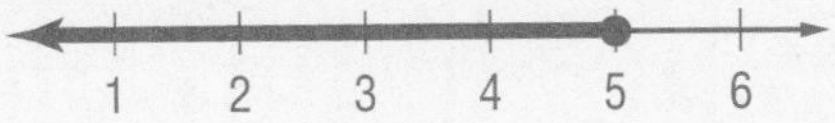

14. 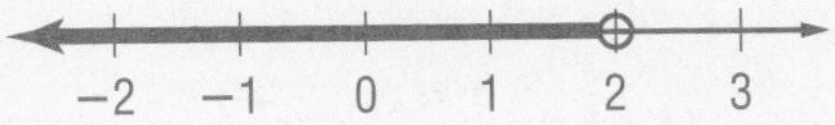

15. 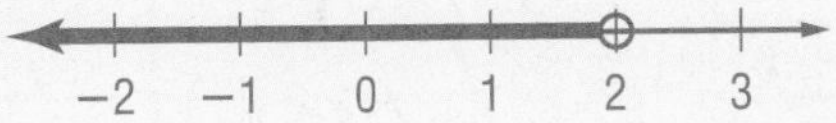

Lesson 4-6

3. 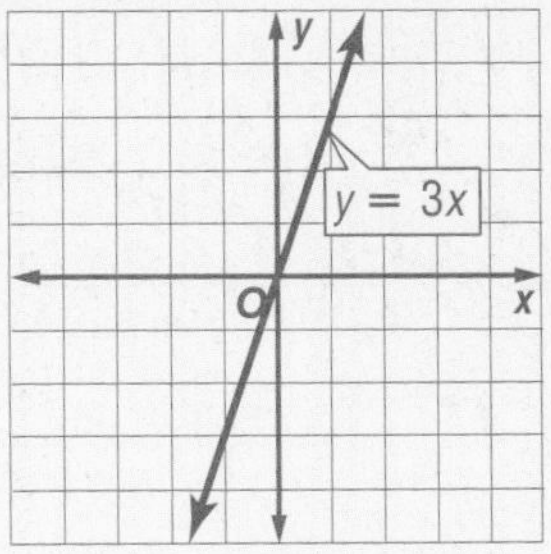

4. 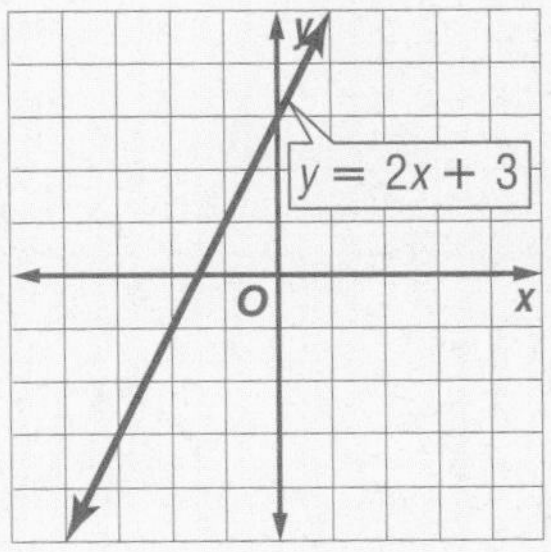

5. 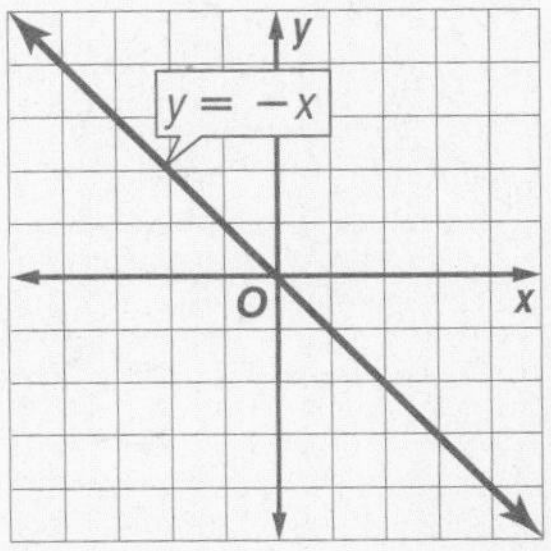

6. 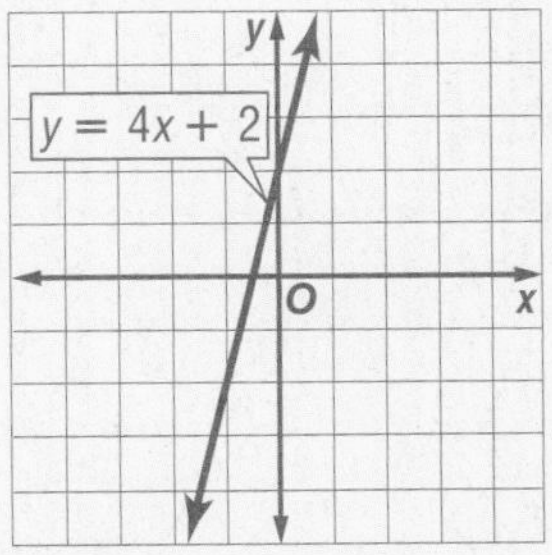

7. 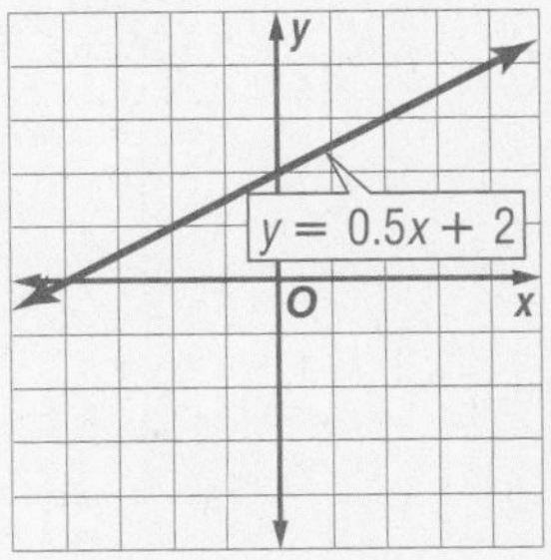

8. 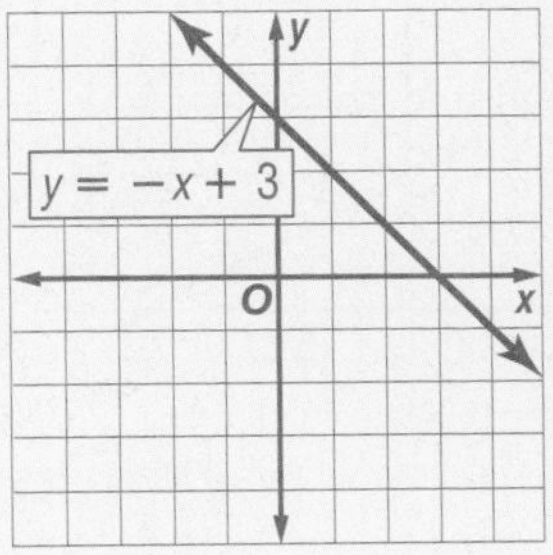

(Continued on p. 574)

Lesson 4-6

9.
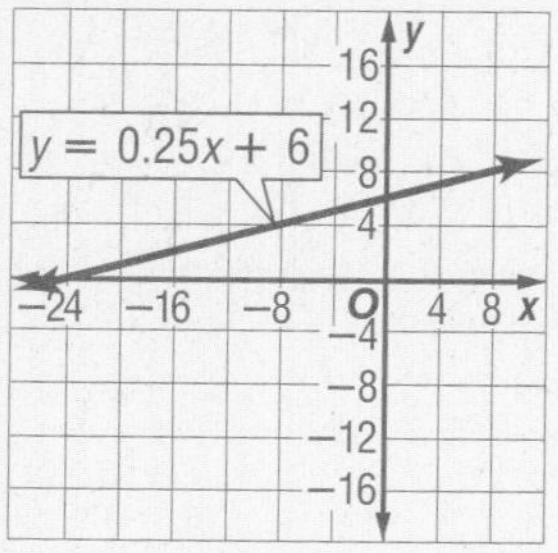

10.
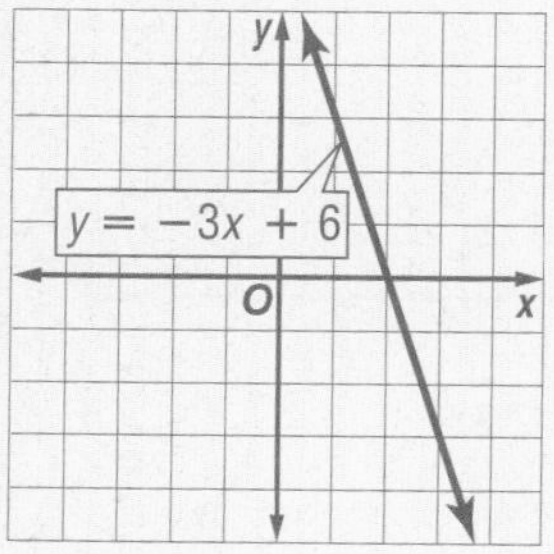

11.
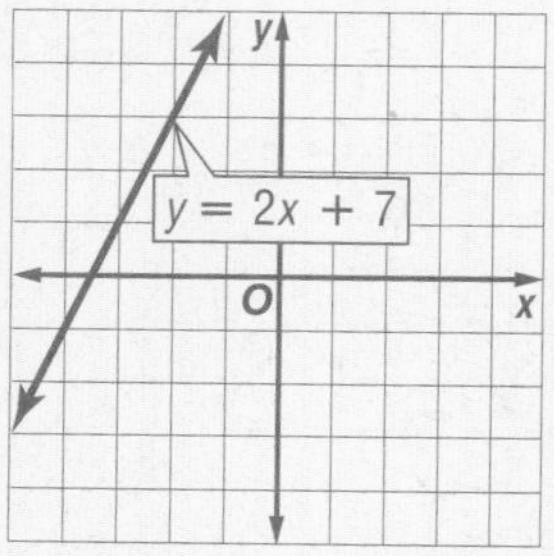

12.
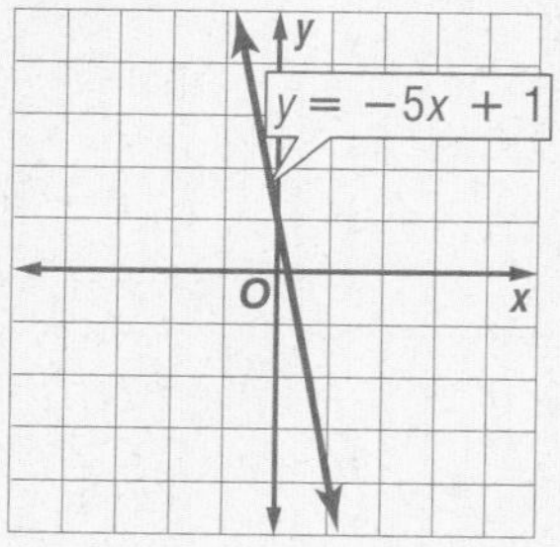

13.
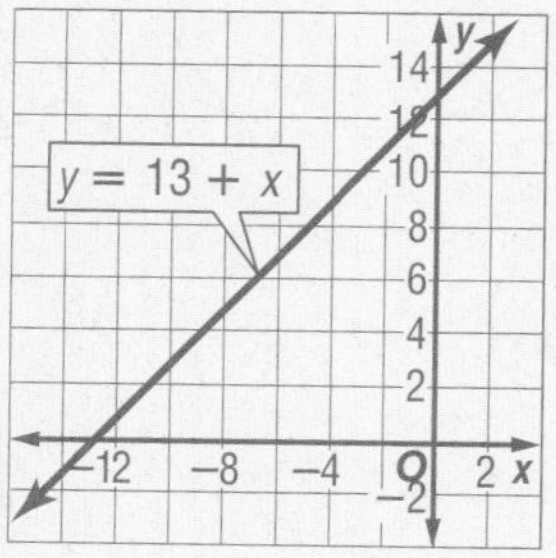

14.
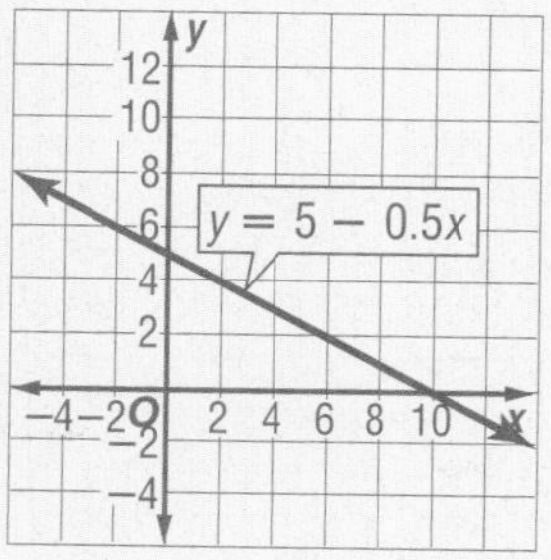

Extra Practice

Lesson 4-7

(Pages 182–185)

Find the slope of the line that passes through each pair of points.

1.
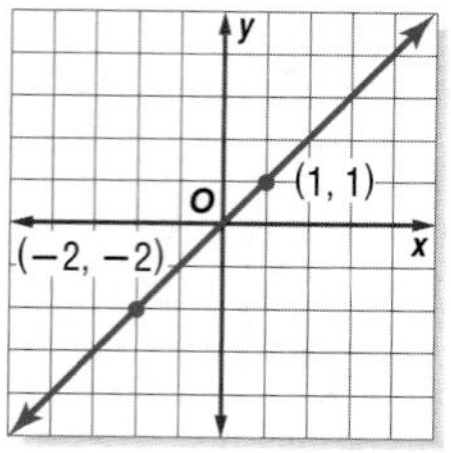

1

2.
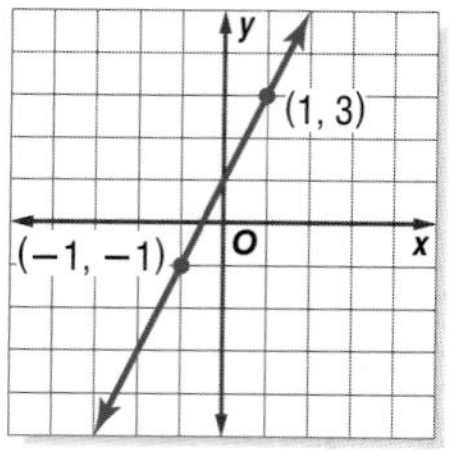

2

3.
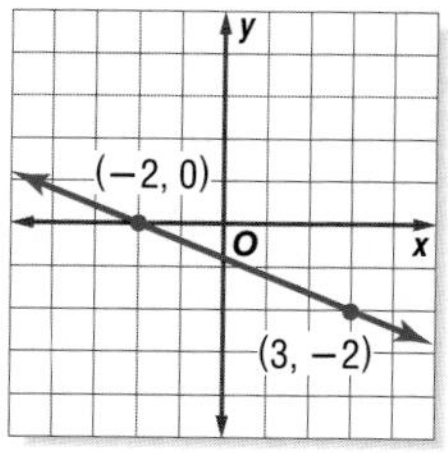

$-\frac{2}{5}$

4. (3, 4), (−2, 0) $\frac{4}{5}$
5. (9, −8), (−4, 1) $-\frac{9}{13}$
6. (−2, −3), (−3, −6) 3
7. (0, 0), (−2, −5) $\frac{5}{2}$
8. (5, 2), (−1, 3) $-\frac{1}{6}$
9. (1, 5), (−2, 4) $\frac{1}{3}$
10. (7, 2), (−4, 3) $-\frac{1}{11}$
11. (2, −6), (9, −5) $\frac{1}{7}$
12. (4, −1), (0, −5) 1

Lesson 5-1

(Pages 197–200)

Determine whether each number is *prime* or *composite*.

1. 32 composite
2. 41 prime
3. 52 composite
4. 21 composite
5. 71 prime
6. 102 composite
7. 239 prime
8. 93 composite
9. 123 composite
10. 251 prime
11. 321 composite
12. 423 composite

Find the prime factorization of each number.

13. 81 3^4
14. 525 $3 \times 5^2 \times 7$
15. 245 5×7^2
16. 1,120 $2^5 \times 5 \times 7$
17. 750 $2 \times 3 \times 5^3$
18. 2,400 $2^5 \times 3 \times 5^2$
19. 914 2×457
20. 975 $3 \times 5^2 \times 13$
21. 423 $3^2 \times 47$
22. 972 $2^2 \times 3^5$
23. 144 $2^4 \times 3^2$
24. 72 $2^3 \times 3^2$

Write each expression as a product of its factors. 25–33. See margin.

25. $35xy$
26. $14a^2$
27. $19bc$
28. $30n$
29. $27cd^2$
30. $37c^2$
31. $4s^2t^2$
32. $72abc$
33. $60p^2qr$

Lesson 5-2

(Pages 203–206)

Find the GCF of each set of numbers.

1. 12, 16 4
2. 63, 81 9
3. 225, 500 25
4. 37, 100 1
5. 32, 240 16
6. 412, 640 4
7. 36, 81 9
8. 140, 350 70
9. 72, 170 2
10. 51, 255 51
11. 48, 72 24
12. 86, 200 2
13. 12, 18, 42 6
14. 24, 56, 120 8
15. 48, 60, 84 12
16. 32, 80, 96 16
17. 14, 49, 70 7
18. 8, 10, 20 2

Find the GCF of each set of algebraic expressions.

19. $18b, 24b$ $6b$
20. $2a, 3a$ a
21. $5n, 5mn$ $5n$
22. $12cd, 24c$ $12c$
23. $30x, 50x^2$ $10x$
24. $15az, 25az$ $5az$
25. $2c, 4ac, 8a$ 2
26. $d, 6c^2d, 12d$ d
27. $10ab, 15bc, 20b^2$ $5b$

For Exercises 15–17, sample answers are given for the function tables.

15.

x	y
2	−2
1	−3
0	−4
−1	−5

$y = x - 4$

16.

x	y
1	12
0	0
−1	−12
−2	−24

$y = 12x$

17.

x	y
2	9
1	8
0	7
−1	6

$y = x + 7$

5-3

(Pages 207–209)

ach fraction in simplest form.

$\frac{1}{2}$	2. $\frac{15}{25}$ $\frac{3}{5}$	3. $\frac{100}{300}$ $\frac{1}{3}$	4. $\frac{14}{35}$ $\frac{2}{5}$	5. $\frac{9}{51}$ $\frac{3}{17}$
$\frac{27}{28}$	7. $\frac{75}{90}$ $\frac{5}{6}$	8. $\frac{24}{40}$ $\frac{3}{5}$	9. $\frac{180}{270}$ $\frac{2}{3}$	10. $\frac{312}{390}$ $\frac{4}{5}$
$\frac{15}{28}$	12. $\frac{71}{82}$ $\frac{71}{82}$	13. $\frac{333}{900}$ $\frac{37}{100}$	14. $\frac{85}{255}$ $\frac{1}{3}$	15. $\frac{84}{128}$ $\frac{21}{32}$
$\frac{2}{3}$	17. $\frac{99}{99}$ 1	18. $\frac{3}{99}$ $\frac{1}{33}$	19. $\frac{44}{55}$ $\frac{4}{5}$	20. $\frac{57}{69}$ $\frac{19}{23}$
$\frac{15}{37}$	22. $\frac{204}{408}$ $\frac{1}{2}$	23. $\frac{5}{125}$ $\frac{1}{25}$	24. $\frac{144}{216}$ $\frac{2}{3}$	25. $\frac{15}{75}$ $\frac{1}{5}$

wo fractions that are equivalent to each fraction. 26–35. Sample answers are given.

$\frac{4}{6}$, $\frac{6}{9}$	27. $\frac{1}{5}$ $\frac{2}{10}$, $\frac{3}{15}$	28. $\frac{2}{7}$ $\frac{4}{14}$, $\frac{6}{21}$	29. $\frac{3}{10}$ $\frac{6}{20}$, $\frac{9}{30}$	30. $\frac{7}{8}$ $\frac{14}{16}$, $\frac{21}{24}$
$\frac{10}{24}$, $\frac{15}{36}$	32. $\frac{3}{8}$ $\frac{6}{16}$, $\frac{9}{24}$	33. $\frac{8}{9}$ $\frac{16}{18}$, $\frac{24}{27}$	34. $\frac{1}{4}$ $\frac{2}{8}$, $\frac{3}{12}$	35. $\frac{2}{11}$ $\frac{4}{22}$, $\frac{6}{33}$

n 5-4

(Pages 210–213)

each repeating decimal using bar notation.

333333… $0.\overline{3}$	2. 0.121212… $0.\overline{12}$	3. 4.3151515… $4.3\overline{15}$
23023023… $7.\overline{023}$	5. 0.544444… $0.5\overline{4}$	6. 18.75484848… $18.75\overline{48}$

each fraction or mixed number as a decimal. Use bar notation if
cimal is a repeating decimal.

0.8	8. $\frac{25}{100}$ 0.25	9. $1\frac{7}{8}$ 1.875	10. $\frac{1}{6}$ $0.1\overline{6}$
0.275	12. $5\frac{13}{50}$ 5.26	13. $\frac{55}{300}$ $0.18\overline{3}$	14. $1\frac{1}{2}$ 1.5
$0.\overline{5}$	16. $2\frac{3}{4}$ 2.75	17. $\frac{9}{11}$ $0.\overline{81}$	18. $4\frac{1}{9}$ $4.\overline{1}$

each decimal as a fraction in simplest form.

6 $\frac{13}{50}$	20. 0.75 $\frac{3}{4}$	21. 0.4 $\frac{2}{5}$	22. 0.1 $\frac{1}{10}$
8 $4\frac{12}{25}$	24. 9.8 $9\frac{4}{5}$	25. 0.91 $\frac{91}{100}$	26. 11.15 $11\frac{3}{20}$

n 5-5

(Pages 216–218)

each ratio as a percent.

out of 100 39%	2. $\frac{23}{100}$ 23%	3. 17:100 17%
per 100 72%	5. 4 to 100 4%	6. 98 in 100 98%

each fraction as a percent.

50%	8. $\frac{2}{5}$ 40%	9. $\frac{60}{100}$ 60%	10. $\frac{17}{20}$ 85%
28%	12. $\frac{1}{20}$ 5%	13. $\frac{8}{100}$ 8%	14. $\frac{7}{7}$ 100%
90%	16. $\frac{1}{100}$ 1%	17. $\frac{50}{50}$ 100%	18. $\frac{49}{50}$ 98%

each percent as a fraction in simplest form.

2% $\frac{3}{25}$	20. 23% $\frac{23}{100}$	21. 1% $\frac{1}{100}$	22. 94% $\frac{47}{50}$
6% $\frac{9}{25}$	24. 4% $\frac{1}{25}$	25. 72% $\frac{18}{25}$	26. 100% 1
5% $\frac{13}{20}$	28. 47% $\frac{47}{100}$	29. 15% $\frac{3}{20}$	30. 48% $\frac{12}{25}$

Extra Practice

Lesson 5-1

25. $5 \times 7 \times x \times y$
26. $2 \times 7 \times a \times a$
27. $19 \times b \times c$
28. $2 \times 3 \times 5 \times n$
29. $3 \times 3 \times 3 \times c \times d \times d$
30. $37 \times c \times c$
31. $2 \times 2 \times s \times s \times t \times t$
32. $2 \times 2 \times 2 \times 3 \times 3 \times a \times b \times c$
33. $2 \times 2 \times 3 \times 5 \times p \times p \times q \times r$

Extra Practice

Extra Practice 575

Lesson 5-6

(Pages 220–223)

Write each percent as a decimal.

1. 42% **0.42**
2. 100% **1.0**
3. 8% **0.08**
4. 20% **0.2**
5. 35% **0.35**
6. 3% **0.03**
7. 62% **0.62**
8. 50% **0.5**
9. 28% **0.28**
10. 87% **0.87**
11. 7.5% **0.075**
12. 87.5% **0.875**
13. 1.8% **0.018**
14. 99.9% **0.999**
15. $85\frac{1}{4}\%$ **0.8525**
16. $24\frac{1}{2}\%$ **0.245**
17. $64\frac{4}{5}\%$ **0.648**
18. $36\frac{3}{4}\%$ **0.3675**
19. $1\frac{1}{5}\%$ **0.012**
20. $2\frac{1}{2}\%$ **0.025**

Write each decimal as a percent.

21. 0.16 **16%**
22. 0.1 **10%**
23. 0.5 **50%**
24. 0.98 **98%**
25. 0.31 **31%**
26. 0.76 **76%**
27. 0.07 **7%**
28. 0.8 **80%**
29. 0.07 **7%**
30. 0.10 **10%**
31. 0.90 **90%**
32. 1.00 **100%**
33. 0.666 **66.6%**
34. 0.725 **72.5%**
35. 0.138 **13.8%**
36. 0.899 **89.9%**
37. 0.256 **25.6%**
38. 0.038 **3.8%**
39. 0.0525 **5.25%**
40. 0.017 **1.7%**

Lesson 5-7

(Pages 224–226)

Find the LCM of each set of numbers.

1. 4, 9 **36**
2. 6, 16 **48**
3. 24, 36 **72**
4. 48, 84 **336**
5. 8, 9 **72**
6. 49, 56 **392**
7. 42, 66 **462**
8. 15, 39 **195**
9. 56, 64 **448**
10. 24, 42 **168**
11. 80, 250 **2,000**
12. 16, 24 **48**
13. 13, 14 **182**
14. 36, 48 **144**
15. 10, 100 **100**
16. 25, 200 **200**
17. 1, 2, 5 **10**
18. 2, 3, 7 **42**
19. 1, 9, 27 **27**
20. 2, 24, 36 **72**
21. 7, 21, 35 **105**
22. 12, 18, 28 **252**
23. 32, 80, 96 **480**
24. 5, 18, 45 **90**
25. 11, 22, 33 **66**
26. 35, 70, 140 **140**
27. 25, 200, 400 **400**
28. 100, 200, 300 **600**

Lesson 5-8

(Pages 227–231)

Find the LCD for each pair of fractions.

1. $\frac{3}{8}, \frac{2}{3}$ **24**
2. $\frac{5}{9}, \frac{7}{12}$ **36**
3. $\frac{4}{9}, \frac{8}{15}$ **45**
4. $\frac{11}{24}, \frac{17}{42}$ **168**
5. $\frac{12}{36}, \frac{15}{42}$ **252**
6. $\frac{25}{27}, \frac{43}{81}$ **81**
7. $\frac{32}{64}, \frac{15}{48}$ **192**
8. $\frac{2}{6}, \frac{14}{15}$ **30**

Replace each ● with <, >, or = to make a true sentence.

9. $\frac{7}{9}$ ● $\frac{3}{5}$ >
10. $\frac{14}{25}$ ● $\frac{3}{4}$ <
11. $\frac{8}{24}$ ● $\frac{20}{60}$ =
12. $\frac{5}{12}$ ● $\frac{4}{9}$ <
13. $\frac{18}{24}$ ● $\frac{10}{18}$ >
14. $\frac{4}{6}$ ● $\frac{5}{9}$ >
15. $\frac{11}{49}$ ● $\frac{12}{42}$ <
16. $\frac{5}{14}$ ● $\frac{2}{6}$ >

Order each set of ratios from least to greatest. **19. $\frac{61}{100}$, 61.5%, 0.65 21. 10.5%, $\frac{7}{50}$, 0.15**

17. 70%, 0.6, $\frac{2}{3}$ **0.6, $\frac{2}{3}$, 70%**
18. 0.8, $\frac{17}{20}$, 17% **17%, 0.8, $\frac{17}{20}$**
19. $\frac{61}{100}$, 0.65, 61.5%
20. $0.\overline{42}$, $\frac{3}{7}$, 42% **42%, 0.42, $\frac{3}{7}$**
21. 0.15, 10.5%, $\frac{7}{50}$
22. $\frac{1}{8}$, 0.81, 18% **$\frac{1}{8}$, 18%, 0.81**

Determine whether each number is rational. Write *yes* or *no*.

23. $4.\overline{28}$ **yes**
24. $-\frac{5}{4}$ **yes**
25. $\frac{0.5}{4}$ **yes**
26. $\frac{1}{7}$ **yes**
27. 0.414114111... **no**
28. 10.987... **no**
29. 7.22... **yes**
30. $3\frac{2}{5}$ **yes**

576 Extra Practice

Extra Practice

6-1

(Pages 240–243)

e. Sample answers are given.

$\frac{6}{8}$ $\frac{1}{2} + 1 = 1\frac{1}{2}$	2. $\frac{3}{9} + \frac{7}{8}$ $\frac{1}{2} + 1 = 1\frac{1}{2}$	3. $\frac{1}{8} + \frac{8}{9}$ $0 + 1 = 1$
$+ 7\frac{6}{7}$ $3 + 8 = 11$	5. $4\frac{2}{3} + 6\frac{7}{8}$ $5 + 7 = 12$	6. $3\frac{2}{3} \times 2\frac{1}{3}$ $4 \times 2 = 8$
3 $1 \cdot 3 = 3$	8. $9\frac{7}{8} - 6\frac{2}{3}$ $10 - 7 = 3$	9. $\frac{3}{7} - \frac{1}{15}$ $\frac{1}{2} - 0 = \frac{1}{2}$
$\frac{7}{8}$ $1 \cdot 1 = 1$	11. $7\frac{1}{4} \div \frac{2}{3}$ $7 \div 1 = 7$	12. $\frac{5}{6} \div \frac{2}{3}$ $1 \div 1 = 1$
$+ 3\frac{1}{8}$ $10 + 3 = 13$	14. $5\frac{1}{3} - 2\frac{3}{4}$ $5 - 3 = 2$	15. $13\frac{7}{8} - 2\frac{1}{3}$ $14 - 2 = 12$
$\frac{3}{8}$ $1 \cdot \frac{1}{2} = \frac{1}{2}$	17. $\frac{1}{9} \div 2$ $0 \div 2 = 0$	18. $\frac{5}{8} - \frac{1}{16}$ $\frac{1}{2} - 0 = \frac{1}{2}$
$+ 4\frac{7}{8}$ $10 + 5 = 15$	20. $\frac{1}{2} \cdot 25$ $\frac{1}{2} \cdot 26 = 13$	21. $35\frac{1}{3} \div 6\frac{3}{4}$ $35 \div 7 = 5$
$\div 3\frac{3}{4}$ $100 \div 4 = 25$	23. $\frac{1}{3} \cdot 38$ $\frac{1}{3} \cdot 39 = 13$	24. $\frac{15}{16} - \frac{3}{4}$ $1 - 1 = 0$
16 $\frac{1}{5} \cdot 15 = 3$	26. $47\frac{1}{2} \div 6\frac{5}{8}$ $49 \div 7 = 7$	27. $\frac{1}{9} \cdot 79$ $\frac{1}{9} \cdot 81 = 9$

6-2

(Pages 244–247)

subtract. Write in simplest form.

$+ \frac{9}{11}$ $1\frac{3}{11}$	2. $\frac{5}{8} - \frac{1}{8}$ $\frac{1}{2}$	3. $\frac{7}{10} + \frac{7}{10}$ $1\frac{2}{5}$
$- \frac{5}{12}$ $\frac{1}{3}$	5. $\frac{2}{9} + \frac{1}{3}$ $\frac{5}{9}$	6. $\frac{1}{2} + \frac{3}{4}$ $1\frac{1}{4}$
$- \frac{3}{12}$ 0	8. $\frac{3}{7} + \frac{6}{14}$ $\frac{6}{7}$	9. $\frac{1}{4} + \frac{3}{5}$ $\frac{17}{20}$
$- \frac{1}{2}$ $\frac{17}{18}$	11. $\frac{5}{7} - \frac{4}{6}$ $\frac{1}{21}$	12. $\frac{3}{4} - \frac{1}{6}$ $\frac{7}{12}$
$- \frac{3}{4}$ $1\frac{7}{20}$	14. $\frac{2}{3} - \frac{1}{8}$ $\frac{13}{24}$	15. $\frac{9}{10} + \frac{1}{3}$ $1\frac{7}{30}$
$+ \frac{2}{9}$ $\frac{34}{45}$	17. $\frac{6}{7} + \frac{6}{9}$ $1\frac{11}{21}$	18. $\frac{3}{7} + \frac{3}{4}$ $1\frac{5}{28}$

te each expression if $a = \frac{2}{3}$ and $b = \frac{7}{12}$.

$+ a$ $\frac{13}{15}$	20. $a - \frac{1}{2}$ $\frac{1}{6}$	21. $b + \frac{7}{8}$ $1\frac{11}{24}$
$- a$ $\frac{5}{24}$	23. $a + b$ $1\frac{1}{4}$	24. $a - b$ $\frac{1}{12}$

n 6-3

(Pages 248–251)

r subtract. Write in simplest form.

$+ 1\frac{1}{3}$ $3\frac{2}{3}$	2. $5\frac{2}{7} - 2\frac{3}{7}$ $2\frac{6}{7}$	3. $6\frac{3}{8} + 7\frac{1}{8}$ $13\frac{1}{2}$
$- 1\frac{1}{4}$ $1\frac{1}{2}$	5. $5\frac{1}{2} - 3\frac{1}{4}$ $2\frac{1}{4}$	6. $2\frac{2}{3} + 4\frac{1}{9}$ $6\frac{7}{9}$
$+ 9\frac{3}{10}$ $17\frac{1}{10}$	8. $3\frac{3}{4} + 5\frac{5}{8}$ $9\frac{3}{8}$	9. $10\frac{2}{3} + 5\frac{6}{7}$ $16\frac{11}{21}$
$\frac{2}{9} - 12\frac{1}{3}$ $4\frac{8}{9}$	11. $6\frac{5}{12} + 12\frac{5}{12}$ $18\frac{5}{6}$	12. $7\frac{1}{4} + 15\frac{5}{6}$ $23\frac{1}{12}$
$+ 4\frac{2}{3}$ $10\frac{19}{24}$	14. $7 - 6\frac{4}{9}$ $\frac{5}{9}$	15. $8\frac{1}{12} + 12\frac{6}{11}$ $20\frac{83}{132}$
$+ 8\frac{1}{4}$ $15\frac{11}{12}$	17. $12\frac{3}{11} + 14\frac{3}{13}$ $26\frac{72}{143}$	18. $21\frac{1}{3} + 15\frac{3}{8}$ $36\frac{17}{24}$
$\frac{1}{7} + 6\frac{1}{4}$ $25\frac{11}{28}$	20. $9\frac{2}{5} - 8\frac{1}{3}$ $1\frac{1}{15}$	21. $18\frac{1}{4} - 3\frac{3}{8}$ $14\frac{7}{8}$
$+ 2\frac{1}{12}$ $3\frac{5}{24}$	23. $2\frac{1}{12} - 1\frac{1}{8}$ $\frac{23}{24}$	24. $10 - \frac{2}{3}$ $9\frac{1}{3}$

Extra Practice 577

Extra Practice

Lesson 6-4
(Pages 254–257)

Multiply. Write in simplest form.

1. $\frac{2}{3} \times \frac{3}{5}$ $\frac{2}{5}$
2. $\frac{1}{6} \times \frac{2}{5}$ $\frac{1}{15}$
3. $\frac{4}{9} \times \frac{3}{7}$ $\frac{4}{21}$
4. $\frac{5}{12} \times \frac{6}{11}$ $\frac{5}{22}$
5. $\frac{3}{8} \times \frac{8}{9}$ $\frac{1}{3}$
6. $\frac{3}{5} \times \frac{1}{12}$ $\frac{1}{20}$
7. $\frac{2}{5} \times \frac{5}{8}$ $\frac{1}{4}$
8. $\frac{7}{15} \times \frac{3}{21}$ $\frac{1}{15}$
9. $\frac{5}{6} \times \frac{15}{16}$ $\frac{25}{32}$
10. $\frac{6}{14} \times \frac{12}{18}$ $\frac{2}{7}$
11. $\frac{2}{3} \times \frac{3}{13}$ $\frac{2}{13}$
12. $\frac{4}{9} \times \frac{1}{6}$ $\frac{2}{27}$
13. $3 \times \frac{1}{9}$ $\frac{1}{3}$
14. $5 \times \frac{6}{7}$ $4\frac{2}{7}$
15. $\frac{3}{5} \times 15$ 9
16. $3\frac{1}{2} \times 4\frac{1}{3}$ $15\frac{1}{6}$
17. $3\frac{5}{8} \times 4\frac{1}{2}$ $16\frac{5}{16}$
18. $\frac{4}{5} \times 2\frac{3}{4}$ $2\frac{1}{5}$
19. $6\frac{1}{8} \times 5\frac{1}{7}$ $31\frac{1}{2}$
20. $2\frac{2}{3} \times 2\frac{1}{4}$ 6
21. $\frac{7}{8} \times 16$ 14
22. $5\frac{1}{5} \times 2\frac{1}{2}$ 13
23. $7 \times \frac{1}{14}$ $\frac{1}{2}$
24. $22 \times \frac{3}{11}$ 6
25. $8\frac{2}{3} \times 1\frac{1}{2}$ 13
26. $4 \times 6\frac{1}{2}$ 26
27. $\frac{1}{2} \times 10\frac{2}{3}$ $5\frac{1}{3}$
28. $\frac{2}{3} \times 21\frac{1}{3}$ $14\frac{2}{9}$
29. $\frac{7}{8} \times \frac{8}{7}$ 1
30. $1\frac{2}{3} \times 1\frac{2}{3}$ $2\frac{7}{9}$
31. $21 \times \frac{1}{2}$ $10\frac{1}{2}$
32. $11 \times \frac{1}{4}$ $2\frac{3}{4}$

Lesson 6-5
(Pages 258–261)

Find the multiplicative inverse of each number.

1. $\frac{2}{3}$ $\frac{3}{2}$ or $1\frac{1}{2}$
2. $-\frac{5}{4}$ $-\frac{4}{5}$
3. 1 $\frac{1}{1}$ or 1
4. 10 $\frac{1}{10}$
5. -23 $-\frac{1}{23}$
6. $-\frac{1}{7}$ -7
7. $\frac{9}{16}$ $\frac{16}{9}$ or $1\frac{7}{9}$
8. 15 $\frac{1}{15}$
9. $1\frac{1}{3}$ $\frac{3}{4}$
10. $-3\frac{3}{4}$ $-\frac{4}{15}$
11. $7\frac{3}{8}$ $\frac{8}{59}$
12. $-\frac{19}{10}$ $-\frac{10}{19}$
13. $-\frac{3}{49}$ $-\frac{49}{3}$ or $-16\frac{1}{3}$
14. $6\frac{2}{5}$ $\frac{5}{32}$
15. $33\frac{1}{3}$ $\frac{3}{100}$
16. $-66\frac{2}{3}$ $-\frac{3}{200}$

Solve each equation. Check your solution.

17. $\frac{a}{13} = 2$ 26
18. $\frac{8}{9}x = 24$ 27
19. $\frac{3}{8}r = 36$ 96
20. $\frac{3}{4}t = \frac{1}{2}$ $\frac{2}{3}$
21. $16 = \frac{h}{4}$ 64
22. $\frac{m}{8} = -12$ -96
23. $\frac{5}{8}n = -45$ -72
24. $10 = \frac{b}{10}$ 100
25. $\frac{1}{7}x = 7$ 49
26. $-5 = \frac{1}{5}y$ -25
27. $\frac{4}{3}m = -28$ -21
28. $\frac{2}{3}z = 20$ 30
29. $\frac{c}{9} = 81$ 729
30. $\frac{m}{9} = 9$ 81
31. $-16 = \frac{4}{9}f$ -36
32. $\frac{15}{8}x = 225$ 120

Lesson 6-6
(Pages 264–266)

Divide. Write in simplest form.

1. $\frac{2}{3} \div \frac{3}{2}$ $\frac{4}{9}$
2. $\frac{3}{5} \div \frac{2}{5}$ $1\frac{1}{2}$
3. $\frac{7}{10} \div \frac{3}{8}$ $1\frac{13}{15}$
4. $\frac{5}{9} \div \frac{2}{5}$ $1\frac{7}{18}$
5. $4 \div \frac{2}{3}$ 6
6. $8 \div \frac{4}{5}$ 10
7. $9 \div \frac{5}{9}$ $16\frac{1}{5}$
8. $\frac{2}{7} \div 2$ $\frac{1}{7}$
9. $\frac{1}{14} \div 7$ $\frac{1}{98}$
10. $\frac{2}{13} \div \frac{5}{26}$ $\frac{4}{5}$
11. $\frac{4}{7} \div \frac{6}{7}$ $\frac{2}{3}$
12. $\frac{7}{8} \div \frac{1}{3}$ $2\frac{5}{8}$
13. $15 \div \frac{3}{5}$ 25
14. $\frac{9}{14} \div \frac{3}{4}$ $\frac{6}{7}$
15. $\frac{8}{9} \div \frac{5}{6}$ $1\frac{1}{15}$
16. $2\frac{1}{2} \div 5$ $\frac{1}{2}$
17. $\frac{7}{8} \div 10$ $\frac{7}{80}$
18. $16 \div \frac{3}{4}$ $21\frac{1}{3}$
19. $22 \div 3\frac{2}{3}$ 6
20. $40\frac{5}{8} \div 4$ $10\frac{5}{32}$
21. $\frac{3}{8} \div 2\frac{1}{2}$ $\frac{3}{20}$
22. $5\frac{1}{2} \div 2\frac{1}{2}$ $2\frac{1}{5}$
23. $3\frac{1}{4} \div 5\frac{1}{2}$ $\frac{13}{22}$
24. $12\frac{5}{6} \div 2\frac{1}{6}$ $5\frac{12}{13}$
25. $7\frac{1}{2} \div 3\frac{1}{2}$ $2\frac{1}{7}$
26. $3\frac{1}{2} \div 7\frac{1}{2}$ $\frac{7}{15}$
27. $4\frac{2}{3} \div 2\frac{1}{3}$ 2

578 Extra Practice

6-7

(Pages 267–269)

ete.

0 lb = ___?___ T **2**

2. 5 T = ___?___ lb **10,000**

3. 5 lb = ___?___ oz **80**

00 lb = ___?___ T **6**

5. $\frac{1}{4}$ lb = ___?___ oz **4**

6. 12 pt = ___?___ c **24**

l = ___?___ pt **24**

8. 24 fl oz = ___?___ c **3**

9. 8 pt = ___?___ c **16**

t = ___?___ qt **5**

11. $2\frac{1}{4}$c = ___?___ fl oz **18**

12. 6 lb = ___?___ oz **96**

al = ___?___ qt **40**

14. 4 qt = ___?___ fl oz **128**

15. 4 pt = ___?___ c **8**

00 ft = ___?___ mi $2\frac{1}{2}$

17. 120 oz = ___?___ lb $7\frac{1}{2}$

18. $9\frac{1}{4}$ gal = ___?___ qt **37**

0 yd = ___?___ mi $4\frac{1}{4}$

20. $12\frac{1}{2}$ lb = ___?___ oz **200**

21. $7\frac{1}{2}$ qt = ___?___ pt **15**

= ___?___ fl oz **25**

23. $2\frac{1}{4}$ mi = ___?___ ft **11,880**

24. $3\frac{2}{3}$ T = ___?___ lb $7{,}333\frac{1}{3}$

6-8

(Pages 270–273)

e perimeter and area of each rectangle.

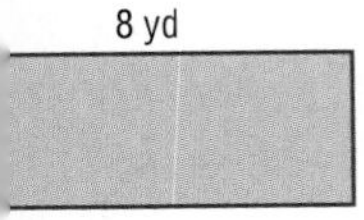

yd; 24 yd^2

2.

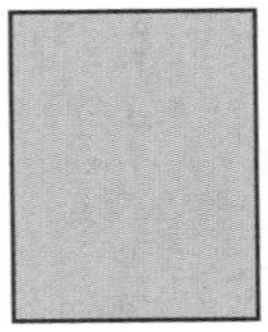

55.4 cm; 189.1 cm^2

3.

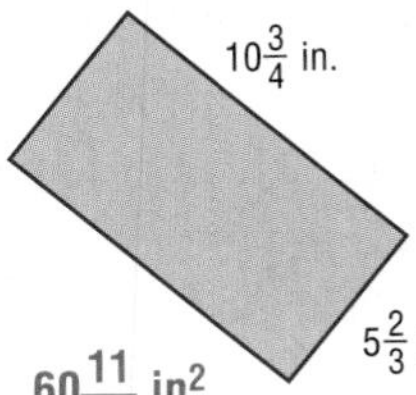

$32\frac{5}{6}$ in., $60\frac{11}{12}$ in^2

80 yd, w = 20 yd **200 yd; 1,600 yd^2**

5. ℓ = 75 cm, w = 25 cm **200 cm; 1,875 cm^2**

5.25 km, w = 1.5 km **13.5 km; 7.875 km^2**

7. ℓ = 8.6 cm, w = 2.5 cm **22.2 cm; 21.5 cm^2**

20.25 m, w = 4.75 m **50 m; 96.1875 m^2**

9. ℓ = 12 ft, w = 3 ft **30 ft; 36 ft^2**

$5\frac{1}{4}$ mi, $w = 2\frac{1}{2}$ mi **$15\frac{1}{2}$ mi; $13\frac{1}{8}$ mi^2**

11. $\ell = 10\frac{2}{3}$ ft, $w = 5\frac{5}{6}$ ft **33 ft; $62\frac{2}{9}$ ft^2**

$12\frac{1}{3}$ yd, $w = 5\frac{2}{3}$ yd **36 yd; $69\frac{8}{9}$ yd^2**

13. $\ell = 3\frac{7}{8}$ in., $w = 1\frac{1}{4}$ in. **$10\frac{1}{4}$ in.; $4\frac{27}{32}$ in^2**

n 6-9

(Pages 275–277)

e circumference of each circle. Use 3.14 or $\frac{22}{7}$ for π. Round to the t tenth if necessary.

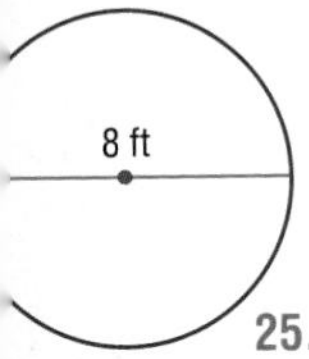

25.1 ft

2.

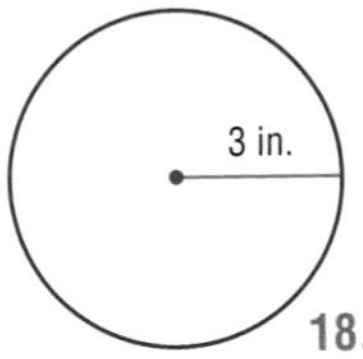

18.8 in.

3.

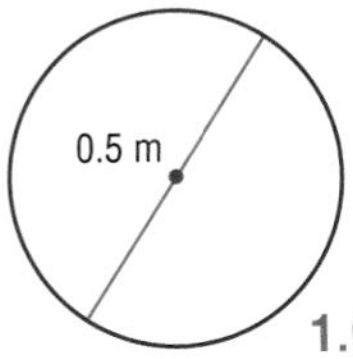

1.6 m

4.

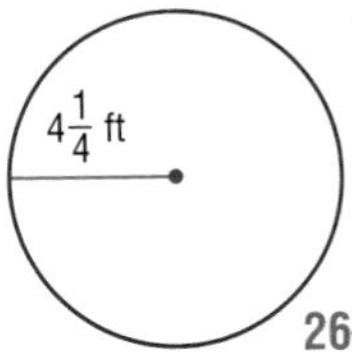

$26\frac{5}{7}$ ft

= 1 m **6.3 m**

6. d = 2 yd **6.3 yd**

7. d = 5,280 ft **16,587.6 ft**

= 350 mi **2,198 mi**

9. r = 1.5 in. **9.4 in.**

10. d = 1.5 in. **4.7 in.**

= 0.5 cm **3.1 cm**

12. d = 6.4 m **20.1 m**

13. r = 10.7 km **67.2 km**

= $\frac{5}{8}$ in. **$3\frac{13}{14}$ in.**

15. $d = 8\frac{1}{4}$ in. **$25\frac{13}{14}$ in.**

16. $r = 2\frac{1}{3}$ yd **$14\frac{2}{3}$ yd**

= $\frac{3}{16}$ in. **$\frac{33}{56}$ in.**

18. $r = 5\frac{1}{2}$ mi **$34\frac{4}{7}$ mi**

19. $d = 42\frac{3}{4}$ ft **$134\frac{5}{14}$ ft**

Extra Practice
Extra Practice
Extra Practice 579

Lesson 7-1

16. 616 = 616
17. 144 > 16
18. 6,500 = 6,500
19. 1,800 > 1,152
20. 500 = 500
21. 1,536 > 6
22. 300 = 300
23. 2,100 > 1,800
24. 224 > 200
25. 120 = 120

Extra Practice

Extra Practice

Lesson 7-1

(Pages 288–291)

Write each ratio as a fraction in simplest form.

1. 45 to 15 $\frac{3}{1}$
2. 64:128 $\frac{1}{2}$
3. 14 to 49 $\frac{2}{7}$
4. 125:25 $\frac{5}{1}$
5. 18 to 81 $\frac{2}{9}$
6. 16:40 $\frac{2}{5}$
7. 120 to 180 $\frac{2}{3}$
8. 32:64 $\frac{1}{2}$
9. 90 to 100 $\frac{9}{10}$
10. 12 weeks out of 15 $\frac{4}{5}$
11. 33 min:60 min $\frac{11}{20}$
12. 10 ft to 8 yd $\frac{5}{12}$
13. 25 pounds to 80 pounds $\frac{5}{16}$
14. 15 minutes:4 hours $\frac{1}{16}$
15. 10 yards to 12 feet $\frac{5}{2}$

Determine whether the ratios are equivalent. Explain. 16–25. See margin for explanations.

16. 14 to 77 and 8 to 44 yes
17. $\frac{48}{16}$ and $\frac{1}{3}$ no
18. 65:13 and 500:100 yes
19. 72 to 90 and 20 to 16 no
20. 250:100 and 5:2 yes
21. $\frac{32}{2}$ and $\frac{3}{48}$ no
22. 3 lb to 60 oz and 5 lb to 100 oz yes
23. 12:25 and 84:150 no
24. 28 in.:5 ft and 40 in.:8 ft no
25. 8 hours to 5 days and 24 hours to 15 day yes

Lesson 7-2

(Pages 292–295)

Find each unit rate. Round to the nearest hundredth if necessary.

1. \$240 for 4 days \$60/day
2. 250 people in 5 buses 50 people/bus
3. 500 miles in 10 hours 50 miles/hour
4. 18 cups for 24 pounds $\frac{3}{4}$ cup/pound
5. 32 people in 8 cars 4 people/car
6. \$4.50 for 3 dozen \$1.50/dozen
7. 245 tickets in 5 days 49 tickets/day
8. 12 classes in 4 semesters 3 classes/semes
9. 60 people in 4 rows 15 people/row
10. 48 ounces in 3 pounds 16 ounces/pound
11. 20 people in 4 groups 5 people/group
12. 1.5 pounds for \$3.00 0.5 pound/dollar
13. 45 miles in 60 minutes 0.75 mile/minute
14. \$5.50 for 10 disks \$0.55/disk
15. 360 miles for 12 gallons 30 miles/gallon
16. \$8.50 for 5 yards \$1.70/yard
17. 24 cups for \$1.20 20 cups/dollar
18. 160 words in 4 minutes 40 words/minute
19. \$60 for 5 books \$12/book
20. \$24 for 6 hours \$4/hour

Lesson 7-3

(Pages 297–300)

Determine whether each pair of ratios forms a proportion.

1. $\frac{2}{5}$ and $\frac{5}{8}$ no
2. $\frac{125}{275}$ and $\frac{150}{300}$ no
3. $\frac{28}{42}$ and $\frac{70}{105}$ yes
4. $\frac{55}{22}$ and $\frac{35}{14}$ yes
5. $\frac{99}{100}$ and $\frac{100}{99}$ no
6. $\frac{85}{17}$ and $\frac{5}{1}$ yes
7. $\frac{100 \text{ ft}}{8 \text{ s}}$ and $\frac{250 \text{ ft}}{20 \text{ s}}$ yes
8. $\frac{\$15}{45 \text{ lb}}$ and $\frac{\$20}{65 \text{ lb}}$ no
9. $\frac{32 \text{ mi}}{6 \text{ h}}$ and $\frac{44 \text{ mi}}{8.25 \text{ h}}$ yes

Solve each proportion.

10. $\frac{u}{72} = \frac{2}{4}$ 36
11. $\frac{12}{m} = \frac{15}{10}$ 8
12. $\frac{36}{90} = \frac{16}{t}$ 40
13. $\frac{g}{32} = \frac{8}{64}$ 4
14. $\frac{5}{14} = \frac{10}{a}$ 28
15. $\frac{k}{18} = \frac{5}{3}$ 30
16. $\frac{120}{150} = \frac{p}{20}$ 16
17. $\frac{15}{w} = \frac{60}{4}$ 1
18. $\frac{81}{90} = \frac{y}{20}$ 18
19. $\frac{14}{s} = \frac{8}{4}$ 7
20. $\frac{h}{3} = \frac{36}{9}$ 12
21. $\frac{50}{8} = \frac{150}{t}$ 24
22. $\frac{45}{8} = \frac{36}{d}$ 6.4
23. $\frac{125}{v} = \frac{20}{5}$ 31.25
24. $\frac{4}{5} = \frac{x}{3}$ 2.4
25. $\frac{45}{75} = \frac{j}{3}$ 1.8
26. $\frac{3}{7} = \frac{21}{d}$ 49
27. $\frac{3}{10} = \frac{z}{36}$ 10.8

n 7-4

(Pages 304–308)

ıap, the scale is 1 inch = 50 miles. For each map distance, find the distance.

ıches **250 mi** | 2. 12 inches **600 mi** | 3. $3\frac{1}{2}$ inches **175 mi** | 4. $2\frac{3}{8}$ inches **$118\frac{3}{4}$ mi**

nch **40 mi** | 6. $6\frac{3}{4}$ inches **337.5 mi** | 7. $2\frac{5}{6}$ inches **$141\frac{2}{3}$ mi** | 8. 8 inches **400 mi**

inches **75 mi** | 10. 3.25 inches **162.5 mi** | 11. 4.75 inches **237.5 mi** | 12. 5.25 inches **262.5 mi**

cale drawing, the scale is $\frac{1}{2}$ inch = 2 feet. Find the dimensions of oom in the scale drawing.

feet by 18 feet **$3\frac{1}{2}$ in. by $4\frac{1}{2}$ in.** | 14. 32 feet by 6 feet **8 in. by $1\frac{1}{2}$ in.**

et by 5 feet **$\frac{3}{4}$ in. by $1\frac{1}{4}$ in.** | 16. 20 feet by 30 feet **5 in. by $7\frac{1}{2}$ in.**

et by 15 feet **2 in. by $3\frac{3}{4}$ in.** | 18. 25 feet by 80 feet **$6\frac{1}{4}$ in. by 20 in.**

n 7-5

(Pages 312–315)

each percent as a fraction in simplest form.

% **$\frac{8}{25}$** | 2. 89% **$\frac{89}{100}$** | 3. 72% **$\frac{18}{25}$** | 4. 11% **$\frac{11}{100}$**

$\frac{1}{100}$ | 6. 28% **$\frac{7}{25}$** | 7. 55% **$\frac{11}{20}$** | 8. 18.5% **$\frac{37}{200}$**

75% **$\frac{91}{400}$** | 10. 25.2% **$\frac{63}{250}$** | 11. 75.5% **$\frac{151}{200}$** | 12. 48.25% **$\frac{193}{400}$**

% **$\frac{13}{200}$** | 14. 1.25% **$\frac{1}{80}$** | 15. 88.9% **$\frac{889}{1,000}$** | 16. $52\frac{1}{4}\%$ **$\frac{209}{400}$**

each fraction as a percent. Round to the nearest hundredth if sary.

56% | 18. $\frac{28}{50}$ **56%** | 19. $\frac{14}{20}$ **70%** | 20. $\frac{7}{10}$ **70%**

100% | 22. $\frac{80}{125}$ **64%** | 23. $\frac{9}{12}$ **75%** | 24. $\frac{4}{6}$ **66.67%**

91.67% | 26. $\frac{9}{16}$ **56.25%** | 27. $\frac{8}{9}$ **88.89%** | 28. $\frac{3}{16}$ **18.75%**

15.63% | 30. $\frac{1}{16}$ **6.25%** | 31. $\frac{8}{15}$ **53.33%** | 32. $\frac{9}{11}$ **81.82%**

n 7-6

(Pages 316–318)

each percent as a decimal and as a mixed number or fraction in est form.

5% **8.95; $8\frac{19}{20}$** | 2. 555% **5.55; $5\frac{11}{20}$** | 3. 480% **4.8; $4\frac{4}{5}$** | 4. 920% **9.2; $9\frac{1}{5}$**

2% **1.22; $1\frac{11}{50}$** | 6. 825% **8.25; $8\frac{1}{4}$** | 7. 0.3% **0.003; $\frac{3}{1,000}$** | 8. 0.42% **0.0042; $\frac{21}{5,000}$**

'8% **0.0078; $\frac{39}{5,000}$** | 10. 765% **7.65; $7\frac{13}{20}$** | 11. 0.99% **0.0099; $\frac{99}{10,000}$** | 12. 1,000% **10; 10**

each decimal as a percent.

350% | 14. 12 **1,200%** | 15. 0.002 **0.2%** | 16. 6.78 **678%**

)056 **0.56%** | 18. 1.95 **195%** | 19. 0.0077 **0.77%** | 20. 0.0102 **1.02%**

.0 **1,400%** | 22. 0.0064 **0.64%** | 23. 16.2 **1,620%** | 24. 44.3 **4,430%**

each number as a percent.

0.4% | 26. $\frac{7}{400}$ **1.75%** | 27. $1\frac{1}{4}$ **125%** | 28. $7\frac{9}{10}$ **790%**

90% | 30. $\frac{1}{500}$ **0.2%** | 31. $12\frac{1}{2}$ **1,250%** | 32. $\frac{1}{1,000}$ **0.1%**

475% | 34. 25 **2,500%** | 35. 900 **90,000%** | 36. $18\frac{2}{5}$ **1,840%**

Extra Practice

Extra Practice 581

Lesson 8-1

13. $0.1 \cdot 40 = 4$ and $3 \cdot 4 = 12$
14. $0.1 \cdot 100 = 10$ and $7 \cdot 10 = 70$
15. $0.1 \cdot 150 = 15$ and $9 \cdot 15 = 135$
16. $0.1 \cdot 70 = 7$ and $7 \cdot 7 = 49$
17. $0.1 \cdot 90 = 9$ and $8 \cdot 9 = 72$
18. $0.1 \cdot 180 = 18$
19. $0.1 \cdot 220 = 22$ and $5 \cdot 22 = 110$
20. $0.1 \cdot 500 = 50$ and $2 \cdot 50 = 100$
21. $0.1 \cdot 400 = 40$ and $8 \cdot 40 = 320$
22. $0.1 \cdot 60 = 6$ and $4 \cdot 6 = 24$
23. $0.1 \cdot 180 = 18$ and $6 \cdot 18 = 108$
24. $0.1 \cdot 450 = 45$ and $2 \cdot 45 = 90$

Extra Practice

Lesson 7-7

(Pages 319–321)

Find each number. Round to the nearest tenth if necessary.

1. 5% of 40 is what number? 2
2. What number is 10% of 120? 12
3. Find 12% of 150. 18
4. Find 12.5% of 40. 5
5. What number is 75% of 200? 150
6. Find 13% of 25.3. 3.3
7. 250% of 44 is what number? 110
8. What number is 0.5% of 13.7? 0.1
9. Find 600% of 7. 42
10. Find 1.5% of $25. $0.38
11. Find 81% of 134. 108.5
12. What number is 43% of 110? 47.3
13. What number is 61% of 524? 319.6
14. Find 100% of 3.5. 3.5
15. 20% of 58.5 is what number? 11.7
16. Find 45% of 125.5. 56.5
17. What number is 23% of 500? 115
18. Find 80% of 8. 6.4
19. 90% of 72 is what number? 64.8
20. What number is 32% of 54? 17.3

Lesson 7-8

(Pages 323–325)

Find each number. Round to the nearest tenth if necessary.

1. What number is 25% of 280? 70
2. 38 is what percent of 50? 76%
3. 54 is 25% of what number? 216
4. 24.5% of what number is 15? 61.2
5. What number is 80% of 500? 400
6. 12% of 120 is what number? 14.4
7. Find 68% of 50. 34
8. What percent of 240 is 32? 13.3%
9. 99 is what percent of 150? 66%
10. Find 75% of 1. 0.8
11. What number is $33\frac{1}{3}$% of 66? 22
12. 50% of 350 is what number? 175
13. What percent of 450 is 50? 11.1%
14. What number is $37\frac{1}{2}$% of 32? 12
15. 95% of 40 is what number? 38
16. Find 30% of 26. 7.8
17. 9 is what percent of 30? 30%
18. 52% of what number is 109.2? 210
19. What number is 65% of 200? 130
20. What number is 15.5% of 45? 7.0

Lesson 8-1

(Pages 334–337)

Estimate by using fractions. 1–36. Sample answers are given.

3. $(4 \cdot 20) + \left(\frac{1}{2} \cdot 20\right)$

1. 28% of 48 $\frac{1}{4} \cdot 48 = 12$
2. 99% of 65 $1 \cdot 65 = 65$
3. 445% of 20
4. 9% of 81 $\frac{1}{10} \cdot 80 = 8$
5. 73% of 240 $\frac{3}{4} \cdot 240 = 180$
6. 65.5% of 75 $\frac{2}{3} \cdot 75 = 50$
7. 48.2% of 93 $\frac{1}{2} \cdot 90 = 45$
8. 39.45% of 51 $\frac{2}{5} \cdot 50 = 20$
9. 287% of 122 $3 \cdot 120 = 36$
10. 53% of 80 $\frac{1}{2} \cdot 80 = 40$
11. 414% of 72 $4 \cdot 72 = 288$
12. 59% of 105 $\frac{3}{5} \cdot 105 = 63$

Estimate by using 10%. 13–24. See margin.

13. 30% of 42
14. 70% of 104
15. 90% of 152
16. 67% of 70
17. 78% of 92
18. 12% of 183
19. 51% of 221
20. 23% of 504
21. 81% of 390
22. 41% of 60
23. 59% of 178
24. 22% of 450

Estimate.

25. 50% of 37 $\frac{1}{2} \cdot 40 = 20$
26. 18% of 90 $\frac{1}{5} \cdot 100 = 20$
27. 300% of 245 $3 \cdot 250 = 75$
28. 1% of 48 $\frac{1}{100} \cdot 50 = 0.5$
29. 70% of 300 $\frac{7}{10} \cdot 300 = 210$
30. 35% of 35 $\frac{7}{20} \cdot 40 = 14$
31. 60.5% of 60 $\frac{3}{5} \cdot 60 = 36$
32. $5\frac{1}{2}$% of 100 $\frac{3}{50} \cdot 100 = 6$
33. 40.01% of 16 $\frac{2}{5} \cdot 15 = 6$
34. 80% of 62 $\frac{4}{5} \cdot 60 = 48$
35. 45% of 119 $\frac{9}{20} \cdot 120 = 54$
36. 14.81% of 986 $\frac{3}{20} \cdot 1{,}000 =$

582 Extra Practice

1 8-2

(Pages 340–343)

ın equation for each problem. Then solve. Round to the nearest f necessary. **1–18. See margin for equations.**

d 45% of 50. **22.5**

2. 75 is what percent of 300? **25%**

o of what number is 2? **12.5**

4. 75% of 80 is what number? **60**

of what number is 12? **240**

6. Find 60% of 45. **27**

d 22% of 22. **4.8**

8. 12% of what number is 50? **416.7**

s what percent of 62? **61.3%**

10. 80 is what percent of 90? **88.9%**

s what percent of 95? **94.7%**

12. $28\frac{1}{2}$% of 64 is what number? **18.2**

d 46.5% of 75. **34.9**

14. What number is 55.5% of 70? **38.9**

5% of what number is 80.5? **100**

16. $66\frac{2}{3}$% of what number is 40? **60**

d 122.5% of 80. **98**

18. 250% of what number is 75? **30**

1 8-3

(Pages 345–347)

table shows the results of a survey tudents' favorite cookies. Predict v many of the 424 students at Scobey h School prefer chocolate chip kies. **253 students**

Cookie	Number
chocolate chip	49
peanut butter	12
oatmeal	10
sugar	8
raisin	3

2. The circle graph shows the results of a survey of teens and where they would prefer to spend a family vacation. Predict how many of the 4,000 teens in the Central School District would prefer to go to an amusement park. **1,800 students**

Vacation Survey

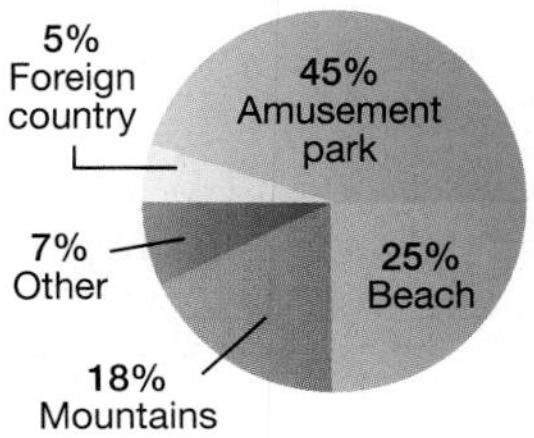

urvey showed that 7 out of 10 first graders received an allowance. ed on that survey, how many of the 800 first graders in the Midtown ool District receive an allowance? **560 first graders**

2000, about 29% of the foreign visitors to the U.S. were from Canada. particular hotel had 150,000 foreign guests in one year, how many uld you predict were from Canada? **43,500 Canadians**

n 8-4

(Pages 350–353)

ach percent of change. Round to the nearest whole percent if ary. State whether the percent of change is an *increase* or *decrease*.

ginal: 450
w: 675 **50%, inc.**

2. original: $1,500
new: $1,200 **20%, dec.**

3. original: 750
new: 700 **7%, dec.**

ginal: 350
w: 420 **20%, inc.**

5. original: 500
new: 100 **80%, dec.**

6. original: $75
new: $50 **33%, dec.**

ginal: 3.25
w: 2.95 **9%, dec.**

8. original: 32.5
new: 44 **35%, inc.**

9. original: 180
new: 160 **11%, dec.**

ginal: $65
w: $75 **15%, inc.**

11. original: 22
new: 88 **300%, inc.**

12. original: 450
new: 445.5 **1%, dec.**

Extra Practice

Lesson 8-2

1. $n = 0.45 \cdot 50$; 22.5
2. $75 = n \cdot 300$; 25%
3. $2 = 0.16 \cdot n$; 12.5
4. $n = 0.75 \cdot 80$; 60
5. $12 = 0.05 \cdot n$; 240
6. $n = 0.60 \cdot 45$; 27
7. $n = 0.22 \cdot 22$; 4.8
8. $50 = 0.12 \cdot n$; 416.7
9. $38 = n \cdot 62$; 61.3%
10. $80 = n \cdot 90$; 88.9%
11. $90 = n \cdot 95$; 94.7%
12. $n = 0.285 \cdot 64$; 18.2
13. $n = 0.465 \cdot 75$; 34.9
14. $n = 0.555 \cdot 70$; 38.9
15. $80.5 = 0.805 \cdot n$; 100
16. $40 = 0.662 \cdot n$; 60.4
17. $n = 1.225 \cdot 80$; 98
18. $75 = 2.5 \cdot n$; 30

Extra Practice

Lesson 8-5

1. $47.70
2. $16.14
3. $179.10
4. $31.03
5. $44.00
6. $14.25
7. $157.50
8. $62.30
9. $41.15
10. $154.06
11. $210
12. $12.54
13. $719.20
14. $112.88
15. $527.12
16. $361.25
17. $12,968.75
18. $52.57

Lesson 8-6

1. $800
2. $33.33
3. $37.50
4. $17.11
5. $28.34
6. $3.34
7. $138
8. $1,584
9. $15.93
10. $135
11. $6.75
12. $1,800
13. $270
14. $380
15. $27
16. $2,315.63
17. $298.99
18. $165.36
19. $91.41
20. $183.33
21. $680.40
22. $202.50
23. $141.59
24. $80

Lesson 8-5

(Pages 354–357)

Find the total cost or sale price to the nearest cent. **1–18. See margin.**

1. \$45 sweater; 6% tax
2. \$18.99 CD; 15% discount
3. \$199 ring; 10% discount
4. \$29 shirt; 7% tax
5. \$55 plant; 20% discount
6. \$19 purse; 25% discount
7. \$150 clock; 5% tax
8. \$89 radio; 30% discount
9. \$39 shoes; 5.5% tax
10. \$145 coat; 6.25% tax
11. \$300 table; 30% discount
12. \$12 meal; 4.5% tax
13. \$899 computer; 20% discount
14. \$105 skateboard; $7\frac{1}{2}$% tax
15. \$599 TV; 12% discount
16. \$425 skis; 15% discount
17. \$12,500 car; $3\frac{3}{4}$% tax
18. \$49.95 gloves; $5\frac{1}{4}$% tax

Find the percent of discount to the nearest percent.

19. regular price: \$72, sale price: \$60 **17%**
20. regular price: \$125, sale price: \$120 **4%**
21. regular price: \$360, sale price: \$280 **22%**
22. regular price: \$90, sale price: \$22.50 **75%**
23. regular price: \$25,000, sale price: \$22,000 **12%**
24. regular price: \$0.99, sale price: \$0.82 **17%**

Lesson 8-6

(Pages 358–360)

Find the interest earned to the nearest cent for each principal, interest rate, and time. **1–12. See margin.**

1. \$2,000, 8%, 5 years
2. \$500, 10%, 8 months
3. \$750, 5%, 1 year
4. \$175.50, $6\frac{1}{2}$%, 18 months
5. \$236.20, 9%, 16 months
6. \$89, $7\frac{1}{2}$%, 6 months
7. \$800, 5.75%, 3 years
8. \$5,500, 7.2%, 4 years
9. \$245, 6%, 13 months
10. \$1,200, 3%, 45 months
11. \$225, $1\frac{1}{2}$%, 2 years
12. \$12,000, $4\frac{1}{2}$%, 40 months

Find the interest paid to the nearest cent for each loan balance, interest rate, and time. **13–24. See margin.**

13. \$750, 18%, 2 years
14. \$1,500, 19%, 16 months
15. \$300, 9%, 1 year
16. \$4,750, 19.5%, 30 months
17. \$2,345, 17%, 9 months
18. \$689, 12%, 2 years
19. \$390, 18.75%, 15 months
20. \$1,250, 22%, 8 months
21. \$3,240, 18%, 14 months
22. \$675, 15%, 2 years
23. \$899, $10\frac{1}{2}$%, 18 months
24. \$1,000, 8%, 1 year

Lesson 9-1

(Pages 370–373)

Use the spinner at the right to find each probability. Write as a fraction in simplest form.

1. P(even number) $\frac{1}{2}$
2. P(prime number) $\frac{1}{2}$
3. P(factor of 12) $\frac{5}{8}$
4. P(composite number) $\frac{3}{8}$
5. P(greater than 10) **0**
6. P(neither prime nor composite) $\frac{1}{8}$

A package of balloons contains 5 green, 3 yellow, 4 red, and 8 pink balloons. Suppose you reach in the package and choose one balloon at random. Find the probability of each event. Write as a fraction in simplest form.

7. P(red balloon) $\frac{1}{5}$
8. P(yellow balloon) $\frac{3}{20}$
9. P(pink balloon) $\frac{2}{5}$
10. P(orange balloon) **0**
11. P(red or yellow balloon) $\frac{7}{20}$
12. P(*not* green balloon) $\frac{3}{4}$

9-2 1–5. See Extra Practice Answer Appendix for tree diagrams. (Pages 374–377)

h situation, make a tree diagram to show the sample space. Then e total number of outcomes.

ing 2 number cubes **36 outcomes**

osing an ice cream cone from waffle, plain, or sugar and a flavor of cream from chocolate, vanilla, or strawberry **9 outcomes**

king a sandwich from white, wheat, or rye bread, cheddar or Swiss ese and ham, turkey, or roast beef **18 outcomes**

ping a penny twice **4 outcomes**

osing one math class from algebra and geometry and one foreign guage class from French, Spanish, or Latin **6 outcomes**

1 9-3 (Pages 378–380)

e Fundamental Counting Principle to find the total number of es in each situation.

osing a local phone number if the exchange is 398 and each of the r remaining digits is different **5,040 outcomes**

osing a way to drive from Millville to Westwood if there are 5 roads t lead from Millville to Miamisburg, 3 roads that connect Miamisburg Hathaway, and 4 highways that connect Hathaway to Westwood **60 outcomes**

sing a quarter, rolling a number cube, and tossing a dime **24 outcomes**

nning the spinners shown below **96 outcomes**

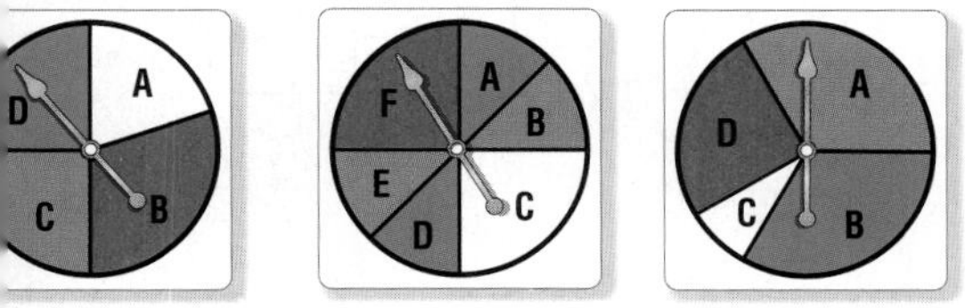

1 9-4 (Pages 381–383)

e value of each expression.

2	2. 7! **5,040**	3. 1! **1**	4. 8! **40,320**
5! **720**	6. 2! · 5! **240**	7. 10 factorial **3,628,800**	8. 1! · 9! **362,880**
8 · 7 **504**	10. 6! · 5! **86,400**	11. 7 · 6 · 5 · 4 **840**	12. 8! · 3! **241,920**

ht runners are competing in a 100-meter sprint. In how many ways the gold, silver, and bronze medals be awarded? **336 ways**

a lottery for which 30 tickets were sold (all to different people), in v many ways can the grand prize, second prize, and third prizes be arded? **24,360 ways**

e-digit locker combinations are assigned using the digits 1–9. In how ny ways can the combinations be formed if no digit can be repeated? **15,120 ways**

w many ways can you arrange the letters in the word *factor*? **720 ways**

Extra Practice

Extra Practice 585

Lesson 9-5
(Pages 387–390)

Tell whether each problem represents a *permutation* or *combination*. Then solve the problem.

1. How many ways can four students be placed in a line? **permutation; 24 ways**
2. In how many ways can six books be displayed on a rack from a collection of 12? **combination; 924 ways**
3. In how many ways can you select three report topics from a total of 8 topics? **combination; 56 ways**
4. How many ways can nine students finish first, second, third, or fourth at a speech competition? **permutation; 3,024 ways**
5. In how many ways can you ride five out of nine roller coasters if you don't care in what order you ride them? **combination; 126 ways**
6. In how many ways can you ride five out of nine roller coasters if you care in what order you ride them? **permutation; 15,120 ways**

Lesson 9-6
(Pages 393–396)

The frequency table shows the results of a fair number cube rolled 40 times.

Face	Frequency
1	5
2	9
3	2
4	8
5	12
6	4

1. Find the experimental probability of rolling a 4. $\frac{1}{5}$
2. Find the theoretical probability of *not* rolling a 4. $\frac{5}{6}$
3. Find the theoretical probability of rolling a 2. $\frac{1}{6}$
4. Find the experimental probability of *not* rolling a 6. $\frac{9}{10}$
5. Suppose the number cube was rolled 500 times. About how many times would it land on 5? **83**
6. Suppose the number cube was rolled 500 times. About how many times would it land on a prime number? **250**

Lesson 9-7
(Pages 398–401)

Determine whether each event is *independent* or *dependent*.

1. selecting and eating a candy from a dish and then selecting another candy **dependent**
2. tossing a coin and then rolling a number cube **independent**
3. choosing a card from deck 1 and then choosing a card from deck 2 **independent**

Find each probability.

4. Two evenly balanced nickels are flipped. Find the probability that one head and one tail result. $\frac{1}{2}$
5. A wallet contains four \$5 bills, two \$10 bills, and eight \$1 bills. Two bills are selected without the first selection being replaced. Find *P*(\$5, then \$5). $\frac{6}{91}$
6. Two chips are selected from a box containing 6 blue chips, 4 red chips, and 3 green chips. The first chip selected is not replaced before the second is drawn. Find *P*(red, then green). $\frac{1}{13}$
7. A bag contains 7 blue, 4 orange, 8 red, and 5 purple marbles. Suppose one marble is chosen and then replaced. A second marble is then chosen. Find *P*(purple, then red). $\frac{5}{72}$

586 Extra Practice

n 10-1
(Pages 413–415)

y each angle as *acute*, *obtuse*, *right*, or *straight*.

2. 3.

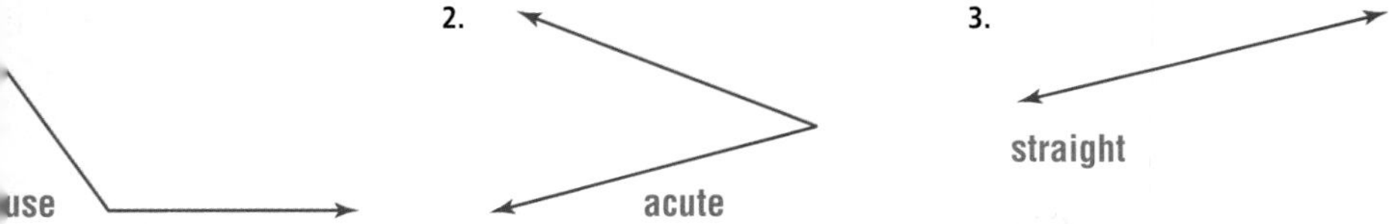

use acute straight

acute 5. 24° acute 6. 110° obtuse 7. 95° obtuse 8. 90° right

an angle having each measurement. Then classify each angle as *obtuse*, *right*, or *straight*. 9–18. See margin for angles.

° obtuse 10. 45° acute 11. 15° acute 12. 180° straight 13. 175° obtuse

acute 15. 52° acute 16. 160° obtuse 17. 80° acute 18. 55° acute

n 10-2
(Pages 418–421)

a circle graph of the data in each table. 1–2. See margin.

Car Sales	
Style	Percent
lan	45%
V	22%
kup truck	9%
rts car	13%
npact car	11%

2.

Favorite Flavor of Ice Cream	
Flavor	Number
vanilla	11
chocolate	15
strawberry	8
mint chip	5
cookie dough	3

n 10-3
(Pages 422–425)

fy each pair of angles as *complementary*, *supplementary*, or *neither*.

complementary 2. supplementary 3.

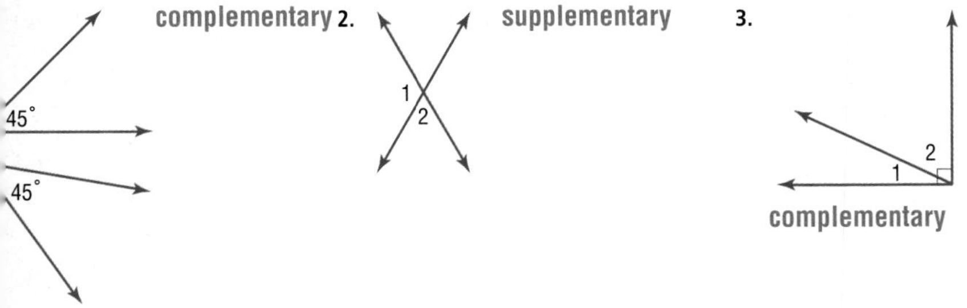

complementary

he value of x in each figure.

40 5. 55 6. 160

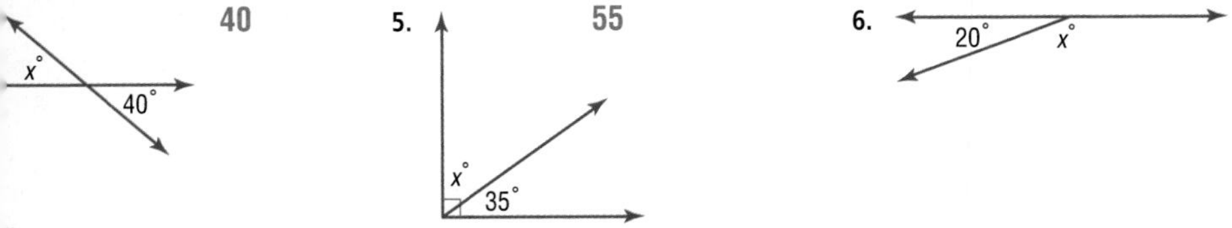

Lesson 10-1

9.–18.

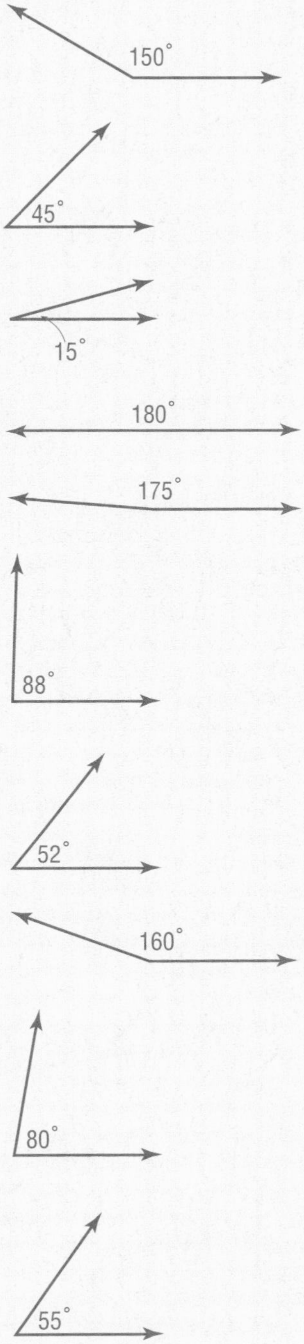

Lesson 10-2

1.

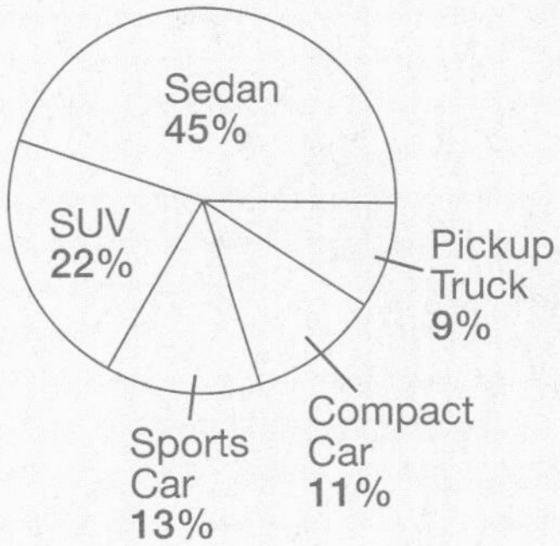

Lesson 10-2

2.

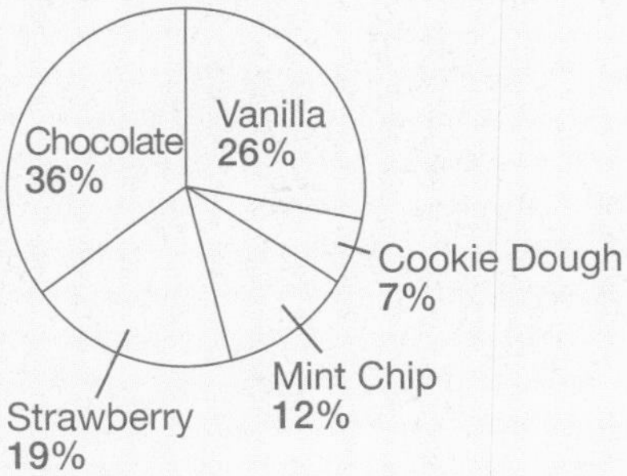

Extra Practice

Lesson 10-8

1\.

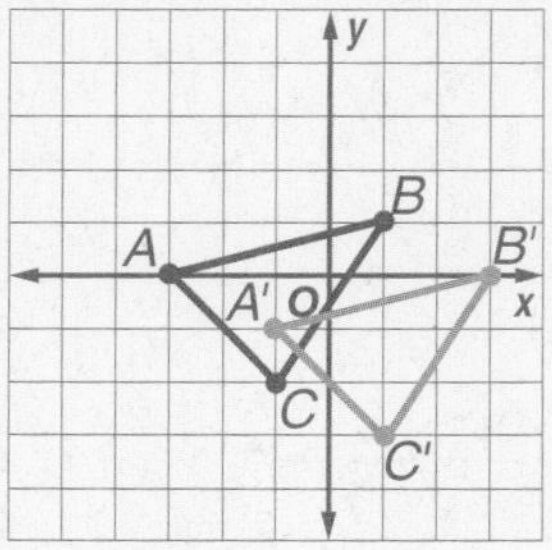

2\.

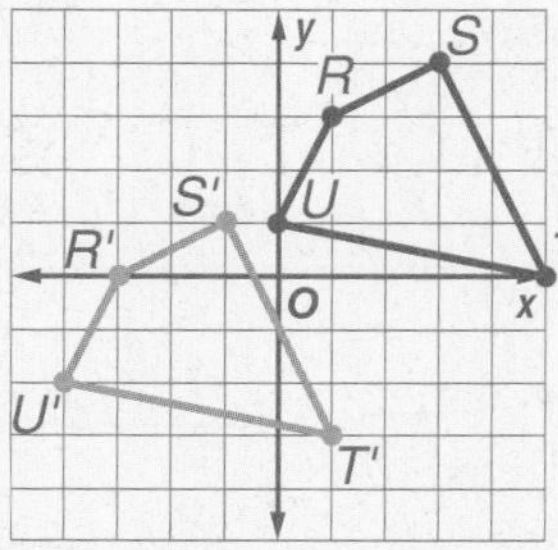

3\.

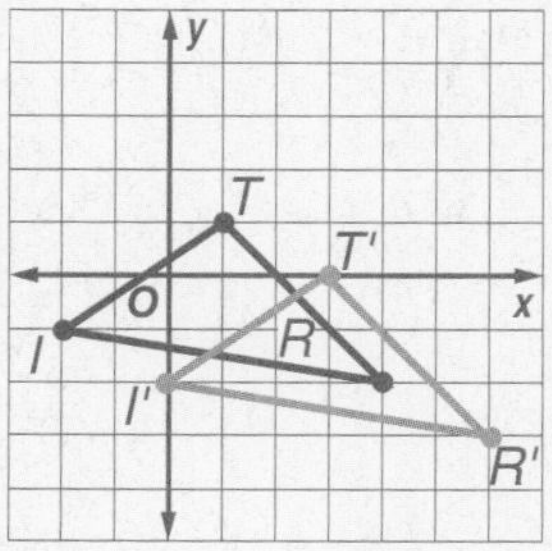

4\.

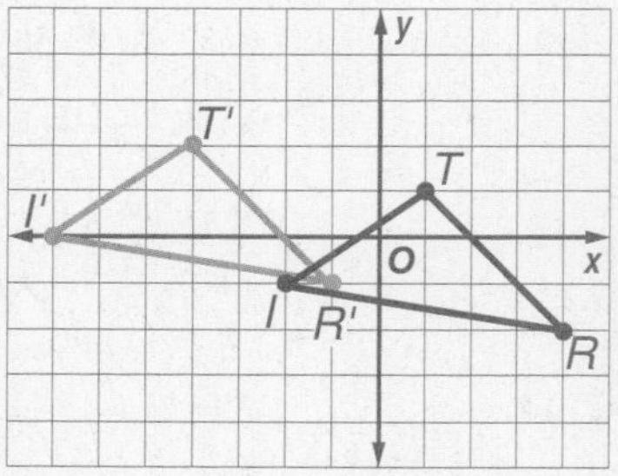

5\.

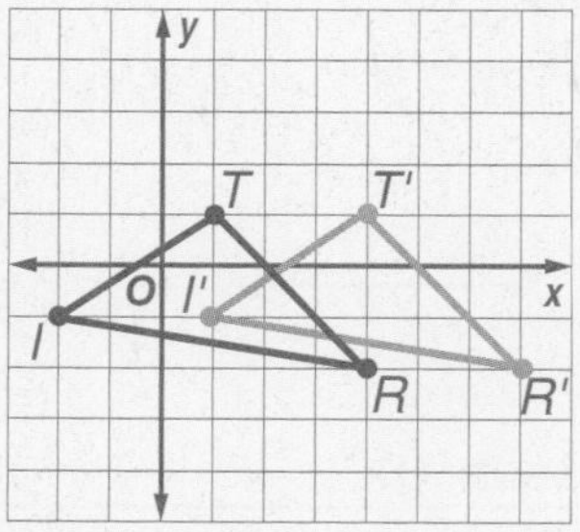

6\.

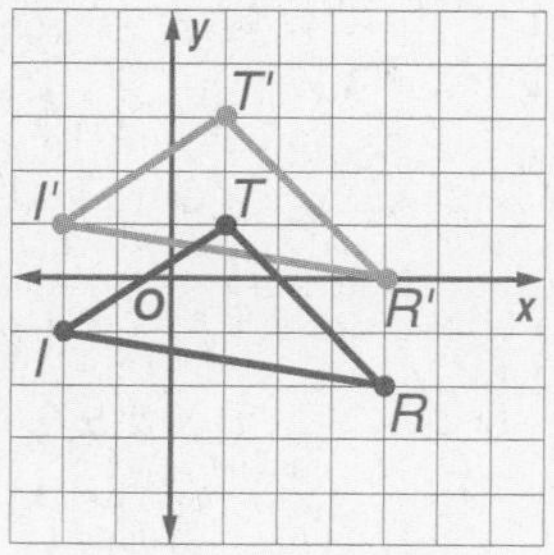

Extra Practice

Lesson 10-4

(Pages 428–431)

Classify each triangle by its angles and by its sides.

1\. 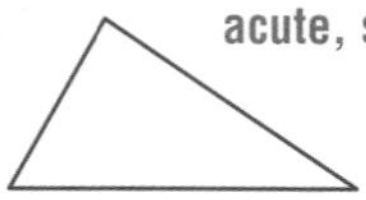 acute, scalene

2\.  right, scalene

3\. 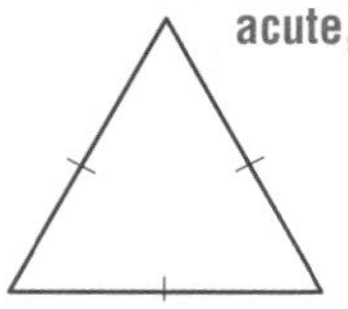 acute, equilat

Find the missing measure in each triangle. Then classify the triangle as *acute, right,* or *obtuse.*

4\. 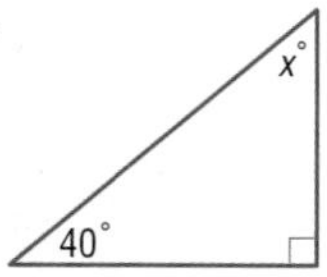

50°; right

5\. 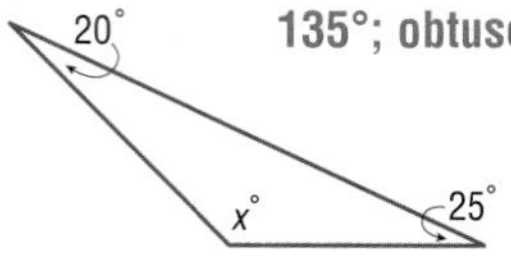

135°; obtuse

6\. 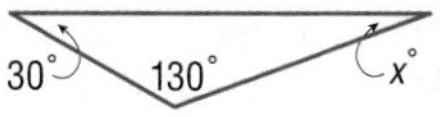

20°; obtuse

Lesson 10-5

(Pages 434–437)

Classify the quadrilateral using the name that *best* describes it.

1\. 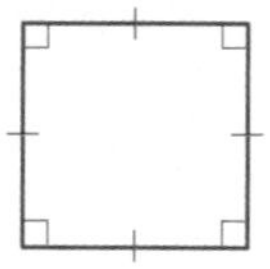 square

2\. 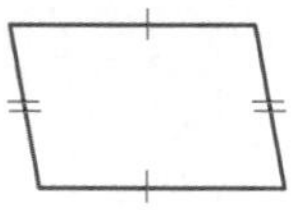 parallelogram

3\. 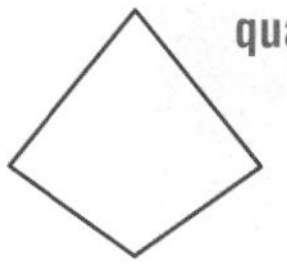 quadrilateral

Find the missing measure in each quadrilateral.

4\. 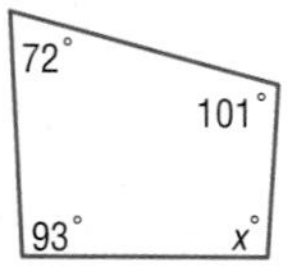

94°

5\. 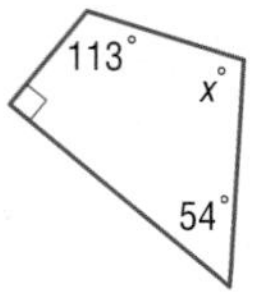

103°

6\. 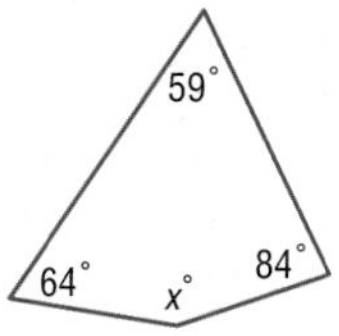

153°

Lesson 10-6

(Pages 440–443)

Find the value of *x* in each pair of similar figures.

1\. 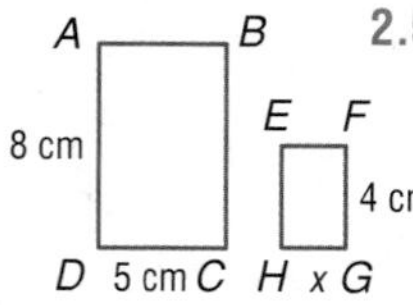

2.5 cm

2\. 8 ft 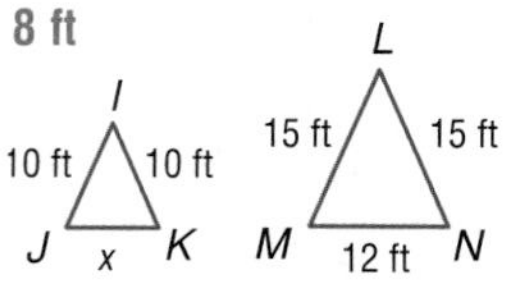

3\. 25 ft 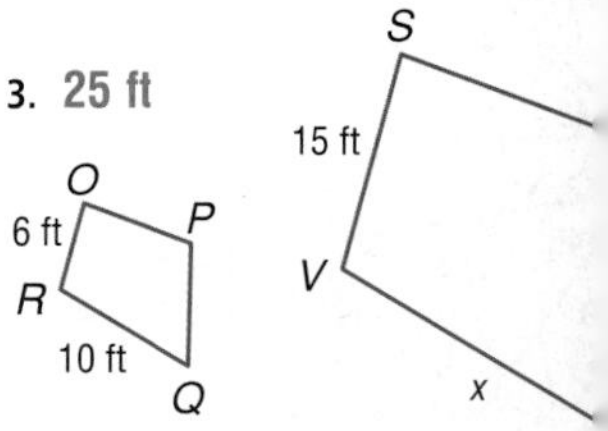

4\. 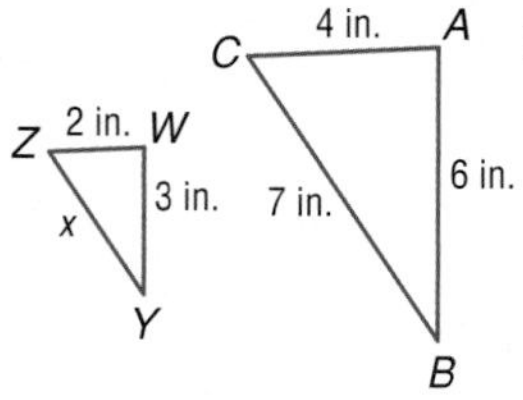

3.5 in.

5\. 40.5 cm 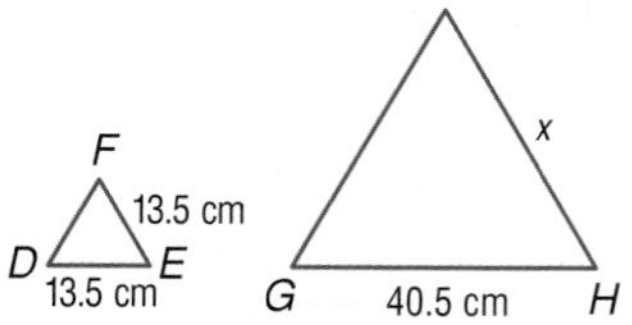

6\. 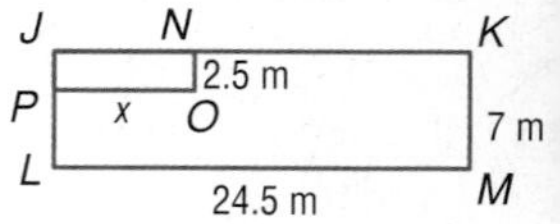

8.75 m

10-7

(Pages 446–450)

ine whether each figure is a polygon. If it is, classify the polygon te whether it is regular. If it is *not* a polygon, explain why.

decagon, not regular

2. 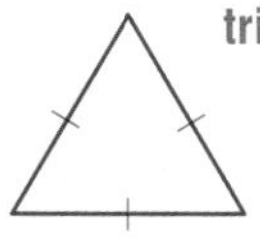 triangle, regular

3. not a polygon; sides are curved 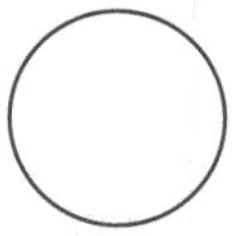

e measure of an angle in each regular polygon. Round to the t tenth degree, if necessary.

gle 60° 5. 30-gon 168° 6. 18-gon 160° 7. 14-gon 154.3°

10-8 1–2. See margin.

(Pages 451–454)

slate $\triangle ABC$ 2 units and 1 unit down.

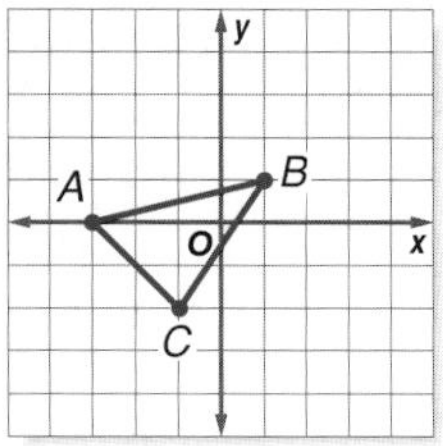

2. Translate quadrilateral *RSTU* 4 units left and 3 units down.

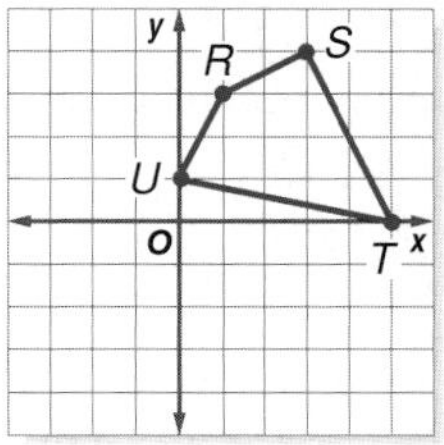

e *TRI* has vertices $T(1, 1)$, $R(4, -2)$, and $I(-2, -1)$. Find the vertices of after each translation. Then graph the figure and its translated image. 3–6. See margin for graphs.

its right, 1 unit down 4. 5 units left, 1 unit up

its right $T'(4, 1)$, $R'(7, -2)$, $I'(1, -1)$ 6. 2 units up $T'(1, 3)$, $R'(4, 0)$, $I'(-2, 1)$

0), $R'(6, -3)$, $I'(0, -2)$ 4. $T'(-4, 2)$, $R'(-1, -1)$, $I'(-7, 0)$

10-9

(Pages 456–459)

ine which figures have line symmetry. Write *yes* or *no*. If *yes*, draw s of symmetry.

 yes

2. 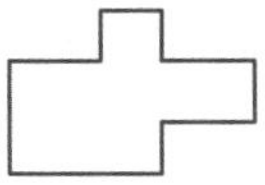 no

3. 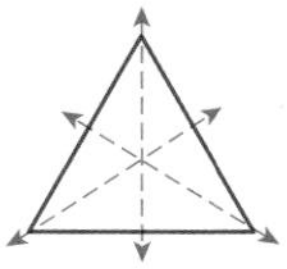 yes

e coordinates of each figure after a reflection over the x-axis. raph the figure and its reflected image. 4–7. See margin for graphs.

rilateral *QUAD* with vertices $Q(-1, 4)$, $U(2, 2)$, $A(1, 1)$, and $D(-2, 2)$ $Q'(-1, -4)$, $U'(2, -2)$, $A'(1, -1)$, $D'(-2, -2)$

gle *ABC* with vertices $A(0, -1)$, $B(4, -3)$, and $C(-4, -5)$ $A'(0, 1)$, $B'(4, 3)$, $C'(-4, 5)$

e coordinates of each figure after a reflection over the y-axis. raph the figure and its reflected image.

llelogram *PARL* with vertices $P(3, 5)$, $A(5, 4)$, $R(5, 1)$, and $L(3, 2)$ $P'(-3, 5)$, $A'(-5, 4)$, $R'(-5, 1)$, $L'(-3, 2)$

agon *PENTA* with vertices $P(-1, 3)$, $E(1, 1)$, $N(0, -2)$, $T(-2, -2)$, and 3, 1) $P'(1, 3)$, $E'(-1, 1)$, $N'(0, -2)$, $T'(2, -2)$, $A'(3, 1)$

Extra Practice

Lesson 10-9

4. 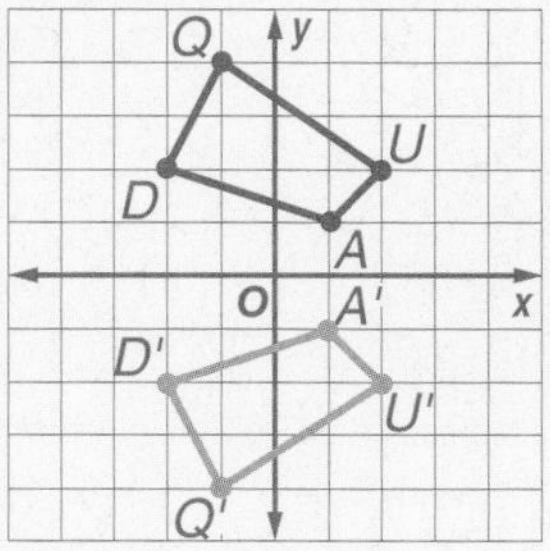

5. 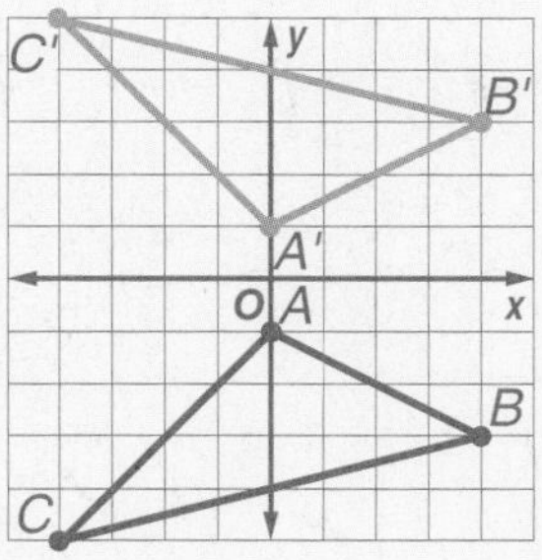

6. 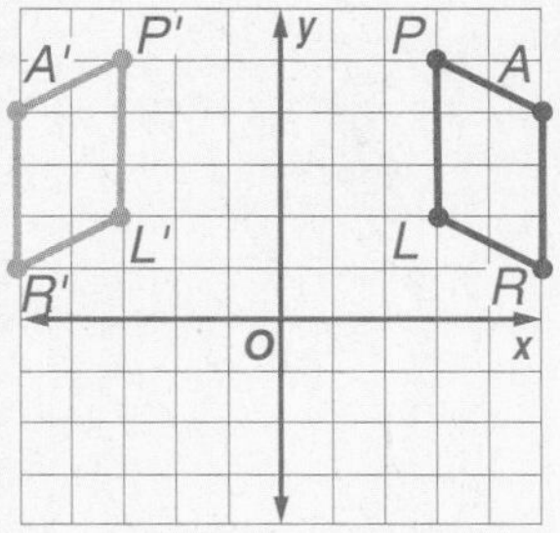

7. 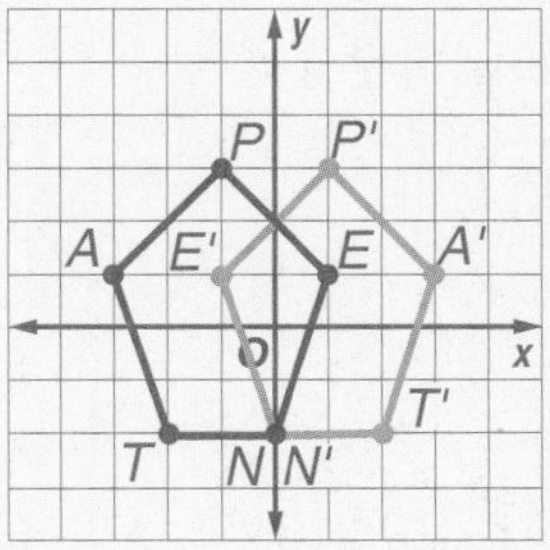

Extra Practice

Extra Practice

Lesson 11-1
(Pages 470–473)

Find the square of each number.

1. 4 **16**
2. 19 **361**
3. 13 **169**
4. 25 **625**
5. 9 **81**
6. 2 **4**
7. 14 **196**
8. 24 **576**
9. 40 **1,600**
10. 50 **2,500**
11. 100 **10,000**
12. 250 **62,500**

Find each square root.

13. $\sqrt{324}$ **18**
14. $\sqrt{900}$ **30**
15. $\sqrt{2,500}$ **50**
16. $\sqrt{576}$ **24**
17. $\sqrt{8,100}$ **90**
18. $\sqrt{676}$ **26**
19. $\sqrt{100}$ **10**
20. $\sqrt{784}$ **28**
21. $\sqrt{1,024}$ **32**
22. $\sqrt{841}$ **29**
23. $\sqrt{2,304}$ **48**
24. $\sqrt{3,025}$ **55**

Lesson 11-2
(Pages 475–477)

Estimate each square root to the nearest whole number.

1. $\sqrt{27}$ **5**
2. $\sqrt{112}$ **11**
3. $\sqrt{249}$ **16**
4. $\sqrt{88}$ **9**
5. $\sqrt{1,500}$ **39**
6. $\sqrt{612}$ **25**
7. $\sqrt{340}$ **18**
8. $\sqrt{495}$ **22**
9. $\sqrt{264}$ **16**
10. $\sqrt{350}$ **19**
11. $\sqrt{834}$ **29**
12. $\sqrt{3,700}$ **61**
13. $\sqrt{298}$ **17**
14. $\sqrt{101}$ **10**
15. $\sqrt{800}$ **28**

Use a calculator to find each square root to the nearest tenth.

16. $\sqrt{58}$ **7.6**
17. $\sqrt{750}$ **27.4**
18. $\sqrt{1,200}$ **34.6**
19. $\sqrt{1,000}$ **31.6**
20. $\sqrt{5,900}$ **76.8**
21. $\sqrt{999}$ **31.6**
22. $\sqrt{374}$ **19.3**
23. $\sqrt{512}$ **22.6**
24. $\sqrt{3,750}$ **61.2**
25. $\sqrt{255}$ **16.0**
26. $\sqrt{83}$ **9.1**
27. $\sqrt{845}$ **29.1**
28. $\sqrt{200}$ **14.1**
29. $\sqrt{500}$ **22.4**
30. $\sqrt{10,001}$ **100.0**

Lesson 11-3
(Pages 479–483)

Find the missing measure of each right triangle. Round to the nearest tenth if necessary.

1. **7.2 ft**

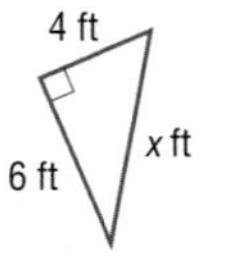

2. **11.3 cm**

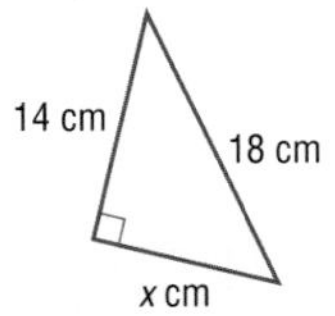

3. **18.7 ft**

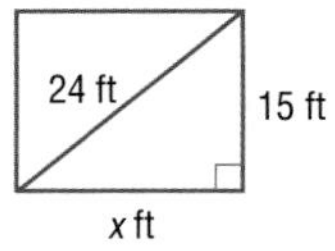

4. $a = 12$ cm, $b = 25$ cm **27.7 cm**
5. $a = 5$ yd, $c = 10$ yd **8.7 yd**
6. $b = 12$ mi, $c = 20$ mi **16**
7. $a = 15$ yd, $b = 24$ yd **28.3 yd**
8. $a = 4$ m, $c = 12$ m **11.3 m**
9. $a = 8$ mm, $b = 11$ mm **13.**
10. $a = 1$ mi, $c = 3$ mi **2.8 mi**
11. $a = 5$ yd, $b = 8$ yd **9.4 yd**
12. $b = 7$ in., $c = 19$ in. **17.7**
13. $a = 50$ km, $c = 75$ km **55.9 km**
14. $b = 82$ ft, $c = 100$ ft **57.2 ft**
15. $a = 100$ m, $b = 200$ m **22**

Determine whether each triangle with the given side lengths is a right triangle. Write *yes* or *no*.

16. 6 mm, 8 mm, 10 mm **yes**
17. 12 ft, 15 ft, 20 ft **no**
18. 300 m, 400 m, 500 m **yes**

Extra Practice

590 Extra Practice

11-4
(Pages 483–485)

e area of each parallelogram. Round to the nearest tenth if ary.

12 m²

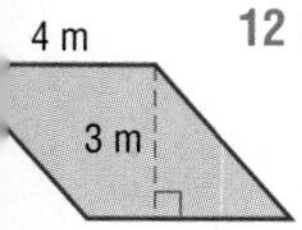

2. **108 m^2**

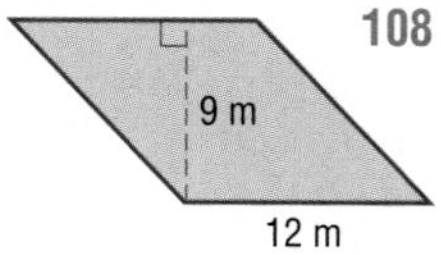

3. **437 ft^2**

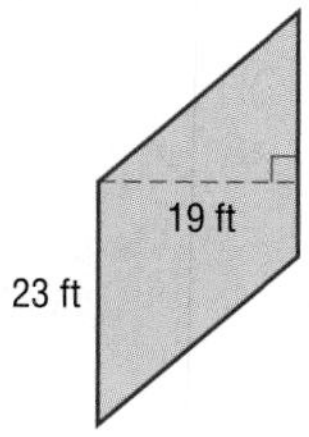

e = 19 m
ght = 6 m **114 m^2**

5. base = 15 ft
height = 12 ft **180 ft^2**

e = 25 yd
ght = 53 yd **1,325 yd^2**

7. base = 135 in.
height = 15 in. **2,025 in^2**

e = 8.2 cm
ght = 5.5 cm **45.1 cm^2**

9. base = 29.3 m
height = 10.1 m **295.9 m^2**

11-5
(Pages 489–492)

e area of each figure. Round to the nearest tenth if necessary.

20 ft^2

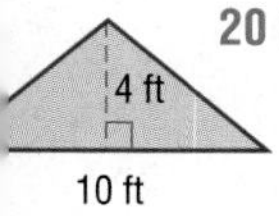

2. **7.5 cm^2**

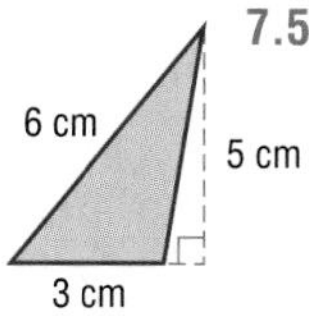

3.

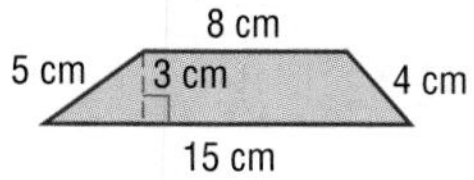

34.5 cm^2

ngle: base = 5 in., height = 9 in. **22.5 in^2**

ezoid: bases = 3 cm and 8 cm, height = 12 cm **66 cm^2**

ezoid: bases = 10 ft and 15 ft, height = 12 ft **150 ft^2**

ngle: base = 12 cm, height = 8 cm **48 cm^2**

ezoid: bases = 82.6 cm and 72.2 cm, height = 44.5 cm **3,444.3 cm^2**

ngle: base = 500.5 ft, height = 254.5 ft **63,688.6 ft^2**

11-6
(Pages 493–495)

e area of each circle. Round to the nearest tenth.

28.3 cm^2

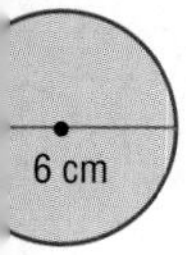

2. **12.6 yd^2**

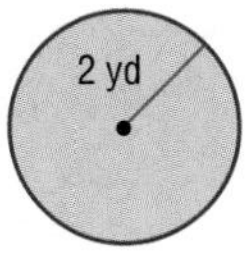

3. **3.1 in^2**

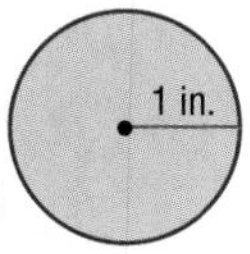

ius = 8 in. **201.1 in^2**
5. diameter = 5 ft **19.6 ft^2**
6. radius = 24 cm **1,809.6 cm^2**

meter = 2.3 m **4.2 m^2**
8. diameter = 82 ft **5,281.0 ft^2**
9. radius = 68 cm **14,526.7 cm^2**

ius = 9.8 mi **301.7 mi^2**
11. diameter = 25.6 m **514.7 m^2**
12. diameter = 6.75 in. **35.8 in^2**

ius = $1\frac{1}{4}$ ft **4.9 ft^2**
14. diameter = $5\frac{2}{3}$ yd **25.2 yd^2**
15. diameter = $45\frac{1}{2}$ mi **1,626.0 mi^2**

nswers are calculated using the π key on a calculator and then rounded.

Extra Practice 591

Extra Practice

Lesson 11-7

(Pages 498–500)

Find the area of each figure. Round to the nearest tenth if necessary.

1. 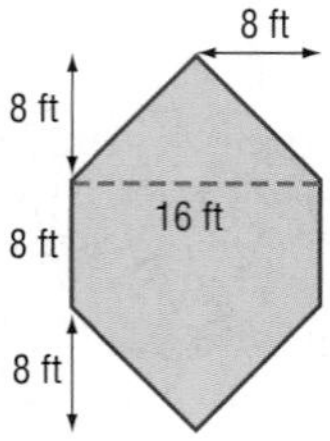

256 ft^2

2. 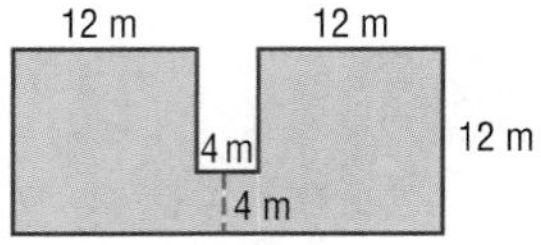

304 m^2

3. 300 in^2 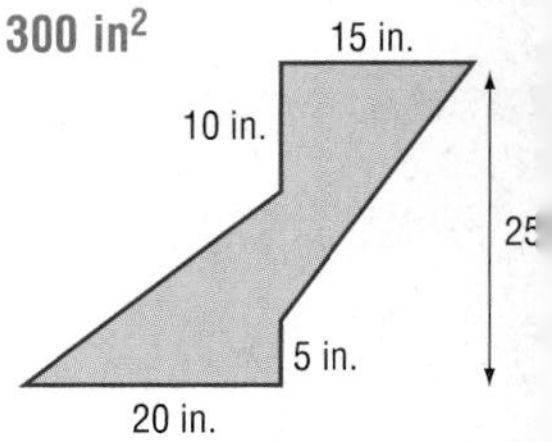

4. 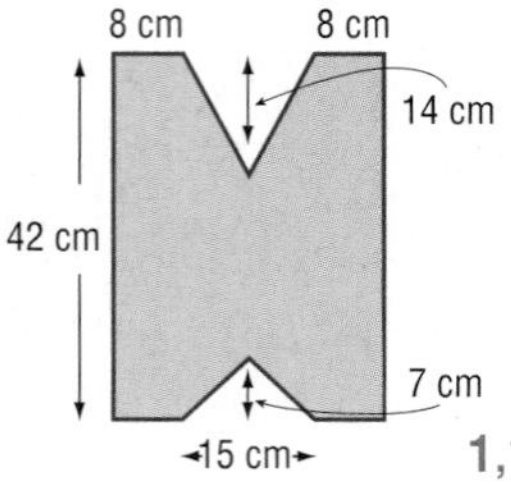

1,144.5 cm^2

5. 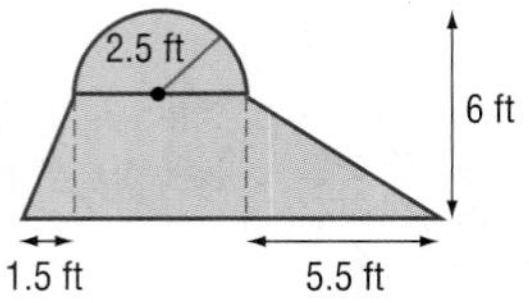

39.6 ft^2

6. 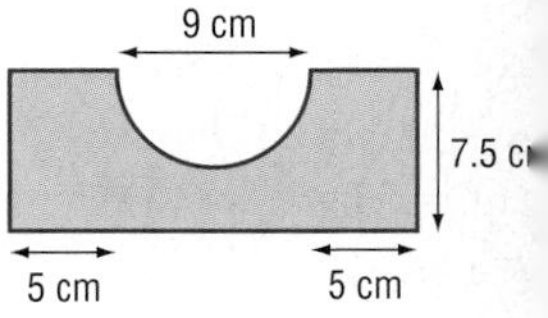

110.7 cm^2

7. 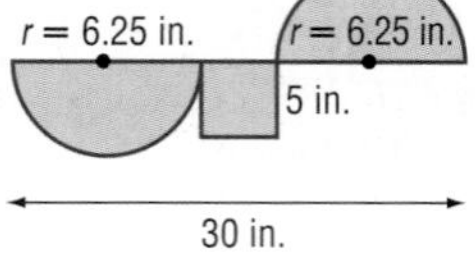

147.7 in^2

8. 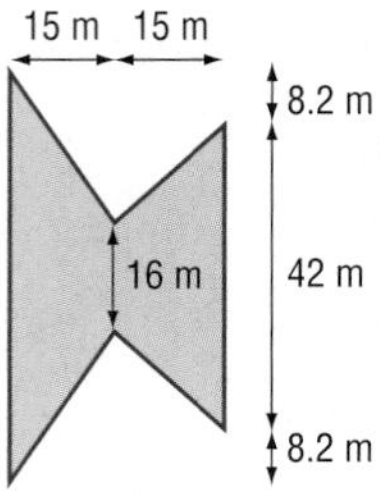

993 m^2

9. 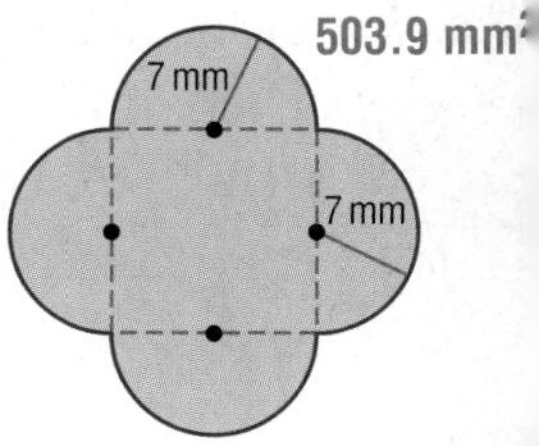

503.9 mm^2

Lesson 11-8

(Pages 501–503)

A randomly dropped counter falls in the squares. Find the probability that it falls in the shaded squares. Write as a percent. Round to the nearest tenth if necessary.

1. 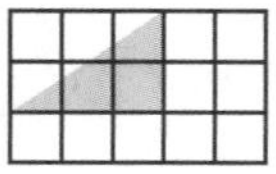 20%

2. 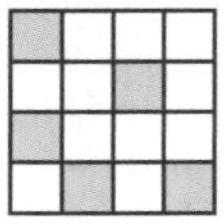 31.3%

3. 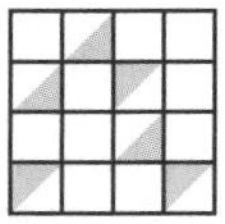 18.8%

4. 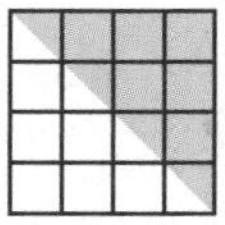 50%

5. 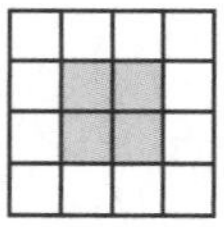 25%

6. 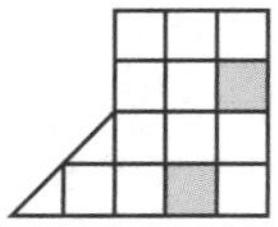 14.3%

7. 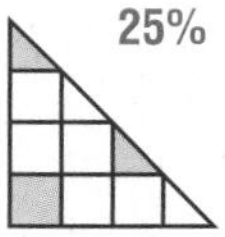 25%

8. 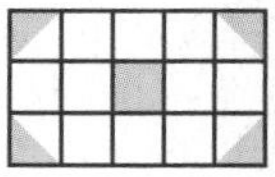 20

9. 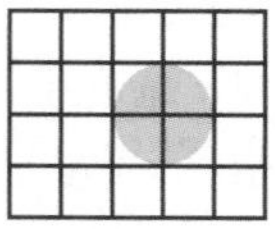 15.7%

10. 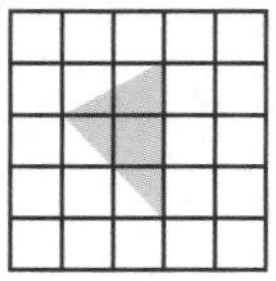 12%

11. 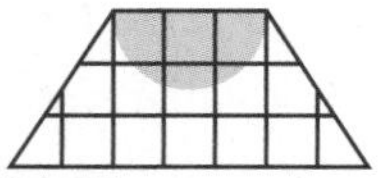 23.6%

12. 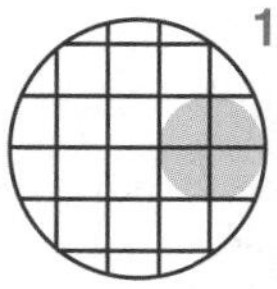 16%

592 Extra Practice

12-1

(Pages 514–517)

top, a side, and a front view of each solid. **1–6. See margin.**

2. 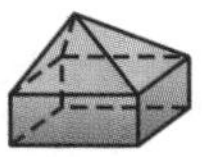

3. 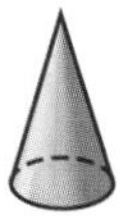

each solid using the top, side, and front views shown. Use ric dot paper.

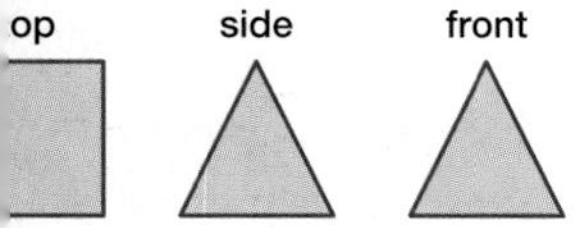

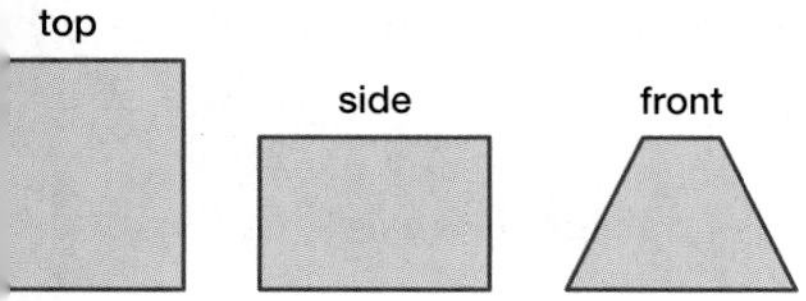

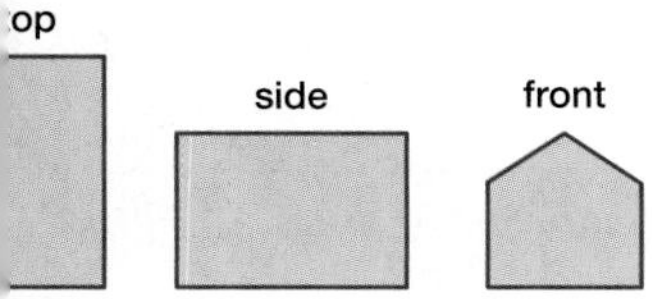

12-2

(Pages 520–522)

e volume of each rectangular prism. Round to the nearest tenth if ary.

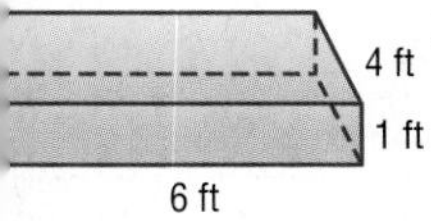

ft^3

2. 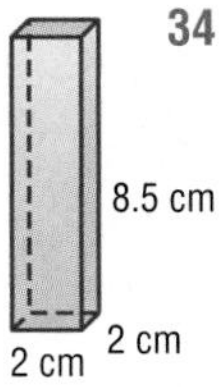

34 cm^3

3. 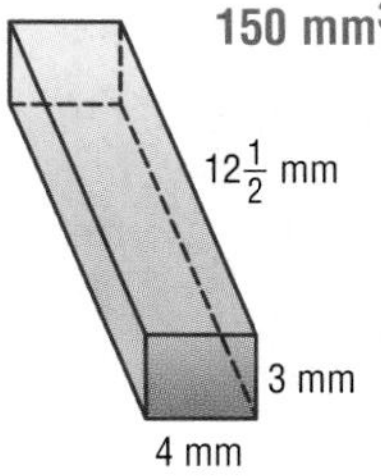

150 mm^3

4. 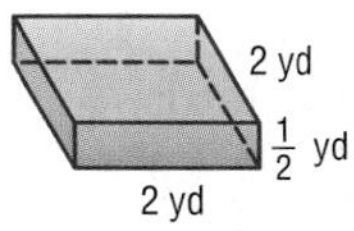

2 yd^3

gth = 3 ft
lth = 10 ft
ght = 2 ft **60 ft^3**

6. length = 18 cm
width = 23 cm
height = 15 cm **6,210 cm^3**

7. length = 25 mm
width = 32 mm
height = 10 mm **8,000 mm^3**

8. length = 1.5 in.
width = 3 in.
height = 6 in. **27 in^3**

gth = 4.5 cm
lth = 6.75 cm
ght = 2 cm **60.8 cm^3**

10. length = 16 mm
width = 0.7 mm
height = 12 mm **134.4 mm^3**

11. length = $3\frac{1}{2}$ ft
width = 10 ft
height = 6 ft **210 ft^3**

12. length = $5\frac{1}{2}$ in.
width = 12 in.
height = $3\frac{3}{8}$ in. **222.8 in^3**

d the volume of a rectangular prism with a length of 3 yards, a width 5 feet, and a height of 12 feet. **540 ft^3**

Lesson 12-1

1. 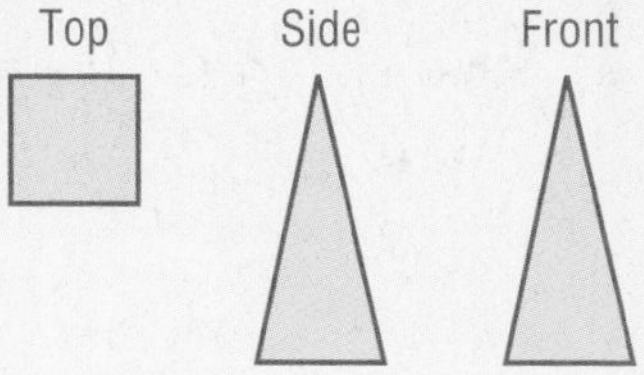

2. 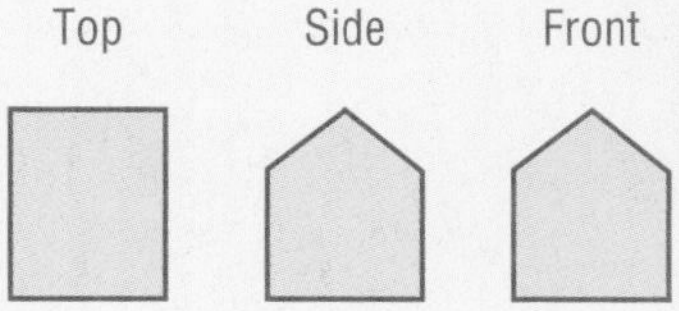

3. 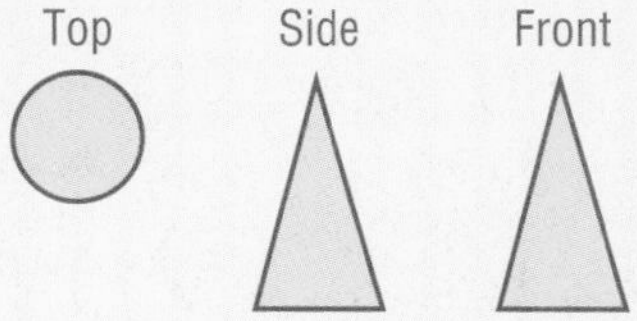

4. 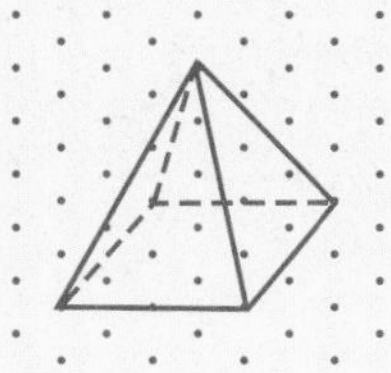

5. 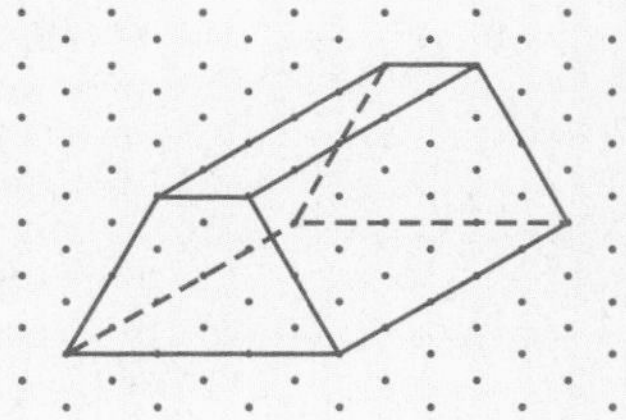

6. 

Lesson 12-3 Answers are calculated using the π key on a calculator and then rounded. (Pages 524–527)

Find the volume of each cylinder. Round to the nearest tenth.

1. 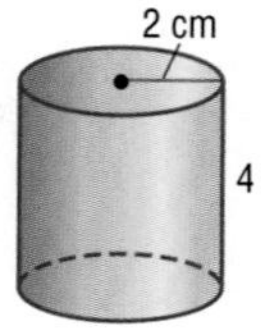

50.3 cm³

2. 

45.9 yd³

3. 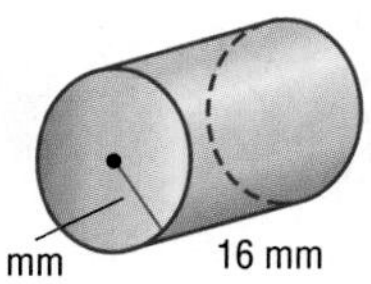

2,827.4 mm³

4. 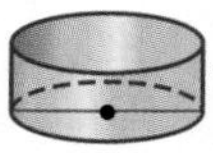

2:

5. radius = 6 in.
height = 3 in.
339.3 in³

6. radius = 8 ft
height = 10 ft
2,010.6 ft³

7. radius = 6 km
height = 12 km
1,357.2 km³

8. radius = 8.5 cm
height = 3 cm
680.9 cm³

9. diameter = 16 yd
height = 4.5 yd
904.8 yd³

10. diameter = 3.5 mm
height = 2.5 mm
24.1 mm³

11. diameter = 12 m
height = 4.75 m
537.2 m³

12. diameter = $\frac{5}{8}$ in
height = 4 in.
1.2 in³

13. diameter = 100 ft
height = 35 ft
274,889.4 ft³

14. radius = 40.5 m
height = 65.1 m
335,460.1 m³

15. radius = 0.5 cm
height = 1.6 cm
1.3 cm³

16. diameter = $8\frac{3}{4}$
height = $5\frac{1}{2}$ in.
330.7 in³

17. Find the volume of a cylinder whose diameter is 6 inches and height is 2 feet. 678.6 in³ or 0.4

18. How tall is a cylinder that has a volume of 2,123 cubic meters and a radius of 13 meters? 4.0 m

Lesson 12-4 (Pages 532–535)

Find the surface area of each rectangular prism. Round to the nearest tenth if necessary.

1. 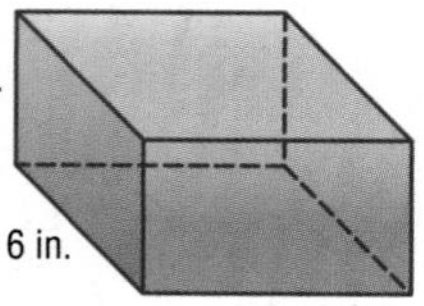

188 in²

2. 

272 cm²

3. 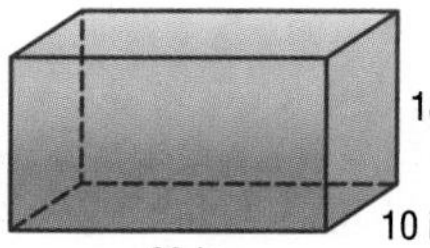

2,152 in²

4. 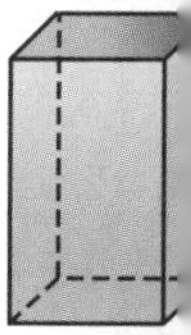

1,72

5. length = 10 m
width = 6 m
height = 7 m
344 m²

6. length = 20 mm
width = 15 mm
height = 25 mm
2,350 mm²

7. length = 16 ft
width = 20 ft
height = 12 ft
1,504 ft²

8. length = 52 cm
width = 48 cm
height = 45 cm
13,992 cm²

9. length = 8 ft
width = 6.5 ft
height = 7 ft
307 ft²

10. length = 9.4 m
width = 2 m
height = 5.2 m
156.2 m²

11. length = 20.4 cm
width = 15.5 cm
height = 8.8 cm
1,264.2 cm²

12. length = 8.5 mi
width = 3 mi
height = 5.8 mi
184.4 mi²

13. length = $7\frac{1}{4}$ ft
width = 5 ft
height = $6\frac{1}{2}$ ft
231.8 ft²

14. length = $15\frac{2}{3}$ yd
width = $7\frac{1}{3}$ yd
height = 9 yd
643.8 yd²

15. length = $4\frac{1}{2}$ in.
width = 10 in.
height = $8\frac{3}{4}$ in.
343.8 in²

16. length = 12.2 m
width = 7.4 mm
height = 7.4 mm
470.6 mm²

17. Find the surface area of an open-top box with a length of 18 yards, a width of 11 yards, and a height of 14 yards. 1,010 yd²

18. Find the surface area of a rectangular prism with a length of 1 yard, a width of 7 feet, and a height of 2 yards. 162 ft² or 18 yd²

594 Extra Practice

…n 12-5 (Pages 538–541)

…e surface area of each cylinder. Round to the nearest tenth.

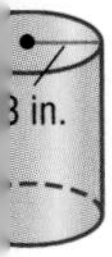

188.5 in^2

2. 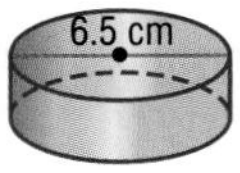

107.2 cm^2

3. 

31.8 m^2

4. 

19.6 ft^2

…nswers are calculated using the π key on a calculator and then rounded.

…ght = 6 cm
…ius = 3.5 cm
….9 cm^2

6. height = 16.5 mm
diameter = 18 mm
1,442.0 mm^2

7. height = 22 yd
radius = 10.5 yd
2,144.1 yd^2

8. height = 6 ft
radius = 18.5 ft
2,847.9 ft^2

…ght = 10.2 mi
…meter = 4 mi
….3 mi^2

10. height = 8.6 cm
diameter = 8.2 cm
327.2 cm^2

11. height = 5.8 km
diameter = 3.6 km
86.0 km^2

12. height = 32.7 m
radius = 21.5 m
7,321.8 m^2

…ght = $2\frac{2}{3}$ yd
…meter = 6 yd
….8 yd^2

14. height = $12\frac{3}{4}$ ft
radius = $7\frac{1}{4}$ ft
911.1 ft^2

15. height = $5\frac{1}{5}$ mi
radius = $18\frac{1}{3}$ mi
2,710.8 mi^2

16. height = $5\frac{1}{2}$ in
diameter = 3 in
66.0 in^2

…n 12-6 (Pages 542–545)

…fy the precision unit of each measuring tool.

0.2 cm

2. 

1 lb

$\frac{1}{4}$ lb

4. 

$\frac{1}{8}$ in.

…each measure using significant digits. 5–8. Sample answers are given.

2.4 lb

6. 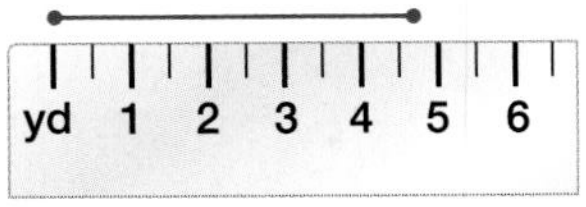

4.7 yd

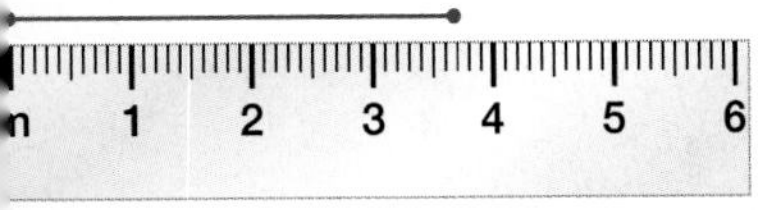

3.65 m

8. 

7.55 kg

Extra Practice

Extra Practice 595

Page 567, Lesson 2-3

1.

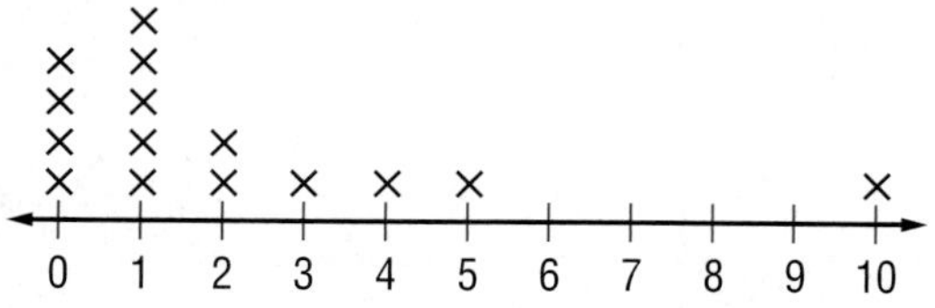

Sample answer: cluster 0–2; gap 5–10; outlier 10

2.

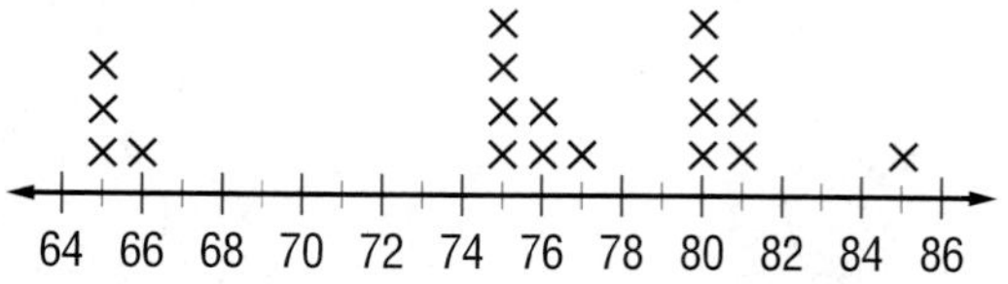

Sample answer: clusters 65–66, 75–77, 80–81; gaps 66–75, 77–80, 81–85; outlier 85

3.

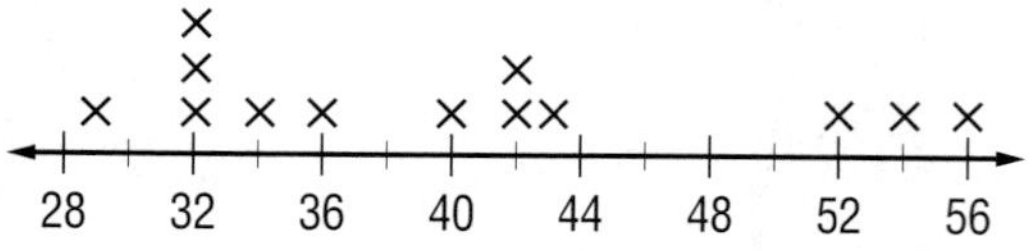

Sample answer: clusters 32–36, 40–43; gaps 36–40, 44–52; outlier none

4.

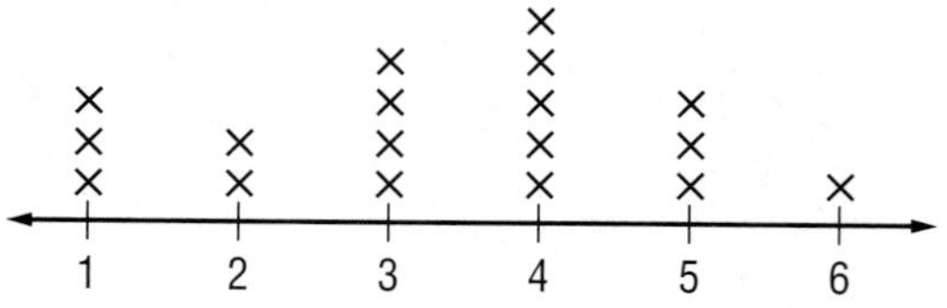

Sample answer: cluster 1–6; gap none; outlier none

Page 568, Lesson 2-6

1a. 1.00, 1.70, 2.20, 3.70, 4.20

1b.

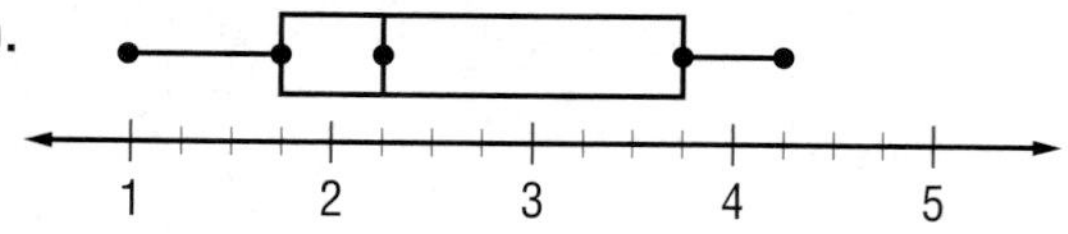

2a. 808, 1,000, 1,492, 1,988, 4,000

2b.

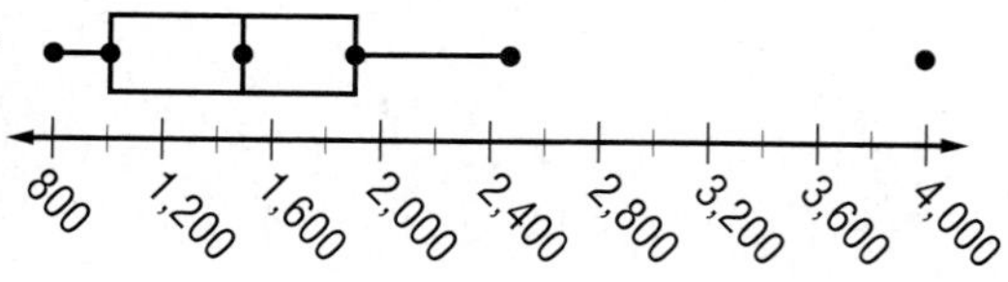

3a. 1,214, 1,362, 1,395, 1,772, 1,831

3b.

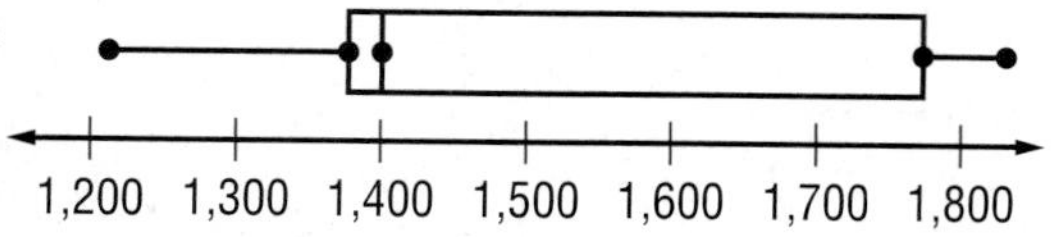

4a. 65, 70.5, 80, 91, 100

4b.

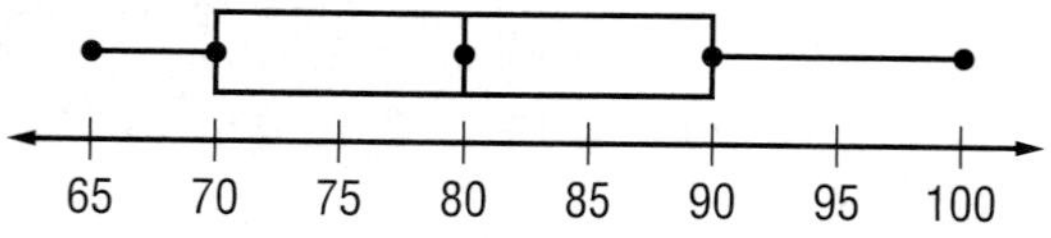

Page 569, Lesson 2-7

1.

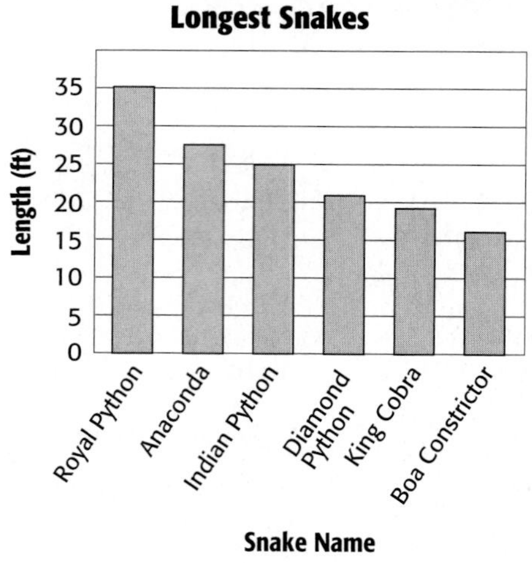

2.

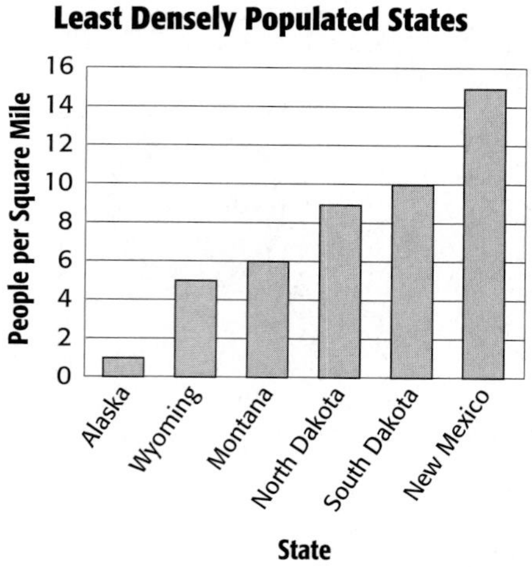

Page 585, Lesson 9-2

1.

Number Cube 1	Number Cube 2	Sample Space
1	1	1-1
	2	1-2
	3	1-3
	4	1-4
	5	1-5
	6	1-6
2	1	2-1
	2	2-2
	3	2-3
	4	2-4
	5	2-5
	6	2-6
3	1	3-1
	2	3-2
	3	3-3
	4	3-4
	5	3-5
	6	3-6
4	1	4-1
	2	4-2
	3	4-3
	4	4-4
	5	4-5
	6	4-6
5	1	5-1
	2	5-2
	3	5-3
	4	5-4
	5	5-5
	6	5-6
6	1	6-1
	2	6-2
	3	6-3
	4	6-4
	5	6-5
	6	6-6

2.

Ice Cream Cone	Ice Cream Flavor	Sample Space
waffle	chocolate	waffle, chocolate
	vanilla	waffle, vanilla
	strawberry	waffle, strawberry
plain	chocolate	plain, chocolate
	vanilla	plain, vanilla
	strawberry	plain, strawberry
sugar	chocolate	sugar, chocolate
	vanilla	sugar, vanilla
	strawberry	sugar, strawberry

3.

Bread	Cheese	Meat	Sample Space
white	cheddar	ham	W, C, H
		turkey	W, C, T
		roast beef	W, C, RB
	Swiss	ham	W, S, H
		turkey	W, S, T
		roast beef	W, S, RB
wheat	cheddar	ham	WH, C, H
		turkey	WH, C, T
		roast beef	WH, C, RB
	Swiss	ham	WH, S, H
		turkey	WH, S, T
		roast beef	WH, S, RB
rye	cheddar	ham	R, C, H
		turkey	R, C, T
		roast beef	R, C, RB
	Swiss	ham	R, S, H
		turkey	R, S, T
		roast beef	R, S, RB

4.

Penny 1	Penny 2	Sample Space
H	H	HH
	T	HT
T	H	TH
	T	TT

5.

Math	Foreign Language	Sample Space
Algebra	French	Algebra, French
	Spanish	Algebra, Spanish
	Latin	Algebra, Latin
Geometry	French	Geometry, French
	Spanish	Geometry, Spanish
	Latin	Geometry, Latin

Mixed Problem Solving

Chapter 1 Decimal Patterns and Algebra

(pages 4–51)

HISTORY For Exercises 1 and 2, use the following information.
In 1932, Amelia Earhart flew 2,026 miles in 14 hours 56 minutes. (Lesson 1-1)

1. How many minutes did she fly? **896 min**
2. To the nearest mile, what was her speed in miles per minute? **2 mi/min**

3. **TRANSPORTATION** An MD-80 aircraft burns 950 gallons of fuel per hour. How much fuel burns during a three-hour flight? (Lesson 1-1) **2,850 gal**
4. **LIGHT** The speed of light is about 67^3 kilometers per second. About how many kilometers per second is this? (Lesson 1-2) **about 300,763 km/s**

SALES For Exercises 5 and 6, use the following information.

A department store is having a back-to-school sale. The table shows the prices of three popular items. **5. 2(37.99) + 3(19.88) + 1(22.50)**

Item	Price
jeans	\$37.99
sweatshirt	\$19.88
polo shirt	\$22.50

Latonia wants to buy 2 pairs of jeans, 3 sweatshirts, and 1 polo shirt. (Lesson 1-3)

5. Write a numerical expression to represent the total cost of the six items.
6. Evaluate the expression to find the total cost. **\$158.12**
7. **FITNESS** You can estimate how fast you walk in miles per hour by evaluating the expression $\frac{n}{30}$, where n is the number of steps you take in one minute. Find your speed if you take 96 steps in one minute. (Lesson 1-4) **3.2 mi/h**

8. **BASEBALL** Last year, Scott attended 13 Minnesota Twins baseball games. This year, he attended 24. Solve $13 + n = 24$ to find how many more games he attended this year than last. (Lesson 1-5) **11 games**

9. **HOT AIR BALLOONS** Miyoki paid \$140 f a four-hour hot air balloon ride over th Bridger Mountains. Solve $4h = 140$ to fi the cost per hour of the ride. (Lesson 1-5) **\$35**

ENTERTAINMENT For Exercises 10 and 11, the following information.
The five members of the Wolff family went an amusement park. They each purchased all-day ride pass and a water park pass, as shown below. (Lesson 1-6) **10. 5(14.95 + 6.50) 5(14.95) + 5(6.50)**

Item	Price
all-day ride pass	\$14.95
water park pass	\$6.50

10. Use the Distributive Property to write t different expressions that represent the cost for the family.
11. Find the total cost of the passes. **\$107.2**

12. **NUMBER THEORY** Numbers that can be represented by a square arrangement o dots are called *square numbers*. The first square numbers are shown below. (Lesso

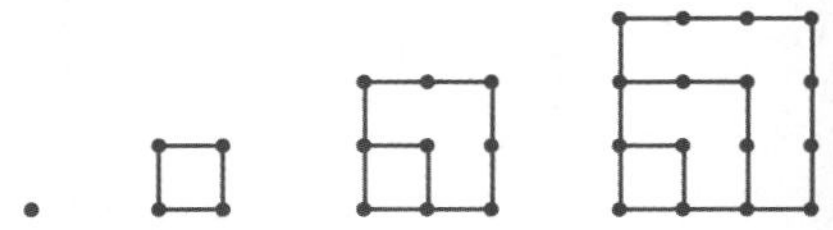

1 2 3 4

Write a sequence formed by the first eig square numbers. **1, 4, 9, 16, 25, 36, 49,**

13. **BIRDS** The world's smallest bird, the bee hummingbird, weighs only about 1.6 grams. How many milligrams is this (Lesson 1-8) **1,600 mg**

14. **BRIDGES** The Akashi Kaikyo bridge in J is 2,003 meters long. What is the length the bridge in kilometers? (Lesson 1-8) **2.00**

15. **SPACE** The Andromeda Spiral is a gala 2,200,000 light years from Earth. Write distance in scientific notation. (Lesson 1-9) **2.2×10^6**

Mixed Problem Solving

596 Mixed Problem Solving

L For Exercises 1 and 2, refer to the table It shows the number of passengers in U.S. airports in one year. (Lesson 2-1)

Airport Passengers (millions)						
80	41	36	34	28	25	37
72	39	36	33	27	61	34
68	37	35	31	25	31	

Source: *The World Almanac*

e a frequency table using the intervals 4, 35–44, 45–54, and so on.

ch interval has the greatest frequency? ain what this means. **1–2. See margin.**

'ING The frequency table below shows ural interstate speed limits for the fifty s. How many states have a speed limit ? (Lesson 2-1) **11 states**

Speed Limit	Tally	Frequency
55	I	1
60		0
65	𝍸 𝍸 𝍸 𝍸	20
70	𝍸 𝍸 𝍸 III	18
75	?	?

urce: *The World Almanac*

'ING For Exercises 4 refer to the table at the t shows the winning ic times for the 's 4 × 100-meter le Relay in swimming. 2)

Year	Time (s)
1976	225
1980	223
1984	224
1988	221
1992	220
1996	219
2000	217

Source: *ESPN Sports Almanac*

e a line graph of the **See margin.**

ict the winning time in vent in 2008. **Sample answer: 215 s**

ION For Exercises 6–9, use the data They are the grams of Carbohydrates en different energy bars. 16, 15, 20, 24, 6, 2, 20, 26, 14, 20, 20, 16, 16 **6. See margin.**

e a line plot of the data. (Lesson 2-3)

t is the range of the data? (Lesson 2-3) **24**

ify any clusters, gaps, or outliers and in what they represent. (Lesson 2-3) **See margin.**

the mean, median, and mode. (Lesson 2-4) **16; 16**

BASKETBALL For Exercises 10–12, refer to the table below. It shows the number of games played by Michael Jordan each year from 1986–1987 to 2001–2002.

Number of Games Played							
82	82	81	82	82	80	78	0
17	82	82	82	0	0	0	60

Source: www.nba.com

10. Find the mean, median, and mode of the data. (Lesson 2-4) **55.625; 80.5; 82**
11. Make a stem-and-leaf plot of the data. (Lesson 2-5) **See margin.**
12. Find the lower extreme, LQ, median, UQ, and upper extreme. (Lesson 2-6) **0; 8.5; 80.5; 82; 82**

VACATION For Exercises 13–15, refer to the box-and-whisker plot below. It shows the number of vacation days for employees at a company. (Lesson 2-6)

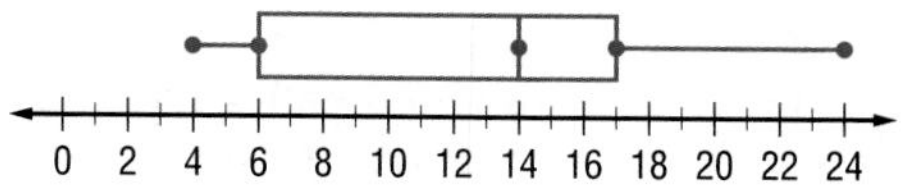

13. What fraction of the employees receive 6 to 14 vacation days? $\frac{1}{4}$
14. What fraction of the employees receive 14 or more vacation days? $\frac{1}{2}$
15. What is the range of the data? **20 days**

16–17. See Answer Appendix.

16. **TOURISTS** The table shows the countries from which the most tourists in the United States came. Make a bar graph of the data. (Lesson 2-7)

Country	Number of Visitors (millions)
Canada	14.6
Mexico	10.3
Japan	5.0
United Kingdom	4.7

Source: Department of Commerce

TRAVEL For Exercises 17 and 18, refer to the data in Exercises 1 and 2. **18. See margin.**

17. Make a histogram of the data. (Lesson 2-7)
18. Would the mean or median best describe the data? Explain your reasoning. (Lesson 2-8)

6.

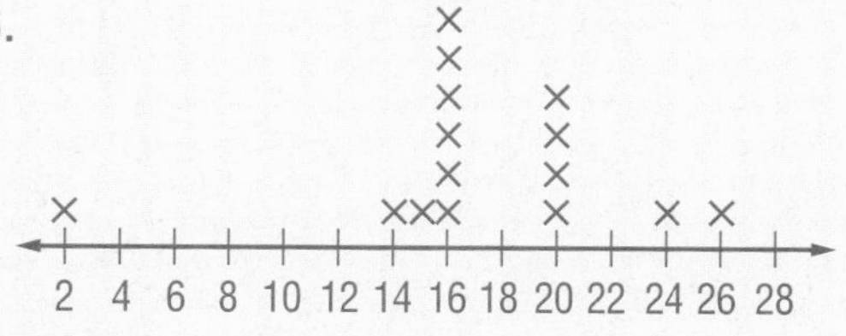

Chapter 2

1.

Passengers (millions)	Tally	Frequency
25–34	𝍸 IIII	9
35–44	𝍸 II	7
45–54		0
55–64	I	1
65–74	II	2
75–84	I	1

2. 25–34 million; More airports had 25 to 34 million passengers in one year than any other number of passengers.

4.

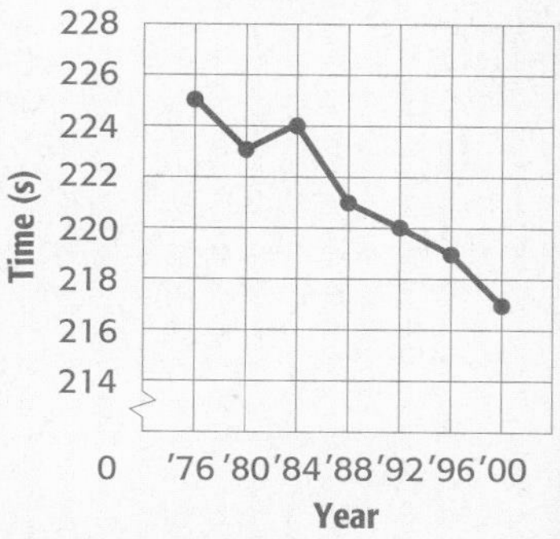

6. See bottom margin.

8. Sample answer: cluster 14–16: most of the energy bars have 14–16 grams of carbohydrates; gaps 2–14 and 16–20: no bars have between 2 and 14 or between 16 and 20 grams; outlier 2: one bar has 2 grams, which is much less than the rest of the bars.

11.

Stem	Leaf
0	0 0 0 0
1	7
2	
3	
4	
5	
6	0
7	8
8	0 1 2 2 2 2 2 2 2

7|8 = 78 games

18. Sample answer: The median best represents the data because the mean is increased by the 4 large outliers.

Mixed Problem Solving

Chapter 3

2. −220, −436, −511, −433, −282, −383, −505, −235, −1,312, −230
3. −1,312, −511, −505, −436, −433, −383, −282, −235, −230, −220
5. Hot Tower (0, 3); Ferris Wheel (0, −4)

Mixed Problem Solving

1. **AIR CONDITIONING** Jacob turned on the air conditioning, and the temperature in his apartment decreased 8 degrees. Write an integer to represent the drop in temperature. (Lesson 3-1) **−8**

EARTH SCIENCE For Exercises 2 and 3, use the table below. It describes the deepest land depressions in the world in feet below sea level. **2–3. See margin.**

Depth (ft)				
220	436	511	433	282
383	505	235	1,312	230

Source: *The Top 10 of Everything*

2. Write an integer to represent each depth. (Lesson 3-1)
3. Order the integers from greatest depth to least depth. (Lesson 3-2)

ENTERTAINMENT For Exercises 4–8, use the diagram below. It shows the locations of several rides at the Outlook Amusement Park. (Lesson 3-3)

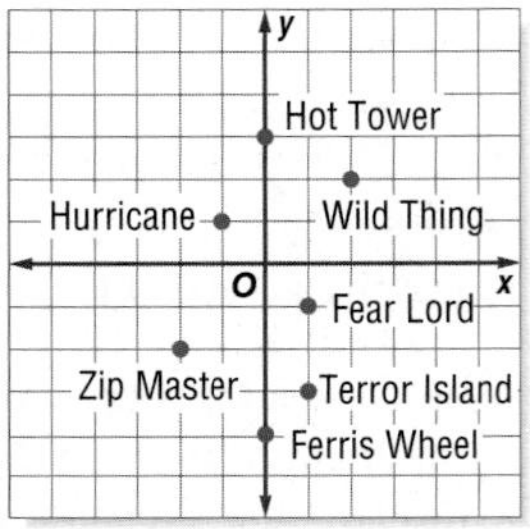

Zip Master

4. Which ride(s) is located in quadrant III?
5. Which ride(s) is located on the y-axis? Name the coordinates. **See margin.**
6. Which ride(s) has coordinates in which the x- and y-coordinates are equal?
7. In which quadrant is the Hurricane located?
8. A new ride is built with a location on the x-axis and 5 units left of the origin. Name the coordinates of this point. **(−5, 0)**

6. Wild Thing, Zip Master 7. quadrant II

9. **CAVERNS** Adriana is 52 feet underground touring the Lewis and Clark Caverns. She climbs a ladder up 15 feet. What is her new location? (Lesson 3-4) **37 ft underground or −37 ft**

10. **RECORDS** The lowest temperature reco in Verkhoyansk, Russia, was about −90 The highest temperature was about 99° What is the difference between these temperatures? (Lesson 3-5) **189°F**

11. **EARTH SCIENCE** The highest and lowes points in California are shown in the ta What is the difference in elevations? (Lesson 3-5) **14,776 ft**

Location	Elevation
Mount Whitney	14,494 ft above sea level
Death Valley	282 ft below sea level

Source: *The World Almanac of the U.S.A.*

12. **TEMPERATURE** Refer to the table below How much colder is outer space than li helium? (Lesson 3-5) **5°C**

Item	Temperature (°C)
dry ice	−108
outer space	−457
liquid helium	−452
bird's body	108

Source: *The Sizesaurus*

13. **RIDES** A glider ride over the Crazy Mountains has a maximum altitude of 12,000 feet, and it descends about 300 f per minute. How long will it take the g to reach an altitude of 6,000 feet? (Lesson **20 min**
14. **POOLS** A swimming pool holds 5,000 gallons of water. A hose can drain the p at the rate of 225 gallons per hour. How much water will still be in the pool afte has been draining for 5 hours? (Lesson 3- **3,875 gal**
15. **TEMPERATURE** A temperature of −89°C was recorded in Antarctica. Use the expression $\frac{9C}{5} + 32$, where C is the temperature in degrees Celsius, to find temperature in degrees Fahrenheit. (Lesson 3-7) **about −128.2°F**
16. **DIVING** In two minutes, a scuba diver ascended from 40 feet below sea level t 8 feet below sea level. By how many fe she ascend per minute? (Lesson 3-7) **16 ft/**

598 Mixed Problem Solving

RISM The Statue of Liberty in New York, ,ʏ York, and the Eiffel Tower in Paris, ıce, were designed by the same person. Statue of Liberty is 152 feet tall. It is 732 shorter than the Eiffel Tower, x. Write an ıtion that models this situation. (Lesson 4-1) $= x - 732$

ONS For Exercises 3, use the table at the ınd the following ıation.

Number of Electoral Votes 2000	
California	54
New York	33
Texas	?
Florida	25
Pennsylvania	23

Source: *The World Almanac*

ʼork has one more al vote than Texas. ylvania has 9 fewer al votes than Texas. -2)

:e two different ations to find the ıber of electoral votes in Texas, n.

ı the number of electoral votes. **32 votes** $= n + 1; 23 = n - 9$

LER COASTERS The track length of ›pular roller coaster is 5,106 feet. The ›r coaster has an average speed of ut 2,000 feet per minute. At that speed, ʼ long will it take to travel its length of 6 feet? Use the formula $d = rt$. (Lesson 4-3) **ıt 2.553 min**

INESS Carla's Catering charges a $25 fee ›rve 15 or fewer people. In addition to fee, they charge $10 per appetizer. You ›aving a party for 12 people and can ıd a total of $85. How many appetizers you order from Carla's Catering? ›n 4-4) **6 appetizers**

NEY Koto works in a hardware store and ıs $6 per hour. She plans on saving her ıey to buy a DVD player that costs $189. te an inequality for the least number of ›rs x Koto needs to work in order to be › to purchase the DVD player. Then solve inequality. (Lesson 4-5) **$6x \geq 189$; $x \geq 31.5$ h**

CS The 26th amendment to the United ›es Constitution guarantees the right to › to citizens who are eighteen years of age ›lder. Write an inequality showing the › a of all voters. (Lesson 4-5) **$a \geq 18$**

8. **GEOMETRY** The formula for the perimeter of a square is $P = 4s$, where P is the perimeter and s is the length of a side. Graph the equation. (Lesson 4-6) **See margin.**

AGES For Exercises 9–12, use the table below. It shows how Jared's age and his sister Emily's age are related. (Lesson 4-6)

Jared's age (yr)	1	2	3	4	5
Emily's age (yr)	7	8	9	10	11

9. Emily is 6 years older than Jared.

9. Write a verbal expression to describe how the ages are related.
10. Write an equation for the verbal expression. Let x represent Jared's age and y represent Emily's age. **$y = x + 6$**
11. Predict how old Emily will be when Jared is 10 years old. **16 years**
12. Graph the equation. **See margin.**

13. **RAMPS** A ramp used to load cargo onto ships is modeled below. Is the slope of the ramp greater than or less than 1? Explain. (Lesson 4-7) **Less than; the slope is $\frac{5}{15}$ or $\frac{1}{3}$.**

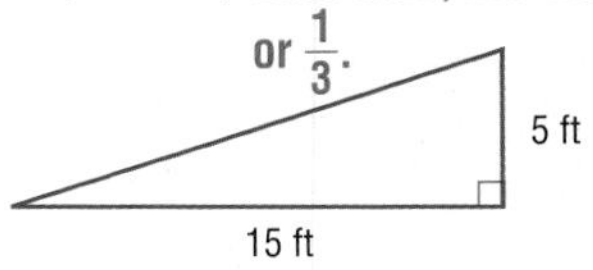

CYCLING For Exercises 14–16, use the table below. It shows the distances traveled by Jamal and Jasmine. (Lesson 4-7)

Hours Traveled	Cycling Distance (miles)	
	Jamal	Jasmine
1	8	11
2	16	22
3	24	33
4	32	44

14. Suppose each of these functions was graphed on a coordinate plane. Without calculating, which line is steeper? Explain. **See margin.**
15. Find the slope of each line. **Jamal, 8; Jasmine, 11**
16. What does the slope of each line represent? **distance traveled per hour**

Mixed Problem Solving

Chapter 4

8.

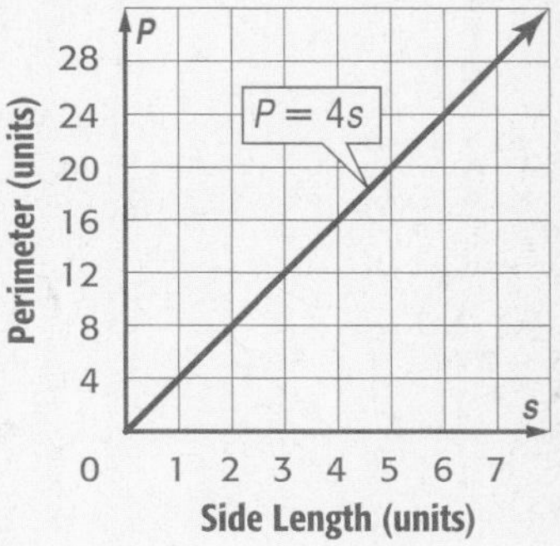

12.

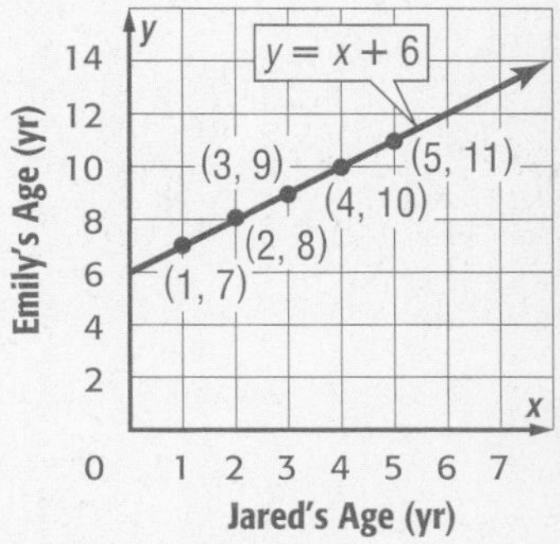

14. Jasmine's; both changes in x are the same, but Jasmine's change in y is 11, which is greater than Jamal's change in y, 8.

Mixed Problem Solving

Mixed Problem Solving

LAND For Exercises 1–3, use the information below.
A section of land is one mile long and one mile wide. (Lesson 5-1) **1.** $2^5 \times 3 \times 5 \times 11$

1. Write the prime factorization of 5,280.
2. Find the area of the section of land in square feet. (*Hint*: 1 mile = 5,280 feet) $27{,}878{,}400$ ft^2
3. Write the prime factorization of the area that you found in Exercise 2. $2^{10} \times 3^2 \times 5^2 \times 11^2$

DECORATIONS For Exercises 4 and 5, use the information below.
Benito is cutting streamers from crepe paper to decorate for a party. He has a red roll of crepe paper 144 inches long, a white roll 192 inches long, and a blue roll 360 inches long. (Lesson 5-2)

4. If he wants to have all colors of streamers the same length, what is the longest length that he can cut? 24 in.
5. How many total streamers can he cut? 29 streamers

OLYMPICS For Exercises 6 and 7, refer to the table below. It shows the medals won by the top three countries in the 2000 Summer Olympics.

Country	Medals		
	Gold	Silver	Bronze
United States	40	24	33
Russia	32	28	28
China	28	16	15

Source: *The World Almanac*

6. Write the number of gold medals that Russia won as a fraction of the total number that Russia won in simplest form. (Lesson 5-3) $\frac{4}{11}$
7. Write the fraction that you wrote in Exercise 6 as a decimal. (Lesson 5-4) $0.\overline{36}$
8. **GEOGRAPHY** Alaska makes up about 0.16 of the total land area of the United States. Write this decimal as a fraction in simplest form. (Lesson 5-4) $\frac{4}{25}$
9. **SPORTS** At Belgrade Intermediate School, 75 out of every 100 students participate in sports. What percent of students do *not* participate in sports? (Lesson 5-5) 25%

ADVERTISING For Exercises 10–13, use the table below. It shows the results of a survey in which teens 13 to 17 years old were asked which types of advertising th pay attention to.

Type of Advertising	Percent of Teens
television	80%
magazine	62%
product in a movie	48%
ad in an E-mail	24%

Source: E-Poll

Write each percent as a fraction in simple form. (Lesson 5-5) **12.** $\frac{12}{25}$

10. television $\frac{4}{5}$
11. magazine $\frac{31}{50}$
12. product in a movie
13. ad in an E-ma
14. **POPULATION** In 2000, about 5.4% of the people in Wyoming lived in Laramie. V this percent as a decimal. (Lesson 5-6) 0.0

GEOMETRY For Exercises 15–17, refer to the grid at the right. (Lesson 5-6)

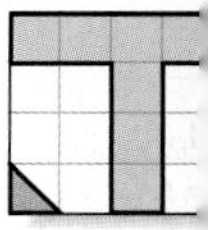

15. Write a decimal and a percent to represent the "T" shaded area. 0.4; 40%
16. Write a decimal and a percent to repres the area shaded pink. 0.05; 5%
17. What percent of the grid is *not* shaded?
18. **FLOWERS** Roses can be ordered in bun of 6 and carnations, in bunches of 15. If Ingrid wants to have the same number roses as carnations for parent night, wh the least number of each flower that sh must order? (Lesson 5-7) 30
19. **WATER** The table at the right shows the fraction of each state that is water. Order the states from least to greatest fraction of water. (Lesson 5-8) Alaska, Wisconsin, Michigan

What Part is W	
State	Fra
Alaska	
Michigan	
Wisconsin	

Source: *The World Almar U.S.A.*

Mixed Problem Solving
600 Mixed Problem Solving

LS A box of instant potatoes contains ups of flakes. A family-sized bowl of toes uses $3\frac{2}{3}$ cups of the flakes. Estimate many family-sized bowls can be made n one box. (Lesson 6-1) **Sample answer: - 4 or 5 bowls**

LL For Exercises 2 and 3, use the table. ws the average annual precipitation for of the driest locations on Earth. (Lesson 6-2)

Location	Precipitation (in.)
Arica, Chile	$\frac{3}{100}$
Iquique, Chile	$\frac{1}{5}$
Callao, Peru	$\frac{12}{25}$

Source: *The Top 10 of Everything*

v much more rain does Iquique get per r than Arica? **$\frac{17}{100}$ in.**

v much more annual rain does Callao get n Iquique? **$\frac{7}{25}$ in.**

FTS Kyle uses $2\frac{3}{4}$ yards of fabric for each t item that he makes. If he uses a 25-yard ic bolt and sells each item for $35, what e greatest amount of money that he can n? (Lesson 6-3) **$315**

RS The star Sirius is about $8\frac{7}{10}$ light years n Earth. Alpha Centauri is half this ance from Earth. How far is Alpha tauri from Earth? (Lesson 6-4) **$4\frac{7}{20}$ light years**

IENCE For Exercises 6 and 7, use the below. It shows the average growth per of hair and fingernails. (Lesson 6-5)

Average Monthly Growth	
hair	$\frac{1}{2}$ in.
fingernails	$\frac{2}{25}$ in.

ve $3 = \frac{1}{2}t$ to find how long it takes hair to w 3 inches. **6 mo**

ve $\frac{2}{25} = \frac{d}{12}$ to find how much fingernails w in 1 year. **$\frac{24}{25}$ in.**

8. **SEWING** Jocelyn has nine yards of cloth to make table napkins for the senior citizens' center. She needs $\frac{3}{8}$ yard for each napkin. Use $\frac{3}{8}c = 9$ to find the number of napkins that she can make. (Lesson 6-5) **24 napkins**

9. **WHALES** During the first year, a baby whale gains about $27\frac{3}{5}$ tons. What is the average weight gain per month? (Lesson 6-6) **$2\frac{3}{10}$ T/mo**

10. **LIFE SCIENCE** An adult has about 5 quarts of blood. If a person donates 1 pint of blood, how many pints are left? (Lesson 6-7) **9 pt**

11. **COFFEE** In Switzerland, the average amount of coffee consumed per year is 1,089 cups per person. How many pints is this? (Lesson 6-7) **544.5 pt**

CHESS For Exercises 12–15, use the chess board below. (Lesson 6-8)

12 in.

12 in.

12. What is the perimeter of the chess board? **48 in.**
13. What is the area of the chess board? **144 in^2**
14. What is the area of each small square? **2.25 in^2**
15. A travel chess board has half the length and width of the board shown. What is the perimeter and area? **24 in.; 36 in^2**

16. **LIFE SCIENCE** A nest built by bald eagles had a diameter of $9\frac{1}{2}$ feet. What was the circumference of the nest? (Lesson 6-9) **29.8 ft**

17. **EARTH SCIENCE** Earth has a diameter of 7,926 miles. Use the formula for the circumference of a circle to approximate the circumference of Earth at its equator. (Lesson 6-9) **about 24,900 mi**

Mixed Problem Solving

Mixed Problem Solving

Mixed Problem Solving 601

1. **SCHOOLS** In a recent year, Oregon had 924 public elementary schools and 264 public high schools. Write a ratio in simplest form comparing the number of public high schools to elementary schools. (Lesson 7-1) $\frac{2}{7}$

2. **TEACHERS** The table shows the ratio of students to teachers for three states. Are any of the ratios equivalent? Explain. (Lesson 7-1)

State	Students : Teachers
Alabama	152 : 10
Kentucky	231 : 15
Montana	76 : 5

Source: *The World Almanac*

yes; Alabama = $\frac{76}{5}$ and Montana = $\frac{76}{5}$

3. **EXERCISE** A person jumps rope 14 times in 10 seconds. What is the unit rate in jumps per second? (Lesson 7-2) **1.4 jumps per s**

4. **FOOD** A 16-ounce box of cereal costs $3.95. Find the unit price to the nearest cent. (Lesson 7-2) **$0.25 per oz**

5. **PHOTOGRAPHS** Mandy is enlarging a photograph that is 3 inches wide and 4.5 inches long. If she wants the width of the enlargement to be 10 inches, what will be the length? (Lesson 7-3) **15 in.**

6. **GEOMETRY** If the ratio of length to width of a rectangle is about 1.6 to 1, then the rectangle is said to be a *golden rectangle*. Determine whether either figure below is golden rectangle. (Lesson 7-3) **yes; 32 ft by 20 ft**

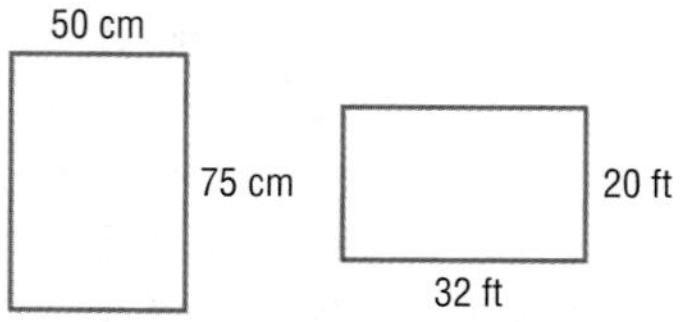

7. **MAPS** Washington, D.C., and Baltimore, Maryland, are $2\frac{7}{8}$ inches apart on a map. If the scale is $\frac{1}{2}$ inch : 6 miles, what is the actual distance between the cities? (Lesson 7-4) **$34\frac{1}{2}$ mi**

8. **MODELS** Ian is making a miniature bec for his daughter's doll house. The actu bed is $6\frac{3}{4}$ feet long. If he uses the scale $\frac{1}{2}$ inch = $1\frac{1}{2}$ feet, what will be the lengt of the miniature bed? (Lesson 7-4) **$2\frac{1}{4}$ in.**

9. **POPULATION** According to the U.S. Cen Bureau, 6.6% of all people living in Flo are 10–14 years old. What fraction is thi (Lesson 7-5) $\frac{33}{500}$

10. **INTERNET** In 2000, 41.5% of U.S. housel had access to the Internet. Write this pe as a fraction in simplest form. (Lesson 7-5)

COINS For Exercises 11 and 12, use the ta below. It shows the fraction of a quarter that is made up of the metals nickel and copper. Write each fraction as a percent. Round to the nearest hundredth if necess. (Lesson 7-5)

Metal	Fraction of Quarter
nickel	$\frac{1}{12}$
copper	$\frac{11}{12}$

Source: *The Top 10 of Everything*

11. nickel **8.33%**

12. copper **91.67%**

13. **FARMS** Approximately 0.0009 of the land in Montana is farmland. Write this as a percent. (Lesson 7-6) **0.09%**

14. **POPULATION** In 2000, 1.5% of the U.S. population was 85 years and older. Wri this percent as a fraction in simplest for (Lesson 7-6) $\frac{3}{200}$

15. **SEEDS** A packet of beans guarantees th 95% of its 200 seeds will germinate. Ho many seeds are expected to germinate? (Lesson 7-7) **190 seeds**

16. **SKIS** Toshiro spent $520 on new twin-ti skis. This was 40% of the money he ear at his summer job. How much did he e at his summer job? (Lesson 7-8) **$1,300**

Mixed Problem Solving

Mixed Problem Solving

602 Mixed Problem Solving

GRAPHY In Washington, about 5.7% of otal area is water. If the total area of hington is 70,637 square miles, estimate number of square miles of water by using . (Lesson 8-1) **Sample answer: 0.1 = 70,000 = 0; $\frac{1}{2}$ × 7,000 = 3,500 mi²**

ULATION According to the U.S. Census eau, 9.4% of the population in the tle-Bellevue metropolitan area is Asian. is area has a population of 2,414,616, nate the number of people who are n. (Lesson 8-1) **Sample answer: $\frac{1}{10}$ · 0,000 = 200,000 people**

ESTS In the United States, forests cover % of the land. If the total land area of the is 3,536,338 square miles, how many re miles are forests? Round to the est square mile. (Lesson 8-2) **873,475 mi²**

ERNMENT Of the 435 members in the House of Representatives, 53 are from fornia and 13 are from North Carolina. he nearest whole percent, what percent e representatives are from California? North Carolina? (Lesson 8-2) **12%; 3%**

For Exercises 5 and 6, use the graph . It shows the results of a survey in people were asked how they determine ng food has been in their freezer. -3)

Frozen Foods

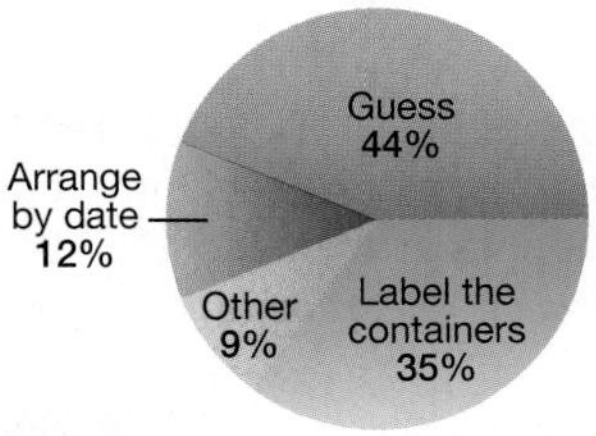

Source: Opinion Research Corp. for Tupperware

city of 56,000 people, predict how many nem guess to determine how long food been in their freezer. **24,640 people**

pose 420 people surveyed said they label r freezer containers. How many people k part in the survey? **1,200 people**

SPORTS For Exercises 7–9, use the table below. It shows the number of participants ages 7 to 17 in the sports listed. (Lesson 8-4)

Sport	Number (millions)	
	1990	2000
in-line skating	3.6	21.8
snowboarding	1.5	4.3
roller hockey	1.5	2.2
golf	23.0	26.4

Source: National Sporting Goods Association

7. What is the percent of change in in-line skaters 7 to 17 years old from 1990 to 2000? Round to the nearest percent and state whether the percent of change is an *increase* or *decrease*. **506% increase**
8. Find the percent of change from 1990 to 2000 in the number of children and teens who played roller hockey. Round to the nearest percent. **47% increase**
9. Which sport had a greater percent of increase in participants from 1990 to 2000: golf or snowboarding? Explain. **snowboarding, with a 187% increase; 15% increase for golf**
10. **MUSIC** Find the total cost of a $15.50 CD if the sales tax is 8%. (Lesson 8-5) **$16.74**

COMPUTERS For Exercises 11 and 12, use the following information.
The Wares want to buy a new computer with a regular price of $1,049. (Lesson 8-5)

11. If the store is offering a 20% discount, what will be the sale price of the computer? **$839.20**
12. If the sales tax on the computer is 5.25%, what will be the total cost with the discount? **$883.26**

BANKING For Exercises 13–16, complete the table below. The interest earned is simple interest. (Lesson 8-6)

	Principal	Rate	Time (yr)	Interest Earned	
13.	$1,525.00	5%	$2\frac{1}{2}$	?	**$190.63**
14.	$2,250.00	4%	?	$337.50	**3.75**
15.	?	3.5%	4	$498.40	**$3,560**
16.	$5,080.00	?	3	$952.50	**6.25%**

Mixed Problem Solving

Mixed Problem Solving

Mixed Problem Solving 603

Mixed Problem Solving

Mixed Problem Solving

1. **DENTISTS** A dental hygienist randomly chooses a toothbrush in a drawer containing 17 white, 12 green, and 5 blue toothbrushes. What is the probability that she chooses a green toothbrush? Write as a fraction in simplest form. (Lesson 9-1) $\frac{6}{17}$

SURVEYS For Exercises 2 and 3, use the table below. It shows the results of a survey in which adults were asked how proud they were to be an American. (Lesson 9-1)

How Proud Are You?

Response	Number
extremely	650
very	250
moderately	60
little/not at all	30
no opinion	10

Source: Gallup Poll

2. If one person participating in the survey is chosen at random, what is the probability that the person is extremely patriotic? Write as a fraction in simplest form. $\frac{13}{20}$
3. If one person participating in the survey is chosen at random, what is the probability that he is *not* moderately patriotic? Write as a fraction in simplest form. $\frac{47}{50}$

RANCHING For Exercises 4–6, use the following information. **4. See Answer Appendix.** For Roger to reach his cattle pasture, he must pass through three consecutive gates. Any of the three gates can be either *open* or *closed*. (Lesson 9-2)

4. Make a tree diagram to show all of the possible positions of the gates.
5. What is the probability that all three gates will be closed when Roger visits this pasture? Write as a fraction. $\frac{1}{8}$ **6.** $\frac{1}{4}$
6. What is the probability that the first two gates are open? Write as a fraction in simplest form.
7. **HIKING** To *run the ridge* in the Bridger Mountains, you can choose two paths up the face of the first peak, four paths across the ridge, and three routes down the final peak. Find the number of ways you can *run the ridge*. (Lesson 9-3) **24 ways**

8. **SKATEBOARDS** World Sports makes skateboards with several different patt on the deck. You can choose one of fou deck lengths and one of six types of wl If they advertise that they have 120 dif skateboards, how many deck patterns there? (Lesson 9-3) **5 patterns**
9. **READING** Mr. Steadman plans to read children's novels to his second graders during the school year. In how many w can he arrange the books to be read? (Lesson 9-4) **40,320 ways**
10. **CRAFTS** Marina has print fabric in pink blue, magenta, green, yellow, and tan. How many different stuffed bears can s make if each bear has only four differe fabrics, and the order of the fabrics is n important? (Lesson 9-5) **15 bears**
11. **FOOD** The graph shows the results of a survey in which 7th graders at Plentyw Middle School were asked to name the favorite fruit.

Favorite Fruits

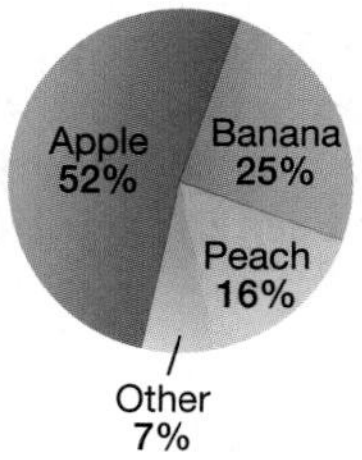

If a 7th grader at the school is randoml selected, what is the probability that th chose bananas as their favorite? Write a fraction in simplest form. (Lesson 9-6) $\frac{1}{4}$

12. **CANDY** A bag has 32 candies as shown in the table. What is the probability of randomly selecting one yellow candy and then one purple candy? Assume that the first candy is not replaced. (Lesson 9-7) $\frac{5}{124}$

Color	Numb
red	10
blue	6
purple	10
yellow	4
green	2

604 Mixed Problem Solving

or Exercises 1 and the diagram of the e American artifact. 0-1)

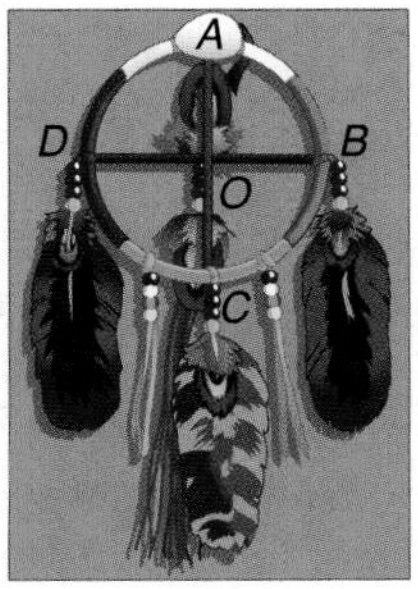

ne a right angle a straight angle.

pose one side of ngle is $\overline{OA}$. cribe the location oint E so that OE is acute. **See margin.**

ple answers: $\angle AOB$; $\angle AOC$

SION For Exercises 3 and 4, use the shown below. It shows the results of a in which families were asked how T.V. channels they watched. (Lesson 10-2)

Channels that Families Watch	
Number	**Percent**
5 or fewer	30%
6–12	33%
13–25	19%
26 or more	14%

fifth category in the survey is *No T.V. or pinion*. What percent of the people veyed were in this category? **4%**

ke a circle graph of the data. **See margin.**

CAL SCIENCE For Exercises 5 and 6, use llowing information.

sics, the *law of reflection* states that the of incidence of incoming light is equal in re to the angle of reflection of outgoing when light strikes a mirror, as shown in agram below. (Lesson 10-3)

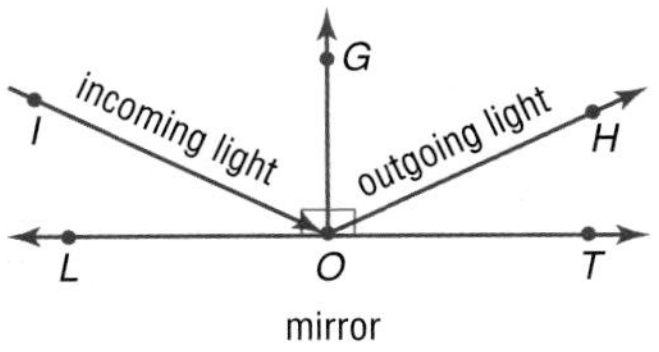

angle of incidence is $\angle IOG$ and the angle eflection is $\angle HOG$. If $m\angle IOG = 65°$, find HOG. **65°**

d $m\angle LOI$. Name another angle with this asure. **25°; $\angle TOH$**

For Exercises 7 and 8, use the figure below.

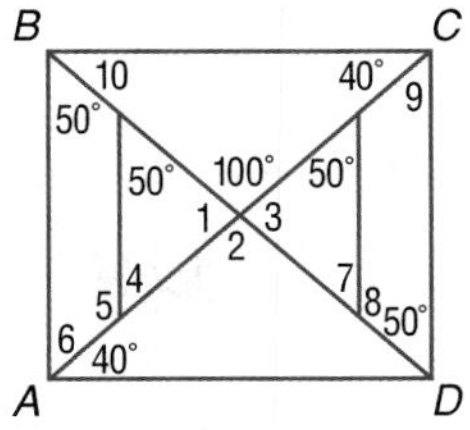

7. Find the measure of each angle numbered from 1–10. (Lesson 10-4)
8. Find the *best* name to classify quadrilateral *ABCD*. Explain your reasoning. (Lesson 10-5)

7–8. See margin.

9. **CRAFTS** Priscilla makes porcelain dolls that are proportional to a real child. If Jody is $4\frac{2}{3}$ feet tall with a 23-inch waist, what should be the waist measure of a doll that is 13 inches tall? Round to the nearest inch. (Lesson 10-6) **5 in.**
10. **ART** Draw a tessellation using two of the polygons listed at the right. Identify the polygons and explain why the tessellation works. (Lesson 10-7) **See margin.**

regular triangles
quadrilaterals
pentagons
hexagons
octagons

For Exercises 11 and 12, use the quadrilateral *MOVE* shown below. **11–12. See margin.**

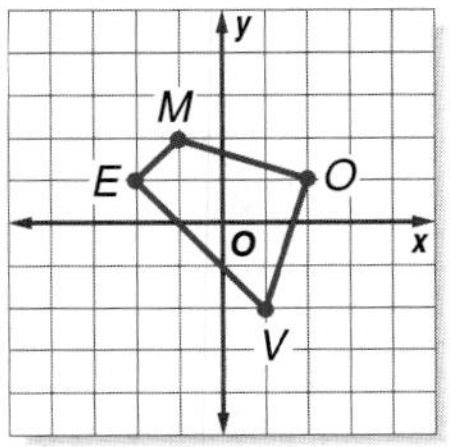

11. Describe the translation that will move M to the point at $(2, -2)$. Then graph quadrilateral $M'O'V'E'$ using this translation. (Lesson 10-8)
12. Find the coordinates of the vertices of quadrilateral *MOVE* after a reflection over the y-axis. Then graph the reflection. (Lesson 10-9)

Mixed Problem Solving

7. $m\angle 1 = 80°$, $m\angle 2 = 100°$, $m\angle 3 = 80°$, $m\angle 4 = 50°$, $m\angle 5 = 130°$, $m\angle 6 = 50°$, $m\angle 7 = 50°$, $m\angle 8 = 130°$, $m\angle 9 = 50°$, $m\angle 10 = 40°$
8. Rectangle; it has four right angles, but the four sides may not be congruent.
10. Sample answer: triangles and quadrilaterals; the sum of the angles where the vertices meet is 360°.

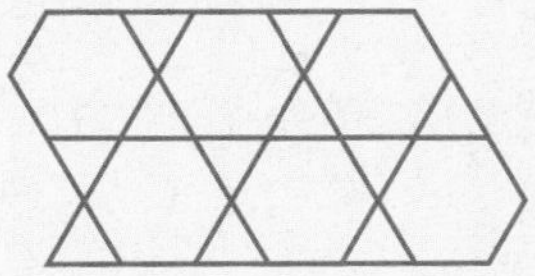

11. 3 units right, 4 units down

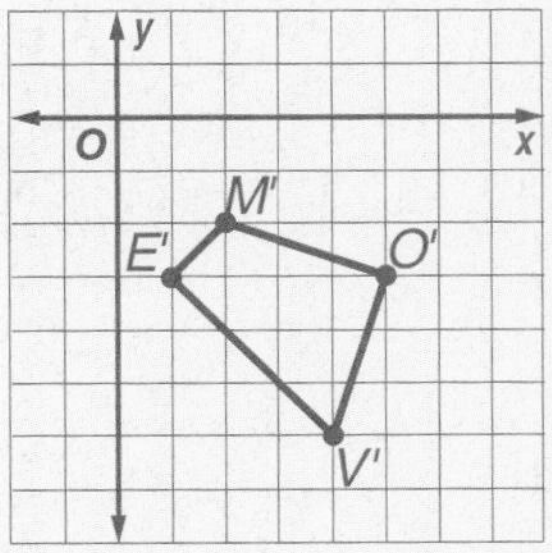

12. $M'(1, 2)$, $O'(-2, 1)$, $V'(-1, -2)$, $E'(2, 1)$

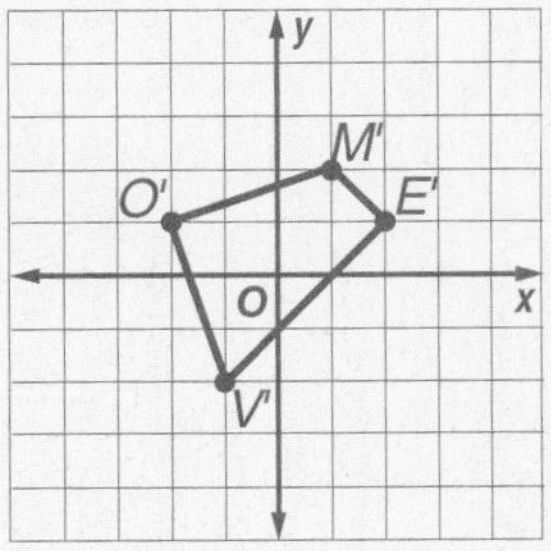

Mixed Problem Solving

Chapter 10

2. Sample answer: on the circle between *A* and *B* or between *A* and *D*

4. Channels that Families Watch

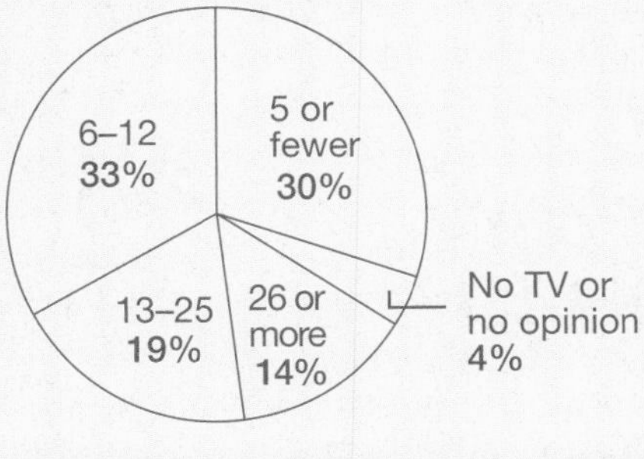

1. **PROPERTY** If a square section of land contains 1,742,400 square feet, what are the side dimensions? (Lesson 11-1) **1,320 ft**

2. **RECORDS** In Japan, a gigantic omelet was made with an area of 1,383 square feet. If the omelet was a square, what would be the side lengths? Round to the nearest tenth. (Lesson 11-2) **37.2 ft**

3. **SOFTBALL** A softball diamond is a square measuring 60 feet on each side.

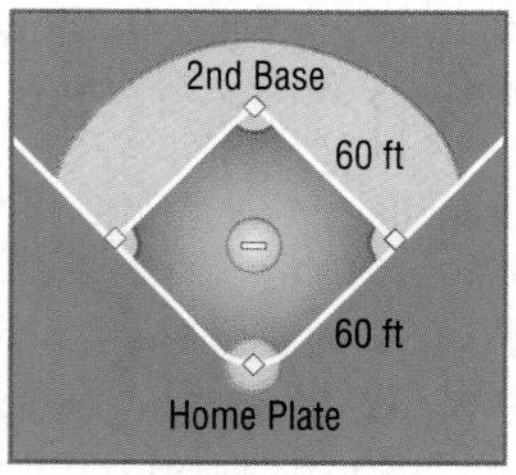

How far does a player on second base throw when she throws from second base to home? Round to the nearest tenth. (Lesson 11-3) **84.9 ft**

4. **BANDS** Mr. Garcia is planning a formation for the middle school band at a football game. The diagram shows the dimensions of the field.

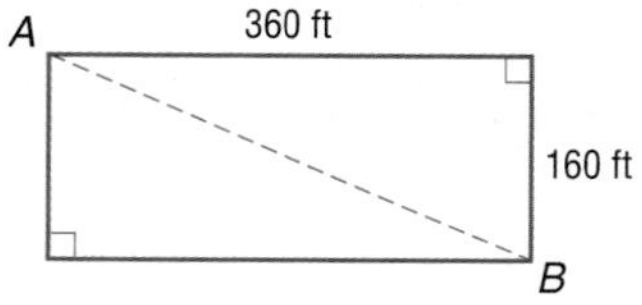

To the nearest foot, what is the distance from *A* to *B*? (Lesson 11-3) **394 ft**

5. **CRAFTS** A quilt pattern uses 25 parallelogram-shaped pieces of fabric, each with a base of 4 inches and a height of $2\frac{1}{2}$ inches. How much fabric is used to make the 25 pieces? (Lesson 11-4) **250 in^2**

6. **FURNITURE** A corner table is in the shape of a right triangle. If the side lengths of the tabletop are 3.5 feet, 3.5 feet, and 4.9 feet, what is the area? Round to the nearest tenth if necessary. (Lesson 11-5) **6.1 ft^2**

7. **PUZZLES** A *tangram* is a square puzzle formed by the seven geometric figures shown at the right. Find the area of each small and large shaded triangle. (Lesson 11-5) **small: 12.5 cm^2, large: 25 cm^2**

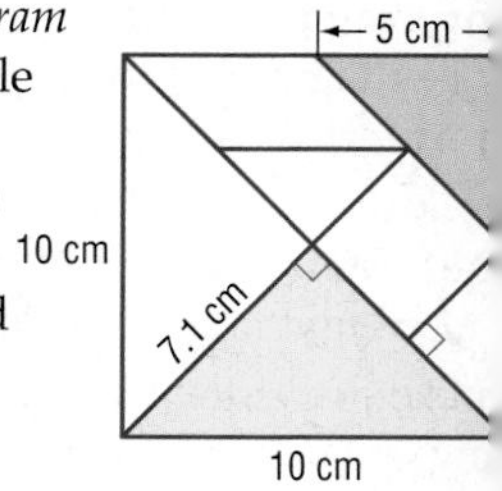

8. **COOKIES** In New Zealand, a giant circu chocolate chip cookie was baked with a diameter of 81 feet 8 inches. To the near square foot, what was the area of the cookie? (Lesson 11-6) **5,238 ft^2**

9. **LANDSCAPING** Find the area of the flower garden shown in the diagram at the right. Round to the nearest square foot. (Lesson 11-7) **3,149 ft^2**

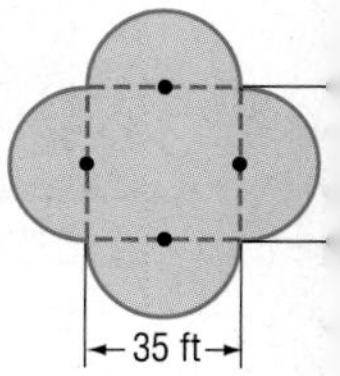

10. **CONSTRUCTION** A design for a deck is shown below. Find the area of the deck. (Lesson 11-7) **82.5 ft^2**

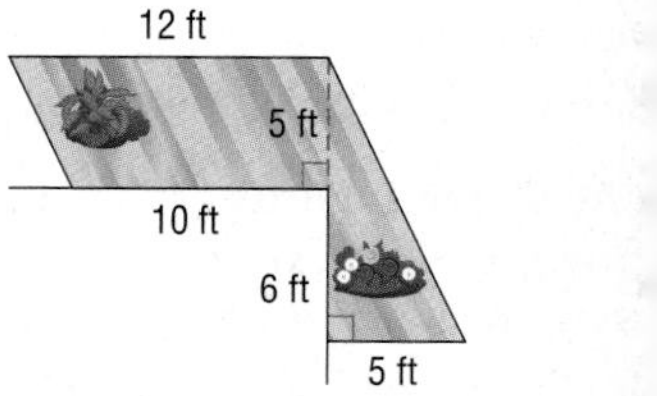

11. **GAMES** At the Moore Middle School carn one game has a dartboard as shown belo

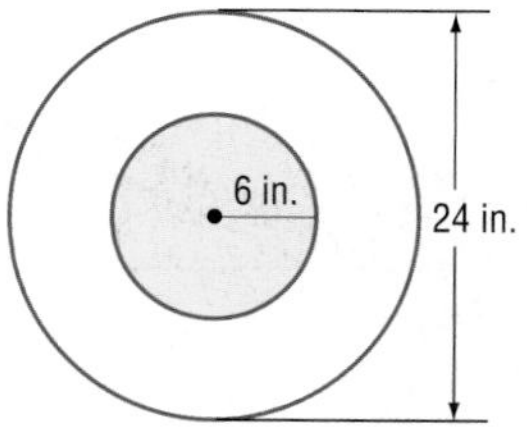

If it is equally likely that a thrown dart land anywhere on the dartboard, find th probability that a dart lands in the shad region. Write as a percent. (Lesson 11-8) **25**

606 Mixed Problem Solving

ORDS According to the *Guinness Book of* *ld Records*, the tallest hotel in the world is 1,053-foot sail-shaped Burj Al Arab in ai, United Arab Emirates.

w possible sketches of the top, side, and t views of the hotel. (Lesson 12-1) **See margin.**

IS For Exercises 2 and 3, use the ing information.
tlantic Ocean has an area of about ,000 square miles. Its average depth is feet. (Lesson 12-2)

he nearest hundredth, what is the rage depth of the Atlantic Ocean in miles? *nt*: 1 mi = 5,280 ft) **2.22 mi**

at is the approximate volume of the ntic Ocean in cubic miles? **74,192,400 mi^3**

R For Exercises 4–6, use the cylinder- d water tank. (Lesson 12-3)

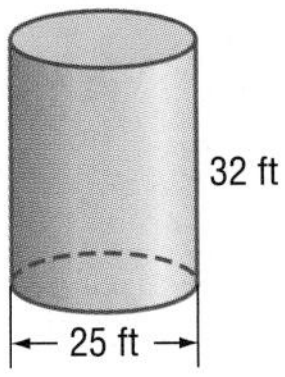

d the volume of the tank. Round to the rest cubic foot. **15,708 ft^3**

gallon of water equals 231 cubic inches water. Find the volume of the water tank allons. **117,504 gal**

ylinder-shaped water tank holds 200,000 ons of water. The height of the tank is 48 . To the nearest tenth of a foot, what is the ius of the tank? **13.3 ft**

For Exercises 7 and 8, use the following information.
Liz is designing some gift boxes. The small box is 6 inches long, 4 inches wide, and 2.5 inches high. The medium box has dimensions that are each 3 times the dimensions of the small box. (Lesson 12-4)

7. Find the surface area of the small box. **98 in^2**
8. What are the dimensions of the medium box? Find the surface area of the medium box. **18 in. by 12 in. by 7.5 in.; 882 in^2**

STORAGE For Exercises 9–11, use the following information.
The two canisters shown below each have a volume of about 628.3 cubic inches. (Lesson 12-5)

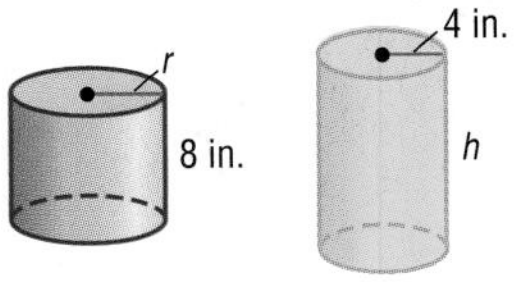

9. What is the radius of the blue canister? Round to the nearest tenth. **5.0 in.**
10. What is the height of the yellow canister? Round to the nearest tenth. **12.5 in.**
11. What is the difference between the surface area of the two canisters? **6.3 in^2**

WATER For Exercises 12 and 13, refer to the water tank shown in Exercise 4. (Lesson 12-5)

12. What is the surface area of the water tank, including the bottom? Round to the nearest tenth. **3,495.0 ft^2**
13. A gallon of paint covers 400 square feet of surface. How many gallons are needed to cover the top and curved surface of the tank? Round to the nearest gallon. **8 gal**

For Exercises 14 and 15, use the following information.
In science class, Yvette and Tyrell use different rulers to measure the length of an earthworm. Using significant digits, Yvette records 13.2 centimeters, and Tyrell records 13.25 centimeters. (Lesson 12-6) **14. 1 cm; The last digit, 2, is estimated, so the precision unit is 1 cm.**

14. What is the most precise unit on Yvette's ruler? Explain.
15. What is the most precise unit on Tyrell's ruler? Explain. **0.1 cm; The last digit, 5, is estimated, so the precision unit is 0.1 cm.**

Mixed Problem Solving

Chapter 12

1. Sample answer:

top side front

Mixed Problem Solving

Page 597, Chapter 2

16. **Tourists in U.S.**

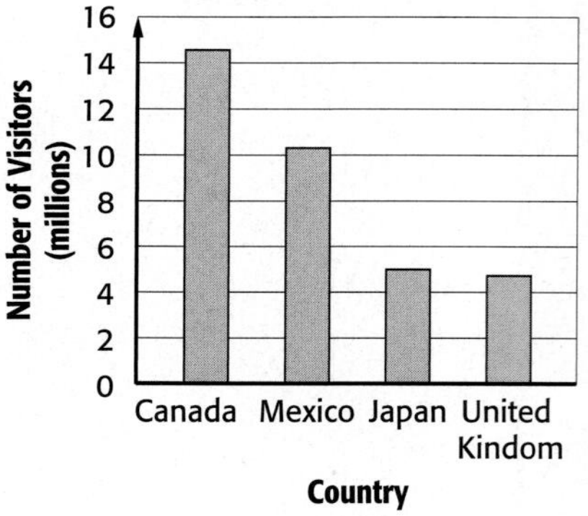

17. **Airport Passengers**

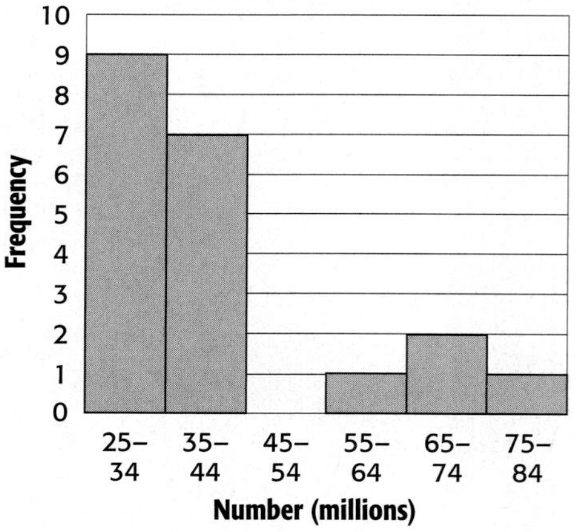

Page 604, Chapter 9

4.

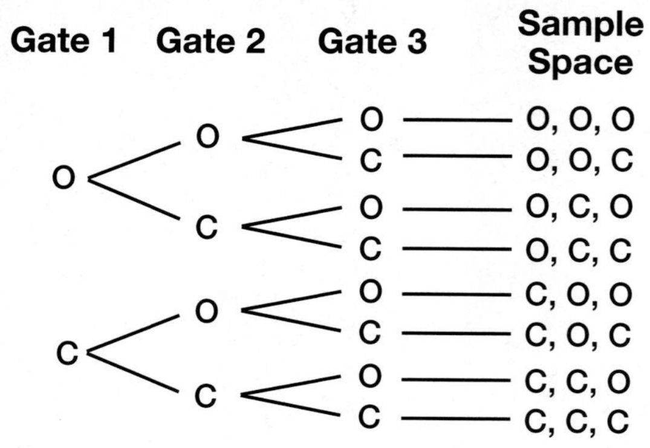

O = Open   C = Closed

NOTES

Glossary/Glosario

Cómo usar el glosario en español:
1. Busca el término en inglés que desees encontrar.
2. El término en español, junto con la definición, se encuentran en la columna de la derecha.

English / Español

A

absolute value (p. 107) The distance the number is from zero on a number line.

valor absoluto Distancia a la que se encuentra un número de cero en la recta numérica.

acute angle (p. 413) An angle with a measure greater than 0° and less than 90°.

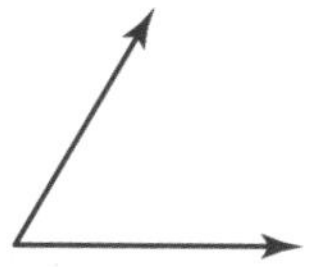

ángulo agudo Ángulo que mide más de 0° y menos de 90°.

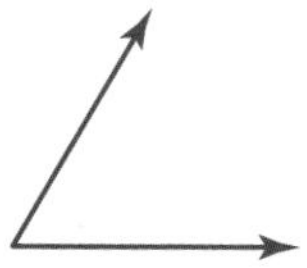

acute triangle (p. 429) A triangle having three acute angles.

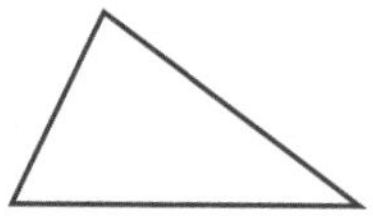

triángulo acutángulo Triángulo con tres ángulos agudos.

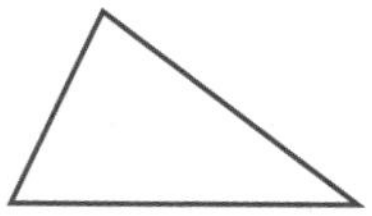

additive inverse (p. 121) The opposite of an integer. The sum of an integer and its additive inverse is zero.

inverso aditivo El opuesto de un entero. La suma de un entero y su inverso aditivo es cero.

algebra (p. 18) The branch of mathematics that involves expressions with variables.

álgebra Rama de las matemáticas que involucra expresiones con variables.

algebraic expression (p. 18) A combination of variables, numbers, and at least one operation.

expresión algebraica Combinación de variables, números y por lo menos una operación.

angle (p. 413) Two rays with a common endpoint form an angle. The rays and vertex are used to name the angle.

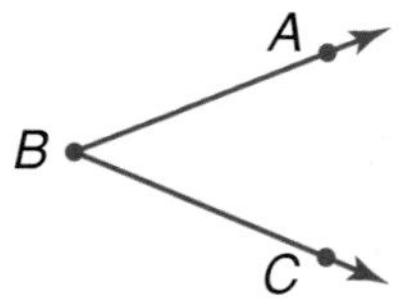

$\angle ABC$, $\angle CBA$, or $\angle B$

ángulo Dos rayos con un extremo común forman un ángulo. Los rayos y el vértice se usan para nombrar el ángulo.

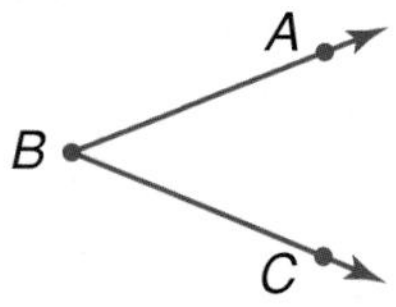

$\angle ABC$, $\angle CBA$, o $\angle B$

area (p. 271) The number of square units needed to cover a surface enclosed by a geometric figure.

área El número de unidades cuadradas necesarias para cubrir una superficie cerrada por una figura geométrica.

arithmetic sequence (p. 34) A sequence in which each term is found by adding the same number to the previous term.

sucesión aritmética Sucesión en que cada término se encuentra sumando el mismo número al término anterior.

Associative Property (p. 31) The way in which three numbers are grouped when they are added or multiplied does not change their sum or product.

propiedad asociativa La manera de agrupar tres números al sumarlos o multiplicarlos no cambia su suma o producto.

B

bar graph (p. 85) A graphic form using bars to make comparisons of statistics.

gráfica de barras Forma gráfica que usa barras para hacer comparaciones estadísticas.

bar notation (p. 211) In repeating decimals, the line or bar placed over the digits that repeat. For example, $2.\overline{63}$ indicates that the digits 63 repeat.

notación de barra Línea o barra que se coloca sobre los dígitos que se repiten en decimales periódicos. Por ejemplo, $2.\overline{63}$ indica que los dígitos 63 se repiten.

base (p. 10) In a power, the number used as a factor. In 10^3, the base is 10. That is, $10^3 = 10 \times 10 \times 10$.

base En una potencia, el número usado como factor. En 10^3, la base es 10. Es decir, $10^3 = 10 \times 10 \times 10$.

base (p. 323) In a percent proportion, the whole quantity, or the number to which the part is being compared.

base En una proporción porcentual, la cantidad total o el número con el que se compara la parte.

base (p. 483) The base of a parallelogram or triangle is any side of the figure. The bases of a trapezoid are the parallel sides.

base La base de un paralelogramo o triángulo es el lado de la figura. Las bases de un trapecio son los lados paralelos

box-and-whisker plot (p. 80) A diagram that summarizes data using the median, the upper and lower quartiles, and the extreme values. A box is drawn around the quartile values and whiskers extend from each quartile to the extreme data points.

diagrama de caja y patillas Diagrama que resume información usando la mediana, los cuartiles superior e inferior y los valores extremos. Se dibuja una caja alrededor de los cuartiles y se trazan patillas que los unan a los valores extremos respectivos.

C

center (p. 275) The given point from which all points on a circle or sphere are the same distance.

centro Un punto dado del cual equidistan todos los puntos de un círculo o de una esfera.

circle (p. 275) The set of all points in a plane that are the same distance from a given point called the center.

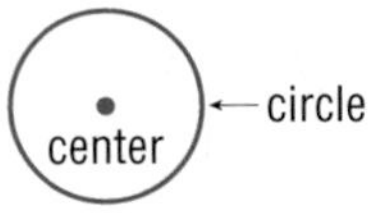

círculo Conjunto de todos los puntos en un plano que equidistan de un punto dado llamado centro.

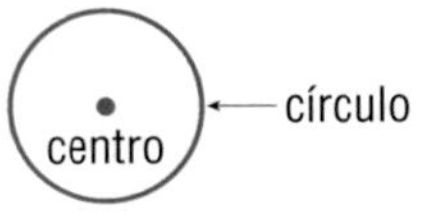

circle graph (p. 418) A type of statistical graph used to compare parts of a whole.

gráfica circular Tipo de gráfica estadística que se usa para comparar las partes de un todo.

circumference (p. 275) The distance around a circle.

circunferencia La distancia alrededor de un círculo.

cluster (p. 65) Data that are grouped closely together.

agrupamiento Datos estrechamente agrupados.

coefficient (p. 19) The numerical factor of a term that contains a variable.

coeficiente El factor numérico de un término que contiene una variable.

combination (p. 387) An arrangement, or listing, of objects in which order is not important.

combinación Arreglo o lista de objetos donde el orden no es importante.

common denominator (p. 227) A common multiple of the denominators of two or more fractions. 24 is a common denominator for $\frac{1}{3}$, $\frac{5}{8}$, and $\frac{3}{4}$ because 24 is the LCM of 3, 8, and 4.

común denominador El múltiplo común de los denominadores de dos o más fracciones. 24 es un denominador común para $\frac{1}{3}$, $\frac{5}{8}$ y $\frac{3}{4}$ porque 24 es el mcm de 3, 8 y 4.

Commutative Property (p. 31) The order in which two numbers are added or multiplied does not change their sum or product.

propiedad conmutativa El orden en que se suman o multiplican dos números no afecta su suma o producto.

compatible numbers (p. 242) Numbers that are easy to compute mentally.

números compatibles Números que son fáciles de computar mentalmente.

complementary angles (p. 422) Two angles are complementary if the sum of their measures is 90°.

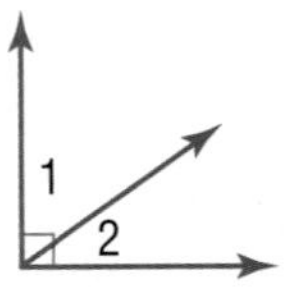

ángulos complementarios Dos ángulos son complementarios si la suma de sus medidas es 90°.

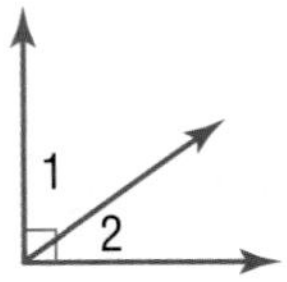

complementary events (p. 371) The events of one outcome happening and that outcome not happening are complementary events. The sum of the probabilities of complementary events is 1.

complex figure (p. 498) A figure made of circles, rectangles, squares, and other two-dimensional figures.

compound event (p. 398) An event consisting of two or more simple events.

composite number (p. 197) A whole number greater than 1 that has more than two factors.

congruent angles (p. 422) Angles that have the same measure.

congruent segments (p. 429) Sides of a triangle having the same length.

constant (p. 19) A term that does not contain a variable.

coordinate grid (p. 112) A plane in which a horizontal number line and a vertical number line intersect at their zero points. Also called a coordinate plane.

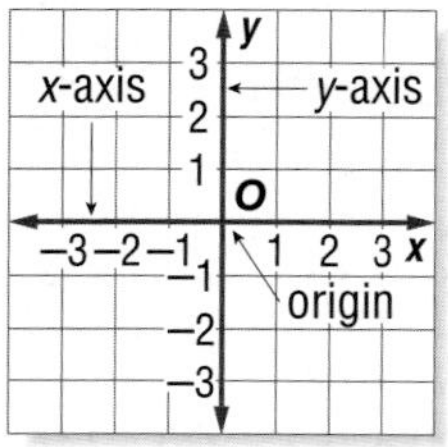

coordinate plane (p. 112) See coordinate grid.

cross product (p. 297) In a proportion, a cross product is the product of the numerator of one ratio and the denominator of the other ratio.

cubed (p. 10) The product in which a number is a factor three times. Two cubed is 8 because $2 \times 2 \times 2 = 8$.

cup (p. 267) A customary unit of capacity equal to 8 fluid ounces.

cylinder (p. 524) A three-dimensional figure with two parallel congruent circular bases.

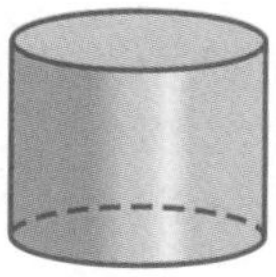

eventos complementarios Se dice de los eventos de un resultado que ocurren y el resultado que no ocurre. La suma de las probabilidades de eventos complementarios es 1.

figura compleja Una figura compuesta por círculos, rectángulos, cuadrados y otras dos figuras bidimencionales.

evento compuesto Un evento que consiste en dos o más eventos simples.

número compuesto Un número entero mayor que 1 que tiene más de dos factores.

ángulos congruentes Ángulos que tienen la misma medida.

segmentos congruentes Los lados de un triángulo que tienen la misma longitud.

constante Término que no contiene una variable.

sistema de coordenadas Plano en el cual se han trazado dos rectas numéricas, una horizontal y una vertical, que se intersecan en sus puntos cero. También conocido como plano de coordenadas.

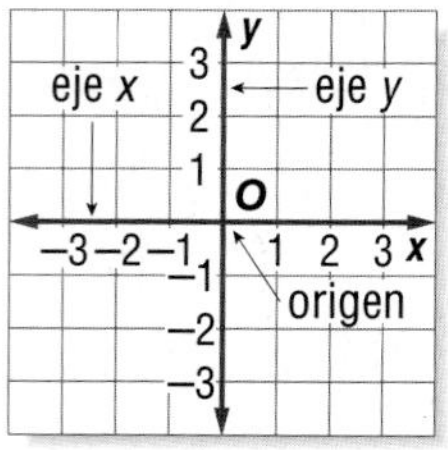

plano de coordenadas Véase sistema de coordenadas.

productos cruzados En una proporción, un producto cruzado es el producto del numerador de una razón y el denominador de la otra razón.

al cubo El producto de un número por sí mismo, tres veces. Dos al cubo es 8 porque $2 \times 2 \times 2 = 8$.

taza Unidad de capacidad del sistema inglés de medidas que equivale a 8 onzas líquidas.

cilindro Figura tridimensional que tiene dos bases circulares congruentes y paralelas.

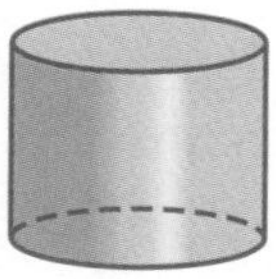

D

data (p. 54) Pieces of information, often numerical, which are gathered for statistical purposes.

datos Información, a menudo numérica, que se recoge con fines estadísticos.

decagon (p. 446) A polygon having ten sides.

defining the variable (p. 25) Choosing a variable to represent an unknown value in a problem, and using it to write an expression or equation to solve the problem.

degrees (p. 413) The most common unit of measure for angles. If a circle were divided into 360 equal-sized parts, each part would have an angle measure of 1 degree.

dependent events (p. 399) Two or more events in which the outcome of one event does affect the outcome of the other event(s).

diameter (p. 275) The distance across a circle through its center.

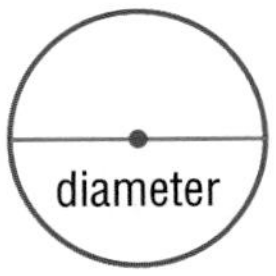

discount (p. 355) The amount by which the regular price of an item is reduced.

Distributive Property (p. 30) To multiply a sum by a number, multiply each addend of the sum by the number outside the parentheses.

domain (p. 177) The set of input values for a function.

E

equation (p. 24) A mathematical sentence that contains an equals sign, =.

equilateral triangle (p. 429) A triangle having three congruent sides.

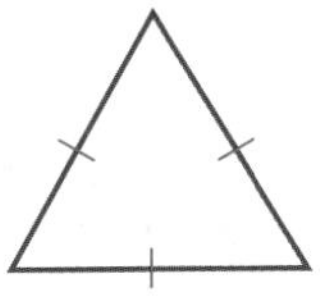

equivalent expressions (p. 30) Expressions that have the same value.

equivalent fractions (p. 207) Fractions that have the same value. $\frac{2}{3}$ and $\frac{4}{6}$ are equivalent fractions.

equivalent ratios (p. 289) Two ratios that have the same value.

evaluate (p. 11) To find the value of an expression.

experimental probability (p. 393) An estimated probability based on the relative frequency of positive outcomes occurring during an experiment.

exponent (p. 10) In a power, the number that tells how many times the base is used as a factor. In 5^3, the exponent is 3. That is, $5^3 = 5 \times 5 \times 5$.

exponential form (p. 11) Numbers written with exponents.

decágono Un polígono con diez lados.

definir una variable El elegir una variable para representar un valor desconocido en un problema y usarla para escribir una expresión o ecuación para resolver el problema.

grados La unidad más común para medir ángulos. Si un círculo se divide en 360 partes iguales, cada parte tiene una medida angular de 1 grado.

eventos dependientes Dos o más eventos en que el resultado de uno de ellos afecta el resultado de los otros eventos.

diámetro La distancia a través de un círculo pasando por el centro.

descuento La cantidad de reducción del precio normal.

propiedad distributiva Para multiplicar una suma por un número, multiplica cada sumando de la suma por el número fuera del paréntesis.

dominio El conjunto de valores de entrada de una función.

ecuación Enunciado matemático que contiene un signo de igualdad, =.

triángulo equilátero Triángulo con tres lados congruentes.

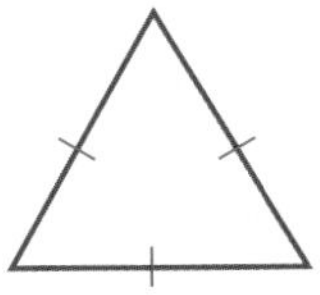

expresiones equivalentes Expresiones que tienen el mismo valor.

fracciones equivalentes Fracciones que tienen el mismo valor. $\frac{2}{3}$ y $\frac{4}{6}$ son fracciones equivalentes.

razones equivalentes Dos razones que tienen el mismo valor.

evaluar Calcular el valor de una expresión.

probabilidad experimental Estimado de una probabilidad que se basa en la frecuencia relativa de los resultados positivos que ocurren durante un experimento.

exponente En una potencia, el número que indica las veces que la base se usa como factor. En 5^3, el exponente es 3. Es decir, $5^3 = 5 \times 5 \times 5$.

forma exponencial Números escritos usando exponentes.

F

factorial (p. 381) The expression $n!$ is the product of all counting numbers beginning with n and counting backward to 1.

factors (p. 10) Two or more numbers that are multiplied together to form a product.

factor tree (p. 198) A diagram showing the prime factorization of a number. The factors branch out from the previous factors until all of the factors are prime numbers.

fair game (p. 374) A game in which players of equal skill have an equal chance of winning.

formula (p. 271) An equation that shows a relationship among certain quantities.

frequency table (p. 54) A table for organizing a set of data that shows the number of pieces of data that fall within given intervals or categories.

function (p. 177) A relation in which each element of the input is paired with exactly one element of the output according to a specified rule.

function table (p. 177) A table used to organize the input numbers, output numbers, and the function rule.

Fundamental Counting Principle (p. 378) Uses multiplication of the number of ways each event in an experiment can occur to find the number of possible outcomes in a sample space.

factorial La expresión $n!$ es el producto de todos los números naturales, comenzando con n y contando al revés hasta 1.

factores Dos o más números que se multiplican entre sí para formar un producto.

diagrama de árbol Diagrama que muestra la factorización prima de un número. Los factores se ramifican a partir de los factores previos hasta que todos los factores son números primos.

juego justo Juego en que jugadores con la misma habilidad tienen igual oportunidad de ganar.

fórmula Ecuación que muestra una relación entre ciertas cantidades.

tabla de frecuencias Tabla que se usa para organizar un conjunto de datos y que muestra cuántas piezas de datos caen dentro de intervalos o categorías dadas.

función Relación en que cada elemento de entrada es apareado con un único elemento de salida, según una regla específica.

tabla de funciones Tabla que organiza las entradas, la regla y las salidas de una función.

Principio Fundamental de Contar Este principio usa la multiplicación del número de veces que puede ocurrir cada evento en un experimento para calcular el número de posibles resultados en un espacio muestral.

G

gallon (p. 267) A customary unit of capacity equal to four quarts.

geometric sequence (p. 34) A sequence in which each term can be found by multiplying the previous term by the same number.

gram (p. 39) A unit of mass in the metric system equivalent to 0.001 kilogram.

graph (p. 106) The process of placing a point on a number line at its proper location.

greatest common factor (GCF) (p. 203) The greatest of the common factors of two or more numbers. The GCF of 18 and 24 is 6.

galón Unidad de capacidad del sistema inglés de medidas que equivale a 4 cuartos de galón.

sucesión geométrica Sucesión en que cada término se puede calcular multiplicando el término anterior por el mismo número.

gramo Unidad de masa del sistema métrico. Un gramo equivale a 0.001 de kilogramo.

graficar Proceso de dibujar o trazar un punto en una recta numérica en su ubicación correcta.

máximo común divisor (MCD) El mayor factor común de dos o más números. El MCD de 18 y 24 es 6.

H

height (p. 483) The length of the segment perpendicular to the base with endpoints on opposite sides. In a triangle, the distance from a base to the opposite vertex.

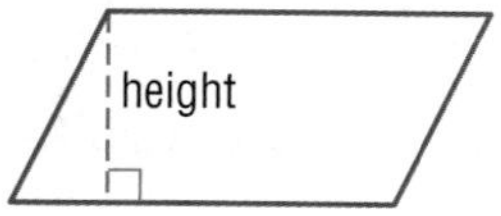

altura Longitud del segmento perpendicular a la base y con extremos en lados opuestos. En un triángulo, es la distancia desde una base al vértice opuesto.

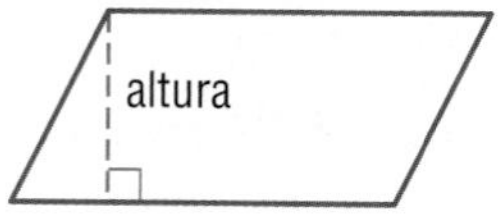

heptagon (p. 446) A polygon having seven sides.

hexagon (p. 446) A polygon having six sides.

histogram (p. 86) A special kind of bar graph in which the bars are used to represent the frequency of numerical data that have been organized in intervals.

hypotenuse (p. 479) The side opposite the right angle in a right triangle.

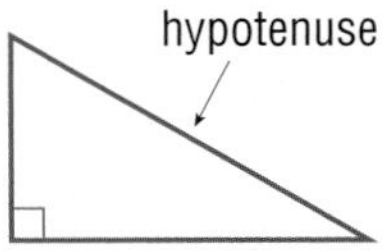

heptágono Polígono con siete lados.

hexágono Polígono con seis lados.

histograma Tipo especial de gráfica de barras que usa barras para representar la frecuencia de los datos numéricos, los cuales han sido organizados en intervalos iguales.

hipotenusa El lado opuesto al ángulo recto en un triángulo rectángulo.

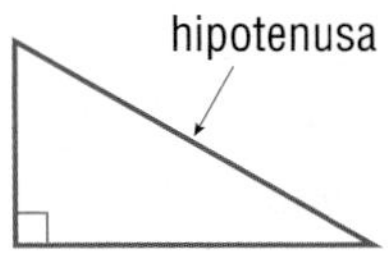

I

Identity Property (p. 31) The sum of an addend and zero is the addend. The product of a factor and one is the factor.

indirect measurement (p. 441) Finding a measurement by using similar triangles and writing a proportion.

integer (p. 106) The whole numbers and their opposites.
$\ldots, -3, -2, -1, 0, 1, 2, 3, \ldots$

independent events (p. 398) Two or more events in which the outcome of one event does not affect the outcome of the other event(s).

inequality (p. 172) A mathematical sentence that contains the symbols $<, >, \leq$, or $\geq$.

interquartile range (p. 81) The range of the middle half of a set of numbers.
Interquartile range = UQ − LQ

interval (p. 54) On a scale, the difference between the greatest and least values in each category.

inverse operations (p. 156) Operations that "undo" each other. Addition and subtraction are inverse operations.

irrational number (p. 476) A number that cannot be written as a fraction.

isosceles triangle (p. 429) A triangle having at least two congruent sides.

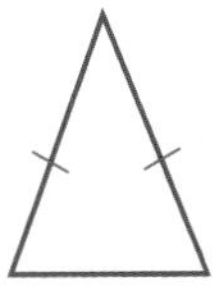

propiedad de identidad La suma de un sumando y cero es el sumando mismo. El producto de un factor y uno es el factor mismo.

medida indirecta Técnica que se usa para calcular una medida a partir de triángulos semejantes y proporciones.

entero Los números enteros y sus opuestos.
$\ldots, -3, -2, -1, 0, 1, 2, 3, \ldots$

eventos independientes Dos o más eventos en los cuales el resultado de uno de ellos no afecta el resultado de los otros eventos.

desigualdad Enunciado matemático que contiene los símbolos $<, >, \leq$ o $\geq$.

amplitud intercuartílica El rango de la mitad central de un conjunto de datos o números. La amplitud intercuartílica = CS − CI

intervalo En una escala, la diferencia entre el valor mayor y el menor en cada categoría.

operaciones inversas Operaciones que se "anulan" mutuamente. La adición y la sustracción son operaciones inversas.

número irracional Número que no se puede escribir como fracción.

triángulo isósceles Triángulo que tiene por lo menos dos lados congruentes.

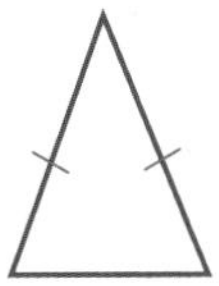

K

kilogram (p. 39) The base unit of mass in the metric system equivalent to 1,000 grams.

kilogramo Unidad fundamental de masa del sistema métrico. Un kilogramo equivale a mil gramos.

L

leaf (p. 76) The second greatest place value of data in a stem-and-leaf plot.

least common denominator (LCD) (p. 227) The least common multiple of the denominators of two or more fractions.

least common multiple (LCM) (p. 224) The least of the common multiples of two or more numbers. The LCM of 2 and 3 is 6.

leg (p. 479) Either of the two sides that form the right angle of a right triangle.

linear equation (p. 178) An equation for which the graph is a straight line.

line graph (p. 60) A type of statistical graph using lines to show how values change over a period of time.

line of symmetry (p. 456) A line that divides a figure into two halves that are reflections of each other.

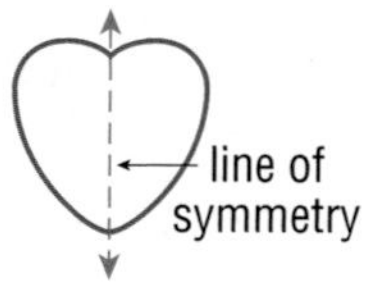

line plot (p. 64) A graph that uses an × above a number on a number line each time that number occurs in a set of data.

line symmetry (p. 456) Figures that match exactly when folded in half have line symmetry.

liter (p. 39) The base unit of capacity in the metric system. A liter is a little more than a quart.

lower extreme (p. 80) The least number of a set of data.

lower quartile (p. 80) The median of the lower half of a set of numbers, indicated by LQ.

hoja El segundo valor de posición mayor en un diagrama de tallo y hojas.

mínimo común denominador (mcd) El menor múltiplo común de los denominadores de dos o más fracciones.

mínimo común múltiplo (mcm) El menor múltiplo común de dos o más números. El mcm de 2 y 3 es 6.

cateto Cualquiera de los lados que forman el ángulo recto de un triángulo rectángulo.

ecuación lineal Ecuación cuya gráfica es una recta.

gráfica lineal Tipo de gráfica estadística que usa segmentos de recta para mostrar cómo cambian los valores durante un período de tiempo.

eje de simetría Recta que divide una figura en dos mitades que son reflexiones entre sí.

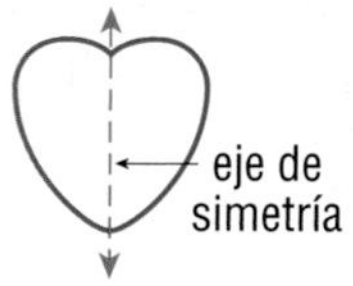

esquema lineal Gráfica que usa una recta numérica y una × sobre un número en la recta numérica, cada vez que el número aparece en un conjunto de datos.

simetría lineal Exhiben simetría lineal las figuras que coinciden exactamente al doblarse una sobre otra.

litro Unidad básica de capacidad del sistema métrico. Un litro es un poco más de un cuarto de galón.

extremo inferior El número menor de un conjunto de datos.

cuartil inferior La mediana de la mitad inferior de un conjunto de datos, la cual se denota por CI.

M

mean (p. 69) The sum of the data divided by the number of items in the data set.

measures of central tendency (p. 69) Numbers that are used to describe the center of a set of data. These measures include the mean, median, and mode.

median (p. 70) The middle number in a set of data when the data are arranged in numerical order. If the data has an even number, the median is the mean of the two middle numbers.

meter (p. 38) The base unit of length in the metric system.

metric system (p. 38) A base-ten system of measurement using the base units: meter for length, kilogram for mass, and liter for capacity.

media La suma de los datos dividida entre el número total de artículos en el conjunto de datos.

medidas de tendencia central Números que se usan para describir el centro de un conjunto de datos. Estas medidas incluyen la media, la mediana y la moda.

mediana Número central de un conjunto de datos, una vez que los datos han sido ordenados numéricamente. Si hay un número par de datos, la mediana es el promedio de los dos datos centrales.

metro Unidad fundamental de longitud del sistema métrico.

sistema métrico Sistema de medidas de base diez que usa las unidades fundamentales: metro para longitud, kilogramo para masa y litro para capacidad.

mode (p. 70) The number or numbers that appear most often in a set of data.

moda Número o números que aparece(n) más frecuentemente en un conjunto de datos.

multiple (p. 224) The product of a number and any whole number.

múltiplo El producto de un número y cualquier número entero.

multiplicative inverse (p. 258) The product of a number and its multiplicative inverse is 1. The multiplicative inverse of $\frac{2}{3}$ is $\frac{3}{2}$.

inverso multiplicativo El producto de un número y su inverso multiplicativo es 1. El inverso multiplicativo de $\frac{2}{3}$ es $\frac{3}{2}$.

N

negative integer (p. 106) An integer that is less than zero.

entero negativo Un entero menor que cero.

nonagon (p. 446) A polygon having nine sides.

enágono Polígono que tiene nueve lados.

numerical expression (p. 14) A combination of numbers and operations.

expresión numérica Combinación de números y operaciones.

O

obtuse angle (p. 413) Any angle that measures greater than 90° but less than 180°.

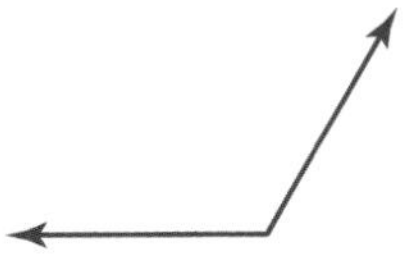

ángulo obtuso Cualquier ángulo que mide más de 90° pero menos de 180°.

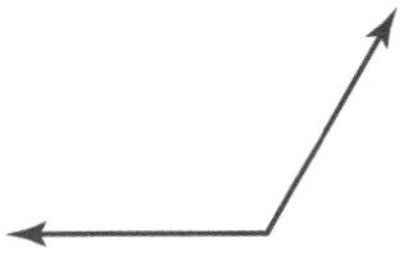

obtuse triangle (p. 429) A triangle having one obtuse angle.

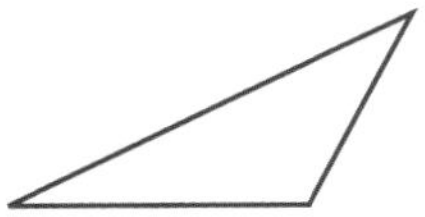

triángulo obtusángulo Triángulo que tiene un ángulo obtuso.

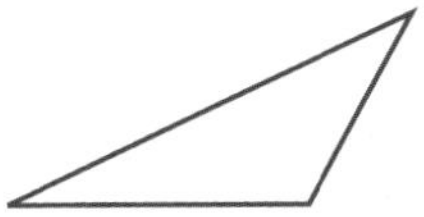

octagon (p. 446) A polygon having eight sides.

octágono Polígono que tiene ocho lados.

opposites (p. 121) Two integers are opposites if they are represented on the number line by points that are the same distance from zero, but on opposite sides of zero. The sum of two opposites is zero.

opuestos Dos enteros son opuestos si, en la recta numérica, están representados por puntos que equidistan de cero, pero en direcciones opuestas. La suma de dos opuestos es cero.

order of operations (p. 14) The rules to follow when more than one operation is used in a numerical expression.
1. Do all operations within grouping symbols first.
2. Evaluate all powers before other operations.
3. Multiply and divide in order from left to right.
4. Add and subtract in order from left to right.

orden de operaciones Reglas a seguir cuando se usa más de una operación en una expresión numérica.
1. Primero ejecuta todas las operaciones dentro de los símbolos de agrupamiento.
2. Evalúa todas las potencias antes que las otras operaciones.
3. Multiplica y divide en orden de izquierda a derecha.
4. Suma y resta en orden de izquierda a derecha.

ordered pair (p. 112) A pair of numbers used to locate a point in the coordinate plane. An ordered pair is written in the form (x-coordinate, y-coordinate).

par ordenado Par de números que se utiliza para ubicar un punto en un plano de coordenadas. Se escribe de la siguiente forma: (coordenada x, coordenada y).

origin (p. 112) The point at which the x-axis and the y-axis intersect in a coordinate plane.

origen Punto en que el eje x y el eje y se intersecan en un plano de coordenadas.

ounce (p. 267) A customary unit of weight. 16 ounces equals 1 pound.

outcome (p. 370) One possible result of a probability event. For example, 4 is an outcome when a number cube is rolled.

outlier (p. 65) A piece of data that is quite separated from the rest of the data. In a box-and-whisker plot, data that are more than 1.5 times the interquartile range from the quartiles.

P

parallelogram (p. 434) A quadrilateral with opposite sides parallel and opposite sides congruent.

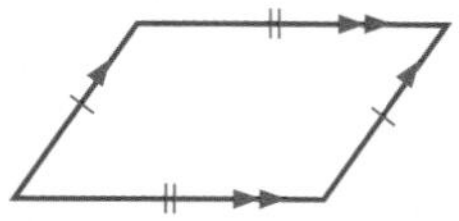

part (p. 323) In a percent proportion, the number that is compared to the whole quantity.

pentagon (p. 446) A polygon having five sides.

percent (p. 216) A ratio that compares a number to 100.

percent equation (p. 340) An equation that describes the relationship between the part, base, and percent.
part = percent · base

percent of change (p. 350) A ratio that compares the change in a quantity to the original amount.

percent of decrease (p. 350) A percent of change when the original quantity decreased.

percent of increase (p. 350) A percent of change when the original quantity increased.

percent proportion (p. 323) Compares part of a quantity to the whole quantity using a percent.

$$\frac{\text{part}}{\text{base}} = \frac{\text{percent}}{100}$$

perfect squares (p. 471) Numbers whose square root is a whole number. 25 is a perfect square because the square root of 25 is 5.

perimeter (p. 270) The distance around a closed geometric figure.

permutation (p. 381) An arrangement, or listing, of objects in which order is important.

pint (p. 267) A customary unit of capacity equal to two cups.

polygon (p. 446) A simple closed figure in a plane formed by three or more line segments.

population (p. 345) The entire group of items or individuals from which the samples under consideration are taken.

positive integer (p. 106) An integer that is greater than zero.

onza Unidad de peso del sistema inglés de medidas. 16 onzas equivalen a una libra.

resultado Uno de los resultados posibles de un evento probabilístico. Por ejemplo, 4 es un resultado posible cuando se lanza un dado.

valor atípico Dato que se encuentra muy separado del resto de los datos. En un diagrama de caja y patillas, los datos que distan de los cuartiles respectivos más de 1.5 veces la amplitud intercuartílica.

paralelogramo Cuadrilátero cuyos lados opuestos son paralelos y congruentes.

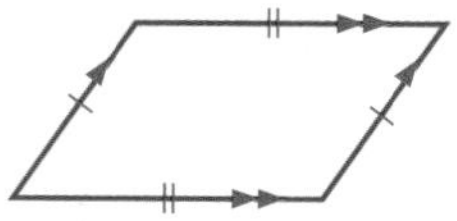

parte En una proporción porcentual, el número que se compara con la cantidad total.

pentágono Polígono que tiene cinco lados.

por ciento Razón que compara un número con 100.

ecuación porcentual Ecuación que describe la relación entre la parte, la base y el por ciento.
parte = por ciento · base

porcentaje de cambio Razón que compara el cambio en una cantidad, con la cantidad original.

porcentaje de disminución Porcentaje de cambio cuando disminuye la cantidad original.

porcentaje de aumento Porcentaje de cambio cuando aumenta la cantidad original.

proporción porcentual Compara parte de una cantidad con la cantidad total mediante un por ciento.

$$\frac{\text{parte}}{\text{base}} = \frac{\text{por ciento}}{100}$$

cuadrados perfectos Números cuya raíz cuadrada es un número entero. 25 es un cuadrado perfecto porque la raíz cuadrada de 25 es 5.

perímetro La distancia alrededor de una figura geométrica cerrada.

permutación Arreglo o lista en que el orden es importante.

pinta Unidad de capacidad del sistema inglés de medidas que equivale a dos tazas.

polígono Figura simple cerrada en un plano, formada por tres o más segmentos de recta.

población El grupo total de individuos o de artículos del cual se toman las muestras bajo estudio.

entero positivo Un entero mayor que cero.

pound (p. 267) A customary unit of weight equal to 16 ounces.

powers (p. 10) Numbers expressed using exponents. The power 3^2 is read *three to the second power,* or *three squared.*

precision (p. 542) The exactness of a measurement which depends on the unit of measure.

precision unit (p. 542) The smallest unit on a measuring tool.

prime factorization (p. 198) Expressing a composite number as a product of prime numbers. For example, the prime factorization of 63 is $3 \times 3 \times 7$.

prime number (p. 197) A whole number greater than 1 that has exactly two factors, 1 and itself.

principal (p. 358) The amount of an investment or a debt.

probability (p. 370) The chance that some event will happen. It is the ratio of the number of ways a certain event can occur to the number of possible outcomes.

properties (p. 31) Statements that are true for any number or variable.

proportion (p. 297) An equation that shows that two ratios are equivalent.

Pythagorean Theorem (p. 479) In a right triangle, the square of the length of the hypotenuse is equal to the sum of the squares of the lengths of the legs. $c^2 = a^2 + b^2$

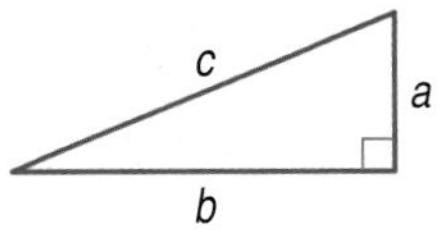

Q

quadrant (p. 113) One of the four regions into which the two perpendicular number lines of the coordinate plane separate the plane.

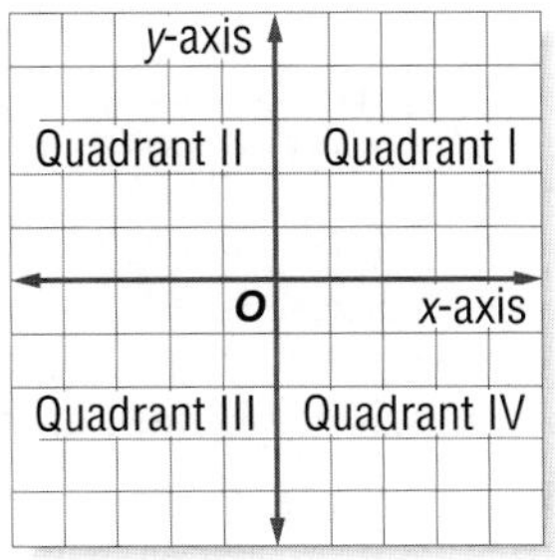

quadrilateral (p. 434) A closed figure having four sides and four angles.

quart (p. 267) A customary unit of capacity equal to two pints.

libra Unidad de peso del sistema inglés de medidas que equivale a 16 onzas.

potencias Números que se expresan usando exponentes. La potencia 3^2 se lee *tres a la segunda potencia* o *tres al cuadrado.*

precisión El grado de exactitud de una medida, lo cual depende de la unidad de medida.

unidad de precisión La unidad más pequeña de un instrumento de medición.

factorización prima Escritura de un número compuesto como el producto de números primos. La factorización prima de 63 es $3 \times 3 \times 7$.

número primo Número entero mayor que 1 que sólo tiene dos factores, 1 y sí mismo.

capital La cantidad de una inversión o una deuda.

probabilidad La posibilidad de que suceda un evento. Es la razón del número de maneras en que puede ocurrir un evento al número total de resultados posibles.

propiedades Enunciados que se cumplen para cualquier número o variable.

proporción Ecuación que afirma la igualdad de dos razones.

Teorema de Pitágoras En un triángulo rectángulo, el cuadrado de la longitud de la hipotenusa es igual a la suma de los cuadrados de las longitudes de los catetos. $c^2 = a^2 + b^2$

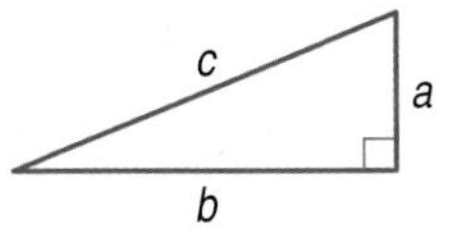

cuadrante Una de las cuatro regiones en que dos rectas numéricas perpendiculares dividen el plano de coordenadas.

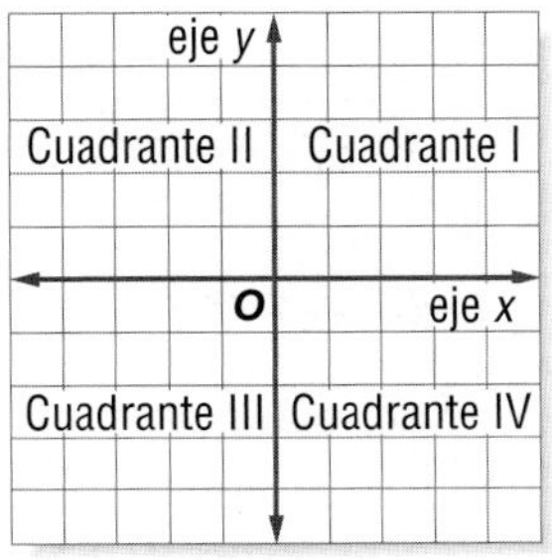

cuadrilátero Figura cerrada que tiene cuatro lados y cuatro ángulos.

cuarto de galón Unidad de capacidad del sistema inglés de medidas que equivale a dos pintas.

R

radical sign (p. 471) The symbol used to indicate a nonnegative square root, $\sqrt{\ }$.

radius (p. 275) The distance from the center of a circle to any point on the circle.

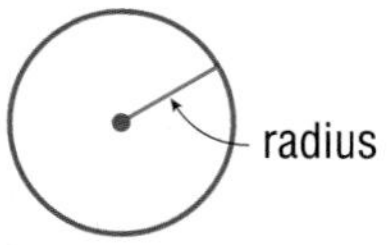

random (p. 370) Outcomes occur at random if each outcome is equally likely to occur.

random sample (p. 345) A sample is called random if the members of the sample are selected purely on the basis of chance.

range (p. 65) The difference between the greatest and least numbers in a data set.

range (p. 177) The set of output values for a function.

rate (p. 292) A ratio that compares two quantities with different kinds of units.

ratio (p. 216) A comparison of two numbers by division. The ratio of 2 to 3 can be written as 2 out of 3, 2 to 3, 2:3, or $\frac{2}{3}$.

rational numbers (p. 229) Numbers that can be written as fractions, including terminating and repeating decimals, and integers.

reciprocal (p. 258) The multiplicative inverse of a number.

rectangle (p. 434) A parallelogram having four right angles.

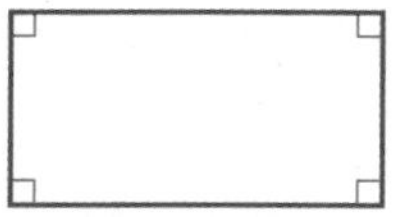

rectangular prism (p. 520) A solid figure that has two parallel and congruent bases that are rectangles.

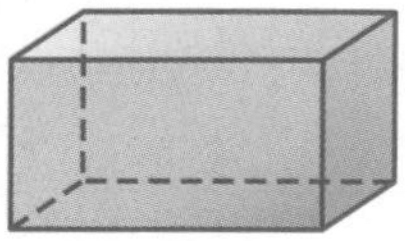

reflection (p. 457) A type of transformation where a figure is flipped over a line of symmetry.

regular polygon (p. 446) A polygon that has all sides congruent and all angles congruent.

repeating decimals (p. 211) A decimal whose digits repeat in groups of one or more. Examples are 0.181818… and 0.83333… .

signo radical Símbolo que se usa para indicar una raíz cuadrada no negativa, $\sqrt{\ }$.

radio Distancia desde el centro de un círculo hasta cualquier punto del mismo.

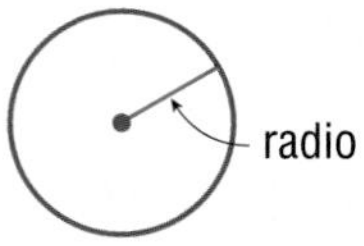

aleatorio Un resultado ocurre al azar si la posibilidad de ocurrir de cada resultado es equiprobable.

muestra aleatoria Se dice que una muestra es aleatoria si sus miembros sólo se eligen aleatoriamente.

rango La diferencia entre el número mayor y el menor en un conjunto de datos.

rango Conjunto de los valores de salida de una función.

tasa Razón que compara dos cantidades que tienen distintas unidades de medida.

razón Comparación de dos números mediante división. La razón de 2 a 3 puede escribirse como 2 de cada 3, 2 a 3, 2:3 ó $\frac{2}{3}$.

números racionales Números que pueden escribirse como fracciones, incluyendo decimales terminales y periódicos y enteros.

recíproco El inverso multiplicativo de un número.

rectángulo Paralelogramo con cuatro ángulos rectos.

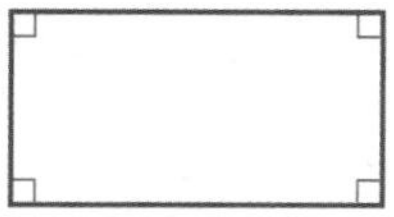

prisma rectangular Figura sólida con dos bases paralelas y congruentes que son rectángulos.

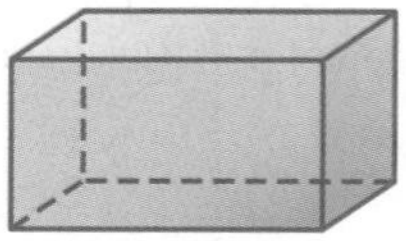

reflexión Transformación en que a una figura se le da vuelta de campana por encima de una recta.

polígono regular Polígono con todos los lados y todos los ángulos congruentes.

decimales periódicos Decimal cuyos dígitos se repiten en grupos de uno o más. Por ejemplo: 0.181818… y 0.83333… .

rhombus (p. 434) A parallelogram having four congruent sides.

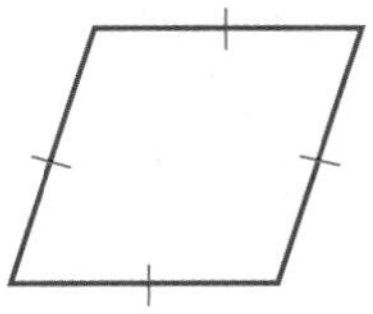

rombo Paralelogramo que tiene cuatro lados congruentes.

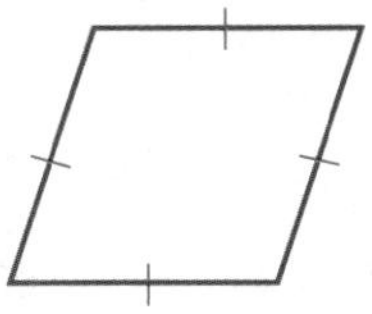

right angle (p. 413) An angle which measures exactly 90°.

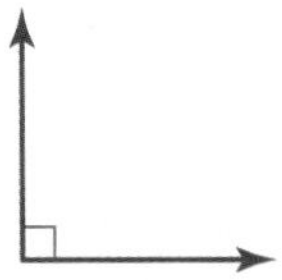

ángulo recto Ángulo que mide exactamente 90°.

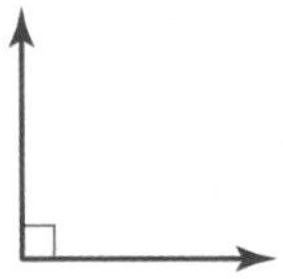

right triangle (p. 429) A triangle having one right angle.

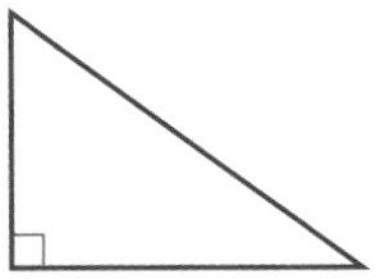

triángulo rectángulo Triángulo que tiene un ángulo recto.

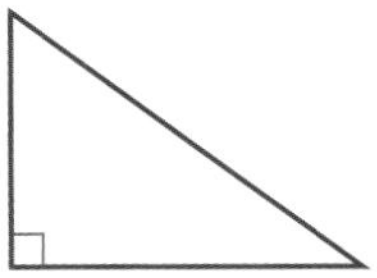

S

sales tax (p. 354) An additional amount of money charged on items that people buy.

impuesto sobre las ventas Cantidad adicional de dinero que se cobra por un artículo comprado.

sample space (p. 374) The set of all possible outcomes of a probability experiment.

espacio muestral Conjunto de todos los resultados posibles de un experimento probabilístico.

scale (p. 54) The set of all possible values of a given measurement, including the least and greatest numbers in the set, separated by equal intervals.

escala Conjunto de todos los posibles valores de una medida dada, el cual incluye los valores máximo y mínimo del conjunto, separados por intervalos iguales.

scale (p. 304) On a map, intervals used representing the ratio of distance on the map to the actual distance.

escala En un mapa, los intervalos que se usan para representar la razón de las distancias en el mapa a las distancias verdaderas.

scale drawing (p. 304) A drawing that is similar but either larger or smaller than the actual object.

dibujo a escala Dibujo que es semejante, pero más grande o más pequeño que el objeto real.

scale factor (p. 305) A scale written as a ratio in simplest form.

factor de escala Escala escrita como una tasa en forma reducida.

scale model (p. 306) A model used to represent something that is too large or too small for an actual-size model.

modelo a escala Réplica de un objeto real, el cual es demasiado grande o demasiado pequeño como para construirlo de tamaño natural.

scalene triangle (p. 429) A triangle having no congruent sides.

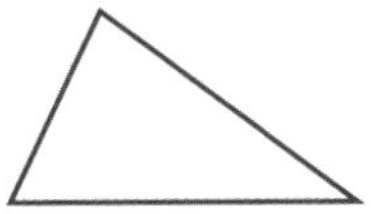

triángulo escaleno Triángulo sin lados congruentes.

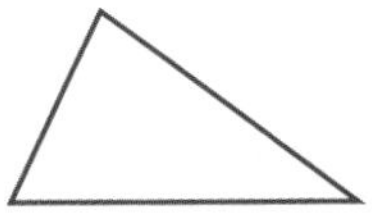

scatter plot (p. 61) In a scatter plot, two sets of related data are plotted as ordered pairs on the same graph.

diagrama de dispersión Diagrama en que dos conjuntos de datos relacionados aparecen graficados como pares ordenados en la misma gráfica.

scientific notation (p. 43) A number written as a product of a number that is at least 1 but less than 10, and a power of ten. For example, $687{,}000 = 6.87 \times 10^5$.

sequence (p. 34) A list of numbers in a certain order, such as 0, 1, 2, 3, or 2, 4, 6, 8.

significant digits (p. 542) All of the digits of a measurement that you know for sure, plus one estimated digit.

similar figures (p. 440) Figures that have the same shape but not necessarily the same size.

simple event (p. 370) One outcome or a collection of outcomes.

simple interest (p. 358) The amount paid or earned for the use of money. The formula for simple interest is $I = prt$.

simplest form (p. 207) A fraction is in simplest form when the GCF of the numerator and the denominator is 1.

slope (p. 182) A number that tells how steep a line is. Slope of a line is found by looking at the change in y with respect to the change in x.

solid (p. 514) A three-dimensional figure. Prisms, pyramids, cones, and cylinders are examples of solids.

solution (p. 24) A value for the variable that makes an equation true. The solution of $12 = x + 7$ is 5.

solving an equation (p. 24) The process of finding a solution to an equation.

square (p. 434) A parallelogram having four right angles and four congruent sides.

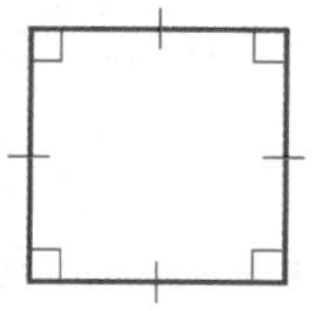

square (p. 470) The product of a number and itself. 36 is the square of 6.

squared (p. 10) A number multiplied by itself; 4×4, or 4^2.

square root (p. 471) One of the two equal factors of a number. The square root of 9 is 3.

standard form (p. 11) Numbers written without exponents.

statistics (p. 54) The study of collecting, organizing, and interpreting data.

stem (p. 76) The greatest place value common to all the data values is used for the stem of a stem-and-leaf plot.

notación científica Números escritos como el producto de un número que es al menos igual a 1, pero menor que 10, por una potencia de diez. Por ejemplo, $687{,}000 = 6.87 \times 10^5$.

sucesión Lista de números en cierto orden, tales como 0, 1, 2, 3 ó 2, 4, 6, 8.

dígitos significativos Todos los dígitos de una medición que se sabe son exactos, más un dígito aproximado.

figuras semejantes Figuras que tienen la misma forma, pero no necesariamente el mismo tamaño.

eventos simples Un resultado o una colección de resultados.

interés simple Cantidad que se paga o que se gana por el uso del dinero. La fórmula para calcular el interés simple es $I = prt$.

forma reducida Una fracción está escrita en forma reducida si el MCD de su numerador y denominador es 1.

pendiente Número que indica el grado de inclinación de una recta. La pendiente de una recta se calcula comparando el cambio en y con respecto al cambio en x.

sólido Una figura tridimensional. Los prismas, las pirámides, los conos y los cilindros son ejemplos de sólidos.

solución Valor de la variable de una ecuación que hace verdadera la ecuación. La solución de $12 = x + 7$ es 5.

resolver una ecuación Proceso de encontrar el número o números que satisfagan una ecuación.

cuadrado Paralelogramo con cuatro ángulos rectos y cuatro lados congruentes.

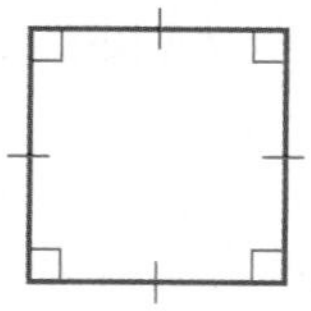

cuadrado El producto de un número por sí mismo. 36 es el cuadrado de 6.

al cuadrado Número multiplicado por sí mismo; 4×4, ó 4^2.

raíz cuadrada Uno de dos factores iguales de un número. La raíz cuadrada de 9 es 3.

forma estándar Números escritos sin exponentes.

estadística Estudio de la recolección, organización e interpretación de datos.

tallo El mayor valor de posición común a todos los datos es el que se usa como tallo en un diagrama de tallo y hojas.

stem-and-leaf plot (p. 76) A system used to condense a set of data where the greatest place value of the data forms the stem and the next greatest place value forms the leaves.

straight angle (p. 413) An angle that measures exactly 180°.

supplementary angles (p. 422) Two angles are supplementary if the sum of their measures is 180°.

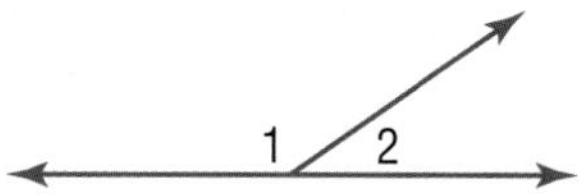

surface area (p. 532) The sum of the areas of all the surfaces (faces) of a three-dimensional figure.

survey (p. 345) A question or set of questions designed to collect data about a specific group of people.

T

term (p. 19) A number, a variable, or a product or quotient of numbers and variables.

term (p. 34) Each number in a sequence.

terminating decimals (p. 211) A decimal whose digits end. Every terminating decimal can be written as a fraction with a denominator of 10, 100, 1,000, and so on.

tessellation (p. 447) A repetitive pattern of polygons that fit together with no holes or gaps.

theoretical probability (p. 393) The ratio of the number of ways an event can occur to the number of possible outcomes.

ton (p. 267) A customary unit of weight equal to 2,000 pounds.

transformation (p. 451) A movement of a geometric figure.

translation (p. 451) One type of transformation where a geometric figure is slid horizontally, vertically, or both.

trapezoid (p. 434) A quadrilateral with one pair of parallel sides.

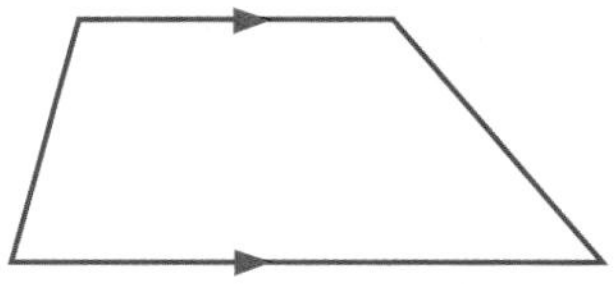

tree diagram (p. 374) A diagram used to show the total number of possible outcomes in a probability experiment.

diagrama de tallo y hojas Sistema que se usa para condensar un conjunto de datos y en el cual el mayor valor de posición de los datos forma el tallo y el segundo mayor valor de posición de los datos forma las hojas.

ángulo llano Ángulo que mide exactamente 180°.

ángulos suplementarios Dos ángulos son suplementarios si la suma de sus medidas es 180°.

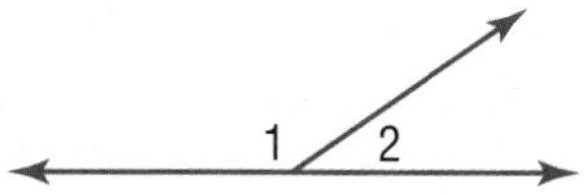

área de superficie La suma de las áreas de todas las superficies (caras) de una figura tridimensional.

encuesta Pregunta o conjunto de preguntas diseñadas para recoger datos sobre un grupo específico de personas.

término Un número, una variable o un producto o un cociente de números y variables.

término Cada número en una sucesión.

decimales terminales Decimal cuyos dígitos terminan. Todo decimal terminal puede escribirse como una fracción con un denominador de 10, 100, 1,000, etc.

teselado Un patrón repetitivo de polígonos que coinciden perfectamente, sin dejar huecos o espacios.

probabilidad teórica La razón del número de maneras en que puede ocurrir un evento al número total de resultados posibles.

tonelada Unidad de peso del sistema inglés que equivale a 2,000 libras.

transformación Movimientos de figuras geométricas

traslación Tipo de transformación en que una figura se desliza horizontal o verticalmente o de ambas maneras

trapecio Cuadrilátero con un único par de lados paralelos.

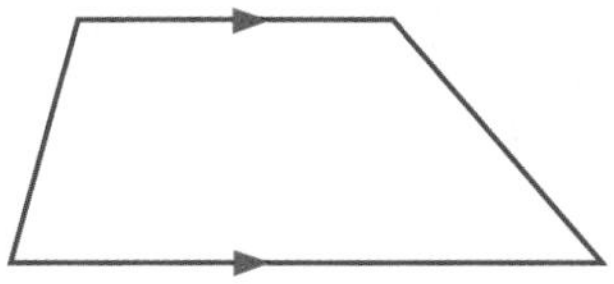

diagrama de árbol Diagrama que se usa para mostrar el número total de resultados posibles en experimento probabilístico.

triangle (p. 428) A polygon that has three sides and three angles.

triángulo Polígono que posee tres lados y tres ángulos.

two-step equation (p. 166) An equation having two different operations.

ecuación de dos pasos Ecuación que contiene dos operaciones distintas.

U

unit rate (p. 292) A rate with denominator of 1.

tasa unitaria Una tasa con un denominador de 1.

upper extreme (p. 80) The greatest number of a set of data.

extremo superior El número máximo de un conjunto de datos o números.

upper quartile (p. 80) The median of the upper half of a set of numbers, indicated by UQ.

cuartil superior La mediana de la mitad superior de un conjunto de números, denotada por CS.

V

variable (p. 18) A placeholder, usually a letter, used to represent an unspecified value in mathematical expressions or sentences. In $3 + a = 6$, a is a variable.

variable Marcador de posición, por lo general, una letra, que se usa para representar un valor desconocido en expresiones o enunciados matemáticos. En $3 + a = 6$, a es una variable.

Venn diagram (p. 203) A diagram that uses circles to show how elements among sets of numbers or objects are related.

diagrama de Venn Diagrama que usa círculos para mostrar la relación entre los elementos en un conjunto de números u objetos.

vertex (p. 413) A vertex of an angle is the common endpoint of the rays forming the angle.

vértice El vértice de un ángulo es el extremo común de los rayos que lo forman.

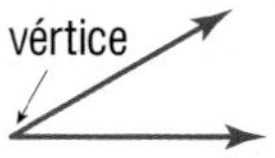

vertical angles (p. 422) Opposite angles formed by the intersection of two lines.

ángulos opuestos por el vértice Ángulos opuestos que se forman de la intersección de dos rectas.

volume (p. 520) The number of cubic units needed to fill the space occupied by a solid.

volumen Número de unidades cúbicas que se requieren para llenar el espacio que ocupa un sólido.

X

***x*-axis** (p. 112) The horizontal number line in a coordinate plane.

eje *x* La recta numérica horizontal en el plano de coordenadas.

***x*-coordinate** (p. 112) The first number of an ordered pair. It corresponds to a number on the x-axis.

coordenada *x* El primer número de un par ordenado. Corresponde a un número en el eje x.

Y

***y*-axis** (p. 112) The vertical number line in a coordinate plane.

eje *y* La recta numérica vertical en el plano de coordenadas.

***y*-coordinate** (p. 112) The second number of an ordered pair. It corresponds to a number on the y-axis.

coordenada *y* El segundo número de un par ordenado. Corresponde a un número en el eje y.

Selected Answers

Chapter 1 Decimal Patterns and Algebra

Page 5 Chapter 1 Getting Started

1. true **3.** 105.8 **5.** 60.64 **7.** 11.3 **9.** 27.69 **11.** 30.8 **13.** 17.01 **15.** 8.15 **17.** 1.1 **19.** 2,450 **21.** 61

Pages 8–9 Lesson 1-1

1. Sample answer: It helps to organize your thoughts and focus on how to approach solving the problem. **5.** 13 E-mails **7.** 1,280,000

9.

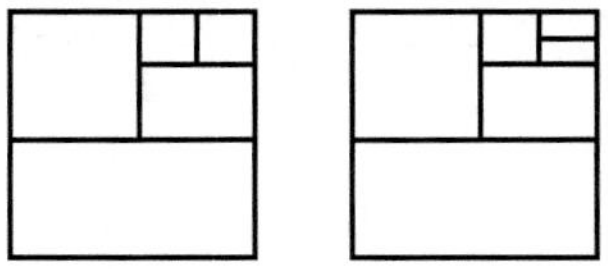

11. 2004 **13.** Sample answer: sweater, gloves, and chocolates **15.** 100 **17.** 625

Pages 11–13 Lesson 1-2

1. Five is used as a factor four times, or $5 \cdot 5 \cdot 5 \cdot 5 = 5^4$. **3.** Sample answer: exponential form: 4^2; standard form: 125 **5.** $6 \cdot 6$ **7.** $8 \cdot 8 \cdot 8 \cdot 8 \cdot 8$ **9.** 3,125 **11.** 5^6 **13.** 1,331 **15.** $4 \cdot 4$ **17.** $9 \cdot 9 \cdot 9$ **19.** $11 \cdot 11 \cdot 11$ **21.** 64 **23.** 625 **25.** 10 **27.** 4,096 **29.** 20,736 **31.** 125,000 **33.** 7^4 **35.** 6^5 **37.** 13^5 **39.** 2^8 **41.** Sample answer: A number taken to the third power is the same as the volume of a cube, or the amount of space inside a cube. **43.** false **45.** false **47.** false **49.** $6^3, 15^2, 3^5, 2^8$ **51.** $8^3, 3^6, 6^4, 5^5$ **53.** $11^3, 5^5, 9^4, 6^5$ **55.** Sample answer: It is easier and less error-prone to write 10^{15} than writing out 15 zeros. **57.** $5^4 \cdot 4^3$ **59.** Sample answer: 390,625 < 534,000 < 823,543, or $5^8 < 534{,}000 < 7^7$ **61.** $5^1 \cdot 10^3$ **63.** C **65.** 3 **67.** false **69.** 3.4 **71.** 3.07 **73.** 33.98 **75.** 3.5

Pages 16–17 Lesson 1-3

1a. $14 - 7$ **1b.** $24 \div 3$ **3.** Cynthia; the first step is to do the division, $24 \div 6$. Yutaka incorrectly multiplied 6 and 2 first. **5.** 5 **7.** 27 **9.** 28 **11.** 53.5 **13.** 58 **15.** 27 **17.** 13 **19.** 13 **21.** 14.2 **23.** 5.8 **25.** 4 **27.** 10 **29.** 2 **31.** 88 **33.** 600 **35.** 195 **37.** 22.8 **39.** 19 **41.** Sample answer: $2(7) + 2(9.5) + 6$; 39 cm **43.** 450 **45.** 296,500 **47.** 57.9 **49.** $72 \div (9 + 27) - 2 = 0$ **51.** 16; The calculator followed the order of operations. Ruby incorrectly added $2 + 2$ first when she should have divided $12 \div 2$ first. **53.** D **55.** Sample answer: $30 - 20$, or about 10 days longer **57.** 121 **59.** 1,024 **61.** $8 \cdot 8 \cdot 8 \cdot 8 \cdot 8$ **63.** $4 \cdot 4 \cdot 4 \cdot 4 \cdot 4 \cdot 4$ **65.** 9 **67.** 18 **69.** 13

Pages 20–21 Lesson 1-4

1. coefficient: 2; constant: -6 **3.** $\frac{x}{y}$ is a division expression; the others are multiplication expressions. **5.** 20 **7.** 21 **9.** 5 **11.** 16 **13.** 5 **15.** 7 **17.** 8 **19.** 12 **21.** 48 **23.** 1.6 **25.** 3 **27.** 32 **29.** 65.2 **31.** 5 **33.** 18 **35.** 3 **37.** Always; ab means a is multiplied by b. **39.** $16n$ **41.** 2, 2.5, 3, 3.5 **43.** about 118 **45.** \$155; \$200; \$245 **47.** Sample answer: $x = 5, y = 2$ **49.** August **51.** 12 **53.** 4 **55.** true **57.** false

Pages 26–27 Lesson 1-5

1. Sample answer: Find the value of the variable that makes a true sentence. **3.** Ivan; $75 - 25 = 50$ is a true statement. **5.** 17 **7.** 54 **9.** 23 **11.** 4 **13.** 23 points **15.** 7 **17.** 38 **19.** 63 **21.** 5 **23.** 11 **25.** 20 **27.** 64 **29.** 20 **31.** 12

33.

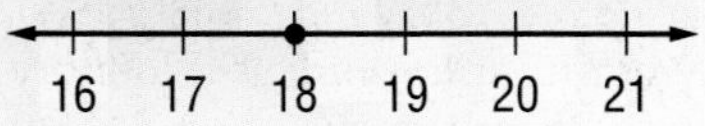

$x = 18$

35.

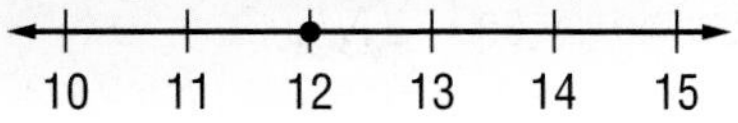

$z = 12$

37. 40 **39.** 25 **41.** 7.5 h **43.** 3,500 mi **45.** $t = 51 - 18$ or $51 = t + 18$; 33 in. **47.** Sample answer: $b = 0$; a is any number **49.** 958 **51.** 31 **53.** 19 **55.** 28 **57.** 150

Pages 32–33 Lesson 1-6

1. $p(q + r)$ **3.** false; $(11 + 18) \times 5 = 145$; $11 + 18 \times 5 = 101$ **5.** $(10)2 + (8)2$; 36 **7.** Commutative Property of Addition **9.** Distributive Property **11.** $2(6) + 2(7)$; 26 **13.** $(11)8 + (3)8$; 112 **15.** $7(8) - 7(6)$; 14 **17.** Sample answer: The second expression, because it is easy to multiply $3 \cdot 400$, $3 \cdot 50$, and $3 \cdot 6$ mentally, and then add. **19.** Associative Property of Multiplication **21.** Commutative Property of Multiplication **23.** Associative Property of Addition **25.** Distributive Property **27.** $239 + 531 = 531 + 239$ **29.** $152 + 0 = 152$ **31.** $y + 5$ **33.** $32b$ **35.** $2x + 6$ **37.** $5c + 6$ **39.** $6(40 + 8) = 6(40) + 6(8) = 240 + 48 = 288$; Distributive Property **43.** A **45.** 8 **47.** 7.1 **49.** $7y$ **51.** 625 **53.** 54 **55.** 20 **57.** 32.4

Pages 35–36 Lesson 1-7

1. Sample answer: Each term in an arithmetic sequence is found by adding or subtracting the same number to or from the previous term. Each term in a geometric sequence is found by multiplying or dividing the previous term by the same number. **5.** Add 2 more than what was added to the previous term; neither. **7.** 324, 972, 2,916 **9.** 6.6, 7, 7.4 **11.** 70, 7, 0.7, 0.07 **13.** + 12; arith. **15.** × 3; geo. **17.** Multiply by 1 more than was multiplied by the previous term; neither. **19.** 20.6, 24.6, 28.6 **21.** 1.3, 1.6, 1.9 **23.** 0.2, 0.04, 0.008 **25.** 1, 3, 6, 10, 15, 21, 28, 36 **27.** 76 **29.** 100 **31.** Distributive **33.** 520 **35.** 3.2

Pages 40–41 Lesson 1-8

1. There are a greater number of smaller units. **3.** Hunter; Arturo multiplied 45.7 by 1,000; he should have divided by 1,000. **5.** 45,000 **7.** 370 **9.** 2,340 **11.** 7.2 **13.** 64,000 **15.** 0.925 **17.** 8.5 **19.** 9,100 **21.** 0.043 **23.** 0.997 **25.** 0.0821 **27.** 57 **29.** 0.169 km **31.** 0.02 km, 3,000 cm, 50 m **33.** 0.06 L, 660 mL, 6.6 kL **35.** 93,000 **37.** 0.04 **39.** 50 cm **41.** 5 cobblers **43.** 3,900,000,000 m **45.** C **47.** + 9; arithmetic **49.** + 1.1; arithmetic **51.** $5(9) + 5(7)$; 80 **53.** $8(7) - 8(2)$; 40 **55.** 800 **57.** 400,000 **59.** 6,500

Pages 44–45 Lesson 1-9

1. Sample answer: It is easier to keep track of the place value. **3.** 3.0×10^6; Sample answer: Chicago has a population of about 3 million people: $3.0 \times 10^6 = 3{,}000{,}000$. **5.** 100,000 **7.** 7.0×10^4 **9.** 1.75×10^7 **11.** 73,000 **13.** 338,000 **15.** 8,980,000 **17.** 798,000 **19.** 990,000,000 **21.** 80,600,000 **23.** 53,400, 5.03×10^5, 4.98×10^6, 4,980,100 **25.** 1.98×10^4 **27.** 9.97×10^6 **29.** 5.05×10^5 **31.** 2.36×10^7 **33.** 1.1×10^6 **35.** 6.08×10^{10} **37.** 54.3 quadrillion **39.** > **41.** < **43.** 2.94×10^{14} mi **45.** B **47.** 1,010 **49.** 2,330

Pages 46–48 Chapter 1 Study Guide and Review

1. b **3.** d **5.** f **7.** $89.40 **9.** 40,353,607 **11.** 324 **13.** 10,000 **15.** 100 **17.** 25 **19.** 20 **21.** 33 **23.** 0 **25.** 11 **27.** 36 **29.** 56 **31.** $965 **33.** 8 **35.** 7 **37.** 108 **39.** 67 **41.** Associative Property of Multiplication **43.** Identity Property of Multiplication **45.** arithmetic; +1.2; 6.8, 8.0, 9.2 **47.** $125, $62.50, $31.25 **49.** 29,000 **51.** 0.007 **53.** 5.9×10^4 **55.** 3.24×10^5 **57.** 1.03×10^6 **59.** 9.1×10^6

Chapter 2 Statistics: Analyzing Data

Page 53 Chapter 2 Getting Started

1. hundredths **3.** greatest to least **5.** thousands **7.** ones **9.** 5.062, 5.16, 5.61 **11.** 0.398, 3.98, 39.8 **13.** 10.26, 1.26, 1.026 **15.** 69.79, 68.99, 68.9 **17.** 86 **19.** 555

Pages 56–57 Lesson 2-1

1. An advantage of using a frequency table is that it organizes data to provide a quick summary. It makes it easier to see quickly the number of times each item of data appears. A disadvantage is that a frequency table may not show the individual pieces of data. **3.** The interval in each category must be the same, 50.

5. Sample answer:

Number of Movie Rentals	Tally	Frequency
0–2	III	3
3–5	卌	5
6–8	II	2
9–11	卌 III	8
12–14	III	3
15–17	II	2
18–20	I	1

7. cereal

9. Sample answer:

Year	Tally	Frequency
1900–1919	II	2
1920–1939	IIII	4
1940–1959	IIII	4
1960–1979	卌 IIII	9
1980–1999	卌 III	8

11. Sample answer:

Chore	Tally	Frequency
wash dishes	卌 III	8
vacuum	卌	5
clean room	卌 卌 II	12
do laundry	卌 I	6
do yard work	IIII	4

13. Sample answer:

Year	Tally	Frequency
1970–1979	II	2
1980–1989	III	3
1990–1999	卌 III	8
2000–2009	卌 II	7

17. Sample answer: 91–110 Calories

19. Sample answer:

Price ($)	Tally	Frequency
11–20	II	2
21–30	IIII	4
31–40	IIII	4
41–50	卌	5
51–60	卌	5
61–70	II	2
71–80	IIII	4
81–90	III	3
91–100	卌 I	6

Most of the scooter prices are spread out evenly among the intervals, with the $91–$100 category containing the greatest number. An advantage of using this scale and interval is that you can get a more detailed distribution. A disadvantage is that it is difficult to see clusters or trends.
21. F **23.** 533,000 **25.** 60,000,000 **27.** 78°F

Pages 61–63 Lesson 2-2

1. Graphs often show trends over time. If you continue the pattern, you can use it to make a prediction. **3.** Tally mark; the other three are ways to display data. **5.** Yes; in the 7th week he will be able to run 1 mile in about 7 minutes. **9.** Sample answer: 210 nations **11.** Sample answer: As the speed increases, the distance required to come to a complete stop increases. **13.** The data do not appear to fall along a straight line. So, birth month and height do not seem to be related. **15.** A **17.** Sample answer: 2; 18–47 **19.** 9^2 **21.** 7^3 **23.** 6.1, 6.2, 6.2, 6.2, 6.4, 6.5, 6.5, 6.7, 6.7, 6.8, 6.8, 6.8, 6.8, 7.0, 7.0

Pages 66–68 Lesson 2-3

1. Sample answer: the greatest and least values, clusters of data, gaps in data, how the data are spread out, the range, and the pieces of data that appear most often **3.** Darnell; 16 is the mode and 20 is the greatest value.

5.

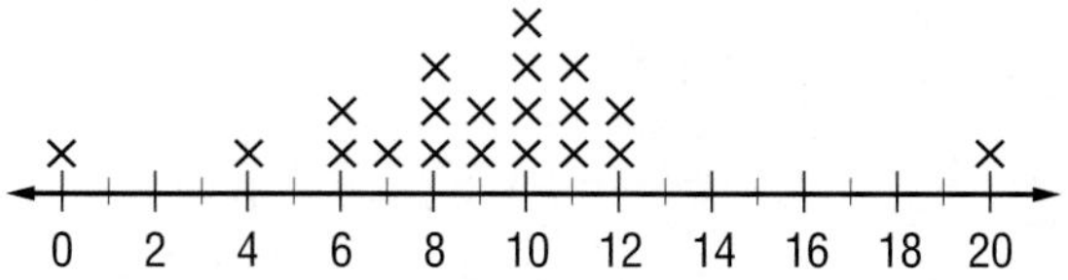

7. Sample answer: Most of the students in the class own 6–12 music CDs. **9.** 104°F **11.** The range would be 29 rather than 37.

13.

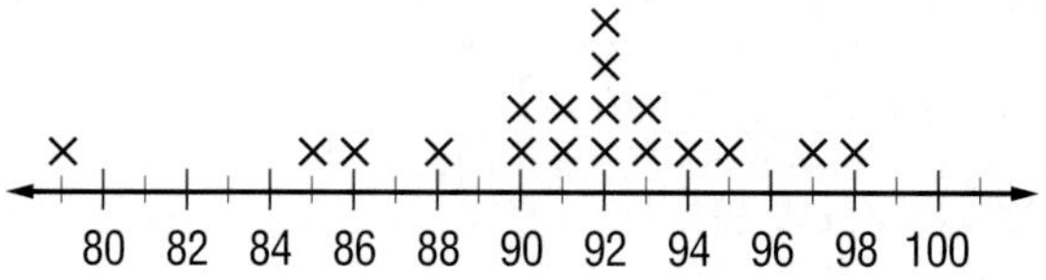

Sample answer: cluster 90–93; gap 79–85; outlier 79
15. Sometimes; clusters may appear anywhere on the line plot. **17.** 5

19.

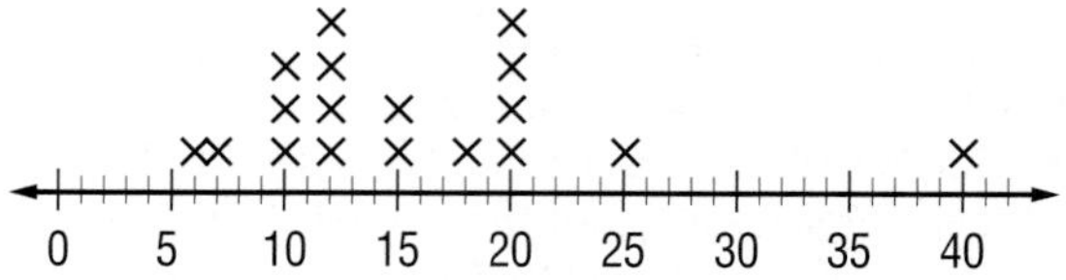

21. Sample answer: Most of the animals represented on the line plot have a life span of 10–20 years. **23.** Sample answer: Bowler B is more consistent because excluding the outlier of 90, the range of scores is 4 points, compared with a range of 10 for Bowler A. Bowler A has scores evenly spread out from 90 to 100. **25.** D **27.** Sample answer: The more it rains, the taller the grass becomes. **29.** 75 **31.** 29 **33.** 47 **35.** 16.1 **37.** 12.6 **39.** 36.4

Pages 71–72 Lesson 2-4

1. Sometimes; if there is an odd number of items, the median is the middle number. If there is an even number of items, the median is the mean of the two middle numbers. **3.** Jared; the median is the 11th term of the ordered data, or 2. The mode is the number that appears most often, or 3. **5.** 52.3; 57; 59 **7.** 8.2; 8; 7 **9.** Sample answer: Any of the three measures could be used to represent the data. The mean is slightly greater than most of the data items, and therefore is a less accurate description of the data. **11.** 8.7; 9.0; 8.0 and 9.3 **13.** $12; $9; $6 **15.** $7.94; $8; $7.50
17. Sample answer: The mean or median best represent the data because half of the items are less than these values and half are greater. Since there are four modes, the mode does not best represent the data.
19. Sample answer: The mean would be most affected; the mode would be the least affected, since it would not change. **21.** 12 **23.** line graph **25.** ones **27.** tens

Pages 78–79 Lesson 2-5

1. Both a stem-and-leaf plot and frequency table show the frequency of data occurring. However, a stem-and-leaf plot shows individual data values and a frequency table shows only the number of data items occurring in each interval.
3. 0, 1, 2, 3
5.

Stem	Leaf
1	6 9
2	1 3 5 5 9
3	1 3 4 5 9
4	2 7 8 *3|4 = 34*

7. 31 **9.** 32.9; 35; 40
11.

Stem	Leaf
1	3 3 5
2	4 8
4	0 1 2 2 5 6 8 8 8 *1|3 = 13*

13.

Stem	Leaf
40	3
41	1
42	
43	3 4 9
44	2 9
45	1 4 4 7 8 9 *43|9 = 439*

15. 26 **17.** $110; $115 and $150
19.

Stem	Leaf
5	1 2 7
6	0 1 2 3 5 7 8 8
7	3 3 3 4 5 6 7 8 9
8	1 2 2 4 5 9
8	1 2 2 4 5 9
9	4 4 7
10	6 8 *9|4 = 94 points*

21. Sample answer: Most of the game point totals are clustered from 60 to 89 points. There do not appear to be any gaps or outliers.
23. Sample answer:

Price ($)	Tally	Frequency
20–29	IIII	4
30–39	~~IIII~~ ~~IIII~~	10
40–49	~~IIII~~	5
50–59	I	1

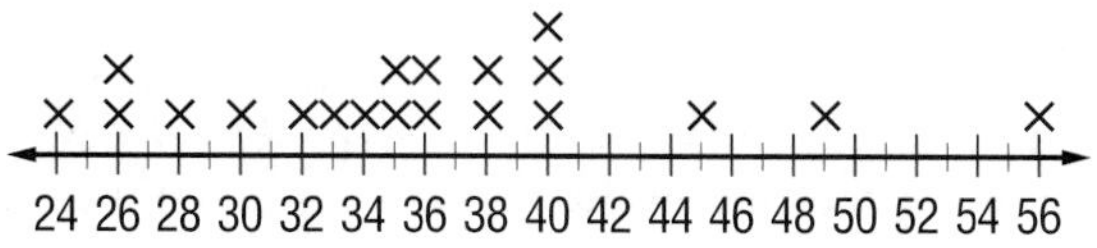

Stem	Leaf
2	4 6 6 8
3	0 2 3 4 5 5 6 6 8 8
4	0 0 0 5 9
5	6 *3|5 = $35*

All three representations show the frequency of data occurring. The frequency table shows intervals of data and is useful in comparing price ranges. The line plot gives a good picture of the spread of the data. The stem-and-leaf plot shows individual prices as in the line plot, as well as intervals, such as $20s, $30s, and so on.
25. 97 **27.** 4.3; 3.6; 3.6 and 2.4 **29.** 21 **31.** 13 **33.** 6, 7.9, 8, 8.4 **35.** 0.09, 0.14, 0.33, 0.7

Pages 82–83 Lesson 2-6

1. Sample answer: In a stem-and-leaf plot, you can see each data item and how the data are clustered. In a box-and-whisker plot, you can immediately see the median and how the data are dispersed. **3.** 270; 350; 380; 588; 1,221
5.

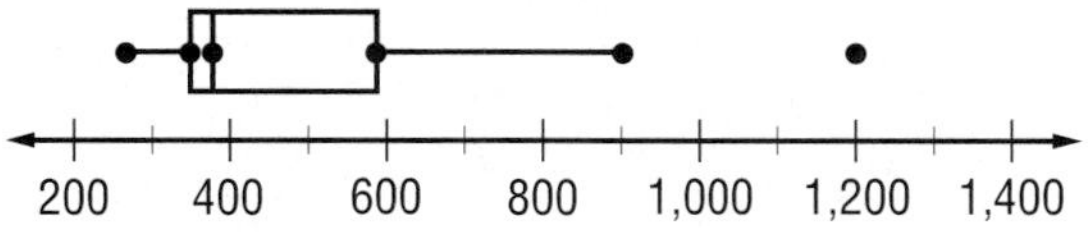

7. 15; 30.5; 45; 51.5; 58 **9.** one-fourth **11.** 0 and 4,600; no outliers **13.** one half **15.** A set in which an extreme value and a quartile are the same has no whisker on that side of the box-and-whisker plot. **17.** 20.5 **19.** 31.6; 32; none
21. Sample answer: 5; 0–19

Pages 87–89 Lesson 2-7

1. Sample answer: Each interval represents a portion of the data set. The number of items in each interval is indicated by the frequency, typically shown along the vertical scale. By adding the frequencies for each interval, you can determine the number of values in the data set. **3.** World Cup Soccer titles of seven countries **5.** The horizontal axis represents the countries and the vertical axis represents the number of wins. **7.** the number of candies found in 50 snack-size packages
9.

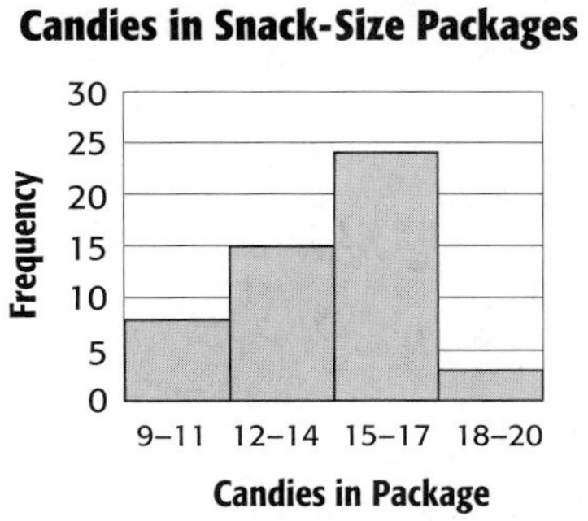

11. Sample answer: The bars are horizontal; the categories are on the vertical axis and the scale is on the horizontal axis.
13. Sample answer: 200,000

15. Sample answer: The histogram is strongly skewed toward the lower end, with a high concentration in the number of states with 10 or fewer representatives. The distribution spreads out from there, with fewer and fewer states as the number per state increases. **17.** Sample answer: Both histograms show that many players have salaries in the lowest interval of the histogram, less than

Selected Answers

$1 million. Both histograms show one, two, or three players in several of the other intervals. The Diamondbacks' salaries show only three players' salaries of more than $7 million, while the Yankees have seven players in that same range. **19.** Sample answer: It is easier to compare two sets of data. **21.** histogram

23.

Stem	Leaf
0	2 5 9
1	0 2 3 6 7
2	3 5 5
3	1 *2|3 = 23*

25. false

Pages 93–95 Lesson 2-8

1. Sample answer: Outliers may distort measures of central tendency; data shown in graphs may be exaggerated or minimized by manipulating scales and intervals. **3.** The cost is increasing over time. In about a year's time, the cost has more than tripled. **5.** Sample answer: The median, because it will not be affected by the high prices.
7. Sample answer: Although both graphs show a decrease in network viewers, in Graph B, a change in the vertical scale makes it appear that the viewers are decreasing more rapidly. **9.** 488.3; none; 150 **11.** median **13.** Yes; all the other test scores were lower than 96. **15.** B **17.** the number of in-line skaters in New York, almost 2 million **19.** $7.57; $7.50; $7.50 and $8.00 **21.** 8,500 **23.** 0.643

Pages 96–98 Chapter 2 Study Guide and Review

1. true **3.** true **5.** false; outlier **7.** true
9. Sample answer:

Hours of Sleep	Tally	Frequency
5	I	1
6	III	3
7	~~IIII~~	5
8	IIII	4
9	III	3

11. Sample answer: 17 million
13.

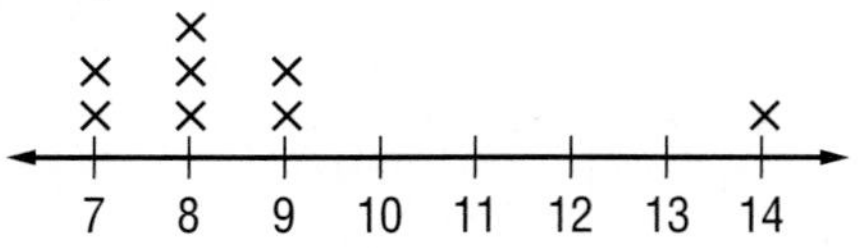

Sample answer: cluster 7–9; gap 9–14; outlier 14
15.

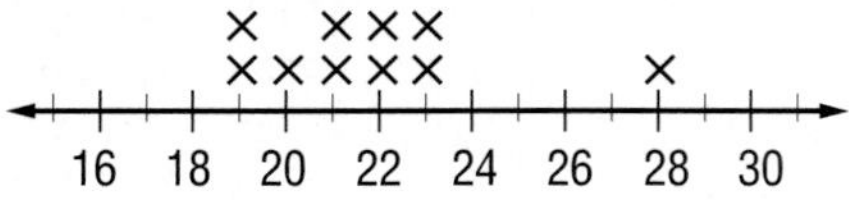

Sample answer: cluster 19–23; gap 23–28; outlier 28
17. 24.4; 24; 18 and 31 **19.** 132.3, 130.5, 136

21.

Stem	Leaf
5	3
6	
7	5 8 8
8	3 5 7 7 9
9	1 2 *7|5 = 75*

23.

Stem	Leaf
20	1 4
21	8 9
22	1 5 9
23	4 *23|4 = 234*

25. 11.5
27.

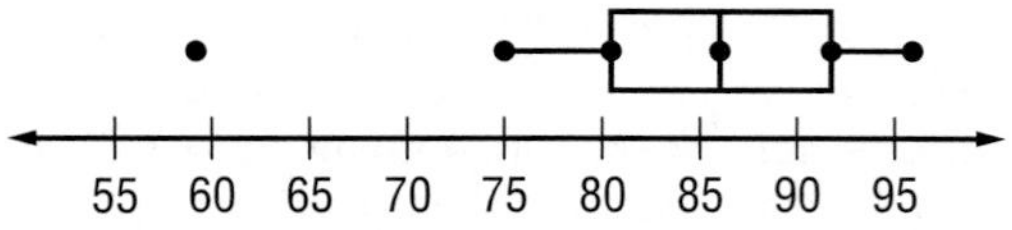

29. 8

Chapter 3 Algebra: Integers

Page 105 Chapter 3 Getting Started

1. false; median **3.** < **5.** < **7.** 123 **9.** 1,730 **11.** 30 **13.** 90 **15.** 7 **17.** 25 **19.** 14; 17 **21.** 100; 34

Pages 107–108 Lesson 3-1

1. Sample answers: temperature, golf scores, elevation, loss of money **3.** −3; The other expressions equal 3. **5.** −11 **7.** 7 **9.** 6 **11.** 9 **13.** 0 **15.** 2008 **17.** −50 **19.** 6 **21.** 9 **23.** 4 **25.** 14 **27.** 0 **29.** 2 protons: 2; 5 electrons: −5
31.

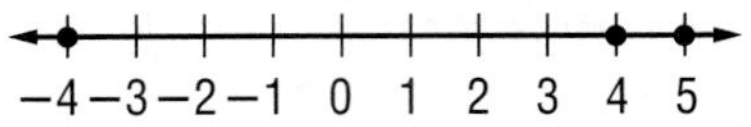

33.

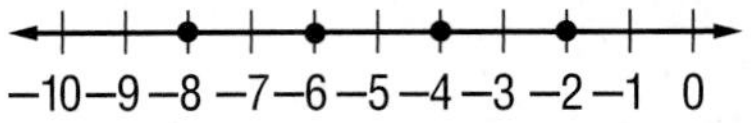

35. true **37.** B **39.** The very high salary increases the mean. **41.** > **43.** < **45.** >

Pages 110–111 Lesson 3-2

1.

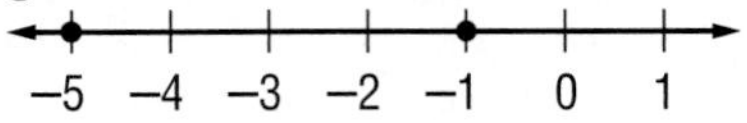

3. right **5.** > **7.** > **9.** 15, −13, 10, −4, 13, −5 **11.** > **13.** > **15.** > **17.** > **19.** > **21.** > **23.** false; Sample answer: 8 > 5 **25.** true **27.** true **29.** Mar. 2, Mar. 6, Mar. 1, Mar. 5, Mar. 3, Mar. 4, Mar. 7 **31.** −7, −6, −4, 1, 3, 5 **33.** They are negative integers. **35.** 7 **37.** 20
39. $x = 2$

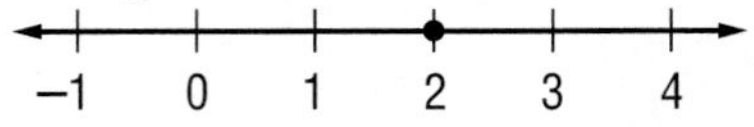

41. $x = 3$

Pages 114–115 Lesson 3-3

1. Sample answer: Point *A* is 1 unit to the right and 2 units down from the origin, in quadrant IV. Point *B* is 2 units to the left and 1 unit up from the origin, in quadrant II.
3. (−2, −4), III **5.** (0, 3), *y*-axis
6–8.

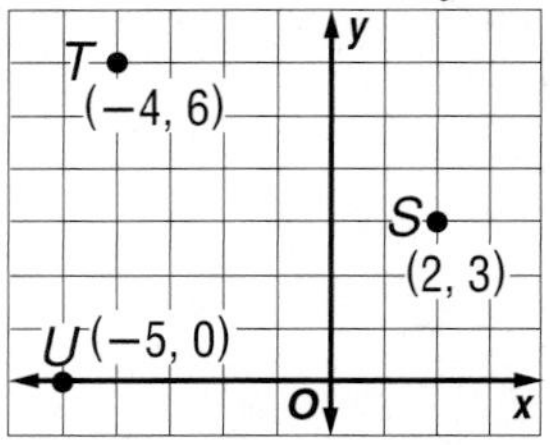

9. (−2, 2), II **11.** (−5, 0), *x*-axis **13.** (2, −2), IV **15.** (3, −5), IV **17.** (2, 2), I **19.** (0, −32)

20–31.

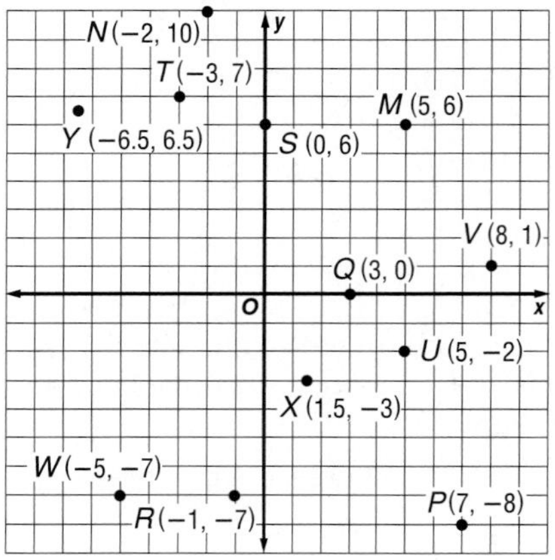

33. Sometimes; both (−2, 0) and (2, 0) lie on the *x*-axis. **35.** China **37.** Children's Zoo **39.** (−3, −2) **41.** Sample answer: On the coordinate plane, I would walk 1 unit right, then 4 units down. **43.** Quadrant I when both x- and y-coordinates are positive; quadrant III when both x- and y-coordinates are negative; the origin when both x- and y-coordinates are 0. **45.** G **47.** $<$ **49.** $>$ **51.** 101 **53.** 384 **55.** 7,470

Pages 122–124 Lesson 3-4

1. $2 + (-7) = -5$ Sample model:

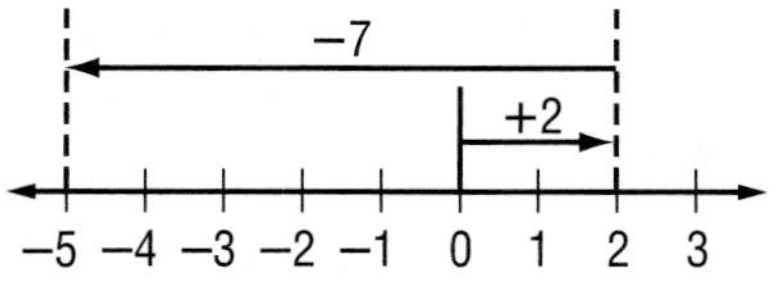

3. Brooke; Javier added correctly but did not apply the sign of the greater absolute value. **5.** −14 **7.** −4 **9.** less since $-\$42 + \$35 = -\$7$ **11.** 0 **13.** −6 **15.** −25 **17.** −14 **19.** 0 **21.** −38 **23.** 8 **25.** −30 **27.** $-3 + (-6)$; −9°F **29.** $-6 + 14$; 8 ft **31.** 4 **33.** −13 **35.** 12 **37.** 0 **39.** −3 **41.** −18 **43.** −12 **45.** Webb; $-14 < -12$ **47.** $y + 2$ **49.** n **51.** $f + (-8)$

53. Sample answer:

$-6 + 9 + (-4)$	$= -6 + (-4) + 9$	Comm. (+)
	$= (-6 + (-4)) + 9$	Assoc. (+)
	$= -10 + 9$	Simplify.
	$= -1$	

55. Sample answer:

$8 + 10 + (-8)$	$= 8 + (-8) + 10$	Comm. (+)
	$= (8 + (-8)) + 10$	Assoc. (+)
	$= 0 + 10$	Additive Inv.
	$= 10$	

57. Sample answer:

$8 + (-9) + 9$	$= 8 + ((-9) + 9)$	Assoc. (+)
	$= 8 + 0$	Additive Inv.
	$= 8$	Simplify.

61. −2; −2; No, addition is associative. **63.** −11 **65.** (0, −2), y-axis **67.** (1, 1), I

69.

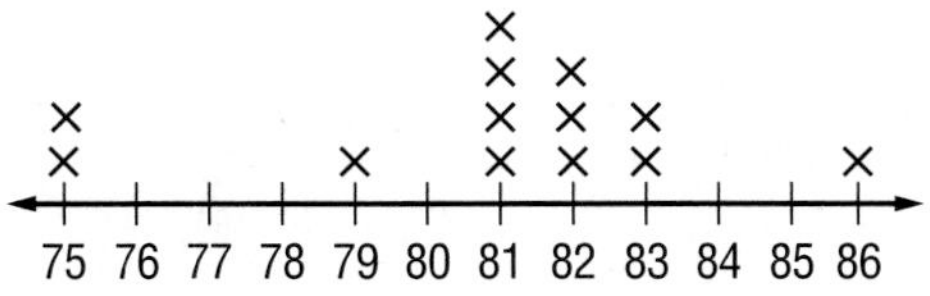

71. 24

Pages 130–131 Lesson 3-5

1. Sample answer: To subtract an integer, add its additive inverse. **3.** Mitsu; Bradley did not add the additive inverse of −19. **5.** −3 **7.** −2 **9.** −8 **11.** 265°F **13.** −10 **15.** 4 **17.** −20 **19.** 23 **21.** 29 **23.** 2 **25.** −6 **27.** −3 **29.** 1 **31.** −4 **33.** 18 **35.** 0 **37.** −12 **39.** 11 **41.** −3 **43.** 8 **45.** 2,139 m **47.** Tiberius; 79 yr **49.** always; $-2 - 3 = -5$ **51.** sometimes; $5 - 2 = 3$, $2 - 5 = -3$ **53.** true **55.** 300°C **57.** −11 **59.** −14 **61.** 70 **63.** 192

Pages 136–137 Lesson 3-6

1. $2(-3) = -6$

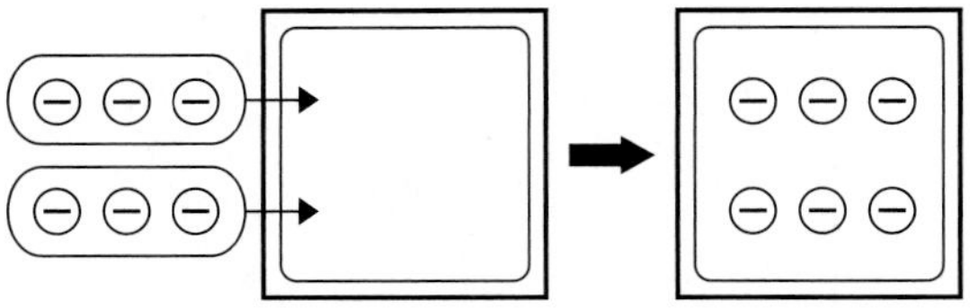

3. negative; Sample answer: $-2(-3)(-4) = -24$ **5.** 45 **7.** $-10a$ **9.** $45c$ **11.** 70 **13.** −875, or 875 feet below sea level **15.** 80 **17.** 170 **19.** 36 **21.** −450 **23.** −220 **25.** −64 **27.** 49 **29.** $-70d$ **31.** $-72f$ **33.** $24h$ **35.** $-10rs$ **37.** −40 **39.** −400 **41.** −300 **43.** −700 **45.** −32, 64; geometric: multiply the previous term by −2

47. Sample answer: new triangle $P'Q'R'$ is on the other side of the origin (in quadrant III) from original triangle PQR (which is in quadrant I).

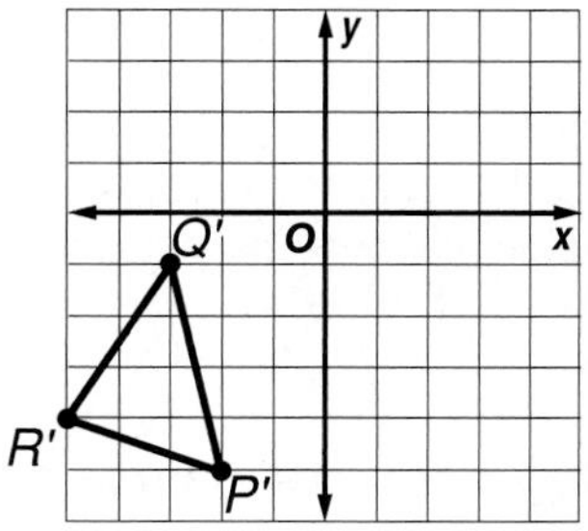

49. when n is an even number **51.** F **53.** 8 **55.** 0 **57.** 2 **59.** 20 and younger, 21–25, 26–30 **61.** 8 **63.** 4

Pages 140–141 Lesson 3-7

1. $-42 \div 7 = -6$, $-42 \div (-6) = 7$ **3.** $-18 \div (-9)$ has a positive quotient; the others have negative quotients. **5.** −8 **7.** −1 **9.** −4 **11.** −2 **13.** 5 **15.** −7 **17.** 4 **19.** 15 **21.** −1 **23.** −5 **25.** −1 **27.** −3 **29.** 2 **31.** 3 **33.** $p = -45 \div 3$; $p = -15$ or 15-yd penalty **35.** −3, 1; Divide the previous term by −3. **37.** −36 sea otters per year **39.** 1, 2, 4, 5, 10, 20, −1, −2, −4, −5, −10, −20 **41.** H **43.** −28 **45.** 60 **47.** 18 **49.** 0.003 g

Pages 142–144 Chapter 3 Study Guide and Review

1. negative **3.** 7 **5.** origin **7.** x-coordinate **9.** negative **11.** −150 **13.** 8 **15.** 11 **17.** 5 **19.** 25 **21.** $>$ **23.** $<$ **25.** $<$ **27.** −32, −23, −21, 14, 19, 25 **29.** 3°, 1°, 0°, −1°, −2° **31.** (−3, −5), III **33.** (0, 3), y-axis

34–37.

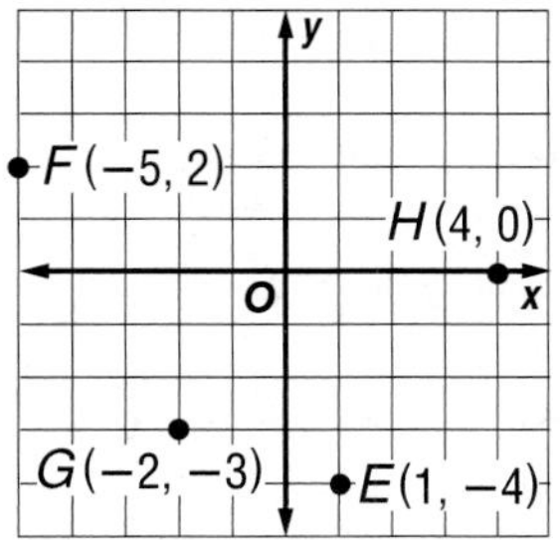

39. −13 **41.** 0 **43.** −13 **45.** 7 **47.** −12 **49.** 35 **51.** 28 **53.** −35 **55.** 5 **57.** −2 **59.** −50 ft per h

Chapter 4 Algebra: Linear Equations and Functions

Page 149 Chapter 4 Getting Started

1. solved **3.** 4 **5.** −11

7.

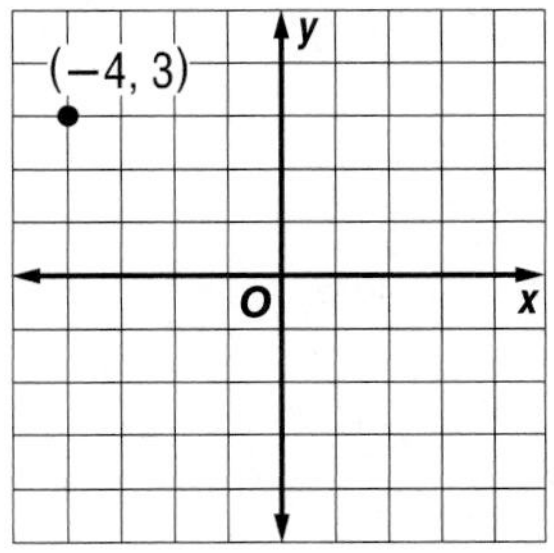

9. −8 **11.** 4 **13.** −11 **15.** 14 **17.** 2 **19.** −2

Pages 151–152 Lesson 4-1

1. Sample answer: 5 dollars more than my allowance is 8 dollars. **3.** $x + 8$ **5.** $10h$ **7.** $n + 4 = -9$ **9.** $a + 2.4 = 35.3$ **11.** $y + 5$ **13.** $n - 6$ **15.** $2f$ **17.** $\frac{z}{-12}$ **19.** $c + 2 = 4$ **21.** $10k = 280$ **23.** $n - 5 = 31$ **25.** $2z + 3 = 15$ **27.** $y + 7$ or $4.5y$ **29.** $x + 2$; $x - 2$ **31.** H **33.** −5 **35.** −8 **37.** −1 **39.** 0

Pages 158–159 Lesson 4-2

1. Addition Property of Equality **3.** $b + 5 = 4$; All of the other equations have solutions equal to 1. This one equals -1. **5.** -2 **7.** 11 **9.** -3 **11.** 7 **13.** -3 **15.** -9 **17.** 17 **19.** 7 **21.** -4 **23.** 61 **25.** -5 **27.** -12 **29.** -5 **31.** 18.4 **33.** 6.4 **35.** -0.68 **37.** $415 = 105 + x$; 310 ft **39.** The value of y must decrease by 3. **41.** 3 **43.** -5 **45.** 9 **47.** 1.2 **49.** 301

Pages 162–163 Lesson 4-3

1. No; $-3(-4)$ equals 12. **3.** Jesse; the variable is multiplied by -5. To solve for x, you need to divide each side of the equation by the entire coefficient, -5. **5.** 2 **7.** -4 **9.** 7 **11.** 6 **13.** -7 **15.** 3 **17.** 14 **19.** -5 **21.** -133 **23.** 10 **25.** 55 **27.** 40 **29.** 2.5 h **31.** 24 in. **33.** 1.5 **35.** 5.4 **37.** 0.8 **39.** 100 s **41.** *Blue Streak* 24.36 ft/s; *Corkscrew* 17.08 ft/s; *Magnum* 42.55 ft/s; *Mean Streak* 32.89 ft/s **43.** C **45.** -10 **47.** -3 **49.** $a + 5$ **51.** $2r$ **53.** -14 **55.** -21

Pages 168–169 Lesson 4-4

1. Sample answer: $3x + 10 = 4$ **3.** 2 **5.** 3 **7.** 6 **9.** 3 **11.** -4 **13.** 4 **15.** 9 **17.** 8 **19.** 11 **21.** -8 **23.** 0 **25.** 3 **27.** 2 **29.** 14 **31.** 2.25 **33.** 2.1 **35.** 28 **37.** 16 **39.** 3 h **41.** Less than; -3°F is about -19.4°C. **43.** C **45.** 7 **47.** 41 **49.** 14 ft **51.** > **53.** <

Pages 174–175 Lesson 4-5

1.

3. Sample answer: $2x < 4$ and $x + 1 < 3$

5.
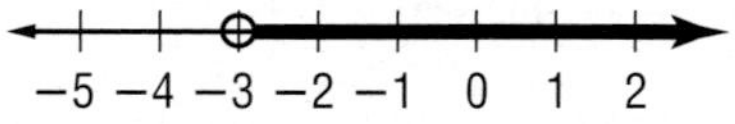

7.
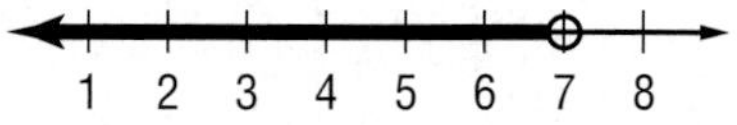

9. $d \geq 4$

11.
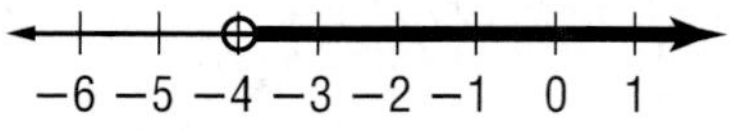

13.
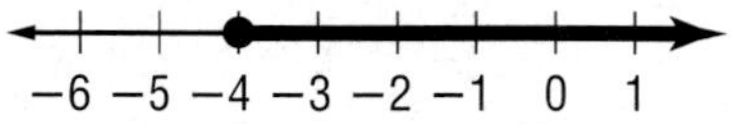

15.
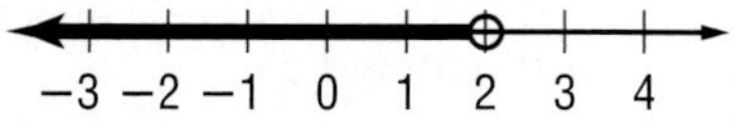

17.
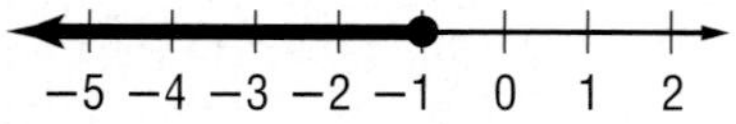

19.
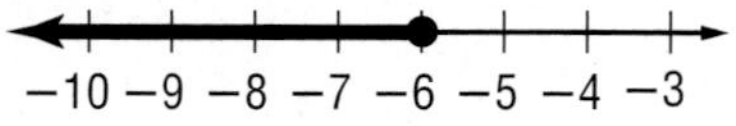

21. $x < -6$ **23.** $g > 7$ **25.** $u \geq 6$ **27.** $y \leq \frac{4}{3}$ **29.** $p \leq -1$ **31.** $d < 0$ **33.** $w + 1 \geq 5$; $w \geq 4$ **35.** $a \geq 16$ **37.** all except baseball-softball **39.** $h > 2{,}000{,}000$ **41.** $j \leq 2.14$, so 1 or 2 pairs of jeans **43.** B **45.** 8 **47.** -13

49.
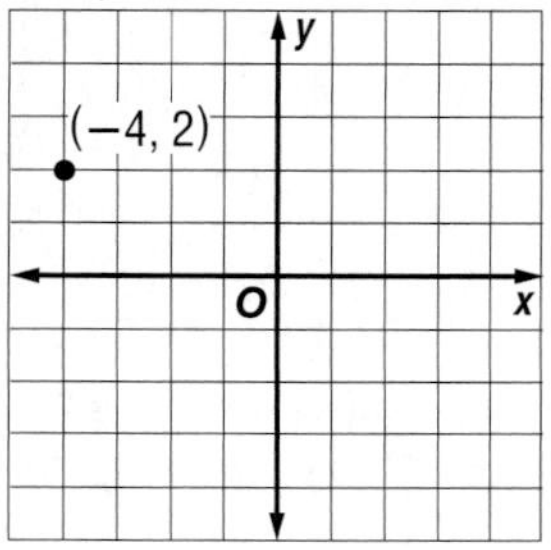

51.
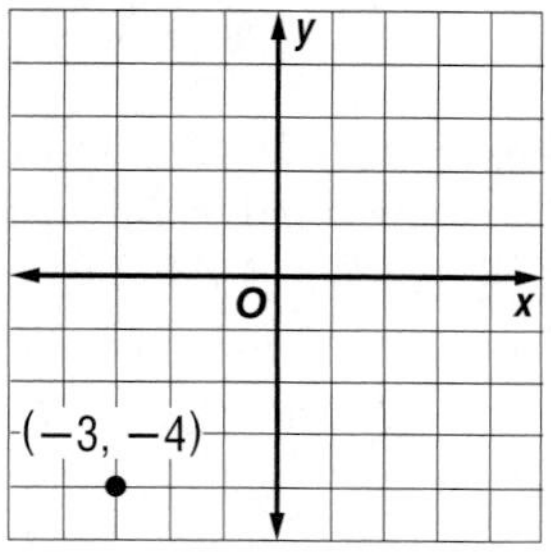

Pages 179–181 Lesson 4-6

1. Sample answer: $y = 2x$

3.

x	$x - 2$	y
1	1 − 2	−1
2	2 − 2	0
3	3 − 2	1
4	4 − 2	2

domain: {1, 2, 3, 4}; range: {−1, 0, 1, 2}

5.
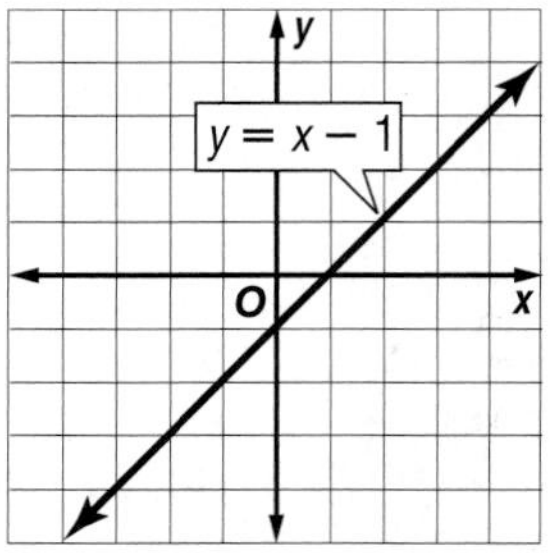

7.
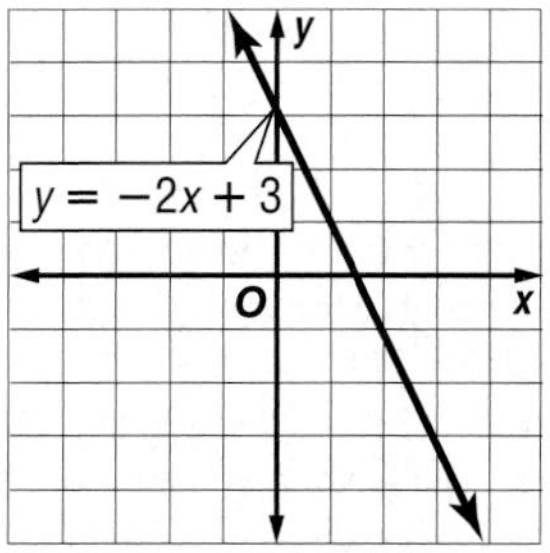

9.

x	$x + 5$	y
1	1 + 5	6
2	2 + 5	7
3	3 + 5	8
4	4 + 5	9

domain: {1, 2, 3, 4}; range: {6, 7, 8, 9}

11.

x	$-6x$	y
−1	−6(−1)	6
0	−6(0)	0
1	−6(1)	−6
2	−6(2)	−12

domain: {−1, 0, 1, 2}; range: {6, 0, −6, −12}

13.

x	$-2x - 2$	y
−1	−2(−1) − 2	0
0	−2(0) − 2	−2
1	−2(1) − 2	−4
2	−2(2) − 2	−6

domain: {−1, 0, 1, 2}; range: {0, −2, −4, −6}

15.
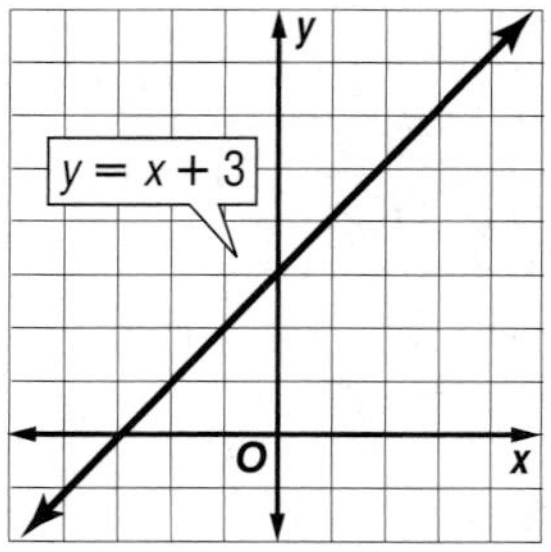

17.
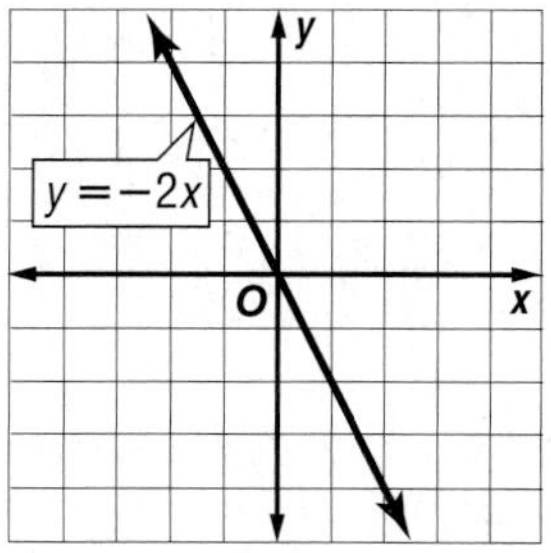

19.
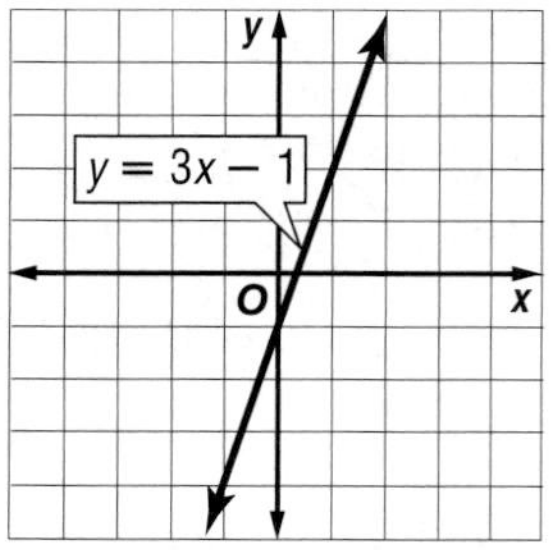

21.
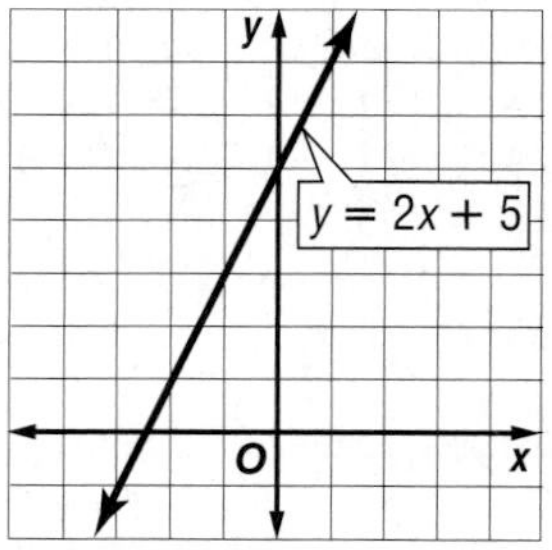

23.
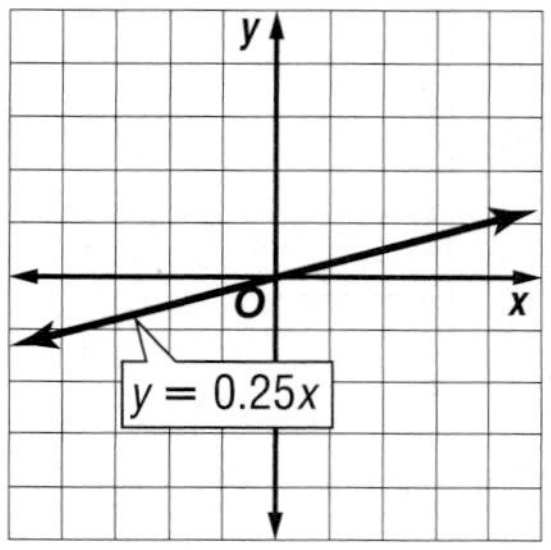

25.
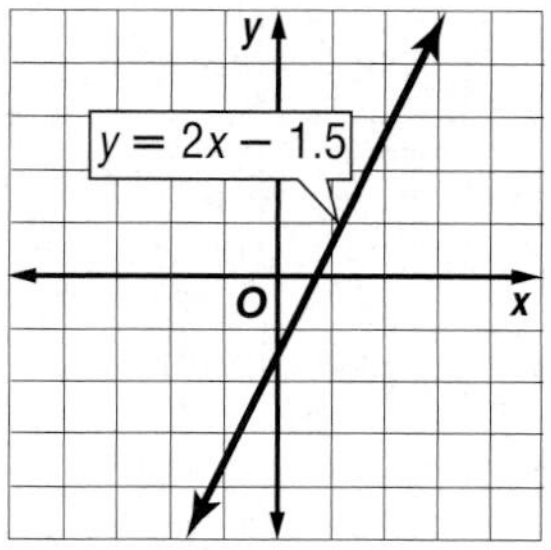

27. $y = x - 5$

x	$x - 5$	y
−1	−1 − 5	−6
0	0 − 5	−5
1	1 − 5	−4
2	2 − 5	−3

29.

x	$2.67x$	y
1	2.67(1)	2.67
2	2.67(2)	5.34
3	2.67(3)	8.01
4	2.67(4)	10.68

31. **Average Defense Spending (per person)**

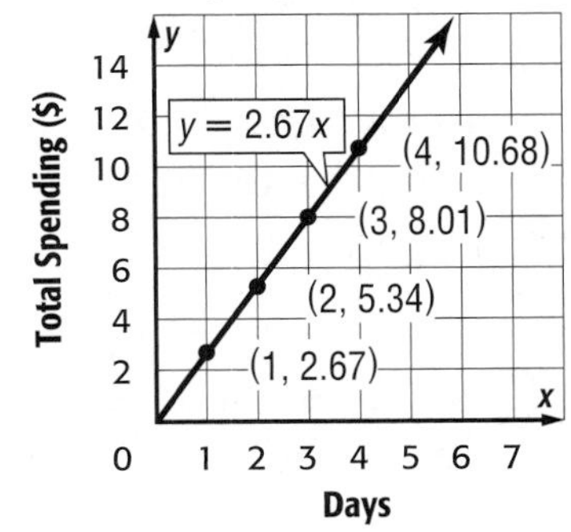

33. $y = 20x$

35. **Rectangles**

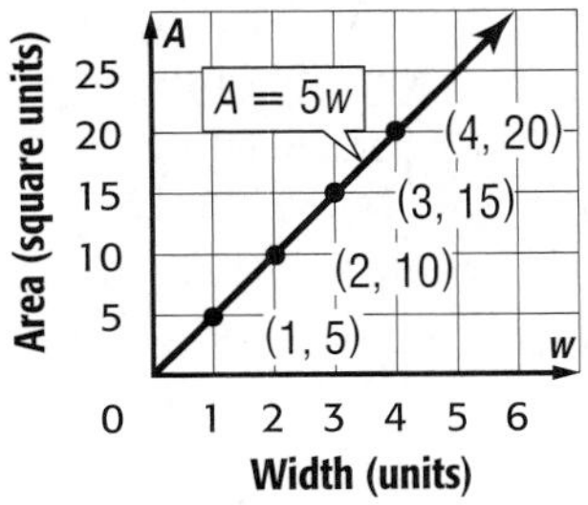

37. $y = x + 2$
39. $y = -4x$

41. nonlinear;

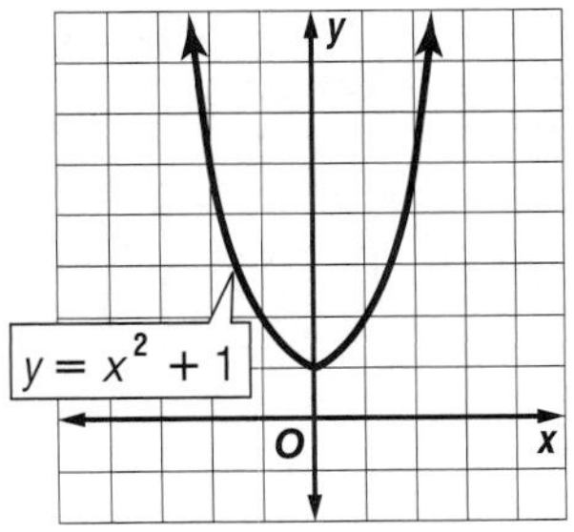

43. linear;

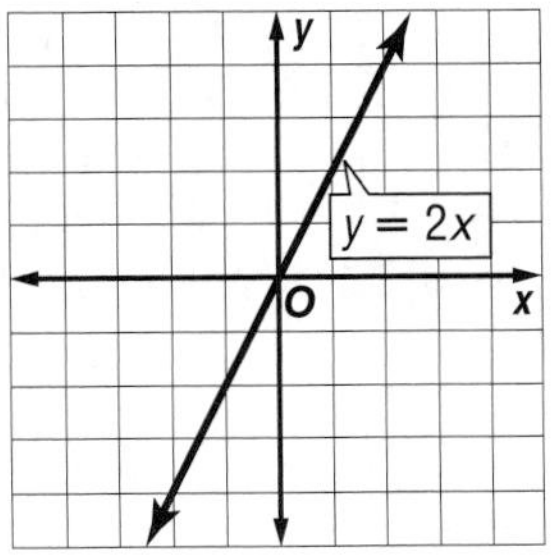

45. nonlinear;

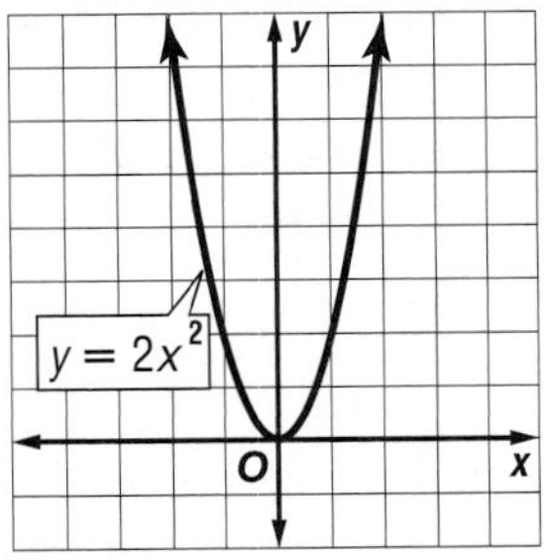

47. D **49.** $x \geq 25$ **51.** -13 **53.** -2 **55.** 3

Pages 184–185 Lesson 4-7

1. the steepness of a line from left to right
3. Sample answer:

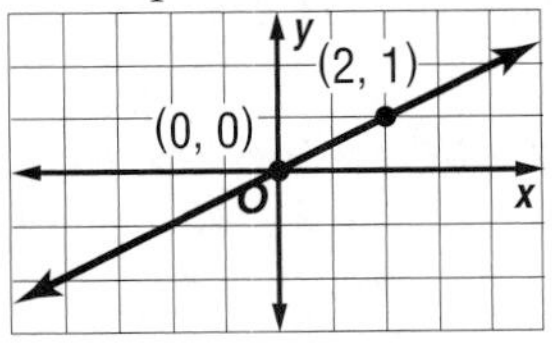

5. -1 **7.** 3 **9.** 1 **11.** -3 **13.** $\frac{1}{3}$ **15.** $-\frac{1}{2}$ **17.** Monica; her earnings are increasing faster.

19. The slope represents the amount of money earned with respect to the number of hours worked. **21.** 1994–1995; The slope is steepest. **23.** The sales remained constant; there is no change in the slope of the line. It is horizontal and the slope is equal to zero.
25. nonlinear

27.

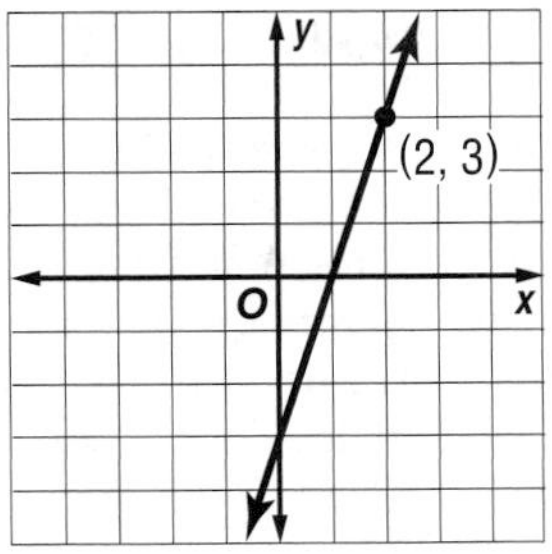

29.

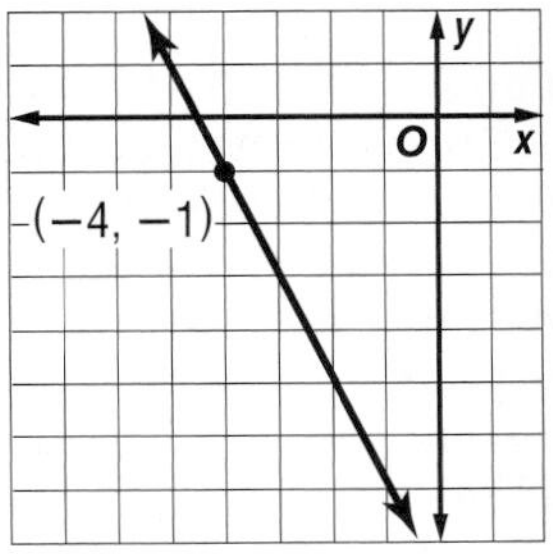

31. C

33.

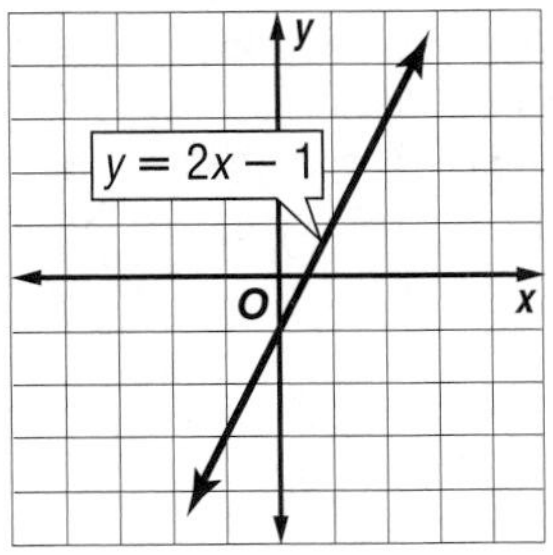

35.

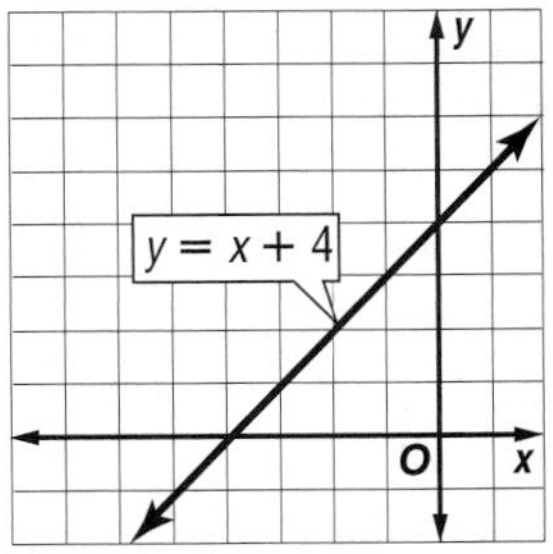

Pages 186–188 Chapter 4 Study Guide and Review

1. true **3.** true **5.** true **7.** false; 7 **9.** false; -6
11. $n + 5$ **13.** $2a$ **15.** $a + 10 = 23$ **17.** $\frac{56}{z} = 14$ **19.** 3
21. -1 **23.** -13 **25.** 23 **27.** -2 **29.** 4 **31.** -12 **33.** -9
35. 11 **37.** 6 **39.** -1 **41.** 9.8 **43.** 3

45. $y > 3$

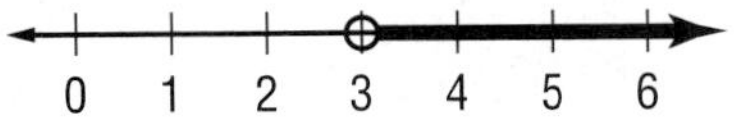

47. $d \leq 5$

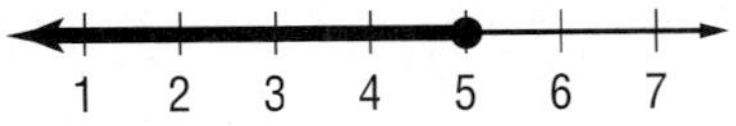

49. $s \leq 11$

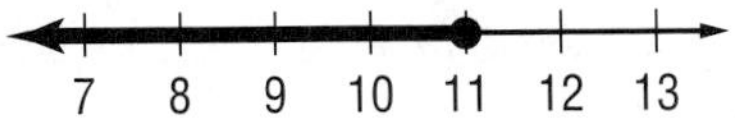

51. $m > -3$

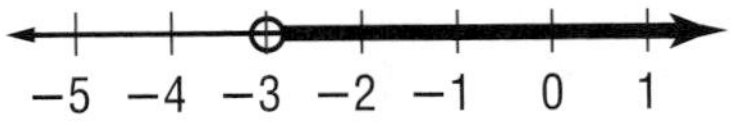

53. $t \geq 2$

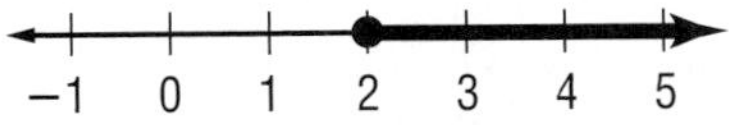

55.

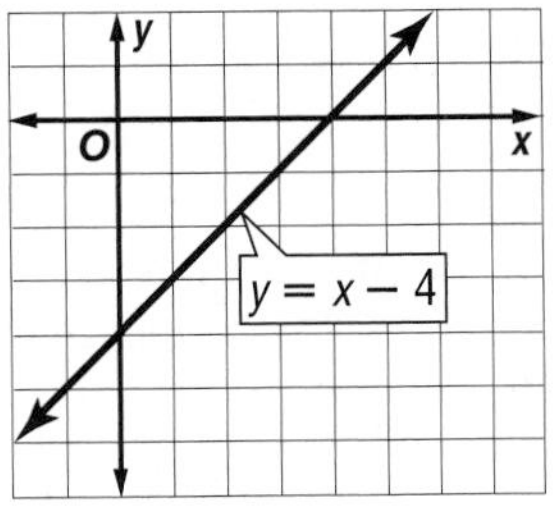

57.

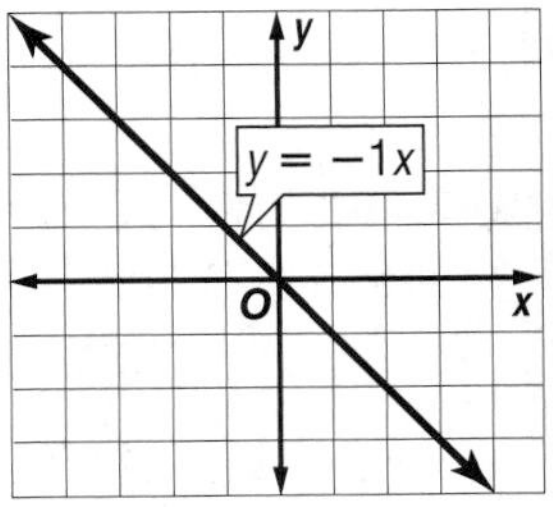

59.

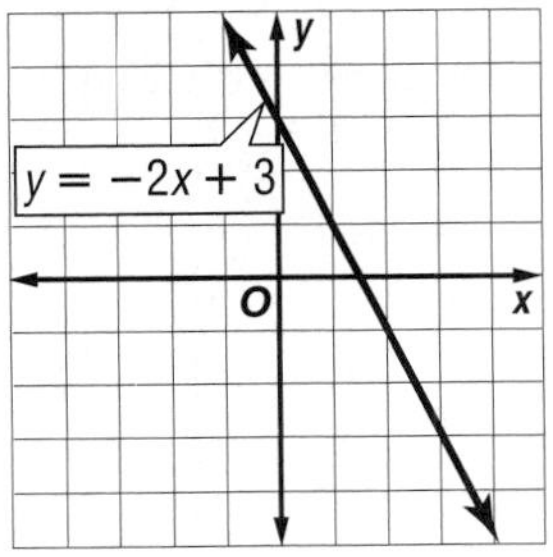

61. $y = 6x$ **63.** 4 **65.** -1

Selected Answers

Chapter 5 Fractions, Decimals, and Percents

Page 195 Chapter 5 Getting Started
1. greater **3.** 0.61 **5.** 0.33 **7.** 5 **9.** none **11.** 0.4 **13.** $2 \times 2 \times 2$ **15.** 7×7 **17.** 4^3 **19.** 1^5

Pages 199–200 Lesson 5-1
1. Sample answer: Both prime and composite numbers are whole numbers greater than 1. A prime number has exactly two factors, 1 and itself. Composite numbers have more than two factors. **3.** Sample answer: 9 is not prime. **5.** prime **7.** prime **9.** $2 \times 3 \times 5$ **11.** $2 \cdot 5 \cdot a \cdot c$ **13.** composite **15.** prime **17.** composite **19.** prime **21.** $2^2 \times 3^2$ **23.** $3^2 \times 11$ **25.** $2 \times 3 \times 5 \times 7$ **27.** $2 \times 3^2 \times 7$ **29.** composite **31.** $2 \cdot 2 \cdot 5 \cdot p \cdot q$ **33.** $7 \cdot 7 \cdot y \cdot y$ **35.** $2 \cdot 2 \cdot 2 \cdot 2 \cdot 3 \cdot a \cdot a \cdot b \cdot b$ **37.** 8 **39.** 7 **41.** 5^2 **43.** Sample answer: 5 and 7, 11 and 13, and 17 and 19 **45.** $2^3 \times 3 \times 5$ **47.** Sample answer: twenty 2-ft-by-3-ft tiles, or twelve 2-ft-by-5-ft tiles **49.** 36 **51.** composite
53.

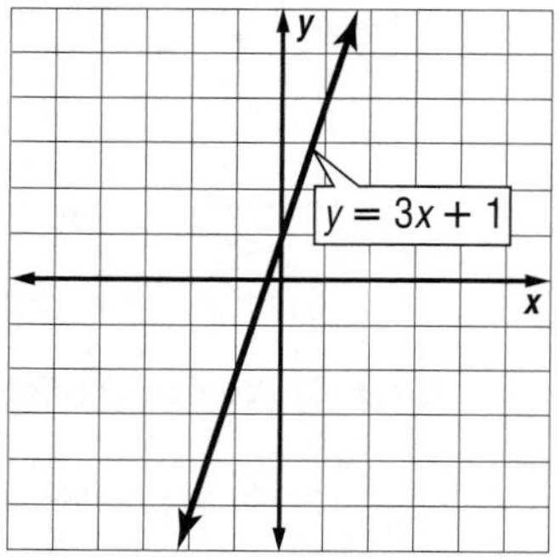

55. negative **57.** zero **59.** 2, 5, 10 **61.** 3, 9

Pages 205–206 Lesson 5-2
1. Sample answer: 32 and 48 **3.** Tiffany is correct; Charles included extra factors of 2 and 3. **5.** 15 **7.** 6 **9.** 6 **11.** 5 **13.** 24 **15.** 8 **17.** 6 **19.** 7 **21.** 5 **23.** never **25.** sometimes **27.** $9b$ **29.** $10x$ **31.** Sample answer: The first building is 9 in. high, 3 in. long, and 4 in. wide. The second building is 9 in. high, 6 in. long, and 5 in. wide. The third building is 9 in. high, 5 in. long, and 5 in. wide. **33.** No; Sample answer: There are a total of 16 pieces if the sandwiches are cut into 4-inch long pieces. **35.** True; their only common factor is 1. **37.** True; their only common factor is 1. **39.** G **41.** composite **43.** composite **45.** $\frac{1}{2}$ **47.** 0.02 L **49.** 24 **51.** 16

Pages 208–209 Lesson 5-3
1. Sample answer: A fraction is in simplest form if the GCF of the numerator and denominator is 1. **3.** Seki; Luther did not divide the numerator and denominator by the GCF. **5.** $\frac{3}{5}$ **7.** $\frac{4}{13}$ **9.** $\frac{5}{7}$ **11.** $\frac{7}{10}$ **13.** $\frac{4}{7}$ **15.** $\frac{6}{7}$ **17.** 1 **19.** $\frac{5}{6}$ **21.** $\frac{6}{10}, \frac{15}{25}$ **23.** $\frac{10}{16}, \frac{20}{32}$ **25.** $\frac{3}{5}$ **27.** $\frac{3}{4}$ **29.** No, because both the numerator and denominator can be divided by 2. **31.** Sample answer:

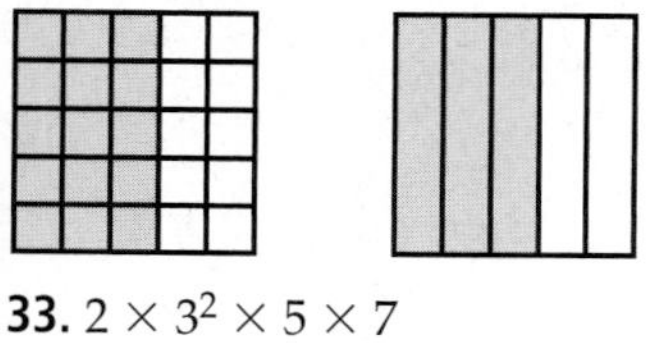

33. $2 \times 3^2 \times 5 \times 7$ **35.** 0.25 **37.** 0.375

Pages 212–213 Lesson 5-4
1. Sample answer: $\frac{1}{2} = 0.5$; $\frac{2}{9} = 0.\overline{2}$ **3.** $0.6\overline{3}$ **5.** $12.\overline{470}$ **7.** 3.625 **9.** $1.8\overline{3}$ **11.** $\frac{1}{10}$ **13.** 0.522 **15.** $5.92\overline{1}$ **17.** $13.1\overline{46}$ **19.** $30.\overline{6841}$ **21.** 0.8 **23.** $0.\overline{72}$ **25.** 0.15 **27.** 6.425 **29.** $0.\overline{714285}$ **31.** $4.\overline{203}$ **33.** $\frac{3}{4}$ **35.** $\frac{1}{5}$ **37.** $5\frac{24}{25}$ **39.** 13; $3\frac{12}{25}$ **41.** $8\frac{9}{10}$; $2\frac{3}{50}$ **43.** 2.3 h **45.** 3.160; Archimedes' **47.** B **49.** $\frac{5}{12}$ **51.** $\frac{2}{7}$ **53.** 12 **55.** $\frac{3}{9}$ **57.** $\frac{4}{12}$

Pages 218–219 Lesson 5-5
1. $\frac{9}{25}$; 36% **3.** 1%; the other ratios equal 10%. **5.** 29.2% **7.** 25% **9.** 80% **11.** $\frac{9}{10}$ **13.** $\frac{3}{4}$ **15.** 75% **17.** 12.2% **19.** $11\frac{2}{7}$% **21.** 99.9% **23.** 20% **25.** 80% **27.** 74% **29.** 26% **31.** 95% **33.** 12% **35.** 76% **37.** $\frac{1}{4}$ **39.** $\frac{9}{20}$ **41.** $\frac{1}{2}$ **43.** $\frac{31}{50}$ **45.** $\frac{17}{25}$ **47.** $\frac{11}{25}$ **49.** $\frac{1}{50}$ **51.** 36% **53.** 17 **55.** D **57.** $0.6\overline{5}$ **59.** $0.\overline{41}$ **61.** 162 **63.** 0.144

Pages 222–223 Lesson 5-6
1. Sample answer: 0.25, $\frac{1}{4}$, 25% **3.** 0.68 **5.** 0.276 **7.** 56% **9.** 8% **11.** 0.012 **13.** 0.7 **15.** 0.01 **17.** 0.022 **19.** 0.3025 **21.** 8% **23.** 60% **25.** 14.5% **27.** 70.25% **29.** 90.5% **31.** 80.8% **33.** 10% to 67% **35.** Z **37.** O **39.** 76.5% **41.** 99% **43.** $\frac{45}{80}$ **45.** G **47.** 72% **49.** 18% **51.** 9.375 **53.** 23 students **55.** 2×5^2 **57.** $2^2 \times 19$

Pages 225–226 Lesson 5-7
1. 3 and 4 are factors of 12; 12 is a multiple (and LCM) of 3 and 4. **3.** Sometimes; the LCM of 3 and 9 is 9; the LCM of 3 and 5 is 15. **5.** 42 **7.** 4 packages of hot dogs and 5 packages of buns **9.** 48 **11.** 72 **13.** 315 **15.** 72 **17.** 420 **19.** 144 **21.** $45n$, where $n = 5, 6, 7, \ldots$ **23.** 5 s; 3 s; 30 s **25.** $2^2 \times 5^2$, or 100 **27.** 2012 **29.** $\frac{17}{25}$ **31.** $\frac{3}{25}$ **33.** $<$ **35.** $<$

Pages 229–231 Lesson 5-8
1. Sample answer: $\frac{2}{15}, \frac{5}{6}$ **3.** 45 **5.** $>$ **7.** $>$ **9.** 0.59, 60%, $\frac{31}{50}$ **11.** yes **13.** 20 **15.** 70 **17.** 36 **19.** $>$ **21.** $<$ **23.** $=$ **25.** $<$ **27.** $>$ **29.** $<$ **31.** $<$ **33.** $>$ **35.** 3 saves out of 4; 3 out of 4 has an average of 0.75; 7 out of 11 has an average of about 0.64. **37.** 19%, $\frac{1}{5}$, 0.23 **39.** 0.615, 62%, $\frac{5}{8}$ **41.** 0.49, 0.5, $\frac{49}{50}$ **43.** Christopher **45.** no **47.** no **49.** yes **51.** Sample answer: $2\frac{3}{4}$ is the same as 2.75. Fill the cup to halfway between 2.7 and 2.8. **53.** $\frac{63}{32}$; the difference of $\frac{63}{32}$ and 2 is the least. **55.** McGwire: $\frac{70}{509} = 0.138$; Sosa: $\frac{66}{643} = 0.103$; McGwire **57.** Sample answer: When the numerators are the same, the fraction with the largest denominator is the least fraction. **59.** 47%, $\frac{4}{7}$, $47\frac{2}{5}$, 47.41, $47.\overline{4}$ **61.** 0.04 **63.** 0.14

Pages 232–234 Chapter 5 Study Guide and Review
1. c **3.** d **5.** a **7.** prime **9.** composite **11.** 2^7 **13.** 5×19 **15.** 9 **17.** 14 **19.** $18w$ **21.** $\frac{4}{5}$ **23.** $\frac{1}{11}$ **25.** $\frac{37}{45}$ **27.** $\frac{1}{20}$

29. 0.875 **31.** $4.\overline{3}$ **33.** $1.\overline{857142}$ **35.** $\frac{11}{25}$ **37.** $\frac{9}{50}$ **39.** $\frac{2}{25}$ **41.** 44% **43.** 40% **45.** $\frac{19}{20}$ **47.** $\frac{4}{25}$ **49.** 0.48 **51.** 0.125 **53.** 61% **55.** 19% **57.** 0.23 **59.** 8 **61.** 24 **63.** 120 **65.** = **67.** > **69.** =

Chapter 6 Applying Fractions

Page 239 Chapter 6 Getting Started

1. Property **3.** 35 **5.** 30 **7.** 21.6 **9.** 0.83 **11.** 7.5 **13.** $2\frac{6}{5}$ **15.** $5\frac{5}{4}$ **17.** $\frac{43}{4}$ **19.** $\frac{22}{5}$

Pages 242–243 Lesson 6-1

1. Sample answer: Models are useful to determine whether fractions are closer to 0, $\frac{1}{2}$, or 1, and to check estimates. **3a.** Yes; the sum of $\frac{1}{2}$ and a fraction greater than $\frac{1}{2}$ is greater than 1. **3b.** Yes; both fractions are greater than $\frac{1}{2}$, so the sum is greater than $\frac{1}{2} + \frac{1}{2}$ or 1. **3c.** No; both fractions are less than $\frac{1}{2}$, so the sum is less than 1. **5–37. Sample answers are given.** **5.** $6 \times 3 = 18$ **7.** $1 - 0 = 1$ **9.** $22 \div 11 = 2$ **11.** $4 + 5 = 9$ **13.** $5 - 3 = 2$ **15.** $3 \cdot 6 = 18$ **17.** $1 + \frac{1}{2} = 1\frac{1}{2}$ **19.** $\frac{1}{2} - 0 = \frac{1}{2}$ **21.** $0 \times 1 = 0$ **23.** $1 \div 1 = 1$ **25.** $\frac{1}{2} \times 14 = 7$ **27.** $25 \div 5 = 5$ **29.** $12 \div 4 = 3$ **31.** $24 \div 3 = 8$ **33.** $-3 \times 9 = -27$ **35.** $-\frac{1}{6} \times (-66) = 11$ **37.** $3 - 1$ or 2 c **39.** It will be greater than the actual quotient. **41.** H **43.** > **45.** < **47.** 12 **49.** 24

Pages 246–247 Lesson 6-2

1. Sample answer:

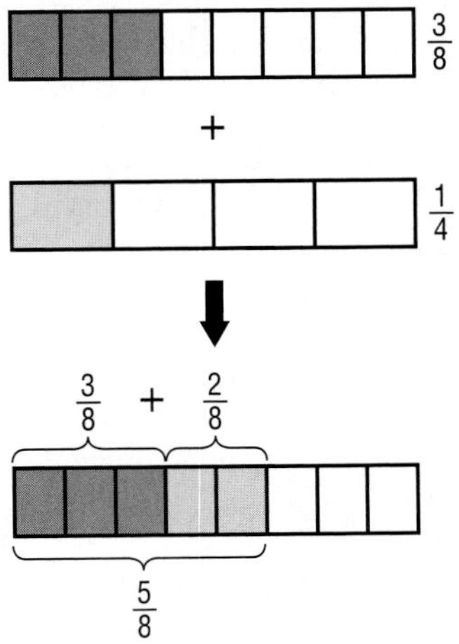

3. Jacinta; Marissa did not rename the fractions using the LCD. **5.** $1\frac{2}{7}$ **7.** $\frac{1}{4}$ **9.** $\frac{1}{4}$ **11.** $1\frac{5}{18}$ **13.** $\frac{8}{15}$ **15.** $\frac{4}{7}$ **17.** $\frac{2}{3}$ **19.** $\frac{2}{3}$ **21.** $\frac{19}{30}$ **23.** $\frac{1}{5}$ **25.** $1\frac{1}{4}$ **27.** $\frac{1}{24}$ **29.** $1\frac{13}{24}$ **31.** $\frac{1}{20}$ **33.** $\frac{11}{15}$ **35.** $\frac{5}{6}$ **37.** $\frac{1}{12}$ **39.** Jon; $\frac{1}{18}$ of the race **41.** Sierra; she has $\frac{1}{4}$ of her allowance left over; Jacob has $\frac{1}{15}$ left over. **43.** C **45.** Sample answer: $1 - \frac{1}{2} = \frac{1}{2}$ **47.** Sample answer: $16 \div 8 = 2$ **49.** friend/family **51.** $4\frac{5}{3}$ **53.** $11\frac{6}{5}$

Pages 250–251 Lesson 6-3

1. Sample answer: A table is $8\frac{1}{2}$ feet long. A tablecloth is $2\frac{1}{3}$ feet longer. How long is the tablecloth? Answer: $8\frac{1}{2} + 2\frac{1}{3}$ or $10\frac{5}{6}$ ft **3.** less than; $7 + 1 < 2 + 7$ **5.** $4\frac{2}{3}$ **7.** $9\frac{13}{20}$ **9.** $8\frac{1}{4}$ **11.** $17\frac{5}{24}$ **13.** $4\frac{5}{8}$ **15.** $9\frac{5}{9}$ **17.** $8\frac{1}{3}$ **19.** $7\frac{5}{12}$ **21.** $7\frac{1}{2}$ **23.** $8\frac{1}{4}$ **25.** $5\frac{1}{15}$ **27.** $6\frac{5}{6}$ **29.** $19\frac{2}{3}$ **31.** $6\frac{3}{5}$ **33.** $4\frac{1}{3}$ **35.** $31\frac{3}{16}$ **37.** $38\frac{5}{8}$ **39.** $6\frac{6}{11}$ **41.** $13\frac{7}{18}$ **43.** 20 ft **45.** H **47.** Sample answer: $1 \div 1 = 1$ **49.** Sample answer: $9 \times 7 = 63$ **51.** > **53.** < **55.** $\frac{19}{8}$ **57.** $\frac{51}{10}$

Pages 256–257 Lesson 6-4

1. Sample answer: $\frac{2}{3}$ and $\frac{4}{5}$ **3.** Less than; $\frac{4}{5}$ of 18 is a fraction of 18. **5.** $\frac{1}{5}$ **7.** $\frac{2}{3}$ **9.** $6\frac{3}{5}$ **11.** $\frac{3}{32}$ **13.** $\frac{1}{6}$ **15.** $\frac{1}{6}$ **17.** $\frac{1}{2}$ **19.** $\frac{5}{36}$ **21.** $4\frac{1}{2}$ **23.** 2,700 votes **25.** $2\frac{2}{3}$ **27.** 16 **29.** 42 **31.** $8\frac{2}{3}$ **33.** $\frac{1}{4}$ **35.** $\frac{7}{40}$ or 0.175 km **37.** Always; sample answer: the numerators of the fractions are less than the denominators, so the product of the numerators will be less than the product of the denominators, resulting in a fraction that is less than 1. **39.** 7/16 **41.** $\frac{6}{7}$ **43.** $1\frac{7}{18}$ **45.** 4 **47.** 3 **49.** −5°F **51.** 62.7 **53.** 2.88

Pages 260–261 Lesson 6-5

1. no; $\frac{8}{3} \neq 24$ **3.** $\frac{2}{3}$, 3; The other pairs of numbers are reciprocals, but these are not. **5.** $\frac{9}{2}$ or $4\frac{1}{2}$ **7.** $\frac{5}{29}$ **9.** −12 **11.** $1\frac{1}{2}$ **13.** 20.5 **15.** $\frac{2}{11}$ **17.** $-\frac{8}{3}$ or $-2\frac{2}{3}$ **19.** $\frac{1}{3}$ **21.** $\frac{5}{22}$ **23.** 36 **25.** 30 **27.** −24 **29.** $-\frac{4}{5}$ **31.** 42.12 **33.** 2.88 **35.** 137.5 mi **37.** Sample answer: Use the Multiplicative Property to multiply each side by 2. Then use the Division Property to divide each side by h. So, $b = \frac{2A}{h}$. **39.** 16 **41.** 9 **43.** $2\frac{2}{3}$ **45.** $\frac{6}{17}$ **47.** $\frac{5}{6}$ **49.** Sample answer: $24 \div 12 = 2$ **51.** Sample answer: $1 \div 1 = 1$

Page 266 Lesson 6-6

1. Sample answer: How many $\frac{1}{4}$-pound hamburgers can be made with 10 pounds of ground beef? **3.** $2\frac{2}{5}$ **5.** $\frac{1}{15}$ **7.** 24 boxes **9.** $\frac{2}{3}$ **11.** $1\frac{1}{6}$ **13.** $\frac{2}{9}$ **15.** $3\frac{3}{4}$ **17.** $\frac{1}{6}$ **19.** $2\frac{7}{15}$ **21.** greater than 1 because $7\frac{1}{6} > 3\frac{2}{3}$ **23.** $\frac{11}{16}$ lb **25.** $\frac{7}{6}$ or $1\frac{1}{6}$ **27.** $\frac{1}{8}$ **29.** $\frac{1}{16}$ **31.** 14 **33.** 16

Pages 268–269 Lesson 6-7

1. Sample answer: A recipe calls for 2 pints of strawberries. How many cups is this? Answer: 4 c **3.** 3 **5.** 3 **7.** 500 **9.** 100 T **11.** 6 **13.** 16 **15.** 9 **17.** 10,560 **19.** $4\frac{1}{2}$ **21.** $5\frac{1}{2}$ **23.** $1\frac{1}{3}$ **25.** $10\frac{3}{4}$ **27.** 82.8 **29.** $\frac{1}{2}$ **31.** $43\frac{3}{4}$ gal

33.

Pounds	Ounces
1	16
2	32
3	48
4	64

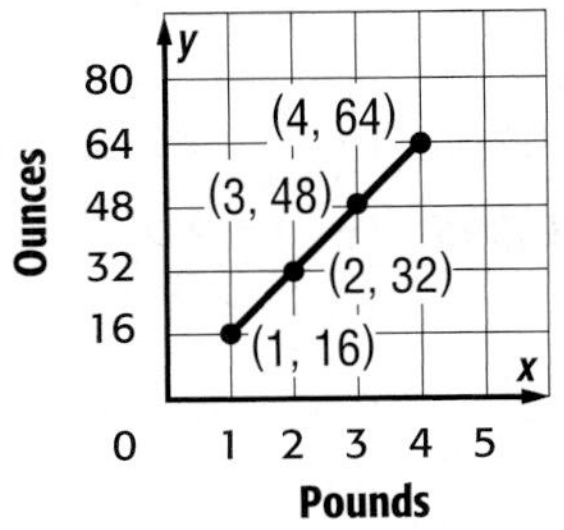

The graph is a straight line. For each x-value increase of 1, the y value increases by 16. **35.** A **37.** 28 **39.** −10.8 **41.** 34 **43.** 19

Pages 272–273 Lesson 6-8

1. Perimeter is the sum of measures, so the units are the same. Area is the product of measures, so the units are squared. **3.** 18 yd; 20 yd^2 **5.** 26 cm; 42 cm^2 **7.** $23\frac{1}{3}$ ft **9.** 82 in.; 414 in^2 **11.** 65.4 m; 183.6 m^2 **13.** $30\frac{1}{5}$ in.; $47\frac{7}{10}$ in^2 **15.** 60 cm; 216 cm^2 **17.** $15\frac{3}{20}$ in.; 14 in^2 **19.** $37\frac{1}{3}$ yd or 112 ft; $81\frac{2}{3}$ yd^2 or 735 ft^2 **21.** The perimeter is doubled, the area is quadrupled. **23.** $35\frac{13}{24}$ ft **25.** 7 boxes **27.** 1.5 acres; 65,010 ÷ 43,560 **29.** The rectangles have the same perimeters; as the lengths and widths get closer, the areas increase. **31.** 36 **33.** A **35.** 10,500 **37.** $7\frac{1}{2}$ **39.** 33 **41.** 44 **43.** $4\frac{5}{7}$

Pages 276–277 Lesson 6-9

1. The circumference increases by 2π for every 1-unit increase in r. **3.** Mya; Aidan incorrectly applied the formula that uses the diameter. **5.** 44 m **7.** 73.5 cm **9.** 11 yd **11.** 50.2 ft **13.** 44 yd **15.** 47.4 m **17.** 12.6 km **19.** 285.7 m **21.** $16\frac{1}{2}$ in. **23.** 785 ft **25.** 2π or 6.28 **27.** 69.1 in. **29.** 17.6 cm; 19 cm^2 **31.** 8 yd

Pages 278–280 Chapter 6 Study Guide and Review

1. diameter **3.** Area **5.** reciprocal **7.** Sample answer: $3 \div 1 = 3$ **9.** Sample answer: $1 \times 0 = 0$ **11.** Sample answer: $\frac{1}{2} \cdot 26 = 13$ **13.** $\frac{1}{6}$ **15.** $\frac{1}{12}$ **17.** $1\frac{3}{5}$ **19.** $\frac{5}{24}$ **21.** $1\frac{1}{10}$ **23.** $7\frac{3}{5}$ **25.** $10\frac{1}{7}$ **27.** $2\frac{4}{15}$ **29.** $15\frac{1}{8}$ **31.** $\frac{6}{35}$ **33.** $\frac{2}{7}$ **35.** $1\frac{3}{4}$ **37.** $1\frac{11}{16}$ mi **39.** $\frac{1}{5}$ **41.** 16 **43.** -22.8 **45.** $\frac{7}{10}$ **47.** $6\frac{1}{2}$ **49.** $3\frac{3}{10}$ **51.** $\frac{3}{22}$ **53.** $\frac{3}{4}$ **55.** 24 **57.** 4 **59.** 3 **61.** $6\frac{1}{4}$ ft; $1\frac{5}{16}$ ft^2 **63.** 11 in.; $2\frac{1}{2}$ in^2 **65.** $13\frac{2}{3}$ ft; $10\frac{1}{2}$ ft^2 **67.** 25.1 yd **69.** 26.4 ft

Chapter 7 Ratios and Proportions

Page 287 Chapter 7 Getting Started

1. ratio **3.** 48.1 **5.** 7.4 **7.** $\frac{1}{5}$ **9.** $\frac{19}{23}$ **11.** $\frac{39}{50}$ **13.** $\frac{3}{50}$ **15.** 24 **17.** $5\frac{1}{2}$ **19.** 6 **21.** $1\frac{1}{2}$

Pages 290–291 Lesson 7-1

1. $\frac{8}{15}$, 8:15, 8 to 15 **3.** 3 to 2; The other ratios equal $\frac{2}{3}$. **5.** $\frac{4}{15}$ **7.** $\frac{11}{4}$ **9.** $\frac{4}{3}$ **11.** yes; $\frac{9}{15} = \frac{3}{5}$ **13.** $\frac{21}{23}$ **15.** $\frac{1}{5}$ **17.** $\frac{15}{7}$ **19.** $\frac{11}{30}$ **21.** $\frac{3}{16}$ **23.** $\frac{5}{1}$ **25.** $\frac{3}{8}$ **27.** yes; $\frac{9}{15} = \frac{3}{5}$ and $\frac{6}{10} = \frac{3}{5}$ **29.** yes; $8:21 = \frac{8}{21}$ and $16:42 = \frac{8}{21}$ **31.** yes; 6 lb:72 oz $= \frac{4}{3}$ and 2 lb:24 oz $= \frac{4}{3}$ **33.** $\frac{1}{13}$ **35.** no **37.** $\frac{2}{5}$; $\frac{7}{3}$; No, the ratio is greater for the dollar bill. **39.** 4; The ratios of successive terms increase by 1. **41.** G **43.** 26 ft; 36 ft^2 **45.** 19 cm; 15 cm^2 **47.** 4.9 **49.** $0.31

Pages 294–295 Lesson 7-2

1. Sample answer: $\frac{30\text{ mi}}{2\text{ h}} = \frac{15\text{ mi}}{\text{h}}$ **3.** a; Sample answer: $\frac{20\text{ ft}}{1\text{ min}} \rightarrow \frac{40\text{ ft}}{1\text{ min}}$ represents an increase in speed since 40 ft/min > 20 ft/min. If the same distance is covered in a longer period of time, then the rate is decreased. Consider $\frac{20\text{ ft}}{1\text{ min}} \rightarrow \frac{20\text{ ft}}{2\text{ min}}$. Since $\frac{20\text{ ft}}{2\text{ min}} = 10$ ft/min, the rate is decreased. **5.** 60 words per min **7.** $0.98 per can **9.** Music Rox; $14.95/CD < $17.99/CD **11.** 152 customers per day **13.** $1.48 per lb **15.** 270 C per serving **17.** 32 mi per gal **19.** 150 for $11.99 **21.** 4,000 gal per min **23.** about 179 people per sq mi **25.** Sometimes; a ratio that compares two measurements with different units is a rate, such as $\frac{2\text{ mi}}{10\text{ min}}$. A ratio that compares two numbers or two measurements with like units is not a rate, such as $\frac{2\text{ c}}{3\text{ c}}$. **27.** human, 72 beats/min; elephant, 35 beats/min **29.** 32 tickets per h **31.** I **33.** $\frac{6}{1}$ **35.** $\frac{10}{7}$ **37.** 56 **39.** $1\frac{1}{3}$

Pages 299–300 Lesson 7-3

1. Compare the cross products. If they are equal, the ratios are equivalent. **3.** $\frac{2}{9}$; The other ratios equal $\frac{2}{3}$. **5.** no **7.** 2 **9.** 21 **11.** 9 **13.** 29.4 **15.** $6\frac{2}{3}$ or about 6.7 oz **17.** no **19.** no **21.** no **23.** 8 kittens **25.** 6 **27.** 6 **29.** 75 **31.** 2 **33.** 63 **35.** 22.5 **37.** 1,040 **39.** 13.5 **41.** $\frac{13}{15} = \frac{x}{100}$; 87% **43.** 4 c **45.** 210 words **47.** Sample answers: 1 and 36; 2 and 18; 4 and 9 **49.** 64 **51.** $\frac{5}{9}$ **53.** $\frac{7}{40}$ **55.** 56 **57.** $1\frac{3}{5}$ **59.** 52

Pages 306–308 Lesson 7-4

3. 200 km **5.** 3.5 cm **7.** $\frac{1}{90}$ **9.** 255 km **11.** 1,205 km **13.** 6 in.; $\frac{1}{240}$ **15.** 12 cm; $\frac{1}{300}$ **17.** 4 in. **19.** $\frac{3}{4}$ in. **21.** $\frac{1}{48}$ **23.** $12\frac{3}{16}$ mi **25.** $\frac{1}{12}$ **27.** 1 in. = 70 mi **29.** D **31.** 40 mi/h **33.** $1.55 per lb **35.** $\frac{11}{40}$ **37.** $\frac{23}{75}$

Pages 314–315 Lesson 7-5

1. Sample answer: Write a proportion using the given fraction and $\frac{n}{100}$. Then solve for n and add a %. Or, write the fraction as a decimal. Then multiply by 100 and add a %. **3.** Sample answer: It is easier to compare percents than it is to compare fractions. **5.** $\frac{3}{16}$ **7.** $\frac{23}{40}$ **9.** 62.5% **11.** 36.36% **13.** $\frac{7}{40}$ **15.** $\frac{31}{500}$ **17.** $\frac{23}{80}$ **19.** $\frac{1}{3}$ **21.** $\frac{13}{16}$ **23.** $\frac{63}{80}$ **25.** 55% **27.** 37.5% **29.** 96.67% **31.** 71.43% **33.** 1.25% **35.** 41.67% **37.** 48% **39.** less than; $\frac{26}{125} = 20.8\%$ **41.** D **43.** 67.5 ft **45.** $1\frac{2}{3}$ **47.** 17.5 **49.** $\frac{9}{100}$ **51.** $\frac{42.5}{100}$

Pages 317–318 Lesson 7-6

1. 170%; 1.7 **3.** $63\frac{1}{2}$; It is the only number not equivalent to $63\frac{1}{2}\%$. **5.** 2; 2 **7.** 0.0015; $\frac{3}{2{,}000}$ **9.** 0.15% **11.** 0.05% **13.** 3.5; $3\frac{1}{2}$ **15.** 6; 6 **17.** 0.006; $\frac{3}{500}$ **19.** 0.0055; $\frac{11}{2{,}000}$ **21.** 2.6; $2\frac{3}{5}$ **23.** 0.0005; $\frac{1}{2{,}000}$ **25.** 850% **27.** 0.9% **29.** 264% **31.** 0.34% **33.** 350% **35.** 0.4% **37.** 162.5% **39.** 0.003; $\frac{3}{1{,}000}$ **41.** about 0.595% **43.** A **45.** $\frac{3}{40}$ **47.** $\frac{1}{16}$ **49.** 6 in. **51.** 0.065 **53.** 0.123

Pages 320–321 Lesson 7-7

1. $\frac{x}{22} = \frac{0.5}{100}$ **3.** Sample answer: Find the number of people surveyed who use a white erasable markerboard to record information during a meeting; 525 people. **5.** 110.5 **7.** 0.1 **9.** 65 **11.** $160 **13.** 51.3 **15.** 12.5 **17.** 0.4 **19.** 0.6 **21.** $\frac{x}{1{,}746} = \frac{22}{100}$; about 384 students **23.** $300.96 **25.** B **27.** 750% **29.** 0.04% **31.** 73.33% **33.** 20 **35.** 4.5

Page 325 Lesson 7-8
1. b **3.** 36% **5.** 0.7 **7.** 25% **9.** 8.6 **11.** 375 **13.** 7.5% **15.** 4.1 **17.** 30 **19.** 0.04 **21.** 60% **23.** 20% of 500, 20% of 100, 5% of 100; Sample answer: If the percent is the same but the base is greater, then the part is greater. If the base is the same but the percent is greater, then the part is greater. **25.** 65% **27.** 31.5

Pages 326–328 Chapter 7 Study Guide and Review
1. e **3.** a **5.** c **7.** g **9.** d **11.** $\frac{4}{3}$ **13.** $\frac{2}{3}$ **15.** yes; $\frac{3}{5} = \frac{21}{35}$ **17.** yes; 27:15 = $\frac{9}{5}$ and 9:5 = $\frac{9}{5}$ **19.** $4.75 per lb **21.** $9.50 per h **23.** $3\frac{1}{2}$ laps per min **25.** 6 **27.** 12.5 **29.** 9 **31.** 94 cm **33.** 27 cm **35.** 5.2 cm **37.** $\frac{27}{500}$ **39.** 12.5% **41.** 17.5% **43.** 0.0075; $\frac{3}{400}$ **45.** 0.005; $\frac{1}{200}$ **47.** 475% **49.** 0.95% **51.** 34.1 **53.** 0.6 **55.** 0.3 **57.** 130.2

Chapter 8 Applying Percent

Page 333 Chapter 8 Getting Started
1. true **3.** 48 **5.** 1,512 **7.** 1.75 **9.** 0.8 **11.** 130 **13.** 216.7 **15.** 0.4 **17.** 0.075 **19.** 39.5%

Pages 336–337 Lesson 8-1
1. Sample answer: Find $\frac{1}{5} \cdot 140$ or find 10% of 140 and multiply by 2; both estimates equal 28. **3.** Ian; Mandy incorrectly changed 1.5% to 1.5, which is 150%.
5–31. Sample answers are given. **5.** $\frac{1}{2} \cdot 160 = 80$ **7.** $\frac{3}{4} \cdot 20 = 15$ **9.** $0.1 \cdot 400 = 40$ and $2 \cdot 40 = 80$ **11.** $(1 \cdot 70) + \left(\frac{1}{2} \cdot 70\right) = 105$ **13.** $0.01 \cdot 80 = 0.8$ and $\frac{1}{2} \cdot 0.8 = 0.4$ **15.** $\frac{1}{4} \cdot 400 = 100$ **17.** $\frac{3}{4} \cdot 280 = 210$ **19.** $\frac{2}{3} \cdot 15 = 10$ **21.** $0.1 \cdot 40 = 4$ and $6 \cdot 4 = 24$ **23.** $0.1 \cdot 80 = 8$ and $8 \cdot 8 = 64$ **25.** $0.1 \cdot 120 = 12$ and $3 \cdot 12 = 36$ **27.** $(1 \cdot 54) + \left(\frac{1}{3} \cdot 54\right) = 72$ **29.** $4 \cdot 12 = 48$ **31.** $0.01 \cdot 400 = 4$ and $\frac{1}{2} \cdot 4 = 2$
33. Sample answer: $0.01 \cdot 3{,}000 = 30$ lb **35.** Sample answer: Find 1% of a number, then multiply by $\frac{3}{8}$. **37.** H **39.** 64.8 **41.** 157.1 **43.** 50 **45.** 357.1

Pages 342–343 Lesson 8-2
1. missing base **3.** $n = 0.88 \cdot 300$; 264 **5.** $3 = 0.12 \cdot n$; 25 **7.** 29.7%; 0.297 **9.** $n = 0.53 \cdot 470$; 249.1 **11.** $26 = n \cdot 96$; 27.1% **13.** $17 = 0.40 \cdot n$; 42.5 **15.** $30 = n \cdot 64$; 46.9% **17.** $1.45 = 0.33 \cdot n$; 4.4 **19.** $n = 0.752 \cdot 600$; 451.2 **21.** $14 = 0.028 \cdot n$; 500 **23.** $230 = n \cdot 200$; 115% **25.** 6% **27.** 42% **29.** 28% **31.** Sample answer: If the percent is less than 100%, then the part is less than the number; if the percent equals 100%, then the part equals the number; if the percent is greater than 100%, then the part is greater than the number. **33.** I **35.** Sample answer: $\frac{3}{10} \cdot 50 = 15$ **37.** 1,312.5 mg **39.** 18 **41.** 16.7% **43.** 31.6

Pages 346–347 Lesson 8-3
1. Sample answer: Randomly select a part of the group to get a sample. Find their preferences, and use the results to find the percent of the total group. **3.** 8.96 million **5.** 1,396 students **7.** about 110 owners **9.** 6,500 teens **11.** 800,000 people **13.** $54 = 0.72 \cdot n$; 75 **15.** 0.3 **17.** 4.5

Pages 352–353 Lesson 8-4
1. Sample answer: The sale price of a T-shirt is $14. If the shirt originally cost $25, what is the percent discount? Answer: 44% decrease **3.** Jada; Miranda did not write a ratio comparing the change to the original amount. **5.** 19% inc. **7.** 25% inc. **9.** 73% inc. **11.** 20% inc. **13.** 50% dec. **15.** 71% dec. **17.** 30% dec. **19.** 18% inc. **21.** 1% inc. **23.** 100% **25.** 21% inc. **27.** 1.2% increase **29.** 41% **31.** 16 to 17; 72% **33.** C **35.** 660 students **37.** $n = 0.21 \cdot 62$; 13.0 **39.** 0.055 **41.** 0.0675

Pages 356–357 Lesson 8-5
1. $6.86 **3.** Sample answer: The regular price of a CD is $17.95. The total cost is $18.98. **5.** $1,338.75 **7.** $47.33 **9.** 9% **11.** $46.40 **13.** $103.95 **15.** $2.58 **17.** $26.53 **19.** $26.07 **21.** 75% **23.** 23% **25.** $131.66 **27.** $600.88 **29.** $366.56 **31.** $57.60 **33.** B **35.** 50% inc. **37.** 73% inc. **39.** 292.3 **41.** 99

Pages 359–360 Lesson 8-6
1. Sample answer: Write the interest rate as a decimal and write the months in years. Then multiply $500 \cdot 0.06 \cdot 1.5$. **3.** $38.40 **5.** $1,417.50 **7.** $23.20 **9.** $21.38 **11.** $330 **13.** $8.54 **15.** $45.31 **17.** $26.37 **19.** $210 **21.** 4% **23.** $627, $655.22, $684.70 **25.** G **27.** 29% inc. **29.** 35% dec.

Pages 362–364 Chapter 8 Study Guide and Review
1. true **3.** false; original **5.** true **7.** true **9.** true
11–17. Sample answers are given. **11.** $\frac{1}{4} \cdot 80 = 20$ **13.** $\frac{3}{4} \cdot 40 = 30$ **15.** $0.1 \cdot 80 = 8$ **17.** $0.1 \cdot 1{,}000 = 100$ and $5 \cdot 100 = 500$ **19.** Sample answer: $\frac{1}{10} \cdot 300 = 30$ households **21.** $39 = 0.65 \cdot n$; 60 **23.** $n = 0.07 \cdot 92$; 6.4 **25.** about 208 **27.** 48% inc. **29.** 20% dec. **31.** about 12% increase **33.** $178.50 **35.** $26.80 **37.** 14% **39.** 10% **41.** $47.50 **43.** $412.50 **45.** $121.50 **47.** $1,360

Chapter 9 Probability

Page 369 Chapter 9 Getting Started
1. simplest **3.** 105 **5.** 52 **7.** 160 **9.** 360 **11.** 336 **13.** 5,040 **15.** $\frac{2}{3}$ **17.** simplified **19.** 5 **21.** 4

Pages 372–373 Lesson 9-1
1. Sample answer: Since 0.7 is greater than 0.5 and less than 0.75, the event is likely to happen. **3.** 0.65, 0.55 are probabilities that are not complementary because $0.65 + 0.55 \neq 1$. The other pairs are complementary. **5.** $\frac{1}{4}$ **7.** $\frac{3}{5}$ **9.** 75% **11.** $\frac{4}{5}$ **13.** $\frac{1}{2}$ **15.** $\frac{1}{10}$ **17.** Sample answer: The complementary event is the chance of no rain. Its probability is 63%. **19.** $\frac{7}{10}$ **21.** $\frac{1}{2}$ **23.** $\frac{7}{10}$ **25.** 0 **27.** $\frac{1}{11}$ **29.** $\frac{9}{10}$ **31.** A **33.** $60 **35.** about 2.5 billion **37.** 0.45 **39.** $1.\overline{7}$ **41.** simplified **43.** $\frac{3}{8}$

Pages 376–377 Lesson 9-2
1. Sample game: Each player tosses a coin 10 times. If it comes up heads, player 1 receives 1 point. If it comes up tails, player 2 receives 1 point. The player with the most points at the end of 10 tosses wins.

3. 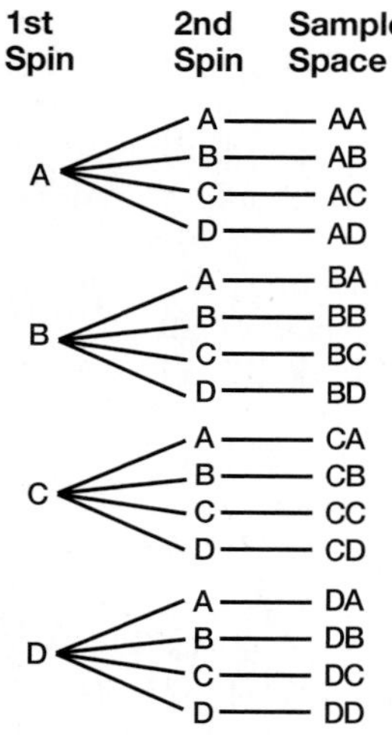

5. $\frac{1}{16}$

7. 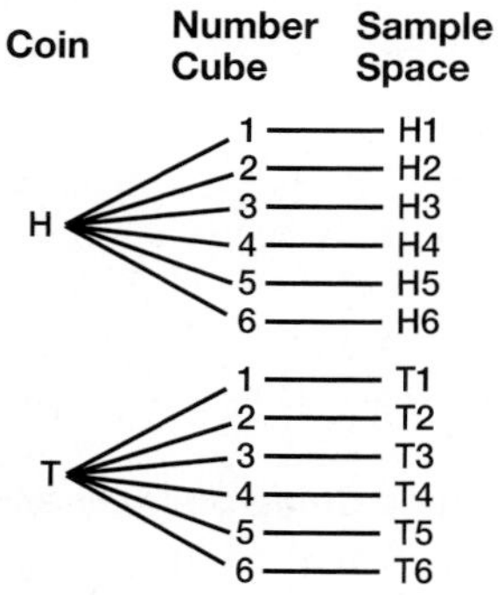

12 outcomes

9. 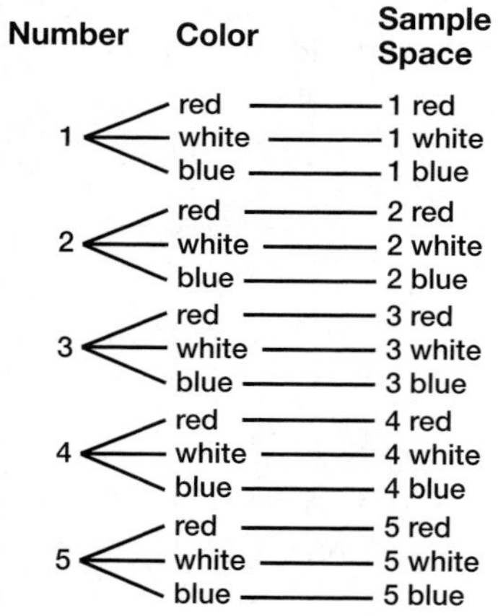

15 outcomes

11. 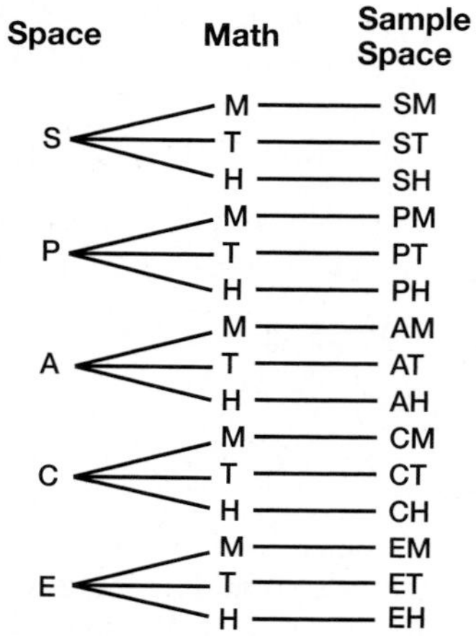

15 outcomes

13. 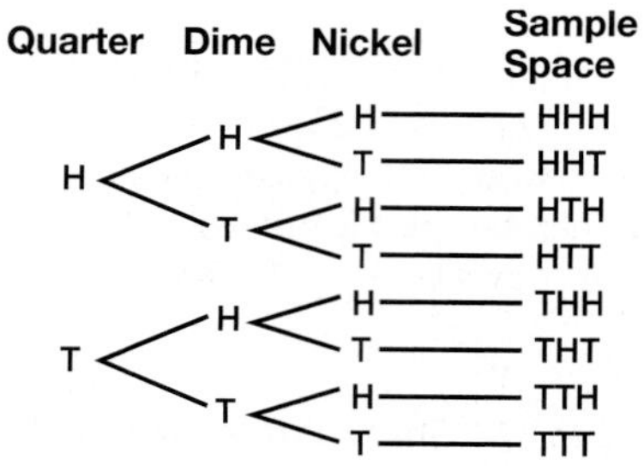

8 outcomes

15. 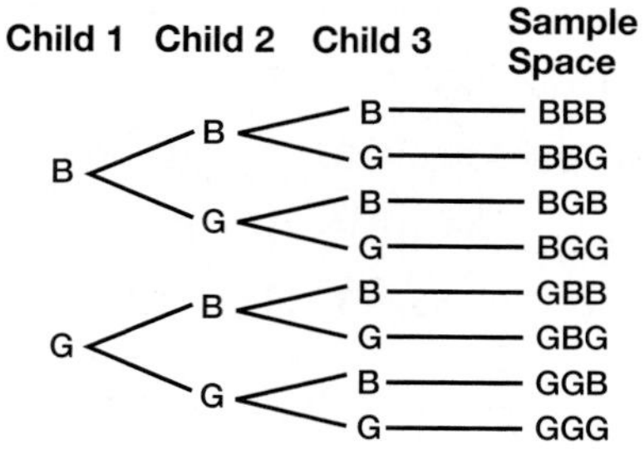

17. $\frac{1}{8}$ **19.** $\frac{4}{24}$ or $\frac{1}{6}$ **21.** P(Player 1) $= \frac{6}{8}$ or $\frac{3}{4}$; P(Player 2) $= \frac{2}{8}$ or $\frac{1}{4}$; Player 2 does not have an equal chance of winning. **23.** C **25.** $\frac{2}{5}$ **27.** $\frac{13}{20}$ **29.** 154 **31.** 176 **33.** 524

Pages 379–380 Lesson 9-3

1. Sample answer: Multiply the number of ways each event can occur to find the total number of outcomes. **3.** Sample answer: Yes, a pair of events can have a number of outcomes that is prime. For example, 11 different pairs of shoes can come in 1 color. $11 \cdot 1$, or 11 outcomes **5.** 8 **7.** 60 **9.** 24 **11.** 12 **13.** 217 **15.** No; the number of selections is 350, which is less than 365. **17.** 2 **19.** 8 **21.** B **23.** $\frac{1}{2}$ **25.** $\frac{3}{4}$ **27.** 6 **29.** 120

Pages 382–383 Lesson 9-4

1. Sample answer: Five factorial has extra factors of 2 and 1, $5! = 5 \cdot 4 \cdot 3 \cdot 2 \cdot 1$. **3.** 6 **5.** 120 **7.** 120 **9.** 144 **11.** 4,320 **13.** 2,880 **15.** 3,024 **17.** 11,880 **19.** $\frac{1}{7}$ **21.** 24 **23.** 96 **25.** 10 **27.** 2.5

Pages 389–390 Lesson 9-5

1. Sample answer: The order of the numbers is important in a lock. **3.** Francisca; Allison found the number of permutations of three people chosen from a group of seven. **5.** permutation; 720 ways **7.** 10 **9.** 36 **11.** combination; 210 **13.** permutation; 6 ways **15.** permutation; 5,040 **19.** A **21.** 120 signals **23.** Sample answer: $0 + 1 = 1$ **25.** Sample answer: $\frac{1}{2} \cdot 20 = 10$ **27.** $\frac{1}{2}$ **29.** $\frac{1}{8}$ **31.** $\frac{1}{2}$

Pages 395–396 Lesson 9-6

1. Sample answer: Both probabilities are ratios that compare the number of favorable outcomes to the total number of possible outcomes. Experimental probability is the result of an experiment. Theoretical probability is what is expected to happen. **3.** $\frac{14}{25}$ **5.** $\frac{3}{16}$ **7.** $\frac{5}{6}$ **9.** $\frac{9}{10}$ **11.** 150 **13.** 14 **15.** $\frac{9}{25}$ **17.** 120 **19.** 1:5 **21.** 2:4 or 1:2 **23.** 1/5 **25.** 720 **27.** $\frac{3}{8}$ **29.** $\frac{7}{8}$

Pages 399–401 Lesson 9-7

1. Sample answer: Events are independent when neither outcome affects the other. Events are dependent when the outcome of the first event affects the outcome of the second event. **3.** Shane; Kimi treated spinning the spinner twice as dependent events. **5.** $\frac{5}{14}$ **7.** $\frac{1}{8}$ **9.** $\frac{2}{87}$ **11.** $\frac{1}{6}$ **13.** $\frac{13}{51}$ **15.** independent **17.** 0.03 **19.** 13.5% **21.** 0.096 **23.** $\frac{1}{2}$ **25.** B **27.** 1 **29.** −27 **31.** 24

Pages 402–404 Chapter 9 Study Guide and Review

1. sample space **3.** experimental **5.** permutation **7.** random **9.** is not **11.** $\frac{1}{2}$ **13.** $\frac{1}{4}$ **15.** $\frac{1}{2}$

17.
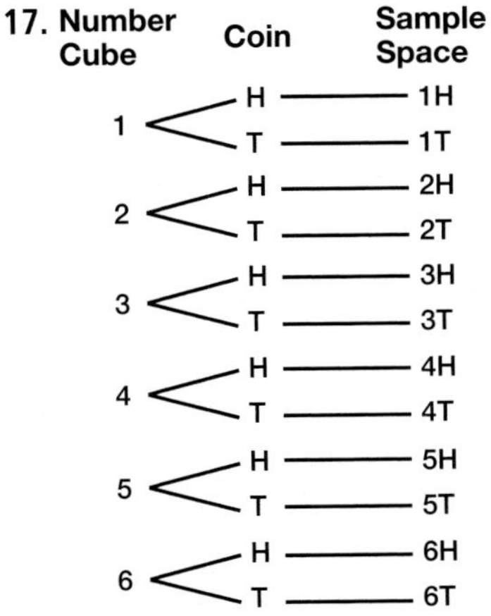

12 outcomes

19.
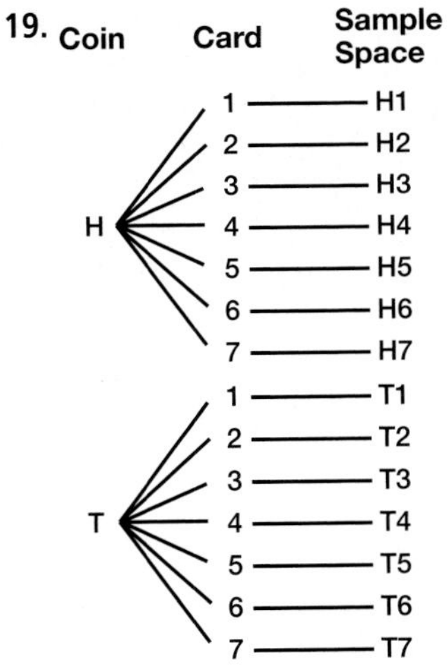

14 outcomes

21. 36 **23.** 20 **25.** 24 **27.** 1,680 **29.** 720 **31.** 60 **33.** 126 groups **35.** 45 ways **37.** $\frac{3}{25}$ **39.** $\frac{1}{5}$ **41.** $\frac{1}{4}$ **43.** $\frac{1}{15}$

Chapter 10 Geometry

Page 411 Chapter 10 Getting Started

1. ratio **3.** 306 **5.** 0.15 **7.** 0.11 **9.** 53 **11.** 131 **13.** 92 **15.** 90 **17.** 54 **19.** 12 **21.** 24 **23.** 8

Pages 414–415 Lesson 10-1

1. Sample answer: An angle with a measure of exactly 90°, which is indicated by a right angle symbol. **3.** obtuse

5.
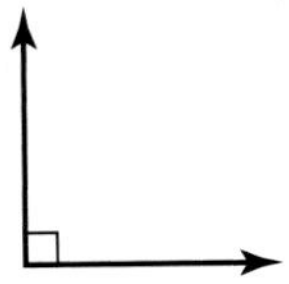
right **7.**

acute

9. acute **11.** obtuse **13.**
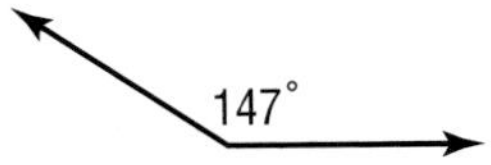

obtuse

15.
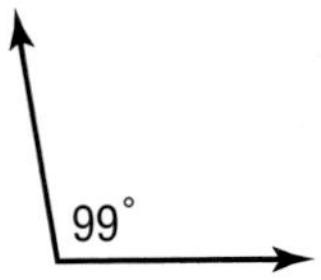

obtuse **17.**
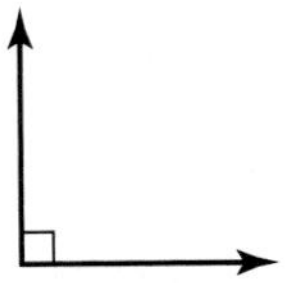
right

19. 150° **21.** Sample answer: acute angles: between their legs and the front of their skis; obtuse angles: between their legs and the back of their skis **23.** 48 times **25.** A **27.** $\frac{1}{4}$ **29.** $\frac{7}{10}$ **31.** 90 **33.** 0.32

Pages 420–421 Lesson 10-2

3. **Favorite Shades of Blue**

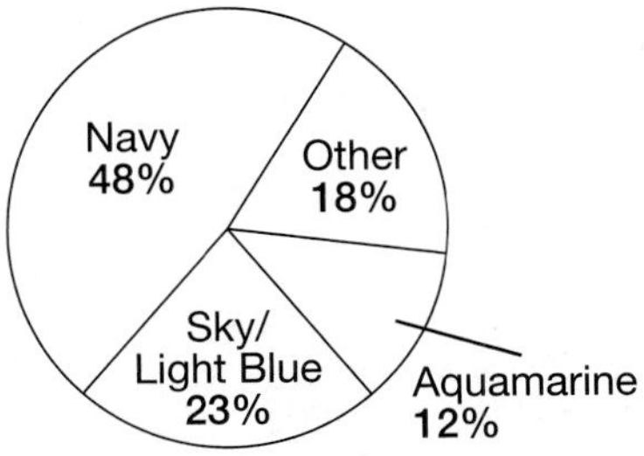

5. grades 1–8

7. **NYC Commuters**

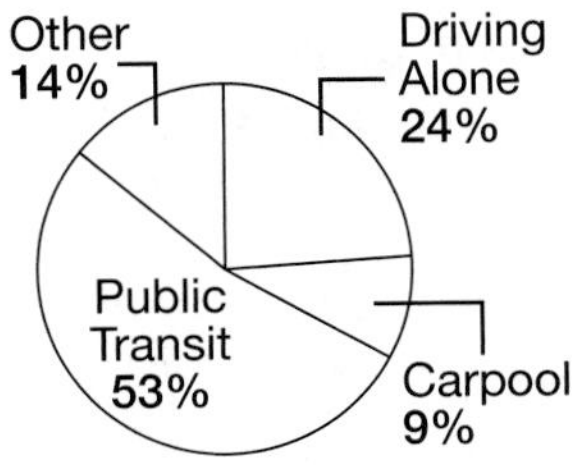

9. Sample answer: A greater percent of commuters in Los Angeles drive alone, while in New York, a greater percent use public transportation. A greater percent of people carpool in Los Angeles than in New York.

11. **U.S. Regions**

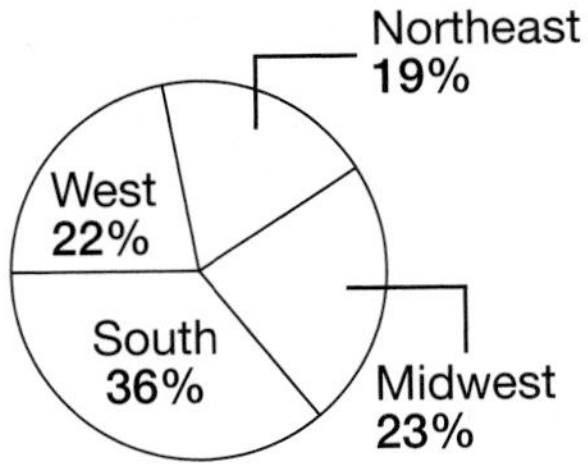

13. yes: 262.8°; no: 90°; not sure: 7.2° **17.** A **19.** acute **21.** right **23.** 72 **25.** 141

Pages 424–425 Lesson 10-3

1. Sample answer: $m\angle 1 + m\angle 2 = 90°$

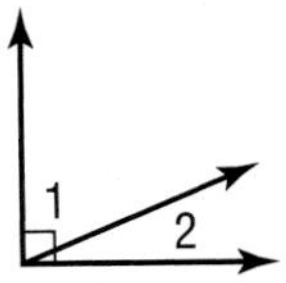

3. supplementary **5.** complementary **7.** 75 **9.** 96° **11.** supplementary **13.** supplementary **15.** complementary **17.** 69 **19.** 42 **21.** 90 **23.** Public Restroom and Subway/train **25.** never; Sample answer: One straight angle measures 180°. **27.** sometimes; Sample answer: Adjacent angles can have the same or different angle measure. **29.** $m\angle E = 39°$, $m\angle F = 51°$ **31.** C **33.** acute **35.** 0.53 **37.** 4.31 **39.** 68 **41.** 101

Pages 430–431 Lesson 10-4

1. An obtuse triangle has one obtuse angle and two acute angles. **3.** 44°; acute **5.** acute, equilateral **7.** right, isosceles **9.** 60°; right **11.** 65°; acute **13.** 37°; right **15.** right **17.** 17° **19.** obtuse, scalene **21.** right, scalene **23.** acute, isosceles **27.** B **29.** supplementary **31.** neither **33.** 60

Pages 435–437 Lesson 10-5

1. Sample answer: A square has all the properties of a rectangle. However, a rectangle does not have four congruent sides, which is a property of a square. **3.** Justin; Venus did not mention the 4 congruent sides of a square. **5.** quadrilateral **7.** 100° **9.** 35° **11.** trapezoid **13.** rhombus **15.** parallelogram **17.** sometimes; Sample answer: A parallelogram can also have angles with different degree measures. **19.** 56° **21.** 67° **23.** 90° **25.** rhombus, rectangle **29.** Yes; A square is a rhombus and a rectangle. **31.** Yes; One pair of opposite sides are parallel, while the other pair of opposite sides are congruent. This figure is called an isosceles trapezoid.

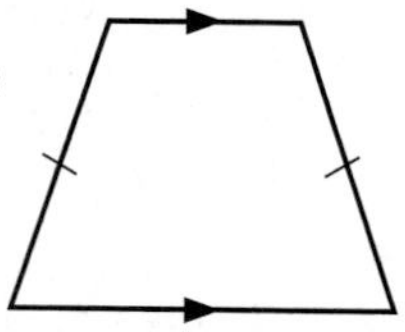

33. B **35.** obtuse, isosceles **37.** acute, equilateral **39.** \$3.78 **41.** 85 **43.** 41.7 **45.** 400 mi **47.** 3 **49.** 7

Pages 442–443 Lesson 10-6

1. Corresponding sides are proportional and corresponding angles are congruent. **3.** 35 m **5.** 48 ft **7.** 12 m **9.** 10.2 mm **11.** yes; $\frac{6}{18} = \frac{11}{33}$ **13.** about 42 ft **15.** 120 m **17.** 1:16 **19.** 90 ft **21.** trapezoid **23.** 69° **25.** 90 **27.** 120

Pages 448–450 Lesson 10-7

1. the point where vertices meet: 45° + 45° + 135° + 135° = 360° **3.** All of the angles of a rhombus do not necessarily have the same measure. **5.** not a polygon, it is not closed **7.** No; 140 does not go into 360 evenly. **9.** regular octagon **11.** not a polygon; the figure has a curved side **13.** regular decagon **15.** 140° **17.** 162° **19.** 135°, 45°, 45° **21.** hexagon and triangle **23.** octagons and squares **25.** 43.2 cm **27.** $37\frac{1}{2}$ yd **31.** Yes; the sum of the measures of angles of any triangle is 180°.

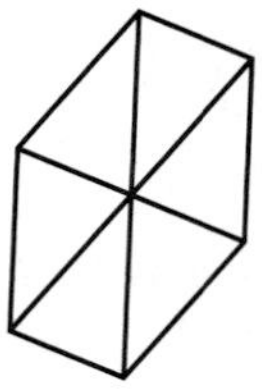

33. I **35.** trapezoid **37.** 32 servings **39.** $3\frac{1}{6}$ **41.** $4\frac{37}{40}$

42–45.

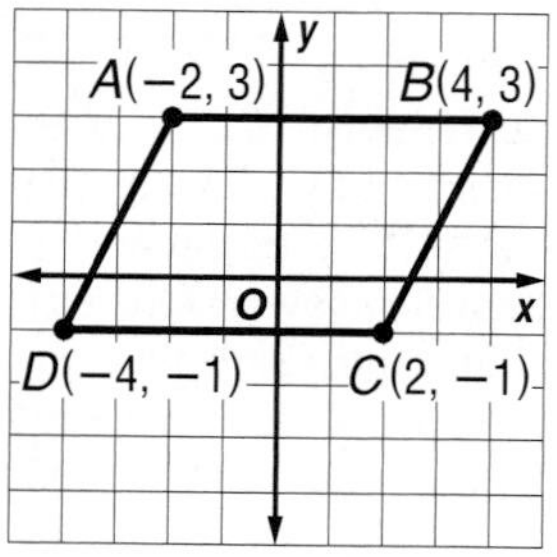

Pages 453–454 Lesson 10-8

1. Sample answer:

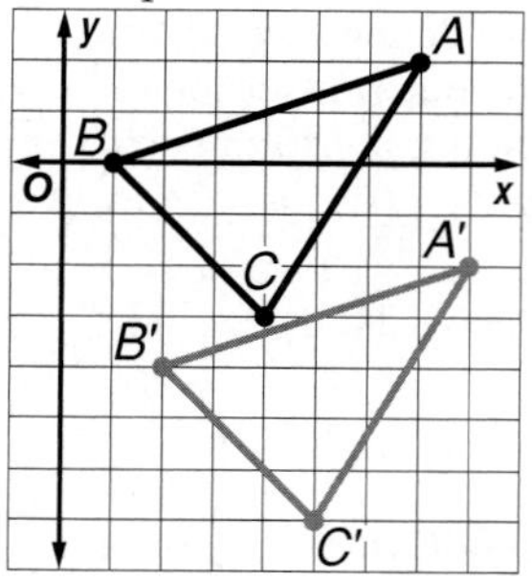

3.

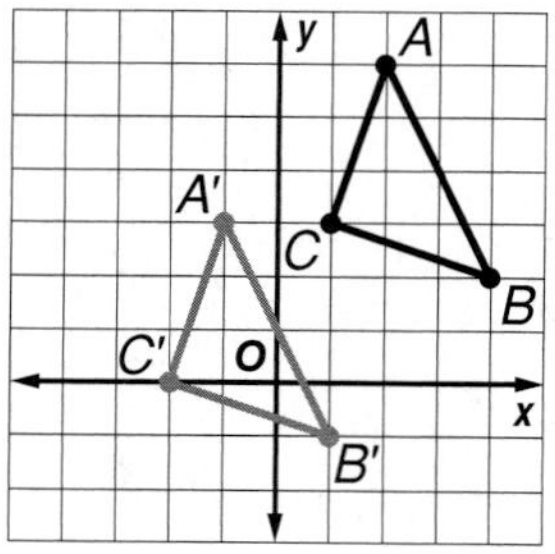

5. $D'(7, 0)$, $E'(4, -2)$, $F'(8, 4)$, $G'(12, -3)$

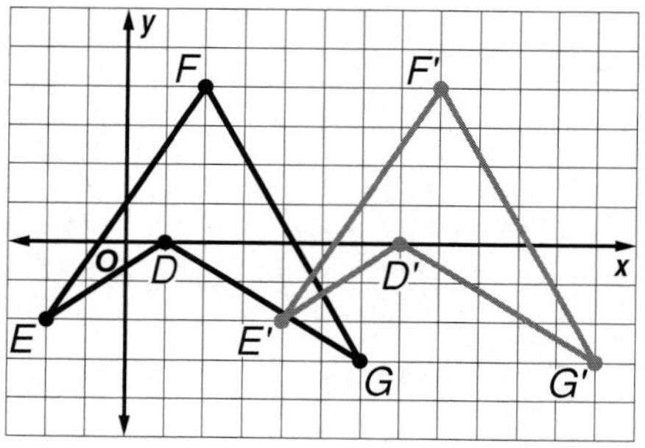

7.

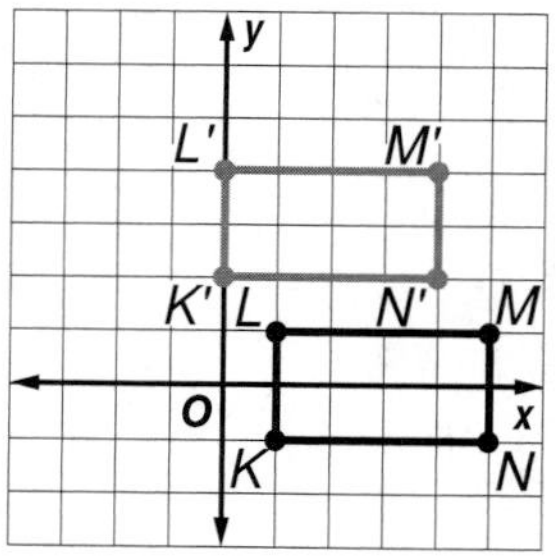

9. $P'(-8, -1)$, $Q'(-3, -3)$, and $R'(-11, 5)$

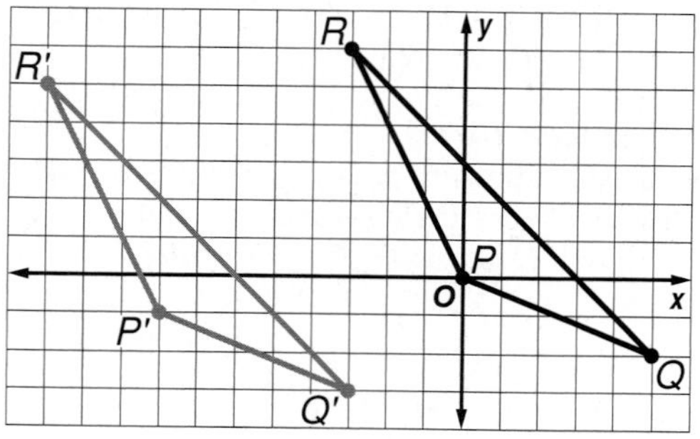

11. $P'(0, -9)$, $Q'(5, -11)$, and $R'(-3, -3)$

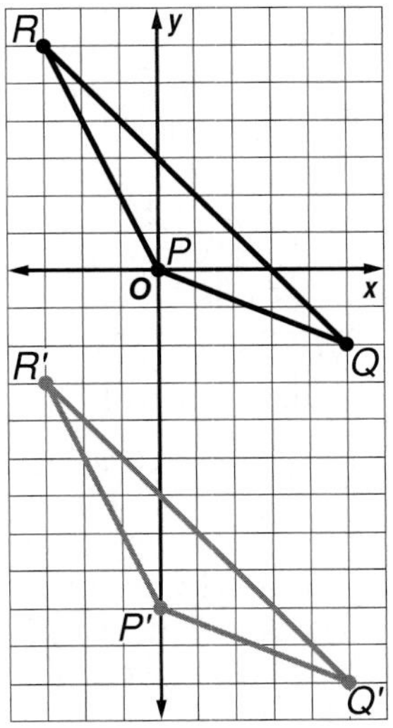

13. Sample answer: there are two main images, the brown horseman and the yellow horseman. Both main horsemen are translated to different parts of the picture. These translations allow for the tessellation of both horsemen. **15.** Sample answer: $\triangle X'Y'Z'$ was translated 5 units left and 2 units down.

17.

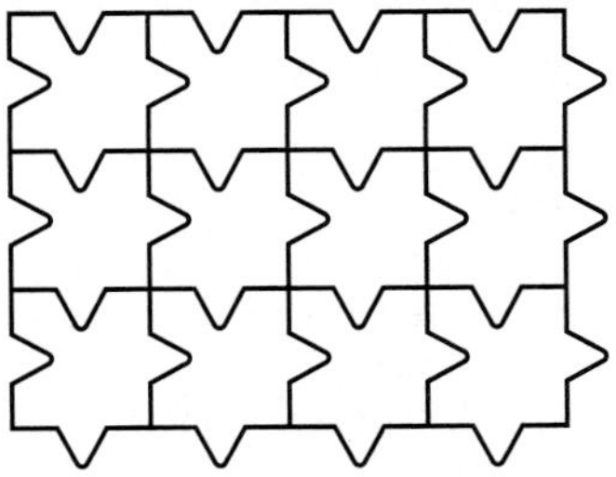

19. C **21.** octagon **23.** yes **25.** no

Pages 458–459 Lesson 10-9

1.

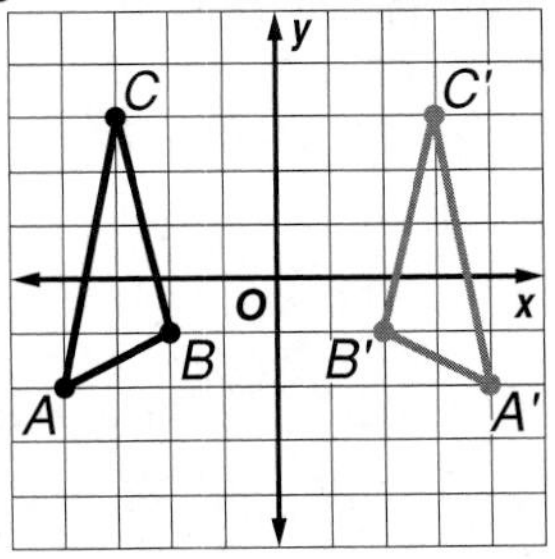

3. no

5. $A'(5, -8)$, $B'(1, -2)$, and $C'(6, -4)$

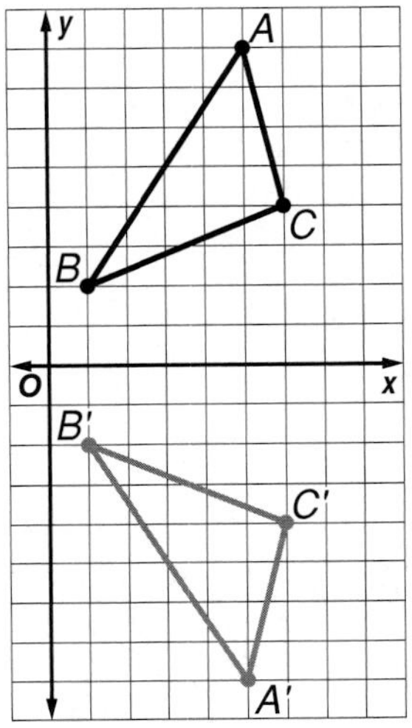

7. yes

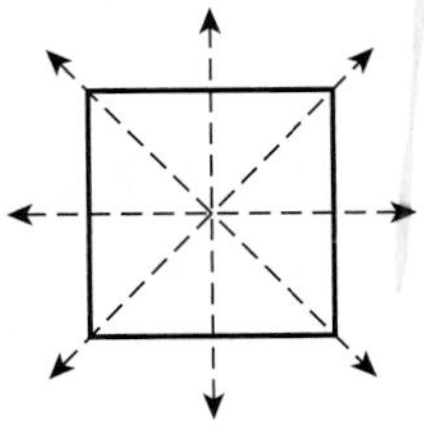

9. yes

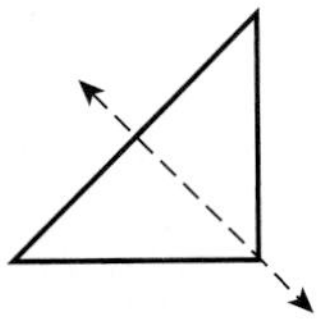

11. $D'(-3, -6)$, $E'(-2, 3)$, $F'(2, -2)$, and $G'(4, -9)$

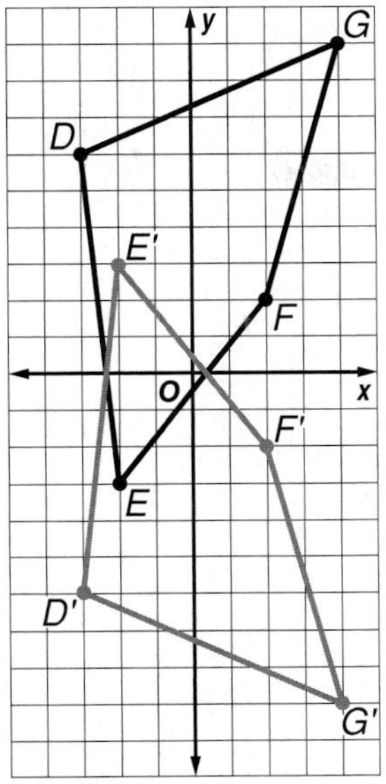

13. $Q'(-2, -5)$, $R'(-4, -5)$, and $S'(-2, 3)$

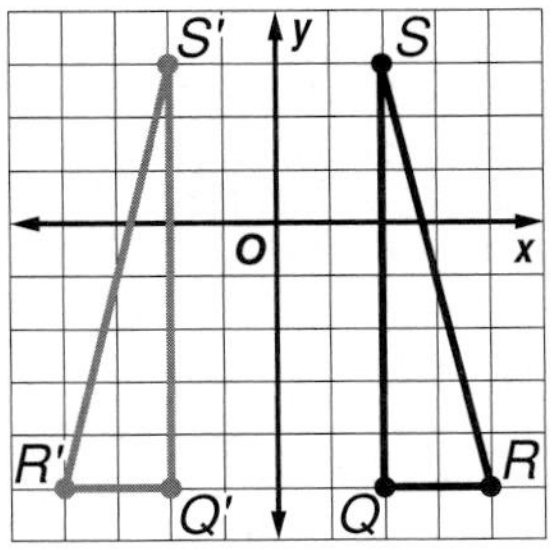

15. vertical line that divides the building in half and a line at the top of the water from the reflection in the water
17. All four flags have line symmetry.

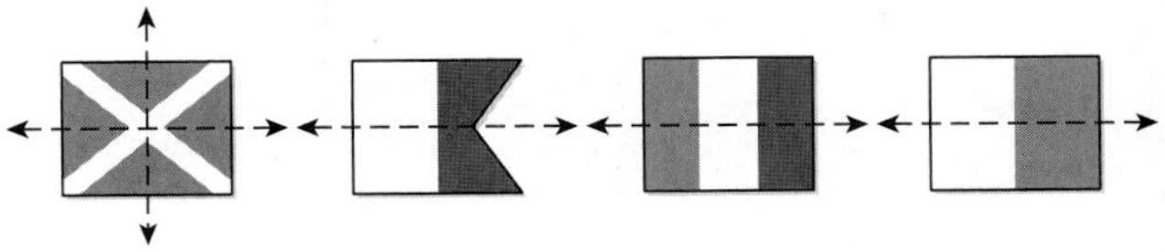

19. $J''(7, -4)$, $K''(-7, -1)$, and $L''(-2, 2)$ **21.** 2 **23.** No, the measure of one angle is 144°, which is not a factor of 360°.
25. Sample answer: $0 \times \frac{1}{2} = 0$ **27.** $6n \leq 18$; $n \leq 3$

Pages 462–464 Chapter 10 Study Guide and Review

1. c **3.** f **5.** a **7.** right **9.** acute

11. **Favorite Soft Drinks**

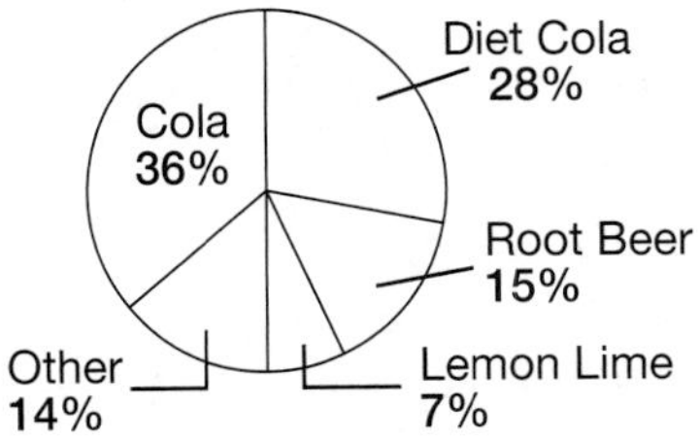

13. 79 **15.** obtuse, scalene **17.** 139° **19.** parallelogram
21. 10 cm **23.** heptagon **25.** 156°
27. $P'(8, -3)$, $Q'(2, -4)$, and $R'(3, 5)$

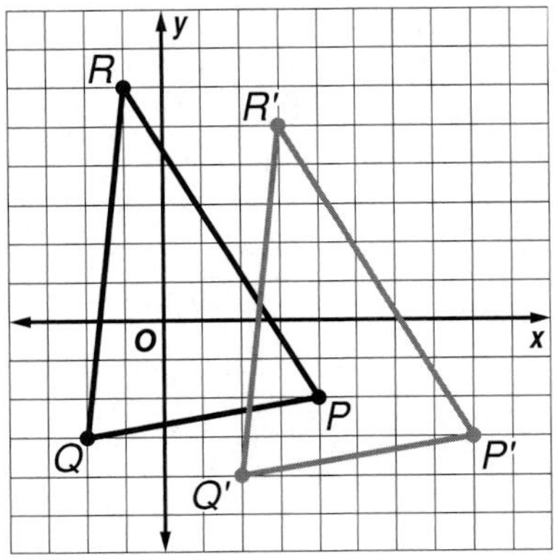

29. $P'(4, -9)$, $Q'(-2, -10)$, and $R'(-1, -1)$

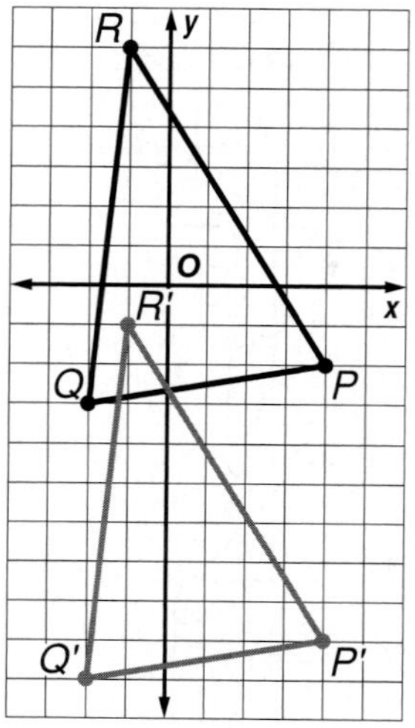

31. $A'(-1, 3)$, $B'(-2, -1)$, $C'(-5, -1)$, $D'(-4, 3)$

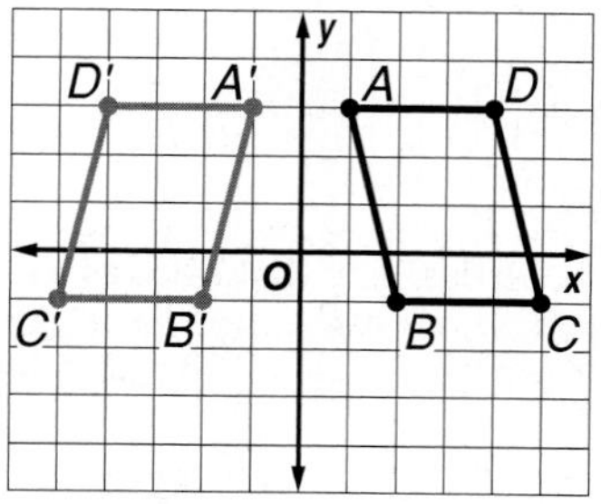

Chapter 11 Geometry: Measuring Two-Dimensional Figures

Page 469 Chapter 11 Getting Started

1. true **3.** false; right **5.** > **7.** = **9.** 64 **11.** 36 **13.** 60 **15.** 112 **17.** $\frac{1}{6}$ **19.** $\frac{1}{2}$

Pages 472–473 Lesson 11-1

1. Both involve multiplying a number by itself. **3.** 116; There are not two equal factors whose product is 116. **5.** 100 **7.** 900 **9.** 11 **11.** 23 **13.** 25 **15.** 49 **17.** 256 **19.** 324 **21.** 2 **23.** 12 **25.** 27 **27.** 35 **29.** 484 **31.** 10, −10 **33.** 1,312 **35.** 202,788 mi^2 **37.** Yes; for example, a garden that measures 30 ft by 30 ft has the same perimeter, but its area is 900 ft^2, which is greater than 500 ft^2. **39.** $\frac{3}{5}$

40–41. **Squares**

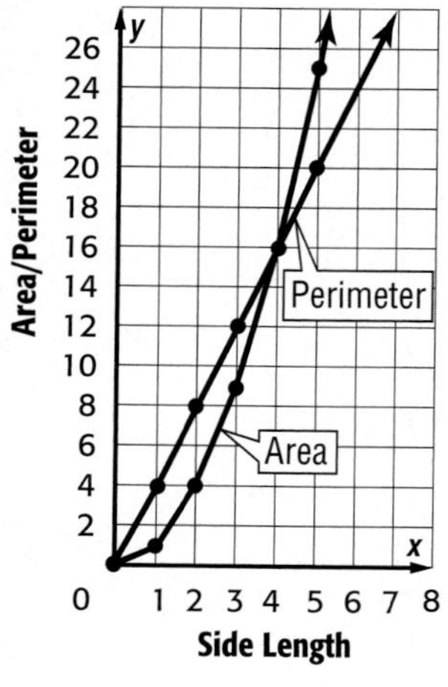

43. when the side length is between 0 and 4; when the side length is 0 or 4 **45.** 55 m^2 greater; The area of an 8-by-8-m square is 64 m^2. **47.** 1,024 cm^2

49. $A'(3, 6)$, $B'(4, 3)$, $C'(1, 2)$

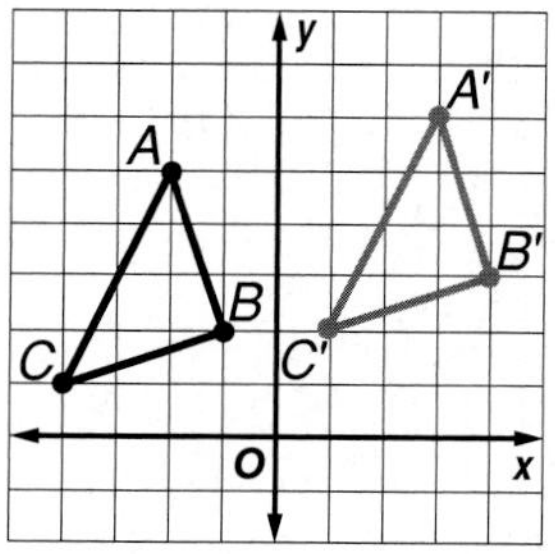

51. > **53.** >

Pages 476–477 Lesson 11-2

1. It cannot be written as a fraction. **3.** 51 is between the perfect squares 49 and 64. Since 51 is closer to 49 than to 64, $\sqrt{51}$ is closer to $\sqrt{49}$, or 7. **5.** 10 **7.** 12 **9.** 7.1 **11.** 21.5 **13.** 3 **15.** 6 **17.** 9 **19.** 12 **21.** 3.9 **23.** 6.6 **25.** 12.6 **27.** 25.4 **29.** $\pi, \frac{14}{3}, \sqrt{87}, 10$ **31.** 3.4 **35.** D **37.** 13 **39.** 28 **41.** 74 **43.** 16

Pages 481–482 Lesson 11-3

1. the lengths of both legs or the length of one leg and the hypotenuse **3.** Devin; the hypotenuse is 16 cm, not x cm as Jamie assumed. **5.** 24.5 in. **7.** yes **9.** 35 in. **11.** 8.0 cm **13.** 25 in. **15.** 23.7 m **17.** yes **19.** no **21.** 19.1 ft **23.** about 10.4 in. **25.** 11.3 in. **27.** 12 **29.** 17 **31.** 213.6 **33.** 28

Pages 484–485 Lesson 11-4

1. $b = 6, h = 3$ **3.** False; if the base and height are each doubled, then the area is $2b \cdot 2h = 4bh$, or 4 times greater. **5.** 1.1 m^2 **7.** 64 in^2 **9.** 312.5 mm^2 **11.** 428.4 mm^2 **13.** 207 in^2 **15.** 180 ft^2 or 20 yd^2 **17.** 1,575 yd^2 **19.** 56.9 cm^2 **21.** 15 in. **23.** C **25.** no **27.** no **29.** 84 **31.** 19.5

Pages 491–492 Lesson 11-5

1. Sample answer: $\frac{1}{2}(10)(20 + 30) = 250$ in^2 **3.** The area of a parallelogram is twice the area of a triangle with the same base and height. **5.** 105.6 m^2 **7.** $1,417.50 **9.** 38.4 mm^2 **11.** 112 m^2 **13.** 161.5 ft^2 **15.** 168 in^2 or 1.2 ft^2

17. Sample answer: 28 in^2 **19.** Sample answer: 7 ft^2

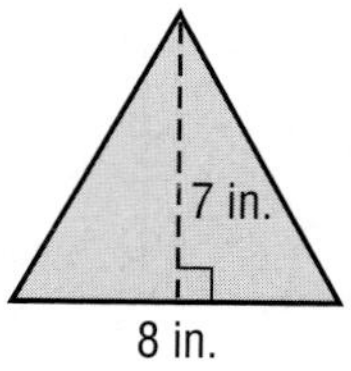

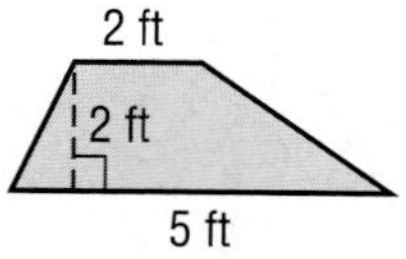

21. about 1,872 mi^2 **23.** 382.4 cm^2 **25.** 3.7 in^2 **27.** 12.4 ft **29.** acute **31.** obtuse **33.** 40.8 **35.** 804.2

Pages 494–495 Lesson 11-6

1. Sample answer:

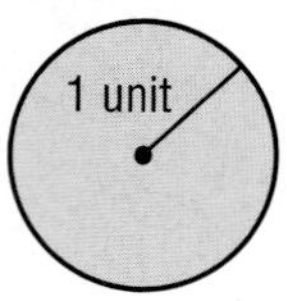

3. 2π; It is about $2 \cdot 3$, or 6; $\sqrt{7}$ is less than $\sqrt{9}$, or 3; 1.5^2 is less than 2^2, or 4. **5.** 254.5 in^2 **7.** 55.4 ft^2 **9.** 452.4 mm^2

11. 201.1 cm^2 **13.** 95.0 ft^2 **15.** 227.0 cm^2 **17.** 113.1 ft^2 **19.** 7.1 cm^2 **21.** 63.6 in^2 **23.** 29.0 m^2 **25.** 43.2 ft^2 **27.** 1 : 2; 1 : 4; No, $\frac{1}{2} \neq \frac{1}{4}$. **29.** 5.9 in^2 **31.** Never; the area is quadrupled because $A = \pi(2r)^2$, or $A = 4\pi r^2$. **33.** 3.7 cm **35.** 120 in^2 **37.** 100.1 m^2 **39.** 113.04 **41.** 150.5

Pages 499–500 Lesson 11-7

1. Sample answer: Separate it into a rectangle and a triangle, find the area of each, and add. **3.** 112 m^2 **5.** 145 m^2 **7.** 58.6 in^2 **9.** 257.1 mm^2 **11.** 510.3 ft^2 **13.** Sample answer: Add the areas of a trapezoid and a small rectangle. **15.** 44 in^2 **17.** 69.4 cm^2 **19.** 176.7 in^2 **21.** $\frac{1}{6}$ **23.** $\frac{1}{3}$

Pages 502–503 Lesson 11-8

1. Sample answer: The probability of an event occurring is the ratio of the shaded area (the number of ways an event can occur) to the entire area (the number of possible outcomes). **3.** 31.3% **5.** 16% **7.** 34.3% **9.** 23.1% **11.** 73.8% **13.** 21.5% **15.** $\frac{4}{25}$ or 16% **17.** 3/10 **19.** 102.1 m^2

Pages 504–506 Chapter 11 Study Guide and Review

1. equal to **3.** 625 **5.** hypotenuse **7.** 36 **9.** 529 **11.** 16 **13.** 2 **15.** 7 **17.** 4 **19.** 7.8 **21.** 21.1 **23.** 13 mm **25.** 7.8 ft **27.** 13.3 m **29.** 99 cm^2 **31.** 135 cm^2 **33.** 36 ft^2 **35.** 187.7 cm^2 **37.** 408.3 in^2 **39.** 1,256.6 ft^2 **41.** 60 m^2 **43.** 182.8 in^2 **45.** 25%

Chapter 12 Geometry: Measuring Three-Dimensional Figures

Page 511 Chapter 12 Getting Started

1. circumference **3.** 90 **5.** 35 **7.** 24.1 **9.** 187.2 **11.** 24 **13.** 32 **15.** 27 **17.** $1\frac{1}{14}$ **19.** 855.3 cm^2 **21.** 113.1 ft^2

Pages 515–517 Lesson 12-1

3–17. Sample drawings are given.

3. top side front

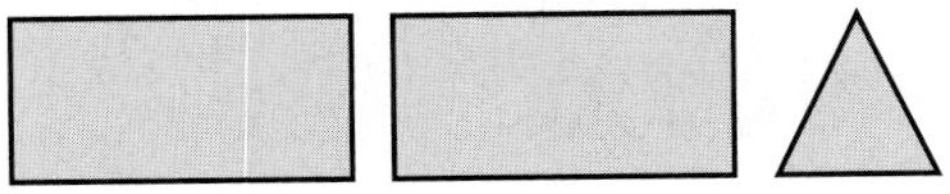

5.

7. top side front

9. top side front

11. top side front

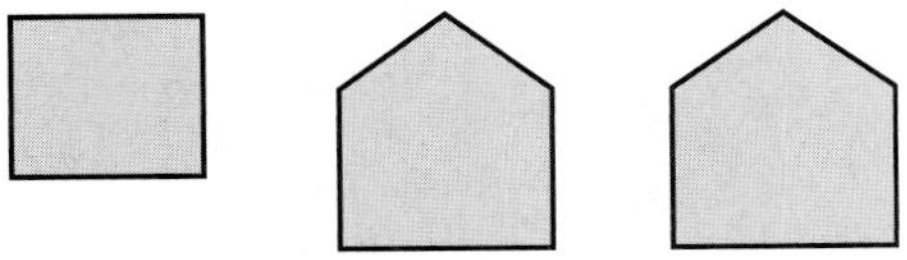

13.

15.

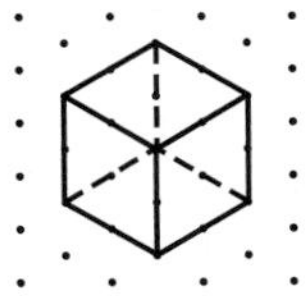

17.

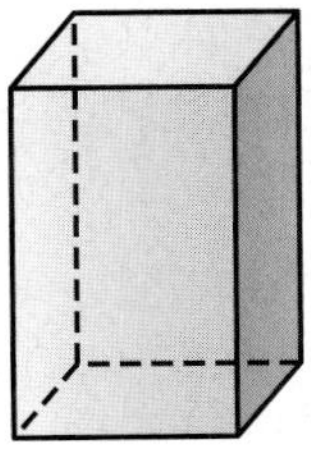

19. cylinder

21. Sample answer:

top side front

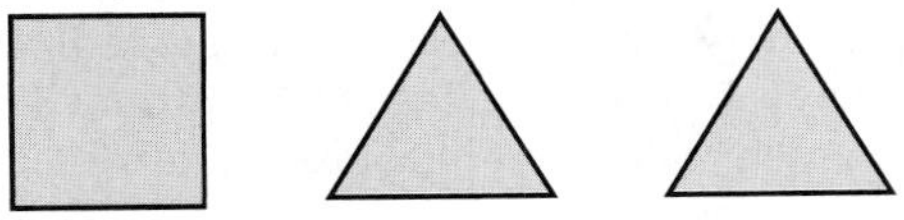

23. Sample answer:

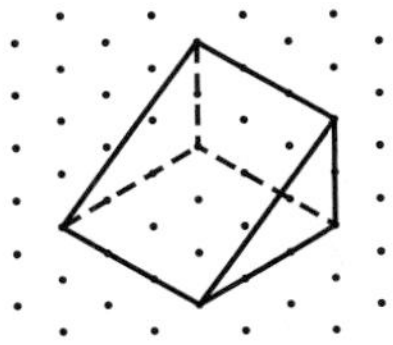

25. G **27.** 96 ft^2 **29.** 112.8 in^2 **31.** 22 **33.** 68

Pages 521–522 Lesson 12-2

1. yd^3 **3.** Ling; the width of Box B is half that of Box A. All three dimensions have to be considered when comparing volumes. **5.** 9.5 cm^3 **7.** 36,000 in^3 **9.** 960 in^3 **11.** 236.3 cm^3 **13.** 27 cm^3 **15.** 2,425.5 ft^3 **17.** 1,000,000 cm^3 **19.** C **21.** Sample answer:

top side front

23. Sample answer: 18 **25.** Sample answer: 72

Pages 525–527 Lesson 12-3

1. Sample answer:

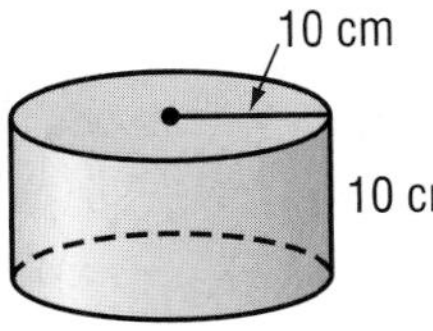

3. 141.4 in^3 **5.** 617.7 ft^3 **7.** 241.5 m^3 **9.** 402.1 in^3 **11.** 167.1 cm^3 **13.** 1.3 m^3 **15.** 2,770.9 mm^3 **17.** 603.2 ft^3 **19.** 603.2 cm^3 **21.** 56.5 in^3 **25.** 1 to 4 **27.** about 41 ft **29.** B **31.** 152.88 m^3

Selected Answers

33. Sample answer: 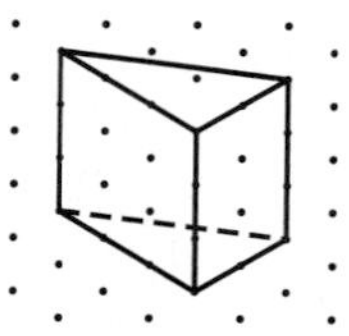 **35.** 572 m^2 **37.** 174 **39.** 43.5

Pages 533–535 Lesson 12-4

1. Find the sum of the areas of the faces. **3.** Surface area measures the area of the faces, and area is measured in square units. **5.** 183.4 cm^2 **7.** 314 cm^2 **9.** 833.1 mm^2 **11.** 125.4 in^2 or $125\frac{3}{8}$ in^2 **13.** 7 m^2 **15.** 1,531.9 ft^3 **17.** surface area $= 6x^2$ **19.** 12 cm^2 **21.** 5 cm long, 4 cm wide, 5 cm high **25.** D **27.** 7 **29.** 72.9 cm^3 **31.** 63 in^3 **33.** 804.2 ft^2 **35.** 145.3 yd^2

Pages 539–541 Lesson 12-5

3. 88.0 mm^2 **5.** 653.5 m^2 **7.** 1,215.8 m^2 **9.** 183.8 ft^2 **11.** 1,120.0 in^2 or $1{,}119\frac{44}{45}$ in^2 **13.** 471.2 m^2 **15.** 172.8 in^2 **17.** Sample answer: prism: volume = 275 in^3, surface area = 237.5 in^2; cylinder: volume = 274.3 in^3, surface area = 211.1 in^2; Both containers hold about the same amount of popcorn. However, the cylinder uses less cardboard. So, the cylinder would be a more cost-effective container. **19.** Sample answer: The areas of both curved surfaces are the same. However, the areas of the top and bottom of cylinder A are greater than the areas of the top and bottom of cylinder B because it has a greater circumference (11 in.).
21. Sample answer:

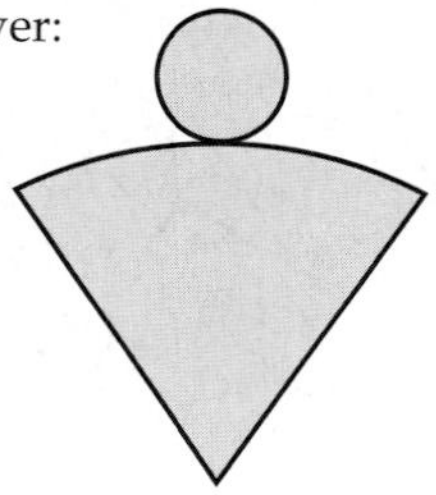

23. G **25.** 112 cm^2 **27.** 326.7 in^3 **29.** 14.0 **31.** 5.4

Pages 543–545 Lesson 12-6

1. Inches, because it is the smallest unit of measure. **3.** 8.6 mi, because it has 2 significant digits. The other measures have 3 significant digits. **5.** Sample answer: 15.65 cm **7.** Sample answer: The measure is precise to the nearest 0.01 second. **9.** 0.1 cm **11.** 2 lb **13.** Sample answer: 13.6 ft **15.** Sample answer: 9.15 cm **17.** Sample answer: 2.45 mL **19.** Neither; they are both given to the nearest tenth of a mile. **25.** B **27.** 502.7 yd^2 **29.** 748.0 in^2 **31.** 45 **33.** 24

Pages 546–548 Chapter 12 Study Guide and Review

1. rectangular prism **3.** cubic **5.** cylinder **7.** smaller
9. Sample answer:

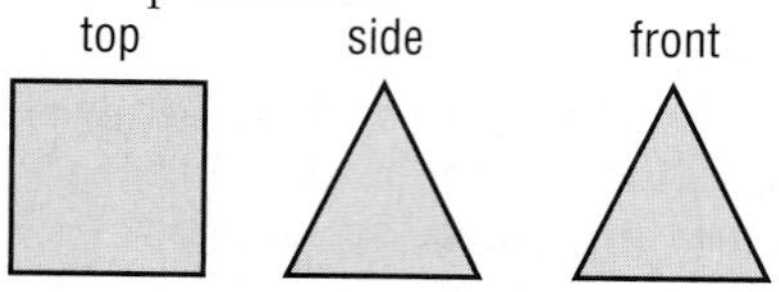

11. Sample answer:

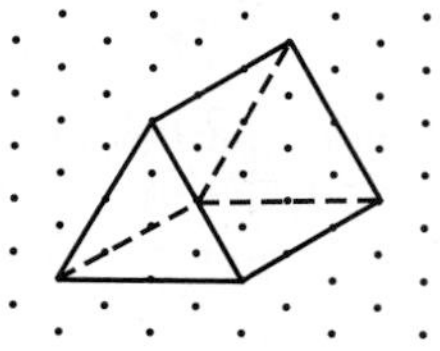

13. 639.6 in^3 or $639\frac{5}{8}$ in^3 **15.** 4,042.4 m^3 **17.** 196.3 in^3 **19.** 531.3 m^2 **21.** 427.3 mm^2 **23.** 69.1 in^2 **25.** 5 min **27.** Sample answer: 0.8 cm

Photo Credits

Cover (tl b)Mark Gallup, (tr)PhotoDisc; **i** Mark Gallup; **iv v x** Aaron Haupt; **xi** Robert Laberge/Getty Images; **xii** Matthias Kulka/CORBIS; **xiii** Jeffrey L. Rotman/CORBIS; **xiv** Aaron Haupt; **xv** Haslin/Sygma/CORBIS; **xvi** PhotoDisc; **xvii** NASA; **xviii** CORBIS; **xix** DUOMO/CORBIS; **xx** CORBIS; **xxi** David Muench/CORBIS; **xxii** Kevin Fleming/CORBIS; **xxiii** John Evans; **xxiv** (t)PhotoDisc, (b)John Evans; **xxv** (t bl)PhotoDisc, (br)John Evans; **xxvi** John Evans; **2–3** Nicole Duplaix/Getty Images; **4–5** Jeff Greenberg/PhotoEdit; **7** CORBIS; **8** Dimitri Lundt/CORBIS; **13** Robert Laberge/Getty Images; **15** Chuck Savage/CORBIS; **19** Aaron Haupt; **22** (l)PhotoDisc, (r)Richard Hutchings/Photo Researchers; **24** Getty Images; **27** Tim Davis/Photo Researchers; **29** John Evans; **31** Brian A. Vikander/CORBIS; **32** Miles Ertman/Masterfile; **39** Aaron Haupt; **40** Phyllis Picardi/Stock Boston; **41** Bob Daemmrich/Stock Boston; **43** Owen Franken/Stock Boston; **52–53** Roger Ressmeyer/CORBIS; **54** Grant Symon/Getty Images; **55** Tim De Waele/TempSport/CORBIS; **57** George Clark, Hattiesburg American/AP/Wide World Photos; **58** John Evans; **59** Matthias Kulka/CORBIS; **60** Kelly–Mooney/CORBIS; **64** Joe Sohm/Photo Researchers; **67** CORBIS; **68** Craig Tuttle/CORBIS; **70** CORBIS; **73** First Image; **75** John Evans; **77 78** Matt Meadows; **79** Aaron Haupt; **80** John E. Kelly/Getty Images; **82** (tl tr)White House Historical Association, (b)Matt Meadows; **83** Jim Corwin/Photo Researchers; **85 86** CORBIS; **87** David Cannon/Getty Images; **94** Don Hebert/Getty Images; **95** Bob Daemmrich/Stock Boston; **102–103** Mike Powell/Getty Images; **104–105** Roger Ressmeyer/CORBIS; **106** Jed Jacobsohn/Getty Images; **109** Mark Tomalty/Masterfile; **111** (l)Timothy Fuller, (c)Thomas Veneklasen, (r)Crown Studios; **113** Geoff Butler; **117** John Evans; **120** Richard Kaylin/Getty Images; **122** Australian Picture Library/CORBIS; **123** Tony Feder/Getty Images; **129** (l)Jeffrey L. Rotman/CORBIS, (r)National Weather Service/Unisys Corp.; **131** Richard T. Nowitz/Getty Images; **132** John Evans; **133** Michelle Garrett/CORBIS; **137** Tim Wright/CORBIS; **139** NASA; **148–149** First Light Associated Photographers; **150** United Features Syndicate; **151 152** PhotoDisc; **158** (t)Michael Zito/SportsChrome, (c)CORBIS; **159** Six Flags, World of Adventure, Aurora, Ohio; **161 162** PhotoDisc; **163** Cedar Point; **164 165** PhotoDisc; **167** Tim Courlas; **169** PhotoDisc; **171** John Evans; **172** Aaron Haupt; **173** PhotoDisc; **176** CORBIS; **177** Aaron Haupt; **179** Brett Froomer/Getty Images; **183** CORBIS; **184** PhotoDisc; **192–193** ThinkStock/SuperStock; **194–195** Neil Rabinowitz/CORBIS; **199** Aaron Haupt; **201** John Evans; **204** Sandy Felsenthal/CORBIS; **206** KS Studios; **208** David Tardif/CORBIS; **210** Haslin/CORBIS SYGMA; **212** Carolyn A. McKeone/Photo Researchers; **215** John Evans; **216** Will & Deni McIntyre/Photo Researchers; **217** Doug Martin; **221** Ralph A. Clevenger/CORBIS; **223** Bob Krist/CORBIS; **227** Aaron Haupt; **228** Getty Images; **231** Ron J. Berard/DUOMO/CORBIS; **238–239** Roy Ooms/Masterfile; **243** Doug Martin; **245** PhotoDisc; **247** M. Angelo/CORBIS; **249** Aaron Haupt; **251** KS Studios; **252** John Evans; **253** Matt Meadows; **254** Richard Hamilton Smith/CORBIS; **257** (t)Joseph Nettis/Photo Researchers, (b)Latent Image; **263** John Evans; **265** Aaron Haupt; **268** Christopher Marona; **273** Charlotte Observer; **274** Matt Meadows; **275** Ben Wood/CORBIS; **277** Circlemakers; **284–285** James Richard/Stock Connection/PictureQuest; **286–287** Joseph Sohm/CORBIS; **289** Elaine Shay; **291** David Young-Wolff/PhotoEdit; **295** Johnny Johnson; **297** Doug Martin; **298** Project Exploration; **299** Aaron Haupt; **300** KS Studios; **301** Aaron Haupt; **302** (l)Ryan McVay/PhotoDisc, (r)KS Studios; **308** Underwood & Underwood/CORBIS; **311** John Evans; **312** Richard Lord/The Image Works; **315** Aaron Haupt; **317** Johnny Johnson; **319** Lynn M. Stone; **321** Matt Meadows; **323** NASA; **332–333** Richard Price/Getty Images; **334** CORBIS; **337** AFP/CORBIS; **338** (l)John Evans, (r)Matt Meadows; **341** Spencer Grant/PhotoEdit; **344** Mary Kate Denny/PhotoEdit; **346** Tim Fuller; **347** CORBIS; **349** John Evans; **351** Charles O'Rear/CORBIS; **352** David Young-Wolff/PhotoEdit; **354** Matt Meadows; **355** KS Studios; **356** Tony Freeman/PhotoEdit; **357** Morton & White; **359** Courtesy George Schweighofer/Currency Quest; **368–369** KS Studios; **370** Aaron Haupt; **371** Matt Meadows; **373** Laura Sifferlin; **375** Tim Fuller; **377** Aaron Haupt; **378** Geoff Butler; **382** DUOMO/CORBIS; **385** John Evans; **388** Geoff Butler; **389** The Supreme Court Historical Society; **391** Laura Sifferlin; **394** (t)KS Studios, (b)Michael Newman/PhotoEdit; **397** William Sallaz/DUOMO/CORBIS; **398** CORBIS; **400** Digital Vision/Getty Images; **408–409** Charles O'Rear/CORBIS; **410–411** Aaron Haupt; **414** Doug Martin; **415** Wally McNamee/CORBIS; **419** Jon Feingersh/CORBIS; **420** Tim Fuller; **423** Bettmann/CORBIS; **428** CORBIS; **434** Jed Jacobsohn/Getty Images; **439** John Evans; **441 443** PhotoDisc; **444** Laura Sifferlin; **447** Aaron Haupt; **448** (l)Tom Stack & Associates, (r)Aaron Haupt; **449** Christine Osborne/CORBIS; **450** EyeWire/Getty Images; **452** Wolfgang Rattay/Reuters/NewsCom; **454** Cordon Art-Baarn-Holland; **458** CORBIS; **459** Mug Shots/CORBIS; **468–469** John & Lisa Merrill/CORBIS; **471** Mark Gibson/H. Armstrong Roberts; **477** Stephen R. Swinburne/Stock Boston; **480** Collin Reid/AP/Wide World Photos; **487** John Evans; **490** David Muench/CORBIS; **496** (l)Laura Sifferlin, (r)Doug Martin; **497** Matt Meadows; **502** Doug Martin; **510–511** LWA-Dann Tardif/CORBIS; **514** Tribune Media Services, Inc.; **518** Laura Sifferlin; **522** Kevin Fleming/CORBIS; **523** Geoff Butler; **526** CORBIS; **529** John Evans; **533** Tim Fuller; **539** Alan Klehr/Getty Images; **540** (t)Ariel Skelley/CORBIS, (b)Geoff Butler; **542** Claudia Kunin/CORBIS; **552** Mark Gallup.

T1 Aaron Haupt; **T5** Mug Shots/CORBIS; **T8** Tim Courlas; **T11** Ariel Skelley/CORBIS; **T16** Steve McAlister Productions/Getty Images.

Photo Credits

Index

A

Absolute value (| |), 107

Acute angle, 413

Acute triangle, 429

Addition
- Associative Property of, 31
- Commutative Property of, 31
- decimals, 559
- fractions, 244
- Identity Property of, 31
- integers, 120–122
- Inverse Property of, 121
- mixed numbers, 248, 249
- phrases indicating, 150
- solving equations, 156

Addition Property of Equality, 157

Additive Inverse Property, 121

Additive inverses, 121

Addressing Individual Needs. *See* Differentiated Instruction

Algebra, 18
- equation, 24–27
- equivalent expressions, 30
- evaluating expressions, 18–21, 123, 129–130, 135–136, 139–140
- functions, 176–181
- graphing inequalities, 172–175
- graphing linear equations, 178, 179
- graphing solutions of equations, 25
- linear equations, 178
- solving equations, 24–27, 154–155, 156–159, 160–163, 166–169, 258–261
- solving proportions, 298
- variable, 18–21
- writing equations, 25, 151, 153, 158, 161, 167
- writing expressions, 150

Algebraic equations. *See* Equations

Algebraic expressions, 18
- evaluating, 18, 19
- factoring, 198
- order of operations, 15
- simplifying, 122, 123
- translating words into, 150

Angles, 413, 414
- acute, 413
- adjacent, 415
- alternate exterior, 427
- alternate interior, 427
- bisecting, 417
- classifying, 414, 423
- complementary, 422
- congruent, 422
- constructing, 416, 417
- corresponding, 427
- drawing, 414
- measuring, 412
- obtuse, 413
- relationships, 422, 423
- right, 413
- straight, 413
- sum in a polygon, 447
- sum in a quadrilateral, 435
- sum in a triangle, 428
- supplementary, 422
- vertex of, 413
- vertical, 422, 427

Applications. T3. *See also* Real-Life Careers; Real-Life Math
- advertising, 380
- age, 165
- agriculture, 208
- airplanes, 262
- allowances, 99, 353
- ambulance, 459
- animals, 6, 49, 67, 85, 86, 153, 161, 257, 267
  - birds, 223, 308, 431
  - bears, 317
  - cats, 347
  - dogs, 212, 383
  - elephants, 253
  - horses, 148
  - insects, 36
  - penguins, 27
  - pets, 77, 247, 447
  - turtles, 161, 257
  - whales, 27, 295
- aquariums, 250
- archaeology, 480
- architecture, 81, 498
- art, 308, 449, 459, 519, 535
- aspect ratio of photographs, 291
- astronomy, 139, 248
- baby-sitting, 162
- baking, 41, 230, 243, 525, 540
- balloons, 223
- behavior of dog owners, 212
- bicycles, 214, 414, 519
- blood pressure, 21, 173
- blueprints, 306, 307, 329
- books, 365
- boxes, 534
- breakfast, 172, 383
- bridges, 83, 306, 307
- buildings, 64, 458
- business, 92, 303
- cameras, 347
- candy, 370, 372, 541
- capacity, 49
- cars, 23, 46, 315, 390
- cellular phones, 46, 88
- cereal, 57, 133, 521, 534
- children's sports injuries, 175
- chores, 339
- civics, 88, 181, 389, 400
- clocks, 413
- clothes, 354, 445
- coins, 133, 217, 375
- collecting/collections, 9
  - coin collecting, 32
  - ornaments, 26
  - souvenirs, 23
- colors, 418
- comics, 514
- commission, 320
- communication, 339, 346, 497
- community service, 519
- computers, 306, 347, 354
- congressional representatives per state, 88
- construction, 242, 249
  - carpentry, 535
  - flooring, 273
  - painting, 235
  - remodeling, 519
- containers, 526
- cooking, 9, 281, 300
- cost of gasoline, 182
- cotton production, 343
- crafts, 266
- credit cards, 359
- crops, 343
- currencies, 300
- data sense, 56, 66, 72, 420
- delicatessen, 376
- design, 436, 539, 548
- diamonds, 226
- dining, 27
- dinosaurs, 268
- discount, 355
- dog ages, 33
- drinking straws, 549
- driving, 174, 477
- earnings, 184
- earthquakes, 163
- education, 216, 353
- elections, 256
- electricity, 218
- elevators, 59, 253
- employment, 99, 162

Index

entertainment, 43, 94, 97, 348
environment, 303
family, 303, 376
fashion, 400
Ferris wheels, 238, 275
fiber-optic cable, 23
firefly flashes, 133
flag day, 281
flags, 257
food, 133, 151, 165, 177, 202, 206, 231, 234, 256, 266, 353, 357, 373, 378, 387, 389, 394, 465
fountains, 294
furniture, 442
games, 14, 202, 303, 368, 374, 377, 380, 452, 502, 503
  cards, 400
  checkers, 388
  chess, 78
  croquet, 392
  darts, 230
  Parcheesi, 371
  video games, 15, 356, 436
gardening, 46, 445, 506, 528
gears, 288
geography, 45, 94, 113, 114, 130, 165, 220, 221, 222, 308, 431, 446, 472
geology, 337
gifts, 396, 533
grades, 98, 253
grilling, 334
health, 19, 21, 173, 425
heart rates, 295
heating, 74
heights, 290
history, 31, 55, 76, 268, 359, 494, 516
  math history, 213
home run totals, 58
household size, 72
indirect measurement, 442
industrial technology, 205
interest, 358, 359
interior design, 500
internet, 180, 203
investing, 358, 360
juice, 253
kites, 162, 240, 428
landscaping, 200, 448, 486, 495
lawn service, 170
light years, 47
loans, 359, 365
lottery, 401
lunch, 383
manufacturing, 449
maps, 112, 304, 306, 307, 453
measurement, 57, 206, 260, 300, 442, 450, 473, 522, 526
meteorology, 130
military, 137
models, 299, 308, 315
model trains, 223
money, 8, 17, 27, 110, 122, 123, 152, 153, 165, 168, 177, 247, 261, 291, 300, 303, 392, 419, 494, 519
monuments, 251
  Mount Rushmore, 286
  Statue of Liberty, 60
movies, 95, 189, 291, 325, 356, 373, 540
moving, 209
music, 23, 72, 194, 208, 209, 219, 276, 294, 339, 355, 383, 492
number names, 45
number theory, 23, 59, 165, 200, 202, 226, 342
nutrition, 69, 72, 77, 80, 294, 299, 343
Olympics, 55, 62, 218, 419, 543
packaging, 199, 256, 549
parade, 204, 405
parking, 209
parks, 270
pasta consumption, 77
patterns, 136, 140, 165, 226, 392
photography, 291, 317, 392
physical fitness, 507
picnics, 225, 392
politics, 421
ponds, 289
Pony Express, 31
pools, 305, 547
population, 13, 151, 295, 318, 336, 339, 353
pottery, 547
presidents, 82, 246
probability, 377, 380, 401, 415, 421, 473, 501
production, 13
profits, 111, 357
rainfall, 83
rain forest, 253
reading, 59, 220
real estate, 425
record temperatures, 66, 168
refreshments, 395
research, 13, 41, 57, 62, 79, 206, 277, 295, 318, 357, 377, 421, 436, 450, 477, 544
reservoirs, 44
restaurants, 30
retail sales, 378
roller coasters, 159, 163
Roman history, 131
safety, 59, 482
salary, 93
sales, 98, 99, 342, 353, 497
sales tax, 354
savings, 202, 262, 358
scale drawing, 318
scaling surface area and volume, 536
school, 62, 71, 202, 380
science, 156
  Earth Science, 45, 59, 120, 129, 133, 137, 140, 202, 253, 392, 415, 497
  Life Science, 33, 133, 233, 257, 295, 298, 336
  Physical Science, 13, 23, 141, 303, 340, 471, 472, 534
  space science, 392
science fair, 202
scooters, 54, 375
seat belts, 339
sewing, 243
shoes, 65
shopping, 78, 175, 189, 253, 310, 325, 339, 403
signal flags, 459
skid marks, 477
snacks, 61, 399
snowfall, 23
space, 70
space shuttle, 526
speed of sound, 13
spending, 180
spinners, 395
sports, 8, 28, 59, 83, 97, 115, 175, 352, 471
  baseball, 20, 231, 342, 392, 396
  basketball, 24, 36, 27, 33, 79, 133, 213, 339, 387, 403, 423, 434, 445
  boating, 21
  bowling, 59, 167
  cycling, 214
  diving, 141, 169
  football, 106, 116, 140, 273
  golf, 123
  hockey, 312
  marathons, 16
  mountain biking, 228, 357
  racing, 9, 13, 54
  running, 62
  sailing, 390, 443
  skateboarding, 123, 179
  skating, 95
  skiing, 184, 268, 415
  snowboarding, 332
  soccer, 45, 212, 389
  softball, 210, 227
  swimming, 61, 329, 382, 519
  tennis, 222
  track & field, 41, 396, 398
  triathlons, 544
  volleyball, 271, 382, 389
static electricity, 108
statistics, 63, 108, 124, 425
statues, 443
stock market, 107, 124, 145, 251, 351
storage, 521
student government, 372, 379, 384, 388
submarine, 123, 136

Index

supplies, 253
surveying, 265
surveys, 312, 319, 405
teachers' salaries, 62
technology, 10, 88, 341, 354, 373
television, 7, 256, 345, 497
temperature, 169
tessellations, 454, 459
tests, 70, 94, 99, 223, 299, 303, 313
textbooks, 86
theater, 497
time, 425
time spent preparing for testing, 17
tourism, 152
towers, 454
traffic, 395, 400, 519
transportation, 92, 382, 392, 522
travel, 32, 153, 162, 165, 253, 437, 445, 482
trees, 291, 351
trucks, scale model, 310
uniforms, 377
utilities, 94
vacation, 260, 339
videos, 9, 165
volunteering, 136
weather, 66, 106, 108, 109, 123, 145, 168, 174, 230, 234, 245
world records, 542
X games, 395
yogurt, 405

Area, 271. *See also* Surface area
and probability, 501, 502
circles, 493, 494
complex figures, 498, 499
parallelograms, 483
rectangles, 271
trapezoids, 490
triangles, 489

Arithmetic sequences, 34

Assessment. *See also* Prerequisite Skills; Standardized Test Practice
Mid-Chapter Practice Test, 28, 74, 116, 170, 214, 262, 310, 348, 384, 438, 486, 528
Practice Test, 49, 99, 145, 189, 235, 281, 329, 365, 405, 465, 507, 549

Assessment, T12, 4E, 52E, 104E, 148E, 194E, 238E, 286E, 332E, 368E, 410E, 468E, 510E
Higher-Level Thinking, 13, 27, 36, 41, 63, 72, 82, 115, 124, 152, 175, 185, 213, 226, 247, 257, 261, 276, 295, 315, 321, 353, 357, 360, 376, 383, 421, 425, 450, 454, 476, 492, 500, 519, 535
Open-Ended Assessment
Modeling, 21, 33, 42, 59, 88, 108, 137, 169, 219, 231, 253, 266, 291, 325, 337, 343, 373, 401, 415, 431, 473, 495, 503, 517, 541
Speaking, 23, 45, 68, 95, 130, 141, 163, 181, 209, 223, 251, 269, 303, 318, 347, 380, 392, 443, 445, 459, 484, 497, 527, 545
Writing, 9, 17, 57, 79, 111, 125, 133, 153, 159, 165, 200, 202, 206, 243, 273, 300, 308, 339, 390, 395, 437, 474, 482, 522

Associative Property
of Addition, 31
of Multiplication, 31

Auditory/Musical Learning Style. *See* Differentiated Instruction

Average, 69, 73

B

Bar graphs, 85, 86

Bar notation, 211

Base. *See also* Percents
for power, 10
parallelogram, 483
trapezoid, 490
triangle, 489

Bellringer
Communication, 6, 38, 58, 69, 120, 128, 150, 172, 207, 216, 220, 252, 258, 264, 297, 312, 334, 350, 358, 381, 387, 434, 440, 470, 489, 514
Hands-On Activity, 10, 34, 43, 60, 76, 85, 112, 132, 156, 166, 182, 197, 203, 248, 267, 292, 316, 345, 378, 391, 413, 428, 451, 475, 498, 518, 532
Problem Solving, 14, 24, 54, 64, 109, 134, 160, 164, 201, 224, 240, 244, 254, 288, 302, 323, 340, 374, 398, 422, 444, 483, 496, 501, 524
Reading Mathematics, 18, 22, 30, 80, 92, 106, 138, 177, 210, 227, 270, 275, 304, 319, 338, 354, 370, 393, 418, 446, 479, 493, 520, 538, 542

Benchmark. *See* Estimating

Bias, 344

Box-and-whisker plots, 80, 81

Building on Prior Knowledge, 4C, 52C, 104C, 148C, 194C, 238C, 286C, 332C, 368C, 410C, 468C, 510C

C

Calculator, 11, 44, 120, 210, 211, 276, 313, 382, 471, 476, 493

Capacity
changing customary units, 267–268
changing metric units, 39
cup, 267
fluid ounce, 267
gallon, 267
kiloliter, 39
liter, 39
milliliter, 39
pint, 267
quart, 267

Careers. *See* Real-Life Careers

Celsius temperature, 140

Center, 275

Centimeter, 38

Central tendency. *See* Measures of central tendency

Circle graphs, 418, 419

Circles, 275
area, 493, 494
center, 275
circumference, 274, 275
diameter, 275
radius, 275

Circumference, 275

Classroom Management
Tips for New Teachers, 25, 114, 167, 204

Closure Property, 266

Clustering, 558

Clusters, 65

Coefficient, 19

Combinations, 386–388
counting, 387

Commission, 342

Common denominator, 227

Commutative Property
of Addition, 31
of Multiplication, 31

Comparing and ordering
decimals, 556
fractions, 227
integers, 109–111
irrational numbers, 477
percents, 228
rational numbers, 227
scientific notation, 45

Compass, 416

Compatible numbers, 242

Complementary angles, 422

Complementary events, 371

Complex figures, 498
area, 498, 499

Composite numbers, 197

Index

Composite shapes. *See* Complex figures

Compound events, 398

Compound inequality, 175

Cones, 541
net of, 541

Congruent angles, 422

Congruent segments, 429

Constant, 19

Constructions
bisecting an angle, 417
congruent angles, 416
equilateral triangle, 432
isosceles triangle, 433
parallel lines, 426
perpendicular lines, 417

Continuity of Instruction, 4C, 52C, 104C, 148C, 194C, 238C, 286C, 332C, 368C, 410C, 468C, 510C

Contractions. *See* Dilations

Coordinate grid. *See* Coordinate plane

Coordinate plane, 112, 113
axes, 112
graphing points on, 113
ordered pair, 112
origin, 112
quadrants, 113
x-axis, 112
y-axis, 112

Coordinates, 112
x-coordinate, 112
y-coordinate, 112

Coordinate system. *See* Coordinate plane

Counting
fundamental principle, 378, 379
permutations, 381, 382
tree diagrams, 375

Course Planning Guide. *See* Pacing

Critical thinking, 9, 13, 17, 21, 27, 33, 36, 41, 45, 57, 63, 68, 72, 79, 83, 89, 95, 108, 111, 115, 124, 131, 137, 141, 152, 159, 163, 169, 175, 181, 185, 200, 206, 209, 213, 219, 223, 226, 231, 243, 247, 251, 257, 261, 266, 269, 273, 277, 291, 295, 300, 308, 315, 318, 321, 325, 337, 343, 347, 353, 357, 360, 373, 377, 380, 383, 390, 396, 401, 415, 421, 425, 431, 437, 443, 450, 454, 459, 473, 477, 482, 485, 492, 495, 500, 503, 517, 522, 527, 535, 541, 545

Cross products, 297

Cubed, 10

Cup, 267

Customary system, 267–269
capacity units, 267
conversions, 267–268
length units, 267
weight units, 267

Cylinders, 524
surface area, 538, 539
volume, 524, 525

D

Daily Intervention. *See* Intervention

Data, 54
bar graphs, 85, 86
box-and-whisker plots, 80, 81
center, 69, 70
clusters, 65
interquartile range, 81
line graphs, 60
line plots, 64, 65
mean, 69
median, 70
mode, 70
outliers, 65
quartiles, 80
range, 65
scatter plots, 61
stem-and-leaf plots, 76, 77

Data Updates. *See* Internet Connections

Decagon, 446

Decimals, 210, 211
adding, 559
comparing and ordering, 556
dividing, 562
estimating, 558
expanded notation, 555
multiplying, 560
place value, 555
repeating, 211
bar notation, 211
rounding, 557
subtracting, 559
terminating, 211
writing as percents, 221

Deductive reasoning, 444

Degrees, 413

Dependent events, 399

Diameter, 275

Differentiated Instruction, T4, T6
Auditory/Musical, 35, 85, 177, 255
Interpersonal, 77, 129, 167, 201, 289, 394, 515
Intrapersonal, 161, 320, 355, 480
Kinesthetic, 25, 60, 121, 157, 245, 302, 371, 423
Logical, 11, 198, 241, 293, 338, 399, 497, 525
Naturalist, 139, 271, 324, 441, 499
Verbal/Linguistic, 15, 55, 217, 265, 335, 418, 434, 476, 539
Visual/Spatial, 22, 69, 113, 204, 304, 341, 381, 447, 490, 518

Dilations, 455

Discount, 355

Distance formula, 161

Distributive Property, 30

Dividend, 139

Divisibility patterns, 554

Division
decimals, 562
dividend, 139
divisor, 139
fractions, 264
integers, 138, 139
mixed numbers, 265
phrases indicating, 150
quotient, 139
solving equations, 259

Division Property of Equality, 160

Domain, 177

E

English-Language Learner (ELL), 4G, 15, 46, 52G, 55, 96, 104G, 142, 148G, 186, 194G, 217, 232, 238G, 265, 278, 286G, 326, 332G, 335, 362, 368G, 402, 410G, 418, 434, 462, 468G, 476, 504, 510G, 546

Equality
Addition Property, 157
Division Property, 160
Multiplication Property, 259
Subtraction Property, 157

Equals sign (=), 24

Equations, 24, 151
addition, 156
checking solutions of, 157, 161
division, 259
linear, 178
models of, 154–155, 160, 166
multiplication, 161
nonlinear, 181
solution of, 24
solving, 24, 25, 156–158, 160, 161, 259
two-step, 166, 167
subtraction, 157
writing sentences as, 151

Equilateral triangle, 429

Index

Equivalent expressions, 30

Equivalent fractions, 207

Estimating
clustering, 558
compatible numbers, 242
front end, 240
percents, 334–336
square roots, 475, 476
with decimals, 558
with fractions, 240

Evaluating, 11
algebraic expressions, 18, 19
expressions with powers, 15
numerical expressions, 14
powers, 11

Evaluation. *See* Assessment

Events. *See* Probability

Expanded notation, 555

Experimental probability, 393, 394

Exponential form, 11

Exponents, 10, 11

Expressions, 18, 19
equivalent, 30
numerical, 14
simplifying, 122, 123
translating words into, 150

F

Factorial, 381
n!, 381

Factoring
algebraic expressions, 198
whole numbers, 197, 198

Factors, 10, 196

Factor tree, 198

Fahrenheit temperature, 140

Fair game, 374

Figures
similar, 440, 441
three-dimensional (*See* Solids)

Find the Error, 16, 26, 40, 66, 71, 78, 122, 130, 151, 162, 168, 174, 205, 208, 222, 246, 250, 276, 290, 294, 336, 352, 389, 399, 435, 481, 494, 521, 539

Flexible Grouping, 7, 31, 66, 92, 109, 133, 150, 183, 211, 225, 228, 249, 252, 267, 317, 346, 358, 378, 391, 429, 445, 451, 471, 493, 533

Flowcharts, 125
symbols, 125

Foldables™ Study Organizer, T8, 8, 11, 16, 20, 28, 32, 40, 62, 71, 74, 94, 114, 116, 122, 130, 136, 140, 151, 158, 162, 168, 170, 184, 199, 205, 212, 214, 218, 222, 225, 229, 242, 250, 256, 262, 290, 299, 306, 310, 314, 324, 342, 348, 352, 356, 372, 384, 389, 395, 400, 430, 436, 438, 442, 448, 472, 481, 486, 491, 516, 526, 528, 534, 540
decimal patterns and algebra, 5
fractions, 239
fractions, decimals, and percents, 195
geometry, 411
integers, 105
measuring figures, 469
percents, 333
probability, 369
ratios and proportions, 287
solving equations, 149
statistics, 53
surface area and volume, 511

Formula, 271

Four-step problem-solving plan, 6–9, 22, 58, 132, 164, 201, 252, 302, 338, 391, 496, 518

Fractions
adding and subtracting, 244, 245
comparing, 227, 228
dividing, 264
equivalent, 207
estimating with, 240
like, adding and subtracting, 244
multiplying, 254, 255
ordering, 227, 228
simplest form, 207
simplifying, 207, 208
unlike, adding and subtracting, 245
writing as decimals, 211
writing as percents, 217, 313, 314

Frequency distributions. *See* Frequency tables

Frequency tables, 54, 55
interval, 54
scale, 54
tally, 54
tally marks, 54

Functions, 177, 178
domain, 177
graphing, 176, 178
input, 177
linear, 178, 179
nonlinear, 181
ordered pairs, 176
output, 177
range, 177
real-world, 179–181
rule, 178
tables, 177–181, 182
writing expressions for, 179–181

Function table, 177

Fundamental Counting Principle, 378, 379

G

Gallon, 267

Game Zone
algebraic expressions, 29
describing solids, 529
factoring integers, 215
finding squares, 439
finding the mean, 75
graphing points on a coordinate plane, 117
multiplying fractions, 263
percent of a number, 349
probability, 385
solving equations, 171
solving proportions, 311
square roots, 487

GCF. *See* Greatest common factor (GCF)

Geometric sequences, 34

Geometry. *See* Angles; Area; Circles; Constructions; Lines; Perimeter; Polygons; Quadrilaterals; Transformations; Triangles

giga, 41

Gram, 39

Graphing
a line using slope, 185
linear equations, 178, 179
on coordinate plane, 113
solutions of equations, 25
solutions of inequalities, 172

Graphing Calculator Investigation, T4
box-and-whisker plots, 84

Graphs
bar graphs, 85, 86
box-and-whisker plots, 80, 81
circle, 418, 419
coordinate plane, 112
histograms, 86
linear equations, 178
line graphs, 60
line plots, 64, 65
misleading, 92

Index

multiple bar, 90
multiple line, 90
number line, 106
scatter plots, 61
stem-and-leaf plots, 76, 77

Greatest common factor (GCF), 203, 204
of algebraic expressions, 204

H

Hands-On Lab
adding integers, 118–119
average student, 73
building three-dimensional figures, 512
changes in volume and surface area, 536–537
circumference, 274
constructing and bisecting angles, 416–417
constructing parallel lines, 426
constructing triangles, 432–433
experimental probability, 397
exploring combinations, 386
exploring factors, 196
exploring sequences, 37
functions and graphs, 176
measuring angles, 412
nets and surface area, 530–531
Pythagorean Theorem, 478
rate of change, 296
rotations, 460–461
sampling, 344
solving equations using models, 154–155
subtracting integers, 126
triangles and trapezoids, 488
using a percent model, 322
wildlife sampling, 301

Hands-On Mini Lab
adding fractions on a ruler, 244
area of a circle, 493
area of a parallelogram, 483
area of a triangle, 489
converting length units, 38
finding a pattern, 18
finding the mean, 69
folding triangles, 428
making a scale drawing, 304
making predictions, 60
modeling division by a fraction, 264
modeling division of integers, 138
modeling multiples, 224
modeling multiplicative inverses, 258
modeling percent of change, 350
modeling percents, 316
modeling primes, 197
modeling subtraction of integers, 128
patterns and sequences, 34
perimeter and area, 470
probability game, 374
similar figures, 440
solving equations using models, 160
tessellations, 451

Height
parallelogram, 483
trapezoid, 490
triangle, 489

Heptagon, 446

Hexagon, 446

Higher-Level Thinking, 13, 27, 36, 41, 63, 72, 82, 115, 124, 152, 175, 185, 213, 226, 247, 257, 261, 276, 295, 315, 321, 353, 357, 360, 376, 383, 421, 425, 450, 454, 476, 492, 500, 519, 535

Histograms, 86–89

Homework Help, 9, 12, 16, 20, 26, 32, 36, 40, 45, 56, 62, 67, 72, 78, 83, 88, 94, 108, 111, 114, 123, 130, 136, 140, 152, 159, 162, 168, 174, 180, 184, 199, 205, 209, 212, 218, 222, 226, 230, 243, 246, 250, 256, 260, 266, 269, 272, 277, 290, 294, 299, 307, 315, 318, 321, 325, 337, 342, 347, 352, 356, 360, 372, 376, 380, 383, 389, 395, 400, 415, 420, 424, 430, 436, 442, 449, 453, 458, 472, 477, 482, 485, 491, 494, 500, 503, 516, 522, 526, 534, 540, 544

Hypotenuse, 479

I

Identity Property
of Addition, 31
of Multiplication, 31

Improper fractions, 563

Independent events, 398

Indirect measurement, 441

Inductive reasoning, 444

Inequalities, 172, 173
graphing solutions, 172
solving, 172, 173
symbols, 172
two-step, 173

Input. *See* Functions

Instructional Strategies, T4–T5

Integers, 106, 107
absolute value, 107
adding, 120–122
different signs, 121
on a number line, 120
same sign, 120
using models, 118–119
Additive Inverse Property, 121
comparing, 109
greater than (>), 109
less than (<), 109
dividing, 138, 139
different signs, 138, 139
same sign, 139
multiplying, 134, 135
different signs, 134
same sign, 135
negative, 106
on a number line, 106, 109
ordering, 110
positive, 106
subtracting, 128, 129
negative, 129
positive, 128
using models, 126–127
zero, 106
zero pairs, 118

Interactive Chalkboard, T5, 4B, 52B, 104B, 148B, 194B, 238B, 286B, 332B, 368B, 410B, 468B, 510B

Interdisciplinary Project. *See* WebQuest

Internet, T5, T7, T15, T17, 4B, 4D, 4E, 4G, 4H, 52B, 52D, 52E, 52G, 52H, 104B, 104D, 104E, 104G, 104H, 148B, 148D, 148E, 148G, 148H, 194B, 194D, 194E, 194G, 194H, 238B, 238D, 238E, 238G, 238H, 286B, 286D, 286E, 286G, 286H, 332B, 332D, 332E, 332G, 332H, 368B, 368D, 368E, 368G, 368H, 410B, 410D, 410E, 410G, 410H, 468B, 468D, 468E, 468G, 468H, 510B, 510D, 510E, 510G, 510H

Internet Connections
msmath2.net/careers, 19, 77 113, 173, 208, 249, 298, 351, 394, 448, 480, 539
msmath2.net/chapter_test, 49, 99, 145, 189, 235, 281, 329, 365, 405, 465, 507, 549
msmath2.net/data_update, 27, 66, 72, 123, 169, 174, 231, 261, 300, 342, 383, 459, 492, 516
msmath2.net/extra_examples, 7, 11, 15, 19, 25, 31, 35, 39, 44, 55, 61, 65, 69, 77, 81, 87, 93, 107, 109, 113, 121, 129, 135, 139, 151, 157, 161, 167, 173, 179, 183, 197, 203, 207, 211, 217, 221, 225, 229, 241, 245, 249, 255, 259, 265, 267, 271, 275, 289, 293, 297, 305, 313, 316, 319, 323, 335, 341, 345, 351, 355, 359, 371, 375, 379, 382, 387, 393, 399, 414, 419, 423, 429, 435, 441, 447, 451, 457, 471, 475, 479, 483, 489, 493, 499, 501, 515, 521, 525, 533, 539, 543

Index

msmath2.net/other_calculator_keystrokes, 84
msmath2.net/self_check_quiz, 9, 13, 17, 21, 27, 33, 36, 41, 45, 57, 63, 67, 72, 79, 83, 89, 95, 108, 111, 115, 123, 131, 137, 141, 152, 159, 163, 169, 175, 181, 185, 199, 205, 209, 213, 219, 223, 226, 231, 243, 247, 251, 257, 261, 266, 269, 273, 277, 291, 295, 299, 307, 315, 318, 321, 325, 337, 343, 347, 353, 357, 360, 373, 377, 380, 383, 389, 395, 401, 415, 421, 425, 431, 437, 443, 449, 453, 459, 473, 477, 482, 485, 491, 495, 500, 503, 517, 522, 527, 535, 541, 545
msmath2.net/standardized_test, 51, 101, 147, 191, 237, 283, 331, 367, 407, 467, 509, 551
msmath2.net/vocabulary_review, 46, 96, 142, 186, 232, 278, 326, 362, 402, 462, 504, 546
msmath2.net/webquest, 3, 95, 103, 185, 193, 277, 285, 401, 409, 545

Interpersonal Learning Style. *See* Differentiated Instruction

Interquartile range, 81

Interval. *See* Frequency tables

Intervention, T10–T11
Daily Intervention, 7, 11, 15, 16, 19, 22, 25, 26, 31, 35, 40, 43, 55, 59, 60, 66, 69, 71, 77, 78, 80, 85, 92, 106, 109, 113, 121, 122, 129, 130, 133, 135, 139, 150, 151, 157, 161, 162, 164, 167, 168, 173, 174, 177, 183, 198, 201, 204, 205, 208, 211, 217, 221, 222, 225, 228, 241, 245, 246, 249, 250, 252, 255, 259, 265, 267, 271, 275, 276, 289, 290, 293, 294, 298, 302, 304, 313, 317, 320, 324, 335, 336, 338, 341, 346, 351, 352, 355, 358, 371, 375, 378, 381, 388, 389, 391, 394, 399, 400, 413, 418, 423, 429, 434, 435, 441, 445, 447, 451, 471, 476, 480, 481, 485, 490, 493, 494, 497, 499, 501, 515, 518, 520, 521, 525, 533, 539, 540, 543
Tips for New Teachers, 7, 129, 198, 253, 422, 441, 480, 539

Intrapersonal Learning Style. *See* Differentiated Instruction

Inverse operations, 156

Inverse Property
of Addition, 121
of Multiplication, 258

Irrational numbers, 476
models of, 475, 476

Irregular figures. *See* Complex figures

Isosceles triangle, 429

K

Kilogram, 39

Kilometer, 38

Kinesthetic Learning Style. *See* Differentiated Instruction

L

Labs. *See* Hands-On Lab and Hands-On Mini Lab

LCD. *See* Least common denominator (LCD)

LCM. *See* Least common multiple (LCM)

Leaf. *See* Stem-and-leaf plots

Learning Styles. *See* Differentiated Instruction

Least common denominator (LCD), 227

Least common multiple (LCM), 224, 225

Leg, 479

Length
changing customary units, 267
foot, 267
inch, 267
metric units, 38
mile, 267
yard, 267

Lesson Objectives, 4A, 52A, 104A, 148A, 194A, 238A, 286A, 332A, 368A, 410A, 468A, 510A

Linear equations, 178, 179

Line graphs, 60
multiple, 63

Line plots, 64, 65

Lines. *See also* Linear equations
parallel, 426
perpendicular, 417
slope, 182, 183

Lines of symmetry, 456

Line symmetry, 456

Liter, 39

Logical Learning Style. *See* Differentiated Instruction

Logical reasoning, 444

Lower extreme, 80

Lower quartile (LQ), 80

LQ. *See* Lower quartile

M

Magnifications. *See* Dilations

Manipulatives, T5, T16, 4B, 52B, 104B, 121, 148B, 161, 194B, 211, 238B, 256B, 332B, 341, 368B, 371, 410B, 414, 452, 468B, 510B, 521

Markup, 357

Mass
changing units of, 39
gram, 39
kilogram, 39
milligram, 39

Mathematical Background, 4C–4D, 52C–52D, 104C–104D, 148C–148D, 194C–194D, 238C–238D, 286C–286D, 332C–332D, 368C–368D, 410C–410D, 468C–468D, 510C–510D

MathPASS: Tutorial Plus, T11, 4B, 52B, 104B, 148B, 194B, 238B, 256B, 332B, 368B, 410B, 468B, 510B

MathScape, T13, 4H, 52H, 104H, 148H, 194H, 238H, 286H, 332H, 368H, 410H, 468H, 510H

Mean, 69
versus median, 93

Measurement. *See also* Customary system; Metric system
angles, 412
area
circles, 493
complex figures, 498–500
converting, 273
parallelograms, 483
rectangles, 271
trapezoids, 488, 490
triangles, 488, 489
circumference, 274–276
effect of changing dimensions, 271, 484, 495, 526, 536–537
indirect, 441
perimeter
irregular figures, 271
rectangles, 270
precision, 542
Pythagorean Theorem, 479–482
significant digits, 542, 543
similar solids, 523
surface area
cylinders, 538–539
rectangular prisms, 532–533
volume
cylinders, 524–525
rectangular prisms, 520–521

Measures of central tendency, 69
mean, 69
median, 70
mode, 70
summary, 71

Index

Median, 70
versus mean, 93

Meeting Student Needs. *See also* Prerequisite Skills; Differentiated Instruction
T8–T9

Mental Math. *See also* Number sense
198, 211, 221, 228, 314, 320

Meter, 38

Metric system, 38, 39
capacity units, 38
conversions
capacity units, 38
length units, 38
mass units, 39
unit conversion table, 38
length units, 38
mass units, 39
unit abbreviations, 39

Mid-Chapter Practice Test, 28, 74, 116, 170, 214, 262, 310, 348, 384, 438, 486, 528

Millimeter, 38

Misconceptions, 19, 43, 59, 80, 106, 135, 164, 173, 208, 221, 259, 275, 298, 313, 351, 375, 388, 413, 485, 501, 520, 543

Misleading statistics, 92, 93

Mixed numbers
adding and subtracting, 248, 249
dividing, 265
improper fractions and, 563
multiplying, 255

Mode, 70

Modeling. *See also* Open-Ended Assessment
T6, T9, 52G, 104G, 148G, 194G, 510G

Multiple, 224

Multiplication,
Associative Property of, 31
Commutative Property of, 31
decimals, 560
fractions, 254, 255
Identity Property of, 31
integers, 134, 135
Inverse Property of, 258
mixed numbers, 255
phrases indicating, 150
solving equations, 259
using dot for, 10

Multiplication Property of Equality, 259

Multiplicative Inverse Property, 258

Multiplicative inverses, 258

N

NAEP 2005, T12, T22, 4A, 52A, 104A, 148A, 194A, 238A, 286A, 332A, 368A, 410A, 468A, 510A

nano, 41

Naturalist Learning Style. *See* Differentiated Instruction

NCTM Principles and Standards, T21–T22, 4A, 52A, 104A, 148A, 194A, 238A, 286A, 332A, 368A, 410A, 468A, 510A

Negative integers, 106

Nets, 530, 531, 541

Nonagon, 446

Number line
absolute value, 107
adding integers, 120–121
comparing rational numbers, 228
graphing inequalities, 172–175
graphing integers, 106
graphing irrational numbers, 228
graphing solutions of equations, 25

Numbers
compatible, 242
composite, 197
exponential form, 11
integers, 106
irrational, 476
perfect squares, 471
prime, 197
rational, 229
squares, 470, 471
standard form, 11, 43

Number sense, 32, 44, 110, 122, 136, 212, 225, 231, 242, 250, 256, 294, 336, 352, 379, 424, 476, 484, 491, 494, 526

Numerical expressions, 14

O

Obtuse angle, 413

Obtuse triangle, 429

Octagon, 446

Odds, 396

Open-Ended Assessment
Modeling, 21, 33, 42, 59, 88, 108, 137, 169, 219, 231, 253, 266, 291, 325, 337, 343, 373, 401, 415, 431, 473, 495, 503, 517, 541
Speaking, 23, 45, 68, 95, 130, 141, 163, 181, 209, 223, 251, 269, 303, 318, 347, 380, 392, 443, 445, 459, 484, 497, 527, 545
Writing, 9, 17, 57, 79, 111, 125, 133, 153, 159, 165, 200, 202, 206, 243, 273, 300, 308, 339, 390, 395, 437, 474, 482, 522

Opposites, 121

Ordered pairs, 112
graphing, 113

Order of operations, 15

Origin, 112

Ounce, 267

Outcomes. *See* Probability

Outliers, 65
box-and-whisker plots, 81, 82
effect on mean, 72

Output. *See* Functions

P

Pacing, T20

Parallel lines, 426
properties of, 427

Parallelograms, 434
area, 483, 484
base, 483
height, 483

Part. *See* Percents

Pascal's Triangle, 392

Pentagon, 446

Percent equation, 340

Percent proportion, 324

Percents, 216, 312, 313
base
in percent equation, 340, 341
in percent proportion, 324
commission, 342
discount, 355
estimating, 334–336
greater than 100%, 317, 318
less than 1%, 317, 318
markup, 357
models of, 216, 217, 322
part
in percent equation, 340, 341
in percent proportion, 324
part/whole relationship, 323
percent equation, 340, 341
percent of a number, 319, 320
percent of change, 350, 351
percent of decrease, 350
percent of increase, 350
percent proportion, 324, 325
sales tax, 354
simple interest, 358, 359, 361
writing as decimals, 220, 221
writing as fractions, 217, 312

Index

Perfect squares. *See* Numbers

Perimeter, 270, 271
rectangles, 270

Permutations, 381, 382
counting, 382

Perpendicular lines, 417

Pi (π), 275
benchmark for, 213

Pint, 267

Place value, 555

Point, on a coordinate plane, 112

Polygons, 446, 447
angle measurements, 447
classifying, 446, 447
decagon, 446
heptagon, 446
hexagon, 446
nonagon, 446
octagon, 446
pentagon, 446
regular, 446
tessellations, 447, 448
translations, 451, 452

Population, 345

Population density, 295

Portfolios, 49, 99, 145, 189, 235, 281, 329, 365, 405, 465, 507, 549

Positive integers, 106

Pound, 267

Powers, 10, 11
base, 10
of ten, 561

Practice Test, 49, 99, 145, 189, 235, 281, 329, 365, 405, 465, 507, 549

Precision, 542

Precision unit, 542

Predictions, 60, 61
using line graphs, 60
using probability, 394
using scatter plots, 61
using statistics, 345

Prerequisite Skills, T3, T10
Adding and Subtracting Decimals, 559
Comparing and Ordering Decimals, 556
Dividing Decimals, 562
Divisibility Patterns, 554
Estimating with Decimals, 558
Getting Ready for the Next Lesson, 9, 13, 17, 21, 27, 33, 36, 41, 57, 63, 68, 72, 79, 83, 89, 108, 111, 115, 124, 131, 137, 152, 159, 163, 169, 175, 181, 200, 206, 209, 213, 219, 223, 226, 243, 247, 251, 257, 261, 266, 269, 273, 291, 295, 300, 308, 315, 318, 321, 337, 343, 347, 353, 357, 373, 377, 380, 383, 390, 396, 415, 421, 425, 431, 437, 443, 450, 454, 473, 477, 482, 485, 492, 495, 500, 517, 522, 527, 535, 541
Getting Started, 5, 53, 105, 149, 195, 239, 287, 333, 369, 411, 469, 511
Mixed Numbers and Improper Fractions, 563
Multiplying Decimals, 560
Place Value and Decimals, 555
Powers of Ten, 561
Rounding Decimals, 557

Prime factorization, 198

Prime numbers, 197
relatively prime, 206
twin primes, 200

Principal, 358

Prisms, rectangular
surface area, 532, 533
volume, 520, 521

Probability, 370, 371
area models, 501, 502
events, 370
complementary, 371
compound, 398
dependent, 399
independent, 398
mutually exclusive, 401
simple, 370
experimental, 393, 394, 397
odds, 396
outcomes, 370
counting, 375, 378
random, 370
sample space, 374
theoretical, 393, 394
tree diagrams, 374, 375

Problem solving, 6–8
four-step plan, 6
list of strategies, 7

Problem-Solving Strategy
act it out, 391
determine reasonable answers, 338
draw a diagram, 302
eliminate possibilities, 252
guess and check, 22
look for a pattern, 132
make a model, 518
make an organized list, 201
solve a simpler problem, 496
use a graph, 58
use logical reasoning, 444
work backward, 164

Products, integers, 134

Professional Development. *See also* Staff Development
4D, 52D, 104D, 148D, 194D, 238D, 286D, 332D, 368D, 410D, 468D, 510D

Program Validation, T18–T19

Project CRISS, T9, 4F, 52F, 104F, 148F, 194F, 238F, 286F, 332F, 368F, 410F, 468F, 510F

Projects. *See* WebQuest

Properties, 30, 31
Addition Property of Equality, 157
Additive Inverse Property, 121
Associative Property, 31
Closure Property, 266
Commutative Property, 31
Distributive Property, 30
Division Property of Equality, 160
of geometric figures, 427, 434
Identity Property, 31
Multiplication Property of Equality, 259
Multiplicative Inverse Property, 258
Subtraction Property of Equality, 157

Proportional reasoning
aspect ratio, 291
circumference, 275
distance formula, 161
indirect measurement, 441
linear functions, 178–179
measurement conversions, 267–268, 38–39
percent proportion, 324
population density, 295
rate of change, 296
ratio, 288–289
scale drawings, 304–306
scale factor, 305
scaling, 304–309
similar figures, 440–443
similar solids, 523
slope, 182–185
solving proportions, 298
speed, 161
unit price, 293
unit rate, 293

Proportions, 297, 298
cross product, 297
identifying, 297
solving, 298

Protractor, 412

Pythagorean Theorem, 479–481
model of, 478

Index

Q

Quadrants, 113

Quadrilaterals, 434, 435
- classifying, 434
- parallelogram, 434
- rectangle, 434
- rhombus, 434
- square, 434
- sum of angles, 435
- trapezoid, 434

Quart, 267

Quartiles, 80, 81
- interquartile range, 81
- lower, 80
- upper, 80

Quotients, integers, 138

R

Radical sign, 471

Radius, 275

Random, 370

Random sample, 345

Range
- of a data set, 65
- of a function, 177

Rate of change, 296

Rates, 292, 293
- distance formula, 161
- population density, 295
- speed, 161
- unit, 292, 293
- unit price, 293

Rational numbers, 229
- comparing and ordering, 228
- identifying, 229
- models of, 229
- percents, 217, 220, 221
- repeating decimals, 211
- terminating decimals, 210–211

Ratios, 216, 288, 289. *See also* Fractions
- comparing, 289
- equivalent, 289, 297
- simplifying, 288
- with units, 289
- writing as percents, 216
- writing in simplest form, 288

Reading, Link to, 30, 34, 64, 69, 120, 182, 210, 258, 340, 387, 413, 451, 498, 520

Reading Math
- addends and sums, 118
- division sentences, 139
- first power, 10
- formula for simple interest, 359
- line segments, 441
- minuends, subtrahends, and differences, 127
- multiplication, 19
- names for variables, 18
- parallel lines, 434
- percent, 217
- percents, 317
- polygons, 446
- probability notation, 398
- quartiles, 81
- regular polygons, 447
- rotations, 460
- scatter plots, 61
- set theory, 107
- square roots, 471
- squaring a number, 470
- subscripts, 490
- theorem, 480
- triangle, 428
- unit prefixes, 39

Reading and Writing Mathematics, T6–T7, 4F, 4G, 9, 15, 17, 46, 52F, 52G, 55, 57, 79, 96, 104F, 104G, 111, 125, 133, 142, 148F, 148G, 153, 159, 165, 186, 194F, 194G, 200, 202, 206, 217, 232, 238F, 238G, 243, 265, 273, 278, 286F, 286G, 300, 308, 326, 332F, 332G, 335, 339, 362, 368F, 368G, 390, 395, 402, 410F, 410G, 418, 434, 437, 462, 468F, 468G, 474, 476, 482, 504, 510F, 510G, 522, 546

Real-Life Careers
- archaeologist, 480
- athletic trainer, 19
- cartographer, 113
- chef, 394
- composer, 208
- graphic designer, 539
- landscape architect, 448
- nutritionist, 77
- paleontologist, 298
- plumber, 249
- stockbroker, 351

Real-Life Math
- animal speeds, 161
- basketball, 423
- bowling, 167
- checkers, 388
- chess, 452
- coins, 217
- communicating by E-mail, 346
- food, 151
- games, 502
- geography, 221, 490
- gift wrapping, 533
- history, 31
- International Space Station, 70
- metric system, 39
- mountain biking, 228
- oceanography, 122
- Olympic Games, 55, 419
- parades, 204
- Parcheesi, 371
- physical science, 471
- ponds, 289
- presidents' native states, 82
- scooters, 375
- skateboarding, 179
- skiing, 268
- surveying, 265
- swimming, 382
- technology use, 341
- television, 7
- video games, 15
- weather, 245

Reasoning, logical
- deductive, 444
- inductive, 444

Reciprocals, 258

Rectangles, 434
- area, 271
- perimeter, 270

Reflections, 457

Regular. *See* Polygons

Repeating. *See* Decimals

Research on program, T18–T19

Resources for Teachers, T4–T5, 4B, 52B, 104B, 148B, 194B, 238B, 286B, 332B, 368B, 410B, 468B, 510B

Rhombus, 434

Right angle, 413

Right triangle, 429
- hypotenuse, 479
- legs, 479
- Pythagorean Theorem, 479–481

Rounding
- decimals, 557
- fractions, 240

Rubrics, 51, 101, 147, 191, 237, 283, 331, 367, 407, 467, 509, 551

Rule. *See* Functions

S

Sales tax, 354

Sample space, 374

Sampling, 301, 344
- bias, 344
- random, 344, 345

Scale, 304. *See also* Frequency tables

Index

Scale drawings, 304, 305
draw objects to scale, 306, 309
equivalent scales, 305
scale factor, 305
spreadsheets and, 309

Scale factor, 305

Scale model, 306

Scalene triangle, 429

Scaling, 304–309

Scatter plots, 61

Scientific notation, 43, 44

Segments, congruent, 429

Sentences, as algebraic equations, 151

Sequences, 34, 35
arithmetic, 34
geometric, 34
terms, 34

Sets
element, 107
intersection, 175
list by roster, 107
list by rule, 229
subset, 107
union, 175
Venn diagram, 203, 229

Significant digits, 542, 543

Similar figures, 440, 441

Similarity
aspect ratio, 291
scaling, 304–309

Similar solids, 523
changes in volume and surface area, 536–537

Simple events, 370

Simple interest, 358, 359

Simplest form, 207
for ratios, 288

Simplifying
expressions, 122, 123
fractions, 207, 208

Slope, 182, 183
constant, 185
graph a line using, 185
negative, 182
positive, 182
rate of change, 296

Solids
drawing, 514
similar, 523

Solution, of an equation, 24

Solving
equations, 24, 25
two-step, 166, 167
equations involving addition and subtraction, 156–158
equations involving multiplication, 160, 161, 259
inequalities, one-step, 173
proportions, 298

Speaking. *See* Open-Ended Assessment

Spreadsheet Investigation
dilations, 455
multiple line and multiple bar graphs, 90
scale drawings, 309
similar solids, 523
simple interest, 361

Square. *See* Numbers

Squared, 10

Square roots, 471
estimating, 475, 476

Squares, 434

Staff Development, T16–T17, 4C, 4D, 4F, 4G, 52C, 52D, 52F, 52G, 104C, 104D, 104F, 104G, 148C, 148D, 148F, 148G, 194C, 194D, 194F, 194G, 238C, 238D, 238F, 238G, 286C, 286D, 286F, 286G, 332C, 332D, 332F, 332G, 368C, 368D, 368F, 368G, 410C, 410D, 410F, 410G, 468C, 468D, 468F, 468G, 510C, 510D, 510F, 510G

Standard form, 11, 43

Standardized Test Practice
Extended Response, 51, 101, 147, 191, 237, 283, 331, 367, 407, 467, 509, 551
Multiple Choice, 9, 13, 17, 21, 23, 27, 28, 33, 41, 45, 49, 50, 57, 59, 63, 68, 72, 74, 79, 83, 89, 95, 99, 100, 108, 111, 115, 116, 124, 131, 133, 137, 141, 145, 146, 152, 159, 163, 165, 169, 170, 175, 181, 185, 189, 190, 200, 202, 206, 209, 213, 214, 219, 223, 226, 231, 235, 236, 243, 247, 251, 253, 257, 261, 262, 269, 273, 277, 281, 282, 291, 295, 300, 303, 308, 315, 318, 321, 325, 329, 330, 337, 339, 343, 347, 348, 353, 357, 360, 365, 366, 373, 377, 380, 390, 392, 396, 401, 406, 415, 421, 425, 431, 437, 438, 443, 445, 450, 454, 459, 465, 466, 473, 477, 482, 485, 486, 492, 495, 497, 500, 503, 507, 508, 517, 519, 522, 527, 535, 541, 545, 549, 550
Short Response/Grid In, 9, 13, 21, 27, 28, 33, 36, 41, 45, 51, 63, 72, 74, 79, 83, 89, 95, 101, 108, 111, 115, 116, 124, 131, 137, 141, 147, 159, 169, 170, 185, 191, 200, 206, 209, 213, 214, 219, 223, 226, 231, 237, 257, 261, 266, 269, 273, 277, 283, 300, 308, 310, 315, 321, 325, 331, 347, 353, 357, 367, 377, 380, 383, 384, 390, 396, 401, 405, 407, 415, 421, 425, 431, 438, 443, 450, 454, 459, 467, 473, 477, 482, 485, 486, 492, 495, 500, 503, 509, 522, 528, 535, 551
Worked-Out Example, 25, 87, 110, 183, 229, 259, 293, 355, 379, 441, 499, 525

Statistics, 54. *See also* Graphs
measures of central tendency (*See* Measures of central tendency)
misleading, 93
population, 345
predicting using, 345, 346
random sample, 345
survey, 345

Stem. *See* Stem-and-leaf plots

Stem-and-leaf plots, 76, 77
leaf, 76
stem, 76

Straight angle, 413

Straightedge, 416

Study Guide and Review, 46–48, 96–98, 142–144, 186–188, 232–234, 278–280, 326–328, 362–364, 402–404, 462–464, 504–506, 546–548

Study Organizer. *See* Foldables™ Study Organizer

Study Skill
making a web, 474
reading math problems, 153
taking good notes, 42, 125

Study Tip
alternate method for converting percents to decimals, 316
area units, 271
canceling units, 289
check, 433
checking solutions, 157
checking solutions of inequalities, 173
check the reasonableness of results, 429
check your answer, 371
choosing a method, 204, 225, 313
clusters, 65
common error, 484
the Commutative and Associative Properties, 31
Commutative Property, 198
compatible numbers, 242

Index

congruent segments, 429
defining the variable, 25
dependent events, 399
diagonals, 447
discount, 355
entering a negative integer on a calculator, 120
equivalent scales, 305
estimating percents, 335
estimation, 293, 521
exponents on the calculator, 11
expressions with powers, 15
factorials on a calculator, 382
finding circumference on a calculator, 276
finding percents mentally, 320
finding the base, 324
finding the median, 70
four-step plan, 7
fraction-decimal equivalencies, 211
fractions, 241
front-end estimation, 240
hypotenuse, 481
input and output, 178
irrational numbers, 476
length of parts of a box-and-whisker plots, 81
look back, 135, 155, 158, 259, 296, 298, 314, 441
mental math, 38, 198, 211, 255
Multiplicative Property of Zero, 135
percents and decimals, 221
reading histograms, 86
reflecting over the *x*-axis, 457
reflecting over the *y*-axis, 457
scale, 306
scientific notation on the calculator, 44
simplest form, 207
simplifying, 255
slope, 183
solving equations, 167
subtraction and negative numbers on a calculator, 129
transformations, 451
translation, 452
unbiased samples, 344
using the percent equation, 340
writing percents, 341
writing positive integers, 106

Subtraction
decimals, 559
fractions, 244
integers, 128, 129
mixed numbers, 248, 249
phrases indicating, 150
solving equations, 156

Subtraction Property of Equality, 157

Supplementary angles, 422

Surface area, 532
cylinders, 538, 539
nets, 530, 531
rectangular prisms, 532, 533

Survey, 345

Symbols
absolute value (| |), 107
factorial (!), 381
is congruent to (≅), 422
is equal to (=), 24
is greater than (>), 109
is greater than or equal to (≥), 172
is less than (<), 109
is less than or equal to (≤), 172
is not equal to (≠), 24
is similar to (~), 440
measure of angle 1 ($m\angle 1$), 428
pi (π), 275

Symmetry, lines of, 456

T

Tally. *See* Frequency tables

Teacher to Teacher, 4C, 52D, 104D, 148C, 194C, 238D, 286D, 332D, 368D, 410D, 468D, 510D

TeacherWorks, T5, T20

Teaching Strategies, 15, 31, 61, 217, 241, 245, 298, 305, 313, 335, 388, 423, 440, 490

Teaching Tips, 37, 73, 84, 90, 118, 126, 154, 176, 196, 274, 296, 301, 309, 322, 344, 361, 386, 397, 412, 416, 426, 432, 455, 460, 478, 488, 512, 523

Technology. T5, T7, T11, T12, T16–T17. *See also* Calculators; Graphing Calculator Investigation; Internet Connections; Spreadsheet Investigation; WebQuest

Temperature, 140

Term, 19

Terminating. *See* Decimals

Terms, of a sequence, 34

Tessellations, 447, 448

Test-Taking Tip
backsolving, 259
drawings, 499
educated guess, 379
eliminate possibilities, 229
eliminating answer choices, 110
elimination, 355
grid-in answers, 293, 441
make a drawing, 183
read a graph, 87
reading, 525
solutions, 25

Theoretical probability, 393, 394

Three-dimensional figures
building, 512–513
cone, 541
cylinder, 524
drawing, 514–517
effect of changing dimensions, 526, 536–537
modeling, 518–519
nets, 530–531
pyramid, 531
rectangular prism, 520–521
similar solids, 523, 526, 536–537
solids, 514
surface area of, 532–535, 538–541
triangular prism, 522
volume of, 520–527

Tips for New Teachers
Classroom Management, 25, 114, 167, 204
Intervention, 7, 129, 198, 253, 422, 441, 480, 539
Teaching Strategies, 15, 31, 61, 217, 241, 245, 298, 305, 313, 335, 388, 423, 440, 490
Using Applications, 39, 107, 110, 293, 382, 419, 471
Using Manipulatives, 121, 161, 211, 341, 371, 414, 452, 521

Ton, 267

Transformations, 451
dilations, 455
identifying, 461
reflections, 456
rotations, 460, 461
translations, 451, 452

Translations, 451, 452
coordinates, 452

Transversal, 427

Trapezoid, 434
area, 490
base, 490
height, 490

Tree diagrams, 374, 375

Triangles, 428, 429
acute, 429
area, 489
base, 489
classifying, 429
equilateral, 429
height, 489
isosceles, 429
obtuse, 429
right, 429
hypotenuse, 479
leg, 479
scalene, 429
similar, 440, 441

Index

Triangular numbers, 36

Tutorials
MathPASS, T11

Two-dimensional figures. *See also* Measurement
complex, 498
constructing triangles, 432–433
decagon, 446
heptagon, 446
hexagon, 446
octagon, 446
parallelogram, 434
pentagon, 446
polygons, 446
quadrilateral, 434–437
rectangle, 434
regular polygon, 446
similar, 440–443
square, 434
tessellation, 447–448
transformations of, 451–461
triangle, 428–431

Two-step equations, 166, 167

U

Unit rates, 292
best buy, 293

Units
customary, 267
converting between, 267, 268
metric, 38
converting between, 38, 39

Unlocking Misconceptions. *See* Misconceptions

Upper extreme, 80

Upper quartile (UQ), 80

UQ. *See* Upper quartile

USA TODAY Snapshots®, 6, 17, 62, 175, 180, 209, 222, 261, 320, 343, 345, 396, 421, 425, 492, 544

V

Variables, 18
defining, 25

Venn diagram, 203

Verbal/Linguistic Learning Style. *See* Differentiated Instruction

Verbal phrases, as algebraic expressions, 150

Vertex, 413

Vertical angles, 422

Visual/Spatial Learning Style. *See* Differentiated Instruction

Vocabulary Builder, T4, T7, 4, 52, 104, 148, 194, 238, 286, 332, 368, 410, 468, 510

Volume, 520
cylinders, 524, 525
rectangular prisms, 520, 521

W

WebQuest, T4, T6–T7, T15, 3, 95, 103, 185, 193, 277, 285, 401, 409, 545

Webs, 474

Weight
changing customary units, 267–268
ounce, 267
pound, 267
ton, 267

Which One Doesn't Belong?, 11, 20, 61, 107, 140, 158, 199, 218, 229, 256, 260, 268, 290, 299, 317, 372, 379, 453, 458, 472, 515, 543

Whiskers. *See* Box-and-whisker plots

Writing Math, 8, 11, 16, 20, 26, 32, 35, 37, 40, 44, 56, 61, 66, 71, 73, 78, 83, 87, 93, 107, 110, 114, 119, 122, 127, 130, 136, 140, 151, 155, 158, 162, 168, 174, 176, 179, 184, 196, 199, 205, 208, 212, 218, 222, 225, 229, 242, 246, 250, 256, 260, 266, 268, 272, 274, 276, 290, 294, 296, 299, 301, 306, 314, 317, 320, 322, 325, 336, 342, 346, 352, 356, 372, 376, 379, 382, 386, 389, 395, 397, 399, 412, 414, 417, 420, 424, 427, 430, 433, 435, 442, 448, 453, 458, 461, 472, 476, 478, 481, 484, 488, 491, 494, 499, 502, 513, 515, 521, 525, 531, 533, 537, 539, 543
See also Reading and Writing Mathematics; Open-Ended Assessment

X

***x*-axis,** 112

***x*-coordinate,** 112

Y

***y*-axis,** 112

***y*-coordinate,** 112

Index